WITHDRAWN FROM STOCK
The University of Liverpool

orange label
QL703. MAC

THE NEW ENCYCLOPEDIA OF MAMMALS

THE NEW ENCYCLOPEDIA OF
MAMMALS

Edited by David Macdonald

Assistant Editor Sasha Norris

This edition published by:
OXFORD
UNIVERSITY PRESS

Great Clarendon Street, Oxford OX2 6DP

Oxford University Press is a department of the University of Oxford. It furthers the University's objective of excellence in research, scholarship, and education by publishing worldwide in

Oxford New York

Athens Auckland Bangkok Bogotá Buenos Aires Cape Town Chennai Dar es Salaam Delhi Florence Hong Kong Istanbul Karachi Kolkata Kuala Lumpur Madrid Melbourne Mexico City Mumbai Nairobi Paris São Paulo Shanghai Singapore Taipei Tokyo Toronto Warsaw

with associated companies in Berlin Ibadan

Oxford is a registered trade mark of Oxford University Press in the UK and in certain other countries

British Library Cataloguing in Publication Data
Data available

Library of Congress Cataloging in Publication Data
Data available

ISBN 0 19 850823 9

All rights reserved. No part of this publication may be reproduced, stored in a retrieval system, or transmitted in any form or by any means electronic, mechanical, photocopying, recording, or otherwise, without the prior permission in writing of the copyright holder, Andromeda Oxford Limited.

You must not circulate this book in any other binding or cover and you must impose this same condition on any acquirer.

Originated in Malaysia by Global Colour Ltd.

Printed in Italy
on acid-free paper
by Milanostampa SpA, Milan

10 9 8 7 6 5 4 3 2 1

Photos page i: *Antelope ground squirrels* Stephen J. Krasemann/Bruce Coleman Collection. page ii-iii: *Polar bear* François Gohier/Ardea London. above: *Spotted dolphins* Jeff Foott/Bruce Coleman Collection.

AN ANDROMEDA BOOK
Planned and produced by
Andromeda Oxford Limited,
11–13 The Vineyard, Abingdon,
Oxfordshire, OX14 3PX
United Kingdom.
www.andromeda.co.uk

Copyright © 2001 Andromeda Oxford Limited

First published 2001

The moral rights of the authors have been asserted

Database right Andromeda Oxford Limited

Publishing Director Graham Bateman
Project Manager Peter Lewis
Editors Tony Allan, Mark Salad
Art Director Chris Munday
Designers Frankie Wood, Mark Regardsoe
Cartographic Editor Tim Williams
Picture Manager Claire Turner
Picture Researcher Vickie Walters
Production Director Clive Sparling
Editorial and Administrative Assistants Rita Demetriou, Marian Dreier, Moira Elliott
Proofreader Lynne Wycherley
Indexer Ann Barrett

Western Gorilla
see page 414

Ibex *see page 570*

For my wife, Jenny,
and Ewan, Fiona and Isobel

Advisory Editors

Professor Hans Kruuk,
Centre for Ecology and Hydrology,
Banchory,
United Kingdom.

Dr Richard Connor,
University of Massachusetts
at Dartmouth, USA.

Professor John Harwood,
Gatty Marine Laboratory,
University of St. Andrew's,
United Kingdom.

Dr Guy Cowlishaw,
Institute of Zoology,
London,
United Kingdom.

Professor Johan du Toit,
Mammal Research Institute,
University of Pretoria,
South Africa.

Professor Jerry O. Wolff,
University of Memphis,
Tennessee, USA.

Dr Christopher R. Dickman,
University of Sydney,
Australia.

Dr Gareth Jones,
University of Bristol,
United Kingdom.

Artwork Panels

Priscilla Barrett
Denys Ovenden
Malcolm McGregor
Michael R. Long
Graham Allen

CONTENTS

Capybara
see page 678

CONTRIBUTORS

WHAT IS A MAMMAL? pp. xvi–xxxi

KB Kate Barlow, British Antarctic Survey

RD Richard Dawkins, University of Oxford, UK

WGK Warren G. Kinzey, City University of New York, USA

DWM David W. Macdonald, WildCRU, University of Oxford, UK

WAM William A. Mason, University of California, Davis, USA

CARNIVORES pp. 2–145

SKB Simon K. Bearder, Oxford Brookes University, UK

CB Colleen Begg, Begg Nature Photography, South Africa

KB Keith Begg, Begg Nature Photography, South Africa

JDSB Johnny D. S. Birks, Universities of Exeter and Durham, UK

LB Luigi Boitani, University of Rome, Italy

TC Tim Caro, University of California, Davis, USA

SC Scott Creel, Montana State University, USA

MC Melanie Culver, Virginia Tech. Fisheries and Wildlife, USA

MJD Michael Daniels, WildCRU, University of Oxford, UK

MD Miguel Delibes, CSIC, Seville, Spain

JMD James M. Dietz, Michigan State University, USA

GWF George W. Frame, Utah State University, USA

EKF Erik K. Fritzell, Oregon State University, USA

DLG David L. Garshelis, State of Minnesota Department of Natural Resources Wildlife Research, USA

EG Eli Gefen, Tel Aviv University, Israel

JLG John L. Gittleman, University of Virginia, USA

MEG Matthew E. Gompper, Columbia University, USA

CH Clare Hawkins, University of Edinburgh, UK

RJ Rodney Jackson, International Snow Leopard Trust, USA

AJTJ A. J. T. Johnsingh, Wildlife Institute of India

WJ Warren Johnson, National Cancer Institute, USA

KK Kaarina Kauhala, Finnish Game and Fisheries Research Institute, Helsinki

RK Roland Kays, New York State Museum, USA

CMK Caroline M. King, University of Waikato, New Zealand

HK Hans Kruuk, Centre for Ecology and Hydrology, Banchory, UK

AL Arild Landa, Institute of Natural Resources, Greenland

SL Serge Larivière, Institute for Wetland and Waterfowl Research, Memphis, USA

AL Andy Loveridge, WildCRU, University of Oxford, UK

TRL T. R. Loughlin, National Marine Mammal Laboratory, USA

RMcD Robbie A. McDonald, The Game Conservancy, UK

JMcN John McNutt, Wild Dog Research Center, Botswana

DWM David W. Macdonald, WildCRU, University of Oxford, UK

JM James Malcolm, University of Redlands, USA

LM Laurie Marker, WildCRU, University of Oxford, UK

FM François Messier, University of Saskatchewan, Canada

MGLM Michael G. L. Mills, Kruger National Park, South Africa

DM Dale G. Miquelle, Hornocker Wildlife Institute, USA

PDM Patricia D. Moehlman, Tanzania Guides, Tanzania

AM Axel Moehrenschlager, Calgary Zoo Animal Health Center, Canada

LN Latika Nath, WildCRU, University of Oxford, UK

SJO'B Stephen J. O'Brien, National Cancer Institute, USA

CP Craig Packer, University of Minnesota, USA

PCP Paul C. Paquet, Conservation Biology Institute, Canada

RAP Roger A. Powell, North Carolina State University, USA

ARR Alan Rabinowitz, Wildlife Conservation Society, Bronx Zoo, New York, USA

AR Anne Rasa, Bonn University, Germany

IR Ian Redmond, Ape Alliance and UK Rhino Group

PRKR Philip Richardson, Africa Wildlife Films, South Africa

JR Jon Rood, National Zoological Park, Virginia, USA

Ocelot
see page 32

JKR James K. Russell, National Zoological Park, Washington DC, USA

CS-Z Claudio Sillero-Zubiri, WildCRU, University of Oxford, UK

PS Philip Stander, Ministry of Environment and Tourism, Namibia

IS Ian Stirling, Canadian Wildlife Service, Edmonton, Canada

AV Arun Venkatamaran, Wildlife Institute of India

CW Chris Wemmer, National Zoological Park, Virginia, USA

RBW Rosie B. Woodroffe, University of Warwick, UK

WCW W. Chris Wozencraft, Lewis–Clark State College, USA

NY Nobuyuki Yamaguchi, WildCRU, University of Oxford, UK

PY Pralad Yonzon, Resources Himalaya, Kathmandu, Nepal

LZ Lu Zhi, Beijing University, China

EZ Erich Zimen, Saarbrücken University, Germany

Sea Mammals pp. 146–289

PKA Paul K. Anderson, University of Calgary, Canada

SSA Sheila S. Anderson, British Antarctic Survey

KB Ken Balcombe, Center for Whale Research, USA

RB Robin Best, Instituto Nacional de Pequisas de Amazonia, Brazil

WNB W. Nigel Bonner, British Antarctic Survey

ILB Ian L. Boyd, British Antarctic Survey, UK

PB Paul Brodie, Bedford Institute of Oceanography, Canada

DPD Daryl Domning, Howard University, USA

JD Jim Darling, West Coast Whale Research, Canada

AWE Albert W. Erickson, University of Seattle, USA

PGHE Peter G.H. Evans, University of Oxford, UK

FHF Francis H. Fay, University of Alaska, USA

RG Ray Gambell, International Whaling Commission, UK

DEG David E. Gaskin, University of Guelph, Canada

CG John Craighead George, North Slope Borough Dept. of Wildlife Management, Alaska, USA

JG Jonathan Gordon, WildCRU, University of Oxford, UK

JH Janice Hannah, International Marine Mammal Association, Guelph, Canada

JH John Harwood, NERC Sea Mammal Research Unit, UK

LH Lex Hiby, Conservation Research Limited, Great Shelford, UK

KMK Kit M. Kovacs, Norwegian Polar Institute, Tromsø, Norway

SDK Scott D. Kraus, New England Aquarium, USA

DML David M. Lavigne, International Marine Mammal Association, Guelph, Canada

BLeB Burney J. Le Boeuf, University of California, Santa Cruz, USA

CL Christina Lockyer, British Antarctic Survey, UK

IAM Ian A. McLaren, Dalhousie University, Canada

CMacL Colin MacLeod, Glasgow University, UK

HM Helene Marsh, James Cook University, Australia

AM Tony Martin, NERC Sea Mammal Research Unit, UK

DKO Daniel K. Odell, University of Miami, USA

JMP Jane M. Packard, University of Florida, USA

VP Vassili Papastavrou, University of Bristol, UK

PP Paddy Pomeroy, NERC Sea Mammal Research Unit, UK

KR Katherine Ralls, Smithsonian Institution, USA

GBR Galen B. Rathbun, California Academy of Sciences, USA

AR Andrew Read, Duke University Marine Laboratory, USA

RR Randall Reeves, University of Quebec, Canada

REAS Robert E. A. Stewart, Department of Fisheries, and Oceans, Winnipeg, Canada

AT Andrew Taber, Wildlife Conservation Society, USA

PLT Peter L. Tyack, North East Fisheries Science Center, USA

MW Michelle Wainstein, University of California, Santa Cruz, USA

LW Lindy Weilgart, Dalhousie University, Canada

RSW Randall S. Wells, Moss Landing Marine Laboratories, USA

HW Hal Whitehead, Dalhousie University, Canada

BW Bernd Würsig, Texas A&M University, USA

Primates pp. 290–433

LB Louise Barrett, University of Liverpool, UK

DB-J Douglas Brandon-Jones, Natural History Museum, UK

RB Redouan Bshary, Max-Planck-Institute, Seewiesen, Germany

DC Dorothy Cheney, University of Pennsylvania, USA

DJC David J. Chivers, University of Cambridge, UK

TC-B Tim Clutton-Brock, University of Cambridge, UK

GC Guy Cowlishaw, Institute of Zoology, London, UK

XD-R Xavier Domingo-Roura, University of Barcelona, Spain

RD Robin Dunbar, University of Liverpool, UK

JEF John E. Fa, Jersey Wildlife Preservation Trust, UK

AHH Alexander H. Harcourt, University of California, Davis, USA

PH Peter Henzi, University of Liverpool, UK

PH Paul Honess, WildCRU, University of Oxford, UK

GJ Gerald H. Jacobs, University of California, Santa Barbara, USA

CHJ Charles H. Janson, State University of New York, USA

PK Peter Kappeler, German Primate Center, Göttingen, Germany

DK Devra Kleiman, National Zoological Park, Washington DC, USA

HK Hans Kummer, University of Zurich, Switzerland

JMacK John MacKinnon, ASEAN Regional Center for Biodiversity, Philippines

KMacK Kathy MacKinnon, World Bank, USA

RDM Robert D. Martin, University of Zurich, Switzerland

RAM Russell A. Mittermeier, Conservation International, USA

KM Katherine Milton, University of California, Berkeley, USA

CN Carsten Niemitz, Free University of Berlin, Germany

RN Ronald Noë, University of Strasbourg, France

RP Ryne Palombit, University of Pennsylvania, USA

FEP Frank E. Poirier, Ohio State University, USA

JIP J. I. Pollock, Duke University Primate Center, USA

TER Thelma E. Rowell, University of California, Berkeley, USA

ABR Anthony B. Rylands, Conservation International, USA

RS Robert Seyfarth, University of Pennsylvania, USA

BS Barbara Smuts, University of Michigan, USA

CvS Carel van Schaik, Duke University, USA

DW David Watts, Yale University, USA

AW Andrew Whiten, University of St. Andrew's, UK

RWW Richard W. Wrangham, Harvard University, USA

PCW Patricia C. Wright, State University of New York, USA

EZ Elka Zimmermann, Hannover Veterinary School, Germany

LARGE HERBIVORES pp. 434–577

RFWB Richard F. W. Barnes, Karisoke Research Center, Rwanda

JB Joel Berger, University of Nevada, USA

CB Cristian Bonacic, WildCRU, University of Oxford, UK

MB Mark Boyce, University of Alberta, Canada

PB Peter Brotherton, English Nature, UK

HGC Hernan Castellanos, Caracas, Venezuela

LC Lynn Clayton, Imperial College, London, UK

TC-B Tim Clutton-Brock, University of Cambridge, UK

RAC Rosemary A. Cockerill, University of Cambridge, UK

TC Tim Coulson, Institute of Zoology, London, UK

DHMC David H.M. Cumming, WWF Southern Africa Regional Programme Office, Zimbabwe

JCD James C. Deutsch, Crusaid, London, UK

ID-H Iain Douglas-Hamilton, Save the Elephants, Kenya

CCD Craig C. Downer, Andean Tapir Fund, USA

GD G. Dubost, National Natural History Museum, France

JdT Johan du Toit, University of Pretoria, South Africa

KE Keith Eltringham, University of Cambridge, UK

WLF William L. Franklin, Iowa State University, USA

VG Valerius Geist, University of Calgary, Canada

MG Morris Gosling, University of Newcastle, UK

SJGH Stephen J. G. Hall, University of Cambridge, UK

SH Simon Hedges, IUCN Asian Wild Cattle Specialist Group, UK

HNH Hendrik N. Hoeck, Kreuzlingen, Switzerland

CMJ Christine M. Janis, Brown University, USA

MVK Marina V. Kholodova, Russian Academy of Science

JK Jonathan Kingdon, University of Oxford, UK

DK David Kitchen, Humboldt State University, USA

KK Karl Kranz, Smithsonian Institution, USA

RK Richard A. Kock, Organisation of African Unity, Nairobi, Kenya

LEBK Loeske E. B. Kruuk, University of Edinburgh, UK

NL-W Nigel Leader-Williams, University of Kent at Canterbury, UK

SL Sandro Lovari, University of Siena, Italy

AAL Anna A. Luschekina, Russian Academy of Science

TSMcC T.S. McCarthy, University of Witwatersrand, South Africa

CRM Christine R. Maher, University of Southern Maine, USA

EJM-G Eleanor J. Milner-Gulland, Imperial College, London, UK

MGM Martyn G. Murray, University of Edinburgh, UK

PN Paul Newton, University of Oxford, UK

NO-S Norman Owen-Smith, University of Witwatersrand, South Africa

KP Katy Payne, Cornell University, USA

RAP Robin A. Pellew, University of Cambridge, UK

MSP Mark Stanley Price, Office of the Advisor for Conservation of the Environment, Oman

CR Craig Roberts, University of Newcastle, UK

KR Keith Ronald, University of Guelph, Canada

DIR Dan I. Rubinstein, Princeton University, USA

SS S. Sathyakumar, Wildlife Institute of India

HS Hezy Shoshani, Cranbrook Institute of Science, USA

TS Tony Sinclair, University of British Columbia, Canada

AS Andrew Spalton, Diwan of Royal Court, Muscat, Sultanate of Oman

AT Andrew Taber, Wildlife Conservation Society, USA

ST Simon Thirgood, University of Edinburgh, UK

RU Rod Underwood, University of Cambridge, UK

RJvA Rudi J. van Aarde, University of Pretoria, South Africa

ACW A. Christy Williams, Wildlife Institute of India

SW Stuart Williams, WildCRU, University of Oxford, UK

PW Peter Wirtz, University of Freiburg, Germany

SMALL HERBIVORES pp. 578–721

GA Greta Ågren, University of Stockholm, Sweden

KBA Kenneth B. Armitage, University of Kansas, USA

CB Claude Baudoin, University of Franche Comté, France

DB Diana Bell, University of East Anglia, UK

RB Robin Boughton, University of Florida, USA

PB Peter Busher, Boston University, USA

TMB Thomas M. Butynski, Kibale Forest Project, Uganda

MC Marcelo Cassini, University of Lujan, Argentina

GBC Gordon B. Corbet, Leven, UK

DPC David P. Cowan, Central Science Laboratory, York, UK

MJD Michael J. Delany, University of Bradford, UK

CRD Christopher R. Dickman, University of Sydney, Australia

JFE John F. Eisenberg, University of Florida, USA

JE James Evans, US Fish and Wildlife Service, USA

JF Julie Feaver, University of Cambridge, UK

THF Theodore H. Fleming, University of Miami, USA

WG Wilma George, University of Oxford, UK

GH Göran Hartman, University of Agricultural Sciences, Sweden

EH Emilio Herrera, Simon Bolivar University, Venezuela

TH Tony Holley, Brent Knoll, Somerset, UK

UWH U. William Huck, Princeton University, USA

Idaho ground squirrels
see page 607

JH Jane Hurst, University of Liverpool, UK

JUMJ Jenny U. M. Jarvis, University of Cape Town, South Africa

PJ Paula Jenkins, Natural History Museum, UK

TK Takeo Kawamichi, University of Osaka, Japan

CJK Charles J. Krebs, University of British Columbia, Canada

TEL Thomas E. Lacher, Texas A&M University, USA

KMacK Kathy MacKinnon, World Bank, USA

DWM David W. Macdonald, WildCRU, University of Oxford, UK

JP James Patton, University of California, Berkeley, USA

GBR Galen B. Rathbun, California Academy of Sciences, USA

DAS Duane A. Schlitter, Texas A&M University, USA

ES Eberhard Schneider, University of Göttingen, Germany

GIS Georgy I. Shenbrot, Ramon Science Center, Israel

PWS Paul W. Sherman, Cornell University, USA

GS Grant Singleton, CSIRO Wildlife and Ecology, Australia

ATS Andrew T. Smith, Arizona State University, USA

DMS D. Michael Stoddart, Hobart University, Tasmania, Australia

RS Robert Strachan, WildCRU, University of Oxford, UK

AT Andrew Taber, Wildlife Conservation Society, USA

RJvA Rudi J. van Aarde, University of Pretoria, South Africa

LW Luc Wauters, University of Insubria, Varese, Italy

JW John O. Whitaker, Indiana State University, USA

JOW Jerry O. Wolff, University of Memphis, USA

CAW Charles A. Woods, Florida State Museum, USA

HY Hannu Ylonen, University of Jyvaskyla, Finland

ZZ Zhang Zhibin, Chinese Academy of Sciences

Insectivores and Marsupials
pp. 722–865

MLA Mike L. Augee, University of New South Wales, Australia

CJB Christopher J. Barnard, University of Nottingham, UK

JWB Jack W. Bradbury, Cornell University, USA

GB Gary Bronner, University of Cape Town, South Africa

LB Linda S. Broome, New South Wales National Parks and Wildlife Service, Australia

AC Andrew Cockburn, Australian National University, Canberra, Australia

CRD Christopher R. Dickman, University of Sydney, Australia

BF Brock Fenton, York University, Ontario, Canada

THF Theodore H. Fleming, University of Miami, USA

GG Greg Gordon, Queensland National Parks and Wildlife Service, Australia

MLG Martyn L. Gorman, University of Aberdeen, UK

TG Tom Grant, University of New South Wales, Australia

PJ Peter Jarman, University of New England, Australia

CJ Christopher Johnson, James Cook University, Australia

GJ Gareth Jones, University of Bristol, UK

MJ Menna Jones, Australian National University, Canberra, Australia

AKL A.K. Lee, Monash University, Australia

DL David Lindenmayer, Australian National University, Canberra, Australia

WJL W. Jim Loughry, Valdosta State University, USA

CMM Colleen M. McDonough, Valdosta State University, USA

RM Roger Martin, James Cook University, Australia

VN Virginia Naples, Northern Illinois University, USA

MEN Martin E. Nicholl, Smithsonian Institution, USA

MAO'C Margaret A. O'Connell, Eastern Washington University, USA

MP Mike Perrin, University of Pretoria, S. Africa

RAR Renee Ann Richer, Harvard University, USA

CR Carlo Rondinini, Institute of Applied Ecology, Rome, Italy

EMR Eleanor M. Russell, CSIRO Wildlife and Rangelands Research Division, Australia

JR Jens Rydell, Gothenburg University, Sweden

JBS Jeremy B. Searle, University of York, UK

JHS John H. Seebeck, Department of Natural Resources and Environment, Victoria, Australia

ES Erik Seiffert, Duke University, USA

PS Paula Stockley, University of Liverpool, UK

JW Jerry Wilkinson, University of Maryland, USA

RDW Ron D. Wooller, Murdoch University, Perth, Australia

AW Andrew Wroot, Royal Holloway College, London, UK

PREFACE

To say that this new edition of **The Encyclopedia of Mammals**, *like its predecessor in 1984, covers all known members of the class Mammalia is an accurate but arid summary of the book. My goal has been to use the newest discoveries of modern biology to weave a thread among this array of 4,600 or so species, giving insight into the lives of animals so intricate in their adaptations that the reality renders our wildest fables dull. To achieve this, in the first edition I strove for a style that would satisfy the highest standards of both science and communication. We were thrilled when the prestigious journal* Science *reviewed the result as "comprehensive, readable and highly entertaining" and the late Sir Peter Scott, the greatest popular naturalist of his day, dubbed it "in a class of its own."*

Twenty years have passed since we began work on that original edition, years that have seen a revolution in the ideas and techniques of biology, and a deluge of new discoveries that make a new edition timely. This time, aided by word processors and the internet (neither of which were in common use two decades ago), we have assembled a fresh team of international researchers to assist in the production of 400,000 words of new text – every entry has been revised, most of them completely re-written.

An important point for the specialist is the way we have grouped orders together. Systematists have made major advances in resolving superordinal relationships in the class Mammalia. For example, while we were preparing proofs for the printer, two benchmark publications concluded that the golden moles and tenrecs belonged with the hyraxes and elephants in the superorder Afrotheria. Where time has allowed, and where unanimity among taxonomists seemed unshakable, we have adapted the layout of the *Encyclopedia* to accord with these new findings, but elsewhere we have retained arrangements that group mammal orders of similar ecological and morphological characteristics. Nonetheless, in every case we have sought to alert the reader to the newest ideas (thus, while the tenrecs and golden moles will be found alongside the Insectivora, not the elephants, there is a "box feature" to explain the new proposal). A different compromise between taxonomy and presentation is illustrated by the rodents. Formerly, this order was divided into three suborders – the Sciuromorpha (squirrel-like), Myomorpha (mouse-like) and Hystricomorpha (cavy-like rodents). However, the new taxonomy is bipartite, with suborders Hystricognathi and Sciurognathi (the second of which includes the mouse-like rodents); without dissenting from the new taxonomy, the visual distinctiveness of the three former suborders made it presentationally convenient for us to retain these groupings.

The *Encyclopedia* begins by presenting the carnivores (order Carnivora) – creatures among which are the epitomes of power, endurance, gentility, and quickness of wit and fang alike. The symbol of majesty is the lion, of tirelessness the wolf, of guile the fox, and of our own ruination of wild places the Giant panda. Other images of savagery, menace, and treachery may be undeserved, but they are no less vivid. There are some who think that to probe the real lives of the King of Beasts and others of his realm is to sully their poetic images. This *Encyclopedia* proves them wrong.

Traditionally, the order Carnivora has been split into two major suborders: the marine carnivores (seals, sea lions, and the walrus) called Pinnipedia, and the terrestrial carnivores, whose members are known collectively as Fissipedia. However, following recent reassessment, it is now broadly accepted that the order Carnivora comprises two distinct monophyletic groups: the suborder Caniformia, meaning "dog-like" (or Superfamily Arctoidea as it is sometimes called), and the suborder Feliformia meaning "cat-like" (or Superfamily Aeluroidea or Feloidea). The Caniformia includes the families Canidae (foxes, jackals, wolves), Ursidae (bears), Procyonidae (raccoons), and Mustelidae (badgers, mink, weasels, etc.), while the Feliformia includes the Viverridae (genets, civets), Hyaenidae (hyenas), Felidae (cats), and Herpestidae (mongooses). The interrelationships of the groups are equivocal; for example, there is much disagreement as to the exact placement of the Red panda, *Ailurus fulgens*, with various authorities allying this problematic species with the ursids, procyonids, and mustelids, and yet others advocating it should be the sole representative of the distinct family Ailuridae.

Even more contentious, however, is just how the pinnipeds fit into this arrangement. Based mainly on morphological evidence, the pinnipeds recently have been included in the suborder Caniformia. Indeed, the view of taxonomists is that placing them in a separate order would make them paraphyletic (see Glossary). General consensus seems to be that the pinnipeds represent a monophyletic clade most closely allied to the Ursidae (although there is even some evidence for a close affinity with the Mustelidae). Certainly, there is no doubt that there are affinities between the marine and terrestrial carnivores (the blood proteins of seals, for instance, are similar to those of bears) but the links are ancient and unclear. Consequently, we have followed the presentationally convenient school of elevating both divisions to ordinal status, dealing in this section with the terrestrial carnivores (Fissipedia) and in the next with the aquatic carnivores (Pinnipedia) and other sea mammals.

Comprising the next part of the *Encyclopedia* are three quite unrelated orders whose members are adapted to life at sea – the whales and dolphins (order Cetacea), seals, sea lions, and the walrus (order Pinnipedia), and the dugong and manatees (order Sirenia). Much more important, these pages bridge the waters between the faunas of our comfortably familiar terrestrial surroundings and the disquietingly alien world of ocean and ice-floe; a world where distances, empty landscapes, and even some of the creatures are so immense as to seem unreal. Superficially, the torpedo-shaped uniformity of marine mammals masks the individual character of each species.

But with closer study, the ways of whale and porpoise, of seal and walrus, spring intimately to life, and in so doing emphasize the subtlety and the frailty of the natural web, and the dependence of the monumental upon the minute.

As members of the order Primates, humans have much in common with all mammals, but our link with other primates, especially the apes, is so immediate that they are uniquely intriguing. There are those, perhaps fearful of the closeness of our relationship, who laugh, mockingly, at the behavior of monkeys and apes, deriding them as a shoddy zoological charade on humanity. In this section, however, the reader encounters creatures with whom we share our roots and, in the apes, in which the similarities between us are at least as compelling as are the differences. The primates are outstanding for their athleticism and intelligence, but above all for their intricate social relationships. Their societies encompass every variant between friendship and feud, and are engrossingly interesting, not only in themselves, but in what they tell us about ourselves.

To the non-scientist the intricacies of classification (the ordering of animals into groups of increasing size and comprehensiveness on the basis of common ancestry) are difficult to comprehend, although there is a tendency to believe that there can be only one correct answer. That this is far from the truth is indicated by changing ideas about the orders Scandentia (tree shrews) and Dermoptera (colugos). Both these groups are small (18 and two species, respectively) and have a long evolutionary history. The latter factor, coupled with an incomplete fossil record, has contributed to their equivocal phylogenetic affiliations in the past. They have separately and variously been allied to other orders such as the Macroscelidea (elephant shrews), Insectivora, Lagomorpha (rabbits and hares), Chiroptera (bats) and, most commonly, the Primates. Today, the tree shrews and colugos are widely treated as separate orders, but their inter-group relationships remain unresolved. Morphological evidence suggested their inclusion in the superorder Archonta, a clade including the Chiroptera and Primates. However, the validity of this clade was doubtful, particularly as regards the relationship of the bats to the other three orders. Indeed, recent genetic studies prove that the bats belong to an entirely different evolutionary branch, and that the tree shrews, colugos, and primates (the so-called Euarchonta) are closely related to the rodents and lagomorphs (the Glires). For convenience, therefore, the tree shrews and colugos have been positioned in the section of the book dealing with primates. The animals have remained the same, but our knowledge and understanding has altered, and will probably continue to do so.

Throughout the animal kingdom shape and form reflect function: size, color, tooth, and claw are all sculptured by evolution, fitting each species to a particular lifestyle. Few mammals, however, flaunt their evolutionary adornments so flagrantly as many of the ungulates, for it is among these that conspicuous antlers and fearsome tusks are commonplace. The functions of bony crowns of the warthog, spreading palmate fingers above the moose's head or violent stilettos on the duiker's forehead doubtless include armament against both predator and rival, and the elegance of their design is a forceful reminder of the power of natural selection.

The ungulates are represented by five extant orders only distantly united by common descent (Perissodactyla, Artiodactyla, Tubulidentata, Proboscidea, and Hyracoidea) . To this eclectic mix could probably be added two further orders of mammals: the Cetacea (whales and dolphins) and the Sirenians. Indeed, there is convincing evidence for a close relationship between the Artiodactyla (odd-toed ungulates) and the Cetacea (together comprising the so-called Cetartiodactyla). The Cetartiodactyla are, in turn, related to the Perissodactyla. The Proboscidea and Hyracoidea, on the other hand, are very closely allied to the Sirenia in a strongly supported group referred to as the Paenungulata (or Subungulata). Included with the Paenungulata, in the clade known as the Afrotheria, is the order Tubulidentata, which contains only one species – the bizarre aardvark. The aardvark has evolved to eat just ants and termites so its diet differs from that of the other "ungulates." In this new view the Ungulata – hoofed grazing animals known colloquially as the ungulates – no longer formally exists in a genetic sense of shared ancestors. Indeed, some of the ungulates – the artiodactyls and perissodactyls – belong within the monophyletic assemblage, the Laurasiatheria, whereas the Paenungulata belong to the other monophyletic assemblage, the Afrotheria.

Two of the other orders of ungulates (Proboscidea, Hyracoidea) include creatures as diverse as towering elephants and their closest surviving relative the diminutive hyraxes. Both of the two large ungulate orders, the odd- and the even-toed ungulates (Perissodactyla, Artiodactyla), include familiar members (horses and cows, respectively). They also include the terrestrial giants – rhinoceroses and hippopotamuses – and the variety of horses, deer, antelopes, and pigs is as great in shape and form as it is in behavior. One thing unites these ungulates (with the exception of the aardvark): their adaptation to herbivory – eating plants.

In a world cloaked in greenery these vegetarians might be thought to have opted for an easy lifestyle. On the contrary, although their "prey" may be neither fleet of foot nor have acute "senses," herbivores have to join battle with a devilish armory of spikes, thorns, poisons and unexpected collaborations, all deployed via life histories which are intricately adapted to outmaneuver the plant's predators. What is more, the cellulose walls that encase plant cells are unassailable to the normal mammalian digestive system, thereby reducing the mightiest elephant to dependence on an alliance with minuscule bacteria in its gut, which have the capacity to digest these resilient tissues. The armory of adaptation and counter-adaptation arrayed between large herbivores and their food has been further refined by the continual pressure on most of them to dodge attacks from their own predators. The result is an overwhelming diversity of form and function.

Three orders of small mammals are covered in the next part of the book – rodents (Rodentia), rabbits and hares (Lagomorpha), and the elephant shrews (Macroscelidea). Their small size makes them difficult to study, but their lifestyles are nevertheless intriguing. Inconspicuous scuttlings of mice and the pestilential image of rats that characterize these creatures in the public's mind belie the extraordinary interest of the rodent order. This group of mammals embraces some 2,000 species, nearly half of all mammals. Rodents are captivating for at least two reasons. First, they are outstandingly successful. Not only does one form or another occupy almost every habitable cranny of the earth's terrestrial environments, but many do so with conspicuous success while every man's hand is turned against them. Their adaptability, and particularly their ability to breed with haste and profligacy, is stunning. Second, rodents are captivating because so many of them surprise us by not conforming at all to our stereotypes – some burrow like moles, others climb, while others hunt fish in streams; in South America some rodents (capybaras) are the size of sheep, while others (maras) are, at a glance, indistinguishable from small antelopes. Their societies, as these pages illustrate, are no less varied than their bodies and give us an insight into a community in microcosm that reverberates with life beneath the grassy swathes of the countryside.

The rabbits and hares (Lagomorpha), share many of the characteristics of rodents: global distribution, small size, incredible breeding capacity, and pestilential propensities. They mostly conform to the "bunny" stereotype, but one group – the pikas – look more like hamsters, and are to be found mostly living on rocky outcrops in North America and Asia. Morphological studies and recent phylogenetic analyses have provided strong support for a sister-group relationship between the rodents and the lagomorphs (the so-called Glires). The elephant shrews (order Macroscelidea) possess characteristics displayed by several other groups – snouts like elephants, the diet of shrews, and the body form of a small antelope with a rat-like tail. This order, confined to Africa, was long included within the Insectivora, though its ordinal rank is now largely unchallenged. However, their relationship with other mammalian orders is less clear and, while they have been closely allied to the lagomorphs and the rodents, recent and extensive molecular studies together with some fossil evidence supports their placement in the Afrotheria, with the aardvark and the Paenungulata.

In its final part, the *Encyclopedia* treats a diverse series of orders (insectivores, xenarthrans, pholidotes, bats, monotremes and marsupials). Here, the reader will find accounts of mammals that fly, burrow, run and swim, that can navigate, hunt and communicate with ultrasounds and echoes quite beyond human perception; some are nearly naked, some blind, some are poisonous; others lay eggs and one, Kitti's hog-nosed bat, weighs just 1.5g (0.05oz) and is the smallest mammal in the world. That most of these species may be unfamiliar makes them all the more intriguing.

The Insectivora (shrews, hedgehogs, moles, and formerly golden moles, tenrecs and others) include small predatory creatures that not only bear "primitive" resemblance to the very earliest mammals, but which, with their tiny size and racing metabolisms, push modern mammalian body processes to the very limit. However, the monophyly of this order has recently been challenged by a deluge of evidence that supports the inclusion of some African members – the golden moles, tenrecs, and even the Madagascan hedgehog – within the Afrotheria. Indeed, moves are well advanced to fragment – indeed to disband entirely – the Insectivora between such orders as Afrosoricida (associated with the Afrotheria) and Eulipotyphla (associated with the Laurasiatheria).

About one-quarter of mammal species are bats (Chiroptera), making them the second largest order of mammals after the rodents. The bats are divided into two suborders, the Megachiroptera (megabats) and the Microchiroptera (microbats). Central to the debate of the origins of bats is whether it is feasible that the power of flight could have evolved twice in the course of mammalian evolution. One corner of the debate suggests that the Megachiroptera are descendants of the Primates, which implies that flight evolved independently in what would consequently represent two unrelated groups of bats. The other, more accepted corner, holds that the chiropterans are monophyletic (with all members having arisen from a common ancestor) implying that the evolution of flight in mammals was a singular event. The latter view has been corroborated by recent genetic studies. Whenever the climate allows their activity, bats have filled the aerial niche. They have done so with such success that within this vast assemblage are to be found plant and fruit eaters, insectivores, carnivores, fish eaters, and even blood eaters. For a group so large, knowledge of their lifestyles is slim, but within these pages will be found insights into how they are adapted to life on the wing and how their societies work.

A taste for ants and termites, lack of teeth, and possession of a long, sticky tongue are the hallmarks of two other orders – the Xenarthra (represented by the anteaters, sloths and armadillos) and Pholidota (pangolins). These two groups have evolved to occupy the "ant-eating" niche in South America and Africa, respectively, and this again emphasizes how evolution molds animals with different origins into similar form. Formerly, both were included in a group known as the Edentata – we retain the umbrella term Edentates for the sake of convenience – but molecular evidence reveals that the two are unrelated: Xenarthra represent a distinct superordinal grouping, while the Pholidota are closely related to the cetartiodactyls, perissodactyls, carnivores, and even bats (the Laurasiatheria).

Many people regard marsupials as a small group of mammals, typically the kangaroos and the koala, inhabiting Australia. In fact, the supercohort Marsupialia actually includes some 300 species, in all manner of shapes and sizes, and represents one of the major surviving subdivisions of the mammals, occurring throughout the Australian region as well as parts of South and North America. Among the marsupials there are species living in almost all of the niches which placental mammals occupy elsewhere.

The two types of mammals, placentals and marsupials, represent a worldwide, evolutionary experiment in the making of mammals: the fact that the results are often strikingly similar animals with quite distinct origins but adapted in the same way to cope with the same niche is compelling evidence for the processes of evolution. Today, it seems that the third such "experiment" in mammal-hood, which produced the egg-laying monotremes (platypus, echidnas) is imperiled, since only three species survive.

In planning this *Encyclopedia* our aim was precise and rather ambitious. Zoologists continue to discover fascinating facts about the behavior and ecology of wild mammals, and to develop new and intriguing ideas to explain their discoveries. Yet so scattered is the information that much of it fails to reach even professional zoologists, far less percolate through to a general readership. Worse still is the erroneous idea that the public will, or can, only accept watered-down science. Our aim, then, has been to gather the newest findings and to present them lucidly, entertainingly, but uncompromisingly. If we have succeeded, these pages will refresh the professional as much as they enthrall the schoolchild. The best way to convey the excitement of discovery is through the discoverers, and for this reason the book is written by the researchers themselves. The passing of two decades since I recruited the first generation of authors has offered a tremendous opportunity in this second edition to combine their original insights with those of a fresh generation of researchers; while some of the original authors have died or been lost from contact, many have contributed afresh – some entries therefore now have two authors, the "founding father" and a new-generation disciple. I am hugely grateful to them all.

As already mentioned, there is no single accepted classification of the animal kingdom, and so we have had to select from the views of different taxonomists. In 1993, Don Wilson and DeeAnn Reeder published a mighty synthesis of the taxonomy and distribution of all mammals, titled *Mammal Species of the World*. Subsequently, in 1997, Malcolm McKenna and Susan Bell published their *Classification of Mammals Above the Species Level*, an update of George Gaylord Simpson's seminal 1945 classification. Both publications are monumental and authoritative works (McKenna and Bell is commendable for its inclusion of the fossil mammals, listing some 5,100 genera). Unfortunately, the differing classifications in the two works makes it practically impossible to attempt to equate the two and come to some kind of middle-ground in this *Encyclopedia*. Consequently, since Wilson and Reeder's work extends to the species level and is more widely available, we opted to use this list as our touchstone. However, this is by no means a straightforward task, partly because new species have been discovered since 1993, partly because new techniques, especially the wizardry of molecular biology and phylogenetic analyses, are ever more frequently reshuffling old classifications, and also because reasons variously scientific,

IUCN CATEGORIES

Ex Extinct, when there is no reasonable doubt that the last individual of a taxon has died.

EW Extinct in the Wild, when a taxon is known only to survive in captivity, or as a naturalized population well outside the past range.

Cr Critically Endangered, when a taxon is facing an extremely high risk of extinction in the wild in the immediate future.

En Endangered, when a taxon faces a very high risk of extinction in the wild in the near future.

Vu Vulnerable, when a taxon faces a high risk of extinction in the wild in the medium-term future.

LR Lower Risk, when a taxon has been evaluated and does not satisfy the criteria for CR, EN or VU.

Note: The Lower Risk (LR) category is further divided into three subcategories: Conservation Dependent (cd) – taxa which are the focus of a continuing taxon-specific or habitat-specific conservation program targeted toward the taxon, the cessation of which would result in the taxon qualifying for one of the threatened categories within a period of five years; Near Threatened (nt) – taxa which do not qualify for Conservation Dependent but which are close to qualifying for VU; and Least Concern (lc) – taxa which do not qualify for the two previous categories.

tactical, or whimsical fuel the eternal debate between those who would split and those who would lump species. Serious readers will want to know what rules of thumb we have brought to bear on this highly technical morass. The example of the Primates serves to illustrate our approach.

Primate taxonomy has been a veritable minefield of mistaken identities, splitting, and lumping. In the first edition we listed 181 species in 52 genera, while Wilson and Reeder (on the basis of a review by Colin Groves) recognized 233 species in 60 genera. However, on assembling all the material provided by our authors, we found their tally was at least 256 species. Some of these additions were strongly felt – for example our bush baby author had himself been involved in describing new species on the basis of their vocalizations – others were probably little more than the whim of a particular non-taxonomic author to follow a particular classification. Between these were many cases where new species have been proposed, often on the basis of new molecular techniques, but not yet thoroughly reviewed, far less universally accepted, by the taxonomic community. What was to be done to balance the books in terms of numbers of species, genera and families? Hoping to minimize both the confusion of readers and the wrath of authors, we opted to take Wilson and Reeder as the baseline, but to follow the lead of those authors who strongly advocated departure from this. However, on the taxonomic panels that introduce each family or order we draw attention in parentheses to any disparities between our species count and that of the 1993 list, and in the species tables, in addition to listing the species on the "established list" (i.e. Wilson and Reeder), we include mention of proposed new species, or subspecies that some believe should be elevated to full species. In short, we provide as much information as possible, making use of the most topical opinions of our authors, but retaining a firm footing on the solid ground of a widely used taxonomy.

In the case of genera, we also erred on the side of inclusiveness. For example, Wilson and Reeder list nine species within the genus *Callithrix*. Yet our author proposed that two of these, along with nine newly described species, be assigned to the genus *Mico*. The species table attempts to reconcile these approaches by referring to *Mico* (*Callithrix*)... . People cross-referring to a text that uses either genus will thus be able to find the species they are looking for, and adherents of both *Callithrix* and the *Mico* schools will find their view acknowledged and (at least by inference) explained.

The reader will encounter four distinct types of entry in the *Encyclopedia*. First, for each order or group of orders, there is a general essay highlighting common features and main variations of the biology (particularly body plan), ecology, and behavior of the group concerned and its evolution. Second, forming the bulk of the *Encyclopedia*, are the accounts of individual species, groups of closely related species, or families of species. The text on these pages covers details of physical features, distribution, evolutionary history, diet and feeding behavior, social dynamics and spatial organization, classification, conservation and relationships with man. Third, the special features – the fruit of authors' "cutting edge" research, offer detailed insights into the social organization, foraging, breeding biology, and conservation of particular species. Finally, the photo stories showcase some stunning sequences of wildlife photography, which cover such diverse topics as the birth of a dolphin and reindeer herding by the Sami.

Throughout the *Encyclopedia* we have added, wherever possible, the IUCN (World Conservation Union) Red List codes for each species. For the second edition, we had the luxury of using the 2000 IUCN Red Lists. However, in the case of primates the 2000 Red List adopts a totally novel and very speciose taxonomy stemming from the work of the IUCN/SSC Primate Specialist Group. In many instances, therefore, matching up the taxonomy appearing in Wilson and Reeder with the 2000 Red List is well-nigh impossible. We have adopted all categories as defined by the IUCN down to and including Lower Risk (in other words, omitting the categories Data Deficient and Not Evaluated). The codes are explained in the accompanying table.

The success of books like this rests upon the integrated efforts of many people with diverse skills. I want to thank all the authors – old and new – for their enthusiastic cooperation. A special pleasure to me has been the involvement of core members of the original cast in the new edition: most especially, Priscilla Barrett and Denys Ovenden have continued their painstaking attention to detail in the new artwork, and Dr Graham Bateman, now Andromeda's publishing director, retained a presiding role in the project. Augmenting their expertise are a number of new stalwarts, notably Dr Sasha Norris, my editorial assistant, whose tireless energies, dedication, and friendship have contributed hugely to this edition. At Andromeda, the book took shape through the industrious efforts of a team of editors and designers led by project manager Dr Peter Lewis and senior art editor Chris Munday. Other invaluable helpers include Lauren Harrington, who scrutinized the taxonomies, and Mike Hoffmann, who updated the Appendix and Glossary.

In the Preface to the first edition I thanked my wife, Jenny, for her tolerance of the avalanches of manuscripts which, for two years, had cascaded from every surface of our home. To tolerate this once was heroic, but to do so twice has been a marvel, and this time I must add my thanks to my children, Ewan, Fiona, and Isobel, who have been surprisingly good-natured despite my peculiar preoccupations with this venture. For two years, we have paddled our editorial boat hither and thither, fearful of being smashed on the rocks of gauche sensationalism or becalmed in the stagnant pools of obsessive detail. Hopefully, the reader will judge that we have charted a course where the current is invigorating and the waters clear.

David W. Macdonald

DAVID W. MACDONALD

DEPARTMENT OF ZOOLOGY AND LADY MARGARET HALL

UNIVERSITY OF OXFORD

What is a Mammal?

MAMMALS COULD BE CATEGORIZED AS a group of animals with backbones, and bodies insulated by hair, which nurse their infants with milk and share a unique jaw articulation. Yet this fails to convey their truly astonishing features – their intricate adaptations, thrilling behavior, and highly complex societies.

A more satisfactory response might therefore be that the essence of mammals lies in their diversity of form and function, and above all their individual flexibility of behavior: the smallest mammal, Kitti's hog-nosed bat, weighs 1.5g (0.05oz), the Blue whale weighs 100 million times as much; the wolf may journey through 1,000sq km (400sq mi), the Naked mole rat never leaves one burrow; this latter species gives birth to litters of up to 28, the orangutan to only one; the elephant, like man, may live three score years and ten, while the male Brown antechinus never sees a second season and dies before the birth of the first and only litter he has fathered. No facet of these varied lives is random; they are diverse but not in disarray. On the contrary, each individual mammal maximizes its "fitness," its ability, relative to others of its kind, to pass on genes, usually by leaving viable offspring.

There are some 4,680 species of mammals (modern molecular analysis has revealed several new ones), among which ancient relationships permit subdivisions into 1,100 or so genera, 139 families, 18 orders, and 2 subclasses. These subclasses acknowledge a 200 million-year separation between egg-laying Prototheria and live-bearing Theria. Moreover, a 90 million-year-old split within the Theria divides the marsupials from the placental mammals. Some splits among the placentals may be almost as ancient.

Even within taxonomy's convenient compartments there is a bewildering variation in the size, shape and life-histories of mammals. Indeed, it is especially characteristic of mammals that even individuals of the same species behave differently depending on their circumstances.

Experiments in Tandem

CONVERGENT EVOLUTION

From ancient starting points very distinct lineages have followed strikingly convergent routes to achieve a mammalian style of existence. Formerly, the marsupials were thought of as "primitive" mammals, alongside the "advanced" placentals. Now, however, it is clear that they are the separate, intriguingly different yet amazingly similar results of an experiment begun more than 90 million

Below Placental mammals are now thought to fall into four groups: Xenarthra, Afrotheria, Laurasiatheria, and Euarchonta plus Glires. Each of the first three almost certainly each stems from a single common ancestor (they are monophyletic), but it remains uncertain whether the Euarchonta and Glires have common or separate (paraphyletic) roots.

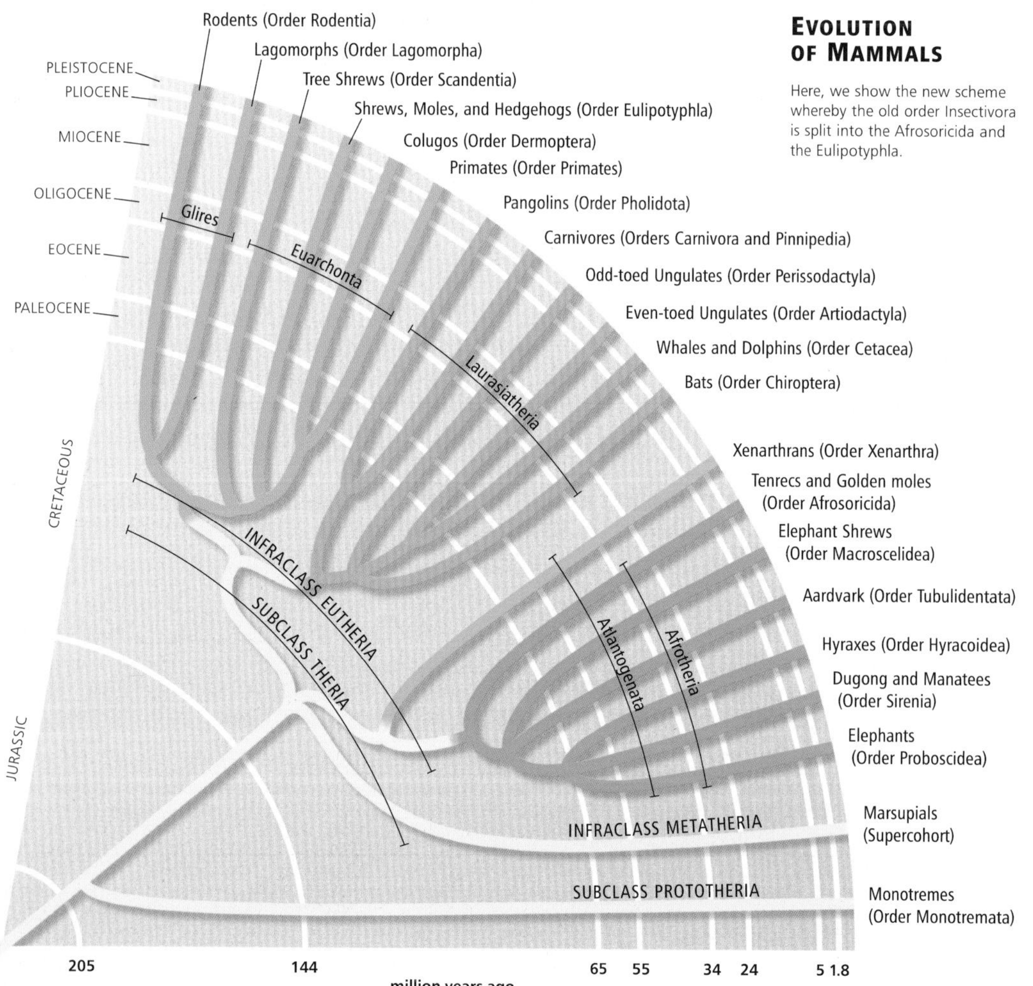

years ago. The ancestral mammal lineage split along the two routes to form these lineages; in the Pliocene both faced the opportunities presented by the creation of continental climates and the emergence of grasslands and both arrived at the same solutions. For example, in tackling the challenges of herbivory, each array has produced both fore- and hindgut fermenters, using completely different microbial communities to achieve the same effects; both have solved problems of tooth wear on abrasive diets in parallel ways; and both evolved toe-reduction and sociality in response to predation risks while foraging in open habitats.

Remarkably, an equally ancient instance of parallel evolution has only recently come to light, which demonstrates that there have been (at least) two great experiments in placental mammals. These mammals began to diversify in the late Cretaceous, at a time when the southern landmass (Gondwana, including Afro-Arabia) was isolated from the northern continent (Laurasia). It has

CLASS: MAMMALIA
2 subclasses; 27 orders; 139 families

SUBCLASS THERIA (Livebearers)

INFRACLASS EUTHERIA (Placentals)

INFRACLASS METATHERIA (Marsupials)

SUBCLASS PROTOTHERIA (Egg-layers)

Left *The red fox* (Vulpes vulpes) *embodies several of the traits that characterize mammalian success – adaptability, opportunism, and intelligence, plus the capacity for intricate social relationships and subtle communication. It also epitomizes the grace and beauty that captivates people's enthusiasm for other members of our Class.*

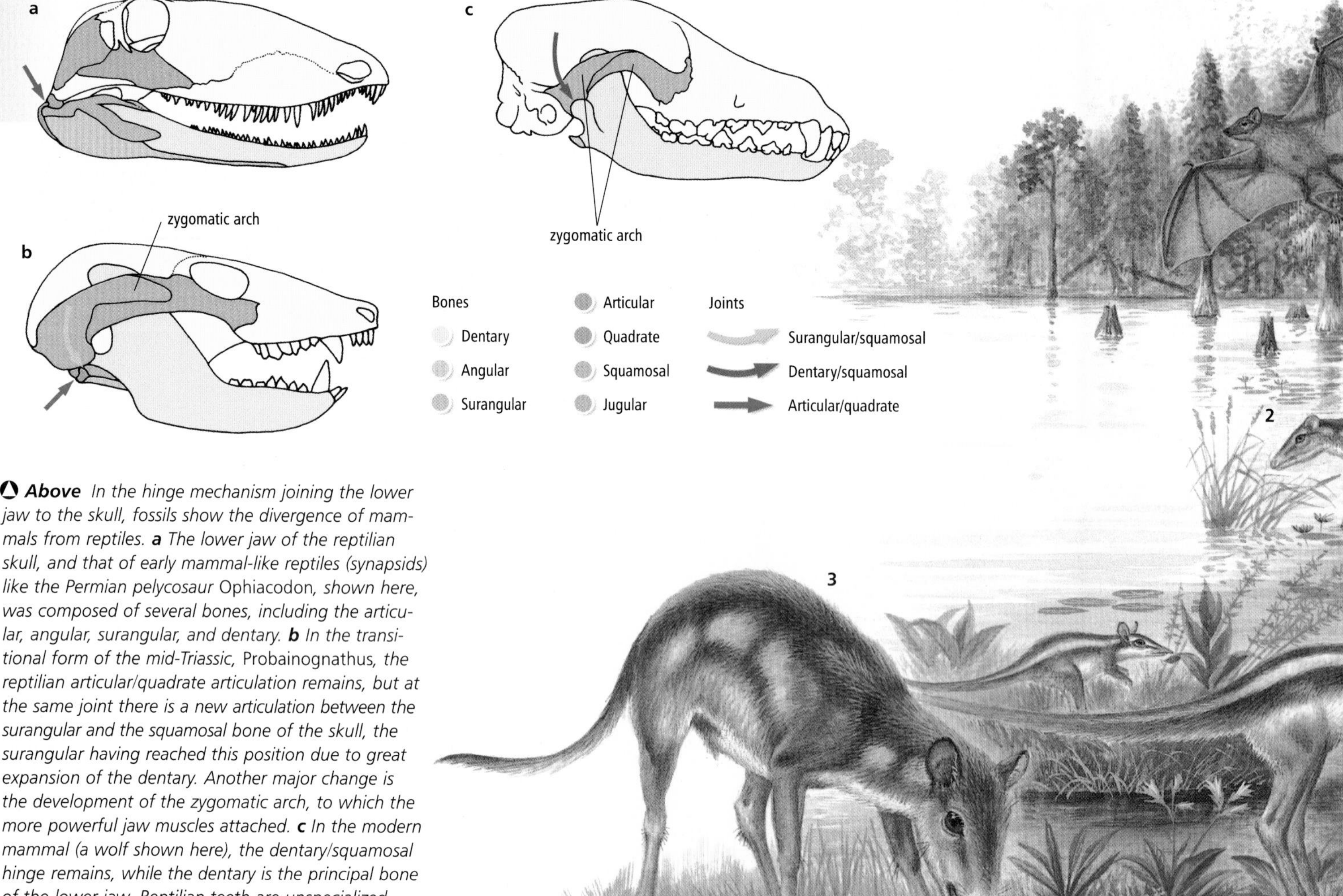

Above *In the hinge mechanism joining the lower jaw to the skull, fossils show the divergence of mammals from reptiles.* ***a*** *The lower jaw of the reptilian skull, and that of early mammal-like reptiles (synapsids) like the Permian pelycosaur* Ophiacodon, *shown here, was composed of several bones, including the articular, angular, surangular, and dentary.* ***b*** *In the transitional form of the mid-Triassic,* Probainognathus, *the reptilian articular/quadrate articulation remains, but at the same joint there is a new articulation between the surangular and the squamosal bone of the skull, the surangular having reached this position due to great expansion of the dentary. Another major change is the development of the zygomatic arch, to which the more powerful jaw muscles attached.* ***c*** *In the modern mammal (a wolf shown here), the dentary/squamosal hinge remains, while the dentary is the principal bone of the lower jaw. Reptilian teeth are unspecialized (homodont), while those of modern mammals are specialized to fulfil different functions (heterodont).*

long been known that ancestral placentals were present in the northern continent at that time and many paleontologists thought that Afro-Arabia acquired its mammals later from lineages that evolved in Laurasia. For example, Paenungulates (the elephants, hyraxes and manatees) have conventionally been aligned with the Ungulates such as cetartiodactyls (the shared lineage leading to whales and cloven-hoofed ungulates) and perissodactyls (the odd-toed ungulates), while tenrecs and golden moles were placed in the order known variously as Insectivora or Lipotyphla alongside moles, shrews, hedgehogs and solenodons. The new revelation is that Afro-Arabia was the cradle of evolution for a separate array of mammals – an ecologically diverse assemblage of endemic African placentals that includes elephant shrews, the aardvark, golden moles and tenrecs, and paenungulates (hyraxes, manatees, and elephants) – collectively dubbed the Afrotheria. From the north, with parallel aquatic, ungulate, and insectivore-like forms, come the Laurasiatheria, namely the cetartiodactyls (artiodactyls and whales), perissodactyls, carnivores, pangolins, bats and Eulipotyphlan insectivores (shrews, moles, and hedgehogs). During their subsequent separation , each lineage has produced parallels: the Gondwanan golden moles versus Laurasian true moles, manatees versus whales, tenrecs versus hedgehogs, and so on. The only failure in this parallelism is that there is no flying Afrotherian.

In many respects Afrotherians are more "primitive" than other placentals. The male reproductive system is more like that of the non-Therian monotremes – for instance, the testes of most Afrotherians are completely abdominal and the penis is extremely long. In a similar vein, Afrotherians generally have a minimally developed thermoregulatory system (they came from warm climates). Indeed, Afrotheria may either be the first branch of the placental family tree, or, together with the South American Xenarthrans (e.g. sloths and armadillos), constitute the sister taxon of all other placentals. Either way, the southern continent, Gondwana, played a much more important role in early placental evolution than the old Laurasian focus had suggested. Some representatives of both lineages – the tenrecs of Gondwana and the solenodons of Laurasia – have changed little from Cretacous placentals. Indeed, the placement of the Afrotherians and Xenarthrans at the bottom of the placental lineage suggests that the first placental mammals were probably terrestrial, not arboreal as previously thought.

Other recent analyses of DNA reveal that South American and Australasian marsupial families are less distinct than was once supposed. The bandicoots (Perameloidea; now only in Australasia) were separated from all other marsupials before most of the South American and other Australasian families separated, and the South American Monito del Monte (Microbiotheriidae) sits among the upper branches of the marsupial tree alongside many Australasian families. The implication is that Australia and South America, joined by Antarctica, shared many marsupial lineages until the continents separated; and that since that time lineages have gone extinct on one continent or the other, so that now no family is shared between them.

Application of DNA studies to mammalian systematics and dating the separation of lineages, and a firmer grasp of the dates of the tectonic separation and amalgamation of land masses, are between them revolutionising understanding of the evolutionary history of mammals.

Above *An array of fossils from the early Eocene (49 m.y.a.) were found at Lake Messel, Germany. This reconstruction shows a community at the dawn of the age of mammals:* **1** Archaeonycteris, *the first known bat;* **2** Messelobunodon, *an ancestral artiodactyl;* **3** Propaleotherium, *an ancestral horse;* **4** Lepticidium, *an insectivore;* **5** Paroodectes, *a miacid;* **6** Eurotamandua, *an anteater; and* **7** Pholidocercus, *a hedgehog.*

From Reptiles to Mammals

Evolutionary Biology

In the Carboniferous period some 300 million years ago, the ancestors of today's mammals were no more than a twinkle in an ancient reptilian eye. The world was spanned by warm, shallow seas and the climate was hot, humid and constant. Among the reptiles of the late Carboniferous, one line heralded the mammal-like reptiles – the subclass Synapsida. The synapsids flourished, dominating the reptilian faunas of the Permian and early Triassic about 280–210 m.y.a. Over millions of years their skeletons altered from the cumbersome reptilian mold to a more racy design that presaged the early mammals. Yet, despite these auspicious beginnings, this was a false start for the mammals. The late Triassic saw the dazzling ascendancy of the dinosaurs, which in the Mesozoic era (225–65 m.y.a.) not only eclipsed the synapsids, but nearly annihilated them through competitive superiority. Inconspicuous mammal-like reptiles survived and evolved during the Triassic (225–195 m.y.a.) into true mammals, of which the first were 5cm (2in) long and nocturnal. Their unobtrusive scuttlings gave little evidence of what was to become the most exciting radiation in vertebrate history when, well over 100 million years later, in the late Cretaceous period, the dinosaurs lumbered into oblivion.

By the Triassic, among synapsids the order Therapsida prevailed and in the fossils of these mammal-like reptiles lie the roots of modern mammals. Over millennia, they developed an expanded temporal skull opening and a corresponding rearrangement of the jaw musculature; a secondary palate appeared, forming a horizontal partition in the roof of the mouth (formed by a backward extension of the maxillary and palatine bones); their teeth became diverse; six of the seven bones of the reptilian lower jaw were reduced in size while the fifth, the dentary, was hugely enlarged; ribs were no longer attached to the cervical and lumber vertebrae, but only to the thoracic ones; the pectoral and pelvic girdles were streamlined, and angles on the heads of femora and humeri altered so that the limbs were aligned beneath, rather than to the side of, the body. These and other changes promoted more effective, agile, and swift working of the body. For example, the false palate forms a bypass for air from the nostrils to the back of the mouth, facilitating simultaneous eating and breathing, made more efficient by the evolution of the diaphragm – a muscular plate separating the chest cavity from the abdomen.

All modern mammals arose from a group called cynodonts. These derived mammal-like reptiles of the middle and late Triassic were somewhat dog-like predators. A cynodont from the mid-Triassic beds of Argentina, *Probainognathus*, is thought to best represent the transition. It retains only the flimsiest articular–quadrate joint and illustrates the development toward the articulation between the dentary and squamosal bones. Moreover, the bones of the old reptilian jaw joint are juxtaposed so as to foreshadow their transformation into the ossicles of the mammalian middle ear apparatus (the articular, quadrate, and stapes become, respectively, the malleus, incus and stapes).

Since soft parts do not fossilize, the history of modern mammals must be traced from fragments of bones and teeth. In addition to their jaw articulation, modern mammalian skulls are distinguished by the entotympanic bone, part of the auditory bulla. Furthermore, mammalian teeth develop only from the premaxillary, maxillary, and dentary bones and are generally diversified in function (heterodont, consisting of incisors, canines, premolars and molars). Typically, placental mammals have two sets of teeth, the milk, or deciduous, set often differing in form and function from the adult set; marsupials replace only some of their teeth. All mammalian teeth consist of a core of bone-like dentine wrapped in a hard case of enamel (largely calcium phosphate). In most mammals the pulp cavity seals, and the tooth ceases to grow, once adult. But the incisors of rodents, the tusks of several species, and the grinding teeth of a few herbivores (such as wombats), all teeth experiencing rapid wear, remain open-rooted and growing.

Mammal Central Heating

Endothermy and its Costs

Two fundamental traits of mammals lie not in their skeletons, but at the boundary of their bodies – the skin. These features are hair and the skin glands, including the mammary glands that secrete milk, and the sweat and sebaceous glands. They may not seem spectacular, and may even have evolved before mammal-like reptiles crossed the official divide. But these traits are associated with endothermy, a condition whose repercussions affect every aspect of mammalian life.

Endothermic animals are those whose internal body temperature is maintained "from within" (endo-) by the oxidation (essentially, burning) of food within the body. Some endotherms maintain a constant internal temperature (homeothermic), whereas that of others varies (heterothermic). The temperature is regulated by a "thermostat" in the brain, situated within the hypothalamus. In regulating their body temperature independent of the environment, mammals (and birds) are unshackled from the alternative, ectothermic, condition typical of all other animals and involving body temperatures rising and falling with the outside (ecto-) temperature. Endothermic and ectothermic animals are sometimes, misleadingly, called warm- and cold-blooded respectively. However, since the major heat source for, say, a lizard is outside its body, coming from the sun, it can have a body temperature higher than that of a so-called warm-blooded animal, but when the air temperature plummets the reptile's body temperature falls too, reducing the ectotherm to compulsory lethargy. In contrast, the internal processes of the endothermic mammal operate independently of the outside environment. This difference is overwhelmingly important because the myriad of linked processes that constitute life are fundamentally chemical reactions and they proceed at rates which are dependent upon temperature. Endothermy confers on mammals an internal constancy that not only allows them to function in a variety of environments from which reptiles are debarred, but also assures a biochemical stability for their bodies. The critical effect of temperature on mammalian functioning is illustrated by the violence of the ensuing delirium if the "thermostat" goes awry and allows the temperature to rise by even a few degrees.

Above *Mammals must expend energy to warm or cool themselves, depending on the vagaries of their surroundings. Elephants lose body heat by flapping their ears, which allows air to flow over blood vessels.*

Endothermy is costly. There are many adaptations involved in minimizing the running costs and the most ubiquitous is mammalian hair. The coat may be adapted in many ways, but there is often an outer layer of longer, more bristle-like, water-repellent guard hairs that provide a tough covering for densely-packed, soft underfur. The volume of air trapped among the hairs depends on whether or not they are erected by muscles in the skin. Hair may protect the skin from the sun's rays or from freezing wind, slowing the escape of watery sweat in the desert or keeping aquatic mammals dry as they dive. Hairs are waterproofed by sebum, the oily secretions of sebaceous glands associated with their roots.

Mammals differ in their body temperatures – e.g. monotremes 30°C (86°F), armadillos 32°C (89.6°F), marsupials and hedgehogs 35°C (95°F), man 37°C (98.6°F) and rabbits and cats 39°C (102.2°F). Some mammals minimize the costs of endothermy by temporarily sacrificing homeothermy: they do not maintain a constant internal temperature. The body temperature and hence metabolic costs of hibernating mammals drop while they are torpid, as do those of tenrecs and many bats during daily periods of inactivity. The body temperature of echidnas fluctuates between 25–37°C (77–99°F), falling much lower during winter torpor.. Because of the huge area for heat loss in their hairless wings, some microbats cannot maintain homeothermy when at rest, but allow their temperature to fall. They get so cold that when they awaken they have to exercise vigorously to raise their temperature before take-off.

The coiled sweat glands in the skin of mammals secrete a watery fluid. When expressed onto the skin's surface this evaporates, and in so doing draws heat from the skin and cools it. Mammals vary in the distribution and abundance of their sweat glands: primates have sweat glands all over the body, in cats and dogs they are confined to the pads of the feet, and whales, sea cows, and golden moles have none. Species with few sweat glands lose heat by evaporation of saliva, either by panting or by licking exposed skin.

Strange Senses and Perfumes

NAVIGATION AND COMMUNICATION

Many mammals have senses quite different to our own. Consider the electromagnetic sensory powers of platypus and echidna, or the ability of elephants to detect low-frequency calls. An obvious example is the tremendous olfactory prowess of most mammals. Less obvious is the barometric sense of the Eastern pipistrelle bat, *Pipistrellus subflavus*. These bats have to conserve every possible calorie of energy and so must avoid fruitless foraging trips. Some nights are almost useless for catching insects, and on these occasions the bats would be better to rest, their metabolisms idling, deep in the shelter of their caves. Yet in these dark recesses there is neither light nor wind nor temperature fluctuation, so how are they to judge the weather outside? It happens that insect activity is greatest on warm nights, and the warmer the night the lower the barometric pressure. Remarkably, bats can read the barometer – the lower the pressure, the more active they are in the roost (at the lowest pressure their metabolic rate increases up to fourfold as they prepare to sortie forth). No-one knows how they sense barometric pressure, but it may be related to the paratympanic (Vitali) organ (which, alone among the mammals, bats possess).

Mammals are unique among animals with backbones in the potency and social importance of their smells. This quality also stems from their skin, wherein both sebaceous and sweat glands become adapted to produce complicated odors with which mammals communicate. The sites of scent glands vary between species: they are aloft the snout in capybaras, on the lower leg in Mule deer, behind the eyes in elephants, and in the middle of the back in hyraxes. It is very common for scent glands to be concentrated in the anogenital region (urine and feces are also socially important odors); the perfume glands of civets lie in a pocket between the anus and genitals – for centuries their greasy secretions have formed the base of expensive perfumes. (Glands around the genitals of Musk deer serve a similar purpose.) Most carnivores have scent-secreting anal sacs, whose function is largely unknown, although in the case of the skunk it is clear enough. The evolution of scent glands has led to a multitude of scentmarking behaviors whereby mammals deploy odors in their environment. Scentmarks, unlike other signals, have the advantage of transmitting long after the sender has moved on. They are often assumed to function as territorial markers, exerting an aversive effect on trespassers, but evidence of this is patchy. Probably most scents convey a plethora of information on the sex, status, age, reproductive condition, and diet of the sender. Many mammals have several scent glands, each of which may send different messages, e.g. the cheek glands of the Dwarf mongoose communicate status, whereas their anal gland secretions communicate individual identity. In labs, almost identical mice have been bred, which differ by only one gene in their so-called Major Histocompatibility Complex (MHC) of genes (which is concerned with the immune system). These mice can nonetheless discriminate each other on the odor of their urine, showing that just one gene difference is sufficient to confer individuality on an odor.

Above *The complicated snout and enormous ears of Bate's slit-faced bat* (Nycteris arge) *are part of the remarkable apparatus of ultrasonic navigation that characterizes the microbats. They also belong to a family, Nycteridae, which is unique in that the tips of their tails are T-shaped.*

Right *Scents and scentmarking in mammals:* ***1*** *Spotted hyena marking with glands on the soles of its feet;* ***2*** *Bighorn sheep sniffing another to determine sex and status;* ***3*** *Tasmanian devil dragging the scent glands around its anus along the ground;* ***4*** *Rhinoceros marking the boundary of its territory with a pile of feces;* ***5*** *Fox using urine to mark out its territory – the scent left will signal the identity of the individual as well as the age of the mark;* ***6*** *Western spotted skunk in "handstand" posture – skunks repel predators with their notoriously powerful scent;* ***7*** *Ring-tailed lemurs rubbing secretions from their forearms onto their tails, which they waft at rivals in "stink" fights.*

Milk and Reproduction

Placental and Marsupial Pregnancy

Mammary glands are unique to mammals and characterize all members of the class. The glands, which are similar to sweat glands, should not be confused with the mammillae or teats, which are merely a way of delivering the milk. Only females' glands produce milk (with the startling exception of the male Dayak fruit bat). Numbers of teats vary from two in, say, primates and the Marsupial mole, to 19 in the Pale-bellied mouse opossum. Generally, a female has twice as many mammaries as the average litter size.

Courtship among mammals varies from force (elephant seals) to elaborate enticement (Uganda kob). Pairings may be ephemeral (Grizzly bears) or lifelong (Silver-backed jackals), and matings monogamous (elephant shrews) or polygynous (Red deer). In all cases fertilization involves intromission, which can last a few seconds (hyraxes) or several hours (rhinoceroses). Each of these variations correlates with the species' niche; for example, among cavy-like rodents duration of intromission is briefest in species which mate in the open, exposed to predators – the males of these species have elaborate penile adornments, perhaps to stimulate cervical contraction in the female.

The three monotreme species, sole survivors of the egg-laying subclass Prototheria, (like the metatherian marsupials, and some of the Insectivora) have a cloaca (a common external opening of urinary and reproductive tracts); and their testes remain in the abdomen (as they do in elephants, rabbits, and some Insectivora). As in birds, only the left ovary of the female platypus sheds eggs into the oviduct, where they are coated with albumen and a shell, and laid after 12–20 days. Echidnas incubate their eggs in a pouch, platypuses keep theirs in a nest where they are incubated for about three weeks. Meanwhile, the embryo is nourished from the yolk. On hatching, the young sucks milk that drains from mammary glands onto tufts of hair on the female abdomen. Monotremes lack nipples.

Among marsupials, eggs are shed by both ovaries into a double-horned (bicornuate) uterus. There the developing embryo spends 12–28 days, gaining nourishment from its yolk sac and "uterine milk" secreted by glands in the uterine walls. The early embryo (blastocyst) rests in a shallow depression in the uterine wall, its vascular yolk sac (chorion) in contact with the slightly eroded wall of the mother's uterus. This point of contact allows limited diffusion between maternal and fetal blood and is called a chorio-vitelline placenta (the chorio-allantoic placenta in bandicoots is discussed below). At birth the marsupial infant is highly altricial (poorly developed), weighing only 0.8g (0.03oz) for the 20–32kg (44–70lb) female Eastern gray kangaroo. Nevertheless, its sense of smell and its fore limbs are disproportionately developed and enable it to work through the thicket of fur on its mother's belly to reach the teats, often in a pouch. The infant attaches to a single teat, which swells so as to plug into the baby's mouth. Marsupials detach from the nipple at about the same weight as that at which a placental mammal of comparable size is born.

Compared to this neat system, the prolonged pregnancies of placental mammals may seem ungainly. Placentals have evolved a chorio-allantoic placenta. This organ facilitates nutritional, respiratory, and excretory exchange between the circulatory system of mother and infants. The mother's enhanced ability to sustain infants in the uterus permits prolonged gestation periods and the birth of more developed (precocial) young. The placenta permits a remarkable liaison between mother and unborn infant. The blastocyst first adheres to the uterus and then, assisted by protein-dissolving enzymes secreted by its outer membrane, sinks into the maternal tissue, reaching an inner layer called the endometrium. The outer membrane of the embryo, the chorion, is equivalent to the one that lines the shell of reptile and bird eggs. Protuberances (villi) grow out from the chorion into the soup of degenerating maternal tissue known as the embryotroph. The villi absorb this nutritious broth. Blood vessels proliferate in the uterus at the site of implantation and the chorionic villi vastly increase the absorbtive surface – a human placenta grows 48km (30mi) of villi. The marsupial bandicoot's placenta lacks villi, and is thus inefficient compared to the placental mammal's version, but nonetheless provides another stunning instance of parallel evolution. Mammalian orders differ in the extent to which the maternal and embryonic membranes of the placenta degenerate to allow mixing of parent and offspring fluids. Among pigs,

Diligent Fathers

In the family life of most mammal species, fathers contribute little directly to the care of their young, but striking exceptions are the South American titi monkeys (genus *Callicebus*), which live in small, monogamous family units usually comprising an adult male, an adult female, and one or two of their immature offspring. These monkeys are notable not just for the remarkable tail-twinning posture **1** adopted by family members as they huddle together in their "sleeping trees" at rest, but also for the prominent role played by the father, who is the infant's primary caretaker and source of emotional security. He will, for example, shift his position to cover and protect it in a heavy rain storm. Whenever there is danger, as from strong winds or a falling tree, the infant moves closer to its father, who is mostly responsible for carrying it **2** until it is old enough to keep up with the adults on its own, at the age of 4–5 months. The father also grooms the infant **3**.

At dawn titi monkeys engage in long bouts of calling, during which members of a pair sit side-by-side and perform a duet that may have taken them a year to perfect together. Puzzlingly, Dusky titi groups move to their highly traditional territorial boundaries and confront each other in vigorous vocal battles, whereas Yellow-handed titis use similar calls from well within the more ephemeral boundaries of their territories. WAM/WGK

lemurs, horses, and whales the chorionic villi simply plug into the maternal endometrium. This is a huge advance on the marsupial system, but nevertheless is 250 times less efficient at salt transfer from mother to fetus than the placentae of most rodents, rabbits, elephant shrews, New World monkeys, and bats. In their cases the maternal and embryonic tissues are so eroded that the fetal blood vessels are bathed in the mother's blood. The great significance of the placenta is that without it the mother's body would reject the baby like any other foreign body. This tolerance of the embryo allows the placental mammals to have longer pregnancies and hence to bear precocial young, although not all do so. The placenta facilitates feeding the embryo during gestation, and milk nourishes it after birth. Yet both have an additional function, namely to transfer the mother's antibodies to her offspring, thus enhancing its immunity to disease. The "afterbirth" of placental mammals is the fetal part of the placenta.

Species differ in the duration of both gestation and lactation, and in their combined length. Gestation length is ultimately constrained by the size of skull that will fit through the mother's pelvis, but where agility, speed or long travels put a premium on the mother's athleticism, then pregnancy will be short compared with the period of lactation, and birth weight of the litter relatively small.

Parental Care and Milk

LACTATION AND GESTATION

Mammals are not the only vertebrates to bear live young (viviparity), but they are unique in that the availability of milk buffers their infants from the demands of foraging for themselves while they are still small and undeveloped copies of their parents. To a large extent a young mammal prospers initially on the strength of its parents' competitive ability, as reflected in the supply of its mother's milk, until reaching an age and size when it can compete more or less on adult terms. In the tree shrew, parental care is entirely nutritional, the mother visiting her infants once every two days solely to suckle them for a few minutes. However, especially where food is elusive, additional parental care eases the transition to adulthood; indeed, since the female can store fat (and scarce minerals) in anticipation of nursing and then convert it to milk, she is free to spend more time with her offspring if necessary. Carnivora carry prey back to their offspring and may (e.g. wolves, African wild dogs) regurgitate for them. Koalas feed on toxic eucalyptus leaves and produce special feces of partially digested and detoxified material on which the weanling feeds, whereas the Two-toed sloth overcomes a comparable problem by nursing for up to two years. Lactation not only prolongs infant dependence and accelerates growth, it detaches the infant mammal from the environment: short-term food shortages are ironed out as the mother continues to lactate, if necessary mobilizing her own tissue, minerals, and trace elements to provide abundant, digestible, and nutritious food for her young. For the young, suckling is hardly arduous, so it can devote more energy to growth than it could if hunting, doubtless inefficiently, for itself. Last but not least, parental care prolongs the young mammal's apprenticeship in complex adult skills.

The evolution of lactation has facilitated a marked increase in the sophistication of mammalian teeth. Once formed, mammalian teeth, encased in a dead shell of enamel, cannot grow in girth (some continue to grow outward). Lactation postpones the time when the teeth must erupt and this may have been a precondition for the evolution of the complex occlusion (fitting together) of cusps of teeth in upper and lower jaws (diphyodonty) that is characteristic of mammalian teeth and necessary for chewing. In a growing jaw such teeth would be thrown out of alignment. The importance of lactation is that it postpones the need for teeth until much of the jaw's growth is complete. As part of this process, mammalian jaws grow quickly; after birth, the growth of a mammal's head suddenly spurts relative to the rest of the body, giving infants their typically big-headed appearance. Furthermore, the growth of jaws and teeth is very resistant to variation, proceeding almost unabated whether the infant is starving or overfed. There are some interesting variations - it takes over 30 years for an elephant jaw to reach full size, but nevertheless the upper and lower teeth are perfectly aligned throughout because their premolars and molars (in both milk and permanent teeth) erupt sequentially (as do kangaroos'), one at a time from the rear, a bigger tooth emerging and

EVOLUTION, SOCIETY, AND SEXUAL DIMORPHISM

Thomas Malthus' 1798 *Essay on the Principle of Population* sowed a seed that germinated in Charles Darwin's mind as the theory of natural selection, in his *Origin of Species* (1859). Malthus observed that, although a breeding pair usually produce more than two offspring, many populations do not grow as fast as this would imply, if at all. Darwin was impressed by the subtlety of species' adaptations and saw that individuals differed in the perfection of their adaptations to prevailing conditions, or "fitness." The variation between individuals arose from the mixing of genetic material involved in sexual reproduction, and from mutation, although the link between these mechanisms and Darwin's theory was not realized until 1900, when Mendel's work was rediscovered.

Since populations do not necessarily grow, many of the young must die, and the variation among individuals facilitates selective death, allowing better adapted individuals to thrive. Traits that confer an adaptive advantage will thus spread, if they are heritable, since those that bear them will become an ever larger part of the breeding population. Natural selection fashions individuals of succeeding generations to be ever better adapted to their circumstances. The characteristics of a species represent the sum of the actions of natural selection on similar individuals.

It is wrong to say that animals behave "for the good of the species" – more accurately, individuals are adapted to maximize their own fitness, which is often equivalent to maximizing the number of their offspring that survive to breed. In fact, selection acts on the genetic material that underlies each individual's traits, and so individuals actually behave in ways that promote the survival of the genes for which they are temporary vehicles – hence biologist Richard Dawkins' now famous term, the "selfish gene." Sometimes an individual helps its relatives, behaving in a way that seems detrimental to its own interests but is on balance beneficial to its genes, and so improves its overall (or "inclusive") fitness.

Individual mammals behave so as to maximize their reproductive success and since the pattern of reproduction is the core of society, adaptations to this end are reflected in the huge variety of mammalian social systems. There is an asymmetry between males and females in this respect: sperm are cheaper to produce than ova, and only female mammals bear the costs of pregnancy and lactation. Thus males may more readily maximize their reproductive success by mating with many females. Females, in contrast, can mother only a relatively small number of young and so maximize their reproductive success by investing heavily in the quality of each and, in particular, securing the best (evolutionarily "fittest") father. Infanticide and helping are striking examples of the lengths to which individuals will go to spread their genes at the expense of their rivals'.

Females are a resource for which male mammals compete. The stringent natural selection that operates between competing males is called sexual selection. It explains why many mammals are polygynous (one male mates with several females) and why males are often bigger than females. A big male defeats more rivals, secures matings with more females and thus sires more offspring; if his size and prowess are passed to his sons they will in turn become successful, dominant males. So, females adapted to behave in a way that enables their sons to prosper will select only the most successful males as mates (but may still try to defray the infanticidal tendencies of other males my mating with them promiscuously). The situation is different if the species' niche is such that a male's reproductive success is affected by the quality of his parental care rather than simply by the quality of his sperm; for example, among canids the survival of young depends on their father providing them with prey, and the male would find it impossible to provide for more than one or perhaps two litters. In this case natural selection favors monogamy and sexual dimorphism is less pronounced. But it is less obvious why sexual dimorphism is especially prevalent among larger species. One possible answer is that energy demands are relatively less on larger species and so they can afford to invest more heavily in muscle and armaments. Of course, this is just one line of logic, linked to particular sets of selective pressures, which explains general trends, but there is variation in all these things: equids are not dimorphic but one male does monopolize a harem.

Infanticide – A Major Factor in Mammalian Sociology

Among the subtleties of mammalian sociology is the importance of infanticide – the killing of young by conspecifics. Infanticide has been recorded in over 100 species from at least five orders and 18 families. Nor is this habit confined to one sex – infanticide by females is known for 25 species from three orders: lagomorphs, carnivores, and rodents. However, the potential motives may differ between the sexes – generally, males may be seeking to eat the babies or increase their mating opportunities, whereas females may be after foraging and nest sites.

Possible benefits of infanticide include: a) securing food (male and female chimpanzees eat their victims); b) eliminating competitors – including those of their future offspring (among common marmosets subordinate non-reproductive female helpers increase the weaning success of infants born to dominants; infanticide by the dominant provides her with a bereaved helper and removes rivals to her own offspring); c) stealing resources (infanticidal female rabbits and Belding's ground squirrels seldom eat their victims but do occupy their burrows); d) sexual selection – infanticide is adaptive for the infanticidal male that destroys a rival's offspring and causes the bereaved mother to stop lactating and come into estrus sooner (e.g. chimpanzees, lions, grizzly bears, and lemmings). Perpetrators reduce the risk of mistakenly killing their own offspring by being able to recognize their odor, or by sparing the offspring of females with which they have mated, or females in places where they have mated (in White-footed mice infanticide is generally by newly-arrived males, since resident males are inhibited from killing pups for the 35–40 days after mating, exactly the time needed for their progeny to mature and disperse); e) Parental manipulation, where parents regulate the sex ratio of a litter; and f) minimizing the risk of accidentally adopting non-kin and so wasting effort (perhaps applicable to the pinnipeds which commonly kill pups separated from their mothers).

Many aspects of mammalian behavior can be interpreted as counter-strategies against infanticide. For example, male deer mice defend territories around females at high (but not low) densities. Female counter strategies to male infanticide include mating with dominants, which are better defenders and would be a greater threat to the offspring had they not sired them. In Hanuman langurs, lions, and meerkats females at risk leave the troop to give birth and shelter their young. The infanticidal race brings costs: female prairie dogs fall victim to infanticide while they are out killing somebody else's pups.

Infanticide may have far-reaching consequences for mammalian society. In species in which males commit infanticide, the most obvious correlate is that females are promiscuous. Thus infanticide may be one of the underlying determinants of some mammalian mating systems. The risk of female infanticide may cause females to become aggressive and to space themselves out territorially, which may in turn may favour delayed emigration of their offspring if the habitat is saturated with territory holders, thus leading to kin-group formation and reproductive suppression and/or synchrony in female reproduction. This will influence a male's ability to monopolize females and thus the likelihood of the species having a monogamous, polygynous or promiscuous mating system.

Right Infanticide is a key element of Hanuman langur societies. This brutality is balanced by equally intriguing acts of cooperation and care. Non-human primates demonstrate the complexity of cooperation, coalition, and conflict that may have been the driving force behind the evolution of the human's unique mental capacities.

Below Infanticide is the main form of infant mortality in certain species (e.g. lions, prairie dogs, and probably many primates). This scheme, devised by biologist Jerry Wolff, illustrates the entanglement of factors that shape mammalian societies, and how many can be affected by the risk of infanticide. For instance, frail young and the phenomenon of bereaved mothers coming into estrus may facilitate male infanticide and have a cascade of effects that fashion male and female behavior and their societies.

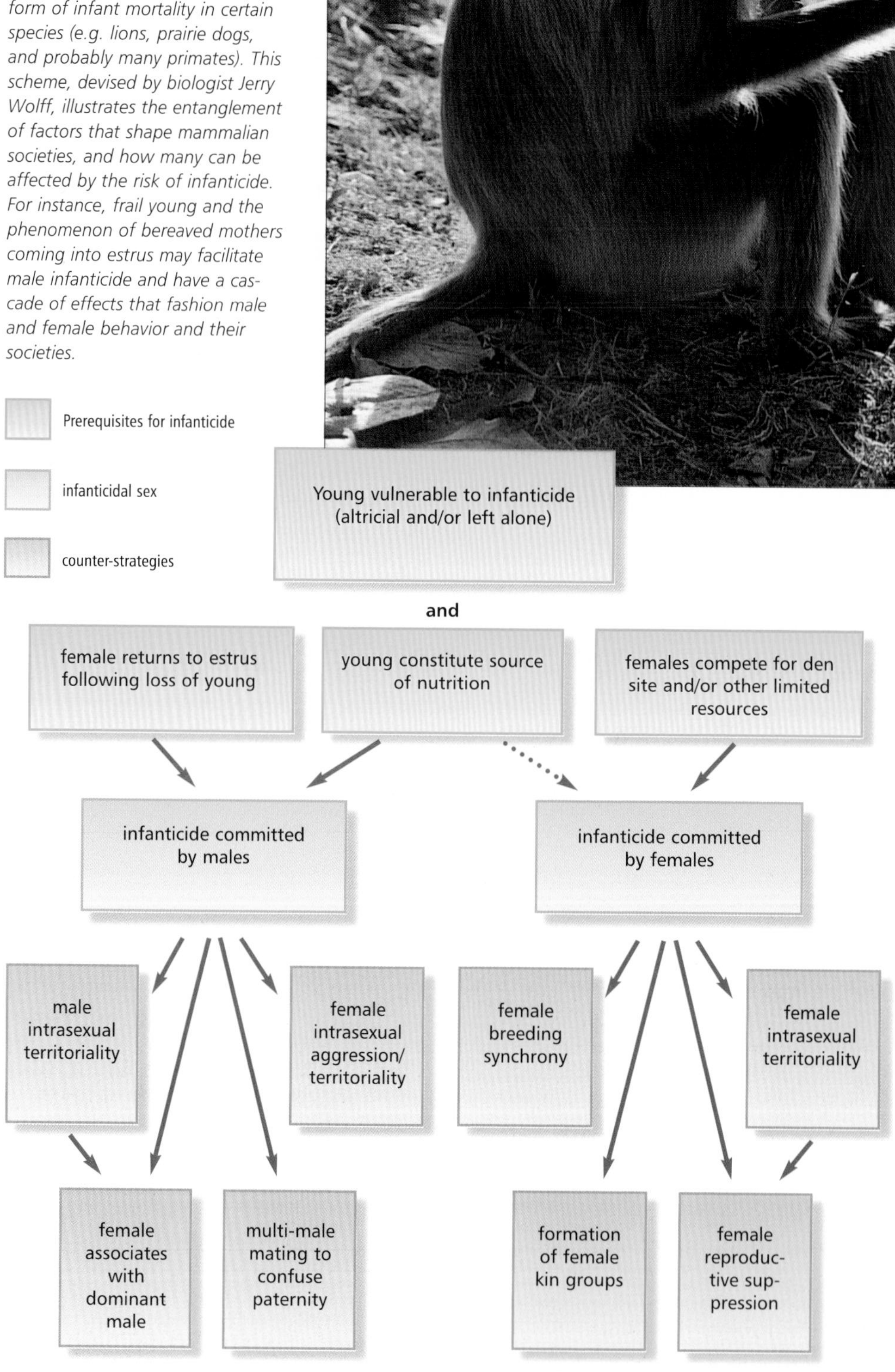

migrating forward along the jaw as the animal grows and as the previous one wears out.

Milk contains water, proteins, fats, and carbohydrates, but in proportions that vary widely between species. Mammals with high-protein milk grow fastest, but the diets of many species preclude their producing protein-rich milk. Pinnipeds have very fat-rich milk: that of California sea lions is 53 percent fat, perhaps because of the need for rapid weight gain prior to immersion in cold seas. Elephant seals born at 46kg (100lb) quadruple their weight in three weeks. Small mammals also grow very fast; Least shrews double in weight by the time they are four days old. The composition of milk may change during lactation: among kangaroos the early milk is almost fat-free, but later it contains 20 percent fat; when the mother nurses two babies of different sizes each teat delivers milk with a fat content appropriate to the stage of development of the infant sucking it.

The timing of breeding in many mammals living in seasonal environments is critical, and often triggered by the effect of daylength on the pineal body of the brain. The costs of pregnancy and, even more, lactation are high and the weaned youngsters will place an additional burden on the food resources within their parent's range. Consequently, in seasonal environments, many species give birth at periods when food is most abundant. This can lead to extreme synchrony in mating time – in the marsupial Brown antechinus all births occur within the same 7–10 days each year! The onset of heat (estrus) in some rodents is triggered by the appearance in their diet of chemicals contained in sprouting spring vegetation.

A difficulty may arise when other factors intervene to make it disadvantageous to mate one gestation period in advance of the optimal birth season. For example, Eurasian badgers give birth in February, but their gestation period of eight weeks would seem to necessitate them mating at a time when they are normally inactive, conserving energy while living on their winter fat reserves. Mammals have evolved some intricate adaptations to resolve this dilemma. In the case of the badger (and some other members of the weasel family, some pinnipeds, some bats, the Roe deer, the Nine-banded armadillo, and the Tammar wallaby) the adaptation is delayed implantation. This interrupts the normal progression of the fertilized egg down the oviduct to the uterus where it implants and develops: instead, the egg, at a stage of division called the blastocyst (where it consists of a hollow ball of cells), reaches the uterus where it floats in suspended animation, encased in a protective coat (zona pellucida) until the optimal time for its development. In the case of the Eurasian badger this means mating any time from February to September. The most protracted delay to implantation is in the fisher – a marten – whose total pregnancy lasts 11 months, the same as that of the Blue whale. The fisher's "true" gestation is two months. Cetaceans (whales and dolphins) have very short pregnancies relative to their body weight. The longest mammalian pregnancy is 22 months in the African elephant. The shortest on record is 12.5 days in the Short-nosed bandicoot.

Most kangaroos and wallabies exhibit nonseasonal embryonic diapause. A female conceives after giving birth (post-partum estrus) but so long as her current infant continues to suckle the new embryo does not implant in the uterine wall. The consequence is the ready availability of a replacement should one infant succumb, with the added advantage of a rapid succession of offspring to be squeezed into good breeding seasons. Embryonic diapause or delayed implantation provide yet another example of an adaptation that has arisen separately in metatherians and eutherians.

A different method of ensuring that birth is at a convenient season is sperm storage. This is used by the Noctule bat and other nontropical members of the families Rhinolophidae (horseshoe bats) and Vespertilionidae (vesper bats). All the males produce sperm in August (thereafter their testes regress). They continue to inseminate females, often while the latter hibernate throughout the winter. The sperm are stored for 10 weeks or more in the uterus, until ovulation in the spring.

Sons, Daughters, and Favoritism

INVESTMENT STRATEGIES

Some mammals treat their offspring differently depending on whether they are sons or daughters. In the two highly polygynous species of elephant seals, male pups are born heavier, grow faster, and are weaned later than their sisters. These differences arise partly because mother elephant seals allow their sons to suckle more than daughters. Similarly, male Red deer calves are born heavier than females, after longer gestation. Thereafter, males suckle more frequently and grow faster and evidently cost their mothers more, since hinds that bear sons are inclined either to breed later in the succeeding season than hinds that rear daughters, or not at all. In these species mothers seem to invest more heavily in sons than in daughters.

The opposite pattern prevails amongst dominant female Rhesus macaques, amongst which a mother that rears a son is more likely to breed the following year than one that has reared a daughter. The implication is that a daughter costs her more, depleting her resources further than does a son. Amongst macaques it seems that part of the extra burden of bearing daughters is, remarkably, that females pregnant with female fetuses are more frequently threatened or attacked by other females than are those bearing male fetuses.

What underlies this favoritism? The answer lies in the limited time, effort, and resources that parents have at their disposal for investment in offspring. Natural selection will favor parents that invest more heavily in one sex of offspring if that investment is later repaid by the production of a larger crop of grandchildren. It is easy to see how just such a process has operated among elephant seals, Red deer, and probably many polygynous mammals. In these species almost all females breed, but only a minority of very dominant males sire the great majority of the young.

Depending on status, males vary hugely in evolutionary fitness from indefatigable studs to reproductive flops. Attaining dominant status depends on a male's size and strength, and these attributes can be greatly influenced by early nourishment. A mother Northern elephant seal that lavishes nourishment on her son is weighting the odds in his favor for the future day when he joins battle to win a harem. If he is victorious his brief but orgiastic reproductive career may secure for his mother up to 50 grandchildren each year for as many as 5 years – an ample return for that extra milk. Since in harem-living species all females breed, largely irrespective of strength, comparable extra investment in daughters' muscle-power would be wasted. Put another way, producing a feeble son that fails to breed is worse than useless, for the parental investment of time and energy is wasted on an evolutionary dead-end. In short, if parents' investment influences their offspring's' future reproductive success, then natural selection will favor those parents that invest more in offspring of the sex in which that contribution has the greatest benefit.

The same principle underlies the opposite result among dominant female macaques. These live in matrilinear groups whose members are linked by a female line of descent. Young males disperse, while mothers form coalitions with their adult daughters who thereby inherit their mother's social rank. The breeding success of a female macaque improves with the strength of the other females in her coalition. An attempt to promote this strength may explain why dominant females allocate extra investment to daughters. The attacks on females bearing female embryos (and later also upon female infants) may arise because mothers in rival coalitions react to these infants as potential competitors of their own daughters. How they perceive the sex of the unborn embryo is an unknown but fascinating twist to the story.

Not only do some species invest more in offspring of one sex than the other, some actually bear more of one sex. Among African wild dogs the sex ratio at birth is biased toward males (59 to 41 percent). This may have evolved because several males are required to rear the offspring of one female, so that parents producing a male-biased litter will thereby secure more grandchildren. In effect, a litter of African wild dogs requires the paternal investment of several "fathers" to survive; thus to gain equal returns (i.e. future descendants) from their investments in sons and daughters, parents of this species may require more sons. Female coypu with ample fat reserves, and hence the opportunity of investing heavily in a litter, selectively abort small, mainly female litters. Later they produce larger litters, with the result that females in peak condition produce more sons. This is beneficial, since sons of females in good condition grow to be stronger and so, in a polygynous society, have a competitive advantage over the less robust sons of less healthy mothers.

Size and the Energy Crisis

ALLOMETRY

To survive, each animal must balance its income of energy with its expenditure. The particular problem for mammals is that their endothermy remorselessly imposes high expenditure. A mammal's body temperature is unlikely to be exactly that of its surroundings, so even when totally inactive the mammalian system must work to maintain its constant temperature and to avoid heat flowing out of or into its body: when at rest, 80–90 percent of the energy "burned" by endotherms is used solely to maintain constant temperature (homeothermy). As summer turns to winter, a mammalian body requires more energy, as it loses heat more quickly to the environment. The heat from the mammal's core is lost through its skin, and as a small mammal grows larger its volume increases faster than its surface area (the surface area of a body increases with the square of its length, whereas its volume increases with its cube). When inactive, the energy costs (and hence requirements) per unit weight of a horse are one-tenth of those of a mouse. This phenomenon, of bodily dimensions varying together but at different rates, is called allometry. A crucial upshot of this increase in surface area (and thus heat loss) relative to weight (and volume) in smaller bodies is that energy consumption rises so steeply with diminishing body size that one of the smallest terrestrial mammals, the Pygmy white-toothed shrew (2–3.5g/0.07–0.12oz), has to eat almost non-stop. The rate at which the body's chemical processes occur and at which it requires energy is called the

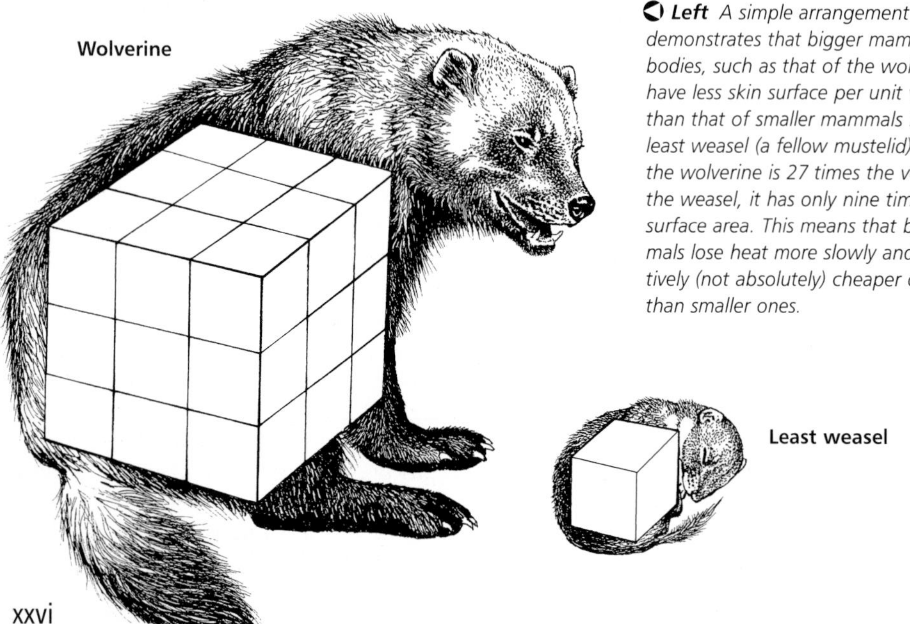

Left *A simple arrangement of cubes demonstrates that bigger mammalian bodies, such as that of the wolverine, have less skin surface per unit volume than that of smaller mammals like the least weasel (a fellow mustelid). Though the wolverine is 27 times the volume of the weasel, it has only nine times the surface area. This means that big mammals lose heat more slowly and are relatively (not absolutely) cheaper on fuel than smaller ones.*

Above *Because the thermal conductivity of water is greater than that of air, marine mammals face a special problem in conserving energy. They all need abundant insulation, a feature amply demonstrated by these hauled-out Atlantic walruses.*

metabolic rate, so fuel-hungry small mammals are said to have a high, or fast, metabolic rate.

Larger mammals score over smaller ones in conserving energy. On the other hand they are at a disadvantage when dissipating heat. Mechanisms for aiding heat loss include the elephants' ears and seals' flippers. The strictures of temperature are reflected in the wide geographical variation in the size of ears of North American hares: the Arctic hare's ears are slightly shorter than its skull, while those of the Antelope jack rabbit of Arizona are vast radiators, twice the length of its skull.

Body Size, the Cost of Living, and Diet

NUTRITION AND METABOLISM

The struggle to maintain body heat is especially acute for marine mammals. Many have evolved to gargantuan proportions to take advantage of a more favorable surface:volume ratio. Thus, the massive Blue whale has a tenfold more advantageous ratio than a small porpoise. This, combined with its far greater depth of insulating blubber, puts the Blue whale at a 100-fold thermal advantage in cold water. Because they lead an energy-expensive life generating adequate heat, smaller whales have even higher metabolic rates than would be predicted from their size. By contrast, fossorial (subterranean) rodents have a lower metabolic rate than to be expected for their size, because of the difficulty of dissipating heat in their humid, windless burrows where sweat cannot evaporate (perhaps relative freedom from predators also allows them to "tick over" more slowly.

If all else were equal, the energy-expensive metabolisms of smaller mammals would force them to eat relatively more than their larger cousins. However, all else is not equal, since foods differ in quantity and availability of energy. Animal tissues, fruits, nuts, and tubers are all rich in readily converted energy, in contrast to most vegetation, where each cell's nutrients are encased in tough cell walls of indigestible (without the aid of microbes) cellulose. The energy contained in a meal of meat is not only greater than that in a comparable weight of foliage, it is easier to digest. Thus, a carnivorous weasel is 26 times more efficient in extracting energy from its food than is its herbivorous prey, the vole. Smaller members of a mammal order tend to sustain their high energy demands by eating richer foods than do their large relatives. The 7kg (15lb) duiker selects buds and shoots

Above *The Gray squirrel has special enzymes that enable it to detoxify acorns. In its native North America this helps it flourish; in Britain, where it is introduced, it is helping it drive the native Red squirrel to extinction.*

whereas the 900kg (2,000lb) Giant eland can survive on coarse grasses; a bush baby eats fruit but the gorilla eats leaves, and the Bank vole eats seeds and roots while the capybara eats grasses. Among carnivores large size facilitates the capture of larger prey, which exempts them from the general rule that quality of diet declines with larger body size.

Diets differ in their availability. "High quality" foods are less abundant than "lower quality" ones. Overall, the abundance of food available to a species depends on which tier of the food chain it tries to exploit: since living things, like other machines, are imperfect, energy is wasted at each link in the chain and so less is available for creatures at the top. This is why the total weight (biomass) of predators is less than that of their prey and why that of herbivores is less than that of their food plants, which in turn are the primary converters of the sun's energy into edible form.

The general rule is that smaller species require more energy per unit weight than do larger ones, and so smaller species are pushed toward more nutritious diets and bigger species can tolerate less nutritious, but often more abundant food. So, for example, the 35g (1.2oz) Bank vole has a higher metabolic rate than the 1.4kg (3lb) Musk rat, even though they are close relatives (both are arvicoline rodents). Many of the species that defy this general rule do so in order to exploit a specialized diet: for example, ant- and termite-eaters, arboreal leaf-eaters, and flying insectivores tend to have slower-than-expected metabolisms. Such mammals are united in their thrifty use of energy by the fact that their diets all preclude the possibility of a consistent and abundant supply of fuel necessary to run a fast metabolism; flying insects are only seasonally available, many tree leaves are loaded with toxins and deficient in nutrients, and quantities of indigestible detritus inevitably adhere together with termites to the anteater's sticky tongue and so diminish the rewards of its foraging. The difficulty of securing and/or processing fuel destines mammals in these niches to an economical "tick-over" metabolism that is sparing in its use of fuel, rather like a slow-running engine.

Other mammals have highly tuned, "souped-up" engines – their metabolisms burning energy even faster than expected for their body sizes. Among these are the seals and sea lions, whales and dolphins, and river and sea otters, which must generate heat to survive in freezing waters.

Quality versus Quantity

K- VERSUS R-SELECTION

Those, generally larger, mammals with a lower metabolic rate cannot grow so fast, and so their embryos have relatively longer gestations and their infants have slower postnatal growth. Litter-weight at birth is a smaller fraction of maternal body weight in larger species. Infants of larger mammals thus require even longer postnatal care because they are born so small relative to adult size. The need for protracted parental care would be increased even further if the overall litter weight was divided into many smaller infants, rather than a few bigger ones, since the smaller infants would need even longer growing times. To minimize this problem larger mammals have smaller average litter sizes. These trends combine, so that mammals with lower metabolic rates have longer intervals between generations and a lower potential for population increase from one generation to the next.

Thus the rate of chemical reactions in the cells of a mammal species has repercussions throughout the species' life-history and even determines the pattern of their population dynamics. Mammal species with fast, expensive metabolisms have a greater capacity for rapid production of young; they are preadapted to population explosions. Viewed against the variety of mammalian sizes from 1.5g (0.05oz) to 150 tonnes, this interaction between size, metabolic rate and reproductive potential raises the intriguing possibility that while some species have evolved a particular size largely in order to overcome the mechanical problems of exploiting a particular niche, others may be a particular size largely due to selection for high reproductive potential of which their size is a secondary consequence. A giraffe has to be tall in order to exploit its treetop food and a Harvest mouse must be small to clamber nimbly aloft grass stalks, and their sizes shackle them respectively to the reproductive consequences of slow and fast metabolisms – by the time a giraffe bears its first, single offspring, a Harvest mouse born at the same time has already been dead four years, and potentially could have left behind more than 10,000 descendants.

On the other hand, mammals like voles and lemmings, with high reproductive potential, are at a great advantage in unstable environments. Those that can breed prolifically and at short notice can take advantage of an unexpected period of bountiful food, and the capacity to breed fast requires the rapid growth permitted by the high metabolic rates typical of small body size. Mammals dependent upon unpredictable resources are therefore generally small.

The key feature of unstable environments is that supply of resources may exceed demand, for example when the few survivors of a harsh period find their food supply is replenished. In these circumstances, survival is no longer so dependent on population density or direct competitive prowess, and an individual will increase its reproductive success by investing more heavily in a larger number of offspring, and by breeding prolifically at the earliest opportunity, while the going is good. Though the term is now less current among scientists, species adapted to these conditions are called r-selected (selected for rate of increase, r).

In a stable environment the situation is very different, because the population will be finely adjusted to the maximum that the environment can sustain, so competition for food and other resources will be intense. In these circumstances the pressure of natural selection increases in proportion to population density. Heightened competition in a saturated environment puts a premium on the competitive prowess of juveniles, and so parents must invest heavily in each offspring, preparing them for entry into the fray. Species adapted to these conditions have smaller litters, emphasizing the quality rather than quantity of infants born at a more advanced (precocial) stage of development, and infants are given more protracted parental care. Such species are said to be K-selected (selected to survive at carrying capacity, K). Clearly, the slower metabolism of larger mammals will push them towards K-selected life-histories. It also makes them less able to recover from persecution and this is why very many endangered mammals are large, slow-breeding species. All else being equal, K-selected mammals produce fewer young per lifetime than do r-selected ones, since not only do they have to invest more in each offspring, but they also have to invest heavily in their own competitive activities, such as territoriality, and in muscle power. The big decision in a mammal's life-history boils down to how to partition energy between reproduction and self-maintenance.

Right *A Polar bear mother nursing her cub. This species is a prime example of a K-selected mammal, whose reproductive strategy is to invest longer parental care in fewer offspring. Thus, Polar bear cubs remain with their mother for well over two years.*

MAMMALS IN CONTEXT

The lives of mammals must be interpreted in the context of the creatures around them. Not only are they preyed upon by, and prey on, non-mammals, but they also compete with them. Take, for example, the two most abundant predators of South Georgia, the Antarctic fur seal, *Arctocephalus gazella*, and the macaroni penguin, *Eudyptes chrysolophus*. Both breed in the austral summer and the major component in the diet of both is Antarctic krill, *Euphausia superba*. Scientists from the British Antarctic Survey have used satellite tracking to follow both species on forays from Bird Island, a tiny island on the northern tip of South Georgia. Most parents of both species travel north or northwest to feed, fur seals ranging to 150km/93mi (on trips of 3-5 days), penguins ranging to 50km/31mi on trips of 1–2 days when feeding chicks (and up to 400km/248mi over 15 days during incubation). Since exploitation of fur seals ceased 60 years ago, the population at South Georgia has grown from a few thousand to more than 3 million. Most of this increase has taken place since the 1970s. Over the same period the estimated population of macaroni penguins has halved to some two and a half million breeding pairs. Could these concurrent changes in numbers be related?

It is a fundamental ecological concept that two species cannot coexist stably when they are in direct competition for some limiting resource. In this case, in the summer, the seals and penguins take similar size-classes of krill from largely overlapping areas of ocean. To raise each pup to weaning, a female fur seal needs to catch over 900kg (1,984lb) of krill, whereas macaroni penguins need only to feed about 23kg (51lb) to fledge their chick. Nevertheless, energetic economics of larger size (female fur seals weigh 40kg/88lb, macaroni penguins 4kg/8.8lb), and the greater range of fur seals when both are rearing offspring, may mean that these seals outcompete the penguins within the latter's more restricted range. The penguins may increasingly be forced to work harder to find sufficient food, or switch to less profitable prey. KB

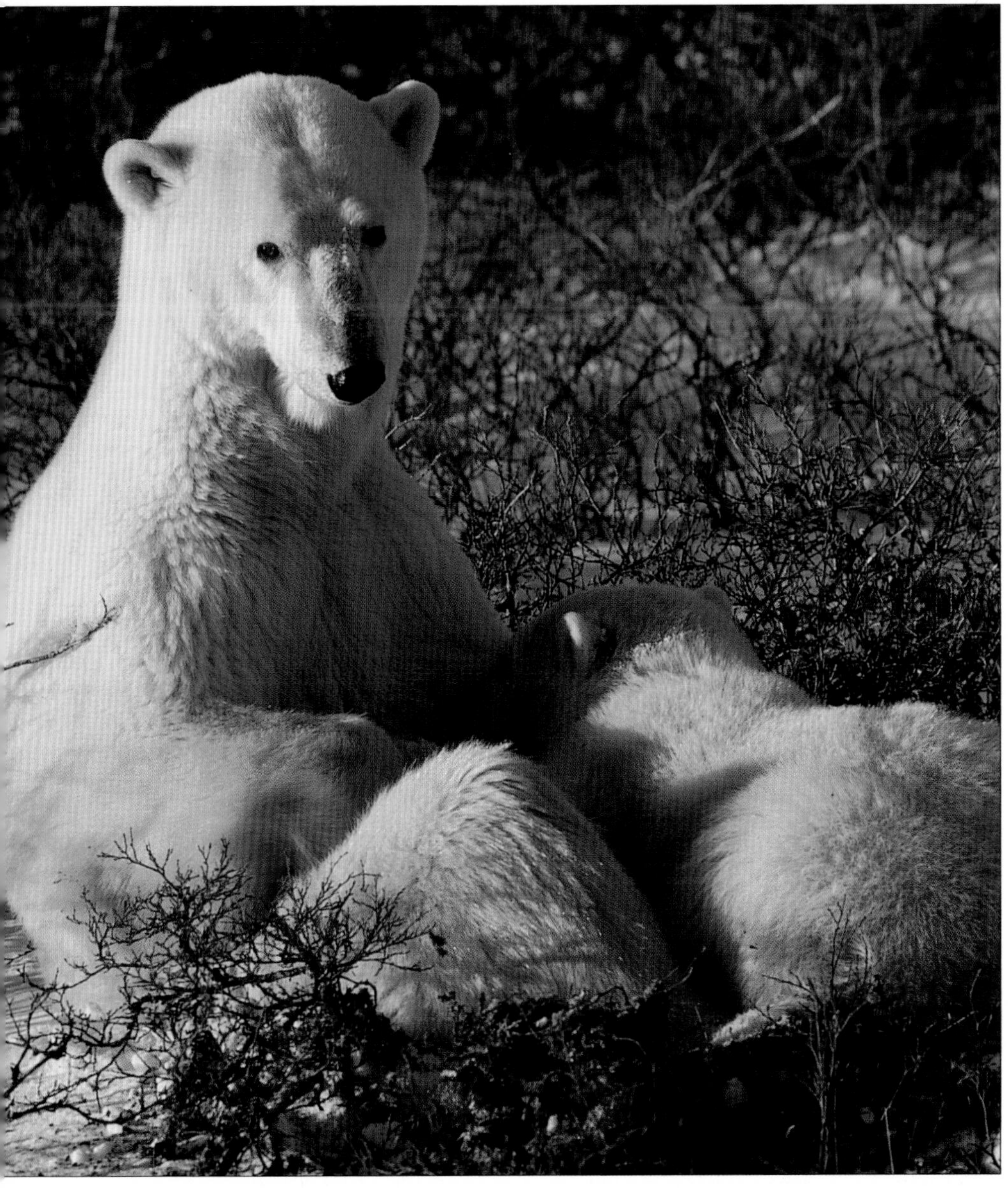

Spendthrifts and Population Explosions

RESPONSES TO UNSTABLE ENVIRONMENTS

The most dramatic illustration of this association between small size, rapid metabolism and great potential for population increase comes from mammals with unexpectedly high metabolic rates for their sizes. Why do arvicoline rodents, rabbits, and weasels spend extra energy on rapid metabolisms when comparably sized marsupials, anteaters and pocket mice maintain homeothermy without recourse to such fuel-hungry "engines"? The answer is that these energy spendthrifts have apparently shouldered the additional burden of meeting extravagant fuel requirements in order to increase their reproductive potential as an adaptive response to unstable environments. The high "cost of living" of r-selected species such as the lemming, compared to the similarly sized (but K-selected) elephant shrew, is thus a tolerable side-effect of a reproductive rate that enables a female to have 12 offspring by the time she is 42 days old. An elephant shrew at best would have two young in 100 days.

Considering mammals of similar size but different metabolic rate, those species whose populations tend to dramatic fluctuations and cycles (e.g. arvicoline rodents and lagomorphs) have more rapid metabolisms than species typified by stable populations (e.g. pocket mice and subterranean rodents). The Arctic hare has an unexpectedly slow metabolism compared to other lagomorphs and its populations do not exhibit the dramatic population cycles typical of the otherwise similar Snowshoe hare. Similarly, Brown lemmings show population cycles with peaks of population density that exceed the troughs 125-fold; the Collared lemming has a lower metabolic rate and shows a maximum of 38-fold variation in numbers (of course, there are added complications – the Brown lemming feeds on abundant mosses and grasses, whereas the Collared lemming feeds on less plentiful dicots and dwarf shrubs). The fluctuations in numbers of voles and lemmings result in huge variation in prey availability for weasels. The weasel's small body-size and even higher than expected metabolic rate enable it to breed twice a year (fast by carnivore standards), which may be an adaptation allowing them to respond as quickly as possible to such a sudden increase in prey numbers. This gives the weasel an advantage over one of its competitors, the stoat, which is broadly similar, but larger and can only breed once a year.

If small mammals can produce many more young, why are any mammals big? Competition drives mammals into countless niches on land, sea, and air and some of these can only be exploited by large species. Large size confers qualities that can be indispensable assets. Such advantages may include (depending on diet and other factors as well as size) the ability to survive on poorer food, to use longevity, efficient storage of metabolites, and memory to "iron out" environmental

Left *The revolution in molecular techniques that has shed new light on mammal evolution has illuminated debates such as the classification of the endangered Red panda (now established as a procyonid).*

fluctuations, to travel farther and faster and hence to exploit widely separated resources, to repel larger predators and to survive colder temperatures. Thus, within an awesome diversity of size, shape and behavior, mammals and their characteristics can be categorized according to a series of trends that shimmer elusively through the cloud of adaptation and counter-adaptation.

An Uncertain Future

Mammal Conservation

The reader may marvel at the stories of adaptation, the process of natural selection, and the beauty of the creatures found in this Encyclopedia, but there is great cause to be fearful for them. Some losses have been dramatic: in the last 40 years, black rhino populations have plummeted from an estimated 100,000 in 1960 to 2,500. According to the 2000 IUCN Red List of Threatened Species, at least 24 per cent of all mammals (or one in four) face a high risk of extinction, either directly or indirectly as a result of human activities. The IUCN lists some 1,130 mammal species in the top three threat categories (up from the 1,096 listed in the 1996 Red List), including 180 species classed as Critically Endangered, 340 as Endangered and 610 as Vulnerable. In addition, 74 species are regarded as Lower Risk: Conservation Dependent, and another 602 species as Lower Risk: Near Threatened.

Indonesia has the largest number of threatened mammals in the world (135 species), followed by India (80), Brazil (75), and China (72); no fewer than seven African countries feature in the top 20 positions. Comparing the number of threatened species against the total mammal diversity for each country, 19 of the top 25 countries (for which the number of threatened species is higher than expected) are island states, among them Australia. Top of this list is Madagascar, which has already lost over 90 percent of its original natural habitat and has more Critically Endangered and Endangered primates than anywhere else. Indeed, the lemurs, which are endemic to Madagascar, are the most threatened family of mammals. Recently, conservationists have identified 25 "biodiversity hotspots," containing more than two-thirds of the world's most endangered mammals. A priority for conservation efforts is to focus on these areas.

The most universal threat to mammals is habitat loss and degradation, followed by invasive exotics, hunting, and trade. Appendix I of the Convention on International Trade in Endangered Species of Wild Fauna and Flora (CITES) lists 219 mammal species threatened with extinction which are or may be affected by trade (another 364 and 56 species are on Appendices II and III, respectively). In Australia, introduced mammals such as Red fox and European rabbit pose a major threat to native species.

According to the Committee on Recently Extinct Organisms (CREO), 83 species of mammals have become Extinct in the last 500 years while at least another four are considered Extinct in the Wild, the most famous example being Przewalski's horse, *Equus przewalskii*. Without the dedicated conservationists of the last century, the tiger, rhinoceroses, elephants, cheetah, and giant panda and many more would certainly already be on this list. While some recently extinct mammals, such as the quagga, *Equus quagga*, have become flagships for conservation efforts, others have simply been forgotten. Of the 83 species listed as Extinct by the IUCN, 16 are marsupials – the thylacine or Tasmanian wolf, *Thylacinus cynocephalus*, disappeared in the 1930s mainly as a result of human persecution.

The first mammalian extinction of the new "millennium" took place only six days after celebrations had ended, with the death of the last known Pyrenean ibex (*Capra pyrenaica pyrenaica*). This subspecies, formerly found in the Spanish Pyrenees, was widespread in the Middle Ages but by the 1990s had fallen to 10. The last animal died in captivity in Ordesa National Park in Spain, on 6 January 2000, when a tree fell on it. This followed the first documented extinction in the 20th century of a primate: Miss Waldron's red colobus monkey, *Procolobus badius waldroni*, which, despite over two decades of warnings, was wiped out by indiscriminate hunting in West Africa.

Though nothing can offset this alarming species loss, on the other side of the "balance sheet," new species of mice and bats are described every year. Sometimes species thought extinct are rediscovered, as in the case of the Bornean bay cat, *Catopuma badia*. This felid, endemic to the island of Borneo, was previously known only from six specimens, five of which were collected between 1855 and 1900. Then, in November 1992, an adult female was trapped on the Sarawak–Indonesia border.

Another recent rediscovery is that of the Indochinese warty pig, *Sus bucculentus*. This species, originally described from two skulls over a century ago, was declared Extinct in 1996. One year later its rediscovery was announced following the retrieval of a partial skull of a juvenile male from indigenous hunters in the Annamite mountains between Laos and Vietnam. Indeed, five species of large mammals have been newly discovered or rediscovered in this region. The Saola, *Pseudoryx nghetinhensis*, first became known to science in 1992 with the discovery of three horns in the Vu Quang Nature Reserve in Vietnam. This reserve also yielded the giant muntjac, *Megamuntiacus vuquangensis*, described in 1994. In fact, a number of new muntjac species have come to light since the early 1990s, including the Truongson muntjac (*Muntiacus truongsonensis*) and *M. putaoensis*. In 1999, a potential new species of camel was discovered in the sand dunes north of the Altun mountains in China's Xinjiang province. DNA tests suggest that the species is at least three per cent different from other camels although their distinctiveness from domesticated Bactrians will only be confirmed by cross-breeding trials. The camels can survive on water that is too salty for other camels. However, this small herd of two-humped camels is threatened by poachers who hunt by laying land mines at the brackish waterholes where they drink. DWM

KIN SELECTION AND RECIPROCAL ALTRUISM

Optimizing conditions for gene survival

ADAPTATION SEEMS "AIMED" AT SURVIVAL, BUT survival of what? Not of the individual, for individual survival is only a means to reproduction, and even personal reproduction is not the whole story. Meerkats and wolves look after younger sisters and brothers, nephews and nieces. Naked mole rats go further, like worker ants even forgoing their own reproduction. Is the goal then survival of the colony, the species, or even the ecosystem? No, these are incidental consequences. We have to think harder, go back to Darwinian first principles.

A sharper consequence of successful individual survival and reproduction is survival of the individual's genes. This is significant, because genes have the unusual potential of immortality. Not the DNA molecules themselves, but the coded information in their nucleotide sequences can survive through unlimited generations. Not all of them do – that is natural selection – but successful sequences will still be here ten million generations hence. The world becomes full of successful genes.

So, having reached the level of the genes, we are finally talking about survival, pure and simple. If adaptations are "aimed at" anything, it is gene survival. But how do genes survive? By building individual organisms as receptacles or vehicles which take action to preserve them. That usually means take action to stay alive and reproduce. But if it is statistically likely that identical genes will be present in certain classes of other individuals, such as sisters or nephews, an individual may work to preserve them too, even at the expense of its own survival and reproduction. "Hamilton's Rule" states that a gene for altruism towards a particular class of relatives will spread if the cost C to the altruist is exceeded by the benefit B to the recipient devalued (multiplied) by the "coefficient of relationship" r (the proportion of genes shared, identical by descent, between them).

Some Values of *r*, the Coefficient of Relationship:

Full sibling (also parent, offspring)	1/2
Half sibling (also grandchild, niece, nephew)	1/4
First cousin	1/8
First cousin once removed	1/16
Second cousin	1/32

When W. D. Hamilton first published the theory in 1964, he coined the term Inclusive Fitness, as a sympathetic gesture to biologists long comfortable with the notion that individuals work to maximize something: Darwinian fitness. Hamilton made the minimal adjustment necessary to accommodate fitness gained through other relatives such as brothers and nieces, as well as through offspring. Inclusive fitness has been informally defined as "That quantity which an individual will appear to maximize, when what is really being maximized is gene survival." Hamilton's own definition seems complicated, but the complications are strictly necessary if you want to focus on individual organisms as maximizing agents. If you are happy to talk, equivalently, about genes maximizing something, that something becomes simply gene survival, and the condition for an altruistic gene's survival is Hamilton's Rule, $rB>C$.

Hamilton's Verbal Definition of Inclusive Fitness: "…the animal's production of adult offspring… stripped of all components…due to the individual's social environment…and augmented by certain fractions of the quantities of the harm and benefit the individual himself causes to the fitnesses of his neighbours. The fractions in question are simply the coefficients of relationship… ." (Hamilton 1964)

Kin selection is not an alternative to "individual selection," nor an unparsimonious complication to be "resorted to" only when individual selection fails. Animals are still maximizing inclusive fitness even when they stick ruthlessly to personal reproduction and never lift a finger for anybody else. Hamilton's Rule is still governing their behavior, but their ecology is such that $rB<C$.

Recipients of altruistic acts and cooperation are not always related to the donors. A supplementary theory is that of Reciprocal Altruism. The cooperation is mutual, and is based on what games theorists call a non-zero sum game. In human games of this kind there is a "banker" and the opportunity exists for the "players" to cooperate with one another at the expense of the banker. Each player brings something to the partnership which the other needs. The ratel or Honey badger (*Mellivora capensis*) has the strength and immunity from stings to open bees' nests, but it cannot find them. A small flying bird, the Honey guide (*Indicator* spp.), can find bees' nests but cannot open them. The ratel eats honey but cannot digest wax. The Honey guide (uniquely among vertebrates) has a special enzyme which enables it to digest wax. These two "players," with complementary needs and skills, cooperate profitably at the expense of the 'banker' (the bees). Honey guides have a special call with which they lead a ratel (or, opportunistically, a man) to a bees' nest. The ratel breaks open the nest and eats the honey, then the bird goes in and eats the wax and larvae.

The theory of reciprocation requires that for each player the cost of giving to the partner should be on average less than the benefit gained from the partner. In the Honey guide/ratel game, the condition is achieved by the complementarity of the needs and skills of the players. In other games, such as the Vampire bat (q.v.) blood sharing scheme, the needs and skills of the players are on average the same, but they alternate on different occasions. Blood meals are hard to find but large when found. On a day when a bat is out of luck, it is saved from starvation by a gift of blood from a lucky, bloated colleague. The favor is repaid when the luck is reversed.

This raises the possibility of cheating. The second plank of reciprocation theory is that cheating must be reliably punished, otherwise natural selection will favor genetic tendencies to take when you need, but not give when your turn comes to pay back. This is modeled by the game of Prisoner's Dilemma (see box). Under the right conditions "Tit for Tat" (or "Reciprocal Altruist") is an evolutionarily stable strategy (ESS). As long as the frequency of reciprocal altruists in the population exceeds a critical number, a mutant cheat does not prosper.

In practice, reciprocal altruism is expected in species that live in stable groupings where repeated encounters with the same individuals are likely. As it happens, these same conditions favor kin-selected altruism, because stable groups are likely to be families. It is often difficult to decide whether cooperation evolves under the influence of one or the other selection pressure. It may often be a mixture of both.

RD

PAYOFF MATRIX	You cooperate	You do not cooperate
I cooperate	We both prosper at the banker's expense	I get the low sucker's payoff. You get the high cheat's payoff
I do not cooperate	I get the high cheat's payoff. You get the low sucker's payoff	Both get the low ("punishment") payoff for not cooperating

The Prisoner's Dilemma *Under the conditions given in the payoff matrix, the strategy Reciprocal Altruist (cooperate, but retaliate against cheats) is evolutionarily stable (ESS). So is the strategy Always Cheat. When two strategies are ESSs, whichever one chances to attain more than a critical frequency in the population will not be invaded by the other. The critical frequency will depend upon detailed circumstances. When the whole population is playing Always Cheat, a player wins and loses on different occasions. On average all score less than when playing Reciprocal Altruist. Here, no player wins, they draw prosperously, and the "banker" loses. It is to be expected that the Reciprocal Altruist ESS will be found in nature quite often.*

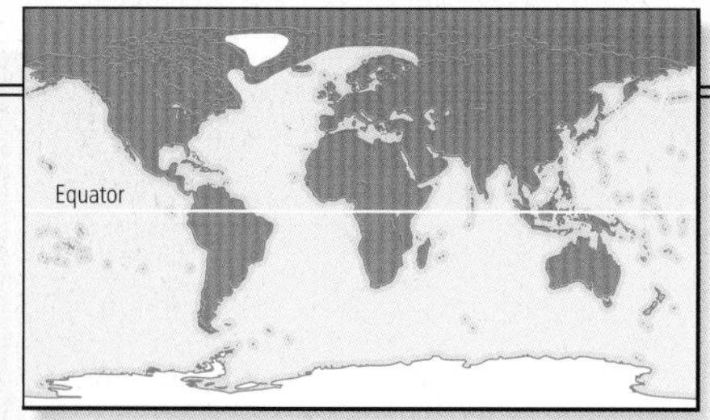

CARNIVORES

CARNIVORES ARE WONDERFULLY DIVERSE IN form, function, and habitat. The Polar bear is up to 25,000 times heavier than the Least weasel; the Giant panda ambles about foraging for bamboo shoots, while the cheetah dashes in pursuit of antelope at up to 96km/h (60mph); the aardwolf licks up termites and the kinkajou lives almost entirely on fruit; the aquatic Sea otter and the arboreal Palm civet rarely touch terra firma.

Little in the outward appearance of carnivores unites them; they may be long and thin, short and fat, powerful or delicate, solitary or sociable, predatory or preyed-upon. They are also numerous. The combined total of 231 terrestrial (fissiped) species would rise to 263 if, as might be evolutionarily more correct, the marine (pinniped) carnivores, which branched off from the fissipeds about 25 million years ago, were included, but here they are treated as a separate order.

So what distinguishes them from other mammals? Ultimately their common lineage rests on a single shared characteristic – they are (or, in the cases of aardwolves and some bears, their immediate ancestors were) equipped with four marvelously engineered carnassial (scissor) teeth. Many species in other orders, past and present, have been meat-eaters, but only members of the Carnivora stem from ancestors whose fourth upper premolar and first lower molar were adapted to shear through flesh. Only the more predatory of the modern species retain this pair of slicing teeth, collectively called carnassials. In species with more vegetarian inclinations, such as pandas, they have reverted to grinding surfaces.

In fact, diet is at the root of one major division within the Carnivora: that between large and small species. Many small carnivores eat invertebrates, but these prey, while easy to catch, are small. The rate at which insects become available, combined with the energy demands of the predators, puts a ceiling of about 20–25kg/44–55lb on the size of carnivore that can be supported in the insect-eating niche (Eurasian badgers and aardwolves being cases in point). Virtually all the larger meat-eaters prey on vertebrates, with one notable exception. The hefty but insectivorous Sloth bear bucks the trend, perhaps because its huge claws, powered by muscular limbs, enable it to rip open termite mounds, making the insects available in sufficient bulk to support a large predator.

Tactics of the Hunters

PREDATORY BEHAVIOR

Many living Carnivora are now adapted to either mixed (omnivorous) or even largely vegetarian diets, but meat-eating has been their specialty in the past. Although easier to digest than plant material, animal prey is harder to catch, and much of the fascination of carnivores lies in the stealth, efficiency, precision, and almost unfathomable complexity of their predatory behavior.

Left *Although they share a carnivorous past, today's carnivores are far from being exclusively meat-eaters. Some, like the Giant panda, have become largely vegetarian; and even voracious hunters like this European common weasel very occasionally eat berries.*

Prey are killed in various ways. Weasels and polecats (Mustelidae), together with civets (Viverridae) and the mongooses (Herpestidae) with which they are closely allied, are generally "occipital crunchers," biting into the back of the head to smash the back of their prey's braincase. Handling prey is dangerous to predators, and this highly stereotyped behavior keeps the victim's armory well out of harm's way. A weasel, for example, will throw itself on its side or back while delivering the killing bite, pushing away the claws and teeth of the struggling prey with powerful thrusts of its legs. When dealing with small prey, small members of the cat family (Felidae) strike at the neck, delving with their highly innervated canine teeth for the crevices that enable them to prize apart the cervical vertebrae. Members of the dog family (Canidae) generally grab for the nape of the neck as they tackle small prey, or pinion it to the ground with their forepaws. This grabbing action is followed by a violent, dislocating shake of the head. Canids tend to immobilize larger prey through shock via throat and nose holds and bites to exposed soft parts that often disembowel the victim.

Left *As a storm gathers in Kenya's Masai Mara National Reserve, a pride of lions settle down to take their ease. Many carnivore species live in groups – a trait that in top predators like lions is usually ascribed to the advantages to be gained from cooperation in hunting.*

The Carnivore Body Plan

ANATOMY AND PHYSIOLOGY

The skeletons of all carnivores, irrespective of whether they walk on the soles of their feet in plantigrade fashion, as bears do, or on their toes (digitigrade) like canids, share an evolutionarily ancient modification of the limbs – the fusion of bones in the foot (see overleaf). For members of the dog family, the development of this "scapholunar" bone might plausibly be interpreted as providing a firm strut for absorbing the shock of landing at the end of a limb adapted for running. However, the fused bones were present in older, now extinct, forest-dwelling carnivores, so perhaps the scapho-lunar bone originally provided a firm basis for flexion at the midcarpal joint in carnivores that needed both to climb well and to grapple with prey.

The advantages of a long stride when running may explain the relatively undeveloped collar bone (clavicle), which is free at both ends and lodged within shoulder muscles. The main function of a large clavicle, as in primates, is to stabilize the lower end of the shoulder blade and to provide attachment for the muscles controlling side-to-side movement of the limbs; neither function is either necessary or desirable for the fore-and-aft swing of the limbs of carnivores, which primarily run down prey.

The remainder of carnivore anatomy is just as varied as the species' diverse lifestyles would lead us to expect. The retractile claws so typical of cats are not common to all carnivores; they are otherwise found only among the Viverridae (some civets and genets). Canids, by contrast, have digging claws that they use, among other contexts, when caching food.

A few unusual reproductive traits are found in carnivores. The penis of members of all families

ORDER: CARNIVORA
9 families; 95 genera; 231 species

CATS Family Felidae p8
37 species in 4 genera
Includes **lion** (*Panthera leo*), **jaguar** (*P. onca*), **leopard** (*P. pardus*), **tiger** (*P. tigris*), **Snow leopard** (*P. uncia*), **Clouded leopard** (*Neofelis nebulosa*), **cheetah** (*Acinonyx jubatus*), **lynx** (*Felis lynx*), **ocelot** (*F. pardalis*), **Margay cat** (*F. wiedi*), **Wild cat** (*F. silvestris*).

DOGS Family Canidae p40
35 species in 10 genera
Includes **jackals** (*Canis adustus, C. aureus, C. mesomelas*), **coyote** (*C. latrans*), **dingo** (*C. dingo*), **wolf** (*C. lupus*), **Arctic fox** (*Alopex lagopus*), **Red fox** (*Vulpes vulpes*), **Swift fox** (*V. velox*), **dhole** (*Cuon alpinus*), **African wild dog** (*Lycaon pictus*).

BEARS Family Ursidae p70
8 species in 5 genera
Includes **Polar bear** (*Ursus maritimus*), **Grizzly** or **Brown bear** (*U. arctos*), **Andean bear** (*Tremarctos ornatus*). The **Giant panda** (*Ailuropoda melanoleuca*), although formerly sometimes separated as the family Ailuropodidae, is now widely accepted as an ursid.

RACCOONS Family Procyonidae p86
19 species in 7 genera
Includes **raccoon** (*Procyon lotor*), **coati** (*Nasua narica*). The **Red panda** (*Ailurus fulgens*) is sometimes treated as a subfamily of the procyonids, and sometimes separated as the monotypic family Ailuridae.

WEASELS Family Mustelidae p96
55 species in 24 genera
Includes **Least weasel** (*Mustela nivalis*), **Black-footed ferret** (*M. nigripes*), **ermine** or **stoat** (*M. erminea*), **Eurasian badger** (*Meles meles*), **Marine otter** (*Lutra felina*), **Sea otter** (*Enhydra lutris*).

SKUNKS Family Mephitidae p96
10 species in 3 genera
Includes **Striped skunk** (*Mephitis mephitis*), **Western spotted skunk** (*Spilogale gracilis*), **Amazonian skunk** (*Conepatus semistriatus*).

CIVETS Family Viverridae p122
35 species in 20 genera
Includes **Palm civet** (*Paradoxurus hermaphroditus*), **genets** (*Genetta* species).

MONGOOSES Family Herpestidae p122
35 species in 17 genera
Includes **Gray meerkat** (*Suricata suricatta*), **Indian mongoose** (*Herpestes auropunctatus*), **Dwarf mongoose** (*Helogale parvula*), **Banded mongoose** (*Mungos mungo*).

HYENAS Family Hyaenidae p140
4 species in 4 genera
Striped hyena (*Parahyaena hyaena*), **Brown hyena** (*H. brunnea*), **Spotted hyena** (*Crocuta crocuta*), **aardwolf** (*Proteles cristatus*).

except the hyenas contains an elongate bony structure known as the baculum or os penis. This penis bone functions to prolong copulation, which may be especially important in species where ovulation is induced by copulation. The shape of the baculum is characteristic for each species. The so-called copulatory tie, which "locks" male and female together during copulation, occurs only in canids and may function in sperm competition.

In the majority of species the cycle of development of the fertilized egg is typical of that found in mammals – the egg develops continuously from fertilization to birth. In some carnivores, principally members of the weasel family, a pause in development occurs – the so-called delayed implantation, which probably times mating and birth at convenient seasons irrespective of the length of gestation.

The senses of the carnivores are all acute. Perhaps most intriguing is their refined ability to use scent not only to find prey (and to escape predators) but as a method of communication. Apart from the signal value of urine and feces, which are deployed at strategic locations, most carnivores have several odorous skin glands. Doubtless these odors convey far more complex information than we can yet confirm. It is already known, for example, that one mongoose can recognize the identity and status of another by scent alone, and it is likely that most carnivores can recognize other individuals of their species from their scentmarks.

Urban Badgers and Imported Stoats

RANGE AND HABITAT

Wild carnivores have a worldwide distribution, with the Arctic foxes of Greenland and the feral cats of subantarctic Marion Island at the extremes of their latitudinal range. Each family is widespread, the dog family most of all. Some species have been introduced by man to areas where they are not native, generally with disastrous consequences: small Indian mongooses shipped to the Caribbean to control rats spread rabies instead; feral cats imported to remote islands for companionship or rat control annihilated flightless birds; stoats and Red foxes, introduced to New Zealand to control rabbits and to Australia to provide sport, actually ended up by posing a threat to the native faunas.

There is one recent and intriguing move that the carnivores have undertaken voluntarily, and that is into the urban environment. The rubbish tips of the Middle East have provided welcome tidbits for jackals since biblical times; skunks and raccoons forage in the suburbs of North American cities, and the Spotted hyenas that wander the

CARNIVORE BODY PLAN

Right The typical dental formula of carnivores is I3/3, C1/1, P4/4, M3/3 = 44: in other words, there are 44 teeth consisting of three incisors, one canine, four premolars, and three molars on each side of the mouth in both the upper and lower jaws. Of particular evolutionary significance are the carnassials – the last upper premolar and the first lower molar – which in carnivores are equipped with sharp tips and high cusps to shear through flesh (the skull of a typical canid is shown right); by contrast, in a primate such as man, they are flattened to provide a grinding surface. There are, however, considerable variations on the typical formula between different Carnivora families. The Gray wolf, for example, lacks one molar on the upper jaw from the standard carnivore complement.

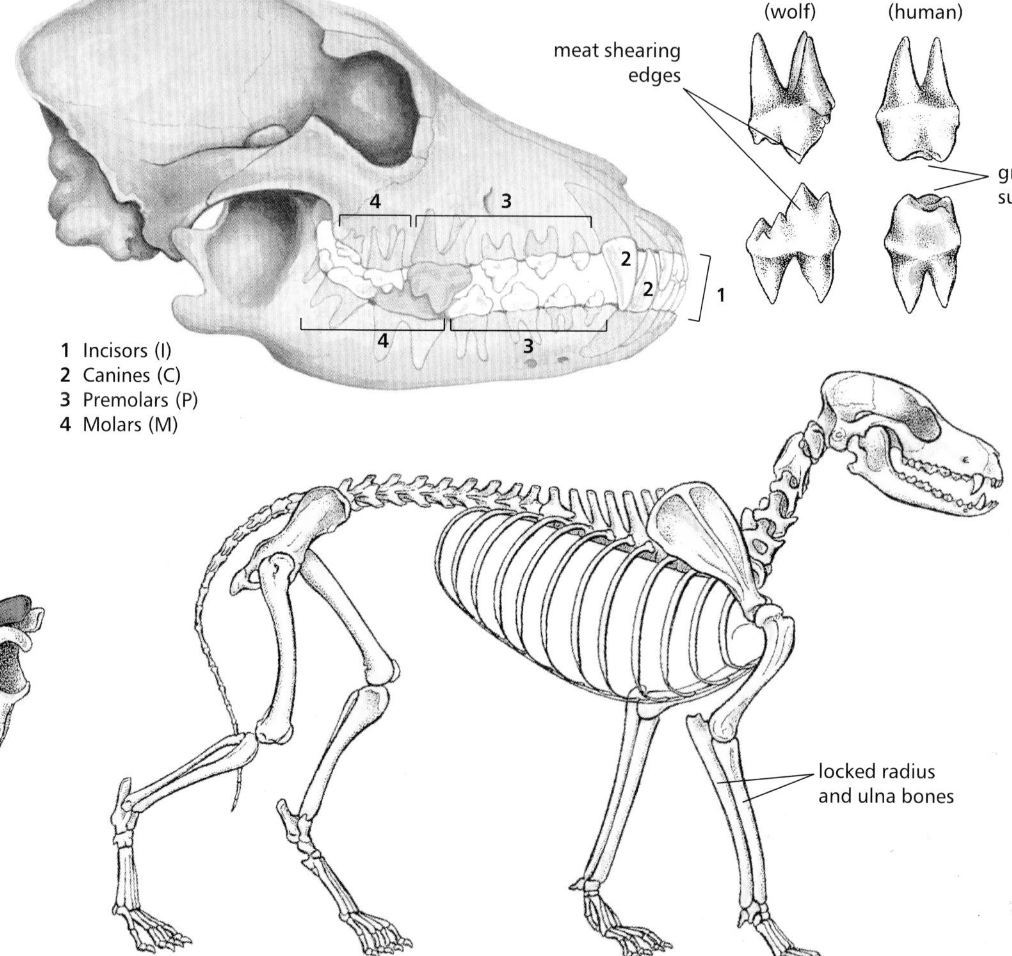

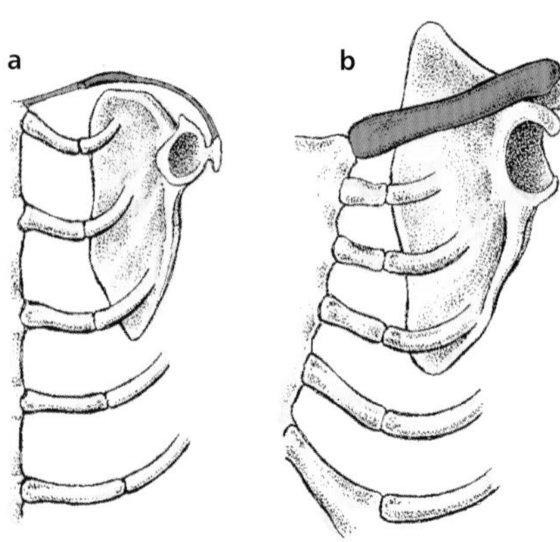

Above The collar bone is reduced in all carnivores in comparison with other mammals. Shown here is the collar bone (clavicle) of a wolf **a**, which appears reduced to a mere sliver (red) suspended on ligaments (blue) when set against the more substantial structure found in Homo sapiens **b**.

Above Most carnivores, like the Gray wolf shown here, have a powerful, agile body and a strong skeleton well adapted for running. In addition to the modified clavicle, wrist, and dentition shown elsewhere in the box, the ulna and fibula (usually the thinner of the two main bones that make up mammals' front and rear lower limbs respectively) are well-developed as an adaptation to the swift pursuit of prey that characterizes members of the dog family. In addition, the ulna is locked to the radius – the other component of the lower front limb – to prevent rotation.

streets of Harar in Ethiopia are widely reported. In recent decades carnivores have been knocking at the gates of the capitals of Europe; Eurasian badgers now occupy setts in London and Copenhagen, and Red foxes are seen by lamplight in the streets of Stockholm, Copenhagen, Paris, London, and many other towns besides. For hitherto unexplained reasons, urban foxes and badgers are most established in the United Kingdom. Red foxes are common throughout the city of Oxford, where the vixen that reared cubs in an automobile factory competes in notoriety with foxes seen in London's Trafalgar Square and Waterloo Rail Station! Oxford's foxes may supplement the rural fox's diet with scraps from bird tables and compost heaps, and with the smaller casualties of road traffic. Only rarely do they raid and overturn dustbins (the culprits are usually dogs).

Right *An American brown bear rears up on its hind legs to scare off an intruder. Grizzlies travel on all fours, walking on the soles of their feet in plantigrade fashion. But they will occasionally take advantage of their great height, to confront a threat or simply to survey the surrounding landscape.*

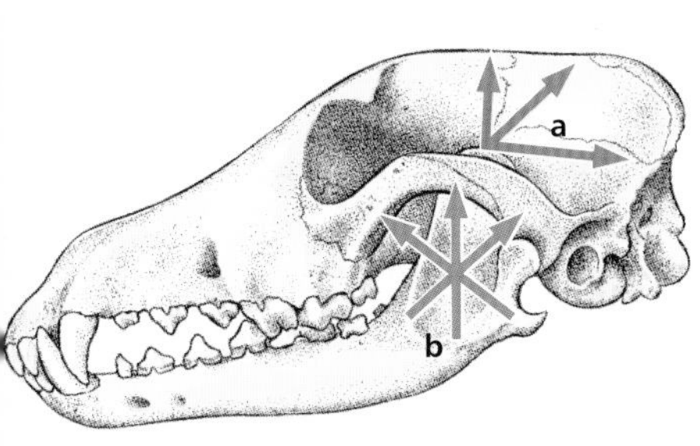

Above *Jaw power is crucial for the capture and tearing up of prey. Shown here are the lines of force exerted by the jaw-closing muscles of the dog. The massive temporalis **a** delivers the power to exert suffocating or bone-splitting pressure, even when the jaws are agape; the rearmost (posterior) fibers of the muscle are most effective when the jaws are open wide. The masseter muscle **b** provides the force needed to cut flesh, and for grinding when the jaws are almost closed.*

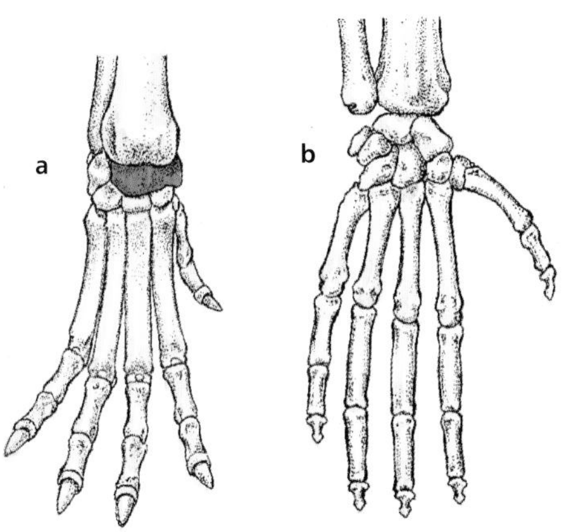

Above *Wrists in which the scaphoid, lunar, and centrale bones are fused to form the scapholunar bone are typical of carnivores **a**; in a primate like man **b**, these bones remain independent.*

Tracking the Past from Teeth

EVOLUTION

The early mammals are known largely from their teeth, since the smallness and fragility of their bodies, and also their forest habitats, have not favored the preservation of complete fossil remains. Consequently, we have only fragmentary knowledge of the origins of mammals 190 million years ago in the Tertiary era, and also of the ancestors of modern mammalian orders about 70 million years ago. Among the ancient carnivorous types a specialized pair of shearing carnassials evolved independently several times, for example in the now extinct order of Creodonts, in which they evolved in different parts of the tooth row from those of modern carnivores.

The most likely forerunners of all living Carnivora are members of the extinct superfamily Miacoidea. These poorly-known forest dwellers had spreading paws, probably indicating a tree-dwelling lifestyle, and carnassials derived from the fourth upper premolar and first lower molar, although the scaphoid and lunar bones were not yet fused. From the miacids the modern carnivore families developed during a fast radiation in the Eocene and Oligocene periods (54–26 million years ago). Doubtless this proliferation of predators mirrored a similar evolutionary explosion of potential prey, which in turn developed from the availability of more diverse vegetable food.

For the last 50 million years or so, the Carnivora have been split between two branches. One, descending from the Viverravines, includes all the catlike families; the other, the dog branch, stems from the Vulpavines. Today the cat branch is represented by four families: the Viverridae (civets), Felidae (cats), Hyaenidae (hyenas) and Herpestidae (mongooses). There is debate about how many families exist in the dog branch: the

Right *The basic division of the Carnivora into dog and cat branches took place more than 50 million years ago with the split between Vulpavines and Viverravines. The seals diverged from the dog branch much later; although they are carnivores in evolutionary terms, they are assigned to the separate order Pinnipedia in this encyclopedia.*

Canidae (dogs), the Ursidae (bears), and the Mustelidae (weasels) and closely-allied Procyonidae (raccoons) are generally accepted, but in addition some taxonomists separate the Ailuridae (Red panda) from the Procyonidae (although the Giant panda is now firmly established as a member of the Ursidae).

The Subtle Roots of Sociality

SOCIAL COOPERATION

Refined though the anatomical specializations of the carnivores may be, and however elegant the details of their behavior, the overwhelming feature of their biology is the subtlety of their societies. The collective strength, coordinated strategy, and awesome effectiveness with which the cooperative hunters overpower their prey has captured the human imagination. But there is much more than cooperative hunting, spectacular though it is, to the societies of carnivores that hunt together, and it is increasingly clear that some carnivores have quite different, yet equally complex, societies whose origins and maintenance have nothing to do with cooperative hunting.

Traditionally, two ideas were advanced to explain why some carnivores go around in groups. Some – such as wolves and lions – hunt together, it was argued, only in order to cooperate in the capture of large, dangerous quarry; whereas others, including some mongooses, were thought to travel together to enjoy greater vigilance for marauding predators, some of which they could collectively repel. Certainly both these are among the selective pressures that have fashioned some carnivore societies, but neither adequately explains why other species live in groups but travel and hunt alone.

There are several species whose shy, nocturnal habits and small prey previously misled people into thinking them strictly solitary or asocial – for example foxes, civets, Brown hyenas, farm cats, and Eurasian badgers. The use of radio-tracking and night-vision equipment have now revealed that in each of these species (and probably many more) several adults may share roughly the same home range, even if the cohabitants meet only infrequently when foraging, and sometimes even den separately. Such species may be said to live in "spatial groups" whose members' home ranges overlap more than would be expected by chance.

One species that shows how other pressures may favor the formation of spatial groups in the absence of concerted hunting or antipredator behavior is the Eurasian badger. In much of rural

EVOLUTION OF CARNIVORES

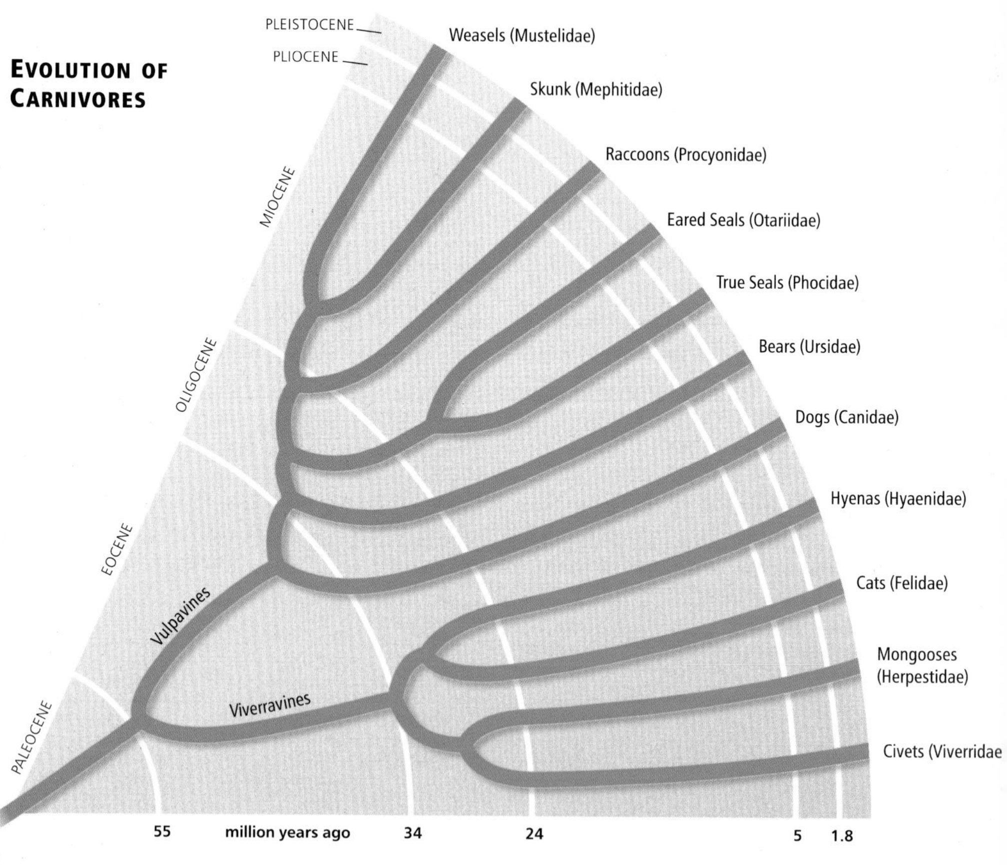

SHARED PARENTHOOD

In some carnivore species, the task of raising young does not fall exclusively on parents. The Black-backed jackal family group below is a case in point. It comprises not just the breeding female ***1*** and her pups (two shown suckling and one begging her to regurgitate food), but also a sibling "helper" ***2*** illustrated in submissive posture towards its father, the breeding male ***3***. "Helpers" bring food to lactating mothers, and guard the young while the parents go off foraging or hunting in pairs.

Although helping behavior is relatively widespread, specific arrangements differ markedly between species. In Banded mongooses and coatis, for example, maternal duties are shared among several breeding females. Lions and domestic cats (but not other felids) commonly live in closeknit groups of related females who share the nursing of the young. Within the dog family, however, nursing coalitions and joint denning of litters are the exception rather than the rule; reproduction is generally the prerogative of a dominant pair, to which subordinate animals defer, either indefinitely or until they accede to dominance or establish a new group. In such circumstances, at least some nonbreeders may help tend the breeders' offspring. In Red foxes, these are invariably female, but in other canids including jackals, males and females are equally likely to be helpers.

Among jackals at least, such behavior has been shown to improve the chances of pup survival, but at first sight it offers little benefit to the helpers themselves; their own potential for reproduction may be delayed for the time they stay with the breeding pair. Several explanations for the custom have been offered. The helpers may acquire practice at parenthood, or benefit subsequently from increased group size. Alternatively, since groups are often composed of kin, they may be investing in infants with which they share almost as many genes as they would with their own offspring. Such is certainly the case with Silverbacked jackals; since their parents are monogamous, helpers are as closely related to their full siblings as they would be to their own young. PDM/DWM

Britain the badger lives in group territories within which 2–10 animals (sometimes as many as 30) den together but forage alone, principally hunting for the earthworm *Lumbricus terrestris*. At night this species of worm only crawls from its burrow when the grass temperature is over 2°C (35.6°F), the air calm, and humidity high. The problem for the badgers is that the worms emerge in different places from one night to the next, depending on slight variations in weather conditions. So, to be sure of finding worms, a badger requires access to a territory large enough to accommodate such a variation of climate, and therefore of worm availability. However, in the night's "good patch," many more worms may be available than one badger can possibly consume, so at no personal cost it can tolerate the presence of others in its territory. From this idea – called the Resource Dispersion Hypothesis (RDH) – flowed the prediction that where patches of earthworms were scattered, badger territories would be bigger, and, independently, that where patches were richer in worms, social groups of badgers would be larger.

Many carnivores depending on patchily dispersed resources adhere to the RDH model: territory size is determined by the dispersion of resources such as feeding sites or water holes, but group size by the amount of a limiting resource – such as food – at each patch. A whole band of coatis in a Mexican forest can share the water holes around which their lives are configured; a mob of meerkats can forage together for beetles in the dung-rich shade of an acacia tree; a group of Spotted hyenas may gorge communally on a large carcass in the moments before it is usurped by scavenging lions. For all these, as for the badger, there may be little cost to tolerating some additional group members so long as food is plentiful – and, furthermore, there may also be positive advantages of group formation, such as alloparenthood or "helping' (see box).

In addition to cooperative hunting, vigilance, and infant care, other advantages of group living are also becoming apparent. Larger groups of coyotes and Golden jackals can better defend their prey from rival groups, as can Spotted hyenas from marauding lions; Dwarf mongooses and meerkats collaborate in the care of ailing group members; coalitions of lion and cheetah males roam together to give themselves a better chance of usurping resident males than they would have alone. The list of advantages is still growing, though always ecological circumstances such as food availability set the limits of feasibility.

Above *Scavenger supreme, a Spotted hyena supplements its diet from a garbage dump in Botswana. Although the spread of cities has had disastrous effects on many mammals, which have lost vital habitat to development, a few species have learned to take advantage of such unwanted by-products of the human presence as refuse and roadkills.*

Troubled Ties to the Human Race

CONSERVATION ISSUES

Man's relationship with carnivores is one of extremes – the Domestic dog and cat are to be found in their millions in all corners of the globe, while some wild species have had their numbers reduced to hundreds and others have been completely annihilated. The dog was one of the first animals to be domesticated, the origins of its close relationship with humans going back some 14,000 years (some molecular evidence suggests as much as 50,000) to a time when people mainly lived in hunter–gatherer societies. It is now generally agreed that the wolf is the ancestor of the Domestic dog, but it is still debated whether dogs were deliberately domesticated to serve as hunters, guards, or scavengers, as sources of food or for warmth at night, or as companions. The ancestry of domesticated cats dates back no more than 4,000 years, and there is no indication that they were domesticated for any practical purposes. Now, both dogs and cats threaten their wild ancestors by crossbreeding with them and by spreading disease.

In common with many groups of mammals, numerous Carnivora species are threatened by man either directly through persecution and exploitation or indirectly through destruction of their habitat. Even if in no immediate danger of extinction, almost all carnivores require conservation in the sense of thoughtful management, since their maligned reputations, as much as their predatory behavior, have turned rural people against them. Despite a generally open verdict on whether predator control is beneficial for either stock protection or disease (rabies) control, more than one Red fox is killed annually per square kilometer (two per square mile) over much of Europe. The onslaught by stockmen on the coyote of North America is notorious, while in the USSR, following an estimated annual loss of 1 million cattle to wolves in the 1920s, a precedent was set for killing up to 40–50,000 wolves annually. Today wolves still attract a bounty in Russia (a female plus pups fetches 200 rubles) at a time when biologists elsewhere are struggling to secure the survival of tiny relict populations – in Egypt of about 30 animals and in Norway of less than 10. Such paradoxes are everywhere: throughout much of Europe the otter is highly protected, but in areas of Eastern Europe where carp ponds produce traditional Christmas fare, otters, growing sleek and populous, are perceived as pests.

As their habitats dwindle and their populations become more fragile, the fate of many species is totally in human hands. Our society must decide whether these fascinating, often strikingly beautiful creatures are to survive or not. The problem lies in our conflicting visions of them; the same Red fox, for example, may be seen by different people as aesthetically stunning, as a rabies vector, as a noble (if inedible) quarry, as a killer of lambs or pheasants, as a "useful" predator on rabbits and rodents, or as a pelt to be harvested. DWM

DISTRIBUTION Worldwide except Australia, Madagascar, and Antarctica. Domestic cat (*F. silvestris catus*) everywhere but Antarctica.

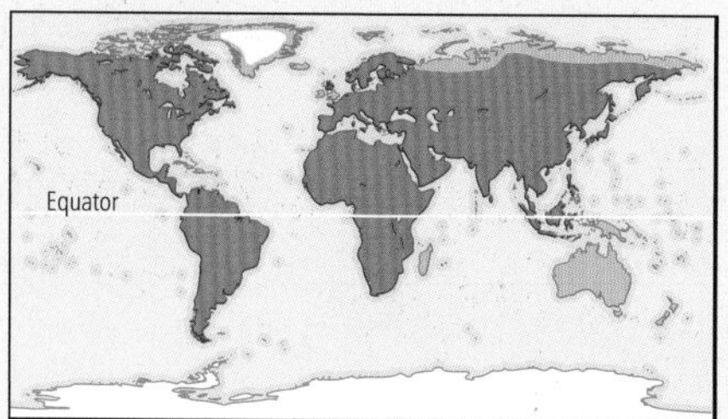

Family: Felidae
37 species in 4 genera

BIG CATS

Seven species in 3 genera:

Lion *Panthera leo*	p10
Tiger *Panthera tigris*	p18
Leopard *Panthera pardus*	p28
Jaguar *Panthera onca*	p30
Snow leopard *Panthera uncia*	p30
Clouded leopard *Neofelis nebulosa*	p30
Cheetah *Acinonyx jubatus*	p22

SMALL CATS p32–39

30 species of *Felis*, including **Lynx** (*F. lynx*), **Bobcat** (*F. rufus*), **Puma** (*F. concolor*), **Ocelot** (*F. pardalis*), **Wild cat** (*F. silvestris*), **Jaguarundi** (*F. yaguarondi*).

Cat Family

FELIDS ARE THE MOST CARNIVOROUS OF THE order Carnivora, and sit at the pinnacle of many food pyramids. The family comprises 37 species (including the Domestic cat) ranging over five continents. Domestication began in the Middle East between 7,000 and 4,000 years ago, as African wild cats (*Felis silvestris lybica*) were tolerated around human settlements because of their ability to prey on rodents infesting grain stores. Worshiped by the Ancient Egyptians, the cat was first introduced to Europe some 2,000 years ago.

The very earliest felids began to evolve during the lower Eocene, some 40 million years ago. Today's feline species descend from an ancestor named *Pseudailurus*, from which a group of large, saber-toothed cats and surviving wild cats emerged from the Oligocene (38–26 million years ago) onward. Saber-tooths disappeared only relatively recently, in the Pleistocene (20–10,000 years ago) at the time of the last ice age, which also saw the extinction of several other mammals.

The cats share several adaptations inherited from their common ancestor, including blunt, flattened faces, large eyes, claws that can be unsheathed, and large, sensitive ears. Their tawny color range and patterning serve as adaptive camouflage, since three-quarters of all cats live isolated existences in dense forests. Pelage or coat display among the living cats varies in pattern – from stripes in tigers to the marbled coats of Clouded leopards, King cheetahs, and Marbled cats – as well as in pigmentation (albino in lions and tigers, black in leopards and jaguars) and hair length. Domestic cat breeds display many coat types and patterns, suggesting that an early ancestor of all cats had the necessary genetic diversity to underpin such variation.

All wild cat species are listed by the IUCN and CITES as threatened in some measure. Their survival is jeopardized by habitat loss, hunting and depredation, and poaching. Awareness of this threat has given rise to conservation initiatives, and the need to halt the extinction of these remarkable animals continues to be a priority. SJO'B

CLASSIFYING CATS

Along with the hyena, mongoose, and civet families, living cats comprise the ailurid, or catlike, side of the Carnivora family tree. The link came to light when a common physiological feature was identified – an ossified segment in the auditory bulla, a part of the inner ear. This feature is not found in the arctoids – the name means bearlike – comprising the raccoons, bears, dogs, and weasels, which make up the other main branch of the order Carnivora. Today, a variety of additional morphological and DNA-based characters have confirmed this historic separation.

Mitochondrial and nuclear DNA analysis have also clarified felid taxonomy at the subfamily level. The gene comparisons cluster the species into three major subfamilies: the ocelot lineage, comprising 7 species; the domestic cat lineage (7 species); and the pantherine lineage (23 species). Genetic analysis also enables the cats to be assorted into eight monophyletic groups (i.e. groups displaying evidence for a recent common ancestor subsequent to divergence from the older common ancestor for all modern cats, *Prionailuris*). Two of the monophyletic groups are supplied by the ocelot and domestic cat lineages. The pantherine lineage, however, splits into six separate monophyletic groups: the *Panthera* genus; the *Lynx* genus; the Asian leopard cat group; the Caracal cat group; the Bay cat group; and the Puma group (which includes the cheetah). Two species, the serval and the Rusty-spotted cat, cannot yet be assigned to any lineage. The eight groups will likely represent a future genus proposal for the Felidae.

SENSES AND REFLEXES

All felids have large eyes with binocular and color vision. In daylight, they see as well as humans, but under poor illumination their sight is up to six times more acute. Their eyes adapt quickly to sudden darkness by rapid action of the iris muscles. The image is further intensified by a reflecting layer, the tapetum lucidum, which lies outside the receptor layer of the retina. Any light that passes through the receptor layer without being absorbed is reflected back again and may stimulate the receptors a second time. The outward sign of this is the "eyeshine" seen when light is shone into cats' eyes at night.

Felids also have sharp hearing, thanks to large ears that funnel sound waves efficiently to the inner ear. The small cats are particularly sensitive to high-frequency sound.

One of the most renowned reflex abilities of cats is always to land on their feet. As a cat falls, the vestibular apparatus of the inner ear, which monitors balance and attitude, acts in conjunction with vision to provide information on orientation. The neck muscles rotate the head to an upright horizontal position, with which the rest of the body rapidly aligns.

Below Cats large and small: **1** Lynx Felis lynx) *and* **2** Bobcat (F. rufus) *are* mall North American felids of similar body shape but with distinctive coloration to suit their different habitats – the plain, brownish-gray coat of the Lynx enables it to remain inconspicuous in dense, moss-laden coniferous swamps, while the black-spotted brown coat of the bobcat camouflages it effectively against the rocks and sagebrush of the arid areas it inhabits; **3** Cheetah (Acinonyx jubatus) – *two subadults are shown play-chasing, an activity that helps hone their skill in running down fast-moving prey;* **4** Tiger (Panthera tigris), *using its impressive roar to warn other tigers that it is in residence;* **5** Jaguar (Panthera onca) *in a characteristic posture of burying its prey with leaf litter;* **6** Leopard (Panthera pardus) *caching its antelope kill up a tree, a strategy sometimes employed by this species to foil scavengers.*

Right In common with other cats, the Jaguar is a highly specialized carnivore.

SKULLS AND DENTITION

Felid skulls are small, with a short face resulting from reduction in the nasal cavity and jaw length. The dental formula is I3/3, C1/1, P3/2, M1/1 = 30, except for the Lynx and Pallas's cat, which lack the first upper premolars and hence have 28 teeth. The molars and premolars are adapted as gripping and tearing carnassials. The upper carnassial tooth (premolar 3) has a dual purpose: it has a sharp cutting edge, but its anterior cusp is relatively broad and is used to crush bones. The canines are large (particularly so in the Clouded leopard) and used for grabbing and killing prey. Jaw mobility is restricted to vertical movements, with the powerful masseter muscle giving a vicelike grip. To compensate for the lack of chewing molars, the tongue is coated with sharp-pointed papillae that retain and lacerate food and rasp flesh off a carcass. Each lineage has its own arrangement of papillae.

Wild cat
9.7 cm

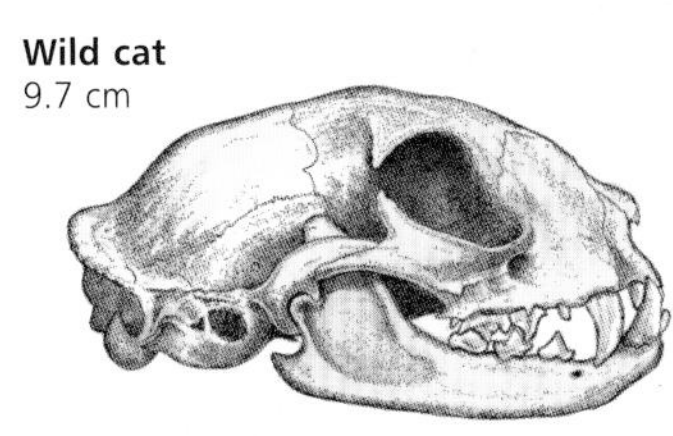

Cheetah
19.7 cm

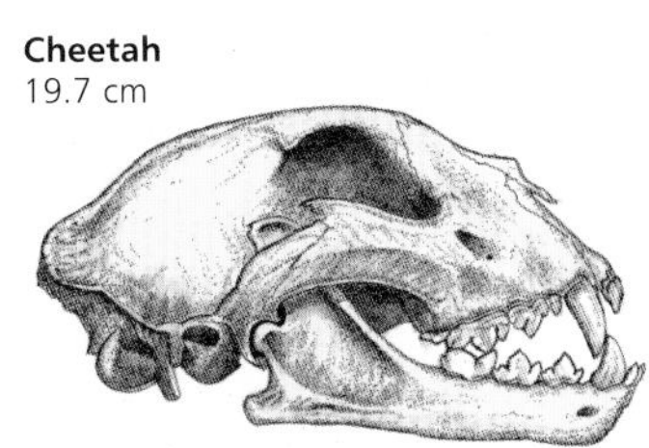

Clouded leopard
23 cm

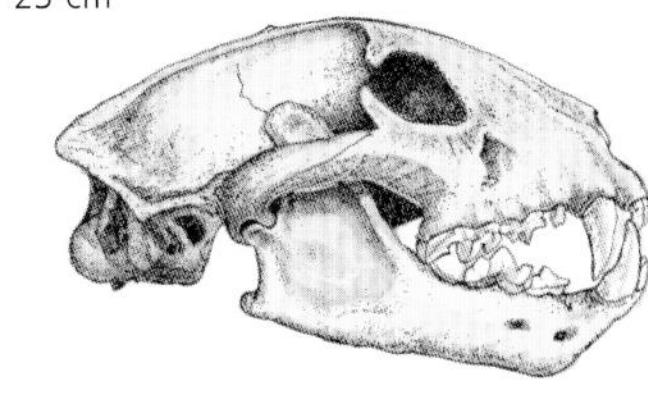

Lion

FOR MILLENNIA, THE LION'S STRENGTH AND *ferocity have earned it the title of "King of the Beasts." Lions were portrayed in the art of ancient Egypt, Assyria, India, and China, even adorning the walls of European caves some 32,000 years ago. Myths of the lion's supernatural powers survive in numerous African cultures: many local peoples believe that the body parts of a lion can cure illness, restore sexual prowess, and provide protection against enemies.*

The powerful image of the creature still lures foreign hunters, who sometimes pay huge sums of money for the opportunity to shoot a male African lion. In the past, local populations were considerably reduced by these trophy hunters, who might shoot as many as a dozen lions in only a few days. However, trophy hunting is regulated today and most tourists prefer to express their fascination for these magnificent animals less destructively, simply by observing or photographing them.

King of the Beasts

FORM AND FUNCTION

Like other members of the cat family, the lion has a lithe, muscular, and deep-chested body. The short and powerful skull and jaws are highly adapted to killing and eating large prey. The upper surface of the tongue is covered with stiff, backward-curved papillae (projections) that are useful both in feeding and in grooming the fur. Lions rely primarily on vision and hearing to find prey. As in most cat species, adult male lions are considerably larger and heavier (30–50 percent) than adult females, presumably because of intense competition between males for mates. Whatever the evolutionary reason for this dimorphism, the males' greater size enables them to monopolize carcasses when feeding with the rest of the pride, and to capture much larger prey than the females.

Although lions have the reputation of being cooperative hunters, this is only common in harsh environments with poor prey availability or during the capture of large, dangerous prey. Cooperation generally occurs when an individual lion has less than about a 10 percent chance of capturing the prey by itself. Here cooperation is practically a prerequisite for a successful hunt. When hunting together, the lions fan out and partially encircle the prey, cutting off potential escape routes.

In many other cases, however, only one or two members of a larger group will attempt to hunt, while the rest of the pride watches them from a safe distance. These non-cooperative hunts are more common when the prey is relatively easy to capture (with a solo hunting success on the order of 20 percent or more). Companions presumably hang back on these occasions because they cannot make much of an impact on the success of the hunt – although they are more than willing to join their successful companions in eating the captured prey!

Above left *Lionesses preparing to attack. They learn from experience to plan the best approach: while many prey simply outrun them, others may inflict fatal wounds. A zebra's kick can break a lion's jaw, dooming the lion to death by starvation.*

FACTFILE

LION

Panthera leo

Order: Carnivora

Family: Felidae

One of 5 species of the genus *Panthera*

5 Subspecies: **Angolan lion** (*P. l. bleyenberghi*), Zimbabwe, Angola, Democratic Republic of Congo; **Asiatic lion** (*P. l. persica*), Gir Forest, India; **Masai lion** (*P. l. massaicus*), Kenya, Tanzania, Uganda; **Senegalese lion** (*P. l. senegalensis*), W Africa; **Transvaal lion** (*P. l. krugeri*), Transvaal.

DISTRIBUTION S Sahara to S Africa, excluding Congo rain forest belt; Gujarat, India (a remnant population in Gir Forest Sanctuary).

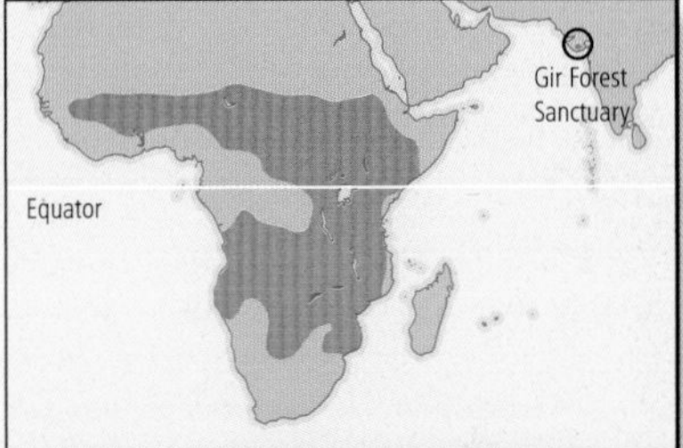

HABITAT Varied, from savanna woodlands of E Africa to sands of Kalahari Desert.

SIZE Head–body length male 1.7–2.5m (8.5–10.8ft) female 1.6–1.9m (5.2–6.2ft); **shoulder height** male 1.2m (4ft) female 1.1m (3.6ft); **tail length** male and female: 60–100cm (2–3.3ft); **weight** male 150–240kg (330–530lb) female 122–182kg (270–400lb).

COAT Light to dark tawny; lighter on abdomen and inner side of legs; back of ears black. Immature animals have a rosette pattern which fades as they mature, although vestiges may remain on lower abdomen and legs of adults. The male's mane varies from platinum blond through reddish-brown to black.

DIET Mainly hoofed mammals such as gazelles, zebras, antelopes, giraffes, and wild hogs, and the young of larger mammals like elephants and rhinos. Will also take smaller prey such as rodents, hares, small birds, reptiles. Females need about 5kg (11lb) of meat a day, males 7kg (15lb).

BREEDING Females are sexually mature at about 36–46 months (24–28 in captivity). 2–4 cubs born at any time of year after gestation 100–119 days. Cubs completely independent by about 2 years 6 months.

LONGEVITY 18 years (up to 25 in captivity).

CONSERVATION STATUS CITES listed and ranked Vulnerable by the IUCN, with numbers declining due to habitat loss. The Asiatic subspecies (*P. l. persica*) is ranked Critically Endangered, while the Cape lion (*P. l. melanochaiatus*), Cape to Natal, and the Barbary lion (*P. l. leo*), formerly found in N Africa, are both listed as Extinct (see box on Barbary lion).

Below *Lionesses hunting as part of a cooperative team will stalk their prey, trying to encircle the herd and cut off escape routes. Because of their relative lack of speed they need to get within close proximity of the prey before launching into the final assault.*

Above *A male lion has a zebra in a suffocating throat bite. Lions are one of the few carnivores to regularly take prey over 250kg (550lb) and healthy adults rather than the young, old, or sick. They kill other predators such as leopards but rarely eat them.*

While lions can reach 58km/h (36mph), their prey can run as fast at 80km/h (50mph), so the lion must use stealth to approach to within 15m (50ft) before charging and either grabbing or slapping their quarry on the flank. Lions do not consider wind direction when hunting, even though they are more successful when stalking upwind. Typically, only about one in four hunts ends successfully. When they do bring down a large animal, the victim is suffocated by a clamping bite to the muzzle or throat.

Males dominate all other lions at a kill, and females dominate the cubs and subadults. Squabbling is common at the dinner table. To protect their share of the kill, lions will clamp down on the carcass with their teeth and smack their pridemates in the face with their paws, sometimes tearing each other's ears in the process. Often the successful hunter is so focussed on throttling the prey that the rest of the pride will manage to consume most of the carcass before it can even start to feed. Adult females require 5–8kg (11–18lb) of meat per day and adult males 7–10kg (15–22lb). But lions only feed irregularly, engorging themselves with huge meals once every 3–4 days – an adult male lion may eat as much as 43kg (95lb) in a single sitting!

Males show early signs of mane growth when they reach full body size at about 2 years of age, but their manes do not reach full size until they are 4–5 yrs old. Mane color generally darkens until the age of 9–10 yrs. Males will lose their manes if they are castrated or badly injured.

A Much-Reduced Range

DISTRIBUTION PATTERNS

Panthera leo was once more widely distributed than today. Archaeological finds testify to its widespread presence in Europe and North America until about 10,000 years ago. Aristotle mentions lions in Greece as recently as 300BC, and the Crusaders frequently encountered lions on their journeys through the Middle East. Lions could still be found in much of the Middle East and northern India up to the early 1900s.

Like all big cats, lions suffer from expanding human populations. Since lions are large and conspicuous and live in more open habitats than their more secretive cousins, they were quickly eliminated from large tracts of land by expanding numbers of agriculturalists and ranchers. Today, lions are largely restricted to a series of national parks and game reserves where they are protected from human persecution.

Big-Game Hunters

DIET

Lions mostly feed on hoofed mammals weighing from 50–500kg (110–1,100lb), but they are also known to eat rodents, hares, birds, and reptiles as well as the occasional elephant and rhino. Lions primarily hunt at night, though they take advantage of dry weather to ambush prey animals drinking at water holes during daylight hours. Females capture most of the small- to mid-sized prey (such as warthog, gazelle, springbok, wildebeest, and zebra). Males specialize in large, slow-moving prey such as buffalo and giraffe.

Lions live in the same habitat as a variety of other carnivores (including leopards, cheetahs, wild dogs, and spotted hyenas) that feed on many of the same prey species. All five carnivore species hunt animals weighing less than 100kg (220lb) such as warthog and gazelle, but only the lion regularly kills species that are larger than about 250kg (550lb) – for example, buffalo, eland, and giraffe. The large-bodied and nocturnal hyena is probably the lion's strongest competitor (with both species showing a similar preference for wildebeest and zebra when they are available), but lions can consistently steal meat from hyena kills, and males are especially persistent scavengers. Lions not only scavenge from cheetah and wild dog kills, but are so aggressive toward them that these species can seldom coexist with lions in the smaller reserves.

RESTORING THE BARBARY LION

Probably due to its proximity to Europe, North Africa (or, to be more precise, Constantine in Algeria) provided the lion that Linnaeus used to classify the species – as *Panthera leo* – in 1758. Sadly, this historical distinction could not save the entire population of lions in North Africa from becoming extinct in the wild less than two centuries later.

Historical records indicate that by the early 18th century lions had all but disappeared from eastern North Africa, and survived only in western North Africa. By the mid-19th century, even that remnant was greatly diminished, due to the wide availability of firearms and a policy of eradication implemented by the Ottoman administration, from which only Morocco (as an independent sultanate) was exempt. The last recorded lion was killed in Algeria in 1893. In Morocco, lions survived until the 1920s.

One population escaped extinction, however. At the same time as they were becoming increasingly scarce in the wild, Barbary lions (BELOW) were presented as tokens of fealty to the rulers of Morocco by the indigenous Berber people. These lions bred in captivity, and aroused great interest among zoologists when King Hassan II transfered his collection from the royal palace to Rabat Zoo in 1970. It transpired that the royal lions' morphology closely approximated historical accounts of the so-called "Barbary lion." In 1998, there were still 52 lions originating from the Moroccan king's collection (24 males and 28 females) in Rabat Zoo and 13 other collections across Europe.

The idea began to take shape of reviving the Barbary lion, but two questions had first to be answered. Firstly, were the royal lions truly representative of Barbary lions? And, more profoundly, was the Barbary lion ever sufficiently distinct to merit singling it out from other surviving lion populations in the first place? An anecdote that the Moroccan sultans' lions had been brought from West Africa on the trans-Saharan trade route raised doubts as to their origins. And newly-developed molecular techniques questioned the Barbary's distinctiveness: these DNA studies suggested that all lions shared a common ancestor relatively recently, perhaps only 50–200,000 years ago. Animals as mobile as lions can travel great distances over a few generations. Even widely separated populations could naturally intermingle over time, crossing the deserts and arid plains that divide them.

To date, these two questions have remained largely unanswered, but advances in biotechnology – in particular, the recovery of ancient DNA from museum specimens – promise new insights. Comparative analysis of mitochondrial DNA and short tandem repeats of nuclear DNA should in time bear fruit. The result will either finally dispel the myth of the Barbary lion, or pave the way for a major restoration project. NY

The Sociable Cat

SOCIAL BEHAVIOR

The lion is the most social of all the felids. The pride is the unit of social life, typically consisting of 3–10 adult females, their dependent offspring, and a coalition of 2–3 adult males. Some prides have been observed to contain as many as 18 adult females and 10 adult males. Unlike a wolf pack or a monkey troop, a lion pride is a fission–fusion society. Lions do not remain in constant contact with their pridemates; rather, each lion may spend days or weeks living on its own or in a small subgroup.

Female pridemates are always close genetic relatives, but they are almost never related to the males in their pride (except in small, isolated populations). The males may or may not be related to each other. If a male reaches maturity without linking with another male, he will seek out another solitary during his nomadic phase and form a pair, and pairs may occasionally recruit a third companion. But most males already have same-aged companions in their natal cohort, and these ready-made coalitions may contain as many as nine or ten brothers and cousins, though three or four is the norm. Larger coalitions have longer tenure in a pride, gain access to more females, and father more surviving offspring per capita than do smaller coalitions.

The resident coalition sires all the cubs born during their tenure, but the pattern of paternity within each coalition is complex. Dominance relations are subtle in male lions, and females often come into estrus synchronously; thus males in small coalitions gain an equivalent access to

Above An adult male lion with a young female. Lionesses are responsible for the vast majority of kills; however, once the prey has been dispatched, the lion will assert its dominance and drive all other animals from the carcass until it has eaten its fill.

Below A lioness carrying a young cub, which is probably no more than 2 months old. At this age the cub will still be feeding on its mother's milk, though within another month or two the cub's diet will also include meat.

mates. However, this equivalence breaks down in larger coalitions, in which many of the males are essentially nonbreeding helpers to their successful male relatives – helpers who can only hope to perpetuate their genetic legacy through an increased production of nephews and nieces.

Given that kinship is so fundamental to lion society, it is ironic that these coalitions of unrelated males are among the most cooperative units in any mammalian species. Both members of an unrelated pair benefit directly from their joint tenure in a pride, and they are highly affectionate to each other and supportive during intergroup encounters. Indeed, their behavior is virtually indistinguishable from pairs of brothers.

Female lions are 30–38 months old before they are ready to become mothers. They can become sexually receptive at any time of year, and estrus lasts 2–4 days with a 2–3 week interval between cycles. Mating pairs copulate about 3 times per hour, but it is not clear whether ovulation is induced by copulation (as in the domestic cat), or if it is spontaneous (as in primates). Gestation is short for such a large mammal – 110 days – and newborn cubs are a relatively tiny 1 percent of adult weight. Litter sizes vary from 1–6, with an average of 2–3. Females live to a maximum age of 18 in the wild, but typically stop breeding by the age of 15.

If females from the same pride give birth within a few months of each other, they will rear their young together, even suckling each other's cubs. Mothers do, however, recognize their young and reserve most of their milk for their own offspring. Cubs start eating meat by 3 months of age but continue to nurse for an additional 3 months. Mortality of cubs can be high; in particularly harsh years, as many as 80 percent may die before their first birthday, although cub mortality can be as low as 10 percent in good years. Cubs show signs of independence by the time they are about 18 months old, and the mother usually gives birth to her next litter around the time they reach their second birthday. However, a female will rapidly resume mating after the death of her current litter.

Both sexes defend the territory. Lions are most persistently cooperative during territorial encounters, and they generally approach the roars of like-sexed intruders. Males range widely to protect their ownership of the pride, while the females safeguard their core area against other groups of females. Males maintain their territory by roaring, urine marking, and patrolling, leaving the females closer to the middle of the pride range. Females, on the other hand, are more cautious than males and more often merely react to the presence of strangers in their territory rather than actively patrolling the range. As they approach adulthood, juvenile females become increasingly active in assisting their mothers in repelling trespassing females; in contrast, subadult males are relatively indifferent to strange females.

Although females are more cooperative during territorial defense than while hunting or rearing their cubs, their behavioral strategies are diverse. Some females always lead the way during an inter-pride encounter, while others always lag behind. But certain females ("friends in need") are more active when their pride is outnumbered and their pridemates need them most. Other females ("fair-weather friends") are most cooperative when their pride safely outnumbers their opponents!

In general, cooperation during an interpride encounter is a numbers game: larger groups dominate smaller ones. Females are most likely to confront strangers when their own group contains at least two more females than the opposing group. Males, on the other hand, will almost always approach an invading coalition unless they are outnumbered by at least three to one.

Territorial intruders usually withdraw immediately on seeing the approach of a resident, although lions may fight and kill a stranger if they have the chance. Most lions probably die from violent intergroup attacks, either following single combat or a gang attack. Most killing bites are directed toward the back of the skull or the spine.

The pride range covers an area of 20–500sq km (8–200sq mi), depending on the size of the pride and the abundance of prey. More than one group of lions may use overlapping areas of their neighboring ranges, but they will avoid straying into each other's central area.

The lion's social system is remarkably similar throughout the species' range. The size and composition of lion prides in the Gir Forest of India are similar to those found in African reserves as diverse as the Serengeti, Kruger, Ngorongoro Crater, and the Kalahari Desert. It seems to be a universal feature of lion society that daughters join their mothers' pride, whereas cohorts of young males venture off together into new prides.

Because lions are so sociable, they are often viewed as paragons of cooperation. However, there is a dark side to the evolution of lion sociality. Males form coalitions to compete against other groups of males: solitary males have little hope of gaining residence in a pride. Females band together to fight their neighbors: solitary females cannot defend a territory against larger prides.

Males lead a more intense life than females. A male coalition can only maintain residence in a pride for a few years, whereas it takes two years for females to rear each set of cubs. On entering a new pride, the males kill the smaller cubs (to whom they are not related), causing the females to be ready to mate within a few days. This infanticide, on average, speeds up the reproduction of the females by several months, at which point they give birth to the cubs of the incoming males. Not all encounters between females and extra-pride males lead to a replacement of the resident coalition. Females are often successful in protecting their cubs against these invasions.

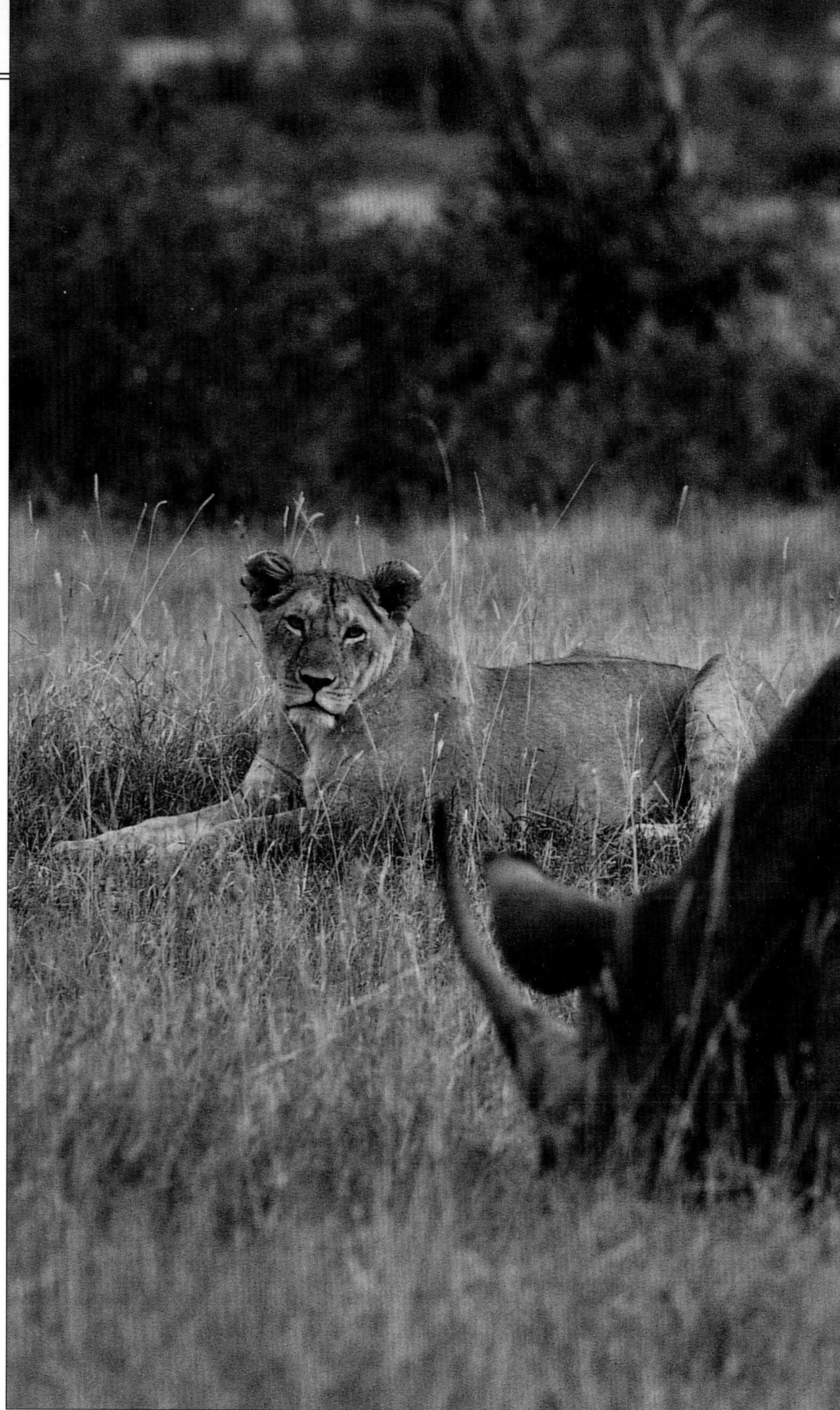

Above *Lionesses and cubs keep close watch on a resting rhinoceros, which, no doubt, is looking back equally warily at them. This stand-off may continue for hours: a healthy adult rhinoceros has the stamina to defend itself, and an experienced hunting lion is cautious about approaching the rhino's deadly horns. There is no guarantee that the cubs will get to eat after a kill and it is not uncommon for cubs to starve. By about 18 months most cubs should be able to secure some food for themselves following a kill.*

Maneaters at Risk

CONSERVATION AND ENVIRONMENT

Instances of humans falling victim to lions are common. Attacks by maneaters are well known in the wild, although they are often perpetrated by injured or aged animals unable to kill their normal food. Man is an easy prey, being neither swift nor strong. Many cases of maneating have followed the extermination by humans of the lions' normal supply of game. For example, at the end of the 19th century, two healthy male lions preyed

regularly on the laborers of the Uganda–Kenya railway, leading to a temporary halt in construction. These "Maneaters of Tsavo" were themselves the victims of human activity, for just a few years earlier Europeans had inadvertently imported a cattle virus that decimated wild ungulates as well as livestock, leaving little for the lions to eat.

While lions are not immediately threatened with extinction, their long-term survival is far from assured. Many lions are still killed illegally. Poachers inadvertently trap lions in snares set for other animals; cattle herders set out poisoned carcasses to eliminate entire prides.

A more significant threat comes from the fact that large areas of land are necessary to support lions and their natural prey. As agriculture spreads, lions are quickly eliminated. The range of the Asian lion was reduced to a single reserve in less than a century, and a similar pattern may soon take place in Africa. Lions survive very poorly outside designated reserves, and only a few hundred animals may still exist throughout West Africa. Indeed there may only be two parks in Africa with numbers greater than a thousand; only a handful of lions survive in each of the many smaller parks. However, the lion's range is expanding in many parts of South Africa that are restoring natural ecosystems for "ecotourism." Lions have been successfully translocated into numerous reserves. Similar efforts are planned to extend the Asian lion into a second reserve in India. Hopefully these efforts will stop the modern-day lion from going the way of the European lion of prehistory. CP

Why Lions Roar

Long-distance vocal communication in African prides

Asked to visualize a roaring lion, most people would probably call to mind the one at the start of Metro Goldwyn Meyer (MGM) films. Ironically, though, the MGM lion is not roaring at all; it is merely snarling at some annoyance just off camera. In a genuine roar, a lion purses its lips, thrusting out its chin and pointing its mouth toward the horizon, its body heaving in rhythm with the exertions of its groans and grunts. Humans, enchanted if unnerved by the grandeur of the sound of the roar, can hear it from 8 km (5 miles) away through the African night. Yet the lion is not roaring for our benefit – so who exactly is the king of the beasts communicating with?

Lions maintain a complex web of social interactions. Pridemates may spread themselves over areas of 50sq km (20sq mi), but enemy prides are always somewhere in the vicinity, and a lone, intruding lion may be killed if it is caught by surprise. It therefore becomes vitally important to keep track of friend and foe. Like many other species in fission–fusion societies, lions communicate among themselves over long distances (probably much further than the range of human hearing). The roar follows a sequence: a slow series of long, low groans, followed by a rapid succession of staccato grunts. Males and females both roar, but the male's voice is louder and deeper.

Lions roar at night when the air is still and they are active, but only if they are in control of a territory. Most young males must endure a phase as nomads when they keep a low profile, avoid resident males, and remain silent whenever the established prides roar to each other at night. But once they have found a spot of their own and marked their new territory, they also begin to roar. Roaring lions answer each other back – and it is this behavior that researchers have exploited by using recorded roars to investigate the meaning behind the calls. Through high-quality sound systems, electronic roars are broadcast to lions in the Serengeti and Ngorongoro Crater, Tanzania, and the lion's responses are measured.

For a lioness, the roars are the sound of reassurance. Mothers with cubs need much luck and hard work to raise them. The males of the pride spend little time in their company; even so, having shared in fathering the litter, they have a direct stake in its survival, and so they patrol the range to protect the cubs from infanticidal strangers. Hearing the roars of her offsprings' fathers resound through the night, a lioness can rest easy in the knowledge that the world is in order. In contrast, the sound of a group of strange males roaring in the immediate vicinity would indicate that something had gone terribly wrong.

Left The lion's roar, which carries far over the open savanna of East Africa, is an unequivocal signal to would-be interlopers that a territory is occupied and will be defended. Though most commonly heard after sundown, it is also made after the pride has devoured a kill.

Whenever roars of the resident males were played to a group of mothers and cubs, the females showed little response. But roars of a neighboring coalition caused them to become agitated and either snarl at the loudspeaker or gather their cubs and run away. And when tapes of the roars of a neighboring group of females were played, the mothers reacted as if to a set of competitors: they confidently approached the loudspeaker, ready to attack.

Furthermore, lionesses were able to tell how many lions were approaching and gauge their response accordingly. When a group of females heard the recording of a single stranger, their reaction depended on how many real lions were present. Although a single lioness would only rarely approach a recording of another singleton, a pair went forward about half the time, and trios almost every time. When the odds were changed by playing the roars of three strangers, trios responded as a single female did to another singleton; a quartet acted in the same way as two lionesses did on hearing one roar; and a quintet as a trio reacting to a lone intruder. The consistency of their response provided strong evidence that lions can count, and that they simultaneously track the number of strangers and companions.

So lions can listen to a roar and tell if it comes from a male or a female, a friend or a foe, and whether they safely outnumber the roaring group. Females may live in prides of 1–18 females, but members of even the largest pride spend much time alone. When such females are temporarily outnumbered, they respond to recorded roars by roaring in their turn to recruit absent partners, which then join them in approaching the loudspeaker.

Lionologists have thus learned much about the responses of lions to each other's roars, but we have yet to learn exactly how they tell which lion is which from the sound of a roaring voice alone. Some lions have a gravelly intonation, others a clear baritone, and individuals can vary their roars according to their motivation: sometimes the slow groans take on a clear sense of urgency, but at other times they can seem no more than half-hearted. Thus this most evocative of sounds in the animal kingdom resonates with meanings that we are only just beginning to understand. CP

Left As well as the territory-marking roar, lions and lionesses have at least eight other vocalizations, each of which is employed for a different purpose. Here, a lioness on a rocky outcrop, or kopje, can be seen calling for her lost cubs.

Above Alert to any approach, a lioness tends cubs of various ages from the pride. On hearing a roar in the vicinity, she will be able to tell instantly whether fight, flight, or simply resting easy is the best course of action.

Tiger

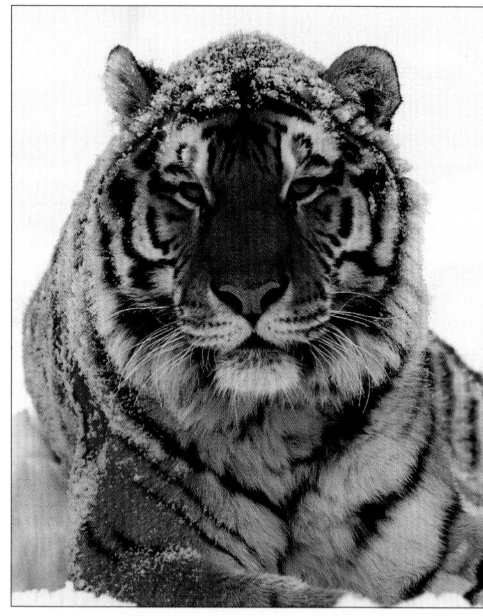

Left *The Amur (or Siberian) tiger is the largest living member of the cat family: the average male of this subspecies may weigh anything between 180–300kg (397–661lb); however, one individual specimen was found that weighed a reported 384kg (845lb).*

FROM THE VILLAINOUS SHERE KHAN IN Kipling's Jungle Book *to the protective "guardian of the west" of Korean mythology, tigers play a more vivid role in the human psyche than perhaps any other animal. Latterly they have become a symbol for conservation, and their survival on the planet has come to represent the human struggle to balance our conflicting needs and desires.*

Generally considered the largest of living felids, the tiger and lion are, in fact, very similar in size. The biggest tigers are found in the Indian subcontinent and Russia, where males average 180–300kg (397–661lb), while more southerly, island populations are the smallest – Sumatran males average 100–150kg (220–330lb).

Built for Killing

Form and Function

The exacting specializations needed to stalk, ambush, and kill prey have been so perfected in the felid family that characteristics other than size and coat color vary little between species. Tigers, like other "great cats," rely on prey generally much larger than themselves, and subsequently have short, heavily-muscled forelimbs with long, sharp, retractable claws, enabling them to grab and hold large prey once contact is made. The skull is foreshortened, increasing the leverage of the powerful jaws. Tigers usually kill their prey with a crushing bite to the back of the neck, though sometimes they exert an unrelenting grip to the throat to strangle it. Completely unique to the tiger, though, are its orange-and-white background coat and black stripes; in fact, individual animals can be identified by their own particular patterns. The white tigers so common in zoos – usually they have chocolate stripes, and so are technically not albinos – are all derived from Mohan, a single male Bengal tiger caught by the Maharaja of Rewa in Madhya Pradesh, India. Tigers that are almost completely black have also been reported from other regions of India, but both white and black animals are exceedingly rare in the wild. While variations in coat patterns exist across the tiger's range (darker coats being associated with the rain-forest jungles of southeast Asia), the vertical striping common to all subspecies provides surprisingly good camouflage that allows tigers to stalk close enough to capture prey in a final, deadly, explosive rush.

Signs of Tiger Country

Distribution Patterns

The sun-dappled, vertical shading patterns of tall grasses, shrubs, and trees that provide the camouflage and cover essential to hunting are perhaps the single common feature linking the variety of habitats in which tigers were once widely distributed. These include the tropical rain forests of the islands of the Sunda Shelf, the tall grasslands and riverine forests of northern India and Nepal, the mixed deciduous, dry evergreen, and dry dipterocarp forests of Thailand, the mangrove swamps of the Sunderbans, the temperate and boreal forests of the Russian Far East, and until recently, the "reed jungles," riparian thickets, and montane forests of the Caspian region.

Dense cover, access to water in hot climates, and high densities of large ungulates are the principle features of tiger habitat. The animals' distribution, as well as much of their behavior and social structure, has always been largely dictated by the availability of deer, wild cattle, and wild pig species, which comprise the large majority of their diet. This has been so since tigers first evolved from *Panthera* stock some 2 million years ago, a development fueled by the Pleistocene radiation of large cervids and bovids in Southeast Asia, which provided a niche for a large-bodied, forest-edge predator.

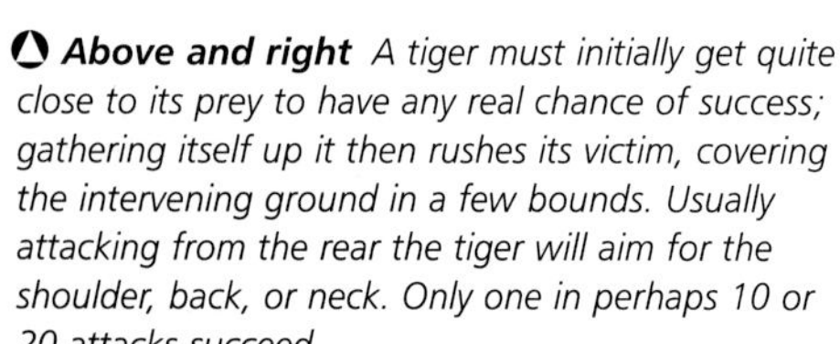

Above and right *A tiger must initially get quite close to its prey to have any real chance of success; gathering itself up it then rushes its victim, covering the intervening ground in a few bounds. Usually attacking from the rear the tiger will aim for the shoulder, back, or neck. Only one in perhaps 10 or 20 attacks succeed.*

Above *A tiger in mid-stride during an attack conveys all the power and agility of this top predator. To seek out prey and defend their home ranges, it is not uncommon for tigers to travel distances of 10–20km (6–12 miles) a day.*

Keeping in Touch at a Distance

SOCIAL BEHAVIOR

Unlike lions and cheetahs, which hunt in open habitats, the tiger is a "stalk and ambush" predator most effective hunting on its own. There is little to gain from hunting cooperatively in a closed environment where prey are scattered, so tigers have a dispersed social system in which social contacts are maintained, but mostly at a distance.

Radio-tracking studies in Nepal, India, and Russia have shown that both males and females occupy territories they defend against intruders of the same sex. Female territories are smaller, and are determined by the amount of food and water needed to survive and raise young. Males attempt to maximize the number of female territories they can defend; their success, and the size of their

FACTFILE

TIGER

Panthera tigris

Order: Carnivora

Family: Felidae

8 subspecies have been recognized, although morphological studies suggest that some may simply represent clinal variations: **Bengal tiger** (*P. t. tigris*), India, Bangladesh, Bhutan, China, W Myanmar, Nepal. **Indo-Chinese tiger** (*P. t. corbetti*), Cambodia, China, Laos, Malaysia, E Myanmar, Thailand, Vietnam. **Sumatran tiger** (*P. t. sumatrae*), Sumatra. **Amur tiger** (*P. t. altaica*), Russia, China, North Korea (unconfirmed). **South China tiger** (*P. t. amoyensis*), China. The **Caspian tiger** (*P. t. virgata*), once found in Afghanistan, Iran, Chinese and Russian Turkestan, and Turkey, is now extinct, as are the **Javan tiger** (*P. t. sondaica*) and the **Bali tiger** (*P. t. balica*).

DISTRIBUTION India, SE Asia, China, SE Russia.

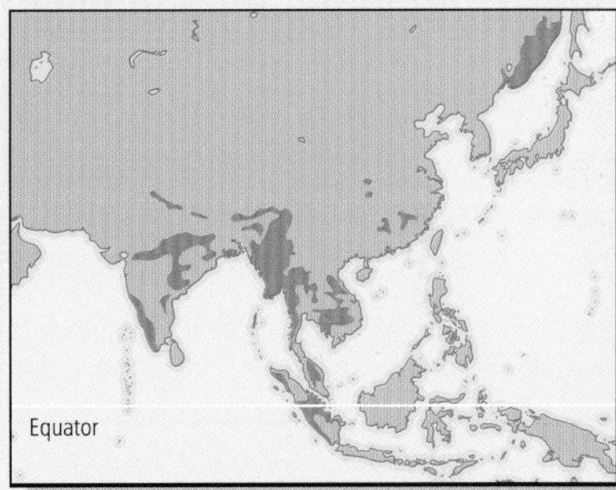

HABITAT Extremely varied, from reedbeds of C Asia to tropical rain forests in Southeast Asia and temperate mixed conifer–deciduous forests of the Russian Far East.

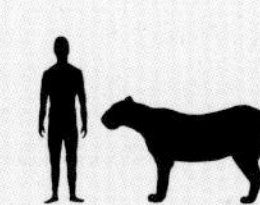

SIZE Head–body length Bengal tiger male 2.7–3.1m (8.8–10.2ft) female 2.4–2.65m (7.9–8.7ft); **weight** male 180–258kg (397–569lb) female 100–160kg (220–353lb).

COAT Black stripes on an orange background on back and sides; undersides mostly white; males have a prominent ruff on the head. Darker pelages are found in tropical rain forests of SE Asia and Sunda Islands; the Amur tiger is paler, with much variation in winter and summer coats. Occasional white tigers (with brown stripes) from C India are the result of a breeding of two animals with a recessive gene; although popular in zoos, they are not of conservation significance.

DIET Mostly large ungulates, including various wild deer, cattle, and pig species. Will also take smaller prey, such as monkeys, badgers, and even fish.

BREEDING Females reach adulthood at 3–4 years, males slightly later, at 4–5 years; usually 2–3 cubs (range 1–7) born at any time of year after a gestation that averages 103 days; cubs independent after approximately 1.5–2 years.

LONGEVITY One individual is known to have survived 15 years in the wild in Royal Chitawan National Park, Nepal; up to 26 years in zoos.

CONSERVATION STATUS Rated Endangered. Numbers declining rapidly due to poaching, habitat loss, and prey depletion. Three of eight recognized subspecies already extinct, and South China tiger on the verge of extinction.

territory, are dependent on their strength and fighting ability. A male will normally retain exclusive breeding rights to females in his territory as long as he can defend them from invaders.

Although this land-tenure system appears to be common to all tigers, the size of home ranges varies depending on prey densities. In Nepal and India, where prey densities are high, female home ranges average only 20sq km (8sq mi), while in the Russian Far East, where the densities are dramatically lower, the ranges cover 470sq km (188sq mi). Transient animals – mostly subadults looking for a home range – move across these territories and along their edges, often in marginal lands, waiting for an opening.

Maintenance of a home range is potentially dangerous – even a victorious fight can leave a tiger wounded and incapable of hunting. Thus, tigers seek ways to advertise their presence and minimize direct conflict. Urine, mixed with anal gland secretions, is sprayed on trees, bushes, and rock faces, and feces and scrapes are left along trails, natural travel routes, and conspicuous places throughout a territory. These signals probably function as "no vacancy" signs, while also conveying additional information to neighbors and intruders; for instance, individuals can probably be recognized by scent. The importance of scent-marking is shown when a tiger fails to visit portions of its home range (usually as a result of death), as the unoccupied territory is usually usurped within a few days or weeks.

Tigers reach sexual maturity at 3–5 years, but it may take longer to establish a territory and begin breeding. This can occur year-round, even in the winter months. Females in estrus advertise their condition by roaring frequently and increasing the rate at which they scentmark. Mating takes place for 2–4 days, during which the male may mount the female as many as 30–40 times a day, with each mating lasting 10–15 seconds. Usually two or three blind and helpless young are born after a pregnancy that averages 103 days. For at least the first month cubs are dependent on their mother's milk, and remain at a single den site, or are moved between sites by their mother, who holds them gently in her teeth by the scruff of the neck.

After one to two months the mother starts to take cubs to kill sites, but they are still left behind during the hunts themselves. Only when they are 6 months old will she begin to introduce them to hunting, stalking, and killing prey. Males are not involved in the rearing of young, though they will on occasion join the family unit, and will even share kills with the tigress and her cubs. When a male acquires another's home range, he will kill the offspring of his competitor, thereby forcing the female of the newly acquired range to come into estrus and give birth to his own offspring sooner.

Cubs remain dependent on their mother for at least 15 months, after which a gradual separation begins. Although the mechanisms are not completely clear, the young tigers either leave on their own or are pushed out, often coincidentally with the timing of the next litter. In both Nepal and Russia, it has been documented that older adult females will occasionally donate all or part of their home range to a daughter, greatly increasing her chance of survival and reproduction.

A Sad Tale of Declining Numbers

CONSERVATION AND ENVIRONMENT

The "Tyger, tyger, burning bright" of William Blake's famous poem is growing dim. Of the eight recognized subspecies, the three smallest and most isolated are now extinct. The Bali tiger was the first to go – the last reliable sighting was in 1939 – followed by the Caspian and Javan tigers, last seen in 1968 and 1979 respectively. Today, South China tigers appear to be on the verge of

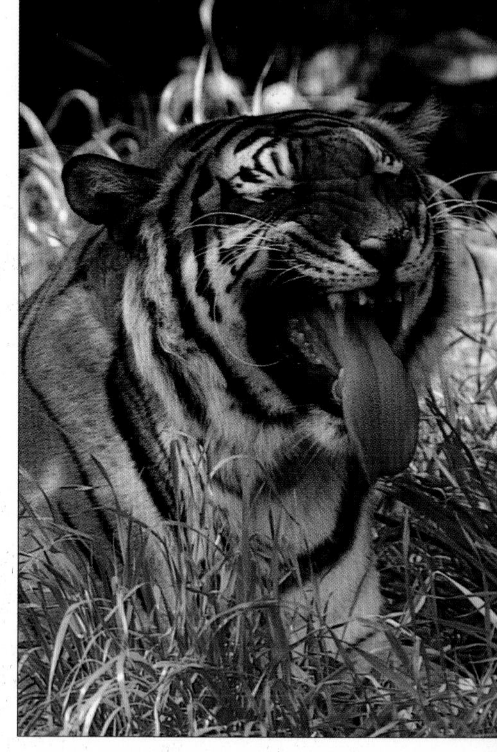

Above *Males can detect the reproductive status of females by sniffing their urine traces while performing "flehmen," an unusual posture in which the head is raised and the lips are curled back into a grimace, allowing olfactory and chemical clues to pass over the naso-vomeral organ positioned in the nostrils.*

TIGERS AND HUMANS

The relationship between tigers and humans is troubled. The world's people, enamored of the tiger's majesty, yearn to save them from the very real threat of extinction. In countries with tigers, people try to balance their own livelihood, sometimes their own lives, with the needs of conservation. In the absence of natural prey tigers will kill cattle. The culprits are usually young, dispersing adults or old, injured, or physically disabled tigers that have been pushed out of territories by younger and stronger animals. Villagers retaliate by poisoning carcasses and putting them out as bait for the troublesome animals. Less often, but too often, tigers kill humans.

Tigers become maneaters for three main reasons – injury, old age, and hunger. In very rare circumstances, cubs pick up the habit from the mother. In India tigers seeing the back end of a moving animal (a human crouching over collecting firewood) discover too late the target was a human being. In these cases, a single blow often results in death and the victim is left uneaten. At other times, the victim either accidentally walks into a tigress with cubs or a tiger at a kill, and is killed in an act of self-defense. The third most common reason for attacks is when cowherds are killed trying to protect their cattle against tigers. In a number of cases these are isolated attacks and the tiger does not turn maneater. In India a tiger is only declared a maneater when it has killed a minimum of three humans.

There is only one place in the world where man-eating is a habit regularly seen in tigers. This is in the Sundarban Tiger Reserve that lies in the delta of the Ganges. This is a unique tiger habitat as it is a mangrove forest. Tigers here have no fear of man, possibly as a result of the historical fact that it is probably the only place in the Indian subcontinent where tigers were not hunted for sport. Living alongside large carnivores has never been easy. Learning to do so is testing human ingenuity, but the price of failure is too great to pay. LN

extinction, and all the others are severely threatened. The best available figures suggest that tiger numbers worldwide declined from perhaps 100,000 at the beginning of the 20th century to approximately 5,000–7,500 at its end. Current estimates of the surviving populations suggest between 3,000 and 4,600 Bengal tigers left in the wild, along with 1,200–1,800 Indo-Chinese tigers, 400–500 Sumatran tigers, and about the same number of Amur tigers; the remnant population of South China tigers is put at only 20–30.

Surviving tigers face three main threats: direct poaching, habitat loss, and prey depletion. The growing demand for tiger bones for traditional Asian medicines, and for skins as trophies, has resulted in ever-growing losses. The greatest exporter of tiger-bone products is China, which distributed over 27 million items from 1990 to 1992. Japan, Hong Kong, and South Korea are the largest importers: South Korea took delivery of nearly 9,000kg (20,000lb) of tiger bone between 1970 and 1993. Attempts to control this illegal trade have had results, but it remains substantial.

The threat to habitat comes from the degradation and fragmentation that ensue as human populations grow. As tiger populations become isolated and reduced, the fragmented remnant populations grow increasingly susceptible to extinction. Inventive land-use plans have been proposed for such countries as Nepal, Thailand, and Russia, seeking to link protected areas via a network of conservation units and ecological corridors that aim to maintain the integrity of entire metapopulations. Such schemes have become more effective with the use of geographic information systems (GIS) that rely on remotely-sensed satellite imagery and allow analyses of multiple elements of the landscape, including forest cover, prey densities, and human impacts. Successful habitat restoration projects in Nepal provide incentives to villagers in return for the protection of communal lands.

Even if habitat is successfully protected, sufficient prey must be present. Many tracts in Asia that would otherwise be suitable are devoid of tigers today because of a shortage of large ungulates. An adult tiger can eat 18–40kg (40–88lb) of meat at a sitting, and must kill 50–75 large ungulates a year. For those ungulate populations to persist, predation and other causes of death should not account for more than 30 percent of their numbers annually, and in many cases densities are not high enough to support tigers. In other parts of the tiger's range, prey only survive in sufficient numbers in protected areas; in the Russian Far East, for example, numbers are often 3–4 times higher inside protected zones. Better controls on ungulate harvest in unprotected areas, along with the elimination of hunting in the protected ones, could benefit both tiger and human populations with the potential for a sustainable harvest.

Ultimately, tigers will survive only if local people find it in their interest to preserve and protect them. Local people in many parts of the tiger's range regard the animal as a compelling and necessary component of their environment, however much they may fear it. Seeking means to meet the needs of tigers and humans in remaining forest tracts will be the key to the animals' survival. DM

Below *In hot locales tigers will spend much of the day near streams or other water and often lie or stand in the water to keep cool. Tigers are proficient swimmers and can cross rivers that are 7–8 km (15–17 miles) wide without difficulty.*

Cheetah

CHEETAHS ARE FAST. THEIR ENTIRE PHYSIQUE is designed for speed: slight build, thin legs, a deep, narrow chest, and a small, delicate, domed skull help them attain speeds of 95km/h (60mph) – greater than those of any other land animal.

Cheetahs are easily distinguished from other cats, not only by their distinctive markings, but also by their loose and rangy frame, small head, high-set eyes, and small, rather flattened ears. Their usual prey are gazelles (especially Thomson's gazelle), impala, wildebeest calves, and other hoofed mammals up to 40kg (88lb) in weight. A single adult may kill every few days; however, females with cubs may need to kill something almost every day. They stalk prey using a concealed approach, followed by a full sprint starting from about 30m (33yd) away. About half the sprints end in prey capture. An average chase, which will last between 20–60 seconds, will cover a distance of 170m (183yd); cheetahs cannot continue a chase for more than about 500m (550yd), which is why hunts invariably fail if there is a large initial starting gap between cheetah and prey . On average, a wild cheetah eats approximately 2kg (4.4lb) a day.

Profiting from Maternal Care

SOCIAL BEHAVIOR

Before parturition, mothers select a lair, a rocky outcrop or marshy area with tall grass, and give birth to one to six cubs each weighing 250–300g (8–11oz). Mothers nurse cubs in the lair, leaving them only to go hunting; males show no parental care. Cubs accompany their mothers from eight weeks onwards, when they are first introduced to solid food. Weaning occurs at three to four months, but cubs remain with their mothers until 14–18 months of age.

Cubs play vociferously with each other and rehearse hunting skills using live prey that the mother brings back to them. However, they are still only amateurs at capturing prey at independence. Littermates stay together after adolescence for a period of six months in order to gain safety in numbers. Each sister then quits the group, while their brothers remain together for life. Adult female cheetahs are solitary except when accompanied by cubs, whereas males either live alone or in small coalitions of two or three individuals.

Right The cheetah's small upper canine teeth – as displayed in this snarl – have correspondingly small roots bounding the sides of its nasal passages. This feature affords the animal an increased intake of air, so allowing it to maintain a relentless, suffocating bite on the throat of its prey.

The Danger from Lions

CONSERVATION AND ENVIRONMENT

Cheetahs show low levels of genetic variation, suggesting that they all descend from a very small population that existed 6,000–20,000 years ago. Genetic monomorphism such as this may result in reduced juvenile survival, because deleterious recessive alleles may be exposed, and in disease susceptibility, because if a disease overcomes the immune system of one individual it can overcome the immune systems of all members of the population. Early work suggested that difficulties in breeding captive cheetahs and an outbreak of

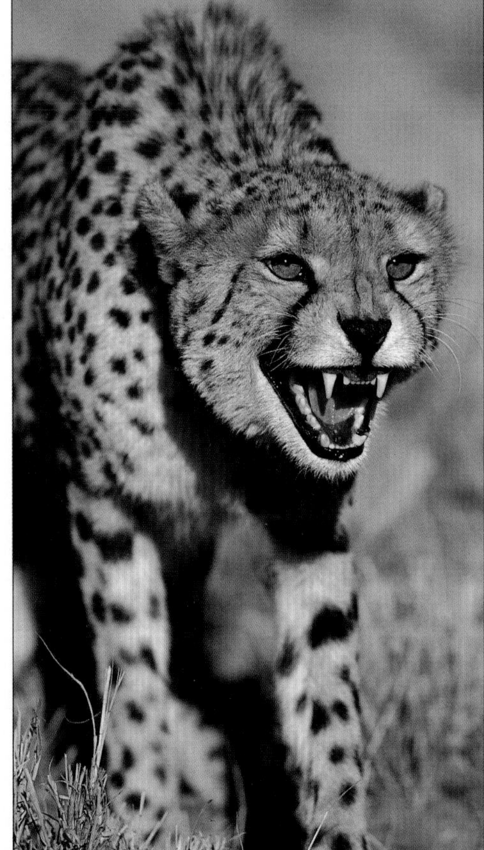

Above Cheetahs, unlike other large cats, have claws that are blunt and straight and cannot be fully retracted. This helps to give them traction while twisting and turning to run down prey such as gazelle. Once the prey is caught, a suffocating bite is applied to the throat. Cheetahs were once tamed for hunting game, for instance by the Mughal emperors of India.

Right Young cheetahs have a thick mantle of smoky gray hair that extends down their nape, shoulders, and back. This mane is pronounced on cubs up to three months old, but dwindles steadily as they mature. The precise function of this hair is unknown, but may mimic the hyena's to deter predators.

CHEETAHS AND GUARD DOGS

The majority of the world's 12,500 remaining wild cheetahs are found outside protected reserves, making them liable to come into conflict with livestock farmers. From 1980–90, more than 6,000 cheetahs were killed by farmers who believed that they either had killed or were going to kill their animals (in fact cheetahs are responsible for less than 5 percent of livestock losses).

To safeguard against cheetahs and other predators, a novel conservation approach in Namibia is reviving the tradition of pitching big cats against big dogs. Anatolian shepherd dogs have been used for 6,000 years to protect sheep flocks on Turkey's arid central plateau. The Anatolian shepherd is large, with excellent eyesight, sharp hearing, and independent behavior.

The dogs are raised with the herd and, true to the nature of canids, protect their charges tenaciously. If they sense danger close at hand, they warn the herd with a loud bark, which also serves to let the predator know of the guardian's presence. If the cat persists in its attack, the dog places itself between it and the herd. Faced by such a large, ferocious guard dog, most cats choose flight over fight. The dogs have proved effective not just against the non-aggressive cheetah, but also against more formidable opponents such as leopards, caracals, jackals, baboons, and even human poachers. LM

FACTFILE

CHEETAH

Acinonyx jubatus

Order: Carnivora

Family: Felidae

2 Subspecies: **African cheetah**, (*A. j. jubatus*); and **Asiatic cheetah**, (*A. j. venaticus*). The **King cheetah**, a mutant form occurring only in S Africa, was once incorrectly described as a separate species, *Acinonyx rex*. Coat: spots along spine joined together in stripes, with small splotches on the body.

DISTRIBUTION Africa, Middle East.

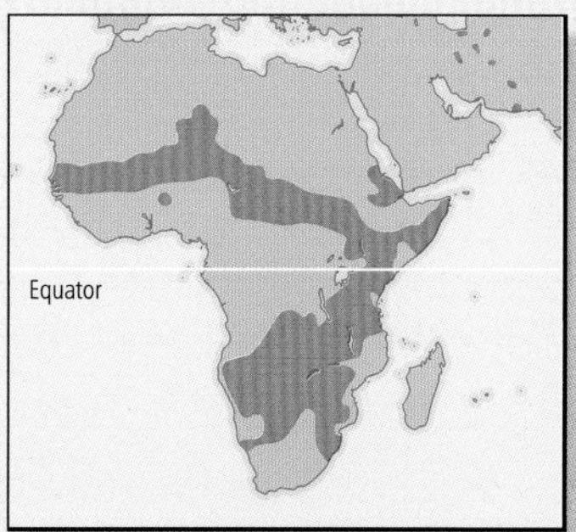

HABITAT Savanna and dry forest.

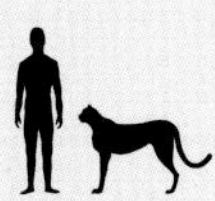

SIZE **Head–body length** 112–135cm (44–53in); **tail length** 66–84cm (26–33in); **weight** 39–65kg (86–143lb). Males are 15 percent larger than females.

COAT Tawny, with small round black spots. Face marked by conspicuous "tear stripes" running from the corner of the eyes; cubs under three months old are blackish, with a mantle of long blue-gray hair on top of the back and neck. Each individual cheetah has a distinctive pattern of spots.

DIET In Africa, mainly middle-sized antelope, Thomson's gazelles, puku, and impala. Cheetahs also feed on hares and newborn gazelles, which they flush while walking through long grass.

BREEDING Females can breed at 24 months and are polyestrous, cycling approximately every 12 days. Males reproduce at 3 years of age.

LONGEVITY Up to 12 years (17 in captivity).

CONSERVATION STATUS Endangered throughout their range, principally as a result of habitat destruction, elimination of their antelope prey by humans, and direct persecution. Between 5,000 and 15,000 cheetahs may remain in Africa, and as few as 200 in Asia, where *A. j. venaticus* is listed as Critically Endangered.

disease in a North American breeding institution might be as a result of this lack of genetic variation, and thus might pose a conservation predicament for the species.

In the wild, however, cheetahs produce cubs rapidly (wild females give birth at intervals of about 18 months, but if the cubs are lost may have another litter much sooner) and no disease outbreaks have been reported. Moreover, breeders have subsequently overcome difficulties in mating and cub-rearing in captivity. As a consequence, current conservation thought regards genetic concerns as unfounded and it may well be that cheetahs are simply less adaptable to the presence of human populations and changes in their environment than other large feline predators, such as the leopard.

For large carnivores, cheetah cubs have an exceptionally high mortality rate. The current focus of attention is on the extent to which cub death as a result of other, larger predators keeps cheetah populations in check. On the Serengeti Plains of Tanzania, for example, lions kill cheetah cubs in the lair so frequently that 95 percent of cubs never reach independence. Across protected areas in Africa, cheetah densities are low where lion densities are high, suggesting such interspecific competition is a common phenomenon.

Cheetahs are therefore "mesopredators," kept in check by larger carnivores. Since elimination of other top predators would have ecosystem ramifications, cheetah conservation may have to rely not just on national parks but also on a patchwork of reserves including pastoral rangelands and farms from which lions and spotted hyenas have already been eliminated. TC

SERENGETI SURVIVAL

Cheetahs' strategies for defending territories

ON THE SERENGETI PLAINS OF TANZANIA, FEMALE cheetahs either live alone or with their dependent cubs, and have enormous home ranges as broad as 800sq km (300sq mi) that circumscribe the annual movements of Thomson's gazelles. By contrast, males live either in lifelong coalitions of two or three animals or else alone. Most of these coalitions are composed of littermates, but approximately 30 percent include an unrelated male. Territorial males, unlike females, do not migrate to follow the prey; however, when there is no food within their territory they will temporarily leave to feed nearby. Superficially, cheetahs resemble lions, in which permanent groups of males, often composed of relatives, jointly defend the pride against other coalitions, but the benefits of pride defence cannot apply to cheetahs since, unlike lions, females cheetahs are both nomadic and solitary.

Adult males exhibit two distinct behavioral tactics. Resident males defend and urine-mark small territories (typically of around 37sq km/14sq mi) although they do not always occupy them year round, whereas non-resident (floater) males roam over huge (777sq km/300sq mi) areas and rarely urine-mark. Non-residents are less relaxed than residents: they sit up and lie alert more often; they exhibit signs of physiological stress, specifically elevated cortisol levels; and they are generally in poorer physical condition, with higher counts of white blood cells and eosinophils (which serve to fight infections), lower muscle mass, and more sarcoptic mange. All male cheetahs start out as floaters when they leave their mother's home range, but whereas some remain non-residents all their lives, others become territorial; however, the opposite course also happens and others that are first encountered as residents end their lives without a territory. These nomadic cheetahs frequently encounter territorial males, who respond aggressively toward them.

Coalitions of males are more likely to obtain a territory than are singleton males (this was true of only 9 percent of 35 singletons observed as opposed to 60 percent of 25 coalitions). The most plausible explanation lies in coalitions' numerical advantage in fights. Fights over territories are an important source of mortality, as males are more likely to die inside or on the immediate borders of territories than outside them, and many males die on territories at the time they are occupied. Males may hold territories for between 4 and 4.5 years. Coalitions are more likely to displace residents from a territory than are singletons, the latter usually acquiring territories by taking over a vacancy. Nevertheless, coalitions hold territories for no longer than singletons; nor do larger groups of males hold larger territories than smaller groups. Thus the key benefit of being a coalition member must be that it gives a male a greater chance of acquiring a territory in the first place.

What are the advantages of territoriality? Greater numbers of females are found on occupied territories than outside territories or on those that are not occupied, and it is four times more common to find males guarding females on occupied territories than elsewhere. Thus territories held by males are female "hotspots." Although the reproductive pay-offs of coalition formation for individual males are not known with accuracy, it appears that males in coalitions can gain prolonged access to areas where females congregate during the course of the year. Thus in both cheetahs and lions, males in coalitions are able to encounter and inseminate greater numbers of females than singleton males. If high densities of females are responsible for group living in both cheetahs and lions, then it should follow that females would be widely dispersed or live at low densities among felids where males are solitary.

HOME RANGES *The maps below show the home ranges of three different cheetah families observed on the Serengeti Plains over the same period. These ranges were overlapped by the home ranges of other cheetah families.*

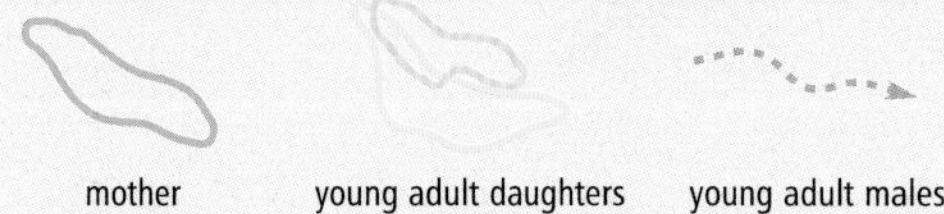

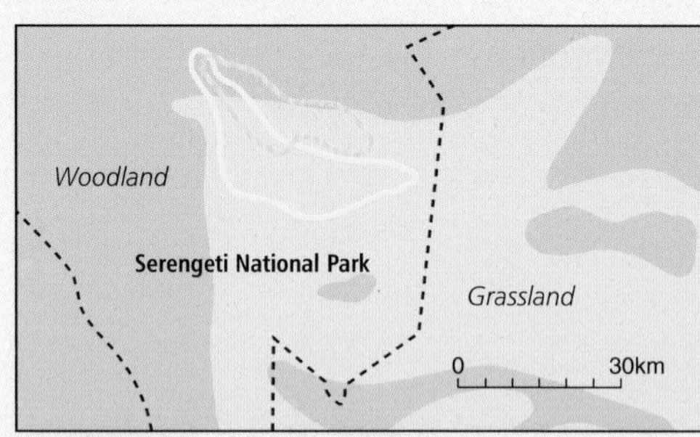

Family 1 *Two female cheetah littermates' home ranges overlap; they also overlap the ranges of several other cheetahs, including their mother and several males. The large range size is a result of their following migratory prey. The home ranges represent the limits of their movement over a full year.*

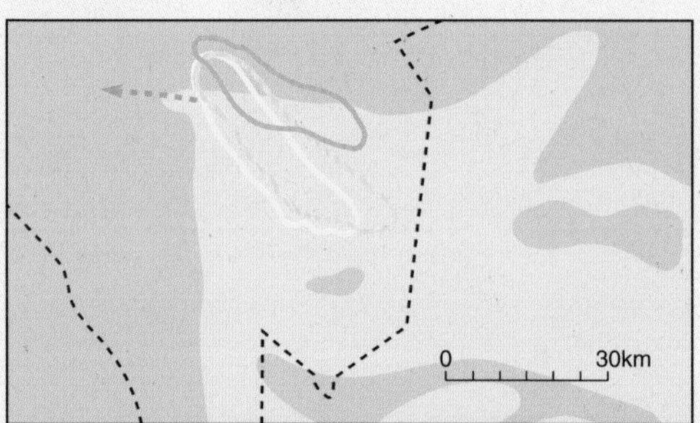

Family 2 *Two adult daughters overlap their mother's home range. The young adult males probably left the area because of aggression from territorial males whose ranges overlap that of their mother. The juveniles will remain nomadic until they can successfully defend a territory.*

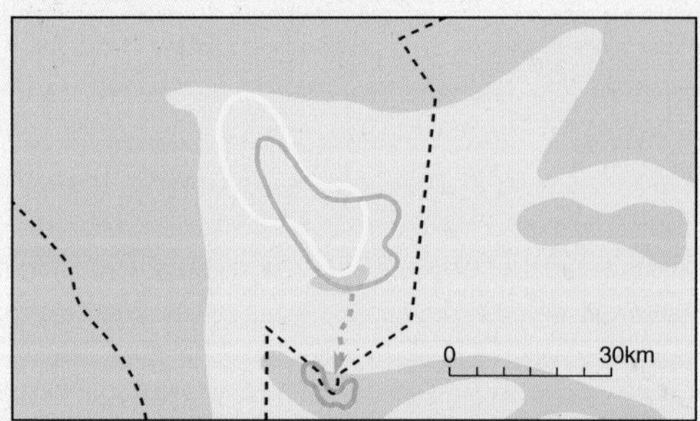

Family 3 *The exclusively grassland home ranges of this mother and daughter overlap considerably. The two young adult males emigrated more than 18km (11mi) from their mother's home territory before successfully ousting two territorial adult males and establishing themselves in their territory.*

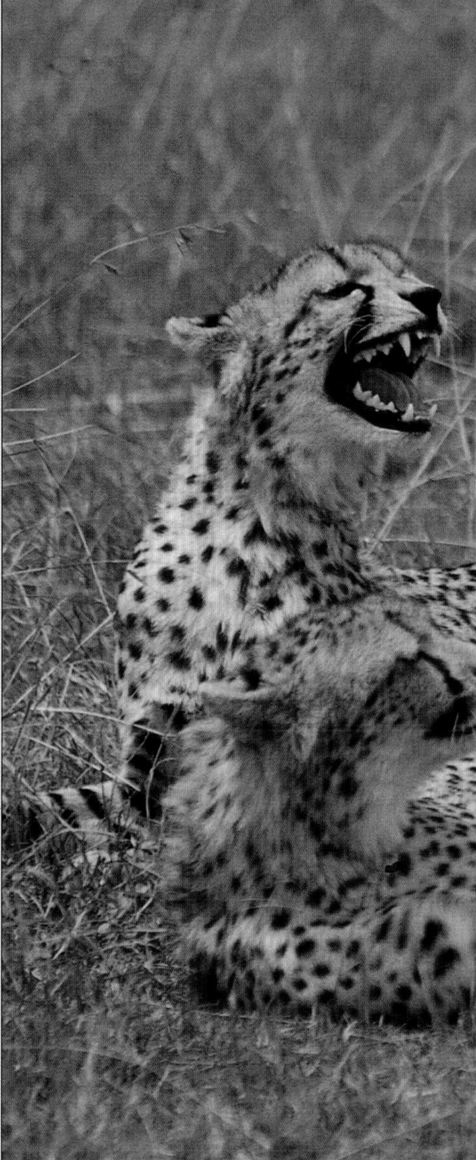

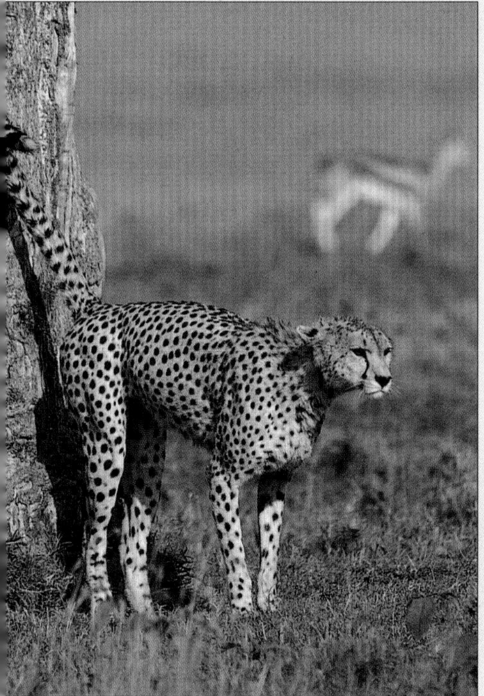

Left *Scentmarking a tree. Territorial males will spray urine onto many prominent landmarks across their range to deter other males. Females in heat scentmark to attract males which, when they discover the scent, hurriedly follow the trail. Nomadic males scentmark only rarely.*

Above *Juvenile cheetahs react defensively to the encroachment of an unfamiliar adult male upon their home range. This group is too young to instigate an attack on this nomadic individual; a territorial adult male, on the other hand, would respond by repelling the interloper.*

And indeed, this does appear to be the case among bobcats, lynx, pumas, leopards, ocelots, snow leopards, and tigers.

It is high female densities and overlapping home ranges together that drive the rare instances of male sociality in felids. The reasons why female cheetahs and all other felids except lionesses and domestic cats live alone are less well understood. One possibility is that among species living in open habitats that usually capture large prey, females are social because large carcasses, which last for some time, risk being seen and stolen by conspecifics, especially in areas where the latter are numerous. Under these circumstances it is more profitable for females to live with relatives and share food rather than relinquish it to non-relatives. These conditions pertain only to lions. An alternative hypothesis is that only among lions are large prey sufficiently numerous to support groups of females living together; females of other species cannot afford to share prey because large prey items (say, one to two times the weight of an adult female) are simply unavailable. A third suggestion, that female lions exhibit sociality for reasons of cooperative defense against infanticidal males, cannot in itself offer sufficient explanation, since other, non-social felids commit infanticide as well.

While the reasons for felids being asocial are not yet resolved, both plausible hypotheses stress the cost of foraging as the main reason preventing the formation of groups in female cats. TC/GWF

LEARNING A PREDATOR'S SKILLS

1 Cheetah cubs playfight with their mother, practicing such hunting skills as the neck bite. Many threats confront cubs as they grow; on average only 1 in 3 reaches adulthood, while in areas like Tanzania's Serengeti Plain, where cheetah cubs fall prey to lions, the rate is closer to 1 in 20. Mothers provide what protection they can, remaining with their families for a year or more after weaning.

2 From the age of 8 weeks, cubs accompany their mother on foraging expeditions, eating from her catches and learning her hunting techniques. However, even though this is a vital part of their education, the presence of boisterous cubs can spoil the mother's chances of catching prey, and put the whole family at risk of starvation.

3 Young cubs investigate a gazelle calf, learning to recognize its smell and behavior. Families sometimes flush out calves from refuges in the long grass; on other occasions mothers will deliberately bring back live prey in order to educate their young. Such experience is very necessary; when they first start hunting on their own, at about 18 months, inexperienced adults often pursue unsuitable prey such as buffalo.

4 Older adolescents corner a fleeing impala calf. As adults, their success in hunting will depend above all on a combination of surprise and sudden acceleration. Although cheetahs are the world's fastest animals, capable of bursts of up to 95km/h (60mph), they have limited powers of endurance and rarely continue a pursuit for more than 500m (545yd). As a result, most chases last for less than 30 seconds, and only about half are successful.

⑤ A juvenile cheetah tackles a warthog piglet – relatively easy prey suitable for a beginner. Later it will graduate to larger kills. Cheetahs' small teeth are not well equipped for tearing flesh, so the animals strangle their prey, cutting off the air supply with a relentless grip. Even then, the catch may not be secure, since cheetahs lose many kills to vultures or lions. To lessen the risk, they normally drag carcasses off to cover, where they can eat their meal undisturbed.

4

Leopard

CONFUSION SURROUNDS THE LEOPARD. EVEN its name originates from the mistaken belief that the animal was a hybrid between the lion (Latin: Leo) and the "pard," or panther. And only just over a century ago, it was still disputed whether leopards and "panthers" were separate species. In fact, the words "pard" and "panther" are vague, archaic terms that have been used to describe several large cats, especially the leopard, jaguar, and puma.

Seven subspecies of this elegant feline hunter have been named. The commonest is the African leopard, which occurs over most of the leopard's range. The other subspecies are small or geographically isolated populations, most of them now critically at risk.

From Leopardskin to Black Panthers

FORM AND FUNCTION

In form the leopard is average among the large cats – slender and delicate compared with the jaguar, but sturdy and solid in comparison with the cheetah. There are various aberrant coat patterns. One of the commonest and most striking alternatives is melanism, which makes the leopard totally black. The recessive gene responsible for this condition is apparently more frequent among leopard populations that live in forests and mountainous regions, and in Asia. In the Malay Peninsula as many as 50 percent of leopards may be black; elsewhere the proportions are much lower. The name "Black panther" is sometimes erroneously applied to such animals in the belief that they are a distinct species. Several other cat species, including the jaguar and serval, also exhibit melanism.

Adaptable Hunters

SOCIAL BEHAVIOR

The leopard is the most widespread member of the cat family, largely thanks to its secretive nature and adaptable foraging behavior. Leopards are highly variable in their diet. Beside scavenging, they catch a great variety of prey species, which includes reptiles, birds, small mammals, medium-sized antelopes, and even occasionally other carnivores, such as Bat-eared foxes and cheetahs. Leopards hunt alone and generally at night, and employ a combination of opportunism, stealth, and speed. They are masters of concealment and are known to stalk to within 2m (6.5ft) of their prey before making a short, fast rush. Small prey is often killed by a bite to the back of the neck. Despite a common misconception to the contrary, leopards do not always store their prey up trees; in fact, most kills are dragged several hundred meters and hidden in thick vegetation. Due to the variety and small size of their prey, leopards avoid intense competition with such fellow-carnivores as lions, tigers, Spotted hyenas, and African wild dogs, which depend on larger prey.

Over most of their range, leopards have no particular breeding season. Females are sexually receptive at 3–7 week intervals, and the period of receptivity lasts for a few days, during which mating is frequent. Litters can contain up to six cubs, but usually consist of one or two blind, furred young weighing 430–570g (15–20oz). The cubs are kept hidden until they start to follow the mother at 6–8 weeks. After the cubs have been weaned, the mother frequently leads them to food but otherwise spends little time with them. When her cubs are 18–20 months old she encourages them to leave and mates again. Leopards are thought to reach sexual maturity at about 2 years' old, the age at which male and female cubs disperse and settle elsewhere, although female cubs frequently settle in an area overlapping their mother's home range.

Leopard females spend nearly half their lives with attendant cubs, but males are almost entirely solitary. The home ranges of leopards are extremely variable in size. High variability in habitats and

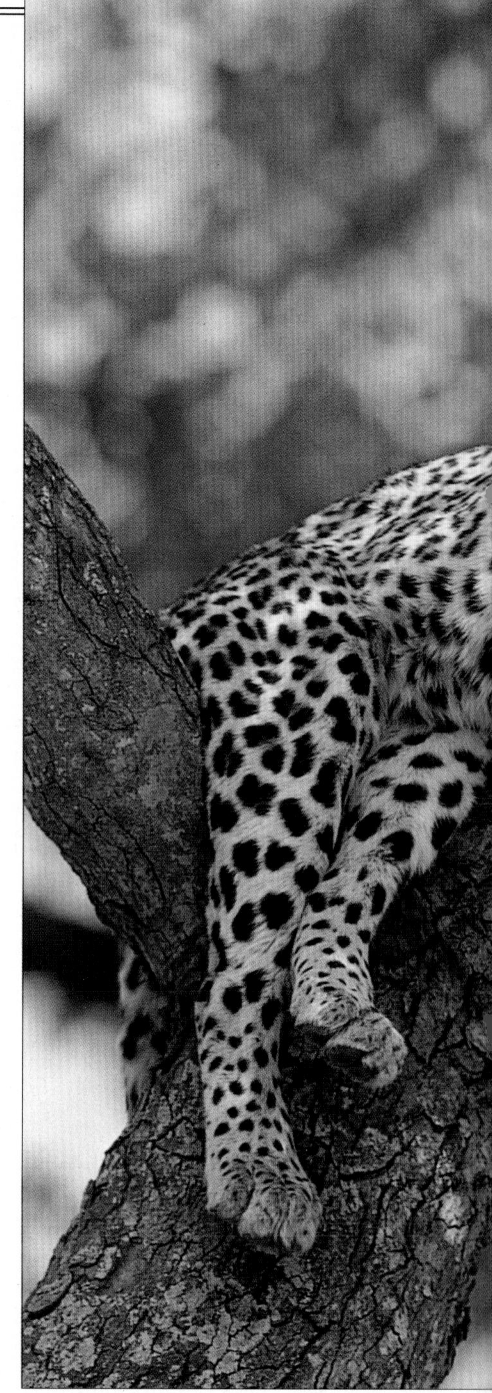

Above *Leopards, like this inhabitant of the Okavango Delta, Botswana, frequently rest in the branches of trees, taking refuge from the heat of the midday sun. Their aptitude for climbing also allows them to evade harassment by other predators; there are many instances of dhole packs treeing leopards in India. Only infrequently will a leopard go to the effort of dragging a kill aloft, to keep it safe from scavengers.*

Above left *An alert leopard partially concealed in undergrowth. The leopard's coat spots provide it with excellent camouflage, especially among trees and bushes, and the prominent, extremely sensitive whiskers (vibrissae) are a hallmark of an animal that hunts by night.*

availability of prey directly affect the density and home ranges of leopards – the fewer prey, the fewer leopards and the larger their home ranges. Females occupy home ranges of 10–290sq km (4–112sq mi) that may overlap by as much as 40 percent with those of other females. Home ranges of males are larger, ranging from 18–1,140sq km (7–440sq mi), with extensive overlap (again as much as 40 percent) between different males. Home ranges of male leopards also overlap extensively with those of females.

The main vocalization is a rough, rasping sound, rather like that of a saw cutting coarse wood; it serves both to proclaim the home-range holder's presence and to make contact between separated individuals. When she is on heat, a female rasps to attract a male, and a mother rasps to call her cubs.

Tourists to the Rescue

CONSERVATION AND ENVIRONMENT

Despite the possible loss of some subspecies, the adaptable and highly successful leopard is likely to continue to thrive in a large part of its present range. Numbers are declining in some areas as a result of habitat loss and of persecution in response to attacks on domestic livestock. However, even in areas where such killings are common, populations appear to be stable.

As well as killing leopards to protect livestock, people also shoot them for sport – in Africa the animal is one of the "Big Five" most highly-rated prey of the sport hunter, the other favored species being the lion, buffalo, elephant, and rhinoceros. The Convention on International Trade in Endangered Species (CITES) allocates annual quotas for trophy hunting to countries that support viable leopard populations. In 1999 a total of 1,635 trophy-hunting quotas were allotted to ten countries, but only 878 leopards were actually hunted, for an average fee of US$2,500.

Leopards are a highly popular attraction for visitors to National Parks, and the growing tourism industry throughout most of their range is making a significant contribution to the survival of the species. A detailed study on leopard-based tourism in Namibia concluded that local communities derived financial benefits amounting to US$2,500 per year by allowing wildlife enthusiasts to view leopards on their land. Thus, in areas where they were once regarded as a nuisance, the animals have now become prized as an important new source of income for the community. PS

FACTFILE

LEOPARD

Panthera pardus

Order: Carnivora

Family: Felidae

1 of 5 species of the genus *Panthera*

7 subspecies: **Amur leopard** (*P. p. orientalis*); **Anatolian leopard** (*P. p. tulliana*); **Barbary leopard** (*P. p. panthera*), Morocco, Algeria, Tunisia; **African leopard** (*P. p. pardus*), Africa except extreme N, Asia; **Sinai leopard** (*P. p. jarvisi*), Sinai; **South Arabian leopard** (*P. p. nimr*); **Zanzibar leopard** (*P. p. adersi*), Zanzibar.

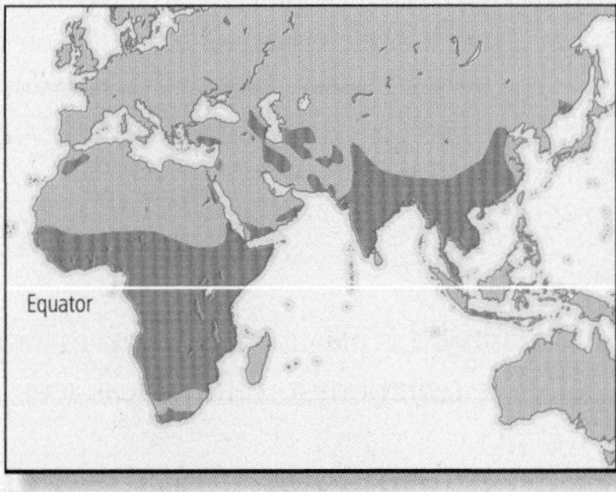

DISTRIBUTION Africa S of the Sahara, and S Asia; scattered populations in N Africa, Arabia, Far East.

HABITAT Most areas with a reasonable amount of cover, a supply of prey animals, and freedom from excessive persecution; from tropical rain forests to barren deserts, from cold mountains almost to urban suburbs.

SIZE Head–body length 100–190cm (40–75in); tail length 70–95cm (28–37in); shoulder height 45–80cm (18–32in); weight 30–70kg (66–155lb). Males are about 50 percent larger than females.

COAT Highly variable, though essentially black spots on a fawn to pale brown background. Typically, the spots are small on the head, larger on the belly and limbs, and arranged in rosette patterns on the back, flanks, and upper limbs.

DIET Reptiles, birds, small mammals, occasionally small carnivores including cheetahs and Bat-eared foxes.

BREEDING Gestation 90–105 days; litter size usually 1–2 (maximum 6).

LONGEVITY Up to 14 years (20 in captivity).

CONSERVATION STATUS The South Arabian, Barbary, Anatolian, and Amur leopards are all Critically Endangered, while the Zanzibar leopard may now be extinct.

Other Big Cats

The jaguar, clouded leopard, and snow leopard live in very different parts of the world, from tropical rain forests to snow-covered mountains. Populations of all three species have declined dramatically in recent decades due to pelt-hunting and extensive habitat loss.

Jaguar

PANTHERA ONCA

The jaguar is the largest cat of the Americas. While similar in appearance to the leopards of Asia and Africa, its behavior and ecological requirements make it the New World equivalent of the tiger.

Although its capacity to roar links it with the big cats, it can more commonly be heard grunting or coughing when hunting, or snarling and growling when threatened (males make a mewing cry during the mating season). The massive head and strong canines are adaptations for crunching skeletal material and cracking open well-armored prey such as turtles.

Jaguars typically prefer dense forest or swamps with good cover, as they normally stalk their prey on the ground; they are also excellent swimmers. While they usually hunt large prey like peccary and deer, over 85 species have been recorded in their diet including monkeys, birds, frogs, fish, and small rodents, as well as domestic livestock.

Jaguars are solitary animals, coming together only in the breeding season. Territories can vary from 25–150sq km (10–60sq mi), depending on prey abundance and diversity. A female can bear up to four young, each weighing 700–900g (25–32oz), although most will raise only two cubs to independence. Cubs are blind at birth, open their eyes after about 2 weeks, and remain with their mother for up to 2 years. Sexual maturity is achieved at approximately 2–3 years for females and 3–4 for males. AR

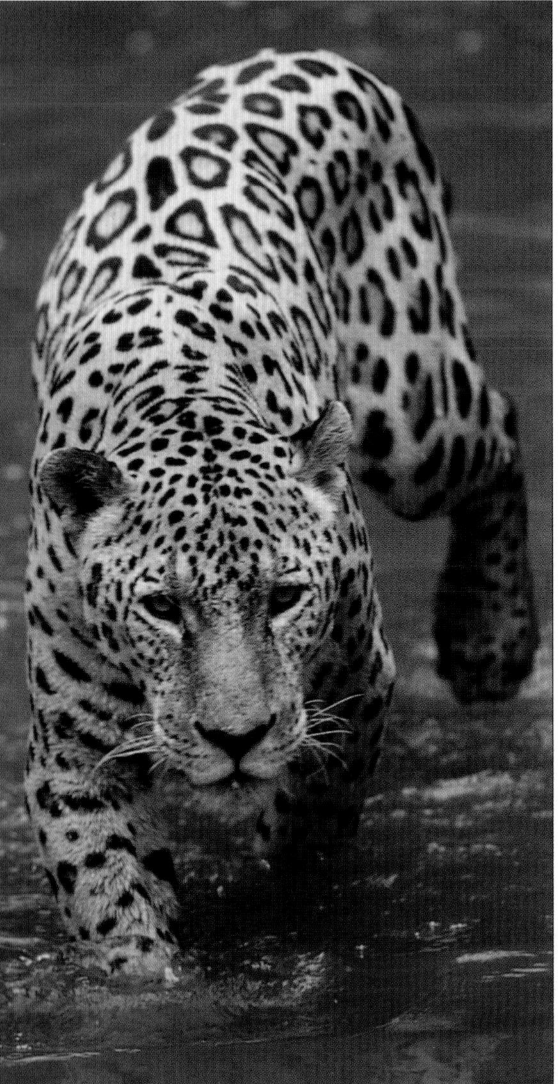

Snow Leopard

PANTHERA UNCIA

Snow leopards are uniquely adapted to extreme conditions. Adaptations for mountain life include a well-developed chest, short forelimbs, and a thick tail nearly a meter (3ft) long that helps them keep their balance; to cope with the cold, they have an enlarged nasal cavity that warms air passing into the body and long body hair with a dense, woolly underfur up to 12cm (5in) thick. Encompassing the Gobi desert in their range, the cats can tolerate temperatures of up to 40°C (104°F) in summer and –40°C (–40°F) in winter.

Blue sheep and ibex are the most widespread prey, but Snow leopards will feed upon a wide range of ungulates, from goats and deer to domestic stock such as subadult yak, sheep, and horses up to three times their weight. Marmots and hares are important food items in summer. A stalker whose camouflage enables it to blend into the rocky background, a Snow leopard must get within 30–40m (100–130ft) of an intended victim before making its final rush.

Home range sizes vary from about 20–40km (12–25mi) in good habitat to 1,000sq km (400sq mi) in prey-sparse Mongolia. Male and female home ranges overlap extensively. Use of a particular area is separated temporally, so that different animals are usually at least 1km (0.6mi) or more apart, except when a female is in estrus.

Snow leopards live alone, except during their mating season from January to March, when their long-drawn-out, wailing calls could be mistaken for those of a yeti, or while females are accompanied by dependent offspring. Common travel lanes are marked with scrapes, feces, and pungent scent-sprays, disclosing the informants' sex and reproductive status.

One to four young are born in spring or early summer in a well-concealed rocky den. Initially, their spots are almost completely black. Opening their eyes on day 7–9, they are quite active by 3 months, but are thought to remain with their mother up to the age of 18–22 months. The interval between births is at least two years.

Snow leopards have become extremely rare in large parts of their range. The primary threats include hunting for coats, bones, and body parts. They are thought to occur in about 120 protected areas, but most of these reserves are too small to support more than a few animals. R

Above *Rare, secretive, and inhabiting some of the most remote terrain on Earth, the Snow leopard has near-mythical status. Until recently, little was known about this magnificent cat, but modern research techniques such as satellite tracking have shed new light on how it has adapted to life on the "Roof of the World."*

Right *A Clouded leopard moving stealthily through the dense Malaysian jungle. Though its local name, meaning "branch-of-tree tiger," suggests it is chiefly arboreal, recent research indicates that it may be more of a habitat generalist than was once supposed.*

Left *The current distribution of the jaguar is barely half what it was in the early part of the 20th century, when its range extended from the southwestern United States to northern Argentina. The killing of jaguars is now banned throughout their range, but illegal hunting for their highly sought-after pelts continues unabated.*

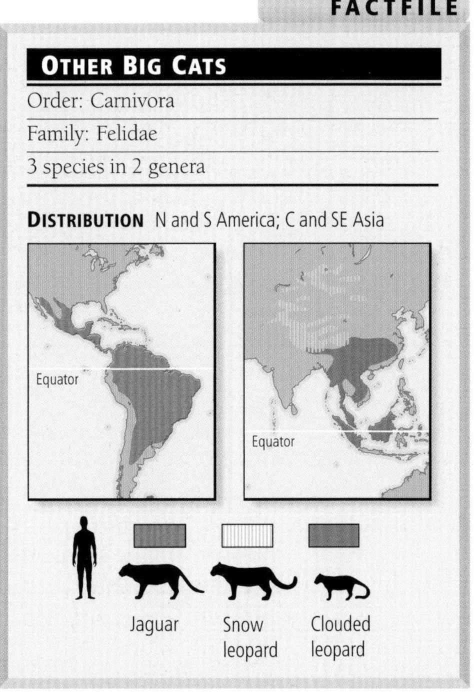

FACTFILE
OTHER BIG CATS
Order: Carnivora
Family: Felidae
3 species in 2 genera
DISTRIBUTION N and S America; C and SE Asia
Equator
Equator
Jaguar
Snow leopard
Clouded leopard

Clouded Leopard

NEOFELIS NEBULOSA

Secretive and reportedly strictly nocturnal, the Clouded leopard is one of the world's most enigmatic felids. Unlike the Snow leopard, it is disinclined to leave scat and scrapes along trails, so its presence is easily missed.

Clouded leopards' tree-climbing skills are regularly observed in captivity, where the animals will run down tree trunks headfirst, clamber upside down along horizontal branches, or hang by their hind paws with the tail providing balance. In the wild, they appear mainly to rest in trees. Early naturalists speculated that their long canines and stocky build were adaptations for relatively large ungulate prey, but the few field sightings and scats examined to date indicate that primates like Proboscis monkeys, Pig-tailed macaques, and gibbons are the primary food items. Other reported prey include birds, small mammals, porcupines, deer, and wild boar, as well as livestock including chickens and goats.

There is very little information on the Clouded leopard's social and reproductive behavior in the wild. One to four blind and helpless young are born after 87–102 days of gestation; their coloration and coat patterning differ from adults' in that the large spots on the sides are completely dark. Cubs open their eyes after 10–11 days, and are quite active by 5 weeks, probably achieving independence by 9 months. Poor breeding success in captivity, due to the notoriously high incidence of males killing females, has now been resolved by "pair-bonding" mates at an early age.

Primary threats to the Clouded leopard are deforestation and hunting; the pelt, bones, and long canines are sold to outsiders and local tribes for medicinal, decorative, and ceremonial purposes. Densities range from one per 4km (2.5mi) in Sabah to one per 14km (9mi) in Thailand, although Borneo is thought to contain the densest population. RJ

Jaguar LR
Panthera onca

SW USA to C Patagonia. Tropical forest, swamps, and open country, including desert and savanna.
HBL 112–185cm (44–73in); TL 45–75cm (18–30in); SH 68–76cm (27–30in): WT 57–113kg (126–249lb); females are on average 20 percent smaller than males.
COAT: basically yellowish-brown, but varying from almost white to black, with a pale chest and irregularly placed black spots on belly; back marked with dark rosettes; lower part of tail ringed with black; a black mark on the lower jaw near the mouth; outer surface of ear pinnae black.
BREEDING: gestation 93–110 days.
LONGEVITY: up to 20 years in captivity.

Snow leopard En
Panthera (Uncia) uncia

Patchy occurrence across the high mountains of C Asia, from the Himalayas to S and W Mongolia and S Russia. Occurs in 12 countries, with China constituting 60 percent of its total range. Alpine steppe, grassland, scrub, and open coniferous forest from 1,800–5,500m (6,000–18,000ft), but as low as 900m (3,000ft) in Russia and the Gobi Desert of Mongolia; avoids dense montane forest. Total population in wild estimated at 4,500–7,500; densities range from over 10 to less than 0.5 per 100km (62mi).
HBL up to 130 cm (51in); TL 80–100cm (31–39in); SH 60cm (24in); WT male: 45–55kg (100–120lb), female 35–40kg (77–88lb).
COAT: long, thick, smoky-gray tinged with yellow, patterned with dark gray rosettes and black spots; tail long, thick, and heavily furred; ear pinnae black-edged; winter coat lighter.
LONGEVITY: to 21 years in captivity.

Clouded leopard Vu
Neofelis nebulosa

India, S China, Nepal, Burma, Indochina to Sumatra and Borneo, possibly also Taiwan (where it may be extinct). Tropical and subtropical primary and secondary evergreen forests up to altitudes of 2,000m (6,500ft); also recorded in Nepal from scrub forest and tall grassland, and from mangrove swamps in Borneo.
HBL 60–110cm (24–43in); TL: 60–90cm (24–35in); SH 80cm (31in); WT 11–20kg (24–44lb).
COAT: ochraceous to tawny or silvery gray, although black and whitish individuals have been reported; large "cloud-shaped" ellipses along the flank strongly edged in black, with solid black ovals and other spots along limbs and belly; back of neck conspicuously marked with two thick black bars; tail encircled with black rings.
LONGEVITY: to 17 years in captivity.

Abbreviations HBL = head–body length TL = tail length SH = shoulder height WT = weight

LR Lower Risk En Endangered Vu Vulnerable

Small Cats

FOR MOST PEOPLE THE TYPICAL SMALL CAT IS the familiar pet of millions of households around the world. Of the many wild species, some do indeed resemble their domestic brethren, both in looks and behavior. But not all members of the family are in fact small – the puma is actually bigger than some "big cats," such as the leopard and the cheetah.

An inability to roar (due to the hardening of the hyoid bone in the throat) has traditionally been used to distinguish the small cats from the big cats of the *Panthera* lineage (i.e. the lion, leopard, tiger, jaguar, and Snow leopard). Yet finding other common traits that link them is difficult; molecular genetic data suggests that they probably derived from at least seven different evolutionary lineages, all descending from common ancestry.

Solitary and Secretive Hunters

FORM AND FUNCTION

Species of the *Felis* lineage range in size from the diminutive Rusty-spotted cat's 1.1–1.6kg (2.4–3.5lb) to the puma's impressive 40–110kg (88–243lb). Family members include the Mountain cat, inhabiting the treeless *altiplano* of the South American Andes; the Sand cat of many of the desert environments of Africa and Asia; the Canadian lynx, common in the wintry reaches of North America; and the Jungle cat from tropical Asia. Molecular genetic studies indicate that, from an evolutionary perspective, the cheetah should also be considered a small cat, although it has commonly been distinguished from the others on morphological and behavioral grounds.

Because they are secretive and difficult to observe, the habits of small cats remain hard to fathom. Even so, some general tendencies have come to light from in-depth studies on certain species – notably the bobcat, wildcat, lynx, and puma. Most are solitary and do not form long-term associations with other individuals. Pumas, for instance, live in isolation, never in social groups of prides or packs. Seclusion and large home ranges are essential for their survival, and for the most part they achieve this by inhabiting remote tracts of jungle, mountain, and swampland. Going against this general trend is the feral domestic cat, which becomes sociable when human food scraps are abundant.

All small cats are almost exclusively carnivorous and kill their own prey, though given the opportunity they will also occasionally eat carrion. The exact makeup of their diets depends largely upon what animals are available. The average size of prey generally correlates well with cat size, measured, for example, by body mass or jaw length. Pumas take large wild ungulates (guanacos, tapirs, bighorn sheep, and moose), in addition to domestic livestock as large as fully-grown horses, burros, and cattle. But since cats are both opportunistic and highly effective hunters, they will also prey on a range of small mammals, lizards, amphibians, birds, and fish. Thus, pumas are also partial to rabbits, and even mice and birds. Likewise in Africa, caracal, which weigh just 10–18kg (22–40lb), commonly take prey as large as antelope (and more rarely springbok and kudu), but they are also excellent birdcatchers and can rely on small rodents for their daily feed.

With the exception of the Black-footed cat, Sand cat, Pallas's cat, and Mountain cat, small cats tend to be associated with either open or closed forest habitats. Most species are not habitat specialists, however, and dwell in a variety of vegetation types. Many species occupy human-altered habitats and are tolerant of limited amounts of human activity; for example, Jungle cats often inhabit the cultivated fields and forest plantations of Asia, as do Tiger cats in parts of Brazil and Rusty-spotted cats in India and Sri Lanka.

Saving Skins

CONSERVATION AND ENVIRONMENT

All small cats, with the exception of the Domestic cat, are managed and regulated to various degrees by national and international laws and treaties. But in practice, the actual level of protection on the ground varies greatly around the world, depending on local densities, traditions, and interactions with humans and their livestock.

The international market in small-cat products, which averaged between 250,000 and 600,000 pelts a year during the 1960s and 1970s, and took an especially heavy toll on the ocelot and Geoffroy's cat, diminished significantly toward the end of the century. As people grew aware of the negative impact of the trade on cats, they became more reluctant to buy fur coats. Consequently, by 1990 the legal market in cat furs had slumped to only a quarter of what it was in 1980, falling from around 450,000 to 100,000 pelts a year, and by the mid-1990s consisted almost totally of lynx and Leopard cat pelts.

Local persecution by sport hunters or farmers protecting livestock is still a threat to small cats, especially when populations are vulnerable for other reasons, such as habitat degradation. Habitat loss causes absolute loss of animals as well as isolation of populations due to changes in terrain that impede movement. This, in turn, reduces the effective population size, as only animals that can reach each other can interbreed. As a case in point, hunting and habitat loss have virtually extirpated pumas east of the Mississippi River, save for

Below *Ten species of small cat, arranged in a west (America) to east (Asia) order reflecting distribution:* ***1*** *Ocelot;* ***2*** *Tiger cat;* ***3*** *Jaguarundi;* ***4*** *and* ***5*** *European and African Wild cat;* ***6*** *Black-footed cat;* ***7*** *Sand cat;* ***8*** *Jungle cat;* ***9*** *Leopard cat;* ***10*** *Asiatic golden cat.*

FACTFILE

SMALL CATS

Order: Carnivora

Family: Felidae

All 30 species of the genus *Felis*. Species include the Wild cat (*F. silvestris*); lynx (*F. lynx*); caracal (*F. caracal*); puma or cougar (*F. concolor*); bobcat (*F. rufus*); ocelot (*F. pardalis*); Margay cat (*F. wiedi*); Geoffroy's cat (*F. geoffroyi*).

DISTRIBUTION N and S America, Eurasia, Africa.

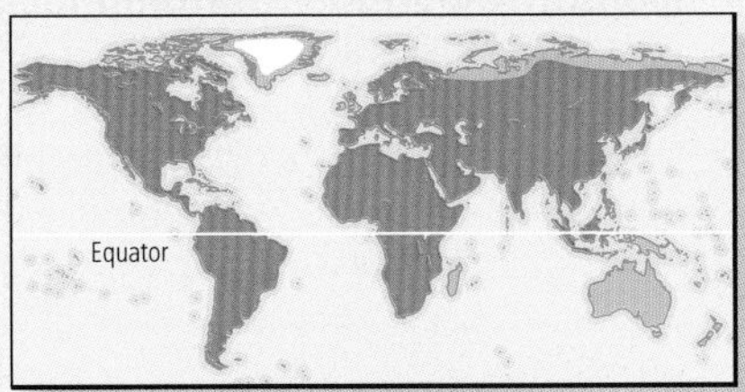

HABITAT From arid regions with sparse cover (Desert cat), through steppe, bush, and savanna (African wild cat), to cool-temperate forest (European wild cat).

SIZE Head–body length from 35–40cm (14–16in) in the Black-footed cat to 105–196cm (41–77in) in the puma; **weight** from 1–2kg (2.2–4.4lb) to 110kg (243lb).

COAT Most often spotted or striped, sometimes uniform; face markings often striped with a black "tear" stripe from the eye.

DIET Mostly small mammals, though the puma will eat wild ungulates and livestock up to fully-grown cattle.

BREEDING Gestation from about 56 days in the Leopard cat to 90–96 days in the puma; litter size from 1–6.

LONGEVITY 12–15 years in the European wild cat.

See species table ▷

▶ ***Right*** *Superficially, the puma resembles a lioness, a likeness that struck Christopher Columbus and Amerigo Vespucci, early European explorers in the Americas. But the puma is in fact very different from the big cats of the Old World. It lives a solitary existence over 110° of latitude, from the Yukon tundra down through the Rockies, Central America, the Andes, and Patagonia to the Straits of Magellan. Also, unlike big cats, it can emit a soft and continuous purr.*

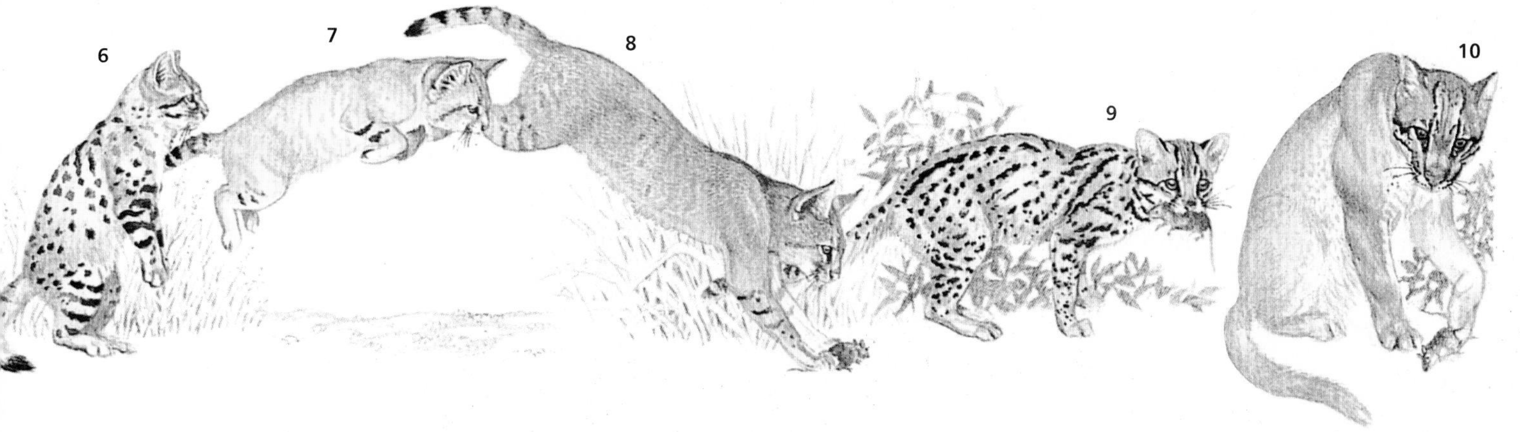

The World's Most Threatened Felid

The Iberian lynx is believed to have evolved from the Cave lynx, a 30-kg (66-lb) felid that roamed continental Europe some 100,000 years ago. Paleontologists think that Pleistocene glaciations isolated a population of these cats between the northern ices and the Mediterranean Sea. The same conditions stimulated the evolution of a new lagomorph species into what we now know as the European rabbit. In response, natural selection shrank the Cave lynx to a size better suited to living off this new delicacy, and gave it a well-camouflaged spotted coat. The new features also reduced competition with the Eurasian lynx, the largest living species of lynx, which was simultaneously spreading into Europe from Asia. While the small Iberian lynx eats rabbits, the bigger Eurasian lynx eats roe deer and chamois.

Throughout the Iberian Peninsula and the south of France, archaeological sites have yielded many remains of Iberian lynx (and its rabbit prey), providing evidence of early human persecution of the cat for its skin and flesh. Its habitat steadily dwindled, as Mediterranean scrubland gave way to arable and livestock farming and forestation. Nevertheless, lynx continued to flourish in the south up to the first half of the 19th century, while at the same time becoming scarcer in the north.

By the early 20th century, the population was restricted to mountainous areas in the southwest of Iberia. Two factors were to hasten the further demise of even this drastically reduced group. First, the lynx were systematically persecuted by the Spanish government. To make matters worse, myxomatosis was introduced from France, decimating rabbit populations. Though given protected status in Spain and Portugal in 1974, lynx numbers did not improve, as they were caught in traps and snares set for other animals. Today, only a few survive; they are still illegally shot, their habitat continues to be degraded, and some animals are killed by cars. Moreover, their rabbit prey is now threatened anew, by viral haemorrhagic disease. Lynx currently inhabit less than 2 percent of the Spanish countryside (as compared to 11 percent in 1960). Small populations live in ten or so isolated locations, and total about 600 animals.

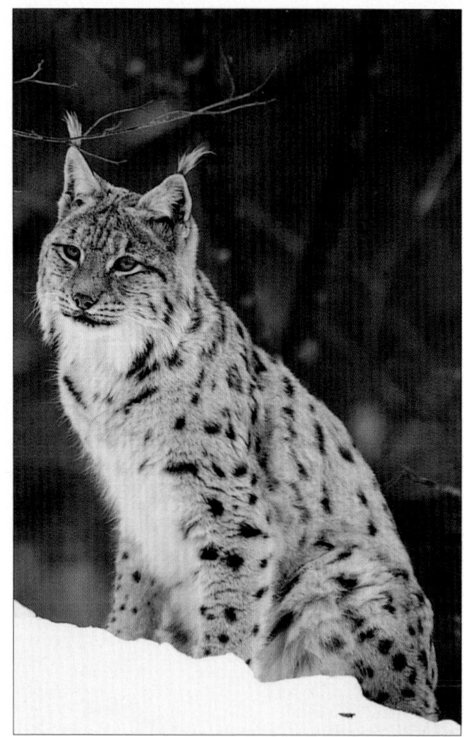

The IUCN considers the Iberian lynx the world's most vulnerable felid, and it seems set to disappear forever in the first decades of the 21st century. European conservationists are desperately concerned that this should not be allowed to happen. MD

a tiny relict population in the Florida cypress swamps. There are fewer than 40 of these so-called "Florida panthers" left, and close inbreeding – a consequence of drastic population reduction over a century of depredation – has led to severe heritable genetic abnormalities. These include kinked tails, holes in the heart, double and single cryptorchidism (small testes), and malformed spermatozoa affecting some 90 percent of the males.

The conservation story for small cats is not all negative, however. The impact of legal hunting and commerce has been most intensively studied in North America, where the exploitation and control of bobcat, lynx, and puma populations as a multimillion-dollar industry is often held up as an example of how cat populations can be managed as a renewable resource. And where it has been necessary to impose an outright hunting ban, such as in the American West, pumas are making a comeback in places like the California Sierra, where the population has doubled in size. Other conservation initiatives include the reintroduction of Texas cougars to Florida. In most of the world, though, the scarcity of reliable information and the lack of adequate local controls accentuate the need for continued and improved protection of cat populations.

Modern research techniques are extending our knowledge of small cat distributions and relative abundances. Automatic camera traps are used to capture images of such elusive species as Golden cats and Leopard cats in Asia, Iberian lynx in Spain, and Kodkod, Ocelot, and Margay in Central and South America. Remote radio tracking of animals and computer-assisted analyses of habitat availability are enhancing the study and monitoring of populations. Molecular genetic techniques are providing information on the evolutionary history of cat populations and data on social structure and behavior. WJ/SJO'B/MC

Left *An ocelot moving through the rain forest of Ecuador. Its striking spotted and striped pelage affords it good camouflage in the dense vegetation of its habitats. Yet the beauty of its coat also made the ocelot the most favored species of the international fur trade, before it was given legal protection.*

Right *The large, tufted ears and slender face clearly distinguish the caracal of Africa and the Near East from other small cats. Caracals are supremely agile hunters that can leap acrobatically into the air to bring down prey, such as guinea fowl, on the wing. They were once tamed and trained by hunters in India and Persia to catch deer and game birds.*

Small Cat Species

Genus *Felis*

FOR AN ACCOUNT OF THE ORGANIZATION of cat lineages, see Cat Family, Classification of Cats. Some authorities split the 30 *Felis* species below between 14 genera, given here in parenthesis.

DOMESTIC CAT LINEAGE

Jungle cat
Felis chaus

Egypt to Indochina and Sri Lanka. Dry forest, woodland, scrub, reed beds, often near human settlements. Preys on rodents and frogs, occasionally birds; most active in day.
HBL 60–75cm; TL 25–35cm; WT 7–13.5kg.
FORM: coat sandy-brown to yellow-gray, sometimes with dark stripes on face and legs and with a ringed tail; young have distinct close-set striped pattern which disappears in the adult; tail short; legs long; ears tapered and tufted, with a light spot at the base.
BREEDING: gestation 66 days; litter 2–5.

Pallas's cat
Felis (Otocolobus) manul
Pallas's cat or manul

Iran to W China. Mountain steppe, rocky terrain, woodland. Preys mainly on rodents.
HBL 50–65cm; TL 21–30cm; WT 3–5kg.
FORM: coat long, orange-gray, with black and white head markings; belly light gray; ears small and rounded, widely separated on a broad head with low forehead; pupils contract to a circle; front premolars missing, giving 28 teeth; eyes face almost directly forward.
BREEDING: Litter 1–5.

Sand cat
Felis margarita

N Africa and SW Asia (Sahara to Baluchistan). Desert. Preys on small rodents, lizards, insects: nocturnal.
HBL 40–57cm; TL 25–35cm; WT 2–2.5kg.
COAT: plain yellow-brown to gray-brown; tail ringed, with black tip; kittens have clear coat markings that usually fade with age; hairy paw pads; head very broad; eyes large and forward on head; ears tapered.

Black-footed cat
Felis nigripes

S Africa, Botswana, Namibia. Steppe and savanna. Preys on rodents, lizards, insects.
HBL 35–40cm; TL 15–17cm; WT 1–2kg.
FORM: coat light brown with dark spots on body and black patches on underside of feet; legs with wide black rings on upper parts; skull broad; ears large; pupils contract to slit; hair on soles of feet.
BREEDING: gestation 63–68 days; litter 2–3.

Wild cat
Felis silvestris

Includes Domestic cat (*F. s. catus*) and African wild cat (*F. s. lybica*). W Europe to India; Africa (*F. s. catus* worldwide – introduced by man). Open forest, savanna, steppe. Preys on small mammals and birds; nocturnal.
HBL 50–80cm; TL 28–35cm; WT 3–6kg (slightly smaller for *F. s. lybica*).
COAT: medium brown, black-striped; *F. s. lybica* is light brown with stripes; *F. s. catus* shows many color forms; females generally paler than males; tail black-tipped.
BREEDING: gestation 68 days; litter 3–6.

Chinese desert cat
Felis bieti

C Asia, W China, S Mongolia. Steppe and mountain. May be endangered; ranked Data Deficient by IUCN.
HBL 70–85cm; TL 30–35cm; WT about 5.5kg.
FORM: coat brownish-yellow with dark spots merging into stripes; underside paler; red tinge on the back; tail ringed; skull broad; ears large; soles of the feet padded with fur.

OCELOT LINEAGE

Ocelot
Felis (Leopardus) pardalis

Arizona to N Argentina. Forest and steppe. Preys on small mammals, birds, reptiles; excellent climber and swimmer; may live in pairs.
HBL 65–97cm; TL 27–40cm; WT 11–16kg.
FORM: coat ocher-yellow to orange-yellow in forested areas, grayer in arid scrubland; black striped and spotted; underside white; tail ringed; eyes brownish; hair curls at the withers to lie forward on upper neck.
BREEDING: gestation 70 days; litter 2–4.

Tiger cat LR
Felis (Leopardus) tigrinus
Tiger, Little spotted cat, Ocelot cat, oricilla

Costa Rica to N Argentina. Forest. Preys on small mammals, birds, lizards, large insects; good climber.
HBL 40–55cm; TL 25–40cm; WT 2–3.5kg.
COAT: light brown with very dark brown stripes and blotches; underparts lighter; white line above the eyes.
BREEDING: gestation 74 days; litter 1–2.

Margay cat
Felis (Leopardus) wiedii
Margay cat or tigrillo

N Mexico to N Argentina. Forest, scrubland. Preys on: rats, squirrels, opossums, monkeys, birds; excellent climber.
HBL 45–70cm; TL 35–50cm; WT 4–9kg.
FORM: coat yellow-brown with black spots and stripes; tail ringed; eyes large, dark brown.
BREEDING: litter 1–2.

Mountain cat Vu
Felis (Oreailurus) jacobita
Mountain cat or Andean cat

S Peru to N Chile. Mountain steppe. Preys on small mammals and birds.
HBL 70–75cm: TL about 45cm; WT 3.5–7kg.
COAT: brown-gray with dark spots, a ringed tail, and white belly; long- and thick-haired, especially on the tail, which appears perfectly round.

Kodkod Vu
Felis (Oncifelis) guigna
Kodkod, Chilean cat, or huiña

C and S Chile, W Argentina. Forest. Preys on birds and small mammals; probably nocturnal.
HBL 40–52cm; TL 17–23cm; WT 2–3kg,
COAT: gray varying to ocher-brown, with dark spots and ringed tail; underside whitish; prominent dark band across the throat, but few markings on the face.

Geoffroy's cat
Felis (Oncifelis) geoffroyi
Geoffroy's cat or Geoffroy's ocelot

Bolivia to Patagonia. Upland forests and scrub. Preys on birds and small mammals; climbs and swims well.
HBL 45–70cm; TL 26–35cm; WT 2–3.5kg.
COAT: silver-gray, through ocher-yellow to brownish-yellow with small black spots.
BREEDING: litter 2–3.

Pampas cat
Felis (Oncifelis) colocolo

Ecuador to Patagonia. Grassland, forest, scrub. Eats small- to mid-sized rodents, birds, lizards, large insects; probably nocturnal. (Previously called *F. pojeros*, from the Spanish *paja*, meaning "straw," because it lives in reed beds.)
HBL 52–70cm; TL 27–33cm; WT 3.5–6.6kg.
FORM: coat long, soft, gray-brown with brown spots (very variable) and with reddish-hue; ears tapered and tufted; eyes yellow-brown: pupil contracts to a spindle.

BAY CAT LINEAGE

Bay cat Vu
Felis (Catopuma) badia
Bay or Bornean red cat

Borneo. Rocky scrub. Preys on small mammals and birds.
HBL about 50cm; TL about 30cm; WT 2–3kg.
FORM: coat: uniform bright reddish-brown, lighter colored on underside; head short and rounded.

Asiatic golden cat LR
Felis (Catopuma) temmincki
Asiatic golden cat or Temminck's golden cat

Nepal to S China and Sumatra. Forest. Lives on rodents, small deer, game birds.
HBL 75–105cm; TL 40–55cm: WT 6–11kg.
COAT: uniform golden-brown; head typically striped with white, blue, and gray.
BREEDING: litter 2–3.

CARACAL LINEAGE

Caracal
Felis (Caracal) caracal
Caracal, lynx, African lynx

Africa and Asia from Turkestan and NW India to Arabia. Wide habitat tolerance. Preys on rodents and other small mammals including young deer, which are either run down or pounced on; mainly active at twilight but will hunt at night in hot weather or by day in the winter.
HBL 55–75cm; TL 22–23cm; WT 16–23kg.
COAT: reddish-brown to yellow-gray; underside white; ears tufted; legs very long; eyes yellow-brown; pupils contract to a circle.
BREEDING: gestation 70–78 days; litter 1–4.

African golden cat
Felis (Profelis) aurata

Senegal to Zaire and Kenya. Forest and dense scrubland. Lives on small mammals and birds.
HBL 70–95cm; TL 28–37cm; WT 13.5–18kg.
FORM: coat chestnut-brown to silver-gray;

ABBREVIATIONS HBL = head–body length TL = tail length WT = weight
Approximate nonmetric equivalents: 10cm = 4in; 230g = 8oz; 1kg = 2.2lb

Ex Extinct
EW Extinct in the Wild
Cr Critically Endangered
En Endangered
Vu Vulnerable
LR Lower Risk

patterning variable in type and extent; eyes brown; tail tapered at tip; ears small and rounded.

LYNX LINEAGE

Lynx
Felis (Lynx) lynx
Lynx, Eurasian or Northern lynx

W Europe to Siberia. Coniferous forest and thick scrub. Preys on rodents, small ungulates; crepuscular.
HBL 80–130cm; TL 5–19cm; WT 8–31kg.
FORM: coat light brown with dark spots; tail black-tipped (coloration and patterning very variable); ear tufts long and black; two tassels on throat; tail short; paws large with thick fur padding; pupils contract to a circle; 28 teeth.
BREEDING: gestation 60–74 days; litter 1–5.

Canadian lynx
Felis (Lynx) canadensis

Alaska, Canada, N USA. Dense forests and tundra. Hunts deer, showshoe hares, birds.
HBL and TL As for Lynx; WT 5–17kg.

Iberian lynx En
Felis (Lynx) pardinus

Spain and Portugal; Woodland and scrubland. Preys on rabbits and gamebirds.
HBL 65–100cm; TL 5–19cm; WT 5–13kg.
FORM: coat marked with black spots on body, tail, and limbs.

Bobcat
Felis (Lynx) rufus
Bobcat or Red lynx

S Canada to S Mexico. Rocky scree, rough ground, thickets, swamp. Preys on rodents, small ungulates, large ground birds; active at twilight.
HBL 62–106cm; TL 10–20cm; WT 6–31kg.
FORM: coat barred and spotted with black on reddish-brown (very variable) basic color; underside white, tail tip black; heavily built with a short tail and short ear tufts.
BREEDING: gestation 60–63 days; litter 1–4.

Marbled cat
Felis (Pardofelis) marmorata

Sumatra, Borneo, Malaya to Nepal. Forest. Preys on rodents, birds, small mammals, insects, lizards, snakes; nocturnal and arboreal.
HBL 40–60cm; TL 45–54cm; WT ca. 5.5kg.
COAT: soft, long fur, light brown with striking patterns of dark brown blotches and spots all over.

ASIAN LEOPARD CAT LINEAGE

Leopard cat
Felis (Prionailurus) bengalensis
Leopard cat or Bengal cat

Sumatra, Java, Borneo, Japan, Philippines, Taiwan. Forest, scrubland, especially near water. Eats rodents, small mammals, birds, which it drops on from above; crepuscular and nocturnal; good swimmer and climber.
HBL 35–60cm; TL 15–40cm; WT 3–7kg.
COAT: base color ocher-yellow to ocher-brown; underside paler; black-spotted coat; white spot between eyes; eyes yellow-brown to greenish-yellow; ears rounded and black with a white spot on back.
BREEDING: gestation about 56 days; litter 2–4.

Iriomote cat En
Felis (Prionailurus) iriomotensis

Iriomote and Ryukyu islands. Sub-tropical rain forest, always near water. Preys on waterbirds, small rodents, crabs, mudskippers; nocturnal and strictly territorial with ranges up to 2sq km (0.8sq mi). May be subspecies of *F. bengalensis*

Flat-headed cat Vu
Felis (Prionailurus) planiceps

Borneo, Sumatra, Malaya. Forest and scrub; near water. Preys on small mammals, birds, fish, amphibians; nocturnal.
HBL 41–50cm; TL 13–15cm; WT 5.5–8kg.
FORM: coat plain reddish-brown, underside white; dark spots on throat, belly, inner sides of legs; ears black with ocher spot at base; tear streaks white; head slightly flattened; legs short; paws small; ears small and rounded; claws not fully retractile.

Fishing cat LR
Felis (Prionailurus) viverrinus

Sumatra, Java, to S China and India. Forest, swamps, marshy areas (dependent on water). Preys on fish, small mammals, birds, insects, and crustacea.
HBL 57–85cm; TL 20–32cm; WT 5.5–8kg.
FORM: coat short and coarse, light brown with dark brown or black spots; tail ringed with black; paws slightly webbed and claws not fully retractile.
BREEDING: gestation 63 days; litter 1–4.

PUMA LINEAGE

Puma
Felis (Puma) concolor
Puma, cougar, Mountain lion, panther

Includes Eastern cougar (*F. c. cougar*), and Florida cougar (*F. c. coryi*). S Canada to Patagonia. Range covers a wider array of ecological zones than any other terrestrial New World mammal: desert, savanna, tropical rain forest, and alpine steppe. Prey ranges from small rodents to fully-grown deer; mainly active at twilight.
HBL 105–196cm; TL 67–78cm; WT 36–103kg.
FORM: coat plain gray-brown to black (very variable); cubs initially dark spotted; head round and small; body very slender; eyes brown; pupils circular; tail tip black.
BREEDING: gestation 90–96 days; litter 3–4.

Jaguarundi
Felis (Herpailurus) yaguarondi
Jaguarundi, jaguarondi, eyra, otter-cat

Arizona to N Argentina. Forest, savanna, scrub. Preys on birds. rabbits, rodents, frogs, fish, poultry; active at twilight.
HBL 55–67cm; TL 33–61cm; WT 5.5–10kg.
COAT: either uniform red or uniform gray, lighter underneath; newborn dark spotted; legs very short; body long and slender; ears small, round; eyes brown; pupil contracts to a slit.
BREEDING: gestation 63–70 days; litter 2–4.

OTHERS

The following two species are of uncertain affiliation.

Serval
Felis (Leptailurus) serval

Africa. Savanna, normally near water. Preys on game birds, rodents. small ungulates; good climber.
HBL 70–100cm; TL 35–40cm; WT 3.5–19kg.
FORM: coat orange-brown with black spots (very variable); slender build with long legs, small head, rather long neck and large, rounded ears; eyes yellowish; pupils contract to a spindle.
BREEDING: gestation about 75 days; litter 1–3.

Rusty-spotted cat
Felis (Prionailurus) rubiginosus

S India and Sri Lanka. Scrub, forest, near waterways and human settlements. Preys on small mammals, birds, insects.
HBL 35–48cm; TL 15–25cm; WT 1–2kg.
COAT: rust colored with brown blotches and stripes.

HOW SOUTH AMERICA'S SMALL CATS EVOLVED

The formation of the land bridge linking North and South America about 3–4 million years ago set in motion an impressive natural evolutionary experiment. As placental mammals moved into South America many novel species arose, filling new niches and outcompeting many of the previous residents. Cat species were among the most successful immigrants; although some, like the Saber-toothed tiger, the American lion, and the American cheetah became extinct, ten South American cat species remain. They include the jaguar, the puma, and the jaguarundi.

The other seven – the ocelot and kodkod, and the Margay, Pampas, Tiger, Geoffroy's, and Andean mountain cats – are small, spotted species that belong to the "ocelot lineage." Although they all arose from a common ancestor, they have since taken divergent evolutionary paths, reflected in their current distributions and patterns of molecular genetic variation. Two of the earliest species to diverge were the Andean mountain and the Pampas cats, followed by the ocelot, Margay and Tiger cats. Within several hundred thousand years, each species had become further divided into distinct, genetically divergent populations or subspecies. The biogeographic distribution patterns of these subunits suggests that both the Amazon river system and the Darien Strait between North and South America constituted important historical barriers to gene flow, reducing dispersal of neotropical cats and leading to recognizable differentiation among isolated groups.

In southern South America, the Andes were probably the major barrier to gene flow. The "youngest" South American cat species, the Geoffroy's cat and kodkod, are for the most part separated geographically by the mountains between Chile and Argentina.

The evolutionary pressures that shaped small cat distribution patterns are still continuing to change, but at a much more rapid pace. During the next few hundred thousand years, the most important changes will all be brought about by human alterations of the environment, increasing or decreasing population sizes and their degree of interconnectedness. WJ

What Makes a Wildcat Wild?

The difficulties of distinguishing wild and domestic species

The Scottish wildcat is an enigma. Its very existence is in question, or at least its identity, as it becomes ever more difficult to tell apart from its domesticated relative, the hearthside tabby. House cats have their ancestry in the wildcat, and the changes undergone by the original wildcats to become the "sophisti-cats" we know and love are in fact very slight. This causes problems, because the Scottish wildcat is considered one of Britain's most charismatic wild animals and there is a desire to preserve it – but what exactly is it?

To answer that question, it is first necessary to look at the early history of human–cat interactions. Before domestication, the wildcat was widely distributed, from Scotland to South Africa and from Portugal to the Caucasus. Huge infestations of rats and mice in grain stores probably attracted many cats to human settlements in Egypt some 4–5,000 years ago. The wide availability of rodent prey would have meant that these cats could live communally rather than in exclusive territories. It is possible that the only wildcats to exploit this new feast had an innate lack of fear of humans, and as such were genetically predisposed by their own personalities to domestication. In other words, the cat may have domesticated itself rather than been domesticated. But though the exact route by which this wild animal adapted to man and his environment remains obscure, it is clear that the Egyptians kept the cat in large numbers and venerated it.

Due to the relatively short timescale involved (mere millennia, as opposed to the millions of years involved in evolution), domestication does not create new species, but rather separate races or "ecotypes." Although humans kept and bred cats, they did not, it seems, interfere too much in how these cats looked. Their gray, tabby, or black-and-white coloration arises from genes they share with their wild relatives.

Wildcats retain a uniform brindle pattern, probably because it is the best camouflage color. Any animals with a different pattern would be less likely to survive to adulthood and reproduce, and so, in the wild population, the genes for different coats remain out of sight at fairly low levels (in other words, they are recessive). Domestic cats, however, are freed from the need to be inconspicuous in order to stay alive, and so can express many different shades and designs.

By the time of the Romans, the Domestic cat had become well established as a pet. It was taken throughout their empire into areas previously occupied solely by the wildcat – and that is where the problems started. For more than 2,000 years, Domestic cats have been escaping the human grasp and venturing into true wildcat territory.

So if cats gone wild are wild cats, what is a wildcat? The answer – and this applies not just in Scotland but throughout the species' range – is that nobody really knows. Domestic cats readily revert to a wild, or "feral," state when they leave human care, and can and do readily interbreed with what we know as the wildcat. Yet the question of what defines a wildcat is of more than academic interest. On Scottish grouse moors, wildcats are highly protected by law, whereas feral domestics are regarded as pests that can legally be shot.

Wildcats have been claimed to have a distinctive coat pattern; to attain a certain size; to have a reduced-size intestine; to exhibit a distinct skull morphology; and to be genetically different. It has also been asserted that wildcats are distinct in their susceptibility to disease, have lower and more seasonal fertility, and are "fiercer" and less social than domestic cats.

Some of the characteristics used to distinguish a wildcat, however, are exactly those that a Domestic cat gone wild might be expected to take on.

Right *The mother of these kittens exhibits certain traits that have traditionally been used to characterize the wildcat – a blunt-tipped, five-ringed tail, a uniformly dull brindle coat pattern, and a robust body. In contrast, the kittens display such "domestic" features as tapered tails, more vivid coat markings, and a slimmer build. Yet none of these characters can alone determine whether an animal is a "true" wildcat or a hybrid.*

The reduced fertility of wildcats may be seen in the wild cat as simply another response to the hardship of life away from human luxury. Conversely, the shorter gut of wildcats is likely to develop in a wild cat trying to reduce the cost of carrying a big stomach around and facing challenges unknown to lazy animals amply provisioned with tinned meat.

Attempting to identify a wildcat population in the face of such conundra has been a challenge. Nonetheless, from a sample of over 300 Scottish "wild-living" cats, one group was found that had significantly longer limbs combined with a shorter gut length and, for the most part, the brindle coat pattern. Many of this group lived at relatively high altitudes, where they may be subject to a harsher environment. So, a wildcat as distinct from the wild cat may yet exist, even if simplistic characters like coat color may be inadequate to describe it.

The results raise an important question: What could maintain a difference between these groups despite interbreeding? The docility and tolerance of the Domestic cat has generally been contrasted with the "fierceness" and untamability of the wildcat, which may have discouraged interbreeding. Yet, intriguingly, it is also possible that these behavioral differences may have a genetic basis, linked to coat color. Although the precise physiological mechanisms remain unexplained, it is thought that production of the coat pigment melanin affects an animal's personality. This may happen by melanin affecting the size and function of glands like the pituitary, adrenal, and thyroid, and their associated behavioral hormones. Adding weight to the argument, certain color types (non-agouti and blotched tabby) are more common in town cats than in rural or feral populations. Tameness, however, is also affected by many environmental factors, such as early exposure to humans or other cats. It is true to say that even Domestic cats, once in the wild where food is scarce, will defend territories and live alone.

In sum, defining a wildcat remains very difficult. There is no doubt that wildcats exist – it is just that we are not sure where to draw the line between them and their feral, interbred, and domestic cousins. But if we cannot define a wildcat, how can we monitor populations, let alone protect them?

One answer to the problem may be to preserve the wildcat for what it is rather than what it looks like. Protecting areas where wildcats are known to have persisted relatively free from habitat disturbance or interference by Domestic cats may be the best way of ensuring not just their own integrity but also that of other wild species whose very identity is being threatened by humans and their animal companions. MJD

Above *An impressive threat display by a wildcat. The wildcat tends to lead a solitary existence, defending an exclusive territory. Its diet include game birds, rodents, and even reptiles and insects.*

Below *The wildcat subspecies of Africa* (F. silvestris lybica) *is the progenitor of the domestic cat. Brought to Europe by the Romans, it interbred with local subspecies* F. s. silvestris *to produce* F. s. catus.

DISTRIBUTION Worldwide, except a few areas (e.g. Madagascar, New Zealand) into most of which the Domestic dog (*Canis lupus familiaris*) has been introduced.

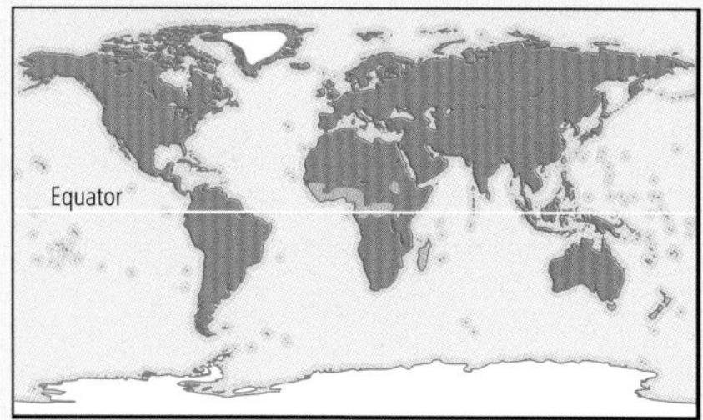

Family: Canidae
35 species in 10 genera

Dog Family

CANIDS EVOLVED FOR FAST PURSUIT OF PREY IN open grasslands, and their anatomy is clearly adapted to this life. Although the 35 species and 10 genera vary in size from the tiny Fennec fox to the large Gray wolf, all but one have lithe builds, long bushy tails, long legs, and digitigrade, four-toed feet with nonretractile claws. The lowslung, stocky Bush dog is the sole exception. The smallest canid, the Fennec, is associated with arid habitats and poor food supplies, while the largest (wild dogs and wolves) are associated with abundant prey. A vestigial first toe (pollex) is found on the front feet in all but the African wild dog. The dingo and the Domestic dog also have vestigial first claws (dew claws) on their hind legs. Other adaptations to running include fusion of wrist bones (scaphoid and lunar) and locking of the front leg bones (radius and ulna) to prevent rotation.

Canids originated in North America during the Eocene (55–34 million years ago), from which five fossil genera are known. Two forms, *Hesperocyon* of North America and *Cynodictis* of Europe, are ancient canids, with civetlike frames. They share a long-bodied, short-limbed physique with the Miacoidea from which all Carnivora evolved. As modern canid features evolved, the family blossomed: there were 19 genera in the Oligocene (34–24 m.y.a. and 42 in the Miocene (24–5 m.y.a.).

The heel of the carnassial teeth in most canids has two cusps, but in the Bush dog, African wild dog, and dhole only one; hence, some taxonomists once classified these genera as the subfamily Simocyoninae, distinct from the subfamily Caninae, containing all other species except the Bat-eared fox (sole member of the Otocyoninae). Members of the three largest genera, *Canis*, *Vulpes*, and *Dusicyon*, are generally more akin to members of their own genus than to those of others, but distinctions between genera can be slight. In descending order, the most atypical body forms are those of the African wild dog, the Bush dog, the Bat-eared fox, the Raccoon dog, the dhole, the Maned wolf, and the Arctic fox. These all occupy single-species genera, but as none is any more closely related to any other than to the bulk of canids, it seems wise to abandon the division of canids into subfamilies. DWM

LIFE IN THE PACK

The most striking feature of the canids is their opportunistic and adaptable behavior. This is most evident in the flexible complexity of their social organization. Remarkably, there is in this respect almost as much variation within as between species. Though African wild dogs, and possibly dholes and Bush dogs, almost always hunt in packs, Gray wolves, coyotes, and jackals feed on prey ranging from ungulates to berries. Partly as a result, they lead social lives that vary from solitary to sociable – Gray wolves may live in isolated monogamous pairs, or in packs of up to 20 members.

These species, and some others like Red and Arctic foxes, live in groups even where large prey does not abound and where they hunt alone. Indeed, there are many other reasons for group living – cooperative defense of territories or large carcasses, communal care of offspring, rivalry with neighboring groups. This is clearly illustrated by the Ethiopian wolf, which lives in packs but almost never hunts cooperatively, its prey being largely rodents.

***Right** African wild dogs crowd around the carcass of an impala.*

1

4

5

2

3

Left and above *Members of the family Canidae:* ***1*** *European wolf howling.* ***2*** *Tibetan wolf cocking its leg.* ***3*** *Gray wolf/husky cross, a common wolf/Domestic dog hybrid.* ***4*** *White-form Arctic fox, seen in its summer coat.* ***5*** *Side-striped jackal playing with a dead mouse.*

OGS UNDER THREAT

or all their adaptability, members of ne dog family cannot escape the indiect threat of habitat destruction. The mall-eared dog and the Bush dog are een so rarely that there are fears for heir futures. The Ethiopian wolf numers less than 500 individuals, the African wild dog from 3,000–5,500 ndividuals, and the Maned wolf only ,000–2,000 in its Argentinian and razilian strongholds. These species are ll endangered. The plight of the sociale canids is especially intense insofar as they are victims of the so-called Allee Effect – that is, at low numbers they enter a downward spiral to extinction. African wild dogs depend on cooperation, so packs with fewer than about five members enter a vortex of decline because they are too small to simultaneously hunt, defend kills, and babysit. Thus, the African wild dogs are even more endangered than their population of 3,000 might suggest, this being equivalent to no more than 600 viable packs across the continent.

KULLS AND DENTITION

Canids have long muzzles, well-developed jaws and a characteristic dental formula of I3/3, C1/1, P4/4, M2/3 = 42, as exemplified by the Gray wolf. Three species depart from this pattern, the Bat-eared fox (= 48), the dhole (= 40), and the Bush dog (= 38). Shearing carnassial teeth (P4/M1) and crushing molars are well developed and, except in the Bat-eared fox, are the largest teeth.

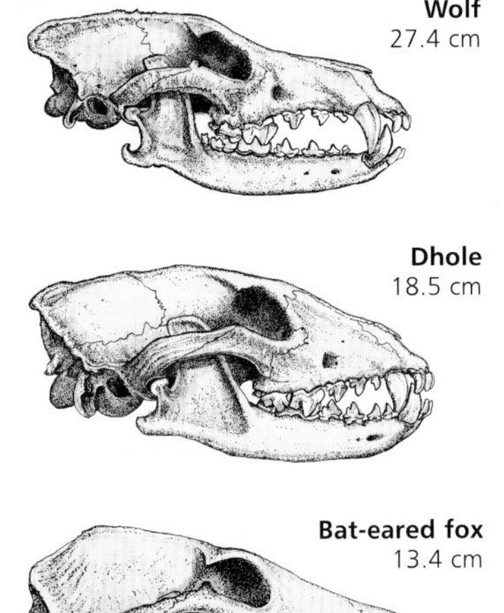

Wolf
27.4 cm

Dhole
18.5 cm

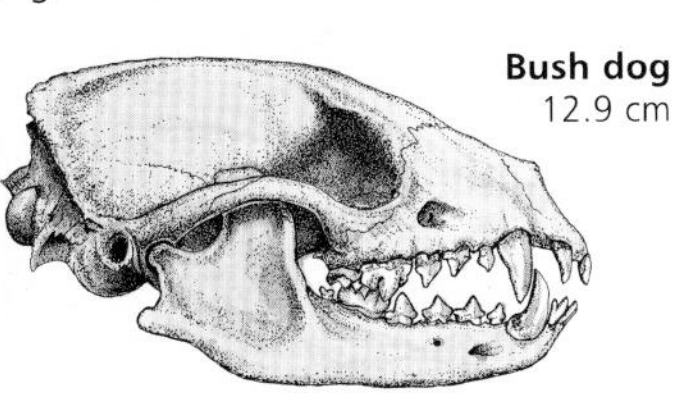

Bush dog
12.9 cm

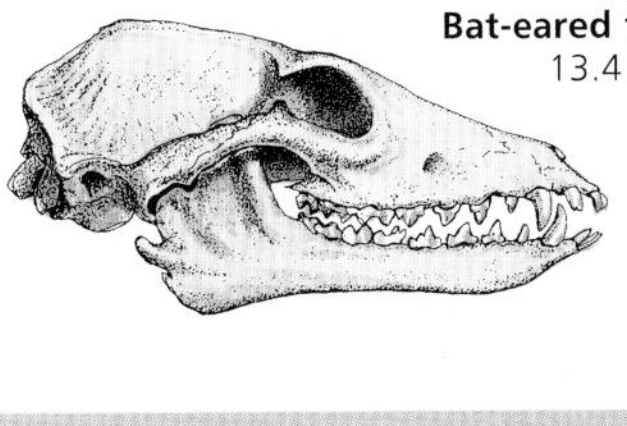

Bat-eared fox
13.4 cm

DOMESTICATION

Various origins have been proposed for Domestic dogs, and doubtless many different canids have been partly domesticated at one time or another. Even so, the wolf is generally accepted as the most likely ancestor of today's Domestic dogs. Domestic dogs are thus known to science as a subspecies of wolf – *Canis lupus familiaris*. The earliest known archaeological indication of domestication comes from a single canine jawbone unearthed at a site in Germany. More foreshortened than that of a wolf, with the teeth more closely packed together, this find is thought to be around 14,000 years old. Other early remains of what are believed to be domestic dogs include a specimen from Coon in Iran, which dates back over 11,000 years. These various discoveries demonstrate that the wolf entered into domestic partnership with man before any other animal species and before the cultivation of plants for food. Indeed, recent molecular evidence suggests that dogs may even have been domesticated as much as 100,000 years ago.

The precise circumstances of domestication have been the subject of considerable speculation. Various theories have been advanced that center on our ancestors' deliberate use of wolves for practical purposes: hunting, guarding, tidying carrion and refuse around settlements, or even as food items. However, it is equally likely that domestication simply came about by accident, with hunter–gatherer societies capturing and raising young wild animals as pets.

Below *The Chihuahua, a variety of domestic dog indigenous to Mexico, was bred by the Aztecs before the Spanish conquest of 1519.*

Wolves

THERE IS NO ANIMAL MORE ENSHRINED IN THE myths of northern cultures than the Gray wolf. Aesop's fables celebrated its cunning, while the story of Romulus and Remus, Rome's legendary founders, was only one of many telling of children raised by wolves. Today some people are seeking to reintroduce the wolf to areas where it had previously been exterminated, while others are doing their best to eliminate it forever.

For thousands of years wild wolves have competed with humans for game and killed farm animals, while the tame wolf has become "man's best friend" – the domestic dog. Little wonder, then, that our relationship with this biggest of canids has been so contradictory. Stories of attacks on humans are widespread, even though shepherds rarely need more than a sturdy stick or a guard-dog to fend off a threatening animal. Wolves can, however, wreak havoc on unprotected livestock; herders in Europe and North America have occasionally reported wolf packs killing dozens of sheep in a single night.

Dogs' Predatory Cousins

FORM AND FUNCTION

From their former omnipresence, wolves are now restricted to the remaining large forested areas in Eastern Europe, a few isolated mountain refuges in the Mediterranean, mountains and semidesert areas of the Middle East, and the wilderness areas of North America, Russia, and China. The largest numbers of wolves are found in Russia, where estimates range from 40,000–60,000 animals. Canada claims about 40,000 wolves, and Alaska has approximately 6,000.

Wolves are relatively massive for canids, and their adaptability to differing climes has resulted in many widely varying forms. Depending on the subspecies, mature male wolves may be 2m (6.5ft) long including the tail, can stand more than 1m (3.3ft) high at the shoulder, and can weigh up to 75kg (165lb). The typical adult wolf, however, is the size of a large German shepherd dog, weighing about 38kg (84lb) and standing 70cm (27.5in) at the shoulder. The smallest wolves inhabit desert and semidesert areas. Forest-dwelling wolves are of medium size, Arctic wolves the largest.

Despite its name, the Gray wolf shows marked variation in color, ranging from white through gray to black; the most common pelage is gray flecked with black. The lightest colored wolves inhabit desert and Arctic regions. In North America and Russia, wolves are often brown or black, but black wolves are extremely rare in Europe. The explanation for this variation remains elusive.

Below *People speak of wolves "baying at the moon," but in fact the animals usually howl to advertise their presence to other wolves. Generally the purpose is to warn off neighboring packs, reducing the risk of a hostile encounter. Lone wolves rarely howl, for to do so could invite attack.*

Below *Although wolves may hunt small prey singly, they rely on numbers to bring down the large animals that form the bulk of their diet. A big carcass will feed a group the size of this pack of Montana timber wolves for days, even though an adult wolf can eat up to 9kg (20lb) of meat in a single feeding.*

THE SECRET LANGUAGE OF WOLVES

Two wolves meet nose to nose; others lie sleeping, or observing pups at play. The wolf pack appears calm, friendly, harmonious. Only the informed observer (and wolves!) will notice the signs indicating the hierarchy and tension: a tail kept somewhat lower and ears held back as a subordinate wolf greets a more dominant pack member **1**; the non-aggressive play face of one pup as it is grabbed by another **2**; a stiff-legged shove as a pup tries to exert authority over one of its contemporaries **3**.

From late fall to late winter, when breeding starts, such interactions become more frequent and purposive. Only one female usually mates, and females may fight vigorously for that right. Even if the dominant female is not challenged, fights may break out among low-ranking females seeking a higher place in the hierarchy. Juveniles and pups may join fights, and some pack members may be driven out. Males may also fight to become the pack sire. Eventually, the dominant pair copulate, generally two or three times a day for approximately 14 days. After that, calm is restored. With the hierarchy fixed for several months, some subordinate pack members may decide to leave. EZ

FACTFILE

WOLVES

Order: Carnivora

Family: Canidae

2 of 9 species of the genus *Canis*. There are up to 32 subspecies of Gray wolf.

DISTRIBUTION N America, Europe, Asia

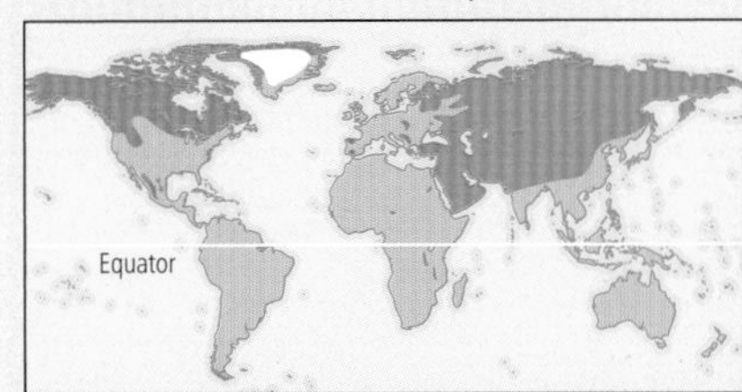

GRAY WOLF *Canis lupus*
Gray, Timber, or White wolf
N America, Europe, Asia, Middle East. Forests, taiga, tundra, deserts, plains, and mountains. Subspecies include the **Common wolf** (*C. l. lupus*), forests of Europe and Asia, medium-sized with short, dark fur; **Steppe wolf** (*C. l. campestris*), steppes and deserts of C Asia, small, with a short, coarse, gray-ocher coat; **Tundra wolves** of Eurasia (*C. l. albus*) and America (*C. l. tundrarum*), large, with coat long and light-colored; **Eastern timber wolf** (*C. l. lycaon*), once the most widespread in N America, but now only in areas of low human population density; smaller, usually gray in color. The **Great Plains wolf** or **Buffalo wolf** (*C. l. nubilas*), white to black in color, which once followed the great herds of bison on the N American plains, is now extinct.
HBL 100–150cm (40–58in); TL 31–51cm (13–20in); SH 66–100cm (26–39in); WT 12–75kg (27–165lb); males larger than females. Coat: usually gray to tawny-buff, but varies from white (in N tundra) through red and brown to black; underside pale. Breeding: gestation 61–63 days. Longevity: 8–16 years (to 20 in captivity).

RED WOLF *Canis rufus*
SE USA. Coastal plains, forests. WT 15–30kg (33–66lb); size intermediate between Gray wolf and coyote. Coat: cinnamon or tawny, with gray and black highlights. Breeding: as Gray wolf. Longevity: as Gray wolf. Conservation status: Critically Endangered, and now probably extinct in wild (but see special feature).

Abbreviations HBL = head–body length TL = tail length SH = shoulder height WT = weight

Sharing Large Spoils

DIET

Wolves consume a wide range of foods, most of which they obtain by killing animals larger than themselves. Large hoofed animals such as moose, elk, deer, sheep, goats, caribou, musk oxen, and bison are the primary prey. Though wolves are quite capable of killing healthy adult animals, field studies show that more than 60 percent of their prey are young, weak, or older animals. The proportion of debilitated prey could in fact be higher than reported, as wolves are keen observers of behavior, able to detect subtle susceptibilities not evident to humans. Healthy, vigorous prey often escape wolf predation by fighting back. On occasion, moose, bison, elk, and deer are able to gain the upper hand and even kill attacking wolves.

Smaller mammals such as voles, beaver, and hares supplement the diet. Where available, fish, berries, and carrion are consumed seasonally. Wolves denning in the Canadian Arctic may subsist on small mammals and birds when their primary prey, caribou, migrate to their summer range. During this time pack cohesion appears to break down, with some members maintaining only a loose affiliation with the denning parents. Wolves occasionally scavenge at refuse dumps and rubbish bins, even when wild prey are available; in Romania and Italy, scavenging occurs in or near towns and cities.

Life in the Pack

Social Behavior

All wolves exhibit similar behaviors, though a degree of variability exists. Communication involves vision, sound, and smell. As with domestic dogs, an elevated tail and erect ears convey alertness and sometimes aggression. Facial expressions, emphasized by the position of the lips and display of the teeth, are the most dramatic form of communication. Vocalizations include squeals, yelps, barks, and howls, which are used to maintain contact over distances of up to 8km (5mi). When alone, very young wolves will deepen their howl to sound more like adults. Scent from urine and possibly fecal matter conveys social status and breeding condition, and serves to advertise territorial occupancy. A gland at the base of the tail may also exude chemicals used in communication.

Studies of captive wolves have shown them to be extremely intelligent and highly gregarious. Though some live solitarily, most live in packs – extended family units of 5–12 members, the exact size depending on the availability of prey. Packs in the Northwest Territories of Canada occasionally coalesce into groups of 20–30 animals when hunting bison.

The pack contains the parents, pups, offspring from other years, and related individuals. The nucleus is the breeding pair, which often mate for life and produce the pack's only annual litter. Most wolves do not breed until at least 22 months of age, though females can conceive at 10 months.

Wolves organize themselves into strict hierarchies within packs, with individual position reflecting status and privilege. Social structure depends on the number, sex, and ages of pack members. Males and females have separate hierarchies in which each animal knows its exact position; interactions between the sexes, however, are more complex, because of breeding relationships. At the top of the hierarchies are the highest-ranking male and female wolves, one of which – male or female – serves as pack leader. Animal behaviorists suggest that the duties of alpha animals range from maintaining pack order to making decisions about where to hunt. Rank positions are not permanent, and contests are most intense during the winter breeding period.

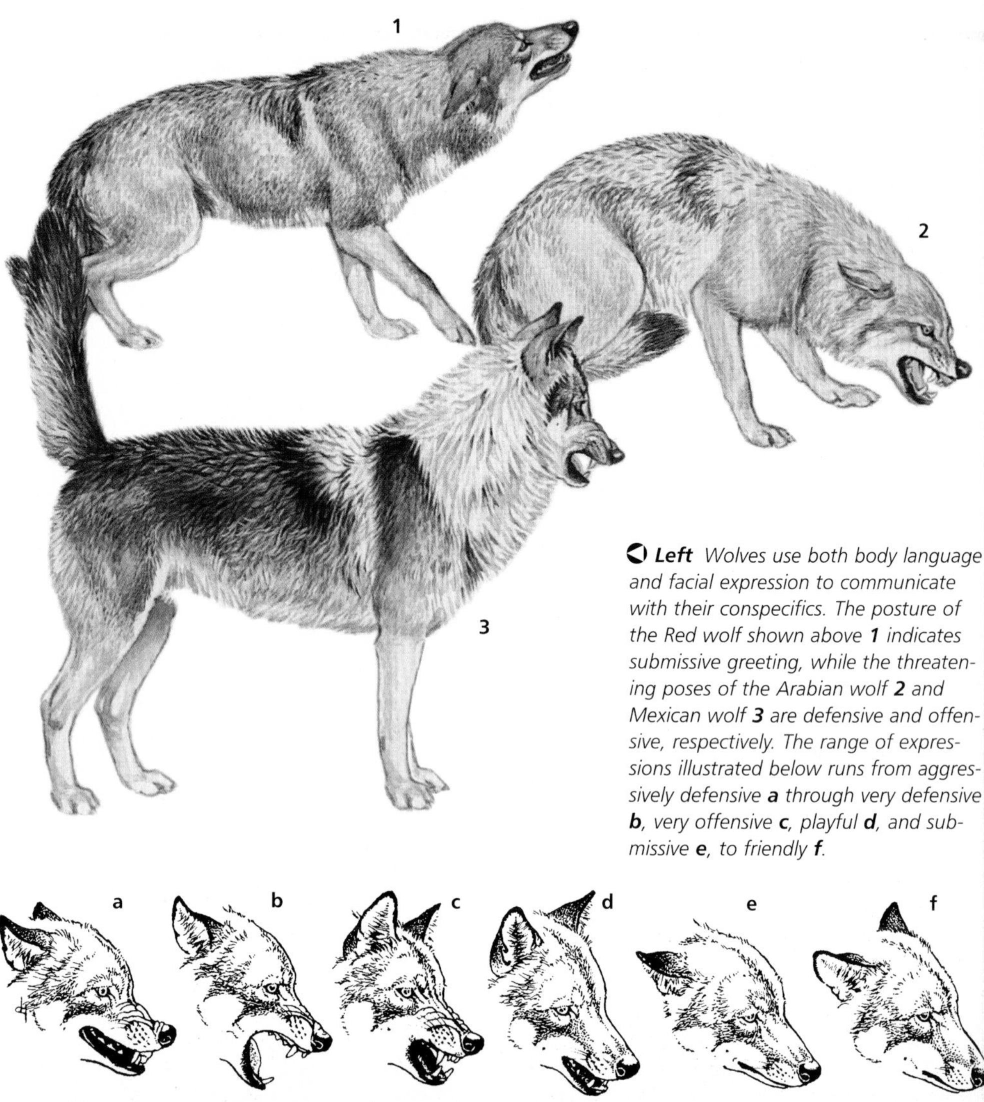

Left Wolves use both body language and facial expression to communicate with their conspecifics. The posture of the Red wolf shown above **1** indicates submissive greeting, while the threatening poses of the Arabian wolf **2** and Mexican wolf **3** are defensive and offensive, respectively. The range of expressions illustrated below runs from aggressively defensive **a** through very defensive **b**, very offensive **c**, playful **d**, and submissive **e**, to friendly **f**.

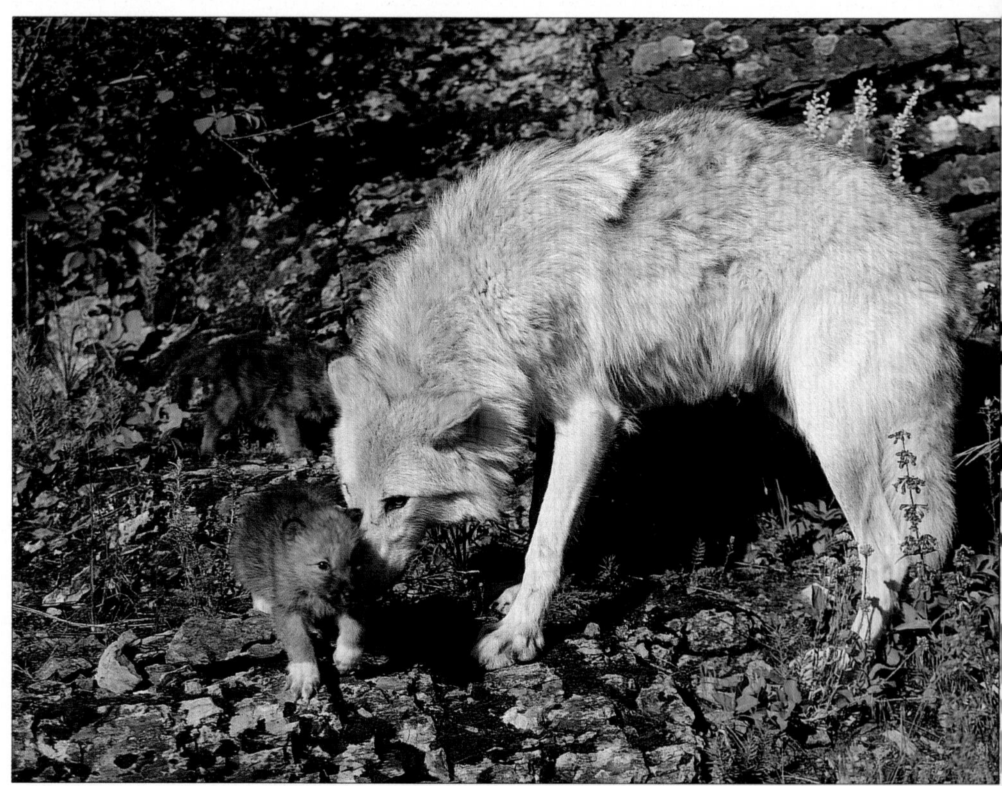

EUROPE'S WOLVES STAGE A COMEBACK

After centuries of persecution that has reduced their numbers to only a few thousand, wolves are making a comeback in isolated areas of Europe. Population sizes currently vary greatly across the continent; six countries have more than 1,000 wolves and eleven more than 500, while eight have populations of fewer than 50 animals. In Italy, where 25 years ago there were only about 100 wolves, about 500 now exist. In Sweden and Norway there were no wolves in the early 1970s, but there is now a population of 60–70 animals, built upon a few dispersing animals from Finland. In the western Alps the last wolf was killed in 1921, but a new population is now growing in size. Wolves from some eastern countries have expanded west, and at the same time Italian wolves have migrated north, dispersing into France, where there is now a population of about 40 animals. Germany, Switzerland, and the Czech Republic are increasingly visited by dispersing animals.

Attitudes have shifted in favor of wolves, and changes in human population density and activity in mountain and rural areas have contributed to the upturn in their fortunes. In most countries they are officially protected, but there are exceptions. Wolves can be a serious problem to farmers, especially where livestock is not defended by the continuous presence of shepherds. Damage done by wolves is compensated in most European countries, and the cost of these schemes is highly variable depending on livestock and terrain types, husbandry techniques, and relative livestock–wolf–wild prey densities. In Spain, Greece, and some East European countries, the wolf is still managed as a game species, with legal hunting seasons. Recolonizing wolves sometimes move back into areas where local ecological and socio-economic conditions now seriously conflict with their presence, as in Norway, France, Switzerland, and northern Italy.

The need for a transnational approach to wolf conservation led to the approval, in 1999, of a pan-European action plan by the Bern Convention that is now being used to coordinate various national conservation schemes. Maintaining and restoring viable wolf populations as an integral part of Europe's ecosystems will be one of the continent's most challenging conservation tasks for the new millennium. LB

Above *European wolf pups practice survival skills by play-fighting. Pups are born in dens, emerging after a few weeks to be taken to a "rendezvous site" – a place near water, marked by trails and matted vegetation. There they can romp at leisure, receiving food and attention not just from their parents but from other pack members as well.*

When packs meet, fights may develop, in which case the common outcome is a dead wolf, lying where it fell with a snarl on its face as testimony to a violent end. Such disastrous encounters are rare, and to reduce their likelihood, packs generally restrict their movement to a relatively exclusive territory of 65–300sq km (25–115sq mi). Its bounds may be 10–20km (6–12mi) across, but only the outer kilometer or so is shared with neighboring packs or lone wolves. This periphery is visited much less often than the rest of the territory, presumably because of the danger of accidentally running into hostile neighbors.

To reduce that risk further, wolves advertise their territories by scentmarking. As packs travel about, the dominant animals urinate on objects or at conspicuous locations about once every three minutes. The density of scent marks in the border regions is twice as high as elsewhere, probably because wolves increase their rate of scent marking when they encounter scent marks left by strangers, and they do this much more frequently on their occasional visits to the edges of their pack territory. This higher density of scent marks – both its own and those of strangers – apparently enables a pack to recognize the periphery and keep from trespassing into even more dangerous areas beyond.

Yet scent marks alone cannot obviate the danger of an unsought encounter between two packs simultaneously patrolling the shared boundary of their territories. In such circumstances packs may howl. Yet this can be a dangerous strategy, for the volume of the howl advertises the pack's strength, and may actually encourage aggression by a larger group. So wolves howl only occasionally, and when they do they call in unison. A neighboring pack will normally only reply if it can match the callers' size or if it is defending a resource such as a fresh kill that it is not prepared to relinquish.

Left *Showing the light coat typical of Arctic-dwelling subspecies, a Tundra wolf bitch nuzzles a pup from her litter. When cubs are initially weaned, they take their first solid food from the mouths not just of the mother but also from other members of the pack. This partially regurgitated food has already been thoroughly chewed, and is easier for the young to digest than raw meat.*

Last of the Wild

CONSERVATION AND ENVIRONMENT

Wild wolves need wild areas to survive. Packs living in highly productive environments such as Yellowstone National Park require exclusive territories of about 150–300sq km (100–200sq mi), while wolves living in the Arctic and dependent on caribou may use areas of 40,000sq km (25,000sq mi) or more. To subsist, a population of wolves requires at least 40 Red deer (or equivalent) per 100sq km (62sq mi). In our human-dominated world, however, such requirements are becoming rare commodities.

Yet the success of efforts to conserve wolves depends as much on social acceptance by their human neighbors as on protecting their biological requisites. In most parts of the world, the goodwill of landowners working in conjunction with governments and conservation organizations is necessary to ensure their survival. Ironically, the species once regarded as a threat to our survival is turning out to be a test of how likely we are to achieve sustainable coexistence with the natural elements that sustain us. PCP

SPECIAL FEATURE

AN ONGOING GENETIC MYSTERY

Are North America's Red wolves mongrels or an ancient species?

CONSERVATION EFFORTS ARE USUALLY PRIORITIZED toward the species most at risk of extinction – and most admired by humans. For a long time the North American Red wolf seemed to fall firmly into both of those categories, but its story is now complicated by unresolved arguments that call its parentage into question, and its survival may hinge on erudite distinctions as to whether it in fact constitutes a separate species at all.

Most wolves belong to the Gray wolf species, of which several types are recognized, including the Timber and Tundra wolves. Zoologists, however, have in the past mostly accepted the Red wolf as another, independent North American species that once roamed throughout much of the eastern United States, ranging from Pennsylvania in the north to Texas in the west. According to the traditional view, aggressive persecution and ongoing habitat loss to human development doomed this wolf in the wild. The loss of indigenous populations of Red wolves was documented from the early 1960s, and by the late 1980s only about 80 of these creatures remained, most of them in captivity. Despite efforts by the US Fish and Wildlife Service to restore the species to the wild, the Red wolf is now the rarest and most endangered canid in the world.

But for some time, there has been an alternative position. Some people have argued that the Red wolf is not a pure species at all. Rather than having evolved over millions of years, Red wolves, in their view, arose when coyotes and Gray wolves, disturbed by human settlement, began to interbreed. The loss of habitat caused previously isolated populations to mix, and the resulting hybrid offspring were the intermediate Red wolves.

Intensive examination of the Red wolf's physical and genetic characteristics over the past few years has produced contradictory and controversial results. Morphological and behavioral studies of skull characteristics, mitochondrial DNA, and microsatellite DNA have supported the Red wolf's status as a unique North American species. This evidence suggests that the Red wolf had an Early Pleistocene origin more than 500,000 years ago, and, importantly, that it was the predecessor from which both modern coyotes and Gray wolves descended. According to this theory, Red wolf numbers began to diminish in the wild after 1940. Subsequently, coyotes and Red wolf–coyote hybrids replaced Red wolves in most of their original range.

Yet there is also genetic evidence for the hybridization theory. The results of most studies conducted in the 1990s tend to support a new hypothesis, according to which Gray wolves and coyotes may even have hybridized repeatedly before European settlement in the south-central United States, though the possibility that the crossbreeding may indeed have occurred only recently and as a result of human-caused environmental changes as previously suggested has not been ruled out.

Another twist in the tale, however, has come from further genetic studies revealing a close relationship between the Red wolf and the wolves living in southeastern Canada that were previously believed to be Gray wolves – a link also confirmed by morphology and the fossil record. Perhaps the most important similarity is their ability to hybridize with coyotes, making both susceptible to genetic swamping. According to one view, a significant portion of the supposed Gray wolves of eastern Canada and the Great Lakes region of the US may in fact be Red wolves or Gray wolf–Red wolf hybrids, while the "coyotes" that now inhabit the eastern US may themselves be Red wolf–coyote hybrids.

This new information has also thrown up a fresh theory to explain the origins of the Red wolf. This theory suggests that Gray wolves, Red wolves, and coyotes may all have descended from a common North American ancestor, with Red wolves and coyotes forming one distinct evolutionary lineage and Gray wolves another. Under this model, the progenitor of all modern wolves migrated to Eurasia 1–2 million years ago, where it evolved

Right *It is virtually impossible to distinguish Red wolves by sight from their far more common relative, the Gray wolf. Only in some individuals does a pronounced red coloration occur. Mostly, as in this fine specimen, the coat is a mixture of russet, cinnamon, gray, and black.*

Below *A mother Red wolf and cub at the captive breeding facility in Washington State. The program to propagate this species and rescue it from extinction was launched in the 1970s. Though most of the 280 offspring born since have stayed in captivity, some have been successfully released back into the wild.*

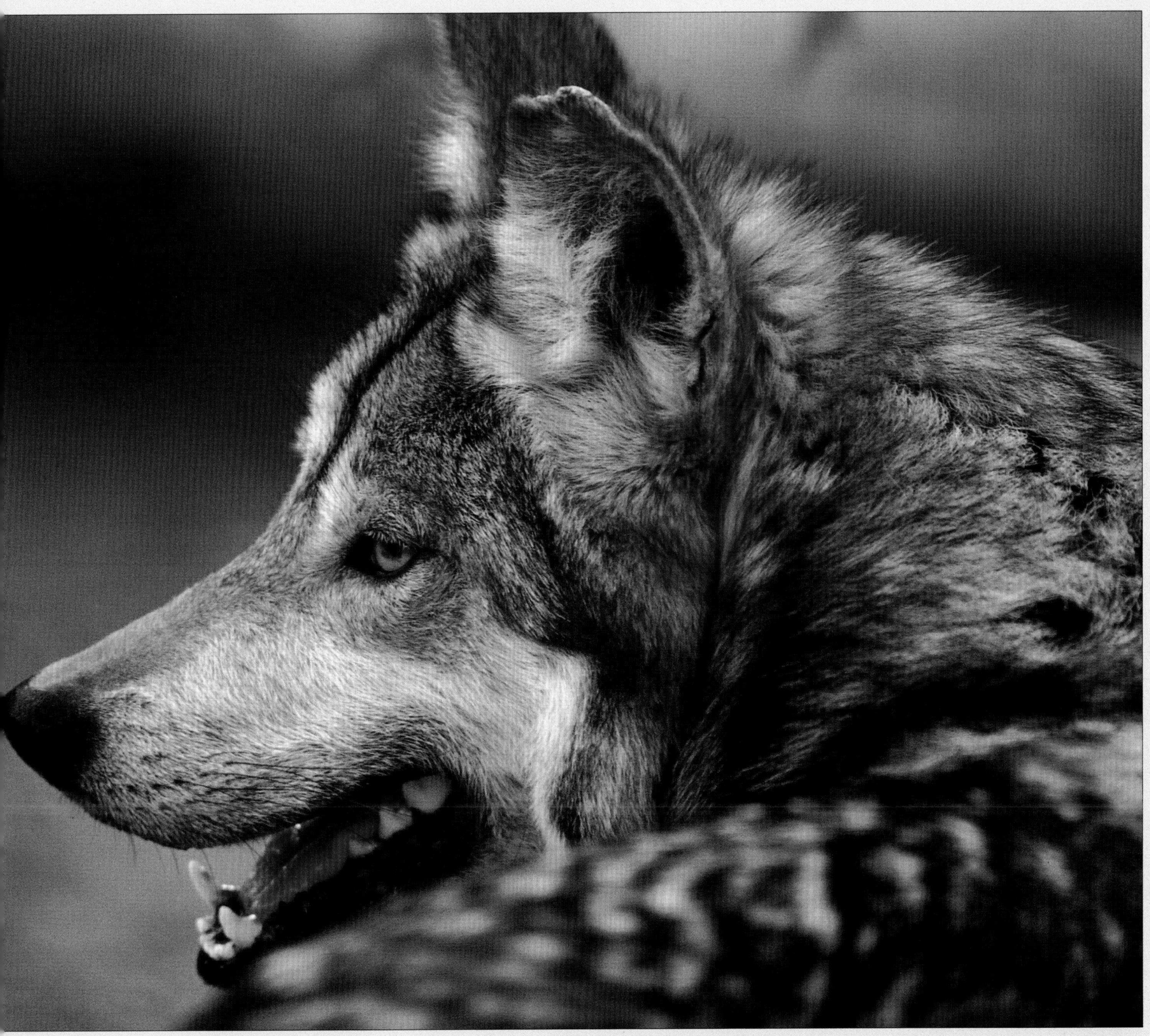

nto the present-day Gray wolf before returning o North America during the Pleistocene – approxinately 300,000 years ago. At about the same ime, the Eastern Canadian/Red wolf and the coyote, which had jointly evolved in North America, diverged.

Clearly, Red and Gray wolf recovery efforts in North America would need to be reevaluated if it turned out that there are extant Red Wolf populations in southeastern Canada. This is a matter of more than academic interest, given that the current Fish and Wildlife Budget for the reintroduction of Red wolves runs at approximately US$4 million annually.

Knowing whether the eastern Canadian wolves are pure Red wolf stock is clearly crucial to the debate. Current thinking is that they are actually a mixed bag, including some pure Gray wolves (old world lupus types) and some pure Red wolves (new world lycaon types), along with Red wolf–coyote and Red wolf–Gray wolf hybrids. Indirectly there may even be Gray wolf–coyote hybrids, for while geneticists believe that direct hybridization between Gray wolves and coyotes probably does not occur, it may be possible for "stepping-stone" Red wolf–coyote hybrids to mate with Gray wolves.

The dilemma for conservationists is to determine whether such hybridization would have occurred had there been no human disturbance. If the Red wolf is not a valid species, the point is moot. If it is a valid species and hybridization occurs as a natural phenomenon, then it is the "process" rather than the "entity" that needs to be protected. On the other hand, if the Red wolf is valid but hybridization is unnatural, then the species should be protected for all the reasons we work to protect biodiversity. The lesson would seem to be that we should give equal consideration to the protection of evolutionary processes as well as to the protection of species. Until now, conservationists have focused only on the protection of species, but that approach will fail in evolutionary time.

Clearly, the evolution of North American wolves is a "work in progress." Although the emerging science of genetics has greatly improved its ability to resolve taxonomic histories and uncertainties over the past few decades, vast gaps still remain in our knowledge of the relatedness of wild wolves. Discovering the recipe for the "canid soup" now found in North America is a priority for conservationists working to protect endangered species.

PCP

Coyote

THE AZTECS NAMED THE WILY CANID THAT *circled their cities and pierced the night air with its yipping howl* coyotl. *This quintessential North American opportunist has adapted to habitats from the Arctic tundra to the city center of Los Angeles; it may live alone or in packs, may eat anything from fruit and insects to mice and antelope, and will even fish or climb trees in search of food.*

Coyotes were long considered solitary animals, but studies have since shown that in some situations they live cooperatively, like wolves. They can interbreed with Domestic dogs, as well as with the Red and, possibly, the Gray wolf (see special feature on the Red wolf); coyote–dog hybrids ("coydogs") are even more liable to attack farm and domestic animals than are coyotes themselves.

Resourceful Predators

FORM AND FUNCTION

The coyote is a medium-sized canid with a rather narrow muzzle, large pointed ears, and long slender legs. Size varies between populations and from one locale to another, and adult males are usually heavier and larger than adult females.

While the geographic ranges of most predators are shrinking, that of the coyote is increasing. A northerly and, particularly, an easterly expansion from the central Great Plains began in the late 19th century, as local populations of the larger canids, the Gray wolf (*Canis lupus*) and the Red wolf (*C. rufus*), were decimated by man.

Like jackals and wolves, the animals are opportunistic predators. Mammals, including carrion, generally make up over 90 percent of their diet. Coyotes normally hunt small prey singly, stalking it occasionally from as far as 50m (165ft) and for as long as 15 minutes. Two or more coyotes may chase larger prey for up to 400m (1,300ft).

Both sexes attain sexual maturity during the first breeding season (January to March) following birth. Females produce one litter a year, averaging six pups per litter. The young are born blind and helpless in a den and are nursed for a period of 5–7 weeks. At three weeks pups begin to eat semi solid food regurgitated by both parents and other pack members of both sexes. Most young disperse in their first year and may travel up to 160km (100mi) before settling down.

***Below** A coyote pack defends a carcass on the edge of its territory. Three pack members **1** are feeding while the dominant male **2** aggressively threatens an intruder **3**, who assumes a defensive threat posture. Another male **4**, backing up his dominant partner, shows less intense aggression. Another trespasser **5** watches for the outcome of the encounter, while other coyotes **6** wait in their own territory for the resident pack to leave the carcass.*

Pack-Dwellers that Often Roam Alone

SOCIAL BEHAVIOR

For pack animals, coyotes spend a surprising amount of time alone, their activity patterns reflecting regional differences in prey abundance. In Wyoming's Grand Teton National Park, coyotes hunting widely distributed rodents were solitary during 77 percent of summer sightings, and all groups consisted of five or fewer members; those living off ungulate carrion, a large and defensible winter food source, were more likely to roam together. Size matters during pack encounters;

Above *After stalking its prey perhaps for as much as 15 minutes, a coyote pounces through long grass for the kill. Although the animals will occasionally cooperate to pursue and bring down large prey such as deer, most hunting is done alone. Coyotes rely mainly on stealth to creep up on small prey, but when necessary they can show a surprising turn of speed; over short distances they have been timed running at up to 64km/hr (40mph).*

at 2.6 members, the winning groups in aggressive encounters were on average twice the size of the losers, which were driven away from precious prey.

Home-range size varies between regions, and it too is affected by food-related activity. Coyotes foraging on abundant fruits, rodents, and rabbits on a Texas ranch only required 3sq km (1sq mi) ranges, while males in Alaska covered 104sq km (40sq mi) when their principal prey, the Snowshoe hare (*Lepus americanus*), was scarce.

Pack life requires social bonding, pup-rearing, and territory defense, and individual roles differ among the pack members. The alpha pair is usually the only one to breed, while juveniles from previous litters act as helpers looking after future generations of pups.

The alpha pair's exclusive grip on breeding provides the other pack members with a tricky dilemma. They may remain within the pack in the hope of one day becoming the dominant breeders, or they can disperse to find a mate – a risky option, since survival rates of dispersers are 25 percent lower than those that stay with the pack. In addition, reproductive success following dispersal is usually small, and many coyotes roam as transients without a mate or territory, covering ranges ten times larger than those of territorial coyotes in search of food.

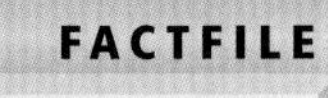

COYOTE

Canis latrans

Coyote or Prairie wolf or Brush wolf

Order: Carnivora

Family: Canidae

One of 8 species of the genus *Canis*

DISTRIBUTION America, from N Alaska to Costa Rica.

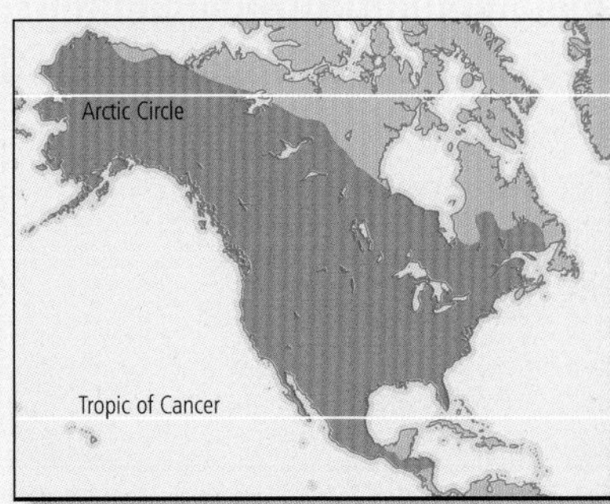

HABITAT Open country, grassland or semi-desert; deciduous and coniferous forests, alpine zones and tundra.

SIZE Head–body length 70–97cm (28–38in); **tail length** 30–38cm (12–15in); **shoulder height** 45–53cm (18–21in); **weight** 8–22kg (17–48lb); males 20 percent heavier than females.

COAT Grizzled buff-gray; muzzle, outer side of ears, forelegs, and feet dull brownish-yellow; throat and belly white; black patches on forelegs, near base of tail, and tailtip.

DIET Opportunistic, taking fruits, insects, rodents, rabbits, small birds, snakes, turtles, poultry, sheep, deer, pronghorn antelope, mountain sheep, carrion, and rubbish.

BREEDING January–March (later in north); both sexes may breed as early as 10 months of age; one 2–5 day estrus each year; gestation 63 days; litter size averages 3.2–6.0, maximum 19.

LONGEVITY Maximum 14.5 years (18 in captivity).

CONSERVATION STATUS Not threatened.

Keeping Numbers in Check

CONSERVATION STATUS

Coyotes are notoriously effective predators of domestic livestock in general and of sheep in particular, and as a result are fiercely persecuted. Eighty-one percent of those observed in a Colorado study, and 57 percent of those in Texas, died as a result of human activities in the form of trapping, shooting, poisoning, or roadkills.

Even so, the effectiveness of the methods used to control them is often questionable.On the Naval Petroleum Reserves of California, 581 coyotes were killed over five years, but the overall population remained unaffected. In Yellowstone National Park, the reintroduction of wolves has proved a more efficient means of reducing coyote numbers, cutting back the population by 50 percent over two winters; moreover, the average pack size of the survivors fell from six to four. AM

AN UNLIKELY HUNTING PARTNERSHIP

While coyotes compete fiercely with foxes over prey carcasses, cooperation does occur with one prairie carnivore, the American badger (*Taxidea taxus*). Coyotes occasionally prey on badgers, and badgers have been known to kill coyote pups in dens. Yet the Navajo long ago noticed that solitary coyotes and badgers sometimes travel and hunt together. As the badger digs into rodent or rabbit burrows, the coyote will wait to catch any fleeing prey. In the course of one observed incident, when humans disturbed a coyote and a badger hunting ground squirrels together, the coyote ran 700m (750yd), waited for the badger to catch up, and then both predators were seen to travel on together. Although the coyote may protect the badger from other predators and lead it to prey carcasses, it nonetheless seems to be the primary benefactor of this unusual relationship.

Jackals

IN ANCIENT TIMES JACKALS WERE A FAMILIAR *sight around abattoirs and funeral grounds. This apparent association with death may have prompted ancient Egyptian priests to incorporate the jackal-headed Anubis into their panoply of gods as the arbitrator of souls and the keeper of the dead.*

In fact the presence of jackals at any place where food might be forthcoming is due more to their dietary opportunism than to any interest in the afterlife. In modern times jackals are commonly sighted in many African game parks and are also sometimes found, as in ancient times, on the outskirts of human settlements.

The Way Canids Used to Be

FORM AND FUNCTION

Jackals are slender, medium-sized canids that show little dimorphism between the sexes. Their long legs and sharp, curved canines are adaptations for hunting small mammals, birds, and reptiles, while their well-developed carnassial teeth are used for shearing tough skin. All species have blunt-clawed feet, and the bones in the forelimbs are fused, making them (along with other canids) poor climbers but excellent runners, able to maintain a slow trot of 12–16km/h (7–10mph) for extended periods while searching for food or surveying their territories. Jackals are nocturnal, with peaks of activity during the early morning and evening; in some areas where human disturbance is low, they may be active during the day.

Present-day jackals are probably very similar to early canids. The three species are not, however, genetically similar, despite the fact that they are very alike in appearance. Analysis of mitochondrial DNA suggests that they all diverged from a common ancestor approximately 6 million years ago. Fossil evidence indicates that Side-striped and Black-backed jackals evolved in Africa, while the Golden jackal very likely had its origins in Europe or Asia. In general the Golden jackal inhabits relatively arid regions and dry grassland (but may be found in forested areas in Asia); the Side-striped prefers moist woodland and riverine forest, while the Black-back shows a preference for dry grassland and *Acacia* woodland.

A group of jackals squabbling over an antelope carcass recently abandoned by lions or hyenas is a common sight in many of Africa's national parks. Even so, jackals are less dependent on carrion than is popularly believed; in many areas scavenged material makes up only 6–10 percent of the diet. They are skillful hunters of small birds and mammals up to the size of a hare or a newborn antelope, and readily eat insects and fruit. Indeed, jackals are adept at locating any rich source of food, so their diet is catholic, varying a great deal between seasons.

Vertebrate prey must be actively pursued, and jackals are expert rodent hunters. They use their acute hearing to pinpoint prey in long grass, then leap into the air to pounce with the forefeet before delivering a killing bite. Side-striped jackals have been observed bumping up against bushes or stamping the ground to disturb locusts, beetles, and grasshoppers, which are subsequently snapped up. When larger mammalian prey such as hares or newborn antelope is available, jackals will often cooperate with their mates in order to make a catch. While one member of a pair chases the prey, the other may harass would-be protectors or else cut off the prey as it tries to zigzag. When hunting gazelle fawns or springhares, jackals that cooperate are two to three times more successful than those that hunt alone.

FACTFILE

JACKALS

Order: Carnivora

Family: Canidae

3 species of the genus *Canis*

DISTRIBUTION Africa, the Middle East, SE Europe, S Asia.

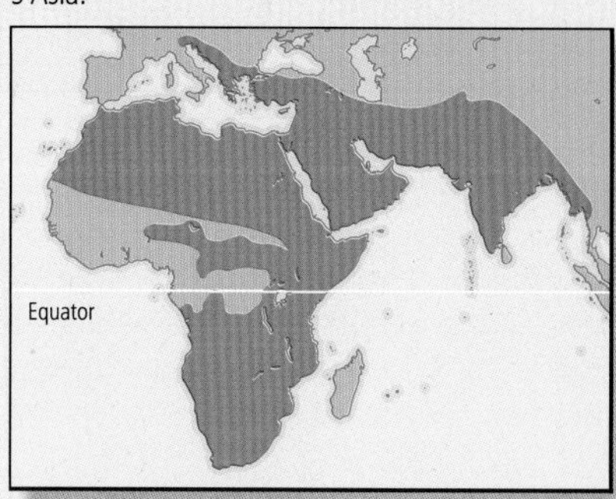

Habitat From arid short grasslands to moist woodlands or dry brush woodlands.
Diet Omnivorous: small to medium-sized mammals, birds, reptiles, carrion, fruit, insects.
Breeding 3–4 pups are born to a monogamous pair after a gestation of approximately 63 days; young jackals mature at around 11 months, sometimes remaining with their parents as helpers for 1–2 years.
Longevity 4–8 years (16 in captivity).

GOLDEN JACKAL *Canis aureus*
Golden or Common jackal
From N and E Africa through to the Middle East, SE Europe, and on into S Asia. HBL 83–132cm (33–52in); SH 40cm (16in); WT 7–15kg (15–33lb). Coat: yellowish to silvery gray, geographically and seasonally variable.

SIDE-STRIPED JACKAL *Canis adustus*
Throughout C Africa, extending into parts of E and W Africa. HBL 96–120cm (38–47in); SH 41cm (16in); WT 6.5–14kg (14–31lb). Coat: fawn and gray, darker on back, with distinctive white horizontal stripe on flanks and a white tail tip.

BLACK-BACKED JACKAL *Canis mesomelas*
Black- or Silver-backed jackal
Two separate populations, one in E Africa, one in S Africa; this species does not occur in C Africa. HBL 96–110cm (38–43in); SH 38cm (15in); WT 7–13.5kg/15–30lb (E Africa), 6.8–9.5kg/15–21lb (S Africa). Coat: marbled black and white back and neck, black tail, rufous brown flanks, head, and legs.

Abbreviations HBL = head–body length SH = shoulder height WT = weight

Protecting the Territory

SOCIAL BEHAVIOR

The monogamous pair bond is the basis of the social group. A pair of jackals defends a territory against other jackal pairs. Together they advertise their territory by leaving urine and feces marks on conspicuous objects. Any unrelated jackal detected within this area is vigorously chased away, especially during the breeding season. Resources within the territory are sometimes plentiful enough to support additional jackals, so subadult and adult young may remain for up to two years before dispersing, after which they set off to find mates and establish their own territory. During this time these young jackals often help to feed and guard the current litter of pups.

In Africa jackal pups are born in a subterranean den or disused termitarium. The time of the year when they are born varies between areas, but usually coincides with food resources being at their peak (during or just after the rainy season). Food to be transported to the den by the adult jackals is swallowed and then regurgitated for the young pups in the den, a practice that prevents loss of food to kleptoparasites such as hyenas that might otherwise steal food carried by mouth. Young jackals are independent at 14 weeks and reach adult size at 9–10 months, although coat markings are not well defined until they are 2 years old.

Jackals are often very vocal. The sound most associated with the Golden and Black-backed jackals is a high-pitched, wavering howl, while the Side-striped jackal calls with a low, hooting noise. A chorus of jackal howls can often be heard in the early evening, as members of a family group relocate one another and advertise their presence to neighboring groups.

Neighbors Not at Risk

CONSERVATION AND ENVIRONMENT

Jackals fall into that small, fortunate group of animals that are able to survive and thrive close to human settlement. Along with their adaptable diet, this accounts for their widespread distribution, despite the fact that Black-backed jackals are known to occasionally kill young sheep and goats and are heavily persecuted in the livestock farming areas of southern Africa, while in some areas they are also killed for their fur.

Jackals' social behavior, with close-knit family groups and aggression toward trespassing animals, makes populations very susceptible to the the spread of the rabies virus, which is transmitted in the saliva and enters a new host's body via bite wounds. Rabies epizootics in jackals occur every 1–8 years and can last for up to 9 years, during which large numbers of jackals are infected and die. They also spread the disease, accounting for around 25 percent of all recorded rabies cases in central southern Africa. AL

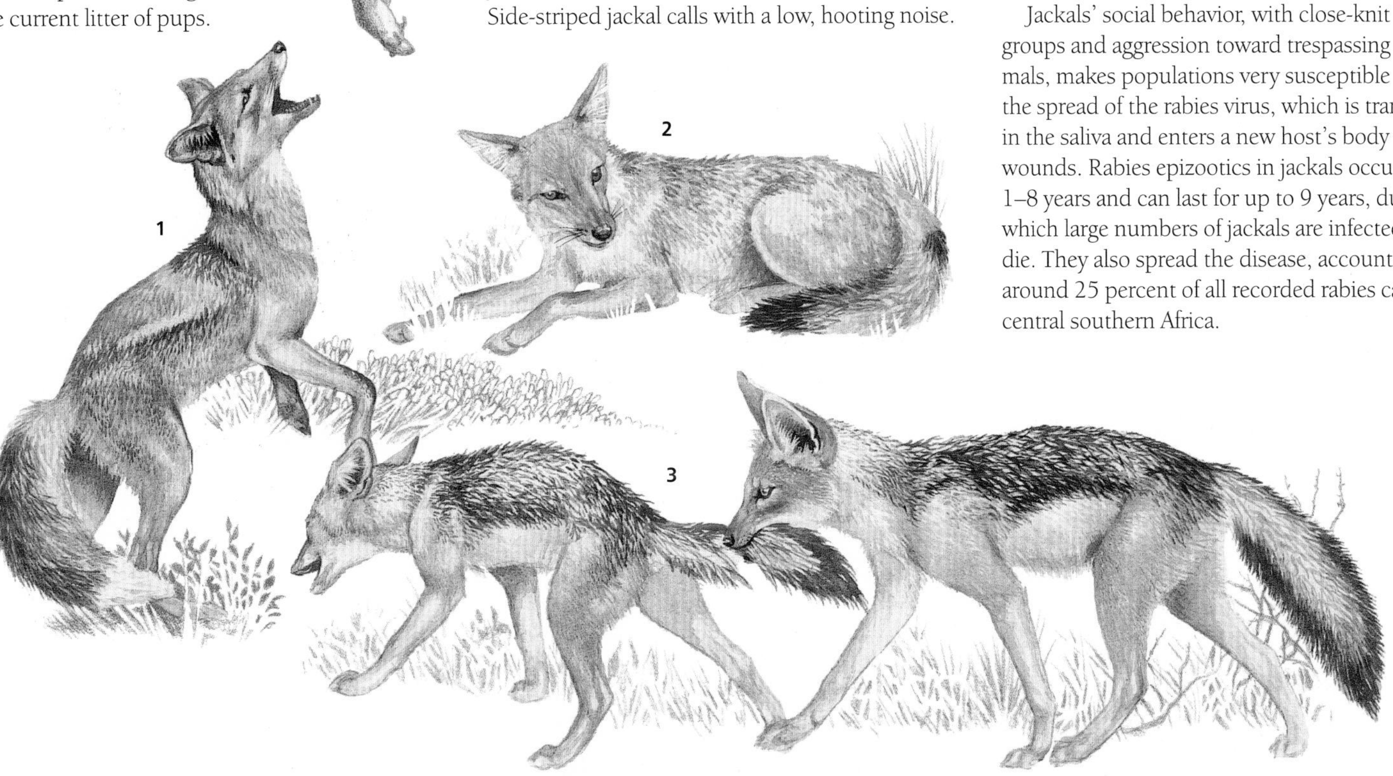

Above The three species of jackal: **1** A Side-striped jackal playing with a dead mouse; **2** A recumbent Golden jackal; **3** Black-backed jackal juveniles engaged in a tail-pulling game: the arched back and laid-back position of the ears reveal the unease of the jackal having its tail pulled.

Right Golden jackal cubs in Israel. Litters can range in size from one to about eight, but the average is two to four. The pups' eyes open after about 10 days. For the first three months of their lives, they live off regurgitated food fed to them from the mouths of their parents, and sometimes also other close kin.

Left An aggregation of Black-backed jackals descend upon a carcass in the Kalahari, South Africa. Although jackals live in pairs, each defending their own home range against outsiders, up to 30 animals may gather together in temporary agglomerations to scavenge off a large prey item such as a lion kill.

Ethiopian Wolf

CLOSE RELATIVES OF THE GRAY WOLVES AND coyotes, Ethiopian wolves probably descend from a late canid arrival to Northeast Africa. Until recently they were notable principally for their precarious status on the verge of extinction, for almost nothing was known about their ecology and behavior, but recent research has revealed a close-knit social life.

The wolf is just one of several mammal species found only in the Ethiopian highlands; others include the African mole rat (*Tachyoryctes macrocephalus*) and the Mountain nyala (*Tragelaphus buxtoni*). Over the last 15,000 years the afroalpine habitats in which they live have become increasingly rare as the African continent has warmed, reducing the available habitat to a tenth or less of its former expanse. Today the wolves cling on in a dozen isolated pockets, the largest of them in the Bale Mountains.

FACTFILE

ETHIOPIAN WOLF

Canis simensis

Ethiopian wolf, Simien fox, or Simien jackal

Order: Carnivora

Family: Canidae

One of 8 species of the genus *Canis*

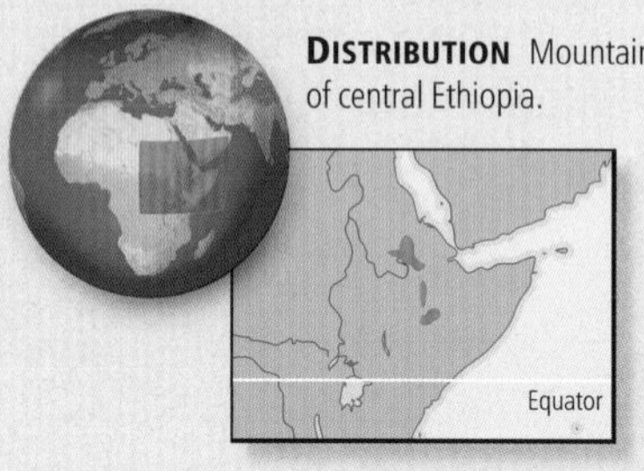

DISTRIBUTION Mountains of central Ethiopia.

HABITAT Grass- and heathland above 3,000m (10,000ft).

SIZE Head–body length 84–100cm (33–40in); tail length 27–40cm (11–16in); shoulder height 53–62cm (21–24in); weight 11–20kg (24–44lb); males on average are about 20 percent larger than females.

COAT Tawny red with pale ginger underfur; chin, inside ears, chest, and underparts white, with distinctive white band around the underside of the neck.

DIET Mostly rats and other rodents.

BREEDING Gestation 60–62 days; litter size 2–6.

LONGEVITY Not known

CONSERVATION STATUS Critically Endangered

A Rapacious Taste for Rodents

FORM AND FUNCTION

Resembling coyotes in shape and size, these medium-sized canids with their distinctive long legs and muzzles are sometimes known as Simien or Simenien foxes. Early explorers and biologists named them Abyssinian wolves, Simien jackals, Red foxes, or Ethiopian jackals, whereas in Ethiopia itself they are variously called *ky kebero* (meaning "red jackal"), *jedalla farda* ("horse's jackal"), or *walgie* ("trickster"). The confusion of names points to the fact that, although they are "true" members of the genus *Canis*, their adaptation to prey exclusively on rodents has resulted in them evolving to resemble oversized Red foxes. Broad, pointed ears, elongated skulls, slender, pointed muzzles, and small, widely-spaced teeth all equip them for seizing small mammals.

In its sparkling red-and-white livery, a lone wolf is conspicuous when it crisscrosses the heathlands hunting the omnipresent afroalpine rodents. Regular targets are the African mole rat, a temptingly large prey that can weigh up to 1,000g (35oz), and several grass-rat species. In the Bale Mountains, rodents account for nearly 96 percent of all prey identified in wolf feces.

A Close-knit Community

SOCIAL BEHAVIOR

The wolves are most active by day, synchronizing their activity with that of above-ground rodents. Although they mostly hunt alone, they adopt the group-hunting behavior of other canids to pursue Mountain nyala calves, reedbuck (*Redunca redunca*), Starck's hare (*Lepus starcki*), and Rock hyraxes (*Procavia habessinica*).

Despite their solitary foraging habits, Ethiopian wolves live in large, close-knit packs that communally share and defend exclusive territories. A pack normally consists of 3–13 adults (mean = 6), including 3–8 related adult males, 1–3 adult females, 1–6 yearlings, and 1–7 pups. The home range will be discrete and relatively small, averaging 6.4km/4mi in optimal habitat but up to 15km/9mi in areas of lower prey biomass. Although it may look deceptively barren, the home range will harbor more than 10,000kg (10 tons) of rats.

Since unoccupied rat-rich habitat is scarce, packs must defend their patches from intruders. Early mornings and evenings are spent patrolling and tandem-marking the boundaries, using raised-leg urinations, defecations, and scratches. Whenever a neighboring pack is encountered, a ritualized aggressive interaction ensues; these are highly vocal, with little contact, and always end with the smaller group fleeing from the larger, which is then able to extend its range.

Healthy wolf populations occupy all available habitat in a given range, at densities of up to 1.2 adults per km (1.9 per mi). Home ranges remain stable over years, altering only after significant demographic changes, often resulting from a rabies epizootic, or the disappearance of a pack.

Males do not disperse, and are recruited into philopatric packs in which the adult sex ratio is biased toward them by 2.6:1. Over half the females disperse at 2 years of age and become "floaters," occupying narrow ranges between pack territories until a breeding vacancy becomes available. Dispersing wolves often have nowhere to go, at worst ending up stranded in agricultural land, so they only move if absolutely necessary.

Above The wolf's rangy, foxlike build is well adapted to hunting rats and other small rodents in its highland habitat. Even so, the animal's survival skills have been sorely tested by human encroachment, and total numbers of surviving adults are now counted in the hundreds rather than the thousands.

Right Like other young canids, Ethiopian wolf pups have a close and enduring bond with their mothers. Only dominant females breed, though other pack members will bring prey to help feed the young after they have weaned – typically at about 10 weeks.

The dominant female of each pack may give birth once a year between October and December (though only about 60 percent breed successfully). All pack members guard the den and bring prey to feed the pups until they are 6 months old. Although cooperative breeding would seem a direct advantage of living in groups, the survival of young is not related to pack size; it seems instead that territorial defense and expansion are the main benefits of group living.

Subordinate females are rarely allowed to breed, but they often assist the dominant female in suckling the pups. Breeding females are typically replaced after death by a resident daughter. This seemingly convenient arrangement could have disastrous consequences if females were to mate with the pack males – in practice, their own father, brothers, or uncles. In fact they avoid the dangers of inbreeding by an unusual mating system that represents a departure from the canid monogamous norm. For a 2–4 week period at the end of the rainy season, most mature females come into estrus more or less synchronously, and a hectic mating season ensues in which females actively seek encounters with neighboring males, while male coalitions comb territorial borders seeking receptive females. As a result, up to 70 percent of matings involve males from outside the pack. In addition, some litters may be fathered by more than one male, further increasing genetic variability. Interestingly, when a female mates within her pack it is almost invariably with the dominant male, but in her liaisons with neighbors she will mate with males of any status.

The Most Threatened Canid

CONSERVATION STATUS

Atypically for canids, Ethiopian wolves are very specific in their needs, and this specialization has led to their present quandary: small populations dependent on a fragmented habitat. They have consequently long been considered rare, and were listed as requiring protection as early as 1938.

More recently, the threat has increased due to the development of high-altitude subsistence agriculture and overgrazing that has trapped the surviving populations in isolated islands of suitable habitat. This fragmentation increases the risk of extinction. Exposure to canid-related diseases and hybridization with Domestic dogs are other threats emerging from increasing contact with people. With no more than 500 adult wolves remaining, this is one of the most endangered large mammalian species.

Although few wolves survive, the task of protecting them is made easier by the need to concentrate on only a handful of populations living in relatively small areas. Conservation actions include the vaccination and sterilization of Domestic dogs within those areas to prevent hybridization, as well as an extensive community education campaign. CS-Z

Foxes

IN AESOP'S FABLE THE CUNNING FOX OUTWITS the stork; the hero of the medieval tales of Reynard the Fox always gets the better of his adversary, the wolf; and in the children's stories of Beatrix Potter, the sly, sinister fox fools the gullible bird Jemima Puddle-duck. The fact that foxes figure as the wily character in the popular tales of many cultures reflects both their wide distribution and their resourceful behavior.

The largest genus of foxes, *Vulpes*, which contains 12 species, is also the most widespread of the canid genera. Moreover, one of its members, the Red fox, is the most widely dispersed and arguably the most adaptable of all carnivores. Together with the Gray wolf, it has the most extensive natural range of any land mammal (with the exception of Man).

Opportunists Par Excellence

FORM AND FUNCTION

Foxes are small canids with pointed muzzles, slender, somewhat flattened skulls, large ears, and long, bushy tails. All species are opportunistic foragers, using hunting techniques which vary from stealth to dash-and-grab.

The similar foraging behavior of different fox species may affect their geographical distribution, as it leads to severe competition for food. Arctic and Red foxes were once thought to be separated by the Arctic species' remarkable tolerance to cold temperatures – its metabolic rate does not even start to increase until –50°C (–58°F), in contrast to the Red fox's, which increases at –13°C (8.6°F). Yet Red foxes are sometimes found in even colder places than Arctic foxes, so the two species are probably separated by food competition. The Red is up to twice as heavy, needs correspondingly more food, and thus in the far north, where prey is sparse, cannot match energy gains with expenditure in the way that the Arctic fox can. However, in areas where both species can subsist, the Red fox's greater size enables it to intimidate the Arctic fox, in effect determining the southern limit of the latter's range.

Direct competition may also have affected the distribution and sizes of *Dusicyon* species. In central and southern Chile both the colpeo and the Argentine gray zorro eat rodents, birds, birds' eggs, and snakes in comparable quantities. However, the two species vary in size with latitude throughout their range. Average body length of the colpeo increases from 70 to 90cm (28 to 35in) and that of the Argentine gray decreases from 68 to 42cm (27 to 17in) with the change in latitude from 34°S to 54°S. Where the two species are of similar size (34°S), the colpeo inhabits higher altitudes of the Andes, so reducing competition. Further south, where the altitude decreases, bringing the species into apparently direct competition, the much smaller size of the Argentine gray predisposes it to hunting smaller prey than the colpeo, and so, again, competition is reduced.

Weighing only 1–1.5kg (2.2–3.3lb), the fennec lives deep in the Sahara and is the smallest of all foxes. The fennec starts to tremble with cold at less than 20°C (68°F), and neatly wraps its tail like a stole around its nose and feet. However, it has an amazing record of its own: it starts to pant only when the temperature exceeds 35°C (95°F), and its jaws open to a full pant only at 38°C (100°F). But when the animal does pant, its resting rate of 23 breaths per minute rockets to a maximum of 690 breaths per minute. A panting fennec curls its tongue up, so as not to waste even a precious drop of saliva. Its butterfly ears constitute 20 percent of its body surface and, when the temperature soars, it dilates the blood vessels in its ears and feet, increasing the amount of heat radiated to the outside. If the air temperature climbs higher than its normal body temperature of 38.2°C (100.8°F), the fennec lets its body heat up to 40.9°C (105.6°F), thus reducing the water it has to "waste" in sweating. The fennec also saves energy by having a metabolism that functions at

FACTFILE

FOXES

Order: Carnivora

Family: Canidae

23 species in 4 genera. Gray foxes: the **Gray** or **Tree fox** (*Urocyon cinereoargenteus*), and the **Island fox** (*U. littoralis*). **Bat-eared fox**: *Otocyon megalotis*. Vulpine foxes: species include **Red fox** (*Vulpes vulpes*), **Swift fox** (*V. velox*), **Arctic fox** (*V. lagopus*). Southern American zorros: species include **Argentine gray zorro** (*Dusicyon griseus*), **Crab-eating zorro** (*D. thous*).

DISTRIBUTION Americas, Europe, Asia, Africa.

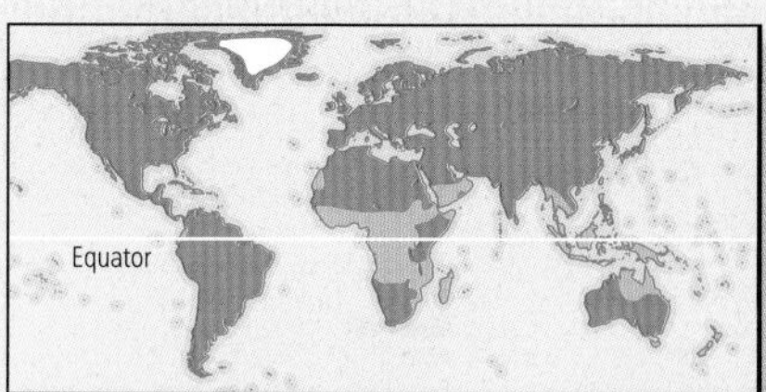

HABITAT Very wide-ranging, from Arctic tundra to city centers.

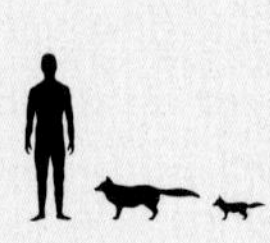

SIZE Ranges from the Fennec fox to the Small-eared dog. **Head–body length** 24–100cm (9.5–39in); **tail length** 18–35cm (7–14in); **weight** 1–9kg (2.2–20lb).

COAT Mostly gray to red-brown, though white or bluish-gray in the Arctic fox in winter.

DIET Diverse, including small mammals, rodents, birds, invertebrates including beetles and earthworms, fish, and fruit when available.

BREEDING Gestation between 51 days (Fennec fox), and 60–63 days (Red fox). Litter size normally 1–6.

LONGEVITY Up to 6 years (to 13 in captivity).

See species table ▷

Above *An Arctic ground squirrel* (Spermophilus undulatus) *clamped firmly in the jaws of a Red fox. Its ability to vary its diet to take whatever prey is locally available has seen the Red fox become one of the most widely distributed species on Earth.*

Right *The coat of the northern Red fox occurs in three color forms, here shown in a group of foxes scavenging from a carcass.* **1** *Two individuals with the vivid, flame-red coloring typical of most high-latitude Red foxes;* **2** *the melanistic ("silver") form; and* **3** *the indeterminate so-called Cross fox. The different forms are probably under complex control of two different genes.*

Left *South American foxes of the genus* Dusicyon. **1** *The Small-eared zorro* (D. microtis)*;* **2** *the Colpeo zorro* (D. culpaeus)*;* **3** *the Argentine gray zorro* (D. griseus)*;* **4** *Azara's zorro* (D. gymnocercus)*; and* **5** *the Crab-eating zorro* (D. thous).

only 67 percent of the rate predicted for such a small animal. Similarly, its heart rate of 118 beats per minute is 40 percent lower than could be expected for its body size.

Canny Hunters

DIET

Apart from the Bat-eared fox, which eats mainly termites, there are no proven differences in species' diets other than those imposed by the limits of available prey. Arctic foxes will take sea birds, ptarmigan, shore invertebrates, fruit, and berries, together with carrion found while methodically beachcombing. They time their shore visits to coincide with the receding tide, when fresh debris is stranded. Red foxes have similarly diverse diets, ranging from small hoofed mammals, rabbits, hares, rodents, and birds to invertebrates such as beetles, grasshoppers, and earthworms. Red foxes have been observed to fish, wading stealthily through shallow marshes. In season, fruit such as blackberries, apples, and the hips of the Dog rose can form as much as 90 percent of their diet.

All vulpine foxes catch rodents with a characteristic "mouse leap," springing a meter off the ground and diving, front paws first, onto the prey. This aerial descent may be a device to counter the vertical jump used by some mice to escape predators. Red foxes catch earthworms that leave their burrows on warm, moist nights by crisscrossing pastures at a slow walk and listening for the rasping of the worms' bristles on the grass. Once a worm is detected, the fox poises over it before plunging its snout into the grass. Worms whose tails retain a grip in their burrows are not broken but gently pulled taut after a momentary pause – a highly effective technique that foxes have in common with fishermen collecting bait.

The few fox species that have been studied in several different habitats have been found to eat whatever food was locally available. Nevertheless, foxes may have some preferences. For example Red foxes, if given the choice, will prefer rodents of the family Microtinae, such as Field voles, to members of the family Muridae, such as Field mice. However, being true opportunists, they will cache even unfavored prey for future use, and have a good memory for the location of these larders.

A Complex Society

SOCIAL BEHAVIOR

Foxes breed once a year. Litter sizes are normally from one to six, the average for the Red fox varying with habitat between four and eight; the maximum number of fetuses found in a Red fox vixen is 12. Vixens have six mammae, or teats. Known gestation periods are 60–63 days for the Red fox and 51 for the fennec. Cubs are generally born in burrows (either dug by the vixen or appropriated from other species) or rock crevices. Litters of Red foxes have been found in hollow trees, under houses, or simply in long grass. Foxes have generally been considered as monogamous, but communal denning has now been recorded for the Indian and the Red fox, and "helpers" at the den

Above *Arctic fox cubs engaged in play-fighting. The coat of this species occurs in two color forms – the white (or polar) and the blue – each with a different winter and summer coloration. These cubs are white forms in winter coat.*

Below *Eight species of fox, depicted in a dash and swipe after a bird, are shown in east-to-west order of distribution (only Gray and Corsac, and occasionally Red foxes, climb trees).* ***1*** *Gray fox* (Urocyon cinereoargenteus)*;* ***2*** *Swift fox* (Vulpes velox)*;* ***3*** *Cape fox* (V. chama)*;* ***4*** *Fennec fox* (V. zerda)*;* ***5*** *Rüppell's fox* (V. rüppelli)*;* ***6*** *Blanford's fox* (V. cana)*;* ***7*** *Indian fox* (V. bengalensis)*; and* ***8*** *Corsac fox* (V. corsac).

TRACKING DOWN DARWIN'S ZORRO

During his voyage on the *Beagle* in 1831, Charles Darwin collected a specimen of the zorro that was later to bear his name. In his journal he wrote of the capture, on Chiloë Island, about 30km (19mi) from the coast of central Chile, of a "fox, of a kind said to be peculiar to the island, and very rare in it, and which is an undescribed species."

Though Darwin's surmise that his zorro was unique has recently been proved correct, the animal's taxonomy was long the subject of speculation. Despite its dark-brown, almost rufous, head color and its relatively short legs, morphological studies found no basic traits to distinguish it from the Argentine gray zorro. The fact that it was restricted to Chiloë Island seemed to corroborate the view that it was an insular subspecies; yet this circumstantial evidence was called into question by the discovery of a mainland population of Darwin's foxes in the 1960s in Nahuelbuta National Park, some 600km (380 mi) to the north.

Conclusive proof had to await the advent of DNA analysis. Because mitochondrial DNA accumulates changes through evolutionary time, populations that are not interbreeding will begin to have molecular haplotypes that differ increasingly from each other and that will eventually distinguish one species from the other.

In the case of Darwin's zorro, sequences from a portion of the mitochondrial DNA genome in several individuals confirmed its uniqueness. They also indicated that it is probably an ancient inhabitant that diverged from a common ancestor with the Argentine gray zorro and colpeo several hundred thousand years ago. A denizen of the dense forest that covered much of South America during the last Ice Age, its population has dwindled as climate change and human activity have drastically reduced this habitat.

Conservation efforts have been galvanized by this discovery. The Nahuelbuta population consists of just 50 animals in less than 5,000ha (12,350 acres), while the Chiloë Island population numbers 500 in 1 million ha (2.5 million acres). Both groups are pressured by human incursion and other threats. Consequently, a number of special study programs have been instituted by the Chilean authorities to promote its survival. WJ

occur in both Arctic and Red foxes and the Crab-eating zorro. Among Red foxes, the proportion of vixens that breed varies greatly between areas, from 30 percent to almost 100 percent.

Foxes have been characterized as solitary carnivores, foraging alone for small prey for which cooperative hunting would be a hindrance rather than an advantage. In this respect their social behavior has been contrasted with that of pack-hunting canids such as wolves. However, with modern radio-tracking studies and night-vision equipment, it has become clear that fox society is complex. In some areas foxes are monogamous; in others they may live in groups, generally composed of one adult male and several vixens. So far, the maximum proven adult group size for Arctic foxes is three and for Red foxes six. There is no evidence of successful immigration of vixens into such groups, so the female members are probably relatives, whereas almost all male offspring emigrate. Dispersal distance varies with habitat, and records of over 200km (125mi) exist. Males generally disperse farther than females.

Although their paths may cross many times each night, foxes within a group may forage mainly in different parts of the territory, with dominant animals monopolizing the best habitat. Range sizes for Red foxes have been found to vary between 10 and 2,000 or more hectares (25–5,000 acres), those of Arctic foxes between 860 and 6,000 hectares (2,100–15,000 acres). Territory area and group size are unrelated. Feces

and urine are left on conspicuous landmarks such as tussocks of grass. These scent marks are distributed throughout the foxes' range, but especially in places visited often. Dominant animals scentmark with urine more than subordinates do, and individuals can distinguish the scent of their own urine from that of strangers. Foxes have paired anal sacs on either side of the anus; these can be evacuated voluntarily, or the secretion may be coated onto feces. Foxes also have a skin gland, 2cm (0.8in) long, on the dorsal surface of the tail, near the base. This "supracaudal" or "violet" gland is covered in bristles and appears as a black spot on the tails of all vulpine foxes. Its function is unknown. There are yet more glands between the foxes' toes. Both males and females may "cock their legs" when urine-marking.

Territory sizes are probably determined by the availability of food and by the mortality rate, which is mainly dependent on man and rabies. Where mortality through hunting is high, few Red foxes survive three years. The oldest vixen known from the wild was the 9-year-old matriarch of a group of four occupying a 40ha (100-acre) territory in Oxfordshire, England. Foxes survive for up to 13 years in captivity.

The Bat-eared Fox – An Insect Eater

Apart from its remarkable ears (among foxes only the fennec's are larger in relation to its body), it is teeth and diet that set this African species apart. Although usually found near large herds of hoofed mammals, such as zebras, wildebeest, and buffalo, the Bat-eared fox is the only canid to have largely abandoned mammalian prey.

The teeth of the Bat-eared fox are relatively small, but it has between four and eight extra molars. On the lower jaw, a steplike protrusion anchors a large muscle used for rapid chewing. The diet includes fruits, scorpions, and the occasional mammal or bird, but 80 percent is insects, and of these most are dung beetles (Scarabidae) and termites, particularly the Harvester termite (genus *Hodotermes*).

These insects abound only where large ungulates are numerous, so the Bat-eared fox remains dependent on mammals for provision of its food despite the insectivorous nature of its diet. Colonies of Harvester termites (which can make up 70 percent of the fox's diet) live underground, but large parties surface to forage for grasses that occur mainly where large ungulates feed. Dung beetles eat the dung of ungulates and lay their eggs in dung balls, which the female beetles bury. Bat-eared foxes eat both adults and larvae, which they locate by listening for the sound of the grub as it gnaws its way out of the dung ball. In this activity, large ears up to 12cm (5in) long serve the fox well.

Bat-eared foxes usually breed in self-dug dens in pairs, but also form small family groups with helpers of either sex and occasional communal nursing when two females breed. The pups (usually numbering between 2 and 5) are born after a 60-day gestation. Juveniles achieve full adult size at four months and, when accompanying adults, may account for reported group sizes of up to 12. Foraging is usually done on an individual basis.

Extensive overlap of foraging areas has been reported. Contact between groups varies from peaceful intermingling to overt aggression. Two or three breeding dens are occasionally clustered within a few hundred meters – probably a response to locally suitable soil or vegetation. Reported home-range sizes vary from 0.5 to 3sq km (0.2–1.2 per sq mi). Population density can sometimes reach 10 per sq km (26 per sq mi), but 0.5–3 per sq km (1.3–7.8 per sq mi) is more usual. JM

Above *The Fennec fox is the smallest of all canids, and is distinguished by its huge ears – the largest in the dog family relative to body size. They effectively act as the animal's "radiators," allowing it to regulate its body temperature in the extreme heat encountered in its desert habitat.*

Like other canids, foxes communicate by means of sound as well as by scentmarking and postural signals. The Arctic fox calls, for example, when an enemy approaches, or during the breeding season. The vocalizations of the Red fox include aggressive yapping and a resonant howl used by young foxes in winter and, more often, in the mating season. Barks, soft whimpers (between vixen and pups), and screams are also part of the Red fox's repertoire.

Victory Against Rabies

CONSERVATION AND ENVIRONMENT

Despite their fabled cunning, foxes, too, have their endangered species. The little-known Small-eared zorro of tropical South America is listed as threatened by the IUCN. The Swift fox, a small 2kg (4.4lb) North American prairie-dwelling fox, was rarely seen in the northern Great Plains between 1900 and 1970, and in Canada seems to have been completely exterminated by hunting and poisoning. In 1928 it disappeared from the province of Saskatchewan and in 1938 from Alberta. However, it has now been successfully reintroduced to the Canadian prairies (where the main threat is now predation by coyotes and Golden eagles, and potential competition from the spread of Red foxes). In Chile, Darwin's zorro is also under threat.

The foxes provide marvelous examples of evolution in action. For example, between 10,000 and 16,000 years ago Gray foxes reached islands offshore of California. A group of three northern islands were at this time a single landmass and were probably reached by rafting or swimming foxes. The three southern islands, however, were never connected either to the northern islands or the mainland, and were only reached by Gray foxes 3,000 years ago. These individuals derive directly from the northern island populations and were most likely transported by Native Americans who arrived in the area some 9,000–10,000 years ago. On all six islands there are miniature replicas of the mainland Gray fox, most of them distinguished as subspecies. Called the Island fox, they weigh only 1.1–2.7kg (2.4–6lb), whereas their mainland counterpart weighs 5kg (11lb). The reason for this is unclear, as their habitats are very similar, but it may relate to diet. The Island fox is mainly insectivorous, whereas the Gray fox eats a wider variety of vertebrate prey found only on the mainland. The three northern islands subspecies have recently declined dramatically. Less than six individuals are now known on San Miguel, for example, from a population of several hundred in 1993. Golden eagle predation and disease have played a part but cannot fully explain the decline. A high-priority plan, including captive breeding, has been enacted to save these species. And yet, less than 200km (124 miles) to the south, on San Clemente Island, another island fox subspecies (which only occurs here) is being killed by the US Navy in an attempt to eliminate predators of the endangered San Clemente Island shrike.

The management and conservation of foxes turn on three main issues: killing for sport or predator control; hunting for pelts; and rabies control. The 1970s saw a revival of the pre-war vogue for fox skins that resulted in huge harvests. In the 1977–78 season, 388,643 Red foxes, 264,957 Gray foxes, and 37,494 Arctic foxes were taken in North America. In 1978, nearly 1 million skins of the Argentine gray zorro were reputedly exported from Argentina. This new market for pelts had repercussions beyond the normal fur-producing nations: Britain and Ireland, where there are no monitoring systems, became fur-trapping nations for the first time in recent history, and it is estimated that 50,000–100,000 Red fox skins were exported in 1980. Soon afterward, however, the vogue passed, and in 2000 British fox skins were again almost valueless.

All fox genera are liable to spread, and to suffer from, rabies. Millions of foxes have been slaughtered in vain attempts to control the disease, but foxes have such resilience that populations can withstand about 75 percent mortality without declining further. Control was revolutionized in the 1980s and 1990s by the use of oral vaccination: baits (pieces of meat impregnated with anti-rabies vaccine) were laid where rabies was endemic. The foxes eating the baits were immunized and hence withdrawn from the pool of susceptibles. The result is that rabies has been widely eliminated from continental Europe – perhaps the greatest 20th-century triumph of wildlife management. Importantly, roughly the same proportion of foxes ate vaccine baits as were formerly killed in attempted lethal control. The difference must lie, then, in the behavior of the foxes that survived: perhaps this was so perturbed by the disruption of their societies that they continued to spread rabies between them, despite their reduced numbers. DWM

Below *The Swift fox was one of many wildlife victims of the rapid westward expansion of population across the Great Plains of North America in the 19th century. As well as suffering habitat loss to agriculture and taking poison bait laid for larger predators, it was hunted for its pelt. Surviving populations in the northern USA were used to reintroduce the Swift fox into western Canada in the 1980s and 1990s.*

Fox Species

FOX CLASSIFICATION IS COMPLICATED BY similarities between imperfectly distinguished species and by fragmentary knowledge of their behavior. The genera are mainly distinguished by the fact that the "brows" formed by the frontal bones above and between the eyes are slightly indented or dished in the genus *Vulpes* and flat in *Dusicyon* (as opposed to convex in *Canis*).

GENUS *UROCYON*

Two species from North America.

Gray fox
Urocyon cinereoargenteus
Gray or Tree fox

Separated as *Urocyon* partly because of longer tail gland; able to climb trees. Central USA to the prairies, S to Venezuela, N to Ontario.
HBL 52–69cm; TL 27–45cm; WT 2.5–7kg.
FORM: canine teeth shorter than average for *Vulpes*; coat gray-agouti; throat white; legs and feet tawny; mane of black-tipped bristles along dorsal surface of tail.

Island gray fox LR
Urocyon littoralis

Islands of W USA.
HBL 59–79cm; TL 11–29cm; WT 1.1–2.7 kg.
FORM: Smaller than Gray fox, but otherwise identical.
CONSERVATION STATUS: LR Conservation Dependent.

GENUS *OTOCYON*

Distinguished from other genera on the basis of unusual dentition.

Bat-eared fox
Otocyon megalotis
Bat-eared or Delandi's fox

Two populations, one from S Zambia to S Africa, the other from Ethiopia to Tanzania. Open grasslands.
HBL 46–58cm; TL 24–34cm; WT 3–4.5kg.
FORM: coat gray to buff, with face markings, tips of ears, feet, and dorsal stripe all black; ears large (to 12cm); teeth weak but with extra molars to give 46–50 teeth in all; lower jaw with step-like angular process to allow attachment of large digastric muscle used for rapid chewing of termites.

GENUS *VULPES*

N and far N of S America, Europe, Africa, and Asia.
FORM: muzzle pointed; ears triangular and erect; tail long and bushy; skull flattened by comparison with *Canis*. Tail tip often different color from rest of coat (eg black or white); black, triangular face marks between eyes and nose.

Indian fox
Vulpes bengalensis
Indian or Bengal fox

India, Pakistan, and Nepal. Steppe, open forest, thorny scrub, and semi-desert up to altitudes of 1,350m.
HBL 45–60cm; TL 25–35cm; WT 1.8–3.2kg.
FORM: coat sandy-orange, with legs tawny-brown; tail black-tipped. Regarded anatomically as the "typical" vulpine fox.

Cape fox
Vulpes chama
Cape or Silverbacked fox

Africa S of Zimbabwe and Angola. Steppe, rocky desert.
HBL 45–61cm; TL 30–40cm; WT 3.6–4.55kg.
FORM: anatomy, especially skull, similar to Indian fox and Pale fox; ears elongate; coat rufous-agouti with silvery-gray back; tail tip black; dark facial mask lacking.

Corsac fox
Vulpes corsac

SE Russian Federation, Turkestan, Mongolia, Transbaikalia to N Manchuria and N Afghanistan. Steppe.
HBL 50–60cm; TL 22–35cm; WT unknown.
COAT: russet-gray with white chin.

Tibetan fox
Vulpes ferrilata
Tibetan or Tibetan sand fox

Probably descended from Corsac fox. Tibet and Nepal. High steppe. (4,500–4,800m).
HBL 67cm; TL 29cm; WT not known.
FORM: head long and narrow; canines elongate; coat pale gray-agouti on body and ears; tip of tail white.

Pale fox
Vulpes pallida

N Africa from Red Sea to Atlantic, Senegal to Sudan and Somalia. Desert.
HBL 40–45cm; TL 27–29cm; WT 2.7kg.
COAT: pale fawn on body and ears; legs rufous; tail tip black; no facial marks; fur short and thin; whiskers relatively long and black.

Rüppell's fox
Vulpes rüppelli
Ruppell's or Sand fox

Scattered populations between Morocco and Afghanistan, NE Nigeria, N Cameroon, Chad, Central African Republic, Gabon, Congo, Somalia, Sudan, Egypt, Sinai, Arabia. Desert.
HBL 40–52cm; TL 25–35cm; WT 1.7kg.
COAT: pale, sandy color; conspicuous white tail tip and black muzzle patches; whiskers relatively long and black.

Arctic fox
Vulpes (Alopex) lagopus
Arctic or Polar or Blue or White fox

Circumpolar in tundra latitudes. Tundra and intertidal zone of seashore.
HBL 55cm (male), 53cm (female); TL 31cm (male), 30cm (female); WT 3.8kg (male), 3.1kg (female).
COAT: very thick with two dichromatic color forms: "white," which is gray-brown in summer; and "blue," which is chocolate-brown in summer; 70 percent of fur is fine, warm underfur. Has remarkable tolerance to cold.

Swift fox LR
Vulpes velox

Central USA from Texas to South Dakota; reintroduced into Canada and Montana. Prairies. Hybrid zone with Kit foxes in New Mexico.
HBL 37–53 cm; TL 22–35 cm; WT 1.9–3.0 kg (male averaging 2.4 kg, female 2.3 kg).
COAT: buff-gray in winter, buff-red in summer, black tail tip, black patches on muzzle sides.
CONSERVATION STATUS: LR Conservation Dependent.

Red fox
Vulpes vulpes
Red or Silver or Cross fox

Formerly divided between 48 subspecies, but in all canids the classification of subspecies is likely to be changed with the advent of molecular techniques. N Hemisphere from Arctic Circle to N African and C American deserts and Asiatic steppes, with natural S limit in Sudan; introduced into Australia. (The N American subspecies *V. v. fulvus* was formerly considered a separate species, *V. fulva*.) Wide-ranging Arctic tundra to European city centers.
HBL 68cm (male), 66cm (female); TL 44cm (male), 42cm (female); WT 5.9kg (male), 5.2kg (female). (*V. v. fulva*: HBL 55–62cm; TL 35–40cm; WT 4.1–5.4kg.)
COAT: rust to flame-red above (with silver, cross, and color phases); white to black below; tip of tail often white.

Kit fox
Vulpes macrotis

NW Mexico and SW USA; population contiguous with exception of the San Joaquin kit fox, *V. m. mutica*, which is geographically isolated in California. Prairies and arid steppe.
HBL 38–50cm; TL 22–30cm; WT 1.8–3kg.
COAT: buff-red; limbs and feet tawny; black tip to very bushy tail.

Blanford's fox
Vulpes cana
Blanford's or Hoary or Afghan fox

Afghanistan, SW Russian Federation, Turkmenistan, NE Iran, Baluchistan; isolated population in Israel. Mountainous regions
HBL 42cm; TL 30cm; WT 1kg.
FORM: coat like miniature Red fox; naked foot-pads for cliff-dwelling lifestyle.

Fennec fox
Vulpes (Fennecus) zerda

Formerly separated as *Fennecus* because of large ears, rounded skull, and weak dentition. N Africa, throughout Sahara, E to Sinai and Arabia. Sandy desert.
HBL 24–41cm; TL 18–31cm; WT 0.8–1.5kg
FORM: coat cream with black-tipped tail; soles of feet furred; ears very large, up to 15cm long; dark bristles over tail gland; whiskers relatively long and black. The smallest fox.

GENUS *DUSICYON*

Restricted to S America.
FORM: anatomy intermediate between *Vulpes* and *Canis*, with the extinct *D. australis* (Falkland Island wolf) most doglike and *D. vetulus* most vulpine. Coat usually gray with tawny grizzling. Skull long and thin; ears large and erect; tail bushy. Biology poorly known.

Colpeo zorro
Dusicyon (Pseudalopex) culpaeus

Andes, from Ecuador and Peru to Tierra del Fuego. (*D. culpaeolus* found in

---

ABBREVIATIONS

HBL = head–body length TL = tail length WT = weight
Approximate nonmetric equivalents: 10cm = 4in; 230g = 8oz; 1kg = 2.2lb

Ex Extinct
EW Extinct in the Wild
Cr Critically Endangered
En Endangered
Vu Vulnerable
LR Lower Risk

Right *Azara's zorro is widely distributed throughout central South America. It has the grizzled gray coat typical of the* Dusicyon *genus.*

Uruguay is similar, but smaller, whereas *D. inca* from Peru is larger; both may be better considered subspecies of *D. culpaeus*, as indeed may Azara's zorro) Mountains and pampas.
HBL 60–115cm; TL 30–45cm; WT 7.4kg.
COAT: back, shoulders grizzled gray; head, neck, ears, legs tawny; tail black-tipped.

Argentine gray zorro
Dusicyon (Pseudalopex) griseus
Argentine gray or Gray or Pampas fox

Distribution as colpeo, but at lower altitudes in Ecuador and N Chile. Plains, pampas, and low mountains.
HBL average 42–68cm; TL 30–36cm; WT 4.4kg.
COAT: brindled pale gray; underparts pale.

Azara's zorro
Dusicyon (Pseudalopex) gymnocercus
Azara's or Pampas fox

Perhaps same species as Colpeo. Paraguay, Chile, SE Brazil, S through E Argentina to Rio Negro. Pampas.
HBL 62cm; TL 34cm; WT 4.8–6.5kg.
COAT: uniform grizzled gray.

Sechuran zorro
Dusicyon (Pseudalopex) sechurae

N Peru and S Ecuador. Coastal desert.
HBL 53–59cm; TL about 25cm; WT about 4.5kg.
COAT: pale agouti without any russet tinges; tail black-tipped.

Hoary zorro
Dusicyon (Pseudalopex) vetulus
Hoary fox or Small-toothed dog

Previously allocated to *Lycalopex* on basis of small teeth. SC Brazil: Minas Gerais, Matto Grosso.
HBL 60cm; TL 30cm; WT 2.7–4kg.
FORM: A small *Dusicyon*, with short muzzle, small teeth, and reduced upper carnassials. Dark line present on dorsal surface of tail.

Darwin's zorro
Dusicyon (Pseudalopex) fulvipes

Chiloë Island and Nahuelbuta National Park, Chile.
WT: c. 2kg.

Crab-eating zorro
Dusicyon (Cerdocyon) thous
Crab-eating fox or Common zorro

Formerly considered sufficiently distinct to be placed in a separate genus, *Cerdocyon*. Colombia and Venezuela to N Argentina and Paraguay. Savanna, llanos, and woodland.
HBL 60–70cm; TL 28–30cm; WT 5–8kg.
COAT: gray-brown; ears dark; tail with dark dorsal stripe and black tip; foot pads large; muzzle short.

Small-eared zorro
Dusicyon (Atelocynus) microtis
Small-eared dog, Zorro negro, or Small-eared zorro

Formerly considered sufficiently distinct to be placed in separate genus, *Atelocynus*. Amazon and Orinoco basins, parts of Peru, Colombia, Ecuador, Venezuela, Brazil. Tropical forests.
HBL 72–100cm; TL 25–35cm; WT 9kg.
FORM: coat dark; ears short (5cm) and rounded; has long, heavy teeth and enlarged second lower molar; gait reputedly catlike.

FOX CLASSIFICATION

Convergence or parallel evolution have selected for similar morphology in many of the "foxlike" species. Morphological characters used in the past to classify these small canids have generated conflicting results, and an excess of genera. The recent classification is based on evolutionary (phylogenetic) associations between these taxa largely revealed by molecular studies.

There are three separate lineages of "foxlike canids." Those closest to the ancestors of all canids are the two gray fox species (*Urocyon*; 4–6 million years old). Although morphologically resembling the classic vulpine fox, they are not genetically related to any of the foxes. The Bat-eared fox (*Otocyon*) is also an ancient canid that is genetically and morphologically distant from all other foxes (Pliocene; 3 million years old). These species form the first lineage.

A second lineage clusters the vulpine foxes. The vulpine lineage is separated into two parts: the Red fox-like and the Fennec-like forms. The Blanford's fox and the Fennec are sister taxa, and are an old radiation (4–4.5 million years). The Red fox-like cluster includes the Kit, Swift, and Arctic foxes and many of the Old World species. The Kit, Swift, and Arctic foxes have only recently evolved (mid-Pleistocene; 0.5 million years), and form a subunit within the Red fox-like clade or cluster.

All the South American species cluster into a third lineage, which is more closely related to the *Canis* radiation than to the other foxes. The Small-eared and Crab-eating zorros are ancestral within that group (3 million years); most other *Dusicyon* species have evolved fairly recently (1.0–2.5 million years). The *Dusicyon* foxes are intermediate in appearance between the popular images of foxes and that of canids such as wolves and coyotes; indeed, in 1868 the taxonomist J. E. Gray called them Fox-tailed wolves. The Falkland Island wolf (*D. australis*) was first described by Darwin, who noted that it fed on the goose *Chloephaga picta*. In their skull and teeth these wolflike foxes were more similar to *Canis* than *Dusicyon*, but their extermination by pelt-hunters by 1880 has sadly relegated them to the status of a biological mystery. Some maintain that *D. australis* was descended from a domestic canid, pointing out that it had a white-tipped tail, rather than the black one of all surviving *Dusicyon* species. Another possibility is that it was related to the genus *Cuon*, of which only the dhole survives (*Cuon* species were common in South America several million years ago). Another South American mystery is *D. hagenbecki*, a foxlike canid known from a single skin found in the Andes. DWM/EG

African Wild Dog

Below A wildebeest falls prey to wild dogs in Tanzania (one dog is wearing a radio-tracking collar). The grabbing of the victim's upper lip is a common tactic: typically another dog will cling onto the tail. Other species hunted by the wild dogs include Greater kudu, Red lechwe, ostrich, springbok, reedbuck, puku, warthog, and several duiker species.

WITH NO MORE SOUND THAN A BREEZE stirring leaves in a dry woodland, a pack of African wild dogs roams tirelessly through its vast home range. The pack moves as one; individuals pause only occasionally to investigate a noteworthy scent detected in passing on a micro-current of warm air. Consummately cooperative, the dogs are described most succinctly by two fundamental behavioral characteristics: cohesive sociality and mobility.

The dogs are also distinguished by their unusual social system, which is the reverse of that found among most other pack-dwelling mammals. Within packs, all the male animals are related to one another, and all the females to each other but not to the males. Females migrate into the pack, whereas males stay with the group into which they were born – a uniquely "back-to-front" arrangement.

FACTFILE

AFRICAN WILD DOG

Lycaon pictus

Wild dog, Painted or Cape hunting dog

Order: Carnivora

Family: Canidae

Sole member of genus

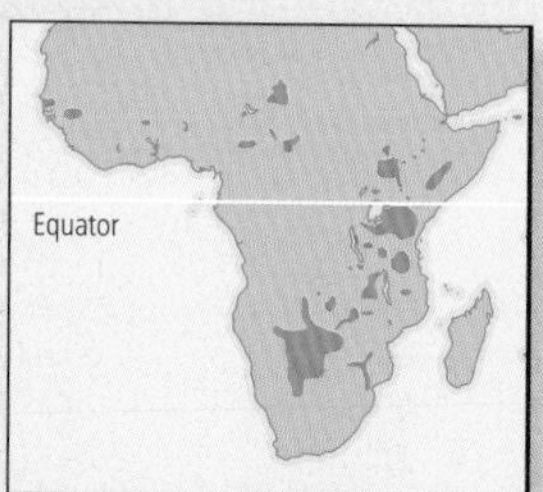

DISTRIBUTION Africa, from the Sahara to S Africa.

HABITAT From semi-desert to alpine zones; savanna woodland probably preferred.

SIZE Head–body length 75–120cm (30–47in); tail length 30–44cm (12–18in); shoulder height 75cm (30in): weight 20–32kg (44–71lb). No variation between sexes. Dogs from S African populations are consistently larger than E African.

COAT Short and dark, with a unique pattern of irregular white and yellow blotches on each individual; muzzle dark; tail white-tipped.

BREEDING Gestation 70–73 days.

LONGEVITY About 10 years.

CONSERVATION STATUS Endangered, with fewer than 5,500 surviving animals.

Keeping Up with the Pack

FORM AND FUNCTION

The African hunting dog is the least typical canid. It is exclusively carnivorous and has a relatively short, powerful muzzle housing an array of shearing teeth, with the last molar poorly developed; unlike other canids, it has lost the fifth digit – the dew claw – on the front feet. The distinctive, large, rounded ears probably aid cooling, but also emphasize the role of vocal signals in social contact. A wild dog will often find itself separated from pack members by some distance soon after the start of a hunt. The pack remain in touch by using soft, distinctive contact calls – "hoo" – that they can detect from as much as 2km (1.2mi) away. Pack members will reassemble after short periods of separation by tracking the direction and distance of the calls.

Despite relying on visual contact to pursue prey, wild dogs hunt in the low light of dawn and dusk, taking advantage of the cool conditions. They also hunt at night during periods of bright moonlight. Packs will rally and move off to hunt twice every day, but an average pack successfully captures prey only twice every three days.

Wild dogs hunt a wide range of species, from rabbits to buffalo calves and zebra. They are best adapted to pursuing medium-sized antelope; in southern Africa, impala constitute as much as 85 percent of wild dog prey, with numerous species making up the remainder. Some packs specialize on prey species that others rarely hunt. In the Serengeti, for example, two packs of dogs have been known to hunt zebra. In northern Botswana warthog and ostrich are relatively common, but only a few packs have the skills to capture them.

A hunt begins with a lethargic rally after the dogs have rested through the heat of the day. They start with a ritual greeting, complete with begging, face-licking, and excited vocalizations from the young dogs. Shortly after the greeting, the pack starts walking away stiffly in a direction usually chosen by one or two of the older dogs, followed by the others spread out in a loose string. In a few minutes they pick up the pace to a steady, near-effortless trot, a gait they can maintain for hours in cool temperatures.

They hunt opportunistically whatever they encounter. Sometimes they sight potential prey before it runs, and decrease the distance by stalking with ears laid flat in plain view of the staring prey. In that case the chase only gets underway when the prey suddenly breaks and runs for cover. It is common for members of the pack to each chase a different impala, and for more than one to

A Hunting Tradition

A zebra stallion trots away from his group of mares and foals as the wild dog pack approaches. The zebra – head lowered, teeth bared, and nostrils flaring – charges at the dogs leading the pack, who turn and flee.

This is usually what happens when African wild dogs and zebras meet. But not always. A minority of wild dog packs will attack zebras. When they do, it is they who charge first, in order to cause a stampede before the stallions can take the initiative. Then they mount a closely coordinated attack, one dog grabbing the chosen victim's tail and another its upper lip, while the rest of the pack disembowel it.

Of 10 wild dog packs recently studied on the Serengeti Plain in Tanzania, only two had the ability to turn the tables on an adult zebra – eight times a dog's weight – and kill it. These two zebra-hunting packs were large, both containing eight or more adult members. Pack size, however, was not the crucial factor: three other packs of similar size ignored zebras and, on one occasion, just four dogs from a zebra-hunting pack were observed to kill a zebra.

Why then do not all African wild dog packs hunt zebras? A clue to the answer seems to lie in the fact that one of the two zebra-hunting packs was known to have hunted zebras for at least 10 years, over three generations. For this pack, zebra-hunting was a tradition, learned by each generation from its predecessor. Other packs studied did not exhibit any such hunting tradition, although some did show a preference for Grant's gazelles over Thomson's gazelles.

Not only zebra hunting but also such knowledge as the location of water, prey concentrations, and range boundaries may be passed on as a tradition. Studies of African wild dogs (and also of some monkeys) indicate that man's previously supposed unique reliance on cultural as against genetic inheritance should rather be viewed as a dramatic extension of a pattern that exists in other social animals. JM

be captured. There are several benefits to hunting as a group, not least an increased rate of capture.

The primary method of capture is to pursue the prey until it tires, enabling a dog to grab it by the flank and pull it to the ground. Over open ground, wild dogs have been recorded running at speeds of 60km/h (37mph). It is common in bushy habitat for a single wild dog to make a kill and then return to fetch the reassembled pack. More often, additional pack members will join in the kill, having kept pace during the chase. Sometimes the entire pack is right behind at the moment of capture. In these circumstances a medium-sized antelope will be dispatched and consumed by all the dogs present. They eat in an exuberance of frenetic excitement and a flurry of dust, which belies the cooperation and precision with which they carve.

Above *This African wild dog, in Botswana, carries a young pup in her mouth. On average only about 5(percent of pups reach adulthood.*

The Caring Carnivore

SOCIAL BEHAVIOR

African wild dogs hunt, rest, travel, and reproduc in packs, which average seven or eight adult mem bers but may include anything from two to more than thirty. Packs of over 50 individuals occur bu rarely endure for more than a few months. When they swell to numbers in excess of 30 with pups from a third or fourth litter, packs usually break up. This normally happens midway through the year, either through fission or, more commonly, vi dispersal of single-sex groups.

Packs are formed when a group of females join an unfamiliar group of males. An average new pac consists of two sisters from one pack – usually

Below and right *Friendly relations prevail in a pack most of the time, as in the ritualized so-called "midday greeting ceremony." All the members of a pack run around excitedly, squeaking and thrusting their muzzles into each other's faces, a gesture that derives from infantile begging (BELOW). At other times adult male packmates and juveniles join in friendly playfights with each other.*

ttermates – and three or four brothers from nother. A "dominant pair," consisting of one nale and one female, emerges immediately to lead he new coalition. Occasionally these new groups ail to coalesce into a functioning pack, whereupon the males and females separate. More often hey thrive and establish a new range.

Wild dogs tend to mate seasonally. Reproducion is monopolized by the dominant male and emale. Some sub-dominant adults never actually aise their own young, cooperating instead to proision the pups produced once a year by the domnant pair. They are often uncles or aunts of the ups, however, so in evolutionary terms they have vested interest in the survival of the youngsters vith which they share their genes. The dogs find a uitable den site, such as a hole dug by an aardark or porcupine, and within a month of the ack forming, a litter of pups is born. Wild dogs ave the largest litters among canids, with an averge of 10 pups.

Below A wild dog pack is on the move. Both dults and pups will play in shallow waters, where here is little chance of being attacked by crocodiles. acks range over enormous distances and a typical vild dog pack may require a range of around 750sq m (290sq miles).

By 4–5 weeks of age, pups begin to eat meat regurgitated for them by the adults in the pack. They are weaned by 10 weeks. At 14–16 weeks, it becomes impossible for the adults to go hunting without some or all of the pups following them. If they go astray on a hunt, they are retrieved by the adults, and are led to each kill as soon as it is made. There, the pups are allowed to eat first until they have had their fill. Only then do the adults allow themselves whatever is left of the carcass.

Physiologically mature at 13–14 months, wild dogs rarely reproduce before two years of age. All dogs surviving to one year remain in their natal pack and assist in the provisioning and protection of younger siblings. Females tend to disperse in their second year, while males are more likely to wait until the following year. When males emigrate from their natal pack they do so in larger sibling groups – the average is three but as many as eight brothers have been recorded dispersing together – and they disperse twice as far as the females.

Although their preferred habitat is woodland and scrub, wild dogs have been recorded in widely varying environments. Where healthy resident prey populations exist, wild dog densities average only one pack per 400sq km (150sq mi), while in less optimal habitats, densities as low as one per 2,000sq km (770sq mi) are not uncommon.

Running for their Lives

CONSERVATION AND ENVIRONMENT

The entire population of wild dogs is estimated to be less than 5,500. They are persecuted by humans and are run over by cars; they are susceptible to epidemic diseases, especially rabies; and the dogs' habitat is shrinking in face of an expanding human population. In addition, lions can pose a serious predation threat.

Wild dogs also suffer from losing food to scavenging hyenas. Dogs normally hunt for around 3.5 hours per day, but must increase this to 12 hours if they lose 25 percent of their food, which would increase their metabolic rate to a physiologically unfeasible 12 times the basal metabolic rate. They need vast areas of land on which to hunt and invariably conflict with the owners of free-ranging, unattended livestock.

A handful of substantial, but largely isolated, populations remain in east and southern Africa, along with several smaller populations throughout sub-Saharan Africa. The future of the species will depend on prevention of direct persecution, as well as on the management of land with their interests in mind. Those countries where African wild dogs do survive and prosper will be able to take pride in their habitat management for a species with very particular needs. JMcN

Dhole

BRANDED AS CRUEL AND WANTON KILLERS *because they often kill by disemboweling their prey, dholes are persecuted throughout their range. Yet they rarely take domestic livestock, and do little harm to humans – only one attack on a person has ever been reported.*

Above *The dhole has a relatively short, broad muzzle. Hunting cooperatively in packs, dholes normally feed together peaceably at kills, such as this large chital stag, but individuals may occasionally take away portions of the carcass to consume undisturbed at a distance.*

Although the social structure of these secretive animals has not been exhaustively studied, it has been established beyond doubt that they are group-living, with cooperative hunting and group care of the young at the heart of their societies. In many respects their lifestyle resembles that of the African wild dog.

Hunters and Scavengers

FORM AND FUNCTION

The dhole differs from most canids in having a thick-set muzzle and one fewer molar tooth on each side of the lower jaw, both adaptations to an almost wholly carnivorous diet. They eat fast: a fawn is dismembered within seconds of the kill and, rather than fight, the dholes will attempt to eat more quickly than their packmates to boost their share. Heart, liver, rump, eyeballs, and other soft parts are eaten first. When water is nearby, dholes drink frequently as they eat, and if not they make for it soon after eating. Unguarded leopard and tiger kills are also scavenged by dholes. In Bandipur Tiger Reserve in southern India, which is rich in prey, each dhole consumes around 1.8kg (4lb) of meat every day.

FACTFILE

DHOLE

Cuon alpinus

Dhole or Asian wild dog or Red dog

Order: Carnivora

Family: Canidae

Sole member of genus

DISTRIBUTION W Asia to China, India, Indochina to Java: rare outside protected areas.

HABITAT Chiefly forests with abundant large ungulates and plentiful water supplies.

SIZE Head–body length 90cm (35in); tail length 40–45cm (16–18in); shoulder height 50cm (20in); weight average 17kg (37lb). Russian populations some 20 percent larger.

COAT Sandy russet above, underside paler; tail black, bushy; pups born sooty brown, acquiring adult colors at 3 months; distinct winter coat in dholes in Russia.

DIET Mammals ranging in size from rodents to deer; also birds, lizards, insects, and wild berries.

BREEDING Gestation 60–62 day. Average litter size 8.

LONGEVITY 8–12 years in the wild; up to 16 in captivity.

CONSERVATION STATUS Of 10 subspecies, two – the East Asian dhole (*C. a. alpinus*), and the smaller West Asian dhole (*C. a. hesperius*) – are listed as threatened. Of three Indian subspecies, only *C. a. dukhunensis*, found S of Ganges, is fairly common; *C. a. primaevus* (Kumaon, Nepal, Sikkim, Bhutan) and *C. a. laniger* (Kashmir, Lhasa) are both rare.

HUNTING STRATEGIES

Dholes are active chiefly during the day, although hunts on moonlit nights are not uncommon. Most hunts involve all adult members of a pack, but solitary dholes often kill small mammals such as chital fawns or Indian hares. Prey is often located by smell. If tall grass conceals their prey, dholes will sometimes jump high in the air or stand briefly on their hind legs in order to spot it.

Dholes have evolved two strategies to overcome the problems posed by hunting in thick cover, both of which depend heavily on cooperation. In the first, the pack moves through scrub in extended line abreast. Any adult capable of killing may begin an attack when it locates suitable prey. If the prey is small it will be dispatched by one dhole. When the prey is larger – for example a chital stag – the sound of the chase and the scream of the prey attract other pack members to assist. It is rare for two large animals to be killed in one hunt.

In the second strategy (BELOW), some dholes remain on the edge of dense cover to intercept fleeing prey as it is flushed out by the other pack members. In thick jungle the chase seldom lasts more than half a kilometer (0.3mi). A nose-hold is frequently used to subdue larger prey such as chital and sambar stags with hard antlers, which are potential lethal weapons. Attack on the rump and flank leads to evisceration; the resultant trauma and loss of blood kills the prey. Small mammals are caught by any part of the body and killed with a single head-shake. Even before their prey is quite dead, dholes start eating. In general they are efficient killers – two or three dholes can kill a deer of 50kg (110lb) within 2 minutes. Interference during the last stages of attack by human observers can prolong the death throes, and this may have helped perpetuate the myth that dholes are particularly cruel hunters.

Caring for the Pups

SOCIAL BEHAVIOR

A dhole pack is an extended family unit, usually of 5–12 animals, and rarely exceeding 20. In a pack observed in Bandipur, the average number of adults was eight, but pups increased this to 16, and there were consistently more males than females. One benefit of pack life is to ward off predators such as leopards and tigers.

Dholes are sexually mature at about 1 year. While mating there is a copulatory tie, holding the pair together for 7–14 minutes. Whelping occurs in November–April, and the average litter size is eight. Before giving birth the bitch prepares a den, usually in an existing hole or shelter on the banks of a streambed or among rocks. In the Bandipur pack, more than three adults took part in feeding both the lactating mother and the pups, which eat regurgitated meat from the age of 3–4 weeks. At this important time the pack stayed much closer to the den in hunting, extending only around 11sq km (4sq mi), which is much smaller than its normal range of 40sq km (15sq mi). Sometimes a second adult (the so-called "guard dhole") stays by the den with the mother while the rest of the pack is away hunting. Pups leave the den at 70–80 days. Thereafter the pack continues to care for the pups, feeding them, escorting them as they play, and letting them eat first at kills. Quick to learn and develop, the pups actively follow the pack at 5 months, and at 8 months participate in kills.

Dhole calls include whines, growls, growl-barks, screams, and whistles; the pups squeak. The whistle is a contact call most often used to reassemble the pack after an unsuccessful hunt. Latrine sites, established at the intersection of trails and roads, may be a major means of communication by smell, serving to warn off neighboring packs at the edge of the home range and to mark how recently an area has been hunted. This ensures efficient use of all parts of the home range, which can be as large as 80sq km (30sq mi).

Taking Refuge in the Reserves

CONSERVATION AND ENVIRONMENT

Until recently, hunters considered dholes as rivals and tried to eradicate them by offering bounties and poisoning their kills; even now, tribal peoples follow packs of dholes and pirate their kills. Occasional poisoning still takes place. However, the main threat now comes from destruction of the dhole's forest habitat through wood-cutting and cattle-grazing, as well as the steady elimination of prey species. The situation is especially serious in Kazakhstan and Siberia, where dholes are often killed by poisoned bait left for Gray wolves. In India the creation of many tiger reserves and national parks has helped conserve the species, and the population is now thought to number between 5,000 and 8,000 animals. AJTJ/AV

Other Dogs

THE WILD DOGS OF EAST ASIA, AUSTRALASIA, *and South America are some of the least-studied canids, and much remains to be learned about them. Some have found a place in the popular imagination: in Brazil, the night cry of the Maned wolf is believed to portend changes in the weather.*

Maned Wolf

CHRYSOCYON BRACHYURUS

The long legs that are the Maned wolf's most distinctive feature are most likely an adaptation to tall grassland habitats. The animals hunt pacas that weigh around 8kg (18lb), but rabbits, small rodents, armadillos, and birds are the most common prey, along with occasional fish, insects, and reptiles. Over half the diet (64 percent) is fruit, particularly *Solanum lycocarpum*, known as "fruta do lobo," or wolf's fruit. The latter may have therapeutic properties against the Giant kidney worm (*Dioctophynia renale*). Individuals hunt alone and at night. Their slow stalk and stiff-legged pounce are reminiscent of the Red fox. Aplomado falcons have been observed to follow and hunt the prey flushed by the pouncing wolf.

Sexually mature at about one year, breeding at two years, females produce one litter per year, usually in June–September. The young reach adult size by about one year. The extent to which males take part in raising the young is not fully known, but males in captivity have been observed to feed pups by regurgitation.

The Maned wolf is classified by the IUCN as Near Threatened, mainly due to habitat loss. In addition, some wolves attack small farm stock, which leads to conflict with farmers. The animals are also occasionally hunted for sport, and are frequently captured for sale to zoos. In Brazil, parts of the Maned wolf's body – and even its feces – are supposed to have medicinal value, or to work as charms. Maintaining the animals in captivity is complicated by the need for enclosures of at least several hundred square meters. JMD

Bush Dog

SPEOTHOS VENATICUS

The Bush dog is one of the least-known canids, and one of the least doglike in appearance. It is a stocky, broad-faced animal with small, rounded ears, squat legs, and a short tail. Bush dogs are elusive, and little is known of their behavior in the wild. Packs of up to 10 animals are reputed to hunt together, often seeking prey larger than themselves. Packs are said to pursue amphibious prey into the water, where the dogs swim and dive with agility. Their teeth are sturdy and highly adapted to their carnivorous way of life. The dental formula (I3/3, C1/1, P4/4, M1/2 = 38) is not shared by any other American canid.

Studies in captivity have shown that males feed their nursing mates and that littermates squabble very little over food in comparison with, say, young foxes. This is probably related to their cooperative hunting in later life. Adults are said to

Far right Since its release into European Russia, the Raccoon dog has colonized parts of Scandinavia, the Baltic states, and Eastern Europe. The pelts of the Raccoon dog are worth little money, although several hundred thousand are used annually.

Left The long legs and red coat of the Maned wolf, here in its grassland habitat, make it a popular exhibit at zoos, for which the animals are sometimes caught. A number of superstitions and beliefs surround the Maned wolf. In Brazil, for example, folklore maintains that it can fell a chicken with its gaze, and that if the left eye is removed from a live Maned wolf it will bring good fortune – though obviously not for the wolf!

Far right An adult Bush dog tending its young cubs. Bush dogs often hunt prey considerably larger than themselves, such as capybaras and rheas. Like wolves, which also hunt cooperatively in later life, Bush dog pups have a tendency to be amicable over the division of food.

Below This male dingo from central Australia is digging out a rabbit from its warren. The dingo has lived in Australia for a very long time and Aboriginal people venerated the first dingoes as Dreamtime cultural heroes; however, they are now commonly viewed as pests and in many areas there are bounty schemes to reduce their numbers.

eep in contact by means of frequent whines; this nay be an adaptation favorable to maintaining roup cohesion whilst foraging in forest under-rowth where visibility is poor. When scent mark-ng, males urinate by cocking their legs at 90°, but emales reverse up to trees and urinate on the runk from a handstand position. DWM

Raccoon Dog

NYCTEREUTES PROCYONOIDES

rimitive canids whose closest relatives are the Gray and Bat-eared foxes, Raccoon dogs are omni-ores with a diet that varies according to the sea-onal availability of fruits, berries, invertebrates, rogs, and reptiles; in most areas, though, small nammals (voles, mice, and shrews) make up bout half of what they eat. They prefer to forage n moist forests with an abundant understory, specially of ferns, and to live near water. When disturbed, they may escape into water, as they are xcellent swimmers.

From the 1920s to the 1950s they were intro-luced into western Russia for fur farming; they vere also deliberately released into the wild, and subsequently spread throughout much of eastern Europe and northwest to Finland. Outside the fur rade, this range expansion has been viewed as ndesirable, since the Raccoon dog carries rabies, and its arrival has further complicated the control of the disease in eastern Europe. There is some evidence that during their winter hibernation – tself a unique feature amongst canids – Raccoon logs may incubate the rabies virus and hence cause the disease to persist from one season to the next in places where fox densities are so low that it night otherwise die out.

Although they are the only members of the amily reputed not to bark, the behavior of Rac-coon dogs is otherwise recognizably canid. Like Bat-eared foxes, they occasionally hold their tails aloft in inverted-U positions during social interac-tions. Subordinate animals apparently do not wag or lash their tails, unlike other canids. However, mating pairs do become tied together as in dogs.

Raccoon dogs are thought to live in pairs or temporary family groups, with males babysitting while females forage. The pups are fed milk until they start to forage for themselves. Radio-tracking studies from Japan suggest that the home ranges of Raccoon dogs overlap widely, and when food is clumped, such as at a fruiting bush, they may feed together amicably. In Finland the mated pair is the basic social unit, and the tie between male and female sharing a home range is strong. KK

Dingo

CANIS DINGO

The origins of the dingo are obscure, although it has lived unchanged in Australia for many thou-sands of years. The first surviving fossil remains are probably from Thailand, and are 5,500 years old; the earliest from Australia are 3,500 years old.

Asian seafarers are thought to have tranported dingoes to Australia, as well as to the islands of Southeast Asia and to Madagascar. The Aborigines may have used the animal for warmth at night, as food, or as a guard, but probably not for hunting.

The flexibility of dingo social behavior parallels that of the coyote or wolf; indeed, the dingo is probably descended from the Indian wolf (*C. lupus pallipes*). Its diet varies between small mammals caught by individual dogs, to large ungulates devoured as carrion or hunted by a pack.

Although the name "dingo" is most commonly applied to the wild canid of Australia, it is also used to describe various little-known dogs such as the New Guinea singing dog and the Malaysian, Siamese, and Filipino wild dogs. The term "pariah dog" is sometimes loosely used to cover both wild dingoes and feral domestic dogs.

In Australia the dingo is economically impor-tant, directly because of attacks on sheep and thus indirectly due to the costs of poisoning, paying bounties, and erecting dingo-proof fencing. Aerial poisoning with baits has greatly reduced dingo numbers, but populations are nonetheless gener-ally robust, except in Queensland where remnants of the pure dingo are likely to be swamped to extinction by feral dog genes. DWM

FACTFILE

OTHER DOGS

Order: Carnivora

Family: Canidae

4 species in 4 genera

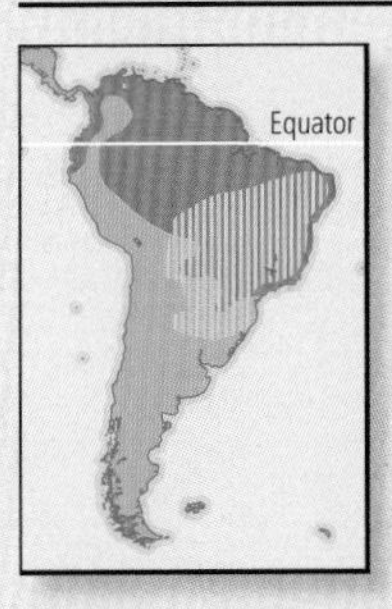

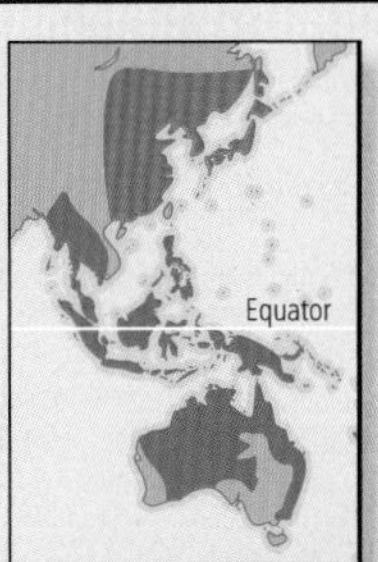

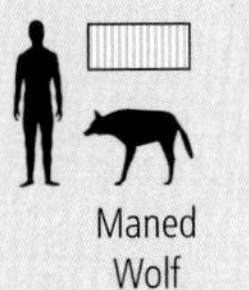

MANED WOLF *Chrysocyon brachyurus*
C and S Brazil, Paraguay, N Argentina, E Bolivia, SE Peru. Grassland and scrub forests. HBL 105cm (41in); SH 87cm (34in); TL 45cm (18in); WT 23kg (51lb); no variation in size between sexes or different populations. Coat: buff-red, with black "stockings," muzzle, and mane; white under chin, inside ears, and at tail tip; pups born black but with white-tipped tail. Breeding: gestation about 65 days; one litter of 2–5 pups a year. Longevity: unknown in wild (12–15 years in captivity). Conservation status: Lower Risk (Near Threatened).

BUSH DOG *Speothos venaticus*
Bush dog or vinegar fox
Panama to Guiana and throughout Brazil. Forests and forest-edge marshland. HBL 66cm (26in); SH 26cm (10in); TL 13cm (5in); WT 5–7kg (11–15lb). Coat: dark brown; lighter fawn on head and nape; chin and underside may be cream-colored or dark. Breeding: gestation reportedly 80 days or more. Longevity: not known. Subspecies: 3. Conservation status: Vulnerable.

RACCOON DOG *Nyctereutes procyonoides*
E Asia, in Far East, E Siberia, Manchuria, China, Korea, Japan, and N Indochinese Peninsula; introduced in Europe. Woodland and forested river valleys. HBL 51–70cm (20–28in); SH 20cm (8in); TL 18cm (7in); WT up to 12kg (26lb). (In Finland: mean in June 5.2kg/ 11.5lb, in October 8.0kg/17.6lb). Coat: long brindled black-brown body fur with black facial mask, sleek black legs, and black stripe on tail. Breeding: gestation 60–63 days. Longevity: up to 7 years. Subspecies: 5–6.

DINGO *Canis dingo*
Australasia, including Indonesia (part), also Malaysia, Thailand, Burma. Ubiquitous, from tropical forest to semiarid regions. HBL 150cm (59in); SH 50cm (20in); TL 35cm (14in); WT 20kg (44lb). Coat: largely reddish-brown with irregular white markings. Breeding: gesta-tion 63 days. Longevity: up to 14 years in captivity; widely subject to bounty schemes in wild.

Abbreviations HBL = head–body length; SH = shoulder height; TL = tail length; WT = weight

DISTRIBUTION Arctic Ocean, N America, Europe, Asia, and S America.

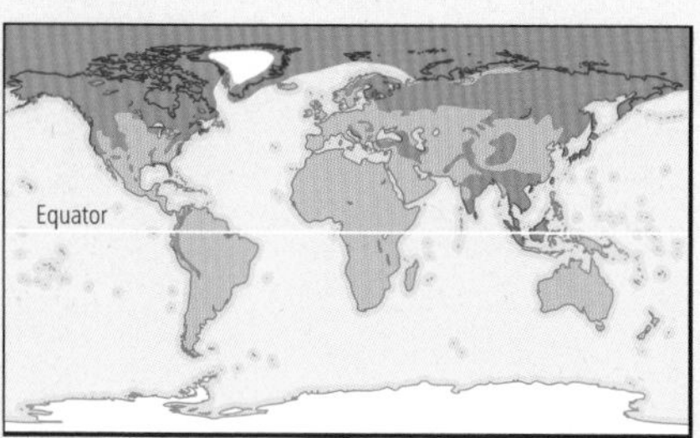

Family: Ursidae
8 species in 5* genera

Brown bear *Ursus arctos*	p72
Polar bear *U. maritimus*	p76
American black bear *U. americanus*	p80
Asian black bear *U. thibetanus*	p84
Sun bear *Helarctos* (or *Ursus*) *malayanus*	p84
Sloth bear *Melursus* (or *Ursus*) *ursinus*	p84
Andean bear *Tremarctos ornatus*	p84
Giant panda *Ailuropoda melanoleuca*	p82

*3 if *Ursus* subsumes the genera *Melursus* and *Helarctos*.

Bear Family

THE BEAR FAMILY (URSIDAE) CONTAINS THE largest terrestrial carnivores – Brown bears and Polar bears – although Brown bears are not especially carnivorous and Polar bears are not really terrestrial (they are effectively marine mammals).

While foraging, bears rely more heavily on strength than on speed. Their build is bulky and muscular, and to support their weight their hind feet are plantigrade (the whole sole touches the ground); this also allows them to stand bipedally. The front feet vary by species from plantigrade to semi-digitigrade (the back of the foot being partly raised off the ground). All species have five toes on each foot, armed with curved, nonretractile claws. Polar bears and most Brown bears in North America do not climb trees; European Brown bears and all other species of bears readily climb to forage or to sleep, although they spend more time on the ground. Surprisingly for a tree-climbing family, and unique among the Carnivora, all members have a short tail; equally uniquely, no bears have facial vibrissae.

Most taxonomists agree that there are eight extant species of bears in three distinct lineages, generally referred to as subfamilies. The Andean bear (Tremarctinae) and Giant panda (Ailuropodinae), each in its own subfamily and genus, have different numbers of chromosomes from the other bears (Ursinae). A puzzling issue for taxonomists is that different species and even different genera of bears can interbreed and produce fertile offspring; however, they do not do so in the wild, even where two species' ranges overlap. There is only one small area, in eastern India, where three species of bears are sympatric.

Bears currently inhabit four continents; they are absent from Australia, Antarctica, and Africa. Although three of the extant species (Andean, Sun, and Sloth) are tropical, bears originated in the northern hemisphere. Increasing seasonality during their evolution created conditions for diversification, and also selected for greater size. Large size increased the volume of food that could be consumed when good sources were found; it also offered protection from carnivores, aided in the killing of big herbivores, and, most crucially, enabled the accumulation of the fat stores needed to survive food shortages and cold weather. DLG

FOOD AND FEEDING

Most bears are generalist omnivores, feeding chiefly on berries and nuts, but also meat, fish, insects, buds, tubers, and leaves. To sustain their large bodies, they rely on a keen sense of smell, color vision, and a good memory to locate concentrated food sources. They seem to possess a detailed mental map of their home range and are able to retrace routes of former foraging excursions, even outside its normal limits.

Despite their reliance on plant matter, ursids are not good at digesting it. They do not have the cellulose-digesting micro-organisms present in the guts of ruminants, so fibrous green vegetation and berry casings often pass through intact. Bear scats (feces) often contain whole fruits that look as though they have just been picked.

The Polar bear is the only ursid that subsists almost entirely on an all-meat diet, preying on Ringed and Bearded seals, as well as young walruses, belugas, and narwhals. Occasionally, they will also scavenge on carcasses.

Above *A Sun bear licking termites from a mound it has broken open.*

Right *Ground squirrels supplement the diet of the Grizzly bear.*

IOW DANGEROUS ARE BEARS?

ears are often thought of as aggres-ve and dangerous animals because ney have the bulk and weaponry easily) kill a person. Yet while it is true that ears do sometimes kill each other, nd breeding males commonly fight, aving gaping wounds and broken eeth, their propensity to attack people greatly exaggerated.

As natural predators, Polar bears IGHT; two subadult males playfight-g) are really the only Ursids that are kely to stalk and kill humans. Attacks re prompted by hunger and not by ne bears having been startled. How-ver, few people live near Polar bears, nd those who do avoid the animals nd carry firearms, with the result that ar more Polar bears are killed by peo-le than vice versa.

Brown bears are the next most nenacing species, but their aggressive-ess varies geographically. Those living Siberia and the interior of North merica (Grizzly bears) can be danger-us, especially when defending their ubs or a carcass. But generally speak-ng, they go out of their way to avoid ontact with humans. Within Europe, rown bears appear to be more ggressive in the eastern countries (in omania 12 people were killed and 94 njured during a recent 8-year period) han in Scandinavia (only a single nprovoked, bear-inflicted human atality during the 20th century).

American black bears, especially uman-habituated campground resi-ents, often bluff-charge but rarely arm people. Given the frequent inter-ctions between these bears and peo-le, serious injuries are remarkably ncommon and fatalities rare (less han 40 across North America in the ntire 20th century). Andean bears are articularly wary and unaggressive oward people, though they will some-imes kill livestock.

Among Asian bears, Giant pandas, taunch vegetarians, pose no threat o people. Sun bears are often feared y local inhabitants because, when larmed, they may suddenly rear up, mit a piercing bark, and swat or lunge t the intruder, but they seldom initiate n attack. In contrast, Asian black ears and Sloth bears, whose behavior nay have been shaped by eons of ncounters with large felids and ungu-ates, are much more prepared to ttack than flee. Over a recent 6-year eriod, Sloth bears in one region of ndia were responsible for 745 human asualties, of which 51 were fatal. Many people consider Sloth bears nore dangerous than tigers.

SKULLS AND DENTITION

Bears have massive skulls, the largest of any carnivores. Differences between their skulls and those of canids, from which they evolved 20–25 million years ago, reflect a shift from a carnivorous to a largely herbivorous diet. Their broad molars with rounded cusps are adapted for crushing and grinding plant matter, and the large surface areas at the back of their skull and mandible provide attachments for correspondingly strong jaw muscles. Polar bears have a carnivorous diet and thus somewhat sharper cheek teeth, but due to their recent ancestry (splitting from brown bears less than one million years ago) their teeth are much less well adapted to shearing than those of other meat-eaters.

Bears have retained large canines for breaking branches, tearing open insect-laden logs, fighting, and – depending on the species – for killing vertebrate prey. The number of premolars varies not only among species but also among individuals within species (dental formula: I3/3, C1/1, P2–4/2–4, M2/3 = 34–42). Anterior premolars tend to be vestigial and widely spaced (less so in Sun bears, Andean bears, and Giant pandas). This creates a gap (known as the diastema) between the molars and canines, which is sometimes used to strip berries or catkins off branches. It also serves to extend the length of the jaw, creating a wider gape. The sloth bear has the most unorthodox dentition, lacking the two upper, central incisors as an adaptation for sucking up termites.

Sloth bear
31.5 cm

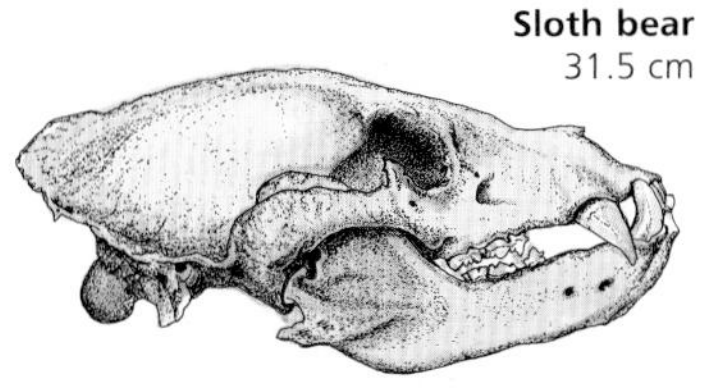

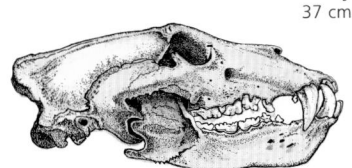

Grizzly
37 cm

BEAR ANCESTORS

Although animals with skulls like bears but the cursorial legs of dogs (subfamily Hemicyoninae) evolved some 25–5 million years ago, the most likely progenitors of today's bears were small carnivores from the same period (subfamily Agriotheriinae). The earliest member of this group, the Dawn bear (*Ursavus elmensis*), was a long-tailed animal like a raccoon, but later members were more akin to modern species in both size and appearance. The group finally died out, but not before radiating into the three present subfamilies. Giant pandas (two species, one of which went extinct less than 1 million years ago) were the first to diverge, followed by the Ursines and Tremarctines.

Without a doubt, the most remarkable early Ursid was the Giant short-faced bear (*Arctodus simus*), which inhabited North America in the Pleistocene (2 million–11,000 years ago). The largest of the Tremarctines, weighing up to 1,000kg (2,200lb), it was a fast, long-legged animal that could bring down bison and camels. Its sole surviving relative in the Tremarctine lineage is the small Andean Bear.

Brown Bear

JUDGING BY THE SCIENTIFIC NAME OF THE species – Ursus arctos *– which means "bear" twice over, first in Latin and then in Greek, the Brown bear might be considered the quintessential example of the Ursidae family. Spanning three continents, it certainly has the distinction of being the most widespread bear on Earth.*

It is tempting nowadays to associate Brown bears with northern climes, in view of their current strongholds in Russia, Canada, and Alaska, but the species' range once extended as far south as North Africa (until the mid-1800s) and central Mexico (until at least the 1960s). In the Middle Ages, they also lived throughout mainland Europe, including all the Mediterranean region as well as the British Isles. At present, due to excessive killing, habitat loss, and population fragmentation from road building, they have a much more fragile distribution. The species is so diverse and widespread that it was once divided into some 232 living and 39 fossil species and subspecies; all of these, including the Grizzly bear of North America (named for its silver-tipped pelage), are now considered a single species.

Unfussy Eaters

FORM AND FUNCTION

The geographic range of Brown bears overlaps that of American and Asiatic black bears, but Brown bears are less restricted to forest. They can live beyond the treeline, up to elevations of 5,000m (16,000ft) – uncommon for either species of Black bear. Like Black bears, Brown bears feed largely on small berries and nuts, but, with their muscular shoulder hump and hefty claws, they are better adapted to digging up small mammals, insects, and roots as well. Their robust jaw muscles also enable them to process more fibrous foods, and thus better survive a diet of excavated plant parts. Their disturbance of the soil may enhance the production of some of these plants. In some places, they feast heavily on aggregations of insects or salmon, or kill and scavenge large-bodied ungulates. They may even limit ungulate densities by taking large numbers of calves. Their aggressive behavior poses a threat to other bears.

Room to Roam

DISTRIBUTION PATTERNS

In Asia, aside from Russia, Brown bears occur in isolated pockets in the Himalayas and the Tibetan Plateau, as well as in the mountainous regions of some Middle Eastern countries. There is even a small population in the Gobi desert. Brown bears and Black bears coexist in many areas, but in at least some of these they tend to use different habitats, or else to share the same habitats at different times of day. Yet they appear unable to coexist on islands; Alaskan islands are home to either Brown bears or American Black bears, but rarely both. Being larger, Brown bears have more extensive home ranges than Black bears; on the mainland, male Brown Bear ranges average 200–2,000sq km (80–800sq miles) and female ranges 100–1,000 sq km (40–400sq miles). Black bear territories are respectively 20–500sq km (8–200sq miles) and 8–80sq km (3–30sq miles). Although home ranges do contract on islands, small islands may simply be too restricted to support Brown bears.

Above *The silver-gray flecked (or "grizzled") coloration of the coat of an adult bear in the Rocky Mountains, Montana, well illustrates why this subspecies* (Ursus arctos horribilis) *is commonly known as the Grizzly. The prominent nose and small eyes indicate that it relies mainly on smell rather than sight.*

Right *Pacific salmon, swimming upstream to reach their spawning grounds, form an important part of the diet of Brown bears on the northwestern seaboard of North America. During this annual glut of easily-caught prey, bears will often gorge themselves, taking a single bite out of a salmon before discarding it and fishing another from the river.*

Iowever, some large islands have Black bears but no Brown bears. For example, Honshu, the largest apanese island, has fossil records of Brown bears, probably displaced by Asiatic black bears arriving ia the Korean land bridge. Brown bears still live on Hokkaido, the northernmost island of Japan, which has no record of Black bears.

Hibernation Strategies

SOCIAL BEHAVIOR

Brown bears exhibit a behavioral trait common o all northern bears. As these evolved from canid ncestors and shifted toward greater reliance on ruits, they ran up against an enormous problem – carcity of winter food. One solution was to fall sleep all winter, like some rodents and bats: that s, to hibernate. A hibernating animal preserves nergy by dramatically reducing its body temperaure (often to near-freezing); small mammals, however, periodically awaken and rewarm, and nay drink, eat cached food, and eliminate wastes. Conversely, frugivorous northern carnivores, like accoons and skunks, grow a thick pelage, put on at, and spend harsh winters dormant in an insuated den, but they retain near-normal body temperature. Denning bears also lower their body emperature only a few degrees, from about 38°C 100°F) to 34°C (93°F), but their heart and respiatory rates decline appreciably, and they exhibit other unique physiological adaptations that must be regarded as true hibernation.

FACTFILE

BROWN BEAR

Ursus arctos

Brown, Grizzly, or Kodiak bear

Order: Carnivora

Family: Ursidae

Subspecies: N. American Grizzly bears (*U. a. horribilis*), Kodiak bears (*U. a. middendorffi*) from Kodiak, Afognak, and Shuyak Islands off Alaska, and Eurasian brown bears (*U. a. arctos*) are all sometimes considered separate subspecies.

DISTRIBUTION NW N America, Scandinavia through Russia to Japan, scattered in S and E Europe, Middle East, Himalayas, China, and Mongolia.

HABITAT Forests, shrubby subalpine to open alpine tundra, and desert or semi-desert.

SIZE **Head–body length** 1.5–2.8m (4.9–9.2ft) **shoulder height** 0.9–1.5m (3–4.9ft) **weight** male: 135–545kg (300–1200lb), occasionally reaching 725kg (1600lb) on Kodiak Island and in coastal Alaska and possibly Kamchatka; female: 80–250kg (175–550lb), and rarely up to 340kg (750lb); extreme variation by season (heaviest in the fall, just before denning) and geographic area (those with access to fish and meat are heavier).

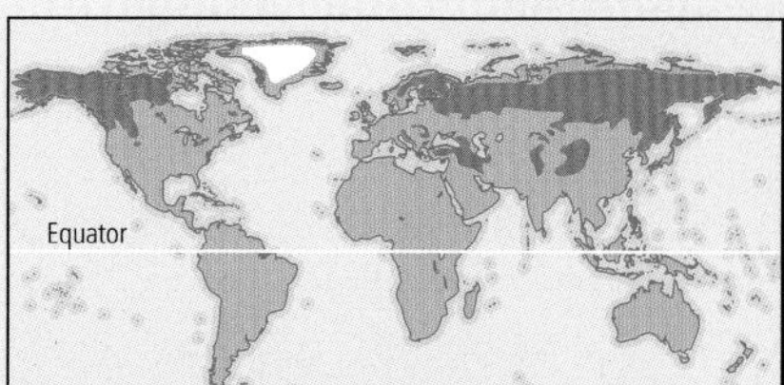

COAT Uniform brown or blonde to silver-tipped (grizzled in interior N America) to near black (E Asia).

DIET Roots, tubers, forbs, grasses, sedges, fruits (berries and nuts), pine seeds, insects, fish, rodents, and ungulates (including livestock).

BREEDING After mating May–July, fertilized eggs develop to blastocyst stage, then delay further development until November, when they implant. 1–4 cubs (typically 2 or 3) 6–8 weeks after that (total gestation 6.5–8.5 months).

LONGEVITY Typically to 25 years, but up to 36 years recorded in the wild (43 in captivity).

CONSERVATION STATUS All populations are CITES Appendix II listed, except Chinese and Mongolian populations, which are Appendix I listed. US populations outside Alaska are on the Endangered Species List.

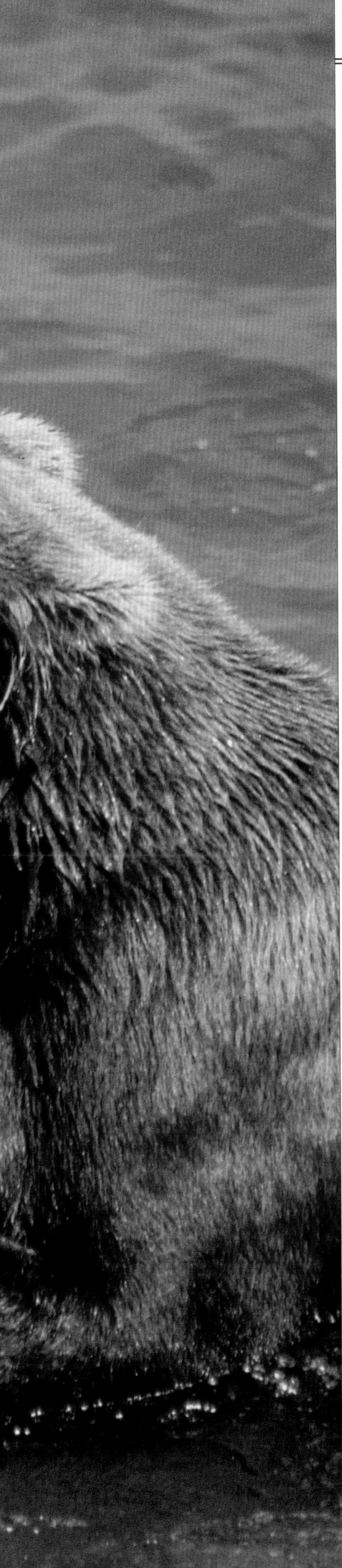

Bears are the only mammals able to survive for half a year or more without eating, drinking, urinating, or defecating. Energy is supplied mainly from stored fat reserves; the fatter the bear at the onset of hibernation, the less lean body mass (muscle) it burns during the winter. Fluids and amino acids are derived by recycling products that are normally excreted as urine, and bone does not deteriorate despite complete inactivity. This process is so efficient that bears rarely die in their dens; starvation-related mortality is more likely to occur in the spring, when their metabolism increases again.

Why bears maintain their body temperature while sleeping away the winter is unclear. Because of their large size, they may not be able to hide as effectively as smaller creatures, and so must be ready to defend themselves if discovered by a potential predator, like a wolf or another bear that has not yet denned. Another possible explanation is that high body temperature is necessary to support the development of cubs, which are born in mid-winter. Winter birthing and nursing by a hibernating mother is a remarkable feat that probably arises from the cubs' need to maximize growth before the next winter, when they too must hibernate.

Cubs are born with their eyes closed, have extremely fine fur, and are smaller in proportion to the size of the mother (less than 1 percent) than is the case with any other placental mammal. Mammalian fetuses have difficulty utilizing fat, because it cannot pass through the placenta; but newborns can absorb it in the form of milk, so the birth of cubs at a very early ("altricial") stage of development gives them the fastest possible access to their mother's stores of fat. Feeding the young in the den takes a very high toll on the mother, however. The burden of producing cubs increases over-winter weight loss by 50–100 percent, which may result in a 40 percent total decline in the mother's body weight during the period of hibernation .

Although bears that live in the tropics do not need to hibernate, because food is available all year round, females of these species still give birth to highly altricial cubs. The mothers stay with their cubs in a den and fast for several weeks or months, but it is unknown whether they hibernate in the same physiological sense as northern bears. It is also uncertain whether all or just some of the bears are induced ovulators (that is, stimulated to release eggs by copulation) and have delayed implantation after fertilization (another northern adaptation).

Left *Though this struggle between two juvenile bears might appear mortal, it is in fact just playful combat, designed to hone the animals' fighting skills. In later life, however, in territorial disputes between males, real fights can result in serious injury to the combatants, or even have a fatal outcome.*

Struggling with Man

CONSERVATION AND ENVIRONMENT

Brown bears have had a long, close, and often fraught relationship with humans, but have never been domesticated. Though generally a peaceable animal that avoids contact with people, the Grizzly bear gained an early reputation for aggression and predation on livestock after its habitat on the Great Plains and in the Rockies was invaded by ranchers, homesteaders, and railroad builders in the 19th century. In just a few decades, the Grizzly population of some 50,000 was decimated.

Recent incidents of Grizzlies maiming or killing people are isolated, but remain a potential problem with increasing recreational use of wilderness areas. Stringent measures have been put in place at campsites throughout the western national parks of North America to prevent scavenging by bears. Bear-proof food containers, garbage disposal units, and incinerators are employed to reduce available food.

A fascinating association, dating back many centuries, exists between Brown bears and the indigenous Ainu people of Hokkaido. The bears were caught as cubs, raised in captivity, fed human food – even being breast-fed by humans – and eventually sacrificed years later in a religious ceremony (which is now banned). Throughout medieval Europe, Brown bears were used in various ceremonies and entertainments. Their ability to walk on their hind legs was exploited in circuses; indeed, the tradition of "dancing bears" still persists in Turkey and Greece.

Because of its wide distribution, covering many remote areas, Brown bear population figures are inexact, but are estimated at 200,000–250,000. Yet this apparently healthy figure belies the fact that many populations are very small and in danger of extirpation. In the United States, excluding Alaska, only five isolated populations survive, totaling about 1,000 animals. Scattered, remnant populations of less than 25 animals each inhabit Spain, France, Italy, and Greece. Bears have been transplanted to augment populations in France, Italy, Poland, and Austria.

Recovery of small populations is hampered by generally low reproduction rates, although these vary geographically with the food supply: the average age of first birthing ranges from 5–10 years, and the interval between successful litters is typically 2–5 years. Twins are most common, but litters can contain 1–4 cubs. Females usually mate from May–July, and remain fertile until well into their twenties. Conflict with humans, their only predators, is exacerbated by the bears' large home ranges. The hunting of Brown bears is legal in Russia (where an annual kill of 4,000–5,000 is permitted), Japan (250), some European countries (total 700), Canada (400), and Alaska (1,100). Such regulated hunting is designed to be sustainable, but illegal killing remains a problem. DLG

Polar Bear

POLAR BEARS ARE THE LARGEST OF THE world's bear species. They have a remarkable ability to devour large amounts of fat rapidly when food is available, and then to fast for protracted periods when it is scarce. They are also metabolically unique in being able to switch from a normal state to a slowed-down, hibernation-like condition at any time of year when there is dearth; other species can only do this in the winter.

Polar bears evolved from Brown bears (*Ursus arctos*) during the late Pleistocene; the oldest known fossil, dating from less than 100,000 years ago, was found in London's Kew Gardens. Their molars and premolars are more jagged and sharper than the flat grinding teeth of other bears, reflecting their rapid evolutionary shift to a carnivorous diet. Despite their very different appearance, Polar and Brown bears are closely related, and in captivitiy can interbreed to produce young, though not in the wild as their habitats do not overlap.

Adaptations for a Harsh Environment

FORM AND FUNCTION

Polar bears live on the ice-covered waters of the circumpolar Arctic. They occur in about 20 populations that interbreed very little. Individual populations vary in size from a few hundred to a few thousand, with a world population estimated at 22,000–27,000. Their preferred habitat is the annual ice near the coastlines of continents and islands, where there are high densities of Ringed seals (*Phoca hispida*), their primary prey. Some Polar bears are also found on the less productive, thick, multi-year pack ice of the central Arctic Ocean. The southern limit of their distribution in winter varies with the southern extent of seasonal pack ice in the Bering, Labrador, and Barents seas. In areas where the ice melts completely in summer, such as Hudson Bay or southeastern Baffin Island, they spend several months on shore, fasting on their stored fat reserves until the waters freeze in the fall. Depending on the area, the sizes of their home ranges may vary from a few hundred to over 300,000sq km (115,000sq mi).

The bears are plantigrade, and have five toes with nonretractable claws on each foot. The forepaws are large and oarlike as an adaptation for swimming, but the toes are not webbed; in the water, the hind legs serve only as rudders. The body is stocky, but lacks the shoulder hump that characterizes the Brown bear. The Polar bear's neck is longer in relation to the rest of the body than in other bears; the ears are small, the fur white, and the skin beneath the coat is black.

Above *A layer of fat and a thick coat protect the Polar bear against the Arctic winter; the only unfurred parts of its body are the foot pads and the tip of its nose, which reveals the black skin underlying the white pelt. Small ears are another adaptation against the bitter cold.*

Right *A Polar bear leaps from an ice-floe off the Canadian coast. Despite their massive bulk, the bears are surprisingly agile. They can swim steadily for several hours, if necessary, and on land may travel 20km (12mi) in a day, though prolonged exertion can cause them to overheat.*

The main food source is Ringed and, to a lesse degree, Bearded seals, which the bears capture as they surface to breathe. They can locate seal breathing-holes under a metre (3.3ft) or more of compacted, wind-blown snow, from measured distances of almost a kilometre (0.6mi) entirely b smell – one of the greatest olfactory feats among mammals. Polar bears also prey opportunistically on walruses, belugas (*Dephinapterus leucas*), narwhals (*Monodon monoceros*), waterfowl, and seabirds. The largest proportion of their annual caloric intake occurs between about late April an mid-July, when newly-weaned Ringed seal pups are abundant, naive, and composed of 50 percen fat by wet weight.

When Polar bears are fasting, they are capable of synthesizing protein and water biochemically and of recycling metabolic byproducts. In Hudso Bay, where the annual ice melts completely by mid-July and does not refreeze until mid-November, pregnant females do not feed for 8 months, during which time they must support themselves as well as their newborn cubs. Polar bears of all other ages and sex classes may occupy temporary dens for weeks on end to conserve energy during periods of particularly cold or inclement weather.

A Lonely Life on the Ice

SOCIAL BEHAVIOR

Polar bears mostly live a solitary existence, with the exception of breeding pairs and family groups At times during the open-water season in summe and fall, a dozen or more adult males may fast together at preferred sites

FACTFILE

POLAR BEAR

Ursus maritimus

Order: Carnivora

Family: Ursidae

HABITAT Sea ice and waters, islands and coasts

SIZE **head–body length** male 200–250cm (6.5–8.3ft) female 180–200cm (5.8–6.5ft); **weight** male 400–600kg (880–1320lb) female 200–350kg (440–770lb), though exceptionally up to 500kg (1100lb) or more.

COAT White, though in practice often yellowish from staining and oxidation of seal oil, particularly in summer.

DIET Mostly Ringed (and to a lesser extent Bearded, Harp, and Hooded) seals; also walrus, beluga, narwhal, small mammals, waterfowl, and seabirds.

DISTRIBUTION Circumpolar Arctic

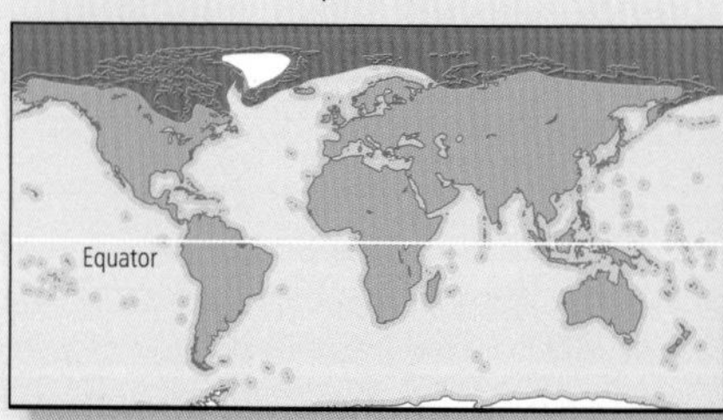

BREEDING Gestation (with delayed implantation) about 8 months; typical litter size 2.

LONGEVITY Females mid–20s to early 30s; males early (occasionally late) 20s.

CONSERVATION STATUS Lower Risk: Conservation Dependent. Once Vulnerable, but numbers have stabilized since the 1973 Agreement on Polar Bear Conservation.

along the coast in cohesive groups; competition for females is precluded at this time because testosterone levels are low, and in addition there is no food to compete for. Larger groups may also tolerate each other's company when feeding off a large food source such as a dead whale.

Polar bears breed in April and May. Since females generally keep their cubs for 2.5 years, they are available for mating only once every three years; competition for females is consequently intense, which probably explains why males are twice the size of females. Females have induced ovulation, which means they must mate many times over a period of several days before ovulation and fertilization can occur, and breeding pairs must remain together for 1 to 2 weeks to ensure successful mating. A female may mate with more than one male if her first suitor is displaced.

Polar bears also have delayed implantation until mid-September to mid-October, depending on latitude. In most areas, the young are born in a snow den some 2–3 months later, with their eyes closed and weighing about 0.6kg (1.3lb) at birth; their body hair is so fine that they have mistakenly been reported as hairless. Cubs are nursed by the female until late March or early April, when they weigh 10–12kg (22–26lb) and are large enough to accompany their mother onto the sea ice.

Females mate for the first time at 4–5 years of age, and continue to have cubs until they die. About two-thirds of the litters are twins, while single-cub litters are next most common. Triplets, while infrequent, are not rare, occuring at frequencies of up to 12 percent in some populations in some years, though usually less than 5 percent. The largest litters and heaviest cubs are born to the heaviest adult females, aged about 8–18, while younger and older females are the more likely to have single-cub litters. There are a small number of documented instances of adult females switching their litters in the wild, as well as of adopting unrelated bears.

Threats to a Top Predator

CONSERVATION AND ENVIRONMENT

The Polar bear is not endangered, though if hunting was not regulated it would be because of its slow rate of population growth. Hunting is currently carried out mainly by indigenous peoples, and the annual harvest worldwide is estimated at around 700 animals. Otherwise, the principal threats to the species are increasing pollution and the prospect of climatic warming. The Agreement on the Conservation of Polar Bears, which came into effect in 1976, requires the bears be managed according to "sound conservation practices," and mandates protection of their habitat. IS

Waiting for the Sea to Freeze Over

❶ *Ringed (and less commonly, Bearded) seals are Polar bears' main prey. They are hunted at breathing holes in the sea ice and at breeding sites where inexperienced seal pups make easy prey. The bears approach with stealth, then rush and plunge after their quarry. To enlarge small holes, they stiffen their fore legs and use their bulk to smash the ice and squeeze the front half of their body through, seizing the seal in their strong jaws and hauling it out. In the Arctic summer, the ice retreats from the coastline and may even melt completely, depriving the bears of their hunting ground.*

❷ *"Laid-back" bears conserving energy. In summer, which the bears spend ashore, they have to survive on their reserves of body fat for four months or more. They may rest up in earth dens, but once the snow arrives in the fall, they head for the coast and wait for the sea to freeze.*

❸ *Aggregations of bears only occur in summer (except when a large food source, say a beached whale, attracts several animals). They are remarkably sociable for a normally solitary species, as there is little competition for food or mates. Play-fights are not uncommon, but each animal remains very aware of its status with respect to other bears.*

4

❹ *As winter draws near and fat reserves dwindle, the bears move north and wait on the peninsula at the mouth of the Churchill River, where the ice usually freezes first. Yet this is the site of the town of Churchill; bears venturing too close – around 100 each season – are darted by the Polar Bear Alert team and "jailed" or flown north by helicopter.*

5

❺ *An inquisitive bear investigates the back of a Tundra Buggy – an outsized bus designed to bring tourists close to bears in safety. For more than 20 years, bear enthusiasts have flocked to Churchill, the self-styled "Polar Bear Capital of the World" during October and November. Baiting the bears with food or fish oil is strictly prohibited, but mobile hotels now offer overnight stays out near the bears, and even the smell of washing-up water is a tantalizing attraction to the most sensitive nose in the animal kingdom. If global warming leads to an earlier melting and later freezing of the ice, the problem of hungry bears congregating around Churchill is set to get worse.* IR

American Black Bear

AMERICAN BLACK BEARS ARE BY FAR THE MOST common bear – in fact, they are probably twice as numerous as all other bears put together, with a total population of around 800,000. They were once even more widely distributed than they are now, inhabiting virtually all of the forested areas of North America, from northern Canada to central Mexico.

During the Pleistocene (2 million –11,000 years ago), Black bears shared North America with several other ursids, but even then they were the most numerous. They increased in size through the Pleistocene, probably in response to the more seasonally extreme weather, until they were as big as today's Brown bears. Through the Holocene (the last 11,000 years), however, their size diminished.

Opportunist Extraordinaire

FORM AND FUNCTION

Black bears are extremely adaptable. They occupy regions as diverse as the hot, dry, shrubby forests of Mexico, the mossy, coniferous rainforests of coastal Alaska, the steamy hardwood swamps of the southeastern USA, and the treeless tundra of Labrador, where they behave like Grizzlies, hunting small mammals and caribou. In the extreme southern parts of their range, food is available year-round, so only pregnant females den; in the southern swamps, they may den high in trees. Conversely, in the far north they tend to choose insulated dens underground or under root masses, where they may remain for over 7 months (the record is 8 months, in Alaska).

The geographical range of these bears extensively overlaps that of Grizzlies. Coat color is not a reliable means of differentiating the two species, as many Black bears are actually brown. An increasing frequency of brown-colored Black bears from east to west across North America, and in the western states from north to south, may be related both to habitat (lighter coat in more open areas) and to mimicry of Grizzlies, which are occasional predators of Black bears. Black bears are distinguishable from Grizzlies by the straight (versus dished) profile of their face and by their lack of a shoulder hump. They also have smaller claws, which are adapted more to climbing than digging.

The Black bear's diet varies with the seasons. In the spring it includes herbaceous vegetation, buds, young leaves, and occasionally carrion (winter-killed ungulates) or nuts left over from the previous fall. Insects and young deer and moose may be added by late spring. By summer, as berries and nuts ripen, these become the main foods. Because of their large size and the need to build up stores of fat before winter, the bears may eat tens of thousands of nuts or berries in a day. Just to maintain weight on an exclusively berry diet, an average Black bear must feed for about 12 hours per day, consuming an average of a berry per second. In certain areas, especially if Grizzlies are absent, Black bears rely heavily on fish, which are nutritionally far more profitable.

Reproduction rates vary with food availability and also geographically: the age of maturity decreases, and litter size increases, from west to east and from north to south. In some areas early-maturing females produce their first cubs at 2 years old, but 4–6 years is more typical. Likewise, very large mothers may produce litters of five or six cubs (six matching the number of mammae), but litters of two or three are more common. Cubs generally remain with their mother for about 17 months, and thus spend their first birthday with her in a winter den. Females usually breed in June or July, following the departure of their offspring; as a result, the nominal interval between litters is two years. Longer intervals can occur if females spend an extra year with their offspring or skip a cycle; conversely, females have the capacity to produce cubs in successive years if all cubs in a litter

Above *A female cinnamon-colored subspecies of the black bear (*U. a. cinnamomum*) adopts a threatening stance toward an approaching male. The geographical variation in coat color among American black bears is very great, especially along the Pacific coast; other subspecies here include the Kermode bear (*U. a. kermodei*), which can have a pur white coat, and the Glacier bear (*U. a. emmonsii*), with bluish coloration.*

Left *In common with all other ursids except the Polar bear, the American black bear favors a mainly herbivorous diet of berries and nuts, along with unde ground roots and tubers. It needs to eat around 5–8k (11–18lb) of food each day. The search for nutrition is especially urgent in the fall, to fatten for the winter. Here, a bear feeds on a bush of chokecherries.*

Right *Two cubs cling to a silver birch tree. Climbing trees is a vital natural aptitude for American black bears, which they exploit in order to obtain foo and to escape danger. Cubs are able to climb shortly after exiting their natal den, at about 12 weeks of age Mothers sometimes leave cubs in "refuge trees" while they go off to forage.*

die before the breeding season. Cubs die from accidents, disease, starvation, or predation, including cannibalism.

As in Brown bears, a mother may abandon a singleton cub – a female that carries on caring for a single cub for two years in the end rears fewer cubs than if she abandons the cub, breeds the next year, and produces three young.

A Burgeoning Population

CONSERVATION AND ENVIRONMENT

American black bears are abundant, in part because they have become tolerant of human presence and, due to their unaggressive nature, people have become increasingly tolerant of them. Moreover, those Black bears able to obtain human-related food, such as grain crops, fruits, garbage, and bird or pet food, benefit reproductively by maturing faster and producing larger litters than bears that have to rely on purely wild foods. On the other hand, humans also constitute their main source of mortality. Across North America, hunters take over 40,000 Black bears annually. Although this accounts for only 5–6 percent of the total population, some local areas sustain annual harvests of over 20 percent. Oddly enough, whereas American black bears are subjected to a higher rate of legal hunting than any other ursid, the kill is so thoroughly regulated by management agencies that this is the only bear species whose numbers are increasing virtually throughout its range. DLG

FACTFILE

AMERICAN BLACK BEAR

Ursus americanus

American black bear or Black bear

Order: Carnivora

Family: Ursidae

Subspecies: **Kermode bear** (*U. a. kermodei*) of coastal British Columbia, which contains some white color-phase individuals; the **Louisiana black bear** (*U. a. luteolus*), and the **Florida black bear** (*U. a. floridanus*). Many other subspecies have been identified, especially in the W part of the range, but experts disagree over their validity.

DISTRIBUTION Throughout Canada, US except central Plains states, N Mexico.

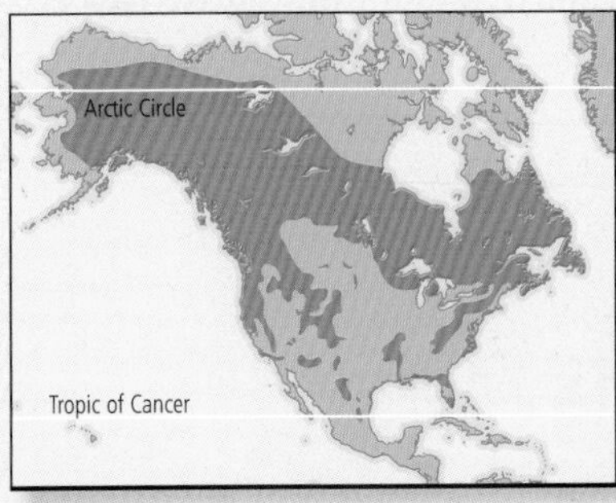

HABITAT Dense forests to more open woodlands and brushlands.

SIZE Varies seasonally and geographically (heavier in N and E). **Head–body length** 1.2–1.9m (4–6.2ft) **Shoulder height** 0.7–1.0m (2.3–3.3ft) **Weight** male: normally 60–225kg (130–500lb), but in some areas where they eat corn, they may weigh up to 300kg (650lb) or more rarely up to 400kg (880lb); female: 40–150kg (90–330lb) but occasionally exceed 180kg (400lb).

COAT Uniform black, brown, cinnamon, or blonde, sometimes with white marking on chest; all-white (non-albino) individuals compose up to 10 percent of some isolated populations in British Columbia, Canada, and also occur rarely elsewhere.

DIET Fruits (berries and nuts), shoots, buds, catkins, insects, some young ungulates, and fish.

BREEDING Mating generally occurs in May–July (stretching into August in S latitudes), uterine implantation of blastocysts in November, and births of 1–6 cubs (typically 2–3) in January (total gestation 6.5–8.5 months).

LONGEVITY Typically to 25 years, but recorded up to 35 years in the wild (same in captivity).

CONSERVATION STATUS Generally not threatened. However *U. a. luteolus* is listed as a threatened subspecies under the US Endangered Species Act and shows some genetic distinctions and *U. a. floridanus* has been considered for federal listing, although its taxonomic status is less clear. American black bear parts (chiefly the gall bladder) are indistinguishable from those of endangered Asian bears, so to control the trade in these, Black bears have been listed under the "copy-cat" clause of CITES, Appendix II.

Giant Panda

SINCE THE FRENCH NATURALIST PÈRE DAVID first discovered the Giant panda in remote western Sichuan in 1869, the animal has become a symbol for people all over the world. They treasure it not only for its distinctive black-and-white coat, gentle demeanor, and mysterious life-history, but also for its rarity. Critically endangered in the wild, it has become an international emblem of wildlife conservation, providing the World Wide Fund for Nature with its logo.

Giant pandas are the focus of a concerted conservation effort, but despite their protected status some are still poached for their skins. Poaching was once punishable by death, and still attracts a heavy jail sentence, of up to 14 years. More often, though, pandas are caught and killed in snares set for other animals, such as musk deer and takin.

The Bamboo Bear

FORM AND FUNCTION

The classification of the Giant panda was controversial for over a century, but recent genetic studies have shown that it represents an early divergence from the bear family. However, the long separation from other bears has made the Giant panda distinctive: it does not hibernate in winter; one of its wrist bones has evolved into a "pseudo-thumb" useful for handling bamboo stems; and when a baby is born, it is remarkably small, usually weighing just 100–200g (3.5–7oz), about 0.001 percent of its mother's weight.

It is also unique in eating bamboo (local people sometimes call it the "bamboo bear"), though it will occasionally eat meat when available. Bamboo provides enough nutrition for survival, but little extra, so pandas in the wild eat for as much as 14 hours a day, consuming 12–38kg (26–84lb) of bamboo – up to 40 percent of their body weight.

In the Qinling Mountains, in Shaanxi Province, a few brown-and-white pandas occur among the normally-colored ones. As a species, the genetic diversity of the panda has been found to be within the normal range, despite the fact that the total population has been reduced and fragmented.

Sex is Not a Problem

SOCIAL BEHAVIOR

The much-publicized failure of zoo-kept Giant pandas to reproduce has given rise to a popular misconception that the panda has trouble breeding. However, field research paints a completely different picture in the wild.

Usually alone or with dependent young, pandas rarely congregate. Each adult has a well-defined home range, which extends to 30sq km (12sq mi) for males and around 4–10sq km (1.5–4sq mi) for females; males' ranges usually include all or part of the ranges of several females. During the mating season,which lasts from March to May, the two sexes come together for a brief period of 2–4 days. Fighting is apt to break out at this time between males competing to court females in estrus. A dominant male often has priority in mating, but subordinate males also have opportunities afterwards.

Gestation takes about 5 months, but pandas have delayed implantation of the blastocysts for between 1 and 3 months. Females give birth once every two or three years from age 4 to at least 20. The young are born in a very immature stage of development, being small, blind, and helpless. The female chooses the base of a hollow tree or a cave in which to give birth and remains there for over a month, carefully tending her cub by cradling it protectively in her massive paw.

A panda cub weans at around 1 year, but will not leave its mother until she becomes pregnant again. After becoming independent, some young settle in ranges that overlap with their mother's; others, especially females, go further afield. Long-term research in the Qinling Mountains showed that the population there has a slow growth rate or is at least stable when habitat destruction and poaching are controlled.

Captivity as such does not kill pandas, but it does seem to lower libido, especially in males; so far, only two captive-born males have successfully mated. Twins are more frequent in captivity than in the wild. Although great progress has been made in captive breeding since 1990, thanks to artificial insemination and improved hand-raising of newborns, the captive population, which totals over 140 pandas, is not yet self-sustaining.

Pandas, People, and Policy

CONSERVATION AND ENVIRONMENT

The threat to the panda comes from humans. China's expanding population has cleared areas for agriculture and timber, reducing panda habitat in Sichuan province by almost 50 percent from 1974–89. Having been eliminated from the subtropical lowlands, panda habitat persists only in high-altitude areas with mixed broad-leaved and coniferous forest with an understory of forbs, saplings, and bamboo. In a national survey in the 1980s, it was revealed that there were about 1,000 pandas remaining in a region of 13,000sq km (5,000sq mi) scattered across six isolated mountain ranges in central and western China. Within the ranges, the habitat is further fragmented into more than 20 smaller patches by cultivation and logging.

The Chinese government has designated 33 Giant panda reserves, and over 50 percent of their habitat now enjoys protected status. Beside safeguarding the species, these conservation programs have many other benefits. The pandas' mountain habitats are vital watersheds that support thousands of people in the immediate area, plus millions more downstream, and the range of plants and animals there is far more diverse than similar ecosystems in other temperate areas of China. LZ

Left Giant pandas feed almost exclusively on bamboo – new leaves and shoots are most nutritious and least fibrous. Every 30–100 years, different types of bamboo flower and die. Pandas traditionally coped with this by switching to other species, but habitat loss has now greatly reduced their range of feeding options.

Below Juvenile pandas engaged in playfighting. Young male pandas occupy a lowly position in the hierarchy and usually do not have a chance to mate until aged 7–8.

FACTFILE

GIANT PANDA

Ailuropoda melanoleuca

Giant panda, Panda bear, Bamboo bear, or panda

Order: Carnivora

Family: Ursidae (but sometimes classified with the Procyonidae)

Sole member of genus

DISTRIBUTION Sichuan, Shaanxi, and Gansu provinces of C and W China.

HABITAT Cool, damp bamboo forests at altitudes of 1,500–3,400m (4,900–11,000ft).

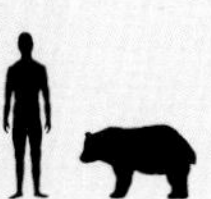

SIZE Shoulder height 70–80cm (27–32in); can measure about 170cm (67in) when standing **Weight** 100–150kg (220–330lb), males being 10 percent larger than females.

COAT Ears, eye patches, muzzle, hindlimbs, forelimbs, and shoulders black, the rest all white.

DIET Primarily bamboo, though pandas in the wild also eat bulbs, grasses, and occasionally insects and rodents.

BREEDING Gestation 125–150 days.

LONGEVITY Usually less than 20 years in the wild (over 30 years in captivity).

CONSERVATION STATUS Endangered.

THE TROUBLE WITH CLONING

Since 1997, scientists from the Chinese Academy of Sciences have been trying to clone Giant pandas, and recently succeeded in introducing the somatic nucleus of a Giant panda cell into a rabbit ovum. This breakthrough has been hailed by some as a potential way of saving the animal in the wild.

Exciting though it is, however, cloning has many problems. One key aspect of conservation is to ensure that animal populations remain genetically diverse when numbers are dwindling – indeed, sex itself may have evolved because individuals that swapped genetic material and thus provided offspring with various chances in the lottery of life were more likely to produce surviving grandoffspring. If all offspring are identical, as in cloned populations, the chance of a single parasite wiping them all out increases. Thus cloning is unlikely to replace sexual reproduction in the conservation field, though it may well be used to restore extinct animals or those whose populations have been reduced to just a few individuals.

Lesser-known Bears

Right *The four species of smaller, lesser-known bears:* ***1*** *the Asian black bear is mainly herbivorous but may, as here, take carrion;* ***2*** *the Andean bear, the only South American ursid, shown climbing a tree in search of fruit;* ***3*** *the Sun bear, here seen eating termites, is the smallest bear of all;* ***4*** *The shaggy-coated Sloth bear makes good use of its long, curved claws and flexible snout to forage for insects and grubs.*

THE BEARS THAT LIVE EXCLUSIVELY OUTSIDE North America and Europe are, with the exception of the Giant panda, not as well-known as their more northerly cousins. All four lesser-known species are probably declining due to illegal hunting and to degradation of their forest habitat.

Asian Black Bear

URSUS THIBETANUS

The Asiatic black bear is often referred to as a cousin of the American black bear because the two are about the same size, shared a common ancestor (about 4 million years ago), and have similar food habits, life histories, and ecologies. They are good tree climbers, and often create tree nests in the process of breaking branches inward toward the trunk to obtain fruits and nuts.

Until recently it was placed in the genus *Selenarctos* (literally "Moon bear," after the prominent, crescent-shaped white mark on its chest, also seen on Sloth and Sun bears). Asiatic black bears of both sexes also share with Sloth bears the ruff of longer hair on their necks. These may both be anti-predator attributes, arising from a long coexistence with tigers: the ruff exaggerates their size, while the white chest marking provides a startling display when the bear stands up.

For some 3,000 years people have hunted these bears for their paws and gall bladder; dried bile is used in traditional Chinese medicine. Japan is currently the only country where hunting is legal, but poaching is rampant across much of Asia. In Pakistan, cubs are taken from the wild and raised to fight bull terriers in exhibitions of bear baiting. The current world population of Asiatic black bears is at least an order of magnitude less than its American counterpart.

Sun Bear

HELARCTOS MALAYANUS

Sun bears are the smallest of bears with the shortest hair, and hence are sometimes called Dog bears. Because of their size and demeanor, they are commonly kept as pets. The name Sun bear and the genus *Helarctos* both refer to the chest marking, which is an ocher or white circle or semicircle. Sun bears coexist with Asian black bears throughout Southeast Asia, and also with Sloth bears in eastern India.

Sun bears are highly arboreal, often sleeping in specially-built nests in trees. Their small size, short snout, and short ears may be arboreal adaptations. They are largely frugivorous but when fruit is scarce they subsist on insects. Sun bears are naturally diurnal, but they become more nocturnal in

Above *In order to get at termites and bees living inside trees, the Sun bear uses its disproportionately large canine teeth and strong jaws to gnaw at the wood. Once it has broken into a cavity, the bear's long tongue comes into its own, probing for larvae, insects, and honey.*

the presence of human disturbance. Unusually for bears, mating and birth occur throughout the year, and gestation varies considerably.

Sloth Bear

MELURSUS URSINUS

The Sloth bear occurs solely on the Indian subcontinent, where it adjoins or slightly overlaps the range of Asiatic black bears; however, the Sloth bear is a more lowland species. The generic name *Melursus* derives from the bear's avid attraction to honey; it will climb trees and endure the stings of swarming bees to obtain honeycombs. Yet honey forms only a small part of its diet, which includes berries during the fruiting season; when these are scarce, it lives on termites, ants, and other insects. Physical adaptations for digging and eating insects include long, slightly curved claws, a broad palate for sucking, the absence of two front upper incisors, and large, protrusible lips – indeed, it was once known as the Lip bear, *U. labiatus*.

The long coat is a puzzling feature on a tropical bear. It may have the same function as the loose clothing worn by people in hot, dry areas with little shade. During the Sloth bear's evolution, the Indian subcontinent was largely grassland and forest, with sparse, scrubby trees.

In a tradition dating back four centuries, the Qalandars, an itinerant people of India, train and exhibit "dancing" Sloth bears as a way of life. They abduct cubs from dens, remove their teeth and claws, and put a rope or ring through the bear's nose or snout to control its movements. Over 1,000 dancing bears still entertain primarily urban and tourist audiences in India.

Above *A female Sloth bear with cubs. Sparse tree cover in this species' native habitat of India is thought to have prompted its habit of carrying its young on its back, and may also explain why it is largely nocturnal. These two behaviors are unique among ursids.*

Andean Bear

TREMARCTOS ORNATUS

Andean bears (or Spectacled bears, from the white markings that often encircle their eyes) together with Sun bears are the only ursids that live on both sides of the equator. Like Sun bears, Andean bears will often climb trees (and vines) for food, build tree nests for sleeping, and are relatively slim and agile; even so, they are the second-largest land mammals in South America. In northern zoos they generally give birth in the period from November to February, but on their home continent birth dates are far more variable.

Andean bears range along the Andean mountain chain, from high altitudes near the snowline (4700 m/15,500 ft) to coastal lowlands (250 m/ 800 ft). They occupy a great diversity of habitats, including cloud forests, rain forests, high-altitude grasslands, and scrub desert. Their diet varies by habitat, but consists mainly of plant matter such as fruits, palm petioles, bamboo, and cactuses. Bromeliads, both arboreal and terrestrial, are their mainstay in most areas, especially during non-fruiting periods. Andean bears are often blamed for killing free-roaming cattle, although some purported kills may simply be the result of scavenging. They are also attracted to cornfields, where farmers may kill them in an attempt to prevent damage to their crops. DLG

FACTFILE

LESSER-KNOWN BEARS

Family: Ursidae

DISTRIBUTION S and SE Asia, from India and Iran to Japan; Andean bear in S America from Venezuela to Bolivia.

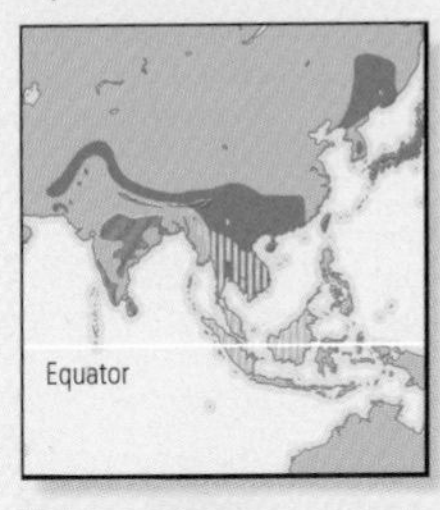

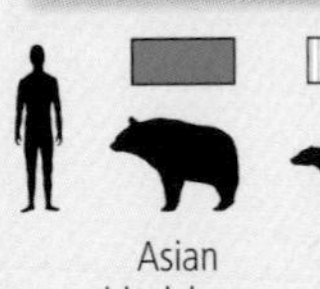

ASIAN BLACK BEAR *Ursus thibetanus*
Asiatic black bear, Moon bear, White-breasted bear
Iran to SE Asia, Taiwan (Formosan black bear), gap in central China, continuing in N China, E Siberia, and Japan; temperate, subtropical, and tropical forests. HTL 1.2–1.9m (3.9–6.2ft); SH 0.7–1.0m (2.3–3.3ft); WT (males) 60–200kg (132–441lb), rarely more than 200kg, (females) 40–140kg (88–309lb). Coat: black with white V on chest, plus white on chin; ruff around neck. Breeding: gestation 6–8 months, possibly shorter in tropics. Longevity: mid-20s in wild (37 in captivity). Conservation status: Vulnerable; Baluchistan bear (*U. t. gedrosianus*) of S Pakistan and Iran is Critically Endangered.

SUN BEAR *Helarctos (Ursus) malayanus*
Malayan sun bear, Malay bear, Honey bear, Dog bear
SE Asia, Sumatra, Borneo, isolated patches in E India and S China; lowland forests. HTL 1.1–1.5m (3.6–4.9ft); SH about 70cm (2.3ft); WT 27–65kg (60–143lb). Coat: short, black, with white, buff, or orange crescent- or doughnut-shaped marking on chest, often extending onto neck. Breeding: gestation 3–8 months. Longevity: to 33 years in captivity (not known in wild). A slightly smaller form in Borneo is considered a separate subspecies (*H. m. euryspilus*). Conservation status: Data Deficient.

SLOTH BEAR *Melursus (Ursus) ursinus*
Honey bear, Lip or Labiated bear, Aswail
India, Nepal, Bhutan, Sri Lanka, and possibly Bangladesh; mainly lowland forests and grasslands. HTL 1.4–1.9m (4.6–6.2ft); SH 60–90 cm (2–3ft); WT (males) 80–145kg (176–320lb), rarely more than 190kg (419lb), (females) 55–95kg (121–209lb). Coat: long, shaggy, black, with white U on chest. Breeding: gestation 4–7 months. Longevity: to 34 years in captivity (not known in wild). A slightly smaller form in Sri Lanka is considered a separate subspecies (*M. u. inornatus*). Conservation status: Vulnerable.

ANDEAN BEAR *Tremarctos ornatus*
Spectacled bear or Ucumari
Andes from W Venezuela to Bolivia–Argentina border; montane forests to high grasslands. HTL 1.3–2.0m (4.3–6.6ft); SH 70–90cm (2.3–3ft); WT (males) 100–175kg (220–386lb), rarely more than 200kg (441lb), (females) 60–80kg (132–176lb). Coat: black with creamy white, biblike marking on chin, neck, and/or chest, and variably white around eyes. Breeding: gestation 5.5–8.5 months. Longevity: to 39 years in captivity (not known in wild). Conservation status: Vulnerable.

Abbreviations HTL = head–tail length SH = shoulder height; WT = weight

DISTRIBUTION N, C, and S America; Red panda in Asia.

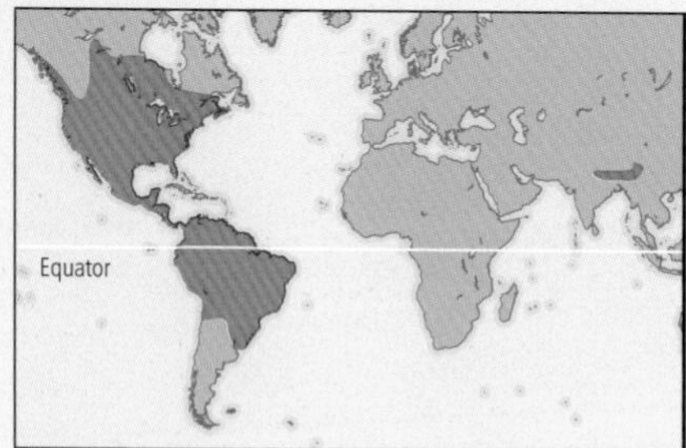

Family: Procyonidae
19 species in 7 genera

RACCOONS AND COATIS Subfamily Procyoninae

12 species in 4 genera
Raccoons 7 species of *Procyon* p88
Coatis 3 species in 2 genera: *Nasua* and *Nasuella* p90
Ringtail and cacomistle 2 species of *Bassariscus* p94

KINKAJOU AND OLINGOS Subfamily Potosinae

6 species in 2 genera
Kinkajou *Potos flavus* p92
Olingos 5 species of *Bassaricyon* p94

PANDA Subfamily Ailurinae

Red panda *Ailurus fulgens* p94

Raccoon Family

THE RACCOON FAMILY COMPRISES ONLY SEVEN genera and 19 species, but its members display a remarkable diversity in form and ecology. This diversity is reflected in the scientific controversy over its classification, which is still a live issue. Here, three subfamilies are recognized – the Ailurinae (with a single Asian genus), the Potosinae, and the Procyoninae (together accounting for the other six genera, all in the Western hemisphere). Much debate concerns the position of the Red panda. Some taxonomists group it with the Giant panda in a separate family, the Ailuropodidae, while others retain it in the Procyonidae.

The raccoon family is descended from the dog family, Canidae. Recognizable *Bassariscus* fossils have been found from 20 million years ago, a time when Europe and North America were one continent. When the continents separated, the family split, with procyonines remaining in the New World and the ailurine ancestors of the Red panda in the Old World.

Procyonids are mid-sized, long-bodied mammals with long tails. The kinkajou is uniformly colored, but the others have distinctive coats, ringed tails, and facial markings that vary from the black mask of the raccoon to white spots in the coatis and cacomistle. The pelage of the Red panda is chestnut brown with a white face.

Procyonids can live for 10–15 years in captivity, but rarely more than 7 in the wild. Females usually breed in the spring of their first year, males from their second year on. The young weigh about 150g (5oz) and are poorly developed at birth. In most species there are three or four young in a litter, but Red pandas produce only one or two and kinkajous usually only one. Females give birth in dens or nests and provide all of the parental care until the young are on their own. All procyonids are nocturnal, except coatis, which are mainly diurnal. The social organization of procyonids has only recently been studied in the wild. Even though some species (e.g. the ringtail) seem to be solitary and others, like the coati, move in large groups, it is likely that all procyonids maintain complex social relationships within and among the sexes.

Some procyonid species are sustaining themselves, while others are now classified as endangered or threatened in some measure. The Red Panda, olingo, and cacomistle are declining, primarily as a result of destruction of their forest habitats. The conservation status of raccoons varies greatly according to species; whereas the Common raccoon continues to expand its range and grow in numbers, insular species are on the endangered list. One, the Barbados raccoon (*Procyon gloveralleni*), has not been sighted since the 1960s and is now classed as extinct by the IUCN. JKR/EKF

1

2

GETTING ABOUT

The feet of procyonids have five toes, with the third toe being the longest, and the animals walk partly or wholly on the soles of their feet, like bears (plantigrade gait). The Red panda has a small extra digit that functions like an opposable thumb. Claws are nonretractile, except that ringtails and Red pandas have semiretractile claws on their forepaws.

Kinkajous have a prehensile tail (a feature that once misled taxonomists into classifying them as primates); in coatis, the tail is semiprehensile and is also used for balance. All procyonids are adept at climbing. Indeed, the kinkajou and cacomistle live exclusively in the tree canopy and almost never come down to the level of the forest floor.

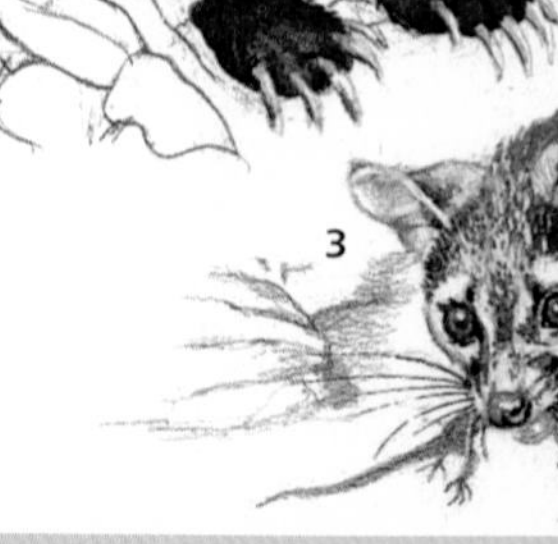

3

KULLS AND DENTITION

ıe teeth of procyonids are general- ed, as befits omnivores. The typical ental formula is I3/3, C1/1, P4/4, 2/2 = 40, but this varies with ›ecies. The kinkajou has only three remolars above and below. The Red anda has three premolars above and ›ur below. Only the cacomistle has ell-developed carnassials. In rac- ›ons the carnassials are unspecial- ed and the molars flat-crowned. In ›atis the molars and premolars are gh-cusped as adaptations to a more sectivorous diet: the canines are ›ng and bladelike and may be used ›r cutting roots while digging.

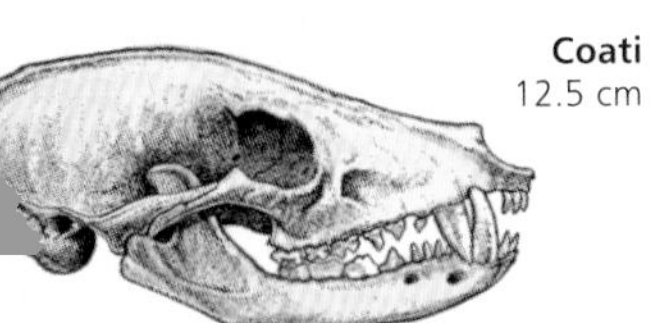

Coati
12.5 cm

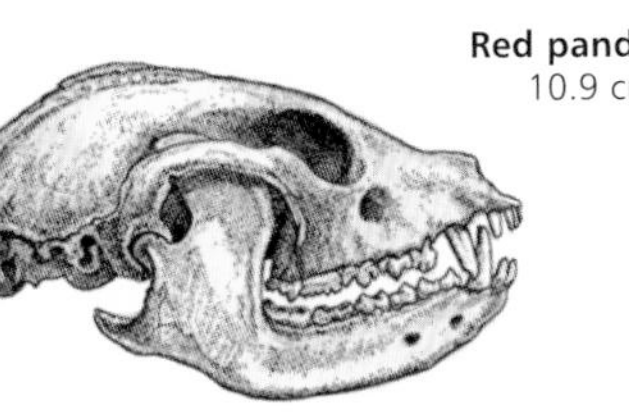

Red panda
10.9 cm

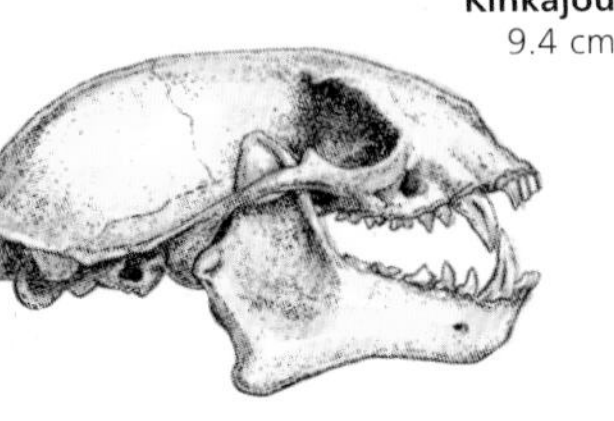

Kinkajou
9.4 cm

Left and below *Various ›embers of the raccoon family.*
A kinkajou licking nectar from flower while holding on to a ›ranch with its prehensile tail.
A coati grubbing among the ›rest-floor vegetation for insects.
A ringtail eating a lizard.

A VARIED DIET

Considering they are members of the order Carnivora, procyonids are not very carnivorous. The kinkajou eats almost no animal prey other than the occasional insect; and the Red panda (ABOVE), is also largely vegetarian, surviving primarily on bamboo shoots, along with fruit, roots, lichen, and acorns. Fruit is in fact a staple for most procyonids, although the other species are generally more omnivorous in their appetites, supplementing their diets with a variety of insects and small animals. Raccoons eat worms, snails, fish, crayfish, and clams in addition to fruits, berries and nuts, while the olingo will take birds and small mammals. The most carnivorous of the procyonids are the ringtail (LEFT) and the cacomistle, both of which are equipped with doglike dentition to catch and eat prey up to the size of rabbits.

Certain variations in external appearance bear witness to the differing diets of the procyonids. In general, their muzzles are pointed, though not markedly so; however, the fruit-eating kinkajou has a shorter-than-average muzzle, as well as a very long tongue for obtaining nectar from flowers (FAR LEFT). At the opposite end of the spectrum is the coati, which has a long, flexible snout that is ideal for probing crevices for insects. The ears of the cacomistle and ringtail are unusually large to help them locate their prey.

Raccoons

r *ACCOONS ARE HIGHLY ADAPTABLE ANIMALS, which have learned how to thrive in human-dominated environments. They are notorious crop marauders and garbage thieves, capable of breaking into, and out of, human-built structures.*

In North America raccoons are so familiar that they are as much a topic of TV cartoon shows as of biological research – and yet surprisingly little is known of their social lives. Humans and raccoons can have a close relationship; young "coons" make enchanting pets, but hormonal and behavioral changes in the maturing animal can try the patience of the most devoted owner.

The Masked Bandit

Form and Function

Physically, raccoons are distinguished by a foxlike face with a black mask across the eyes, a stout build, and a ringed tail. Adults typically weigh from 5–8kg (11–18lb), but weights vary geographically and seasonally – and males tend to be 20–30 percent bigger than females.

Raccoons may become sexually mature during their first spring, but some do not breed until the second season. The peak of breeding occurs from February to April, depending on latitude, with 4–6 young born nine weeks later. Yearling females may reach puberty later than adults, and some adult females will even breed a second time if their first litter dies, so the breeding season may be extended into the autumn.

While raccoons are excellent climbers, radio-tracking reveals that in addition to tree hollows they often use ground burrows, brushy "nests," old buildings, cellars, log piles, and haystacks as dens. Shelter from potential predators and the elements is the only criterion for a suitable nest site.

In their feeding habits raccoons are opportunistic omnivores, adapting their diet to the available foodstuffs. In most areas they forage at night near lakes, streams, or marshy areas, where fish, crayfish, snails, clams, aquatic insects, and other potential foods are sought. They travel to upland areas as well, especially to exploit fruits, berries, nuts, and seeds. They also consume insects, stored grain, and earthworms. Fresh sweetcorn (maize) is a particular delicacy, taken just before it is ready to be picked for human consumption.

Raccoons are mainly nocturnal, focusing their activities between sunset and midnight. In the northern United States and southern Canada, they become inactive during the winter months, although they are not true hibernators. They will remain in the same den for a month or more unless nighttime temperatures rise above freezing. Communal denning is common; up to 23 raccoons have been reported in a single den, and these large winter congregations are most likely to be socially structured.

FACTFILE

RACCOONS

Order: Carnivora

Family: Procyonidae

All 7 species of the genus *Procyon*

DISTRIBUTION
N, C, and S America

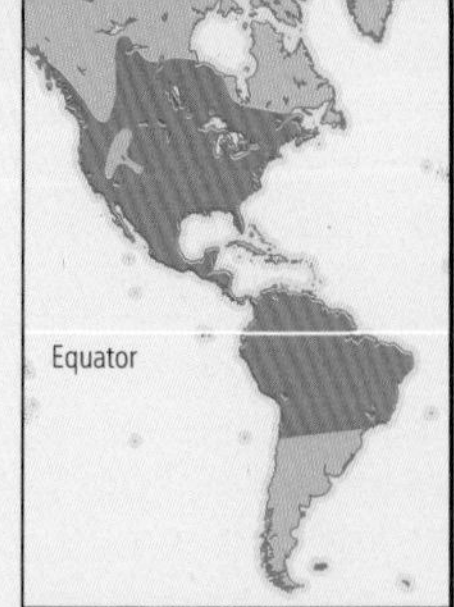

COMMON RACCOON *Procyon lotor*
S Canada, USA, C America; introduced in parts of Europe and Asia. The commonest species, occupying diverse habitats. HTL 55cm (22in); TL 25cm (10in); WT usually 5–8kg (11–18lb), sometimes up to 15kg (33lb), females being about 25 percent smaller than males. Coat: usually grizzled gray but sometimes lighter and more rufous (albinos also occur); tail with alternate brown and black rings (usually 5); black face mask accentuated by gray bars above and below, black eyes, and short, rounded, light-tipped ear pinnae. Breeding: gestation 63 days. Longevity: 13–16 years (over 17 years recorded in captivity).

TRES MARÍAS RACCOON *Procyon insularis*
María Madre Island, Mexico. Coat: shorter, coarser, lighter-colored than *P. lotor*. Endangered.

BARBADOS RACCOON *Procyon gloveralleni*
Barbados. Coat: darker than *P. lotor*. Extinct.

CRAB-EATING RACCOON *Procyon cancrivorus*
Costa Rica south to N Argentina. Coat: shorter, coarser, more yellowish-red, and with less underfur than *P. lotor*; hair on nape of neck directed forward; tail longer than that of *P. lotor.*

COZUMEL ISLAND RACCOON *Procyon pygmaeus*
Cozumel Island, Yucatán, Mexico. The smallest raccoon, often weighing only 3–4kg (6.6–8.8lb). Coat: lighter than *P. lotor*. Endangered.

GUADELOUPE RACCOON *Procyon minor*
Guadeloupe. Coat: paler than *P. lotor*. Endangered.

PROCYON MAYNARDI
Nassau Island, Bahamas. Previously considered conspecific to *P. lotor*. Endangered.

Abbreviations HTL = head–tail length WT = weight

*◐ **Left** Common raccoons confront one another on a tree branch. This widespread species now ranges from Panama to the southern portions of the Canadian provinces, only being absent from parts of the inter-mountain western United States. It has spread by adapting to human-induced environmental changes, using old buildings for winter den sites, and feeding on stored grain.*

Group Living

SOCIAL BEHAVIOR

The raccoon's dextrous forepaws give it not just its English name (derived from the Algonquian *arakun*, which roughly translates as "he who scratches with his hands"), but also the German *Waschbär* or "washing bear," and its scientific name *lotor* (from *lavere*, meaning "to wash" in Latin, which is also the root of the word "lotion"). The rubbing, dunking, and manipulating actions are not truly washing, but are associated with finding and catching aquatic prey. The behavior is innate, and displacement "feeling" actions by captive animals are common even in the absence of water.

Raccoons are certainly polygynous and may be promiscuous, but their precise social organization is unknown. Several females, usually related to one another, have overlapping home ranges that are circumscribed by the larger home ranges of one or more males. Females form "consortships" with between one and four males during the breeding season (though in one group under observation, most – 62 percent – associated only with only one male). Heavier males dominate access to the females.

Individual adult males, or a group of males, occupy distinct, territory-like areas ranging in size from 50 to 5,000ha (125–12,500 acres); in general, most raccoons regularly roam over about 500ha (1,500 acres) in a year. Males within a group may travel together at times, but scars appear during the breeding seasons, suggesting competition for mates. A study of consortships among members of a marked raccoon population suggested that one male out of 5–8 in the group did 50–60 percent of the mating with the 10–12 females in the group's area, but secondary males probably secured occasional matings.

Social relationships are probably established and maintained through the animals' diverse postures, vocalizations, and scents. For example, 13 distinct "calls" have been identified. At communal latrines, one or more raccoons regularly deposit feces at the same prominent location, such as the crotch of a tree, a backyard woodpile, or a corner of an attic, though the significance of this behavior is unknown.

City Slicker and Hunters' Quarry

CONSERVATION AND ENVIRONMENT

While several of the lesser-known raccoon species are considered endangered by the IUCN and one – the Barbados raccoon (*Procyon gloveralleni*) – is now listed as extinct, there are few threats to the Common raccoon's long-term survival. Disease, in particular distemper, can temporarily reduce populations, but they quickly rebound. The sport of "coon hunting" with specially trained hounds is popular on chilly nights from September to December in eastern North America. Periods of high fur prices can cause an upsurge in trapping and hunting, which may account for over 4 million raccoons annually. Many others die on the roads. In rural Illinois, for example, trained observers recorded one dead raccoon for every 286km (178 miles) traveled, and mortality densities are much higher in urban areas

The Common raccoon is the major rabies vector in the southeastern United States, and in 1997 accounted for half of all cases of rabies from wild animals reported in the country as a whole. It also often hosts the Raccoon roundworm (*Baylisascaris procyonis*), an intestinal parasite that does not harm the raccoon itself but has caused death in domestic animals and, in two reported cases, in small children. Despite these drawbacks, raccoons are popular both as quarry for sportsmen and with city dwellers as urban wildlife. EKF

Above *Displaying clearly the black eye-mask shared by all species, a Common raccoon reaches for rosehips. The extremely diverse diet of raccoons includes various freshwater fish and crustaceans, all kinds of fruit and vegetables, and even earthworms and insects.*

Right *Raccoons are a frequent sight in the backyards of urban North America, where they come in search of food scraps from ash cans, especially at night. Most homeowners discourage them, not only in order to avoid mess and noise, but also for fear that they will infect their pets with rabies (of which they are principal carriers).*

Coatis

THE EARLY EUROPEAN NATURALISTS WHO FIRST observed coatis distinguished as many as 30 different species, based on the animals' habits and coat colors, but modern taxonomists have reduced the range to only three. The confusion is understandable, because coatis are exceptional in the breadth of variation in their behavior and morphology: in fact, males and females are so distinct that they often act as if they were indeed different species!

The differences have their roots in coati social organization, which is very distinctive – the males are solitary, while females live in highly organized groups termed "bands." Behavioral interactions are extremely complex, involving cooperation reminiscent of primate sociality; for example, band members often groom each other, nurse each other's offspring, and jointly fend off potential predators.

Long Snouts for Snuffling

FORM AND FUNCTION

Coatis are easily recognized by their slender tails and their long and flexible snouts, which protrude beyond the end of the lower jaw. The coati nose is dense with sensory receptors, giving the animals an acute sense of smell. Numerous muscles add flexibility to the tip of the snout, allowing coatis to poke and pry into crevices while seeking prey. Active by day, they spend much time moving through the underbrush and leaf-litter in quest of food, actively brushing aside leaves or hastily digging to expose tidbits, usually in the form of an invertebrate or fruit.

Although coatis inhabit some arid regions, for example in the southwestern United States, they are more commonly denizens of tropical forests. Invariably, they have a highly synchronous mating season. All females in a band, and among bands within a population, come into estrus at the same time, and mating occurs within a period of just two to four weeks. While males may attempt to monopolize access to females by guarding bands, studies of the genetic make-up of offspring and of their potential parents have revealed that different females within a group may nonetheless mate with different males.

The synchronized mating season results in a baby boom some 2–3 months later. Prior to giving birth, females leave the band and become solitary, producing litters of 3–4 young, most of which will not survive to reproduce due to the combined effects of predation, disease, and starvation. Females and their infants remain independent of the band for five or six weeks, until the young are old enough to travel. The timing of the mating season varies between regions according to the availability of food.

Although taxonomically carnivores, coatis are in fact omnivorous, with fruit and arthropods dominating their diet. Reliance on fruit, along with the pressures of predator avoidance, have helped fashion the coati's social structure of solitary males and band-living females. These bands may be quite large – up to 25 or more. In one study in Panama, average band size was found to be 15 individuals, although the number varied seasonally with the availability of food and the comings and going of pregnant females departing from and then returning to the band.

Above *The social bonds between adult female coatis, often shown by mutual grooming* ***1****, bring benefits to all the members of a band, but mostly to the juveniles. The adults and subadults surround the juveniles, and also keep a watch for predators* ***2****. Adults will cooperate to chase a predator away vigorously. Juveniles also get nearly as much grooming* ***3*** *from their mother's allies, which are even allowed to nurse, as they do from her. They also spend much time playing together* ***4****.*

Left *A long, mobile snout and strong front claws are tools of the trade for the Ringtailed coati, which snuffles around in the litter on the forest floor for insect prey, and excavates the surface soil to grub out lizards and spiders from their burrows. Males forage quietly alone, and catch more lizards and rodents than females and young, which forage in groups.*

THE INDIVIDUAL IN THE COATI BAND

Which animals are card-carrying members of a coati band? One might assume that the bands would be made up of blood relatives, but in fact genetic fingerprinting indicates that they also commonly contain some unrelated individuals. An extended field study in Panama has revealed that the unrelated coatis are more likely to be targets of aggression from other members, forced to live on the outskirts of the band, where they are perhaps more susceptible to predators. Such costs may be severe, but in the final analysis it seems that it is better to gain some of the benefits of membership, such as increased access to food, rather than to be left out in the cold altogether.

Solitary Males and Sociable Females

SOCIAL BEHAVIOR

Typically among mammals, adolescents of one sex or the other disperse in search of territory and breeding opportunities as they approach sexual maturity, thereby avoiding mating with close relatives and competing with kin. Coatis accomplish this in an unusual way: females rarely leave the band in which they were born, but early in their third year males disperse from their natal band without vacating its territory – rather, they linger as interlopers, continuing to use their natal home range. They avoid interbreeding by leaving the range during the intense, synchronous mating period to search for unrelated mates elsewhere, returning after several weeks at the end of the breeding season. They can thus continue to live close to home while at the same time minimizing competition with their female kin.

Home ranges of solitary males may overlap by as much as 72 percent, and neighbors are often relatives. The ranges are not generally defended, though access to especially rich food patches, such as fruiting trees, may provoke aggression, usually short-lived. The ranges of bands of females and immature offspring may overlap by up to 66 percent, though core areas tend to be exclusive.

During intra-band conflicts, individuals quickly come to each other's aid, forming coalitions. The extent to which different band members display these traits is influenced by factors such as age, class, and genetic relationships (see box).

COATIS

Order: Carnivora

Family: Procyonidae

3 species in 2 genera

DISTRIBUTION Southern N America, C and S America.

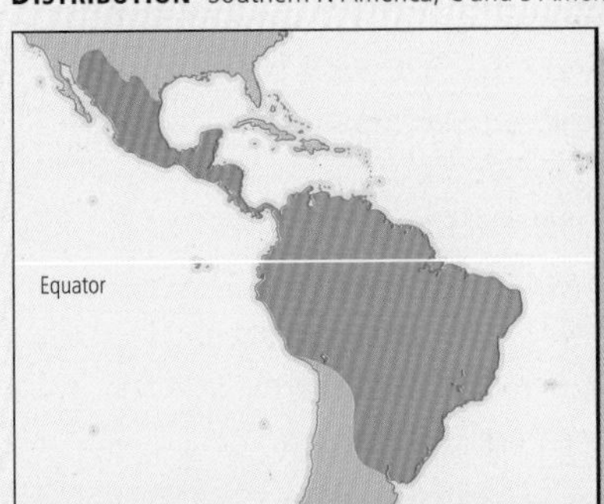

Habitat Wide-ranging, including tropical lowlands, dry high-altitude forests, oak forests, mesquite grassland, and on the edge of forests.
Gestation 77 days
Longevity 7 years (to 14 years in captivity)

RINGTAILED COATI *Nasua nasua*
Forests of S America, E of Andes, S to N Argentina and Uruguay. HTL 80–130cm (32–50in), somewhat more than half being tail; WT male 4–5.6kg (9–12lb), female 3.5–4.5kg (8–10lb). Coat: tawny-red with black face; a small white spot above and below each eye and a large one on each cheek; white throat, belly; black feet, black rings on tail.

WHITE-NOSED COATI *Nasua narica*
SE Arizona, Mexico, C America, W Colombia, and Ecuador. Size: as Ringtailed coati. Coat: gray or brown with silver grizzling on sides of arms and a white band round the end of the muzzle; otherwise as in Ringtailed coati.

MOUNTAIN COATI *Nasuella olivacea*
Mountain forests of Ecuador, W Venezuela, and Colombia. HTL 70–80cm (28–32in), tail shorter than the body. Coat: olive brown, with black muzzle, eye rings, and feet; tail with black rings.

Abbreviations HTL = head–tail length WT = weight

Isolated Populations at Risk

CONSERVATION AND ENVIRONMENT

Most coatis are currently safe from extinction – indeed, in many areas they are among the most common species. There are a few points of concern, however. One important but little-studied subspecies, *Nasua narica nelsoni* – possibly even a separate species – on Cozumel Island, Mexico, is under threat from the tourist industry; and the mountain coati (*Nasuella olivacea*), which is restricted to inaccessible high-altitude habitats in the Andes of western Venezuela, Colombia, and Ecuador, seems distressingly susceptible to local deforestation and land conversion. The IUCN currently categorizes it as Data Deficient. MEG/JLG

Kinkajou

Misled by the kinkajou's prehensile tail, big, forward-facing eyes, and propensity for fruit, early taxonomists declared it a primate, giving it the name Lemur flavus. *Yet twisted into the double helix of DNA within the animal now known as* Potos flavus *lies the truth – the kinkajou is not a monkey. Little else about these nocturnal fruit-eaters betrays their true taxonomy, and it took intricate anatomical and genetic sleuthing to uncover their true, Carnivoran ancestry. The kinkajous' closest relatives are in fact the coatis, raccoons, and ringtails of the family Procyonidae.*

In appearance and habits, the kinkajou is probably most like the olingo (see Other Procyonids); the two animals even sometimes forage together. Though there are differences in diet – the olingo eats a wide range of insects, mammals, and birds, while the kinkajou limits itself to sweet foods – they are so broadly similar that some taxonomists have accorded them a separate subfamily.

Acrobats of the Tree Canopy

FORM AND FUNCTION

The kinkajou's most monkeylike trait is its prehensile tail. This is used as a safety line, latched to secure branches as the 2–3kg (4–7lb) nocturnal animal scampers from one dark rainforest tree canopy to another. Kinkajous can also hang upside-down by their tails while using their dexterous forepaws to manipulate food.

The security offered by this tail, and the dangers of the forest floor at night, combine to confine the kinkajou's activities to the canopy, with the result that they are rarely found in the diet of terrestrial predators such as jaguars. Kinkajous are too nocturnal for eagles, too large for Neotropical forest owls, and too arboreal for potential mammalian predators. This low predation risk is evident not only in their fearless unconcern with terrestrial human observers, but also in their indifference to moonlight. Furthermore, kinkajous have a low reproductive rate (females typically have a litter size of one, and give birth no more than once per year), suggesting that their populations could not sustain a high predation rate.

While avoiding predators in the treetops, kinkajous also take advantage of the easily accessible foodstuffs there. Their diet consists of 90 percent fruit and 10 percent leaves and nectar. Panamanian kinkajous eat no animal prey, although in other parts of their range insects may be an important food. Kinkajous eat a wide variety of fruit species (at least 78 in central Panama), and they prefer fruits that are fleshy and sweet. With the exception of fruit bats, very few mammals are as dedicated frugivores as the kinkajou. Among primates, spider monkeys, chimpanzees, and orangutans are considered fruit specialists, but fruit rarely exceeds 70 percent of their total diet.

A Monkey-like Lifestyle

SOCIAL BEHAVIOR

Low predation risk frees the kinkajou from the pressure to be gregarious, since the advantages of strength in numbers to detect and fend off predators are obviated. Individuals typically forage alone at night like most carnivores – 80.4 percent of feeding bouts feature singletons – but still regularly meet up with other members of their social group, like many primates. This solitary group life is an adaptation to reduce feeding competition. Competition is less important in very large fruiting trees, thus providing opportunities for animals to meet up in an arboreal picnic (in large trees, 35.3% of feeding bouts occur in groups). Social groups also convene at day dens.

Above *Kinkajous can move relatively swiftly within a tree, using their long, short-haired prehensile tail to secure themselves to one branch while they swing and reach for another. The tail is one of the features that distinguish the creature from the olingo, which has a long, bushy, non-prehensile tail.*

Left *Kinkajous have dexterous front feet that aid them in both manipulating food and negotiating their way around the rainforest canopy. The rear feet are longer than the front feet.*

FACTFILE

KINKAJOU

Potos flavus

Order: Carnivora

Family: Procyonidae

Sole member of genus

DISTRIBUTION E Central and South America, from S Mexico to Brazil.

HABITAT Tropical forest.

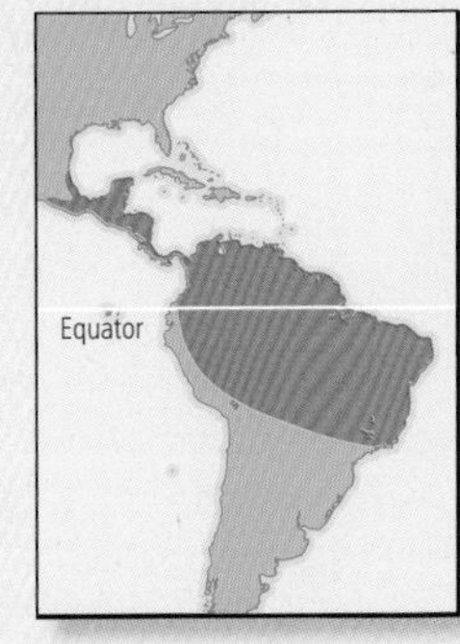

SIZE Head–body length 42–57cm (16.5–22.5in); tail length 40–56cm (16–22in); weight 1.4–4.6kg (3–10lb).

COAT Short and brown, with a tawny coloration.

DIET Mainly fruit, but also honey, flowers, insects, and small vertebrates.

BREEDING Gestation 112–118 days.

LONGEVITY In captivity, up to 32 years.

CONSERVATION STATUS Not under threat.

Below *Being a nocturnal creature the kinkajou tends to spend its days asleep in tree hollows, although it will sometimes emerge on hot days to lie out on a branch. This kinkajou, known in Venezuela as a Cuchi-Cuchi, displays the long, narrow, extrudable tongue that is probably used for probing flowers in pursuit of nectar or for obtaining honey. The kinkajou has a tendency to be solitary when feeding, but will be more social when food is plentiful.*

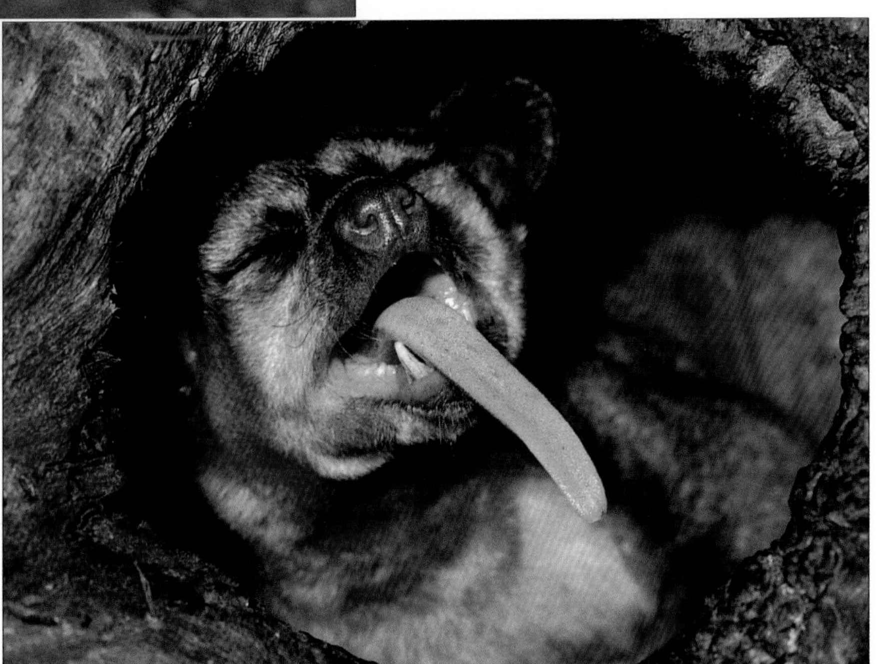

Between dawn and dusk kinkajous seek refuge inside tree holes or thick palm fronds, and up to five animals will squeeze into one of these hideouts. Once reunited with the group, kinkajous show a range of social behavior including allogrooming, group feeding, and play with juveniles. On average, grooming bouts last 6.4 minutes, but can stretch on for up to 28 minutes. Although all combinations of group members participate in this grooming, bouts between adult and subadult males are most common.

Kinkajou social groups consist of a breeding female with her juvenile (up to 1 yr old) and subadult (1–3 yr old) offspring, accompanied by two adult males. These males can be related and are generally amicable, but occasionally squabble, especially around breeding time. One male dominates these brief fights and monopolizes the breeding by defending receptive females, thus obtaining 91.7 percent of the copulations. Inheritance in kinkajou social groups may be patrilineal, since males are the more social sex and dispersal is female-biased. Evidence of this bias comes not only from the dispersal of marked females, but also from a microsatellite genetic study that showed males to be more closely related to their male neighbors than females were to their female ones. Clearly, females leave home more often than males. In addition, not all breeding females join these social groups, and some raise offspring apart from a group structure as "single mothers."

Adult kinkajous have home ranges of 30–50ha (75–125 acres), with strict territorial boundaries between groups. Kinkajous have a unique set of scent glands on their chin, throat, and chest that are probably used to mark territorial boundaries. "Single-mum" kinkajous, excluded from group membership, live along the borders of neighboring social groups and overlap slightly with the group males, but never with the group females. Exactly why pairs of males coexist in each social group is puzzling, but may be related to the costs of marking and defending a territory large enough to completely contain one group female and also overlap partially with at least one "single mum." In this respect, kinkajou behavior is similar to male coalitions that overlap solitary females in other carnivores such as the cheetah and raccoon.

Other aspects of kinkajou sociality have no parallels in related species. Yet even though their combination of a fission–fusion social life, male–male sociality, and female dispersal is unique among carnivores, it is characteristic of spider monkeys and chimpanzees. Although group membership in these primates involves much more extrovert socializing, it may be telling that these are two of the most frugivorous primates and that both enjoy relatively low predation risk. Given these similarities, it is easy to see why early naturalists mistook the kinkajou for a primate, and why local people still refer to it as *mono de la noche* or "monkey of the night." RK

Other Procyonids

THE RED PANDA OF ASIA AND THE NOCTURNAL species of Central and South America are among the least-known procyonids. Shy and retiring, they live on mixed diets: the Red panda mainly on bamboo, the American species on varying combinations of fruit, insects, and small mammals.

Red Panda

AILURUS FULGENS

The Red panda is the sole species of the subfamily Ailurinae, in the Procyonidae (raccoon family), although it has also been associated with the Ursidae (bear family) and the Ailuropodidae (Giant panda) on the basis of morphological, anatomical, biochemical, and paleontological evidence. The animal's taxonomic placement has not stabilized since it was first reported in 1827, even though it was the only panda known until 1869, when the Giant panda was first discovered.

The pelage of the Red panda is made up of long, coarse hairs and a mat of very dense underwool to keep them warm and dry in a cold, moist environment. The soles of its feet are covered by a dense mat of white hair unlike anything seen in other tropical mammals; the nearest parallel is with Arctic species like the polar bear (*Ursus maritimus*). There are five well-developed toes on each foot that are widely separated and have curved, semiretractile claws. Red pandas share the Giant panda's radial sesmoid – its famous "thumb" – though in their case it is not as large.

The stomach is simple with no cecum, as is typical of carnivores. A series of minute pores located between the plantar pads may secrete scents. Paired anal glands are present in both sexes, adjacent to the anal openings; they are dark green, with a characteristic, pungent odor. The animals' vocal repertoire includes squeals, twitters, and a peculiar "quack–snort"; non-vocal sounds are hisses, jaw clapping, and grunts.

Although folivores, Red pandas may forage on the ground for roots, succulent grasses, fallen fruits, insects, and grubs. Their primary diet is leaves and shoots of bamboo. When feeding, they grasp bamboo stems by the forepaws, shearing the leaves off with their mouths; terminal leaves are nipped off with the incisors.

Although Red pandas are active at all times of the day, they are primarily crepuscular, with other peaks of activity around midday and at about 11 p.m. On average they are awake for only 56 percent of the day, probably as an adaptation to their low-quality diet; the figure increases marginally in the fall, when favored fruits are widely spaced, necessitating additional traveling, and rises to 63 percent in winter, when the pandas mate. In a two-year study in Nepal's Langtang National Park, Red panda home ranges were found to overlap between sexes and between males, but seldom between females. The size of the ranges varied from 1.4–11.6sq km (0.5–4.5sq mi); male ranges were larger than females', especially during the mating season. PY

Olingos

BASSARICYON SPECIES

The five olingo species are very similar to kinkajous in external appearance and habits. They all have long bodies and tails and short legs, and are nocturnal fruit-bearers; they sometimes even forage together, usually in groups of several kinkajous to one or two olingos.

On closer examination, though, important differences emerge. Olingos have long muzzles and bushy, non-prehensile tails. In addition, they are slightly smaller than kinkajous, and are subordinate to them; the bigger animals will chase them out of feeding trees (especially flowering balsas) in the low-fruit season. Reports from captivity also suggest that olingos have a more carnivorous diet, eating large insects, small mammals, and birds as well as fruit, grain, and nuts; their dentition is certainly better suited for meat-eating.

Olingos are strictly nocturnal, and, although relatively common, are rarely seen. They are thought to live at relatively low densities, with only one male and one female generally found in any single area. The home range of one olingo male was found to be 37.5ha (93 acres), and the animal traveled about 1,400m (1,500yd) in a single night.

Above *A pair of ringtails. This species was formerly reared as a companion and mouser in prospectors' camps in the early American West – hence its alternative name of "Miner's cat."*

Above left *The striking facial features of the Red panda, which lives in temperate forests with bamboo undergrowth from Nepal through Sikkim (India), Bhutan, northern Myanmar (Burma), and part of southern China. The Red panda's restricted ecological range makes it especially susceptible to the destruction of mountain forest ecosystems, and its extinction in the Himalayas is now a serious threat.*

Right *An olingo (species* Bassaricyon alleni*) from the rain forest of western Brazil. Olingos are found from sea level up to an altitude of 2,000m (6,500ft), and spend the day in nests made from dry leaves in hollow tree trunks, emerging at night to hunt for insects and small mammals.*

Ringtail and Cacomistle

BASSARISCUS ASTUTUS/B. SUMICHRASTI

Both the ringtail and its slightly larger relative, the cacomistle, have doglike teeth that reflect their predatory nature. The two species also share long legs, lithe bodies, and long, bushy, ringed tails. They have foxlike faces, and their ears are larger than in other procyonids: the ringtail's are rounded, while the cacomistle's are tapered. The ringtail has semiretractile claws; the cacomistle's are non-retractile

Although one of the smallest procyonids, the ringtail is the most carnivorous, eating rodents, insects, and birds as well as fruit and other vegetable matter. The cacomistle spends much more time in trees than the ringtail. One cacomistle telemetry study found that the animals move about 2.5km (1.5mi) per night and have home ranges of 16–22ha (40–54 acres), and they may be important seed dispersers because they eat a good amount of (preferably ripe) fruit. RK/JKR

FACTFILE

OTHER PROCYONIDS

Order: Carnivora

Family: Procyonidae

8 species in 3 genera

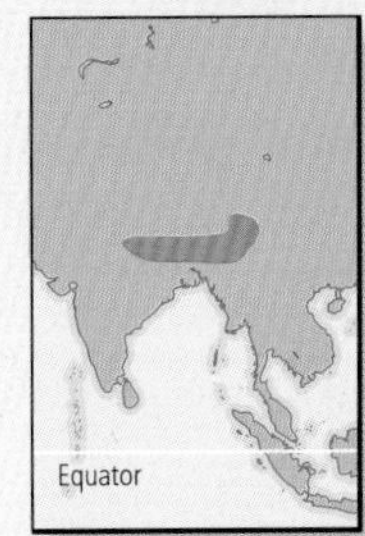

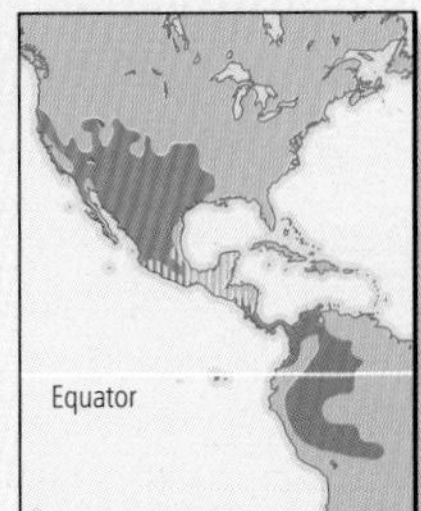

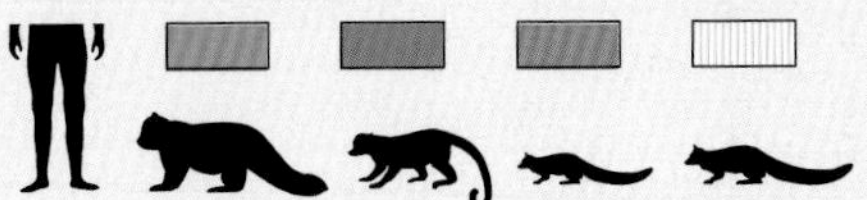

Red Panda Olingos Ringtail Cacomistle

RED PANDA *Ailurus fulgens*

Red or Lesser panda

Himalayas to S China. Favors remote, high-altitude bamboo forests. HBL 50–60cm (20–24in); TL 30–50cm (12–20in); WT 3–5kg (6.6–11lb). Coat: soft, dense, rich, chestnut-colored fur on the back; limbs and underside darker; variable amount of white on face and ears. Breeding: gestation 112–158 days; litter size 1–2 young; newborn young weigh between 110–130g (3.9–4.6oz). Longevity: up to 14 years. Subspecies: *A. f. fulgens*, Himalayas from Nepal to Assam; *A. f. styani*, N Burma and S China. Conservation status: Endangered.

OLINGOS *Bassaricyon* spp.

C America and NW South America (*B. gabbii*); Amazonia (*B. alleni*). Tropical rain forest at about 1,800m. HBL 42–47cm (17–19in); TL 43–48cm (17–19in); WT about 1.6kg (3.5lb). Coat: gray-brown, long, loose hair with blackish hues above, yellowish below and on insides of the limbs; yellowish band across neck to back of ears; tail with 11–13 black rings, often indistinct. Breeding: gestation 73–74 days. Longevity: more than 15 years in captivity. *Bassaricyon lasius* (Costa Rica) and *B. pauli* (Panama) are probably subspecies of *B. gabbii*, and *B. beddardi* (Guyana) a subspecies of *B. alleni*. Conservation status: *B. gabbii* is currently listed as Lower Risk: Near Threatened; *B. lasius* and *B. pauli* are both classed as Endangered.

RINGTAIL *Bassariscus astutus*

Ringtail, Civet cat, Miner's cat, or Ring-tailed cat

W USA, from Oregon and Colorado S, and throughout Mexico. Dry habitats, especially rocky cliffs. HBL 31–38cm (12–15in); TL 31–44cm (12–17in); WT 0.8–1.1kg (1.8–2.4lb). Coat: overall gray or brown; white spots above and below each eye and on cheeks.

CACOMISTLE *Bassariscus sumichrasti*

C America. Dry forests. HBL 38–50cm (15–20in); TL 39–53cm (15–21in); WT 0.9kg (2lb). Coat: as for ringtail. Conservation status: Lower Risk (Near Threatened).

Abbreviations HBL = head–body length; TL = tail length; WT = weight

DISTRIBUTION All continents except Antarctica and Australia, though three species have been introduced to New Zealand.

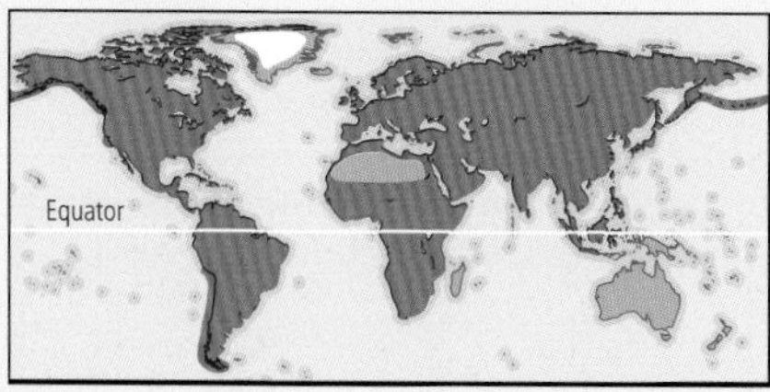

Weasel and Skunk Families

MUSTELIDS AND MEPHITIDS ARE CLOSELY allied; indeed, some taxonomists still consider the Mephitids – the skunk family – as a Mustelid subfamily. Although between them the two families make up the largest carnivore grouping and have successfully spread over five continents, much remains to be learned about their biology. A new species, the Colombian weasel, was discovered as recently as 1978, and even some of the commonest species have so far avoided scientific scrutiny. There is still an active debate about the mustelids' evolutionary origins, and the membership of their various subfamilies may yet be revised completely.

Even though they range in size from the Least weasel, which, at less than 50g (2oz) is the world's smallest carnivore, to the much larger Sea otter, which can exceed 45kg (100lb), mustelids share an energetically demanding body shape. Typically they are long and thin, with short legs, and are suited to moving in confined spaces, such as the burrows of their rodent prey. Their slender form gives mustelids a distinct advantage over the prey, but at the same time provides them with a large surface area, which means that they lose proportionally great amounts of energy as radiated body heat. To keep up with this rapid energy loss they have high metabolic rates and high food requirements. The smallest weasels must eat about a third of their weight in prey daily, and in lactating females the figure can reach two-thirds.

Although some members of the mustelid family can be very common locally, the IUCN classifies 17 species and 10 subspecies as threatened, and further 9 species and 4 subspecies as vulnerable. Overexploitation and direct control by humans have often precipitated declines in mustelid populations. Habitat loss and the elimination of prey animals, as in the case of the Black-footed ferret, also cause grave problems. As international law increasingly encourages protection for carnivores and prevents trade in endangered species, the future of many mustelids seems more secure, but urgent measures are still needed to conserve the few species that remain on the brink of extinction. RMcD

FOOD AND FEEDING

Mustelids are highly active foragers that seek and pursue their prey relentlessly in all their refuges. Several mustelids, such as stoats, martens, and tayras, are agile climbers and can snatch birds and squirrels from their nests. To avoid periods of food shortage, mustelids adopt a strategy of surplus killing. This means that when prey is plentiful they kill as much as possible – much more than they can eat immediately. The excess is then dragged away and cached for consumption later, when prey may be scarce.

The diet of smaller mustelids, such as weasels, stoats, and polecats, consists mainly of burrow-dwelling small mammals, such as mice, voles, and rabbits. Otters typically eat fish and aquatic invertebrates, while in Britain the Eurasian badger, one of the larger mustelids, lives on a diet almost entirely composed of earthworms and plant roots. The wolverine, sometimes known as the "glutton" because of its reputedly insatiable appetite, can kill animals as large as reindeer. Fur trappers exploit the fondness of some mustelids, especially martens, for sweet berries and sometimes bait their traps with jam. Skunks eat a wide range of food, including small rodents and snakes, insects and other invertebrates, as well as fruit and green vegetation.

KULLS AND DENTITION

lustelid and mephitid skulls tend to e long, flattened, and more or less vedge-shaped, tapering to the muz- e. Skull size and the number and daptations of teeth vary widely. Most nembers of the subfamily Mustelinae, or example the common weasel, have dental formula I3/3, C1/1, P3/3, M1/2 34, with prominent, sharp canines nd cutting carnassials. In the wolver- ne (I3/3, C1/1, P4/4, M1/2 = 38) the eavy premolars and powerful jaws an crush even thick bones. The dental ormula of the honey badger is I3/3, 1/1, P3/3, M1/1 = 32, while in the urasian badger it is I3/3, C1/1, P4/4, 11/2 = 38 – the largest number of eeth in the family. In otters of the enus *Lutra* the arrangement is I3/3, 1/1, P3–4/3, M1/2 = 34–36.

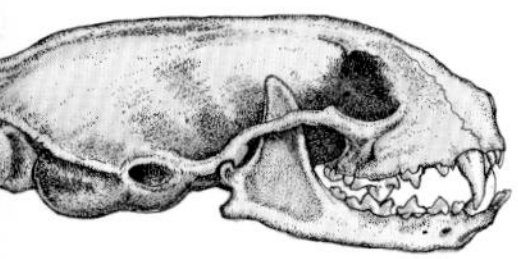

uropean common weasel
.4 cm

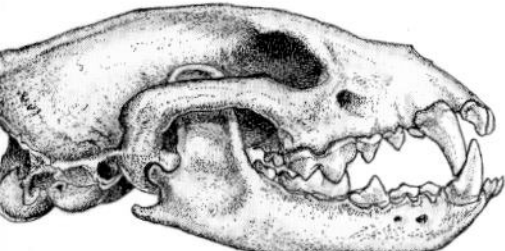

olverine
6.8 cm

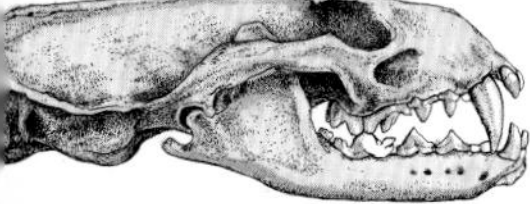

tter (*Lutra*)
2 cm

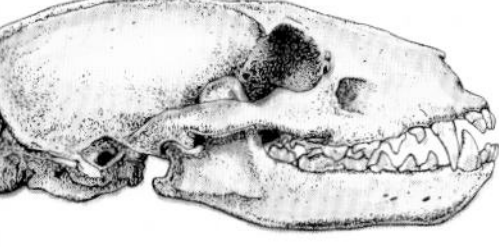

urasian badger
3 cm

Left *Various members of the easel family, shown in characteristic ctivities:* ***1*** *Indian smooth-coated otter holding a shell in its dextrous, heavily webbed forepaws;* ***2*** *European weasel dragging a mouse;* ***3*** *American mink with a rabbit;* ***4*** *Eurasian badger forag- ing;* ***5*** *European polecat in its winter coat;* ***6*** *Pine marten, having successfully caught a bird;* ***7*** *Wolverine following a scent trail across the ground.*

CHEMICAL COMMUNICATION

Mustelids and mephitids are generally solitary animals. Only five species of otter, the Eurasian badger, and the Spotted skunk exhibit any degree of group living. Typically, they lead unsocial lives after leaving their mother's care. Meetings with conspecifics are usually hostile, except during the mating season when a temporary truce exists between the sexes. Despite this, they have evolved a complex system of social communication based on chemical signals, which are deposited in conspicuous places in their urine, feces, and skin secretions. Specialized anal glands also make the acrid, musky-smelling secretions that are used in displays of aggression, threat, and fear. Although all mustelids possess these anal glands, the ability to expel a fine spray of foul-smelling liquid at an intruder is most pronounced in skunks (BELOW). Dominant mustelids mark out their territories and keenly defend their patch from interlopers of the same sex. However, male territories usually enclose those of several females, and in the breeding season males will do their utmost to defend their access rights to these potential mates.

Weasels, Mink, and Polecats

The Least weasel has been described as "the Nemesis of Nature's little people." This identification with the ancient Greek goddess of retribution seems appropriate, since the whole group of lithe, sleek-bodied carnivores of which it is the smallest member are formidable hunters. They often track animals to their burrows and take on prey substantially bigger than themselves.

The mustelids may be small and much persecuted by man, but they are awe-inspiring predators with the potential to alter the entire ecology of a region. All are fiercely carnivorous and extremely strong for their size – a weasel is easily capable of running at speed carrying a load of meat equivalent to half its own weight, which no lion could do.

Mousetraps with Teeth

FORM AND FUNCTION

While most mustelids weigh less than 2kg (4.4lb), the wolverine is ten times that weight. Many species exhibit pronounced differences in size between males and females. Knowledge of the group is very uneven. The three Northern Hemisphere weasel species (the Least weasel, the ermine or stoat, and the Long-tailed weasel) are known in some detail, but at the other end of the scale there is no reliable information at all about some of the tropical species. Yet, as a group, they are among the most common and widespread carnivores in the world.

The northern weasels have long, slim bodies and short legs, a flat-topped, sharp-faced, almost triangular head, and short, rounded ears. All mustelid species have the habit of sitting up on their haunches for a better view than they can get from on all fours.

Mink have small ears and long, bushy tails. The coat provides important insulation against low water temperatures because, like otters, the animals have little subcutaneous fat. The fur has two components: long guard hairs each surrounded by 9–24 underfur hairs that are one-third or half the length. There are two molts each year: the thick, dark winter coat is shed in April, to be replaced by a much flatter and browner summer coat. The summer molt occurs in August or September, and the winter coat is in its prime condition by late November. Northern subspecies have darker fur than southern forms.

The European mink may be very similar in appearance to the American mink, but it is not closely related. DNA analysis has shown that the two species are as far apart from each other as it is possible to be while both remaining within the single subfamily Mustelinae, and they do not interbreed. European mink do, however, interbreed with the European polecat.

Prolific Killers

DIET

All mustelids are terrestrial hunters. They will take whatever small rodents, rabbits, birds, insects, lizards, and frogs are locally available. Most eat nothing but meat, and their teeth are highly adapted to killing and cutting up prey. They do not feed indiscriminately; members of each species choose different items from the menu on offer. What is more, they are strategic killers. The Marbled polecat uses different techniques to dispatch different types of prey: a bite to the head or neck for large prey, but to the thorax for small. Fleeing prey are bitten dorsally, while those turning to defend themselves get bitten in the throat. Only one polecat, the rare Black-footed ferret of the western prairies of the USA, is a true specialist depending entirely on Prairie dogs.

The small Least or Common weasels are relentless hunters of mice and voles, especially Field voles and lemmings, which they can follow along their runways in thick grass and under snow. This habit leaves the prey little option, for if they run into the open, they are attacked by avian predators such as kestrels. Despite their small size, Least weasels can kill young rabbits, but it is probably not worth their while if they can find enough small rodents and birds. The larger weasels, such

◐ **Left** *A Long-tailed weasel. Like the ermine, this 'orth American species – one of the larger mustelids – /as once widely hunted for its pelt. However, changes ı fashion have seen a sharp decline in this trade.*

◐ **Above** *Polecats, like this European polecat, can be distinguished from weasels primarily by their coat color (they are never brown and white). In general, they are also more stockily built.*

FACTFILE

WEASELS, MINK, AND POLECATS

Order: Carnivora

Family: Mustelidae

Subfamily: Mustelinae

23 species in 7 genera. Species include: Least or European common weasel (*Mustela nivalis*), European polecat (*M. putorius*), ermine or stoat (*M. erminea*), American mink (*M. vison*), European mink (*M. lutreola*), grison (*Galictis vittata*).

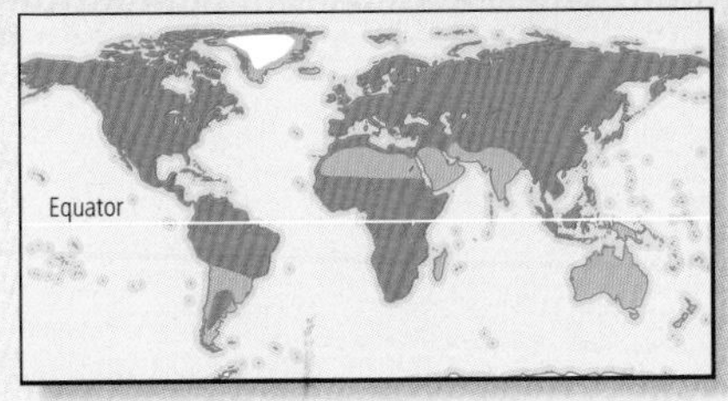

DISTRIBUTION Widespread from the tropics to the Arctic, in the Americas, Eurasia, Africa; introduced in New Zealand.

HABITAT Very varied, from forests to mountains, farmland, semi-desert, steppe, and tundra.

SIZE Ranges from Least weasel males to grison males. **Head–body length** 15–55cm (6–22in); **tail length** 3–21cm (1–8.5in); **weight** 30g–3.2kg (1oz–7lb).

COAT Weasels are brown above, white or yellow below in summer, entirely white in northern populations in winter. Polecats sport various colors, often having bold black and white markings. Mink have thick, glossy, black or brown coats, darker in winter; the European species is distinguished from the American by a white patch on its upper lip.

DIET Small rodents, rabbits, birds, insects, lizards, frogs.

BREEDING Gestation 35–45 days, extended by delayed implantation in ermine, American mink, and Long-tailed weasel.

LONGEVITY Average less than 1 year for the Least weasel in the wild, though up to 10 in captivity. Larger species are probably longer-lived; mink and polecats are thought to live to 6 years or more in the wild (mink to 12 in captivity)

See species table ▷

WHY ARE MALE WEASELS BIGGER THAN FEMALES?

Male mustelids are usually much larger than females. This is especially evident in the smaller species, such as the Least weasel, where the female is only half the weight of the male. The reason for this remarkable sexual dimorphism has perplexed scientists for generations, and a number of theories have been proposed.

The simplest explanation is that the size difference enables males and females to eat different prey and so avoid competition with each other, particularly when food is scarce. A predator will always try to catch the largest prey animal possible, to get the best return for its effort; males can catch prey too big for females to handle, although there is still a large overlap in their choices.

Another explanation gives quite different reasons for the varying body sizes of males and females, seeing any difference between the sexes in diet as a consequence of the dimorphism rather than as its purpose. For a female who raises young on her own, small size has benefits, for she must spend long hours foraging to meet her own physiological needs as well as producing milk for her offspring. If she was the size of the male, she would need to catch and consume an extra vole a day, and that would mean spending longer away from her litter, which require her for warmth and protection. Males are promiscuous and territorial – they compete with other males for access to females and for the best territories. For them, fighting ability and mating success increases with size, so sexual selection has favored the development of large individuals.

as the ermine and Long-tailed weasel, will chase any rodents that expose themselves above cover, but gain a better return in energy from concentrating on larger prey such as rabbits and water voles.

Mink take a wide variety of prey, including small mammals, frogs, fish, and crayfish, as well as some insects, worms, and birds. The diet of both mink species is very similar when in the same area, only the European mink takes more frogs, the American more small rodents; water vole numbers have declined steeply since its appearance in Britain. The animals' eyesight is not particularly well adapted to underwater vision, and fish are often located from above before the mink dives in pursuit. Mink rely heavily upon their sense of smell when foraging for terrestrial prey.

Rabbits and rodents normally reproduce at a much faster rate than their mustelid predators. In habitats occupied year-round by a wide range of predators, many hungry mouths await the annual new crop of young rabbits and rodents. In the harsher climates of the far north, however, typically occupied by various species of voles and lemmings, ermine and Least weasels are often the only resident overwintering predators. When they are rare, the rodents can increase rapidly under the snow, and the mustelids cannot reproduce fast enough to catch up. By the time they do, the rodents are abundant, and predation makes no difference to their numbers. Predation is usually heaviest when rodent numbers are already in decline for some other reason (for example, voles in peak-year populations usually mature later, breed for a shorter season, and disperse more widely). Then the hungry weasels have no other option but to pursue the last few rodents right into their nests. When this happens, predation can accelerate, or even prolong, the decrease in rodent numbers; but in turn, the weasels cannot breed when rodent numbers are too low. Before long the weasels become rare or locally extinct and their toll on the rodents ceases, permitting the rodents to increase again. This "on/off" weasel predation is one plausible explanation for the well-known population cycles among northern voles and lemmings.

Evidence that the mustelids might be responsible for the changing fortunes of their prey came from a study in Finland. Mammalian (Least weasel and ermine) and avian (Tengmalm's owl and kestrel) predators were removed from some areas but not others. In areas of removal, voles and lemmings did not suffer the characteristic dip in numbers. Furthermore, it was the formidable Least weasel that seemed to have the single greatest impact, accounting for 40 percent of all vole predation. Removal of all predators except the Least weasel did not stop the decline in vole numbers.

In other habitats, mustelids seldom have much effect on prey numbers, except where the prey is

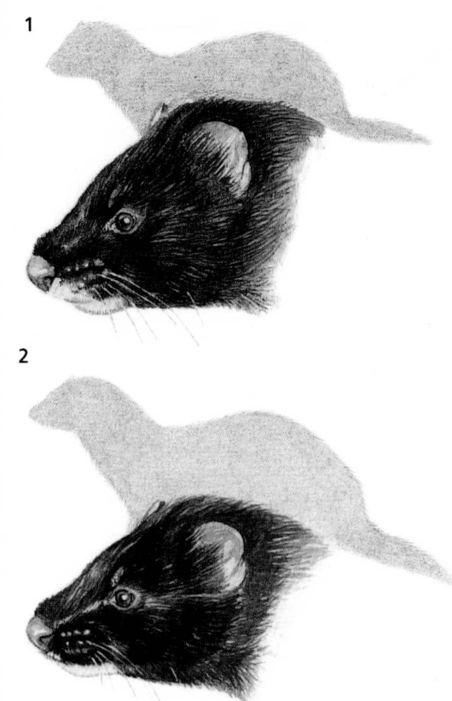

Above The European mink **1** always has a white patch on its upper lip. American Mink **2**, now naturalized in Europe, are larger, and most lack the white patch. However, 10–20 percent not only possess the patch, but it may be as large as that of the European species. In such animals, only study of the skeleton can guarantee correct identification.

Left An American mink displays the dense pelt that has led the animals to be trapped for their fur for centuries, and has made them the object of commercial breeding programs since 1866. Fur farmers prefer this species to the European variety because of its hardiness, and the range of mutant colors it produces.

THE VERSATILE MINK

The carnivorous diet of mink includes crayfish, crabs, fish, small burrowing mammals, muskrats, rabbits, and birds. This range of prey – hunted in water, on land, in swamps, and down burrows – is considerably greater than that of more specialized mustelids, such as otters and weasels.

For mink, the so-called "broad niche" that they occupy carries both costs and benefits. The costs arise when mink compete with more specialized predators. Since a specialist is better adapted than a generalist to exploiting certain prey, the generalist fares the worse when those prey become the object of competition; for example, in times of absence or scarcity of other prey groups, mink may depend heavily on fish, bringing them into direct rivalry with better-equipped otters. On the benefit side, mink have such a wide choice of prey that they can normally turn to alternatives if one type of prey becomes scarce; such an option is closed to specialists.

The adaptability of mink is reflected in the variety of habitats in which they thrive – from the arctic wastes of Alaska to the swamps of the Florida Keys; from inland lakes and rivers to the wave-battered rocks of the Atlantic coast. JDSB

isolated and has no safe refuge. Ermine introduced to an island off Holland in 1931 bred rapidly, and by 1937 had wiped out a plague of water voles. Likewise, in an 8.5-ha (21-acre) enclosure on a New Zealand farm, ferrets and feral domestic cats almost exterminated a dense rabbit population (up to 120 per ha or 48 per acre) in only three years.

Mustelids will themselves adapt to changes in the prey they live on. When American mink first arrived in the territories of the European species in Belarus, the invaders were much larger than the native species, with imported males averaging a weight of 1,907g (4.2lb) and the females 895g (2lb) to the natives' 854g (1.9lb) and 537g (1.3lb) respectively. But six years later the two had converged, with American males averaging 1,180g (2.6lb) as against the European males' 977g (2.2lb) and the females' 677g (1.5lb). The invading species had apparently adapted its size to the local requirements and resources, while only the larger of the native mink had survived the attacks of the invaders.

Below An American mink – one of many now breeding in England – drags its latest prey, a moorhen, back to its burrow. Birds are only a secondary element in the mink's diet; small mammals, fish, frogs, and crayfish are all more often taken by this agile and versatile waterside predator.

Solitary Hunters

SOCIAL BEHAVIOR

Weasels hunt largely underground or under snow, and are active day and night. Some larger species are more often nocturnal. Usually, males and females have separate home ranges; these may overlap between sexes, but never within the same sex. Residents avoid each other whenever possible; when they do meet, females are subordinate to males, except when with young. The grison, and possibly the African striped weasel, may be more sociable, as they are reputedly seen in groups (of unknown composition); and, in captivity, a male and a female, or two male grisons, may safely be kept in one cage for long periods. However, four grisons observed together in captivity showed no evidence of the mutual grooming characteristic of social animals. Most mustelids, sociable or not, may be seen in parties of females and young just before the dispersal of the litters.

Home ranges are smaller in habitats rich with prey, in the non-breeding season, and in the smallest species. For example, male Least weasels in Britain have ranges of 1–25ha (2.5–62 acres), while those of European polecats in Russia are ten times this size.

Both mink species are solitary and territorial. Individuals defend linear territories of 1–4km (0.6–2.5mi) along rivers or the shores of lakes by scentmarking and overt aggression. Each territory contains several waterside dens plus a core foraging area. Marshland territories cover up to 9ha (22 acres), while those on a rocky coastline, such as Vancouver Island or the Scottish west coast, are only 0.7km (0.4 miles) long, reflecting the abundance of rockpool crabs and fish. Female mink ranges are about 20 percent smaller than those of the males quoted above. Mink scentmark with feces coated by secretion from glands at the end of the gut (the so-called proctodeal glands), employing an "anal drag" action, and by secretions from glandular patches on the underside of the throat and chest, deposited by ventral rubbing.

Breeding habits are known in detail for only a few mustelid species. In mink, males leave their territories as the mating season approaches (February to March in the northern hemisphere) and travel long distances in search of females. Both sexes are promiscuous, mating with several partners in a single season or in a single estrus. Experiments in mink farms have shown that the last male to mate fathers most of the kits. Fighting is common between rutting males.

In the ermine and the Long-tailed weasel, implantation of the fertilized egg is delayed for almost a year after mating. American (but, intriguingly, not European) mink also practice delayed implantation, though in their case only 7–30 days elapses. In the Marbled polecat, the date of birth depends on the weather preceding implantation. It seems that this delay allows females to be flexible about when their young are born, and time it to the best conditions possible. In other species this is apparently not the case.

Below *The ermine or stoat, here seen carrying its young, eats a variety of food, including game birds and poultry (and their eggs). As a result, they are widely exterminated by gamekeepers.*

Young mustelids are born blind and thinly furred, in a secure nest often borrowed from prey species and lined with fur from previous meals. The eyes open at 3–4 weeks in the Least weasel, 5–6 weeks in the larger weasels and polecats. The young chew on meat at 3–5 weeks, although lactation lasts about 6–12 weeks (8–10 in mink), and the family stays together some weeks after weaning. Young mink disperse from the natal territory at 3–4 months of age, the males to a greater distance, often 50km (31 miles) or more from their starting-point. Sexual maturity is reached at 10 months.

The larger mustelid species, and males of the ermine and Long-tailed weasels, first breed as 1-year-olds. Young females of these species mate when still in or only just out of the nest, with the male that holds the territory where their mother resides, but cannot produce a litter until they, too, are 1 year old. If food is plentiful, both sexes among Least weasels may mature at 3–4 months, and adults can have a second litter, which gives these species the capacity to respond more rapidly to population increases in small rodents than any other carnivore.

Friend or Foe?

CONSERVATION AND ENVIRONMENT

Small mustelids have long been trapped and shot by gamekeepers. The temporary extinction of the European polecat in England was probably due to intensive gamekeeping on the great sporting estates of the late 19th century. A hundred years later, ermine and Least weasels still thrive, and European polecats are returning to England from their refuges in Wales and Scotland.

Mink occupy a special position, owing to the economic value put upon their fur. The American species originally occurred only in North America, but it is now farmed extensively, supporting a

Above *A Least or European common weasel. This species coexists with its larger relative, the stoat, by taking smaller prey, such as voles and mice.*

Right *These mostly southerly species share the same body plan as the ermine or Least weasel but tend to have black, not brown, as the predominant coloration, or to be larger.* **1** *North African banded weasel* (Poecilicitis libyca). **2** *African striped weasel (Poecilogale albinucha).* **3** *Marbled polecat* (Vormela peregusna). **4** *The skunklike zorilla* (Ictonyx striatus), *which appears to threaten to stink-spray.* **5** *Little grison* (Galictus cuja) *and* **6** *European polecat* (Mustela putorius) *in winter coats, both in upright sniffing/lookout stance.* **7** *Patagonian weasel* (Lyncodon patagonicus) *in typical flattened weasel posture.* **8** *Black-footed ferret* (Mustela nigripes) *at Prairie dog burrow.*

'ade that was worth US$6 billion in 1996. There 're now fur farms throughout the Americas, urope, and Asia, and inevitably animals have scaped and established the species in the wild in ll these areas. Some 20,000 American mink have lso been released in Russia and neighboring tates to support fur trapping, and in 1999 alone round 13,000 were freed from British fur farms y animal rights protesters, causing problems for ocal wildlife.

Besides exploiting their fur, humans have tradi- ionally assigned mustelids another role, as allies n the war against rodents and rabbits. Each year, nustelids can make very heavy inroads into local opulations of crop pests. Farmers have translo- ated weasels and polecats to places where they vere absent to help rid houses and farms of odents – or at least prevent outbreaks. Mustelids vere particularly valued in Europe before the ntroduction of the domestic cat in the 9th centu- y. The Least weasel, ermine, and feral ferret were leliberately introduced into New Zealand in 1884 o control a plague of European rabbits on the new sheep pastures. Unfortunately, the initiative lid not succeed; both species, particularly rmine, turned instead to eating the vulnerable native ground-nesting birds, and both, particularly errets, carried bovine TB. Small mustelids are now onsidered worse pests in New Zealand than they ver were on the game estates of their homeland.

While a rampant fur trade has historically taken ts toll on mustelid populations, and despite local fforts to control their spread, most members of he weasel–polecat group are in no immediate langer. Yet one species, the Black-footed ferret, lid very nearly disappear in the wild at the end of he 20th century, due to its dependence on Prairie logs, a species humans have eliminated from arge areas of farmland. When the last known colony died out in North Dakota in 1974, the fer- ets were thought extinct, but a fresh colony of some 30 individuals was found in Wyoming in 1987 and taken into captivity. This group was to orm the basis of a successful captive-breeding and reintroduction program. By the year 2000 here were 300 Black-footed ferrets in captivity and around 200 once more living wild in the USA.

The situation with mink is very different; here he threat is one of a conflict between species. The American mink is widely regarded as a pest and a hreat to native prey species in countries where it s now naturalized; it is still expanding in Europe and Asia. Ironically, one victim is the European mink, with which it competes aggressively. Being arger, it usually succeeds in expelling or killing its smaller rival.

As a result, the long-term survival of the Euro- pean species is severely threatened, with the last remnant populations declining fast. A rescue pro- ject is now under way to protect European mink on an island in the Baltic from which the Ameri- can species has been excluded. CMK/HK

Weasel, Mink, and Polecat Species

FIVE SPECIES ARE SUFFICIENTLY DISTINCT to be placed in separate monotypic genera, and there are just two species of grison (*Galictis*). The other 16 species belong to the genus *Mustela*, although some authorities recognize fewer species. Figures for size and breeding are mostly very approximate or unknown. Males of most species are much heavier than females.

Least weasel
Mustela nivalis
Least or European common weasel

N America, from Arctic to about 40°N; Europe from Atlantic seaboard (except Ireland), including Azores, Mediterranean islands; N Africa and Egypt, E across Asia N of Himalayas; introduced in New Zealand. In Sweden, two subspecies are recognized: the one described by Linnaeus, *M. n. nivalis* in N and C Sweden, is smaller, shows less sexual dimorphism, normally has white winter fur (but not always), and has more white on underside when compared to *M. n. vulgaris* in S Sweden and the rest of W Europe, which retains summer coat in winter and also has a brown spot on cheek; the two forms interbreed and produce fertile offspring, both in captivity and across a narrow hybrid zone in C Sweden. Some taxonomists regard the American subspecies, *M. n. rixosa*, as conspecific with the Swedish *M. n. nivalis*, and that in turn as specifically distinct from *M. n. vulgaris*. Others see the ability of the two forms to interbreed (see above) as firm evidence that they are both only subspecies of *M. nivalis*.
SIZE: very large variation, from small northern form – in America HBL (male) 15–20 cm; TL 3–4 cm; WT 30–70g – to the largest forms in S beyond range of ermines – e.g. Turkmenistan: HBL (male) 23–24 cm; TL 5–9cm; WT to 250g.
COAT: brown above, white below, turning entirely white in winter except in W Europe and S Russia; no black tip to tail.
BREEDING: gestation 34–37 days; litter size 4–8; may produce 2, even 3, litters a year during vole plagues.

Ermine
Mustela erminea
Ermine, Stoat, or Short-tailed weasel

Tundra and forest zones of N America and Eurasia, S to about 40°N, including Ireland and Japan, but not Mediterranean region, the semi-deserts of Kazakhstan and Mongolia, or N Africa; introduced in New Zealand.
SIZE: very large variation, especially in N America; largest in N (male HBL 24cm; WT 200g), smallest in Colorado, where Least weasel is absent (male HBL 17cm; WT 60g); Russian races 130–190g; British and New Zealand races up to 350g.
COAT: brown and white; prominent black tip to tail, even in white winter coat.
BREEDING: delayed implantation of 9–10 months, from early summer mating in one year to whelping in spring of the next, which cannot be shortened by abundance of food; active gestation about 28 days; litter size 4–9, sometimes up to 18.

Long-tailed weasel
Mustela frenata

N America from about 50° N to Panama, extending through northern S America along Andes to Bolivia.
HBL (male) 23–35cm; TL 13–25cm; WT (male): 200–340g, (female): 85–200g.
COAT: and breeding as in *M. erminea*; S races have white facial markings and yellow underparts.
BREEDING: as for *M. erminea*.

Tropical weasel
Mustela africana

E Peru, Brazil. Formerly placed by some in separate genus, *Grammogale*.
HBL (male) 31–32cm; TL 20–23cm; WT unknown.
COAT: reddish-brown above, lighter below, with median abdominal brown stripe; black tail tip indistinct; foot soles naked.

Colombian weasel En
Mustela felipei

A new species from highlands of Colombia first described in 1978. Formerly placed by some in separate genus, *Grammogale*.
SIZE: of only 2 males measured, HBL 21–22cm; TL 10–11cm; WT unknown.
COAT: blackish-brown above, orange-buff below; no black tail tip; short ventral brown patch (not stripe); feet bare and webbed.

European polecat
Mustela putorius

Forest zones of Europe, except most of Scandinavia, to Urals. Includes the **ferret** (*M. p. furo*), a domesticated form of *M. putorius* (and possibly also of *M. eversmannii*) bred in captivity as early as the 4th century BC. Among ferrets, albinos and white or pale fur are common. Introduced into New Zealand, where there is now a large populations of feral animals.
HBL (male) 35–51cm; TL 12–19cm; WT 0.7–1.4kg.
COAT: buff to black, with dark mask across eyes.
BREEDING: gestation 40–43 days; litter size typically 5–8.

Steppe polecat
Mustela eversmannii

Steppes and semi-deserts of Russia, Kazakhstan, and Mongolia, to China.
HBL (male) 32–56cm; TL 8–18cm; WT about 2kg.
COAT: reddish-brown, darker below and on feet and face mask; ears and lips white.
BREEDING: gestation 36–42 days; litter 3–6, occasionally up to 18.

Black-footed ferret En
Mustela nigripes

W prairies of N America, only within range of Prairie dogs (*Cynomys*); was rated extinct in the wild, but maintained in captivity and being locally reintroduced; derived from *M. eversmannii*, invading N America during Pleistocene (2 million –11,000 years ago).
HBL (male) 35–41cm; TL 11–13cm; WT 0.9–1kg.
COAT: yellowish with dark facial mask, tail tip, and feet.
BREEDING: gestation 42–45 days; litter 3–4.

Mountain weasel
Mustela altaica

Forested mountains of Asia, from Altai to Korea and Tibet.
HBL (male) 22–29cm; TL 11–15cm; WT 350g.
COAT: dark yellowish to ruddy-brown, with creamy-white throat and ventral patches; paler in winter, but not white; white upper lips and chin shading to adjacent darker areas (cf. *M. sibirica*).
BREEDING: gestation 30–49 days; litter usually 1–2, but up to 7–8.

Kolinsky
Mustela sibirica
Kolinsky or Siberian weasel

European Russia to E Siberia, Korea, China, Japan, and Taiwan (see also *M. lutreolina*).
HBL (male) 25–39cm; TL 15–21cm; WT 650–800g.
COAT: dark brown, paler below; may have white throat patch; paler in winter, but not white; dark facial mask with white upper lips and chin sharply contrasting with surrounding darker fur; tail thick and bushy.
BREEDING: gestation 35–42 days.

Yellow-bellied weasel
Mustela kathiah

Himalayas, W and S China, N Burma.
HBL 23–29cm; TL 17cm; WT unknown.
COAT: deep chocolate-brown (rusty-brown in winter), yellow below; may have white spots on forepaws and whitish throat patch, chin, and upper lips; tail long-haired, at least in winter.

Back-striped weasel Vu
Mustela strigidorsa

Nepal, E through N Burma to Indochina
HBL (female) about 29cm; TL 15cm; WT unknown.
COAT: deep chocolate-brown (paler in winter), with silvery dorsal streak from base of skull to tail root, and yellowish streak from chest along abdomen; upper lip, chin, and throat whitish to ocherous; tail bushy; feet naked at all seasons.

Barefoot weasel
Mustela nudipes

SE Asia, Sumatra, Borneo.
COAT: uniform bright red with white head; feet naked at all seasons.

Indonesian mountain weasel En
Mustela lutreolina

High altitudes of Java and Sumatra. Probably derived from *M. sibirica* stranded on the islands at the end of the Pleistocene. Some authorities consider *M. lutreolina* and *M. sibirica* as one species.
FORM: the few known specimens are similar in size and color to European mink/ *M. lutreola* (russet-brown, no face mask, variable white throat patch); but skull similar to *M. sibirica*.

American mink
Mustela vison
American or Eastern mink

Originally N America, but now naturalized throughout Europe, C and E Asia, and in southern S America. Fourteen subspecies.
SIZE: HBL (male) 34–54cm, (female) 30–45cm; TL (male) 15–21cm, (female) 14–20cm; WT male and female 0.5–1.5kg.
COAT: thick, glossy, black or brown; winter coat darker.
BREEDING: gestation 34–70 days (delayed implantation).

ABBREVIATIONS HBL = head–body length TL = tail length WT = weight
Approximate nonmetric equivalents: 10cm = 4in; 230g = 8oz; 1kg = 2.2lb

Ex Extinct
EW Extinct in the Wild
Cr Critically Endangered
En Endangered
Vu Vulnerable
LR Lower Risk

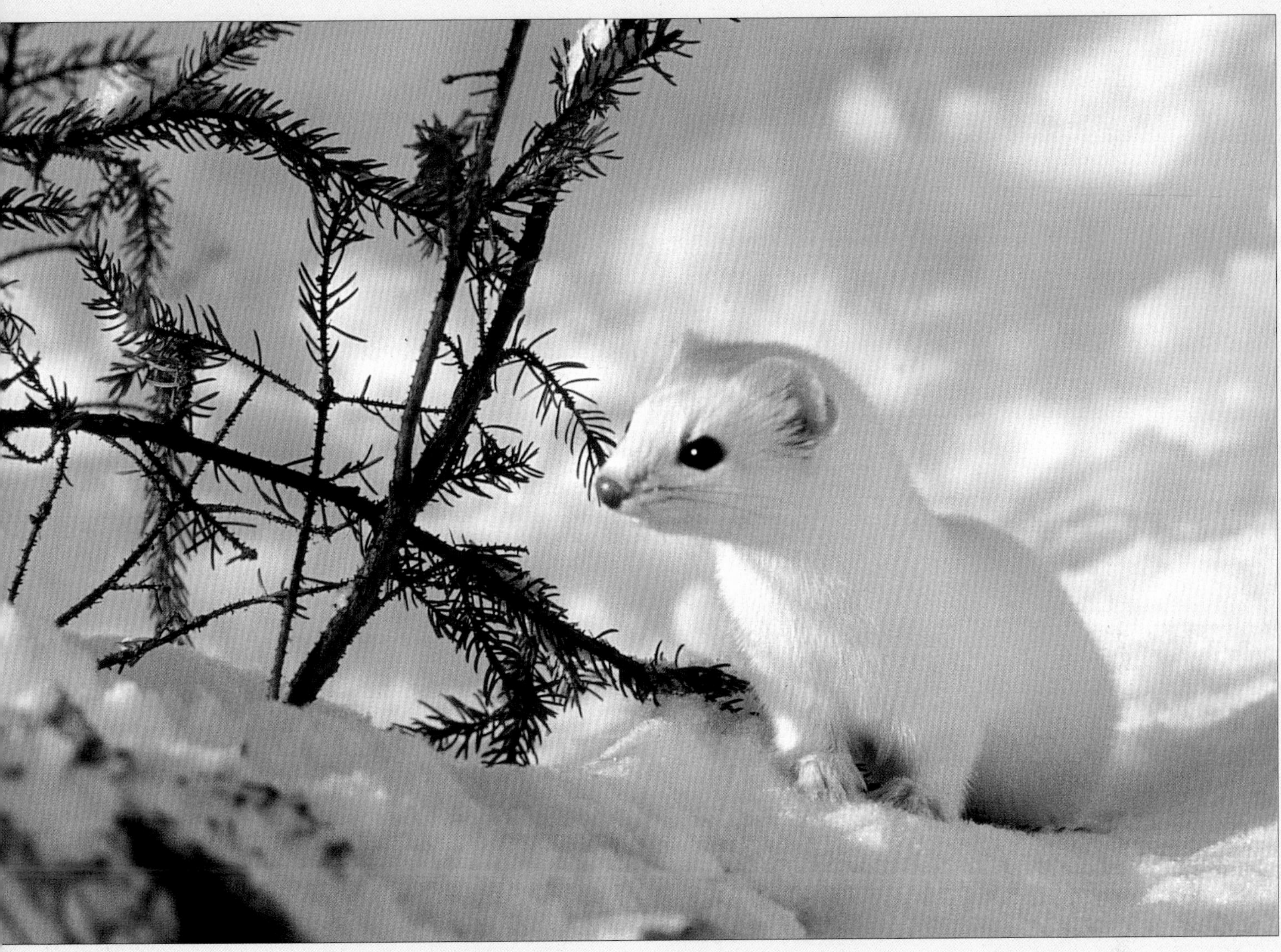

Above *The ermine takes on its distinctive white winter coat by molting for around one month in the fall. Its winter pelage is thicker and longer to help it preserve body heat in the cold.*

European mink En
Mustela lutreola

Originally throughout W, C and E Europe (not Britain), but now virtually confined to a few, declining populations in E Europe and Spain. 7 subspecies.
SIZE: HBL (male) 28–43cm, (female) 32–40cm ; TL (male) 12–19cm, (female) 13–18cm; WT slightly less than American mink.
COAT: similar to that of American mink but with white patch on upper lip.
BREEDING: gestation 35–42 days (no delayed implantation).
LONGEVITY: up to 6 years or more (to 12 in captivity).

Marbled polecat
Vormela peregusna

Steppe and semi-desert zones from SE Europe (Romania) E to W China, Palestine and Baluchistan.
HBL (male and female) 33–35cm ; TL 12–22cm; WT about 700g.
COAT: black, marked with white or yellowish spots and stripes; face like European polecat (*M. putorius*).
BREEDING: gestation 56–63 days (but up to 11 months with delayed implantation); litter size typically 4–8.

Zorilla
Ictonyx striatus
Zorilla, African polecat, striped polecat

Semiarid regions throughout Africa S of Sahara.
HBL (male and female) 28–38cm; TL 20–30cm; WT 1.4kg.
COAT: black, strikingly marked with white; hair long and tail bushy.
BREEDING: gestation 36 days; litter 2–3. Young weigh c.15g at birth.
LONGEVITY: 13 years in captivity.

North African banded weasel
Poecilictis libyca

Semi-desert fringes of the Sahara, from Morocco and Egypt to N Nigeria and Sudan; closely related to *Ictonyx*, possibly same genus.
HBL (male and female) 22–28cm; TL 13–18cm; WT (male) 200–250g.
COAT: black, marked with variable pattern of bands and spots.
BREEDING: gestation unknown; litter 1–3.

African striped weasel
Poecilogale albinucha

Africa S of Sahara.
HBL (male and female) 25–35cm; TL 15–23cm; WT 230–350g.
COAT: black with 4 white and 3 black stripes down back; tail white.
BREEDING: gestation 32 days; litter 1–3.

Grison
Galictis vittata
Grison or Huron

C and S America, from Mexico to Brazil, up to 1,200m altitude.
HBL (male and female) 47–55cm; TL 16cm; WT 1.4–3.2kg.
FORM: face, legs, and underparts black; back and tail smoky-gray, with white stripe across forehead; feet partly webbed.
BREEDING: gestation unknown; litter probably 2–4.

Little grison
Galictis cuja

C and S America, at higher altitudes than *G. vittata*.
HBL (male and female) 40–45cm; TL 15–19cm; WT about 1kg.
COAT: as *G. vittata*, but back is yellowish-gray or brownish.

Patagonian weasel
Lyncodon patagonicus

Pampas of Argentina and Chile.
HBL (male and female) 30–35cm; TL 6–9cm; WT unknown.
FORM: top of head creamy-white; back grayish; underparts brown; only 28 teeth.

Martens

ONE MIGHT SUPPOSE THAT THE NORTH American porcupine's fearsomely sharp quills present an impregnable defense. However, one species of marten, the fisher, has succeeded where other predators have failed, and developed a unique method of killing these porcupines. Foraging both in trees and on the ground, the fisher, like its fellow martens, is a highly adapted and efficient hunter.

Left *A fisher surveys the world warily from a rock. Despite its name, fish form only a small part of this North American species' diet, which – typically for the marten family as a whole – consists chiefly of small mammals such as mice and squirrels, together with fruit, eggs, and insects.*

Martens were a distinct group within the weasel family by the Pliocene (7–2 million years ago). Skeletal characteristics of these ancestral martens show that the three present-day subgenera – *Martes* (pine martens), *Charronia* (yellow-throated martens), and *Pekania* (fisher) – were already distinguishable by that time. Evolution of the separate species within the first two subgenera began in the Pleistocene from 2 m.y.a.

Agile and Graceful Predators

FORM AND FUNCTION

Martens are medium-sized carnivores, only moderately elongated in shape, with wedge-shaped faces and rounded ears that are larger than in some mustelids. While the bushy tail serves as a balancing-rod, the large paws with haired soles and semiretractile claws are also great assets to these semiarboreal animals. Martens leap from branch to branch effortlessly, and are among the most agile and graceful members of the weasel family. Pathways, in trees or on the ground, may be marked with scent from the anal glands as well as with urine.

The Stone, Beech, or House marten inhabits coniferous and deciduous woodlands from Europe to Central Asia and is often found near human habitations. Its large white throat patch extends onto its forelegs and underside. It is very similar to those species restricted to northern coniferous forests – the Pine marten, sable, Japanese marten, and North American marten – which, however, have smaller bibs. Differences between these four (which some authorities prefer to classify as a single species) are graded from Europe eastward to North America: for example, the Pine marten is the largest and the North American marten the smallest. Two species from southern Asia, the Yellow-throated marten and Nilgiri marten, have striking yellow bibs and are again sometimes considered one species. The fisher, from coniferous and mixed woodlands of North America, is the largest species and lacks the throat patch of other martens.

Fishers are typical martens in most respects. They are not named for their skill at catching fish or "fishing" bait out of traps, as is commonly supposed. Rather, the name is thought to derive from old English ("fiche"), Dutch, and French words for the European polecat and its pelt. Their special technique for killing porcupines involves considerable guile and patience. The arrangement of quills on a porcupine protects it from an attack to the back of the neck, where most carnivores attack, but its face is not protected. Fishers stand low to the ground and can thus direct an attack to a porcupine's face, yet are big enough to inflict damaging wounds. A fisher circles the porcupine on the ground, taking advantage of any chance to bite its face. The porcupine attempts to keep its back and tail toward the fisher and to seek protection for its face against a log or tree. If the fisher delivers enough solid bites to the porcupine's face, the porcupine suffers shock or is unable to protect itself. The fisher then overturns the porcupine and begins feeding on its unquilled belly. Killing a porcupine is long, hard work and a successful kill may take over half an hour. Depending on how many scavengers share the kill, the fisher may have enough food for over two weeks. Where porcupines are common, they may make up a quarter of a fisher's diet.

The tayra (*Eira barbara*) is regarded as a marten-like mustelid, but is considerably larger than most true martens, and exhibits different social behavior. It lives in forests from Mexico to Argentina and on the island of Trinidad. Its diet includes

Left *At full stretch, a Beech marten leaps across a pool of water. Beech martens primarily feed in trees, but may also hunt rats and mice around farms, even sometimes denning in attics. They also inhabit rocky areas, sometimes nesting in crevices or cairns – hence their alternative name of Stone marten.*

birds, small mammals, and fruits – it can cause substantial damage to banana crops. Though broadly similar in size and form to fishers, there are some important physiological differences. Tayras are less elongate in shape, have longer legs, and are less sexually dimorphic in body size than the martens. Finally, the tayra has a metabolic rate that is lower than might be expected for a mammal of its size, while the fisher is known to have a slightly elevated metabolic rate.

Hunters that are Hunted

SOCIAL BEHAVIOR

Martens are generally solitary. They are polygamous, and mating normally occurs in late summer (early spring in the fisher). During early spring, litters of 1–5 sparsely furred, blind, and deaf kits are born. The kits are weaned around 2 months of age. They are able to kill prey by 3–4 months, shortly before they leave their mothers. While martens are intolerant of other members of their species and are territorial toward members of their own sex, tayras are fairly tolerant of conspecifics and are often found in pairs, as well as larger, probably family, groups.

Because several of the martens, especially the sable and fisher, have been valued for their fur, hunting and trapping pressure on marten populations has sometimes been very high. This pressure, in combination with the destruction of the conifer and conifer–hardwood forests preferred by these species, has led to a decline in some populations. At present, however, only the Nilgiri marten of southern India is considered vulnerable. RAP

Above *A Pine marten stands guard over a freshly-killed Red squirrel. Martens are opportunistic hunters that dispatch prey with a bite to the back of the neck.*

FACTFILE

MARTENS

Order: Carnivora

Family: Mustelidae

Subfamily: Mustelinae

Martens: 8 species of genus *Martes*

Tayra: sole species of genus *Eira*

DISTRIBUTION: Asia, N America, Europe (genus *Martes*), C and S America (genus *Eira*).

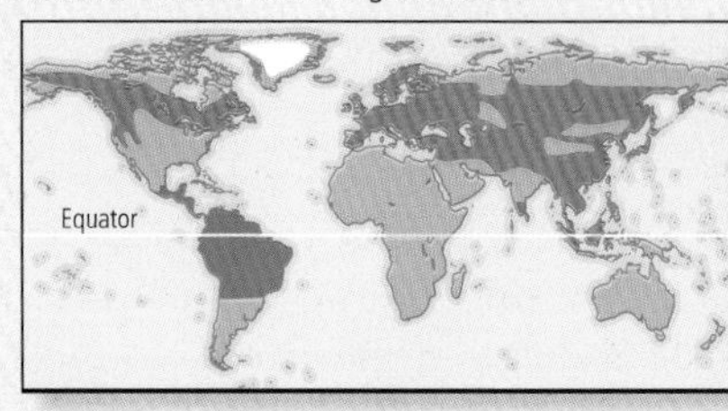

Martens   Tayra

HABITAT Coniferous and some deciduous forests, plus tropical mountain zones; Stone marten in urban areas.

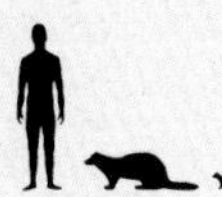

SIZE Head–body length 30–75cm (12–45in); tail length 12–45cm (4.7–18in); weight 0.5–6kg (1–13lb). Males are 30–100 percent heavier than females in all Martes species.

COAT Generally soft, thick, brown, with feet and tail darker, sometimes black, and often a pale throat patch or bib; tail bushy; soles of feet furred.

DIET Small mammals, birds, fish, insects, and fruit.

BREEDING Gestation 8–9 months (including 6–7 month delay in egg implantation) in most N species.

LONGEVITY To 10–15 years in most species.

CONSERVATION STATUS Nilgiri marten is Vulnerable; others not threatened.

For data on Tayra, see table below

Pine marten
Martes martes

C and N Europe, W Asia.
HBL 40–55cm (16–22in); TL 20–28cm (8–11in); WT 0.9–2kg (2–4lb).
COAT: chestnut-brown to dark brown, bib creamy-white.

Japanese marten
Martes melampus

Japan, Korea.
HBL 30–45cm (12–18in); TL 17–23cm (7–9in); WT 0.5–1.5kg (1–3lb).
COAT: yellow-brown to dark brown, bib white/cream.

Yellow-throated marten
Martes flavigula

SE Asia to Korea, Java, Sumatra, Borneo.
HBL 48–70cm (19–28in); TL 35–45cm (14–18in); WT 1–5kg (2–11lb).
COAT: dark brown to yellow-brown, bib yellow to orange, legs and tail dark brown to black.
BREEDING: gestation variable 5–6 months with 3–4 months' delayed implantation.

American marten
Martes americana

Northern N America to Sierra Nevada and Rockies In Colorado and California.
HBL 30–45cm (12–18in); TL 16–24cm (6–9in); WT 0.5–1.5kg (1–3lb).
COAT: golden-brown to dark brown, bib cream to orange.

Sable
Martes zibellina

N Asia, N Japanese islands.
HBL 35–55cm (14–22in); TL 12–19cm (5–7in); WT 0.5–2kg (1–4lb).
COAT: dark brown, yellowish bib not always clearly delineated; tail short.
BREEDING: egg implantation and gestation 1 month longer than above species.

Stone marten
Martes foina
Stone, Beech or House marten

S and C Europe to Denmark and C Asia.
HBL 43–55cm (17–22in); TL 22–30cm (9–12in); WT 0.5–2kg (1–4lb).
COAT: chocolate-brown, underfur lighter than *M.zibellina*. White bib often in 2 parts.

Fisher
Martes pennanti
Fisher, Pekan or Virginian polecat

Northern N America to California (Sierra Nevada) and W Virginia (Appalachians).
HBL 47–75cm (19–30in); TL 30–42cm (12–17in); WT 2–5kg (4–11lb).
COAT: medium to dark brown; gold to silver hoariness on head and shoulders; legs and tail black; variable cream chest patch.
BREEDING: gestation 11–12 months (implantation delayed 9–10 months).

Nilgiri marten Vu
Martes gwatkinsi
Nilgiri or Yellow-throated marten

Nilgiri mountains of S India. Smaller than *M. flavigula*, but coat similar.

Tayra
Eira barbara

C and S America, Trinidad.
HBL 90–115cm (35–45in); TL 35–45cm (14–18in); WT 4–6kg (9–13lb).
FORM: dark brown to black; yellow to white throat patch. Soles of the feet are not furred. Strong, nonretractile claws.
DIET: Vegetable matter, especially fruit, more important in diet of Tayra than in that of martens.
BREEDING: gestation 63–67 days (no delayed implantation); litters average 3 kits.

Abbreviations HBL = head–body length TL = tail length SH = shoulder height WT = weight  Vulnerable

Wolverine

THE WOLVERINE IS ONE OF THE LEAST-KNOWN large carnivores of northern Eurasia and America. Long ago, Native American mythology portrayed the animal as a trickster hero and a link to the spirit world. In the 18th century the famous Swedish taxonomist Carolus Linnaeus was uncertain whether to classify it as a weasel or a dog. Although we now know that wolverines are members of the weasel family, they remain mysterious in many ways.

Wolverines are the largest mustelids. Their powerful build means that they are able to tackle prey many times their own size. They have even been known to drive bears and cougars from their kills and drag moose carcasses for several kilometers to a hidden cache or den.

Tough Predators

FORM AND FUNCTION

Wolverines' heads are broad and rounded, with small eyes and short ears. Their pelage ranges from dark brown to black, and most have a light silver facial mask and a buff stripe running down from their shoulders and crossing their rump just above the tail.

Communication takes place through a variety of short-range vocalizations, and via scentmarking techniques that include urination, defecation, and abdominal rubbing. Wolverines have well-developed anal musk glands and, when traveling, will periodically stop to leave scent on a tussock or willow shrub. During the breeding period, females frequently mark in the den area before leaving to search for food.

The wolverine's Holarctic distribution spans Alpine, tundra, and northern taiga habitats – its thick, glossy fur and mode of living are well-suited to such harsh conditions. Large paws and a plantigrade gait facilitate movement through deep snow, while the paws also enable them to dig tunnels through the snow to food caches; they can stay out of sight in these for several days.

In winter, wolverines either kill or scavenge reindeer and caribou, which they locate with their extremely keen sense of smell. Good use is made of carcasses that cannot be eaten in one sitting – they are dismembered, and the pieces hidden in widely-dispersed caches in soil, snowfields, crevices, streams, or marshes. The prey may be exhumed up to six months later, by which time a female may have dependent cubs. In summer, wolverines also feed on birds, smaller mammals, plants, and the remains of reindeer calves or other prey killed by lynx, wolves, and bears.

Above *The wolverine's strong teeth and jaw muscles give it an extremely powerful bite, allowing it to gnaw frozen meat and crush bones as thick as the femur of a moose to get at the marrow.*

Lone Rangers of the North

SOCIAL BEHAVIOR

Wolverines mate between April and August, but the semi-developed eggs are not implanted in the womb to develop until the following winter. This delay ensures that females give birth at an optimal time for rearing cubs, typically in late February or early March. Litter size averages two or three cubs, and the 80–90g (3oz) young are born with a complete coat of white fur. Weaned at 9–10 weeks, they begin to travel with their mothers by late April and grow rapidly, reaching adult size by early winter and finally becoming independent at around 12 months. Female wolverines attain sexual maturity at about 15 months, but few 2-year-old females produce litters. Males have to wait a little longer; although they are potentially sexually mature at 14 months of age, in practice they do not mate until 3–4 years of age. Some females have been reported to have litters at 10 years of age, but generally reproduction ceases at 8. Males become reproductively senescent at 5–7.

Females give birth in a den, usually located among rocks in a tunnel dug through the overlying snow. Den sites tend to be used again and again, although the specific structure may change from year to year. Females may move den site between parturition and cub weaning. Snowmelt, human disturbance, parasites in the den, or threat from predators may all occasion a relocation.

The home ranges of wolverines cover an enormous area. Females with cubs have the smallest, varying from 40 to 100sq km (15–40sq mi).

lales' ranges are usually two to three times larger, nd can spread over as much as 1,500sq km 500sq mi). Subadults and very old non-breeders nay roam over several thousand square kilometers n their search for a place to settle. Range use eems to vary with season; generally, wolverines witch to lower altitudes in winter, probably in rder to take advantage of the greater availability f small prey and carrion in the low-lying areas.

Retreat and Decline

CONSERVATION AND ENVIRONMENT

listorical records indicate that wolverine distribu-on, both in America and Eurasia, has contracted orthward (and up to ever higher altitudes) as result of persecution, deforestation, human ncroachment, and urbanization. At one time ney were found as far south as southern Norway, weden, Finland, the Baltic States, and northeast-rn Poland, along with much of Russia. In North merica, there were once wolverines in Washing-on State, Oregon, Colorado, and California, as vell as in the Canadian provinces of Quebec and Ontario. They are now no longer present in any of nese areas.

Wolverines are hunted and trapped for sport nd for fur in some states of North America and n Russia. In North America they are rarely associ-ted with depredation of livestock, largely because heir range does not overlap much with that of domestic sheep (and where it does, the sheep are usually herded). Yet the killing of livestock is one of the main reasons advanced for their control, and historical decline, in Fennoscandia, where compensation has been claimed for more than 12,000 sheep in Norway and 18,000 domestic reindeer in Norway, Sweden, and Finland.

Over most of their distribution, wolverines occur in association with semi-domestic and wild reindeer and caribou, suggesting that these form an important, if not crucial, part of their diet. As the deer are highly sensitive to human disturbance, increasing problems for the wolverine seem likely in the future.

Although the various countries in which wolverines now live each have their own classifica-tions of the animal's conservation status, it may broadly be considered as vulnerable over most of its range, and endangered where it exists in small and partly isolated populations. The overall trend is one of retreat and declining populations, with increases in only a few areas of Fennoscandia and North America. AL

Below Somewhat bearlike in appearance, the wolverine is far smaller than any ursid. Yet, for all its compact size, it is an adept hunter that can dispatch substantial prey like reindeer and caribou. Because its large feet act like snowshoes, spreading its weight, it can run over soft, deep snow that will slow down its heavy, hoofed quarry. Its supposedly ravenous appetite led early writers to name it the "glutton."

FACTFILE

WOLVERINE

Gulo gulo

Order: Carnivora

Family: Mustelidae

Subfamily: Mustelinae

Sole species of the genus

DISTRIBUTION Circumpolar, in N America and Eurasia.

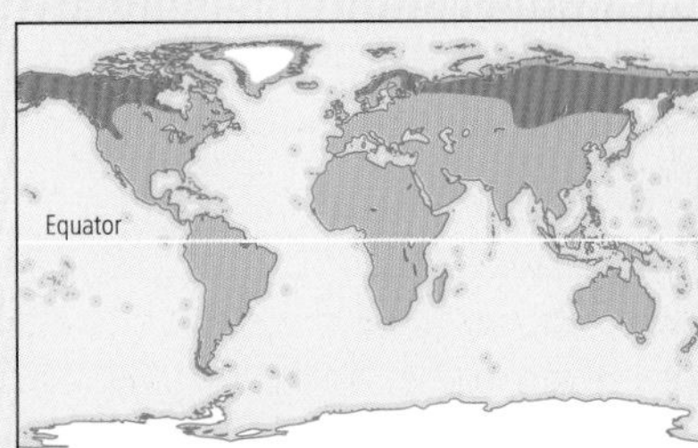

HABITAT Arctic and subarctic tundra and taiga.

SIZE Head–body length up to 95cm (37in); tail length 20cm (8in); weight up to 25kg (55lb): males average 15kg (33lb), females 10kg (22lb).

DIET Mainly reindeer or caribou; in summer, also birds, small mammals and plants.

COAT Long, dark brown to black; lighter band along flanks to upper side of bushy tail.

BREEDING Total gestation about 9 months; delayed implantation; development from implantation to delivery about 30–50 days. Litter size: 1–5.

LONGEVITY To 17 years.

CONSERVATION STATUS Vulnerable over most of its range; isolated populations endangered.

Otters

OTTERS ARE THE ONLY TRULY AMPHIBIOUS *members of the weasel family. All are similar in general appearance, but different species show striking behavioral variations. The majority forage mostly in water, but spend much of their time resting on land; the Sea otter, however, rarely comes ashore.*

The densely-packed underfur – around 70,000 hairs per sq cm (450,000 to the sq in) – and long guard hairs of otters trap a layer of insulating air while the animals are underwater. Because marine otters lack the insulating fat layers of seals and sealions, they need to have regular access to fresh water in order to wash their coats and maintain its insulating quality.

Life in the Fast Lane

FORM AND FUNCTION

Otter bodies are elongated, sinuous, and lithe, built for vigorous swimming. In most species the limbs are short and the paws are webbed. The tail is fully haired, thick at the base and tapering to a point, and in some species horizontally flattened. There are numerous stiff whiskers (vibrissae) around the nose and snout, and in tufts on the elbows. These tactile hairs are very sensitive and are used in locating prey. The ears are small and round and, like the nostrils, are closed under water. Most otters have claws; clawless species of the *Aonyx* and *Amblonyx* genera rely solely on their acute sense of touch, especially when searching under stones and debris for crabs, which they retrieve with their long fingers.

Otters eat mainly fish, but most species also catch frogs, crayfish, and crabs, as well as some birds and small mammals. The prey are often sluggish, bottom-living species like eels, though faster fish are also sometimes taken when not active.

Otters have an unusually high metabolic rate, probably in order to meet the special demands of their habitat. Water rapidly conducts heat away from the body, which creates high energy demands on animals active within it. Individual Eurasian otters need the equivalent of up to 15 percent of their body weight per day in fish, depending on the temperature of the water. Sea otters feed even more voraciously, although quantities are difficult to estimate. To catch high-calorie prey, the animals employ a high-risk foraging strategy that expends a lot of energy. This makes them vulnerable to changes in prey availability; for instance if, in water of 10°C, an otter cannot catch at least 100g (3.5oz) of fish per hour, it will not survive. Otters (other than sea otters, which almost never leave the water) spend three to five hours per day fishing (though a female with cubs will spend up to eight). Most species fish and hunt on their own, but some, like the Giant, Smooth-coated, Short-clawed, North American, and Spot-necked otters, will hunt in groups.

Denizens of the World

DISTRIBUTION PATTERNS

The Eurasian otter still has the largest geographical range, formerly even larger; before its extinction in Japan in the 1970s, it spanned from there to Ireland and from Siberia to Sri Lanka. It does not overlap with the similar Hairy-nosed otter, but may be found in the same rivers as Smooth-coated and Short-clawed otters. Similarly, in the New World, *Lontra* species generally do not overlap, but the North American river otter occurs in the same habitats as the Sea otter, and the Neotropical river otter shares rivers with the Giant otter. In Africa the two clawless otters do not overlap, but they are sympatric with the Spot-necked otter.

One New World species, the Marine otter, inhabits the rough waters off the western coast of South America from Peru to Cape Horn. It is a fish- and crab-eater, with unusually coarse fur. This species and the Sea otter of the northern Pacific live exclusively in the sea (while several others use the sea as well as fresh water).

The Hairy-nosed otter has a nose entirely covered in fur, like the Giant otter of Brazil, and is a spot-necked species. It lives in streams and lakes in Southeast Asia, but little is known of its ecology and status and it appears to be rare.

FACTFILE

OTTERS

Order: Carnivora

Family: Mustelidae

Subfamily: Lutrinae

13 species in 7 genera: North American river otter (*Lontra canadensis*), Marine otter (*L. felina*); Southern river otter (*L. provocax*), Neotropical river otter (*L. longicaudis*), European river otter (*Lutra lutra*), Hairy-nosed otter (*L. sumatrana*), Spot-necked otter (*L. maculicollis*), Cape clawless otter (*Aonyx capensis*), Congo clawless otter (*A. congicus*), Short-clawed otter (*Amblonyx cinereus*); Smooth-coated otter (*Lutrogale perspicillata*), Giant otter (*Pteronura brasiliensis*), Sea otter (*Enhydra lutris*).

DISTRIBUTION Widespread in subpolar regions excluding Australasia and Madagascar.

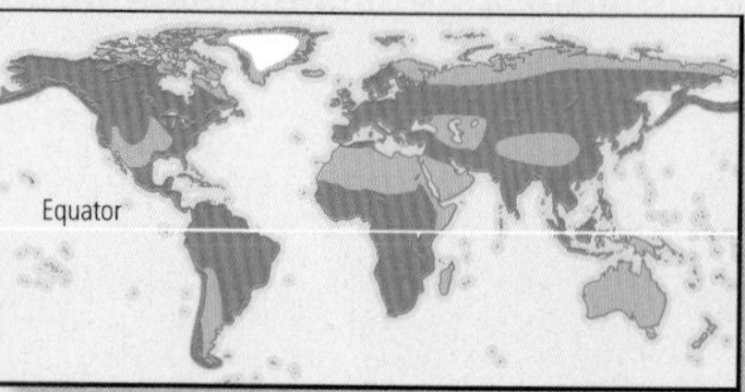

HABITAT Aquatic (including marine), and terrestrial.

SIZE **Head–body length** ranges from 41–64cm (16–25in) in the Oriental short-clawed otter to 96–123cm (38–48in) in the Giant otter; **tail length** from 25–35cm (10–14in) to 45–65cm (18–26in); **weight** from about 5kg (11lb) to up to 30kg (66lb). The shorter Sea otter may attain 45kg (88lb).

COAT Brown; darker above, chest, throat, and underside usually lighter.

DIET Primarily fish, though most species also eat frogs, crayfish and crabs, and may also take birds and small mammals.

BREEDING Gestation mostly 60–70 days, but up to 12 months with delayed implantation in some species. Litter size in most species from 1–4.

LONGEVITY Up to 20 years.

See species table ▷

Above left *A group of Short-clawed otters; this small Asian species lives in loosely organized family units comprising around 12 individuals. Although Short-clawed otters in the wild feed predominantly on crustaceans, frogs, and mollusks, and take relatively few fish, captive animals have been trained to catch fish by fishermen in Southeast Asia.*

Right *This European river otter clearly displays the short, dense fur that is essential in keeping the animal well-insulated in its watery element. People came to prize the particularly fine coat of this and other Lutra species, and hunted them for many centuries with specially bred otter hounds.*

The Spot-necked otter inhabits streams and lakes of sub-Saharan Africa. In East African lakes it is diurnal, traveling in groups of up to 20 and eating mostly small cichlid fish. Farther south it is nocturnal, living in streams, but never in the sea.

The Indian smooth-coated otter is larger and more heavily built than the Eurasian otter, and it displaces this species when they meet. Like the Giant otter, it has a flattened tail and thickly webbed paws, which are nonetheless remarkably agile in manipulating and retrieving small objects. The shortened face and domed skull house broad molars, and it feeds mostly on rather large fish, often caught cooperatively. In Sumatra it also occurs along the coast, in mangrove swamps.

The genus *Aonyx* contains two large African species, the Cape and Congo clawless otters. Their forepaws are not webbed, but have strong fingers that they manipulate with monkeylike dexterity – indeed, the thumb can be opposed when picking up or holding down objects. The hindfeet have a small web, and the middle toes have claws used for grooming. The fingers are used to probe mud and crevices. The broad cheek teeth are well adapted to grinding the tough carapaces of crustacea.

The Short-clawed otter is the smallest species. It lives in streams and rice paddies of south Asia. Its forefeet are only partially webbed, and have stubby, agile fingers tipped with tiny, vestigial claws that grow like short pegs on the top of bulbous fingertips. Like their clawless relatives, these diminutive otters use their sensitive forepaws constantly to search for prey by touch alone.

The Giant otter, which can measure over 1.8m (5.9ft) overall and weigh up to 30kg (66lb), is probably the rarest otter. Its annual life cycle is closely linked to the rise and fall of the water level during the rainy season from April to September. Giant otters have disappeared over much of their former range as a result of pelt hunting. Almost 20,000 skins were exported from Brazil alone in the 1960s; the trade declined significantly when bans came into effect in the early 1970s. Poaching is still widespread, but habitat degradation now presents the greatest threat.

The Sea otter of the north Pacific is unlike other members of the subfamily. Not so slender and with a relatively short tail, it can weigh even more than the Giant. Its large, rounded molars are perfectly adapted to crushing sea urchins, abalones, and mussels. Following near-extinction, Sea otters are now thought to number 150,000, most inhabiting coasts from Prince William Sound in Alaska to the Kurile Islands in Russia, though a southern subspecies of some 2,000 lives off California. Genetic evidence suggests that, despite suffering severe population bottlenecks due to hunting, major loss of genetic variability was avoided.

Although the Eurasian, Smooth-coated, and Short-clawed otters may be sympatric in Asia, the are adapted to different diets, so competition between them is limited. Where they occur together, the Smooth-coated takes large fish, the Eurasian small fish and frogs, and the Short-clawed mostly freshwater crabs. The Eurasian otter tends to avoid the larger Smooth-coated one

In Alaska the North American river otter and the Sea otter occupy the same coasts. They divide the habitat, Sea otters diving to depths of over 30m (100ft) – the record dive is 97m (318ft) – to retrieve molluscs and medium-sized fish, while river otters forage for small, bottom-dwelling fish on the nearshore.

From Sociable to Solitary

SOCIAL BEHAVIOR

The social behavior of otters ranges from solitary to group-living. Sea otters may occur in "rafts" of several hundreds (this may be an anti-predator protection against sharks and killer whales). In contrast, the Eurasian otter is rather solitary; females and cubs share ranges with other females within which each family defends an exclusive core area. Male otters have much larger ranges that overlap with those of several females. Home (group) ranges of these otters are very large; along sea coasts female ranges are 5–14km (3–8.7mi)

Below *The Giant otter preys on slow-moving fish on stream- c lake-beds. Like other otters, it eat its catch immediately, either at th water's surface or, when dealing with unwieldy prey, on land.*

◐ ***Left** A variety of otter characteristics: **1** the Short-clawed otter is hand-oriented and always reaches out with its forelimbs to take food; **2** the Spot-necked otter, in contrast, is mouth-oriented, stretching for food with its neck and body. The presence or absence of webbing and claws distinguishes many otter species. Shown here are the forepaws of: **a** the Short-clawed otter; **b** the Cape clawless otter; **c** the Giant otter; **d** the Indian smooth-coated otter; **e** the Spot-necked otter; and **f** the North American river otter. **3** Smooth-coated otter using its heavily webbed, but highly dextrous, forepaws to hold a shell up to its mouth. **4** Sea otter cracking a shell on a stone on its chest. **5** All otters are characterized by a broad, flat, streamlined skull; shown here is the head of a North American river otter.*

A Tool-using Carnivore

In order to survive, the Sea otter needs to eat fully 20–25 percent of its body weight each day, and has developed some unique skills to ensure a regular and plentiful supply of food.

Even compared to other otters, the Sea otter shows remarkable dexterity. Most food items are collected by picking them off the bottom or from kelp stalks, but when digging for clams, the otter kicks with its hind flippers to stay close to the bottom while digging rapidly with its forepaws in a circular motion and rooting with its head. It will remain submerged for 30–60 seconds and return to the same hole on three or more successive dives, to enlarge the hole laterally with each dive and retrieve any clams it finds.

The Sea otter is the only mammal apart from primates reported to use a tool while foraging. To dislodge abalone they grasp a stone between their mitten-like forepaws and bang it against the edge of the abalone shell. It may require three or more dives to dislodge an abalone; the same stone may be used through 20 or more dives.

Food items are almost always brought to the surface for consumption, from depths of up to 40m (130ft). The Sea otter may then place a stone on its chest and use it as an anvil on which to open mussels, clams, and other shell-encased prey. The stone is carried to the surface in a flap of skin in the armpit and the food item in the forepaws. The stones are usually flat and about 18cm (7in) in diameter. When pounding, the arms are raised to about 90° to the body and the mollusc brought down forcefully, so that the hard shell strikes the stone. An uninterrupted series of 2–22 or more blows at about 2 per second is usually enough to crack the shell. It is then bitten and the contents extracted with the lower incisors, which project forward. The otter may roll over in the water between bites to jettison debris and to keep its fur clean. In sandy or muddy areas, where stones are not available, the otter uses one clam or mussel as a hammer and another as the anvil.

Wild Alaskan Sea otters, unlike captive individuals and members of the Californian population, rarely use the anvil technique, probably because they feed on prey that they can crush in their teeth, such as crab, snails, and fish. Sea otters also prey on sea urchins; it is the dye in the urchin's shell that causes the purple color of some Sea otter skeletons. TRL

long, but inland the average female range stretches from 20km (12.4mi) up to a maximum of 40km (25mi). For males the average is 35km (21.7mi), with a maximum of 84km (52.2mi). Males and females associate only very briefly during the mating season, and males take no part in rearing the cubs. The habitat of males is generally different from that of females, with males spending most of their time in large rivers and along exposed sea coasts, females more along small streams and in sheltered bays.

The female Eurasian otter is a devoted mother. Remaining with her for around a year, the cubs learn to fish through lessons in which she releases live prey for them to re-catch. Even so, fish-catching is an art that the young may take 18 months or more to perfect. The extended period of juvenile dependency, the small numbers of cubs (from one to three, or rarely four) and the short average life span (4 years) of Eurasian otters results in a low rate of population growth and makes them more vulnerable to extinctions.

The North American river otter has similar-sized ranges to the Eurasian species, but in this species males may go around in "bachelor" groups of 12 or more individuals. Similarly, clawless otter males are found in groups in the Cape.

The Sea otter presents a further social variation: after mating, the sexes separate into coastal resting areas that average 40ha (100 acres) in males and 80ha (200 acres) in areas occupied by females with their young; these areas are distinct, but may be close to areas of about 30ha (75 acres) patrolled by single males. Female Sea otters from Prince William Sound are sexually mature between 2 and 5 years of age and go on bearing young until they are 15 years old. Young females tend to produce more female offspring, while older females produce equal numbers of each sex.

The playfulness of otters is often remarked upon, but this may be more characteristic of captive otters than their wild counterparts. However, wild otters will tunnel through a snow drift or sometimes slide down a mudbank, seemingly for entertainment; juveniles may rough-and-tumble ashore or chase each other in the water.

Otters are very vocal, with a large repertoire of calls. The most commonly-heard call of the Eurasian otter is a high-pitched whistle, used between mother and cubs. When fighting, otters caterwaul like cats; they make a huffing sound when alarmed, and during play-fighting their twittering noises carry a long way.

All species except one produce scent marks. Only the Sea otter (which rarely comes ashore) lacks the paired scent glands at the base of the tail which give otters their typical fishy, musky smell. The most common scentmark is the "spraint" (feces) left at conspicuous points along banks. Spraints are usually closely associated with feeding territories; most likely, they communicate to other otters that nearby foraging sites are being exploited and they therefore assist the otters in spacing out their activities. Otters do not mark their territorial boundaries with scent.

Social species also use communal latrines, where the urine and feces are thoroughly mixed and trampled into the substrate by stomping with the hind paws (and, in the case of the Giant otter, by kneading with the forepaws). Giant otter marking-sites are particularly striking: a pair or a group may clear and scentmark a site 8m (26ft) long by 7m (23ft) wide, during a bout of feverish activity that leaves the area denuded of vegetation and smelling of scent, feces and urine. The strong, dank odor can be detected near any site which has been visited within the previous several weeks. Urine may be dribbled during vegetation marking, when a Giant otter pulls down armfuls of leaves, rubbing its body over them. Giant otters trampling the vegetation cover their fur with the scent they are themselves spreading. Later, while resting, they rub themselves against the ground and each other until there is a composite scent characteristic of a pair or even a group. The significance of these scentmarking behaviors is still unknown.

Above *Spot-necked otters playfully chasing a terrapin in the clear waters of Lake Tanganyika, East Africa. Such play serves to reinforce social bonds, and helps young otters perfect their hunting techniques. It also generates a vital commodity: body heat.*

Below right *Sea otter mothers habitually carry their young pups on their chests, where they nurse and clean them meticulously. Grooming makes the pups' fur so buoyant that they cannot sink, even if they try to dive; pups only start to dive when they are two months old.*

Decline and Revival

CONSERVATION AND ENVIRONMENT

Particularly in Europe, otters were once widely persecuted for their fur, and to prevent depredation of fish stocks. The European river otter is now absent from many countries it once inhabited, including the Netherlands, Belgium, and Switzerland. Yet even after otter hunting was banned, further decline resulted from pollution of rivers and prey by agricultural chemicals – mainly organochlorines and polychlorinated biphenyls (PCBs). By the end of the 1950s in the British Isles, significant otter populations were to be found only in Scotland; however, in the 1980s and 1990s, with harmful pesticides having been outlawed, the otter began to return to some of its former habitats in England, sometimes as a result of reintroduction programs. Indeed, so successfully has the otter managed to reestablish itself in certain areas that it appears to have stemmed the rise of the American mink. Elsewhere, however, otter survival is still jeopardized by hunting; for example, this is the major cause of decline of South American and Asian species, such as the Southern river otter, the Marine otter, and the Hairy-nosed otter.

A dramatic example of recovery was that of the Sea otter population in Alaska, severely depleted by overhunting for pelts in the late 19th and early 20th centuries. A hunting ban agreed between the USA, Britain, Russia, and Japan in 1911 saw numbers steadily increase from a low-point of around 1,500 animals to some 150,000. Yet, as a stark reminder of how human agency can suddenly devastate even a thriving population, the grounding of the tanker *Exxon Valdez* in Prince William Sound in 1989 – triggering the world's worst oil spill – killed 5,000 Sea otters (plus an untold number of North American river otters). HK

Otter Species

orth American river otter
ontra canadensis
orth American river otter or Canadian otter

anada, USA including Alaska.
BL 66–107cm (26–42in); TL 32–46cm
13–18in).
ORM: coat very dark, dusky brown
bove, almost black to reddish-black or
ccasionally grayish-brown; lighter, silvery
r grayish on belly; throat and cheeks sil-
ery to yellowish-gray; feet well-webbed,
laws strong.
REEDING: March–April; gestation 10–12
nonths, with delayed implantation; litter
–5 (average 2–3).

arine otter En
ontra felina

oast and coastal islands of Chile and
eru; exterminated in Argentina.
BL 57–59cm (22–23in); TL 30–36cm
12–14in).
ORM: coat dark brown above, with
nderside a lighter fawn color; feet well-
vebbed, claws strong.
REEDING: no data on season, may be
ecember–January; gestation 60–70 days;
tters average 2.

outhern river otter En
ontra provocax

rgentina, Chile.
BL 57–70cm (22–28in); TL 35–46cm
14–18in).
ORM: coat dark to very dark, burnt
mber above; underside a lighter cinna-
non color; claws strongly webbed.
REEDING: unknown.

eotropical river otter
ontra longicaudis

and S America, from Mexico to Argenti-
a. Taxonomy in the process of revision;
ere considered to include *L. annectens*,
. platensis, *L. incarum*, *L. enudris*, *L. insu-
aris*. *L. repanda*, and *L. latidens*.
BL 50–79cm (20–31in); TL 38–57cm
15–22in).

FORM: coat cinnamon-brown to grayish-brown on back, sometimes with one or more lighter (buff or cream) spots or patches; claws strong, webbing present.
BREEDING: season varies with locality; gestation unknown; litter 1–4 (average 2–3).

European river otter Vu
Lutra lutra
European or Eurasian river otter

Most of Eurasia S of tundra line, N Africa.
HBL 57–70cm (22–28in); TL 35–40cm (14–16in).
FORM: coat brownish (lighter in Asian races); throat brown to cream; feet well-webbed, claws strong; tail cylindrical, thick at the base.
BREEDING: nonseasonal; gestation 61–65 days; litter 2–5 (average 2–3).

Hairy-nosed otter Vu
Lutra sumatrana

Sumatra, Java, Borneo, Thailand, Vietnam, Malaysia.
HBL 50–82cm (20–32in); TL 35–50cm (14–20in).
FORM: coat very dark brown above, underside very slightly paler; throat sometimes white; feet well-webbed, claws strong.
BREEDING: unknown.

Spot-necked otter Vu
Lutra maculicollis

Africa S of Sahara; absent only from desert areas, such as Namibia.
HBL 58–69cm (23–27in); TL 33–45cm (13–18in).
FORM: coat very dark or raw umber above; underside slightly lighter; throat and/or groin usually with irregular patches and spots of cream-white (buff-yellow in juveniles); webbing to near tips of toes, claws strong.
BREEDING: season variable; gestation unknown: litter 1–4.

Smooth-coated otter Vu
Lutrogale perspicillata

Discontinuous: Iraq (Tigris river), lower Indus, India, SE Asia, Burma, SW China, Malay Peninsula, Sumatra, Borneo.
HBL 66–79cm (26–31in); TL 41–51cm (16–20in).
FORM: coat raw umber to smoky gray-brown; throat, cheeks very light gray or almost white; feet quite large; webbing well-developed and thick; claws strong; tail tapered, but with slight flattening at sides.
BREEDING: October or December; gestation 63–65 days; litter 1–4 (average 2).

Short-clawed otter LR
Amblonyx cinereus

India, Sri Lanka, SE Asia, Indonesia, Borneo, Palawan Islands, S China.
HBL 41–64cm (16–25in); TL 25–35cm (10–14in).
FORM: coat burnt umber to dusky brown; throat noticeably lighter, whitish to grayish; feet narrow, webbed only to about last joint of toes; claws blunt, peglike, rudimentary.
BREEDING: nonseasonal; gestation 60–64 days; litter 1–6 (average 2–3).

Cape clawless otter
Aonyx capensis

Africa S of 15°N, from Senegal to Ethiopia S to the Cape; absent only from desert regions of Namibia.
HBL 73–95cm (29–37in); TL 41–67cm (16–26in).
FORM: coat dark brown above, sometimes frosted with white or grizzled hair tips; cheeks and neck lighter brown; forefeet virtually unwebbed, looking like pinkish hands; hindfeet similar, but with some webbing; no claws on fingers, inner or outer toes; 3 middle toes have short, peg-like claws.
BREEDING: season variable; litter 1–4.

Congo clawless otter
Aonyx congicus

African forest streams and rivers of the Congo basin.
FORM: in almost all aspects very similar to the Cape clawless otter, but cheeks and neck white, small differences in molars. Ecology and behaviour unknown.

Giant otter En
Pteronura brasiliensis

In all countries of S America except possibly Chile, Argentina, and Uruguay.
HBL 96–123cm (38–48in); TL 45–65cm (18–26in).
FORM: coat very dark burnt umber, groin never spotted; chin, throat, and chest usually marked with cream-colored patches, blotches or spots; muzzle, lips, and chin often whitish or spotted white; claws and webbing very well developed – webs reach to tips of toes and fingers; feet very large, fleshy; tail lance-shaped, widest at mid-point.
BREEDING: non-seasonal; gestation 65–70 days; litter 1–5 (average 2).

Sea otter En
Enhydra lutris

Kurile, Aleutian Islands, Alaskan coast and Gulf of Alaska; re-introduced into parts of former range along Pacific coast of N America (notably California) and Russia.
HBL 55–130cm (22–51in); TL 13–33cm (5–13in).
FORM: adult coats dark brown above with head straw-colored; juveniles uniform dark brown; cubs fawn-colored at birth, tail somewhat flattened, short and not markedly tapered; muzzle with thick whiskers; forefeet small with no obvious toes; hindfeet very large, flipperlike.
BREEDING: season variable, but may be only 4–5 months; gestation about 9 months, with delay in implantation; litter 1, rarely 2.

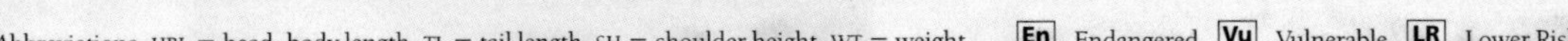
Abbreviations HBL = head–body length TL = tail length SH = shoulder height WT = weight En Endangered Vu Vulnerable LR Lower Risk

Badgers

Badgers are stockily-built mustelids that are widely distributed in the northern hemisphere. They originated in the forests of Asia during the Tertiary period (the genus Meles *emerging about 2 million years ago), but have diversified into a broad array of habitats, including open plains, semi-desert, and cultivated land. While the primitive ferret badgers resemble their arboreal ancestors, the other badgers are determinedly terrestrial, rooting and digging to find their prey on the ground.*

The badgers' name reflects their adaptation: the word is believed to be derived from the French word *bêcheur*, meaning "digger." The European, American, hog, and stink badgers are all well-designed for burrowing, being compact, short-legged, well-muscled, and possessed of long claws. Additionally, the front toes of the Indonesian stink badger are joined together back to the claw roots in what appears to be a further adaptation for digging. American badgers are such efficient diggers that, when threatened in the open, they are able to burrow their way to safety and disappear from view in the space of a minute. By contrast, the more primitive ferret badgers are considerably less heavily built, with elongated bodies rather like those of martens.

Designed for Digging

FORM AND FUNCTION

All of the badgers inhabit burrows, but the European badger digs the most complex dens (setts), which are passed on from generation to generation. Some setts are known to have persisted for hundreds of years. The largest sett to have been excavated was estimated to consist of 879m (0.54 miles) of tunnels, with 129 entrances. Constructing this sett – which would have been carried out by generations of badgers over many years – involved the removal of 62 tonnes (61 tons) of soil. Other badgers dig simpler burrows, though the American badger spends so much of its foraging time digging for rodents that it may well rival the European badger in its enthusiasm for excavation.

All of the badgers have striking coloration on the face, from the bold stripes of the European badger to the curious "mask" of the ferret badgers. The role that these markings play is not known for any species, but they may be some form of warning coloration. Badgers are known for their strength and ferocity when threatened – as well as their pungent scent – and their distinctive markings may serve to remind potential predators of past unpleasant encounters. Nevertheless, badgers are victims of predation – in Thailand, Hog badgers are an important prey of leopards, and the European badger was probably formerly prey to the wolves and bears that would also have inhabited woodland regions.

Out of Asia

DISTRIBUTION PATTERNS

Badgers evolved from an ancestral, marten-like stock in the forests of Asia, and the ferret, hog, and stink badgers still inhabit that region. European badgers are more generalist in their habitat requirements, and consequently more widely distributed. They are found across a huge swathe of territory that extends from Ireland to Japan and takes in dry Mediterranean islands, the boreal forests of Scandinavia and Russia, the semi-deserts of Israel and Jordan, and various city suburbs. They even occur in parts of Russia as far north as the Arctic Circle. European badgers reach their highest densities, however, in patchworks of deciduous woodland and pasture in the British Isles. The American badger, by contrast, is more restricted to open plains.

FACTFILE

BADGERS

Order: Carnivora

Family: Mustelidae

Subfamilies: Melinae, Mellivorinae, Taxidiinae

10 species in 6 genera

DISTRIBUTION Africa, Eurasia, N America

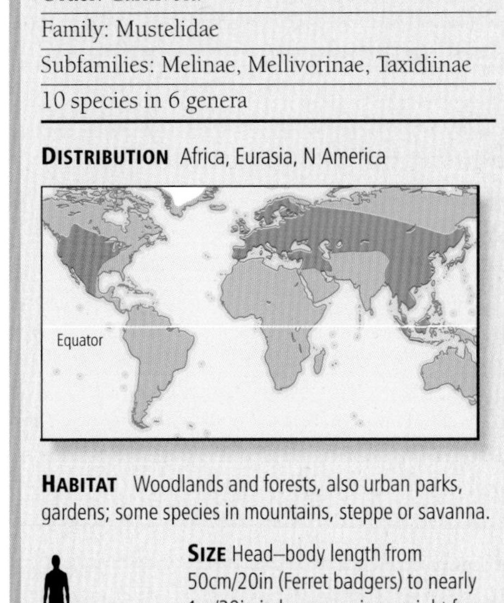

HABITAT Woodlands and forests, also urban parks, gardens; some species in mountains, steppe or savanna.

SIZE Head–body length from 50cm/20in (Ferret badgers) to nearly 1m/39in in large species. weight from 2kg (4.4lb) up to 12kg (26lb).

COAT Gray-black, dark or pale brown. Facial and/or dorsal stripes in all species.

DIET Insects, earthworms, fruit, and vegetables.

BREEDING Gestation 3.5–12 months, including period of delayed implantation.

LONGEVITY To 25 years in captivity (not known in wild).

See species table ▷

Left *An American badger emerging from its sett; can be identified by its reddish-grey upperparts and hite facial stripe. Powerful shoulders and long claws ake the American badger, which inhabits the North merican continent from Mexico to the southern aches of Canada, particularly proficient at digging.*

Above *Regular black-and-white facial markings characterize the European badger, here seen foraging in its woodland habitat. The black stripe extends from the nose, around the eye, to the ear on either side of the face. Badgers will often follow the same routes, creating a number of well-worn pathways.*

Adaptable Eaters

DIET

/ith the exception of the American badger, all ıdgers are generalist omnivores. They feed on a ıriety of insects and other invertebrates, as well ; small vertebrates, cereals, and tubers. In the :itish Isles, European badgers are so dependent pon earthworms that they have been referred as "worm specialists" and may eat several hun- red in a night; elsewhere their diets focus on ifferent prey, including rabbits in southern Spain ıd insects and olives in Italy. European badgers' ng claws and thick skin make them one of the w species able to kill and eat hedgehogs – as a sult hedgehogs are rare or absent where badgers e abundant, and avoid the scent of badgers hile foraging.

In contrast with other badger species, American ıdgers are specialist predators and their diet kes in burrowing rodents, including Prairie ogs, ground squirrels, and gophers. Where prey abundant – such as in the environs of Prairie dog "towns" – American badgers are sometimes found in foraging associations with coyotes, a partnership that appears to be genuinely beneficial to both species.

The American and European badgers both have a seasonal cycle of fat deposition which reflects the availability of food. In temperate regions, both vertebrate and invertebrate prey are often very scarce during the winter. Both badger species deposit fat during the autumn, and rely almost exclusively on stored fat to survive through the winter. In European badgers, this cycle is under hormonal control and persists even when badgers have free access to food at all times. In the very cold periods of midwinter, American badgers enter a state of torpor and may not emerge above ground for as long as two months. European badgers do not hibernate in the British Isles, but they do become extremely inactive and may spend several days, possibly even weeks, in their den; in the colder parts of northern Europe this period may be greatly extended.

Flexible Families

SOCIAL BEHAVIOR

Most badgers are fiercely solitary. Young American badgers, for example, disperse away from their mothers' range as soon as they reach independence, and males and females come together only to mate. Males do not defend territories; instead, they range widely in search of receptive females. Females' ranges also overlap to some extent. Both sexes move dens continually – one female used 50 dens within a home range of 7.5sq km (2.9 sq mi).

Solitary living seems to have been the habit of badgers' ancestors, but one species, the European badger, has evolved a tendency to live in groups. In undisturbed populations European badgers are territorial, marking the edges of their ranges with a variety of scent marks, and occasionally getting involved in serious fights on territory boundaries. Puzzlingly, on other occasions they seem tolerant of trespassing neighbors. Within a territory, however, most interactions are amicable, with mutual grooming and scentmarking. This may be a very basic example of the reciprocity that is thought to underpin mammalian societies: a grooming session between two badgers involves quickfire swapping. If one reneges on this tit-for-tat deal for even a few seconds, the cooperation breaks down. On cold days, two or three badgers will often sleep together in the same nest chamber.

European badger groups form because cubs remain in the territory where they are born – in the British Isles as many as 80 percent of young badgers never leave their natal group. This means that groups can grow large – up to 27 badgers may share a single territory and one or more setts.

The reason for group living among European badgers is unclear, but may be related to weather and food supply. Badgers only form groups in temperate areas with relatively high rainfall. Scarcity of food in cold or dry regions, such as northern Scandinavia or southern Spain, causes badgers to live alone or in pairs in large home ranges (4–5sq km/1.5–1.9sq mi). In contrast, in the lush pastures and woodlands of southern England, territories may be as small as 20ha (50 acres) and still support a group of four or more adults.

Because of the way badger groups form, almost all members are close relatives. As a result, there may sometimes be no suitable unrelated mating partners within the group and, during the mating season, both males and females may visit neighboring territories to mate with nonrelatives. Some badgers move permanently to other groups. In several populations, only males move – usually to neighboring territories – while females spend their entire lives in the group where they are born. Elsewhere, both sexes may disperse to other groups. Females, in particular, may disperse as coalitions of two or three relatives, and there is some evidence to suggest that newcomers displace resident females from their territories.

In southern England up to four females may raise cubs within a group, though at higher latitudes breeding is restricted to a single female, perhaps because food is more limited. The mating system is intriguing, with as many as six males mating with one female during a single estrus, sometimes apparently with minimal animosity between males in the queue (a waiting male has been seen grooming the male engaged in mating!). Another, possibly related, surprise in badger groups is that it is extremely difficult to detect any stable social hierarchy among the adults.

Pregnancy begins in midwinter, when badgers are very inactive. This may explain why European badgers have an unusual reproductive cycle known as "delayed implantation": females can conceive at any time between February and October, but development of the fertilized eggs is suspended until midwinter, when the embryos implant in the walls of the uterus and gestation begins again. American badgers have a similar system, and this phenomenon is also thought to occur in Hog badgers. In ferret badgers, by contrast, breeding is less highly synchronized among females and, though only one litter of cubs seems to be born each year, there is little evidence of delayed implantation.

Because both European and American badgers undergo pregnancy during midwinter, when they are very inactive or torpid, mothers must sustain both themselves and their fetuses almost entirely on stored fat. Probably as a result, newborn badger cubs are much smaller, relative to their mothers' body size, than most other newborn mammals. A 9.5kg (21lb) American badger, for example, produces a litter of two cubs each weighing less than 100g (4oz), when she would be expected to produce a litter weighing about 1kg (2.2.lb) in total. This low-cost pregnancy must be paid for later: suckling badger cubs place a higher energy demand on their mothers than do other young mammals. This reproductive strategy of the temperate-zone badgers resembles that of bears, which also produce relatively small cubs after a winter pregnancy.

THE VORACIOUS HONEY BADGER

The Honey badger, or ratel, is famous for its foraging association with a small bird, the Honey guide. These birds are adept at finding the nests of wild bees, but cannot break into them. So, they attract a ratel with a distinctive song, and wait while it breaks open the nest with its powerful claws. Once the badger has had its fill of the honey inside, the bird feasts on the larval bees and wax left behind.

A recent study in the southern Kalahari in South Africa revealed that the ratel has a diverse diet – 59 prey species, ranging from bee larvae to 3m-long pythons. Honey badgers dig out 85 percent of their prey, but will also scale tall Camelthorn acacias in search of raptor chicks. While gerbils, lizards, and geckos are the most common prey, larger prey, especially snakes, make up over half the biomass eaten.

This study also found that other creatures benefit from the ratel's strength and foraging abilities. Thus, Palechanting goshawks (*Micronisus gabar*) and Black-backed jackals (*Canis mesomelas*) pounce on rodents and reptiles flushed out by the badger. The slow badger is powerless to prevent these hangers-on, and gains no advantage from their company. An added twist is that the young of both goshawks and jackals are prey of the Honey badger. CB/KB

Above *Between two and five European badger cubs (usually three) are born at the end of winter, an stay below ground in the sett for their first 8 weeks. This early birth probably allows cubs to become independent at a time of high food availability, in the ear summer, and gives them as much time as possible to gain weight to survive the following winter.*

Right *The tailless Malayan stink badger or teledu and the long-tailed ferret badgers are all found in Southeast Asia, and overlap in Borneo and Java. The ferret badgers are the smallest species, and the only ones to climb trees.*

Subfamily Melinae

Eight species in four genera, from India, SE Asia, the Far East, and Europe.

Teledu
Mydaus javanensis
Teledu, Malayan or Indonesian stink badger

Mountains of Borneo, Sumatra, Java, and N Natuna Islands. Terrestrial, burrowing.
HBL 37–51cm (15–20in); TL 5–8cm (2–3in); WT 1.4–3.6kg (3–8lb).
FORM: coat dark brown or blackish, with white crown to head and either white stripe down back or row of white patches.

Palawan stink badger Vu
Mydaus marchei

Palawan and Busuanga NE of Borneo. Terrestrial; burrowing.
HBL 32–46cm (13–18in); TL 1.5–4.5cm (0.6–1.8in); WT 3kg (7lb).
FORM: coat dark brown to black above, brown below, yellowish cap and back streak, fading at shoulders; muzzle off-white; anal region hairless and pale-skinned.

Hog badger
Arctonyx collaris

Forest zones from Peking in N, throughout S China and Indochina to Thailand, and Sumatra. Terrestrial; burrowing.
HBL 55–70cm (22–28in); TL 12–17cm (5–7in); WT 7–14kg (15–31lb).
FORM: coat back yellow, gray, or blackish; ears and tail white; feet and belly black. Dark facial stripes through eyes.

Ferret badgers
Four species in the genus *Melogale*

Indian ferret badger (*M. personata*), India, Nepal, Burma; Oriental ferret badger (*M. orientalis*), Java, Bali, and SE Asia; Chinese ferret badger (*M. moschata*), China, Taiwan, Assam, Burma, and SE Asia; Everett's ferret badger (*M. everetti*), Borneo. All terrestrial; burrowing.
HBL 33–43cm (13–17in); TL 15–23cm (6–9in); WT 2kg (4lb).
COAT: upper parts pale to dark brown, with white or reddish dorsal stripe; belly paler; face with conspicuous black and white or yellowish pattern.

European badger
Meles meles
European or Eurasian badger

Woodland and steppe. N Europe up to S Scandinavia, European Russia up to Arctic Circle, S to Palestine, E to Iran, Tibet, and S China. Terrestrial; burrowing.
HBL 67–81cm (26–32in); TL 15–20cm (6–8in); WT 12kg/26lb (male), 10kg/22lb (female).
FORM: Coat gray-black above, black below. Head and ear tips white, black facial stripe from snout through eyes to behind ears. Male's neck broad and thick. Tail narrow, pointed, white or pale in male, broader, grayer in female.

Subfamily Mellivorinae

Single genus and species from Africa and S Asia.

Honey badger
Mellivora capensis

From open, dry savanna to dense forest in Africa, from Cape to Morocco in W and Ethiopia, Sudan, and Somalia in E; Arabia to Turkmenistan, Nepal, and India. Terrestrial, in burrows or among rocks.
HBL 60–70cm (24–28in); TL 20–30cm (8–12in); WT up to 12kg (26lb).
COAT: upper parts from head to tail white (extent of mantle variable, may be absent), sometimes with gray or brown tinge; sides, underparts, and limbs pure black. Young rusty brown above.

Subfamily Taxidiinae

Single genus and species from N America.

American badger
Taxidea taxus

From SW Canada and N central USA, S to Mexico. Terrestrial; burrowing.
HBL 42–64cm (17–25 in) in females, 52–72cm (20–28in) in males; TL 10–16cm (4–6in); WT 3.5–12kg (8–26lb). Females smaller than males.
COAT: upper parts grayish to reddish, underparts buff; feet dark brown to black. Central white facial stripe from nose at least to shoulders; black patches on face and cheeks. Chin, throat, and mid-ventral region whitish.

Abbreviations HBL = head–body length TL = tail length WT = weight. Vu Vulnerable

Honey badgers bear cubs throughout the year fter a gestation of six to eight weeks. A single cub born, and 53 percent die before reaching matu-ty. The cub is totally dependent on its mother or 14–16 months. At any one time, the females within the home range of a male are in different eproductive states and males rely on scentmark-ng to find receptive females. Females routinely nark by token urination in foraging holes and only nark at latrines when they are in estrus, to adver-se this fact to the males that constantly patrol trines and liberally mark each site with feces, rine, and secretions from their anal scent glands.

Honey badgers are thought to be induced ovu-tors; once in estrus, they need frequent copula-ons over a long period to ensure fertilization. Iales are not territorial, but will guard an strous female by sequestering her in a bur-ow and stopping her from leaving for up to iree days while mating occurs. Moreover, iere is a dominance hierarchy among iales; prominent scars or calluses on the acks of 62 percent of adult males are iought to result from repeated male onflict, and signal sexual experience.

A Mixed Picture

CONSERVATION AND ENVIRONMENT

he Palawan stink badger and Everett's ferret bad-er are the only badger species facing serious con-rvation threats at present, although the Javan rret badger and several other endemic badger ibspecies are also locally threatened. These for-st-dwelling species may suffer habitat loss from widespread deforestation. Other, less threatened, species also come into conflict with man. In particular, American badgers have declined markedly, apparently as a result of indiscriminate poisoning and destruction of their rodent prey.

As vectors of bovine tuberculosis, European badgers come into serious conflict with farmers, particularly in the British Isles. To combat the disease, badger culls have been conducted in Britain and Ireland, though their effectiveness is in doubt, since it is impossible to estimate the number of cattle infected specifically by badgers. What is clear is that badger culls inflame public opinion. Nevertheless, badgers receive substantial legal protection and their numbers have increased steadily in recent years, sometimes to very high densities – as many as 29 adults per sq km. Intriguingly, maximum growth occurs where they have long been protected, suggesting that something other than legislation is behind the increase: one possibility is climate change, with wet summers producing more earthworms and hence better cub survival. The fact that European badgers thrive in the highly modified habitats of the agricultural and suburban landscape is eloquent testimony to their adaptability. RBW/DWM/CB/KB

Skunks

THE PUNGENT AROMA OF SKUNK MUSK irritates even the least sensitive of human noses, but for the animals themselves it is the smell of safety. Skunks are one of the few groups of mammals that rely on chemicals to defend themselves.

Skunks' black-and-white fur patterns have a similar function to the yellow-and-black stripes on the abdomen of a wasp; they are a so-called "aposematic signal" designed to warn off predators. When threatened, skunks raise their tails, stamp their feet, hiss, or even perform bluff charges and handstands – all to give their assailants every opportunity to avoid a shower of foul-smelling musk. Predators that ignore the warnings and continue to harass the skunk experience the full potency of a mixture of sulfur, butane, and methane compounds.

Chemical Warfare on Legs

FORM AND FUNCTION

Skunks are omnipresent throughout North, Central, and South America, except in the far north of Canada. They are in their element where there are both fields and woods, although they will also frequent suburban and urban areas as long as there is food and shelter. Resting in burrows, culverts, or under buildings, urban skunks emerge at night to feed on garbage and dig up lawns and gardens. This, together with the fact that humans and their pets may receive a dose of the skunks' noxious spray, can make them unwelcome tenants.

Skunks normally hit targets accurately within 2m (6.5ft), but the smell may be detected by the human nose for up to 1km (0.5mi) downwind. The musk, which causes intense irritation and temporary blindness if it gets in the eyes, is stored

Right *An Eastern spotted skunk grubbing for foo[d]. No two animals of the genus* Spilogale *have the sam[e] pattern of spots and stripes.*

Below *Black and white markings are common to all skunks, although the exact pattern varies markedly between, and even within, species.* **1, 2** *Two forms o[f] the Striped skunk, the commonest species.* **3** *Hog-nosed skunk foraging with its long, naked snout.* **4** *Western spotted skunk; all spotted skunk species, which are smaller than most other skunk types, can augment their defensive strategy by adopting a characteristic "handstand" posture, balancing on their forepaws and advancing toward their would-be attacker.* **5** *White-backed form of the Hooded skunk, in the black-backed variant, the wide white stripes appear on the animal's sides.*

■ two glands on either side of the anus, flanked ■ muscles that force it out through two small nip■es. Enough musk is stored to allow 5–6 shots. ■though musk can be replenished within 48 ■ours, skunks rely on it for survival and so are ■areful not to waste it.

In build, the animals are intermediate between ■easels and badgers. They have long claws on the ■ont feet that they use to dig for food, or to exca■ate burrows in which to rest, sleep through the ■inter, and give birth and raise young.

All the family are efficient grubbers and ■ousers. Insects and rodents are their staple diet; ■ey especially relish underground grubs and lar■e. Frogs, salamanders, snakes, and bird eggs are ■so favored, as are carrion and human garbage. ■ost prey are located by sound or smell but rarely ■ sight, since skunks' distance vision is poor; ■etails more than 3m (10ft) away elude them, ■hich is why they are so vulnerable to road traffic. ■ northern latitudes, stocking up on body fat is ■nportant for hibernation, and in late summer and ■ll a skunk will double its spring weight in order ■ survive the winter and reproduce in the spring.

Solitary Ranges, Communal Dens

SOCIAL BEHAVIOR

■kunks are solitary for most of the year. In the ■orth, they den communally, with as many as 20 ■nimals – typically, a single adult male along with ■everal females – bedding down together for up to ■ months. Breeding occurs in early spring, after ■hich all animals become solitary again. Soon ■ter emergence in March, females start preparing ■aternity dens. Young Striped skunks are born in ■id-May, and remain dependent until late July. By August, the young have reached adult size, and disperse to fend for themselves. For most of the year females occupy home ranges of 2–4sq km (0.8–1.5sq mi), each overlapping greatly with those of other females. Male home ranges may reach over 20sq km (8sq mi), and also overlap with each other. Males do not care for, and may even kill, young skunks, so females defend maternity dens aggressively against intruding males.

The Risk from Rabies

CONSERVATION AND ENVIRONMENT

Skunks are important vectors of rabies throughout North America; a variant of the virus specific to the family occupies the largest geographic area of all North American rabies variants. The epidemiology of skunk rabies is poorly known, and the current lack of an efficient oral vaccine hampers management of the disease. Skunks excrete the virus in their saliva, and infect other animals by biting. Transmission among skunks is probably helped by their gregarious winter habits; periodic, localized outbreaks of rabies effectively limit their numbers. Their preference for farm buildings as denning sites increases the risk of transmission to domestic animals. Although animals infected with skunk rabies may die from the disease, infection and transmission rates to other species are much lower than to other skunks.

Perhaps thanks to the skunk's chemical deterrent, coyotes, domestic dogs, badgers, foxes, and Great horned owls rarely kill them. Humans, on the other hand, are thought to cause the deaths of half of all the skunks born each year – as a result of roadkills, shooting, or poisoning. Less than 10 percent of skunks live more than 3 years. SL/FM

FACTFILE

SKUNKS

Order: Carnivora

Family: Mephitidae

10 species in 3 genera

DISTRIBUTION
N, C, and S America

STRIPED SKUNK *Mephitis mephitis*
S Canada, USA, N Mexico. Commonest species, often found in suburban and urban areas. Dens in burrows or under buildings. HTL 68cm (27in); WT 1.5–6kg (3–13lb). Coat: black, with forking white stripes on back; white patch on head, white head stripe. Breeding: mates in February–March; true gestation 62–66 days; 3–9 young born in April–May. Longevity: most often less than 3 years in the wild, 8–10 years in captivity.

HOODED SKUNK *Mephitis macroura*
SW USA. Prefers rocky canyons and SW desert habitats. Rare, more secretive than *M. mephitis*. Coat: three color forms – black with fully white back; black with 2 thin white stripes on each upper side; or mixture of both. HBL 31cm (12in); WT 0.9kg (2lb). Breeding: mates in March–April; true gestation 63 days; 3–6 young born in May–June.

SPOTTED SKUNKS
Three species of the genus *Spilogale*
Western spotted skunk (*S. gracilis*), W USA to C Mexico; Eastern spotted skunk (*S. putorius*), SE and C USA to E Mexico; Pygmy spotted skunk (*S. pygmaea*), W and SW Mexico. All readily climb trees, den in crevices, burrows, or under buildings. HTL 40cm (16in); WT 0.5kg (1lb). Coat: black with 4–6 broken white stripes or spots; hair silkier than in other genera. Breeding: all bear 2–6 young. *S. putorius* and *S. pygmaea* mate in February–April; true gestation 50–65 days. *S. gracilis* mates in late summer; implantation is delayed until March–April; young are born in May.

HOG-NOSED SKUNKS
Five species of the genus *Conepatus*
Western hog-nosed skunk (*C. mesoleucus*), S USA, Nicaragua; Eastern hog-nosed skunk (*C. leuconotus*), E Texas, E Mexico; Amazonian skunk (*C. semistriatus*), S Mexico, N Peru, E Brazil; Andes skunk (*C. chinga*), Argentina, Bolivia, Chile, Paraguay, Peru; Patagonian skunk (*C. humboldtii*), S Chile, Argentina. Diverse habitats, but prefer rugged terrain; den in rock crevices and burrows. HTL 60cm (24in); WT 1.5–2kg (3–4.5lb). Coat: black, with large white band on back and white tail; distinguished by lack of white head stripe, and by bare, elongated snout. Breeding: mate in February; gestation 60 days; 2–4 young born in April–May.

Abbreviations HBL = head–body length WT = weight
HTL = head–tail length

DISTRIBUTION S Italy, France and Iberian peninsula; Africa, Madagascar, Middle East to India and Sri Lanka, much of C and S China; SE Asia to Sulawesi and the Philippines.

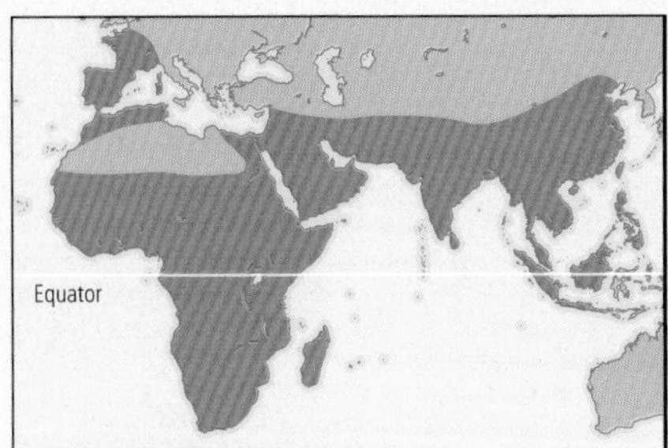

CIVETS AND GENETS
Family Viverridae p124–131

35 species in 20 genera, including:

Sulawesi palm civet (*Macrogalidia musschenbroekii*), **African palm civet** (*Nandinia binotata*), **Banded palm civet** (*Hemigalus derbyanus*), **Otter civet** (*Cynogale bennettii*), **Common genet** (*Genetta genetta*), **African linsang** (*Poiana richardsoni*), **falanouc** (*Eupleres goudotii*), **fanaloka** (*Fossa fossa*), and **fossa** (*Cryptoprocta ferox*).

MONGOOSES Family Herpestidae p132–139

35 species in 17 genera, including:

Ring-tailed mongoose (*Galidia elegans*), **Dwarf mongoose** (*Helogale parvula*), **Banded mongoose** (*Mungos mungo*), and **Gray meerkat** (*Suricata suricatta*).

Civets, Genets, and Mongooses

THE VIVERRIDS AND HERPESTIDS ARE LARGE carnivore families, but their natural distribution is confined to the Old World. These families include all species called civets, linsangs, genets, and mongooses. The mongooses were once assigned to the Viverridae along with the civets and genets, but most authorities now treat them as a separate family – the Herpestidae. Several subfamilies have at various times been raised to family status.

Viverrids and herpestids so closely resemble the ancestors of carnivores, the Miacoidea, that their fossils are almost indistinguishable from these early Eocene relatives. The tooth structure and skeletal morphology has barely changed over some 40 to 50 million years.

Perhaps the modern families are simply a continuation of this old lineage. However, in spite of their primitive dentition, viverrids and herpestids have a highly developed inner ear and so present an evolutionary mosaic of primitive and advanced features, making their systematic position uncertain. They are sometimes placed between the weasel and the cat families.

Viverrids and herpestids vary greatly in form, size, gait (from digitigrade to near-plantigrade), and habits. Most civets and genets look like spotted, long-nosed cats, with long, slender bodies, pointed ears, and short legs. However, the binturong (or "Bear cat") resembles a wolverine in build but has long black hair, a very long, thick, prehensile tail and a catlike head; the African civet is somewhat doglike in habits and appearance; the fossa so closely resembles a cat that some scientists once regarded it as a primitive member of the cat family; the Otter civets could pass as long-nosed otters; while the falanouc looks like a mongoose with an elongated nose and a bushy tail like a tree squirrel. Mongooses vary less in gross form: they have long bodies, short legs, and small rounded ears.

There is a tendency for males to be slightly larger than females, with the notable exception of the binturong, where females may be 20 percent larger. Females have one to three pairs of teats and males possess a baculum (stiffening bone in the penis). Vision and hearing are excellent. WCW

OOD AND FEEDING

vets, genets, and mongooses tend to omnivorous. Most species eat small ammals, birds, reptiles, insects, eggs, nd fruit. Palm civets, however, feed most exclusively on fruit, nuts, and ulbs. Mongooses generally consume ss fruit than civets. Some species, articularly those of the genus *erpestes*, are sufficiently carnivorous nd predatory to have been introduced to some areas to curb the popula- ons of rodents and poisonous snakes. ne disadvantage of such introduc- ons has been that the mongooses ave proceeded to decimate the other una in the region.

SKULLS AND DENTITION

Skull forms and dentition vary considerably among viverrids and herpestids. The facial part of the skull is long, the canine teeth relatively small, and the carnassials relatively unprominent. The normal dental formula is I3/3, C1/1, P4/4, M2/2 = 40, but the number of molars and premolars may be reduced. Skulls of genets are catlike (shown here is the Common genet, *Genetta genetta*), while those of civets are similar, but more heavily built. In mongooses, such as *Herpestes* species, the skull is about the same length, but more robust. The skulls of Palm civets are also sturdily built, like those of mongooses, with the binturong having a particularly domed form; the principally vegetarian diet of this species is reflected in its flattened, straight canine teeth, reduced premolars (with one lower pair missing) and molars, and peglike upper incisors.

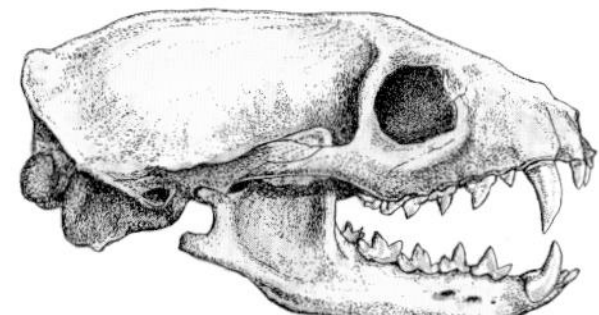

Mongoose (*Herpestes*)
9.7 cm

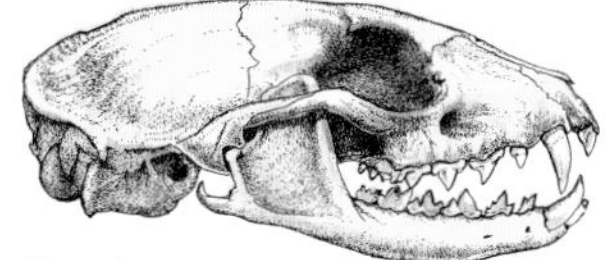

Genet
9.6 cm

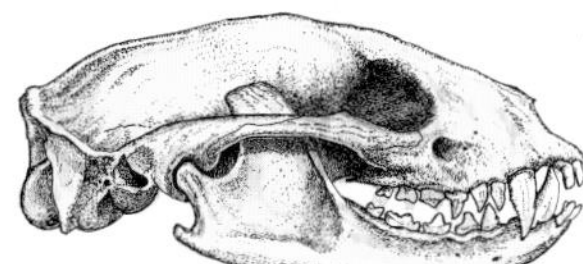

Binturong
13.8 cm

◐ ***Left*** *Scorpions are among the favored prey of the Yellow mongoose* (Cynictis penicillata), *a species that is widespread across southern Africa. Heavily carnivorous, it will also eat small snakes and lizards, birds' and reptiles' eggs, and rodents.*

◑ ***Right*** *An African linsang, a member of the civet and genet family, raiding a bird's nest. The linsang's varied diet includes nestlings, insects, nuts, and other plant matter.*

MONGOOSE OR CIVET?

Mongooses differ from members of the civet and genet family in a variety of morphological, behavioral, and genetic characteristics. Most civets have long tails equal to or even exceeding their body length, while those of mongooses average only half to three-quarters of their body length. Whereas civets have five toes on each foot, partly or totally retractile claws, webbing between toes, and pointed ears projecting above the profile of their heads, all mongooses have four or five toes on each foot, nonretractile claws, reduced or absent webbing between their toes, and rounded ears located on the side of the head, which rarely protrude above the head's profile.

The coat of civets is generally spotted or striped, the ear flaps have pockets (bursae) on the lateral margins, and most have a perineal (civet) gland situated near the genitals. Although true mongooses (subfamily Herpestinae) are uniformly colored, lack ear bursae and the perineal gland, Malagasy mongooses (subfamily Galidiinae) have ear bursae, some have perineal glands, and three of the four species are variously striped. Finally, one key morphological distinction between the families relates to the musk-containing anal glands. In all mongooses, these and the associated sacs are well developed, forming a pouch outside the anus proper. The anal scent glands of civets and genets are far less prominent.

The two families also exhibit different living habits and hunting strategies. In the main, viverrids tend to inhabit densely forested regions; herpestids generally live in more open terrain. Civets are primarily nocturnal and tree-dwelling (though there are a few terrestrial and semiaquatic species), while the majority of mongooses are terrestrial, and the family divides between nocturnal and diurnal species. Moreover, in marked contrast to the solitary civet, some mongoose species, such as the meerkat (LEFT), have highly developed social systems.

Expert opinion is divided on whether the fossa, the catlike predator of the Madagascar rain forests, should be assigned to the Viverridae or Herpestidae – here, it is included in the former group.

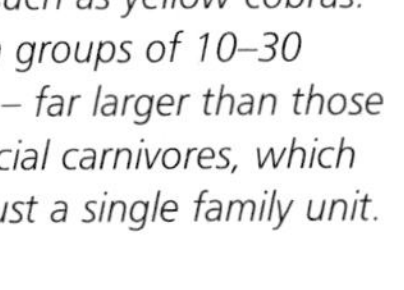

◐ ***Left*** *Alert meerkats on the lookout in the Kalahari desert, South Africa. Like some other mongoose species, meerkats will band together to drive off predators such as yellow cobras. They live in groups of 10–30 individuals – far larger than those of most social carnivores, which comprise just a single family unit.*

Civets and Genets

In the humid night air of the West African rain forest, a loud, plaintive cry is repeated, like the hooting of an owl; then another series of cries penetrates the dark from a kilometer away. These are African palm civets, the best known of some 35 species of a group including the civets, linsangs, and genets. This diverse assemblage of mostly catlike carnivores displays a wide range of lifestyles and coat markings. Primarily nocturnal foragers and ambush killers, they usually rest in a rock crevice, empty burrow, or hollow tree during the day. They are solitary animals, only occasionally forming small maternal family groups.

These animals are best known to us as the source of commercial civet oil (see box). Because many are inhabitants of tropical forests, a number of species find themselves in an ever-dwindling, threatened habitat – whether this be as a result of increasing deforestation in Southeast Asia or land clearance for cattle ranching in Madagascar. The IUCN lists as Endangered the Otter civet and the Falanouc, with several other species ranked Vulnerable. In contrast, the Common palm civet has become so plentiful and habituated to humans that it is frequently found living in and around villages and coffee plantations, and in some areas is even considered a pest.

Variations on a Theme

FORM AND FUNCTION

The seven southern Asian and one African specie of palm civets (subfamily Paradoxurinae) are characterized by the possession of semiretractile claw and a perineal scent gland lying within a simple fold of skin. The vocal African palm civet spends most of its time in the forest canopy, where it feeds chiefly on fruits of trees and vines, and occasionally on small mammals and birds. Adult male occupy home ranges of over 100ha (250 acres) and regularly scentmark trees on the borders of their territory.

All other species of Palm civets are confined to the forests of Asia. They are skillful climbers, aide by their sharp, curved retractile claws, usually naked soles, and partly fused third and fourth toes, which strengthen the grasp of the hindfeet. The Common palm civet is one of the most widespread. Like the Masked palm civet (which, unusually, has no body markings except for the head), it probably forages on the ground for fallen fruit and for animals. The tails of both species are only moderately long and are used to brace the animal during climbing. Most other species have longer tails. They seem to spend more time foraging in trees. The massively muscular tail of the binturong is prehensile (uniquely among viverrids) and, along with the hindfeet, is used to grasp branches while the forelimbs pull fruiting branches to the mouth. Binturongs have also been reported to swim in rivers and catch fish. The Sulawesi, Giant or Brown palm civet has very flexible feet with a web of naked skin between the toes. It is an acrobatic climber and lives in steep forested ravines and ridges of central and northeastern Sulawesi.

There are five species of banded palm civet and otter civet (subfamily Hemigalinae), all confined to the rain forests of Southeast Asia. The Banded palm civet, the best known, is named for the broad, dark vertical bands on its sides. Very carnivorous, it forages at night on the ground and in trees for lizards, frogs, rats, crabs, snails, earthworms, and ants, resting during the day in holes in tree trunks. One to three young are born to a litter and they begin to take solid food at the age of 10 weeks. Hose's palm civet resembles the Banded palm civet in body form, but is distinguished by skull characters and by its uniform blackish-brown color pattern. Owston's banded civet has spots as well as band markings, and seems to be a specialist on invertebrate foods; stomachs of the few specimens now in museums were found to contain only worms.

FACTFILE

CIVETS AND GENETS

Order Carnivora

Family Viverridae

35 species in 6 subfamilies and 20 genera. The Paradoxurinae subfamily comprises 7 species in 5 genera, including the Sulawesi palm civet (*Macrogalidia musschenbroekii*). The subfamily Nandiniae contains only the African palm civet (*Nandinia binotata*). The subfamily Hemigalinae comprises 5 species in 4 genera, including the Banded palm civet (*Hemigalus derbyanus*). The Viverrinae subfamily comprises 19 species in 7 genera, including the Common genet (*Genetta genetta*) and the African linsang (*Poiana richardsoni*). The subfamily Euplerinae comprises 2 species in 2 genera, the falanouc (*Eupleres goudotii*) and the fanaloka (*Fossa fossa*). The fossa (*Cryptoprocta ferox*) is the sole member of the subfamily Cryptoproctinae.

DISTRIBUTION Africa and Madagascar, SW Europe (Spain and France), Near East, Arabian peninsula, India, SE Asia to Borneo and the Philippines.

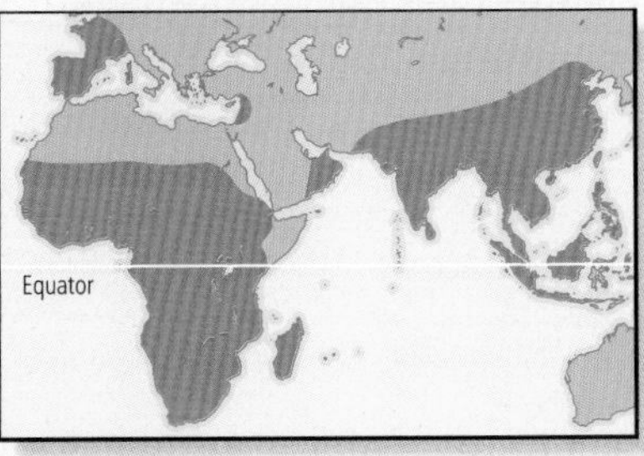

HABITAT Rain forests to woodlands, brush, savanna and mountains: chiefly arboreal, but also on ground and by riverbanks.

SIZE Head–body length from 33cm/13in (African linsang) to 88cm/35in (Sulawesi palm civet); **tail length** from 32cm (13in) to 55cm (22in); **weight** from 600g/1.3lb (Spotted linsang) to 13kg/29lb (African civet).

Though generally shorter than the Sulawesi palm civet or African civet, some fossas can weigh up to 20kg (44lb).

COAT Various textures; some monochrome species, but dark spots, bands, or stripes on lighter ground, and banded tail frequent.

DIET Small mammals, birds, reptiles, insects, eggs, and fruit. Palm civets feed almost exclusively on fruit, nuts, and bulbs.

BREEDING Gestation 70 days in genets, 80 in African civet, 90 in Palm civets.

LONGEVITY To about 20 years (15 years for Masked palm civet and 34 for genet recorded in captivity).

See species table ▷

◀ ***Left*** *Keen eyesight and a lithe build makes genets, like this Small-spotted genet, highly effective predators. Combining speed with stealth, they stalk their prey in a series of dashes punctuated by short pauses.*

◀ ***Above right*** *A binturong foraging for fruit in a tree. This arboreal species tends to move slowly and cautiously. Despite their somewhat ferocious appearance, binturongs are reportedly quite easy to tame.*

The Otter civets are the unorthodox members of the group. They have smaller ears, a blunter muzzle, a more compact body and a short tail. There are two species – the Otter civet and Lowe's otter civet – that differ in details of coloration and teeth. Their dense hair, valve-like nostrils, and thick whiskers equip Otter civets for life in water, catching fish. The toes are less webbed than in otters, which the Otter civets nevertheless resemble in habit and appearance. Otter civets are fine swimmers and can also climb trees.

The African civet is the largest and best known of the true civets and linsangs (subfamily Viverrinae). It lives in all habitats from dry scrub savanna to tropical rain forest, foraging exclusively on the ground by night and resting by day in thickets or burrows.

The African civet's Asian relatives probably share many features of its natural history. The similar Large Indian civet, Large-spotted civet, and Malay civet share eye-catching black and white body and back stripes, and have a rather doglike body plan, with a crest of long hair over the spine that can be raised to make the animal seem larger if threatened. Malay civets and Large Indian civets also have the latrine habit of the African civet.

The Small Indian civet has a narrower head and a genet-like build. It lacks a spinal crest, the body spots and tail rings are less well defined than in a genet, and the coloration is drab. A skillful predator of small mammals and birds, it stalks its prey like a cat. But it is also an opportunist that feeds on insects, turtle eggs, and fallen fruit. It is not a particularly good climber and often lives on cultivated land and near rural villages.

The linsangs are among the rarest, least known, and most beautiful members of the group. All three species are small, quick, trimly built, secretive forest animals with darkly-marked torsos and banded tails. They depend almost entirely on small vertebrates for food. The stomachs of four Banded linsangs were found to contain remains of squirrels, Spiny rats, birds, Crested lizards, and insects. In all likelihood they live alone, and there is good evidence to suggest that the African linsang builds leafy nests in trees. The two Asian species, the Spotted linsang and Banded linsang, apparently sleep in nests lined with dried vegetation under tree roots or hollow logs. They lack civet glands and the second upper molar. In common with the genets, all three linsangs have fully retractile claws.

Above *An African civet at rest during the day. At night, it is an omnivorous feeder. Small mammals such as gerbils, springhares, and spiny mice make up the bulk of its diet, but it will also eat carrion. Ground birds like francolins and guinea fowl are sometimes caught; reptiles, insects, and fruit complete the menu.*

Right *Representative civets and linsangs, shown in characteristic activities:* ***1*** *African linsang* (Poiana richardsoni) *feeding on a nestling;* ***2*** *Banded palm civet* (Hemigalus derbyanus) *eating a lizard;* ***3*** *Oriental or Malayan civet* (Viverra tangalunga) *with dorsal crest erect;* ***4*** *Common palm civet* (Paradoxurus hermaphroditus) *scenting the air;* ***5*** *Binturong* (Arctictis binturong) *foraging for fruit, while grasping a branch with its prehensile tail.*

The Fishing or Aquatic genet or Congo water civet is a rare and little-known inhabitant of streams and small rivers of the Central African forest block. It feeds on fish and possibly crustaceans, and local people report that it even occasionally eats cassava left to soak in streams by people preparing the vegetable for human consumption. It is not particularly specialized for swimming, but probably uses its naked palms to locate fish lurking under rocks and undercut river banks, grabbing them with its retractile claws and killing them with its sharp teeth.

Because of their nocturnal habits and cryptic coat patterns, genets have been studied mainly in the laboratory rather than in the wild. There are 11 species of these medium-sized, long-bodied, and short-legged carnivores, from Africa, Arabia, the Near East, and southwest Europe. They all have rows of dark spots along the body, or stripes, which are denser on the upper surfaces, on a light brown or gray background. The tails are ringed and about as long as the body. They have a long face, and pointed muzzle with long whiskers, largish ears, binocular vision, fully retractile claws and five toes on all four feet. In Africa they occupy all habitats except desert, but they prefer areas of dense vegetation. In Spain, Common genets are widespread even in high mountains.

Common genets may have been imported to Europe as pets by the Moors in the Middle Ages, or they may be a remnant population left after the Gibraltar land bridge was broken. A size variation, with smaller specimens in the north of the range, suggests that the distribution of the Common genet is natural. However, the genet population of the Balearic Islands is definitely the result of human introduction. The subspecies from Ibiza (*G. genetta isabelae*) is smaller than the other European forms; it closely resembles the Feline genet subspecies *G. felina senegalensis* found in Senegal, West Africa.

Three unusual viverrids occur on Madagascar The rare and secretive falanouc inhabits wet and low-lying rain forest from the eastern-central region up to the northwest of the island. It is on of two species in the subfamily Euplerinae, the other being the similarly-sized fanaloka. The falanouc's specialized teeth are used to seize and hold earthworms, slugs, snails, and insect larvae Loath to bite in self-defense, it employs its sharp claws (long on the forefeet) for this purpose.

The fanaloka resembles a small spotted fox in build and gait, but its relationship to other viverrids is uncertain. Its anatomy differs in many respects from the Banded palm civets with whic it was once grouped and it is known only from Madagascar, far away from its presumed south Asian relatives. Fanalokas are nocturnal and live exclusively in dense forests, frequenting remote ravines and valleys for preference. Also, unlike most other viverrids, they live in pairs. They are not particularly good climbers, but rely on heari and vision to find their favored food – rodents, frogs, mollusks, and sand eels. Despite its scientic name (*Fossa fossa*), the fanaloka should not be confused with the much larger fossa.

The fossa (see box overleaf) is the largest Malagasy carnivore and large individuals exceed in weight all other members of the civet–mongoose family. Its catlike head (large frontal eyes, shortened jaws, and rounded ears) and general appearance prompted its early classification as a felid. The fossa's resemblance to cats, however, is a result of independent (convergent) evolution. It quite different from all other viverrids and is the only member of the subfamily Cryptoproctinae.

Civet Oil and Scentmarking

The term "civet" derives from the Arabic word *zabad*, which denotes the unctuous fluid (and its odor) obtained from the perineal glands of most viverrids, except mongooses. The gland is associated with the genitalia, but it differs in anatomy between species: in Palm civets, it is a simple long fold of skin that produces only a thin film of the scented secretions, while in the three *Viverra* species and in the African civet, it is a deep muscular pouch that may accumulate several grams of civet a week. Genet scent has a subtle, pleasing odor, but that of the true civets (*Viverra*) is powerful and disagreeable; the scent of the binturong is reminiscent of cooked popcorn. Civetone, which has a pleasant, musky odor, is probably the most widespread component, but other compounds, such as scatole, often impart a fetid odor to the secretion.

Scentmarking is important in viverrid communication, but the method differs between species. The binturong spreads its scent passively while moving about as the gland touches its limbs and vegetation. Some species scentmark by squatting and then wiping or rubbing the gland along the ground or on a prominent object. The Large Indian civet elevates its tail, turns the pouch inside out and presses it backward against upright saplings or rocks. Like most other species, genets scentmark while squatting, but they also leave scent on elevated objects by assuming a handstand posture,

The close association of the perineal gland and the genitalia suggests that the secretion may have sex-related functions. Indeed, civetone may "exalt" volatile compounds from the reproductive tract of females on heat. Perineal gland scent may also carry information on sex, age, and individual identity.

Civet has long played a key role in the perfume industry – it was imported from Africa by King Solomon in the 10th century BC. Once refined, it is cherished within the perfume industry because of its odor, its ability to exalt other aromatic compounds, and its longlasting properties. Civet oil also has medicinal applications, being used to reduce perspiration, as an aphrodisiac, and as a cure for certain skin disorders. Since the development of synthetic chemical substitutes, the collection of civet oil is not as vital as it once was to the industry. Nevertheless, several East African and Oriental countries still ship large quantities of civet oil each year and, in some instances, civets have been introduced to supply a primitive economic base for poor areas. The animals are kept in small cages, restrained by manhandling, and the scent scraped out with a special spoon.

1
2
3
4
5

Above *The three Malagasy viverrids:* ***1*** *the falanouc is mongoose-like, has an elongated snout and body and nonretractile claws, and feeds mainly on invertebrates;* ***2*** *the fanaloka or Madagascar civet, is more foxlike in appearance, has retractile claws and feeds primarily on small mammals, reptiles, and amphibians;* ***3*** *the fossa has a catlike head and retractile claws for capturing its prey.*

Fruit, Birds, and Bats

DIET

All Palm civets eat a wide variety and a vast quantity of fruit, as well as some rodents, birds, snails and scorpions, which supplement the protein in a diet that is otherwise high in carbohydrates. In Java, the Common palm civet eats the fruits of at least 35 species of trees, palms, shrubs, and creepers. Some fruits harmful to humans are eaten without ill effects. The seed of the Arenga palm, for example, has a prickly outer pulp, but it is consumed in large quantities, passing through the digestive tract undamaged. The Small-toothed palm civet has small, flat-crowned premolars that seem to be adaptations for a diet of soft fruit. Nearly all Palm civets are notorious banana thieves, and the Common palm civet is also known as the "Toddy cat" for its fondness for the fermented palm sap (toddy) which over much of southern Asia is collected, in bamboo tubes attached to palm trunks, for human consumption.

Genets are almost all carnivorous, though insects and fruit also form a regular part of their diet. One exception is Johnston's genet, which is thought to be largely insectivorous. In Africa, small mammals such as rodents make up the major part of the Common genet's diet, while in Spain lizards are also a favored item. Feces analysis of Common genets in Spain shows that passerine birds comprise up to half the diet in spring and summer, fruit is important during autumn and winter, rodents (mainly wood mice) are taken all year round, and insects from spring to autumn, though they make up only a small part of the diet. Some genets stalk frogs at the side of rain pools. One Forest genet has been observed regularly to take bats as they leave their roosts, while in West Africa Forest genets are known to feed on the profuse nectar of the tree *Maranthes polyandra*; the flowers are bat-pollinated, so perhaps the initial attraction was to the bats rather than the nectar.

Below *A Large spotted genet. Like all genets, this southern African species is well adapted to an arboreal way of life. Excellent binocular vision allows it to judge distances accurately as it jumps from branch to branch or pounces on its prey.*

Versatile Predators

SOCIAL BEHAVIOR

Dominant male Palm civets use many kilometers of boughs and vines to patrol their home ranges once every 5–10 days. Subordinate males – usually smaller, immature, or aged animals – occupy small areas within the range, but avoid dominant males and traverse their ranges at irregular intervals. Eventually, however, when one of these subordinates matures, the dominant male's priority to mate with a female in heat is challenged. Often fatal wounds are inflicted in the ensuing fight and the vanquished sitting tenant, weakened by deep bites, retreats to the ground where he dies or is killed by a leopard or other predator.

One to three females live within the home range of a dominant male and he visits the home range of each for several days as he makes his rounds. For most of the year, females do not tolerate a male staying in the same tree. But in the long rainy season males and females keep track of one another's whereabouts by calling in the darkness. Mating takes place in June over several days, as the pair roost in the same tree. Three months later, 1–3 young are born in a secluded tangle of vines. They are weaned six months later and reach sexual maturity shortly after the second year. Females share their home ranges only with daughters less than two years old. Male offspring emigrate shortly after weaning.

African civets almost always defecate at dung heaps (middens or "civetries") near their route of movement. Civetries are normally less than 0.5sq m (5sq ft) in area and there is evidence that they are located at territorial boundaries serving as contact zones between neighbors. Fruit trees and shrubs are often scentmarked with the perineal gland, and so to a lesser extent are grass, dry logs, and rocks. Female African civets are sexually mature at one year and may produce as many as two litters a year after a gestation of about 80 days. One to three young are born, probably in a secluded thicket. The mother nurses them until they are 3–5 months old. They begin to catch and eat insects before weaning, and the mother summons them with a chuckling contact call when she wishes to share a rodent or bird kill.

Genets are adapted to living in trees, but they often hunt and forage on the ground. Their coloration helps them avoid detection, especially on moonlit nights. In Spain, Common genets' activity starts at or just before sunset and ends shortly after dawn. They are inactive for a period during the night, and occasionally there is no morning activity. In daytime the genets only stir to move from one resting site to another. In the Serengeti, the Feline genet, unlike other small carnivores of

THE FASCINATING FOSSA

Found only in the forests of Madagascar, the fossa is a civet whose livelihood is lemurs. Lemurs account for over half its diet – no mean feat, given the intelligence, agility, and relatively low population density typical of these primates! It is unusual for walking on the whole foot (plantigrade gait), not on the toes as most viverrids do. To aid its arboreal lifestyle, the fossa's tail is as long as its body (70cm/28in), but it is equally at home on the ground, where it may take snakes, Common tenrecs, and Guinea fowl.

It is not only the fossa's eating habits that set it apart. Fossa mating is also a strange affair. A single female will exclusively occupy a site high in a tree, below which a number of males congregate. The female mates with several of these, sometimes copulating incessantly over a long period (up to 165 minutes). Mating sites are easy to locate from the vocalizations of the males as they fight and call up to the female. After a week of this behavior, a new female arrives. She takes over the site and, like her predecessor, mates with the males gathered there.

The prolonged mating period is all the more surprising when one considers the structure of the male genitalia. The penis is supported by a bone and covered with hard, backward-pointing spines. When erect, these spines stick out, and are probably responsible for the lock seen between mating fossas (BELOW), similar to that between domestic dogs.

Fossas also undergo a strange developmental phase unknown in other mammals. During their adolescence, female fossas take on some distinctly masculine characteristics not seen in adult females. The clitoris grows disproportionately, develops a small supporting bone, and becomes covered with spines. In addition, the cream fur of the underparts is colored by a bright orange secretion. In adult fossas, this secretion is seen only in males. This temporary "masculinization" has not been identified in any other female mammals, and it remains unclear whether it is a mysterious adaptation or an unusual side-effect of pubescent hormonal changes.

Madagascar became separated from continental Africa some 165 million years ago, and the fauna there evolved in isolation from their counterparts on the mainland. This long seclusion may provide the sole explanation for the fossa's peculiarity. On the other hand, it could be that the fossa's supposedly unique attributes are waiting to be discovered in other civet species. Both the Binturong (*Arctictis binturong*) and the Common palm civet (whose revealing Latin name is *Paradoxurus hermaphroditus*!) are said to be difficult to sex, although in their cases the confusion apparently lies with their scent glands, which can make males appear female. With the fossa, this interesting possibility remains to be explored. CH

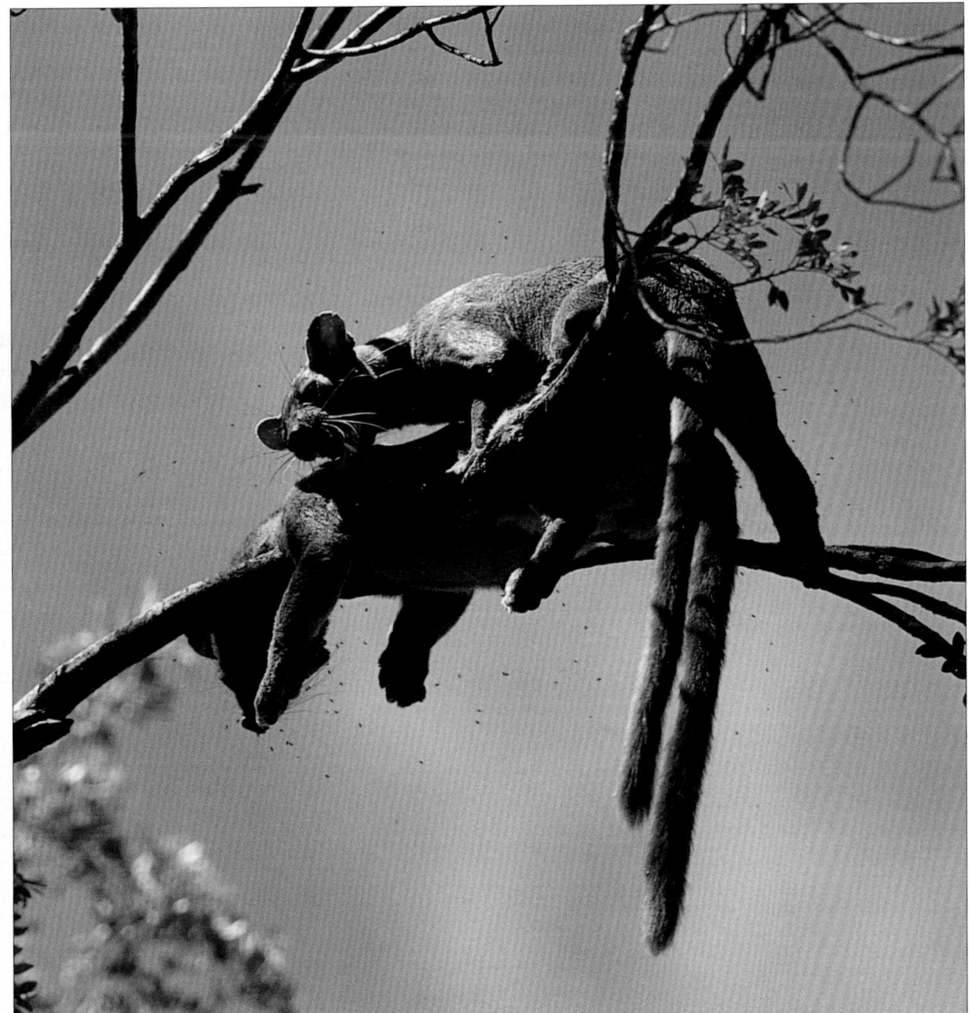

the plain, is relatively inactive at dawn and dusk and most active around midnight. This pattern of activity may reduce competition for similar foods.

Genets breed throughout the year, but in many areas there are seasonal peaks (e.g. April and September for Common genets in Europe). Two to four young are born, after a gestation of about 70 days, in a vegetation-lined nest in a tree or burrow. They are blind at birth and about 13.5cm (5in) long. Their eyes open after eight days and they venture from the nest soon after. They are weaned after six months, although they take solid food earlier. Genets are thought to become independent after one year and are sexually mature after two. Scentmarking – by feces, urine and perineal gland secretion – plays a key role in genet social life, allowing animals to tell the identity, familiarity, sex, and breeding status of conspecifics. Males, and to a lesser extent females, show a seasonal variation in the frequency of marking, with an increase before the breeding season.

Genets are mostly solitary. In Africa, population densities of 1–2 per sq km (2.5–5 per sq mi) have been reported, with home ranges as small as 0.25 sq km (0.1sq mi). Females appear to be more territorial than males, which wander farther. Population density is much lower in Spain. Two male Common genets have been tracked by radio over home ranges of 5sq km (2sq mi), moving quickly over long distances, up to 3km (1.9mi) in an hour.

Genets make good pets; in Europe they were kept as rat catchers until they were superseded by the modern domestic cat in the Middle Ages. In parts of Africa, however, they are regarded as a pest because of their attacks on poultry.

The falanouc has a solitary, territorial lifestyle, with a brief consort between mates and a longer mother–young bond that dissolves before the onset of the next breeding season. The base of the tail serves as a fat storage organ for the cold, dry months of June and July when food is in short supply. Subcutaneous fat is deposited in April and May. The falanouc's reproductive pattern deviates from that of most carnivores: a single offspring is born in summer. The newborn's well-developed condition suggests that mother and young are highly mobile shortly after birth. An animal born in captivity had open eyes at birth and was able to follow its mother and hide in vegetation when only two days old. It did not take solid food until nine weeks old, but weaned quickly thereafter. These traits enable the young to stay close to the mother while she roams far and wide for food. Falanouc young develop locomotory and sensory skills very early, but grow and mature at a slightly slower pace than other similar-sized carnivores.

The fanaloka mates in August and September and gives birth to one young after a three-month gestation. The young are born in a physically advanced state. The eyes are open at birth and in a few days young are able to follow the mother. The baby is weaned in 10 weeks. CW

Civet and Genet Species

THERE ARE SIX SUBFAMILIES OF VIVERRIDS. Some authorities assign the genus *Cryptoprocta* to the family Herpestidae.

AFRICAN PALM CIVET

Subfamily Nandiniiae

African palm civet
Nandinia binotata
African palm civet or Two-spotted palm civet

From Guinea (including Fernando Póo Island) to S Sudan in the north, to Mozambique, E Zimbabwe and C Angola in south. Arboreal.
HBL 50cm; TL 57cm; WT 3kg.
COAT: a uniform olive brown with faint spots; 2 cream spots on the shoulders vary geographically in size and intensity.

PALM CIVETS

Subfamily Paradoxurinae

7 species in 5 genera
Semiarboreal/arboreal; nocturnal. Teeth specialized for fruit or mixed diet; carnassial teeth weakly to moderately developed; two relatively flat-crowned molars in upper and lower jaws of all species. Perineal scent glands in both sexes of all species, except *Arctogalidia*, where it is lacking in the male. In *Nandinia* the gland is in front of the sex organs. Claws semiretractile.

Small-toothed palm civet
Arctogalidia trivirgata
Small-toothed or Three-striped palm civet

Assam, Myanmar, Thailand, Malayan and Indochinese Peninsulas, China (Yunnan), Sumatra, Java, Borneo, Riau-Lingga Archipelago, Bangka, Bilitung, N Natuna Islands. Arboreal.
HBL 51cm; TL 58cm; WT 2.4kg.
COAT: more or less uniform, varying from silvery to buff to dark brown, sometimes grizzled on head and tail; 3 thin, dark-colored stripes on back, often from base of tail to shoulders; white streak down middle of nose; tail sometimes has vague dark bands; tip sometimes white.

Common palm civet
Paradoxurus hermaphroditus
Common palm civet or Toddy cat

India, Sri Lanka, Nepal, Assam, Bhutan, Myanmar, Thailand, S China, Malaya, Indochina, Sumatra, Java, Borneo, Ceram, Kei Islands, Nusa Tenggara (Lesser Sunda Islands) as far E as Timor, Philippines. Semiarboreal.
HBL 54cm; TL 46cm; WT 3.4kg.
COAT: variable from buff to dark brown depending on locality; usually black stripes on back and small to medium spots on sides and base of tail; face mask of spots and forehead streak; spots variable both locally and geographically; tail tip sometimes white.

Golden palm civet
Paradoxurus zeylonensis

Sri Lanka. Arboreal.
HBL 51cm; TL 46cm; WT 3kg.
COAT: brown, golden-brown or rusty; spots and stripes barely visible; nap of hair forward on neck and throat; tip of tail sometimes white.

Jerdon's palm civet Vu
Parodoxurus jerdoni

S India, Palni and Nilgiri hills. Tranvancore and Coorg. Arboreal.
HBL 59cm; TL 52cm; WT 3.6kg.
COAT: deep brown, or brown to black with silver or gray speckling; nap of hair as *P. zeylonensis*.

Masked palm civet
Paguma larvata

India, Nepal, Tibet, China (N to Hopei), Shansi, Taiwan, Hainan, Myanmar, Thailand, Malaya, Sumatra, N Borneo, S Andaman Is.; introduced to Japan. Arboreal.
HBL 63cm; TL 59cm; WT 4.8kg.
COAT: uniform grayish or yellowish-brown to black depending on geographic origin; face dark, but may be marked with a light frontal streak or spots under the eyes and in front of the ears; tip of tail sometimes white.

Sulawesi palm civet Vu
Macrogalidia musschenbroekii
Giant civet or Brown palm civet

NE and C Sulawesi (Celebes). Semiarboreal.
HBL 88cm; TL 62cm; WT 4.2kg.
COAT: uniform brown with vague darker spots on either side of midline and faint light-colored rings on the tail; hair lighter above and beneath eyes. Cheek teeth of upper jaw arranged in parallel rather than diverging rows.

Binturong
Arctictis binturong
Binturong or Bear cat

India, Nepal, Bhutan, Myanmar, Thailand, Malaysia and Indochina, Sumatra, Java, Borneo and Palawan. Arboreal.
HBL 77cm; TL 73cm; WT 7.6kg (females 20 percent larger).
COAT: black with variable amount of white or yellow restricted to hair tips (yellowish or gray binturongs always have black undercoats); hair long and coarse; ears with long black tufts and white margins. Tail heavily built, especially at the base, and prehensile at the tip.

BANDED PALM CIVETS

Subfamily Hemigalinae

5 species in 4 genera
Nocturnal. Second molar large with many cusps. Perineal scent glands present in all species, but not as large as in other subfamilies. Claws semiretractile.

Banded palm civet
Hemigalus derbyanus

Peninsular Myanmar, Malaya, Sumatra, Borneo, Sipora and S Pagi Islands. Semiarboreal.
HBL 53cm; TL 32cm; WT 2.1kg.
COAT: pale yellow to grayish-buff, with contrasting dark brown markings; face and back with longitudinal stripes; body with about 5 transverse bands extending halfway down flank; tail dark on terminal half, with about 2 dark rings on the base.

Hose's palm civet Vu
Diplogale hosei

Borneo, Sarawak (Mt Dulit to 1,200m). Semiterrestrial.
Dimensions unknown.
COAT: uniform dark brown with gray eye and cheek spots; chin, throat, and backs of ears white; belly white or dusky gray.

Owston's banded civet Vu
Chrotogale owstoni
Owston's banded or Owston's palm civet

N of Indochinese Peninsula. Terrestrial.
HBL 55cm; TL 43cm; WT unknown.
COAT: similar to Banded palm civet, but with only 4 transverse dark-colored dorsal bands and with black spots on neck, torso and limbs.

Otter civet En
Cynogale bennettii
Otter civet, Water civet

Sumatra, Borneo, Malayan and Indochinese Peninsulas. Semiaquatic.
HBL 64cm; TL 17cm; WT 4.7kg.
COAT: uniform brown, soft dense hair, with faint grizzled appearance; front of throat white or buff-white. First three upper premolars unusually large with high, compressed and pointed crowns; remaining cheek teeth broad and adapte[d] for crushing. Ears small, but designed to keep out water. Feet naked underneath.

Lowe's otter civet
Cynogale lowei

N Vietnam. Semiaquatic.
COAT: dark brown above, white to dirty white below from cheek to belly; tail dar[k] brown. Other features as *C. bennettii*.

TRUE CIVETS, LINSANGS & GENET[S]

Subfamily Viverrinae

19 species in 7 genera
Teeth specialized for an omnivorous diet shearing teeth well developed; two mola[rs] in each side of upper and lower jaws of a[ll] except *Poiana* and *Prionodon* (one upper molar). Perineal gland in all genera excep[t] *Prionodon*, presence not certain in *Poian[a]* and *Osbornictis*. Soles of feet normally hairy between toes and pad.

African linsang
Poiana richardsoni
African linsang or oyan

Sierra Leone, Ivory Coast, Gabon, Camer[o]oun, N Congo, Fernando Póo Island. Arboreal; nocturnal.
HBL 33cm; TL 38cm; WT 650g.
COAT: torso spotted, stripes on neck; tai[l] white with about 12 dark rings and light-colored tip. Claws retractile.

Spotted linsang
Prionodon pardicolor

Nepal, Assam, Sikkim, N Myanmar, Indo[-]china. Semiarboreal; nocturnal.
HBL 39cm: TL 34cm; WT 600g.
COAT: light yellow with dark spots on torso and stripes on neck; tail with 8–9 dark bands alternating with thin light bands. Claws retractile.

Banded linsang
Prionodon linsang

W Malaysia, Tenasserim, Sumatra, Java, Borneo. Semiarboreal; nocturnal.
HBL 40cm; TL 34cm; WT 700g.
COAT: very light yellow with 5 large transverse dark bands on back; neck stripes broad with small elongate spots and stripes on flank; tail with 7–8 dark bands and black tip. Claws retractile.

ABBREVIATIONS	HBL = head–body length TL = tail length WT = weight	Ex Extinct	En Endangered
	Approximate nonmetric equivalents: 10cm = 4in; 230g = 8oz; 1kg = 2.2lb	EW Extinct in the Wild	Vu Vulnerable
		Cr Critically Endangered	LR Lower Risk

mall Indian civet
'iverricula indica
mall Indian civet, rasse

China, Myanmar, W Malaysia, Thailand, umatra, Java, Bali, Hainan, Taiwan, ndochina, India, Sri Lanka, Bhutan; ntroduced to Madagascar, Sokotra and Comoro Islands. Terrestrial; nocturnal/ repuscular.
HBL 57cm; TL 36cm; WT 3kg.
COAT: light brown, gray to yellow-gray vith small spots arranged in longitudinal tripes on the forequarters and larger spots n the flanks; 6-8 stripes on the back; eck stripes not contrasting In color as in *'iverra* and *Civettictis*; 7–8 dark bands on ail and tip often light. Claws semiretrac-tile; skin partially bare between toes and oot pads.

Malayan civet
'iverra tangalunga
Oriental or Ground civet, tangalunga

Malaya, Samatra, Riau-Lingga Archipelago, Borneo, Sulawesi, Karlinata, Bangka, Buru, Ambon and Langkawi Islands, and Philippines. Terrestrial: nocturnal/crepus-ular.
HBL 66cm; TL 32cm; WT 3.7kg.
COAT: dark with many close-set small black spots and bars on torso, often form-ng a brindled pattern; crest, which can be erected, of black hair from shoulder to nidtail; black and white neck stripes that pass under throat; white tail bands, inter-upted by black crest and black tip. Claws emiretractile.

Large Indian civet
'iverra zibetha

N India, Nepal, Myanmar, Thailand, Indo-hina, Malaya. S China. Terrestrial; noctur-nal/crepuscular.
HBL 81cm: TL 43cm; WT 8.5kg.
COAT: tawny to gray with black spots, osettes, bars and stripes on torso, neck vith black and white stripes that pass under the throat; erectile spinal crest of black hair from shoulder to rump; tail vith complete white bands and black tip. Claws semiretractile.

Large-spotted civet
'iverra megaspila

S Myanmar, Thailand, formerly the coastal listrict and W Ghats of S India. Indo-hina, Malay Peninsula to Penang. Terres-trial; nocturnal/crepuscular. Population in he W Ghats of India (which the IUCN classes as a separate species, *V. civettina*) is now regarded as Critically Endangered, due to hunting and habitat loss from farm-ng; fewer than 250 individuals survive.
HBL 76cm; TL 37cm; WT 6.6kg.
COAT: grayish to tawny with small indis-tinct black or brown spots on the oreparts; large spots on the flanks often fusing into bars and stripes; pronounced black and white neck stripes; spinal crest of erectile black hair from shoulder to rump, bordered on either side by a longi-tudinal row of spots; tail with 5–7 white bands, most of which do not circle the tail completely, and black tip. Claws not retractile, soles of feet scantily haired between toes and foot pads.

African civet
Civettictis civetta

Senegal E to Somalia in N, through C and E Africa to KwaZulu–Natal, Transvaal, N Botswana and N Namibia in S. Terrestri-al; nocturnal/crepuscular.
HBL 84cm; TL 42cm; WT 13kg.
COAT: grayish to tawny, torso marked with dark brown or black spots, bars and stripes (degree of striping and distinctness of spots geographically variable); spinal crest from shoulders to tail; tail with indis-tinct bands and black tip. Claws not retractile; soles of feet bare between toes and foot pads.

Aquatic genet
Osbornictis piscivora
Fishing genet or Congo water civet

Kisangani and Kibale-Ituri districts of Zaire. Semiaquatic; nocturnal.
HBL 47cm; TL 37cm; WT 1.4kg.
COAT: uniform chestnut-brown with dull red belly; chin and throat white; tail uni-form dark brown and heavily furred. Claws semiretractile.

Common genet
Genetta genetta
Small-spotted genet or European genet

Africa (N of Sahara), Iberian Peninsula, France, Palestine. Open or wooded coun-try with some cover.
HBL 40–50cm TL 37–46cm; WT 1–2.3kg.
COAT: grayish-white with blackish spots in rows; tail with 9–10 dark rings and white tip; prominent dark spinal crest.

Feline genet
Genetta felina

Africa S of the Sahara except for rain forest; S Arabian Peninsula. Open or wooded country with some cover.
HBL 40–50cm; TL 37–47cm; WT 1–2.3kg.
COAT: light gray to brownish-yellow with blackish spots in rows; tail with 9–10 black rings and white tip; prominent spinal crest of black hairs; hind legs with gray stripe.

Forest genet
Genetta maculata (formerly *G. pardina*)

Southern part of W Africa, C Africa, S Africa (except Cape region). Dense forest.
Dimensions as *G. genetta*.
COAT: grayish to pale brown, more heavily spotted than Common genet; tail black with 3–4 light rings at base and tip dark or light; spinal crest short and can be erected. Relatively long-legged.

Large-spotted genet
Genetia tigrina
Blotched genet or Tigrine genet

Cape region of S Africa. Woodland and scrub.
Dimensions as *G. genetta*.
COAT: brown-gray to dirty white with large brown or dark spots; tail relatively long with 8 or 9 black rings and dark tip. Relatively short-legged.

Servaline genet
Genetta servalina
Servaline genet or Small-spotted genet

C Africa, with restricted range in E Africa. Forest.
HBL 42–53cm; TL 41–51cm; WT 1–2kg.
COAT: ocherous and more evenly covered with small blackish spots than other genets; underparts darker; tail with 10–12 rings and white tip. Face relatively long.

Giant genet
Genetta victoriae
Giant or Giant forest genet

Uganda, N Zaire. Rain forest.
HBL 50–60cm; TL 45–55cm; WT 1.5–3.5kg.
COAT: yellow to reddish-brown, very heavily and darkly spotted; tail bushy with 6–8 broad rings and black tip; dark spinal crest; legs dark.

Angolan genet
Genetta angolensis
Mozambique genet, Hinton's genet

N Angola, Mozambique, S Zaire, NW Zambia, S Tanzania. Forest.
Dimensions as *G. genetta*.
COAT: largish dark spots in 3 rows each side of erectile spinal crest; tail very bushy with 6–8 broad rings and dark tip; neck striped; hind legs dark with thin gray stripe. Relatively long-haired.

Abyssinian genet
Genetta abyssinica

Ethiopian highlands, Somalia. Mountains.
HBL 40–46cm; WT 0.8–1.6kg.
COAT: very light sandy-gray with black horizontal stripes and black spotting; tail with 6–7 dark rings and dark tip; back with 4–5 stripes; black spinal crest poorly developed.

Villier's genet
Genetta thierryi
Villier's genet or False genet

W Africa. Forest and Guinea savanna.
Dimensions and coat as for *G. abyssinico* but with chestnut or black spotting which is poorly defined; tail with 7–9 dark rings (first rings rufous) and dark tip.

Johnston's genet
Genetta johnstoni
Johnston's genet or Lehmann's genet

Liberia.
COAT: ground color yellow to grayish-brown; black erectile spinal stripe; rows of large dark spots on sides; tail with 8 dark rings. Skull larger than other genets with overlapping distribution, but greatly reduced teeth suggest insectivorous diet.

MALAGASY CIVETS

Subfamily Euplerinae
2 species in 2 genera

Falanouc En
Eupleres goudotii

East C to NW Madagascar. Rain forest. Terrestrial; solitary.
HBL 48–56cm; TL 22–25cm; WT not known.
COAT: light to medium brown; whitish-gray on underside. Snout elongated. First premolars and canines short, curved back-ward and flattened, for taking small soft-bodied prey. No anal or perineal gland.

Fanaloka Vu
Fossa fossa
Fanaloka, Madagascar or Malagasy civet

Madagascar. Dense rain forest. Terrestrial; nocturnal; lives in pairs.
HBL 47cm; TL 9.5cm; WT 2.2kg.
COAT: brown with darker brown dots in coalescing longitudinal rows; faint dark banding on upperside of short tail. Per-ineal gland absent.

Subfamily Cryptoproctinae

Fossa En
Cryptoprocta ferox

Madagascar. Rain forest. Arboreal; noc-turnal.
HBL 70cm; TL 65cm; WT 9.5–20kg.
COAT: reddish-brown to dark brown. Head cat-like with large frontal eyes, short jaws, rounded ears; tail cylindrical. Car-nassial teeth well developed, upper molars reduced (formula I3/3, C1/1, P3/1, M1/1 = 32). Claws retractile. Feet webbed. Anal gland well developed, perineal gland absent.
BREEDING: seasonal, mating in September and October. Gestation 3 months; 2–4 young, WT 80–100g, born in tree or ground den. Physical development slow; the eyes do not open for 16–25 days and solid food is not taken for 3 months. They are weaned by 4 months and growth is complete at 2 years, although they are not sexually mature for another two years.
CW/WCW

Mongooses

SLENDER, AGILE CARNIVORES THAT PREY ON a wide range of invertebrates and small vertebrates, mongooses are distributed throughout sub-Saharan Africa and southern Asia. Because their diets vary from purely insectivorous to highly carnivorous, these Old World animals fill niches and have lifestyles that are occupied in the New World by the members of several orders, including Carnivora, Insectivora, and Rodentia.

The mongoose family are close evolutionary relations of the civet and genet family; indeed, there are some taxonomists who classify both families jointly as Viverridae. Even so, there are a significant number of differences between the two groups. Features specific to the mongoose family include nonretractile claws, four or five toes on each foot, reduced or absent webbing between the toes, and rounded ears – located on the side of the head – that rarely protrude above the head's profile. In contrast to the civets, which are solitary creatures, some mongooses have quite highly developed social systems; and in addition, while civets are for the most part both nocturnal and tree-dwelling, mongooses are more commonly terrestrial, while some species are active during daylight hours.

Short Legs and Long Tails

FORM AND FUNCTION

Often the most abundant carnivores in the locations they inhabit, mongooses are agile and active terrestrial mammals. The face and body are long, and they have small, rounded ears, short legs, and long, tapering, bushy tails.

Most mongooses are brindled or grizzled, and few coats are strongly marked. No species have spots (unlike civets and genets), and few have shoulder stripes; the feet or legs, and tail or tail tip, are often of a different hue. The Banded mongoose and the suricate have darker transverse bands across the back. Among the four Madagascar mongooses, two species have stripes that run along the body and one has a ringed tail. Considerable color variation occurs, even within the same species; for example, the Slender mongoose is gray or yellowish-brown throughout most of its range, but in the Kalahari desert it is red, and there is also a melanistic (black) form. Variations usually correlate with soil color, suggesting that camouflage is important for survival.

Most mongooses have a large anal sac containing at least two glandular openings. Scentmarking with anal and cheek glands can communicate the sex, sexual receptivity, and individual and pack identity of the marker.

FACTFILE

MONGOOSES

Order: Carnivora

Family: Herpestidae

35 species in 2 subfamilies and 17 genera. The Galidiinae subfamily (Madagascan mongooses) comprises 5 species in 4 genera, including the **Ring-tailed mongoose** (*Galidia elegans*). The Herpestinae subfamily (African and Asian mongooses) comprises 30 species in 13 genera, including the **Dwarf mongoose** (*Helogale parvula*), **Banded mongoose** (*Mungos mungo*), and **Suricate or Gray meerkat** (*Suricata suricatta*).

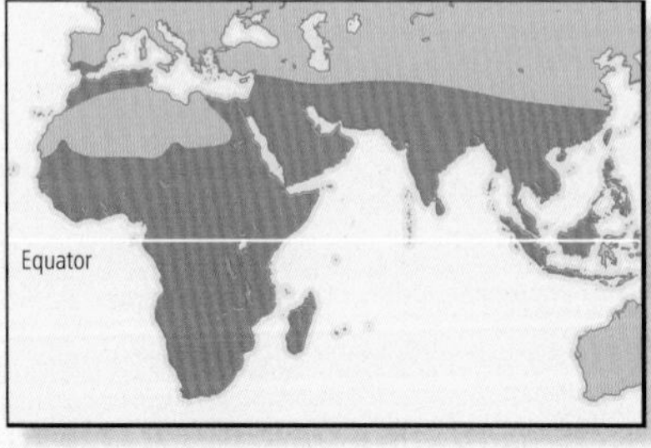

DISTRIBUTION Africa and Madagascar, SW Europe, Near East, Arabia to India and Sri Lanka, S China, SE Asia to Borneo and the Philippines; introduced in W Indies, Fiji, Hawaiian Islands.

HABITAT From forests to open woodland, savanna, semi-desert, and desert; chiefly terrestrial, but also semiaquatic and arboreal.

SIZE **Head–body length** from 24cm/9.5in (Dwarf mongoose) to 58cm/23in (White-tailed mongoose); **tail length** from 19cm (7.5in) to 44cm (17in); **weight** from 320g (11oz) to 5kg (11lb); some Egyptian mongooses larger in total length.

COAT Long, coarse, usually grizzled or brindled. Color ranges from dark-gray through brown to yellowish or reddish; some genera have bands or stripes.

DIET Insects and other invertebrates, small vertebrates, birds' eggs, crabs, fish, occasionally fruit and other vegetable matter. Some species (notably of the genus Herpestes) kill and eat venomous snakes.

BREEDING Gestation mostly about 60 days, but 42 in the Small Indian mongoose, 105 in the Narrow-striped mongoose.

LONGEVITY To about 10 years (19 in captivity).

See species table ▷

Mongoose lifespans can be long; one female Dwarf mongoose in Serengeti National Park lived 13 years. In practice, however, few animals in the wild live out their natural span, and only 25 of 846 dwarf mongooses in the same study attained their seventh birthday.

Niche Opportunists

DISTRIBUTION PATTERNS

Mongooses are mostly restricted to the Old World, inhabiting a broad band straddling both sides of the Equator from western Africa to Southeast Asia, though they have also been introduced into the West Indies, Fiji, and the Hawaiian islands. They have adapted to a variety of habitats, ranging from forests and open woodland to savanna and deserts.

Among communally-dwelling mongooses, the location of specific populations can be influenced by denning opportunities. In African grasslands with hard soils, for example, where termite mounds are often the only available refuge, several species prefer to den whenever possible in the ventilation shafts of the large mounds built by *Macrotermes* termites. Consequently, these mongooses are often found only where such termite mounds are available, rather than where other aspects of habitat, such as vegetation type, are right for them.

In contrast, woodlands, rocky habitats, and soft soils provide numerous retreats, and mongooses are often the most abundant carnivores in such regions. In the Serengeti study, Dwarf mongooses were found to attain densities up to 8 adults per sq km (21 per sq mi) and the combined densities of Dwarf, Banded, and Slender mongooses could exceed 20 adults per sq km (52 per sq mi). At least three other species (White-tailed, Large gray, and Marsh mongooses) also used the area, so total mongoose density probably exceeded 25 adults per sq km (65 per sq mi).

Right *Some representative mongoose species.* ***1*** *White-tailed mongoose* (Ichneumia albicauda), *the largest of all the mongooses;* ***2*** *Bushy-tailed mongoose – Kenyan subspecies* Bdeogale crassicauda omnivora *– sniffing the air in a typical mongoose "high-sit" posture;* ***3*** *Ring-tailed mongoose* (Galidia elegans) *in fast, active trot;* ***4*** *Dwarf mongoose* (Helogale parvula) *adult shown feeding beetle to juvenile;* ***5*** *Selous' mongoose* (Paracynictis selousi) *in a low, sitting posture;* ***6*** *Narrow-striped mongoose* (Mungotictis decemlineata)*;* ***7*** *Egyptian mongoose* (Herpestes ichneumon) *preparing to break open an egg by throwing it between its legs onto a rock;* ***8*** *Marsh mongoose* (Atilax paludinosus) *scentmarking a stone by using the "anal drag" method.*

1
2
3
4
5
6
7
8

Digging for Dinner

DIET

Ever since Rudyard Kipling recounted the duel between Riki-tiki-tavi and the cobra, it has been a common assumption in the West that mongooses feed mainly on snakes, but in fact there are probably not enough of these within the range of any species to predominate in its diet. Most mongooses are actually opportunists feeding on small vertebrates, insects, and other invertebrates, and also occasionally on fruits. The structure of their feet reflects the diet, for the long, nonretractile claws are an adaptation for digging; the mongoose sniffs along the surface of the ground and, when it finds an insect, either snaps it up from the surface or unearths it from its underground home.

From Solitary to Social

SOCIAL BEHAVIOR

Most mongooses, like most carnivores, live alone, though mothers form stable groups with their dependent offspring. The Egyptian mongoose, a resident of woodlands and savannas, provides a good example of their typical social organization. Each female has a territory that is defended against others of her sex. Males also defend territories, which are larger and generally overlap the ranges of several females. Adult males and females rarely travel together, and rarely interact outside of mating periods. Male mongooses make little or no contribution to parental care during gestation, lactation, or the rearing of weaned young.

Feeding on prey like mice, snakes, and lizards requires stealth, which explains why most mongooses are solitary; if two mongooses are easier to detect than one, then hunting together carries a cost. Solitary mongooses habitually take prey small enough to be easily killed by a single hunter, so there is little scope for cooperative hunting to yield benefits.

In a minority of mongoose species, however, circumstances favor group living. In contrast to the 3.5kg (7.7lb) Egyptian mongoose, Dwarf mongooses weigh only 350g (12oz), and are vulnerable to a much broader set of predators. Thus, there is a strong premium for Dwarf mongooses to be aware of nearby activity on the ground and in the air, particularly as foraging for beetles and larvae requires that they keep their heads down to shift through soil and litter. One obvious solution to these conflicting demands is to have more than one pair of eyes, and Dwarf mongooses accomplish this by living in groups. Pack sizes range from 2–21 adults, with an average of nine.

Other species that have adapted to communal living include Gray meerkats and Banded mongooses. Gray meerkats, like Dwarf mongooses, are targets for a long list of larger carnivores and raptors. When a group is foraging, one meerkat will climb to an exposed vantage point and scan for predators, sometimes remaining at its post for more than an hour. If a predator is detected, the sentinel gives a loud alarm call.

Social mongooses have developed a vocabulary of calls to inform group-mates whether the threat comes from the ground or from the air, so they can react accordingly. To avoid a raptor, mongooses sprint headlong to the nearest hole; to avoid a larger carnivore, they will still sprint for cover, but the dash is less heedless. There are good reasons for this behavior. First, raptors approach far more quickly than a mongoose's top speed, and nothing less than a full-tilt sprint is likely to get an animal to cover in time. Yet over short distances a mongoose can run almost as fast as a larger carnivore and so has some leeway to pick a path that will avoid the attacker. Another factor is that, while terrestrial predators may have only intermittent sight of their quarry, raptors have a relatively uninterrupted view, making route-picking less profitable when fleeing from birds of prey.

Social species share common features: they are small, diurnal, occupy open habitats such as grasslands or tree-savannas, and eat insects as opposed to vertebrate prey. Solitary species, in contrast, are generally larger, and typically occupy woodland habitats; many are nocturnal or crepuscular. The small mice or birds that they live on are usually found in much lower densities than insect prey, making their diet much less easy to share.

Not all mongooses neatly fit this pattern, however, and the exceptions reveal the intricacy of social evolution. For example, the White-tailed mongoose is the largest member of the family, and it is strictly nocturnal (and thus is rarely seen, even where it is common) – traits that are typically associated with solitary species in which each individual defends a territory. Yet White-tails feed primarily on termites and, congruently with their hybrid ecology, display a hybrid social system. In

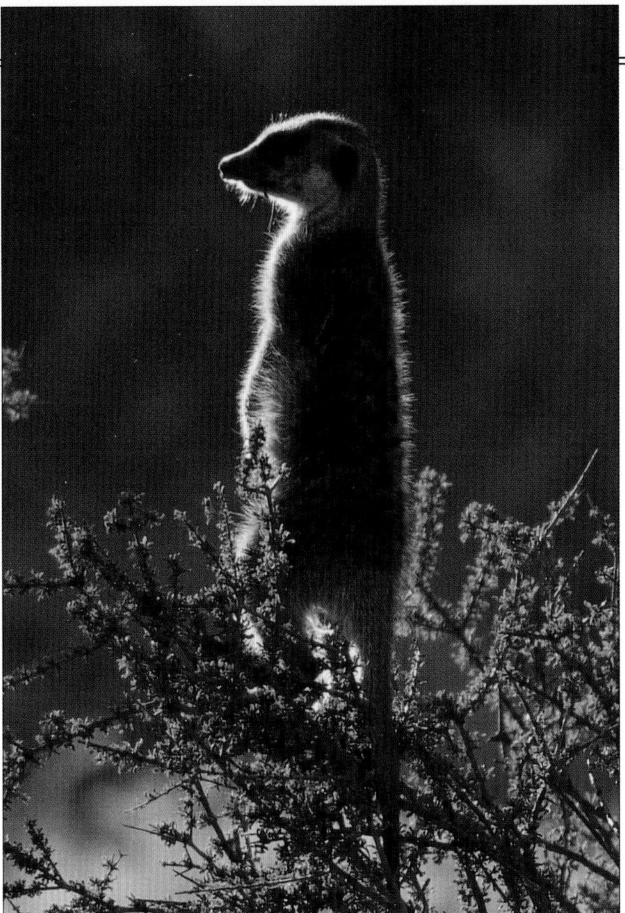

Left A Marsh mongoose ngrossed in eating a mussel at the ater's edge in the Serengeti. espite a complete lack of webbing etween the toes, the Marsh monoose is the most aquatic of all the pecies; it is often found in the cinity of estuaries and marshes.

Right A suricate, or meerkat, anding watch from its vantage oint in a thorn bush. A call from uch a sentinel will cause a meerkat oop to scatter instantly and take over. Troops typically contain 2–3 mily units comprising a male, a male, and 2–5 young.

Below A pack of Banded monooses in Namibia. They employ a utual defense strategy; for example, when a jackal approaches they ill move toward it as a tightly knit unch, thus creating the appearance f a single large animal from which e jackal will retreat.

the western Serengeti, while individuals were almost always seen alone, radio-tracking showed that several females were in fact using and defending almost identical home ranges. The species, it seems, is solitary in its foraging and movements, but social in its use of space. A simple experiment showed that removing all of the insects on the ground in a small plot had no obvious effect on the number available for hungry mongooses later that same night. Thus, two White-tailed mongooses foraging independently within the same home range would be unlikely to reduce the availability of food for one another, thereby permitting coalitions of females to form with no cost.

The benefits of group living for White-tailed mongooses come when the young are due to leave home. Females settle with their mother, but males must leave. Instead of heading off alone, they gang together in coalitions of 2–4 animals. As the chances of dying at this time may exceed 50 percent, especially in densely populated areas where young males must search far and wide for a homestead, grouping is likely to carry survival benefits.

A Bird–Mammal Alliance

Death in the talons of a raptor is not uncommon in mongoose society. Subordinate males in a group take it in turns to act as guards to warn of a raptor's approach. But the Dwarf mongooses in the Commiphora woodland of Kenya's Taru desert have an unusual ally: they are accompanied as they travel by Yellow-billed, Red-billed, and Decken's hornbills (*Tockus flavirostris*, *T. erythrorhynchus*, and *T. deckeni*) that also keep watch from the air.

In the morning, the birds wait in trees around the termite mound until the mongooses emerge, shortly after which they all set off foraging together. Should the mongooses be late, the birds will call into the ventilation shafts, bringing the animals to the surface almost immediately.

Once underway the advantage to both the birds and the mongooses becomes apparent. The birds catch grasshoppers and flying insects disturbed by the mongooses. In return, the hornbills are very efficient sentries, and being more numerous than the mongoose guards (around 3–4 birds per guard) usually spot threatening raptors first. Hornbills give their warnings by flying vertically into the trees and calling, causing the mongooses to scatter and hide as they do when their own guards shout a warning. Extraordinarily, hornbills warn for small raptor species that are not dangerous to themselves but which are predators on baby mongooses. The birds do not call when they spot small raptors that are not mongoose enemies. Again, the bird's strategy replicates that of the mongoose guards.

When 12 or more birds are present, the mongooses frequently post no guards at all. By relying on the hornbills, subordinate mongoose males can spend more time foraging and at the same time reduce their chances of being predated. Although they may lose food to the hornbills, mongooses profit from the birds' superior vigilance abilities: a uniquely understood case of mutualism between birds and mammals. AR

An Ecological Success Story

CONSERVATION AND ENVIRONMENT

Mongooses are a widespread and successful group. No species is known to be in danger of extinction, but the most vulnerable are likely to be the four Madagascar mongooses, as a result of destruction of their habitat.

The Small Indian mongoose, Yellow mongoose, and suricate have all been targets of persecution, but even so all are still abundant. The two latter species, both from southern Africa, have been shot or gassed in their burrows as rabies carriers. The Small Indian mongoose, which was first introduced into the West Indies in the 1870s (and to the Hawaiian Islands in the 1880s) in an attempt to control rats in the sugarcane plantations, is now considered a pest because of its attacks on chickens and native fauna; it has also been implicated with rabies. Although it is sometimes accused of causing the extinction of native birds and reptiles, this has not been proved. On many islands this mongoose is still an important predator on harmful rodents, and its economic status should be considered separately on each one. SC

Mongoose Species

MONGOOSES ARE DIVIDED INTO TWO subfamilies: the Galidiinae (Madagascar) and the Herpestinae (other parts of Africa and Asia). Herpestinae species are uniformly colored, lack ear bursae and perineal scent gland, and females have two or three pairs of teats. Three of the five Galidiinae are striped; all have ear bursae, some a perineal gland, and females have a single set of teats.

AFRICAN & ASIAN MONGOOSES

Subfamily Herpestinae

Marsh mongoose
Atilax paludinosus
Marsh or Water mongoose

Gambia E to Ethiopia, S to S Africa. Nocturnal; solitary; semiaquatic.
HBL 50cm; TL 35cm; WT 3.5kg.
FORM: coat dark brown to black; no webbing between toes; bare heel pad present.

Bushy-tailed mongoose
Bdeogale crassicauda

Mozambique, Malawi, Zambia, Tanzania, Kenya.
HBL 43cm; TL 24cm; WT 1.6kg. Nocturnal; solitary.
FORM: coat dark brown to black; Kenya subspecies lighter in color; four (not five) toes on each foot.

Jackson's mongoose
Bdeogale jacksoni

C Kenya and SE Uganda.
HBL, TL, and WT as for *B. nigripes*.
FORM: considered by some as conspecific with *B. nigripes*.

Black-legged mongoose
Bdeogale nigripes

Nigeria to N Angola, C Kenya, SE Uganda. Nocturnal; solitary.
HBL 60cm; TL 37cm; WT 3kg.
FORM: coat light gray to brown, with yellow to white-tipped guard hairs; fur dense and short; belly and tail white; chest and legs black; four (not five) toes on each foot.

Alexander's mongoose
Crossarchus alexandri
Alexander's or Congo mongoose

Zaire, W Uganda, Mt Elgon, Kenya. Diurnal; group-living.
HBL 40cm; TL 27cm; WT 1.5kg.
FORM: coat mainly brown with black feet; nose elongate and mobile.

Angolan mongoose
Crossarchus ansorgei

N Angola, SE Zaire. Diurnal; group-living.
HBL 32cm; TL 21cm; WT about 1kg.
FORM: coat brownish, with black legs and tail tip; nose not elongate.

Kusimanse
Crossarchus obscurus
Dark mongoose, kusimanse, or Long-nosed mongoose

Sierra Leone to Cameroon. Diurnal; group-living.
HBL 35cm; TL 21cm; WT 1kg.
FORM: coat dark brown to black; nose elongate and mobile.

Yellow mongoose
Cynictis penicillata
Yellow mongoose or Red meerkat

S Africa, Namibia, S Angola, Botswana. Diurnal; lives in pairs or family groups.
HBL 31cm; TL 22cm; WT 0.8kg.
FORM: coat orange-yellow, speckled with gray; chin and tail tip white; ears large.

Pousargues' mongoose
Dologale dybowskii
Pousargues' or Dybowski's or African tropical savanna mongoose

NE Zaire, Central African Republic, S Sudan, W Uganda. Diurnal.
HBL 27.5cm; TL 20cm; WT 0.4kg.
COAT: dark brown grizzled with tan; feet and legs black.

Dwarf mongoose
Helogale parvula (Includes *H. hirtula.*)

Ethiopia to northern S Africa, W to N Namibia, Angola, and Cameroon. Diurnal; group-living.
HBL 24cm; TL 19cm; WT 0.32kg.
FORM: coat varies from grayish-tan to dark brown with fine grizzling; bare heel pad.

Short-tailed mongoose
Herpestes brachyurus (Includes *H. hosei.*)

Malaysia, Sumatra, Java, Philippines. Nocturnal/crepuscular; solitary.
HBL 49cm; TL 24cm; WT 1.4kg.
COAT: brown to black, with guard hairs banded with brown to red; legs black.

Egyptian mongoose
Herpestes ichneumon
Egyptian mongoose or ichneumon

Most of Africa except for the Sahara, C and W African forest regions, and SW Africa; Israel, S Spain, and Portugal. Mainly diurnal; solitary.
HBL 57cm; TL 50cm; WT 3.6kg.
COAT: grizzled gray, with black tail tuft.

Indian gray mongoose
Herpestes edwardsii

E and C Arabia to Nepal, India, and Sri Lanka. Diurnal; solitary.
HBL 43cm; TL 39cm; WT 1.5kg.
COAT: gray to light brown, finely speckled with black.

Indian brown mongoose
Herpestes fuscus

S India and Sri Lanka. Solitary.
HBL 38cm; TL 30cm; WT 1.6kg.
FORM: coat blackish-brown to sandy-gray, speckled with black; feet darker than head and body.

Long-nosed mongoose
Herpestes naso

SE Nigeria to Gabon and Zaire. Nocturnal: solitary.
HBL 55cm; TL 41cm; WT 3kg.
FORM: coat dark blackish-brown grizzle with buff; crest of hairs from nape to shoulders; nose distinctly elongate.

Small Indian mongoose
Herpestes javanicus (Includes *H. auro unctatus.*)
Small Indian mongoose or Javan gold-spotte mongoose

N Arabia to S China and Malay Peninsu Sumatra; Java; introduced into W Indie Hawaiian Islands, Fiji. Diurnal; solitary.
HBL 39cm; TL 26cm; WT 0.8kg.
COAT: either light brownish-gray, speck with black to dark brown (in arid region or red speckled with black and gray (we tropical regions).

Cape gray mongoose
Herpestes (Galerella) pulverulentus

S Angola, Namibia, S Africa. Crepuscular/nocturnal; solitary.
HBL 34cm; TL 34cm; WT 0.75kg.
FORM: coat brownish-gray to gray, spec led with black (black and red forms known); tail with dark brown to black ti feet dark brown; bare heel pad.

ABBREVIATIONS HBL = head–body length TL = tail length WT = weight.
Approximate nonmetric equivalents: 10cm = 4in; 230g = 8oz; 1kg = 2.2lb

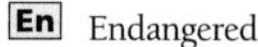
En Endangered Vu Vulnerable

Left An Indian gray mongoose squares up to an Indian cobra. Mongooses are popularly believed to be immune to snakebites, but this is not the case; they are merely fast and agile enough to avoid being bitten.

nder mongoose
rpestes (Galerella) sanguineus

ica S of Sahara. Diurnal; solitary.
. 35cm; TL 30cm; WT 0.6kg.
RM: coat varies from gray, yellowish, or dish-brown to red; tail tip black; bare el pad.

rpestes (Galerella) swalius

nd Central Namibia
, TL, and WT: as for *H.sanguineus.*
RM: considered by some to be conspe- c with *H. sanguineus.*

rpestes (Galerella) flavescens

nd C Namibia, S Angola
, TL, and WT: as for *H. sanguineus.*
RM: considered by some to be conspe- c with *H. sanguineus.*

ab-eating mongoose
rpestes urva

hina, Nepal, Assam, Burma, Indochi- se peninsula, Taiwan, Hainan, Sumatra, rneo, Philippines. Nocturnal; solitary.
L 50cm; TL 32cm; WT 3.4kg.
AT: dark brown to gray; legs black; ite strip from mouth to shoulder; tip of light-colored. May include *H. palustris.*

Ruddy mongoose
Herpestes smithii

India, Sri Lanka. Nocturnal; solitary.
HBL 45cm; TL 40cm; WT 1.9kg.
COAT: light brownish-gray to black, speckled with white and red; feet dark brown; tail tip black.

Stripe-necked mongoose
Herpestes vitticollis

S India, Sri Lanka. Diurnal and crepuscular; solitary.
HBL 54cm; TL 35cm; WT 2.9kg.
FORM: coat grizzled-gray, tipped with chestnut; legs and feet dark brown or black; tail tip black; black stripe from ear to shoulder; bare heel pad.

Meller's mongoose
Rhynchogale melleri

S Zaire, Tanzania, Malawi, Zambia, C and N Mozambique. Nocturnal; solitary.
HBL 47cm; TL 38cm; WT 2.6kg.
COAT: reddish-brown, grizzled with tan; legs dark brown.

Liberian mongoose En
Liberiictis kuhni

Liberia. Diurnal; group-living.
HBL 42cm; TL 20cm; WT 2.3kg.
COAT: blackish-brown with dark brown stripe, bordered by two light brown stripes from ear to shoulders.

Gambian mongoose
Mungos gambianus

Gambia to Nigeria. Diurnal; group-living.
HBL 32cm; TL 19cm; WT 1.8kg.
COAT: grizzled-gray and black; black stripe on side of neck contrasting with buff-white throat.

Banded mongoose
Mungos mungo

Africa S of Sahara, excluding Congo and SW Africa. Diurnal; group-living.
HBL 34cm; TL 22cm; WT 1.8kg.
COAT: brownish-gray; feet dark brown to black; tail with black tip; dark brown bands across back.

White-tailed mongoose
Ichneumia albicauda

Sub-Saharan Africa except for the C and W African forest regions and SW Africa; S Arabia. Nocturnal; solitary.
HBL 58cm; TL 44cm; WT 4kg.
COAT: gray, grizzled with black and white; tail usually white but may be dark in some areas; legs and feet black.

Selous' mongoose
Paracynictis selousi
Selous' mongoose or Gray meerkat

Southern Africa from Angola to northern S Africa. Nocturnal; solitary.
HBL 45cm; TL 35cm; WT 1.7kg.
FORM: coat brown, speckled gray-black; tail white-tipped; four toes on each foot.

Suricate
Suricata suricatta
Suricate, meerkat, Gray meerkat, or stokstertje

Angola, Namibia, S Africa, S Botswana. Diurnal; group-living.
HBL 29cm; TL 19cm; WT 0.9kg.
COAT: tan to gray; broken brown bands on back and sides; head and throat grayish-white; eye rings, ears, and tail tip black.

MALAGASY MONGOOSES

Subfamily Galidiinae

Ring-tailed mongoose Vu
Galidia elegans

Madagascar. Diurnal; live in pairs.
HBL 37cm; TL 27cm; WT 0.9kg.
FORM: coat light tan to dark red-brown; 5–7 dark bands on tail; bare heel pad.

Broad-striped mongoose Vu
Galidictis fasciata
Broad-striped mongoose or Madagascar banded mongoose

Madagascar. Nocturnal; live in pairs.
HBL 35.5cm; TL 29cm; WT unknown.
COAT: gray-brown with broad, longitudinal stripes from nape to beyond tail base.

Giant-striped mongoose En
Galidictis grandidieri

Spiny desert of SW Madagascar.
HBL, TL, and WT unknown.

Narrow-striped mongoose En
Mungotictis decemlineata

W Madagascar. Diurnal; live in pairs.
HBL 34cm; TL 27cm; WT 0.8kg.
FORM: coat brownish-gray with speckling on back and sides; 10–12 narrow, reddish-brown to dark brown longitudinal stripes on back and sides; bare heel pad.

Brown mongoose Vu
Salanoia concolor

E Madagascar. Diurnal; live in pairs.
HBL 24cm; TL 16cm; WT 0.8kg.
COAT: reddish-brown, speckled with black and tan. JR/WCW/SC

TOUGH AT THE TOP

The costs of dominance in Dwarf mongoose packs

OUR UNDERSTANDING OF EVOLUTION BY NATURAL selection emphasizes that individuals who survive and reproduce better than their contemporaries will pass on the traits that favored their success. Conversely, the traits of individuals that do not survive and breed will disappear. Ultimately, the number of offspring, grand-offspring, and great-grand-offspring is the currency by which evolutionary success and failure are measured.

At first sight this solid underpinning of evolutionary biology would not appear to explain why some animals, including mongooses, wild dogs, and wolves, raise the young of other individuals within their group while they themselves bear no offspring. A new understanding of why this strategy just might pay dividends has come, perhaps oddly, from analysis of the animals' droppings.

Among many social mongooses, dogs, and hyenas, groups are structured by a dominance hierarchy within each sex, and the only individuals assured of breeding are the alpha (or dominant) male and female. The other animals may be prevented from breeding by the sophisticated internal workings of their hormones. Research carried out on captive animals has shown that stress, when sufficiently intense and prolonged, can shut down reproduction. Stress increases the secretion of the hormone cortisol by the animal's adrenal gland, and cortisol in turn reduces the production of sex hormones such as testosterone and estrogen which make an animal's reproductive processes tick. Unpleasant physical stressors such as cold or hunger can increase cortisol secretion, but so too can psychological stressors such as fighting, and particularly losing a fight. Since subordinate animals, by definition, lose more fights than they win, chronic social stress might explain why subordinates do not breed in many social species.

This notion is sometimes known as the "psychological castration" hypothesis. Dominant animals, so the theory goes, are rarely subject to aggression from their group mates, and thus experience little social stress. Subordinates, on the other hand, are burdened with constant aggression and social stress, provoking cortisol secretion that throws a wrench into their reproductive physiology.

In the past, this hypothesis has been difficult to test in the wild. The problem is that determining hormone levels involves getting hold of some of an animal's blood, which in turn means catching, restraining, and anesthetizing it. The animal meanwhile believes it has been caught by a large predator (which it has!) and quickly increases the production of cortisol preparing it for flight or fight – which makes it difficult to get a measure of its normal hormone levels.

Now, however, methods have been developed to measure the levels of stress and reproductive hormones in urine and feces. The idiosyncratic ups and downs of an individual's hormones can be tracked without harming or affecting the animal at all, but simply by collecting these natural byproducts of everyday life.

This new tool has yielded many new findings on the dwarf mongooses of Serengeti National Park. Living in groups of 4 to 5 adults of each sex, with all adults cooperating to raise the pack's offspring, dwarf mongooses provide an excellent opportunity for weighing up the benefits and drawbacks of having children of one's own. One to four litters are born in each rainy season, with 2–3 offspring per litter. Genetic data reveal that 85 percent of these young are born to the alpha (dominant) female, and 76 percent are fathered by the alpha male. In other words, the alphas have a near-monopoly on being parents to the young raised by the group.

Yet, against expectations, the new data has shown that it is the dominant individuals that have exceptionally high stress hormone levels, while those of the subordinates are altogether more relaxed. Dominants are therefore breeding successfully despite levels of cortisol that are consistently more than double those of their underlings. Subsequent studies of free-living African wild dogs, wolves, several birds, and a number of primates have identified the same pattern. It now appears that, for many animals, social stress is a cost of being dominant, not a consequence of being subordinate.

Above *Young dwarf mongooses foraging for food on a termite mound. Packs are organized along hierarchical lines, and are always headed by a mature female; on her death, the pack will likely disperse.*

Left *Dwarf mongooses, particularly animals of the opposite sex, often perform mutual grooming. This species displays one of the most highly developed forms of social organization among mammals.*

This surprising result helps explain why subordi-ates tolerate their status. If, in addition to mono-olizing reproduction, dominants benefited from aving low stress-hormone levels, then social ominance would be a one-way street, with all the enefits accruing to the alphas. The new endocrine ata suggest, however, that the costs and benefits f dominance flow both ways. Dominant animals aintain their reproductive advantage only by ithstanding the costs of a prolonged increase in ress-hormone levels.

In dwarf mongooses, age is a good predictor of ominance rank – so good that an uninformed uess that the oldest animal is dominant proves orrect 8 times out of 10. But this fact only raises e question of why a large, vigorous mongoose in s prime should accept the situation, rather than ggressively challenging an older animal that may e smaller and in poorer condition. Dominance is ought to evolve so that individuals who are con-sting a resource can avoid the costs of fighting. If the outcome can be predicted fairly reliably in advance, then the outcome of "respecting" dominance is the same as if the fight actually occurred, with the prize (usually food or a mating opportunity) going to the dominant animal. This argument rests on the assumption that the dominant animal would win if challenged to a serious fight. In mongooses, as in most carnivores, dominants are involved in agonistic (confrontational) interactions at higher rates than subordinates.

However, if social dominance carries the hidden cost of tolerating chronically high stress-hormone levels, then being dominant may compromise survival. Raised cortisol levels have undesirable side effects: they make energy production less efficient, impair digestion, raise blood pressure, and even suppress the immune system. Thus far, there is little evidence to show that dominant individuals have shorter lifespans than subordinates, but this too raises a tricky problem. If individuals that become dominant are in better condition or of high genetic quality, then they may start with survival prospects that are better than average, but that are reduced by these costs. So the two effects may cancel each other out, resulting in no difference in death-rates between the groups.

It has proved very difficult to document this sort of fitness trade-off in the wild, and many biologists would argue that controlled experimental studies in captivity are necessary to resolve the problem. In the meantime, it is clear that the physiological consequences of social status are more complex and interesting than was originally thought. SC

DISTRIBUTION Africa except Sahara and the Congo basin. Middle East to Arabia, India, and S Nepal.

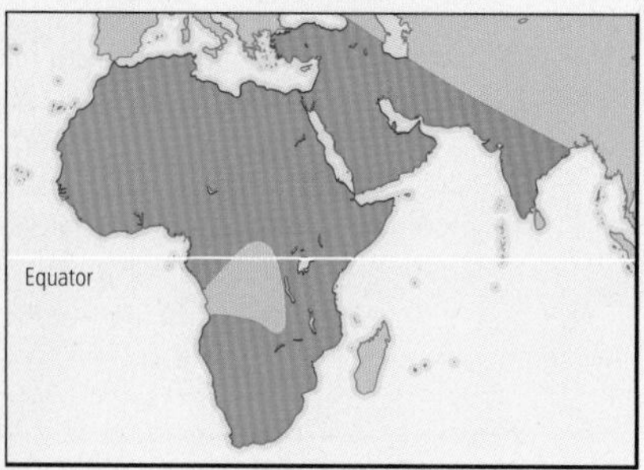

Order: Carnivora. Family: Hyaenidae. 4 Species in 4 genera

STRIPED HYENA *Hyaena hyaena*
5 subspecies: *H. h. barbara*, NW Africa; *H. h. dubbah*, NE Africa; *H. h. syriaca*, Syria, Asia Minor, Caucasus; *H. h. hyaena*, India; and *H. h. sultana*, Arabia. HTL 1.2–1.45m (4–5ft); SH 66–75cm (2.2–2.5ft); WT 26–41kg (57–90lb). Coat: long-haired, pale gray to beige, with 5–9 vertical body stripes, black throat patch. Diet: mainly mammalian carrion, also invertebrates, eggs, wild fruits, and organic human waste; may kill small mammals and birds. Breeding: 1–4 (usually 3) cubs born throughout the year after 90-day gestation; cubs weaned at 10–12 months; females sexually mature at 2–3 years. Conservation status: Lower Risk.

BROWN HYENA *Parahyaena brunnea*
Brown hyena or Beach or Strand wolf
Widespread in S Africa, particularly in the W; also into S Angola. Mainly inhabits arid regions. HTL 1.26–1.61m (4.1–5.3ft); SH 72–88cm (2.4–2.9ft); WT 28–49kg (62–108lb). Coat: shaggy and dark brown to black, with neck and shoulders white; white horizontal stripes on legs. Diet: as Striped hyena; eats about 2.8kg (6.2lb) of food daily. Breeding: 1–5 cubs born throughout the year after 90-day gestation; cubs weaned at 12 months and leave den at 18 months; females sexually mature at 2–3 years. Conservation status: Lower Risk.

SPOTTED HYENA *Crocuta crocuta*
Spotted or Laughing hyena
Sub-Saharan Africa, except the Congo rain forests and far S. Inhabits desert to fringes of tropical rain forest; highest densities in savannas. HTL 1.3–1.85m (4.3–6ft); SH 70–95cm (2.3–3.2ft); WT male: 45–62kg (100–137lb), female: 55–82.5kg (121–182lb). Coat: short-haired, sandy, ginger or dull grey to brown, dark spots on back, flanks, rump, and legs. Diet: scavenges and hunts medium to large hoofed mammals; eats 3–6kg (7–13lb) of meat daily. Breeding: usually 1–2 cubs born throughout the year after a gestation of about 90 days; cubs weaned at 12–16 months; females sexually mature at 2–3 years. Conservation status: Lower Risk.

AARDWOLF *Proteles cristatus*
2 subspecies: *P. c. cristatus*, S Africa, and *P. c. septentrionalis*, E Africa. Mainly open, grassy plains with 100–600mm (4–24in) annual rainfall. HTL 0.85–1.05m (2.8–3.4ft), SH 0.45–0.5m (1.5–1.6ft); WT 8–10kg (17.6–22lb). Coat: long-haired, yellowish-white to rufous, paler on throat and underparts; 3–5 vertical black stripes on body, 1–2 diagonal stripes on fore- and hindquarters; irregular horizontal stripes on legs. Diet: mainly termites. Breeding: seasonal breeder; 1–4 cubs born after 90-day gestation; cubs weaned at about 4 months, when they leave the den. Conservation status: Lower Risk.

Abbreviations HTL = head–tail length SH = shoulder height WT = weight

Hyena Family

HYENAS HAVE NEVER ENJOYED A GOOD PRESS. They are often regarded as ugly, cowardly, and sinister, though they are really none of these. In fact they are among the most interesting and intelligent of animals, with an extraordinary social system. Although there are only four living species in four genera, they show a remarkable diversity of ecological adaptations and social behavior.

Rituals that Maintain Social Bonds

FORM AND FUNCTION

The three true hyenas – the Spotted, Brown, and Striped – are doglike animals with large heads and powerful forequarters. Their phenomenally strong teeth and jaws enable them to crush all but the largest bones of heavy animals such as zebra and kudu in order to extract the nutritious marrow. The hydrochloric acid in their stomachs is highly concentrated, making them more efficient than any other mammal at extracting nutrients from bone. In contrast to the forequarters, the hindquarters are poorly developed, which gives rise to their backward-sloping appearance. The Brown and Striped hyenas (and the aardwolf) have long hair, which they raise in situations of conflict, an adaptation that serves to make them look larger and more intimidating. The Spotted hyena has short fur.

The female Spotted hyena's genitalia are almost indistinguishable from the male's – hence the myth that they are hermaphrodites. The clitoris is large, like a penis, and the vaginal labiae are fused to form pseudo-testes. Females give birth through the penislike clitoris. These organs are displayed in a ritualized meeting ceremony during which two hyenas stand head to tail, raise the inner hin leg, and mutually sniff and lick each other's erec ed "penis." These rituals maintain social bonds and hierarchies. In the other species the sexual organs are normal, and meeting ceremonies involve mutual anal sniffing.

A Vigorous Life within the Clan

SOCIAL BEHAVIOR

Hyenas illustrate the generalization that ecology, and particularly feeding ecology, has a profound effect on social behavior. Whereas the Brown hyena always moves around on its own, Spotted hyenas are frequently encountered in groups. Because carrion is easy to locate by smell, there i no need for Brown hyenas to cooperate when fo aging. More importantly, most of the food items they find provide a meal for only one individual, so foraging in company would lead to strife. Wh little is known of Striped hyenas suggests their lifestyle is similar to the Brown hyena's.

The most obvious explanation for why Spotte hyenas forage in groups is that, through coopera tion, they can improve their hunting success. This, however, is only true up to a point. In the Ngorongoro Crater large groups are necessary to pull down zebra, but in the southern Kalahari a single hyena is just as successful as several at pulling down a gemsbok calf, which is their maj prey in this region. A more important reason for group hunting is that the hyenas often feed on large food items that are capable of providing a meal for several animals at once. A closer look at

Below *The feeding behavior of two species of the hyena family:* **1** *Spotted hyena pack cooperatively hunting down a zebra;* **2** *two Brown hyena juveniles play as an adult approache with its kill, a Bat-eared fox.*

Food and Feeding

Strong jaws and teeth, a superbly efficient digestive system, and the ability to cover large distances at night make hyenas the most efficient mammalian scavengers. Carrion makes up easily the bulk of the diets of Brown and Striped hyenas; however, this will be supplemented by invertebrates, wild fruits, eggs, and the occasional small prey animal. In this sense, these hyenas are nature's waste-disposal system; in Africa's southern Kalahari, only 4.2 percent of Brown hyenas' diet is made up of animals that they have killed themselves.

The Spotted hyena (LEFT), in contrast, is as efficient at hunting as it is at scavenging. Spotted hyenas hunt by running down their prey over distances of up to 3km (2mi) at speeds approaching 60km/h (37mph). Their targets are usually the young of large antelope, such as gemsbok and wildebeest, but they are also capable of pulling down prey as large as zebra and even buffalo adults. The proportions of scavenged versus killed food in the diet of individual clans varies significantly from area to area; in South Africa's Kruger National Park, they kill less than 50 percent of their food, a proportion that rises to about 70 percent in the southern Kalahari and to about 90 percent in Tanzania's Ngorongoro Crater. Spotted hyenas often bury food in muddy pools; when they are hungry they will return to this hidden cache.

The diet of the aardwolf differs greatly from that of the three true hyenas: even in captivity they appear reluctant to consume meat. The aardwolf's diet is composed mainly of termites and insect larvae, for which they forage at night.

Skulls and Dentition

The skulls of hyenids are robust and long. All members of the family have a complete dental formula of I3/3, C1/1, P4/3, M1/1 = 34 (among carnivores only members of the cat family have fewer teeth). However, in the aardwolf the cheek teeth are reduced to small, peg-like structures, spaced widely apart and often lost in adults, leaving as few as 24 teeth.

The more massive skulls of hyenas have relatively short jaws which give a powerful grip. Hyena skulls suggest two distinct trends of adaptation. That of the Spotted hyena is specialized for crushing large bones and cutting thick hides, which other large carnivores are unable to consume and digest. The bone-crushing premolars are relatively large and the carnassial teeth are used almost solely for slicing or shearing.

In Brown and Striped hyenas the corresponding premolars are smaller and the carnassials do the crushing, chopping, and shearing. These differences relate to the smaller species' dependence on a wider range of food, including insects, wild fruit, and eggs as well as carrion and prey.

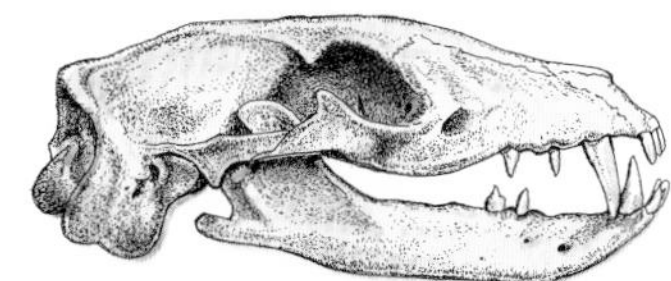

Aardwolf
13.5 cm

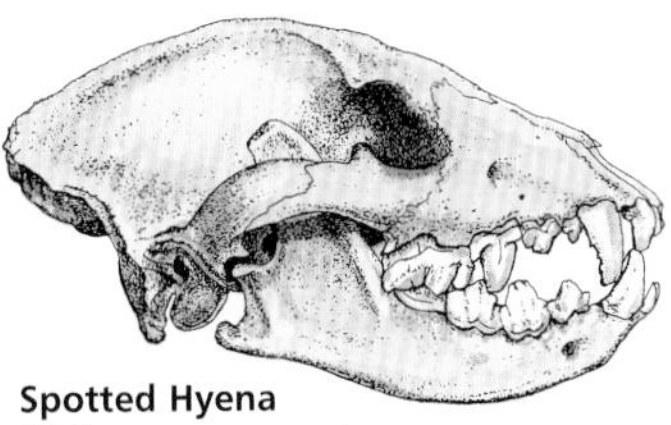

Spotted Hyena
26.3 cm

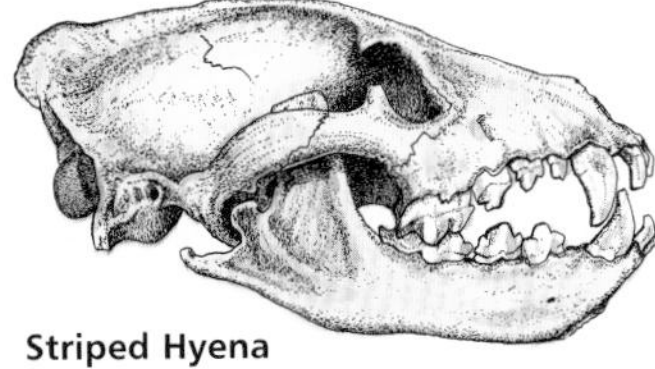

Striped Hyena
23.7 cm

Scentmarking and Greeting

Hyena social systems are complex. The true hyenas have elaborate meeting ceremonies and efficient long-range communication by scent. In the aardwolf, only scentmarking approaches this complexity.

A unifying feature of hyenas is the anal pouch, used in a unique form of scentmarking called pasting. Striped and Spotted hyenas deposit a single creamy paste; however, the Brown hyena produces a lipid white secretion and a black paste. A Brown hyena places its anal gland on a grass stalk **1** and moves forward, leaving a blob of white paste. As the anal gland retracts, a black smear is deposited above the white blob **2**. There may be as many as 15,000 active pastings in a territory. Thus intruders are quickly made aware of the owner's presence. The white paste, which has a short-lived smell, warns intruders off; the black paste probably provides information about foraging sites to clan members.

All three species turn the anal pouch inside out during encounters with other animals; however, in Spotted hyenas, which do not present the anal region during meeting ceremonies, this is linked to aggression. The selective advantage of reducing tension has, it seems, resulted in two types of display during meeting ceremonies. In Brown and Striped hyenas these involve erection of the dorsal crest **3**, sniffing of the head and body, inspection of the anal pouch, and ritual fighting. In Spotted hyenas it includes mutual sniffing and licking of the genital area and erect penis or clitoris as the animals stand head to tail with one hind leg lifted **4**. This display differs from the state of the sex organs at mating; it is conspicuous in cubs, and the Spotted hyena that begins contact is often lower in the dominance hierarchy. The function is one of appeasement.

the composition of these groups shows that they are more likely to comprise individuals that are closely related to each other than would be expected if the group was selected randomly fror the members of the clan. Kin selection is the process whereby animals favor their close relative over more distantly related or unrelated individuals, with whom they share fewer genes. In this case, the advantage is that, if the group has a successful hunt, then the rewards of each individual animal's efforts will be shared by its relatives rather than by unrelated animals.

The clan is the unit of hyena society. Striped, Brown, and Spotted hyenas all live in clans. Members of a clan share a territory and defend it against neighbors. Spotted hyena clans are ruled by a strict, female-dominated, linear hierarchy, in which the highest-ranking of the males is subordinate to the lowest-ranking female. Males leave their natal clan at subadulthood, work their way into a new clan, and then ascend up the male hierarchy to get mating opportunities. Females usually remain in their natal clan and inherit thei mother's rank. Brown hyena clans are more egalitarian. Some males and females leave the clan at subadulthood, others stay with their natal clan fc an extended period, perhaps for life. Males that leave their clans become nomadic, or join a new clan. Both nomadic males and immigrants mate with females.

Even when moving alone, Spotted hyenas maintain some direct contact with the rest of the clan, responding to sounds that are only audible to humans with the aid of an amplifier and headphones. Calls that the unaided human ear can

Above This close-up of Striped hyenas shows nale nibbling the neck of a female. In the rather ıgthy ritual fighting that is a component of the eeting ceremony of the Striped hyena, the throat or ck of a subordinate hyena is often bitten and held shaken.

Below In the Chobe National Park, Botswana, ɔack of Spotted hyenas gather around the carcass an elephant. Group feeding is frequently noisy, but ely involves serious fighting; the animals will gorge ite rapidly and may consume as much as 15kg 3lb) of flesh in a sitting.

hear include whoops, yells, and a kind of demented cackle, from which this species derives its alternative name of Laughing hyena. Whoop calls, in particular, are well suited to long-range communication, since they carry over several kilometers; each call is repeated a number of times, which helps the listener locate the caller, and each hyena has a distinctive voice. Woodland Spotted hyenas of the Timbavati Game Reserve in the Transvaal, South Africa, will frequently approach or answer tape-recordings of their companions; but if recordings of strange hyenas are played, the residents often arrive in groups at the run, calling excitedly, with manes raised, tails curled high, and anal glands protruding. Infant hyenas will answer the pre-recorded whoops of their mothers, but not those of other clan hyenas.

The size of the clans varies between the species – Spotted hyena clans tend to be larger than the others – and also within species under different ecological regimes. In the southern Kalahari, a Spotted hyena clan may have as few as five members, whereas in the Ngorongoro Crater a clan may be 80 strong. Brown hyena clans may comprise no more than a female and her latest litter of cubs, or as many as 15 members. So too the size of the clan territory can vary widely. In the Ngorongoro Crater the 80 Spotted hyenas may live in a 40sq km (15.5sq mi) territory, whereas the five Kalahari hyenas might live in a territory of up to 1,000sq km (400sq mi). The dispersion pattern of food is largely responsible for territory size variation. In the Ngorongoro Crater the density of wildebeest and zebra provides enough food to support a large clan in a small area. In the arid southern Kalahari, where the hyenas may have to travel 50km (30mi) or more to find a kill, the territories are much larger and the clans smaller.

All hyenas keep their young in dens, usually composed of a series of underground holes. Only the cubs can fit into a den, which provides excellent protection during the long periods that the adults are away. The denning period is extended, lasting about 18 months. In a Spotted hyena clan all the females keep their cubs at a large communal den. Mostly there will be only one breeding female in a Brown hyena clan, although sometimes two or even three females will breed, and then a den will be shared.

Spotted hyenas are unusual amongst carnivores in that cubs are born with their eyes open and their teeth erupted. Litter mates, which are normally twins, engage in high levels of aggression within minutes of birth, which quickly leads to the establishment of a dominance hierarchy between siblings, allowing the dominant cub to control access to maternal milk. Sometimes this aggression will lead to the death of the smaller cub. This appears most likely to happen when resources are in short supply and probably insufficient to sustain two cubs.

The manner in which hyenas raise their young varies. The communally-denning Spotted hyenas do not feed their cubs meat until at about 9 months, when the cubs are able to travel with their mother to kills. Until then the cubs are wholly reliant on the mothers' very rich milk. The young are not fully weaned until 15 months of age. Brown hyena cubs are also not weaned until they are more than a year old. However, their milk diet is supplemented from about three months of age with food that all clan members carry back to the den.

Why these differences in cub rearing? Once again we need to look to the feeding habits of the two species for the answers. If a Brown hyena acquires a large piece of food, it can eat part of it, then carry the rest off to the den. Should five Spotted hyenas acquire a similar food item, there

would obviously not be anything left to take bac to the den. Additionally, at a Brown hyena den there is usually one litter of cubs of equal size. A a Spotted hyena den there may be many cubs of unequal ages and the older ones would domina the food. Finally, in many areas, the large distances that Spotted hyenas move from the den would make carrying food back a fairly dauntin task. Not that Brown hyenas do not sometimes perform heroic feats of food carrying. One instance involved a Brown hyena carrying a carcass weighing about 7kg (15lb) for 15km (9 miles) to a den. As the Brown hyenas are related to the cubs at the den, which are their offspring, half-sibs, or cousins, food carrying is another example of kin selection.

Victims of Ancient Prejudice

CONSERVATION AND ENVIRONMENT

Although none of the hyena species are endangered, several populations are threatened, often because of unnecessary persecution brought about by misconceptions, myths, and a generall negative attitude on the part of their human neighbors. Throughout the Arabian peninsula a North Africa, the Striped hyena is loathed as a grave robber and therefore severely persecuted through poison-baiting and trapping. The Brow hyena is enjoying slightly better fortunes, as farme

in its southern African home are slowly changing their attitudes. It is safe in some of the large protected areas, such as the Namib-Naukluft Park, the Kgalakgadi Transfrontier Park, and the Central Kalahari Game Reserve. The Spotted hyena, being predatory, is the most likely member of the family to come into conflict with local people by killing livestock, and its IUCN status is therefore Lower Risk: Conservation Dependent, in contrast to the Lower Risk: Least Concern status of the other two species. It is particularly common in many of the large national parks and other protected areas in eastern and southern Africa. MGLM/SKB

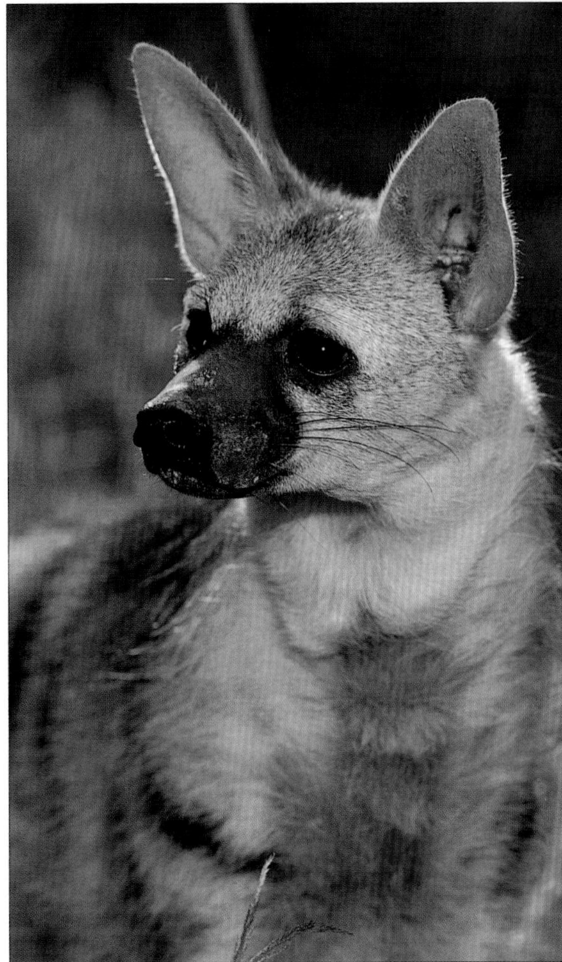

The Aardwolf

PROTELES CRISTATUS

The aardwolf, which resembles a miniature Striped hyena, has poorly developed jaws, and its molar teeth are reduced to tiny pegs in testimony to its insect diet. Paleontological and genetic studies show that it belongs to the hyena family, having diverged from other hyenas 15–32 million years ago. In contrast, the Spotted hyena separated from the Brown and Striped hyenas 10 million years ago and the Brown and Striped split six million years ago.

The aardwolf is a highly specialized eater, feeding predominantly on termites and in particular on one local species of Nasute (snouted) harvester termite of the genus *Trinervitermes*; these termites are nocturnal, as a result of poor pigmentation that leaves them unable to tolerate direct sunlight; they also produce a chemical that deters many other predators. In the cold winter months in South Africa, when the Nasute harvester termites become inactive, the aardwolf switches to the Diurnal pigmented harvester termite (*Hodotermes mossambicus*). In the rainy season in East Africa, it also broadens its diet slightly and includes termites of the genera *Odontotermes* and *Macrotermes*. Unlike other ant- or termite-eating animals, which dig out their prey from their underground nests, the aardwolf uses its long tongue to lick up insects. The Snouted harvester termites forage in dense columns and the aardwolf may consume as many as 200,000 in one night. Very occasionally, aardwolves will also take small mammals, nestling birds, and carrion.

The aardwolf has a different type of social system from the other hyenas. In terms of spatial organization, it appears to be monogamous; a pair share a territory of 1 to 4sq km (0.4–1.5sq mi) with their most recent offspring. The size of the territory is determined by the density of termites. An intruding aardwolf may be chased away up to 400m and fights will take place if the intruder is caught. However, the territorial system may be relaxed somewhat at times when food is short and individuals from other territories may forage in the same area at the same time.

Aardwolves are seasonal breeders. During the mating season, which occurs in July in South Africa, the appearance of monogamy is abandoned, as highly promiscuous, dominant males make forays into neighboring territories and mate with their neighbor's female. Thus the male aardwolves, which diligently help in rearing young by guarding the den, particularly against jackals, may be cuckolded, unwittingly protecting another male's offspring.

An aardwolf family group may have 10 dens scattered throughout their territory; they use only one or two at any time, but will change dens as often as every four to six weeks.

Cubs are usually born in litters of 2–4; they are born with their eyes open, but are otherwise helpless. For the first 6–8 weeks the cubs will remain in the den, during which time the male may spend six hours a day keeping watch on a den while the female forages. At about 12 weeks the cubs begin foraging, accompanied by at least one adult; by 16–20 weeks they will forage alone for much of the night. By the start of the next breeding season the cubs will be wandering far beyond their parent's territory; most of the subadults will have emigrated from the area by the time the next generation of cubs is beginning to forage.

In terms of conservation, the aardwolf is classified by the IUCN as Lower Risk: Least Concern. In South Africa, however, it is susceptible to insecticide poisoning aimed at locusts. PRKR

Above *A Spotted hyena suckles cubs in the Kruger* *tional Park, South Africa. Although hyena young* *raised in communal dens, care of cubs is the sole* *ponsibility of the mother. These cubs will not be* *ly weaned until they are about 15 months old, per-* *os even older; at no point during this period will* *od be carried back to the den for the cubs, as* *curs with both the Brown and Striped hyenas.*

Above right *A captive aardwolf. The external* *s of the aardwolf are very large, and it appears that* *e animal locates its prey mainly through the use of* *und. The aardwolf also has a long tongue, covered* *a sticky saliva and large papillae, which helps it lick* *insects.*

Left *In the Kalahari National Park, this Brown* *ena engages in an aggressive display by raising its* *rsal crest. This is often done in the ritualized fighting* *at forms part of the Brown hyenas meeting ceremony,* *t may also be done to intimidate an enemy by mak-* *g an individual animal look bigger.*

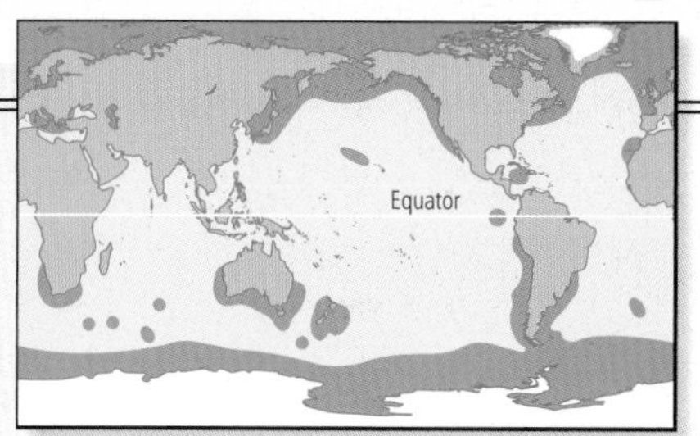

SEALS & SEA LIONS

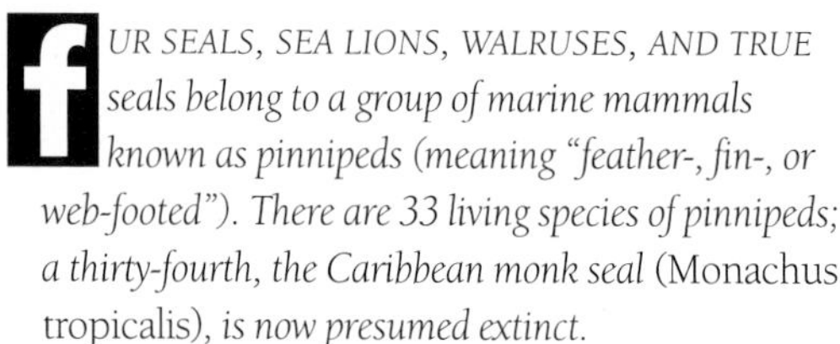

f UR SEALS, SEA LIONS, WALRUSES, AND TRUE seals belong to a group of marine mammals known as pinnipeds (meaning "feather-, fin-, or web-footed"). There are 33 living species of pinnipeds; a thirty-fourth, the Caribbean monk seal (Monachus tropicalis), *is now presumed extinct.*

All pinnipeds have streamlined, spindle-shaped bodies with limbs that are modified as flippers. They have a relatively large adult body size, ranging from small female fur seals that may weigh less than 30kg (66lb), to massive adult male Southern elephant seals, which can weigh over 4 tonnes.

The pinnipeds have long been recognized as aquatic relatives of terrestrial mammals belonging to the order Carnivora. Unlike other marine mammals the pinnipeds have never severed their links with the land. For this reason, pinnipeds have been aptly described as "amphibious carnivora." Although they spend much of their lives in the water, they must still return to land (or ice) in order to give birth.

The pinnipeds divide into three families: the Otariidae or eared seals, including nine species of fur seal and five of sea lion; the Odobenidae, comprising only one species, the walrus; and the Phocidae or earless seals, including 18 species of true or hair seals.

The phocid seals can be easily distinguished because their hindflippers extend behind the body and cannot be brought forward in order to

1b

1a

2a

3a

2b

3b

alk. In water, propulsion is provided by alternat-
g strokes of the hindflippers. The phocids also
k external ears. All phocids have a thick layer of
sulating subcutaneous fat, termed blubber.

In contrast, otariids are capable of bringing eir hind flippers forward and under their bodies order to walk. In water, propulsion is provided the fore flippers, while the hind flippers play no tive role. Otariids possess a flap of skin, a rudi-entary pinna with cartilaginous support, near e external ear opening. Their pelage is generally icker than that of true seals, and is made up of merous, long guard hairs overlying a thick, oolly underfur. The otariid pelt is essentially aterproof and traps air within its structure to ovide insulation underwater. Otariids generally ve less blubber than phocid seals.

The one surviving odobenid, the walrus, hibits a combination of phocid and otariid aits. Its flippers are triangular in shape like those phocid seals, but the walrus can bring them rward in the manner of otariids. When moving ound on ice, the walrus will often drag its ndquarters with the hindflippers trailing hind, in a manner similar to phocid seals. hen in water, the hindflippers provide propul-on. Also as in phocids, the walrus ear lacks an vious pinna. Unlike either of the other two milies, however, the walrus is essentially naked, pelt being reduced to scattered hairs over the dy surface. The walrus is the only pinniped in hich males have descended testes that are quite vious during the breeding season. Both sexes ve elongated canine teeth.

A Single Ancestor

ORIGINS AND EVOLUTION

here has been a great deal of debate about the igins, evolution, and phylogeny of pinnipeds id, hence, about their taxonomy and historical ogeography. Disagreement has focussed largely n whether seals are monophyletic or diphyletic. wealth of new information, however, rejects the tter hypothesis in favor of the former.

According to the monophyletic view, all pin-peds evolved from a single ancestor, an arctoid rnivore. This lineage gave rise to the earliest nown pinniped, *Enaliarctos mealsi*, which first pears in the fossil record about 23 million years go. Among the latter group, modern otariids represent a conservative family that was the first to split off from the main line of pinniped evolution. In other words, walruses and phocid seals share a more recent common ancestor and, consequently, are more closely related.

Acceptance of the monophyletic theory has several important implications regarding pinniped taxonomy. Most obviously, the superfamily names Otarioidea and Phocoidea disappear entirely from the classification scheme. Within the order Carnivora, *Enaliarctos*, together with extant pinnipeds, comprise a monophyletic group that is termed the Pinnipedimorpha. However, the old name Pinnipedia – used formerly to denote the pinnipeds either as an order unto themselves or as a suborder within the Carnivora – has been retained for all here. Taxonomists are still determining which specific taxonomic level, if any, this term should be reapplied to.

Left *The characteristic features of the three fami-s of pinnipeds.* ***1*** *The Cape fur seal* (Arctocephalus usillus), *a typical eared seal, showing* ***a*** *the scroll-like ternal ear flaps and thick fur; the male shown here as a particularly thick mane. On land* ***b*** *eared seals pport themselves on their fore flippers and bring eir rear flippers beneath the body.* ***2*** *The walrus* Odobenus rosmarus), *showing* ***a*** *its distinctive tusks, hich on land* ***b*** *are often used as a lever.* ***3*** *The arbor seal* (Phoca vitulina), *a typical true seal, show-g* ***a*** *its sleek hair and its lack of external ear flaps. ue seals are cumbersome on land* ***b***.

Cold Water Specialists

PRESENT-DAY DISTRIBUTION

Generally, pinnipeds are inhabitants of polar and subpolar seas in both hemispheres. The greatest concentrations are found in subarctic regions of the North Atlantic and North Pacific, and in the Southern Ocean around Antarctica.

ORDER: CARNIVORA (PINNIPEDIA)

3 families: 21 genera: 33 species

EARED SEALS Family Otariidae p160

14 species in 7 genera

Species include: **Antarctic fur seal** (*Arctocephalus gazella*), **California sea lion** (*Zalophus californianus*), Northern fur seal (*Callorhinus ursinus*), **Steller sea lion** (*Eumetopias jubatus*), **South American fur seal** (*Otaria flavescens*)

WALRUS Family Odobenidae p174

1 species: walrus (*Odobenus rosmarus*)

TRUE SEALS Family Phocidae p180

18 species in 13 genera

Species include: **Mediterranean monk seal** (*Monachus monachus*), **Northern elephant seal** (*Mirounga angustirostris*), **Leopard seal** (*Hydrurga leptonyx*), **Weddell seal** (*Leptonychotes weddellii*), **Harbor seal** (*Phoca vitulina*), **Baikal seal** (*Pusa sibirica*), **Gray seal** (*Halichoerus grypus*), **Harp seal** (*Pagophilus groenlandicus*), **Bearded seal** (*Erignathus barbatus*), **Crabeater seal** (*Lobodon carcinophagus*), **Hooded seal** (*Cystophora cristata*)

Below *Eared seals are highly social animals, and may often congregate in considerable numbers. The almost balletic grace with which they glide through the water is demonstrated by this large group of California sea lions.*

Modern pinnipeds are largely confined to productive seas where surface temperatures do not exceed 20°C (68°F) at any time of the year. The one major oceanic region from which pinnipeds are virtually absent is the Indo-Malayan western Pacific, a region with a continuous history of warm temperatures.

The only pinniped species that do not live in regions simultaneously characterized by relatively low environmental temperatures and relatively high levels of productivity are the monk seals. Of the three species known in modern times, two – the Mediterranean monk seal and the Hawaiian monk seal – currently number about 500 and 1,400 animals respectively. Their numbers have been declining in recent decades. The third, the Caribbean monk seal, apparently became extinct sometime in the last 50 years.

Above *A Weddell seal basking on a rocky Antarctic islet. This most southerly of all seal species has the long, broadly-webbed, and mobile fore flippers that characterize all southern true seals.*

Wide Dispersal

Historical Zoogeography

If all pinnipeds arose from a single ancestor in the north Pacific, the question arises as to how modern pinnipeds came to be so widely dispersed. Briefly, the fossil record indicates that the otariids remained in the Pacific basin for millions of years, moving northward and westward to occupy coastal regions on both sides of the ocean. Later, with the closure of the Central American Seaway (an ancient waterway that separated North and South America), ancestors of modern fur seals moved into the south Pacific from the west coast of South America. From here they gained access to the South Atlantic and moved northward until they were limited by warmer waters. Other otariids dispersed to southern Africa and throughout the Southern Ocean, where they encountered suitable breeding habitats on the dispersed subantarctic islands. More recently, ancestors of modern sea lions appear in the fossil record on both sides of the Pacific basin. In a relatively short period of geological time, they apparently crossed the equator and dispersed into the south Pacific, following the fur seals to the west and, later, east coasts of South America, and to Australia and New Zealand, where they still exist.

The historical zoogeography of the phocid seals is more speculative, because of a limited fossil record. Phocid ancestors must have moved fron their center of origin in the North Pacific throug the Central American Seaway and into the Atlan basin, where they appear in the fossil record of t eastern seaboard of the United States some 15 million years ago. Whether the ancestors of the Hawaiian monk seal made their way to Hawaii without first entering the Caribbean remains a mystery. What is known, however, is that this species retains some characteristics that are mor primitive than those found in the earliest fossil phocids from the eastern United States.

Regardless, the phocid seals continued to mo northward and eastward to occupy both sides of the North Atlantic, where they continue to resid Later, ancestors of the Ringed seal probably mov eastward into the Paratethys Sea (covering the ar now occupied by the Black, Caspian, and Aral seas, and intermittently linked to the Mediterran ean) from which they entered the Arctic Ocean and dispersed throughout the circumpolar regio More recently, some of these seals moved southward along river systems to reach Lake Baikal in Siberia. A number of phocid lineages presumabl reentered the Pacific basin via the Arctic Ocean.

While most of the northern phocids became increasingly adapted to cold, productive subpola and polar conditions, the monk seals tried a diffe ent experiment, occupying warmer regions such as the Caribbean and, later, the Mediterranean Sea. Interestingly, the monk seal lineage also gave rise both to the elephant seals and to the Antarct phocids. The ancestors of at least some of these seals probably reentered the Pacific basin and, possibly, like the otariids, moved southward alon the west coast of South America around to the ea coast of South America, subsequently dispersing

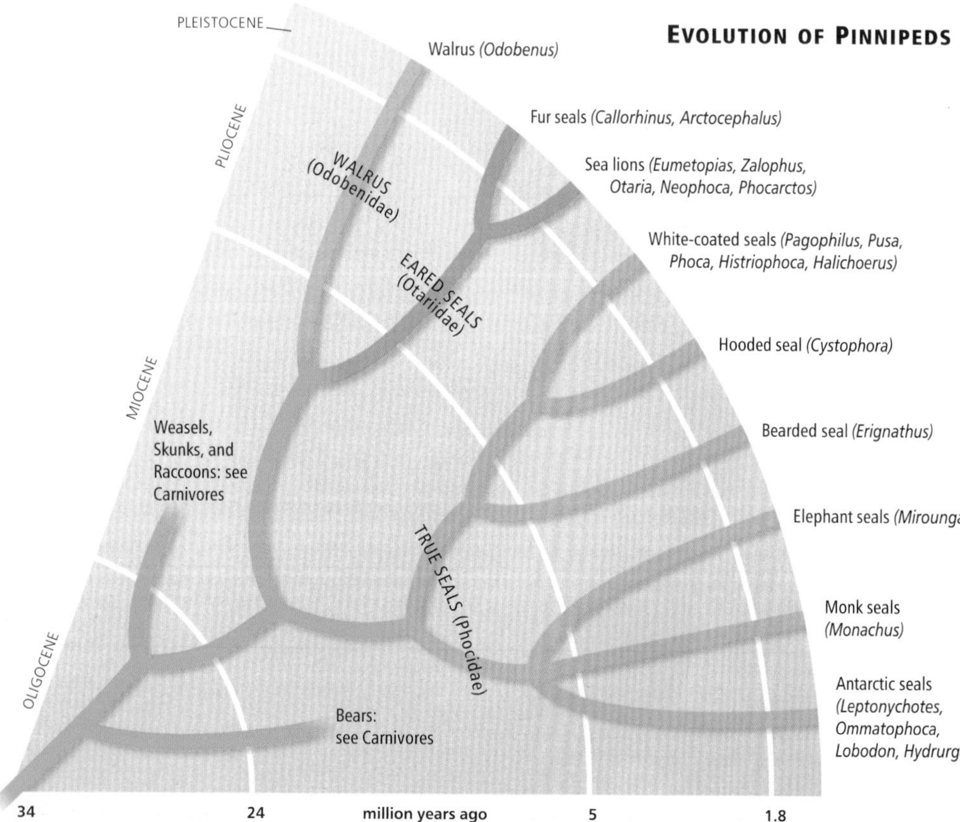

PINNIPED BODY PLAN

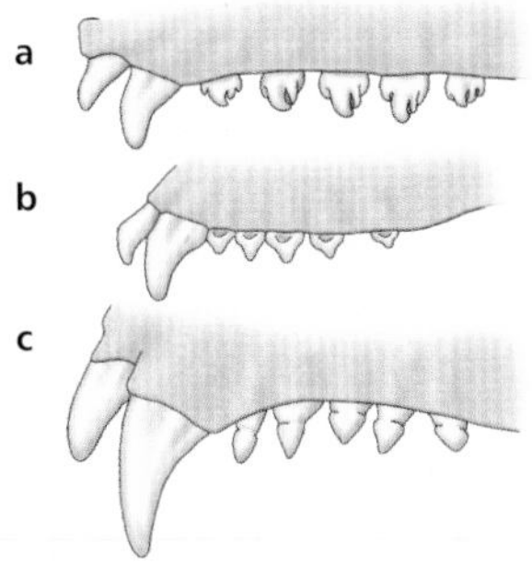

a

b

c

Left *The teeth of pinnipeds are more variable in number than those of most land carnivores. The teeth of the Crabeater seal **a** have quite elaborate cusps that leave only small gaps when the jaws are closed. This has to do with its habit of feeding almost exclusively on the small crustacean, krill. In contrast, the Weddell seal **b**, feeding on fish and bottom-dwelling invertebrates, has far simpler teeth. The dental formula of both Crabeater and Weddell seals is I1/1 or 2, C1/1, P5/5, while that of the South American sea lion **c** is I3/2, C1/1, P5/5.*

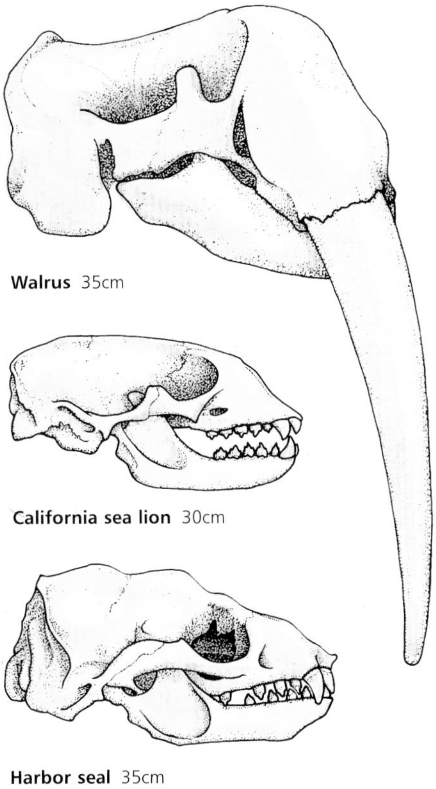

Walrus 35cm

California sea lion 30cm

Harbor seal 35cm

Above *Skulls of the walrus, California sea lion (an eared seal), and Harbor seal (a true seal). The skulls of true and eared seals are generally similar, except for the region behind the articulation of the lower jaw.*

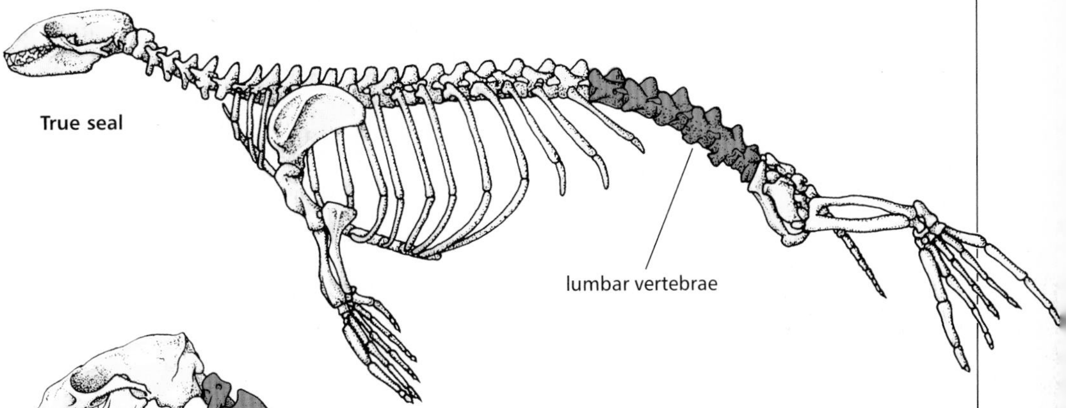

True seal

lumbar vertebrae

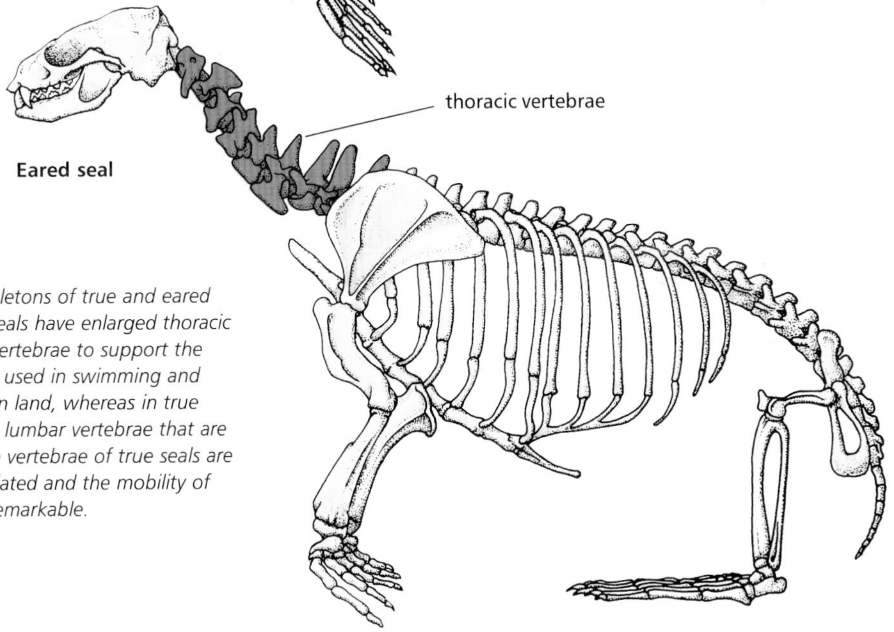

thoracic vertebrae

Eared seal

Right *Skeletons of true and eared seals. Eared seals have enlarged thoracic and cervical vertebrae to support the large muscles used in swimming and locomotion on land, whereas in true seals, it is the lumbar vertebrae that are enlarged. The vertebrae of true seals are loosely articulated and the mobility of the spine is remarkable.*

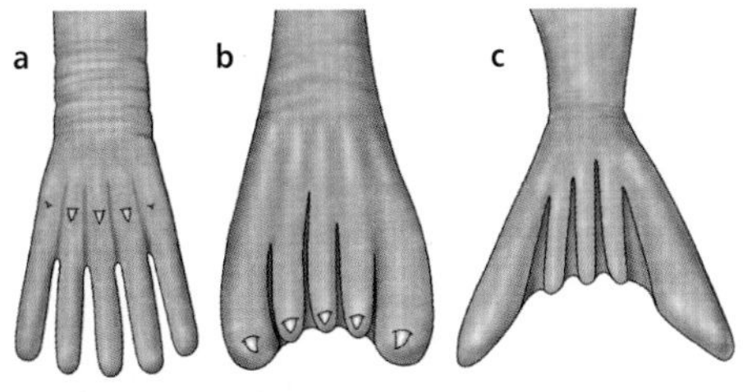

a b c

Above *Hindflippers of **a** a sea lion, **b** a Harbor seal, and **c** an Elephant seal. In sea lions, there are cartilaginous extensions to the digits, and the nails are reduced to nonfunctional nodules, some distance from the edge. The Harbor seal, in common with all northern true seals, has large claws, but these are reduced in southern seals, such as the elephant seals, which have fibrous tissue between the digits, which increases the flipper's surface area.*

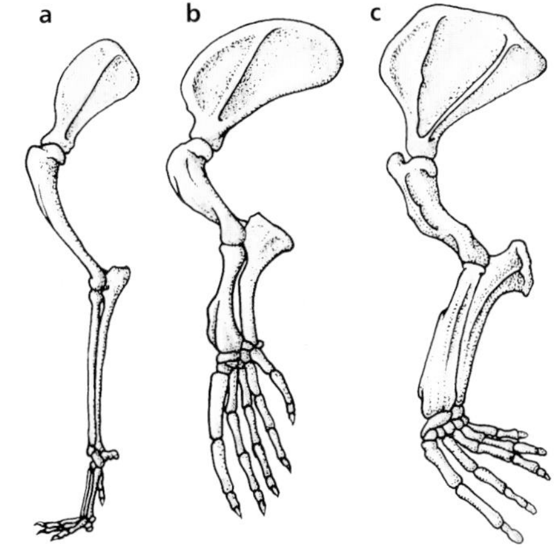

Left *Fore limbs of pinnipeds and carnivores contrasted. Compared to the greyhound **a**, both true seals **b** and eared seals **c** show a broadening and elongation of the digits. In eared seals, the digits decrease in length serially from the first. The fore flippers of true seals are more variable: the fifth digit of northern true seals is not much shorter than the first but in the monk seals, shown here, the fifth is considerably shorter, while the other four are of similar length.*

throughout the islands of the Southern Ocean to Antarctica. Alternatively, it is also possible that at least some of the Antarctic phocids reached Antarctica via the Atlantic Ocean.

The oldest walruses are also known from the north Pacific. Members of the modern walrus lineage may have crossed the Central American Seaway into the Caribbean and dispersed northward in the north Atlantic. According to this view, the surviving genus (*Odobenus*) returned to the Pacific via the Arctic Ocean within the last 1 million years. Alternatively, *Odobenus* arose in the north Pacific and moved through the Arctic Ocean and into the North Atlantic. Both hypotheses account for the modern distribution of *Odobenus*.

Unique Adaptations

Sensory Organs

Pinnipeds, like other marine mammals, evolved from terrestrial ancestors that apparently first invaded coastal waters in response to new ecological opportunities, such as abundant food resources. But, unlike other marine mammals, the pinnipeds remained dependent on solid substrates in order to give birth and to nurse the young. The resulting dichotomous life style has only been accomplished through an evolutionary modification of the terrestrial mammalian plan. Pinnipeds thus exhibit a unique set of traits that permit an air-breathing, warm blooded, mammal that evolved initially on land to feed successfully and live for extended periods in cold, productive oceans. Such a feat of adaptation required numerous adjustments, the sum total of which have made the pinnipeds the unique creatures they are.

An important area of adaptation was sensory organs. Mammals' highly developed senses initially evolved to function on land, but the pinnipeds' lifestyle required them to function efficiently in water while retaining some capabilities in air as well. Various evolutionary modifications were required to meet these conflicting needs.

For pinnipeds, vision is probably the most important source of sensory information, and they generally have large eyes. The outer window of the eye, the cornea, protects the delicate inner parts. The cornea is constantly lubricated by tears from lacrimal glands, which protect it from salt water or sand. Unlike those of terrestrial mammals, pinnipeds' eyes lack tear ducts to drain away the tears.

The pinniped eye must function not only in air and in water, but also over a wide range of light intensities – this can extend from the blinding brightness of the pack ice on a sunny day to the gloomy darkness of a deep dive. In the former situation, the corneas of ice-dwelling pinnipeds can tolerate levels of ultraviolet radiation that would damage the human cornea.

In air, the typical mammalian eye focuses a clear image on the retina, a layer of light-sensitive cells lining the back of the eyeball. The image is focused by two "lenses," the outer cornea, and the lens inside the eye. In water, however, the cornea effectively disappears (much as a glass bead apparently vanishes when placed in a glass of water) because it has virtually the same refractive index as water; this is why most mammals cannot see well underwater. The pinnipeds have – like cetaceans – compensated for the "loss" of the cornea underwater by evolving a large, almost spherical, fishlike lens to increase the focusing power of their eyes.

Above *The eyes of pinnipeds are perfectly adapted for a life spent seeking out underwater prey. The pupils of this Northern elephant seal pup are hugely dilated as it moves through murky reedbeds. Due to their lack of any tear ducts, when seals are on land or ice, fluid simply drains onto the pelage around their eyes, making them look as though they are crying.*

Such a modification of the eye for underwater vision might be expected to result in impaired vision out of the water. In air, with both the cornea and the large lens bending light rays, seals do tend to be shortsighted, at least under some light conditions. However, their myopia is minimized in bright light by the iris diaphragm, which controls the size of the pupil. In bright light, the pinniped pupil closes to a narrow, vertical slit with a small pinhole at the top. Light entering the eye passes through this pinhole, reducing the ability of the cornea and lens to bend it and, in the process, minimizing the degree of myopia experienced.

In dim light the pinniped pupil opens to form a wide circle in order to let in as much light as possible. In so doing, both the cornea and the large lens become involved in focusing the image. Thus, in dim light above the water's surface, pinnipeds tend to be quite myopic and, because of the large, curved cornea, may suffer the additiona blurred vision associated with astigmatism.

Pinnipeds' retinas have also evolved to cope with the dimly-lit conditions they often encounte underwater. As in nocturnal mammals, the retina contain large numbers of very sensitive cells, known as rod photoreceptors, that provide increased sensitivity at the expense of the perceptio of fine detail and color. Further, the light-sensitiv pigments in the rods of pinnipeds tend to be mo sensitive to the color of their underwater environments. Thus, coastal species, like Harbor seals, have eyes that are most sensitive to green light; deep divers like Southern elephant seals, which inhabit the bluer waters of the open ocean, have blue-sensitive visual pigments. To enhance the sensitivity of their eyes, pinnipeds, like cats and other nocturnal mammals, also have a reflective layer behind the retina, called the tapetum lucidum. Light passing through the retina hits th tapetum and is reflected back, so that it has two chances of being detected by the photoreceptors.

Pinnipeds also retain a small number of photoreceptors – called cones – that in humans and other mammals provide acute color vision in bright light. As a consequence, there is some evidence that seals and walruses are capable of limited color vision.

Just as an eye that evolved on land does not nction very well underwater without modifica- on, so the ears of terrestrial mammals are not itirely appropriate for hearing in water. Sound aves no longer enter the ear via the air-filled iditory canal; rather, they are transmitted through e bones of the skull and reach the inner ear from l directions, resulting in distortion and the loss directional hearing.

Through the processes of evolution, the pin- ped ear has been modified to function under ater. Structural changes in the inner ear serve to nplify sound reception, and the bone housing e inner ear has become isolated from all but one the other bones of the skull, reducing the nount of sound that can reach it through the anium. Other modifications of the temporal one of the skull ensure that the ears remain fectively separated on either side of the head, permitting good directional hearing underwater.

Pinnipeds also have specialized tissues lining the auditory canal and the middle-ear cavity that automatically adjust the pressure within the ear when an animal dives. Additionally, evolutionary changes in the relative size of the internal parts of the ear prevent their overstimulation in water, where sound travels five times faster than in air.

As a result of such adaptations, the pinniped ear actually works better in water than in air. On land, their hearing range is similar to that of humans, only not quite as sensitive. Under water, however, pinniped hearing is superior, extending to higher frequencies than humans can register

Pinnipeds' possession of relatively high-frequency hearing, and the observation that the vocal repertoire of some species includes pulsed clicks, has led a number of biologists to conclude that seals and walruses, like some toothed whales and bats, use echolocation as an additional source of information. Although there is some laboratory evidence that seals may be able to discriminate between objects using clicks, the importance of this ability in the wild is unknown.

All pinnipeds have well-developed whiskers, called vibrissae. These are arranged in horizontal rows along both sides of the snout, and they vary in number and form between species. Some phocid seals (including Harp and Hooded seals) have whiskers that are "beaded" in structure, whereas those of several other pinnipeds (including monk

Below *Seals' and sea lions' mystacial vibrissae – the whiskers on the side of the nose – are highly sensitive. Each is set in a follicle surrounded by a connective tissue capsule well supplied with nerve fibers, indicating that they may be used to detect slight water movements caused by prey.*

and Bearded seals) lack the beads and are smooth.

The pinnipeds' vibrissae are well supplied with nerves, blood vessels, and muscles, and seem well-equipped to provide sensory information about the external world. While their precise functions are not well understood, there is some evidence that they provide tactile information in the same way as the vibrissae of cats and dogs do. The whiskers seem, for example, to aid ice-breeding seals in locating the circular holes they use to obtain access to the surface of the ice. On land or ice, they are used to investigate objects or other animals. It has also been suggested that they are sensitive to low-frequency, waterborne vibrations, and may function to detect the movement of fish or other aquatic organisms, especially when visibility is poor. Below the surface, they may also help the seal to judge speed, and in addition they serve as communication devices; for example, they are often pulled forward and held erect during aggressive encounters. Since there is considerable variation in the length, diameter, and surface structure of pinniped vibrissae, and in the angle at which they protrude from the nose, there may be some functional differences among species in the kinds of information they provide.

The region of the brain responsible for the sense of smell (the olfactory lobes) is small in pinnipeds, but even so odors remain important in their social life. Some pinnipeds, like Harp seals, recognize their own offspring primarily by smell, and male fur seals and sea lions use odor to determine the reproductive condition of females. In water, however, they keep their nostrils tightly closed and the sense of smell ceases to function.

From Krill to Penguins

Food and Feeding

When the pinnipeds first appeared, about 25 million years ago, they underwent a rapid species radiation, perhaps in response to the appearance of increased food stocks. These might have been associated with an increase in upwelling processes (caused by climatic events or movements of the Earth's crust), bringing nutrients to the surface and increasing oceanic productivity. Upwelling is common along western coastlines at high latitudes and at current divergences, and it is in such places that we find large numbers of pinnipeds.

Pinnipeds generally are opportunistic feeders, able to take a variety of prey. Seals surface to eat large prey, but smaller prey are eaten underwater.

Not all pinnipeds are generalist feeders. Some, such as the Crabeater seal, are extreme specialists. The Ringed seal feeds extensively on crustacea in its Arctic habitat; the Southern elephant seal and the Ross seal feed largely on squid; the walrus and the Bearded seal eat mainly bottom-dwelling invertebrates, such as clams. Some pinnipeds feed on warm-blooded prey. Many sea lions commonly take birds, while some prey on the offspring of other seals.

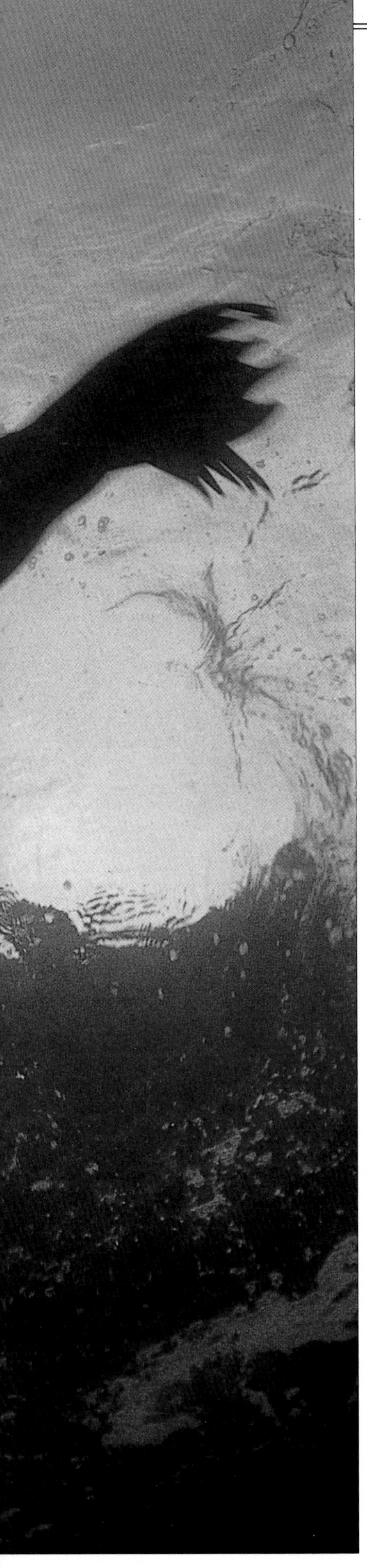

Steller sea lion

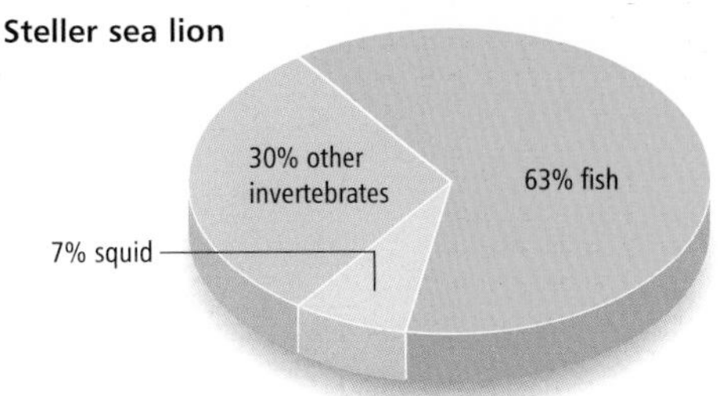

Crabeater seal

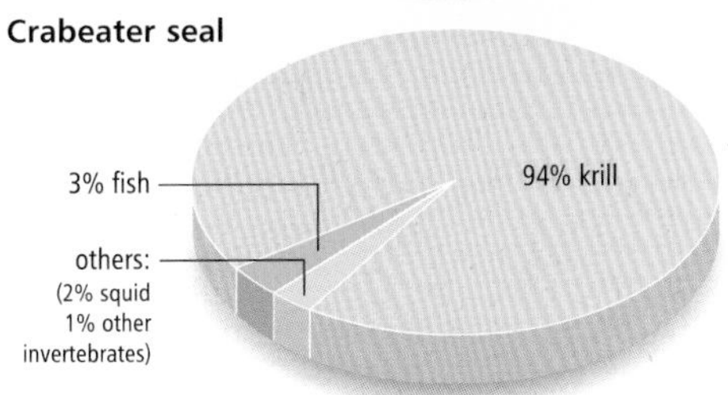

Leopard seal

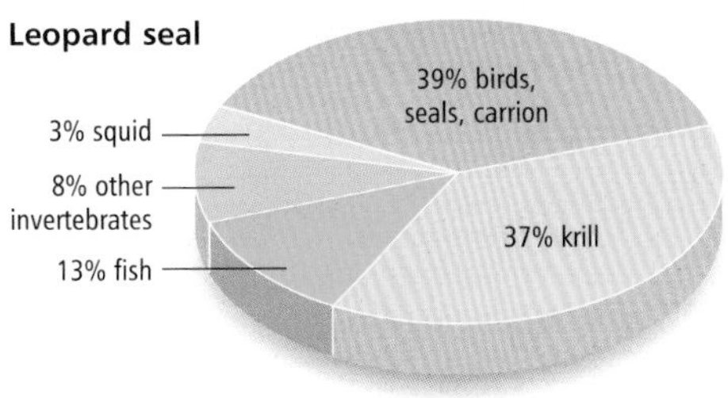

Above *The contrasting composition of seal and sea lion diets, from an opportunist general predator, the Leopard Seal, to a specialist, the Crabeater.*

Left *Twisting its lithe body acrobatically, a California sea lion grasps a spiny sea urchin in its jaws. This generalist feeder consumes a wide variety of fish, squid, and invertebrates.*

The jaws and teeth of pinnipeds are adapted for grasping prey rather than for chewing it. Most prey is swallowed whole, although pieces may be torn off a large item. Plankton feeders, like Crabeater or Ringed seals, have elaborately cusped teeth through which water can be strained out of the mouth before the prey is swallowed. In many seals, even krill eaters, the cheek teeth are reduced.

The pinniped stomach is simple and aligned with the long axis of the body, which may assist in engulfing large prey. The small intestine is often very long – in an adult male Southern elephant seal it can reach 202m (660ft) – in comparison with the human small intestine (about 7m (23ft) long). The cecum, colon, and rectum are relatively short in seals.

Little is known about the feeding rates of seals in the wild. Activity and water temperatures can affect these greatly. The Northern fur seal has been calculated to require 14 percent of its body weight in food daily for maintenance alone. Juveniles need proportionately more food than adults to allow for growth and greater heat loss. Most pinnipeds can undertake prolonged fasts in connection with reproductive activities or molting. The blubber layer, which acts as a food store as well as insulation, is very important in this respect.

Insulation by Fur and Fat

HEAT CONSERVATION

Because sea water is always colder than blood temperature, and heat is lost much more rapidly to water than to air, seals need adaptations to avoid excessive loss of heat. One very effective way of minimizing heat loss is to reduce surface area. The streamlining of the seal's body goes some way toward achieving this. Another important method is to take advantage of the relationship between surface and volume: for bodies of the same shape, larger ones have relatively less surface area than small ones. As a result, there are no small pinnipeds, in contrast to the numerous small rodents, insectivores, or land-based carnivores.

Another way to control heat loss is to insulate what surface there is. The layer of air trapped in the typical mammalian hair coat is an effective insulator on land. In water, however, a hair coat is much less effective, since the air layer is expelled as the hair is wetted; even so, by retaining a more or less stationary layer of water against the surface of the body, it still has a significant effect.

Yet one group of pinnipeds, the fur seals, have developed an improved method of insulation. The coat of all pinnipeds consists of a great number of units, each composed of a bundle of hairs and a pair of associated sebaceous (oil) glands. In each hair bundle, there is a long, stout, deeply-rooted guard hair and a number of finer, shorter fur fibers. In true seals and sea lions only a few (1–5) of these fibers grow in each bundle, but in the fur seals they comprise a dense mat of underfur. The fine tips of the fur fibers and the secretions of the sebaceous glands make the fur water-repellent.

Fur is an effective insulator, but it has the disadvantage that if the seal dives the air layer in the fur is compressed by half its thickness for each 10m (33ft) of depth, reducing its efficiency accordingly. As a result, seals have developed another mode of insulation in the form of the thick layer of fatty tissue, or blubber, beneath their skin, which also provides energy during fasting and lactation. Fat is a poor conductor of heat, and in air a blubber layer is about half as effective an insulator as an equal thickness of fur. In the water, however, the blubber insulation is reduced to about a quarter of its value in air, but its effectiveness is unaffected by depth. Seals commonly have in excess of 7–10cm (3–4in) of blubber, which effectively prevents heat loss from the body core. True seals have thicker blubber than eared seals.

In cold conditions, loss of heat from the flippers, which do not have an insulating covering, is minimized by reducing the flow of blood to them, only enough circulation being maintained to prevent freezing. Beneath the capillary bed, there are special shunts (passages) between the arterioles and the venules known as arterio-venous anastomoses, or AVAs. By opening the AVAs, more blood can be circulated through the superficial layers and heat can be lost when necessary.

Insulation that is effective in the water will also be effective in air. Skin surface temperatures, of course, may be much higher than the air temperature if the sun is shining. Most seals can thus easily tolerate cold climates, since they can endure almost any air temperature, and water never becomes much colder than about –1.8°C (28°F).

Not all pinnipeds live in cold climates, however; for those in temperate or tropical regions (mainly eared and monk seals) disposing of excess heat when out of the water can be a big problem. Monk seals in Hawaii avoid dry beaches on sunny days.

Fur seals can suffer severely from heat stress after periods of activity, for they can only lose heat across the surface of their naked flippers. To do this, the AVAs are dilated, so that more blood is diverted through the superficial layer, and heat is then radiated away from the black surface of the flippers. This may be aided by spreading the flippers widely, fanning them, or urinating on them.

True seals have AVAs over the whole of the body. Blubber contains blood vessels, and a true seal can divert blood through to the skin surface to lose heat. Conversely, the system can be used to gain heat from radiation in bright sunshine, even at very low air temperatures. Walruses also have AVAs over their body surface.

Every pinniped must periodically renew its hair covering and the superficial layer of its skin. In eared seals, molting is a relatively lengthy process. They shed the underfur fibers first; in fur seals some of these are retained in the hair canal. Guard hairs are molted shortly afterwards, but not all are lost at each molt.

Among true seals, however, molting is a far more abrupt process (evident in the great flakes of skin shed by the elephant seal). In order that the necessary growth of new hair may take place, the blood supply to the skin has to be increased, which also entails an increase in heat loss. Because of this, most seals stay out of the water for much of the duration of the molt, and some species, for example elephant seals, may gather together in large heaps, conserving heat by lying in contact with their neighbors.

Above *During the annual molt, which can last from just three days to well over a month, the skins of Southern elephant seals come to resemble an aerial view of the icefloe-bound waters they inhabit! Uniquely among seals, elephant seals shed the superficial layer of skin in large flakes, along with their rather scanty hair.*

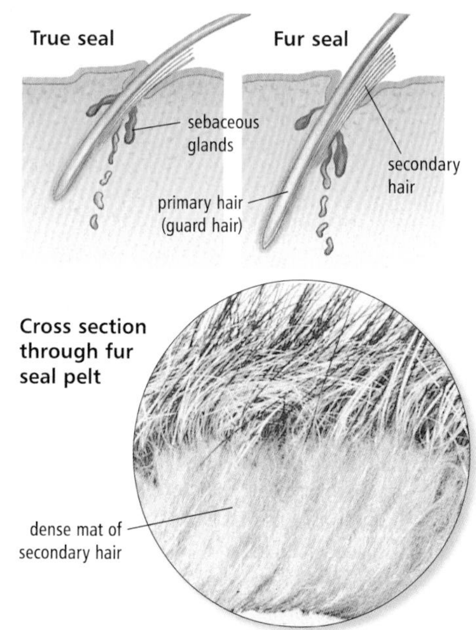

Above *Hair bundles of a true seal and a fur seal. In true seals, the primary (guard) hairs are accompanied by only a few secondary hairs, but in fur seals there may be 50 such fibers to each guard hair, giving a fiber density of up to 57,000 per sq cm. This extremely dense mat of fur is supported by the shafts of the guard hairs, as shown in the pelt section above.*

A Vulnerable Time

Reproductive Strategies

The adaptations that fit pinnipeds supremely we for life at sea often render them clumsy and vulnerable on land, to which they must resort to reproduce. Because they are more vulnerable to predation when they are ashore, they have had to develop various strategies to ensure their safety during the period of birth and the dependence o the young.

Typically, pinnipeds breed in the spring or earl summer. After a period of intensive feeding, they assemble at the chosen breeding site. Most true seals breed on ice (though monk seals, elephant seals, the Harbor seal, and most Gray seal popula tions are exceptions), which none of the eared seals do. Walruses are also ice-breeders.

Often the males arrive on the breeding grounc a few days or weeks before the females and take up territories ashore. The breeding females, carry ing fetuses that were conceived the previous season, come ashore only shortly before giving birth Birth is a speedy process in all pinnipeds, as the pup forms a convenient, torpedo-shaped packag that can slip out with equal facility either head or tail first. A single young is produced at a birth; twins are extremely rare, and are probably never reared successfully. The newborn pup is covered with a specialized birth coat, or lanugo. This is more woolly than the next coat, and is often of a different color. In most true seals, the lanugo is molted after two or three weeks, although Harbo and Hooded seals shed theirs while in the womb In eared seals, the lanugo is usually shed after tw or three months.

A few hours usually elapse before the mother first feeds her pup. The suckling period can be as short as a few days in some ice-breeding true seal such as the Hooded seal, while in other true seals

Above *South American sea lions commonly assem-* *on breeding beaches from October to December.* *peak of activity is reached in January, as the bulls* *outly defend their groups of cows against competi-* *rs. As many as 18 females may form a male's harem,* *ough, as here, the number is usually far smaller.*

is longer – about three weeks in Gray seals, and x weeks in Ringed seals. Many true seal mothers not feed themselves at all while suckling their ıps. At the end of lactation, the mother comes to season and is mated, weans her pup abruptly, ıd then deserts it. Subsequently, there is only sual contact between mother and pup.

In eared seals, the mother–pup association sts longer. About one week after the birth, the male comes into season and is mated by the earest dominant bull. She then departs on a ries of feeding excursions, between which she turns to feed her pup. The pup may be independent after 4–6 months, but often continues to ceive some milk from its mother until the arrival the next pup, or even beyond it.

In both eared and true seals, the fertilized egg itially develops only enough to form a hollow ıll of cells known as a blastocyst. It then lies dormant in the womb until the main period of feeding the previous pup is completed, usually at out four months after birth. After this, the blastocyst implants in the wall of the womb, develops placenta, and begins to develop normally. This elayed implantation may serve to let seals combine birth and mating into a single period.

Some seals remain near their breeding grounds roughout the year, but most disperse, either cally or, in the case of certain species, migrating r thousands of miles. During this period the als are building up reserves to see them through e next breeding season.

Exploitation and Recovery

Seals and People

Seals and humans have had a close relationship since primitive man first spread into the coastal regions where seals were abundant. Seals were ideally suited to hunter–gatherers, in part as a result of the modifications that fitted them to an aquatic life. They were large enough for the pursuit and killing of a single animal to provide an ample reward, yet not so big that there were major risks involved. Their furry skins made tough, waterproof garments to keep out the elements. The layer of blubber beneath the skin could be eaten as food with the rest of the carcass, but could also be burned in a lamp to provide light and warmth.

Stone Age hunters have left records of their association with seals in the form of engravings on bones and teeth. In the Arctic, the indigenous populations developed a culture that was largely dependent on seals and walruses for its very survival. They hunted whatever species were available, developing a complex technology of harpoon and kayak to do so. Seal carvings figure prominently in Eskimo art.

Native American peoples, from British Columbia southward, hunted seals and sea lions. At the extreme tip of South America, in Tierra del Fuego, the Canoe Indians hunted fur seals; when European seal hunters all but exterminated these, the Indians starved.

Subsistence sealing, as practiced by primitive communities or by crofter-fishermen in Europe up to the 20th century, made relatively little impact on seal stocks. However, the large aggregations of breeding monk seals and South American fur seals encountered by European explorers in the 15th and 16th centuries were particularly vulnerable to overexploitation, and some populations were rapidly wiped out. Other large aggregations of pinnipeds, such as the vast herds of Harp seals breeding on the ice off the east coast of Canada, required more sophisticated hunting techniques and so were not targeted until the early part of the 18th century.

Eared seals have suffered as much, if not more, than the true seals. The Northern fur seal was heavily exploited on the Pribilof Islands, in the Bering Sea off Alaska, from 1786 onward, with perhaps 2.5 million animals being killed by 1867. The banning of pelagic sealing in 1911 under the North Pacific fur Seal Convention allowed the population to recover, but an attempt to increase the potential harvest from the population by killing adult females between 1956 and 1962 brought a steady decline in numbers that has yet to be reversed.

In the Southern Hemisphere. fur sealing was combined with the hunting of elephant seals for their oil. Elephant seal stocks recovered in the first half of the 20th century to form the basis of a properly controlled and lucrative industry in South Georgia between 1910 and 1964. There is no commercial elephant sealing today. The Antarctic fur seal, almost exterminated in the 19th century, has now regained its former abundance.

DML/KMK/WNB

Right and Below *Pinnipeds have long played a central role in the lives of Arctic peoples. The first seal harpooning is an important rite of passage for an Eskimo boy. All parts of a kill are used; the hides may be sewn together to cover a family boat, or umiaq, while the walrus tusk has been carved into a pipe depicting whales, walruses, and seals.*

MUGSHOTS FROM THE DEEP

Computer photo-identification of marine mammals

WATCHING AND ENJOYING A TELEVISION SOAP opera would be impossible if we could not instantly recognize the individual characters involved by their faces. In the same way, any study of social behavior in other animals depends on being able to spot individuals within a group or population. While animals may all look the same to the untrained eye, some biologists have come to be as familiar with their study species as we all are with our own social groups. By using a variety of naturally-occurring features (ranging from general body shape, posture, and gait to such details as scars or ear-tears) researchers have been able to identify, for example, several hundred individual elephants and scores of African wild dogs.

Not all researchers can develop such an intimate knowledge of their study species, however, and so catalogs containing sketches or photos of the animals are used for comparison against photographs taken by individual workers in the field. Every new sketch or photo then initiates a search through the collection to find out which, if any, of the individuals it matches.

Photographs of Humpback and Right whales in the North and South Atlantic, and Sperm whales off the Azores and Galapagos islands, have been used to build photo-identification catalogs containing thousands of individual whales. Finding a match among such a large number of possible contenders is difficult, and so computer software has been developed both to speed the search and to reduce the risks of either failing to find a match or of making an erroneous one. Avoiding mistakes is particularly important if the catalog is being used to estimate population size, calving intervals, or migration rates, and even more so for endangered species. Thus, recent results from the North Atlantic Right whale catalog suggest a worrying decline in reproductive rate; it is important to be able to show that such an observation is real and not the result of missing matches.

An obvious way to speed the search for a match is to classify individuals into broad categories, such as "male" or "immature." A further refinement is to group according to appearance, for instance "dark" versus "light" or "boldly patterned" versus "finely patterned." If a match still proves elusive, a verbal description of the pattern on a defined section of the animal's body may help – say, "notches" or "scallops" along the trailing edge of a fin or fluke. Animals' bodies are divided into a number of zones, each of which can be classed according to the presence or absence of a given characteristic, and the search restricted accordingly. Some allowance is made for the fact that animals, like people, can look very different in different photos.

There is a limit, however, to what can be done purely by classification. To restrict the search still further, the patterning on a particular body-part is scanned into a computer file. Numerical descriptions of the patterns are then compared automatically in order to provide a similarity index between the new photo and each one in the whole catalog.

One type of numerical description is a "feature vector" – a list of summary statistics derived from the gray-shade values scanned from the image. A digital image is made up of small dots or pixels, each of which is assigned a gray-shade value from, say, 0 for black to 255 for white. This is the same approach used by a computer converting scanned text to a word-processing document, or in robot vision, where the statistics are chosen to distinguish one class of pattern from another, such as the letter "a" from the letter "b" or a chair from a cup.

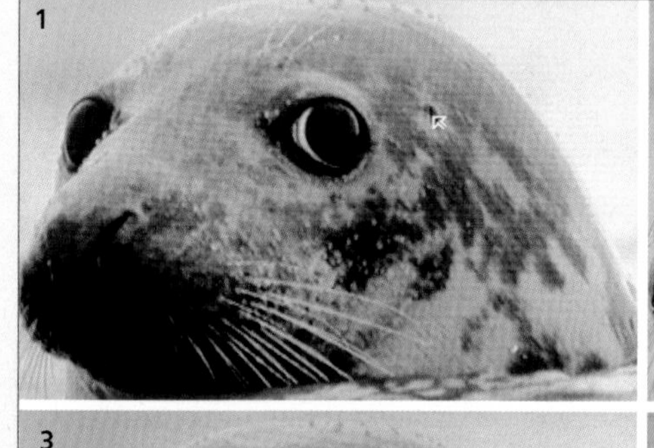

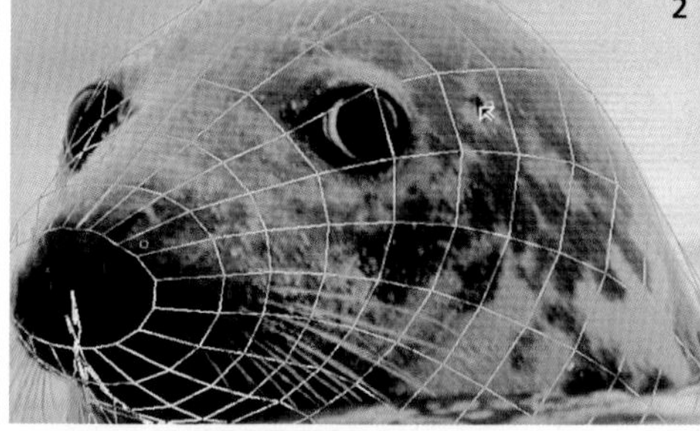

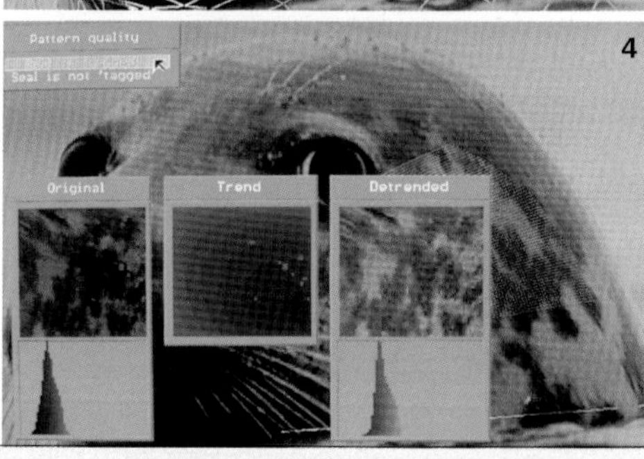

The problem in individual recognition is different, however: the pattern we are searching for is identical to, not just in the same class as, the one scanned from the new image, but it may be one among thousands. Then again, we need to scan just a small subregion from a larger region of continuous pattern, and it is not possible to design a feature vector that can deal with large errors in the location of the subregion.

As an alternative way of deriving similarity indices a system of "template matching" has been developed. This involves looking for patterns where the dark and light patches can be lined up with those on the new image. Furthermore, a 3-D surface model is used to locate the subregion, and

Above *The stages of the photo-identification process:* ***1*** *close-up portrait of a Gray seal, digitized and displayed on a computer screen;* ***2*** *a three-dimensional model of the seal's head is aligned on top of the original image of the Gray seal;* ***3*** *the 3-D model zones in on an area of distinctive patterning on the side of the seal's head, from which an array of gray-shade values will be scanned;* ***4*** *the array of gray-shade values from the patterned area is displayed and corrected for any shading resulting from the angle of the sun when the photograph was taken – the array will be used to initiate an automatic search for the individual seal in the photolibrary.*

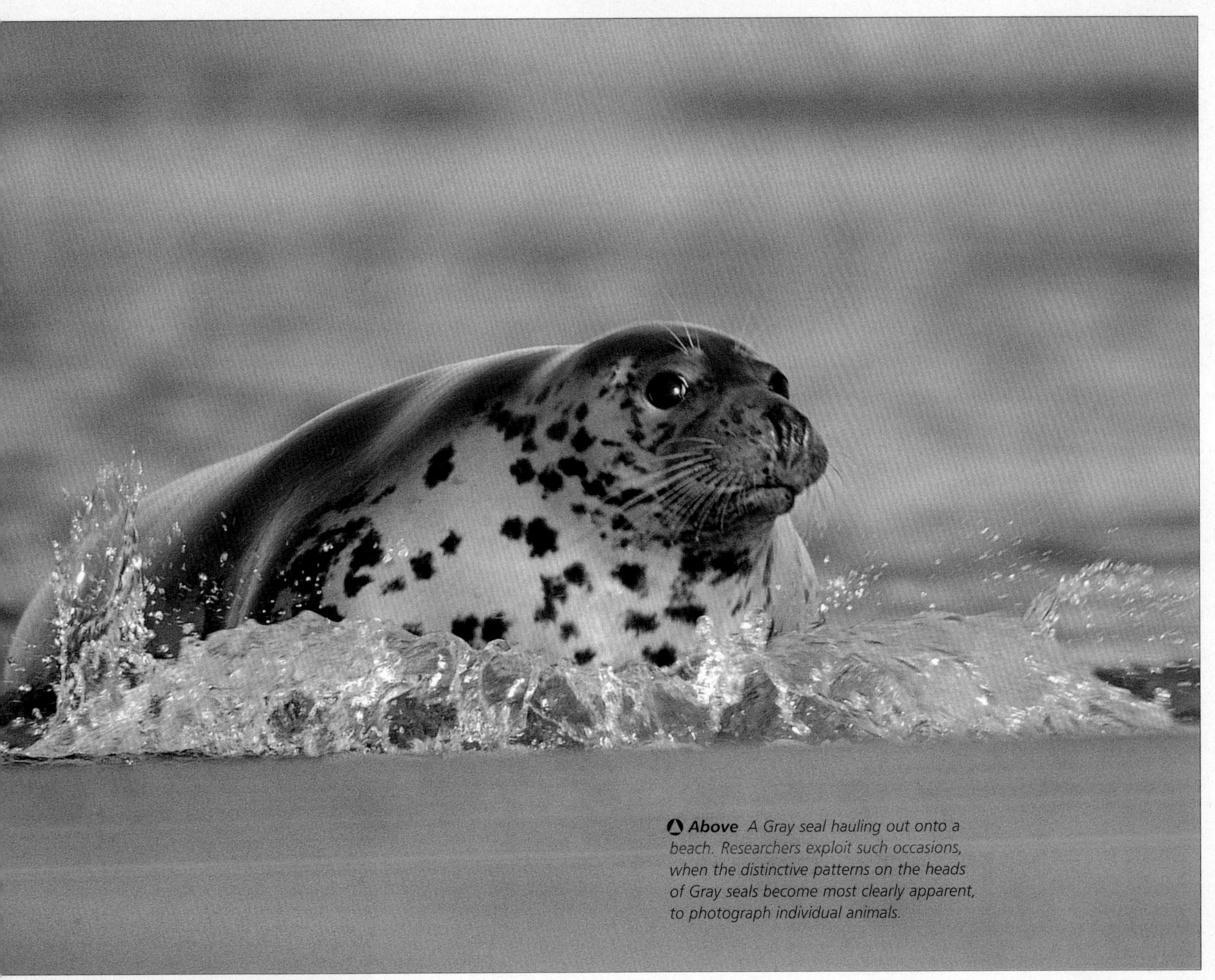

Above *A Gray seal hauling out onto a beach. Researchers exploit such occasions, when the distinctive patterns on the heads of Gray seals become most clearly apparent, to photograph individual animals.*

o ensure that the same area of the pattern is canned from each photograph.

Recognition has to be based on a small subre-ion, because it is impossible to photograph an ndividual from exactly the same viewpoint each me. A subregion must be used that will be visible n any photo, or at least any taken from a wide ange of angles. Gray seals, for example, are most asily photographed when swimming in the sea ear their "haul-out" sites. At these times they ften raise their heads to look at objects up and own the beach, thereby exposing a region of pat-ern on the side of the head – it extends roughly rom the ear to the corner of the eye and down to he level of the mouth. Of course the head has two ides, so the individual can be represented by two uch "fingerprint" regions – or by only one if the hotographer failed to get photographs of both ides of the head before the seal swam away.

Back in the laboratory, the photo is digitized and isplayed on a computer monitor. Then a 3-D urface model is displayed on the same screen and uperimposed on the image of the head by lining up the model's ears, eyes, and nose with those on the image. The model can also bend and twist its neck to mimic the posture of the seal in the photograph. In this way the subregion of the image that is to be scanned is identified as the area lying within a certain region of the model, once the two are accurately aligned. The scanning of the subregion also allows for the curvature of the head, the sample points being more closely spaced where the camera angle is more oblique to the surface.

The array of gray-shade values scanned from a new photo is compared to the library of arrays scanned from all previous photographs to generate similarity indices. Often the index obtained between the new array and one scanned from a previous photo of the same seal is so high that a match can be identified immediately. In any case, all new photographs corresponding to all arrays scoring beyond a threshold similarity index are checked by eye to confirm a match. Images are never recorded as showing the same seal unless they pass this final scrutiny, so the risk of a false-positive match is negligible.

From this new process, a catalog of some 4,500 North Sea gray seals has been compiled. The same method has been used to establish Irish and Baltic Sea catalogs that currently hold at least 1,500 seals each. These are all mature female and young male seals – mature males are too dark for individual recognition via pelage patterns. The identifications have made it possible to derive local population estimates and migration rates from each catalog.

Thus, preliminary results from the Baltic suggest that the population is still severely depleted by a combination of earlier hunting and pollution, and is still well below the level it was at the beginning of the 20th century, as some researchers had anticipated. Results from the Irish Sea show an influx of seals during the breeding season, with many seals moving to Cornwall and the Scilly Isles in the summer. As the catalogs expand, the estimates will be refined. Other results appear almost by accident; for example, one female seal was seen off the Irish coast with opaque lenses in her eyes. When she showed up in Wales, two months later, she was completely blind – but still alive! LH

SPECIAL FEATURE

DEEP DIVERS

Adaptations that allow sea mammals to forage the ocean depths

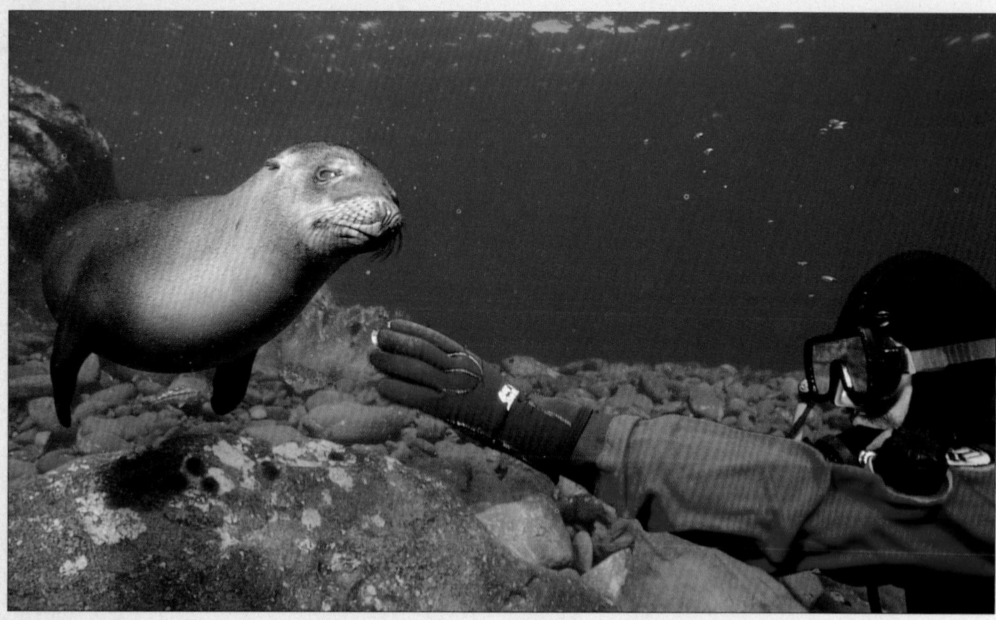

Above *The underwater environment is fundamentally hostile to mammals. Without an aqualung, humans can survive for six minutes at most, whereas some seals can stay underwater for nearly two hours.*

IF YOU ENROLL FOR A DIVING COURSE, IT IS LIKELY that you will be asked to sign a liability waiver form. Listed there will be an alarming catalog of ailments that might befall not just a human but also any other mammal adapted to life on land once it ventures underwater. Nitrogen narcosis (causing loss of consciousness at depths of 100m/330ft); high pressure nervous syndrome (causing convulsions and tremors at 500–1,000m/1,650–3,300ft); oxygen toxicity, causing convulsions; extreme hypoxia, causing blackout; decompression sickness (also known as "the bends"): all are possible problems associated with diving at depth. Yet mammals such as seals and whales appear to have evolved their way out of these hazards.

Many mammals forage in or around water, but only two orders, the Cetacea (whales and dolphins) and the Sirenia (dugongs and manatees), have become adapted to an exclusively aquatic lifestyle. In addition, the seals, sea lions, and walruses of the suborder Pinnipedia live most of their lives in the water and only have to leave it occasionally. All these groups are highly adapted to life underwater, but they are still air-breathers that must return to the surface to fill their lungs. To allow them to dive without the encumbrances of either a human-style oxygen tank or any of the physical distresses listed above, these mammals have evolved a set of superb adaptations.

Step one for a diving seal is to expel most of the air from the respiratory system. After submergence, the lungs collapse under the pressure of the water and any residual air is pushed into the bronchial tubes, to exit through the mouth or nose. This mechanism of lung collapse means that, by the time the animal reaches a depth of about 60m (200ft), it has no free gas in its body. It is thus not only incompressible, but also avoids "the bends," which are caused by free gas bubbles appearing in the blood as the diver ascends.

Once they have submerged, all diving mammals must have enough oxygen stored in their bodies to supply their metabolic requirements for the whole of the dive. To address this problem, their circulatory systems are finely tuned to ensure that what oxygen there is reaches the brain at all times. More

Above *While many seals can dive for protracted periods, they need a relatively long recovery time – thus, a dive of 45 minutes for certain species might entail a surface recovery time of nearly an hour. However, elephant seals are able to dive repeatedly for long periods with only a couple of minutes' recovery time between dives.*

Left *Although it may not seem a good idea to collapse the lungs, it is an essential facility for seals to function freely at depths. Once they are collapsed, dive duration is dependent only on the amount of oxygen that can be stored in the rest of the body.*

of the blood, and more red cells within it, contain increased amounts of hemoglobin, which binds with oxygen and transports it around the body. Significantly, they are also able to store oxygen in their muscles, using large amounts of myoglobin, which binds strongly to the life-giving gas.

Another crucial adaptation is to limit the need for oxygen while submerged by reducing or completely stopping circulation of blood to the limbs and non-critical organs. This "dive response" is produced by constriction of the blood vessels and by slowing the heart rate – in some cases from 60–80 beats per minute at the surface to as little as 4 beats per minute during parts of dives. Even the heart rate of fetuses slows down in parallel with that of their mothers, so that they too contribute to the parsimonious use of oxygen.

These physiological changes allow mammals to perform extraordinary diving feats. At present elephant seals are thought to be the champion divers, but it seems likely that even their abilities will be eclipsed as more information about sperm and beaked whales becomes available. Female elephant seals dive, on average, for 30–40 minutes at a time, and these dives are separated by only 1–2 minutes spent at the surface replenishing oxygen supplies. When at sea (where they spend 10 months of the year), Elephant seals spend 90 percent of their lives submerged. If necessary, they can stay under water for up to 2 hours with apparent ease, enabling them to catch fish at great depths.

Normally elephant seals feed at depths of 300–600m (1,000–2,000ft), but in extreme cases they may dive as far as 1,500m (5,000ft) or more. Their large size helps, for increased body size results in reduced metabolic costs per unit of body mass. Since oxygen storage capacity increases in proportion to body mass, bigger marine mammals have a larger oxygen store both in absolute terms and also, more importantly, relative to their rate of oxygen use. Yet even female fur seals, which have the body size only of a large dog, can remain submerged for more than 5 minutes and may dive to over 200m (650ft).

There is much scientific debate about how these animals, whose basic, functional anatomy and physiology are very similar to those of most other mammals, can make such prolonged dives. Clearly, diving mammals have exploited all the flexibility of mammalian form and function to allow them to gain access to the wealth of food that is available for them in the oceans. Yet even so there are limits beyond which they cannot go, presumably set mainly by the amount of oxygen they can store in their bodies. We also know, however, that some animals can use anaerobic metabolism while diving, in which case the physiological limit is set by factors such as the acid/base buffering of blood and muscle in the face of the rapidly increasing lactic acid concentrations that are a by-product of such a metabolism, even though marine mammals appear to be more resistant to high lactic acid than humans. Anaerobic metabolism is 18 times less efficient than aerobic metabolism, and it usually involves long recovery times, so it may be used only in extreme circumstances, to escape acute dangers. Otherwise, it would need to provide the diver with access to a very rich food source indeed to balance out the relatively high energy cost required to fuel the process. ILB

Eared Seals

ON A SANDY BEACH, A HUGE, MANED BULL *seal throws back its head and bellows to proclaim its mastery. Around it are its harem of up to 80 females; beyond them are further groups, each in turn presided over by a similar beachmaster. These are eared seals, and all 14 species are gregarious, social breeders.*

The surviving eared seal species comprise the fur seals and sea lions. As a group, they are distinguished from the true seals by their use of the foreflippers as the principal means of propulsion through the water. Generally, most sea lions are larger than most fur seals, and have blunter snouts; those of the fur seals tend to be sharp. The flippers of sea lions are also usually shorter than those of fur seals. The most obvious difference between the two, though, is the underfur, which is abundant in the fur seals and sparse in sea lions. Nevertheless, it is clear that the two fur seal genera, *Callorhinus* and *Arctocephalus*, are less closely related to each other than *Arctocephalus* is to the sea-lion genera, and the division of fur seals and sea lions that is sometimes made into the subfamilies Arctocephalinae and Otariinae is unjustified.

Big Males, Small Females

FORM AND FUNCTION

Because their hind limbs have not been greatly involved in aquatic locomotion, eared seals have retained a useful locomotory function on land, and they are comparatively agile. Circus sea lions can be trained to run up a ladder; more to the point, a bull fur seal can gallop across a rocky beach in pursuit of a rival. On broken terrain, a fur seal can move faster than humans can run.

Eared seals are a more uniform group than the true seals, both in appearance and behavior. In all eared seals, males are much larger than females – up to five times heavier in the Northern fur seal, a disparity in size that is rivaled among mammals only by the Southern elephant seal, a true seal species in which the male may be up to four times heavier than the female. Also, successful breeding males maintain a harem during the breeding season, a strategy known as polygynous breeding, as opposed to monogamous breeding, in which a male mates with only one female.

Most eared seals are generalist feeders, there being few specialists of the kind found among true seals. No eared seal populations have adopted a freshwater existence, as have several true seal species, including the Baikal seal, and some Ringed seal and Harbor seal subspecies.

In evolutionary terms, the earliest known otariid is *Pithanotaria*, known from several localities in California from 12–13 million years ago. *Pithanotaria* was a small animal, at 1.5m (5ft) in length about the same size as the Galápagos fur seal, the smallest living otariid. It had a uniform dentition and bony processes about the eye sockets, both characteristic of modern otariids.

By 8 million years ago there were otariids in the North Pacific that showed an increase in body size and were clearly sexually dimorphic, males being larger than females. Except for slight differences in some of the limb bones and the retention of double-rooted cheek teeth, these forms could easily be taken for modern sea lions. The Northern fur seal, *Callorhinus*, diverged from the main otariid stem about 6 million years ago, and soon afterwards otariids dispersed southwards to the Southern Hemisphere. There is no evidence that any otariid managed to follow other early pinnipeds through the Central American Seaway into the North Atlantic.

From 6 million to maybe 2 or 3 million years ago, there was little diversification in the otariid stock, which remained very similar to the existing genus *Arctocephalus*, the southern fur seals. In the past 2 million years, however, there was a sudden acceleration in the rate of size increase, as well as

FACTFILE

EARED SEALS

Order: Carnivora (Pinnipedia)

Family: Otariidae

14 species in 7 genera: southern fur seals (*Arctocephalus*, 8 species); Northern fur seal (*Callorhinus ursinus*); South American sea lion (*Otaria flavescens*); Steller sea lion (*Eumetopias jubatus*); California sea lion (*Zalophus californianus*); New Zealand sea lion (*Phocarctos hookeri*); Australian sea lion (*Neophoca cinerea*).

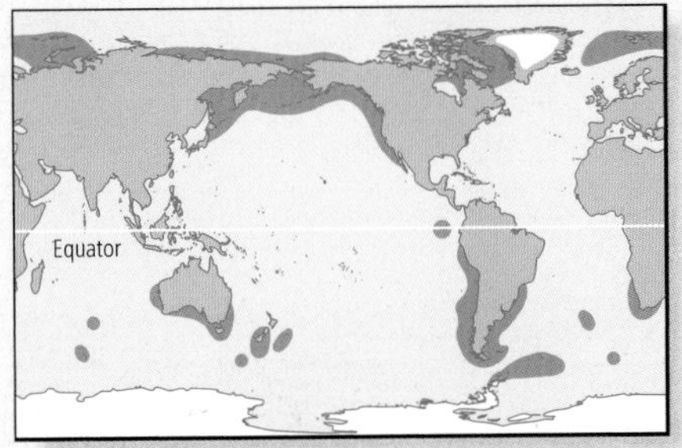

DISTRIBUTION N Pacific coasts from Japan to Mexico; Galápagos Islands; W coast of S America from N Peru round Cape Horn to S Brazil; S and SW coasts of South Africa; S coast of Australia; South Island, New Zealand; oceanic islands circling Antarctica.

HABITAT Generally coastal on offshore rocks, islands, and beaches; occasionally estuarine and freshwater rivers.

SIZE **Head–tail length** ranges from about 1.2m (3.9ft) in the Galápagos fur seal to 2.8m (9.3ft) in the Steller sea lion; **weight** from about 30kg (66lb) in the Galápagos fur seal to 566kg (1,250lb) in the Steller sea lion. Within species, males are larger than females.

COAT Distinguished from true seals by the underfur, which is abundant in the fur seals, sparse in the sea lions.

DIET Mostly generalist feeders, taking fish, krill, invertebrates such as rock lobster, and occasionally warm-blooded prey, principally penguins; in addition, sea lions sometimes eat fur seal pups.

BREEDING Gestation 11–12 months, including a 3–4 month period of suspended development (delayed implantation). The Australian sea lion is an exception: gestation lasts 18 months including a 5–6 month period of delayed implantation. Lactation lasts from 4 months to 3 years.

LONGEVITY About 20 years

See species table ▷

Left *In its defense of a small mating territory against rival males, a New Zealand fur seal gives a full-throated roar. Competition for space on the crowded breeding grounds is so intense that males cannot leave their patch even to hunt for food. Instead, they go hungry, sometimes fasting for as long as 70 days.*

Below *The hind flippers of eared seals are less adapted for swimming than those of true seals, remaining closer in form to the rear limbs of the terrestrial mammals from which pinnipeds evolved. As a result, eared seals can move relatively easily on land, using the flippers to support their weight.*

in the development of single-rooted cheek teeth, and generic diversification. The existing five sea-lion genera appeared from the arctocephaline stock within the last 3 million years or so.

The 14 living species of eared seal are found today on North Pacific coasts from Japan to Mexico; on the Galápagos Islands and on the western coast of South America, from northern Peru round Cape Horn to southern Brazil; on the south and southwest coasts of southern Africa; on the southern coast of Australia and South Island, New Zealand; and on the oceanic island groups circling Antarctica. These locations tend to be cool- rather than cold-water areas (although the Northern fur seal, the Steller sea lion, and, particularly, the Antarctic fur seal all occur in regions of near-freezing water). There are no ice-breeding eared seals.

Sea Hunters

DIET

Eared seals concentrate in areas where rising currents carry nutrients to the surface, feeding on a variety of open-sea and sea-bottom organisms, both fish and invertebrates – whatever food is most abundant and easy to catch. Many take their food, such as rock lobsters and octopus, from the bottom of the sea. Australian fur seals have been caught in traps and trawls at a depth of 120m (400ft), but eared seals are generally relatively shallow feeders in comparison with true seals.

Sometimes eared seals turn to warm-blooded prey. At Macquarie Island, the New Zealand sea lion feeds largely on penguins, and some Southern fur seals, often subadult males, also take these birds. Steller sea lions occasionally take young Northern fur seals. There are also observations on record of Southern sea lions attacking South American fur seals. The motivation for such attacks appears to differ for subadults and adults, with subadult males using captured fur seals as female sea-lion substitutes, guarding them from others and copulating with them, whereas adult sea lions hunt fur seals as food.

The Antarctic fur seal is one of the few specialist feeders. It lives largely on Antarctic krill, which is the only food found to be taken by the breeding females.

The amount of food consumed by eared seals is not easily determined. It varies from species to species, of course, and small animals need proportionately more food than large ones.

Coping with the Crowds

SOCIAL BEHAVIOR

Eared seals are all more or less social animals, tending to live in groups and to gather in aggregations, which may be very large during the breeding season. At its peak, the breeding haul-out of Northern fur seals at the Pribilof Islands must have represented one of the largest aggregations of large mammals to be seen anywhere in the world at the time. As already noted, male eared seals are polygynous, maintaining a harem of very many females, as do some other socially-breeding pinnipeds, notably the elephant seals. The fact that similar behavior has evolved in both the eared and true seals is believed to be related to the basic facts of life for pinnipeds, involving birth on land and offshore marine feeding.

Because pinnipeds have limited mobility on land, they seek out specially advantageous sites where the absence of terrestrial predators allows them to breed successfully. Such sites are relatively rare and space is often restricted, factors that tend to bring the females together. The males are more widely spaced because of the aggression they display towards each other. This tendency of the females to clump and the males to space out means that some males will be excluded from a position among the females, while the females will be drawn to the more successful males.

Such behavior is believed to favor large size in the males, for two reasons. Firstly, males need to be powerful to defend their territories, and to have impressive features that they can use in threat displays and in courtship. Secondly, a successful male cannot relinquish his territory to feed in the water until he has mated with as many females as possible; to maximize the number requires a lengthy period of fasting, with reliance on a large store of blubber for energy (also, large animals need less energy per unit weight than do small animals). Thus the larger bulls will tend to produce more offspring than smaller bulls.

As a consequence of this course of evolution, an otariid breeding beach is a lively scene. Bulls patrol their territorial boundaries, displaying frequently to neighbors. Most encounters between neighboring territorial bulls go no further than display and threat, but actual fights are frequent when a newcomer attempts to establish a place on the beach. The development of a tough hide and a massive mane over the forequarters does something to lessen injuries, but even so serious wounds are common, and it is not unusual for a bull to die from his wounds. Many pups are trampled to death in these battles, but few if any will be related to the bulls that are involved. Because of the strain imposed by intense activity in territorial encounters and the long period of fasting (for example, 70 days in the case of the bull New Zealand fur seal), few males are able to occupy dominant positions on a breeding beach for more than two or three seasons.

The breeding beaches are loud with barks and grunts, and as a result different sounds have evolved to communicate various emotions. Adult southern sea lion males produce at least four calls: high-pitched yelps and barks, both made during aggressive interactions, such as attack and retreat displays, when establishing territories, or during fights; growls when interacting with females; and exhalations after agonistic encounters. Females call to their pups, for example after birth or when separated from them, and grunt when fighting with other females. Pups call back in response to their mothers, and also vocalize when hungry or searching for mother. Several of these calls are complex enough to be individually distinct, such that each sea lion has a different voice.

The seals have an annual round that is well exemplified by the Antarctic fur seal. During the winter, from May to October, the adults are at sea and little is known of this phase of their life. In late October the breeding bulls begin to come ashore to establish territories. At this stage there little fighting, as there is ample space on the beach. Later, as the beach fills up, boundary disputes are frequent and there is much fighting. The first cows arrive ashore 2–3 weeks later, pregnant from the previous year's mating. By the first week in December, 50 percent of the pups have been born, 90 percent of these within a three-week period. Cows come ashore about two days before giving birth. For the next six days the mother stays with her pup, suckling it at intervals, and coming into heat again eight days after the birth. During this time the bulls are very active, fighting with neighbors and endeavoring to accumulate more cows in their territories. Though they cannot actively collect cows, they do their best to prevent those already in the territory from leaving. This interception of cows that show signs of leaving brings the bull into contact with all the females

Below An Australian sea lion bull protects its territory from the attentions of an encroaching neighbor. Males usually seek to scare off intruders with aggressive displays, but real fighting can break out, especially when a newcomer tries to establish its presence.

Below *Representative sea lions and fur seals. All species are sexually dimorphic, males being larger and generally darker in color than females. Sea lions have broader muzzles than fur seals and lack underfur:* ***1*** *A male California sea lion* (Zalophus californianus). ***2*** *Female Steller sea lion* (Eumetopias jubatus). ***3*** *Female South American sea lion* (Otaria flavescens). ***4*** *Male New Zealand sea lion* (Phocarctos hookeri). *Fur seals have thick coats that cause overheating on land:* ***5*** *Female South American fur seal* (Arctocephalus australis). ***6*** *Male Northern fur seal* (Callorhinus ursinus).

coming into heat when they become restless. Copulation follows once a receptive cow has been detected, and very soon after being mated the cow departs to sea on a feeding excursion.

There then follows a lactation period of about 117 days, during which the cow comes back from time to time in order to suckle her pup between feeding trips. There are on average about 17 of these feeding/suckling episodes, about twice as long being spent at sea feeding as on shore suckling. While the cows are away, the abandoned pups move to the back of the breeding beach, where they lie about in groups. A cow returning from a feeding trip comes back to the beach where she left her pup and calls for it with a characteristic pup-attraction call. The pup answers with its own call, which is recognized by the mother. She confirms recognition by smelling the pup, then leads it to a sheltered place, often on top of a clump of tussock grass, to feed it.

Eventually, weaning takes place. Surprisingly, this occurs in the Antarctic fur seal by the pup taking to the sea, so it is simply not present for the female's final return. Pups tend to take to the sea in groups, so weaning is more synchronized than births. Consequently, late-born pups tend to have a shorter lactation period, and lower weaning weights, than those born earlier. Perhaps this grouping of pups is an anti-predator strategy, since Leopard seals are known to kill young fur seals.

The abruptly terminated lactation period of the Antarctic fur seal is not typical of eared seals. The only other species that shows it, the Northern fur seal, is also migratory, abandoning its breeding places completely in the winter. Most other eared seals will continue to feed their pups until the arrival of the next, alternating suckling and feeding trips. Some, indeed, will go further and can be seen suckling a pup and a yearling, or even with a two-year-old as well, as recorded, for example, in northern populations of Steller sea lions. The Galápagos fur seal is another species that may suckle its pup for 2–3 years, and it has been shown that the presence of an unweaned yearling or two-year-old inhibits the birth of a younger sibling. When a pup is born in these circumstances, it almost always dies in the first month if its older

Right *South American sea lions crowd a breeding ground, or "rookery," on the coast of Peru in the mating season. Although there can be as many as 18 females in a male's territory, the average for this species is less than three – an unusually small number.*

Below *On the Galápagos Islands, a California sea lion mother sniffs a new-born pup; in weeks to come, scent will be one of the main ways by which she will recognize her offspring on the crowded breeding grounds. Pups can swim at birth, and can walk with some degree of coordination within 30 minutes.*

sibling is a yearling, or in about 50 percent of cases if it is a two-year-old.

Australian sea lion females do not give birth until 18 months after mating. Studies of 96 free-ranging animals from South Australia have shown that the fertilized egg remains dormant in the female's reproductive tracts for between 3.5 and 5 months, after which concentrations of certain hormones increase and the blastocyst reactivates and implants. Gestation will then continue for 14 months, the longest post-implantation period recorded in any pinniped.

In California sea lions, mitochondrial DNA analysis has shown that populations from the Gulf of California do not frequently interbreed with those from the Pacific coast, suggesting that females are philopatric, returning to the beaches where they were born.

Right *With their hind flippers serving as rudders, California sea lions perform underwater acrobatics as they search for food on a shallow seabed off the Mexican coast. Water quality is vital to seals' well-being, as blubber can store pollutants along with body fat.*

Left *On the coast of southern Australia, a recently-born Australian sea lion pup practices its vocalizing skills. Each pup has its own distinctive call, by which its mother will be able to recognize it even among a crowd of fellow youngsters on a busy rookery.*

Below *In Namibia, a half-grown Cape fur seal pup feeds on its mother's milk. She will suckle it for about a year in all, though for at least half that time it will also be foraging effectively on its own behalf, learning the hunting skills it will need as an adult.*

On the Mend?

CONSERVATION AND ENVIRONMENT

Many of the eared seals were hunted almost to extinction in the 19th century, but the general picture thereafter has been one of recovery. The Juan Fernandez fur seal, once considered extinct, was rediscovered in 1965; the Guadalupe fur seal, thought to have been exterminated in 1928, was refound in 1954. The Antarctic fur seal has made the most dramatic recovery of all, from near-extinction to a population of 700,000–1 million.

The South American fur seal has the longest continuous record of exploitation of any fur seal, with the first skins being exported from Uruguay in the 16th century; the seals on Isla de Lobos are still exploited. The Northern fur seal was heavily exploited at the Pribilof Islands from 1786 onward, and was seriously depleted by overexploitation in the course of the 19th century. Then from 1891 on a number of protective measures were adopted, culminating in 1911 when the Treaty for the Preservation and Protection of Fur Seals banned pelagic sealing.

While sealing continued on the breeding grounds, the Northern fur seal quickly recovered from its depleted state. Following the seal hunts of 1956–62, however, which involved the killing of females, the population began a steady decline that led to it once more being declared "depleted" in 1988 under the US Marine Mammal Protection Act. Since the mid-1980s, hunting on the Pribilof Islands has been restricted to a subsistence kill by native Aleuts, which has averaged less than 2,000 seals per year. Despite the reduced kill, however, the population has yet to show any significant signs of recovery.

An increasing threat to the marine environment comes from chemical pollution. Many studies are showing concentrations of pollutants incorporated within the bodies of fur seals and sea lions. Steller sea lions from the Gulf of the Farallones in central California have been found to have elevated levels of the trace elements copper (91.0mg/kg dry weight) and selenium (4.1mg/kg). Similarly, in the coastal waters off Hokkaido, Japan, concentrations of the plastic pollutant butyl tin and a number of organochlorine compounds, which accumulate in blubber, were found to be elevated in Steller sea lions. One particular group of organochlorines, the polychlorinated biphenyls (PCBs) were found at concentrations of 5.7 to 41 $\mu g.g^{-1}$ (lipid weight) in males and from 0.57 to 16 $\mu g.g^{-1}$ in females in Steller sea lions from Alaska and the Russian Bering Sea. DDT and its metabolites are also still a problem, as are chlordane products. The residue levels of these toxins increases with age in males, whereas in females they decrease sharply after maturity, suggesting the transfer of organochlorines in large quantities during lactation to offspring. Residue levels of PCBs and DDTs in the liver of male Steller sea lions from the Bering Sea have been found to be significantly lower than those from Alaska.

Steller sea lion numbers in the United States have declined by about 75 percent since the early 1970s. The first general, range-wide survey of Steller sea lions was completed in 1989, with a total of 68,094 adults and juveniles (non-pups) counted. This figure included 47,960 in Alaska (70 percent of the total), 10,000 animals in Russia (15 percent), 6,109 in British Columbia (9 percent), 2,261 in Oregon (3 percent), and 1,764 in California (also 3 percent). At the time, the world

ɔpulation was estimated to be 116,000 animals, · somewhere between 39 and 48 percent of the ł0,000–300,000 that had been estimated 30 ars previously. In Alaska, Steller sea lions are ɔw classified under the U.S. Endangered Species ːt as Threatened in the eastern portion of their nge and as Endangered in the western portion. ırious additional hypotheses have been put forard to explain the reasons for the collapse in the askan population, including competition with e large local pollack fishery, by-catch in fishing ar, indiscriminate shooting, and oceanographic ange resulting in a reduction in food availability. ɔw rates of survival among juveniles, apparently sociated with increasing food shortages, appears to be the main factor involved in this alarming population decline.

Another pinniped that is currently in peril is the New Zealand sea lion. Pups are born at only five sites, four of which are in the subantarctic Auckland Islands and the other at Campbell Island in the same region. The highly localized and historically reduced distribution make this species vulnerable. The principal source of human-caused mortality has been the incidental by-catch of sea lions by trawl fisheries operating in their habitat. Iin recent years, steps have been taken to limit the level of by-catch and so provide increased protection for the species.

Some seals – especially the California, South American, and Steller sea lions – are still harassed by fishermen, but apart from unexplained declines such as that of the South American sea lion in the Falklands (from 300,000 in the 1930s to 30,000 in recent years), most eared seal populations are now in a fairly secure position.

The work of the Marine Mammal Center in California demonstrates a more positive side of the human–pinniped relationship. It gives help to sick and injured animals, responding to growing humanitarian concerns over animal welfare. From 1984 to 1990, its operatives successfully treated almost half of the 786 California sea lions and 18 Northern fur seals brought to them, mostly by concerned members of the public. DML/WNB

Eared Seal Species

THE 14 SPECIES OF EARED SEAL ARE commonly divided into two groups: the fur seals, assigned to 2 genera, *Callorhinus* and *Arctocephalus*, and the 5 genera of sea lion, each of them monotypic (containing a single species). In fact *Callorhinus* also consists of just one species, the Northern fur seal, which some authorities consider more closely related to the sea lions than to the *Arctocephalus* furred seals, which it superficially resembles.

FUR SEALS

9 species in 2 genera

Northern fur seal Vu

Callorhinus ursinus

Northern or Alaskan fur seal

5 stocks recognized: Pribilof Islands, E Bering Sea; Commander Islands, W Bering Sea; Robben Island, Sea of Okhotsk; Kuril Islands (Kamennye Lovuski Islands and Srednev Rocks), W Pacific; San Miguel Island, California. Newer sites have been established at Bogoslof Island in the Aleutians, and at Castle Rock, California. Individuals have been found NE along the Arctic coast to Amundsen Gulf and SW to Shandong, China. Except for the San Miguel population, which probably remains in Californian waters all year round, the males leave the breeding islands from August to October; little is known about their winter whereabouts, but they are thought to stay near the Aleutian Islands. Females, pups, and juveniles leave by late November and migrate as far S as S California in the E and Japan in the W. The Northern fur seal population has been declining since 1956. Total population size was estimated in 1993 at about 1.2 million: 900,000 at the Pribilof Islands; 225,000–230,000 at the Commander Islands; 55,000–65,000 at Robben Island; 50,000–55,000 at the Kuril Islands; 4,000 at San Miguel Island.
HTL male 210cm, female 100–142cm; WT male 180–270kg, female 30–50kg.
FORM: adult males are dark brown to black, with gray guard hairs on the back of the mane. Females and subadults are silvery-gray above and reddish-brown below, with paler chest. Pups are black at birth, with lighter color on belly and face. Muzzle short, giving a characteristic profile. Long pinnae and hind flippers.
DIET: principal prey items include small schooling fish (anchovy, herring, walleye pollock) and squid.
BREEDING: females reach sexual maturity at 2–5 years; males reach the height of their reproductive status at about 9. Births from June to early August. Male pups are 55–63cm long and weigh 5.4–5.9kg, female pups 53–62cm and 4.5–4.8kg. Mating occurs about 1 week after birth. Lactation about 4 months. Adults are most often seen at sea alone or in pairs, but sometimes in groups of 3 or more.
LONGEVITY: about 25 years

Antarctic fur seal

Arctocephalus gazella

Antarctic or Kerguelen fur seal

Found on islands S of the Antarctic Convergence, but rarely below 65°S. Main breeding colony at South Georgia; much smaller groups in S Shetland Islands, S Orkney Islands, S Sandwich Islands, Bouvetoya, Marion Island, Heard Island, Macquarie and McDonald Islands, Crozet Island, Prince Edward Islands, and Kerguelen, where the species was once very abundant (hence its alternative name). Vagrants to Tierra del Fuego, Mar del Plata (Argentina), and the Juan Fernández Islands. Population, based on estimated 1990–91 pup production, 1.3–1.7 million.
HTL male 165–200cm, female 115–149cm; WT male 90–210kg, female 25–55kg.
FORM: males are dark grayish-brown to charcoal, with frosting on the guard hairs of the mane. Females and juveniles are medium gray, occasionally darker above. Often have a creamy throat and chest; creamy coloration extends onto the sides and back, with a pale area on the flanks to the hindflippers. Light areas on the face, muzzle, and ears. A pale form, yellowish or off-white, occurs infrequently (1 in 2,000). Pups are black. Forehead convex, muzzle short and broad. Postcanine teeth often no more than buttons of enamel.
DIET: krill, fish, and squid, depending on location. Some of the seals eat penguins, although these do not appear to be important in the diet
BREEDING: males and females reach sexual maturity at 3–4 years, although males probably do not mate until 6–10. Seals spend the austral winter at sea, returning to the breeding places in October. Peak pupping occurs in early December. Pups measure 60–73cm and weigh 4.5–6.5kg at birth. Lactation about 17 weeks, with the pups molting towards the end of this period. Male pups weigh about 17.8kg at weaning, females about 14.7kg. Adult females and males leave the breeding grounds in April, to resume an aquatic existence until the next breeding season. Subadults and some adults remain at the rookeries all year.
LONGEVITY: males 15 years, females 23.

Guadalupe fur seal Vu

Arctocephalus townsendi

Little is known about this species. Breeding range now restricted to Guadalupe Island, Baja California, Mexico. Wanders as far N as the Farallon Islands and Sonoma County, California, and S around Cabo San Lucas, Mexico, into the Gulf of California. Formerly found on islands off the coast of California. Population of about 6,000 animals, estimated from a 1987 count of 3,259 animals.
HTL male about 180cm, female 120cm; WT male about 100kg, female unknown.
FORM: adult males are dark grayish-brown to grayish-black. Tips of the guard hairs of the mane may be light. Much of the head and back of the neck may appear tan. Females are dark gray-brown to grayish-black and paler below. Chest and underside of the neck can be creamy gray. Head of the male is large, and the muzzle is long and pointed.
DIET: not well studied, but have been observed to eat squid and lanternfish.
BREEDING: occurs from mid-June to July or August, although most pups are born from mid to late June. Females mate 7–10 days after giving birth. Pups are weaned between 8–11 months. Little is known about their annual distribution, but it appears that males, juveniles, and females without pups may live at sea at different times during the year. Males are absent from the rookeries in the winter.

Juan Fernández fur seal Vu

Arctocephalus philippii

Rookery sites at Robinson Crusoe Island, and at Alejandro Selkirk and Santa Clara Islands in the Juan Fernández group. Also found on San Felix Island and San Ambrosio, Chile; vagrants to Punta San Juan, Peru. Nearly exterminated by hunters by the 19th century, and only "rediscovered"

Right *Resting its head sleepily on a rock on the Namibian coast, a Cape fur seal displays the vestigial pinna, or projecting outer ear, from which the eared seals take their name.*

ABBREVIATIONS	HTL = head-to-tail length WT = weight	En Endangered
	Approximate nonmetric equivalents: 1m = 3.3ft; 10kg = 22lb	Vu Vulnerable

in 1965; total population estimated at about 12,000 in 1990–91.
HTL male about 200cm, female 140cm; WT male about 140kg, female 50kg.
FORM: males are dark blackish-brown with silver tips on the longer guard hairs of the mane. Females are gray-brown to dark brown on the back and paler below, especially on the chest and the underside of the neck, which can be creamy gray. May have light areas on the face as well. Forehead convex, with a long muzzle and a prominent snout. Postcanine teeth large and unicuspid.
DIET: fish and cephalopods
BREEDING: occurs from mid-November to end of January, with peak pupping from late November to early December. At birth, male pups weigh 6.8kg and are 68cm long; females 6.2kg and 65cm. When on shore, prefer rocky and volcanic shorelines with boulders, grottoes, and caves.

Galápagos fur seal Vu

Arctocephalus galapagoensis

Confined to the Galápagos Archipelago, they breed on several islands, notably Isabela and Fernandina in the W and Pinta, Marchena, Genovesa, and Wolf in the N. Prefer rocky beaches with boulders and lava, and seek shelter from the sun under ledges and between boulders. Population size is approximately 40,000, based on census data from 1978 and 1988–89, although numbers may have declined as a result of the 1982–83 El Niño event.
HTL male up to 150cm, female up to 130cm; WT male 60–75kg, female about 30kg. The smallest and least sexually dimorphic of the eared seals.
FORM: adults are dark brown above; some may have lighter hues. The muzzle is pale; in males, this color can extend to the face. Females and subadults have a pale, grayish-tan chest and a rusty-tan belly. The mane of the male contains guard hairs only slightly longer than on the rest of the body, and may have a grizzled appearance. Pups are blackish-brown and may have grayish areas around the mouth and nose. Forehead flat in profile; the muzzle very short, with a pointed snout. Postcanine teeth small and unicuspid.
DIET: mainly mycotophids, bathylagids, and small cephalopods.
BREEDING: females reach sexual maturity at 3–4 years. Territorial males are 7–10 years old. Births mid-August to mid-November, with a peak in late September through early October. Pups weigh between 3–4.5kg at birth, and are weaned from 1–3 years. Mating occurs about 1 week after birth. The breeding regime is affected by the high temperatures to which the fur seals are exposed: females tend to seek out shaded places or caves in lava rocks, while males ensure their territories have access to the sea, so that they can cool off in the water when overheated.

South American fur seal

Arctocephalus australis

2 subspecies inhabit coastal waters of South America: *A. a. gracilis* is found on rookeries from Isla Lobos de Tierra, Peru, S to Rocas Abtao, Chile; from Isla Chiloë, Chile, S to Isla de Los Estados, Argentina; and from Isla de Lobos, Uruguay, N to Recife dos Tôrres, Brazil; vagrants to Pacific coast of Colombia and to Juan Fernández Islands. *A. a. australis* is found throughout the Falkland Islands. Estimated 1993 population for the mainland was 307,000. The 1993 Falkland population was estimated at 15,000–16,000.
HTL male up to 200cm, female up to 150cm; WT male 160kg, female 60kg. Fur seals from the Falkland Islands are said to be larger than those from the mainland.
FORM: adult males are blackish-gray with grayish tinges. Adult females and subadults are dark brown to grayish-black above and paler below. Lighter areas on the underside of the lower neck, sometimes on muzzle and around ears. Muzzle moderately long and flat-topped; noticeable forehead and long pinnae. Pups are dark at birth, but may have paler areas on face and belly. Postcanines medium-sized and usually tricuspid, though in some specimens the lateral cusps are missing.
DIET: a variety of cephalopods, fish (e.g. anchovies, horse mackerel), crustacea (e.g. rock lobster, krill), lamellibranchs, and sea snails.
BREEDING: births occur from mid-October to December depending on location, and mating takes place about 1 week after the pup is born. Pups weigh between 3.5–7kg at birth; males are 60–65cm long, females 57–60cm. Lactation 8–24 months, depending on location. Females reach sexual maturity at 3 years, males at 7. Most females remain near the rookeries throughout the year; little is known about the seasonal movements of males and juveniles.

Cape fur seal

Arctocephalus pusillus

Cape or South African fur seal, Australian fur seal

Two subspecies: *A. p. pusillus* (Cape or South African fur seal), distributed along the coasts of South Africa and Namibia from Cape Cross S to the Cape Peninsula and E to Algoa Bay and Black Rocks, Cape Province; and *A. p. doriferus* (Australian fur seal), distributed in the coastal waters of SE Australia, with rookeries located from Victoria to Pedra Blanca, Tasmania, and New South Wales. *A. p. pusillus* vagrants occasionally found to Baia do Quicombo in Angola and SE to Marion Island; *A. p. doriferus* vagrants have been found to Port Stephens, New South Wales. The Cape fur seal population was estimated at about 2 million in the early 1990s. The Australian population was estimated to be between 30,000–50,000 in 1991.
HTL *A. p. pusillus*: male 200–230cm, female 120–160cm; *A. p. doriferus*: male 200–225cm, female 125–170cm. WT *A. p. pusillus*: male 200–360kg, female 40–80kg; *A. p. doriferus*: male 220–360kg, female 36–110kg.
FORM: males are dark gray or brown on the back and lighter on the belly; mane becomes lighter with age. Females are brownish-gray on the back and light brown on the belly. Paler on the muzzle, lower jaw, and face. Pups are black at birth and turn grayish-fawn with a pale throat. Forehead convex, muzzle relatively long, but heavy. Postcanine teeth robust and tricuspid, with prominent anterior and posterior cusps. Teeth and skull very similar to those of the Australian sea lion. In many respects, including its large size, *Arctocephalus pusillus* is closer to the sea lions than to the fur seals.
DIET: pelagic schooling fish such as maasbanker, pilchard, hake, and Cape mackerel; also squid and cuttlefish. Can also bottom-feed, taking octopus.
BREEDING: females reach sexual maturity at 3–6 years, males at 4–5, although they probably do not mate until they reach 9–12. Births take place from late October to early January, with peak pupping in the first week of December, with some variation between colonies. Pups of *A. p. pusillus* measure 60–80cm at birth and weigh 5–6kg. Pups of *A. p. doriferus* measure 64–81cm and weigh 5–12.5kg for males, 62–79cm and 4.5–10kg for females. Lactation about 12 months or longer, with a small proportion of young being suckled for a second year and, possibly, a third. Mating occurs about 6 days after pupping. Neither subspecies are migratory, but move locally within their ranges.

New Zealand fur seal

Arctocephalus forsteri

Two populations exist: one along the S coast of Australia, and the other around New Zealand. The Australian population is found on rookeries from Eclipse Island, W Australia, E to Maatsuyker Island off th[e] end of Tasmania. The New Zealand pop[u]lation is found on rookeries on Cape Pa[l]liser, off the North Island, along the W coast of South Island, and on Stewart Island, Solander Island, the Snares, Au[ck]land Islands, Campbell Island, Antipo[des] Islands, Bounty Island, and Chatham Island. Disperses N to both coasts of th[e] North Island, N to Three Kings Islands and S to Macquarie Island. Has been found as far as New Caledonia off the E coast of Australia. A 1989–90 survey o[f] seals in South and Western Australia resulted in a population estimate of 27,200. The New Zealand population i[s] estimated at 30,000–50,000. Both pop[u]lations are thought to be increasing.
HTL male 145–250cm, female 125–150cm; WT male 120–185kg, fe[male] 25–50kg.
FORM: adult males are dark brownish-gray on the back and paler below. The muzzle is also pale. Females are genera[lly] paler on the underside of the neck and chest. Pups are blackish with a pale mu[z]zle, molting to a coat similar to the adu[lt] 2–3 months. Forehead slightly convex, muzzle moderately long, with a somew[hat] bulbous, pointed nose. Postcanine tee[th] are small and unicuspid.
DIET: barracuda, octopus, squid, and small fish. Birds, including penguins, a[re] occasionally eaten.
BREEDING: females reach sexual matu[rity] at 4–6 years. Males are able to defend t[er]ritories at 10–12, although they reach s[ex]ual maturity several years earlier. Births from mid-November to January, with p[eak] pupping from late November to mid-December. Pups measure 40–45cm at birth; males weigh an average of 3.9kg, females 3.3kg. Mating occurs about 9 d[ays] after pupping. Lactation lasts 10–11 months. Few males remain on the rook[eries] throughout the winter.

Subantarctic fur seal

Arctocephalus tropicalis

Subantarctic or Amsterdam Island fur seal

Found on islands to the N of the Antar[ctic] Convergence: Tristan da Cunha, Goug[h] Island, Amsterdam Island, Saint Paul Island, Prince Edward Islands, Marion Island, Crozet Island, Heard Island, an[d] Macquarie Island. Vagrants reported fr[om] such diverse locations as South Georgi[a], Brazil, W and SE coasts of South Africa, Madagascar, Comoros, S Australia, Ne[w] Zealand, and Juan Fernández Islands. Combined population estimates from 1987 and 1990 amount to at least 310,000 animals.

ABBREVIATIONS

HTL = head-to-tail length WT = weight
Approximate nonmetric equivalents: 1m = 3.3ft; 10kg = 22lb

Ex Extinct
EW Extinct in the Wild
Cr Critically Endangered
En Endangered
Vu Vulnerable
LR Lower Risk

male 180cm, female 145cm; WT
e 165kg, female 55kg.
RM: the only fur seal to show a clear
or pattern; brown to dark gray, with yel-
chest and throat area, most apparent
nales; dark belly. Pale or yellow face to
ind the ears. Males are usually darker
n females. Adult males have a conspic-
s crest of longer guard hairs on the top
he head and a thick mane. The fore-
d is only slightly convex, the muzzle
rt and narrow. Ear pinnae are not par-
larly prominent. Postcanine teeth
all and unicuspid.
T: fish, cephalopods, and euphausiids;
guins are sometimes taken.
EEDING: females reach sexual maturity
–6 years, males at 4–8, although the
er are not likely to breed before 8.
ths occur from November to January,
h peak pupping in December. Pups are
m long and weigh about 5kg at birth.
tation lasts 9–10 months. Mating
urs about a week after pupping. Most
he population, except for females with
ps, spend most of the winter and spring
ea.
ngevity: males 18 years; females 23.

SEA LIONS

pecies in 5 genera

eller sea lion **En**

metopias jubatus

ller or Northern sea lion

nabits the N Pacific from the Bering
ait S to Hokkaido (Japan) and the
annel Islands off S California. Impor-
t breeding colonies in the Kuril Islands,
mchatka, on islands in the Sea of
hotsk, the Aleutian Islands, and on the
skan–Canadian coastline. Some breed
San Miguel, California. Vagrants to Her-
el Island in the Beaufort Sea and to
ngsu, China. World population estimat-
at about 110,000 in 1989, roughly
e-third of the population estimate from
1960s. The population is declining
pecially in the Gulf of Alaska, Aleutian
ands, and off Russia.
L male 282cm, female 228cm; WT
ale 566kg, female 263kg. Steller sea
ns are the largest otariids.
RM: both sexes light to reddish-brown,
ghtly darker on the chest and belly.
ult males have distinctive manes, but
t as conspicuous as in South American
a lions. The head and muzzle are large,
d the male has an obvious forehead.
ps dark to blackish-brown, molting to a
hter pelage at about 6 months.
ET: a wide variety of fish (e.g. walleye
llock, herring, rockfish), cephalopods
oth squid and octopus), bivalve mol-
sks, shrimps, and crabs. Some young
ales eat Northern fur seal pups.
REEDING: females reach sexual maturity

at 3–6 years, males at 3–7, although they are not socially mature until about 10. Females arrive on the rookeries from mid-May through June, giving birth about 3 days after arriving. Peak pupping occurs in late June. Pups are about 100cm long and weigh 16–23kg at birth. Lactation 12–36 months, depending on location or population. Occasionally a female will suckle a newborn and a yearling (and, rarely, a 2-year old as well). Usually seen at sea in groups of 1–12 animals. Individuals, especially males, range widely after the breeding season. Males apparently move N to feed and do not return to the rookeries until the following spring.

California sea lion

Zalophus californianus

California or Galápagos sea lion

Three subspecies recognized: *Z. c. californianus*, the California sea lion, includes two geographical divisions, Pacific coast and Gulf of California. The Pacific population ranges from Solander Island, Canada, and the Strait of Georgia, S to Baja California, including 2 offshore islands, Islote Zapato and Rocas Alijos. The Gulf of California population ranges throughout the gulf and includes rookeries from Roca Consaf S to Los Islotes. *Z. c. wollebaeki*, the Galápagos sea lion, inhabits islands in the Galápagos Archipelago; a small rookery has also been established off the coast of Ecuador. *Z. c. japonicus* once occurred in the East Sea and the coastal waters of Japan, but is now considered extinct. Population size for *Z. c. californianus* was estimated at 204,000–214,000 in 1999, for *Z. c. wollebaeki* at 30,000 in 1979.
HTL male 240cm, female 180cm; WT male 275kg, female 91kg.
FORM: males are generally a dark chestnut brown, although some males do not darken and remain lighter on head and muzzle, and on the sides, hind, and belly; females and juveniles are tan. Mane not as well developed as in the Steller or South American sea lions. Sagittal crest in males creates a high, domed forehead.
DIET: includes fish (anchovy, Pacific hake, jack mackerel, rock fish) and squid.
BREEDING: females reach sexual maturity at 4–5 years; males unknown. *Z. c. californianus* gives birth from May through July; births in *Z. c. wollebaeki* extend over a long season from May to January. Pups measure about 75cm and weigh 7–9kg at birth. Lactation 6–12 months, but females are occasionally seen suckling yearlings. *Z. c. californianus* migrates N from the rookeries in August and September and returns from March to May. Female movements are unknown, but they are believed to remain near the rookeries year round. *Z. c. wollebaeki* apparently stays around the Galápagos Archipelago all year.

South American sea lion

Otaria flavescens (O. byronia)

South American or Southern sea lion

Found along the coast of South America, with rookeries from Zorritos, Peru, S to Tierra del Fuego and Isla de los Estados, Argentina, and N to Recife des Tôrres, Brazil. Stragglers reported from Bahia (Brazil) on the Atlantic side, and Colombia, Panama, and the Galápagos Islands on the Pacific side. South American sea lions have been found in estuaries and freshwater systems. Early population estimates from the 1950s and 1960s were around 273,500. Estimates from the different areas in 1973 and 1982 add up to about 155,000, although this may not cover the total population.
HTL male 260cm, female 200cm; WT male approximately 300kg, female 150kg.
FORM: males dark brown to brownish-orange, darkening with age; mane and underside may be lighter, the face darker. Females and subadults are yellow to brownish-orange; color may be patterned and have different hues. The muzzle is blunt and upturned at the end and the ears are small and lie close to the head. These characteristics are magnified in the male, the most "lionlike" of the sea lions in appearance. Pups are black above and grayish-orange below, molting to dark brown at about 4 weeks.
DIET: includes fish (rock fish, South Pacific hake, herring), cephalopods, and crustaceans; ducks and penguins are reportedly sometimes taken.
BREEDING: females reach sexual maturity at 4 years, males at 5–6. Time of birth differs depending on location, occurring from September to March, with a peak in January. Male pups at birth are about 79–85cm long and weigh 13–15kg; females are 73–82cm and weigh 10–14kg. Lactation lasts 6–12 months; rarely a female will suckle both a newborn pup and a yearling.

Australian sea lion

Neophoca cinerea

Inhabits coastal waters of W and S Australia; found on island rookeries from Houtman Abrolhos (Western Australia), S and E to the Pages Islands, South Australia. Vagrants N to Shark Bay, Western Australia. Population estimates were formerly in the range of 3,000–5,000 animals, but more recent estimates are in the range of 10,000–12,000.
HTL male 200–250cm, female 130–180cm; WT male 250–300kg, female 70–110kg.
FORM: adult females silver-gray above, creamy yellow beneath, fading to brownish; adult males dark blackish-brown, with a mane of longer, coarser hairs over the shoulders. Newborn pups are chocolate brown with a pale fawn crown. Pups molt to juvenile coat similar to that of adult females at about 2 months.
DIET: cephalopods; fish including salmon, whiting, and sharks; sometimes penguins.
BREEDING: females reach sexual maturity at 4.5–6 years, males at 6 or more. Different breeding locations have different pupping seasons. Newborn pups are 62–68cm long and weigh 6.4–7.9kg. Females prefer sandy beaches and smooth rocks for pupping, and are known to move inland. Mating occurs about 10 days after birth. Lactation often continues until the next pup is born, with females suckling young up to three-quarters of their own length. Adults form relatively small breeding colonies of about 200 animals. Considered non-migratory.

New Zealand sea lion **Vu**

Phocarctos hookeri

New Zealand or Hooker's sea lion

Lives in coastal waters of New Zealand and on nearby subantarctic islands. Main colonies occur on Enderby, Dundas, and Figure-of-Eight Islands in the Auckland group; there are also small rookeries on the Snares and Campbell Island. Stragglers can occasionally be seen at Macquarie and Stewart Islands, and on the South Island of New Zealand to the Otago Peninsula. The older population estimate of approximately 5,000 animals was increased to 10,000–15,000 in 1992.
HTL male 250–350cm, female 200cm; WT male 300–450kg, female 160kg.
FORM: adult males dark blackish-brown all over, with well-developed, dark manes; adult females and subadult males and females are silver-gray on the back, pale yellow or creamy white on the belly, above the ears to the eyes, and down the sides of the muzzle. Pups are dark brown, with cream markings on the top of the head, extending down the nose and over the top of the head to the nape of the neck. Pups molt at about 2 months.
DIET: squid, octopus, small fish, prawns, crayfish, crabs; penguins are also occasionally taken.
BREEDING: females reach sexual maturity at 3–4 years, males at 5. Females give birth on open, sandy beaches from early December to early January, the bulls having arrived in late October and early November, and the cows from late November to December. Pups are sexually dimorphic at birth, the males weighing on average 7.9kg and females 7.2kg. Males weigh up to 3kg more than females at 20 days of age. Females mate 7–10 days after giving birth. Lactation lasts nearly 1 year; females can sometimes be seen suckling young estimated to be yearlings or older. New Zealand sea lions do not seem to be migratory, but disperse widely during the nonbreeding season. DML/JH

The Fight to Mate

Breeding strategy of California sea lions

EVENLY SPACED BURSTS OF BUBBLES RISING TO the surface offshore of a California sea lion rookery (pupping/mating site) indicate that, below, a male is patrolling his territory. These, the best-known of all sea lions thanks to their performances in circuses and oceanaria, sometimes have territories that are mostly in the water, and males bark to warn intruders of their territorial boundaries.

The California sea lion is currently found in the eastern North Pacific from British Columbia to Baja California, and in a separate population on the Galápagos Islands, near the Equator. Male sea lions are larger than females and maintain territories on the rookeries during the breeding season, which lasts from May to August in California. Each male mates with as many females as possible. A successful adult male must defend his stretch of beach from all other males in order to maximize his mating success. Fighting occurs during the establishment of territories, but is soon reduced to ritualized boundary displays. These performances include barking, head shaking, oblique stares, and lunges at the opponent's flippers. Displays are most likely to occur on territorial boundary lines and can be used to plot the locations of individual territories on the rookery.

Ritualized fighting and large size are the two most important factors enabling the males to stay on their territories for long periods of time without feeding. For a male to maximize the numbers of his offspring, he must remain on the rookery for as long as possible, and ritualized fighting uses less energy than does actual combat. Large size is not only an asset in combat but also confers a lower rate of energy expenditure and the ability to store abundant blubber. The blubber serves as a layer of insulation when the sea lion is in cold water, and is its only source of food when it is on its territory.

Also important in the male sea lion's reproductive strategy is the timing of territory occupation. Ideally, territories should be occupied when the greatest number of receptive females are present. On average, there are 16 females for every territorial male, and 2 females for every pup. In the Nort ern fur seal, the females are receptive about five days after they have given birth, and males estab lish their territories before the females arrive on t rookeries in the Bering Sea. But in the California sea lion the females are not receptive until about 21 days after they have given birth. The fact that

Above *Two male California sea lions confront ch other underwater at the boundary between eir adjoining territories. Such encounters usually ke the form of hostile displays rather than fights.*

Above left *In a rookery at Monterey in lifornia, a group of females clusters around a terri- ial male. Males only hold their territories for an erage of 27 days, in which time they will mate with many females as they can.*

Left *Oblique aggression, as displayed by male lifornia sea lions. Once they have initially estab- hed their territories, the animals use ritualized ges- res to maintain the boundaries. These displays take e place of fighting, allowing males to save energy r maintenance. The order in which the individual stures are performed varies, but a typical sequence* **1** *head-shaking and barking as the males approach e boundary, followed by* **2** *oblique stares inter- ersed with lunges, and* **3** *more head-shaking and rking. During lunges, the males try to keep their re flippers as far as possible from each other's ouths to avoid being bitten. The thick skin on the est helps fend off potentially serious blows.*

Right *A newborn pup. Young animals run the k of being trampled to death on the rare occasions hen actual fighting does break out between rival lls in the crowded rookeries.*

male California sea lions only hold their territories for an average of 27 days means that it is counter-productive for them to establish territories before the females arrive, and in fact they do not even begin to set up territories until after the first pups are born. The number of territories on a rookery increases gradually, reaching a peak about five weeks after the peak of the pupping period.

The weather also affects the sea lions' breeding strategy. Temperatures of more than 30°C (86°F) occur during the breeding season, and while this is generally favorable to the pups, which have not yet fully developed their ability to regulate their body temperature, the territorial males may suffer. All sea lions have only a limited ability to regulate body temperature on land, and normally cool off by entering the water. But for a territorial male to do this is to risk losing his territory. Therefore a successful territorial male must have access to water as a part of his territory. During hot weather, territories without direct access to water cannot be defended.

Sometimes territories are mostly in the water. This often occurs at the base of steep cliffs where there is a little beach providing just enough room for females to come ashore and give birth. It is at these points that the males are most likely to patrol their territories, barking underwater. It is possible that a male with a large portion of his territory under water would have an energetic advantage over one with most of his territory on land. What is certain is that any advantage a male can gain in order to leave more offspring will be exploited to the full. DKO

Walrus

IF SAILORS OF OLD THOUGHT MANATEES WERE mermaids, one can only speculate what they made of their first walrus sighting. Walruses are ponderous and ungainly on land, but agile and powerful in the sea. Heavily mustached above gleaming tusks and capable of a wealth of sounds from grunts and roars to a delicate, bell-like rasp, the animals make a dramatic first impression.

Walruses have long been revered by the indigenous peoples of the Arctic, who see many human attributes in them. Highly social and gregarious, slow to mature and reproduce, fiercely protective of their young, vocally communicative, and long-lived – walruses invite human empathy.

Sea Lions with Tusks

FORM AND FUNCTION

Early descriptions of walruses compared the animals to pigs, partly because of their tendency to huddle together, sometimes one on top of another, but also in recognition of their sparsely haired, rotund bodies, about as big around as they are long. Outwardly they are most similar to sea lions, except for their squarer heads and long tusks. Male walruses have a pair of highly inflatable air sacs in their throat, which have the dual purposes of producing special sounds during courtship and acting as floats while the walrus is resting at sea.

Measuring the mass of an animal as large and remote as a walrus is difficult, and generally there are too few data to generate reliable mass growth curves. Moreover, mass varies with season and reproductive status as well as with age, and there are also differences among locations, reflecting the size of the walrus population and the relative availability of prey. Fewer than 15 Atlantic walruses have been weighed; asymptotic masses range from about 635–650kg (1,400–1,430lb) for females and 850–1,750kg (1,870–3,850lb) for males. Based on a large sample, the mature mass of Pacific walruses is estimated at about 850kg (1,870lb) for females and 1,225kg (2,700lb) for males. Three males at Svalbard also weighed about 1,230kg (2,700lb). In the absence of direct data, the mass of walruses has been estimated from length measurements, sometimes combined with girth. Based on this approach, walrus asymptotic mass ranges from 640–720kg (1,400–1,580lb) for females and 900–1,115kg (1,980–2,450lb) for males.

When on land or ice, walruses stand and walk on all four limbs. The heels of the hind flippers are brought under the rump for support and the toes are turned forward and outward. The palms of the fore limbs support the trunk, and the "fingers" are turned outward and back. Walruses have considerable reach and flexibility in their flippers, managing to scratch most parts of their body with them. In water, the walrus propels itself almost exclusively by means of the hind limbs; the fore limbs are used mainly as rudders.

Walruses have remarkable skin. It is 2–4cm (0.8–1.6in) thick and thrown into creases and folds at every joint and bend of the body. This thick skin protects the animal against injury from the tusks of other walruses and when hauling out on sharp ice or rough rocks. Everywhere but on the flippers, the skin is covered by coarse hair about 1cm (0.4in) long, which imparts a furry or velvety texture to the body surface of females and young males; adult males (bulls) tend to be sparsely haired and to have nearly bare, knobbly skin on the neck and shoulders. That knobbly skin is up to 5cm (2in) thick, for added protection, and imparts a distinctive appearance that clearly separates the bulls from all other animals. The folds of the skin can be infested by blood-sucking lice that seem to cause some irritation, for walruses often rub and scratch their skin.

The walrus's nearest living relatives are the fur seals, with which it evolved from bearlike ancestors, the Enaliarctidae, in the North Pacific Ocean

Above *Taking to the water from a haul-out on the askan coast, a group of walruses create the marine uivalent of a traffic jam. Walruses are naturally gre-rious, and males and females gather in herds that n number hundreds or even thousands of animals.*

oout 20 million years ago. Early walruses were milar in appearance to modern sea lions, and om about 5–10 million years ago they were the ost abundant and diverse pinnipeds in the acific Ocean. Some of those early forms were sheaters, but others had already changed their et to mollusks and other bottom fauna. They adually also altered in appearance and behavior. It was probably in this connection that the change from forelimb (Otariid-style) to hind limb propulsion took place, and the tusks began to enlarge.

Between 5 and 8 million years ago, some of these bottom-feeding walruses with tusks made their way into the North Atlantic Ocean through the Central American Seaway, an open channel across what is now Costa Rica and Panama. They flourished in the North Atlantic. The fate of those in the North Pacific is more controversial. It was long thought that the population there died out, to re-emerge about 1 million years ago when Atlantic walruses recolonized the Pacific via the Arctic Ocean. Recent discoveries of *Odobenus*

FACTFILE

WALRUS

Odobenus rosmarus

Order: Carnivora (Pinnipedia)

Family: Odobenidae

Sole member of genus. 2 or 3 subspecies: **Atlantic walrus** (*O. r. rosmarus*), C Canadian Arctic to Kara and Barents Seas. **Pacific walrus** (*O. r. divergens*), Bering and Chukchi Seas. The **Laptev Sea walrus** is sometimes also considered a separate subspecies (*O. r. laptevi*).

DISTRIBUTION Arctic seas, from C Arctic Canada and Greenland to N Eurasia and W Alaska.

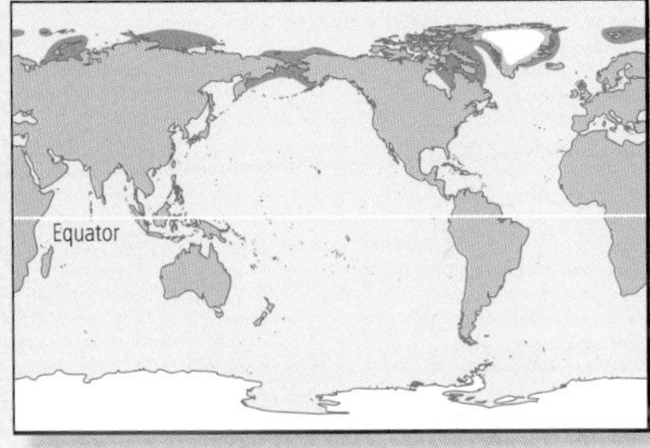

HABITAT Chiefly open water and pack ice over continental shelf.

SIZE Head–body length male: 3.1–3.2m (10.2–10.5ft), female: 2.7m (8.9ft) on average, with marked regional variations; smallest walruses are in Hudson Bay, where average lengths are 2.9m/9.5ft (males), 2.5m/8.2ft (females); in contrast, females from Foxe Basin average 2.8m/9.2ft. **Weight** male: 795–1,210kg (1,750–2,670lb), female: 565–830kg (1,250–1,835lb). **Tusk length** also varies, with the largest tusks, in the Pacific walrus, averaging about 55cm (22in) in adult males and 40cm (16in) in females.

FORM Skin cinnamon-brown to pale tawny, darkest on chest and abdomen; immature animals are darker than adults. Surfaces of flippers hairless; black in young animals, becoming brownish to gray with age. Hair sparse on neck and shoulders of adult males.

DIET Mainly mollusks; occasionally seals and sea birds.

BREEDING Gestation 15–16 months, including 4–5 months of delayed implantation. Litter size: 1.

LONGEVITY 40 years or more.

CONSERVATION STATUS Not considered endangered as a species, though numbers have been seriously reduced as a result of past hunting.

remains in Japan, however, have challenged this view. They indicate that walruses existed in at least the western Pacific into the middle Pleistocene, and these animals may have been the origin of the modern Pacific walrus (*O. r. divergens*). The second scenario is consistent with genetic differences found between the two subspecies of walrus, but the origins of the two groups remain uncertain.

Not Just in Shallow Seas

DISTRIBUTION PATTERNS

It was long thought that walruses were limited to shallow seas, but new evidence from satellite-linked data loggers indicates that they can dive to at least 180m (600ft). While still shallow compared to a number of other marine mammals, this is a considerable advance on the 80–100m (300ft) previously considered their maximum limit. Most walrus dives are not that deep because their main prey is limited by depth, but the animals can and do cross deeper water between feeding areas.

Pacific walruses have been reported out of their normal range as far east as central Canada, while

WHY DO WALRUSES HAVE TUSKS?

In both male and female walruses, the upper canine teeth develop into great tusks. These have many functions, serving both as icepicks and defensive weapons, but their primary role lies in signifying the bearer's status in walrus society. In any herd, the largest walrus with the largest tusks tends to be the dominant one. Simply by posturing to display the size of its tusks, the dominant animal can move unchallenged into the most comfortable or advantageous positions, displacing shorter-tusked subordinates. If the dominant walrus encounters another with tusks of comparable size, however, their confrontation may escalate from visual displays to stabbing duels **1**, until eventually one of the combatants concedes defeat by turning away **2** and withdrawing. Such contests can involve both sexes, but are most intense among bulls in the breeding season.

The social value of the tusks extends beyond the competition for dominance. By their size and shape, the tusks convey much information about the sex and age of their bearer. For about the first year and a half after birth, young walruses have no visible tusks. Their canine teeth do not emerge through the gums until 6–8 months of age, and they are covered by the ample upper lip for another year thereafter. As a result, any small animal with no tusks is immediately recognizable to all the others as being young and dependent, and its larger companion can therefore tentatively be identified as an adult female. At all ages, the tusks of females are usually shorter, more slender, more curved, and rounder in cross section than those of males. In old age, the tusks of animals of both sexes tend to be stout, but shortened and blunted by fracture and abrasion. A human observer can identify the sex and approximate age of a walrus from its tusks, and other walruses are probably at least as perceptive.

Occasionally, a walrus emerging from the water onto an ice floe will use its tusks as a fifth limb, jabbing the points into the ice to help heave the body forward **3**. Tales of this behavior led an 18th-century zoologist to give the walrus the generic name *Odobenus*, a contraction of the Greek words *odontos* and *baenos*, meaning literally "toothwalk." Walruses also use their tusks as head-rests **4** and to break breathing-holes in the ice.

Left *A male Pacific walrus slides into the water, ealing the tough, wrinkled hide that is one of the cies' most distinctive features. In older males, the n can be up to 5cm (2in) thick around the head and k, where it provides protection against injuries.*

antic walruses have occasionally been seen off coasts of the Netherlands, the British Isles, d, less often, France and Spain. Recent sightings walruses in the Gulf of St. Lawrence, Canada, ay also be extralimital excursions from farther rth, or tentative first steps toward recolonizing s former walrus habitat.

Specialists in Seafood

DIET

alruses today feed primarily on bivalve mollusks, clams, cockles, and mussels that abound on e continental shelves of northern seas. They also t about 40 other kinds of invertebrates from the afloor, including several species of shrimp, abs, snails, polychaete and priapulid worms, topuses, sea cucumbers, and tunicates, as well a few fishes. Some walruses eat seals, probably th as scavengers and as predators.

To locate their food, walruses probably rely ore on touch than on any other sense, for they ed in total darkness during the winter, and in urky waters or at depths where light penetra- on is poor for much of the rest of the year. Their nse of touch appears to be most powerfully veloped on the front of their snout, where the in skin and about 450 coarse whiskers are high- sensitive. The central vibrissae are the most tac- e, and can discriminate objects about the size of ncil erasers.

The upper edge of the snout is armored with ugh, cornified skin, and is used for digging out small clams and other invertebrates in the bottom mud. Those buried deeper in the sediments are excavated by jetting water into their burrows. The ability of walruses to squirt large amounts of water from their mouth is well known to zoo keepers.

The old hypothesis that walruses dig out clams with their tusks is now known to be incorrect. The tusks are primarily social organs, like the antlers of deer and the horns of sheep (see box). They become worn not from digging but from being dragged along in the mud and sand while the walrus moves forward excavating its prey.

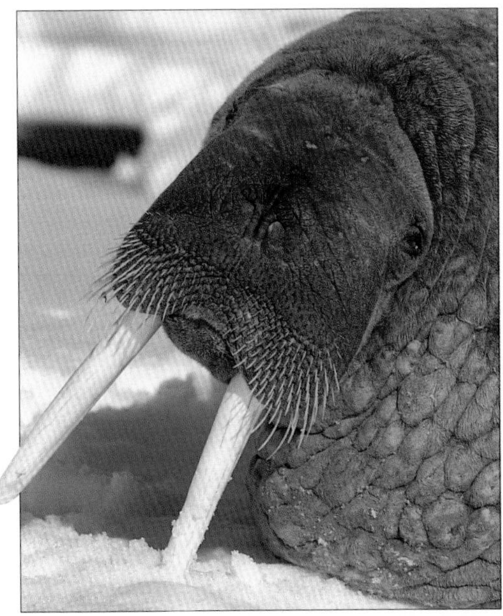

Above *On Svalbard in the Arctic Ocean, a bull walrus displays the mustache-like vibrissae that play an important part in feeding. In the inky depths, the animals use these sensitive whiskers to "feel out" mollusks and other seabed prey.*

Caring for the Calves

SOCIAL BEHAVIOR

A few females first breed at 4 years of age, and some as late as 10; the average is 6 or 7 years. Males mature more slowly; for most, full physical development requires about 15 years. Only then can a male achieve full social maturity or dominance. The bulls compete for mates, and only those that are large enough in body and tusk are able to compete successfully.

Mating takes place during the coldest part of the winter in January–February, probably in the water. There is little information on breeding behavior; most observations of Pacific walruses have been made on pack ice, and those of their Atlantic cousins mostly at polynya (areas of permanently open water surrounded by sea ice). These habitat differences may be responsible for divergences noted between the subspecies.

Among Pacific walruses, the adult females (cows) and young congregate in traditional breeding areas, gathering into relatively small herds that travel and feed together. Several such herds may coalesce when they haul out onto the pack ice to rest between feeding bouts. Each herd is accompanied by one to several bulls, which mostly remain in the water. These bulls engage incessantly in vocal displays, consisting of set sequences of repetitive clicks, knocks, and bell-tone rasps and strums underwater, and shorter sequences of "clacks" and whistles at the surface. Like the songs of birds, these repetitive calls probably serve to attract mates and to repel potential competitors. The bell-like sound is apparently produced by using one of the inflatable sacs in the throat as a resonance chamber, and is used only in sexual display. The normal sounds of walruses are barks of variable pitch.

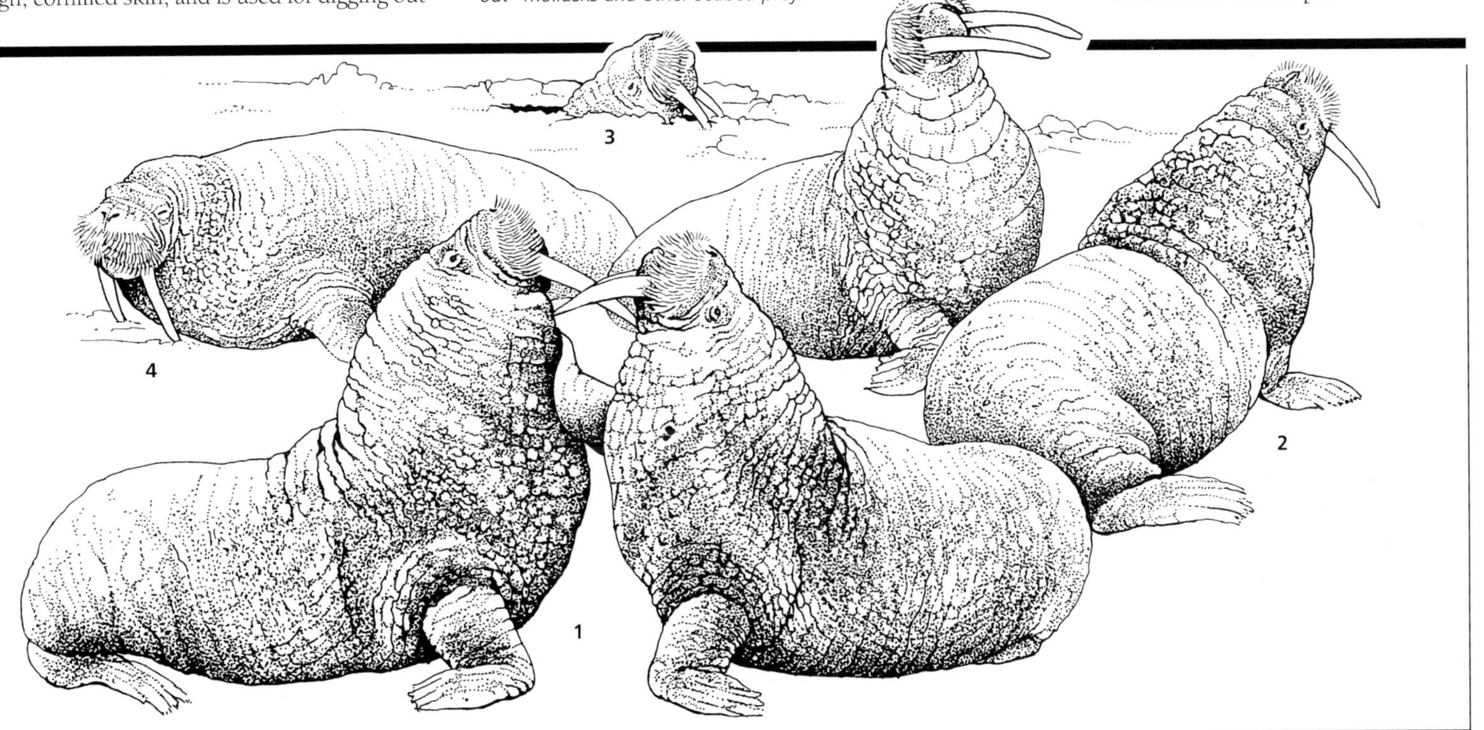

Atlantic walruses at polynya are less mobile, and probably limited in number by the available food resources. Males make similar vocal displays, but can interact with all their competitors to develop stable dominance relationships and to defend access to groups of females. This female-defence polygyny contrasts with the lek-like system reported for Pacific walruses, in which males appear to display to a group of females they cannot defend. The difference in breeding systems may be a result of increased male competition and decreased habitat stability among Pacific walruses.

The female gives birth to a single calf in the spring of the following year, usually in May. The long pregnancy means females breed only once every two years at most, and the gap becomes longer with age. For this reason, the walrus has the lowest rate of reproduction of any pinniped, and one of the lowest among mammals in general. Twins are uncommon.

At birth the calf is about 1.1m (3.6ft) long and weighs 50–65kg (110–145lb). It has a short, soft coat of hair, pale grayish flippers, a thick white mustache, and no visible teeth. It feeds only on milk for the first 6 months, then begins to eat some solids.

By the end of its first year, the calf has approximately tripled in weight, and developed tusks 2.5cm (1in) long. For another year, the calf remains with its mother, while gradually developing benthic feeding ability and becoming more independent. At 2–3 years it separates from the mother. Some females are accompanied by both very young and older calves.

After weaning, the young walruses continue to associate and travel with the adult females. Gradually, over the next 2–4 years, the males break away, forming their own small groups in winter or joining with larger herds of bulls in the summer. Seasonal segregation of sexes occurs to some degree in all walrus populations, but it is particularly apparent in the Bering–Chukchi region and at Svalbard/Franz Josef Land. In the Bering–Chukchi area, most of the bulls congregate in separate haul-out and feeding zones in the Bering Sea during the spring, while the cows and most of the immature animals migrate north into the Chukchi Sea. They remain separated throughout the summer, then, as the cows migrate south in the fall, the bulls apparently meet them in the vicinity of the Bering Strait and accompany them to the wintering–breeding areas in the Bering Sea. The immature males spend the winter in other parts of the pack ice, outside the breeding areas.

Mixed herds are common in Foxe Basin, and both mixed and segregated herds are found in the Canadian high Arctic. Segregation by age and sex diffuses the impact on the food supply and reduces potential conflict between adult and adolescent males in the breeding season. Why some groups segregate more than others is not known.

Threats on the Arctic Horizon

CONSERVATION AND ENVIRONMENT

Walruses are still valued by the northern peoples of North America, Russia, and Greenland as a major source of food and other materials, just as they have been for thousands of years. Farther south, in Europe, Asia, and North America, however, the main interest in walruses has been for the ivory from their great white tusks, only second in size and quality to those of elephants.

Walrus populations throughout the Arctic were severely depleted during the 18th, 19th, and 20th centuries by commercial hunters from Europe and North America, who sought the animals principally for their tusks, skins, and oil. The North Atlantic population was the first to be depleted and was reduced to the lowest numbers. The maritime stock of eastern Canada was extirpated in the 1800s, and walruses at Svalbard nearly met the same fate. Most other stocks of the Atlantic subspecies have also been reduced.

Population estimates for Atlantic walruses are incomplete or out of date, but the subspecies is definitely reduced from pre-exploitation times. The size of the Pacific walrus population is also uncertain. Joint American/Soviet estimates from 1975–90 suggested at least 200,000–250,000 animals. Arriving at a more precise figure is difficult for several reasons: the gregarious nature of walruses, which means that surveys find many animals if they find any at all; haul-out patterns that vary, making it impossible to know what portion of a population is visible at any one time; and the animals' wide distribution, which means that more remote groups may be missed.

Walruses are killed for subsistence through most of their range, though in all areas this hunting is regulated, whether by quota, seasons, or other methods. In Norway they are completely protected. In Russia, Canada, and Greenland, aboriginal subsistence is now recognized as the first consideration after that of the survival of the walruses themselves. The task of managing stock can be difficult, not only because of the international nature of many populations but also because of a growing awareness, through genetic and stable isotope studies, that populations may be finely subdivided, so that what appears to be a single stock may in fact be two or more groups with different probabilities of being hunted.

Oil contamination can have an impact on walruses, both directly through ingestion of oil residues during benthic feeding and also indirectly by reducing prey populations. The physiological reactions of walruses to oil and the aromatic hydrocarbons associated with them are unknown.

In general, walruses have low levels of organochlorine contamination, reflecting the low lipid

Left *Half submerged on an ice floe, two walruses take their ease in the still waters of North Hudson Bay. Even though the Hudson Bay population are among the smallest walruses, males still average an impressive 800kg (1,750lb).*

Right *Turning pink in the sun's heat, basking male walruses jam an inlet on Round Island, off Alaska. Walruses usually prefer to haul out on ice, but when that option is not available, they will settle instead for remote stretches of shoreline where they are unlikely to be disturbed.*

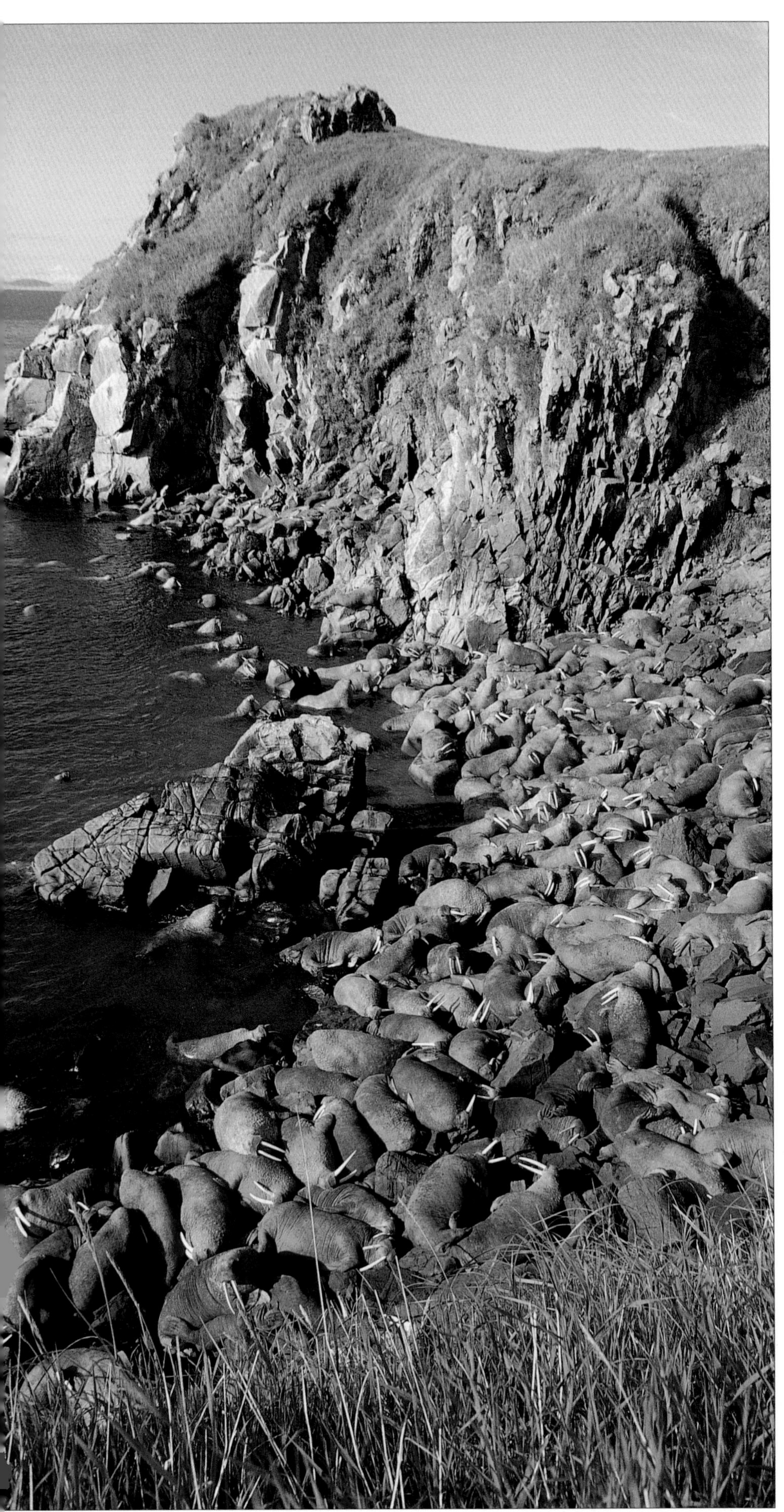

content of their prey; the highest levels are found in seal-eating walruses, for the contaminants accumulate in the seals' fat. Mercury levels are generally lower in walruses than in other marine mammals, in the range of 1 to 2 ($\mu g.g^{-1}$ wet weight) for liver. Seal-eaters also had slightly elevated hepatic lead levels, at around ~0.15 ($\mu g.g^{-1}$ wet weight) compared to 0.07–0.11 ($\mu g.g^{-1}$ wet weight) in walruses from Foxe Basin and Alaska. Lead in walrus teeth from Foxe Basin appears to be mostly natural, without a significant anthropogenic contribution. Cadmium levels range from ~6 to 11 ($\mu g.g^{-1}$ wet weight) in liver; research on the levels found in the teeth of Foxe Basin walruses has shown that they have not changed from pre-industrial times.

Walruses are vulnerable to noise pollution, being particularly sensitive to the sound of aircraft. Several haul-outs have been abandoned in areas where human activities have increased, though it is difficult to distinguish an abandonment caused by a population decline from one resulting from a shift in distribution. Fishing vessels can also cause problems by taking food that would otherwise be available to walruses, generating noise pollution, and destroying bivalve habitat.

Although walruses once ranged into southern waters, suggesting they can tolerate warmer climes than those of their present distribution, global warming may nonetheless affect them indirectly in negative ways. Low calf production in Alaska has been associated with poor ice years there, and the distribution of walrus prey may also change with increasing sedimentation or changing sea temperature. Walrus distribution may shift with respect to the human populations that rely on them.

Walruses are host to a number of diseases and parasites. Walrus viruses include calcivirus and a morbillivirus similar to phocine distemper virus. Several bacterial infections have been identified, often in association with physical trauma to tusks, eyes, or flippers. The effects of the viral infections on the animals are unknown, but the bacterial infections are usually accompanied by generally poor body condition, lethargy, and death; *Brucella* bacteria, to which the animals are known to be prone, cause reproductive failures in other mammals. The most common ectoparasite on walruses is an anopluran louse. Internal parasites include a number of nematode worms, and acanthocephalans. One nematode, *Trichinella*, causes sickness and death in humans.

Natural mortality also occurs through misadventure. There have been several instances of mass death by trauma among Pacific walruses, probably resulting from stampedes of hauled-out walruses in the course of which many individuals are overrun and trampled by others. Falls, ice-entrapment, and injuries from conspecifics are other sources of mortality. Predators, aside from man, include killer whales (*Orcinus orca*) and polar bears (*Ursus maritimus*). REAS/FHF

True Seals

DRAGGING ITSELF PONDEROUSLY ACROSS THE ice, a true seal is transformed once it reaches the sea, where it can plunge to 600m (2,000ft) and stay submerged for an hour. True seals' lack of agility on land is more than offset by their grace in the water.

Despite the extreme refinement of their physiology in respect to diving, the true seals of the Phocidae family are still not fully emancipated from their terrestrial ancestors of some 25 million years ago. The tie to land or ice for birth and nurture of the young sets the pattern of their lives.

Built for Diving

FORM AND FUNCTION

Unlike eared seals, true seals swim by powerful sideways movements of their hindquarters. The trailing hind limbs are bound to the pelvis so that the "crotch" is at the level of the ankles and the tail scarcely protrudes. The long, broadly-webbed feet make very effective flippers but are useless on land. The fore limbs, unlike those of eared seals, are not strongly propulsive; they are buried to the base of the hand, and are used for steering in the water and, sometimes, to assist in scrambling over land or ice. The northern true seals have evolved more powerful arrangements of muscle attachments along the spine, whereas Antarctic species have longer, more mobile fore flippers.

Respiration and circulation in true seals are adapted to one overwhelming purpose; that of spending long periods of time underwater. The Weddell seal is a supreme diver, with descents of 600m (2,000ft) to its credit. Even so, it has to cede preeminence to the elephant seals; the longest dive yet recorded, lasting for 120 minutes was made by a Southern elephant seal, while the current depth record of 1.5 km (almost 1 mile) is held by a Northern elephant seal.

Many profound adaptations have evolved to permit such feats, including some that affect the seals' eyesight. Sight is probably important for locating and capturing prey underwater, and the pupils of pinnipeds are adapted for optimal use in their respective foraging environments. The pupils of individual species dilate relative to the level of brightness they are accustomed to encountering in the areas of sea in which they hunt; for example, the Northern elephant seals, which dive into deep, dimly-lit waters, have a wider range of pupil dilation than the Harbor seal, which inhabits shallower waters.

Anatomically, the 18 living species of true seal fall into two subfamilies, both of which further subdivide into three distinct tribes. In the souther

Left Although seals are equipped with physiological adaptations that enable them to make long dives, the physical exertion of staying submerged for a prolonged period takes its toll. After such dives, seals may require a recovery period longer than the dive itself. Here, a Harp seal surfaces through a hole in the ice.

Below A Weddell seal swims with agility in the sea that shares its name. This most southerly of all mammal species lives on the land-fast ice around the coast of Antarctica. It uses its excellent underwater vision to hunt cod – its favored prey – and also, by following radiating cracks in the ice, to locate breathing holes.

:als, or Monachinae, these are the tropical awaiian and Mediterranean monk seals (another ionk seal, the Caribbean, was declared extinct in 996); the Northern and Southern elephant seals; id the Antarctic species (the Crabeater, Leopard, oss, and Weddell seals). The northern seals, or hocinae, are divided between the Bearded seal :rignathus barbatus); the Hooded seal (*Cystophora istata*); and a third tribe containing the Baikal, aspian, Gray, Harbor, Harp, Ringed, and Spotted :als.

Although now largely found in high latitudes of oth the Northern and Southern hemispheres, ie true seals probably originated in warmer aters, where the monk seals still live. All of the orthern species breed on ice, with the exception f the Harbor seal, which breeds as far south as aja California (the Gray seal can breed on land or :e). Of the southern seals, the Northern and outhern elephant seals breed respectively from :alifornia to Mexico and in the temperate to sub- ntarctic parts of the Southern Ocean. The four ntarctic seals breed on ice, and occur generally outh of the Antarctic Convergence at 50–60°S.

The most obvious differences between species re in size and the relative sizes of the sexes. Some opulations of Ringed seals reach weights of only bout 45kg (100lb), whereas a fully-grown male outhern elephant seal may be more than 50 mes heavier. In most species, males and females re of similar size, but in the other Monachinae, specially the monk, Leopard, and Weddell seals, :males tend to be larger than the males, whereas he males of some northern seals – the Gray, looded, and elephant seals – are hugely larger han the females; these large males also have eavy, arched skulls and nasal protuberances for ggressive displays. The size disparity is most narked in the Southern elephant seal, in which he male can be more than seven times the weight of the female.

Marine Opportunists

DIET

Most seals have a diet of relatively small or soft ood, and so the array of premolars and molars ound in terrestrial carnivores, adapted for cutting nd crushing, is reduced to rows of uniform teeth, isually five.

Where several species inhabit the same area, ome differentiation is apparent. In the Okhotsk nd Bering Seas, the Ringed seal breeds on land- ast ice or heavy pack (drifting) ice and feeds on mall fish and planktonic crustaceans, while the Spotted and Ribbon seals use somewhat lighter

▲ ***Above*** *The marked disparity in size between large males and much smaller females – here, a Northern elephant seal pair on a mating beach – does not apply to all species of true seal.*

FACTFILE

TRUE SEALS

Order: Carnivora (Pinnipedia)

Family: Phocidae

18 species in 13 genera: monk seals (genus *Monachus*, 2 species); Crabeater seal (*Lobodon carcinophagus*); Weddell seal (*Leptonychotes weddelli*); Leopard seal (*Hydrurga leptonyx*); Ross seal (*Ommatophoca rossii*); elephant seals (*Mirounga*, 2 species); Bearded seal (*Erignathus barbatus*); Hooded seal (*Cystophora cristata*); Harbor and Spotted seals (*Phoca*, 2 species); Ribbon seal (*Histriophoca fasciata*); Ringed, Baikal, and Caspian seals (*Pusa*, 3 species); Harp seal (*Pagophilus groenlandicus*); and Gray seal (*Halichoerus grypus*).

HABITAT Land-fast ice, pack ice, offshore rocks and islands, beaches and rocky inlets.

SIZE Head–tail length ranges from 1.3m (4.3ft) in the Ringed seal to 4.2m (13.8ft) in the male Southern elephant seal; weight from 68kg (150lb) in the Ringed seal to 2,200kg (4,850lb) in the male Southern elephant seal.

DISTRIBUTION Generally in polar, subpolar, and temperate seas, except for the monk seals of the Mediterranean and Hawaiian regions.

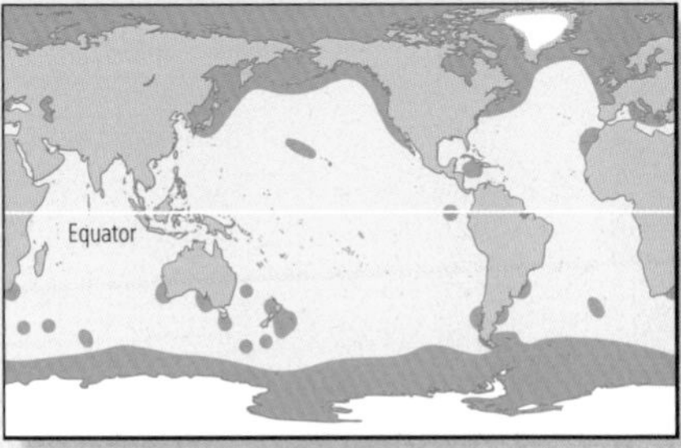

COAT Streamlined and lacking the underfur that distinguishes the eared seals. Some species have spotted or banded coloration.

DIET Fish, shellfish, cephalopods; Leopard seals prey on penguins and other seals.

BREEDING Gestation 10–11 months, including 2–4 months of suspended development (delayed implantation). Lactation lasts from 4 days to 2.5 months.

LONGEVITY About 25 years; maximum lifespans of 43 (Ringed seal) and 46 (Gray seal) recorded in the wild.

See species table ▷

pack ice and feed respectively on shallow-water fish and deep-water fish and squid. The Bearded seal, which also inhabits this region, is unique among true seals in feeding almost entirely on bottom-dwelling mollusks and shrimps; its teeth wear down quite early in life.

Under fast ice around Antarctica, the Weddell seal eats fish. In the pack ice, the Ross seal subsists on deep-water squid, the Leopard seal eats large quantities of seals and penguins (see box), and the Crabeater seal (see The Krill-eating Crabeater) feeds mostly on krill, which it strains through its many-pointed teeth.

Strategies for Reproduction

SOCIAL BEHAVIOR

Since the discovery, in the mid-20th century, of a way of determining animals' ages from layers in the true seals' teeth, the basic patterns of their growth, reproduction, and survival rates have been extensively documented. Ages of sexual maturity vary rather unexpectedly, being later in small species like the Ringed and Caspian seals than in the large Antarctic species or the huge elephant seals. Early maturity may be a disadvantage in species like the Harbor and Ringed seals that disperse in complex, near-shore environments where land (or ice) predation is a threat and where learning about surroundings is essential for safe reproduction. Although Gray and elephant seals of both sexes are fertile when quite young, males are incapable of securing mates until they are much larger, some years later.

Although species differences remain, Baikal, Ringed, Harp, Harbor, and elephant seal females have all been shown to mature earlier in populations reduced by exploitation. This effect has been attributed to an increase in the amount of food available per animal, resulting in faster growth rates in juveniles and an earlier onset of maturity. A remarkable decrease in the mean age of first reproduction among female Crabeater seals, from more than 4 years in 1945 to less than 3 in 1965, may, for example, have been associated with a vast "release" of its krill food base through the reduction in the number of the great whales at that time as a result of overexploitation.

The Harbor seal (*Phoca vitulina*) has one of the broadest geographic distributions of any pinniped, stretching from the Baltic across the Atlantic and Pacific Oceans to southern Japan. Although individuals may travel several hundred kilometers on annual feeding migrations, Harbor seals are generally believed to be philopatric, returning to the same areas each year to breed. Genetic studies of mitochondrial DNA from seals from 24 localities bear this out. Populations in the Atlantic and Pacific Oceans are significantly differentiated, as are the east and west coast populations in each ocean. Within these four regions, populations that are geographically farthest apart are generally the most differentiated and often do not share genotypes, or differ in genotype frequency. However, genetic discontinuities may exist even between neighboring populations, such as those on the coasts of Scotland and eastern England or the eastern and western shores of the Baltic. Evidence from genetic studies of mitochondria are consistent with an ancient isolation of populations in both oceans, due to the development of polar sea ice. In the Atlantic and Pacific, populations appear to have been colonized from west to east, with the European populations showing the most recent common ancestry, perhaps as a result of recolonization from Ice Age refugia – areas relatively unaffected by climate change – after the last glaciation

Reproductive seasons may be set by females becoming receptive at optimal times for either rearing the young or for fostering their independence; males are often potent long before and long after. Occasionally, newborns occur as much as six months outside the normal season; these

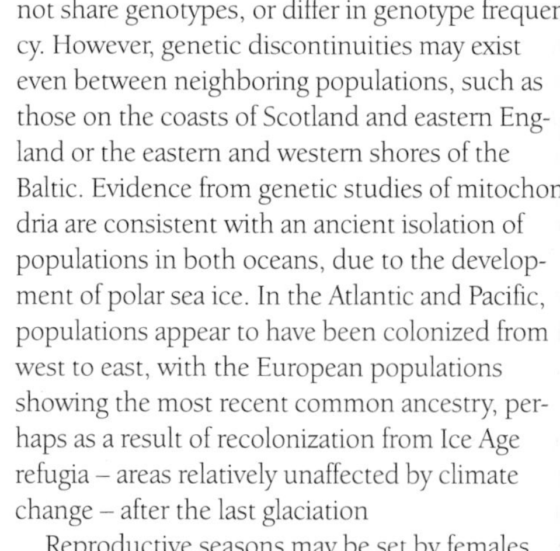

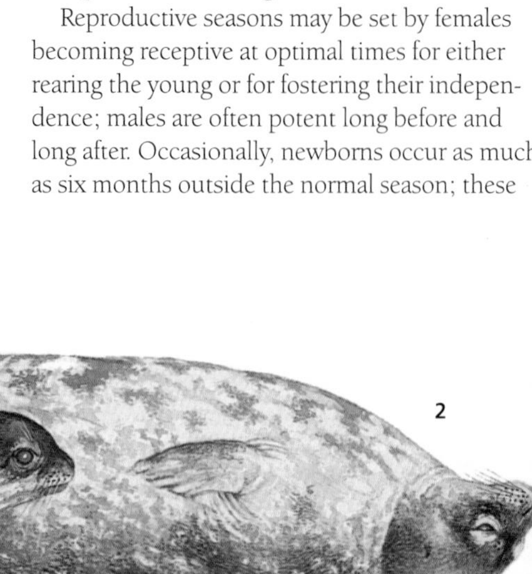

Below True seals from both the northern and southern oceans. The first six are southern seals, the last six are northern seals. **1** *Ross seal* (Ommatophoca rossii); **2** *Weddell seal* (Leptonychotes weddellii); **3** *Crabeater seal* (Lobodon carcinophagus); **4** *Leopard seal* (Hydrurga leptonyx); **5** *Southern elephant seal* (Mirounga leonina); **6** *Hawaiian monk seal* (Monachus schauinslandi); **7** *Hooded seal* (Cystophora cristata) *with inflated black hood* **a** *and red nostril bladder* **b**; **8** *Ringed seal* (Pusa hispida); **9** *Gray seal* (Halichoerus grypus); **10** *Harp seal* (Pagophilus groenlandicus); **11** *Bearded seal* (Erignathus barbatus); **12** *Ribbon seal* (Histriophoca fasciata).

ve been attributed to young mothers whose cles have not been set. Most females of a species produce at about the same time, although popations at higher latitudes may be later than the st. Gray seals show marked regional differences timing and choice of breeding sites, and treme local variability also occurs among Harbor als in western North America, where quite proxate populations may differ by up to four onths, perhaps as a result of "drift" in this relaely nonseasonal region.

Mean lactation periods are 1–2 weeks in pack- e seals and up to 12 weeks in the Ringed and ikal seals, which suckle their young in "snow ves" on the fast ice. The differences seem to be lated to the relative stability and protectiveness the nursery in which the young are raised. Pups Weddell seals on fast ice, as well as pups of Haror and monk seals on land, are weaned when ey are around 5–6 weeks old, whereas those of e elephant and Gray seals (in which the males mate on land with as many females as they can) are weaned at 3–4 weeks, perhaps as an evolutionary response to preemptive males. Pups of most species increase in weight on average 2.5–3.5 times during the lactation period, although the Baikal seal, with a lactation of 8–10 weeks, is reputed to increase its weight 5.5 times.

The blubber of females is transferred to the pup in the form of very rich, fatty milk. The fat content of Harp seal milk increases from around 23 percent at the start of lactation to more than 40 percent by the end, with a corresponding decline in water content. Since the female fasts while lactating, the decreasing water content may be important in maintaining her water balance. The layer of blubber below the surface of pinnipeds' skin serves as insulation as well as acting as an energy reserve; Harbor seals maximize the effectiveness of their blubber in this respect by seasonal changes in their girth, blubber volume, and thickness.

Although lactation is physiologically demanding of nutrients, many species not only of true seals but of bears and baleen whales as well fast for much or all of it. Great body mass allows these species to store large amounts of fat and protein relative to their rates of milk production. Mammals that lactate while fasting may lose up to 40 percent of their initial body weight. The production of milk entails the loss of up to one-third of body fat and 15 percent of body protein in several seal species, greatly depleting the mother's resources; thus, Southern elephant seal mothers lose, on average, 35 percent of their postpartum mass during lactation and 40 percent during the whole breeding period, with levels of expenditure apparently determined largely by female mass at parturition. Initially their milk is 70 percent water, but in the first 20 days the composition changes rapidly, becoming "half fat" (actually 52 percent), while the water content drops to 33 percent. Northern elephant seal pups gain mass quickly; in the four weeks for which they nurse, average body mass goes up from an average of 42kg (93lb) to 127kg (280lb)! Mothers depress their metabolism as an adaptation for energy conservation during long-term fasting.

In species that breed colonially, harassment by males may affect the length of the breeding season. For example, male Gray seals have been shown disturb late-pupping mothers significantly more than peak-pupping ones. Late-pupping mother spent 22 percent less time suckling and had pu that were 16 percent lighter on weaning than equivalent-sized mothers that gave birth at the peak of the season. Reduced weaning mass may lead to a lower chance of survival, and so such harassment may well play a role in synchronizin reproduction.

Pups are occasionally adopted, and some ma pups of Northern elephant seals may sometime take advantage of the tolerance of unrelated females to steal milk from them, thereby gainin

xtra weight. In Southern elephant seals, the more paced out the seals are on the breeding beaches, he less aggression occurs between females and he less likely it is that mothers will be separated rom their pups. Smaller, younger females are nore likely to be separated from pups than older, arger females.

Of 35 observed separations among Harbor eals, 68 percent occurred on the same day as, or vithin one day of, a storm, suggesting that this is he primary cause of separations. Seal pups have mazing homing abilities, and can often find their ookeries if separated at sea. Of 75 Northern elehant seal pups that were captured and released

THE RINGED SEAL'S SNOW LAIRS

The Ringed seal, *Pusa hispida*, is the smallest of the marine Arctic pinnipeds, and is one of only two species adapted to life on the land-fast sea ice. The animals' habitat lies at latitudes subject to extreme low temperatures, and is characterized by a stable ice platform that forms in early winter. The small body size of the adults and the semi-altricial pups are unusual adaptations to cold, allowing the seals to use shelters that they construct in the snow overlying their breathing holes. These small snow caves hide them from predators, especially Polar bears and Arctic foxes. It appears that pups still wearing their lanugo, or birth coat, can withstand the Arctic cold without shelter, but pups that have been wetted become hypothermic and require shelter to regain a stable body temperature. Since female seals actively swim away with their pups from attacks on their birth lairs by foxes and bears, both the physical and the thermal protection of alternate lairs under the snow are important for pup survival. Weddell seals resident in the land-fast ice of the Antarctic are the Ringed seals' southern ecological counterparts, but their large body size represents a more typical cold-adaptive strategy for polar seals. SSA

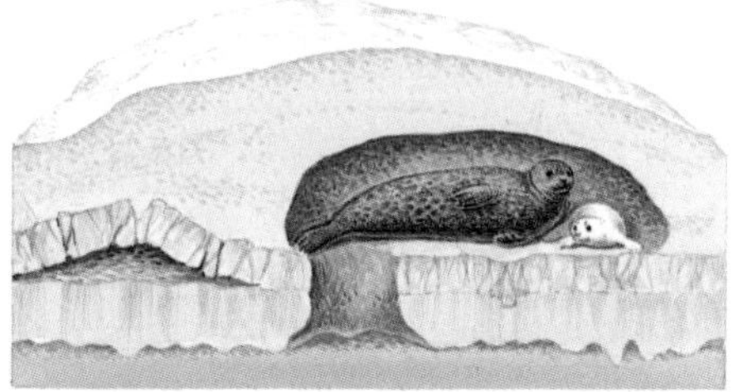

Above As the winter pack ice thickens and snow builds up around pressure ridges in the ice, pregnant Ringed seals dig upwards with their fore flippers from the water below, to create a lair or snow-cave. Here the pup is born and spends its first one or two months. This excavation behavior is unique among seals.

Left Except for individual animals that are occasionally sighted in the connecting river systems, Baikal seals are largely confined to Lake Baikal, in Siberia. Large aggregations of this species only form at certain times of the year, such as during the summer or spring at feeding areas.

Below The Hooded seal has not one but two forms of bizarre nasal display. The lining of one nostril can be forced out through the opposite nostril to form an inflated red bladder. Alternatively, the whole of the black hood, which is an enlargement of the nasal cavity, can be blown up.

100km (60mi) away from their rookeries, 75 percent found their way home, many by a direct route, and by traveling an average of 39km (24 mi) a day. In colonially breeding species, such as the Gray seal, where separations and reunions occur routinely, females may be able to recognize their own pups by vocalizations.

Copulation takes place on land in elephant and Gray seals and normally in the water among all other species. In all species, mating evidently occurs soon after, and sometimes even shortly before, pups are weaned, so gestations last 10–11 months. However, the period of active embryonic growth is only 6.5–8 months. This delay in implantation and growth of the embryo has the consequence that males compete for females when they are localized and restrained by maternal duties; at the same time it adjusts the rate of fetal growth to the feeding and physiological capacities of the females.

Although some true seals have been reported to mate with only a single partner each year, males of all species are probably promiscuous whenever they have the opportunity. In species that aggregate to breed, dominant males can either actively control access to specific groups of females, as is the case in elephant seals, or defend a position in an area where there is likely to be a particularly high density of receptive females, as is the case for Gray seals. Females may incite male contests by vocalizing loudly during attempted matings; this is probably done to assess male fitness, providing a more dominant male with the opportunity to displace the original one. Although mating behavior varies considerably between and within species, in all cases only a fraction of sexually mature males are successful. In extreme instances, such as the Southern elephant seal, the effective sex ratio on the breeding beaches may exceed 100 females to 1 male.

Such extreme differences in the reproductive success of males have fascinating repercussions for the breeding strategies of mother seals. When a mother is in bad condition, giving birth to a male pup may be counterproductive, as he may grow up small and have little or no success with females. A smaller mother with a male fetus may benefit from terminating her pregnancy and allocating the resources she saves to her own growth. She could then give birth to, and raise, a larger pup in the subsequent season. Female Southern elephant seals must be large – at least 300 kg (660lb) – before they have their first pup, but small females (i.e. under 380kg/840lb) rarely produce male pups. This may be because male pups are 14 percent heavier than females at birth, and this may pose an unacceptably high cost on a small female and prejudice her future chances of reproduction. There is equal maternal investment for pups of both sexes in the Gray seal, despite the sexual dimorphism that exists in the species.

Underwater territoriality has been suspected or confirmed in some species that mate in the water.

Above *A group of Northern elephant seals resting on a beach during the molting season. The species makes two annual migrations, one for breeding and one for molting, which normally takes place during July and August. Why the seals travel vast distances to molt remains a mystery.*

Above *A Gray seal on the hunt for fish. The diet of this species includes some invertebrates, such as crabs, but the bulk of its prey consists of bottom-dwelling fish. Its consumption of salmon and cod has caused the Gray seal to be persecuted – sometimes quite vigorously – by fishermen.*

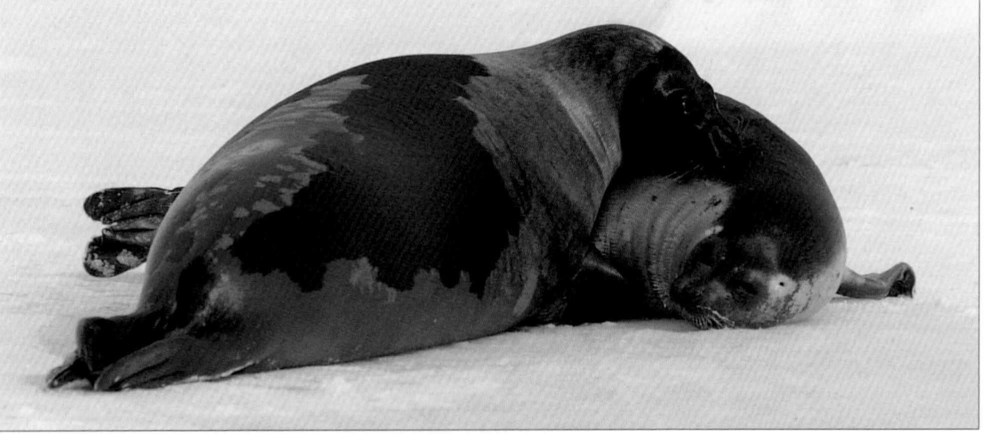

Left *Because of delayed implantation, female Har[p] seals mate within a few days of giving birth to the pu[p] they conceived the previous year. Mating occurs before the newborn pup is fully weaned, a process that takes less than a fortnight.*

THE KRILL-EATING CRABEATER

The world's most abundant sea mammal

ANTARCTIC PACK ICE IS THE MOST EXTENSIVE but dynamic habitat in the world, its annual range varying seasonally from 4 million sq km (1.6 million sq mi) in the late summer up to 22 million sq km (8.5 million sq mi) in late winter. The animal most suited to this dramatic environment is the Crabeater seal. Crabeaters are almost certainly the most abundant large mammal species in the world after humans (there are 6 billion of us), probably numbering between 7 and 14 million individuals; former proposals of a population of 15–40 million are likely to have been overestimates.

Krill is the keystone species of the Antarctic marine food web. It grazes on algal blooms at the edge of the pack ice, and also on algae released from the ice as it is broken up by the perpetual grinding of wind and waves. Krill is the main food of seabirds, whales, and many seals, including, of course, the Crabeater, which sieves it from the water through highly modified teeth.

By inhabiting the extensive pack ice, Crabeaters can exploit this resource in ways not open to competitors such as the whales, which feed mainly in open water. The seals rest on ice floes as a means of escaping predation, and perhaps also to save the energy needed to keep warm in water that is perpetually close to freezing. Ice, seals, and krill all drift together, which means that Crabeaters have constant access to food simply by slipping through the brash ice filling the cracks between the floes and into the deep, still waters beneath. The seals have been observed to cross almost half Antarctica's circumference in this way in just a few months. As a result, Crabeaters are now thought to form a single large population, all members of which can interbreed.

The dynamic nature of the pack-ice habitat also influences the seals' social structure. During the breeding season from October to November, mothers stay together with their new-born, furry, milk-coffee-colored pups for up to 3 weeks, during which time the pup grows from 25kg (55lb) to 110kg (243lb) solely on the mother's milk. Unlike most other seal species, Crabeaters are not colonial breeders; they travel in mock-family groups in which the male is not necessarily the father. The male is typically very aggressive and persistent, seeking to separate the female from the pup. The females tend to be slightly larger, however, averaging 235kg (518lb) as against the males' 225kg (496lb), and so they are able to drive the male away from the pup if need be. The females' aggression towards the male fades as the pup grows, and the pup is weaned at the point when the male becomes dominant. Little is known about the seals' mating behavior, although it seems likely that the males wait for the female to come back into estrus in the last few days of weaning the pup. Usually family groups remain 1–2km (0.6–1.2mi) apart, but occasionally other males will approach across the ice, in which case a fight lasting as much as six minutes may ensue. Old male Crabeaters are heavily scarred, probably as a result of fights during this mate-guarding phase of breeding.

Leopard seals and Killer whales are the main predators of Crabeater seals. Pups, particularly, are at risk, and seem to die at a very high rate in their first year of life; in contrast, adults often bear the scars of close encounters with death, but may go on to live to 40 years. The seals grow up to the age of 5–7, after which they breed, although weight is more important than age in determining whether a female gets pregnant; growth rate and food abundance are therefore likely to influence the rate at which the seals reproduce. The near-extinction of the largest baleen whales (which also eat krill) in the early 20th century may have left more food for Crabeaters, so it is possible this highly abundant seal owes at least some of its success to the otherwise sad demise of the whales. ILB/RML

Below *Crabeater seals recumbent on an icefloe in the Antarctic peninsula. Although they are abundant, there are fears that recovering whale populations could impact upon krill abundance and so reduce Crabeater seal numbers.*

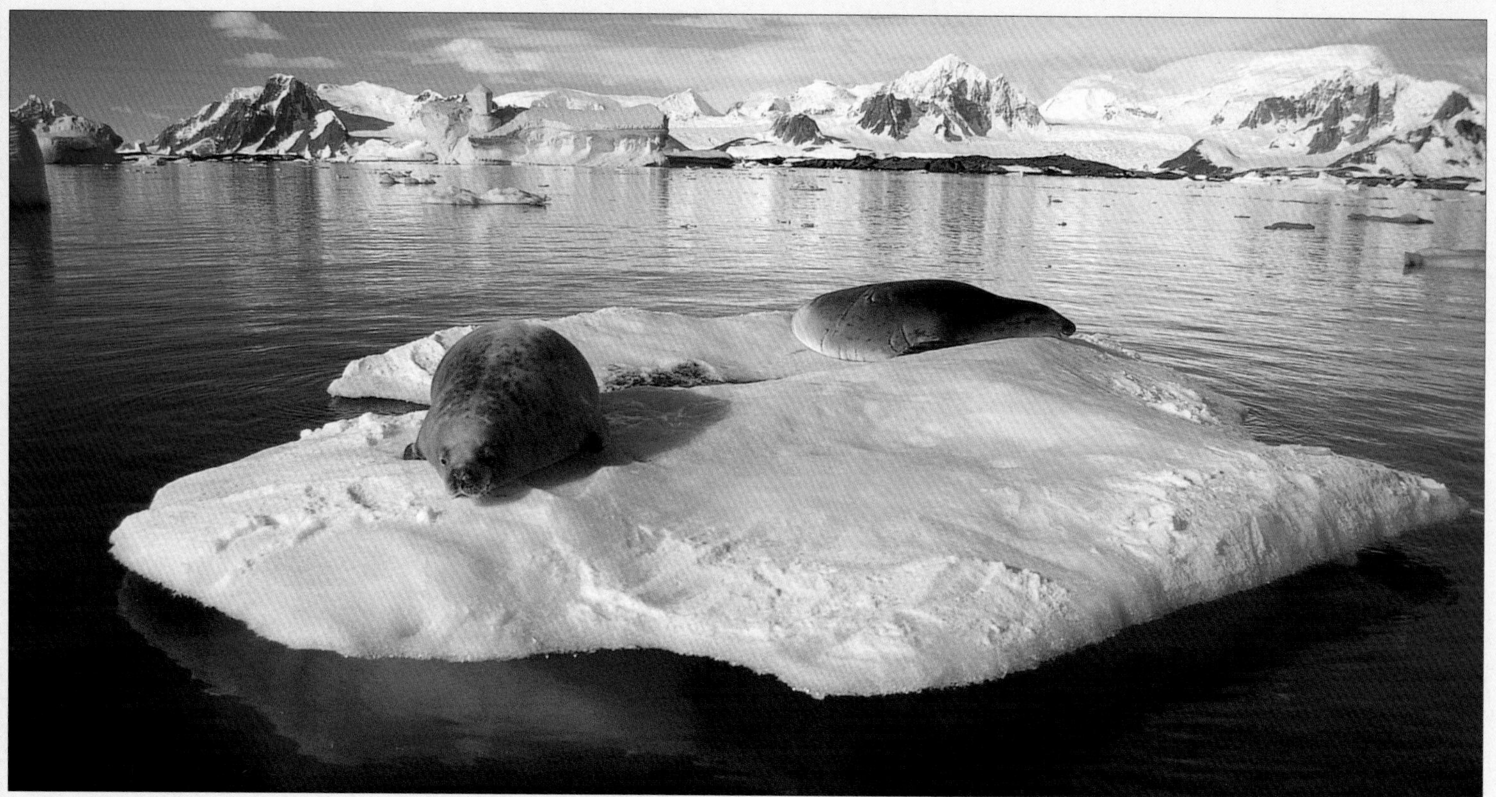

Hunter of the Southern Ocean

The largest of the Antarctic seals, the Leopard seal is the only one that regularly preys on warm-blooded animals. It is an opportunistic predator, feeding on a wide variety of prey, including krill (37 percent), fish (13 percent), and squid (3 percent) as well as penguins (25 percent), other seabirds (5 percent), and seals (9 percent). It only hunts in the water; on land it is cumbersome.

Only larger and older Leopard seals appear to take large prey. Although spectacular, predation on penguins in the vicinity of rookeries is seasonal, and appears to be the speciality of just a few seals, since the mobility of swimming penguins makes them hard to catch.

The Leopard seal is unique among true seals in habitually feeding on other seal species. Such an attack has never been witnessed, but the high frequency (55 percent) of scarring on Crabeaters, together with frequent finds of seal remains in Leopard seal stomachs, indicates a fairly high level of predation. The majority of the seals preyed upon are young animals, but freshly scarred older seals attest to the fact that all age classes are vulnerable. In addition to the Crabeater, prey species include the Weddell seal, the pups of elephant seals, fur seals, and presumably also the Ross seal. Unlike the other true seals (but like the eared seals), the Leopard seal has elongated fore flippers that give it an advantage in speed and maneuverability.

The scarring on Crabeater seals as a result of Leopard seal attacks consists of slashes up to 30cm (12in) long, often in parallel pairs, coursing tangentially across the body. In the past these scars were mistakenly thought to be caused principally by Killer whales, but it now appears that they come from the evasive rolling action that often enables Crabeaters to escape from Leopard seals. When a Crabeater is caught, only its skin and attached blubber are eaten. AWE

***Below** The skull of the Leopard seal is characteristically elongated. This individual displays its well-developed canine teeth. The complicated cheek teeth are used for filtering krill.*

Male Weddell seals display in, and aggressively defend, narrow stretches of water up to 200m (650ft) long under females congregated with young along ice cracks. Individual male Ringed seals may use ice holes as much as 1km (0.6mi) apart, excluding other males but not females. Male Harbor seals "sing" underwater during the breeding season from specific locations that are often visited by breeding females. Many of these animals have bite wounds, indicating that these locations are actively defended. Although Crabeater, Gray, Hooded, and Spotted seals are said to form "families" of male and female with young, the males are merely waiting for the females to become receptive for mating. The females are aggressively defended, but may be abandoned once mating has occurred, as the males leave to seek other matings.

Adult Northern elephant seals go to sea twice each year for a total of around 8 months, during which time they range widely in the northern Pacific. New tracking technology has demonstrated that the species (and individuals) return to the same foraging areas after breeding and molting, documenting the first double migration for any animal. The seals dived continually to depths of 250–550m (820–1,800ft) during both migrations, and the males traveled distances of at least 21,000 km (13,000mi) during the 250 days they were at sea. (The equivalent figures for females were18,000km/11,200mi and 300 days.) These are the longest annual migrations yet recorded for individual mammals. The double migrations are a

ɔnsequence of the seals' requirement to return ı land twice each year, to molt and to breed, .though the reasons why seals favor molting sites n the California Channel Islands over island and ɔntinental beaches nearer their foraging areas are ɔt known.

Spending so much time at sea necessitates eeping in the water. Northern elephant seals can ɔ for up to 25 minutes without breathing when ıey are asleep underwater, and can rise to the ırface to breathe without apparently waking up.

The social behavior of most true seal species as been little studied outside of the breeding eason. Species may be basically solitary, aggregat-ıg merely because of clumped food resources or esting places, or they may interact in truly social ays. Harbor seals in Québec, Canada, have been hown to reduce their individual rates of scanning ɔr danger when they are assembled in larger roups. It is thought that the navigational skills of Iarp seals may be enhanced when they migrate in erds. And, finally, weaned young Crabeater seals re believed to gather together for greater protec-on from Leopard seals.

Grounds for Hope

CONSERVATION AND ENVIRONMENT

)nly the monk seals are truly endangered as a roup, although isolated populations or sub-pecies of other seals are rare or in decline. Many emale Ringed and Gray seals in the Baltic Sea ave reproductive abnormalities that have been ttributed to the effects of pollution. Lake Saimaa in Finland contains a small population of around 200 individuals of a unique Ringed seal sub-species, which is vulnerable to entanglement in fishing gear and to loss of breeding habitat. The distinctive Harbor seal of the Kurile Islands, which numbers some 5,000 individuals, is protected in Russian territorial waters, but some are killed in northern Japan.

By contrast, most populations of seals are probably stable or increasing, some after heavy exploitation in the past. A particularly striking example of recovery is the Northern elephant seal, which has increased from fewer than 100 in 1912 to more than 100,000 today.

Several seal populations around the world come into conflict with fishermen, who look upon them as competitors. However, in most cases there is little scientific evidence to suggest that the seals are having a significant impact on fish stocks or fisheries. For example, a study of the interac-tions between Harp seals and cod trawlers in Canada found that seals damaged only 0.002 per-cent of the catch, although much higher levels of damage have been recorded in gill-net fisheries for cod and salmon.

Seals are also killed for subsistence and for their skins by aboriginal peoples in northern regions and by coastal peoples elsewhere. In common with the body parts of many other mammals, seal bones, organs, and pelts are used in traditional Chinese medicine. Genetic sequencing techniques have also identified pinniped penises among such trade items. The trade in body parts of some seal species is legal, although individual species are protected against hunting. Unfortunately, because the species origin of body parts is not easily identi-fiable without genetic sequencing, the lucrative market for these organs could lead to the poach-ing of protected species. DML/IAM

Above *Ecotourism, in this case to photograph Harp seal pups in Canada, offers another way for local communities to gain an income from seal populations in addition to hunting.*

Below *The Southern elephant seal molts between January and May; the molt may last for as little as a few days or as much as a month and more. Sandy beaches are prime sites; here, a bull Southern elephant seal uses its flexible flippers to throw sand over itself in an effort to reduce its body temperature.*

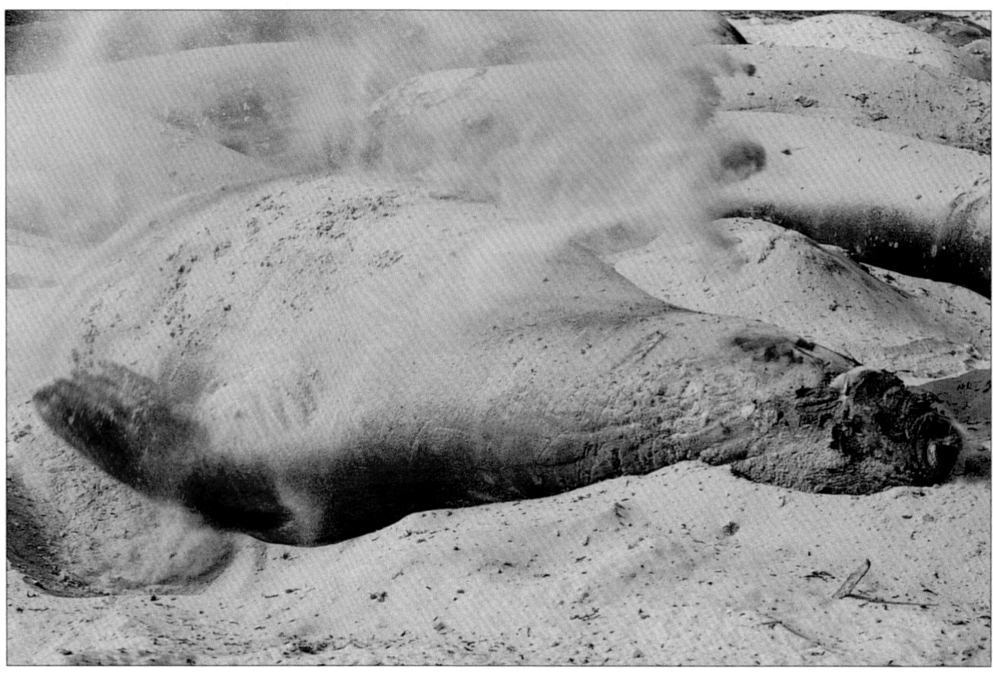

True Seal Species

THE 18 EXTANT SPECIES OF TRUE SEAL OR phocid include the monk seals (genus *Monachus*) and the elephant seals (genus *Mirounga*), as well as the Harbor and Spotted seals of the genus *Phoca*. Some authorities also list the Harp and Ribbon seals as *Phoca* species, and also regard the genus *Pusa*, linking the Ringed, Baikal, and Caspian seals, as a *Phoca* subgenus. A nineteenth species, the Caribbean or West Indian monk seal (*Monachus tropicalis*) was last seen in 1952 and was declared extinct by the IUCN in 1996, although at the time of writing it still appears in CITES Appendix 1 as an endangered species.

Hawaiian monk seal En
Monachus schauinslandi

Inhabits sandy beaches and surrounding waters of the NW Hawaiian Islands, also known as the Leeward Chain: Kure Atoll, Midway Atoll, Pearl and Hermes Reef, Lisianski Island, Laysan Island, French Frigate Shoals (FFS), Necker and Nihoa Islands; a small number of animals also inhabit the main Hawaiian Islands. Subpopulations breeding on Kure, Midway, Pearl, and Hermes appear to be increasing, whereas those at FFS (by far the largest subpopulation), and at Lisianski and Laysan Islands, are in decline. Numbers of seals at Necker and Nihoa, while small, also appear to be increasing. Trends on the main islands are unknown. Total population numbers c.1,400 animals, declining at approximately 5 percent per year, apparently as a result of decreased food availability (especially at FFS), possibly related to competition with fisheries; entanglement in fishing gear and other human disturbance; and mobbing behavior, where females and immature seals are killed by sexually aggressive males.
HTL male 210cm, female 230cm; WT male 170kg, female 205kg.
FORM: adult coat is silvery-gray on back, fading to cream on throat, chest, and belly; additional light patches may also be found on body. Over time, coat looks brown above and yellow below; males, and some females, turn dark brown or black with age. Pups are born with long, black lanugo (fetal hair that is retained after birth), which is shed at about 6 weeks. The Hawaiian monk seal still retains a number of anatomical features that are more primitive than those found in fossil relatives dated 14–16 million years ago.
DIET: fish, cephalopods, and crustaceans, including lobster.
BREEDING: females reach sexual maturity at c.4–8; age of sexual maturity in males unknown. Births occur throughout year, but are most common between February and August. Peak pupping occurs on beaches from March to early April. Pups weigh 14–17kg at birth, are nursed for 5–6 weeks, and are weaned at 60–75kg. Adults tend to be solitary, both on land and in the water.
LONGEVITY: 20–25 years

Mediterranean monk seal Cr
Monachus monachus

Historically widespread throughout Mediterranean, NW coast of Africa, and Black Sea, but now the most endangered of all pinniped species. Present today in the NE Atlantic, in remnant colonies along the coast of W Sahara/Mauritania (the largest remaining colony), and on the Desertas Islands, Madeira. In the Mediterranean, they occur mainly in the Ionian and Aegean Seas, and in the Cilician Basin off the S coast of Turkey. A few individuals (probably no more than 2–5) may still remain in the Black Sea. The species is still reported to occur off the coasts of Croatia, Italy, Cyprus, Libya, Algeria, and Morocco but numbers are very low (probably no more than 50 animals in total). Total world population is probably only 400–500. Major threats include: loss of habitat from development, mass tourism, and cliff collapse; competition with fishing industry for prey, incidental catch in fishing gear, and killing by fishermen; disease and toxic algal blooms.
HTL male 240cm, female 238cm; WT male 315kg, female 300kg.
FORM: adult coat usually brown or gray on back, lighter on belly. White patch common on underside of belly; other irregular light patches not uncommon. Older males tend to be black, with an irregular white patch on the belly. New born have black, wooly lanugo, white or yellow patch on belly, the shape of which can be used to determine sex.
DIET: fish and octopus, in inshore waters.
BREEDING: males and females thought to reach sexual maturity at 5–6 years. Females give birth to a single pup weighing 16–18kg. Births may occur in any month, but most are born in September–October. Virtually all pupping today occurs in caves and grottoes in sea cliffs. Lactation 5–6 weeks, but may be as long as 17 weeks at Turkish sites. Males possibly polygynous, mating in water; social structure unknown, although somewhat gregarious in breeding caves.
LONGEVITY: 20–30 years.

Northern elephant seal
Mirounga angustirostris

Colonies exist in California on Santa Barbara, San Nicolas, San Miguel, Santa Rosa, Año Nuevo, and Southeast Farallon Islands, and on the mainland at Ano Nuevo Point and Point Reyes. Colonies also exist in Mexico on Guadalupe, San Benito, and Cedros Islands. A few births also occur on Natividad, San Martín, the Coronado Islands, and on San Clemente Island. Outside breeding season, females range along coasts of Oregon and Washington, while adult males migrate as far N as Gulf of Alaska and Aleutian Islands. Hunted almost to extinction in the 19th century; only 8–20 individuals thought to remain by end of century. Following protection, numbers increased until, by 1992, world population was estimated at c.125,000 animals.
HTL male up to 500cm, female up to 300cm; WT male 1.8–2.7 tonnes, female 350–900kg.
FORM: coat gray, tan, or brown, although males are generally darker. With age, bulls become corrugated and heavily scarred in the thick-skinned neck region, and develop an inflatable snout or proboscis (hence "elephant"); they also become pale on face, proboscis, and head, pinkish on chest. Canines very large, cheek teeth peglike, but sometimes with cusps and double roots. Newborn pups have black lanugo that is replaced by a light gray or silver coat at about 3 weeks.
DIET: skates, rays, sharks, hake, shrimp, euhausiids, octopus, whiting, and crab, but predominately deepwater squid.
BREEDING: females reach sexual maturity at 2 years, although they probably do not have their first young until 4. Males reach sexual maturity at 6–7, but usually do not mate successfully until 9–10. Births occur on islands from mid-December through March. Male pups are c.153cm long and weigh 36kg at birth; females pups are 147cm and 31.5kg. Lactation 27 days. At the start of the breeding season, females gather on beaches after smaller number of large males establish dominance hierarchies by visual and vocal displays, and sometimes by physical combat. A single male may mate with up to 80 females in a season. The annual molt begins with juveniles in April and ends with adult males in August. After breeding season, adults spend most of their time offshore diving for food.
LONGEVITY: males 14 years; females up to 18.

Southern elephant seal
Mirounga leonina

Almost circumpolar in distribution in the hemisphere, generally N of the pack ice. Breeds mostly on islands on both sides of the Antarctic Convergence, in 3 separate groups (possibly subspecies): 1) S Georgi Falkland, Gough, S Shetland Islands, islands and mainland of Antarctic Peninsu la, Patagonia N to Valdes Peninsula; 2) Ke guelen, Heard, Marion, and Crozet island 3) Macquarie and Campbell islands. Ano er group, on Juan Fernández Island, was wiped out by hunting in the 19th century Scattered births found elsewhere, N to Namibia, Oman, and the Comoros Island South Island of New Zealand, Tasmania, Tristan da Cunha. The species migrates to the Antarctic mainland and has strayed N to S Australia, Mauritius, Rodriguez Islanc Uruguay, Peru, and the North Island of New Zealand. World population estimate in 1991 at 700,000–800,000. Some bree ing populations in the Antarctic portion o the Indian and Pacific Oceans have been declining in recent years; Atlantic population appeared stable or increasing in 199
HTL male up to 620cm, female 266cm; WT male 3.7 tonnes, female 350–800kg. Mal are the largest of all pinnipeds.
FORM: coat light to dark silvery-gray or brown in adults and juveniles. Newborn with black lanugo, changing to a short, silvery-gray coat at c.3 weeks of age. Adult males have thick, scarred neck shield; pro boscis is shorter and skull generally more massive than in Northern elephant seal; large canine teeth with small incisors and even smaller, peglike postcanines.
DIET: mostly cephalopods, some fish.
BREEDING: females reach sexual maturity at 3–4 years, depending on location, male at about 4, although few successfully bree before the age of 10. Births occur on shore and, occasionally, on shore ice, in September and October. Male pups weigh c.45kg female 40kg. Pups are nursed for c.23 day At weaning, males weigh c.119kg; females c.112kg. Breeding behavior similar to that of Northern elephant seals, but single mal may defend "harem" of up to 50 females, "share" much larger aggregations, in whic the sex ratio may reach 300 females to eac male. After breeding, Southern elephant seals disperse widely. They molt ashore in January–April.
LONGEVITY: males generally less than 20 years (only 1.2 percent reach 13 on Macquarie Island), females about 14.

ABBREVIATIONS	HTL = head–tail length WT = weight Approximate nonmetric equivalents: 1m = 3.3ft; 10kg = 22lb	Ex Extinct EW Extinct in the Wild Cr Critically Endangered	En Endangered Vu Vulnerable LR Lower Risk

·abeater seal
bodon carcinophagus

stributed throughout pack ice of Antarc- a, following advancing and retreating ice. ıs strayed as far N as Heard Island, South ·ica, New Zealand, S Australia, and South ıerica. The most numerous of all pinni- ds; census data from between 1968 and 83 suggest a world population of 11–12 llion, although significant declines in nsity were noted in some areas.
L both sexes 257cm; WT male 225kg, male 235kg.
·RM: coat uniform; usually light gray, metimes darker, becoming paler overall th age; irregular patches of spots and gs on the sides, flippers, and around the ›per insertions. Immature seals darker, d may be somewhat mottled. Often heav- scarred on the back and sides from acks by Leopard seals, and around the e and flippers from fights with other abeaters during the breeding season. ·wborn pups have a soft, woolly, grayish- own lanugo; flippers may be darker. ults have a long, slender body and a rela- ely long, thin muzzle. Cheek teeth elabo- ely multicuspid for straining acroplanktonic food.
ET: principal, and often only, prey is krill. ay also feed seasonally on fish such as tarctic silver fish.
EEDING: males and females reach sexual aturity from 2.5–6 years, depending on od availability. Females give birth to a sin- pup, approximately 120cm long and ighing 20–40kg, on pack ice in Septem- r–December, peaking in October. Wean- g occurs between 14 and 21 days, when e pups weigh about 110kg. Males join males with pups to form "triads" during e whelping season. After the pups are eaned, males mate with females, and pairs ay stay together for as long as 1–2 weeks erwards. Throughout the rest of the year, nerally found alone or in small groups, hough large groups have been observed.
NGEVITY: approximately 20 years.

eddell seal
ptonychotes weddellii

rcumpolar in distribution, the most utherly of all seals, breeding in fast ice ound Antarctica and on islands N to S eorgia. Has strayed as far N as Uruguay, an Fernández Island, N New Zealand, S stralia, Kerguelen Islands, and Isla ocha, Chile. Although world population s been reported to be over 750,000 ani- als, no good population estimate actually ists.
L male 250–290cm, female 260–330cm; T both sexes approximately 400–450kg, with wide seasonal fluctuations.
FORM: coat variable: dark silvery-gray, darker above, mottled with black, gray, and whitish blotches. Lighter patches over the eyes and around the lower muzzle. Newborns have gray or light brown woolly lanugo with a dark line down the back. Canines and two large, protruding incisors often worn from "sawing" at ice holes.
DIET: mostly fish (some large), cephalopods, crustaceans.
BREEDING: females reach sexual maturity from 3–6 years, males from 7–8. Females haul out along cracks in the pack ice, and occasionally on islands, to give birth. Pupping occurs from mid-September to early November depending on latitude. Pups are c.120cm long and weigh 22–25kg at birth. Lactation lasts 6–7 weeks, and pups weigh c.113kg at weaning. Adult males vocalize, display, and fight under the ice along the cracks to mate in underwater "territories," or may simply prevent access to breathing holes by other males. Weddell seals are not particularly gregarious and remain a discrete distance when hauled out.
LONGEVITY: approximately 25 years.

Leopard seal
Hydrurga leptonyx

Polar and subpolar waters of the S hemisphere, from the edge of the pack ice to Antarctic continent, as well as on and near many subantarctic islands. Heard Island in the S Indian Ocean has a year-round population. Present on South Georgia, Kerguelen, Macquarie, and the Falkland Islands; also occur at Campbell and Auckland Islands. Has strayed as far N as Sydney, Australia, Rarotonga in the Cook Islands, to S Africa, Tristan da Cunha, and N Argentina. Data from between 1968 and 1983 indicate world population of at least 300,000–500,000.
HTL male 250–320cm, female 241–338cm; WT male 200–450kg, female 225–591kg.
FORM: coat silver to dark gray above, paler below, with light and dark spots most noticeably on throat, shoulders, sides, and belly. Newborn with long, soft lanugo resembling adult pattern. Appearance is often described as reptilian. Bodies elongate, heads large and jaws massive, with large canines, and massive postcanine teeth.
DIET: krill, penguins, fish, squid, and other seals.
BREEDING: females reach sexual maturity at 3–7 years, males at 2–6. Births in September to January, peak in November–December on pack ice, occasionally on islands. Lactation said to last c.4 weeks. Males are not seen with females and young on the ice; their whereabouts at this time is unknown. Female pupping rate of 47–61 percent per annum low for pinnipeds. Usually solitary on the ice and at sea. Often found in association with other pinnipeds (particularly Crabeater and Antarctic fur seals) and large colonies of penguins.
LONGEVITY: 26 years or more.

Ross seal
Ommatophoca rossii

Patchily distributed around the Antarctic continent, but most abundant in the King Haakon VII and Ross Seas. Has strayed to Heard Island and S Australia. World population said to number c.200,000 in 1984.
HTL male 168–208cm, female 196–236cm; WT male 129–216kg, female 159–204kg.
FORM: coat dark gray to chestnut above, silvery-white below, with light and dark flecks where the two colors meet; light and dark chestnut or chocolate stripes from chin to chest, and sometimes along sides of neck. Lanugo of newborn reportedly has pattern similar to adult coat. A thick-necked, short-muzzled species with wide-set eyes and long fore and hind flippers. Incisors and canines sharp and recurved to secure slippery prey, but cheek teeth small, barely piercing the gum.
DIET: primarily cephalopods (squid), but some fish and krill as well.
BREEDING: females reach sexual maturity at 3–5 years, males at 2–7. Births occur on pack ice in November. Newborn pups are approximately 105–120cm long and weigh 27kg. Lactation estimated at c.28 days. Make trilling, sirenlike calls or chugging sounds. Thought to be largely solitary, although very little is known of this species.
LONGEVITY: approximately 20 years.

Bearded seal
Erignathus barbatus

Circumpolar distribution in Arctic and subarctic waters, S to S Labrador, S Greenland, N Iceland, White Sea, Hokkaido, and the Alaska Peninsula. Has strayed to Tokyo Bay, Cape Cod, and N Spain. Two subspecies have been proposed: *E. b. barbatus* (Atlantic basin) and *E. b. nauticus* (W Canadian Arctic to central Siberia). Bearded seals are solitary and inhabit areas of relatively shallow water and moving ice. When ice is not available, they will haul out on land and gravel beaches. Incomplete estimates from early 1980s suggest 450,000 animals inhabit the Pacific.
HTL both sexes 210–250cm, with females slightly longer than males; WT 200–360kg.
FORM: coat varies from silver blue-gray to chocolate brown above, with white or cream colored patches on the body. Births on pack ice mid-March to mid-May. Newborn lanugo is brown or grayish, with light patches on the face and shoulders.
DIET: includes bottom-dwelling crustaceans (crabs and shrimps), mollusks (clams and whelks), polychaetes, cephalopods (squid and octopus), and a variety of fish including Arctic and saffon cod, flounders, sculpins.
BREEDING: pups are born weighing c.33kg and measuring c.130cm, and can enter the water within hours. Weaning occurs between 18 and 24 days at a weight of 78–110kg. Four mammae (as in southern seals, but unlike the 2 found in other northern seals); large body, disproportionately small head, and square fore flippers; abundant, long, densely-packed whiskers. Skull with deep jaw; teeth often worn down or missing in older animals. Females reach sexual maturity between 3 and 6 years; males between 6 and 7. Males "sing" underwater during the breeding season, perhaps to attract females or to defend underwater territories.
LONGEVITY: approximately 25 years.

Harbor seal
Phoca vitulina
Harbor or Common seal

A nonmigratory species occurring near islands and along coastal regions of both N Atlantic and N Pacific oceans over a wide range of latitudes. 5 subspecies recognized: *P. v. vitulina* in NE Atlantic, from Murmansk along coast of Barents Sea (rare), Faeroes, Svalbard, and Iceland, to the outer Baltic, UK S to Channel coast of Ireland, N France, to N Portugal (rare); *P. v. concolor* in the NW Atlantic, from S and W Greenland to Admiralty Inlet on Baffin Island in the Canadian Arctic, to Hudson Bay, S along Atlantic coast of North America to Cape Cod, Massachusetts, and, rarely, to N Florida; *P. v. richardsi* in the NE Pacific, from the Aleutian Islands, Bristol Bay, and the Pribilof Islands, Alaska, S to the central W coast of Baja California, Mexico; *P. v. stejnegeri* in the NW Pacific from the Aleutian Islands, Commander Islands, S and E Kamchatka, the Kurile Islands, and along the coast of NE Hokkaido, Japan; and *P. v. mellonae* landlocked in the area of Lacs des Loups Marins, the freshwater "Seal Lakes" of N Quebec, Canada. Total world population is probably around 500,000 animals.
HTL male 150–186cm, female 120–169cm; WT male 55–170kg, female 45–142kg; size varies with subspecies.
FORM: coat color and pattern variable. Light to dark gray or brown base coat, with overlay of spots, rings, and blotches on adults. Pups usually shed whitish lanugo before birth; otherwise born with "adult" coat. Doglike face, with teeth set obliquely in jaws.

DIET: includes herring, flatfish, cod, walleye, squid, octopus, shrimp.
BREEDING: females sexually mature at 3–4 years, males at 5–6. Females give birth to a single pup on land from January to October, depending on location. Pups weigh 8–12kg at birth and are highly precocial, sometimes entering the water and swimming within 5 minutes of birth. Pups are nursed for 4–6 weeks and are weaned at up to 30kg. Harbor seals haul out, often in groups, on islets, rocks, sandbars, sometimes ice, generally in inshore waters. Most often alone or in small groups at sea.

Spotted seal
Phoca largha
Spotted or Largha seal

Pack ice of N Pacific. 8 separate breeding populations: Gulf of Laotung; Peter the Great Bay; Tatarskiy Strait; Sakhalin Island to N Hokkaido; N Sea of Okhotsk; Karaginskiy Island in Kamchatka; NW Bering Sea; and SE Bering Sea. In spring, found as far S as Fujian, China; Shikoku, Japan; and the E Aleutian Islands. In summer, ranges N into the Chukchi Sea and to Herschel Island, Yukon. World population said to number c.400,000.
HTL male 150–170cm, female 140–160cm; WT male 85–110kg, female 65–115kg.
FORM: typically light or silver gray, darker above, with many small, uniform, oval dark spots. Born with whitish lanugo. Teeth set straight in jaws, unlike closely related Harbor seal.
DIET: fish, cephalopods, crustaceans.
BREEDING: females reach sexual maturity at 3–4 years, males at 4–5. Births in January to mid-April; pups weigh 7–12kg and measure 75–90cm. Lactation 28 weeks, but varies with stability of pack ice. During breeding season, male, female, and pup form "triads," scattered widely on ice. Moves to coasts in summer, resting on land and sometimes entering rivers.
LONGEVITY: approximately 35 years.

Ringed seal
Pusa (Phoca) hispida

There are 5 subspecies, 3 inhabiting marine waters and 2 freshwater lakes: *Phoca hispida hispida* (Arctic Ocean) inhabits shorefast ice and stable pack ice throughout Arctic Ocean and Bering Sea. Found S to James Bay, Strait of Belle Isle, Kap Farvel, Barents Sea off N Norway, White Sea, Karaginskiy Island in Kamchatka, and N Bristol Bay. Also found in freshwater Nettilling Lake and in the Koukdjuak River on Baffin Island. Vagrants as far as the Azores, Germany, Portugal, New Jersey, and S California. *P. h. botnica* (Baltic ringed seal) found throughout N Baltic Sea, S to Stockholm, Sweden, and Riga, Latvia. *P. h. ochotensis* inhabits the W, N, and NE parts of the Sea of Okhotsk, S to the N coast of Hokkaido and Cape Lopatka, Kamchatka; vagrants found as far as Jiangsu, China and Shikoku and Kyushu, Japan. *P. h. ladogensis* (Ladoga ringed seal), found almost entirely in freshwater Lake Ladoga in Russia, although seals are said occasionally to transit the Neva Reka to the Gulf of Finland. *P. h. saimensis* (Saimaa ringed seal) inhabits the freshwater Lake Saimaa, Finland. World population believed to be 6–7 million.
HTL 99–157cm; WT *P. h. ladogensis* 32–56kg; *P. h. saimensis* 45–100kg; *P. h. hispida* 45–107kg; *P. h. botnica* 60–140kg. Males are slightly larger than females.
FORM: coat highly variable: silvery to dark gray below, darker on back, with rings on sides and back. Born with white lanugo. Replaced by mostly unspotted coat, silver on the belly and darker gray on the back, by 6–8 weeks. Short muzzle, with fine, cuspid teeth.
DIET: wide variety of fish and planktonic crustaceans.
BREEDING: females reach sexual maturity at 3–7 years, males at 5–7 depending on population. Births March–April in snow lairs over breathing holes in fast ice, for protection against cold temperatures and predators, especially polar bears; exception is *P. h. ochotensis* in the Sea of Okhotsk, which breeds on unstable sea ice. Pups are born approximately 50–65cm long and weigh only 4–5kg. Lactation mostly between 39 and 45 days on fast ice; approximately 21 days for *P. h. ochotensis*, 5–8 weeks for *P. h. saimensis*. Many adults remain in the same localized area year round. Haul out on ice, rarely land, and are wary, scanning for predators.
LONGEVITY: 30–40 years.

Baikal seal
Pusa (Phoca) sibirica

Freshwater Lake Baikal in Siberia, N of Mongolia. On rare occasions enters rivers that drain into the lake. Population approximately 60,000–70,000 (1990).
HTL both sexes 122cm; WT c.50–130kg.
FORM: coat silver-gray above, with lighter sides and belly; unspotted. Born with white lanugo, shedding to a silver-gray coat at 4–6 weeks. Skull foreshortened, with large eye sockets. Fore flippers and claws larger and stronger than in Ringed and Caspian seals.
DIET: fish, including *Comephorus dybowski, C. baikalensis*, and sculpins.
BREEDING: females reach sexual maturity at 2–5 years, males at 4–7. Pups are born in lairs from mid-February to late March, weighing c.4kg and measuring 64–66cm. Lactation 8–10 weeks. Adult molt begins in late May or early June and may continue into July. Most Baikal seals are solitary.
LONGEVITY: males 52 years, females 56.

Caspian seal Vu
Pusa (Phoca) caspica

Found in the landlocked, saline Caspian Sea, sometimes entering rivers. During breeding season, they concentrate at shallow N end of lake, which freezes over. During summer, they move to water in the deeper middle and S portion. Population said to number approximately 500,000–600,000 but concerns about its status persist.
HTL both sexes c.130–140cm; WT 50–60kg.
FORM: gray back with lighter gray sides and belly; dark spots on backs of males, lighter spots on females. Newborn has white lanugo, shed at c.3 weeks to reveal short, dark hair. Skull like that of Ringed seal.
DIET: wide range of small fish (gobies, sculpins, clupeids), crustaceans, shrimps.
BREEDING: females reach sexual maturity at 4–5 years, males at 6–7. Aggregations of females give birth to single pups weighing c.5kg in late January to early February on pack ice N of Kulaly Island. Lactation 28–35 days. After ice melts, Caspian seals are occasionally seen on islets and rocks.
LONGEVITY: approximately 35 years.

Gray seal
Halichoerus grypus

Separated into 3 distinct populations based on distribution, time of breeding, and body size: NW Atlantic (from Labrador to Nantucket, Massachusetts), NE Atlantic (from C France to Kola Peninsula, Russia), and Baltic. Most of the NW Atlantic population gives birth in the Northumberland Strait, and on sandy Sable Island, Canada. In NE Atlantic, most pupping occurs around British Isles, especially off N and NW coasts of Scotland. Smaller colonies off Norway, Russia, Faeroe Islands, Iceland, France, Netherlands, and Germany. World population approximately 300,000. Baltic population is considerably reduced from its historic size, but now increasing.
HTL NW Atlantic male 235cm, female 200cm; NE Atlantic male 195cm, female 165cm; WT NW Atlantic male 300–350kg, female 150–200kg; NE Atlantic male 170–310kg, female 103–180kg.
FORM: males dark brown, gray, or black, with small blotches. Female is lighter, with dark spots and blotches on gray, tan, or yellowish background. Pups are born with white lanugo, molting to a muted version of the adult pelt between 2 and 4 weeks of age. Snout elongate and arched, especially in adult males, which also develop a heavy, scarred neck region. Large, peglike teeth.
DIET: a wide variety of fish (including sand lance, cod, flatfish, silver hake, herring, and mackerel), and invertebrates (cephalopods and crustaceans).
BREEDING: females become sexually mature at 3–5 years, males at c.3–6, although they may not mate until 8 years older because of competition with older males for females. Gray seals are gregario forming large aggregations during the pu ping, breeding, and molting seasons. In t NE Atlantic population, most births occu on land between September and Novemb Most pupping in the Baltic occurs on ice February and March. In the NW Atlantic, pupping takes place on both land and ice between late December and early Februa At birth, pups are 90–105cm long and weigh 11–20kg, gaining up to 2kg per da throughout the 16- or 17-day nursing per od. Male gray seals are considered polygy nous, monopolizing access to groups of females rather than defending territories controlling the movement of females.
LONGEVITY: males approximately 25 yea females up to 35 years.

Harp seal
Pagophilus groenlandicus (Phoca groenlandica)

Three discrete and highly migratory popu lations breed off NE Newfoundland and i Gulf of St. Lawrence off E Canada; off the coast of Greenland, around the island of Jan Mayen; and in the White Sea off the coast of Russia, summering in waters off E Canadian Arctic Archipelago, W and E Greenland, and in the Kara and Barents Seas. Extralimital sightings have been reported from the Mackenzie Delta, as far south as Virginia on the E coast of the US and the southern North Sea. World popu tion at end of 20th century numbered approximately 7 million.
HTL both sexes (for NW Atlantic population) 170cm; WT 130kg in early March (slightly larger elsewhere).
FORM: coat silvery-white or gray, with da spots in juveniles. Black, wishbone, or "harp-shaped" markings on silvery-white backs of mature animals, tending to devel op most quickly in males at onset of sexu maturity; some females never develop full harps and remain spotted. Black head on adults, ending behind the external ear openings and under the chin. Newborn have white lanugo. A relatively slender, active species, with small, cuspid teeth.
DIET: invertebrates such as amphipods ar shrimp, fish including Arctic cod, capelin polar cod, sand lance, herring, and the occasional Atlantic cod.
BREEDING: sexual maturity at 4–6 years. Gregarious females give birth to a single pup, weighing c.11kg, on newly-formed pack ice (whelping patches) in mid-Febru ary to early March in the Gulf of St. Lawrence and White Sea, slightly later off Newfoundland, and from mid-March to April off Jan Mayen. Females nurse pups a fat-rich milk for c.12 days. Pups are weaned at c.38kg, when females move on to mate with waiting males in preparation

the next year's pupping season. Adults ‑m all 3 populations soon move N, where y once again congregate on ice (molting ‑ches) to undergo the annual molt. The 3 ‑pulations then migrate N to separate ‑nmer feeding grounds.
‑NGEVITY: up to 30 years.

‑bon seal
‑striophoca (Phoca) fasciata

‑iberian and Chukchi Seas, SE to Bristol ‑, Alaska, and Unalaska Island, SW along coast of Kamchatka as far as N Hokkai- Japan, including the Sea of Okhotsk S ‑atarskiy Strait. Has been found as far as ‑dova, Alaska, and Morro Bay, Califor- . Not well studied. World population ‑3,000–240,000, based on estimates ‑m 1982 and 1979 respectively.
‑ 155–165cm, occasionally up to ‑0cm; WT 70–80kg, occasionally up to ‑0kg.
‑RM: males are dark brown to black, with ‑ite to yellowish bands around the neck, ‑pers, and hips. Females are lighter, and bands less distinct. Born with long, ‑itish lanugo, replaced with a coat that is ‑e-gray on the back and silver-gray on the belly after the first molt, at approximately 4–5 weeks. Skull foreshortened, with small teeth, large eye sockets.
DIET: fish (including pollock, eelpout), shrimps, cephalopods, bottom invertebrates; the young are said to eat krill.
BREEDING: females said to reach sexual maturity between 2 and 5 years, males between 3 and 6. Most pups are born on pack ice in early to mid April. Pups weigh 9–10kg at birth, are nursed for 3–4 weeks, and are weaned at 27–30kg. Males are not seen on the ice with females and pups. Adults are pelagic, remaining offshore for rest of year.
LONGEVITY: approximately 20 years.

Hooded seal
Cystophora cristata

Two stocks recognized; one in NW Atlantic, one in Greenland Sea. There are 4 major breeding areas: in the Gulf of St. Lawrence, on the "Front" E of Newfoundland, in Davis Strait (between Greenland and N Canada), and on the "West Ice" near Jan Mayen. Hooded seals are highly migratory; adults from the "West Ice" are know to occur regularly off the W coast of Ireland and young seals have been recorded as far south as Portugal. Seals from the NW Atlantic have been recorded in Puerto Rico and California. World population currently estimated at 500,000–700,000 individuals.
HTL male c.250cm, female 220cm; WT male up to 435kg, female up to 350kg.
FORM: both sexes silvery-gray, with scattered, irregularly-sized black blotches and spots. Dark colored fore and hind flippers and head to behind the eyes. Fetuses have a light gray lanugo that is molted prior to birth. Pups, called bluebacks, are born with blue-gray backs, silvery-gray or yellowish sides and belly, a dark mask, and dark flippers. Bluebacks molt to "adult" coat at c.14 months. Mature males have an inflatable "hood" on top of head; they can also inflate their nasal septum through one or both nostrils as a red "balloon." Skull heavy, with 2 lower, 4 upper incisors, large canines, and peglike cheek teeth.
DIET: deepwater fish (redfish, Greenland halibut) and squid. Weaned pups feed on capelin, polar cod, amphipods.
BREEDING: females reach sexual maturity at c.3 years, males at c.5. Generally solitary except during the spring breeding season and when molting, which occurs from June to August in the Denmark Strait off Greenland. Loose aggregations of females give birth to single pups weighing c.20kg on heavy, drifting pack ice in the second half of March. Hooded seals have the shortest lactation of any mammal – 4 days. Pups are nursed with milk containing over 50 percent fat; they grow at c.5kg per day, and are weaned at an average weight of 40kg. During lactation, female and pup are attended by one or more aggressive males competing for mating privileges. Spectacular displays involving the "hood" and nasal septum are used to fend off other males and to impress the female. Large males generally mate with several females during breeding season.
LONGEVITY: 30–35 years.

DML/JH

Above *Found only in Lake Baikal and the surrounding river system, the Baikal seal is restricted to a freshwater environment. In the summer they haul out and bask in the southern reaches of the lake, as here on the Ushkani Islands.*

THE WORLD'S LARGEST SEAL HUNT

The continued killing of northwest Atlantic Harp seals

Above and below *Images of the clubbing to death of Harp seals, such as this from the 1970s, inflamed public opinion and led to hunting restrictions by the mid-1980s. It is not hard to see why th annual seal hunt provoked such a vehement reaction; the very young pups or "whitecoats" are the epitome of innocence and vulnerability. But althou their slaughter was stemmed, the killing of older se pups was allowed to continue, and reached record levels in the late 1990s.*

FOR MANY YEARS, THE COMING OF SPRING focused international attention on Canada's east coast and its annual seal hunt. The spectacle of photogenic, white-coated Harp seal pups (which made up the bulk of the catch), killed within days of their birth and while still nursing, was more than many people could bear. The resulting outcry over more than a decade led, in 1983, to a temporary two-year ban on the import of whitecoat pelts by the European Union. That ban was extended in 1985 for an extra four years, and made indefinite in 1989. The weight of public opinion also led a Royal Commission on Seals and Sealing in Canada to note, in 1986, that the clubbing of whitecoats was "widely viewed as abhorrent both in Canada and abroad," and to recommend cessation of the commercial whitecoat hunt. In late 1987, that recommendation was implemented by the Canadian government.

With the killing of whitecoats virtually ended, many people assumed that Canada's seal hunt was over. But nothing could be further from the truth. The hunt for northwest Atlantic Harp seals not only continued both in Canada and off west Greenland as it had for centuries but, by the late 1990s, it was recording levels of killing that had not been seen since before the birth of the anti-sealing movement in the late 1960s and the subsequent introduction of quotas in 1971 to limit the size of Canada's hunt.

In Canada, the impetus for an increased hunt began in 1995, when new federal subsidies were introduced to encourage sealers to kill more seals. It continued in 1996 when Canada's fisheries minister raised the total allowable catch to 250,000 and, for the second year in a row, provided additional subsidies to encourage sealing, ostensibly to benefit depleted cod (*Gadus morhua*) stocks. During the hunt the following spring, more than 242,000 Harp seals were landed in the largest hunt since 1970. Then, in December 1996, the fisheries minister increased the total allowable catch again, this time to 275,000 animals. Landed catches increased in response, reaching 264,202 in 1997 and 282,070 in 1998, before dropping back to 244,552 in 1999. The majority – in fact, more than 80 percent – of the catch in these years was still made up of pups, but they were no longer whitecoats. Rather, they were the so-called "ragged-jackets" (weaned pups approximately 2–3

Right *One of the justifications advanced by proponents of culling is that the abundance of Harp seals has severely depleted stocks of cod; however, there is no scientific evidence to support this claim. "Ragged-jacket" pups, characterized by their partially molted coats (seen here), and older "beaters" have borne the brunt of recent killing.*

eeks old, in the process of shedding their white, eonatal coats) and "beaters" (fully molted pups ore than 3 weeks of age).

Over the same period, Greenland's heavily subdized and unregulated summer hunt for the ame animals increased exponentially, reaching 0,000 or more by 1996. Taken together, Canada nd Greenland's landed catches for the period 996–99 averaged more than 300,000 animals er year. The last time average landed catches ere so high – between 1950 and 1969 – the seal opulation declined by 50 percent or more.

To fully appreciate the magnitude of human npact on the northwest Atlantic Harp seal poputiion, one must look beyond the landed catch tatistics provided by Canada and Greenland. To et an estimate of the total number of animals illed, animals taken by sealers but neither landed or recorded in the official catch statistics must lso be counted, as must seals killed incidentally commercial fisheries. When all these figures are dded together, some 400,000–500,000 Harp eals were killed in the northwest Atlantic each ear between 1996 and 1999. The hunt for these eals remains the largest of a marine mammal opulation anywhere in the world.

In response to this situation, the anti-hunt lobby has redirected its protest to the killing of ragged-jackets and beaters, either with clubs or with rifles. It also highlights the cruelty of large numbers of older animals shot and wounded and left to die a lingering death at sea.

There is also scientific concern about whether the hunt in recent years has achieved the Canadian government's objective of maintaining the size of the Harp seal population. There is evidence, for example, that the levels of killing between 1996 and 1999 were higher than the population could sustain. The government's own most recent estimates, following a 1999 aerial survey of pup production, indicate that the population leveled out in the mid-1990s and is now declining slightly, a result of the increased hunts in recent years.

While Canada's landed catch in 2000 fell to 91,602 – apparently as a result of reduced subsidies and a lack of new markets for most seal products – Greenland's unregulated hunt continued unabated. It now lands some 100,000 Harp seals each year. Plus, for every seal landed, another is estimated to be killed but not landed, meaning that some 200,000 Harp seals are now being killed in the Greenland summer hunt alone.

On the other side of the current debate are those who would like to see even more seals killed. They argue that the northwest Atlantic Harp seal is an abundant species; the Canadian government currently puts its population at just over 5 million. With stocks of many fish (particularly Atlantic cod) dwindling in the northwest Atlantic, largely as a result of overfishing, there have in recent years also been renewed calls for culling Harp seals, on the grounds that the seals are impeding the recovery of fish stocks and that fewer seals would mean more fish for commercial fishers. Opponents of this view contend that Harp seals rarely eat Atlantic cod, and so have little impact on their recovery. Moreover, and perhaps most importantly, the complexity of feeding relationships in the ocean is such that a Harp seal cull might actually be counterproductive, reducing rather than increasing the availability of Atlantic cod and other commercially important species.

As we enter the new millennium, yet another expert panel has been assembled by the Canadian government to make recommendations about future hunts or culls of the Northwest Atlantic harp seal population. The controversy, it seems, is set to continue. DML

COUNTING THE COSTS OF MOTHERHOOD

Maternal investment strategies among pinnipeds

LIKE ALL OTHER PREDATORS, THE THREE FAMILIES of pinnipeds – the true seals (Phocidae), eared seals (Otariidae), and walruses (Odobenidae) – must convert the prey they eat into offspring if they are to ensure their own genetic survival. But pinniped females have the low reproductive rates characteristic of long-lived mammals, usually giving birth to only a single young annually. In addition, they must leave the water to give birth, often in places far from their feeding grounds. This separation of reproduction and feeding is the single most important factor determining the varying reproductive patterns of the different species, which do not always follow neat phylogenetic divisions; for example, a few true seals, Harbor seals among them, have foraging patterns during the reproductive period that are closer to those of some eared seals than to their fellow phocids. Whatever strategy they pursue, though, female pinnipeds must allocate expenditure for breeding from resources acquired during a feeding schedule that varies from season to season, and this involves difficult trade-offs.

Mothers invest heavily in their pups, both in the form of the milk they feed them and of the energy they expend on defending them against other adults and external predators. Almost all pinnipeds breed in a highly synchronized annual cycle and, aided by delayed implantation, whelping occurs at the same time each year. Until recently, it was thought that the true seals supported their lactation solely from stored fat, foregoing foraging until the pup was weaned, whereas eared seals foraged as they fed their pups: a distinction summed up in the catchy adage that true seals were "capital breeders" and eared seals "income breeders." The difference could be convincingly attributed to body size, the larger true seals having a greater capacity to store fat. Sadly, this convenient distinction has recently been tarnished by the discovery that several true seals also forage during lactation, among them both large- and medium-sized ice-breeders (Weddell and some Harp seals) as well as the small, land-breeding Harbor seal.

Even so, size remains a critical factor in the maternal investment equation. Female true seals do indeed tend to be larger than eared seals (the respective median weights are 220kg/485lb as against 40–50kg/c.100lb), and for the most part they do employ a fasting strategy and usually

Above *A Cape fur seal mother reclines luxuriously* *ı her back while feeding a good-sized pup. Typically* *r the eared seals, lactation in this species lasts 12* *onths or more; a few pups are suckled for a sec-* *ıd, or even a third year.*

Left *Round as a barrage balloon, this Harp seal* *other exemplifies the true seal strategy of building* *ɔ fat reserves in advance of a birth. She will nurse* *er pup for as little as 10–12 days, during which time* *s weight will more than double, while she goes* *ithout food. After the young seal is weaned,* *other and pup will have little further contact.*

bstain from feeding during the brief but energy-tensive lactation period, which lasts as little as ɔur days in Hooded seals. Pups grow quickly (a ooded seal pup will double its birth weight in ıose four days!) and are weaned abruptly. To chieve these remarkable growth rates, milk fat ɔntent can be as high as 60 percent.

In contrast, female otariids – the eared seals – st for only 5–12 days from stored reserves. After ıat, they return to sea repeatedly to forage etween suckling bouts in the course of an extended lactation period of anything from 4 months (in the Galápagos sea lion) to 24 (in Galápagos and South American fur seals). The pups of eared seals generally grow more slowly than their true seal counterparts, and are less chubby at weaning, when they weigh only 30 percent of the mother's mass, in comparison to the 40 percent characteristic of true seals. Walrus maternal care is less well known, but pups always accompany their mothers during feeding trips and are typically suckled in the water.

Within most pinniped species there is a huge range in the size of adult females – the largest are often twice the mass of the smallest. The relationship between a mother's expenditure (the obvious and dramatic use of body reserves to raise a pup) and her investment (changes in her fitness resulting from such expenditure) is complicated by this variation in the mothers' size, since a hefty mother has to invest relatively less of her bulk in raising a baby of a given size than a diminutive mother does. This intra-species variation also tends to confuse comparisons between species in terms of relating body size to reproductive behavior.

One way of measuring the cost of reproduction, however, is in terms of the penalty the mother pays in reduced fecundity and survival. For example, Galápagos fur seals still accompanied by pups from previous seasons have a greatly reduced chance of bearing a pup in the following one (45 percent as compared to 90 percent). In Antarctic fur seals, previous pregnancy is associated with lowered survival rates as well as reduced fecundity – a female that has never given birth has three times the chance of reaching age 15 than has a female giving birth each year, and 48 percent of females do not give birth in the year following a pregnancy. Among the true seals, the birth rate of Weddell seals that produced pups in one year falls by 5 percent the following year.

These costs can differ with circumstances, as illustrated by two populations of Northern elephant seals in California. At the crowded site of Año Nuevo, females that first gave birth aged 3 and 4 had approximately half the weaning success of companions that postponed breeding until they were older. In a low-density colony on the Farallon Islands, however, no evidence of reduced survival in younger breeders was found; in fact in these surroundings females that bred early and survived had greater lifetime reproductive success than later breeders. Even so, females were less likely to pup in the year following a first birth, and the longer they postponed their first pregnancy, the greater their fertility was in subsequent years.

Evidence that the costs of reproduction are affected by the mother's size (as distinct from her age) comes from Gray seals. On North Rona off the Scottish coast, female Grays tend to skip breeding in years following higher than average expenditures. In any one year, mothers have similar rates of pupping success, irrespective of their body size. Over two-year periods, however, those that expended more than the average amount of bodyweight (estimated at 40 percent of postpartum mass) in the first year had reduced pupping success (an 0.85 rate as against the average of 0.93) in the second year. Smaller-sized females tended to have smaller expenditures (36 percent of postpartum mass), and most of them (90 percent) returned to pup the following year.

Indeed, these two types of female are thought by some researchers to indicate the existence of two different strategies within the one species: at one extreme are the females that "live fast and die young," breeding early and often, while at the other are females that start to reproduce late and breed less regularly, but live longer. It remains to be seen which strategy has the higher pay-off in terms of surviving offspring. PP

THE RAREST SEALS

Monk seal conservation

MONK SEALS ARE THE ONLY PINNIPEDS THAT live in warm, subtropical seas. Extremely vulnerable to human interference, one of the three species – the Caribbean monk seal – is probably already extinct. Reports of seals in the Caribbean can usually be traced to escaped California sea lions; however, some sightings do appear to be true seals, although these could be vagrant Hooded or even Southern elephant seals. The Mediterranean monk seal has been reduced to less than 500 animals, divided into many small groups distributed over a wide area. There are still about 1,300 Hawaiian monk seals divided among several breeding colonies but the population is slowly declining.

Aristotle's description of the Mediterranean monk seal is the first historical record of a pinniped, and sightings of this species may have given rise to the ancient myth of the sirens, temptresses whose enchanting song lured passing ships to their destruction on the rocks. By contrast, in the real world, it is people who are increasingly responsible for the untimely deaths of seals. The Mediterranean Sea and its beaches are used intensively for recreation, fishing, and waste disposal. Seals are deliberately killed by fishermen or die accidentally after being caught in fishing gear. Mediterranean monk seals, which are highly sensitive to such intrusion, no longer use open beaches but haul out and give birth mainly in caves, which can collapse. This occurrence is not as unlikely as it might seem – since the population around the Cabo Blanco peninsula in the Western Sahara was discovered in 1923, the colony has occupied at least eight caves, five of which have collapsed. A mass mortality in this region in 1997 killed two-thirds of the population. The exact cause is unknown, but a viral epidemic or poisoning by toxic algae are the most likely candidates. Apart from those at Cabo Blanco, the largest remaining colonies of Mediterranean monk seals around the Aegean Sea. One major problem hampering the conservation of Mediterranean monk seals has been the difficulty of coordinating the management plans of the ten or more countries where colonies are found.

Hawaiian monk seals breed on eight of the nine Northwest Hawaiian Islands, a low-lying, fragmentary chain of coral atolls and rock islets that extends over 1,600km (1,000mi) northwest from the main Hawaiian Islands (where a small population also exists). Sealing expeditions in the 19th century reduced the number of seals, while expeditions for guano, feathers, and whales further disturbed their habitat. Later, during World War II, the establishment of manned US Naval and Coast Guard stations led to major declines in monk seal populations at Midway Island, French Frigate Shoals, and Kure Atoll, as humans and their dogs began to encroach on the beaches. Female seals, in particular, were frightened off and began to give birth on exposed offshore sandbars, where many of the pups died.

Unlike the Mediterranean monk seal, the fate of the Hawaiian monk seal resides in the hands of a single country, the United States, which in 1976 added it to its list of endangered species. The National Marine Fisheries Service (NMFS), runs a substantial research and monitoring program on the species. Over time, the defense installations at monk seal breeding sites have fallen into disuse. All major sites except Kure, which is a Hawaii State Seabird Sanctuary, are now part of the US National Wildlife Refuge system. Moreover, most sites are protected from casual visitors, though properly supervised ecotourists are now allowed to visit Midway Island to observe monk seals and breeding sea birds.

Populations are currently increasing at Pearl and Hermes Reef, Kure, and Midway, and remain stable at Laysan and Lisianski Islands. The largest population, at French Frigate Shoals, increased dramatically after the Coast Guard station there was closed in 1979, but then declined again by about 55 percent from 1989 to 1997, due to yet another problem that besets monk seals – insufficient food. Many pups are starving, while others weigh less than is optimal when they are weaned making their chances of subsequent survival very poor. Only 8 to 25 percent of weaned pups survived to age 2 during the years from 1989 to 1997, compared to at least 80 percent between 1984 and 1987. The same problems exist to a

Below *Like all monk seals, Hawaiian monk seals are particularly sensitive to disturbance. The human population increases that have taken place elsewhere may have caused them to be confined to the relatively unpopulated Leeward Islands.*

sser extent at Laysan (juvenile survival 30 to 70 ercent in the period 1989–97). Some of the ecline in prey availability may be due to a fall in ıe general productivity of the local waters, which ›nds to vary according to an approximately 10- ear cycle of changing oceanographic conditions. s of 2000, productivity seemed to be improving t lower trophic levels of the oceanic ecosystem round the Hawaiian Islands, which could have ositive consequences for the seals.

There are a number of other problems affecting lawaiian monk seal colonies. These include con- ict with local lobster fishermen over common shing grounds (lobster are an important part of ıe monk seals' diet), predation from Galápagos nd tiger sharks (especially at French Frigate hoals), and the hazard presented by the growing olume of marine debris. In 1999 alone, 25 monk eals were found entangled in underwater nets nd fishing gear, and many more probably died nnoticed. Several local clean-up efforts are under- vay to combat this menace, but the problem is international, since debris can drift for thousands of miles before snagging on Hawaiian reefs.

To make matters worse for an already beleaguered group, monk seal populations are given to "mobbing," an aggressive mating behavior pattern in which groups of adult males chase, attack, and sometimes even kill adult females or immature seals of both sexes. Mobbing can occur when chance events cause the death of many females, with the result that males predominate and come into violent conflict over the remaining females. In 1994, NMFS biologists decided to remove 22 culprit males from Laysan, where the problem was especially acute, and release them around the main Hawaiian Islands. This radical strategy appeared to pay off, and has now been adopted more widely. Yet despite such efforts, as well as other initiatives on shark control and fishery protection, the long fight to save the monk seals from extinction is far from won. KR

Left *Owing to their geographical location, Mediterranean monks seals have been in documented contact with humankind longer than any other pinniped. Although they are now protected throughout most of their range, many are still killed annually as a result of conflict with fishermen.*

Below *Hawaiian monk seals tend to remain inactive during the daytime and feeding is carried out at night; this lifestyle is possibly an attempt to avoid overheating in the temperate waters around Hawaii, for the Hawaiian monk seals have the same blubber content as their polar counterparts.*

STORY

Life and Death on the Beaches

1 A bull elephant seal basking on a beach at the Año Nuevo State Reserve, California, displays its huge proboscis. This trunklike appendage, which gives the seals their name, takes eight years to become fully developed. During the breeding season it is inflated and used as the principal organ of sexual display.

2 A bull Northern elephant seal (Mirounga angustirostris) *forces its attentions on a female nursing two young. From the moment that they arrive on the breeding beaches, females are hounded by would-be mates. Bulls not only pursue receptive females; they also attempt to mate with females not capable of being impregnated, including those that are already pregnant or nursing. Courtship in elephant seals is direct and aggressive. Without preliminaries, a male moves directly to the side of a female, puts a foreflipper over her back, bites her on the neck, pulls her to him, and attempts copulation. If the female protests or attempts to move away, as is usually the case, the male pins her down by slamming the full weight of his head and forequarters on her back one or more times, and bites her more vigorously.*

In these sexual disagreements, the outcome is usually bad for the female. Mounting is harmful for pregnant females and interrupts feeding and causes mother–pup separation in nursing mothers. Moreover, in their eagerness to drive off other bulls, males sometimes crush pups in their path; up to 10 percent of those born in any one season may die in this way.

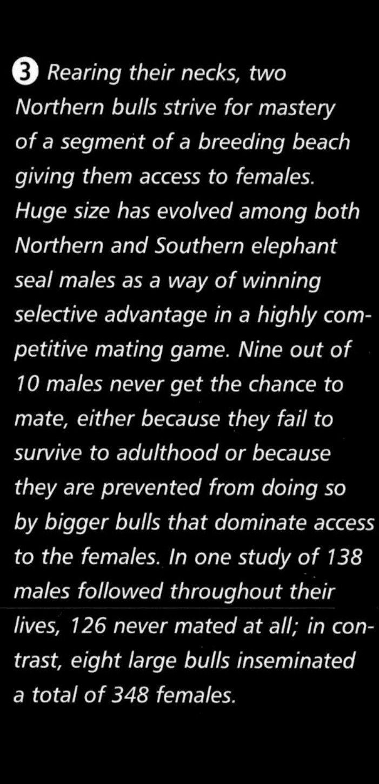

❸ *Rearing their necks, two Northern bulls strive for mastery of a segment of a breeding beach giving them access to females. Huge size has evolved among both Northern and Southern elephant seal males as a way of winning selective advantage in a highly competitive mating game. Nine out of 10 males never get the chance to mate, either because they fail to survive to adulthood or because they are prevented from doing so by bigger bulls that dominate access to the females. In one study of 138 males followed throughout their lives, 126 never mated at all; in contrast, eight large bulls inseminated a total of 348 females.*

❹ *A male Southern elephant seal at Grytviken on South Georgia shows the scars of battle. The head and neck bear the brunt of the attacks, and scar tissue here toughens and thickens the skin. Males of this species weigh up to 3.7 tonnes, dwarfing the females, which are on average only one-third to one-tenth their size.* BJLeB/MW

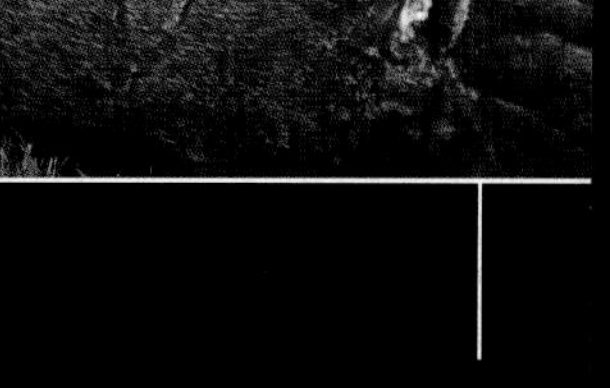

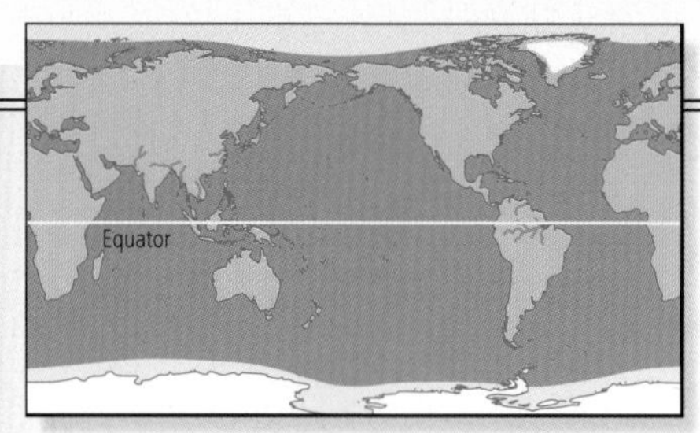

WHALES & DOLPHINS

MAMMALS EVOLVED ON LAND, AND MOST people still generally think of them as terrestrial animals. Yet over two-thirds of the surface of the planet is covered with water, so it is hardly surprising that in the course of evolutionary time some of them should have taken advantage of the expansive aqueous niche. Whales and dolphins have subsequently adapted to the physical properties of water and evolved in ways that would not have been possible had they remained on land; only supported by water's dense buoyancy could the largest whales have reached their present size.

Adaptation to the watery environment has given the whales and dolphins a fishlike appearance, and it was not until 1758 that, thanks to the great Swedish biologist Carolus Linnaeus, they were in fact recognized as mammals. It took even longer to reveal the adaptations that have enabled them to spend their entire lives in oceans and seas; animals that encircle the whole world underwater and feed as deep as 1,500m/4,900ft (even 3,000m for sperm whales) have inevitably been difficult for land-dwelling humans to study. Technology – so much the enemy of whales in other ways – is, however, now helping us to understand their requirements and their complexity.

In fact, whales are clearly distinguished from fish by a number of specifically mammalian features. They are warm-blooded, breathe air through lungs, and give birth to living young that are suckled on milk secreted by the mammary glands of the mother. Unlike most land mammals, they do not have a coat of hair for warmth, as this would impede their progress through the water, reducing the advantage gained by the streamlining of the body. Of the three marine mammal orders – besides Cetacea, to which whales and dolphins belong, these comprise Pinnipedia (seals, sea lions, and walruses) and Sirenia (dugongs and manatees) – it is the whales and dolphins that are most specialized for life in the water; pinnipeds must return to land or ice to breed.

One half of the order Cetacea comprises the generally small dolphins and porpoises. They belong to the suborder Odonteceti, or toothed whales, which in total accounts for 72 of the 85 cetacean species. The toothed whales feed mainly on fish and squid, which they pursue and capture with jaws usually containing a large array of teeth. In contrast, most of the great whales belong to the suborder Mysticeti, the baleen whales; these have a system of horny plates – the baleen – in place of teeth, using them to filter or strain planktonic organisms and larger invertebrates, as well as schools of small fish, from the sea.

Body Shape, Locomotion, and Diving

ANATOMY AND PHYSIOLOGY

The largest animal ever to have lived on this planet is the Blue whale. It reaches a length of 24–27m (80–90ft) and weighs 130–150 metric tons, equivalent to the weight of 33 elephants – the largest terrestrial mammal. Such an enormous body could only be supported in an aquatic medium, for on land it would require limbs so large that mobility would be greatly restricted. Though populations have been severely reduced by overhunting, the Blue whale still survives today.

Despite their size and weight, whales and dolphins are very mobile, their streamlined bodies

Left A mother and calf Bottlenose dolphin diving to feed in the Red Sea. These dolphins are often seen in captivity, as they appear to be an intelligent species that can learn complex tasks quickly. Some people even claim to have seen wild Bottlenoses using their tail to flip fish onto a beach before retrieving them.

ideally adapted to fast movement through water. The head is elongated in comparison to other mammals, and passes imperceptibly into the trunk, with no obvious neck or shoulders. Rorquals, river dolphins, and the white whales have neck vertebrae that are separate, allowing greater flexibility; the remainder of the species have between two and seven fused together.

Whales' skeletons still bear signs that their ancestors were once land mammals with four legs, but hind legs have now entirely disappeared from the outside of the body. In fact, few body parts protrude to create drag. Instead of ears, cetaceans merely have two tiny openings that lead directly to the organs of hearing. The male's penis is completely hidden within muscular folds, and the teats of the female are housed within slits on either side of the genital area. The only protuberances are a pair of horizontal fins or flippers, a boneless tail fluke, and, in many species, an upright but boneless dorsal fin of tissue that is firm and fibrous.

In most of the toothed whales the jaws are extended as a beaklike snout, behind which the

Left The Pygmy killer whale (Feresa attenuata) *is predatory species, about which little is known. They ave occasionally been known to prey upon the offpring of some species of dolphins, and are found rincipally in the coastal waters off Japan, Hawaii, nd South Africa.*

Above *The tail flukes, powered by huge back muscles, are the great whales' sole source of propulsion. The Northern right whale, in common with other species, usually only raises its tails flukes clear of the water before making a deep dive. One speculation is that this species may use its tail as a sail in high winds.*

ORDER: CETACEA
14 families; 40 genera; 85 species

TOOTHED WHALES
Suborder: Odontoceti

RIVER DOLPHINS Families Lipotidae, Iniidae, Platanistidae, and Pontoporiidae p218
5 species in 4 genera in 4 families
Yangtze river dolphin (*Lipotes vexillifer*), **Amazon dolphin** (*Inia geoffrensis*), **Ganges dolphin** (*Platanista gangetica*), **Indus dolphin** (*P. minor*), and **La Plata dolphin** (*Pontoporia blainvillei*).

DOLPHINS Family Delphinidae p220
At least 36 species in 17 genera
Includes **Pantropical spotted dolphin** (*Stenella attenuata*), **Common dolphin** (*Delphinus delphis*), **Killer whale** (*Orcinus orca*), **Pilot whale** (*Globicephala melas*), **Risso's dolphin** (*Grampus griseus*), **White-beaked dolphin** (*Lagenorhynchus albirostris*).

PORPOISES Family Phocoenidae p236
6 species in 3 genera
Includes **Harbor porpoise** (*Phocoena phocoena*) and **Finless porpoise** (*Neophocaena phocaenoides*)

BELUGA AND NARWHAL Family Monodontidae p240
2 species in 2 genera: **beluga** (*Delphinapterus leucas*) and **narwhal** (*Monodon monoceros*).

SPERM WHALE Family Physeteridae p244
1 species *Physeter catodon (macrocephalus)*

PYGMY SPERM WHALES Family Kogiidae p244
2 species in 1 genus: **Pygmy sperm whale** (*Kogia breviceps*) and **Dwarf sperm whale** (*K. simus*).

BEAKED WHALES Family Ziphiidae p250
At least 20 species in 6 genera
Includes **Northern bottlenose whale** (*Hyperoodon ampulatus*), **Blainville's beaked whale** (*Mesoplodon densirostris*), and **Cuvier's beaked whale** (*Ziphius cavirostris*).

BALEEN WHALES
Suborder: Mysticeti

GRAY WHALE Family Eschrichtiidae p256
1 species *Eschrichtius robustus*

RORQUALS Family Balaenopteridae p262
8 species in 2 genera
Includes **Blue whale** (*Balaenoptera musculus*), **Fin whale** (*Balaenoptera physalus*), **Minke whale** (*Balaenoptera acutorostrata*), **Humpback whale** (*Megaptera novaeangliae*).

RIGHT WHALES Family Balaenidae p270
3 species in 3 genera
Includes **Bowhead whale** (*Balaena mysticetus*) and **Northern right whale** (*Balaena glacialis*).

PYGMY RIGHT WHALE Family Neobalaenidae p270
1 species *Caperea marginata*

Note As with several other mammalian orders, the taxonomy of the cetaceans is in considerable flux. For example, the classification established in 1993 accepted 11 genera, 41 families, and 78 species.

forehead rises in a rounded curve or "melon." Unlike the baleen whales (or any other mammal) they possess a single nostril; the two nasal passages, which are separate at the base of the skull, join close below the surface to form a single opening – the blowhole. In extreme cases, one passage is devoted to sound production, leaving the other as the sole breathing tube. The blowhole typically takes the form of a crescent-shaped slit protected by a fatty and fibrous pad or plug. Efficient adaptation means that the slit is closed by water pressure, but can be opened by muscular action when the whale surfaces to breathe. The skull bones of the nasal region are usually asymmetrical in their size, shape, and position, although porpoises and the La Plata dolphin have symmetrical skulls.

The baleen or whalebone whales differ from toothed whales in a number of ways. They are generally much larger, and the baleen apparatus takes the place of teeth in the mouth. The baleen grows as a series of horny plates from the sides of the upper jaw, occupying the position of the upper teeth in other animals. Baleen whales feed by straining large quantities of water containing plankton and larger organisms through these plates. The paired nostrils remain separate, so that the blowhole is a double hole forming two parallel slits that are close together when shut. Other features specific to baleen whales are single-headed ribs and a breast bone (sternum) composed of a single bone articulating with the first pair of ribs only. All baleen whales have symmetrical skulls, though the size and shape varies between species.

Like all mammals, cetaceans are warm-blooded, using part of the energy available to them to maintain a stable body-core temperature, in their case of 36–37°C (97–99°F). The sea is a relatively cool environment, with temperatures usually below 25°C (77°F), and cetaceans lack fur coats, instead insulating the vital organs that lie just inside their skin with blubber – a layer of fat up to 50cm (20in) thick in the Bowhead whale. Larger species have a distinct advantage over smaller ones, having less surface area over which to lose heat in relation to their body mass. This may explain why the smaller dolphins do not occur a[t] very high latitudes. The liver is also an importan[t] fat store, and in some species there are significan[t] quantities of fat (up to half the body's total) in th[e] form of oil laid down in the skeletal bones.

Whales and dolphins rely upon controlled bloodflow from the body core to the skin and appendages for heat regulation. Unlike land mam[-]mals, whose insulation typically overlies vascula[r] circulation to the skin, the insulating blanket of cetacean blubber is penetrated by massive, contorted spirals of blood vessels called *retia mirabili[a]* (mainly arteries, but also including some thin-walled veins) that usually form blocks of tissue o[n] the inner dorsal wall of the thoracic cavity and o[n] the extremities or periphery of the body. These vessels function as countercurrent heat exchangers, maintaining a heat differential between oppo[-]sitely-directed flows and so increasing the amou[nt] of heat transferred. Such heat exchangers in the flippers, flukes, and fins serve to conserve body

DOLPHIN AND WHALE BODY PLANS

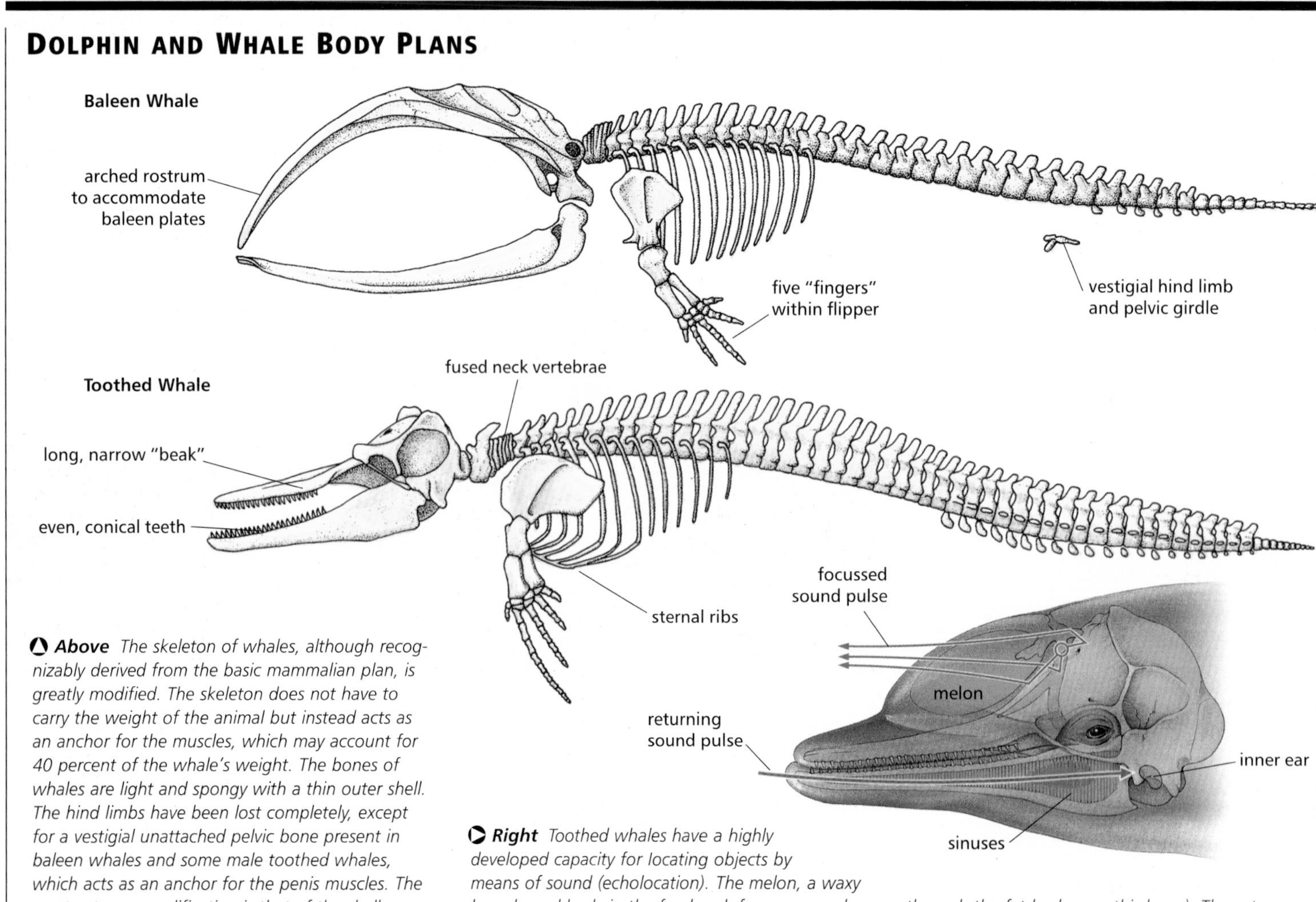

Above The skeleton of whales, although recognizably derived from the basic mammalian plan, is greatly modified. The skeleton does not have to carry the weight of the animal but instead acts as an anchor for the muscles, which may account for 40 percent of the whale's weight. The bones of whales are light and spongy with a thin outer shell. The hind limbs have been lost completely, except for a vestigial unattached pelvic bone present in baleen whales and some male toothed whales, which acts as an anchor for the penis muscles. The most extreme modification is that of the skull, which is greatly extended in both baleen and toothed whales. The loss of teeth in baleen whales and associated changes have produced a skull with a grotesque form, unlike that of any other animal.

Right Toothed whales have a highly developed capacity for locating objects by means of sound (echolocation). The melon, a waxy lens-shaped body in the forehead, focuses sounds produced in the nasal passages (more focussing is achieved by reflections off the skull and air sacs). Returning sound waves are channeled to the inner ear through oil-filled sinuses in the lower jaw (and through the fat body over this bone). The extreme sensitivity of this system is assisted by the isolation of the inner ear from the skull by means of a bubbly foam. Sound is thus very precisely channeled without the interference of extraneous resonances.

Γhe baleen whales have not yet been shown to : echolocation as toothed whales and dolphins and may instead rely on sight to locate the arms of plankton on which they feed. However, y can emit very loud pulses of sound at low fre- encies (from 15Hz to 30kHz) and for short rations that, as revealed by SOSUS (SOund rveillance System) arrays deployed by the US vy in deep waters, are audible over tens to hun- :ds of kilometers across the deep-ocean chan- ls. The low frequencies of Blue and Fin whale nes have very long wavelengths, from 50m at Hz to100m at 15Hz. A typical Blue whale call ts for 20 seconds and in water is approximately km in length. If these sounds were used for nolocation, they would not be able to discrimi- te targets finer than those wavelengths. They ay therefore be used primarily to detect gescale oceanic features like continental elves, or possibly the sharp differences in water nsity associated with cold-water upwellings.

Baleen whale sounds may also serve a commu- :atory function, particularly if individual calls ry and are recognizable. In toothed whales and lphins, which are often social, signature whis- s provide opportunities for individual recogni- n, and these might lead to the evolution of ographic or pod-specific dialects.

Evolution

CETACEAN ANCESTRY

ie origins of present-day cetaceans are poorly iderstood, although the introduction of new olecular techniques in the 1990s has led to an :tensive re-evaluation of relationships on the sis of a combination of fresh molecular and orphological information. However, whereas orphology indicates that cetaceans are most osely related either to Artiodactyla (even-toed ngulates) or to Perissodactyla (odd-toed ungu- tes), molecular data suggest cetaceans and hip- opotamid artiodactyls may form extant sister xa, with Ruminantia (antelope, deer, and kin) eing the next most closely related.

Mammals recognizable as cetaceans first appear ; fossils in rock strata from the early Middle ocene. The first was *Pakicetus*, of the Protoceri- ae, an elongated aquatic animal with reduced nd limbs and a long snout. Classified within a parate suborder called Archaeoceti, such mam- ials flourished during the Eocene epoch, but ost were extinct before the end of the Oligo- ene, and none survived beyond the Miocene. lowever, even by the late Middle and early Upper ocene, they had become so specialized that they ould not have been the ancestors of modern etaceans.

Looking further back in time, members of the errestrial suborder Mesonychia may have given se to the archaeocetes (and thence to all other etaceans) at the end of the Cretaceous and then olonized the sea during the Paleocene. These are

EVOLUTION OF WHALES

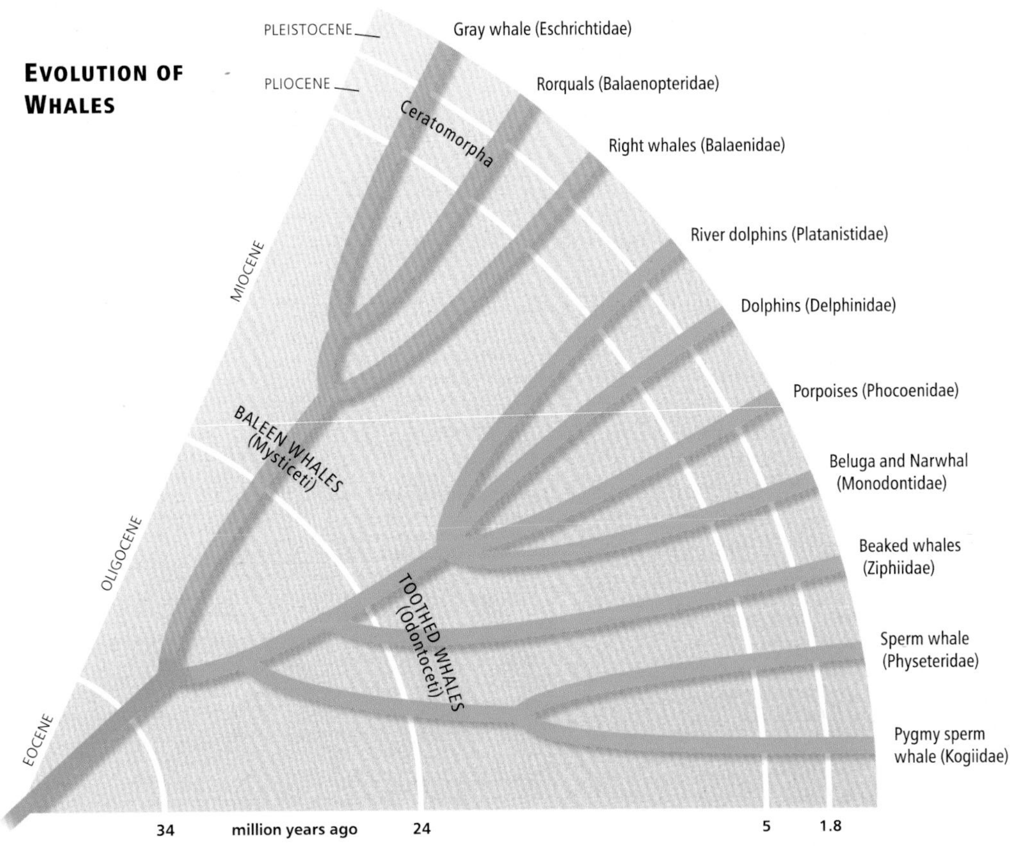

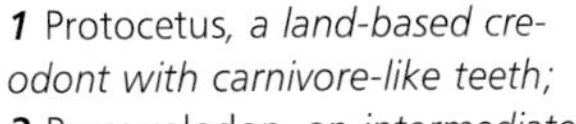

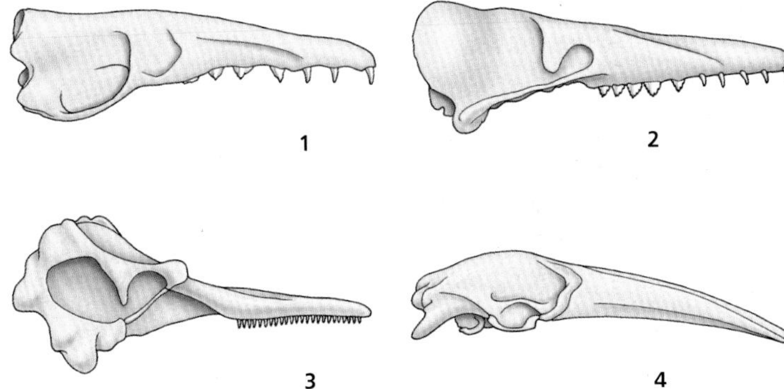

Right *The modifications to bone structure and teeth that have been involved in the evolution of the whale skull.* **1** Protocetus, *a land-based creodont with carnivore-like teeth;* **2** Prosqualodon, *an intermediate form;* **3** *The Bottlenose dolphin, a modern toothed whale, showing the uniform teeth;* **4** *a rorqual, showing the most extreme modification of the bone structure, loss of all the teeth, and their replacement by baleen (not shown).*

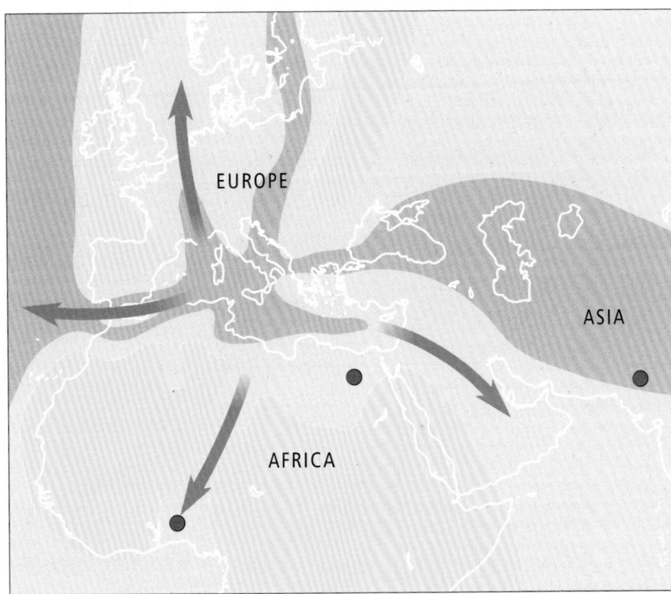

Left *The Tethys Sea was an enormous trough in the earth's crust. It formed a seaway that separated Europe and Africa, continued across the Middle East, and then southeast to Myanmar (Burma). It is the probable center of whale evolution: dispersal routes show how modern distribution came about.*

original extent of the Tethys Sea
expansion of the Tethys Sea during the Eocene
source of the remains of the oldest known whales
possible dispersal route

thought to share with the suborder Arctocyonia (the likely ancestors of present-day ungulates and their relatives) ancestry from the Condylarthra (otherwise known as creodonts). The Mesonychidae resemble the archaeocetes in a number of skull and dental characteristics, and although the similarities are not all clear-cut, they appear at present to be the most likely ancestors of the Cetacea.

Most of the early archaeocete remains come from the Mediterranean–Arabian Gulf region, which during the Paleocene formed a narrow arm of the western part of the ancient Tethys Sea. It was probably here that populations of terrestrial creodonts started to colonize marshes and shallow coastal fringes during the late Paleocene, exploiting niches vacated at the end of the Cretaceous by the vanishing plesiosaurs, ichthyosaurs, and other reptiles. As population pressure on resources intensified during the Eocene, we can speculate that selection would have favored adaptations for the capture of fast-moving fish rather than the sluggish fish and freshwater and estuarine mollusks that previously formed their main diet.

This was the age of mammals, with massive adaptive radiation into many species; such rapid evolution may help to explain the sparse fossil record for this period, with forms developing relatively specialized cetacean characters quite quickly. During the Eocene, the warm waters of the western Tethys Sea were dominated by the archaeocetes, but as the climate started to deteriorate and the Tethys Sea enlarged during the Oligocene, they probably declined in density and abundance, and by the Miocene were being entirely replaced by odontocetes and mysticetes. The progressively more aquatic mode of life resulted in a backward and upward shift of the external nostrils, and the development of ways to seal them against water. The long, mobile neck, functional hind limbs, and, eventually, most of the pelvic girdle were all lost, together with any remaining coat, and horizontal tail flukes developed for propulsion. The body became more torpedo-shaped to provide greater streamlining, and a dorsal fin developed for hydrodynamic control and temperature regulation.

Most modifications of the archaeocete skull towards an odontocete (toothed whale) form involved the telescoping of the front of the skull, which probably paralleled the development of acoustic scanning as a means of location, thereb aiding the capture of agile prey. At the same time various specialized organs developed – notably the melon, spermaceti organ, and nasal diverticula. Their functions are thought to relate to the pr cessing of sounds and, in the case of species like the Sperm whale, possibly also to diving. Later, sexual selection may have favored greater develo ment of some of these in males of species in which competition for mates is important.

The teeth of archaeocetes were differentiated into incisors, canines, and grinding teeth. These teeth either became modified during the Eocene, forming the long rows of sharp, uniform teeth ty ical of present-day toothed whales, or later, main in the Oligocene, gave rise to the baleen plates: the remarkable feeding structures that evolved from the curved transverse ridges of the palate. Most present-day baleen whales still have the

Above Small prey for large mammals; individual Antarctic krill grow to a maximum size of just 7.5cm (3in); however, they exist in staggering numbers, totaling some 500–750 million tonnes. Krill is the principal food of the large baleen whales, particularly the Blue whale, but also the Fin, Sei, and Minke whales.

Left Humpback whales "bubble net" feeding. One or two Humpbacks work together, swimming up beneath a shoal, all the while spiraling and blowing bubbles. The fish will swim in through the barrier of bubbles to join their companions but will not swim out, so the "net" effectively concentrates the shoal.

Below Feeding strategies vary from species to species of dolphin. Some herd shoals in an apparently cooperative manner, while others feed independently. Here, a Bottlenose dolphin is seen chasing a shoal of baitfish. Bottlenose dolphins have also been seen to temporarily disable fish by knocking them out of the water with their powerful tail.

developmental precursors of teeth (buds) in the early stages of fetal development, a further indication of their common ancestry with the toothed whales, which is also supported by anatomical and chromosomal evidence.

The earliest true toothed whales were the squalodonts, a group of short-beaked whales with triangular, sharklike teeth. These were possibly most abundant in the late Oligocene to early Miocene, dispersed throughout the Southern Hemisphere, but by the middle Miocene they were being superseded by representatives of families with living relatives. In particular, the *Ziphiidae* (beaked whales) can be traced back to a squalodont ancestor, and so can other groups including the *Physeteridae* (Sperm whales) and, in fresh and brackish waters, the *Platanistidae* (river dolphins), *Delphinidae* (true dolphins), and *Phocoenidae* (porpoises). Sperm whales, which have a much more marked asymmetry of the skull than any other odontocetes and a quite distinct chromosomal structure, almost certainly diverged early in the history of the line, around the mid-Oligocene. However, recent molecular evidence refutes the idea that they may be more closely related to baleen whales than to other toothed whales.

While the toothed whale skull was becoming modified to contain acoustic apparatus, the baleen whale skull became adapted to a different lifestyle. The upper margin of the "forehead" of the skull underwent considerable forward extension, probably mainly to combat the stresses on the skull and jaws imposed by the wide opening and closing of the mouth at irregular intervals as the animals feed.

A Catholic Diet

FOOD AND FEEDING

From the predatory Killer whales, which feed on other cetaceans, Sea otters, and pinnipeds, through fish- and cephalopod-pursuing dolphins to Humpback and Blue whales sieving krill, cetacean diets encompass much of what is on offer from the sea. Many small toothed whales appear to have generalist diets, opportunistically taking in a range of shoaling open-sea fishes, but the extent to which diets overlap between species within a region remains poorly known. Among the baleen whales, the thickness and number of baleen plates is related to the size and species of prey taken. Thus the Gray whale, a highly selective seabottom feeder, has a shorter, stiffer baleen and fewer throat grooves (usually 2–3) than the rorquals (with 14–100), and is thereby adapted for "scouring" the sea bottom. In the rorquals, the baleen is longer and wider. In the Blue whale, the plates may reach a width of nearly 0.75m (2.5ft); in the other rorquals they are correspondingly narrower, and this dictates the diet of each. In the Right and Bowhead whales the baleen is extremely long and fine and these whales feed on the smallest planktonic invertebrates of any of the baleen whales.

Whereas baleen whales and some toothed whales, such as the Sperm whale, Northern bottlenose whale and Harbor porpoise, may tend to feed independently, a number of small toothed whales appear to herd shoals cooperatively by a combination of breaching and fast surface-rushing in groups. Communication between individuals presumably is carried out by vocalization (high-pitched squeaks, squeals, or grunts) and perhaps also by particular types of breaching. These latter activities often seem to be quite complex, but until we can routinely follow marked individuals (preferably below water), we cannot be sure of the extent of cooperation between individuals.

Ecology and Behavior

LIFESTYLE AND BREEDING

The different evolutionary courses that the baleen and toothed whales have taken have strongly influenced their respective ecologies. Generally speaking, the ocean areas with the highest primary productivity (quantities of phytoplankton), and hence the most fish and squid dependent upon this, are close to the Poles (where 150–250mg of carbon are fixed from light energy per square meter daily in a season), whereas at tropical latitudes productivity is relatively low (though rich upwellings of nutrients do occur patchily – high in coastal waters but at rates of 50–100mg of carbon per square meter per day in open waters where there is no upwelling). Polar regions show great seasonal variations, and during the summer the rapid increase in temperature, sunlight, and daylength, and the relatively stable climatic conditions, allow phytoplankton – and hence zooplankton and higher organisms such as fish and squid – to build up to very high densities.

During the 120-day period of summer feeding, the great baleen whales probably eat about 3–4 percent of their body weight daily. Present-day whale populations in the Southern Ocean (including all species of baleen whales) consume about 40 million metric tons of krill each year. Before these whale populations were exploited by man, the figure may have been as high as 200 million tons. Thus, during part of the year (about four months in the Southern Ocean, but often more than six months in the North Pacific and North Atlantic, where productivity is lower), the great whales migrate to high latitudes to feed, and here they may put on as much as 40 percent of their body weight as blubber. During the rest of the year, feeding rates may be reduced to about a tenth of the summer value, which results in much of the blubber store being used by the time the whales return to the feeding grounds.

Why should the great baleen whales use up the food they have stored in their blubber to migrate to regions nearer the Equator where there is little food? There are several speculative answers, yet no definitive one. Many smaller cetacean species spend all the year at high latitudes and appear to be perfectly capable of rearing their young in this relatively cool environment. Breeding at high latitudes in winter would be constrained by lack of food, but in summer primary productivity is very high and water temperatures are also much more favorable.

One explanation why the larger whales travel to warm waters to breed is that the growth rates required for the young to attain anything like the large size of their parents, together with the energy intake required by mothers to sustain both themselves and their calves, would require a

Above *A Killer whale beaches on the coast of Patagonia, Argentina, and snatches a South America[n] sea lion. This species has fewer teeth than most toothed whales, but they are large and strong for seizing large fish, squid, and other marine vertebrate[s]. The Killer whale does not chew its food but swallows it whole or in large chunks, which it tears off.*

Below *One of the functions of breaching is to communicate with other whales, but it may also serv[e] to panic or stun shoals of fish, making it easier to fee[d] on them. Humpback whales are very adept at breaching and have been observed to leap clear of the wate[r] belly-up, reenter headfirst, and circle underwater to return to the starting position.*

ger period of high productivity than is available polar summer. This proposition is supported the observation that plankton has a short sea- of abundance, whereas fish and squid are ilable year-round, and the great whales that lertake extensive migrations are mostly plank- -feeders (although some migratory Humpback ales depend more on fish). Secondly, the wer may be historical. The earliest fossil ains of baleen whales, from about 30 million rs ago, occur in low latitudes of the North antic. With the juxtaposition of continental d masses by tectonic plate movement and inges in sea temperature during the Cenozoic they radiated and dispersed toward the poles. with some long-distance migrant bird species, present-day movements of the great whales y be partly a vestige of earlier times, in their e of an epoch when high productivity was to be nd in more equatorial regions. For this to ly, however, the energetic costs of latitudinal g-distance travel would need to be small, for ection pressures would otherwise have elimi- ed this behavioral trait. A third explanation is t whales avoid breeding in polar and subpolar s because they experience high predation pres- es there, due to the higher density of Killer ales. A full understanding of whale migration l eludes science. However, it is worth noting t some large baleen whales, such as the Fin ale, do not necessarily show strong latitudinal grations.

Whereas baleen whales feed chiefly on zoo- nkton, the toothed whales feed largely on her fish or squid. All three prey groups have mparable energy values, weight for weight, and hough this takes no account of differences in otein, fat, or carbohydrate contents, daily feed- rates are comparable across groups. Body size pears to be the factor that determines whether not toothed whales move out of high latitudes winter; smaller species have relatively higher ding rates, irrespective of diet (the smallest ecies put on 8–10 percent of body weight daily, mpared with 3–5 percent for the largest), but al daily intake for a smaller individual is obvi- sly proportionately lower.

The smaller cetaceans may be found at most itudes. Even though they range over large areas for example, the home range of the Pantropical otted dolphin in the North Pacific appears to be 0–480km (200–300mi) in diameter – they do t tend to make strong north–south migrations.

Although the differences in migratory habits nnot be entirely attributed to diet, other tacean features do appear to be related to what e animals eat. Plankton- and fish-feeding species have gestation periods of between 10–13 onths whatever their size, whereas squid-feeders ve periods of the order of 12–16 months. The nger gestation may reflect the relative food val- s (protein, fat, and carbohydrate amounts, rather than simply energy values) of the different prey, or it may relate to the relative seasonal availabilities of those prey. Amongst large whales, lactation periods are relatively longer in the squid-feeding Sperm whale than in plankton-feeding baleen whales. Among smaller cetaceans, the pattern is similar, though less clear-cut: squid-feeding species have lactation periods varying from 12–24 months, whereas in fish-feeding species they are generally around 10–12 months.

The social systems of whales may also be influenced by diet. In the plankton-feeders such as the Right and Bowhead whales, males may compete with one another to mate with a single female, but spend most of their time either singly or in pairs; small groups of usually less than 10 individuals occur primarily at feeding concentrations or during long-distance movements. We have little information on the social systems of rorquals, but male Humpback whales seasonally advertise themselves to females, singing long songs and breaching repeatedly; they form "floating leks," in which males consort temporarily with females for mating purposes, rebuffing or displacing other males. On the other hand, at least some of the squid-feeding whale species form stable groups, often matriarchal with young of both sexes, to which adult males attach themselves for short periods for mating purposes; other groups can comprise bachelor males, which may also travel alone. This system is exemplified by the Sperm whale. The pilot whale may be more akin to the Killer whale, with males remaining in their natal group, in contrast to the Sperm whale, in which males disperse. Microsatellite DNA studies of the Long-finned pilot whale indicate that the species travels in matrifocal kin groups, with males mating with two or more females though rarely with members of their own pod. The Killer whale, which has a mixed diet of squid, fish, marine birds, and mammals, also forms very stable, matrilineal social groups, each typically consisting of an older mature female, her male and female offspring, and the offspring of the second generation's mature females. Mature males remain with the pod into which they were born, and movement or exchange of individuals among pods has not been documented. Genetic studies suggest that male Killer whales do not mate with closely-related pod members but rather with members of other pods in ephemeral encounters. Most fish-feeders, however, have a rather fluid social system, with mixed groups or family units (which may simply be mother–calf pairs) that aggregate on the feeding or mating grounds, and also during long-distance movements. Individuals come and go so that the group is not stable, though it may have a constant core. In several species studied, it seems that there is no stable pair bond (long-term associations being mainly between female kin), and both sexes are promiscuous.

Above *A pair of Bottlenose dolphins mating. The mating season varies according to the geographical location of the dolphins and the waters that they inhabit. The gestation period is usually about one year. The newborn may not be fully weaned until they are 18 months old, but will begin taking some solids from about 6 months.*

Although almost certainly rare, hybridization between Fin and Blue whales does occur, despite an evolutionary separation of 5 million years or more. Hybridization between more distantly related mammalian species may not be excluded, but it is probable that the Blue and Fin whales are nearly as different in their mitochondrial DNA sequences as hybridizing mammal species can be – as different, in fact, as chimps and humans.

Location and Life History

DISTRIBUTION

Cetaceans are not randomly distributed over any region but instead seem to favor certain oceanographic features such as upwellings (where food concentrations tend to occur), or undersea topographic features such as continental shelf slopes (which may serve as cues for navigation). Breeding areas for most cetacean species (particularly

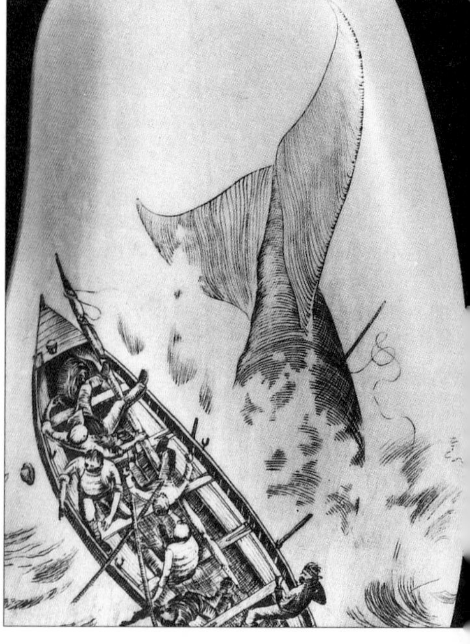

Above *Whaling was a far more dangerous activi before the advent of explosive harpoons and factory ships. This scene of early whalers being tossed abou in their small boat by a harpooned whale is an exam ple of scrimshaw, the art of engraving pictures onto whalebone or whale ivory – in this instance, a Sperm whale's tooth.*

Above left *Dolphins figure widely in classical art and literature – including works by Aristotle, Aesop, and Herodotus – often as rescuers of people in peril at sea. This Greek bowl, made in the 5th century BC, depicts an incident involving Dionysus, the god of wine. Attacked by pirates, the god retaliated by tran forming his assailants into the benign form of dolphi*

Below *Pictorial representations of whales exist as far back as the Stone Age. However, following the introduction of scientific classification by Linnaeus, works such as this 18th century engraving – intende primarily as a typological aid – became widespread.*

small toothed whales) are very poorly known, but are better known for some of the large whales.

Gray and Right whales seem to require shallow coastal bays in warm waters for calving, whereas balaenopterids such as Blue, Fin, and Sei whales possibly breed in deeper waters further offshore. The former group thus has more localized calving areas than the latter.

During the period of mating, some cetacean species congregate in particular areas. These may be the same warm-water areas as those in which calving occurs during the winter months, or they may be on feeding grounds at high latitudes during the summer, as with many small toothed whales. Mating is usually seasonal, but in a number of gregarious dolphin species sexual activity has been observed during most months.

The lengths of the gestation and lactation periods naturally dictate the frequency with which a female may bear young. Cetaceans give birth to single young. In the smaller fish-feeding species (such as porpoises), a female may reproduce every year; among the large plankton-feeding whales, the period is every alternate year (in some species perhaps only every 3 years); in squid-feeding species, females bear young every 3–7 years; while in the Killer whale, with its mixed diet, females reproduce at intervals of 3–8 years. Furthermore, many species do not reach sexual maturity for a number of years (4–10 years in plankton- and fish-feeders, 8–16 in squid feeders and in the Killer whale). It is thus not surprising that most species are long-lived (12–50 years in the smaller species, but 50–100 years in the large baleen whales like the Bowhead and the Sperm, as well as the Killer whale).

Natural mortality rates seem to decrease in different whale species as their size increases, with those for juveniles being somewhat higher than the ones for adults. Current estimates are 9–10 percent per annum for Minke whales; 7.5 percent for Sperm whales; and 4 percent for Fin whales. The long maturation in squid-feeders probably results from their need for a long period to learn efficient capture of the relatively difficult and agile squid prey. There is evidence for differences in annual mortality rates between sexes in some

ecies, for instance Killer whales (3.9 percent in ult males and 0.5–2.1 percent in adult females) d Sperm whales (6–8 percent in males and 5–7 rcent in females).

Whales and Man

AN HISTORICAL OVERVIEW

an has interacted with whales for almost as long we have archaeological evidence of human tivity. Carvings showing whaling activities have en found in Norse settlements from 4,000 years o, and Alaskan Eskimo middens 3,500 years d contain the remains of whales. It is quite pos- le, of course, that at this time whales were not much actively hunted as taken primarily when tering nearshore waters to strand. However, th the likely seasonal abundance of whales in e polar regions as the oceans warmed after the eistocene, it would be surprising if these early nters had not actively exploited them.

At about the same time (3,200 years ago), the ncient Greeks incorporated dolphins into their lture in a nonconsumptive way, for they appear frescoes in the Minoan temple of Knossos in ete, and many Greek myths refer to the ani- als' altruistic behavior. One describes how the ic poet and musician Arion was set upon by the ew of the boat in which he traveled from Italy. hen they threatened to kill him, he asked the vor of playing one last tune, which was so sweet at it attracted a school of dolphins. Seeing them, ion leapt overboard and was carried to safety on lolphin's back. The Greek philosopher Aristotle 84–322BC) was the first to study cetaceans in y detail, and even though some of his informa- n was incorrect and contradictory, many of his tailed descriptions of their anatomy clearly indi- te that he had dissected specimens.

The earliest record of regular whaling in Europe mes from the Norsemen of Scandinavia, in ound 800–1000AD. Basque seafarers were also nong the first to exploit whales; by the 12th cen- ry, there are accounts of extensive whaling in e Bay of Biscay. Early fisheries probably concen- ted on Right and Bowhead whales, since they e slow-moving and float after death due to their gh oil content. A Gray whale population that ce existed in the North Atlantic was probably nted to extinction by the early 18th century.

From the Bay of Biscay, whaling gradually read northward up the European coast and ross to Greenland. By the next century the utch and then the British started whaling in Arc- waters. During the 17th century, whaling from stern North America was also getting under y. All through this period, the whalers used nall sailing ships and struck their prey with har- ons hurled from rowing boats. The whales were en towed ashore to land or ice floes, or cut up d processed in the sea alongside the boat. In ntrast, whaling in Japan, which developed ound 1600, used nets and fleets of small boats.

Above *Small-scale whaling, like this Pilot whale hunt on the Faeroes, still persists in some places. Both the USA and Russia, for example, are permitted limited catches by indigenous peoples of the Arctic Circle, for whom whaling has long been a traditional way of life.*

As vessels improved, whalers started to pursue other species, notably the Sperm whale. In the 18th and 19th centuries, the whalers of New England (USA), Britain, and Holland moved first southward in the Atlantic and then west into the Pacific around Cape Horn, and eastwards into the Indian Ocean around the Cape of Good Hope. In the first half of the 19th century, whaling started in South Africa and the Seychelles. By this time, the Arctic whalers had penetrated far into the icy waters of Greenland, the Davis Strait, and Svalbard, where they took Bowhead, Right, and, later, Humpback whales. Whaling for Right whales also started up in the higher latitudes of the Pacific off New Zealand and Australia, and from 1840 onward for Bowheads in the Bering, Chukchi, and Beaufort seas.

Overhunting brought about the collapse of whaling in the North Atlantic by the late 1700s, and in the North Pacific during the mid-1800s. Sperm whaling flourished until about 1850, but then declined rapidly. The situation worsened after 1868, when a Norwegian, Svend Foyn, developed an explosive harpoon gun and steam-driven vessels replaced the sailing ships. Both these innovations had a significant impact on the remaining great whales, allowing ships to pursue even the fast-moving rorquals. By the late 1800s, populations of Bowheads and Right whales had declined markedly. British Arctic whaling ceased in 1912.

By the end of the 19th century, whalers were concentrating on the Pacific and the waters off Newfoundland and the west coast of Africa. Then, in 1905, they discovered the rich Antarctic feeding grounds of Blue, Fin, and Sei whales. In 1925 the first modern factory ship began working in the Antarctic and the industry ceased to be land-based. As a result Antarctic whaling expanded rapidly, with 46,000 whales taken in the 1937–38 season, until, yet again, populations declined to commercial extinction. The largest and hence most valuable of the rorquals, the Blue whale, dominated the catches in the 1930s, but had declined to very few by the mid-1950s, and was eventually totally protected in 1965. As these populations declined, attention turned to the next largest rorqual, and so on down.

Sperm whales continued to be taken following the population collapse in the 1850s, though with a world catch of only about 5,000 annually until 1948. Catches then increased quite rapidly, with about 20,000 a year being taken, mainly in the North Pacific and the Southern Hemisphere, until 1985, when the species became protected.

Until the middle of the 20th century, the whaling industry was dominated by Norway and the United Kingdom, with Holland and the United States also taking substantial shares. After World War II, however, these nations abandoned deep-sea whaling, and the industry was largely taken over by Japan and the Soviet Union, although many nations continued coastal whaling. Data from the former Soviet Union that has recently come to light show, for example, that from 1951–71 at least 3,368 Southern right whales were taken, even though the species had officially been under international protection since 1935.

Originally, the most important product of modern deep-sea whaling was oil; that of baleen whales was used in margarines and other foodstuffs, and that of Sperm whales in specialized lubricants. From about 1950, however, chemical products and meal for animal foodstuffs became increasingly important, although meat from baleen whales was also highly valued for human consumption by the Japanese. The Soviet Union, on the other hand, used very little whale meat and instead concentrated upon Sperm whales for their oil. By the late 1970s, whale catches in the Antarctic were yielding 29 percent meat, 20 percent oil, and 7 percent meal and solubles.

Over the last 30 years public attention and sympathy has increasingly turned towards the plight of whales. People watching whales off the eastern United States and in the lagoons of California were impressed by their confiding nature and fascinating behavior. At the same time, most of the great whales were continuing to decline in number as a result of overexploitation. The International Whaling Commission was set up in 1946 to regulate whaling activities, but it remained generally ineffective because the advice of its scientific committee was often overruled in the interests of short-term commercial considerations.

In 1972, the US Marine Mammal Act prohibited the taking and importing of marine mammals and their products except under certain conditions, such as by some native peoples, Inuits, and Aleuts for subsistence purposes or for the making of native handicrafts. In the same year, the United Nations Conference on the Human Environment called for a 10-year moratorium on whaling. The latter was not accepted by the International Whaling Commission, but continued publicity and pressure from environmental bodies and concern expressed by many scientists over the difficulties in estimating population sizes and maximum sustainable yields, finally had an effect. In 1982, a ban on all commercial whaling was agreed upon, and this took effect from 1986. Following the conservationist stance taken by the majority of IWC member nations during the 1980s and 90s, a separate management organization called NAMMCO (North Atlantic Marine Mammal Commission) was formed alongside the IWC, bringing together those countries like Norway, the Faeroes, Iceland, Greenland, and parts of Canada that wished to continue whaling in the North Atlantic. Within the IWC itself, Norway continued taking Minke whales for scientific purposes and, from the late 1990s, restarted an annual commercial take of a few hundred animals. Japan continues to lobby for the resumption of commercial whaling.

With the largely global ban on whaling, a nu ber of large whale species have shown signs of recovery. Humpback whales in the western No Atlantic were estimated to number 10,600 in 1999, with the well-studied summering popula tion in the Gulf of Maine growing at a rate of 6. percent per annum. The eastern North Pacific Blue whale population is also showing some encouraging signs of recovery, with 2,000 anim estimated in the 1990s and a trend toward increasing population size over several years. However, the status of some baleen whale popu tions is causing great concern, due to their over scarcity and various associated problems, inclu ing human-induced mortality. All populations of Northern right whales are seriously endangered there are little more than 300 left in the northwestern Atlantic, and only a few tens in the east ern North Atlantic. The Bowhead whales of the Okhotsk Sea and various parts of the eastern Ar tic, Gray whales in the western North Pacific, an Blue whales in several areas all also remain at pe ilously low numbers.

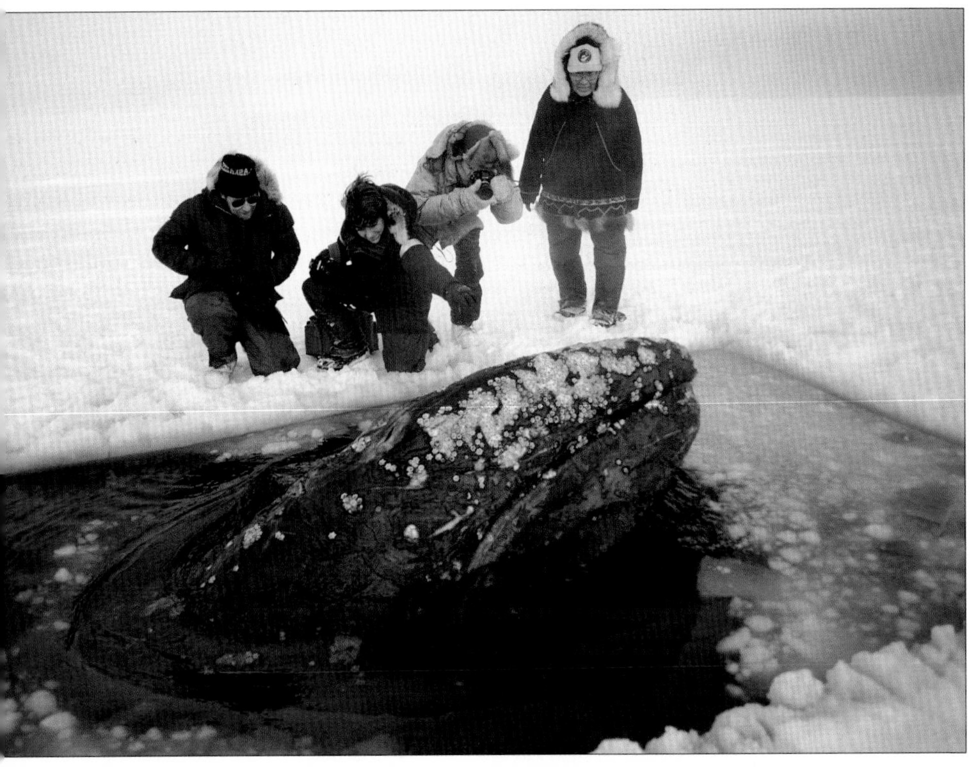

Above *Without human intervention there is still ificant danger for the world's whale populations. man concern for these creatures no longer simply ends to attempts to curtail whaling; operations are ied out to help whales that have become stranded, - as is the case with this Gray whale at Barrow, ska – trapped in pack ice.*

Left *Iceland, which withdrew from the Inter- ional Whaling Commission in the 1990s, is one of andful of countries that still pursues whaling activi- in the North Atlantic. At a processing plant in lfjördhur, the carcass of a huge Fin whale is stripped ts flesh in readiness for its eventual resale as meat.*

Below *A diver swims alongside a Right whale off coast of Argentina. Despite the massive size of the ht whale – average length is about 14m (46ft), rage weight around 22,000kg (48,400lb) – they st on a diet of tiny crustaceans, which are mostly y a few millimeters in diameter. They feed by swim- g along, mouth open, near the surface.*

Changing Marine Environments

NEW THREATS

The story of man's often unhappy relationship with whales does not end here. Even if we terminated commercial whaling forever, or acquired sufficient knowledge to manage whale populations in a sustained manner, cetaceans would continue to face a variety of other threats of mankind's making. The marine environment is being modified in many parts of the world as human populations increase and become more industrialized, making greater demands upon the sea either by removing organisms for food or by releasing toxic waste products. Acoustic disturbance comes from seismic testing during oil and gas exploration, from loud military sonar, and from motor-vessel traffic. Although it is difficult to determine the longterm consequences, short-term negative reactions have been observed in several species, and there is some evidence that extended exposure to loud sounds may cause hearing damage. Besides emitting sounds, vessels may also pose a direct threat of physical damage through collisions. The most endangered of all the great whales, the Northern right whale, which numbers just 300 animals, faces the threat of ship strikes off the eastern seaboard of the United States. Another new danger is that of high-speed ferries, which are being introduced into many parts of the world and are already resulting in collisions with slower-moving cetaceans like Sperm and Pilot whales.

Toxic chemical pollution, particularly from heavy metals, oil, and persistent chemicals, may also have serious harmful effects in enclosed bodies of water such as the Baltic, Mediterranean, and North Seas, making coastal species such as the Harbor porpoise particularly vulnerable. Sometimes high pollutant burdens have been linked with disease and mass die-offs, as during the early 1990s when more than 1,000 Striped dolphins died of a morbillivirus in the Mediterranean. In that instance the animals were found to have large concentrations of PCBs (polychlorinated biphenyls) in their tissues, which affect the immune system, reducing resistance to disease.

Destruction of suitable habitat by the building of coastal resorts, breakwaters that change local current patterns and encourage silting, and dams that regulate water flow in rivers, all pose additional threats. Species most vulnerable are usually those that are rare and localized in distribution, like the vaquita in the Gulf of California or the baiji (*Lipotes vexillifer*) on the Yangtze River in China.

The greatest threat of all for many species may come from commercial fishing. In some instances, independent observers aboard vessels have been able to estimate incidental takes, and comparisons with population estimates have shown that the mortality from these activities is unsustainable. Every year, the world's fisheries draw up an estimated 27 million metric tons of nontarget marine life that is simply discarded back into the sea, dead or dying. This figure represents a quarter of the global catch, and a wide variety of species are affected. On the continental shelf of Europe and North America, the Harbor porpoise is particularly at risk. Further offshore, the more pelagic dolphins like the Striped, Common, Spinner, and Spotted varieties are sometimes taken in large numbers. Even the great whales fall victim to particular fisheries, adding further pressure upon rare and endangered species.

Direct competition with fisheries for food may represent a further pressure, though this is one that is very difficult to disentangle from other effects upon the marine ecosytem. As particular prey species become scarce through overexploitation, others may take their place and the structure of marine communities alters. Although many cetacean species have catholic diets and appear able to switch prey species, we have little understanding of the longterm consequences for them.

Finally, increasing industrialization has contributed to widespread climate changes. These may have a variety of consequences upon cetacean populations. Sea temperature rises are likely to affect the composition of cetacean species in a particular region. With the melting of portions of the ice cap, the rise in water levels will affect the marine communities of shallow seas. And the greater frequency of storms may result in less stable climate systems elsewhere and the breakdown of plankton frontal systems, on which many cetaceans depend. All these pressures are unlikely to evoke the same passions as direct killing by humans, but nevertheless they will have to be addressed if these magnificent creatures are to continue to grace our seas. PGHE

DISTRIBUTION All oceans

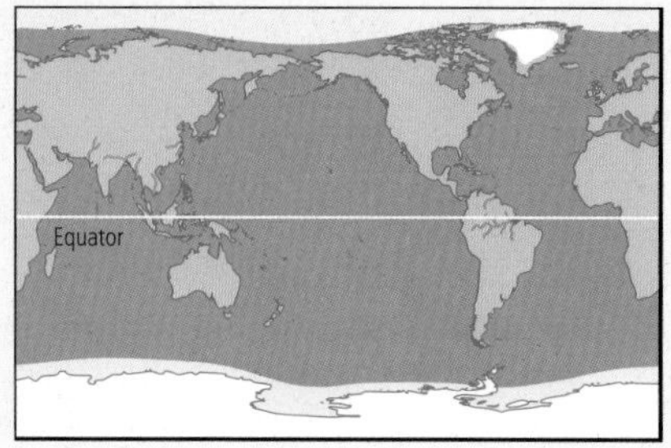

Suborder: Odontoceti
At least 72 species in 34* genera and 10 families

*35 if *Lagenorhynchus acutus* becomes *Leucopleurus acutus*, as has recently been recommended.

Toothed Whales

NEARLY 90 PERCENT OF CETACEAN SPECIES belong to the suborder Odontoceti – the toothed whales. The great majority are relatively small dolphins and porpoises, usually less than 4.5m (15ft) in length. Most toothed whales have a prominent, beaklike snout, behind which the forehead rises in a rounded "melon." As their name suggests, they always bear teeth, however rudimentary, in the upper or lower jaws, or both, at some stage of their lives. The teeth look alike, have a single root, and last a lifetime (monophyodont dentition).

The members of the Platanistidae family (river dolphins) are the most primitive living cetaceans. They all have a long beak and broad, short flippers. Their eyes are very small – some species are almost blind – so they navigate by echolocation.

The family Physeteridae contains the Sperm whale, at 18m (59ft) the largest of the toothed whales. Its Pygmy and Dwarf relatives are smaller, with a less massive head and a distinct dorsal fin.

The beaked whales (family Ziphiidae) are named for their distinctive, long beak. Most are rarely seen, some being known only from stranded specimens.

The remaining three families – Monodontida Phocoenidae, and Delphinidae – are all closely related, probably having diverged in the middle Miocene. The two species of Monodontidae, th narwhal and beluga, are confined to the norther oceans, particularly the Arctic. They are relativel small whales, and neither has a dorsal fin.

The family Phocoenidae (the porpoises) radia ed and dispersed from the tropics into temperat waters of both hemispheres during the Miocene Pliocene (about 7 million years ago). All are sma species with a blunt snout, no beak, and spade-shaped teeth.

Most of the Delphinidae family are characterized by functional teeth in both jaws, a melon w a distinct beak, and a dorsal fin. The largest, the Killer whale, is unusual in having rounded flippe in lacking any beak, and in preying on other sea mammals. Some dolphin genera occur in both hemispheres away from the tropics; another gro has almost identical populations in each hemisphere, plus a smal ler form in the tropics; other have a cosmopolitan distribution; and still other a restricted pantropical distribution. PGI

CAPTIVE DOLPHINS

Many people have only seen toothed whales in the flesh in dolphinariums or "marine parks," which have proliferated since the 1970s. These commercial enterprises justify keeping dolphins and orcas captive on the grounds of scientific research and public education. Yet conservationists argue that captivity cannot recreate the complex social lives of cetaceans in the wild and that, even in well-run dolphinariums, enforced daily interaction with people places the animals under unnatural stress. There is particular concern about confined "petting pools," where dolphins can be touched and fed all day long. It is beyond dispute that life expectancy is far lower among captive whales and dolphins. With the growth of ecotourism, more people are being encouraged to go and see dolphins in the wild, and so learn more about the animals within their natural habitat.

1

2

3

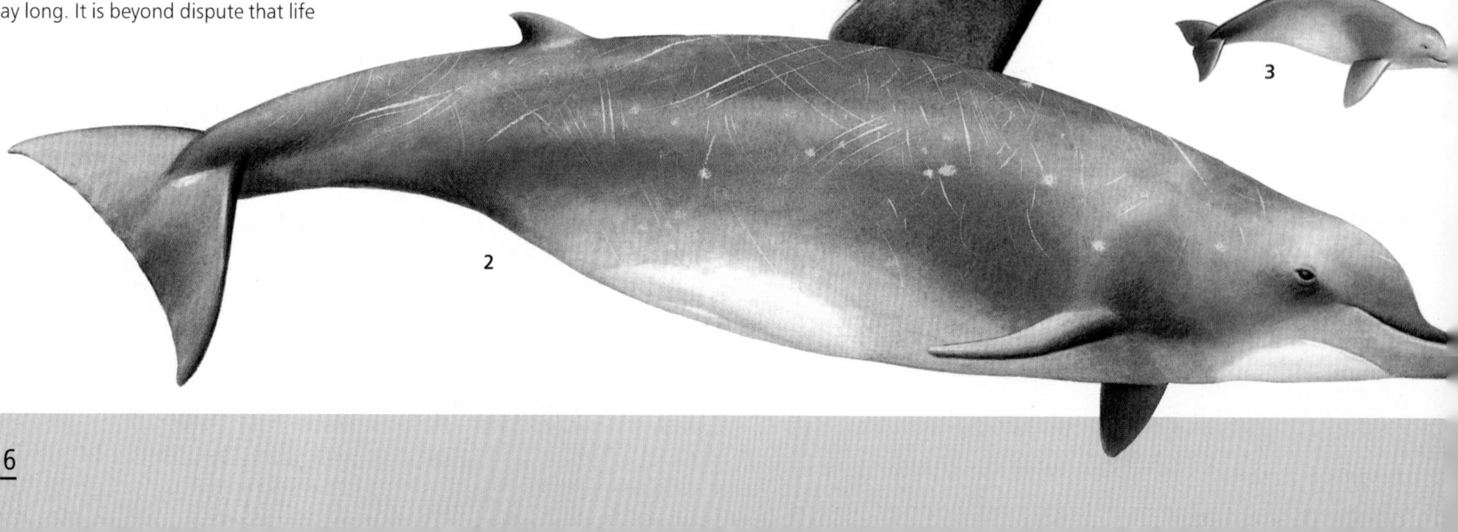

ɒthed Whale Skulls

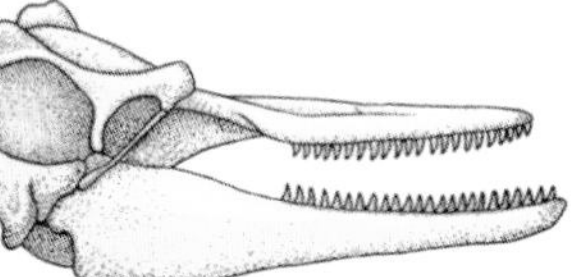

Bottlenose dolphin
60cm

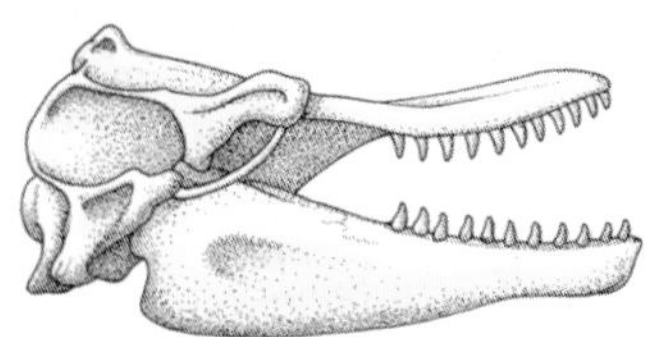

Killer whale
120cm

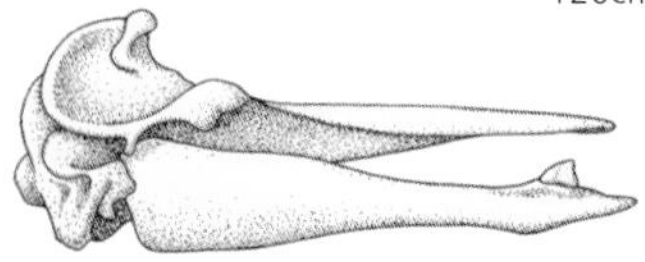

Gervais's beaked whale
120cm

ɔothed whales evolved from a ter-
ʳial carnivorous ancestor of modern
ulates, the skulls telescoped, result-
ɪn a long, narrow "beak" and a
ʳement of the posterior maxillary
e to the supra-occipital region (top
ne skull). These changes were asso-
ed with the development of echo-
ting abilities and the modification
ne teeth for catching fish. The teeth
ne toothed whales' ancestors were
erentiated into incisors, canines,
molars, as in modern carnivores,
the ideal dentition for fish-eaters is
ng row of even, conical teeth, and
is in fact roughly the pattern in all
thed whales.

Bottlenose dolphins are classic fish-eaters, with many small, sharp teeth. In the Killer whale (RIGHT) which feeds on mammals as well as fish, the number of teeth is greatly reduced (10–14 on each side of both jaws). The beaked whales, which eat squid, have become almost toothless; *Mesoplodon* species, like Gervais's beaked whale, have a single pair of teeth in the lower jaw.

***ɛelow** Representative species of ʳhed whale, showing their great ʳrsity of form and size: **1** Bottle- e dolphin* (Tursiops truncatus)*; ɜird's beaked whale* (Berardius bairdii)*; **3** Ganges dolphin* (Platanista gangetica)*; **4** Narwhal* (Monodon monoceros)*; **5** Sperm whale* (Physeter macrocephalus)*; **6** Dall's porpoise* (Phocoenoides dalli)*;*

Toothed Whale Nostrils

The nostrils of toothed whales in general show a migration backward and toward the top of the skull, compared to land mammals. The narwhal and Pygmy sperm whale are typical. The Sperm whale is unusual, in that its development of the huge spermaceti organ made this pattern unworkable and so two existing passages were displaced forward; the right nasal passage lies below the spermaceti organ.

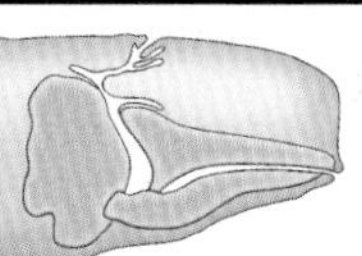

Narwhal

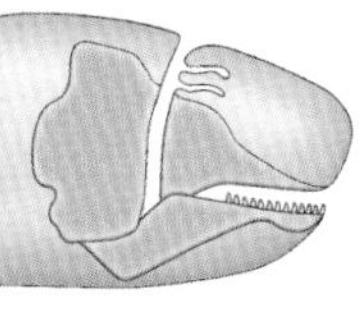

Pygmy sperm whale

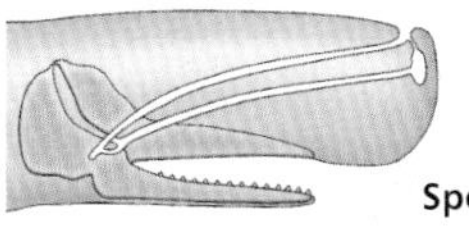

Sperm whale

6

5

4

River Dolphins

Dolphins are usually thought of as inhabiting the vastness of the oceans, so the idea of them navigating a narrow channel through the Amazonian rain forest, in a muddy, sunbaked watercourse crossing the countryside of Bangladesh, or in the clear, cool flow of a Himalayan foothill stream may be unexpected. And yet these are all circumstances to which river dolphins are adapted.

The term "river dolphin" has conventionally been applied to a group of longbeaked, many-toothed small cetaceans: the Yangtze River dolphin; the Ganges and Indus river dolphins; the Amazon dolphin; and the La Plata dolphin. These dolphins may not have a common ancestry, but have converged due to their similar niches. All but the La Plata dolphin, which lives in the estuary of the River Plate and along the Atlantic coast of South America between northern Argentina and central Brazil, are confined to freshwater environments.

Several other small cetacean species have populations that inhabit both marine waters and freshwater systems, although they are not true river dolphins. The species in question are the tucuxi (*Sotalia fluviatilis*) of Central and South America, the Irrawaddy dolphin (*Orcaella brevirostris*) of Southeast Asia, and the Finless porpoise (*Neophocaena phocaenoides*) of the Far East. Although there is no conclusive proof, these populations of "facultative" river cetaceans probably remain permanently in fresh waters and do not regularly migrate into and out of the sea. Here, they are treated mainly in the entries on Dolphins and Porpoises, respectively.

Flipper in Fresh Waters

FORM AND FUNCTION

In addition to their long, narrow beaks and numerous teeth, river dolphins exhibit some extreme characteristics in their morphology and sensory systems. Amazon dolphins, for example, are the only modern cetaceans with differentiated teeth; those in the front half of the jaw are conical, typically for dolphins, but in the back they are flanged on the inside portion of the crown. Amazon dolphins have a diverse diet that includes not only many species of softbodied fish but also spiny and hardbodied prey like crabs, armored catfish, and even small turtles, so they probably use their modified rear teeth to crush and soften prey items. It is a misconception that all river dolphins are blind, but the atrophied eyes of the Ganges and Indus dolphins lack a lens, leaving these species unable to resolve images. The most they can do is perceive the presence or absence of light. The other species within the group also have reduced vision, but not nearly to the same degree; in fact, the Amazon dolphin has quite good vision.

The sonar abilities of all the river dolphins are highly developed, enabling them to detect objects (including prey) in the murky conditions prevailing in much of their environment. The Ganges and Indus dolphins are known to be side swimmers; in other words, after surfacing to breathe, they immediately roll onto one side, "feeling" for the bottom with a flipper while scanning the area ahead of them with constant clicking sounds.

Social or Solitary?

SOCIAL BEHAVIOR

Groups of river dolphins rarely exceed 10 in number, and it is not unusual to encounter solitary individuals. The relationships of associated individuals have not been studied, so the composition, structure, and nature of river dolphin societies are completely unknown. Calves, of course, remain close to their mothers for several months. Although densities, or the numbers of animals present in a given area of water surface, can be a good deal higher for river dolphins than for most marine dolphins, this is probably due mainly to the differences of scale between rivers and the open sea. Small bands of Tucuxis or Irrawaddy dolphins will typically surface in synchrony and within one or two body lengths of one another. In contrast, river dolphins proper will often app to surface and dive independently, while also c veying the impression that animals across a wi area are engaged in similar activities and are so how linked.

River dolphins are found mainly in confluences (where rivers or streams converge), at sharp bends, along sandbanks, and near the downstream ends of islands. Such areas offer both deep water and hydraulic refuge, the latter in the form of eddy countercurrent systems, where dolphins can reduce their energy expenditure but still avoid being swept downstream by the main current.

5

Below and left *The five species of "true" river dolphins. Despite their widely separated habitats, they are very similar in appearance, differing primarily in skin color, length of beak, and number of teeth:* ***1*** *Amazon dolphin;* ***2*** *La Plata dolphin;* ***3*** *Ganges dolphin;* ***4*** *Indus dolphin;* ***5*** *Yangtze river dolphin.*

A Lethal Proximity to Humans

CONSERVATION AND ENVIRONMENT

:r dolphins everywhere live in danger of inci-.tal capture in fishing gear (especially gill nets), . of the transformation and fragmentation of ır habitat to serve human needs. Levels of toxic .taminants found in river dolphin tissue – for mple, organochlorine pesticides, polychlori-ed biphenyls (PCBs), and organotins – have ed concern about their implications for health reproduction. The animals' close proximity to .rces of pollution (such as sewage outfalls, fac-y discharge, and agricultural runoff) and their .tive inability to metabolize contaminants make m especially vulnerable. Dams also impede ir natural movements; subpopulations upriver barrages have declined steadily, and some are ady extinct. In Asia especially, much of the er impounded above barrages is diverted to gate fields and supply homes and businesses, s directly reducing the dolphins' habitat.

The Yangtze river dolphin is the most critically langered cetacean in the world. Discovered by stern science only in 1918, it was still common l widely distributed along the entire length of Yangtze River in the 1950s. However, from the rt of China's "Great Leap Forward" (the rapid ustrialization of the state) in the autumn of 58, the dolphins were hunted intensively to ovide meat, oil, and leather. Today, although y are legally protected, they continue to be ed accidentally in fisheries, and to die from col-ons with powered vessels and exposure to derwater blasting. Their figures are thought to mber only in tens. The Finless porpoises that are much of the Yangtze river dolphin's histori-range have also suffered a rapid decline, hough several hundred of this species, possibly en a thousand and more, survive.

Perhaps only a few thousand Ganges dolphins survive, and fewer than 1,000 Indus dolphins. Tribal people in remote reaches of the Ganges and Brahmaputra rivers still hunt dolphins for food, and fishermen in the Subcontinent use their oil to lure a highly prized species of catfish. The Irrawaddy dolphins in the Mekong and Mahakam rivers are in grave danger of extinction. The situation is somewhat brighter in South America, where two river dolphins, the Amazon and the tucuxi, are still quite numerous and widely distributed. RR

RIVER DOLPHINS

Order: Cetacea

Families: Platanistidae, Lipotidae, Iniidae, Pontoporiidae

5 species in 4 genera

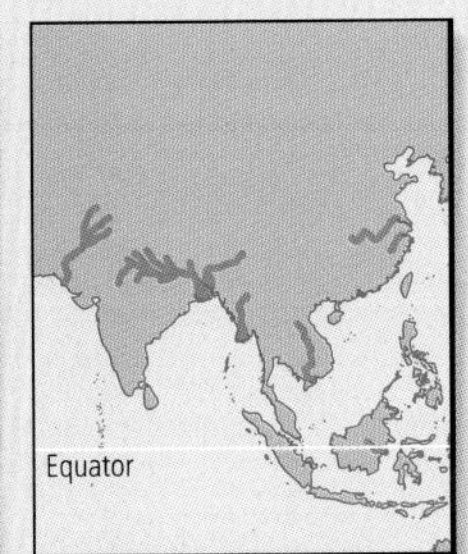

DISTRIBUTION SE Asia, S America

GANGES DOLPHIN
Platanista gangetica
Ganges dolphin or susu
India, Nepal, Bangladesh, in Ganges-Brahmaputra-Meghna river system. HTL 210–260cm (83–102in); WT 80–90kg (175–200lb). Skin: light grayish-brown, paler beneath. Gestation: 10 months. Longevity: over 28 years. Conservation status: Endangered.

INDUS DOLPHIN *Platanista minor*
Indus dolphin or bhulan
Pakistan, in Indus river. Size, coat, diet, gestation, and (probably) longevity: as Ganges dolphin. Conservation status: Endangered.

YANGTZE RIVER DOLPHIN *Lipotes vexillifer*
Yangtze River dolphin, Whitefin dolphin, or baiji
China, in Yangtze and lower Fuchunjian Rivers. HTL 230–250cm (91–98in); WT 135–230kg (300–510lb). Skin: bluish-gray, white underneath. Gestation: probably 10–12 months. Conservation status: Critically Endangered.

AMAZON DOLPHIN *Inia geoffrensis*
Amazon dolphin or boto
S America, in Amazon and Orinoco river systems. HTL 208–228cm (82–90in) (Orinoco), 224–247cm (88–97in) (Amazon); WT 85–130kg (190–285lb). Skin: dark bluish-gray above, pink beneath; darker in Orinoco. Gestation: probably 10–12 months. Conservation status: Vulnerable.

LA PLATA DOLPHIN *Pontoporia blainvillei*
La Plata dolphin or franciscana
Coast of E South America, from Ubatuba to Valdes Peninsula (not in La Plata river). HTL 155–175cm (61–69in); WT 32–52kg (70–115lb). Skin: light, warm brown, paler beneath. Gestation: 11 months. Longevity: more than 16 years.

Abbreviations HTL = head–tail length WT = weight

Dolphins

FROM THE DOLPHINS THAT RESCUED THE lyric poet Arion from pirates in Greek mythology to the eponymous Killer whale hero of the 1993 Hollywood movie Free Willy, *dolphins have always had a special appeal for humankind. It has been argued that their intelligence and developed social organization are equaled only by the primates – perhaps even only by humans – while their general friendliness and apparent lack of aggression have been compared favorably with our own.*

This anthropocentric view has in recent years required modification, as it has become apparent, for example, that aggression is a not uncommon element of dolphin behavior. Even so, the more we find out about dolphins' learning abilities, social skills, and life below the waves, the more we marvel at the great variety of behaviors and social structures they exhibit as different populations or species adapt to local conditions.

Agile and Intelligent

FORM AND FUNCTION

The family Delphinidae is a relatively modern group, having evolved during the late Miocene, about 10 million years ago. They are the most abundant and varied of all cetaceans.

Most dolphins are small to medium-sized animals, with well-developed beaks and a central, sickle-shaped dorsal fin that curves backwards. They have a single, crescent-shaped blowhole on top of the head, with the concave side facing forwards, and have functional well-separated teeth in both jaws (anything from 10 to 224, though most have 100–200). The majority of delphinids have a forehead melon, although this is indistinct in some species, for example the tucuxi, and entirely absent in *Cephalorhynchus* species; the melon is pronounced and rounded to form an indistinct beak in Risso's dolphin and the two species of pilot whales, and tapered to form a blunt snout in Killer and False killer whales. Killer whales also have rounded, paddle-shaped flippers, whereas the pilot whales and False killer have narrow, elongated flippers.

The wide variation in color patterns between species can be categorized in a number of ways. One classification recognizes three types: uniform (plain or evenly marked), patched (with clearly demarcated pigmented areas), and disruptive coloration (black and white). Color differentiation helps individuals to recognize one another, and colors may also help conceal hunters from their prey. Dolphins feeding at depths where the ligh dim are often uniform, while surface feeders ten to be countershaded (dark above, light below) s that they blend into the background when lit fr above. The color patterns of some species may also act as anti-predator camouflage: saddle patterns afford protection through their counterligh ing effect, while spotted patterns blend in with sun-dappled water. Criss-cross patterns have bo countershading and disruptive elements.

Dolphins, like other toothed whales, rely grea ly on sound for communication. Their sounds range from a narrow band and modulated whist to trains of clicks of 0.2 kHz into the ultrasonic range around 80–220 kHz; these appear to be used for tracing prey by echolocation, and possibly also for stunning it. Although different whistles have been categorized and associated with particular behaviors, there is no evidence of a la guage with syntax.

Dolphins can perform quite complex tasks ar are fine mimics capable of memorizing long routines, particularly where learning by ear is

Left *Bottlenose dolphins mainly inhabit tropical and subtropical waters. This individual clearly displays the short beak that is characteristic of the species. Bottlenoses also commonly have the white patch seen here on the tip of the lower jaw.*

involved. In some tests they rank with elephants. Bottlenose dolphins can generalize rules and develop abstract concepts. Dolphins have large brains relative to body size – adult Bottlenose dolphins weighing 130–200kg (290–440lb) may have a brain of about 1,600g (3.5lb) weight; in comparison, the brains of humans weighing 36–95kg (80–210lb) range from 1,100–1,540g (2.4–3.4lb). They also have a high degree of folding of the cerebral cortex, comparable with that found in primates. These features are considered to be indications of high intelligence.

Brain tissue is metabolically expensive to produce, and is therefore unlikely to evolve unless there is a strong benefit. Several different functions have been ascribed to the large brains exhibited by some cetaceans. (Not all species have large brains, however; mysticetes have relatively small brains). One suggestion is that processing acoustic information requires greater "storage" space than visual information; another explanation is that cetaceans may simply need larger brains to do the same tasks that land mammals achieve with smaller ones. A third hypothesis is that brain power plays an important part in social evolution, and that extended parental care, cooperation with conspecifics in feeding and defence, alliance formation, and individual recognition with individual-specific social bonding may together have favored cerebral development.

Dolphins' oft-cited lack of aggression has been exaggerated. Bottlenose species, and perhaps also spinners, develop dominance hierarchies in captivity, in which aggression is manifested by directing the head at the threatened animal, displaying with an open mouth, or

Left *Bottlenose dolphins leaping. Dolphins may use this behavior to herd fish and for sexual display, but also sometimes leap just for the fun of it. Such displays of graceful agility have helped secure dolphins a special place in the human imagination.*

Below *Juvenile Atlantic spotted dolphins in the Atlantic waters around the Bahamas. The patterns on many dolphins may help to conceal them from their prey or from predators, as light playing near the surface blends with the dolphin's coloration and breaks up its outline.*

FACTFILE

DOLPHINS

Order: Cetacea

Family: Delphinidae

At least 36 species in 17 genera, including common or saddleback dolphins (*Delphinus*, 3 species); spinner, spotted, and striped dolphins (*Stenella*, 5 species); white-sided and white-beaked dolphins (*Lagenorhynchus*, 5/6 species); southern or piebald dolphins (*Cephalorhyncus*, 4 species); hump-backed dolphins (*Sousa*, 3 species); bottlenose dolphins (*Tursiops*, 2 species); right whale dolphins (*Lissodelphis*, 2 species); pilot whales (*Globicephala*, 2 species).

DISTRIBUTION All oceans

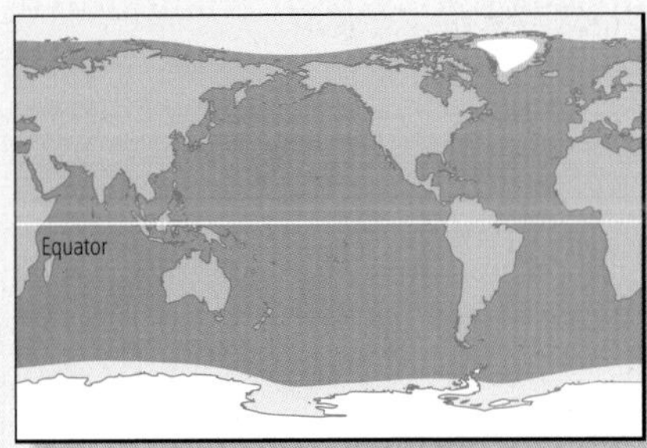

HABITAT Generally over the continental shelf, but some open sea.

SIZE **Head–tail length** from 1.2m (3.5ft) in Heaviside's dolphin to 7m (23ft) in the Killer whale; **weight** from 40kg (88lb) to 4.5 tonnes (4.4 tons) in the same two species.

FORM Snout beaklike (versus blunt in porpoises) and teeth spade-shaped (versus cone-shaped in porpoises); body form slender and streamlined. Pectoral and dorsal fins sickle-shaped, triangular, or rounded; dorsal fin positioned near middle of back, except in right whale dolphins, which lack a dorsal fin.

DIET Primarily fish or squid; Killer whales also eat other marine mammals and birds.

BREEDING Gestation 10–16 months (13–16 months in the Killer whale, False killer whale, pilot whales, and Risso's dolphin; the rest 10–12 months).

LONGEVITY Up to 50–100 years (Killer whale).

See species table ▷

clapping of the jaw. Fights have also been observed in the wild, in which scratches and scrapes have been inflicted by one individual running its teeth over the back of another; and some species, for example Bottlenose dolphins, are known to attack individuals of other, smaller species (for example, Spotted and Spinner dolphins); they have even been known to kill Harbor porpoises.

Different Diets, Different Forms

DIET

Differences in diet between dolphin species are reflected in their body shapes and dentition. For example, those species that feed primarily upon squid generally have the most rounded foreheads, blunt beaks, and (often) reduced dentition.

Killer whales, whose diet also includes marine mammals and birds, have particularly enlarged foreheads. One theory is that this is an adaptation for receiving and focusing sound signals to obtain an accurate picture of the location of agile, fast-moving prey. The other members of the family feed primarily upon fish: all appear to be opportunist feeders, probably catching whatever species they encounter within particular size ranges. Some – for example, bottlenose and hump-backed dolphins – are primarily inshore feeders, though they also prey on both bottom-dwelling and open-sea fish. Other species, such as members of the genera *Stenella* and *Delphinus*, are more pelagic, feeding further out to sea on shoaling fish – both those found close to the surface such as anchovy, herring, and capelin, and those located at great depths, including lantern fish.

Most dolphins are partial to squid and even shrimps. These overlaps make it difficult to determine the extent of competition among species. One way to avoid having to share food is to avoid other dolphins with similar dietary requirements. In the tropical eastern Pacific, Spotted dolphins feed largely on open-sea fish near the surface, whereas the related Spinner dolphin feeds at deeper levels; and the two may also feed at different times of the day.

Deeper-ocean dolphins tend to travel in herds of up to 1,000 or more, whose members may cooperate in the capture of shoaling fish (see Cooperative Killers). Inshore species usually form smaller herds of 2–12 individuals, possibly because they are feeding on prey at low densities. Offshore, dolphin schools may spread out to form a band that may be anything from 20m (66ft) to several kilometers wide. Small groups of 5–25 tend to coalesce within the larger assemblage. The dolphins often follow underwater escarpments or other landmarks, and they also exploit tidal currents to ensure efficient journeying. When shoaling fish are present in large numbers, the dolphins come together in feeding activity that may sometimes appear frenzied but in fact involves cooperation to herd the shoal into a tight ball, through which the dolphins weave picking off mouthful after mouthful.

Radio-tracking studies have shown that dolphins may have home ranges varying in size from about 125sq km (48sq mi) in the Bottlenose to 1,500sq km (580sq mi) in the Dusky dolphin. Successive generations of Bottlenose dolphins have been observed occupying the same range for more than 28 years. Individual movements of more than 1,800km (1,140mi) a year have been recorded in Spotted dolphins, and are probably not uncommon for open-sea species.

Home Is Where the Group Is

SOCIAL BEHAVIOR

Although most species have an open social structure, with individuals entering and leaving the herd over periods of time, some, such as pilot whales and the Killer whale, seem to have more stable group membership. Genetic data from Long-finned and observational data from Short-finned pilot whales suggest that pods are composed primarily of related females and their offspring, although one or more unrelated adult males may associate with the group when there are opportunities

mating. Grown-up offspring of both sexes stay h their mothers, although adult sons may ›ve among pods to mate before returning to ir natal pods. Herds of bottlenose dolphins m to comprise family groups of male, female, 1 calf, or else mother–calf pairs, which may ;regate to form larger herds, some of which may segregated by sex and age. Among bottlenoses re is evidence of strong male–male bonds. The mals' mating system is as yet imperfectly derstood, but generally appears to be promis- ous. The frequent scars observed in some ›cies appear to be most prevalent on males, ;gesting male–male competition for mating ess to females. Polygyny may also occur, but, atever the mating system, the male–female and le–calf bonds are relatively weak.

Sexual behavior occurs throughout the year, hough there is usually a peak in calving during : summer months, even in lower latitudes. One f is born, and it remains with the mother for ›eral months; she will continue to feed it with lk for up to 3.5 years. Many species therefore ›ed at minimum intervals of 2–3 years (rising to 8 years in the case of Killer and pilot whales). e age of sexual maturity probably ranges from 7 years (in Commerson's, Spinner, and Com- on dolphins) up to 16 years in the male Killer ale, with most species breeding at about 8–12 ars of age.

Many species undertake seasonal migrations search of food; these are usually offshore– shore movements, although they can be lati- dinal. If discrete calving areas exist, they have ›ely been identified, although they may be in deeper offshore waters where there is less turbulence from coastal currents. Adults and young in several species may then move into shallower waters to exploit prey that aggregate around reefs and seamounts.

Dolphins are gregarious, though the large herds of 1,000 or more generally only occur during long-distance movements or among concentrations at major food sources. In most cases, herd membership is fluid; individuals may enter or leave the group over a period of weeks or months, with only a minority remaining within it for a longer time. There is rarely any indication of the stable, well-developed social organization typical of primates, although in a few species such as the Killer whale, kin relations can last a lifetime. It has not been easy to determine the extent to which dolphins cooperate with one another in infant rearing and prey capture, but even if we accept that some of the more gregarious species do so, such behavior is also found among primates, carnivores, and birds.

Above *The characteristic poses of 13 dolphin species.* ***1*** *Bottlenose dolphin* (Tursiops truncatus); ***2*** *Rough-toothed dolphin* (Steno bredanensis); ***3*** *Atlantic white-sided dolphin* (Lagenorhynchus acutus); ***4*** *Atlantic spotted dolphin* (Stenella frontalis); ***5*** *Common dolphin* (Delphinus delphis); ***6*** *Northern right whale dolphin* (Lissodelphis borealis); ***7*** *Dusky dolphin* (Lagenorhynchus obscurus); ***8*** *Atlantic humpbacked dolphin* (Sousa teuszii); ***9*** *Melon-headed whale* (Peponocephala electra); ***10*** *Commerson's dolphin* (Cephalorhynchus commersonii); ***11*** *False killer whale* (Pseudorca crassidens); ***12*** *Killer whale* (Orcinus orca); ***13*** *Risso's dolphin* (Grampus griseus).

Right *A heavily scarred male Risso's dolphin. Prominent weals like this are caused by fights over feeding territories or mating, which may sometimes be prolonged and vicious.*

The Problem with Gill-nets

CONSERVATION AND ENVIRONMENT

Large herds of dolphins sometimes concentrate at feeding areas, and conflict can result if these coincide with human fisheries. Many dolphins are caught and drowned in gill-nets. Inshore species such as Dall's and Harbor porpoises are most at risk, but in the late 1960s and early 1970s the eastern Pacific purse-seine tuna fishery annually caused the death of between 150,000 and 500,000 individuals – mainly Spinner and Atlantic spotted but also Common dolphins. Subsequently, the numbers later fell thanks to the introduction of various methods of making the nets more conspicuous to dolphins, including the application of lines of floats at the water's surface, as well as of panels through which captured dolphins could escape. By the end of the century, the numbers of animals killed accidentally each year had fallen to about 3,000 (all from non-US vessels, since the United States ceased this fishing activity in the region from 1995).

Yet incidental capture in fishing gear continues to be a worldwide problem. Bottom-set gill-nets in the North Sea kill several thousand Harbor porpoises a year, a larger number than the local populations can sustain. Mitigation measures such as "pingers" (acoustic alarms) have been shown to reduce kills substantially in some circumstances, most recently off Denmark and the UK. Such techniques are now being applied more widely, but they will not be effective in every situation.

Less obvious threats to dolphins come from inshore pollution by toxic chemicals and disturbance from boats. A recent study of stranded porpoises in the UK found that those with high pollutant burdens were significantly more likely to be diseased, and the same factors may well be at work in Common and Bottlenose dolphins in the western Mediterranean and off southern California. The rise in recreational traffic in coastal waters poses threats to species like Bottlenose dolphins that share those waters, whilst the introduction of high-speed ferries in many parts of the world has led to collisions with pilot whales.

Dolphin hunting is not widespread. It continues off the coasts of Japan and South America and on a small scale off many tropical islands. Until recently, large numbers of Common dolphins were taken in the Black Sea (Turkish catches were 40,000–70,000 annually until 1983, when the practice was outlawed, although poaching still continues in a minor way). Finally, direct competition for particular fish species may be an important potential threat as man increasingly exploits the marine environment for food. PGHE

The Mystery of Mass Strandings

Dolphins have long been found stranded on dry land; in fact, up to the 1970s, virtually all information about cetaceans came from carcasses found on coasts. Most such animals have died at sea and been washed ashore, but some strand while alive. Such incidents are most conspicuous when groups of animals strand together, a phenomenon that is particularly prevalent among pilot and False killer whales.

In some cases, mass strandings can be directly related to a particular cause. From 1989–92, several hundred Striped dolphins that washed ashore along the Mediterranean coasts of Spain, France, Italy, and Greece were found to have a morbillivirus infection, but also very high levels of PCBs (polychlorinated biphenyls). It has been suggested that the animals, weakened by food shortage, mobilized their blubber reserves, where the pollutants were stored in an inert state, thus lowering their resistance to disease. Mass mortality from disease has been reported along the eastern seaboard of North America in a number of species, particularly Bottlenose dolphins.

The cause of mass strandings of live animals is rarely known. While disease and old age may explain many individual cases, they are obviously unlikely to lead to an entire herd coming ashore, so the explanation may be that most of the stranded animals were simply following a leader – usually an experienced older animal (which among species with matriarchal societies would normally be a female). So far as we know, mass-stranding species tend to form fairly stable herds; they are also pelagic, so they are less likely to be familiar with shallow coastal areas and more likely to become disorientated. Other theories that have been put forward to explain the phenomenon include infection of the inner ear by nematode parasites, disturbing the animals' balance or echolocation abilities; the effects of upsetting sounds, such as underwater explosions or magnetic disturbances; and disorientation after the animals have followed prey into unfamiliar or shallow waters.

Below *Stranded Long-finned pilot whales. Pilot whales appear to strand in large numbers more frequently than any other whale; however, this may just be a result of their abundance.*

SPECIAL FEATURE

SLEEK SPINNERS

School life of the Spinner dolphin

THE SLENDER DOLPHINS LEAP HIGH OUT OF THE water, twisting and spinning rapidly around their longitudinal axes. The movement instantly identifies these as Spinner dolphins, inhabitants of oceans throughout the tropics and subtropics.

Spinner dolphins of Hawaii rest and socialize during the day within protected bays and along shallow coastlines, in tight groups of usually 10–100 animals. At night, they move into deep water 1km (0.6mi) or more offshore, and dive to 100m (330ft) or more to feed. When this occurs, the group spreads out.

The groups are ephemeral. During the night, many individual dolphins change their companions, so by the time that they head shoreward at dawn, group membership is partially reshuffled. This shifting is not random, however, as small subgroups of 4–8 animals, possibly related, stay together for up to four months and possibly longer, changing group affiliations at will. Ties between some dolphins last throughout their lives.

The dolphins may range up to 100km (60 mi) along the coast daily, though subgroups have a preferred "home area." The shallow water is usually calmer than the open ocean, making resting and socializing easier, and deepwater sharks that prey on dolphins are rarer and more easily detected.

It is likely that Spinners recognize their group mates – and possibly more distant individuals – far from their home area. When Spinners meet after a long separation, surface leaping, spinning, and tailslapping takes place. There is also much vocalizing, which may be part of a greeting ceremony.

Hawaiian Spinner dolphins have the advantage of the shoreline for protection. In deep waters, Spinners mix with Atlantic spotted dolphins, and adopt a mutual defense strategy: Spinner dolphins feed mostly at night, Spotted dolphins feed mostly during the day. Each species helps the other while it is resting by guarding against the danger of surprise attack by large, deepwater sharks.

Deepwater Spinners may cover several thousand kilometers in a few months. It is not known whether their social affinities are as transient as those of their Hawaiian relatives. Perhaps the open-ocean school, which travels together and may number 5,000–10,000 animals, has its coastal equivalent in the population of many of the interchanging groups of the Hawaiian coast.

The open-ocean dolphins also associate with Yellowfin tuna in the tropical Pacific. The tuna may benefit from the excellent echolocation abilities that dolphins use to help find and identify prey. Since the tuna often swim below the dolphins, movements, such as breaking ranks in the face of a shark attack, may be easily detected by the dolphins, so the benefits may be mutual.

In the past four decades, a downside of such associations has emerged for both species. Dolphins surface to breathe, and so are easily spotted by fishermen, who set their nets around dolphin–tuna schools. In the past, many dolphins used to drown after becoming entangled in the tuna nets. Over the past 20 years, tuna fishermen have adopted special nets and practices to reduce the threat to dolphins. A panel of finer mesh is used in that part of the net furthest from the fishing boats, where the fleeing dolphins used to get snared and drown as the net was being tightened (pursed). The dolphins can thus now escape over the net rim, while the tuna usually dive and are retained in the net. However, things can still go wrong, especially if a net canopies above dolphins attempting to get to the surface to breathe.

Helpfully, one further aspect of dolphin behavior has also been recognized. Dolphins caught in tuna nets often lie placidly as if feigning death (although the rigidity may be due to shock). Such dolphins were previously thought to have drowned, and were hauled up onto the deck of the vessel, where they did indeed die. Now the divers who monitor the nets signal to colleagues in small boats stationed around the net to manually help such unmoving animals over the net rim. Although some animals may be missed, the divers do their best to release dolphins entangled in the mesh – a good example of humans attempting to live in harmony with other animals. BW/RSW

Below A school of Hawaiian Spinner dolphins. Following legislation in the USA, deaths of these dolphins in fishing bycatch fell from about 180,000 in 1972 to just over 100 in 1993.

Dolphin Species

DOLPHIN TAXONOMY IS COMPLICATED BY the fact that several genera, such as *Delphinus*, *Stenella*, *Sousa*, and *Sotalia*, include species that are virtually indistinguishable in physical appearance. Controversy remains, but genetic analysis has added weight to the morphological differentiation of at least two species of Common dolphin that live together but have different lengths of beak. Two separate species of spotted dolphin are now recognized – the Pantropical and the Atlantic – while Spinner dolphins have recently been divided into the Long-snouted and the Clymene varieties.

As more studies are conducted on populations inhabiting remote regions of the world, evidence for further species will no doubt come to light, and it is likely that there will be further splits within genera. For example, at present only two species of the genus *Tursiops* are recognized, *T. truncatus* and *T. aduncus*; the latter distinguished both genetically and externally by its spotted undersurface. Within these groupings, however, some inshore and offshore populations show both morphological and genetic differences, and variation between populations in the Indo-Pacific is by no means clear-cut. The nomenclature adopted below follows Rice (1999).

Commerson's dolphin
Cephalorhynchus commersonii
Commerson's or Piebald dolphin

Cool waters of southern S America and Falkland Islands, possibly across Southern Ocean to Kerguelen Island.
HTL 1.3–1.4m; WT c.50kg.
FORM: dark gray on back but with large white-pale gray cape across front half, extending down across belly, leaving only small black area around anus; white area also on throat and chin, so that dark gray frontal region is confined to forehead. Snout and broad band across neck region to flippers. Rounded black flippers and centrally placed low rounded dorsal fin. Short, rounded snout with no melon and very short beak. Small, stout, torpedo-shaped body.

Black dolphin
Cephalorhynchus eutropia
Black, Chilean, or White-bellied dolphin

Coastal waters of Chile.
HTL c.1.6m; WT c.45kg.
FORM: black on back, flanks, and part of belly but with three areas of white, variable in extent, on throat, behind flippers, and around anal area; thin, pale gray or white margin to lips of both jaws; sometimes pale gray area around blowhole. Low triangular dorsal fin, centrally placed with longer leading edge and blunt apex. Short, rounded snout with no melon and very short beak. Small, stout, torpedo-shaped body, with keels above and below tail stock.

Heaviside's dolphin
Cephalorhynchus heavisidii

Coastal waters; W coast of southern Africa.
HTL 1.2–1.7m; WT c.40kg (max. 74kg).
FORM: black on back and flanks, white belly, extending upwards as three lobes, two on either side of the flipper, and one from anal region up along flanks to tail stock. Small, oval-shaped black flippers, and centrally-placed, low, triangular dorsal fin. Short, rounded snout with no melon and no distinct beak. Small, fairly stout, torpedo-shaped body.

Hector's dolphin En
Cephalorhynchus hectori

Coastal waters of New Zealand.
HTL 1.2–1.4m; WT c.40kg.
FORM: pale to dark gray around anus. Rounded black flipper and centrally placed low, rounded dorsal fin. Short, rounded snout with no melon and a short beak. Small, stout, torpedo-shaped body, narrowing at tail stock.
CONSERVATION STATUS: North Island subpopulation Critically Endangered.

Rough-toothed dolphin
Steno bredanensis

Offshore waters of all tropical, subtropical, and warm temperate seas.
Male HTL 2.3–2.4m; WT c.140kg. Female HTL 2.2–2.3m; WT c.120kg.
FORM: dark gray to dark purplish-gray on back and flanks, and white throat and belly; pinkish-white blotches on flanks round to belly; often scarred with white streaks. Centrally-placed, sickle-shaped dorsal fin. Long, slender beak not clearly demarcated from forehead, white/pinkish-white on both sides, including one or both lips and tip of snout. Slim, torpedo-shaped body, keels above and below tail stock.

Atlantic hump-backed dolphin
Sousa teuszii

Coastal waters and river systems of W Africa. Possibly a form of Indo-Pacific hump-backed dolphin which it closely resembles, differing in having fewer teeth and more vertebrae.
HTL about 2.0m; WT c.100kg. Shape and coloration variable.
FORM: dark gray-white on back and upper flanks, lightening on lower flanks to white belly; young uniformly pale cream. Small but prominent triangular, centrally-placed dorsal fin, sickle-shaped in young, becoming more rounded later. Rounded flippers. Long slender beak with slight melon on forehead. Stout, torpedo-shaped body, with distinct dorsal hump in middle of back (on which is dorsal fin) and similar marked keels above and below tail stock.

Indo-Pacific hump-backed dolphin
Sousa chinensis

Coastal warm waters of E Africa to Indonesia and S China.
HTL 2.0–2.8m; WT c. 85kg. Shape and coloration variable.
FORM: dark gray-white on back and upper flanks, usually lightening on lower flanks to white belly; adults may develop spots or speckles of yellow, pink, gray, or brown; young uniformly pale cream. Small but prominent triangular, centrally-placed dorsal fin, sickle-shaped in young, becoming more rounded later. Rounded flippers. Dorsal fin and flippers may be tipped white. Long slender beak (with white patch on tip in some individuals) with slight melon on forehead. Stout, torpedo-shaped body, with distinct dorsal hump in middle of back (on which is dorsal fin) and similar marked keels above and below tail stock.

Indian hump-backed dolphin
Sousa plumbea

Coastal waters; E Africa to Thailand. Size and form as for *S. chinensis*, but coloration darker. Regarded by some as synonym for *S. chinensis*.

Tucuxi
Sotalia fluviatilis

Orinoco and Amazon river systems and coastal waters of NE South America and E Central America.
HTL 1.4–1.8m; WT 36–45kg.
FORM: coloration variable geographically and with age; medium to dark gray on back and upper flanks with brownish tinge, lighter gray sometimes with patches of yellow-ocher on lower flanks and belly; two pale gray areas sometimes extend diagonally upwards on flanks. Coloration lightens with age, sometimes becoming cream-white. Small, triangular, centrally placed dorsal fin. Relatively large spoon-shaped flippers. Pronounced beak (mid to dark gray above, light gray-white belly) and rounded forehead. Small, stout, torpedo-shaped body. Orinoco population (previously recognized as a separate species, *S. guianensis*) generally darker, sometimes with a brownish band exte[nd]ing from anal area diagonally upwards over flanks to leading edge of dorsal fin.

Bottlenose dolphin
Tursiops truncatus

Atlantic and temperate N Pacific. Coastal waters of most tropical, subtropical and temperate regions.
HTL 2.3–3.9m; WT 150–200kg.
FORM: usually dark-gray on back, light gray on flanks (variable in extent), grad[ing] to white or pink on belly. Some spotting may be present on belly. Centrally plac[ed] tall, slender, sickle-shaped dorsal fin. Robust head with distinct, short beak, often with white patch on tip of lower j[aw]. Stout, torpedo-shaped body, with mod[er]ately keeled tail stock.

Indian Ocean bottlenose dolphin
Tursiops aduncus

Indo-Pacific and Red Sea.
HTL 2.3m; WT 150 kg.
Form as for *T. truncatus*, though sometimes darker in coloration. Regarded by some as synonym for *T. truncatus*.

Atlantic spotted dolphin
Stenella frontalis

Subtropical and warm temperate Atlan[tic].
HTL 1.9–2.3m; WT c. 100–110kg.
FORM: coloration and markings variab[le] with age and geographically; dark gray black on back and upper flanks, lighter gray on lower flanks and belly (sometim[es] pinkish on throat); white spots on upp[er] flanks, dark spots on lower flanks and belly, absent at birth but enlarging with age; spotting also decreases away from coasts of N America; distinct dark gray-black area (or cape) on head to dorsal fi[n]; pronounced pale blaze on flanks, slant[ing] up on to back behind dorsal fin. Slende[r] sickle-shaped, centrally-placed dorsal f[in]; pale, medium, or dark gray or pinkish f[lip]pers. Long, slender beak, both upper a[nd] lower lips white or pinkish, and distinc[t] forehead. Slender to relatively stout (in coastal populations) torpedo-shaped b[ody] with marked keel below tail stock (som[e]times also one above tail stock).

ABBREVIATIONS	HTL = head–tail straight–line length WT = weight	Ex Extinct	En Endangered
	Approximate nonmetric equivalents: 1m = 3.3ft; 10kg = 22lb	EW Extinct in the Wild	Vu Vulnerable
		Cr Critically Endangered	LR Lower Risk

Above *A school of Southern right ʼhale dolphins off the coast of Peru. ʼnis species' common name comes ɔm the fact that, like the Right whale, has no dorsal fin.*

ɪntropical spotted dolphin LR
ɫenella attenuata

ɾopical Pacific, Atlantic, and Indian ▪cean. Size and form as for *S. frontalis*, ɹt with black circles around eyes and a ɾoad black stripe from origin of flipper to ɔrner of mouth (which fades as spotting ▪creases), giving a banded appearance to ɪe light gray sides of the head.
ONSERVATION STATUS: LR Conservation ▪ependent.

pinner dolphin LR
ɫenella longirostris

ɾobably in all tropical oceans. At least ɔur different races.
TL 1.7–2.1m; WT c. 75kg.
ORM: dark gray, brown, or black on back; ghter gray, tan, or yellowish-tan flanks, ɪd white belly (purplish or yellow in ɔme populations); distinct black to light ɾay stripe from flipper to eye. Slender ɾect to sickle-shaped centrally-placed dorɪl fin, often lighter gray near middle of n; relatively large black to light gray flipers. Medium to long, slender beak and istinct forehead. Slender to quite stout, ɔrpedo-shaped body, which may have ɪarked keels above and below tail stock.
ONSERVATION STATUS: LR Conservation ɔependent.

Clymene dolphin
Stenella clymene
Clymene or Short-snouted spinner dolphin

Warm temperate to tropical Atlantic.
HTL 1.8–2.1m; WT c. 75kg.
FORM: dark gray back, lighter gray flanks, and white belly. Slender, sickle-shaped dorsal fin; relatively large dark to light gray flippers. Short beak and distinct but sloping forehead. Slender to moderately stout torpedo-shaped body, which may have marked keels below and occasionally above tail stock.

Striped dolphin LR
Stenella coeruleoalba
Striped, Euphrosyne, or Blue-white dolphin

All tropical, subtropical, and warm temperate seas, including Mediterranean.
HTL 2.0–2.4m (male), 1.85–2.25m (female); WT 70–90kg (rarely to 130 kg).
FORM: dark gray to brown or bluish gray on back, lighter gray flanks, and white belly; two distinct black bands on flanks, one from near eye down side of body to anal area (with short secondary stripe originating with this band, turning downwards towards flippers) and the other from eye to flippers; most have additional black or dark gray fingers extending from behind dorsal fin forward and about halfway to eye; black flippers. Slender, sickle-shaped, centrally-placed dorsal fin. Slender, long beak (but shorter than Common dolphin) and distinct forehead. Slender, torpedo-shaped body.
CONSERVATION STATUS: LR Conservation Dependent.

Common dolphin
Delphinus delphis
Common or Saddleback dolphin

All tropical, subtropical, and warm temperate seas, including Mediterranean and Black Seas. Usually offshore waters.
HTL 1.8–2.2m (male), 1.7–2.1m (female); WT 80–110kg (male), 70–100kg (female).
FORM: coloration variable; black or brownish-black on back and upper flanks; chest and belly cream-white to white; hourglass pattern of tan or yellowish tan on flanks becoming paler gray behind dorsal fin where it may reach dorsal surface; black stripe from flipper to middle of lower jaw. and from eye to base of beak; flippers black to light gray or white (Atlantic population). Slender, sickle-shaped to erect dorsal fin, centrally placed. Long, slender beak and distinct forehead; slender, torpedo shaped body.

Long-beaked common dolphin
Delphinus capensis

Eastern N Pacific. Size and form as for *D. delphis*, but with a longer beak and one or two gray lines running longitudinally on lower flanks. Only recently recognized as a separate species from *D. delphis*.

Arabian common dolphin
Delphinus tropicalis

Arabian Sea, Gulf of Aden, and Persian Gulf to the Malabar Coast of India and the South China Sea. Dimensions and form as for *D. capensis*, of which some authorities consider it a separate population.

Fraser's dolphin
Lagenodelphis hosei
Fraser's or Sarawak dolphin

Warm waters of all oceans.
HTL 2.3–2.5m; WT 160–210 kg.
Form: medium-dark gray on back and flanks, white or pinkish-white belly; two parallel stripes on flanks: upper cream-white, beginning above and in front of eye, moving back and narrowing to tail stock; lower more distinct, dark gray-black from eye to anus; sometimes also a black band from mouth to flipper; white throat and chin, but tip of lower jaw usually black. Small, slender, slightly sickle-shaped dorsal fin, pointed at tip, centrally placed. Very short rounded snout with short beak. Fairly robust torpedo-shaped body, with marked keels above and below tail stock.

Peale's dolphin
Lagenorhynchus australis
Peale's or Black-chinned dolphin

Cold waters of Argentina, Chile, and the Falkland Islands.
HTL 2.0–2.2m; WT c. 115kg.
FORM: dark gray-black on back, white belly; light gray area on flanks from behind eye to anus, and, above this, a narrow white band behind the dorsal fin extending backwards, enlarging to tail stock; thin black line running from leading edge of black flipper to eye. Centrally-placed, sickle-shaped fin. Rounded snout with short black beak and torpedo-shaped body.

White-beaked dolphin
Lagenorhynchus albirostris

Temperate and subpolar waters of N Atlantic, mainly on continental shelf.
HTL 2.5–2.8m (male), 2.4–2.7m (female); WT 300–350kg (male), 250–300kg (female).
FORM: dark gray or black over most of back, but pale gray-white area over dorsal surface behind fin (less distinct in young individuals); commonly dark gray-white blaze from near dorsal surface, behind eye, across flanks and downward to anal area; white belly. Centrally-placed, tall (particularly in adult males), sickle-shaped dorsal fin. Rounded snout and short beak, often light gray or white. Very stout torpedo-shaped body, with very thick tail stock.

Atlantic white-sided dolphin
Lagenorhynchus acutus

Temperate and subpolar waters of N Atlantic, mainly along shelf edge and beyond.
HTL 2.1–2.6m (male), 2.1–2.4m (female); WT 215–234kg (male), 165–182kg (female).
SKIN: black on back, white belly, gray flanks but with long white oval blaze from below dorsal fin to area above anus; directly above but originating slightly behind the front edge of white blaze is an elongated yellow band extending back to tail stock. Centrally-placed, sickle-shaped dorsal fin, relatively tall, pointed at tip. Rounded snout with short, black beak; stout, torpedo-shaped body with very thick tail stock (particularly in adult males), narrowing close to tail flukes. Recently reassigned by some authorities to a separate genus, *Leucopleurus*.

Dusky dolphin
Lagenorhynchus obscurus

Circumpolar in temperate waters of S Hemisphere.
HTL 1.8–2.0m; WT c. 115kg.
FORM: dark gray-black on back, white belly; large gray area (varying in intensity) on lower flanks. extending from base of beak or eye backwards and running to anus; light gray or white areas on upper flanks extending backwards from below dorsal fin as two blazes which generally meet above anal region and end at tail stock. Centrally-placed, sickle-shaped dorsal fin, slightly more erect and less curved than rest of genus, commonly pale gray on posterior part of fin. Rounded snout with very short black beak and torpedo-shaped body.

Pacific white-sided dolphin
Lagenorhynchus obliquidens

Temperate waters of N Pacific.
HTL 1.9–2.0m; WT about 150kg; male slightly larger than female.
FORM: dark gray or black on back, white belly, large, pale gray oval area on otherwise black flanks in front of fin above flipper and extending forward to eye, which is encircled with dark gray or black; narrow, pale gray stripe above eye running along length of body and curving down to anal area where it broadens out; pale gray blaze also sometimes present on posterior part of centrally-placed, sickle-shaped dorsal fin. Rounded snout with very short black beak and torpedo-shaped body.

Hourglass dolphin
Lagenorhynchus cruciger

Probably circumpolar in cooler waters of Southern Ocean.
HTL 1.6–1.8m; WT c. 100kg.
FORM: black on back, white belly, two large white areas on black flanks forward of dorsal fin to black beak and backward to tail stock, connected by narrow white band; area of white variable in extent. Centrally-placed, sickle-shaped dorsal fin, usually strongly concave on leading edge. Rounded snout with very short black beak and torpedo-shaped body.

Northern right whale dolphin
Lissodelphis borealis

Offshore waters of temperate N Pacific.
HTL 2.1–3.1m; WT c. 70kg.
SKIN: black on back and flanks, extending down around navel; white belly, in some individuals extending up flanks around flipper so that only the tips are black; otherwise, flippers all black. Juveniles light gray to brown on back and flanks. Dorsal fin absent. Rounded snout with distinct beak and white band across bottom of lower jaw. Small, slender, torpedo-shaped body, with marked keel above tail stock.

Southern right whale dolphin
Lissodelphis peronii

Offshore waters of Southern Ocean, possibly circumpolar.
HTL 1.8–2.3m; WT c. 60kg.
FORM: black on back and flanks, white belly extending upwards to lower flanks behind flippers and forward across forehead in front of eyes so that entire back is white; flippers all white. Dorsal fin absent. Rounded snout with distinct beak. Small, very slender torpedo-shaped body, with underside of tail fluke white.

Risso's dolphin
Grampus griseus
Risso's or Gray dolphin

All tropical and temperate seas.
HTL 3.5–4.0m (male), 3.3–3.5m (female); WT c.40kg (male), c.35kg (female).
FORM: dark to light gray on back and flanks, palest in older individuals, especially leading edge of dorsal fin, so that head may be pure white; many scars on flanks of adults; white belly enlarging to oval patch on chest and chin. Long, pointed black flippers; tall, centrally-placed, sickle-shaped dorsal fin (taller and more erect in adult males). Blunt snout, rounded with slight melon. No beak. Stout, torpedo-shaped body, narrowing behind dorsal fin to quite narrow tail stock.

Melon-headed whale
Peponocephala electra
Melon-headed whale or Many-toothed blackfish

Probably all tropical seas.
HTL 2.3–2.7m (males slightly larger than females); WT c. 160kg.
FORM: black on back and flanks; slightly lighter on belly; areas around anus, genitals, and lips pale gray or white. Pointed flippers. Sickle-shaped, centrally-placed dorsal fin. Rounded head (though snout slightly more pointed than in *Feresa attenuata*) with slightly underslung jaw, and slender body with slim tail stock.

Killer whale **LR**
Orcinus orca
Killer whale or orca

All oceans.
HTL 6.7–7.0m (male), 5.5–6.5m (female); WT 4,000–4,500kg (male), 2,500–3,000kg (female).
FORM: black on back and sides, white belly extending up the flanks and a white oval patch above and behind the eye; indistinct gray saddle over back behind dorsal fin. Rounded, paddle-shaped flippers and centrally placed dorsal fin, sickle-shaped in female and immatures, but very tall and erect in male. Broad, rounded head and stout, torpedo-shaped body.
CONSERVATION STATUS: LR Conservation Dependent.

Pygmy killer whale
Feresa attenuata

Probably all tropical and subtropical seas.
HTL 2.3–2.7m (male), 2.1–2.7m (female); WT 150–170kg (male and female).
SKIN: dark gray or black on back, often lighter on flanks; small but conspicuous white zone on underside from anus to tail stock and around lips; chin may be entirely white. Flippers slightly rounded at tip. Sickle-shaped, centrally-placed dorsal fin. Rounded head with underslung jaw; slender body.

False killer whale
Pseudorca crassidens

All oceans; mainly tropical and warm temperate.
HTL 5.0–5.5m (male), 4.0–4.5m (female); WT c.2,000kg (male), c.1,200kg (female).
FORM: all black except for a blaze of gray on belly between the flippers, which have a broad hump on the front margin near middle of flipper. Tall, sickle-shaped dorsal fin just behind midpoint of back, sometimes pointed. Slender, tapered head, underslung jaw, and long, slender body.

Long-finned pilot whale
Globicephala melas
Long-finned pilot whale or Pothead whale

G. m. melas temperate waters of N Atlantic; *G. m. edwardi* all waters of all seas in S Hemisphere.
HTL 5.5–6.2m (male), 3.8–5.4m (female); WT c.3,000–3,500kg (male), 1,800–2,500kg (female).
FORM: black on back and flanks, with anchor-shaped patch of grayish-white on chin and gray area on belly, both variable in extent and intensity (lighter in younger animals). Some have gray dorsal fin. Long sickle-shaped flippers, and fairly low dorsal fin slightly forward of midpoint, with long base, sickle-shaped (in adult females and immatures) to flag-shaped (in adult males). Square, bulbous head, particularly in old males, with slightly protruding upper lip and robust body.

Short-finned pilot whale **LR**
Globicephala macrorhynchus

All tropical and subtropical waters but with possible separate form in N Pacific.
HTL 4.5–5.5m (male), 3.3–3.6m (female); WT c.2,500kg (male), c.1,300kg (female).
FORM: black on back, flanks, and most of belly, with anchor-shaped gray patch on chin and gray area on belly (lighter in younger animals). Long, sickle-shaped flippers, and fairly low dorsal fin, slightly forward of midpoint, with long base, sickle- to flag-shaped. Square, bulbous head especially in old males, with slightly protruding upper lip and robust body.
CONSERVATION STATUS: LR Conservation Dependent.

Irrawaddy dolphin
Orcaella brevirostris

Coastal waters from Bay of Bengal to N coast of Australia.
HTL 2–2.5m; WT c.100kg.
FORM: blue-gray on back and flanks, lighter gray on belly. Stout, torpedo-shaped body and tail stock; robust, rounded head with distinct melon but no beak; small, sickle-shaped dorsal fin with rounded tip, slightly behind centre of back. Sometimes regarded as a member of the Monodontidae family.

SPECIAL FEATURE

A Dolphin's Day

Moods of the Dusky dolphin

THE SLEEK, STREAMLINED DOLPHINS WERE LEAPING around the Zodiac rubber inflatable at Golfo San José, off the coast of southern Argentina. When the divers entered the cool water, a group of 15 dolphins repeatedly cavorted under and above them, approaching the humans to within an arm's length and showing no fear of these strangers from another world.

These were Dusky dolphins, whose playful behavior indicated that they had recently been feeding and socializing, for Dusky dolphins have different moods, and will not interact with humans when they are hungry or tired.

Season and time of day also affect their behavior. These small, round dolphins feed on Southern anchovy during summer afternoons. Nighttime is spent in small schools of 6–15 animals not more than about 1km (0.6 mi) offshore. When danger approaches in the form of large sharks or killer whales, they retreat close inshore to hide in the turbulence of the surfline.

In the morning, the dolphins begin to move into deeper water 2–10km (1–6 mi) out, line abreast, each animal 10m (33ft) or more from the next, so that 15 dolphins may cover a swath of sea 150m (500ft) or more wide. They use echolocation to find food, and, because they are spread out, they can sweep a large area of ocean. When a group locates a school of anchovy, individuals dive down and physically herd it to the surface by swimming around and under the fish in an ever-tightening formation.

The marine birds that gather above the anchovy to feed, and the leaping of the dolphins around the periphery of the school, indicate what is going on to human observers as far as 10km (6 miles) away. Other small groups of dolphins, equally distant, will also see such activity, and move rapidly toward it.

The newly-arrived dolphin groups are immediately incorporated into the activity; the more dolphins present, the more efficiently they are able to corral and herd prey to the surface. Five to ten dolphins cannot effectively herd prey and so give up quickly, but groups of 50 have been seen to feed on average for 27 minutes. By mid-afternoon groups of as many as 300 dolphins (normally scattered in 20 to 30 small groups covering a total area of about 1,300sq km/500 square miles) may come together to feed for 2–3 hours. There is much social interaction in such a large group, and considerable sexual activity, particularly toward the end of feeding. Mixing in large groups allows individuals a wider choice of mates, and thus avoids the problems of inbreeding. Both males and females will mate with more than one partner, but females do seem to be particular about the partners they will allow close.

Play must wait for resting and feeding to end, but it is nonetheless a crucial activity. In order for dolphins to function effectively while avoiding predators, hunting for food, and cooperatively herding prey, they must know each other well and communicate efficiently. Socializing helps to bring this about. Toward the end of feeding, they swim together in small, ever-changing subgroups, with individuals touching or caressing each other with their flippers, swimming belly to belly, and poking their noses at each others' sides or bellies. At this time, the dolphins will readily approach a boat, ride on its bow wave if it is moving, and swim with divers in the water.

In the evening, the large school splits into many small groups once again, and the animals settle down near shore to rest, the mood changing abruptly to quiescence once again. Although there is some interchange of individuals between small groups from day to day, some of the same dolphins travel together on subsequent days. Indeed, some Dusky dolphins have been observed to stay together for as long as 12 years.

Occasionally in summer, and for much of the winter, anchovy are not present, so Dusky dolphins feed in small groups on squid and bottom-dwelling fish, mainly at night. Such prey does not occur in large shoals, so the feeding dolphins stay in small groups and remain in somber mood. Winter and hunger do not make for a playful dolphin, but in summertime the living is easy and the dolphins are high! BW

Above *A Dusky dolphin performing an exuberant leap. After feeding on shoals of anchovy, Dusky dolphins habitually leap acrobatically and play together. This is perhaps the most important time for these highly social animals, since play reinforces the bonds between individuals that make them an effective hunting unit.*

Left *A group of Dusky dolphins in close formation. When hunting for anchovy, the dolphins will space themselves more widely apart, in order to sweep the maximum area of ocean with their echolocating capacity.*

How Dolphins Keep in Touch

Acoustic communication in the undersea environment

PLACE A HYDROPHONE IN THE SEA CLOSE TO A group of dolphins and you will almost certainly hear a rich variety of clicks, whistles, and cries. Dolphins are among the most vocal of mammals: they have a well-developed echolocation system for which they produce the high-frequency clicks, and they also communicate acoustically, primarily using tonal sounds such as whistles.

The importance of vocal communication for this group is explained by the animals' biology and by the physical characteristics of the medium in which they live. Most dolphins are highly social animals, interacting and coordinating their behavior with many different individuals as well as maintaining longterm relationships. They also spend their lives moving rapidly though an extensive, featureless environment that transmits light poorly – the visual range is typically of the order of tens of meters or less – but through which sound is conducted more efficiently than any other form of energy. To maintain the social organization that is so important for their survival, dolphins need to communicate over considerable distances, and sound is the only efficient means for them to achieve this.

Some of the most characteristic dolphin vocalizations are narrow-band whistles. All dolphins, with the exception of a few inshore species whose social organization is unknown, produce whistles. Some whistles have characteristic, distinctive patterns of frequency modulation that are unique to individual animals; they sweep up and down the frequency scale in a specific way. These are called "signature whistles," and they seem to serve as individual identification signals.

Signature whistles are usually produced by dolphins when they are out of visual contact with other members of their group, and they may function primarily as contact calls. Dolphins are adept at vocal learning, an ability that is relatively uncommon in mammals. A dolphin may imitate the signature whistles of other individuals in its group, perhaps as a way of gaining their attention. A dolphin establishes its own signature whistle by the time it is 2 years of age, and it then remains fixed. Males may copy the whistles of their mothers, which serves to prevent incest.

Acoustic identification by humans of individuals or groups has been possible in several dolphin species and seems to occur at the level that makes most sense biologically. For example, the smaller dolphins typically exhibit a complex fission–fusion type of social organization, with the composition of groups changing from day to day, which makes it important to develop an ability to recognize individuals and to behave appropriately to them, based on past experience. The social structure of the largest member of the dolphin family, the Killer whale, however, is quite different. It lives in extremely stable social groups, and in this species it is the groups, or pods, that have characteristic vocalizations. Each pod has a unique dialect (a particular repertoire of calls), and the degree of similarity in dialects between pods generally reflects their genetic relatedness and the extent to which they spend time together.

Relatively little has been discovered about the way that dolphins use sound to communicate in the wild. However, specific calls that may coordinate cooperative feeding have been identified in some species, and correlations have also been demonstrated between activity states and the sort of vocalizations that dolphin groups produce.

In captivity, dolphins have readily learned a variety of artificial languages. Controversially, some studies have suggested that they use rules of syntax and grammar to understand novel sentences. This capacity suggests they may have evolved a sophisticated communication system for use in the wild.

Another intriguing aspect of dolphin communication is that they seem able to eavesdrop on the echolocation calls of other dolphins. At the very least, dolphins can probably "hear" how successfully other individuals within auditory range (which may extend to 1km/0.6mi or more) are feeding. Recent work with captive animals has shown that at shorter ranges dolphins can actually "analyze" the echoes from the echolocation clicks of other individuals, using sonar in a bistatic mode. Dolphins can also produce sound by nonvocal means, for example when they slap the water with their flukes or crash back down after a leap.

Although acoustics is undoubtedly the dominant sense used by these animals, other senses, such as vision, touch, and taste, are also important. Vision, for example, may be useful for coordinating movements at short range, and this may be responsible for the bold patterns and coloration on the heads and flanks of many dolphins. Characteristic body postures and movements, signifying behavioral states such as threat or submissiveness, have also been identified in both captive and wild dolphins. Dolphins have no sense of smell, but they are able to taste chemicals in the seawater through which they and their companions swim. Chemicals released in urine and feces may be an important way of communicating breeding condition or feeding success. JG

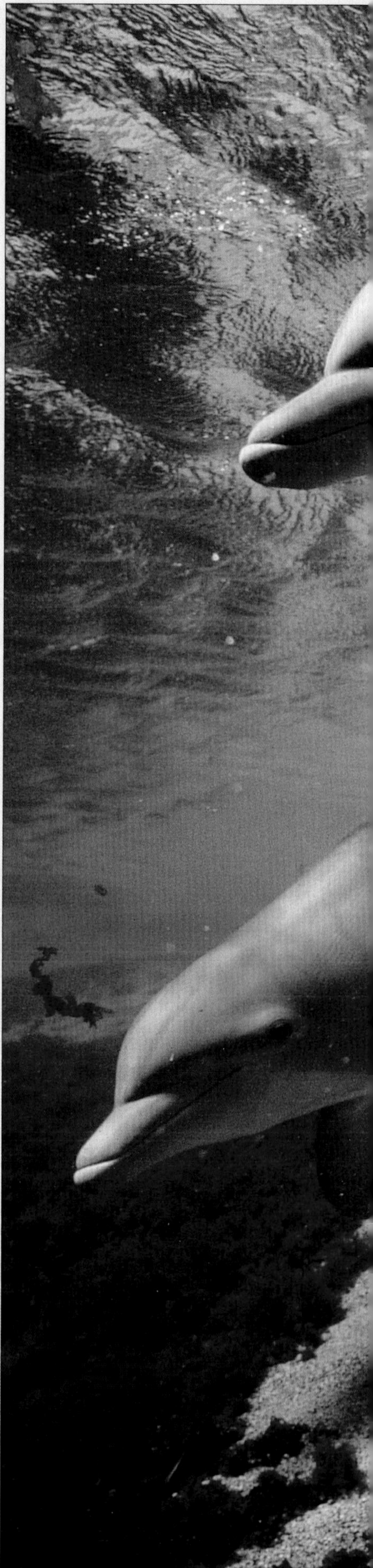

Right *Sociable and intelligent, dolphins have developed complex auditory signals to keep in touch in an undersea world in which sound is the best medium for long-distance communication.*

COOPERATIVE KILLERS

The hunting strategies of Killer whales

AS THEY APPROACH THE ROCKY POINT, AROUND which the tide is flowing rapidly toward them, the pod of 20 Killer whales is spread in line abreast, each one about 50m (160ft) apart. The whales are swimming slowly near the surface, occasionally rising to breathe and slap the surface with their long, oval flippers and large tail flukes. Underwater, the slaps sound like muffled gunshots against a background of squeals and ratchetlike clicking sounds. Then comes a long, wavering whistle, punctuated by a honk like that from a squeeze-bulb horn at an Indian bazaar, and the whales converge methodically on their prey – a school of several thousand Pacific pink salmon they are herding between the rock and the roaring current. For several minutes the whales have the fish loosely but effectively trapped. One by one, they pick out and swallow several 3kg (6.6lb) salmon from the periphery of the school. Then the whales seem to lose interest in the hunt and instead roll lazily in the water and casually "spy-hop" to look around at the boatfuls of salmon anglers floating near the point. With another underwater whistle and a honk, all the whales simultaneously submerge, to reappear five minutes later in a close-knit group beyond the point, upcurrent and far away from the fishermen. They remain in a close group, swimming slowly and silently for two hours toward the next rocky point, where the cooperative hunt is repeated.

Killer whales, or orcas, are the largest members of the dolphin family. The adult males, which may attain 9m (29ft) in length, are recognizable by their upright dorsal fins, up to 2m (6.6ft) tall – the largest of any whale. Females are slightly smaller, with the dorsal fin typically about 70cm (2ft) tall. The shape of the fin varies between individuals and populations as a consequence both of damage and of genetic influences.

The body pigmentation pattern is strikingly black on the back and white on the belly, with a white patch above the eye and a gray "saddle-patch" on the back behind and below the dorsal fin. The variations in fin shape and saddle pattern have permitted individual whales to be indentified and studied across many different regions of the world, and this, along with DNA evidence, has given us an unusally detailed insight into the orca's underwater world.

Killer whale pods are comprised of mothers and their offspring, which remain together for generations. The adult males in the pod are typically sons, brothers, and uncles of the other pod members, not harem bulls as was once believed. Killers look for mates outside of their own family or matriline. The cohesion of these family groups over very long periods of time, in conjunction with changing prey distributions, has caused the development of ecotypes or groups of specialized whales adept at catching particular types of prey.

Killer whales are capable of eating a wide range of prey, but pods tend to feed mainly on locally abundant resources. This can affect the dynamics of hunting, and alters the optimum pod size and even the body shape of the whales themselves. The so-called "transient" Killer whales of North America, which feed on seals and other marine mammals, travel in small groups (average size three), often capture prey individually, and are physically larger than the piscivorous "resident" Killer whales that specialize in salmon. The herring eating Killer whales of Norway often forage in very large groups, with many pods represented in the coordinated "carousel" encirclement of tens of thousands of herring, separated off from many millions in the fish school. The solitary whale that slides out on a beach in Argentina to grab a baby seal is yet another ecotype.

Left *A group of Killer whales converges on a shoal of Pacific salmon, herding them by a combination of effective cooperation and underwater sounds. When trapped between the shore and the pod, the salmon are picked off one by one until the Killers' hunger is satiated.*

Killer whale females usually become sexually mature in their early teens, and they live from 50 to 100 years. Males mature later and die a little

unger. An adult female can bear a single calf ce every three years from her teens to about age . Gestation lasts for 15–17 months, and nursing proximately one year. Females give birth to out five viable calves in their reproductive life- ne, although obviously not all young survive to aturity. After the age of 40, the female whale ay have social status as babysitter and teacher to ung whales in the pod.

Some Killer whale pods may travel many hun- eds of kilometers to keep up with the move- ents of prey species, while others maintain atively restricted ranges where food is abundant year. As top predators, Killer whales are not numerous, but the appearance of several pods of these mobile hunters in an area where prey is locally or seasonally abundant can give the false impression that their population is quite large.

A distinct disadvantage of being at the top of the food chain in today's world is that contaminants may bioaccumulate in prey species and ultimately affect the predator itself. In the Pacific Northwest of North America, both the resident and "transient" Killer whales have been found to have some of the highest tissue PCB (polychlorinated biphenyl) levels in the world, and there is a corresponding decrease in reproduction rates and in the general survivability of their populations. KB

Above *A pod of Killer whales surfaces close to shore in the San Juan Islands, off the coast of Washington on America's Pacific littoral. Such groups are typically composed of mothers and their offspring, though here the tall dorsal fin in the foreground indicates the presence of an adult male – perhaps a brother or uncle of the other whales.*

STORY

BIRTH OF A DOLPHIN

❶ *One hour before giving birth a heavily pregnant Bottlenose dolphin* (Tursiops truncatus) *swims in the waters of an Italian dolphinarium. Like all cetaceans, dolphins carry their young for a long time, 12–14 months in the case of Bottlenoses. Single calves are usual with fewer than 1 percent of births involving twins, and there are no records of two calves being reared successfully. In cool European waters, Bottlenose dolphins only give birth in the midsummer months, but in warmer waters this can be all year round.*

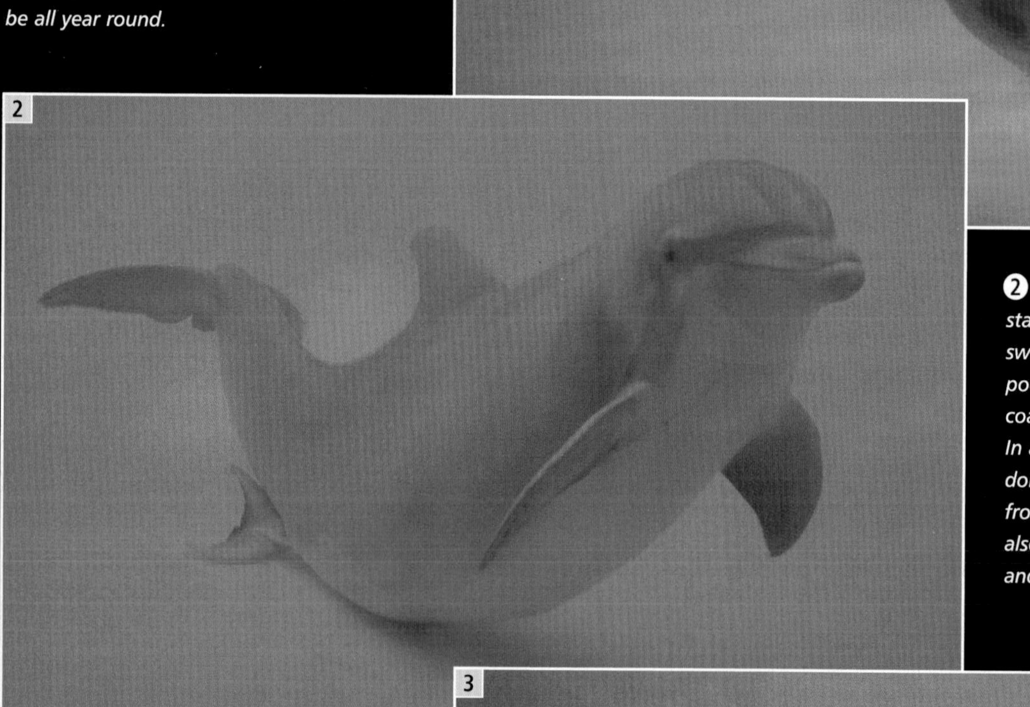

❷ *Half an hour before delivery, the calf has started to emerge and the mother stops swimming and rests on the bottom of the pool. In the wild, dolphins choose warm coastal shallows as favorite calving spots. In attendance will be one or more female dolphin "helpers," who protect the mother from predators. After the birth they will also help to lift the calf to the surface and nurture it.*

❸ *Calves are born tailfirst, pushed backward by the mother's powerful muscular contractions. Calf and mother are linked by a short umbilical cord, which has built-in weak spots that allow it to break spontaneously after the birth. Backward birth is rare among terrestrial mammals, but is advantageous to cetaceans, protecting the young for the longest possible time from the airless underwater environment. This tailfirst delivery gave rise to various legends; some seafarers claimed that young calves swam back into their mothers' wombs for protection, while others contended that they extruded their tails before birth in order to practice swimming.*

5

4 Seconds after birth the newborn calf breaks free as the umbilical chord is pulled taut and breaks. A stream of blood and amniotic fluid clouds the water behind the mother. At birth, dolphin calves measure 100–135cm (39–53in) long and weigh 10–20kg (22–44lb), about 10 percent of their mothers' weight.

5 The calf takes its first breaths within seconds of the birth, making its way instinctively toward the surface to do so. Even so, the mother will help it on its way, gently nudging it upward with her snout or a flipper. Though initially unsteady in the water, it will be able to swim well within about half an hour.

6

4

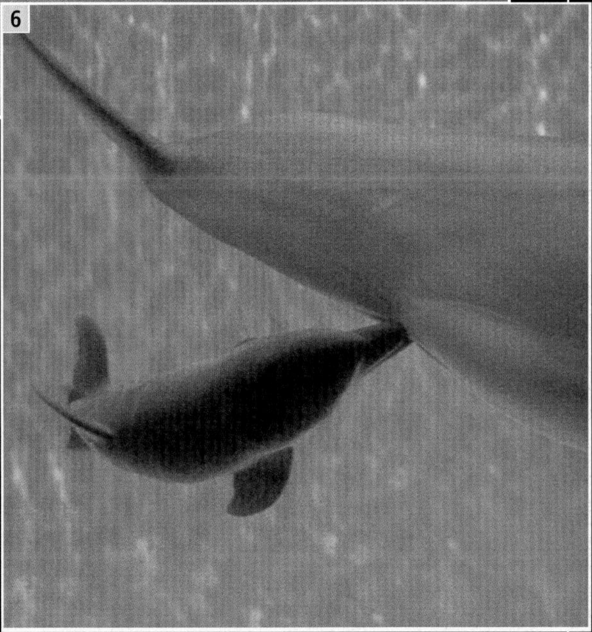

6 Two hours after birth, the calf suckles its mother's fat-rich milk from one of the two nipples situated on either side of the genital slit. Lactation lasts for 18–20 months in all, although some solid food will be taken from six months. The mother–calf bond lasts for 3–6 years and sexual maturity is reached at around 5–14 years for females and 9–13 years for males. Calves are normally born at two-year intervals.

Porpoises

MANY PEOPLE ON THE COASTS OF NORTH America and northern Europe get their first sight of live cetaceans in the shape of porpoises – typically Phocoena phocoena, *the Harbor porpoise, glimpsed as a small, elusive, dark object from a ferry or a vantage point on shore. Yet we still have a great deal to learn about these coastal animals – a matter of concern, as their numbers may be declining rapidly.*

The true porpoises comprise the Phocoenidae, one of ten families of the suborder Odontoceti, the toothed whales. The family is made up of six species, unified by their morphology and small size. Phocoenids are closely related to the true dolphins of the family Delphinidae, but porpoises and dolphins have separate ancestries – that is, they are phylogenetically distinct. Both families evolved from a common ancestor about 10 million years ago, but since then have evolved significant differences in many aspects of their biology. In terms of their behavior and anatomy, porpoises and dolphins are as different as cats and dogs.

Small, Plump, and Wet

FORM AND FUNCTION

In anatomy, the true porpoises are a rather uniform group. Compared to other cetaceans, they are very small; no member of the family exceeds 2.5m (8ft) in length. Their small size presents porpoises with a considerable thermal challenge – how to stay warm in a cold, highly conductive environment. Species inhabiting cooler parts of the world, such as the Harbor and Dall's porpoises, have met this challenge with a rounded shape and small extremities, which minimize their surface area, and a thick, highly insulating layer of blubber that reduces heat loss.

Like other toothed whales, porpoises breathe through a single blowhole positioned slightly to the left of the midline of the skull. Their foreheads are characterized by a melon – a lipid-rich tissue lying just above the anterior portion of the cranium (the domed forehead) that focuses the sound used in echolocation. Their tail stocks are laterally compressed into keels, which reach an extreme in the Dall's porpoise, and their flukes are separated by a distinct notch.

Porpoises are different from dolphins. They a lack the rostrum, or beak, that is characteristic c most dolphins, and all – with the exception of t Finless porpoise – have small, triangular fins. Al but the Dall's porpoise possess several rows of functionally mysterious skin swellings on the leading edge of their dorsal fins (or dorsal ridge the case of the Finless porpoise). Porpoise teeth are laterally compressed and spatulate, or spade shaped, in contrast to the conical teeth of dolphins. Both dolphins and porpoises use their

Right *As is the case with dolphins, contact with fishing enterprises can be dangerous for porpoises. Here researchers tend a Harbor porpoise that has be rescued from a herring weir in the Bay of Fundy, Canada. Accidental capture is responsible for the deaths of thousands of porpoises each year.*

Below *A Harbor porpoise speeding along the surface off Shetland, Scotland. Sleek and fast when swimming in the open sea, this species is, however, more prone than others to stranding, which they us ally do singly on the sloping shelves of sandy beache or on mudflats.*

th to grasp prey, rather than for shearing or ewing, but it is not known why the two families ve evolved different dental morphologies. Por- ise skulls have prominent "bosses" – protruber- ces – on the premaxillae. The heads of adults ture many apparently juvenile characteristics, ch as short rostra, large, rounded braincases, d the delayed fusion of cranial sutures. Paedo- orphosis, the retention of juvenile features in adult form, is common to all porpoises, but reasons for it remain a mystery.

Rivers, Coasts, and Open Oceans

DISTRIBUTION PATTERNS

e oldest fossils of porpoises date from the late ocene (about 10–12 million years ago), and are und in the North Pacific, so this may be the dle of this group of animals. The first true dol- ins appeared at about the same time, and the ogenitors of both families – the Kentriodontidae disappeared soon after. Porpoiselike fossils have en found in late Miocene strata of southern lifornia, but in Europe and the Southern Hemi- here phocoenid fossils occur only from the ocene, 4 or 5 million years ago.

The six species of modern porpoise appeared a v million years ago, but in that relatively short ace of time they have evolved to take advantage a number of different environments. The Fin- s porpoise ranges along the tropical coasts of Indo-Pacific region and is found in estuaries as ll as in the heart of major river systems, includ- g the Yangtze in China. Harbor porpoises, rmeister's porpoises, and vaquitas are primarily coastal species; the latter has the most limited geographical range of any cetacean, being entirely restricted to the northern part of the Gulf of California. The Dall's and Spectacled porpoises are creatures of the open ocean, found respectively throughout the North Pacific and the circumpolar Southern Ocean.

The geographical locations of many cetacean species has been heavily influenced by the periodic changes in temperature that occur in the world's oceans. This has caused closely related species of porpoise to become separated by warmer waters on either side of the tropics, so that matched, closely related species exist in both the northern and southern hemispheres. Such species are said to exhibit antitropical distributions, because recent forms now separated in the temperate waters of the two hemispheres originated from a common ancestor that crossed the equator during cool periods. The vaquita, for example, is more closely related to the Burmeister's porpoise of South American waters than to the Harbor porpoise that is found in the nearby coastal waters off California. It is likely that porpoises first colonized the Gulf of California during a cool glacial period during the Pleistocene, when the ancestors of Burmeister's porpoise crossed the equator only to become isolated once the waters warmed again. Similarly, the isolation of Harbor porpoises in the Black Sea probably occurred during a warming period, after porpoises initially penetrated the Mediterranean from the Atlantic during a cool period. Porpoises are absent from the Mediterranean today.

FACTFILE

PORPOISES

Order: Cetacea

Family: Phocoenidae

6 species in 3 genera

DISTRIBUTION Most major oceans

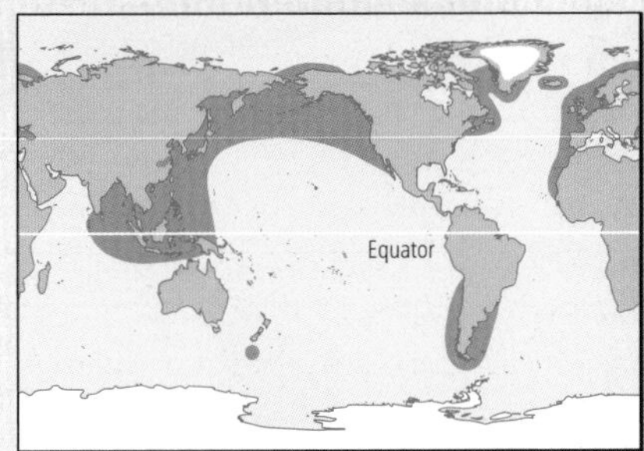

HARBOR PORPOISE *Phocoena phocoena*
Coastal temperate N Atlantic and N Pacific, Baltic, and Black Sea. HTL 1.4–2.0m (4.5–7ft); WT 40–80kg (88–176lb). Skin: dark gray cape with mottled lateral surfaces, subtle eye patch, distinct flipper stripes, dark chin patch, white abdomen; triangular dorsal fin. 20–29 teeth in each row. Conservation status: Vulnerable.

BURMEISTER'S PORPOISE *Phocoena spinipinnis*
Coastal S America from Peru to S Brazil. HTL 1.6–2.0m (5–7ft); WT 60–100kg (132–220lb). Skin: uniformly dark gray dorsal and lateral surfaces, subtle dark eye patch and flipper stripe, light gray abdomen; low dorsal fin placed far back on body. 11–25 teeth in each row.

VAQUITA *Phocoena sinus*
Upper Gulf of California. HTL 1.3–1.5m (4–5ft); WT 35–50kg (77–110lb). Skin: dark gray cape, pale gray lateral field, distinctive dark eye ring, lip patches, and flipper stripe, white abdomen; tall, triangular dorsal fin. 16–22 teeth in each row. Conservation status: Critically Endangered.

SPECTACLED PORPOISE *Phocoena (Australophocaena) dioptrica*
Southern Ocean. HTL 1.8–2.3m (6–7.5ft); WT 100–180kg (220–397lb). Skin: black dorsal and lateral surfaces, sharply demarcated from white abdomen, black eye patch surrounded by thin white line, black lips; dorsal fin triangular and dimorphic, larger in mature males. 16–25 teeth in each row.

DALL'S PORPOISE *Phocoenoides dalli*
N Pacific Ocean. HTL 1.7–2.4m (5.5–8ft); WT 100–220kg (220–485lb). Skin: dorsal and lateral surfaces black, white lateral blaze extending from urogenital region to level of dorsal fin or to flippers, light frosting on flukes and dorsal fin; dorsal fin slightly falcate and dimorphic, canted forward in mature males. 21–28 teeth in each row. Conservation status: Lower Risk – Conservation Dependent).

FINLESS PORPOISE *Neophocaena phocaenoides*
Coastal Indo-Pacific waters from the Persian Gulf to Indonesia and N to Japan. HTL 1.4–2.0 m (4.5–6.5ft); WT 30–80kg (66–176lb). Skin: light gray, with pale abdomen; lacks dorsal fin. 12–23 teeth in each row.

Abbreviations HTL = head-to-tail length WT = weight

Below *The six species of porpoise:* **1** *Vaquita* (Phocoena sinus)*;* **2** *Burmeister's porpoise* (Phocoena spinipinnis)*;* **3** *Finless porpoise* (Neophocaena phocaenoides)*;* **4** *Dall's porpoise* (Phocoenoides dalli)*;* **5** *Spectacled porpoise* (Phocoena dioptrica)*;* **6** *Harbor porpoise* (Phocoena phocoena).

Echolocating Prey

DIET

All porpoise species feed principally on small, schooling fish. Harbor porpoises and Burmeister's porpoises take lipid-rich herring, anchovies, and capelin, supplementing this oily diet with small, demersal prey living close to the seabed, such as juvenile cod and similar fish. Harbor porpoises are capable of diving to depths of more than 200m (650ft) in search of their prey. Young harbor porpoises learn to forage independently by taking euphausiids, the prey of the fish pursued by their mothers. Finless porpoises frequently also eat crustaceans and squid. Dall's porpoises exploit small fish and squid that comprise the "deep scattering layer," an aggregation of small animals that moves up and down in the water column each day in response to changes in light levels. This species may feed most intensively at night, when their prey migrate vertically toward the surface. With porpoises, there is no evidence for cooperative foraging of the kind observed in many dolphin species. When prey density is low, porpoises feed far apart, but they can assemble rapidly as soon as large concentrations of prey are located.

Sound production has been studied only in the Harbor porpoise. Harbor porpoises produce sequences of clicks at two major frequencies: one near 2 kHz, and the other at around 130 kHz. The use of echolocation in prey detection and capture is not well understood, although it seems hunting also involves passive listening (for the noises made by fish) and looking. The prey are not chewed but swallowed whole.

Captive porpoises are often maintained on daily rations of 5–8 percent of body mass, although the amount varies depending on the caloric density of the food and the activity level and reproductive status of the porpoise.

Ghosts of the Sea

SOCIAL BEHAVIOR

In general, porpoises are shy, unobtrusive animals that occur alone or in small groups. Most phocoenids are difficult to sight, let alone follow, except in the calmest conditions, and many coastal residents are unaware of their presence. The only species readily attracted to moving vessels is the Dall's porpoise, a fast and boisterous swimmer that generates fans of spray, known as "rooster tails," visible for hundreds of meters. Harbor and Finless porpoises are much more subtle in their surface behavior, although the former may make horizontal leaps partly clear of the surface in rough seas or when chasing prey. Harbor porpoises occasionally lie motionless at the surface on glassy calm days. Vaquitas are particularly sh animals, and very few researchers have ever seen the species alive in the field.

Porpoises live in fission–fusion societies, in which animals come together when necessary b there is no evidence of long-term social bonds. Much, however, remains to be discovered about their social life. The only lasting association occ between a nursing female and her calf. Mothers with calves may band together, but are often seg gated from other animals. The observed presenc of small Harbor porpoises with mother–calf pair early in the summer may indicate that some old calves stay near their mothers for short periods after weaning. No species of porpoise is known strand en masse, like the delphinids and Sperm whales, which form stable, longterm association

Harbor porpoise females are larger than male an unusual feature among mammals, though les so among cetaceans. Male Harbor porpoises pos sess extremely large testes, which reach 4 percen of total body mass during the breeding season b ress considerably at other periods. This suggests that sperm competition, in which several males mate with a single female and the battle for paternity is fought within the female's reproductive tracts, may play an important role in the reproduction of this species. Many porpoises exhibit considerable reproductive seasonality, in which ovulation, conception, and parturition occur during brief intervals of a few weeks, usually in the late spring or early summer.

Observations of live and stranded Harbor porpoises in the Moray Firth, Scotland, indicate that porpoises are occasionally attacked and killed by Bottlenose dolphins; similar findings have also been reported from North America. The reasons for this unusual behavior are unclear.

The Threat from Fishing Nets

CONSERVATION AND ENVIRONMENT

Coastal areas are as attractive to porpoises as they are to humans, and for many of the same reasons. Our seaside activities, however, are causing many problems for the world's cetaceans.

Many porpoises live in waters used intensively by humans and are affected adversely by certain human activities. For example, the habitat of Finless porpoises in the South China Sea is being lost irrevocably or degraded by reclamation, dredging, and coastal pollution. Up to 40,000 Dall's porpoises are taken by harpoon for their meat each year off the coast of northern Japan; such numbers are unlikely to be sustainable. Intense vessel traffic and other anthropogenic sounds reduce the amount and quality of habitat suitable for Harbor porpoises, as witnessed by carefully controlled experiments in which porpoise densities have been shown to decline sharply following the introduction of such sounds.

It is the accidental capture of porpoises by fishermen, though, that poses the greatest threat. All species of porpoise, even the Spectacled porpoise of the remote Southern Ocean, are taken as "by-catch" in commercial fisheries. It is hard to fathom why, with such a sophisticated system of echolocation, porpoises become entangled in fishing nets. The most common type of fishing gear involved in these interactions is the gill-net, which is hung near the water's surface to capture fish swimming in the water column, or near the bottom to entangle bottom-dwelling species. Both configurations are deadly to porpoises, which either do not detect the nets, do not perceive them as dangerous, or else make navigational mistakes that result in their entanglement.

The number of porpoises killed in this manner can be enormous. Every year Danish gill-net fisheries in the North Sea alone account for the deaths of around 7,000 Harbor porpoises. The number of porpoises taken by vessels of other countries in this area is unknown. Widespread conservation efforts are being undertaken by researchers and fishermen to resolve the problem of by-catches. One possible solution is the use of high-frequency alarms, known as pingers, that fishermen can place on their nets to warn porpoises of their deadly presence. AR/DEG

THE VULNERABLE VAQUITA

The vaquita, the smallest of all cetaceans, has an extremely limited geographical range. The only population of this shy, unassuming odontocete is cut off at the very top of the 600 mile blind-alley that is formed by the Gulf of California, Mexico. The species is believed to have evolved from a population of Burmeister's porpoises – its closest relative – that became isolated in the Upper Gulf after a period of cooler water conditions in the Pleistocene came to an end. In its very restricted domain, the vaquita now faces a desperate struggle for survival. Several factors are to blame. Firstly, productivity of all species in the Upper Gulf has fallen as more and more water is diverted from the Colorado River for irrigation and domestic use on its course through the western USA, so reducing its flow into the head of the Gulf. Vaquita numbers may have sunk to such a low level (around 600, according to estimates at the end of the 1990s) that inbreeding depression may further be reducing its viability. Yet everyone agrees that the principal cause of its drastic decline is fisheries' by-catch. Mortality in fishing gear, a threat to small cetaceans throughout the world, arises in this case not from industrial-scale fishing but from the sheer density of numbers of artisanal gill-net fishermen. The difficulties facing the Mexican government in saving the vaquita are immense: not only are very limited data available on the species' biology and its seasonal distribution, but the urgent need to change and regulate fishing practices must be balanced against protection of peoples' livelihoods in this economically disadvantaged region. JG

Beluga and Narwhal

BELUGAS AND NARWHALS, THE so-called "white whales", are among the most social of all cetaceans. A large aggregation of brilliant white belugas in an Arctic bay makes an impressive enough sight, but a procession of hundreds or even thousands of narwhals moving along the coast is truly awe-inspiring. White whales are known to have lived in temperate seas in prehistoric times, but now exclusively occupy cold Arctic waters.

The narwhal's skin coloration is striking in itself: small patches of gray-green, cream, and black pigmentation seem painted onto its body by short strokes of a stiff brush. Even more astonishing, though, is the renowned spiraled tusk, thrust above the water as the male breaks the surface. Not only does this seem disproportionately long – 3m (10ft) of tusk on a 5m (15ft) whale – but it is also curiously off-center, protruding from the left upper lip at an awkward angle and pointing down. To crown these oddities, the tails of older males appear to be put on back-to-front!

Fat for Insulation

FORM AND FUNCTION

Narwhals and belugas are similarly shaped, though the beluga is slightly smaller. One feature that is peculiar to the beluga is its neck; unlike most whales, it can turn its head sideways to a near right-angle. The beluga has no dorsal fin – hence its scientific name, *Delphinapterus* or "dolphin-without-a-wing" – although there is a ridge along the back from mid-body to the tail; a true fin might lose body heat, and would be at risk of getting damaged in ice.

In both species, males are about 50cm (20in) longer than females, and their flippers increasingly turn upward at the tips with age. The flippers of belugas are capable of a wide range of movements, and appear to serve an important function in close-quarters maneuvering, including very slo reverse swimming. In aging male narwhals, the shape of the tail changes; the tips migrate forw giving a concave leading edge when viewed fro above or below. Both species have thick layers o blubber to provide insulation against the near-freezing water in which they live, but the belug is so fat that the head (which of necessity is less well-endowed with blubber) always looks too small for its body.

Below *From June to September, belugas congre gate in hundreds and thousands at their traditional birthing grounds in wide river estuaries. The immen numbers involved in migratory patterns may be judg from this aerial view of belugas at the mouth of the Cunningham River, on Somerset Island.*

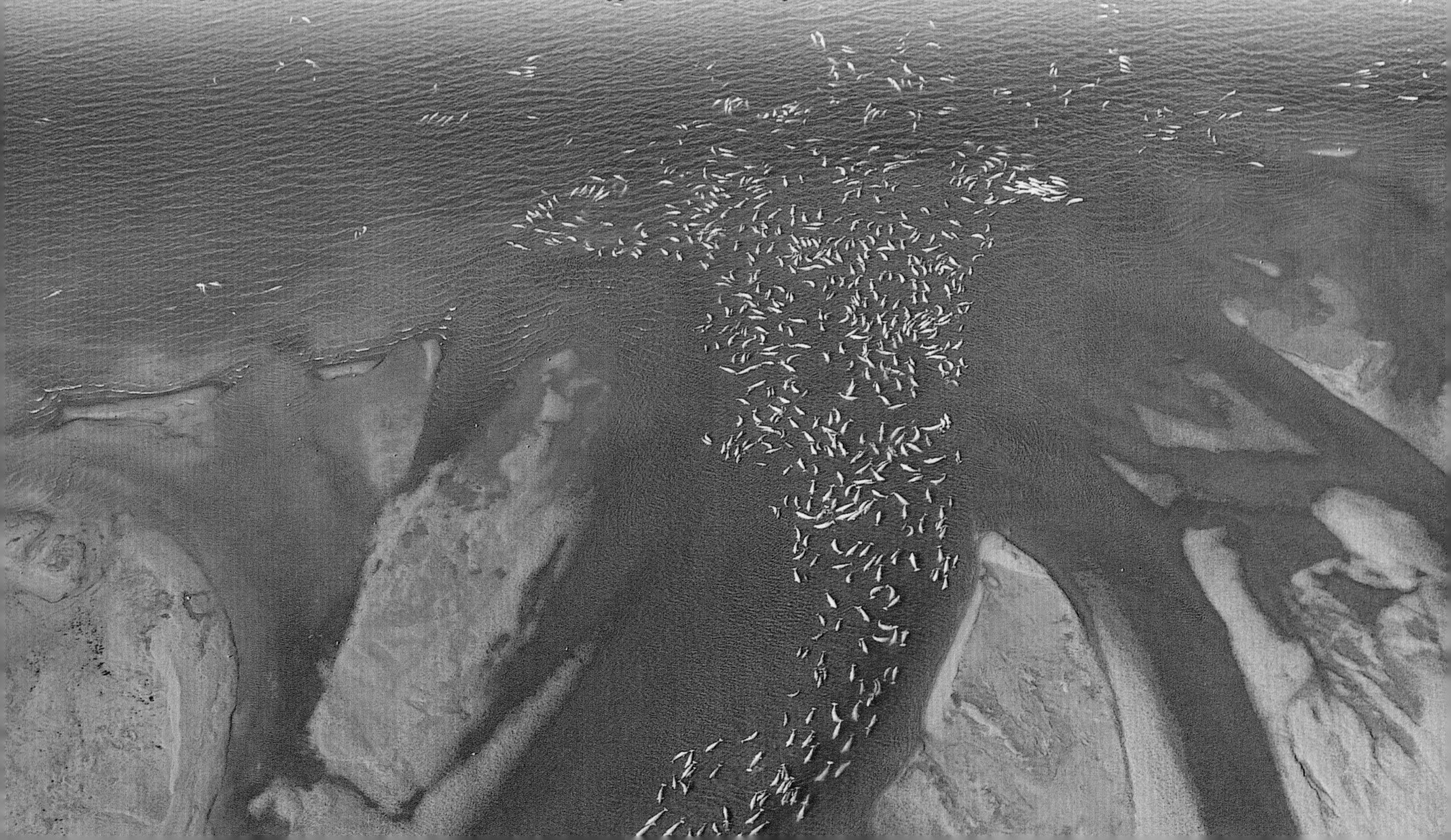

Left A beluga and its calf. Suckling may last for two years, during which time the mother and calf are almost insepara-ble. Newborn belugas are brown, and the skin gradually lightens through gray, as in this one-year-old, to white.

FACTFILE

BELUGA AND NARWHAL

Order: Cetacea

Family: Monodontidae

2 species in 2 genera

DISTRIBUTION Northern circumpolar

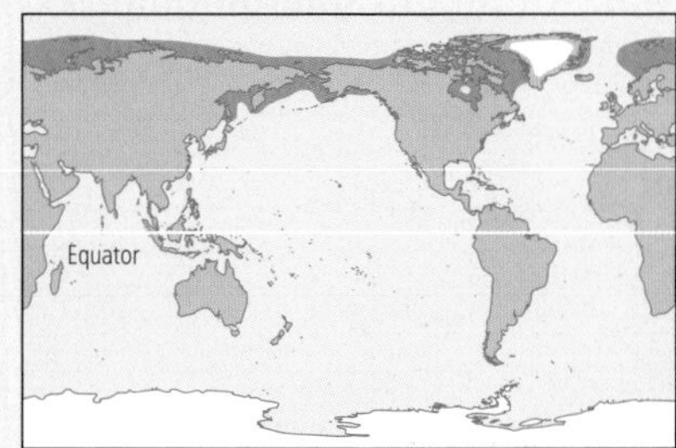

BELUGA *Delphinapterus leucas*
Beluga or White whale
N Russia and N America, Greenland, Svalbard. Cold waters, usually near ice; offshore and coastal; estuaries in summer. HTL 300–500cm (10–16ft), WT 500–1,500kg (1,100–3,300lb). Adult males c.25 percent longer than females, and nearly double their mass. Skin: adults white or yellowish; young slate-gray, becoming medium gray at 2 years and white on maturity. Diet: mostly benthic; schooling fish, crustacea, worms, mollusks. Breeding: gestation probably 14–15 months. Longevity: 30–40 years. Conservation status: Vulnerable.

NARWHAL
Monodon monoceros
N Canada and Russia, Greenland, Svalbard. Cold waters, invariably in or near sea ice; mainly offshore, but often fjords and coasts in summer. HTL 400–500cm (13–16.5ft); WT 800–1,600kg (1,760–3,520lb); male tusk length 150–300cm (5–10ft); males larger than females. Skin: mottled gray-green, cream, and black, whitening with age, beginning with the belly; young dark gray. Diet: Arctic cod, flatfish, cephalopods, shrimps. Breeding: gestation probably 14–15 months. Longevity: 30–40 years.

Abbreviations HTL = head–tail length WT = weight

The narwhal has only two teeth, h of which are non-functional. In female, these grow to about 20cm a) in length but never emerge from gums; in the male, the left tooth ntinues to grow to form the tusk. A tiny nority of males – less than 1 percent – produce n tusks, while a similar proportion of females e single tusks. The purpose of the tusk has n the subject of various theories but it seems be simply a secondary sexual characteristic that ys a role in establishing dominance in social life l breeding.

Belugas are capable of a wide range of bodily l facial expressions, including an impressive uth gape displaying 32–40 peglike teeth that t one another. The surfaces can be heavily rn – sometimes so much so that they cannot be ctive for grasping prey. This – and the fact that teeth do not fully emerge until well into the ond or third year – suggests that their prime ction may not be for feeding. Belugas com-nly clap their jaws together to make drumming nds, and the teeth may contribute to these; y are also used visually in showing-off displays.

In contrast to the narwhal, the beluga is a high-ocal animal, producing moos, chirps, whistles, l clangs that long ago earned it the nickname "sea canary." Some of the sounds it makes are ily heard through the hulls of boats and even ove water; underwater, the din from a herd is niniscent of a barnyard. In addition to its vocal l echolocation skills, the beluga also uses on for both communication and predation. e versatility of its expressions suggests the likeli-od of subtle social communication.

***Above** The development of facial features and expressions in the beluga. Adult belugas have a very pronounced forehead melon, but this is slow to devel-op: in newborn calves **a** it is almost absent; in year-lings **b** the melon is quite large but the beak undevel-oped; maturity **c** is reached at 5–8 years. The beluga's mouth and neck are highly flexible; belugas communi-cate with each other a great deal by sound and facial expression; in repose the beluga seems to our eyes to be smiling. **1** Besides clicks and bell-like tones, belu-gas produce loud reports by clapping their jaws together. Belugas are versatile feeders and the pursed mouth **2** is believed to be used in bottom-feeding.*

***Right** A pod of narwhals off the coast of Baffin nd, northern Canada. On migratory journeys such his, pods may coalesce into schools of up to 2,000 mals, all swimming near the surface. However, en foraging for food, narwhals are capable of div- to great depths to reach the ocean floor.*

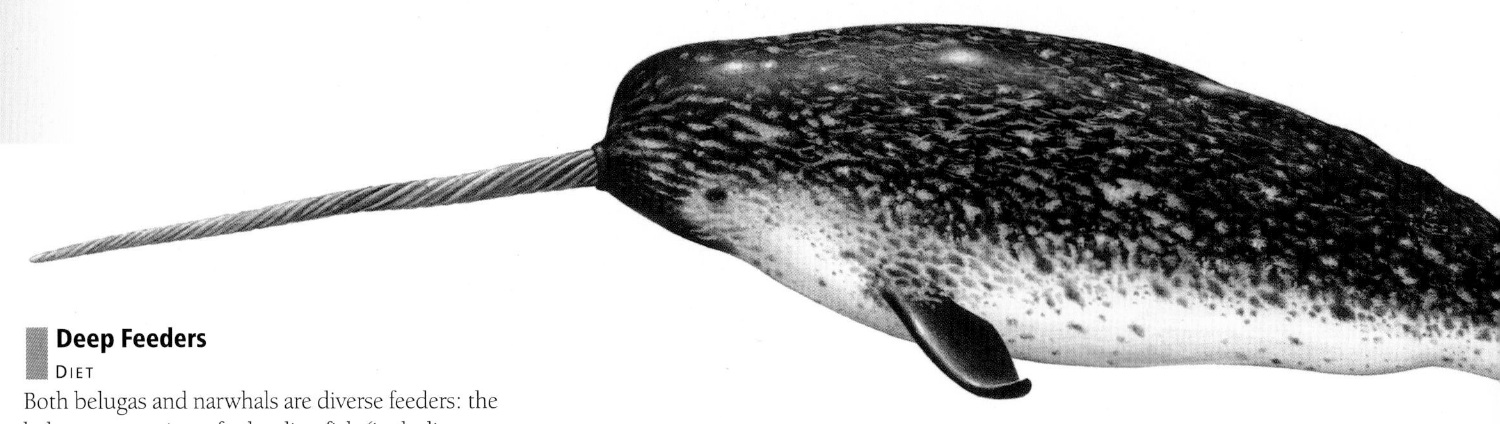

Deep Feeders

DIET

Both belugas and narwhals are diverse feeders: the beluga on a variety of schooling fish (including cod), as well as crustacea, worms, and sometimes mollusks, and the narwhal on cephalopods, arctic cod, flatfish, and shrimps. Belugas find most of their prey on the seabed at depths of up to 500m (1,650ft), and the narwhal similarly feeds at depth, though not necessarily on the bottom. Both species are capable of dives in excess of 1,000m (3,300ft), and are limited only by their breath-hold capacity, which is normally 10–20 minutes during deep dives but can exceed 20 minutes in exceptional circumstances. The highly flexible neck of belugas permits a wide visual or acoustic sweep of the bottom, and they can produce both suction and a jet of water to dislodge prey. The narwhal, with no functional teeth at all, can, like the beluga, probably suck up its dinner. The male and the tuskless female have similar diets, so the tusk appears to play no part in feeding – indeed, it can only be an impediment, obstructing the mouth of the whale as it approaches prey.

Migrating Whales

SOCIAL BEHAVIOR

The beluga and the narwhal are probably similar in their growth and reproduction, although more is known about the beluga. Females become sexually mature at around 5 years, males after 8, but there are differences between populations. Judging by the sexual dimorphism, dominant males probably mate with many females. Births mostly occur in early summer as the sea ice begins to break up. Many populations occupy estuaries in July, but this is probably not linked to breeding, for few calves are born in these sheltered areas. Single calves are the norm, with twinning an extremely rare event. A strong bond is established between mother and calf immediately after birth, and physical contact is maintained so closely that the calf seems attached to the mother's side or back. The mother may provide milk for more than two years, after which she will again be pregnant. The complete reproductive cycle of gestation and lactation takes three or more years.

Narwhals sometimes move from the offshore pack ice into fjords during mid-summer, but spend less time in shallow waters than do belugas. Both species are extremely social and sometimes appear in the same fjord together, but such co-occurrence seems coincidental and does not normally result in any obvious interaction. Animals are rarely seen alone or even in small groups, so the occasional beluga or narwhal observed in temperate European waters is abnormal socially as well as geographically.

Aggregations of hundreds or thousands of whales are common, often covering areas of many square kilometers. The aggregations behave as one entity, even though, when seen from the air, they clearly comprise many smaller, tightly-knit groups, usually containing whales of a similar size and/or sexual status. Females with calves come together, as do groups of large adult males. It is known that these male groups often stay together for several months, and maybe even longer.

Satellite telemetry has also revealed much about the migrations of belugas and narwhals. Both species spend most of the year offshore in areas dominated by sea ice, but sometimes in open areas within the pack ice called polynyas. Narwhals may remain offshore all year, or may enter fjords for brief periods in July or August. Most beluga populations frequent estuaries in summer, but individual whales do not remain in them for long. In the Canadian Beaufort Sea, belugas pause in their eastward migration in the immense Mackenzie Delta for a week or so, then continue on to deeper waters. The area can be likened to a highway service station; even though hunters and observers see hundreds of whales there every day for more than a month, there is in fact a continuous turnover of individuals, and tens of thousands of belugas visit during this period. In some areas such as Svalbard, estuaries are not available and belugas head for glacier fronts instead. What estuaries and glaciers have in common is that each is a source of fresh water. At this time of year, the belugas undergo a molt, in which their old yellow skin is shed to reveal the new, gleaming white skin underneath. Fresh water absorbed through the epidermis may hasten the shedding process, which certainly occurs very rapidly, aided by the whales rubbing themselves on the gravel of the seabed.

Above *The narwhal's extraordinary tusk results from its left tooth growing out into a counter-clockw spiral. Like antlers in deer, the tusk's sheer size may serve to indicate an animal's prowess; it is also used as a weapon, as evidenced by the accumulation of scars on aging narwhals and the occasional tusk-tip found embedded in a skull.*

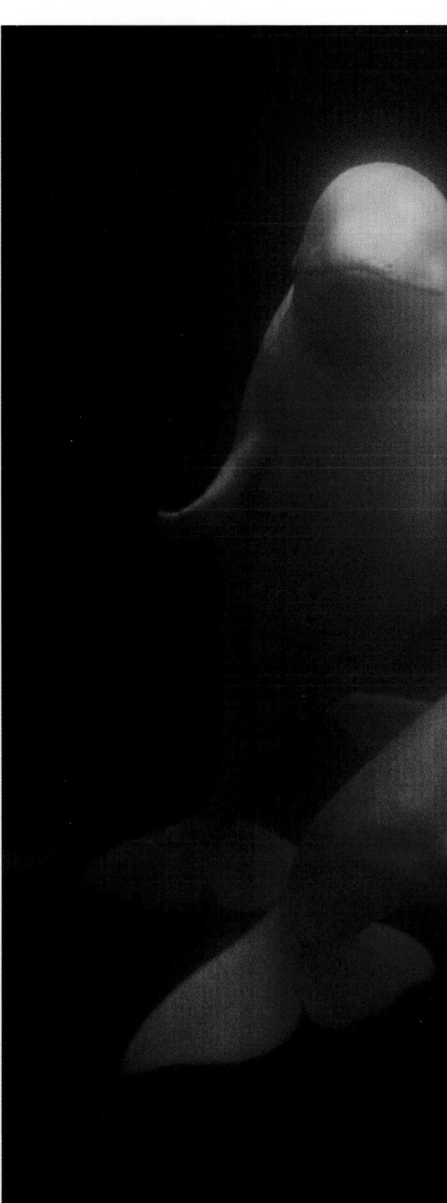

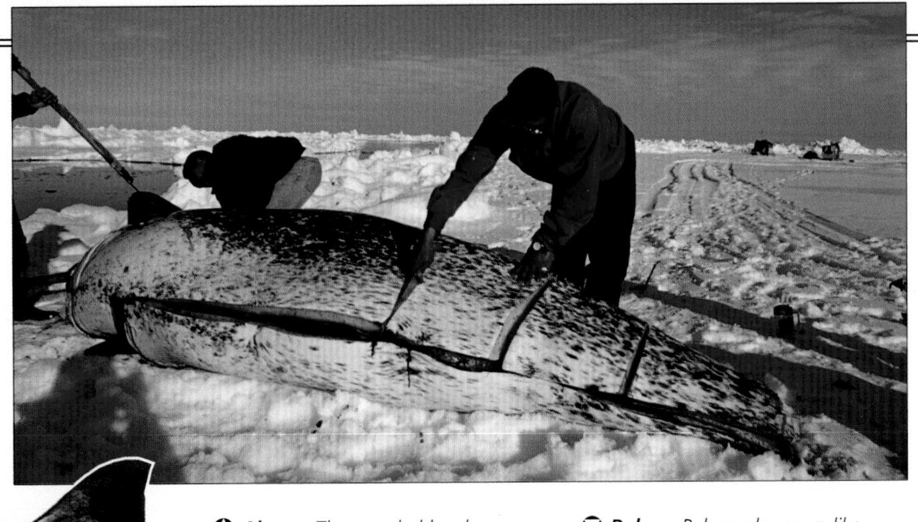

Above *The narwhal has long been hunted by the indigenous Eskimo peoples of the Arctic. Every part of the animal is used – its meat and skin (which is rich in vitamin C) are eaten, while its sinews are dried and turned into stout cord. Yet the high prices commanded by the narwhal's tusk have raised concerns about overexploitation of the species.*

Below *Belugas loom up like ghostly apparitions in the murky waters of Hudson Bay. The bulbous frontal "melon" is used for echolocation, in which the beluga sends out acoustic signals to judge its distance from objects by the frequency at which the sound waves bounce back. This is vital for navigating and finding prey in dark or muddy waters.*

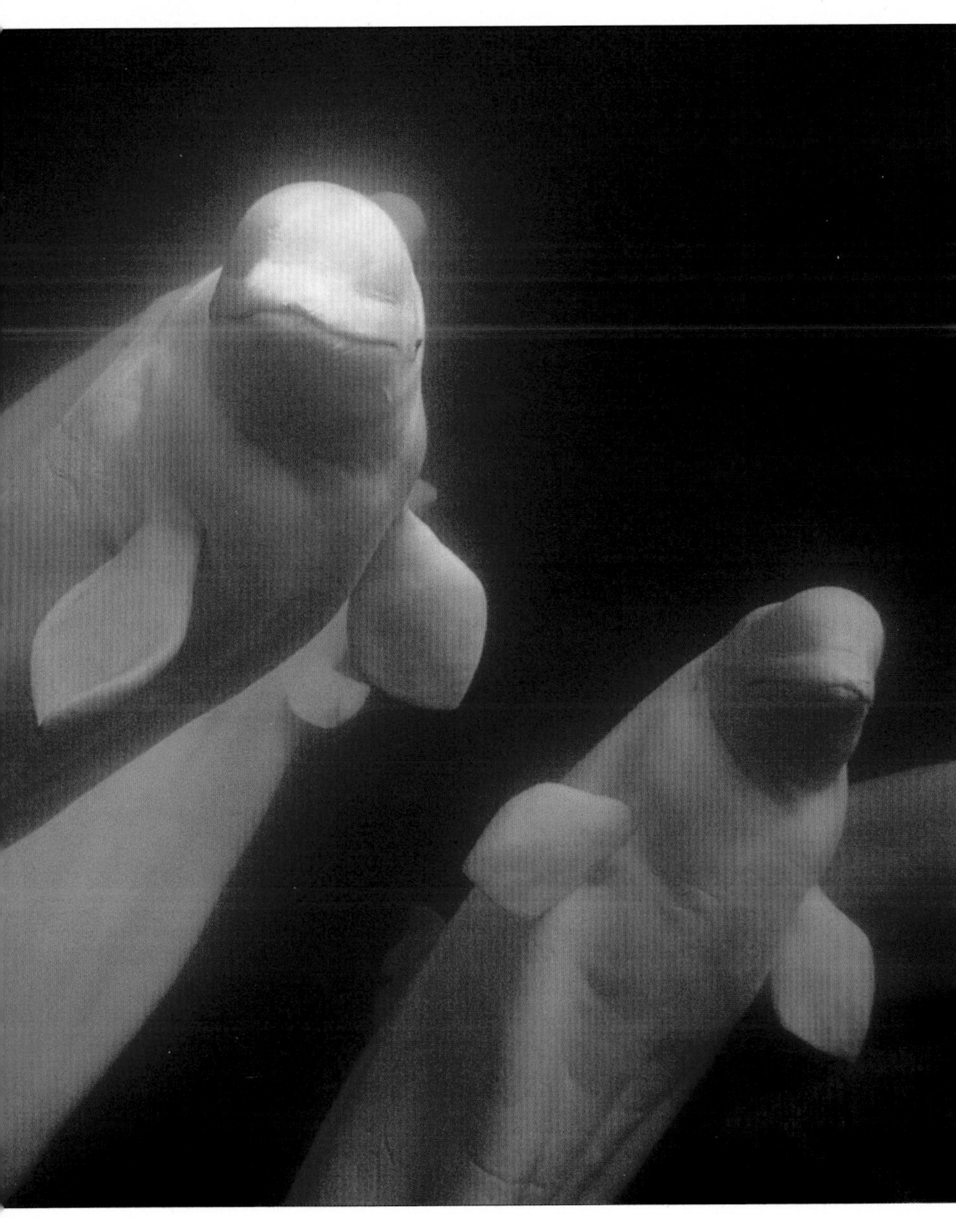

Easy Pickings for Hunters

CONSERVATION AND ENVIRONMENT

Belugas return faithfully to their summer grounds along migration routes that are probably learned in infancy. This fidelity can have unfortunate consequences, for the whales will go on returning to a given site even if overexploited by hunters until the population is extinguished. Furthermore, the 30 recognized populations of belugas are so inflexible in their preference for familiar routes and breeding grounds that they will not recolonize areas where the population has been wiped out. One such location is Ungava Bay in Labrador, where belugas were formerly abundant but are now rarely encountered.

The predictable migratory behavior of belugas has made the species particularly vulnerable to exploitation. In the 18th and 19th centuries American and European whalers would force mass strandings of hundreds of belugas in order to "top up" their cargo of whale oil rendered from their primary quarry, the Bowhead. Aboriginal peoples also hunted the whales, though historically they took relatively small harvests which were probably sustainable; today's Eskimo hunters are more mechanized (with high-velocity rifles, explosive harpoons, and motorboats) and serve greater human populations, and so have the potential seriously to deplete white whale stocks. Currently, belugas are thought to number in excess of 100,000 worldwide, but individual populations range from healthy (in the tens of thousands) to effectively extinct following excessive hunting. Combined annual catches are now in the hundreds or low thousands, with the precise number varying from year to year.

The beluga's seasonal affinity for shallow coastal areas also makes it vulnerable to more indirect modern-day threats. The pollution of prey, and habitat degradation through oil exploration and the building of hydroelectric dams, are presently of greatest concern, while global warming and its effects on sea ice may be a looming problem. Though belugas and narwhals have survived periodic fluctuations in Arctic ice extent within even the recent geological past, the currently predicted rate of change in this region is exceptional and will demand rapid adaptation if these pagophilic (ice-loving) species are to continue to thrive.

The most recent estimate of narwhal numbers is in the order of 25,000–30,000, consisting of three putative stocks. In contrast to the beluga, this species avoids nearshore waters for most of the year, thus reducing the threat from industrial development, but hunting occurs and is not evenly distributed. The narwhal's tusk has always been the prime cause of its persecution, from medieval times – when it was reputed to be the horn of the mythical unicorn, endowed with magical properties – up to the present day, when it is much in demand from private collectors and museums. AM/PB

Sperm Whales

IMMORTALIZED IN HERMAN MELVILLE'S NOVEL Moby Dick, Sperm whales run to extremes. They are the largest of the toothed whales, have the biggest brains on Earth, are very sexually dimorphic (males weigh three times as much as females), and make possibly the deepest and longest dives of any creature in the animal kingdom.

Long ago, sailors thought that the regularly-spaced clicking noises they heard through the hulls of their ships originated from what they called "carpenter fish," because they sounded like hammers tapping. They were in fact listening to the sounds of Sperm whales. As for the name "Sperm whale," it seems to have come from whalers misinterpreting the nature of the oily substance known as spermaceti found in the whales' massive foreheads.

Sound in the Deep Ocean

FORM AND FUNCTION

The ancient family of Physeteridae seems to have separated from the main odontocete line early in cetacean evolution, about 30 million years ago. The single extant species, the Sperm whale – *Physeter catadon (macrocephalus)* along with the far smaller Dwarf and Pygmy sperm whales (family Kogiidae) – all share a barrelshaped head, a long, narrow, underslung lower jaw with uniform teeth, paddleshaped flippers, and a blowhole which is displaced to the left side. The *Kogia* genus emerged much later, around 8 million years ago.

The Sperm whale's great, square forehead, situated above the upper jaw and in front of the skull, makes up a quarter to a third of the animal's length. It holds the spermaceti organ, an ellipsoid structure contained within a sheath of connective tissue. Both the organ itself and the tissue surrounding it are dense with spermaceti – a semi-liquid, waxy oil. Air sacs bound both ends of the spermaceti organ. The skull and air passages that surround the organ are highly asymmetrical. The two nasal passages have become very different in both form and function: the left is used for breathing, the right for sound production.

Why the Sperm whale carries such large, unwieldy headgear is not clear. One suggested reason is that it helps focus the clicks which serve in echolocating food in the murky depths of the deep sea. Sperm whales also use the clicks for communication, being the most vocal of the three Physeteridae species.

The rodlike lower jaw of the Sperm whale contains 20–26 pairs of large teeth, while the Dwarf sperm whale may have 8–13 and the Pygmy sperm whale 10–16 pairs. These teeth do not seem to be required for feeding, as well-fed Sperm whales have been caught lacking teeth or even lower jaws. Moreover, the teeth do not erupt (emerge) until the whales reach sexual maturity. None of the species normally have teeth in the upper jaw, and if they do, the teeth usually do not erupt. *Kogia* teeth are thin, very sharp, curved, and lack enamel.

Except for the head and tail flukes, the skin of the Sperm whale is corrugated to create an irregular, undulating surface. The low dorsal fin may be topped by a rough, whitish callus, especially in mature females.

Sperm whales make repeated, deep foraging dives, on average to about 400m (1,300ft) and for about 35 minutes, though they can reach over 1,000m (3,300ft) and last for more than an hour. Between dives, the whales surface to breathe, on average for about 8 minutes. The whales descend nearly vertically, raising their flukes straight out of the water.

For both sexes, squid form the bulk of the diet. Female Sperm whales spend about 75 percent of their time foraging. Although they go for smaller prey than the males, they will occasionally take on Giant and Jumbo squid, and scars from the squids' sucker marks on their heads bear testimony to their undersea battles. Prey items recovered from male Sperm whales tend to be larger versions of the same species eaten by and recovered from females, though the males also eat rather more fish, including sharks and rays.

In the Pygmy and Dwarf sperm whales, the head is more conical and much shorter in relation to the overall body length. The two *Kogia* species look rather sharklike, with an underslung mouth and sharp teeth, plus a bracket-shaped mark on the side of the head resembling gill slits. While they mainly eat squid and octopus, *Kogia* have a flattened snout, and benthic fish and crabs have been recovered from their stomachs, suggesting that they may be bottom-feeders at least some of the time. Their diet is otherwise not dissimilar to that of the Sperm whale.

Global Voyagers

DISTRIBUTION PATTERNS

Few animals on earth are as widely distributed as Sperm whales. They occupy waters from near both poles to the equator. The two sexes seem to be geographically segregated for much of the year, with females and juveniles inhabiting warmer waters at latitudes of less than about 40°, while the males move o to higher latitudes as they age and grow. The largest males are found close to the edge of the Arctic and Antarctic pack ice. In order to mate, they must migrate to the tropics, where the females abide.

Genetic studies indicate that all Sperm whal populations are broadly similar. Mitochondrial DNA, transmitted only through the mother, shows no geographical structure at scales of les than an ocean basin. Nuclear DNA, half of whic comes through the wide-ranging males, is even more geographically homogeneous – there are significant differences between Sperm whale po ulations within oceans, and whatever difference

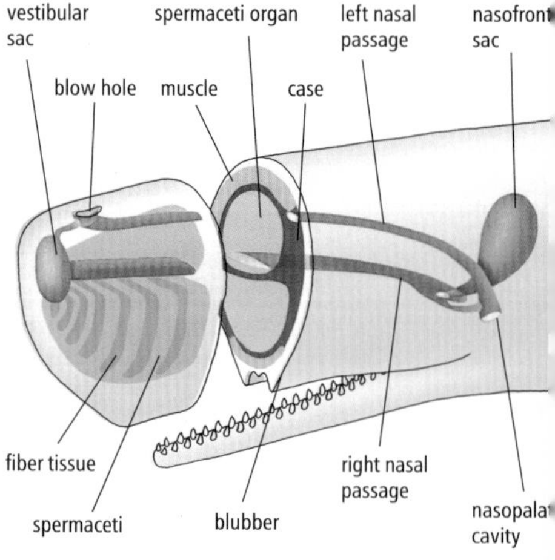

Above *Cutaway drawing of the head of a Sper whale, showing its anatomical features. The whale alter its buoyancy by cooling or warming the huge volume of spermaceti in the upper part of its hea It achieves this by regulating the flow of water through its nasal passages.*

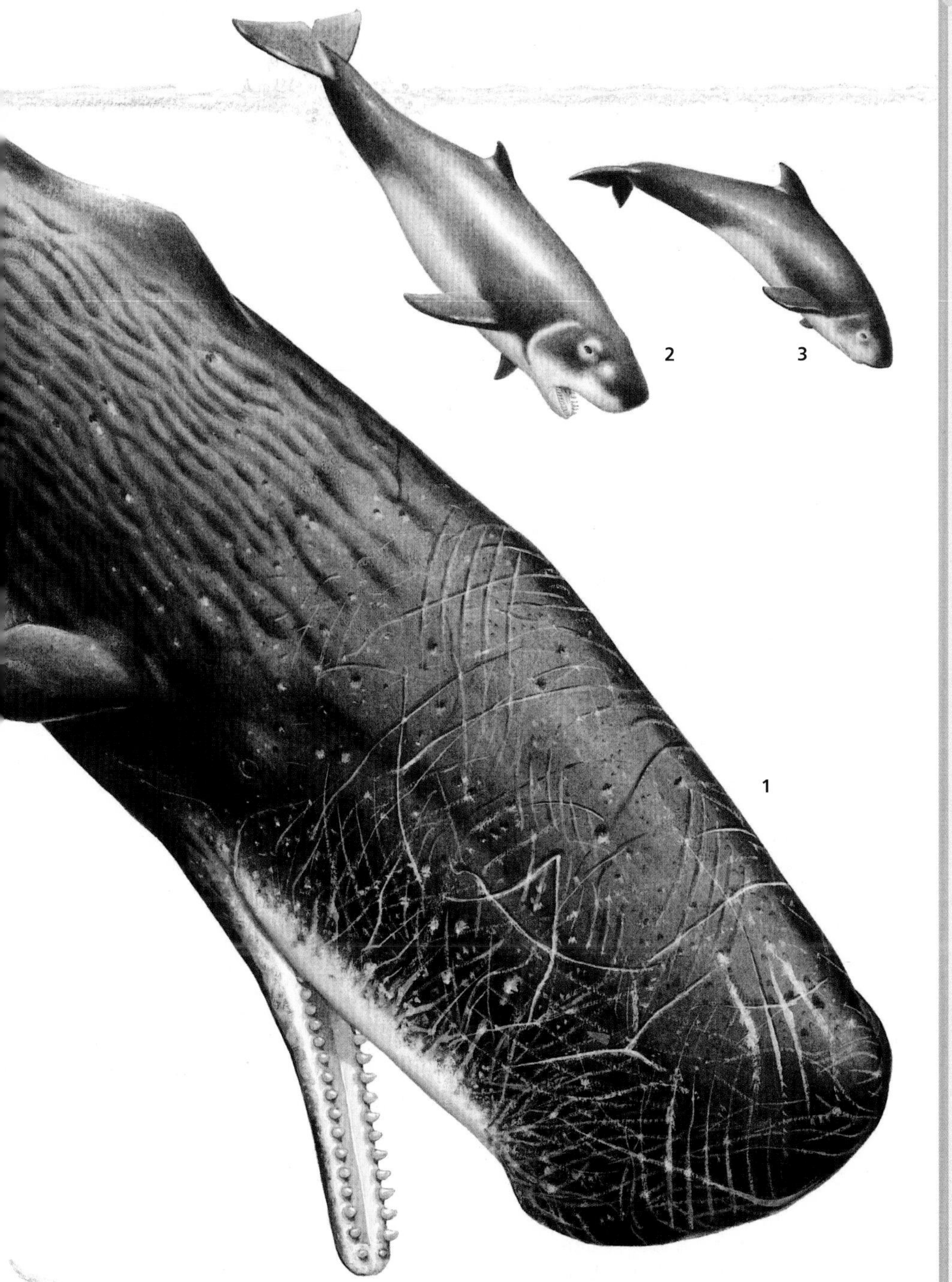

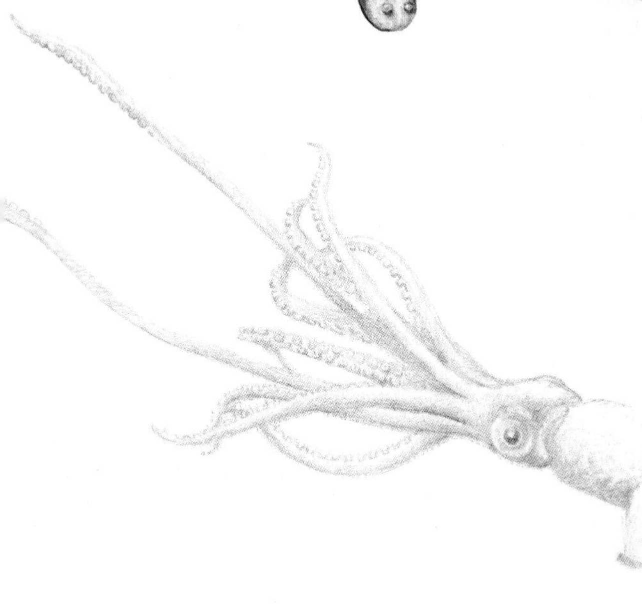

Above *The three species of Sperm whale:* ***1*** *Sperm whale* (Physeter catodon) *diving for squid, which it sometimes catches at extraordinary depths near the ocean floor;* ***2*** *Pygmy sperm whale* (Kogia breviceps)*;* ***3*** *Dwarf sperm whale* (Kogia simus).

FACTFILE

SPERM WHALES

Order: Cetacea

Families: Physeteridae and Kogiidae

3 species in 2 genera

DISTRIBUTION Worldwide in tropical and temperate waters to latitudes of about 40°; mature male Sperm whales to polar ice edge.

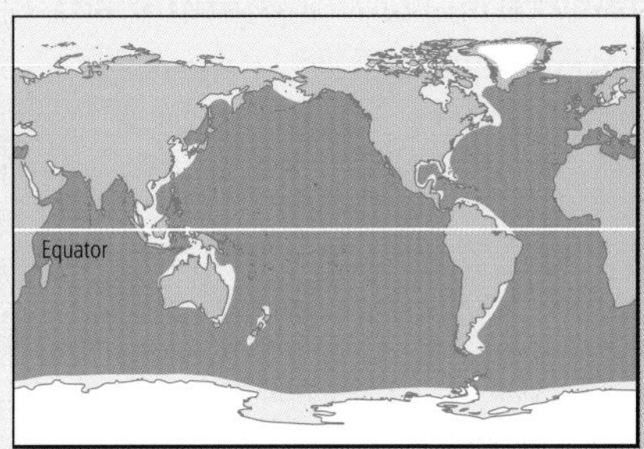

HABITAT Mainly deep waters (over 1,000m/3,300ft) off the edge of the continental shelf. Juvenile and immature *Kogia* may inhabit shallower, more inshore waters over the outer part of the continental shelf.

SPERM WHALE *Physeter catodon (macrocephalus)*
Cachalot, Spermacet whale, Pot whale, sea-guap
HTL male 16m (52.5ft), max. 18m (59ft); female 11m (36ft), max.12.5m (41ft). WT male 45 tonnes, max. 57 tonnes; female 15 tonnes, max. 24 tonnes. Skin dark gray, but often with white lining to mouth and white patches on belly; wrinkled, except for head and flukes. Breeding: females mature sexually around age 9; in males, puberty extends from ages 10–20, though they do not participate actively in breeding until roughly age 30. Single calves born in summer after 14–15 months' gestation; calf care is prolonged, with lactation lasting 2 years or more. Longevity: at least 60–70 years. Conservation status: Vulnerable.

PYGMY SPERM WHALE *Kogia breviceps*
Pygmy or Lesser sperm whale, Lesser cachalot
HTL male 4 m (13ft); female max. 3m (10ft). WT 318–408kg (700–900lb). Skin: bluish gray on back, shading to lighter gray on the sides, and white or pinkish on belly; lighter mark – the "bracket" or "false gill" – on side of head. Breeding: mating believed to occur in summer; gestation period 9–11 months, with births in the spring. Calves, which are about 1m (3.3ft) long at birth, are nursed for about a year; females may give birth two years in a row. Longevity: may be 17 years or more.

DWARF SPERM WHALE *Kogia simus*
Dwarf sperm whale or Owen's pygmy whale
HTL 2.1–2.7m (7–9ft). WT 136–272kg (300–600lb). Skin: bluish-gray on back, shading to lighter gray on the sides, and white or pinkish on the belly; lighter "bracket" or "false gill" on side of head. Breeding: calves are somewhat smaller at birth than those of *K. breviceps*. Longevity: unknown.

Abbreviations: HTL = head–tail length WT = weight

do exist are between oceanic populations. They live in deep water, usually deeper than 1,000m (3,300ft) and far from land. The edges of the continental shelf seem to suit them well.

Pygmy and Dwarf sperm whales are also cosmopolitan. Pygmy sperm whales are found in deep waters of temperate, subtropical, and tropical seas. The Dwarf sperm whale occurs in somewhat warmer water.

The two *Kogia* species spend a substantial amount of time lying quietly on the surface exposing the back of their heads, with their tails hanging down limply. Pygmy sperm whales are timid and slowmoving. They will not swim toward boats themselves, but are easy to approach as they lie motionless at the surface. They come up to breathe in a slow and deliberate manner, and have an inconspicuous blow. When startled or distressed, *Kogia* may discharge a reddish-brown intestinal fluid that perhaps helps them to escape from predators such as large sharks and killer whales, in the manner of an octopus's ink. *Kogia* also have eyes adapted for functioning in the poor light of the deep sea.

Very little is known of the reproductive strategy of the Pygmy or Dwarf sperm whales. No sexual dimorphism is apparent in either species. This is in marked contrast to the Sperm whale, which is radically sexually dimorphic; the size of the mature male seems to confer a reproductive advantage. *Kogia* may therefore have a different mating system.

Female Sperm whales attain physical maturity at roughly 30 years of age, but males continue to grow until they are about 50 years old. Large, mature, male Sperm whales (late 20s or older) make the journey from pole to equator, where they roam between groups searching for receptive females with whom to mate. It is not known whether males return annually or biennially. The time spent with each group varies from a few minutes to several hours. Breeding males mostly avoid one another, much like bull elephants in the rutting condition known as "musth." Occasionally, though, they may fight, as evidenced by deep scars on the heads of some mature adults. The spacing of these scars leaves little doubt that they are produced by other males' teeth.

Pygmy sperm whales may bear a calf two years in succession, i.e. they can be pregnant and nursing simultaneously. In contrast, Sperm whales give birth roughly once every 5 years or so, following a a gestation period that is not accurately known, but is within the range of 14–15 months. Reproductive rates in females decline with age.

Right *A family group comprising mother, daughter, and new calf swimming together near the Azores. Sperm whales are about 4m (13ft) long at birth and may weigh as much as 1,000kg (2,200lb). Sperm whale calves eat solid food by one year of age, but continue to suckle for several years.*

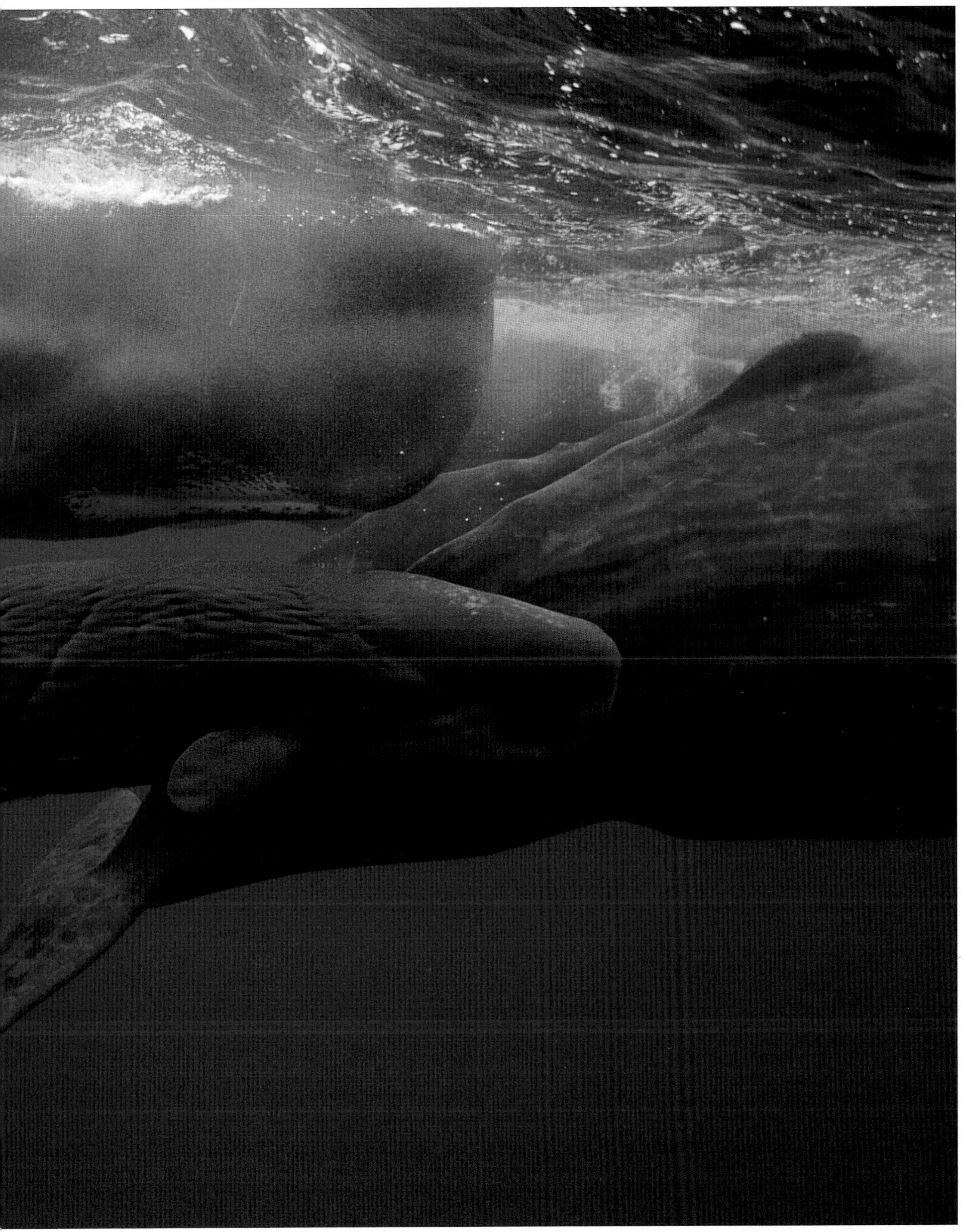

Care in a Cetacean Community

SOCIAL BEHAVIOR

Female Sperm whales are exceedingly gregarious. Their social life is based on the family unit, which consists of about 12 permanently attached, closely related females and their young. Two or more units may join together for a period of several days, forming a cohesive group of about 20 whales, perhaps to enhance foraging efficiency or at least to reduce interference among different units foraging in the same area.

Male Sperm whales, in contrast, leave the natal unit when they are roughly 6 years old to form "bachelor schools." As males age, they are found in progressively smaller aggregations. Mature males' associations with other males rarely last longer than a day, although males have been found stranded together on beaches, suggesting social relationships are not entirely absent.

It is the potential of other whales to act as babysitters which seems to draw female Sperm whales to affiliate. Young calves are apparently unable to make prolonged dives with their mothers to the depths to feed. Left alone at the surface, they would be vulnerable to attacks by sharks or killer whales. Members of groups containing young calves seem therefore to stagger their dives, so th[at] some adults are at the surface at all times. In add[i]tion to this communal care within family units, there is strong, though not definitive, evidence f[or] females suckling calves who are not their own.

Communal group defense against predators extends to protecting other adults too. Clusterin[g] tightly together, the whales coordinate their effo[rts] in a "marguerite" formation, with their heads together at the center and their bodies radiating out like the petals of a flower. Alternatively, they may adopt a "heads out" formation. The former strategy makes use of the whales' flukes as defense; the latter, their jaws.

On occasion, individual whales will even plac[e] themselves at risk to help others. In one well-observed incident off California, Sperm whales under attack by killer whales were seen to leave the comparative safety of the marguerite formati[on] to "rescue" another whale which had become se[p]arated and was being badly mauled by the killers.

Female Sperm whales regularly gather at the

CHAMPION DIVERS

Sperm whales may be considered to be the champion divers among all aquatic mammals. They have been accurately recorded by sonar as diving to 1,200m (3,936ft), and Sperm whale carcasses have been recovered entangled in cables from 1,140m (3,740ft), where they had probably been feeding on the bottom-dwelling squid that form the bulk of their diet. One of two bulls observed diving for one to nearly two hours each dive, was found – on capture – to have in its stomach two specimens of Scymodon, a small, bottom-living shark. The depth of water in the area was about 3,200m (10,500ft), suggesting an amazing diving ability. The fact that Sperm whales dive right to the seabed for food is borne out by the discovery of all manner of objects in their stomachs, from stones to tin cans, suggesting that they literally shovel up the bottom mud.

Bulls are the deepest and longest divers, although females may dive to 1,000m (3,280ft) for more than one hour. Juveniles and calves dive for only half this time, to about 700m (2,300ft). Females often accompany young whales, and this, rather than an inability to dive deeper, may be what limits their diving range. However, the gregariousness and caring behavior within the nursery school means that young calves may temporarily be adopted by other females, thus enabling the mother to dive deeper for food than she might otherwise be able to do.

If diving as a group, Sperm whales appear to remain close and do almost everything together. They recover quickly from a long, deep dive, and dive again after only 2–5 minutes. After several long dives, they reach their physiological limit and recover by lolling on the surface for many minutes.

The descent and ascent rates are astonishing. The fastest recorded averaged 170m (550ft) per min in descent and 140m (450ft) per min in ascent. The adaptations that enable the Sperm whale to perform these prodigious feats are largely similar to those of other cetaceans but more efficient. For example, the muscle in Sperm whale can absorb up to 50 percent of the total oxygen store – at least double the proportion in land mammals, and significantly more than in baleen whales and seals.

A unique feature of the Sperm whale is the vast spermaceti organ, which fills most of the upper part of the head, and is thought to be an aid to buoyancy control. The theory is that the nasal passages and sinuses which permeate the organ can control the cooling and warming rate of the wax, which has a consistent melting point of 29°C (84.2°F). As the whale dives from warm surface waters to the colder depths, the flow of water into the head passages is controlled to quickly cool the head wax from the whale's normal body temperature of 33.5°C (92.3°F). As a result, the wax solidifies, shrinking as it does so, increasing the density of the head, and thus assisting the descent. On ascending, the blood flow to the capillaries in the head can be increased, so warming the wax slightly and increasing buoyancy to provide lift to the exhausted whale.

CL

Above *The power of the Sperm whale is dramat[i]cally apparent as this pod forges ahead in formation. The dorsal hump, slightly suggestive of a submarine, [is] prominent here, and the individual on the right is demonstrating its oblique blow.*

Right *The "marguerite" formation, in which mem[]bers of a pod will encircle an injured Sperm whale while it remains alive. Such supportive behavior used to be disastrous for the whales, allowing them to be picked off one-by-one by whalers.*

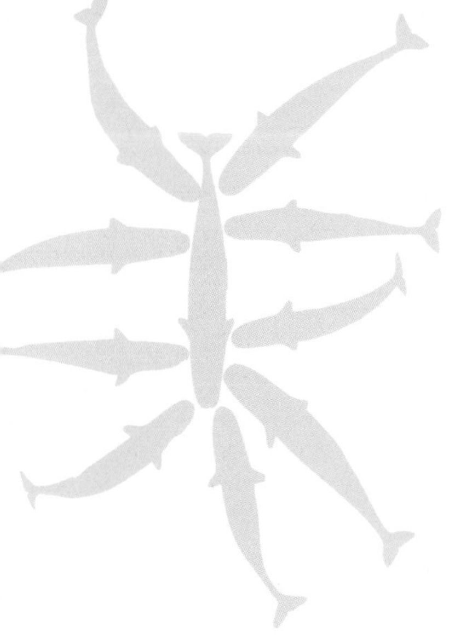

rface to rest or socialize for several hours a day. ey may lie parallel to each other, in a behavior metimes called "logging" (because of their semblance to immobile logs), or twist and turn the water, rolling and touching each other. The hales also perform breaches (leaps from the ater), lobtails (hitting the water with their tail kes), and spy-hops (raising the head vertically t of the water). Females and immatures breach lobtail at an overall rate of about once an hour. owever, breaches and lobtails are almost always ustered into bouts, which often coincide with e start or end of periods of surface socializing.

During these social times, Sperm whales often emit "codas" (stereotyped, patterned series of around 3–20 clicks). Reminiscent of our Morse code, codas last around 1–2 seconds, and can often be heard as exchanges, or "dialogs," between individuals. Codas are a form of communication. So, one whale may emit "click-click–pause–click," and another may respond with "click-click-click-click-click." Two whales can emit the same coda almost simultaneously, forming a duet that sounds like an echo. Groups of females have distinctive repertoires of roughly 12 commonly-used codas ("dialects"), and coda repertoires also vary geographically. Coda repertoire is probably culturally transmitted, passed down from mother and family unit to offspring.

More commonly, Sperm whales produce precisely-spaced echolocation clicks – called "usual" clicks – that are iterated at about two clicks per second. Also in the repertoire are streams of clicks, known as "creaks" because together they make a creaking sound. These are used in social situations among codas or in foraging, perhaps to home in on potential prey. Ringing "slow" clicks, produced around once every six seconds, are characteristic of large, breeding males. It is thought that slow clicks may advertise a breeding male's presence, size, and/or fitness, repel other males, attract females, or be bounced off other whales to aid the originator in his echolocation. Sperm whales are a notable exception to other social, toothed whales in that their sounds consist almost entirely of clicks (rather than whistles, as made by dolphins).

Kogia appear less sociable than Sperm whales. Pygmy sperm whales live either alone or in groups of up to six animals, while up to ten Dwarfs may coexist. In contrast to Sperm whales, male Dwarf sperm whales may associate in groups with females and their calves, and groups of immatures also form. All three species in the family Physeteridae are known to strand, but the Pygmy sperm whale is particularly prone, being one of the most commonly stranded cetacean species. In fact, much of what is known of *Kogia* comes from data collected at strandings.

Past Exploitation, Present Risk

CONSERVATION AND ENVIRONMENT

Estimates for the global population of Sperm whales range from 200,000 to 1.5 million. Sperm whales are listed as Vulnerable on the IUCN Red List, and the International Whaling Commission's 1988 moratorium has forbidden commercial whaling. *Kogia* are probably also rare, but their populations have not been properly censused.

Historically, however, the Sperm whale's contribution to human civilization has been prodigious. The huge animals' spermaceti oil and blubber largely fueled the Industrial Revolution. A second wave of whaling took place in the 20th century, using mechanized catcher vessels and explosive harpoons, resulting in the deaths of up to 30,000 Sperm whales each year. Due to their much greater size and proportionally larger spermaceti organs, large males were the main targets of the hunters, who by now marketed the oil as a high-grade industrial lubricant. This practice continues in the southeast Pacific, where large males are now rare and calving rates are so low as to put the longterm survival of this population in doubt.

According to the International Whaling Commission's Sperm whale model, populations grow at the tiny rate of less than one animal per year, even in ideal conditions. Sperm whales also die from entanglement in fishing gear, choking on plastic bags, and collision with ships. Chemical pollution is also evident in their blubber; levels of contaminants in Sperm whales are intermediate between inshore odontocetes – the highest – and baleen whales.

Because they rely so heavily on sound for all aspects of their lives, and because sound often carries further at depth, noise pollution is considered to be an additional threat. Shipping, underwater explosions, seismic exploration, oil drilling, military sonars and exercises, and oceanographic experiments all contribute to the increased undersea noise levels in the modern world. Sperm whales react markedly to such threats; for example, in response to sonar during the US invasion of Grenada in 1983, Sperm whales fell silent and presumably ceased foraging. They have also been seen to respond in a similar way to the sounds of a seismic vessel at work, even though it lay several hundred kilometers away. LW/HW

Beaked Whales

3

The beaked whales are one of the less well-known families of large mammals. Two new species have been described since 1990, and at least one probable species (referred to as Mesoplodon *sp. 'A') that matches no currently known species has been observed in the wild. Several species have yet to be seen alive and one* (M. bahamondi) *is only known from cranial remains.*

4

Beaked whales were long thought to be impossible to identify to species level at sea, but recent research has shown that this may not be true. Three species have been subjected to commercial whaling: Baird's and Cuvier's beaked whales, which were taken in small numbers off Japan, and the Northern bottlenose whale, of which about 50,000 were taken in the northern North Atlantic in the late 19th and early 20th centuries.

5

Dwellers in the Depths

FORM AND FUNCTION

The whales' family name comes from the distinctive beak found in all species; it derives from the Greek *xiphos*, meaning sword – hence Ziphiidae, or "sword-nosed whales." In all species there is no crease between the beak and the forehead as is found in other beaked cetaceans, notably some dolphins. The beak, which is long and slender in species such as Sowerby's beaked whale (*Mesoplodon bidens*), is short and poorly defined in Cuvier's beaked whale (*Ziphius cavirostris*). All beaked whales have one or two pairs of teeth that protrude from the closed mouth in adults to form tusks. In all but the genus *Berardius*, these tusks are sexually dimorphic and only erupt beyond the gums in adult males. Shepherd's beaked whale (*Tasmacetus shepherdi*) is the only species that has retained any teeth other than these tusks, so in most species juveniles and adult females are functionally toothless. This reduction in teeth is thought to relate to specialization on a diet of squid and to the use of suction-feeding for prey capture. The tusks appear to be used as weapons, and in most species adult males are heavily scarred with tusk-marks. The position and shape of the tusks vary considerably between species, particularly in the genus *Mesoplodon*, and are commonly used to identify species.

There is little variation in body form between Ziphiidae species other than in dentition, the shape of the forehead, and the length of the beak. These are medium-sized whales, varying in length from under 4m (13ft) in the Pygmy to over 12m (40ft) in Baird's beaked whale. The body is generally robust and spindle-shaped, with the widest girth at the midpoint. The pectoral fins are relatively small; in the genus *Mesoplodon*, they fit into small depressions in the side of the body when not being used for maneuvering. The dorsal fin is small and set two-thirds of the way from beak to tail. The tail fluke is broad relative to other cetaceans and lacks a central notch. Between the mandibles are two throat grooves, which are characteristic of the beaked whales; these converge anteriorly but do not meet, and are thought to function during suction-feeding.

Sexual dimorphism in most species is restricted to the development of tusks, although in the bottlenose whales the forehe[ad] also changes shape due to the development of t[wo] bony crests on the front of the skull that are among the largest masses of solid bone found in any mammal species. In most species there is lit[tle] or no sexual dimorphism in body size, although the bottlenose whales and Cuvier's beaked wha[le] males are larger, while for the four-toothed whal[es] the opposite holds true.

Life Beyond the Edge

SOCIAL BEHAVIOR

Most beaked whales live in oceanic waters at or beyond the shelf edge. They seem to be particularly common in the deep waters around oceani[c] islands, such as the Azores or the Canaries, and around submarine features such as canyons, gul[lies], and seamounts. Beaked whales probably fo[r]age at or close to the seabed in water depths of u[p] to 1,000m (3,300 ft) or more, and may be amo[ng] the deepest and longest divers of all whales. Div[es] lasting 20–30 minutes are commonly recorded, and Northern bottlenose whales have been recorded diving for up to 80 minutes or more.

Currently the three best-known species of beaked whale are Cuvier's, Blainville's, and the Northern bottlenose whale. Northern bottlenos[e] are widely distributed in the North Atlantic, but have only been studied at one location; all other information about them comes from past whalin[g]. The single population studied is thought to be resident year round in an area off the east coast [of] Canada, close to a deep gully on the seabed. All ages and sexes in the area have been recorded, and some individuals have been followed over many years. The average group size for this area i[s]

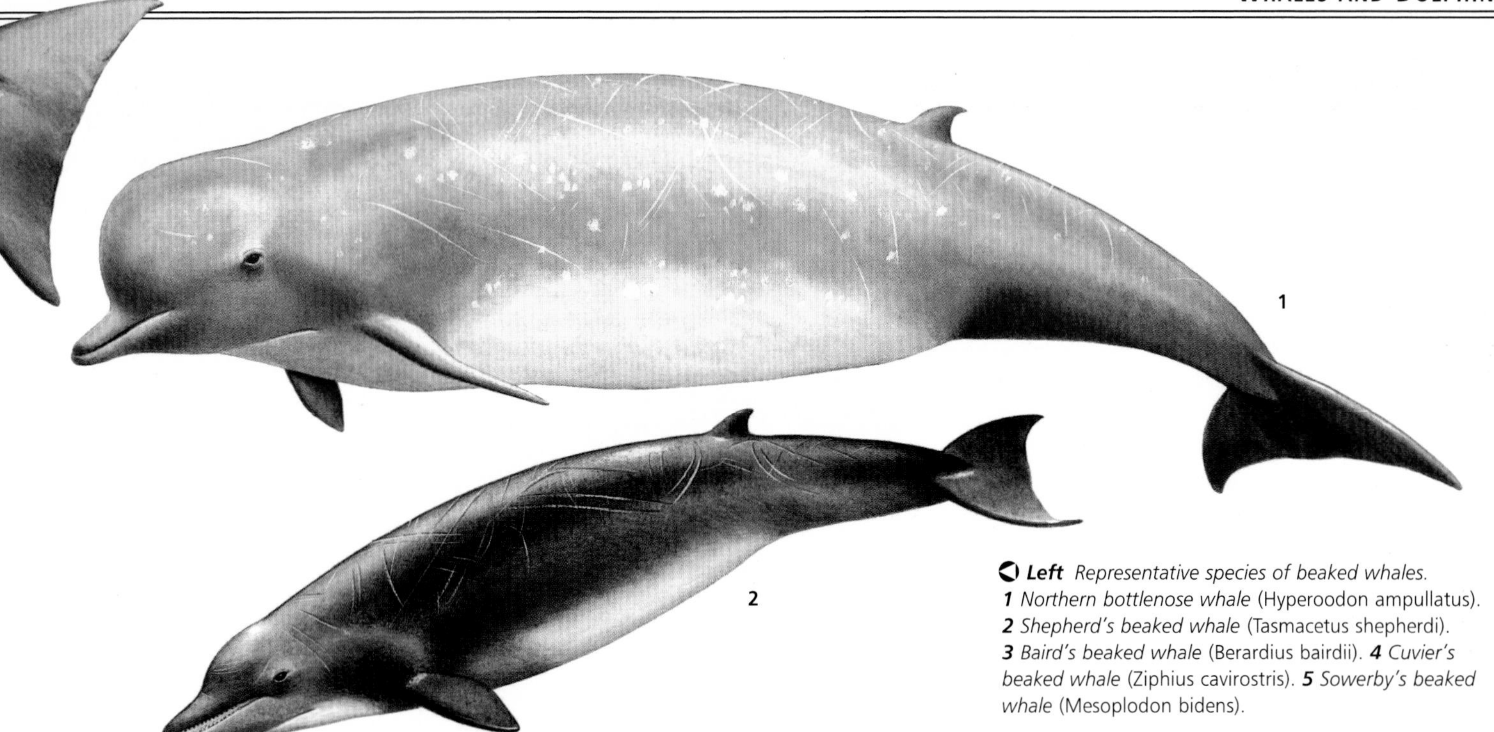

Left *Representative species of beaked whales.* **1** *Northern bottlenose whale* (Hyperoodon ampullatus). **2** *Shepherd's beaked whale* (Tasmacetus shepherdi). **3** *Baird's beaked whale* (Berardius bairdii). **4** *Cuvier's beaked whale* (Ziphius cavirostris). **5** *Sowerby's beaked whale* (Mesoplodon bidens).

ır animals, and although groups may consist of ages, there is some segregation between the :es. Longterm associations between adult males ve been recorded. In other areas, particularly ther north, the whales are migratory, moving rth in spring and south in late summer. Group es of up to 20 animals have also been recorded. Cuvier's beaked whales are generally seen in ters more than 200m (660ft) deep, and are quently recorded at 1,000m (3,300ft) or more. htings at sea commonly occur around sea-ounts, escarpments, and canyons. Groups of up seven animals have been observed, but pods of 4 may be more common. In some areas, indi-ual animals have been recorded several times se to the same localities.

Although Blainville's beaked whales are found roughout the warm temperate and tropical ters of the world's oceans, most information them currently comes from the northern hamas. Here the whales are commonly recorded in groups of up to seven at depths of 200–1,000m (660–3,300ft). Individual animals are commonly resighted both within a single year and over several years, especially near bottom features such as deepwater canyons. Groups generally consist of a number of adult females and juveniles, and it is rare for more than one adult male to be present. The whales of this species are probably polygamous, with adult males moving between groups of adult females.

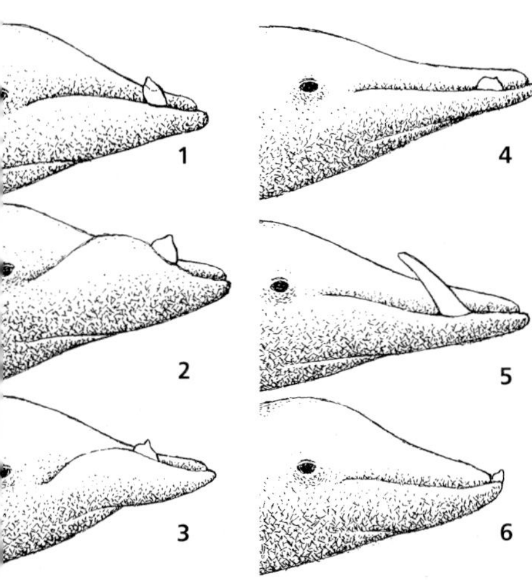

No Escape from Pollution

CONSERVATION AND ENVIRONMENT

The lack of information about living beaked whales means that little is known about their status or any threats to their conservation. In the past their preference for deeper waters may have isolated them from impacts that affected more coastal species, but this is starting to change. Noise pollution has been linked to a number of mass strandings since the mid-1980s, and increasingly biocontaminants are being recorded from their blubber. Stranded beaked whales often have plastics, in the form of bags or sheets, within their stomachs, and in some cases this may have been the cause of death. With increased exploitation of deepwater fisheries around the world, beaked whales have also become increasingly susceptible to bycatch in fishing gear, and may also suffer from depletion of prey if future expansion of the fisheries continues. CMacL

Left *Beaked whale species of the genus* Mesoplodon *have a single tooth, the shape and jaw position of which clearly distinguishes them from one another.* **1** *Hubb's beaked whale* (M. carlhubbsi). **2** *Blainville's beaked whale* (M. densirostris). **3** *Ginkgo-toothed beaked whale* (M. ginkgodens). **4** *Gray's beaked whale* (M. grayi). **5** *Strap-toothed whale* (M. layardii). **6** *True's beaked whale* (M. mirus).

FACTFILE

BEAKED WHALES

Order Cetacea

Family: Ziphiidae

At least 20 species in 6 genera: four-toothed whales (*Berardius*, 2 species); bottlenose whales (*Hyperoodon*, 2 species); genus *Mesoplodon* (13 or 14 species); Longman's beaked whale (*Indopacetus pacificus*); Shepherd's beaked whale (*Tasmacetus shepherdi*); and Cuvier's beaked whale (*Ziphius cavirostris*).

DISTRIBUTION All oceans

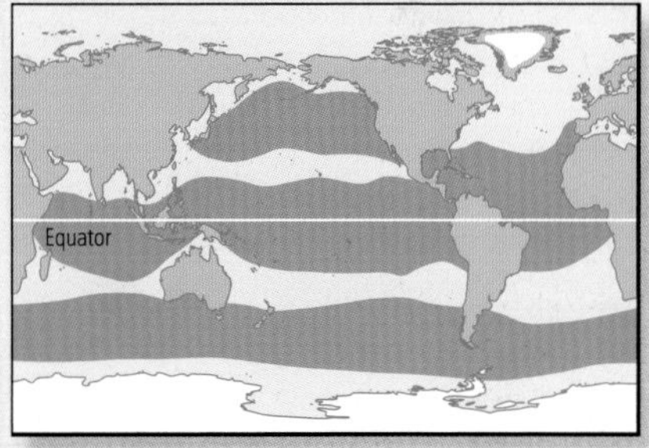

HABITAT Deep waters of shelf edge, slope, and abyssal plain.

SIZE Head-to-tail length 3.4–12.8m (11.5–42ft.); weight 1–15 tonnes.

SKIN Colors range from off-white through various shades of gray and brown; often marked by scarring.

DIET Deep water squid, fish, and some crustaceans.

BREEDING Gestation thought to last 12 months in the Northern bottlenose,17 months in Baird's beaked whale.

LONGEVITY Unknown for most species, though Northern bottlenose whales have been recorded up to 37 years of age.

See species table ▷

Beaked Whale Species

THE BEAKED WHALES WERE ONE OF THE first groups to split from the lineage of living cetaceans and they are now the second largest family in the order after the dolphins. The ancestors of the surviving species can be traced back as far as the Miocene, about 20 million years ago. Much remains to be learned about them, and new species are still being identfied. The limitations of current knowledge extends to the whale's conservation status; all species not otherwise listed below are classified by the IUCN as Data Deficient.

GENUS *BERARDIUS*

The largest of the beaked whales. Unlike other beaked whales, four-toothed whales have two pairs of teeth that erupt to form tusks in adults. The anterior pair, found at the tip of the lower jaw, are larger and triangular in shape, while the posterior pair, separated from the front teeth by a short gap, are smaller and peglike.

Arnoux's beaked whale LR
Berardius arnuxii

Arnoux's or Southern beaked whale, Southern four-toothed whale, Southern giant bottlenose whale, New Zealand whale

Circumpolar in the colder waters of the southern oceans, from 34°S to the ice edge.
HTL c.9m; WT 7–10 metric tons.
SKIN: bluish gray, sometimes with a brownish tint; flippers, flukes, and back darker, underside lighter; old males are dirty white from head to dorsal fin. Adult males and females may have extensive scarring.
BREEDING: gestation thought to last 17 months.
CONSERVATION STATUS: LR Conservation Dependent.

Baird's beaked whale LR
Berardius bairdii

Baird's beaked whale, Northern or Giant four-toothed whale, Northern giant or North Pacific bottlenose whale

N Pacific, from 24°N off California to 63°N; most sightings occur N of 35°N.
HTL up to 12.8m (female), 11.9m (male); WT 11–15 tonnes.
SKIN: bluish gray, sometimes with a brownish tint; flippers, flukes, and back darker, underside lighter; white blotches around the navel and anus; old males are dirty white from head to dorsal fin. Adult males and females may have extensive scarring.
BREEDING: gestation thought to last 17 months.
CONSERVATION STATUS: LR Conservation Dependent.

GENUS *HYPEROODON*

Bottlenose whales, characterized by a short, well-defined beak and a bulbous forehead. Adult males have two large bony crests on the skull that may function as weapons or protective shields. A single pair of pear-shaped teeth are situated at the tip of the lower jaw; these erupt, although only barely, in adult males. The Northern bottlenose whale is the better known of the two species, but their biology is thought to be similar.

Northern bottlenose whale LR
Hyperoodon ampullatus

Northern or North Atlantic bottlenose whale, flathead

N Atlantic, from 77°N to Cape Verde Islands in the E and from Davis straits to Cape Cod in the W. Also recorded in W Mediterranean and North Sea.
HTL up to 9.8m (male), 8.7m (female); WT 7.5 tonnes (male), 5.8 tonnes (female).
SKIN: younger animals are dark dorsally and light ventrally, becoming lighter with age; males develop a white area on the forehead that increases in size as they get older. Males have relatively few of the scars observed in other beaked whales.
BREEDING: gestation estimated at 12 months.
LONGEVITY: to 37 years (male), 27 years (female).
CONSERVATION STATUS: LR Conservation Dependent.

Southern bottlenose whale LR
Hyperoodon planifrons

Southern or Antarctic bottlenose whale, Flathead

S hemisphere, from 30°S to N edge of Antarctic icesheet; may be most common between 58°S and 62°S. May also occur in tropical Indo-Pacific.
HTL 7.45m (male), 6.94m (female); WT c.8 tonnes (male), 6 tonnes (female).
SKIN: younger animals are dark dorsally and light ventrally, becoming lighter with age; males develop a white area on the forehead that increases in size as they get older. Males have relatively little scarring.
CONSERVATION STATUS: LR Conservation Dependent.

GENUS *INDOPACETUS*

A single species, whose general morphology is unknown, though there are probably two conical teeth at the tip of the lower jaw.

Longman's beaked whale
Indopacetus pacificus

Longman's, Pacific, or Indo-Pacific beaked whale

A little-known species, known until the late 1990s only from two skulls, one from Queensland, Australia, the other from Somalia; however, recent reexamination of reports of a previously unidentified whale from the tropical Indo-Pacific, similar in morphology to a bottlenose, suggests they may refer to this species.
HTL estimated at 7–7.5m from length of skulls.

GENUS *MESOPLODON*

Characterized by a conservative body plan with little variation between species, except in the shape and position of the single pair of teeth that give this genus its name (*Mesoplodon* = armed with teeth in the middle of the jaw); these vary from small cones at the tip of the lower jaw to 30cm-long tusks set midway along the length of the gape. In addition, there is some variation in body size and the relative length of the beak. When observed in the wild, group size varies from 1–15 animals.

Mesoplodon species "A"

The designation of this species is based on about 30 sightings in the eastern tropical Pacific. These may represent living records of *M. bahamondi*, though until a stranded animal is recovered this remains speculation, and until a type specimen to compare with others is available, the species will remain unnamed.
HTL estimated at 5–5.5m.
SKIN: two distinct color forms have been observed: one (possibly adult male) with a broad, pale swathe contrasting sharply with an otherwise dark body; the other (possibly females and juveniles) uniformly gray-brown.

Mesoplodon bahamondi

Described in 1995 and known only from cranial remains. Thought to occur only off the Pacific coast of S and C America. No other information currently available, but may be the same species as *Mesoplodon* sp. "A".

Sowerby's beaked whale
Mesoplodon bidens

Sowerby's or North Sea beaked whale

Subpolar to warm temperate waters of Atlantic; most northerly sighting 71°3[illegible] N; one possible record from E Mediterranean, one from Gulf of Mexico. Commonly strands on coasts around the N[illegible] Sea, where it may become trapped due unfamiliarity with shallow waters and [illegible] of suitable food.
HTL c.5m; WT 1–1.3 tonnes. Identified the presence of two teeth, which are approximately triangular in shape and erally flattened, located in approximate the middle of the lower jaw.

Andrew's beaked whale
Mesoplodon bowdoini

Andrew's, Splay-toothed, Bowdoin's, or Deepcrest beaked whale

Cool temperate waters of Australia and New Zealand. Known only from a limit number of stranded specimens, and ha never been seen alive.
HTL 4–4.7m; WT 1–1.5 tonnes.
SKIN: generally dark all over, with a wh tip to the beak and lower jaw. Identifie the presence of two large laterally-flatte teeth in raised sockets about 20cm fro tip of lower jaw.

Hubb's beaked whale
Mesoplodon carlhubbsi

Hubb's or Arch-beaked whale

Temperate waters of N Pacific, from Bri Columbia to California; in W, most records are from Japan. Distribution thought to correlate with the deep subarctic current system.
HTL c.5m; WT 1–1.5 tonnes.
SKIN: Generally dark all over, with a wh tip to the beak and a white area around the blowhole. Identified by the presenc two large, laterally–flattened teeth set b from the tip of the lower jaw.

Blainville's beaked whale
Mesoplodon densirostris

Blainville's or Dense beaked whale

The most widely distributed *Mesoplodo* species, found in warm temperate and tropical waters of all the world's oceans Commonly observed in water depths o 200–1,000m, particularly around mari canyons.
HTL c.4.5m; WT c.1 tonne.
SKIN: young are dark on top and lighte below; adults are dark all over, althoug

ABBREVIATIONS HTL = head–tail length WT = weight

Approximate nonmetric equivalents: 1m = 3.3ft; 10kg = 22lb

Ex Extinct
EW Extinct in the Wild
Cr Critically Endangered
En Endangered
Vu Vulnerable
LR Lower Risk

Left *A group of Blainville's beaked whales* (Mesoplodon densirostris). *This is one of the smaller species of beaked whale.*

actual color can vary from brown to k gray. Adult males are frequently ·ked by a tangled mass of furrows and ite scars, which may cover top of head l back as far as dorsal fin. Identified by sence of a stepped lower jaw; in adult les, two large, almost conical teeth pro-de from highest point of the step and visible above the animal's head.

rvais's beaked whale
·soplodon europaeus
vais', Antillean, or Gulf Stream beaked whale

rm temperate and tropical N Atlantic; ge may also stretch S across equator, as re have been sightings from Ascension and.
. c.5m; WT 1–1.5 tonnes.
N: Generally dark on top and lighter ow. Adult males may have linear scar- g caused by the teeth of conspecifics. ntified by presence of a single pair eeth set 7–10cm back from tip of ver jaw.

nkgo-toothed beaked whale
·soplodon ginkgodens
kgo-toothed or Japanese beaked whale

nge poorly defined due to rarity of andings and sightings; occurs in ific and Indian Oceans and recorded m California, Mexico, Japan, Taiwan, lapagos Islands, Sri Lanka, Indonesia, atham Islands, and New South Wales.
. 4.7–5.2m; WT 1.5–2 tonnes.
N: dark all over, with white spots and tches around the navel. Few or none of linear scars characteristic of adult les of other beaked whales species. ntified by the presence of a single pair eeth in lower jaw set back from the tip l shaped like leaves from the ginkgo tree.

Gray's beaked whale
Mesoplodon grayi
Gray's, Scamperdown, or Southern beaked whale

Cold temperate waters of the southern oceans.
HTL 4.5–5.6m; WT 1–1.5 tonnes.
SKIN: dark bluish gray, brownish gray, or black on the upper surface, with a paler underside; beak, throat, and front of forehead may be white. Identified by the presence of a single pair of laterally-flattened teeth, shaped in profile like a flattened onion, about 20cm from tip of lower jaw.

Hector's beaked whale
Mesoplodon hectori
Hector's or New Zealand beaked whale, Skew-beaked whale

Southern oceans, particularly New Zealand, but also Argentina, Falkland Islands, and South Africa; thought to inhabit cooler waters than *M. grayi*. Also recorded in N Pacific, but genetic analysis suggests these specimens may be a separate, and therefore new, species.
HTL 4–4.5m; WT 1–2 tonnes.
SKIN: dark gray or brownish gray back and upper sides, paler or white underneath. Males may have many pale linear scars caused by the teeth of other males. Identified by the presence of a single pair of laterally-flattened teeth close to the tip of the lower jaw.

Strap-toothed whale
Mesoplodon layardii
Strap-toothed whale, Strap-toothed beaked whale, or Layard's beaked whale

Circumpolar in S hemisphere between 30°S and 65°S; one of the most commonly recorded species of *Mesoplodon*.
HTL up to 6.2m; WT up to 3 tonnes.
SKIN: highly contrasting dark and light pattern: dark face mask, white throat and upper back, dark body and tail, white patch around genital slits. Identified by the presence of a single pair of teeth set well back in the lower jaw; in adult males these can reach over 30cm in length, and may curl up and cross over the upper jaw, restricting the animal's gape.

True's beaked whale
Mesoplodon mirus
True's or Wonderful beaked whale

Two separate populations, one in the warm temperate N Atlantic, the other in the S Indian Ocean.
HTL 4.9–5.3m; WT 1–1.5 tonnes.
SKIN: dark gray or bluish-gray back, with mottled gray or pale undersides; dark patch around eye; in the S hemisphere population, the rear third of the body is paler. Identified by a single pair of teeth with an oval cross-section situated at the tip of the lower jaw.

Pygmy beaked whale
Mesoplodon peruvianus
Pygmy, Peruvian, or Lesser beaked whale

The smallest species of beaked whale, first described in 1991. Seems to be restricted to warmer waters of Pacific.
HTL 3.4–3.7m.
SKIN: uniformly dark gray above, with paler underside. Identified by the presence of a single pair of teeth in the lower jaw which are set on a raised area of bone; the teeth, even in adult males, are very small relative to other species.

Stejneger's beaked whale
Mesoplodon stejnegeri
Stejneger's, Saber-toothed, Bering Sea, or North Pacific beaked whale

Confined to the cold temperate waters of the N Pacific, Sea of Japan, and S Bering Sea. Reported to have blue-green colored milk, a feature found in only one other mammal.
HTL 5–5.3m; WT 1–1.5 tonnes.
SKIN: black, dark gray, or brown above, with lighter-colored undersides; white tip to the end of the beak. Identified by the presence of a single pair of large and tusk-like laterally-flattened teeth set about 20cm back from the tip of the lower jaw.

GENUS *TASMACETUS*

A single species, marked by a long, slender beak with two large, sexually dimorphic conical teeth at the tip of the lower jaw. 26–27 smaller, conical, nonsexually dimorphic teeth line the rest of the lower jaw, and 19–21 similar teeth are present in the upper jaw. This is the only genus of beaked whale with functional teeth in the upper jaw.

Shepherd's beaked whale
Tasmacetus shepherdi
Shepherd's beaked whale, Tasman beaked whale, or Tasman whale

Known from about 20 specimens, all recorded in S hemisphere.
HTL c.7m.; WT 2–3 tonnes.
SKIN: dark brown back and sides with creamy white underparts; two lighter diagonal stripes on either side, and a light patch above flippers; top of head may be lighter than dorsal surface of the body.

GENUS *ZIPHIUS*

A single species, marked by a short, stubby, indistinct beak, which resembles that of a goose in profile, with two conical teeth at the tip of the lower jaw.

Cuvier's beaked whale
Ziphius cavirostris
Cuvier's beaked whale, Goose-beaked or Goosebeak whale

Widest distribution of any beaked whale species, found throughout the world's oceans from the cold temperate regions to equatorial waters.
HTL up to 7m; WT 2–3 tonnes.
SKIN: dark brown, tan, or gray on the back and sides, paler underneath; males develop a lighter area on the head and upper back, which may extend as far back as the dorsal fin in the oldest animals. Adult males are frequently marked with long, pale, linear scars, caused by the teeth of other males during fights.

DISTRIBUTION All oceans

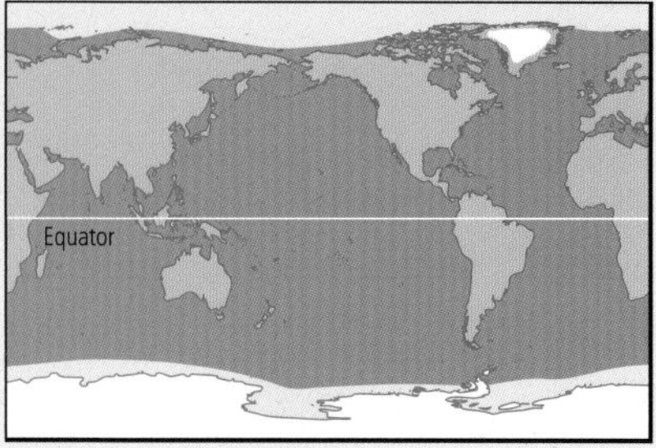

Suborder: Mysticeti
13 species in 6 genera and 4 families

GRAY WHALE Family Eschrichtiidae p256
1 species *Eschrichtius robustus*

RORQUALS Family Balaenopteridae p262
8 species in 2 genera
Blue whale *Balaenoptera musculus*
Fin whale *B. physalus*
Sei whale *B. borealis*
Bryde's whale *B. edeni*
Balaenoptera brydei
Northern Minke whale *B. acutorostrata*
Southern Minke whale *B. bonaerensis*
Humpback whale *Megaptera novaeangliae*

RIGHT WHALES Family Balaenidae p270
3 species in 2 genera
Northern right whale *Eubalaena glacialis*
Southern right whale *E. australis*
Bowhead whale *Balaena mysticetus*

PYGMY RIGHT WHALE Family Neobalaenidae p270
1 species *Caperea marginata*

Baleen Whales

THE SUBORDER MYSTICETI, OR BALEEN WHALES, contains relatively few living species, but they make up for their paucity by their size – the Blue whale is the largest animal ever to have lived. In baleen whales, teeth appear only as vestigial buds in the embryo. In their place, there has evolved a new structure, the baleen plates, which are totally unconnected to teeth. These strain the small fish and crustaceans that make up their diet. Blue whales strain krill – tiny zooplanktonic organisms – in huge quantities. The baleen plates are fringed with bristles, and the food organisms are dislodged from the baleen by the tongue.

The Gray whale is the sole member of the family Eschrichtiidae, and is confined to the North Pacific (a North Atlantic population only became extinct in the early 18th century). It has a narrow, gently arched rostrum (beak), two – rarely, four – short throat grooves, and no dorsal fin.

The slim, torpedo-shaped rorquals of the *Balaenoptera* genus are linked in the family Balaenopteridae with the Humpback whale, the sole representative of *Megaptera*. All the family members have a series of throat pleats that expand as they ingest water filled with plankton, and then contract to force the water over the baleen plates. This leaves the plankton stranded on the fibrous mat that forms the frayed inner edges of the plates. The Humpback differs from other rorqua in being rather stoutly built, with fewer and mu coarser throat grooves and a pair of very long, robust flippers. The head, lower jaw, and flipper front edges are covered with irregular knobs, an the trailing edge is indented and serrated. All ba aenopterids except the Blue whale will also feed on fish; indeed, fish are the staple diet of the Minke whale, the smallest species.

The three members of the family Balaenidae have massive heads that take up as much as one third of their total body length. In these species, the rostrum is long and narrow, and arches upward, though the lower jaw bones do not. This leaves a space that is filled by the huge lower lip that rise from the lower jaw; these in turn enclo the long, narrow baleen plates that hang from th edges of the rostrum. All have seven neck verteb fused into a single mass, a feature they share wit the sole member of the family Neobalaenidae, th Pygmy right whale (which alone has a dorsal fin)

With the exception of the tropical Bryde's whale, rorquals are found throughout the world oceans. The Bowhead has a restricted Arctic dis bution; the Northern right whale occurs only in the North Atlantic and the Southern right whale only in the South Atlantic; while the Pygmy righ whale is confined to the Southern Ocean. PGI

TYPES OF BALEEN

Although the basic principle of filter feeding through the use of the baleen plates is the same in all the mysticete whales, there are significant variations in physiology and feeding technique. The two extremes are typified by the Northern and Southern right whales and rorquals such as the Sei whale. The Right whales have a narrow rostrum and long baleen plates. They feed by skimming the surface of the sea, collecting food organisms that they then dislodge with their tongues. The Sei whale, on the other hand, has a wide rostrum with short baleen plates. It gulps down huge mouthfuls of seawater and then raises its tongue to force the water back through the baleen plates, leaving the krill behind. See also Rorqual Feeding Habits.

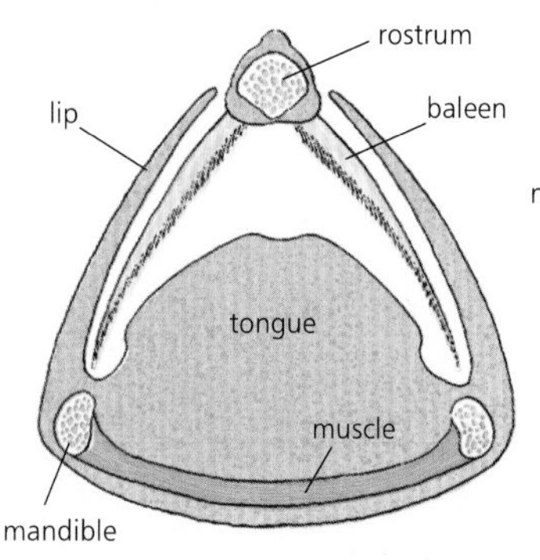

Northern and Southern right whales

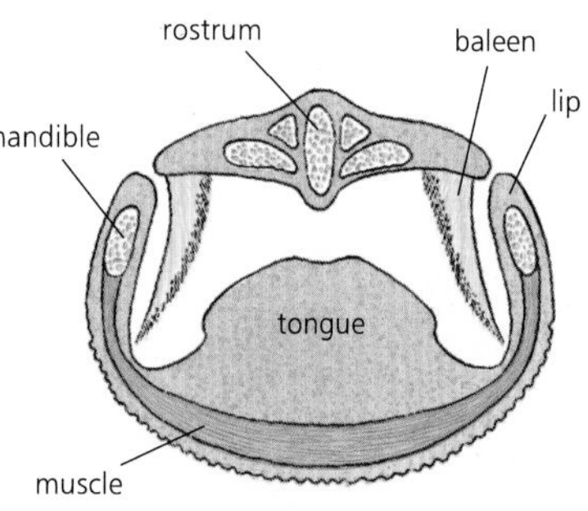

Sei whale

Above *Gray whale baleen. This bottom-feeding species has baleen that is much heavier and shorter than the surface filter-feeders, enabling it to plow through seafloor sediment in search of crustaceans, worms, and mollusks.*

E ADVANTAGE OF SIZE

st baleen whales are characterized heir huge size. Large bodies are e to store more fat; moreover, large y mass is a distinct advantage for m-blooded animals when it comes etaining heat in freezing oceans. iough whales' bodies ostensibly sent a huge surface area over which t loss might occur, the key factor in intaining body heat is the surface-volume ratio, which grows more favorable with increased bulk. Insulation for sea mammals is provided by a subcutaneous layer of fatty tissue – blubber. In the Bowhead whale, an Arctic species, this can be up to 50cm (20in) thick. However deep a whale dives, the blubber ensures that its body-core temperature stays stable at around 36–37°C (97–99°F). Of course, this can also cause them problems with cooling down!

Since oxygen storage capacity increases in proportion to body mass, larger whales can store more oxygen both in absolute terms and relative to their rate of oxygen use. Little wonder, then, that the largest whales are also the deepest divers.

Below The extremes of baleen whale body mass are represented by the Pygmy right whale and the Blue whale. The Blue whale exceeds in size even the largest dinosaurs ever to have lived on Earth.

my right whale

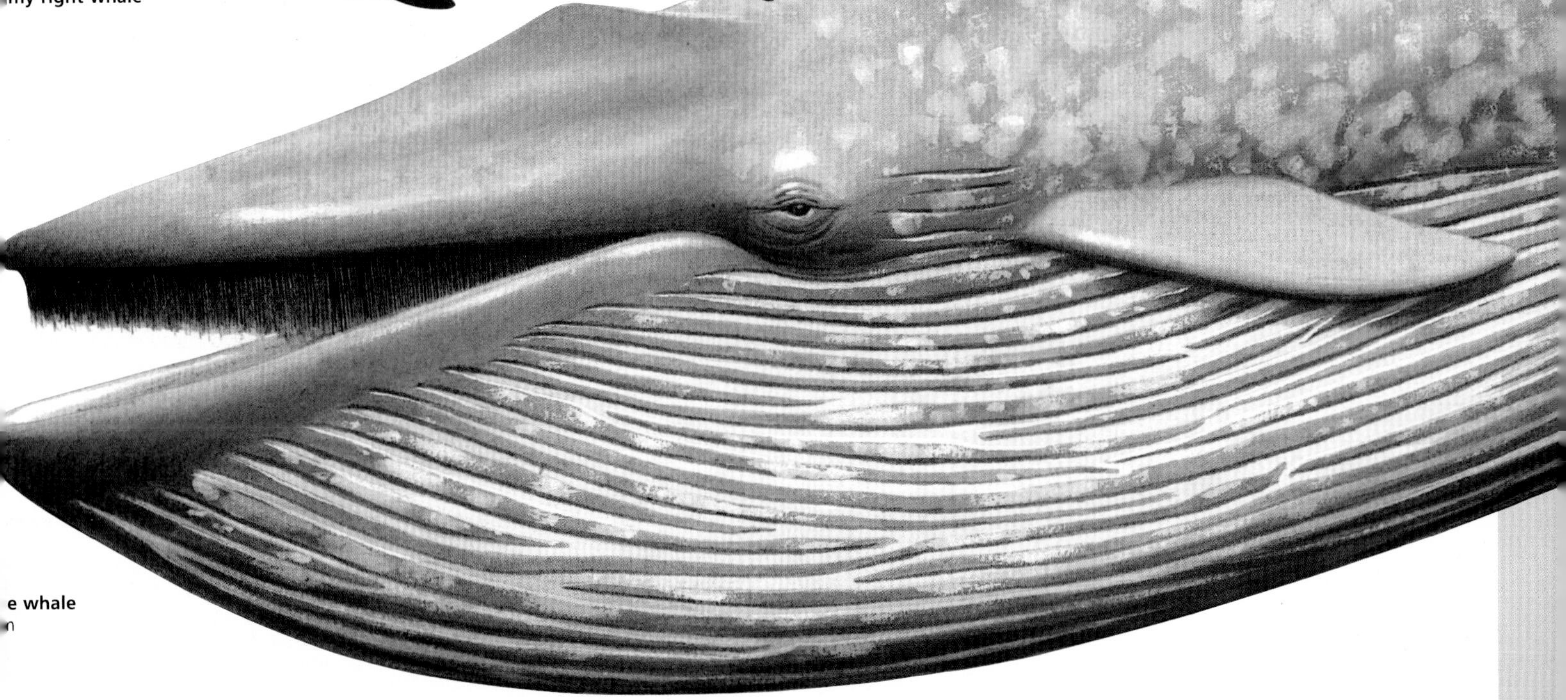

e whale
n

RIGINS AND EVOLUTION

e baleen whales are thought to have olved in the western South Pacific, iere rich zooplankton deposits in e Oligocene strata, together with e occurrence of fossils of the earliest ysticete forms (early cetothere ances-rs of the whales) suggest that these ay have favored the evolution of leen and a filter-feeding mode of . From here they may have dispersed o the Pacific and Indo-Pacific regions ong lines of high productivity during e late Cenozoic, although the rquals appear to have been originally stributed in the warm, temperate rth Atlantic.

SKULLS OF BALEEN WHALES

Baleen whales exhibit a more extreme modification of the skull than toothed whales – so much so that it is at first hard to believe that such bones could support the head and enclose the brain of these creatures. The principal modifications are the extension of the jaws, the upper of which (the rostrum) supports the baleen plates; forward movement of the supraoccipital (back of the skull) region over the frontal; and the consequent merging of the rostral and cranial bones. In the Bowhead whale, the rostrum has a pronounced curve to accommodate the long baleen plates. In rorquals such as the Humpback whale, the rostrum is broader and only gently curved. The Gray whale is a bottom-feeder, which "plows" the ocean floor, and its jaws are consequently shorter and thicker than in the other species, the upper supporting short, stiff baleen.

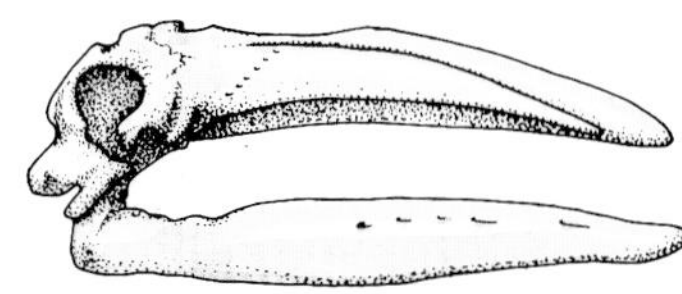

Gray whale
240 cm

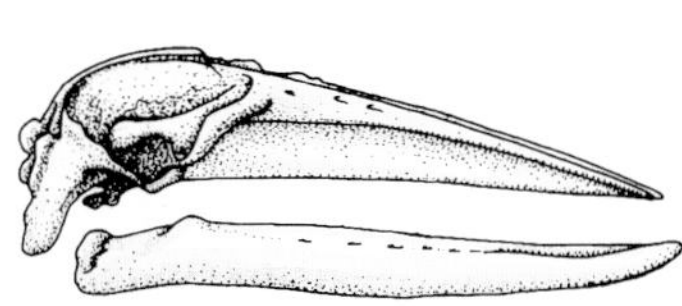

Humpback
350 cm

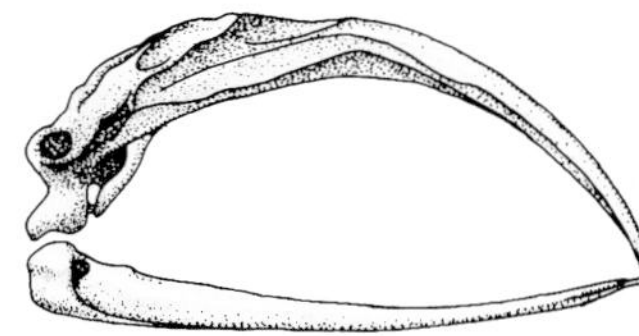

Bowhead whale
500 cm

Gray Whale

GRAY WHALES ARE THE MOST COASTAL OF THE baleen whales, and are often found within one kilometer of shore. This preference for coastal waters, and the accessibility to humans of the breeding lagoons in Mexico, make them among the best-known cetaceans. Thousands of people watch the "Grays" swimming past the shores of California.

Each fall and spring, gray whales migrate along the western coast of North America on their yearly passage between summer feeding grounds in the Arctic and winter calving areas in the protected lagoons of Baja California, Mexico. Their migration can be the longest of any mammal; some individuals may swim as far as 20,400km (12,675mi) yearly from the Arctic ice-pack to the subtropics and back.

Big and Barnacled

FORM AND FUNCTION

The Gray whale averages about 12m (40ft) in length, but can reach up to 15m (50ft). The skin is a mottled dark to light gray in color, and is one of the most heavily parasitized among cetaceans. Both barnacles and whale lice (cyamids) live on it in great abundance, barnacles particularly on top of the whale's relatively short, bowed head, around the blowhole, and on the anterior part of the back; one barnacle and three whale lice species are unique to the Gray whale and have not – so far – been found anywhere else. Albino Grays have been sighted, though they would appear to be extremely rare.

Gray whales lack a dorsal fin, but do have a dorsal ridge of 8–9 humps along the last third of the back. The baleen is yellowish-white, and is much heavier and shorter than in other baleen whales, never exceeding 38cm (15in) in length, no doubt because Gray whales strain bottom sediments to get at their prey, whereas other whales merely strain the water column (see Deep Harvest box). Under the throat are two longitudinal grooves about 2m (6.6ft) long and 40cm (16in) apart. These grooves may stretch open and allow the mouth to expand during feeding, thus enabling the whale to take in more food.

While migrating, Gray whales swim at about 4.5 knots (8km/h), but they can attain speeds of 11 knots (20km/h) under stress. Migrating Grays swim steadily, surfacing every 3–4 minutes to blow 3–5 times. The spout is short and puffy, and is forked as it issues from both blowholes. The tail flukes often come out of the water on the last blow in a series as the whale dives.

Above *Gray whale mother and calf. The young Gray whales are smooth and sleek compared to thei barnacled elders. Calves are usually born between la December and early February and at birth can weigh up to 500kg (1,100lb), at which point they still lack the blubber necessary to withstand the freezing tem peratures of Arctic waters.*

Below *Like tiny, bejeweled grottoes, barnacle clusters surround the blowhole of a Gray whale. Mo of the great whales are host to barnacles, but Gray whales are particularly heavily encrusted. It is in and around these clusters of barnacles that whale lice live – small, pale, spidery creatures usually about 2.5cm (1in) in length.*

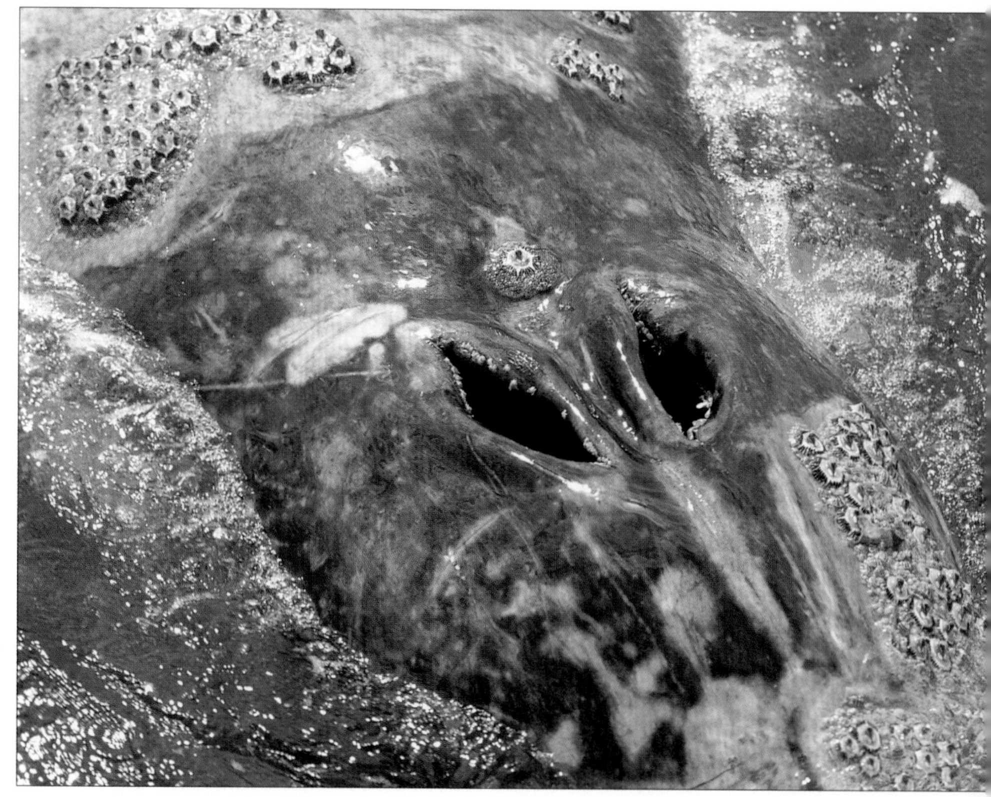

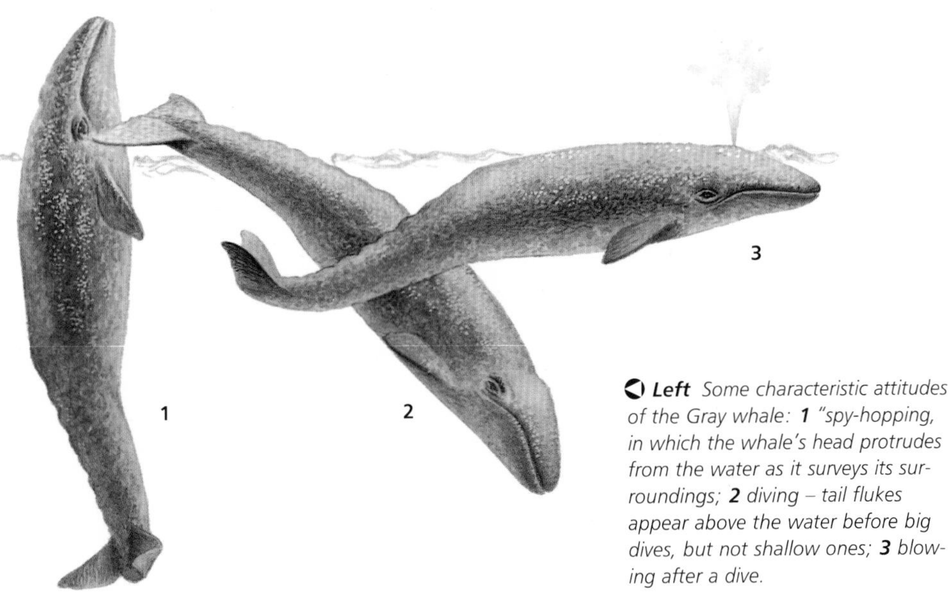

Left *Some characteristic attitudes of the Gray whale:* ***1*** *"spy-hopping, in which the whale's head protrudes from the water as it surveys its surroundings;* ***2*** *diving – tail flukes appear above the water before big dives, but not shallow ones;* ***3*** *blowing after a dive.*

The Gray whale's sound repertoire includes ınts, pulses, clicks, moans, and knocks; in the ;oons of Baja California calves also emit a low, onant pulse that attracts their mothers. But in ay whales sounds do not appear to have the mplexity or social importance of those proced by other cetaceans. The exact significance most of their communications is not known.

Along the Pacific Coast

DISTRIBUTION PATTERNS

present there are only two stocks of Gray ıales: the Californian, and the separate Western cific stock. Gray whales once inhabited the orth Atlantic but disappeared in the early 1700s, obably due to whaling.

The Californian Gray whale calves during the nter in lagoons, such as Laguna Ojo de Liebre d Laguna San Ignacio, on the desert peninsula Baja California, Mexico. They summer in the orthern Bering Sea near Saint Lawrence Island d north through the Bering Straits into the ıukchi Sea, almost to the edge of the Arctic pack e. A small portion of the population summers ong the North American coast from northern alifornia to Alaska. The only known current mmer ground of the Western Pacific stock is ar Sakhalin Island in the Okhotsk Sea. This opulation, numbering as few as 100 animals, igrates southward each fall past the east and est shores of Japan to unknown calving grounds.

Calving Grounds to Feeding Grounds

SOCIAL BEHAVIOR

ay whales reach puberty at about 8 years of age ange 5–11 years), when the mean length is 1.1m (36ft) for males and 11.7m (38ft) for males; they attain full physical maturity at about). Like the other baleen whales, females of the ecies are larger than males, probably to satisfy e greater physical demands of bearing and nursing young. Females give birth in alternate years to a single calf about 4.9m (16ft) long, after a gestation period of just over a year.

Gray whales are adapted to migration, and many aspects of their life history and ecology reflect this yearly movement from the Arctic to the subtropics. The majority of the Eastern Pacific, or Californian, population spends from May through November in Arctic waters.

At the start of the Arctic winter their feeding grounds begin to freeze over. The whales then migrate to the protected lagoons, where the females calve. The calves are born within a period of 5–6 weeks, with a peak occurring about 10th January. At birth the calves have coats of blubber that would be too thin for them to withstand cold Arctic water, though they thrive in the warm lagoons. For the first few hours after birth the breathing and swimming of the calf are uncoordinated and labored, and the mother sometimes has to help the calf to breathe by holding it to the surface with her back or tail flukes. The calves are nursed for about seven months, beginning in the confined shallow lagoons, where they gain motor coordination and perhaps establish the mother–young bond necessary to keep together on the migration north to the summering grounds where they are weaned. By the time the calves have arrived in the Arctic, they have built up thick insulating blubber coats from the milk of the nursing females. In the lagoons and off southern California, the calves stay close to and almost touching their mothers; but by the time they reach the Bering Sea in late May and June they are good swimmers and may be seen breaching energetically away from their mothers.

Since the migration route follows the coast closely, the whales may navigate simply by staying in shallow water and keeping the land on their right or left side, depending on whether they are migrating north or south. At points along the migration route Gray whales are often seen "spy-hopping." To spy-hop, a whale thrusts its head straight up out of the water and then slowly sinks back down along its horizontal axis. This contrasts with the breach, where a whale leaps half way or more out of the water and then falls back on its side, creating a large splash. It is possible that Gray whales spy-hop to view the adjacent shore and thus orient their migration.

Mating and other sexual behavior have been observed throughout the range at all times of year, but most conceptions occur within a three-week period during the southward migration, with a peak around mid-December. Gray whale sex may involve as many as five or more individuals rolling and milling together, but when conception occurs is unknown. Some authors have speculated that the extra animals are necessary to hold the mating

FACTFILE

GRAY WHALE

Eschrichtius robustus

Gray whale, California gray whale, or devilfish

Family: Eschrichtiidae

Sole member of genus

DISTRIBUTION Two stocks: E Pacific, or Californian, from Baja California along Pacific coast to Bering and Chukchi seas; W Pacific from S Korea to Okhotsk Sea.

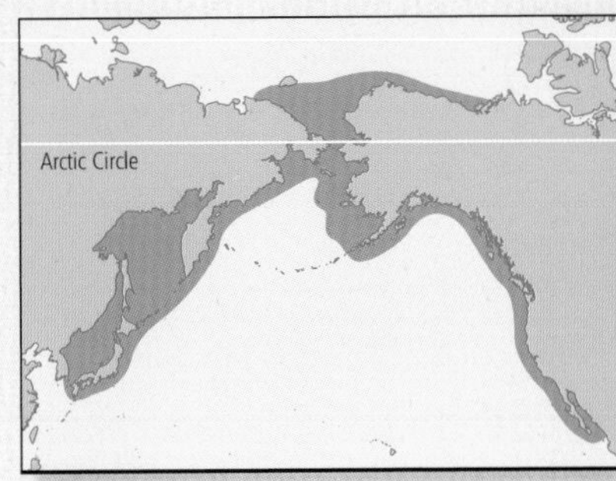

HABITAT Usually in coastal waters less than 100m (330ft) deep.

SIZE **Head–tail length** male 11.9–14.3m (39–47ft), female 12.8–15.2m (42–50ft); **weight** male 16 tonnes, female (pregnant) 31–34 tonnes.

SKIN Mottled gray, usually covered with patches of barnacles and whale lice, no dorsal fin, but low ridge on rear half of back. Two throat grooves. White baleen. Spout paired, short, and bushy.

DIET Bottom-dwelling amphipods and a variety of planktonic invertebrates.

BREEDING Gestation 13 months; a single calf is born in alternate years.

LONGEVITY Sexually mature at 8 years, physically at 40; maximum recorded lifespan 77.

CONSERVATION STATUS Classified generally as Lower Risk: Conservation Dependent, though the Northwest Pacific stock is Critically Endangered.

Above *As part of the mating ritual, male and female Gray whales caress one another as they swim along. Provided that the female does not rebuff the male's advances, a fleeting congress will occur; although this only lasts between 10–30 seconds, it will occur repeatedly.*

Right *Pinkish in color, the Gray whale's penis, which is pliable, is some 2m (6ft) in length and approximately 20cm (8in) broad at the base. When the penis is flaccid, it folds up in an "S" shape into an abdominal groove.*

pair together; if so this would class as an extreme example of cooperation.

The migration off California occurs in a sequence according to reproductive status, sex, and age-group. Heading south, the migration is led by females in the late stages of pregnancy, presumably responding to a physiological imperative to give birth in warm water; all other whales probably make the migration for social purposes connected with mating. Next come the recently impregnated females who have weaned their calves the previous summer. Then come immature females and adult males and finally the immature males. The migration north is led by the newly pregnant females, perhaps hurrying to spend the maximum length of time feeding in the Arctic to nurture the fetus developing inside them. The adult males and nonbreeding females follow, then immature whales of both sexes, and finally, meandering slowly, come females with newborn calves.

Observers have noticed changes in the sizes of groups as the migration progresses. In the early part of the southward migration, single whales predominate, presumably mostly females carrying near-term fetuses, and almost no whales are in groups of more than six. These leading whales swim steadily, seldom deviating from the migratory path, which suggests that they are hurrying south to give birth. During the remainder of the migration, groups of two predominate, but there may be as many as 11 in one group in the middle of the procession. These later whales seem to have a tendency to loiter more en route, particularly toward the end of the migration.

In the calving grounds, the males and subadults are concentrated in the areas around the lagoon mouths where much rolling, milling, and sexual play can be seen, while the mothers and calves seem to use the shallower portions deep inside the lagoons. In the Arctic, 100 or more Gray whales may gather to feed in roughly the same area.

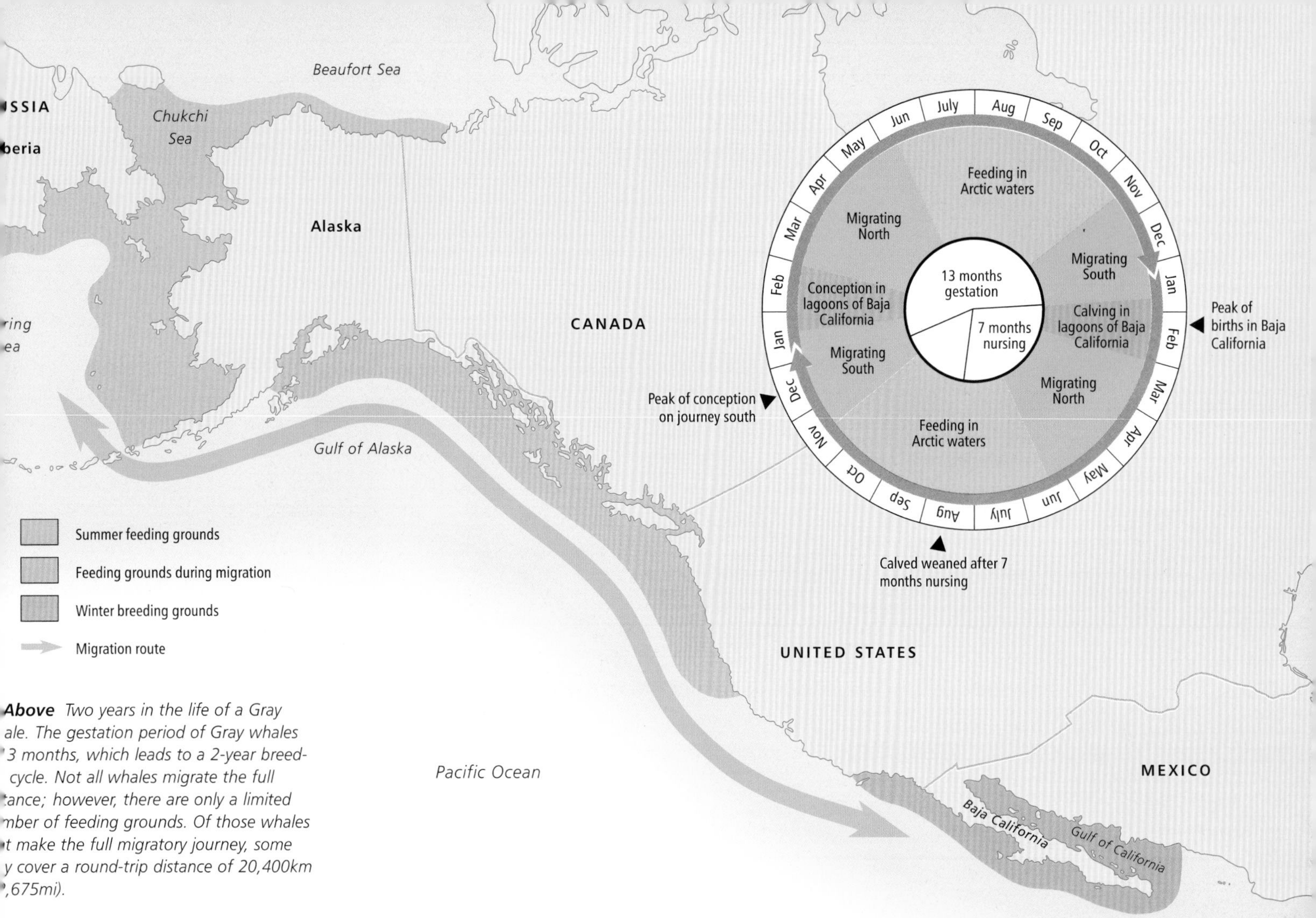

SSIA

beria

Beaufort Sea

Chukchi Sea

Alaska

ring ea

CANADA

Gulf of Alaska

UNITED STATES

Pacific Ocean

MEXICO

Baja California

Gulf of California

July | Aug | Sep | Oct | Nov | Dec | Jan | Feb | Mar | Apr | May | Jun

Feeding in Arctic waters

Migrating North

Migrating South

Conception in lagoons of Baja California

Migrating South

13 months gestation

7 months nursing

Calving in lagoons of Baja California

Migrating North

Feeding in Arctic waters

Jan | Dec | Nov | Oct | Sep | Aug | July | Jun | May | Apr | Mar | Feb

Peak of births in Baja California

Peak of conception on journey south

Calved weaned after 7 months nursing

Summer feeding grounds

Feeding grounds during migration

Winter breeding grounds

Migration route

Above *Two years in the life of a Gray ale. The gestation period of Gray whales 3 months, which leads to a 2-year breed- cycle. Not all whales migrate the full ance; however, there are only a limited mber of feeding grounds. Of those whales t make the full migratory journey, some y cover a round-trip distance of 20,400km ,675mi).*

Some individuals do not make the entire migra- n north. For example, off the coast of British lumbia some individuals stay in the same area ding for the eight or nine months between rth and southbound migration, and some indi- ual whales have been recorded returning to the ne location each summer. Similar small, sum- r resident populations occur from northern lifornia to Alaska. These residents seem to lude both sexes and all age groups, including nales with calves. This is perhaps an alternative ding strategy to making the full migration, but s one that only a few whales can afford, since ding areas south of the northern Bering Sea are obably rare and can, therefore, only support a ction of the population.

The only known nonhuman predator on the ay whale is the Killer whale. Several attacks have en observed, most often on cows with calves, sumably in an attempt to get at the relatively fenseless calf. Killer whales seem to attack par- ularly the lips, tongue, and flukes of the Grays, areas that may most readily be grasped. Adult ays accompanying calves will place themselves otectively between the attackers and the calves. hen under attack, Grays swim toward shallow ter and kelp beds near shore, where Killer ales seem hesitant to enter. Gray whales pond to underwater playback of recordings of ler whale sounds by swimming rapidly away or taking refuge in thick kelp beds.

Below *When young Gray whales are first born they are not especially coordinated and sometimes the mother will hold the calf to the surface on her back to help it breathe; however, as the calf gets a little older swimming onto its mother's back becomes a game.*

Hostages to the Hunt

CONSERVATION AND ENVIRONMENT

The Western Pacific Gray whale currently exists in very small numbers, and is one of the most endangered whales. Heavy whaling pressures in the first third of the 20th century, and sporadic whaling since, are undoubtedly responsible for this sad decline.

For thousands of years, Eskimo, Aleut, and Native American whaling tribes took Gray whales from the Californian stock in the northern part of its range. In the 1850s Yankee whalers began killing Gray whales, both in the calving lagoons and along the migration route, in such numbers that by 1874 one whaling captain, Charles Scammon, was predicting that the Californian Gray would soon be extinct. By 1900, with the Californian stock reduced to a tiny remnant population, whaling virtually ceased. It resumed in 1913, continuing sporadically until 1946, when the International Whaling Commission (IWC) was formed. This body prohibited a commercial take of Gray whales, although the Soviet Union was permitted an aboriginal catch for the Eskimo people living on the Chukchi peninsula. At present Russia has an annual quota of 140 Gray whales. The United States also has a hunt quota of five whales a year, taken by the Makah people of Washington State.

Since the decline of whaling, the Eastern Pacific, or Californian, stock has made a steady recovery. The present population is estimated at approximately 25,000, with some indications that its growth is beginning to level off. JD/AT

Right *A Gray whale calf breaching. Not to be confused with the "spy-hop," in which the head is thrust out of the water, a breach occurs when a whale leaps half-way – or more – out of the water and then falls back onto its side.*

DEEP HARVEST

Gray whales are adapted to exploiting the tremendous seasonal abundance of food that results as the Arctic pack ice retreats in spring, exposing the sea to the polar summer's 24-hour daylight and thus triggering an enormous bloom of micro-organisms in the water from the surface down to the sea floor. While present in the Arctic from May to November, Gray whales store enough fat to sustain them virtually without feeding through the rest of the year, when they migrate to calve in warm waters while their summer feeding grounds are covered with ice. By the time they return to the feeding grounds, they may have lost up to one-third of their body weight.

The whales feed in shallow waters between 5–100m (15–330ft) deep; the diet comprises amphipods and isopods (both orders of crustaceans), polychaete worms, and mollusks that live on the ocean floor or a few centimeters into the bottom sediment. The gammarid amphipod *Ampelisca nacrocephala* is probably the species that is most commonly taken. To feed, Gray whales dive to the sea floor, turn on their side (usually the right), and plant the side of their head on the bottom. By suction they obtain a mouthful of invertebrate prey along with sand and mud sediments. They filter out the food by forcing the mixture through their baleen, trailing long plumes of sand and mud behind as they surface. The food items are swallowed; they take a few breaths, and dive again. As Gray whales feed, they leave pits, or bites, in the ocean floor. Some scientists have speculated that they may thus effectively plow the sea floor, possibly increasing its productivity in subsequent years.

Although Gray whales are primarily known as bottom-feeders, they may also prey on a variety of small planktonic organisms present in the water column, such as mysids and crab larvae, herring eggs, and even small fish. Most feeding occurs on the summer grounds, but if the opportunity arises, it may also occur during migrations, or even on the winter grounds.

Several species of sea bird associate with feeding Gray whales, including Horned puffins, Glaucous gulls, and Arctic terns. These birds apparently feed on crustaceans that escape through the baleen during the straining process while the whales are surfacing. The discovery of this association answered the perplexing question of how large numbers of bottom-dwelling invertebrates, from beyond the birds' diving depth, got into their digestive tracts.

Rorquals

WITHIN THIS GROUP OF WHALES IS THE largest animal that has ever lived: the giant Blue whale, which weighs up to 150 tonnes, the weight of 25 six-tonne male African elephants. The rorquals also include one of the most tuneful and agile whales, the Humpback, which not only produces eerie and wide-ranging sounds but also performs remarkable acrobatics, leaping from the water, sometimes upside down.

The name "rorqual" comes from the Norwegian, and literally means "furrow whale" – a reference to the longitudinal folds of skin below and behind the mouth that are a distinctive feature of these species. Many rorquals travel great distances across the world's oceans on annual migrations to and from breeding grounds in the tropics and feeding grounds in the polar regions. The larger species have been hunted intensively over the past 100 years, and their numbers have consequently been severely reduced.

Giants of the Deep

FORM AND FUNCTION

Sleek and streamlined in form, rorquals have a series of grooves or folds of skin that, in all species except the Sei whale, extend from the chin backward under the belly to the navel. During feeding these grooves expand, allowing the mouth to increase considerably in volume. Photos of dead whales show the throats sagging, contributing in the past to a falsely clumsy picture of these creatures that was a grotesque distortion of their sleek underwater reality.

Animals in the southern hemisphere are a little bigger than those in the northern, and in all species the female grows to be slightly bigger than the male. The head occupies up to a quarter of the body length and, except in the Humpback whale, has a distinct central ridge running forward from the blowhole to the snout; the Bryde's whale also has additional ridges on either side of this one. In all species, the lower jaw is bowed and protrudes beyond the end of the snout.

Below *This Southern minke whale swimming in the waters around Australia's Great Barrier Reef belongs to the dwarf form of the species, which averages just 7m (23ft) in length. Its dimensions are modest in comparison with the gigantic size of the Blue whale; one female Blue taken in Antarctic waters early in the 20th century measured 33.58m (110ft).*

Above Silhouetted against the setting sun in ...kan waters, a Humpback whale dives in search of ... the main component of northern Humpbacks' ...s; those in southern oceans are predominantly ...-eaters. The tail flukes are so distinctive in shape ... coloration that they can be used to identify indi...al whales.

The flippers are lancetlike and narrow in all but ... Humpback whale, where they are scalloped ...at least the leading edge and form almost a ...d of the body length. The dorsal fin is set far ...k. The tail flukes are broad, with a conspicu... indentation in the middle, spreading especial...widely in the Humpback whale. The blow as ... whale exhales consists of a single spout from a ...ible blowhole on the top of the head; the ...ght and shape vary between species.

Across the Seven Seas

DISTRIBUTION PATTERNS

...e, Fin, Sei, Minke, and Humpback whales are ...nd in all the world's major oceans. They spend ... summer months in polar feeding grounds and ... winter in more temperate breeding grounds. ...mpback whales swim close to coasts during ...ir migrations, but the other rorquals tend to be ...re oceanic. Bryde's whale occurs only in tem...ate and warm waters, generally near shore in ... Atlantic, Pacific, and Indian Oceans.

In the southern hemisphere, Blue whales start ...migrate ahead of the Fin and Humpback ...ales, with Sei whales following some two ...nths later. Within each species, age and sex ...ermine the distribution of individuals. Older ...mals and pregnant females tend to migrate in ...vance of the other classes, with sexually imma...e whales at the rear of the stream. In all species, ... bigger, older animals tend to travel closer to ... poles than the younger whales.

In contrast to the other rorqual species, Blue and Minke whales occur right up to the ice-edge. Fin whales do not go quite that far, while Sei whales are much more sub-Antarctic in their distribution. This succession is not so clearly apparent in the northern hemisphere, where a more complex pattern of land masses and water currents has a distorting effect.

The various species of rorquals are thought to be divided across the world's oceans into different populations that do not generally interbreed. Yet genetic evidence, and the recovery of marked whales, indicate some interchange, at least between the northern and southern hemispheres. Most populations tend to be spread widely across the oceans, though the Humpback, which breeds in coastal waters, tends also to be more concentrated around feeding grounds.

A Life of Great Migrations

SOCIAL BEHAVIOR

The life cycles of the Blue, Fin, Sei, Minke, and Humpback whales are very closely related to the pattern of seasonal migrations. In both hemispheres the whales mate in low-latitude warm waters during the winter months, and then migrate to their respective polar feeding grounds, where they spend 3–4 months feeding on the rich plankton that constitutes their diet. After this period of intensive feeding, they migrate back to the temperate zone once again, where females give birth to a single calf 10–12 months after mating. Conceptions and births may occur at almost any time of the year, but the greatest reproductive activity is confined to relatively short peak periods of 3–4 months.

The newborn calf is about a third of the mother's length and 4–5 percent of her weight. It accompanies its mother on the spring migration, swimming 3,200km (2,000mi) or more towards the polar seas. During this time it lives on its mother's rich milk, which has a fat content of up to 46 percent, as compared to the 3–5 percent in human and cows' milk. The mother squirts the milk into the mouth of the calf with the aid of the muscles surrounding her mammary glands as the baby holds on to one of her two nipples. The young grow quickly on this high-energy diet, gaining as much as 90kg (200lb) a day, so that in six to seven months a Blue whale calf will add about

FACTFILE

RORQUALS

Order: Cetacea

Family: Balaenopteridae

8 species in 2 genera: *Balaenoptera* (7 species), including the **Blue whale** (*B. musculus*) and **Fin whale** (*B. physalus*); and *Megaptera* (1 species), the **Humpback whale** (*M. novaeangliae*).

DISTRIBUTION All major oceans

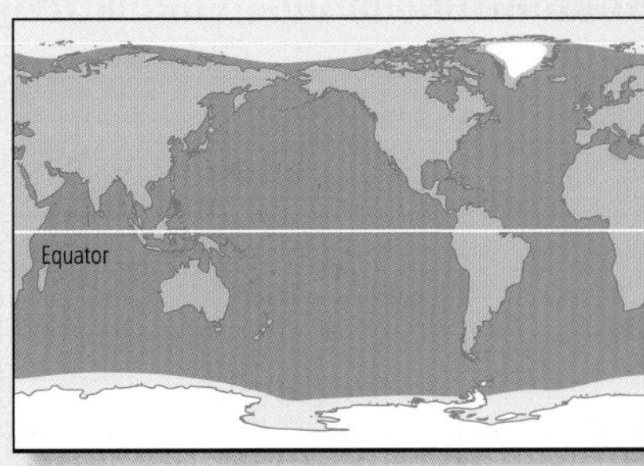

HABITAT All species but *B. edeni* and *B. brydei* migrate between summer feeding grounds in the polar regions and winter breeding grounds in temperate waters.

SIZE **Head–tail length** ranges from 9m (30ft) in the Northern minke whale to 27m (90ft) in the Blue whale, the world's largest animal. **Weight** ranges from 9 tonnes to 150 tonnes in the same 2 species. In all species, females grow slightly larger than males.

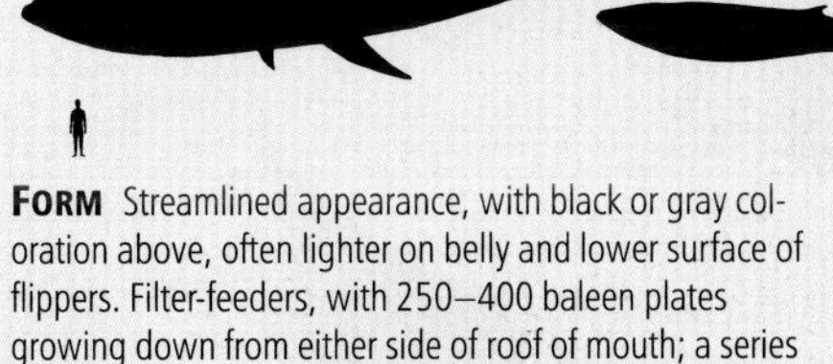

FORM Streamlined appearance, with black or gray coloration above, often lighter on belly and lower surface of flippers. Filter-feeders, with 250–400 baleen plates growing down from either side of roof of mouth; a series of folds or grooves that expand during feeding extend from the chin back under the belly. Tail flukes are broad and conspicuously indented in the middle.

DIET Krill, copepods, and fish, in varying proportions; Bryde's whale eats mainly fish, the Blue whale almost exclusively krill.

BREEDING Single calf born 10–12 months after mating. In most species, there is a 2-year gap between pregnancies.

LONGEVITY From 45 years in the minkes up to 100 or more in the bigger species.

See species table ▷

17 tonnes to its 2.5–tonne birth weight. Calves are weaned at 7–8 months, when they are abou 10m (33ft) long.

The age at which rorquals start to reproduce has changed over recent years due to a fascinati interaction between whale biology and human predation. Fin whales born before 1930 becam sexually mature at around 10 years, but the me age has subsequently declined to about 6 years Sei whales caught before 1935 did not reprodu until around 11 years, but they are now ready a just 7 in some areas. As for the southern Minke whales, their age of maturity has dropped by a massive 8 years, from 14 to 6.

The most likely explanation for these change that individual whales now have access to increased amounts of food because of the mass reductions in whale stocks, permitting the survivors to grow more quickly. Because the start c reproduction is closely related to body size, fast growth means that the whales reach the critical size at a younger age.

Left *Seen from behind, a Blue whale exhibits th twin "nostrils" that jointly form the blowhole in this in the other rorqual species. When the whale exhale a single jet of spray emerges from both orifices. The size and shape of the spout varies between species; experienced whale-watchers use this clue to distinguish different types of rorqual from a distance.*

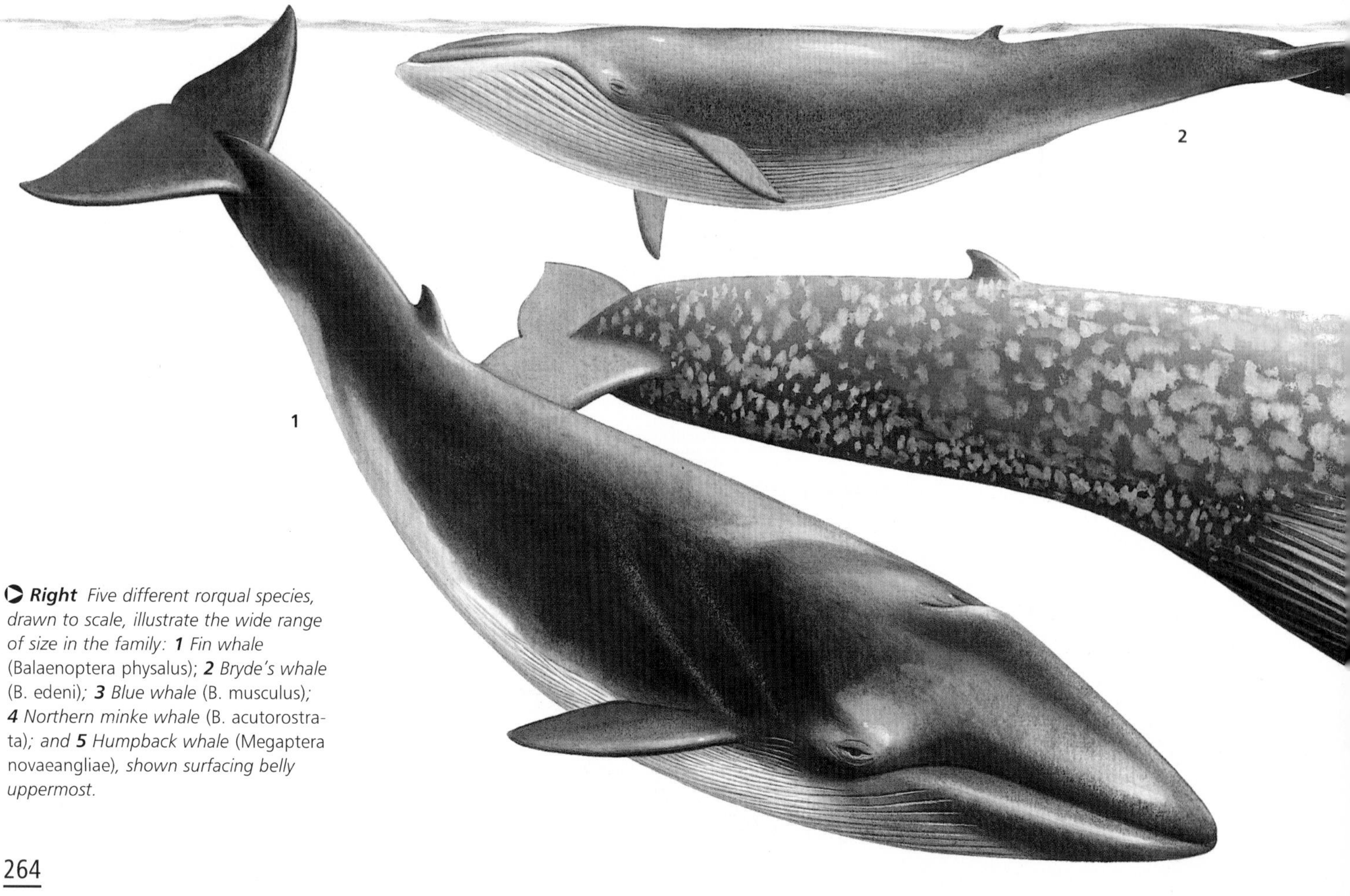

Right *Five different rorqual species, drawn to scale, illustrate the wide range of size in the family:* ***1*** *Fin whale* (Balaenoptera physalus); ***2*** *Bryde's whale* (B. edeni)*;* ***3*** *Blue whale* (B. musculus)*;* ***4*** *Northern minke whale* (B. acutorostrata)*; and* ***5*** *Humpback whale* (Megaptera novaeangliae)*, shown surfacing belly uppermost.*

Fin whale En

Balaenoptera physalus

Polar to tropical seas. 2 subspecies: *B. p. physalus*, N Atlantic and N Pacific, HTL 24m (80ft); WT 70 tonnes; and *B. p. quoyi*, S Hemisphere, HTL 27m (90ft), WT 80 tonnes.
FORM: skin gray above, white below, asymmetrical on jaw; flipper and flukes white below; baleen plates: 260–470, blue-gray with whitish fringes, but left front white; throat grooves: 56–100.

Sei whale En

Balaenoptera borealis

Polar to tropical seas. 2 subspecies: *B. b. borealis*, N Atlantic and N Pacific oceans, HTL 19m (60ft), WT 30 tonnes; and *B. b. schlegellii*, S.Hemisphere, HTL 21m (70ft), WT 35 tonnes.
FORM: skin dark steely-gray, white grooves on belly; baleen plates: 320–400, gray-black with pale fringes; throat grooves: 32–62.

Bryde's whale

Balaenoptera edeni

Coastal waters of E Indian Ocean and W Pacific.
HTL 11m (36ft); WT 20 tonnes.
FORM: skin dark gray; baleen plates: 250–370, gray with dark fringes; throat grooves: 47–70.

Balaenoptera brydei

Offshore tropical and warm temperate waters worldwide.
HTL 15m (50ft); WT 26 tonnes.
FORM: as for *B. edeni*, of which *B. brydei* was formerly regarded as a subspecies.

Blue whale En

Balaenoptera musculus

Polar to tropical seas. 3 subspecies: Northern blue whale (*B. m. musculus*), N Atlantic and N Pacific, HTL 24–27m (80–90ft), WT 130–150 tonnes; Southern blue whale (*B. m. indica*), S hemisphere, HTL 27m (90ft), WT 150 tonnes; and the Pygmy blue whale (*B. m. brevicauda*), S hemisphere, particularly in the southern Indian Ocean and South Pacific, HTL 24m (80ft), WT 70 tonnes. FORM: skin mottled bluish-gray (silvery-gray in *B. m. brevicauda*), flippers pale beneath; baleen plates: 270–395, blue-black; throat grooves: 55–88 (76–94 in *B. m. brevicauda*).

Northern minke whale LR

Balaenoptera acutorostrata

Polar to tropical seas. 2 subspecies: *B. a. acutorostrata*, N Atlantic, and *B. a. scammoni*, N Pacific.
HTL 9m (30ft); WT 9 tonnes.
FORM: skin dark gray above, belly and flippers white below; white or pale band on flippers; pale streaks behind head. Baleen plates: 230–350, yellowish-white, some black. Throat grooves: 50–70.
CONSERVATION STATUS: LR Near Threatened.

Southern minke whale LR

Balaenoptera bonaerensis

Polar to tropical seas of the S hemisphere.
HTL 11m (36ft); WT 10 tonnes.
FORM: as for *B. acutorostrata*, though often without pale band on flippers. There also appears to be a "Dwarf" minke whale in the lower latitudes of the S hemisphere, resembling the N Pacific population.
CONSERVATION STATUS: LR Conservation Dependent.

Humpback whale Vu

Megaptera novaeangliae

Polar to tropical seas.
HTL 16m (50ft); WT 65 tonnes.
FORM: skin black above, grooves white, flukes with variable white pattern below; baleen plates: 270–400, dark gray; throat grooves: 14–24.

Abbreviations HTL = head–tail length WT = weight En Endangered Vu Vulnerable LR Lower Risk

◐ ***Left*** *Sei whales* (Balaenoptera borealis), *not shown below, are similar in appearance to Bryde's whales. The two are most easily distinguished by the ridges on the head; there are three in Bryde's whale* ***a****, only one in the Sei* ***b****.*

Behemoths at Risk

CONSERVATION AND ENVIRONMENT

The future of the rorquals now depends largely on the success of the measures taken in recent years to protect them from overhunting. There is evidence of increasing numbers in some species, but because of their very low birthrates it will be decades before there is full recovery. Left to themselves, whale populations may double in size in 10–20 years, depending on reproduction rates; but the Antarctic Blue whale still probably numbers only 5–10 percent of its original abundance, and it does not reproduce very quickly.

There is also increasing concern about alterations in, and degradation of, the marine environment due to climate change and pollution. While it is unlikely that increases in water temperature would have a direct effect on whales, which are insulated from their immediate environment by their blubber, the food organisms on which they depend, such as krill and fish, may well move in response to these environmental alterations and to changes in the ocean currents. Similarly, it is possible that depletion of the ozone layer over the polar regions will allow more ultraviolet radiation to enter the surface waters, changing the productivity of the areas of the oceans that have long been frequented as feeding grounds by the whales. Direct pollution in the form of harmful chemicals, as well as of nondegradable objects such as plastic bags, bottles, and other waste that can be swallowed and block the whales' food tracts, is a cause for growing concern, as is sound pollution, which impinges on their sensing and communication capacity. To these hazards must be added the risks of entrapment in fishing gear and collisions with vessels in increasingly busy shipping lanes. RG

__Right__ Showing one of its two flippers – the longest of any cetacean species – a Humpback whale breaches majestically before returning to the water with a resounding splash. As with other species, breaching seems to serve two main functions: stunning or panicking fish shoals and communicating information to other herd members.

BIG APPETITES, SMALL PREY

The biggest animals in the world keep themselves alive by eating some of the smallest. Rorquals are filter-feeders. They have huge, sievelike baleen plates growing from the roofs of their mouths, which gather the tiny plants and animals in the marine water column. The whale opens its mouth widely, engulfing the plankton in a large volume of water **a**. The water is then sieved through the spaces between the baleen plates as the mouth closes; the previously expanded throat region tightens up and the tongue is raised. The food material is held back on the bristles lining the inner edges of the baleen plates before being swallowed **b**.

Sei whales can also feed by skimming through patches of plankton-rich water with their mouths half open. The head is normally raised a little above the surface, so that water and food are sieved continuously through the baleen plates. When enough has been collected, the whale closes its mouth and swallows the food.

The bristles fringing the baleen plates vary in texture between the species, as do the shapes and sizes of the plates, thus determining which food organisms each different species can capture. The Blue whale has rather coarse baleen bristles, and feeds almost exclusively on shrimplike food, especially krill. This organism is the basic sustenance for all the baleen whales in the Antarctic, but a wider range of nutriment appears in other areas, especially in the northern hemisphere.

Fin whales, with their medium-texture baleen bristles, eat mainly krill and copepods, with fish third in importance, but there is considerable variation by area and season. Sei whales, which have much finer baleen fringes, are primarily copepod feeders, but krill and other crustaceans are also consumed. Minke and Humpback whales feed mainly on fish in the northern hemisphere and krill in the south, while Bryde's whales are more exclusively fish eaters, with only a few crustacea in their diet.

The fish taken by these whales are generally schooling species, and include herring, cod, mackerel, capelin, and sardines. The Humpback and Minke whales have a characteristic lunging action which may serve to scare and concentrate the prey fish as the whale circles them before shooting up vertically with its mouth open to engulf the food. Humpback whales may also "herd" their prey by releasing a circle of bubbles that whirl to the surface around them.

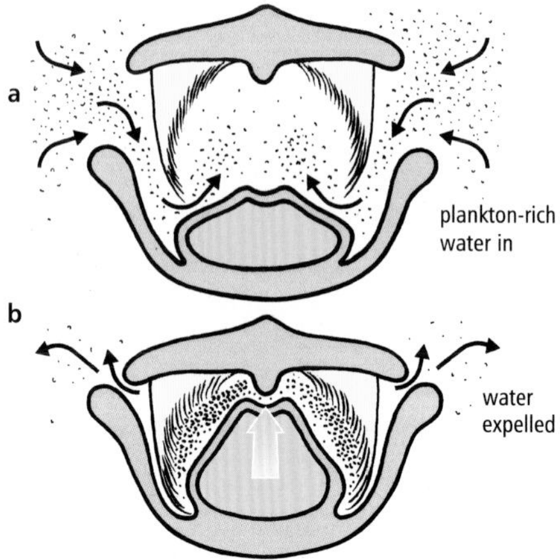

It is estimated that a large Blue whale weighing 100 tonnes has to eat 4 tonnes of krill every day during the summer feeding season. The first chamber of the three-part stomach can hold about one tonne at a time, so the whale has to fill this about four times a day. The stored energy represented by the oil and fat laid down in blubber under the skin and around the internal organs fuels the animal during the rest of the year when very little food is consumed. Over the course of the whole year the whale eats roughly 1.5–2 percent of its own body weight daily.

SPECIAL FEATURE

NEW LIGHT ON THE SINGING WHALE

What fresh research techniques have shown about the Humpback

THE 1990S WITNESSED GREAT ADVANCES IN scientific research on cetaceans in general, and on the Humpback whale in particular. The new techniques that have been applied to studying the fascinating behavior of this leviathan fall into three main categories: genetic analysis of its DNA; electronic tagging for satellite tracking; and acoustic monitoring by deep-sea hydrophone arrays.

Increasingly sophisticated techniques of genetic analysis have had a profound influence on several different fields of research. In broad-based studies of populations, they can be used both to determine their structure and to assess the long-term consequences of particular behavioral patterns across different populations over evolutionary time. More narrowly, DNA analysis can confirm the sex of whales and throw new light on paternity and matrilineal relationships. During the early 1990s, biologists photoidentified and biopsied several thousand Humpbacks in the North Atlantic, a significant percentage of the entire population. From these samples, an accurate picture is emerging of which calves are likely to have been sired by which males. Formerly, whale paternity was shrouded in mystery and conjecture, as no scientist had ever managed to observe Humpback copulation at first hand!

One particular line of genetic research has confirmed and deepened our awareness of Humpbacks' migratory patterns. This has been achieved by taking samples of mitochondrial DNA. Mitochondria are tiny energy-processing structures that occur inside cells of the body. They are present in eggs but not in sperm, so they are inherited only from mother to offspring, never coming down the male line. Mitochondria contain DNA, and the mitochondrial DNA of whales from different feeding grounds is distinct enough to suggest that there is low gene flow across these subpopulations. Nuclear DNA, which is inherited from the father as well as the mother, does not show the same specificity to feeding ground. Therefore, the genetic analyses corroborate evidence from sightings, which indicates matrilineal fidelity to feeding areas; but while sighting data are limited to time scales of a decade or two, the genetic data show that the pattern has prevailed for thousands of years. This result is especially striking, since Humpback whales from the same tropical breeding grounds only migrate once with their mother; even so, the genetic data suggest that they then continue to travel thousands of kilometers to the same feeding ground annually for the rest of their life.

Modern methods of satellite tracking have proved invaluable in plotting the movement of Humpbacks over thousands of kilometers of ocean. Fitting whales with electronic tags not only allows scientists to follow their migratory routes over vast distances and protracted periods, but also to gather information on such features as dive profiles. Tagging has also yielded a mass of data on local movements within feeding and breeding grounds.

The seismic shift in superpower relations in the 1990s brought an added bonus to researchers tracking Humpback whales. During the Cold War, the US Navy secretly installed an extensive array of hydrophones on many ocean basins to follow the movement of Soviet submarines. Data collected by these listening devices were telemetered back to central sites, enabling sound sources to be monitored over enormous ranges. This system, which was called SOSUS (SOund SUrveillance System), was declassified and made available to biologists in the 1990s. With its aid, the sounds of Humpbacks and other whales can be picked up at a distance of hundreds of kilometers.

What these diverse research methods have all served to emphasize is the far-reaching effects of the annual migratory cycle on baleen whales. It influences every aspect of Humpback life and behavior, beginning most fundamentally with their biology; the annual cycle actually selects for large size in the whales. In the summer Humpbacks feed intensely, and then must live off their energy stores for the rest of the year as they migrate thousands of kilometers to breed and calve. Not only can large whales store more energy; they can also swim long distances more efficiently, and they have lower metabolic costs of thermoregulation. The extra energetic demands of pregnancy and lactation in females mean that they must store even more energy, and this has selected for adult females to grow larger than adult males.

The ability to lactate after months of fasting is very unusual among mammals, and is a critical feature of the baleen whale reproductive cycle. A pregnant female must bring her calf to term while fasting during migration. After she gives birth in the tropics, she must suckle the calf on her energy reserves for up to half a year before she returns to the feeding area the next summer. The calf grows rapidly during this period in order to wean at a length of 8–9m (26–30ft) within a year.

Two further key areas of research interest – also intimately linked with the migratory cycle – are feeding and mating. In the North Atlantic, Humpback whales return to discrete feeding areas such as Iceland or Norway, though within each of these the whales may shift their distribution to match the changing distribution of prey. Over the years, resighting data of naturally marked whales has suggested that Humpbacks from many feeding grounds mix and interbreed on the tropical breec ing grounds, but then reliably migrate back to their mother's feeding area.

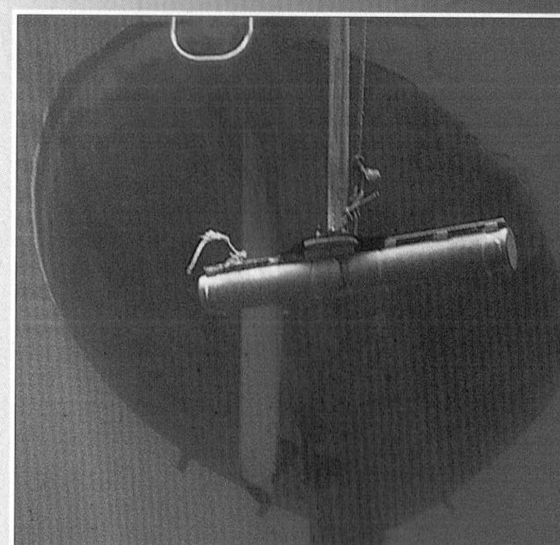

Above Most Humpback recordings are made from small inshore boats, but the songs can also be tracked by deep-sea SOSUS arrays. Research shows that the whales sing over large expanses of ocean and that their songs can be heard at great ranges.

During the winter, Humpback whales migrate to tropical waters where they mate and give birth Female humpbacks seek out calm, protected waters in the lee of an island or in a shallow bank Males congregate near these areas to mate with the females. One mating strategy is to compete directly with other males for access to a female; this involves one male humpback escorting a female, trying to prevent rivals from approaching her, while the other males swim close by, mounting challenges. Such intense competition usually only lasts for a few hours at most, during which time challengers may change roles with principal escorts several times. The winner does not seem t stay with the female for long, for females have been seen with other males on the very next day;

Above *A Humpback mother and* *lf. Modern genetic analysis has proved* *at offspring return to the mother's* *eding ground for their entire lives.*

on the feeding grounds, associations between dividuals other than mother and calf are brief. enetic analyses of successive calves of the same male have proved that females mate with differ-nt males in different breeding seasons.

The other male mating strategy is to sing. Dur-g the breeding season, lone males repeat long, omplex songs, each lasting for about 10 minutes, an unbroken series. They can sing incessantly for ver 24 hours. Analysis of the songs reveals that e sound sequences of which they are composed e generated by a hierarchical model. Each song onsists of a series of themes repeated in a specific der; the themes in turn are made up of phrases peated a variable number of times.

Most males within a single population perform ry similar songs. However, the song changes adually but continuously over time, so that ngs from different years are quite distinct from one another, and there are no common sounds in songs recorded a decade apart. Usually the songs from different populations of Humpbacks are independent, though biologists noted a fascinating exception to this rule off the east coast of Australia, the site of long-running song research. One year, several whales were recorded singing a song totally different from that of the rest – yet very like songs heard off Australia's west coast. It appeared that one or two west-coast Humpbacks had migrated east, a rare but not unknown phenomenon. The west-coast song gradually gained ascendancy and eclipsed the original east-coast song within two years. Both the gradual evolution of songs and this sudden adoption of a new one show that Humpback males learn the detailed acoustic structure of their songs, and that whales from different populations do not sing different songs because of different genotypes, but rather because they imitate what they hear. In other words, Humpback song is a form of culture.

Once singing Humpbacks join with others, their song ends. When a singer joins a female, behaviors associated with mating (e.g. rolling and flippering) have been seen. Yet when another male is met, aggressive interactions often ensue, after which the original singer or the joiner will resume singing once the whales have split up again. These sightings suggest that males sing to attract females, and that other males compete with singers. Humpback males do not appear to compete over territory, for daily photo-ID of whales in the same area shows a very low incidence of resightings; rather, the whales move on, borne on ocean currents. Unlike terrestrial animals, many of which defend a specific geographic location, Humpbacks seem to maintain more coordination with respect to one another than to any particular place. PLT

Right Whales

THE RIGHT WHALE OWES ITS NAME TO WHALERS. It was so called because it was the "right" whale to hunt – it swam slowly, floated when killed, and had a high yield of baleen and oil. No other whale was hunted to such precariously low levels as this species. Even today, after decades of protection from industrial whaling, the North Atlantic and North Pacific populations both number less than a few hundred animals.

Right whales worldwide are at risk from humans. They share with us a preference for coastal waters, where they give birth to their young. This leads the most vulnerable members of their populations into the most crowded habitats in the world's oceans. In the North Atlantic, these habitats are filled with ships and fishing gear, which kill Right whales at a rate that threatens their survival. Due to human-induced mortality and declining birth rates, this population appears to be in a state of decline, and recent population models predict extinction within two centuries. In the North Pacific, insufficient information is available to make any assessment of either population size or growth rate. Southern hemisphere Right whales, however, appear to be doing relatively well, numbering perhaps 6,000 animals and growing at a rate of 6–7 percent per year.

Big Heads and Curved Jaws

FORM AND FUNCTION

Right whales, Bowhead whales, and Pygmy right whales share certain characteristics that distinguish them from the rorquals. These include an arched rostrum (upper jaw), giving a deeply curved jawline in profile in contrast to the nearly straight line of the rorqual mouth; this trait is most pronounced in the Bowhead whale, in which the head may make up as much as 40 percent of total body length. Other distinctive features include long, slender baleen plates instead of relatively short ones, and a complete absence of throat grooves in the large species (the Pygmy has two), compared to many in all the rorquals. There are also a number of marked differences in cranial features: in particular, the upper jawbone is narrow in Right whales and broad in rorquals. In all three Right whale species, the head is large in proportion to the rest of the body. While the two large species are exceptionally bulky in comparison to the rorquals, the Pygmy right whale is relatively small and slim; unlike the other two species, it has a small, triangular dorsal fin.

Right whales are uniquely identifiable from Bowheads as well as other cetacean species through the patches of thickened skin, called "callosities," that form along the rostrum, above the eyes, and along the lower jaw. The largest patch, found on the snout, was called the "bonnet" by old-time whalers. Colonies of whale lice (species of the genus *Cyamus*) live within these outgrowths. Callosities are slightly larger in males than females, and may be employed in competition for females. They are also useful to scientific observers, for they form unique patterns that can be used to identify individual whales without the need to capture or contact them. Thanks to them identification catalogues have been drawn up for every Right whale population currently being studied; from these it is possible to tell such facts as how long each animals lives, when it reproduces, and how it migrates.

Right whales make low-frequency sounds, and studies have shown that they have at least two types of call – contact calls between widely separated individuals, and calls used by females to attract mates – though in practice their repertoire is probably much broader. Unlike humpbacks,

Below *Male Northern right whales gather round a female in estrus during the breeding season. While there have been reports of inter-male aggression at this time, there are also reliable accounts of reproductive cooperation, with one male supporting a female while another male mates with her.*

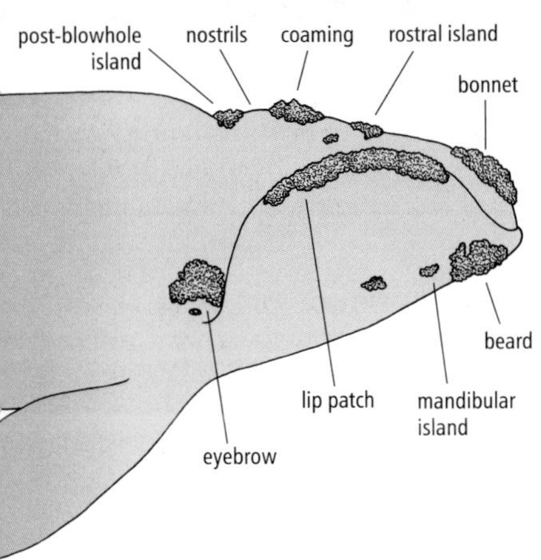

Above *Surfacing off Argentina's Valdez Peninsula, a Southern right whale mother and calf display the "bonnets" that are among the species' most distinctive features. Consisting of layers of hardened skin protruding from the upper jaw, the bonnets are usually infested with barnacles and other parasites.*

Left *The bonnet is in fact only one of several groups of callosities displayed by most Right whales. They are present from birth, but their exact function is still unknown. Observers have noted that males have more of them than females, leading some people to suggest that they might serve as weapons used to graze other males' skin during breeding contests.*

ich produce repeated "song" sequences, Right ales make many single and grouped sounds in 50–500Hz range. Feeding whales also make a riable 2–4kHz low-amplitude noise that has en traced to water rattling across their partially-posed baleen plates. Nothing is known of vocal-tions in the Pygmy right whale, whereas those the Bowhead are simple and change over time.

A Long-Distance Annual Round

DISTRIBUTION PATTERNS

e migratory habits of the Right whale are not lly known. In the North Atlantic, the coastal ters of Florida and Georgia are the primary nter calving ground, but the distribution of n-calving animals at this time is unknown. out two-thirds of the North Atlantic population regularly seen in the spring, summer, and fall in e Gulf of Maine, but genetic and sightings data ggest the existence of a second, as-yet unidenti-d summering habitat for this subspecies.

In the southern hemisphere, Right whales are own to congregate in winter for calving in the astal waters of South Africa, Argentina, Australia, and New Zealand's sub-Antarctic islands. Exactly where they feed or spend the summer is only partly known, although Right whales have been spotted occasionally around the Antarctic.

In the north Pacific, scattered sightings suggest that remnant populations of Right whales occur seasonally in the Gulf of Alaska and the Sea of Okhotsk. Although rare winter sightings of Right whales along the west coast of the USA and in Hawaii have been reported, no information on wintering habitats, calving grounds, or migratory routes is available. Combined data suggest that calves are born in the winter and that mating, leading to conception, occurs during late fall to early winter in the respective hemispheres. Therefore, the "mating activity" observed in all areas during the rest of the year may be more related to social behavior than actual reproduction.

The annual migratory cycle of the Bowhead is best seen in the Bering Sea–Beaufort Sea population, which is by far the largest surviving stock of this species, and the best-known in the Arctic. Distributions are closely connected with seasonal changes in the position and extent of ice-free

FACTFILE

RIGHT WHALES

Order: Cetacea

Families: Balaenidae and Neobalaenidae

4 species in 3 genera, *Eubalaena*, *Balaena*, and *Caperea*; some authorities, though, include the Right whales and the Bowhead within a single genus, *Balaena*. The Pygmy right whale (genus *Caperea*) is sometimes treated as a subfamily, but here we consider it as the sole representative of its own family, the Neobalaenidae.

DISTRIBUTION Arctic and temperate waters

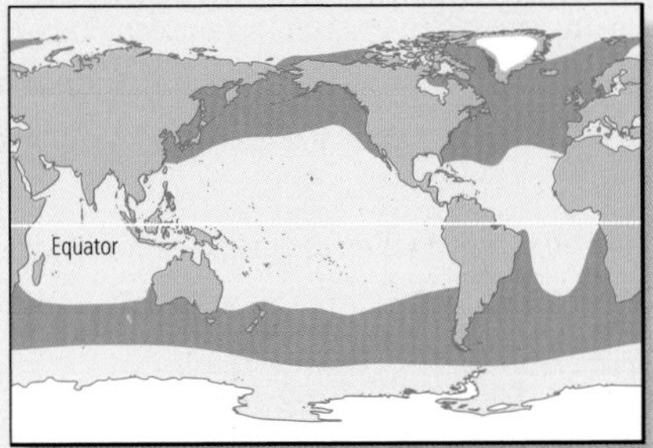

NORTHERN RIGHT WHALE *Eubalaena glacialis*
Temperate waters of N hemisphere; in the Atlantic, recorded as far S as Florida. HTL Up to 18m (59ft), average adult about 15m (49ft): WT 50–56 tonnes; N Pacific animals larger than other species by 5–10 percent. Form: body black, with white patches on the chin and belly, sometimes extensive; head and jaws characteristically bear individually distinctive patterns of rough, thickened skin patches called callosities that are heavily infested with crustacean parasites known as cyamids; baleen is black, and up to 2.5m (8ft) in length. Breeding: gestation 12–13 months. Longevity: one known female lived at least 65 years. Conservation status: Endangered.

SOUTHERN RIGHT WHALE *Eubalaena australis*
Temperate waters of S hemisphere; in the Atlantic, recorded as far N as S Brazil. Length, Form, Breeding, and Longevity: as for Northern right whale. Conservation status: Lower Risk – Conservation Dependent.

BOWHEAD WHALE *Balaena mysticetus*
Arctic Basin, with winter migration into Bering and Labrador Seas. HTL 3.5–20m (11–66ft), average adult about 17m (56ft); WT probably 60–80 tonnes. Form: body black, except for white or ocherous chin patch; no callosities; baleen narrow, dark gray to blackish, up to 4m (13ft) in length. Breeding: gestation 10–11 months. Longevity: evidence (e.g. old harpoon heads) points to extreme age. Conservation status: Lower Risk – Conservation Dependent, though the Svalbard–Barents Sea stock is Critically Endangered, while the Okhotsk Sea and Baffin Bay subpopulations are Endangered.

PYGMY RIGHT WHALE *Caperea marginata*
Circumpolar in S temperate and sub-Antarctic waters; not a true Antarctic species. HTL 2–6.5m (7–21ft), average adult about 5m (16ft): WT about 3–3.5 tonnes. Form: body gray, darker above and lighter below, with some variable pale streaks on the back and shoulders, and dark streaks from eye to flipper; baleen plates relatively long for its size, whitish with dark outer borders. Breeding: gestation probably 10–11 months.

Abbreviations HTL = head–tail length WT = weight

areas. The route and timing of migration each year are dictated by the development of open channels (leads) between the ice floes from the northern Bering Sea eastward to the Amundsen Gulf in spring and summer.

Bowheads winter in the Bering Sea, particularly around St. Lawrence and St. Matthew Islands, and it is here that the calves are born. Mating occurs in the spring during the first stage of the migration. Aerial and satellite photographs reveal that the ice in the northern Bering and southern Chukchi Seas develops fractures in April, which first open to Cape Lisburne and then Point Barrow. The leads are relatively close to shore, so that most of the population passes Barrow on their way into the Beaufort Sea. Beyond Barrow, however, the winds and current circulation open large offshore leads, and the eastward migration shifts further from land. Whales reach Cape Bathurst and the Amundsen Gulf as early as May. The slow breakup of coastal ice east of Alaska normally prevents Bowheads utilizing the Mackenzie Delta and Yukon shore in any numbers until the second half of July.

Eskimo hunters state that there is segregation by age and sex during this migration (as in Australian Humpback whales). The migration certainly takes place in "pulses," with animals during May and June straggling in a column along the whole length of the route from Barrow to southwestern Banks Island. The return migration to the Bering Sea in late summer–early fall tends not only to be rather rapid (according to old whaling records) but also further offshore, and hence less easy to observe.

Skimming the Plankton Swarms

DIET

Right whales, Bowheads, and Pygmy right whales all feed primarily on copepods, but the North Atlantic Right whale also takes juvenile krill, and occasionally other swarming planktonic larvae. The Southern Ocean population appears to eat adult krill regularly as well.

Bowhead feeding is usually associated with restricted belts of high productivity in arctic areas, such as the edge of the plume from the Mackenzie River, where nutrient enrichment and water clarity are both optimal for active photosynthesis by phytoplankton, resulting in turn in relatively high zooplankton production. Both Bowhead and Right whales generally feed by skimming with their mouths open through concentrations of zooplankton; this is in contrast to the feeding methods of most rorquals (other than Sei whales), which tend to gulp patches of highly concentrated fish or krill and filter the food through their baleen plates. In most northern hemisphere feeding grounds, Right whales usually feed at depth, diving for 8–12 minutes at a time, though they will occasionally feed at the surface when prey concentrations are dense. Right whales have occasionally been observed to feed side by side.

Right whales and Bowheads probably need 1,000–2,500kg (2,200–5,500lb) of food per day, and the Pygmy right whale perhaps 50–100kg (110–220lb); little is known of its feeding habits, except that two animals taken by the Russians had stomachs full of copepods.

Seeking Company in the Ocean Waste[s]

SOCIAL BEHAVIOR

Right whale social structure is poorly understoo[d]. Recognizable individuals may be seen alone at some times of day, and with one or more groups later that day or on other days. When large grou[ps] of whales are seen within a few kilometers of eac[h] other, however, it is most likely merely in respon[se] to concentrations of food. These groups are prob[ably] not behaviorally comparable to pods of dolphins or toothed whales.

The most tightly-linked social pairing is that between mother and calf. These two can remain within one bodylength for the first 6 months of the calf's life. Weaning occurs at 10–12 months, and mothers and their offspring are rarely seen together again. Breaching behavior (leaping from the surface) and lobtailing (slapping the water with the flukes) occur frequently in this species; these behaviors may allow the whales to indicate their location, especially when surface noise

Above *Different Right whale species have quite* *tinct body forms. The Northern right* (Eubalaena cialis) ***1*** *is distinguished by its huge baleen and* *gue, its deeply arched lower jaw, and its callosities.* *vheads (*Balaena mysticetus*)* ***2*** *have an even more* *onounced curve to the lower jaw, but no callosities.* *e Pygmy right whale* (Caperea marginata) ***3*** *has a* *rsal fin and only moderate bowing of the lower jaw.*

Below *A Southern right whale calf* (Eubalaena stralis) *accompanies its mother in surface waters off* *agonia. The mother–calf pair are inseparable for* *first six months after birth, but rarely link up again* *ce the infant has been weaned, at about 1 year.*

means that vocalizations cannot be heard.

The long reproductive cycle (3 years or more between births) means that less than one-third of the adult females in a given area may be receptive to males each year. Females appear to solicit males acoustically, and then make mating difficult, either by swimming away or by lying on their back with both flippers in the air so that the genital region is inaccessible. Males compete for access to the female through active pushing and by displacing one another, and it appears that females mate with many males. Males also probably employ sperm competition – male testes are both the largest in the world (over 800kg) and the largest relative to body size in the baleen whales.

A single 4–5m (13–16ft) calf is born to a Right whale mother in the winter months. Cows give birth every 3–5 years after a 12–13 month gestation, and suckle the young for 10–12 (or sometimes as long as 17) months. By the time it is 1 year old, the rapidly-growing calf is already 8–9m (26–30ft) long. Young females reach sexual maturity around 9 years of age, although unusually precocious whales have been observed with their first calf as early as age 6.

The number of calves born to each mature North Atlantic Right whale female has declined significantly from 1980–99. Since the mid-1990s, the interval between calves for an individual female has increased from 3.67 years (1980–92) to over 5 years. The population growth rate appears to be substantially lower than among southern Right whale populations off Argentina and South Africa. Many factors may be reducing success rates, including inbreeding depression, competition for food from other species, climatic changes resulting in reduced food availability, disease, biotoxins, and the harmful (but not lethal) effects of toxic contaminants.

Natural mortality rates in the North Atlantic animals have been estimated at 17 percent for the first year of life, and about 3 percent for the next 3 years. Adult mortality rates are apparently very low – only three adults are known to have died of natural causes in this population since 1970. Longevity remains unknown, although at least one North Atlantic female has been sighted since 1935 (making her over 60 years old). About 7 percent of the North Atlantic population have scars from Killer whale attacks, but while this may cause some deaths, anecdotal reports of orca/Right whale encounters suggest that Right whales are more than adequately capable of defending themselves. No fatal diseases or epizootics have been reported for this species, although lesions thought to be related to illness have been reported. Three species of cyamid lice have been found on northern Right whales, *Cyamus ovalis*, *C. gracilis*, and *C. erraticus*. None of these appear to have a long-term effect upon the whales, even though they apparently live on the animals' sloughing skin.

Approximately 38 percent of all mortality in this species is due to collisions with large ships, with an additional 8 percent due to entanglements in fishing gear. Almost 60 percent of all Right whales in the North Atlantic display scars from entanglement in fishing gear at some time in their lives. Vertical lines from lobster and crab pots and groundfish gill-nets appear to be primarily responsible. Extensive efforts are underway in the US to develop alternative fishing methods to reduce kills of Right whales.

The Pygmy right whale resembles the Right whale in its preference for relatively shallow water at some times of year, and there is speculation that mating occurs during this inshore phase. Nevertheless, Pygmy right whales have been seen during most months of the year in all the regions from which the species has been reported, so it may be that there are localized populations and limited migrations. There, however, the resemblance to their larger distant cousins ends; no deep diving for long periods has been noted, despite an earlier suggestion that the peculiar flattening of the underside of the ribcage might indicate that Pygmy right whales spent long periods on the bottom. In addition, there is none of the exuberant tail-fluking or lobtailing and breaching characteristic of the larger species.

The Pygmy right swims relatively slowly, often without its dorsal fin breaking the surface, and the whole snout usually breaks clear of the water at surfacing – behavior similar to that of the Minke whale, with which it is easily confused. The respiratory rhythm of undisturbed animals is regular, with rather less than one blow per minute in a sequence of about five ventilations with dives of 3–4 minutes between them. The Pygmy right's general behavior has been characterized as "unspectacular" – another feature, coupled with its small size, that has contributed to the lack of recorded observations of the species. It appears to be present in sub-Antarctic and southern temperate zones right around the globe – areas of low population density and relatively little land mass where there are few researchers on hand to note the whales' behavior or attributes.

Above *Shedding rivulets of water, a Southern ri whale leaps from the South Atlantic within sight of Argentinian coast. The whales breach more often in inshore waters than in the open ocean, perhaps as a way of communicating in areas where human-induc disturbance makes acoustic signaling difficult.*

Below *A Bowhead from America's eastern popu tion basks on the surface off Baffin Island during th annual southward migration. Each spring the whale travel north from waters off the Gulf of St. Lawrenc to feeding grounds in the high Arctic, returning in t fall to spend winter in slightly more temperate clime*

Saving the Survivors

CONSERVATION AND ENVIRONMENT

remaining Bowhead and Right whales stocks remnants of much larger populations. The ear- st Right whale hunting, by the Basques, began r a thousand years ago, and the methods devel- ed then provided the basis for much of the rldwide whaling industry up until the factory- p technology of the 20th century. The last ht whaling by Europeans and Americans took ce in the early 1900s in the Atlantic. Right ales were finally given international protection 1935, although illegal Soviet whaling in the 60s is now known to have resulted in the aths of a few thousand Right whales in the ath Atlantic, Indian, and Pacific Oceans.

Historically, much of the hunting was carried t in southern-hemisphere calving grounds and s. Traditionally, whalers would attempt to take alf first, to draw the mother in for an easy kill. e carcasses were hauled ashore, or into the shal- s, and the baleen cut out. If the oil was taken, blubber was stripped and cut into pieces to be ndered down" in large cast iron "try pots."

The future outlook for Right whales is mixed. athern-hemisphere populations appear to be reasing, and marine parks and regulations are place to protect the animals around South ica, Argentina, New Zealand, and Australia. e North Pacific population is currently known, but new studies should provide more ormation soon. Since this population does not pear to have a particular coastal calving ground, l the known summer areas are mostly offshore, se whales may be less vulnerable than other pulations to conflicts with human activities. In North Atlantic, increasing human-induced aths and decreasing calving rates are driving the pulation to extinction. Even so, most earchers and conservationists believe that, if aths from shipping and fishing gear can be pped, the North Atlantic Right whales could r time make a comeback. SDK/DEG

THE ALASKA BOWHEAD HUNT

For thousands of years the Eskimo peoples of Alaska have hunted the Bowhead whales that follow the inshore lead in the pack ice eastward each spring, traditionally using ivory or stone-tipped harpoons and sealskin floats as their weapons. For much of that time the whale population in the region numbered 10,000–20,000 or more, and the effect of the hunt was negligible. In the 19th century, however, American and European commercial whalers reduced eastern Bowhead populations to a few hundreds, and the Bering Sea population as a whole to perhaps a few thousands, in a matter of decades.

Commercial whaling ceased in 1915, although a low-level Eskimo hunt persisted. Yet the eastern population did not regain its former levels. Factors such as predation by Killer whales, inbreeding depression, and (rarely) suffocation under ice may all have been implicated in the failure. But it was also suggested that changes in Eskimo hunting methods – since the 1880s, grenade-tipped darting guns have been in use – may have played their part.

In 1977, the Scientific Committee of the International Whaling Commission (IWC) recommended that the hunt cease. The Eskimos vehemently insisted that it was materially and culturally necessary for their own welfare, and a small quota was allowed by the IWC, to be administered by the Alaska Eskimo Whaling Commission. Improvements in hunting methods – for example the replacement of black-powder grenades by a modern explosive, penthrite – have also greatly increased the efficiency (and hence humaneness) of whale killing.

Studies have subsequently shown that the western (Bering/Chukchi/Beaufort Sea) population of Bowheads is increasing at an annual rate of 3 percent even with the Eskimo harvest, leading biologists to conclude recently that the present population (8,200 in 1993) can permit the nutritional needs of the Eskimo people for approximately 56 whales annually to be met comfortably. So, the situation looks bright both for the Eskimo hunters and for the Bowheads. CG

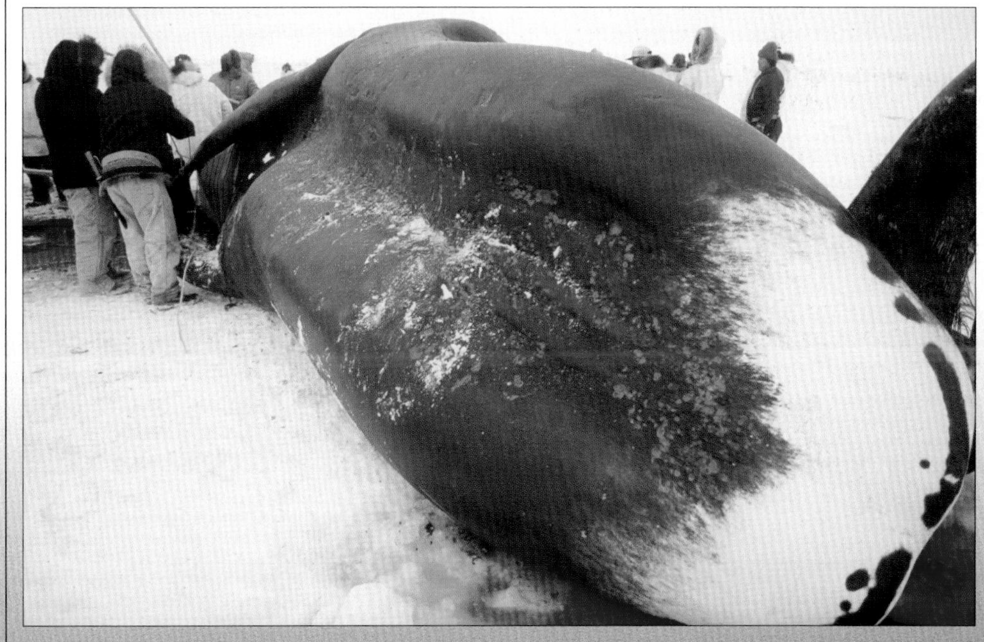

WHALEHUNTING TO WHALEWATCHING

The role of ecotourism in sustainable whale management

THE INTERNATIONAL WHALING COMMISSION (IWC) decided in 1982 that all commercial whaling should be suspended from 1986. In 1983 the Convention on International Trade in Endangered Species (CITES) listed the great whales on Appendix 1, thus prohibiting international trade and bringing CITES into line with the IWC. Despite these decisions, however, both Norway and Japan continue whaling for primarily commercial purposes. Norway filed a formal objection to the IWC moratorium decision, and is therefore not bound by it; Japan uses the loophole of scientific research to kill whales in the Southern Ocean and North Pacific. The moratorium was adopted in part because it was recognized that the existing regulatory mechanisms were inadequate. The IWC needed breathing space to design a better system for managing any future whaling, should the moratorium be lifted.

The whaling issue is unique in that whales have a special status in international law, both as highly migratory species and specifically as whales. All nations can contribute to decisions regarding their future by joining the IWC, which is a global intergovernmental convention. The IWC was set up to ensure that any use of whales should be sustainable, both to ensure the "proper conservation of whale stocks" and for the "orderly development of the whaling industry." In reality, "sustainable" has proved a very difficult term to pin down, and the IWC has been pivotal as a forum for deciding what is needed to ensure sustainability in practice.

No single group of mammals has been exploited as ruthlessly as the great whales. To take one example, only a few hundred of an original population of 250,000 Blue whales now remain in the Southern Ocean. Most populations of great whales have shown no signs of recovery, partly because there is insufficient data to detect trends in abundance and partly because insufficient time has elapsed since they were protected. As early as 1946, at its inception, the IWC recognized that "the history of whaling has seen overfishing of one area after another and of one species of whale after another to such a degree that it is essential to protect all species of whale from further overfishing." However, the Commission has so far largely failed to achieve this objective.

The world is at odds over whaling. Some countries, such as Australia and the United Kingdom, have argued that commercial whaling meets no pressing human need, and therefore oppose it on the grounds that it is cruel and unnecessary. A number of other nonwhaling countries are instead pressing for the IWC to develop means to ensure that any future whaling, if permitted, would be strictly managed. As a first step, catch limits would be calculated under IWC supervision and not just by the whaling nations themselves. Then a mechanism to prevent illegal whaling, provisionally known as the Revised Management Scheme, would need to be agreed and implemented.

Above *As ecotourism continues to grow in popularity, the financial worth of wildlife can chan[ge] dramatically – causing communities that previously hunted a resource to have an interest in conservin[g] it. Here tourists watch a Gray whale.*

Left *The crew of a Norwegian vessel haul aboa[rd] the carcass of a Minke whale in the Barents Sea. Despite assurances to the contrary from nations sti[ll] active in whaling, there is evidence that protected species of whale are still culled.*

Meanwhile, Norway and Japan avoid IWC decisions and are pressing to take more whales without international regulation.

The mechanism for calculating catch quotas, known as the Revised Management Procedure, i[s a] good example of sustainability in practice. Permi[t]ted catch sizes depend on the quality of the dat[a] on whale numbers and the level of depletion of the population. Thus, if the population has alrea[dy] been reduced by whaling, catches are set at zer[o] until that population recovers. If there is poor da[ta] on the size of a population, the catch limits are lowered, under the precautionary principle that [it] is better to err on the side of safety. The burden [of] proof is put on those who wish to exploit the resource.

Whaling's sorry history demonstrates that stri[ct] measures would be required to prevent past mis

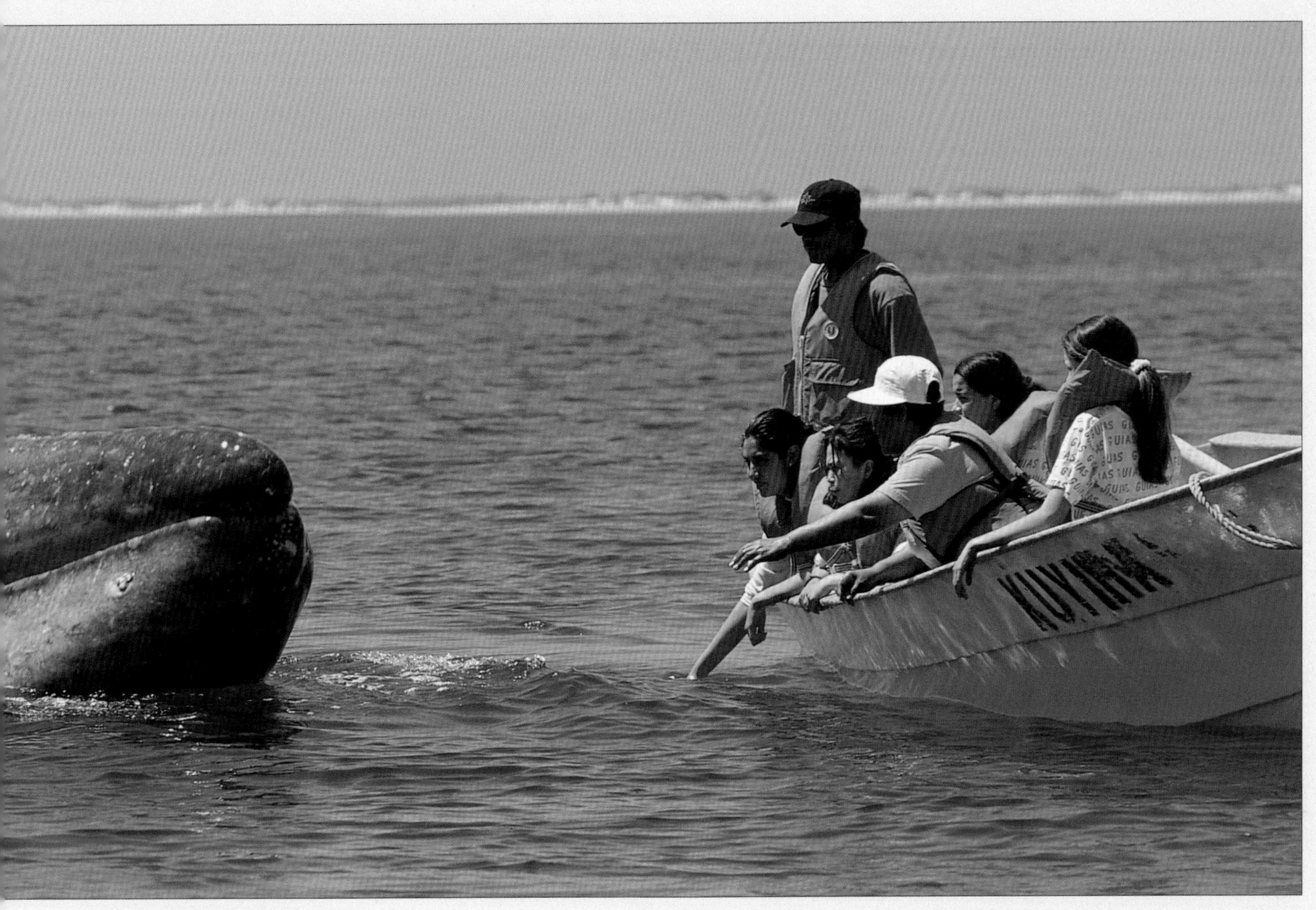

‹es from being repeated. More often than not in ·lier days, IWC catch limits were not respected. the 1960s and 1970s, the whaling ships of the ·mer Soviet Union broke every rule in the book, ·tching protected species and supplying false ·ta to the international community. The falsifica-·n of records, which continued for almost 30 ars, has only recently come to light, thanks to ·search by Russian scientists, some of whom were ·emselves involved in whaling operations when ·unger. So, for instance, it is now known that in ·e 1960s a Soviet factory ship illegally killed sever-·hundred Right whales in the Okhotsk Sea, ·spite the fact that the species had been protect-· since 1935. It is known too that Japan's whal-·g ships also falsified data, and it is likely that ·her countries did the same.

Nations that want to continue whaling argue ·at such problems are now past history. However, ·JA analyses of whale meat recently on sale in ·pan has shown that the scientific culling of ·nke whales is in fact providing a cover for the ·e of meat from other species, including such ·otected ones as the Gray, Humpback, Sei, and ·erm whales. It is impossible, however, to deter-·ne where exactly the meat is coming from ·cause of insufficiencies in Japan's regulation of · internal market. It is also unclear how much ·hale meat is being smuggled around the world.

The fact is that, without proper regulation, any whaling, even of abundant species, also puts endangered species at risk. Both Norway and Japan argue against the international supervision that could be provided by DNA monitoring of markets and fully international inspection and observation schemes, stating that the industry can be regulated nationally instead. Despite the present lack of an agreed scheme to regulate whaling, both Norway and Japan submitted proposals to the CITES meetings in 1994, 1997, and again in 2000 to allow international trade in some populations of whale, all of which failed. In that respect, the present discussions within the IWC mirror those of 30 years ago. It is clear that if commercial whaling were ever to be conducted on a genuinely sustainable basis, the costs of preventing overfishing would be substantial.

One option for commercially exploiting whales that has much more potential for sustainability involves watching rather than killing them. By 1998, whalewatching was an activity conducted for profit in 87 countries, with over 9 million participants and an annual turnover of more than US$1 billion. In many countries whalewatching has become a substantial source of foreign currency: many isolated communities, such as Husavik, a small fishing town in northern Iceland, and Kaikoura in New Zealand, have seen their fortunes transformed by it. One of the advantages of whalewatching is that it allows money to circulate widely within local communities, unlike the whaling industry, which concentrates profits in the hands of a few.

Even so, whalewatching itself has the potential to cause problems; for instance, there are concerns that whales could be disturbed if boats approach them too closely or too fast, and there is a risk of collisions. The rapid growth of the industry has resulted in concerns that it should be conducted in a sustainable, humane, and equitable manner, and the IWC has produced a framework to assist in the establishment of rules and guidelines. While recognizing that the rules have to be developed on a case-by-case basis, a number of general principles have been put forward: thus, the whales should be allowed to control the nature and the duration of the interactions; whalewatchers' boats should be designed and maintained to minimize risks; and the industry as a whole needs to be managed, for example to limit approach distances and numbers of vessels.

Whatever turn the whaling debate takes, and regardless of whether the IWC moratorium is eventually lifted, it is likely that the issues of defining genuine sustainability, ensuring appropriate regulation, and agreeing mechanisms to enforce compliance will remain hotly debated. VP

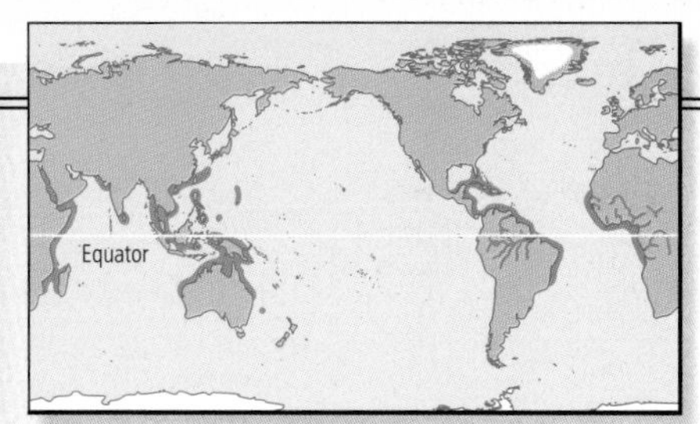

DUGONG & MANATEES

ALTHOUGH SIRENIANS HAVE STREAMLINED body forms like those of other marine mammals that never leave the water, they are the only ones that feed primarily on plants. This unique feeding niche is the key to understanding the evolution of the order's form and life history, and possibly explains why it contains so few species.

Sirenians are descended from terrestrial mammals that once browsed the shallow, grassy swamps of the Paleocene, some 60 million years ago. These herbivores gradually became more aquatic, yet their closest modern relative remains a land mammal, the elephant.

Current theories suggest that, during the relatively warm Eocene period (55–34 million years ago), a sea cow (*Protosiren*) that was the ancestor of the modern dugong and manatees fed on the vast seagrass meadows found in shallow tropical waters of the west Atlantic and Caribbean. After the global climate cooled during the Oligocene (34–24 million years ago), the seagrass beds retreated. The manatees (family Trichechidae) appeared during the Miocene (24–5 million years ago), a geological period that favored the growth of freshwater plants in nutrient-rich rivers along the coast of South America. Unlike the seagrasses, these floating mats of river grass contained silica, an abrasive defense against herbivores, which causes rapid wearing of the teeth. To counter this deterrent, manatees have an unusual adaptation that minimizes the impact of wear: throughout their lives, worn teeth are shed at the front and are replaced at the back (see box opposite).

Today, there are only four sirenian species: one dugong and three manatees. A fifth, Steller's sea cow, was exterminated by humans in the mid-1700s. Adapted to the cold temperatures of the northern Pacific, Steller's sea cow was a specialist, feeding on kelp, the dense marine algae that became abundant after the retreat of the seagrass beds (see An Extinct Giant box).

Large, Slow, and Docile

FORM AND FUNCTION

Sirenians are non-ruminant herbivores, like the horse and elephant but unlike sheep and cows, and they do not have a chambered or compartmentalized stomach. The intestines are extremely long – over 45m (150ft) in manatees – and between the large and small intestines there is a large mid-gut cecum, with paired, blind-ending branches. Bacterial digestion of cellulose occurs in this hind part of the digestive tract and enables the four species to process the large volume of relatively low-quality forage they require to obtain adequate energy and nutrients; this amounts to 8–15 percent of their body weight daily.

Sirenians expend little energy: for manatees, about one-third that of a typical mammal of the same weight. Their slow, languid movements are said to have reminded early mariners of mermaids – sirens of the sea. Although capable of rapid movement when pursued, they have little need for speed in an environment without humans, having few other predators. Living in tropical waters, sirenians can afford to have a low metabolic rate, because they expend little energy on regulating body temperature. Sirenians also conserve energy by virtue of their relatively large body size.

Manatees have the typical sirenian body form and are distinguished from the dugong mainly by their large, horizontal, paddle-shaped tails, which move up and down when they swim. They have only six neck vertebrae; all other mammals have seven. The lips are covered with stiff bristles, and there are two muscular projections that grasp and

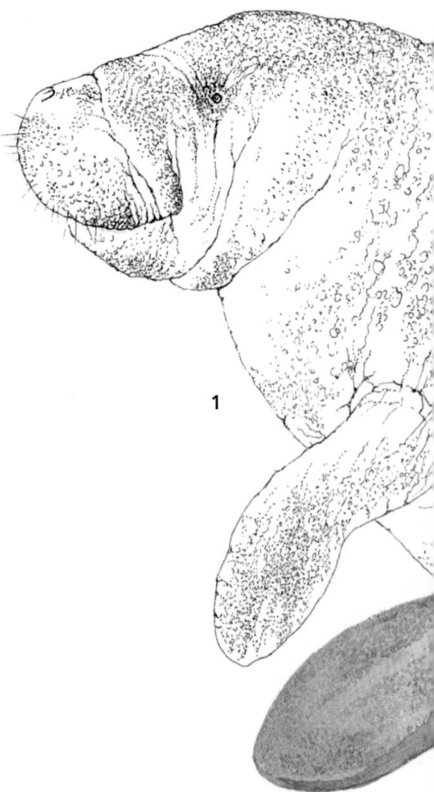

SKULLS AND DENTITION OF SIRENIANS

Adult dugongs of both sexes have only a few, peglike molar teeth, located at the back of the jaws. Juveniles also have premolars, but these are lost in the first years of life. Adult males also have a pair of "tusks": incisor teeth that project through the upper lip a short distance in front of the mouth and behind the disk. The uses to which these stubby tusks are put are not clear, but it is thought that the males may use them to guide their slippery mates during courtship.

A unique feature of manatees is a constant horizontal replacement of the molar teeth. When a manatee is born, it has both premolars and molars. As the calf is weaned and begins to eat vegetable matter, the mechanical stimulation involved in chewing starts a forward movement of the whole tooth row. New teeth entering at the back of the jaw push each row forward through the jawbone until the roots are eaten away and the tooth falls out. This type of replacement is unique to manatees.

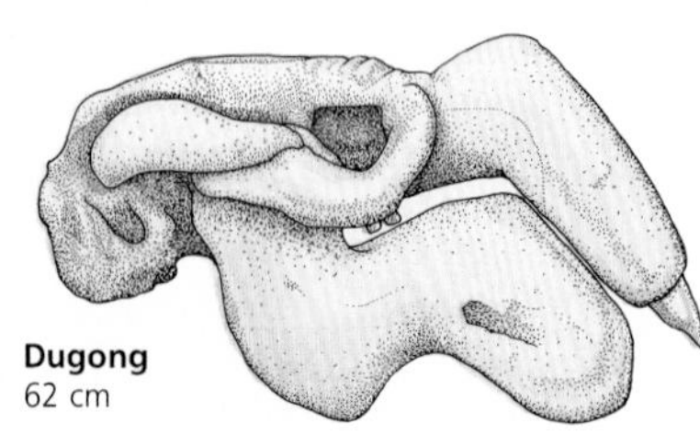

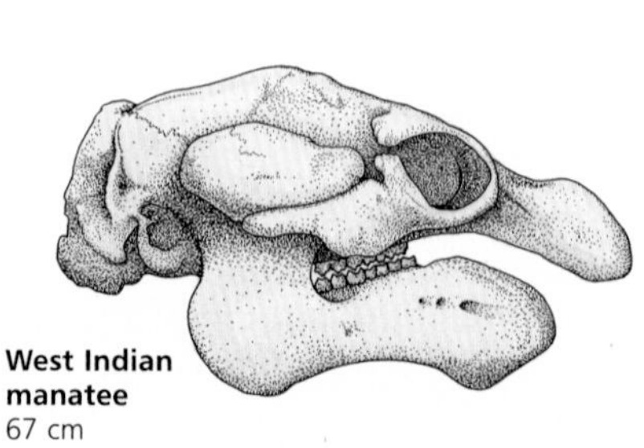

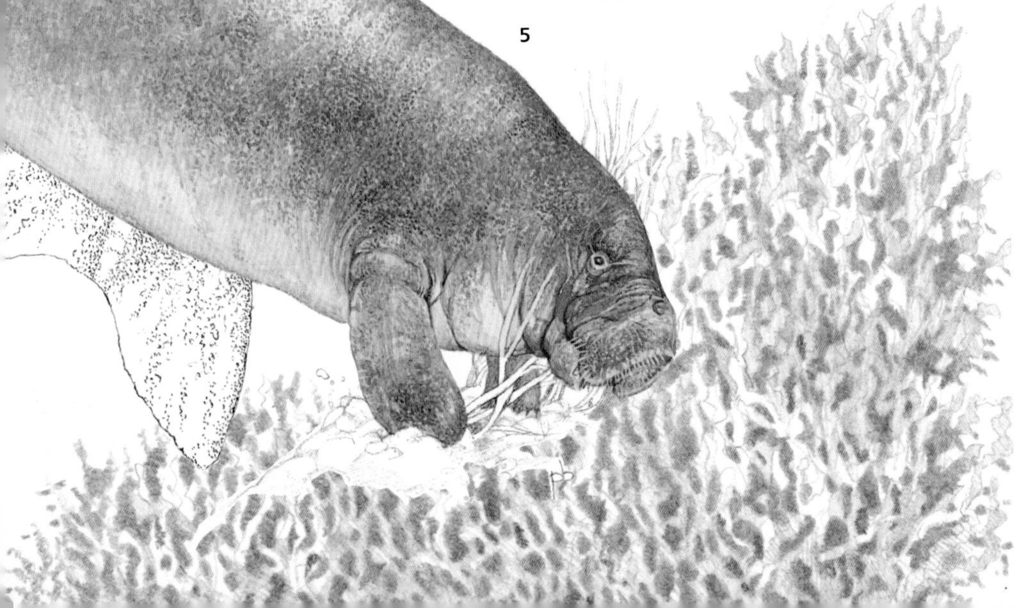

ORDER: SIRENIA

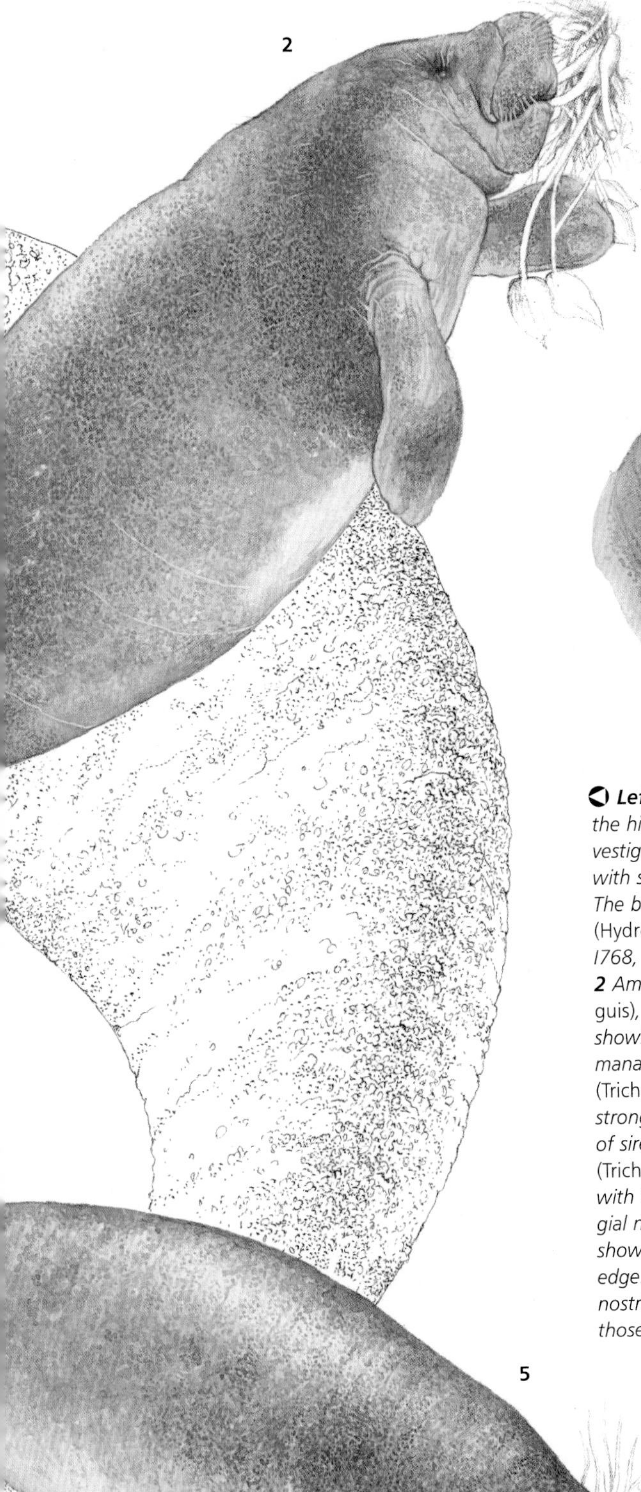

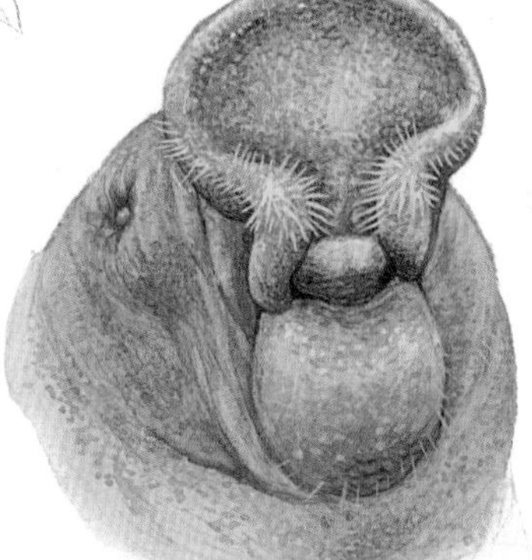

2 families; 2 genera; 4 species

Distribution Tropical coasts of E Africa, Asia, Australia, and New Guinea; SE North America, Caribbean, and N South America; River Amazon; W African coast (Senegal to Angola).

Habitat Coastal shallows and river estuaries.

Size Head–tail length from 1–4m (3.3–13ft) in the dugong, 2.5–4.6m (7.7–15ft) in manatees; **weight** 230–900kg (500–2,000lb) in the dugong, 350–1,600kg (770–3,550lb) in manatees.

Diet Water plants – sirenians are the only mammals that have evolved to exploit plant life in coastal waters.The dugong is a seabed grazer, feeding primarily on sea grasses and some algae. The manatees browse on a variety of submerged and floating plants, including large quantities of water hyacinths in Florida and of mangroves in W Africa, where manatees are said also to depend heavily on riverbank growth. They also ingest some crustaceans along with the vegetation, and have been reported to eat fish entangled in nets.

WEST INDIAN MANATEE *Trichechus manatus*

West Indian or Caribbean manatee

SE North America (Florida), Caribbean, and N South America on Atlantic coast to C Brazil. Shallow coastal waters, estuaries, and rivers. 2 subspecies – *T. m. manatus* and *T. m. latirostris* – have been proposed for the North and South American coastal population and the Caribbean populations respectively, but such a division is probably not justified because detailed comparative studies of the two groups have not yet been made. Head–tail length 3.7–4.6m (12.1–15.1ft); weight 1,600kg (3,500lb). Skin: gray-brownish and hairless; rudimentary nails on fore flippers. Breeding: gestation approximately 12 months. Longevity: 28 years in captivity, probably longer in the wild. Conservation status: Vulnerable.

WEST AFRICAN MANATEE *Trichechus senegalensis*

West African or Senegal manatee

W Africa (Senegal to Angola). Other details, where known, are similar to those of the West Indian manatee. Conservation status: Vulnerable

AMAZONIAN MANATEE *Trichechus inunguis*

Amazonian or South American manatee

Amazon river drainage basin in floodplain lakes, rivers, and channels. Head–tail length 2.5–3m (8–10ft); weight 350–500kg (770–1,100lb). Skin: lead-gray with variable pink belly patch (white when dead); no nails on fore flippers. Breeding: gestation not known, but probably similar to that of the West Indian manatee. Longevity: greater than 30 years. Conservation status: Vulnerable.

DUGONG *Dugong dugon*

Dugong or Sea cow or Sea Pig

SW Pacific Ocean from New Caledonia, W Micronesia, and the Philippines to Taiwan, Vietnam, Indonesia, New Guinea, and the N coasts of Australia; Indian Ocean from Australia and Indonesia to Sri Lanka and India, the Red Sea, and S along the African coast to Mozambique. Coastal shallows. Head–tail length 1–4m (3.3–13ft); weight 230–900kg (500–2,000lb). Skin: smooth, brown to gray, with short sensory bristles at intervals of 2–3cm (0.8–1.2in). Breeding: gestation 13 months (estimated). Longevity: to around 60 years. Conservation status: Vulnerable.

Left *Sirenians have only fore flippers, the hind limbs having been lost, leaving a vestigial pelvic girdle; the head is large, with small eyes and tiny ear openings. The biggest species was Steller's sea cow* (Hydrodamalis gigas) ***1****, extinct since 1768, which had a tough, barklike skin.* ***2*** *Amazonian manatee* (Trichechus inunguis)*, feeding on floating vegetation and showing the rounded tail typical of all manatees.* ***3*** *West African manatee* (Trichechus senegalensis) *displaying the strong bristles on very mobile lips typical of sirenians.* ***4*** *West Indian manatee* (Trichechus manatus) *carrying vegetation with its flippers. This manatee has vestigial nails.* ***5*** *Dugong* (Dugong dugon) *showing the tail with a concave trailing edge. The dugong has no nails, and its nostrils are placed further back than those of manatees.*

pass the grasses and aquatic plants that they feed on into the mouth.

The eyes of manatees are not particularly well-adapted to the aquatic environment, but their hearing is good, despite the tiny external ear openings. They seem particularly sensitive to high-frequency noises, which may be an adaptation to shallow water, where the propagation of low-frequency sound is limited. The hearing abilities of manatees and other marine mammals may have also been shaped by ambient and thermal noise curves in the sea.

Being unable to hear low-frequency noises may be a contributing factor to the manatees' inability to effectively detect boat noise and therefore avoid collisions. They do not use echolocation or sonar, and may bump into objects in murky waters; nor do they possess vocal cords. Even so, they do communicate by vocalizations, which may be high-pitched chirps or squeaks; how these sounds are produced is a mystery.

Taste buds are present on the tongue, and are apparently used in the selection of food plants; manatees can also recognize other individuals by "tasting" the scent marks left on prominent objects. Unlike toothed whales, they still possess the brain organs involved in smell, but since they spend most of their time underwater with the nose valves closed, this sense may not be used.

Manatees explore their environment by touch, using their highly-developed muzzles and muscular lips.The tactile resolving power of their bristle-like hairs is lower than that of pinnipeds, but compares well with that of the trunk of Asian elephants. This increases grazing and browsing efficiency and maximizes the potential of the manatee as a generalist feeder.

Manatees can store large amounts of fat as blubber beneath the skin and around the intestines, which affords some degree of thermal protection from the environment. Despite this, manatees in the Atlantic Ocean generally avoid areas where temperatures drop below 20°C (68°F). The blubber also helps them to endure long periods of fasting – up to six months in the Amazonian manatee during the dry season, when aquatic plants are unavailable.

The dugong grows to a length of 3m (10ft) and

weight of 400kg (880lb). In contrast to the three anatee species, all of which spend time in fresh ater to varying degrees, it is the only extant lant-eating mammal that spends all its life at sea. nlike those of manatees, its tail has a straight or ightly concave trailing edge. A short, broad, unklike snout ends in a downward-facing exible disk and a slitlike mouth.

Dugongs appear to "chew" vegetation, mainly ith rough, horny pads located in the roof and oor of the mouth. While their preferred feeding ode is piglike, rooting out carbohydrate-rich hizomes (underground storage roots) from the eabed, the name "sea cow" is not always a nisnomer. At Shark Bay, Western Australia, low inter temperatures drive the herds from their summer feeding grounds and their choicest food plants. After a migration of over 160km (100mi) to the warmer waters of the western bay, they feed during the winter months by browsing the terminal leaves of *Amphibolis antarctica*, a tough-stemmed, bushlike seagrass.

In both feeding modes, the dugong's foraging apparatus is the highly mobile, horseshoe-shaped disk at the end of its snout. In the disk, laterally-moving waves of muscular contraction sweep away overlying sediments, while stiffer bristles scoop up exposed rhizomes and any leaves that may remain attached. A meandering, flat-bottomed furrow is left behind on the seabed as evidence of a dugong's passage. Foraging dugongs rise to the surface to breathe every 40–400 seconds; the deeper the water, the longer the intervals become.

Isolated Survivors

DISTRIBUTION PATTERNS

The four sirenian species are geographically isolated. The dugong's range spans 40 countries from east Africa to Vanuatu, including tropical and subtropical coastal and island waters between about 26° and 27° north and south of the Equator. Their historical distribution broadly coincides with the tropical Indo-Pacific distribution of seagrasses. Outside Australia, the dugong probably survives through most of its range only in relict populations separated by large areas where it is close to extinction or even extinct. The degree to which dugong numbers have dwindled and their range has fragmented is unknown.

The West African and West Indian manatees have been isolated for long enough to become

Above *Displaying its rubbery, almost seal-like skin, an Amazonian manatee basks in the shallows. The Amazonian is the smallest of the three manatee species and is the only one that occurs exclusively in fresh water. Other distinctive features are flippers that usually lack nails and an elongated snout.*

Left *Fearlessly approaching the photographer, a West Indian manatee demonstrates the curiosity that is a feature of all sirenian species. One reason why they may wish to explore unfamiliar newcomers from close up is that their eyesight is poor; touch and hearing are more important weapons in their sensory arsenal.*

Below *A dugong cruises the Pacific shallows in search of seagrass to graze on. Dugongs are less bulky than manatees, and can most easily be distinguished from them by the shape of the tail; this is rounded or fan-shaped in the manatee, but in the dugong is indented to form a shallow V.*

Below *The mother–child bond, exhibited here by a West Indian manatee and her calf, is easily the strongest social tie in the sirenian world. Cows give birth to a single offspring once every other year, and the young stay with them for 12–18 months, learning about choice feeding areas and annual migration routes.*

distinct since their supposed common ancestor migrated to Africa across the Atlantic Ocean. Eac can occupy both saltwater and freshwater habitats. The Amazonian manatee apparently becam isolated when the Andes mountain range was uplifted in the Pliocene 5–2 million years ago, changing the river drainage out of the Amazon basin from the Pacific to the Atlantic Ocean. Am: zonian manatees are not tolerant of salt water an occupy only the Amazon River and its tributaries

Despite the manatee's ability to move thousands of kilometers along continental margins, genetic studies have revealed strong population separations between most locations. These findings are consistent with tagging studies which indicate that stretches of open water and unsuitable coastal habitats constitute substantial barrie to gene flow and colonization. Conversely, withi Florida and Brazil, manatees are genetically more similar than might be expected, which may be explained by recent colonization into high latitudes or bottleneck effects. Adult survival probabilities in these areas appear high enough to maintain growing populations if other traits such as reproductive rates and juvenile survival are als sufficiently high. Lower and variable survival rate on the Atlantic coast are a cause for concern.

Grazing the Shallows

DIET

Sirenians have few competitors for food. In terres trial grasslands there are many grazers and browsers, requiring a complex division of resources; but the only large herbivores in seagra meadows are sirenians and sea turtles. Marine plant communities are low in diversity compared with terrestrial communities and lack species wit high-energy seeds, which facilitate niche subdivision among herbivores in terrestrial systems. It is not surprising that dugongs and manatees dig int the sediments when they feed on rooted aquatics over half of the mass of seagrasses is found in the rhizomes, which concentrate carbohydrates. In contrast, the cold-blooded sea turtles subsist by grazing on the blades of seagrasses without disturbing the rhizomes, and appear to feed in deeper water. Thus even the herbivorous sea turtles probably do not compete significantly for food taken by the sirenians.

As aquatic herbivores, manatees are restricted to feeding on plants in, or very near, the water. Occasionally, they feed with their head and shoulders out of water, but normally they consume floating or submerged grasses and other vascular plants. They may eat algae, but this does not forn an important part of the diet. The coastal West Indian and West African manatees feed on seagrasses growing in relatively shallow, clear marine waters, and also enter inland waterways to feed o freshwater plants. Amazonian manatees are surface feeders, browsing floating grasses (the murky

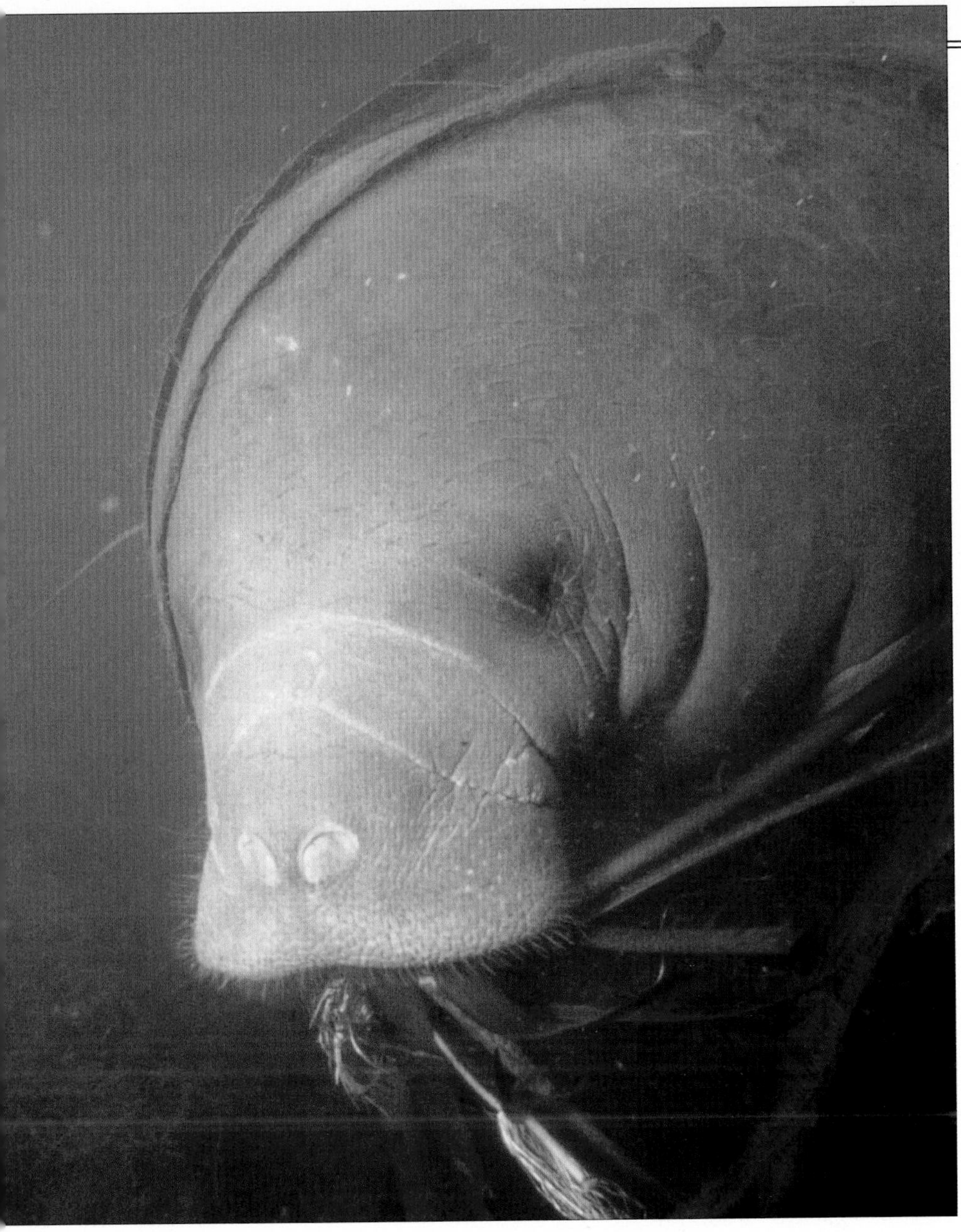

Above *The sensitive vibrissae (whiskers) on a manatee's protruding upper lip play an essential part in feeding. With their aid, an animal can feel its way through clumps of floating vegetation or root for nutrients on the seabed, much as pigs do on land.*

Left *A West Indian manatee feeds on aquatic plants off Florida. As the downward-pointing snout might suggest, this species spends some of its time grazing on the seabed, in contrast to the Amazonian manatee, which is almost entirely a surface feeder.*

nazon waters inhibit the growth of submerged uatic plants). The habit of surface feeding may plain why the downward deflection in the snout Amazonian manatees is much less pronounced an in bottom-feeding West Indian and African anatees. Some 44 species of plants and 10 ecies of algae have been recorded as foods of the 'est Indian species, but only 24 for the Amazon- n manatee.

Many of the food plants on which the manatees aze have evolved anti-herbivore protective echanisms – spicules of silica in the grasses, and nnins, nitrates, and oxalates in other aquatics – at reduce their digestibility and lower their food lue. Microbes in the manatees' digestive tract ay be able to detoxify some of the plants' chemi- il defenses.

Dugongs feed on seagrasses – marine flowering ants that sometimes resemble terrestrial grasses id are distinct from seaweeds. Seagrasses grow n the bottom in coastal shallows, and dugongs enerally feed at depths of 2–6m (6–20ft). The od they most prefer is the carbohydrate-rich rhi- omes of the smaller seagrass species.

The Cow–Calf Bond

SOCIAL BEHAVIOR

The large sirenian body size, dictated by the requirements of nutrition and temperature regulation, is associated with traits seen in other large mammalian herbivores as well as large marine mammals. The life span is long – ages of 30 or more have been recorded in captivity – and the reproductive rate is low. Females give birth to a single calf after about a year's gestation, calves stay with the mother for 1–2 years, and sexual maturity is delayed for 4–8 years. Consequently, the potential rate of population increase is low. It is possible that rapid reproduction brings no advantage where the renewability of food resources is slow and there are few predators.

Manatees are extremely slow breeders: at most they produce only a single calf every two years, and calves may be weaned from 12–18 months. Although young calves may feed on plants within weeks of being born, the long nursing period probably allows them to learn the necessary migration routes, foods, and preferred feeding areas from their mother.

In highly seasonal environments such as the Amazon, and probably also at the northern and southern limits of their distribution, the availability of food dictates when the majority of manatee females are ready to mate, and this, in turn, results in a seasonal peak in calving. The reproductive biology of male manatees is poorly known, but it is not uncommon for a receptive female to be accompanied by 6–8 males and to mate with several of these within a short time. Direct observation and radio-tracking studies have shown that manatees are essentially solitary, but occasionally form groups of a dozen or more.

Comparatively little is known about the behavior and ecology of dugongs, for they are not easily studied. The waters in which they are found are generally turbid, and they combine shyness and curiosity in a way that frustrates close observation. When disturbed, their flight is rapid and furtive; only the top of the head and nostrils are exposed as they rise to breathe. When underwater visibility is adequate and they are approached cautiously, they will come from 100m (330ft) or more to investigate a diver or a small boat, probably alerted at first by their extremely keen underwater hearing. Normal behavior stops until their curiosity is satisfied; then they swim off, frequently on a zigzag course that keeps the intruder in view with alternate eyes.

The dugongs' curiosity suggests that, as adults at least, they have few predators, although attacks by Killer whales and sharks have been recorded. Dugongs have smaller and less complexly structured brains than whales and dolphins, and their greater tendency to approach and investigate

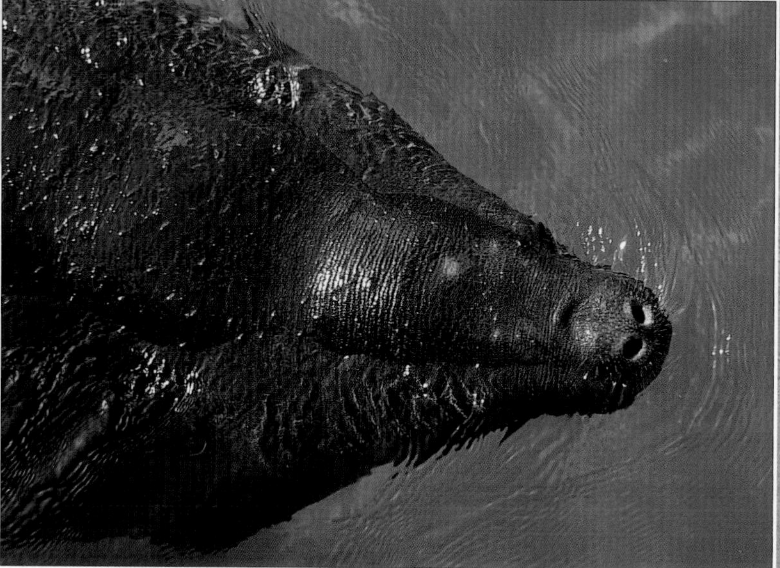

Left *A manatee raises its nostrils to the water's surface to take in air. Manatees normally breathe at intervals of less than a minute, though dives of more than 15 minutes have been recorded.*

objects visually is consistent with a lack of echolocation apparatus. Known dugong calls include chirps, trills, and whistles, which may signal danger and maintain mother–young contact. Large size, tough skin, dense bone structure, and blood that clots very rapidly to close wounds seem to be an adult dugong's main means of defense.

Dugongs sometimes form large herds, but are more often encountered in groups of less than a dozen; many individuals may be solitary. Gender is difficult to determine in the wild, but the groups generally appear to include one or more females with calves. In some habitats, larger numbers may gather to exploit rich seagrass resources; in these situations, herds 60–100 strong may "cultivate" the beds, cooperating to maintain the seagrass in a maximally nutritious state

Radio tagging has shown that dugongs are for the most part relatively sedentary, inhabiting home ranges of a few dozen square kilometers. Sometimes, however, they make return trips over distances of hundreds of kilometers. The function of these excursions remains obscure.

Tropical environments make long mating seasons possible, and dugongs may mate over a period of 4–5 months. Females become sexually mature at 10–17 years, and give birth to a single calf after a gestation of about 13 months. Few births have been observed, but it seems that the females seek out shallows at the water's edge. The calf keeps close contact with its mother for up to 2 years, suckling from a single teat in the axilla of each flipper while lying beside her and taking refuge behind her back in the presence of danger. Although females can become pregnant while lactating, inter-birth intervals average from 3–7 years. Females may live to over 60 years.

Right *In Florida, a pair of manatees share a snack at a favorite feeding spot. Although the big animals are not highly social, they tolerate one another's company without showing aggression, and enjoy apparent play activities that include nuzzling and "kissing."*

Left West Indian manatees lack cohesive social organization, with th exception of the mother–calf bond. **1** Other assemblages are either agg gations at locations where resource such as food or warm water are co centrated, or else are ephemeral an have no consistency in composition, as is the case with mating herds an all-male cavorting groups. Despite t lack of continuity in social grouping the animals often exhibit social inte tions that are characterized by simp gestures, such as physical contact a "kissing." **2** Even when alone, mana tees can communicate with one ano via "rubbing posts" **3**, prominent objects on which they deposit tas and odors that other manatees detect chemically. The big ma mals sometimes relax by lyin their backs on the seabed

Lambs to the Slaughter

CONSERVATION AND ENVIRONMENT

Docility, delicious flesh, and a low reproductive capacity are not auspicious characteristics for animals in the modern world. The dugong and manatees have all three, and are consequently among the most threatened of aquatic mammals.

All three species of manatees are considered by the IUCN to be vulnerable as a result of both historic and modern overhunting for their meat and skins; they are also at risk from more recent threats such as pollution and high-speed pleasure craft. One study estimated that a 10 percent increase in adult mortality or reproduction would drive the Florida population to extinction over a 1,000-year time scale, whereas a 10 percent decrease in adult mortality would allow slow population growth. They are protected under the Convention on International Trade in Endangered Species of Fauna and Flora (CITES), and legally in most countries where they exist.

In Costa Rica, local residents blame an apparent decline in their numbers on illegal hunting, high levels of toxicants in coastal waters, ingestion of plastic banana bags, and increased motorboat traffic. Badly managed "eco-tourism," and environmental degradation have also played a part. A study of manatee carcasses in Florida revealed that most deaths there were due to human interaction, especially captures and watercraft collisions. When a manatee does die of natural causes, it is usually a dependent calf.

Amazonian manatees have been commercially exploited for their meat and hide since 1542, and

AN EXTINCT GIANT

The only close relative of the dugong to survive into historic times was Steller's sea cow. This giant marine grazer was the largest of all the sirenians, with a body length of up to 7.5m (25ft) and a weight of 4.5–5.9 tonnes. The sea cow was unique among mammals in having no phalanges. The pectoral flipper had a stumpy, densely bristled termination that was described by a contemporary observer as "hoof-like." Seacows were apparently unable to submerge, and instead used these appendages to support themselves against the rocks while they fed on kelp in the surfzone. Fossil evidence suggests that some 100,000 years ago their range extended along northern Pacific coasts from Baja California up through the Aleutian Islands as far as Japan.

As with the dugong, inshore feeding habits made Steller's sea cow vulnerable to hunters in small boats. As a result, it was probably much preyed upon by native peoples, and by the time it first came to the attention of the West the total population had already been severely reduced, possibly to as few as 1,000–2,000 animals. At the time of its discovery, by a shipwrecked Russian exploring party in 1741, it was restricted to the shores of two subarctic Pacific islands, each 50–100km (30–60mi) long. The survivors of the shipwreck themselves discovered the sea cow's vulnerability and edibility out of dire necessity, and over the next quarter of a century parties of fur-hunters found it convenient to winter on the two islands in order to exploit this easily accessible food supply. By 1768 no more of the giants could be found.

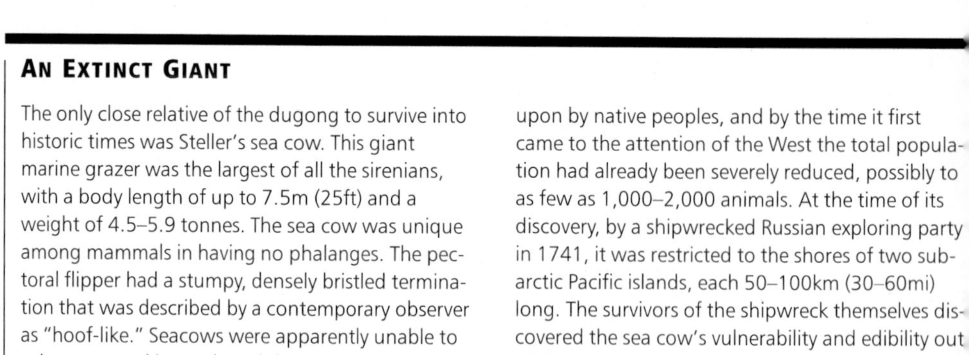

In later times, several isolated dugong populations seem to have suffered similar fates. Even though there are still a few surviving herds as numerous as those of Steller's sea cow were at the time of its discovery, the story of the extinction of the species' huge relative nevertheless has an obvious moral for the present day. PKA

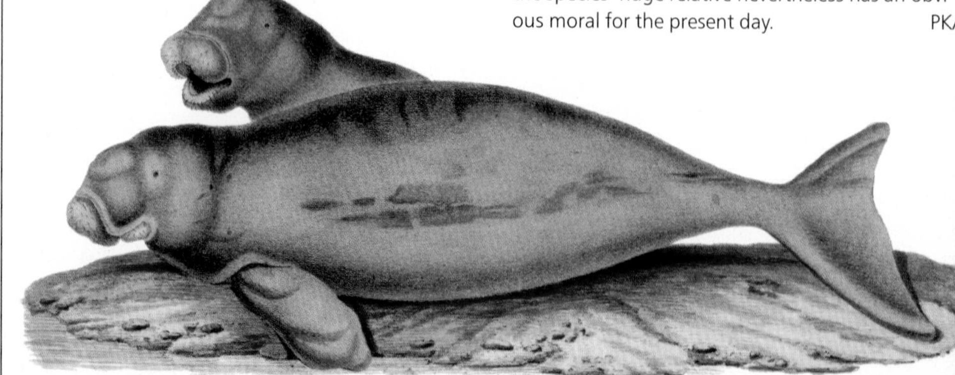

y are now considered an endangered species. ıough they have been legally protected since '73, exploitation for meat has in fact continued hout any practical restrictions. Manatees are) hunted in Peru, and are sometimes taken inci- tally in fishing gear.

Another important factor in manatee deaths are oms of toxic algae that may be exacerbated by lution. During a period of several weeks in the ing of 1996, over 200 manatees were found d or dying in Florida's coastal waters or on the ches of the west coast. At the same time, high sities of a dinoflagellate which produces a ent neurotoxin that binds to manatee brain s were observed in the same coastal areas. Yet ther threat is morbillivirus, which may cause l infections and possibly also more insidious cts on the immune system or reproduction.

mproving the manatee's lot will require proac- management. Scientific research, rescue, and abilitation also have their place. Boat-free es provide sanctuaries for manatees in Florida, l are an effective management tool. If speed l boating regulations are effectively enforced in key coastal counties, Florida's manatees uld be able to coexist indefinitely with human eational requirements, whereas if regulation is unsuccessful, the population is likely to decline slowly toward extinction. Control of hunting and management of tourism may also make the manatee's future more certain elsewhere.

Above *In Florida's Homosassa Springs State Wildlife Park, a conservation worker feeds vitamin supplements to an eager clutch of manatees. The species' long-term survival in the state now depends on the effectiveness of management programs.*

In recent times, people have found a non-lethal employment for manatees that they seem only too happy to fulfill: that of clearing weeds from irrigation canals and the dams of hydroelectric power stations. It is thus possible that the animals' gentle, herbivorous lifestyle might yet help them to survive in an aggressive world.

As a marine creature, the dugong has had less direct contact with people than the riverine manatees, but even so the relationship has not been a happy one. Dugongs have traditionally been hunted by coastal peoples in most of their range. In recent times, the increase in human populations and the growing availability of nylon gill-nets and boats with outboard motors have led to the decimation of dugong populations outside of Australia and the Persian Gulf. PKA/JMP/GBR/DPD/RB

Above *Caught in a fish-trap off Sulawesi (Indonesia), a young dugong uses its sensitive snout to investigate its prison. Over the centuries hunting has driven the dugong from much of its former range.*

GRAZING THE SEAGRASS MEADOWS

Feeding strategy of the Dugong

A PERSON WANDERING THE INTERTIDAL SEAGRASS meadows of northern Australia will likely notice long, serpentine furrows devoid of all vegetation. These are the feeding trails of the dugong, a large marine mammal that uproots whole seagrass plants, including their roots and rhizomes. Dugongs prefer small, delicate, "weedy" seagrasses that are low in fiber, yet high in available nutrients and easily digested – mainly species from the genera *Halophila* and *Halodule*. Experiments simulating dugong grazing indicate that their feeding alters both the species composition and nutrient qualities of seagrass communities, causing them to become lower in fiber and higher in nitrogen. In effect, dugongs are like farmers cultivating their crops. If these animals were to become locally extinct, the seagrass meadows would, in turn, deteriorate as dugong habitat.

Over most of its range, the dugong is known only from incidental sightings, accidental drownings, and the anecdotal reports of fishermen. However, within Australia, intensive aerial surveys have produced a more comprehensive picture of dugong distribution, which is now known to extend from Moreton Bay in Queensland, on the east coast, around to Shark Bay in Western Australia. These same surveys show dugongs to be the most abundant marine mammal in the inshore waters of northern Australia, numbering some 85,000 individuals. What is more, this figure is probably an underestimate, since some areas of suitable habitat have not been surveyed and the mathematical correction for animals that cannot be seen in turbid water is conservative. In other words, Australia is the dugong's last stronghold.

More than 60 individual dugongs have been tracked using satellite transmitters. Most of their movements have been localized to the vicinity of seagrass beds and are dictated by the tide. At localities where the tidal range is large, dugongs can gain access to their inshore feeding areas only when the water depth is at least one meter (3.3ft). In areas with low tidal amplitude or where seagrass grows subtidally, dugongs can generally feed without making significant local movements. However, at the high-latitude limits of their range, dugongs make seasonal movements to warmer waters. While overwintering in Moreton Bay, many dugongs frequently make round trips of 15–40km (9–25mi) between their foraging grounds inside the bay and oceanic waters, which are, on average, up to 5°C warmer. Dugongs also apparently relocate within Shark Bay itself, moving from west to east, where the water is warmer. Some travel for long distances; for example, in the Great Barrier Reef region and the Gulf of Carpentaria, several dugongs have been recorded making trips of 100–600km (62–372mi) over just a few days. Many of these movements were return trips. One plausible explanation for such long journeys is that dugongs are checking the status of the seagrass beds in their region. Many seagrass meadows arise and promptly disappear again, for no apparent reason. Sometimes hundreds of kilometers of seagrass may be lost after storms or flooding.

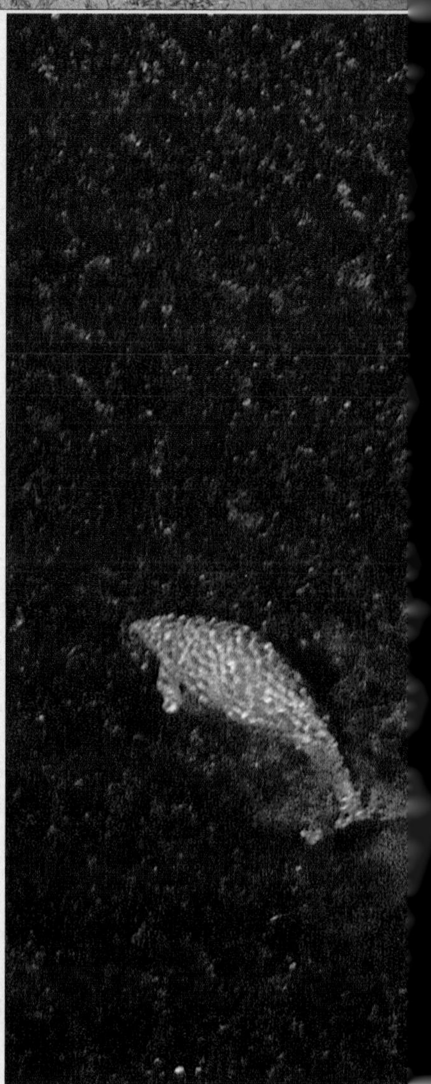

Above and Right *Seagrasses grow abundantly in the shallow waters that surround northern Australia. This plentiful supply of vegetation attracts great numbers of dugongs. Stirring up the substrate as they browse, these sedate marine grazers appear when seen from the air almost like combine harvesters moving slowly across a field of crops.*

Dugongs are long-lived, with a low reproductive rate, long generation time, and a high investment in each offspring. On the basis of annual growth rings in its tusks, the oldest dugong was estimated to be 73 years old when she died. Females give birth aged 10–17, and the period between successive births varies from three to seven years. The gestation period is around13 months, and the (almost always single) calf suckles for at least 18 months. Dugongs start eating seagrasses soon after they are born and grow rapidly during the suckling period. Population simulations indicate that a dugong population is unlikely to be able to increase by more than 5 percent annually. This makes the dugong highly susceptible to overexploitation by indigenous hunters or to incidental drowning in fishing nets. Consequently, they are classified as Vulnerable to global extinction. HM

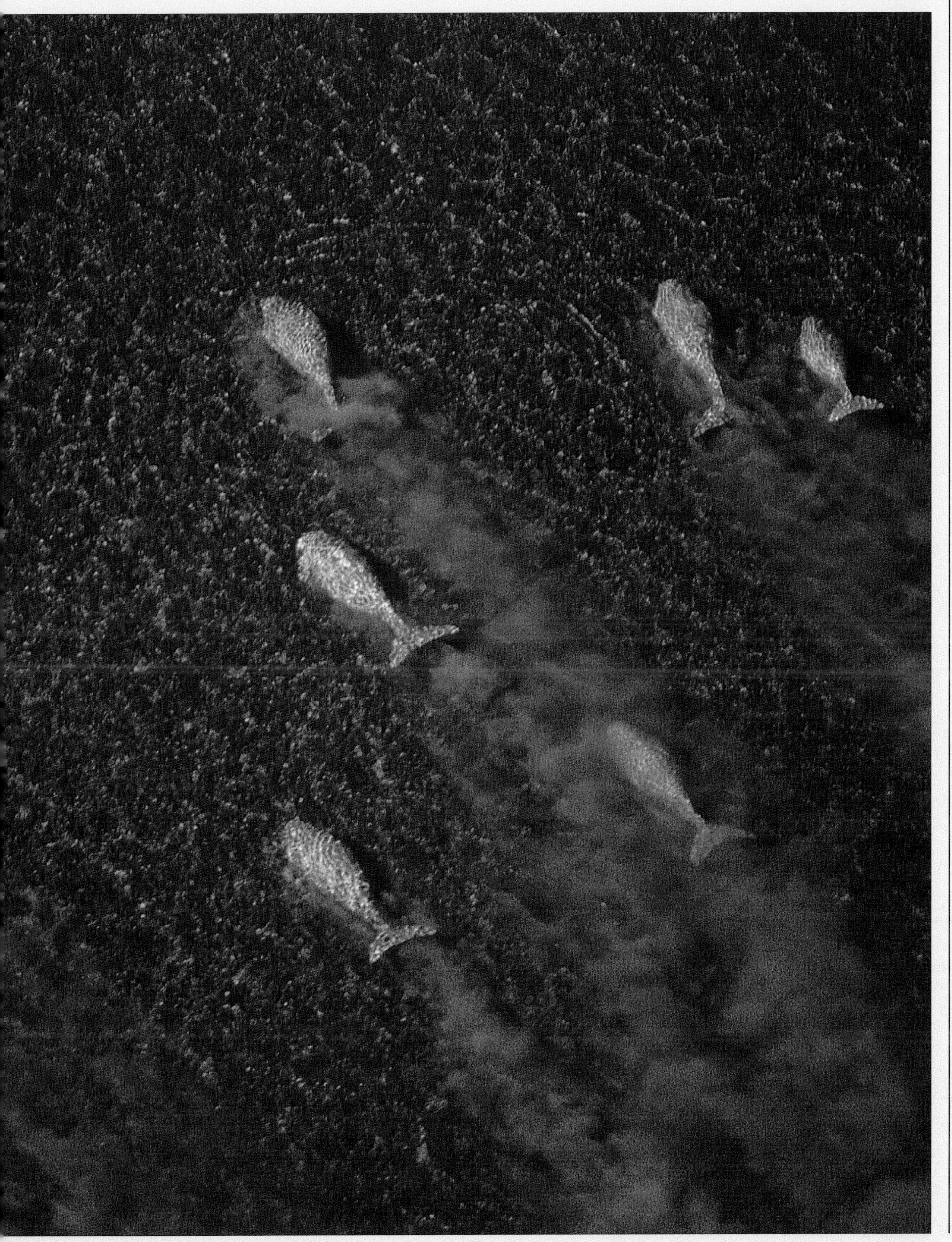

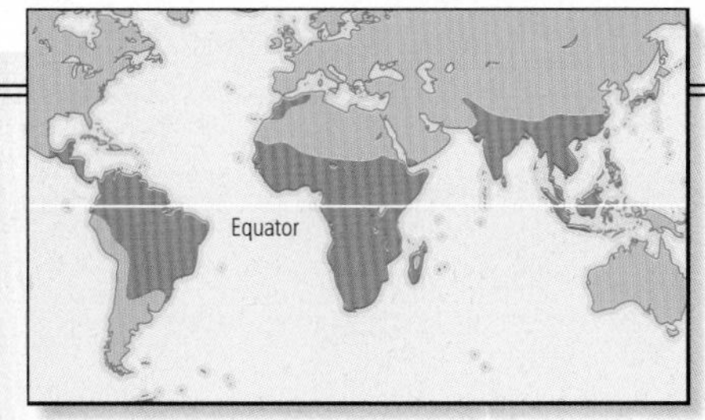

PRIMATES

The technical description of a group of mammals often fails signally to convey the thrilling physical presence of the creatures concerned. This is rarely more true than of the ponderous definition of primates offered by the English biologist Saint George Jackson Mivart in 1873: "unguiculate, claviculate, placental mammals, with orbits encircled by bone; three kinds of teeth, at least at one time of life; brain always with a posterior lobe and calcarine fissure; the innermost digits of at least one pair of extremities opposable, hallux with a flat nail or none; a well-developed caecum; penis pendulous; testes scrotal; always two pectoral mammae." Yet while this list goes no way toward capturing the sheer charisma of primates, and makes no mention of the intriguing social behavior that is perhaps their most striking feature, it remains largely accurate. Modern biologists would add only two further defining features: firstly, a foreshortened muzzle plus flattened face, associated with reliance on vision at the expense of olfaction, leading to efficient stereoscopic color vision; and, secondly and crucially, a larger brain, with elaboration and differentiation of the cerebral cortex.

This is a very general blueprint; in practice, not all primates exhibit all of these traits. Furthermore, those listed are not all unique to primates: for example, many other mammals have clavicles (collarbones), three kinds of teeth, and a pendulous penis. Indeed, anatomically, primates are remarkably generalist mammals.

The primates divide into two suborders: strepsirhines and haplorhines. In strepsirhines, the muzzle (or "rhinarium"), the area of bare skin around the nostrils, is prominent, moist, and glandular, whereas in haplorhines, it is considerably suppressed and hairy. The strepsirhines are more typical of ancestral primates, with longer snouts, a better developed sense of smell, and smaller brains. They comprise three major subgroups: the bush babies of Africa (the Galagonidae), the angwantibo/loris/potto group of the Old World (the Loridae), and the lemurs of Madagascar (a diverse assemblage of five families: the Lemuridae, Cheirogaleidae, Indriidae, Megaladapidae, and Daubentoniidae). Notably, strepsirhines are absent from the Americas.

The haplorhines are also subdivided into three: the platyrrhines and catarrhines (names denoting

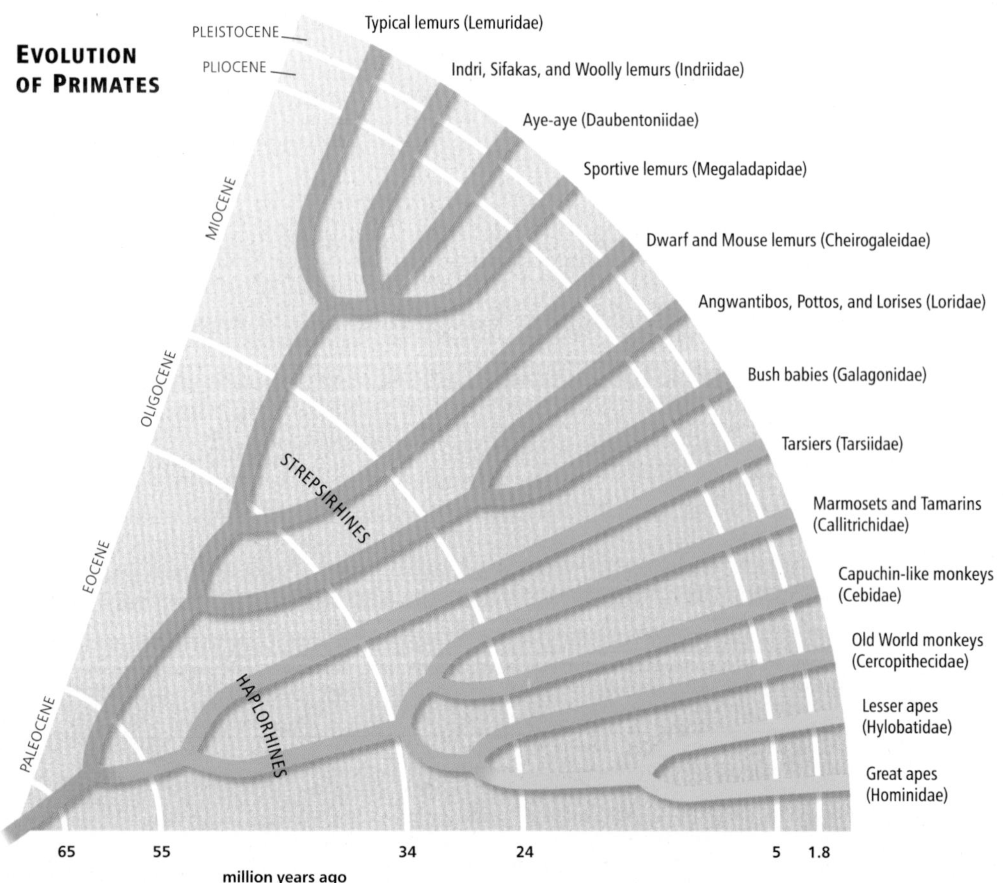

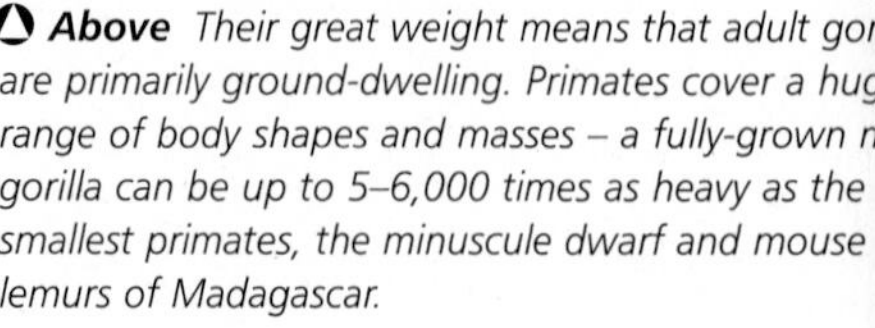

Above *Their great weight means that adult gor are primarily ground-dwelling. Primates cover a hug range of body shapes and masses – a fully-grown n gorilla can be up to 5–6,000 times as heavy as the smallest primates, the minuscule dwarf and mouse lemurs of Madagascar.*

*Top Posture and gait. Some lemurs **1** walk on all* *s; the tree-dwelling lifestyle of most species is re-* *ted in the longer hind limbs for leaping. The tarsier* *a "vertical clinger and leaper" adapted for leaping* *ween vertical trunks. Most arboreal monkeys like* *Diana monkey **3**, have well-developed hind limbs and a long tail for balancing; baboons and other ground-dwellers have forelimbs at least as long as their hind limbs. Among apes, the knuckle-walking gorilla **4** is the most terrestrial. The gibbons **5** arm-swing under branches on their long arms and, like the chimpanzee **6**, may walk upright, thus freeing the hands.*

ORDER: PRIMATES
13 families; 64 genera; 256 species

STREPSIRHINES
Suborder Strepsirhini

TYPICAL LEMURS Family Lemuridae p310
10 species in 4 genera
Includes **Ring-tailed lemur** (*Lemur catta*).

SPORTIVE LEMURS Family Megaladapidae p314
7 species in 1 genus
Includes **Northern sportive lemur** (*L. septentrionalis*).

DWARF AND MOUSE LEMURS
Family Cheirogaleidae p316
13 (7) species in 5 (4) genera
Includes **Greater dwarf lemur** (*Cheirogaleus major*).

INDRI, SIFAKAS, AND WOOLLY LEMURS
Family Indriidae p320
6 (5) species in 3 genera
Includes **indri** (*Indri indri*).

AYE-AYE Family Daubentoniidae p322
1 species *Daubentonia madagascariensis*

ANGWANTIBOS, POTTOS, AND LORISES
Family Loridae p324
7 (6) species in 5 (4) genera
Includes **Slow loris** (*Nycticebus coucang*), **potto** (*Perodicticus potto*).

BUSH BABIES Family Galagonidae p324
17 (11) species in 4 genera
Includes **Garnett's galago** (*Otolemur garnettii*).

TARSIERS, MONKEYS, AND APES
Suborder Haplorhini

TARSIERS Family Tarsiidae p332
5 species in 1 genus
Includes **Western tarsier** (*Tarsius bancanus*)

MARMOSETS AND TAMARINS
Family Callitrichidae p334
38 (26) species in 6 (4) genera
Includes **Common marmoset** (*Callithrix jacchus*), **Emperor tamarin** (*Saguinus imperator*)

CAPUCHIN-LIKE MONKEYS Family Cebidae p344
47 (58) species in 11 genera
Includes **night monkeys** (*Aotus* species), **titi monkeys** (*Callicebus* spp.), **spider monkeys** (*Ateles* spp.).

OLD WORLD MONKEYS
Family Cercopithecidae p356
87 (81) species in 18 genera
Includes **Savanna baboon** (*Papio hamadryas*), **mandrill** (*Mandrillus sphinx*), **Hanuman langur** (*Semnopithecus entellus*).

GIBBONS OR LESSER APES Family Hylobatidae p398
11 species in 1 genus
Includes **siamang** (*Hylobates syndactylus*) and **Lar gibbon** (*H. Lar*).

GREAT APES Family Hominidae p406–420
7 (5) species in 4 genera
Includes **Western gorilla** (*Gorilla gorilla*), **Common chimpanzee** (*Pan troglodytes*), **Bornean orangutan** (*Pongo pygmaeus*).

Note Primates are one of the groups whose taxonomy is in the greatest flux. A comprehensive revision in 1993 recognized 13 families, 60 genera, and 233 species. To illustrate the divergence between the tallies according to our authors and those in the 1993 established taxonomy, the latter are given in parentheses above and in the Factfile panels throughout.

the shape of their nostrils), and the tarsiers. The platyrrhines comprise the New World monkeys – the marmosets and tamarins (Callitrichidae) and the cebids (Cebidae), including capuchins, uakaris, howler monkeys, spider monkeys, titi monkeys, and the night monkeys. Catarrhines include the Old World monkeys (the Cercopithecidae) – the baboons, colobus monkeys, guenons, leaf monkeys, macaques, and mangabeys – and also the apes – the great apes (Hominidae), and the lesser apes or gibbons (Hylobatidae) of Southeast Asia. The tarsiers (Tarsiidae) are a relatively small group, made up of just five species of nocturnal insectivores from Southeast Asia.

Older classifications divided the primates into the suborders prosimians (i.e. the strepsirhines and tarsiers together) and anthropoids (i.e. the haplorhines minus the tarsiers). While there is still debate over which classification is the more appropriate, recent research on tarsiers tends to vindicate the strepsirhine/haplorhine model. Tarsiers appear to share with the simian (or anthropoid) primates not only a common, more recent evolutionary history, but also certain anatomical traits – notably the hairy rhinarium and an upper lip that is not cleft, plus features of the internal structure of the nose and ear as well as those of the blood supply to the brain and in the placenta. Adopting this model has the effect of pushing back to about 55 million years ago the first known haplorhine ancestors of the anthropoids. While the earliest known anthropoid is *Eosimias* from the late Eocene of China (about 42 million years ago), the predecessors of tarsiers and anthropoids were most likely the omomyids, from the early Eocene. Their oldest antecedent is *Tei hardina americana*. Two other mammalian grou both at different times regarded as Primates, are here considered as separate: the tree shrews (or Scandentia) and the colugos (order Dermopter

Most primates live in tropical forest between 25°N and 30°S. Among the exceptions are five Old World monkeys whose ranges occur entire outside the tropics. Two are macaques: the Barbary macaque, found in North Africa and introduced to Gibraltar; and the Japanese macaque, which occurs on Hokkaido and Honshu. The r are three closely-related species of langur: the Y nan, Guizhou, and Sichuan snubnosed monke of China. The reason the primates are restricted the tropics is probably related to their diet, whi consists largely of fruit, shoots, or insects – iten that are scarce during winter in temperate regio

PRIMATE BODY PLAN

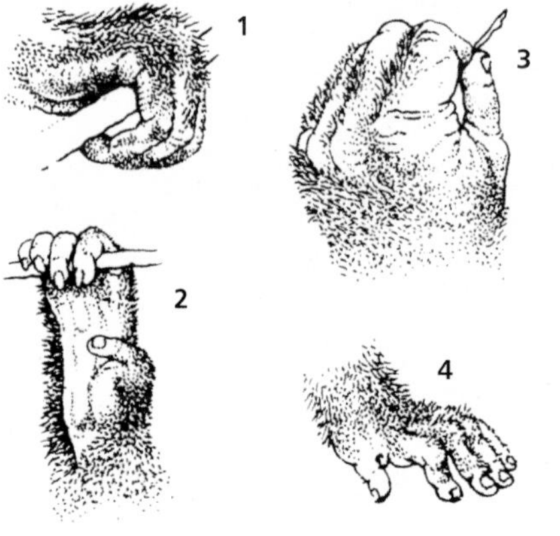

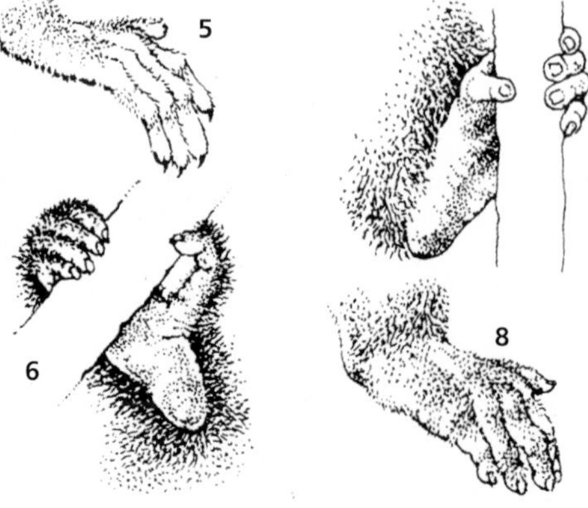

Above The structure of primate hands and feet varies according to the ways of life of each species. **1** Hand of a spider monkey, showing the much reduced thumb of an arm-swinging species. **2** Gibbon: short opposable thumb well distant from arm-swinging (brachiating) grip of fingers. **3** Gorilla: thumb opposable to other digits. allows precision grip. **4** Macaque: short opposable thumb in hand adapted for walking with palm flat on ground. **5** Tamarin: long foot of branch-running species with claws on all digits except big toes for anchoring (all other monkeys and apes have flat nails on all digits). **6** Siamang and **7** Orangutan; broad foot with long grasping big toe for climbing. **8** Baboon: long slender foot of ground-living monkey.

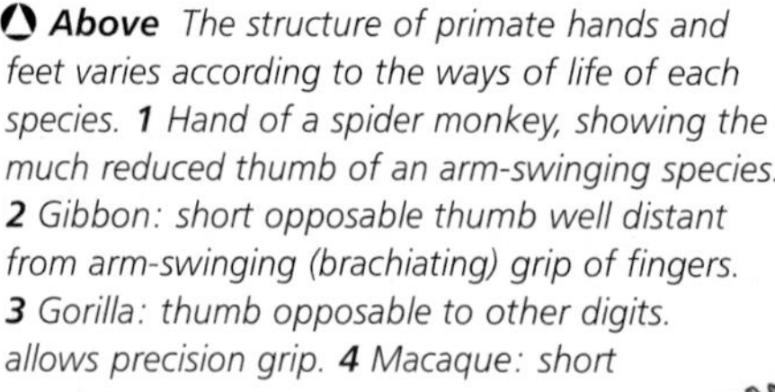

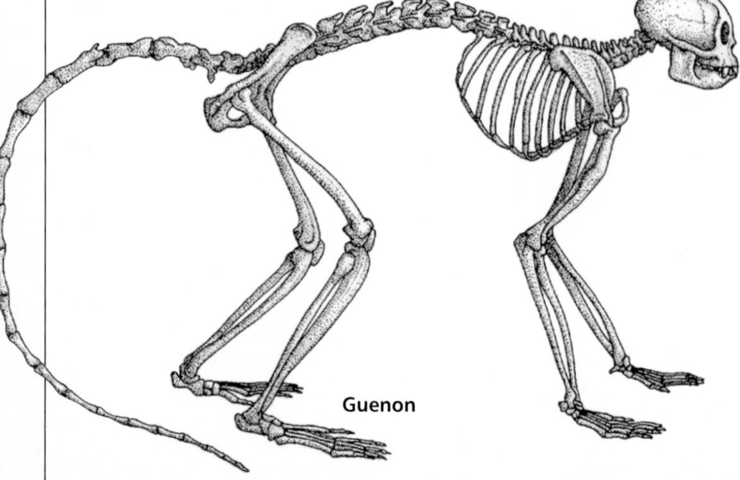

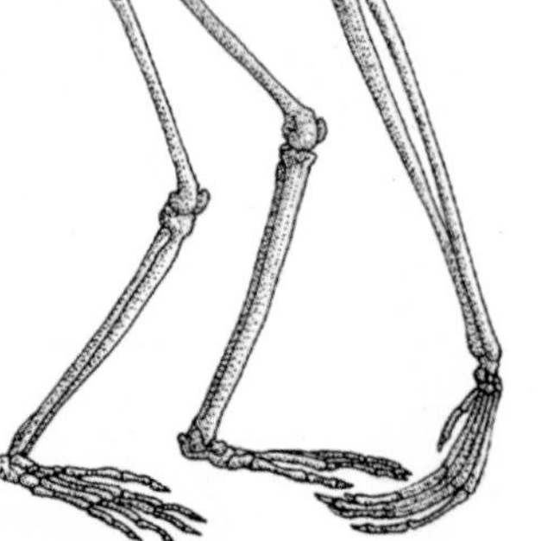

__Below__ Skeletons. The quadrupedal lemurs and most monkeys like the guenons retain the basic shape of early primates – long back, short, narrow rib cage, long, narrow hip bones, and legs as long as – or longer – than the arms. Most live in trees and move about by running along or leaping between branches. Their long tail serves as a rudder or balancing aid while climbing and leaping. Ground-living monkeys, such as the baboons, generally have more rudimentary tails.

Neither apes nor the slower-moving prosimians have tails. In the orangutan and other apes. the back is shorter, the ribcage broader and the pelvis bones more robust – features related to a vertical posture. Arms are longer than legs, considerably so in species, such as the gibbons and orangutan, that move by arm-swinging (brachiation). Further dexterity of the hands has accompanied the development of the vertical posture in apes, some of which (as more rarely some monkeys) may at times move about bipedally like humans.

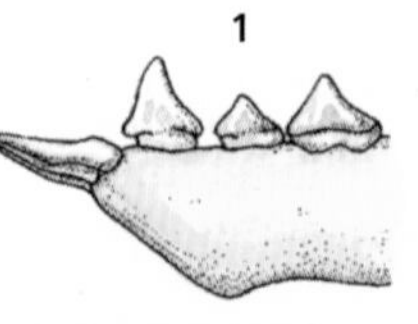

__Below__ Insectivorous precursors of primates had numerous teeth with sharp cusps. In strepsirhines such as the lemur __1__, the first lower premolar is almost canine-like in form, while the crowns of the lower incisors and canines lie flat to form a toothcomb, as in bush babies, which is used in feeding and grooming. In leafeating monkeys of the Old World, such as Presbytis *__2__, the squared-off molars bear four cusps joined by transverse ridges on the large grinding surface that helps break up the fibrous diet. In apes like the gorilla __3__ the lower molars have five cusps and a more complicated pattern of ridges.*

Small Families, Big Brains

ANATOMY AND PHYSIOLOGY

rimates are medium-sized mammals. They are rger than most insectivores and bats, but smaller an most ungulates and cetaceans. They range in ody mass from about 30g (1oz), as in Gray ouse lemurs, to over 150kg (331lb), as in male orillas (an adult man weighs about 80kg/176lb). s in other mammals, the body mass of a primate an important determinant of its life history: larg- primates reproduce more slowly but live longer.

In the fast lane, mouse lemurs begin breeding one year of age, give birth to two offspring weighing 6.5g/0.2oz) following a gestation peri- d of only two months, and will produce another ter ten months later. The longest recorded life pan is 15 years. In contrast, female gorillas do not egin to breed until ten years of age, give birth to a ngle offspring (weighing 2.1kg/4.6lb) following a station period of nine months, and will not pro- uce another offspring for another four years. orillas can live for up to 40 years. Primates there- re show considerable diversity in their reproduc- ve patterns, with the sole exception of litter size; ey normally only have between one and two oung at a time.

A fundamental difference between primates d other mammals is that primates have unusu- y slow life histories. In other words, their rates birth, growth, and death are only about one- arter to one-half of those recorded in other ammals of similar body mass. The reasons for is difference are unclear, but may relate to brain ze. Brain tissue is one of the most energetically pensive tissues in the body, so the high meta- lic demands of large primate brains may reduce e metabolic energy available for other areas of owth or reproduction.

One important consequence of the slow rate of production in primates is their relatively strong edisposition to infanticide. Male primates will ten attempt to kill infants that have been sired other males, because a lactating female cannot nceive further offspring until her present infant s been weaned. Males who have access to such males are reluctant to delay any mating opportu- y, because they may have only a narrow time ndow in which to breed. While at their peak ysical strength and able to fight off competitors, ales do all they can to secure genetic immortali- In male Hanuman langurs, this period compris- only 800 days in a life span of 20 years.

Body size varies not only between different ecies of primates but also between males and males of the same species. Males are generally ger than females, although there is much varia- n in this pattern, related to the type of mating

Right *A young Sumatran orangutan. Only four ch offspring will be born to a female orangutan over years of fertility, thanks to the species' extremely w rate of breeding.*

Above *A Spot-nosed monkey* (Cercopithecus nictitans). *Guenon species such as this forage high in the canopy of the West African forest for their preferred diet of leaves, fruit, and insects.*

system involved. When a single male can potentially monopolize exclusive access to a large number of estrous females, competition between males becomes intense: because large size is advantageous in combat, it leads to selection for large male body mass. Male Hanuman langurs, for example, can sometimes monopolize access to groups of over 20 females, and can grow to a size 160 percent larger than that of females. In contrast, males that normally mate with a single female will often be no different in size, as in the case of gibbons. Unusually, lemurs display very little sexual dimorphism, despite the fact that they share similar patterns of social organization with other primates. This strongly suggests that the mating systems that underpin lemur social systems may be quite different to those encountered in other primates.

It is not only size that has altered in the face of primates' fierce competition for paternity. Formidable weaponry in the form of huge canines are commonly used in fights and aggressive displays. In addition, males have developed more subtle ways to get ahead in the mating game. In species where fertile females are mated in short succession by several males, the males grow large testes to produce greater numbers of sperm. The more sperm a male produces, the greater the likelihood that he will be the one to fertilize the female's ova by swamping the sperm of other males already in the female's reproductive tract.

Sexual competition affects not just male morphology. In many primates, estrous cycles are accompanied by a swelling of the female perineal skin. This swelling produces a large, turgid, and often brightly-coloured ornament that can be visible from several hundred meters, reaching its maximum size around the time of ovulation. These swellings are most common in species that exist in large multimale–multifemale social groups. Males find the sexual swelling attractive, and the bigger the swelling the more attention the female gets. Competition between males for mating access is most intense at this time, ensuring that females are likely to be fertilized by the strongest, high-quality males. In addition, because swellings also attract matings by other males even when they are not fully developed (in other words, when there is still a low probability of conception), they help a female to minimize the risk of future infanticide by those males: even a remo chance of paternity can be sufficient to discourag them from killing her future infant. The use of swellings as signals has become so important tha female Gelada baboons, whose perineal skin is rarely in view due to the sitting posture they customarily adopt while foraging, have developed a patch of skin on their chests with a fringe of vesicles that serves the same function: the vesicles swell and change color during the estrous cycle.

A Complex Ecology

Diet and Lifestyle

Despite being predominantly creatures of the tropical forest, primates also occupy temperate forest, savanna, desert, montane, and coastal regions. The order contains terrestrial (ground-dwelling) species as well as arboreal (tree-dwellin ones; nocturnal species as well as diurnal ones; and specialized insectivores as well as fruiteaters (frugivores) and leafeaters (folivores). Fifty-five percent of all species – including guenons, spide monkeys, and gibbons – are active by day, eat fruit, and live in trees. Another 20 percent, amor them colobus monkeys, howler monkeys, and langurs, eat leaves instead of fruit. The remainde split about evenly between nocturnal, arboreal insect-eaters, such as tarsiers, lorises, and bush babies; nocturnal, arboreal frugivores – for exam ple, dwarf lemurs; and diurnal, terrestrial fruit-eaters, including baboons, macaques, and chimpanzees.

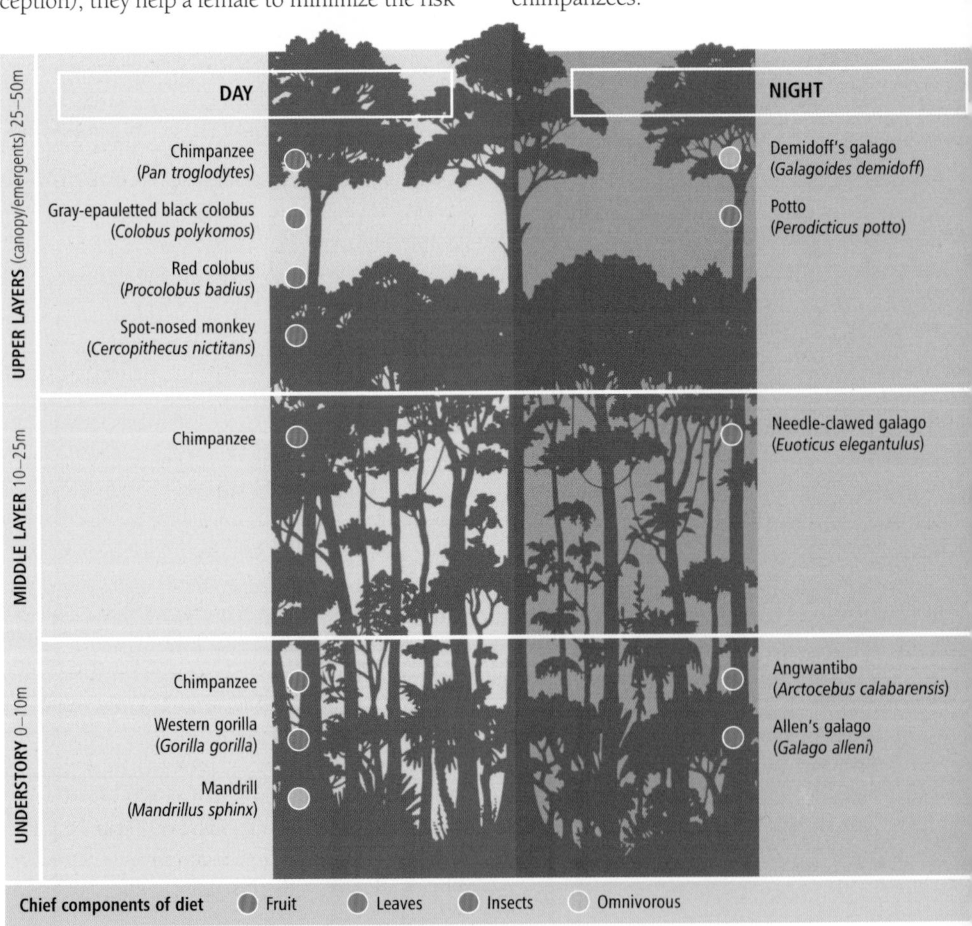

Above *Savanna baboons are equipped with very ong hands, which they use to uproot the bulbs and omes of flowering plants such as water lilies. These culent storage organs help the monkeys survive iods when other food is scarce.*

Left *Feeding ecology. Many primate species share same forest by splitting up their environment and educing competition for food. In this "share out" means of natural selection, differences in feeding es, kinds of diet, and levels where the animals age are all important. In the West African rain ests, five different species of lorisids commonly for- by night: the diminutive Demidoff's galago feeds inly on insects in the upper levels, which are used by the fruit-eating potto. The fruiteating Southern edle-clawed galago uses the middle and lower lev- while Allen's galago is mostly confined to forest or shrubs, as is the insect-eating angwantibo. ong the day-active species in the same forest is the nivorous mandrill, which obtains most of its food m the ground or shrub layer, as does the leaf-eating rilla. The chimpanzee, a fruiteater, uses all levels of forest. Also in the upper levels, the Red colobus nkey has to feed on many species for its diet of shoots, flower buds, and flowers, whereas the ay-epauletted black colobus also eats mature leaves d ranges less widely.*

There are, however, several exceptions to this schema. The Gelada baboon, for example, mainly eats grass. The Sportive lemurs, in contrast, live in trees and eat leaves, but do so by night. In addition, several members of the Lemuridae family, such as the Brown lemur and the Bamboo lemurs, are now cathemeral (active by both day and night), extending their activity period into the daytime possibly as a result of the extinction of large, predatory eagles from Madagascar, which would previously have posed too much of a threat for activity during the daytime.

The ancestral primates ate insects, were awake at night, and lived in the forest canopy. Living on the ground, eating fruit and leaves, and being awake by day came later. The night monkeys of the New World have returned from diurnal ancestry to being creatures of the night, and are now, along with tarsiers, the only nocturnal haplorhines.

Thanks to their insular seclusion, the lemurs have been able to rebel against the strepsirhine norm of nightlife, insect-eating, and treehouses. When the ancestral lemurs arrived in Madagascar about 40 million years ago, probably by rafting from Africa on vegetation, they encountered an island impoverished in species, with many vacant ecological niches. So, the ancestral lemurs underwent an adaptive radiation that produced over 40 different species.

The size of a primate is closely related to its lifestyle, with good reason. Many predators are active during the day, so diurnal species (including most haplorhines) tend to be larger than nocturnal species (most strepsirhines). The typical mass of a female haplorhine is 6kg (13lb), compared to 1kg (2.2lb) for a female strepsirhine. Similarly, terrestrial species tend to be larger than arboreal species, because they face a risk of predation from both ground-dwelling and aerial predators. Thus, the terrestrial baboons, mandrills, and drills have a greater body mass (a typical female weighing 14kg/31lb) than the principally arboreal guenons, where an average female weighs just 4kg (8.8lb).

Finally, body size is another factor that constrains diet. Most of the large primates are folivorous, because only foliage is available in sufficient abundance to meet all their daily requirements. Although foliage is a low-quality food with extremely poor nutritional value, large primates can cope with it because their metabolic energy demands are relatively low. In contrast, smaller primates, with relatively high metabolic demands, do not have this luxury. These primates are obligate insectivores. In between these two extremes are the frugivores and a number of generalist species that consume a mixture of fruit, foliage, and insects.

Body size and diet also play an important role in determining species abundance. All else being equal, larger animals need larger home ranges in which to find food to sustain themselves. Among folivorous strepsirhines, for example, the 3.3-kg (7-lb) Golden-crowned sifaka ranges across 9–12ha (17–30 acres), while the 600-g (21-oz) Weasel sportive lemur can make do with only 1.5ha (4 acres). The kind of food eaten also affects the size of territory required. Garnett's greater bush baby is not much larger than the Sportive lemur but, due to its preference for high protein insects, requires as much space for its activities as the sifaka. These patterns in home range area are inevitably reflected in the population densities of these species: the Weasel sportive lemur is more abundant than either the Golden-crowned sifaka or Garnett's galago.

Leafy greens are not always the easiest foods to digest, so primates are equipped with a range of adaptations to process foliage, fruit, and insects. Molar teeth with shearing crests help grind the plant food. An enlarged digestive tract houses the bacteria needed to break down plant cell walls, either in the stomach (for instance, in colobus monkeys and langurs), the cecum (Bamboo lemurs and Sportive lemurs), or the colon (howler monkeys and gorillas). In contrast, frugivores are characterized by relatively large incisors, since fruits require more preparation than leaves before they can be chewed, and simple digestive tracts, because there is little cellulose to be broken down in fruit. Insectivores possess a combination of these features; they have shearing molars, to crush the hard skins of beetles, but simple guts.

Some primates eat the gums and resins secreted by certain tree species, employing special incisors to gouge holes in the bark and claws to scrape out their semiliquid prize. The gum-eating Southern needle-clawed bush baby, the Fork-marked lemur, and the marmosets may also have an enlarged cecum, suggesting that bacteria in the gut may play a part in helping to digest the gum.

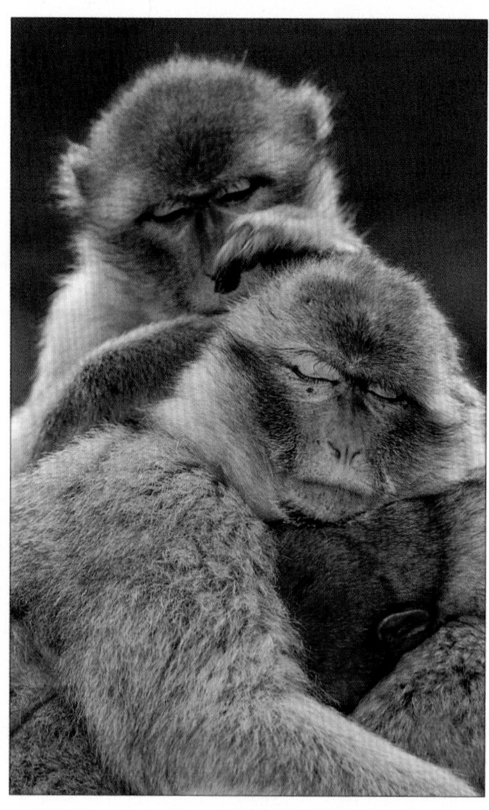

Above *Grooming is a highly characteristic aspect of primate life that helps cement social relationships.*

Below *Types of primate society. Most primates live in groups. Gibbons live in monogamous pairs, up to four immature offspring remaining with their parents. In bush babies, the ranges of several related females and their offspring overlap; males occupy larger areas that include the ranges of several females. Gorilla groups comprise a single breeding male with a harem of females together with their young. Yellow (or Common) baboons live in multimale troops with several breeding males and many breeding females, while harem groups of Gelada baboons aggregate into larger bands that exclude nonbreeding mature males. In chimpanzees, communities of unrelated females with individual home ranges but considerable overlap, are monopolized by groups of related breeding males.*

Clearly, this insectivore–frugivore–folivore schema simplifies a lot of important diversity a flexibility. In addition, since the annual diet of most primates will encompass leaves, fruits, an insects at some point, these descriptions of dietary strategy only reflect the relative importance of each primary component. In this resp the tarsiers are a special case, as the only prima to restrict their diet exclusively to animal prey.

Killing and eating other vertebrates is also a facet of chimpanzee and baboon life. Most hun ing is carried out by males and targets small un lates and other primates. Chimpanzees have developed hunting to a sophisticated level, reg larly undertaking complex cooperative hunts w great success. At Gombe, Tanzania, chimpanze predation of Red colobus monkeys has been su ciently heavy to have halved the size of the colobus groups that are hunted. This is unusua predation is something all primates have to be wary of, but the executioner is not normally another primate.

Snakes, eagles, and – especially in the case o terrestrial primates – big cats are the primate's worst enemies. In a vervet monkey population Amboseli, Kenya, predation normally accounts 50 percent of all mortality, but in one year a sin resident leopard was responsible for 70 percent all deaths. Moreover, predators may play an important role in limiting the total size and growth of primate populations: this is probably true for the Mouse lemurs of the Beza Mahafaly Reserve, Madagascar, where Barn owls harvest a quarter of the population annually.

To defend themselves, primates have respon ed with a range of morphological and behaviora defenses. Small primates, such as the nocturnal strepsirhines, tend to forage alone and rely on crypsis (hiding) to avoid detection. Diurnal primates, though, have responded by forming groups. Individuals in groups gain a range of antipredator benefits, including several extra se of eyes, which brings improved detection of

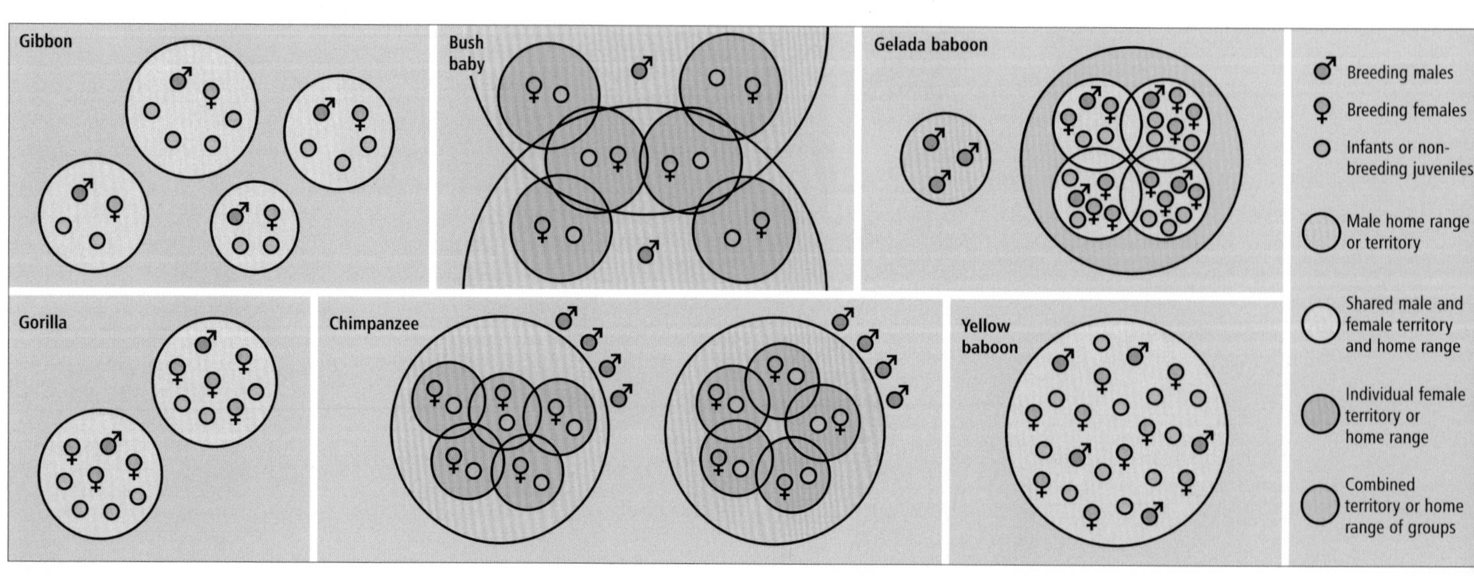

›roaching predators (the "detection effect"). : chances of any one individual ending up dead also reduced (the "dilution effect"), and the ıup can gang together to retaliate and mob dators (the "deterrence effect"). In baboons l some other species, such as Red colobus nkeys, males will form coalitions against poten- predators: these coalitions can be so devastat- that there are at least four recorded cases of ›oons killing leopards (their primary predator). ese benefits tend to increase with group size, so mals like terrestrial baboons, living riskily on ground, tend to cohabit with at least 30 other mals, while the arboreal guenons only congre- e in groups of 20. Primates in many communi- take further advantage of this fact when the ial groups of two or more species gather to for- together in "polyspecific associations."

To minimize the risk of predation, baboons will de off the quality of food in a habitat against its ociated risk of predation, sometimes preferring avoid foraging in food-rich areas if to do so uld entail a high risk of predator attack. Similar- arboreal primates such as capuchin monkeys d Long-tailed macaques will prefer to forage h in the canopy rather than on the forest floor, ere they are more vulnerable to predators. Even restrial species such as baboons will be reluc- it to stray far from tall trees or cliff faces, which y will use as a refuge if attacked by a predator.

If primates cannot avoid entering high-risk situ- ons, they will become more vigilant and per- m alarm calls when predators are detected. rvet monkeys have a different warning for big s, eagles, and snakes, and listening vervets pond differently to each, respectively either eing to the trees, looking upward and fleeing to shes, or standing bipedally and scanning the ound. Each response is the best escape strategy m the predator in question.

Seeking Safety in Numbers

SOCIAL COOPERATION

ciality is perhaps the primates' most striking aracteristic. Almost all monkeys and apes live in oups, as do the majority of strepsirhines. More- er, while group membership in many other cial animals may vary from hour to hour as indi- luals come and go, primate groups generally y together.

The driving force behind the evolution of pri- ate sociality appears to have been predation risk. ving with other members of one's own kind has disadvantages – competition for food means ving to work harder for every mouthful – but e reduced risk of being eaten makes the extra ort worthwhile.

The species with the largest group size is the elada baboon. It grazes on the open terrain of e montane grasslands of the Ethiopian high- nds, and can live in stable groups of 150 mem- rs, sometimes collecting in herds of up to 600.

Above *Ring-tailed lemurs live in social groups that can number over 20 individuals. Uniquely among mammals, adult females dominate the hierarchy.*

When social groups can monopolize a food resource such as a fruiting tree by aggressively repelling other smaller groups, large group size can actually benefit the individual members' search for food. Among Wedge-capped capuchins at Hato Masaguaral (Venezuela), for example, females in larger groups have a higher birth rate due to their enhanced nutritional status resulting from better access to fruiting trees. Such cases, however, are the exception rather than the rule; in most primate populations, female birth rates are higher in smaller groups where feeding competition is lower.

Group size is only one aspect of primate society; the composition of the group is also important. The linchpin of primate societies is the female and her offspring. In haplorhines and several lemurs (such as the Bamboo lemurs, Brown lemur, sifakas, and indri), mothers carry their infants with them. In the nocturnal strepsirhines and remaining lemurs, mothers "park" their infants at a nest.

Females have a choice of either foraging together or avoiding one another. Where females do not forage with other females, they either forage alone or in the company of a single adult male. Where they forage alone, males usually defend extensive territories that overlap the smaller home ranges of several females. Subordinate males are excluded from access to females, and adopt home ranges of their own, often in less favorable habitat. The orangutan and many of the nocturnal strepsirhines exhibit this pattern.

When females forage in the company of a single male, the basic social unit is typically the monogamous pair defending a shared territory. Since it is usually in the evolutionary interest of a male to fertilize as many females as possible, and therefore to defend a territory encompassing that of several females, the survival of monogamy in primates remains something of a mystery. The most likely explanation is that the males provide an essential service in the rearing of offspring, such as in protecting them from predators or infanticidal males. Providing such care, though costly in reducing the number of different females a male can impregnate, may increase a male's chances of leaving surviving offspring. There are several examples of monogamous primates, including the indri, titi monkeys, night monkeys, and gibbons.

When females forage together, they may be accompanied by one or several adult males. The precise number appears to depend on how easy it is to prevent other males from gaining access to the females. A single adult male can defend a small number of females from other males in a unimale group, or "harem." This type of social organization is shown by most guenons, most species of colobus and leaf monkey, some species of baboon, and, among the apes, the gorilla. Those males which are unable to enter these groups often associate with one another to form all-male bands. Many arboreal species that live in unimale groups, such as the forest guenons and leaf monkeys, defend territories. In contrast, the unimale groups of several terrestrial species, such as Gelada baboons, mandrills, and snubnosed monkeys, are not territorial and will aggregate into larger groups when predation risk is high or food resources are abundant.

However, once the number of females in a group exceeds about 6–12 (the precise number is variable), it becomes more difficult to exclude other males from joining, so the group becomes multimale. The capuchins and howler monkeys of the New World and the macaques and Savanna baboons of the Old World, among other species, usually live in such multimale groups. Territoriality is uncommon in these big groups because the extensive ranging areas they require are difficult to defend effectively.

Unsurprisingly, given the behavioral flexibility of primates, there are several variants on these basic types of social organization. Among the marmosets and tamarins, for example, group composition is highly dynamic: females may mate monogamously, polygynously – with several females mating with the same, usually high-quality male – or polyandrously, with one female mating with several males; and although several adult females may live together in a group, only one will normally breed. An alternative social model is that of the chimpanzees and spider monkeys. Their unstable groups vary constantly in size, but they belong to larger communities that are stable. In chimpanzee communities, females appear to have individual home ranges that overlap widely, and they will collect in groups at particularly rich feeding sites. Bands of related males monopolize breeding access to communities of females, maintaining hostile relations with the males of neighboring communities.

Patterns of Dominance

Hierarchies and Coalitions

Dominance hierarchies are one of the hallmarks of primate sociality. Dominance is determined by the ability to win fights, with the strongest individuals usually ranking highest. Importantly, high-ranking individuals of either sex typically have more offspring, because they can obtain better access to those resources that limit reproduction. These are usually food patches in the case of females, and fertile females in the case of males.

The influence of high dominance rank is greatest where these resources are clumped together and can be monopolized by a single powerful individual (via so-called "contest" competition). Fruit growing in dense clusters on the same tree one such resource. In the wild, female dominanc rank is therefore particularly pronounced and obvious in frugivorous species such as baboons, macaques, and chimpanzees. In contrast, where resources are widely dispersed and cannot be monopolized ("scramble" competition), rank is less important. This is typical of folivorous specie such as colobus monkeys and gorillas, in which female dominance hierarchies are often shallow and poorly-defined.

A similar pattern can be seen across male primates: male dominance rank appears to be relatively less important in those species that breed
seasonally (such as squirrel monkeys and several
macaque species), because females all become
ready to mate simultaneously and a single male
cannot monopolize them all.

Dominance rank is not always solely dependent on individual fighting ability. Primates often form alliances, or friendships with other males, to enhance their power through cooperation. These appear to be largely "negotiated" through mutual grooming, and can have a strong influence on

Above and right *Social interactions. Like many other Old World monkeys, both Gelada (ABOVE) and Hamadryas baboons usually live in harem groups. In the Gelada baboon the principal social bonds are between females (indicated by solid lines in the diagram); the breeding male mates with the females but otherwise interacts with them relatively little (broken lines). In the Hamadryas baboon it is the dominant male that maintains cohesion of the group, by continually herding his females together, and social bonds between females are relatively weak.*

primate politics through the network of coalitions that arise. The Old World monkeys exhibit this most strongly.

The social groups of Old World monkeys tend to consist of related females and unrelated males (since males disperse from their natal group at maturity, but females remain in their natal group for life). The close family relations among these females helps strong social bonds to develop, such that family members support one another in aggressive conflicts. Consequently a female's rank can depend on the family, or "matriline," to which she belongs, to the extent that all members of high-ranking matrilines outrank all members of low-ranking ones. In addition, due to power struggles within matrilines in some populations of baboons and macaques, individual rank is inversely related to priority in order of birth (see Figure).

In contrast, in some primate species such as the Red colobus monkey and the Common chimpanzee, females disperse while males usually stay in their natal group. In general, most (perhaps even all) of the Cercopithecines have male dispersal, while most other primate species have mixed sex-dispersal. Female-only dispersal (as in some Red colobus monkeys and chimpanzees) is almost certainly the most uncommon arrangement. In these species, adult females tend to be relatively aloof, while males form close friendships.

Yet cooperative relationships are not solely restricted to relatives. In some baboon populations, for example, a male baboon that wishes to gain access to a fertile female – by displacing the dominant male that is guarding her – may attempt to enlist one or more nonrelated males to help him. Coalitions such as this can be shortlived, and may have little impact on an individual's position in the dominance hierarchy. While it is usually only the most powerful male in the coalition that finally gains "ownership" of the estrous female, other males seem willing to help because there is always a slim chance that they might also gain some limited access to the female.

Similarly, females sometimes prefer to form coalitions with nonrelated males. The development of such coalitionary bonds, or "friendships," between males and females is believed to play an important role in preventing infanticide. Male friends have often previously mated with the female in question, and thus protect the infant, possibly their own offspring, from danger. One way males help is to carry the infant out of the way of potentially infanticidal males. There are many potential benefits from male–female bonds across all types of primate societies, but where females carry their infants with them – in monogamous, unimale, or multimale groups – the reduced risk of infanticide appears to be the key factor.

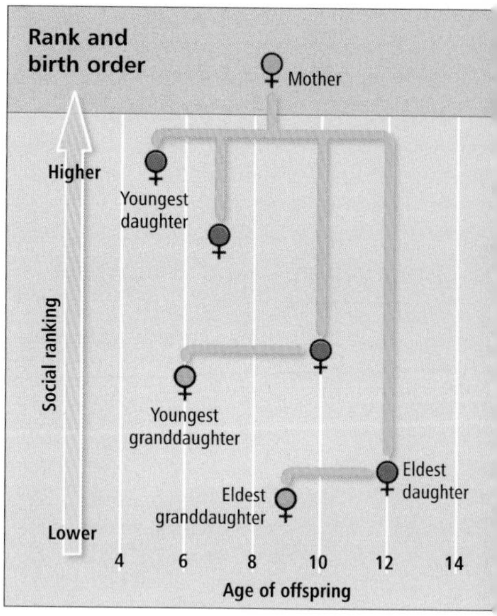

The complex network of social relationships within primate troops requires an elaborate and accurate system of communication. Among the nocturnal strepsirhines, olfactory communication usually plays a major role, and individuals mark their territories with urine, feces, or the secretion of specialized glands, which can signal the animal's sex, reproductive status, and individual identity. However, a system of communication based on smell is both too slow and too limited to meet the needs of the gregarious primates, which have evolved elaborate visual and vocal signaling systems, based on continuously varying signals that can convey subtle nuances of meaning.

Above *Rank and birth order. Dominance hierarchies of males and females are found in most primate groups. Among many baboons and macaques males leave the group after adolescence, while females stay. Social bonds between these related females maintain the cohesion of the troop, which is therefore matrilinear. The rank of different members of the same matriline may depend on birth order: daughters may rank immediately below their mother but above older sisters. In due course, a daughter may mate and found her own matriline. All members of a matriline share the same ranking in relation to members of other matrilines.*

TACTICAL DECEPTION AND "MINDREADING"

Primates have long been known for their intelligence, but the most taxing everyday problems they face in the wild are often caused by the sophisticated social skills of their companions. Deception and counter-deception offer some of the most graphic examples of this social complexity (see Why Primates have Big Brains).

Tactical deception occurs when an animal borrows a behavior from its normal context and deploys it in another situation, so misleading its companions. For example, when baboons spot a predator or another group of baboons, they typically show overt signs of interest, rearing up on their hind legs, craning their necks, and staring into the distance. Sometimes they will give alarm calls too. Other baboons quickly tune in to this, peering in the same direction to find out the cause of the disturbance. At other times, however, a baboon being harassed by others will distract them by using these signals to raise a false alarm. While its pursuers are busy checking to see if there is a predator approaching, the first animal slips away.

Identifying tactical deception is a real challenge for scientists, because by its very nature it is subtle and relatively rare – a monkey simply cannot get away with indiscriminately "crying wolf." Nevertheless, researchers have gradually accumulated a catalog of tactical deception techniques, the diversity of which is a tribute to the primate mind. Distraction may be achieved not only through the ploys outlined above, but also by leading others away from a prime food source only to double back, or even by apparently disingenuous grooming.

Distraction, however, is by no means the only tactic. Concealment is another major ploy, again taking several forms, from simply hiding prized food items to the more subtle ability to temporarily hide interest in a choice, half-hidden fruit until potential competitors have left the scene. More complex tactics involve fooling individuals into serving as "social tools" with which to manipulate others. For example, juveniles have been seen to act as if an adult in possession of desirable food had attacked them when in fact they had not, so inciting a higher-ranking ally of the juvenile to drive off the innocent individual, leaving the food to the youngster.

Deceivers may be punished if they are found out. In one experiment with free-ranging Rhesus monkeys, desirable food items were dropped in front of monkeys while they were alone. About half the time they would produce a food-call, attracting others, but on other occasions they remained silent and kept their prize secret. If other monkeys chanced on the scene, they were much more likely to direct aggression at an animal that had stayed quiet than at one that had called.

Coupled with other complexities of social maneuvering and alliance formation, behaviors like these have led primates to be described as having a "Machiavellian intelligence." Since the social scheming of young children soon comes to rely on some understanding of how other individuals' minds work, we may wonder if the same is true of primates. Experiments show that, in apes at least, there is some understanding of intention (distinguishing accidental acts from deliberate ones) and attention (discriminating plausible targets another individual is looking at). Abilities like these may represent the early stages of the reading of others' minds that we humans obsessively engage in in so much of our own everyday lives. AW

The Risk of Extinctions

CONSERVATION ISSUES

The primates are more widely threatened with extinction than most mammalian orders, with almost half of all species presently at risk, including all the great apes and most of the lemurs. Within the last 1,000 years alone, 15 primate species, representing eight unique genera, became extinct, following the arrival of humans on Madagascar. Globally, many species are now classed as dangered or Critically Endangered: some of ·se, for example the Golden lion tamarin and · Moloch gibbon, may well disappear over the xt 50 years.

What is often forgotten when primate conser- ion is under discussion are the great benefits ·y bring to human society. In Kibale Forest ganda), for example, primates disperse the ·ds of more than one-third of all forest trees; ly 42 percent of these tree species play an portant role in local human communities, pro- ling resources such as wood (for fuel and furni- ·e), food, medicine, and fodder. If the primates come extinct, those trees – and perhaps the ole forest – will disappear.

There are two basic threats to primate species: bitat disturbance and hunting. Habitat distur- nce is caused primarily by agricultural and estry activities. It threatens species through the odification, fragmentation, and overall loss of eir environments. Of these processes, outright ss is the most serious: between 1981 and 1990, estimated 8 percent of the world's remaining opical forests were destroyed. While primates n cope with both habitat modification and frag- entation, they need certain factors to remain, ch as food trees; and it is also vital that "islands" habitat should be connected by corridors of itable territory so that populations can inter- ingle. If species are to be preserved in habitats odified by humans, these factors must be con- dered. Biology dictates which species are affect- l most by disturbance. Small generalist species, ch as Blue monkeys and Redtail guenons, often ow the greatest tolerance, because their flexible ets allow them to switch to different food urces in the absence of others. Small size also anslates to a high reproductive rate that allows e population to bounce back quickly from large- ale mortality.

People usually hunt primates for food, though me species are caught and traded alive for bio- edical research, or are otherwise killed for their ins or organs (the latter primarily for use in tra- tional Eastern medicines). Hunting has its most vere impact when it becomes commercial: the ushmeat trade in Africa currently poses a very rious threat to many primate populations. unting can drive populations to extinction even undisturbed habitat, although habitat distur- ance further exacerbates the impact of hunting; r example, logging activities can open up previ- usly inaccessible forests to hunters and settlers, has happened in northern Congo. Relative to her mammals, primates are particularly hard-hit y hunting, as a result of their slow reproductive tes. Large, conspicuous primates such as drills nd gorillas tend to be the most vulnerable to xtinction from hunting because they provide a t of meat and are easy to find, while their low irth rates can quickly be overwhelmed by high unting losses.

Most primate conservation is aimed at protection of the animals within their habitats. Recently the notion of sustainable harvesting has become fashionable among conservation biologists. This allows for regulated hunting of animals by local communities, thereby placing an economic value on the wildlife as a way of winning support for its continued presence in an area. This, however, is generally a poor option for primates, given their slow reproductive rates. Conservation through other uses – especially tourism – has more potential. Some captive breeding and reintroductions are carried out for primates, but the expense of these programs usually makes it more cost-effective to focus on the protection of existing wild populations. GC/TC-B

Above *A family group of Cotton-top tamarins – a species in which older offspring delay their own breeding by staying in the family unit and caring for younger siblings.* ***1*** *While a father carries on his back one of his infants, it is groomed by an older offspring helper,* ***2****. Meanwhile, another helper* ***3*** *receives the twin of the first infant* ***4*** *from its mother* ***5****, who has been suckling it.*

WHY PRIMATES HAVE BIG BRAINS

The role of neocortex size in social interaction

THE BRAIN – WITH ITS EXTENSION, THE SPINAL cord – provides one of the defining characteristics of the vertebrates (animals with backbones). Yet not all vertebrates have brains of similar size. We ought not to be surprised to discover that an elephant has a larger brain than a shrew: after all, the elephant has more muscle mass to manage, and the intricacies of maintaining body temperature and blood flow through such a massive body impose much greater demands on the central computer in the brain. But when we plot brain size against body size, we find that the various groups of vertebrates lie on different planes. Fish and reptiles have the smallest brains for body size; above them lie birds and most mammals; and above these yet again, on a plane all their own, are the primates, which have the largest brain-for-body size of any animals. A primate weighing 5kg (11lb) has a brain that is three times the size of that of a 5kg non-primate mammal, which in turn has one 30 times as big as that of a similarly-sized reptile.

Thus, mammals have unusually large brains and, within the mammals, primates have by far the largest relative brain sizes. Why do we need such big brains? Conventional wisdom has always assumed that they are necessary to handle the problems of survival, but recent evidence has suggested that the world whose complexities they help us navigate is not so much the physical as the social one.

Traditionally, the brain has been seen as the place where information acquired through our sensory organs (eyes, ears, noses, skin) is analyzed and interpreted. Brains, it was argued, are there to help us make sense of the world as we direct our bodies toward satisfying our physiological needs (finding food, water, and shelter) and avoiding being eaten by other creatures. But in that case why should mammals, and especially primates, need larger brains than, say, fish?

The one respect in which primates do seem to differ from other creatures sufficiently to explain such a divergence is in the complexity of their social lives. In this regard, consider the example of Mel. Mel is a young, female Chacma baboon living in the Drakensberg Mountains of South Africa. The mountains contain few bushes and trees to provide the fruit-based diet that baboons prefer. Instead, these mountain baboons make use of the roots and tubers that grasses and other plants use to store nutrients below ground. But digging these up is no easy task; most require the strength of an adult to get them out of the ground.

Mel was observed watching a young adult male named Paul as he struggled to dig a large and succulent bulb out of the ground. At the critical moment when he finally wrenched it free, Mel let out an almighty yell of the kind that young baboons usually give when being attacked by a much older and larger animal. Mel's mother, who had been feeding out of immediate view some distance away, rushed to her daughter's defense. She took in the scene at a glance, put two and two together, and came up with the wrong interpretation: it was obvious to her that Paul must have attacked Mel. In response, she attacked Paul so fiercely that, in his surprise, he dropped the bulb and fled. Mel nonchalantly picked up the abandoned bulb and sat down to enjoy lunch.

Although we need to be cautious in interpreting anecdotes, scientists have been impressed by the fact that examples of this kind are commonly reported from studies of monkeys and apes, but only rarely from other species of mammals (including strepsirhines like lemurs and galagos). Maybe primates need larger brains to manage the computations involved in these more sophisticated kinds of social interaction.

One way of assessing the validity of this argument is to ask whether species with larger brains live in more complex societies. However, it is not total brain size but the size of a crucial component, the neocortex, that turns out to be most important in this respect. This thin sheet of cells, just a few millimeters thick, forms the outer sheath of the mammalian brain, and has grown disproportionately large in the brains of primates, where it represents a new addition tacked onto the older, reptilian-type brain. More importantly, the neocortex is the part of the brain linked with creative thinking – all those processes we associate with consciousness. And it is among the primates tha the neocortex reaches its fullest development. In fact, the evolution of the primate brain could be called the story of the evolution of the neocortex

Comparing the size of the neocortex with mea sures of social complexity such as group size give interesting results. As group size increases, the number of relationships that each individual has to keep track of grows exponentially. And relative neocortex size turns out to correlate much more strongly with social group size than, say, with ecc logical variables such as home range size or diet. It seems that in order to survive well in groups above a certain size, primates need at least a min mum amount of brain power.

Other aspects of social complexity (including the size of grooming cliques, males' exploitation of alternative strategies for gaining access to females with whom to mate, and the amount of social play in juveniles) have now also been show to correlate with relative neocortex size. More importantly, although total brain size in primates correlates with the length of gestation and lactation (as it does in all mammals), the relative size o the neocortex has been shown to correlate best with the length of the juvenile period. In other words, neocortex size seems to be closely associated with the length of time over which socialization and education take place – as it were, the time it takes for the software programs to be loaded into the brain's computer.

The implications of this "social brain hypothesis" are not limited to primates, Evidence is also now emerging to suggest that a similar relationship between neocortex size and social complexity may occur in other groups of mammals, including advanced insectivores, carnivores, the whale family, and possibly the bats. The scaling of the relationship may be different between these groups, but the same general pattern seems to apply – suggesting that it is mammals' social life that is most intimately related to the size and complexity of their brains.

RD

Right The fact that primate juveniles engage in social play more than non-primate juveniles may be related to their larger neocortex.

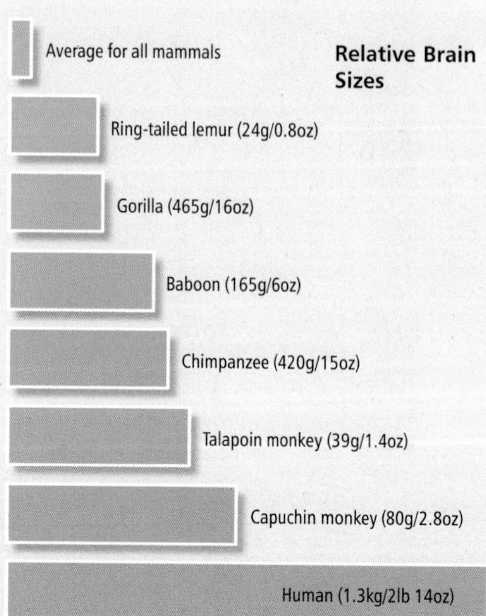

Left The degree of flexibility in the behavior of a species is related to both absolute and relative brain size. It is no surprise that, in terms of absolute brain weight (figures in brackets), the great apes are closest to man. However, as shown in the diagram, when the comparison is based on an index that allows for the influence of body size on the size of the brain (i.e. relative brain size), it is the versatile Capuchin monkey that turns out to be the closest to man.

SEEING IN COLOR

The evolution of trichromatic color vision

COLOR VISION CANNOT BE TAKEN FOR GRANTED. The capacity to recognize, say, red apples or skin pimples can have huge implications for food or mate selection. Yet not all mammals see color in the same way humans do.

To support color vision, an animal's eyes must have various cells called cone photoreceptors, each containing a different type of photopigment (light-sensitive molecules). Most human eyes have three types of cone photopigment (labeled S, M, and L in the diagram below) and these, in conjunction with an appropriately organized nervous system, yield color vision that is trichromatic (made up of combinations of three primary colors). The result is an impressive capacity that, among other things, permits people to discern some two million different surface colors.

The color vision of humans, Old World monkeys, and apes is distinctly different from that of most other mammals. Many, including probably most cat and dog species, have only two types of cone photopigment (an S pigment similar to that shown and a second, single pigment with maximum sensitivity somewhere in the range encompassed by M and L in the diagram). This results in a much diminished, dichromatic color vision formally similar to that of people with red–green color-blindness. Dichromats see many fewer distinct hues than trichromats, confusing colors that are quite distinct to the rest of us. Still other creatures (e.g., some marine mammals and rodents) have only a single class of cone photopigment, and so have no color vision at all. Comparisons of the genes that encode the photopigment proteins (called "opsins") and direct studies of animal color vision have recently provided some insights into why such large differences exist and how primates might have gained their unique color vision.

While many bird, reptile, and fish orders include representatives of four different cone-opsin gene families, only two of the four appear in contemporary non-primate mammals. The other two may have been lost during early mammalian evolution, though precisely why they disappeared is unclear; one possibility is that color vision is of little use in low light, so early nocturnal mammals would not have profited much by possessing it. What is certain is that primate color vision has been elaborated from the basic mammalian plan of two types of cone photopigment. In humans, instead of a single pigment from the M/L range, there are two cone pigments M and L (see diagram); the genes that encode for them are found side by side on the X-chromosome and are virtually identical in structure. A gene found on chromosome 7 specifies the third type of photopigment (S).

Gene-sequence comparisons suggest that the M and L opsin genes arose from a gene duplication that occurred in the early catarrhine primate line some 30–40 million years ago, giving these early primates trichromatic color vision. This gene and photopigment arrangement has been passed down to all the succeeding catarrhines, accounting for the trichromacy of contemporary catarrhines and the near-identity of color vision in all members of this group.

Color vision in New World monkeys is a very different matter. Here variation rules. In nearly every species from this group there are both dichromats and trichromatic individuals. All this comes about because of a photopigment polymorphism that causes individual variation in the M and L cone pigments. All monkeys share a common S-pigment but some individuals get only one pigment from the M/L range, giving them dichromatic vision; others have two pigments from this range and are trichromats. Unlike their Old World cousins, most New World monkeys have only a single pigment gene on their X-chromosomes. This means that the only way they can achieve two different pigment genes is to have two X-chromosomes and be heterozygous. This is only possible for females, so only heterozygous female New World monkeys can have trichromatic color vision.

This complex arrangement in New World monkeys may have been a path that the Old World monkeys trod on their way to universal trichromatic color vision. In other words, it is possible that our catarrhine ancestors may have been polymorphic in the manner of New World monkeys before gene duplication happened. Supporting that possibility is the discovery that, relatively recently (certainly no more than 13 million years ago), New World howler monkeys underwent a gene duplication that changed them from having polymorphic color vision akin to most other platyrrhine monkeys to having universal trichromatic color vision nearly identical to that of catarrhine primates.

A special case among New World monkeys is the genus *Aotus* (the night monkeys). Like many mammals, they have a single type of M/L cone pigment. However, they also lack an S pigment, and thus have no color vision. The reason is that in

Left *The ancestors of the night monkeys (genus* Aotus*) of South and Central America are thought to have been day-active and to have possessed dichromatic color vision. And yet the night monkeys themselves have no color vision at all; it may be that the transition to becoming nocturnal rendered the ability to differentiate between certain colors unnecessary.*

Right *A foraging Squirrel monkey. Almost all species of New World monkeys contain both individuals that have trichromatic color vision and those that do not. Some biologists think that the former group are far better equipped for locating brightly-colored fruit in dense foliage.*

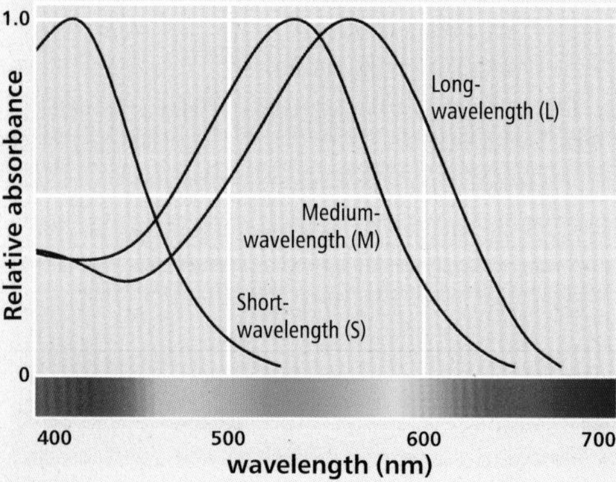

Left *Absorption spectra for the three cone pigments found in the eyes of Old World monkeys and apes (including humans). The presence of these pigments, plus the appropriate nervous system wiring, allows for trichromatic color vision. Primates or other mammals with dichromatic color vision all possess the S-cone photopigment, but only one or other of the M and L pigments. This means that they are unable to perceive a full range of colors over the middle- and long-wavelength part of the spectrum.*

their case the gene that normally specifies S-cone opsin contains fatal mutations: this arrangement suggests that an ancestor had a functioning second cone pigment and dichromatic color vision.

The discovery of variations in color vision among primates raises questions about the costs and benefits of having it in the first place. Primate trichromacy allows for acute discriminations among green, yellow, orange, and red targets. In the past it has been claimed that such a capacity is particularly useful for detecting yellow, orange, and red fruits from among foliage background. This would suggest that trichromatic New World monkeys should be at a considerable advantage relative to their dichromatic peers, and indeed one recent study suggests that trichromatic monkeys are the more efficient foragers.

But if trichromatic color vision provides an advantage, why have all New World monkeys not become uniformly trichromatic? The answer may be nothing more complicated than that opsin gene duplications are quite rare events that have happened only twice in primate evolution – once in catarrhine history, and again during the evolution of howler monkeys.

Color vision requires an elaborate nervous system organization, so there are costs as well as benefits associated with this adaptation. Under those lifestyle circumstances in which color vision has lowered utility – for example, in dim-light environments – the selective pressure to maintain these adaptations may lower and color-vision capacity may be lost.

In conclusion, the color vision of haplorhines follows two general patterns – either two X-chromosome pigment genes and routine trichromatic color vision, or a single X-chromosome pigment gene with polymorphic variation producing rampant individual variations in color vision. It is not possible to characterize the color vision of the more primitive strepsirhine primates so simply. Some strepsirhine species resemble many non-primate mammals in having only two cone pigments and dichromatic color vision; alternatively, at least some nocturnal strepsirhines (e.g., bush babies) appear to lack color vision entirely for the same reason as the night monkeys. But it has recently also been discovered that still other strepsirhine species have photopigment polymorphisms in the fashion of platyrrhine monkeys, thus allowing individuals to have either dichromatic or trichromatic color vision. Understanding of strepsirhine color vision is in its infancy, but it may well prove crucial in illustrating how our early primate ancestors began to escape from the confines of dichromatic color vision typical of most mammals. GJ

GROOMING AND FAMILY LIFE

Exchanging services among female monkeys

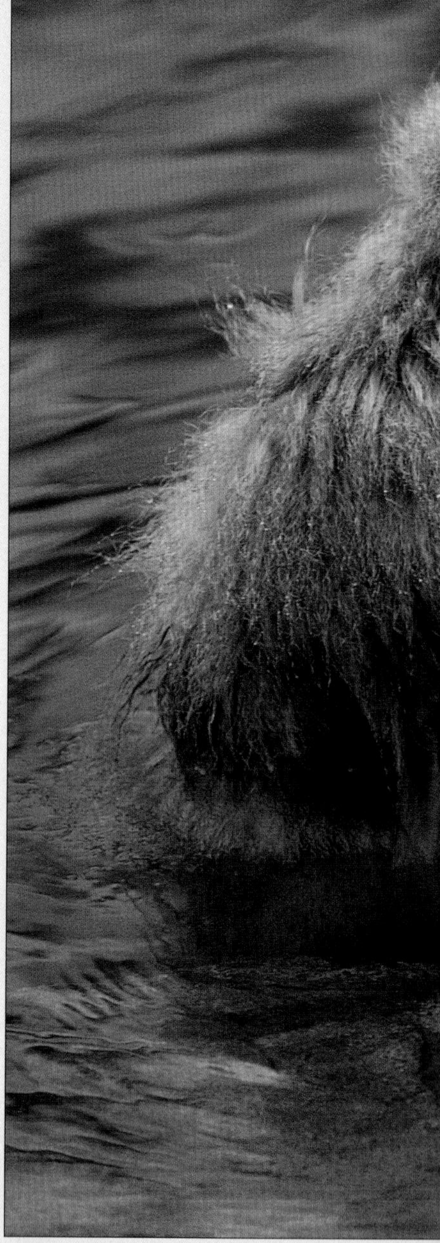

A QUINTESSENTIAL FEATURE OF MONKEY LIFE IS the effort that animals put into grooming other members of their troop, combing through their fur and picking out dirt, flakes of skin, and ectoparasites such as lice. This behavior has contributed to their reputation as the most social of animals, but there may be more to grooming than just sociality and hygiene. A new view suggests that primates picking ticks from the fur of their fellows may be illustrating the principle of exchange that drives the world's economies and stock markets.

Old World cercopithecine monkeys, and South American species such as the capuchins and squirrel monkeys, form social groups consisting of a permanent core of related adult females and their dependent offspring, with one or more unrelated and transient adult males. Living in groups provides protection against predators, and possibly against males intent on killing young monkeys, but it also entails one major disadvantage: the need to compete for food. As a result, it has been assumed that females are very much "reluctant partners," forced by external pressures to spend their lives among competitors, and that grooming evolved as a mechanism to deal with this problem.

According to this argument, grooming allows females to build up relationships that become beneficial at times of crisis. Whenever a female faces aggression from another, her grooming partner should – according to the theory – come to her aid and help her defeat the opponent. Social groups are said to consist of distinct subsets of alliance partners who reciprocate support for one another against more dominant animals.

For such a strategy to work, females ought to target their grooming efforts toward individuals most likely or able to support them. Getting dominants whose status is respected or feared by other group members on side would be especially useful, so grooming should be directed toward them. One mechanism to achieve this would be to invest time in grooming alliance partners in order to forge a "bond of trust"; those who put effort into a relationship could be deemed worthy of support and unlikely to cheat on their partners by not responding to their cries for help. In this view, grooming is the glue that cements strategic relationships.

Such notions have been widely accepted but have proved hard to confirm. The theory implies that low-ranking animals should spend much time grooming dominants, but in many species such as capuchins and Bonnet macaques, the reverse is actually true: dominant females put more effort into grooming low-ranking females. In Weeper capuchins (*Cebus olivaceus*), for example, for every minute of grooming that high-ranking females receive from lower-ranking females, they give a minute and a half in return.

More importantly, there is actually little evidence that adult females form coalitions against one another. Savanna baboons in southern Africa, for example, never form coalitions, yet they still make time to groom. In one unusual population living in the Drakensberg Mountains, grooming relationships persist even though competition for food is practically nonexistent: the small, dispersed food items eaten by the animals, such as the underground corms of *Watsonia socium* and the flowers of *Moraea stricta*, are just not worth fighting over. Yet these animals nonetheless place great emphasis on grooming; indeed, the patterns of troop sizes found in the mountains are directly related to time constraints on females' ability to groom each other. This suggests that grooming itself is a valuable service; females groom simply to ensure that they get groomed back.

Grooming is not only valuable for an animal's hygiene, but also brings advantages of an altogether more hedonistic nature. The "happy hormone," beta-endorphin, is produced by animals enjoying a bout of grooming, making the occupation a highly pleasurable experience. Female monkeys are compulsive groomers – you could almost say that they are addicted – and the stress-reducing properties of the activity may help ease the tensions of group life. This is important for females, since high levels of stress can reduce their ability to conceive. Low-ranking female Gelada baboons, for example, take 3–5 months longer to fall pregnant than their higher-ranking counterparts, with the

ult that they produce offspring at a slower rate n dominant females.

While the alliance theory assumed grooming ; a function rather like money in our society – ing no inherent value of its own but being used a means of acquiring other things that are valu- e – this new idea proposes that the value of oming lies in grooming itself. Put simply, oming is not a "currency" that is used to buy port from other animals but a "commodity" t is bartered with other individuals.

This view of female monkeys as traders, not col- orators, carries important implications for our derstanding of their sociality. In its light, groom- can be seen as a commodity traded in the arketplace" of the group, while grooming rela- nships are based on how many other individuals interested in acquiring that commodity. The s of supply and demand will then determine "price" that individuals must pay to obtain

Above *Grooming is vital for removing parasites, and the dynamics of who grooms whom reveals much about the social forces in primate societies, such as the matrilineal groups of these Japanese macaques.*

Left *Although there are many theories to explain why monkeys will groom one another, it is clear that there is pleasure to be had – thanks to the release of hormones – for the monkey being groomed.*

grooming, just as in a human economic market. Differences in the amount of time that individuals spend grooming each other (known as "pay-off asymmetries") may therefore reflect an individual's standing in the marketplace, and their ability to command a good price for their services.

One example of this kind of pay-off asymmetry is found in the grooming of dominant animals by subordinates. Chacma baboon females living in habitats where food is contested are prepared to give more grooming to dominant females than they receive in return, which may secure them tolerance at choice feeding sites. Because tolerance in such a situation is a more valuable commodity than grooming, subordinates are prepared to pay more for it.

Patas monkey females also use grooming to buy access to infants, as do their baboon counterparts. Newborn infants are very attractive to other monkeys, and females in particular are very keen to inspect and fondle them. Both baboons and Patas females groom mothers who are reluctant to part with their vulnerable offspring; and, having done so, they are allowed to handle the youngsters in return. Even here, however, market forces prevail: low-ranking baboon females have to groom for much longer periods than high-ranking females before being allowed to handle an infant. On this view, social relationships within a primate group are determined by the relative bargaining power of any two individual females. LB/PH

DISTRIBUTION Madagascar, Africa, S of Sahara, S India, Myanmar (Burma) to Indonesia and Philippines.

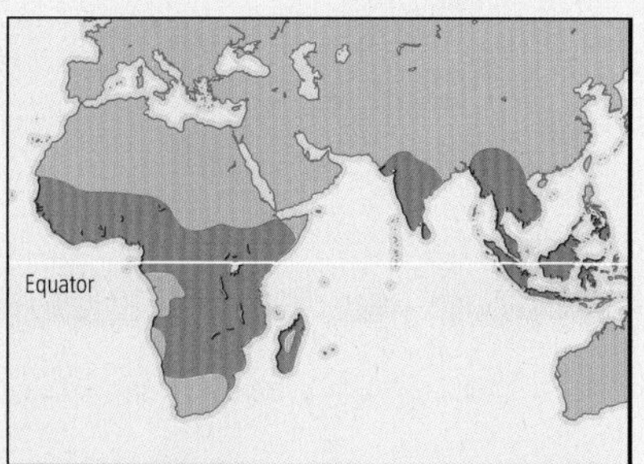

Suborder: Strepsirhini
61 species in 23 genera and 7 families

TYPICAL LEMURS Family Lemuridae p310
10 species in 4 genera

SPORTIVE LEMURS Family Megaladapidae p314
7 species in the genus *Lepilemur*

DWARF AND MOUSE LEMURS
Family Cheirogaleidae p316
13 (7) species in 5 (4) genera

INDRI, SIFAKAS, AND WOOLLY LEMURS
Family Indriidae p320
6 (5) species in 3 genera

AYE-AYE Family Daubentoniidae p322
1 species *Daubentonia madagascariensis*

BUSH BABIES, LORISES, AND POTTOS
Families Galagonidae and Loridae p324
24 (17) species in 9 (8) genera

Note The established 1993 classification recognized 47 species and 21 genera (in parentheses).

Strepsirhines

THE STREPSIRHINES OCCUR ONLY IN THE OLD World: lemurs in Madagascar; bush babies and pottos in Africa; and lorises in South India and Southeast Asia, Malaysia, and Indonesia. All species live in tropical arboreal habitats and rarely, with the notable exception of the Ring-tailed lemur (*Lemur catta*), spend much time on the ground. Strepsirhines move in a variety of ways, including vertical clinging and leaping plus quadrupedalism in the bush babies and lemurs, and quadrupedal walking and climbing in the lorises and pottos. Most have a generalized diet, which may include animal protein (mostly invertebrates), fruit, leaves, gum, flowers, nectar, and pollen. However, in the group there are dietary specialists like the leaf-eating lemurs of the genus *Lepilemur*, the bamboo-eating lemurs (genus *Hapalemur*), and some gum-eating specialists such as the galagos *Galago moholi* and *Euoticus elegantulus* and the Fork-marked lemur (*Phaner furcifer*).

Several new forms, some of which may be new species, have recently been discovered, for example the Golden potto (*Arctocebus aureus*). The Golden-brown mouse lemur (*Microcebus ravelobensis*) and the Rondo galago (*Galago rondoensis*) have also been added to the tally of recognized species. A number of factors have aided scientific understanding – including a growing tendency of researchers to venture into strepsirhines' habitats at night, the use of new tools to determine species identity, and advances in vocalization analysis and genetics. Even so, many tracts of forest still remain to be explored for new species.

While it was formerly convenient to classify bush babies (galagos), lorises, pottos, and lemu along with the tarsiers as prosimians ("forerunr of the simians" or "pre-monkeys"), because of their superficially similar appearance, we now know that this is phyletically incorrect. A more objective classification groups the lemurs, galag lorises, and pottos together as strepsirhines anc the tarsiers, monkeys, lesser and great apes together as haplorhines (see Primates).

Yet while significant differences in the structu of the eye and in dentition (see below) make a powerful case for classifying primates as strepsirhines or haplorhines rather than prosimians or anthropoids, some features still link tarsiers anc strepsirhines. These include one anatomical feature and certain aspects of their behavior. Tarsie and strepsirhines share the same structure (bicc nuate) of the uterus, as opposed to the simplex structure seen in the other haplorhines. And in their behavior, particularly their grouping patter tarsiers and strepsirhines exhibit less complexit than the monkeys and apes. Apart from some o the diurnal lemurs, strepsirhines are mostly solitary feeders, occasionally forming small groups; very similar pattern is found in tarsiers. This apparent lack of complexity can be partly attributed to their nocturnal lifestyle, where maintaining group cohesion is more difficult than durin the day, due to visual constraints. But it may als reflect an origin for the tarsiers before the evolution of this degree of complexity in the ancestor of the other haplorhines.

SKULLS AND DENTITION

Strepsirhines have moderately developed jaws, relatively large eye sockets, and a medium-sized, rounded braincase. There is a complete bony bar on the outer margin of the eye socket (orbit). This bar provides support for the outer side of the eye. As in primates generally, the orbits are directed forward for binocular vision. Orbits are relatively larger in nocturnal strepsirhines than in day-active species (for example, some of the larger Malagasy lemurs).

Strepsirhines have relatively simple jaws and teeth compared to many mammals. The two halves of the lower jaw remain unfused throughout life. The molars are squared off in both upper and lower jaws and typically have four cusps, though tarsiers still have primitive three-cusped molars in the upper jaw. The molar cusps are generally low, but sharper, higher cusps are found in insect- or leaf-eating species. The common full dental formula of I2/2, C1/1, P3/3, M3/3 = 36 is found in dwarf lemurs (Cheirogaleidae), most true lemurs (Lemuridae) and all bush babies (Galagonidae) and lorises (Loridae), all of which have a toothcomb in the lower jaw. The Sportive lemurs (Lepilemuridae) differ from this basic formula in lacking the upper incisors, while members of the mostly day-active indri group (Indriidae) have lost the canines from the toothcomb and a premolar from each side of both jaws. The aye-aye (dental formula: I1/1, C0/0, P1/0, M3/3 = 18) only retains its powerful, continuously growing incisors and a set of relatively tiny cheek teeth. Its powerful, forward-curving, bevel-edged incisors are used to tear through the husks of hard fruit.

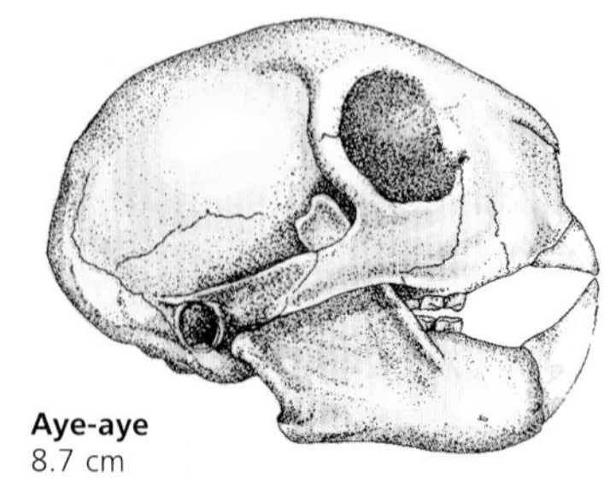
Aye-aye
8.7 cm

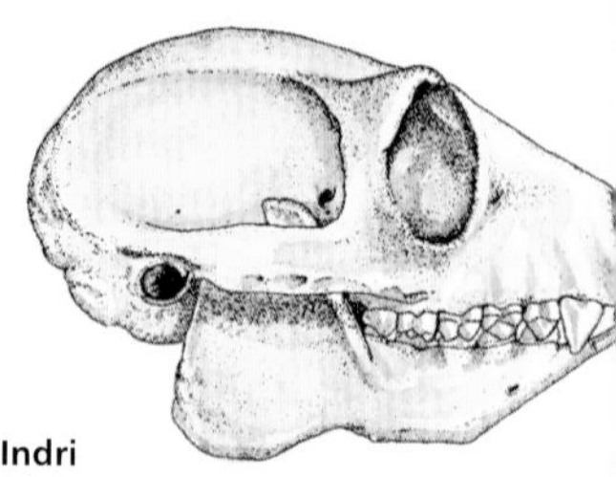
Indri
10.3 cm

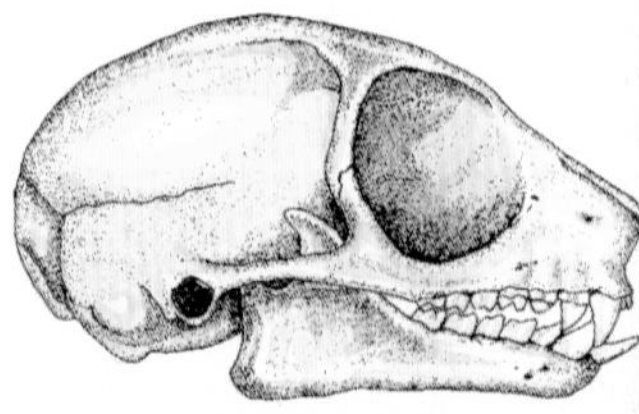
Needle-clawed bush baby
4.7 cm

EING IN THE DARK

h the exception of some diurnal urs, such as the Ring-tailed lemur, eyes of all strepsirhine species are racterized by the presence of the etum lucidum. This crystalline layer ind the retina, which is found in ny nocturnal mammals, improves sitivity in dim light conditions or in nplete darkness by reflecting light k through the retina. In addition, re are some other features of the na that are consistently different ween the haplorhines and strep- ines, such as the presence of a pit e fovea) and yellow spot (macula ea). These occur in the central visual a of the retina, are associated with rnal vision, and are found in all hap- ines (except the nocturnal night nkeys, *Aotus* spp.), but are absent he strepsirhines. Another feature of eye that distinguishes the hap- ines from the strepsirhines con- ns the structure of the orbit. All epsirhines have a bony bar around outer edge of the orbit. In hap- ines, however, including the tar- s, the eye is more thoroughly tected. They have a wall or septum t affords total bony closure to the k of the orbit; this separates the ball from any distorting effect of large chewing muscles.

Right *The large eyes of the cturnal Slender loris* (Loris digradus).

MUR DIVERSITY

ring their long isolation on the nd of Madagascar, lemurs have olved a spectacular diversity of life- tory adaptations, rivaling that of the monkeys and apes combined. s diversity manifests itself in many ferent ways.

Firstly, there is great variation in nur size; still inhabited by the small- known primate, the Pygmy mouse nur, Madagascar was also once me to some of the largest ones, like e recently extinct Giant sloth lemurs, ich weighed more than 200kg 40lb) and were bigger than male rillas. Most lemurs are exclusively ive at night, although diurnal activi- has evolved at least twice indepen- ntly in this group, and some species read their activity in irregular bursts er both day and night. Lemurs have olved nearly all dietary adaptations own from other primates; some spe- lize on fruit, others on leaves, and others on gum or small prey. Their productive rates and adaptations are equally diverse. Small mouse lemurs begin reproducing within their first year of life and give birth annually to 2–4 helpless young that spend their first weeks "parked" in nests and tree holes. Sifakas and indris, on the other hand, do not reproduce until they are 4 or 5 years old, when they begin giving birth, usually only every other year, to a single offspring that is carried around by the mother.

Left *Typifying the diversity of lemur form among extant species are **1** the nocturnal, arboreal Hairy-eared dwarf lemur and **2** the diurnal, largely terrestrial Ring-tailed lemur. (Not to scale.)*

Finally, the radiation of the lemurs has given rise to all the main types of social system found among primates. In most species, individual adults spend their nocturnal active period alone, but they may share daytime shelters with conspecifics. In comparison with other primates, a relatively large number of lemur species live in pair-bonded family units, but permanent groups consisting of several males and females have also evolved at least twice. Lemurs that exhibit different combinations of these life-history traits have colonized habitats ranging from evergreen rain forest to spiny desert forest, and as such they offer an exciting opportunity for evolutionary comparisons with other primates and mammalian radiations. PK

Typical Lemurs

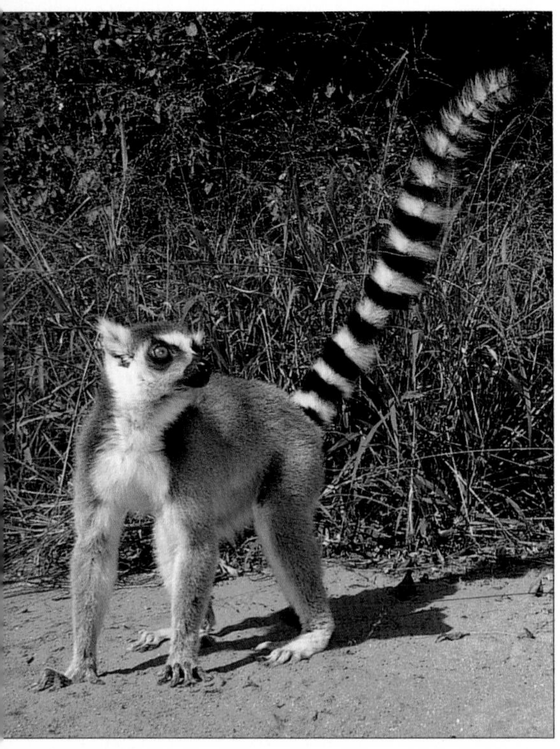

ONLY ON MADAGASCAR CAN LEMURS BE found (with a few species also introduced into the Comoro Islands to the northwest). Over 50 known species, of which some 35 are still extant, represent the endpoints of an adaptive radiation that originated from a single successful colonization event during the Paleogene, some 40 million years ago. Colonization may have begun with a pregnant female crossing from East Africa on a raft of vegetation.

The family Lemuridae includes the lemurs best known to science and the general public. Ring-tailed lemurs, in particular, have been studied for four decades, and are found in most zoos. They are responsible for the common name of all their close relatives, because their howls reminded early biologists of the cries of the Roman *lemures*, or spirits of the dead. The Ring-tailed lemur exhibits a social structure that is unique among mammals: all adult females completely dominate all males of their group. The Lemuridae are found in virtually all Madagascan forests. Ring-tailed lemurs are restricted to the arid south and southwest, but a mountain population has also been recently described. Bamboo lemurs inhabit the eastern rain forests. Two isolated populations of bamboo lemurs live in the humid northwest and in dry western forest, respectively. Ruffed lemurs are rainforest inhabitants that can be found all along the east coast. True lemurs (*Eulemur* species) have colonized all forest types with five different species. Especially noteworthy are brown lemurs, whose six subspecies are found distributed all around the island, except for the arid south. All the Lemuridae are good climbers and spend most of their time in trees; only Ring-tailed lemurs spend a lot of time traveling on the ground in the scrubby dry forests they inhabit.

A 24-hour Lifestyle

FORM AND FUNCTION

Ring-tailed lemurs are cat-sized, with dense gray fur, white underparts, a black face mask, and a long, upright, black-and-white striped tail. They move quadrupedally, and spend more time on the ground (about one third) than any other lemur species. Their closest relatives are the bamboo or gentle lemurs, which are dark brown, orange, or ash gray, and the size of a small cat. The true lemurs are the largest group in this family. They are all approximately cat-sized, quadrupedal, and have long tails, but their coat coloration is extremely variable within and between species. Most taxa have distinct facial patterns composed of black, white, or orange stripes, patches, or beards. In addition, this is the only lemur genus in which males and females differ in coat color and facial pattern (sexual dichromatism). Ruffed lemurs are the largest living members of this family. They are also quadrupedal, with long, bushy tails, and they come in two spectacular color morphs. Both have black faces, tails, hands, and feet, but the body of one subspecies has a very variable black and white pattern, while the other sports a brilliant chestnut-red.

Most true lemurs exhibit an activity pattern that is virtually unique among primates and rarely encountered among other arboreal tropical mammals. They are cathemeral, which means that their activity is spread over several bursts, distributed over both day and night. Ring-tailed, and perhaps ruffed lemurs, in contrast, are predominantly active during the day. All of these species lack trichromatic color vision, characteristic of other day-active primates. Cathemeral activity may either represent a stage in the evolutionary transition from night to day activity, or an almost unique adaptation of these lemurs, shared only by one species of South American night monkey, *Aotus azarae*.

The diversity of reproductive patterns exhibited by this family offers a glimpse of possible stages in the evolution of primate life histories. Ruffed lemurs give birth to litters of 2–4 little-developed

Left *With its extravagant, boldly hooped tail, the Ring-tailed lemur is the most instantly recognizable o all lemurs. But although it is relatively widespread an abundant, conservationists are concerned that it is rapidly losing ground to human encroachment, and i being overhunted.*

Below *Representatives of the family Lemuridae:* ***1a*** *and* ***b*** *Black lemur* (Eulemur macaco), *male and female coloration;* ***2a*** *and* ***b*** *Mongoose lemur* (Eulemur mongoz), *male and female coloration;* ***3*** *Bamboo lemur* (Hapalemur griseus) *rubbing scent on a branch;* ***4*** *Brown lemur* (Eulemur fulvus) *markir its tail with the scent glands on its wrists;* ***5*** *a subspecies, the White-fronted brown lemur* (E. f. albifror *male – the female has gray rather than white fur around the face);* ***6*** *Black-and-white morph of the Ruffed lemur* (Varecia variegata) *engaged in anogen scentmarking.*

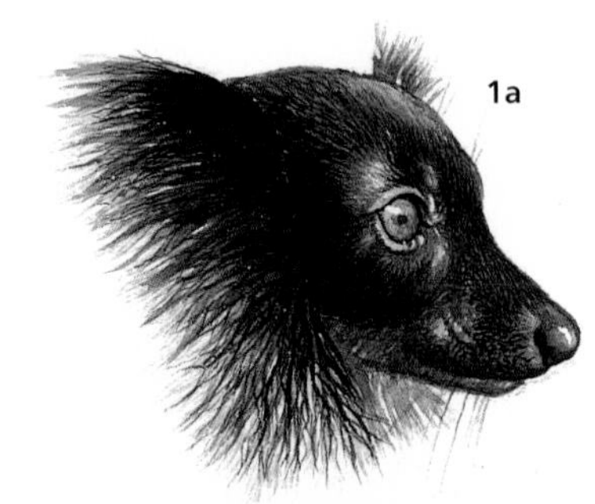

ıng after 100 days in the womb. For the first ee weeks following birth, they are left in a self-ıstructed nest, and later carried in the mouth to ıer sites where they are also guarded by males. ey are never carried on the mother's body, but tead develop so rapidly that by four months y are as agile and mobile as adults. Some bam-ɔ lemurs also build nests for their typically sin-offspring, which are born after a gestation iod of about 140 days and carried either in the ıuth or clinging to the fur. Finally, Ring-tailed d true lemurs usually give birth to single infants er about 120 days of gestation (but they can ɔduce twins under good conditions). These ecies carry their young on their body from birth; y are weaned at 4 months, and become sexual-nature in their second or third year of life. ngevity in the wild is unknown, but in captivity larger species live for well over 20 years. Males d females do not differ in average body size.

Generalists and Specialists

DIET

The diet of the Lemuridae can be either omnivorous or specialized. Bamboo lemurs, as their name implies, are extremely specialized, in that they all feed on bamboo. In areas where all three species occur together, they even use different parts of the same plant; *Hapalemur griseus* concentrates on leaves, *H. aureus* on new shoots, which contain concentrations of cyanide that are lethal to most mammals, and *H. simus* on the soft pith, which it obtains by tearing apart the wooden bamboo poles with its powerful jaws. Ring-tailed lemurs prefer the fruit of the introduced tamarind tree, while ruffed lemurs are the most frugivorous of all lemurs. True lemurs have more omnivorous diets, including locally available fruit but also flowers and leaves during other seasons. They have also been observed eating various arthropods and some small vertebrates.

As regards supplementing the diet of other species, members of this lemur family are commonly preyed upon by the fossa (*Cryptoprocta ferox*; see Carnivores: Civets, Genets, and Mongooses) where it still occurs. In addition, young lemurs may fall victim to the larger birds of prey; to warn their offspring, adults utter alarm calls at the appearance of raptors.

FACTFILE

TYPICAL LEMURS

Order: Primates

Family: Lemuridae

10 species in 4 genera: true lemurs (*Eulemur*, 5 species); bamboo or gentle lemurs (*Hapalemur*, 3 species); Ring-tailed lemur (*Lemur catta*); and Ruffed lemur (*Varecia variegata*).

DISTRIBUTION Coastal areas throughout Madagascar.

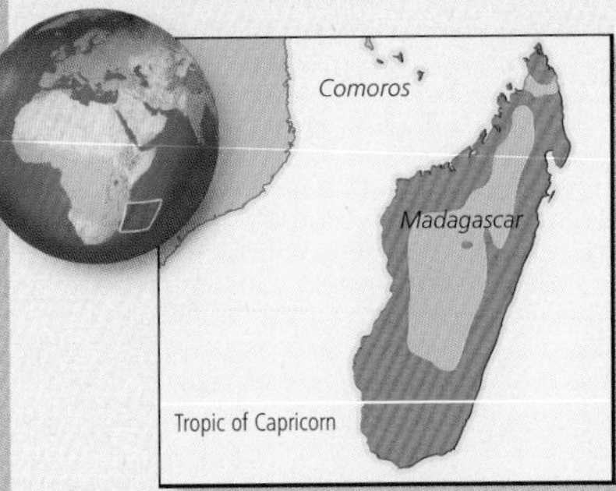

HABITAT Forest habitats throughout the island (True lemurs). Dry, spiny forest (Ring-tailed lemur); rain and humid forests (Ruffed and bamboo lemurs).

SIZE **Head–body length** ranges from 25–45cm (10–18in) in bamboo lemurs to 50–55cm (20–22in) in Ruffed lemur; **tail length** from 32–48cm (13–19in) to 60–65cm (24–26in); **weight** from 0.7–2.5kg (25–88oz) to 3–4.5kg (106–159oz) in same species.

COAT Short and dense. Pelage variable among species; includes black, maroon, gray, brown, and black-and-white. Long tail; short ears; prominent muzzle.

DIET For Ring-tailed, Ruffed, and true lemurs, mainly fruit, plus leaves, flowers, or occasional animal prey. Bamboo lemurs specialize on bamboo.

BREEDING Females sexually mature at 1.5–2.5 years. In most species, usually a single offspring born after gestation of 126–140 days. The Ruffed lemur has a high rate of twin births, and is unusual in leaving its young in the nest.

LONGEVITY Not known in the wild, but a Brown lemur survived in captivity for over 35 years.

See species table ▷

3

5

4

6

Submissive Males

SOCIAL BEHAVIOR

All Lemuridae species live either in pairs or in small groups of adult males and females that rarely exceed 20 members. Pair-living, rare among primates and other mammals, has evolved in all four genera. Some ruffed and brown lemurs live in fluid associations, in which community members form subgroups, whose composition typically changes on a daily basis. The groups of the more gregarious species vary in size and make-up, but contain on average the same number of adult males and females. Most other primates live in groups with more females, since aggression among males for access to females leads to excess male deaths and a higher proportion of females in the population. Lemur males are the same size or smaller than females, and do not have enlarged canines. The unique behavioral trait of female dominance displayed by some species is likely to be related to these traits in males. All adult females can elicit submissive behavior from all adult males, often without aggression. Evolutionary reasons for this sex-role reversal are still unknown.

Top *The red color morph of the Ruffed lemur* (Varecia variegata ruber). *This resident of Madagascar's east coast is the largest of the typical lemurs and is mostly active at dusk.*

Above *Immediately after its birth, the infant Ri tailed lemur clings to its mother's underside; after t weeks, however, it switches position and begins to ride on her back.*

ɔroup-living lemurs provide an opportunity ɔompare patterns of social behavior with haplo- ıe primates, where life in groups evolved later l independently. Behavioral studies show that ıur groups are not glued together by social ıds among closely-related females, even though y typically remain in their natal groups for life. ly mothers and daughters seem to support each er; more distantly related females may even ːt each other from groups. Ring-tailed lemurs ·e formalized dominance relations and a clear ɪarchy, also within the sexes, whereas Brown ıurs have no behavioral mechanisms to signal ɔgnition of individual asymmetries in power.

Most typical lemurs are very vocal, using calls ılert others to predators or to communicate h neighboring groups. They also communicate scent, secreted in special scent glands on the ɔat, hands, and anogenital region. The degree erritoriality varies greatly among species, and ɔears to depend mainly on the local availability l distribution of key resources. Like monkeys l apes, Lemuridae often groom each other to ɪent social bonds. In contrast to monkeys, wever, they use their lower incisors, which tilt ward at an angle of about 45°, as a specialized othcomb" to remove excess fur and parasites m their grooming partner.

The pair-living lemurs appear to have monogamous mating systems. Group-living species are more promiscuous. During the short annual breeding season, when each female is in estrus only for a single day, males fight ferociously with each other. In the end, females end up mating with virtually all resident males, and sometimes even with visitors from neighboring groups. Preliminary genetic paternity analyses have revealed that it is often a single, dominant male who ends up siring most infants.

Common and Rare

CONSERVATION AND ENVIRONMENT

The Lemuridae include both some of the rarest and some of the commonest lemurs. The Golden and Greater Bamboo lemurs are considered Critically Endangered. These and many other species suffer from destruction of their remaining habitat. The bamboo lemurs also have the disadvantage of minuscule geographical ranges. A Critically Endangered subspecies of Lesser bamboo lemur is the focus of one of Madagascar's most intense conservation programs, but the survivors number only in the hundreds. True lemurs are commonly hunted and also face the steady reduction of their habitats, but they are currently still abundant, except for some rare subspecies. PK

THE BYGONE WEALTH OF MALAGASY LEMURS

It is thought that around 50 lemur species were extant on Madagascar when humans arrived some 2,000 years ago. Yet within a few centuries, hunting, habitat destruction, and microclimatic change made many of them extinct. So relatively recent is their demise that their remains have not even had time to form fossils; rather, they are known from intact bones and skeletons in caves.

The extinct lemurs spanned a wide range of physical and ecological types. Body size ranged from forms as small as a mouse to others larger than orangutans or gorillas. All types of locomotion were represented: quadrupedal terrestrial, quadrupedal arboreal, slow-grasping climber, vertical clinging and leaping, and brachiation (swinging by the arms); and, very likely, the larger-bodied types tended to be highly gregarious leaf-eaters, active chiefly in the daylight hours, like some monkeys in Africa today. It is now widely agreed that the Indriidae alone included (together with the present-day forms) the baboon-like *Archaeolemur*, the gelada-like *Hadropithecus*, the enormous and robust *Archaeoindris*, and the large brachiating *Palaeopropithecus*. Competing with *Archaeoindris* in body size was *Megaladapis*, koala-like in proportions but with a skull over a foot (30cm) long in one species. JIP

Ring-tailed lemur Vu
Lemur catta

Deciduous forests, gallery forests, arid bush/forest. Diurnal. Feeding largely arboreal, but animals often travel on ground. Births in August–November.
HBL 39–46cm (15–18in); TL 56–63cm (22–25in): WT 2.3–3.5kg (5–7.7lb).
COAT: back gray, limbs and belly lighter, extremities white; top of head, rings about eyes and muzzle black; tail banded black and white.

Brown lemur
Eulemur fulvus

Seven subspecies from E and W coasts of Madagascar and from Comoro Islands. In all types of forest, except in dry S and SW. Some populations diurnal, others sporadically active through day and night. Some subspecies becoming very rare. Sometimes included in *E. macaco*.
HBL 38–50cm (15–20in), TL 50–60cm (20–24in); WT 1.9–4.2kg (4.2–9.3lb).
COAT: colors highly variable depending on subspecies, and often between sexes.
CONSERVATION STATUS: One subspecies (White-collared brown lemur) is classed as Critically Endangered, and two others (Sanford's brown lemur and Collared brown lemur) as Vulnerable.

Mongoose lemur Vu
Eulemur mongoz

NW Madagascar, and Moheli and Anjouan in Comoro Islands. Moist forest, deciduous forest and secondary growth, likes nectar, also eats flowers, fruit and leaves. Most populations nocturnal and arboreal.
HBL 32–37cm (13–15in); TL 47–51cm (19–20in); WT 2–2.2kg (4.4–4.8lb).
COAT: males gray, with pale faces, red cheeks and beards. Females have browner backs, dark faces, white cheeks and beards.

Black lemur Vu
Eulemur macaco

West of N Madagascar. Humid forests. Primarily arboreal, daytime and dusk activity.
HBL 38–45cm (15–18in); TL 51–64cm (19–25in); WT 2–2.9kg (4.4–6.4lb).
COAT: highly dichromatic: males uniformly black; females light-chestnut brown, with darker faces and heavy white ear tufts.

Crowned lemur Vu
Eulemur coronatus

Extreme north. Dry forests and, recently, high-altitude moist forest. Active in day and at dusk. Sexually dichromatic.
HBL 32–36cm (13–14in); TL 42–51cm (17–20in); WT c.2kg (4.4lb).
COAT: males have medium-gray backs, lighter limbs and underparts; faces whitish, with V-shaped orange marking above forehead; crown of head black. Female upperparts and head cap lighter in color.

Red-bellied lemur Vu
Eulemur rubriventer

Medium to high altitudes in E coast rain forest. Poorly known but probably diurnal, limited to highest strata of forest.
HBL 36–42cm (14–17in); TL 46–54cm (18–21in); WT unknown.
COAT: upperparts chestnut-brown, tall black, face dark. Males have reddish-brown underparts, females whitish.

Ruffed lemur En
Varecia variegata

Sparsely distributed in E coast rain forest. Largest of Lemuridae. Poorly known: probably fruit-eating, in upper strata of forest. At least 2 subspecies recognized, primarily on basis of coat color.
HBL 51–56cm (20–22in); TL 56–65cm (22–26in); WT 3.3–4.5kg (7.3–9.9lb).
COAT: both have long, dense fur, especially around neck. Prominent muzzle, face covered by short hair.

Bamboo lemur LR
Hapalemur griseus

E coast humid forest and two isolated populations along W coast. Essentially limited to bamboo forests and reed beds.
HBL 27–40cm (11–16in); TL 32–40cm (13–16in); WT 0.7–1kg 1.6–2.2lb).
FORM: rounded head with small, furred ears, short muzzle and woolly coat.
CONSERVATION STATUS: LR Near Threatened. Three subspecies, one of which (Alaotran bamboo lemur) is classified as Critically Endangered and another (Western Lesser bamboo lemur) as Vulnerable.

Golden bamboo lemur Cr
Hapalemur aureus

Extremely limited range in central SE coast humid forest.
HBL 28–45cm (11–18in); TL 24–40cm (9–13in); WT 1–1.5kg (2.2–3.3lb).
COAT: coloration more reddish-brown than *H. griseus*.

Greater bamboo lemur Cr
Hapalemur simus

Extremely limited range in central SE coast humid forest.
Total length c.90cm (35in).
COAT: gray to gray-brown with lighter underparts. Larger and more heavily built than *H. griseus*. RDM/PK

Abbreviations HBL = head–body length TL = tail length WT = weight

Cr Critically Endangered En Endangered Vu Vulnerable LR Lower Risk

Sportive Lemurs

SPORTIVE LEMURS ARE SOME OF THE SMALLEST specialized leafeaters among mammals. They appear to be able to thrive on such a low-energy diet only by drastically reducing their metabolic rates to levels far below those predicted for mammals of their size.

As a by-product of this metabolic adaptation, sportive lemurs are among the least active primates, and have very small home ranges. Their common name is thus misleading; it is in fact based on their defensive posture, which is vaguely reminiscent of that of a boxer.

Lazy Leapers

FORM AND FUNCTION

Sportive lemurs are medium-sized lemurs, with an average body mass of less than 1kg (2.2lb). Their pelage is dense and short, and includes various shades of brown and gray, with great variation among individuals. The color of the ventrum varies from gray to white. They have a prominent, moist muzzle, big eyes, and medium-sized ears. The tail is thin and about the same length as the body (each about 25cm/10in). They typically rest in a vertical posture and locomote between tree trunks in deliberate, short leaps, maintaining their vertical orientation. Sportive lemurs are active at night, spending the day in hollow trees or liana tangles, though they have occasionally been observed sunbathing or moving short distances just outside their shelters during the day. They are strictly tree-dwelling and do not come down to the ground.

Sportive lemurs are energy savers. Once they exit their tree holes, they may only visit a few feeding trees during the course of a night. They can spend hours ingesting leaves on a single tree, interspersed by long periods of inactivity and digestion. They combine reduced activity with metabolic rates at least 50 percent below levels predicted for their body mass. Unlike some mouse and dwarf lemurs, however, they do remain active year-round.

The reproductive patterns of sportive lemurs remain extremely poorly known. After about 130 days of gestation, a single infant is born. Mothers initially carry infants in their mouths and "park" them in the vegetation next to where they forage. Later on, infants are also carried on the mother's back. It is thought that juveniles become independent after one year and reach sexual maturity at around 18 months. Reproductive rates and life spans are not known. Sexual size dimorphism is lacking. Males of some species have larger cani than females, but their relative testis size is extremely small.

Because sportive lemurs exhibit a number o morphological similarities with recently extinct giant lemurs of the genus *Megaladapis*, some taxonomists consider them as a subfamily of the Megaladapidae.

Above *Small ears and a very pronounced muzzle mark this animal out as a Gray-backed sportive lemur. This species is also commonly named for the small island of Nosy Bé off the northwest coast of Madagascar, one of its last strongholds.*

Leafeaters of the Malagasy Forests

DISTRIBUTION PATTERNS

Sportive lemurs are found in all Malagasy forest including eastern rain forest, northern humid fo est, western dry forest, and southern spiny fore often in very high densities. As far as is known, two species overlap in their geographical range but some hybrid zones may exist. They occupy the middle and upper forest layers. Their diet is mainly based on leaves. They can tolerate high concentrations of secondary plant compounds, and species inhabiting deciduous forests may

FACTFIL

SPORTIVE LEMURS

Order: Primates

Family: Megaladapidae

7 species of the genus *Lepilemur*

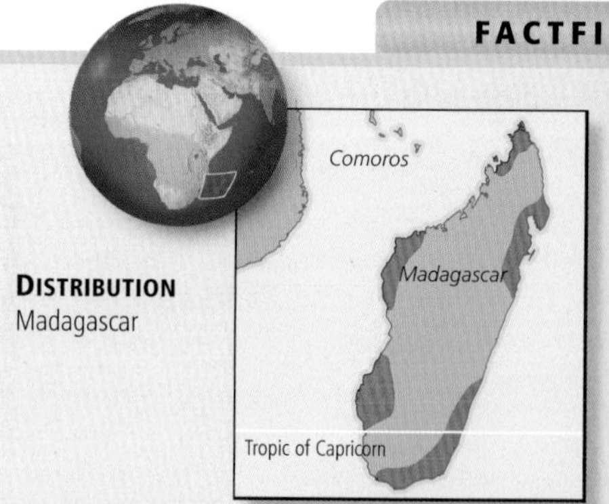

DISTRIBUTION Madagascar

Size HBL 25–35cm (10–14in); TL 22–30cm (9–12in); WT 0.5–1kg (1–2lb). No sexual dimorphism.

Diet All sportive lemurs are folivores that occasionally take flowers and fruit. Low metabolic rates and gut specializations allow them to survive on this low-energy diet.

Breeding Reproductive biology is poorly known. Sexual maturity at around 18 months. Single infants are born September–December after 130–135 days' gestation.

Longevity Unknown in wild. Up to 12 years in captivity.

WEASEL SPORTIVE LEMUR *Lepilemur mustelinus*
Xerophytic, gallery, deciduous, and humid forests. Form: short face, moist rhinarium, prominent ears, dense, woolly fur. Conservation status: Lower Risk – Near Threatened.

SMALL-TOOTHED SPORTIVE LEMUR
Lepilemur microdon
Small-toothed or Light-necked sportive lemur
E Madagascar. Conservation status: Lower Risk – Near Threatened.

WHITE-FOOTED SPORTIVE LEMUR *Lepilemur leucopus*
S gallery forests and *Didierea/Euphorbia* bush. Form: small-bodied, upperparts medium gray, underparts gray-white, tail light brown. Ears large. Conservation status: Lower Risk – Near Threatened.

RED-TAILED SPORTIVE LEMUR *Lepilemur ruficaudatus*
SW *Didierea/Euphorbia* bush and gallery forests. Form: back light brown, underparts paler, reddish tail, pale face and throat. Ears large. Conservation status: Lower Risk – Near Threatened.

MILNE-EDWARDS'S SPORTIVE LEMUR
Lepilemur edwardsi
WC deciduous forests. Form: similar to *L. ruficaudatus*, but coat darker, especially on upper part of back, gray-brown face, underparts gray, ears large. Conservation status: Lower Risk – Near Threatened.

GRAY-BACKED SPORTIVE LEMUR *Lepilemur dorsalis*
NW moist forests. Form: small-bodied, coat medium to dark brown above and below, face dark, ears small. Conservation status: Vulnerable.

NORTHERN SPORTIVE LEMUR
Lepilemur septentrionalis
Extreme N deciduous forests. Form: coat gray on upperparts and crown, rump and hind limbs paler, tail pale brown, underparts and face gray. Conservation status: Vulnerable.

Abbreviations HBL = head–body length TL = tail length WT = weight

◐ ***Left*** *Milne-Edwards's sportive lemurs customarily spend the day resting together in groups of two or three in a hollow tree, before dispersing at night to feed. Although they are relatively abundant, this habit makes them very vulnerable to hunters, who break open the tree trunk and kill the animals for food.*

even subsist on dry leaves during the dry season. Sportive lemurs have long intestines with a large blind-ending sac (cecum) and bacterial microflora that assist them in breaking down food matter – the cellulose in leaves – that is hard to digest. Some individuals have been observed to ingest feces, presumably to take up such bacteria. Sportive lemurs are important prey for the fossa (*Cryptoprocta ferox*), the largest Malagasy carnivore (see Carnivores: Civets, Genets, and Mongooses), and also for the Long-eared owl (*Asio madagascariensis*). The diurnal Harrier hawk (*Polyboroides radiatus*) and large boas over 2m (6.5ft) in length – for example *Sanzinia madagascariensis* and *Acrantophis* sp. – have been observed to take sportive lemurs out of their daytime tree holes.

The Loneliest Lemurs

SOCIAL BEHAVIOR

Sportive lemurs are typically solitary during their nocturnal activity. In some species, adults rarely associate or interact. They also spend the day alone in tree holes or other shelters. In Milne-Edwards's and Red-tailed sportive lemurs, pairs of adult males and females may by associated for at least part of the year. They move independently in overlapping home ranges of about one hectare (2.5 acres), and may share a sleeping tree. Brief, nocturnal encounters may include exchanges of allogrooming. Sportive lemurs use loud vocalizations, especially early in the night, to signal their presence to conspecifics. Occasional branch- and tree-shaking displays may also serve a territorial function. The social behavior of sportive lemurs is the least well understood of all the primate families. Their mating systems remain completely unstudied.

Under the Ax

CONSERVATION AND ENVIRONMENT

Sportive lemurs are still abundant in many areas. Population densities can be very high in habitats that remain intact. They will tolerate secondary forests, and are even found in some plantations. However, Northern and Gray-backed sportive lemurs have very small remaining geographical ranges, which puts them at the greatest risk of extinction. Sportive lemurs suffer more from human hunting activities than any other lemur taxon. When threatened, instead of fleeing, they try to defend themselves inside their tree holes, which makes them easy victims for people armed with axes. It has been estimated that several thousand sportive lemurs end up in the cooking pot every day! PK

Dwarf and Mouse Lemurs

MEMBERS OF THIS FAMILY HAVE RETAINED many primitive characteristics, and so are our best living models for ancestral primates. At the same time, several species have very specialized physiological adaptations that allow them to hibernate for several months during the cool, dry season.

The dwarf lemur family includes the smallest known primate: the Pygmy mouse lemur, which, at 30g (1oz), weighs little more than a house mouse. It is one of six new species discovered in the 1990s, several of which have extremely small ranges and are at risk of going extinct before we even understand the rudiments of their biology.

Hibernating Dwarfs

FORM AND FUNCTION

The mouse and dwarf lemurs are the smallest primates on Madagascar. They have elongated bodies with short arms and legs, and run and jump quadrupedally. The head is small, with prominent eyes, a moist muzzle, and (in most species) large, sparsely furred ears. The tail is long and can be used to store fat reserves. All species live in trees, bushes, and lianas. *Mirza* and *Microcebus* also briefly descend to the ground, catching small animal prey in the leaf litter to supplement their otherwise frugivorous diet. All dwarf and mouse lemurs are only active at night. Their vision is improved by a tapetum of light-reflecting crystals behind the retina, an adaptation found in many other nocturnal mammals.

Microcebus, *Cheirogaleus*, and perhaps *Allocebus*, are unique among primates in that they spend days, weeks, or months when climatic conditions are unfavorable in a state of torpor or hibernation. They reduce their metabolic rates dramatically and lower their body temperatures close to ambient temperatures (as low as 15°C/59°F) to maximize energy savings. Hibernating species nearly double their weight during the plenteous warm season in preparation for their long inactivity. Curiously, among Gray mouse lemurs only females hibernate, whereas males remain active throughout the dry season. *Mirza* and *Phaner*, on the other hand, are active year-round.

All dwarf and mouse lemurs develop rapidly, and females reach sexual maturity within their first year of life. All members of this family, except *Phaner*, produce litters of 2–4 poorly developed infants after 2–3 months of gestation. Infants are initially "parked" in safe sites like tree-hole nests. Females of the hibernating species lactate even when putting on fat reserves for the upcoming dry season. In general, females tend to be larger than males, but sexual dimorphism may fluctuate during the course of the year. Males of some promiscuous species, such as *Mirza* and *Microcebus*, have huge testes for their body size; a single testis exceeds the brain in both volume and mass!

A Complex Network of Niches

DISTRIBUTION PATTERNS

Members of this family have colonized all forest habitats on Madagascar. Coquerel's dwarf lemurs are limited to the dry deciduous forests of the west, and Hairy-eared dwarf lemurs are confined to the rain forests along the east coast, but the other genera have representatives inhabiting both types of habitat. Some of them, especially mouse lemurs, do well in modified forests, and also occur

FACTFILE

DWARF AND MOUSE LEMURS

Order: Primates

Family: Cheirogaleidae

13 (7) species in 5 (4) genera: mouse lemurs (*Microcebus*, 4 species); dwarf lemurs (*Mirza*, *Allocebus*, *Cheirogaleus*, and *Phaner*, 5 species).

DISTRIBUTION Madagascar

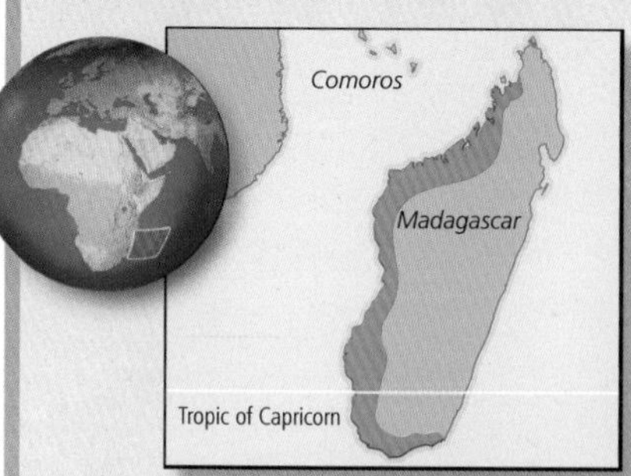

HABITAT Throughout the island, including E rain forest, W dry deciduous forest, and S spiny forest.

SIZE **Head–body length** ranges from 9–11cm (4in) in the Pygmy mouse lemur (*M. myoxinus*) to 22–30cm (9–12in) in the Fork-marked lemur (*P. furcifer*); **weight** ranges from 24–38g (0.8–1.3oz) to 350–500g (12–18oz) in the same species.

COAT Fur short and dense; mostly gray-brown above, white to cream below, depending on species.

DIET Mostly fruiteaters. Mouse and Fork-marked lemurs also feed on gum and tree sap, mouse lemurs hunt small invertebrates, and Coquerel's dwarf lemurs hunt small reptiles and other vertebrates.

BREEDING 2–4 young born after 2–3 months' gestation. Fork-marked lemur bears a single infant.

LONGEVITY Rarely more than 5 years in the wild (in captivity, more than 10).

See species table ▷

Right *The huge eyes of the Brown mouse lemur are equipped with a light-reflecting layer of cells behind the retina that give it excellent night vision as it goes about its lone foraging activities in the trees. Its extremely varied diet includes leaves, gum, flowers, spiders, insects, and sometimes even small reptiles.*

Left *Twin infant Coquerel's dwarf lemurs – a not uncommon sight in this species, whose litters more often consist of twins than of individuals. Although their survival is threatened by habitat destruction, Coquerel's dwarf lemurs can breed throughout the year, and populate some areas quite densely.*

in plantations and close to human settlements. In most forests, two or more members of this family occur together; in central western Madagascar, five species from three genera share the same habitat.

Mouse and dwarf lemurs have evolved several dietary strategies. The most diverse diet of fruit, small arthropods, and gum is eaten by *Microcebus* species, while fork-marked lemurs are gum specialists, with a long tongue and specialized teeth to chisel holes in tree bark to stimulate the flow of tree sap. Fat-tailed dwarf lemurs appear to prefer fruit, which are abundant during their short annual period of activity. Coquerel's dwarf lemurs are also frugivorous, but spend most of their active period looking for animal prey, including arthropods and small vertebrates such as snakes and chameleons. During the dry season, they also feed on sweet secretions of homopteran larvae. All members of this family eat flowers and nectar, and so act as pollinators for some plant species.

The way species use the habitat is also distinctive. Fork-marked lemurs, for example, are mainly active in the canopy, whereas mouse lemurs tend to use the lowest forest layers. Diurnal resting sites also vary among species. Dwarf and mouse lemurs compete over hollow trees suited for their long periods of inactivity. Coquerel's dwarf lemurs build spherical leaf nests high up in the canopy, which may be taken over by fork-marked lemurs.

Because of their small size and typically high population densities, mouse and dwarf lemurs are quantitatively important prey items for several predators. Nocturnal raptors (owls) alone can take up to 30 percent of a mouse lemur population per year. Endemic civets and mongooses and large snakes also prey on these small lemurs.

Encounters in the Night

SOCIAL BEHAVIOR

Despite their very similar life histories and co-occurrence in many habitats, dwarf and mouse lemurs have evolved diverse social systems. Most activity is carried out alone, embedded in some higher-order social complexity. Gum sites and animal prey can only be fed upon by one animal at a time, which necessitates solo feeding. In the better-studied solitary species, such as *M. muri*[…] and *M. coquereli*, females occupy stable home ranges that overlap with those of close relatives[…] with so many mammals, males tend to have lar[…] home ranges than females, especially during th[…] breeding season. During their nocturnal activity[…] they use various acoustic and olfactory signals, including ultrasound, to coordinate their activi[…] with neighbors. Direct encounters occur infrequently in the night, and individuals may chase[…] groom, or ignore each other. Among fork-marke[…] and Fat-tailed dwarf lemurs, pairs defend territo[…]ries, using spectacular vocal duets and fecal ma[…] respectively, to advertise their presence.

During the day, mouse and dwarf lemurs bu[…] their own nests or use tree holes or other shelte[…] Patterns of use of these shelters provide a glimp[…] of additional social complexity that ongoing research has only begun to reveal. In Coquerel'[…] dwarf lemur, for example, individual adults typically sleep alone, although occasional pairs of females have been observed. Fat-tailed and fork[…] marked lemurs, on the other hand, rest with a

MOUSE LEMURS

Members of the *Microcebus* genus, mouse lemurs include most of the smaller members of the Cheirogaleidae family. Taxonomically, their status is in flux. A brown variety of *M. murinus* – *M. rufus* – was declared a separate species in 1977. The period 1995–2001 witnessed a great increase in the number of recognized species : the Pygmy mouse lemur (*M. myoxinus*) was distinguished from *M. murinus* in 1995, and in the following year the Golden-brown mouse lemur (*M. ravelobensis*) was described. By 2001, a combination of morphological studies and analysis of mitochondrial DNA undertaken by researchers from the University of Antananarivo and the German Primate Center had revealed four further species from the dry forests on Madagascar's west coast: Northern rufous mouse lemur (*M. tavaratra*), Berthe's mouse lemur (*M. berthae*), Sambirano mouse lemur (*M. sambiranensis*), and the Gray-brown mouse lemur (*M. griseorufus*).

Gray mouse lemur
Microcebus murinus

NW, W, and S Madagascar, in forest fringes and secondary vegetation; nocturnal.
HBL 10–20cm (4–8in); TL 10–20cm (4–8in); WT 30–70g (1–2.5oz).
COAT: fur on back gray to gray-brown; white to cream below; large membranous ears.
BREEDING: gestation 60 days; litter size 2–3.
LONGEVITY: to 15 years in captivity.
CONSERVATION STATUS: not immediately threatened, but affected by habitat destruction.

Brown mouse lemur
Microcebus rufus

E Madagascar, in forest fringes and secondary vegetation; nocturnal.
HBL 10–20cm (4–8in); tL 10–20cm (4–8in); WT 30–70g (1–2.5oz).
COAT: fur on back brown; white to cream below; medium-sized membranous ears.
BREEDING: gestatión 60 days.
LONGEVITY: 12 years in captivity.
CONSERVATION STATUS: not immediately threatened, but affected by habitat destruction.

Pygmy mouse lemur En
Microcebus myoxinus

W C Madagascar.
HTL 18–22cm (7–8.5in); WT 24–38g (0.8–1.3oz).
COAT: fur on back reddish-brown with an orange tinge. Formerly included within *M. murinus*.

Golden-brown mouse lemur En
Microcebus ravelobensis

NW Madagascar, in the Ankaranfantsika Nature Reserve; nocturnal. WT 60g (2oz). Similar in appearance to the Pygmy mouse lemur, but of slighter build, and with longer tail and more pointed nose.
DIET: Primarily insectivorous.

DWARF LEMURS

Five species in four genera are now recognized. Coquerel's lemur once assigned to *Microcebus*, but now sometimes assigned a genus of its own, *Mirza*. In most areas two or more species of dwarf lemurs – in the west, up to five – occur sympatrically.

Coquerel's dwarf (or mouse) lemur Vu
Mirza (Microcebus) coquereli

Disjunct distribution in well-established coastal forests of W and NW Madagascar; nocturnal.
HBL 20–23cm (8–9in); TL 30–33cm (12–13in); WT 300–350g (11–12oz).
COAT: fur on back gray-brown to pale brown, yellowish below; tip of tail darker than rest of fur; large, membranous ears.
BREEDING: gestation 86 days; litter 1–2.
LONGEVITY: up to 15 years in captivity.

Hairy-eared dwarf lemur En
Allocebus trichotis

NE Madagascar; very restricted distribution in primary rain forests; nocturnal.
HBL 6–13cm (2–5in); TL 10–20cm (4–8in); WT 30–70g (1–2.5oz).
COAT: fur on back pale brown; white to cream below; ears short, but with pronounced tufts of long hair.
BREEDING and LONGEVITY: unknown.

Greater dwarf lemur
Cheirogaleus major

E Madagascar in well-established secondary forests and primary forests: nocturnal.
HBL 20–25cm (8–10in); TL 20–28cm (8–11in): WT 140–400g (5–14oz).
COAT: fur fairly short and dense; gray-brown above, white to cream below; black ring around each eye.
BREEDING: gestation 70 days.
LONGEVITY: 15 years in captivity.

Fat-tailed dwarf lemur
Cheirogaleus medius

NW, W, and S Madagascar, in well-established secondary forests and primary forests; nocturnal.
HBL 20–25cm (8–10in); TL 20–28cm (8–11in); WT 140–400g (5–14oz).
COAT: fur short and dense; pale gray above, white to cream below.
BREEDING: gestation 61 days.
LONGEVITY: 18 years in captivity.

Fork-marked lemur LR
Phaner furcifer

Disjunct distribution in W, NW, and NE Madagascar, in well-established forests in coastal regions; nocturnal.
HBL 22–30cm (9–12in); TL 28–37cm (11–15in); WT 350–500g (12–18oz).
COAT: fur gray-brown on back, white to cream below; conspicuous dark line running over back of head and forking to join up with dark rings surrounding the eyes.
BREEDING and LONGEVITY: unknown.
CONSERVATION STATUS: LR – Near Threatened; three subspecies ranked as Vulnerable.

RDM/PK

Abbreviations HBL = head–body length HTL = head–tail length TL = tail length WT = weight   En Endangered   Vu Vulnerable   LR Lower Risk

Top *For the first few weeks of their lives, the tiny spring of dwarf and mouse lemurs are carried ound in their mothers' mouths. When the female s to go searching for food, she "parks" her infant – e, a young Gray mouse lemur – at a secure ation.*

Above *Nectar forms an important part of the diet all dwarf and mouse lemurs. In feeding from plants h as epiphytic orchids that grow in tree forks, they p spread pollen, which sticks to their thick fur, from wer to flower.*

partner of the opposite sex, with which they form stable pair bonds. In further contrast, Gray and perhaps other mouse lemurs often sleep in larger groups comprising members of both sexes. More than 15 individuals have been seen sharing a tree hole, although groups of 2–5 are most common. To what extent these various social arrangements reflect different mating systems is still poorly known, but promiscuous matings appear to be common, even among pair-living species.

In the Dark

CONSERVATION AND ENVIRONMENT

Mouse and dwarf lemur populations can maintain extremely high densities. Thus, species with large remaining geographical ranges are generally not at immediate risk of extinction. However, some, such as the Hairy-eared dwarf lemur, are extremely rare, and their geographical distribution and natural histories remain virtually unstudied. Similarly, the known distribution of some of the newly described species of mouse lemur are very limited and are therefore vulnerable to habitat destruction. Some mouse lemurs appear rather resilient to habitat modifications and continue to use secondary forest, forest fragments, and plantations, but it has recently been shown that they cannot maintain viable populations in such habitats. So far, conservation projects have generally not focused on these little brown lemurs, partly because their biogeography and ecology remain among the most poorly known of all primates. PK

LEMUR DIALECTS

Humans around the world speak in dialects that not only reflect regional and social background but also often have an effect on social acceptance. Over the years, animals as diverse as bees, frogs, birds, and mammals have also been reported to employ dialects, though such findings remain controversial. Now, however, Malagasy mouse lemurs, the smallest primate taxon on Earth, have been shown to use them.

The Gray mouse lemur (*Microcebus murinus*), inhabits the western Malagasy deciduous forest, clustered in forest patches with suitable rearing sites such as safe tree holes. Long-term capture-recapture and radio-tracking data as well as genotyping of 161 individuals with 8 microsatellite markers suggest that a population within a forest patch may consist of a dispersed network of individual neighborhoods, each with about 35 resident individuals. Young females seem to stay with or close to their mother (female philopatry), whereas young males migrate.

Individuals within this dispersed society use a rich array of different sounds in their social interactions. Many of their high-pitched calls, which range from 10–36kHz (human hearing range is 0.02–20kHz),can only be heard with the aid of modern bat recording and computer analysis equipment. Analysis of the results has provided fresh insights into the lemurs' secretive vocal signaling system. As in other social mammals, the calls of the mouse lemur reveal the sex and identity of the individual. But not only are individual animals' calls different; whole neighborhoods of animals have been shown to communicate via different dialects.

During the breeding season, reproductively active males emit a trill call – the most complex in the whole repertoire – as part of the mating display. Between 13–35kHz in pitch and lasting 0.3–0.9 seconds, the call consists of an ordered sequence of broadband frequency modulated syllables, like that of bird song. The call may be repeated up to 1.5 times a minute, depending on the motivation of the sender and on how courted females and competing males respond. Males of the same locality emit individually distinct trill calls within an overall theme that are quite distinct from those of neighboring localities with which they appear to interbreed.

Furthermore, the lemurs seem to consciously manipulate the differences in dialect. Laboratory experiments have shown that young males during play will produce early attempts at the trill call that are highly variable, but when they reach sexual maturity, the sound crystallizes into a form more similar to that of its companions than to non-companions. Adult males that are transferred from their home to a new location will shift their call to resemble that of their neighbors.

The tricky question is why dialects exist at all. One answer from sociobiology is that individuals migrating into a new neighborhood and showing an open vocal system might experience less aggression from residents and win greater social acceptance. In that case, dialects in human and non-human primates may share a similar biological function. EZ

Indri, Sifakas, and Woolly Lemurs

This family includes the largest lemurs still living. One of them, the indri, exhibits striking similarities with Asian gibbons: it is the only lemur without a tail, and pairs of adults defend territories with loud song duets that can be heard over several kilometers. The other members of the family, sifakas and avahis, also have striking vocalizations.

Some of the most endangered lemurs are found in this group; indeed, most of its known members have become extinct since the arrival of humans on Madagascar about 2,000 years ago. At least ten species of closely-related Giant sloth lemurs were extirpated within the last millennium.

Nocturnal Leapers

FORM AND FUNCTION

The members of this family vary in activity and size. Avahis, or Woolly lemurs, are medium-sized, nocturnal primates, weighing around 1kg (2.2lb). Sifakas and indri, in contrast, include the largest living lemurs, exceeding 7kg (15lb) in weight, and are active strictly during the day. They also differ greatly in appearance. Woolly lemurs have a thick, dense coat, varying from reddish-brown through olive to gray, with pale underparts. The tail is longer than the body and can be rolled up on the ventral side, as it can in sifakas. The head is rounded, with short ears and big eyes. Sifakas are extremely colorful, with different species and subspecies showing distinctive patterns. The extent of variation ranges from completely white (in the romantically named Silky sifaka) to completely black (Perrier's sifaka); there are also species with orange, maroon, or gray patches on their head, arms, or legs, which are otherwise black or white. In all species the face is naked and black, surrounding a prominent muzzle. The indri has dense black and white fur, but the relative proportions of the two colors vary among individuals and populations. The indri also has a naked, black face, with ears that are more prominent than the sifakas'. It is the only lemur with a vestigial tail.

Indri, sifakas, and woolly lemurs have body plans that support a highly specialized mode of locomotion: vertical-clinging-and-leaping. Powerful legs, about one-third longer than their arms, propel them between trees while allowing them to keep an upright body posture. The larger species can cover up to 10m (33ft) with such leaps. There is a general tendency in this family for the females to be slightly larger and heavier than the males.

All members of this family have females that have only one offspring at a time, and these then develop much more rapidly than the young of comparably-sized primates. From birth, infants are carried across the belly, later switching to a jockey position on their mother's back. Breeding is seasonal, followed by a gestation period of more than five months in sifakas and the indri. Female sifakas do not reach sexual maturity before 4 years; indri females may be as old as 9 before they start to reproduce. Infant and juvenile mortality are high (around 50 percent during the first 2 years), whereas individuals over 2 years survive well from year to year. The oldest age achieved by these lemurs in the wild is not known, but the larger species probably exceed 20 years. The life histories of woolly lemurs remain essentially unknown.

Folivores of the Tree Canopy

DISTRIBUTION PATTERNS

As a group, the Indridae have a wide distribution. Woolly lemurs are divided into an eastern rainforest and a western dry-forest species. *Propithecus verreauxi* has colonized the dry south and west with several subspecies, while *P. diadema* is found in the moist east and north. *Propithecus tattersalli*

Above A mother Western woolly lemur and her offspring. The young of this rainforest species are b in the dry season and carried around for many mon

Right To traverse open ground, the sifaka (pictu here is a Verreaux's sifaka) performs a unique, skipping "dance" on its hind legs, holding its front legs outstretched for balance.

FACTFIL

INDRI, SIFAKAS, & WOOLLY LEMURS

Order: Primates

Family: Indriidae

6 (5) species in 3 genera: woolly lemurs (*Avahi*, 2 (1) species); sifakas (*Propithecus*, 3 species); and indri (*Indri indri*).

DISTRIBUTION Madagascar

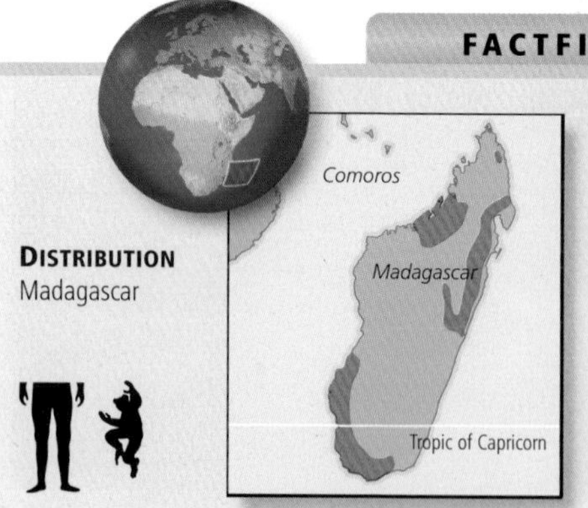

Diet Leaves, fruit, and flowers, in different proportions and according to seasonal availability.

Breeding Sifakas and indri are sexually mature at 4–5 years. Gestation period is about 5 months. Single infants are carried by clinging to mother's fur from birth onward, and are weaned after 4–6 months. Breeding is seasonal, but interbirth intervals can be 2 or 3 years.

Longevity Not known in wild, but some individual sifakas have been followed for well over 10 years.

EASTERN WOOLLY LEMUR *Avahi laniger*
E rain forest. HBL 25–30cm (10–12in); TL 31–37cm (12–15in); WT 0.7–1.3kg (1.5–3lb). Coat: thick, grayish-brown fur and whitish underparts, with a long, thin tail. Conservation status: Lower Risk – Least Concern.

WESTERN WOOLLY LEMUR *Avahi occidentalis*
Disjunct areas of NW dry forest. HBL, TL, WT, and coat: as for *A. laniger* (which some authorities still include it within). Conservation status: Vulnerable.

VERREAUX'S SIFAKA *Propithecus verreauxi*
N and SW spiny and dry forests. HBL 45–50cm (18–20in); TL 41–60cm (16–24in); WT 3–7.5kg (7–17lb). Coat: short, dense fur, predominantly white but with brown spots; top of head brown. Conservation status: Vulnerable.

GOLDEN-CROWNED SIFAKA *Propithecus tattersalli*
NE rain forest. HBL 45–50cm (18–20in); TL 41–60cm (16–24in); WT 3–7.5kg (7–17lb). Coat: mainly white, but with bright orange head and underparts. Conservation status: Critically Endangered.

DIADEMED SIFAKA *Propithecus diadema*
N and E rain forest.S and W spiny and dry forests. HBL 45–50cm (18–20in); TL 41–60cm (16–24in); WT 3–7.5kg (7–17lb). Coat: whitish-gray with orange lower limbs; some subspecies entirely black, or black with white rump. Conservation status: Endangered.

INDRI *Indri indri*
E rain forest. HBL 60–70cm (24–28in); TL 4–5cm (1.5–2in); WT 6–7.5kg (13–17lb). Coat: black and white, with geographical and individual variation in distribution and proportions. Only lemur to have a vestigial tail. Conservation status: Endangered.

Abbreviations HBL = head–body length TL = tail length WT = weight

; restricted to a few forests in the northeast, /hereas the indri occurs throughout the northern art of the eastern rain forest. All species prefer to emain high in the tree canopy, and will only ven-ure down very occasionally to eat soil or to cross pen space.

Sifakas and indri are vegetarian; leaves are their aple food, but they also eat seasonal flowers and uit. Unlike other diurnal lemurs, sifakas and ndri do not come to the ground to drink. For dry-rest sifakas, this means no water (except for rare ew) for up to nine months each year, including e energetically stressful time of lactation.

Females in Control

SOCIAL BEHAVIOR

Voolly lemurs and indri live in nuclear families. he age at which the young leave the exclusive mily home range varies, so parents may live with eir offspring for several years. To defend the ome range, woolly lemurs use a distinctive call 'aya-hee") which is responsible for their vernacu-r and generic name (*Avahi*). Indri employ loud alls in long duets that resemble those of gibbons. ndri females dominate their mates, which typical- have to wait lower in the trees until the female as finished feeding up in the canopy. Females so lead most of the group's expeditions around s range, and often sleep separately from males.

Sifakas live in small groups of variable size and omposition. These may contain 1–13 individu-s, with a mean of five. There are usually equal umbers of males and females. Females typically emain in the group into which they were born, ut males transfer between groups several times uring their lives. Females dominate males during raging and other activities. Groups can be quite spersed while traveling, so that most socializing ccurs when the animals are having extended sts, for which they come together in the same ees. Sifakas also spend the night high up in tall ees, safe from their main predator, the fossa *Cryptoprocta ferox*). In some populations, sifakas e very possessive about their territory – behavior at is thought to be related to a scarcity of food. lsewhere, neighboring groups regularly meet to ngage in mutual threats and chases, or peaceful, mporary associations. Sifakas are comparatively lent, even though their vernacular name derives om their contact call ("Shee-fak").

Mating strategies of sifakas are poorly under-ood. Both sexes undertake multiple matings, but e males have minute testes, smaller even than ose of the tiny mouse lemurs. This suggests that perm competition is probably not an important echanism of reproductive competition. There is me evidence that dominant males, who also ave bigger testes, physiologically suppress repro-uctive function in rivals, perhaps by means of emical signals from their scent glands. Killing of ependent young (infanticide) by adult males has so been observed in several populations.

The Surviving Minority

CONSERVATION AND ENVIRONMENT

The majority of known members of this family were driven to extinction during recent centuries; they are now known only from subfossil remains. Some of the surviving taxa are already extremely rare, and have small and fragmented remaining ranges. *Propithecus tattersalli*, for example, is Critically Endangered, and has received the highest conservation priority rating because no part of its range is legally protected. It has one of the smallest populations and most limited distributions of all lemur species. Perrier's and Silky sifakas are similarly threatened, with remaining populations totaling probably less than 1,000 individuals. The overwhelming threat to sifakas and the indri is habitat loss to slash-and-burn agriculture and tree-felling. Hunting is less of a problem, despite their large size, because in many areas killing them is still considered taboo by the local people. PK

Aye-aye

Left The aye-aye eats coconuts by gnawing a ho and extracting the milk and pulp with its long third finger – prior to gnawing it may even tap the cocor with its third finger, possibly to ascertain how much milk it contains. They appear to be frequent visitors coconut plantations.

ONE OF THE MOST DISTINCTIVE MAMMALS, the aye-aye is the only living member of its family. It has such unique and unusual morphological adaptations that it was at first classified as a rodent. Its appearance is so bizarre that the natives of Madagascar for long considered it an omen of bad luck that must be killed on sight.

As a result, the aye-aye was thought to be nearly extinct only a few years ago. However, recent field research has shown that it is more widely distributed than previously feared, and intense conservation efforts provide at least some hope for its continued survival.

Percussion Foragers

FORM AND FUNCTION

The aye-aye's appearance is unique. The coat is thick and dark brown, with long, bright guard hairs; only the face is pale, with distinctive dark rings around the big eyes. The long tail is very bushy, and the head is relatively massive and rounded. Aye-ayes run and jump quadrupedally. They have big, naked ears, and elongated digits with long, curved nails. They are the largest nocturnal primate.

Equally distinctive are the improbably long middle fingers, which are used in combination with permanently growing incisors as foraging tools to chisel holes in dead wood and extract insect larvae. Aye-ayes also use their middle fingers as drumsticks to knock on wood; the returning echoes serve, via echolocation or a cutaneous sense, to locate larvae hidden inside the wood. Only one other animal is known to share these strange adaptations: the New Guinean marsupial *Dactylopsila palpator* – one of the Petauridae, or gliders – mirrors the aye-aye's use of an empty ecological niche occupied elsewhere by birds (the woodpeckers).

Aye-ayes build nests from twigs and leaves high up in the forest canopy where they spend the day. Each animal uses several nests, and they swap these in such a way that the same treehouse will typically have different occupants on successive nights. Their undeveloped, small, single infants are born after about 170 days of gestation, and spend the first two months of their lives in a nest. They take at least seven months to reach weaning age. As in other lemurs, adult males and females do not differ in size.

Rangers of the Night

SOCIAL BEHAVIOR

Aye-ayes were once thought to be restricted in their distribution to a few lowland rain forests along Madagascar's east coast. Sightings are difficult, because these nocturnal animals live in low-density populations. However, intense recent censusing has revealed their presence in other rain forests along the east coast, in several localities in humid northwestern forests, and even in deciduous forests of the west. They are not confined to

FACTFIL

AYE-AYE

Daubentonia madagascariensis

Order: Primates

Family: Daubentoniidae

Sole member of genus

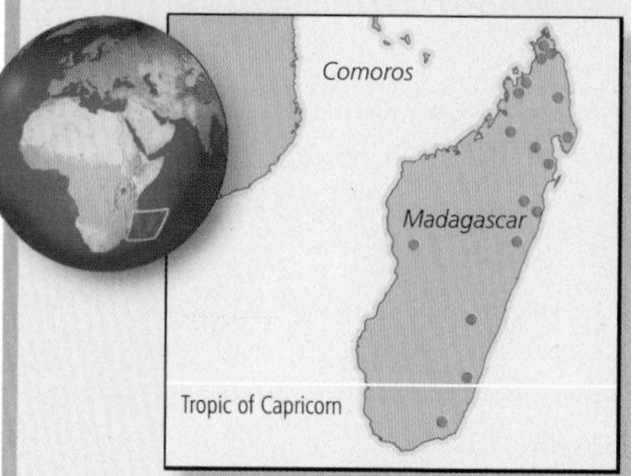

DISTRIBUTION E and NW Madagascar. One isolated population in W of the island.

HABITAT Humid and rain forest; W population in dry forest.

SIZE Head–body length 30–40cm (12–16in); **tail length** 45–55cm (18–22in); **weight** 2.4–2.8kg (5–6lb). No sexual dimorphism.

COAT Long, shaggy, dark gray-brown fur, with long guard hair with white tips. Face and underparts creamy or grayish. Eyes surrounded by dark rings. Big, leathery ears. Tail long and very bushy. All digits elongated. A quadrupedal climber and leaper, with legs slightly longer than its arms.

DIET Excavated insect larvae and inner parts of nuts; coconuts, mangoes, and other planted fruit are also taken where available.

BREEDING Sexual maturity after 2–3 years. No fixed breeding season. Single infant is "parked" in a nest during the first months of its life.

LONGEVITY Not known

CONSERVATION STATUS Endangered. Once thought to be among the most threatened of all mammals, but intensified research has revealed its presence over a large area. Lives at very low densities and is threatened by habitat destruction and hunting. One larger sister species went extinct in recent centuries.

Above *Aye-ayes do not have opposable thumbs; ›wever, the first toe – the only one to have a flat ther than clawlike nail – is opposable. They can hang om a branch by their hind feet, leaving their hands ee for feeding.*

imary forest, and are also found in mangroves, iickets, and cultivated areas, such as plantations. ye-ayes prefer to move in trees, but will come to ie ground to traverse forest gaps.

Gnaw marks made by their powerful incisors hen opening nuts or fibrous fruits are a reliable idirect sign of their presence. Most of their noc-urnal activity is devoted to traveling and foraging, id they may cover several kilometers per night.

Detailed observations of the social behavior id organization of aye-ayes have only been made on a population introduced to the island of Nosy Mangabe, where the animals live at higher densities than on mainland Madagascar. Existing studies have revealed that aye-ayes feed alone by night and sleep alone by day. Individual animals may converge at rich feeding sites, but females are invariably intolerant of each other, using urine, scent marks, and vocalization to advertise their presence within their large home ranges, which may extend to over 30ha (75 acres). Males occupy overlapping home ranges that are up to four times larger than those of the female. Females may become receptive only once every two or three years, and at any time of the year, which is exceptional among lemurs. They advertise their estrus with loud vocalizations, and will mate with several of the males that are attracted by their cries.

A Global Conservation Priority

CONSERVATION AND ENVIRONMENT

As the only surviving representative of an entire and unusual primate family, the aye-aye has been assigned one of the highest global conservation priorities. This rating is due to its taxonomic status and uniqueness. However, actual population numbers are higher than those of several other lemur species, primarily because of the aye-ayes' relatively wide geographical distribution. Where they still occur, large areas need to be preserved to maintain viable populations because the relatively large home range results in a low population density. The survival of aye-ayes is jeopardized by habitat loss, as well as from killing by superstitious people and farmers whose plantations they have raided. PK

Bush Babies, Lorises, and Pottos

BUSH BABIES – ALSO KNOWN AS GALAGOS – lorises, and pottos are nocturnal. They resemble the lemurs, with whom they share a common ancestor, but none confines its activity to the daylight hours, for this niche is occupied on mainland Africa and in Asia, where they live, by monkeys and apes.

Besides being active by night, all members of this group are relatively small tree-dwellers that feed alone. While they retain features similar to those of the earliest primates, they cannot be said to be "primitive" as they possess many characteristics that are highly specialized and evolved.

Leaping Galagos and Creeping Lorises

FORM AND FUNCTION

All Loroidea species' digits have nails except for the second digit of each foot, which has a "toilet" claw used for grooming. Unlike the haplorhine primates (monkeys, apes, and tarsiers), loriforms do not show true opposability and only possess whole-hand grasping. This means that while they can grasp, they cannot touch their thumb to each of the other same-hand fingers separately. True opposability is important for tasks involving fine manipulation, such as food handling.

Their dental formula (I2/2; C1/1; P3/3; M3/3 = 36) is similar to that of some of the earliest strepsirhine primates (a category that combines bush babies, lorises, and pottos with lemurs), with the lower incisors and incisor-like canines procumbent (projecting forward), forming a toothcomb-like arrangement that plays an important role in grooming, and feeding off insects or gums. On the underside of the tongue is a cartilaginous brush (the sublingua), the points of which fit between the teeth of the toothcomb and serve to clean out particles.

Trying to tell the sexes apart can be difficult, as the females possess a large, well-developed clitoris, which can be mistaken for a penis. It is particularly easy to confuse subadult males as they may not have an obvious scrotum. In all loriforms the penis has an elongated penis bone (baculum); from the skin cells of the glans emerge backward-pointing spines, hardened with keratin and less than 2mm (.08in) in length. These form part of a "lock and key" system that holds copulating couples together longer, thus increasing the likelihood of a successful mating. The bacula and spines of galagos are sufficiently varied to distinguish the species; they may also help prevent cross-species matings. Among the lorisines, the spines are large in the angwantibo, but reduced to keratinized plates or papillae in pottos and the Slow, Slender, and Pygmy lorises. In primates these spines are generally related to mating systems, being longer and more complex in species with non-gregarious mating systems and less so in those with monogamous and polygynous systems. In some galago species "mating chases" occur, where several males pursue a female in estrus that may mate with more than one male. The spines may provide a clear selective advantage for the first male, prolonging copulation and so giving a head start to his sperm. In females the labia majora can be large, resembling a scrotum, and in the middle of the estrus cycle the vaginal epithelium becomes thickened and hardened.

In locomotion and general body form, the loriforms fall into two groups: the leaping galagos and the slow-moving, creeping lorises and pottos. All four limbs of lorises and pottos are of approximately equal length, and they have extremely short or vestigial tails. They move on all fours, climbing or walking along branches, crossing from tree to tree by stretching between terminal branches. The angwantibo, which specializes in living in areas of forest where there are fallen trees, may be found walking along the forest floor, but otherwise lorises and pottos never leave the tree tops. For security in the lofty heights, they possess an extremely powerful grip that can be maintained for at least a full day. This is made possible by the presence of a specialized arrangement of blood vessels in the wrists and ankles, known as the *rete mirabile*, which keeps the necessary muscles supplied with oxygen and nutrients and removes metabolic products (e.g. lactic acid) that may otherwise cause muscle cramps and damage.

In primates a clear relationship exists between modes of locomotion and the relative lengths of the fore- and hind limbs. The majority of primates are quadrupedal, with limbs of approximately equal length and this is the case with the lorisines such as the potto and the Slender loris. In contrast, the galagos are specialist leapers, with hind limbs longer than their forelimbs. They also have bushy tails 1.2–1.8 times longer than their bodies. Very long back legs propel the Senegal galago on 5m (16ft) jumps, while a tail up to 30.3cm (1ft) long acts as a stabilizer. It is common for members of this group to "urine wash" (urinating on their hands and feet); while many theories have been advanced to account for this behavior, such as thermoregulation and scentmarking, the most credible explanation is that it improves grip.

Like the lemurs, the loriforms have retained a strong sense of smell from their early mammal

FACTFIL

BUSH BABIES, LORISES, & POTTOS

Order: Primates

Superfamily: Loroidea

Families: Galagonidae (bush babies or galagos – 17 (11) species in 3 (4) genera) and Loridae (lorises and pottos – 7 (6) species in 5 (4) genera).

DISTRIBUTION Warm areas of Africa, S India, Sri Lanka, and SE Asia.

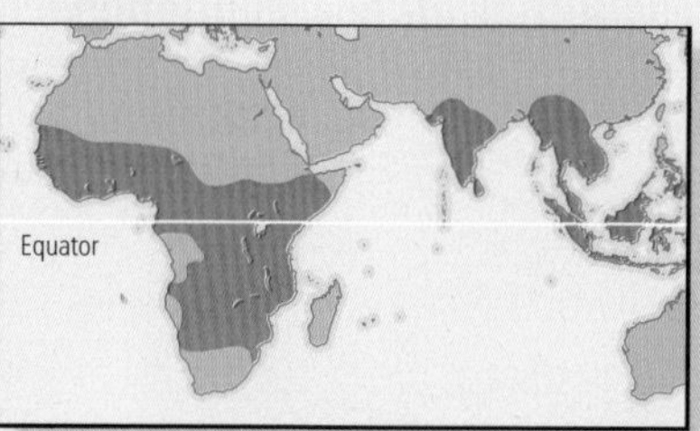

HABITAT A variety of arboreal habitats, from savanna woodland, thickets, and dry and coastal forests to rain forest, including plantations and mixed groves of tree crops.

SIZE Head–body length from 10.7cm (4.2in) in the Rondo galago to 30.7cm (12.1in) in the Thick-tailed galago; **tail length** from 18.4cm (7.2in) to 42cm (16.5in); **weight** from 60g (2oz) to 1130kg (40oz).

FORM Coat thickly furred, in various shades of gray or reddish brown. Galagos are more slimly built than lorises and pottos, with longer arms and legs and lengthy, almost bushy tails.

DIET Fruits, gums, nectar, insects, eggs, and various small prey, including birds, bats, and rodents. Some bush babies subsist largely on gums, while the angwantibo preys chiefly on caterpillars.

BREEDING Gestation ranges from 111 days (Demidoff's galago) to 197 days (potto).

LONGEVITY Up to 26 years (potto).

See species table ▷

Right *Bush babies moving about in trees; they a[re] agile leapers with long hind limbs and bushy tails tha[t] they use for balance when jumping:* **1** *Thick-tailed galago* (Otolemur crassicaudatus); **2** *Demidoff's gala[go]* (Galagoides demidoff); **3** *The needle-clawed bush babies (here,* Euoticus elegantulus) *have needle poin[ts] that help to grip on trees on the nails of all digits except the thumb and the big toe.*

Lorises and pottos are slow-moving climbers with a strong grip, opposable first digit, and no tail. **4** *The Slender loris* (Loris tardigradus) *has a particularly mobile hip joint for climbing:* **5** *Slow loris* (Nycticebu[s] coucang); **6** *Potto* (Perodicticus potto); **7** *Angwantib[o]* (Arctocebus calabarensis).

1
4
5
6
7
3
2

ancestors. To aid in this, they have an elongated muzzle with an area of moist, glandular, naked skin (the rhinarium) around the nostrils. Within the nasal cavity are scroll-shaped bones (the ethmoturbinals), covered with nasal epithelium, that are better developed and more numerous than in haplorhines. Adding to the profile of animals whose sensory life is dominated by chemical sense is the presence of a vomeronasal or Jacobsen's chemosensory organ. This lies between the nose and mouth cavities, and is linked to the rhinarium by a groove passing along the the palate.

Large eyes help loriforms to see at night. They have a crystalline layer at the back of their eyes between the retina and choroid layer that reflects light back through the retina, increasing the stimulation of its photoreceptors and allowing vision at low light levels. In contrast to the majority of diurnal primates, which recognize ripening fruit by color, these nocturnal forms have little or no color vision due to the very low proportion of rod to cone photoreceptors in the retina. While the pottos and lorises have relatively small pinnae (external ears), those of galagos are large and can move independently of one another; they can also be folded flat to avoid damage. Their sense of touch is centered in the fleshy pads on their hands and feet, which contain touch-sensitive Meissner's corpuscles. These tiny receptors, found on fingerprint ridges and in the clusters of vibrissae on the head, ankles, and wrists, are unique to primates.

Above *Although, as a rule, Slow lorises proceed along branches at a relatively sedate pace, they can strike with great speed when hunting, lunging forw[ard] to seize their prey with both hands.*

Left *The batlike ears of the Senegal, or Lesser, galago* (Galago senegalensis) *enable it to track the movements of insect prey in the dark. It is also capa[ble] of snatching insects from the air as they fly by.*

Hunting Insects by Night

DIET

All loriforms eat some animal protein, whether in the form of invertebrate or vertebrate prey. Galagos often detect insects by sound before locating them visually. They are capable of catching insects on the wing by hand, snatching them while grasping a branch with their feet. The smallest galagos tend to eat proportionately more animal protein (about 70 percent in the case of Demidoff's galago and 50 percent in Garnett's galago). The energy value of invertebrate food is high, and most species will select invertebrates over other food types if given a choice. However, larger species of over 350g (12oz) adult body weight find it difficult to gather sufficient invertebrate prey. As a result, they supplement their diet with fruit, which can be found in large quantities. Eating vertebrate prey is generally rare, though some species have been known to take small birds and reptiles.

The loriforms have two idiosyncrasies in their feeding habits: they consume unpalatable or poisonous invertebrates, and they eat gum. They detect slow-moving or stationary insects by scent, swallowing even poisonous millipedes and caterpillars with irritating or poisonous hairs. The angwantibo deals with hairy caterpillars by holding the head in its mouth and rubbing the hairs off with its hands. Slow lorises and pottos have a slow metabolism, with a basal metabolic rate about 40 percent lower than might be predicted for their size; this gives them time to neutralize undesirable or toxic chemicals in the gut before they are absorbed and take effect. Perhaps the most specialized gum-eater is the Southern needle-clawed galago, whose diet consists of approximately 75 percent gum, harvested from lianas and trees. The Mohol and Senegal galagos also eat considerable amounts of gum (up to 50 percent of their diet). Gums contain long-chain sugars that can only be digested through bacterial fermentation in the gut. Loriforms that eat gum have an enlarged cecum (a bag joining the large intestine which houses fermenting symbiotic bacteria). In the Southern needle-clawed galago, the cecum is five times larger than might be expected for its body size.

Dispersed Communities of Foragers

SOCIAL BEHAVIOR

Loriforms are solitary for considerable periods during the night, but while foraging they frequently meet conspecifics with whom their home ranges overlap. In such cases the two animals sometimes touch, engaging in mutual grooming or else they may send postural signals to one another. Mostly, though, they communicate through smells and sounds. Both galagos and lorisines have a variety of glandular areas of skin depending on species, these are under the chin, on the inside of the arm near the elbow, on the chest, or around the genitals. Exudates from th[e] glands, together with scents in excreta, are used [to] mark territories, self-anoint, and scentmark mem[]bers of the opposite sex. In lorises and pottos, vocalizations are mainly used for mother–infant interactions and to signal alarm and aggression.

In galagos, which rely more heavily on vocal communication, each species has a repertoire of ten or more loud calls. Many are complex, other[s] simple and repetitive. These play a part in attrac[t]ing mates, repelling rivals, and indicating alarm. The calls vary in frequency and rhythm, and ma[y] be combined or graded to reflect the mood of th[e] caller. Certain galago calls have been shown to b[e] species-specific, and analyzing them has proved [a] valuable tool for exploring species diversity and discovering new species.

While most diurnal primates are gregarious, t[he] nocturnal galagos, lorises, and pottos can broadl[y] be described as solitary foragers living in socially complex, but dispersed communities. Individua[ls] of many species, including the potto, sleep alone during the day. In other species, including many agos, males sleep alone, whereas females with ung sleep in groups of up to ten; in a few ecies (for example the Zanzibar galago), sleeping oups may be of mixed sex. Sleeping sites, of ich there may be several in an individual's nge, vary from branches sheltered by dense, tan-ed vegetation, and flat leaf-nests to the complex herical leaf-nest made by Demidoff's galago.

Most galago species live in polygynous social stems with home ranges that may overlap when suits the occupants. Typically, males have larger me ranges, with dominant males and adult males excluding others of their kind. A single minant male will overlap the ranges of several ature females and their subadult female off-ring (formed into a matriarchy). Non-breeding, badult males may be tolerated within the terri-ries of dominant males. At times of mating, the stem of territories may break down. Mating, and erefore birthing, occurs year-round in some ecies, while others show a seasonal pattern of ce or twice a year, depending on climate; for stance, Slender lorises mate from April to May d from October to November.

Loriforms are long-lived for their size; Allen's ago can reach 12 years and pottos have lived to e age of 26. They also have lengthy gestation riods, ranging from 111 days in Demidoff's ago to 191 days for the potto. Most mothers oduce singletons or sometimes twins, although e Thick-tailed galago may have triplets. Usually ere is just one litter per year, but some species ve two. To move the young between nests or to ark" them while feeding, mothers carry them in eir mouths by their flank fur; alternatively, the ung may clasp onto the mother. Weaning age in agos varies from 53 days to 140, but is longer lorisines: from 115 days in the angwantibo to 6 onths in the Slow loris and potto. Female off-ring typically stay within their mother's territory til they become reproductively active and estab-h a range of their own. At puberty males leave e natal area to establish their own territories.

Sharing the Fate of the Forests

CONSERVATION AND ENVIRONMENT

me galago species, such as the Mohol, Senegal, d Thick-tailed galagos, are widespread and in serious danger of either local or general extinc-n. Others, as well as all lorises and pottos, have ore restricted distributions, and of these the ost endangered species are those restricted to pical forest habitats that are themselves under reat. These include several putative new species ot yet tallied in most lists), such as the Matun-, Rondo, and Mountain galagos, which are lim-d to a few coastal, montane, or lowland forests East Africa, and the two angwantibo species, ose equatorial rainforest habitat in West Africa at risk. In addition, loriforms have been trapped limited numbers for research and the food or t trade.

The future of the South and East Asian lorises also looks bleak, as their tropical forest habitat continues to be destroyed. As with all other primates, the loriforms are listed in CITES Appendix II. Among the African species, the Allen's, Somali, Zanzibar, and Northern and Southern needle-clawed galagos are (along with those mentioned above) deemed at greatest risk (as is the putative new species, Grant's galago). Securing their future is hampered by the fact that the true extent of species diversity in the galagos and the lorisines (especially the Southeast Asian forms) still remains unclear and needs to be clarified before effective conservation measures can be implemented. PH

Above *Generally, the nocturnal pottos – like this juvenile from the Ituri Rainforest Reserve in the Democratic Republic of Congo – stay high up in the tree canopy, rarely descending below 10m (33ft). Their diet is mainly fruit-based, although insects and some small vertebrates, such as birds, are also eaten. Young male pottos will leave their mother's home range when they are about six months old. Bony processes project from vertebrae between the potto's shoulder blades to help form a "shield" on the nape and neck.*

Bush Baby, Loris, and Potto Species

THE LAST YEARS OF THE 20TH CENTURY saw a major revision of bush baby taxonomy. As recently as the 1980s, only 6 galago species were described, but now the established view is that there are 11 species of Galagonidae (17 if "upgraded" subspecies – marked † below – or newly discovered species – marked †† – are recognized).

Diversity within the pottos and the lorises (family Loridae) may have been similarly underestimated; though no-one yet knows for sure, there may be as many as six species of potto and twelve of loris.

The reappraisal has been prompted by new genetic evidence, but also by computerized analysis of galago vocalizations. It is easy to see how valuable this latter approach is in the study of small-bodied, forest-living, nocturnal species that can be hard to see and that rely primarily on vocal communication. Vocalization studies have confirmed what is already clear to the ear – that there are significant differences in certain call types between different species – and have already helped unearth at least one new species, the Rondo galago. The technique is now being used to establish the exact relationship between the Matundu and Zanzibar galagos (particularly the Kenyan coastal form of the latter), and to determine the number of separate species that may be "hidden" within what is currently listed as Allen's galago.

BUSH BABIES OR GALAGOS

Family Galagonidae

17 species, all in Africa. All have very large eyes adapted to nocturnal vision; long tail with tuft; large, membranous ears that can be folded up; long hind limbs adapted to leaping. Most construct nests.

Southern needle-clawed galago LR
Euoticus elegantulus
Southern needle-clawed or Elegant galago

W and C Africa. Sleeps in forest branches, sheltered by dense vegetation; with *Euoticus pallidus*, among the only galagos not to make nests or take refuge in holes in trees.
HBL 30cm; TL 29cm; WT 300g.
FORM: coat an attractive reddish color, with darker line down back, ash-gray underside and flanks; tail gray, always with white tip; nails elongated to form line point; branches etc. may be held by pad on last joint of digits as in other galagos, or by digging in nails; heavier muzzle than other galagos, which, with large, golden eyes, give it a striking appearance.
DIET: 75 percent gums, with corresponding adaptations – second premolar in form of canine, elongated dental comb, longer intestine.
LONGEVITY: to 15 years in captivity.
CONSERVATION STATUS: LR Near Threatened.

Northern needle-clawed galago LR
Euoticus pallidus
Northern needle-clawed or Pale galago

W and C Africa. Details as for *Eoticus elegantulus*, though with a more northerly distribution.
CONSERVATION STATUS: LR Near Threatened.

Allen's galago LR
Galago alleni

W and C Africa. Inhabits lower storey of rain forest, leaps from trunk to trunk and between bases of lianas; sleeps in hollows.
HBL 20cm; TL 25cm; WT 260g.
COAT: quite thick, smoky-gray, with reddish flanks, thighs, and arms; underside pale gray.
DIET: fruits, small prey.
BREEDING: mating and birth are not seasonal; one, usually single, birth a year; gestation 133 days; newborn weighs 24g.
LONGEVITY: to 12 years in captivity.
CONSERVATION STATUS: LR Near Threatened.

Gabon galago †
Galago gabonensis

W and C Africa. Previously considered subspecies of *Galago alleni,* from which it is distinguished by tail coloration, which may range from dark to silver-gray, with or without a 1–6cm white tip.

Somali galago
Galago gallarum

NE African woodland. Formerly considered subspecies of *Galago senegalensis*.

Eastern needle-clawed galago
Galago matschiei
Eastern needle-clawed or Matschie's galago

E African forests, in Great Rift Valley in Rwanda, Burundi, E Uganda.
HBL 16cm; TL 23cm; WT 250g.
COAT: very dark.

Mohol galago
Galago moholi
Mohol or Southern lesser galago

Arid woodland or forest in southern Africa from Angola to Tanzania. Sleeps in tree hollows. Formerly considered a subspecies of *Galago senegalensis*.

Senegal galago
Galago senegalensis
Senegal or Lesser bush baby

Arid woodland from Senegal to Kenya; sleeps in tree holes.
HBL 16cm; TL 23cm; WT 250g.
FORM: coat paler gray than Allen's galago, with longer ears; long back legs are used to make jumps of up to 5m.
DIET: very varied, including small prey, Acacia gum, fruits, nectar; gums provide the basic food in dry periods.
BREEDING: reproduces twice a year; often twins, exceptionally triplets; gestation 123 days; newborn weighs 12g.
LONGEVITY: to 14 years in captivity.

Demidoff's galago
Galagoides demidoff
Demidoff's or Dwarf bush baby or galago

W and C Africa: Gabon, Central African Republic, Uganda, W Tanzania, Burundi, DRC, Congo; Senegal, S Mali, Upper Volta, SW Nigeria, Dahomey to Senegal; very small area on coast of Kenya. In tropical rain forest, in thick foliage and lianas, crowns of tallest trees, beside tracks, in former plantations; makes complex, spherical leaf nests for sleeping.
HBL 12cm; TL 17cm; WT 60g; smallest lorisid and, with *Microcebus murinus* of Madagascar, the smallest of all primates.
COAT: gray-black to reddish, depending on the individual and age (darker in young animals).
DIET: small insects, fruits, and gums.
BREEDING: in Gabon, usually a single young per year (1 in 5 births twins); shortest gestation of any galago species, at 111 days; newborn weighs 7–12g; infants weaned at 53 days, reaching adult size at 2–3 months and puberty at 8–9 months.
LONGEVITY: unknown in wild; to 12 or more years in captivity.

Grant's galago †
Galagoides granti

SE Africa. Sleeps in tree holes. Once considered a subspecies of *Galago senegalensis*.

Right *A Thick-tailed galago. Th[e] loud croaking noise made by this species is said to resemble the soun[d] of a child crying.*

Thomas's galago †
Galagoides thomasi

W and C Africa. Formerly considered s[ub]species of *Galagoides demidoff*. Distinguished by its long, bushy tail, up to 1 times head–body length.

Mountain galago †
Galagoides orinus

East African forests. Formerly consider[ed] subspecies of *Galagoides demidoff*.

Rondo galago ††
Galagoides rondoensis

E African forests. Discovered by comp[uter] analysis of galago vocalizations.
HBL 10.7cm; TL 18.4cm; WT 60g, mak[ing] this the smallest galago.

Matundu galago ††
Galagoides udzungwensis

E African forests.

Zanzibar galago
Galagoides zanzibaricus

E African forests. Sleeps in tree hollows. Formerly considered subspecies of *Gal[ago] senegalensis*.
BREEDING: births seasonal, from Febru[ary] –March and August–October.

Thick-tailed galago
Otolemur crassicaudatus

Southern Africa, from Angola to Tanzar[nia]. Dense dry and gallery forests; sleeps in leaf-nests.
HBL 30.7cm; TL 42 cm; WT 1,130g, ma[k]ing it the largest of the galagos and the least adept at leaping.
COAT: gray.
DIET: similar to *Galago senegalensis*, bu[t] prey items often bigger (birds, eggs, sm[all] mammals, reptiles).

Garnett's galago
Otolemur garnettii

SE Africa. Formerly considered subspec[ies] of *Otolemur crassicaudatus*.
BREEDING: infants weaned at 140 days.

ABBREVIATIONS

HBL = head–body length TL = tail length WT = weight
Approximate nonmetric equivalents: 2.5cm = 1in; 230g = 8oz.
† = 'new' species. See p291

Ex Extinct · EW Extinct in the Wild · Cr Critically Endangered · En Endangered · Vu Vulnerable · LR Lower Risk

LORISES AND POTTOS

mily Loridae

very short, limbs more equal in length n in bush babies, adapted for climbing. nds and feet developed into pincers, h opposable thumb and much reduced ex finger. Does not construct nests.

lden potto LR
tocebus aureus

stricted to equatorial forests of meroon, Nigeria, Gabon, and Congo. merly considered a subspecies of *Arcto-us calabarensis*, but smaller and more nder. WT 210g.
NSERVATION STATUS: LR Near Threat-d.

gwantibo LR
tocebus calabarensis

stricted to equatorial forests of meroon, Nigeria, Gabon, and Congo, ng in areas of forest where there are fall-trees; in Gabon, in wet, low forest rich ianas, also in former scrubby plantations; only species of the family that does not live exclusively in the tree canopy.
HBL 24cm; TL 1cm; WT 200–500g.
FORM: slender and light compared to the heavier, compact potto. Coat light reddish.
DIET: chiefly caterpillars.
BREEDING: mating and birth occur year-round; gestation 135 days; one young born weighing 24g with white-spotted coat; mother mates a few days after birth, so weaning occurs a few days before the next birth, at about 115 days.
CONSERVATION STATUS: LR Near Threatened.

Slender loris Vu
Loris tardigradus

Restricted to forests in S India and Sri Lanka.
HBL 24cm; no tail; WT 300g.
FORM: coat gray or reddish; build similar to angwantibo, but larger; eyes surrounded by two black spots separated by narrow white line down to nose.
BREEDING: mates seasonally, in April–May and October–November.

Slow loris
Nycticebus coucang

More widespread than Slender and Pygmy loris, in forests from E India to Vietnam, Malay peninsula, W Indonesia, and the Philippines.
HBL 30cm; TL 5cm; WT 1.2kg.
FORM: anatomy and lifestyle similar to potto, but coat color more varied; ash-gray, darker dorsal line divides on head into two branches that surround eyes.
BREEDING: one infant, 45g; weaning occurs at 6 months.

Pygmy loris Vu
Nycticebus pygmaeus

Restricted to forests in Laos, Cambodia, Vietnam, and S China.

Potto
Perodicticus potto

In tropical forests of W African coast from Guinea to Congo and from Gabon to W Kenya.
HBL 32cm; TL 5cm; WT 1.1kg.
FORM: large, muscular, compact; coat reddish-brown to blackish; ears often yellowish within; bony processes project from vertebrae between shoulder blades to help form "shield" on nape and back.
DIET: chiefly fruits, some gums; small, often irritant prey items (birds, bats, rodents) eaten whole.
BREEDING: gestation, at 197 days, the longest of any of the two families; a single young, born dark-colored in Gabon, speckled with white in Côte d'Ivoire, weighing 50g; weaning occurs at 6 months; young attain puberty at about 1 year.
LONGEVITY: up to 26 years in captivity.

Martin's false potto ††
Pseudopotto martini

Restricted to equatorial forests of Cameroon, Nigeria, Gabon, and Congo.

DISTRIBUTION Chiefly within the tropics, in S and C America, Africa, and Asia. Mostly forest-dwellers; some species in grasslands.

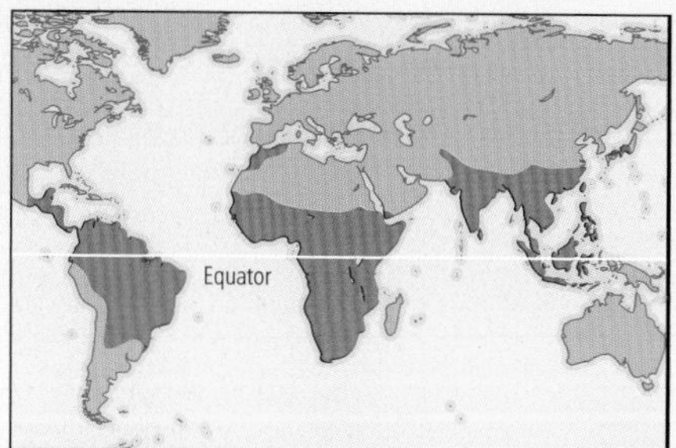

4 families: Cercopithecidae, Cebidae, Callitrichidae, Tarsiidae
178 species in 36 genera

TARSIERS Family Tarsiidae p332
5 species in the genus *Tarsius*

NEW WORLD MONKEYS

Marmosets and tamarins (family Callitrichidae)
38 (26) species in 6 (4) genera p334
Capuchin-like monkeys (family Cebidae)
47 (58) species in 11 genera p344

OLD WORLD MONKEYS Family Cercopithecidae

Baboons, drills, mandrills, mangabeys, guenons, and macaques (subfamily Cercopithecinae)
45 (47) species in 11 genera p356
Colobus and leaf monkeys (subfamily Colobinae)
42 (34) species in 7 genera p380

Note The established 1993 classification recognized 170 species and 34 genera (in parentheses).

Monkeys and Tarsier[s]

RANGING IN SIZE FROM THE PYGMY MARMOSET (120g/4oz) to the mandrill (32kg/70lb), monkeys are medium-sized mammals that eat mainly fruit and foliage and, for the most part, live in diurnal social groups. Traits that distinguish them from lower primates include the structure of their skulls and placenta, and the relatively large size of their brains; they also show an increased reliance on sight over sound and smell, and greater flexibility in their mode of locomotion. They now divide into two geographically separate lineages – the New World monkeys, sometimes known as Platyrrhines, and the Old World monkeys, or Catarrhines (see opposite).

In turn, the New World monkeys split into two families: the marmosets and tamarins (the Callitrichidae) and the capuchin-like or cebid monkeys (the Cebidae). Marmosets and tamarins are all small (typically about 350g/12oz) and live principally in small groups. All 38 species are diurnal and arboreal.

The cebid monkeys are more diverse, comprising six main groups. These are the howler monkeys, relatively large at 5–7kg (11–15lb); night monkeys, active by night; titi monkeys, and the sakis and uakaris, which are all broadly similar in social organization, diet, and body size (around 1–3kg/2–7lb), but occupy a diurnal niche; cap[u]chin and squirrel monkeys; and spider and wo[olly] monkeys. These last two types live in mixed-se[x] groups that can exceed 50 individuals.

The Old World monkeys are probably the m[ost] recent of major primate groups. Although they belong to a single family, the Cercopithecidae, comprises the two largest primate subfamilies: the Colobinae (42 species), which are mostly tree-dwellers living in small groups, and the Cercopithecinae (46 species). Sometimes refer[red] to as "typical monkeys," the cercopithecines liv[e] in societies that generally consist of families of philopatric females spending their lives close to the area where they were born, together with th[eir] offspring and also one or more attached adult males. With the exception of the Asian macaqu[es], most of the cercopithecine species come from sub-Saharan Africa; in contrast, the colobines a[re] predominantly Asian, with just nine African species among them.

Tarsiers were formerly grouped alongside the lemurs, bush babies, lorises, and pottos in the suborder Prosimii. However, more modern clas[si]fications tend to assign the tarsiers and the anth[ro]poids (monkeys and apes) together to the suborder Haplorhini GC/TC-

SKULLS AND DENTITION

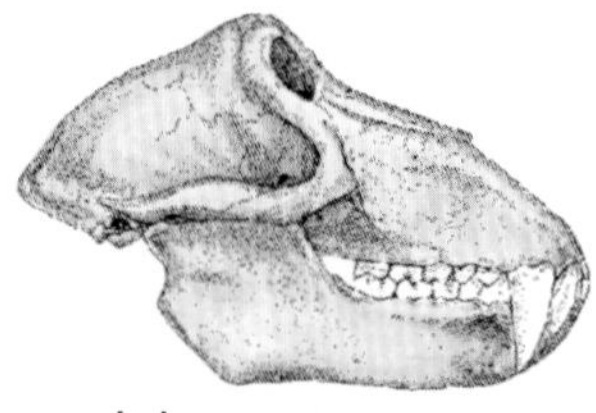
Savanna baboon
19.6 cm

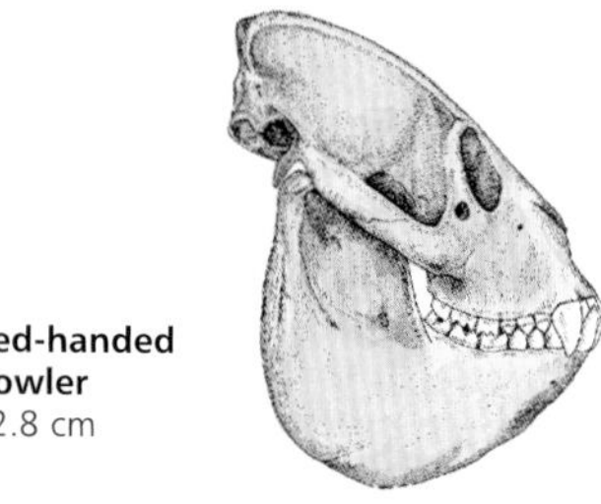
Red-handed howler
12.8 cm

White-footed tamarin
4.5 cm

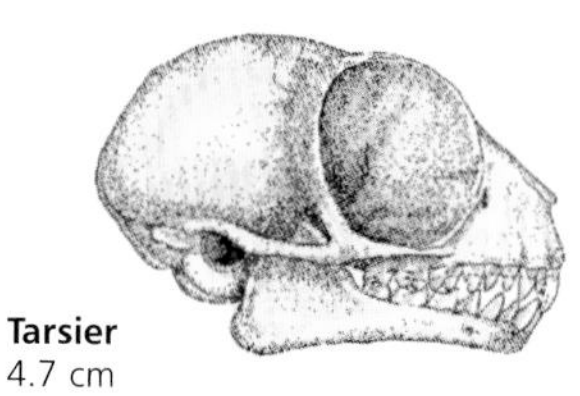
Tarsier
4.7 cm

Monkeys tend to have large brains, so their skulls are characterized by a large, globular braincase. All monkeys and tarsiers have eyes directed forward for binocular vision and contained in bony sockets produced by a virtually complete plate behind the orbit. The frontal bones of the forehead become fused together early in life. The two halves of the typically deep lower jaw are fused together at the midline.

Monkeys have spatulate (shovel-shaped) incisors, conspicuous canines, and squared-off molar teeth, which typically have four cusps. In marmosets, the lower incisors are as tall as the canines. Among the New World monkeys, the cebid monkeys all have a formula of I2/2, C1/1, P3/3, M3/3 = 36, while both marmosets and tamarins have I2/2, C1/1, P3/3, M2/2 = 32, and are the only primates to have reduced the dental formula by loss of molar teeth. As in other respects, Goeldi's monkey is intermediate, in that it has the full dental formula of the cebid monkeys but also has tiny third molars in both upper and lower jaws. In Old World monkeys (as in apes and man) there is a single dental formula that has been attained by losing premolars: I2/2, C1/1, P2/2, M3/3 = 32. The molars of Old World monkeys are distinctive in having four cusps joined in pairs by transverse ridges (bilophodonty).

The dental formula of tarsiers is I2/1, C1/1, P3/3, M3/3 = 34. The large upper incisors are pointed and the upper canines are relatively small. Corroborating the classification of tarsiers with the haplorhines, this dentition varies markedly from the overall strepsirhine pattern, which is characterized by a toothcomb, two rather than four incisors on the lower jaw, and a spac[e] or medial gap, between the front inc[i]sors of the upper jaw. Tarsiers' cheek teeth are adapted to deal with insect[s], their main prey.

Sexual dimorphism, often marked [in] monkeys, is less pronounced in New World species, among which howler monkeys are the most extreme case: their proclivity for leafeating is reflect[ed] in the depth of the lower jaw. In the [Old] World baboons, the male can be nea[rly] twice as heavy as the female, and the skull is accordingly much larger.

*Right Members of three families of monkeys: **1** Geoffroy's tamarin* (Saguinus geoffroyi); ***2** Humboldt's woolly monkey* (Lagothrix lagotricha); ***3** Mandrill* (Mandrillus (Papio) sphinx), ***4** Patas monkey* (Erythrocebus patas) *this final species is the fastest primate of all, reaching speeds of up to 55km (35mph); its long legs and short digits are adaptations for speed.*

TELLING OLD FROM NEW

Old and New World monkeys differ greatly in appearance – a distinction recognized in the names "catarrhine" and "platyrrhine." The terms derive from the shape of the nose – a reliable way of distinguishing between the two groups. The platyrrhines **a**, or New World monkeys, have nostrils that are wide open and far apart, while in the catarrhine (Old World) species **b** they are narrow and close together. Other differences between the two populations include the evolution of a prehensile tail in the larger New World monkeys but not in their Old World counterparts, and the development of ischial callosities – hard pads on the lower side of the buttocks for sitting – in Old but not New World monkeys.

FOOD AND FEEDING

Among the New World monkeys, marmosets and tamarins are tree-dwellers that forage on fruit, insects, and gums. The cebids have mixed tastes. The howler monkeys are leafeaters, but all the rest subsist mostly on fruit, supplemented by leaves and insects; some species will even extend their diet to eggs and small prey such as frogs, snails and lizards. Sakis and uarakis possess special teeth for processing hard-cased seeds and fruit.

The Old World species have even more diverse eating habits. The cercopithecines are omnivorous though predominantly fruit-eating, while the colobines are specialized leaf-eaters, though they will vary their diet with fruit and flowers when available. They are the primate equivalent of ruminants, equipped with large forestomachs containing the bacteria that digest the cellulose in the leaves.

Tarsiers (LEFT) have a unique diet among primates, consisting solely of animal protein (small vertebrates, insects, and even poisonous snakes). Like bush babies and many lemurs, they hunt this prey first by listening with independently moving ears, and then making a sudden lunge that involves finely-tuned hand-to-eye coordination.

Tarsiers

TARSIERS ARE EXTRAORDINARY ANIMALS IN *many respects. Relative to their body weight, they have the biggest eyes of any mammal; thus, each eye of the Western tarsier is around 16mm (0.6in) in diameter and, at 3g/0.1oz, weighs slightly more than its brain! Tarsiers also give birth to the largest babies relative to the parent's body weight.*

On Sulawesi, tarsiers are abundant in some localities, but all are vulnerable. In Malaysia and Indonesia tarsiers are protected by law. Some people have tried to keep the animals as pets, but in the absence of appropriate live food, they usually die within days. A much greater threat is escalating habitat loss.

Champion Leapers

FORM AND FUNCTION

All tarsier species are of a similar size, except for the Pygmy tarsier, which, judging from museum material, is thought to weigh less than 100g (3.5oz). Tarsiers are buff-gray or ocher, sometimes beige or sand-colored. The coat is softer than velvet. These vertical-clingers-and-leapers are able to jump more than 40 times their own length. In its entirety, the hind limb is about twice as long as the head-and-body length, and the thigh, lower leg, and foot are each about equal in length. As early as 4 months old (only 6 weeks after weaning), the feet reach adult size. The three species from Sulawesi all sport long, scaly tails. In the other two species, only the distribution of hairs indicates how such scales were once arranged. The tufts of hair on the tails of Sulawesi tarsiers (except for one sparsely-tufted specimen from Selayar) are about 13–15cm (5.1–5.9in) long and bushy; those of the Philippine and the Western tarsiers measure less than 10cm (3.9in). The second and third toes of tarsiers are equipped with a so-called "toilet claw" for grooming.

The tarsier's fingers are long and slender. In *Tarsius spectrum* and *T. dianae*, the third finger is only about 15 percent shorter than the humerus, while in *T. bancanus* it is even some 10 percent longer. The fingers form a very effective cage to trap swift insect prey in the darkness of a forest night.

The likely explanation for the tarsier's huge eyes, especially relative to those of other nocturnal primates, lies in its route to night-living. Ancestral tarsiers were probably diurnal and had lost their nocturnal predecessors' Tapetum lucidum (a reflecting layer that maximizes the light-capturing capacity of many nocturnal mammals). Thus, when tarsiers secondarily adapted to night-life, they solved the low-light problem in a different way, evolving bigger eyes to maximize light-gathering capacity. In the tarsier's retina there is a central groove or fovea – a structure unique to higher primates, and responsible for pin-sharp vision. Together with other characters like the haplorhine nose, the common structure of the fovea suggests that tarsiers are closely related to South American monkeys like the Owl monkey and marmosets.

Crossing Wallace's Line

DISTRIBUTION PATTERNS

While fossil relatives of tarsiers have been found in Asia, Europe, North America, and Africa, modern tarsiers are restricted to a few islands in Southeast Asia. Most animals in this region are distributed either on one side or the other of Wallace's Lin[e], a zoogeographic border that separates the cent[ers] of evolution for the Eurasian and Australian fau[nas]. However, *Tarsius* is unusual in that its distr[ib]ution crosses Wallace's Line. Based on comparisons with other members of the Wallacean fauna, this may indicate that tarsiers hav[e] lived in the region for more than 40 million yea[rs].

The Western and Philippine tarsiers are the s[ole] representatives on their respective islands. At le[ast] three species occur on Sulawesi, with the Spect[ral] tarsier occupying the northern peninsula, and Dian's tarsier (only described in 1991) and the Pygmy tarsier living on the central mainland. More species may remain to be discovered on other islands. While the three Sulawesi species [are] closely related, the Philippine and the Western species differ sufficiently that assigning them to a separate genus has been seriously discussed.

FACTFIL[E]

TARSIERS

Order: Primates

Family: Tarsiidae

5 species of the genus *Tarsius*. The tarsiers on the islands of Sangihe, Selayar, and Togian may yet be judged separate species.

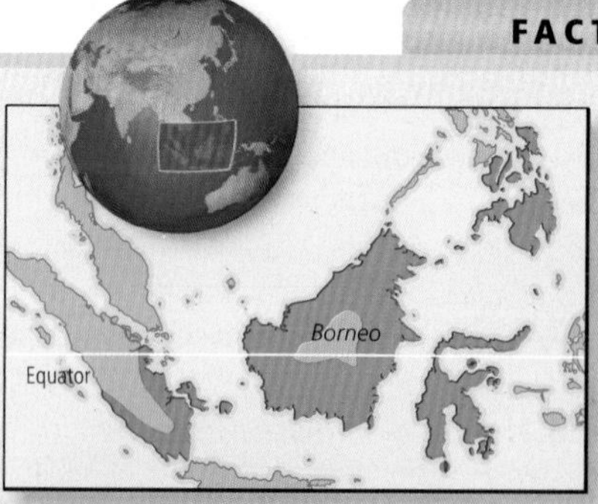

DISTRIBUTION Islands of SE Asia

Diet Mainly insects; also invertebrates, snakes, birds.

PHILIPPINE TARSIER *Tarsius syrichta*
SE Philippine islands (Mindanao, Bohol, Samar etc.). Rain forest and shrub. Crepuscular and nocturnal. HBL 11–12.7cm (4.3–5in); TL 21–25cm (8.3–10in); WT female 110–119g (3.9–4.2oz), male 131–138g (4.6–4.9oz). Coat: gray to gray-buff; face more ocher; tail tuft very sparse and short. Longevity: in captivity 8–12 years.

WESTERN TARSIER *Tarsius bancanus*
Borneo, Bangka, S Sumatra. Primary and secondary rain forest, shrubs, plantations. Lives in territorial pairs; crepuscular and nocturnal. HBL 11.5–14.5cm (4.5–5.7in); TL 20–23.5cm (7.9–9.2in); WT females 107–127g (3.8–4.5oz), males 122–134g (4.3–4.7oz). Coat: buff, brown-tipped; tail tuft well developed but not bushy. Longevity: 8 years in the wild.

SPECTRAL OR EASTERN TARSIER *Tarsius spectrum*
N peninsula of Sulawesi (Minahassa). Primary and secondary rain forest, mangroves, shrubs, plantations. Lives in pairs or polygynous groups with offspring; crepuscular and nocturnal. HBL 9.5–14 cm (3.7–5.5in); TL 20–26cm (7.9–10.2in); WT females 102–114g (3.6–4oz), males 118–130g (4.2–4.6oz). Coat: gray to gray-buff, may be darker than species from other islands; tail tuft long and bushy, scalelike skin on tail. Conservation status: Lower Risk – Near Threatened.

DIAN'S TARSIER *Tarsius dianae*
C Sulawesi. Primary and secondary rain forest at 700m (2,300ft) above sea level; plantations. Lives in family groups with one adult pair; crepuscular and nocturnal. HBL 11.5–12cm (4.5–4.7oz); TL 21.5–23cm (8.5–9in); WT female 110g (3.9oz). Coat: woolly, grayish buff, whitish on upper lip; no brown on lateral side of thigh; finger and toenails dark and keeled; tail tuft long and bushy. Conservation status: Lower Risk – Conservation Dependent.

PYGMY TARSIER *Tarsius pumilus*
C Sulawesi. Montane forests. Crepuscular and nocturnal. HBL 9.5–10.5cm (3.7–4.1in). Coat: color similar to other species, but fur more curly.

Abbreviations HBL = head–body length TL = tail length WT = weight

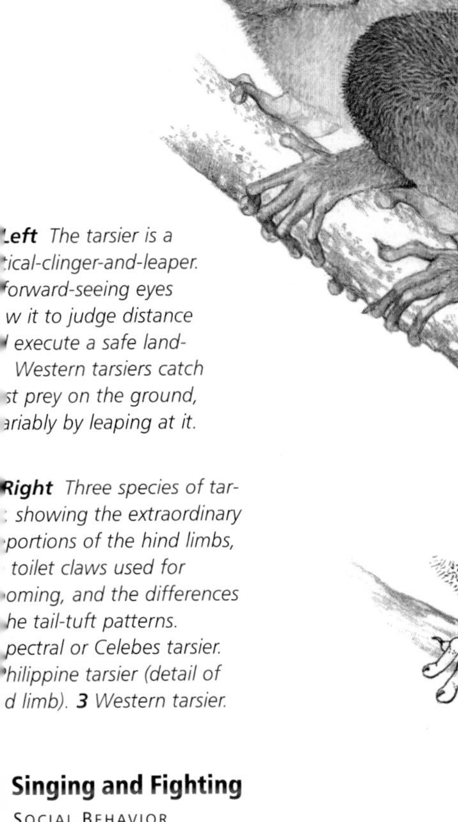

Left *The tarsier is a* *ical-clinger-and-leaper.* *orward-seeing eyes* *w it to judge distance* *execute a safe land-* *Western tarsiers catch* *st prey on the ground,* *ariably by leaping at it.*

Right *Three species of tar-* *showing the extraordinary* *portions of the hind limbs,* *toilet claws used for* *oming, and the differences* *he tail-tuft patterns.* *pectral or Celebes tarsier.* *hilippine tarsier (detail of* *d limb).* ***3*** *Western tarsier.*

Singing and Fighting

SOCIAL BEHAVIOR

siers are exclusively insectivorous and carnivo-is. In the Western and Spectral tarsiers, moths, usts, beetles, and cicadas together make up ore than half of their prey. In the Western tarsier tebrates represent less than 1 percent of food ms. A Western tarsier can catch and kill a bird ger than itself. Venomous snakes are also some-nes eaten. Spectral tarsiers also collect prey from aves. Tarsiers may eat about 10 percent of their n body weight every 24 hours and drink several nes per night.

Western and Spectral tarsiers are sexually ature when aged about 1 year, but in the West-n species, some young males may delay repro-ctive maturity for an as-yet unknown period til they establish a home range of their own.

Courtship in Western tarsiers involves much asing around, sometimes with soft vocaliza-ns. However, during mating, which takes place a tree, the pair are silent. The gestation period is out 190 days in the Spectral and Western tar-ers. In Western tarsiers, births occur throughout e year. In the Spectral tarsier, births are more fre-ent around April and May, but some occur in ovember and December.

A single young, about 25 percent of its moth-'s weight, is born fully furred and with its eyes en. It can climb at just 1 day old. Both Spectral d Philippine tarsiers have been seen carrying their infants by mouth. Western and Spectral tarsier mothers can leap between trees with their infant slung under their belly.

In the Western tarsier, home ranges are generally occupied by a pair. Spectral tarsiers may live in small groups comprising one or two adult females, with an average group size of 2.8 individuals. The large testes of the males in all three species suggest a promiscuous mating sys-tem. Tarsiers scentmark their home ranges with urine and a secretion from a skin gland on the chest. Each of the three species from Sulawesi has a specific social call. They are most vocal in the morning near their sleeping site, which is marked by scents. Males and females use their very differ-ent but equally high-pitched voices to perform beautiful duet songs, which are so loud that they can be heard 100m (110yds) away. Almost as soon as the sun rises, female Spectral and Dian's tarsiers start to sing. The song of Dian's tarsier is characterized in a long middle section by a contin-uous series of fast notes in both sexes, whereas in the Spectral tarsier the females repeat a series of phrases, the single calls being much slower. While captive Spectral tarsiers utter these typical calls, Philippine tarsiers do not, and Western tarsiers seem never to perform duets. Based on the evalua-tions of social duetting among Dian's tarsiers, a pair dominates a small group, even when a third female or male, or else another pair, contribute to the social song.

The average home range size of the Spectral tar-sier is barely 2 hectares (4.9 acres) per group. In Dian's tarsier the range is some 0.6–1.6ha (1.5–4 acres), while that of the Western tarsier is thought to be around 2ha. Western tarsiers appear to be rather belligerent, judging by the frequency of scars and injuries. CN

Marmosets and Tamarins

WITH THEIR FINE, SILKY COATS, LONG tails, and a wide array of tufts, manes, crests, mustaches, and fringes, marmosets and tamarins are the most diverse and colorful of the New World primates. The name marmoset itself is thought to derive from an old French word for "grotesque figure."

These diminutive, squirrel-like monkeys of the tropical American forests share several features that are very unusual among primates: a variable breeding system; sexes that look similar; multiple births (usually of twins); extensive care of the young by the father and other group members; and social groups as large as 20. Adult offspring may remain in the group and help care for their younger siblings. Marmosets (but not tamarins) are, uniquely, specialized gum-eaters.

Smaller is Better

FORM AND FUNCTION

The diminutive marmosets and tamarins differ anatomically from other, larger monkeys of the New World. The possession of modified claws rather than nails on all digits except the big toe sets them apart, as does the presence of two, as opposed to three, molar teeth on either side of each jaw. Their propensity to give birth to twins further distinguishes them. These features, together with their simple uterus and the lack of a rear inner cusp on the upper molars, have led to the suggestion that they have evolved to take advantage of an insect diet and, as part of this process, have become smaller in the course of their evolution. Goeldi's monkey is also believed to have undergone such "phyletic dwarfism," but shares traits with the other New World monkeys – for example, single offspring and three molar teeth in each jaw – while being small and having claws rather than nails like the marmosets and tamarins.

Different Groups with Different Needs

DISTRIBUTION PATTERNS

Both the Pygmy marmoset and Goeldi's monkey are restricted to the upper Amazon, in Brazil, Peru, southern Colombia, and northern Bolivia. The Pygmy marmoset is very particular about living in riparian and seasonally flooded forest, where their preferred tree-gum sources are most abundant. Here populations reach up to 40–50 groups per sq km (104–130 groups/sq mi), while in less appealing secondary forest or away from rivers, they occur in lower densities of 10–12 groups per sq km (26–31 groups/sq mi). The rarer Goeldi's monkey has a preference for dense, scrubby undergrowth, especially in bamboo forests, and so exists in patches of suitable vegetation that may be isolated from one another by several kilometers.

The marmosets can be divided into two major groups, those (which some believe merit their own genus, *Mico*) of the Amazon south of the Rio Amazonas (though *M. melanurus* extends into the Chaco of Paraguay and eastern Bolivia), and those (which all agree are in the genus *Callithrix*) typically found in the Atlantic forest, though with two species (*C. jacchus* and *C. penicillata)* extending into the central savanna (*cerrado)* and the thorn scrub and forests (*caatinga)* of northeast Brazil. The tamarins (*Saguinus*) are distributed widely through Amazonia north of the Rio Amazonas and, in the south, west of the Rio Madeira. Several new species are likely to change the established list. One Saddleback tamarin, *S. f. fuscicollis*, extends its range east across the upper Madeira, where it is sympatric with a marmoset species (previously called *Callithrix emiliae*, but now being redescribed), while the Black-handed tamarin (*S. niger*) occurs south of the Amazonas at its mouth. Only three species – the Cottontop, Geoffroy's, and the Silvery-brown tamarin – occur outside Amazonia, in northern Colombia and Panama.

The monkeys inhabit a wide variety of forest types, from tall primary rain forest with secondary growth patches to semideciduous dry forest, including the drier forests of northern Colombia and Central America, plus gallery forest and forest

1 2 3

Above and right *1 Emperor tamarin* (S. imperator). *2 Mustached tamarin* (S. mystax). *3 Cotton-top tamarin* (S. oedipus). *4 Goeldi's monke* (Callimico goeldii) *in "arch-bristle" offensive threat posture used within the troop. 5 Geoffroy's tamarin* (Saguinus geoffroyi) *scentmarking a branch with glands situated around its genitals. 6 Silvery marmos* (Mico (Callithrix) argentata). *7 Buffy tufted-ear marmoset* (Callithrix aurita). *8 Black tufted-ear marmose* (C. penicillata). *9 Black-tailed marmoset* (M. melanurus) *presenting its rear with tail raised as an offensive threat posture used between members of different troops. 10 Golden-headed lion tamarin* (Leontopithecus chrysomelas) *using elongated fingers to probe a bromeliad for insects. 11 Santarém marmos* (Mico (Callithrix) humeralifer). *12 Pygmy marmoset* (Cebuella (Callithrix) pygmaea) *gouging tree for gum and sap. 13 Red-bellied tamarin* (S. labiatus) *markin with chest glands. 14 Saddle-back tamarin* (S. fuscicollis) *marking with glands above pubic area. 15 Mottle faced tamarin* (S. inustus).

FACTFIL

MARMOSETS AND TAMARINS

Order: Primates

Family: Callitrichidae

38 (26) species in 6 (4) genera: Amazonian marmosets (*Mico*, 11 (–) species); E Brazilian marmosets (*Callithrix*, 6 (9) species); Pygmy marmoset (*Cebuella (Callithrix) pygmaea)*; tamarins (*Saguinus*, 15 (12) species); lion tamarins (*Leontopithecus*, 4 species); Goeldi's monkey (*Callimico goeldii*).

DISTRIBUTION S Central and N South America

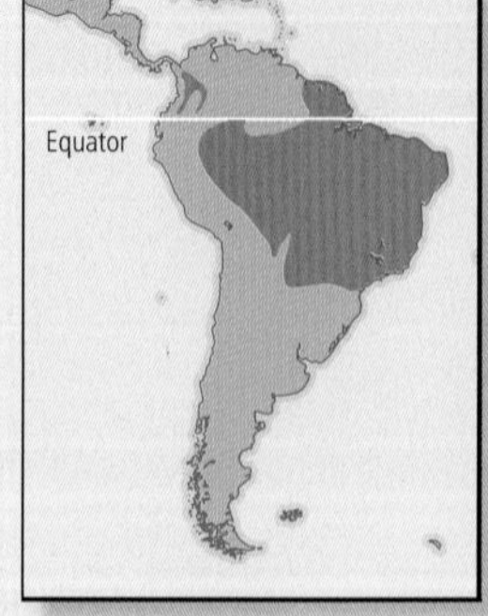

HABITAT Chiefly tropical rain forest; also gallery forest and forest patches in savanna.

SIZE **Head–body length** ranges from the tiny Pygmy marmoset, at 17.5–19cm (7–7.5in), to the lion tamarins, 34–40cm (13–16in); **tail length** from 19cm (7.5in) to 26–38cm (10–15in) in the same species; **weight** from 120–190kg (4.2–6.7oz) to 630–710g (22.2–25oz). Most species weigh between 260 and 380g (9.2–13.4oz).

COAT Fine, silky, often colorful. Many species have ear tufts, mustaches, manes, or crests.

DIET Fruits, flowers, nectar, plant exudates (gums, saps, latex), along with animal prey including frogs, snails, lizards, spiders, and insects.

BREEDING Gestation 130–170 days

LONGEVITY Unknown in wild (7–16 years in captivity)

See species table ▷

4
5
6
7
8
9
10
11
12
13
14
15

patches in savanna regions in Amazonia, the Chaco of Bolivia and Paraguay, the *cerrado* of central Brazil, and the *caatinga* of northeast Brazil. Mature forest with patches of secondary growth provides homes for 3–5 groups per sq km (8–13 groups/sq mi). Successional river-edge forest arising from changes in the courses of rivers, often found around many of the Amazonian tributaries, is also dense with these monkeys. There they feed on the small and juicy fruits of colonizing trees, and sleep among bushy vegetation and liana tangles. Thus they avoid forest hawks and the predatory, martenlike tayra, while being able to capture more of their own preferred animal prey.

The four species of lion tamarin survive in widely-separated remnant lowland forests in southeastern Brazil, in low densities of 0.5–1 group per sq km (1–3 groups/sq mi). Although they exploit forest in even quite early stages of succession, they depend on tall, mature forest for their sleeping holes, which are dug out by woodpeckers, and for sufficient animal-prey foraging sites, especially bromeliad epiphytes and leaf-litter piles in vines and palm-tree crowns.

Above *An Emperor tamarin* (Saguinus imperato *displays the small but sharp teeth it uses to eat the stick insects, mantises, and other orthopterans that make up the major part of its diet.*

The Forest Menu

DIET

Marmosets and tamarins eat fruits, flowers, nectar, plant exudates such as gums, saps, and latex, and animal prey, including frogs, snails, lizards, spiders, insects – especially grasshoppers, beetles, and stick insects – and even fledglings and bird's eggs. Exudates are a key resource for the marmosets, but the tamarins (*Saguinus*) and lion tamarins (*Leontopithecus*) eat them opportunistically, especially the large quantities sometimes exuded from wounds in wild cashew trees (*Anacardium*) and *Inga*, as well as the beans of leguminous trees. They are not leafeaters, although they do occasionally eat leaf buds. The fruits are usually small and sweet, and such genera as *Pourouma*, *Ficus*, *Cecropia*, *Inga*, and *Miconia* are particularly important. Recent research has demonstrated the significant role that marmosets and tamarins play in seed dispersal. Many of their fruits (drupes or arilate seeds) have small seeds that are swallowed with the pulp and pass through the digestive tract unharmed after 1–3 hours, well away from the parent plant. In some cases the seeds ingested may even be quite large, up to 1.5cm (0.6in) in length, with many seeds in the digestive tract at the same time. Although seed-swallowing is generally accepted to be a way of digesting pulp or arils which are hard to remove from the seed (and as such a mechanism by which the plant can promote seed dispersal), researchers have also indicated that the very size of the seeds, which undoubtedly stretch the intestine in their passage, may play a role in dislodging intestinal parasite especially acanthocephalan hookworms.

The two groups spend 25–30 percent of the active time foraging for animal prey, searching through clumps of dead leaves, among fresh leaves, along branches, and peering and reachir into holes and crevices in branches and tree trunks. Marmosets also exploit the insects disturbed by army-ant (*Eciton*) swarm raids. The Pygmy marmoset spends 67 percent of its feedi time tree-gouging for gums. Spiders and insects are also important for this species, while fruit is too scarce in its tiny home ranges to be anythin more than an occasional part of its diet. This co trasts with other, primarily fruit-eating marmose which also partake of flowers, animal prey, and, particularly at times of fruit shortage, gum or ne tar. No marmoset species share the same forest, possibly due to rivalry for plant exudates, thoug several tamarin species coexist by exploiting dif ent levels, sites, and prey. The marmosets and the Mustached and Emperor tamarins glean the foliage for small insects, particularly orthoptera while Lion and Saddle-back tamarins break rott wood or masses of peaty soil to uncover larger insects, dedicating much of their foraging time t investigating crevices and holes in treetrunks an branches. To help with this mode of food-findin lion tamarins have longer hands and fingers tha other callitrichids. Differences in foraging techniques and the prey exploited are believed to be the reason why Saddleback tamarins and the th Mustached tamarins (*S. imperator*, *S. mystax*, an

MONKEYS THAT EAT GUM

The Pygmy marmoset and the larger marmosets mainly eat gum. (Gums are a trees' defense system against damage to its bark, usually by wood-boring insects.) Although the Fork-crowned lemur uses its toothcomb to scrape up gums that have already been exuded, these monkeys are the only primates that regularly gouge their own holes. The amount of gum recovered is usually quite small, and they spend only 2 minutes at most at any one hole. Unlike the tamarins **a**, the marmosets **b** have relatively large incisors, beyond which the canines barely protrude. The lower incisors lack enamel on the inner surface, while on the outer surface the enamel is thickened, producing a chisel-like effect. Marmosets anchor the upper incisors in the bark and gouge upwards with these lower incisors. The holes they produce are usually oval and 2–3cm (1in) across at most, but favored trees may be riddled with channels as long as 10–15cm (4–6in). The small size and clawlike nails of

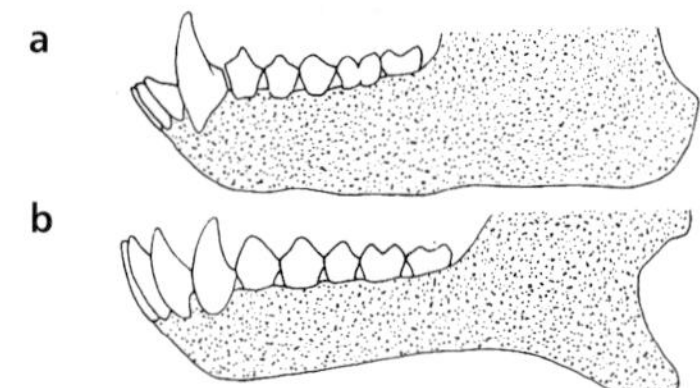

callitrichids are important adaptations that enable them to cling to tree trunks . Gums are particularly important for the Pygmy marmosets, which have such small home ranges that they cannot depend on fruits all year round, and the same may be true of the dwarf marmoset, *C. humilis*. However, they are also a significant dietary supplement for all marmosets, particularly at times of fruit shortage. For such species as the Common and the Black tufted-ear marmosets they are vital, allowing them to survive in the relatively harsh environments of northeast and central Brazil.

labiatus) can be sympatric and, where they are, rm mixed-species groups. The Mustached marins are larger (at around 450g/16oz) than e Saddle-back tamarins (around 350g/12oz), d forage at higher levels in the forest. A number reasons for this association have been put forrd, including improved resource use in locating d finding fruit trees, benefits in predator detecn, and the likelihood of Saddle-back tamarins nefiting from the insects that fall from the upper ers of the forest as a result of the activities of the ustached tamarins above them.

Life in an Extended Family

SOCIAL BEHAVIOR

armosets and tamarins generally live in groups 4–20 individuals. Pygmy marmoset groups nge in size from 2–9, including a reproductive male, a mate, and her offspring, and sometimes e or two unrelated adults. Marmosets (*Mico* and *llithrix*) generally occur in larger groups than e tamarins, probably as a result of having access an additional food source in gums. Groups are mprised usually of a single breeding pair and eir offspring, but in the larger groups there may two or more generations of offspring who main in the group even when reaching adultod, as well as unrelated subadults and adults at have migrated. Different species have home nges of vastly different sizes. Pygmy marmosets quire 0.1–0.3ha (0.2–0.7 acres), *Saguinus*, *Mico*, d *Callithrix* live in 10–40ha (25–100 acres), ile lion tamarins have home ranges that can be large as 200ha (500 acres). Groups visit approxately one-third of their range each day, traveling to 2km (1.2mi).

The home range is defended, and it is in this tivity that the idiosyncratic hair styles of the different monkeys come into their own. Calling often ecedes a showy confrontation. Marmosets raise eir tail, fluff their fur, and display the rump and stinct white genitalia. Lion tamarins raise their anes, and tamarins fluff their fur and tonguek. Territorial scentmarking is also important. armosets use chest (chest-rubbing) and suprabic glands (sprawling) for such marking. A ird type of scentmarking, using the glands ound the genitals, is used when members of a oup wish to communicate with each other. Marosets often take tree-gouging as an opportunity mark, and sometimes to urinate in, the holes ey gouge, possibly as a means of maintaining minance relations, and taking advantage of the t that other group members will feed at the site d therefore be forced to smell their scents.

The most obvious and prolonged social behav within a group is mutual grooming, most freently between the adult breeding pair and their est offspring. The time spent on grooming, rest (often huddled together), and play depends the time of year and the amount of food availle, which affects the time that has to be spent

Right *From its forest perch, a Cotton-top tamarin* (Saguinus oedipus) *keeps a wary eye out for possible predators. Highly arboreal like all the callitrichids, Cotton-tops spend their days moving below the canopy in search of food, particularly ripe fruit, nectar, and insects; they will rarely travel much more than 1–2km (1mi) in all. At twilight they return to a sleeping tree, where they spend the night resting in broad forks among the branches. A typical troop will number about 6–9 animals.*

Above *The Maués marmoset* (Mico (Callithrix) mauesi) *is named for the River Maués, a southern tributary of the Amazon in Brazil's Amazonas province. A denizen of the deep forest, the species was only discovered in 1985 and was first described in 1992.*

foraging. When food is abundant, it may be as much as a couple of hours, generally around midday or in the early afternoon. Other forms of communication involve a limited number of facial expressions, specific postures, and patterns of hair-erection, as well as complex, graded, high-pitched and birdlike vocalizations. The monkeys have characteristic long-calls, which play an important role in helping group members maintain contact.

In the wild, there is generally only one breeding female in each group. In the tamarins and lion tamarins, twins are born once a year, but the marmosets tend to produce two litters annually, at intervals that may be only a few days longer than the gestation period. This is probably related to their tree-gouging and gum-feeding, which guarantee a year-round food source sufficient to sustain pregnancy. When tamarins lack fruit they switch to nectar, which, although rich in sugar, is only available in small quantities. In some areas where Saddleback tamarins are able to raid pygmy marmoset gum-feeding sites, they breed twice a year. In many studies, the breeding female has been seen to copulate with more than one group male, and more than one adult male carries the young, indicating a polyandrous mating system. Reproduction is suppressed in other female members of the group as a result of behavioral domination by the reproductive female and by the effects of chemicals (pheromones) in the scentmarks from her genital glands. With the exception of the lion tamarins and Goeldi's monkey, marmosets and tamarins can suppress ovulation altogether.

The reason for this suppression, and probably also for the polyandrous mating of the breeding female, would seem to be related to the need for helpers to carry offspring. In Goeldi's monkey groups, there may be two breeding females, each producing a single offspring once a year, and there have been recorded cases of two (and in one instance of three) females breeding in marmoset and tamarin groups. Reasons for this behavior can in some cases be directly linked to situations where the dominance position for breeding is in dispute, with the oldest daughter taking over from her mother; in others the familial relations are not known, but the situation is apparently linked to instability in group membership, such as at the death of the breeding male.

Newborn marmoset and tamarin twins weigh a massive 19–25 percent of the mother's weight, a considerably higher proportion than in any other primates except tarsiers. The infants are carried rather than being left in a nest, and the help of the male parent or even parents (the twins are dizygotic) and other group members in such transportation is considered one of the key adaptations which allows for the high reproductive rate. Marmoset infants are completely dependent for the first two weeks, but by two months can travel independently, catch insects or rob them from other group members, and spend extended pe ods in play, wrestling, cuffing, and chasing eac other. They reach puberty at 12–18 months, ai adult size at 2 years of age. Among the tamarin lion tamarins, and Goeldi's monkey, whose ne born are smaller (about 9–15 percent of the mother's weight), males help carry the young from the age of 7–10 days. Tamarins mature slightly slower than marmosets, becoming ind pendent at two and a half months.

All group members carry the young and sur der food morsels to the young and the breedin female. This form of cooperative breeding wou appear to be unique among primates. The adu helpers that stay in the family group may gain breeding experience while waiting until suitab habitat becomes available for them to breed th selves, or until they get the chance to breed in neighboring group. Once established as breed callitrichids have a higher reproductive potenti than any other primate. In suitable conditions, female marmoset can produce twins about onc every five months.

The Dangers of Habitat Destruction

CONSERVATION AND ENVIRONMENT

The variety of species is greatest in the central a upper Amazon region. There the animals occur limited distributions between even quite small river systems, and an amazing eight new specie have been discovered during the last ten years. fourth lion tamarin, the Black-faced lion tamari was also discovered in 1990. Marmosets and tamarins used to be wrongly considered carrier yellow fever and malaria, and were persecuted i consequence; until 1970–73, they were captur and exported in large numbers, particularly for zoological gardens and for biomedical research. While the small size of many of the callitrichids means that they are not hunted for food as the larger platyrrhines are, their limited geographic distributions makes them extremely vulnerable extensive habitat destruction. This is the case fo the endangered Buffy-headed (*Callithrix flavicep* and Buffy-tufted-ear marmosets (*Callithrix aurit* which are naturally rare, occupying small range in montane areas in southeast Brazil where the forests have been widely destroyed. The lion tamarins of the southeast of Brazil are likewise a a particular disadvantage in that they evidently require mature lowland forest, and, except for t Golden-headed lion tamarin (whose estimated numbers are 4,000–6,000), the fragmented po ulations in the wild each total less than 1,000 a mals. The Pied tamarin (*Saguinus bicolor*) is the most threatened of any of the Amazonian callitrichids. It has a minute distribution around Manaus, limited to a radius of 35–40km (20–2 miles) from the city. Curiously, it is now becom evident that the Red-handed tamarin (*Saguinus midas*) is posing an even more serious threat by replacing it on the periphery of its range. A

armoset and Tamarin Species

SMALL SIZE AND LOW POPULATIONS armosets and tamarins, and their e forest habitat, make them difficult als to study. As recently as 1996 a species, the Dwarf marmoset, was vered in a small patch of forest on the bank of the lower Aripuanã river in zonas state, central Brazil. Other dis- ries of new species may well follow. venty-six callitrichid species appeared e last established list of mammalian es, in 1993. Since then, the tendency een toward a more speciose taxono- 12 species not listed in 1993 are ed † below.

MARMOSETS

nera *Cebuella, Callithrix,* l *Mico*

phological and genetic studies from 980s onwards recognize the genus *ella* and align it with *Callithrix*. More- recent discoveries suggest as many new species of the *Mico (Callithrix)* ge, which are not yet included in the ies tally of the established list, but be at the next major revision. n primary tropical rain forest mixed secondary growth, gallery forest, and st patches. All but two species (*C. pyg- a* and *M. melanurus*) are restricted to il. General data for the group are as ws: HBL 17–21cm; TL 19–29cm; WT –350g. COAT: variable, some species wing subspecific variation from dark to pletely pale or white; varying degrees artufts. BREEDING: gestation 150–151 ; litter size 2.

fy-tufted-ear marmoset En
ithrix aurita

nant forests in SE Brazil in states of de Janeiro, São Paulo, and Minas ais.
AT: whitish or buffy eartufts; forehead erous to whitish; front of crown tawny ale buff; side of face and temples k; back agouti to dark brown or black ated, as in Common marmoset; tail ed, underparts black to ocherous.

ffy-headed marmoset En
lithrix flaviceps

nnant forests in SE Brazil in states of irito Santo and Minas Gerais.
AT: eartufts, crown, side of face and eks ocherous; back grizzled striated uti; underparts yellowish to orange; distinctly banded.

Geoffroy's marmoset Vu
Callithrix geoffroyi

Remnant forests in SE Brazil in states of Minas Gerais and Espírito Santo, between the Rios Jequitinhonha and Doce.
COAT: blackish-brown, with elongated black eartufts; forehead, cheeks and vertex of crown white; underparts dark brown; tail black, lightly ringed.

Common marmoset
Callithrix jacchus

NE Brazil, W, and S from the Rio Parnaíba; probably formerly restricted to the N of the Rio São Francisco, but today ranging S (introduced) in the states of Bahia, Minas Gerais, Espírito Santo, and Rio de Janeiro, and even in the suburbs of Buenos Aires, Argentina. An adaptable species in the N Atlantic coastal forest and gallery forest and forest patches in the *caatinga* (dry thorn scrub) and *cerrado* (bush savanna).
COAT: body mottled gray-brown; crown blackish with a white blaze on forehead; elongated white eartufts; tail ringed gray and white.

Wied's black-tufted-ear marmoset
Callithrix kuhlii

Atlantic forest in S Bahia, Brazil, between the Rios de Contas and Jequitinhonha.
COAT: body mottled blackish with red-dish-brown bases on the hairs on lower back, flanks, and outer thighs; ringed tail; black face with offwhite cheek patches and a white patch on the forehead between the eyes; in juveniles the crown is black, turning grayish-brown when adult; distinctive reddish-brown bases to hairs on outer thighs and flank; tail black, lightly ringed.

Black tufted-ear marmoset
Callithrix penicillata

S central Brazil, in the states of Goiás, Bahia, Minas Gerais, and São Paulo. Seasonal semideciduous and deciduous forest patches and gallery forest in the Cerrado (bush savanna).
COAT: body mottled gray; black, pencil-like eartufts; ringed tail; black face, with a white patch on the forehead between the eyes.

Pygmy marmoset
Cebuella (Callithrix) pygmaea

Upper Amazonia in Colombia, Peru, Ecuador, N Bolivia, and Brazil. Prefers floodplain forests and natural forest edge. The smallest living monkey. The pygmy marmoset was formerly placed in its own genus, *Cebuella*, but morphological and genetic studies in the 1990s have resulted in some authorities placing it in the genus *Callithrix*.
HBL 17.5cm; TL 19cm; WT 120–190g.
COAT: tawny agouti; long hairs of head and cheeks form a mane; tail ringed. Two subspecies: *C. p. pygmaea*, from the N of the Rio Solimões and W of the Rio Japurá-Caquetá in Brazil, Colombia, Ecuador, and Peru, has an ocherous belly; *C. p. niveiventris*, from the S of the Rio Solimões in Brazil, Peru, and Bolivia, has a sharply contrasting whitish chest, belly, and inner surface of arms and legs.
DIET: mainly exudate, insects, and spiders, but also some fruits.
BREEDING: gestation 136 days.

Silvery marmoset
Mico (Callithrix) argentata

E Brazilian Amazon, S of Rio Amazonas, E of the lower Rio Tapajós, extending through the lowlands of the lower Rios Xingú and Tocantins.
COAT: predominantly silvery white; pink face and ears; no eartufts; black tail.

Golden-white tassel-ear marmoset † Vu
Mico (Callithrix) chrysoleucus

A small area of the Brazilian Amazon between the Rio Amazonas and the S bank tributaries Uraria-Canumã.
COAT: face pink; fur near-white; long white eartufts; rest of body pale golden to orange.

Snethlage's marmoset †
Mico (Callithrix) emiliae

Upper reaches of the Rios Irirí and Xingu in the Brazilian Amazon.
COAT: very similar to *M. argentatus*, but darker with a blackish crown and grayish-brown back; no eartufts; tail black.

Santarém marmoset
Mico (Callithrix) humeralifer

Brazil, between the Rios Maués-açú and Tapajós, S of the Rio Amazonas. Prefers secondary forest and dense terra firme forest.
COAT: pigmented face; long silvery eartufts; mantle mixed silvery and black; back with pale spots and streaks; tail silvery ringed with black; white hip patches.

Dwarf marmoset †
Mico (Callithrix) humilis

From a very small area W of the lower Rio Aripuanã, S of the Rio Solimões in Brazil. Prefers dense terra firme (unflooded) forest and secondary forest. This remarkable marmoset, discovered only in 1996, weighs about 165–200g when adult, and as such is intermediate in size between the Pygmy marmoset and the remaining members of the genus.
HBL 15cm; TL 25cm; WT 200g.
COAT: upper and outer parts of the body and limbs are olive brown; a black, triangular crown on the head; white rim above eyes extending to temples; sides of neck and shoulders are grayish white; black tail.
DIET: Exudate, insects and small animal prey, fruits.

Aripuanã marmoset †
Mico (Callithrix) intermedius

Between the Rios Roosevelt and Aripuanã of the Rio Madeira basin in the Brazilian Amazon. In tall, dense, mature forests, but more abundant in secondary-growth forest.
COAT: face pink, very reduced eartufts; upper chest and back creamy white; rump and base of tail dark brown, underparts orange; pale thigh stripe as in *M. melanurus*; tail offwhite, variably marked at the base and tip with brown.
DIET: fruits, gums (especially when fruits are scarce at end of wet season), small animal prey including frogs and lizards.

Golden-white bare-ear marmoset † Vu
Mico (Callithrix) leucippe

Brazilian Amazon, S of the Rio Amazonas, between the Rios Cuparí and Jamanxim, E bank tributaries of the Rio Tapajós.
COAT: head and body silvery or creamy white; tail pale, with orange or gold tip; no eartufts.

Marca's marmoset †
Mico (Callithrix) marcai

Known only from three museum specimens from the W bank of the mouth of Rio Roosevelt, S of the Rio Amazonas in Brazil.
COAT: dark face, with a small whitish patch between the eyes; dark brown crown, paler brown on nap and mantle, speckled with white on the middle and lower back; thighs ocherous; underparts reddish; tail dark brown.

ABBREVIATIONS

HBL = head–body length TL = tail length WT = weight
Approximate nonmetric equivalents: 2.5cm = 1in; 230g = 8oz; 1kg = 2.2lb
† = 'new' species. See p291

Ex Extinct
EW Extinct in the Wild
Cr Critically Endangered
En Endangered
Vu Vulnerable
LR Lower Risk

Maués marmoset †
Mico (Callithrix) mauesi

A small range bounded by the Rio Maués-açú (on the W bank), Uariá, and Abacaxis, S of the Rio Amazonas in Brazil. Observed in dense primary forest and forest edge. Discovered in 1985 and first described in 1992.
HBL 19.8–22.6cm; TL 34–38cm; WT 315–405g.
COAT: general appearance silvery-brown; notable for its erect, "neatly-trimmed" eartufts, with the hairs growing from the border of the pinna and unlike any other marmoset; face thinly-haired and pinkish, with whitish tips to the hairs on the cheeks; shoulders and back marbled dark brown and white; light silvery hip patch as seen in *M. humeralifer*; thighs and legs silvery gray; underparts buffy with orange tint; tail black, with faint banding of silvery gray.

Black-tailed marmoset †
Mico (Callithrix) melanurus

Brazilian Amazon, S of Rio Amazonas, into E Bolivia and N Paraguay.
COAT: predominantly dark brown, darker on lower back than mantle; black face, sometimes mottled white around the nose and mouth; black ears, no eartufts; pale hip and thigh patches; black tail.

Black-headed marmoset † Vu
Mico (Callithrix) nigriceps

Brazilian Amazon, E of the Rios Jiparaná and Madeira and W to the Rio dos Marmelos. Abundant in marginal and disturbed forest habitats. First described in 1992, it is considered threatened due to its small range in an area undergoing colonization and development.
HBL 19–22cm; TL 31–32cm; WT 330–400g.
COAT: black crown and face, cheeks blackish to grayish brown; no eartufts; mantle grayish brown/silvery; lower back drab brown, forelimbs, hips, and upper thighs pale orange; ventrum yellow to orange; tail black (dark brown at base).

TAMARINS

Genus *Saguinus*

In tropical, evergreen, primary forests, secondary-growth forest, and semideciduous, dry forests.
HBL 19–21cm; TL 25–29cm; WT 260–380g.
COAT: color very variable.
DIET: fruit, flowers, insects, spiders, snails, lizards, frogs, plus plant exudates such as gums and nectar when readily available.
BREEDING: gestation period averages 140–170 days; litter size 2.

Above Geoffroy's marmoset (*Callithrix geoffroyi*) combines a mo[nkey's] face with an almost catlike bo[dy].

Black-mantle tamarin
Saguinus nigricollis

Upper Amazonia, N of Rio Amazonas, in S Colombia, Ecuador, Peru, and Brazil.
COAT: head, neck, mantle, and forelimbs black-brown; hairs around mouth gray; lower back, rump, thighs, and underparts olivaceous or buff-brown. Two subspecies, *S. n. nigricollis* and *S. n. hernandezi* (latter Vulnerable), between Rios Caquetá and Caguán (Colombia), have blackish mid-dorsal band on lower back and tail base.

Graell's black-mantle tamarin †
Saguinus graellsi

Upper Amazonia between the Rios Napo and Pastaza, extending N to S Colombia. Formerly included in *S. nigricollis*.
COAT: head, neck, mantle, forelimbs blackish to blackish brown, crown pale brown, tail black, except for the basal portion, which is brownish.

Saddle-back tamarin
Saguinus fuscicollis

Upper Amazonia, W of Rio Madeira to S of Rio Amazonas and S of Rios Japurá-Caquetá and Caguán in Brazil, Bolivia, Peru, and Ecuador. Eleven subspecies: *S. avilapiresi, crandalli, cruzlimai, fuscicollis, illigeri, lagonotus, leucogenys, melanoleucus, nigrifrons, primitivus,* and *weddelli.*
COAT: extremely variable among subspecies occurring between major as well as minor river systems. All have trizonal back coloration which, except in palest forms, divides a distinct rump, saddle, and mantle; the cheeks have white hairs.

Golden-mantle saddle-back tamarin
Saguinus tripartitus

Amazonian Peru and Ecuador on the right bank of the Río Napo, W from the mouth of the Río Curaray, extending to the left bank of the Río Napo above its confluence with the Río Santa Maria, and N to the Río Putumayo on the Colombian border.
COAT: striking golden-orange to creamy mantle, sharply defined from the black fur on the head and the gray-to-golden marbled back; head and facial skin black; a grayish or white chevron on the forehead between the eyes; arms orange-colored; chest, belly, and inner parts of the arms and legs have orange-colored fur; tail is predominantly black, with a reddish-orange patch by the rump.

Red-bellied tamarin
Saguinus labiatus

Three subspecies: *S. l. labiatus* S of Rio Ipixuna between Ríos Purus and Made[ira] in Brazil, Bolivia, and extreme SE Peru; *S. l. rufiventer* S of Río Solimões betwe[en] the Ríos Madeira and Purus, S to the R[io] Ipixuna; *S. l. thomasi* between Rios Jap[urá] and Solimões in Brazil.
COAT: crown with golden, reddish, or [cop]pery line, and a black, gray, or silvery s[pot] behind; mouth and cheeks covered by thin line of white hairs; back black, ma[r]bled with silvery hairs; throat and upp[er] chest black; underparts reddish or ora[nge].

Mustached tamarin
Saguinus mystax

E Peru, W Brazil, NW Bolivia, possibly Colombia. With *S. labiatus* and *S. impe[ra]tor*, may form a subgroup of related tax[a].
COAT: black crown; white area around mouth; well-developed but not particu[lar]ly long mustache; black underparts.

peror tamarin
uinus imperator

azonia in extreme SE Peru, NW Bolivia, NW Brazil.
AT: gray; elongated mustaches; crown ery-brown; tail reddish-orange. Two species: the Bearded emperor tamarin, *subgrisescens*, has a small white beard ing in the Black-chinned emperor arin, *S. i. imperator*. These are quite inct tamarins and considered by some e separate species.

-handed tamarin
guinus midas

azonia, N of Rio Amazonas, E of Rio gro in Brazil, Surinam, Guyana, and nch Guiana.
AT: black face; middle and lower back ck, marbled with reddish or orange rs; hands and feet a striking golden nge.

ck-handed tamarin †
guinus niger

f the mouth of the Rio Amazonas in the te of Pará, Brazil. Formerly included in *nidas*.
AT: similar to *S. midas*, but darker and h black hands and feet.

ttle-faced tamarin
guinus inustus

of Rio Amazonas between Rio Negro l Rio Japurá in Brazil, extending W into lombia.
AT: uniformly black, parts of face with- pigment, giving mottled appearance.

ed tamarin En
guinus bicolor

of Rio Amazonas between the Rio Negro d Rio Urubú, Brazil.
AT: head from throat to crown black d bare; tail blackish to pale brown above d reddish to orange below; large ears; ite forequarters; brownish-agouti dquarters; reddish underbelly.

re-face tamarin †
guinus martinsi

azilian Amazon, N of the river, between Rios Uatumã and Nhamundá (*S. m. hraceus*) and the Rios Uatumã and lower epecurú (*S. m. martinsi*). Formerly luded in *S. bicolor*.
AT: both forms have bare, black faces; *m. ochraceus* is a paler brown, with fore- arters paler than hindquarters, *S. m. rtinsi* is a more uniform brown color; in th the tail is brown above, orange below, h orange tip.

tton-top tamarin En
guinus oedipus

V Colombia.
AT: crest of long, whitish hairs from ehead to nape flowing over the shoulders; back brown; underparts, arms, and legs whitish to yellow; rump and inner sides of thighs reddish-orange; tail reddish-orange towards base, blackish towards tip.

Geoffroy's tamarin
Saguinus geoffroyi

NW Colombia, Panama, and Costa Rica.
COAT: skin of head and throat black with short white hairs; wedge-shaped, mid-frontal white crest, sharply defined from reddish mantle; back mixed black and buffy hairs; sides of neck, arms, and upper chest whitish; underparts white to yellowish; tail black, mixed with reddish hairs toward base, black at tip.

Silvery-brown tamarin Vu
Saguinus leucopus
Silvery-brown or White-footed tamarin

N Colombia between Ríos Magdalena and Cauca.
COAT: hairs of cheeks are long, forming upward and outward crest; forehead and crown covered with short, silvery hairs; back dark brown; outer sides of shoulders and thighs whitish; chest and inner sides of arms and legs reddish-brown; tail dark brown with silvery-orange streaks on undersurface.

Lion Tamarins
Genus *Leontopithecus*

In non-overlapping and minute distributions in remnants of mature lowland forests of the Atlantic forest of Brazil. HBL 34–40cm; TL 26–38cm; WT 630–710g. The largest of the callitrichids, and all threatened with extinction. The lion tamarins are notable for their consistent use of tree holes as sleeping sites and their long, thin hands and fingers, useful for probing between the leaves of bromeliad epiphytes for their prey.
COAT: long hairs of crown, cheeks, and sides of neck form an erectile mane, with variable pelage patterns of black and red-dish-gold.
DIET: fruits, flowers, frogs, lizards, snails, insects, plus plant exudates (gums and nectar) when readily available.
BREEDING: gestation period 128 days; litter size 2.

Golden lion tamarin Cr
Leontopithecus rosalia

Occurs in Atlantic coastal forest remnants in Rio de Janeiro state, SE Brazil. It has been the subject of a major conservation program lead by the Smithsonian Institution since 1983, including captive breeding, reintroduction of captive-bred animals, the translocation of groups from threatened to safe forests, and environmental education in the region of its last stronghold. About 600 Golden lion tamarins are left in the wild.
COAT: entirely golden-red.

Black lion tamarin Cr
Leontopithecus chrysopygus

In six forest patches in the state of São Paulo in SE Brazil. After an interval of 65 years during which the species had not been seen or heard of, Adelmar F. Coimbra-Filho, Brazilian pioneer primatologist, rediscovered it in 1973 in a forest reserve in the far W of the state, now the Morro do Diabo State Park. Less than 1,000 remain in the wild.
COAT: black, with golden rump and thighs; black face.

Golden-headed lion tamarin En
Leontopithecus chrysomelas

The cocoa-growing region in the S of Bahia state, NE Brazil, between the Rio Jequitinhonha and Rio de Contas. While currently less threatened than the other lion tamarins, agriculture, the timber industry, and cattle-ranching are resulting in the widespread demise of its remaining forest. Less than 6,000 are believed to survive in the wild.
COAT: black with a golden mane, forearms, and rump.

Black-faced lion tamarin Cr
Leontopithecus caissara

In coastal forest of SE Brazil (including sandy-soil forests, called *restinga*), on the mainland in N Paraná state extending into the far SE of São Paulo state, and on the island of Superagüi. It may be a variation or subspecies of the Black lion tamarin. First described in 1990. Total population estimated at 350 or fewer.
COAT: black face, head, forearms, and tail, with a golden back.

Genus *Callimico*

Goeldi's monkey
Callimico goeldii

Upper Amazon in Brazil, N Bolivia, Peru, Colombia, and Ecuador, in dense undergrowth of terra firme and bamboo forests. HBL 25–31cm; TL 25.5–32cm; WT 390–670g.
COAT: black or blackish-brown, with bobbed mane, tiered pair of lateral tufts on back of crown, thick side-whiskers extending below jaws.
DIET: fruits, insects, spiders, lizards, frogs, and snakes.
BREEDING: gestation period 150–160 days; litter size 1.

THE GOELDI'S MONKEY MYSTERY

Goeldi's monkey, *Callimico goeldii*, is a small (less than 1kg/2.2lb), rare occupant of bamboo forests in the upper Amazon. Its size, and the clawlike nails on all the digits but the big toe, clearly align it with the marmosets and tamarins. For many years it was believed to be the closest living relative of these monkeys' common ancestor, retaining certain primitive characteristics such as a third molar (other marmosets and tamarins have two), and giving birth to single young when marmosets and tamarins have twins. The third molar and singleton offspring are otherwise typical of the cebids.

As a result, some authorities in the past placed *Callimico* in a family all of its own, the Callimiconidae. However, recent research has changed our thinking in this respect. In 1978, immunological studies indicated that *Callimico* was closely related to the *Mico* and *Callithrix* marmosets. This at the time was merely found to be odd, and was ignored. However, more recent studies in molecular genetics, sequencing DNA in both nuclear genes and mitochondria, have clearly confirmed the close phylogenetic relationship between *Callimico* and *Mico/Callithrix*. This calls for some rethinking about the morphological and reproductive aspects that differentiate it from marmosets as well as from the tamarins.

In the case of the teeth, it is now acknowledged that not just *Callimico* but all of the marmosets and tamarins are undergoing an evolutionary process called dwarfism; they are tending to become smaller, thereby accounting for the loss of the third molar as the jaw shortens. The third molar of *Callimico* is very small and apparently non-functional, and it would seem that it just has not quite lost it yet.

The reproductive aspect is more problematic. *Cebuella*, *Mico*, *Callithrix*, *Saguinus*, and *Leontopithecus* are the only anthropoid primates to regularly produce twins, and they do so in a way that is unique among mammals. The twins are not identical, but they share the same placenta with extensive mixing of the blood of the two foetuses – a phenomenon called chimerism. The fact that *Callimico* produces single offspring, while the more distantly related tamarins produce twins just like the marmosets, indicates only two possibilities. The first is that the marmosets and tamarins evolved twinning with chimerism independently. This is highly unlikely, as no other mammal has evolved a similar system and it hardly seems probable that two closely related primates would do so independently. The second, more convincing, hypothesis is that a common ancestor evolved chimeric twinning, but *Callimico* has reverted to producing singletons. The crucial question of what ecological factors caused *Callimico* to revert to single offspring remains unanswered. ABR

ON THE BRINK OF EXTINCTION

Saving the lion tamarins of Brazil

THE JESUIT ANTONIO PIGAFETTA, CHRONICLER of Magellan's voyage around the world, referred to them as "beautiful, simian-like cats similar to small lions." Though not taxonomically correct, the description, based on observations made in 1519, captures well the first impression made by a Golden lion tamarin (*Leontopithecus rosalia*) – one of the most strikingly colored of all mammals and, unfortunately, one of the most endangered. The three other lion tamarin species – the Black-faced (*L. caissara*), Golden-headed (*L. chrysomelas*), and Black (*L. chrysopygus*) – are all similarly threatened with extinction.

The Golden lion tamarin itself has a magnificent, reddish-gold coat and a long, back-swept mane that covers the ears and frames the dark, almost bare face. The Black-faced tamarin – only discovered in 1990 – is also mostly gold, although the head is dark and it also has a black tail, hands, and feet. The other two species have black bodies, though with a golden mane and hindquarters respectively. All four species are found only in remnant patches of Brazil's Atlantic Forest, a belt of evergreen woodland in eastern Brazil now reduced to perhaps 5 percent of its original extent, with small extensions in Paraguay and northern Argentina. The east coast was the first part of Brazil to be colonized, almost 500 years ago, and it is now the most densely inhabited region of the country, containing the enormous conurbations of São Paulo and Rio de Janeiro.

Lion tamarins have always been restricted to low-altitude forests, usually below 300m (1,000ft). Because they are easy to clear for agricultural and pasture land, such forests are usually the first to disappear in the tropics. Habitat destruction in Rio de Janeiro state further increased in the 1970s when the Niterói Bridge was constructed across Guanabara Bay, allowing rapid access to beaches northeast of Rio and facilitating access to remaining portions of the lion tamarins' range. More recently, plantation owners in Bahia have taken to cutting down the remaining dense forest on their cacao estates in order to create pasture and diversify their crops. In addition, landless squatters have been invading public and private land in Rio de Janeiro and São Paulo states and cutting down forest as well as exploiting the remaining flora and fauna.

Added to this major problem of habitat destruction has been the live capture of Golden lion tamarins for pets and zoo exhibits (the other three species have been less affected by this commerce). The trade was declared illegal in the early 1970s, and since that time the number of animals captured for such purposes has diminished substantially. Even so, Golden lion tamarins still appear from time to time in the local markets of eastern Brazil. The monkeys also suffer number loss because of natural predators, such as ocelots, tayras, and even domestic dogs as well as snakes, hawks, and eagles.

All these factors have combined to result in critically low numbers surviving in the wild. The largest population is of Golden-headed lion tamarins in Bahia, where probably a few thousand remain. However, the forests in which they live are rapidly disappearing. Another protected area, the similarly-sized Serra do Conduru State Park, was set up in 1997 within the geographic distribution of the species, but since that time no lion tamarins have been recorded there. There may be 1,000 Black lion tamarins surviving in forest fragments in São Paulo state, with the largest cluster in the Morro do Diabo State Reserve, which covers 37,000ha (145sq mi) and smaller numbers in the 2,170ha (5,400-acre) Caetetus Reserve. Here too, however, there are problems with squatters that are degrading and cutting down the few habitat patches that remain.

Golden lion tamarins number about 800 individual animals. They are grouped in two main reserves – Poço das Antas, covering 6,500ha (16,000 acres), and the 2,400ha (6,000-acre) Uniao, which was only created in 1998 – as well as on about 15 tiny forest fragments on private land. The species is under severe threat from continued deforestation, however, much of which is undertaken to create weekend beach properties. Only about 40 percent of the Poço das Antas reserve is forested, and that area is blighted by almost annual fires that retard regeneration of secondary growth; in addition, the reserve is bisected by a railway line used to carry toxic chemicals. A now-unused dam, built 20 years ago, flooded part of its lands.

Perhaps as few as 250 Black-faced lion tamarins survive on an island reserve in Paraná, in southern Brazil, and along the mainland coast in São Paulo state. Other potential sanctuaries in the region, including the 150,000ha (580sq mi) Jacupiranga State Park, are largely deforested. These tamarins are threatened by the development of tourism, the harvesting of palm hearts, and further deforestation to provide more land for cattle ranching.

Efforts to save the lion tamarins began with the work of a leading primatologist, Dr. Adelmar Coimbra-Filho, in the 1960s. He not only

Above right *The Golden lion tamarin has endured high demands for live capture; however, there are captive breeding programs and recent reintroduction schemes appear to have been successful.*

Left *Golden-headed lion tamarins grooming. The only reserve created for this species, the 7,100ha (17,500 acre) Una Biological Reserve, has been plagued by squatters.*

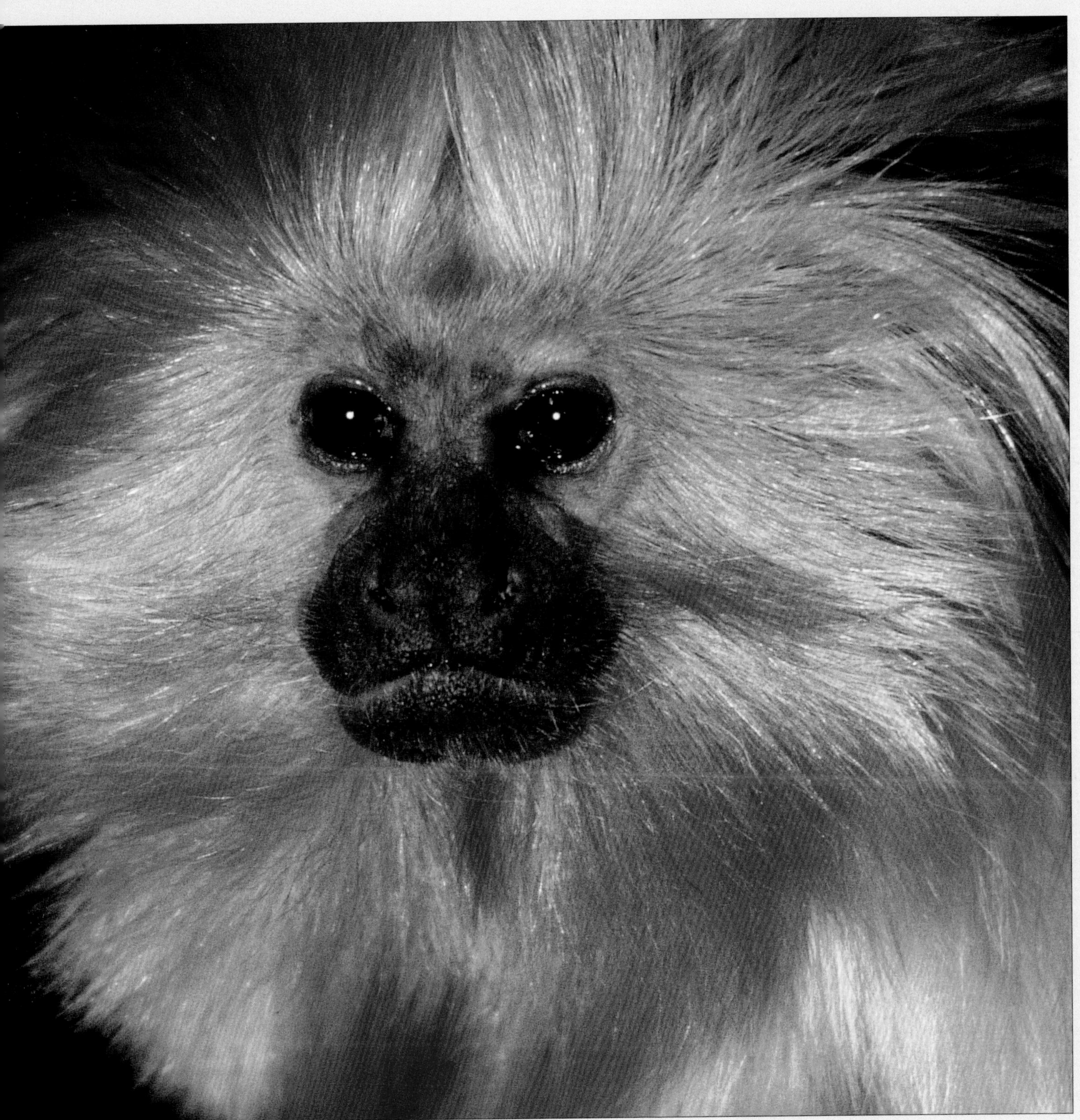

discovered the Black lion tamarin in 1973, but so founded the Rio de Janeiro Primate Center, here endangered Brazilian primates are bred.

In the 1970s the number of Golden lion marins held in zoos was dropping as rapidly as in e wild; however, researchers at the National Zoogical Park in Washington, DC, developed techques for managing and breeding the animals in ptivity. This led to the development of a plan, verseen by the Brazilian Golden Lion Tamarin ssociation, that integrated both zoo and field rategies to implement innovative genetic and demographic management of the captive population. The plan sponsored pioneering longterm field studies of the species' behavior and ecology, and undertook ground-breaking programs to educate local communities in Brazil about conservation of the species, as well as providing increased habitat protection for the tamarins and corridors between surviving forest fragments. The success of the project is vouched for by the fact that nearly 40 percent of the current wild population of Golden lion tamarins derives from a reintroduction program involving zoo-born animals. Conservation of the Black, and more recently of the Black-faced, lion tamarins will similarly involve multiple reintroductions and translocations to tackle the genetic isolation of the few remaining populations.

Yet despite the international efforts to preserve them, the future prospects for all four species remain bleak. The Golden, Black, and Black-faced lion tamarins in particular are still among the 25 most endangered primate species in the world, and without constant vigilance and protection any one could become extinct in the wild within the next decade. DGK/RAM

Capuchin-like Monkeys

The monkeys of the New World have evolved into an extraordinary array of ecological, social, and anatomical types, many of them unique. The family Cebidae includes the world's only nocturnal monkey, some of the world's brainiest nonhuman primates, and the only primates with prehensile tails.

New World monkeys encompass a huge variety. There are cebid species that occur only in a small and isolated mountain range (the Yellow-tailed woolly monkey) or a single small island (the Coiba Island howler), while others have spread throughout tropical South America (for instance, the night monkeys and the Brown capuchin). Their social organizations range from strict monogamy to large polygamous groups. Yet despite their broad range of adaptations, cebids share some common features, set apart from other primates by the wide form of the nose (specifically, of the septum that separates the nostrils), the absence of cheek pouches, and a shared tooth formula (I2/2, C1/1, P3/3, M3/3 = 36).

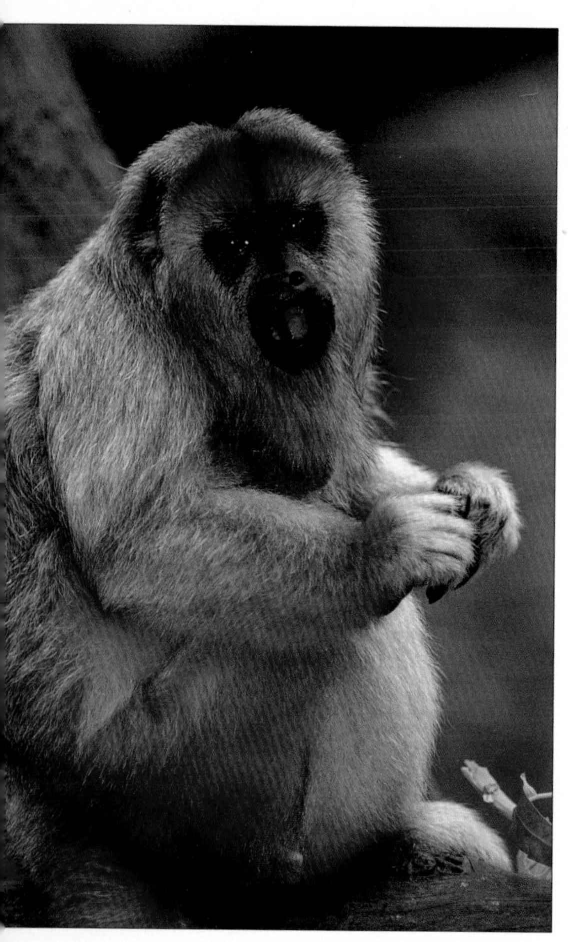

Toilers in the Treetops

FORM AND FUNCTION

Cebid monkeys are mostly found in tropical and subtropical evergreen forest, although some have adapted to elevations as high as 3,000m (9,900ft) and to forests with marked dry seasons. They live almost exclusively in trees, but some species will descend to the ground to play (White-fronted capuchin), to look for food, or to travel between patches of woodland. Nevertheless, unlike the Old World primates, none of the cebids show obvious specializations for life on the ground.

All the cebid monkeys that have been tested have some color vision, usually fairly acute, although they have poor sensitivity to the red end of the spectrum. The genetics of color vision in cebids is quite different from Old World monkeys, and one odd consequence is that males in some species are always colorblind, while only a fraction of the females are.

Species such as the squirrel monkeys, which move by leaping, have thighs that are shorter in relation to the lower leg than species that clamber, such as the howler monkeys; this adaptation allows more force to be exerted in the jump. Others, like spider monkeys, that swing hand over hand below branches have relatively long arms and extra mobility in the shoulder joint. The very flexible, prehensile tail of the larger cebids serves as a fifth limb, used to grasp branches for safety and to hang from when feeding near the tips.

Stiff Competition

DIET

Each species has a diet that matches, and may be limited by, its anatomy, in particular the forms of the jaws, teeth, and gut. Leafeating Mantled howler monkeys have relatively broad, flat teeth, very deep lower jaw bones, and a relatively large gut occupying one third of their body volume, because they must process large amounts of plant material. The insect-eating squirrel monkeys have sharp, narrow teeth and a short, simple gut occupying less than one sixth of its body.

New World monkeys also use behaviors to extend the range of foods they can eat. Like saki monkeys, capuchins eat very hard seeds, but they cannot open them with their teeth; instead, they resort to primitive tool use, banging one seed against another to crack the shell. Species that rely on scarce large fruiting trees, such as spider monkeys, have excellent spatial memory and can navigate directly from any location to the nearest fruit patch. Leafeating species, including howler monkeys, descend to the ground to eat claylike soils

__Above__ A Common squirrel monkey female with infant. Males of this species play no part in rearing th young, but other female "friends" will help carry and watch over an infant while its mother forages. Fema that do not have their own young are the ones most able to help a mother, and often are her own offspring from previous years.

__Left__ A female Black-and-gold howler monkey. Th is one of the few cebid species to exhibit sexual dimo phism in coat coloration. Whereas the female sports an olive-buff color, the male has an all-black coat.

CAPUCHIN-LIKE MONKEYS

Order: Primates

Family: Cebidae

47 (58) species in 11 genera: night monkeys (*Aotus*, 2 (10) species); titi monkeys (*Callicebus*, 13 species); squirrel monkeys (*Saimiri*, 4 (5) species); capuchin monkeys (*Cebus*, 4 species); saki monkeys (*Pithecia*, 5 species); bearded sakis (*Chiropotes*, 2 species); uakaris (*Cacajao*, 2 species); howler monkeys (*Alouatta*, 6 (8) species); spider monkeys (*Ateles*, 6 species); woolly monkeys (*Lagothrix*, 2 species); and the muriqui (*Brachyteles arachnoides*).

DISTRIBUTION America (Mexico) S through S America to Paraguay, N Argentina, S Brazil.

HABITAT Mostly tropical and subtropical evergreen forests, from sea level to 1,000m (3,280ft).

SIZE **Head–body length** from 25–37cm (10–15in) in male Squirrel monkey to 46–63cm (18–25in) in Woolly spider monkey; **tail length** from 37–45cm (15–18in) to 65–74cm (26–29in); **weight** from 0.6–1.1kg (1.3–2.4lb) to 12kg (26lb) or more. Males often larger than females, but not always.

COAT White, yellow, red to brown, black; patterning mostly around head.

DIET Fruits, nuts, seeds, leaves, insects, occasionally small mammals.

BREEDING Gestation from about 120 to 225 days, depending on genus.

LONGEVITY Maximum in 12–25 year range for most species.

See species table ▷

at appear to help bind the poisonous comunds found in many tropical tree leaves.

The distinctive differences in form and ecology ong species are closely related to body size. e smallest and lightest species can most easily p from branch to branch without risk of injury, ich may explain their lack of a prehensile tail. e small species also inevitably have small, weak vs, which limit the size and hardness of the its they can eat. Small size also means a relaely fast metabolism and thus a high demand scarce energy-rich and protein-rich foodstuffs, ch as insects and ripe fruits. Finally, the small t and short digestive times of the smaller cebid onkeys are adequate for easily digested foods, t not adapted for tough plant materials like ature leaves. Large monkeys, of course, have the opposite advantages and disadvantages: high risk of falling, strong jaws, relatively low energy demand, and long digestive times.

Although cebid monkeys coexist with many nonprimate competitors, they compete most directly with other cebid monkeys. It is common for as many as five species to feed on one tree species, or even in the same tree, in which case physical clashes determine who has precedence. Usually the smaller or less agile species are evicted, and this vulnerability may have contributed to their specialized lifestyles – for instance, titi monkeys are able to eat green fruit before it is palatable to the larger monkeys, while night monkeys feed at night in trees that are dominated during the day by big species (see box). Squirrel monkeys find safety in numbers too great for larger monkeys, living in smaller groups, to chase out of a tree.

Perhaps as a response to feeding competition between species, even the larger cebids show a number of fairly distinct ecological types. Bearded sakis are specialized for opening the seeds inside fruits, whereas uakaris, which eat many of the same foods as other medium-sized monkeys, are found only in a restricted habitat – flooded swamp forest – where other cebids are absent. Howler monkeys are the only species able to survive for appreciable periods on a diet purely of leaves.

Different species within one genus are likely to compete severely because they are more likely to be similar in anatomy and ecology. The idea that competition for food has influenced the evolution of cebids is supported by the fact that in any one place it is rare to find more than one species of the

same genus. Where there is an overlap, such as between the Brown capuchin and other capuchin species, there is some evidence that they have reduced competition by eating less similar foods where they coexist than where they occur alone.

There is a growing body of evidence to suggest that some cebid species may use plants selectively for purposes other than eating. White-faced capuchins in Costa Rica were seen to apply to their skin material from at least four plant genera (*Citrus*, *Clematis*, *Piper*, and *Sloanea*). The first three contain secondary compounds with known anti-insect and/or medicinal properties; ethnographic records indicate that indigenous peoples throughout the New World use these plants in similar ways.

Societies Shaped by the Search for Fo

SOCIAL BEHAVIOR

Each species of cebid monkey also has its distinc tive social structure. Most smaller cebids in the 0.7–1.5kg (1.5–3.3lb) range, including the nigh monkeys, titi monkeys, and probably also sakis, are monogamous, maintaining the pair bond bo by friendly interactions between the adult coupl and by active aggression toward other members of the same species and sex. In monogamous species, juveniles may stay with their parents for year or two after the birth of the next infant, so that groups of up to five animals may be seen.

Exceptionally among the smaller species, the squirrel monkeys live in large groups, typically containing 30–40 animals, with up to a dozen reproductive females and several adult males. Squirrel monkey species are unusual among primates for the diversity of social systems they sho So, for example while female Bolivian squirrel monkeys stay in their natal group and interact closely with relatives, Red-backed females migra upon becoming adult and exhibit few friendly interactions among each other or with males – whereas in Common squirrel monkeys, females interact mostly with males.

All the larger cebids (2–9kg/4.4–20lb) typically live in groups of at least five animals and are more or less polygamous. At low population densities, howlers and capuchins often live in harems in which one male monopolizes 1–3 females. At higher population densities, these same species occur in larger groups of 7–20 animals that include several adult and subadult males. Such large, multimale groups are normal for bearded sakis, uakaris, woolly monkeys, and muriquis. Spider monkeys have variable social groupings in which individuals daily join and leave subgroups of changing size and composition.

In these larger species, females seem to prefer to mate with the dominant male or males, but will solicit and mate with subordinate or subadult males as well. However, genetic analyses show that in Brown capuchin monkeys the dominant male sires nearly all the offspring, even though he accounts for only about half the matings. Conversely, it appears that in woolly and Woolly spider monkeys mating success is not related to dominance. Instead, a female will mate with a number of males, and success may depend simply on the volume of sperm produced by the male; indeed, among the New World monkeys, Woolly spider monkeys have the largest testes relative to their body size.

Group size seems to depend to a large extent on the productivity and abundance of the food. Most species that live in small groups feed on small, scarce, and scattered resources, such as insects and small fruit crops in vines. Large-group species typically use productive but sparse clumps of resources, such as fruits on large fig trees. The exception to this pattern are the night monkeys, which live in small monogamous groups but feed in large, clumped food trees.

How a species uses its home range also depends to a large extent on how its food resources are distributed. Titi monkeys, for example, feed in small, scattered food trees and use their home range in a very even fashion, visiting each part of it every few days. Bolivian squirrel monkeys and White-fronted capuchins depend on large, clumped, or rare trees, and their large groups travel their home ranges in an uneven way, spending most of their time in a small sector until the fruit there is exhausted, then moving on to find new sources.

Above *Representative species of capuchin-like* *onkeys, showing their movement among the trees.* *most species, males are slightly larger than the* *males (shown here); in a few species, there are color* *fferences between the sexes.* ***1*** *Guianan saki* (Pithecia *thecia) – the male has a mat of buff hair on top of* *e head.* ***2*** *Red uakari* (Cacajao rubicundus). *Both* ***3*** *e Dusky titi* (Callicebus moloch) *and* ***4*** *the Common* *uirrel monkey* (Saimiri sciureus) *move chiefly by leap-* *g and have tails that are not prehensile.* ***5*** *Brown* *puchin* (Cebus apella). ***6*** *Black-and-gold howler* *onkey* (Alouatta caraya) *– as in other larger cebids,* *tail is prehensile, naked beneath the tip for better* *ipping. In* ***7*** *the Black-handed spider monkey* (Ateles *eoffroyi) the tail may be used for picking up small* *jects, such as items of food.* ***8*** *Humboldt's or* *nokey woolly monkey* (Lagothrix lagotricha).

SOCIAL STATION IN BROWN CAPUCHINS

As a group of Brown capuchins spreads out in search of food, it is possible to predict which position each monkey will take up as it forages. Each individual's position depends upon its foraging success when food is scarce. If there is squabbling at a food tree, it is the dominant male and those he tolerates that almost always feed first and longest. Subordinates must wait until his entourage has left the tree.

All animals in a foraging group try to be close to food and safe from predators. Individuals in the center are generally safest, for they benefit from the watchfulness of their neighbors, whose presence also shelters them from direct attack. However, food availability may be highest at the leading edge of a moving group, where new resources are being discovered. Best of all is the position just behind the leading edge, for its occupant can parasitize the efforts of the 8–12 members who scout ahead. This is the dominant's favored spot; he is the only group member that obtains more food by "scrounging" the discoveries of others than by finding his own.

Group members keep a wary eye on the dominant male and adjust their own position accordingly. He is very tolerant of infants and young juveniles, nearly all of whom he will have sired, and these tend to follow him, just behind the center of the group. Tolerated females, plus their older offspring, often form the leading edge, getting the best access to food, but standing in the line of predator fire. Females and males not tolerated by the dominant male are found on the periphery of the group, at the trailing edge, or even further away. Their priority seems to be to avoid aggression, even if this means spending more time looking for food and a greater risk of attack by predators.

These patterns of home-range use in turn affect how neighboring groups of a species interact. Species that exploit their home range in an even way are usually territorial, and may defend their range with loud dawn calls, as in the case of titi and howler monkeys. At the opposite extreme, different groups of Bolivian squirrel monkeys seem to overlap completely, and usually show no overt aggression toward each other even when feeding in the same fruit tree.

Regardless of the extent of competition for food between groups, feeding competition between members of the same group is often intense. Within their groups of 3–15 or more, Brown capuchins often fight over food (see box), and individuals that cannot win aggressive confrontations suffer markedly reduced feeding success. Fighting is most common in small fruit trees, or when food is scarce. At these times some group members may forage alone or in smaller groups.

However, there are also compensating benefits to large group size. For instance, individuals in large groups probably suffer less predation; the more eyes and ears there are to watch for predators, the lower the chance of a successful predator attack. Throughout Central and South America there are a variety of large hawks and eagles, some of them specialized to eat monkeys. In southeastern Peru, each capuchin group is attacked by an eagle once every two weeks on average, and there are less serious threats up to several times a day. Brown capuchins are so wary that they consistently take alarm at harmless birds that fly by.

Other advantages to group life include communal discovery and defense of food trees. Although individuals of many species know in advance what trees will be in fruit and where they are, some portion of their diet comes from sources that they find by chance. In either case, extra individuals in the group can increase the pool of food available to the group as a whole by sharing knowledge about fruit trees; at least one species, the Brown capuchin, has very distinctive loud whistles which are given when a group member finds a rich food source. Larger capuchin groups also usually win fights for possession of fruit trees.

Relationships between males and females in cebid species range from friendly to occasionally antagonistic. Males can perform useful services fc female group members, not least because they ar more vigilant and often detect and alert the other to predators or other monkey groups. Among spi der and some squirrel monkeys, some males form subgroups that move independently of females except when a female is ready to mate.

Several cebid monkeys actively form mixed groups with members of other species. A great variety of stable associations have been recorded, notably between squirrel monkeys and capuchin squirrel monkeys and uakaris, even capuchins an spider monkeys. Although both species in such a group may benefit from reduced predation risk, research in Peru and Panama suggests that the benefits are often one-sided. For instance, it is almost always the squirrel monkeys that join up with the capuchins, and capuchins usually determine the direction of movement of the combinec group. Because capuchins have far smaller home ranges than squirrel monkeys – 1–2sq km (0.4–0.8sq mi) compared with over 4sq km (1.5sq mi) – they are probably better at locating fruiting tree

Some social displays are common to many, if not all, cebid genera. Most species show some form of threat behavior to other members of the same species, and even toward other species and predators. The usual posture is with the mouth opened and teeth bared to expose the canines. Not surprisingly, this display seems to be absent in titi monkeys, which have small canines, and night monkeys, which fight at night. Along with the open mouth display, a monkey often shakes a nearby branch or tries to break it off.

Grooming is the most common friendly behavior shown by cebids. Often an individual will approach another and lie down in a characteristic posture on the branch next to it, or even right on top of its feet. The groomer usually grooms areas difficult for the other monkey to reach or see. An individual grooms for only a few minutes at a time, then turns around and solicits grooming from its partner: the pair usually exchange grooming bouts several times; the sequence stops wher one of the partners refuses to return grooming.

The behavior of infant monkeys differs from that of adults in many ways. They are clearly less coordinated and less able to match appropriate behavior to particular occasions. Some juvenile

Above right *The unmistakable white facial mask of the male Guianan saki. Like many other New Worl monkeys, Guianan sakis live in monogamous pairs or small family groups.*

Left *A Black uakari in the Brazilian rain forest. Together with its relative, the Red uakari, it is the only short-tailed primate inhabiting the New World.*

behavior is almost entirely absent from the adult repertoire, and vice versa. It seems that many adult behavior patterns have to be learned before an individual can perform them correctly. Since their food supply is assured, infants spend much of their time at play, either exploring their environment or play-fighting with other juveniles.

Much still remains to be learned about cebid monkeys. A number of species are known only from museum specimens. Opinions on the division of the family into subfamilies, for example, are constantly changing. Recent evidence from DNA sequences suggests that the night monkeys, squirrel monkeys, and capuchins may be more closely related to the marmoset family than to other cebid species. Only a few species have been studied for over five years, yet longterm studies are essential to understanding the behavioral flexibility and development of individuals and the relationship between environment and behavior.

Rain Forests in Retreat

CONSERVATION AND ENVIRONMENT

Nearly one-third of cebid species are listed as Endangered or Vulnerable. In the Amazon basin, spider monkeys and woolly monkeys are shot for food and have been eradicated from many areas. These species are especially vulnerable because of their slow maturation and low reproductive rates, as well as their dependence on mature rain forest – currently being felled at a rate of over 40,000 hectares (100,000 acres) a day in some areas.

Smaller cebids, like the titis, night monkeys, and squirrel monkeys, can adapt more easily to habitat loss and change. But even for these, the impending destruction of mature rain forest throughout South America will prevent scientists from ever observing them in a truly natural state. Species that are restricted to tiny ranges or are naturally fairly scarce may become extinct before their unique adaptations can be studied. CHJ

MONKEYS IN THE MOONLIGHT

Inhabiting the forests of much of South America, night monkeys (*Aotus* spp.) are the only truly nocturnal monkey species. With their enlarged eyes, the monkeys have excellent night vision. On moonlit nights they make spectacular 3–5m (10–16 feet) leaps from tree to tree, adeptly locating fruit trees and catching insects even in very low light levels.

Morphological evidence suggests they were not always nocturnal. In particular, the structure of the eye indicates that the ancestor of *Aotus* was exclusively day-active and had some color vision (whereas Night monkeys have no color vision at all). So why has a day monkey evolved into nocturnal species?

The answer may lie in the need for safe access to food sources. Other small South American monkeys are hunted by diurnal raptors, and large monkeys, particularly capuchins, chase smaller monkeys from fruit trees. Night monkeys avoid these two problems by only venturing out punctually 15 minutes after sunset, when daytime predators and competitors have roosted. The only nocturnal predators big enough to eat a night monkey are jaguars and ocelots, which are not sufficiently agile to catch them in the trees.

However, in the open forests of the dry *chaco* of Paraguay and Argentina, where daytime predators and competitors are rare but nocturnal Great horned owls (*Bubo virginianus*) are common, night monkeys have partially reverted to daytime activity. They sleep on open branches and are active for 1–3 hours in daylight. In fact, in the cold *chaco* winter, during times of no moonlight, groups travel nearly as far by day as in the night. Yet much of these groups' behavior, such as fighting in the moonlight, does not change and is typical of the night monkey in all habitats. PCW

Above *The Northern night monkey* (A.trivirgatus).

Capuchin-like Monkey Species

THE CONSTANT REVISION OF THE TAXONOMY of the cebids is a reflection of how little is known about these New World monkeys. Many species have not been studied since they were first discovered, and their distributions are known only from a few collecting localities. The scientific criterion for distinguishing a species – whether it can interbreed extensively with other animals – cannot be applied to many cebids, since their ranges do not overlap. In other cases, there are important differences in the genetic material between forms that cannot easily be distinguished by color. New forms are still being discovered and described, and forms that were once deemed subspecies have been elevated to full species status by some authors. In most cases, these species do not overlap with other species in the genus and thus form a "superspecies" group. In the following accounts, species names appear in bold, but not every species is given a separate entry. Where too little is known about a species, it appears within a group entry on its superspecies complex – the present author recognizes two superspecies groups, the titi monkeys.

At the higher taxonomic level, recent evidence from DNA sequences supports an earlier suggestion that the cebids actually have two origins. In this new view, capuchins (*Cebus*) and squirrel monkeys (*Saimiri*) are the closest relatives of the marmosets and tamarins, with the night monkeys *Aotus* a possible primitive member of this group. Species in this group all depend largely on insects for protein. The remaining New World monkey species form a distinct group of more omnivorous and folivorous species. The treatment used here is the more usual one of keeping the marmosets and tamarins as a separate family and uniting the remaining monkeys, awaiting more definitive molecular data for a resolution of the evolutionary relationships of this primate radiation.

NIGHT MONKEYS

Genus *Aotus*

Found in many habitats from dry to wet, from sea level to over 3,000m (9,840ft). Eats mainly fruits, leaves, and insects; flowers eaten seasonally, as are vertebrates and eggs. Mainly nocturnal. Monogamous, in groups of 2–5 individuals. Sexes similar in size. Gestation 120–133 days; litter size one, rarely two; birth frequency one per year, births seasonal in some areas.

Northern night monkey
Aotus trivirgatus
Northern or Gray-necked night monkey, Owl monkey, douroucouli

From 81°W in Panama S to N bank of Amazon in Peru and Brazil except Guianas, most of Venezuela and E Brazil. Consists of 5 chromosomally distinct, but superficially similar forms – *A. brumbacki* VU, *A. lemurinus* VU, *A. hershkovitzi*, *A. trivirgatus*, and *A. vociferans* (some authorities regard these as a superspecies group).
HBL 29.6–42.0cm; TL 25–44cm; WT 0.83–1.27kg.
COAT: grizzled brown or gray on back, back of head, and limbs; underside buff-white; head distinctive, with triagonal white patches above large eyes, and three black stripes, between and either side of eyes, converging on crown.

Southern night monkey
Aotus nigriceps
Southern or Red-necked night monkey, Owl monkey, douroucouli

From S bank of Amazon in Peru and Brazil to N Argentina except E and SE Brazil. Consists of five chromosomally distinct, but superficially similar forms – *A. azarai, A. infulatus, A. miconax, A. nancymae, and A. nigriceps* (some authorities regard these as a superspecies group). Chiefly nocturnal, but partly diurnal at far southern limit of range.
HBL 29.6–42.0cm; TL 25–44cm; WT 0.78–1.1kg.
COAT: grizzled brown or gray on back, back of head, and limbs; underside orange; head distinctive, with triagonal white patches above large eyes, diamond–shaped spot between eyes on forehead, and black stripes outside and behind the ocular white patches.

TITI MONKEYS

Genus *Callicebus*

In disturbed to mature, moist to rain forest; from sea level to about 1,000m (3,280ft). Prefer understory up to 10m (33ft). Feed on fruits, often unripe, insects and leaves; may ingest dirt from leaf-cutter ant mounds. Coat dense and long. Canines short relative to other cebids. Males do not consistently weigh more than females. Diurnal. Monogamous, in groups of 2–6. Gestation 155 days; litter size always one; births once a year, usually in early rainy season.

Dusky titi
Callicebus moloch (superspecies group)

S of the Amazon river in C and E Amazonia to S Bolivia. In this group are gathered five species, including one with a tiny distribution (*C. cinerascens*), and the rest with abutting geographical distributions with little or no overlap (*C. hoffmannsi, C. donacophilus, C. brunneus*, and *C. moloch*). Favor swampy, flooded, disturbed habitats along rivers.
HBL 31–45cm; TL 36–53cm; WT 0.7–1.2kg.
DIET: mainly fruit, but dusky titis eat more leaves than the Yellow-handed titi, and can live on leaves and green fruits alone for short periods. Insects form about one-sixth of the diet, but often are conspicuous noxious or defended prey (such as larvae of ants).
COAT: gray to red-brown; belly same as back, or distinctly paler orange (*C. hoffmannsi*); forehead does not contrast with rest of face. Tip of tail usually paler than remainder.

Chestnut-bellied titi
Callicebus cupreus (superspecies group)

N and S of Amazon river in Peru and Brazil, west of Rio Negro, east to Andes and south to N Bolivia. Prefers lowland forest along riverbanks. This group contains up to six species, two with abutting distributions (*C. dubius* VU and *C. cupreus*), one (*C. caligatus*) that overlaps broadly with *C. cupreus*, and three species with minute geographical ranges which may be relict species associated with this group (*C. olallae, C. modestus*, and *C. oenanthe* VU). Diet not known in detail for any species in this group.
HBL 30–41cm; TL 33–48cm; WT 1.2kg.
COAT: buff, grizzled to agouti or brown, usually with a contrasting belly in shades of orange to red–brown; the forehead contrasts with the rest of face in all the species except the Bolivian relicts *C. modestus* and *C. olallae*. The tail is generally uniform in coloration.

Yellow-handed titi
Callicebus torquatus
Yellow- or White-handed titi, Widow monkey, Collared titi

N of Amazon and Rio Marañon (Peru) to Orinoco and the Guianas; S of Amazon to Rio Purus in Brazil. Prefers unflooded, mineral-poor, sandy-soil forests. Overlaps with several species in the Dusky and Chestnut-bellied titi superspecies grou[ps].
HBL 23–36cm; TL 42–49cm; WT about 1.1–1.5kg.
DIET: eats fewer leaves than Dusky titi, and forages more actively for insects.
COAT: red-brown to black; forehead fa[ce], throat, and tip of tail may have contras[ting] white-cream to orange fur; fur on hand[s] pale cream; forearms, legs, and feet bla[ck]; tail dark brown to black.

Masked titi
Callicebus personatus

E coast of Brazil from Bahia S to São Pa[ulo] (no overlap with other species).
HBL 31–42cm; TL 42–56cm; WT 0.97–1.65kg.
DIET: mostly fruit, with leaves about o[ne-]fifth of total; no insect feeding known.
COAT: medium brown with buffy tips o[f] hairs; belly may be more red-brown; e[n]tire head fur contrasting dark brown or bla[ck]; hands and feet always black; tail reddis[h] brown mixed with black.

SQUIRREL MONKEYS

Genus *Saimiri*

Moist to wet, disturbed to mature upla[nd], river edge, and mangrove forest. Diurna[l]. Polygamous, in groups usually of 30–4[0] individuals. Litter size one. Birth freque[n]cy one per year.

Common squirrel monkey
Saimiri sciureus

Amazonia and Guyanan Shield N of the Amazon river from the Atlantic ocean to the Andes mountains, S of the Amazon only in E Brazil. From sea level to about 2,000m (6,560ft). Close relatives of this species include *S. vanzolinii* VU, with a tiny geographic range encompassed by *sciureus*, although the two types can be distinguished by their chromosomes. Complex vocalization system with up to 20 distinct call types.
HBL 28–37cm (female), 27–37cm (male); TL 38–45cm (female), 37–45cm (male); WT 0.65–1.25kg (female), 0.55–1.15kg (male). Males not markedly larger than females.
DIET: mostly fruit and insects, but may survive on insects.
COAT: short, generally olive on back; underside paler yellow; limbs, hands, a[nd] feet orange-yellow. Face, throat, and ear white, with contrasting black muzzle.

ABBREVIATIONS	HBL = head–body length TL = tail length WT = weight	Ex Extinct	En Endangered
	Approximate nonmetric equivalents: 2.5cm = 1in; 230g = 8oz; 1kg = 2.2lb	EW Extinct in the Wild	Vu Vulnerable
		Cr Critically Endangered	LR Lower Risk

◐ **Left** *In common with other spider monkeys, the Black-handed spider monkey* (Ateles geoffroyi) *is a supremely agile acrobat of the rain forest, which uses its long limbs and prehensile tail to swing swiftly through the trees.*

hite fur over the eye forms a pointed othic" arch. Crown dark green to gray to ıck.
EEDING: gestation 170 days; births fair- synchronous within a group, but birth aks do not follow strict seasonality.

olivian squirrel monkey
imiri boliviensis

Amazonia S of the Amazon river to SE livia. From sea level to about 1,000m 280ft) elevation.
L 31cm; TL 36cm; WT 0.7–0.9kg male), 0.96–1.1kg (male). Males com- te for dominant status and are larger an females; during the breeding season, males become "fatted" with a marked increase in size about the shoulders and neck.
DIET: prefers fruit in large figs and other canopy trees when available, but can survive on scattered small fruit sources and insects. Females form aggressive alliances with relatives to gain access to fruit trees.
COAT: short, generally green-gray on back, darker in females. Underside paler white. Face, throat, and ears white, with contrasting black muzzle. White fur over the eye forms a rounded "Roman" arch. Crown and sideburns black.
BREEDING: gestation 155–170 days; births very seasonal and synchronous within a group.

Red-backed squirrel monkey En
Saimiri oerstedii

C America from Costa Rica to Panama (80°W). Near sea level.
HBL 27cm; TL 36cm; WT 0.6–0.79kg (female), 0.75–0.95kg (male). Males strive for dominant status and are larger than females; during the breeding season, males become "fatted" with a marked increase in size about the shoulders and neck.
DIET: mostly scattered fruit and insects.
COAT: appearance similar to Common squirrel monkeys, but with a black crown.
Breeding: gestation period unknown; births very seasonal and synchronous within a group, probably as a defense against predators attacking the recent newborns.

Golden-backed squirrel monkey
Saimiri ustus

C Amazonia S of the Amazon river to N Bolivia, overlapping with neither *S. sciureus* or *S. boliviensis*. Formerly considered a form of *S. sciureus*. From sea level to about 400m (1,310ft).
HBL 31cm; TL 44cm; WT 0.71–0.88kg (female), 0.62–1.2kg (male).
DIET: chiefly fruit and insects, but may survive on insects.
COAT: short, generally golden on back; underside white; limbs, hands, and feet yellow. Face white with contrasting black muzzle. Ears less hairy and more golden-colored than other *Saimiri*. White fur over the eye forms a pointed "gothic" arch. Crown grey to golden-brown.
BREEDING: gestation period unknown.

CAPUCHIN MONKEYS

Genus *Cebus*

In many habitats from dry to rain forest, from sea level to over 2,500m (8,200ft), also riverine forests, swamps, mangroves, seashores. Prefer understory to mid-canopy, but often come to ground to forage or play. Mostly eat ripe fruits and insects, but may use unripe fruits, vegetation, flowers, seeds, roots, other invertebrates (snails, spiders), young and adult vertebrates (especially nestling birds), cultivated crops. Fur of medium thickness and length. Tail prehensile. Males bigger than females in body and relative canine size. Diurnal. Polygamous, in groups of 3–30 with 1 to several males. Adult male/female sex ratio in groups is about 1:1 in small groups, up to 1:3 in large groups. Complex social behavior, with frequent grooming and formation of aggressive alliances within and between the sexes. Allo-parenting mostly takes the form of carrying of offspring. Intricate sexual courtship usually initiated by females. Although not strictly territorial, groups are usually aggressive toward each other. Gestation about 150 days; litter size one (twins very rare); births usually once every 2 years, but may be faster if previous infant dies. Births often seasonal, in early rainy season, but not as well synchronized as the squirrel monkeys. Slow to mature, with first reproduction in females typically at age 5–6 years. Extremely long-lived for their size, 44 or more years in captivity. Two species groups exist, one including only *C. apella*, which overlaps with members of the other group; the other three species do not overlap or barely meet.

Brown capuchin
Cebus apella
Brown, Tufted, or Black-capped capuchin

Throughout South America E of the Andes in subtropical and tropical forest, except Uruguay and Chile. Includes *C. xanthosternos* CR, which some primatologists regard as a separate species. Prefers moister forests than other capuchins, ranges to higher elevations (2,700m/ 8,860ft). congregates in groups of 3–15 animals, but can reach 35 in Argentina.
HBL 33–48cm (female), 32–56.5cm (male); TL 38–47cm (female), 38–56cm (male); WT 1.9–3.0kg (female), 1.9–4.8kg (male).
DIET: vegetation and larger fruits.
COAT: coarser than other species, generally light to dark brown, paler on belly; extremities of limbs always black; facial pattern is highly variable, but usually pale skin shows through sparse whitish fur on face, surrounded by distinctive black sideburns extending to chin; black cap forming a downward pointing triangle on forehead, usually dividing forehead. Hairs may extend to form more or less distinct paired "tufts" on crown.

White-faced capuchin
Cebus capucinus
White-faced or White-throated capuchin

C America from Belize south to N and W Colombia. In dry to wet forests up to 2,100m (6,900ft), and in mangroves. Group size 10–24.
HBL 32–40.5cm (female), 33–46cm (male); TL 42–45.5cm (female), 40–50cm (male); WT 2.6–4.1kg (female), 3.2–5.5kg (male).
COAT: fur on upper limbs, sides of body and belly pale cream to white, grading to black on back and extremities of limbs; distinct cap nearly black, rest of facial fur white; longer forehead and crown hairs on older individuals may form a ruff.

Weeper capuchin
Cebus olivaceus
Weeper or Wedge-capped capuchin

S America N of Amazon and N and E of Rio Negro in Brazil through the Guianas to central Venezuela. Habitat similar to *C. capucinus*. Group size 10–33. May contain a newly-described species, *C. kaapori*.
HBL 39cm (female), 38–46cm (male); TL 45cm (female), 44–49cm (male); WT 2.9–3.2kg (female), 3.0–4.5kg (male).
COAT: similar in pattern to White-faced capuchin, but less contrast between light and dark colors on body, rarely becoming black; crown patch narrow, coming to distinct point on forehead.

White-fronted capuchin
Cebus albifrons

W and S of range of Weeper capuchin, from Venezuela and Brazil W of Rio Negro and Tapajós throughout upper Amazon to Bolivia; also W of Andes in Ecuador, and in N Colombia and W Venezuela; Trinidad. May prefer less disturbed, moister forests than other capuchins, but is found along river edges in savanna habitat. Group size 7–30.
HBL 33–42cm (female), 35–44cm (male); TL 41–50cm (female), 40.5–49.5cm (male); WT 1.55–2.2kg (female), 2.7–3.3kg (male).
COAT: very similar to Weeper capuchin, but never dark brown to black; cap paler and broad, covering most of top of head; limb extremities and tail tip usually paler, unlike other *Cebus*.

SAKI MONKEYS

Genus *Pithecia*
Prefer mature high-ground forest, but also in disturbed forests and savannas; rarely swamp forests. From sea level to about 500m (1,640ft). Generally prefer upper levels of forest including the canopy. Diet consists of fruits, including mature and immature seeds, and some leaves.
Coat is coarse and long, especially on tail, which is not prehensile. Gap between canines and premolars. Males slightly larger than females. Diurnal. Probably monogamous or in extended family groups of 2–7 individuals. Groups are often spread apart, so that only 1–2 individuals are seen feeding in one location. Gestation 163–176 days (known only for Guianan saki); litter size one; birth interval one year.

Guianan saki
Pithecia pithecia
Guianan or White-faced saki

N of Amazon, E of Rios Negro and Orinoco, in lowland and montane forests. Prefer lower and middle canopy levels, unlike other sakis.
HBL 33–34cm (female), 33–37.5cm (male); TL 34–43.5cm (female), 35–44.5cm (male); WT 0.78–1.75kg (female), 0.96–2.5kg (male).
COAT: one of the few cebids with different coat color in males and females. Male black except for white to reddish forehead, face, and throat; female brown to brown-gray above, paler below, with white to pale red-brown stripes from eyes to corners of mouth.
BREEDING: birth peaks in December–April.

Monk saki
Pithecia monachus
Monk or Red-bearded saki

Upper Amazonia N of Amazon river in Brazil, Peru, Colombia, and Ecuador.
HBL 40.5cm (female), 40–44cm (male); TL 40–50cm (female), 45–47cm (male); WT 2.2kg (female), 2.5kg (male).
DIET: more fruit pulp than seeds, which form about one-third of diet.
COAT: grizzled brown-gray hands and feet paler; beard and underside reddish; males have mat of buff-colored hair on forehead and crown.

Bald-faced saki
Pithecia irrorata

W and central Amazonia S of the Amazon river to N Bolivia and E Peru.
HBL 38–42cm (female), 39–41cm (male); TL 47–55cm (female), 47–50cm (male); WT 2.2 kg (female), 2.9kg (male).
COAT: black with silvery-white tipped hairs, producing overall gray color.

Buffy saki
Pithecia albicans

Between lower Rios Juruá and Purus in Brazil, restricted to primary flooded and unflooded forests.
HBL 36.5–40.5cm (female), 40–41cm (male); TL 40.5–45.5cm (female), 42–44cm (male); WT 3.0kg.
DIET: seeds comprise nearly half of diet, followed by fruit pulp. Young leaves and flowers form one-sixth of diet.
COAT: black back and tail. Limbs, ventral fur, and head ruff buff to red.

Equatorial saki
Pithecia aequatorialis

Distribution broadly overlaps that of *P. monachus*, but is somewhat more to W. Formerly considered a subspecies of *P. monachus*. Preference for riverine and flooded forests.
HBL, TL, and WT not available.
COAT: generally black with short white tips. Face white, in contrast to rest of coat, similar to the Guianan saki, but underparts are orange-brown, sexes do not differ in color, and the distributions of the two species do not overlap.

BEARDED SAKIS

Genus *Chiropotes*
From near sea level to mountain forests. Prefer upper canopy, rarely in understory. Feed primarily on seeds, often unripe, and fruits; leaves eaten rarely. Coat of medium length, dense except for bushy tail, which is not prehensile except in infants. Canines and premolars separated by a space. Sexes similar in size. Diurnal. Probably polygamous; in groups of 8–30 individuals, with more than one male. Gestation about 5 months; litter size one; birth frequency unknown, but births seasonal in some areas at start of rainy season.

Bearded saki En
Chiropotes satanas
Bearded or Black saki

N of Amazon, E of the Rio Negro and S of Rio Orinoco into the Guianas, and S of Amazon from Rio Tocantins E to Atlantic. Prefers unflooded, mature forests, but found in montane savanna and blackwater swamp forests.
HBL 38–41cm (female), 40–48cm (male); TL 37–42cm (female), 40–42cm (male); WT 1.9–3.3kg (female), 2.2–4.0kg (male).
DIET: seeds form two-thirds of diet, followed by fruit.
COAT: mostly black, with back and shoulders light yellow-brown to dark brown; distinctive swollen temples covered with fur, and beards roughly as long as the face.

White-nosed saki
Chiropotes albinasus

S of Amazon between Rios Xingu and Madeira, to near Bolivia. In both high ground and blackwater swamp forests.
HBL 41–47cm; TL 38–45cm; WT 2.2–2.8kg (female), 2.7–3.3kg (male).
DIET: more fruit and less seeds than that of *C. satanas*.
COAT: entirely black except for pink nose covered with short white hair; temporal swellings and beard less developed than Bearded saki.

UAKARIS

Genus *Cacajao*
Apparently prefer lowland flooded forest, if not limited to it. Eat mostly seeds (two-thirds of diet), fruit, flowers, leaves, and some insects. They travel in the upper canopy, but will descend to or near the ground to pick up fallen seeds and fru[it]. Coat long and coarse. Tail very short, [c]about one-third body length. Canines [and] premolars separated by a space. Males slightly larger than females. Diurnal. P[rob]ably polygamous in groups of 15–30 o[r] more, with several adult males. Gestat[ion] about 6 months; litter size one; birth f[re]quency unknown in wild, births seaso[nal] in captivity (semi free-ranging conditi[ons]

Red uakari
Cacajao calvus
Red or White uakari

N of Amazon from Rio Japura (Brazil) [W] to Andean foothills; S of Amazon from Juruá (Brazil) to Rio Huallaga (Peru). T[he] Red uakari occurs nearly always in blac[k]water flooded forest, while the white s[ub]species prefers whitewater flooded fore[st].
HBL 54–57cm (female), 54–56cm (mal[e]); TL 14–16.5cm (female), 15.5–18.5cm (male); WT 2.9kg (female), 3.45kg (ma[le]).
COAT: entirely white to chestnut-red; f[ore]head and crown sparsely haired or bal[d]; facial skin pink to scarlet, may become paler if diseased.

Black uakari
Cacajao melanocephalus
Black or Black-headed uakari

N of Amazon, E of Rio Japurá to Rios Negro and Branco. Blackwater swamp [for]est, upland forests on sandy soils; near [sea] level to 600m (1,970ft) elevation.
HBL 30–50cm; TL 12.5–21cm; WT 2.4–4.0kg.
COAT: head and neck black; back, hindlimbs and tail chestnut-brown to y[el]lowish; facial skin black, fur extends ov[er] crown to forehead.

HOWLER MONKEYS

Genus *Alouatta*
In modified and undisturbed dry to rai[n] forests, wooded savannas and gallery forests out to mangrove forests. From s[ea] level to about 2,500m (8,200ft), but us[u]ally below 1,000m (3,280ft). Prefer low to middle canopy level, but descend to understory to feed, and travel on groun[d] when necessary. Prefer to eat fruit, but with high proportion of unripe fruit. Di[et] 40 percent or more mature or young leaves. Fur medium length, sparse on belly. Tail is prehensile. Male larger tha[n] female. Diurnal. Polygamous, in groups of usually 4–11, with 1 or more males. Gestation 180–194 days; litter size: one (twins rare); birth interval: 1–2 years. Births not markedly seasonal, or with tw[o]

ABBREVIATIONS HBL = head–body length TL = tail length WT = weight
Approximate nonmetric equivalents: 2.5cm = 1in; 230g = 8oz; 1kg = 2.2lb

Ex Extinct
EW Extinct in the Wild
Cr Critically Endangered
En Endangered
Vu Vulnerable
LR Lower Risk

h peaks per year. Male takeovers of ıps often associated with infanticide by ıigrant male(s). Both sexes emigrate n natal group, unlike most other poly-ous species, in which only one sex ıally males) migrate.

xican black howler
uatta pigra
ican black or Guatemalan howler

atán Peninsula (Mexico), N temala, Belize.
50–54cm (female), 60–64cm (male);
4–67cm (female), 66.5–71cm (male);
6.4kg (female), 11.4kg (male).
AT: black, base of hairs reddish-brown.

ntled howler
uatta palliata

m S Veracruz, Mexico, S to N tip of ombia on both sides of Andes, contin-g on W side to Ecuador. Contains the atened form *A. coibensis*, which lives y on Coiba Island, and has been sug-ed as a separate species. Typically ti-male groups of larger size than most er howler species, up to 21 individuals.
48–63cm (female), 51–68cm (male);
6–66cm (female), 55–61cm (male);
3.1–7.6kg (female), 4.5–9.8kg (male).
AT: black or brown, mixed on back h golden brown; flank fringe yellow-wn.

d howler
uatta seniculus

m N Colombia and Venezuela S to azon, including the Guianas: S of Ama-from Andes east to Rio Madeira, S to C ivia. The Bolivian population is some-es accorded species status as *A. sara*. oitat generalist, from mangroves to annas, woodlots, gallery forests, and mary rain forest.
48–57cm (female), 51–63cm (male);
52–68cm (female), 57–68cm (male);
4.2–7.0kg (female), 5.4–9.0kg (male).
ET: red howlers have been observed eat-soil more often than other species.
AT: orange-brown, paler below; beard nales is darker than rest of coat.

d-handed howler
uatta belzebul

Amazon from Rio Madeira E to antic ocean. Uni-male groups of 2–8 ividuals.
46–57cm (female), 58–60cm (male);
58–69cm (female), 56–66cm (male);
4.85–6.2kg (female), 6.5–8.0kg ale).
AT: black or blackish-brown with yel-to reddish hands, feet and tip of tail.

Brown howler
Alouatta fusca

E coast of Brazil from Bahia S to Rio Grande do Sul. In modified and undisturbed forest, often associated with *Araucaria* ("monkey puzzle") trees.
HBL 45–49cm (female), 54–59cm (male);
TL 52–57cm (female), 48.5–67cm (male);
WT 4.1–5.0kg (female), 5.3–7.15kg (male).
COAT: black to brown to dark red, paler below; females paler than males.

Black-and-gold howler
Alouatta caraya

S Brazil, Paraguay, E Bolivia, and N Argentina. Reaches the highest densities of any cebid species on islands in the Paraná river in Argentina.
HBL 50cm (female), 60–65cm (male); TL 55–60cm (female), 62–65cm (male); WT 3.8–5.4kg (female), 5.0–8.3kg (male).
COAT: one of the few species with different coat colors in males and females. All black in male, olive-buff in female.

SPIDER MONKEYS

Genus *Ateles*

Mature, tall, moist to rain forest. From sea level to 2,300m (7,545ft). Prefer middle to upper canopy, often moving into emergent trees. Move, suspended by hands and tail, hand over hand below branches, as well as quadrupedally. Eat fruit and leaves; ripe fruit usually accounts for more than 75 percent of the annual diet. Tree bark and wood are eaten regularly in small amounts for unknown reasons. Fur short, not dense. Diurnal. Polygamous, in groups or communities of about 20, although generally only 2–8 individuals move together at one time. Larger subgroups form when fruit is more abundant. Males tend to travel in groups more than females and cooperate in defending territorial boundaries against other communities. Females leave their natal group, males do not. Gestation 225 days; litter size one; birth frequency: one every 2–4 years. Births not notably seasonal.

Black-faced black spider monkey
Ateles chamek

S of Amazon, in upper Amazonia from Rio Ucayali E to Rio Juruá and S to headwaters of Rio Madeira (Bolivia).
HBL 40–52cm (female), 45cm (male); TL 80–88cm (female), 82cm, (male); WT 7.0kg.
COAT: all glossy black. Face black, muzzle pink.

Black spider monkey
Ateles paniscus

N of Amazon, E of Rios Negro and Branco through the Guianas.
HBL 49–62cm (female), 52–58cm (male);
TL 64–93cm (female), 72–85cm, (male);
WT 6.5–11.0kg (female), 5.5–9.2kg (male).
COAT: all glossy black, face pinkish.

Long-haired spider monkey Vu
Ateles belzebuth
Long-haired or White-bellied spider monkey

From N Colombia S; from upper Orinoco drainage and headwaters of Rio Negro, in upper Amazonia to N Peru.
HBL 34–58cm (female), 42–50cm (male);
TL 61–88cm (female), 69.5–82cm (male);
WT 7.5–10.4kg (female), 7.3–9.8kg (male).
COAT: black, or dark brown with paler brown to white underparts, hindlimbs, and base of tail; white to yellow-brown triangular patch on forehead present in about one-third of individuals.

White-whiskered spider monkey En
Ateles marginatus

S of Amazon river from Juruá E to Atlantic ocean.
HBL 34–51cm (female), 50cm (male); TL 61–77cm (female), 75cm (male); WT 5.8kg (female).
COAT: black, with white patch on forehead and white line connecting ears to chin.

Brown-headed spider monkey
Ateles fusciceps

From S Panama through N Colombia, on W side of Andes to Ecuador.
HBL 45–55cm (female), 37–59cm (male);
TL 60–81cm (female), 63–72cm (male);
WT 8.8kg (female), 8.9kg (male).
COAT: all black with white chin whiskers, or brownish-black with olive to yellow-brown crown cap.

Black-handed spider monkey Vu
Ateles geoffroyi

C America from Veracruz (Mexico) to central Panama.
HBL 34–52cm (female), 38–49.5cm (male); TL 70–84cm (female), 59–82cm (male); WT 6.0–8.9kg (female), 7.4–9.0kg (male).
COAT: variable golden brown, red to dark brown; hands and feet black.

GENUS *BRACHYTELES*

Muriqui Cr
Brachyteles arachnoides
Muriqui or Woolly spider monkey

Moist rain forest of SE Brazil from Bahia to São Paulo, in several relict populations. Two species may exist, a northern one (*B. hypoxanthus*) and a southern one (*B. arachnoides*), differing in several aspects of morphology, genetics, and social structure. From sea level to 1,000m (3,280ft). Lives in multimale troops of 8–45, but these often divide into two foraging parties. Diurnal.
HBL 54–60cm (female), 58–61cm (male);
TL 74–84cm (female), 67–69cm (male);
WT 9.5kg (female), 12.1kg (male).
DIET: fruit, leaves and flowers; muriquis ingest more leaves than spider monkeys but less than howler monkeys in same area.
COAT: uniform pale gray to brown, sometimes with extensive yellow in males. Fur thick and dense, similar to woolly monkeys (*Lagothrix*).
BREEDING: gestation period unknown; single infant born.

WOOLLY MONKEYS

Genus *Lagothrix*

Usually in mature moist to rain forest from sea level to 3,000m (9,840ft). Distribution discontinuous and patchy, perhaps because of competition from spider monkeys. Primarily eats ripe fruit, but also leaves and other vegetation. Prefers canopy level of forest. Fur moderately long and very dense. Sexes similar in length, but males often heavier. Diurnal. Polygamous, in groups of 5–43 or more, with several males. Male dominance status is determined by age, but males do not compete strongly for mating, and a receptive female may mate with several males. Gestation 223 days; litter size one; births every 1.5–2 years, not obviously seasonal.

Humboldt's woolly monkey Vu
Lagothrix lagotricha
Humboldt's or Smokey or common woolly monkey

From N Colombia throughout Amazonia E to Rio Negro (N of Amazon) and Rio Tapajós (S of Amazon), S to N Bolivia.
HBL 46–58cm (female), 46–65cm (male);
TL 62–72cm (female), 53–77cm (male);
WT 3.5–6.5kg (female), 3.6–10.0kg (male).
COAT: body gray to olive-brown, to dark brown to black; head often darker, almost black.

Yellow-tailed woolly monkey
Lagothrix flavicauda

Peru, in departments of Amazonas, San Martin, and La Libertad, in mid-elevation forest (1,700–2,700m/5,580–8,860ft).
HBL 52cm (female); TL 56cm (male); WT 10.0kg.
COAT: medium brown except for yellow underside to tail at tip around naked prehensile portion; fur on nose and mouth whitish.

LEAFEATERS OF THE NEW WORLD

Diet and energy conservation in howler monkeys

FIRST-TIME VISITORS TO THE NEOTROPICS OFTEN emerge from the forest in great excitement, claiming to have heard a lion or some other huge beast roaring nearby. It comes as quite a surprise for them to learn that this terrifying vocalization is being produced not by a large carnivore but rather by a 7–9kg (15–20lb) Neotropical monkey called, in deference to its loud cry, a "howler" monkey. Far from being dangerous predators, howler monkeys are placid, tree-dwelling vegetarians whose diet is made up of a wide variety of young tree and vine leaves, flowers, and tropical forest fruits. One might wonder why such monkeys should need to create such a raucous hullabaloo.

The calls produced by howler monkey species (genus *Alouatta*) are among the loudest made by any animal. Under some conditions, a howler's call can be heard for over 1.6km (1mi). Charles Darwin suggested that, within vertebrate species, the male that makes the loudest calls would attract the most females by advertising his strength. Darwin's explanation has proved accurate for some frogs, but thus far there is little evidence to support this theory for howler monkeys.

Another suggestion is that the calls announce a troop's rights to food trees in its home range. This does seem correct, but the evolutionary motivations behind the howling are complex and demand a look at the monkeys' feeding behavior, their social lives, and the energetic constraints of leafeating. These relationships have been revealed in recent studies of the Mantled howler monkey (*A. palliata*) of Central and South America.

Above *Howler monkeys can spend over half of their waking hours resting – an important strategy for the conservation of energy; most of the remainder of their time is spent eating, with only a tiny amount given over to social activities. Howling has an important role in energy-saving strategies.*

Below *A Brown howler monkey foraging among the foliage of the Caratinga Reserve in eastern Brazil. Young, tender leaves are a more efficient source of energy than older ones. The ability to see in color, which certain New World monkeys possess, is important in differentiating between similarly hued foliage.*

Howler monkeys have the widest geographical distribution of any New World monkey – from southern Mexico to northern Argentina. They generally make up the highest percentage of the primate biomass in the areas they occupy, testifying to the success of their ecological niche. One key factor is their ability to use leaves as a major dietary component. Tropical trees generally do not shed their leaves seasonally, but rather keep them year-round. In tropical forests leaves also tend to be more abundant than alternative foods such as ripe fruits. In general, a leafeating primate faces fewer problems in finding food.

Since leaves are relatively abundant, one might wonder why other monkey species – spider monkeys, capuchin monkeys, night monkeys – are not also leaf specialists. Observations show that most Neotropical monkeys do not eat large quantities of leaves, and some eat no leafy material at all. But leaves, while omnipresent, have a major drawback: they are low in nutrients and high in cellulose and hemicellulose, the structural carbohydrates that compose much of the leaf cell wall. Mammals do not have enzymes capable of digesting this material, so leafeating monkeys are filling their stomachs in large part with indigestible bulk. Though young leaves can be a good source of protein, they are low in sugars and fats. To be a successful leafeater, a primate must find some way of circumventing these problems.

In Old World tropical forests there are many different leafeating monkeys, placed together in the subfamily Colobinae. All colobine monkeys are alike in that they have a highly specialized, sacculated stomach, similar in many respects to

Right *The hyoid bone and lower jaw of the ɔwler monkey is enlarged – here, a male Black ɔwler – enabling them to make their loud calls.*

e cows. Special bacteria in one or more sections the colobine stomach are capable of digesting e cellulose and hemicelluose of leaf cell walls. In is digestive process, called fermentation, energy- ch gases (also known as volatile fatty acids) are oduced that can be absorbed by the monkey nd used to help fuel its daily activities. It is only rough the intervention of these specialized bac- ria that colobine monkeys or other mammals can otain energy from plant cell walls.

Unlike the colobines, howler monkeys do not ave a sacculated stomach. Rather, they have a mple, acid stomach, akin to that of humans. But owlers do have two enlarged sections – in the ecum and colon respectively – in which fermenta- e bacteria that break down leaf cellulose and emicellulose are found. Like colobines, howlers e the gases produced as an energy source.

In general, hindgut fermenters such as howler onkeys are not as efficient as stomach fer- enters such as colobines at obtaining energy om leafy foods. To improve their returns from afeating, howler monkeys must feed very selec- ely, choosing young leaves that can be rapidly rmented. Howlers also eat ripe fruits and flow- s, but they can live for weeks only on leaves, pro- ding the leaves are of sufficiently high quality.

Even with their selective feeding regime, howler onkeys must still watch their energy expenditure, nce the daily energy returns from fermentation e limited. Members of a howler troop typically end more than 50 percent of the day resting or eeping. The monkeys have a short range, moving ly some 400m (1,300ft) per day; total home nge size for a troop of 15–20 animals is only ound 31ha (77 acres). Thus all potential food urces in a home range lie within an average y's travel. In contrast, a fruiteating spider mon- y has a home range of 300ha (1,000 acres) or ore, and an average daily range of well over 1km .6mi). This is because ripe fruit is far less abun- nt than young leaves.

As part of their strategy of energy conservation, wlers also show a division of labor between the xes. Males help settle disputes and defend troop embers from predators. In addition, they use eir powerful calls to defend the troop's rights to portant food trees in their home range. All of ese activities spare females from having to carry t these tasks. Females can then invest more ergy in reproduction and care of the young.

Howler males produce their dramatic howl by awing air through a cavity within an enlarged oid bone in the throat. The hyoid of all male wlers is considerable larger than that of all males; in addition, that of the Red howler (*A. seniculus*) is large relative to males of *A. palliata*. The size of the hyoid influences the type of howl produced; for example, in *A. seniculus*, the call resembles a long-drawn-out deep moan, whereas in *A. palliata* it is more like a true howl.

In all howler species all male members of each troop give a "dawn chorus" that is then answered by males from all other troops within earshot. A howler troop does not have an exclusive territory, but shares parts of its home range with neighbor- ing troops. By howling each morning and again whenever the troop moves to a new feeding site, one troop is able to inform neighboring troops about its location throughout the day.

When two howler troops meet there is a con- siderable uproar, with animals – particularly adult males – expending much energy in howling, leap- ing, running, and, at times, even fighting. Female troop members can become scattered or lost, and valuable time that could be spent eating or resting is wasted. Thus, rather than patrolling the territory borders constantly or getting into energetically costly squabbles between troops, the monkeys howl in order to let the other groups know where they are.

There is a dominance hierarchy between differ- ent troops, apparently based on the fighting abili- ties and coordinated behaviors of the adult males. By listening to the howls, a weaker troop knows the location of a stronger troop and can avoid meeting it or traveling to a food source that it would not be permitted to use, thereby saving energy. What is more, the stronger troop also ben- efits, since its members do not have to expend valuable energy and expose themselves to poten- tial danger by defending preferred food trees. Howling therefore functions to help troops space themselves as efficiently as possible, allowing the monkeys to surmount the low-energy problems associated with leafeating. KM

Guenons, Macaques, and Baboons

ACTIVE AND GREGARIOUS, NOISY, IMITATIVE, and curious, the cercopithecines are the "typical monkeys" of legend, well known because their distribution and ways of life bring them into contact with people. Many are adaptable generalists, able to take advantage of the wastefulness or generosity of human neighbors to make a living, or to take their share from unwilling human hosts by skillful theft from unharvested crops or food stores.

In the past, when laboratory scientists referred to "the monkey," they nearly always meant the Rhesus macaque, a cercopithecine that has long borne the brunt of our invasive curiosity. It is only in recent years that zoologists have begun to appreciate the great variety among the cercopithecines, and to gain fresh insights into the dynamics of group life and the possibility of recent, ongoing, and future change in this young and probably most rapidly evolving subfamily of primates.

Colorful Sexual Displays

FORM AND FUNCTION

Cercopithecine monkeys have the same dental formula as man: I2/2, C1/1, P2/2, M3/3 = 32. They possess powerful jaws, with an arrangement of muscles designed to give an effective "nutcracker" action between the back teeth. With the exception of some of the smaller guenons, the face is rather long, the "dog-faced" baboons being the most extreme examples.

In baboons, mangabeys, and macaques, bright colors occur on patches of bare skin on the face, rump, and, in the gelada, the chest. The red face color of some macaques depends on exposure to the sun: monkeys in the wild have red faces, while those kept indoors have gray-white ones. The black face color of the Patas monkey also fades indoors, just like a human tan. Skin color also depends on gonadal hormones, so that among adults (but not juveniles) the bright red color is brighter still in the mating season, and castration

FACTFILE

GUENONS, MACAQUES, & BABOONS

Order: Primates

Family: Cercopithecidae

Subfamily: Cercopithecinae

45 (47) species in 11 genera: macaques (*Macaca*, 15 (16) species); Savanna baboon (*Papio hamadryas*); mandrills (*Mandrillus*, 2 species); gelada (*Theropithecus gelada*); mangabeys (*Cercocebus* and *Lophocebus*, 4 species); guenons (*Cercopithecus*, 18 (19) species); vervet (*Chlorocebus aethiops*); Allen's swamp monkey (*Allenopithecus nigroviridis*); Patas monkey (*Erythrocebus patas*); and talapoin (*Miopithecus talapoin*).

DISTRIBUTION Throughout Asia except high latitudes, including N Japan and Tibet; Africa S of about 15°N (plus Barbary macaque in N Africa).

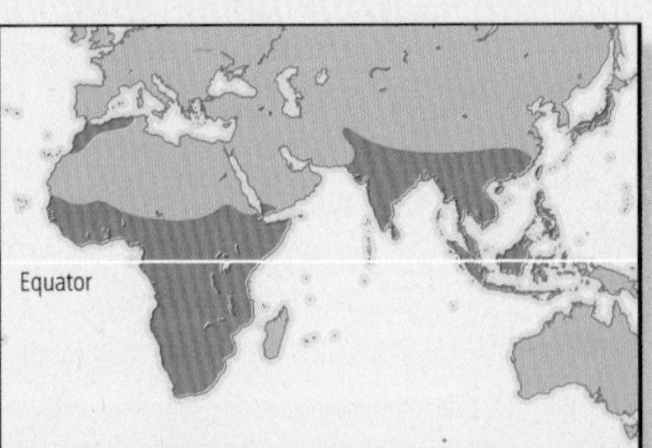

HABITAT From rain forests to mountains snow-capped in winter; also in savanna, brush.

SIZE Ranges from the talapoin to drill and mandrills. Head–body length 34–70cm (13–28in); tail length 12–38cm (5–15in); weight 0.7–50kg (1.5–110lb).

COAT Long, dense, silky, often (especially in males) with mane or cape. Colors generally brighter in forest-dwelling species. Skin color on face and rump important in other species.

DIET Primarily fruit, but may also include seeds, flowers, buds, leaves, bark, gum, roots, bulbs, as well as snails, crabs, fish, lizards, birds, and small mammals.

BREEDING Most species conceive in a limited mating season. Gestation 5–6 months.

LONGEVITY 20–31 years, depending on species.

See species table ▷

Left *One of the distinctive features of the Lion-tailed macaque is the ruff of grayish hair on either si[de] of its face. These inhabitants of the wet forests of southern India are primarily tree-dwelling and spend only a tiny amount of their lives on the ground.*

ıses the color to fade. Mandrills have patches of ·e skin as well as red, but the blue is a structural or. A similar bright blue color appears in the otal skin of several guenons at adolescence. e formation of this requires the presence of the le hormone testosterone; however, once med it is stable, and does not disappear if the le is castrated.

Female baboons, mangabeys, some macaques, d some guenons develop a swelling around the va, which in certain species can extend up to root of the tail and round onto the thighs. In rmal animals, these swellings increase in size ring the first, follicular phase of the menstrual cle, and decrease after ovulation. These ellings provide for a clear identification of a nale that is about to ovulate, and adult males l not normally attempt to copulate with a nale unless she has a swelling. The exact site d patterning of the swelling differs from species species, and indeed individual animals are recnizable by their swellings. The swellings tend to rease in size in successive menstrual cycles, so t captive monkeys that are not breeding reguy may grow enormous swellings that can constitute as much as 10 or 15 percent of their total body weight. In the wild, adult females rarely undergo more than one or two successive cycles before conceiving anew, and the process of having a baby prevents them from coming into season again and consequently they have smaller swellings. Rhesus macaques and some other species show a rather different pattern of swelling. The adolescent Rhesus female swells around her tail root and along her thighs, with the swelling reaching a maximum just before menstruation in the luteal phase of her cycle. This pattern ends when the female is fully mature and ready to conceive. Similarly, adolescent Patas females show vulval swelling, whereas adults usually do not: however, sometimes these swellings develop in females that are returning to reproductive activity after a long interval.

In the wild, changes that occur during the course of pregnancy in some species are even more conspicuous than the changes that are related to the menstrual cycle. Thus the black, naked skin over the ischia on the rump of baboons loses its pigment and becomes a glowing red. The vulva of Vervet monkeys and mangabeys also reddens when they are pregnant, sufficiently to be easily recognized at a distance. Since these kinds of sexual signals are more common among terrestrial species than arboreal species, it seems likely that they evolved to encourage males to join and stay in the social groups in order to provide protection against predators.

Newborn cercopithecines have short, velvety fur, often of a color that contrasts with the pelage of the adult: olive-brown Stump-tailed macaques have primrose-yellow infants, for instance, while yellow-gray baboons have black infants. The infants are even more conspicuous because they have little or no pigment on the naked skin of the face, feet, perineum, and ears (which seem to be almost adult-sized). The newborn male baboon has a bright red penis, while infant male macaques have a very large, empty scrotum. (Infant guenons, however, are very difficult to sex even when you have them in your hand.) It is while sporting their distinct coat that the young are the focus of attention for other members of the group. They frequently inspect the newborn's genitals, perhaps to determine the sex of the new group member so that their interaction with it is appropriate. In the third month, the natal coat begins to be replaced by a juvenile coat, which is usually a fluffier version of the adult pattern, less brightly colored or clearly marked.

Left *The unmistakable facial shape and coloration of a male mandrill; the distinctive outgrowths on either side of the nasal ridge are at their most pronounced in adult males. The mandrill is the largest of all monkeys – a fully grown adult can reach a weight of over 36kg (79lb).*

Below *A vervet* (Chlorocebus aethiops) *with young in Tanzania. Young vervets are nursed until the arrival of the next offspring. Vervets are the most widespread of the guenons. Unlike the other guenons, which are primarily forest-dwelling, vervets live in many local variants throughout the African savanna.*

Ecotype-casting by Species

Distribution Patterns

The cercopithecines are a modern group that originated in Africa, their macaquelike forebears appearing in the fossil record toward the end of the Miocene, around 10–15 million years ago. Macaques spread north and east, to Europe in the Pliocene (5–1.8 million years ago) and to Asia by the Plio-Pleistocene (2 million years ago). During the same period they died out in Africa south of the Sahara, or perhaps evolved and radiated into the modern African genera – the baboons, guenons, and mangabeys. The Barbary ape appears to be somewhat separated from the Asian macaques in terms of molecular genetics as well as some anatomical details, perhaps retaining some more primitive features.

The radiation of the guenons in Africa is very recent. The emergence of distinct species (a process known as speciation) seems to have occurred during the last main glaciation, when Africa as a whole was colder and much drier, so that the forests retreated to a few scattered areas in equatorial Africa – along the east and west coasts, and around Mt. Cameroon and the Ruwenzoris. When the climate became wetter and warmer about 12,000 years ago, the forests spread, and with them their now distinct monkeys. In the same way, in Asia, cold, dry periods associated with glaciation restricted the habitats of macaques and separated populations, which differentiated. In Asia the picture is more complicated, because variations in sea level allowed macaques to move from one island of Southeast Asia to another.

In Africa, the history of baboon species is closely associated with the history of grasslands. Between 5 and 2 million years ago, the most common baboons were relatives of today's gelada, specialized to harvest grasses on the lush grasslands of sub-Saharan Africa. Following a dramatic drying and cooling of the climate around 2 million years ago, these grasslands were raised up into the upper levels of the mountains and replaced by the drier grasslands we associate today with Africa's savannas. The Savanna or "Common" baboon, once probably a forest-edge species, spread with the advance of wooded savannas, and eventually replaced the geladas everywhere but in their Ethiopian highland stronghold.

All macaques are placed in a single genus – *Macaca* – that occupies the whole of Asia except the high latitudes. Most areas are occupied by a single species with characteristics appropriate to local conditions. The one surviving macaque in Africa, the Barbary, has thick fur and no tail, traits which help it to survive the snowy winters of the Atlas mountains. Similar adaptations are also useful to the Japanese macaque of northern Japan, and the Tibetan macaque in the high mountains of Tibet. The sturdily-built Rhesus monkey of the Himalayan foothills, northern India, and Pakistan is replaced in southern India by the smaller, more lightly built, and longer-tailed Bonnet macaque, which, further south still in Sri Lanka, gives way to the rather similar Toque macaque.

The equatorial rain forest has enough niches to be home to two species of macaque. Thus, the arboreal Lion-tailed

***Below** Drills and baboons, the largest of the monkeys (adult males shown). **1** Mandrill* (Mandrillu sphinx)*; **2** drill* (M. leucophaeus)*; **3** gelada* (Theropithecus gelada)*, showing bare patches on the nec and chest. Various subspecies of baboon, showing color distinctions: **4** Hamadryas baboon* (Papio hamadryas hamadryas)*; **5** Guinea baboon* (P. h. papio)*; **6** Yellow baboon* (P. h. cynocephalus) *of lo land East and Central Africa; **7** Olive baboon* (P. h. anubis) *of highland East Africa; **8** Chacma baboon* (P. h. ursinus) *of southern Africa.*

:aque lives in the forest of southwest India ve the heads of the more terrestrial Bonnet :aque, and in Sumatra and Borneo the small, ıg-tailed macaque lives in the same forests as heavily-built, short-tailed, more ground-living -tailed macaque. On the island of Sulawesi lebes), a number of different species (between ıd seven, according to different authors) of :k, stump-tailed macaque has evolved to occu- :ach of the sprawling peninsulas.

:harismatic and conspicuous, the baboons live rywhere in Africa where there is water to drink. :y have heavy, doglike muzzles, and limb modi- .tions that allow them to walk long distances the ground; Savanna baboon troops commonly /el as much as 5km (3mi) a day, while the madryas baboon subspecies averages 13km ni). The Savanna or Common baboon is the re widespread, and individuals from different ions are sufficiently different in appearance to 'e been given separate species status in the past; they are now generally regarded as subspecies. They live in grass- and bushland, and along the edge of forests.

The Hamadryas baboon subspecies replaces all the Savanna baboon subspecies in the deserts of northeast Ethiopia and on the Arabian peninsula: male Hamadryas baboons have red faces and rumps instead of black, and a long cape of gray fur, and consequently look very different.

Along a narrow boundary zone in Ethiopia, however, the two different subspecies hybridize.This hybrid zone probably came about as a result of Savanna baboons extending their range up into the more desertlike habitats favored by the Hamadryas baboons. The hybridization itself, however, is principally due to Hamadryas males entering Savanna baboon troops and mating with the females. Hamadryas troops are much

FRIENDSHIPS BETWEEN THE SEXES

Forming enduring relationships in an Olive baboon troop

THE OLIVE BABOONS OF EAST AFRICA ARE HIGHLY social primates, living in large troops that occupy home ranges of up to 40sq km (15.4sq mi). The 30–150 troop members remain together all the time, traveling, feeding, and sleeping as a cohesive unit. The enduring relationships formed by females with one another, with young, and with adult males lie at the heart of baboon society.

Females and their grown-up female offspring typically maintain close bonds, whereas males leave one by one to join new troops. These associations among maternal kin produce a network of social relationships extending down through three generations and out to include first cousins. Within the kin group (or matriline), each adult female ranks below her mother. When a troop of Olive baboons takes a break during the day, relatives often gather around the oldest female in the family to rest and groom. At night, family members usually sleep huddled together, and they will come to one another's aid if a member of the group is threatened by another baboon. Although females compete for food and status, serious fights among them are extremely rare, status being routinely expressed by such submissive gestures as fear grins and raised tails. When conflict does erupt between females, they often reconcile afterwards by grunting at one another or, less commonly, by touching or embracing.

A male immigrant unfamiliar to members of his new troop must therefore penetrate a dense network of relatives and friends. He usually begins by cultivating a relationship with an adult female who is not nursing a young infant. He will follow her about, smacking his lips, grunting softly, and making the friendly "come hither" face – a distinctive expression in which the ears are flattened back against the skull and the eyes are narrowed – whenever he catches her eye; and, if she permits, he will groom her. After many months, the male may succeed in establishing a stable bond with the female. If so, their relationship will serve as a kind of passport, allowing him to gradually extend his ties to her friends and relatives. Even when past this initiation stage, incoming males have to maintain the goodwill of the females. Male–female bonding is not immediately related to sexual activity, for although females will mate and interact with many different males when they are in estrus, they spend most of their adult lives either pregnant or nursing, at which times they will not mate.

Such male–female special relationships, or "friendships," have been documented in every well-studied troop of savanna baboons. In one large troop of Olive baboons at Gilgil in Kenya, most of the 35 pregnant or nursing females had a friendship with just one, two, or three of the 18 adult males (a 1:2 adult sex ratio is typical of savanna baboons). While foraging for grasses, bulbs, roots, leaves, and fruits, three out of four remained within 5m (16ft) of "friends," while avoiding close encounters with all other adult males. In addition, nearly all amicable interactions between non-estrous females and males, including 98 percent of grooming, involved friends.

Since a female is only about half the weight of an adult male (9–16kg/20–35lb versus 22–35 kg/49–77lb) and lacks his long, razor-sharp

Above *A female Olive baboon with baby being groomed in the Masai Mara National Park, Kenya. Often females will exchange grooming services for the opportunity to handle another female's infant.*

Below *When Olive baboon troops take a break to rest, the main group may well subdivide into smaller, related subgroups.*

chance of forming a consortship with a female as a non-friend. While in some savanna baboon populations the dominant male gets up to 80 percent of the fertile matings, in this particular Olive baboon troop at Gilgil, lower-ranking, long-term resident males consorted more than twice as often as more dominant newcomers during estrous cycles leading to conception.

Individual personality traits are important in determining who pairs up with whom, and it is not surprising that friendships vary considerably from one couple to another. Some friends spend a lot of time together but rarely touch, while others groom frequently, embrace when their paths cross, and huddle together at night. Friendships change as each baboon matures through different phases of life. Some dissolve: a female may forge a new bond during a series of sexual encounters; a male may gain in dominance status and drop his adolescent friend for a new relationship with the influential alpha female. But many friendships persist for years, settling into a pattern of comfortable, relaxed familiarity. In general, males who interact with females and infants frequently are more likely to remain in a troop for many years.

Baboons are often portrayed as a highly competitive species, but that is just one side of their nature. Natural selection has also favored in them a capacity to develop close and enduring bonds with a few others. The fact that humans share this capacity with other primates suggests that it is an ancient and fundamental aspect of our evolutionary heritage. BS

nes, she cannot defend her infant against cks by males. Infanticide by recent male new-ers has been observed in all three subspecies avanna baboons. Field experiments using play-ks of Chacma baboon vocalizations, however, w that males are highly attuned to a female d's screams for help, especially if she has a ng infant and especially if her scream is paired a playback of an aggressive vocalization by a nt male immigrant. The male is much less likely to attend to the screams of a nursing mother who is not his friend. For a female, friendships with males may mean the survival of her offspring.

What males stand to gain from friendships is not always so obvious. Often, the male is known to have mated with the female around the time she conceived, in which case he may be protecting his own infant. Yet this is not always the case. If he is not the father, he may be aiming to be so next time around, for on average, a friend has twice the

Left *The Stump-tailed macaque, with its distinctive red facial mask, inhabits the montane forests of Southeast Asia. It is now under threat from habitat loss over parts of its range, and has been classified as Vulnerable by the IUCN.*

Right *Facial expressions in macaques (adult males illustrated).* **1** *Barbary macaque* (Macaca sylvanus)*: lip-smacking with infant;* **2** *Moor macaque* (M. maura)*: open-mouth threat;* **3** *Bonnet macaque* (M. radiata)*: canine display yawn;* **4** *Long-tailed macaque* (M. fascicularis)*: fear grin;* **5** *Stump-tailed macaque* (M. arctoides)*: open mouth threat;* **6** *Pig-tailed macaque* (M. nemestrina)*: approach pout-face – can precede copulation or attack or even grooming another individual;* **7** *Rhesus macaque* (M. mulatta)*: aggressive stare.*

more difficult for males from outside to enter, because they consist of male brotherhoods that act in alliance to keep strange males out.

Two species occupy the forest floor of west-central Africa: the drill and the mandrill. In Cameroon the two species live in the forests, one on either side of the Sanaga River. The largest of the baboons, they are mainly black and have short tails. Their niche is perhaps parallel to that of the Pig-tailed macaque, largest of the macaques.

The gelada is a long-haired species that inhabits the cool highland regions of Ethiopia, where its diet is based exclusively on grass. The gelada is unique in being the only true grazer among the primates. Because grasses are much more abundant than the fruiting trees favored by other cercopithecines, large numbers of gelada can congregate in one place. Herds of up to 500 animals – the largest naturally-occurring groups of primates – are not an uncommon sight.

The four species of mangabeys (*Cercocebus* and *Lophocebus* species) have always been generally considered to be lightly-built, long-tailed baboons, but while two conform to this description, the other two are actually more closely related to the drill and the mandrill. Mangabeys live exclusively in closed-canopy forests. The Gray-cheeked mangabey and the Black mangabey are highly arboreal, while the Agile mangabey and the White mangabey usually move on the forest floor. Arboreal and terrestrial species are able to co-exist in the same forest: Gray-cheeked and Agile mangabeys, for example, both occur in the forest reserve of Dja in southern Cameroon.

The guenons are grouped mainly in the genus *Cercopithecus*, though some of the more eccentric species have been given generic status. The many species of *Cercopithecus*, recognized by differences in coat color, are regional variations of perhaps half-a-dozen ecotypes (forms of a species fitted to a particular ecological role). Although most forests accommodate only one example of an ecotype, the richest habitats may be home to four or five guenon species of different ecotypes. For example, Blue monkeys and Sykes' monkeys are similar in build and habits, and tend not to occur together. In contrast, Blue monkeys and Redtail monkeys differ in size and feeding habits, and so commonly live side by side in the same forests.

All the other species of *Cercopithecus* are forest species, including the most widespread Spot-nosed/Blue monkey (*C. nictitans/mitis*) group. These large guenons eat leaves, living wherever there is a patch of closed-canopy forest. Next come the Redtail guenons (the *C. cephus/C. ascanius* group), smaller monkeys that seem to require a more layered canopy to the forest, with tangles of creepers, possibly because they provide cover from such predators as eagles, to which the monkeys' small size makes them vulnerable. They are followed by the *C. mona* group, smaller still and more insectivorous. These three species commonly form associations of more than one species, for example in the rain forests of Gabon. De Brazza's monkey inhabits wet patches of forest, especially those rich in palms. The ground-living L'Hoest's monkey can inhabit quite high-altitude forest.
Where overlap occurs, some

›ridization of species is known; for example, ltail and Blue monkeys hybridize in the Kibale est of western Uganda.

Four other guenon genera have only a single cies each. The most widespread guenon is the vet (also called the Grivet or Green monkey), ich has as many as 16 local variants throughout anna Africa. It never strays far from water, nding much of its time in the *Acacia* trees that river banks. The Patas monkey is set apart by skeletal adaptations to ground-living. The gest of the guenons, its long legs, orange-colored t, white mustache, and fast, bounding run oss open grasslands won it its alternative name he Hussar monkey – the sight apparently ninded 19th-century travelers of a light cavalry iment charging. It lives in the open *Acacia* odlands and scrub of drier, more seasonal areas rth of the equatorial forest. Its habitat is often acent to that of the vervets, and, although ch larger, patas avoid vervets when they meet.

The talapoin is the smallest of all the Old World nkeys, and lives in floodplain forests of west-ntral Africa. Allen's swamp monkey, as its name gests, frequents swamps, in the Congo basin. h are placed in separate genera, partly because females have a perineal swelling, unlike nales of other guenons.

Dietary Opportunists

DIET

Cercopithecines are primarily fruiteaters, but they are also opportunistic feeders. Their diet includes most things that are edible in any forest: seeds, flowers, buds, leaves, bark, gum, roots, bulbs and rhizomes, insects, snails, crabs, fish, lizards, birds, and mammals – anything that is digestible and not poisonous is fair game. Most food is caught or gathered with the hands. Selection and preparation of food is learned from observation, initially of the mother. In this way, local traditions in food preference develop. Adult baboons will prevent juveniles from eating unfamiliar food. On the other hand, in experimental situations, juveniles have come to recognize new food items and to devise their own preparation methods; other juveniles and adult females have learned from these young entrepreneurs, but adult males do so less readily. This transmission of information is a crucial function of group living: the troop is primarily an educational establishment.

Species that live near water use aquatic foods. Japanese macaque troops living by the seashore have recently incorporated seaweeds into their diet. The Crab-eating macaque (an alternative name for the Long-tailed macaque) is so called for good reason; Savanna baboons living on the coast of southern Africa take shellfish off the rocks; and talapoins are said to dive for fish.

Where several species live in association, for example in a West African forest, the smaller species tend to eat more insects, especially active types like grasshoppers, while larger species eat more caterpillars, and more leaves and gum.

Monkey Alliances

Groups of two or more species of monkey can sometimes be observed traveling and feeding together. In some cases, these groups form more by accident than intent, and last for only a matter of hours before the individual species part company. This usually happens because one species travels faster or farther than the other. The temporary associations that form between gelada and Hamadryas baboons on the Ethiopian highlands are of this kind. A relatively long-lived kind of "polyspecific association" is, however, quite common between Blue and Red-tailed monkeys in East Africa and between several of the small- to medium-sized forest monkey species in West Africa. These groups probably form to provide protection against predators such as eagles, while avoiding the ecological competition that would otherwise occur if two groups of the same species foraged together.

Right *Bonnet macaques are both arboreal and terrestrial, which allows them to share territory with more exclusively tree-dwelling macaque species.*

Mangabeys have powerful incisors, used to open hard nuts, which are inaccessible to guenons.

Patas monkeys are adapted for running in grassland between patches of *Acacia* woods where they feed on fruit, leaves, and gum, as well as on insects and some small vertebrates. Their diet is not very different from that of guenons living in forests. Baboons, in contrast, have a diet that includes large quantities of grass, which accounts for their long jaw supporting a large area of molar teeth for chewing the tough leaves. Their very powerful, spoon-shaped hands are strong enough for digging, and they subsist through severe dry seasons by digging up the rhizomes and the leaf-base storage bulbs of grasses and lilies. Geladas too have relatively large and high-crowned molars, to enable them to grind down the fiber in grasses without wearing down their teeth too fast. By using a rapid pinching movement of the thumb and forefinger of each hand alternately, they can crop the sward as closely as sheep. Baboons also take small, mammalian herbivores, kill and eat the kids of gazelles which are left hidden in the grass, and hunt hares. In a simple form of cooperative hunting, a group of baboons will spread out to set moving and head off small prey, almost like beaters, although the prey is shared only reluctantly by the eventual captor.

Whenever people and monkeys come into contact (which is relatively frequently, as their ecological requirements are rather similar), the diet of the monkeys expands to include human provisions. The monkeys' behavior is clear evidence that learning plays an important role in how they acquire food. They time their arrival at feeding stations to coincide with the arrival of food, and raid crops when people are predictably absent, for instance in heavy rainstorms or during the siesta. Baboons will enter a field where women are working, and even chase them away, but avoid men, who are usually armed. Likewise, talapoins will crowd quite close to people who are washing or fishing at a river in the forest, but avoid people setting out to hunt. All these instances suggest a sophisticated appreciation of human behavior.

Late Developers

SOCIAL BEHAVIOR

Cercopithecine monkeys are slow to mature, slow to reproduce, and live long. In well-fed captive populations, the Rhesus macaque usually conceives first at 3½ years of age and gives birth at 4, but perhaps 10 percent will mature a year earlier, and another 10 percent a year later than that. Patas, the largest guenons, usually conceive at 2½ years (range 1½–3½ years). They are thus the fastest-maturing cercopithecine monkeys so far recorded. On the other hand, females of the smallest cercopithecine, the talapoin, do not conceive until 4–5 years old, and the same is true of other forest guenons, such as Sykes' monkey and De Brazza's monkey. The effects of food availability on these developmental stages is dramatic. Captive baboons, for example, may conceive at 3½ years, but baboons in a deteriorated natural environment, as at Amboseli, Kenya, may not do so until 7½ years. Vervets at Amboseli conceive first at about 5 years, while in captivity they conceive at about 2½ years. This difference has n been noticed in forest guenons or Patas monkeys. Males begin to produce sperm at about the same age as the females of their species first conceive, but they are not fully grown and are still socially immature. As a result, males are several years older than females before they begin to breed.

Most cercopithecines conceive during a limit mating season. In high latitudes, mating occurs the fall. In the wet tropics, guenon conceptions

:ur in the dry season, while in dry country, :as conceive during the rains. Baboons and ngabeys breed at any time of the year, but stres- s, such as a drought that causes the death of ·eral infants, may have the effect of synchroniz- ; the next pregnancies of the mothers. Never- ·less, it is climate that drives breeding seasonality ·ough its effects on food supply. Mating may last several months, as in vervets, or be concentrat- into a few weeks, as in Patas monkeys and talapoins, in which case the female probably ovulates only once in the annual season. Individual females generally mate in bouts lasting several days; and females that have already conceived may continue to show bouts of receptivity. Male Rhesus macaques also show seasonal changes, with reduced testosterone levels and testis size in the non-breeding summer months.

Courtship is minimal, since mates are usually familiar with each other, and is confined to signals which indicate immediate readiness to mate. The courtship of Patas monkey females is more elaborate: a crouching run with tail tip curled, chin thrust forward, lips pouted. The patas female also puffs out her cheeks, and often holds her vulva with one hand or rubs it on a branch at the same time. Consortships are frequently formed, a pair remaining close together and at the edge of their troop for hours or days. In some species a single mount may lead to ejaculation, in others several

JUST LIKE US?

The limits of the human–monkey analogy

WATCH ANY DOCUMENTARY ON THE SOCIAL behavior of monkeys, and you are tempted to conclude that they are remarkably similar to humans. Monkeys live in complex Machiavellian societies, have emotions and facial expressions that look like ours, and certainly seem to undertake planned, premeditated actions. But it is equally clear that monkeys aren't just furry humans: they don't have language, build machines, tell stories, or write books. Their intelligence lies in unknown territory, somewhere between that of rote, unthinking automata and fully evolved human beings. What, exactly, makes them different from us? How should we characterize a mind that is so much like ours, and yet so different?

If the subject under scrutiny were humans, language would be our guide. Language provides a window on the mind by revealing how we think. Monkeys don't have language but do communicate, and for many years scientists have used the vocal communication of primates to study how they think and how they see the world.

Many animals can distinguish their own kin from others. Humans, however, go one step further and recognize the kin relations that exist in families other than their own. Nonhuman primates do the same, as shown by field playback experiments on vervet monkeys (*Cercopithecus aethiops*) and baboons (*Papio hamadryas*). Both species live in groups containing matrilineal families (mothers and their offspring) arranged in a linear dominance hierarchy. This means that all of the members of family A rank above all the members of family B, and so on. In one experiment involving baboons, researchers waited until two adult females (say, B and D) were resting close together with their offspring elsewhere. Then, from a hidden loudspeaker, they played the aggressive and submissive calls of two other group members and filmed the females' responses. When the playback sequence mimicked a fight between two individuals who were unrelated to the females, neither responded. If it mimicked a fight between an unrelated individual and the close relative of female B, D looked at B. And if it mimicked a fight between the close relatives of both B and D, the two females looked at each other. Moreover, in the ensuing 15 minutes, female B was much more likely to be aggressive toward D. In much the same way, a fight between the members of two vervet monkey families will often prompt a fight between two other, previously uninvolved, members of the squabbling matrilines.

Monkeys also appear to recognize other individuals' relative ranks. When subjects are played a sequence of calls that mimics a fight in which animal D, for example, threatens the higher-ranking animal B, they take a strong interest in this apparent violation of the dominance hierarchy. A monkey in the middle of the hierarchy, then, does more than just divide other individuals into those who rank higher or lower than herself; she appears to have an actual rank order in her mind.

Like humans, therefore, monkeys seem to view their social groups not just as a random collection of individuals but as a network of social relationships with a ranked, matrilineal structure. Moreover, there is good reason for natural selection to have favored such sophisticated social knowledge. Although many animals form alliances with close family members, only nonhuman primates appear to recruit their alliance partners strategically, seeking out allies that outrank both them and their current opponents. Strategic choice requires an encyclopedic knowledge of group members and their relationships – a mental feat that, in a group of 80 baboons, is enormously eased if one has a structure within which to organize the information.

Do the vocalizations of monkeys also share some of the attributes of human language? At first sight, this would appear to be the case. The alarm calls of vervet monkeys, for example, seem to function

Right *Vervet monkeys perch in the branches of a tree in the Moremi Game Reserve, Botswana. Vervets, along with certain other monkeys, differentiate between predators when giving their alarm calls.*

Below *The adaptability of vervets' alarm signal behavior does not extend to varying their calls to warn of particular hazards to their offspring.*

ıost like human words. Alarm calls are produced ıt encode specific information about the threat-ng animal. Leopard calls cause other group ·mbers to run into trees. A different alarm call, en for eagles, causes listeners to look up and run o bushes, while the "snake" alarm call makes ıers stand on their hind legs and peer into the rounding grass. How might one explore the ·ntal events that take place in a monkey's mind ring the seconds after an alarm call is heard?

In the rain forest of West Africa, Diana monkeys *ercopithecus diana*) also give different alarm ls for leopards and eagles. Normally females ıt hear the growl of a leopard, the shriek of an gle, or a male Diana monkey's eagle alarm res-nd with a chorus of their own predator-specific ls. However, if females hear a male's leopard alarm call, and then, five minutes later, hear the growl of a leopard, they do not respond to the growl and the same applies to an eagle. Yet the females respond with a chorus of alarms if, after hearing one type of alarm, they hear the call of a different predator. Apparently, female Diana monkeys group leopard growls and males' leopard alarm calls into the same functional class, classifying calls not according to whether or not they sound alike, but according to the sort of danger they represent.

In other respects, however, there are profound differences between monkeys' vocalizations and human language. Unlike humans, monkeys appear not to invent new "words" for new circumstances; they do not modify their calls, nor do they learn new ones. Equally important, monkeys never alter their behavior or their vocalizations to inform or teach those who are ignorant, or to comment on events or objects that aren't present. In terms of their vocalizations, then, monkeys seem to be unaffected by the mental states of their companions. As an example, vervet monkeys do not inform their offspring about which species are dangerous and which are harmless. Although adults give alarm calls to species that prey on them, they often fail to warn their offspring of predators that prey exclusively on juveniles. Similarly, baboon mothers respond to their infant's "lost" calls only if they themselves are also separated from the group. Like humans, monkeys have mental states like thoughts, beliefs, and emotions. But unlike humans, monkeys seem not to have mental states about the mental states of others. RS/DC

mounts precede it. Copulating pairs are sometimes harassed by juveniles, especially the female's own offspring, whose interests are presumably served by delaying the advent of a sibling with whom the mother's attentions will have to be shared. The disruption may be enough to make repeated mounting necessary before ejaculation is achieved.

A single infant is born after a gestation of 5–6 months (twins are very rare). Labor is much shorter than in humans, and females give birth wherever they happen to be at the time. The newborn infant is furred, and its eyes are open; it often grasps at the mother's hair with its hands even before the legs and feet have emerged. Infants cling to the mother's belly immediately, and can usually support their own weight, although the mother typically puts a hand to the infant's back, supporting it as she moves about during the first few hours. The newborn usually has the nipple in its mouth and uses it to support its head even when it is not nursing. Most monkeys are born at night, in the tree where the mother sleeps, and she eats the placenta and licks the infant clean before morning. (None of the cercopithecines ever makes a nest.) Patas monkey infants, on the other hand, are usually born on the ground and during the daytime. It seems that the timing of births is subject to selection by predation pressure, for Patas monkeys sleep at night in low trees, where they are vulnerable to predators. Although females giving birth attract little interest from other group members, the baby, once born, becomes a focus of attention for other adult and juvenile females, who may compete for the mother's permission to touch and even look after it.

Changes in maternal care match the growing ability of the infant to move independently. Nursing becomes infrequent after the first few months, but usually continues until the next infant is born. This usually occurs after about a year in most macaques, vervets, Patas monkeys, and talapoins, and after two years or more in forest guenons like the Blue monkey. In baboons, the birth interval varies, probably depending on food availability, between 15 and 24 months. If an infant is stillborn or dies, the birth interval may be shortened, although not significantly so in species that breed seasonally.

The mother–daughter bond typically lasts for life; in contrast, the mother's bond with her sons lasts only until sexual maturity, when the young adult males of most species leave their natal group and enter another one or become solitary for a while. Beyond infancy, the bond is seen in the frequency of grooming or sitting together, and in mutual defence. Juveniles also form bonds with their siblings, and where hierarchies are in evidence, a female may rank just below her mother but just above her older sisters. Males lose their inherited rank when they leave the troop, but a young male may join the same new troop as his older brother, who helps with his introduction.

When more than one female is receptive in a troop, there is some tendency for males to prefer older females, which also have rather longer periods of receptivity. Similarly, females tend to prefe older males. Baboons may have preferred mates, spending time together even when the female is not receptive. In species where a single adult ma lives with a group of females, other males who u to that point have been living a solitary life may join the troop when several females are receptive thus providing the females with a choice of mate In polygynous Red-tail monkeys, males survive less well than females, so those that do make it t adulthood contribute more of their genes to the subsequent generations.

The basic unit of cercopithecine social organization is the matriline, in which daughters stay with their mothers as long as they live, while ma usually leave the natal group around adolescenc The Hamadryas baboon seems to be an exceptio in this species, males stay together in brotherhoods, and it is the females who more often mo

Below *Small and medium-sized cercopithecines.* ***1*** *Gray-cheeked mangabey* (Lophocebus albigena)*: western race with double crest;* ***2*** *Allen's swamp monkey* (Allenopithecus nigroviridis)*:* ***3*** *The Mustached monkey* (Cercopithecus cephus)*:* ***4*** *Talapoin* (Miopithecus talapoin)*: the smallest Old World monkey;* ***5*** *Sooty mangabey, a geographical race of the White mangabey* (Cercocebus torquatus)*;* ***6*** *Patas monkey* (Erythrocebus patas).

:ween groups (See A Male-dominated Society). ıereas some years ago it seemed possible to ntify species-typical group sizes, home ranges, ɫ social organization, recent research has ·ealed considerable variation within a species, ɫh in time and across habitats. Under favorable ɪditions, a single founding female may be sur- ·ed by several daughters, each now the head her own matriline, and all still within a single op. In harsh conditions, survival may be so low ɪt the matrilinear organization can only be :ected after several years of study.

An upper limit to troop size may be determined constraints on foraging; macaque troops fre- ently begin to split into smaller troops, mainly ɪng matrilines, but only after troop numbers ve reached sizes far larger than occur naturally. ɪmbination or fusion of troops is rarer, but can :cur in poorer habitats when mortality reduces a ɔup below the size at which it is viable. Howev- the smaller of the two troops invariably does s well out of the merger: its members often end at the bottom of the social hierarchy, and may t survive well. Group living has its costs, with ger groups experiencing more feeding competi- n, and thus having to travel longer distances to d enough food.

In general, troops live within a defined home ıge. While guenons and mangabeys may defend erritory against adjacent troops, baboons and

Above *A group of Patas monkeys slake their thirst at a waterhole. The Patas monkey's long legs, long feet, and short, strong digits are all special adaptations that assist in the process of running in this fastest-moving of all primates; they may attain speeds of up to 55km/h (35mph).*

4

5

6

Patas monkeys often live in ranges that are too large to defend, and the ranges of neighboring groups often overlap considerably as a result. The range is the "property" of the females which form the permanent nucleus of the troop, and in Blue and Redtail monkeys it is the females, with juveniles, who are most often involved in boundary disputes. The male troop members are more transitory. They may remain in a troop for periods ranging from a few weeks to as long as 2–3 years, but rarely more. Adult males make loud calls which are highly species-specific and (where they have been studied) also characteristic of an individual male. The loud calls may serve as a rallying point, but also advertise to rivals that there is a male resident in the group.

Cercopithecines have been categorized into one-male and multimale group species. Male baboons, mangabeys, and macaques will tolerate each other's presence in a troop; nonetheless, a small troop may still include only a single fully adult male. Males living together in a troop will establish a hierarchy based on the outcome of competitive interactions. The rank order is not very stable, but changes with age, or as males join or leave the troop. When there are fewer than four adult males in the group, the dominant male is able to monopolize most of the matings, providing not too many females are receptive at the same time – when there are more, the dominant male cannot keep all his rivals away from the females, and matings are more widely distributed.

Vervets and talapoins are unusual among the guenons in that they also live in multimale groups, but Patas monkeys and most forest guenons have one-male troops. In these species, the male's tenure may be as short as 2–3 years, and, in a mating season when many females are receptive, other adult males will join the troop and also copulate.

In a number of baboon and guenon species, newly-arrived adult males have been seen to kill infants they find in the troop. Though not necessarily typical of all populations all the time, this behavior

Above and left *Parental behavior in baboons. In the photograph a male Savanna baboon carries an infant on his back in the Krug National Park, South Africa. In the illustration a female baboon begins the process of encouragi independence in her offspring through a series c training exercises.* **1** *the mother rises, turns her back, and withdraws contact from the infant baboon;* **2** *the mother moves a few steps away from the infant, with the infant following her;* **3** *the mother turns around again to reassure the infant that it is not being abandoned; the infant stops;* **4** *older infant baboons, who have already gained some measure of independence will be rejected more frequently until full independence is finally achieved.*

sufficiently widespread to suggest that it is part ' the male's reproductive strategy (See Why Pri- ates Kill their Young).

Adult male guenons who are not members of troop are usually found alone, although male ıtas monkeys will form small temporary parties. ı captivity, more than one male will live happily ›gether, but only providing there are no females resent. Talapoins live in very large multimale oops, but outside the mating season the males ve in a subgroup whose members interact mainly ith each other and only very rarely with females. amadryas baboons and geladas have "harem" roups within their troops. Each harem consists of –6 adult females and their young with a single dult male, though occasionally there may be two dult males; bachelor males live in a peripheral ıbgroup. Baboons move in procession, usually ith adult males at the front and rear, adult males also towards front and rear (including ıose carrying infants), and juveniles towards the enter of the column.

Martyrs to Medical Research

CONSERVATION AND ENVIRONMENT

ll forest-living monkeys may be considered ıreatened, because tropical forests are being estroyed at such a high rate. (Fully 8 percent of ıe world's forest cover was lost between 1980 nd 1990, and each year 100,000 sq km/38,500 q mi of forest is felled in the Amazon Basin one.) In addition, monkey populations are ways at risk because of their slow reproductive ıte. Where monkeys are considered a delicacy, ıe introduction of guns and the increased com- ıercialization of hunting have further taken their oll on populations. As crop-growing areas are extended, the displaced monkeys raid crops: modern, cash-oriented economies are less tolerant of such theft than traditional societies. Monkeys also share many diseases with people, tuberculosis among them. Monkeys have been shown to carry yellow fever, to which they are extremely susceptible, and baboons carry asymptomatic schistosomiasis. There have been occasional suggestions that monkeys should be exterminated to control disease, but probably no actual attempts to do so.

For several years it seemed as if the increasing demand for monkeys for use in medical research – up to 130,000 Rhesus monkeys were shipped from India to the USA each year during the 1950s and 1960s – would bring about the extinction of the most commonly used species. The animals also suffered high mortality rates in trapping and shipping. Latterly, however, a decline in the research industry, increasing efforts to breed monkeys in captivity for research, and awareness of the need to handle newly-caught animals carefully have reduced this threat. As a result, between 1989 and 1994 the number of Old World monkeys dispatched from the four main exporting countries fell from about 29,000 to about 18,000 per year. Yet the conservation of monkey species is principally a matter of preserving the ecosystems in which they live in large enough patches to allow viable populations to survive. Successful management will depend upon controlling human encroachment.

RD/TER

Above *A tense stand-off takes place between two groups of Long-tailed macaques. The prize at stake in this altercation between the rival factions is leadership of the troop.*

MONKEY KILLERS

Monkeys, some of which prey on smaller creatures, themselves fall prey to other animals. Some of the largest eagles feed mainly on forest monkeys. The African Crowned hawk eagle (BELOW), which often hunts in pairs, has a special technique that is lethal for clever monkeys. One swoops and perches among a troop of monkeys, which respond by approaching to mob it. Meanwhile, with the monkeys' attention diverted, its mate swoops from behind and snatches an unsuspecting victim. Forest monkeys use a special alarm call when an eagle flies over, and respond by diving into thick cover. The Crowned hawk eagle can, however, fly through forest and hunt on the forest floor, and has been seen to kill a near-adult male mandrill, the largest of all the cercopithecines. In the Kibale forest, Uganda, monkeys accounted for 84 percent of the prey taken by Crowned hawk eagles, while a single pair was estimated to remove about 8 percent of the guenon population each year at Makokou in Gabon. In open country, the Martial eagle may prey on vervets and baboons. Vervets are also prey to pythons, which wait in ambush for them at the base of trees. These monkeys have quite distinct alarm calls to warn of the approach of snakes, leopards, and eagles. Monkeys are probably only incidental food items for carnivores; nonetheless, these species may still be major predators of monkeys. Leopards, for example, are the main predators of baboons, attacking them both by day and by night, with kill rates that may be as high as 75 percent of attacks launched. Lions are the second most important baboon killer. Other primates can also be significant predators of monkeys. Baboons themselves occasionally take vervets at Amboseli in Kenya, and chimpanzees eat baboons, Red colobus, and guenons at Gombe in Tanzania. Monkey is also the preferred meat of some people in West and Central Africa and in Southeast Asia.

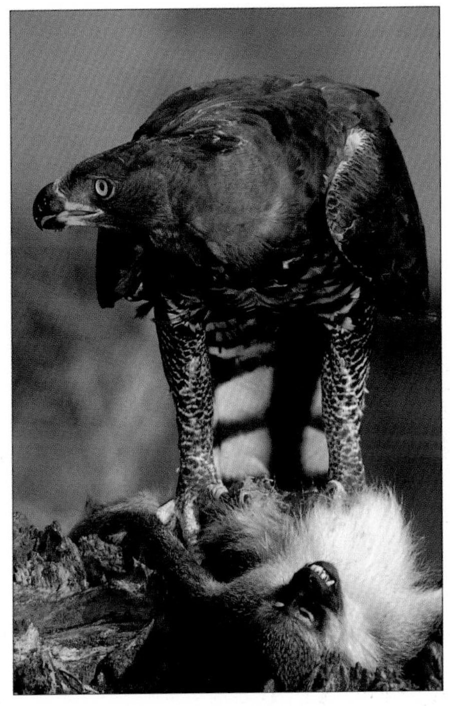

Guenon, Macaque, and Baboon Species

THE CERCOPITHECINE SUBFAMILY OF OLD World monkeys actually includes other groups beside the guenons, macaques, and baboons, notably the baboonlike drills and mandrills and the mangabeys. The group as a whole also takes in such well-known species as the vervets (classed as guenons) and the Rhesus monkeys (macaques).

MANGABEYS

Genus *Cercocebus* and *Lophocebus*

Medium-sized monkeys closely related to the baboons, but restricted to forests. The brownish species (Agile and White mangabeys) are considered to be closely related to each other, and rather distantly separated from the blackish species (Gray-cheeked and Black mangabeys), for which the genus name *Lophocebus* has been proposed. The first two derive from a drill/mandrill ancestor, while the *Lophocebus* pair are thought to share a common ancestor with the baboon/gelada group. All have tails longer than their bodies. Females are smaller than males, but not as markedly as in guenons. Large, strong incisor teeth allow mangabeys to exploit hard seeds which are not accessible to guenons, with which they share habitats. Pregnancy lasts about 6 months, and there is no evidence of breeding seasonality. Infants are the same color as adults. Mangabeys live in large groups that include several males. They are very vocal, and the adult male has a dramatically loud long-distance call (the "whoop-gobble" of the Gray-cheeked mangabey), while the adult females of a group also perform loud choruses.

Agile mangabey LR
Cercocebus galeritus
Agile, Crested, or Tana River mangabey

Cameroon and Gabon, Kenya and Tanzania. Includes *C. agilis*. The recently discovered E populations are scattered and separated from the W ones by thousands of kilometers. Rain forest and gallery forest. Terrestrial.
HBL 44–58cm; WT 5.3kg (female), 9.6kg (male).
COAT: dull yellowish-brown; hair on top of head forms crest.
DIET: palmnuts, seeds, leaves.
CONSERVATION STATUS: LR Near Threatened.

White mangabey LR
Cercocebus torquatus
White, Collared, Red-capped, or Sooty mangabey

Senegal to Gabon. Primary rain forest. Terrestrial.
HBL 66cm; WT 6.2kg (female), 11kg (male).
FORM: coat gray; geographical races have color variants – a white collar in Ghana, a red cap in Cameroon. Females have a cyclic vulval swelling.
DIET: palmnuts, seeds, fruit, leaves.
CONSERVATION STATUS: LR Near Threatened.

Gray-cheeked mangabey
Lophocebus albigena

SW Cameroon to E Uganda. Primary moist, evergreen forest. Arboreal.
HBL 44–58cm (female), 45–62cm, (male); WT 6kg (female), 8.3kg (male).
FORM: coat black, with some brown in long shoulder hair; short hair on cheeks grayish; hair on head rises to single (E races) or double (W races) crest. Female has bright pink cyclic vulval swelling.
DIET: fruit and seeds, also flowers, leaves, insects, and occasional small vertebrates.

Black mangabey LR
Lophocebus aterrimus

Democratic Republic of Congo (DRC). Rain forest. Arboreal. Considered by some authorities as a subspecies of *L. albigena*.
HBL 71cm; WT 5.8kg (female), 7.8kg (male).
COAT: black.
DIET: fruit, seeds.
CONSERVATION STATUS: LR – Near Threatened.

GUENONS

Genus *Cercopithecus*

These African long-tailed monkeys are mainly forest-dwelling. Both sexes have brightly-colored coats, but patterns are more pronounced in males. Infants have dark or dull-colored coats, with pink faces at birth that darken later. Tails are considerably longer than bodies. Males are much larger than females, the difference being greater in larger species; females range from two-thirds to half the weight of males. Taxonomy is complex: several species groups are recognized, with species from one group replacing each other geographically in the guilds of guenons present in each forest. These species groupings are indicated in the species entries below. Other species occur in suitable habitat over a very large area without obvious racial differentiation. Social organization is varied and is described by species. Adult males give loud, species-specific distance calls. Groups of different species may travel together for long periods.
BREEDING: gestation periods are estimated at around 5 months. Breeding is seasonal where known. Typical birth intervals vary between 1 and 3 years, and first births occur between 3 and 7 years, variation being attributed to species and habitat differences in different cases.

Redtail monkey
Cercopithecus ascanius
Redtail or Coppertail monkey, Schmidt's guenon

NE and E Democratic Republic of Congo (DRC), S Uganda, W Kenya, W Tanzania, SW Rwanda. One of the *C. cephus* species group. Mature rain forest and young secondary forest. Arboreal. Groups often have only one adult male.
HBL 41–48cm; WT 2.9kg (female), 3.7kg (male).
COAT: yellow-brown, speckled, with pale underparts; limbs gray; tail chestnut-red on lower end; face black, bluish around eyes, with white spot on nose and pronounced white cheek fur.
DIET: insects, fruit, leaves, flowers, buds.

Mustached monkey
Cercopithecus cephus

S Cameroon to N Angola. Rain forest. Arboreal. Groups may include only one adult male.
HBL 48–56cm; WT 2.9kg (female), 4.3kg (male).
COAT: red-brown agouti with dark gray limbs and back; lower part of tail red; throat and belly white; face black with blue skin around eyes, white mustache bar, white cheek fur.
DIET: fruit, insects, leaves, shoots, crops.

Red-eared monkey Vu
Cercopithecus erythrotis
Red-eared or Sclater's monkey

S Nigeria and W Cameroon. One of the *C. cephus* species group. Includes *C. sclateri*, an endangered species first identified in the wild in 1988 and sometimes thought to be a hybrid between *C. erythrotis* and *C. erythrogaster*. Rain forest. Arboreal.
HBL 36–51cm; WT 2.9kg (female), 3.6k (male).
COAT: brown-agouti with gray limbs; p of the tail red; face blue around eyes, n and eartips red, cheek fur yellow.
DIET: fruit, insects, shoots, leaves, crop

Red-bellied monkey
Cercopithecus erythrogaster

SW Nigeria. Rain forest. Arboreal. A lit known species, similar to *C. petaurista* in *C. cephus* species group.
HBL 46cm; WT 2.4kg (female), 4.1kg (male).
COAT: brown-agouti; face black, throat ruff white; belly variable from reddish t gray.
DIET: fruit, insects, leaves, crops.

Lesser spot-nosed monkey
Cercopithecus petaurista
Lesser spot-nosed or Lesser white-nosed monkey

Sierra Leone to Benin. One of the *C. cep* species group. Rain forest. Arboreal.
HBL 36–46cm; WT 2.9kg (female), 4.1k (male).
COAT: greenish-brown agouti; underpa white, lower part of tail red; face black with white spot on nose, prominent wh throat ruff and white eartufts.
DIET: fruit, insects, shoots, leaves, crop

Owl-faced monkey
Cercopithecus hamlyni
Owl-faced or Hamlyn's monkey

DRC to NW Rwanda. Rain and montan forest. Arboreal. Lives in small groups w a single male.
HBL 56cm; WT 3.4kg (female), 5.5kg (m
COAT: olive-agouti with darker extremities; scrotum and perineum bright blue, face black with yellowish diadem and th white stripe down nose.
DIET: fruit, insects, leaves.
CONSERVATION STATUS: LR Near Thre ened.

L'Hoest's monkey
Cercopithecus lhoesti

Mt. Cameroon and E DRC to W Uganda Rwanda. Close relatives include *C. preus* (Preuss's monkey). Montane forest. Terr trial. Lives in small groups with a single adult male.
HBL 46–56cm; WT 3.5kg (female), 6kg (male).

ABBREVIATIONS	HBL = head–body length TL = tail length WT = weight Approximate nonmetric equivalents: 2.5cm = 1in; 230g = 8oz; 1kg = 2.2lb	Ex Extinct EW Extinct in the Wild Cr Critically Endangered	En Endangered Vu Vulnerable LR Lower Risk

Left *De Brazza's monkey sports a distinctive white beard and an orange "diadem" on its brow.*

ORM: coat dark gray-agouti with chest-ut saddle; underparts dark. The E form as a striking white bib, while the W form less strikingly marked, with small bib, ght gray cheek fur, and whitish mustache narkings. Tail hook-shaped at end.
IET: fruit, leaves, insects.
ONSERVATION STATUS: LR Near Threat-ned.

lue monkey
ercopithecus mitis
lue, Silver, Golden, or Samango monkey

W Angola to SW Ethiopia, down hrough southern Africa. *C. mitis* is eplaced in W Africa by the closely similar *. nictitans*. Rain forest and montane bam-oo forest. Arboreal. Lives in medium-ized groups of about 20–40, often with nly a single adult male.
IBL 49–66cm; WT 4.3kg (female), 7.9kg male).
FORM: coat gray-agouti, with geographic variants often given subspecific rank. The Blue monkey (*C. m. stuhlmanni*) has a bluish-gray mantle, black belly and limbs, dark face with pale yellowish diadem; Silver (*C. m. doggetti*) and Golden (*C. m. kandti*) monkeys, from W Uganda, Rwanda, and E DRC, are variants with lighter and yellowish mantles respectively. The samango of southern Africa is a drab rusty-gray.
DIET: fruit, flowers, nectar, leaves, shoots, buds, insects; prey includes wood owls and bush babies.

Sykes' monkey
Cercopithecus albogularis

E of Rift Valley, from Somalia S to E Cape Province in S Africa; also the islands of Phylax, Zanzibar, Mafia.
COAT: chestnut saddle; pronounced white ruff. Other details as *C. mitis*, of which it is sometimes considered a subspecies.

Spot-nosed monkey
Cercopithecus nictitans
Spot-nosed or Greater white-nosed monkey or hocheur

Sierra Leone to NW DRC. Rain forest. Arboreal. Habits similar to Blue monkey, which replaces it to the west.
HBL 44–66cm; WT 4.3kg (female), 6.7kg (male).
COAT: dark olive-agouti; belly, extremities and tail black, face dark gray with white spot on nose.
DIET: fruit, leaves, shoots, insects, crops.

Mona monkey
Cercopithecus mona

Senegal to W Uganda. Rain forest. Arboreal. Lives in fairly large groups, which may contain a single, or more than one, adult male. Its name derives from the moaning contact call of the female. Similar to Crowned guenon, which replaces it to the S.
HBL 46–56cm; WT 5.1kg (male).
COAT: back brown-agouti, rump and underparts white; upper face bluish-gray, muzzle pink; hair round face yellowish, with dark stripe from face to ear.
DIET: fruit, leaves, shoots, insects, crops.

Crowned guenon
Cercopithecus pogonias

S Cameroon to Congo basin. Forest. Arboreal.
HBL 46cm; WT 2.9kg (female), 4.3kg (male). Similar in habits to Mona monkey.
COAT: brown-agouti with black extremities; lower part of tail black; belly and rump yellow; face blue-gray with pink muzzle; prominent black line from face to ear, and median black line from forehead forming crest; fur yellow between black lines.
DIET: fruit, leaves, shoots, insects, crops.

De Brazza's monkey
Cercopithecus neglectus

Cameroon to Ethiopia, Kenya to Angola. Swamp forest. Semiterrestrial. Lives in small groups, usually a pair with offspring. Freezes when alarmed.
HBL 41–61cm; WT 4.1kg (female), 7.4kg (male).
COAT: gray-agouti with black extremities; tail black; white stripe on thigh; rump white; face black with white muzzle; long white beard and orange diadem; scrotum blue.
DIET: fruit, leaves, insects.

Diana monkey En
Cercopithecus diana

Sierra Leone to SW Ghana. Forest. Arboreal. A wide-ranging species of the high canopy, living in medium-sized groups with a single adult male. This species may be allied to the Vervet monkey.
HBL 41–53cm; WT 3.9kg (female), 5.2kg (male).
COAT: gray-agouti and chestnut back; extremities and tail black; white stripe on thigh; rump fur red or cream in different races; face black, surrounded by white ruff and beard.
DIET: fruit, leaves, insects.

Wolf's monkey
Cercopithecus wolfi
Wolf's or Dent's monkey

A little-known species from DRC, NE Angola, W Uganda, Central African Republic. Arboreal. Includes *C. denti*.
HBL 45–51cm; WT 2.9kg (female), 3.9kg (male).

Campbell's monkey
Cercopithecus campbelli

A little-known species from Gambia to Ghana.
HBL 36–55cm; WT 2.7kg (female), 4.5kg (male).

Dryas monkey
Cercopithecus dryas
Dryas or Salongo monkey

A little-known monkey from DRC.
WT 3kg (male).

Suntailed monkey Vu
Cercopithecus solatus

Most recently discovered Old World primate, first described from forests of Gabon (W Africa) in 1986. Little known about it.
WT 3.9kg (female), 6.9kg (male).

VERVET

Genus *Chlorocebus*

A single species separated from *Cercopithecus* through its lack of certain cranial specializations. The vervet also displays some terrestrial adaptations similar to those of *Erythrocebus* but not present in *Cercopithecus*.

Vervet
Chlorocebus aethiops
Vervet, grivet, Savanna, or Green monkey

Senegal to Somalia, E Africa down to southern Africa. Savanna, woodland edge, never far from water and often on banks of water courses. Semiterrestrial. Groups usually include several adult males. Closely related to the Diana and Patas monkeys.
HBL 46–66cm; WT 3kg (female), 4.3kg (male).
FORM: coat yellowish- to olive-agouti, underparts white, lower limbs gray, face black with white cheektufts and browband; eyelids white; scrotum bright blue, penis and perineal patch red. Geographical races have been recognized within their vast range and given specific status according to detail of color and pattern of cheektufts, but there is also variation of these characters within one troop.
DIET: fruit, leaves, flowers, insects, eggs, nestlings, rodents, crops.

SWAMP MONKEY

Genus *Allenopithecus*

A single species, separated from *Cercopithecus* because females have periodic (perineal) swelling.

Allen's swamp monkey LR
Allenopithecus nigroviridis

E Congo and W DRC. Swamp forest. Habits unknown.
HBL 41–51cm; TL 36–53cm; WT 3.2kg (female), 6.1kg (male).
COAT: green-gray agouti with lighter underparts; hair flattened on crown.
DIET: fruit, seeds, insects, fish, shrimps, snails.
CONSERVATION STATUS: LR Near Threatened.

TALAPOIN

Genus *Miopithecus*

A single species, separated from *Cercopithecus* because females have cyclic perineal swelling. Talapoins live in large groups of 70–100, including many adult males. They are sharply seasonal breeders, mating in the long dry season and giving birth 5½ months later. Infants are colored like adults except for the pink face, which darkens after about 2 months. The juvenile period is long, with first births occurring at 5 or 6 years.

Talapoin monkey
Miopithecus talapoin

S Cameroon to Angola. Wet and swamp forest, and alongside water courses. Arboreal.
HBL 34–37cm; TL 36–38cm; WT 1.1kg (female), 1.4kg (male).
COAT: greenish-agouti; underparts and inner sides of limbs pale; scrotum blue; face gray with dark brown cheek stripe.
DIET: fruit, insects, flowers, crops.

PATAS

Genus *Erythrocebus*

A single species separated from *Cercopithecus* because of its long limbs and adaptations for running. Patas monkeys live in moderately-sized groups, usually with a single adult male. They are seasonal breeders, mating in the wet season and giving birth 5 months later. Infants are light brown with pink faces, which darken by 2 months. The juvenile period is short, with first births occurring at 3 years or even earlier.

Patas monkey
Erythrocebus patas
Patas, Military, or Hussar monkey

Senegal to Ethiopia, Kenya, Tanzania. Terrestrial. Thought to be closely related to the Vervet monkey.
HBL 58–75cm; TL 62–74cm; WT 5.8kg (female), 10.6kg (male).
COAT: shaggy, reddish-brown; underparts, extremities, and rump white; scrotum bright blue; penis red; face black, with white mustache; cap brighter red, with black line from face to ear.
DIET: *Acacia* fruit, galls, and leaves, other fruit, insects, crops, tree gum.

MACAQUES

Genus *Macaca*

Heavily built, often partly terrestrial monkeys. The coat is generally dullish brown, but the naked skin on face and rump may be bright red; some species have sexual swellings. Tails are mostly shorter than body length or totally absent, depending on species. Males are somewhat larger than females. Diets are eclectic, with fruit the most common item. Breeding is seasonal for the most part; mating takes place in the fall and birth occurs in the spring, after about 5 months' gestation. Infants have a distinctively-colored soft natal coat which is replaced after about 2 months. Macaques live in fairly large groups that may include several adult males. Females generally remain in their natal group for life, but males emigrate at adolescence and thereafter live alone, in small groups of males, or in other groups with females for varying periods of time.

Stump-tailed macaque Vu
Macaca arctoides
Stump-tailed or Bear macaque

E India to S China and Vietnam. Forest, particularly montane. Terrestrial and arboreal.
HBL 50–70cm; TL 1–10cm; WT 8.4kg (female), 12.2kg (male).
FORM: coat dark brown; face naked, dark red and mottled; rump also naked and dark red. No perineal swelling.
DIET: fruit, insects, young leaves, crops, small animals.

Assamese macaque Vu
Macaca assamensis

N India to Thailand and Vietnam. Forest. Terrestrial and arboreal.
HBL 53–68cm; TL 19–38cm; WT 6.9kg (female), 11.3kg (male).
COAT: varying shades of yellowish to dark brown; face and perineum naked, red in adult.
DIET: fruit, insects, young leaves, crops, small animals.

Formosan rock macaque
Macaca cyclopis
Formosan rock or Taiwan macaque

Taiwan. Terrestrial and arboreal.
HBL 56cm; TAIL moderately long; WT 4. (female), 6kg (male).
COAT: dark brown.
DIET: fruit, insects, young leaves, crops small animals.

Long-tailed macaque
Macaca fascicularis
Long-tailed, Crab-eating, or Cynomolgous macaque

Indonesia and Philippines to S Burma. Forest edge, swamp, banks of water cou es and coastal forest. Terrestrial and arb real.
HBL 38–65cm; TL 40–66cm; WT 3.6kg (female), 5.4kg (male).
FORM: coat varying shades of brown (grayish or yellowish or darker); unders paler; face skin dark gray; prominent fri of gray hair round face. No perineal swelling.
DIET: fruit, insects, young leaves, crops, small animals.
CONSERVATION STATUS: LR Near Thre ened.

Japanese macaque
Macaca fuscata

Japan. Forest. Terrestrial and arboreal.
HBL 47–60cm; TL 7–12cm; WT 8kg (female), 11kg (male).
FORM: coat brown to gray; face and rum skin naked, red in adult. No perineal swelling.
DIET: fruit, insects, young leaves, crops, small animals.

Rhesus macaque
Macaca mulatta
Rhesus macaque or Rhesus monkey

India and Afghanistan to China and Viet nam. Forest, forest edge, and outskirts o towns and villages. Terrestrial and arboreal.
HBL 47–64cm; TL 19–30cm; WT 5.4kg (female), 7.7kg (male).
FORM: coat brown with paler underside face and rump naked, red in adult. No perineal swelling.
DIET: fruit, insects, young leaves, crops, small animals.
CONSERVATION STATUS: LR Near Threatened.

ABBREVIATIONS HBL = head–body length TL = tail length WT = weight
Approximate nonmetric equivalents: 2.5cm = 1in; 230g = 8oz; 1kg = 2.2lb

Ex Extinct
EW Extinct in the Wild
Cr Critically Endangered
En Endangered
Vu Vulnerable
LR Lower Risk

-tailed macaque Vu
caca nemestrina

dia to Indonesia. Wet forest. Includes *agensis*, found only on the Mentawai nds off Sumatra. Terrestrial and real.
47–60cm; TL 13–24cm; WT 6.5kg ale), 11.2kg (male).
RM: coat varying shades of brown, with r underside and darker brown areas nd face. Females have large cyclic per- l swelling.
T: fruit, insects, young leaves, crops, ll animals.

nnet macaque
caca radiata

dia. Forest, forest edge, and outskirts owns and villages. Terrestrial and arbo- . HBL 35–60cm; TL 48–69cm; WT kg (female), 6.7kg (male).
RM: coat grayish-brown with paler lerparts; hair on head grows out in orl from central crown. No perineal lling.
T: fruit, insects, young leaves, crops.

n-tailed macaque En
caca silenus

dia. Wet forest. Terrestrial and arboreal. 46–61cm; TL 25–38cm; WT 6.1kg nale), 8.9kg (male).
RM: coat black with gray around face, outstanding ruff; tail with slight tuft at Females have cyclic perineal swelling.
T: omnivorous.

que macaque Vu
caca sinica

Lanka. Wet forest, edges of water- rses, scrub. Terrestrial and arboreal.
43–53cm; TL 47–62cm; WT 3.2kg male), 5.7kg (male).
RM: coat reddish or yellowish-brown h paler underparts; hair on top of head ws out from central crown. No perineal elling.
T: fruit, insects, young leaves, crops.

rbary macaque Vu
caca sylvanus
bary macaque, Barbary ape, or Rock ape

Algeria and Morocco; managed popula- n on Gibraltar. Mid- to high-altitude for- , also scrub and cliffs. Terrestrial and oreal.
L 50–60cm; TAIL absent; WT 11kg male), 16kg (male).
RM: coat yellowish-gray to grayish- own, with paler underparts; face dark sh-colored. Females have dark gray-red clic perineal swelling.
DIET: fruit, young leaves, bark, roots, occasionally invertebrates.

Tibetan macaque LR
Macaca thibetana
Tibetan, Père David's, or Tibetan stump-tailed macaque

Tibet to China. Montane forest. Semiterrestrial.
HBL 60cm; TL 6cm; WT 9.5kg (female), 12.2kg (male).
COAT: brown.
DIET: omnivorous.
CONSERVATION STATUS: LR Conservation Dependent.

Moor macaque En
Macaca maura

Sulawesi. Forest.
HBL 66cm; TAIL absent; WT 6kg (female), 9.7kg (male).
FORM: coat brown or brownish-black; ischial callosities large and pink. Females have cyclic perineal swelling.
DIET: omnivorous.

Celebes macaque En
Macaca nigra

Sulawesi. Forest.
HBL 55cm; WT 5.5kg (female), 9.9kg (male).
FORM: coat black, with prominent pink ischial callosities; face black, prominent ridges down side of nose; hair on head rises to stiff crest; tail absent. Females have cyclic pink perineal swelling.
DIET: omnivorous.

Tonkean macaque LR
Macaca tonkeana

Sulawesi. Forest.
HBL ca. 60cm; WT 9kg (female), 14.9kg (male).
FORM: coat black, lighter brown rump, cheeks; ischial callosities prominent; tail absent. Females have cyclic pink perineal swelling.
DIET: omnivorous.
CONSERVATION STATUS: LR Near Threatened.

BABOONS

Genus *Papio*

The classification of baboons is controversial and several systems have been proposed. Here the Savanna or Common baboon is considered to be one species containing five races previously considered separate species. Baboons live in large groups. The Hamadryas baboon subspecies has a hierarchical group structure based on the one-male unit or harem. Savanna baboons have more informal groups including several adult males. Breeding: gestation c.185 days; breeding not seasonal. Birth intervals vary around 2 years, depending on the food supply, and first births occur when females are from 4 to 8 years old. Infants have a black natal coat and pink skin for the first 2 months.

Savanna baboon LR
Papio hamadryas
Savanna or Common baboon

West Africa to Ethiopia and Somalia; Saudi Arabia and S Yemen, south to S Africa and Angola. Savanna woodland and forest edge; rocky desert and subdesert, with some grass and thorn bush (Hamadryas baboon subspecies *P. h. hamadryas*). Terrestrial.
HBL 56–79cm; TL 42–60cm; WT 9.9–13.3kg (female), 16.9–25.1kg (male).
FORM: coat gray-agouti, with longer hair over shoulders, especially in adult males; shiny black patch of bare skin over hips; face naked and black with prominent lateral ridges on the long muzzle, especially in adult males. Females have cyclic perineal swelling. First 3 or 4 tail vertebrae fused in adult, giving hook-shaped base to tail. Coat color varies geographically in subspecies once accorded specific status. The lowland E and C African Yellow baboon (*P. h. cynocephalus*) is, as its common name suggests, yellowish. Similarly, the highland E African Olive baboon (*P. h. anubis*) is olive-greenish. The southern African Chacma baboon (*P. h. ursinus*) is dark gray. The Guinea baboon (*P. h. papio*) is brown with red naked skin on rump, and a brownish-red face. The Hamadryas baboon (*P. h. hamadryas*) is characterized by its silver-gray coat and red naked skin on face and rump. The nose shape also varies geographically, the Olive baboon having a pointed nose extending beyond the mouth a little, while the Yellow and Chacma subspecies have retroussé noses.
DIET: grass, fruit, seeds, bulbs and tubers, insects, hares and young ungulates, crops.
CONSERVATION STATUS: LR Near Threatened.

MANDRILL

Genus *Mandrillus*

Large, terrestrial forest-dwelling baboons confined to the rain forests of W Central Africa. Ranges do not overlap, with drill in N and mandrill in S.

Drill En
Mandrillus leucophaeus

SE Nigeria and W Cameroon. Rain forest. Terrestrial.
HBL 70cm; TL 12cm; WT 12.5kg (female), 20kg (male). Females much smaller than males.
COAT: dark brown, with blue to purple naked rump; face black with white fringe of hair around it. Muzzle long, with pronounced lateral ridges along it.
DIET: fruit, seed, fungi, roots, insects, small vertebrates.

Mandrill Vu
Mandrillus sphinx

S Cameroon, Gabon, Congo. Rain forest. Terrestrial.
HBL 80cm; TL 7cm; WT 12.9kg (female), 36.1kg (male). Females much smaller than males.
COAT: olive-brown agouti with pale underparts; blue to purple naked rump in adult males, duller in females and juveniles. Face very brightly colored in adult male, with red median stripe on muzzle, ridged side of muzzle blue, beard yellow. Females and juveniles similarly colored but duller.
DIET: fruit, seeds, fungi, roots, insects, small vertebrates.

GELADA

Genus *Theropithecus*

A single species, the only survivor of an important fossil group once widespread throughout Africa from the Mediterranean to S Africa. Now confined to high-altitude habitats above 1,700m (5,800ft) in NW Ethiopia. Commonly referred to as a baboon, but very different in vocal and visual communication patterns from *Papio*. Geladas live in large herds, within which adult males have harems of several females. Other males live in bachelor groups at the periphery.

Gelada LR
Theropithecus gelada
Gelada or Gelada baboon

NW Ethiopia. Grassland. Terrestrial.
HBL 50–74cm; TL 32–50cm; WT 11.7kg (female), 19kg (male).
FORM: coat brown, fading to cream at end of long hairs; mane and long cape over shoulders; naked area of red skin around base of neck, surrounded by whitish lumps in the female which vary in size with the menstrual cycle. Rump of both sexes also red and naked, and rather fat. Muzzle with concave upper line, longitudinal ridges along side of snout. Upper lip can be everted, used in flash display.
DIET: grass, roots, bulbs, seeds, fruit, and insects.
CONSERVATION STATUS: LR Near Threatened.

A Male-dominated Society

The Hamadryas baboons of Cone Rock, Ethiopia

It is an hour before sunset in the semi-desert landscape of the southern Danakil plain in Ethiopia. A long column of Hamadryas baboons – brown females and young interspersed with large, gray-mantled males – is crossing a dry river bed and threading its way up a gravel slope. The troop is making for a cliff where it will pass the night in safety from leopards. Suddenly, a male near the front runs back along the column at full speed. A female separates from the last group in the line and hurries toward him, seemingly aware that she has severely tested his tolerance by lagging too far behind. Upon reaching her, the male delivers a shaking bite to the back of her neck. Squealing, she follows him closely up to the cliff where his other females are waiting. He then leads his family to a ledge where they settle for grooming.

The neck bite is an extreme form of the threat that Hamadryas males habitually employ in herding their females. Four-fifths of the troop's adult males own harems, which range in size from one to 10 females and average about two. The sex ratio is not skewed, but males only form pair bonds long after sexual maturity, at 10–12 years, while females do so at 4–6 years. Whereas males of other baboon species consort with only one female at a time and only for hours or days when she is in heat, 70 percent of Hamadryas pair bonds last longer than three years and continue uninterrupted through the periods of pregnancy and lactation, when the female is not sexually accessible. This demonstrates that primate pair bonds are not necessarily sexually motivated.

The troop living at Cone Rock in northeastern Ethiopia numbers several hundred and is organized into groups of four levels – harem families, clans (numbering some 15–30), bands (each containing around 65–90 individuals), and finally the troop. Members of the same group interact about 10 times more often with each other than they do with outsiders that belong to next larger grouping.

The reproductive career of a young Hamadryas male is successful only if he can win control of a harem. His difficulty is that all reproductive females in a troop belong to a male, who will fight any encroaching rival. Moreover, experiments have shown that males have an inhibition against stealing females from another male, even from a weaker one. This ensures peace, but prevents young males from gaining access to mature females of their own. To get around this problem, young males belonging to different bands at Cone Rock have adopted quite different strategies. A subadult male in Band 1 will attach himself to a family with a small daughter. She is still immature, her father is not very keen to defend her, and other grown males disregard her. The subadult male takes to closely preceding the juvenile female whenever she follows her mother, to instill in her the idea that she is actually following him. And indeed, after weeks of gentle maneuvering, she does begin to tag along with him. Using this skillful technique, the subadult can eventually prize her away from her family. In stark contrast to this subtle approach, males in Band 2 always bide their time until they are young adults, and then abduct a juvenile female suddenly and by force. Clearly, then, divergent styles of social behavior exist even within the same troop. By acquiring a pre-pubertal female the subadult male avoids competition from adult males. Once mature she will become his first mate, whereupon the inhibition of other males works in his favor.

Aging males, however, no longer benefit from others' inhibition. Fully grown followers in one band were seen to attack the three old leaders of their clan and take some or all of their females by force. Within weeks, the defeated males lost weight and their gray hair changed to the brown color of females. Their appearance indicated a drop in testosterone levels. Of the three defeated leaders, only one lived on for several years in his clan, where his influence was mostly felt in the clan's decisions on travel directions. The other two disappeared: they either left the troop or died.

It appears that the male's reproductive career depends on his association with his clan and band. Clan and band males obtain most of their females from one another and cooperate in their defense against outsiders. In fact, a male remains in his

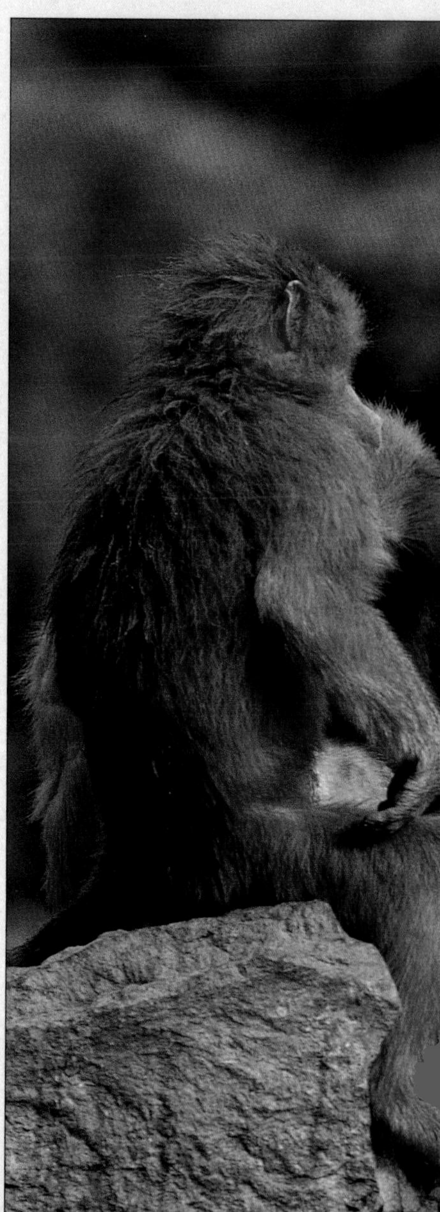

Right *In Hamadryas baboons grooming is the most time-intensive social activity. Aggression between females almost invariably concerns getting the attention of the male and females will fight each other for the right to groom him.*

Left *Baboons partition their time so as to be able to engage in a variety of important activities during the day. Some of these activities can bring them into conflict with other members of the group:* ***1*** *an aggressive encounter between males;* ***2*** *juvenile exploring;* ***3*** *foraging;* ***4*** *female grooming male;* ***5*** *infants playing;* ***6*** *female presenting to male (not shown) during courtship.*

natal clan to the end of his reproductive life. He first becomes one of its followers and finally, with luck, one of its harem leaders. When two mothers of juvenile sons were taken over by males from outside the clan, the sons actually left their mothers and rejoined their fathers' clans – an extremely unusual preference in a juvenile primate. Females, however, are often transferred to other clans or even bands. The Hamadryas baboons of Cone Rock are exceptional in living in societies that appear to be driven by relationships between related males rather than females, which is the norm in most related primates.

Nevertheless, females are far from being mere chattels in Hamadryas society. Experiments have revealed that females demonstrate definite preferences for certain males, and that rival males heed these preferences. Thus, in choice tests, the less a female favored her present male, the more likely it was that a rival would overcome his inhibition and steal her in subsequent tests of inhibition.

The adult males of a band interact in selecting the direction of the day's foraging trip before they set off. Toward the end of the morning social session near the cliff, individual males will leave the periphery of the resting band, only to sit down again after a few meters, facing outward. Other individuals may then follow. Protuberances grow out of the periphery of the group in several directions and withdraw again if followers are lacking. Males approach and look at each other. Eventually one protuberance gains ascendancy over the others and the band departs in its direction. After a few hundred meters it splits up, but around noon it reassembles at the waterhole that lies in the direction of the original joint departure. Waterholes can be dangerous places and seem to require the full strength of numbers. The band tends to appoint one of them as a meeting place hours in advance. HK

MONKEYS IN THE SNOW

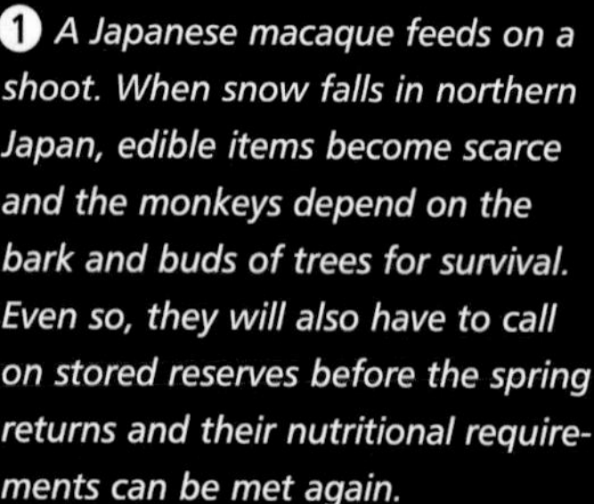

❶ *A Japanese macaque feeds on a shoot. When snow falls in northern Japan, edible items become scarce and the monkeys depend on the bark and buds of trees for survival. Even so, they will also have to call on stored reserves before the spring returns and their nutritional requirements can be met again.*

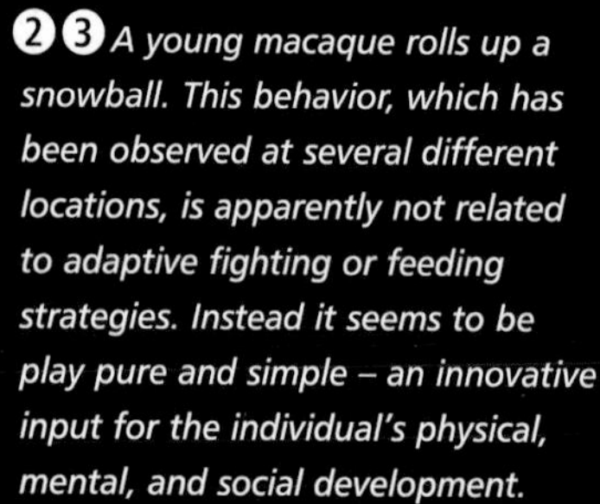

❷❸ *A young macaque rolls up a snowball. This behavior, which has been observed at several different locations, is apparently not related to adaptive fighting or feeding strategies. Instead it seems to be play pure and simple – an innovative input for the individual's physical, mental, and social development.*

Grooming not only helps p the skin clean of debris d ectoparasites but also serves reinforce social bonds. Such avior is common not just ong related animals but also nin the larger framework of the op. Japanese macaque troops primarily matrilineal, although y may contain both male and nale members; the females tend stay for life, however, while the es are more transitory, passing n one group to another.

⑤ *A young macaque snuggles in its mother's long hair against the winter cold. Females protect their single offspring fiercely in the first few months after birth. Even though the young are weaned by about 1 year of age, maternal support continues for much longer; female offspring take 3 years to become active adult members of their matrilineal kingroups, while males stay even longer before dispersing to other groups in search of mating opportunities.*

⑥ *Macaques bask in the warm thermal waters of a pool at Jigokudani, near Nagano in the mountains of eastern Honshu, Japan's central island. The monkeys live farther north than any other nonhuman primates and, unlike their cousins in more tropical regions, they experience four distinct seasons annually. Winters are particularly harsh – temperatures can fall as low as -15°C (5°F) in January – and in these bleak times, when the mating season has finished and nomadic males are back in the troop, it is a pure sensual pleasure to relax in a hot bath.* XD-R

Colobus and Leaf Monkeys

The species of the subfamily Colobinae are long-tailed, tree-dwelling monkeys that almost all depend on leaves as the main constituent of their diet. Anatomically, their most distinctive feature is an unusual stomach that enables them to digest leaves more efficiently than any other primates.

The colobines includes species as diverse as the long-nosed, pale-faced Proboscis monkey, anatomically a ground-dweller but trapped by island climate change in mangrove and rain forest, and the Red-shanked douc monkey, whose pale, ethereal facial skin and white, rufflike whiskers recall a chorister who is about to don his cassock over a gray, black, and red school uniform. More flamboyant is the black guereza with its long-haired white mantle, while the endearing, wizened faces of the simple-nosed Asian leaf monkeys have earned one of their genera the name *Presbytis*, meaning "old woman" in Greek.

Slender Leafeaters

Form and Function

The Old World monkeys (family Cercopithecidae) are all anatomically alike, and few consistent differences distinguish the two subfamilies. The colobus and leaf monkeys of the Colobinae branch are principally distinguished by the absence of cheek pouches, and the presence of large salivary glands and a complex sacculated stomach (partitioned into sacs). Colobine molar teeth have high, pointed cusps; the inside of their upper molars and the outside of the lower ones are less convexly buttressed than in cercopithecines. The inner surface enamel on the lower incisors is thicker, and there is a lateral process on the lower second incisor. The molars tend to erupt precociously in relation to the incisors. Underjet (protrusion of the lower incisors beyond the upper incisors) is common in colobines, but rare in cercopithecines.

The majority of living colobines are more slenderly built than cercopithecines. The exceptions are the *Nasalis* and *Pygathrix* species, which include some of the largest, although not the heaviest, of monkeys. Their fore- and hindlimbs are more equal in length than in other living species, indicating that at some point in their history they occupied a more terrestrial habitat. The

Right *The Red sureli inhabits the island of Borneo in Southeast Asia. It is a diurnal rainforest dweller; however, it is also occasionally found in areas of secondary woodland.*

FACTFILE

Colobus and Leaf Monkeys

Order: Primates

Family: Cercopithecidae

Subfamily: Colobinae

42 (34) species in 7 genera: Genus *Nasalis* (2 species); snub-nosed and douc monkeys (*Pygathrix*, 6 (5) species); surelis (*Presbytis*, 8 species); langur and leaf monkeys (*Semnopithecus* and *Trachypithecus*, 15 (10) species); black colobus monkeys (*Colobus*, 5 (4) species); red colobus monkeys (*Procolobus*, 6 (5) species).

DISTRIBUTION S and SE Asia; equatorial Africa.

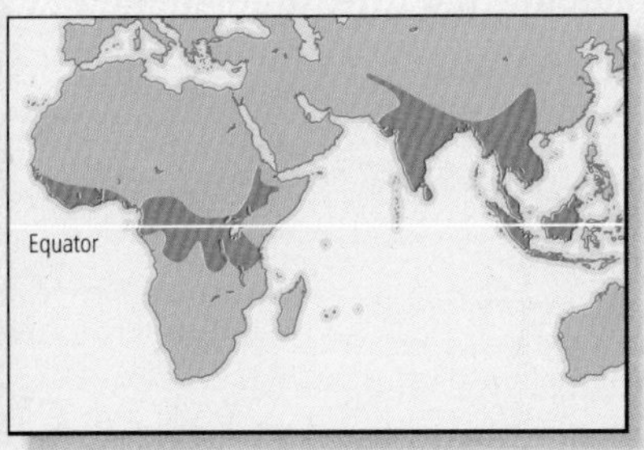

HABITAT Chiefly forests; also dry scrub, cultivated areas, and urban environments.

SIZE **Head–body length** from 41–78cm (16–31in) in Hanuman langur to 43–49cm (17–20in) in Olive colobus; **tail length** from 69–108cm (27–43in) to 57–64cm (23–25in); **weight** from 5.4–23.6kg (12–52lb) to 2.9–5.7kg (6.4–12.5lb), both in same species.

COAT Very varied between species, especially on the head, where crests, fringes, or whorls may be present.

DIET Leaves, fruit, flowers, buds, seeds, and shoots; a few species sometimes eat insects.

BREEDING Gestation 140–220 days, depending on species.

LONGEVITY About 20 years (29 in captivity).

See species table ▷

Above A male Proboscis monkey, clearly displaying its characteristic pendulous nose. By contrast, females of this species have snub noses. In younger monkeys the nose is turned upward, only becoming long and pronouncedly bulbous as the animal ages. The Proboscis monkey is the largest of all the Asian colobines, with males weighing almost twice as much as females.

trend to thumb loss, barely evident in the snub-nosed monkeys, the Proboscis monkey, and the fossil genus *Mesopithecus*, progresses through other Asian colobines until, in the *Colobus* genus itself (the word means "docked" in Greek), it is either absent or reduced to a small tubercle, sometimes with a vestigial nail. The vulnerability of the thumb to injury during leaping presumably outweighs the adaptive advantages of its retention. The ischial callosities are separate in females and contiguous in males, except in male *Pygathrix* and male *Procolobus* species, in which the callosities are separated by a strip of furred skin.

Coat color variation in newborn colobines can be quite dramatic but is usually similar across different species of the same genus. Dental and visceral anatomy, and external features such as the position of the ischial callosities, further confirm the distinctiveness of some genera. At species level, colobines are discriminated chiefly by coat color, but also by vocalization and hair style, especially on the head, which may sport crests, fringes, whorls, or partings.

The Asian Connection

DISTRIBUTION PATTERNS

The present stronghold of the colobine subfamily is Asia, with four genera and 31 species, compared to only the two genera and 11 species that are found in Africa. The fossil representation of colobines is, however, strongest in Africa, and two of the earliest fossil genera, *Mesopithecus* and *Dolichopithecus*, are European. Asian species range from about latitude 35.5°N at the Afghanistan–Pakistan frontier to the Lesser Sunda island of Lombok, stopping short of Sulawesi and the Philippines. *Pygathrix* and *Nasalis* species inhabit southern China, eastern Indochina, the Mentawai Islands, and Borneo, but, intriguingly, they are absent from the intervening areas of the Malay peninsula and Sumatra.

The living African species are distributed from the Gambia on the west through the Guinea forest belt and the central African forest to Ethiopia in the east, with outlying populations in other parts of East Africa as well as on the islands of Bioko (in the Gulf of Guinea) and Zanzibar (off the Eastern coast of Tanzania). Fossil African colobines inhabited both northern and southern Africa. Borneo is home to the greatest modern concentrations of colobine species, with six species altogether, although not more than five are to be found in any one part of the island; and in northeastern Indochina and West and Central Africa, each with three species.

Tracking Ecological Change Through Species Distribution

The colobines have much to tell us about the long-term instability of the rainforest environment. For example, one south Indian leaf monkey shows an uncanny resemblance to another that is now restricted to inhabiting a single island off north Vietnam. The similarities between the two are so great that at one time the distributions of the two must have joined up, indicating that, 200,000 years ago, most of present-day India must have been covered with rain forest. Shortly thereafter, the sudden severe drought accompanying the onset of a glaciation evidently reduced rainforest distribution to just a small proportion of its present extent. Now it is confined to the southwestern and extreme northeastern areas of the subcontinent.

Primate distributions in Sumatra indicate that this event was the earlier and more drastic of two deforestations that have moulded modern rainforest distribution. The rain forest recovered much lost ground during the interglacial period, only to lose most of it again at the beginning of the most recent period of glaciation, which was about 80,000 years ago. Earlier glaciations over the previous two million years undoubtedly also influenced primate distributions, though their effects are no longer readily apparent; even so, it is not unreasonable to assume that the rain forest has undergone a prolonged series of major expansions and contractions.

In Africa, the Gray- and the White-epauletted black colobus monkeys are closely related, but they are geographically separated by the guereza, which is probably an interglacial or postglacial descendant of the former species, filling the hiatus created by central African glacial deforestation. This hiatus is also evident in the distribution of red colobus species, which are absent from the area of west-central Africa occupied by the Satanic black colobus. Further gaps occur in their east African distribution. All Colobus species, along with many other animals, are subject to a well-known geographic hiatus in West Africa at the Dahomey Gap.

This climatic background provides a fascinating insight into a phenomenon first observed in South American mammals, but probably widespread elsewhere. Termed "metachromism," it is the tendency for a unidirectional evolution of coat and skin color characterized by progressive pigment dilution. Many of the Asian colobines, whose central area of distribution vanished during the glaciations, are relatively darkly pigmented; indeed, some are almost entirely black. Their descendant species are gray, almost invariably succeeded by brown posterity; in some cases the sequence persists through red to virtually white. Originally taken as a longterm irreversible process, this is in fact quite probably a short-term genetic effect produced by the relatively rapid periods of population dispersal that accompanied post-glacial rainforest regeneration. The distribution of the differently-colored colobines may equate to zones of forest regrowth, indicating that climatic warming occurred in phases rather than as a gradual, uninterrupted event.

A Fondness for Foliage

DIET

The essential feature of the colobine stomach is the segregation of its sacculated, expanded upper portion from the lower acid region. The upper chamber contains a neutral medium essential for the fermentation of foliage by anaerobic bacteria. The enlarged salivary glands probably assist in neutralizing acid seepage from the lower stomach. Because the food is of low nutritional value, the colobine stomach is made to accommodate large volumes; stomach contents can constitute more than a quarter of the adult colobine body weight, and as much as half in a semi-weaned infant. The symbiotic bacteria in colobine stomachs enable them to digest leaves more efficiently than other primates. Energy can be released from denatured cellulose (a major component of all leaves); and many toxins can be deactivated, allowing colobines to consume items fatal to other primates. There are also bacteria in the colobine gut that serve to recycle urea, an adaptation that may allow the Hanuman langur to survive in very aric regions, as it does in artiodactyl ruminants.

Plants defend themselves against being eaten by producing toxins in their leaves, which reach their highest concentrations on nutrient-poor soils, where the plants are less able to replace leaves. In forests on good soil, leaves are nutritio and easy to digest, and colobines eat leaves princ pally of common trees. This is the case, for exam ple, with the guereza and Red colobus in Kibale Forest, Uganda. On poor soil, colobines are oblig ed to be more selective. They substitute other plant parts, particularly seeds, for many commor leaves, as, for instance, in the Satanic black colobus in Cameroon, and the Banded and Bornean Red sureli.

A growing proclivity for leaves and other plant parts, less susceptible than fruits to seasonal varia tion in availability, enabled ancestral Old World monkeys to enter open woodland and savanna habitats that were inhospitable to more frugivorous hominoids. As the climate and vegetation

Above Hanuman langurs foraging for food. In ·mmon with all other langurs, Hanumans are almost :clusively vegetarian. While foraging – an activity)stly undertaken in the early morning or late after-on – groups may cover an area of many square kilo-:ters. Though they are mainly tree-dwelling, they are ·ite well adapted to living in areas where forestation scarce, traveling across the ground on all four feet.

Right A Mitered sureli. Group size for this species erages around 15 individuals. In captivity they can long-lived, one individual reaching 18 years of age. ·ants are born white with a dark stripe extending wn their backs to the end of their tails. Adult coat loration may vary considerably.

changed, the cercopithecine diet became more varied, while colobines specialized on leaves. Today, where known, the diet of all colobines includes leaves. The general preference is for young rather than mature leaves, and some species may be unable to cope with the latter. Leaves are so far the only natural dietary item recorded for the barely-studied Barbe's leaf monkey. Fruits form part of the diet of all the remaining investigated species. The guereza, for example, eats about 15 percent fruits and seeds in comparison to 81 percent leaf material. Most species eat flowers, buds, seeds, and shoots. The Hanuman langur, the Banded sureli, the Javan Grizzled sureli, the better-studied leaf monkeys, the guereza, and the Red colobus have all been observed eating soil or termite clay. Golden leaf monkeys specifically eat salty earth or sand, and the Bornean Grizzled sureli churns up the mud at salt springs, presumably to eat. Invertebrates comprise a small proportion of the diet of some surelis, of the Hanuman langur, of the Dark-handed red colobus, and of the Hooded black leaf monkey (and probably also of other leaf monkeys). A Bornean Grizzled sureli has even been observed raiding a bird's nest for its eggs or nestlings. Hanuman langurs and red colobus monkeys eat insect galls and fungi, and African colobines eat lichen and dead wood. The Yunnan snub-nosed monkey predominantly feeds on arboreal lichens. Pith occurs in the diets of the Hanuman langur and the Red sureli, and roots in those of the Hanuman langur and the Mitered sureli. The latter unearths and consumes cultivated sweet potatoes. The Hanuman langur eats gum and sap, and can digest with impunity quantities of the strychnine-containing fruit of *Strychnos nux-vomica* that would be fatal to a Rhesus macaque. It also eats repulsive and foul-smelling latex-bearing plants such as the ak (*Calotropis*). Water is generally obtained from dew, moisture in the diet, or rainwater in treetrunk hollows.

Solemn-faced Monkeys

SOCIAL BEHAVIOR

Female colobines reach sexual maturity at about 4 years of age, males at 4–5. Copulation is not restricted to a distinct breeding season, but there tends to be a birth peak, with weaning synchronized to coincide with the seasonal abundance of solid food. Sexual behavior is usually initiated by the female. A receptive female Proboscis monkey will purse the lips of her closed mouth when eyeing a male. If he returns her glance, she (like the female *Semnopithecus*) rapidly shakes her head. The Proboscis male responds by assuming a pout-face and either approaches the female or she him, presenting her anogenital region. A Hanuman langur female may hit the male, pull his fur, or even bite him if he fails to respond to her advances. A female Red-shanked douc will characteristically adopt a prone position, and view the male over her shoulder. He in turn may signal his arousal by intently staring at the female, and then gazing at a suitable copulation site. Soliciting in female *Colobus* species is similar, but emphasized by tongue-smacking. During copulation, douc and *Colobus* females remain prone, whereas Proboscis monkey and *Semnopithecus* females adopt the normal cercopithecid quadrupedal stance. The Proboscis female continues to headshake, and both partners show the mating pout-face.

At birth, infants are about 20cm (8in) long and weigh about 0.4kg (14oz). The eyes are open and the infant can cling to its mother strongly enough to support its own weight, although Olive colobus infants are carried in the mother's mouth. Body hair is shorter, more downy, and commonly of a different color than in adulthood. The skin and ischial callosities are usually paler than in maturity, though infant facial skin is darker in the Proboscis monkey and the Red-shanked douc. Births are single or, rarely, twin. Except for the Banded sureli and the red colobus species, parental care in all studied species involves acquiescence by the mother to other females borrowing her offspring. Soon after birth, the infant is frequently handled and carried up to 25m (75ft) from its mother. A mother may suckle her own and another infant simultaneously, and one has even been observed carrying three infants. "Babysitting" females tend to abruptly abandon their charges, leaving the mother to retrieve the screaming infant. In *Semnopithecus* species, active rejection of the young has

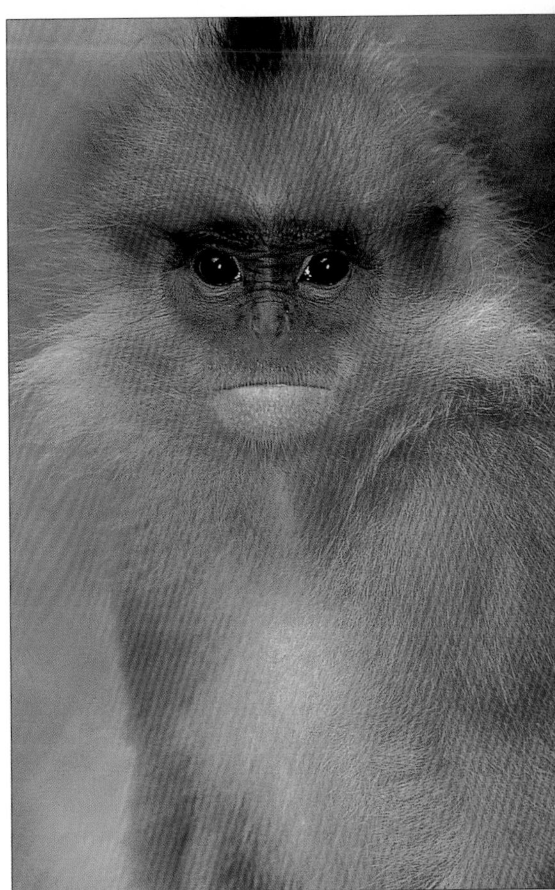

been observed as early as at about 5 weeks. This may accelerate the infant's independence, enhancing its chances of surviving the high infant mortality that often accompanies the violent replacement of the adult male in the one-male troops characteristic of some populations (see Why Primates Kill their Young). It may also allow the mother to concentrate on time-consuming foraging. Transition from the neonatal to the adult coat occurs at 5–10 months; adult size is attained at about age 5.

As compared with macaques, colobine species show a generally lower level of aggressive, socio-sexual, vocal, and even gestural interactions in their social relationships; their general demeanor has been described as "grave and serious," prompting the generic name *Semnopithecus* (Greek for "solemn ape"). One reason may be connected to feeding behavior. Tree-dwelling animals whose food is evenly distributed through the forest and requires prolonged periods of sedentary feeding have less need for tight troop coordination, and so facial expressions are kept to a minimum. Entry into, or movement within, a feeding tree is characterized by meticulous care to avoid close encounters with companions already stationed there. Once a monkey takes up a feeding position, its very nature, facing towards the periphery of the tree where the bulk of the food is located, enables the colobine to feed for long periods with minimal interaction with its neighbors. Although a peak period is common in the morning and late afternoon, the low nutritional content of the monkeys' diet usually necessitates feeding on and off throughout the day, further reducing opportunities for complex social behavior.

Group sizes range from solitary animals, usually males, to groups of over 120 langurs (possibly a temporary aggregation of troops seeking water). Groups of 200–300 – even of over 600 in the case of Golden snub-nosed monkeys – have been reported, though these were probably aggregations of smaller family units. Groups of 60 or so have been reported for the Proboscis monkey, the Red-shanked douc, and the Western red colobus, and troops of up to 40 for most genera. Average troop sizes, though, are lower, ranging from 3.4 in the Mentawai sureli which, with the simakobu and some Javan Grizzled sureli populations, is frequently monogamous, to 37 in one Hanuman langur population. In the simakobu and the White-epauletted black colobus, the figure is about five; in the Hooded black, Purple-faced, and Capped leaf monkeys, most of the surelis, and the guereza, it ranges from six to nine; and in most of the remaining species from 10 to 18. The exception is the Dark-handed red colobus, where the typical group size is probably 50.

The representation of adult males in mixed-sex troops is roughly proportional to their size, and is equaled or exceeded in number by adult females. The Black-shanked douc in eastern Cambodia is

invariably encountered in (presumably mixed-sex) pairs, and the Bornean Grizzled sureli often in "family parties" of three, but some troops number at least twelve. All-male bands occur in the Proboscis monkey, the Guizhou snub-nosed monkey, and many of the simple-nosed Asian species, but are extremely rare in *Colobus* species.

Where assessed, home range in most species is about 30ha (74 acres), ranging up to 64ha (158 acres) for the Capped leaf monkey and about 130ha (320 acres) for the Proboscis monkey anc the Dark-handed red colobus. In the Hooded black leaf monkey, range sizes vary from 6–260h (15–642 acres), and in the Hanuman langur fro 5–1,300ha (12–3,200 acres). Grizzled and Banc ed surelis, Purple-faced, Silvered, and Dusky leaf monkeys, and some Hanuman langur populatio are all considered territorial on the basis of defense and exclusive use of at least the major pa of their home range. Purple-faced leaf monkey

Above Red-shanked douc monkeys with a small ant. Found in the tropical forests of Cambodia, os, and Vietnam, this strikingly-colored creature is rnal and tree-dwelling. Despite the fact that they an Endangered species, they are still captured for e on the pet market.

troops often temporarily desert their home range in order to attack an adjacent troop; while the adult male Purple-faced leaf monkey has such a fastidious sense of territory that it has been seen to chastize fellow troop members for transgressing into other territories. Other Hanuman langur populations, and other species such as the Mentawai sureli, the Hooded black, and the Capped leaf monkey, have exclusive core areas that include important sleeping and feeding trees, and that occupy 20–50 percent of the home range; in one Western red colobus troop studied in the Gambia, the figure reached 83 percent. Within its home range the guereza has a preferred area from which other troops are chased, but not permanently excluded. In contrast, three Dark-handed red colobus troops were found to have very extensive home-range overlap. Relations between these three troops were usually aggressive; they involved only the adult and subadult males, and no matter where an encounter occurred within their home ranges, one dominant troop usually supplanted the other, although the participation of particular males seemed to be a factor in whether their group prevailed. Other red colobus troops entered the area very infrequently and were usually evicted immediately.

Surelis, langurs, and black colobus monkeys are all particularly noted for their loud calls, which are generally most intense and most contagious at dawn. They may also call during the day, especially when a troop is getting ready to move or night-time sleeping positions are being finalized; some species also call during the night. These calls are believed to advertise and defend both territories and mates. During an inter-troop encounter, or sometimes when a predator is detected, the calls may be preceded or accompanied by a dramatic leaping display in which the protagonist plunges down from the heights and then ascends and repeats the performance, with branches swaying and cracking around him.

Population densities are very variable, both within and between species. In the surelis, they range from 3 to 48 animals per sq km (7–125/sq mi); in the simakobu and the leaf monkeys, from 8–220 (20–570/sq mi); in the Hanuman langur, from 3–904 (7–2,340/sq mi); and in the colobus species, from 30–880 (70–2,280/sq mi). Among Hanuman langurs, lower densities are most typical of populations living in open grassland or agricultural land, where troops have large home ranges. Intermediate densities are associated with populations living close to towns and villages, while higher densities are found in forest-dwelling populations, where the troops have small, sometimes overlapping, home ranges.

Decimation by Hunting

Conservation and Environment

In 1819, a ship's crew visiting Da Nang, Vietnam, killed more than 100 Red-shanked doucs between 5am and breakfast-time. Apparently unmolested by the local people, doucs lacked respect for firearms and were even drawn to their deaths by the cries of their wounded companions. By 1831, depredations by Europeans were such that doucs had learned to fear gunfire. In 1974, after depletion of habitat and the ravages of the Vietnam War, only 30–40 doucs were found in a 10-week period in one vicinity. In 1993, however, several multimale–multifemale groups, ranging in size from

Above A female Olive colobus. This species, ch, unusually, carries its newborn offspring around 's mouth, inhabits the lower strata of the rain est. Its diet is exclusively vegetarian, comprising ds, fruit, and foliage, with young leaves being its ferred food.

Left A Hanuman langur leaping its way across the nganga river in India. Hanumans are renowned for ir great agility. In the trees they make horizontal os of up to 5m (16ft); longer leaps, involving some of height, can extend over fully 10m (33ft) – an ressive feat for an animal less than 1m (3ft) tall.

Below White-epauletted black colobus monkeys quite similar in appearance to guerezas, but may distinguished from the latter by their lack of a hy tail. Infants of this species are born pure white, take on adult markings by the time they are 3–4 nths old.

6–17, were found there, amounting to a total of rather less than 300 animals.

The Western black colobus was seriously threatened by the European fur trade at the end of the 19th century, and the guereza continues to suffer from a tourist demand for goods made from its pelt. Other species hunted for their beautiful coats are the Hooded black leaf monkey and the Golden snub-nosed monkey. The coat of the latter was said to protect the wearer from rheumatism, but fortunately only officials and the Chinese Imperial family were allowed to wear it. Both species are now protected, and the former is said to have increased in numbers since 1960, although it is subjected to severe poaching in the extreme north of its range. Most species are hunted by local people for their flesh, which in many species is reputed to have medicinal value, as are the bezoar stones (stones from the intestine with supposedly magical healing properties) reportedly found in all the Bornean colobines except for the Banded sureli and the Silvered leaf monkey. In Africa, the escalation of the bushmeat trade may already have exterminated one Western red colobus subspecies, and even populations in officially protected forests are being destroyed.

In Asia, the most insidious threats to the future survival of colobines are the wider availability of firearms and the destruction of habitat for timber and cultivation that, in Indonesia, has recently led to the deliberate setting of forest fires. It is essential that at least part of the monkeys' remaining habitat should be protected by reserves; the need is particularly urgent in Côte d'Ivoire, the Mentawai Islands, and northern Vietnam.

Apart from hunting and habitat destruction, interactions between colobines and man are minimal, except where they live side by side, as in the case of Sumatra, where relations are amicable even though the monkeys frequently damage crops. The best-known co-habiter is the Hanuman langur, which in some areas obtains 90 percent of its diet from crops. Like all animals, though, it goes unmolested by Buddhists, and is venerated by Hindus. In the great Sanskrit epic, the *Ramayana*, the monkey-god Hanuman was Rama's chief helper in the search for his kidnapped wife Sita. The story tells that while in Sri Lanka, Hanuman stole the mango, previously unknown on the mainland. For this theft he was condemned to be burnt, and his face and paws were permanently blackened by the scorching they received. Many Hindus regularly feed langurs. The exasperation felt by those whose produce is sacrificed to the monkeys, or whose shops are plundered by them, was epitomized by the town that once dispatched a wagonload of langurs to a distant railway station in a desperate attempt to be rid of them. DB-J

CHINA'S ENDANGERED MONKEYS

Mainland China has three endangered snub-nosed colobine species, all unknown in the West before the late 19th century. The most endangered is the Yunnan or Black snub-nosed monkey (*Pygathrix bieti*), now reduced to 1,000–1,500 animals inhabiting Yunnan province and Tibet. The better-protected Guizhou snub-nosed monkey (*P. brelichi*) is more geographically restricted, inhabiting only Fanjing Mountain and its vicinity in Guizhou province. It probably numbers between 800 and 1,200 animals.

The Golden snub-nosed or Golden monkey (*P. roxellana*) is the most numerous and widely distributed species, with 8,000–10,000 animals inhabiting Gansu, Hubei (especially Shennongjia), Shaanxi, and Sichuan Provinces. Golden monkeys inhabit mountain forests in some of the largest troops known for arboreal primates; assemblages of over 600 animals have been reported. In ecologically disturbed areas, troops may number 30–100 animals. Larger troops are organized around polygynous subgroups, each of a single adult male, five adult females, and their offspring. There are also peripheral and solitary males, but within a troop adult females outnumber adult males. Males defend the troop against predators (chiefly Yellow-throated martens). They are primarily leafeaters, supplementing their diet with fruit, pinecone seeds, bark, insects, birds, and birds' eggs. Because they lack cheek pouches and eat leaves, they feed often and in large quantities. In zoos, they are fed 850–1,380g (2–3lb) daily. There are difficulties in providing a balanced diet, and they fare poorly in captivity.

Humans have valued Golden monkeys' decorative shoulder and back hair (up to 10cm/14in long) for making coats for over 1,000 years, and herbal medicines are made from the meat and bones. But the major threat to all three species' survival has been destruction of their limited habitat.

In recent decades, the Chinese government has taken steps to preserve these primates. Hunting was banned in 1975, but lack of funds and staff mean that it persists, as does forest destruction. Preservation areas have been established, although only 2 of the 13 subpopulations of the Yunnan snub-nosed monkey fall completely within a reserve. Births are increasing in preservation areas, allowing grounds for guarded optimism, but longterm studies are urgently needed. FEP

Colobus and Leaf Monkey Species

THE SUBFAMILY COLOBINAE COMPRISES 42 individual species in 7 genera (the 1993 established taxonomy recognizes 34 in 7; species not included there are marked † below). They extend throughout equatorial Africa, the Indian subcontinent and Southeast Asia (including Java, Sumatra, Borneo, and many of the smaller islands). At least half of the individual species are threatened, chiefly through loss of habitat to forest clearance and agriculture.

GENUS *NASALIS*

Thickset build: macaque-like limb proportions, skull shape, and coat color; nose prominent.

Proboscis monkey En

Nasalis larvatus

Borneo, except C Sarawak. Tidal mangrove, nipapalm–mangrove, and (mainly riverine) lowland rain forest. Swims competently.
HBL 54–76cm; TL 52–75cm; WT 20.3kg (male), 9.8kg (female).
COAT: crown reddish-orange, with frontal whorl and narrow nape extension flanked by paler cheek and chest ruff; rest of coat orange-white or pale orange, richer on lower chest, variably suffused with gray, flecked with black and reddish on shoulders and back: white triangular rump patch adjoining white tail. Newborn face vivid blue; adult face pinkish beige. Pendulous, tongue-shaped nose in adult male; penis reddish-pink; scrotum black.

Simakobu En

Nasalis (Simias) concolor

Simakobu, Pig-tailed snub-nosed monkey, Pagai Island langur

Mentawai Islands, Indonesia. Rain forest and mangrove forest.
HBL 45–55cm; TL 10–19cm; WT 7kg.
COAT: blackish-brown, pale-speckled on nape, shoulder, and upper back; white penile tuft. Face skin black, bordered with whitish hairs. Tail hairless, except sparse hair at tip. 1 in 4 individuals are cream-buff, washed with brown.

GENUS *PYGATHRIX*

Large, with arms only slightly shorter than legs; face short and broad; shelflike brow ridge; region between eyes broad; nasal bones reduced or absent; nasal passages broad and deep; small flap on upper rim of each nostril.

Golden snub-nosed monkey Vu

Pygathrix (Rhinopithecus) roxellana

Golden, Orange, Snub-nosed, or Roxellane's monkey, Moupin langur

In Chinese provinces of Gansu, Hubei, Shaanxi, and Sichuan. High evergreen subtropical and coniferous forest and bamboo jungle, snowclad for more than half the year; migrates vertically biannually.
HBL 47–78cm; TL 48–88cm; WT 16.8kg (male), 9.9kg (female).
COAT: upperside and tail dark brown or blackish, darkest on nape and sagittal ridge on crown; underside, tail tip, and long (10cm) hairs scattered over shoulders whitish orange; legs, chest band, and facial margin richer orange; orange suffusion increases with age in adult male. Muzzle white; area around eyes and nose pale blue. Newborn gray, changing at four weeks to creamy white.

Yunnan snub-nosed monkey En

Pygathrix (Rhinopithecus) bieti

Yunnan or Black snub-nosed monkey

Yunling mountain range (26.5°–31°N), Yunnan, and Tibet. High coniferous forest (3,000–4,000m), frostbound for 280 days per annum. Included by some in *P. roxellana*.
HBL 51–83cm; TL 49–72cm; WT 15kg (male), 9kg (female).
COAT: forehead, inner forearm, inner leg, and ring around belly gray; belly itself white; paws, sagittal crest, much of tail, and some upper lip hairs blackish; long yellowish-gray hairs with black-brown tips scattered on shoulders; rest of coat grayish-black above and white below. Adult males generally blacker than adult females, with longer white flank hair. Lips deep pink; area around eyes pale pink; rest of muzzle and sides of nose white. Newborn bright white, becoming gray over several months.

Guizhou snub-nosed monkey En

Pygathrix (Rhinopithecus) brelichi

Guizhou, Gray, or Oxtailed snub-nosed monkey

Fanjing Mountain, Guizhou Province, China. Semi-evergreen subtropical forest. Included by some in *P. roxellana*.
HBL 64–73cm; TL 85–97cm; WT 14.6kg (male), 7.8kg (female).
COAT: upperparts grayish-brown, pale gray on thigh; tail, paws, forearm, and outside of shank blackish; tail tip and blaze between shoulders yellowish-white; nape and vertex whitish-brown, blackish at front and sides; forehead, inner knee, and inside of upper arm orange; belly pale yellowish-gray; midline parting from forehead to crestless vertex; tail hairs sometimes long with midline parting. Adult males more brightly colored than females, with white skin on prominent nipples. Lips and eyelids pink; rest of face bluish white.

Tonkin snub-nosed monkey Cr

Pygathrix (Rhinopithecus) avunculus

Tonkin or Dollman's snub-nosed monkey

Central N Vietnam. Subtropical forest and bamboo jungle.
HBL 51–65cm; TL 66–92cm; WT 13.8kg (male), 8.3kg (female).
COAT: upper side blackish; brown between shoulders (occasionally white-sprinkled); nape and rear of crown brown or yellowish-brown, with narrow, blackish-brown border at front and sides; paws blackish-brown; pale orange of underside sometimes restricted to throat and chest band, with rest yellowish-white, which encircles face and most of elbow, ankle and hip; tail blackish-brown with whitish-yellow or orange-gray hair tips; tail hairs without parting, or long with midline parting or helical parting; crown hairs flat. Tail tip whitish in adults. Pale blue around eyes, violet in between; lips deep pink; rest of muzzle blue-black. Penis black; scrotum white-haired pale bluish white.

Red-shanked douc monkey En

Pygathrix nemaeus

Red-shanked douc or Cochin China monkey

S Laos and C Vietnam. Tropical rain and monsoon forest.
HBL 52–64cm; TL 57–68cm; WT 12.6kg (male), 8.9kg (female).
COAT: fingers and toes black; lips, cheeks, and throat, inside of thigh, perineum, tail, and small triangular rump patch white; white areas surrounded by black, often with intervening deep orange band, most evidently between throat and chest; crestless crown, upper arms, and trunk between black areas black-speckled gray. Penis reddish-pink; scrotum white. Skin of muzzle white; ears, nose, and rest of face orange. Northern subspecies has white forearm and deep orange-red shank; southern one (probably a distinct species) has gray forearm and gray shank.

Right Silvered leaf monkeys. A[c]... have a distinctive, shaggy, silvered [c]... but infants are even more striking, being born with an orange coat th[at] persists for the first three months.

Black-shanked douc monkey †

Pygathrix nigripes

Black-shanked or Black-footed douc mon[key]

E Cambodia and S Vietnam. Tropical [rain] forest; gallery and monsoon forest. In[clud]ed by some in *P. nemaeus*.
HBL 50–72cm; TL 59–77cm; WT 13.1[kg] (male), 8.7kg (female).
Distinguished from *P. nemaeus* by pala[te] anatomy and black-speckled gray fore[arm] and blackish shank. Blackish areas expanded; orange reduced. Penis red; scrotum and inside of thigh blue. Faci[al] skin blue, with reddish-yellow tinge o[n] muzzle.

SURELIS

Genus *Presbytis*

Forearm relatively long; brow ridges u[sual]ly poorly developed or absent; bridge [of] nose convex; muzzle short; 5th cusp [of] lower 3rd molar usually reduced or absent; cusp development on upper 3[rd] molar variable; broad, underjetted low[er] incisors have tubercle on inner surface; sagittal crest; coat of newborn whitish

Mentawai sureli

Presbytis potenziani

Mentawai or Red-bellied sureli

Mentawai Islands, Indonesia. Rain an[d] mangrove forest.
HBL 44–58cm; TL 50–65cm; WT 5.4–7.3kg.
COAT: small ridgelike crest; upperside tail blackish; pubic region yellowish-white; brow band, cheek, chin, throat, upper chest, and sometimes tail tip whitish; rest of underside and sometim[es] collar reddish-orange, brown, or occas[ion]ally whitish-orange.

Grizzled sureli

Presbytis comata

Grizzled or Gray or Sunda Island sureli

N and E Borneo, W Java, N Sumatra. T[rop]ical rain forest; riparian forest in grassl[and;] rubber plantations; fruit gardens. Previ[ously] (incorrectly) named *P. aygula*. Cl[ose] relatives include *P. hosei* and *P. thomasi*.

Continued ▷

ABBREVIATIONS	HBL = head–body length TL = tail length WT = weight		
	Approximate nonmetric equivalents: 2.5cm = 1in; 230g = 8oz; 1kg = 2.2lb	Ex Extinct	En Endangered
	† = 'new' species. See p291	EW Extinct in the Wild	Vu Vulnerable
		Cr Critically Endangered	LR Lower Risk

HBL 42–60cm; TL 55–84cm; WT 5–8kg.
COAT: paws and sometimes shank black; crown black, gray, or brown, with central N Bornean subspecies sexually dichromatic in extent of white trim; rest of upperside pale gray speckled with black or brown; underside white.

Fuscous sureli †
Presbytis fredericae

C Java, Indonesia. Tropical rain forest; araucaria plantations. Usually included in *P. comata*.
HBL 45–58cm; TL 57–70cm.
COAT: blackish brown above, with long, paler guard hairs and blackish-brown cummerbund and pectoral tract on white underside, or dark gray above with gray pectoral tract on white underside; fingers and toes variably white.

White-fronted sureli
Presbytis frontata

E, SE, and C Borneo. Tropical rain forest.
HBL 42–60cm; TL 62–79cm; WT 5.6–6.5kg.
COAT: paws, cheeks, and brow blackish; forearm, shank, and sometimes tail base and crest blackish-brown; trunk pale grayish-brown, yellowish below; tail yellowish, speckled with dark gray. Hairline raised and crest raked forward over pale-skinned forehead.

Banded sureli LR
Presbytis femoralis

NW Borneo, Malay Peninsula, Singapore, C Sumatra (including Batu Islands). Rain forest; swamp; mangrove swamp.
HBL 43–60cm; TL 62–83cm; WT 5.9–8.1kg.
COAT: dark brown to black, with variable amounts of white on underside. Included by some in *P. melalophos*. A Bornean subspecies is orange with black upperparts and white underparts.
CONSERVATION STATUS: LR Near Threatened.

Pale-thighed sureli †
Presbytis siamensis

Riau Archipelago, S Malay Peninsula, E C Sumatra, Great Natuna Island. Tropical rain and swamp forest. Included by some in *P. femoralis*.
HBL 41–61cm; TL 58–85cm; WT 5–6.7kg.
COAT: limb extremities and brow blackish; outside of thighs grayish white; rest of upperparts and tail pale grayish or blackish brown; underparts white; horizontal fringe radiates from 0–2 whorls at front end of crest.

Mitered sureli LR
Presbytis melalophos
Mitered or Black-crested sureli or simpai

SW Sumatra. Tropical rain forest, villages.
HBL 42–59cm; TL 53–81cm; WT 5.8–7.4kg.
COAT: back (occasionally broad midline band only) brown-red to pale orange, or gray, variably suffused with black/gray; brow red to whitish, often circumscribed by blackish, brown, or gray crest and brow hairs that may extend to ear; underside white, sometimes suffused with yellow or orange, especially at chest and limbs.
CONSERVATION STATUS: LR Near Threatened.

Red sureli
Presbytis rubicunda
Red or Maroon sureli

Karimata Island and Borneo, except C Sarawak and lowland NW Borneo. Rain forest.
HBL 45–55cm; TL 64–78cm; WT 5.2–7.8kg.
COAT: blackish-red or reddish-orange, paler underside, variably more black or brownish on tail tip and paws; horizontal fringe radiates from 0–2 frontal whorls; 1–2 nape whorls sometimes present.

LANGURS AND LEAF MONKEYS

Genus *Semnopithecus* and *Trachypithecus*

Brow ridge shelflike and newborn coat black in species of genus *Semnopithecus*. In other species (genus *Trachypithecus* – considered by some a subgenus of *Semnopithecus*), brow ridges resemble raised eyebrows; females have pale pubic patch; newborn coat orange (sometimes white) suffused with gray, brown, or black.

Malabar langur †
Semnopithecus hypoleucos

SW India, between W Ghats and coast to 14°N. Evergreen forest; cultivated woodland; gardens.
HBL 61–70cm; TL 85–92cm; WT 8.4–11.5kg.
COAT: paws and forearm blackish; legs blackish or grayish-brown; tail blackish or dark gray at base, tip blackish, yellowish-gray, or white; midline of back dark brown or dark gray; flanks and rear of thighs pale yellowish-gray; crown orange-gray or yellowish-white; throat and underside orange-white. Frontal whorl.

Hanuman langur LR
Semnopithecus entellus
Hanuman or Common or Gray langur

S Himalayas, from Afghanistan frontier to Tibet between Sikkim and Bhutan; India NW of range of Malabar langur to Aravalli Hills and Kathiawar, and NE to Khulna province, Bangladesh; N, E, and SE Sri Lanka. Allegedly in W Assam. Forest; scrub; cultivated fields; village and town centers (0–4,080m).
HBL 41–78cm; TL 69–108cm; WT 5.4–23.6kg.
COAT: upperparts gray or pale grayish-brown, often tinged with yellow; in Bengal, gray almost replaced by pale orange; crown, underparts, and tail tip white or yellowish-white; paws often, and forearms occasionally, blackish or brownish; in Sri Lanka and SE India, crown usually crested. Frontal whorl.
CONSERVATION STATUS: LR Near Threatened.

Purple-faced leaf monkey
Trachypithecus (Semnop.) vetulus
Purple-faced leaf monkey or wanderoo

SW, C, and N Sri Lanka. Forest; swamp; rocky, treeless, coastal slopes; parkland. Formerly known as *S. senex*.
HBL 47–70cm; TL 62–92cm; WT 4.3–10kg.
COAT: brown, darkest at limb extremities and sometimes with yellow to brown tail tip and crown, or blackish with pale brown crown and yellowish tail tip; white to yellow throat and sideways-directed whiskers; rump patch sometimes almost white or yellow; tail base, thighs, and back sometimes gray-speckled.

Hooded black leaf monkey Vu
Trachypithecus (Semnop.) johnii
Hooded or Leonine or Gray-headed black leaf monkey, or Nilgiri langur

W Ghats of India and Cat Ba Island, N Vietnam. Evergreen and riverine forest; deciduous woodland; on Cat Ba, stunted, treeclad limestone hills.
HBL 49–71cm; TL 69–97cm; WT 9.8–13.6kg.
COAT: yellowish vertex grades (further in Cat Ba) through brown to glossy brown-tinged black of rest of body; gray speckle of short-haired rump sometimes extends to thigh and tail.

White-rumped black leaf monkey † Cr
Trachypithecus (Semnop.) delacouri
White-rumped or Pied or Delacour's black leaf monkey

N Vietnam. Tropical monsoon forest on limestone mountains. Sleeps in caves. Included by some in *S. francoisi*.
HBL 57–58cm; TL 82–86cm.
COAT: glossy black, with white from nape whorl to end of mouth; sharply demarcated white area on hindpart of back and outside of thighs; mid-tail hairs long; pointed coronal crest.

White-browed black leaf monkey †
Trachypithecus (Semnop.) laotum

C Laos. Limestone mountains, some vi[rtu]ally treeless. Included by some authori[ties] in *S. francoisi*.
HBL 46–54cm; TL 81–90cm.
COAT: glossy black with pale hairs on c[heeks] and throat; white from nape whorl to e[nd] of mouth, including forehead and cove[ring] most of back of head; pointed coronal crest.

Bar-headed black leaf monkey †
Trachypithecus (Semnop.) hatinhen[sis]
Bar-headed or Ha Tinh black leaf monkey

E C Laos, C Vietnam. Tropical monsoo[n] forest. Included by some in *S. francoisi*.
HBL 50–67cm; TL 81–87cm; WT 8–9kg.
COAT: glossy black with white from behind ear to end of mouth; pointed coronal crest.

White-sideburned black leaf monkey
Trachypithecus (Semnop.) francoisi
White-sideburned or Francois' black leaf mo[nkey]

Guangxi and Guizhou, China, NE Vietnam. Tall riverside crags; tropical monsoon forest on limestone mountains. Bivouacs in caves. Includes the variably albinotic *T. leucocephalus*.
HBL 47–67cm; TL 77–90cm; WT 5.7–9.5kg.
COAT: glossy black with white from ear [to] end of mouth; pointed coronal crest; 2 nape whorls.

Ebony leaf monkey †
Trachypithecus (Semnop.) mauritius
Ebony or Moor or Negro leaf monkey

S W Java (Indonesia), E C Laos, W C Vietnam. Forests. Previously included i[n] *S. auratus*.
HBL 55cm; TL 73–82cm.
COAT: glossy black, tinged with brown, especially on underside and cheeks; mi[d]dle of paw sometimes slightly white.

Spangled leaf monkey
Trachypithecus (Semnop.) auratus
Spangled or Ebony or Moor or Negro leaf mon[key]

Bali, Bangka, Belitung, S Borneo, C and [E] Java, Lombok, Serasan, S Sumatra and Riau–Lingga archipelago. Forest; plantations. Includes *S. cristatus*.
HBL 43–65cm; TL 61–87cm; WT 5–8kg.
COAT: glossy black, tinged with brown, especially on underside and cheeks; ha[ir] tips variably sprinkled with white; unde[r]side sometimes gray.

ABBREVIATIONS

HBL = head–body length TL = tail length WT = weight
Approximate nonmetric equivalents: 2.5cm = 1in; 230g = 8oz; 1kg = 2.2lb
† = 'new' species. See p291

Ex Extinct
EW Extinct in the Wild
Cr Critically Endangered
En Endangered
Vu Vulnerable
LR Lower Risk

vered leaf monkey
chypithecus (Semnop.) cristatus

nd N Borneo, Cambodia, coastal W laysia, Riau archipelago, N Sumatra, S C ıiland, S Vietnam. Forest swamp; bam-›; scrub; plantations; parkland; villages. . 43–59cm; TL 63–84cm; WT 4.9–6.2kg. AT: brown, brownish-gray or blackish-›wn, darker on paws, tail, and forehead; or masked by grayish or yellowish hair ; groin and underside of tail base yel-ish; coronal crest variably developed.

rbe's leaf monkey †
chypithecus (Semnop.) barbei

:hina, N Indochina into Burma. Forest. merly divided between *S. cristatus* and *hayrei.* _ 43–60cm; TL 62–88cm; WT 4.6–8.7kg. AT: gray to blackish-brown; paws and brow black or blackish-brown, upper n and (sometimes) underside, leg, tail, ›e, or back suffused with silvery gray or low; coronal crest variably developed.

sky leaf monkey
chypithecus (Semnop.) obscurus

sky or Spectacled leaf monkey

pura (NE India), adjacent Bangladesh, 5han, and lowland SW and S Burma, lay Peninsula and neighboring small ınds (not Singapore). Forest; scrub; ntations; gardens. Includes *T. phayrei* rt). . 42–68cm; TL 57–86cm; WT 4.2–10.9kg. AT: dark gray to blackish-brown; nape er, occasionally yellowish-white; back dline usually paler and sometimes with nge sheen; elbow, legs, and tail root en paler than back, occasionally pale yish-yellow; paws and eyebrows black blackish-brown; underside yellowish-›wnish to blackish-gray, dark brown, or casionally pale orange; coronal crest ually present in NW; frontal whorl in an subspecies.

pped leaf monkey
chypithecus (Semnop.) pileatus

pped or Bonneted leaf monkey

ngladesh and Assam E of Jamuna and anas rivers, N and highland W Burma. rest; swamp; bamboo. Overlaps with sky leaf monkey in Tripura and adjacent ngladesh. L 49–76cm; TL 81–110cm; WT)–14kg. AT: upperparts gray, darkest on anterior back and occasionally tinged with nge; eyebrow and (occasionally) fore-ad black; paws and tail root black or rk gray; paws sometimes partially ange-white; cheeks and underside gray, ıitish to orange; crown hairs semi-erect d project over cheek hairs. Scrotum portedly absent.

Golden leaf monkey En
Trachypithecus (Semnop.) geei

Bhutan and Assam W of Manas river. Forest; plantations.
HBL 49–72cm; TL 71–94cm; WT 9.5–12kg.
COAT: orange-white; underside, rear of back, and (sometimes) cheeks orange; blackish hair tips on cap; faint gray tinge on forearm and shank, sometimes on rear of back and upperside of tail; crown hairs semi-erect and project over cheek hairs. Pubic skin pale. Scrotum reportedly absent.

BLACK COLOBUS MONKEYS

Genus *Colobus*

Stomach 3-chambered; larynx large; sac below hyoid bone; facial skin black.

Satanic black colobus Vu
Colobus satanas

Bioko, E and SW Cameroon, Equatorial Guinea, NW Gabon, probably W Congo. Forest; montane shrub; meadows.
HBL 58–72cm; TL 60–97cm; WT 6–11kg.
COAT: entirely glossy black; crown hairs semi-erect and forward-directed on forehead; newborn coat brown or pale gray.

White-epauletted black colobus
Colobus angolensis

White-epauletted or Angolan black or Black-and-white colobus

NE Angola, SW, C and NE Democratic Republic of Congo (DRC), SW Uganda, W Rwanda, W Burundi, W and E Tanzania, coastal S Kenya, possibly Malawi, vagrant in NW Zambia. Forest; woodland; maize cultivation.
HBL 47–66cm; TL 63–92cm; WT 5.9–11.3kg.
COAT: glossy black with white or whitish-gray cheeks, throat, longhaired shoulder epaulettes, and tip (or occasionally major part) of tail; pubic region, and occasionally narrow brow band and chest region, white; brow fringe, frontal whorl, or nape parting sometimes present. Newborn coat white.
CONSERVATION STATUS: Generally not threatened, though the DRC and Tanzanian populations are classed as Data Deficient and one subspecies – *C. a. ruwenzorii* – is listed as Vulnerable.

Gray-epauletted black colobus LR
Colobus polykomos

Gray-epauletted black colobus or King or Western black-and-white colobus

Guinea to W Côte d'Ivoire. Forest; scrub–woodland in Guinea savanna.
HBL 57–68cm; TL 94–110cm; WT 6.7–10.1kg.
COAT: glossy black; tail white; face border and throat sprinkled with white, extending to longhaired shoulders. Point of nose reaches or protrudes beyond mouth. Newborn coat white.
CONSERVATION STATUS: LR Near Threatened.

White-thighed black colobus † Vu
Colobus vellerosus

White-thighed black colobus or Ursine black-and-white colobus

W Côte d'Ivoire to W Nigeria, with hiatus at Dahomey Gap. Forest.
HBL 57–70cm; TL 76–100cm; WT 7–10kg.
COAT: glossy black; callosity border, face border, throat, and tail white; white either sparsely sprinkled on shoulders or on outside of thighs, occasionally both. Point of nose reaches or protrudes beyond mouth. Newborn coat white.

Guereza
Colobus guereza

Guereza, or White-mantled or Magistrate black colobus, or Eastern black-and-white colobus

N Congo, E Gabon, Cameroon, E Nigeria, Central African Republic, NE DRC, NW Rwanda, Uganda, S Sudan, Ethiopia, W Kenya and adjacent Tanzania. Forest; woodland; wooded grassland.
HBL 45–70cm; TL 52–90cm; WT 5.4–14.5kg.
COAT: glossy black; face and callosities surrounded by white; U-shaped white mantle of varying length on sides and rear of back; outside of thighs variably whitish; tail variably bushy and whitish or yellowish from tip towards base. Albinism common on Mt. Kenya. Point of nose nearly touches mouth. Newborn coat white.

RED COLOBUS MONKEYS

Genus *Procolobus*

Equatorial Africa. Limb proportions similar to *Pygathrix*; stomach 4-chambered; larynx small, no sac below hyoid. Sexual swelling in female and sometimes in immature male. Male skull usually with sagittal crest. Most, or all, species included by some authorities in genus *Colobus*.

Western red colobus En
Procolobus badius

Western red or Bay colobus

Senegal, Gambia, to SW Ghana. Forests, savanna woodland, savanna.
HBL 47–63cm; TL 52–75cm; WT 5.5–10kg.
COAT: crown, (sometimes) forehead, back, outside of upper arm, outside of thighs, and tail gray or blackish; pubic area white; rest of body whitish-orange to orange-red.
CONSERVATION STATUS: Several subspecies are Endangered, and one – *P. b. waldroni*, in Côte d'Ivoire and Ghana – is Critically Endangered.

Preuss' red colobus En
Procolobus preussi

W Cameroon. Lowland rain forest.
HBL 56–64cm; TL 75–76cm.
COAT: crown and back pale-stippled dark gray; cheek, flank, outside of leg reddish-orange; tail blackish-red; underparts whitish-orange.

Pennant's red colobus En
Procolobus pennantii

Bioko and Niger delta. Forest.
COAT: upperparts reddish orange, tinged with blackish brown on paws, crown, shoulders, and sometimes entire back; shank, upper arm, and tail tip brownish orange; cheek and underside grayish to yellowish white, sometimes extending onto front outside of forearms.

Red-crowned red colobus †
Procolobus tholloni

Red-crowned or Thollon's or Tshuapa red colobus

DRC, S of R Zaïre to Lake Tanganyika and Ituri Forest. Forest.
HBL 50–64cm; TL 58–75cm.
COAT: upperparts dark orange, sometimes tinged brown or gray on rump, flank, and extremities; dark red on crown, shoulder, tail tip, and (usually) hands; underside pale yellow, often tinged with gray, brown, or orange.

Dark-handed red colobus En
Procolobus kirkii

Dark-handed or Kirk's red colobus

E Congo, DRC N of R Zaïre, SW Uganda, Rwanda, Burundi, Tanzania, Zanzibar, and lower Tana river, Kenya. Forest. Subspecies include *P. k. rufomitratus.*
HBL 45–65cm; TL 58–77cm; WT 5.2–11.3kg.
COAT: paws blackish brown to black, occasionally pale gray; shoulder, flank, limbs, and tail tip often similar, sometimes with red on shoulder and white on limbs; brow orange or white, variously tinged with black, red, gray, or brown; cheek and underside white or yellowish-white, occasionally tinged with gray or orange; rest of body orange or pale gray, variously tinged with black, red, gray, brown, yellow, white, or black-flecked.

Olive colobus LR
Procolobus verus

Olive or Van Beneden's colobus

Sierra Leone to SW Togo; C Nigeria, S of Benue River. Forest; abandoned cultivation.
HBL 43–49cm; TL 57–64cm; WT 2.9–5.7kg.
COAT: upperparts black-stippled grayish-orange, grayer toward limb extremities; pale gray below; occasionally diluted gray leaving more orange above and more white below. Shorthaired sagittal coronal crest flanked by whorl on either side.
CONSERVATION STATUS: LR Near Threatened.

WHY PRIMATES KILL THEIR YOUNG

Incidences of infanticide in monkey and ape species

ONE OF THE MOST ARRESTING EXAMPLES OF aggression in the animal kingdom is infanticide, the killing of dependent infants by conspecifics. Infanticide is not something discussed over polite dinner tables, but it has been widespread even among humans, practiced at some level in the majority of cultures, whether hunter–gatherer, horticulturalist, or agrarian, though apparently less so today than in the past. Indeed, it may have evolutionary implications that extend as far as those of the institution of marriage itself.

Descriptions of infanticide in animals date back to the time of the ancient Greeks, but it was only in the 1960s that the practice was documented among non-human primates, in the Hanuman langurs of south Asia. Hanumans often live in groups comprising one breeding male, several adult females, and their offspring of various ages, which include nursing infants. The resident male periodically dies or is actively ousted from the group by a band of bachelor males – when this occurs the new male that replaces him may then attempt to kill some of the infants in the group.

Infanticide has since been documented in many primates, among them several lemur species, howler monkeys, leaf monkeys, the guenon monkey group, and Savanna baboons, as well as among Mountain gorillas and common chimpanzees. Infanticide is usually, though not always, carried out by males, and happens most often when new males move into a group. On a day-to-day basis, infanticide is rare and not easily witnessed – only about 60 episodes have been well described to date – but its apparent rarity should not obscure its significance. In primates such as Red howler monkeys, Mountain gorillas, and Chacma baboons, infanticide is a major source of infant mortality, accounting for 25–38 percent of infant deaths. In other words, about 13 percent of the infants born in Mountain gorilla and Red howler monkey groups are killed by males.

Infanticide and why it happens have always been highly controversial topics, and they remain hotly debated today. One idea is that infanticide is an aberrant behavior resulting from overcrowding or other unusual circumstances – the social pathology hypothesis. And yet infanticide has been seen to follow male replacement even in low-density populations of Hanuman langurs.

An alternative explanation recognizes a critical aspect of female primate biology. Lactation and the nursing of infants suppress ovulatory cycles in females for long periods. Pregnant mothers, or those with young infants, cannot be impregnated by an incoming male. Males usually have a limited window of opportunity to breed, because other males are always eager to usurp their dominance. Killing infants works as a strategy to return mothers to the regular cycle of fertility much sooner than would normally be the case if their infants survived to weaning. For example, in Chacma baboons there is usually a gap of about 18 months separating the birth of one (surviving) infant and the time when the mother conceives her next infant. If the infant dies, however, the mother typically resumes cycling immediately and becomes pregnant again within 5 months. Thus, infanticide can potentially speed up fertilization of females. Males may benefit in other ways by removing offspring not their own, such as by reducing competition for food, but this does not appear to be the main motivation, as males rarely attack either recently-weaned young or young juveniles fathered by the previous resident that are independent of their mothers.

Another idea claims that infanticide is an accidental side-effect of aggression surrounding male replacement. This argument posits that infants are simply more likely than other individuals to be fatally "caught in the middle" of the attacks a new male launches to establish his residency or dominance in the group. However, DNA extracted from the feces of Hanuman langurs in Ramnagar, Nepal, confirm that infanticidal males do not kill erratically but specifically target infants sired by other males. Further genetic work is needed to examine whether infanticidal primate males sire the females' subsequent infants.

Primate infanticide has been documented mainly in single-male groups, but recent studies have

Above *Before reaching maturity Hanuman males* *ave or are driven out of the troop and join other* *attached males: these bands watch breeding* *oops with an eye to usurping the alpha male.*

Left *Among Chacma baboons infanticide is a fair-* *frequent occurrence, since male leadership of a* *oup changes quite rapidly. Here – in the Chobe* *ational Park, Botswana – an adult carries away a* *ad infant.*

so revealed that infanticide takes place in multi-ale settings. For example, among the Chacma aboons of Botswana, which live in societies of -10 adult males, a couple of dozen adult males, and many youngsters, a new male immi-rant into the group who attains the alpha posi-on may attempt to kill infants sired before his rival, and will then mate with the mothers when ney resume cycling. About one-third to one-half f new alpha males commit infanticide in this way. nce alpha males have greater sexual access to females, they are more likely to benefit from the removal of infants.

This evidence would suggest that male infanticide is an adaptive reproductive strategy for the baboons. However, it is much less common among the closely-related Olive baboons of East Africa. The reasons for the difference are unclear, but one factor seems paramount: alpha male Chacma baboons have a very brief tenure of only about 7 months, whereas Olive baboons may retain their status for 1–4 years. The latter thus have longer to get a female pregnant, while the former must sire their offspring quickly. All this points to infanticide being a chance for males to optimize their reproductive options.

However, infanticide remains a perplexing enigma in another primate with multimale societies – the chimpanzee. No single hypothesis unambiguously explains the patterning of infanticide in chimps. All three of the hypotheses mentioned above may apply, plus the additional possibility that infants are exploited as a source of food. In some observed cases, for example, males killed infants in neighboring communities, gaining no obvious reproductive advantage as the mothers failed to transfer to the killers' group. In reported cases among humans, too, infant killing seems to have little to do with male competition for mating opportunities. The decision to commit infanticide is usually made by the infant's mother or father, which implicates parental control of reproduction or fertility as an underlying cause. Even when infanticidal maltreatment is perpetrated by a step-parent – as notoriously portrayed in literature and folklore – it may be more likely motivated by decreased willingness to invest resources in others' offspring than in any direct mating advantage gained by removing children.

Infanticide not only has fascinating origins; its repercussions are also far-reaching. One compelling possibility is that the cohesive social bonds that exist between males and females evolved as a deterrent to infanticide. Around the time a female Chacma baboon gives birth, she typically seeks out a particular adult male within the group and establishes a "friendship" with him. She sticks closely by the chosen male, following him around incessantly, grooming him much more than he grooms her, and allowing him alone to touch and handle her baby. Why do nursing mothers associate with a male in this way? The answer, in Chacma baboons at least, may be to gain protection from infanticide, for a female's male friend is more likely to actively defend her offspring than other group males. Moreover, infanticidal attacks made by a new alpha male seem more likely to fail when male friends step in, and conversely are more likely to succeed when these friends are absent. In a recent study, for example, direct intervention by male friends was observed in all infanticidal attacks by new alpha males in which the infants escaped injury; however, male friends were absent in two-thirds of the attacks in which infants were severely or fatally wounded.

Whether these males are the fathers of the infants they protect is currently unknown, but will no doubt become clear in time through genetic data. If they are not, their "friendly" behavior may increase the chances that female friends will prefer them as sires of their future offspring.

Although a seemingly negative and "antisocial" behavior, infanticide could ultimately account for the evolution of the ostensibly positive social attachments and "pair bonds" shared by males and females in some species. The possibility even exists that infanticide explains female–male bonds in humans and their ancestors. RP

FRUITFUL COOPERATION

Interspecies associations in an African forest

IT MAY NOT COME AS A SURPRISE WHEN individuals of the same species help each other out, but when different species come together to cooperate it becomes more worthy of attention. In the Taï National Park, in southwest Cote d'Ivoire, all possible combinations of seven different monkey species – three species of colobus and three guenons, plus Sooty mangabeys – can be seen keeping company. One can even encounter all seven together in a single aggregation of more than 250 individuals. The key species in this association system is the Diana monkey (*Cercopithecus diana*), highly appreciated by human observers for its splendid looks and by other monkeys for its alertness. Olive colobus and Diana monkeys form permanent attachments, while those between Red colobus and Diana monkeys are more transient. Observations of these mixed species groups have yielded some insights into how and why they might come together. In both cases stable combinations of partner groups share a common range.

Western red colobus monkeys (*Procolobus badius*) are fairly large, with bright red bellies and slate-colored backs. They live in large groups of around 75 members, and feed high up in the canopy on young leaves, blossoms, buds, and unripe fruit. Olive colobus (*Procolobus verus*) are half the size of their red cousins and live in groups often of less than 10 individuals. They have a diet similar to the Red colobus, but avoid the niche of the larger species by feeding on small trees, low in the forest. The Diana monkeys have a group composition typical for guenons: a single male with 5 to 10 females and some immatures. They search for fruit and insects in all layers of the canopy.

The primary function of both associations is to improve defenses against the four major monkey predators found here: Crowned hawk eagles (*Stephanoaetus coronatus*), leopards (*Panthera pardus*), chimpanzees (*Pan troglodytes verus*), and humans (*Homo sapiens*). Group living is the primary defense strategy of most diurnal primates against predators – the problem is that every new animal in a group potentially means less food per individual, which may outweigh the benefits. The solution is to associate with a species that has a different diet – and colobus monkeys fit well with Diana monkeys because their diets hardly overlap.

Two elements are important in a defense system based on safety in numbers. The first is early warning: when alerted in time, each individual can follow the best strategy to lower its risk, depending on its location and on the predator approaching. (Information on the type of predator is contained in the alarm calls of the different species; the monkeys can apparently extract the relevant data from the alarm calls of other species too.) Then there is dilution: the more potential prey, the lower the chance of any single individual being attacked.

The value of Diana monkeys as sentinels became apparent when a scientist wrapped in a leopard-print cloth approached mixed groups. It was almost always a Diana which warned first – even when the Dianas were outnumbered by members of other species, or were further away from the danger. Dianas' high mobility and preference for foraging at the outside of the tree crown contributed to their success as lookouts.

There are, however, costs to congregating with other species, particularly when each prefer quite different types of food. As a consequence, Red colobus only choose to associate at times of the year when they are at high risk from predation, or when food is favorably distributed. Their particular bane are cooperatively-hunting chimpanzees, which kill many more Red colobus than Diana monkeys; so the Red colobus most often take the initiative in forming associations with Dianas during the chimpanzees' hunting season, and much less from September through November when the chimpanzees do not hunt.

Above *Cooperation is essential during certain tasks, especially such activities as drinking at a water hole, which is a dangerous time because it requires the animals to leave the cover of the forest canopy and come out into more open ground. These Red colobus and vervets – one of the many permutation of interspecies association to be found among African monkeys – take turns at drinking and keeping watch for predators.*

Left *The Diana monkey's watchful eye and general alertness are valuable contributions to the formation of interspecies associations: the occurrence of such mixed species groups may depend on many variables, such as food abundance or prevalence of predators.*

Playing the sounds of chimpanzees through a peaker induced the Red colobus immediately to pproach their Diana partner group. In addition, roups stayed together longer when they heard he screams of chimpanzees in the early morning han when they heard other (or no) sounds.

Diana monkeys profit from the presence of the ed colobus through the effects of dilution. Their roup size quadruples when a Red colobus group oins, so the chances of any individual being aught are reduced. The colobus, which generally orage higher, also improve the warning system for rowned eagles approaching over the canopy.

The Olive colobus, in contrast, are less useful, ontributing almost nothing to the safety of the iana monkeys. These uniformly brown monkeys re extremely good at hiding at the first alarm, vhich they rarely give themselves. They sit frozen a thicket while the other monkeys scurry around. Their small size and small groups may be adaptations to this cryptic lifestyle, though perhaps their style of living does not allow a higher total biomass per group. Unlike Red colobus, which come and go, Olives never leave their Diana sentinels, and must find their food within the limited area where the Diana monkeys search.

On the island of Tiwai in Sierra Leone, Red and Olive colobus and Diana monkeys all live relatively free from predators. In spite of this, the Olive colobus still follow the Diana monkeys around, but the Red colobus mix with them no more than would be expected on the basis of chance encounters. It would seem that, in the course of evolution, the tendency to associate with Diana monkeys developed into an unconditional, "hard-wired" strategy in Olive colobus, while Red colobus only seek out other monkeys when the safety provided by their own ranks does not suffice. RN/RB

Above *The benefits of the association between the Olive colobus and Diana monkeys appears to favor the Olive colobus; however they have little dietary overlap, so the cost is not especially high.*

DISTRIBUTION E India to S China and south through SE Asia to Malay Peninsula, Sumatra, Java, Borneo, W and C Africa.

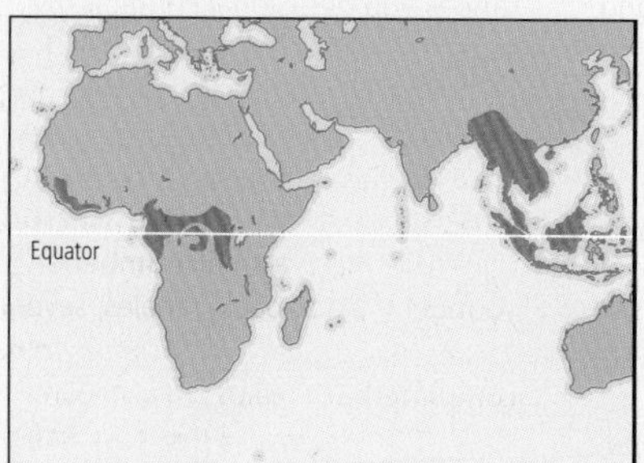

Families: Hylobatidae, Hominidae
18 species in 5 genera

GIBBONS OR LESSER APES
Family Hylobatidae p398

11 species in the genus *Hylobates*

GREAT APES Family Hominidae

7 (5) species in 4 genera

Common chimpanzee *Pan troglodytes*	p406
Bonobo *P. paniscus*	p406
Western gorilla *Gorilla gorilla*	p414
Eastern gorilla *G. beringei*	p414
Bornean orangutan *Pongo pygmaeus*	p420
Sumatran orangutan *P. abelii*	p420
Human *Homo sapiens*	

Note The 1993 established list of mammalian species recognized 16 species in 5 genera (in parentheses). Following a meeting of leading primatologists in Orlando, USA, in 2000, there has been a move to elevate former great ape subspecies to new species.

Apes

THE APES ARE OUR CLOSEST RELATIVES, AND together with *Homo sapiens* comprise the superfamily Hominoidea, the largest living primates. The great apes in particular are remarkably similar to human beings in their reproductive biology, and in all apes maternal care of infants is prolonged, ranging from some 18 months in gibbons to almost three years in great apes. There is a fairly sharp distinction between the medium-sized lesser apes (gibbons and siamangs; family Hylobatidae) and the much larger great apes (gorillas, chimps, and orangs; family Hominidae). Fossil relatives of apes are known from the early Miocene, some 20 million years ago. In fact, the lesser apes may have become distinct as long ago as the early Oligocene (about 34 million years ago), but as yet no convincing direct fossil relatives have been found.

The apes have no tail and their fore limbs, prominent in locomotion, are longer than the hind limbs. The chest is barrel-shaped (rather than flattened from side to side as in monkeys), and a modified wrist structure with interposition of a cartilage meniscus between the end of the ulna and the wrist bones permits greater mobility. Gibbons display the most spectacular pattern of movement, known as "true brachiation," in which the body is swung along beneath the branches with the arms taking hold alternately. The great apes are far less athletic in their movement through the trees. The orangutan, the largest living arboreal mammal, moves slowly and deliberately, suspending the great weight of its body from all four limbs in a quadrumanous (four-handed) progression. In contrast to their Asiatic relatives, the great apes of Africa usually travel along the ground. Both chimpanzees and gorillas "knuckle-walk," with the knuckles of the hands providing the points of contact with the ground (orangutans, especially old, heavy males, may descend to the ground and show "adducted fist-walking" involving the outer margins of the hands – a more rudimentary pattern of forelimb support). Chimps spend between one-quarter and one-third of their time in the trees, but gorillas only about 10 percent.

The lesser apes, like most monogamous mammals, exhibit very little difference in body size between the sexes, although in some species there are differences in male and female coat coloration. In chimpanzees there is relatively little size difference, but in both orangutans and gorillas there is extreme sexual dimorphism in body size. RD

SKULLS AND DENTITION

Apes have relatively well-developed jaws, a flattened face with forward-pointing eyes, and a globular braincase, reflecting the fairly large brain in relation to body size. As in monkeys, each eye is enclosed within a complete bony socket, formed by development of a plate of bone that separates the orbit from the jaw musculature behind it. The dental formula is identical to that of Old World monkeys and humans – I2/2, C1/1, P2/2, M3/3 = 32 – with spatulate (shovel-shaped) incisors and squared-off cheek teeth bearing relatively low cusps that reflect the predominance of plant food in the diet. The canine teeth are prominent, and the rear edge of each upper canine hones against the front edge of the anterior lower ("sectorial") premolar. The lower jaw is fairly deep, and the two sides of the jaw are fused at the front as in monkeys and humans. Similarly, in common with monkeys and humans, the frontal bones of the skull are fused at the midline.

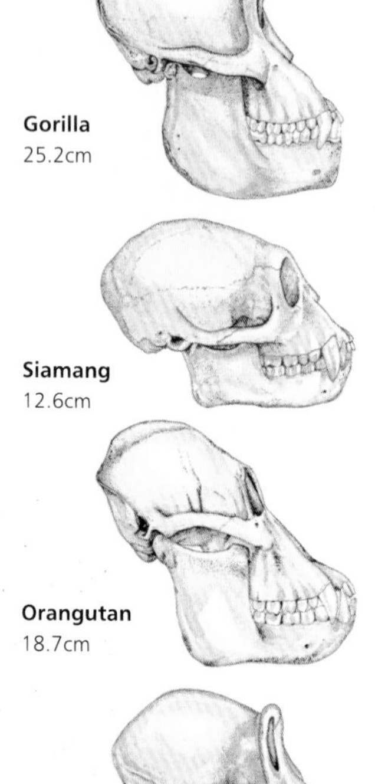

In association with their much smaller body size, the skulls of lesser apes are relatively lightly built, and there is relatively little difference between males and females (sexual dimorphism) in body size. In the much heavier great apes, the skull is robustly built and sexual dimorphism in body size is reflected in the heavier build of male skulls. Indeed, the particularly powerfully developed jaw and neck muscles of male gorillas and orangutans have required the development of substantial midline (sagittal) and neck (nuchal) crests on the skull to provide additional surfaces for the attachment of muscles. The facial area of the orangutan's skull slopes back far more than in other great apes.

Below **1** *Female Sumatran* [o]ngutan (Pongo abelii) *grasping* [b]ranch *with its hooklike hands;* [2 M]oloch *gibbons* (Hylobates mol[och]) *giving alarm calls;* **3** *Common* [chi]mpanzee (Pan troglodytes) *with* [a y]oung *duiker it has killed;* [4 ...]male *Western gorilla* (Gorilla [go]rilla) *with its harem.*

Ape Social Systems

Patterns of social organization among the apes cover almost the entire range known for primates. The lesser apes are all essentially monogamous, and a mated pair may remain together for life. The interval between births in gibbons is typically around 3 years, sexual maturity is reached at about 6 years of age in both sexes, and sexually mature adults emigrate from the parents' home range. As a result, the maximum size of gibbon groups is about five individuals. Orangutans feed in ones or twos, adult males typically avoid one another, and the only common social unit is that of the mother and her offspring, although adult females and/or immature orangutans may form small temporary groups. Yet the orangutan does have some kind of social system, broadly comparable to the "dispersed societies" of many nocturnal lemur and loris species, based on overlapping ranges and occasional contacts. By contrast, the great apes of Africa have quite well-defined social groups, gorillas are essentially harem-living, and groups average about 12 members, typically consisting of a single mature silverback male, a small number of younger ("black-back") males, several adult females, and immature animals. Lone silverback males are also common. In chimpanzees there are at least two different levels of grouping. The fundamental social unit is a territorial community of 40–80 individuals, including numerous adults of both sexes, but it is rare to find the whole community together in one place. Clear-cut territorial behavior seems to be restricted to the lesser apes and the chimpanzees. Gibbons are well-known for their loud territorial calls. In chimps, territorial demarcation is more discreet, although encounters between members of neighboring communities can lead to death.

Food and Feeding

The apes are all predominantly vegetarian, although chimpanzees and orangutans eat some animal food as well. Orangs have access to a wide selection of fruit (ABOVE) in times of abundance. Only gorillas are predominantly leaf eaters rather than fruit eaters. In the remaining apes, particularly in the gibbons and siamang, their "suspensory" movements while feeding are linked to adaptations for feeding in the terminal branches. Although animal food represents a small part of its diet, the chimpanzee's feeding activity has attracted much interest because of its possible relevance to the emergence of hunting behavior in early humans. In both East and West Africa, chimpanzees prey on fellow primates (e.g., Red colobus monkeys and baboons) and other medium-sized mammals (e.g., bushpigs); individuals (especially males) show a limited tendency to hunt cooperatively and they share food. Chimpanzees (primarily females) also feed on termites, using a "tool" – a carefully selected twig.

Gibbons

Left A siamang vocalizing; the calls, or songs, of this species are made louder as a result of the sound waves resonating across the vastly inflated throat sac, in much the same way as the sound hole on an acoustic guitar amplifies the plucked strings.

CONTRARY TO POPULAR BELIEF, THE APES AND man do not share an ancestor that habitually swung through the trees by its arms. While all apes have long arms and mobile shoulders and stand erect, only the gibbons have developed powerful propulsive abilities in their upper limbs. Climbing and feeding in an upright posture (sitting or hanging) with strong arms and a long reach would be the ancestral positional behaviors common to all apes.

Gibbons are characterized by their spectacular arm-swinging form of locomotion (brachiation) and habitual erect posture, which are key adaptations for their unique suspensory behavior. They utter loud and complex calls of considerable purity in a stereotyped manner, which capture the spirit – both exuberant and melancholic – of the jungles of the Far East. These beautiful calls, mainly performed as duets, may serve to develop and maintain pair bonds and to exclude neighboring groups from the territory of the monogamous family group. Altogether, then, the gibbons' key attributes (monogamy, territoriality, a frugivorous diet, "suspensory" behavior, and elaborate songs) form a blend that is unique among primates.

Successful Apes

FORM AND FUNCTION

While the great apes have developed sexual dimorphism in body size, adult male and female gibbons are more or less the same size. They are relatively small, slender, and graceful apes with very long arms, longer legs than one might expect, and dense hair; they are more efficient at bipedal walking than the great apes, and do so on any firm support, such as branches too large to swing beneath, not just on the ground as is commonly supposed. Coat color and markings, especially on the face, clearly distinguish the species and, in some cases, age and sex. Some species have developed throat (laryngeal) sacs, which enhance the carrying capacity of calls. These calls, especially those of the adult female, provide one of the easiest ways of identifying species.

In terms of diversity and abundance, the gibbons are the most successful of the apes. From an adept climbing and fruiteating ancestor, they have diversified throughout the forests of Southeast Asia over the last million or so years, maintaining the same body form and size (with the chief exception of the siamang) for hanging to feed from the terminal branches and for brachiating through the forest canopy. It was the frequent periods of isolation in different parts of the Sunda Shelf during the Pliocene changes in sea level (5–1.8 million years ago) that led to the differentiation of gibbons into present species.

A Life in the Trees

DISTRIBUTION PATTERNS

The gibbons are distributed throughout the mainland and islands of Southeast Asia forming the Sunda Shelf and, being almost exclusively arboreal, depend on the evergreen tropical rain forests. It seems that about one million years ago the ancestral gibbon spread down into Southeast Asia to become isolated in the southwest, northeast, and east (the Asian mainland would have been uninhabitable during the early glaciations). These three lineages respectively gave rise to the siamang, the

FACTFIL

GIBBONS

Order: Primates

Family: Hylobatidae

11 species of the genus *Hylobates*

DISTRIBUTION Extreme E of India to far S of China, S through Bangladesh, Burma, and Indochina to Malay Peninsula, Sumatra, W Java, and Borneo.

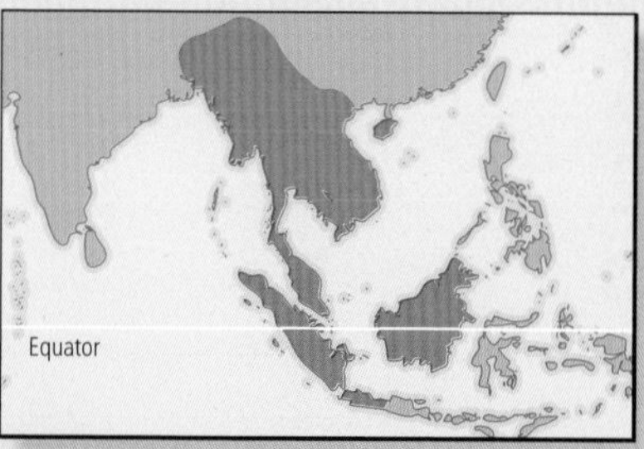

HABITAT Evergreen rain forests (SE Asia) and semi-deciduous monsoon forests (mainland Asia).

SIZE Head–body length 45–65cm (18–26in) in most species; weight 5.5–6.7kg (121–14.7lb). The siamang has a head–body length of 75–90cm (30–35in); weight 10.5kg (23lb). Sexes similar in size.

COAT Color distinguishes species (and sometimes sexes and age group within a species), as do calls.

DIET Ripe, pulpy fruit, leaves, some invertebrates.

BREEDING Gestation 7–8 months.

LONGEVITY In wild, siamang to 25–30 years, Lar gibbon to 25 years or more. In captivity, up to 40 years.

See species table ▷

Left *A female Lar gibbon* (Hylobates lar) *with infant. This species has a long weaning period, which is not completed until the offspring is at least 18 months old. Lar gibbon numbers are in decline across much of their range as a result of deforestation in areas such as mainland Malaysia, where – in recent years – vast areas of forest have been burned to clear ground for farming.*

rested gibbons, and the rest. The greatest ıanges occurred subsequently in the eastern oup, which spread back toward the Asian mainnd during the interglacial periods, giving rise st to the Hoolock gibbon (and the Kloss gibbon the west), then to the Pileated, and finally, durg and since the last glaciation, to the Agile and ar, with Müller's and Moloch gibbons evolving ı Borneo and Java, respectively. Some authorities ɔw place the Hoolock gibbon in its own subnus, due to its large body size and fewer chroosomes. The range of gibbons has contracted ɔuthward over time – 1,000 years ago, according Chinese literature, it extended north to the Yelw River. Oddly, gibbons are sexually dichromatic ross the north of their range (males are mainly ack, females buff or gray), black in the southest, very variable in color in the center of their stribution, and tending to gray in the east.

Sexually dichromatic taxa occur in seasonal forests, which offer better visibility for visual signals.

The gibbon species are often separated from each other by seas and rivers, except for the much larger siamang, which is sympatric with the Lar gibbon in peninsular Malaysia and with the Agile gibbon in Sumatra. Although otherwise similar in size and shape, as a result of their common adaptation to a particular forest niche, they are readily identified by coat color and markings and by song structure and singing behavior. The siamang used to be placed in a separate genus, but the gibbons are best considered as monogeneric, with the siamang as one subgenus (*Symphalangus*; 50 chromosomes, diploid number), the crested gibbon of the northeast as the second (*Nomascus*), with 3–4 subspecies spread from north to south across the seas and rivers of Indochina (52 chromosomes), the Hoolock gibbon in the northwest (38 chromosomes) as the third (*Bunopithecus*), and the Lar gibbon group as the fourth (*Hylobates*) in the center and the east (44 chromosomes). The crested gibbons are as different from the Lar group as the siamang and the Hoolock gibbon. The Hoolock gibbon is sexually dichromatic, like the crested and pileated gibbons. The Kloss gibbon was once called the Dwarf siamang because it is also completely black. The "gray" gibbons are the Moloch and Müller's gibbons. The discovery of a large and ancient hybrid zone in the center of Borneo means that Müller's gibbon may have to be considered as a subspecies of the Agile gibbon. The most widely distributed and variable in color are the Agile and Lar gibbons, which can both be described as polychromatic, although the Lar gibbon in Thailand shows extreme dichromatism, which is apparently not related to sex, but rather a response to seasonal, semideciduous forest.

Year-round Fruit

DIET

Gibbons generally show a preference for small scattered sources of pulpy fruit, which brings them into competition more with birds and squirrels than with other primates. Unlike the monkeys, which feed in large groups and can more easily digest unripe fruit, gibbons eat mainly ripe fruit; they also eat significant quantities of young leaves and a small amount of invertebrates, an essential source of animal protein.

The structural complexity of the gibbons' habitat buffers the effects of any limited seasonality. Within as well as between plant species (climbers as well as freestanding trees) fruiting occurs at different times of year, ensuring year round availability of fruit. Since such plant species rely on animals for dispersal of seeds, this is an important example of co-evolution between plants and animals. Recent studies in Indonesian Borneo and Bangladesh have provided clear evidence of the efficienc of gibbons in seed dispersal, an activity that help promote the natural regeneration of forests. Seed that have passed through their guts are more like to germinate than those that have not, as digestion seems to weaken the seed coat.

About 35 percent of the daily active period of 9–10 hours is spent feeding (and about 24 percent in travel). Feeding on fruit occupies about 6 percent of feeding time, and on young leaves 30 percent, except for the siamang, which eats 44 percent fruit and 45 percent leaves, and the Kloss gibbon (72 percent fruit, 25 percent animal matter, and virtually no leaves). The larger the propor tion of leaves in a species' diet, the relatively large are the cheek teeth and their shearing blades; the voluminous cecum and colon indicate an ability to cope with (and even ferment) the large leaf component of the diet in these simple-stomache animals. Fruit, even small ones, are picked by a precision grip of thumb against index finger, which permits unripe fruit to be allowed to ripen

Left An alternative name for the Lar gibbon is the White-handed gibbon – a distinctive feature clearly displayed by this fruiteating individual. The Lar's coat is not uniform across the species, but males and females in any one population will have similar coloration.

Crested black gibbon En
Hylobates concolor
Crested, Concolor, or White-cheeked gibbon

Tonkin, W and C Yunnan, N Vietnam between Black and Red rivers and NW Laos. 4 subspecies.
HBL 45–64cm (18–25in); WT 5.7kg (12.6lb).
COAT: male black with more or less whitish (or yellowish or reddish) cheeks; female buff or golden sometimes with black patches; infant whitish.
CALLS: male grunts, squeals, whistles; female – rising notes and twitter, sequence of about 10 seconds.

Northern and Southern white-cheeked crested gibbons
Hylobates leucogenys

S Yunnan, N and S Laos, NW and C Vietnam. 2 subspecies.
HBL 45–64cm (18–25in); WT 5.7kg (12.6lb).

Yellow-cheeked crested gibbon
Hylobates gabriellae
Yellow-cheeked crested or Buff-cheeked gibbon

S Laos, S Vietnam, E Cambodia.
HBL As for *H. leucogenys*.

Siamang LR
Hylobates syndactylus

Malay Peninsula, Sumatra.
HBL 75–90cm (30–35in); WT 10.5kg (23lb).
COAT: male, female, and infants black; throat sac gray or pink.
CALLS: male screams; female – bark-series lasting about 18 seconds.
CONSERVATION STATUS: LR Near Threatened.

Hoolock gibbon
Hylobates hoolock
Hoolock or White-browed gibbon

Assam, Burma, Bangladesh.
HBL 45–64cm (18–25in); WT 5.5kg (female; 12.1lb), 5.6kg (male; 12.3lb).
COAT: male black, female golden with darker cheeks, both with white eyebrows; infant whitish.
CALLS: male – di-phasic, accelerating, variable: female – similar to but lower than, and alternating with, male's.

Kloss gibbon Vu
Hylobates klossii
Kloss gibbon, Mentawai gibbon, or beeloh (incorrectly: Dwarf gibbon, Dwarf siamang)

Mentawai islands, W Sumatra (Siberut, Sipora, N and S Pagai).
HBL 45–64cm (18–25in); WT 5.8kg (12.8lb).
COAT: overall glossy black, in male, female, and infant – the only gibbon with such coloration.
CALLS: male – quiver-hoot, moan; female – slow rise and fall, with intervening trill or "bubble" or not, sequence lasts 30–45 seconds.

Pileated gibbon Vu
Hylobates pileatus
Pileated or Capped gibbon

SE Thailand, Cambodia W of Mekong.
HBL 45–64cm (18–25in); WT 5.5kg (male; 12.1lb), 5.4kg (female;11.9lb).
COAT: male black with white hands, feet and head-ring: female silvery-gray with black chest, cheeks, and cap: infant gray.
CALLS: male – abrupt notes, di-phasic with trill after female's; female – short, rising notes, rich bubble: 18 seconds.

Müller's gibbon LR
Hylobates muelleri
Müller's or Gray gibbon

Borneo N of Kapuas, E of Barito rivers.
HBL 45–64cm (18–25in); WT 5.7kg (male; 12.6lb), 5.3kg (female;11.7lb).
COAT: mouse-gray to brown, cap and chest dark (more so in female), pale face-ring (often incomplete) in male.
CALLS: male – single hoots; female – as Pileated gibbon but notes shorter; sequence 10–15 seconds.
CONSERVATION STATUS: LR Near Threatened.

Moloch gibbon Cr
Hylobates moloch
Moloch or Silvery gibbon

W Java.
HBL 45–64cm (18–25in); WT 5.9kg (13lb).
COAT: silvery-gray in male and female, all ages: cap and chest darker.
CALLS: male – simple hoot; female – like Lar gibbon at first, ends with short bubble; 14 seconds.

Agile gibbon LR
Hylobates agilis

Malay Peninsula, Sumatra (most), SW Borneo.
HBL 45–64cm (18–25in); WT 5.9kg (13lb).
COAT: variable (but same in both sexes in one population), light buff with gold, red, or brown: or reds and browns: or brown or black; white eyebrows and cheeks in male, brows only in female.
CALLS: male – di-phasic hoots; female – shorter than Moloch , lighter-pitched, rising notes to stable climax; 15 seconds.
CONSERVATION STATUS: LR Near Threatened.

Lar gibbon LR
Hylobates lar
Lar or White-handed or Common gibbon

Thailand, Malay Peninsula, N Sumatra.
HBL 45–64cm (18–25in); WT 5.3kg (female; 11.7lb), 5.7kg (male; 12.6lb).
COAT: variable (but same in both sexes in one population); Thailand: black or light buff; white face-ring, hands, and feet. Malaysia: dark brown to buff. Sumatra: brown to red, or buff.
CALLS: male – simple and quiver hoots; female – longer notes than Moloch gibbon, climax fluctuates, duration (Thailand) 18, (Malay Peninsula) 21, (Sumatra) 14–17 seconds.
CONSERVATION STATUS: LR Near Threatened.

Abbreviations HBL = head–body length WT = weight   Cr Critically Endangered   En Endangered   Vu Vulnerable   LR Lower Risk

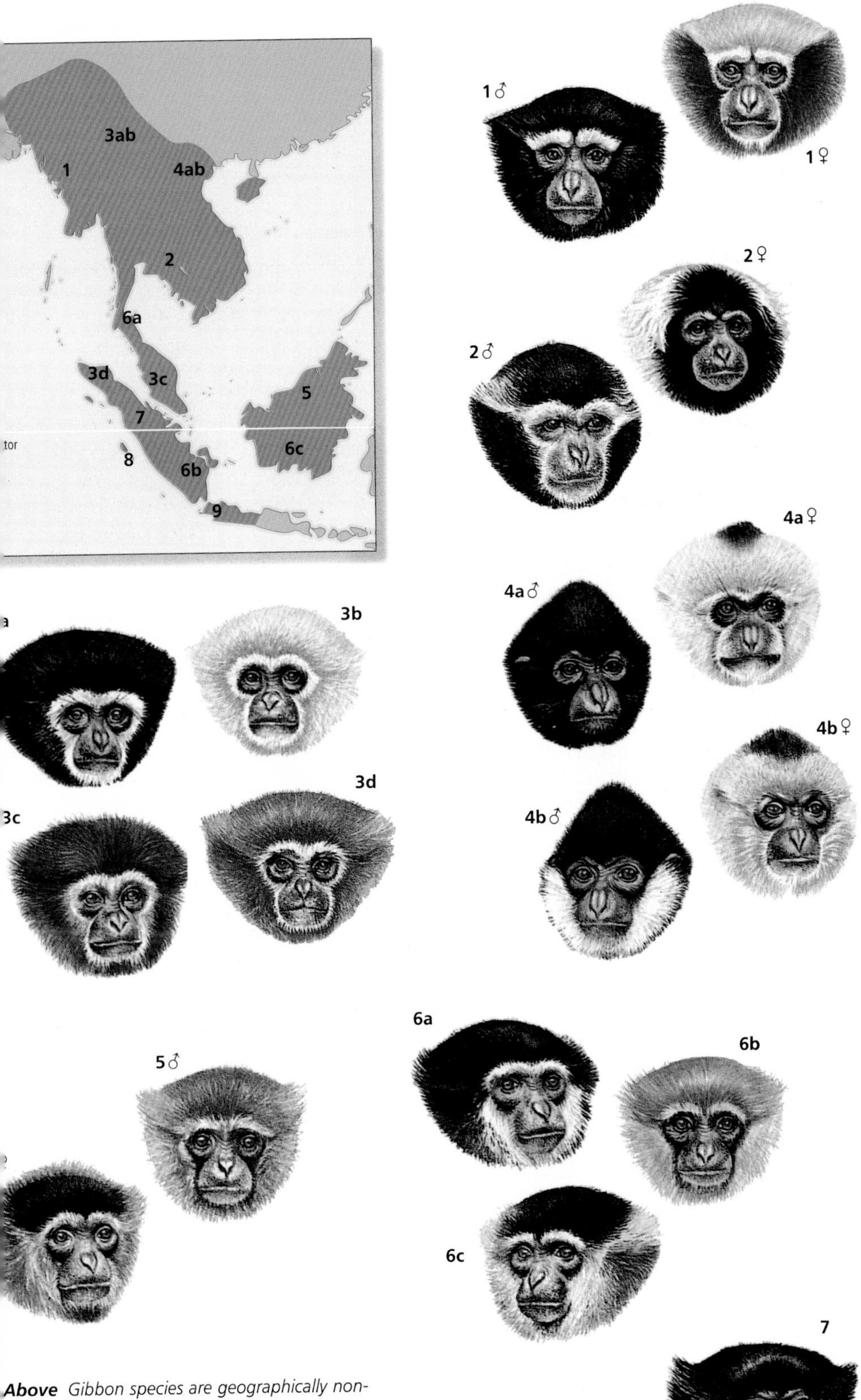

Above *Gibbon species are geographically non-*
rlapping, with the exception of the siamang, which erlaps both Lar and Agile gibbons. Within most ecies (not in the siamang, or Kloss or Moloch gib-ns), coat color can vary according to either or both nder or geographical population. ***1*** *Hoolock gib-n;* ***2*** *Pileated gibbon;* ***3*** *Lar gibbon (sexes similar in ne population):* ***a*** *Thailand, dark phase,* ***b*** *Thailand, ht phase,* ***c*** *south of Malay peninsula,* ***d*** *northern matra;* ***4*** *Crested black gibbon:* ***a*** *black-cheeked ase,* ***b*** *white-cheeked phase;* ***5*** *Müller's gibbon; Agile gibbon: forms in* ***a*** *Malay peninsula,* ***b*** *south- Sumatra,* ***c*** *southwest Borneo;* ***7*** *siamang;* ***8*** *Kloss bon;* ***9*** *Moloch or Silvery gibbon.*

Accomplished Singers

SOCIAL BEHAVIOR

The adult pair of a gibbon family group tend to produce a single offspring every 2–3 years, so that there are usually 2 immature animals in the group, but sometimes as many as 4. Thus, copulation is not seen very often; it is usually dorso-ventral, with the female crouching on a branch and the male suspended behind, but rarely the animals will copulate facing each other. The gestation period lasts 7–8 months and the infant is weaned early in its second year. The siamang is unusual in the high level of paternal care of the infant; the adult male takes over daily care of the infant at about one year of age, and it is from him that it gains independence of movement (by three years of age). Juveniles of either sex are relatively little involved in group social interactions. By about six years the immature animal appears fully grown, and, as a subadult, tends to interact with siblings in a friendly manner, with the adult male in both friendly and aggressive ways, and avoids interaction with the adult female. Conflict with the adult male helps to ease the now socially mature animal out of the group by about eight years of age.

Subadult males often sing alone, apparently to attract a female, but they may also wander in search of one. Thus, either sons or daughters may end up near their parents, although sons perhaps more often. It is clear, however, that the first animal that comes along is not necessarily a suitable mate for life; it may take more than one attempt to find a compatible partner.

The siamang is unusual among gibbons in the high cohesion of the family group throughout daily activities – group members are 10m (33ft) apart on average, and rarely is one animal separated from another by a distance of more than 30m (100ft). In other gibbons the family feeds together only in the larger food sources; for the rest of the day foraging is undertaken individually across a broad front of about 50m (165ft); they come together occasionally to rest and groom and, in some cases, to sleep at night.

Social interactions are infrequent; there are few visual or vocal signals, even in siamang, despite their "expressive" faces and complex vocal repertoire. Grooming is the most important social behavior, both between adults and subadults, and between adults and young; play, centered on the infant, is the next most common.

The most dramatic and energetically costly social behavior is singing, which mostly involves the adult pair, but extends to a chorus of the whole group as the young learn their roles. While it is most commonly explained as a means of communicating between family groups such matters as territorial advertisement and defense, some also believe that singing is crucial not only in forming a pair bond, but also in maintaining and developing it. The elaborate duet that has evolved in most gibbon species is given for 15 minutes a day on

average and ranges in frequency from twice a day to once every five days, according to the species and to factors relating to fruiting, breeding, and social change. The Kloss and possibly the Moloch gibbon do not have duets; however, female Kloss gibbons have an astonishing "great call" and male solos are given at or before dawn in Kloss, Lar, Agile, Müller's, and – perhaps – Moloch gibbons.

This near-daily advertisement of the presence of a group and its determination to defend the area in which it resides is augmented by confrontations at the territorial boundary about once every five days, for an average 35 minutes. Altogether there are perhaps five levels of territorial defense: calls from the center, calls from the boundary, confrontation across the boundary, chases across the boundary by males, and, very rarely, physical contact between males.

The duet song bouts of most gibbon species conform to the same basic pattern: an introductory sequence while the male and female (and young) "warm up," followed by alternating sequences of organizing (behaviorally and vocally) between male and female, and of "great calls" by the female, usually with some vocal contribution from the male, at least at the end as a coda. Only in the Kloss gibbon are the songs of male and female completely separated into solos. In Lar, Agile, Müller's, and Crested gibbons, male and female contributions are integrated sequentially into the duet, whereas in Hoolock and Pileated gibbons and in the siamang, the male and female call together at the same time, even during the female's great call.

There are usually 2–4 gibbon family groups, each of four individuals, to each square kilometer (0.45sq mi) of forest with a total body weight of 45–100kg per sq km (40–90lb/sq mi); however the number of groups in such an area may range from one to six. These groups travel about 1.5km a day (siamang, Pileated, and Müller's gibbons have mean annual day ranges of about 0.8km) around a home range usually of 30–40 hectares (74–99 acres), but about 15 hectares (37 acres) for Lar in Thailand and for Moloch gibbons, and about 60 hectares (148 acres) for Lar gibbons in Malaysia sympatric with siamang. Most gibbon species defend about three-quarters of this home range (25 hectares) as the group's territory. (About 90 percent of the home range is defended by Moloch and Müller's gibbons, and only about 60 percent by siamang and Kloss gibbons.) It is difficult to define territorial boundaries in siamang, however, since disputes are rare; it seems that they use their much louder calls to create a "buffer zone" between territories. Though twice the size of other gibbons, siamang live in smalle home ranges, moving about less and eating larg amounts of more common foods such as leaves.

Above *A pair of Crested black gibbons* (Hyloba concolor)*, showing the sexual dichromatism of this species. The male has a dark coat, sometimes (as h embellished with white cheeks, hence the alternativ name of White-cheeked gibbon. In striking contrast the female is golden-colored, with dark patches of*

Revered Forest Spirits

CONSERVATION AND ENVIRONMENT

The rapid clearance of tropical evergreen rain forests in Southeast Asia places the gibbons' future in great peril. In 1975 there were estimat to be about 4 million, but fears were already bei voiced about the ability of some species to surv in viable numbers. In the intervening years, logging has displaced thousands of gibbons annua (millions in all) and resulted in a high rate of att tion. Nowadays, there is a reluctance to produc mates of population size, for fear of misinter- ation. It is clear, however, that the Kloss, loch, and some of the Crested gibbons are e to extinction. Thus, over 25 years, the loch gibbon has declined from about 20,000 ess than 1,000. Pileated and Hoolock gibbons also near-extinct over large parts of their range, their actual status is impossible to define.

The highest priority must be given to protect- adequate areas of suitable habitat, ideally 200–)sq km (77–193sq mi) in lowland forest and ersheds. In the long term it would be wise, for nomic reasons alone, to maintain rain forests ier than to clear them. As a further immediate), displaced animals should be rounded up and irned to unpopulated forests, or used to estab- breeding centers, preferably in the countries origin, for conservation research (focusing, say, nutrition and breeding), and for education.

As forest dwellers, gibbons pose no threat to nans as pests or as effective carriers of disease. eed, because of their resemblance to man (lack ail, upright posture, intelligent expression), y are venerated as benign spirits by the indige- is forest peoples of Indonesia and the Malay insula. Such societies tend not to hunt them; ier, it is newcomers to the forest who are ponsible for the indiscriminate slaughter of cies, including gibbons (though, fortunately, y remain elusive to most hunters). DJC

Right and below *Müller's gibbons: right, nging through the forests of Kalimantan, Borneo; ow, bedraggled after the rain. In common with all other gibbons, this species faces an uncertain ure, as urban expansion and the need for more icultural land prompt further encroachments into raditional habitat.*

SPECIAL FEATURE

DEFENSE BY SINGING

Great calls and song bouts of the gibbons

Below *Launched into mid-air, a female Kloss gibbon brings her great call to a climax, and serves notice on others to steer clear of her territory. She may also run upright along branches tearing off leaves and be joined by other family group members shaking branches to enhance the display and giving calls of their own, like the clinging infant.*

THE SONG BOUTS OF GIBBONS ARE AMONG the most spectacular sounds of the tropical forests of Asia. Both males and females perform complex song bouts from the forest canopy that can be heard up to several kilometers away. The songs and song bouts are highly sexually dimorphic, males and females singing very different songs, and they appear to serve the two sexes differently.

The solo performances of male gibbons commence before sunrise and reach their conclusion at dawn. What begins as a series of soft, simple warblings gradually develops over the next 20–40 minutes into a bout of increasingly loud and elaborate songs. In the Agile gibbon, the final songs of the bout are twice as long as the earlier songs and contain almost double the number of notes (3.5 seconds and 7.6 notes, in contrast to 1.8 seconds and 4.2 notes); in the Kloss gibbon, these complex songs are called "trills."

Females usually wait to sing until around mid-morning, when they launch into a shorter and much less variable singing session about 10–20 minutes in length. Females simply repeat the same song over and over again, but despite the repetition the performance is deeply impressive: known as the "great call," it comprises some 6–80 notes performed over a period of 7–30 seconds. Perhaps the most outstanding of all great calls are those of the Kloss gibbon, which have been described as "probably the finest music made by any wild land mammal."

Males sing in bed. Their predawn chorus begins before they emerge from their tree platform 30m (100ft) above the ground; and although their repertoire is very diverse, their delivery is relatively low-key. Females, on the other hand, are "drama queens." They perform by daylight, selecting tall trees from which to enact a vigorous physical display involving a great deal of swinging between branches. Rotten branches crash to the forest floor in a display that reaches its climax at the crescendo of the great call. It is not only possible to tell the sexes and the species of gibbons from their songs but also to distinguish individual singers. Females typically differ in the components and duration of their great call, and in the range and number of notes produced.

So why do gibbons sing? Primarily, it seems, to alert other members of the species to their presence and location. More specifically, the performance of male song bouts appears to be related to the density of the population, and therefore to the number of young males who are maturing and seeking mates. Although it was once thought that males sang to defend their mate's feeding territory, it now seems that they are actually singing to defend her from the attentions of unmated males. Mated males sing more frequently – about once every 2 to 4 days – when there are many other gibbons about to threaten their relationships, whereas in areas where the sparsity of other males means competition is low they do not sing at all.

'he male's song is not without its costs. Males
less in cold weather, or when their available
t is lacking in energy-rich sugars. In addition,
nated males looking to find a mate may be able
valuate the strength and quality of their mated
's – and so their ability to defend their mates
n attempted takeovers – simply by listening to
song bouts.

'emale song bouts are strongly related to the
ree to which the singer's territory is threatened
eighboring females wishing to steal valuable
t. Thus, female singing appears to advertise the
ger's presence in an attempt to prevent pilfer-
. Females normally perform song bouts about
e every 1–3 days, but will drop this to 2–5 days
truders only put in an appearance about once
rtnight. In contrast, singing occurs daily when
y are threatened every other day.

Recognizing an opponent is crucial to gibbon
ss levels. If intruders are neighbors, they will
e their own territory nearby and so are proba-
only visiting to gain temporary access to a
npting tree in fruit. Serious fights (and the atten-
nt risks of injury) are normally only worth enter-
into with gibbons from outside the area, which
y well be looking for a territory of their own.

n many gibbon populations, males accompany
nales in song to create complex duets. While the
les' contribution varies between species, they
en sing along with – or immediately following –
female great call. This male song is called the
da." The degree of coordination and synchro-
ation of vocalizations between the sexes during
ets improves with practice, so the quality of the
et is often a good indication of how long a cou-
have been together.

Some authorities believe duetting helps estab-
and maintain the pair bond. However, this
ms unlikely, given that some species (e.g., Kloss
l Moloch gibbons) do not normally duet even
ugh all have close pair bonding. It now seems,
her, that duets are performed to advertise the
r bond in those populations where territorial
ursions are frequent. By joining his mate in a
et, the male signals his presence and support for
to the neighbors. This may help both to reduce
risk of territorial encounters and to prevent
gression from escalating if such encounters
take place. GC

Above *Male and female Moloch gibbons screaming together at their neighbors during a territorial dispute. This Javanese species does not duet – indeed, the male Moloch gibbon does not normally sing at all.*

Right *From his sleeping tree, nale Kloss gibbon completes a ut of singing that may have ed for up to two hours. The ertoire of males of this species xtremely varied.*

Chimpanzees

MOST OF THE SCIENTIFIC COMMUNITY now consider chimpanzees and bonobos to be our closest living relatives. Genetic evidence indicates that we last shared a common ancestor with them about 6 million years ago, after the lineage that led to modern gorillas had diverged.

Bonobos diverged from chimpanzees about 1.5 million years ago, presumably when part of the ancestral chimpanzee population crossed to the south side of the Congo River and became isolated there. Bonobos are restricted to lowland tropical rain forests, including those on the edges of the southwestern African savanna in what is now the Democratic Republic of Congo. Chimpanzees are also mostly rainforest dwellers, but occupy a much wider geographic range and a broader range of habitats, including montane forest, seasonally dry forests, and savanna–woodland areas, where their population density is very low.

Humankind's Nearest Relations

FORM AND FUNCTION

The number of recognized species and subspecies of chimpanzee has varied greatly over time. Until recently, the consensus was that all chimpanzees belonged to a single species that included three subspecies, but now chimpanzee taxonomy is again in flux. Because chimpanzees are so close to us evolutionarily and because their behavior resembles ours in striking ways, they have been the favorite point of comparison in scenarios of early hominid evolution and in attempts to understand the biological roots of our own behavior. However, recent research on bonobos has revealed important contrasts between the two ape species that must be taken into account in comparing them to ourselves.

Both species have bodies well adapted for arboreal activities. Their arms are considerably longer than their legs, their fingers much longer than those of humans, and their shoulder joints are highly mobile. These and other features of their skeletons and musculature enable them to hang from tree branches by their arms and make them adept at climbing tree trunks and lianas and at clambering in the crowns of trees. Indeed, both species do most of their feeding in trees, and they also sleep in trees at night, in nests that they construct by breaking and folding branches. However, they travel mostly on the ground, where they engage in quadrupedal "knuckle-walking," a form of locomotion they share with gorillas. Their bodies show many adaptations to this practice; for example, where the radius, a forearm bone, meets the bones of the wrist, it has a ridge that preven the wrist from buckling when the knuckles are bearing the body's weight.

Bonobos are also known as "pygmy chimpanzees," but this is a misnomer. They have a more slender build than chimpanzees and their skulls are shaped somewhat differently, but the weigh as much as at least the smallest subspeci of chimpanzees. Both chimpanzees and bonob can stand upright and often do so when climbi or reaching for food, but their bipedal locomoti is awkward compared to our own.

Chimpanzees and bonobos have brains that, at 300–400cc (18–24cu in), are large both in absolute terms and relative to their body size. They do notably well on problem-solving tasks laboratory settings, and both have shown some capability to engage in symbolic communicatio in captivity when given intensive training, or at least extensive learning opportunities. They use variety of vocal and visual signals to communica in the wild, but none of these seem to involve t use of symbols. Both species are highly adept at predicting and manipulating the behavior of ot ers, be they conspecifics or human researchers. Some evidence suggests that this is because the can understand that others have desires and knowledge states like their own – that is, that t possess a "theory of mind" somewhat like that humans. However, this claim is currently a topi of debate.

Male chimpanzees and bonobos are 10–20 percent larger than females and are considerabl stronger. They also have larger canine teeth, wh are their main weapons. Otherwise, males and females have similar body proportions.

Starting at adolescence, females develop periodic swellings of the skin around their genitals. These initially occur at irregular intervals and ca last for many weeks, but by adulthood, females have regular menstrual cycles of about 35 days chimpanzees and 40 days in bonobos, and the swellings are confined to about 12–20 days in t middle of each cycle. Females with swellings are estrus (in heat), and are both interested in male sexual advances and motivated to approach ma and to initiate sex with them. In the wild, femal chimpanzees give birth to their first infants whe about 13 years old. Infants develop slowly and a

Left *A group of young chimpanzees resting afte a bout of energetic play. Chimpanzees have a length infancy and adolescence. The protracted weaning period of 4 years may be a result of the long gaps between pregnancies.*

FACTFILE

CHIMPANZEES

Order: Primates

Family: Hominidae

2 species of the genus *Pan*

DISTRIBUTION W and C Africa

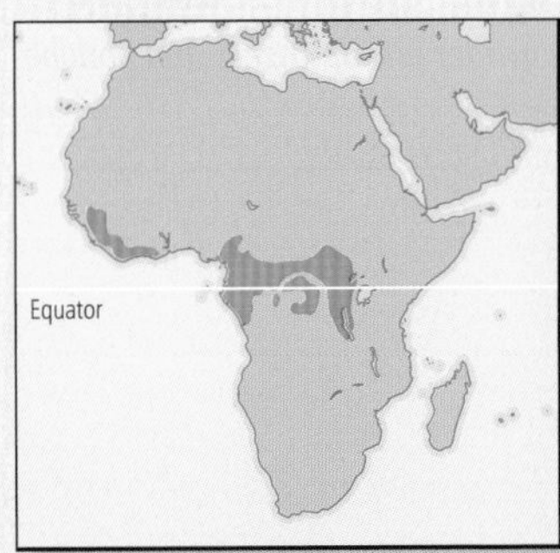

COMMON CHIMPANZEE *Pan troglodytes*
Four subspecies: **Western** or **Masked chimpanzee** (*P. t. verus*; considered by some as a separate species), **Black-faced chimpanzee** (*P. t. troglodytes*), **Long-haired chimpanzee** (*P. t. schweinfurthi*), and one yet to be given a common name (*P. t. vellerosus*). W and C Africa, N of River Congo, from Senegal to Tanzania. Humid forest, deciduous woodland, or mixed savanna; presence in open areas depends on access to evergreen, fruit-producing forest; from sea level up to 2,000m (6,500ft).

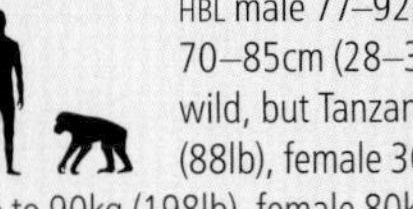

HBL male 77–92cm (30–36in), female 70–85cm (28–33in); WT poorly known in wild, but Tanzanian male weighed 40kg (88lb), female 30kg (66lb); in zoos, male up to 90kg (198lb), female 80kg (176lb). Form: coat predominantly black, often gray on back after 20 years; short white beard common in both sexes; infants have white "tail-tuft" of hair, lost by early adulthood; baldness frequent in adults, typically a triangle on forehead of males, more extensive in females; skin of hands and feet black, face variable from pink to brown or black, normally darkening with age. Breeding: gestation 230–240 days. Longevity: 40–45 years. Conservation status: Endangered.

BONOBO *Pan paniscus*
Bonobo, Dwarf, or Pygmy chimpanzee
C Africa, confined to Democratic Republic of Congo between Rivers Congo and Kasai. Humid forest only, below 1,500m (4,900ft). HBL male 73–83cm (29–33in), female 70–76cm (28–30in); WT (rarely measured in wild) male 39kg (86lb), female 31kg (68lb); body lighter in build than Common chimpanzee, including narrower chest, longer limbs, and smaller teeth. Form: coat as Common chimpanzee, but face wholly black, with hair on top of head projecting sideways; white "tail-tuft" commonly remains in adults. Breeding: gestation 230–240 days. Longevity: unknown. Conservation status: Endangered.

Abbreviations HBL = head–body length WT = weight

Left *The chimpanzee's long arms, which reach to just below the knee when it is standing erect, are invaluable assets that enable it to move swiftly and with great agility around its forest habitat.*

Below *The diet of chimpanzees varies according to the time of year and the abundance of particular foods. In excess of half of all food intake is comprised of fruit and much of the remainder is vegetative matter. Meat is eaten, though it probably accounts for 5 percent or less of the diet. Of the many insects consumed, termites are the most important. Here, a chimpanzee eats fruit directly from the tree.*

not weaned until about 4 years old, and intervals between surviving births average 5–6 years. Compared to other primates, male chimpanzees and bonobos have testes that are enormous relative to the overall size of their bodies, and they are capable of copulating with females at high rates. Males reach their full adult size at about 16 years, but may be fertile well before this.

Dietary Differences

DIET

Chimpanzees and bonobos are generally active from dawn to dusk, a 12–13 hour period in thei equatorial habitats, and feed for at least half of t time. Both mostly eat fruit, supplemented with leaves, seeds, flowers, pith, bark, and other plan parts. Chimpanzees eat from as many as 20 plar species a day and 300 a year. Fruit abundance ir their habitats can vary dramatically over the course of a year. On some days they eat little oth than the fruit of a single species that happens to be available in large quantities. They eat leaves year-round, but consume these and other non-fruit items more when fruit is not abundant. Bonobos seem to rely somewhat more on plant stems and pith than chimpanzees do, and their habitats may offer more steadily available fruit supplies; these differences have important conse quences for their social lives.

Chimpanzees and bonobos also eat foods of animal origin, including insects like termites an meat from a variety of vertebrates; for chimpanzees, ingestion of animal prey can take up as much as 5 percent of feeding time. Chimpanzee hunt more often than bonobos do and take a wider range of prey, including monkeys, bushpi forest antelope, and various small mammals. Monkeys are their most common prey items, an red colobus monkeys are the main prey species wherever they occur with chimpanzees. Chimpanzees mostly hunt in groups, and males hunt more than females. Small antelopes are the mos common prey for bonobos, which are not know to hunt monkeys, and they capture most prey individually and opportunistically, rather than during group hunts.

Hunting success varies among chimpanzee communities for various reasons. Capturing mo keys in the canopy of tall, primary forest is more difficult than catching them in areas where the canopy is low and broken, and chimpanzees in areas with both kinds of forest are more likely to hunt in areas with broken canopies. The numbe of hunters and the extent to which they coopera also influences the outcome, and hunts are mor likely to be successful when many males participate and when they coordinate their attacks wit each other. For red colobus hunts, success rates different habitats range from 50–80 percent, qu high values compared to those of most carnivor

Hunting frequency varies over time. In at leas some habitats, hunts are more common when fruit is abundant; at these times, males often for large parties and can easily gain the energy need to pursue monkeys, and they may commonly tr el several kilometers looking for red colobus groups to hunt.

Feeding is mostly an individual affair, but me eating is a conspicuous exception. Male chimpanzees sometimes fight over carcasses immedi ately after captures, and high-ranking males

whether some female chimpanzees associate with members of two neighboring communities. All communities have more or less friendly social relationships with each other, but relations between communities are antagonistic, much more so in chimpanzees than in bonobos. Within a community, individuals travel and feed in parties that vary in size and composition, and may rarely or never all be together. Party size depends very much on the availability of food, especially fruit. Parties are larger when fruit abundance is generally high, and large parties also gather at large fruiting trees like giant figs, but individuals form smaller parties when fruit is scarcer as a way to reduce competition for food. Males also form large parties, independently of fruit availability, when they are with estrous females. Average party size is slightly higher for bonobos (6–15) than for chimpanzees (3–10), and party size varies less in bonobos than in many chimpanzee populations, presumably because food abundance varies less in their habitats.

Male chimpanzees are more sociable than females, who may often be alone with their dependent offspring. Sex differences in sociability are less pronounced in bonobos. Male chimpanzees move about more widely than females and generally use their community's entire range. Females with dependent offspring tend to limit their activities more to core areas in restricted parts of the community range, although the extent of sex difference seems to vary among habitats. Estrous females travel more widely, usually accompanied by many males. We know less about range use in bonobos, but the sex differences seem to be less pronounced.

'asionally steal meat from subordinates. How-r, peaceful begging for and sharing of meat is nmon. Much sharing involves possessors owing others to take pieces of carcasses, but netimes possessors actively give meat to others. les are most often the possessors in chimizees, and they share mostly with other males l especially with their allies and with their main oming partners. Reciprocity in meat sharing urs between males.

Females often get some meat from males. rous females may be more successful than oth-, but claims that females trade sex for meat e not been substantiated. Males sometimes te with females during meat-sharing episodes, the presence of estrous females does not ays make males more likely to hunt, and meat ring may have little effect on male mating sucs. Bonobo females possess animal carcasses relely more often than chimpanzees, and they en control the enormous fruit of *Treculia* trees, most commonly shared plant food. In contrast chimpanzees, most food sharing in bonobos urs between females. Chimpanzees sometimes o share *Treculia* fruit; as with meat, males most-hare with each other, and especially share with es and major grooming partners.

Above Central African bonobos eating melon. The diet of the bonobo is broadly similar to that of the chimpanzee, though bonobos consume less vertebrate prey and when they do it is most commonly small antelope caught opportunistically.

Below Although there are many similarities in the way bonobos and chimpanzees are raised, there are also significant differences. For example, young adult male bonobos stay in their mother's groups, whereas their chimpanzee counterparts join male groups.

Aggression and Peacefulness

SOCIAL BEHAVIOR

impanzees and bonobos have fission–ion societies. All individuals belong to nmunities that have 15–150 members l that seem to be socially bounded, hough some uncertainty exists about

Above and right *Facial expressions are more varied in higher primates than in any other animal. The chimpanzee has a particularly sophisticated repertoire of expressions, which it uses to convey a wide range of social signals:* ***1*** *play face, characterized by a relaxed, open mouth and the upper teeth fully concealed behind the top lip; the juvenile chimp* ABOVE *is seen making this type of unthreatening face while wrestling with its playmate;* ***2*** *pout, an expression that is commonly used when begging for food;* ***3*** *display face, with the facial hairs erect; this hostile expression is used during an attack or on other occasions when the chimpanzee needs to display aggression;* ***4*** *full open grin, an indication of intense fear or some other form of excitement;* ***5*** *horizontal pout, intended to show submission, for example, while whimpering after being attacked;* ***6*** *fear grin, shown during an approach to – or from – a higher-ranking chimpanzee.*

Male chimpanzees and bonobos remain with their natal communities for life. In contrast, females typically transfer into neighboring ones adolescence, before they have started to reproduce. Adult females occasionally transfer also, although this appears to be rare. Immigrant fem chimpanzees face some aggression from residen females as they try to establish core areas in the new communities, and are dependent on males protect them against this harassment. Immigrar female bonobos seem to face less aggression, an make efforts to develop social ties to particular r ident females that may assist them to gain tolerance within the group.

Notable differences in social relationships ex between chimpanzees and bonobos. Chimpanzees have a male-bonded society. Males espe cially associate with each other. Most form dominance relationships, which can lead to do nance hierarchies, although these may not be apparent in communities with many males. Co petition for high dominance rank is striking. Ho ever, males also have many friendly interactions. Grooming between them is common, and they

OOL USE

himpanzees use a wide variety of tools in many ontexts. The utensils employed include fly whisks, af sponges and wipes, twig probes used to extract one marrow, and branches and stones that serve as ammers and anvils.

Two kinds of food (social insects and fruit) are ommonly obtained with tools, although different opulations vary in their use of them. Most social isects have potent defenses that are overcome by ne use of sticks or soft stems. For instance, chim-anzees prepare smooth, strong wands 60–70cm 24–28in) long for feeding on Driver ants: they lower ne wand into an open nest, wait for ants to crawl p it, then sweep them off into their mouths before ne ants have time to bite. They strip grass stems to nake them supple, poking them into holes on ter-nitaria: soldiers bite the stem, and cling on long nough for the chimpanzee to extract and eat them BELOW LEFT). Sticks are also used to enlarge holes, so hat honey or tree-dwelling ants can be reached.

A second food type eaten with tools is fruits with hells too hard to bite open. Sticks or rocks weighing p to 1.5kg (3.3lb) are used to smash these fruits, ometimes against a platform stone that serves as an nvil. Platform stones have been found with a worn, ounded depression, suggesting they have been sed for centuries.

Tools are not used only when feeding. Adult nales elaborate their charging displays by hurling sticks, branches, or rocks (BELOW RIGHT) of 4kg (8.8lb) or more: in a long display as many as 100 rocks may be thrown, and other individuals have to watch out to avoid being hit. On at least one occasion, missile use has been observed in a hunting context: a male hit an adult pig from five meters (over 16ft), startling it so that it ran off and allowed the chimpanzee to seize its young.

Young chimpanzees need years of observational learning and practice to become skilled at using some tools, especially hammers and anvils. Separate populations differ considerably in their use of them – the most striking example of variation in socially learned traditions, or "culture," in chimpanzees. Bonobos also use tools, but fewer kinds, and less frequently. Until recently, chimpanzees and bonobos were thought to be the only apes to make and use tools routinely in the wild, but we now know that some orangutans commonly manufacture and use them to extract food that would otherwise be hard to obtain. DW/RWW

oom more with each other than with females or an females do with each other. Reconciliation er conflicts is also more common between ales than between females. Some males form iances against others that are important in com-tition for rank, and an alpha male may owe his sition to support from allies. Females do not utinely develop strong social bonds with each her or with particular males. Some females have minance relationships, but they do not form minance hierarchies. All males are dominant to females.

In contrast, bonobo females associate more th each other and develop strong social bonds, spite the fact that they usually are not close rel-ves. Grooming is common between some females, and females also commonly rub their genitals together as a way to reduce tension and maintain affiliative relationships. Females sometimes form coalitions against males and can get the males to behave submissively, and males commonly defer to females during feeding, rather than trying to take feeding spots or food items. Females also support their adult sons in competition with other males and can influence sons' dominance ranks; female chimpanzees play no role in male competition for dominance in the wild. Male bonobos also groom with each other regularly, and grooming may be more common between male pairs than between most female pairs, but they apparently do not form alliances like those seen in chimpanzees.

Male chimpanzees also cooperate with each other in aggression between communities, which takes two forms. When parties from neighboring communities meet during the course of normal activities, they usually display excitedly and may charge at and chase each other, although those on one side may instead flee silently if they are outnumbered. Sometimes males patrol the boundaries of their ranges and even make incursions into those of neighbors. Patrollers are conspicuously silent and wary and are looking for neighbors. If they hear or meet some, their response depends largely on relative numbers: they quietly leave, or even flee, if outnumbered or if the opposing party is not much smaller than theirs, but they attack when they greatly outnumber their opponents. Attacks are severe and can be fatal; males are known to have killed adult and adolescent males, infants, and even nonfertile adult females.

Lethal aggression by male coalitions is unusual in mammals; among primates, it occurs only in chimpanzees and humans. Why it occurs in chimpanzees is not entirely clear. Possibly success in intercommunity competition allows males access to more females with whom they can mate, but it may also be important because it gives females already in the males' community access to more and better food, and thereby improves their reproductive success. Patrolling and successful territory defense also helps to protect females against the threat of infanticide by outside males.

When parties from neighboring bonobo communities meet, they also usually display at each other and may engage in chases. However, sometimes encounters are peaceful – something unknown in chimpanzees – and boundary patrols and serious attacks have not been seen. This difference from chimpanzees may result mostly from the fact that bonobos typically travel in larger parties, so opportunities for groups to make low-risk attacks on much smaller groups or on solitary individuals would be rare.

Chimpanzee mating behavior is complex and highly variable. Most mating is opportunistic: estrous females copulate repeatedly with most or all adult males in a community, and often with immature males. Mating with many males presumably confuses paternity, and could give females some insurance against infanticide by males. However, as females approach the end of their estrus, and ovulation (which usually occurs 1–3 days before their sexual swellings deflate) becomes more likely, high-ranking males sometimes try to guard them and to prevent them from mating with other males. Attempts to monopolize mating involve considerable aggression, mostly aimed at the females themselves. Mating success is positively correlated with male dominance rank, and high rank seems to confer some advantages with regard to fertilization, but we have little information on paternity. Sometimes a male can persuade a female to go on a consort (a temporary courting association) with him, during which they try to avoid other males and may stay on their own for several weeks. Pregnancy lasts about 7.5 months. Once a female conceives, she does not resume regular cycling for 4 years or more if her infant survives.

Bonobo females have longer estrous periods than chimpanzee females. They are more likely to show estruslike behavior during pregnancy, and they start having sexual swellings within a year after giving birth. Female sexual behavior is thus even less closely tied to conception in bonobos than in chimpanzees, which may help to explain why neither mate-guarding nor infanticide have been seen in this species. Sexual behavior is generally much more common in bonobos than in chimpanzees and serves many social functions. Genital–genital rubbing and copulations are particularly common during postconflict reconciliation and during food sharing.

Hunting, male bonding, cooperative intergroup raiding, and the making and use of tools (see box) are all traits that chimpanzees share with humans and that might also have characterized our last common ancestor. If so, we need to explain why male bonding and intergroup raiding are absent in bonobos. The answer could be that a combination of a more reliable and evenly distributed food supply, along with an elaboration of female sexuality, removed chances for low-cost raiding, reduced the intensity of male mating competition, and made females less vulnerable to male aggression.

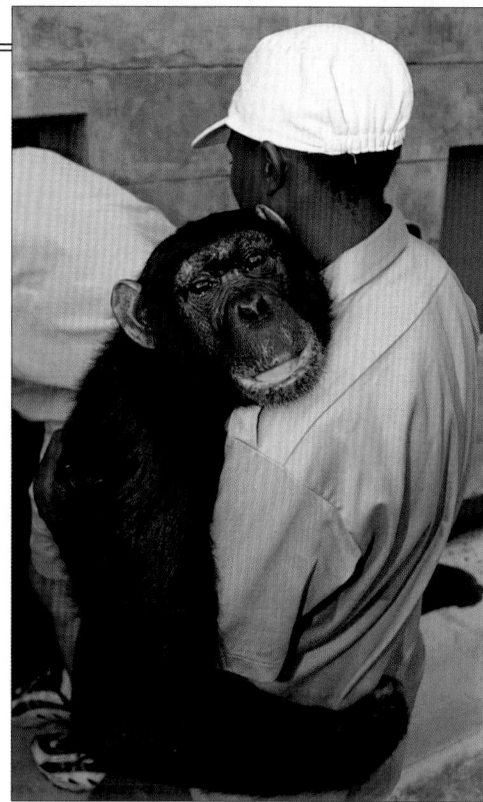

Above *As well as captive individuals there are many captive colonies – managed with the hope of producing a self-sustaining group. This sedated chimpanzee is being released into the wild in a new area.*

Below *A chimpanzee group crossing the Conkouati Lagoon in West Africa. In the course of foraging, chimpanzee groups may cover distances of up to 15km (9mi) each day.*

A Precarious Future

CONSERVATION AND ENVIRONMENT

Whether we will ever have convincing answers the unresolved questions about chimpanzee behavior is uncertain because of the precarious futures that the two species face. Considerable tracts of chimpanzee habitat remain, but most a under threat from logging, conversion to farmland, and other forms of human encroachment. Bonobos are far less common than chimpanzee in the wild and face similar pressures. Both ape are hunted to supply the bushmeat trade (see feature opposite), and both also fall victim to snares that have been set to catch other animals Even populations in national parks are not nece sarily well protected from hunting, and the pop lations in some protected areas may be too sma to be viable. The upheaval created by military co flict has exacerbated the threat to these apes in many areas.

The threats from habitat loss and hunting are particularly menacing because chimpanzees and bonobos reproduce slowly and cannot live at hi population densities. Large numbers of chimpa zees exist in captivity, and their long-term welfa has become a major issue in western countries and Japan. There is no viable captive breeding population of bonobos. The survival of these ap in the wild will depend on continued, intensive conservation efforts and, ultimately, on a resolu tion of the problems that lead to habitat destruc tion and unsustainable hunting. DW/RW

SPECIAL FEATURE

THE BUSHMEAT TRADE

Taking mammals to the marketplace

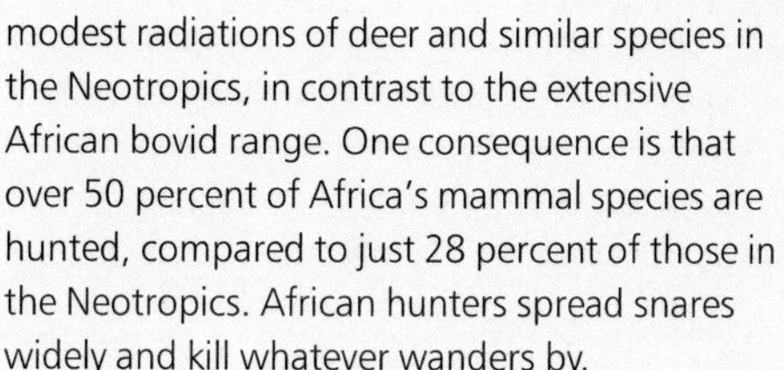

Right *Body parts of a dismembered gorilla are displayed by African hunters prior to sale at a bushmeat market. Estimates of the magnitude of this trade, which have considered its impact on the Congo Basin, showed that a rural human population of 24 million, living in 1.8 million sq km (700,000sq mi) of tropical forest, can consume around 579 million animals every year. This equates to about 4 million tons of dressed mammal meat. Such figures are alarming enough, but may still underestimate the problem, since they take no account of consumption by urban households.*

MANY ISSUES IN TROPICAL WILDLIFE CONSERVATION have overshadowed that of the bushmeat trade, but this has been a mistake. The loss of wildlife through the removal of wild animals for the dinner plate is in some areas proving to be a greater threat even than deforestation. In non-forested areas, too, the bushmeat trade is a serious threat.

Many millions of Africans, ranging from subsistence foragers and farmers to urban dwellers of different social backgrounds, continue to consume large quantities of bushmeat. Ethnic groups such as the Mvae, Yassa, and Kola of Cameroon eat more meat than the average European or North American, and animals as diverse as tortoises and elephants are eaten. But such traditions can only be sustained when the animals are able to reproduce fast enough to replenish their numbers.

Many factors have come together to increase the death toll. To begin with, there are simply more people. In the Congo Basin, the human population has doubled since the 1920s and is likely to double again in the next 30 years. New roads built for the logging industry, together with the increased availability of firearms, have also both played a part. Some roads take hunters deep into the forest, while others carry them far afield to sell their wares at distant markets. New roads maintained by the logging industry can greatly reduce transportation times and distances for those selling bushmeat; the Mouloundou–Yakadouma road in eastern Cameroon, for example, has shortened a two-week journey to less than five hours.

Everywhere demand is increasing, for reptiles, birds, but most of all for mammals. Studies estimate that between 23–56 million individual mammals are killed every year in rural Brazilian Amazonia alone, providing 145,000–356,000 tons of meat. The species affected includes over 50 different taxa, particularly mammals, birds, and reptiles. Between 6.4 and 15.8 million individual mammals are harvested annually, providing 60–148 tons of meat. In the Amazon region, the Collared peccary is the most sought-after prey.

On average, African forests contain around 3,000kg (6,600lb) of potential mammal bushmeat per square kilometer (8.5 US tons per square mile), while Neotropical South American forests offer only half this amount. In Africa, the mammals are mainly ground-dwelling (80 percent), while in America, they inhabit the trees and forest canopy (50–90 percent). Another difference is that African mammals are on average bigger and easier to trap in snares. The Lowland tapir is the only forest mammal over 50kg (110lb) in weight in the Neotropics, whereas 13 species in African forests are in this size class. This is partly a function of the modest radiations of deer and similar species in the Neotropics, in contrast to the extensive African bovid range. One consequence is that over 50 percent of Africa's mammal species are hunted, compared to just 28 percent of those in the Neotropics. African hunters spread snares widely and kill whatever wanders by.

Larger-bodied mammal species are particularly susceptible to over-harvesting because of their generally low reproductive rates and densities. Yet they also tend to be targeted more frequently; from the hunter's point of view, larger animals better compensate for the cost of bullets. Another factor is the risk hunters must run to make a kill. In African forests, the most sought-after prey include the large-bodied red and black-and-white colobus monkeys, because these monkeys – which are vocal (especially the red colobus), live in large groups in the upper canopy, and are generally slower-moving than other primates – pose no threat. Because hunters target large game in forests near long-established communities, populations of these animals are often much reduced near human habitation.

There is still much to be learned about bushmeat use in the tropics, and there is a need for adequate baseline data that can be incorporated into policies and programs for natural resource management, biodiversity conservation, and rural development. On the basis of current information, however, it is predicted that, under current levels of exploitation, most primates and ungulates may become extinct in less than three decades. Geographically restricted species, such as the mandrill and the White-throated guenon among the primates and the zebra and Jentink's duikers among the ungulates, are considered particularly at risk. On the plus side, the available information also indicates that game populations can readily bounce back after exploitation, so long as hunting pressure is not too heavy and large neighboring tracts of undisturbed forest can buffer and replenish hunted areas.

Strategies that reduce incentives for, and the impact of, bushmeat hunting (especially for markets) will need to take into account both law enforcement and the economics of the trade. Various authors have suggested plans that span reduction of local supply, quota systems, local and government enforcement, and limitation of commercial supply. However, much remains to be done in linking community participation, traditional control systems, and land tenure reforms. The complex political context within which the trade occurs makes it difficult to put solutions in place. Some progress in advancing sustainable management of wildlife has been made in recent years; however, there are still too few of these, and in some regions they have probably come too late.

For the bushmeat trade to achieve sustainability, supply and demand would have to be balanced either by increasing the supply to hunters – which is not a practical proposition – or by decreasing consumer demand. Because part of the pressure on resources comes from the increased access provided to bushmeat hunters by the timber industry, most conservation organizations have suggested regulating hunting within concessions and banning companies from transporting hunters or bushmeat. It will be just as important, though, to reduce consumer demand by increasing the availability and lowering the price of alternative food. Providing viable substitutes would benefit local people and lower consumption of bushmeat. JEF

Gorillas

THE GORILLAS ARE THE LARGEST LIVING primates and, along with the two species of chimpanzee, the apes most closely related to man. Indeed, fossils and biochemical data indicate that the chimpanzees and gorillas are more closely related to humans than they are to the orangutans.

The apes are probably the most intelligent land animals apart from humans, at least as judged by our standards. They can learn hundreds of "words" in deaf-and-dumb sign language, and even string some together into simple, grammatical, two-word "phrases." Nevertheless, the gorillas' formidable appearance, great strength, and chest-beating display allowed hunters to give them an otherwise unfounded reputation for ferocity. But, in Rwanda, thousands of tourists every year approach on foot to within a few meters of totally wild gorillas. Threatened by man alone, adult males are dangerously aggressive only when fighting one another over females, or defending their family groups against predators and hunters.

The Greatest Apes

FORM AND FUNCTION

Gorillas differ from the chimpanzees – their close relatives – in being far larger, and in the different proportions of their bodies (longer arms relative to leg size, shorter and broader hands and feet) and a different color pattern. The build is much heavier, and many of the proportional differences are connected with this. In particular, the gorilla requires far larger teeth (especially the molars) to process the amounts of food that are needed to sustain its huge bulk. These in turn are worked by much bigger jaw muscles, particularly the temporal muscles, which in male gorillas meet in the midline of the skull, where they are attached to a tall, bony crest – the sagittal crest. A small sagittal crest may occur in female gorillas or in chimpanzees, but a large one, meeting a big shelf of bone (the nuchal crest) at the back of the skull, is a distinctive characteristic of male gorillas and considerably affects the external shape of the head.

In addition, male gorillas have canine teeth that, in relation to body size, are bigger than those of chimpanzee males, and bigger also than those of the female; the canines are especially large in G. *b. diehli* and biggest of all in *G. b. beringei*. With these canines, male gorillas inflict serious wound on each other as well as on predators. In evolutionary terms, this probably explains why the teeth are so large: the males that won the females and that best defended them, were the ones with the most formidable weapons. The male's skull crest is thus bigger than the female's both because, requiring more food generally, he requires the muscle ability to crush coarser food, and also because he benefits from larger muscles (and hence larger muscle attachment areas) to increase the wounding power of his huge canine. The gorilla has small ears, and the nostrils are

Below *A female Mountain gorilla eating a stingir nettle. The bulk of East African gorillas' diets is made up of leaves and other vegetation, rather than fruit. Gorillas never stay long enough at one feeding site t strip it completely; rather, they crop the vegetation, leaving enough growth for quick rejuvenation to occ*

FACTFILE

GORILLAS

Order: Primates

Family: Hominidae

2 (1) species of the genus *Gorilla*

DISTRIBUTION
Equatorial Africa

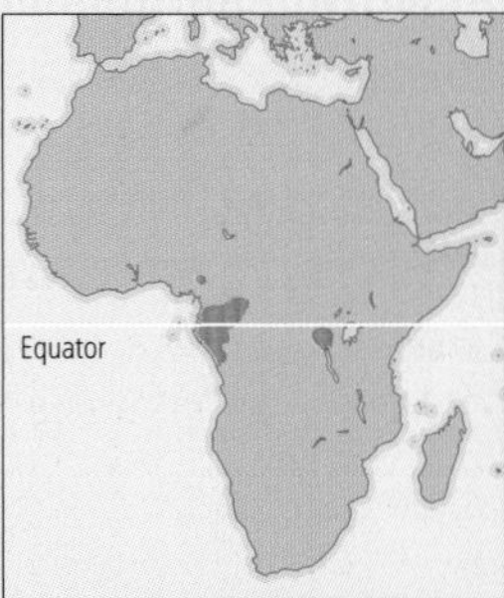

WESTERN GORILLA *Gorilla gorilla*
Two subspecies: **Western lowland gorilla** (*G. g. gorilla*) and **Cross River gorilla** (*G. g. diehli*). Cameroon, Central African Republic, Congo, Gabon, Democratic Republic of Congo (DRC), Equatorial Guinea, Nigeria. Swamp, montane, and secondary tropical forest. HT male 170cm (67in), occasionally up to 180cm (71in), female up to 150cm (60in); WT male 140–180kg (310–400lb), female 90kg (200lb). Coat: brown-gray; males with broad, silvery-white saddle extending to rump; *G. g. gorilla* has a distinctive red forehead, especially on male. Hair short on back, long elsewhere; skin jet black almost from birth. Breeding: gestation 250–270 days; only one young born at a time (if twins, only one usually survives). Longevity: about 35 years in the wild (50 years in captivity). Conservation status: *G. g. gorilla* best assessed as Vulnerable (though currently listed as Endangered on the IUCN Red List); *G. g. diehli*, however, is Critically Endangered, with just a few hundred individuals on the Nigeria–Cameroon border.

EASTERN GORILLA *Gorilla beringei*
Three subspecies: **Mountain gorilla** (*G. b. beringei*), **Eastern lowland gorilla** (*G. b. diehli*), and a third from Bwindi Impenetrable Forest yet to be named. E DRC, Rwanda, and Uganda. Swamp, montane, and secondary tropical forest; *G. b. beringei* at altitudes of about 1,650–3,790m (5,400–12,400ft). Coat: black, turning gray with age; males with broad, silvery-white saddle on back, but not extending as far as in *G. gorilla*. Hair short on back, long elsewhere, particularly on *G. b. beringei*; skin jet black almost from birth. Breeding and Longevity: as for *G. gorilla*. Conservation status: *G. b. diehli* best assessed as Vulnerable, though *G. b. beringei* is undoubtedly Endangered, being reduced to a population of about 600.

Abbreviations HT = height WT = weight

Above *Male gorillas have an extraordinarily powerful build: fully grown adults like this "silverback" Mountain gorilla from the Parc National des Volcans in Rwanda can have a chest circumference equal to their own height. Their great size, combined with the low nutritional content of much of their diet, means that gorillas have to spend much of the day feeding.*

rdered by broad, expanded ridges (naval wings) t extend to the upper lip.

Gorillas are mainly terrestrial and quadrupedal ney walk on the soles of their hind limbs and knuckles of their fore limbs. However, in West ica, where there are far more fruiting trees than East Africa, adults – including the massive les – spend an appreciable time feeding on fruit h up trees. Lightweight individuals will even ng from tree to tree by their arms (brachiation), d the young play in the trees. While they prefer it, including figs where they occur, gorillas fall k on leaves, pith, and stems in areas or times ow fruit availability. The species of sedges, bs, shrubs, and vines that make up the gorilla's back diet grow best in swamp, montane, and ondary forests, where the open canopy allows nty of light to reach the forest floor.

The gorilla's large size and folivorous habits an that the animals must spend long hours ding every day to maintain their body weight. is prevents them from regularly traveling long tances, so while home ranges of gorilla groups ver from 5–30sq km (2–12sq mi), the normal e of travel is only 0.5–2km (0.3–1.2 miles) per . Both home range and daily distance are less eastern than in western Africa, because the East ican forests contain fewer species of fruiting e. The Eastern gorillas thus have a more foliage- h diet than does the western species, and travel s far each day to find it.

The small distance traveled per day means that rillas cannot be territorial. The circumference of n a range as small as 5sq km (2sq mi) is 8 km (5mi), or at least four times the usual daily travel distance. The range is thus effectively indefensible. Consequently, the home ranges of neighboring gorilla groups overlap to a great extent. Indeed, even the most heavily used core areas of a group's ranges can overlap.

Gorillas normally feed during the morning and afternoon, and rest for an hour or two around midday. Like all the great apes, they make "nests" at night – platforms or cushions of branches and leaves pulled and bent under them. These nests keep the gorillas off the cold ground or support them in a tree for the night, and also prevent them sliding down steep slopes – a real danger. These benefits are particularly useful in eastern Africa, where the nests and the dung in and around them are extremely useful for censusing gorillas – an animal's size can be judged from the size of its dung bolus, and the number of animals in family groups from the number of nests. In western Africa, gorillas often do not build nests.

Gorillas do not have a distinct breeding season. Births are usually single, as in all primates weighing more than 1kg (2.2lb). Twins are very rare, and when they do occur they are usually so small at birth (and the mother, who has to carry the infants for the first few months of life, finds it so hard to care for them) that at least one almost always dies. Newborn gorilla babies weigh 1.8–2kg (4–4.4lb), and their grayish-pink skin is sparsely covered with fur. They begin to crawl after about nine weeks, and can walk from 30–40 weeks. Gorillas take longer to wean than humans – from 2½ to 3 years – and females give birth at about four-year intervals. However, a 40 percent mortality rate in the first 3 years of life means that a surviving offspring is produced only about once every 6–8 years in a female's breeding life. Female gorillas mature sexually at 7–8 years of age, but usually do not start to breed until they are 10 or so. Males mature a little later but, because of competition for mates, very few start to breed before 15–20 years of age.

Mountain and Lowland

DISTRIBUTION PATTERNS

The two species of gorilla are found in two widely separated areas of Africa. The western and eastern populations were probably originally separated first by the massive inland Congo Basin lake of the Miocene, and then, from about 5 million years ago, by the gradual drying of the region and the retreat of forests to higher areas. Subsequently, gorillas have not spread back into the central Congo Basin, either because they have not had time or because the heavily-shaded primary forest there does not allow the growth of sufficient ground vegetation to support such a large, predominantly terrestrial animal.

Although they are found over only a small area of Africa, gorilla habitat includes a wide range of altitudes, from sea level in west Africa to 3,790m (12,400ft) in the east. Oddly enough, it is the Mountain gorilla (*G. b. beringei*), found in the extreme east and at the highest altitude, that is the best-known form; the behavior of Western gorillas is relatively little known, because dense vegetation prevents their being easily observed. Generalizations about gorillas, especially concerning their behavior, thus have to be read in the knowledge that most of the information comes from an extreme population.

Life in the Harem

SOCIAL BEHAVIOR

Of all the great apes (family Hominidae), the gorilla shows the most stable grouping patterns. The same adult individuals travel together for months and usually years at a time. It is because gorillas are mainly foliage-eating that they can afford to live in these relatively permanent groups. Foliage, unlike fruit generally and especially the ripe fruit that the ape gut requires, comes in large patches that can in turn support large groups of animals. In west Africa, where fruit forms a far higher proportion of the gorilla's diet than in the east, gorilla groups tend much more often to split into temporary subgroups than they do in east Africa, as animals range far apart searching for the relatively scarce ripe fruit.

Gorilla groups can include up to 30–40 animals, but more usually number 5–10. An average group in the east (Rwanda, Uganda, and the eastern DRC) contains about three adult females, four or five offspring of widely different ages, and one fully adult male – called the "silverback" because of its silvery-white saddle. It used to be thought that groups in west Africa averaged about five animals, though it is possible that observers were actually counting subgroups of larger agglomerations. Nevertheless, it seems that in the west, groups containing in excess of 10 animals are rare, whereas in the east, groups with 15–20 members are not uncommon, and groups of over 30 animals have been recorded.

Adult females in any one silverback's harem are mostly unrelated, and the social ties that exist between them are weak. In this respect they differ from many Old World monkeys, though not the chimpanzees. In these ape species, females often leave the group in which they were born, usually at puberty, to join other groups. Thus, in contrast to many other primates, it is the bonds between each individual female and the silverback, rather than bonds between the females, that hold the group together.

The attractiveness of the silverback to the females is most apparent during the midday rest period, when the whole group gathers around the dominant male; youngsters play, and adults sleep or groom one another or their infants. Mutual grooming, which keeps fur free from dirt and parasites, is an expression of affinity that helps establish and maintain cooperative partnerships in many species, but it is not as frequent in gorillas as in other social primates. Mothers groom offspring, and immature animals and adult females sometimes groom the silverback; but while adult female kin groom one another, grooming among unrelated females in gorilla groups is very rare. The reason why adult female gorillas do not need grooming as a mechanism to inhibit aggression or to maintain cooperative partnerships, and also do not show obvious differences in status, seems to lie in the relative abundance of food, which reduces competition.

Three-quarters of young female gorillas eventually quit the group in which they were born. They do so because there is little to be gained from staying, and also to avoid inbreeding, for the father may still be breeding when they mature. When they leave, they invariably go immediately to a nearby silverback, who is never more than 200–300 meters away. Usually, however, they do not stay with the first male to whom they transfer. What ultimately determines a female's choice of male is probably a combination of the quality of the habitat in the male's range and the male's fighting ability. Fighting ability is important, because it is the silverback that protects the female and her offspring against predators and against other males – a serious potential threat against which females would otherwise have little defence, being so much smaller than the males and usually in the company of non-relatives unwilling to risk themselves on their behalf.

About a third of infants that die are killed by a male that is not their own father. The most likely explanation for this infanticide is that, once the infant is killed, the mother ceases to lactate (nurse), and is therefore ready to mate again far sooner than if she kept the offspring. The male who kills a one-year-old infant can thus reproduc[e] two years earlier than if he had allowed the female's offspring to live. In most species with high rates of infanticide, such as lions or hanuma[n] langurs, the killers are males who have entered th[e] group and ousted the resident male. In gorillas, the resident seems usually to be powerful enoug[h] to protect females and their infants: the infanticides occur when a female joins a male who is no[t] the father, either because her resident male died [or] was killed, or because she moved to the new mal[e] with her infant. However, such moves are rare when the resident is still alive: it seems to be the case that once a female has found a powerful ma[le] and started to breed with him, she usually stays with him for life.

About half of all males leave the group of thei[r] birth (their natal group) at puberty. They initially travel alone or with other males – sometimes for years – until they acquire females from other groups and establish their own harem. What determines whether a male remains in his natal group or leaves it is probably the degree of acces[s] to fertile females that he is allowed by the dominant silverback. This in turn is decided by the degree of dominance the older male exerts, and how many females there are in a group. If the dominant male is in his prime and the group is small, the subordinate male will find it difficult t[o] mate, and he will leave. When the father is old, [or] there are many females, the young male could have substantial access, and be more likely to sta[y]. Exactly how much monopolization or sharing of females goes on among the males will not be cle[ar] until DNA paternity results appear. With half th[e] males leaving the group of their birth, a little ove[r] a third of groups in both east and west Africa co[n]tain two males.

Males attracting females apparently rely on demonstrating fighting prowess as a means of luring them away from their family group. Lone males appear to be prepared to work harder for females than do leaders of established groups, a[nd] therefore pose more of a threat to their peers. When they meet, two silverbacks will perform elaborate acts to advertise their strength – the famous chest-beating display may be accompanied by hoots, barks, and roars, tearing of vegeta[-]tion, and sideways dashes – all of which are designed to intimidate the rival male.

It appears that, once a silverback has successfully established a breeding harem, he will stay

Right *The intelligence for which the gorilla is renowned seems etched on the pensive expression o[f] this magnificent Lowland male. It was once thought that gorilla groups contained only one adult male, b[ut] around one-third of groups in both East and West Africa have been found to host two full-grown male[s].*

with it for life. With some males having almost permanent access to females, while others have none, competition for females is intense. Fights between leaders and lone males can be severe enough to result in the death of the loser – usually the lone male. The frequency with which fights occur varies enormously with the density of gorillas and the number of males without females, but a resident male might expect at least one lethal encounter in his lifetime, and a fight maybe once a year. Clearly, large size is an advantage in these fights, and in the displays that precede them. Intermale competition is thus almost certainly one explanation for the sexual dimorphism (difference of body form between the sexes) in body size, canine size, and jaw musculature shown by the gorilla. In this, the gorilla matches most other polygynous mammal species (those that mate with several females). Large, fearsome males can mate and guard more females than can smaller, less impressive males. Certainly the gorilla is one of the most sexually dimorphic of all primates, the males being almost twice as large as the females (and, moreover, differently colored).

Right After a morning spent feeding, gorillas habitually take a rest in the middle of the day. At this time members of the group will gather around the silverback **1**. Females with infants **2** tend to be closest, while females without infants **3** stay in the background on the periphery of the group. While a subadult male **4** is merely tolerated, infants **5** will play close to the silverback, remaining within his sphere of protection.

Pressures of Encroachment

CONSERVATION AND ENVIRONMENT

The number of gorillas surviving in the wild today is not accurately known from censuses, but can be estimated from reasonable assessments of average densities and from the amount of habitat remaining. The best available estimate (1996) indicates at least 112,000 Western gorillas, over 10,000 Eastern lowland gorillas, and just a few hundred Mountain and Cross River gorillas. In sum, the world gorilla population is around 125,000. Gabon and the Republic of the Congo, which have large expanses of forest and low human populations that are only increasing slowly, hold three-quarters of the western population (over 40,000 gorillas each), and thus over two-thirds of all gorillas.

Throughout the gorilla's range, the forests on which it depends for survival are being cut down for timber and to make way for agriculture. Formerly, deforestation was not a significant problem because the density of the human population was low enough to practice shifting agriculture, and the abandoned fields, with their regenerating secondary-growth forest, provided abundant food supplies for the gorilla. Yet human populations across the continent as a whole are now increasi[ng] rapidly – during the second half of the 20th century they more than tripled, and almost quadrupled in countries where gorillas live – and as the human population grows, so agricultural land-us[e] becomes permanent.

Another threat is posed by hunting. The demand for bushmeat in western Africa, where wild game is preferred to the meat of domesticated animals, now accounts for the deaths of thousands of animals annually, including hundreds o[f] apes (see feature on The Bushmeat Trade). Although legislation to control the hunting and capture of gorillas exists in all nine African countries that have wild gorilla populations, enforcement of wildlife protection laws here – as elsewhere in the world – is often poor.

At the present rates of deforestation and of human population increase, the only gorillas remaining in 150 years' time will be those in protected national parks. If we estimate one gorilla for every 2sq km (0.8sq mi), some 27,000 migh[t] survive, but no single population would numbe[r] more than 2,600. Nevertheless, there are ground[s] for hope. The Virunga population of Mountain gorillas, in the war-ravaged region of the DRC, Uganda, and Rwanda has remained at several hundred for decades. What is more, many of the

Left Gorilla groups like this tend to lead an insula[r] existence, yet gorillas are not averse to social contact Although displays of aggression are common when groups meet, sometimes groups will just ignore each other or will temporarily mingle.

Right *A young mountain gorilla playing in the ees. Even the largest males will sometimes clamber round in the trees, especially in areas where fruit is bundant. However, it is only the younger, lighter orillas that can swing freely from branch to branch.*

npoverished African countries are doing relatively more to save their natural heritage than their nmensely richer Western counterparts.

In all countries where gorillas are present, it is xtremely difficult for governments to fund major onservation programs, given other, more urgent riorities. Long-term ecological stability is usually acrificed for short-term economic gain, often ith the active encouragement of international evelopment organizations. However, the gorilla's ppeal to tourists may turn out to be its saving race – perhaps the revenue earned from visitors ay persuade local populations to leave the gorilla nd its habitat intact. Ultimately, local people ust learn to value gorillas and the forests in hich they live. For this reason, the funding of rograms for conservation education and for the evelopment of tourism are crucial, especially in reas in which agriculture generates more assured, r more equally distributed, income than tourism as it can do.

Such measures, however, represent a rather esperate rearguard action. In the long run, the ell-being both of the gorilla and of its human eighbors must depend on preventing continued nternational exploitation of Africa's forests and on ncreasing the productivity of the agricultural land ow in use. AHH

Orangutans

THE ELUSIVE ORANGUTANS, ASIA'S ONLY GREAT apes, are now restricted to the steamy jungles of Borneo and Sumatra. The two species stand out among primates in many ways. They are the world's largest arboreal – and slowest-breeding – mammals. They are considered social recluses, yet have a very peculiar sex life and establish local traditions reminiscent of nascent human cultures.

Orangutans (Malay for "forest people") deliberately clamber around in the trees. Too heavy to jump, they cross canopy gaps by swinging a tree back and forth until they can grab on to the next one, always holding on with at least two limbs. This is made possible by their long arms and short legs (30 percent shorter than the arms), which can move freely in many directions, and by their long, hook-shaped hands and feet. With the exception of adult Bornean males, which may spend up to 5 percent of their time on the ground (probably because tigers – the apes' main predators – are now extinct in Borneo), they almost never descend to the forest floor. Incapable of knuckle-walking like the African apes, they keep their hands and feet curled up when moving on the ground.

The Great Red Apes

FORM AND FUNCTION

Molecular studies suggest that the orangutan clade separated from the one that gave rise to the African apes and humans around 14 million years ago (m.y.a.). The late Miocene (12–9 m.y.a.) *Sivapithecus* apes of South Asia were very similar, and are widely considered to be ancestral to the extant orangutans. Pleistocene (ca. 1 m.y.a.) orangutans of giant size are known from Indochina, and subfossil orangutans (ca. 40,000 years ago) about 30 percent larger than modern specimens from caves in Sumatra and Borneo. During the Pleistocene, smaller forms than the extant ones also occurred on Java. It is possible that the earlier orangutans were more terrestrial, but the elaborate arboreal adaptations of living orangutans testify to a long history in the forest canopy.

The red apes' jaws are massive, and their big, flat cheek teeth have wrinkled cusps and thick enamel – the perfect anatomy for tearing open woody fruits or branches with termite nests, grinding hard seeds, and pulling off tree bark. At least once a day, these big apes construct a sleeping platform by breaking and folding several branches and braiding branches and twigs on top. During rain, they add an extra layer to make a waterproof cover.

Staying Close to Home

DISTRIBUTION PATTERNS

Because orangutans are large animals with correspondingly huge appetites, their densities are usually low (around 1 animal per sq km, or 2.6/sq mile), but in fertile river valleys and especially swamp forests, densities may reach 7 per sq km (18/sq mi). Sumatran animals live at higher densities than their counterparts in Borneo, and are also found at higher altitudes. In the absence of hunting, densities are determined by fruit productivity, especially of fruit with fleshy pulp. There is more fleshy fruit in valleys than on slopes or ridges, more in the lowlands than in the mountains, and generally more in Sumatra, which is geologically active, than in Borneo.

Orangutans find travel arduous, usually covering less than 1km (0.6mi) per day. Nonetheless, adult females exploit home ranges of several hundred hectares, and males' ranges may reach several thousand hectares. Neither sex is territorial, and ranges overlap widely, although smaller animals may avoid the company of dominants. Daughters tend to stay near their mothers' ranges when sexually mature, but males may roam around for years before they settle.

Right *The only truly arboreal ape, the orangutan also the largest animal that lives in the forest canopy. swings from tree to tree on its long arms, grasping anches with its hooklike hands and feet. Two nests e made each day: one for a short nap, and a sturdier e for night time.*

Left *A 16-year old female orangutan with 9-month d infant, probably the first she has given birth to. male orangutans bear three or four young in a life-ne. The young are reared exclusively by the mother, maining dependent until about 10 years of age.*

An Insatiable Appetite for Fruit

DIET

rangutans have a huge capacity for food and ill sometimes spend a whole day sitting in a ngle fruit tree, gorging. About 60 percent of all od eaten is fruit – hundreds of species, both pe and unripe. The apes favor fruits with sugary r fatty pulp. Where available, the rather bland uits of large strangling fig trees are the staple od, because they come in huge crops and are asily harvested and digested. Orangutans also gularly eat young leaves and shoots, inverte-rates, and occasionally mineral-rich soil; very ccasionally they capture and devour vertebrates ıch as the Slow loris. When ripe fruit is scarce, ey feed on seeds, or rip off large chunks of ark from trees and lianas. It is especially in those an times that their robust dentition stands em in good stead. When little juicy fruit is avail-ble, orangutans drink water from tree holes;

FACTFILE

ORANGUTANS

Order: Primates

Family: Hominidae

2 (1) species of the genus *Pongo*: Bornean orang-utan (*P. pygmaeus*), Sumatran orangutan (*P. abelii*).

DISTRIBUTION Once widespread in SE Asia and Indochi-na, now confined to N Sumatra and most of lowland Borneo.

HABITAT Lowland and hilly tropical rain forest, including dipterocarp and peat-swamp forest. Orangs are tree dwellers and largely solitary.

SIZE **Head–body length** male: 97cm (38in) female: 78cm (31in); **height** male: 137cm (54in) female: 115cm (45in); **weight** male: 60–90kg (130–200lb) female: 40–50kg (88–110lb).

FORM Coat sparse, long, coarse red hair, ranging from bright orange in young animals to maroon or dark chocolate in some adults. Face bare and black, but pinkish on muzzle and around eyes of young animals. Males have pronounced cheek pads of fatty tissue and a throat pouch. Teeth and jaws are relatively massive for opening and grinding shells and nuts. Sumatran species is thinner, has paler coat, longer hair and longer face than the Bornean. Arms can be up to 2m (6.6ft) long and are used to swing between trees.

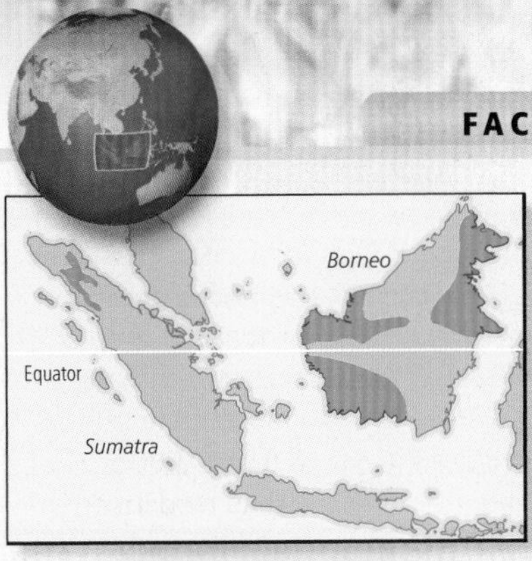

DIET Fruit (e.g. durians, rambutans, jackfruits, lychees, mangosteens, mangoes, figs). Young shoots, bark, insects, woody lianas. Occasionally bird eggs and small vertebrates.

BREEDING Females sexually mature at about 10 years of age remaining fertile until about 30. Single offspring born every 3–6 years after a gestation of 235–270 days. Young are weaned at about 3 years and become completely inde-pendent between 7–10 years.

LONGEVITY Up to about 35 years in the wild, up to 60 in captivity.

CONSERVATION STATUS Critically Endangered (*P. abelii*), Endangered (*P. pygmaeus*). Populations have suffered through habitat destruction and, in the past, collection for zoos and the pet trade.

the ape dips in a hand and sucks the droplets of water as they fall from its hairy wrist.

In some Sumatran swamps, orangutans make stick tools to extricate seeds from the large *Neesia* fruit, which protects its seeds in a mass of stinging hairs. They also use tools to extract honey from the nests of stingless bees, or to probe for ants or termites in treeholes. All individuals in the tool-using populations have the knack, although they vary in the rate of use. In intriguing contrast, members of other populations, sometimes only separated from the tool-users by a river, do not have this ability. Such local tool-use traditions are very similar to those found in wild chimpanzees.

Learning from the Neighbors

SOCIAL BEHAVIOR

Orangutans are slow-growing, slow-breeding, long-lived animals. Their leisurely life history is probably an adaptation to living in habitats with a low rate of unavoidable mortality and periods of very low food abundance. In the wild, females reach puberty at age 10, but do not give birth for another five years. Infants are carried continuously for the first year, and whenever the mother travels until they are about 4 years old. Mothers are very patient with their young, who sleep in their mother's nest until they are weaned at about three years old. Even after that, youngsters often associate with their mother. The interval between births averages eight years. In the wild, females live for about 45 years, and thus have the chance to produce at most four surviving young over a lifetime, perhaps the lowest total of all mammals.

Males become sexually mature ("subadult") at about 12 years old, and over time develop the full regalia of adulthood. Fully adult males are about twice the size of females, with cheek flanges of fibrous tissue that broaden the face, a big throat pouch, and long, capelike hair on their arms and back. They also become capable of producing the booming "long call." There is much variation in how soon these secondary sexual characteristics appear: the fastest-developing subadult males may reach full adult status in less than a decade, while others seem to be arrested for two decades or more before finally growing flanges. This developmental arrest, which probably represents an adaptive mating strategy, may be more common in Sumatra, where the ratio of subadult to adult males in the populations is three times higher than in Borneo.

Orangutans are rather solitary animals, particularly those in Borneo. Adults generally travel and forage independently. Offspring gradually become

Right *The male orangutan is an impressive sight. His size, cheek flaps, throat pouch, and long hair enhance his display in confrontations with other males who try to encroach too far into his range. The throat pouch produces vibrations that can be heard more than a mile away. Orangutans in captivity can become obese, making the flaps and pouch even larger.*

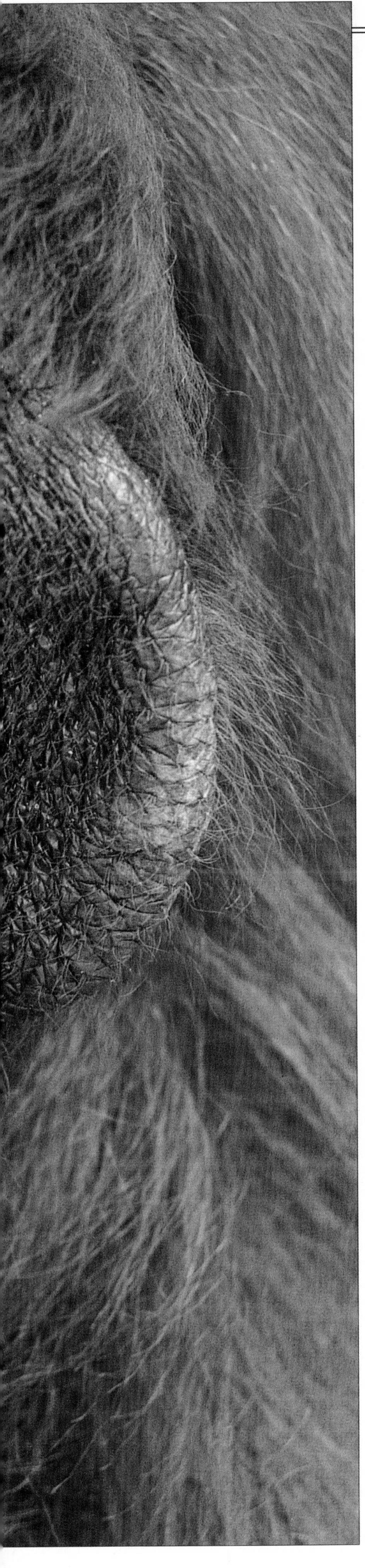

more independent after weaning. By adolescence, males have usually broken ties with their mothers, but females return frequently. Juveniles and adolescents sometimes play together for a few hours, or even travel around in pairs or tag onto family units. When several adult orangutans meet, as when drawn to the same fruit tree, they hardly interact socially and depart separately after eating.

Sumatran orangutans are more sociable. All classes of animal, except low-ranking adult males, are social and travel together. Sumatran orangutans also eat more fruit and invertebrates, and less bark, than those in Borneo, and also have a monopoly on tool-use. These differences ultimately stem from their higher population densities, which in turn reflect the greater productivity of the habitat. In more productive habitats, the costs of traveling and feeding together are lower, and animals can reap the social benefits associated with group life, such as learning tool-using skills.

Orangutans recognize individually all the animals whose ranges they regularly overlap, and form social relationships. Females form clusters of preferential associations, which also tend to be reproductively synchronized. Although subadult males occasionally bond, male relationships are largely competitive. The "long calls" of adult males, emitted several times per day, keep subordinate males away, but when adult males meet they indulge in violently aggressive displays, sometimes leading to chases and fights on the ground. Adult males tolerate subadult males so long as they maintain a respectful distance

Subadult males try to associate with potentially receptive females whenever the opportunity arises, but a female that is ready to conceive seeks out the local dominant adult male, which generally succeeds in preventing most subadult males from mating with her. The nonpreferred males, adult or subadult, therefore tend to force matings when they encounter a lone female, often viciously biting the fiercely resisting female to restrain her.

Female-initiated consortships with the dominant adult male last a few days in Borneo, but may last weeks in Sumatra. Probably related to this is the fact that over half of Sumatran matings are cooperative, whereas some 90 percent of those in Borneo are forced. The benefit to the female of her costly mate preference is still unsolved; it may be her way of securing good genes for her offspring, but probably involves some form of protection by the resident dominant adult male.

Among primates, captive orangutans are among the top scorers in comparative intelligence experiments. Wild orangutans rely on their intelligence to develop complex feeding techniques, sometimes involving the use of tools, which give them access to foodstuffs not available to most other rainforest denizens. They are also excellent imitators, quickly learning skills from others, including the use of tools. Their propensity to mimic the behaviors of others rather than invent variants

Above Orangutans get much of their liquid from eating fruit, but are also partial to water. This Sumatran orang is in the Gunung Leuser reserve, Indonesia.

anew has given rise to local traditions. At different sites, orangutans employ different nest-building techniques, make different vocalizations, and have different ways of handling food.

At Risk of Extinction

CONSERVATION AND ENVIRONMENT

Humans have been a predator and competitor of the orangutan ever since anatomically modern humans invaded Southeast Asia some 40,000 years ago. Hunting by humans is now widely held responsible for the extinction of this ape in most of its former range. Subsistence hunting in historic times is probably responsible for the patchy distribution on both Borneo and Sumatra.

The orangutan now faces extinction in the wild. Orangutans are sensitive to selective logging, and disappear altogether when logging becomes intensive. Most forest outside reserves is being readied for farming or has disappeared already. Thus, the only effective way to protect the orangutan is to preserve as much of its habitat as possible within nature reserves and national parks.

Malaysia and Indonesia have both set up major reserves . However, economic and political turmoil in the late 1990s in Indonesia, which holds over 90 percent of wild orangutans, saw the onset of illegal logging in protected areas. This development came in the wake of devastating forest fires in Borneo, which is becoming increasingly sensitive to long, El Niño-related droughts.

Orangutan numbers are down by over 92 percent compared with a century ago, and fell by fully half in northern Sumatra between 1993 and 2000. Remaining populations are restricted to small islands that will remain separated, since orangutans are poor colonizers. Hence, serious protection and active management of the remaining forests is needed to avert the orangutan's extinction in the wild. CvS/JMacK

PHOTO STORY

Last Chance for the Orang?

❶ A captive orangutan at a rehabilitation center. Throughout their range, on the islands of Sumatra and Borneo, hundreds of animals are held in such facilities, where they are prepared for release into the wild. As the original habitats of orangutans dwindle, the fate of the whole species may rest on these programs. Yet the survival rates of released orangs may be as low as 12 percent.

❷ Young orphaned orangutans a brought to the centers after their parents have been killed by logge Up to the age of 7 or 8, they will happily congregate in small group Rehabilitation programs encourag this social tendency, in the hope that it will foster vital survival skill The young orangs are not releasec until they are at least 5 years old.

❸ A keeper and his charge intera at the Wanariset Rehabilitation Project in eastern Borneo, the largest of the centers. Opinions differ on how much human contac the captive animals should be exposed to. Great care is given to disease prevention, for fear that the animals will contract human infections such as tuberculosis anc hepatitis and transmit them to the wild. All new arrivals are vaccinate and quarantined.

4 The skills that a young orangutan acquires from its mother over seven years of dependency are many and varied. It not only learns to navigate through the forest with great agility, but is also taught to find up to 400 different types of foods, including durians, tree bark, and Pandanus *leaves. Released animals struggle to exploit their habitat with the same efficiency.*

5 A juvenile orangutan rehabilitated into the Meratus Mountains, a remote tract of forest in central Borneo. Ex-captives are now no longer set free into areas that are still populated by wild orangs, as this put too great a strain on limited resources. Because of the difficulties in evaluating the progress of released animals, it is too early to say whether these programs have been successful. Certainly, some released orangs have bred in the wild.

6 Commercial logging (much of it illegal) and the forest fires that devastated Indonesia in the late 1990s are at the root of the alarming decrease in orangutan populations. One study in the Leuser region of Indonesia showed that their numbers had almost halved over a period of just six years.

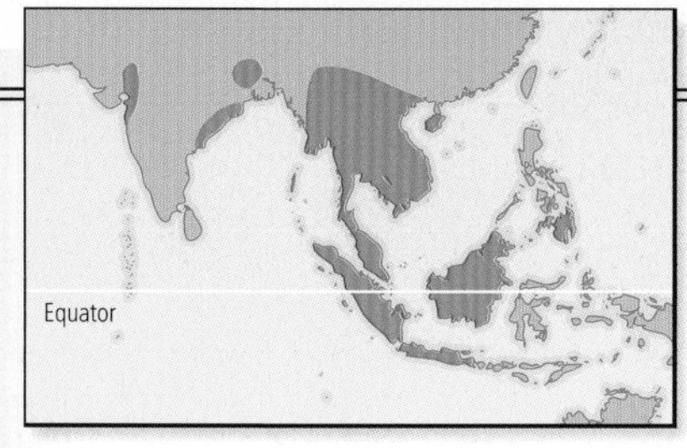

TREE SHREWS

Tree shrews are small, squirrel-like mammals found in the tropical rain forests of southern and southeastern Asia. The first reference to them was in 1780 by William Ellis, a surgeon who accompanied Captain Cook on his exploratory voyage to the Malay Archipelago. It was Ellis who coined the name "tree shrew," a somewhat inappropriate term, as they are quite different from the true shrews (Soricidae) and, as a group, are not particularly well adapted for life in trees – indeed some tree-shrew species are almost completely terrestrial in habit. Most tree-shrew species are semiterrestrial, rather like European squirrels, a resemblance that is underlined by the fact that both tree shrews and squirrels are covered by the Malay word "tupai," from which the genus name Tupaia is derived. At first classified as insectivores, tree shrews were then thought to be primates for 50 years, but recent research distinguishes them from both orders and they are here considered as the only members of the order Scandentia.

None of the six genera covers the entire geographical range of the order, although the genus *Tupaia* is the most widespread. The greatest number of species is found on Borneo, where 10 of the 18 recognized species occur. This concentration is partly a consequence of the large size of the island and the resulting wide range of available habitats, but it is also possible that Borneo was the center from which the adaptive radiation of modern tree-shrew species began.

More Squirrel than Shrew

FORM AND FUNCTION

Tree shrews are small mammals with an elongated body and a long tail, which – except in the case of the Pen-tailed tree shrew – is usually covered with long, thick hair. Their body fur is dense and soft. They have claws on all fingers and toes; the first digits diverge slightly from the others. Their

Below *Representative species of tree shrews.* ***1*** *The largely arboreal Pygmy tree shrew* (Tupaia minor) *"sledging" along a branch and leaving a scer from its abdominal gland.* ***2*** *The arboreal Pen-tailed tree shrew* (Ptilocercus lowii) *holds a captured insect between both hands while devouring it; both its hands and feet are relatively larger than those of the other tree shrews. As well as being the only noc turnal species, it is the only living representative of t subfamily Ptilocercinae.* ***3*** *The Philippine tree shrew* (Urogale everetti) *is the largest species: it spends mc of its time at ground level rooting through debris.* ***4*** *The semi-terrestrial Common tree shrew* (Tupaia g *"chinning" to leave a scent trail from the sternal glar which is located on its chest.* ***5*** *The Terrestrial tree shrew* (Lyonogale tana) *is a large-bodied species: lik the Philippine tree shrew, it too finds most of its foo in litter on the forest floor, rooting with its snout and turning over objects such as stones.* ***6*** *The mainly arboreal Northern smooth-tailed tree shrew* (Dendrogale murina) *snatching an insect from the air with both hands.*

outs range from short to elongated. The ears ve a membranous external flap, which varies in e from species to species and is usually covered th hair. In the Pen-tailed tree shrew the ear flaps bare and larger than those found in other tree rews, doubtless because this nocturnal species ies more heavily on its hearing to find insect ey and avoid predators at night. The species that mainly arboreal, such as the Pygmy tree shrew, small, have short snouts, more forward-facing es, weakly developed claws, and tails that are nger than their combined head-and-body ngth. Terrestrial species, such as the Philippine e shrew, are large, have elongated snouts, well-veloped claws that are suited to rooting after sects, and tails that are shorter than their head-d-body length. Their terrestrial habits permit eater body size with less need for a long tail to lance in the trees.

In most anatomical features, tree shrews show little obvious specialization, although there are some unusual features in the skull and dentition (see right). Except for the Pen-tailed tree shrew, they generally have laterally placed eyes, which are moderately large relative to body size and permit good vision. There is a well-developed subtongue (sublingua).

Hand-to-Mouth Existence

DIET

Like European squirrels, most tree shrews spend more time foraging on the ground than in trees. They scurry up and down tree trunks and across the forest floor with characteristic, jerky movements of their tails, feeding on a wide variety of small animal prey (especially arthropods, their principal food) as well as on fruits, seeds, and other plant matter. All species – except the most

ORDER: SCANDENTIA

Family: Tupaiidae. 18 species in 6 genera and 2 subfamilies

Distribution NW India to Mindanao in the Philippines, S China to Java, including most islands in the Malayan Archipelago. Mainly in tropical rain forest.

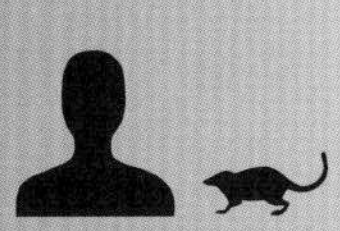

Size Ranges from Pen-tailed tree shrew, with **head–body length** 10–14cm (4–5.5in), **tail length** 9–13cm (3.5–5.1in), **weight** 50g (1.8oz) to Terrestrial tree shrew, with **head–body length** 17–23cm (6.7–9in), **tail length** 14–24cm (5.5–9.4in), **weight** 300g (11oz).

SUBFAMILY TUPAIINAE

Genus *Tupaia* 11 species, including the Pygmy tree shrew (*T. minor*), Belanger's tree shrew (*T. belangeri*), Common tree shrew (*T. glis*)
Genus *Anathana* 1 species, the Indian tree shrew (*A. ellioti*)
Genus *Urogale* 1 species, the Philippine tree shrew (*U. everetti*)
Genus *Dendrogale* 2 species, the Smooth-tailed tree shrews
Genus *Lyonogale* (sometimes included in *Tupaia*) 2 species, including the Terrestrial tree shrew (*L. tana*)

SUBFAMILY PTILOCERCINAE

Genus *Ptilocercus* 1 species, the Pen-tailed tree shrew (*P. lowii*)

See species table ▷

SKULLS AND DENTITION

The tree-shrew skull is that of a strictly quadrupedal mammal. The *foramen magnum* (the opening through which the spinal cord passes) is directed backward, whereas in the more fully arboreal living primates it is typically directed obliquely downward. The skull is longest in terrestrial species that root in leaf-litter. Compared here in dorsal view are the skulls of ground-living Philippine and Terrestrial tree shrews **a**, **b**, the semi-arboreal Common tree shrew **c**, and the tree-dwelling Pygmy tree shrew **d**. Common features include the postorbital bar and an inflated auditory bulla. The dental formula is I2/3, C1/1, P3/3, M3/3 = 38. The canine teeth are relatively poorly developed; the primitive sharp-cusped molars reflect an insectivorous diet, and the forward-projecting lower incisors are used in feeding and grooming.

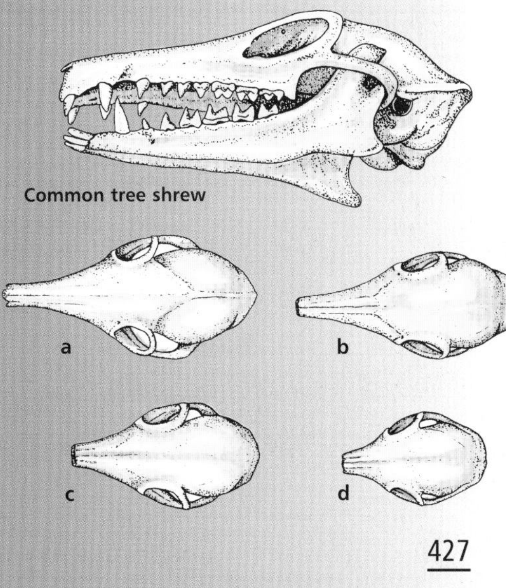

Insectivores, Primates – or Neither?

Misleadingly named, and originally placed in the order Insectivora, tree shrews would probably have remained in obscurity but for the suggestion made by the anatomist Wilfred Le Gros Clark in the 1920s that these relatively primitive mammals might be related to the primates. Comparing the structure of the skull, brain, musculature, and reproductive systems, he concluded that the tree shrews should be regarded as the first offshoot from the ancestral primate stock. This interpretation was accepted by George Gaylord Simpson, who included the tree shrews in the order Primates in his influential 1945 classification of the mammals.

Thereafter, many studies of tree shrews were conducted in the hope of clarifying the evolutionary history of the primates (including man), and they became widely regarded as present-day survivors of our primate ancestors. Today's consensus is that tree shrews are not specifically related to either primates or insectivores, but represent a quite separate lineage in the evolution of the placental mammals.

This view has been reached by several means. The first objection to the "primate" interpretation is that tree shrews may have come to resemble primates through entirely separate (convergent) evolution of certain features because of similar functional requirements. For instance, primates are typically arboreal, while insectivores (e.g. shrews, hedgehogs, moles, tenrecs, etc) are typically terrestrial, so it is possible that tree shrew/primate similarities evolved through convergent adaptations for arboreal life. Various apparently primate-like features – shortening of the snout, forward rotation of the eye-sockets, and greater development of the central nervous system associated with large eyes – are largely confined to the most arboreal of the tree-shrew species, such as the Pygmy tree shrew. So, these special characters cannot reliably be regarded as vestiges from a common ancestral stock of tree shrews and primates, particularly as the ancestral tree shrew was probably closest to the modern semi-terrestrial species.

The second, more important, objection is based on a distinction between similarities shared because they derive from a specific ancestral stock and similarities shared merely because of retention of characters from the more ancient ancestral stock of placental mammals. This second kind of similarity does not itself indicate any specific relationship between tree shrews and primates. A particularly good example is the cecum, a blind sac in the digestive tract at the junction of the small and large intestines, housing bacteria that assist in the breakdown of plant food. Tree shrews and primates typically possess a cecum, whereas it is absent from insectivores. However, the cecum is widespread among mammals, both placentals and marsupials, and is even present among reptiles. It therefore seems likely that the cecum was already present in the earliest placental mammals, so its retention provides no evidence whatsoever for a specific ancestral connection between tree shrews and primates; it is the insectivores that are specialized.

It has also emerged that in some respects tree shrews are very different from primates, particularly in reproductive characters. The development of the placenta in tree shrews is quite unlike that in any primate species, and the offspring are born in a naked, helpless condition that contrasts markedly with the advanced condition of newborn primates. In tree shrews, parental care is extremely limited and also far removed from the elaborate parental care of the primates.

A major difficulty in reconstructing the evolution of tree shrews has been the lack of convincing fossil evidence, a gap now filled to some extent by the discovery in Indian Miocene Siwalik deposits of tree-shrew fossils (*Palaeotupaia sivalensis*) dating back some 10 million years. These support the interpretation that tree shrews derived from a semi-terrestrial ancestral form with moderate development of the snout, but they are far too recent to indicate early relationships.

It is probably wisest to regard the tree shrews as an entirely separate order of mammals that branched off very early during the radiation of placental mammal types. This interpretation is also supported by a wealth of biochemical evidence from immunological cross-reactions of proteins and from comparison of sequence data for proteins and DNA.

It now seems that, rather than being survivors from early primate stock, the tree shrews may well be closer to the common ancestors of placental mammals in general.

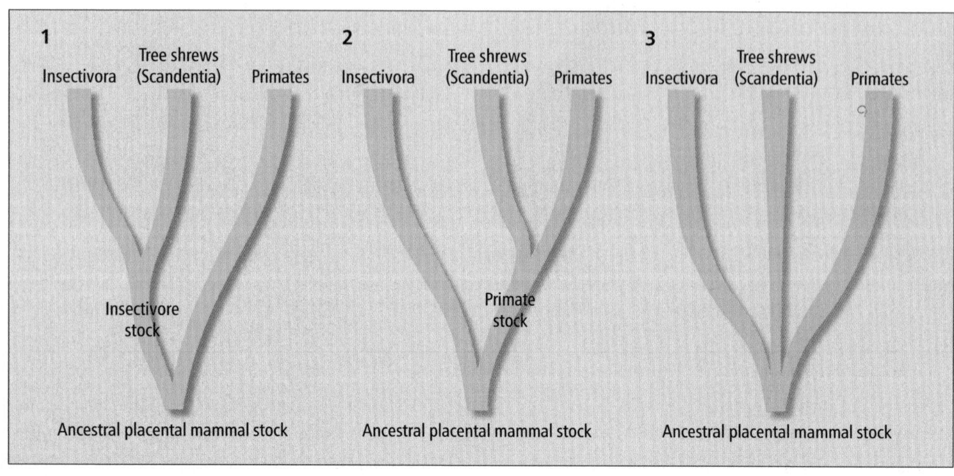

Below *Three possible lines of tree-shrew descent.* ***1*** *from Insectivore stock;* ***2*** *from primate stock;* ***3*** *from common placental mammalian stock.*

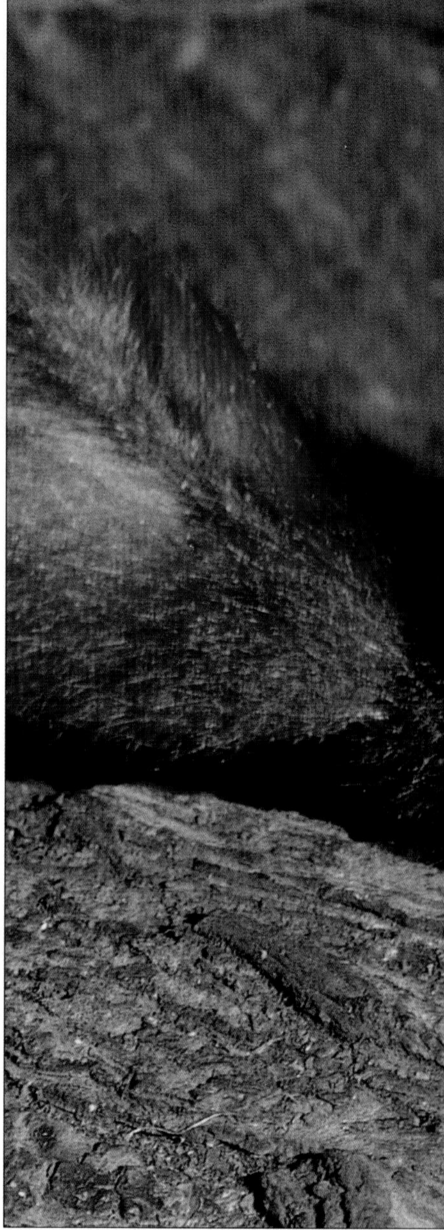

arboreal – spend a great deal of time rooting in leaf litter with the snout and hands. Tree shrews typically prefer to catch food with their snouts, and only use their hands when food cannot be reached otherwise. However, flying insects may caught with a rapid snatch of one or both hands and all tree-shrew species hold food between the front paws when eating. The larger tree shrews probably eat small vertebrates in the wild, for example small mammals and lizards, as in captivity they have been seen to overpower adult mice and young rats and to kill them with a single bite to the neck.

Leaving the Young Alone

SOCIAL BEHAVIOR

Most tree shrews (like squirrels) are exceptions to the general rule that small mammals are nocturnal. The Pen-tailed tree shrew, however, is exclusively nocturnal, and many of its distinctive

tures (larger eyes with a reflecting tapetum, ge ears, long whiskers, gray/black coloration) n be attributed to this difference. It has been ggested that the Smooth-tailed tree shrews ght be intermediate in exhibiting a crepuscular ttern, with peaks of activity at dawn and dusk, t little is known of their natural behavior.

Most tree-shrew species nest in tree hollows ed with dried leaves. The gestation period is –50 days, according to species, and the 1–3 ung are born without fur, with closed ears and es. The ears open after about 10 days and the es in 3 weeks.

Tree shrews are unusual among placental mam-als for the extremely restricted nature of their rental, especially maternal, care. Laboratory idies have shown that in at least three species e Pygmy, Belanger's, and Terrestrial tree shrews) e mother gives birth to her offspring in a sepa-te nest which she visits to suckle the young only once every two days. (Early attempts to breed tree shrews in captivity failed largely because only one nest was provided.) Fieldwork has now confirmed that the Terrestrial tree shrew shows this pattern under natural conditions. The mother's visits are very brief (5-10 minutes), and in this short space of time she provides each infant with 5–15g (0.2–0.5oz) of milk to provision it for the next 48 hours. The milk contains a large amount of protein (10 percent), which permits the young to grow rapidly, and an unusually high fat concentration (25 percent), which enables them to maintain their body temperature in the region of 37°C (98.6°F) despite the absence of the mother from the nest. However, the infants are relatively immobile in the nest, so the milk contains only a small proportion of carbohydrate (2 percent) for immediate energy needs.

In all three shrew species so far studied, the infants have been found to stay in the nest for about a month, after which they emerge as small replicas of the adults. The young continue to grow rapidly, and sexual maturity may be reached by the age of 4 months. Between the birth of her offspring and their eventual emergence from the nest, the mother spends a total of only one-and-a-half hours with the infants, and shows no toilet care during her brief suckling visits. Indeed, maternal care in tree shrews is so limited that if an infant is removed from its nest and placed just alongside, the mother will completely ignore it. She only recognizes her offspring in the nest because of a scent mark which she deposits on them with her sternal gland; if the scent is wiped off she will devour her own infants!

Above *Holding food between the hands while eating is a behavior common to all of the tree shrews. The Terrestrial tree shrew from the East Indies has hands with particularly strong claws.*

Above *A male Belanger's tree shrew. Tree shrews can be aggressive toward others of the same gender, and male home ranges are not thought to overlap greatly. However, a male's home range may overlap with the home range of more than one female.*

Below right *The pair bond between tree shrews appears to be relatively permanent. As well as marking out territorial areas, animals living in pairs have been observed to scentmark each other, and mothers will scentmark either the nest that contains their young or the young themselves.*

The number of pairs of teats is characteristic of each species and is directly linked to the typical number of infants in a litter (one, two, or three pairs of teats broadly corresponding to one, two, or three offspring).

Tree shrews tend to breed over a large part of the year, although a definite seasonal peak of births has been reported in some cases. The short gestation period and rapid maturation of the offspring mean that tree shrews can breed rapidly if the conditions are right, and they are able to colonize new areas quite quickly.

Of all the species, the Common tree shrew has been most closely observed in the wild, although some work has also been done on the Terrestria tree shrew. This Common tree shrew forms loos social groups typically composed of an adult pa and their apparent offspring. The members of each group occupy all or part of a common hom range covering approximately one hectare (2.5 acres), but they usually move around independently during the daytime, predominantly on o near the forest floor. The home range of each group seems to be defended as a territory, as there is very little overlap between adjacent hom ranges and fights have been observed on their boundaries.

Tree shrews have a rather limited range of cal All species, when surprised in the nest or durin attacks on other tree shrews, produce a hoarse, snarling hiss with the mouth held wide open. Infants produce a similar sound when disturbe in the nest. A variety of squeaks and squeals is produced during fights, culminating in really piercing squeals when one combatant is beaten. *Tupaia* species also produce a persistent chatteri call when mildly alarmed, and there is some evidence that this acts as a mobbing call announci the presence of potential predators.

Inconspicuous but not Invulnerable

CONSERVATION AND ENVIRONMENT

Tree shrews are relatively inconspicuous mammals, and their interactions with humans are ve limited. They may be pests in fruit plantations, and they are occasionally found in and around human habitations, but they do not appear to occupy any significant place in the human econ my or in mythology. Nevertheless, they have not been spared the effects of human activity, particularly where it threatens sensitive habitat, and several of the rarer species are now at risk as a result of habitat loss. RD

COMMUNICATING BY SCENT

Tree shrews engage in extensive scentmarking. The details vary from species to species, but in all cases the process involves special scent glands, urine, and sometimes even feces. Belanger's tree shrew possesses two glandular areas on the ventral surface of the body: the sternal gland is used in "chinning," the tree shrew standing with stiffened legs and rubbing the gland over the object to be marked; the abdominal gland is used in "sledging," in which the tree shrew slides down a branch while pressing its abdomen against the surface. Tree shrews also scentmark by depositing droplets of urine, and the Terrestrial tree shrew performs a kind of dance in which the hands and feet are impregnated with urine previously deposited on a flat surface. In captivity, the products of these scentmarking activities accumulate to form a fatty, pungent, orange-yellow crust. Captive tree shrews also deposit feces in specific places in the cage, suggesting that droppings may play a role in territorial demarcation in the wild.

ree Shrew Species

bfamily Tupaiinae

ve in daytime (*Dendrogale* perhaps at ight). Eyes laterally placed. Five genera.

GENUS *TUPAIA*

alay Peninsula, Indonesia, Philippines, ochina. Semiterrestrial or arboreal; lium-sized (160g/6oz) or small g/1.5oz).
M: conspicuous cream or buff shoul- stripes always present; snout short arboreal forms) or slightly elongated niterrestrial forms); canine teeth mod- ely well developed; ear flaps small; ales have 1, 2, or 3 pairs of teats.

anger's tree shrew
aia belangeri

ochina north of 10°N, Myanmar, NE ia, China. Semiterrestrial.
RM: tail equal to length of head and ly combined; weight 160g/6oz; coat ges from olivaceous to very dark brown ve, and from creamy-white to orange- below; females have 3 pairs of teats.

mmon tree shrew
aia glis

lalay Peninsula, Sumatra and surround- islands.
RM: habit, weight, coat, and tail features *T. belangeri*; females have 2 pairs of teats.

ng-footed tree shrew En
aia longipes

neo.
RM: habit, weight, coat, and tail features *T. belangeri*; females have 3 pairs of teats.

Montane tree shrew
Tupaia montana
Montane or Mountain tree shrew

N Borneo (mountains).
FORM: habit, weight, coat, and tail features as *T. belangeri*; females have 2 pairs of teats.

Nicobar tree shrew En
Tupaia nicobarica

Nicobar Islands. Semiterrestrial to arboreal.
FORM: tail longer than length of head and body combined; weight and coat as *T. belangeri*; females have 1 pair of teats.

Painted tree shrew
Tupaia picta

N Borneo (lowlands).
FORM: habit, weight, coat and tail as *T. belangeri*, except for dark stripe down back; females have 2 pairs of teats.

Palawan tree shrew Vu
Tupaia palawanensis

Philippines.
FORM: habit, weight, coat, and tail features as *T. belangeri*; females have 2 pairs of teats.

Rufous-tailed tree shrew
Tupaia splendidula

SW Borneo, NE Sumatra.
FORM: habit weight, coat, and tail features as *T. belangeri*; females have 2 pairs of teats.

Pygmy tree shrew
Tupaia minor

Borneo, Sumatra, S Malay Peninsula and surrounding islands. Arboreal.
FORM: tail length greater than head–body length; weight 45g (1.5oz); coat olivaceous above, off-white below; females have 2 pairs of teats.

Indonesian tree shrew
Tupaia javanica
Indonesian or Javan tree shrew

Java, Sumatra, Bali, and Nias.
FORM: habit, weight, coat, tail and teats as *T. minor*.

Slender tree shrew
Tupaia gracilis

N Borneo and surrounding islands.
FORM: habit, weight, coat, tail, and teats as *T. minor*.

GENUS *ANATHANA*

Indian tree shrew LR
Anathana ellioti

India south of the Ganges. Semiterrestrial. Medium sized (160g/6oz).
FORM: coat brown or gray-brown above, buff below; shoulder stripe tight buff or white; pale markings around eyes; tail equal in length to head and body combined; snout short; canine teeth poorly developed; ear flaps well developed; females have 3 pairs of teats.
CONSERVATION STATUS: LR Near Threatened.

GENUS *UROGALE*

Philippine tree shrew Vu
Urogale everetti

Mindanao, Dinigat, and Siargao. Terrestrial. Large (350g/12oz).
FORM: coat dark brown above, yellowish or rufous below; shoulder stripe pale; tail much shorter than length of head and body combined, and covered in closely set rufous hairs; snout elongated; second pair of upper incisors enlarged, third pair of lower incisors reduced; females have 2 pairs of teats.

GENUS *DENDROGALE*

Arboreal. Small (50g/2oz).
FORM: no shoulder stripes present; tail slightly longer than length of head and body combined, covered with fine smooth hair; snout short; ear flaps large; females have 1 pair of teats.

Southern smooth-tailed tree shrew Vu
Dendrogale melanura
Southern or Bornean smooth-tailed tree shrew

N Borneo.
FORM: coat dark brown above, pale buff below: facial streaks inconspicuous, but orange-brown eye rings prominent; claws sharp.

Northern smooth-tailed tree shrew
Dendrogale murina

E Thailand, Vietnam, and Cambodia.
FORM: coat light brown above, pale buff below; dark streak on each side of face running from snout to ear and highlighted by paler fur above and below; claws small and blunt.

GENUS *LYONOGALE*

Terrestrial. Large (300g/11oz).
FORM: conspicuous black stripe along back: shoulder stripe pale; tail bushy and shorter than length of head and body combined; snout elongated; canine teeth well developed; claws robust; females have 2 pairs of teats.

Terrestrial tree shrew
Lyonogale tana

Borneo, Sumatra and surrounding islands.
FORM: coat dark red-brown above, orange-red or rusty-red below; front of dorsal stripe highlighted by pale areas either side; shoulder stripe yellowish; claws robust and elongated.

Striped tree shrew
Lyonogale dorsalis

NW Borneo.
FORM: coat dull brown above, pale buff below; shoulder stripe creamy buff or whitish; claws less robust and shorter than in *L. tana*.

Subfamily Ptilocercinae

Nocturnal. Eyes forward-facing, giving binocular vision. One genus.

GENUS *PTILOCERCUS*

Pen-tailed tree shrew
Ptilocercus lowii

S Malay Peninsula, NW Borneo, N Sumatra and surrounding islands. Arboreal. Small (50g/1.8oz).
FORM: coat dark gray above, pale gray or buff below; dark facial stripes running from snout to behind eye; no shoulder stripe present; tail considerably longer than combined length of head and body; covered for entire length with scales, except for tuft of hairs at tip; snout short; upper incisors enlarged; ear flaps large, membranous, and mobile; females have 2 pairs of teats.

En Endangered   Vu Vulnerable
LR Lower Risk

EE SHREW CLASSIFICATION

the classification of tree shrews, phasis is placed on coat coloration tterns, tail length and shape, the m of the ears, the development of snout and claws, and the number teats in females.
Tupaia contains the least specialized e shrews; their wide geographical tribution is associated with consider- le speciation – at least 11 distinct ecies are recognized.
The single species of *Anathana* curs in India south of the Ganges, iich divides it from *T. belangeri* to the st, from which it differs in having rel- vely larger ear flaps and more com- x molar tooth cusp patterns.
The two predominantly ground- ng species here placed in the genus onogale have elongated snouts and oust claws on the forefeet.
The single known species of *Urogale* the largest of living tree-shrew species d the sole representative on the and of Mindanao. The dentition is striking in that the second pair of incisors in the upper jaw are prominent and caninelike –dwarfing the canines – while in the lower jaw the third pair of incisors are greatly reduced.

The two species of the genus *Dendrogale* are distinguished by their fine tail fur, diminutive body size, and characteristic facial markings.

The Pen-tailed tree shrew (*Ptilocercus lowii*) is placed in a separate subfamily. It is the only nocturnal tree shrew and appears to be almost exclusively arboreal. The eyes are relatively forward-facing, giving binocular overlap. In the upper jaw the anterior incisors are distinctively enlarged. Pen-tailed tree shrews lack a shoulder stripe, in contrast to all others except *Dendrogale* species.

The Smooth-tailed tree shrews are in many ways intermediate between the other tree shrews and *Ptilocercus*, for example in the sparseness of the hair on their tails.

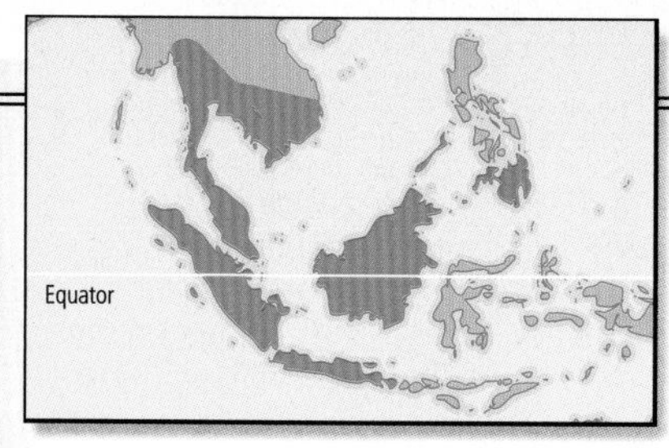

COLUGOS

The curious colugos or flying lemurs of the rain forests of Southeast Asia are neither true fliers nor true lemurs. They are gliding mammals belonging to a distinct order, the Dermoptera ("skinwings"), which includes only one known family, but may have contained another, namely, the extinct Plagiomenidae from the late Paleocene and early Eocene of North America, 60–70 million years ago.

Two species of colugo are known, the Malayan colugo and the slightly smaller and more primitive Philippine colugo. The taxonomic relationships of the colugos are obscure; at different times they have been included with the insectivores and the bats, and to confuse matters further for many years they were known by the family name Galeopithecidae ("cloaked monkeys").

Nature's Hang Gliders

FORM AND FUNCTION

Without doubt the colugo's most distinctive characteristic is the gliding membrane or patagium, which stretches from the side of the animal's neck to the tips of the fingers and toes and continues to the very tip of the tail. About the size of domestic cats, they are so arboreal that a colugo on the ground is almost helpless. The head is broad, somewhat like a greyhound's in appearance, with short, rounded ears and a blunt muzzle. Colugos have large eyes, as befits nocturnal animals, and their stereoscopic vision gives them the depth perception necessary for judging accurate landings following a glide.

Colugos have limbs of equal length, with strong, sharp claws that anchor the animal as it rests on a trunk or hangs beneath a branch, rather in the manner of South American sloths. With the wide patagium (flight membrane) and nonopposable thumbs, colugos are clumsy climbers, ascending tree trunks in a series of slow "hops." Nevertheless, this is a small price to pay for the mobility the gliding membrane affords these animals in moving between trees through the tall rain forest. No other gliding mammal has such an extensive flight membrane. With the patagium outspread, the colugo assumes the shape of a kite, and can execute controlled glides of 70m (230ft) or more with little loss of height.

The colugos' remarkable gliding mechanisms are effective for long-distance travel, but make them vulnerable to fast-flying birds of prey. It is probably because of this vulnerability that both colugos and flying squirrels are largely nocturnal. Colugos spend the day in holes or hollows of trees, or hanging beneath a bough or against a tree trunk with the patagium extended like a cloak. In coconut plantations, colugos may even curl up in a ball among the palm fronds.

A female colugo usually gives birth to a single young – rarely two – after a gestation of 60 days. Lactating females with unweaned young have been found to be pregnant, so it is possible that births may follow in rapid succession. The infant is born in an undeveloped state, like a marsupial, and until it is weaned it is carried on the belly of the mother, sometimes even when she "flies." The patagium can be folded near the tail into a soft, warm pouch for carrying the young. When the female rests beneath a bough with the patagium outstretched, the infant peers out from an exotic hammock. Young colugos emit ducklike cries, and the cry of the adult is said to be similar, although it is only rarely used.

Specialized Stomach

DIET

The colugo's diet seems to consist mainly of leaves, shoots, buds, and flowers, and perhaps also soft fruits. Animals have also been seen licking sap from tree trunks. When feeding, the colugo pulls a bunch of leaves within reach with its front foot, then picks off the leaves with its strong tongue and lower incisors. The stomach is spec ized for ingesting large quantities of leafy vegeta tion, and has an extended pyloric digesting region, the part near the exit to the intestines. Like many arboreal mammals, colugos probably obtain sufficient water from their food and by licking wet leaves.

Hunted by Eagles

CONSERVATION AND ENVIRONMENT

Like many other rainforest species, colugos are threatened by the loss of their forest habitats through logging and conversion to agricultural use. The Malayan colugo is regarded as a pest o coconut plantations, because it eats budding coconut flowers. In addition to habitat destruction the Philippine colugo is threatened by hun ing for its soft fur and its meat, which is considered a local delicacy. Its natural predators include one of the world's rarest birds, the Phili pine monkey-eating eagle (*Pithecophaga jefferyi*); has been estimated that colugos constitute up t 90 percent of the eagles' diet. This suggests that colugos may actually be more common and mo diurnal than is generally supposed. However, th secretive and nocturnal lifestyle make it difficult to assess their status accurately. Nevertheless, as with many other rainforest mammals in Southea Asia, their best hope for long-term survival will l the provision of adequate protection by the esta lishment of reserves within their range. KMa

__Right__ The dappled gray and brown coloration of the Malayan colugo provides a highly effective camouflage against tree bark.

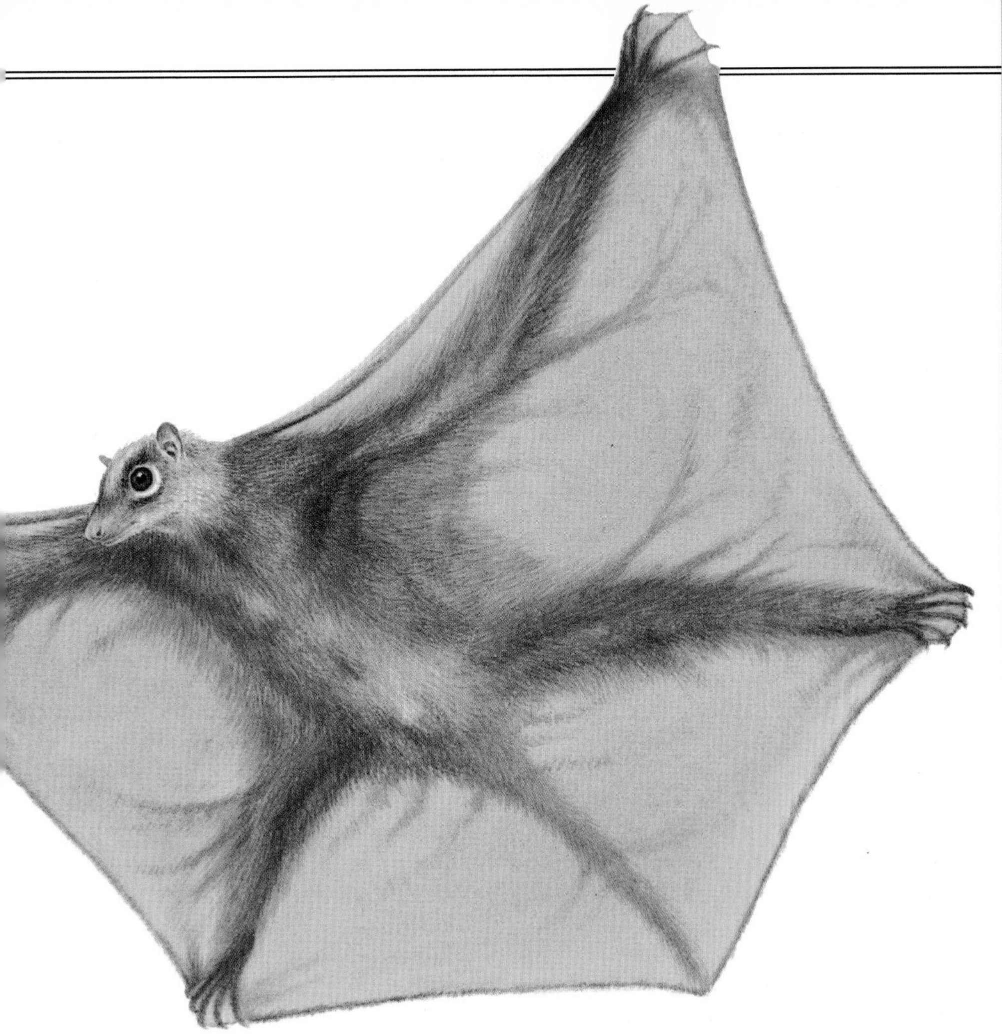

◀ ***Left*** *The Philippine colugo in "flight" with its gliding membrane (patagium) fully outstretched. During a measured glide between trees of 136m (450ft), one colugo was seen to lose only 12m (39ft) in height.*

ORDER: DERMOPTERA

Family: Cynocephalidae. 2 species of the genus *Cynocephalus*

Distribution SE Asia

MALAYAN COLUGO *Cynocephalus variegatus*

Malayan colugo or Malayan flying lemur

Malaya, Thailand, Tenasserim, Sumatra, Borneo, Java, and adjacent islands. Tropical rain forest and rubber plantations. HBL 34–42cm (13–16.5in); TL 22–27cm (8–11in); "wingspan" 70cm (28in); WT 1–1.75kg (2–4lb). Coat: upper surface of flight membranes mottled grayish-brown with white spots (an effective camouflage on treetrunks), underparts paler; females tend to be more gray, males more brown or reddish; females are slightly larger than males. Breeding: one young after 60 days' gestation. Longevity: not known.

PHILIPPINE COLUGO *Cynocephalus volans*

Philippine colugo, Gliding or Flying lemur

Philippine islands of Mindanao, Basilan, Samar, Leyte, Bohol. Forests in mountainous and lowland regions. HBL 33–38cm (13–15in); TL 22–27cm (8–11in); WT 1–1.5kg (2–3.5lb). Coat: darker and less spotted than Malayan colugo. Breeding and Longevity: as Malayan colugo. Conservation status: Vulnerable.

Abbreviations HBL = head–body–length
TL tail–length WT weight

Right *A female Malayan colugo with* *ing. The young, which are weaned at* *out 6 months old, are tiny when born,* *ighing only about 35g (1.2oz) and do* *reach adult size until 2–3 years old.*

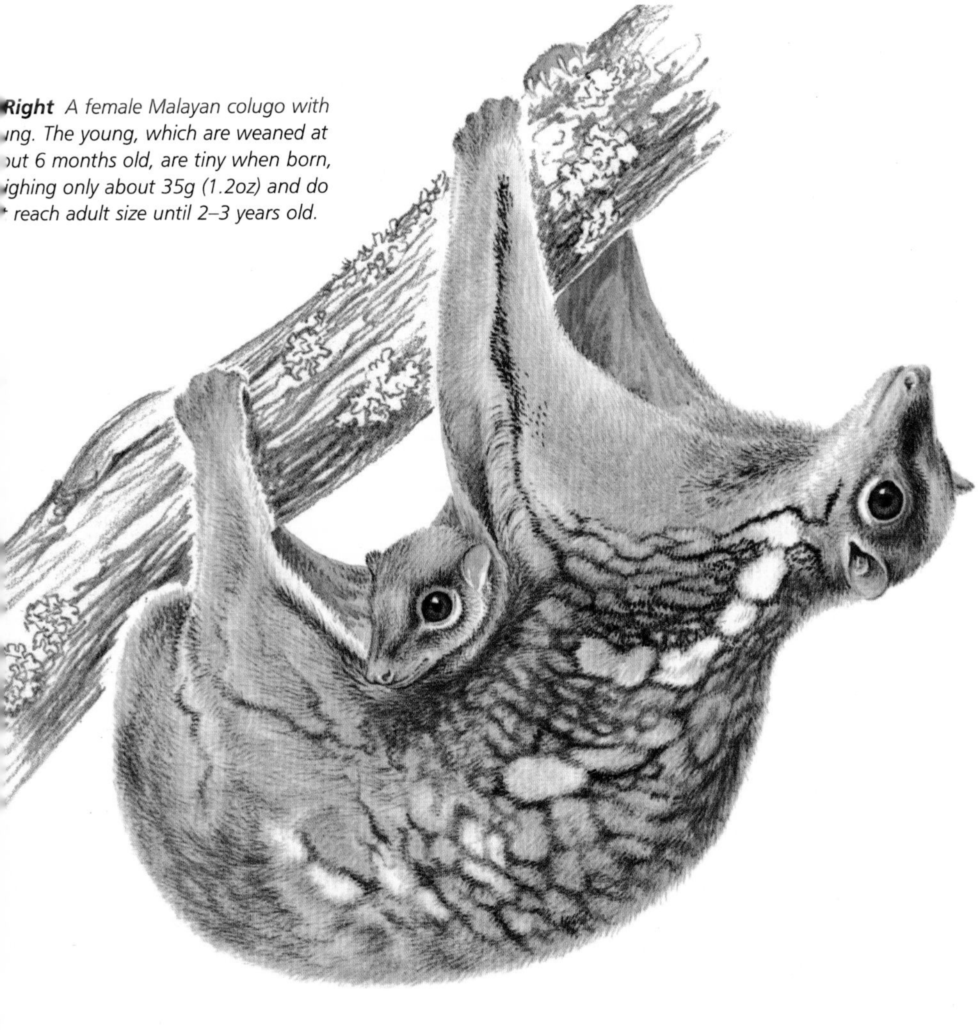

DENTITION OF COLUGOS

Colugos are herbivores with teeth that are highly specialized and unlike those of any other mammal. Like ruminants, they have a gap at the front of the upper jaw, with all the upper incisors at the side of the mouth, but the second upper incisor has two roots, a feature unique among mammals. The most interesting aspect of their dentition, however, is that the two pairs of lower incisors are comblike, with as many as 20 comb tines arising from one root. The function of these unique tines is not fully understood; they may be used as scrapers to strain sap, or for grooming the animal's fur.

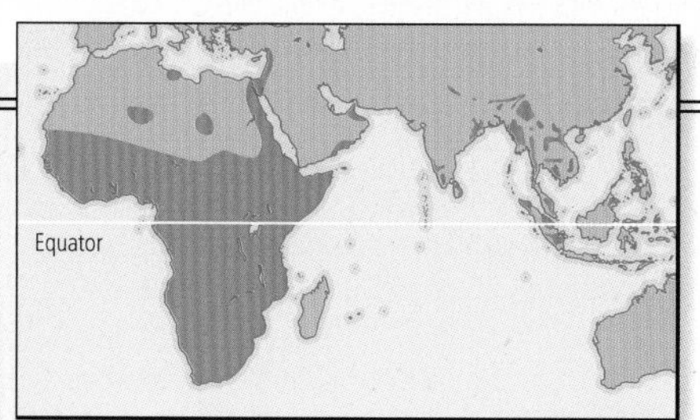

SUBUNGULATES

THE HUGE ELEPHANT, THE RELATIVELY SMALL hyrax, and the aquatic dugongs and manatees all look markedly different from one another. And yet, zoologists believe that these three orders of animals are more closely related to each other than they are to other mammals, and so have grouped them together in the superorder Subungulata.

This surprising relationship between such diverse mammals has been corroborated by research into the animals' molecular biology and anatomy. Indeed, it has even emerged that the sea cows, which belong to the order Sirenia (see Dugong and Manatees) may be more closely related to the elephants than are the hyraxes. The grouping of sea cows and elephants is called the Tethytheria, which also includes an extinct group of amphibious mammals, the desmostylians.

Another type of mammal that many biologists include within the Subungulates is the aardvark (order Tubulidentata). This grouping is preferred by molecular systematists, though anatomists and paleontologists are more skeptical. All of these mammals (with the exception of the sea cows) have an initial appearance and early diversification in Africa, and this may reflect the early isolation of a common ancestor on this continent.

The three orders of Subungulates proper – the Hyracoidea (hyraxes), Sirenia (sea cows), and Proboscidea (trunked mammals, today represented only by the elephants) – share certain derived anatomical characters, including details of the structure of the skull, wrist, and placenta. They have short, nail-like hooves, and share the peculiar feature of a styloglossus muscle, running between the base of the skull and the tongue, that is forked at its insertion into the tongue. The females have two teats between the forelegs (hyraxes have an additional two or four on the belly); the testes remain in the body cavity close to the kidneys. The orders are also related biochemically.

Out of Africa

SUBUNGULATE ORIGINS

The earliest ungulates, the Condylarthra, appeared in the early Paleocene, some 65 million years ago. They were the ancestors of the modern ungulates, the Perissodactyla (odd-toed ungulates) and Artiodactyla (even-toed ungulates). The Subungulata (also called the Paenungulata) evolved in Africa, during this continent's isolation, from a Paleocene offshoot of the Condylarthra. By the early Eocene (about 54 million years ago), the Subungulata had separated into the three distinct orders.

The Tubulidentata (represented only by the aardvark) diverged early on during the Paleocene from the Condylarthra, and specialized in feeding on termites and ants. Despite its resemblance to anteaters, the aardvark is not related to them at all. Fossil evidence suggests that members of this order spread from Africa to Europe and Asia during the late Miocene (about 8 million years ago), but three of the four genera recognized are now extinct.

The Hyracoidea proliferated about 40 million years ago, but spread no further than Africa and southern Eurasia (including China). Today, their only Eurasian stronghold is around the eastern Mediterranean. Some of them became as large a tapirs, but now only the smaller forms remain. The decline of the Hyracoidea during the Miocene, 25 million years ago, coincided with radiation of the Artiodactyla, against whom it is likely that they were unable to compete.

The Proboscidea were a very successful orde that went through a period of rapid radiation in Africa and then spread across the globe except f Australia and Antarctica. The most obvious feat of the Proboscidea is, of course, their large size. Associated with this are their flattened soles, elo gated limb bones, and the modifications to the head and associated structures. Modern elepha are only a remnant of a previously vast diversity proboscideans, many of which survived into th Pleistocene. Elephants are distinguished by hav ing upper tusks only and a dentition adapted fo eating grass. The deinotheres, known from Afric and Asia, had lower tusks only, and appear to have been specialized browsers with tapir-like molars. The mastodons, which spread to North America, were also browsers and, like elephants, eventually lost their

Right *Evolution of the elephant. Beginning with the small, tapir-like* Moeritherium *1, in the early Oligocene (34 million years ago), proboscideans became a large widespread group by the mid-Miocene (around 15 million years ago).* Trilophodon *2 was one of a family of long-jawed gomphotheres found in Eurasia, Africa, and North America from the Miocene to the Pleistocene (24–1.8 million years ago).* Platybelodon *3 was a shovel-tusked gomphothere found in the late Miocene (about 12–5 million years ago) of Asia and North America. The Imperial mammoth* (Mammuthus imperator) *4, the largest ever proboscidean, flourished in the Pleistocene of Eurasia, Africa, and North America. Unlike the earlier forms, it had high-crowned teeth, like those of the modern Savanna elephant* (Loxodonta africana) *5.*

er tusks. Gomphotheres had pig-like molars, icating an omnivorous diet. Most forms kept ir lower tusks, and they spread not only to rth America but also into South America.

Highly Specialized Teeth

DENTITION

subungulates show specializations of the lars (grinding teeth) and incisors, and all have their canines. The elephant's upper incisors e become its characteristic tusks, and the same h have been enlarged in both the hyraxes and dugong (while manatees have no front teeth at Elephants and hyraxes have transverse ridges their molars, while those of the sirenians are ondarily simplified. Hyraxes have molars very ch like those of the primitive horses, which is y the earliest known horse was originally called *acotherium* (having been mistaken for a hyrax). wever, the other subungulates are more special-l. The molars of the dugong are ever-growing.

Elephants have very high-crowned molars, and both elephants and manatees have taken advantage of the fact that mammalian molars erupt in sequence from the back of the jaw in a way so as to prolong the effective life of their dentition. Elephants delay the timing of molar eruption so that only one tooth is in use in the jaw at any one time. Each tooth has been greatly enlarged to the size of the entire back of the jaw; when one tooth wears down, the erupting tooth behind moves in to take its place. Elephants have six teeth in each jaw half (upper and lower, left and right), which represent the normal mammalian complement of three milk molars and three true molars, but when the last tooth has been worn down the elephant is all out of dentition. In contrast, manatees do not enlarge the individual teeth, but have added to the normal mammalian complement, and have a virtually never-ending supply of erupting teeth throughout their life (at least 20 in each jaw half), erupting from the back of the jaw and falling out of the front when worn down. The aardvark's dentition is unique, in that the front teeth are lacking and the peg-like molars and premolars are formed from columns of dentine. CMJ/RFWB

4

5

ORDERS: PROBOSCIDEA, HYRACOIDEA, TUBULIDENTATA

3 orders; 6 genera; 15 species

A further order, the aquatic Sirenia, is considered by some authorities to belong to this grouping. It is discussed only peripherally here (see Dugong and Manatees for main entry).

ELEPHANTS Order Proboscidea p436

3 species

Savanna elephant (*Loxodonta africana*), **Forest elephant** (*L. cyclotis*), **Asian elephant** (*Elephas maximus*)

HYRAXES Order Hyracoidea p448

11 species in 3 genera

Abyssinian hyrax (*Procavia habessinicus*), **Cape hyrax** (*P. capensis*), **Johnston's hyrax** (*P. johnstoni*), **Kaokoveld hyrax** (*P. welwitschii*), **Western hyrax** (*P. ruficeps*), **Ahaggar hyrax** (*Heterohyrax* antineae), **Bruce's yellow-spotted hyrax** (*H. brucei*), **Matadi hyrax** (*H. chapini*), **Eastern tree hyrax** (*Dendrohyrax validus*), **Southern tree hyrax** (*D. arboreus*), **Western tree hyrax** (*D. dorsalis*)

AARDVARK Order Tubulidentata p452

1 species

Aardvark (*Orycteropus afer*)

Note Even more remarkable than the ancestral linkage of the elephants, hyraxes, aardvark, and sea cows is the recent discovery that all these are linked to elephant shrews, tenrecs, and golden moles in the Afrotheria – a supraordinal assemblage of mammals that share a common African ancestor (see Preface, What is a Mammal?, and box p722). Thus, the "ungulates" are actually made up of two convergent stocks – these Afrotherians and their Laurasiatherian counterparts, which include both the artiodactyls and perissodactyls.

Below *Hyraxes, which superficially resemble large rodents, are in fact most closely related to elephants. Their range once spread as far as China, but they are now restricted to Africa and the Middle East.*

Elephants

MODERN ELEPHANTS ARE THE LARGEST extant land mammals; they have the biggest brains in the animal kingdom, live as long as humans, are able to learn and remember, and are adaptable as working animals. The antecedents of today's elephants are Phosphatherium, the earliest proboscidean, which lived around 58 million years ago, and Moeritherium (named for Egypt's Lake Moeris, near which its remains were discovered), from around 34 m.y.a.

For millennia, elephants' great strength has been exploited in agriculture and warfare, and even today, notably in the Indian subcontinent, they are still important economically and as cultural symbols. But demand for elephant tusks, the main source of commercial ivory, has brought about a drastic decline in elephant populations over the past 150 years. Now they are threatened by rising human populations competing for space across most of their range.

Big Bodies, Big Brains

FORM AND FUNCTION

Though African and Asian elephants are ecologically very similar, there are physical and physiological differences between the two (see below). In addition to visible distinctions, the African species have one more pair of ribs than their Asian counterpart (21 versus 20).

Of the African elephants, the Savanna species is better understood than the Forest elephant, since it is far easier to study behavior in the open grasslands of East Africa than in dense forest habitats. Savanna bulls are also the biggest and heaviest elephants of all. The largest known specimen, killed in Angola in 1955 and now on display in the Smithsonian Institute in Washington DC, weighed 10 tonnes and measured 4m (13.1ft) at the shoulder. Body size continues to increase throughout life, so that the biggest elephant in a group is also likely to be the oldest.

The characteristic form of the skull, jaws, teeth, tusks, ears, and digestive system of elephants are all part of the adaptive complex associated with the evolution of large body size (see box). The skull, jaws, and teeth form a specialized system for crushing coarse plant material. The skull is disproportionately large compared with the size of the brain and has evolved to support the trunk and heavy dentition. It is, however, relatively light, due to the presence in the cranium of interlinked air cells and cavities.

The tusks are elongated upper incisors. They first appear at birth and grow throughout life, so that by the age of 60, a bull's tusks may each weigh 60kg (132lb). Such massive "tuskers" have always been prime targets for ivory and big-game hunters, with the result that few remain in the wild today. Elephant ivory is a unique mixture of dentine and calcium salts, and a transverse section through a tusk shows a regular diamond pattern not seen in the tusks of any other mammal. The tusks are used in feeding for such purposes as prizing off the bark of trees or digging for roots, in social encounters as instruments of display, and also as weapons.

The upper lip and the nose of elephants have become elongated and muscularized to form a trunk. Unlike other herbivores, an elephant can reach the ground with its mouth. The fact that early proboscideans did not evolve an elongate neck may have been due to the weight of their heavy cranial and jaw structures. Besides enabli elephants to feed from the ground, the trunk is also used for feeding from trees and shrubs, bre ing off branches, and picking leaves, shoots, an fruits. Further uses include drinking, greeting, caressing and threatening, squirting water and throwing dust, and the forming and amplifying vocalizations. Elephants drink by sucking water into their trunks, then pouring it into their mouths; they also throw water over their backs cool themselves. At times of water shortage they sometimes spray themselves with water that the have stored in a pouch in the pharynx. The trur can also serve as a snorkel, enabling an individu to breathe if submerged, for instance during a river crossing. Elephants use their trunks to rub an itchy eye or to scratch an ear. Trunks are also employed to gesture at enemies, to throw objec or to use tools such as sticks to scratch the skin

The most common elephant vocalization is a growl emanating from the larynx (what hunters used to call the "tummy rumble"). This sound c carry for up to 1 km (0.6m), and may be used a a warning, or to maintain contact with other ele phants. When feeding in dense bush, members of a group monitor each other's positions by low growls with a strong infrasonic component (see box). They vocalize less frequently when the bu is more open and the group members can see o another. The trunk is used as a resonating cham ber to amplify bellows or screams so as to conve a variety of emotions. New evidence suggests th another organ, high in the trunk and known as the alinasal cartilage, may also modify sounds. This cartilage divides the bony opening where th trunk begins (the external nares) and can be use to direct the flow of air. The loud trumpeting of elephants is mainly used when they are excited, surprised, about to attack, or interacting in play.

Visual messages are conveyed by changes in posture and the position of the tail, head, ears, and trunk, which besides the other uses describ earlier has a secondary value in communication Though powerful enough to lift whole trees, it is also an acutely sensitive organ of smell and touc Smell plays an important part in social contacts within a herd and in the detection of external threats. As to touch, the trunk's prehensile two fingerlike lips, endowed with fine sensory hairs, can pick up very small objects. In addition, elephants often touch each other using the trunk;

African elephants

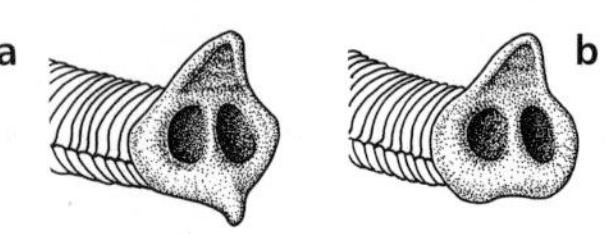

Asian elephant

Above *African and Asian elephants compared. The African elephant is bigger, with a concave back and larger ears. Its trunk has two processes ("lips")* **a** *rather than the Asian's one* **b**. *Whereas both sexes of both African species generally have tusks, these are mainly confined to male Asian elephants.*

mother will continually use hers to guide her ıfant. When elephants meet, they often greet ıch other by touching the other's mouth with e tip of their trunk.

The brain of an elephant weighs 3.6–4.3kg –9.5lb) in females and 4.2–5.4kg (9.3–12lb) males. It has a temporal lobe that is highly con-ıluted – even more so than in humans – thus creasing the active surface area. Its size may be sociated with the need for storage space for formation, for elephants need to differentiate entities, to record memories of other elephants' havior, and to store experiences of droughts, ıngerous places and situations, and promising feeding sites. Some of their social behavior suggests that they possess the mind tools to imagine what other elephants are feeling. Given their long lives, there is a survival premium for families led by matriarchs that make the right decisions on movements in times of peril or drought. All these factors favor the development of intelligence.

Besides their role in communication, the elephant's large ears act as radiators to prevent overheating, always a danger for animals with large, compact bodies. They are well supplied with blood, and can be fanned to increase the cooling flow of air over them; on hot, windy days, elephants sometimes spread their ears to let the

Left *An African elephant cow and calf. The huge ears, which give the elephant such a distinctive frontal appearance, function as radiators for cooling its bulky body by creating a vast surface area over which heat is lost.*

FACTFILE

ELEPHANTS

Order: Proboscidea

Family: Elephantidae

3 species in 2 genera

DISTRIBUTION Africa S of the Sahara; S Asia.

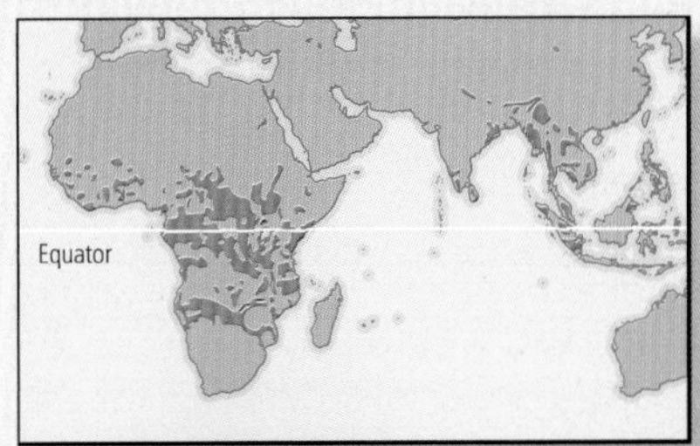

SAVANNA ELEPHANT *Loxodonta africana*
Sub-Saharan E and C Africa. Savanna grassland. HBL male 6–7.5m (20–24.5ft), female 0.6m (2ft) shorter; SH male 3.3m (10.8ft), female 2.7m (8.9ft); WT male up to 6 tonnes, female 3 tonnes. In its densest places, the skin is 2–4cm (0.8–1.6in) thick and sparsely endowed with hair. The Savanna elephant typically has only four toes on the forefoot and three on the hindfoot. Breeding: gestation averages 656 days. Longevity: 60 years (more than 80 in captivity). Conservation status: Endangered.

FOREST ELEPHANT *Loxodonta cyclotis*
C and W Africa; dense lowland jungle. HBL, SH, and WT as for Savanna elephant. Tusks straighter and ears more rounded than Savanna elephant. In common with Asian elephants, five toes on the forefoot and four toes on the hindfoot. A subspecies of the Forest elephant, the Pygmy elephant (*L. c. pumilio*), HBL 2.4–2.8m (7.9–9.2ft); WT 1,800–3,200kg (3,970–7,050lb), occurs from Sierra Leone to DRC. Conservation status: Endangered.

ASIAN ELEPHANT *Elephas maximus*
Indian subcontinent and Sri Lanka, Indochina, parts of peninsular Malaysia, Thailand, and SE Asian islands. Evergreen and dry deciduous forests, thorn scrub jungle, swamp and grassland; from sea level up to 3,000m (10,000ft). Three subspecies: *E. m. maximus* on Sri Lanka, *E. m. sumatranus* on Sumatra, and *E. m. indicus* on mainland Asia. HBL 5.5–6.4m (18–21ft); SH 2.5–3m (8–10ft); WT male 5.4 tonnes, female 2.7 tonnes. Skin: dark gray to brown, sometimes marked with flesh-colored blotches on the forehead, ears, and chest. Breeding: gestation 615–668 days; normally a single calf is born, weighing about 100kg (224lb). Longevity: 75–80 years in captivity. Conservation status: Endangered.

Abbreviations HBL = head–body length SH = shoulder height WT = weight

breeze blow over them. Observation of blood vessels on the medial side of the ear show that when the ambient temperature is cool the vessels are normal and do not protrude beyond the skin, but when temperatures are high, the blood vessels dilate and rise above the skin. Elephants also have a keen sense of hearing and communicate mainly through vocalizations, especially the Forest species.

The massive body is supported by pillarlike legs with thick, heavy bones – this is called graviportal stance. The bone structure of the forefeet is semi-digitigrade (a horse has a digitigrade stance, with the heel raised high off the ground), whereas the hind feet are semi-plantigrade (a man has a plantigrade posture, with the heel on the ground). Elephants usually meander, but at other times walk quickly at about 3–4km/h (2–2.5mph). A charging elephant is said to reach 40km/h (25mph); at this speed over short distances it would easily outstrip a human sprinter, but it is doubtful whether it has ever been measured accurately.

Elephants have a nonruminant digestive system similar to that of horses. Microbial fermentation takes place in the cecum, which is an enlarged sac at the junction of the small and large intestines.

Elephants spend at least three-quarters of their time searching for and consuming food. In the wet season, Savanna elephants eat mainly grasses, plus small amounts of leaves from a wide range of trees and shrubs. After the rains have ended and the grasses have withered, they turn to the woody parts of trees and shrubs. They will also eat large quantities of flowers and fruits when these are available, and they will dig for roots, especially after the first rains of the season.

Above The elephant's dextrous trunk serves many purposes, including plucking succulent leaves and shoots from high branches. Leaves supplement the Savanna elephants' main fare of grass, whereas they form the staple diet of Forest elephants.

Asian elephants eat a varied diet that include up to 100 species of plant; however, more than 85 percent of it derives from between 10 and 25 favorite foodstuffs. Crops also play a part, as the elephants live in landscapes dominated by agric ture. As cultigens like cereal and millet are essen tially grasses that have been selected by humans for their high protein and other nutrient conten it is hardly surprising that elephants generally fi them more attractive than wild grasses.

Because of their large body size and rapid rate of throughput, all elephants need large amounts of food: an adult requires 75–150kg (165–330l of food a day, or over 50 tonnes per year. Less th half of this is thoroughly digested. Elephants rel on the gut microflora to help them with digestio babies, having no microflora of their own, ingest them by eating the dung of older family membe

In addition, elephants require 80–160 liters (20–40 US gallons) of fluid every day, an amou they can imbibe in less than 5 minutes. In the d season they may dig holes in dry riverbeds with their trunks and tusks to find water.

Matriarchs and Musth Bulls

SOCIAL BEHAVIOR

Most of the information on elephants' ranges nc comes from radio-tracking, which has been usec in Africa since 1969. In addition, improved radi collars using global positioning technology were deployed at the end of the 20th century, allowin very accurate positions to be collected every hou

Cumulative average daily movements vary greatly from one elephant to another. In a study in Kenya, an elephant living in a well-watered

WHY ELEPHANTS GREW BIG

At their evolutionary peak, elephants living and extinct spread to all parts of the globe except Australia, New Zealand, and Antarctica, and until the Pleistocene (about 2 million years ago), elephants occupied a range of habitats from desert to montane forest throughout their range. This wide radiation was related to their most outstanding feature: the evolution of large body size.

In order to understand their evolutionary success, one must first consider the early large-herbivore community. The first sizable mammalian herbivores in Africa were perissodactyls – ancestors of the horse – which arose in the late Paleocene, about 58 million years ago. They remained dominant until the coming of the ruminant artiodactyls – predecessors of the antelopes – in the middle Eocene, some 46 million years ago. It is likely that while each perissodactyl species ate a wide range of plants, taking the coarser parts, the ruminants ate a narrower range, taking the softer parts.

The earliest proboscideans appeared in the late Paleocene; the elephantids arose in the late Miocene, at a time when the highly successful ruminants were continuing to evolve and to colonize new ecological niches. As non-ruminants, members of the Elephantidae were able to feed on plant foods which were too coarse for the ruminants, but this brought them into competition with the perissodactyls.

For a given digestive system, differences in metabolic rate enable a large animal to feed on less nutritious plant parts than a smaller one can. There was, therefore, a strong selective pressure for the elephantids to increase their body size and so reduce the competition with perissodactyls. The most nutritious plant parts, such as leaf shoots and fruits, are produced only at certain seasons, and even then may be sparse and widely scattered; but coarse plants are more abundantly distributed both in space and time. By permitting the digestion of the latter, the elephants' evolutionary strategy thus enabled them to feed on plant parts which were not only abundant but also available all the year round. In particular, it enabled them to feed on the woody parts of trees and shrubs, and so to tap a resource which other mammalian herbivores could neither reach nor digest. At the same time, they remained able to eat rich plant parts such as fruits whenever these were available. This catholicity enabled elephants to thrive in a wide range of habitats. RB

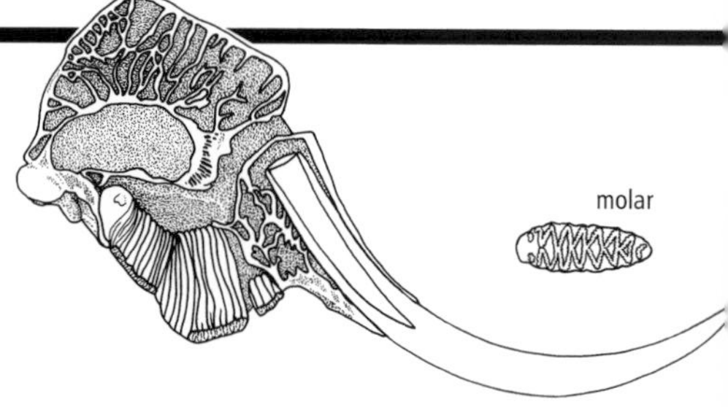

Above The elephant's skull is massive, comprising 12–25 percent of its body weight. It would be even heavier if it were not for an extensive network of air-cells (diploe). the dental formula is I1/0, C0/0, P3/3, M3/3. The single upper incisor grows into the tusk and the molars fall out at the front when worn, being replaced from behind. Only one tooth on each side, above and below, is in use at any one time.

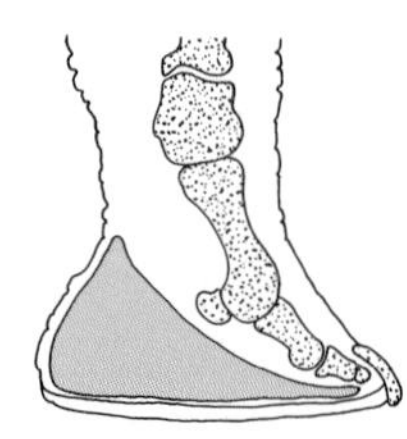

Left The elephant's foot is broad and the phalanges (fingers and toes) are embedded in a fatty matrix (green). The huge weight of the animal is spread so well that it leaves hardly any tracks.

Above *Herds of elephants often lift their trunks h into the prevailing breeze and use their acute se of smell to gain advance warning of any roaching threat.*

Below *Despite its thickness – in places, up to 4cm – the elephant's skin is highly sensitive and requires frequent bathing, massaging, and powdering with dust to keep it free from parasites and diseases.*

forest averaged a mere 3km (1.9mi) a day, while others living in an arid northern area covered 12km (7.5mi) a day. Typically, an elephant will meander over a cumulative daily distance of 7–8km (4–5mi).

A striking feature of elephant movements is the behavior known as "streaking." This involves a spell of relatively rapid motion – typically at 3–4 km/h (2–2.5mph) – with a strong directional component, usually down corridors connecting distinct segments of the animals' range. Streaking, which is fairly rare and often takes place at night, may serve to get elephants rapidly across dangerous areas from one safe haven to another.

Elephants also respond speedily to sudden rainfall, and may travel up to 30km (19mi) to reach a spot where an isolated shower has fallen, in order to exploit the lush growth of grass that soon follows. Likewise in the forest they may make long journeys to find rare trees in fruit. When elephants venture into dangerous feeding zones such as farmlands, they tend to do so only at night. Elephants appear to be able to learn which areas are safe and will venture right up to the edge of a protected area before turning back at the border. Repeated movements often carve broad "elephant roads" cutting through even the densest jungles, which may subsequently be used by many other species including humans.

Some elephant ranges have surprisingly complex structures. By enclosing the outside points of a range, the rough area can be calculated for comparative purposes. Within this area there may be discrete segments, linked by elongated corridors and empty zones where the elephants never venture. Ranges as small as 10sq km (4sq mi) have been described in one forest in Tanzania, while others of up to 18,000sq km (7,000sq mi) were observed in a desert area in Namibia. In a study in Kenya, the home ranges of Savanna elephants in a woodland and bushland habitat were found to average 750sq km (290sq mi) in an area of abundant food and water, and 1,600sq km (617sq mi) in a more arid area. Detailed studies of the ranging behavior of the Forest species were initiated at the beginning of the 21st century, and first results indicate that the ranges can be as much as 60km (37mi) across, far larger than previously thought. Radiotelemetry studies of Asian elephants have revealed that female groups in India range over 180–600sq km (70–230sq mi) or more, while males typically use an area of 160–400sq km (60–150sq mi).

Elephants live in groups and display complex social behavior. The advantages of living in society lie in group defense, the teaching of the young,

Communicating over Long Distances

Elephants make calls that are rich in infrasound – sound below the range of human hearing. Forest elephants make calls as low as 5Hz, two octaves below the lowest sounds that humans ordinarily hear (c. 20–20,000Hz). Many of the calls of Asian and of African Savanna elephants have fundamental frequencies of 14–20Hz. A person standing close to a calling elephant may hear a soft rumble, but at even a short distance calls that are perfectly audible to elephants are not perceived by humans.

Elephants use their calls in long-distance signaling, since very low-frequency sound loses exceptionally little energy in traveling. Playback experiments show that calls at 119dB (SPL) are powerful enough to elicit responses from other elephants as far away as 4km (2.5mi) in the middle of the day. In the evening, distances increase as a result of atmospheric conditions. On a clear savanna night (as 80 percent are in the southern African dry season), a temperature inversion forms. A layer of hot air is trapped under cold air, forming a ceiling at 300m (1,000ft) or below that reflects earth-generated sounds back down to the earth, enhancing sound propagation at or near ground level, especially at low frequencies. One model projects that at such times the calling area for loud elephant calls expands tenfold, from 30 to 300sq km (11.5 to 115sq mi).

Long-distance communication appears to be crucial to reproduction in free-ranging elephants. Adult males and females move about independently most of the time. When a female comes into estrus, she will emit a series of powerful, low-pitched calls to announce her condition. Reproductive males then gather rapidly from varying distances and all directions. It is essential for them to detect her condition quickly, because the estrous periods are extraordinarily short and rare – on average, according to data from Amboseli Park, Kenya, encompassing only 2–4 days every 5 years.

Elephants' powerful infrasonic calls also have long-distance functions in other circumstances. In times of excitement, distress, or separation, family members listen for one another's calls and aggregate. Noisy events that result in aggregation include aggression or calf rescue as well as courtship and mating. Listening males and females are selectively attracted to the site, to help in or to take advantage of the situation. Related families sometimes coordinate their movements for weeks at a time over distances of several kilometers without actually meeting. This appears to be how they avoid exhausting the resources on which they depend while remaining available to one another for purposes of mutual support. KP

Below *As well as long-distance calling, close-quarters interaction – here, a bull tests a cow's receptivity – is a key part of the reproductive cycle.*

and enhanced mating opportunities. Female elephants live in family units, which typically consi[st] of closely-related adult cows and their immature offspring. The adults are either sisters or else mothers and daughters. A typical family unit ma[y] contain two or three sisters and their offspring, [or] one old cow in company with one or two adult daughters and their offspring. When the female calves reach maturity, they stay with the family unit, where they may then breed. As the family grows in size, a subgroup of young adult cows w[ill] gradually separate to form their own unit. As a result, related family units are sometimes separat[ed], but usually move together in a coordinated fashion. Groups of 2–4 related families that trav[el] around together are known as kinship groups or bond groups.

The oldest female, the matriarch, leads the fa[m]ily unit. The social bonds between the members [of] the family are very strong. In times of danger, the family forms a defensive circle with calves in the middle and adults facing outward. The matriarc[h] or other adult females will check the bearing of the threat, generally a human being. Usually she

ill retreat, but sometimes she may advance to confront the danger, spreading her ears and giving nt to trumpeting and thunderous growling. threat charge, with ears outstretched, is often nough to deter an aggressor. But occasionally, the harge is followed through. Unfortunately, this efensive behavior exposes the matriarch to dan-er, so she often is first to fall victim to poachers, aving the rest of the family leaderless.

If a family member is shot, wounded, or tran-uilized, the rest of the group may well come to s aid. With much noise and excitement, they will y to lift the animal to its feet and carry it away, ith family members supporting it on either side.

In Asian elephants too, the basic social struc-ure is a family group of 2–10 females and their ffspring, with an average size of 6.7 animals. ntensive studies conducted in Rajaji National ark in northern India have shown that groups of hree or less adult females and their associated ffspring are highly stable, spending up to 90 per-ent of the time together. Such groups will meet p from time to time with other groups (maybe lated) when grazing in open grasslands or near waterholes, though the intense greetings described in African elephants have not been documented, and the bigger groupings seem to be merely transitory.

In contrast to their sisters, young male elephants leave or are forced out of the family on reaching puberty. Adult bulls tend to associate with one another in small groups that constantly change in number and composition; they also spend short periods alone. The traditional picture has been one of weak associations and little cooperative behavior between bulls, but recent research in Kenya shows that short-term intensive associations may be repeated time and again, with intervening periods of separation. In northern Botswana, small bull populations of several hundred individuals maintain close associations by focusing their activities on waterholes, often provided by the local Department of Wildlife.

Male Asian elephants start to move away from the family group at around 6–7 years of age. Fully-grown adult males are solitary and are rarely seen with female groups unless a cow is in estrus. When they reach the age of 20, mature males start coming into a phase called "musth," in preparation for the rigors of competing for females and mating. Musth (a Hindi or Urdu word meaning "intoxicated") aptly describes this extreme physiological condition, in which the levels of testosterone in the blood may increase by a factor of twenty or more. The animals typically display agonistic or aggressive behavior. In the Rajaji National Park study, the biggest adult males came into musth during the period when most of the females would have been in estrus. Fully-grown adult males (up to 35 years) remain in musth for about 60 days, during which time they wander widely in search of females in estrus.

African elephants also experience musth, though in a less pronounced form. In Kenya's Amboseli Park, where social behavior has been more intensively studied than elsewhere on the continent, the bulls normally do not come into

Above *Young African bull elephants tussling playfully. In later life, the sparring skills they learn may be brought to bear in earnest during the "musth" phase, when adult bulls fight for access to females.*

musth until nearly fully grown at 29 years old. Musth usually lasts for 2–3 months and tends to coincide with periods of high rainfall.

Musth bulls are more likely to be involved in fights than others, and occasionally this can result in the death of one of the contenders. During musth, elephant bulls dramatically decrease their food intake, burning up accumulated fat reserves. Musth bulls emit signals that notify other elephants of their state. The temporal glands between the eye and the ear swell up and discharge a viscous aromatic secretion. They discharge a continuous dribble of urine containing soluble pheromones. The posture of musth bulls is also quite distinct; they carry their heads much higher than normal, and their ears are held high and spread out. There is also a characteristic vocalization, the "musth rumble," which consists of low, pulsating growls somewhat like a low-revving diesel engine.

The purpose of musth seems to be to temporarily increase the status of a bull and to help it win fights; even a small bull in musth will normally prevail over a larger, non–musth bull. Cows are in estrus only 2–4 days during the estrous cycle that lasts about 4 months if females are not pregnant, or every 4–5 years if she conceived, had a successful delivery, and raised the young. During estrus, bulls must be able to locate the females rapidly. Musth bulls travel longer daily distances than other bulls. Females in estrus attract musth bulls with a very loud infrasonic call. It appears that females prefer large bulls to small ones and musth bulls to non-musth. They exercise choice as to which bull to mate with; if a female does not want a particular individual, she will run away, and even if he catches up with her she will refuse to stand still long enough for him to mate successfully.

Female estrus usually comes on during the rains, and bulls of the highest rank come into musth at this time. The presence or absence of larger musth bulls apparently affects the age at which a bull comes into musth and the length of time it stays in the condition, apparently through an intimidation effect. In one population in South Africa, the introduction of older bulls subdued the highly aggressive behavior of young bulls that had been killing the resident rhinos.

Cows reach sexual maturity at about 10 years of age, but this may be delayed for several years

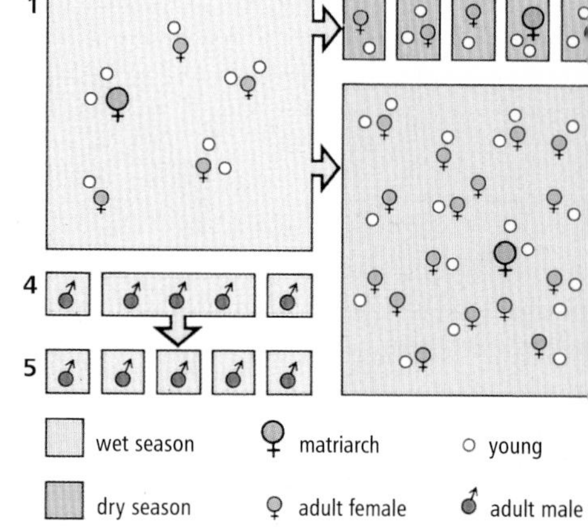

Above A typical elephant family unit is made up of closely related cows (with one dominant cow, the matriarch) and their offspring **1**. When food is scarc **2**, the family groups tend to split up to forage. In th wet season **3** family units may merge to form group of 50 or more. Bulls leave at puberty to join small, loose herds **4** or live alone **5**.

Below A large group of elephants of varying age and comprising a number of family groups assemble at the onset of the wet season.

luring drought or periods of high population density. Once a cow starts to breed, she may produce an infant every three or four years, although this period too can be extended when times are bad. Females are most fecund from the age of 25 to 45.

Most elephant populations show an annual reproductive cycle that corresponds to the seasonal availability of food and water. During the dry season when food is scarce, cows may cease to ovulate. When the rains break and the food supply improves, a period of 1–2 months of good feeding is needed to raise the female's body fat above the level necessary for ovulation. Thus females are in heat during the second half of the rainy season and in the first months of the dry season.

Elephants have an exceptionally long gestation period, averaging 630 days or sometimes not far short of 2 years, which means that the calves are born early in the wet season, when conditions are optimal for their survival. In particular, abundant green food helps ensure that the mother will lactate successfully during the early months. First-year survival of calves has usually been assumed to be around 70–80 percent, but recent data from studies tracking 13 mothers with calves over a two-year period suggests that more than 95 percent of offspring in fact survive past their first year.

At birth, the African elephant weighs about 120kg (265lb). The lengthy gestation is followed by a long period of juvenile dependency. The infant suckles (with its mouth, not its trunk) from the paired breasts between the mother's forelegs, which are roughly the size and shape of large human breasts. (Both elephants and humans have nipples, not teats, an error often found in the literature.) The young elephant grows rapidly, reaching a weight of 1 tonne by the time it is 6 years old. The rate of growth decreases after about 15 years, but growth continues throughout life; males grow faster than females.

During a birth, other cows may collect around the new calf, and so-called "midwives" may assist by removing the fetal membrane. Others may help the infant to its feet, marking the start of a joint family responsibility for the young of a group. Young female elephants called "allomothers" play a key role in bringing up young elephants. They increase the calves' chances of survival by their efforts, and at the same time gain experience for the time when they too will become mothers.

The supposed existence of elephant graveyards is a myth, although it is possible that old elephants whose days are numbered may congregate on riverbanks to feed on the lush vegetation. Some countries have also seen elephant killing-fields, where poachers have left dead elephants strewn across the landscape. This happened, for example, in the Murchison Falls National Park in Uganda, which used to be home to 8,000 elephants; they were killed for their ivory by poachers, many of them soldiers of Idi Amin's army, who reduced the population's numbers to less than 100 in the early 1980s.

A real mystery of elephant life lies in the way the living treat the dead. They exhibit an extraordinary interest in elephant carcasses and bones, and will spend hours sniffing the remains and investigating them with their trunks, picking up some bones and putting them in their mouths or carrying them on their heads. So far there is no satisfactory scientific explanation for this behavior, although it seems that elephants are able to identify particular individuals from the smell of the remains, suggesting a level of understanding that remains a challenge to science.

Stemming the Ivory Trade

AFRICAN ELEPHANT CONSERVATION

Man's relationship with elephants is beset by contradictions. On the one hand, they are regarded with awe and fascination; while on the other, whole populations are eradicated in pursuit of land or ivory. Even though elephants do have natural predators – the young are often killed by lions, hyenas, or crocodiles – by far their most dangerous enemy is humankind.

As far back as classical antiquity, North African elephant populations were dwindling rapidly. They finally disappeared during the Dark or Middle Ages. The Arab ivory trade, which started in the 17th century, precipitated a further rapid decline among elephant populations in both West and East Africa. The colonial era accelerated the process by opening up previously inaccessible areas and introducing modern technology, notably high-powered rifles; in Africa, the destruction of elephants reached a peak between 1830 and 1900. Today, continuing deforestation and the encroachment of roads, farms, and towns into former elephant habitats threaten both African and Asian elephants by restricting their range, cutting off seasonal migration routes, and bringing them into more frequent conflict with people.

Perceptions of the main conservation issues involved have changed radically since the 1960s, when the debate centered on local overpopulation of elephants in protected areas. In the 1970s and 1980s the concern was that elephants were being decimated by poaching for ivory. It used to be thought that the worldwide economic recession of the early 1970s encouraged investors to switch to ivory as a wealth store, but studies by the Ivory Trade Review Group have since demonstrated that it was rather the increase in the buying power of ordinary Japanese citizens, coupled with a desire

Above *In inaccessible parts of the Indian subcontinent, such as Nepal and the Andaman Islands, Asian elephants still play a key role in forestry, carrying logs or clearing vegetation. Here, a* mahout *– an elephant owner and driver – guides his animal through dense undergrowth in Chitwan NP, Nepal.*

Left *In the dry season, when food is at a premium, elephants become extremely destructive, stripping and eating bark and even felling trees to get at leaves and twigs. Raiding elephants can also devastate planted crops, which has led to conflict with some African farmers.*

Above *Ivory poaching was the scourge of elephant ›pulations at the end of the 20th century, and is still ‘e. Here, tusks seized from poachers are destroyed.*

Right *Asian elephants still feature prominently in ‚ligious and cultural festivals, such as this night pro- ›ssion through Kandy, on the island of Sri Lanka.*

› possess ivory seals as status symbols, that iggered the upsurge in killing by stimulating a ‹uge increase in the price of ivory between 1969 ıd 1973. The result was a dramatic decline in ephant numbers; for instance, the Kenyan popu- tion fell from 167,000 in 1970 to 60,000 by 980 and to about 22,000 by 1989. In 1979 a ‘ide-ranging survey estimated that 1.3 million ephants survived in Africa, but by 1989 the ıumber had fallen to 609,000, suggesting that ıore than half of Africa's elephants had been lost › ivory poaching in a single decade. In response, ıe movement to ban the trade reached a climax ı 1989, when a meeting of the Convention on ıternational Trade in Endangered Species of Wild lora and Fauna (CITES) voted in favor of a mora- ›rium on all trade in elephant products.

There has since been much controversy over ‘hether or not the ivory ban has been effective in ›wering elephant killing. In places where moni- ›ring is possible, mortality has been found to ave dropped in the 1990s compared to the previ- us decade, though critics of the ban attribute this ffect merely to better policing. What is certain, owever, is that ivory prices have dropped since 989, reducing the incentive for poaching.

Even so, southern African countries, whose wn elephant populations increased, contend that they are unnecessarily penalized by the ban. Throughout the early 1990s, they sought to reopen a limited ivory trade with Japan, and in 1997 succeeded in having restricted exports reestablished. However, policymakers in East, Central, and West Africa feared that even this small-scale revival might herald a return to the days of uncontrolled slaughter. To forestall this posssibility, all parties at the 2000 CITES conference resolved that ivory trading should be suspended until an adequate system of monitoring illegal killing had been put in place.

Since the original ivory trade ban of 1989, conflict between elephants and rising human populations has played an increasing role in conservation thinking, for while the elephant population outside of the southern countries more than halved in the ensuing two decades, the human population more than doubled. Attitudes toward elephants remain contradictory, varying from the sheer animosity felt by many agriculturalists within elephant ranges, through the indifference shown by most local people, to the tolerance generally exhibited by certain pastoralist peoples such as the Masai and the Samburu – and finally, of course, to the adoration of safari tourists. The key to the animals' survival would seem to lie in extending tolerant attitudes and in educating people to recognize elephants as a valuable natural resource, at best as cultural assets or at least as a potential means of livelihood.

Even though protecting elephants is costly, many conservationists regard the elephant as a so-called "keystone species" – namely, one that pays dividends by benefiting other animals within its ecosystem. Thus, elephants play a pivotal role, dispersing seeds, transforming savannas into grasslands, distributing nutrients in their dung, providing water for other species by digging waterholes, supplying food for birds by disturbing insects and small animals while walking in tall grass, and even alerting small animals to approaching predators. Moreover, since larger species require greater quantities of food and water and larger home ranges than smaller species, an area large enough to support an elephant will automatically support several other species.

In 1995, the African Elephant Specialist Group analyzed all available data to reach the following estimates. Some 285,246 individuals definitely survived, while a further 101,285 probably and 171,892 possibly did so; the existence of an additional 22,752 animals on top of these figures was considered speculative. Although the data from similar general population estimates is too imprecise to accurately identify trends, there is a wealth of repeat-count data for individual populations in the savannas that suggests major declines in East, Central, and West Africa, compared to increases in parts of Southern Africa, such as Botswana, Zimbabwe, and South Africa. One of the best examples of a continuously declining elephant population is that of Eritrea. The most recent count puts their numbers at just 2–8 individuals; immediate efforts must be made to save Africa's second most northerly relict elephant population.

Working Elephants

ASIAN ELEPHANT CONSERVATION

Like their African relatives, Asian elephants also now occupy a much-reduced range. The total estimated population of Asian elephants in the wild is 37,000–57,000, occurring across an area of some 500,000sq km (193,000sq mi). In addition, there are about 15,000 elephants in captivity. It was captive elephants that allowed the British to open up forests in the region for logging in colonial times. Subsequently, the decline of the timber industry across south Asia, Indochina, and Southeast Asia has led to a decline in numbers, though some countries such as Myanmar still have a sizable population of 5,000 or so timber elephants.

Habitat loss and poaching are still the main threats to Asian elephants. The tusked males alone are targeted by the poachers, which radically affects breeding patterns. In Periyar Tiger Reserve in south India, for example, where poaching has left only 1 male for every 100 females, fewer than one-third of adult females are accompanied by calves less than 5 years old; in Rajaji National Park, in contrast, the sex ratio is 1 male for every 2 females, and more than 90 percent of the adult cows are accompanied by a calf. Elephant–human conflict mostly arises on the borders between elephant habitats and inhabited lands; in India about 300 people and 200 elephants per year die through poaching, crop protection incidents, or accidents. ID-H/RFWB/HS/ACW/AJTJ

AN ELEPHANT'S EARLY YEARS

❶ Only an hour old, an elephant calf pulls itself to its feet and takes its first tentative steps. An African elephant weighs about 120kg (265lb) at birth, the Asian elephant about 100kg (220lb). The calf remains closely dependent on its mother for its first 10 years of life, and continues learning social and survival skills until it is about 17. Elephant society is close-knit and matriarchal, and in addition to its mother, a calf has the support and attention of an extended family of aunts, cousins, and subadult males.

❷ The newborn calf will be immediately encouraged to suckle from its mother's teats, which are positioned between her front legs. Milk is the most important food for the first two years of a calf's life, and the mother does not have another calf during this time. Milk continues to be part of the calf's diet until it is 3 or 4 years of age. If the calf's mother dies, the calf usually starves; other females will not suckle it at the expense of their own offspring.

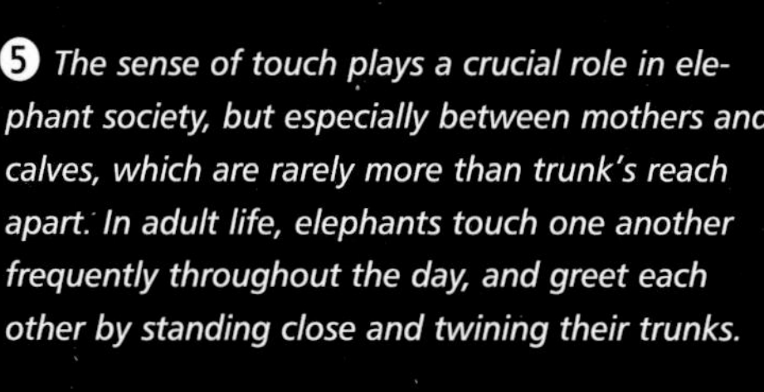

⑤ *The sense of touch plays a crucial role in elephant society, but especially between mothers and calves, which are rarely more than trunk's reach apart. In adult life, elephants touch one another frequently throughout the day, and greet each other by standing close and twining their trunks.*

③ ④ *Play is an essential part of the physical and social development of elephant calves. Spreading its ears wide, a four-month-old calf threatens an imagined enemy. Early morning is a favorite time for playing. Chasing and head-to-head sparring are the preferred games for males, while running through tall grass, chasing birds, throwing sticks, and attacking imaginary enemies are those of female calves. Both sexes, especially juveniles, seem to enjoy rough and tumble games where one elephant clambers over another.*

6

⑥ *Baby elephants learn to eat by putting their trunks inside their mothers' mouths to take food. From the age of 6 months a calf will supplement its milk diet with vegetation and will sample food it sees other herd members eating. Elephants know instinctively which foods are good and which are poisonous, although they can sometimes ingest a toxic meal by accident – a poisonous mushroom growing in the grass, for instance.*

Hyraxes

INTRIGUINGLY, THE ANCIENT HEBREW AND Phoenician word for the hyrax, shaphan, *(meaning, literally, "the hidden one") was responsible for a geographical misnomer. Some 3,000 years ago, Phoenician seamen sailing west through the Mediterranean sighted a coastline teeming with animals they thought were hyraxes, familiar to them from their homeland in the Levant. Accordingly, they called it* Ishaphan *("Island of the Hyrax"). The Romans later modified this to* Hispania, *root of the Spanish* España. *But the animals the sailors saw were in fact rabbits, and so Spain is actually misnamed.*

During the Pliocene – approximately 7–2 million years ago – hyraxes were both widespread and diverse: they radiated from the southern parts of Europe to China in the east, and one fossil form, *Pliohyrax graecus*, was probably aquatic. However, in the present day they are confined to Africa and the Middle East.

The rock hyraxes have the widest geographical and altitudinal distribution, while the bush hyraxes are largely confined to the eastern parts of Africa. Both are dependent on the presence of suitable refuges in "kopjes" – rocky outcrops – and cliffs. As their name suggests, the tree hyraxes are found in arboreal habitats of Africa, but in the alpine areas of the Ruwenzori Mountains they are also rock dwellers. The Eastern tree hyrax might be the earliest type of forest-living tree hyrax, being a member of the primitive fauna and flora of the islands of Zanzibar and Pemba.

Small but Solidly Built

FORM AND FUNCTION

The odd appearance of the hyrax has caused even further confusion. Their superficial similarity to rodents led Storr, in 1780, mistakenly to link them with guinea pigs of the genus *Cavia*, and he thus gave them the family name of Procaviidae or "before the guinea pigs." Later, the mistake was discovered, but the group was given the equally misleading name of hyrax, which comes from the Greek and means "shrew mouse."

Hyraxes are small and solidly built, with a short, rudimentary stump for a tail. Males and females are approximately the same size, reachin a maximum length of about 60cm (24in) and rarely weighing more than a little over 4kg (9lb). The feet are ill-equipped for digging, but have rul bery pads containing numerous glands that exud sweat when the animal is running, which greatly enhances their climbing ability; additionally, part of the underside of the foot can be retracted, which creates something akin to a suction pad to afford the animals extra grip. Species that live in arid and warm zones have short fur, while tree hyraxes and the species that is found in alpine areas have thick, soft fur. Hyraxes have long, tactile hairs at intervals all over their bodies, probabl to aid in orientation when they are in dark fissure and holes. They also have a dorsal gland, which is surrounded by a light-colored circle of hairs that stiffen whenever the animal is excited. The function of this gland is unknown; however, it is thought that it might be important for intraspecific odor communication.

Fossil beds in the Fayum, Egypt, show that 40 million years ago hyraxes were the most importar medium-sized grazing and browsing ungulates. At the time there were at least six genera (though there could have been many more), which range in dimensions from their present size to those th were the size of tapirs. With the first radiation of the bovids (a more advanced mammal) in the course of the Miocene – about 25 million years ago – hyrax populations were considerably reduced and, correspondingly, hyraxes became much less diverse; the animals survived only among the rock and tree habitats into which the bovids did not go.

Modern-day hyraxes retain some primitive features, notably an inefficient feeding mechanism that involves cropping food with the cheek teeth or molars instead of with the incisors, as is the case with modern hoofed mammals. Additionally hyraxes also possess short feet and have poor body-temperature regulation.

Left *Found from southeastern Egypt to northeastern South Africa, Bruce's yellow-spotted hyrax obtains most of its food from browsing. Dominant males will stand guard on high rocks and keep watch while the rest of the group feeds below, though they have also been observed keeping watch while the rest of the group bask on the rocks. If danger approaches, they give a shrill alarm call.*

Coping with a High Fiber Diet

DIET

raxes consume a wide variety of plants. Rock raxes feed mainly on grass, which is a relatively arse material, and therefore have hypsodont ntition – high crowns with relatively short ots, whereas both the browsing bush hyraxes d the tree hyraxes have a diet that is based on ter food and therefore have a brachydont dentin, which comprises short crowns with relatively g roots.

Hyraxes do not ruminate. However, their gut is mplex, comprising three separate areas of crobial digestion, and their ability to digest fiber efficiently is similar to that of ruminants. Their kidneys are efficient enough to allow them to exist on only a minimal moisture intake. In addition, they have a high capacity for concentrating urea and electrolytes, and excrete large amounts of undissolved calcium carbonate. As hyraxes have the habit of consistently urinating in the same place, the crystallized calcium carbonate forms deposits, which whiten the cliffs. These crystals were used as medicine both by several South African tribes as well as by Europeans. The crystals were used for the treatment of such ailments as epilepsy, hysteria, St. Vitus's dance, injuries, and as an abortivum.

KOPJE COHABITANTS

he dense vegetation of the Serengeti and Matobo National Parks, in Tanzania and Zimbabwe respecively, supports two species of hyraxes – the graybrown bush hyrax and the larger dark brown rock hyrax – living together in harmony. Whenever two or more closely related species live together permaently in a confined habitat, at least some of their basic needs must differ, otherwise one species will eventually exclude the other. In the case of these two pecies, the division of food resources is the key facor that allows them to cohabit successfully – thus while the bush hyrax browses on leaves, the rock hyrax feeds mainly on grass. Both species consume plants that are poisonous to most other animals.

When bush and rock hyraxes occur together, they ve in close contact. In the early mornings they hudlle together after spending the night in the same holes. They use the same urinating and defecating places. Births tend to be synchronous, and they how cooperative behavior. Newborns are greeted and sniffed intensively by members of both species. The juveniles associate and form a nursery group; they play together with no apparent hindrance, as play elements in both species are alike. Most of their vocalizations are also similar, including sounds used in threat, fear, alertness, and contact situations. Such a close association has never been recorded between any other two mammal species except among primates and whales.

Even so, bush and rock hyraxes do differ in key behavior patterns. Firstly, they do not interbreed, because their mating behavior is different and they also have different sex-organ anatomy: the penis of the bush hyrax is long and complex, with a thin appendage at the end, arising within a cuplike glans, while that of the rock hyrax is short and simple. The male territorial call, possibly intended as a "keep out" signal, is also different.

Below *A mixed group, comprising Johnston's hyrax and Bruce's yellow-spotted hyrax bask on rocks together in the Serengeti National Park.*

FACTFILE

HYRAXES

Order: Hyracoidea

Family: Procaviidae

11 species in 3 genera

DISTRIBUTION Africa and the Middle East.

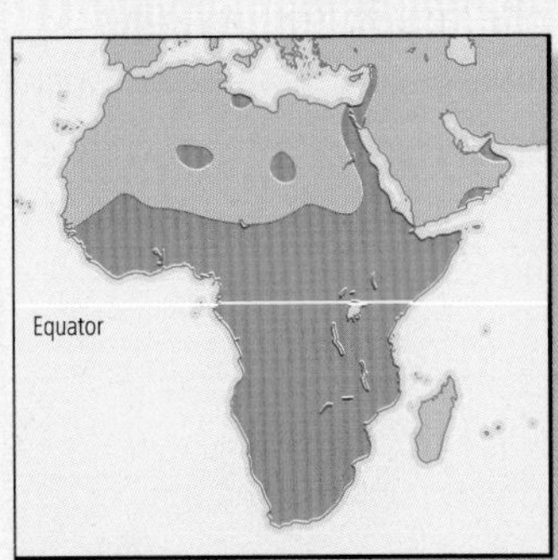

ROCK HYRAXES OR DASSIES Genus *Procavia*
Abyssinian hyrax (*P. habessinicus*); **Cape hyrax** (*P. capensis*); **Johnston's hyrax** (*P. johnstoni*); **Kaokoveld hyrax** (*P. welwitschii*); **Western hyrax** (*P. ruficeps*). SW and NE Africa, Sinai to Lebanon and SE Arabian Peninsula. Rock boulders in vegetation zones ranging from arid to the alpine zone of Mt. Kenya (3,200–4,200m/ 10,500– 13,800ft). Active in daytime. HBL 44–54cm (17–21in); WT 1.8–5.4kg (4–12lb). Coat: light to dark brown; dorsal spot dark brown in Cape hyrax, yellowish-orange in Western hyrax. Breeding: gestation 212–240 days. Longevity: 9–12 years.

BUSH HYRAXES Genus *Heterohyrax*
Ahaggar hyrax (*H. antineae*); **Bruce's yellow-spotted hyrax** (*H. brucei*); **Matadi hyrax** (*H. chapini*). SW, SE to NE Africa. Rock boulders and outcrops in different vegetation zones in E Africa, sometimes in hollow trees; active in daytime. HBL 32–47cm (12.5–18.5in); WT 1.3–2.4kg (2.9–5.3lb). Coat: light gray, underparts white, dorsal spot yellow. Breeding: gestation about 230 days. Longevity: 10–12 years. Conservation status: *H. antineae* and *H. chapini* are both listed as Vulnerable.

TREE HYRAXES Genus *Dendrohyrax*
Eastern tree hyrax (*D. validus*) Kilimanjaro, Meru, Usambara, Zanzibar, Pemba, and Kenyan coast, **Southern tree hyrax** (*D. arboreus*) SE and E Africa, **Western tree hyrax** (*D. dorsalis*) W and C Africa. Evergreen forests up to about 3,650m (12,000ft); Southern tree hyrax among rock boulders in Ruwenzori. HBL 32–60cm (12.5–24in); WT 1.7–4.5kg (3.7–9.9lb). Coat: long, soft, and dark brown; dorsal spot light to dark yellow, from 20–40mm/ 0.8–1.6in long (Eastern tree hyrax) to 40–75mm/ 1.6–3in (Western tree hyrax). Breeding: gestation 220–240 days. Longevity: 10 years plus. Conservation status: *D. validus* is listed as Vulnerable, as is the South African subpopulation of *D. arboreus*.

Note The 1993 established list of mammal species regarded the Abyssinian, Kaokoveld, Johnston's and Western hyrax as synonymous with *P. capensis*, and the Matadi hyrax as a subspecies of *H. brucei*.

Abbreviations HBL = head–body length WT = weight

Gregarious Group-Dwellers

SOCIAL BEHAVIOR

To make up for poor physiological control of body temperature, hyraxes huddle together and bask in the sun. They are also relatively inactive and have low metabolic rates, which enables them to exist in very dry areas that provide food of poor quality. A crucial requirement, however, is shelter that provides them with relatively constant temperature and humidity.

Hyrax social organization varies in relation to living space. On kopjes smaller than 4,000sq m (43,000sq ft), both rock and bush hyraxes live in cohesive and stable family groups consisting of 3–7 related adult females,1 adult territorial male, dispersing males, and the juveniles of both sexes. Larger kopjes may support several family groups, each occupying a traditional range. The territorial male repels all intruding males from an area largely encompassing the females' core area (the average for bush hyraxes is 27 animals in 2,100sq m/ 22,600sq ft, and for rock hyraxes 4 animals to 4,250sq m/45,750sq ft).

The females' home ranges are not defended and may overlap. Rarely, an adult female will join a group and will eventually be incorporated. Females become receptive about once a year, and a peak in births seems to coincide with rainfall. Within a family group, the pregnant females all give birth within a period of about three weeks. The number of young per female bush hyrax varies between 1 and 3 (mean 1.6), and in rock hyraxes between 1 and 4 (mean 2.4). The young are fully developed at birth, and suckling young of both species assume a strict teat order. Weaning occurs at 1–5 months and both sexes reach sexual maturity at about 16–17 months. Upon sexual maturity females usually join the adult female group, while males disperse before they reach 30 months. Adult females live significantly longer than adult males.

There are four classes of mature male: territorial and peripheral, plus early and late dispersers. Territorial males are the most dominant. Their aggressive behavior toward other adult males escalates in the mating season, when the weight of their testes increases a record twentyfold. These males monopolize receptive females, and show a preference for copulating with females over 28 months of age. A territorial male monopolizes "his" female group year-round, and repels other males from sleeping holes, basking places, and feeding areas. Males can fight to the death, although this is probably quite rare. While his group members feed, a territorial male will often stand guard on a high rock and be the first to call in case of danger. The males utter the territorial call all year round.

On small kopjes, peripheral males are unable settle, but on large kopjes they can occupy areas on the edge of the territorial males' areas. They live solitarily, and the highest-ranking takes over female group when a territorial male disappears. These males show no seasonality in aggression b call only in the mating season. Most of their mating attempts and copulations are with females under 28 months – older males probably do not bother mating with females this young as first pregnancies tend to have a higher mortality rate.

Right A Cape hyrax with its offspring. The young are usually born in a crevice, in a litter that may vary in size between a single young and perhaps half a dozen, though most commonly will comprise 2–3. The young are born with hair and become mobile quite soon after birth, usually within 24 hours or so.

Left A Southern tree hyrax eating *Acacia* leaves. Although one genus is commonly referred to as the "tree hyraxes," all hyraxes are adept at climbing and those from other genera will sometimes bask in trees. However, the tree hyraxes' arboreal dependence puts it most seriously at risk from deforestation.

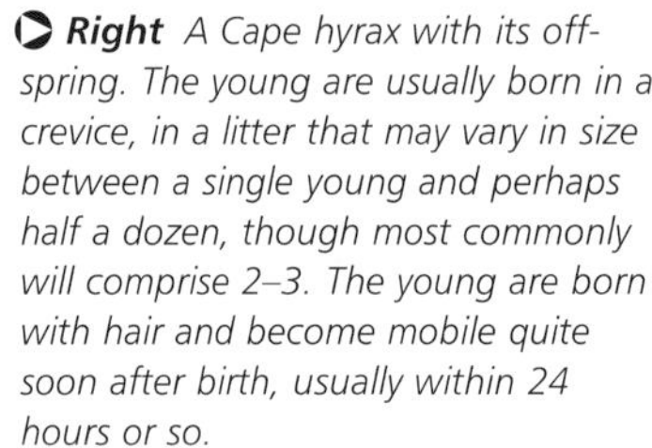

Right Mating in hyraxes is both brief and vigorous. The anatomy of the penis varies quite significantly between the three genera. In rock hyraxes it is short, simply built, and elliptical in cross section; in tree hyraxes it is similarly built and slightly curved; in bush hyraxes, illustrated here, it is long and complex: on the end of the penis, and arising within a cuplike glans penis, is a short, thin appendage, which has the penis opening. **1** The male presses the penis against the vagina; **2** violent copulation takes place, in which the male leaves the ground; **3** the female moves forward, causing the male to withdraw.

The early dispersers include the majority of the venile males. These leave their birth sites at etween 16–24 months old, soon after reaching xual maturity. The late dispersers leave a year ter, but still leave before they are 30 months old. efore leaving their birth sites, both early and late spersers have ranges that overlap with their others' home ranges. They disperse in the mat- g season to become peripheral males. Almost o threatening, submissive, or fleeing behavior as been observed taking place between territorial ales and late dispersers.

Individual rock and bush hyraxes have been bserved to disperse over a distance of at least 2km (1.2mi). However, the greater the distance a dispersing animal has to travel across the open grass plains, where there is little cover and few hiding places, the more the animal is at risk, either of predation or as a result of its inability to cope with temperature stress.

Long-term observations in Serengeti, Matobo (Zimbabwe), and Israel show that hyrax colonies fluctuate, and that small colonies are prone to extinction. Preliminary DNA analysis of rock and bush hyraxes in the Serengeti suggests that there is very little genetic variation among kopjes. Dispersing occurs among closely-located kopjes (around 10 km/6mi distance).

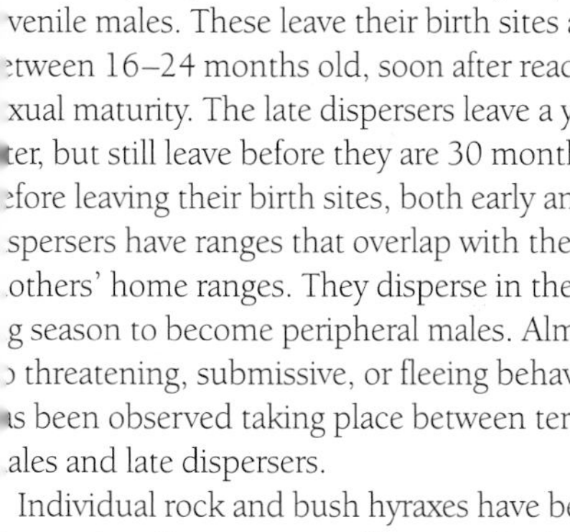

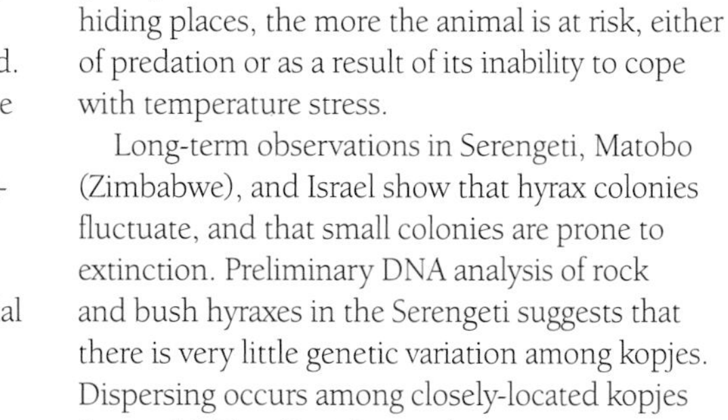

Enemies Within and Without

CONSERVATION AND ENVIRONMENT

The most important predator of hyraxes is the Verreaux eagle, which feeds almost exclusively on them. Other predators are the Martial and Tawny eagles, leopards, lions, jackals, Spotted hyenas, and several snake species. External parasites such as ticks, lice, mites, and fleas, and internal parasites such as nematodes, cestodes, and anthrax, also probably play an important role in hyrax mortality. In Kenya and Ethiopia it was found that rock and tree hyraxes might be an important reservoir for the parasitic disease leishmaniasis.

The Eastern tree hyrax is extensively hunted for its fur in the forest belt surrounding Mt. Kilimanjaro; it takes 48 pelts to yield one rug. Because African forests are disappearing at an alarming rate, the tree hyraxes are probably the most endangered of all the hyrax species, especially in small areas of woodland such as those on the islands of Zanzibar and Pemba and in the Usambara Mountains of Tanzania. No recent information is available on the status of *Heterohyrax antineae* in Algeria or *H. chapini* in Congo, but their status is deemed Vulnerable. HNH

Aardvark

FEW PEOPLE HAVE HAD THE FORTUNE OF A *close encounter with the aardvark, one of Africa's most bizarre and specialized mammals. Nocturnal and secretive, this ant- and termite-eater is the only living member of the order Tubulidentata. Thanks to its elusiveness, it is also one of the least-known of living mammals.*

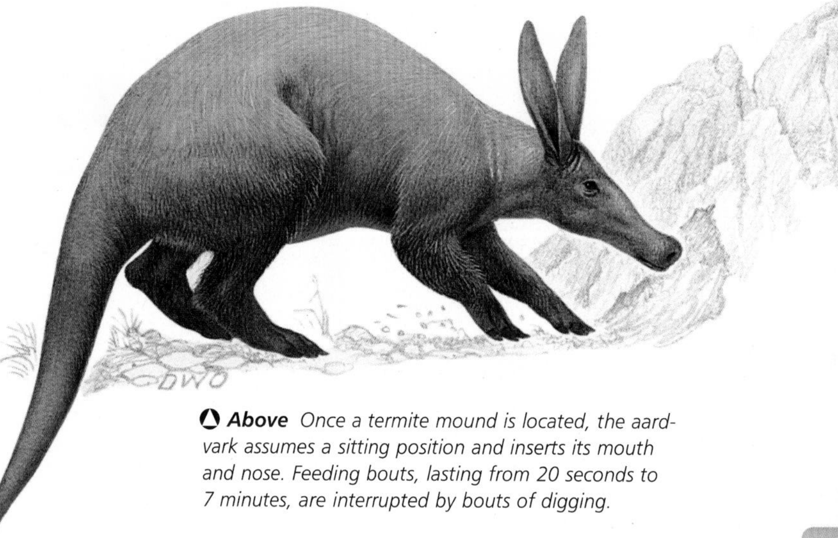

Above *Once a termite mound is located, the aardvark assumes a sitting position and inserts its mouth and nose. Feeding bouts, lasting from 20 seconds to 7 minutes, are interrupted by bouts of digging.*

Aardvark means, literally, "earthpig" in Afrikaans, but the resemblance to a pig is purely superficial. The animal has large ears, a coarse-haired body, and a tubular snout – but one much longer than any pig's, and shaped to the requirements of a diet as specialized as the typical pig's is omnivorous.

Adaptations for Ant-eating

FORM AND FUNCTION

Many adaptations in the aardvark's anatomy equip it for eating ants and termites. With its tubular snout close to the ground, it can make sweeping motions 30cm (12in) wide, following a zigzag course sniffing for food. The same route may be used on consecutive nights. Pushing the end of its soft nose firmly into a suitable patch of ground, the aardvark inevitably sucks up some soil, which is prevented from entering the lungs by a filter of dense hairs in its nostrils. A wall between the nasal slits is equipped with a series of thick, fleshy processes, probably used for sensing the jittery prey.

When an aardvark finds the prey that it is seeking, it will rapidly dig a V-shaped furrow with its forefeet. Its short, powerful limbs have four digits on the front feet and five on the back, all equipped with long, sharp-edged, spoon-shaped claws that make light work of the digging. A sticky, round, long, thin tongue and well-developed salivary glands act like flypaper to ensure aardvarks get the most out of every termite mound. Their stomachs have a muscular pyloric area, which functions like a gizzard, grinding up the food. Aardvarks therefore do not need to chew their food, and so have no need for incisors or canine teeth; instead, their continuously growing, open-rooted cheek teeth consist of two upper and two lower premolars and three upper and three lower molars in each half- jaw. The cheek teeth differ from those of other mammals in that the dentine is not surrounded by enamel but by the thin, bonelike tissue known as cementum.

Both male and female aardvarks have anal scent glands that emit a pungent, yellowish secretion, probably to advertise social status. Large ears signal the animal's acute hearing, but their typical posture during foraging is vertically upward, not earthward. This suggests that aardvarks do not strain to hear the sounds of termites, but rather listen out for predators.

In order to get enough nourishment, aardvarks readily eat more than 50,000 insects in a night. In one study, they were found to consume 21 different species of ants, plus 2 species of termite (the latter mostly in the winter). Aardvarks share their habitat with a variety of other termite- and ant-eating animals, including hyenas, jackals, vultures, storks, geese, pangolins, Bat-eared foxes, and aardwolves, but all of these also take other prey, so reducing competition.

Nighttime Foragers

SOCIAL BEHAVIOR

Aardvarks are almost exclusively nocturnal and solitary. They normally emerge from their burrows shortly after nightfall, though they may come out late in the afternoon in winter. Individuals tracked by radio in South Africa's arid Karoo plateau were found to be most active in the early evening, from 8pm to midnight, though the moon was apparently not a factor in their decision to venture forth. Their home ranges are well-organized, with a system of burrows that provide very efficacious refuges for animals seeking shelter from storms or hungry predators. The animals in the Karoo study foraged over distances varying from 2–5km (1.2–3mi) each night, covering an average distance of 550m (1,800ft) an hour without stopping to rest. Their nightly route was circular, but did not encompass their entire home range. Other studies suggest that aardvarks may range as far as 15km (9mi) during a 10-hour foraging period, covering up to 30km (19mi) in a night.

In the Karoo survey, home ranges for both males and females varied from 133–384ha

FACTFIL

AARDVARK

Orycteropus afer

Aardvark, antbear, or Earth pig

Order: Tubulidentata

Family: Orycteropodidae

18 subspecies have been listed, but most may be invalid; there is insufficient knowledge of the animal for firm conclusions to be drawn.

DISTRIBUTION Africa S of the Sahara, excluding deserts.

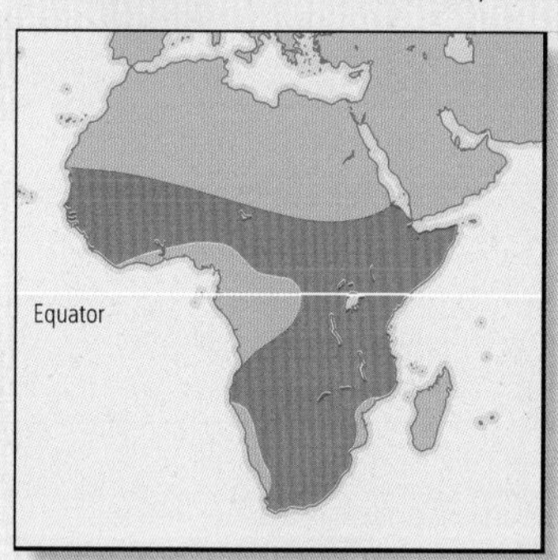

HABITAT Mainly open woodland, scrub, and grassland; rarer in rain forest; avoids rocky hills.

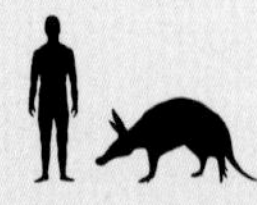

SIZE Head–body length 105–130cm (41–51in); **tail length** 45–63cm (18–25in); **weight** 40–65kg (88–143lb). Both sexes same size.

COAT Pale, yellowish gray, with head and tail off-white (the gray to reddish-brown color often seen results from staining by soil, which occurs when the animal is burrowing). Females tend to be lighter in color.

DIET Mainly ants and termites.

BREEDING Gestation 7 months.

LONGEVITY Up to 10 years in captivity.

CONSERVATION STATUS Not threatened

(328–949 acres), and overlapped considerably with those of adjacent animals. Individuals spent half their time in a core area that took up 25–33 percent of the range. The proportion of time spent above ground feeding was higher in summer than in winter. In winter, aardvarks may remain active above ground for an average of 5 hours, regularly returning to their burrows before midnight, but in summer their nightly sorties may last 8–9 hours.

Aardvarks make burrows of three main types: those dug when looking for food; larger, temporary sites, scattered through the home range, that may be used for refuge; and permanent refuges in which the young are born. The latter are deep and labyrinthine, up to 13m (43ft) long, and usually have more than one entrance. Aardvarks like to change the layout of their homes regularly, and can, if necessary, dig new burrows with considerable speed, disappearing below ground within 5–20 minutes. In the Karoo study, burrowing was estimated to affect 0.05 percent of the total land area per year, and burrows were used for from 4 to 38 consecutive nights, after which they were abandoned. Only mothers and their young share burrows. Droppings are deposited in shallow digs throughout their range and covered with soil.

Little information is available on reproduction, but a single offspring is born weighing around 2kg (4lb), probably just before or during the rainy season, when termites are readily available. The infant first ventures out of the burrow to accompany its mother when only 2 weeks old. It will start digging its own burrows at about 6 months, but may stay with the mother until the onset of the next mating season.

Specialists Ill-fitted for Change

CONSERVATION AND ENVIRONMENT

Due to their specialized food preferences, aardvarks are extremely vulnerable to habitat changes. Intensive crop farming may reduce their density, but increased cattle herding benefits them, as trampling creates the right conditions for termites. However, little progress can be made in formulating management policies until more is known about the behavior and ecology of the animals.

Aardvarks are killed not just for their flesh, which is said to taste like pork, but also for their teeth, worn on necklaces by members of the Margbetu, Ayanda, and Logo tribes of the Democratic Republic of the Congo to prevent illness and as good-luck charms. The animal's bristly hair is sometimes reduced to a powder that is regarded as a potent poison when added to the local beer. It is also believed that the harvest will be increased when aardvark claws are put into baskets used to collect flying termites for food. RJvA

Left The aardvark's long, tubular ears give it vital early warning of predators. Once it has dug far down into the soil, it can fold its ears flat, so preventing them from filling with earth and impairing its hearing.

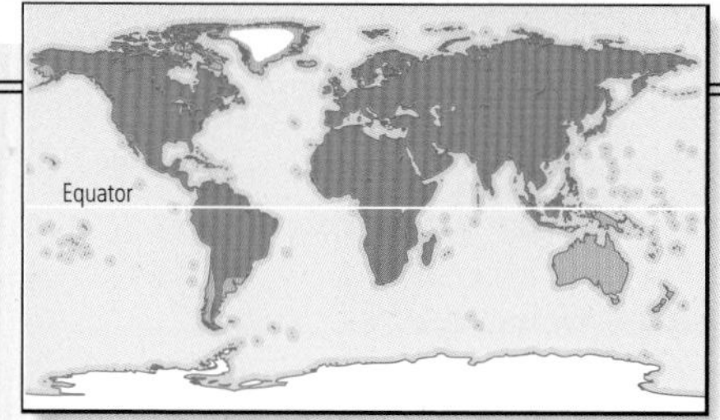

HOOFED MAMMALS

THE DESIGNATION "UNGULATE" IS A GENERAL term applied to all those groups of mammals that have substituted hooves for claws in their evolution. This character appears to follow from a commitment to a terrestrial, herbivorous lifestyle, with rapid locomotion. Ungulates are relatively large animals, none less than 1kg (2.2lb) in body mass. Most are exclusively herbivorous, and they comprise around 80 percent of species of terrestrial mammals over 50kg (110lb). Living ungulates mainly belong to the orders Perissodactyla and Artiodactyla, which diverged from a common ancestor among the primitive "condylarths" some 65 million years ago. (Other ungulates include the subungulates – elephants, hyraxes, and sea cows.)

Despite the superficial similarities between horses and antelopes, rhinos and hippos, tapirs and pigs, the former of each pair belongs to the Perissodactyla (odd-toed ungulates) and the latter to the Artiodactyla (even-toed ungulates). The similarities between them have largely come about due to convergent evolution (natural selection favoring similar body plans in animals evolving from different ancestry). Recent molecular evidence has confirmed that the groupings of these animals based on their morphology is broadly correct and that the two orders are closely related, although their relationship with the subungulates is probably more distant than previously assumed. Intriguingly, there is continued debate about whether whales should be included within the order Artiodactyla (as preferred by molecular biologists) or as the sister taxon of the Artiodactyla (Cetacea; as preferred by morphologists).

Body Shape and Senses

ANATOMY AND PHYSIOLOGY

Despite the variety of bodily shapes and adornments, from a 1kg (2.2lb) chevrotain to a 3-tonne hippopotamus, and from the ponderous rhinoceros to the graceful horse, there is an underlying common theme to the two lineages of modern ungulates. Ungulates generally have long-muzzled heads, held horizontally on the neck, barrel-shaped bodies carried on legs of roughly equal length, and small tails. Their skin is quite thick and tends to carry a coat of hairs (which may be air-filled for insulation) rather than soft fur. The primitive mammalian limb pattern ends in a foot with five digits, all of which are placed on the ground in standing and during locomotion. All ungulates have reduced the number of toes, and none retains the first digit. Ungulates have also lengthened the metapodials (the long bones in the fleshy parts of human hands and feet), and these bones are compressed with movement of the limb mainly in the backward and forward planes. Ungulate "hands and feet" are used only for locomotion and are incapable of grasping. The term "unguligrade" refers to the ungulate posture, in which the animal has evolved to balance on the tips of its toes (like a ballerina on pointe). The long legs and tiptoe stance of ungulates bestow both speed and endurance. The pronghorn, or American antelope (*Antilocapra americana*), is probably the fastest living ungulate, clocked at speeds of up to 86km/h (53mph). Asiatic asses (*Equus hemionus*) are almost as fast at sprinting, and have a sustained galloping speed of 50km/h (31mph).

As the names imply, odd- and even-toed ungulates differ in the type and degree of modification of their feet, and in the number of toes. Even-toed ungulates (artiodactyls) have two or four toes on each foot. The animal's weight is supported on an axis that runs between the third and fourth digits. In ruminants (cud chewers) and camels the third and fourth metapodials are elongated and fused together: the two toes then appear to emerge from a single bone, resulting in a "cloven-hoofed" appearance of the foot. Odd-toed ungulates (perissodactyls) have a single toe or three toes together bearing the weight of the animal, with the axis of the limb passing through the middle (third) digit. All modern species have lost the fifth digit in the hind foot. The metapodials are unfused and relatively short in three-toed forms such as rhinos and tapirs. The evolutionary equivalent in perissodactyls to the long-legged, cloven-hoofed limb of artiodactyls is the elongated single metapodial seen in horses.

The earliest horses were three-toed ungulates, but since the late Eocene (around 40 m.y.a.) horses have borne their weight on the single third toe, with ligaments rather than a fleshy pad for support. The next step toward modern horses is seen in post-Eocene fossils, in which the second and fourth digits are minimized, but would have provided support for the foot in galloping and jumping. These fossil horses did possess a fully-formed hoof. In addition, the metapodials were greatly elongated (although not fused together) to form a long, slender lower limb. All living equid species have reduced these side toes to proximal splint bones, and bear their entire weight at all times on an enlarged single hoof. Single-toed (monodactyl) horses date from the late Miocene (c. 10 m.y.a.).

Ungulates have good but not exceptional hearing, small ears that can be rotated to detect the direction of a sound, an apparently very good sense of smell, and many have excellent eyesight (e.g., horses). The eyes function well by day and night, and they have binocular vision, especially in open-country species, allowing the animals to judge distance and speed accurately. Ungulate communication involves mainly visual signals and sounds, heavily complemented by the use of scent in forest-dwelling species (although less so in open-country species). Perissodactyls lack the diversity of scent glands found on the feet and faces of many artiodactyls. Instead they rely more on vocal communication, with the frequent production of a large variety of sounds; ruminant

ORDERS: PERISSODACTYLA AND ARTIODACTYLA

13 families; 88 genera; 212 species

ODD-TOED UNGULATES
Order Perissodactyla – perissodactyls

HIPPOMORPHS Suborder Hippomorpha

HORSES, ZEBRAS, AND ASSES
Family Equidae – equids p468

7 (9) species in 1 genus
Includes **African ass** (*Equus asinus*), **Domestic horse** (*E. caballus*), **Plains zebra** (*E. burchellii*).

CERATOMORPHS Suborder Ceratomorpha

TAPIRS Family Tapiridae p474

4 species in 1 genus
Includes **Brazilian tapir** (*Tapirus terrestris*).

RHINOCEROSES Family Rhinocerotidae p476

5 species in 4 genera
Includes **Black rhinoceros** (*Diceros bicornis*) and **Indian rhinoceros** (*Rhinoceros unicornis*).

EVEN-TOED UNGULATES
Order Artiodactyla – artiodactyls

SUOIDS Suborder Suina

WILD PIGS AND BOARS Family Suidae p484

13 (16) species in 5 genera Includes **Wild boar** (*Sus scrofa*) and **warthog** (*Phacochoerus africanus*).

PECCARIES Family Tayassuidae p488

3 species in 2 (3) genera
Includes **Collared peccary** (*Tayassu/Pecari tajacu*).

HIPPOPOTAMUSES Family Hippopotamidae p490

2 (4) species in 2 genera
Hippopotamus (*Hippopotamus amphibius*) and **Pygmy hippopotamus** (*Hexaprotodon liberiensis*).

TYLOPODS Suborder Tylopoda

CAMELS AND LLAMAS
Family Camelidae p496

6 species in 3 genera Includes **Bactrian camel** (*Camelus bactrianus*) and **guanaco** (*Lama guanicoe*).

RUMINANTS Suborder Ruminantia

CHEVROTAINS Family Tragulidae p500

4 species in 3 genera
Includes **Water chevrotain** (*Hyemoschus aquaticus*).

MUSK DEER Family Moschidae p502

4 species in 1 genus
Includes **Musk deer** (*Moschus moschiferus*).

DEER Family Cervidae p504

38 (43) species in 16 genera
Includes **Red deer** (*Cervus elaphus*), **reindeer** (*Rangifer tarandus*), and **moose** (*Alces alces*).

GIRAFFE AND OKAPI Family Giraffidae p520

2 species in 2 genera **Giraffe** (*Giraffa camelopardalis*) and **okapi** (*Okapia johnstoni*).

PRONGHORN Family Antilocapridae p528

1 species **pronghorn** (*Antilocapra americana*).

CATTLE, ANTELOPE, SHEEP, AND GOATS
Family Bovidae p530

123 (137) species in 47 (45) genera
Includes **bison** (*Bison bison*), **gemsbok** (*Oryx gazella*), and **ibex** (*Capra ibex*).

Note The 1993 established taxonomy recognized 13 families, 87 genera, and 238 species (given in parentheses above).

Below Africa hosts the greatest diversity of hoofed nmals. In often arid landscapes, bodies of water as focal points, attracting a variety of different ulate species (along with their predators). In Etosha National Park, Namibia, zebra (odd-toed ungulates of the equid family) and eland (even-toed ungulates belonging to the bovid family) gather together to drink at a waterhole.

odactyls such as deer and antelope are virtually te in comparison with the variety of snorts, innies, and other vocalizations of horses. Peris- lactyls also produce a much greater variety of ial expressions than do artiodactyls, in part cause their mode of feeding (using the lips her than the tongue) has bestowed on them a ater complexity of facial musculature.

Processing Cellulose

FOOD, FEEDING, AND DIGESTION

ungulates are predominantly terrestrial herbi- es, feeding on leaves, flowers, fruits, or seeds rees, herbs, and grasses (although pigs and ccaries are characteristically more omnivorous, d may include roots, tubers, and animal flesh in their diet). With rare exceptions, all artiodactyls and perissodactyls stand on the ground to feed; even the amphibious hippopotamus grazes on land. They cannot use forelimbs to manipulate food but must take it directly from the plant, or off the ground if it has fallen, with their lips, teeth, and tongue, and these structures are appropriately modified. The tongue of giraffes and okapis is so long and flexible that they can even employ it to clean their eyes.

Even though ungulates eat plants, not all plants are equal in terms of the nourishment they provide. Plant foliage tends to be abundant in carbohydrates such as sugars and starches, which are easily digested sources of energy, but it is low in fat and frequently in protein as well. A low-fat diet does not seem to be a problem for ungulates and many have lost the gall bladder, which in other mammals produces bile salts that emulsify and break down fats. From proteins, however, come amino acids, from which the body constructs itself. The most abundant source of vegetable protein is seeds, but these are generally small and widely dispersed, so herbivores can have a difficult time obtaining enough dietary protein. This applies particularly to the smaller ungulates, with their higher rates of nutrient turnover. Very occasionally, small antelopes such as duikers will even catch and eat birds as an additional source of protein, while pigs will readily consume carrion. The carbohydrates in vegetation essentially occur in two forms: in the liquid cell contents, and in the fibrous cell wall. The cell wall is indigestible to most animals. To deal with this, ungulate dentition is adapted to grinding; the cheek teeth (molars and premolars) smash the cell wall to release the digestible contents. The back of an ungulate's mouth functions like a mill, with large, flat, square molars that reduce plant matter to fine particles. In conjunction with this, the jaw musculature and skull configuration are both modified so that the lower tooth rows can be moved across the upper with a sweeping transverse grinding motion. By contrast, most other mammals use a more up-and-down motion that simply cuts and pulps food. Some ungulates (mainly grazers) have high-crowned (hypsodont) cheek teeth that are made to last a lifetime of continual abrasion. There is typically a gap (called the diastema) between the milling molars and the plucking incisors. The original mammalian function of canines was as instruments for stabbing or piercing, but among ungulates only the pig family still use them as such. In perissodactyls the canines have been lost or, as in the case of horses, retained only in the males as small pegs for use in display. In ruminant artiodactyls the lower canines have been modified to form part of the lower incisor row. The upper canines are retained in just a few species, in which they are unusually elongated, for example in deer whose males either have small antlers or none at all (such as Musk deer, Chinese water deer, and muntjacs).

Some artiodactyls, such as pigs, prefer eating nonfibrous vegetation like fruits and tubers. These species do not digest cellulose and have a monogastric digestive system resembling that of most other mammals. Other ungulates ingest large quantities of cellulose, however, and thus require a means to digest it. Microorganisms (bacteria and protozoa) are employed for this task within the gut, transforming cellulose into products that can then be absorbed and utilized. Where and how these microorganisms are employed in the gut separates the perissodactyls from those artiodactyls that eat fibrous vegetation. A "ruminant" artiodactyl (a member of the suborder Ruminantia or Tylopoda) has a fermentation chamber situated within a complex, multi-chambered stomach. This complex stomach allows food to be differentiated, with digested food passing through a sieve-like structure to the posterior "true" stomach and the intestines, whereas undigested food is set aside to ferment, from whence it is regurgitated to the mouth to be chewed again. This is called rumination, or "chewing the cud." This fermentation system is very efficient, making maximum use of the available cellulose during the extended period (up to four days) that each mouthful of food is retained in the gut. A simpler form of forestomach fermentation that does not involve rumination is seen in the more herbivorous of the suine artiodactyls, such as peccaries and hippos.

In perissodactyls the hindgut areas of the cecum and colon (large intestine) are the sites of fermentation. Although the processes are biochemically identical to those in ruminants, the digestion of cellulose is less efficient, as the food is only retained for about half the time: a horse will process grass in around 48 hours, as opposed to around 80 hours in a cow. Moreover, while the cow is 80 percent efficient in terms of cellulose digestion, a horse is only about 50 percent efficient. This difference means that the horse has to consume a greater quantity of food per day, which it can do by virtue of its faster passage rate. An analogy is to compare a corner grocery store with a supermarket: the corner store ("ruminant") has a low rate of sales, but recovers a relatively high profit from each item sold; a supermarket ("hindgut fermenter") has a high rate of sales, and so can reduce its profit margin on each of the same items.

Ruminants are able to utilize nutritious young, short herbage, and thus require lower daily intake, surviving where food is nutritious but in short supply. Hindgut fermenters are at an advantage where food is abundant but of limited quality and high in fiber. Desert inhabitants such as the oryx and camel, or Arctic tundra inhabitants such as reindeer or musk oxen, are examples of large herbivores that are successful by virtue of the ruminant digestive system. In habitats such as the tropical savannas of Africa, where both types of animal coexist, there is a partitioning of cropping. For example, zebras (perissodactyls) eat the poorer-

THE UNGULATE BODY PLAN

Right Ungulates have evolved two very diffe[rent] systems for dealing with the indigestible cellulos[e] [of] their highly fibrous food: hindgut fermentation [and] rumination. In the hindgut fermenters (perissodactyls), cell contents are completely digested in [the] (simple) stomach, and then pass to the cecum a[nd] the large intestine, where the cellulose of the pl[ant] cell walls is fermented by microorganisms. Ruminants have a more complex digestive syster[m] and retain food in the gut for much longer. Foo[d] initially passes to the first stomach chamber (the rumen) where it is all fermented by microorganis[ms] and is then regurgitated to be re-chewed. This r[e-]chewing (rumination) regulates food particle size and the small particles eventually pass through t[he] second (reticulum) and third (omasum) chamber[s to] the fourth chamber (the true stomach or aboma[sum]). Surplus microorganisms accompany the fi[ber] particles and provide a source of protein. Digest[ion] is completed in the abomasum, and nutrients are absorbed in the small intestine. Some additional fermentation and absorption occurs in the cecu[m].

Below Ruminants must pay a price for their highly efficient digestive system: it takes a long t[ime] for food to pass through their guts. In large herb[i]vores this price outweighs the advantages of effi[]cient digestion, because large animals have to ea[t a] large volume of food per day and do not have ti[me] to find the scattered items of high-quality food t[hat] smaller species can. They must therefore accept l[ow-]quality food that can be gathered in large quanti[ties], but the ruminant system takes so long to proces[s] such food that a greater net intake of nutrients [is] achieved with a simple gut and fast throughput. Hippos are the largest artiodactyls, and despite h[av]ing some forestomach fermentation, they have a simple stomach, do not chew the cud, and have a faster rate of food passage than ruminants. Sm[all] animals eat less per day than large ones, but the rate at which they burn energy per kilogram of b[ody] mass is much faster. Very small ungulates must t[hus] process their food relatively rapidly, and cannot c[ope] with the slow passage rate required by the rumi[nant] system. This is why small ruminants (e.g. chevrotains), select a high-quality diet that needs little f[er]mentation. Ruminants therefore have a range of sizes outside of which they cannot operate.

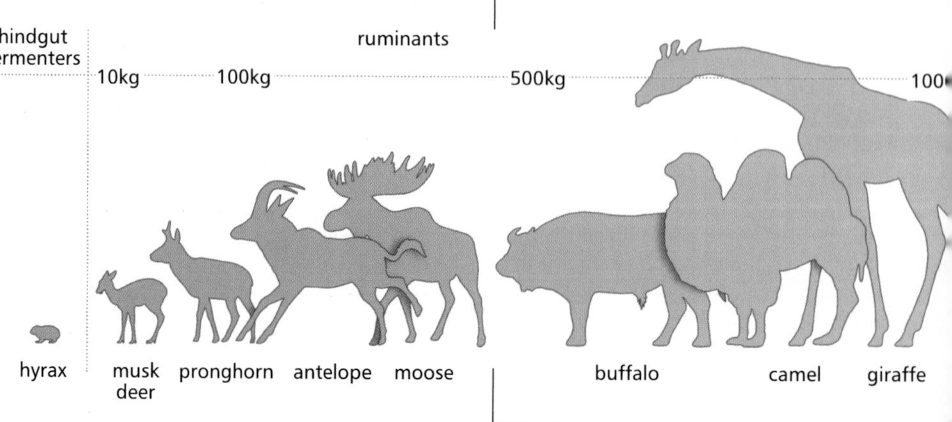

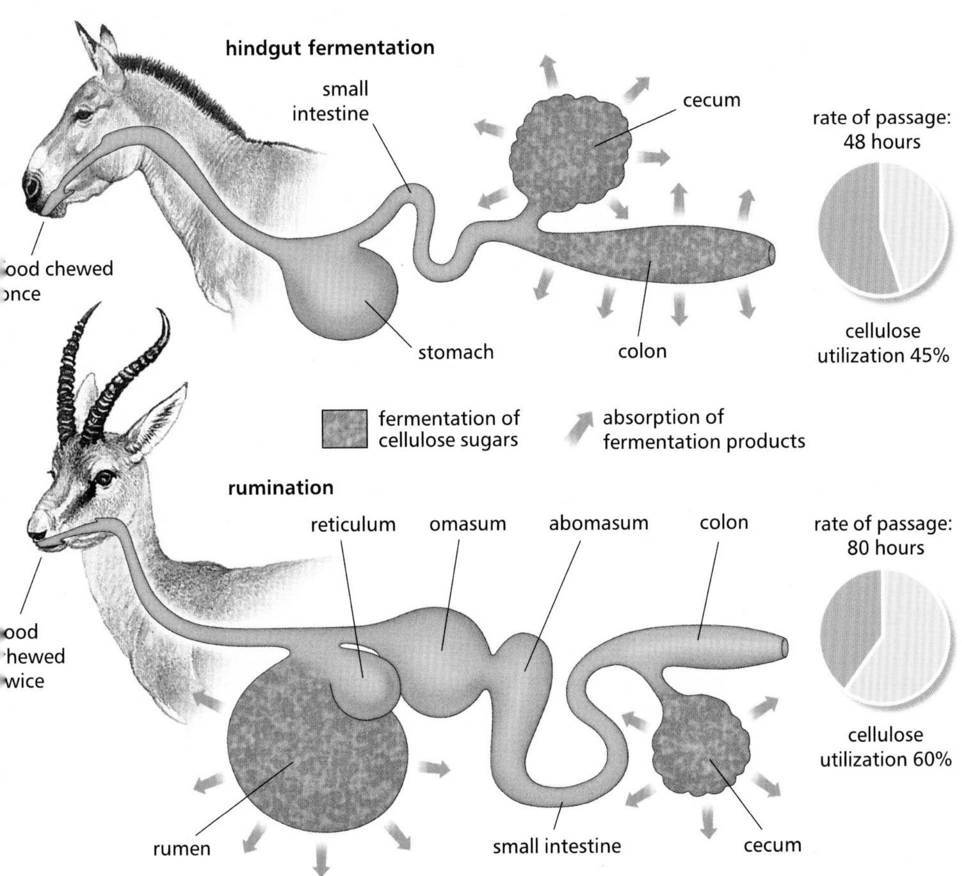

Below Primitive herbivorous mammals have
olars with separate, low, rounded cusps (buno-
nt) rather like our own, designed to pulp and
ish relatively soft food, as seen in omnivorous
gs **a**. Fibrous vegetation is tough, and browsing
d grazing ungulates have modified these simple
sps into high-relief shearing ridges (lophs). In
rissodactyls, such as the rhinoceros **b**, the ridges
n in a primarily crosswise direction across the
oth in a "lophodont" pattern. Horses **c** have mod-
ed this original pattern into a highly complex and
ded one. In ruminant artiodactyls like the deer **d**,
e ridges are crescent-shaped and run in a primarily
ngthwise direction in a "selenodont" pattern.

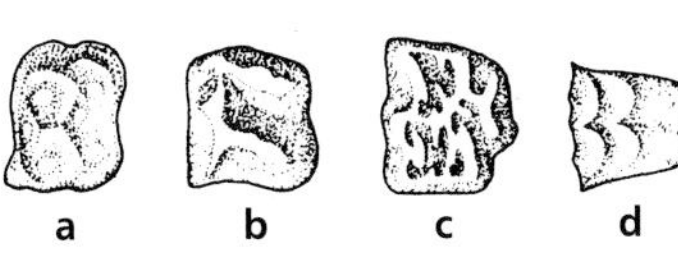

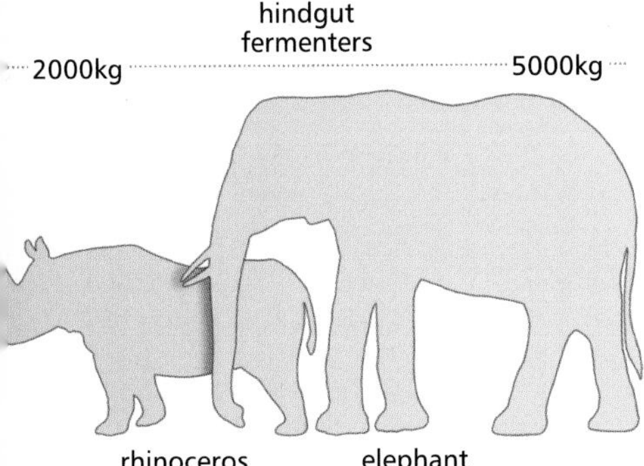

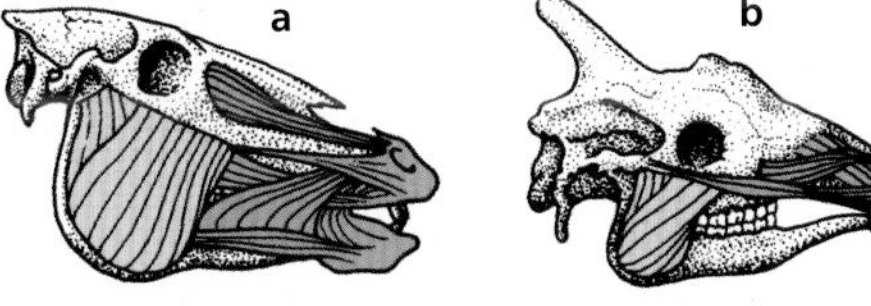

Above The different modes of feeding in perissodactyls and ruminant artiodactyls are reflected in the size of the jaw and musculature. Non-ruminant grazers, such as the horse **a**, have to consume large quantities of tough, fibrous food, and consequently the lower jaw is very deep to accommodate the very large masseter muscle, used for grinding food with the cheek teeth. Ruminants, such as the giraffe **b**, consume less food per day, and much of their chewing is of the already half-digested cud; consequently the lower jaw and the masseter muscle are much less pronounced. Most perissodactyls (with the exception of some rhinos) retain both sets of incisors and use the upper lip extensively in feeding, like the horse **a**. Ruminant artiodactyls, like the giraffe **b**, have lost the upper incisors and make extensive use of a prehensile tongue rather than the upper lip in food gathering. The resulting differences in facial musculature mean that perissodactyls have a much greater variety of facial expressions, used to communicate with each other, than do artiodactyls.

Below In the hoofed mammals, the primitive mammalian foot **a** has been modified in various ways. The toes are reduced in number and the long bones (metapodials) much extended. The foot is held lifted, with only the tips of the toes on the ground (unguligrade). The joint surfaces are restricted so that the limbs cannot be rotated or moved in or out of the body to any great extent. The prime movement is thus fore and aft, which facilitates fast running at the expense of climbing and digging. In the generalized ungulate feet **b–c**, one or two digits are lost, the metapodials somewhat elongated, the tarsal bones more ordered: **b** tapir, **c** pig, **d** peccary, **e** chevrotain. Rhinos **f** and hippos **g** have feet specialized for weight-bearing (graviportal), with short digits and a spreading foot in which the side toes touch the ground when standing. The most drastic modifications occur in hoofed mammals adapted to a fast-running (cursorial) existence. In camels **h** and in pecoran ruminants **j**, **k**, the metapodials (digits 3 and 4) are long and fused for most of their length into a single bone. In the horse **i** there is only a single digit, number 3 (digits 2 and 4 are retained as vestigial splint bones). Deer **j** retain the side toes (remnants of digits 2 and 5), whereas these have been entirely lost in the camel **h** and pronghorn **k**.

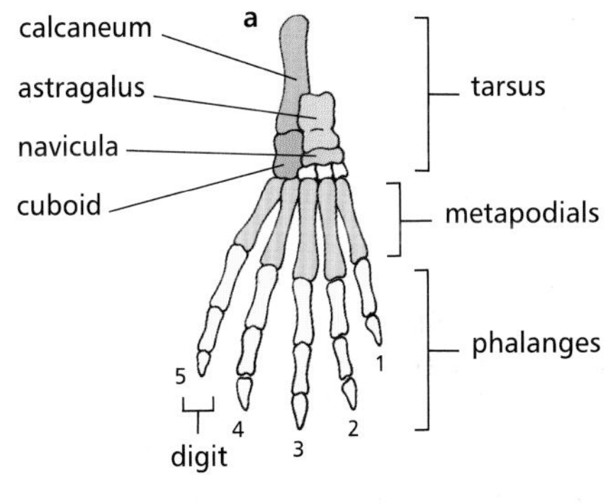

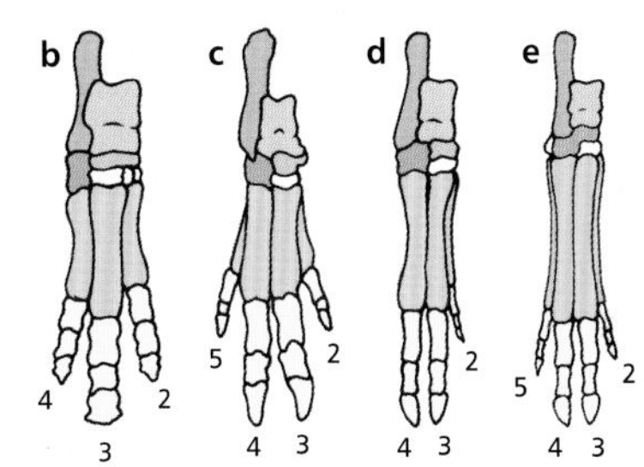

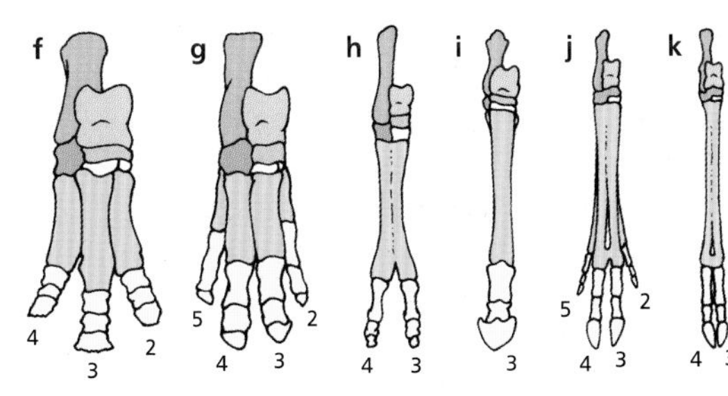

quality old foliage at the top of the grass stand, while gazelles and wildebeest (artiodactyls) eat the higher-quality new foliage that is uncovered by the zebras' feeding.

Absorption of the products of protein digestion is also different between ruminants and hindgut fermenters. The protein in the food of ruminants is initially fermented to ammonia by the stomach microorganisms; this ammonia is transported in the blood to the liver, where it is converted to urea, and then transported back to the stomach, where it is used for additional food by the microorganisms. The ruminant itself absorbs protein only in the form of microorganisms that spill over into the rest of the digestive system. In this way ruminants efficiently recycle urea – normally a waste product excreted in the urine – and thus conserve water. This digestive arrangement results in another pivotal ecological difference between ruminants and other ungulates: desert ruminants such as the oryx and camel need drink only occasionally, whereas desert hindgut fermenters, such as asses, must drink daily to produce sufficient water to make urine.

The ability of ruminants to recycle urea and "farm" microorganisms in their guts are adaptations with potentially far-reaching implications. There are many more species of ruminants than there are of hindgut-fermenting perissodactyls. This is in part because ruminants are free from the burden of having to obtain all essential amino acids from their diets (getting it instead from their in-house bacteria); they can specialize on a narrower range of plants, and thus subdivide available resources among species in a finer fashion. On the other hand, perissodactyls can make maximum use of fruit in their diet because most of the nutrients, particularly sugars, are absorbed before the region of fermentation is reached. Ruminants ferment fruit in the fore-stomach, and in doing so lose much of its nutritional value. The smallest ruminants such as chevrotains and duikers, however, have only small rumens in which minimal fermentation occurs, so they make better use of fruits and hence are freed from the need to eat much fibrous food.

Myriad Variations

HOOFED MAMMAL ORIGINS

Perissodactyls and artiodactyls first radiated in the Northern Hemisphere in the early Eocene, 54 million years ago. The earliest members of these orders were relatively small mammals, the size of a cat or a small dog, and had teeth suggestive of a browser/frugivore/omnivore type of diet. At this time Africa was isolated from Eurasia, and North America was isolated from South America, although a limited amount of interchange of animals was possible between the northern continents. The climate at higher latitudes was warmer than today, with tropical forest reaching northward to within the Arctic Circle.

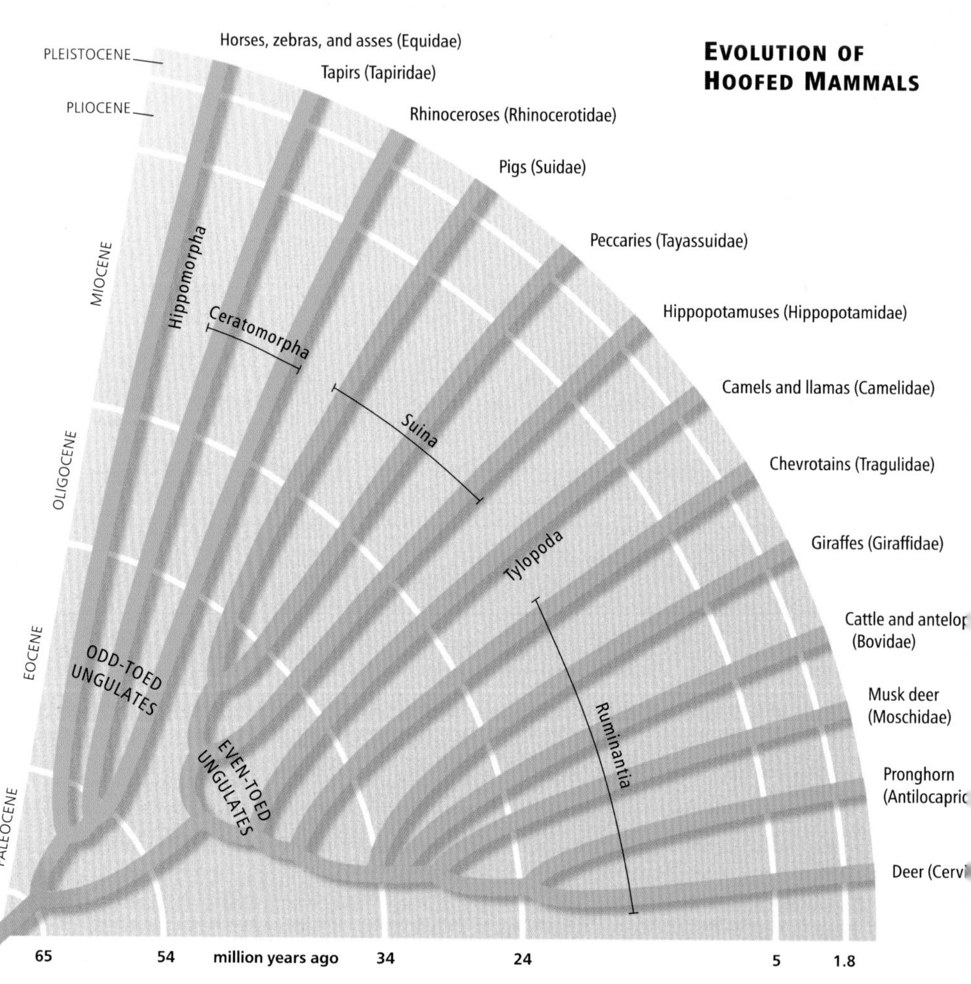

Above *The ancestors of modern hoofed mamma (condylarths) first appeared at the start of the Paleocene, 65 million years ago. During the Eocene epoch, more modern types of ungulates appeared in North American and Eurasia, including the first artiodactyls and perissodactyls, and rapidly diversified, though more archaic forms remained present.*

Eocene perissodactyls had a wide range of body sizes, ranging from about 5kg (11lb) to about 1,000kg (2,200lb), and were the first ungulates to adopt a diet of relatively fibrous vegetation (although they were all browsers, as grass had not yet evolved). They were initially known from both Eurasia and North America. An early, highly

Left Ungulates from the Eocene and Oligocene. Odd-toed ungulates predominated in these early forms: **1** Hyracotherium, *the first known horse, from the early Eocene – its name derives from its misidentification, via its dentition, as a hyrax-like form;* **2** Uintatherium, *a large herbivore from the middle Eocene (belonging to the extinct order Dinocerata);* **3** Brontops, *a late Eocene North American brontothere;* **4** Merycoidodon, *a late Eocene and Oligocene oreodont (tylopod artiodactyl).*

erse group was the tapiroids, which would ntually give rise to the modern tapirs and rhi-s. Many kinds of horses existed primarily in rth America. One interesting family was the ntotheres, which became extinct at the end of Eocene and left fossils in North America and a. Their teeth suggest a mixed diet of leaves l fruit; later species resembled White rhinos t with Y-shaped, bony horns on their noses.

In contrast to this blossoming perissodactyl ersity, early Eocene artiodactyls were all small ıder 5kg/11lb) and either omnivorous or else y ate soft browse and fruits, a tendency ained today in most pigs. There was never, wever, an equivalent radiation of omnivorous issodactyls. Artiodactyls with teeth indicative more fibrous browsing appeared only in the er part of the mid-Eocene. Then, during the late cene (from around 45–34 m.y.a.), the global nate changed dramatically. Plummeting tem-atures in the higher latitudes and increasingly sonal climates in North America and Eurasia ced the ungulates to adapt their foraging habits accommodate seasonal fluctuations in forage antity and quality. Arboreal herbivores depen-nt on the year-round availability of fruit, such as mates, disappeared from the northern conti-nts at this time. A considerable increase in body e occurred among the artiodactyls as the avail-lity of leafy food became seasonally restricted. e lineage (pigs and their relatives) remained gely omnivorous, while the other (ruminants) came specialized in various forms of herbivory. th this initial increase in body size, it was possi-for the ruminant artiodactyls to evolve their aracteristic foregut site of fermentation to utilize nore fibrous diet. Increased body size allowed one disadvantage of being a ruminant (i.e. the bility to avoid fermenting all food, with a con-uent loss of nutrients) to be outweighed by a ater gut capacity and lower energy demand per unit of animal mass. A rumen may also have been advantageous in enabling the recycling of urea and detoxification of plant secondary compounds.

Artiodactyl Ascendancy

LATER EVOLUTION

The appearance and predominance in the higher latitudes of seasonal types of vegetation thus marked the time when ruminants became more diverse and abundant, and began to replace peris-sodactyls as the dominant ungulates. Nevertheless the evolutionary radiation of the ruminant artio-dactyls had little effect on the equids, which were able to specialize on diets too coarse for ruminants. Indeed, equids enjoyed a second burst of diversifi-cation in the late Eocene, as did rhinos, which are first known from the middle Eocene of North America and Asia. In contrast the tapiroid radia-tion was badly affected, despite the persistence of the modern tapir lineage to the present day.

The subtropical woodlands of the Northern Hemisphere during the Oligocene (34–24 m.y.a.) were good habitat for the omnivores and larger browsing ungulates, but the dominant forms were not those of today. This epoch was the peak of rhino diversity in the Northern Hemisphere, including large hippo-like forms and small pony-sized forms, as well as the "true" rhino lineage that survives. In Asia, one branch of the originally pony-sized rhinos produced the giant "giraffe rhino" *Paraceratherium* (also known as *Indricotherium* or *Baluchitherium*) which, at around 12 tonnes, was the largest land mammal that ever lived. Other common Oligocene ungulates included entelo-donts, which were giant (up to bison-size) pig-like animals (but not true pigs) with enormous heads and large canines and powerful premolars, sug-gesting they fed partly on carrion. Oreodonts, native to North America only, were incredibly common for their time but looked like nothing in existence today; they combined a generally pig-like (or hyrax-like) body with deer-like teeth, clearly specialized for browsing rather than omnivory.

Less common in the Oligocene, and retaining a smallish body size (about that of a goat), were the ungulates that would rise to prominence later in the Cenozoic: horses and camels in North Ameri-ca; true pigs and primitive ruminants in Eurasia. Meanwhile Africa, which was still isolated from the northern continents, contained not only early ele-phants but also a huge diversity of hyraxes. Most of these hyraxes would later disappear, perhaps victims of competition from the ruminants and rhinos that invaded from Eurasia in the Miocene.

The early Miocene (23–16 m.y.a.) saw a gradual climatic change in the northern latitudes; temper-atures increased overall, reaching a maximum in the middle Miocene around 14 m.y.a., and the subtropical woodland of the Oligocene was replaced by a more open savanna type of habitat, with the spread of grasslands. Many of the Oligo-cene ungulates now became extinct, with their places taken by species more closely related to the present day horses, camelids, and ruminants.

During the early Miocene the Eurasian ungu-lates had invaded Africa, and both continents were home to a diversification of ruminants. The original stock of small, hornless animals, rather like today's mouse deer, now diversified into the modern horned families of larger body size, including giraffes, bovids, and cervids. The evolu-tion of horns, first seen in the males only, was probably related to the Miocene habitat changes. With the replacement of forest by more open woodland or savanna, the habitat structure was now suitable for the males to set up territories and fight with each other over access to females and choice feeding areas.

Some ruminants invaded North America in the early Miocene, including the ancestors of the pronghorn (antilocaprids), and the deer-like dromomerycids. But the major players on the North American stage were the horses and the camelids. Camelids diversified into a variety of body types mimicking the Old World ruminants, including gazelle-like forms and giraffe-like forms.

While bovids are the most diverse and geographically abundant of the pecoran families today, in the Miocene they were confined to the Old World (Africa and Eurasia). Indeed, the modern situation of equids and bovids coexisting is a relatively recent (late Miocene) development. Equids originally radiated in North America, and bovids in the Old World; they first met when Miocene three-toed horses migrated to the Old World. (The modern, single-toed *Equus* made its journey later, at the start of the Pleistocene 1.8 m.y.a.) Pigs and giraffes, both exclusively Old World forms, were also more diverse in the Miocene than they are now. Giraffes included low-level mixed feeders (sivatheres) with shorter necks, higher-crowned teeth, and moose-like palmate horns. Sivatheres may even have survived into historical times, as animals resembling them appear on archeological artifacts.

Following the warm peak in the middle Miocene, global temperatures fell during the rest of the Cenozoic to the relative low of the present day, and the world also became steadily more arid. Although savanna-like habitats first appeared in the early Miocene, this was only in the higher latitudes, and even there true grazers were not apparent until later in the epoch, around 12 m.y.a. In contrast, present-day African savannas are a fairly recent development, probably absent until the Pleistocene or late Pliocene, 3–1.8 m.y.a. Equids were at their most diverse in North America at the start of the late Miocene (10 m.y.a.), when there were around a dozen genera sharing the same habitats. Most were medium-sized grazers or mixed feeders, like modern zebras, but there was also a large persistently browsing form and a gazelle-sized grazing form. It seems likely that the North American late Miocene savannas were more arid than the African ones of today, resulting in a type of tough vegetation more suitable for sustaining a diversity of equids than grazing ruminant artiodactyls

The diversity of the endemic North American ungulates declined in the Plio-Pleistocene, when the savanna habitat gave way to prairie. However, the larger camelids and one-toed grazing equids retained a moderate diversity and abundance of species, with representatives of both families also dispersing southward when the Americas became connected in the late Pliocene. Then suddenly, at the end of the Pleistocene, almost all of the endemic North American ungulates disappeared. A combination of climate change and hunting pressure by the recently immigrant humans was the likely cause. These events left the pronghorn as the only survivor of the late Tertiary radiations, and also left us with a falsely scant impression of the evolutionary success of equids and camelids. In the Americas the majority of present-day ungulates represent a relatively recent (Plio-Pleistocene) invasion of North America by deer and bovids, while the bovids had yet to reach South America by the time the Spanish conquistadors introduced livestock, in the 16th century.

In contrast to the dramatic changes in North America, the Old World retained a diversity of tropical and subtropical habitats throughout the Plio-Pleistocene, enabling a relatively large number of bovid, giraffid, and equid species to persist in the extensive savannas of Africa and southern Eurasia. But during the Pleistocene Ice Ages the savannas retreated into the tropics, shrinking the geographical range and diversity of the tropical-habitat giraffids, while those of the temperate-habitat deer expanded, including giant forms like the Irish "elk." Both horses and rhinos were abundant as temperate and cold-adapted forms in the Pleistocene of Eurasia, but rhinos are now totally absent from higher latitudes, and the true wild horses succumbed to human interference (including domestication!) during historical times.

Grazers and Browsers

ECOLOGY AND BEHAVIOR

Ungulates spend most of their time engaged in activities that are concerned in some way with e ing, breeding, or avoiding being eaten. As regarc eating, ungulates indulge either in grazing, mea ing that they feed on monocotyledonous plants (grasses), or browsing, meaning that they feed o dicotyledonous plants (woody plants and forbs) Most ungulates are specialized for membership either one of these two "guilds," although some switch between grazing and browsing dependin on the relative availability and quality of each fo type. Such species, of which impalas and goats a typical examples, are termed "mixed feeders." T separation into browsers and grazers, with some overlap for mixed feeders, is an important factor contributing to the coexistence of numerous ungulate species (often more than 20) in some African savanna areas, for example.

Browsing has been a common occupation of ungulates throughout their evolution, and is presently practiced by many species including tapirs, giraffes, and deer. Indeed, the original die of all perissodactyls and ruminant artiodactyls probably consisted of the leaves of dicotyledono trees and shrubs. In the middle Miocene (c. 15 m.y.a.) there was an ecological change, however,

4

5

6

Left *Ungulates from the Miocene, Pliocene, and Pleistocene:* **1** Aepycamelus, *a middle Miocene giraffe-like camel;* **2** Cranioceras, *a late Miocene deer-related dromomerycid ruminant;* **3** Synthetoceras, *a "slingshot-horned" late Miocene protoceratid;* **4** Capromeryx; *a Pliocene pronghorn;* **5** Bison antiquus, *a Pleistocene ancestor of the modern bison;* **6** Coelodonta, *the Woolly rhinoceros of the Pleistocene.*

Below *Giraffes are exclusively browsers, favoring* *e leaves and shoots of trees such as acacia and* *rrh* (Commiphora). *Their mouths, lips, and tongues* *ve become highly specialized to deal with a variety* *browse, be it spiny or thornless.*

which shifted the emphasis of ungulate diets. Grasses emerged as abundant land plants, comprising an excellent and abundant source of nutritious, although cellulose-rich, vegetation.

As a food resource, grass differs from browse (the foliage of dicotyledonous plants) in three important ways. The first is that it occurs close to the ground in large, continuous expanses, while browse usually occurs in scattered clumps (trees and shrubs) at various heights above ground. The second is that grass plants have short roots, and so the availability of green grass is very closely linked to the moisture content of the topsoil, which is in turn closely linked to rainfall. Trees, on the other hand, have deep roots and so maintain green leaves for longer in dry periods. These two differences explain why grazing ungulates commonly occur in large herds, feeding side by side, sometimes migrating in a seasonal pattern in response to the tight links that bind grass growth to rainfall. Migratory grazers often form particularly large herds, such as wildebeest in the Serengeti, for example, which migrate in a herd of about 1.5 million animals. Browsers do not form large herds because they would be unable to keep in contact with each other while moving through trees, and if they did they would always be competing with each other for access to each individual favored tree or shrub. Also, browsing does not lend itself to migration; woody browse is far less affected by short-term weather patterns like rain or snowfall than grass. Finally, the third major distinction between grass and browse is that grasses do not display the complexity of chemical and physical adaptations that have evolved among woody plants as defenses against herbivory. Browsers have to contend with the toxic and taste-repellent properties of resins, phenolics, tannins, and alkaloids in their food, even after negotiating the thorns, prickles, and spines that many woody plants have in addition to their chemical armories. Although some grasses do contain defensive hydrocarbon compounds, the biggest challenges grazers have to contend with while feeding are maximizing the ratio of grass-leaf to stem in each mouthful, and avoiding tooth-wear. Grass leaves are the most nutritious parts of the plant but they also contain abrasive silica crystals, and because grass grows close to the soil surface it is often dusty, which makes it even harder on the teeth. To offset this problem grazing ungulates generally have high-crowned cheek teeth that accommodate a relatively large amount of tooth-wear through life.

Browsing ungulates have to learn which plants are more toxic than others. In fact a goat can learn after only one encounter which plant in its recent feeding experience made it ill. Then, if its foraging options are limited, it will learn after only about three sample meals how much of that plant it can ingest per day without any ill effects. This requirement – to keep the concentration of each plant defensive compound below its toxicity threshold in the animal's body – explains a key aspect of browsing ungulate behavior. Browsers move from tree to tree, taking relatively small meals from each to spread their toxin load; if they filled their guts with leaves from only one tree, they would overdose on at least one type of defensive compound.

In addition to the use of different food classes by grazers and browsers, coexisting ungulate species also tend to have very different body sizes. This provides another means by which common food resources are partitioned among species. The larger-bodied ungulates need to eat more per day, but can tolerate a wider range in diet quality (usually expressed in terms of fiber content); the smaller species can fill their guts quickly while being selective for the highest quality plant parts available. It is thus common to see mixed-species herds of ungulates in African savannas, for example, where small gazelles follow behind the much larger zebras and wildebeest. By trampling and mowing down tall, coarse grass, exposing green leaf and shoot material for the gazelles to pluck with their narrow muzzles, the larger species facilitate feeding opportunities for the smaller ones. Nature is never fair, however, and the larger species usually lose in the end. When small grazers move onto a patch where large grazers are feeding, their highly selective removal of the best plant parts steadily reduces the overall quality of food available in that patch. The bigger animals do not drive the smaller animals away but simply move on, drawing the smaller ones behind them across the plains in what is known as a "grazing succession," through which the sequence is repeated each year. The same principle applies among browsers, but here the bigger animals are displaced vertically rather than horizontally. It explains why, for example, giraffes prefer feeding above the heights reached by smaller browsers, even though the lower levels of the tree canopy are equally available to them.

Despite the inequities in their feeding efficiencies, small ungulates do not enjoy a net ecological advantage over their larger relatives. For one thing, they have limited social lives. This is because the social behavior of an ungulate species is determined by its body size, diet, and the structure of the habitat in which it occurs. Also, the relative advantages and disadvantages of group feeding are affected by the threat of being eaten. Small species are vulnerable to many more predators than are large ungulates, and often cannot avoid predation by flight or self-defense. They seek to avoid detection, and most are cryptically colored. They eat buds and berries, which represent scattered and scarce items of high protein content. These are best sought alone, preferably over familiar ground. The logistics of all this mean that berry-eaters cannot occur in herds, since the foraging horde would soon deplete all the available resources. Many small artiodactyls thus forage singly or in pairs throughout a so-called "resource-defended territory." They live in tropical forest or woodland where such food is readily available, along with plentiful cover against predators. For all animals of this type, contacts and conflicts between members of the same species are few. They tend to be solitary or monogamous in their reproductive behavior, and exhibit little difference in appearance or body size between the sexes. Among artiodactyls, this type of social system is seen among chevrotains, small forest deer such as pudu, and small forest antelopes such as duikers; among perissodactyls, it is exhibited by tapirs and browsing rhinos (for example, the Black, Javan, and Sumatran rhinos).

Larger perissodactyls and artiodactyls can tol ate food of a lower protein content, and are bett able to avoid predators by fleeing from them. Th can thus forage in open habitats on abundant fo like leaves and grasses. Living in groups benefits these open-country ungulates, since cohesion a communication between animals aids early pred tor detection and reduces the chance that any o individual will be the victim of an attack. The greater availability of food prevents these animal from interfering with each other's feeding.

Nearly all large ruminant artiodactyls are foun in herds, ranging from bands of two to three females in the majority of deer species to the vas groups of individuals of both sexes among bison The exceptions to herd-forming among the large ruminants are specialist forest browsers like the okapi and moose, which seek refuge in water or dense forest to escape predators. Grazing periss dactyls also form herds. Equids usually form

Above *Cape buffalo sparring; the horns of this ·ge bovid species are used in contests to determine ·ating rights, but also provide effective defense ·ainst predators. A herd of buffalo may often coop- ·ate to drive off a lion.*

·oups of a dozen or so, consisting of females ·d their offspring and a single stallion, although ·ese smaller groups can sometimes aggregate ·to much larger herds. White rhinos form mixed- ·x groups of up to a dozen individuals.

Larger artiodactyls have two distinct types of ·rds, with either fixed or temporary member- ·ip. Fixed-membership herds usually comprise ·ose relatives, and show group defense against ·edators. Both sexes are present and males estab- ·h a dominance hierarchy to determine mating ·hts for the females. Although the females in ·ch herds usually have horns, male dominance ·pends on size, and in such species the males are usually much larger than the females (up to twice female body mass), and may continue to grow throughout their lives. Such species are mainly large-bodied, open-habitat bovids, such as bison, buffalo, and musk oxen.

In some artiodactyl species that have open-membership herds, mating rites are tied to ownership of a territory, while in other species mating is tied strictly to rank. Territories are often just mating areas that can support only one male, if that, rather than an area where females congregate to take advantage of some resource such as food or water. Thus females will enter and leave the male's territory, and the most successful males are those that hold the most attractive "property." Males must challenge other males for the rights to a territory, and rarely do they hold it for long, as defending it is exhausting. It is in these species that the differences between males and females reach a zenith. Males have elaborate horns or antlers to engage in continual territorial combat. Most medium to large-sized deer and antelope belong to this category. Some species, such as the migratory wildebeest, have this type of social organization, yet appear to lack clearly marked territories that are fixed in locality. In others, such as the fallow deer, a "lek" system is seen, in which receptive females visit the holders of highly-contested, close-packed mini-territories that have no other function than as places for males to advertise to females and to mate. Territory size varies both with the type of territory (they are smaller if used just for breeding) and with the size of the animal (larger animals have larger territories). The seasonal breeding territory of a male kob weighing 100kg (220lb) is only about one square kilometer (0.4sq mi), while a 250kg (551lb) Greater kudu – a non-territorial male ruminant – has a year-round home range of about 20sq km (8sq mi).

The typical perissodactyl social system (as exemplified by equids) differs from the typical ruminant artiodactyl system in that a single male consorts with a particular group of females. The female group is a harem, with strong social bonding among individuals. Fights between males are less common than in artiodactyls, as the male horse neither defends a fixed territory, nor does he have to contend continually with other adult males within the herd. Males without harems (usually younger males) form roving bachelor herds of more variable composition. A modified version of this social system is seen in the wild ass, Grevy's zebra, and the White rhino, where the males additionally defend a territorial area. Interestingly, no perissodactyls have pronounced differences in size and appearance between sexes, in contrast to the larger ruminant artiodactyls.

Most ungulate females mature relatively quickly for their size (female equids can conceive as early as their second year), have long gestation periods (among equids, around 11 months), and usually bear only one, large, well-developed juvenile at a birth. Pigs are an exception in producing several piglets in a litter. All artiodactyl juveniles can see, hear, call out, and stand to suckle soon after being born. All females (except, again, pigs) make no nest but usually seek seclusion just before giving birth. Where cover is available, many ungulate mothers hide their young as best they can while they go off to feed, and the calf lies immobile until her return. In contrast some of the largest, open-country, herd-forming species, for example wildebeest (artiodactyl) and horses (perissodactyl), have young that can get to their feet and start moving remarkably soon after birth; these young can run at near adult speed within half an hour of birth.

Despite relatively slow reproduction, artiodactyls are not very long-lived for their size; 10 is an advanced age for impalas, and 15 for red deer. Perissodactyls are longer-lived; up to 35 years in equids and 45 years in rhinos. As the most abundant large mammals, ungulates form the staple diet of most of the great terrestrial carnivores. For those that do not fall victim to predators, starvation is an annual threat, since populations of many of them, as dominant herbivores in their communities, number close to the yearly carrying capacity of the plant community. Although they have evolved a considerable capacity for storing energy in their body tissues, a coincidence of bad events for an individual, such as a delayed onset of a rainy season while suckling a calf or defending a territory, could tip the balance. Nutritional stress predisposes an animal to predation, parasites, pathogens, and environmental anomalies such as cold snaps. For geriatric animals, which have survived the rigors of life, the sad fate of starvation often awaits once their abrasive diet has finally reduced their grinding molars to eroded stumps that are useless for feeding.

An Invaluable Resource

Ungulates and Humans

The genus *Homo* emerged near the peak of the bovid radiation, and for the past 2 million years has endured the Pleistocene and associated global changes alongside that family. With hindsight, it is a chastening thought to realize that humans have, in one way or another, been largely responsible for the dwindling numbers of ungulate species throughout this period. The abundance of ungulates in most large mammal communities, their size, social organization, and ecology, even their antlers and horns, have all exposed them to damaging interactions with humans.

The human ancestors whose remains have been found in East Africa lived among a community of ungulates similar to that of the same area today. One crucial difference is that then there were a number of now-extinct giant forms. For example the buffalo *Pelorovis* may have weighed twice as much as the living Cape buffalo. Such large ungu lates were at minimal individual risk from most predators. But humans were different; though n ther strong nor fast, they excelled in three traits: inventive intelligence, coordinated group huntir and the use of artificial weapons, especially proje tiles. Through their use, humans became the mo generalized large predators in the community, al to hunt all prey in a wide variety of circumstance Coordinated groups of humans were also able to gang up on other large predators, like the large cats, and steal their prey, using noise, fire, and projectiles to drive them off. From the glacial and interglacial epochs in Eurasia comes a wealth of evidence that early human societies depended heavily on ungulate meat for food. Not only are there bones of the animals they dined on, but th hunters also left their own record in the form of art: carvings, clay models, and, most dramatic o all, vast cave-murals depicting many of the animals. It seems predictable, then, that the huntir of ungulates should have become a means of winning status within a society; indeed, sport ar trophy hunting are still widely regarded as legitimate ways for a person to enhance their position

A crucial event in the relationship between ungulates and man was domestication. Some 15 ungulate species have been domesticated. This has certainly occurred independently with differ ent species; thus, domestication of South Ameri can camelids is quite separate from any Old Wo domestication. Domestic ungulates tend to live i closed-membership herds and are non-territorial characteristics that lend themselves to herding and tending by humans. In the Old World, shee and goats were domesticated by 7500BC (perha much earlier), from mouflon and wild goat respe tively. The pig was domesticated from wild boar by 7000BC in the Middle East, but was probably domesticated independently at other centers in i wide range in Asia at other dates.

Cattle were domesticated by 6500BC from wi cattle or aurochs in Europe and the Near East, although the cattle of India and East Asia may have been domesticated independently from rather different stock. The arrival of the domestic artiodactyl in the lives of humans probably occurred at about the same time as domesticatio of wheat, barley, and the dog, and preceded by 2,000–4,000 years the domestication of donkey horses, elephants, and camels.

In South America, domestic breeds of llamas appeared between 4000 and 2500BC, some 2,000 years after the domestication of corn in th region. The temporarily settled conditions of a

Left *A juvenile White-tailed deer buck signaling deference to a dominant buck by licking. Prominent antlers or horns are hallmarks of species where confl over territory ownership is a constant fact of life.*

imitive, crop-growing society would have been eal for the domestication of captured ungulates; deed, domestication would be essential to sure meat for the new agriculturalists, as hunt- g wild migrating herds requires nomadism.

Horses and donkeys were the last of the com- on livestock animals to be domesticated and ey have been the least affected by human anipulation and artificial selection. Both these uids were probably originally domesticated in ia. The sole existing wild horse, Przewalski's rse, is the same species as the domestic horse; ewise, the African wild ass (originally also own from Asia) is the same species as the mestic donkey. The first domestic horses peared at about 6000BC in the Ukraine, when ey may have initially been used for food. It was t until about 2000BC that the widespread use the horse as a means of transport came about, using a revolution in human mobility and in the chnology of warfare. Indeed, the spread of the arrior Indo-European-speaking peoples such as e Hittites and the Aryans across Eurasia was cilitated by the use of horse-drawn vehicles.

Some domesticated ungulates remained con- ned to the areas in which they were domesticat- l (the yak and water buffalo being examples) but ttle, sheep, goats, horses, donkeys, and camels e typical of species that dispersed rapidly and idely through their associations with humans. omesticated ungulates formed the basis for dis- nct nomadic pastoral cultures, in which people lied almost entirely on their animals, not on ops. These people lived typically on milk from eir stock, supplemented occasionally by meat or en by blood taken without killing the beast. This actice persists today among the Masai people of enya, who bleed their cattle from the jugular in and mix the blood with milk to make a ink that is their prime source of protein.

So successful was the early association tween livestock and people that pas- ralism spread throughout the savanna, eppe, and semidesert lands of Eurasia d Africa. Then, with the advent of ips, livestock quickly spread to the mericas and Australia as well as emingly unlikely places like Green- nd, Iceland, and remote islands ound the world. The cumulative fect has been the gross alteration of large expanses of the Earth's surface through the continual and combined effects of intensive grazing and browsing. Pastoralism everywhere has diminished wild ungulates to the point where natural communities remain only where physical barriers or the risk of disease have kept out humans and their stock. Indeed, livestock now constitutes over 90 percent of the ungulate biomass in African savannas. This alarming figure means that about four species of introduced domesticated ungulate (usually either cattle or camels, plus sheep, goats, and donkeys) have supplanted the 20 or more species that made up the indigenous ungulate biomass in any particular region.

In some areas, such as southeastern Zimbabwe, the cattle density is now about three times greater than would be expected from the long-term average density of buffalo or any other indigenous ungulate species under natural conditions. Extermination of large predators, control of parasites and diseases, and especially the artificial supply of water from boreholes have created this situation. It could be argued, then, that in terms of their present numbers and distribution, the domesticated species are the most successful of all ungulates through evolutionary time, and the suite of traits that favor domestication represent a winning evolutionary "strategy." Domesticated species are now wholly dependent on man, however, and it probably will not be very much further forward along the evolutionary timescale before the success of this "strategy" is severely tested.

CMJ/JdT

Above *A traditional herder in Bhutan milking a Yak, a domesticated indigenous species. In many parts of the world, biodiversity is seriously threatened by a huge influx of non-native domesticated livestock that monopolizes the resources needed by wild ungulates.*

Right *Almost all horses, even ose living wild, are varieties of the mestic horse. There are over 200 cognized breeds, most of which (like ese Austrian Haflinger ponies) have en produced by natural selection ther than by selective breeding pro- ams. Paradoxically, the only gen- nely wild horses still in existence e examples of Przewalski's horse sident in zoos.*

Distribution Africa, Asia, S and C America.

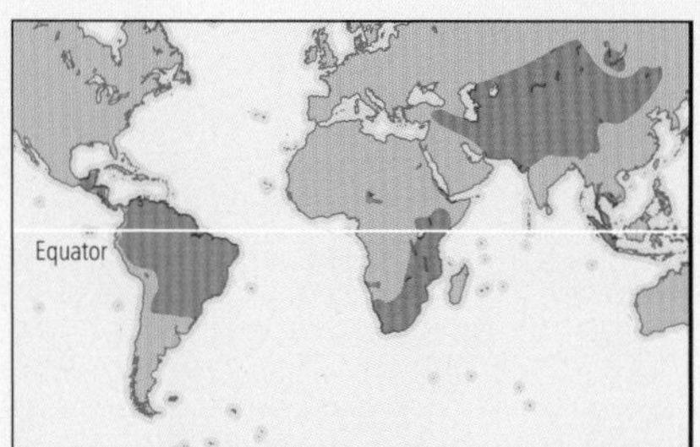

Order: Perissodactyla
16 species in 6 genera and 3 families

Horses, Zebras and Asses
Family Equidae p468
7 (9) species in the genus *Equus*

Tapirs Family Tapiridae p474
4 species in the genus *Tapirus*

Rhinoceroses
Family Rhinocerotidae p476
5 species in 4 genera

Odd-toed Ungulates

The odd-toed ungulates or perissodactyla are a small order of mammals, with the equids – the horses, plus the closely related asses and zebras – as their only widespread members. The other members are animals not usually associated with the graceful, open-country horse: the ponderous rhinos of Africa and Asia, and the elusive tapirs of Malaysia and South America. Yet details of the anatomy of these animals, such as the dentition and the limb structure, together with overall similarities in their behavior and physiology, can be shown to unite them all in a single order.

Today perissodactyls apparently run a poor second to the artiodactyls or even-toed ungulates in terms of numbers of species, geographical distribution, variety of form, and ecological diversity. Yet the perissodactyls were the dominant ungulate order during the early Eocene (from 55 million years ago), and their subsequent decline is more likely to have been due to climatic factors than to direct competition with artiodactyls.

Perissodactyls first appeared in the latest Paleocene around 58 m.y.a., and by the early Eocene 3 million years later they represented a diverse assemblage of species. Equids are first known from North America in this period. The modern families of tapirs and rhinos emerged by the late middle Eocene, derived from several more primitive "tapiroid" taxa. Tapirs and rhinos are closely related, and are often grouped together as the suborder Ceratomorpha. Rhinos are characteristically heavy-bodied and show adaptations of the limbs for weight-bearing. In contrast, from the start of their evolution, equids showed modifications for running, with a progressive tendency to lengthen the limbs and reduce the lateral digits.

Tapirs were originally a North American grou[p] and have only inhabited South America and Southeast Asia for around the past 3 million yea[rs]. Until the end of the Pleistocene 10,000 years ag[o] they persisted in North America and Europe. Living tapirs are medium-sized, and live mainly in tropical and subtropical forest in Malaysia and Central and South America.

Rhinos probably had their origins in Asia, but were common in North America from the late middle Eocene to the early Pliocene (46–4 millio[n] years ago). By the middle Oligocene (c. 30 m.y.a.) they displayed great diversity, from large, hippo-like forms to pony- and tapir-sized species. Modern types first appeared in Africa at the start of th[e] Miocene (24 m.y.a.), and were common throughout Eurasia until the end of the Pleistocene. Surviving rhinos, uniformly large, occur today only i[n] Africa and Southeast Asia.

Today's equids are medium-sized, and are all specialist grazers. They occurred on all continen[ts] except Australasia during the Pleistocene, but ar[e] now found only in Africa and parts of Asia. Feral horse and ass populations flourish in Europe, Au[s]tralia, western North America, and South Americ[a].

The center of equid evolution was North Ame[r]ica, although relatives of the North American lineage were present in Eurasia in the Eocene. True equids (genus *Equus*) did not appear in the Old World until the Pleistocene (1.8 m.y.a.), but earl[i]er invasions of Eurasia were made by a three-toe[d] browsing horse (*Ancitherium*) in the early Miocen[e] and of Eurasia and Africa by a three-toed grazing horse (*Hipparion*) in the late Miocene. *Hipparion* survived into the Pleistocene and was found for [a] time alongside true equids. Equids went extinct in North America only about 10,000 years ago. CMJ

▽ **Below** *Species of odd-toed ungulates, illustrating the great diversity that exists within this small order:* ***1*** *Black rhinoceros* (Diceros bicornis)*;* ***2*** *Mountain tapir* (Tapirus pinchaque)*;* ***3*** *African ass* (Equus asinus).

OMESTICATION OF HORSES

ople killed horses for food long fore they ever thought of riding em; indeed, hunting may have spelt e demise of the native North Ameri- n equids about 10,000 years ago. It as at least 5,000 years later that the st serious attempts were made to me horses, probably on the vast rasian plains north of the Black and spian seas. By that time, donkeys had already been domesticated in North Africa and the Middle East, and were being put to work in the city-states of Mesopotamia as beasts of burden and alongside oxen for plowing. The first renowned horsemen were the Scythians, steppe warriors whose development of the bridle and possibly also the stirrup turned the horse into a formidable instrument of war.

Horses of the type that were first domesticated continued to exist outside of human control until very recent times. The pony-sized tarpan survived in eastern Europe until the late 19th or early 20th century, while Przewalski's horse (*Equus przewalskii*) was last seen in its native habitat near the Altai Mountains in Mongolia in 1968; it is now classed as Extinct in the Wild, and only survives in zoos. The large feral populations of horses that still occur in several parts of the world – notably, the mustangs of the American West and wild horses in the Australian outback – are descendants of domesticated breeds that escaped into the wild.

Below *A group of feral horses in New South Wales, Australia.*

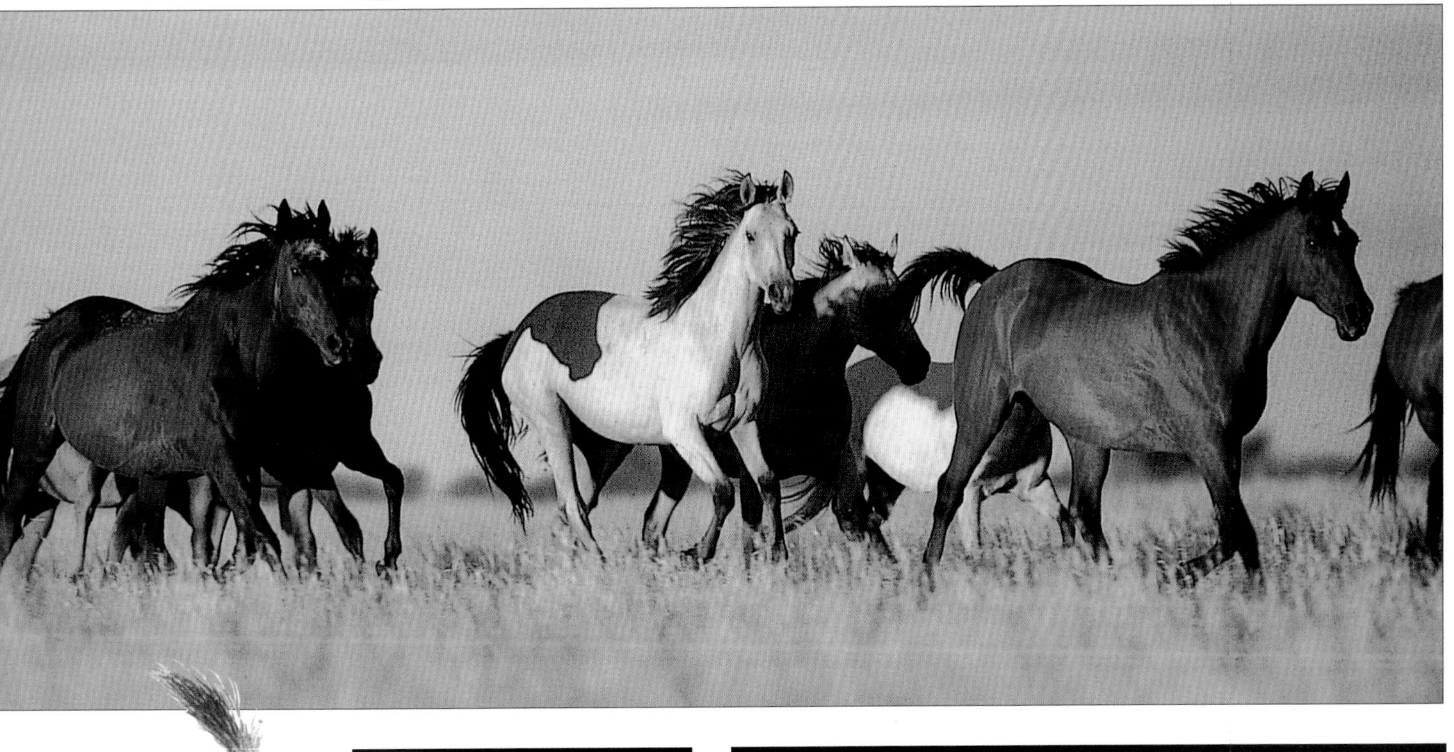

EXTINCT FAMILIES

Three original perissodactyl families are now extinct. The most primitive (taxonomically speaking) were the rhinolike brontotheres, which lived in North America and Asia in the Eocene. The tapirlike paleotheres, from the Eocene of Eurasia, were related to the horses.

The Chalicotheriidae were distantly related to the tapirs and rhinos. Later Tertiary chalicotheres were largish animals, with secondarily substituted claws for hooves. Chalicotheres were first known from the Eocene of North America and Asia, becoming extinct in North America in the mid-Miocene, but surviving into the Pleistocene in Asia and Africa. Conceivably, they survived even longer; chalicothere-like animals, with horselike heads and bearlike feet, are depicted on Siberian tombs of the 5th century BC. The so-called "Nandi bear," a mysterious animal allegedly sighted from time to time in the Kakamega forests of Kenya, may also be a surviving chalicothere.

SKULLS OF PERISSODACTYLS

Odd-toed ungulates may have low- or high-crowned cheek teeth, but they all tend to have molarized premolars, with the molar cusps coalescing to form cutting ridges. Tapirs retain the full mammalian dental formula, and have a fairly generalized skull shape with the exception of the retraction of the nasal bones for the attachment of the proboscis. Rhinos are characterized by an especially deep and long occipital region (rear of the skull), associated with the large mass of neck musculature needed to hold up their massive heads, and by their projecting and wrinkled nasal bones, to which the keratinous horns are attached. Rhinos' incisors are reduced to two uppers and one lower in each jaw half. The skulls of equids show a great elongation of the face, the posterior position of the orbits (to allow room for the very high-crowned cheek-teeth), a complete postorbital bar, and an exceedingly deep and massive lower jaw.

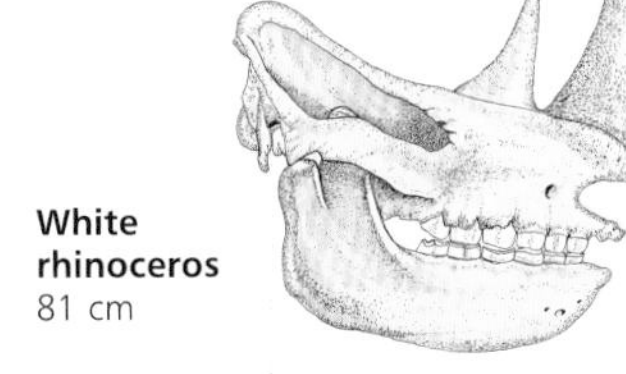

White rhinoceros 81 cm

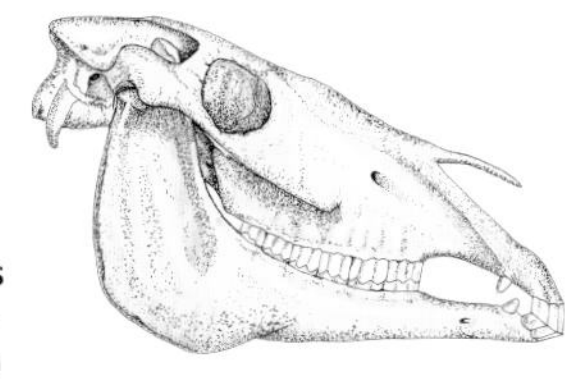

Plains zebra 54 cm

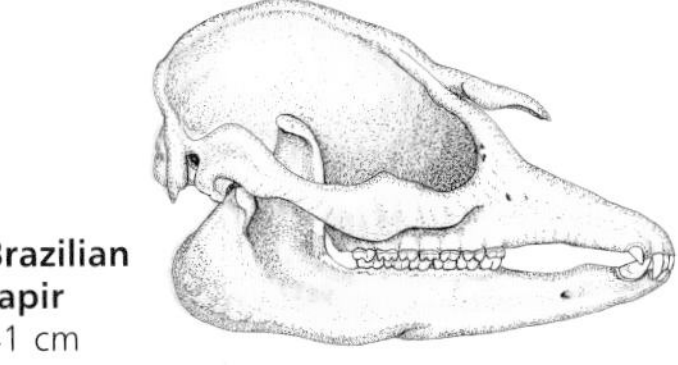

Brazilian tapir 41 cm

Horses, Zebras, and Asses

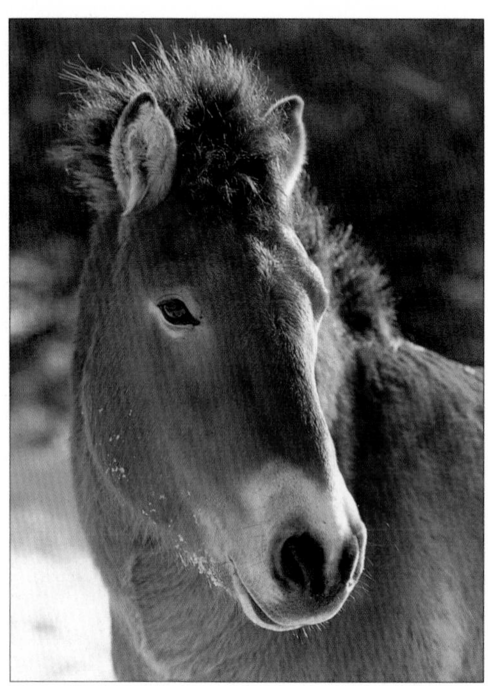

HERDS OF HORSES, ASSES, OR ZEBRAS thundering across wide, open plains capture the imagination like few other sights. They conjure up an impression of power and grace, wildness and freedom. Yet such scenes may soon become mere memories, if the conservation initiatives that have been implemented to stabilize precarious equid populations are not taken further. Moreover, by preserving the landscapes that wild equids inhabit, these conservation efforts are also benefiting numerous other threatened species that share the equids' ecosystems.

Slender-legged Herbivores

FORM AND FUNCTION

Few taxonomic groups have such a richly documented evolutionary history as the equids. The earliest of the horse-like ancestors, *Hyracotherium*, appeared in the Eocene, about 54 million years ago. This small, dog-sized mammal browsed on small shrubs and had low-crowned teeth without the complex enamel ridges of modern equids. Its hind feet had already lost two toes and its fore feet one, but they were still covered with soft pads. Although these small ancestors appeared throughout the Northern Hemisphere, the subsequent evolution of horse-like mammals was centered in North America, with periodic migrations to Eurasia, Africa, and South America. By the Oligocene, 34 m.y.a., large three-toed equids like *Mesohippus* and *Miohippus* appeared. Not only did they bear much of the increased weight on the middle digit: their premolars became more molar-like. But it was only when grasses appeared, in the Miocene 20 m.y.a., that equids began to radiate. Side toes were further reduced, and continuously-growing teeth with high crowns, complex grinding ridges, and cement-filled interstices evolved with the opening up of the habitat and the emergence of tougher forage. By about 5 m.y.a., one-toed, horse-like grazers appeared, first in the form of *Pliohippus* and then as *Equus*, today's only surviving genus.

At this point, the center of evolution shifted from the New World to the Old, where the modern species of zebras and asses appeared. As environments changed, populations became isolated, adapting to local circumstances and giving rise to most of today's living species.

The first species to diverge from the earliest true horses were the asses. They were followed by the zebras, with the Grevy's zebra emerging before either the Mountain or Plains zebra or the quagga (*E. quagga*, which was driven extinct at the end of the 19th century). The only species that probably did not originate in geographic isolation was the ancestor of the Domestic horse, which is thought to be directly descended from some mutant Przewalski's horses. Although all species continued to thrive throughout the Pleistocene, horses had vanished from the American plains by the end of the most recent ice age, perhaps as a result of the arrival of humans from Asia or of changing climates, or maybe a combination of both. Horses were re-introduced to the Americas, with the arrival of European colonists.

All equids are medium- to large-sized herbivores with long heads and necks and slender legs that bear the body's weight only on the middle digit of each hoofed

Top left Przewalski's horse was only discovered in the 1870s by the Russian explorer Nikolai Przewalski in western Mongolia. None have been found in the wild there since 1968.

Right Plains zebras drinking at a lake. Though this African species is not in particular danger, virtually all free-ranging populations of wild equids worldwide are threatened by human activities.

THE ZEBRA'S STRIPES

Several theories have been put forward to explain the function of zebras' conspicuous markings. The notion that the stripes are camouflage is contradicted by zebras' behavior; they gather in great numbers at exposed sites and never try to hide from predators. Likewise, the claim that lions are "dazzled" by a zebra herd's patterning is also not borne out in reality. Finally, the idea that stripes evolved to deter harmful flies disregards the fact that insects are no hazard in most zebra habitats.

Research into vision and brain physiology suggests that the stripes have a social function, promoting group cohesion. Zebras appear actively to seek out the visual stimulation of this patterning. In particular, stripes are thought to stimulate grooming; significantly, the three zebra species' coats are most alike on the neck and shoulders, the preferred grooming area for equines.

Biologists believe that the ancestor of all equines was striped. Selective pressure helps maintain crisp patterns on the social zebra herds of tropical Africa. But among equids that do not live at high density, or where cold climates require shaggy winter coats and annual molts, stripes have become redundant or impractical. JK

ot, providing a springy movement. They possess
oth upper and lower incisors that clip vegetation
nd a battery of high-crowned, ridged cheek-teeth
at are used for grinding grasses.

The four features that seem to have contributed ost to equid ecological success and broad geoaphic range are the springy gait, grinding teeth, rge size, and the lock position in their legs that nables them to rest. This stay apparatus allows e legs to lock without muscular contraction, hich dramatically lowers the energetic costs of sting, feeding, and scanning for predators – l-important but time-consuming activities. By ecreasing relative, or weight-specific, nutritional emands, large size not only enables equids to rage as generalists, but also reduces the cost of ansport, allowing them to range economically.

Horses' ears are moderately long and erect, but n be moved to locate sounds and to send visual gnals. A mane covers the neck; it falls to the side Domestic horses, but stands erect on the other ecies. All equids have long tails, covered by long air in horses and by short hair only at the tip in ses and zebras. The most striking distinguishing ature between species is coat color; zebras have e most dramatic livery, while the coats of horses d asses are more uniform in color.

The species differ somewhat in size, males eing generally 10 percent larger than females. ales also possess large canines, suggesting that ale–male combat for mating opportunities has een sharpened by sexual selection.

FACTFILE

HORSES, ZEBRAS, AND ASSES

Order: Perissodactyla

Family: Equidae

7 (9) species in 1 genus

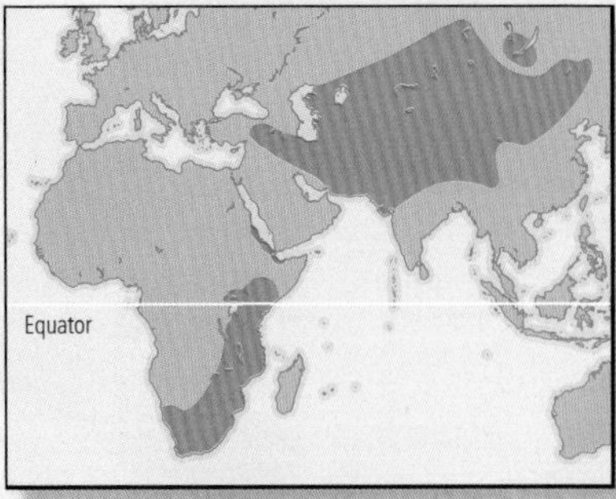

Distribution
E Africa; Near East to Mongolia

PRZEWALSKI'S HORSE *Equus przewalskii*
Przewalski's, Asiatic, or Wild horse
Mongolia near Altai Mts. Open plains and semidesert. HBL 210cm: TL 90cm; WT 350kg. Coat: dun on sides and back, yellowish-white on belly; dark brown, erect mane, legs grayish on inside; thick-headed, short-legged, and stocky. Considered the true wild horse; some authorities regard it as part of *E. caballus*. Conservation status: Extinct in the Wild.

DOMESTIC HORSE *Equus caballus*
Domestic or Feral horse
N and S America and Australia. Open and mountainous temperate grasslands, occasionally semideserts. HBL 200cm; TL 90cm; WT 350–700kg. Coat: sandy to darkish brown; mane falls to side of neck. Dozens of varieties. Feral forms are thick-headed and stocky; domestic breeds, with worldwide distribution, are slender-headed and graceful-limbed.

AFRICAN ASS *Equus asinus*
Sudan, Ethiopia, and Somalia. Rocky desert. HBL 200cm; TL 42cm; WT 275kg. Coat: grayish, with white belly and dark stripe along back; Nubian subspecies has shoulder cross, Somali subspecies leg bands. The smallest equid, with the narrowest feet. Subspecies: 3. Conservation status: Critically Endangered.

ASIATIC ASS *Equus hemionus*
Syria, Iran, N India, Tibet. Highland and lowland deserts. HBL 210cm; TL 49cm; WT 290kg; larger than the African species. Coat: reddish brown in summer, becoming lighter brown in winter; belly is white and has a prominent dorsal stripe. The most horse-like of the asses, with broad, round hoofs. Subspecies: 4; some authorities regard the kiang and onager as separate species. Conservation status: Species listed as Data Deficient, though the Syrian subspecies is extinct, and the Indian wild ass and the onager are classed as Endangered.

PLAINS ZEBRA *Equus burchellii*
Plains or Common zebra
E Africa. Grasslands and savanna. HBL 230cm; TL 52cm; WT 235kg. Coat: sleek with broad vertical black and white stripes on body, becoming horizontal on haunches. Always fat-looking; short-legged and dumpy. Subspecies: 3.

MOUNTAIN ZEBRA *Equus zebra*
SW Africa. Mountain grasslands. HBL 215cm: TL 50cm: WT 260kg. Coat: sleeker than the Plains zebra, with narrower stripes and a white belly; also thinner, with narrower hoofs and a dewlap under the neck. Subspecies: 2. Conservation status: Endangered.

GREVY'S ZEBRA *Equus grevyi*
Grevy's or Imperial zebra
Ethiopia, Somalia, and N Kenya. Subdesert steppe and arid bushed grassland. HBL 275cm; TL 49cm; WT 405kg. Coat: narrow, vertical black and white stripes on body, curving upward on haunches; belly is white and mane prominent and erect. Mule-like in appearance, with long, narrow head and prominent, broad ears. Conservation status: Endangered.

Abbreviations HBL = head–body length TL = tail length WT = weight. Approximate nonmetric equivalents: 50cm = 20in; 100kg = 220lb.

Horses' moods are often indicated visually by changes in ear, mouth, and tail positions. Smell helps individuals keep track of the movements of neighbors, since urine and feces bear social odors. Males use the "flehmen" or lip-curl response to assess the sexual state of females. Yet most social contact takes place through sound. In horses and the Plains zebra, foals whinny when separated from their mothers and mothers nicker to warn foals of danger. Males often nicker to declare their interest in a female, and squeal to warn competitors that further conflict is imminent. In fact, in horses, the squeals of dominant and subordinate males differ. Those of dominants are 50 percent more drawn-out, and tend to end with broad notes rather than with the pure-tone whistles characteristic of subordinates. But the key difference occurs at the onset of the squeal; dominants can instantly hit high-frequency notes, whereas subordinates cannot. This suggests that dominants can exhale air more forcefully than subordinates; sound thus becomes an indicator of superior aerobic capacity, warning other males not to escalate bravado into a real fight. In asses and Grevy's zebra, males often bray when fighting or calling to each other over long distances.

Grasslands, Savannas, and Deserts

Distribution Patterns

The Plains zebra occupies the lushest environments of any wild equid – the grasslands and drier savannas of East Africa, from Kenya to the Cape. The Mountain zebra is confined to two mountainous regions of southwest Africa where vegetation is abundant. The remaining species live in more arid environments with sparse vegetation. Przewalski's horse inhabited the semi-arid deserts of Mongolia, while the Asian wild ass is found in the most arid deserts of central Asia and the Near East. The African wild ass, the least horselike of

Left** Representative species of horses, asses, and zebras.* ***1 *Przewalski's horse* (Equus przewalskii), *the ancestor of all domestic horses, showing the stallion's bite threat.* ***2*** *A female African ass* (Equus asinus), *showing the kick threat, with its ears held back.* ***3*** *A male onager, a subspecies of the Asiatic ass* (Equus hemionus), *adding to a dung pile, as a territorial mark.* ***4*** *The kiang, the largest subspecies of Asiatic ass, showing the flehmen reaction after smelling a female's urine.* ***5*** *A young male Mountain zebra* (Equus zebra) *presenting a submissive face to an adult male. Note the dewlap and the grid-iron rump pattern.* ***6*** *A male Plains zebra* (Equus burchellii) *driving mares in a characteristic low-head posture, with ears held back.* ***7*** *A female Grevy's zebra* (Equus grevyi) *in heat and showing the receptive stance, with hind legs slightly splayed and tail raised to one side.*

ıe equids, roams the rocky deserts of North frica. Generally, the ranges of the species do not verlap, the only exception being Grevy's and the lains zebra, which coexist in the thorny scrub of orthern Kenya.

High-Fiber Foragers

DIET

ll equids forage primarily on fibrous foods. lthough horses and zebras feed mainly on grasss and sedges, they will eat bark, leaves, buds, uits, and roots, all of which are common fare for ıe asses, when these are sparse. In equids the ıicroorganisms such as bacteria and protozoa ıat break down plant cell walls are housed in the ac-like cecum. Food is fermented after passing ırough the stomach, so the passage rate of vegeıtion is not limited as it is in the ruminant grazers, nabling equids to consume large quantities of ɔw-quality forage, and to sustain themselves in ıore marginal habitats and on diets of lower qualy than ruminants. Even when vegetation grows apidly, equids forage for about 60 percent of the day, or up to 80 percent when conditions worsen.

Water plays a critical role in shaping the daily, seasonal, and ranging activities of equids. Individuals of all species need to water once a day, nursing young foals more frequently. In fact, the specific watering needs of females of different reproductive states generate different ranging and association patterns among the arid and mesic-adapted species. In arid areas the best feeding sites often become widely separated from watering points. As a result, lactating and non-lactating females of the asses and Grevy's zebra are forced to split off from the herd for parts of the day. In the more mesic areas, food and water are never so far apart.

The interactions that equids have with other species are varied. Ranges rarely overlap, but when sympatry does occur, as happens for some populations of Grevy's and Plains zebras, competition for forage appears, but it is subtle. Since microhabitat differences separate the lactating females of these species – only lactating Grevy's females remain on the grazing lawns near water – these females have a foraging refuge. Although these areas are limited in expanse, nursing females are able to rear their young in them free from competition. Their more numerous, non-lactating counterparts, however, must search more widely for forage, and in doing so come into contact with Plains zebras, which are more efficient at foraging as well as even more abundant. Since both species are drawn to the best swards of grass, the bite rate of non-lactating Grevy's females is sufficiently depressed to lower bodily condition and reproduction, thus ultimately restricting the species' geographic range to the most arid grasslands.

The relationship of equids to similarly-sized ruminant grazers can involve coexistence, competition, or even facilitation. The relationship between Plains zebras, wildebeest, and Thomson's gazelles typifies the diverse nature of these interactions.

When grasses are growing, slight differences in clipping styles and mouth size provide enough dietary separation to minimize competition. But as habitats dry out and grass growth ceases, the zebras' need for large amounts of forage drives them from the higher plains to the moister, more dangerous valleys, where tough forage is plentiful. As they consume the taller, more fibrous material, the softer, more nutritious green leaves appear. By opening up the habitat, zebras facilitate coexistence with the more selective ruminants that need higher-quality vegetation. Nevertheless, under conditions of severe drought or overgrazing, forage can become so limited that wildebeest, Thomson's gazelles, and the other bovids exploit their more efficient rumen fermentation system to outcompete the equids. This argument has been used to explain why grazing equids declined in abundance during the Pleistocene when similarly-sized bovids radiated. Yet little direct evidence supports this hypotheses, and a more modern view is that the evolutionary replacement may have been more the result of changing vegetation characteristics than comparative fermentation efficiency.

Tight-knit and Fluid Groups

SOCIAL BEHAVIOR

Equids are highly social and exhibit two basic patterns of social organization. In one, typified by the two horse species as well as the Plains and Mountain zebra, adults live in groups with a permanent membership, consisting of a male and a few females that remain in the same harem throughout their adult lives. Each harem has a home range that overlaps with those of neighbors. The second social system, typified by the asses and Grevy's zebra, involves more ephemeral adult associations rarely lasting longer than a few months. Temporary aggregations of one or both sexes are common, but most adult males live alone in large territories. For Grevy's zebra these vary in size from 2–10sq km (0.8–3.9sq mi), but for asses they measure up to 15sq km (5.8sq mi). Within territorial boundaries, which are marked with large piles of dung, owners get exclusive mating access to receptive females that roam through them. In both systems, surplus males live together in bachelor groups.

Social systems involving temporary groupings and solitary territorial males occur in drier habitats where resources are distributed patchily and where water and the best feeding areas become separated. With lactating females unable to associate on a continuing basis with their non-lactating counterparts, the social cohesion so evident in the mesic-adapted species becomes disrupted. Given that receptive and sexually active females populate both types of female groupings, males compete for control of exclusive areas that either control the major travel routes to and from water if they are dominant, or are far from water, containing the best forage if they are not. Only larger, more evenly distributed resources in which food and water occur near each other allow females of differing reproductive states to travel together in permanent groups, thus enabling harems to form.

Mammal groups are typically formed of close female kin, as daughters remain with their mothers, but in equids groups are composed of non-relatives, since both sexes leave their natal area. Females emigrate when they become sexually mature at about two years old, and neighboring harems, or bachelor males, attempt to steal then Males disperse by the fourth year to form bachel associations. Only after a number of years in suc associations are males able to defend territories, steal young females, or displace established hare males. Less able males leaving bachelor groups often form alliances with other similarly ranked males, thereby gaining some mating success.

The relationship between males and females that develop in the harem-forming species is alsc special. Typically, harems form when closed mem bership groups of females coalesce because some foraging or anti-predator benefits accrue from doing so. In horses and Plains zebras, however, females come together because males provide material benefits: increased time for foraging, greater protection for the young, and less sexual harassment. Minimizing strife creates a social sta bility that is vitally important for female welfare, ultimately allowing them to increase their reproductive success. Since males vary in their ability provide these benefits, females are choosy, and may vote with their feet by changing groups.

In Plains zebra societies, harem groups themselves often coalesce to form herds, sometimes joining with 50 or more bachelor males. This multi-leveled social organization is more characteristic of primates like Gelada or Savanna baboo than it is of ungulates. Why such complex societies form is not fully understood, although food resources, intensity of predation, and the pressures that males face in coping with bachelor ma sexual encroachments all seem to play a part.

Mares usually bear only one foal. Only in the Grevy's zebra does gestation last more than a yea since females come on heat 7–10 days after bearing a foal, mating and birth occur during the sam season, coinciding with renewed vegetation growth. Reproductive competition for receptive females is keen among males. It begins with visu displays of head-shaking, neck-arching, or foot-stamping, or with ritualized bouts of defecating and sniffing, or even with squealing. But it is not uncommon for contests to escalate, with animals pushing, rearing, and biting at one another's necks, tearing at "knees," or kicking their hind legs into their opponents' faces and chests. In contrast, amicable activities like mutual groomin cement female relationships. There are, however, dominance hierarchies among females also, and high rank confers benefits, including first access t water and superior vegetation.

The young are up and about within an hour of birth. They start grazing within a few weeks, but are generally not weaned for 8–13 months. Females can breed annually, but most miss a year because of the strains of rearing foals.

Left A group of kiangs. Human settlement of thei habitat on the remote Tibetan Plateau threatens the future security of this subspecies of Asiatic ass.

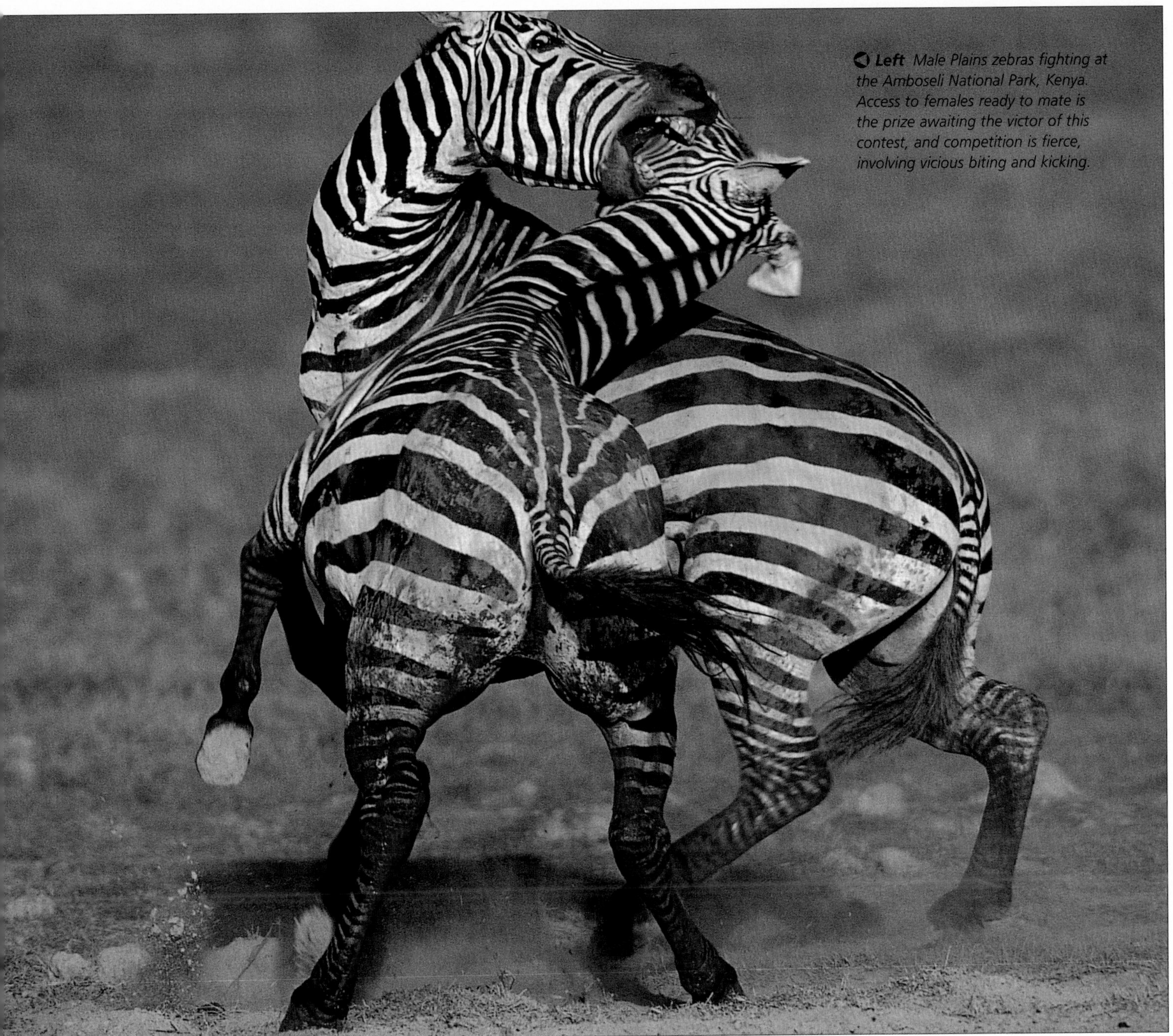

Left *Male Plains zebras fighting at the Amboseli National Park, Kenya. Access to females ready to mate is the prize awaiting the victor of this contest, and competition is fierce, involving vicious biting and kicking.*

Redesigning the Human Relationship

CONSERVATION AND ENVIRONMENT

›f the 20 equid subspecies identified at the start ˈ the 20th century, 3 are now extinct and 13 ıreatened. No Przewalski's horses have been ›en in their natural habitat since 1968, and only ›out 200 animals, scattered among the world's ›os and some reintroduced populations, survive.

Many feral populations of Domestic horses are ›eated as vermin. Selective removals, adoption ›rograms, and female fertility control are pursued ›ggressively to preserve rangelands and minimize ›mpetition with livestock minimized. As for ›bras, only the Plains zebra is plentiful and occu-›ies much of its former range, but even it is under ›reat from humans. Mountain zebra populations ›e already small, yet protected in national parks, ›ut those of Grevy's zebra have fallen drastically, ›nce their beautiful coats fetch high prices.

Even in the Serengeti, which supports large opulations of Plains zebras, numbers fell by some 40 percent during the 1990s. The estimated annual off-take of about 13 percent is clearly not sustainable. On unprotected lands the problem is often worse, since growing human populations bring settlements, crops, and livestock ranching.

If the cycle of poverty and political instability can be broken, wildlife viewing and limited cropping may point the way forward for sustainable zebra management, generating income as an incentive to protect the species. Models have shown that, for most zebra populations, 6 percent of the animals can be harvested per year to maintain stable numbers. Equids are resilient, upping their birth rate and lowering their age of sexual maturity when populations are depressed, so increasing effective genetic population sizes. A further source of genetic diversity comes from the fact that most adult males breed. With their powerful hold on people's imagination and their good fortune in inhabiting areas of natural beauty, wild equids will hopefully prove a major faunal attraction. DIR

GREVY'S ZEBRAS AND PASTORALISTS

In northern Kenya, the number of Grevy's zebras has shrunk drastically since 1977, largely as a result of increased competition with pastoralists and their livestock. Overgrazing has reduced perennial grasses, causing the Grevy's to venture further and more often in search of food. Access to water is also affected; in protected areas, the zebras drink in the middle of the day to avoid predators, but in pastoralist areas waterholes are monopolized by people and their animals during daylight, forcing the zebras to drink at night. This makes them more vulnerable to predation, especially the foals. Lacking the energy to travel as far and as fast as their mothers, foals are left behind in unprotected "kindergartens" some distance from water. At dusk and dawn, the young are particularly susceptible to attack. One solution might be to create enough protected areas to produce young that will survive and then colonize the more dangerous areas inhabited by people. SW

Tapirs

THE FOUR EXTANT TAPIR SPECIES ARE REMNANTS of an ancient lineage that evolved some 55 million years ago, following the eclipse of the dinosaurs. Predating both the horse and rhinoceros families, the versatile, prehensile trunk of the tapir is considered the key to its evolutionary success.

These "living fossils" most closely resemble the ancient extinct order Condylarthra, transitional between insectivores and advanced ungulates. As the Earth has become both drier and colder, grassy habitats have favored the rise of Artiodactyla herbivores, with their more efficient digestive systems, but formerly many species of tapirs populated such areas as North America, Europe, Asia, and northern Africa. These included the giant *Megatapirus*, fossil remains of which have been found in Pleistocene deposits in China's Sichuan province.

Land Animals that Love Water

FORM AND FUNCTION

Tapirs depend on both lowland and montane tropical forest habitats, and frequently enter rivers and lakes. Those species that inhabit lowland rain forests (*T. terrestris*, *T. bairdii*, and *T. indicus*) are especially amphibious, being equally at home on dry ground or in flooded forests and rivers, while even the Mountain tapirs are not averse to a bath. Instinctively, all tapirs take refuge in water when they are being pursued by predators, which include jaguars, tigers, pumas, and Andean bears – ingeniously, the Brazilian tapir of the Amazon may submerge itself in deep water to force a jaguar clinging to its back to release its hold. This species is known regularly to walk on river beds, searching for favored aquatic plants. Giant anacondas and crocodiles are reported to attack and swallow Brazilian tapirs whole, particularly smaller or young animals. Jungle-dwelling tapirs often deposit their feces in water, either in an effort to eliminate their scent trail or to reduce biting flies. For the same reasons, tapirs also wallow in mud. Humans are tapirs' chief predators and often use hunting dogs to track down their quarry.

Lower-elevation tapirs are mainly nocturnal and crepuscular. By contrast, Mountain tapirs tend to emerge more during the day, yet will still take a midday siesta when the heat of the sun is at its most intense. They sleep principally between midnight and dawn, resuming their activity in the early morning. All tapirs have a tendency to be increasingly nocturnal when facing heavy persecution by humans.

The coat of tapirs is designed for camouflage. In all species, young animals less than one year old sport whitish spots and stripes, set on a rufous to dark brown background. The giant Malayan tapir has a white saddle with dark extremities and forequarters, which break up its outline, especially at night. The Mountain tapir has a thick, woolly coat of dark brown to coal-black hair, which enables it to blend in with the strong shadows that occur in montane cloud forests or in the intensely bright, open tracts of the *páramo* (the treeless moorland typical of the northern Andes).

Tapirs are myopic, but their acute senses of hearing and smell are used to good effect in locating food and detecting predators. On the precipitous terrain of the Andes, their keen sense of balance allows them to tread very sure-footedly; on steep slopes with gradients of between 60° and 70°, tapirs can easily elude human researchers who are trying to observe them. Throughout the forest, well-trodden trails connect the tapirs' major water sources, salt licks, and feeding and sleeping areas.

Salt licks are favored spots for tapirs to congregate around at mating time. Males engage in combat for access to females, after which Mountain tapirs may be seen spending time together as isolated pairs.

Reciprocal Relationships with Plants

DIET

All parts of ferns, horsetails, palm fruits and hearts, coarse leaves of tropical trees, and bromeliads (and their berries) are eaten, and tapirs act as major seed dispersers for a substantial percentage of the plants that they consume; in one study area in Ecuador's Sangay National Park, 33 percent of the vascular plant species inventoried and 42 percent of those consumed by Mountain tapirs produced seeds that germinated in tapir feces. Baird's tapir is known to consume 34kg (75lb) of forage in a single night, the majority of which rapidly becomes top soil mulch in a humid rain forest; indeed, this tapir species has been closely linked with the successful reproduction and dispersal of large, seeded fruiting trees such as *Raphia faedigera* in Central America, and avocado and palm species also seem to have similarly mutual relationships with tapirs. Tapirs also open up clearings in the rain forest for the germination of new plants.

Forest Benefactors Turned Victims

CONSERVATION AND ENVIRONMENT

Coming into estrus every 2–3 months, mother tapirs have a long gestation period of 13 months and generally give birth to a single young, which stays with its mother for 1–2 years. Tapirs are also generally solitary and sparsely distributed, with individual home ranges of just a few square kilometers. Their large size and clear trails make them easy targets for hunters, who sometimes lure them by imitating their whistles. They are also slow to recover their numbers; according to one study, the Brazilian tapir goes into a decline after a 20 percent decrease in a viable population from hunting.

Hunting and international trade in tapirs are illegal but continue nevertheless. Habitat destruction is rife, exposing these shy mammals to disease, shortage of mates, inbreeding, and persecution by hunters for meat, hides, hunting trophies, and ingredients for folk medicine. The long-term solution may lie in the restoration of degraded habitat through sustainable agriculture projects such as terracing, as well as through the revival of the traditional, nutritious quinoa grain (*Chenopodium* spp.) in the diet of Andean people in place of livestock dependence. CC

Left *The four extant species of tapirs:* ***1*** *Baird's tapir* (Tapirus bairdii) *shown with young;* ***2*** *Mountain tapir* (T. pinchaque)*;* ***3*** *Brazilian tapir* (T. terrestris)*;* ***4*** *Malayan tapir* (T. indicus).

2

3

4

FACTFILE

TAPIRS

Order: Perissodactyla

Family: Tapiridae

4 species of the genus *Tapirus*

DISTRIBUTION C and S America, SE Asia, in moist tropical forests and grasslands from sea level to nearly 5,000m (16,500ft).

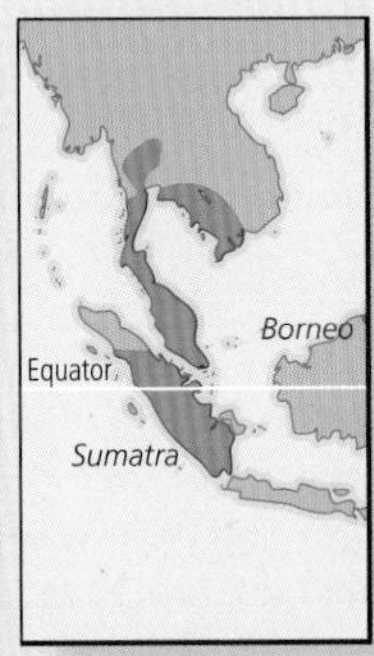

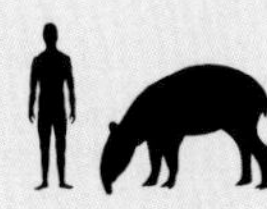

Size Head–body length 180–250cm (71–98in); **tail length** 5–10cm (2–4in); **shoulder height** 75–120cm (29–47in); **weight** 150–300kg (330–660lb).

BRAZILIAN TAPIR *Tapirus terrestris*

South American, Lowland, or Amazonian tapir

E of Andes from N Colombia to S Brazil, N Argentina, and Paraguay, including Amazonia and Orinoco tropical forest basins. Lowland rain forest and lower montane forest; sea level to c.1,700m (5,600ft). Coat: dark to reddish brown to whitish gray dorsally, paler below; short, sparse, bristly; low, narrow, muscular mane from forehead to shoulders. Breeding: gestation 390–400 days. Longevity: 30 years. Conservation status: Lower Risk – Near Threatened.

MOUNTAIN TAPIR *Tapirus pinchaque*

Mountain, Andean, or Woolly tapir

Andes mountains in Colombia, Ecuador, extreme NW Peru. Mid- and high-elevation cloud forests to above treeline in bushy páramo grasslands, also to Alpine meadow; observed from 1,400–4,700m (4,600–15,400ft), commonly 2,000–4,500m (6,500–14,800ft). Coat: thick and woolly, reddish dark brown to coal black in color; white lip, toe, and, typically, upper ear fringes; skin thinner than other tapirs. Breeding: gestation 393 days. Longevity: estimated at 25 years. Conservation status: Endangered.

BAIRD'S TAPIR *Tapirus bairdii*

Baird's or Central American tapir, Mountain cow

S. Mexico through C America and S to Gulf of Guayaquil. Lowland forests, swamps, flooded meadows, lower- to middle-elevation montane forests. Coat: lower chest and cheek whitish or grayish; body reddish brown; white ear fringes; short, thick mane proceeding from forehead, not as peaked as *T. terrestris*. Breeding: gestation 13 months. Longevity: 30 years. Conservation status: Vulnerable.

MALAYAN TAPIR *Tapirus indicus*

Malayan or Asian tapir

S Burma, Thailand, Sumatra, formerly Borneo. Dense primary rain forests; river and lake banks. Coat: middle part of body white, fore- and hindparts black ("disruptive coloration"). Breeding: gestation 390–395 days. Longevity: 30 years. Conservation status: Vulnerable.

Rhinoceroses

Rhinos have short, stout limbs to support the massive weight. The three toes on each foot give their tracks a characteristic "ace-of-clubs" appea ance. The Indian rhino has an armor-plated look produced by the prominent folds on its skin and its lumpy surface. The White rhino has a prominent hump on the back of its neck, containing t ligament supporting the weight of its massive head. In both the White and the Indian rhino, adult males are notably larger than females, whil in the other rhino species both sexes are of simil size. The Black rhino has a prehensile upper lip grasping the branch ends of woody plants, while the White rhino has a lengthened skull and broa lips for grazing the short grasses that it favors. In color, the two species are not notably different,

Left *Surpassed in size only by the Asian and African elephants, the White rhino is the world's thir largest land mammal. Males of this species can weig over 2 tonnes.*

WITH THEIR MASSIVE BULK, LEATHERY skin, and prominent horns, rhinoceroses look at first sight as though they would be more at home with dinosaurs than with the mammals that superseded them. Indeed, this impression is not entirely false – rhinos do have a very ancient ancestry.

Together with the elephants and hippopotamuses, rhinos are surviving representatives of a biological category that was once far more abundant and diverse – the "megaherbivores." Many forms of rhinos existed during the Tertiary era 40 million years ago, while Europe was home to the Woolly rhino as recently as the last Ice Age, just 15,000 years ago; and although the extinct rhinos varied in their possession and arrangement of horns, they were generally large. Of the five surviving species, two are on the brink of extinction, while the other three are increasingly threatened.

The Mammalian Dinosaur

FORM AND FUNCTION

The rhinoceros (from the Greek, meaning "nose horn") is named for its most distinctive feature. Unlike the horns of antelopes, cattle, or sheep, the rhino's have no bony core, but consist instead of an aggregation of keratin fibers perched on a roughened area on the skull. Black and White rhinos (the two African species) and the Sumatran rhinoceros all have two horns in tandem, with the front one usually the larger; Indian and Javan rhinos have a single horn on the end of their snout.

FACTFILE

RHINOCEROSES

Order: Perissodactyla

Family: Rhinocerotidae

5 species in 4 genera

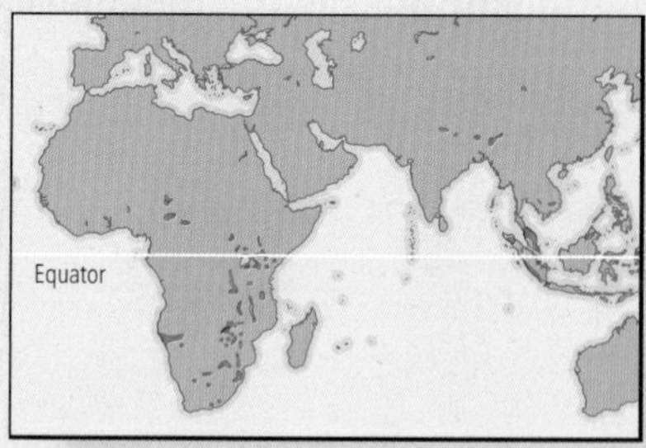

DISTRIBUTION Africa and tropical Asia

BLACK RHINOCEROS *Diceros bicornis*
Black or Hooked-lipped rhinoceros
Africa, from the Cape to Kenya. From montane rain forest to arid scrubland; browser; more nocturnal than diurnal. HBL 2.86–3.05m (9.4–10ft); HT 1.43–1.8m (4.7–5.2ft): TL 60cm (2ft); AH 42–135cm (1.4–4.4ft); PH 20–50cm (8–20cm); WT 0.95–1.3 tonnes. Coat: gray to brownish gray, varying with soil color; hairless. Breeding: 1 young born after gestation of 16–17months. Longevity: 40 years. Conservation status: Critically Endangered.

WHITE RHINOCEROS *Ceratotherium simum*
White or Square-lipped rhinoceros
S and NE Africa. Drier savannas; grazer; both diurnal and nocturnal. HBL male 3.7–4m (12–13ft), female 3.4–3.65m (11.1–12ft); HT male 1.7–1.86m (5.6–6.1ft), female 1.6–1.77m (5.2–5.8ft); TL 70cm (28in); AH male 40–120cm (16–47in), female 50–166cm (20–65in); PH 16–40cm (6–16in), WT male up to 2.3 tonnes, female up to 1.7 tonnes. Coat: neutral gray, varying with soil color; almost hairless. Breeding: 1 young born after a gestation of 16 months. Longevity: 45 years. Conservation status: Lower Risk – Conservation Dependent, though the northern population in the Democratic Republic of Congo is Critically Endangered.

INDIAN RHINOCEROS *Rhinoceros unicornis*
Indian or Greater one-horned rhinoceros
India (Assam), Nepal, and Bhutan. Floodplain grasslands; mainly a grazer; diurnal and nocturnal. HBL male 3.68–3.8m (12.1–12.5ft), female 3.1–3.4m (10.2–11.2ft); HT male 1.7–1.86m (5.6–6.1ft), female 1.48–1.73m (4.9–5.7ft); TL 70–80cm (28–31in); HORN 45cm (18in); WT male 2.2 tonnes, female 1.6 tonnes. Coat: gray, hairless. Breeding: 1 young born after a gestation of 16 months. Longevity: 45 years. Conservation status: Endangered.

JAVAN RHINOCEROS *Rhinoceros sondaicus*
Javan or Lesser one-horned rhinoceros
SE Asia. Lowland rain forests; browser; diurnal and nocturnal. HT up to 1.7m (5.6ft); WT up to 1.4 tonnes. Coat: gray, hairless. Conservation status: Critically Endangered.

SUMATRAN RHINOCEROS *Dicerorhinus sumatrensis*
Sumatran or Asian two-horned or Hairy rhinoceros
SE Asia. Montane rain forests; browser; diurnal and nocturnal. HBL 2.5–3.15m (8.2–10.3ft); HT up to 1.38m (4.5ft); AH up to 38cm (15in); WT up to 0.8 tonnes. Coat: gray, sparsely covered with long hair. Breeding: 1 young born after a gestation of 7–8 months. Longevity: 32 years. Conservation status: Critically Endangered.

Abbreviations: HBL = head–body length HT = height TL = tail length AH = anterior horn PH = posterior horn WT = weight

d the popular names applied to them most)bably arose from the local soil color tinting the ;t specimens seen.

Rhinos have poor vision, and are unable to ect a motionless person at a distance of more an 30m (100ft). The eyes are placed on either le of the head, so that to see straight in front the imals peer first with one eye, then with the ıer. Their hearing is good, their tubular ears iveling to pick up the quietest sounds. Howev- they rely most of all on their sense of smell for owledge of their surroundings: the volume of e olfactory passages in the snout exceeds that of e entire brain!

When undisturbed by humans, rhinos can metimes be noisy animals: a variety of snorts, ffing sounds, roars, squeals, shrieks, and honks ve been described for various species. Snorts, nks, and roars serve largely to maintain spacing tween animals, while squeals are used by calves eking to nurse; White rhino bulls also squeal ıen seeking to block cows from leaving their ter- ories. Loud shrieks or bleats are used defensive- while a squeak-pant is made by bulls chasing other animal. Soft hic-pants are emitted by bulls ıring courtship.

One peculiarity of rhinos is that, as in ele- ıants, the testes do not descend into a scrotum. ıe penis, when retracted, points backwards, so at both sexes direct urine to the rear. Females)ssess two teats located between their hind legs.

Five Species, Three Tribes

DISTRIBUTION PATTERNS

ıe five surviving species of rhino represent three stinct lineages, or tribes, within the family ıinocerotidae. The Sumatran rhino is the sole ırviving representative of the Dicerorhinini, hich originated in the Miocene some 20 million ars ago. The Indian and Javan rhinos – the ıinocerotini – shared a common ancestor in dia about 10 million years ago; the extinct 'oolly rhinoceros was the terminal form of a dif- rent lineage within this tribe. The two African)ecies – the Dicerotini – had an African ancestry ıring the middle Miocene, about 14 million ars ago. The two modern genera diverged in the rly Pliocene, some 5 million years ago.

Modern distribution of species has been radi- lly affected by hunting and habitat loss. In par- cular, the Black rhino's range has dwindled astically. As recently as the 1960s, it roamed vir- ıally all of sub-Saharan Africa except tropical rests, inhabiting mountainous regions of Kenya ıp to 3,000m/10,000ft), stony desert in Mali and amibia, and scrubland from Zambia to Mozam- que. Now, their distribution is very patchy, and ıost survive only within guarded game reserves.

Right *Black rhinos use their thick, hooked, prehen- le lips to pluck woody stems, thickened leaves, and eedy browse.*

Plants by the Kilo

DIET

All rhinoceroses are herbivores dependent on plant foliage, and they need a large daily intake of food to support their great bulk; the wet weight of the grass in the stomach of one White rhino cow that died of peritonitis totaled 72kg (159lb), or 4.5 percent of body mass, and probably represented roughly one day's food intake. With their large size and hindgut fermentation they can tolerate a relatively high-fiber diet, but they prefer more nutritious leafy material when available. Both African species have no incisors or canines, but use their lips (differently configured in each) to graze or browse. Asian species still have incisors, and the Sumatran rhino bears canines, in both cases for fighting rather than gathering food. The broad lips of White rhinos give them a large area of bite, enabling them to glean enough food from the grassland areas that they favor for much of the year. During the dry season, once the short grasses have been depleted, they turn to the shaded areas under trees where buffalo grass (*Panicum maximum*) predominates, and finally resort to stands of tall grasses, dominated by red grass (*Themeda triandra*). Black rhinos use their prehensile upper lips to pick woody plants. Favorite browse includes many of the *Acacia* species, as well as euphorbias, including succulent forms with milky latex. Forbs are also an important part of the diet, but little grass is eaten. Among fruits eaten are the enormous woody "sausages" of *Kigelia africana*.

Below *The White rhino has a wide, unhooked upper lip that predisposes it to grazing, rather than browsing. It suffered heavily from overhunting, but conservation measures have seen its numbers rise.*

Indian rhinos use their prehensile upper lip to gather tall grasses and shrubs, but can fold the tip away when feeding on short grasses. They tend to favor tall grasses, especially *Saccharum* species, but woody browse comprises about 20 percent of their diet during the winter period. They also seek out fallen fruits of *Trewia nudiflora*. Both the Javan and Sumatran rhinos are exclusively browsers, often breaking down saplings to feed on leaves and shoot ends. They also eat selected fruits, as do African Black rhinos and, to a lesser extent, Indian rhinos. Fruits consumed by Sumatran rhinos include *Garcinia mangostana* and *Mangifera indica*.

All rhinos are dependent upon water, drinking almost daily at small pools or rivers when these are readily available. In captivity, White rhinos will drink up to 80 liters per day, though they must get by with less in the wild. In arid conditions, both African species can survive for periods of 4–5 days between waterhole visits.

Rhinos also depend on waterholes for wallowing. Indian rhinos in particular spend long periods lying in water, while the African species more commonly coat themselves in mud. While the water may provide some cooling, the mud coating probably serves mainly to protect them against biting flies (though rhinos have thick hides, the blood vessels lie just below a thin outer layer).

Sociable Solitaries

SOCIAL BEHAVIOR

For large, long-lived mammals like rhinos, life-history processes tend to be protracted. When about five years old, female White and Indian rhinos undergo their first sexual cycles, and by 6–8 years they will bear their first calf, after a gestation period of 16 months. In the smaller Black rhino,

Right *The five species of rhinoceros:* ***1*** *Indian rhinoceros* (Rhinoceros unicornis)*;* ***2*** *Javan rhinoceros* (Rhinoceros sondaicus)*;* ***3*** *Sumatran rhinoceros* (Dicerorhinus sumatrensis)*;* ***4*** *Black rhinoceros* (Dicer bicornis)*;* ***5*** *White rhinoceros* (Ceratotherium simum)

females breed about a year younger than this. A single birth is the rule. Intervals between successive offspring can be as short as 22 months, but more usually vary between 2 and 4 years. The babies are relatively small at birth, weighing only about 4 percent of the mother's weight (65kg/143lb in White and Indian rhinos; 40kg/88lb in the Black rhino). Females seek seclusion from other rhinos around the time of the birth. White rhino calves are able to follow their mothers three days after birth. Indian rhino mothers sometimes forage as far as 800m (2,600ft) from their calves. Indian and White rhino calves tend to run in front of the mother, where she is well able to offer protection, while those of Black rhinos, which inhabit thicker bush, usually run behind. White rhino mothers stand protectively over their offspring should danger threaten.

Males mature at about 7–8 years of age in the wild; but they are prevented from breeding until they can claim their first territories, or attain dominant status, at about 10 years.

Births may take place in any month of the year, but among the African rhinos conceptions tend to peak during the rains, so that most babies are born during the early part of the dry season. The mother's milk supply helps nourish them through this difficult time. Although White rhino calves begin to nibble on grass when they are about 3 months old, nursing continues beyond one year of age.

Adult rhinos are mostly solitary, except for the association between a mother and her most recent offspring. This lasts until the calf is driven away just prior to the birth of the next offspring, typically at around 2–3 years of age. However, immature animals of both sexes, or adult females without young, may join up to form pairs or larger groups, and in White rhinos such temporary herds can on occasion number 10 or more animals. Trios consisting of a female accompanied by a small calf plus a larger adolescent are not uncommon in White rhinos, but generally the female is not the mother of the older calf. Adult females without calves are also quite tolerant of associations by other young animals. Adult males of all species are usually solitary, apart from temporary associations with estrous females.

White and Indian rhinos typically have home ranges of 10–25sq km (4–10sq mi), but can occupy 50sq km (20sq mi) or more in low-density populations. Ranges of Black rhino females vary from about 3sq km (1.2sq mi) in forest patches to nearly 90sq km (35sq mi) in arid areas. Female home ranges in all species overlap extensively, and there is no indication of territoriality. White rhino females commonly engage in friendly nose-to-nose meetings that may lead to gentle horn-rubbing, but Indian rhino females generally respond aggressively to any close approach. However, subadults of both species approach adult females, calves, and other immature animals for nose-to-nose meetings, and sometimes playful wrestling bouts.

Males of all species are given to vicious fights that inflict gaping wounds. Both African species jab at one another with upward thrusts of their front horns. Black rhinos have the highest incidence among mammals of fatal intraspecies fighting: almost 50 percent of males and 33 percent of females die from wounds. Why they are quite so bellicose is not known; in any event, rhino populations with high mortality rates recover only slowly. Asian rhino species attack by jabbing open-mouthed with their lower incisor tusks, or, in the case of the Sumatran rhino, with the lower canines.

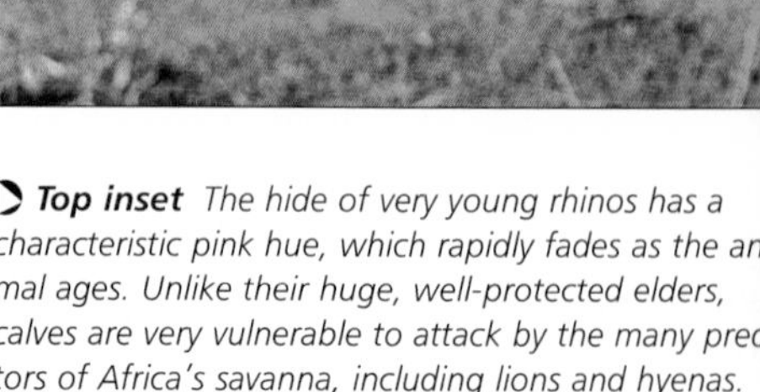

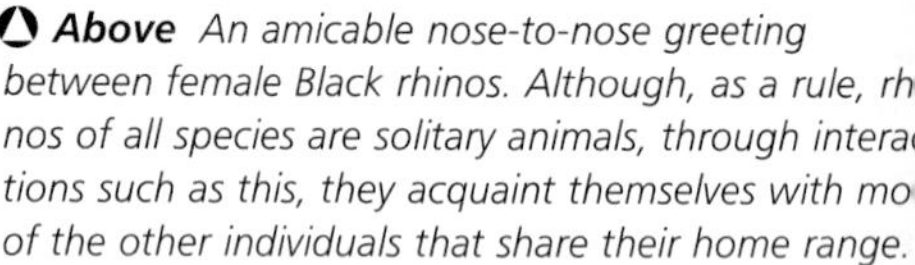

Top inset *The hide of very young rhinos has a characteristic pink hue, which rapidly fades as the an mal ages. Unlike their huge, well-protected elders, calves are very vulnerable to attack by the many prec tors of Africa's savanna, including lions and hyenas.*

Above *An amicable nose-to-nose greeting between female Black rhinos. Although, as a rule, rh nos of all species are solitary animals, through intera tions such as this, they acquaint themselves with mo of the other individuals that share their home range.*

Left *When confronting one another to assert thei rights over territory or breeding partners, rhinos repe the same gestures over and over before one gives way:* ***1*** *horns forced against one another;* ***2*** *wiping the horns on the ground.* ***3*** *A dominant male proclaims his mastery by spray-urinating, while the subor dinate male retreats. Only dominant males display thi particular behavior.*

Black rhinos have a reputation for unprovoked gression but their charges are usually blind shes designed to see off intruders. Indian rhinos quently make aggressive rushes when disturbed, d occasionally attack the elephants used as servation platforms in some of the sanctuaries ey occupy.

In contrast, the White rhino is mild and inof- nsive by nature, and despite its large size is easily ghtened off. Often, a group of White rhinos cluding subadults will stand in a defensive for- ation with their rumps pressed together, facing tward in different directions. While this forma- n may be successful in protecting young ani- als against attacks from carnivores such as lions d hyenas, it is useless against armed humans.

Recently a number of White rhinos have been lled by elephants in parks where orphaned ung elephants saved from culling operations ve been introduced, perhaps as a result of the ale elephants misdirecting their sexual or aggres- e behavior on reaching adolescence.

A Precarious Existence

CONSERVATION AND ENVIRONMENT

An interesting case study in endangerment and conservation is the Black rhino. From 1970 to the late 1990s, Black rhino populations were mercilessly slaughtered, plummeting from an estimated 100,000 individuals to less than 3,000. Protection was legislated by CITES in the mid-1970s, but it was ineffective. Poaching occurred for the sole purpose of selling horns on the black market, with exports going mainly to Yemen and Asia. In Yemen, the horns are crafted into the handles of daggers (*jambiya*) worn by men as status symbols; in Asia, they are ground up for use in traditional medicine. Efforts to curb the slaughter varied, from traditional methods, such as working with local villagers, instituting educational programs, and enlisting more guards, to the more drastic policy of translocating rhinos to safer areas within (and even outside) Africa. Controversial remedies included shooting poachers on sight (in Zimbabwe) and dehorning (in Zimbabwe, Namibia, and Swaziland). The shoot-to-kill policy proved ineffective (though some 150 poachers were killed), while dehorning met with biological and economic obstacles.

While it is encouraging that even today Black rhinos still occur in huge reserves, some in excess of 20,000sq km (such as in South Africa's Kruger and Namibia's Etosha national parks), these, and virtually all other populations are now fenced. The sole exception is the Black rhino population in the rugged Namib Desert, where densities may be as low as 0.002 per square kilometer.

Rhino survival, then, is now dependent in the first instance on armed protection and the flow of tourist revenue. Rhinos co-exist with other fauna in parks and reserves, but it is unclear how, in the long term, conservationists will solve the problems posed by genetic restrictions and overcrowding (including competition with elephants). Moreover, while the market for Chinese traditional medicines remains buoyant, there will be a constant threat of poaching. NO-S/JB

DISTRIBUTION Worldwide, except Australia and Antarctica.

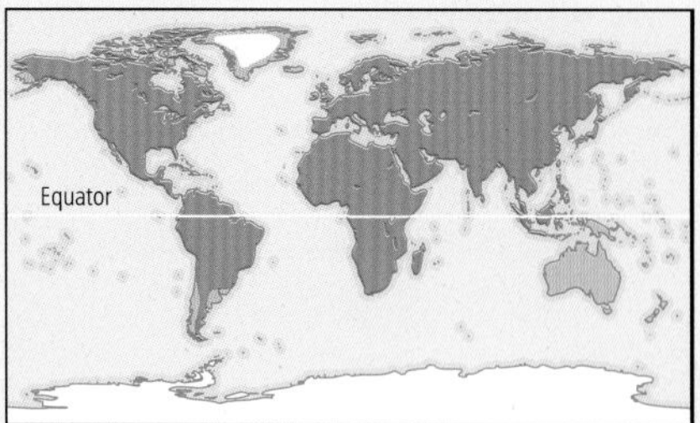

Order: Artiodactyla
196 species in 82 genera and 10 families

Note The 1993 established taxonomy recognized 220 species, 81 genera, and 10 families (given in parentheses above).

Even-toed Ungulates

THE ARTIODACTYLA OR EVEN-TOED UNGULATES are the most spectacular and diverse array of large, land-dwelling mammals alive today. Living in all habitats from rain forest to desert and on all continents except Australasia and Antarctica, they dominated the mammal communities of the savannas where human beings first arose. Now, however, humans control the environments on which their dwindling communities depend, ironically destroying their range by overstocking with domesticated forms of the same animals.

The order Artiodactyla comprises two rather different types of animals, linked by similarities of anatomical structure. The suines or suoids (suborder Suina) include the pigs, peccaries, and hippos. Suines are primarily omnivorous, retaining low-crowned cheek teeth with simple cusps, large tusk-like canines, short limbs, and four toes, although some lineages may be more "progressive" in the modification of these features.

In contrast, ruminant artiodactyls, comprising the suborders Tylopoda (camels and llamas) and Ruminantia (deer, giraffes, and bovids), are specialized herbivores that have evolved multi-chambered stomachs and the habit of chewing the cud to digest fibrous herbage. They have cheek teeth with ridges rather than cusps, and these teeth may be high-crowned (hypsodont). They also show a tendency to elongated limbs and a reduced number of functional toes, down to two from the four found in primitive artiodactyls.

The only living tylopods are the camels and llamas, which are low-level feeders taking a mixture of browse, herbs, and grass. Camels, true to their reputation, are found today in arid steppes and deserts in Asia and North Africa (and are als[o] found as feral animals in similar habitats in Australia), while llamas are known from the high-altitude plains of South America. However, came[ls] were first known from late-middle Eocene Nort[h] America (46 million years ago), and survived the[re] until the end of the Pleistocene (10,000 years ago). They were not known in Asia until the late Miocene (around 6 million years ago), nor in South America until the early Pleistocene (aroun[d] 1.8 million years ago).

The suborder Ruminantia is divided further into the traguloids and the pecorans. Living trag[u]loids are the chevrotains (family Tragulidae), ver[y] small deer that are found today in Old World tro[p]ical forests. Pecorans are all the remaining ruminant families, characterized by their bony horns.

The antelopes and cattle are today the most successful and diverse of the pecoran families, with a wide range of body forms and sizes and many different feeding adaptations. Most notabl[y] this is the only one to have evolved large-sized, open-habitat, specialized grazers, such as the bison and the wildebeest. Bovids have been primarily an Old World group, reaching their peak [of] diversity in the Plio-Pleistocene of Africa. They d[id] not invade North America until the Pleistocene, and were never native to South America. CMJ/J[...]

Below and above right *Artiodactyl diversity:* **1** *Bull hippopotamuses fighting;* **2** *Wild boar foraging;* **3** *male Mule deer displaying its forked antlers;* **4** *the Saola, only discovered in the Laos–Vietnam border region in 1992;* **5** *male Ibex lock horns in combat.*

VERSIFICATION

:iodactyls first appeared in North ıerica and Eurasia in the early cene, 54 million years ago. At that ıe they were represented by small, ɔrt-legged animals, whose simple-ɔped teeth suggest they lived on a ɔnkeylike diet of soft herbage. The three present-day suborders – Suina (pigs and their relatives), Tylopoda (camels and llamas), and Ruminantia (deer and bovids) – had all made their appearance by the late-middle Miocene, some 46 million years ago. Now the order has diversifed to include species as dissimilar as the elegant Dama gazelle, the imposing Bactrian camel, the shaggy-coated yak, and the fiercely-tusked warthog and babirusa.

5

SKULLS

The skulls of ruminants like the Roe deer and water buffalo have a shorter occiput, a complete postorbital bar, a big gap between front and back teeth (diastema), an incisiform lower canine, and the loss of the upper incisors (partially retained in camelids). The molars may be low- or high-crowned, with the cusps coalesced into cutting ridges. Ruminants resemble equids in the position of the orbits, but never develop the deep and massive lower jaw that characterizes horses.

The skulls of pigs like the babirusa are characterized by a deep occiput (rear of the skull), an incomplete postorbital bar, a short space between the front and back teeth, and the retention of large upper and lower canines. The teeth are usually low-crowned and simple-cusped (except in some specialized grazers such as the warthog). The upper incisors are never entirely lost; the upper canines are characteristically curved up to form tusks in pigs, but point down in peccaries and hippos.

e deer
cm

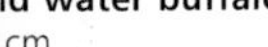

ld water buffalo
cm

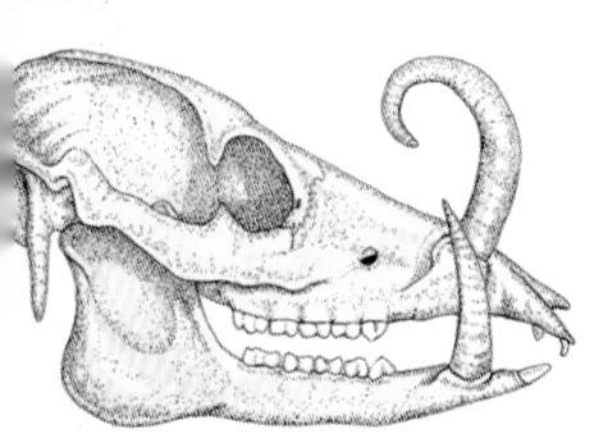

birusa
.5 cm

ARTIODACTYL HORN TYPES

Pecorans include all those ruminants possessing bony horns. The five living families are distinguished primarily by their horn type or antlers, which appear to have evolved independently among the different families. The family Moschidae (musk deer) remains primarily without horns, although the males have saber-like upper canines. Giraffids are tree browsers, found today in African tropical forests (the okapi) or savannas (the giraffe). The deer are primarily small- to medium-sized browsers found today in temperate and tropical woodlands in Eurasia, North America, and South America (they invaded the Americas during the Pliocene). Pronghorns have always been an exclusively North American family. Only one species now remains, a medium-sized, open-habitat browser on the western prairie; but there was a large diversity of antilocaprid species during the Miocene and Pliocene

Of the horned artiodactyls, the giraffe **1** has simple, unbranched postorbital horns covered by skin (in males only). The Roe deer **2** has branched postorbital antlers (horn-like structures of naked bone) that are shed yearly; in this and other cervids, except for the reindeer, antlers are found in males only. The pronghorn **3** has unbranched postorbital horns with a forked, keratinous cover that is shed annually; horns are present in both sexes of this species, but are smaller in females. Bovids such as the Common eland **4** have unbranched postorbital horns that may be spiraled, coiled, or twisted, with a permanent keratinous cover; in some species horns are present in the males only, while in others females have smaller horns.

Wild Pigs and Boars

WHAT WILD PIGS LACK IN GRACE AND beauty they make up for in strength, adaptability, and intelligence. They are admirably adapted to range the forests, thickets, woodlands, and grasslands that they haunt in small bands, to duel for position and mates, to fend off predators, and to enjoy a catholic diet.

Wild pigs are the most generalized of the living, even-toed, hoofed mammals (artiodactyls). They have a simple stomach, four toes on each foot, and, in three of the genera, a full complement of teeth. For the most part, they live in forests or woodlands and are active at night, which minimizes the amount of contact with people. Around 9,000 years ago, with the rise of settled agriculture, the Wild boar began to be domesticated; there are now many different breeds and crossbreeds of domestic pig.

Keen Senses

FORM AND FUNCTION

The living wild pigs are medium-sized artiodactyls characterized by a large head, short neck, and powerful but agile body, with a coarse, bristly coat. Their eyes are small and their expressive ears fairly long. The prominent snout carries a distinctive set of tusks (canine teeth) and ends in a mobile, disk-like nose perforated by two nostrils. The structure of the snout, tusks, and facial warts is intimately linked to diet, mode of feeding, and fighting style. The tasseled tail swats flies and signals mood.

Large canine teeth, molars with rounded cusps, and a prenasal bone supporting the nose are key features of wild pigs; in warthogs, the first and second molars regress and disappear as the third

Right The Bushpig is notorious for raiding cultivated crops in Africa. As a carrier of swine fever, it has also periodically been the subject of control campaigns.

Wild boar
Sus scrofa
Wild boar or Eurasian wild pig

Europe, N Africa, Asia, Sumatra, Japan, Taiwan. Introduced into N America. Feral domestic pigs in Australia, New Zealand, N and S America. Broad-leaved woodland and steppe. Active daylight and twilight.
HBL 90–180cm (35–71in); TL 30–40cm (12–16in); WT 50–200kg (110–440lb).
COAT: brownish-gray bristles, short, dense winter coat; no facial warts.
BREEDING: gestation 115 days; 6 pairs of mammae.
LONGEVITY: 15–20 years.

Pygmy hog Cr
Sus salvanius

Himalayan foothills of Assam. Tall savanna grassland. Nocturnal and crepuscular.
HBL 58–66cm (23–26in); TL 3cm (1.2in).
COAT: blackish-brown bristles on gray-brown skin; no facial warts.
BREEDING: gestation 100 days; 3 pairs of mammae.
LONGEVITY: 10–12 years.

Javan warty pig En
Sus verrucosus

Java, Bawean, Madura (extinct). Forest, lowland grasslands, and swamps.
HBL 90–160cm (35–63in); WT up to 185kg (408lb).
COAT: red or yellow hair with black tips; marked facial warts.
BREEDING AND LONGEVITY: probably as for Wild boar.

Bearded pig
Sus barbatus

Malaysia, Sumatra, and Borneo. Tropical forest, secondary forest, and mangroves.
COAT: dark brown to gray, with distinctive white beard on cheeks; marked facial warts.
BREEDING AND LONGEVITY: probably as for Wild boar.

Philippine warty pig Vu
Sus philippensis

Luzon, Mindoro, Samar, Leyte, and Mindanao islands.

Sulawesi wild boar Cr
Sus celebensis

Sulawesi and adjacent islands.

Indochinese warty pig
Sus bucculentus

Recently rediscovered in Vietnam from a single young animal.

The 1993 established list of mammals recognized 3 further *Sus* species: *S. cebifrons*, *S. heureni*, and *S. timoriensis*.

Babirusa Vu
Babyrousa babyrussa

Sulawesi, Togian, Sulu, and Burn Islands. Tropical forest.
HBL 85–105cm (33–41in); TL 27–32cm (11–13in); WT up to 90kg (198lb).
COAT: sparse, short, white or gray bristles; marked facial warts.
BREEDING: gestation 125–150 days; 1 pair mammae. Longevity: up to 24 years.

Red river hog
Potamochoerus porcus

W Africa and Congo basin. Forest and moist savanna woodlands.
HBL 100–150cm (39–59in); TL 30–40cm (12–16in); WT 50–120kg (110–265lb).
COAT: bright rufous with distinct white dorsal stripes, long white whiskers, and ear tufts; facial warts in male.
BREEDING: gestation 127 days; 3 pairs mammae.
LONGEVITY: 10–15 years.

Bushpig
Potamochoerus larvatus

E and S Africa. Forest, moist savanna woodlands, and grasslands.
COAT: red-brown to dark gray, with or without white masks and ear tufts; facial warts in male.

Giant forest hog
Hylochoerus meinertzhageni

Congo basin, parts of W and E Africa. Tropical forest, and intermediate zone between forest and grassland.
HBL 130–210cm (51–83in); TL 30–45cm (12–18in); WT 130–275kg (287–606lb).
COAT: brown and black bristles; facial warts.
BREEDING: gestation 149–154 days; 3 pairs mammae.

Common warthog
Phacochoerus africanus

Sub-Saharan Africa. Savanna woodland and grasslands. Active daytime.
HBL 110–135cm (43–53in); TL 40cm (16in); WT 50–100kg (110–220lb).
COAT: black or white bristles on gray skin; pronounced facial warts.
BREEDING: gestation 170–175 days; 2 pairs mammae.
LONGEVITY: 12–15 years.

Desert warthog
Phacochoerus aethiopicus

Arid steppes of Somalia and NE Kenya. Extinct in the Cape Province of South Africa, where it was first described. Presence in Somalia only recently established. Smaller than Common warthog.

Abbreviations HBL = head–body length WT = weight TL = tail length Cr Critically Endangered En Endangered Vu Vulnerable

FACTFILE

WILD PIGS AND BOARS

Order: Artiodactyla

Family: Suidae

13 (16) species in 5 genera: **pigs, hogs, and boars** (*Sus*, 7 (10) species); **warthogs** (*Phacochoerus*, 2 species); **Red river hog and bushpig** (*Potamochoerus*, 2 species); **babirusa** (*Babyrousa babyrussa*); and **Giant forest hog** (*Hylochoerus meinertzhageni*).

DISTRIBUTION Europe, Asia, E. Indies, Africa; introduced into N and S America, Australia, Tasmania, New Guinea, and New Zealand.

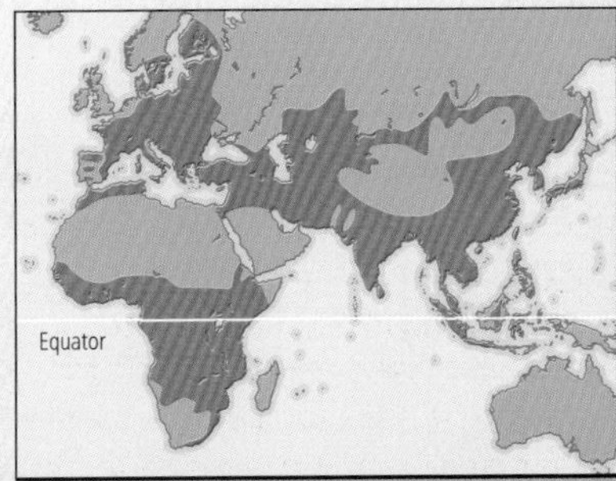

HABITAT Mostly forests and woodlands

SIZE **Head–body length** from 58–66cm (23–26in) in the Pygmy hog to 130–210cm (51–83in) in the Giant forest hog; **weight** from 6–9kg (13–20lb) to 130–275kg (285–605lb) in the same 2 species.

COAT Some wild pigs are almost naked, but more typically the skin is sparsely covered with coarse bristles.

DIET Omnivorous, including fungi, roots, bulbs, tubers, fruits, snails, earthworms, reptiles, young birds, eggs, small rodents, and carrion.

BREEDING Gestation period ranges from 100 days in the Pygmy hog to 175 in the Common warthog. Litter size varies for 1–12.

LONGEVITY Typically 15–20 years; up to 24 in the babirusa.

nolar elongates to fill the tooth row. Wild pigs valk on the third and fourth digits of each foot, vhile the smaller second and fifth digits are usualy clear of the ground. All male pigs are larger than emales, with more pronounced tusks and warts.

Smell, hearing, and vocalization are well develpped, and family groups communicate incessantly py squeaks, chirrups, and grunts. A loud grunt nay herald alarm, while rhythmic grunts characerize the courtship chant which, in warthogs, ounds like the exhaust of a two-stroke engine.

Snouting for Plants and Insects

DIET

Although adult males are solitary, other pigs forage in family parties, feeding on a wide range of plants (fungi, ferns, grasses, leaves, roots, bulbs, nd fruits), as well as rooting in litter and moist arth to take insect larvae, small vertebrates (frogs

Above *Though the warthog's upper tusks are the more impressive, the smaller, sharper lower tusks form its main weapons. In fights, the grotesque facial warts protect it against the incurving tusks of its opponent.*

and mice), and earthworms. The diet of bushpigs in the Knysna forest of South Africa, for example, was found to consist of underground plant parts such as tubers and corms (40 percent), leaves (30 percent), fruit (13 percent), animal matter (9 percent), and fungi (8 percent).

The Giant forest hog and the warthog are more specialized herbivores. The forest hog browses and grazes in evergreen pastures and forest glades, seldom digging with its snout. Warthogs, however, feed almost entirely on grasses, plucking the growing tips with their unique incisors or lips, or scooping grass rhizomes out of the sunbaked savanna soils with the tough upper edge of their nose. In Zimbabwe during the peak of the dry season, grass rhizomes may form as much as 85 percent of the warthog's diet; during the rains, grass leaves make up as much as 95 percent, while grass seeds form 60 percent of it toward the rains' end.

Sounders for the Sows

SOCIAL BEHAVIOR

While sexual maturity is attained at 18 months, males may only mate successfully on reaching physical maturity at about 4 years old. In the moist tropics breeding occurs throughout the year, but in temperate and tropical areas of marked seasonality, mating takes place in the fall and sows farrow the following spring. The young are born either in a grass nest constructed by the mother or (in the case of warthogs) in a hole underground, and weigh between 500 and 900g (18–32oz) at birth. The piglets remain in the nest for about 10 days before following their mother. Each piglet has its own teat. Weaning occurs at about three months, but young pigs remain with their mother in a closely-knit family group until she is ready to farrow again. After farrowing in isolation, young sows may rejoin her to form larger matriarchal herds, or "sounders," that may include several generations.

There are several courtship rituals. A chanting boar will nudge the sow's flanks, sniff her genital region, indulge in lateral displays, and repeatedly attempt to rest his chin on her rump. Bushpig, warthog, and pigs of the genus *Sus*, and possibly others, produce lip-gland pheromones, which may be dispersed while chanting; the *Sus* species also produce a salivary foam which induces a receptive sow to adopt a mating stance. Mating may last for up to 10 minutes; the boar's spiral penis fits into the sow's grooved cervix, in which a plug comprised of coagulated seminal fluid forms after copulation. This plug probably serves to block the entry of sperm from other boars.

Socially, boars tend to be solitary or to live in bachelor groups, while females form matriarchal sounders comprising one or more adult sows wit[h] their young of various ages. When these mother–daughter groups get too large, they fragment into smaller kinship units or "clans." These consist of a number of related sounders with overlapping home ranges where feeding grounds, water holes, wallows, resting sites, and sleeping dens are shared. Wild pigs appear to be nonterritorial, wit[h] home ranges of about 25ha (62 acres) in Pygmy hogs, 1–4sq km (0.4–1.5sq mi) in warthogs and 10–20 sq km (4–8sq mi) in Wild boars. Pigs hav[e] various means of keeping the home range well marked, by secretions from lip glands, pre-orbital glands (warthog and Giant forest hog), or foot glands (bushpig). The mating system appears to be a roving dominance hierarchy among the male[s] within a clan area.

In northeastern Borneo, the bearded pig is noted for the periodic mass migrations it undertakes following population explosions stimulated by successive years of unusually high forest fruit production. The last major migration, involving

The Bizarre Babirusa

The babirusa of Sulawesi is one of the most enigmatic mammals of Asia. An aberrant pig, it has been assigned a subfamily of its own, the Babirousinae, on account of its extraordinary tusks and complex stomach. Its bizarre, curling tusks have long fascinated observers, inspiring demonic Balinese masks and causing the animals to be given as gifts to visiting diplomats by early Sulawesi rulers. These tusks are in fact the babirusa's upper canine teeth, which grow vertically up through the skin of the snout and curve around towards the forehead. The sharp-tipped lower canines also grow far out from the jaw.

The function of the tusks has been the subject of speculation for more than a century. Present only in adult male animals, they are probably secondary sexual characteristics functioning in male–male competition. In advanced combat, adult male babirusa rear up on their hind legs in a "boxing" posture, the ultimate purpose of which might be to break off the opponent's upper tusk, hence reducing its ability to compete for mates.

Babirusa social organization comprises solitary adult males and matriarchal sounders of one or a few adult females and their immature young, usually in groups of up to five. Babirusa often use mud wallows, and rub their bodies on nearby trees afterwards. They are largely frugivorous but also consume leaves, grasses, and animal material. The species has a slow reproductive rate, producing usually one or two piglets that are born in a nest.

Endemic to Sulawesi and a few neighboring islands, the babirusa, which weighs 100kg (220lb), is today seriously endangered. Its population in the wild is down to just a few thousand, and this figure is dwindling due to loss of its rain-forest habitat and to illegal poaching for its meat. Although the babirusa is fully protected by law, between 5 and 15 animals are sold weekly in local markets in Manado, North Sulawesi, where pig meat is a popular delicacy. Babirusa are trapped by hunters in string leg snares. Regular checkpoints to examine dealers' vehicles, greater law enforcement with swift prosecution for poachers, and public education campaigns among hunters, dealers, and consumers provide the solutions to this problem, and these are being implemented by local authorities. However, the rapid range contraction the species has undergone is clear from historical records: in 1869 Alfred Russell Wallace observed babirusa near Manado itself, whereas today dealers must travel many hundreds of kilometers to purchase the meat of this extraordinary species. LC

Below A Wild boar sow with her young; their distinctive, striped coloration is for camouflage, and fades with age. Wild boars roam freely in forests in many parts of Europe and are highly prized as game animals.

million or more pigs, occurred in 1983–84 and ıvolved a round journey of about 200km (125 ıi). Excessive logging has reduced the fruit crop ıd with it these periodic eruptions.

An Enduring Relationship with Humans

CONSERVATION AND ENVIRONMENT

Iumans have reared pigs for meat in Europe and sia since neolithic times, the Wild boar having een domesticated separately in Europe, India, hina, and Malaya between about 8,000 and 1,000 years ago. In parts of Indonesia, there are ill well-developed pig-based cultures in which ıe animal plays a central symbolic role; pig body arts are used to divine the future, and annual rituals such as hunts are enacted. A domesticated ırm of the Sulawesi wild boar occurs in the hilippines, Sulawesi, and neighboring islands.

While in some areas wild species like the Pygmy og are being threatened by habitat loss, elsewhere ral pigs have become a pest threatening native oecies. The population of the Javan warty pig is ow critically low, and survival will depend on abitat protection and vigorous captive-breeding rograms. To thrive in the wild, it will need to be anslocated to protected islands that exclude the Vild boar as a competitor. The position of the frican wild pigs and the widely-distributed Wild oar is presently satisfactory. DHMC

Above *The different fighting styles of wild pig species:* ***1*** *in the Giant forest hog, contact is made with the toughened top of the head;* ***2*** *in the bushpig, snouts are crossed, sword-like, and are protected by warts;* ***3*** *Wild boars slash at each other's shoulders, which are protected by thickened skin and matted hair.*

Peccaries

DEEP IN THE AMAZONIAN RAIN FOREST A herd of over a hundred piglike creatures files toward a salt lick at a disorderly trot. The lick is in a small clearing, which soon fills with the dark, white-chinned animals snuffling their noses in the mineral-rich water and soil. A cacophony of low adult grunts and high-pitched mother-attraction calls from the young fill the air, along with a strong musky smell. In the background, loud cracks are heard as peccaries break open hard palm nuts with their strong jaws.

Such a scene has played itself out in the forest for many thousands of years. Only today there may well be a new twist. Suddenly the sound of an explosion rips the air. An Amerindian hunter, perched in a tree, fires repeatedly into the group with an old shotgun. Soon five White-lipped peccaries lie dead. The herd wheels and turns, following an old, grizzled veteran back into the forest. This is a typical daily scene throughout lowland forest areas in today's Neotropics, where hunting and habitat loss are putting the long-term survival of peccaries in question.

Above *White-lipped peccaries are probably the most wide-ranging of the three species, with home ranges that are considerably larger than those of the other two species. Here, a juvenile White-lipped peccary wanders through grassland.*

Pigs of the Forest

Form and Function

Peccaries are medium-sized animals similar to pigs, but with long, slender legs. They evolved in the Western Hemisphere in the Oligocene era, while true pigs evolved in the Eastern Hemisphere. Long-extinct peccary taxa have been found in North America, Europe, and Asia. Scientists knew one of the modern species – the Chacoan peccary – only from the fossil record until its surprising discovery alive in the wild in 1972.

The three surviving species differ in both color and size. The Collared peccary is smallest and is distinguished by a white collar; the White-lipped peccary is darker, and has white lips and cheeks. The Chacoan peccary is large and dark, but with a white collar like the Collared peccary; several distinctive features – longer legs, a much larger head, more developed dental crowns, eyes located further to the rear of the head, a longer and higher snout – justify its placement in a separate genus, *Catagonus*. In all three species, both sexes are similar in size.

Peccaries are omnivorous, but have a particular taste for fruits (particularly palms), seeds, roots, stems, and vines. Occasionally, they also eat insects, other invertebrates, carrion, and even small mammals. The main foods of the Chacoan peccary are cacti, which are also important seasonally for Collared peccaries in some areas.

Protecting the Herd

Social Behavior

Female Collared peccaries are sexually mature at 33–34 weeks, males at 46–47 weeks. First reproduction for White-lipped and Chacoan peccaries seems to come after the second year in the wild. Copulation lasts just a few seconds and is not preceded by elaborate courtship displays. Females may mate with several males, with adult males establishing a hierarchy within the group to limit mating by subordinates. In Collared peccaries, the young are nursed for about 50 days, up to a maximum of 74. Females of this species do not nurse young other than their own, but communal promiscuous suckling has been observed in White-lipped peccaries.

Peccaries are gregarious. Chacoan peccaries live in groups of 2–10 adults and young, and Collared peccaries in herds of anything from 6 to 50 individuals. White-lipped peccary herds typically number around 100, but vary from 50 to a maximum of perhaps 400 (now extremely rare as a result of overhunting). Large herds may divide into subgroups while foraging, rejoining to move from site to site. Amerindian hunters report that White-lipped peccary groups follow an old leader; certainly, groups of this species are frequently led by adults when traveling, followed by females wi[th] young, then juvenile males, and lastly the old an[d] lame. Cohesion and recognition within groups is reinforced by mutual marking with a gland on th[e] back above the tail. Individuals will stand side-by-side and back to front, vigorously rubbing their cheeks on each other's glands.

Collared and Chacoan peccary groups occupy stable home ranges with limited overlap, suggesting territoriality. Ranges of the Collared peccary vary in size from 30–800ha (75–2,000 acres), while those of Chacoan peccary have been estimated to average 1,100ha (2,700 acres). Collare[d] and White-lipped peccaries mark tree trunks and other objects within the range with secretions from the rump gland. The core area, which is pre[f]erentially used, is also marked by dung piles. White-lipped peccaries live in large home ranges covering 22–110sq km (8.5–42sq mi), though they may also be nomadic or migratory in some circumstances.

The main natural predators of peccaries are ountain lions and jaguars. South American peas- ts claim that the jaguar only kills White-lipped ccaries that stray from the group. Certainly ien White-lipped peccaries confront either man predators or dogs, one or several individu- will stay behind to confront the threat at con- lerable risk to themselves, while the rest of the oup escapes. Such "stay-behinds" are typically ales, but females may return to defend a fallen ngener. If a predator gets very close before tection, Collared peccaries will disperse in all rections to confuse the assailant, while emitting rm calls. Chacoan peccaries, in contrast, tend stand their ground and face the danger head ; this may be a good strategy against large cats, it puts the whole group at risk from human inters with shotguns.

Falling Foul of the Bushmeat Trade

CONSERVATION AND ENVIRONMENT

ccaries are among the most important prey ems for indigenous hunters throughout Latin nerica. Hundred of thousands are killed annual- not just for subsistence but also to fuel a con- lerable commercial harvest of meat and hides. any Amazonian cities and towns, far from abun- nt sources of cattle, have large bushmeat mar- ts in which peccaries figure prominently.

Beyond direct hunting, clearance of forest for crops and pastures is reducing peccary habitat throughout their distributions. The regions they inhabit face some of the world's highest deforestation rates; more than a million hectares of natural habitat are lost each year. Given their large ranges and group sizes, White-lipped peccaries are particularly threatened in this respect, as they require huge areas of wilderness to survive. In contrast, Collared peccaries seem more adaptable to modifications of the landscape, so long as enough forest islands persist; but they are often targeted by farmers, who consider them pests because they destroy crops. Most parlous of all is the state of the Chacoan peccary, now listed as Endangered as a result of overhunting, habitat destruction, and possibly disease.

If these fascinating species are to survive, management plans will be needed to conserve both the animals themselves and the habitats they depend on. Efforts to manage peccary harvests sustainably are under way at many sites, but most of the projects are only just starting and it is unclear how many will succeed in the long term. Yet there are hopeful signs; a highly effective scheme is in place in Loreto, in Amazonian Peru, while in Texas, New Mexico, and Arizona, hunting of Collared peccaries for sport has been sustainably managed for decades. HGC/AT

FACTFILE

PECCARIES

Order: Artiodactyla

Family: Tayassuidae

3 species in 2 (3) genera

DISTRIBUTION SW USA to N Argentina

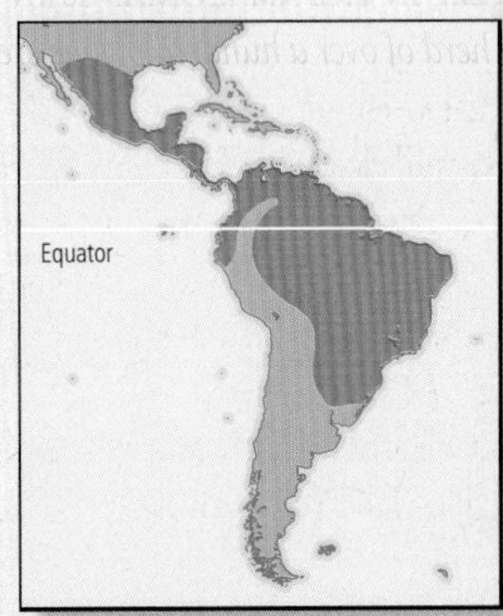

COLLARED PECCARY *Tayassu (Pecari) tajacu*
SW USA to N Argentina. Wet and dry tropical forest, tropical wooded savannas, thorn scrub, chaparral, woodlands. HBL 78–100cm (31–39in); SH 40–49cm (16–19in); TL 2–6cm (0.8–2.4in); WT 16–35kg (35–77lb). Coat: grizzled gray with darker back; limbs blackish; whitish collar runs diagonally from middle back to chest; juveniles russet-colored, collar diffuse. Breeding: gestation 145 days; litter size 1–4, usually 2. Longevity: 16 years in the wild (up to 24 in captivity).

WHITE-LIPPED PECCARY *Tayassu pecari*
SE Veracruz state, Mexico, to N Argentina. Wet tropical forest, dry tropical forest, tropical wooded savanna, thorn scrub. HBL 90–135cm (35–53in); SH 56cm (22in); TL 3–6cm (1.2–2.4in); WT 27–40kg (60–88lb). Coat: dark brown to black, with whitish bristles on lips, chin, and throat; juveniles reddish-brown to gray. Breeding: gestation 158 days; litter size 1–4, usually 2. Longevity: 15 years in the wild (up to 21 in captivity).

CHACOAN PECCARY *Catagonus wagneri*
Gran Chaco (N Argentina, SE Bolivia, W Paraguay). Dry thorn forest with isolated savannas. HBL 93–106cm (37–42in); SH 52–69cm (20–27in); TL 3–10cm (1.2–3.9in); WT 30–43 kg (66–95lb). Coat: grizzled dark gray-brown, with faint whitish collar band from middle back to chest; juveniles tan and black with diffuse collar. Breeding: gestation about 5 months; litter size: 1–4, usually 2. Longevity: at least 9 years in the wild. Conservation status: Endangered.

Abbreviations HBL = head–body length SH = shoulder height TL = tail length WT = weight

Left Young Collared peccaries remain dependent on their mothers for about 24 weeks. Lactation lasts 6–8 weeks, although the young eat solid food within 3–4 weeks. Both parents and other group members help to care for the young, especially in the face of danger, as when threatened by a predator.

Hippopotamuses

HIPPOS ARE AN EXAMPLE OF THE UNUSUAL *phenomenon of the species pair: two closely-related species that have adapted to different habitats (other instances are Forest and Savanna elephants, and buffaloes). The larger hippo inhabits grassland, the smaller forest.*

The hippo is also unusual for the way in which its life is compartmentalized and segregated: daytime and nighttime activities, and breeding and feeding, take place in very different habitats: daytime activities in water, nighttime activities on land. This division of life zones is thought to be associated with their unique skin structure, which causes a high rate of water loss when the animal is exposed to the air, making it necessary for hippos to spend most of the day in water. Indeed the rate of water loss in hippos is several times greater than that found in other animals, losing about 12mg of water from every 5sq cm of skin every 10 minutes, about 3–5 times the rate in man.

Amphibious Ungulates

FORM AND FUNCTION

The common hippo has a large, barrel-shaped body poised on rather short legs that seem too light to support it; in fact, for most of its life the hippo is buoyed up by water. The eyes, ears, and nostrils are placed on top of the skull, allowing the hippo to see, hear, and breathe while in the water. Less aquatic in its habits, the Pygmy hippo has eyes more to the side of its head; in addition, its feet are less webbed and bear shorter lateral toes, and it has sloping forequarters, which facilitate passage through undergrowth, where the common hippo's back is more or less parallel to the ground. In both species, the lower jaw is hinged far back on the skull, giving a huge gape; in the common hippo, this may extend to 150°, where the human mouth can barely manage 45°.

Both hippo species are active mainly at night, but whereas the common hippo is found on the savannas, the Pygmy hippo is a creature of the forest. The common hippo spends the day in water:

THE HIPPO IN MYTH

Hippos have long had a special place in people's consciousness. They featured in the theology of the ancient Egyptians in the form of the goddess Taweret, part hippopotamus and part human. They were also known to Pliny the Elder, who described the animal's anatomy and habits more or less correctly, except for endowing it with a long black mane, presumably to increase the equine resemblance – the name "hippopotamus" being derived from the ancient Greek for "river horse." He also erred in believing that hippos proceed to their grazing grounds backwards so as to avoid being followed. To some modern Africans, such as the Valley Bisa of Zambia, the hippo is a totem animal and consequently is not hunted.

One of the folk tales about the Pygmy hippo is that it is not suckled in the normal way, but licks a secretion from its mother's skin. This myth is probably based on at least superficial observation, for the teats, though present, are inconspicuous, and the animal does produce skin secretions that become worked into a foamy lather after exertion.

Above *Because hippos have a specific gravity that greater than that of water, they can quite easily walk long river beds. Typically, they will stay underwater or about five minutes at a time.*

Left *The location of the eyes, nostrils, and ears make it easy for a hippo to be able to see, smell, and ear while remaining largely submerged.*

river, a lake, or a muddy wallow. At dusk it merges to graze 3–4km (1.9–2.5mi) inland. To ave energy during the wet season, some hippos, usually male, will rest in the temporary wallows hat appear on the savanna rather than go all the vay back to permanent water; these entrepreneurial individuals are thus able to extend their grazing ange up to 10km (6.2mi).

Sometimes Pygmy hippos occupy a burrow or ole in a riverbank. They apparently do not excavate these themselves, relying instead on the activities of other species such as the Cape clawless otter, *Aonyx capensis*, or the Spot-necked otter, *Lutra maculicollis*, which make burrows between he roots of trees that are then enlarged and eroded by the flow of the river waters.

The hippos have traditionally been classified with the pigs and peccaries in the suborder Suina of the order Artiodactyla, but recent studies of mitochondrial DNA provide strong evidence of a closer relationship with whales. That whales are related to artiodactyls is now generally accepted, but the group closest to them has not previously been identified. It now seems that hippos occupy that position, and that the divergence between whales and hippos took place about 54 million years ago. This does not mean that one is descended from the other but only that they had a common ancestor. The fossil records of both whales and hippos are poor, so we may never know what that ancestor looked like.

Exclusively in Africa

DISTRIBUTION PATTERNS

The common hippo, the larger of the two species, is distributed throughout sub-Saharan Africa, with most populations found in countries in the east and south. The smaller Pygmy hippo is confined mainly to Liberia, with a few groups in the neighboring countries of Sierra Leone, Guinea, and Côte d'Ivoire. A separate subspecies of Pygmy hippo occurs some 1,800km (1,100mi) from Liberia in the region around the Niger Delta in Nigeria; never very numerous, there is some doubt whether it still exists.

Although they are now confined to Africa, hippos used to have a much wider distribution. Extinct species occurred throughout Europe and Asia, though not in Australia or the Americas.

Abstemious Eaters

DIET

Considering their huge size, hippos are relatively frugal in their eating habits. They consume only 1 to 1.5 percent of their body weight daily – about half that of comparable mammals like the White rhino.The mean dry weight of the stomach contents of hippos culled in Uganda was found to be just 34.9kg (76.9lb) for males and 37.7kg (83lb) for females. Wallowing in warm water all day is, it would seem, a highly energy-efficient way of life.

The common hippo feeds almost exclusively on grass, but some dicotyledons (forbs) may be ingested incidentally; the animal does not feed on aquatic vegetation to any extent. There are isolated reports of carnivory by hippos, usually in the form of scavenging, though sometimes the "prey" is killed; there is even one reported case of cannibalism. The Pygmy hippo has a more varied diet, consisting of fallen fruits, ferns, dicotyledons, and grasses. The animals leave the water at night to feed on fruit, fern leaves, and sometimes grass on the forest floor. Rather than using teeth to bite food, hippos nip it with their thickened lips.

Hippos rely on an unusual alimentary canal to break down the tough cellulose which makes up much of their grassy diet. The stomach probably consists of four chambers (although some authorities maintain there are only three). These function like those of ruminants such as cattle and antelopes; microorganisms housed within a

FACTFILE

HIPPOPOTAMUSES

Order: Artiodactyla

Family: Hippopotamidae

2 species in 2 genera

DISTRIBUTION Sub-Saharan Africa

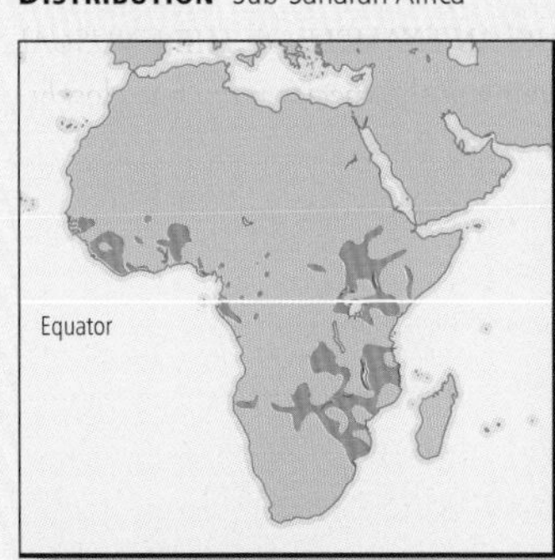

HIPPOPOTAMUS *Hippopotamus amphibius*
W, C, E, and S Africa. Short grasslands (at night); rivers, wallows, lakes (by day). HBL 3.3–3.45m (10.8–11.3ft); SH 1.4m (4.6ft); WT male 1.6–3.2 tonnes, female 1.4 tonnes. Skin: upper part of body gray-brown to blue-black; lower part pinkish; albinos are bright pink. Breeding: gestation about 240 days. Longevity: about 45 years (49 in captivity). Conservation status: not currently threatened, though numbers are declining; one subspecies, *H. a. tschadensis*, is listed as Vulnerable.

PYGMY HIPPOPOTAMUS *Hexaprotodon liberiensis*
Liberia and Côte d'Ivoire; a few also in Sierra Leone and Guinea. Lowland forests and swamps. HBL 1.5–1.75m (4.9–5.7ft); SH 75–100cm (30–39in); WT 180–275kg (397–605lb). Skin: slaty green-black, shading to creamy gray on the lower part of the body. Breeding: gestation 190–210 days. Longevity: about 35 years (42 in captivity). Conservation status: Vulnerable; the Niger Delta subspecies, *H. l. heslopi*, is Critically Endangered.

Note The 1993 established list of mammals included the long-extinct Madagascan species *Hexaprotodon madagascariensis* and *Hippopotamus lemerlei*.

Abbreviations HBL = head–body length SH = shoulder height WT = weight

fermentation "vat" produce the enyzmes that are needed to break down cellulose. Hippos do not "chew the cud" – in other words, regurgitate the partially-digested stomach contents for a second chew in the mouth – but they do enjoy some of the advantage of rumination and so are classed as "pseudoruminants."

Gregarious but Not Social

SOCIAL BEHAVIOR

The social lives of hippos are not very well-known. Around 10 percent of males are territorial; true to their amphibious nature, they defend not a patch of land but a few hundred meters of river or lake shore. They will tolerate other males in their territory, provided they behave submissively, but will expect to exert exclusive mating rights over all the females within it. If a bachelor male does not obey the rules and instead challenges the sovereignty of the territory holder, serious fights can break out. These are often bloody affairs that may result in the death of one of the contestants; most of the damage is done by the animals' razor-sharp lower canines, which may reach a length of 50cm (20in). Hippos also practice "muck-spreading," which occurs when the tail is vigorously wagged during defecation, scattering the feces far and wide. This may have some social significance, since males have been recorded defecating over each other. Most probably, though, muck-spreading has an orientation function, since bushes along the game trails to the grazing grounds are regularly sprayed.

Females gather into "schools," but these do not appear to be social groupings. There is no bonding between the females, and although they usually return to the same stretch of water each morning, they leave the water to graze on their own, unless accompanied by calves.There is also evidence that they change territories regularly. The common hippo is therefore in the odd position of being gregarious but asocial.

The Pygmy hippo is asocial and is usually found alone, except when mating or with a calf. It is not known for sure whether the animals are territorial, but the general consensus is that they are not. It seems likely that males wander over a home range that overlaps with those of other males, and that includes a number of female ranges within its borders. When adult Pygmy hippos are found together, they usually turn out to be a male consorting with a female prior to mating. Such courtship is not found in the common hippo, in which mating is peremptory to a degree, with the male forcing its attention on the female. In both species, mating occurs in the water, where most daytime activity is concentrated, although the Pygmy hippo also mates on land.

The hippos' aquatic habits are partly dictated by the structure of their skin. In both species of hippos this is very thick (up to 35 mm/1.4in in the common hippo), consisting of a thin epidermis (less than 1.0mm/.04in) containing many nerve endings overlying the dermis, a dense sheet of collagen with fibers arranged in a matted but regular fashion that gives it great strength. There is a coarse network of blood vessels deep within the dermis, but no sebaceous (i.e. true temperature-regulating) glands. This means that hippos cannot sweat, so water is essential for body cooling. Unlike many large grassland mammals, which allow the body temperature to rise through

Above *While males can exhibit some tolerance toward each other, aggression does sometimes break out, usually – as in many other animals – in connection with mating rights. Most mature males will have some scarring as a result of battles. Early stages of aggression are characterized by posturing and splashing water, however these fights can be lethal. Dominant males may maintain a territory for up to eight years.*

bsorbing the sun's heat by day and dissipating it uring the cool night, the common hippo keeps s body temperature constant to within 1°C 1.8°F) largely by immersing itself in water. Tem- erature control may work the same way in the ygmy species, whose skin physiology appears to e similar.

Hippo skin is hairless apart from a few bristles n the tail and around the mouth, and is thought o be very sensitive. Hippos are grayish-black in olor, merging to pinkish-brown in places, while ygmy hippos are uniformly black.

The strange myth that hippos secrete blood derives from a copious secretion produced from large glands under the skin surface that may act as the hippos' own sun cream. It appears to turn from colorless to red-brown on exposure to air. The secretion may also have antibacterial properties, as severe wounds inflicted by battling males heal up cleanly and quickly.

The reproductive organs of the hippo follow the usual mammalian plan, except that the testes descend only partially; there is no scrotum, making it difficult to tell the sexes apart. The female is peculiar in two particulars: the upper part of the vagina is marked by a number of transverse ridges, and there are two large diverticula projecting from the vestibule of the reproductive tract. The function of both these features remains unknown.

The common hippo produces a resonant call, starting with a high-pitched squeal followed by a series of deep, bass rumbles, that can be heard over long distances. Most of the calls are made in the air, but recent research has shown that common hippos call underwater as well. The function of the calls, which are not heard while the animals are grazing at night, is unknown, though all common hippos make them. Schools indulge in what has been called "chain chorusing," when a group of hippos bellows in unison, to be answered in turn by neighboring herds, so that the chorus moves along a river rather like a Mexican wave; one such chain was recorded as traveling 8km (5mi) downstream in 4 minutes. Possibly the chorus serves as an advertisement to females, on the basis that the louder the call, the bigger the group and therefore the stronger the territorial male controlling it. The calls might also serve as an indication to nonterritorial males of the relative strength of a territory's holders.

Underwater calls sometimes take the form of clicking noises, which may serve to announce the presence of a calling hippo in murky waters; there is no evidence to suggest that they are employed for the purpose of echolocation. There is clear anatomical evidence, however, that underwater sounds are picked up through the jawbone, so that amphibious calls can be detected simultaneously with the ears above the water's surface and with the jaw under it.

Gestation lasts for around 8 months (6.5 in the Pygmy), which is short for such a large animal, increasing the frequency of births. Births can take place on land, but mainly occur in the water.

Hydro-Engineers of the Okavango

The fan-shaped, gently sloping Okavango Delta of semi-arid northern Botswana is southern Africa's largest wetland. The Okavango river feeds 4,000sq km (1,540 sq mi) of permanent wetland, which can extend to 12,000sq km (4,600 sq mi) when flooded; within this area are numerous islands. Some 96 percent of annual inflow and rainfall is lost by evapotranspiration.

The Okavango is hippo heaven, for the creatures are amphibious, grazing on the islands by night and inhabiting the deep backswamps by day. But the hippos in turn have a major role to play in shaping the landscape they inhabit, for their activities crucially affect the way water flows around the Okavango. As they move between feeding and resting sites, the hippos creates trails that they follow day after day; their regular passage trims marginal vegetation and also keeps the bed clear, for while other large mammals in the area, such as elephants (*Loxodonta africana*) and buffaloes (*Syncerus caffer*), create land paths linking the islands, hippo trails follow low ground and thus remain in water, with side branches leading to the islands. These trails are 2m (6.6ft) wide and as deep as the swamp.

The water flows along these hippo trails, leaking out sideways to supply the permanent swamps. The channels are dense with plant growth that separates them from the less well-vegetated backswamps. Over time, the trails widen through erosion and eventually, collapse, and the main flow of water is forced to take an alternate route, usually through the path of least resistance: another hippo trail. The ground plan of the Okavango is thus hippo-made.

These hippo-derived water diversions may lead to further astonishing changes in the landscape. Hippo trails often lead to lakes, and as they bring the water with them, the accompanying influx of sediment may fill up and close the lake. When the floods come, it is along hippo trails that they flow into fringing land, subsequently creating seasonal swamps. It is likely that the geomorphic functioning of the Okavango would be radically different in the absence of hippos. TSMcC

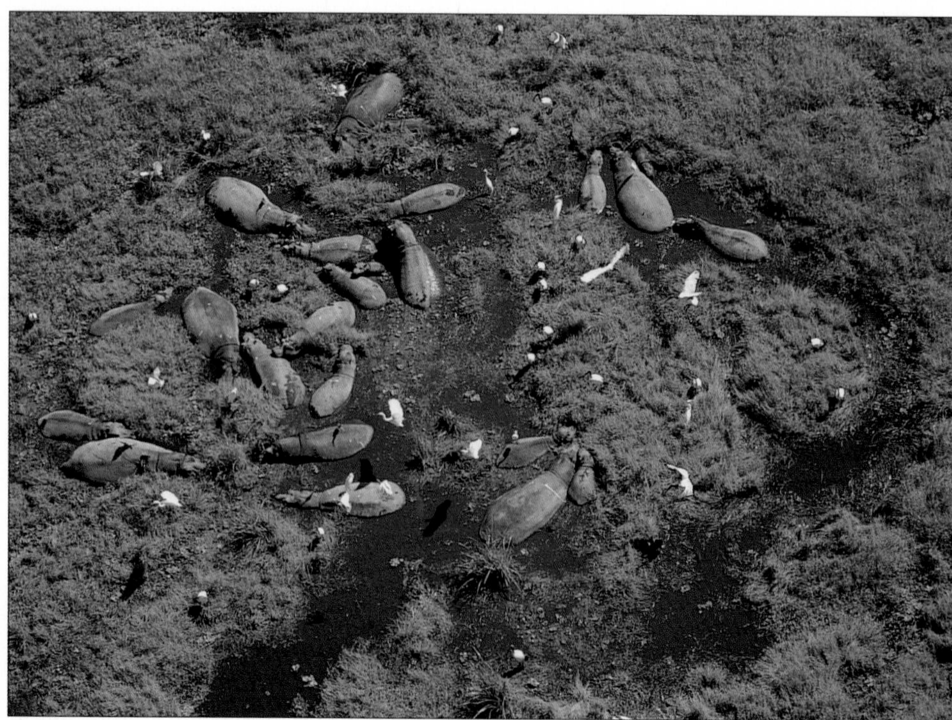

Suckling is amphibious too, and continues for about a year until the calf is fully weaned, a process that begins in both species at between 6 and 8 months. After they are weaned, calves remain with their mothers until fully grown, at about 8 years of age.

The similarity between the reproductive physiology of the two species suggests that the small stature of the Pygmy hippo is a recent evolutionary event. Certainly there are fossil relatives of the Pygmy hippo that are full-sized and relatives of th common hippo that are small.

Easy Prey for Poachers

CONSERVATION AND ENVIRONMENT

The Pygmy hippo is mainly confined to dense rai forest. This fact, coupled with its solitary, nocturnal behavior, makes it difficult to spot, let alone to estimate its numbers. The total population size, therefore, is unknown and estimates vary from a few to many thousands; but these larger estimate have been derived from extrapolations from small sample counts and may not be reliable.

The biggest threat to the survival of the Pygmy hippo is deforestation. Another threat is hunting, which certainly occurs, although it is difficult to assess its significance. The best hope for the species' survival is in the Sarpo National Park in Liberia, which provides suitable habitat and wher there have been some recent sightings. The status

More is known about the numbers of common hippos than of the Pygmy species, thanks to a continent-wide census carried out in 1988–89. This varied in accuracy from country to country, but the best estimate produced a total of about 174,000 hippos for the whole of Africa. Of these 86,400 were in southern Africa, 79,500 in eastern Africa, and only 7,700 in western Africa. The country with the most hippos was Zambia (40,000), followed by the Democratic Republic of the Congo (30,000). In 18 of the 34 countries with hippos, numbers were thought to be declining, while increases were reported in only 2 countries. Populations were said to be stable in another 7 countries, and no opinions were offered on the situation in the remainder.

The survey made it clear that West Africa is the region where conservation concerns are most urgent, for hippos are not only rare there, but are split into small subpopulations of a few hundred animals at the most. West Africa generally provides a less congenial environment for hippos.

The main threat to the common hippo probably comes from hunting rather than habitat loss, although the latter can be locally significant. Because of their habit of congregating into large herds by day, hippos are easy prey for poachers but, by the same token, this makes them easier to guard. Hippos are hunted for their meat, although their large canines and incisors are also prized as trophies; many items on sale that appear to be carved from small elephant tusks in fact derive from hippo canines. The ban on the sale of elephant ivory was followed by an increase in hippo teeth in circulation.

Hippos sometimes arouse hostility by raiding crops, and also occasionally by killing people. The crop most often attacked is rice, which is closely related to the hippos' natural grass food, but even if crops are not eaten, they may be destroyed simply by a hippo passing through the fields.

Farmers protect themselves against hippos mainly by shouting and rattling tins, but this can be dangerous. Shooting is an option, but it has to be carried out by the authorities and not by the farmers themselves, and is often ineffectual due to bureaucratic delay. In Malawi, for instance, an unacceptably high number of hippos escape wounded – 651 out of 928 targeted to be killed. Apart from animal-welfare concerns, these wounded hippos can be extremely dangerous. People most at risk from hippo attacks are fishermen in small canoes, which can easily be overturned by a rampaging animal.

The common hippo, like the Pygmy, does very well in zoos and breeds prolifically, although no single zoo keeps very many specimens. Cooperation between institutions will therefore be necessary if the captive population is to be kept free from genetic problems. It should be possible to maintain hippos indefinitely in captivity, for most of the present zoo specimens are captive-born. KE

Above The foraging range of hippos is curtailed by the amount of time that they can spend away from the water. The use of wallows, in which the hippos can cool down, increases this distance.

Below A Pygmy hippopotamus shows its razor-sharp lower canines in an impressive threat display. This West African species is in severe danger of extinction from habitat loss.

of the subspecies in Nigeria needs to be established, but it is possible that it is already extinct.

Pygmy hippos do well in zoos, but although they breed regularly, they do not do so to their maximum capacity. This may be partly due to the zoos' habit of keeping them as monogamous pairs, which is unnatural and probably results in reduced fertility. Nevertheless, the captive population has doubled within the last 25 years. Almost all present-day zoo specimens have been born in captivity, suggesting that it should be possible to maintain the species in zoos indefinitely even if it goes extinct in the wild, provided that due attention is paid to potential genetic problems.

The common hippo, although similar to the Pygmy in many ways, has a quite different ecology. Its conservation depends on it having access to grasslands as well as to water. Grazing densities of hippos in Uganda have been found to vary from 9.4–26.5 hippos per sq km (24.3–68.6 per sq mi) in the Murchison Falls National Park to about 28 hippos per sq km (72.5 per sq mi) in Queen Elizabeth National Park. The latter densities are very high, and there is evidence that they may lead to overgrazing and habitat destruction. The perceived overabundance led to a culling operation in the early 1960s in Queen Elizabeth National Park, when 7,000 hippos out of 21,000 were shot, resulting in an improvement in the vegetation and a more balanced community of large mammals.

Camels and Llamas

THE MEMBERS OF THE CAMEL FAMILY ARE among the principal large, herbivorous mammals of arid habitats; having evolved to cope with life in near-desert and desert conditions, they have made a crucial contribution to both human survival and development in these environments.

The domesticated, one-humped dromedary of southwestern Asia and north Africa, and the two-humped Bactrian camel, which is still found wild in the Mongolian steppes, are well-known, but four more species in the New World are also classed as camelids: the domesticated llama and alpaca and the wild guanaco and vicuña.

A Unique Ungulate

FORM AND FUNCTION

The camel differs from other hoofed mammals in that the body load rests not on the hooves but on the sole-pads, and only the front ends of the hooves actually touch the ground. The South American camelids are adapted to arid and steppe environments with altitudes that range from sea level up to 4,600m (15,100ft), and the pads of their toes, which are not as wide as those of camels, are movable to assist them on rocky trails and gravel slopes. The split upper lip, the long, curved neck, and the lack of tensor skin between thigh and body – which makes the legs appear very long – are characteristic features of camels. They move at an ambling pace. Camels are unique among mammals in having elliptical blood corpuscles. They all have an isolated upper incisor, which in the males of the South American camelids is hooked and sharp-edged, as are the tusk-like canines that are found in the upper and lower jaws. Camels have horny callosities on the chest and leg joints.

Dromedaries and Bactrian camels are particularly hardy. They consume a wide variety of plants over expansive home ranges, eating food that other mammals avoid, such as thorns, dry vegetation, and saltbush. They can endure long periods without water (up to 10 months if not working), and so are able to graze far from oases. When they do drink, they can take in 136 liters (30 gallons) within a very short time. They conserve water by producing dry feces and little urine, and allow their daytime body temperature to rise by as much as 6–8°C (11–14.5°F) during hot weather to diminish the need for evaporative cooling by sweating, although sweat glands occur over most of their body for use when necessary; the density of sweat glands in the skin is in fact the same in camels and man. Camels do sweat and the moisture is efficiently evaporated. They have simple, coiled, tubular sweat glands, associated with primary hair follicles all over the body except for the upper lip, external nostrils, and perianal region. These are more deeply situated than in most mammals. It is common to feel moisture beneath a camel's harness after even mild exercise.

Various adaptations help camels to survive in extreme conditions. Their nostrils can be closed t keep out blowing sand. The nostril cavities reduce water loss by moistening inhaled air and cooling exhaled air. The hump serves as a store for energy rich fats, enabling them to go for long periods without food. Thick fur and underwool provide warmth during the cold desert nights and some daytime insulation against the heat. They also conserve water by minimizing their body's exposure to sunshine and concentrating their urine.

Camels, Llamas, and Cammas

EVOLUTION AND DISTRIBUTION PATTERNS

Among the earliest of the even-toed hoofed mammals, the camelids first appeared in the late Eocene. The camel family originated and evolved during 40–45 million years in North America, with key dispersals to South America and Asia occurring only 2–3 million years ago.

The relationships between the South American camelids are still being studied. Fertile offspring

Left *Representative members of the camel family:* ***1*** *the Bactrian camel;* ***2*** *the Arabian camel, or dromedary, which like the Bactrian has two-toed feet that spread the weight on sand or snow;* ***3*** *the alpaca, which is raised for its wool;* ***4*** *the llama, a traditionally important beast of burden in South America;* ***5*** *the vicuña, from the high Andes.*

·e produced by all possible pure and hybrid ·atches between the four, and they all have the ·ame chromosome number (74), yet there are ·onetheless major differences between wild ·cuñas and guanacos, in the growth of their ·cisor teeth and in their behavior. The long-held ·ew that both the llama and alpaca are descended ·om the wild guanaco has recently been chal- ·nged by the suggestion that the fine-woolled ·paca is the product of cross-breeding between ·ıe vicuña and llama (or guanaco). Studies based ·n molecular techniques have added more heat to ·ıe controversy, and it seems that the most robust ·onclusion is that they are all closely-related ·ecies still in a process of early differentiation ·om common ancestors.

Based on wool length and body size, there are ·vo widely recognized breeds each for the llama ·ıd alpaca. The suri alpaca is well known for both ·raight and wavy wool fibers. Today no llamas or ·pacas live independently of man in their Andean ·omeland. Recently, the close phylogenetic

FACTFILE

CAMELS AND LLAMAS

Order: Artiodactyla

Family: Camelidae

6 species in 3 genera (including 3 domesticated species)

Distribution SW Asia, N Africa, Mongolia, Andes

Size Height ranges from 86–96cm/34–38in (shoulder height) in the vicuña to 190–230cm/75–91in (height at hump) in the dromedary and Bactrian camel; weight from 45–55kg (99–121lb) to 450–650kg (1,000–1,450lb) in the same species.

LLAMA *Lama glama*
Andes of C Peru, W Bolivia, NE Chile, NW Argentina. Alpine grassland, and shrubland from 2,300–4,000m (7,545–13,125ft). Two breeds: chaku, ccara. HBL 120–225cm (47–88in); SH 109–119cm (3.6–3.9ft); WT 130–155kg (285–340lb). Coat: uniform or multicolored white, brown, gray to black. Breeding: gestation 348–368 days.

ALPACA *Lama pacos*
Andes of C Peru to W Bolivia. Alpine grassland, meadows, and marshes at 4,400–4,800m (14,435–15,750ft). Two breeds: huacaya, suri. HBL 120–225cm (47–88in); SH 94–104cm (3.1–3.4ft); WT 55–65kg (121–143kg). Coat: uniform or multicolored white, brown, gray to black; hair longer than llama. Breeding: gestation 342–345 days.

GUANACO *Lama guanicoe*
Andean foothills of Peru, Chile, Argentina, and Patagonia. Desert grassland, savanna, shrubland, and occasional forest up to 4,250m (13,950ft). Four questionable subspecies. SH 110–115cm (3.6–3.8ft); WT 100–120kg (220–265lb). Coat: uniform cinnamon brown with white undersides; head gray to black. Breeding: gestation 345–360 days.

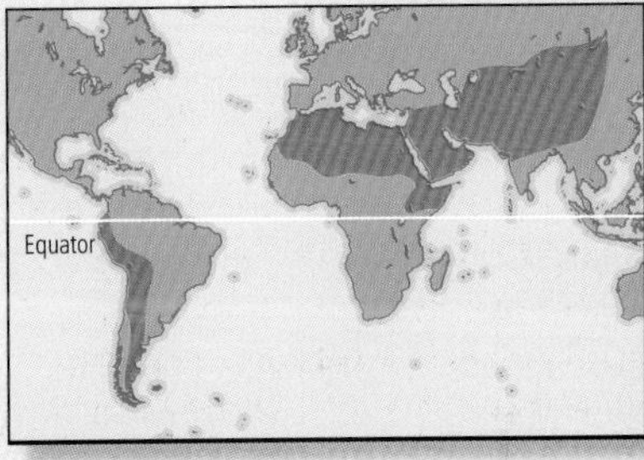

VICUÑA *Vicugna vicugna*
High Andes of C Peru, W Bolivia, NE Chile, NW Argentina. Alpine puna grassland at 3,700–4,800m (12,100–15,750ft). 2 subspecies: Peruvian, Argentinean. HBL 125–190cm (4.1–6.2ft); SH 86–96cm (2.8–3.1ft); WT 45–55kg (100–120lb). Coat: uniform rich cinnamon, with or without long white chest bib; undersides white. Breeding: gestation 330–350 days. Conservation status: Lower Risk – Conservation Dependent.

DROMEDARY *Camelus dromedarius*
Dromedary, Arabian or One-humped camel
SW Asia and N Africa; feral in Australia. Deserts. Height at hump 190–230cm (6.2–7.5ft); WT 450–650kg (990–1,430lb). Coat: short, but longer on crown, neck, throat, rump, and tail tip; color variable from white to medium brown, sometimes skewbald. Breeding: gestation 390–410 days.

BACTRIAN CAMEL *Camelus bactrianus*
Bactrian or Two-humped camel
Mongolia. Steppe grassland. Height at hump 190–230cm (6.2–7.5ft); WT 450–650kg (990–1,430lb). Short ears; small, conical humps; small feet, no chest or leg callosities. Coat: uniform light to dark brown; short in summer with thin manes on chin, shoulders, hindlegs, and humps; winter coat longer, thicker, and darker in color. Breeding: gestation 390–410 days. Conservation status: Endangered.

Abbreviations HBL = head–body length SH = shoulder height WT = weight

relationship of the Old and New World branches of the camel family was confirmed by the artificial production, in Dubai, of a cross between a camel and a llama (dubbed a camma).

The center of South American camelid domestication may have been the Lake Titicaca pastoral region, or perhaps the Junin Plateau to the northwest, where South American camelid domestication has been dated at 4,000–5,000 years ago.

Some biologists believe that the dromedary was first domesticated in central or southern Arabia before 2000BC, others as early as 4000BC, while some place it between the 13th and 12th centuries BC. From there, they spread to North Africa, and later to East Africa and India.

The Bactrian camel was domesticated independently of the dromedary, probably before 2500BC at one or more centers in the plateaus of northern Iran and southwestern Turkestan. From here the animals spread east to Iraq, India, and China.

Today there are approximately 21.5 million camelids in the world, with around 7.7 million in South America. The domestic llamas and alpacas are far more numerous than the wild guanacos and vicuñas. Llamas (3.7 million) are slightly more abundant than alpacas (3.3 million), and guanacos (875,000) are much more common than vicuñas (250,000). Most alpacas and vicuñas are in Peru, the majority of South America's llamas are in Bolivia, and nearly all the guanacos are found in Argentina and Chile. Overall, the numbers of alpacas, vicuñas, and guanacos are increasing due to the value of their wool, whereas llama numbers are falling.

Some 90 percent of the world's 14 million camels are dromedaries, with 63 percent of all camels living in Africa. Sudan (2.8 million), Somalia (2.0 million), India (1.2 million), and Ethiopia (0.9 million) have the largest populations, while Somalia has the highest density (3.14 per sq km). Numbers of camels have declined over recent decades in some countries, due in part to the forced resettlement of nomads, though the worldwide camel population has remained quite stable.

The once vast range of the Bactrian camel has contracted severely, although some remain in Afghanistan, Iran, Turkey, and Russia. Most of the camels in Mongolia and China are Bactrians, and there are probably less than 1,000 wild Bactrians in the trans-Altai Gobi Desert.

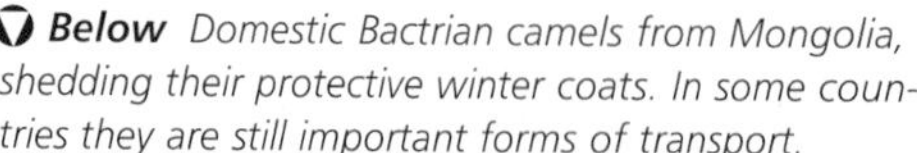

Above *Llamas, once a vital beast of burden on the Andean altiplano, can carry 25–60kg (55–132lb) across rugged terrain for 15–30km (9–18mi) a day.*

Below *Domestic Bactrian camels from Mongolia, shedding their protective winter coats. In some countries they are still important forms of transport.*

Feeding, Breeding, and Sleeping

SOCIAL BEHAVIOR

Camels have a very long gestation period, ranging from 10 to 16 months. The South American camelids are seasonal breeders and mate lying down on their chests – copulation lasts 10–20 minutes. They are induced ovulators, give birth to only a single offspring in a standing position, and neither lick the newborn nor eat the afterbirth. The newborn is extremely active and follows its mother within 15–30 minutes of birth. Births are

Above *Guanacos are protected in Chile and Peru, ıt not in Argentina, where tens of thousands of adult ɔd juvenile guanaco pelts are exported annually. This oup with young are in the Torres del Paine in Chile.*

sually in the morning, and female guanacos in ıtagonia synchronize their births. The female ɔmes into heat again 24 hours after birth, but sually does not mate for two weeks after the terus has fully recovered. Considering the harsh ıvironment and long gestation period (11 ıonths), their reproductive year is extremely ght, which explains why they mate so early after arturition. Most female llamas first breed as 2-ːar-olds, though some may breed earlier. The ıme applies to vicuñas and guanacos in the wild. here is no information on the social organization f the domesticated llama and alpaca, because on-breeding males are castrated. However, cir-ımstantial evidence suggests a territorial system, ith males maintaining a harem.

The vicuña is strictly a grazer in alpine puna ·assland habitats between 3,700–4,800m 2,000–15,700ft). They are sedentary and non-ıigratory, with year-round defended territories. ɔpulations are divided into male groups and fam-y groups. The territories are defended by the ter-torial male and are occupied by the male, several males, and offspring less than one year old. The ːrritory is in two parts – a feeding territory and a sleeping territory in higher terrain. The feeding territory is a predictable source of food; mating and birth occur in these territories, which accommodate the advantages of group living.

The guanaco is both a grazer and browser, and can occur in either sedentary or migratory populations. It is flexible in its habitat requirements, occupying desert grasslands, savannas, shrublands, and sometimes even forest, and ranges in elevation between 0–4,250m (0–13,900ft). Its territorial system is similar to that of the llama and alpaca, but the number of animals using the feeding territory, unlike in the vicuña, is not related to forage production. Vicuña and guanaco young are expelled from the family group by the adult male, as old juveniles in the vicuña and as yearlings in the guanaco. Camels mate throughout the year, but a peak of births coincides with the season of plant growth. Copulation again takes place lying down, and they may be induced ovulators.

Camelids and Humans

CONSERVATION AND ENVIRONMENT

The camelids' robustness as pack animals and their products of meat, wool, hair, milk, and fuel have been indispensable to successful human settlement of temperate and high-altitude deserts. The Inca Empire's culture and economy revolved around the llama, which provided the primary means of transporting people and goods. After the arrival of the Spanish conquistadors in the early 16th century and the destruction of the Inca social system, camelid numbers went into sharp decline, largely as a result of unregulated shooting and the introduction of sheep.

Today, in the Andes the llama is still used as a pack animal and for meat, though it is has been eclipsed by sheep as a livestock animal and by modern forms of transport. Llamas have been introduced into some countries in small numbers for recreational backpacking, wool production, and as pets. In the Andes, the woolly alpaca is replacing the llama as the most important domestic camelid. Nowadays, alpacas and llamas are broadly distributed around the world. They are considered pets in the United States, but Australia, New Zealand, and some European countries are using them as the basis of a new fine-fiber farming system for arid and marginal land.

At the time of the Spanish conquest, there were millions of guanacos in South America, perhaps as many as 35–50 million on the Patagonian pampas alone. Today, although its numbers are greatly reduced, the guanaco is still the most widely distributed camelid.

The vicuña has suffered a drastic decline over the centuries. A population believed to number millions in the 1500s fell to 400,000 in the early 1950s, and then to less than 15,000 in the late 1960s. The vicuña was first placed on the Red List of Endangered Species by the IUCN in 1969. After having been given protected status, its numbers are slowly rising again, with a population now estimated at around 250,000. A program of sustainable use is currently in place in Peru.

Dromedaries have been important beasts of burden in the western Sahara for thousands of years, but they are especially valued for the produce they yield. Their milk is often the principal source of nourishment for desert nomads. Such peoples also use camels as work animals for moving camp and carrying water. They are generally ridden in the cool of the early morning or late in the night beneath a full moon. Camels can walk 30km (19mi) a day at a leisurely pace to allow for feeding and resting; on a cool night they can double this distance. Yet, though camels still occupy a special position in social customs and rituals in the Arab world, their practical role has been largely usurped by automobiles. The future of the camel in Saharan countries depends on the fate of the nomadic peoples. If nomads are urged to settle, the vast stretches of arid land will fall into disuse. However, if traditional ways are fostered, the desert will continue to serve as an important resource for millions of people and camels.

In India, Sudan, and other countries in Asia and Africa, camels are still widely employed in transport and agriculture, and as a source of food. For example, rural communities in western Rajasthan use camel carts, which can transport six to eight people with ease. CB/WLF

Chevrotains

The chevrotains or mouse deer are diminutive inhabitants of Old World tropical forests. They are intermediate in form between pigs and deer, although the smallest species is no bigger than a rabbit.

During the Oligocene and Miocene (34–5 million years ago), chevrotains had a worldwide distribution, but today the four species are restricted to the jungles of Africa and Southeast Asia. In form they have remained virtually unchanged in 30 million years of evolution.

Primitive Ruminants

FORM AND FUNCTION

Chevrotains are among the smallest ruminants – hence the name "mouse deer" – with the Lesser mouse deer the most diminutive, at a mere 2kg (4.4lb). All species have a cumbersome build, with short, thin legs and limited agility. Males are generally smaller than females: in the Water chevrotain, they weigh 20 percent less on average. The coat is usually a shade of brown or reddish brown, variously striped or spotted depending on the species. These are shy, secretive creatures, mostly active at night. They inhabit prime tropical forest, the three Asiatic species showing a preference for rocky habitats. The Water chevrotain is a good swimmer. Their diet consists mainly of fallen fruit, along with some foliage.

Chevrotains are ruminants, which is to say that their gut is modified to ferment their food; even so, they exhibit many non-ruminant characteristics, and are best regarded as primitive examples of the type, providing a living link between ruminants and non-ruminants. They have a four-chambered stomach, although the third chamber is poorly developed. Other anatomical features they share with the ruminants are a lack of upper incisor teeth, incisor-like lower canines adjoining a full set of upper incisors, and only three premolar teeth. Typically for forest ruminants, they are solitary; in addition, they bear a single young, and the females ingest the placenta after birth.

The characteristics they share with non-ruminants include having no horns or antlers. The males possess projecting, continually growing upper canines (these are peg-like in the females) as well as premolars with sharp crowns; in addition, all four toes are fully developed. They also share some behavioral patterns with pigs, including simple sexual behavior involving prolonged copulation and a lack of visual displays, so that communication is limited to smells and cries. In addition, they lack specialized scent glands below the eyes or between the toes, and have a porcine habit of lying down rump first, then retracting their forelimbs beneath them. Of the four species, the Water chevrotain is the most pig-like (i.e. primitive), since its forelimbs lack a cannon bone and the skin on the rump forms a tough shield that protects the animal against canine teeth.

A Shy and Solitary Lifestyle

SOCIAL BEHAVIOR

The reproductive biology of chevrotains is poorly understood. In the Water chevrotain and Greater mouse deer, breeding occurs throughout the year, with only one young per litter. Weaning occurs at 3 months old in both species, and sexual maturity is achieved at 10 months and 4–5 months respectively. Water chevrotains hide their young in undergrowth soon after birth, and maternal care is limited to nursing. In the Greater and Indian spotted chevrotain, mating occurs within two days of birth. In the Water chevrotain, the only mating display is a cry that is made by the male, akin to that of pigs, which brings the courted female to a standstill for copulation, which typically lasts for 2–5 minutes. Water chevrotains are solitary, except during the mating season, and mostly show no interest, aggressive or otherwise, in other members of their species. Communication takes place by means of both calls and scent

Water chevrotains possess anal and preputial glands, and this species marks its home range with urine and feces, the latter impregnated with anal gland secretions. All species have a chin gland; in the Water chevrotain and Indian spotted chevrotain, it is rudimentary, but in males of the other species it produces copious secretions that they use to mark a mate's or male antagonist's back during encounters, Among Water chevrotain, the heaviest and oldest animals are dominant; owing to their solitary life, however, there is no established hierarchy. Fighting between males is reduced to a short rush, each antagonist biting his opponent with his sharp canines.

Right *The well-developed, projecting upper canine teeth of the Greater mouse deer can be clearly seen here; in males these canines grow continually and are enlarged into tusks, which are the primary weapon used in fighting.*

Below *The Lesser mouse deer, the smallest chevrotain. As well as being prey for a number of predators, the animals are also hunted as a food source by various indigenous peoples. Some chevrotains are even kept as pets since, despite their shy disposition, they are not difficult to tame.*

Spacing behavior is only known for Water
ıevrotains. They live mainly on land, but will
:treat to water when danger threatens. Home
ınges cover 23–28ha (60–70 acres) for males
ıd 13–14ha (32–35 acres) for females, and they
.ways border a watercourse at some point. Home
ınges of adult females do not overlap; neither do
ıose of males, although they do overlap those of
males. There is no evidence of territorial defense.
ɔpulation density varies from 7.7–28 per sq km
!0–72 per sq mi) for the Water chevrotain in
abon; for the Indian spotted chevrotain in Sri
anka, it averages 0.58 per sq km (1.5 per sq mi).

Because of their small size, chevrotains are
:latively easy prey for various predators, includ-
ıg large snakes, crocodiles, eagles, and forest-
welling cats. As with many other tropical forest
ɔecies, the survival of these four living fossils will
epend both on conservation of their habitat and
ı strict regulation of hunting. GD

FACTFILE

CHEVROTAINS

Order: Artiodactyla

Family: Tragulidae

4 species in 3 genera

Distribution
Tropical rain forests of Africa, India, and SE Asia.

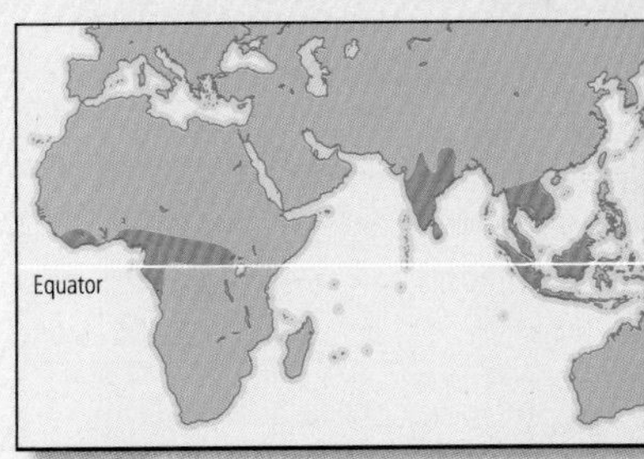

WATER CHEVROTAIN *Hyemoschus aquaticus*
C and W Africa. Tropical rain forest; nocturnal. HBL 70–80cm (28–32in); SH 32–40cm (13–16in); TL 10–14cm (4–5.5in); WT 8–13kg (18–29lb). Coat: blackish red-brown with lighter spots, arranged in rows and continuous side stripes; throat and chest have white herringbone pattern. Breeding: gestation 6–9 months. Longevity: 10–14 years.

INDIAN SPOTTED CHEVROTAIN *Moschiola meminna*
Indian spotted chevrotain or Spotted mouse deer
Sri Lanka and India. Tropical rain forest, particularly rocky; nocturnal .HBL 50–58cm (20–23in); SH 25–30cm (10–12in); TL 3cm (1.2in); WT about 3kg (6.6lb). Coat: reddish brown, with lighter spots and stripes; more or less similar to Water chevrotain. Breeding: gestation probably 5 months.

LESSER MOUSE DEER *Tragulus javanicus*
Lesser mouse deer or Lesser Malay chevrotain
SE Asia. Tropical rain forest and mangroves; nocturnal. HBL 44–48cm (17–19in); SH 20cm (8in); TL 6.5–8cm (2.5–3.2in); WT 1.7–2.6kg. (4–5.7lb). Coat: more or less uniform red, with characteristic herringbone pattern over the fore part of the body. Breeding: gestation probably 5 months.

GREATER MOUSE DEER *Tragulus napu*
Greater mouse deer or Greater Malay chevrotain
SE Asia excluding Java. Tropical rain forest; mainly nocturnal. HBL 50–60cm (20–24in); SH 30–35cm (12–14in); TL 7–8cm (2.8–3.2in); WT 4–6kg (9–13lb). Coat: more or less uniform red. Breeding: gestation 5 months. Conservation status: 1 subspecies, the Balabac chevrotain (T. n. nigricans), is Endangered.

Abbreviations HBL = head–body length SH = shoulder height TL = tail length WT = weight

Musk Deer

MUSK PRODUCED BY MALE MUSK DEER *influences estrous cycles in females and so aids reproduction. Unfortunately for the deer, this natural substance became one of the most sought-after animal products. For centuries, musk was exploited in the manufacture of perfumes; latterly, its use has been confined to indigenous medicines.*

Together with increasing habitat loss, the continuing demand for musk and the high price it fetches – a single kilogram (2.2lb) sells for US$65,000 – have led to local extinctions and declining numbers across the deer's former distribution range.

Diminutive and Delicate

FORM AND FUNCTION

Musk deer are not true deer, but primitive ruminants that differ from the cervids in not having antlers or facial glands. Other distinctive physiological idiosyncrasies include the possession of a gall bladder, caudal gland, and musk gland, and having only one, rather than two, pairs of teats.

The prominent canines can be as long as 7–10cm (3–4in) in males. As in the Chinese water deer and the muntjac, the teeth move in their sockets, which helps feeding and cud-chewing and also lessens the risk of breakage in combat with conspecifics. In females, the upper canines are small and never protrude below the lip of the lower jaw.

The body is stocky, and the head small, with long, harelike ears. Hind legs that are 5cm (2in) longer than their forelegs, a massive back end, and a curved spine give musk deer a characteristic, bounding gait. The central hooves are long and pointed, and the lateral hooves (dew claws) are enlarged. The dense, wavy coat contributes to the deer's thickset appearance.

Only males possess the caudal and musk glands. Set below the tail, the caudal gland has pores on either side, which exude a yellow, viscous secretion with a strong-smelling odor. The musk gland is situated near the navel; it has an outer glandular region producing immature musk, and a central sac in which the secretion matures into a powerfully scented, granular, reddish-brown substance.

Musk deer are well adapted for life at high altitudes. They move about easily on rough, steep, hazardous terrain such as snow-clad mountain slopes, usually walking or jumping and only rarely trotting. The enlargement of the hooves minimizes sinking in soft snow, and their hair has air-filled compartments for insulation.

Right *The long, protruding canines of the male Musk deer give it a lugubrious expression. The Musk deer is thought to be similar to the ancestors of the antler-bearing deer.*

Browsers that can live on a poor-quality diet, musk deer feed on leaves of woody plants, forbs, lichen, mosses, ferns, and grasses. Oak (*Quercus semecarpifolia*), wintergreen (*Gaultheria*), and rhododendron leaves are commonly taken in winter, as well as lichen (*Usnea*) and mosses. Indeed, in Russia, musk deer mainly eat lichen, and in Nepal they will climb trees to get to it (they also use this trick to escape predators). In deep snow (up to 1m/3.3ft), musk deer feed on vegetation on lower or fallen branches of trees.

Musk deer are believed to have originated in the New World, where one fossil has been found dating from the early Miocene (23 million years ago). In the Pleistocene, from about 1.8 million years ago, they diversified in the tropical forests of Asia and Europe. The four modern taxa inhabit the forested mountains of central and east Asia, being found in Afghanistan, Bhutan, China, India, Korea, Myanmar, Pakistan, Russia, and Vietnam.

In the Himalayas, they live in forested areas between 2,500m (8,200ft) and the treeline (3,000–4,300m/9,840–14,100ft). Unlike other ungulates of the region, which need to migrate to lower elevations in winter, musk deer flourish all year round in the subalpine and upper temperate habitats (up to 2,500m).

Musk deer are sedentary animals, seldom straying from small home ranges of 15–30ha (30–75 acres). They rest in dense undergrowth or in montane bamboo thickets during the day, and although they feed in open alpine meadows, they rarely stray far from cover. They tend to be nocturnal or crepuscular.

Communicating by Smell

SOCIAL BEHAVIOR

Musk deer use defecation and the scent glands to communicate by smell; sight and sound are less important. Defecation sites dispersed throughout the home range are used most often during the rut. Males stand downhill of the latrine in a semi-squatting position, scrape the ground with their forefeet, release fresh pellets, and then cover them with mud, old pellets, or litter. Scraping adds visual signals to the animal's odor and, when performed after defecation, impregnates the animal's feet with the scent of feces. When they urinate, males also leave pink marks smelling of musk in the snow. Female urine is amber and odorless.

Some defecation sites are used only by one individual, while others are shared, which corresponds to the degree of overlap between animals ranges. Musk is synthesized seasonally before the rut, so it may influence estrous cycling, like the analogous preputial gland in other mammals. To add to the range of color and smell, males paste secretions from the caudal gland throughout their territory on the stems of dried forbs or bushes, particularly bamboo.

All this olfactory posting may help keep other males away from the territory. Small or weak males will normally avoid entering a strong male's territory, but if they do come close fights may break out. The best territories encompass the home ranges of two or more females.

ATURAL MUSK

Ausk comes from the male's preputial gland, which has both an outer region, where "early" mmature) musk is produced, and a central sac or storing and maturing the secretion. The sac is roughly spherical pod, some 3cm (1.2in) in iameter, which is almost smooth on one side and overed on the other with dark hair surrounding n orifice. Grain musk is the dried secretion that mpties from the pod. The sticky, reddish-brown rains have a slightly bitter taste and a penetrating, ersistent odor. The smell derives from muscone, ketone identified as 3-methyl cyclopentadeca-one. The scent is so intense that just one part in ,000 can be readily detected. Curiously, the per-ume loses its odor when kept in hermetically ealed containers, but revives when small quanti-ies of ammonia or some other alkali are added.

Musk's subtle pervasiveness has long ensured ts popularity in perfume manufacture. It is mployed as a fixative, usually in the form of a incture to improve the perfume as it matures. Ausk is also used in soaps, sachet powders, and nothproofing agents. Musk's role in indigenous nedicine has an even longer history. In India and China, it was used in the Ayurveda and Unani nedical systems as early as 3,500BC. Today it is sed as a sedative to treat asthma, epilepsy, hyste-ia, and other nervous disorders, and has also een shown to be effective against bronchitis, neumonia, and typhoid.

Musk deer temporarily pair up during the rut. The young are born in spring after a gestation of about 6 months. Usually a single fawn is born, but twins are not uncommon. To safeguard them from predators, their mothers hide their fawns in vegetation, but stay in the vicinity to nurse them when necessary. After only 45–50 days, the fawns are independent. When they sense danger, Musk deer make a hissing sound and stot (leap vertically).

A Lucrative but Lethal Trade

CONSERVATION AND ENVIRONMENT

Poaching and widespread habitat destruction are the principal threats to Musk deer. About 300kg (660lb) of musk is traded each year, costing the lives of 24,000 deer. Most of the supply comes from China, India, and Nepal, and is exported to Japan and Southeast Asia, where it is used in stimulants for the heart and central nervous system.

Musk can be obtained without killing the deer, and in China, farming takes place on a large scale. However, it has not stopped poaching, since the quality of musk from captive animals is poor. Moreover, poaching is far more lucrative than tending animals in captivity.

A particular problem on the Indian subcontinent is the destruction by grazing livestock of understory vegetation. About 70 percent of potential musk deer habitat south of the Greater Himalayas has already been lost. However, some conservation work is underway. Several protected areas have been created, and great efforts are being made to enforce anti-poaching measures. SS

FACTFILE

MUSK DEER

Order: Artiodactyla

Family: Moschidae

4 species of the genus *Moschus*

DISTRIBUTION Asia

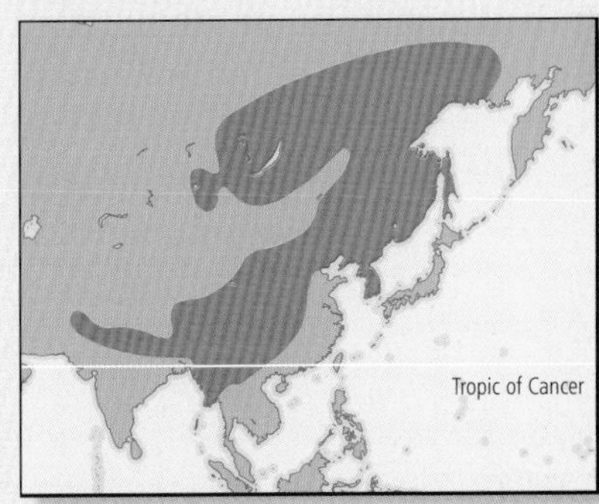

Habitat Mainly forested and alpine scrub habitats in mountains; mostly crepuscular or nocturnal.

Size Head–body length 80–100cm (31–39in); tail length 4–6cm (1.5–2in); height 50–70cm (20–28in); weight 13–15kg (29–33lb).

Coat Grayish-brown to yellow-brown or dark brown. Tips of ears dark brown or ivory-yellow. Double white throat-stripes. Genital region white. Texture of hair soft to coarse. Tail naked except for a tuft of hair at the tip.

Diet Leaves of woody plants, forbs (herbs other than grasses), ferns, lichens, mosses, and grasses.

Longevity To 20 years in captivity

HIMALAYAN MUSK DEER *Moschus chrysogaster*
Himalayan or Alpine musk deer
Greater Himalaya of N Afghanistan, N Pakistan, N and NE India, Nepal, and C Tibet to C China. Breeding: females are sexually mature within their first year, or at the latest by 1.5 years of age; mating takes place during late autumn and early winter; usually a single fawn is born after a gestation period of 180–200 days; fawns are completely independent by about 1.5 months. Conservation status: Lower Risk – Near Threatened.

SIBERIAN MUSK DEER *Moschus moschiferus*
E Siberia, N Mongolia, and N China W to Kansu and Korea. Breeding: as for *M. chrysogaster*, but litter size is typically 2. Conservation status: Vulnerable.

DWARF MUSK DEER *Moschus berezovskii*
Dwarf or Forest musk deer
S and C China including Anhwei, N Vietnam. Breeding: as for *M. moschiferus*. Conservation status: Lower Risk – Near Threatened.

BLACK MUSK DEER *Moschus fuscus*
W Yunnan, SE Tibet, N Myanmar. Conservation status: Lower Risk – Near Threatened.

Left Suckling in Musk deer is unusual in that the young touches its mother's hind leg with a raised fore leg. This gesture is similar to that seen in the courting behavior of other hoofed mammals such as the gerenuk. The female Musk deer usually gives birth to one fawn, which is very small in comparison with its mother, weighing only around 600–700g (20–25oz). After about one month of suckling, the fawn is ready to leave its hiding place and accompany its mother.

Deer

THE ANTLERS OF MALE DEER DISTINGUISH *them from other ruminants. The bony, hornlike antlers are typically regrown and shed each year. This process requires considerable energy and nutrients, particularly in large-bodied deer species, as antlers grow with positive allometry – that is, as body size increases, antler size increases at a greater rate. In the largest living deer, the moose, antlers on large bulls may exceptionally exceed 30kg (66lb) in weight, spanning up to 2m (6.5ft). Even so, they would have been dwarfed by those of the extinct Irish elk, which were twice as heavy and extended up to 4m (13ft) across, accounting for about 10 percent of the animal's total body weight.*

Deer are similar in appearance to other ruminants, particularly antelopes, with graceful, elongated bodies, slender legs and necks, short tails, and angular heads. They have large, round eyes, which are placed well to the side of the head, and triangular or ovoid ears that are set high on the head. The size of deer ranges from the moose to the Southern pudu, which weighs only about 1 percent of the moose's 800kg (1,750lb). Firstly deer were a source of food, then sport for huntsmen; more recently deer have been regarded as creatures of interest in their own right and knowledge of them has broadened.

The Significance of Antlers

FORM AND FUNCTION

Antlers arise from the skin covering the frontal bones of male fawns, and are induced by male hormones. The physiological control of antler growth, however, is not just complex but differs significantly between species. Antlers grow annually, usually in summer, within a sensitive, highly vascularized skin called velvet. Growing antlers are thus warm to the touch and very sensitive, judging from the response of antler-growing males to flies landing on the velvet. The antlers finish growth and turn to bone in about 140 days in large-bodied species, at which time the velvet begins to dry and crack, apparently in response to the release of male hormones. The males rub it off by striking the antlers against shrubs and small trees. This damages the bark, staining the antlers dark with the sap that flows from it. In addition, the antlers may be used to horn the soil, which also colors them.

After the mating season, the reduction in sex hormones lead to antler shedding. A layer of bone-dissolving cells invades their base, weakening their attachment to the skull until they drop off. In New World deer, antlers tend to be shed in early or mid-winter and normally do not start growing until spring; in Old World species, they may be retained until late winter or spring. Here antler growth starts right after shedding.

A number of deer species, particularly in the tropics, have mere spikes or buttons, while one, the water-deer, has no antlers at all. However, males of this small species have long, sharp upper canines – primitive weapons that characterized deer some 30 million years ago in the early part of their long evolutionary history. Biologists have speculated whether this tusked deer is a relict species from the distant past, or if it has lost its antlers and has reverted to a primitive condition. Knifelike tusks are typical weapons of solitary territory-defenders.

Below *Representative species of deer:* **1** *Moose* (Alces alces); **2** *Roe deer* (Capreolus capreolus); **3** *Chital* (Axis axis); **4** *Reeves's muntjac* (Muntiacus reevesi); **5** *Sika deer* (Cervus nippon); **6** *Tufted deer* (Elaphodus cephalophus); **7** *Père David's deer* (Elaphurus davidianus); **8** *Water deer* (Hydropotes inermis).

FACTFILE

DEER

Order: Artiodactyla

Family: Cervidae

38 (43) species in 16 genera and 4 subfamilies. Species include: **muntjacs** (7 species of genus *Muntiacus*); **Fallow deer** (*Dama dama*); **chital** (*Axis axis*); **Red deer** (*Cervus elaphus*); **wapiti** (*Cervus canadensis*); **Sika deer** (*Cervus nippon*); **Roe deer** (*Capreolus capreolus*); **moose** (*Alces alces*); **reindeer** (*Rangifer tarandus*); **huemuls** (2 species of genus *Hippocamelus*); **Red brocket** (*Mazama americana*).

DISTRIBUTION N and S America, Eurasia, NW Africa (introduced to Australasia).

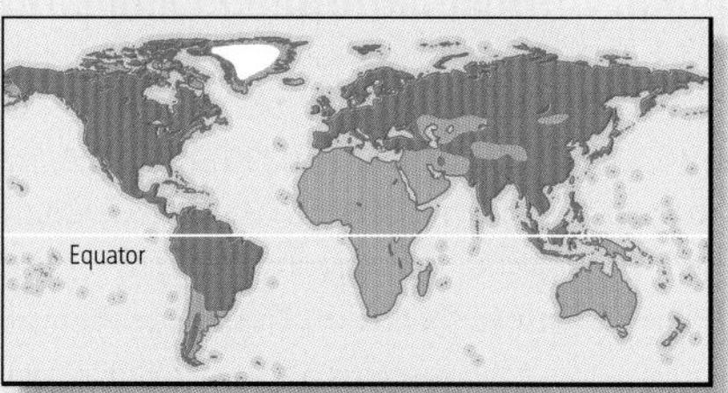

HABITAT Mainly forest and woodland, but also Arctic tundra, grassland, and mountain regions.

SIZE Shoulder height from 38cm (15in) in the Southern pudu to 230cm (90in) in the Moose; **weight** from 8kg (17.5lb) to 800kg (1,750lb) in the same two species.

COAT Mostly shades of gray, brown, red, and yellow; some adults and many young are spotted.

DIET Herbivorous: shoots, young foliage herbs, grasses, lichens, fruits, mushrooms.

BREEDING Gestation from 24 weeks in Water deer to 40 weeks in Père David's deer.

LONGEVITY Average lifespan of caribou in the wild is 4.5 years. In captivity, many deer species live to 20 years or more.

See species table ▷

There are other small deer armed with both small antlers and tusks: the muntjacs of southeast Asia. Among muntjacs, the species with the largest antlers are the most gregarious and widely distributed forms, with males apparently sharing common territories that are large but poor in resources. In contrast, the more solitary species have large tusks but only small or diminutive antlers. These species defend small but resource-rich territories embedded within contiguous ranges of the large-antlered muntjacs. Molecular genetics indicates that some of the small-antlered, large-tusked muntjacs derived from large-antlered, small-tusked forebears, suggesting that large tusks gained them evolutionary priority in the pugnacious defense of small, resource-rich territories. That, in turn, makes it likely that evolution may have occasionally reversed itself to permit antlers to be lost in favor of tusks in species defending such territories.

Intriguingly, though, large antlers may result not from males having fiercer fights but from females trying to get their offspring to grow up faster. In all but a few species of deer, fawns hide alone after birth while the mother goes off to feed. Hiding is a very successful way of avoiding predators, but it becomes hard or impossible in open habitats, particularly for big fawns. If the young can rise to their feet soon after birth and soon start running with their mother, however, the problem is solved.

In deer species that have adapted to open plains that provide little or no cover, the female therefore needs to input large amounts of nutrients into a growing youngster during gestation and lactation to encourage speedy development. Just as importantly, however, she needs to select a partner with good genes for growth. Large antlers are a reliable signal to the female that a male has been able to find enough food to fund these "luxury" tissues. And while such a male's sons

Above *A male Reeves's muntjac drinking. Indigenous to China, but introduced to England, it is rarely found any significant distance from water. This deer species possesses small antlers as well as long upper canine teeth.*

An Unusual Association

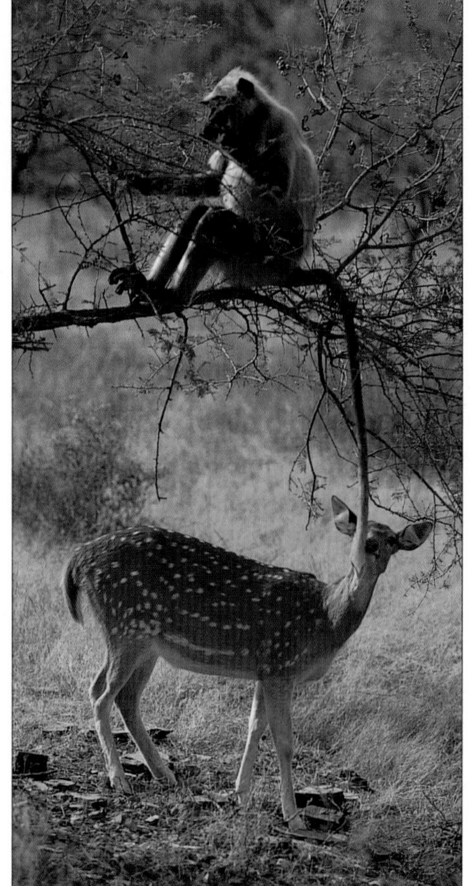

A herd of large, spotted deer forage below a tree in an Indian jungle. Above and among them, a troupe of gray monkeys feed. One buck approaches a female monkey head-on. With a lunge, the monkey momentarily grabs the deer's antler, and the deer backs away. The deer is a chital (*Axis axis*) and the monkey a Common langur (*Semnopithecus entellus*), both common species in deciduous forest on the Indian subcontinent.

In Central India, throughout the year except during the monsoon season (July–October), chital can be found in attendance on langur troupes for as much as 8 hours in 24; a herd has been seen to approach a single troupe as many as 17 times in a single day. The reason is not hard to find: the chital browse upon foliage dropped from the trees by feeding langurs, particularly the leaf blades which the langurs reject, having eaten the stalks. A troupe of 20 langurs drops an estimated 1.5 tonnes of foliage each year, of which 0.8 tonnes is suitable forage for chital. This gleaned waste is particularly valuable for chital between November and June, when grass is sparse. During the monsoon season, when other food is more plentiful, langur/chital associations decline dramatically.

Chital initiate the association, locating feeding langurs either by their movement, scent, or perhaps even by the smell of broken foliage, usually from 50–75m (165–245ft) away. Once located, they move quickly to take up station below the monkeys.

At first sight, the relationship might seem one-sided, with the chital gaining all the advantage. However, chital and langurs respond to each others' alarm calls, so early detection of predators such as tigers and leopards probably benefits both. Also, the sensory abilities of the two species are complementary, langurs having acute vision and chital a good sense of smell.

By providing a concentrated and abundant food supply, this association allows chital to congregate in denser herds than is normal for this opportunistic herbivore. Chital grazing in the loose herds that are typical over most of their range are rarely overtly aggressive towards each other, but where they are close-packed below langurs, large bucks in particular often threaten does and young. In consequence, such bucks probably eat more of the langur-created waste, and are thus at a competitive advantage to other bucks without access to these gleanings.

Some langurs come to the ground to glean items dropped by their companions: thus a mother and her young might descend to take up small fruit discarded by an adult male. On the ground, male langurs may react aggressively towards chital. Chital also occasionally feed on the waste dropped by fruit-eating birds and on wind-felled foliage, so their opportunistic feeding pattern may have pre-adapted them to associating with langurs. The association is probably mutually beneficial, but asymmetrical in that the chital benefit considerably more than the langurs. PN

may themselves be expected to grow large antlers his daughters will give birth to larger babies and produce richer milk.

The connection linking antlers, the size of the young at birth, and the richness of the female's milk extends into other areas, for example speed the fastest species have the largest antlers, the biggest babies, and the richest milk. Even more convincing evidence of the link is the fact that th males of such species spend more time showing off their fine adornments to females than to othe males. Huge antlers thus, indirectly, appear to be nature's paradoxical way of ensuring the safety o young deer in open grasslands and tundra. It is unsurprising, therefore, that the now-extinct Iris elk, which had the largest antlers on record, was also probably the fastest deer ever known.

Antlers make good fossils, fortunately for biolo-gists trying to piece together the deer's past histo-ry. Deer can be identified by the mid-Tertiary, some 30 million years ago, in the shape of small, tusked forms from tropical climes. Antlers appeared from the early Miocene about 20 millio years ago, but tusks also remained. As antlers became bigger and more complex, tusks became smaller or vanished altogether. Antlers now bega tracking evolutionary innovation – that is, they increased in size and complexity as deer evolved to cope with harsher climates. The smallest and simplest antlers are found on tropical deer, while the largest and most complex are on Pleistocene species from subarctic, alpine, and periglacial environments.

Even now, the number of tines (prongs) goes up as the environment becomes more challenging. In the Old World, the smallest and most primitive tropical resource defenders (muntjacs)

ve only one or two prongs, while gregarious, ge-bodied forms from warm regions (Rusa, is, or Swamp deer) have three. Antlers are four-onged in temperate-zone species like the Fallow d the Sika, five-pronged in forms from cold tem-rate climates (most Red deer), and six-pronged alpine, subarctic forms such as the wapiti and me extinct giant deer.

Antlers can only grow big if sufficient nutrients available, but soils from the tropical regions m which deer originally derived are usually low minerals, let alone the calcium and phosphate sential for antler growth. Antler growth also mands a high-protein diet. These nutritional mands have tied deer to fertile localities, such flood plains of large rivers.

Beginning about 5 million years ago in the early ocene, increasingly larger glaciations generated ore and better habitats for deer. Glaciers grind ck into highly fertile dust that is then widely deposited by water (silt) and wind (loess). Deer basically exploiters of nutrient-rich, disturbed, mature ecosystems, and the severe ecological sruptions of the ice ages greatly favored their read and subsequent evolution. Large-antlered er thus appeared repeatedly, beginning with the nor glaciations late in the Pliocene and continu-g into the major glaciations of the Pleistocene m about 1.8 million years ago. There were ge-antlered deer in the Pleistocene pampas of nat is now Argentina, and also similar radiations Europe and Asia, of which the most spectacular ere the giant Megacerines, and in particular the sh elk (*Megaloceros*). These were very successful ring the mid-Pleistocene, but became extinct, ve for the Fallow deer (*Dama*), during the last glaciation. Among extant species, the deer with the largest antler mass relative to body size is the reindeer or caribou. It is also the most highly evolved runner among extant deer, and lives in huge groups on open tundra landscapes. Reindeer mothers bear a single, large, well-developed fawn and produce the richest milk of any deer; the males have showy antler displays in courtship.

Above *A group of male Mule deer, displaying their brown-gray winter coloration (their coats take on a reddish hue in summer). This New World species, which occurs in North and Central America, can run at speeds of up to 64km/h (40mph).*

Old World, New World

DISTRIBUTION PATTERNS

Deer divide into two groups, once recognized as subfamilies, that can be roughly labeled as Old and New World species. Old World deer retain only the upper portions of the second and fifth

Below *Representative species of deer (continued):* ***1*** *Southern pudu* (Pudu pudu)*;* ***2*** *Pampas deer* (Ozotoceros bezoarticus)*;* ***3*** *Marsh deer* (Blastocerus dichotomus)*;* ***4*** *Peruvian huemul* (Hippocamelus antisensis)*;* ***5*** *Red brocket* (Mazama americana)*;* ***6*** *White-tailed deer* (Odocoileus virginianus)*;* ***7*** *Reindeer* (Rangifer tarandus).

metapodia (middle hand or foot bones), the New World ones only the lower portions. There are other differences also, for example in penis placement and the manner of shedding antlers. Circumstantial evidence suggests that the New World deer evolved in temperate forests in the late Tertiary about 5 million years ago, in what are now arctic latitudes of North America and eastern Siberia. The Old World deer originated in Asia.

Only the New World deer go into South America. One species, the White-tailed deer, spans more than 70° of latitude, stretching from the borders of the Arctic Circle in North America to about 18° south of the Equator in South America. Here too are found the smallest of all deer, the tiny pudu and brockets, the ecological counterparts of the muntjacs of southern Asia.

Both Old and New World have marsh-adapted deer, and both also once had grotesque ice-age giants. In this respect, the Old World's Irish elk was surpassed in body size by the Broad-fronted elk, a New World deer of the mid-Pleistocene roughly 1 million years ago.

Making the Most of Minerals

DIET

On the whole, deer favor easily digestible foods and eat broadly similar diets. Many species, both large and small, are "concentrate selectors," feeding on shoots, young foliage, sprouting herbs and grasses, young twigs, lichens, fruit, mushrooms, and even decaying vegetation such as natural ensilage. These species invariably have a small rumen and a rapid digestion. A few tropical forms, such as the chital or sambar, have shifted more onto green grasses and evolved the ever-growing cheek teeth typical of grazers. None have become specialists in digesting coarse-fibered forage like many bovids – not even the Plains-adapted wapiti or reindeer. Wapiti are mixed feeders that require a lot of high-quality forage but also mix in some tough-fibered grasses. Reindeer feed on succulent tundra vegetation in summer and soft lichens in winter. Because antler growth requires large amounts of minerals, deer are limited to high-quality vegetation, and have left feeding on grasses, which are relatively low in bone-building minerals, to bovids and other grazers.

Sparring for Position

SOCIAL BEHAVIOR

Deer with complex antlers use them not only in combat but, more importantly, in sparring. Like human martial arts, these contests are played to rules and serve to advertise the strongest animals without the need for bloodshed; the locking together of horns allows males to use their physical strength to determine fights without resorting to wounding one another.

The largest male may solicit sparring matches from the smallest and, provided the lesser male accepts the invitation, both may engage in protracted, even exuberant antler-play. Through such activities males form "friendships"; sparring partners may feed, rest, and move together, and the smaller may enjoy the protection of the larger against more dominant males.

Surprisingly, the rules of sparring differ greatly between species. In muntjacs, large males bond with smaller to defend a common territory. In Mule deer, smaller sparring partners help the larger protect a clan of females, and may breed when several does are in heat and the larger buc cannot protect and breed them all. The activity is innate, as antlerless fawns occasionally try to spar. Sparring allows deer to live peaceably in groups, in which interactions between individua occur much more frequently than when animals live alone.

Group living has many advantages, not least in reducing the risk of predation. The larger the group, the less likely it is that any given individu will be killed, and the chance of survival is even greater for animals surrounded by other herd members than for those on the group's edges. Ir addition, the larger the herd, the greater the likel hood that a weak individual will trail in flight an become a victim when predators give chase, allowing the remaining herd members to escape For a healthy adult ungulate, the herd is consequently a very safe place to live.

Clearly, members of a herd benefit by not drawing the attention of predators, and one way of doing so is to minimize body injuries and bleeding. It has been suggested that this may ha been a factor in the evolution of antlers: by settling contests by trials of strength rather than by wounding (as in territorial animals with daggerli weapons), the risk of bloodshed was reduced.

Living together with males, females may suffe from competition for food; to counter this, they often evolve the armament and external appearance of the male they must defeat in competitior So, reindeer females have horns that serve them well in chasing away yearling males that, instead of digging their own craters in the snow to get at the lichen beneath, parasitize those of females ar their calves. In contrast, female caribou living in forests where lichens grow on trees lack antlers a they have no need to defend food resources.

Wapiti tend to live on open plains, and the females are much closer to males both in body mass and external appearance than in forest-adapted species.They also have subcutaneous "hornlets" on the skull, and on rare occasions grow small antlers.

Roe deer are highly specialized cervids that exploit the summer's productivity pulse directly for reproduction, so that the bucks become territorial in spring. That demands ready-grown antlers, which consequently sprout in winter anc are shed in the fall; the American pronghorn has convergently evolved a similar system. Uniquely for ungulates, Roe deer also have delayed implan tation; eggs are fertilized during the rut in late summer, but do not implant until January. From

Left A male Roe deer and fawn browsing. This species is very well adapted for the cold climates tha it inhabits; thanks to a system of delayed implantatic newborn Roe deer do not have to face the rigors of winter weather and scant food supplies.

➤ ***Right*** *A Red deer stag in March. Deer are unique among ruminants in their annual cycle of antler growth and shedding. The process is under the control of the sexual and growth hormones. Fur-covered skin (velvet) carries blood to the growing antlers. In the fall the velvet dries out, and the deer thrash their antlers against vegetation to remove it. In the winter the antlers are shed.*

then on the embryo grows to term, so the fawn is born when the vegetation greens up in spring.

Communicating by scent appears to be a highly significant activity among deer. As a rule of thumb, small-bodied species mark the ground or vegetation with urine, feces, or glandular secretions, while large-bodied deer mark their own bodies. The huge frontal glands of muntjac males are used to check off the ground beside a new scent. Once a spot has been marked, it receives little attention on subsequent visits, and so more time can be devoted to investigating new tracks or scents. Large deer such as moose and Red deer use urine to impregnate long hairs on their body, notably the neck mane in Red deer and wapiti and the long, dangling bell of the moose. Old World deer males may advertise vocally during the rut, but also spray their bodies copiously with urine, paw urine-soaked rutting pits, and rub their body and neck in these. Moose, after pawing a rutting pit, will lower their heads close to the pit and splash urine-soaked mud on the bell, which thereby becomes a scent-dispersing organ. The bull's scent is extremely attractive to females. Some New World deer, moose and reindeer included, urinate on their hind legs; those of White-tailed and Mule deer bear large brushes with specialized, scent-capturing hairs that are flared during aggressive encounters with rivals. Deer possess a diversity of skin glands, whose function is in need of further scientific investigation.

The Red Deer Rut

Rutting Red deer stags fight vigorously for the possession of harems. Usually a challenging stag approaches a harem-holder and the two roar at each other for several minutes **1**. Then the contest progresses to a parallel walk **2**, the stags moving tensely up and down until one of them turns and faces his opponent, lowering his antlers **3**. The other stag turns quickly and they lock antlers and push strenuously against each other, each trying to twist its opponent round and to gain the advantage of slope **4**. When one succeeds in pushing the other rapidly backwards, the loser disengages and runs off **5**. Many contests do not reach the fight stage, but end with the challenger withdrawing after the roaring contest or the parallel walk.

A mature stag will fight about five times over the period of the rut, risking considerable danger. In a study on the Isle of Rhum, Scotland, about 23 percent of mature stags were injured during the rut. Even harem-holders that won fights often lost hinds to young stags lurking on the edges of the harem for an opportunity to steal them. At such a price, it is not surprising that stags are reluctant to fight unnecessarily. Unequal contests are less common than well-matched ones, suggesting that stags are able to assess their chances of winning.

Playback experiments have showed that stags attempt to outdo their rivals in roaring frequency, and the roaring rates of individual stags correlate with fighting success. RAC

Adaptability under Pressure

CONSERVATION AND ENVIRONMENT

Deer have not flourished in well-established, species-rich, mature ecosystems, but have succeeded in exploiting ecological opportunities created by major disturbances such as floods, fires, and storms. This tendency has enabled them to benefit from the big landscape changes brought on by glaciations and deglaciations, by megafaunal extinctions, and in recent times by human disturbances. Where predators are very active, deer may move into towns, farms, and suburbs; thus, in North America, White-tailed deer have invaded cities to become a significant pest.

Historically, the Red deer was the royal deer of Europe, much esteemed for its sporting qualities throughout the ages. White-tailed and Mule deer are currently the most important big game animals in North America, as the Roe deer is in Europe.

Some deer species tame easily. Reindeer were apparently the primary food source of our late-Ice-Age ancestors, and became a domesticated form postglacially, while Red and Sika deer were not uncommon table-fare for our parent species Homo erectus. The largest living deer, the moose, was occasionally domesticated during the Middle Ages, and served as a remarkably able mount and beast of burden. However, it is difficult to keep in good health in captivity, as it requires herbs of high nutritional value and a high diversity of browse.

Sika deer, Red deer, and wapiti have been kept in deer farms, a practice that began in China, where the growing velvet antlers were valued for medicines. Deer farming is, however, bad for wildlife conservation, since escaped deer can genetically pollute wild populations and are a conduit for livestock diseases.

In terms of conservation, many north-temperate species are doing reasonably well, but some tropical forms are not. Swamp-adapted deer that require fertile plains are severely affected by habitat loss to intensive agriculture. Others, particularly in South America, are susceptible to predation by dogs, as they seem to have little ability to deal with a pack-hunting, cursorial predator. We have derived great benefits historically from deer and, with attention to their needs, we shall continue to reap great rewards in the future. VG

Left The tangled mass of horn in this clash between reindeer looks as if it must end in bloodshed; yet this is ritualized sparring, which merely serves as a test of strength.

Below A female moose in water. As well as being speedy runners – reaching speeds of up to 56km/h (35mph) – moose are quite adept swimmers. This enables them to feed on water vegetation, in pursuit of which they will occasionally submerge themselves completely.

Deer are the principal prey of many large carni-]ores, and in the ongoing battle for survival evolu-]on has provided them with multiple security]rategies. Saltors (jumpers) depart quickly, then]odge out of sight and hide. Cursors (runners)]n with speed and endurance. There are cliffhop-]ers, such as the huemuls (*Hippocamelus* spp.),]d at one time there was a large, methodical]ck-climber, the extinct *Navahoceros* from the]orth American Rockies.

Among the saltors are some species, such as]e Mule deer, that seek to place steep walls or]gh obstacles between themselves and pursuing]redators; these deer are equipped with enlarged]rs and eyes that allow them to detect predators] a great distance. The Moose uses its large body]ze and very long legs to trot gracefully, with little]stly bodylift, over low obstacles that force small-]odied predators into very costly jumping, tiring]em out as well as slowing them down. Some]ecies such as the West Chinese and Tibetan red]eer have large haunches for jumping on steep]ountains or over high shrubs. Since each form] escape requires somewhat different landscapes]d substrates for success, antipredator adapta-]ons also segregate sympatric deer species ecolog-]ally – in other words, different escape strategies]ut separate species into different localities within]e same landscape, thereby minimizing competi-]on for food.

Deer Species

DEER SPLIT INTO TWO MAIN GROUPS: OLD World species (the primitive Muntiacinae and the Cervinae) and New World species (the antlerless Hydropotinae and the antlered Capreolinae). These terms denote their origin, not their current distribution.

Species that do not appear in the 1993 established taxonomy are marked † below. That listing also included eight species not described here: *Muntiacus atherodes*; *Dama mesopotamica*; *Cervus alfredi*, *C. mariannus*, and *C. schomburgki*; *Capreolus pygargus*; *Mazama bricenii* and *M. nana*. This aggregated to 43 cervid species overall.

SUBFAMILY MUNTIACINAE

Indian muntjac
Muntiacus muntjak

India, Sri Lanka, Tibet, SW China, Burma, Thailand, Vietnam, Malaya, Sumatra, Java, Borneo (introduced in England). Woodland and forest with good undergrowth. 15 subspecies.
SH 57cm (male), 49cm (female); WT 18kg. Antlers usually single spikes; antler pedicels extend down face; AL 17cm; PT 2–4. Upper canines present, and in males form tusks 2.5cm long.
COAT: dark chestnut on back toning to almost white on belly, darker in winter than in summer.
BREEDING: gestation about 180 days.

Reeves's muntjac
Muntiacus reevesi
Reeves's or Chinese muntjac or Barking deer

E China, Taiwan (introduced in England). Habitat, antlers, and canines as Indian muntjac. 2 subspecies. SH 41cm.
COAT: chestnut, with darker limbs and black stripe along nape; rufous patch between pedicels on forehead; chin, throat, and underside of tail white.
BREEDING: 210 days.

Hairy-fronted muntjac — Vu
Muntiacus crinifrons
Hairy-fronted or Black muntjac

E China. Habitat, antlers, and canines as Indian muntjac.
SH 61cm; WT 18–20kg; AL 6.5cm, with small projection from inner side of base.
COAT: blackish brown, with lighter head and neck. Tail longer than other muntjacs.

Fea's muntjac
Muntiacus feae

Thailand, Tenasserim. Similar to Indian muntjac, but coat darker brown, with yellow hairs up to center of the pedicels.

Roosevelt's muntjac
Muntiacus rooseveltorum

N Vietnam. Intermediate in size between Indian and Reeves's muntjac. Some authorities treat Roosevelt's muntjac as conspecific with Fea's muntjac.
COAT: redder than Reeves's muntjac. In other respects same as Indian muntjac.

Gongshan muntjac
Muntiacus gongshanensis

S China, Tibet, Myanmar, and N Thailand. Productive evergreen lowland forests.
SH 57–61cm (female), 49–52cm (male); WT 18–20kg.
COAT: dark chestnut brown. May be conspecific with the Hairy-fronted muntjac, which it resembles in external appearance and size; the two species also share very similar r-DNA sequences and similar karyotypes, 2n=9 and 2n=8 respectively. Also closely related to Fea's muntjac. Although fairly recent phylogenetically, this group of large, long-tailed muntjacs have small, dagger-like antlers, hidden in a tuft of long, reddish hair.

Giant muntjac †
Muntiacus vuquangensis

Highlands of Laos, Vietnam, and Cambodia. Old-growth evergreen forests, but may also venture into second growth.
SH 67cm; WT 35–40 kg.
COAT: grizzled tan to gray-brown, dark dorsal line on the neck; relatively short tail. Males have light knee-patches. Antlers substantial and the antler pedicels unusually stout; AL 17–28cm. Despite its large body and antler size and wide distribution, this species was only described by science in 1994. Initially placed in its own genus, *Megamuntiacus*, subsequent work on its DNA properties has assigned it to *Muntiacus*. It is part of a genetic clade containing the small mountain muntjacs *M. truongsonensis*, *M. putaoensis*, and *M. rooseveltorum*, with which it is largely sympatric.
BREEDING: gestation unknown; may hybridize with the Indian muntjac in captivity.

Tufted deer
Elaphodus cephalophus

S, SE, and C China, NE Burma. Forest up to 4,570m. 3 subspecies.
SH 63cm. Antlers and pedicels smaller than in muntjacs; antlers unbranched, and often completely hidden by tufts of hair which grow from forehead. Upper canines present, and in male form tusks 2.5 cm long.
COAT: deep chocolate brown, with gray neck and head; underparts, tips of ears, and underside of tail white. Young have spots on back.
BREEDING: gestation about 180 days.

SUBFAMILY CERVINAE

Fallow deer
Dama dama

Europe, Asia Minor, Iran (introduced to Australia, New Zealand). Woodland and woodland edge, scrub. 2 subspecies.
SH 91cm (male), 78cm (female); WT 63–103kg (male), 29–54kg (female). Antlers branched and palmate; AL 71cm.
COAT: typically fawn, with white spots on back and flanks in summer, grayish brown without spots in winter; rump patch white, edged with black; black line down back and tail; belly white; color highly variable and melanistic (black), white and intermediate forms common. Young spotted in some color varieties.
BREEDING: gestation 229–240 days.

Chital
Axis axis
Chital, Axis deer, or Spotted deer

India, Sri Lanka (introduced to Australia). Forest edge, woodland. 2 subspecies.
SH 91cm; WT 86kg; AL 76cm; PT 6; beam of antler curves back and out in lyre shape.
COAT: rufous fawn; white spots on back. No seasonal color change. Young spotted.
BREEDING: gestation 210–225 days.

Hog deer
Axis porcinus

N India, Sri Lanka, Burma, Thailand, Vietnam (introduced to Australia). Grassland, paddyfields. 2 subspecies.
SH 66–74cm; WT 36–45kg; AL 39cm; PT 6.
COAT: yellowish brown with darker underparts, the metatarsal tufts lighter colored than the rest of leg; young spotted. Build low and heavy, with short face and legs.
BREEDING: gestation about 180 days.

Kuhl's deer — En
Axis kuhlii
Kuhl's or Bawean deer.

Bawean Island.
SH 68cm; AL unknown; PT 6; pedicles long.
COAT: uniform brown, tail long and bushy; young unspotted.

Calamian deer — En
Axis calamianensis

Calamian Islands. Similar to Hog deer, except face and ears shorter, with white patch from lower jaw to the throat and white mustache mark.

Thorold's deer — Vu
Cervus albirostris

Tibet. SH 122cm. Antlers branched; AL unknown; PT 10.
COAT: brown with creamy belly, white nose, lips, chin, throat, and white patch near ears; winter coat lighter in color than summer. Ears are narrow and lance-shaped. Hooves are high, short, and wide, like those of cattle. Reversal of hair direction on withers gives the appearance of a hump.

Swamp deer — Vu
Cervus duvaucelii
Swamp deer or barasingha

N and C India, S Nepal. Swamps, grassy plains. 2 subspecies. SH 119–124cm; WT 172–182kg; AL 89cm; PT 10–15.
COAT: brown, with yellowish underpar in the hot season stags become reddish the back. Young spotted.

Red deer
Cervus elaphus
Red deer, maral, hangul, shou, Bactrian dee or Yarkand deer

Scandinavia, Europe, N Africa, Asia Mir Tibet, Kashmir, Turkestan, Afghanistan (introduced to Australia and New Zealand). Woodland and woodland edg open moorland in Scotland. 12 subspec
SH 122–127cm in *C. e. hanglu*, though s and weight vary with subspecies and lo ity; females are smaller than males. Ant branched, varying with subspecies and locality, eg AL 123cm, PT 20 in *C. e. hippelaphus*; AL 89cm, PT 12 in *C. e. scoticu*
COAT: red, liver, or gray-brown in sumr becoming darker and grayer in winter; rump patch yellowish, often with a dark caudal stripe; young spotted.
BREEDING: gestation 230–240 days.
CONSERVATION STATUS: 3 subspecies are listed as Endangered.

Wapiti †
Cervus canadensis
Wapiti or elk (elk in America only)

Western N America, Tien Shan Mountai to Manchuria and Mongolia, Kansu, Chi Grassland, forest edge; upward summe movement in mountains. 13 subspecies Some regard it as subspecies of *C. elaph*
SH 130–152cm; WT 240–454kg; female smaller than males. Antlers branched; A 100cm; PT 12.
COAT: in summer, light bay with darker head and legs; in winter, darker with gra underparts; rump patch light-colored; young spotted.
BREEDING: gestation 249–262 days.

Eld's deer
Cervus eldii
Eld's deer or thamin

Manipur, Thailand, Vietnam, Hainan Island, Burma, Tenasserim. Low, flat, marshy country. 3 subspecies.
SH 114cm. Antlers: the long brow tine a the beam form a continuous bow-shape curve, with forks at the end of the beam AL 99cm (does not include brow tine).
COAT: in summer, red with pale brown underparts; in winter, dark brown with whitish underparts; white on chin, arou eyes and margins of ears; females lighte colored than males; young spotted. Wal on the undersides of its hardened paste as well as its hooves.

Sika deer
Cervus nippon
Sika or Japanese deer

Japan, Vietnam, Formosa, Manchuria, Korea, N and SE China (introduced to

New Zealand). Forest. 13 subspecies. SH 65–109cm; WT 48kg. AL 28–81cm; PT 6–8, depending on race; top points tend to be palmate.
COAT: in summer, chestnut-yellowish brown with white spots on sides and a white caudal disc edged with black; in winter, gray brown with spots less obvious; young spotted.
BREEDING: gestation 217 days.
CONSERVATION STATUS: 5 subspecies are Critically Endangered, and 2 Endangered.

Rusa deer
Cervus timorensis
Russ or Timor deer

Indonesian Archipelago (introduced to Australia, New Zealand, Fiji, New Guinea). Parkland, grassland, woods, and forest. 6 subspecies.
SH 98–110cm (male), 86–98cm (female). Antlers branched; AL 111cm; PT 6.
COAT: brown, with lighter underparts and under tail; young unspotted.

Sambar
Cervus unicolor

From Philippines through Indonesia, S China, Burma to India and Sri Lanka (introduced to Australia and New Zealand). Woodland: avoids open scrub and heaviest forest. 16 subspecies.
SH 61–142cm: WT 227–272kg. Antlers branched, with long brow tine at acute angle to beam and forward-pointing terminal forks: AL 60–100cm; PT 6.
COAT: dark brown, with lighter yellow brown under chin, inside limbs, between buttocks, and under tail; females and young lighter-colored than males; young unspotted.
BREEDING: gestation 240 days.

Père David's deer [Cr]
Elaphurus davidianus

Formerly China, never known outside parks and zoos.
SH 120cm. Antlers branched, but with tines pointing backwards. AL 80cm.
COAT: bright red with dark dorsal stripe in summer, iron gray in winter; tail long; hooves very wide. Young spotted.
BREEDING: gestation 250 days.

SUBFAMILY HYDROPOTINAE

Water deer [LR]
Hydropotes inermis
Water deer or Chinese water deer

China, Korea. Swamps, reedbeds, grasslands. 2 subspecies.
SH 52cm (male), 48cm (female); WT 11–14kg (male), 8–11kg (female). Antlers lacking in both sexes. Upper canines present, and in males form tusks 7cm long.
COAT: reddish brown in summer, dull brown in winter: young dull brown, with light spots each side of midline.
BREEDING: gestation 176 days.
CONSERVATION STATUS: LR Near Threatened.

SUBFAMILY CAPREOLINAE

Mule deer [En]
Odocoileus hemionus
Mule or Black-tailed deer

Western N America, C America. Grassland, woodland. 11 subspecies.
SH 102cm; WT 120kg. Antlers branch dichotomously: PT 8–10.
COAT: rusty red in summer, brownish gray in winter; face and throat whitish, with black bar round chin and black patch on forehead; belly, inside of legs, and rump patch white; tail white with black tip or, in Black-tailed subspecies, with black extending up outer surface. Young spotted.
BREEDING: gestation 203 days.
CONSERVATION STATUS: 1 subspecies listed as Endangered.

White-tailed deer
Odocoileus virginianus

N and C America, northern parts of S America. In N America, conifer swamps in winter, woodland edge from spring to fall; in S America, valleys near water in dry season, higher in rainy season. 38 subspecies.
SH 81–102cm; WT 18–136kg. Antlers branch from main beam, ie not dichotomously; beams curve forward and inward; basal snag longer than in Black-tailed deer; PT 7–8, fewer in southern races.
COAT: reddish brown in summer, gray brown in winter; throat and inside ears with whitish patches; belly, inner thighs, and underside of tail white; young spotted. Metatarsal gland on hock shorter (2.5cm) than on Black-tailed deer (7.5–12.7cm).
BREEDING: gestation 204 days.
CONSERVATION STATUS: 1 subspecies listed as Endangered.

Roe deer
Capreolus capreolus

Europe, Asia Minor, Siberia, N Asia, Manchuria, China, Korea. Forest, forest edge, woodland, moorland. 3 subspecies.
SH 64–89cm; WT 17–23kg. Antlers branched; PT 6.
COAT: in summer, foxy red, with gray face, white chin, and a black band from mouth to nostrils; in winter, grayish fawn, with white rump patch and white on throat and gullet; young spotted. After the rut in July–August, implantation is delayed until December and kids are born in April–June.
BREEDING: gestation 294 days (including 150 days pre-implantation).

Moose
Alces alces
Moose or elk (elk in Europe only)

N Europe, E Siberia, Mongolia, Manchuria, Alaska, Canada, Wyoming, NE USA (introduced to New Zealand). 6 subspecies.
SH 168–230cm; WT 400–800kg (females are about 25 percent smaller than males). Antlers large, branched, and palmate; PT 18–20. Shoulders humped, muzzle pendulous, and the "bell" hangs from throat.
COAT: blackish brown with lighter brown underparts, darker in summer than in winter; naked patch on the muzzle is extremely small; young unspotted.
BREEDING: gestation 240–250 days.

Reindeer
Rangifer tarandus
Reindeer or caribou

Scandinavia, Svalbard, European Russia from Karelia to Sakhalin Island, Alaska, Canada, Greenland and adjacent islands. Many reindeer in Europe and Scandinavia are domestic. Woodland or forest edge, all year for some races, but others migrate to Arctic tundra for summer. 9 subspecies.
SH 107–127cm (male), 94–114cm (female); WT 91–272kg (male). Antlers present in both sexes, smaller and with fewer points in the female; branched, with a tendency for palmate top points; the brow points are palmate in the males; males shed antlers November–April, females May–June; AL up to 147 in males; PT up to 44 in males.
COAT: brown in summer, gray in winter, white on rump, tail, and above each hoof; neck paler, chest and legs darker; males have white manes in the rut; only deer with no naked patch on muzzle; young unspotted.
BREEDING: gestation 210–240 days.

Marsh deer [Vu]
Blastocerus dichotomus

C Brazil to N Argentina. Marshes, floodplains, savannas.
SH unknown; WT unknown; similar in size to small Red deer. AL 60cm; the brow tine forks, and there is a terminal fork on the beam which gives PT 8.
COAT: rufous in summer, duller brown in winter; lower legs dark; black band on the muzzle; young unspotted.

Pampas deer [LR]
Ozotoceros bezoarticus

Brazil, Argentina, Paraguay, Bolivia. Open, grassy plains.3 subspecies.
SH 69cm; WT unknown; AL unknown; PT 6.
COAT: yellowish brown, with white on the underparts and inside the ear; upper surface of the tail dark; hair forms whorls on the base of neck; young spotted.
CONSERVATION STATUS: LR Near Threatened.

Chilean huemul [En]
Hippocamelus bisulcus
Chilean huemul or Chilean guemal

Chile, Argentina, High Andes.
SH 91cm: WT 60-65kg. Antlers branched; AL 28cm; PT up to 4.
COAT: dark brown in summer, paler in winter; young unspotted. Ears mulelike.

Peruvian huemul
Hippocamelus antisensis
Peruvian huemul or Peruvian guemal

Peru, Ecuador, Bolivia, N Argentina, High Andes. Weight slightly less than and has antlers similar to Chilean huemul.
COAT: similar to, but paler than, Chilean huemul. Young unspotted.

Red brocket
Mazama americana

C and S America, from Mexico to Argentina. Dense mountain thickets. 14 subspecies.
SH 71cm: WT 20kg. Antlers simple spikes; AL 10–13cm.
COAT: red brown, with grayish neck and white underparts; tail brown above, white below. Young spotted.
BREEDING: gestation 220 days.

Brown brocket
Mazama gouazoupira

C and S America, from Mexico to Argentina. Mountain thickets, but more open than Red brocket. 10 subspecies.
SH 35–61cm; WT 17kg. Antlers simple spikes; AL 10–13cm.
COAT: similar to Red brocket but duller brown. Young spotted.
BREEDING: gestation 206 days.

Little red brocket [LR]
Mazama rufina

N Venezuela, Ecuador, SE Brazil. Forest thickets. 2 subspecies.
SH 35cm. Antlers simple spikes; AL 7cm.
COAT: dark chestnut, with darker head and legs. Preorbital glands extremely large. Young spotted.
CONSERVATION STATUS: LR Near Threatened.

Dwarf brocket
Mazama chunyi

N Bolivia and Peru, Andes.
SH 35cm. Antlers simple spikes.
COAT: cinnamon-rufous brown with buff throat, chest, and inner legs, and white underside to tail; whorl on nape, supraorbital streak and circumorbital band lacking; smaller and darker than Brown brocket. Tail shorter than in other brockets. Young spotted.

Southern pudu [Vu]
Pudu pudu

Lower Andes of Chile and Argentina. Deep forest.
SH 35–38cm; WT 6–8kg. Antlers simple spikes: AL 7–10cm.
COAT: rufous or dark brown, with paler sides, legs, and feet; young spotted.
BREEDING: gestation 210 days.

Northern pudu [LR]
Pudu mephistophiles

Lower Andes of Ecuador, Peru, Colombia. Deep forest. Slightly larger than Southern pudu. 2 subspecies.
AL unknown; antlers simple spikes.
COAT: reddish brown, with almost black head and feet. Young spotted.
CONSERVATION STATUS: LR Near Threatened.

RAC/VG

ABBREVIATIONS

SH = shoulder height TL = tail length WT = weight AL = antler length
PT = number of points on both antlers
Approximate nonmetric equivalents: 10cm = 4in; 1kg = 2.2lb

Ex Extinct
EW Extinct in the Wild
Cr Critically Endangered
En Endangered
Vu Vulnerable
LR Lower Risk

SPECIAL FEATURE

SEX RATIO MANIPULATION IN RED DEER

Why do some hinds regularly have more sons than daughters?

ON AVERAGE, FEMALE MAMMALS USUALLY give birth to equal numbers of male and female young. This strategy makes evolutionary sense for each individual mother, for if mothers consistently produced greater numbers of male offspring, the males would outnumber the females and hence face greater competition for mates – which would mean that the average male had a worse chance of reproduction than the average female. This would favor any mother who had predominantly female young, and so the pendulum of advantage would swing back and forth until it stabilized at an equal sex ratio. Thus sex ratios are mostly 1:1. However, there are situations in which it may benefit a mother to produce sons rather than daughters, or vice versa. The potential advantages of this manipulation of the sex ratio are elegantly illustrated in remarkable results from studies of Red deer.

In Red deer, a hind (female) is likely to produce several young – one a year throughout her 10- to 11-year life. For stags, life is much more of a lottery; some have dozens of offspring, others none at all. On the island of Rhum off the west coast of Scotland, one successful male managed to father 53 offspring. In contrast, 9 percent of males had just two children, 19 percent one, and 35 percent died childless. To make matters worse, around half the young die before they reach adulthood. The reason for this skew is that Red deer have a polygynous mating system in which the successful males defend groups of females – harems – against the attentions of the many.

A male's ability to outperform in the mating game depends on his "quality": his body and antler size, his fighting prowess, his roaring ability, and his potential for coping with defending females rather than feeding in preparation for the winter months ahead. Thus it pays a female to produce male young only when she can be assured of having a good-quality baby. Rather than run the risk of producing poorer-quality male offspring, hinds will produce female calves, which despite their physical inferiority will always have a reasonable chance of reproducing. Of course, all individuals in the present generation of any animal population are the product of successfully reproducing parents and grandparents. A poor-quality male calf may well represent an evolutionary dead end for the mother's lineage. In situations where the offspring is likely to be in relatively poor condition, genes for producing a poor-quality male will most likely die out, while those for producing a female will be more likely to spread.

One of the factors affecting the performance of offspring in Red deer is the dominance status of the mother. Red deer hinds are social: they live in

Above *During the rut, Red deer stags with harems must defend them against rival males. A roaring contest usually heralds a fight, but sometimes one stag can assert dominance by vocalization alone.*

Right *Female Red deer also have fights to establish dominance hierarchy. The increased access to prime feeding spots enjoyed by dominant hinds is important in producing stronger offspring.*

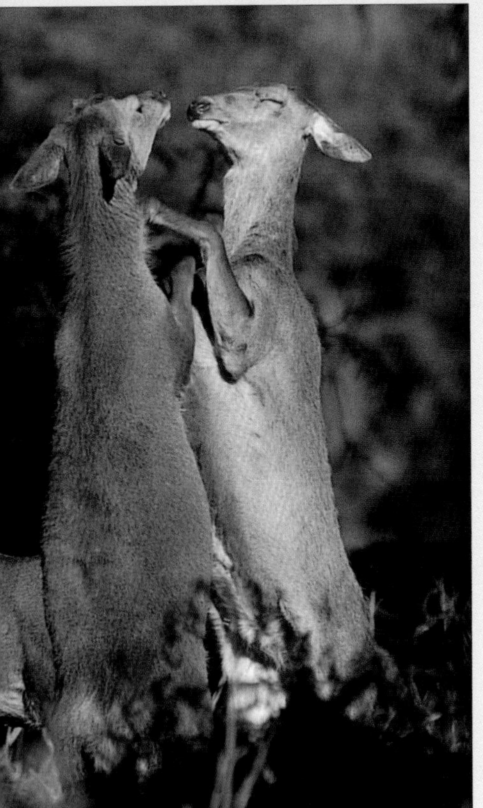

Above *A calf will remain concealed for the first few weeks of its life, emerging only to suckle. Lactation may last 7 months, so the mother's ability to find prime grazing sites is vitally important.*

large groups that are hierarchical in structure. The dominance hierarchy is manifest in interactions between individuals. Dominant hinds are, on average, approximately 7 percent heavier than subordinate hinds, and their offspring are both 14 percent more likely to survive their first year of life and also more likely to be dominant themselves. It follows that if a mother is herself dominant, her young are likely to have what it takes to compete well in the rigors of the rut – so she should produce sons. Similarly, if she is subordinate, her offspring will be smaller and weaker, and so she should produce daughters.

Remarkably, results from a longterm study of Red deer on Rhum exactly fitted this pattern. The most dominant females produced approximately 65 percent sons, whereas the least dominant produced only 35 percent. This difference apparently occurs at conception, and may be associated with levels of hormones in the mother, although the precise mechanism is still unknown. The effects are also small in magnitude: we only observe a bias of 15 percent. This serves as a reminder that the more usual chromosomal mechanisms that determine the sex of offspring may constrain any adaptive manipulation of offspring sex ratio.

To complicate the story further, other factors without any apparent evolutionary benefits also affect the sex ratio of Red deer calves. If mothers are stressed during pregnancy – for example by high population densities and hence low food availability – they may lose the fetus. It turns out that they are more likely to lose a fetus if it is male, so high levels of environmental stress are associated with births of fewer males. On Rhum, an increase in either population density or winter rainfall to 1.25 times their average value reduces the proportion of males born the following spring by about 3 percent. Under these harsher conditions, there is no difference between dominants and subordinates in the sex ratio of their offspring, suggesting that the adaptive manipulation described above requires exactly the right environmental conditions to function. Why males are more prone to dying as juveniles remains a mystery, as does perhaps the most remarkable feature of this entire system: exactly how does the female decide the sex and the fate of an embryo without modern technology? LEBK/TC-B

SEXUAL PARTNERS ON PARADE

The lek mating system in Fallow deer

FROM THE MOMENT OF CONCEPTION, FEMALE mammals invest more in offspring than males do. They usually end up caring for the young, while males merely try hard to demonstrate to females that they are worthy to be their fathers. Perhaps the strangest way that evolution has devised for males to show their worth is on leks – sexual display areas in which males compete to attract mates. Instead of defending resources that attract females or defending the females themselves, males that breed on leks defend small mating territories that contain few or no resources other than themselves. The territories are typically clustered together in groups of anything up to 100; between clusters, other males defend isolated territories or look hopefully for receptive females.

Leks are rare in mammals, and are found mainly among the African antelopes. The discovery in the 1980s that leks occurred in some populations of European Fallow deer launched a decade of research into the circumstances in which they occur and the reasons why males and females mate on them.

In ungulates, the habit of lekking is almost universally variable within species. This is particularly the case with Fallow deer, which exhibit extreme plasticity in male mating strategies, both between and within populations. The diverse strategies may be viewed as a continuum arranged with regard to the degree of territoriality displayed during the rut. At one extreme, males forage for receptive females and defend them in harems of up to 50. Others defend isolated mating territories, varying in size from a few square meters to several hectares, which may contain resources valuable to females; these deer also seek to attract females with vocal and olfactory displays. Finally, at the other extreme of the territoriality continuum, are leks.

Fallow bucks on leks defend tiny territories consisting of little more than a few square meters of soil trampled bare of vegetation. The typical Fallow deer lek consists of 10–20 mature males, although bigger leks of up to 50 males have been observed. To the human observer, leks are often a scene of frantic activity, as males desperately try to attract and retain receptive females on their territories.

The most likely explanation for this apparently bizarre behavior lies in a link with population density. In Fallow deer as in other ungulates, increases in density are often associated with increased male territoriality during the rut. In populations where there are fewer than 10 adult males per sq km (4 per sq mi), Fallow bucks tend to forage for females or defend mobile harems. Males start to defend isolated territories as density increases, and in the densest populations, with more than 40 males per sq km (15 per sq mi), they aggregate together and defend territories on leks. In some populations, increases in density have resulted in a shift from isolated territories to leks, providing convincing proof of the link.

But why do male Fallow deer defend territories on leks? Males on leks fight frequently, often up to ten times per day, and fights may result in serious injury or death. The stress of fighting, and the competition from males desperate to get a place on the lek, mean that tenure of lek territories can be very short, often as little as two or three days. Nevertheless, a lek is the place to be for bucks eager to encounter does. Successful males on leks mate more than five times as often as males in the same populations away from leks, which more than compensates for their shorter tenure of the territories. In most lekking populations, more than 80 percent of matings occur on leks rather than on single territories or in mobile harems. This is because females – and particularly those in estrus – actively travel to and settle on leks. Males on lek territories may also retain estrous females more effectively than do males on single territories. Because females tend to move to adjacent territories when disturbed, females that enter a lek will circulate between territories rather than leave the lek. Thus males on leks usually have more females to mate with than males adopting alternative mating strategies elsewhere.

If male Fallow deer mate on leks because they attract and retain estrous females, then the interesting question becomes, "Why do estrous females mate on leks?" In their attempts to explain female preferences for so doing, most researchers have pointed to the genetic benefits: females choose males that will provide them with offspring that have better survival rates, or with sons that will be more attractive to females. Evidence in support of this "good genes" hypothesis usually involves the demonstration that, within leks, the majority of females mate with the most competitive males. Studies of lekking Fallow deer demonstrate that females choose males on the basis of either their body size, their vocalization rates, or their location in central territories

on the lek, and that females could be using any of these cues to chose good-quality males.

However, even if females do gain genetic benefits from mating on leks, it does not necessarily follow that mate choice is the primary cause of lek formation. An alternative view is that females gain direct benefits from mating on leks through either increased mating rates or as a result of the decreased risk of predation. Studies of Fallow deer and other lekking ungulates have shown that estrous females are harassed by nonterritorial males four times more frequently in mixed-sex herds and on single territories than on lek territories. Experiments have demonstrated that female Fallow deer are more attracted to males with groups of females than to single males, and that this preference is restricted to estrous females – anestrous females do not show it. So females may be seeking protection from harassment by associating thus; the added security, along with the convenience of the lek for deer of both sexes looking for sexual partners, may explain the evolution of this strange phenomenon. ST

Below *A pair of bucks fighting on a lek. At most, males are only able to maintain their position on the lek for a few days at a time, and sometimes even only for a few hours.*

Above *Fallow deer mating on a lek. Male Fallow deer are capable of mating from about 18 months of age; however, they usually do not breed until they are at least four years old.*

PHOTO STORY

Reindeer Migration

1 Reindeer sleds lead the herd on its summer migration. Reindeer naturally migrate to exploit seasonal differences in forage availability. Tundra vegetation has only a short summer growing season, but is highly nutritious, and flies are less of a nuisance in open, windswept areas. In winter, the reindeer migrate to areas where they can dig through the snow for lichens and other vegetation. The nomadic Sami have little altered this natural migratory pattern.

2 Harnessing a reindeer to a sled. The Sami live in northern Norway, Sweden, Finland, and Russia. Initially they followed wild reindeer herds to hunt them, and probably invented the first ski. The need to protect wild reindeer from predators prompted the first attempts at domestication around 500AD. No wild reindeer (and few wolves) now remain in northern Fenno-Scandia. The herds traditionally have provided the Sami with most of their needs: meat; leather and fur for clothes, shoes, and sleeping bags; and antlers and bone for carving. Herded reindeer are only semi-domesticated, but draft and milking animals are fully tamed for regular handling.

3 A herder lassoes a reindeer. Reindeer are the only cervid in which both sexes and all calves carry antlers, around which the first lasso is thrown from some 20 meters (66ft). With a strong and lively animal, the first lasso is wound round a post or tree, while a second lasso is used to trap the hind leg, allowing the reindeer to be drawn in and handled as required. Reindeer belonging to different herders are identified by marks cut in the ears of calves, which allows the herder to pick out his animals from a circling throng in the corral.

4 *A herder in a boat leads his reindeer to summer pasture on an island off the Norwegian coast. Reindeer are well adapted to their Arctic habitats. Splayed hooves and strong legs help them to walk on and dig in snow, to swim, and to migrate long distances. Long interdigital hairs and a glandular secretion prevent snow clogging between the toes. A well-insulated coat of hollow hairs and air pockets also enables them to swim high in the water. Sami herders exploit these adaptations, and have also applied new technologies to herding, such as walkie-talkies and snow scooters.*

5 *Reindeer are confined in corrals at key times in the annual cycle. A summer round-up enables calves to be marked. Reindeer are collected again in autumn for the main period of slaughter, before the herd is split into smaller units for the winter. The capacity of the wintering grounds is the limiting factor on the numbers of animals carried by each Sami pasture unit. The Chernobyl disaster of 1986 has seriously affected sales of reindeer meat, as radioactive fallout accumulates in long-lived lichens, and in turn in the reindeer as they feed on them.* NL–W

5

3

Giraffe and Okapi

TOWERING ABOVE THE GROUND AT A HEIGHT of some 4–5m (13–16ft), a group of giraffes in savanna woodland looks from a distance like a flotilla of disembodied heads and necks moving through the tree canopy. Giraffes are the tallest living animals, and their height, combined with their bold color patterning and unusual body proportions, make them quite unmistakable.

The overwhelming impression giraffes create is one of elongation, an effect emphasized by the short length of the body in relation to the pronounced length of the neck. This peculiarity is all the more marked in comparison with the animal's only close relative, the okapi, an elusive forest-dweller that is closer to a horse in size and form.

High-level Browsers

Form and Function

The functional significance of the giraffe's vertically elongated form, it would appear, is to improve access to foliage in the tree canopy. Through evolutionary time, competition from smaller-bodied browsers presumably made it beneficial for ancestral giraffids to utilize fully the upper levels of their browsing height ranges during critical periods of resource shortage. Taller individuals were less likely to fall victim to starvation, and so more of them survived to pass their genes on to future generations. The giraffe's neck is a textbook example of an "adaptation" that has come about through natural selection, even though it has never actually been demonstrated that giraffes feed more efficiently at high levels.

Support for this notion is growing, however, from experiments underway in South Africa. It is now well established that smaller-bodied herbivores need less food per day, and so can afford to carefully select what they eat; they also have narrower muzzles with which to get in among the foliage and select the prime leaves and buds. The smaller herbivore size-classes also include species that occur in comparatively high local densities, so the food they leave for the larger animals is often suboptimal. Above the heads of the smaller herbivores, however, is a wealth of unexploited plant material. To draw an analogy, elongated necks provide giraffes with the advantages that apple-picking devices provide for people.

Yet being so tall carries some costs. A giraffe's heart is about 2m (6.5ft) above its hooves, and 3m (10ft) below its brain. This means that there is a high column of blood in each leg weighing down on the capillaries in the tissues lower down. In all other living animals the result would be filtration edema, in which the blood is forced outward through the capillary walls. Giraffes, however, have an especially thick sheath of tight skin around the slender lower half of each leg, in order to maintain high extravascular pressure and so counteract the pressure in the blood vessels. The effect is the same as that generated by the "G-suits" worn by fighter pilots, which are designed to prevent blood rushing to the legs and causing blackouts during rapid acceleration to higher altitude.

Another anomaly in the giraffe's circulatory system is that the pumping force required to drive blood up to the brain (215mm hg at heart level) is almost twice that of an average cow. The blood pressure in their brain, however, is maintained at about 90mm hg, which is no higher than in most large mammals. One amazing aspect is that, when a giraffe lowers its head to drink, blood could be expected to rush down its long neck into its head and burst into the brain, causing it literally to explode. In fact the blood vessels in the giraffe neck include an intricate pressure-regulating mechanism involving the carotid rete, a meshed network of narrow vessels, derived from multiple subdivisions of the carotid arteries, which lies at the base of the brain. Furthermore, the idiosyncratic, splay-legged, knees-bent posture of drinking giraffes serves to bring the chest lower to the ground and reduce the height difference between heart and brain.

Left *A Rothschild's giraffe calf. As it develops, the giraffe's curious anatomy becomes even more pronounced, its neck growing proportionally longer.*

FACTFILE

GIRAFFE AND OKAPI

Order: Artiodactyla

Family: Giraffidae

2 species in 2 genera

DISTRIBUTION Sub-Saharan Africa.

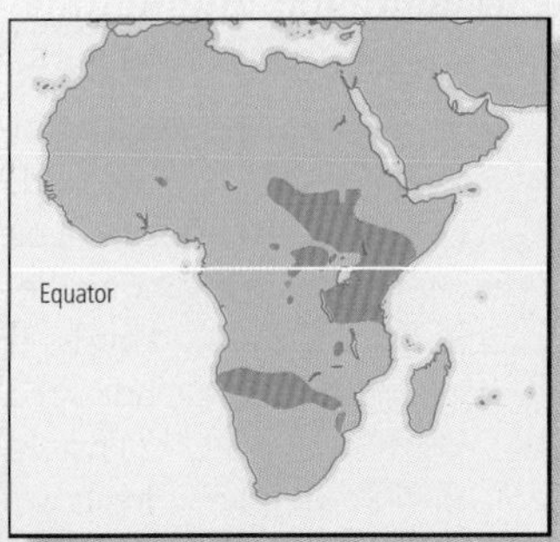

GIRAFFE *Giraffa camelopardalis*
9 subspecies: **West African giraffe** (*G. c. peralta*), **Kordofan giraffe** *(G. c. antiquorum)*, **Nubian giraffe** (*G. c. camelopardalis*), **Reticulated giraffe** (*G. c. reticulata*), **Rothschild giraffe** (*G. c. rothschildi*), **Masai giraffe** (*G. c. tippelskirchi*), **Thornicroft giraffe** (*G. c. thornicrofti*), **Southern African giraffe** (*G. c. capensis*/*G. c. angolensis*). Africa S of the Sahara. Open woodland and wooded grassland. HBL male 3.8–4.7m (12–15ft); TL (excluding tassel) male 80–100cm (31–39in); HT to horn tips male 4.7–5.3m (15–17ft), female 3.9–4.5m (13–15ft); WT male 800–1,930kg (1,765–4,255lb), female 550–1,180kg (1,213–2,601lb). Coat: patches of variable size and color (usually orange-brown, russet, or almost black) separated by a network of cream-buff lines. Breeding: gestation 453–464 days. Longevity: 25 years (to 28 in captivity). Conservation status: Lower Risk – Conservation Dependent; still common in E Africa, where populations are even expanding in the S, but increasingly rare in W Africa.

OKAPI *Okapia johnstoni*
N and NE Democratic Republic of Congo. Dense rain forest; prefers dense undergrowth in secondary forest; often recorded beside streams and rivers. HBL 1.9–2m (6–7ft); TL (excluding tassel) 30–42cm (12–17in); HT 1.5–1.7m (5–6ft); WT 210–250kg (465–550lb). Coat: velvety dark chestnut, almost purplish black, with conspicuous transverse black and white stripes on hindquarters, lower rump, and upper legs; the female is redder in color. Breeding: gestation 427–457 days (in captivity). Longevity: 15 or more years (in captivity). Conservation status: Lower Risk – Near Threatened.

Abbreviations HBL = head–body length TL = tail length HT = height WT = weight

Above right *Although the giraffe's great height* *nables it to forage easily in the tree canopy and so* *void food competition with other browsers, it must* *play its forelegs widely in order to drink at water-* *oles. In this ungainly posture, the animal is vulnerable* *o attack by predators. Fortunately, it gets most of its* *noisture from food and only rarely needs to drink.*

The structural design of the giraffe neck is very similar to that of a crane on a building site. Seven greatly elongated cervical vertebrae (the same number of neck bones found in all mammals) are strung with "cables" (tendons and muscles) to an anchor point in the shoulder hump (vertically extended dorsal processes of the thoracic vertebrae). Within this structure are the pipes that carry air to and from the lungs and food to and from the rumen. Giraffes are the largest living ruminants and, like others in this group of species, they chew their food more than once. When food is being regurgitated up the long esophagus for remastication, a distinct, tennis-ball-shaped lump is visible in the neck, rather like an elevator rising up a tall building. The long neck imposes an even greater challenge to respiration, because the trachea extends over 1.5m (5ft) and is about 5cm (2in) in diameter, thus containing about 3 liters of air. This "dead space" between lungs and nostrils is filled with a mixture of inhaled and exhaled air that has to be moved back and forth in addition to the air that actually enters the lungs. To overcome this difficulty, giraffes have to breathe more frequently than would be expected of animals of their size; the respiration rate of an average adult giraffe at rest is more than 20 breaths a minute, while in humans it is 12–15.

The irregular blotches on a giraffe's coat mimic the dappled combination of light and shade that is found in savanna woodlands. Coloration and patterning vary considerably among giraffe subspecies – the Reticulated giraffe has a particularly striking coat – and to some extent among sex and age classes within populations. In southern Africa, the bulls darken to almost black on their rumps and withers as they get older. Moreover, the coat of each giraffe functions as a kind of distinctive fingerprint. Some field researchers have used these patterns, especially on the neck, to identify and monitor individual giraffes by reference to a photographic catalog.

The Shrinking Savanna

DISTRIBUTION PATTERNS

Giraffes are indigenous to the entire savanna biome of sub-Saharan Africa, but have lost well over 50 percent of their geographical range in historical times. This is mostly as a result of overhunting and habitat loss, combined perhaps with episodic outbreaks of cattle-borne diseases. In West Africa the giraffe distribution is now restricted to one region in the Sahel (the southern fringe of the Sahara), with its center near Niamey in southwestern Niger. On the other hand, in southern Africa the giraffe distribution has begun to expand again in recent years, with the trend toward replacing cattle with game on privately-owned commercial ranchland. Despite this general increase in numbers, there remains a puzzling absence of giraffes in the Zambezi Valley between Zimbabwe and Zambia, which could be a relic of the rinderpest panzootic at the beginning of the 20th century.

Searching for Succulence

DIET

Giraffes are pure browsers, in that they feed almost exclusively on dicotyledonous plants (trees, shrubs, and forbs), with grazing occurring in the wild only very rarely, when new, succulent green grass is available after good rains. Throughout their distribution, giraffes utilize various acacia tree species for their staple browse, although the diet includes a high diversity of species from genera such as *Combretum*, *Commiphora*, *Grewia*, *Boscia*, and many others. Most giraffe feeding studies have been conducted in southern and eastern Africa, where a typical annual giraffe diet includes browse from 40–60 different species of woody plant. Giraffes are selective feeders, browsing on high-quality plant parts such as fresh leaves, shoots, fruits, and flowers whenever these are available. Large body size means giraffes must eat a lot of food, but they can tolerate a relatively wide range in its quality. Thus they can roam freely in undisturbed savanna and usually maintain a diet that is well above that required for maintenance. They can also survive lean months by including in their diets fibrous leaves of drought-hardy trees such as *Colophospermum mopane*, as well as twigs, leaf litter, and dried pods. Being ruminants, giraffes are able to improve the digestibility of their forage by remastication, and they also have a unique ability to ruminate while walking, which allows more time for feeding.

Left *Although the markings on these Reticulated giraffes may look similar to the untrained eye, each is, in fact, highly individual. Reticulated giraffes have the clearest, most distinctive patterning of all subspecies.*

Since plants in savanna ecosystems are scattered and bring forth leaves seasonally, giraffes move large distances, especially in the dry season. In southern and eastern Africa, they may wander across 300–600sq km (115–230sq mi), while in the Sahel region of Niger individuals of both sexes have been found to range over 1,500sq km (580sq mi). During dry seasons, all but the plants along water drainage lines lose their leaves, forcing giraffes to gather in the low-lying areas where there is adequate soil moisture to keep trees in leaf. In more plentiful times they roam freely, and so their distribution is more random. In Kruger National Park their use of vegetation in watercourses and riparian fringes begins to increase when the average monthly rainfall over the preceding four months falls below 60mm (2.4in); if it sinks under 20mm (0.8in), these strips of greenery become the most frequently used of all habitats.

The giraffe's mouth is perfectly adapted to its diet. Its long tongue plucks spiny shoots, which it crushes with its molar teeth. Smooth shoots are pulled obliquely through the gap between molars and canines and stripped of leaves as the stem is dragged over the incisors and canines, which are laterally flattened. Giraffes can feed at any level up to 5m (16.5ft) above the ground, but in practice this depends upon the availability of mature trees that have not been broken down by elephants. Bulls typically feed above the heads of cows and young. This feeding height difference occurs both because bulls are bigger than cows and also because the bulls feed more often with their necks stretched vertically at 180° to the forelegs – cows prefer to feed with the neck inclined at 135°. So consistent is this difference in feeding posture that it can be used to sex giraffes from a distance, but the reasons for it have yet to be unraveled.

The answer may be related to the other activities that make demands on the daily lives of bulls. While females will feed for more than half the time (55 percent) in any 24-hour cycle, the equivalent figure for males is only 43 percent, and while females need to ingest 58kg (128lb) of food in that time, males must take in 66kg (145lb). In addition to needing more food, males also need extra time for searching out mates and for engaging in male–male dominance contests, so they have tight restrictions on their feeding time. It could thus be that intersexual competition for food causes bulls to feed above the levels used by cows and subadults. Another idea is that males like to hold their heads high to keep an eye out for rivals and potential mates.

On a diet containing 70 percent water, giraffes rarely need to drink, but they will do when clean water is available. Their peculiar tendency to chew on old bones and to eat soil is probably related to a lack of minerals in their diet in some areas. Giraffes sometimes die from choking or botulism poisoning as a result of chewing and swallowing bones that have been discarded by scavengers.

Above *To facilitate its feeding on spiny foliage, the giraffe has a 46cm/18in-long, powerful tongue that is as dextrous as a monkey's forearm. It is also equipped with highly mobile, muscular lips.*

Necking and Clubbing

SOCIAL BEHAVIOR

Giraffe social behavior is relatively uncomplicated. Cows, subadult males, and juveniles associate in groups that seldom include more than 20 animals. The composition of any one group is unstable in that individual animals join and leave at will, with the only apparent social bonds being between cows and their unweaned calves. Mature bulls roam widely (typically 20km/12 miles in a day) in search of cows in estrus, and so are frequently alone. The largest dominant bulls monopolize mating. If a subordinate needs to be displaced, the dominant male will stand tall with his neck held vertically and strut with legs stiffened towards his opponent. The winner sometimes chases the loser for a short distance and then stands in a tall display posture until the loser has disappeared, after which he returns to consort with the cow in estrus.

Bulls always check the reproductive status of cows when they encounter them, after soliciting a urine sample by sniffing and nuzzling at each cow's genitalia. A solicited cow halts and delivers a short stream of urine onto the bull's tongue, which he tests in the vomero-nasal organ on the roof of his mouth, while assuming the distinctive flehmen posture – a scent-tasting grimace, with lips pulled back and mouth slightly open. Successful mating is usually only achieved by large males aged 8 or more, while females may first conceive when aged between 4 and 5. Gestation lasts more than a year (15 months), so giraffes have no breeding season. A typical cow will usually conceive again about five months after the birth of a 100 -kg (220-lb) calf and could produce 5–10 calves in her lifetime (up to 25 years). Calves are usually born singly and twins are very rare. The calves grow about 8cm (3in) per month, and may suckle for up to 18 months, so although as much as 50 percent of calves will die in their first six months, some mothers are simultaneously gestating one calf and feeding another. Losses of young giraffes are mainly to such predators as lions and hyenas, although leopards and wild dogs may also kill newborns.

During the first few weeks of its life, a giraffe calf will rest alone in the shade while its mother forages nearby, returning regularly to nurse. Life becomes more sociable for the youngster after this early stage, and it is common to see "nursery groups" of up to 10 calves together in the company of two or three adults. While giraffe mothers appear neglectful, their high visual acuity enables them to spot danger from as far as about 1 km (0.6 mi). When a calf is under attack, its mother will often attempt to defend it by standing over it and kicking violently at the attackers with both front and back legs – this is an effective tactic, although a determined lion pride will almost always win in the end.

Subadult males engage in "necking" contests, during which both animals entwine their necks, wrestling and butting for 30 minutes or more in an apparently amicable manner. This appears to result in a dominance hierarchy being established within each age-cohort of males, and provides young males with an opportunity to develop their neck muscles and hone their jousting skills. When they are mature, these skills will be of crucial importance. Giraffe bulls seldom engage in violent aggression, but when they do it is impressive. They swing their heads at each other like clubs, delivering resounding blows aimed to injure the opponent in the underbelly. A well-placed blow can knock a 1,500kg (1.5 ton) bull off balance, and if the two bulls' heads collide, then injuries to eyes, horns, and jaws result. It is remarkable that the giraffe brain is able to tolerate the forces involved in swinging the head repeatedly and violently through an arc of about 3.5m (11.5ft), as well as the concussion that occurs on each impact. Not surprisingly, the skull of the giraffe bull is heavily armored with progressively-deposited layers of bone, which in older animals form conical lumps above the eyes and around the horns. As a bull ages, the total mass of its head increases at a rate of 1kg (2.2lb) per year, exceeding 30kg (66lb) in 20-year-old specimens. In cows, head mass levels off at about 17kg (37.5lb) from 8 years onwards. Some researchers argue that neck-wrestling and head-clubbing are such bizarre and apparently costly means of establishing male dominance that the giraffe's neck might in fact have a sexually selected origin – that is, the

adaptive value of a long and muscular neck may lie more in securing mating opportunities for males than in enhancing access to browse for males and females. It is impossible to determine the validity of this hypothesis, since the evolutionary history of the giraffe lies far behind us, but one immediately obvious problem is the fact that giraffe cows also have long necks and they do not use them at all for fighting.

Giraffes use their excellent vision and height advantage to maintain contact with one another. Vocalizations – at least in the audible range of humans – are rare, and occur only in stressful situations such as when predators threaten. Then adults may bellow or snort, while calves may sometimes bleat. A group of giraffes all standing staring in the same direction, ears forward, is a sure sign that a predator is being monitored. Lions are the main predators of giraffes and are able to pull down even adult bulls. The lions' strategy is to drive their victim through broken ground and thus force it to slow down or lose balance.

Due to the size of the carcass, a lion pride may feed off a giraffe kill for several days, and in Kruger Park giraffe meat makes up 43 percent of the lion diet – more than that of any other prey species. Giraffe males appear to be especially vulnerable to lion predation, with the female:male sex ratio of giraffe kills attributed to lions in the Kruger Park being about 1:1.8. Wandering alone between groups in search of mating opportunities means single males lack the advantage of group vigilance. This contributes to a skewed sex ratio among adults; there are almost twice as many females as males (5:3) in both the Serengeti and Kruger national parks.

Above *Although not notably belligerent :reatures, bull giraffes are extremely powerful, ınd use various tactics to try and knock an ıdversary off balance. For example, as here, ɔne giraffe may maneuver its neck under its ival's leg. Another very effective measure is o deliver a well-aimed head-butt to the ınderbelly.*

Right *"Necking" is ritualized fighting indulged in mainly by young bulls to determine dominance. They use their horns and heads as butts; the necks are slowly intertwined, pushing from one side to the other, like a bout of arm-wrestling. Only the strongest adult bulls get to mate with the females in an area.*

A Case for Game Ranching?

CONSERVATION AND ENVIRONMENT

The shrinking distribution of giraffes in West Africa and in the northern part of their range in East Africa is a cause for concern. The root of the problem is overhunting, combined with habitat degradation in the Sahel, where trees are chopped down for firewood and to make browse accessible to livestock. It is interesting to note that giraffes are among the very last indigenous large mammals to survive in the Sahel. While all other browsers lost out in the competition with domestic livestock, giraffes were able to feed above their heads and so to survive.

In their eastern and southern ranges, the giraffe populations are stable and even expanding due to their popularity on privately-owned commercial game ranches. There is great potential for further expansion, as the giraffe is an ideal candidate for integration into livestock-based animal production systems. In African savannas, pastoral societies commonly keep more animals than can be naturally sustained by the grass layer, resulting in the encroachment of various spinescent woody plant species. These plants are giraffe food, so giraffes could be a wild alternative or adjunct to cattle.

The problem is that giraffes cannot be easily herded, they have a contempt for fences, and one slaughtered giraffe instantly yields a relatively vast amount of meat in one place and time. A rural household cannot consume or cure all this meat quickly enough, and it is difficult for an individual pastoralist to butcher the carcass efficiently and transport all the meat to market before it spoils. Giraffes therefore fall under communal ownership if they occur on communal rangelands and the best short-term strategy for individual farmers is simply to slaughter a giraffe whenever the chance arises. This is the "tragedy of the commons," perhaps the most central and universal dilemma that conservationists and rural development workers are grappling with across the globe. JdT

Okapi

OKAPI JOHNSTONI

The okapi remains one of the major zoological mysteries of Africa, so little is known about its behavior in the wild. The species was only officially "discovered" in 1901 by Sir Harry Johnston, the British explorer, whose interest had been aroused by the persistent rumors of a horse-like animal, living in the forests of the Belgian Congo, that was hunted by the pygmies. Okapi is the name that the pygmies gave to this creature.

Its present distribution is confined to the rain forest of northern Democratic Republic of Congo (DRC), and up to the Uganda border and the Semliki river in the East. About 5,000 of the estimated 30,000 remaining wild okapi live in the Okapi Wildlife Reserve in the Ituri Forest, a remote region that was badly affected by armed conflict in 1997–98.

Like most forest-dwelling large mammals, the okapi is very localized, but in areas of suitable habitat is relatively common. Local population densities of 1–2.5 per sq km (0.4–1 per sq mi) have been suggested. Suitable habitat comprises dense secondary forest where the young trees and shrubs increase food availability. Riverside woodland, and particularly clearings and glades where the penetration of light encourages the production of low browse, is favored.

Characteristic giraffe-like features include skin covered horns and lobed canine teeth, together with a long, extendable black tongue, which is used to gather food into the mouth. Only the males are horned, although females sometimes have a variable pair of horn sheaths.

Differences in behavior and ecology between the two giraffid species arise from differences in the habitat in which they live: open savanna in the giraffe, forest in the okapi. Early reports suggested that okapis, especially the males, were nomadic, though in fact their regular use of tracks linking favorite feeding areas suggests a more sedentary way of life. Unlike giraffes, okapis have glands on their feet, and they have also been reported to mark bushes with urine. These factors, together with their solitary existence, imply a social system based on male territoriality.

In captivity, heat lasts for up to a month, during which time the male consorts closely with the female. Females advertise their condition by urine marking and by calling, although at other times okapis are silent. The exceptionally long period of heat may serve to allow enough time for the male to locate the female and to overcome the species' solitary inclinations before mating. The initial approach and contact phases of okapi courtship are marked by more female aggression and male dominance display than in giraffes, presumably to overcome the defensive reactions of typically solitary individuals. Male courtship behavior is like that of antelopes, including flehmen (lip-curl), display of the white throat patch, leg kicking, and head tossing. Male encounters, particularly in the presence of a female in heat, are frequently aggressive, the ritualized neck fight being reinforced by periodic charging and butting with the horns.

Calves are born with non-adult proportions, having a small head, a short neck, and long, thick legs. The conspicuous mane is much reduced in the adults. Calves remain hidden for the first few weeks. Vocalization is important to maintain contact between the mother and the offspring. In captivity, calves are usually weaned by six months. The horns of males develop between the first and third years. Leopards are the principal predator of calves. Like giraffes, female okapis defend their offspring by kicking with their feet.

The okapi has had protected status since 1933. However, hunting remains a real problem. Subsistence hunting is unlikely to endanger the status of the species, but this is not the case with the bushmeat trade, which now threatens all the larger forest mammals, including the okapi, in the more accessible parts of its habitat. RA

Left *An adult okapi with young. The strong contrast of the zebra-like stripes of the back legs and flanks are probably important for calf imprinting and as a "follow me" signal.*

Giraffes and Plant Defenses

Interactions between giraffes and Acacia trees

A close ecological relationship exists between giraffes and the *Acacia* trees on which they browse. For millions of years, an evolutionary "arms race" has been fought involving adaptations and counter-adaptations on each side. *Acacia* browse forms the staple giraffe diet but *Acacia* trees have physical and chemical defenses against excessive browsing. Sharp thorns and prickles stab, hook, and rip at giraffe noses, lips, and tongues, while some *Acacia* species also have a flat-topped architecture (e.g. the umbrella thorn, *Acacia tortilis*) that prevents giraffes from gaining access to new shoots on the upper canopy. *Acacia* thorns are particularly long and densely packed in the height zone that giraffes have access to, while above this height, where no browsers go, the tree is more palatable and user-friendly. Chemical defenses include various plant compounds such as tannins, which are repellent in taste, are toxic, and can give giraffes indigestion. Various physiological adaptations enable giraffes to overcome these defenses and although they are not all understood they include highly viscous saliva and specialized liver function, together with an ability to discriminate accurately between foliage containing different concentrations of defensive compounds. This ability develops as the young giraffe samples different types of foliage while being weaned. Trial-and-error sampling in small quantities, eating what mother eats, and associating distinctive smells and tastes with the consequences of previous meals, all contribute to the cumulative development of a young giraffe's dietary preferences.

Where giraffe densities are high, such as in the central region of the Kruger National Park (2.5 giraffes per sq km), the impacts of their browsing can be clearly seen on knobthorns (*Acacia nigrescens*), which are favored food plants. They become pruned into characteristic cone and hourglass shapes, producing a topiary effect similar to that achieved by gardeners for ornamentation. Ironically, trees that are most heavily browsed are the least able to defend themselves. Knobthorns exposed to high browsing pressure (around 40 percent of shoots freshly browsed) produce leaves with about half the condensed tannin concentration of unbrowsed trees, probably because they are forced to allocate most available resources to growing back their lost leaves, with little spare for chemical defense. The tendency for giraffes and other browsers, especially impalas, to concentrate their browsing on localized patches of knobthorns is analogous to the tendency for grazing ungulates to feed in mixed herds on "grazing lawns."

Knobthorns come into profuse bloom with creamy-white bottle-brush flowers in the late dry season. At this time of year, when tree foliage is least abundant, the flowers constitute a particularly appealing food source for giraffes. During the six-week flowering peak giraffes move from tree to tree selecting this floral delicacy, which makes up almost one-quarter of their diet during this period. Effective defenses against giraffe browsing are conspicuously absent from knobthorn flowers, which is surprising in view of the fact that knobthorn leaves are comparatively well defended. The exciting possibility remains that giraffes are in fact performing a vital service for the *Acacia*: pollination. On an evolutionary scale, the loss of some flowers from a superabundant and highly ephemeral bloom may be compensated for by the dispersal of pollen, which brushes onto the hairs of the giraffe's head and neck during feeding. Most flowering plants exploit the flying abilities of insects, occasionally birds or bats and rarely small non-flying mammals such as rodents, marsupials, or primates. The latter cannot fly between trees and usually do not travel very far in a day. An average giraffe, however, gets dusted with pollen (and no doubt transfers some) each time its large, furry head brushes through the canopy 4m (13ft) above ground, while visiting at least 100 flowering knobthorns per day, covering 20km (13mi) in the process. The co-evolved relationship between the knobthorn and the world's tallest animal is clearly a prime candidate for further studies on pollination biology. JdT

Above *The distinctive parasol shape of the umbrella thorn* (Acacia tortilis) *restricts giraffes to browsing on the lower canopy.*

Pronghorn

THE PRONGHORN IS THE ONLY LIVING SPECIES of the Antilocapridae, one of just two families of mammals found exclusively in North America (the other is the Aplodontidae). Pronghorns look and behave like African gazelles, which is due to evolution working similarly on these groups. Traditionally they were considered Bovidae due to the similarities, but recent morphological and molecular evidence has reclassified them as a separate family.

Antilocapridae once included two subfamilies: Merycodontinae, and the Antilocaprinae, but the first became extinct by the early Pliocene, about 5 million years ago. Antilocapra is unique among living ruminants in that it has permanent, unbranched horncores, coupled with a branched horn sheath that is shed annually. Current subspecies differ slightly in size, color, and structure, with darker, larger forms living in the north, and paler, smaller forms found in the south.

Long-Distance Runners

FORM AND FUNCTION

The pronghorn is renowned in both folklore and in fact for its speed, endurance, and curiosity. Speeds of up to 86km/h (55mph) have been recorded, and it can maintain 70km/h (45mph) for over 6.4km (4mi). The animals approach and inspect moving objects, even predators, from long distances. Early settlers exploited this curiosity to attract them within gunshot range by flagging (tying handkerchiefs to poles and waving them in the air). Flagging is now illegal, but it contributed to the pronghorn's decline from 40–50 million in 1850 to just 13,000 by 1920.

Pronghorns have stocky bodies with long, slim legs. Their long, pointed hooves are cloven and cushioned to take the shock of a stride that may reach 8m (27ft) at a full run. Both sexes have black horns. The buck's horns have enlarged, forward-pointing prongs below backward-pointing hooks, but does' horns are small. Their large, protruding eyes allow a 360° field of vision. Long, black eyelashes act as sun-visors. The tan-and-white body and neck are related to the pronghorn's use of open prairie and brush lands, and they are important in communication. Bucks have a black facemask and two black patches beneath the ear. These black markings are used in courtship and dominance displays. Scent is important, too. Males have nine skin glands (2 beneath the ears, 2 rump glands, 1 above the tail, 4 between the toes) and females six (4 between the toes, 2 rump). Rump glands produce alarm odors, and scent from glands beneath the ears is used to mark territories and during courtship.

Pronghorn teeth are adapted to selective grazing of a variety of forbs, shrubs, and grasses; they grow continually in response to wear, thus providing an even surface for grinding rough vegetation.

The Politics of Rutting

SOCIAL BEHAVIOR

Both sexes mature at 16 months, but females occasionally conceive at 5 months. Only dominant males breed, and this usually delays a male's first breeding until 3–5 years of age. Twins are common on good range, and females produce 4–7 eggs, but up to four fertilized eggs may implant. When this occurs, the tip of the embryo in the upper compartment of the womb grows and pierces the fetal membranes of the embryo lower in the womb, causing intrauterine siblicide. The 2–3 week rut occurs from late August until early October, depending on latitude. Fawns are born in late May to early June and weigh 3.4–3.9kg (7.5–8.6lb) at birth.

At two days old, a fawn can run faster than a horse, but it lacks the stamina needed to keep up with a herd in flight; therefore, it hides in vegetation until 21–26 days old. For the first few months, the fawns are vulnerable to predators, including coyotes and golden eagles. Fawns interact with their mothers for only 20–25 minutes each day; even after an older fawn joins a nursery herd, this pattern continues. Does nurse, groom, distract predators, and lead their fawns to food and water. Nursing and grooming continue for about 4–5 months, but development of aggressive and sexual behavior (mounting their mothers) causes weaning of males 2–3 weeks earlier than females. Adult bucks show no direct paternal care.

In territorial populations, male pronghorns establish territories (0.2–4.3sq km/ 0.1–1.7sq mi), in early March, and defend them until the end of the rut even though mating only occurs in the fall. Bachelor herds range across 5.1–12.9sq km (2–5sq mi), but the pickings they have access to are not as rich as those of the territorial males. Whether or not a population of pronghorns is territorial seems to depend on rain. Where it is wet (more than 40cm/16in of annual precipitation), pronghorns are territorial, scentmarking, calling, and challenging intruders more fiercely than in arid zones. The extra food available may mean males have more energy for these activities.

Either way, males defending territories containing the most succulent and protein-rich forage get most of the matings (as much as 95 percent of matings occur on territories and, in one study, 8 males out of a population of 40 got 70 percent of these), and some females will return throughout their life to the same territory to find a mate. Does and fawns will wander over 6.3–10.5sq km (2.4–4sq mi) areas encompassing the territories of several males. Therefore, 30–40 percent of all mating may occur on only the best one or two territories in a given area. A top male on a high-quality territory might hold it for 4–5 years before being replaced, and might sire from 15–30 percent of all fawns for a given year. In a given male ageclass, as few as 35 percent of the bucks out of 30–50 males may sire all of that cohort's offspring. As the rut approaches, males associate more frequently with a group of females and defend that group from other males, vigorously chasing away intruders. In arid zones, males reso[rt] to defending groups of females to gain matings rather than territories.

Back from the Brink

CONSERVATION AND ENVIRONMENT

Conservation efforts have brought pronghorns back from near extinction to a population of over 1 million. While this is low compared to their former abundance, it is a significant recovery and provides excellent recreational opportunities for

…unters, photographers, and nature-watchers. …owever, pronghorns today are threatened by oil …xploration, strip-mining for coal, and habitat loss …ue to expanding human populations. Roads built …r oil fields open up otherwise inaccessible areas … poachers, while fences, particularly around key …aling areas, exclude pronghorns and reduce …vorable habitat. Interestingly, although prong…orns can jump these fences, they are loath to do …, which may be related to their evolutionary his…ry on the Great Plains, where there were no ver…cal obstacles. DK/CRM

Above *A group of pronghorns on the move in the snowy, rocky landscape of Yellowstone National Park. The distances that are traveled for foraging varies with the season and the availability of food supplies; during the winter daily travel will be at least four times as far as during the summer.*

Below *Fights between rival territorial males follow quite ritualized patterns. The beginning of such an encounter is characterized by a staring match; if that fails to resolve the dispute, angry vocalizations and chasing will ensue. Occasionally, when all of this fails, fights will occur.*

FACTFILE

PRONGHORN

Antilocapra americana

Order: Artiodactyla

Family: Antilocapridae

Subfamily: Antilocaprinae

4 subspecies: *A. a. americana*, *A. a. peninsularis*, *A. a. mexicana*, *A. a. sonorensis*.

DISTRIBUTION W USA and Canada, parts of Mexico.

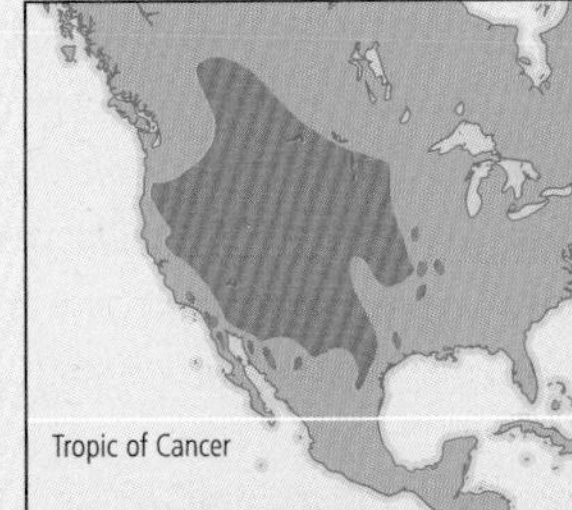

HABITAT Open grass and brush lands, rarely open coniferous forests; active in daytime to twilight.

SIZE **Head–tail length** 141cm (55in); **shoulder height** 87cm (34in); **weight** 47–70kg (103–154lb); **horn length** (female) 4.2cm (1.5in); (male) 43cm (17in). Females are 10–25 percent smaller than males, depending on the population.

COAT Upper body tan; belly, inner limbs, rectangular area between shoulders and hip, shield and crescent on throat, and rump all white. Face mask and subauricular gland on mature males black.

DIET Forbs, shrubs, grasses, and other plants including cacti and domestic crops.

BREEDING Gestation 252 days

LONGEVITY 9–10 years (to 12 years in captivity)

CONSERVATION STATUS Of the 4 subspecies, *A. a. peninsularis* is listed as Critically Endangered and *A. a. sonorensis* as Endangered. *A. a. mexicana* is Lower Risk – Conservation Dependent.

Wild Cattle and Spiral-horned Antelopes

WILD CATTLE ARE LINKED WITH THE FOUR-horned and spiral-horned antelopes in the subfamily Bovinae. At first glance, there seems to be little in common between the three tribes, but cladistic and phenetic analyses have revealed their structural similarities, and for the most part they share a lack of territoriality in their behavior.

Both wild cattle and spiral-horned antelopes evolved from animals resembling the present-day Four-horned antelope and the nilgai (tribe *Boselaphini*). Numerically, the wild cattle are now dwarfed by their domesticated cousins – there are well over a billion Domestic cattle in the world, but only a few dozen Koupreys – and many of the surviving species are under threat. The spiral-horned antelopes have fared somewhat better in recent times, even under heavy poaching pressure, partly as a result of their shy, elusive nature, which has made them one of the least studied antelope groups.

Taxonomic Conundrums

FORM AND FUNCTION

The wild cattle include the largest living members of the Bovidae. Although their evolution from boselaphine ancestors is not in dispute, it is nevertheless unlikely that they form a monophyletic group (a group of species or other taxa descended from a single ancestral species), and a set of parallel evolutions is possible or even likely. Indeed, some biologists think a single tribal name should be used for the Boselaphini and Bovini.

Present-day domestic cattle are descended from the aurochs, which used to range from the Atlantic to the Pacific, and from the northern tundras to India and North Africa (the species was doubtless independently domesticated in several places, and some authorities think that the aurochs of India belonged to a distinct race, or even species). But as agriculture spread, their numbers declined and the last one died in 1627. Bantengs, gaurs, yaks, and water buffaloes have also been domesticated, and with domestication came experimentation. Hybrids between cattle and yaks are economically important in Nepal and China. Attempts have been made in North America to develop a breed based on a bison–cattle cross, called the cattalo, but the male progeny of these hybrids have proved to be infertile. In many places, domesticated animals have been allowed to revert to the wild. In the Northern Territory of Australia, there are thousands of feral water buffaloes and bantengs roaming free, descendants of animals left behind when military settlements were abandoned in the 19th century.

While evolutionary history and tribal affiliation may be of little interest to non-specialists, other taxonomic problems are of more practical importance. For example, American bison are conventionally split into two subspecies, the Wood and the Plains bison. However, recent evidence suggests that environmental factors can account for the differences in coat between the two forms, and genetic data from their mitochondrial DNA also suggests that the species should not be divided. This has serious implications for conservationists

FACTFILE

WILD CATTLE & SPIRAL-HORNED ANTELOPES

Order: Artiodactyla

Family: Bovidae

Subfamily: Bovinae

At least 24 species in 9 genera and 3 tribes. Tribe Bovini (wild cattle) include **Water buffalo** (*Bubalus arnee*); **American bison** (*Bison bison*); **yak** (*Bos mutus*). Tribe Boselaphini (four-horned antelopes) include **nilgai** (*Boselaphus tragocamelus*). Tribe Strepsicerotini (spiral-horned antelopes) include **Common eland** (*Taurotragus oryx*), **Greater kudu** (*Tragelaphus strepsiceros*), and **Mountain nyala** (*T. buxtoni*).

DISTRIBUTION N America to Europe, Africa, China, S and SE Asia, Indonesia, and Philippines.

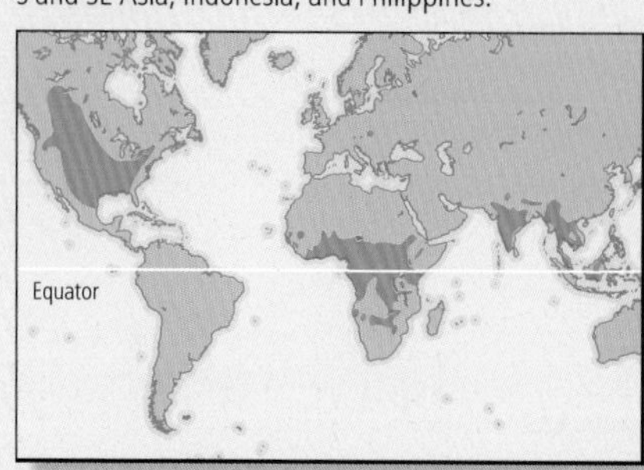

HABITAT Ranges from alpine tundra to tropical forest, sometimes near water.

SIZE **Head–body length** ranges from 1m (3.3ft) in the Four-horned antelope to 2.4–3.4m (8–11ft) in the African buffalo; **weight** from 20kg (44lb) in the Four-horned antelope to 1.2 tonnes in the gaur.

COAT Varies with habitat from the thick, shaggy hair of the mountain-dwelling Yak to the short pelage of the antelopes of the African savanna. Color in the antelopes usually mimics that of their surroundings.

DIET Mostly grasses for the wild cattle; leaves, buds, flowers, fruits, bark, roots, and seed pods for the spiral-horned antelopes.

BREEDING Gestation times average 250–300 days in wild cattle (340 days in the African buffalo) and from 220–270 days in spiral antelopes.

LONGEVITY Wild cattle can live for up to 30 years in captivity, though few reach even half that age in the wild. Among the Boselaphini, nilgai can live for 10 years in the wild, but over 20 in captivity. Spiral-horned antelopes have a maximum longevity in captivity of 20–25 years.

See species table ▷

***Right** An African buffalo attended by oxpecker birds. As well as picking parasites from the hide of the animal, oxpeckers make a hissing sound when alarmed, which may alert the buffalo to danger.*

cause Wood bison are considered to be Endan-red. Furthermore, other evidence has been ken by some biologists to indicate that American d European bison are in fact the same species nd should be placed in the genus *Bos*, not *son*). This would also have serious conservation plications, since the European bison is current- listed as Endangered in the IUCN's Red List, t treating the two bison as one species would most certainly lead to its downlisting – and list-g by CITES or the IUCN determines the extent which trade in a species can occur and how any resources are likely to be made available for conservation.

Unfortunately, bison are not the only problem oup within the Bovini; there is also the question how many anoa taxa there are. The anoas are nall buffaloes that occur only on the Indonesian and of Sulawesi. Conventionally two species ve been recognized, the Lowland and the Mountain anoa, both of which are classed as Endangered by the IUCN. Distinguishing between the two forms has always proved challenging, however, and recent genetic and biochemical research has cast doubt on the validity of the split. Yet the significance of this research is difficult to judge, because the anoas sampled were all zoo animals, some of which may have been descended from hybrids. More recently, other researchers on Sulawesi have suggested that there may in fact be a third type of anoa.

To add to the taxonomists' problems, a new species of bovid came to light in Indochina in the 1990s. The Vu Quang ox or saola (meaning "weaving spindle," from the shape of its horns) came to the attention of scientists in 1992. It differs significantly from all other bovid genera in appearance, morphology, and DNA sequencing. The species was subsequently named *Pseudoryx nghetinhensis*. On the basis of a preliminary analysis of mitochondrial DNA extracted from a smoke-dried skin sample, as well as other, morphological evidence, it was placed in the subfamily Bovinae. More recent molecular analyses have demonstrated unambiguously that this attribution was correct.

A second supposed "discovery" of a hitherto unknown Indochinese species presents a cautionary tale for the overzealous researcher. Zoologists traveling in Vietnam and Cambodia in 1993 found sets of unusual, spiral horns for sale in rural markets and concluded that they came from an elusive bovid long rumored to live in the forest, and known in the Khmer language as the Khting vor. The horns were found to match ones collected by hunters in the same region in the 1930s, which were identified at the time as coming from Kouprey. Although local anecdotal evidence abounded on the Khting vor, no biologist ever sighted the animal. Nevertheless, the "new"

species was given the scientific name *Pseudonovibos spiralis*, and assigned variously by taxonomists to the subfamilies Caprinae or Bovinae. However, close examination of horn samples by French researchers in 2000 found that they were forgeries, made by heating and twisting cattle horns.

The spiral-horned antelopes are essentially sexually dimorphic. The males bear arms for battle, an adaptation to their life in dominance hierarchies with other males with little or no territoriality. The females are more generic in form.

The spiral horn – a layer of keratin (rudimentary horn) over a bony core – is swept back with a dramatic corkscrew in some species, following the plane of the face. It varies in length among species, from 20cm (8in) in the bushbuck to nearly 1m (3.3ft) in the Giant eland or Greater kudu. The size and length is partly a function of body size, but the twisting is a throwback, possibly to four-horned ancestors in Asia. The fusing of the two sets of horns may have been the mechanism behind the spiral, the main driving force being keratin on the frontal ridge of the skull growing at a different rate to the main horn, producing a corkscrew effect. When a kudu is seen running through bushland, with its head extended and horns lying along its back without ever snagging or jarring against the thorny brush, the design makes sense.

When the antelopes are attacked, the horn becomes a formidable weapon, capable of defending almost every part of the body. In confrontations with other males, it accentuates body size and is an important factor in establishing dominance. In contrast, the shorter, straighter, and more robust horns of the Common eland are more suited to the open and better adapted to the close combat involved in dealing with savanna predators such as lions.

The antelopes possess superb hearing and sharp eyes, but have a surprisingly poor sense of smell. More graceful in appearance than their bovine cousins, they have deep, narrow chests with high heads and long, tapering legs, distinctive white stripes on the face, neck, and chest, and socks above the hooves; the crests and dewlaps are more prominent in males. The rich coat color of many of these bovids, varying from red to almost black in the bushbuck, mimics the lush vegetation and dappled light of the forest, riverine thicket, or swamp areas where they live. The gray-coated kudus are entirely lost against the leafless scrub characteristic of *Acacia* woodland for much of the year, as is the Mountain nyala amongst the heathers and junipers 3,500m (11,500ft) up. The eland's fawn coat blends superbly with the brown and yellow mosaic of grassland colors.

Running is not a priority for the antelopes, for which camouflage is often more useful than prolonged flight at speed. The eland, however, is a superb long-distance runner, even though fully-grown adults weigh well over half a tonne.

Excepting deserts, the antelopes have representatives in almost every major habitat on the African continent, though the transition zone between woodland and savanna is the niche they have most successfully exploited. The different species have distinct environmental preferences. Sitatungas seek out reedbeds; Lesser kudus favor the arid acacia woodland of the Horn of Africa, while the Greater kudu is most often found in wooded hill country throughout South, East, and Central Africa; the nyala inhabits thick vegetation in sub-Saharan Africa and bushy low lying zone in southeastern Africa, but the Mountain nyala is restricted to the montane mosaic of pasture, forest, and scrub in a single zone in Ethiopia. The bongo, nocturnal in its habits, hides among shrubs and low vegetation in glades and on forest margins, retreating in daylight to deep-canopy forest; it is a surprisingly large ruminant to occupy such a habitat.

Grazing, Browsing, and Wallowing

DIET

The wild cattle are predominantly grazers, although browse is an important food source for some species, especially during periods of limited grass availability; and all are ruminants. Rumination was a key adaptation in herbivore evolution, allowing cellulose, the main constituent of plant material, to be converted into digestible carbohydrates more effectively than by any other method. The ruminant stomach is very versatile and exists in many forms, each specialized to a particular diet. This versatility has allowed the ruminants to occupy a greater diversity of ecological niches than any other group of herbivores.

The Bovini's broad array of incisors and massive molars enable even the largest species to consume and process sufficient quantities of coarse vegetation. Their tongues are prehensile, permitting more efficient foraging, since large bunches of grass can be torn off at one time. Bovine species are less efficient at feeding on very short grass, however, and may be outcompeted in such areas by smaller herbivores. All species are more or less water-dependent, lacking efficient means of conserving liquid and so needing to drink regularly (although yak and bison can meet their water requirements by eating snow).

Several species also wallow, most notably the Water buffalo. Where large pools are not available, Water buffalo make use of smaller depressions, and they can rapidly churn the ground to a muddy consistency. Wallowing has a thermoregulatory function; buffaloes have fewer sweat glands per unit area of skin than true cattle and evaporative cooling through sweating is of little importance. Mud is more effective than clean water for cooling animals because the water in it evaporates slowly, prolonging the cooling effect. Some reports suggest that wallowing is essential for the water buffaloes' survival, but work on feral animals in

Last Remaining Survivors

Although ancestral members of the Boselaphini tribe may have been the evolutionary progenitors of the entire Bovinae subfamily, only two species presently survive. One of these, the Four-horned antelope or chousingha, which is found in India and Nepal, is now listed as Vulnerable, with less than 10,000 animals remaining. The nilgai or bluebuck is faring somewhat better. It has two centers of distribution, one in eastern Pakistan, India, and Nepal, and the other (an introduced population) in southern Texas. Numbering about 8,000–9,000 individuals, this latter population may well contain as many animals as the entire Asian community.

The nilgai is a medium-sized antelope, which is about 2m (6.5ft) long and stands 1.4m (4.5ft) high at the shoulder. Some individual males may weigh up to 300kg (660lb). The forelegs are noticeably longer than the hind limbs, which gives it a distinctive, sloping profile. In its natural habitat, it haunts forests and low jungle, only occasionally venturing onto open plains. Like all antelopes, it is a fast runner and has been recorded as achieving speeds of up to 48km/h (30mph) when in flight.

Even so, the nilgai is reputed to tame easily. In India, it was traditionally regarded as a close relative of the sacred cow – hornless female nilgais in particular, have a cattle-like appearance – and it long benefited from some of the immunity from persecution that Hinduism confers on that animal. Under those circumstances it thrived, and was reported to be surprisingly at ease in the presence of people. However, throughout the 20th century its numbers declined as a result of increased hunting and habitat loss, and it is now regarded as Conservation Dependent.

Below *Some representative species of wild cattle and spiral-horned antelopes (all animals shown are males).* **1** *Nilgai* (Boselaphus tragocamelus) *running;* **2** *Auroch, from which all present-day domestic cattle are descended;* **3** *Common eland* (Taurotragus oryx; *East African race);* **4** *Wild water buffalo* (Bubalus arnee) *– this species has the longest horns of any bovid;* **5** *American bison* (Bison bison) *bellowing during a broadside threat to another male.*

Australia does not support this conclusion; instead the results indicate that when buffaloes are deprived of wallows, they behave more like cattle and seek shade. Wallowing has the additional function of providing a thick coating of mud on the body that protects the animals from insect bites; indeed, this probably accounts for the buffaloes' notable resistance to tick-borne diseases.

Leaves, buds, flowers, fruits, bark, roots, and even seed pods provide the selective browse taken by the four-horned and spiral-horned antelopes. Only the eland and Mountain nyala show any tendency to exploit grasslands, but they are more picky than most true grazers. They too are highly dependent on the remarkable ruminant stomach to survive the vagaries of climate despite living at relatively low population densities. The efficient digestion it provides has also allowed them to occupy high-altitude niches where few other large mammals in Africa survive.

Herds and Aggregations

SOCIAL BEHAVIOR

The wild cattle mostly live in herds, although this is less true of the tamaraw, while the anoas and the saola are more or less solitary. In the herd-dwellers females spend their lives in groups of stable composition that consist of cows, their offspring, and perhaps further generations. Young bulls generally leave at about 3 years. These groups are small in the gaur, the banteng and the forest-dwelling African buffalo, typically containing up to 12 animals. The American and European bison usually live in larger groups of 20–60. In all these species, temporary aggregations may form, particularly in the breeding season or on particularly favorable feeding grounds. The Cape subspecies of the African buffalo lives in particularly large herds that in some areas may contain more than 1,000 animals. Adult bulls are mostly solitary or live in loose associations known as "bachelor herds," but associate with the maternal herds when females are likely to be in estrus.

Males need to be able to assess each other's fighting potential in order to avoid dangerous fights between adversaries of comparable status. Their facial musculature does not permit a wide range of expressions, so posture and movement are used as guides instead. The few contests that escalate to combat may result in serious injuries.

The horns of bovids are very effective antipredator defenses. A herd of Cape buffalo is quite capable of attacking lions if threatened; the buffalo herd is such an effective defensive unit that blind, lame, or even three-legged individuals may continue to thrive within it. Tigers, however, frequently kill fully-grown gaur, and solitary African buffalo bulls often fall victim to lions.

The advantages of living in herds are in fact n[ot] restricted to defense against predators. Herds ar[e] also effective ways of sharing information about the best places to feed. African buffaloes are eve[n] thought to "vote" by adopting a special posture that indicates the way that the herd should hea[d] following a period of resting and rumination.

The spiral-horned antelopes have evolved fle[xi]ble social lives. Sitatungas, bushbucks, nyalas, a[nd] Lesser kudus have foregone the expense of territ[o]ry defense despite normally remaining resident i[n] areas ranging from 1–500ha (2.5–1,250acres) i[n] extent. Elands, Greater kudus, Mountain nyalas and bongos are more fluid in their movements; eland home ranges stretch to 1,500sq km (600 s[q] mi) or more. Associations of animals of both se[xes] and all ages are generally loose (except for the ti[e] between mother and calf) and usually familial o[r] a result of seasonal factors and food availability.

Aggregations form particularly when vegetati[on] is lush and green shoots are abundant. On rare

Below A nyala male, with female and young. Males of this species usually avoid one another; her[e] the male goes through his display ritual, raising his crests and fringes to make himself look larger.

occasions, eland herds can number several hundred animals, usually as a defensive strategy when feeding in open habitat. Young males will often form loose herds, but older males are mostly solitary, and in some species – like the kudus and the nyalas – they actively avoid one another; if they are forced together by the presence of an estrous female, conflicts will arise.

Male nyalas have the most expressive display forms amongst bovids. They blow themselves up like puffer fish, with fringes and crests raised and the dark gray body presented broadside to the rival. The kudu and Giant eland tend to emphasize height and horn in strutting displays, while the Common elands use their extraordinary bulk to intimidate their rivals. Male bushbucks and elands form clear dominance hierarchies through highly ritualistic sparring, which is less lethal and destructive than that of their bovine cousins. As with bovines, they do not possess specialized scent glands and rely on visual clues for communication. Courtship is a simple matter of the peripatetic pursuit of estrous females by the males. When contact is made, the males vocalize with a variety of bleats and clicks, and badger the female until copulation is accepted. Loose bonds can form between the courting couple at this time.

Gestation varies from 6 to 9 months, increasing in line with the species' body size. Birth peaks occur to coincide with the flush of new vegetation from the rains. For several weeks, the female must daily leave her young one hiding alone while she goes off to feed. It is common for the young to subsequently form crèches. Longevity also varies with body size, from 10 to 20 years on average.

Mixed Fortunes

CONSERVATION AND ENVIRONMENT

Some of the wild cattle species were once exceedingly numerous. In 1898, a traveler in Tibet reported that "on one green hill there were, I believe, more yak visible than hill," while the American bison probably once numbered in the tens of millions. By contrast, there are some species that were probably always rather rare. The gaur and banteng need grassy glades in forests, which were probably uncommon before man started practicing shifting agriculture.

Most wild cattle and buffalo species are now seriously threatened. Tamaraw, kouprey, and khting vor are all thought to be teetering on the brink of extinction. True Wild water buffalo may already have been lost through interbreeding with domestic and feral livestock. Hunting and the illegal trade in trophies threaten banteng and gaur; and the anoas are threatened by hunting for their meat and by habitat loss. Even wild yaks in the remote wastes of the Tibetan Plateau are not safe; they are threatened by commercial hunting for meat, by diseases spread by domestic livestock, and by interbreeding with domestic yaks. While African buffaloes are much more numerous, they have

Above *The astonishing horns of the male Greater kudu can reach lengths of over 1m (3ft), from head to tip, even when measured in a straight line; they are perhaps 30 percent longer measured along the curves.*

still been greatly reduced in number by hunting and habitat loss, and in several southern localities may never have recovered from the devastating rinderpest epidemic of the 1890s. For these reasons, the survival of most of the wild Bovini can only be assured in properly protected reserves. Indeed, the European bison died out in the wild in 1919, and the present thriving populations in Poland and the Caucasus were built up from zoo stocks that were released back into the wild.

There is also a need to conserve the rare and vanishing breeds of Domestic cattle. Together with the wild species, these comprise genetic banks that may well be a valuable source of new varieties and hybrids.

Even with the dramatic changes that have occurred in the African environment this century, the spiral-horned antelopes are faring better than many of their open-country cousins. In some cases where bush encroachment or land-use change has followed upon mismanagement of grazing or farming lands, they have actually recolonized lands from which they had been driven.

Through boredom or greed, younger African farmers sometimes bring automatic rifles to bear on wild ruminants, to the dismay of many older members of the community, who remember a time when such animals were only exploited as a food reserve in periods of drought and livestock death. Yet except for the eland, which has been known to jump a 2m/6.6ft-high fence to raid vegetable crops, the antelopes rarely clash with man, since they do not compete with livestock for food.

A greater threat comes from the bushmeat trade. The scale of bushmeat consumption is

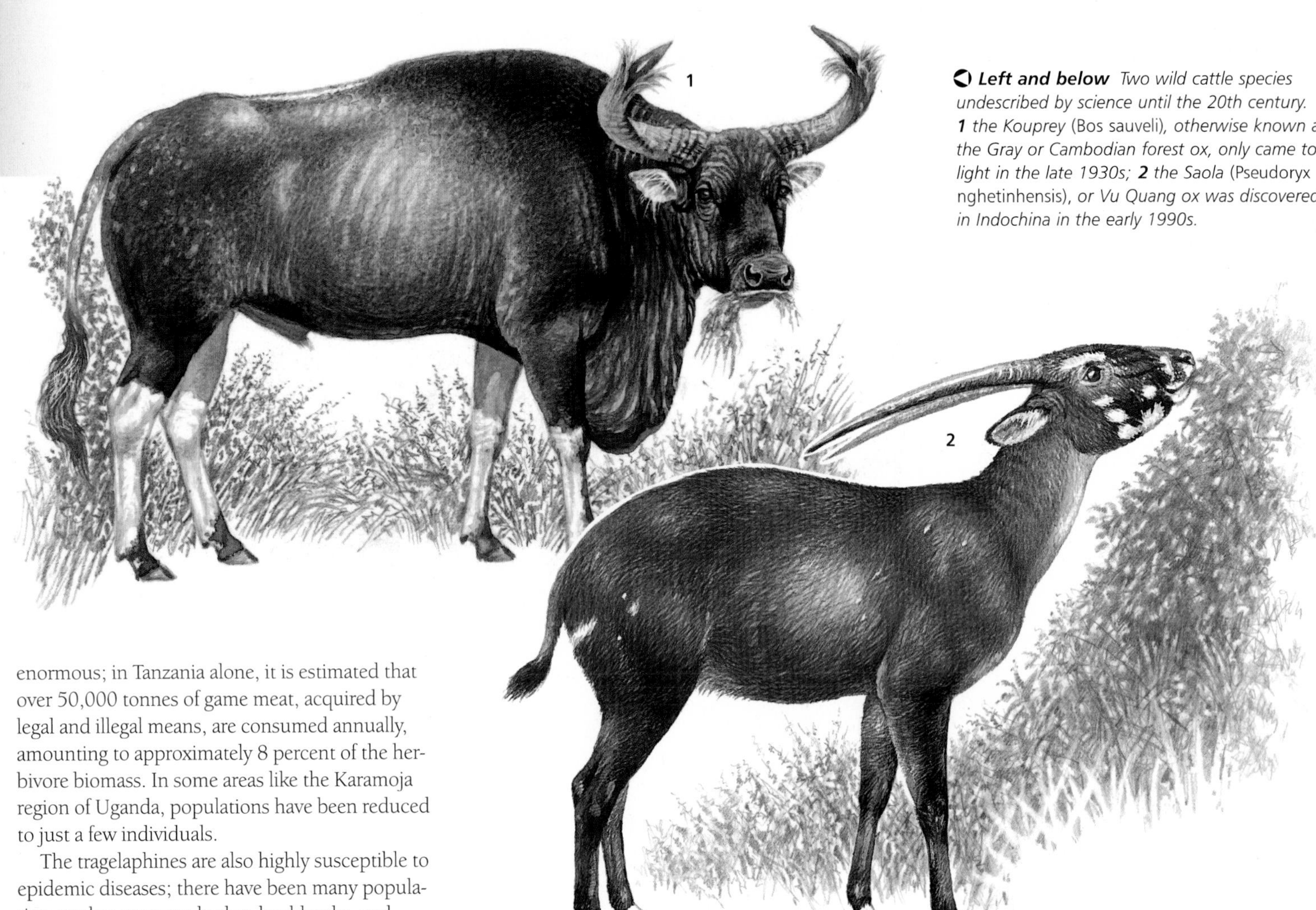

Left and below *Two wild cattle species undescribed by science until the 20th century.* ***1*** *the Kouprey* (Bos sauveli), *otherwise known as the Gray or Cambodian forest ox, only came to light in the late 1930s;* ***2*** *the Saola* (Pseudoryx nghetinhensis), *or Vu Quang ox was discovered in Indochina in the early 1990s.*

enormous; in Tanzania alone, it is estimated that over 50,000 tonnes of game meat, acquired by legal and illegal means, are consumed annually, amounting to approximately 8 percent of the herbivore biomass. In some areas like the Karamoja region of Uganda, populations have been reduced to just a few individuals.

The tragelaphines are also highly susceptible to epidemic diseases; there have been many population crashes amongst kudus, bushbucks, and other tragelaphines as a direct result of such outbreaks. In recent years, for example, Lesser kudus have died in their thousands in East Africa from disease alone; anthrax, introduced by cattle, virtually wiped out the population in Mago National Park, Ethiopia, in 1999, and it was estimated that 90 percent of the species in the Tsavo ecosystem died from rinderpest in 1994–95. Fortunately, numbers recover quickly as long as enough survive to reproduce and the habitat is intact, but this added pressure is contributing to the downward trend in populations in Africa.

The bushbuck remains one of the most abundant antelope species in Africa, with probably over 300,000 individuals spread through 35 countries. Sitatungas are more restricted in distribution and in habitat preference, and so are more vulnerable. Significant populations of over 100,000 animals remain in much of Central Africa, but are likely to lose out as humans divert more water resources for their own ends.

Nyala exist only in Southern Africa. With a population in excess of 26,000 individuals, many in well-protected areas, the species' future is fairly secure, although conservation-dependent. Mountain nyala are restricted to the Bale and Arsi mountain ranges, east of the Rift Valley in the Ethiopian highlands. From near extinction in the early 1990s in the wake of political upheaval, they have won increasing tolerance from the human population, which has led to a recovery in numbers from a few hundred to perhaps 2,000 individuals. Their limited range and close association with humans and cattle make disease and resource competition their biggest threats.

The world's 8,000 or so remaining Lesser kudus are confined to the Horn of Africa. They live among harsh, dry *Acacia melliphora* bushland, which reduces the risk of decline as a consequence of human development, at least in the short term. However, rinderpest, a cattle disease, remains a major threat. The Greater kudu has the ambivalent honor of being a trophy animal shot for sport, which has enhanced its status among landowners, who are able to make a living from incoming hunters. Space has thus been made for this species, whose population in Africa remains around 300,000. The situation in Eastern and West-Central Africa is less secure as a result of disease, poaching, and habitat destruction.

The Common eland is not as common as it once was, having lost over half its former range since 1970. Nonetheless, it remains viable, with over 50,000 head. The Giant eland population is far more vulnerable; fewer than 5,000 individuals remain, spread in fragmented pockets across West-Central Africa. Ironically, safari hunting may be the short-term key to the species' survival, as it is one of the most prized trophy animals in Africa.

Bongo populations are poorly monitored and numbers are unknown, but the species' range is still quite extensive across 10 countries in West-Central Africa, with a small extant population in the Aberdares and Mount Kenya in East Africa. This mountain subspecies is endangered and may become extinct.

Conservationists remain divided as how best to conserve these prized species, whether they are valued for their meat, photographic or hunting tourism, or for their intrinsic value. Some say the answer lies in total protection away from human intrusion; others recommend community participation in conservation efforts, even if this takes the form of sustainable use and consumption. There probably is no single solution in any one country, and many avenues must be tried. Since the fate of these beautiful creatures lies in the hands of some of the poorest and most volatile communities in the world, every effort must be made through education and economic support to improve the lot of both the human and animal population. Only then can we hope to see the antelopes survive into the future. TS/RK/SH

Wild Cattle and Spiral-horned Antelope Species

WILD CATTLE, FOUR-HORNED ANTELOPES, and spiral-horned antelopes comprise three tribes that initially appear quite diverse; however, scientific analyses have shown that there are many structural similarities between the tribes.

TRIBE BOSELAPHINI

Both species in peninsular India. Relicts of the stock from which the Bovini arose. They have primitive skeletal, dental, and behavioral characteristics; females not horned.

Genus *Tetracerus*

Four-horned antelope Vu
Tetracerus quadricornis
Four-horned antelope or chousingha

Wooded, hilly country near water. HBL 100cm; TL 13cm; HT 60cm; HL male: front pair 2–4cm, rear 8–10cm; WT 20kg. COAT: male dull red-brown above, white below, dark stripe down front of each leg; old males yellowish; females brownish-gray. Facial and foot glands.

Genus *Boselaphus*

Nilgai LR
Boselaphus tragocamelus
Nilgai, Blue bull, or bluebuck

Thinly wooded country. Rather horselike. HBL 200cm; TL 45–50cm; HT 120–150cm: HL male: 15–20cm; WT male: 240kg; female: 120kg.
COAT: coarse iron-gray; white markings on fetlocks, cheeks, ears; tuft of stiff black hairs on throat; young bulls and cows tawny.
CONSERVATION STATUS: LR Conservation Dependent.

WILD CATTLE

Tribe Bovini

N America, Africa, Europe, Asia, Philippines, and Indonesia. Stout bodies, low, wide skulls, and smooth or keeled horns in both sexes, splaying out sideways from the skull. No facial or foot glands. Sexual maturation and frequency of calving depend to a degree on health and feeding, but all those which have been studied can achieve maturity at 2 years and produce a calf a year for most of their life. Males may also be able to fertilize at 2 years, but are usually prevented by social factors. Longevity is around 15–20 years; this, and most of the body dimensions quoted, may seldom, if ever, be achieved in most wild populations now.

Genus *Bubalus*

Stout, dark-colored bodies, hair generally sparse. Horn cores triangular in cross section, except in Mountain anoa. No boss of horn in either sex; no dewlaps or hump.

Wild water buffalo En
Bubalus arnee (bubalis)
Wild water buffalo, Asian buffalo, carabao, or arni

Very widespread (Asia, S America, Europe, N Africa) as domestic form, of which there are two groups of breeds, the River and the Swamp buffaloes. Latter is feral in N Australia. True Wild water buffalo are thought to survive in India, Nepal, Bhutan, and Thailand; free-living water buffalo of uncertain provenance and/or pedigree occur in Sri Lanka, Burma (Myanmar), Indochina, Indonesia, and Malaysia. Near (and in) large rivers in grass jungles and marshes, riparian and other forest types. HBL 240–280cm; TL 60–85cm; HT 160–190cm; WT male: 1,200kg; female: 800kg. FORM: coat slaty black, legs dirty white up to hocks and knees; white chevron on lower neck. Flexible fetlock joints make the buffalo nimble in the mud; horns heavy, backswept.
BREEDING: gestation 300–334 days.

Lowland anoa En
Bubalus depressicornis

Endemic to the Indonesian island of Sulawesi. A miniature water buffalo; little-known, but reportedly has more of a requirement for undisturbed forest than the other species of Asian wild cattle. Mainly solitary, unlike other wild cattle.

Below *A yak in the rocky terrain of the Nepalese Himalayas. The yak's dense, hairy coat helps it to conserve heat in this inhospitable environment.*

HBL about 180cm; TL about 40cm; HT about 86cm: HL male: 27–37cm, female: 18–26cm.
FORM: coat black, sparse hair and white stockings; young woolly, brown. Horns flat and wrinkled.
BREEDING: gestation 275–315 days.

Mountain anoa En
Bubalus quarlesi

Endemic to the Indonesian island of Sulawesi. Adults look like juvenile Lowland anoa. Little-known, but reportedly has more of a requirement for undisturbed forest than the other species of Asian wild cattle. Mainly solitary, unlike other wild cattle.
HBL about 150cm; TL about 24cm; HT about 70cm; HL 15–20cm.
FORM: legs same color coat as body, which is brown–black; woolliness persists into adulthood. Horns smooth and circular.
BREEDING: gestation 275–315 days.

Tamaraw Cr
Bubalus mindorensis
Tamaraw or tamarau

Found on the island of Mindoro (Philippines). Less solitary than anoas, tamaraws occur in small groups. Reputedly nocturnal and aggressive, they were once relatively tame and could be seen grazing in the open during the day; hunting pressure led them to become increasingly nocturnal.
HT 100cm; HL about 35–50cm; WT male: about 274kg, female: less than 260kg.
FORM: coat dark brown-grayish black; horns stout.
BREEDING: gestation about 317 days.

Genus *Bos*

Dark, short-haired coat (excepting yak and some Domestic cattle). Horn cores circular or oval in cross section. No boss of horn in either sex; dewlaps and humps well developed in some species.

Banteng En
Bos javanicus
Banteng, tsaine, tembadau

Isolated populations in Burma (Myanmar), Thailand, and Indochina, and on the islands of Borneo, Java, and Bali (domesticated as Bali cattle). Feral in N Australia. Quite thick forest with glades. Very like Domestic cow in general proportions.
HBL (mainland race) 190–225cm; TL 65–70cm; HT 160cm; WT 600–900kg.
COAT: adult bulls dark chestnut, black in the Javan and Bornean races; young bulls and cows reddish brown; all have white band around muzzle, white patch over eyelids, white stockings, and white rump patch; males have bald patch between horns, and a dorsal ridge and dewlap.
BREEDING: gestation 285–300 days.

Gaur Vu
Bos frontalis
Gaur, Indian bison, or seladang

Significant populations in India, smaller populations in Nepal, Bhutan, Bangladesh, Burma (Myanmar), China, Thailand, Indochina, and W Malaysia. Tropical forest with glades, savanna forest.
HBL 240–300cm; TL 70–105cm; HT 170–200cm; HL (male) up to 80cm: WT male: 800–1,225kg, female: about 700kg.
FORM: adult bulls shiny black with gray boss between the horns and white stockings; young bulls and cows dark brown, also with white stockings. Huge head, deep massive body, and sturdy limbs. Hump formed by long extensions of the vertebrae; small dewlap below the chin and a large one draped between the forelegs; the horns sweep sideways and upward.
BREEDING: gestation about 270 days.

Yak Vu
Bos mutus (grunniens)

Wild yak survive in the alpine tundra and cold desert regions of the N Tibetan Plateau, Xinjiang, and Qinghai – almost uninhabited mountainous regions of China at 4,000–6,000m altitude.
SH male: 203cm, female: 156cm; HL male: 80cm, female: 51cm; WT male: 821kg, female: 306kg (all figures come from a 1938 report).
FORM: shaggy fringes of coarse hair about body, with dense undercoat of soft hair; coat blackish brown with white around the muzzle. Massively built, with drooping head, high humped shoulders, straight back, and short sturdy limbs.
BREEDING: gestation probably 255–304 days.

Common cattle
Bos taurus

Present-day common cattle are descended from the aurochs or urus, which died out in 1627. All cattle are completely interfertile, but the skull and blood proteins of the humped zebu cattle are different from those of the humpless breeds; perhaps the domestications were of two races of the aurochs. The hump of the zebu, unlike those of the other species mentioned in this table, is composed of muscle and fat, and is not simply the result of long processes on the vertebrae. Domestic cattle are feral in many places, the most notable being Chillingham Park, NE England. Cattle exist in many sizes and forms. There are long-horned and polled (hornless) breeds.
SH 180–200cm; WT 450–900kg.
BREEDING: gestation around 283 days.

Kouprey Cr
Bos sauveli
Kouprey, Gray ox, or Cambodian forest ox

Not described by scientists until 1937. Probably extinct outside of Cambodia, and almost extinct there. Formerly occurred in forests and wooded savannas of Indochina.
HBL 210–225cm; TL 100–110cm; HT 150–200cm; WT 700–900kg.
FORM: old bulls black or very dark brown, though they may have grayish patches on body; cows and young bulls gray, with underparts lighter and chest and forelegs darker; both sexes have white stockings. Bulls have very long (over 40cm) dewlap hanging from neck; dorsal ridge not well developed. Horns in female are lyre-shaped; in males, horns curve forward and round, then up; horn tips are frayed.

Genus *Pseudoryx*

Saola En
Pseudoryx nghetinhensis
Saola or Vu Quang ox

Discovered by science through horn samples in Vu Quang Nature Reserve in1992; first live animal captured in 1994. Dense forest in Annamite Mountains of WC Vietnam and Laos, from elevations of 200–2,000m (660–6,600ft).
HBL 150–200cm; SH 80–95cm; HL 32–50cm; TL 13–15cm; WT c.100kg.
COAT: reddish brown to dark brown, blackish legs, with black stripe running the length of the back. Buff stripes on face in both male and female, white patches on muzzle and throat area.

Genus *Syncerus*

Physiological and hair characteristics, as well as behavioral traits, differentiate this genus from *Bubalus*. Genetic data also su[pport] the separation of the two genera.

African buffalo
Syncerus caffer

Africa S of the Sahara. Generally considered as two subspecies: the Cape buffalo (*S.c. caffer*) and the Forest buffalo (*S. c. nanus*), with intermediate forms. The former lives in savannas and woodlands, th[e] latter in forests nearer the Equator.
HBL 240–340cm (Cape buffalo), 220cm (Forest buffalo); TL 75–110cm (Cape buffalo), 70cm (Forest buffalo); HT 135–170cm (Cape buffalo), 100–120cm (Forest buffalo); HL 50–150cm (Cape buffalo), 30–40cm (Forest buffalo); WT male: 680kg, female: 576kg (Cape buffalo), male: 270–320kg, female: 265kg (Forest buffalo).
FORM: coat reddish brown in Cape buffalo, brownish-black in Forest buffalo; heavy bodied, thick-necked; horns triangular in cross section; heavy boss between horn bases.
BREEDING: gestation 340 days.
CONSERVATION STATUS: LR Conservation Dependent.

Genus *Bison*

Reddish brown–dark brown coat; long shaggy hairs on neck and head, shoulders and forelegs, and a beard. Short and broad skull, dorsal hump formed by processes of the vertebrae. Smooth horns, circular in cross section, about 45cm long. The two species are completely interfertile. Some authorities, on the basis of molecular DNA, chromosomal, immunological, blood typing, and protein sequence data, have suggested that both should be included in the genus *Bos*.

American bison
Bison bison
American bison or buffalo

Widely considered to be two subspecies, the Plains bison (*B. b. bison*) and the rather larger and darker Wood bison (*B. b. athabascae*), which lives further N; the validity of these subspecies is, however, questionable. N America. Grassland, aspen parkland, coniferous forests. Associated now with prairies, but inhabited forests as well before its virtual extermination last century. Now mainly in parks and refuges.

ABBREVIATIONS
HBL = head–body length TL = tail length SH = shoulder height HT = height
HL= horn length WT = weight
Approximate nonmetric equivalents: 10cm = 4in; 1kg = 2.2lb

Ex Extinct
EW Extinct in the Wild
Cr Critically Endangered
En Endangered
Vu Vulnerable
LR Lower Risk
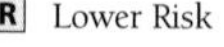

L 380cm; TL 90cm; HT male 195cm; WT ale: 818kg, female: 545kg.
EEDING: gestation 270–300 days.
ONSERVATION STATUS: LR Conservation ependent.

ropean bison En
son bonasus
ropean bison or wisent

came extinct in the wild in 1919 (Russ-–Polish border), but was re-established Bialowieza Primeval Forest in Poland, d later in the Caucasus and elsewhere in ssia. Mixed woods with undergrowth d open spaces.
L 290cm; TL 80cm; HT 180–195cm; WT ale: 800kg.
EEDING: gestation 254–272 days.

SPIRAL-HORNED ANTELOPES

Tribe Strepsicerotini

stricted to Africa. Medium to large body e, more slenderly built than the Bovini, th long necks and deep bodies. Adult ales larger than females; sexes also differ markings and horn structure. Horns in ales only in most species. No distinct cial or foot glands.

enus *Tragelaphus*

tatunga LR
agelaphus spekeii
atunga or Marshbuck

vamps, reedbeds, and marshes of the ctoria, Congo, and Zambezi–Okavango er systems.
L male: 150–170cm, female: 135–5cm; TL 20–25cm; HL 45–90cm (male); T male: 80–125kg, female: 50–60kg.
RM: coat shaggy and slightly oily; males lighter and redder in color than e yellowish to dark-gray-brown males; ots and up to 10 white stripes on the dy; dorsal crest runs the length of the dy and is erectile; white patches on roat, white spots on cheeks. Long, layed hooves distribute weight, allowing e animal to walk through mud without nking into the ground.
ONSERVATION STATUS: LR Near Threatned.

yala LR
agelaphus angasii

verside thicket and dense vegetation in Africa.
BL male: 210cm, female: 179cm; TL ale: 43cm, female: 36cm; SH male: 2cm, female: 97cm; HL 65cm (male); T male: 107kg, female: 62kg.
OAT: shaggy, dark gray-brown, particularly along the underside of the body and throat; usually several poorly marked white vertical stripes; long, conspicuous erectile crest, brown on the neck and white along the back; legs orange; white chevron between the eyes; horns lyre-shaped with a single complete turn, black with whitish tips; females much redder with clearly marked white stripes, short coats, a less obvious chevron, and generally resemble bushbuck females.
CONSERVATION STATUS: LR Conservation Dependent.

Bushbuck
Tragelaphus scriptus

Locally throughout Africa S of the Sahara, except for the arid SW and NW regions, in a wide range of habitats whose common feature is dense cover.
HBL male: 115–145cm, female: 110–130cm; TL male: 20–24cm, female: 20–24cm; HL 25–57cm (male); WT male: 30–75kg, female: 24–42kg.
COAT: short, varying from bright chestnut to dark brown, with white transverse and vertical body stripes being either clearly marked, broken, or reduced to a few spots on the haunches; black band from between the eyes to the muzzle; white spot on cheek, two white patches on the throat; adult males are darker than females and young, especially on the forequarters, and the erectile crest is more prominent.

Mountain nyala En
Tragelaphus buxtoni

Highland forest and heathland of the Arusi and Bale Mountains in Ethiopia.
HBL 190–250cm; HL up to 80cm.
FORM: shaggy, grayish-brown coat; about 4 ill-marked, vertical white stripes; white chevron between the eyes; two white spots on the cheeks; two white patches on the neck; short white mane continued as a brown and white crest. Horns one to one-and-a-half fairly open turns. In many ways resembles the Greater kudu more closely than the nyala.

Lesser kudu LR
Tragelaphus imberbis

Thicket vegetation in Ethiopia, Uganda, Sudan, Somalia, Kenya, and N and C Tanzania.
HBL 160–175cm; TL 26–30cm; HL 60–90cm; WT male: 90–110kg, female: 55–70kg.
FORM: coat sleek and short haired, brownish-gray, with 11–15 clearly marked vertical white stripes; head darker, with incomplete white chevron between the eyes; two white patches on the neck; male's dorsal crest extends forward into a short mane; tail bushy; small but clear spots on cheeks; reddish tinge to legs; female slightly more reddish than male. Horns 2–3 open spirals.
CONSERVATION STATUS: LR Conservation Dependent.

Greater kudu LR
Tragelaphus strepsiceros

Woodland, especially in hilly, broken ground, in E, C, and S Africa.
HBL male: 190–250cm, female: 190–220cm; TL 37–48cm; HL 100–180cm (male); WT male: 190–215kg, female: 120–215kg.
COAT: short, blue-gray to reddish brown; 6–10 vertical white body stripes; white chevron between eyes; up to 3 white cheek spots; dorsal crest extended by mane along whole body; fringe of hairs from chin to base of neck in males; females and young redder than males.
CONSERVATION STATUS: LR Conservation Dependent.

Bongo LR
Tragelaphus eurycerus

Discontinuously distributed in lowland forest in E, C, and W Africa; found outside this habitat in S Sudan, in small populations in montane or highland forest in Kenya, and in the Congo.
HBL 220–235cm; TL 24–26cm; HL 60–100cm; WT male: 240–405kg, female: 210–253kg.
FORM: coat bright chestnut red, much darker in adult males; dark muzzle, white chevron between eyes, about 2 white cheek spots; lower neck and undersides darker, whitish crescent collar at base of neck; black and white spinal crest, and many narrow but clear white vertical stripes on the body; contrasting black and white markings on the legs. Horns present in both sexes, heavy and smooth with an open spiral of one to one-and-a-half turns. Tail long and tufted at tip.
CONSERVATION STATUS: LR Near Threatened.

Genus *Taurotragus*

Common eland LR
Taurotragus oryx
Common or Cape eland

Nomadic grassland and open woodland; may have at least visited all but the most arid of these habitats in E, S, and C Africa in the past; now found only in game reserves and ranches in some areas.
HBL male: 250–340cm, female: 200–280cm; TL 54–75cm; SH male: 135–178cm, female: 125–150cm; HL male: 60–102cm, female: 60–140cm; WT male: 400–950kg, female: 390–595kg.
FORM: coat light tan, darkening to gray in old males; a few light stripes on the forequarters (not found in adults of the S populations); black and white leg markings; black tuft on end of tail; black stripe along back, merging into short mane; adult males develop a tuft of frizzy hair on their foreheads. Ears have much smaller, more horse-like pinnae than those of other spiral-horned antelopes.
CONSERVATION STATUS: LR Conservation Dependent.

Giant eland
Taurotragus derbianus

More of a woodland species than the Common eland, and found in small populations in W and C Africa, with larger and more secure populations in E Africa, particularly Sudan.
HBL male: 290cm, female 220cm; HT male: 150–175cm, female: 150cm; TL 55–78cm; HL male: 80–110cm, female 80–125cm; WT male: 450–900kg, female: 440kg.
FORM: reddish brown coat, becoming slate gray in adult males; 12–15 body stripes, white chevron between eyes, white cheek spots, black stripe along back merging into a short mane, black collar around neck, with contrasting white patches on either side. The collar emphasizes the dewlap, which begins under the chin and finishes above the base of the neck. Horns more strongly keeled than in the Common eland, but also more slender and longer, even in the males.
CONSERVATION STATUS: Western subspecies (*T. d. derbianus*) classified as Endangered; Eastern subspecies (*T. d. gigas*) as Lower Risk: Near Threatened.

SJGH/RU

Managing Bison in Yellowstone

Preserving the final remnant of the Great Plains herds

THE GREAT PLAINS OF NORTH AMERICA WERE once home to vast herds of bison (*Bison bison*), up to 90 million at their height. Many native peoples – notably the Cheyenne, Sioux, Crow, Comanche, Blackfoot, Assiniboin, Mandan, Shoshone, Kiowa, and Gros Ventre – depended on these animals, not only killing them for their meat, but also using their hides for shelter and clothing. But the westward expansion of European settlement and the advent of the railroad from the mid-19th century onward had a devastating effect on Plains societies. A deliberate strategy of extirpating indigenous hunting cultures to make way for ranches and farms involved the wholesale slaughter of bison. In one 3-year period alone (1872–74), some 3 million were shot. So effective was the policy that by 1900 only a few isolated herds remained in the wild, totaling barely 1,000 animals.

Part of this sad, remnant population was the small herd of bison in Yellowstone National Park, which numbered fewer than 50 animals. In 1902 this group was supplemented by 21 semi-domesticated Plains bison. Additional animals were subsequently introduced from Montana and Texas, until by 1917 there were three introduced animals for every wild bison in the park. By the 1950s, the semi-domesticated herds were completely mixed into the wild population. Today, the Yellowstone ecosystem hosts the only unfenced, free-ranging bison herd in the conterminous United States.

Until 1967, the Yellowstone herd was managed essentially as livestock, with culling programs to keep the population low to minimize the effects on range vegetation. Beginning in 1968, however, the National Park Service embarked on an experimental program to allow natural ecological processes to govern the distribution and abundance of the bison, along with all other ungulate herds in the Park. In the years that followed, the bison population grew to reach approximately 4,000 animals by 1995; there were also another 400 animals in Grand Teton National Park, south of Yellowstone.

Bison are highly gregarious; outside of the breeding season, typically only adult males will be found by themselves. During the rut in July, males fight for dominance, and only about one male in three breeds. Cow bison have a single, tan-coloured calf in May. The animals have nomadic patterns of movement and foraging. In Yellowstone National Park they may migrate to high-elevation grasslands during summer, but are forced into lower elevation rangelands in winter when snows become too deep. Dispersal appears to play an important role in regulating the population.

As the herd grew during the 1980s and 1990s, increasing numbers of bison dispersed out of the park altogether, onto National Forest and private lands, especially during winters with heavy snowfall. Dispersing bison met fierce resistance from livestock growers and supporting agriculture agencies, largely because the park's bison are known to host brucellosis, a disease caused by the bacterium *Brucella abortus* that results in abortions in bison and cattle. (People pasteurize milk to avoid contracting this seldom-fatal disease, known in its human form as undulant fever.) Livestock growers had good reason to be concerned, for the economic consequences of contagion could be disastrous; strict rules enforced by the United States Department of Agriculture require infected herds to be slaughtered, and states hosting infected cattle risk losing their brucellosis-free status, requiring expensive testing prior to interstate shipment of stock. As a consequence, fears of contagion led to the slaughter of 1,100 dispersing bison by state and federal officials during the severe winter of 1996–97. Ironically, cattle were the original source of the disease, which was first diagnosed in Yellowstone's bison in 1917.

Above *Although the Yellowstone herd is free-ranging and unfenced, some human intervention is necessitated by issues such as cattle disease.*

Left *In winter bison forage by using their massive heads to clear snow, thereby creating craters in which they can graze.*

Right *Bison calves, which may weigh up to 30kg (66lb) at birth, are generally not weaned until they are 8–12 months old.*

In response to the problem, a bison manageme plan has been designed to protect livestock interes while also safeguarding the herd within the park. Research concentrates on finding an effective vacc that can be used to reduce the prevalence of bruce losis. Currently inoculation is with the vaccine RB5 administered using biobullets (plastic, dissolving b lets containing the vaccine) shot from an air gun ir the legs of calves. Meanwhile, corrals and holding pens have been constructed at the park boundarie to capture dispersing bison for slaughter. Many bi are seriously injured during transportation to slaug terhouses, goring each other with their horns. Tho bison that evade capture have been shot.

The largest terrestrial mammals in the western hemisphere, bison are clearly important components of the Yellowstone ecosystem, even though the impact of a naturally regulated herd on vegetation is still not fully understood. Bison are primarily grazers, foraging on grasses and sedges, and grazed areas in Yellowstone National Park have been shown to have a 36–85 percent higher production of grass than ungrazed areas, partly stimulated by the high mineral content of bison feces and urine, which contain 90 percent of the phosphorus and 65–95 percent of the nitrogen from ingested foods.

In addition to the plant–herbivore interactions, bison fall prey to wolves (*Canis lupus*) and Grizzly bears (*Ursus arctos*). Wolves were reintroduced to the park in 1995, and have proved to be effective predators of bison, even though these constitute only a small part of their diet; over 90 percent of the animals the wolves have killed have been easier prey. For Grizzlies, scavenging winter-killed bison can be a crucial resource on emerging from winter hibernation. Finding a bison carcass in spring can help a female Grizzly ensure the survival of cubs that might otherwise starve. MB

Duikers

ALMOST ANYWHERE IN AFRICA SOUTH OF THE Sahara desert, the observant traveler may catch a glimpse of one of many small to medium-sized antelopes disappearing into dense forest or thickets. These are the duikers – an Afrikaans word meaning "divers," so named for their habit of diving into cover when disturbed.

As their Afrikaans name suggests, duikers are shy creatures that do their best to avoid encountering humans. Their bodies are adapted to a fugitive lifestyle: the short forelegs, longer hind legs, and arched body shape all help them to slip easily through dense vegetation.

Fugitives of the Forests

FORM AND FUNCTION

The sexes are similar in appearance, although females are often up to 4 percent longer than males. Both usually possess short, conical horns, but these are occasionally lacking in females. The coat is often reddish, but in some species is blue-gray, black, or striped. Duikers have the largest brains relative to body size of any antelopes.

Duikers are uncommon as fossils, but the living species retain many features found in the fossil remains of early bovids, and are sometimes considered to be the most primitive living African antelopes.

Although some species show a preference for open country with scattered brush cover, many duikers live secretively in dense forest, which has made their natural history difficult to study. The Yellow-backed duiker may be active by both night and day, but others, including the Blue duiker, are active only during the day, and some, such as the Bay duiker, only at night. Duikers are primarily browsers, and require high-quality food because of their relatively small size. They feed on leaves, fruits, shoots, buds, seeds, and bark. One of their food-finding strategies is to follow groups of frugivorous birds and monkeys, feeding on fallen fruit. Surprisingly, they sometimes stalk and capture small birds and rodents, and occasionally also eat insects and even carrion. In a study of Blue duikers, ants made up over 10 percent of the stomach contents by dry weight.

Pairing Up

SOCIAL BEHAVIOR

Duikers are not gregarious, and are usually seen either alone or in pairs. In the wild the Blue duiker is truly monogamous; pairs seem to mate for life and reside in small, stable territories that are actively defended against other members of the same species by both male and female. In captivity, male Maxwell's duikers are highly intolerant of other males and drive them off, while females dispatch intruding females. Observations on captive animals suggest that several other duiker species may also be monogamous.

In monogamous social systems, which are comparatively rare in mammals, males sometimes provide a great deal of care for the young, and the young remain in the social group for extended periods, during which they often help to care for their younger siblings. This pattern is known as

▶ **Left** *The large scent gland below the eye* ɪn *be clearly seen on this Red forest duiker* ɔm *South Africa. The timing and frequency* ƒ *marking, done by rubbing the gland on an* ɔject, *varies from species to species.*

Below *Representative species of duikers.* Common duiker (Sylvicapra grimmia) ɖuiking" *(fleeing) into the undergrowth;* Yellow-backed duiker (Cephalophus ɪlvicultor) *displaying the erect yellow rump* atch; **3** *male Zebra duiker* (C. zebra) *mark-* ɪg *a sapling with the scent gland, located* ɪst *beneath the eye;* **4** *Jentink's duiker* ̈. jentinki) *suckling a young calf;* **5** *male* Maxwell's duiker (C. maxwelli) *about to rub* ʒ *scent gland on another male as a prelude* ɔ *fighting;* **6** *Red-flanked duiker* (C. rufilatus) ▪ *the process of eating a bird.*

FACTFILE

DUIKERS

Order: Artiodactyla

Family: Bovidae

Subfamily: Cephalophinae

17 (19) species in 2 genera: **forest duikers** (*Cephalophus*, 16 species), including **Blue duiker** (*C. monticola*), **Bay duiker** (*C. dorsalis*), **Maxwell's duiker** (*C. maxwelli*); and **savanna or bush duikers** (*Sylvicapra*, 1 species), comprising solely the **Common duiker** (*S. grimmia*).

DISTRIBUTION Africa S of Sahara

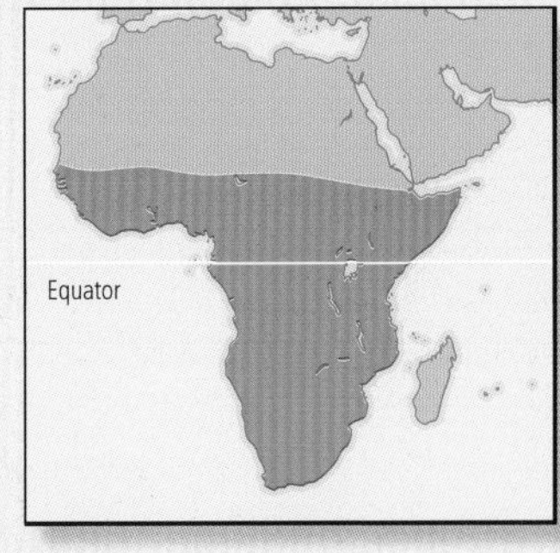

HABITAT Mainly dense forest thickets

SIZE **Head–body length** ranges from 59–72cm (22–29in) in the Blue duiker to 115–145cm (45–57in) in the Yellow-backed duiker; **tail length** from 7–12.5cm (3–5in) to 11–18cm (4–7in), and **weight** from 4–6kg (9–13lb) to 45–80kg (100–176lb), both in the same 2 species.

COAT Often reddish, but in some species blue-gray, black, or striped.

DIET Leaves, fruits, shoots, buds, seeds, and bark; also occasionally insects, small birds, and rodents.

BREEDING Gestation 7.5–8 months in Red-flanked and Bay duikers.

LONGEVITY Unknown in the wild; 10–15 years in captivity.

See species table ▷

obligate monogamy. However, in other monogamous species, males leave most responsibility for the care of the young to the female, and the young disperse from the social group before the birth of their younger siblings (facultative monogamy). In species exhibiting obligate monogamy, such as marmosets and tamarins, mates remain near one another and usually feed and rest together. The Blue and Maxwell's duiker appear to be facultatively monogamous; while males and females feed apart, they commonly rest close together and defend a mutual territory.

Duikers bear only a single calf, and females conceive again soon after giving birth. Calves of all species spend little time with their mothers during the first few weeks of life, remaining well-hidden in vegetation. Young Blue duikers reach sexual maturity at about 1 year of age and leave their parents during the second year of life to attempt to find their own mates and territories. Larger species probably reach sexual maturity at an older age. Calves of some species, such as the Zebra and Maxwell's duikers, resemble their parents in color, while those of other species have distinctive juvenile coats. Young Jentink's and Bay duikers are very similar in appearance, with a uniform dark-brown coat unlike that of the adults of either species. The yellow rump patch of the Yellow-backed duiker does not begin to appear until about 1 month of age, and is not fully developed until about 10 months.

Duikers possess large scent glands beneath each eye. The structure of these glands differs from that of the preorbital glands found in other antelopes. The secretion, which may be clear or bluish in some species, is extruded through a series of pores instead of via a single large opening. In captivity, many duiker species rub these glands on fences, trees, and other objects in their enclosures. This has usually been interpreted as territorial marking. This behavior is very frequent in some species: in captivity, male Maxwell's duikers may mark as often as six times during a 10-minute interval.

Maxwell's duikers also press these glands on the glands of other individuals, first on one side and then on the other. This behavior is called mutual marking. Very forceful mutual marking has been observed in captivity between males as a prelude to fighting. A much more gentle form is often seen between males and females, not just in this species but also in Red and Blue duikers.

Above *Engaged in grooming, a pair of Zebra duikers reveal the pronounced black stripes that give the species its name. Living in lowland forests from western Sierra Leone to the central Côte d'Ivoire, the animals are now listed as Vulnerable by the IUCN.*

Below *With ears pricked to pick up sounds of danger, a Common duiker in South Africa's Kruger National Park shows the alertness these shy animals rely on to survive.*

In Demand for Bushmeat

CONSERVATION AND ENVIRONMENT

The rarest species is Ader's duiker, which is classified as Endangered; however, most *Cephalophus* species are now regarded by the IUCN as being at some degree of risk. The principal threat comes from subsistence hunting – the so-called bushmeat trade. The meat of these animals is greatly prized by humans, and duikers are easily dazzled at night by lights, which makes them relatively easy to either shoot or capture. Some species are also caught by driving them into nets.

In many areas duikers are now the main component of the trade in wildlife species for food (see Primates: The Bushmeat Trade), and evidence suggests that they are often harvested at unsustainable rates. Perhaps surprisingly, the smaller duikers are also in demand by trophy hunters, who seek full-mounts for their collections. KR/KI

)uiker Species

.OST DUIKERS ARE FOREST-DWELLERS OF .e genus *Cephalophus*, a name that refers • the crest of long hair between their ɔrns. The genus *Sylvicapra* includes only single species, the Common duiker, hich is the only species typically found . savanna and open bush country.

FOREST DUIKERS

;enus *Cephalophus*

.orns usual in both sexes, and in same lane as forehead. Ears short and rounded.

laxwell's duiker LR
'ephalophus maxwellii

'igeria west to Gambia and Senegal. Low-.nd forest and adjacent gallery forest.
BL 55–90cm (22–35in); TL -10cm(3–4in); WT 8–9kg (18–20lb).
ORM: coat from gray-brown to blue-gray, ith much variation. Horns sometimes bsent in females.
'ONSERVATION STATUS: LR Near Threat-.ned.

lue duiker
'ephalophus monticola

'igeria to Gabon, east to Kenya and south) S Africa. Lowland forest.
BL 55–72cm (22–28in); TL 7–12.5cm 3–5in); WT 4–6kg (9–13lb).
.OAT: color variable; similar to Maxwell's uiker but may be bluer.

lack-fronted duiker LR
'ephalophus nigrifrons

;ameroon to Angola and east through)emocratic Republic of Congo (DRC) to .enya. Lowland, gallery, montane and ıarshy forests.
BL 85–107cm (33–42in); TL 10–15cm 4–6in); WT 13–16kg (29–35lb).
.OAT: reddish to dark brown, with red-ish brown to black stripe from nose to orns.
.ONSERVATION STATUS: LR Near Threat-ned. One subspecies – *C. n. rubidus* – is sted as Endangered.

led-flanked duiker LR
'ephalophus rufilatus

enegal to Cameroon east to Sudan and Jganda. Gallery forests and forest edges.
BL 60–70cm (24–28in); TL 7–10cm 3–4in); WT 9–12kg (20–26lb).
'ORM: coat reddish yellow to reddish rown, with dark dorsal stripe from nose ɔ tail. Horns regularly lacking in female.
.ONSERVATION STATUS: LR Conservation)ependent.

entink's duiker Vu
'ephalophus jentinki

.iberia and W Côte d'Ivoire. Lowland for-st only. A very rare species.
BL about 135cm (53in); TL about 15cm 6in); WT up to 70kg (154lb).
COAT: head and neck black, shoulders white, back and rump a grizzled "salt and pepper."

Yellow-backed duiker LR
Cephalophus sylvicultor

Guinea-Bissau east to Sudan and Uganda, south to Angola and Zambia. Wide variety of forest types and open bush.
HBL 115–145cm (45–57in); TL 11–18cm (4–7in); WT 45–80kg (99–176lb).
COAT: blackish brown except for yellow rump patch.
CONSERVATION STATUS: LR Near Threat-ened.

Abbott's duiker Vu
Cephalophus spadix

Tanzania. High montane forest.
HBL 100–120cm (39–47in); TL 8–12cm (3–5in); WT up to 60kg (132lb).
COAT: dark chestnut brown to black.

Zebra duiker Vu
Cephalophus zebra

Sierra Leone, Liberia, and Côte d'Ivoire. Lowland forest.
HBL 85–90cm (33–35in); TL about 15cm (6in); WT 9–16kg (20–35lb).
COAT: reddish brown with 12–15 black transverse stripes.

Black duiker LR
Cephalophus niger

Guinea east to Nigeria. Lowland forests.
HBL 80–90cm (31–35in); TL 12–14cm (5–5.5in); WT 15–20kg (33–44lb).
COAT: brownish black to black.
CONSERVATION STATUS: LR Near Threat-ened.

Ader's duiker En
Cephalophus adersi

Zanzibar, coastal Kenya, and Tanzania.
HBL 66–72cm (26–28in); TL 9–12cm (3.5–5in); WT 6.5–12kg (14–26lb).
COAT: tawny-red with white band on rump.

Red forest duiker LR
Cephalophus natalensis

Somalia south to Zimbabwe and Mozam-bique. Lowland, montane forest.
HBL 70–100cm (28–39in); TL 9–14cm (3.5–5.5in); WT 10.5–12kg (23–26lb).
COAT: orange-red to dark brown.
CONSERVATION STATUS: LR Conservation Dependent.

Peter's duiker LR
Cephalophus callipygus

Cameroon and Gabon east through Cen-tral African Republic and DRC. Lowland forest.
HBL 80–115cm (31–45in); TL 10–16.5cm (4–6.5in); WT 15–24kg (33–53lb).
COAT: light to dark reddish brown.
CONSERVATION STATUS: LR Near Threat-ened.

Weyn's duiker LR
Cephalophus weynsi

DRC, Uganda, Rwanda, and W Kenya. A very poorly known species which may be a distinct species or a form of either Peter's duiker or the Red forest duiker.
CONSERVATION STATUS: LR Near Threat-ened.

Bay duiker LR
Cephalophus dorsalis

Guinea-Bissau east to DRC and south to Angola. Lowland forest.
HBL 70–100cm (28–39in); TL 8–15cm (3–6in); WT 19–25kg (42–55lb).
COAT: brownish yellow to brownish red; dorsal black stripe from nose to tail.
CONSERVATION STATUS: LR Near Threat-ened.

White-bellied duiker LR
Cephalophus leucogaster

Cameroon south and east into DRC. Low-land forest.
HBL 90–100cm (35–39in); TL 12–15cm (5–6in); WT 12–15kg (26–33lb).
COAT: light to dark reddish brown with white chin, throat, and belly.
CONSERVATION STATUS: LR Near Threatened.

Ogilby's duiker LR
Cephalophus ogilbyi

Sierra Leone East to Cameroon and Gabon. Lowland forests.
HBL 85–115cm (33–45in); TL 12–15cm (5–6in); WT 14–20kg (31–44lb).
COAT: reddish orange with white stock-ings; dorsal black stripe from shoulder to tail.
CONSERVATION STATUS: LR Near Threat-ened.

SAVANNA OR BUSH DUIKERS

Genus *Sylvicapra*

Horns usually in males only and directed upward. Ears longer and more pointed than in *Cephalophus*.

Common duiker
Sylvicapra grimmia

Subsaharan Africa except DRC. Savanna and open bush.
HBL 80–115cm (31–45in); TL 10–22cm (4–9in); WT 10–18kg (22–40lb).
COAT: sandy-tan.

En Endangered Vu Vulnerable
LR Lower Risk

HBL = head–body length TL = tail length
WT = weight

(The 1993 established list of mammalian species also recognized the following species: Harvey's duiker (*C. harveyi*) and the Ruwenzori duiker (*C. rubidus*)).

Below *Studies of Blue duikers have revealed that a substantial part of their diet is composed of insects.*

Grazing Antelopes

THE GREAT GRASSLANDS OF SUB-SAHARAN Africa, stretching to the north, south, and east of the Congo basin, are home to the highest biomass of mammals on earth. Here species of the extraordinary grass family, the Graminae, characterized by growth from the bottom of the plant, co-evolved with their equally specialized consumers, the grazing antelopes. Grouped together under this name are three separate tribes of animals: the Reduncini of the wetlands and tussocky grassland, including the kob and the reedbucks; the Alcelaphini of moist grassland and open woodland, among them the impala, the hartebeest, and the wildebeest; and the Hippotragini, or horselike antelopes of the arid lands.

Not so much a single habitat as a whole family of habitats, the grasslands vary from the swamplands of the Nile Sudd and the Okavango Delta through the open savannas of East Africa to the bushlands and woodlands of southern Africa and the semidesert of the Sahel and Kalahari. Nowhere stable, the structure of these habitats is constantly altering under the influence of changing rainfall patterns, fire, and the huge and varied impact of grazers and browsers themselves. Within these grasslands, mammalian biomass increases with increasing rainfall, and in the wettest areas it is only the intense grazing and browsing pressure of the herbivores, combined with frequent burning, that prevents the habitat transforming into closed forest or dense bush.

Three Themes with Variations

FORM AND FUNCTION

The first of the three tribes of grazing antelopes, the wetland antelopes or Reduncini, all inhabit relatively moist and productive habitat, from low-lying wetlands to montane grasslands. All are dependent on a supply of high-quality green forage for most of the year. Perhaps because of their frequently waterlogged habitat, none of the Reduncini have highly developed scent glands, and most communicate via unusual, highpitched whistles that may serve both as alarm calls for predators and as mating calls in males.

The smallest of the Reduncini, the reedbucks, may most closely resemble the ancestors of the group. The Mountain reedbuck, found in three disjunct highland populations, subsists on a coarser diet than the rest of the tribe. Home ranges are relatively small, ranging from 10 to 15ha (25–37 acres), and population densities are usually relatively low, although the Kenyan population reaches 11 animals per sq km (28/sq mi).

The Bohor reedbuck and Southern reedbuck are small lowland species inhabiting moist grass- and swamplands respectively north and south of the Congo basin. Like the Mountain reedbuck's, their home ranges are small, although they may reach high density in some areas – for example, 16.6 animals per sq km (33/sq mi) in one Zululand population. Like most small antelope species, their antipredator defense is hiding rather than herding.

The largest of the Reduncini are the fabulously shaggy waterbucks, a species that at first sight would appear more comfortable in North American forests than in the wet tropical grasslands they

Left *The imposing, lyre-shaped horns immediately identify this animal as a male impala. Impalas are the most widespread of the grazing antelopes.*

A Living Relic

The hirola is the sole surviving species of a once-abundant group of antelopes; it has a high conservation priority as a living relic elucidating the evolutionary relationships of the Alcelaphini. It is now confined to small areas of Kenya and Somalia. Over the last 30 years, competition with cattle, disease, and poaching have seen hirola numbers fall from 20,000 to less than 2,000. The Somalian population may now be extinct. Some were introduced to Tsavo National Park in southern Kenya in the 1970s, but their numbers have not risen above 100 animals.

Hirola are grazing antelopes that live in arid, open scrubland. Grass is their staple diet, but they also eat broad-leaved herbs in the dry season. They appear to be attracted to areas that are heavily used (but not overgrazed) by livestock kept under traditional Somali husbandry practices. This habit puts them at risk from diseases like rinderpest and tuberculosis.

Seasonal breeders, hirola mate mostly during the long rains in February–March, and give birth at the start of the short rains in October–November. Females live in groups averaging 8 animals. They separate from the group to calve; young animals are very vulnerable for the first few weeks of life.

In high-quality habitat, males are believed to defend territories, where they try to mate with females that come to feed. Territories are marked with feces and secretions from well-defined facial glands. Non-territorial males form groups of 2–38 animals. In Tsavo, breeding males accompany groups of females over ranges of up to 40sq km (15sq mi). The herd males may exert firm control over their harems, leading them into new habitat or herding them from the rear. Males may thus switch between territory and harem defense, depending on ecological factors and population density. MG

tually inhabit. Their diet is high in protein, hich probably accounts for their high water take – an unusual feature for antelopes – and e fact that they are never found further than a v kilometers from standing water. Because of eir relatively large size they also require a high roughput of vegetation, and so they frequently pplement their grazing diet with reeds, browse, d even aquatic vegetation. Their waterside habi- t opens them up to predation, but in Uganda's ueen Elizabeth Park they have been shown to be nongst the least preferred prey of lions, perhaps cause of the smelly, oily secretion that coats eir fur, especially in males. They are sedentary imals, with female home ranges varying from st 0.3sq km (0.12sq mi) in one dense popula- on around Lake Nakuru in Kenya to 6sq km .3sq mi) in Uganda.

The kob, lechwe, and puku fit between reed- ick and waterbuck in size, but are characterized large home ranges; they can also reach extraor- narily high local population densities in the oist green savanna and seasonally inundated vampland they inhabit. The kobs are found rth of the Congo River, while the closely related iku and lechwe live to the south. All are highly ecialized on rich, green grass, and their ranging havior adapts to the spatial and temporal fluctu- ions in this valuable resource. The most seden- ry species, the puku, graze on predictable umps of grass and sedge in ancient oxbows and lleys. Uganda kob range up to a distance of

Above *Lechwe – seen here in Chobe National Park, Botswana – are members of the Reduncini tribe, all of which favor wetland habitat.*

40km (25mi) as local cloudbursts and fires cause brief flushes of green grass in the northern part of Queen Elizabeth Park. Finally, white-eared kob in the Sudan follow a mass migration across hundreds of kilometers, tracking the rainfall and seasonal inundation of the Nile Sudd. In the 1980s their population was estimated at 800,000 and the migration equaled that of the Serengeti wildebeest in the spectacle it provided. All the kobs and lechwes herd together as an antipredator strategy, and only males have horns.

The second tribe, the Alcelaphini, comprises the wildebeest and their kin. All except the impala are ungainly creatures with shoulders that are higher than their rumps, long faces, and stumpy horns, which are found on both sexes. Yet they are the most successful inhabitants of the classic African savanna. The various species range from the grasslands north of the Congo through the belt of grass to the east of the river's basin down to the woodlands of southern Africa. None are as tied to water or green grass as the Reduncini, and all except the impala, which supplements its graze with browse, are relatively large.

Topis are selective grazers found only in moist savanna areas, often together with kobs or reedbucks. The various hartebeest species are less selective grazers of medium and long savanna

FACTFILE

Grazing Antelopes

Order: Artiodactyla

Family: Bovidae

Subfamily: Hippotraginae

23 species in 11 genera

Distribution Africa, Arabia

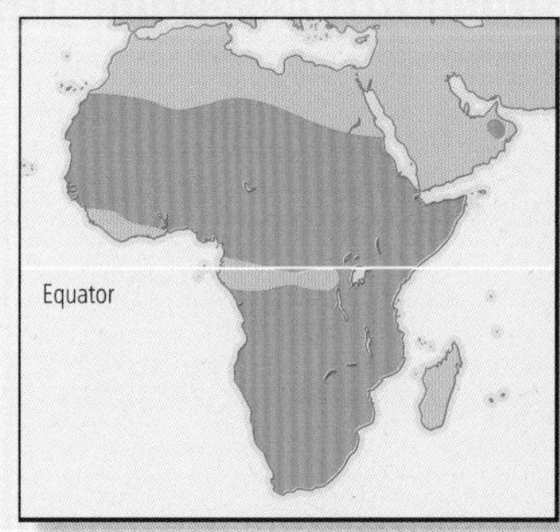

Habitat Dry and wet grasslands up to 5,000m (16,400ft)

Size Shoulder height from 65–76cm (26–30in) in the Mountain reedbuck to 126–145cm (49–57in) in the Roan antelope; weight from 23kg (51lb) in the Gray rhebok to 280kg (620lb) in the Roan antelope.

Coat Varied, from smooth and sleek in the impala to relatively shaggy in the waterbuck. Some species, including wildebeest, oryx, and horse antelopes, have manes.

Diet Grasses, herbs, aquatic plants; some species also browse on bulbs, seedpods of legumes, and tubers.

Breeding Usually only one young born after a gestation of 210–280 days.

Longevity Reedbucks have lived for up to 18 years in captivity, hartebeests to 20 years, gnus to 21 years, and Sable antelopes to 22 years.

See species table ▷

grassland, with medium-sized female home ranges of 3.7–5.5sq km (1.4–2.1sq mi) and moderate densities averaging 22 per sq km (57/sq mi) in Nairobi National Park, Kenya. Wildebeest inhabit the open grassland and woodland of eastern and southern Africa. Though most famous for the spectacular migratory populations of the Serengeti ecosystem and of Botswana, wildebeest may also be sedentary, as in Tanzania's Ngorongoro Crater.

Impalas are the most widespread species of grazing antelope, found throughout eastern and southern Africa. Although classed as Alcelaphini, molecular evidence suggests they are an outgroup less closely related than the other species, and indeed they lack the high shoulders and long snout of the rest of the tribe; in some respects they resemble the smaller, more delicate kob. Also unusually for an Alcelaphini species, they vary their diet between grass and browse: in the Sengwa research area of northwestern Zimbabwe, their diet was found to change from 94 percent grass in the wet season to 69 percent herbs and woody browse in the dry. Impalas are particularly associated with savanna woodland, riverside strips, and zones between different vegetation types. By inhabiting varied habitat and altering their diet seasonally, they are able to obtain relatively high-quality food throughout the year in small home ranges of 0.5–4.5sq km (0.2–1.7sq mi), and are never migratory.

The third tribe, the elegant Hippotragini, are Africa's greatest arid-land specialists. All species are large, with long straight or single-curved horns on both sexes. Females invariably range over huge areas in tight social groups of 5–25 animals, and population densities are never high. In addition to grass, most are able to take some browse, including seedpods of legumes, bulbs of succulents,

Above *Behavior of grazing antelopes.* ***1*** *Southern reedbuck* (Redunca arundinum) *in the "proud posture."* ***2*** *Defassa waterbuck* (Kobus ellipsiprymnus defassa) *showing the dominance display.* ***3*** *Gray rhebok* (Pelea capreolus) *in the alert posture.* ***4*** *Uganda kob* (Kobus kob thomasi) *in the head-high approach to a female during the mating season.* ***5*** *Addax* (Addax nasomaculatus) *performing the flehmen test after sampling a female's urine.* ***6*** *Sable antelope* (Hippotragus niger) *presenting horns, a male dominance display.* ***7*** *Roan antelope* (Hippotragus equinus) *in the submissive posture.* ***8*** *Coke's hartebeest* (Alcelaphus busephalus cokii) *showing the submissive posture of a yearling.* ***9*** *A territorial male impala* (Aepyceros melampus) *roaring during the rutting season.* ***10*** *A male bontebok* (Damaliscus pygargus) *initiating butting by dropping to his knees.* ***11*** *Gemsbok* (Oryx gazella) *showing the ritual foreleg kick during courtship.* ***12*** *Topi* (Damaliscus lunatus) *in the head-up approach to a female.* ***13*** *Blue wildebeest* (Connochaetes taurinus) *in the ears-down courtship approach*

ıbers, and wild melons. The females' horns may e used to exclude non-relatives from their tight ɔcial groups or to compete for scarce resources.

The driest country of all is inhabited by the ddax, a large white antelope with magnificent piralling horns that barely survives in the water-ss areas of the Sahara. The semi-arid areas of the ahel are home to the Scimitar oryx (which was ɔrmerly present north of the Sahara as well). The rabian oryx, the only non-African member of the ımily, which once roamed across the Arabian and nai peninsula, has recently been reintroduced to Oman and Saudi Arabia. Roan and sable antelopes re the least dry-adapted of the tribe, though they hare the same body type and social structure.

The Sociobiological Imperative

SOCIAL BEHAVIOR

More than any other group, the grazing antelopes illustrate the central dogma of sociobiology – that females follow resources, while males follow females. There is a simple theoretical justification for this generalization: because females must bear and feed young, whereas males supply only sperm, there will always be far more males available to mate at any given time than females. Female success at reproduction thus depends primarily on access to resources, while that of a male depends on how many females he can gain access to. Over time, females are selected to track resources, and males are selected to track females.

Reduncine Leks

Annually, during one of the two rainy seasons, in clearings interspersed around the bushland of Queen Elizabeth National Park, Uganda, an extraordinary scene is played out. Evenly spaced around a vast area of closely grazed turf are 65 adult kob males. A few are lying down or grazing, but most are standing erect, threatening other males or displaying to females that are resting in clusters or wandering between territories. As two new females enter the clearing, the chests of many males rise and fall almost in unison, as each produces a series of short, shrill whistles.

A kob territorial breeding ground, or lek, has probably existed on exactly the same spot for centuries. The curious lekking system, in which males hold tiny territories devoid of resources clustered on a traditional mating ground and females leave the herds in which they usually live to walk to the lek to mate, is known from seven species of mammals. It poses two key questions: first, why is this system present in these few species and no others; and second, in circumstances where females appear free to choose mating partners, how do they make that choice?

The mating strategy of the reduncines is determined by the behavior patterns of females, who move widely and erratically in huge, unstable herds, tracking unpredictable rainfall. Thus, for males to defend a fixed territory would be a waste of time, if all the females are several kilometers away looking for suitable grazing. And, because females' social groups are not cohesive, harem defense, the chosen strategy of the Hippotragini, is not an option either. Thus, reduncine males have little option but to stake out a display site and advertise for mates.

Females benefit in several ways from the lek mating system. First, they gain security since the short grass on leks restricts the cover available to predators; in any case, lions usually pick off males on the edge of leks, not females in the center. Secondly, leks enable them to avoid harassment; Kafue lechwe females that attempt to mate outside of leks are often chased and harassed by juvenile males. Finally, the females have a far wider choice; from the throng of competing males, they can select the most attractive, namely those with the best genes to pass on to their offspring.

But how do lekking females arrive at their choices? Studies of Uganda kob have shown that they focus their attention on territories, not individuals. Males often change territories, and when they do, it is the popularity of the territory that predicts how well any individual will do, not the animal's own success on the last territory it held. Moreover, the single characteristic of a territory that seems to attract females most strongly is its previous popularity with other females.

What makes other females' preferences so appealing is still not certain, but there are two main hypotheses. The first is that females access the males that are most popular (and so, presumably, genetically superior) by copying the mate choices of females around them; the second, that the places where many other females have chosen to stay must be safe from predators. Either way, reduncine kobs are a sight to behold! JCD

Based on this generalization, we can divide the grazing antelopes into four basic breeding systems. The first of these, which could be categorized as "male resource–defense territories," is found where females tend to be sedentary within small home ranges and moderate population densities. These species include the puku, where females have access throughout the year to good-quality forage in ancient oxbows; the waterbuck, whose large size and moist habitat enable females to find high-protein forage throughout the year; impalas, where switching between graze and browse makes a sedentary lifestyle possible; and hartebeests, which have a tolerance for poor-quality grass that enables them to stay put even in moderate dry seasons. In all these species females have small home ranges, and strong, mature males can effectively monopolize access to them by defending territories that include all or a substantial part of several females' home range. To the

Above Generally short-grass grazers, the migratory attern of wildebeest apparently reflects the availabili- / of suitable forage, water, and minerals. During their pectacular migrations, many wildebeest fall prey to rocodiles at river crossings.

casual observer it may look as if male waterbucks or impalas are defending a harem of females, but actually they are defending territories with clear boundaries. If the females leave the area, the male will relinquish control of them.

The reedbucks exemplify a second pattern only slightly different from the first, for although they are often found in apparently monogamous pairs, their mating system actually resembles that of the resource-defending waterbuck and puku.

Females, relying on hiding for predator defense and requiring relatively digestible food because of their small body size, live fundamentally solitary lives. Males defend resource-based territories, but frequently these only overlap the territory of a single female.

The third breeding system, harem defense, is found in most Hippotragini species. In these arid-adapted antelopes, the female home ranges are vast – far too large for a male to effectively control.

But females live in tight, stable kin groups of 5–25 animals, perhaps to enable them to defend scarce resources from unrelated individuals. The most common male reproductive strategy is thus to defend the female group directly, rather than protect a territory. The one exception is the Sable antelope, which lives in a lusher habitat where female movements are smaller and more predictable, and males revert back to defending territories.

Finally, there are several species or subspecies in which females live at high densities, range widely and unpredictably over large areas in search of large patches of high-quality forage, and do not form stable or cohesive kin groups. These include the Uganda kob, the White-eared kob, the Kafue lechwe, and perhaps some migratory populations of wildebeest. In these species, males often defend tiny, resourceless territories, clustered together on conventional breeding grounds or "leks." Sexually receptive females leave the large, mixed-sex herds in which they usually live and walk up to several kilometers to the nearest lek. There they remain for 12–24 hours and mate a number of times with several different males.

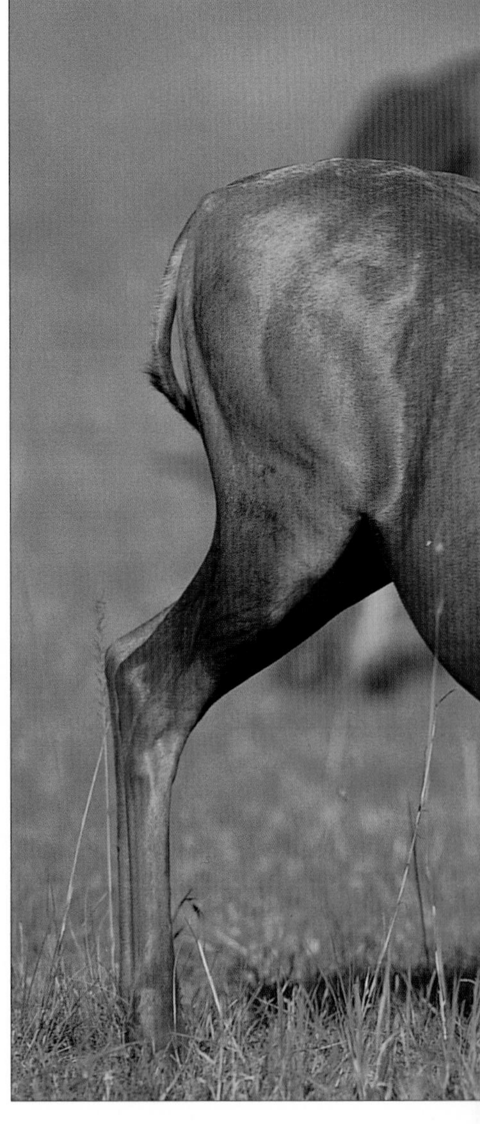

A High-Density Waterbuck Population

At Lake Nakuru National Park, Kenya, the density of waterbuck reaches as much as 100 animals per sq km (250 per sq mi) in some areas. The park average of 30 animals per sq km (75 per sq mi) is so much higher than the more typical 1–2 animals per sq km (2.5–5 per sq mi) found elsewhere that there are grounds for suspecting the Nakuru waterbuck differ in social behavior from other populations.

Male waterbuck usually occupy territories larger than 100ha (250 acres), but at Lake Nakuru the severity of competition probably explains why territories are smaller – ranging from 10–40ha (25–100 acres) – and average duration of territory ownership is shorter (at about 1.5 years). At any time, only a tiny proportion (about 7 percent) of adult males hold a territory, and only 20 percent of males surviving to prime age ever become owners of a territory.

Some 53 percent of the territories at Lake Nakuru contain one or more "satellite males," adult males subordinate to the territory holder and tolerated by him. Satellite males have access to the whole territory, and some have access to two adjacent territories as well. While tolerating his satellite(s), a territory owner will threaten and expel other adult males. Adult males that are neither territory holders nor satellites (i.e. over 80 percent of adult males) unite with young and juvenile males in bachelor herds, which rarely enter the territories of dominant males.

Adult and young males attempting to enter a territory are often confronted and repelled by the satellite male instead of by the territory owner. Satellites apparently share in the defense of the territory, saving the owner energy and decreasing his risk of being wounded in a fight.

When a receptive female is in the territory, usually only the territory holder copulates with her. Occasionally, however, a satellite male will manage to copulate with a receptive female while the territory holder is not close by. Waterbuck territories are situated along rivers and lakeshores where the grass is noticeably greener, so by being inside a territory satellite males also gain access to better resources than bachelor males.

The biggest advantage for the satellite male is probably his improved chance of becoming a territory owner himself. In 5 of 12 observed cases of change of territory ownership, a satellite became the new owner, either of the territory it had already occupied as a satellite or of one adjacent to it. It can be calculated that the average probability of gaining possession of a territory is about 12 times greater for a satellite male than for a bachelor. PW

Topi have been reported to lek-breed in dense populations in Akagera National Park, Rwanda, and the Masai Mara, Kenya, where their ranging behavior resembles that of the Uganda kob. Little is known about the breeding system of migratory wildebeest. Some observers report that males establish small, clustered territories, similar to kob leks, along the path of the migration, intercepting females briefly on their journey. The difficulties of recognizing individuals in such a huge migratory flow have so far precluded reliable data-gathering.

One other aspect of breeding seasons has intrigued observers. Why are some species seasonal breeders and others not? As with many other aspects of the reproductive system, seasonality appears to be a flexible adaptation of females to their habitat. At one extreme are species inhabiting southern African savannas with a single, highly predictable rainy season each year, such as puku. These species usually time reproduction so that lactation – the time when females are expending most energy – occurs just after the rains, when food is optimally available. At the other extreme are species such as Mountain reedbuck, which

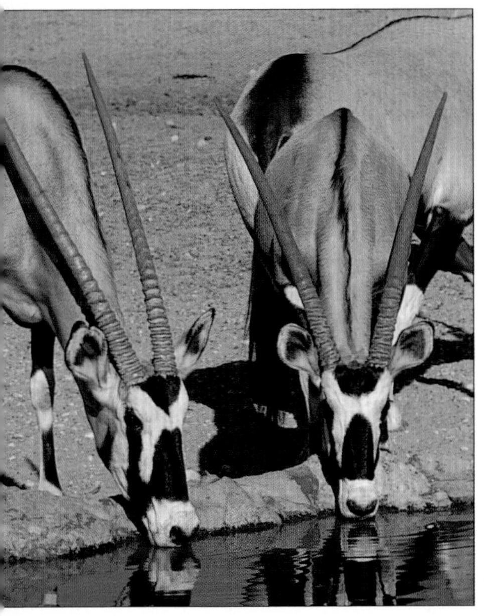

◐ **Below** *The gemsbok is reasonably secure across most of its range. However, all of the other larger grazing antelopes are seriously threatened.*

◐ **Above** *Young male topi wrestling. Expelled from a dominant male's territory when aged around one year, they challenge older bulls at age 3 or 4.*

tend to breed throughout the year. In the middle are species inhabiting partially seasonal habitats, such as kob in Uganda, where rain falls all year round but peaks in October and March. Here females breed year-round, but with seasonal peaks matching the heaviest rainfall. It seems that these species, rather than timing lactation optimally, become sexually receptive whenever they achieve sufficient body condition to predict the probable success of a reproductive attempt.

Food on the Hoof

CONSERVATION AND ENVIRONMENT

Almost all grazing antelope species except waterbuck are popular prey for humans. Because of their high potential reproductive rates, however, a serious risk of extinction has tended to arise in only two circumstances: where the species naturally lives at a very low population density, or where habitat destruction is the primary culprit.

By far the most vulnerable group are the Hippotragini, which share a variety of characteristics predisposing them to extinction. They are large, they live in arid habitats with huge home ranges at extremely low population densities, and they reproduce relatively slowly. They are difficult to preserve in small reserves, and are easily threatened by hunting pressure. Often they share their habitat with humans who face frequent food shortages. One species, the Blue buck (*Hippotragus leucophaeus*), which lived in the southwestern Cape in South Africa, is already extinct, and another, the Arabian oryx, was extinct in the wild until recent efforts at reintroduction. The addax and Scimitar oryx are also in danger of imminent extinction, with the Scimitar already extinct north of the Sahara. All the other species save the gemsbok, which is relatively well-protected in southern African reserves, face some degree of threat.

The Alcelaphini are generally less threatened, except where humans have interfered with migration patterns or degraded habitat badly. In Botswana, the division of grazing land into fenced ranches has drastically reduced the wildebeest population by preventing migration. In Kenya, the Hunter's hartebeest, which lives at low densities in relatively dry savanna, is critically threatened by hunting and competition with cattle for forage.

The Reduncini or wetland antelopes, with their high reproductive rates and frequent preference for swampy areas of less value to humans, are also generally secure, although many local populations have been extinguished in the last fifty years. One exception is the lechwe; several subspecies in Zambia are threatened by hydroelectric schemes and, to an even greater extent, by poaching.

The high population densities and reproductive rates of many species of Reduncini and Alcelaphini would seem to lend themselves to controlled harvesting by humans; failure to implement sustainable management appears to stem from political rather than biological causes. Controlled harvesting was attempted in the Toro Game Reserve of Uganda in the 1960s and 1970s, but widespread civil unrest thereafter led to indiscriminate slaughter, and the local population was almost driven extinct. This pattern has been repeated for many species across the continent.

In the meantime, effective maintenance of wild populations tends to be associated with high-volume tourism, as in the Serengeti. Much will depend on the success of future park planning and management in Africa, which in turn requires balancing the ethical and economic demands of competing interests. On the one hand, high-volume tourism can impact negatively on wildlife habitats, sensitive species, and local people; on the other, limiting visitor numbers in the interest of conservation may limit park revenues and tourist income, although experience suggests that high quality–low volume tourism can be a particularly sustainable combination. JCD/MGM

Grazing Antelope Species

TRIBE REDUNCINI

Africa. Antelopes of wetlands, tall or tussock grassland. Horns on male only; long hair on sides and neck except in kob.

Genus *Redunca*

Light and graceful animals characterized by whistling and high, bouncing jumps. Females of similar size to males. Glandular spot occurs beneath ears. Three species.

Southern reedbuck LR
Redunca arundinum

Africa north to Tanzania and west to Angola. Southern savannas. Largest reedbuck. Two subspecies
HBL 134–167cm; TL 26–27cm; SH 84–96cm; HL 30–45cm; WT 50–80kg.
COAT: light buff to sandy brown; pale zone at base of horns; black and white markings on forelegs.
CONSERVATION STATUS: LR Conservation Dependent.

Mountain reedbuck LR
Redunca fulvorufula

Cameroon, Ethiopia and E Africa, S Africa. Montane grasslands up to 5,000m. Smallest reedbuck. Three subspecies, each confined to its own highland area, including Chanler's mountain reedbuck (*R. f. chanleri*), Ethiopia and E Africa.
HBL 110–136cm; TL 20cm SH 65–76cm; HL 14–38cm; WT 30kg.
FORM: coat soft, woolly and gray; prominent eyes and sockets.
CONSERVATION STATUS: LR Conservation Dependent.

Bohor reedbuck LR
Redunca redunca

Senegal east to Sudan and south to Tanzania. Northern savannas. Seven subspecies, including the Sudan form (*R. r. cottoni*), which has long, splayed horns.
HBL 100–130cm; TL 18–20cm; SH male 75–89cm, female 69–76cm; HL 20–41cm; WT 45kg.
FORM: coat light buff, with strongly marked forelegs; horns forward-hooked.
CONSERVATION STATUS: LR Conservation Dependent.

Genus *Pelea*

Gray rhebok LR
Pelea capreolus
Gray rhebok, Vaal ribbok, or rhebuck

S Africa. High plateaux. SH 76cm; HL 20–29cm; WT 23kg.
FORM: coat brownish gray, soft and woolly; gracefully built, with long slender neck; tail short and bushy; no gland patch beneath the ears; horns short, straight, and almost vertical.
CONSERVATION STATUS: LR Conservation Dependent.

Genus *Kobus*

Medium- to large-sized animals with a relatively heavy gait. Males have ridged horns and females are smaller than males. Five species.

Waterbuck LR
Kobus ellipsiprymnus

S Africa north to Ethiopia and S Sudan, west to Senegal. Savanna and woodland near to permanent water. Two taxonomic groups. The Common waterbuck, Ellipsiprymnus group, with a white ellipse on the rump, is restricted to the east of the Rift valley for most of its range extending from 6°N to 29.5°S. Subspecies: 4. The Defassa waterbuck, Defassa group, with a white blaze on the rump is predominantly reddish in color. North and west of the Rift valley. Nine subspecies.
HBL 177–235cm; TL 33–40cm; SH 125cm; HL 55–99cm; WT 170–250kg.
FORM: coat dark gray or reddish; large and shaggy with heavy gait.
CONSERVATION STATUS: LR Conservation Dependent.

Kob LR
Kobus kob

Gambia east to Sudan and Ethiopia, Uganda. Low-lying flats and gently rolling hills close to permanent water. Ten subspecies, including Buffon's kob (*K. k. kob*), Senegal to NW Nigeria; White-eared kob (*K. k. leucotis*), of S Sudan, whose males are predominantly black.
HBL 160–180cm; TL 115cm; SH male 90–100cm, female 82–92cm; HL 40–69cm; WT 77kg. Male robust and thick-necked with lyre-shaped horns; females more slender and graceful.
COAT: reddish, with white underside and throat chevron; legs with black markings.
CONSERVATION STATUS: LR Conservation Dependent.

Lechwe LR
Kobus leche

Botswana, Zambia, SE Democratic Republic of Congo (DRC). Floodplains and seasonally inundated grasslands. Three subspecies.
SH 99cm; HL 50–92cm; WT male 125kg, female 70kg.
FORM: coat long and rough, bright chestnut to black with white underparts. Hooves long and pointed. Horns long, thin, and lyre-shaped.
CONSERVATION STATUS: LR Conservation Dependent – Red lechwe (*K. l. leche*), Botswana and Zambia. The Black lechwe (*K. l. smithemani*), NE Zambia and SE DRC, and Kafue lechwe (*K. l. kafuensis*), Kafue River, Zambia, are classified as Vulnerable.

Nile lechwe LR
Kobus megaceros
Nile or Mrs Gray's lechwe

Sudan, W Ethiopia. Swamps in the vicinity of the White Nile. Sobat, Baro and Gilo Rivers.
SH 94cm; HL 687cm; WT 86kg.
FORM: coat long and rough, blackish chocolate with white underparts and white patch on shoulders; female uniformly reddish fawn; hooves long, pointed and splayed; horns long and spread out.
CONSERVATION STATUS: LR Near Threatened.

Puku LR
Kobus vardonii

Southern DRC, Botswana, Angola, Zambia, Malawi, Tanzania. Margins of lakes, swamps and rivers, and on floodplains. Two subspecies: puku (*K. v. vardonii*) and Senga kob (*K. v. senganus*).
HBL 126–142cm; TL 28–32cm; SH 77–83cm; HL 40–54cm; WT male 77kg, female 66kg.
FORM: coat fairly long and shaggy, bright golden yellow; horns short, thick, and less lyre-shaped.
CONSERVATION STATUS: LR Conservation Dependent.

TRIBE ALCELAPHINI

Africa. Grazers of open woodland, mois grassland and the zone between these t habitats. Characteristically with high pc ulation density. Except for impala, tribe characteristics include horns on both sexes, a long face, elevated shoulder anc sloping hindquarters, preorbital glands and glands on forefeet only. The female slightly smaller than the male. Impala h commonly been classed with the gazelle (Antelopini) although affinity with the kobs and reedbuck (Reduncini) has alsc been suggested. Evidence now suggests that impala are an early offshoot of the Alcelaphini, having diverged at the start that tribe's history as an independent u

Genus *Beatragus*

Hirola [
Beatragus (Damaliscus) hunteri
Hirola or Hunter's hartebeest

E Kenya and S Somalia. Grassy plains between dry acacia bush and coastal for est.
HBL 120–200cm; TL 30–45cm; SH 100–125cm; HL 55–72cm; WT 80kg.
FORM: coat uniform sandy to reddish tawny, with white spectacles; horns long and lyre-shaped on short bony base (pe cel); female smaller, with lighter horns.

Genus *Damaliscus*

Characterized by a long head but frontal region not drawn upward into a bony pedicel. Slope of body less exaggerated than that of hartebeest. Horns thick and ridged.

Topi
Damaliscus lunatus
Topi, tsessebe, sassaby, tiangs damalisc, korrigum, or Bastard hartebeest

Senegal to W Sudan, E Africa through to Africa. Green grassland of open savanna and swampy floodplains. Seven subspecies, including tsessebe (*D. l. lunatus*) from Zambia southwards; *D. l. jumela* an *Coastal topi* (*D. l. topi*) in NE DRC and E Africa; tiang (*D. l. tiang*) in NW Kenya, V Ethiopia, and S Sudan.
HBL 170cm; TL 43cm; SH 124cm; HL 35–60cm; WT male up to 170kg, female 130kg.
COAT: bold pattern of black patches (notably on face) on glossy mahogany re fawn on lower part of legs.

ABBREVIATIONS			
HBL = head–body length TL = tail length SH = shoulder height	Ex Extinct	En Endangered
HL = horn length WT = weight	EW Extinct in the Wild	Vu Vulnerable
Approximate nonmetric equivalents: 10cm = 4in; 1kg = 2.2lb	Cr Critically Endangered	LR Lower Risk

ONSERVATION STATUS: LR Conservaon Dependent. The Korrigum subspecies). l. korrigum), Senegal to W Sudan, is ulnerable.

ontebok LR
)amaliscus pygargus
ontebok or blesbok

Africa. Open grassland. Two subspecies. H 84–99cm; HL 35–51cm; WT 59–100kg. ORM: coat rich, purplish chestnut rown, darker on the neck and hindquarers; white face patch, rump, belly, and wer legs; horns rather small in simple rm.
ONSERVATION STATUS: LR Conservaon Dependent – blesbok (*D. p. phillipsi*). he bontebok (*D. p. pygargus*), is ulnerable.

Genus *Alcelaphus*
haracterized by a long narrow hammerhaped head, heavy hooked and ridged orns and pronounced slope of back. Two pecies.

Hartebeest LR
Alcelaphus buselaphus
artebeest or kongoni

Coarse grassland and open woodland. enegal to Somalia, E Africa to S Africa. welve subspecies, including: Western artebeest (*A. b. major*), Senegal and uinea; Lelwel hartebeest (*A. b. lelwel*), S udan, Ethiopia, N Uganda and Kenya: ora hartebeest EN (*A. b. tora*) E, Sudan, thiopia; Jackson's hartebeest (*A. b. jacksoni*), E Africa, Rwanda; Coke's hartebeest A. b. cokii), Kenya, Tanzania; Cape or Red artebeest (*A. b. caama*), S Africa, Namiba, Botswana, W Zimbabwe.
IBL 195–200cm; TL 30cm; SH 112–130cm; L 45–70cm; WT male 142–183kg, female 26–167kg.
ORM: coat uniform sandy fawn to bright eddish, lighter on hindquarters, someimes with black markings on legs; frontal egion of head drawn up into a bony pediel; horn shape diagnostic of races.
ONSERVATION STATUS: LR Conservaion Dependent. Swayne's hartebeest A. b. swaynei) E, Ethiopia, Somalia, is /ulnerable.

Lichtenstein's hartebeest LR
Alcelaphus (Sigmoceros) lichtensteinii

Tanzania, SE DRC, Angola, Zambia, Mozambique, Zimbabwe.Open woodland. HBL190cm; TL 46cm; SH 124cm; HL 45–52cm; WT male 160–205kg, female 165kg.
FORM: coat bright reddish, with fawn flanks and white hindquarters; dark stripe down front legs; frontal region of skull does not form pedicel.
CONSERVATION STATUS: LR Conservation Dependent.

Genus *Connochaetes*
Characterized by massive head and shoulders with mane on neck and shoulders, beard under throat and long tail reaching nearly to the ground. Both sexes horned. Two species. Name derives from the Hottentot *t'gnu*, which mimics the typical loud, bellowing snort.

Black wildebeest LR
Connochaetes gnou
Black wildebeest or White-tailed gnu

S Africa. Open grass veld.
SH 115cm; HL 53–74cm; WT 150–180kg.
FORM: coat dark brown to black; tail white; face covered by brush of stiff upward-pointing hairs, tuft of hair between front legs; horns descend forward, then point upward.
CONSERVATION STATUS: LR Conservation Dependent.

Blue wildebeest LR
Connochaetes taurinus
Blue wildebeest or Brindled gnu

Northern S Africa to Kenya just south of the equator. Moist grassland and open woodland. Five subspecies, including Blue wildebeest (*C. t. taurinus*), Zambia and to the south and west; Cookson's wildebeest (*C. t. cooksoni*), Luangwa valley, Zambia; White-bearded wildebeest (*C. t. mearnsi*), Tanzania, Kenya.
HBL 194–209cm; TL 45–56cm; SH 128–140cm; HL 40–73cm; WT male 230kg, female 160kg.
FORM: coat slaty to dark gray; tail black; horns curve downward laterally and then point upward and inward; bovine appearance.
CONSERVATION STATUS: LR Conservation Dependent.

Genus *Aepyceros*

Impala LR
Aepyceros melampus

S Africa to Kenya, Namibia to Mozambique. Open deciduous woodland, especially near water. Six subspecies.
HBL male 142cm, female 128cm; TL 30cm; SH male 91cm; female 86cm; HL 49cm; WT male 80kg, female 45kg.
FORM: coat light mahogany, with fawn flanks and white undersurface of the belly; black vertical stripes on tail and thighs, and black glandular tuft on the fetlocks; straight-backed, light-limbed, and graceful – the quintessential antelope; horns, lyre-shaped, on male only.
CONSERVATION STATUS: LR Conservation Dependent. The Black-faced impala (*A. m. petersi*), E and SW Angola, NW Namibia, is Vulnerable.

TRIBE HIPPOTRAGINI

Horse-like antelopes of dry country and savanna. Africa, Arabia. Females only marginally smaller than males and both sexes with well-developed horns. Six species.

Genus *Hippotragus*
Characterized by heavily ringed horns curving backward. Well-developed mane of stiff hairs. Three species.

Roan antelope LR
Hippotragus equinus
Roan or Horse antelope

Gambia to the Somali arid zone, C Africa to S Africa, but rarely east of the Rift Valley. Thinly treed grasslands. Six subspecies.
HBL 190–240cm; TL 37–48cm; SH 126–145cm; HL 55–99cm; WT male 280kg, female 260kg.
FORM: coat sandy fawn to dark reddish with white underparts; contrasted black and white face markings; ears long with tuft of hair on tip.
CONSERVATION STATUS: LR Conservation Dependent.

Sable antelope LR
Hippotragus niger

C Africa from Kenya to S Africa, Angola to Mozambique. Woodland and woodland-grassland edges. Four subspecies.
HBL 197–210cm; TL 38–46cm; SH 117–140cm; HL 50–164cm; WT male 260kg, female 220kg.
FORM: adult females and young bulls have rich russet coat with pale underparts, darkening to black; mature bulls black with white bellies; calves uniformly dun; forehead narrow and horns scythe-like.
CONSERVATION STATUS: LR Conservation Dependent. The Giant sable antelope (*H. n. variani*), Angola, is Critically Endangered.

Genus *Oryx*
Characterized by short mane, hump over shoulder and large hooves.

Scimitar oryx EW
Oryx dammah
Scimitar or White oryx

Formerly over most of N Africa, now regarded as extinct in the wild.
SH 119cm; HL 102–127cm; WT 204kg.
FORM: coat pale, with neck and chest ruddy brown and brownish markings on the face; scimitar-shaped horns.

Gemsbok LR
Oryx gazella
Gemsbok, oryx, or Beisa oryx

Discontinuous distribution. Namibia, Angola, Botswana, Zimbabwe, S Africa, and Tanzania north to the Ethiopian coast. Seasonally arid areas. Five subspecies, including Gemsbok (*O. g. gazella*) in SW Africa; Beisa oryx (*O. g. beisa*) in the Horn of Africa; and Fringe-eared oryx (*O. g. callotis*) in Kenya and Tanzania.
HBL 153–170cm; TL 47cm; SH 120cm; HL 65–110cm; WT male 200kg female 162kg.
FORM: coat fawn with white underparts, black and white markings on the head; black line down throat and across flanks; black tail; horns straight and long.
CONSERVATION STATUS: LR Conservation Dependent.

Arabian oryx En
Oryx leucoryx
Arabian or White oryx

Formerly Arabian peninsula, Sinai peninsula. Presently reintroduced into Oman. Stony semidesert. About two-thirds the size of gemsbok.
SH 81–102cm; HL 38–68cm; WT 65–75kg.
Form: coat white, with black markings on the face; legs dark chocolate brown to black; tawny-colored line across flanks; horns nearly straight.

Genus *Addax*

Addax Cr
Addax nasomaculatus

Formerly throughout entire Sahara. Presently remnant populations in Mauritania, Mali, Niger, Chad, S Algeria, W Sudan. Sandy and stony desert far from water.
SH 100–110cm; HL 76–109cm; WT 81–122kg.
FORM: coat grayish white with white rump, belly and limbs; chestnut wig on forehead; tail long with black tip; horns long and spirally twisted. MGM

SPECIAL FEATURE

SOLVING THE MASAI MARA MYSTERY

What drives the great ungulate migrations?

ACROSS SNOW-CLAD BOREAL FORESTS AND THE Arctic tundra, over the burning hot plains of central Asia and crocodile-infested rivers in tropical Africa, millions of hoofed animals get on the move every year. What impels them to make their spectacular and risky journeys? Is each migration unique, or might we expect similar factors to underlie the long-distance treks of caribou in Canada, gazelles in the Eastern Steppe of Mongolia, and tiang and kob in southern Sudan?

To answer these questions, scientists have had to contend with the confusion generated by huge numbers of nearly indistinguishable animals traveling over vast annual ranges. Even today with the help of satellites and laptop computers, the combination of remote location and expensive and sometimes unreliable radio-tracking technology continues to hamper progress.

Perhaps the most famous of all migrations is that of the White-bearded wildebeest, which takes place in association with Burchell's zebra and Thomson's gazelle in the Serengeti National Park, Tanzania, and adjoining protected areas. Here scientists have been studying the migratory ecosystem since the late 1950s, utilizing a battery of techniques that include light aircraft surveys, radio-tracking, vehicle counts and observations, immobilization and sampling from live animals, tame animal feeding studies, rumen contents analysis of shot animals, carcass autopsy of natural kills, forage and water analysis, and computer simulation. Slowly, a picture of the causation of animal movements has begun to emerge.

The annual cycle takes herds of up to 1.5 million wildebeest out of the short-grass plains in the southeast of the Serengeti as the rainy season draws to a close. The animals first move northwest toward the wetter margins of Lake Victoria and then northward into the Masai Mara, where they remain over the driest part of the year. When the rains finally return, new grass springs up all over the ecosystem. Strangely, the wildebeest choose this precise moment to return south to the short-grass plains, leaving the northern pastures at their peak in forage quality.

It has been suggested that the wildebeest keep on the move in order to reduce the risk of predation from the estimated 3,000 lions and 9,000 spotted hyenas that share their ecosystem. The large predators are unable to follow the migration freely because they are restricted to the vicinity of dens during their breeding season. This has the added effect of limiting the population size of predators, since for part of the year they are dependent on resident game only. In support of this hypothesis, it has been found that predators limit the numbers of resident ungulates in the Serengeti well below the carrying capacity of their habitats, whereas migratory populations reach much higher population densities and become food-limited.

This explanation, however, has two flaws. Firstly it has been shown that hyenas commute up to 60km (40mi) from their dens on a regular basis, enabling them to keep in touch with the migrating animals for a greater part of their annual journey than had previously been thought. Secondly, it cannot explain why resident ungulates do not also migrate to escape the predator shadow.

A number of other causes for migrations have been put forward. Plagued by biting flies, caribou have been known to swim out to offshore islands. In Africa, tsetse flies can be an intense irritation in woody locations and have been blamed for the migrations of the Serengeti herds. Probably, however, they are responsible for localized movement patterns only. Likewise, the influence of saturated black cotton soils on movement probably has only localized significance in Serengeti.

The most long-standing explanation for migratory movements has supposed that animals take advantage of differences in the quality and availability of forage plants, as well as the important minerals they contain. By setting up fenced enclosures on the wet- and dry-season ranges in the Serengeti and clipping grass samples twice per month, researchers have been able to compare the concentrations of sodium, calcium, phosphorus, protein, and energy in forage through the annual migratory cycle and to contrast them with the minimum requirements of wildebeest. For most of the tests, it turned out that either range would be a suitable habitat for wildebeest throughout the year. There was no difference in the digestible energy of grass leaf in the two areas; protein levels averaged 3 percent lower on the dry season range, but never reached levels likely to constrain food intake; and sodium and calcium concentrations were, similarly, sufficient to meet the requirements of lactating wildebeest on both ranges.

Phosphorus levels, on the other hand, were below the required level in all samples gathered from the dry-season range, but remained well above minimum required levels on the wet season range. As the major constituents of bone, phosphorus and calcium are particularly vital for growth. But whereas natural pastures usually contain enough calcium for cattle or sheep, a deficiency in phosphorus occurs extensively throughout the world, including in parts of the Serengeti. Phosphorus deficiency causes impairment of fertility, appetite, milk yield, and growth, together with bone and teeth abnormalities and increased mortality in grazing livestock.

Above *Migratory wildebeest crossing the Serengeti plains in Tanzania. Vast herds of White-bearded wildebeest make the annual migration. The animals often perish in large numbers on the journey, particularly at river crossings.*

In order to further investigate the role of mineral deficiency in the wildebeest migration, serum and urine samples were collected from immobilized animals on wet- and dry-season ranges. It was found that the level of serum phosphate in animals immobilized on the dry-season range were less than half the critical minimum levels, reaching the same depressed level as that

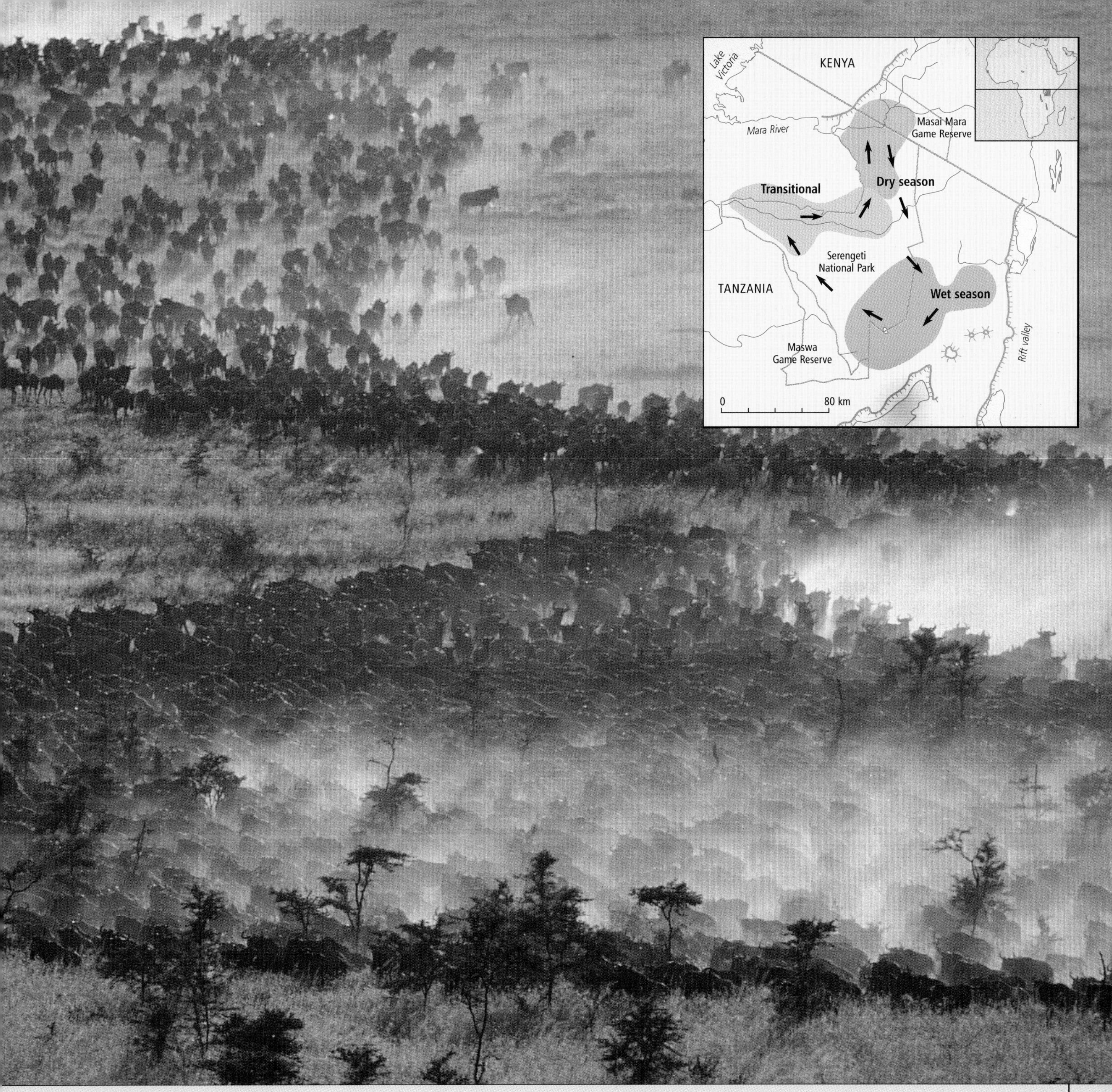

bserved in experimentally deprived sheep, which xhibited a craving for bone, bird feces, and other aturally-occurring sources of phosphorus.

We can now begin to answer the question, Why migrate?" The wildebeest leave the short-rass plains as the rains fail, responding to the lack f food, the shortage of water on the porous vol-nic soils, and the increasing salinity of the few maining pools. They move to the dry-season nge, where there is an abundance of tall grass ith occasional patches of new growth thrown up y passing rain storms, and where fresh water is vailable from permanent rivers. But during this eriod they build up a phosphorus deficiency. Vhen the rains return, the herds respond quickly, oving south onto the short-grass plains to take advantage of the green flush of mineral-rich grasses. By migrating in search of food, water, and minerals, the herds also escape the predator shadow and their numbers increase; but the escape from predators is more likely to be a consequence of the migratory habit than a cause of it. That said, it is certainly true that the larger the migratory herds, the further they have to range in order to find green grass, and in this way, all three causes – animal, vegetable, and mineral – contribute to the overall migratory habit of the wildebeest.

The resident antelopes apparently make up for the general deficiency in phosphorus by feeding in smaller group sizes. They tend to concentrate on easily recognizable hot spots in the dry-season range that have higher mineral concentrations than the surroundings. At an even finer scale, they select from mineral-rich forage on microspots such as termitaria.

The migratory and residential species are well adapted to their different lifestyles. A comparison of two alcelaphines, the hartebeest and the wildebeest, shows that the residential hartebeest has a narrow muzzle and pointed face for selective feeding in tall grasses, a low metabolic rate that is suited to a low-quality diet, and breeding that is conditional on reaching a minimum threshold in fat reserves. The migratory wildebeest, on the other hand, has a high aerobic capacity for use in its tireless rocking canter, together with a wide muzzle suited to cropping on short green grass, and a pattern of annual breeding. MGM

A SPECIALIST FOR EXTREMES

Saving the Arabian Oryx

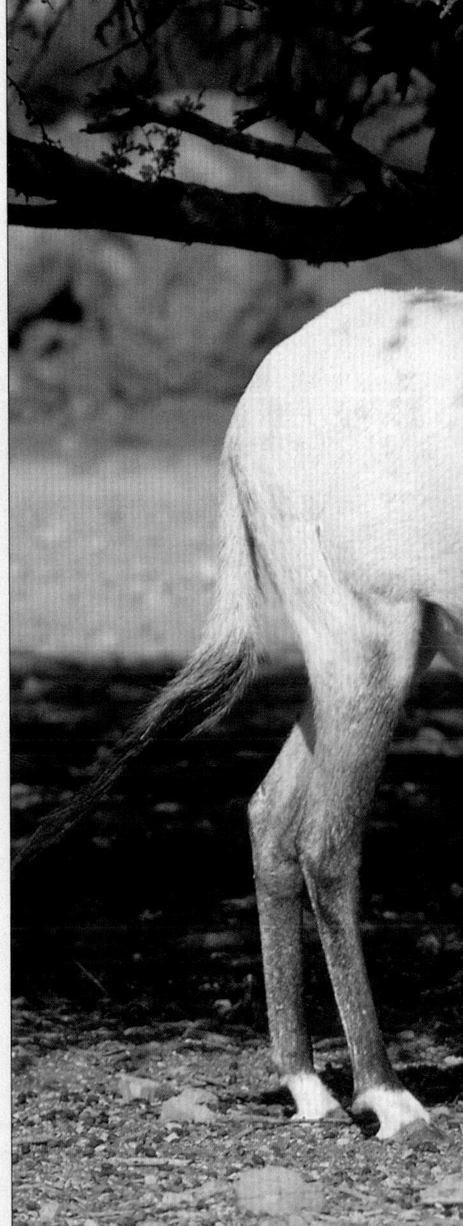

THE DESERT ENVIRONMENT OF CENTRAL ARABIA is one of cruel extremes: summer shade temperatures peak at 50°C (122°F), while winter minima drop to 6–7°C (43–45°F), with strong, cold winds. Throughout the year, the temperature varies by 20°C (36°F) each day. Over large areas, rain may not fall for years and the vegetation lies dormant or in seed; natural surface water is short-lasting after rain. Sand storms can reduce visibility to a few meters for days on end. Resources of food, water, shade, and shelter are sparse and scattered, making exacting demands on the animals living here. Of all the oryx species – gemsbok in the Kalahari, beisa in the East African Somali desert, and Scimitar oryx in the Sahara – none equals the Arabian oryx as a desert specialist.

Sparse food has selected for a small size of 65–95kg (145–210lb), only a third of the weight of the gemsbok in the Kalahari, which, although a barren landscape, is well vegetated in comparison with central Arabia. The Arabian oryx's hooves are shovel-like and splayed, with their large surface areas fully in contact with the ground as an adaptation to sandy surfaces. The oryx is not a great runner, but it can walk for hours on end to reach a favorable habitat, and treks of 25–30km (15–19mi) in one night are not unusual. The oryx often combines trekking with the unavoidable chore of rumination, but if the herd encounters a patch of good grazing, they switch to feeding at once.

The unpredictable nature of rainfall, and thus of grazing, is reflected in the ever-changing and dynamic nature of the oryx herds. In good times herd size may reach 30 animals, but in poor conditions it is typically five; two females, following calves, and an adult male. Many males are solitary and hold large overlapping territories through which females wander in search of improved grazing. While females usually move to new rain-fed grazing, territorial males may continue to hold their drought-stricken territories.

A low frequency of aggressive interactions allows animals to share scattered shade trees, under which they may spend eight of the daylight hours in the summer heat. Under shade trees the oryx excavate scrapes with their fore hooves, so that they lie in cooler sand and reduce the surface area exposed to drying winds. They fine-tune their heat regulation carefully by seeking shade earlier on hot days and not venturing out until a cooler evening breeze blows. Through behavioral avoidance of excessive heatload, precious water is not lost by panting to cool down.

The members of a feeding herd spread out until neighbors are 50–100m (165–330ft) apart, but constant visual checks, especially in undulating terrain, ensure that the animals keep in touch. Cohesion is helped by strong synchronization of activity within the herd. Changes of direction when feeding are usually initiated by the dominant female. She will start in the new direction, then stop and look over her shoulder at the others until more or, gradually, all start to follow her. Singly, oryx search for a herd and can recognize and follow fresh tracks in the sand. Moreover, as oryx are visible to the naked eye at 3km (2mi) in sunlight, the white coat may have evolved in part as a flag to assist herd-location in a open environment where merging with the environment is less necessary. Non-human predators such as the Arabian wolf (*Canis lupus arabs*) and Striped hyena (*Hyaena hyaena*) have never been abundant.

Historically, the Arabian oryx ranged through Arabia, up through Jordan and into Syria and Iraq. It was a prized trophy and source of meat fo[r] Bedu tribesmen who, hunting on foot or from

Above right *Grazing Arabian oryx take advantage of the shade offered by a tree, a rare occurrence in their sparsely vegetated habitat.*

Left *The Arabian oryx's largely white coat has evolved to reflect solar radiation. On winter mornings, the animals' coats take on the look of suede as they erect their hairs to absorb the sun's warmth, which is retained at night by the thicker winter coat. Blacker winter markings also increase heat absorption.*

Far right *Female oryx and their calves became the particular target of well-armed, motorized gangs of illegal hunters in Oman in the late 1990s. Though poaching has now been stemmed, it remains a serious threat to re-establishing this species in the wild.*

amel with primitive rifles, evidently did not signifi-antly deplete the populations. But from 1945 on, notorized hunting and automatic weapons aused a severe contraction of the animal's range nd numbers, and it became extinct in the wild in 972, leaving just a few specimens in private col-ections in Arabia and in the World Herd in the JSA. This grew out of the 1962 Operation Oryx, rganized by the far-sighted Fauna and Flora reservation Society to ensure the survival of the pecies in captivity.

By 1982, ecological and social conditions in entral Oman were deemed right for the release of carefully developed herd, with the long-term aim f re-establishing a viable population in the wild. urther immigrants joined the first ten oryx in 984. At first the released herds grew slowly, as he desert was hit by drought in 1983–86. How-ver, after rainfall in 1986, the oryx left their sup-plementary feed and water for the last time, rigid erd structures broke down, and a wild population began to develop. Further immigrants were released, and the population grew quickly to reach 100 by 1990. Although some old and young oryx died, the population survived extended drought in 1990–1992, the first real test of its fitness. Further immigrants brought the total number of released oryx to 44, and by 1996 the population comprised an estimated 400 animals, utilizing over 16,000sq km (6,000sq mi) of desert.

Sadly, in February 1996 – 14 years after the first release, and 24 years after the oryx had been exterminated in the wild – poachers again began taking oryx. Rangers found the tracks of motorized hunters, who had chased a herd and captured two calves. Over the following months armed poachers removed further animals, many of which died before reaching markets outside Oman. Poaching continued through 1997 and 1998, but was brought to a halt by early 1999. No oryx has been taken since then, and by January 2000 the wild population numbered over 100, with nearly 50, mostly rescued from the wild, awaiting release.

What has happened in Oman demonstrates that, for the oryx, the challenge of today is neither captive breeding nor reintroduction but to protect the species in the wild. Measures have been implemented locally and regionally to ensure that there is no return to poaching and illegal trade. Meanwhile, a second wild population has now been established in Saudi Arabia. Saved from a second extinction in the wild, this desert specialist continues to roam free in the Arabian Peninsula. MSP/AS

Gazelles, Dwarf Antelopes, and Saigas

THE GAZELLES, DWARF ANTELOPES, AND saigas make up a tantalizingly contrasting group, containing some of the most abundant and the rarest, the most studied and the least known of the hoofed mammals. They occupy habitats from dense forest to desert and rocky outcrops, and their range spans three zoogeographic regions from the Cape to eastern China.

Typically slim, slender-legged, and big-eyed, the gazelles and dwarf antelopes have long been considered among the most elegant of all mammals. Now, however, they are taxonomically linked with the saiga, whose sheeplike head and relatively stumpy body can initially seem clumsy and ungainly – though not to anyone who has seen these extraordinarily swift animals moving at speed.

Horn and Hoof

FORM AND FUNCTION

The gazelles of the tribe Antilopini are all very similar in shape. Most have ringed (annulated), S-shaped horns that are used in intrasexual combat rather than against predators. The diversity of horn shapes across species is thought to reflect differences in fighting styles; their complex configuration allows combatants to join horns and wrestle, making fights trials of strength rather than lethal contests involving lasting physical damage. The annulations ensure that the horns lock during combat, preventing them from sliding through and causing injury.

Gazelles are generally pale fawn above and white beneath. An exception is the blackbuck of India, in which males develop conspicuous black upperparts at about 3 years of age. The two-tone coloration of gazelles probably acts as counter-shading to obscure the animal from predators and minimize detection. The gerenuk and dibatag are particularly slender, with extended necks and legs.

Gazelles live in more open habitat than dwarf antelopes and so rely more on visual signals. Some species, such as the springbok and the Thomson's and Red-fronted gazelles, have conspicuous black sidebands. Three functions have been suggested for these bands: they may act as a visual prompt to keep the herd together; they might function as alarm signals when members are fleeing; or they may serve to break up the outline of individuals in a herd. All species have white buttocks with at least some black on the tip of the tail, and dark bands that are more or less distinct.

A feature of the gazelles is their habit of "stotting" or "pronking": jumping repeatedly with the legs straight and rigid and landing on all four limbs together. Stotting may communicate alarm; it might also serve to give the animal a better view of a predator, or to confuse or intimidate a potential aggressor, thereby discouraging pursuit. Stotting is most pronounced in the springbok, which takes its name from the custom; it also erects the line of white hairs on its back at the same time.

The senses of sight and hearing are well-developed in gazelles, and are reflected in the large orbits and ear cavities of the skull. Unlike dwarf antelopes, most females possess horns, the exceptions being the Goitered gazelle, the blackbuck, the gerenuk, and the dibatag. One possible reason for their evolution is that they permit females to better defend food resources, particularly in the dry season or in winter.

The only common features of the dwarf antelope tribe are, firstly, their small size, with females 10–20 percent larger than males; then their tendency to live alone or in pairs on territories; and finally, their well-developed (especially preorbital) glands for scentmarking. Only males have horns, with the sole exception of one East African subspecies of klipspringer (*O. o. schillingsi*); in its case females have evolved horns apparently because they are advantageous in the frequent fights needed to maintain their territories.

Unusually among hoofed mammals, female dwarf antelopes are larger than the males. It is advantageous for the females to be as large as possible within the constraints of their habitats, since

FACTFILE

GAZELLES AND DWARF ANTELOPES

Order: Artiodactyla

Family: Bovidae

Subfamily: Antilopinae

32 (38) species in 3 tribes and 14 genera. Species include **springbok** (*Antidorcas marsupialis*); **Thomson's gazelle** (*Gazella thomsoni*); **Grant's gazelle** (*Gazella granti*); **steenbok** (*Raphicerus campestris*); **Kirk's dik-dik** (*Madoqua kirkii*); **saiga** (*Saiga tatarica*).

DISTRIBUTION Africa, Middle East, Indian subcontinent, China, Tibet, Russia.

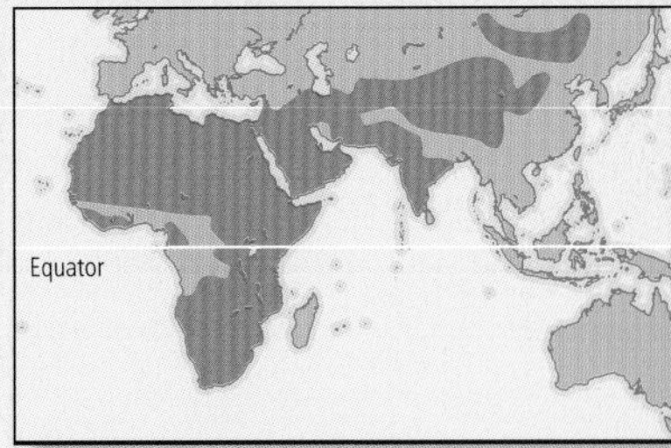

HABITAT Varied, from dense forest to desert, steppe and rocky outcrops.

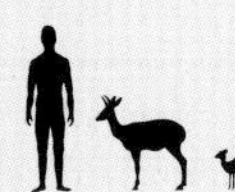

SIZE Head–body length from 45–55cm (18–21.5in) in the Royal antelope to 145–172cm (57–68in) in the Dama gazelle; **weight** from 1.5–2.5kg (3.3–5.5lb) to 40–85kg (88–188lb) in the same 2 species.

COAT Short, from coarse to silky, marked in some species by conspicuous patches and stripes.

DIET Various combinations of grasses, herbs, shoots, leaves, buds, shrubs, fruits.

BREEDING Gestation 150–225 days (gazelles); c.180 days (dwarf antelopes); 140–145 days (saiga). Most gazelles and dwarf antelope species bear single young, but saigas tend to bear twins.

LONGEVITY Springbok and blackbuck over 18 years in captivity; dwarf antelopes 13–14 years; saigas 10–12 years (in the wild).

See species table ▷

Above left *The elongated nose is a prominent* *ature of dik-diks (genus* Madoqua*), and is especially* *onounced in this species (Guenther's dik-dik).* *nother characteristic of dik-diks is the tuft of hair on* *e forehead, which sometimes conceals the male's* *rns. The preorbital scent gland can be clearly seen.*

Above *The horns of the male steenbok are smooth rather than ridged. Males of this species fight quite ferociously and even in captivity are kept separated from each other.*

Below *Older Southern Africans still remember a time when the herds of springbok were so large that it took several days for them to pass their farms.*

they have the additional burden of raising young, and it is also advantageous for calves to be large. On the other hand, there is less selection pressure for large size on dwarf antelope males as the level of competition between them is lower than in other antelope groups; they tend either to live alone or else to form lifetime pair-bonds, and so rarely need to fight with competing males.

Those species that live in dense cover, such as the Pygmy antelope, have a crouched, hare-like appearance, with long, powerful hind legs, an arched back, and a short neck. This body form is suited to rapid movement through thick vegetation. The other species live in more open habitat, where detection of predators by sight is more important, and they have a more upright posture, with a long neck and a raised head.

The exception is the klipspringer, which has an arched back, enabling it to stand with all four limbs together to take advantage of small patches of level rock. Its physical appearance is unusual; it has a thick coat of lightweight, hollow-shafted hair to protect the body against the cold as well as against physical damage from the bare rock. It has peglike hooves, each with a rubbery center and a hard outer ring to gain purchase on the rock.

The long, mobile proboscis of the dik-diks acts as a heat-exchanger, cooling the blood going to the brain. This is one of a series of adaptations (including excreting very dry feces and the most concentrated urine of any ungulate) that allow dik-diks to survive without access to open water, should the need arise.

In comparison to the gazelles, the saiga is not conventionally beautiful. Its body is clumsy, heavy-gauge, and stumpy, supported on thin, short legs. It has an inflated, trunklike nose, with disjointed longitudinal grooves and thin, weirdly-plated skin. When the animal is at rest, the trunk hangs down, but it swings from side to side when running. However, the nose is useful, particularly in winter when temperatures can plunge to –40°C; in such conditions, the extended nose preheats chilled air before it enters the respiratory paths.

It also serves a purpose when the animal runs. An unhurried saiga can appear clumsy, but a herd in flight seems to skim the ground, individuals occasionally leaping to get a better view. The animals are among the fastest in the world, touching speeds up to 70–80km/h (45–50mph). They run with a stiff gait, with their heads held low, generating much dust, and it is here that the snout helps, acting as a respirator to sift air from the dust.

Unlike most deer or antelope, which dodge and jink when fleeing danger, saigas normally run in a straight line. They do so because their shoulder joints move only forward and backward, allowing them faster movement on flat ground with low vegetation. For all its evolutionary advantages, though, this mode of movement offers no protection against motorized poachers, who pursue saigas in a straight line until they drop. Saiga are also good swimmers, and can cross rivers as wide as the Volga.

The saiga's beautiful, large black eyes see poorly, so hearing and smell are the primary senses. Their wide nostrils can detect the smell of the ground after rain, allowing a herd to change direction to exploit freshly-watered pastures. Males have horns 22–40cm (9–16in) long that serve as weapons during the rut. Females usually lack horns, which are unusual in being composed of epidermal cells with a bony core; this gives them a yellowish, translucent appearance. They are lyrate in shape, with black tips in young animals.

Desert, Plain, and Mountain

DISTRIBUTION PATTERNS

The range of the gazelle tribe is extensive, from the springbok in the Cape to the Dorcas gazelle in India and Przewalski's gazelle in eastern China. The species in north Africa, the Horn of Africa, and Asia inhabit arid and desert regions where their distribution is very patchy; they occur in low densities, with populations separated by geographical barriers such as the Persian Gulf and the Red Sea. This has led to many variations in form. A good example is the Dorcas gazelle: in Morocco it is pale-colored with relatively straight parallel horns, but as its range extends eastward it goes through various changes until in India it is reddish, smaller in size, and with lyre-shaped horns.

The dwarf antelope tribe (Neotragini) is also very varied in habitat, ranging from the Pygmy antelope in the dense forests of central Africa to the oribi in open grass plains adjacent to water, and from the dik-diks in arid bush country to the klipspringer on steep, rocky crags.

Although fossil evidence indicates that the saiga once inhabited the steppe all the way from the Crimea to Alaska, they now live only in Kazakhstan (where there are three separate populations), in the Russian republic of Kalmykia, and in Mongolia. The Mongolian populations are quite different from the others, and are considered a separate subspecies; smaller and with shorter, straighter horns, they are also behaviorally distinct, in that they do not migrate.

Right Species of dwarf antelopes and gazelles: **1** *Thomson's gazelle* (Gazella thomsonii); **2** *Kirk's dik-dik* (Madoqua kirkii) *horning vegetation;* **3** *Royal antelope* (Neotragus pygmaeus), *the smallest antelope;* **4** *oribi* (Ourebia ourebi) *marking a stem with its ear gland;* **5** *steenbok* (Raphicerus campestris) *scentmarking with the preorbital gland;* **6** *blackbuck* (Antilope cervicapra) *in territorial display pose;* **7** *Slender-horned gazelle* (Gazella leptoceros). **8** *Tibetan gazelle* (Procapra picticaudata); **9** *Dama gazelle* (Gazella dama), *the largest gazelle;* **10** *Goitered gazelle* (Gazella subgutturosa); **11** *springbok* (Antidorcas marsupialis) *pronking;* **12** *dibatag* (Ammodorcas clarkei) *in alarmed posture;* **13** *klipspringer* (Oreotragus oreotragus) *defecating on its territorial boundary;* **14** *beira* (Dorcatragus megalotis).

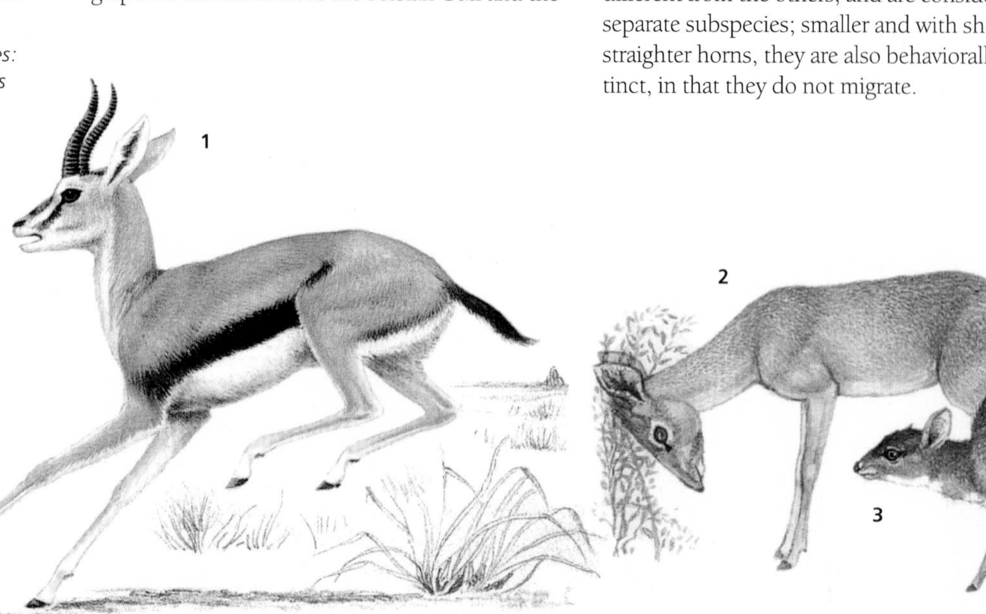

12
10
11
9
8
7
6
4
5

Browsing and Grazing

DIET

Gazelles are mixed feeders, though they mainly browse, taking grass, herbs, and woody plants in varying degrees depending on their availability. Generally, they eat whatever is greenest, so in the early spring when the grass flushes they will turn to browse. Thomson's gazelle is almost entirely a grazer, up to 90 percent of its food being grass. The gerenuk is a browser that feeds on young shoots and leaves of trees, particularly acacias up to a height of 2.6m (8.5ft). To reach the higher leaves, gerenuk have long legs and necks, and rear up on their hind legs while resting their forelegs against supporting branches.

Dwarf antelopes' small size is linked to their diet. All species except the oribi, which is a grazer, are "concentrate selectors," taking easily-digested vegetation low in fiber, such as young green leaves, buds, and fruit. A smaller body size increases the metabolic requirement per kilogram of bodyweight, so a small herbivore has to assimilate proportionally more food than a large herbivore. In all ruminants, the rate of food intake is limited by the length of time food remains in the rumen, but in dwarf antelopes this limit is so low that they have to choose vegetation of especially high nutritive quality.

In its diet, the saiga is a generalist grazer, feeding on more than 150 different items including all parts of forbs, low shrubs, and grasses. Saigas are also able to digest some plants that are poisonous for livestock; in winter they eat rough, dry twigs and shriveled leaves.

Defending Territories

SOCIAL BEHAVIOR

In all gazelles, males establish territories, at least during the breeding season, from which they actively exclude other mature males while the females are receptive. Four types of groupings occur: single territorial males; female groups with their recent offspring, usually associated with a territorial male; bachelor groups comprising nonbreeding males without territories; and mixed groups of all sexes and ages, which are common outside the breeding season. There is a certain amount of mixing of the groups. Males mark their territories with urine and dung piles and, in most species, with secretions from preorbital glands; in territorial male gerenuks, the rate of deposit is 2.24 preorbital scentmarks per hour, and in Thomson's gazelles 3.13, each on territories containing 110–120 marking sites. Subordinate males also add to the dung piles when they move through the territories; they are tolerated by the territory holder as long as they remain subordinate and do not approach females with any intent.

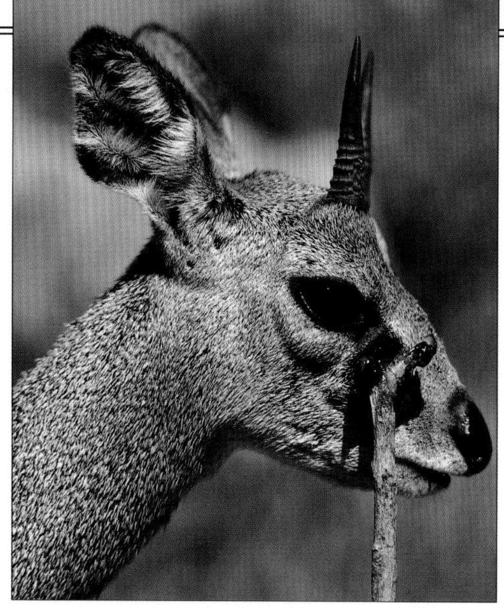

Above *A klipspringer marking a twig with its orbital scent gland. This species will similarly mark hundreds of sites within its territory.*

At the lowest level of intensity, threat displays start with head raised, chin up, and horns lying along the back. This performance reaches its most advanced form in the Grant's gazelle; two males will stand antiparallel to one another with their chins up and heads turned away, and then, at the same instant, will both whip their heads around to face each other. If the situation worsens, the first escalation takes the form of a head-on approach, with chin tucked in and horns vertical. Next comes a posture in which the head is lowered to the ground and the horns point towa the rival. Some gazelles, such as the Sand and Mountain gazelles of Saudi Arabia, indulge in "air-cushion" fights involving a series of head-on charges in which the contestants stop about 30c (12in) apart, although Sand gazelles in particula will escalate these rituals into vigorous physical fighting if opponents do not retreat.

Fights consist of opponents locking horns wit their heads down on the ground, and then pushing and twisting. At any time, an individual can submit by moving away. Fights are sometimes interrupted by a brief ritual in which combatants pretend to graze before resuming the contest. Sparring bouts are very common between bachelor males; these do not result in change of status but doubtless provide experience. Intense, wounding fights do occasionally occur between neighboring territorial males. Territorial males will herd females with a chin-up display, but if females decide to leave a territory the male will not attempt either to stop them or to retrieve them once they have crossed the boundary.

Gazelles have access to more food than dwarf antelopes, and so can afford to maintain many females in their territories. Since there are thus fewer breeding males per female, there is increased competition for the right to establish territories. Evolution has therefore favored large size in males, which in *Gazella cuvieri* are up to 57 percent larger than the females of the species.

In temperate and cold zones, gazelles breed seasonally, so that the births coincide with the vegetation flush in spring or the early rains. Females leave the herd to give birth alone. Fawns lie out for the first few weeks of life. Once they can run sufficiently well, they join the group.

Dwarf antelopes have a wide range of predator including cats like the leopard and caracal, and even large pythons; youngsters are also sometime eaten by large birds of prey and baboons. Genera ly they rely on concealment to escape detection and capture. Their first response on detecting a threat is to freeze, and then, if the predator approaches closer, to run away. Grysbucks tend

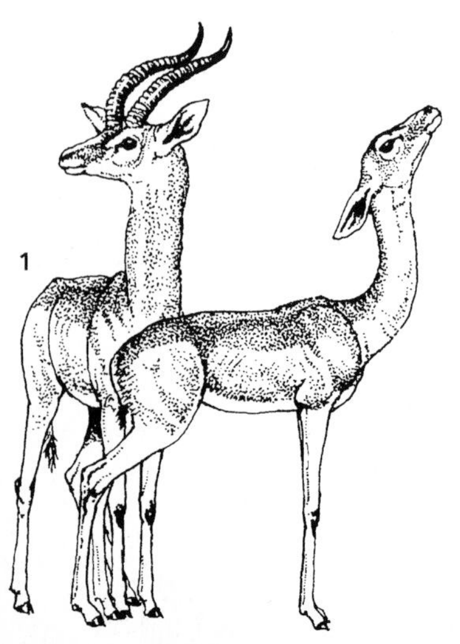

to dash away and then suddenly drop down to lie hidden from sight. In tall grass, oribi lie motionless to escape detection, but if the grass is short they flee, often with a stotting action like that of gazelles. The oribi, the klipspringer, and the dik-diks all have a whistling alarm call. Both male and female klipspringers call together in a duet when they detect predators, standing only a few meters apart. This deters pursuit, possibly because it informs the predator that it has been spotted.

Dwarf antelopes usually live alone or in monogamous pairs. The males play little or no part in raising the young. In most cases the pair-bond is lifelong, although if a mate dies the remaining individual will find a new partner. The oribi is the only member of the tribe known to sometimes form stable groups, though other oribi populations are predominantly monogamous.

The antelopes need to have an intimate knowledge of their habitat in order to locate scattered food items and to be able to hide from predators. It is therefore not surprising that they tend to stay in the same area and defend their territories against intruders. There is growing evidence that the females of the Neotragini tribe, unusually for antelope, are also territorial to some extent.

Both males and females attack intruders, typically of the same sex. Despite being fiercely territorial, male dwarf antelopes rarely make contact during their fights, which are highly ritualized. Dik-diks, for example, have an erectile crest of hair between their ears that they raise to threaten opponents while thrashing bushes with their horns; and if the contest escalates, it is to an "air-cushion" fight. In other species, an intruder will often run away from a territory-owner as far as the boundary of its own territory, where it will turn and fight. This may happen several times, giving the impression that males are taking turns to chase each other.

Territories are demarcated with scentmarks and, in some species, with piles of feces. These marks warn intruders that an area is occupied; they tend to be situated along animal trails or boundaries. Dwarf antelopes have particularly well-developed scent glands. All species have a preorbital gland in front of the eye, the only exception being the beira, which has a single gland in between the hooves; in contrast, the oribi has six different glands. The animals daub preorbital gland secretions repeatedly on particular stems; on klipspringer territories, visible black beads of secretion may be visible at as many as 840 separate sites. Pedal glands on the hooves (absent only in klipspringers) mark the ground along frequently-travelled pathways. Female dik-dik and klipspringers usually decide where the pair should scentmark within a territory, reinforcing the notion that they are territorial in their own right. Males scentmark at a higher rate than females

Right The gerenuk is unique among antelopes in being able to maneuver at full stretch on its hind legs. By browsing at heights, it occupies a distinct niche.

Below Courtship in the gerenuk involves four stages: **1** the female raises her nose in a defensive gesture, while the male displays by a sideways presentation of the head; **2** the male marks the female from behind on the thigh; **3** the male performs the foreleg kick, tapping the females hind legs with his fore leg; **4** the male performs the "flehmen" or lip-curl test, sampling the female's urine to determine whether she is ready to mate.

4

(leaving 3.4 marks per bout as against the females' 2.1), and they will also overmark female marks: this is important in advertising the paired status of the female.

Births occur throughout the year, but there are peaks that coincide with the vegetation flush that follows the early rains. In equatorial regions, where there are two rainy seasons a year, two birth peaks occur. Young are born singly. Newborn young hide for the first few weeks after birth, and their mothers come to suckle them, calling with a soft bleat. In dik-diks, females disperse at 6–8 months of age, shortly before their first estrus, perhaps to avoid being mated by their fathers. Females become sexually mature at 6 months in the smaller species and 10 months in the larger. Male dik-diks disperse much later, waiting until a territory becomes available. As male offspring become sexually mature, they are increasingly threatened by the territorial male. They respond with submissive displays, and eventually leave the territory at about the age of 9–15 months.

For all but the Mongolian saigas, migration is a way of life. Saigas spend the winter in desert areas, where the snow is not so thick and vegetation is relatively plentiful. They move up to 1,000km (620mi) north to the steppe areas in the spring to take advantage of the rich grazing, retracing their steps in the autumn. These migration routes were also followed by nomadic pastoralists in pre-Soviet times. Saigas also move to avoid bad weather, food shortages, predators, and even poachers, their evasive meanderings taking them as far as 40km (25mi) in a single day. Their nomadic habits are advantageous in a patchy environment, such as the semidesert, where food resources can be highly localized.

Herd size varies seasonally; the huge, mixed migrating herds break up into smaller groups of tens to thousands of animals when they reach their destinations. Reproduction also affects the herds' composition; saigas break their journey to the summer pastures in late April–early May to give birth in huge aggregations. More than 80 percent of females give birth in a 10-day period, swamping predators (mainly wolves) with the sheer number of offspring. The sound of these birth areas as mothers and calves communicate with one another resembles the noise inside a football stadium.

In the mating season (December), the saigas adopt a new look; both sexes are distinctive in long, light-colored wool, while males grow lengthy "whiskers" under the eyes and a dark collar around the neck. Then the herds organize into harems. The males bellow as they fight for possession of females, sometimes hurting each other badly in the process. Defending harems of between 5 and 50 females is tiring work for the successful males, who lose weight and condition. They are easy prey for predators and poachers, and suffer heavy losses in harsh winters. Meanwhile, the mated females begin a gestation that lasts 140–145 days.

Like their nomadism, saigas' potential for very rapid population growth is an adaptation to their unpredictable environment. The semi-arid rangelands where they live are climatically harsh, with temperatures reaching lows of –40°C (–72°F) in winter and highs of +30°C (86°F) in summer. Drought and snowy winters can cause mass mortality, but saiga populations bounce back quickly because of their high fecundity. Females reach sexual maturity at 7–8 months and – unusually among ungulates – routinely give birth to twins.

THE TRADE IN SAIGA HORNS

Saiga horns are prized in traditional Chinese medicine as a treatment for serious illnesses such as strokes, often in combination with rhino horn. The trade is far from new; in the 19th century, it was so intensive that it caused a population collapse that led contemporary observers to tell of saiga herds being reduced to a few thousand individuals. The populations in Kazakhstan and Kalmykia declined so drastically that one of the several species of parasitic blowflies that live on the animals was extirpated, having nowhere else to live. The Soviet period saw strong conservation measures and a dramatic recovery of the species, peaking at around 1 million animals in Kazakhstan alone. Now intensive, uncontrolled hunting once again threatens the species.

The fact that the trade in horns targets only males has worrying implications for the saiga's social and reproductive behavior. Fewer males are available, defending larger harems of females. These harems can be so numerous that with the best will in the world it becomes impossible for the male to impregnate all of his wives – perhaps as many as 700 of them. Females thus go without bearing offspring for an entire season, which for an animal that lives only 3–4 years is a large part of its potential reproductive career. This problem was acute in Kalmykia in the 1999–2000 season. AAL/MVK/EJM-G

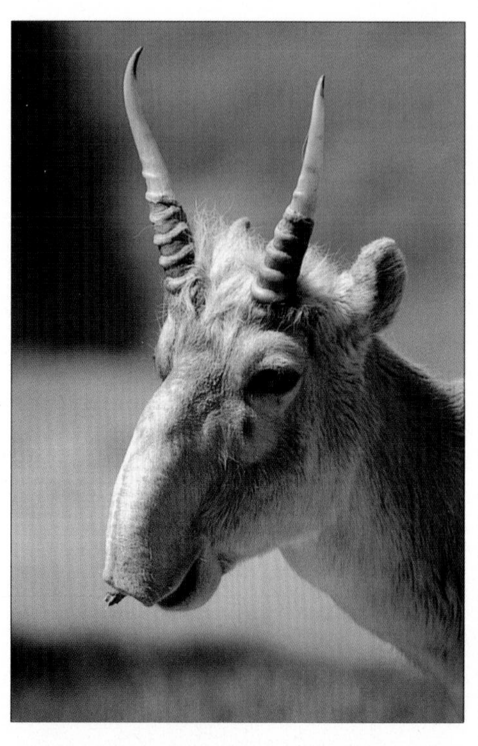

Shrinking Ranges

CONSERVATION AND ENVIRONMENT

Many species of gazelle have been greatly reduced in numbers and range by man, and a large proportion are listed as either Vulnerable or Endangered in the IUCN Red List. Those that inhabit North Africa, the Horn of Africa, and Asia have suffered most, because they live in arid habitats at densities of less than 1 per sq km (2.6 per sq mi) where they can easily be hunted from vehicles. Domestic sheep and goats also compete for the same food plants, and access to springs has been denied them by human activity. Przewalski's gazelle and two subspecies of mountain gazelle (*G. g. acaciae* from Israel and *G. g. muscatensis* from Oman) are critically endangered. In Asia, cultivation has removed many areas of winter range, so that the vast winter aggregations of some species are now rare or absent.

The range of the dwarf antelopes, which are confined to Africa, has almost certainly been affected by habitat disturbance, since many species prefer the secondary growth that invades disturbed areas, notably from slash-and-burn cultivation. According to the current Red List, the western subspecies of klipspringer (*O. o. porteousi*) is endangered, while Piacentini's dik-dik, the beira, and a subspecies of oribi (*O. o. haggardi*) are vulnerable. Another subspecies of oribi, *O. o. keniae*, has recently become extinct.

Despite their hardiness, saigas suffer from various manmade and natural afflictions. The last few years have seen climate change that has left Kalmykia increasingly arid, and drought during lambing has hit hard. The breakup of the Soviet Union has also had major repercussions, causing previously successful conservation and management programs to collapse through lack of funds; at the same time, high levels of poverty and unemployment have encouraged poaching. The collapse of the rural economy in Kazakhstan has led not just to an emptying of the steppe of livestock but also to the re-emergence of foot-and-mouth disease and other infections, putting the susceptible saiga population at risk. Such episodes were common before vaccination began.

Conservation efforts have been limited in recent years because of the socio-economic problems in the region. Smallscale experimental captive breeding programs are under way in Kalmykia and China, but these need to be complemented by large-scale antipoaching measures for in situ conservation of this remarkable beast.
CR/PB/AAL/MVK/EJM-G

Gazelle, Dwarf Antelope, and Saiga Species

THE SUBFAMILY ANTILOPINAE USED TO consist of two tribes – the Antilopini, or gazelles, and the Neotragini, or dwarf antelopes. Recently, however, the Saigini, comprising a single East Asian species, the saiga, has been reassigned to it from the Caprinae or goat antelopes, largely on the strength of phenetic and phylogenetic analysis. In addition, a revision of dik-dik taxonomy has increased the number of these East African dwarf antelope species from three to four.

GAZELLES

Tribe Antilopini

Slender body and long legs. Males larger than females. Fawn upperparts, white underparts, typically with gazelline facial markings of dark band on blaze, white band on either side, dark band from eye to muzzle, and white around eyes. Horns generally S-shaped, annulated. Usually both sexes have horns, though female horns are shorter and thinner. Tail black-tipped. Populations typically comprise territorial males, at least during the breeding season, female groups with their offspring, and bachelor groups of nonterritorial males. Will migrate in response to seasonal changes in vegetation and climate, forming large, mixed-sex aggregations during the winter or dry seasons. In seasonal parts of their range, birth peaks coincide with vegetation flush in spring or early rains. Mostly diurnal. Mixed feeders, but mostly browse.

Genus *Gazella*

Subgenus *Nanger*

Large gazelles. White rump and buttocks, tail white with black tip. 3 species.

Dama gazelle En
Gazella dama

Sahara, from Mauritania to Sudan. Desert. Occurs singly or in small groups. In rainy season moves N into Sahara, in dry season back to Sudan. Very rare.
HBL 145–172cm; TL 22–30cm; SH 88–108cm; HL 33–40cm; WT 40–85kg. Neck and legs long for a gazelle.
FORM: neck and underparts reddish brown, sharply contrasting with white rump, underparts, and head; white spot on neck. Horns sharply curved back at base, relatively short.

Soemmerring's gazelle Vu
Gazella soemmerringii

Horn of Africa, N to Sudan. Bush and acacia steppe. Occurs in small groups of 5–20.
HBL 122–150cm; TL 20–28cm; SH 78–88cm; HL 38–58cm; WT 30–55kg.
FORM: pale fawn on head, neck, and underparts; facial markings very pronounced. Short neck; long head; horns sharply curved inward at tips.

Grant's gazelle LR
Gazella granti

Tanzania, Kenya, and parts of Ethiopia, Somalia, and Sudan. From semidesert to open savanna. Lives in small groups of up to 30. Preorbital gland not used.
HBL 140–166cm; TL 20–28cm; SH 75–92cm; HL 45–81cm; WT 38–82kg.
FORM: black pygal band; heavily built; horns long, variable according to race.
CONSERVATION STATUS: LR Conservation Dependent.

Above *For dik-diks, marking at fecal piles becomes something of a "dung ceremony," with the male scraping dirt over the female's deposits before making his own contribution.*

Subgenus *Gazella*

7 species of small gazelle. Except *G. thomsoni*, occur in small groups.
HBL 70–107cm; TL 15–26cm; SH 40–70cm; HL 25–43cm; WT 15–32kg.
COAT: white on underparts and buttocks, not extending to rump.
BREEDING: gestation 5.5 months.

Mountain gazelle LR
Gazella gazella

Arabian peninsula, Palestine; extinct over much of its range. Semidesert and desert

scrub in mountains and coastal foothills.
COAT: upperparts fawn, pygal and flank bands distinct.
BREEDING: seasonal.
CONSERVATION STATUS: LR Conservation Dependent.

Dorcas gazelle Vu
Gazella dorcas
Dorcas gazelle or jebeer

From Senegal to Morocco, and W through N Africa and Iran to India. Semidesert plains.
COAT: upperparts pale; pygal and flank bands indistinct.

Slender-horned gazelle En
Gazella leptoceros

Egypt E into Algeria. Mountainous and sandy desert.
FORM: upperparts very pale; ears large; hooves broadened; horns long, only slightly curved.

Red-fronted gazelle Vu
Gazella rufifrons

From Senegal in a narrow band running E to Sudan. Semidesert steppe.
FORM: reddish upperparts, narrow black band on side with reddish shadow band below, contrasting with white underparts; horns short, stout, only slightly curved.

Thomson's gazelle LR
Gazella thomsonii

Tanzania and Kenya, and an isolated population in S Sudan. Open, grassy plains. Very abundant; occurs in large herds of up to 200, with aggregations of several thousand during migration. Largest of the subgenus.
FORM: upperparts bright fawn, broad conspicuous dark band on the side; well-pronounced facial markings; horns long, only slightly curved; small and slender in females.
DIET: mixed feeder, but predominantly a grazer.
CONSERVATION STATUS: LR Conservation Dependent.

Speke's gazelle Vu
Gazella spekei

Horn of Africa. Bare, stony steppe. Little known.
FORM: upperparts pale fawn, broad dark sideband; small, swollen, extensible protuberance on nose.

Edmi En
Gazella cuvieri

Morocco, N Algeria, Tunis. Semidesert steppe.
COAT: upperparts dark gray-brown, dark sideband with shadow band below; facial markings pronounced.

Subgenus *Trachelocele*

Goitered gazelle LR
Gazella subgutturosa

From Palestine and Arabia E through Iran and Turkestan to E China. Semidesert and desert steppe. In Asia, forms winter aggregations of several thousand at lower altitudes to avoid snow; in summer, the aggregations disperse, with females going further than males.
HBL 38–109cm; TL 12–18cm; SH 52–65cm; HL 32–45cm; WT 29–42kg.
FORM: upperparts pale, pygal and sidebands indistinct; facial markings not pronounced, fading to white with age. Horns arise close together and curve in at tips; female mostly hornless. Male larynx forms conspicuous swelling. Stocky body, relatively short legs.
CONSERVATION STATUS: LR Near Threatened.

Genus *Antilope*

Blackbuck Vu
Antilope cervicapra

Indian subcontinent. From semidesert to open woodland.
HBL 100–150cm; TL 10–17cm; SH 60–83cm; WT 25–45kg.
FORM: upperparts and neck dark brown to black in adult males, contrasting with white chin, eyes, and underparts; immature males and females are light fawn. Horns long and spirally twisted; females hornless.

Genus *Procapra*

Gazelle-like. Pale fawn upperparts, white rump and buttocks. 3 species.

Tibetan gazelle LR
Procapra picticaudata

Most of Tibet. Plateau grassland and high-altitude barren steppe.
HBL 91–105cm; TL 2–10cm; SH 54–64cm; HL 28–40cm; WT 20–35kg.
FORM: face glands and inguinal glands absent; horns S-shaped, not curving in at tips.
CONSERVATION STATUS: LR Near Threatened.

Przewalski's gazelle Cr
Procapra przewalskii

China, from Nan Shan and Kukunor to Ordos Plateau. Semidesert steppe.
FORM: same size as Tibetan gazelle; horns curve in at tip.

Mongolian gazelle LR
Procapra gutturosa

Most of Mongolia and Inner Mongolia. Dry steppe and semidesert.
HBL 110–148cm; TL 5–12cm; SH 30–45cm; WT 28–40kg.
FORM: small preorbital glands and large inguinal glands present.
CONSERVATION STATUS: LR Near Threatened.

Genus *Antidorcas*

Springbok LR
Antidorcas marsupialis

Southern Africa W of Drakensberg Mountains and N to Angola. Open, arid plains. Very gregarious; used to migrate in vast herds of tens of thousands. Has characteristic pronking gait, in which white hairs on back are erected.
HBL 96–115cm; TL 20–30cm; SH 75–83cm; HL 35–48cm WT 25–46kg.
FORM: bright reddish fawn upperparts, dark sideband contrasting with white underparts; face white, with dark band from eye to muzzle; buttocks and rump white, and line of white erectile hairs in fold of skin along lower back. Horns short, sharply curved in at tip.
DIET: mixed feeders, taking predominantly grass.
CONSERVATION STATUS: LR Conservation Dependent.

Genus *Litocranius*

Gerenuk LR
Litocranius walleri

Horn of Africa S to Tanzania. Desert to dry bush savanna. Occurs singly or in small groups. Does not form migratory aggregations.
HBL 140–160cm; TL 22–35cm; SH 8–105cm; HL 32–44cm; WT 29–52kg.
FORM: reddish brown upperparts, back distinctly dark, white around eyes and underparts. Very long limbs and neck, with small head and weak chin. Tail short with black tip. Horns stout at base, sharply curved forward at tips; females hornless.
DIET: browses on tall bushes by standing on hind legs; feeds delicately on leaves and young shoots. Independent of water.
CONSERVATION STATUS: LR Conservation Dependent.

Genus *Ammodorcas*

Dibatag Vu
Ammodorcas clarkei

Horn of Africa. Grassy plains and scrub desert. Occurs singly or in small groups. Able to stand on hind legs to browse. Holds tail erect in flight. Does not form migratory aggregations.
HBL 152–168cm; TL 30–36cm; SH 80–88cm; HL 25–33cm; WT 23–32kg.
FORM: upperparts dark reddish gray, contrasting with white underparts and buttocks; head with chestnut gazelline markings; tail long, thin, and black; long neck and limbs. Horns curve backward at base, then forward at tip as in reedbuck; females hornless.

DWARF ANTELOPES

Tribe Neotragini

Small, delicate antelopes, females slightly larger than males. Crepuscular or nocturnal. Live singly or in small family groups. Males territorial. Territories marked with dung piles, preorbital and pedal glands; scent glands well-developed. Underparts are white or buffish, horns short and straight; females are hornless, except in one subspecies of klipspringer. Diet consists of young leaves and buds, fruit roots, tubers, fallen leaves, green grass (except for oribi). Dwarf antelopes breed throughout the year, with birth peaks in early rains; gestation about 6 months.

Genus *Neotragus*

Horns smooth, short, inclined backward. Tail relatively long. Back arched, neck short. 3 species.

Royal antelope LR
Neotragus pygmaeus

Sierra Leone, Liberia, Côte d'Ivoire, Ghana. Dense forest. Occurs singly or in pairs; shy, secretive, little-known.
HBL 45–55cm; TL 4–4.5cm; SH 20–28cm; HL 2.5–3cm; WT 1.5–2.5kg; smallest horned ungulate.
COAT: back brown, becoming lighter and bright reddish on flanks and limbs, contrasting with white underparts; tail reddish on top and white underneath and at tip; head and neck dark brown.
CONSERVATION STATUS: LR Near Threatened.

ABBREVIATIONS

HBL = head–body length HTL = head–tail length TL = tail length SH = shoulder height
HL = horn length WT = weight
Approximate nonmetric equivalents: 10cm = 4in; 1kg = 2.2lb

Ex Extinct
EW Extinct in the Wild
Cr Critically Endangered
En Endangered
Vu Vulnerable
LR Lower Risk

ygmy antelope LR
leotragus batesi

E Nigeria, Cameroon, Gabon, Congo, W ganda, Zaire. Dense forest. Predominant- solitary. Often moves into plantations nd recently disturbed land at night. emales have overlapping home ranges. BL 50–58cm; TL 4.5–5cm; SH 24–33cm; L 2–4cm; WT 3kg.
OAT: shiny dark chestnut on the back, ecoming lighter on the flanks; tail dark rown.
ONSERVATION STATUS: LR Near Threat- ned.

uni LR
leotragus moschatus

atchily distributed from Kwazulu-Natal hrough Mozambique and Tanzania to enya. Coastal, riverine, and montane for- st with thick undergrowth. Occurs singly r in small groups, with never more than ne adult male per group. Males grate heir horns on tree trunks and scentmark vith preorbital glands.
IBL 58–62cm; TL 11–13cm; SH 30–41cm; IL 5–13cm; WT 4–6kg.
ORM: dark brown, slightly freckled; ransparent pink ears; horns strongly nnulated at base.
ONSERVATION STATUS: LR Conservation Dependent.

Genus *Madoqua*

Erectile hairs on forehead. Horns in males. arge preorbital scent gland. Tail minute. Underparts white. Males and females form ifelong pairs and defend territories. Larger emporary groupings commonly occur at sites of food abundance, such as under lowering trees. Diurnal. 4 species.

Salt's dik-dik
Madoqua saltiana
Salt's, Swayne's, or Phillips' dik-dik

Horn of Africa. Arid evergreen scrub in oothills and outliers of Ethiopean moun- ains, particularly in disturbed or over- grazed areas with good thicket vegetation.
HBL 52–67cm; TL 3.5–4.5cm; SH 34.5–40.5cm; HL 4–9cm; WT 3–4kg.
FORM: thick coat, back gray and speckled, flanks variable gray to reddish; legs and forehead and nose bright reddish; white rings round eyes; nose slightly elongated.

Piacentini's dik-dik Vu
Madoqua piacentinii
Piancentini's or Silver dik-dik

E Somalia. Semi-arid scrub. Size and coat similar to Salt's dik-dik.

Guenther's dik-dik
Madoqua guentheri

N Uganda, E through Kenya and Ethiopia to the Ogaden and Somalia. Semi-arid scrub.
HBL 62–75cm; WT 3.5–5cm; SH 34–38cm; HL 4–9cm; WT 4–5.5kg.
FORM: back and flanks speckled gray; reddish nose and forehead; nose conspicuously elongated.

Kirk's dik-dik
Madoqua kirkii
Kirk's or Damara dik-dik

Tanzania and S half of Kenya (Kirk's); Namibia and SW Angola (Damara). Over 2,000km (1,250mi) separate these two ranges, and recent genetic work suggests that they may be separate species.
HBL 60–72cm; TL 4.5–5.5cm; SH 35–43cm; HL 4–9cm; WT 4.5–6kg.
FORM: whitish rings round eyes; nose moderately elongated; the Damara dik-dik has rubbery pads on its hooves, and is restricted to hard or rocky ground.

Genus *Oreotragus*

Klipspringer LR
Oreotragus oreotragus

From the Cape to Angola, and up the E half of Africa to Ethiopia and E Sudan; also found in two isolated massifs in Nigeria and Central African Republic. Well-drained rocky outcrops. Gait a stilted, bouncing motion. Occurs in pairs, sometimes accompanied by 1 or 2 offspring. Very well-developed preorbital glands. Aggregations occur at favorable feeding sites. Lacks pedal glands.
HBL 75–90cm; TL 6.5–10.5cm; SH 43–51cm; HL 6–16cm; WT 10–15kg
FORM: yellowish coat, speckled with gray; ears round, conspicuously bordered with black; black ring above hooves; tail minute; back conspicuously arched. Fur is thick, coarse, and brittle, loosely rooted and lightweight, giving the animal a stocky appearance. Hooves peglike; horns smooth, nearly vertical.
CONSERVATION STATUS: LR Conservation Dependent.

Genus *Raphicerus*

Coat reddish, large white-lined ears, tail minute, horns smooth and vertical. Diurnal or crepuscular. 3 species.

Steenbok
Raphicerus campestris

From Angola, Zambia, and Mozambique S to the Cape, and in Kenya and Tanzania. Open, lightly wooded plains. Usually seen singly or in loosely-maintained pairs. Has home range of some 0.62sq km.
HBL 70–95cm; TL 4–6cm; 45–60cm; HL 9–19cm: WT 10–15kg.
FORM: reddish fawn; large white-lined ears; shiny black nose; horns smooth and vertical; tail minute. Preorbital gland used very little for territorial marking.

Sharpe's grysbuck LR
Raphicerus sharpei

Tanzania, Zambia, Mozambique, Zimbabwe. Woodland with low thicket or secondary growth. Mainly nocturnal and cryptic.
HBL 61–75cm; TL 5–7cm; SH 45–60cm; HL 3–10cm; WT 7.5–11.5kg.
FORM: reddish brown, speckled with white on back and flanks; large white-lined ears; horns small, smooth, and vertical; tail small; back slightly arched.
CONSERVATION STATUS: LR Conservation Dependent.

Cape grysbuck
Raphicerus melanotis

Restricted to the S Cape. Not as red as Sharpe's grysbuck, but otherwise similar.

Genus *Ourebia*

Oribi LR
Ourebia ourebi

Range patchy and extensive. E half of southern Africa, Zambia, Angola, and Zaire, and from Tanzania N to Ethiopia, and W to Senegal. Grassy plains with only light bush, near water. Lives mostly in pairs or small family groups, though aggregations do occur at favorable feeding sites. Scent glands and marking very well developed in males. Territories large. Commonly runs with stotting gait. Grazer. Diurnal.
HBL 92–140cm; TL 6–15cm; SH 54–67cm; HL 8–19cm; WT 14–21kg.
FORM: reddish fawn back and flanks, contrasting conspicuously with white underparts; forehead and crown reddish brown; black glandular spot below ear; tail short, with black tip. Ears large and narrow; horns short, vertical, slightly annulated at base.
CONSERVATION STATUS: LR Conservation Dependent.

Genus *Dorcatragus*

Beira Vu
Dorcatragus megalotis

Somalia and Ethiopia, bordering the Red Sea and the Gulf of Aden. Stony, barren hills and mountains. Very rare and little-known. Diurnal.
HBL 70–85cm; TL 14–20cm; SH 52–65cm; HL 7.5–12.5cm; WT 15–26kg.
FORM: gazelle-like; large for a neotragine. Gray coat, finely speckled on back and flanks; distinct dark band on sides; underparts yellowish. Ears very large; tail long and white; horns widely separated, curving slightly forward; rubbery hooves adapted to rocky habitat.

SAIGA

Tribe Saigini

Saiga LR
Saiga tatarica

N Caucasus, Kazakhstan, SW Mongolia, Zinjiang (China). Inhabit cold, elevated, and arid steppe. Migrate annually in huge herds of up to 200,000 individuals, although normal herd size is 30–40.
HTL 123–146cm (male), 108–125cm (female); HT 69–79cm (male), 57–73cm (female); WT 32–51kg (male), 21–41kg (female).
FORM: coat sandy; horns amber and translucent. Only males have horns. Head has grotesque proboscis used to filter out dust during the summer migrations and to heat the air during the icy winters. Nose increases in size in males during rut, at which time hair tufts below the eyes become impregnated with sticky, smelly, preorbital gland secretions; carpal and inguinal glands also present; ears and tail short; neck with short mane.
DIET: mainly grazers, although they may also feed on small herbs and lichens.
BREEDING: gestation 145 days; twinning common, especially after the first birth. Population greatly expanded in recent years.
CONSERVATION STATUS: LR Conservation Dependent. One subspecies, the Mongolian saiga (*S. t. mongolica*), is listed as Endangered.

The 1993 established list of mammalian species recognized five further species of *Gazella*: the Arabian gazelle (*G. arabica*), the Indian gazelle (*G. bennettii*), the Queen of Sheba's gazelle (*G. bilkis*), the Red gazelle (*G. rufina*), and the Saudi gazelle (*G. saudiya*). In addition, the chiru (*Pantholops hodgsoni*)– here accorded its own subfamily, Panthalopinae, within the Goat Antelopes – was included within the subfamily Antilopinae.

Goat Antelopes

THE CAPRIDS ARE RENOWNED AS GAME animals – the heads of giant sheep and bighorns are prized trophies, and permits to hunt these species may fetch hundreds of thousands of dollars.

The goat antelopes and their descendants (subfamily Caprinae) blossomed into a great diversity of species during the Ice Ages, which began some 4 million years ago. They came to occupy a variety of extreme environments – hot deserts, arctic barrens, or alpine plateaus – while also surviving in their evolutionary home, the humid tropics.

Resource Defenders and Grazers

FORM AND FUNCTION

The basic goat-antelope body plan dates back to the late Miocene. However, as the goat antelopes advanced into extreme climates, they abandoned the ancestral body plan, grew in size, and diverged in appearance. The end products were giants such as the Musk ox, the giant rams, the long-horned ibex, and the bull-headed takin, as well as many extinct Ice Age forms.

The goat antelopes and their kin are usually stocky, gregarious bovids. The primitive forms tend to have small horns and strongly patterned coats; the true sheep and goats are characterized by long, curving horns. The Musk oxen and their relatives were all large-bodied with large horns. There is no consistent explanation for the diverse body colors among these species, which range from pure white to pure black. Like other social ungulates, gregarious caprids have strong pelage and color markings on head and tail poles. The females have smaller bodies and horns. At one extreme (the most primitive), there are so-called resource defenders, which live in small, well-defined areas of productive, diverse habitat that supply all their needs and can be easily defended against members of the same species. At the opposite extreme (the most recently evolved) are the grazers, which live in less productive, climatically severe landscapes. They are highly gregarious, roam widely, and have retained very few attributes of their tropical ancestors. They have large horns, engage in a variety of "sporting engagements," and the sexes may be strikingly different. Species of goat antelopes are also adapted to a whole range of lifestyles in between these extremes.

Most resource defenders are dark in color, with male and female being similar in appearance. As small territories are easier to defend, such species are small-bodied and have a wide diet. Male and female serow look alike – dark colored, with long, tasseled ears, a long neck-mane, and large preorbital glands for scentmarking. They are aggressive and successfully fight off predators with their daggerlike horns. Their teeth are typical of browsers, with lower crowns, and their legs are less specialized for climbing than those of their advanced relatives. Their home ranges are well structured, with trails, "horn-rubs" on saplings, and dung heaps. Serow are widely distributed throughout Southeast Asia and vary in color between regions. They are largely confined to moist, shrubby, or timber-covered rocky outcrops.

The two extreme forms of rupicaprids are the Chamois and the Mountain goat. Both are adapted to cold, highland terrain and both have glands on the back of the head. But there the similarities end. The Mountain goat is a massive, ponderous rock-climber that only reverts to resource defense under severe winter conditions. It clings closely to cliffs and can live successfully even in the coastal ranges of Alaska. The Chamois is relatively small-bodied and gregarious. Only large males may usurp pockets of rich habitat in summer to fatten for the rut.

Several lines of goat antelopes became adapted to grazing; thus arose the advanced caprids (sheep and goats) and the Musk ox, as well as their many extinct relatives. Among caprids, the most primitive form – and closest to the ancestral rupicaprids – is the tahr. In morphology and behavior it stands between the rupicaprids and the caprids. It is closely wedded to cliffs and has broad food habits.

The Barbary sheep exemplifies the next stage in evolution. "New designs" are expected to arise at the periphery of the ancestor's range when colonizing a new habitat. Although a new species adapts to the new physical and biotic characteristics of the environment it colonizes, changes in its external appearance arise, initially as a result of social selection during colonization. Because food is abundant during colonization, competition shifts from food to mates. Moreover, the few colonizing individuals reproduce rapidly. Thus social competition and rapid selection quickly generate larger bodies, stouter weapons, and better bluffing abilities. As expected, the Barbary sheep has a bigger head and horns than the tahr, although head-clashing is clearly prevalent. Barbary sheep are similar in appearance to goats, but biochemical evidence suggests they are more closely related to sheep. This odd caprid stands at the dividing point between two great radiations – those of the sheep and true goats.

Although precise details of sheep and goat evolution are obscure, we do know that both genera moved into cold mountains, increasing in body and horn size. Some species arose locally in response to ecological opportunities and it is likely that there were hybridizations and extinctions.

Goats became specialized for cliffs, while sheep inhabited the open dry lands close to cliffs. These differences arise from different ways of evading predators: sheep escape by running and by clumping, while goats are more specialized at bounding among rocks, a terrain that hampers predators. In appearance, sheep and goat males

Below and right *Representative species of goat antelopes:* ***1*** *Goral* (Nemorhaedus goral)*;* ***2*** *Himalayan tahr* (Hemitragus jemlahicus)*;* ***3*** *Urial* (Ovis orientalis)*;* ***4*** *Barbary sheep* (Ammotragus lervia)*,* ***5*** *Wild goat* (Capra aegagrus)*;* ***6*** *Japanese serow* (Capricornis crispus)*;* ***7*** *Chamois* (Rupicapra rupicapra)*,* ***8*** *Ibex* (Capra ibex)*;* ***9*** *Mountain goat* (Oreamnos americanus)*;* ***10*** *Argalis* (Ovis ammon)*;* ***11*** *Takin* (Budorcas taxicolor)*;* ***12*** *Musk ox* (Ovibos moschatus).

10 11 12

9 7 8

6

5

3 4

FACTFILE

GOAT ANTELOPES

Order: Artiodactyla

Family: Bovidae

27 (34) species in 2 subfamilies and 11 (10) genera. Species include: **chamois** (*Rupicapra rupicapra*); **Mountain goat** (*Oreamnos americanus*); **Musk ox** (*Ovibos moschatus*); **Barbary sheep** (*Ammotragus lervia*); **ibex** (*Capra ibex*); and **chiru** (*Pantholops hodgsoni*).

DISTRIBUTION Asia, C Europe, N and C America, N Africa. The chiru inhabits the Tibetan plateau.

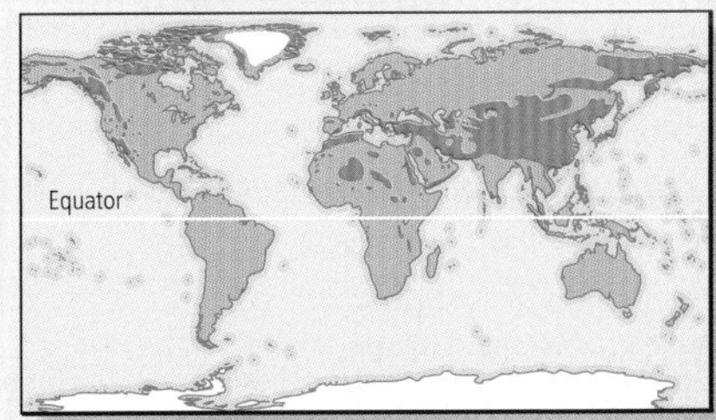

HABITAT Steep terrain, from hot deserts and moist jungles to Arctic barrens. The chiru lives on Alpine steppe.

SIZE Male **head–tail length** ranges from 1.06–1.17m (42–46in) in the goral to 2.45m (8ft) in the Musk ox; **height** 69–78 cm (27–31in) to 1.5m (5ft), and **weight** 28-42kg (62–92lb) to 350kg (770lb).

COAT Very diverse, from pure white in Mountain goats and Dall's sheep to pure black in Musk oxen and Stone's sheep. Gregarious forms tend to have strong frontal and rear markings.

DIET Grasses, herbs, leaves, shoots, fungi, sedges, lichens.

BREEDING Gestation 150–160 days in the urial, 180 days in the chiru, 250–260 days in the goral.

LONGEVITY 6 years in the urial; 9 years in Blue sheep and American bighorn sheep (up to 24 years in captivity).

See species table ▷

differ considerably but females do not. Male goats have long, scimitar-shaped, knobby horns as well as a chin beard. Male sheep lack a chin beard but have massive curling horns; primitive sheep may have long hair on their cheeks and neck. Goats retain a tail and never have a large rump patch, while sheep have short tails and large rump patches. Male goats – but not male sheep – spray themselves with urine. Where sheep and goats occur together they thus occupy different habitats, but this is not the case when one occurs without the other. In the absence of goats, sheep move to cliffs and become like ibex in body shape and size. Some true goats, like the Caucasian tur and Blue sheep, evolved horns not unlike those of sheep and lost their strong body odor. Goat and sheep seem flexible enough to avoid competition, while remaining ecologically interchangeable. In both genera, as horns increase in length and thickness, coat and coloration become less showy. Goats and sheep reach ever higher altitudes and latitudes, growing larger, and delivering ever harder blows in combat. They culminate in the Argalis and the Siberian form of the Ibex, from the high mountains of central Asia.

Sheep had two great radiations from primitive 58 chromosome urial sheep: the first generated the 54/52 chromosome sheep (mouflon and American bighorn sheep), the second the giant argalis with 56 chromosomes. The extreme geographic forms, the bighorns and argalis, are very similar in social behavior and appearance. Asiatic and American sheep differ in ways of dealing with predators. The American sheep resemble ibex and cling to the vicinity of cliffs. Compared to Asiatic sheep, American sheep, which do not bear twins and have a longer gestation, have much lower reproductive rates. While they do live longer, as populations, they cannot respond to decimation by a rapid buildup of numbers.

Widespread Habitats

Distribution Patterns

The goats had less success in geographic expansion than sheep. They stuck closely to cliffs, penetrating as far as the Ethiopian highlands to the south, and Europe to the west. In mountains without adjacent dry, grassy foothills goats apparently were more successful than sheep. The geographic expansion of goats and sheep is such that, where they are found together, primitive goats are paired with primitive sheep, and advanced goats with advanced sheep. The most primitive form is probably the markhor, while the most advanced is the ibex. Species most distant from Asia Minor differ most from the wild goat.

The Caprini are one radiation from the Rupicaprini that adapted to open landscapes and extreme climates, along with the Musk oxen radiation and probably also that of the takin. These latter evolved frontal combat and head-butting, hierarchical social structures, male groups, and gregarious behavior. Many species went extinct, making them difficult to interpret. Only the most highly evolved of the Musk oxen remains, and even this species survives only in Arctic North America and Greenland. The takin is still a mysterious animal, which, like the goats, has dispensed with a collection of odoriferous glands in favor of a general body odor. Like the outsize rupicaprid it appears to be, it still clings to forest cover, but like an open-country form with social tendencies, it will also flock to large, open alpine meadows.

Submission and Aggression

Social Behavior

Rupicaprids and advanced caprids display different social behaviors. In the former group, resource defense obliges individuals to lead solitary, territorial lives. But among the advanced caprids, individuals band together in cooperative herds to

Below *The Ibex, which belongs to the tribe Caprini, is superbly well adapted to rugged terrain and extreme alpine and desert climates. It is found in such habitats across a wide range, from mountainous regions of Europe to the Far East.*

Below *Musk oxen have developed a simple yet effective mode of defense; when a predator is sighted the adults crowd around the young animals. Given their size, strength, and formidable horns, this strategy deters most predators, yet at the same time makes them easy prey for armed hunters.*

reduce predation. Since grassland is productive, and all of its productivity is available for feeding, grazers can live in large flocks. Animals in the center of the herd are virtually untouchable. With grazing comes mobility, but it also means having to process abrasive grasses. This has led to an increase in tooth size among grazers.

Goat antelopes have also evolved diverse strategies for courtship and combat. At one extreme is the rupicaprid Mountain goat. The males of this species, which are subordinate to females for all but two weeks in the mating season, crawl submissively in courtship. Unlike many other caprids, Mountain goat males do not butt heads, but jab each other with their short, sharp, horns. Yet fights are very rare, because of the risk of serious injury. Their primary defense is a very thick skin.

At the opposite extreme are bighorns, advanced caprids with horns that can exceed 13 percent of a male's weight. The horns act as both weapons and shields. Moreover, as display organs, they are meant to impress rivals and females. Large horns signal an ability to find abundant, high-quality food – a vital trait to pass on, especially to female offspring, who must produce plentiful, rich milk.

In cliffside habitats bighorns consort with only one female at a time, since this terrain limits visibility and maneuverability. By contrast, the giant sheep herd harems, living as they do in open terrain, where visibility and maneuverability are assured. Mountain goats also consort in pairs – unless heavy snowfalls press them into small herds. Then rutting males defend as many females as possible against rivals. Heredity and environment thus both ensure a diversity of mating tactics.

Below The tahr is widely dispersed geographically at low latitudes: in the Arabian peninsula, the Nilgiri hills of southern India, and the Himalayas. Its species are thus separated by huge distances, indicating their great evolutionary age. During the Pleistocene, tahrs roamed as far west as western Europe.

CHAMOIS SOCIAL LIFE

The social system of chamois is extremely flexible. Females are usually resident and tend to gather in flocks. Males stay with the mother's flock until they are sexually mature at 2–3 years old. They then live nomadically until full maturity (aged 7–9), when they become attached to a definite area. Their solitary lives are broken by occasional sojourns with the female flocks. The younger males that mix freely with such flocks are expelled by older males during the rut, which reaches its peak in mid-November. Harem holders try hard to defend females on heat, but still have to concede some mating opportunities to peripheral males.

With the first heavy snowfalls, the flocks disperse and move to the woodland winter ranges, where they forage for scattered food resources. This can be a time of heavy mortality. Females bear a single kid in May–early June at precipitous, isolated sites. Large groups will not reform until mid-June.

A large repertoire of threat displays usually forestalls direct aggression among chamois. However, when conflict does ensue, the loser quickly adopts an extreme submissive posture. Not only is this designed to mollify the victor; it also effectively prevents it from using its backward-pointing horn tips to gore its rival. SL

Management of Stocks

CONSERVATION AND ENVIRONMENT

To early man in the Pleistocene, caprids were of little economic significance as game animals. However, in the Neolithic (eight thousand years ago), sheep and goats were domesticated in the Near East. These animals quickly assumed a supreme importance. Their value lay in their ability to exploit poor-quality land, their low maintenance costs, and the valuable products they yield.

Domestic sheep and goats arose in the same region. The mouflon gave rise to Domestic sheep, while Domestic goats came from the common Wild goat. Neither Ibex nor advanced sheep contributed to domestication, although this development would have been feasible. In modern times, caprids enjoy mixed fortunes. In America, bighorn sheep numbers have been increased by various conservation strategies. Desert bighorns are susceptible to diseases of livestock and to disturbances. In Asia, due to the high reproductive rates of Old World wild sheep and goats, the situation is a happy one in some regions, but in most others, such as the Himalayas, it is desperate. Severe competition from domestic stock and overhunting has eliminated them from many areas. VG

Goat Antelope Species

THE 26 SPECIES OF THE CAPRINAE, THE goat antelope subfamily, are coupled here with the chiru (*Pantholops hodgsoni*), a single species assigned its own subfamily, the Panthalopinae, regarded as a sister taxon of the Caprinae and of similar age.

The 1993 established taxonomy recognized three more species of *Nemorhaedus* and four more species of *Capra*.

Subfamily Caprinae

26 species in 3 tribes and 11 genera.

TRIBE RUPICAPRINI

Strong affinity for steep slopes and shrub or tree cover. Defend resources using short, sharp horns. Sexes similar in weight and appearance. Tail short. Four teats. Young follow mother straight after birth.

Mainland serow Vu

Capricornis sumatraensis

Tropical and subtropical E Asia.
Male HTL unknown; HT up to 107cm; WT up to 140kg. Female HTL unknown; HT up to 90cm; WT up to 100kg.
COAT: very variable, often dark, with long (40cm) white or brown neck mane; hair coarse, long, and bristly; short beard from mouth to ears. Males' horns stouter than females', 15–25cm long; ears long, lance-shaped; large preorbital and foot glands.

Japanese serow LR

Capricornis crispus

Japanese or Taiwanese serow

Japan, Taiwan. In Japan, adapted to low temperatures and snowy climates; dwarfed island form on Taiwan. Smaller and lighter colored than Mainland serow, with conspicuous cheek beards in both sexes.
CONSERVATION STATUS: LR Conservation Dependent.

Goral LR

Nemorhaedus goral

Goral, Red or Common goral

N India and Burma to SE Siberia and S to Thailand. Very steep cliffs in dry climates; Manchurian form in moist, snowy climate.
Male (Manchuria) HTL 106–117cm; HT 69–78cm; WT 28–42kg. Female HTL 106–118cm; HT 50–75cm; WT 22–35kg.
COAT: highly variable, ranging from foxy red to dark gray to white; throat patch white; body hair long; neck mane short in male; horns black, up to 23cm long in males, 20cm in females. Tail longer than in serow, with terminal tuft; tail longer in Manchurian form than in southern races. Rudimentary preorbital glands.
BREEDING: gestation 250–260 days in Manchurian, shorter in southern forms.
CONSERVATION STATUS: LR Near Threatened; Vulnerable in Far East.

Chamois

Rupicapra rupicapra

European Alps, Caucasus, Carpathian, and Tatra Mts; NE Turkey; Balkans; introduced to New Zealand. Alpine forest and meadow. 6 subspecies.
HTL 125–135cm; HT 70–90cm; WT 30–50kg (male), 24–42kg (female).
COAT: pale brown in summer, dark brown/black with white rump patch and white to yellow facial stripes in late fall/winter. Occipital glands present in both sexes. Tail short.
BREEDING: gestation 160–170 days.
CONSERVATION STATUS: 2 subspecies are Critically Endangered.

Pyrenean chamois LR

Rupicapra pyrenaica

NW Spain (Cantabrian Mts); Pyrenees, C Apennines. 3 subspecies.
HTL; HT and WT as for Chamois.
COAT: reddish beige in summer, dark brown with wide yellowish patches on throat, neck sides, shoulders, hindlimbs in late fall/winter.
BREEDING as for Chamois.
CONSERVATION STATUS: LR Conservation Dependent. One subspecies is Endangered.

Mountain goat

Oreamnos americanus

SE Alaska, S Yukon, and SW Mackenzie to Oregon, Idaho, and Montana; introduced in other N American mountain areas. Steep cliffs and edges of major glaciers.
Size very variable regionally: male HTL up to 175cm; HT up to 122cm: WT up to 140kg. Female HTL up to 145cm; HT up to 92cm; WT up to 57kg. In southern forms, WT to 70kg (male), 57kg (female).
COAT: yellowish-white, with long underwool and longer guard hairs that elongate into a stiff mane on the neck and rump and into "pantaloons" on legs; molts in July; horns black, 15–25cm long, thicker in males. Massive legs with very large hooves. Tail short.
BREEDING: gestation about 180 days.

TRIBE OVIBONINI

Grotesque, giant rupicaprids from arctic and alpine environments, with long-haired, massive squat bodies carried on short, stout legs. Sexes similar in appearance, but females only 60 percent of males' weight. Horns relatively larger than in rupicaprids, but smaller than in caprids, curved for frontal attacks. Single young born after gestation of 8–8.5 months. Tail short. Four teats.

Takin Vu

Budorcas taxicolor

Takin or Golden-fleeced cow

W China, Butan, Burma. High alpine and subalpine bamboo forests on steep rugged terrain.
Size variable: Male HTL 170–220cm; HT 100–130cm; WT up to 350kg, female to 250kg.
COAT: normally light, long, and shaggy, golden in one race; face dark in bulls, but only nose dark in cows and calves. No localized skin glands, but oily, strong-smelling substance with burning taste secreted over whole body.

Musk ox

Ovibos moschatus

Alaska to Greenland. Arctic and tundra near glaciers; adapted to extreme cold; mixed feeder, restricted to areas with thin snow.
Size variable, increasing from north to south: Male (south) HTL 245cm; HT 145cm; WT 350kg; male (Greenland) HTL 180cm; HT 110–130cm; WT up to 650kg (in captivity).
COAT: black with light saddle and front, possibly bleaching in spring, dense and long, with hair strands up to 62cm long. Foot glands present; preorbital glands large and secrete copiously in bull in rut.

TRIBE CAPRINI

Advanced caprids with large differences in weight and external appearance between sexes. Specialized for frontal combat, with hierarchical social system in males. Spatial segregation of sexes outside the mating season. Horns with annual growth. much larger in males: length of display hair and horn mass inversely related. Teeth strongly molarized. Teats 2–4. Strong preference for open areas and grazing. Highly gregarious.

Genus *Hemitragus*

Most primitive and rupicaprid-like in fo[rm] and behavior of the Caprini. Both sexes horned; those of males stouter, longer, a[nd] heavier, up to 44cm. Prefer steep, arid, o[r] cold cliffs. Males grow thick, heavy skin. Not malodorous; lack foot, groin, preorbital or tail glands. Females about 60 pe[r]cent of male weight. Gestation: 180 days.

Himalayan tahr

Hemitragus jemlahicus

Himalayas. Introduced to New Zealand.
HTL 130–170cm; HT 62–100cm; WT may exceed 108kg in males.
COAT: variable over body from copper to black; hair relatively long, with long-haired, shaggy neck ruff in males; coat shorter in females.

Nilgiri tahr E

Hemitragus hylocrius

S India.
HT 62–100cm
COAT: dark, almost black, short-haired, with silver saddle in males; mane in male short and bristly: females grayish brown with a white belly.

Arabian tahr E

Hemitragus jayakari

Oman.
HTL, HT, WT: not known; is said to be the smallest species.
COAT: brownish with dark dorsal stripe, black legs, white belly, and long hair abo[ve] jaw and withers.

Genus *Ammotragus*

A single species.

Barbary sheep V

Ammotragus lervia

Barbary sheep or aouydad

N Africa. Mountains, particularly high desert. 4 subspecies.
Male HTL 155–165cm; HT 90–100cm; W[T] up to 140kg; females 50 percent lighter.
COAT: short with long-haired neck ruffs and "pantaloons" on front legs in both sexes; beard and rump patch absent. Tail long, flat, naked on underside. Callouses on knees. Subcaudal glands present. No smell on males. Male horns up to 84cm, female up to 40cm. Ears slim and pointe[d].
BREEDING: gestation 160 days.

ABBREVIATIONS HTL = head–tail length HT = height WT = weight
Approximate nonmetric equivalents: 10cm = 4in; 1kg = 2.2lb

Ex Extinct
EW Extinct in the Wild
Cr Critically Endangered
En Endangered
Vu Vulnerable
LR Lower Risk

;enus *Pseudois*

. single species.

3lue sheep LR
'seudois nayaur
lue sheep or bhara

Iimalayas, Tibet, E China.
ew measurements available: male HT 1cm; WT 60kg; females WT 40kg.
:OAT: bluish with light abdomen, legs trongly marked: lacks beard and shin lands; rump patch small; tail naked on nderside. No callouses on knees. Horns urve rearward, cylindrical, up to 84cm ɔng in males, tiny in females. Ears pointd and short.
BREEDING: gestation 160 days.
.ONGEVITY: average 9 years, up to 24.
:ONSERVATION STATUS: LR Near Threatned.

;enus *Capra*

Iales have chin beards and are strongmelling. No preorbital, groin, or foot lands; anal glands present. Both sexes ave callouses on knees and long, flat tails vith bare underside. Ears long, pointed, xcept in alpine forms. Rump patch small. :liff-adapted jumpers, highly gregarious. 3ody size varies locally. Females 50–60 ercent of male weight. In males, horns ncrease in length and weight with age. 'emale horns 20–25cm in all species. 3reeding coat of male becomes more colorful with age in most species. Gestation: 150 days in small-bodied species, up to 170 days in large-bodied species. Twins common. Average life expectancy about 8 ears. Many races, exact number unsettled.

Wild goat Vu
Capra aegagrus
Wild goat or bezoar

Greek Islands, Turkey, Iran, SW Afghanstan, Oman, Caucasus, Turkmenia, Pakistan and adjacent India; Domestic goat worldwide. 4 subspecies, including the Domestic goat (*C. a. hircus*).
Size highly variable: male HTL (Crete) unknown; HT unknown; WT 26–42kg: male (Persia) up to 90kg, female up to 45kg.
COAT: old males very colorful compared to females or young males. Horns flat and scimitar-shaped, 58–126cm long.

Spanish goat LR
Capra pyrenaica
Spanish goat or Spanish ibex

Pyrenees. 4 subspecies.
Male HTL 130–140cm; HT 66–70cm; WT 65–80kg. Female HTL 100–110cm; HT 70–75cm; HT 35–45kg.
COAT: similar to Wild goat in color. Horns differ from those of ibex or goats, up to 75cm long.
CONSERVATION STATUS: LR Near Threatened.

Ibex
Capra ibex

C Europe, Afghanistan and Kashmir to Mongolia and C China, N Ethiopia to Syria and Arabia. Extreme alpine or desert.
Male HTL (Siberian) 115–170cm; HT 65–105cm; WT 80–100kg. Female HTL 130cm; HT 65–70cm; WT 30–50kg.
Alpine ibex may be larger than this (WT up to 117kg): Nubian and Walia ibex are smaller.
COAT: less colorful than in Wild goat; uniformly brown in alpine races; chin beards smaller than in goats. Horns massive and thick, but much more slender than in Caucasian turs; maximum length 85cm (Alpine), 128–143cm (Siberian).

Markhor En
Capra falconeri

Afghanistan, N Pakistan, N India, Kashmir, S Uzbekistan, Tajikistan. Woodlands low on mountain slopes.
Male HTL 162–168cm; HT 86–100cm; WT 80–110kg. Female HTL 140–150cm; HT 65–70cm; WT 32–40kg.
COAT: diagnostic due to long neck mane in male, pantaloons, and strong markings; female does not have display hairs. Horns twisted, maximum length 82–143cm. The largest goat. Dentition more primitive than in ibex and turs.

East Caucasian tur Vu
Capra cylindricornis

E Caucasus. Mountains.
Male HTL 130–150cm; HT 79–98cm; WT 65–100kg. Female HTL 120–140cm; HT 65–70cm; WT 45–55kg.
COAT: uniformly dark brown in winter, red in summer; chin beard very short, up to 7cm long. Horns cylindrical and sharply backward winding, as in *Pseudois*; maximum length 103cm.

West Caucasian tur En
Capra caucasica

W Caucasus. Mountains.
Size and form as East Caucasian tur except that horns are similar to those of Alpine ibex, but more massive and curved. Skull form diagnostic and different from ibex. Chin beard up to 18cm.

Genus *Ovis*

Characterized by the presence of preorbital, foot, and groin glands. No offensive smell on males. Tail short; rump patch small in primitive, large in advanced species. Two teats on females. Horns in both sexes, except a few mouflon populations where females lack horns; female horns normally very small. Horns increase in mass from urials to argalis sheep. In the latter, horns form up to 13 percent of the male's body mass. In mouflons the horns may wind backwards, otherwise horns wind forward. Horn mass in large mature rams: urials 5–9kg; Altai argalis 20–22kg, rarely more; Snow sheep 6–8kg; Stone's and Dall's sheep 8–10kg; bighorns up to 12kg. Horns used as weapons in combat. Prefer grazing. Highly gregarious, with sexes segregated except at mating season.

Urial Vu
Ovis orientalis

Kashmir to Iran. Particularly in rolling terrain and deserts.
Male HTL 110–145cm; HT 88–100cm; WT 36–87kg.
COAT: color variable, usually light brown; males with whitish cheek beards and light-colored long neck ruff; rump patch diffuse; tail thin and long for a sheep. Horns relatively light, forward-winding. Long-legged, fleet-footed.
BREEDING: gestation 150–160 days.
LONGEVITY: 6 years.

Argalis Vu
Ovis ammon

Pamir to Outer Mongolia and throughout Tibetan plateau. Cold, high alpine and desert habitats.
Male HTL 180–200cm; HT 110–125cm; WT 95–140kg (180kg in Altai argalis); largest sheep.
COAT: light brown with large white rump patch and white legs; size of neck ruff inversely related to size of horns; horns up to 190cm long and 50cm in circumference.
BREEDING: twinning common.

Mouflon
Ovis musimon

Asia Minor, Iran, Sardinia, Corsica, Cyprus; widely introduced in Europe. Cold and desert habitats.
Male HTL 110–130cm; HT 65–75cm; WT 25–55kg.
COAT: dark chestnut brown with light saddle; rump patch distinct; lacks a cheek bib but possesses dark ruff; tail short, broad, and dark. Face of adults becomes lighter with age. Smallest wild sheep.

Snow sheep LR
Ovis nivicola
Snow sheep or Siberian bighorn

NE Siberia. Extreme alpine and arctic regions, particularly cliffs. 4 races.
Male HTL 162–178cm; HT 90–100cm; WT 90–120kg; females 60–65kg.
COAT: dark brown, with small, distinct rump patch and broad tail; some races light colored.
CONSERVATION STATUS: LR Conservation Dependent.

Thinhorn sheep
Ovis dalli
Thinhorn, Stone's, Dall's, or White sheep

Alaska to N British Columbia. Extreme alpine and arctic regions, particularly cliffs.
Male HTL 135–155cm; HT 93–102cm; WT 90–120kg.
COAT: white in 2 subspecies, black or gray in the Stone's sheep; large, distinct rump patch in latter.

American bighorn sheep LR
Ovis canadensis
American bighorn sheep or Mountain sheep

SW Canada to W USA and N Mexico. Alpine to dry desert. 7 races.
Male HTL 168–186cm; HT 94–110cm; WT 57–140kg, depending on locality; female 56–80kg.
COAT: light to dark brown, lacks ruff or cheek beards; large rump patches. Body stocky, as in ibex.
BREEDING: gestation 175 days.
LONGEVITY: average 9 years but to 24.
CONSERVATION STATUS: LR Conservation Dependent.

Domestic (Soay) sheep
Ovis aries

Domestic sheep are thought to have derived from *O. orientalis*. The Soay is a subspecies of Domestic sheep that has reverted to feral status. Orkney Islands and St Kilda, Scotland; now largely restricted to North Ronaldsay.
SH male 56cm (22in). WT male 36kg (79lb), female 25kg (55lb)
COAT: brown with white belly; horns strong with single whorl in rams; ewes horned or polled; tail short.

Subfamily Panthalopinae

A single species.

Chiru En
Pantholops hodgsoni
Chiru or Tibetan antelope

Tibet, Tsinghai, Sichuan (China), Ladak (N India). High Tibetan plateau. The fact that the chiru is an only species and has been forced to adapt to an extreme environment suggests a low competitive ability in this subfamily.
HTL 170cm; HT 90–100cm; WT up to 40kg in males.
COAT: sandy, horns black, 50–72cm long; rump patch large; face of male dark. Tail and ears short; nasal sacs the size of pigeon eggs when inflated; auxillary glands present.
BREEDING: gestation about 180 days.
CONSERVATION STATUS: severely endangered by poaching for its exceptionally fine-fibered underwool, which is turned into high-priced garments . VG

AGE, SEX, AND THE WEATHER

Modeling population dynamics among Soay sheep

THE POPULATION OF SOAY SHEEP ON HIRTA – the largest island in the St. Kilda group, an uninhabited archipelago off the west coast of Scotland – has been the focus of an intensive study that began in 1985. Over 95 percent of sheep living in approximately one-third of the 638-ha (1,576-acre) island have been tagged by scientists and can be individually recognized. Data on birth dates and weights, phenotype, death dates, summer weights, gut parasite loads, and genetic information including relatedness between individuals is now known for over 2,500 individuals. The number of sheep in the study population and the weather throughout the study have both been carefully recorded. The total number of sheep living on the whole island has also been counted annually from 1955 to the present day. Probably more is known about this population than is known for any other free-living, unmanaged population of organisms. Analyses of these data have led to many discoveries about animal behavior, population genetics, evolutionary biology, and the way populations change in size from year to year.

To understand why the population fluctuates in size requires basic knowledge of Soay sheep natural history. Lambs are born in April, close to the time that the new grass starts growing. During the summer, sheep concentrate on feeding and laying down fat that will help sustain them through the winter. Mating occurs during the rut in autumn and all animals – including lambs – take part. The most common form of death is starvation, exacerbated by large numbers of parasitic worms living in the animals' guts. Death will occur in late winter if an animal has exhausted its fat reserves. Consequently, any factors that influence how large an animal's fat reserves are, and how quickly it is depleted during winter, affects whether that animal survives or dies. Population size, winter weather, and an animal's sex and age are all associated with the size and rate of loss of the fat reserves.

The sheep population fluctuates in size dramatically, with up to 60 percent of animals dying in some years. Death occurs at the end of winter and the beginning of spring, with the majority of males that die (58.3 percent) dying in March, and the majority of females (47.0 percent) dying in early April. The large population die-offs are followed by years of rapid population growth – the population can recover its losses in two years.

Recent research has uncovered why the population experiences highs and lows – findings that are likely to be applicable to other animal populations. There are three components that influence whether the population stays the same size, increases, or decreases: the size of the population, its structure, and the weather during the winter. When the population size is large, competition for high-quality grazing is greater than when the population is small, so animals have smaller fat reserves at the end of summer and are more susceptible to starvation. Consequently, the population is more likely to decline when it is large than when it is small. Also, animals die faster in stormy winter weather than in periods of sunshine, as they have to expend more energy to keep warm. Population size and winter weather interact – if the population is large, animals have small fat reserves and are more vulnerable to bad winter weather than when the population size is smaller and they have larger fat reserves.

Population structure also plays a part, because the effects of winter weather and population density on survival and reproduction differ both between males and females and between animals of different ages. Different classes of animals invest different amounts of energy in growth and reproduction at different times of the year. For example, males invest energy in reproduction in the autumn during the rut, when they stop feeding and try to mate with as many females as possible. Females, in contrast, feed throughout the mating season, investing energy in reproduction during late gestation and lactation at the end of the winter and during the early spring. Females therefore enter the winter in better condition than males, as they have accumulated a larger reserve of fat. A consequence is that females between the ages of 2 and 6 have, on average, a 92 percent chance of surviving the winter, compared to 84 percent for equivalently aged males.

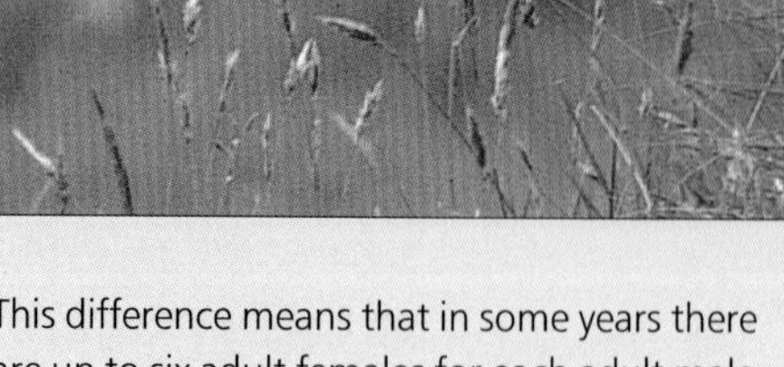

This difference means that in some years there are up to six adult females for each adult male.

Animals of different ages also respond to density and weather in contrasting manners. Lambs and individuals older than 6 years are significantly more likely to die in years when the population is large and winter weather is bad, than adults in their prime. This is because lambs have to use the fat reserves they have built up in the preceding summer for growth, survival, and reproduction (female Soay sheep attempt to breed in their first year of life), while prime-aged adults are already fully grown. At a population size of 200 individuals, female lambs have a 90 percent chance of survival, but when the population reaches 500 their chances fall to 20 percent.

Above *Soay sheep are a subspecies of the mestic sheep (*Ovis aries*), which descended from ouflon-like animals about 10,000 years ago. The eep on Hirta have reverted to feral status. Crucial being able to make accurate predictions about creases and declines in population size is an under-anding of the gender and age balance of the oup. If the percentage of young (*INSET*) is greater an the percentage of adults, the overall population likely to decline if there is a harsh winter.*

How do these differences affect the way the population changes in size from one year to the next? Because winter weather and population size influence survival prospects in different ways among males and females and among individuals of different ages, the structure of the population changes independently of the total population size. For example, the proportion of adult males in the population varies between 3.8 percent and 16.8 percent, and the proportion of adult females from 16 to 47 percent.

The way the population fluctuates is therefore the result of a complex interaction. A population of 350 animals consisting of 55 percent adult males and females in a poor winter is predicted to stay the same size the following year. In contrast, a population of 350 animals that consists of only 35 percent adult males and females in a poor winter is predicted to decline to 220 individuals. All three factors – population size, structure, and the winter weather – are so important that if one is taken out of the mathematical model, it becomes impossible to predict future fluctuations.

The population of Soay sheep on Hirta is one of only a handful of populations that allow the detailed analyses described above to be undertaken. However, our findings suggest that an understanding of the way mammal populations fluctuate over time can only be achieved by incorporating differences in the ways that animals of different ages and sexes respond to population density and weather into models. TC

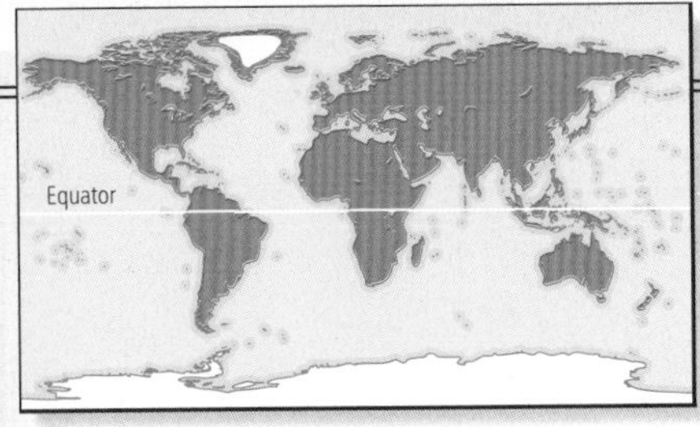

RODENTS

RODENTS HAVE INFLUENCED HISTORY AND *human endeavor more than any other group of mammals. Over 42 percent of all mammal species belong to this one order, whose members live i almost every habitat, often in close association with humans. Frequently this association is to people's disadvantage, since rodents consume prodigious quantities of carefully stored food and spread fatal diseases. It is said that rat-borne typhus has had a greater influence upon human destiny than has any single person, and in the last millennium rat-borne diseases have taken more lives than all wars and revolutions put together.*

It is nevertheless testimony to the entrepreneuria spirit of rodents that they have thrived in human-dominated environments from which so many other animal groups have been extinguished. Fur thermore, many rodent species have an importar function in ecosystems, and are therefore highly beneficial to man. Rodents play key roles in main taining the relationship between plants and fung Many fungi form mycorrhizal (mutually beneficia associations with the roots of plants that increase the ability of the plants to extract nutrients and water from the soil by many thousands of times. So important is this relationship that many plant simply cannot survive without the fungi, and vice versa. One of the most important groups of myco rrhizal fungi are the truffles, which are related to mushrooms but form their fruiting bodies under ground. Truffles and truffle-like fungi rely on animals to dig up the fruiting bodies and to disperse the spores, either in the wind or in the animal's feces after eating the fungus. When gaps form in forests, for example, small mammals deposit fungal spores in their feces in the gaps, thereby bring ing the fungi to the places where plant seeds are germinating. In the forests of North America and Australia, it is believed that this three-sided relationship between plants, fungi, and small mammals – including rodents – is vital for ecosystem

Right *A giant among rodents, the capybara can weigh up to 66kg (146lb) – fully 10,000 times the weight of the smallest mice. The animals share the close-knit social life and herbivorous diet of many other rodent species, but are unusual in their semi-aquatic lifestyle.*

Left *On the alert for food, a Eurasian harvest mouse* (Micromys minutus) *investigates a wheat crop in summer. Rodents are supreme opportunists; their rapid rate of reproduction and wide-ranging diets ha made them among the most successful mammals.*

nction. The role of rodents in other ecosystems, ːe grasslands, is not so well studied but is proba- y just as important.

In South America, Africa, and Asia, some larger ecies are also an important source of protein, ing trapped or deliberately bred for food. ecies in the latter category are the guinea pigs in uth America, and the grasscutter rats and the lible dormouse (*Myoxus glis*) in Africa. Among her rodents, hamsters and gerbils are popular ts in the western world, with the prairie dog ecoming increasingly common in the United ates. Moreover, rats, mice, and guinea pigs today ay an indispensable role in the testing of drugs d in biological research. Rodents are also key- one species in many ecosystems, providing an nportant food source for many species of medi- m-sized carnivores and birds of prey. For exam- e, in agricultural ecosystems it is not unusual to ave 50–100kg (110–220lb) of rodents per ha istributed over tens of thousands of hectares. his provides a huge variety of food for predators ving in agricultural systems or in neighboring rest habitats. The Black-footed ferret (*Mustela gripes*), which is the subject of much human urturing through a reintroduction program after aving become Extinct in the Wild, relies on a iet of prairie dogs for its survival.

Equipped for Gnawing

FORM AND FUNCTION

odents occur in virtually every habitat, from the igh arctic tundra, where they live and breed nder the snow (for example, lemmings), to the ottest and driest of deserts (gerbils). Others glide om tree to tree (flying squirrels), seldom coming own to the ground, or else spend their entire ves in underground networks of burrows (mole- ts). Some have webbed feet and are semi-aquatic nuskrats), often undertaking complex engineer- ng programs to regulate water levels (beavers), hile others never touch a drop of water through- ut their entire lives (gundis). Such species can erive their water requirements from fat reserves.

Most rodents are small, weighing 100g (3.5oz) r less. There are only a few large species; the iggest of them, the capybara, may weigh up to 6kg (146lb).

The term "rodent" derives from the Latin verb *odere*, which means "to gnaw." All rodents have haracteristic teeth, including a single pair of azor-sharp incisors. With these teeth, the rodent an gnaw through the toughest of husks, pods, nd shells to reach the nutritious food contained within. Gnawing is facilitated by a sizable gap, nown as the diastema, immediately behind the ncisors, into which the lips can be drawn, so seal- ng off the mouth from inedible fragments dis- odged by the incisors. Rodents have no canine eeth, but they do possess a substantial battery of nolar teeth by which all food is finely ground. Convoluted layers of enamel traverse these often massive and intricately structured teeth. The pattern made by these layers is often of taxonomic significance. Most rodents have no more than 22 teeth, although one exception is the Silvery mole-rat from Central and East Africa, which has 28. The Australian water rat has just 12. Since rodents feed on hard materials, the incisors have open roots and grow continuously throughout life. They are constantly worn down by the action of their opposite number on the other jaw. If the teeth of rodents become misaligned so that they are not automatically worn down during feeding, they will continue to grow and may eventually end up piercing the skull.

Most rodents are squat, compact creatures with short limbs and a tail. In South America, where there are no antelopes, several species have evolved long legs for a life on the grassy plains (e.g., maras, pacas, and agoutis), and show some convergence with the antelope body form. A very variable anatomical feature is the tail (see panel overleaf and Squirrel-like Rodents).

Modern taxonomists divide rodents into two suborders. For convenience, we present these in three sections, the squirrel-like and the mouse-like forms – both part of the suborder Sciurognathi – and the porcupine and cavy-like forms, the suborder Hystricognathi. Rodents were formerly split into three suborders on the basis of jaw musculature. The main jaw muscle is the masseter, which not only closes the lower jaw on the upper, but also pulls the lower jaw forward, so creating the unique gnawing action. In the extinct Paleocene

ORDER: RODENTIA
28 families: 431 genera: 1,999 species

SCIUROGNATHS Suborder Sciurognathi

SQUIRREL-LIKE RODENTS p588

284 species in 56 genera in 5 families
Includes **beavers** (Family Castoridae); **squirrels** (Family Sciuridae); **springhare** (Family Pedetidae).

MOUSE-LIKE RODENTS p616

1,477 species in 312 genera in 5 families
Includes **rats, mice, voles, gerbils, hamsters,** and **lemmings** (Family Muridae); **jumping mice, birch-mice,** and **jerboas** (Family Dipodidae).

HYSTRICOGNATHS Suborder Hystricognathi

CAVY-LIKE RODENTS
Suborder Hystricognathi p668

238 species in 63 genera in 18 families
Includes **New World porcupines** (Family Erethizontidae); **cavies** (Family Caviidae); **capybara** (Family Hydrochaeridae); **agoutis and acouchis** (Family Dasyproctidae); **chinchillas** and **viscachas** (Family Chinchillidae); **octodonts** (Family Octodontidae); **tuco-tucos** (Family Ctenomyidae); **cane rats** (Family Thryonomyidae); **Dassie rat** (Family Petromuridae); **coypu** (Family Myocastoridae); **Old World porcupines** (Family Hystricidae); **gundis** (Family Ctenodactylidae); **African mole-rats** (Family Bathyergidae).

Note There is some divergence in the numbers of species and genera between the following accounts and the established 1993 list of mammalian species. The interested reader is referred to the Appendix, which contains a comprehensive list of rodent species based on the 1993 taxonomy.

rodents, the masseter was small and did not spread far onto the front of the skull. In the squirrel-like rodents the lateral masseter extends in front of the eye onto the snout; the deep masseter is short and used only in closing the jaw. In the cavy-like rodents it is the deep masseter that extends forward onto the snout to give the gnawing action. Both the lateral and deep branches of the masseter are thrust forward in the mouse-like rodents, providing the most effective gnawing action of all, with the result that they are the most successful in terms of distribution and number of species.

Most rodents eat a range of plant products, from leaves to fruits, along with small invertebrates, such as spiders and grasshoppers. Many northern rodents such as the Field vole (*Microtus agrestis*) eat the bark of woody trees in times of food scarcity due to high populations. In Field voles the toxic effects of bark seem to be neutralized by specially secreted enzymes in the stomach, which allow dietary flexibility in times of famine. A few species are specialized carnivores; for example, the Australian water rat (*Hydromys chrysogaster*) feeds on small fish, frogs, and mollusks.

To facilitate bacterial digestion of cellulose rodents have a relatively large cecum (appendix) that houses a dense bacterial flora. After the food they have eaten has been softened in the stomach, it passes down the large intestine and into the cecum. There the cellulose is split by bacteria into its digestible carbohydrate constituents, but absorption can only take place higher up the gut, in the stomach. Therefore rodents practice refection – reingesting the bacterially-treated food taken directly from the anus. On its second visit to the stomach the carbohydrates are absorbed and the fecal pellet that eventually emerges is hard and dry. It is not known how rodents know which type of feces is being produced. The rodent's digestive system is very efficient, assimilating as much as 80 percent of the ingested energy.

Evolution of Rodents

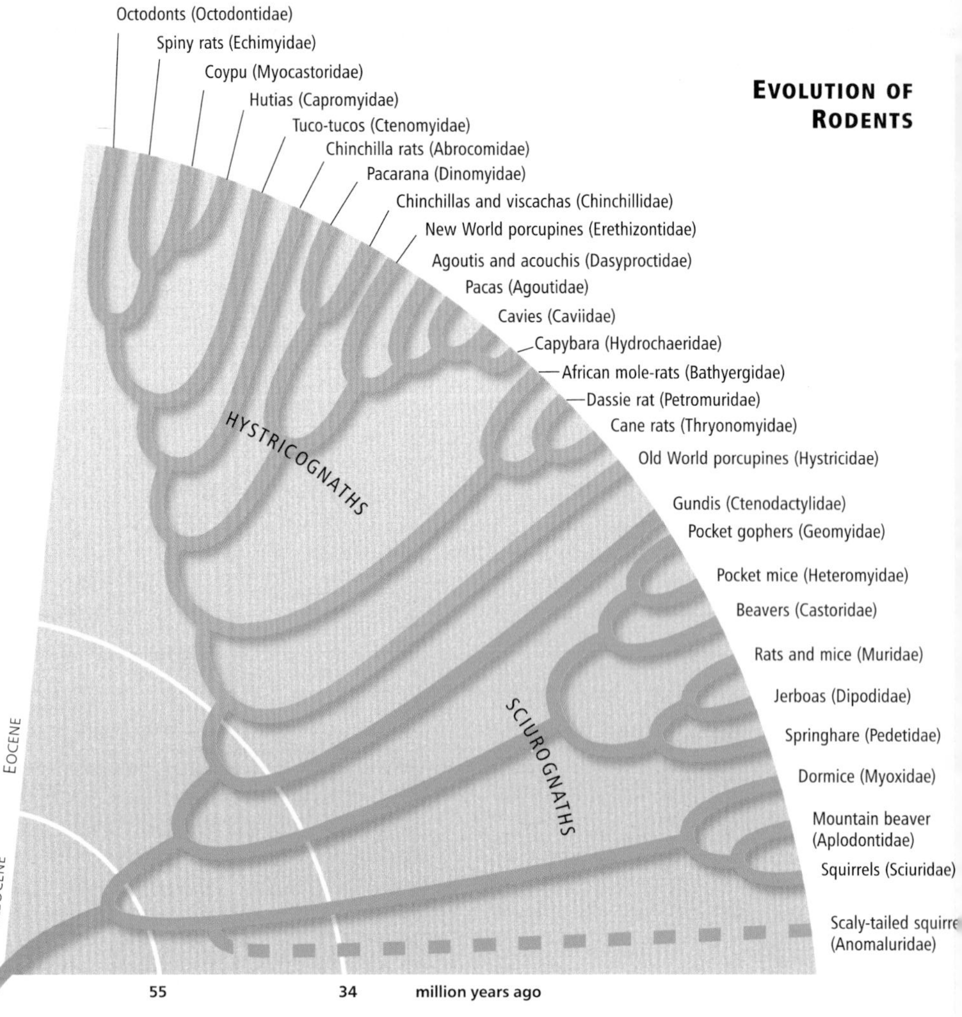

All members of at least three families (hamsters, pocket gophers, pocket mice) have cheek pouches. Fur-lined folds of skin projecting inwards from the corner of the mouth, these may reach back to the shoulders, and can be everted for cleaning. They are used for carrying provisions, and rodents equipped with them often build up large stores – up to 90kg (198lb) in Common hamsters.

Rodents are intelligent and can master simple tasks for obtaining food. They can be readily conditioned, and easily learn to avoid fast-acting poisoned baits – a factor that makes them difficult pests to eradicate. Their sense of smell and their hearing are keenly developed. Nocturnal species have large eyes; in addition, all rodents have long, touch-sensitive whiskers (vibrissae).

Above *In evolutionary terms, rodents are quite young and so retain large, untapped stocks of genetic variability. This evolutionary tree, compiled by biologist Rodney Honeycutt, is based on relationships revealed by molecula techniques. Thus the lengths of the branches are proportional to genetic similarity, and not necessarily to time of separation, which is large ly unknown thanks to limited fossil evidence.*

Left *A South African ground squirrel nibbles on a melon. Manual dexterity is particula ly well developed in squirrels, though other rodents also make good use of their front paws for digging, grooming, and gathering food and nesting materials.*

RODENT BODY PLAN

Below *The skeleton of a Roof rat.* (Rattus rattus) *exhibits typical rodent features in its squat form, short limbs, plantigrade gait ((meaning that the animal walks on the soles of its feet), and long tail.*

Right *Skull of the Roof rat. Clearly shown are the continuously growing, gnawing incisors and the chewing molars, with the gap (diastema) left by the absence of the canine and premolar teeth. All mouselike rodents lack premolars, but the squirrel- and cavy-like rodents have one or two on each side of the jaw.*

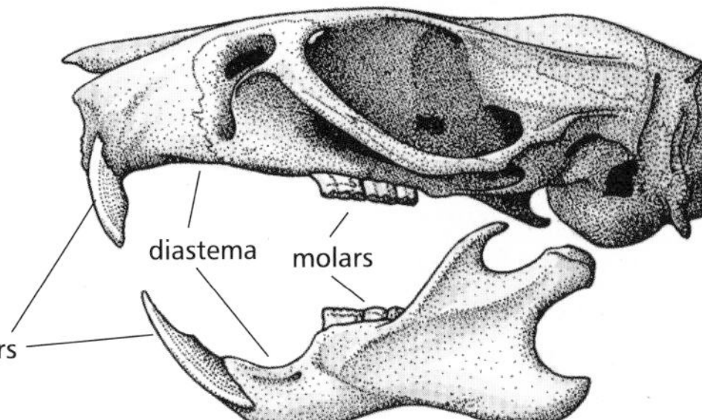

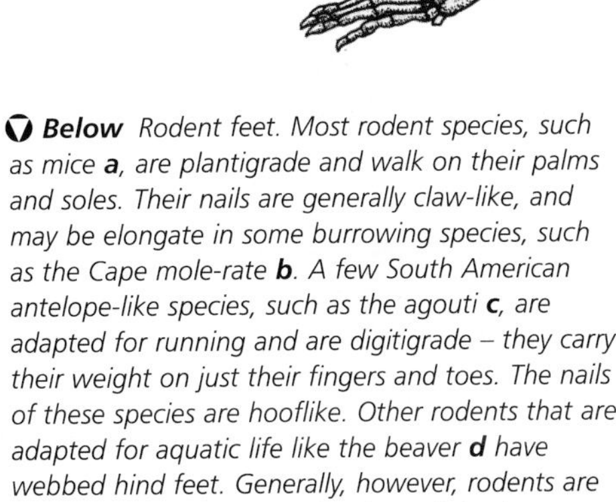

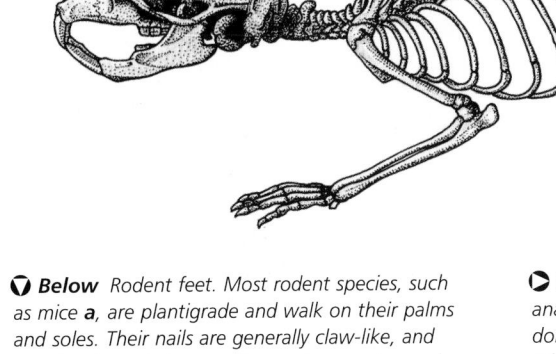

Below *Rodent feet. Most rodent species, such as mice* **a**, *are plantigrade and walk on their palms and soles. Their nails are generally claw-like, and may be elongate in some burrowing species, such as the Cape mole-rate* **b**. *A few South American antelope-like species, such as the agouti* **c**, *are adapted for running and are digitigrade – they carry their weight on just their fingers and toes. The nails of these species are hooflike. Other rodents that are adapted for aquatic life like the beaver* **d** *have webbed hind feet. Generally, however, rodents are not fleet of foot, though the desert-living kangaroo mice and jerboas can bound across sand at speeds of up to 48 kilometers per hour (30mph).*

Right *The rodent tail is a very variable anatomical feature. In the beaver* **a** *it is flattened dorso-ventrally and aids in rapid underwater swimming; in the muskrat* **b** *the flattening is lateral and the tail is used as a rudder. Hopping mice, jerboas* **c**, *and kangaroo mice* **d** *have very long tails, usually with a tuft of hair at the end to act as a balancing organ. In a few species, such as the European harvest mouse* **e**, *the tail is truly prehensile and functions as a fifth limb, allowing incredible gymnastics to be performed among the grass heads. In arboreal and gliding species, such as the Southern flying squirrel* **f**, *it is bushy so as to provide both counterbalance and drag anchor. Some hamsters* **g** *have only a tiny visible tail, and most cavy-like rodents have no visible tail at all.*

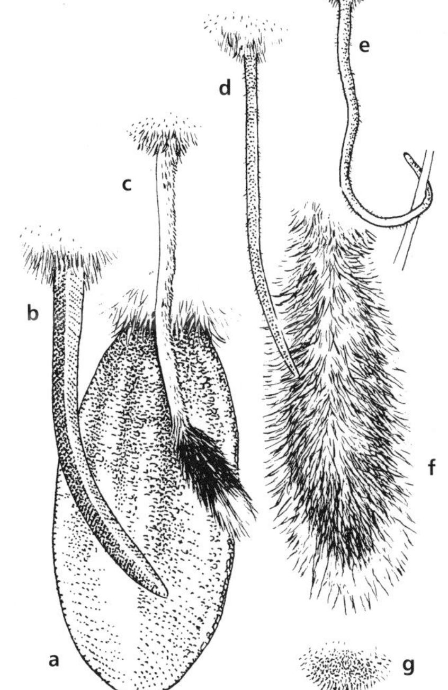

a

b

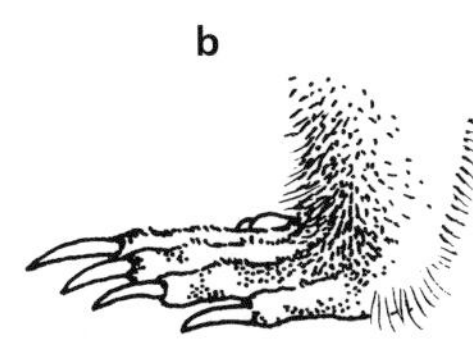

c

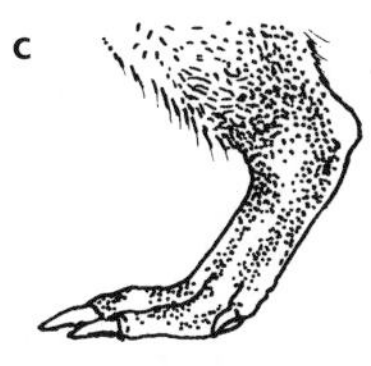

d

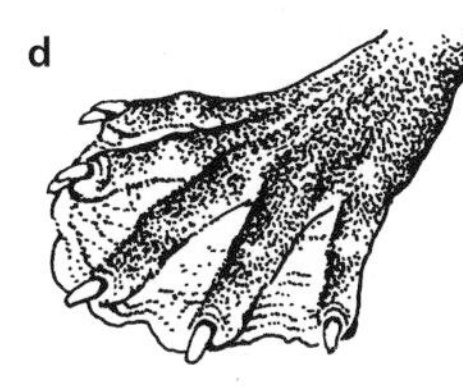

The Prehistory of Rodents

EVOLUTION AND RADIATION

Almost two-thirds of all rodent species belong to just one family, the Muridae, with 1,303 species at present, although numbers change constantly as new species are identified. Its members are distributed worldwide, including Australia and New Guinea, where it is the only terrestrial placental mammal family found (excluding dingoes, introduced approximately 4,000 years ago, and modern introductions such as the rabbit). The second most numerous family is that of the squirrels (Sciuridae), with 273 species distributed throughout Eurasia, Africa, and North and South America.

The fossil record of the rodents is pitifully sparse, partly because finding small bones requires very careful looking. Rodent remains are known from as far back as the late Paleocene era (57 million years ago), by which time all the main characteristics of the order had already developed. The earliest rodents apparently belonged to an extinct sciuromorph family, the Paramyidae.

During the Eocene era (55–34 million years ago) there was a rapid diversification of the rodents, and by the end of that epoch it seems that leaping, burrowing, and running forms had evolved. By the Eocene/Oligocene boundary (34 million years ago) many families recognizable today were already occurring in North America, Europe, and Asia, and during the Miocene (about 20 million years ago) the majority of present-day families had arisen. Subsequently the most important evolutionary event was the appearance of the Muridae from Europe in the Pliocene epoch (5–1.8 million years ago). At the start of the Pleistocene, they entered Australia, probably via Timor, and then underwent a rapid evolution. At the same time murids invaded South America from the north once it was united to North America by a land bridge, with the result that there was an explosive radiation of New World rats and mice across South America.

How Rodents Interact

SOCIAL BEHAVIOR

Rodents are often highly social, frequently living in huge aggregations. Prairie dog townships may contain more than 5,000 individuals. The solitary way of life appears to be restricted to those species that can defend food resources against intruders. These include some that live in arid grasslands and deserts – hamsters and some desert mice – and also species such as the North American red squirrel (*Tamiasciurus hudsonicus*), which lives in northern coniferous forests and stores cones in large central caches called middens.

The Norway or Brown rat (*Rattus norvegicus*) is a miscreant species that originated in Southeast Asia but has spread right around the globe in company with humans. Its social structure is central to the species' ecology and hence to the effectiveness of control measures. Socially dominant rats gain feeding priority and greater reproductive access and success. Social pressures force subordinate male rats to migrate into less favorable areas, resulting in a strongly unbalanced sex ratio (with more females than males) in breeding areas. Larger male rats tend to win dominance contests with smaller rats but, strangely, retain their higher social status long after subsequently losing body weight to levels at or below younger rats in the group. This means that dominance tends to be age-related, with the dominant alpha male often being smaller than many of his subordinates. Rats, it seems, respect their elders. Larger rats tend to accept the status quo because the costs of aggression are too great relative to the value of the contested resource. Moreover, dominant males cannot strictly control access to receptive estrous females. Sometimes, lower-ranking males actually achieve more matings.

Indications are that male Norway rats undergo scramble competition for mates. In naturalistic enclosures, a string of up to seven males may pursue a receptive female whenever she leaves her burrow. To test if female rats actually selected mates in this mad scramble or whether they mated promiscuously by choice, female rats were placed in a central arena surrounded by cubicles in which males were housed (a "rodent invertabrothel"). A circular passage gave the female access to the males but was too narrow for the males to pass. In such conditions, females formed enduring bonds with a single male, but also mated promiscuously. Even solitary rats seem to be profoundly affected by the knowledge that others of their species are in the vicinity. When female captive Heerman's kangaroo rats were entirely separated from males by wooden barriers rather than by clear plastic barriers, their estrous cycles were immediately doubled in length.

Rodent behavior is as adaptable as every other aspect of rodent biology. Having alert and active senses, rodents communicate by sight, sound, and smell. House mice have a sophisticated system of scent communication (See "A Scent-based Information Superhighway"). Kangaroo rats, however, tap-dance to talk. Three species living in the same area in California were each found to have a different rhythm when drumming on the desert floor. The Desert kangaroo rat thumped every 0.2–0.3 seconds, while the Giant kangaroo drummed long footrolls that could last for 100 drums at 18 drums per second; the Banner-tailed thumped footrolls at 3–38 drums per second. Even more intriguing is that individual Bannertails seemed to have their own signature rhythms. The sounds travel seismically through the ground to the ears of listening kangaroo rats hidden in burrows. Rats in burrows respond with carefully-timed drumming which does not overlap with those of the above-ground drummer. Drumming provides useful information about spacing and serves to reinforce territorial ownership. Female rats drumming at males tend to be saying "go away." Mother bannertails drum vigorously at snakes. Bannertails moving into a new neighborhood were found to alter their drums to be different from their neighbors.

While the social systems of some rodents have been well studied (See African Mole Rats), the habits of most of the 2,000 or so species are still a mystery awaiting investigation. Some insights have been achieved. Female Gunnison's prairie dogs mate with several males whereas they should be able to gain all the sperm necessary to fertilize their eggs from a single male. However, the probability of getting pregnant increases from 92 percent to 100 percent if they mate with three males instead of one, and females who mate with more males tend to have larger litters.

Some of the best-known visual displays are seen in the arboreal and the day-active terrestrial species. Courtship display in tree squirrels may be readily observed in city parks in early spring. The male pursues the female through the trees, flicking his bushy tail forward over his body and head when he is stationary. The female goads him by running slightly ahead, but he responds by uttering plaintive sounds similar to those that infants make to keep their mothers close. These sounds stop the female, allowing the male to catch up. Threat displays are dramatic in some species.

Rodents make considerable use of vocalizations in their communication. North American red squirrels and ground squirrels use a wide range of calls to advertise their presence to neighbors, to defend territories, and to sound the alarm when predators are detected. In other rodents, the sounds are far above the range of human hearing (at about 45kHz).

Rodents communicate extensively through odors produced by a variety of scent glands. Males tend to produce more and stronger odors than females, and young males are afforded a measure of protection from paternal attack by smelling like their mothers until they are sexually mature.

ODOR AND RODENT REPRODUCTION

Every phase of rodent reproduction, from initial sexual attraction to mating and the successful rearing of young, is influenced, if not actually controlled, by odor signals. Male rats are attracted to the urine of sexually receptive females, and sexually experienced males are more strongly attracted than naive males. Furthermore, if an experienced male is presented with the odor of a novel mature female alongside that of his mate, he prefers the novel odor. Females, on the other hand, prefer the odor of their mate to that of a stranger. The male's reproductive fitness is, it seems, best suited by seeking out and impregnating as many females as possible. The female, however, needs to produce many healthy young, so her fitness is maximized by mating with the best-quality male – one who has already proved himself. In gregarious species like the House mouse, a dominant male can mate with 20 females in 6 hours if their cycles are synchronized.

In young females, the odor of male urine accelerates not only the peak of sexual receptivity but also the onset of sexual maturity; it also brings sexually quiescent females into breeding condition. The effect is particularly strong from the urine of dominant males, while urine from castrated males has no such effect. It would appear, then, that the active ingredient – a pheromone – is made from, or is dependent upon the presence of, the male sex hormone testosterone. Male urine has such a powerful effect that if a newly-pregnant female is exposed to the urine odor of an unfamiliar male, she will resorb her litter and come rapidly into heat. In contrast, the odor of female urine has either no effect on the onset of sexual maturity or else slightly retards it. DMS

Above *Disturbed by an intruder, a North American porcupine displays its quills. The species is found in forests across most of the USA, northern Mexico, and Canada. Porcupines have quite poor eyesight, move slowly, and cannot jump, but nonetheless frequently climb trees to enormous heights in search of food.*

Left *Threat displays are very dramatic in some rodent species.* ***1*** *When slightly angry the Cape porcupine raises its quills and rattles the specialized hollow quills on its tail. If this fails to have the desired effect the hind feet are thumped on the ground in a war dance accompaniment to the rattling. Only if the threat persists will the porcupine turn its back on its opponent and charge backwards with its lethal spines at the ready.* ***2*** *Slightly less dramatic is the threat display of the Kenyan crested rat. This slow and solidly built rodent responds to danger by elevating a contrastingly colored crest of long hairs along its back, and in so doing exposes a glandular strip along the body. Special wick-like hairs lining the gland facilitate the rapid dissemination of a strong, unpleasant odor.* ***3*** *Finally, the little Norway lemming stands its ground in the face of danger and lifts its chin to expose the pale neck and cheek fur, which contrast strongly with the dark upper fur.*

Controlling Spiraling Populations

RODENTS AND HUMANKIND

With their high powers of reproduction and ability to invade all habitats, rodents are of great economic and ecological importance. Most rodent species are pregnant for just 19–21 days, mate again within 2 days of giving birth, and the young begin breeding at 6 weeks of age; theoretically, a single breeding pair of mice can generate 500 mice in 21 weeks. In good conditions, rodent numbers can soar, up to 1,000–2,000 per hectare (400–800 per acre). Aperiodic outbreaks occur repeatedly in the House mouse on farmland in Australia, in rat species in the uplands of Laos, and among grassland species such as the Common vole.

Of approximately 2,000 rodent species, about 200–300 are economically important, and some of these occur worldwide. A telling example of their economic impact comes from Southeast Asia, where rodents are economically the most important pre-harvest pest. In Indonesia, rodents cause annual losses of around 17 percent of rice production; if these could be halved, there would be enough rice to provide 70 percent of the energy requirements of an extra 17.5 million people. In Vietnam, rodent damage to rice affected 63,000ha (155,000 acres) in 1995, rising to more than 700,000ha (1.7 million acres) in 1999. In 1998, an estimated 82 million rats were killed under bounty schemes. Over 5 million rat tails were returned from January to September in the province of Vinh Phuc alone, where the authorities estimate that there are more than 10 million rats and only 1.1 million people.

For farmers in mountainous regions of Laos, rodents are the pest problem they currently have least control over. It is not unusual for a family to lose more than 70 percent of its crop to rodents; if this occurs to one crop it is a major cause for concern, but if it happens to two crops in a row, the situation becomes catastrophic. And the impact of rodents does not stop once the crop is harvested; they also consume and contaminate significant amounts of stored grain. It is estimated that post-harvest losses are of a similar magnitude to those that occur pre-harvest.

Other than rice, tropical crops damaged by rodents include coconuts, maize, coffee, field beans, oil palm, citrus, melons, cocoa, and dates. Every year rodents consume food equivalent to the world's entire cereal and potato harvest; it has been estimated that a train 5,000km (3,000mi) long – as long as the Great Wall of China – would be needed to haul the take.

The challenge of managing the rodent impact in Southeast Asia is complicated because there are at least 15 major pest species, each with its own peculiarities. The variables include their level of tolerance to a commensal life with humans; their breeding ecology; use of habitat; social behavior; feeding behavior; climbing, swimming, or burrowing abilities; physiological tolerances to climatic conditions (including periods when water is scarce or only present at high levels of salinity); and responses to major disturbances such as fire, new cultivation practices, or floods. Although the use of rodenticides offers short-term respite from the depredations of most species, more environmentally benign and sustainable approaches to pest management demand good knowledge of the ecology of the particular species to be controlled.

The most universal pests are the Norway rat, the Roof rat, and the House mouse. House mouse populations undergo spectacular eruptions in Australian cereal-growing areas, with devastating economic, environmental, and social consequences. Squirrel-like pests include the European Red and Gray squirrels, but in terms of the damage they cause, these are relatively minor. Other problem species include the gerbils, the multi-mammate rats, and the Nile rat, all of which devastate agricultural crops in Africa. In North America as well as eastern and western Europe, voles are prominent pests: they strip bark from trees, often killing them, and consume seedlings in forest plantations or fields. When vole populations peak (once every 3–4 years) there may be 2,000 voles per hectare (almost 5,000 per acre). Other major pests include the Cotton or Cane rat and Web-footed marsh rat (*Holochilus brasiliensis*) from, principally, Latin America; the Polynesian rat (*Rattus exulans*) of the Pacific Islands and Southeast Asia; the Bandicota rats of the Indian subcontinent and Malaysian Peninsula; the Ricefield rat (*Rattus argentiventer*), Lesser ricefield rat (*Rattus losea*), and Philippines ricefield rat (*Rattus tanezumi*) in Southeast Asia; and the prairie dogs, marmots, pikas, Brandt's vole (*Microtus brandti*), Chinese zokor (*Myospalax fontanierii*), and ground squirrel of the Mongolian and Californian grasslands.

The characteristics that enable these animals to become such a problem are a simple body plan adapted to a variety of habitats and climates; opportunistic feeding behavior; gnawing and burrowing habits; and high reproductive potential. Many rodents of the arctic tundra and taiga undergo population explosions every 3-4 years (for instance, the Norway lemming in Europe and the Brown and Collared lemmings in North America) The population density builds up to a high level and then dramatically declines. Several theses have been advanced to explain the decline, none of them wholly satisfactory; for example, it was long held that disease (tularemia or lemming fever found in many rodent populations) was the root cause of the decline, but it now seems more likely that it simply hastens it. Other suggestions are that the rodents become more aggressive at high density, leading to a failure of courtship and reproduction, or that the decline is due to the action of predators or the impoverishment of the forage. Objective observation of lemming behavior at high density shows it to be adaptive, providing the lemmings with the best chances for survival (see Voles and Lemmings).

In Britain, Norway rats may live in fields during the warm summer months when food is plentiful, and they seldom reach economically important numbers; but after harvest, with the onset of cold weather, they move into buildings. Also in Britain, and some other western European countries, the Long-tailed field mouse (*Apodemus sylvaticus*), normally only a pest in the winter when it may for example nibble stored apples, has learned to locate, probably by smell, pelleted sugar-beet seed. The damage it causes, which possibly went unnoticed before the advent of precision drilling, can lead to large barren patches in fields of sugar beet and sometimes necessitates complete re-sowing.

Although rodents usually consume about 15 percent of their body weight in food per day, much of the damage they do is not due to direct consumption. Three hundred rats in a grain store can eat 3 tonnes of grain in a year; but every 24 hours they also contaminate the grain with 15,000 droppings, 3.5 liters (6 pints) of urine, and countless hairs and greasy skin secretions. In sugar cane, rats may chew at the cane directly, consuming only a part of it; the damage, however, may cause the cane to fall over, so the impact of the sun's rays is reduced and harvesting impeded. In addition, the gnawed stem allows microorganisms to enter, reducing the sugar content. Apart from the value of the lost crop itself, a 6 percent drop in sugar content represents an equivalent reduction in the return on investment on land preparation, fertilizers, pesticides, irrigation water, management, harvesting, and processing.

Structural damage attributable to rodents results, for example, from the animals burrowing

*) **Left** A Brown rat* (Rattus norvegicus) *cares for a elpless infant. Although the young are born naked nd blind, they are quick developers; within 15 days hey are fully furred, and after another week are veaned and ready to leave the nest. Newborns will hemselves be ready to start breeding within 90 days.*

*) ▽ **Above and below** Competition for food is a najor cause of conflict between rodents and humans. pecies like the Roof rat* (Rattus rattus; *below) are amed marauders of stored vegetables and grains; and ats will also attack fruit on the tree, as a damaged gyptian orange (above) testifies. In Asia, rodents rou-inely consume 5–30 percent of the rice crop, some-imes devastating areas of 10,000 hectares (25,000 cres) or more; they have also been held responsible or destroying annually 5–10 percent of China's stored rain – enough to feed up to 100 million people.*

THE IMPACT OF RODENT-BORNE DISEASES

Rodent pests are involved in the transmission of more than 20 pathogens, including bubonic plague (transmitted to man by the bite of the rat flea), which was responsible for the death of 25 million Europeans from the 14th to the 17th century. Rats also transmit debilitating chronic diseases. In the late 1970s, 80 percent of the inhabitants of the capital of one developing Asian country were seropositive for murine typhus. Forty percent of those admitted to hospital were diagnosed as having fever of unknown origin; at least some of these, and maybe most, were probably suffering from murine typhus. The impact of the disease on the economy of the country is impossible to determine – and the same country also suffered frequent outbreaks of plague.

Apart from plague, which persists in many African and Asian countries as well as in the USA (where wild mammals transmit the disease, killing fewer than 10 people a year), murine typhus, salmonella food poisoning, leptospirosis, and the West African disease Lassa fever, to mention just a few, are all potentially fatal diseases transmitted by rats.

In the late 1990s, however, attention has tended to focus on rodent-borne hemorrhagic fever viruses. From 1995–2000, at least 25 "new" hantaviruses and arenaviruses were identified, all of them associated with rodents from the family Muridae. The hantaviruses cause pulmonary ailments in the New World (50 percent of infected humans die) and fever with renal ailments in the Old World (approximately 200,000 human cases each year in Asia, with 1–15 percent mortality). The arenaviruses cause South American hemorrhagic fever in the New World (mortality in humans is 10–33 percent) and Lassa fever in the Old World (100,000–300,000 human infections each year in West Africa, causing 5,000 deaths annually). Each virus is normally associated with a specific rodent host. Infection is passed on to humans via rodent urine, feces, or saliva.

After a period in which much work was done on describing and understanding the degree of rodent viral diversity that can generate human infections, the focus for the control or prevention of rodent-borne hemorrhagic disease is now switching to understanding the biology and ecology of the host–disease association.

▷ **Right** *A German engraving shows the bizarre protective clothing worn by a doctor to treat victims of bubonic plague in Nuremberg in 1656.*

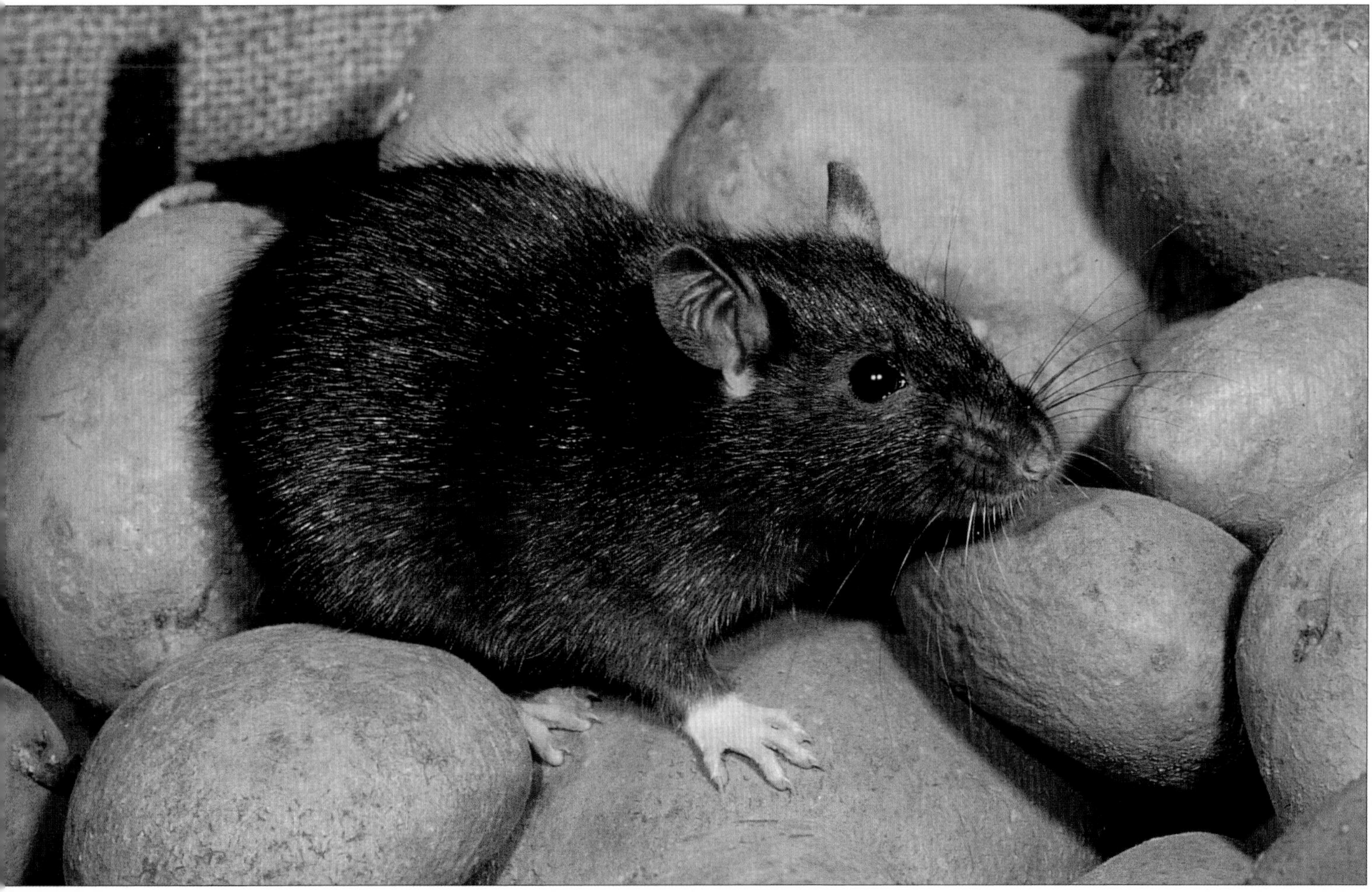

into banks or sewers, or under roads. The effects include subsidence, flooding, and even soil erosion in many areas of the world. Gnawed electrical cables can cause fires, leading to enormous economic impact. Rodents also gnaw through electric wires in the insulated walls of modern poultry and pork units, causing malfunctions of air-conditioning units and subsequent severe economic losses.

Apart from the immediate economic costs that the hordes may bring to farmers, high densities can have a profound effect upon the ecological balance of an entire region. Firstly, considerable damage is often inflicted on vegetation, from which it may take several years to recover. Secondly, predators increase in numbers in response to the abundance of rodents, and when the rodents have gone they turn their attention to other prey. Eruptions of Long-haired rat (*Rattus villosissimus*) populations over hundreds of thousands of square kilometers in northwestern Queensland and the Northern Territory of Australia lead to a feeding bonanza for Letter-winged kites, dingoes, foxes, and cats, and these predators in turn increase dramatically in numbers. In the mid-1990s there were so many cats living in the area that after the rat populations rapidly declined there was grave concern for other native fauna – in some areas many trees were literally a ball of cat fur. The situation was so desperate that the army was called in to help eradicate the feral cat population.

The simplest method of controlling the impact of rodents is to reduce harborage and available food and water. This, however, is often impossible. At best such "good housekeeping" can prevent rodent numbers building up, but it is seldom an effective method for reducing existing populations. One imperative is to develop management techniques based on an understanding of the ecology and behavior of the pest species concerned. The same principles of management apply in controlling the impact of rodents in fields, stores, or domestic premises.

To reduce existing populations of rodents, predators – wild (for example, birds of prey) or domestic (cats or dogs) – have relatively little effect. Their role may lie in limiting population growth. It is a widely accepted principle that predators do not control, in absolute terms, their prey, although the abundance of prey may affect the numbers of predators. Mongooses were introduced to the West Indies and Hawaiian Islands and cobras to Malaysian oil palm estates, both to control rats. The rats remain, however, and the mongooses and cobra are themselves now considered pests, the one a reservoir for rabies, the other a direct risk to people. Even the farm cat will not usually have a significant effect on rodent numbers: the reproductive rate of rodents keeps them ahead of the consumption rate of cats!

One of the simplest methods for combating small numbers of rodents is trapping. Few traps, however, are efficient: most simply maim their

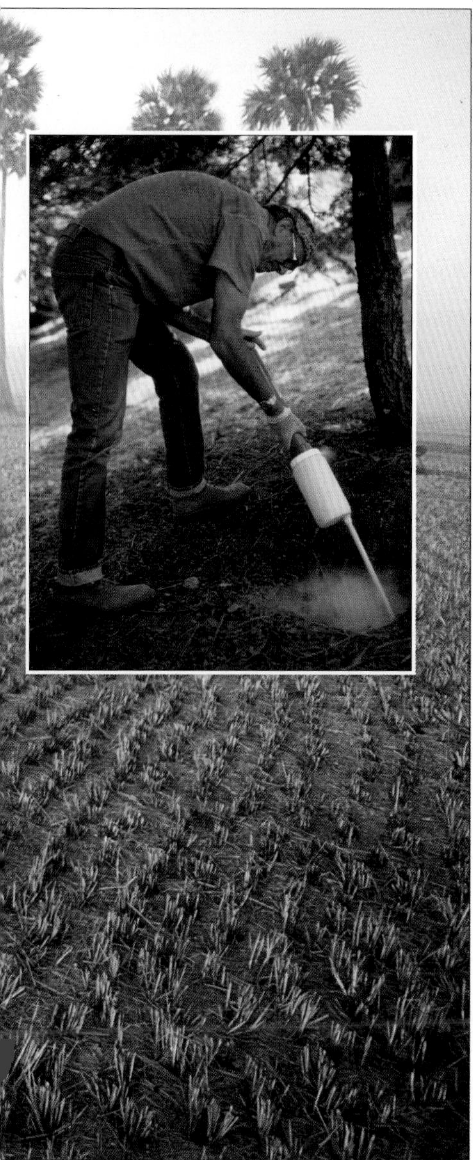

Above *Indian villagers dig out rats' nests in their fields. For farmers, the battle against rodent damage is worldwide and never-ending. Over the centuries poison has been a favorite weapon, but the use of powerful pesticides can come at a cost, as the toxins may also harm non-targeted species, and there is also a risk of contaminating human food supplies. In the fight against rodent-borne maladies, an alternative approach is to dust nests with insecticides to kill disease-bearing parasites, like the exterminator dusting a ground squirrel's den in California (inset).*

Left *A Barn owl heads for the nest with a Field mouse clasped firmly in its beak. Many terrestrial predators as well as birds of prey rely on a plentiful supply of fast-breeding rodents for a large part of their diet. Programs aimed at the local eradication of pest species can have unintended knock-on effects higher up the food chain by depriving these species of their prey.*

victims. One promising method of physical control is the use of multiple capture traps placed at the base of fences (25 x 25m) that enclose early-planted crops (lure crops). These "trap-barrier" systems, which remain in place for the duration of the crop, have significantly reduced the impact of the Ricefield rat in lowland irrigated rice crops in Indonesia, Malaysia, and Vietnam. In Malaysia, 6,872 rats were caught in one night and over 44,000 rats in a 9-week period. In Indonesia and Vietnam, yield increases from surrounding crops have ranged from 0.3 to 1.0 tonne per ha (0.1–0.4 tonnes an acre), representing a 10–25 percent increase in production. The disadvantage of this approach is that it is labor-intensive and requires a coordinated community approach.

The oldest kind of rodenticide – fast-acting, non-selective poison – appears in the earliest written record of chemical pest control; the Greek philosopher Aristotle described the use of strychnine as early as 350 BC. However, fast-acting poisons such as strychnine, thallium sulfate, sodium monofluoroacetate (Compound 1080), and zinc phosphide have various technical and ecological disadvantages, including causing long-lasting poison shyness in sub-lethally poisoned rodents. They also represent a hazard to other animals.

Since 1945, when warfarin was first synthesized, several anticoagulant rodenticides have been developed. These compounds decrease the blood's ability to clot, and consequently bring about death by internal or external bleeding. The first generation of anticoagulants required multiple feeds of the poison, and were initially effective at controlling susceptible species such as the Norway rat. However some Norway rats have since acquired genetic resistance to the substances, while other species, such as the Roof rat, had a natural resistance from the start. These inadequacies have led to the development of more potent "second generation" anticoagulants, with active ingredients such as brodifacoum and bromadiolone, that only require single feeds and generally have a high kill rate for most rodent species. However, there are concerns about the risk to non-target species from these more potent chemicals, which are more persistent in the environment and which accumulate in predators as they eat more and more poisoned rodents until the predators themselves succumb to the poison.

Apart from the choice of toxicant, the timing of rodent control and the coordinated execution of a planned campaign are important in serious control programs. The most effective time to control agricultural rodents, for example, is when little food is available to them and when populations are low (probably just before breeding). "Avant-garde" methods of rodent control involving such relatively novel means as chemosterilants, ultrasonic sound, or electromagnetism are sometimes suggested, but none can yet claim to be as effective as anticoagulant rodenticides.

Species at Risk

CONSERVATION AND ENVIRONMENT

Not all rodents have thrived with the spread of humans. At least 54 species have become extinct in the last two centuries, and another 380 currently face a similar fate. At greatest risk are 78 critically endangered species that have small, isolated populations (often less than 250 individuals) that are continuing to decline. For some of these, such as Margaret's kangaroo rat and the Brazilian arboreal mouse, habitat protection offers hope that extinction will be averted. For others, such as the Bramble Cay mosaic-tailed rat, the future is bleak. This stocky rodent occurs on only one sparsely vegetated coral cay, 340m (1,100ft) long and 150m (500ft) wide, at the northern tip of Australia's Great Barrier Reef. Although its population numbers several hundred individuals, the rat is declining inexorably as the tides erode the coral and threaten to inundate the land.

At slightly less risk are the 100 or so endangered species. These may have total populations of up to 2,500, often scattered among several locations that are at risk of disturbance. Two species of Central American agoutis fall within this category, as do six species of Mexican woodrats.

Active management is assisting the survival of some endangered species. For example, Greater stick-nest rats numbered less than 1,500 individuals in 1990 and were restricted to Franklin Island off the southern coast of Australia. Successful translocations of captive-bred animals to three new islands and also to three large enclosed areas on the Australian mainland allowed the total population to double within nine years. Programs of captive breeding and habitat management currently benefit over 20 endangered species, including Vancouver marmots, Stephens' kangaroo rats, and Shark Bay mice.

While still threatened, almost 200 species classed as Vulnerable face less risk of imminent extinction than their endangered relatives, and may achieve populations up to 10,000 individuals. Examples include the Plains rat and Dusky hopping mouse of central Australia, the Utah prairie dog, and Menzbier's marmot of Tien Shan.

Some rodents are elusive, making it difficult to confidently identify their status. Arboreal species such as the Prehensile-tailed rat and the South American climbing rats do not readily enter traps, while others, such as the southern Australian Heath rat, so resemble other common species that they are easily overlooked. In many instances a lack of recent survey work makes status assessment impossible. The New Britain water rat and the Orange and Mansuela mosaic-tailed rats of Ceram are known only from one or two specimens collected in the early 20th century. Whether abundant or extinct, these rodents will remain known only from museum specimens until intrepid biologists foray back to the sites where they were originally collected. GS/CRD/DMS

DISTRIBUTION Worldwide except for Australia region, Polynesia, Madagascar, southern S America and some desert regions.

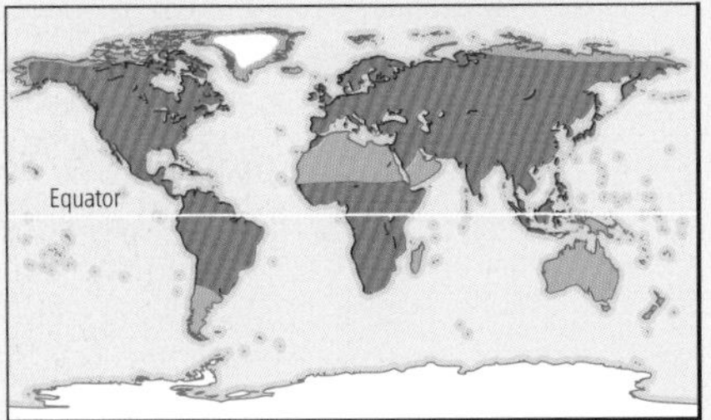

SCIUROGNATHS (part only) Suborder Sciurognathi
284 species in 56 genera and 5 families

Squirrel-like Rodents

SQUIRRELS ARE PREDOMINANTLY SEEDEATERS and are the dominant arboreal rodents in many parts of the world. However, in the same family there are almost as many terrestrial species, including the ground squirrels of the open grasslands – also mostly seedeaters – as well as the more specialized and herbivorous marmots.

Although they may be highly specialized in other respects, members of the squirrel family have a relatively primitive, unspecialized arrangement of the jaw muscles and therefore of the associated parts of the skull, in contrast to the mouse-like ("myomorph") and cavy-like ("hystricognath") rodents, which have these areas specialized in ways not encountered in any other mammals. In squirrels the deep masseter muscle is short and direct, extending up from the mandible to terminate on the zygomatic arch. Because this particular feature is shared by some smaller groups of rodents, notably the Mountain beaver and the true beavers, these families have been grouped together with the squirrels in the suborder Sciuromorpha.

These families appear to have diverged from each other and from other rodents very early in the evolution of rodents and have very little in common other than the retention of the "sciuromorph" condition of the chewing apparatus.

A further primitive feature retained by these rodents is the presence of one or two premolar teeth in each row, giving four or five cheekteeth in each row instead of three as in the murids.

The superficially similar scaly-tailed squirrels of the African rain forests are only distantly relatec and are now placed within a separate suborder, Anomaluromorpha, which also includes the aberrant springhare of the African plains. GB(

›IFFERENT STROKES...

‹here is a great deal of variation in the ›hape and function of the tail among ›e families of squirrel-like rodents. ›ne highly specialized adaptation is ›ound in the various species of flying ›quirrel (RIGHT), in which the tail is sed in conjunction with the gliding ›embrane to control the precise angle f descent. In another group of "fly-›g" squirrel-like rodents – the scaly-›ailed squirrels of the tropical African ›ain forests – the tail aids not only glid-›g but also climbing; overlapping ›cales on its underside give the squirrel ›urchase as it lands and grips onto the ›ark of trees. The semi-aquatic beaver mploys its broad, flat tail both as a ›eans of propulsion and as a rudder. Vhen slowly patrolling on the lake's ›urface, its tail moves from side to ›de. However, when swimming fast ›nderwater (INSET) the beaver propels ›self with powerful up-and-down ›hrusts of the tail. The long, bushy tail ›f the kangaroo-like springhare acts as counterbalance, helping it maintain quilibrium while hopping along at ›peed. When the springhare is at rest, nd standing on its hind feet, it braces ›self with its tail. The familiar, bushy ›ails of squirrels perform different roles ›epending on the habitat of the indi-idual species – serving as warm cloaks ›r the denizens of northern forests, nd as parasols to shade Cape ground ›quirrels from the fierce heat of the ›un in the Kalahari desert.

›HAPING THE ENVIRONMENT

›mong squirrel-like rodents, the ›eaver is renowned for its skill in alter-›g the landscape of its habitat. It cuts ›own trees with its strong incisors, ›ses the felled material to build dams ›nd lodges, and excavates canals with ›s forepaws.

Somewhat less spectacular is the ›ay that smaller sciuromorphs such as ›quirrels shape their environment, and ›et this devastation is far more signifi-ant, thanks to the prolific and wide-pread nature of the squirrel family. ›ed squirrels, which live predominant-› in conifer plantations, cause much ›amage by feeding on young shoots. ›heir cousins the Gray squirrels special-›e in peeling the bark from relatively young trees, up to 30–40 years old. The reasons for doing this are twofold: some bark is taken as a soft lining for nests; but the majority is stripped to gain access to the sweet, tasty sap beneath – a valuable dietary supplement when other food is in short supply (for example, in mid-summer). This destructive activity is especially prevalent where squirrel populations are dense and low-status males are forced to search widely for food during lean times. Bark-stripping presents the forestry industry with a serious problem; though trees are remarkably resilient, if their bark is stripped off entirely around the trunk and the soft tissue gnawed away, the vascular bundles that transport water, sugar, and nutrients through the tree are severed and it dies.

Crop-raiding is also a cause of conflict between humans and squirrel-like rodents. Ground squirrel species feed principally on low-growing vegetation.

› ***Left and above*** *A Red squirrel, ›rom Europe and Asia, and (above) a ›ouglas squirrel from western North ›merica both display the classic squir-›el profile – a long body and bushy tail.*

SKULLS AND DENTITION

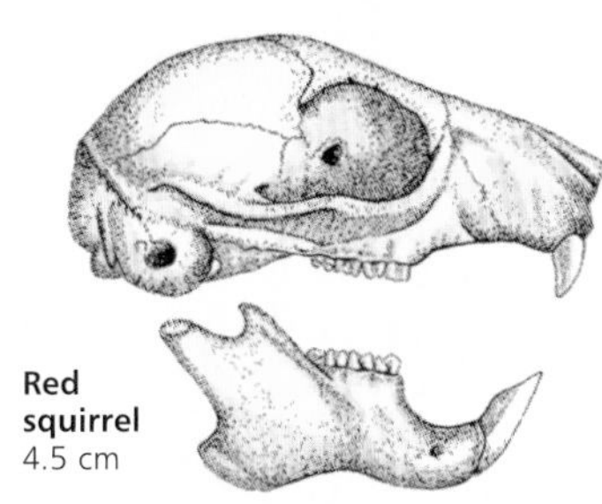

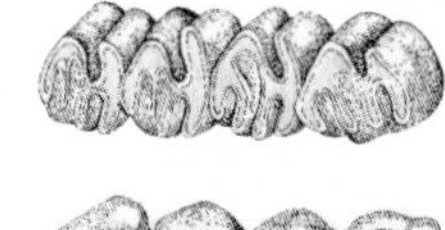

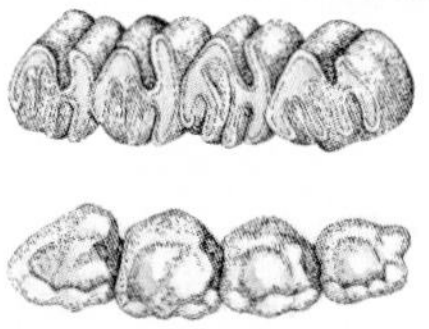

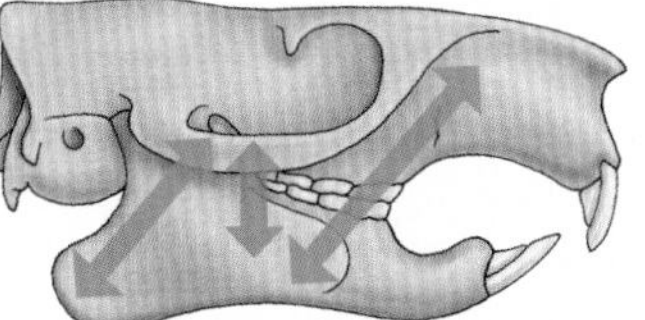

Skulls of squirrels show few extreme adaptations, although those of the larger ground squirrels, such as the marmots, are more angular than those of the tree squirrels. Most members of the squirrel family have rather simple teeth, lacking either the strongly projecting cusps or the sharp enamel ridges found in many other rodents. In the beavers, however, there is a pattern of ridges, adapted to their diet of bark and other fibrous and abrasive vegetation and convergent with that found in some unrelated but ecologically similar rodents like the coypu.

The primitive jaw musculature is characteristic of squirrel-like rodents. The lateral masseter muscle (blue) extends in front of the eye onto the snout, moving the lower jaw forward during gnawing. The deep masseter muscle (red) is short and used only in closing the jaw. Shown above is the skull of a marmot.

Beavers

FEW WILD ANIMALS HAVE HAD AS GREAT AN influence on the world's history and economics as the beaver. Exploration of the North American interior by Europeans was stimulated in large part by the demand for beaver pelts, used for hats and clothing. Records of the Hudson's Bay Company show an annual catch of 100,000 animals. So lucrative was this fur trade that conflicts erupted over access to trapping areas – notably the series of French and Indian Wars in the 18th century, which culminated in British control over the whole of northern North America.

The former importance of beavers may be judged from the fact that they are portrayed on the coats of arms of cities as distant as Härnösand in Sweden and Irkutsk in Siberia. In North America, many indigenous peoples valued the beaver both as a resource and as a spiritual totem. Central to the religion of the Montagnais of Quebec was a benevolent Beaver guardian spirit, while the tribe west of Lake Athabasca in Alberta were named for the main river of their homeland: *Tsades*, or "River of Beavers" (now called the Peace River).

Heavyweights of the Rodent World

FORM AND FUNCTION

Biologically, the beaver's large incisor teeth, flat, scaly, almost hairless tail, and webbed hind feet with a split grooming claw on the second digit are all distinctive, as is the internal anatomy of the throat and digestive tract. Beavers display a rich variety of construction, communication, and social behaviors that set them apart from other mammals. Like humans, beavers live in family groups, have complex communication systems, build homes (lodges and burrows), store food, and develop transportation networks (ponds linked by canals). Furthermore, they too change their environment to suit their needs; the dams they build promote ecological diversity, increase wetlands, affect water quality and yields, and help shape landscape evolution.

After the capybaras of South America, beavers are the heaviest rodents in the world; adults average 20–30kg (44–66lb), up to a maximum recorded weight of 45.5kg (100lb). Beavers always live near water. The animals scull slowly along the surface of lakes, using side-to-side movements of the tail for steering and propulsion. By contrast, when swimming fast or diving, beavers moves their tails up and down in synchrony with powerful thrusts of the hind feet.

Beavers may appear slow and awkward on land, waddling on large, pigeon-toed rear feet, short front legs, and trim forefeet. Yet they can put on a turn of speed if alarmed, outpacing both predators and researchers as they gallop for the water.

A History of Comings and Goings

DISTRIBUTION PATTERNS

The North American beaver ranges over most of the continent from the Mackenzie River delta in Canada south to northern Mexico. By the late 1800s, beavers had effectively been extirpated locally over much of this range, especially in the eastern United States, but state and federal wildlife agencies have reestablished many populations through translocations and reintroductions. The animals have also been introduced into Finland, Russia (to the Karelian Isthmus, the Amur basin, and the Kamchatka peninsula), and in many central European countries, including Germany, Austria, and Poland. The largest population of North American beavers in Europe is today in southeastern Finland, numbering an estimated 10,500. North American beavers were also introduced in

FACTFILE

BEAVERS

Order: Rodentia

Family: Castoridae

2 species of the genus *Castor*

DISTRIBUTION N America, Scandinavia, W and E Europe, C Asia, NW China, Far E Russia, S America.

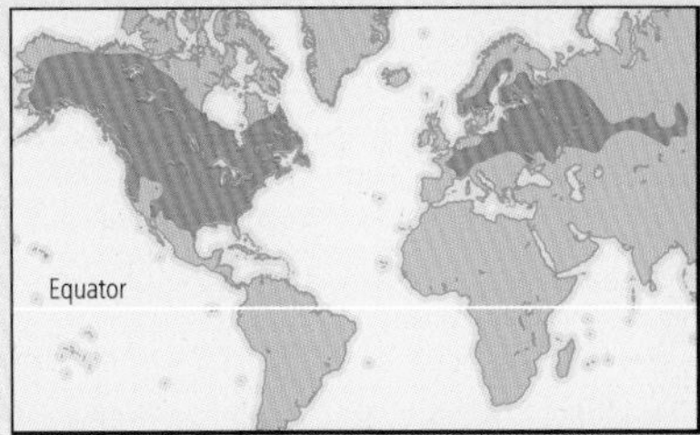

Habitat Riparian, semiaquatic wetlands associated with ponds, lakes, rivers, and streams.
Diet Wood (especially aspen), grasses, roots.

NORTH AMERICAN BEAVER *Castor canadensis*
North American or Canadian beaver
N America from Alaska E to Labrador, S to N Florida and Tamaulipas (Mexico); introduced to S America (Tierra del Fuego, Argentina), Europe, and Asia. HBL 80–120cm (32–47in); TL 25–50cm 10–20in); SH 30–60cm (12–23in); WT 11–30kg (24–66lb); no difference between sexes. Coat: yellowish brown to almost black; reddish brown is most common. Underfur is dense and dark gray. Breeding: gestation about 105–107 days. Longevity: 10–15 years. 24 subspecies are recognized.

EURASIAN BEAVER *Castor fiber*
Eurasian or European beaver
NW and C Eurasia, in isolated, but increasing, populations from France E to Lake Baikal and Mongolia; also in far-eastern Russia (Khabarovsk). Other details as for *C. canadensis*. 8 subspecies are recognized.

Abbreviations HBL = head–body length TL = tail length SH = shoulder height WT = weight

Above *Though somewhat ungainly when out of water, the semiaquatic beaver* (Castor canadensis) *is perfectly adapted to a wetland environment. Its dense, luxuriant pelage made the beaver vulnerable to hunting by fur trappers, and it was exterminated in many countries. However, aided by reintroduction programs, it is now making a comeback.*

Left *Battling against the current, a beaver feeds on the bark of a silver birch branch. When a beaver dives, it shuts its nose and ears tight, while a translucent membrane covers its eyes and the back of the tongue prevents water from entering its throat. This effective, watertight adaptation allows beavers to gnaw and carry sticks underwater without choking; but the comic aspect of closing the lips behind the front teeth is much beloved of cartoonists when depicting the animals.*

South America (to Tierra del Fuego) in 1946. While 24 subspecies are recognized, local exterminations, translocations, and reintroductions have altered the purity of many of these.

The Eurasian beaver was once found throughout Europe and Asia, but only isolated populations have survived the long association with humans. By the early 1900s, only eight relic populations totaling an estimated 1,200 individuals remained, in France, Germany, Norway, Belarus, Russia, Ukraine, Mongolia, and China respectively. Reintroduction and translocation programs began in most European countries in the early to mid 1900s, and many continue at the present time. The Eurasian beaver population is growing (the current estimate is 500,000–600,000 animals) and its range is expanding throughout Europe and Asia. Eight subspecies are recognized, but translocations and reintroductions have altered their historic distributions.

The earliest direct ancestor of the Eurasian beaver was probably *Steneofiber* from the Oligocene, about 32 million years ago. The genus *Castor*, which originated in Europe during the Pliocene (5–1.8 m.y.a.), entered North America while the continents were still connected; thus the present-day North American beaver is thought to be evolutionarily younger than the Eurasian beaver. During the Pleistocene, 10,000 years ago, the two species coexisted with giant forms that weighed 270–320kg (600–700lb), for example *Castoroides* in North America and *Trogontherium* in Eurasia. The two present species are externally similar in size and coloration and are indistinguishable in appearance, but they differ in cranial morphology and chromosome number (*C. fiber* has 48 chromosomes, *C. canadensis* 40). The North American beaver is believed to be derived from the Eurasian beaver because it has fewer chromosomes. It is thought that the reduced

Right *The beaver's lodge is a large, conical pile of logs and branches, sited on the bank or isolated in the middle of a lake. Lodges average 3–4m (10–13ft) in diameter, with rooms measuring 1–2m (3.3–6.6ft), and always incorporate a living chamber above water level. Sometimes there is also a dining area nearer the water.*

Above *Beaver kits are extremely precious – females produce only one small litter each year. At birth, they have a full coat of fur and open eyes, and are able to move around inside the lodge. All family members share in bringing the kits solid food, with the adult male most actively supplying provisions.*

number of chromosomes in *C. canadensis* resulted from the fusion of 8 chromosome pairs in the ancestral *C. fiber*, and that this difference prevents the two species from interbreeding.

Choosy Generalists

DIET

Beavers are generalist herbivores whose diet varies seasonally. In spring and summer they feed on relatively non-woody plants (leaves, roots, herbs, ferns, grasses, and algae), but they turn to trees and shrubs, especially in the fall. Aspen is the preferred tree species, but birch, maple, oak, dogwood, and fruit trees are also taken. In many regions, shrubs such as willow and alder form the bulk of the diet. Wood is not easy to digest; to cope with it, beavers have special microbial fermentation in the cecum, and digestion occurs twice to extract the maximum nutritive advantage.

Beavers are only able to survive the harsh winters of northern latitudes by caching food. Woody stems are stored underwater, where they are safe from other browsers. The beavers can then swim under the ice to fetch the food without having to leave the safety of the pond.

Many European beavers do not cache food (in some populations, only 50 percent of families do so). Instead, they venture onto land to find food in winter. Another strategy that beavers use to survive at this time is to live off the fat stored in their tails.

How Colonies Work

SOCIAL BEHAVIOR

Beavers live in small, closed family units, which are often referred to as "colonies." An established colony contains an adult pair, young of the current year (called "kits"), yearlings born the previous year, and possibly one or more subadults from previous breeding seasons. There are typically four to eight beavers in a family. The inclusion of offspring older than 24 months usually occurs in high-density populations, and these young adults generally do not breed. Under high-density conditions, the normal family structure has been reported to change, with more than one reproductively active adult female present.

The main beaver predator, beside humans, is the wolf, for which beavers are an important prey item. Other large carnivores, such as lynx, coyotes, wolverines, bears, and foxes, may also kill them.

Beavers are unusual among mammals in that they exhibit long-term monogamy. The mated pair occupies a discrete, individual territory, and the relationship usually lasts until one adult dies. The family unit is exceptionally stable, thanks to a low birth rate (one litter annually of 1–5 young in the Eurasian and up to 8 in the North American beaver), a high survival rate for all age–sex classes, and long-term parental care, with the young usually staying in the colony for 2 years. Both Eurasian and North American beavers have an average litter size of 2–3 kits, and the dominance hierarchy within the family is by age class, with adults dominant to yearlings and yearlings to kits. A sex-based hierarchy, with the adult female dominant to all other family members, has occasionally been reported. Physical aggression is rare, although beavers in dense populations are reported to have more tail scars – a sign of fighting, usually with outsiders over territory boundaries.

Mating occurs in winter, usually in the water but also in the lodge or burrow. Kits are born in

THE BEAVER'S 29-HOUR DAY

For most of the year, beavers are active at night, rising at sunset and retiring to their lodges at sunrise. This regular daily cycle is termed a circadian rhythm. During winter at northern latitudes, however, when ponds freeze over, beavers stay in their lodges or under the ice, because temperatures there remain near 0°C (32°F), while air temperatures are generally much lower. Activity above ice would require a very high use of energy.

In the dimly lit world of the lodge and the water surrounding it, light levels remain constant and low throughout the 24-hour day, so that sunrise and sunset are not apparent. In the absence of solar "cues," activity, recorded as noise and movement, is not synchronized with the solar day. The circadian rhythm breaks down, and "beaver days" become longer, varying in length from 26 to 29 hours. This type of cycle is termed a "free-running circadian rhythm."

te spring, which may coincide with the dispersal f 2-year-olds away from the family territory. hough they can swim within a few hours, the its' small size and dense fur make them too uoyant to submerge easily, so they are unable to ive down the passage out of the lodge. Kits nurse or between 6–8 weeks, although they may begin ating solid food before they are weaned.

While the kits are very young, the adult male nay spend more time in territorial defense patrolling and scent marking the family territory) while the female is more involved in care and ursing. The kits grow rapidly, but require many nonths of practice to perfect their ability to con-truct dams and lodges. As yearlings, they partici-pate in all family activities, including construction. Dispersal of the young adults usually occurs in the pring of their second year, when they are approxi-mately 24 months old; they head off to set up ter-itories of their own, usually within 20km (12mi), lthough they may travel as far as 250km (155mi).

One of the ways in which beavers communi-cate is by depositing scent, usually at the borders of the family territory. Both species of beaver centmark on small waterside mounds made of nud and vegetation dredged up from underwater. Eurasian beavers also mark tufts of grass, rocks, and logs, as well as directly onto the ground. The scent, produced by the substance known as cas-oreum (from the castor sacs) and secretions from he anal glands, is pungent and musty. All mem-bers of a beaver family participate, but the adult male marks most frequently. Scentmarking is most ntense in the spring, and probably serves to con-vey information about the resident family to pass-ng strangers and to neighbors. Another mode of communication is to slap the tail against the water. Adults do this more often than younger ani-mals, to let an approaching interloper know they have been spotted. The slap often elicits a response from the stranger, which enables the beaver to gauge what level of threat is posed. Beavers also communicate through vocalizations (hissing and grunting), tooth sharpening, and posture.

Of the various construction activities in which beavers engage, canal-building is the least com-plex and was probably the first that the animals developed. They use their forepaws to loosen mud and sediment from the bottom of shallow streams and marshy trails, pushing it out of the way to the sides. The resulting channels enable the beavers to stay in the water while moving between ponds or to feeding areas. This behavior occurs most often in summer when water levels are low.

Beavers are efficient excavators and usually dig multiple burrows in the family area. A burrow may be a single tunnel or a whole maze, hollowed into the bank from a stream or pond and ending in one or more chambers. In many habitats, beaver fami-lies use burrows as the primary residence. Alterna-tive riverside accommodation is provided by the beaver equivalent of a log cabin – the lodge.

Above *Of all rodents, the beaver is one of the best-adapted for movement in water. Its torpedo-shaped body is hydrodynamically efficient and its fur is waterproof. It uses its broad tail for propulsion and steering, and also gains thrust from its webbed feet.*

Meat, Fur, and Medicines

CONSERVATION AND ENVIRONMENT

Historically, beavers have provided humans with meat, fur, and medicinal products, and they con-tinue to supply the first two to this day. From 200,000–600,000 pelts are collected in Canada each year, while the harvest in the United States is 100,000–200,000. Scandinavian countries take around 13,000 pelts, and Russia some 30,000. However, the global fur market is depressed, and there is little economic incentive for trappers.

Another beaver product of value to humans – especially in the past – was castoreum, a complex substance consisting of hundreds of compounds, including alcohols, phenols, salicylaldehyde, and castoramine. This substance is produced in the animal's castor sacs. As far back as 500–400 BC Hippocrates and Herodotus note its efficacy in treating diseases of the womb. Later writers, such as Pliny the Elder and Galen, mention it as a reme-dy for cramps and intestinal spasms. Down the ages, castoreum has also variously been suggested as a treatment for sores, ulcers, earache, constipa-tion, and even as an antidote to snake venom. Modern studies have remarked on its similarity to the synthetic drug aspirin, which is derived from salicylic acid; castoreum also contains salicin, which the beavers obtain by eating the bark of wil-low and aspen. Although it no longer appears in the conventional pharmacopoeia, it is still used by homeopathic practitioners, and also serves as a base for perfume.

Beavers are today established once more in Bel-gium and the Netherlands, and a reintroduction program is planned in Britain. As with most such initiatives, there is controversy over the potential effect on land used by humans. North American beavers in Europe add to an already charged situa-tion. Currently, large populations exist in Finland and northwestern Russia and there is no consen-sus on how the two species are competing for the limited wetland resource. New techniques make it possible to distinguish the two species in the field, either by using genetic analysis of hair root cells or else by comparing anal-gland secretions.

Beavers and humans conflict when beavers make wetlands out of agricultural fields. In the midwestern and eastern United States, where beavers were once nearly extirpated, they have made a tremendous comeback, and beaver–human conflict is growing, a trend that is likely to continue. Beavers are keystone species in wetland habitats, and it remains for humans to acknowl-edge their environmental contributions and devel-op strategies that allow both humans and beavers to share the landscape. PB/GH

PHOTO STORY

The Construction Worker Rodent

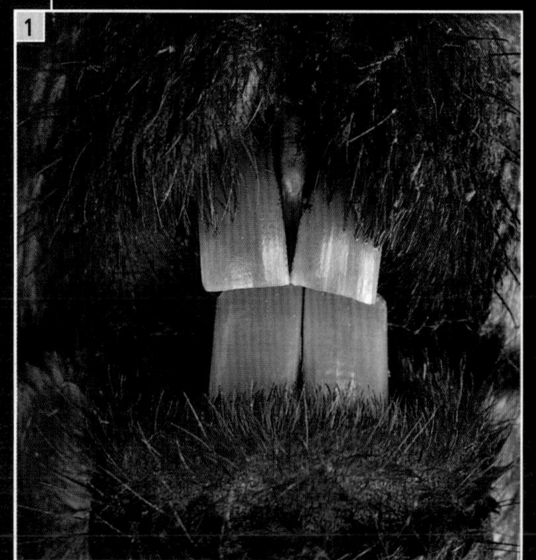

3 *Although many beaver barrages are built across streams under 6m (20ft) wide and in water less than 1m (3.3ft) deep, the animals keep adding material until some dams end up over 100m (330ft) long and 3m (10ft) high. Dam-building is at its most intense in the spring and fall, although construction may go on throughout the year.*

1 *Just as a lumberjack's ax is hardened with iron, so the front enamel of beaver teeth is reinforced. The softer inner surface wears down more rapidly, creating a sharp, chisel-like edge that makes cutting down trees for food and building easier. In common with all rodents, beavers have large incisors that grow as fast as they are worn down.*

2 *Trees felled by beavers; although the damage caused by gnawing may appear terminal, the trees that beavers favor (aspens, poplars, cottonwoods, and willows) are characterized by rapid growth, and beaver "pruning" often stimulates reinvigorated growth the next spring.*

4 *In areas where beavers and humans coexist, beavers' dam-building activity (especially their propensity to block culverts and cause road flooding) is seen as a hazard. But the ecological advantages are often overlooked; by slowing the flow of rivers and streams, dams boost sediment deposition – a natural filtration system that removes potentially harmful impurities from the water. In addition, the large areas of wetland that dams create bring other benefits, such as reduced erosion damage and greater biodiversity.*

5 *The mud, stones, sticks, and branches that beavers use to construct their dams make for a very robust structure, behind which a substantial pond will form. By impounding a large body of water, they effectively surround their home with a wide moat, which increases their security from predators. Moreover, the bigger the lake, the more access the beavers have by water to distant food items. Several lakes and lodges may be the work of a single colony (so, simply counting these in any one area is not a reliable way of estimating the local beaver population). Colonies are forced to move on when the accumulation of sediment in a pond becomes too great. Abandoned, silted-up beaver ponds form the basis of rich new ecosystems; they develop into wetland meadows whose soil, rich in decaying plant matter, supports reeds and sedges, and eventually even large trees.*

Mountain Beaver

THE MOST PRIMITIVE OF ALL PRESENT-DAY rodents, Mountain beavers live only in southwestern Canada and along the west coast of the USA, where they inhabit some of the most productive coniferous forest lands in North America.

Not to be confused with the flat-tailed stream beaver (genus *Castor*), Mountain beavers are strictly terrestrial animals. They are secretive and nocturnal, spending much of their time in burrows, where they sleep, eat, defecate, fight, and reproduce; they even do most of their traveling underground. Consequently, they are seldom seen outside zoos.

FACTFILE

MOUNTAIN BEAVER

Aplodontia rufa

Mountain beaver, boomer, or sewellel

Order: Rodentia

Sole member of family Aplodontidae

DISTRIBUTION USA and Canada along Pacific coast

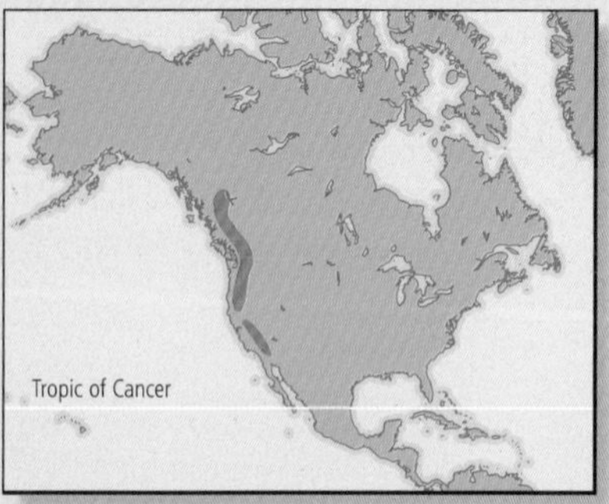

HABITAT Coniferous forest

SIZE Head–body length 30–41cm (12-16in); tail length 2.5–3.8cm (1–1.5in); shoulder height 11.5–14cm (4.5–5.5in); weight 1–1.5kg (2.2–3.3lb). Sexes are similar in size and shape, but males weigh slightly more than females.

COAT Young in their year of birth have grayish fur, adults blackish to reddish brown, tawny underneath.

DIET Leafy materials, grasses, roots.

BREEDING Gestation 28–30 days

LONGEVITY 5–10 years

CONSERVATION STATUS Lower Risk: Near Threatened; two subspecies – the Point Arena and Point Reyes Mountain beavers – are listed as Vulnerable.

Built for Burrowing

FORM AND FUNCTION

Mountain beavers are medium-sized, bull-necked rodents with a round, robust body and a moderately flat and broad head that is equipped with small, black, beady eyes and long, stiff, whitish whiskers (vibrissae). Their incisor teeth are rootless and continuously growing, as are their premolars and molars. Their ears are relatively small, and covered with short, soft, light-colored hair. Their short legs give them a squat appearance; the tail is short and vestigial. The fur on their back presents a sheen, whereas white-to-translucent-tipped long guard hairs impart their flanks with a grizzled effect. A distinctive feature is a soft, furry white spot under each ear.

Mountain beavers can be found at all elevations from sea level to the treeline, and in areas with rainfall of 50–350cm (20–138in) per year and where winters are wet, mild, and snow-free, summers moist, mild, and cloudy. Their home burrows are generally in areas with deep, well-drained soils and abundant fleshy and woody plants. In drier areas Mountain beavers are restricted to habitats on banks and to ditches that are seasonally wet or have some free-running water available for most of the year.

The distribution of Mountain beavers is in part explained by the fact that they cannot adequately regulate the temperature of their bodies and must therefore live in stable, cool, moist environments. Nor can they effectively conserve body moisture or fat, which prevents them from either hibernating or spending the summer in torpor. Mountain beavers require considerable amounts of both food and water, and they must line their nests well for insulation. They satisfy most of their requirements with items obtained from within 30m (100ft) of their nest. Water is obtained mainly from succulent plants, dew, or rain.

Unlike any other rodent or rabbit, Mountain beavers extract fecal pellets individually with their incisors when defecating and toss them on piles in underground toilet chambers. However, they also share with other rodents and rabbits the habit of reingesting some of the pellets.

Mountain beavers are strictly vegetarians. They harvest leafy materials such as fronds of Sword fern, new branches of salal and huckleberry, stems of Douglas fir and Vine maple, and clumps of grass or sedge, and also seek out succulent, fleshy foods, such as fiddle heads of Bracken fern and roots of False dandelion and Bleeding heart. If these foods are not eaten immediately they will be stored underground for subsequent consumption.

Most food and nest items are gathered above ground between dusk and dawn and consumed underground. Decaying, uneaten food is abandoned or buried in blind chambers; dry, uneaten food is added to the nest.

Life below the Forest Floor

SOCIAL BEHAVIOR

It used to be thought that all Mountain beavers were solitary except during the breeding and rearing season, but recent studies using radio tracking have demonstrated that this view is inadequate. Some Mountain beavers spend short periods together in all seasons. Neighbors, for example, will share nests and food caches, or a wandering beaver will stay a day or two in another beaver's burrow system. Sometimes a beaver's burrow will also be occupied, in part if not in its entirety, by any of a range of other animals, such as salamanders, frogs, or deer mice.

Unlike many rodents, Mountain beavers have a low rate of reproduction. Most do not mate until they are at least 2 years old, and females conceive only once a year, even if they lose a litter. Breeding is to some extent synchronized. Males are sexually active from about late December to early March, with aggressive older males doing most of the breeding. Conception normally occurs in January or February. Litters of between 2–4 young are born in February, March, or early April. The young

Left *Mountain beavers seldom appear above ground during daylight hours. This rare image shows well the squat, thickset appearance of this most elusive of North American mammals.*

land. Both sexes disperse in the same manner. Dispersing young travel mainly above ground: they become very vulnerable to predators, for example owls, hawks, and ravens, coyotes, bobcats, and man. Those living near roads are liable to be killed by vehicles.

The Forester's Foe

CONSERVATION AND ENVIRONMENT

Mountain beavers are classed as non-game mammals, and so are not managed in the same way as game animals like deer and elk or furbearing animals such as beavers and muskrats. They can pose a huge threat to young conifers planted for timber; the damage they cause affects about 111,000ha (275,000 acres). Almost all damage occurs while they are gathering food and nest materials.

Although some of the Mountain beavers' forest habitat has given way to urban development and agriculture, they range over about as extensive an area now as they did 200–300 years ago. They are probably more abundant now than they were in the early 20th century, thanks to forest logging practices. Mountain beavers do not appear to be in any immediate danger of extermination from man or natural causes. JE

e born blind, hairless, and helpless in the nest. ney are weaned when they reach about 6–8 eeks old and continue to occupy the nest and urrow system with their mother until late sumer or early fall. Juveniles disperse and leave their atal site to find a territory and burrow system of eir own.

Once on its own, a young beaver may establish new burrow system or, more commonly, restore abandoned one. New burrows may be within)0m (330ft) of the mother's burrow, or up to km (1.2mi) away, the distance depending on opulation densities and on the quality of the

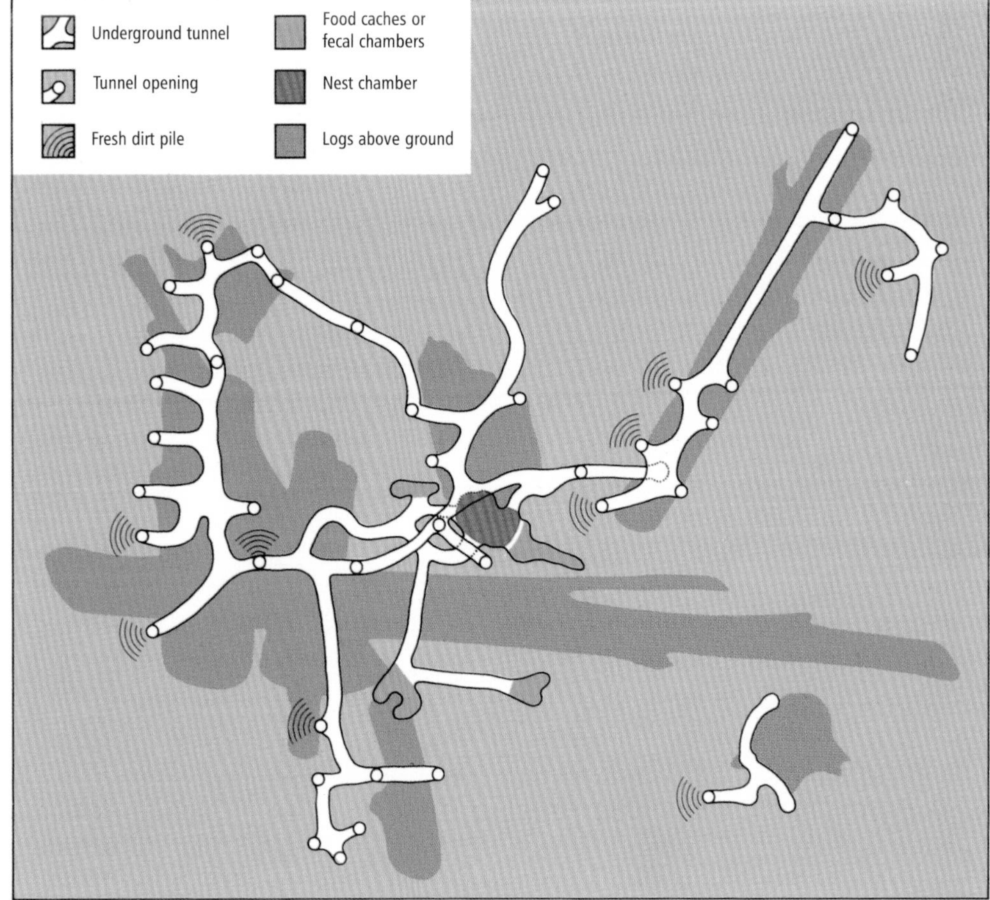

Right *The burrow system of a Mountain beaver. ch burrow system consists of a single nest chamber d underground food caches and fecal chambers hich are generally close to the nest. Most nests are out 1m (3.3ft) below ground in a well-drained, me-shaped chamber, although some may occur 2m .6ft) or more below ground. Tunnels are generally –20cm (6–8in) in diameter and occur at various vels; those closest to the surface are used for travel; ep ones lead to the nest and food caches. Burrow enings occur every 4–6m (13–20ft) or more, depend- g on vegetative cover and number of animals occu- ing a particular area. Densities vary from 1 or 2 ountain beavers per hectare (2.5 acres) in poor habi- t to 20 or more per hectare in good habitat. Up to per hectare have been kill-trapped in reforested ear-out areas, but such densities are rare. Individual rrow systems often interconnect.*

Squirrels

TEMPERATE-ZONE SQUIRRELS ARE A LITTLE like daffodils: they appear suddenly in early spring, add life to the habitat for a few months, and then disappear again. For the ground-dwellers, disappearance signals the start of hibernation, for many of these mammals spend at least half their lives in dormancy: they are active above ground for only 4–6 months and hibernate for the rest of the year in grass-lined nests deep underground.

Because they are relatively unspecialized, squirrels have been able to evolve different body forms and habits that fit them for life in a broad range of habitats, from lush tropical rain forests to semi-arid deserts, and from open prairies to town gardens. This successful family includes such diverse forms as the ground-dwelling and burrowing marmots, ground squirrels, prairie dogs, and chipmunks; the arboreal and day-active tree squirrels; and the nocturnal flying squirrels. The scaly-tailed squirrels (Anomaluridae) are also treated here, though they are taxonomically distinct from true squirrels.

Squirrel Groups and Genera

Pygmy squirrels: *Exilisciurus, Glyphotes, Myosciurus, Nannosciurus, Sciurillus*
Dwarf squirrels: *Microsciurus, Prosciurillus*
Giant squirrels: *Ratufa, Rubrisciurus*
Flying squirrels: *Aeretes, Aeromys, Belomys, Biswamoyopterus, Eupetaurus, Glaucomys, Hylopetes, Iomys, Petaurillus, Petinomys, Pteromys, Pteromyscus, Trogopterus*
Giant flying squirrels: *Petaurista*
Beautiful squirrels: *Callosciurus*
Sun squirrels: *Heliosciurus*
Groove-toothed squirrels: *Rheithrosciurus, Syntheosciurus,*
Palm squirrels: *Epixerus, Funambulus, Menetes, Protoxerus*
Long-nosed squirrels: *Hyosciurus, Rhinosciurus*
Rock squirrels: *Sciurotamias*
Red-cheeked squirrels: *Dremomys*
American red squirrels: *Tamiasciurus*
Tree squirrels: *Sciurus, Sundasciurus*
Striped squirrels: *Funisciurus, Lariscus, Tamiops*
African bush squirrels: *Paraxerus*
Chipmunks: *Tamias*
Ground squirrels: *Ammospermophilus, Atlantoxerus, Xerus (Geosciurus), Spermophilus, Spermophilopsis*
Prairie dogs: *Cynomys*
Marmots: *Marmota*

For full species list see Appendix ▷

Diggers, Climbers, and Gliders

FUNCTION AND FORM

Squirrels are instantly recognizable, with their cylindrical bodies, bushy tails, and prehensile limbs. They have short forelegs, with a small thumb and four toes on the front feet, and longer hind legs with either four (woodchucks, *Marmota monax*) or five toes (ground squirrels and tree squirrels) on the hind feet. Most are diurnal, conspicuous, active, and – often – amusingly clever. They exhibit a wide array of body sizes and behaviors, from ground-dwelling, fossorial species (marmots, ground squirrels, prairie dogs) to arboreal tree squirrels and nocturnal flying squirrels. Their abilities to live in contrasting habitats and to forage opportunistically are the bases for their widespread distribution and numerical abundance, both in species and populations.

Squirrels have large eyes, surrounded by light-colored rings, that are placed on the sides of their heads, affording them a broad field of vision. Keen eyesight enables them to recognize dangerous predators from nondangerous conspecifics at

Above The Harris' antelope squirrel inhabits bur-
ws that it has excavated. Within its range it may
ve several burrows, one of which will have a nest.
e species is diurnal and quite conspicuous.

Left The Red squirrel, which lives in woodland
oss Europe and Central Asia, has distinctive tufting
its ears. It is hunted in Russia for its dark brown
nter pelage.

at distances. Tree and flying squirrels and chip-
nks also have large ears; some, like Red squir-
s and Tassel-eared squirrels (*Sciurus aberti*),
ve conspicuous ear-tufts. All squirrels have
uch-sensitive whiskers (vibrissae) on the head,
t, and the outsides of the legs.

Squirrels have the usual arrangement of teeth
rodents: a single pair of chisel-shaped incisors
each jaw, a large gap in front of the premolars,
d no canine teeth. Their incisors grow continu-
sly and are worn back by use; the cheek
th are rooted and have abrasive chewing sures. The lower jaw is movable, and the lower
cisors can operate independently. Some chipmunks and ground squirrels have internal cheek pouches for carrying food.

Ground-dwelling squirrels are heavy-bodied, with powerful forelimbs and large scraping claws for digging, whereas arboreal squirrels have lighter, longer bodies, less muscular forelimbs, and sharp claws on all toes. Tree squirrels descend tree trunks head first, turning their hind feet backward and sticking the claws into the bark to act as anchors. Their bushy tails are multifunctional: they serve as a balance when the squirrel runs and climbs, as a rudder when it jumps, as a flag to communicate social signals, and as a wrap-around blanket when the animal sleeps. All squirrels have soft pads on the soles of their feet, affording them a better grip of the substrate and food items. When feeding, squirrels squat on their haunches and hold the morsel in their forepaws. Footpads of desert-living Long-clawed ground squirrels (*Spermophilopsis leptodactylus*) are furry, which insulates them as they scurry over hot sand; in addition, fringes of stiff hairs on the outside of the hind feet serve to push away sand during burrowing.

FACTFILE

SQUIRRELS

Order: Rodentia

Family: Sciuridae

273 species in 50 genera

DISTRIBUTION Among the most widespread of mammals, found worldwide except for in Australia, Polynesia, Madagascar, S South America, and the Sahara desert (Africa and Arabia).

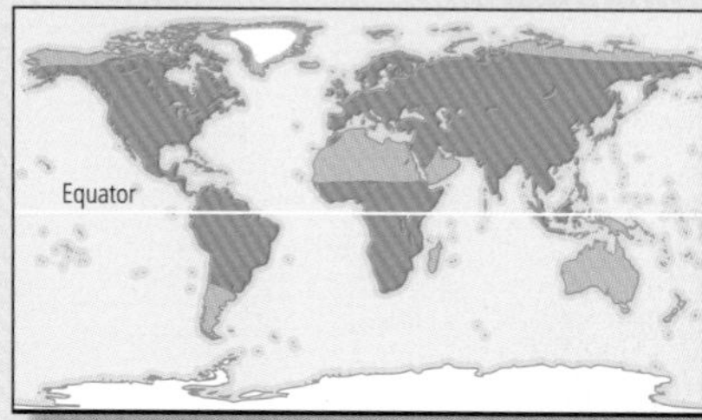

HABITAT Various, from tropical rain forests to temperate and boreal coniferous forests, tundra and alpine meadows to semiarid deserts, and cultivated fields to city parks. Some species are arboreal and nest in tree branches or cavities; others are terrestrial and excavate subterranean burrows.

SIZE **Head–body length** ranges from 6.6–10cm (2.6–3.9in) in the diminutive African pygmy squirrel, *Myosciurus pumilio* to 53–73cm (20.8–28.7in) in the Alpine marmot, *Marmota marmota*; **tail length** from 5–8cm (2–3in) to 13–16cm (5.1–6.3in); **weight** from 10g (0.35oz) to 4–8kg (8.8–17.6lb), in the same two species.

COAT Squirrels come in many colors. Most squirrels molt twice per year. In northern areas a soft, fine summer coat alternates with a stiff, thick winter coat. There are no sexual dimorphisms or age variations in coat texture or colors.

DIET Tree and flying squirrels eat nuts, seeds, fruits, buds, flowers, sap, and occasionally fungi; ground-dwelling squirrels eat grasses, forbs, flowers, bulbs, and especially seeds (*Spermophilus* means "seed loving"). Most species will also eat insects, birds' eggs and nestlings, and small vertebrates if they are available.

BREEDING In most species, females mature sexually before males, usually reproducing by age 1 year. Males are polygynous and, in some species, females also mate with multiple partners, resulting in litters of mixed paternity. Most ground-dwelling squirrels, flying squirrels, and northern populations of tree squirrels bear one litter in late spring; temperate-zone tree squirrels and chipmunks often have another summer litter. Litters typically contain 1–6 (up to 11) pups; larger species have smaller litters.

LONGEVITY Ground and tree squirrels live 2–3 years on average and 6–7 years maximum; larger species, such as Yellow-bellied marmots (*Marmota flaviventris*), live 4–5 years on average and 13–14 years maximum. Females usually live longer than males.

CONSERVATION STATUS Many species are threatened or endangered due to loss of habitat, introduction of exotic species, and harassment. Red squirrels are listed as Near Endangered in the UK and N Italy due to exclusion by the introduced Gray squirrel.

1

2

8

Flying squirrels, like other gliding mammals such as flying lemurs and flying phalangers, have a furred, muscular membrane (or "patagium") that extends along the sides of the body and acts as a parachute when the animal leaps. The patagium stretches from the hind legs to the front limbs, and is bound in front by a thin rod of cartilage attached to the wrist. Once airborne, the squirrel steers by changing the position of its limbs and bushy tail and by varying the tension in the patagium. Flying squirrels descend in long, smooth curves to the base of tree trunks, where they brake by turning the tail and body upward. The larger flying squirrels can glide for 100m (330ft) or more; smaller species cover much shorter distances. Gliding is an economical way to travel and facilitates quick escape from flightless tree predators such as pine martens. When on trees, however, flying squirrels are hindered in their movements by the membrane, which may explain why they are all nocturnal – to evade keen-sighted birds of prey.

The Winter Larder

DIET

Squirrels are primarily vegetarians. Tree squirrels favor nuts and seeds, but will also eat leafy greenery. Gray squirrels in the United Kingdom and the Malabar giant squirrels (*Ratufa indica*) of Indian rain forests will also debark branches and feed on the underlying growing tissue; Horsetail squirrels in Malayan rain forests specialize on bark and sap from the boles of trees. In addition, insects are an important dietary component for some species. Gray and Red squirrels in broadleaf woodlands feed on caterpillars in the spring and early summer when other nutritious foods are scarce. Several tropical squirrels only eat insects; indeed, the Long-nosed squirrel (*Rhinosciurus laticaudatus*) has incisors modified into forceps-like vices perfect for grabbing such small prey. Most tree and flying squirrels will also eat birds' eggs and nestlings.

Ground-dwelling squirrels are similarly omnivorous. Although they feed primarily on plant parts, they will in addition eat insects, birds' eggs, carrion, and, occasionally, each other's young. Typically, no one plant makes up more than 10 percent of a species' diet. Early in the growing season, ground squirrels seek out new grass shoots, which they pull up and eat blade by blade, starting at the base. As the season progresses, flowers and unfurling leaves of forbs are their mainstays. In late summer, species that hibernate concentrate on energy-rich bulbs and seeds of grasses and

7

Left *Representative species of squirrels:* **1** *Southern flying squirrel* (Glaucomys volans) *gliding from a nest hole in a tree trunk;* **2** *Prevost's squirrel* (Callosciurus prevostii)*;* **3** *African pygmy squirrel* (Myosciurus pumilio) *descending a tree head-first;* **4** *Abert or Tassel-eared squirrel* (Sciurus aberti)*;* **5** *American red squirrel* (Tamiasciurus hudsonicus) *hanging by its hind legs;* **6** *Indian giant or Malabar squirrel* (Ratufa indica)*;* **7** *Asiatic or Siberian chipmunk* (Tamias sibiricus) *with filled cheek pouches;* **8** *Alpine marmot* (Marmota marmota) *in vigilant upright posture giving alarm whistle;* **9** *Shrew-faced ground or Long-nosed squirrel* (Rhinosciurus laticaudatus) *foraging for termites;* **10** *Geoffroy's or Western ground squirrel* (Xerus erythropus) *with its tail arched and fluffed, an indication of anxiety.*

legumes, and on nuts. Sometimes foraging activities affect the local vegetational structure: prairie dogs bite off and discard tall plants around their burrows to enhance their field of view, and this constant cropping discourages all but the fastest-growing plants. Most squirrels get all the water they require from ingested plant materials; however, tree squirrels regularly visit sources of drinking water, especially during hot, dry summers.

Because of the seasonal nature of flowering and fruiting in temperate forests, the squirrels that live there depend on different foods at different times of year. Another way of ensuring a constant food supply is to cache seeds and nuts, which can be unearthed and eaten during the winter. From July, when the first fruits mature, until the following April, Gray and Red squirrels in broadleaf woodlands depend on fresh and buried nuts. A poor mast crop can have disastrous consequences in terms of high overwinter mortality (particularly of subadults), and may even prevent breeding the following spring. Adult Red squirrels that recover many cached pine cones and beechnuts are more likely to survive the next spring and summer than those that must rely only on buds, shoots, and flowers; and females are more likely to produce a spring litter if they feed heavily on cached seeds than if they feed mainly on other vegetation.

Tree squirrels within forests may act as foresters; by burying seeds and nuts safely in the soil they promote regeneration of the trees. Squirrels of all ages hoard nuts and seeds; even juveniles exhibit characteristic burying behaviors. As poaching of squirrels' larders is not uncommon, some scatter their stores widely to avoid theft by neighbors. Individual Red and Gray squirrels may bury hundreds of nuts in a season, some close to the tree that bore them but others as much as 30–60m (100–200ft) away. And, although individuals can smell nuts buried as deep as 30cm (12in) below the surface, many are never retrieved. Burying behavior and failed retrieval result, inadvertently, in tree-seed dispersal. There has been a long co-evolution of tree squirrels with many coniferous and mast-producing trees, such as pine, oak, and beech. This may seem odd, as squirrels' teeth can gnaw through hard shells and destroy the embryo of any nut they discover; but trees in fact benefit, by producing crops so abundant that many buried nuts are never recovered, enabling them to germinate.

Douglas squirrels (*Tamiasciurus douglasii*) and American red squirrels (*T. hudsonicus*) do not scatter-hoard. Instead they cut unopened pine and spruce cones off trees and cache them in a single larder, often located in a hollow stump or under a log. Larders may contain up to 4,000 cones in Jack pine forests, and as many as 18,000 in White spruce forests. The food contained in these so-called "middens" is essential for surviving the long winters in the boreal forest, so each squirrel defends a territory surrounding 1–3 middens. Males and females defend similar-sized territories: 0.4–0.5ha (about 1 acre) in spruce forests, and 0.6–0.7ha (1.5 acres) in Jack pine forests. Territories are defended year round: females rarely leave home, and males depart only for a few days in the spring when neighboring females are sexually receptive. The squirrels warn away potential rivals by screeching and rattling calls; if an intruder persists, they will resort to chases and physical combat. Territorial defense is most intensive close to the midden, where the stakes are highest, and most evident in the autumn, when new cones are ready to harvest and dispersing juveniles are attempting to establish territories of their own.

The area defended around "primary" middens varies in size between years. When cone crops are good, juvenile recruitment is high, because territories of adults shrink and there is plenty to eat. Young, transient animals can temporarily reside between territories centered on primary middens. When cone crops fail, however, juvenile recruitment drops due to starvation and lack of living space. Interestingly, female American red squirrels may in very good years leave their territory and its middens to one or two sons or daughters. Being

Above *A White-tailed prairie dog eating a flower. Prairie dogs live in colonies or "towns." Populations of these animals fell during the 20th century, mostly as a result of poisoning programs; indeed, some prairie dog species have been reduced by several hundred million.*

Below *The Least chipmunk* (Tamias minimus) *is the most broadly distributed chipmunk in North America. In colder areas of the continent, such as Canada, this species may be dormant for more than half the year (October to April).*

rger, older, and more experienced, the mother is ore likely to compete successfully for a new ter-ory than her young would be, so leaving her ter-ory to her offspring increases their likelihood of rvival and reproduction.

North American flying squirrels also hoard food d guard their larders throughout the year. These rders are hidden in tree cavities or underground, d are marked by sweat and sebaceous gland cretions. Larders of Southern flying squirrels Glaucomys volans), primarily containing acorns, ay be defended by a single individual or, in win-r, by aggregations of up to 15 squirrels. Hud-ing together reduces daily energy expenditure about 30 percent, and squirrels in a group are ss likely to be taken by predators than an indi-dual living on its own.

In coniferous forests, Northern flying squirrels (*G. sabrinus*) do not have access to acorns, so

Above *The Arizona gray squirrel is a tree squirrel found in forested areas of Arizona, New Mexico, and Mexico. Tree squirrels are not hibernators, but in very cold weather they will take to their nests, only leaving when it becomes necessary to find food. The diet is mainly nuts, seeds, fruit, young shoots, and buds.*

lichens and hypogeal (subterranean) fungi are the predominant winter food. Nests are constructed in tree cavities or attached to branches, and are composed almost entirely of certain special arboreal lichens that lack acids and other secondary compounds. The lichen-covered walls then serve not only as insulation but also as winter food caches. By spring, the squirrels have almost literally eaten themselves out of house and home!

Photos and documentary film of tree squirrels curling up in their winter nests have led to the popular misconception that all squirrels hibernate. Tropical and desert-dwelling species may remain active all year round, and even holarctic tree squirrels stay alert, putting on limited fat reserves in the autumn and relying on the constant availability of high-energy nuts and seeds to get them through the winter. Their winter activity pattern represents a balance between short, frenetic feeding bouts (of about 3–4 hours per day) to meet their energy demands – about 400–700 kJ/24–42kcal per day for a Red squirrel weighing 300–350g (10–12oz) – and long resting periods in the nest.

Slightly flattened and spherical in shape, the drey is made of small branches, twigs, and grasses (or a tree cavity may be used), and is thickly lined with dry grass, moss, and fur. It is so well insulated that, in freezing conditions when the owner is inside, the internal temperature is about 20°C higher than the air temperature outside. During a squirrel's extended resting periods, its body temperature decreases from about 41°C to 39°C, which further reduces energy expenditure. Nonetheless, the animal cannot afford to spend more than two days inside its cozy drey without foraging outside, even in bad weather.

Ground-dwelling squirrels living in temperate and northern climates spend the winter in subterranean burrows. They prepare for this period of forced inactivity either by storing food in their dens or by accumulating fat on their bodies; some will undertake both activities. Food-storing species typically undergo long bouts of torpor that are interspersed with brief periods of activity. Siberian chipmunks (*Tamias sibiricus*) store seeds, acorns, buds, and mushrooms, each in a different burrow compartment. They can carry up to 9g (0.3oz) of grain in their cheek pouches for distances in excess of 1km (0.6mi), and an individual animal may store as much as 2.6kg (5.7lb) of winter food.

Most fat-storing ground squirrels and marmots spend the winter months in hibernation, sometimes in groups of close relatives. The last animal into the den (among Alpine marmots, *Marmota marmota*, this is usually an adult male) plugs the entrance hole with hay, earth, and rocks, for insulation and safety from predators.

Physiologically, hibernation is a state of suspended animation during which a squirrel's metabolism slows down to one-third of its normal rate, and its heartbeat, body temperature, and respiratory rate plummet. When the temperature outside is below freezing, a hibernating Alpine marmot's body temperature will drop to 4.5–7.5°C (40–45.5°F), and the animal may breathe only once or twice per minute. Every 2–4 weeks, hibernators awaken to defecate and urinate; by the time they emerge properly from hibernation in the spring, they will have lost more than half their pre-hibernation body weight. At that point they may feed on the bulbs and seeds that they stored away the previous fall. Such stores are essential when late spring storms keep the animals below ground.

Some ground-dwelling hibernators also become torpid during the summer (estivation) if the vegetation withers away due to heat and drought. In such circumstances it is much better to be safely out of sight doing nothing rather than busily scurrying around on the hot surface, especially when there is no food to be found there. When entering estivation, the animals conceal and insulate the mouth of their den with plugs of grass and sand.

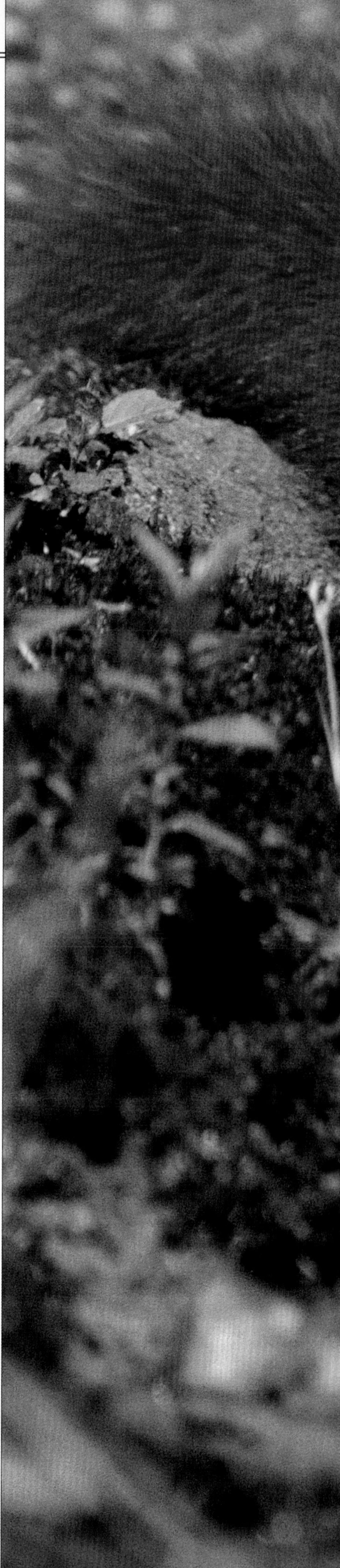

Right *A Eurasian red squirrel* (Sciurus vulgaris) *– a species that occurs throughout Europe and northern Asia – drinking from a pond. Most species of squirrel get most of the water they require from their food; however, tree squirrels will visit watering spots quite frequently, especially during hot periods.*

Scaly-tailed Squirrels

The tropical and subtropical forests of the Old World are inhabited by an interesting array of gliding mammals. In tropical Africa this niche is filled by the scaly-tailed squirrels, members of the Anomaluridae family (7 species in 3 genera – see Appendix for full list), which apparently share only a distant evolutionary relationship with true squirrels.

With the exception of the Cameroon scaly-tail, which does not fly, the anomalures' most distinctive feature is a capelike membrane, stretched between their four limbs, that lets them glide from branch to branch in the depths of the African rain forest. The membrane is supported at the front by a rod of cartilage extending from the elbow or the wrist; at the rear it attaches to the ankles. Spread out in flight, it forms a rough square that permits the squirrels to soar over distances that can exceed 100m (330ft). Although this adaptation resembles the patagia of the flying squirrels and the Australian flying possums, these species too are evidently not closely related; instead, they have developed along convergent lines to fill similar ecological niches in different parts of the world.

The anomalures are squirrel-like in form, with a relatively thin, short-furred tail. They take their name from an area of rough, overlapping scales near its base. The scales help the squirrels get a purchase on trees when they land from a glide, and also provide grip for climbing trunks.

The ecology and behavior of scaly-tailed squirrels are poorly known, because they live in areas that are rarely visited by outsiders and are nocturnal in their habits. Even so, such species as the Lord Derby's Zenker's squirrels are relatively common, and do sometimes come into contact with humans.

The scant information available about anomalure reproduction indicates that females may have two litters of 1–3 young per year. At birth babies are large, well-furred, and active, and their eyes are open. Female pygmy scaly-tail squirrels apparently leave the colonies to bear their single young alone.

Except for Lord Derby's scaly-tailed flying squirrel, these interesting but poorly-known rodents depend entirely on primary tropical forest for their existence. To the extent that African primary forests are being destroyed, they are endangered, for they require for their survival a mature forest habitat with hollow trees in which they can nest. For want of more detailed information on their condition, the IUCN currently lists most species as Near Threatened. THF

Above *A White-tailed antelope squirrel on a Joshua tree. This species, which is found in Baja California and New Mexico, has stable dominance hierarchies. The breeding season for this species is from February to June.*

Below *Cape ground squirrels sunning themselves in the Kalahari Gemsbok Park. The species lives in social groups in burrows that may have anything up to 100 openings. Colonies average about 5–10 members, but can number as many as 25–30 individuals.*

Living in Clusters

SOCIAL BEHAVIOR

All squirrels raise young. Ground-dwelling hibernators reproduce very early in the spring, presumably to maximize the time their young have to grow and fatten before winter returns. Males typically emerge first and wait for the females, who mate soon after emergence. The exact timing of emergence and mating depends on the severity of the preceding winter, and it differs greatly between species, among populations of the same species at different altitudes and latitudes, and within the same population under different weather conditions. For example, Belding's ground squirrels (*Spermophilus beldingi*) living at 2,200m (7,200ft) emerge and mate 5–8 weeks before conspecifics living just 800m (2,600ft) higher and 15km (9mi) away; however, the timing of emergence at 3,000m (9,800ft) may itself vary by 5–6 weeks, depending on snow depth and spring weather.

Squirrels' mating systems are diverse. In some species, such as Belding's ground squirrels, males defend small mating territories which females visit when they are receptive. The males of other species such as Thirteen-lined ground squirrels (*S. tridecemlineatus*) search for widely scattered females, and wait in line to mate with them. At the opposite extreme, some marmots mate while still submerged in their winter burrows. Sometimes, as in the case of Idaho ground squirrels (*S. brunneus*), males guard females closely before and after mating. And, in most marmots, many ground squirrels, and Black-tailed prairie dogs, males vigorously defend territories surrounding the burrows of several females, who mate exclusively with the territory-holder.

In tree squirrels and chipmunks, receptive females attract males using chemical signals and vocalizations, then lead the males on long, spectacular mating chases that may last from 4 to 10 hours. Often numerous males simultaneously chase the same female (3–5 in Red squirrels, up to 10 in Gray squirrels, and 9–17 in Beautiful squirrels). By running away but remaining conspicuous (rather than hiding), receptive females force males to compete, enabling the females to compare their suitors' stamina and fighting abilities. In general, the dominant male stays closest to the female and accounts for 80–90 percent of copulations. After mating, he guards the female for up to an hour. In some species males also hinder further mating with physical barriers of coagulated sperm and seminal fluids. However, females occasionally solicit matings with males other than the alpha (up to 4 or more in Beautiful squirrels), and females often remove copulatory plugs to facilitate

remating, as in the case of Gray and Fox squirrels (*Sciurus niger*). Multiple mating may enable females to hedge their bets against the possibility of the dominant male being infertile. Ground squirrel females also mate with multiple males after removing copulatory plugs. Genetic analyses of Arctic (*S. parryii*), Belding's, and Thirteen-lined ground squirrel litters indicate that mixed paternity is frequent, though the first male to mate with a female (usually the dominant male) sires the majority of each litter.

The young are gestated for 3–6 weeks and then suckled for slightly longer again, especially in tree and flying squirrels. Litter sizes range from 1 in the Giant squirrel (*Ratufa macroura*) and Giant flying squirrel (*Petaurista elegans*) to 2 in beautiful squirrels, 2–5 in most tree squirrels and marmots, 4–6 in prairie dogs and chipmunks, and 5–11 in ground squirrels. In many species, middle-aged females in good condition produce larger litters and heavier pups than younger or older females and than females in poor condition. Across species, litter sizes vary inversely with female body size and degrees of sociality: bigger, more social species have smaller litters.

Young squirrels are born naked, toothless, and helpless, with skin over their eyes. They develop rapidly. Young Red squirrels begin to sprout hair at 10–13 days, and are fully furred by 3 weeks; lower incisors appear at 22 days, upper ones at 35 days; eyes open at 30 days; self-cleaning begins around 35 days. Juveniles take their first solid food and begin venturing from the nest at about 40 days. At 8–10 weeks juveniles are independent, although they remain near their mother

Above *Idaho ground squirrels. The male is on guard against rivals and stays close to the female with whom he has just mated. Since females of this species are widely dispersed and locating them is a time-consuming activity, to ensure he fathers some young each season, the male adopts the strategy of keeping competitors away rather than looking for more females. The females are only sexually receptive for a few hours, just after they emerge from hibernation.*

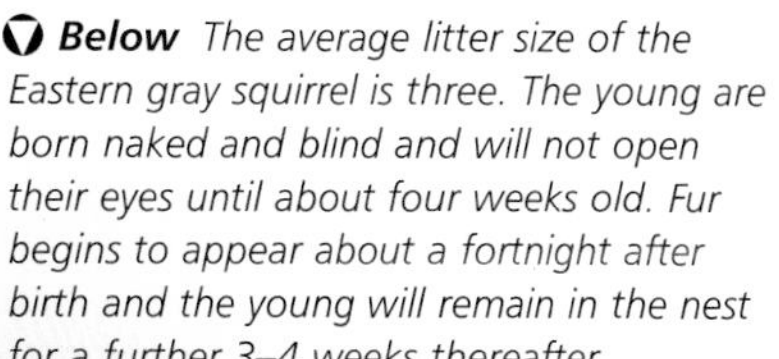

Below *The average litter size of the Eastern gray squirrel is three. The young are born naked and blind and will not open their eyes until about four weeks old. Fur begins to appear about a fortnight after birth and the young will remain in the nest for a further 3–4 weeks thereafter.*

and may still share her nest. The ontogeny of young flying squirrels is similar.

Interestingly, young ground-dwelling squirrels develop much more rapidly: chipmunk and ground squirrel pups are independent by 3–4 weeks, and marmot pups by 6 weeks. These differences probably relate to each species' ecology. So, for ground-dwellers it is dangerous to remain in the maternal nest too long because pups can be trapped by digging predators, whereas for young tree and flying squirrels it is risky to leave the nest too soon because climbing about in trees requires considerable balance and coordination.

In most tree squirrels, ground squirrels, and chipmunks, parental care is provided solely by the mother, but marmots, prairie dogs, and certain other ground squirrels live in family groups. The basic social unit is a cluster of female kin, with daughters spending their lives in the group they were born into (natal philopatry). Females display nepotism (kin-assisting behavior) by sharing and jointly defending territories, and giving warning calls when predators approach (see the Role of Kinship special feature page).

In some species, one or more males append themselves to these female clusters; they sire the next generation and, sometimes, help care for the young. Alpine marmots live in just such mixed-sex colonies, ranging in size from 2 or 3 to over 50 animals. One large colony may occupy an immense burrow system. Colony members scent-mark their territory with substances secreted from cheek glands, and chase unrelated intruders, accompanied by tooth gnashing and calling. While the young marmots play, other members of the family stand guard. Juveniles hibernate and live together in the parents' burrow for the summer or two after they are born. Black-tailed prairie dogs also live in mixed-sex family groups, and these too are aggregated into immense colonies. Each female has her own burrow, but families cooperate to defend their shared territory, to nurse hungry pups, and to warn each other of danger.

Among tree squirrels, social behavior often varies between populations of the same species according to seasonal and spatial variation in seed and nut crops. Among Red squirrels, when food availability is stable in time and space, site-fidelity is high and home-range size is small (about 2–4ha/5–10 acres in females and 6–8ha/15–20 acres in males). Females defend territories (especially the core area around the nest) against other females, and dominant individuals will not move unless a territory with greater food abundance becomes available nearby. Males do not defend territories, but neighboring males that share foraging areas differ in social status: heavy, old males dominate the rest. Both sexes exclude dispersing juveniles and subadults from territories, and these young become "floaters," searching over large areas for a vacant home range. In high-quality woodlands in central Europe, Red squirrel

Niche Separation in Tropical Tree Squirrels

For two or more species to live in the same habitat, their use of resources must be sufficiently different to avoid the competitive exclusion of one species by another. Such lifestyle differences as ground living or tree dwelling, daytime or nocturnal activity, and insectivory or frugivory are obvious types of ecological separation. Sometimes, however, squirrel species occurring naturally in the same habitat appear to utilize the same food resources. Do they actually occupy different niches?

This situation is illustrated by squirrels found in the lowland forests of West Malaysia. Of the 25 species found there, 11 are nocturnal and 14 diurnal. The latter can be divided into terrestrial, arboreal, and climbing categories, with different species making different use of the various forest strata. For example, the Three-striped ground squirrel and Long-nosed squirrel feed on the ground or around fallen trees, whereas the Slender squirrel is most active on tree trunks at lower forest levels. Plantain and Horse-tailed squirrels travel and feed mainly in the lower and middle forest levels, but nest in the upper canopy. The three largest species live highest in the canopy.

Malaysian squirrels show considerable divergence in food choices when food is abundant, but considerable overlap when it is scarce; then, all species rely heavily on bark and sap. Unlike African or temperate forest species of comparable size, none of the smaller species, except *Sundasciurus hippurus*, are seed specialists. The Three-striped ground squirrel feeds on plant and insect material, and the Long-nosed squirrel is an insectivore; diets of these species overlap with those of tree shrews more than with other arboreal squirrels. Horse-tailed squirrels feed mainly on bark and sap, and most beautiful squirrels feed opportunistically on a variety of plant materials and insects. The larger flying squirrels eat a higher proportion of leaves than the smaller species, which eat mainly fruit.

The three largest species of diurnal squirrels diverge less than the smaller species in the lower forest levels. All three are fruiteaters, but the Pale giant squirrel (*Ratufa affinis*) primarily uses the middle canopy levels and prefers leaves. The Black giant squirrel (*Ratufa bicolor*) and Prevost's squirrel (*Callosciurus prevostii*) are often seen feeding together at the same fruit trees, but the latter eats a smaller range of fruit, and the two species also have different foraging patterns. The giant squirrel is larger, giving it a competitive advantage, but it cannot move as far or as fast as the smaller Prevost's squirrel; the latter can afford to spend less time feeding each day and is able to travel farther to scattered food sources, consuming them before the giant squirrels arrive. KMacK

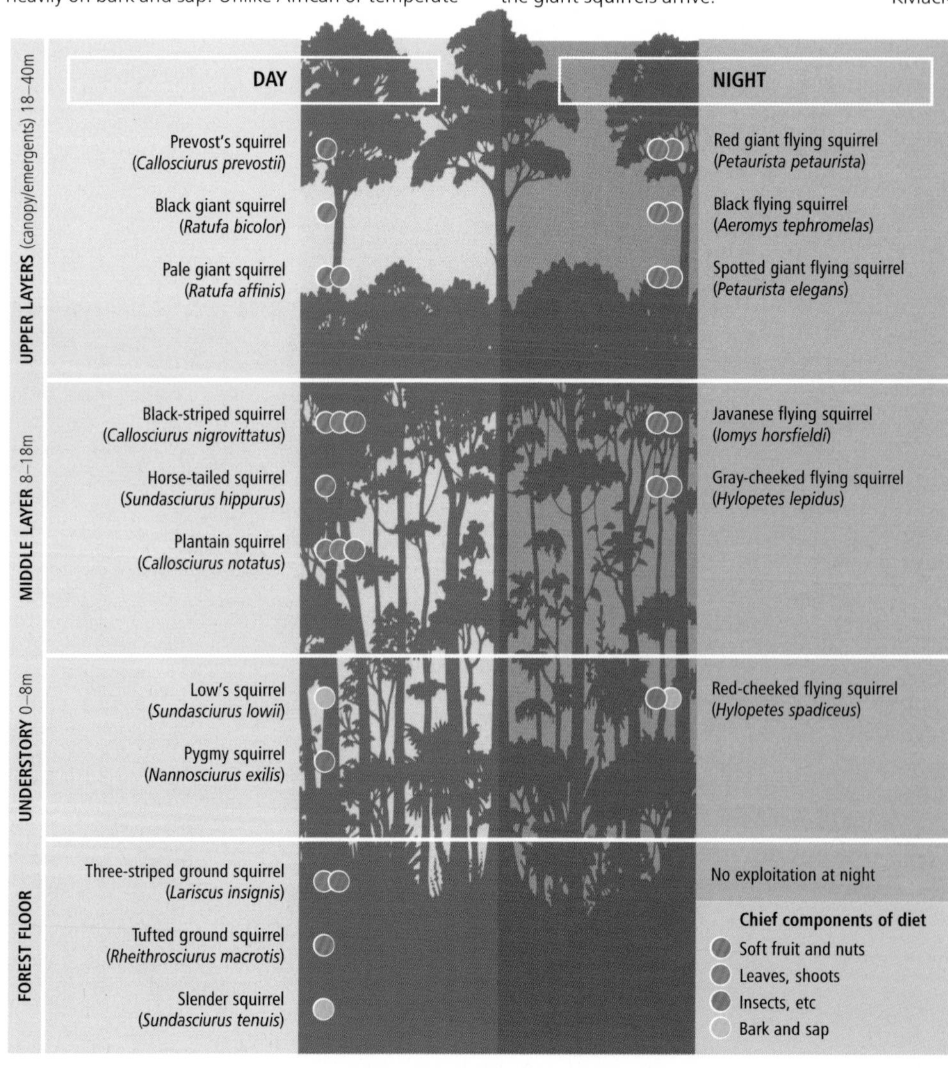

umbers vary little between years (numbering .7–2.2 squirrels per hectare, or 0.3–0.9 an acre). y contrast, in boreal coniferous forests in which ine and spruce seed abundance varies over time nd between areas, both males and females fre- uently abandon territories when food becomes carce. When the spruce cone crop fails, adult nales disperse after mating, and adult females fter they wean their litter. Since reproductive ates and adult and juvenile survival are directly elated to the food supply, Red squirrel numbers vithin an area can vary between years by an order f magnitude of 0.03 –0.3 squirrels per hectare 0.01–0.1 squirrels an acre).

Gray squirrel populations also fluctuate annual- y in size, and density varies with habitat (e.g., –10 squirrels a hectare, or 1–4 per acre). These lifferences can be traced to variations in acorn nd hickory-nut production, through their effects n the animals' reproductive rates, litter sizes, nd, ultimately, survival. Population density is letermined largely by the recruitment or other- vise of locally-born young, but interactions etween resident squirrels and juveniles drive lispersal behavior. Adult males aggressively chase uveniles of both sexes (but especially males), vhereas females allow daughters to remain on neir home range. For females, living together vith relatives leads to communal nesting and ther forms of amicable behavior.

Squirrels, Predators, and Man

CONSERVATION AND ENVIRONMENT

Squirrels are prey for many carnivores, including badgers, weasels, foxes, coyotes, bobcats, and (feral) cats on the ground, and hawks, eagles, falcons, and owls from the air. Sometimes squirrel populations are decimated by infectious diseases, particularly bubonic plague and tularemia.

Yet the greatest threat to squirrel populations worldwide is undoubtedly human population expansion, and the associated habitat fragmentation and loss due to the growth of cities, changes of land use practices in rural areas (e.g., from rangeland to intensive agriculture), tree harvesting, fire suppression, and the introduction of non-native plants. Squirrels are hunted for pelts and meat and for sport, and both hunting and poisoning is still largely unregulated – if not actually encouraged – by governmental agencies that consider squirrels primarily as pests and carriers of disease. Little notice has been taken of the ecological and aesthetic losses associated with the dwindling size of many squirrel species, or of the benefits their activities bring. For example, the burrowing activities of ground-dwelling squirrels aerate the soil and bring nutrients to the surface, promoting plant growth, while the health of many forest ecosystems depends critically on tree squirrels' role as seed dispersers and controllers of insect outbreaks. PWS/LW

Above *The Northern striped palm squirrel is quite social and several may be observed in the same tree. Found in parts of Iran, India, Nepal, and Pakistan. Females can produce several litters per year.*

Below *In its woodland domain, the Malabar or Indian giant squirrel is very agile, rarely descending to the ground. However, along with all other species in the genus* Ratufa *, it is listed as Vulnerable, largely as a result of extensive recent habitat loss.*

THE ROLE OF KINSHIP

The annual round in Belding's ground squirrels

BELDING'S GROUND SQUIRRELS ARE SOCIAL rodents inhabiting grasslands in the cold deserts of the far western United States (California, Nevada, Oregon, and Idaho). They are active above ground most of the day, going to subterranean burrows for refuge from predators and inclement weather and to spend the night. Primarily vegetarians, they forage on forbs and grasses, and are particularly fond of flower heads and seeds; but they will also eat birds' eggs, carrion, and, occasionally, each other's young.

A population of Belding's ground squirrels at Tioga Pass, high in the central Sierra Nevada of California (3,040m/9,945ft), has been studied for more than two decades. The animals are active only from May to October, hibernating for the rest of the year. Adult males emerge first each spring, often tunneling through several meters of accumulated snow to do so. Once the snow melts the females emerge, and the annual cycle of social and reproductive behavior starts – a complex and intriguing mix of competition and cooperation.

Females mate about a week after emergence. Although they are only receptive for a single afternoon, they take full advantage of this opportunity, typically mating with at least three (and sometimes up to eight) different males. Genetic analyses have revealed that about two-thirds of litters are fathered by more than one male, so that pups develop *in utero* among full- and half-brothers and sisters. Although a female's first mate is the predominant sire of her offspring, some litters are sired by as many as four different males.

In order to mate successfully, males defend small, exclusive territories. In the presence of receptive females, they threaten, chase, and fight with each other. Virtually every male sustains physical injuries during the mating period, some of them serious. The heaviest and most experienced males usually win such conflicts, and receptive females remain near those males. Dominant males copulate with multiple females (up to 13 in a season), but more than half of all males mate only once or not at all in any given year.

After mating, each female digs her own nest burrow in which she will rear her young. Females produce only one litter per season. Gestation lasts about 24 days, lactation 27 days, and the mean litter size is five pups. Females shoulder the entire parental role; indeed, some males do not even interact with the young, because by the time weaned juveniles begin to emerge above ground in July, they have already gone back into hibernation. Adult females hibernate early in the fall; and finally, when it starts snowing, the year's young begin their first long, risky winter underground.

The hibernation period, lasting 7–8 months, is a time of heavy mortality in which two-thirds of the juveniles and one-third of the adults perish. Most die because they deplete their stores of body fat and freeze to death; others are eaten by digging predators such as badgers and coyotes. Males live 2–3 years on average, whereas females typically live 3–4 years; the males apparently die younger due to infection from wounds incurred during fights over females or from the increased exposure to predators resulting from their higher levels of mobility, leading to more rapid aging (senescence). The sexes also differ markedly in their respective tendencies to leave the area where

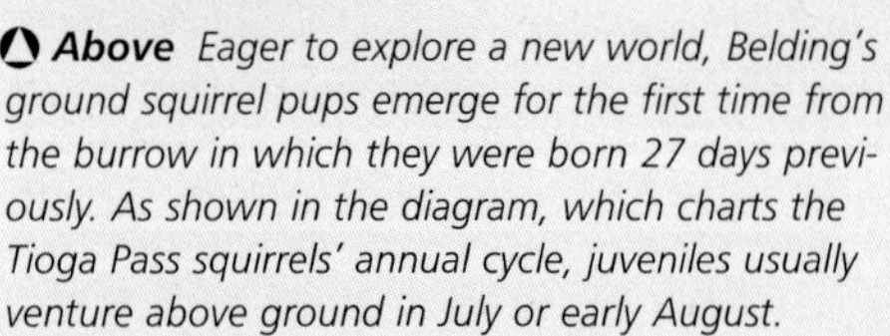

Above Eager to explore a new world, Belding's ground squirrel pups emerge for the first time from the burrow in which they were born 27 days previously. As shown in the diagram, which charts the Tioga Pass squirrels' annual cycle, juveniles usually venture above ground in July or early August.

Left A female ground squirrel gathers grass to line a nest. Typically, burrows are 5–8m (16–26ft) long, 30–60cm (12–24in) below ground, and have multiple surface openings. It can take more than 50 loads of dry grass to line a single nest.

Above *Keeping watch for the benefit of the kin community in which she lives, a female squirrel notices the approach of a predator. She gives a repeated alarm cry, even at the risk of drawing the intruder's attention to herself.*

hey were born. Juvenile males begin dispersing oon after they are weaned and never return ome, whereas females seldom disperse, staying lose to the natal burrow and interacting with naternal kin throughout their lives.

The ground squirrels' matrilocal population tructure has set the stage for the evolution of epotism (the giving of favors to family members) mong females. This manifests itself in four main vays. First, close relatives (mothers and offspring nd sisters) seldom attack or fight each other vhen establishing nests; females with living kin hus obtain residences with less expenditure of me and energy, and lower danger of injury, than hose without kin. Second, close relatives share ortions of their nesting territory and permit each ther access to food and hiding places on these efended areas. Third, close kin help each other to vict distantly related and unrelated squirrels from ach other's territories. Fourth, females give warn-ıg cries when predatory mammals approach.

At the sight of a badger, coyote, or weasel, ome females stand erect and utter staccato alarm rills. A greater proportion of callers than non-allers are attacked and killed, so giving such calls is dangerous. However, not all individuals take the same risks. The most frequent callers are old, lactating, resident females with living offspring or sisters nearby, whereas males and immigrant females with no relatives in the area call infrequently. Callers thus apparently behave altruistically – as if they were trading the risks of exposure to predators for the safety and survival of dependent kin.

The squirrels also vocalize when predatory hawks swoop at them, but these calls sound quite different: they are high-pitched whistles, each containing only a single note. Interestingly, these calls also apparently have a different function. Upon hearing a whistle, all the other squirrels scamper for cover, and their rapid flight inadvertently benefits the caller, both by creating pandemonium, which can confuse the predator's focus, and a scurrying group, in which there is numerical safety. So where the staccato alarm trill promotes the survival of offspring and other relatives, the single-note whistle call promotes self-preservation. The implication is that Belding's ground squirrels have different sounds for different degrees of danger.

Another important manifestation of nepotism is cooperative territory defense by related females. During gestation and lactation, such females will exclude others from the area surrounding their nest burrow. Territoriality functions to protect the helpless pups from predation by other squirrels, for when territories are left unattended, even briefly, unrelated females or yearling males sometimes attempt to kill the young.

In the males' case, infanticide is motivated by hunger, since they typically eat their victims. Females seldom do so; in their case, the killing is triggered by the loss of their own litters to predators. After such a loss, they often leave the burrow that proved unsafe and move to a more secure site – and if there are young already present there, they will try to kill them, thereby reducing future competition for themselves and their daughters.

Females with close relatives as neighbors lose fewer offspring in this way than do females without neighboring kin, because groups detect marauders more quickly and expel them more rapidly; also, pups are defended by their mother's relatives even when she is temporarily away from home foraging. In sum, living in extended families is a strategy for survival and reproduction among female Belding's ground squirrels. PWS

THE ROOTS OF MARMOT SOCIALITY

The advantages of living in extended family groups

MARMOT SOCIALITY BEGINS WITH HIBERNATION. All 14 species of marmots live in the northern hemisphere, mainly in mountainous areas where food is unavailable during the winter. When the weather is fine, they lay down fat to keep them alive during a hibernation that may last 9 months. Some species, including the Black-capped, Long-tailed, and Steppe marmots, live in environments so harsh that they mate, initiate gestation, and even give birth before emerging above ground.

When active, marmots use energy at a rate 8–15 times greater than during hibernation. To meet this heavy energy demand, they store copious amounts of body fat – comprising about 30 percent of body mass in the case of *Marmota bobak*, rising to 53 percent in *M. olympus*. Larger animals can store more fat and, relative to smaller animals, burn it more slowly. Marmots are the largest true hibernators, and the average adult body mass at the time of immergence ranges from 3.4kg (7.5lb) for *M. flaviventris*, the smallest species, to 7.1kg (15.7lb) for *M. olympus*, the largest. Even with the advantage of large body size, the demands on energy for reproduction, growth, activity, and hibernation are so great that individual females of at least 10 species, including *M. camtschatica*, *M. caudata*, and *M. menzbieri*, cannot accumulate sufficient resources for annual reproduction, which they skip, sometimes for two or more years.

To reach full size, marmots need to grow for a long time. Only one species, the woodchuck (*M. monax*), has an active season sufficiently long (at 5 months or more) to allow youngsters to mature and reproduce when 1 year old. Young woodchucks disperse and hibernate singly. The woodchuck is the only non-social species of marmot. Adult female woodchucks live alone, and males defend a home range that includes one or more females. Widespread forest–meadow-edge habitat enables the young to establish independence.

Marmots of most species live with their parents until they are 2 years old. The Yellow-bellied marmot (*M. flaviventris*) requires one additional year. While male Yellow-bellies disappear from the natal area to make their own way in the world, half of the females remain. Thus mother–daughter–sister assemblages of up to five are formed into matrilines. One male will enter an area containing more than one matriline and defend it.

Male Olympic (*M. olympus*) and Hoary (*M. caligata*) marmots typically live with two females and their young. The females look after the young until they are 2 years old, and do not produce another litter during this time. Adolescent young disperse at age 2 and breed one year later. All members of the group hibernate together in a burrow.

The social lives of other marmot species involve living in extended family groups in which a dominant pair shares the home range with young of various ages that may stay at home beyond the age when they could be reproducing for themselves. This apparently paradoxical behavior would be explicable if the youngsters stood a chance of inheriting the parents' territory. In fact, however, if a member of the dominant pair were to die and a son or daughter inherited the territory, the new owner could end up mating with its own mother or father. To avoid this inbreeding, which could lead to physical inadequacies in the young, yearling Yellow-bellied marmot males always disperse away from home, while half of the yearling females remain. The young females do not normally end up mating with their father, as he does not usually hold ownership of the territory long enough – for only 2.2 years on average. A female is in her third year of residency before she can breed, and her breeding career lasts an additional 2.95 years. Only about 17 percent of the males reach an age that makes reproduction likely. Male mortality is higher than that of females; at age 2, when many males are still in the dispersal stage, about 50 percent die, as opposed to only about 30 percent of the two-year-old females. Some level of inbreeding does occur in family groups, but there is no evidence of deleterious effects. In addition, relatedness, and hence inbreeding, is reduced when a male or female from some other family replaces the territorial dominant.

For Yellow-bellied marmots, family living seems to bring all kinds of benefits. Animals in larger groups survive better and have more offspring. Group defense of resources ensures that they never go hungry, while the many extra pairs of eyes increase the chances of spotting a predator before it gets within striking distance. The rate of survival increases from about 0.6 for a single female to about 0.8 in groups of three or more. Net reproductive rate increases from 0.5 for a single female to 1.15 in groups of three.

Setting out to find a new home as a young marmot is a treacherous business, and waiting until 2 years or older increases the chances of success. Survival rates are significantly greater in *M. caudata* and *M. vancouverensis*, which disperse at age 3 or older, than in *M. flaviventris*, which disperses at 1 year old. However, staying at home with their parents results in reproductive suppression. In Gray marmots (*M. baibacina*), most breeding is carried out by older females, and young females only get an opportunity when the population has been reduced. Reproductive suppression may be so pervasive that even when a dominant female Alpine

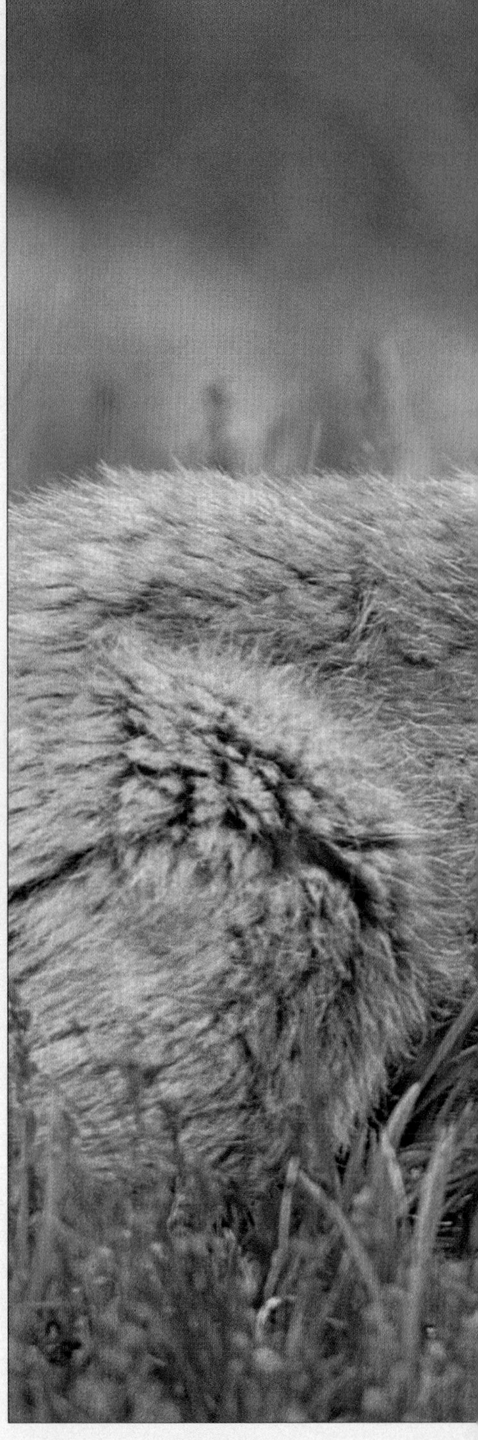

Above *A juvenile Olympic marmot approaches an adult marmot in grassland, Olympic National Park, Washington. Olympic marmots, the largest marmot species, are very tolerant and sociable and such face-to-face encounters are very common. The dispersal of young is a relatively slow process and appears to be determined by the young themselves rather than as a result of adult aggression driving them out.*

marmot does not produce any young, as in the case of *M. marmota*, the subordinate females will still fail to reproduce.

Any animal that spends its entire life caring for others without having offspring of its own will die without passing its genes on to the next generation, and the genes responsible for causing such caring behavior will die with the childless carer. Thus such indiscriminate "caring genes" cannot evolve. Relatives, however, share genes, so those that cause individual animals to care for close kin and thus increase their likelihood of successfully rearing more young with more of those "relative-caring" genes can flourish. In biology, this principle is known as "kin selection." Individuals can afford to adopt an altruistic attitude toward family members because they share sufficient genes for their efforts not to be wasted. This is especially the case if having a family of one's own involves potentially dangerous activities, such as searching for a new home alone.

One reason why young marmots stay at home may be to help raise their siblings. Alpine marmots cuddle up to the young, preventing heat loss that would otherwise burn up fat reserves and hence cause weight loss, which in turn would reduce their chances of survival. When subordinate full siblings are present, juvenile mortality is about 5 percent, but this figure increases to about 22 percent when they are not present.

Even so, the benefits of kin selection hardly compensate fully for failing to produce young of one's own. A further reason why marmots stay at home when they could be establishing their own territories may be that there is simply nowhere for them to go. Habitats may become saturated, so that there are no territories available for a young family. In such circumstances they will stay at home, gaining experience by helping to raise their siblings and waiting for an opportunity to finally strike out on their own, richer in experience and still alive thanks to their parents' tolerance. KBA

Springhare

SCATTERED THROUGH THE ARID LANDS OF East and South Africa are numerous flood plains and fossil lake beds. After the rainy season grass is superabundant here, but for much of the year it is too short and sparse for large grazers to forage efficiently. The result is an unused food supply or, to use the ecological term, an empty niche. To fill it requires an animal small enough to use the grass efficiently, yet large and mobile enough to travel to it from areas that can provide shelter from the weather and from predators. These are attributes of the springhare.

In spite of its adaptiveness, the springhare still faces formidable problems. In terms of its size it is small enough to be killed by snakes, owls, and mongooses, and yet big enough to be attractive to large predators, including lions, wild dogs, and man. Many of the animal's specialized physical and behavioral features are adaptations, not just for an arid environment, but also for the efficient avoidance of predators.

Squirrel, Hare, or Kangaroo?

FORM AND FUNCTION

There is little evidence of the springhare's origins. Some people believe that its closest living relatives are the scaly-tailed flying squirrels of the genus *Anomalurus*. The springhare, except for the bushy tail, actually resembles a miniature kangaroo; its hind legs are very long, and each foot pad has four toes, each equipped with a hoof-like nail. Its most frequent and rapid type of movement is hopping on both feet. Its tail – slightly longer than its body – helps to maintain balance while hopping. The tail has a thick brush at the end that is of a darker coloration. The front legs, which are only about a quarter of the length of the hind legs, have long, sharp claws that are curved for use in digging burrows. The head is rabbit-like, with large ears and eyes and a protruding nose; sight, hearing, and smell are well-developed. When pursued by a predator, a springhare can leap 3–4m (10–13ft). If captured, it will try to bite the predator with its large incisors and to rake it with the sharp nails of its hind feet.

Though springhares are herbivorous, they occasionally eat mineral-rich soils and accidentally ingest insects, such as locusts and beetles. They are very selective grazers, preferring green grasses high in protein and water. However, they also eat bulbs and roots. If necessary, a springhare can go for at least 7 days without drinking by reducing food intake and producing concentrated urine and very dry feces.

Keeping Out of the Way of Predators

SOCIAL BEHAVIOR

Springhares are nocturnal; shunning the midday sun means they avoid losing water in the arid regions they inhabit. Normally they forage within a maximum of 250m (820ft) of their burrows but occasionally they have been observed to travel much farther away. While foraging, they are highly vulnerable to predators because they are completely exposed to detection. On nights with a full moon they appear to be particularly at risk, and venture only an average of 4m (13ft) to the feeding area; in contrast, on moonless nights they travel some 58m (190ft). When they are above ground springhares spend about 40 percent of their time in groups of 2–6 animals, presumably because a group is more efficient at detecting danger.

Springhares spend the hours of daylight in burrows located in well-drained, sandy soils. Burrows lie about 80cm (31in) deep, have 2–11 entrances, and vary in length from 10–46m (33–151ft). Each burrow is occupied either by a single springhare or by a mother and an infant. Burrows provide considerable protection against the arid environment and also against predators. Some predators, such as snakes and mongooses, can enter burrows, however, so springhares often block entrances and passageways with soil after entering. When predators do enter, the tunnels and openings in the burrow system provide many escape routes. The absence of chambers and nests within the burrow suggests that springhares do not rest consistently in any one location – probably another precaution against predators.

In springhare populations the number of males equals the number of females. There is no breeding season, and about 76 percent of the adult females may be pregnant at any one time. Adult females undergo about three pregnancies per year, usually resulting in the birth of a single, large, well-developed infant; twins are born, but this is a rare occurrence.

Newborn springhares are well furred and are able to see and move about almost immediately. However, they are confined to the burrow and are totally dependent on milk until they are half-grown, at which time they can become completely active above ground. Immature springhares

Above *With its short front paws and hopping motion, the springhare resembles a miniature kangaroo. It is a vital resource for the indigenous Bushmen of southwest Africa, who eat its meat, make its skin into garments, use its long tail sinew as thread, and even smoke its fecal pellets.*

usually account for about 28 percent of all the individuals that are active above ground.

Although the reproductive rate of springhares is surprisingly low, there are two distinctive advantages to the animals' reproductive strategy. First, the time and energy available to the female for reproduction is funneled into a single infant, resulting apparently in low infant and juvenile mortality. When the juvenile springhare first emerges from its burrow, its feet are 97 percent, and its ears 93 percent, of their adult size: it is almost as capable of coping with predators and other environmental hazards as a fully-grown adult. Second, in having to provide care and nutrition for only one infant, the mother is subject to minimal strain.

A Tale of Declining Numbers

CONSERVATION AND ENVIRONMENT

Springhares are generally common where they occur, despite the fact that they are frequently hunted by man. In the best habitats there may be more than 10 springhares per hectare (4 per acre), and this can include areas where domestic stock have grazed the forage to a suitable height. However, when arid, ecologically sensitive areas are overgrazed by domestic stock, as occurs in the Kalahari Desert, springhare densities are lower.

Springhares are of considerable importance to man as a source of food and skins. In Botswana they are the most prominent wild animal in the human diet, and a single band of bushmen may kill more than 200 annually. They can be a significant pest to agriculture, feeding on a wide variety of crops including corn, sweet potatoes, and wheat. Partly as a result, there has been an upsurge in hunting activity that, together with habitat loss, is reckoned to have reduced springhare numbers by at least 20 percent in the 1990s alone. The animals are now classed as Vulnerable by the IUCN. TMB

FACTFILE

SPRINGHARE

Pedetes capensis

Springhare or springhass

Order: Rodentia

Family: Pedetidae (sole member of family)

DISTRIBUTION Kenya, Tanzania, Angola, Zimbabwe, Botswana, Namibia, South Africa.

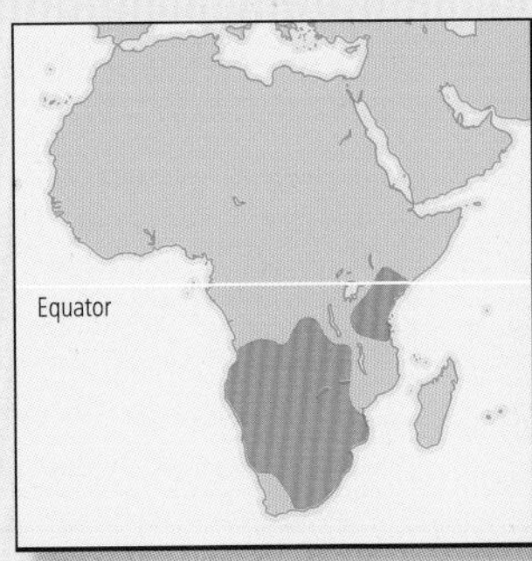

HABITAT Flood plains, fossil lake beds, savanna, other sparsely vegetated arid and semi-arid habitats on or near sandy soils.

SIZE Head-body length 36–43cm (14–17in); tail length 40–48cm (16–19in); ear length 7cm (3in); hind foot length 115cm (6in); weight 3–4kg (6–9lb). (Dimensions are similar for both female and male.)

COAT Upper parts, lower half of ears, basal half of tail yellow-brown, cinnamon, or rufous brown; upper half of ears, distal half of tail, and whiskers black; underparts and insides of legs vary from white to light orange.

DIET Herbivorous, with a preference for protein-rich green grasses.

BREEDING Gestation about 77 days.

LONGEVITY Unknown in wild; more than 14 years in captivity.

CONSERVATION STATUS Vulnerable

Left *Some typical postures of the springhare:* ***1*** *sitting down to groom;* ***2*** *hopping – when the springhare is in motion, it holds its tail (thought to be an aid to balance) either horizontally or curled up;* ***3*** *foraging on all fours;* ***4*** *standing on its hind legs.*

DISTRIBUTION Worldwide except for Antarctica.

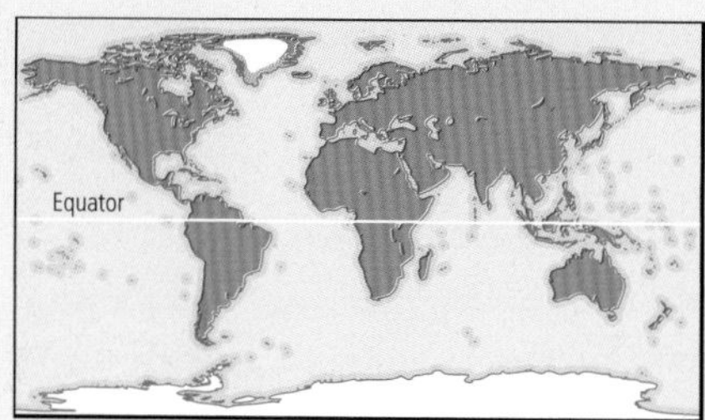

SCIUROGNATHS (part only) Suborder Sciurognathi
1,477 species in 312 genera and 5 families

Mouse-like Rodents

MORE THAN A QUARTER OF ALL SPECIES OF mammals can loosely be described as mouse-like rodents. Although today they join the squirrels in the suborder Sciurognathi, they once occupied their own order, the Myomorpha. They are very diverse – so much so that they are difficult to describe in terms of a typical member. However, the Brown rat and House mouse are fairly representative, both in size and overall appearance. The great majority are small, terrestrial, prolific, nocturnal seedeaters. The justification for believing them to comprise a natural group derived from a single ancestor is debatable but lies mainly in two features: the structure of the chewing muscles of the jaw and the structure of the molar teeth.

Most mouse-like rodent species belong to the mouse family. The minority groups are the dormice, the jerboas and jumping mice,the pocket mice, and the pocket gophers. These represent early offshoots that have remained limited in species numbers and also somewhat specialized, the dormice being arboreal and (in temperate regions) hibernating, the jerboas adapted for the desert. The murids have undergone more recent and much more extensive changes (adaptive radiation) beginning in the Miocene epoch, i.e. within the last 24 million years. Some of the resultant groups are specialists, for example the voles and lemmings, which are adapted to feeding on grass and other tough but abundant vegetation. Yet many species have remained versatile generalists, feeding on seeds, buds, and sometimes insects, all more nutritious but less abundant than grass.

In ecological terms, most mouse-like rodents may be classified as "r-strategists," that is, they are adapted for early and prolific reproduction rather than for long individual life spans ("k-strategists"). Although this applies in some degree to most rodents, the rats and mice show it more strongly and generally than, for example, their nearest relatives, the dormice and the jerboas. GBC

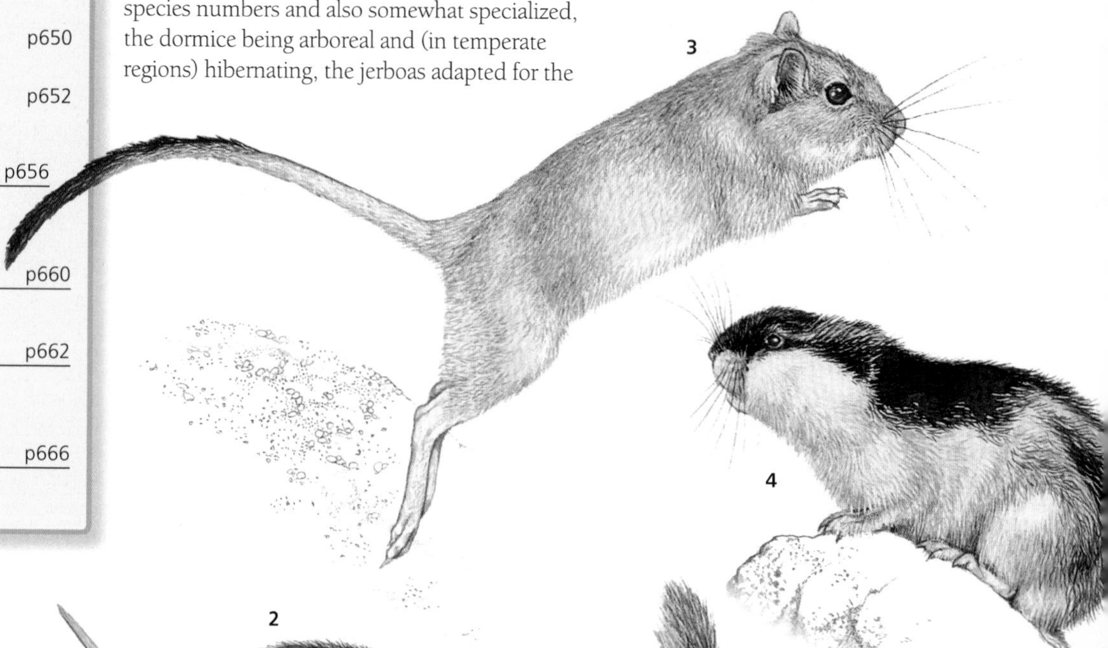

3

4

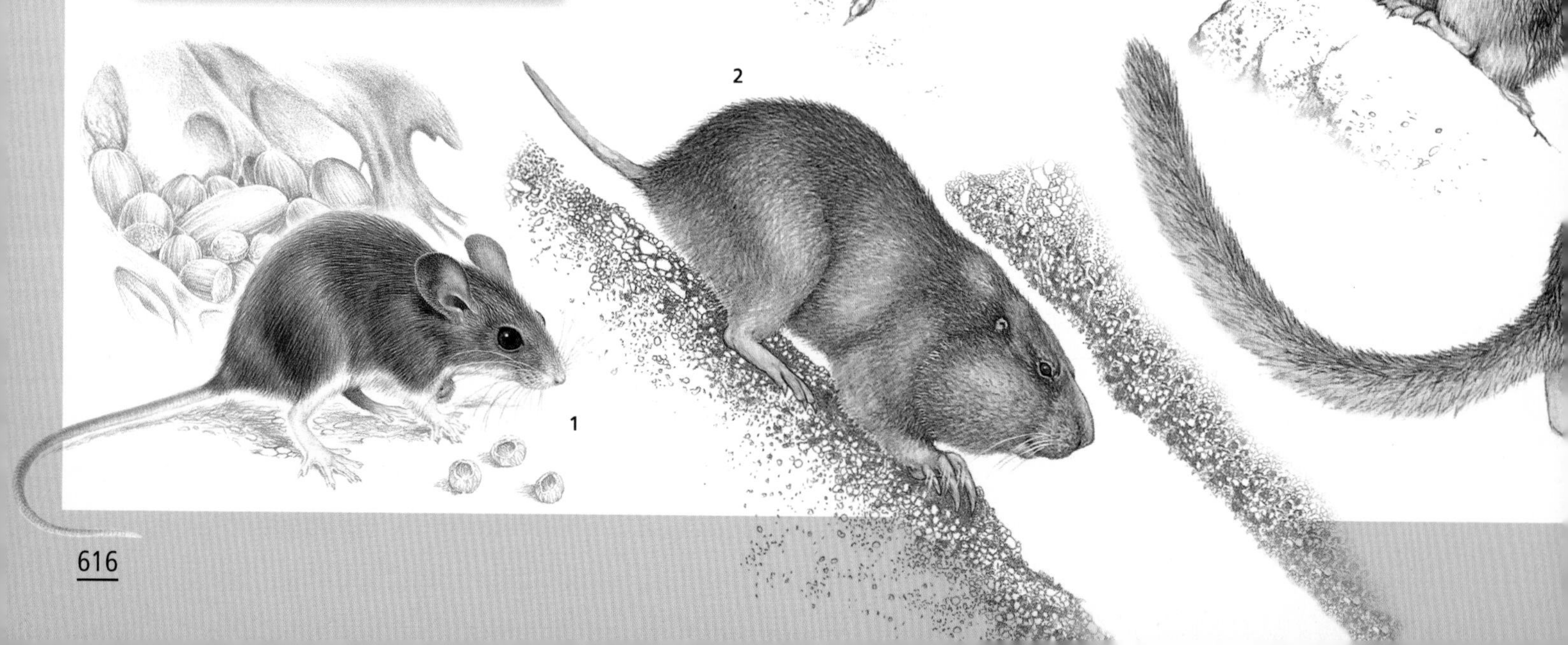

2

1

GREAT PROLIFERATORS

The greatest diversification of species in the entire evolution of mammals has occurred in the mouse family, which has over 1,000 living species. Its members are found around the world in almost every terrestrial habitat. Those that most closely resemble the common ancestor of the group are probably the common mice and rats found in forest habitats world-wide. These are versatile animals, predominantly seedeaters but capable of using their seedeating teeth to exploit many other foods, such as buds and insects.

From such an ancestor many more specialized groups have arisen, capable of exploiting more difficult habitats. Most gerbils (subfamily Gerbillinae) have remained seedeaters but have adapted to hot arid conditions in Africa and Central Asia. The hamsters (subfamily Cricetinae) have adapted to colder arid conditions by perfecting the arts of food storage and hibernation; the voles and lemmings (subfamily Arvicolinae) and the superficially similar African swamp rats (subfamily Murinae) have cracked the problem of feeding on tough herbage, opening up fresh possibilities for expansion.

SKULLS AND DENTITION

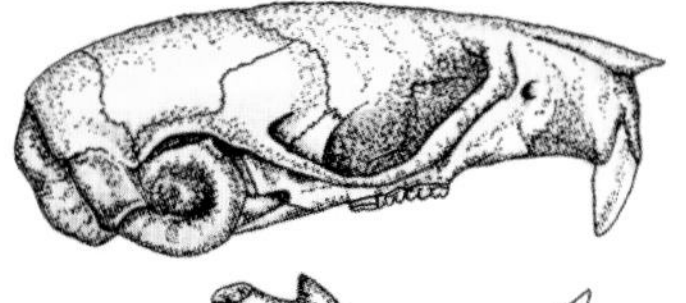

Harvest mouse 1.6 cm

European hamster

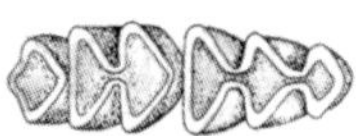

Libyan jird

Harvest mouse

Norway lemming

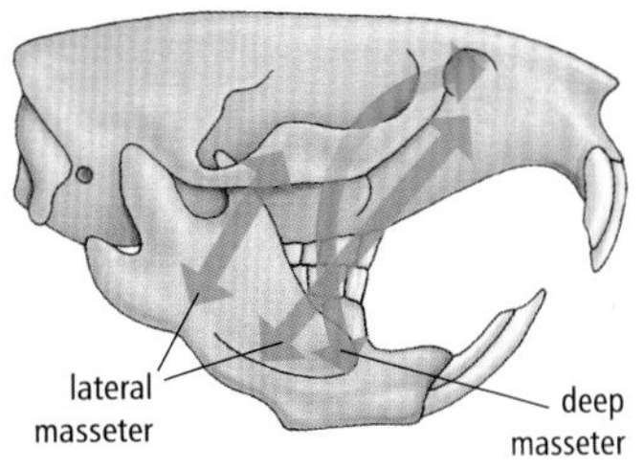

Most mouse-like rodents have only three cheekteeth in each row, but they vary greatly in both their capacity for growth and the complexity of the wearing surfaces. The most primitive condition is probably that found in the hamsters – low-crowned, with rounded cusps on the biting surface arranged in two longitudinal rows and no growth after their initial eruption. The rats and mice of the subfamily Murinae, typified by the Harvest mouse, have developed a more complex arrangement of cusps, forming three rows, while retaining most of the other primitive characters. These two groups are often treated as separate families, the Cricetidae (so-called cricetine rodents) and Muridae (so-called murine rodents) respectively.

The gerbils have high-crowned but mostly rooted teeth, in which the original pattern of cusps is soon transformed by wear into a series of transverse ridges. The voles and lemmings take this adaptation to a tough, abrasive diet – in this case mainly grass-even further by having teeth with high crowns and complex shearing ridges of hard enamel that continue to grow and develop roots only late in life or, more commonly, not at all.

The other feature that principally distinguishes the mouse-like rodents from other rodent groups is the structure of the jaw muscles. In the mouse-like rodents, both the lateral (blue in the diagram LEFT) and deep (red) masseter muscles are thrust forward, providing a very effective gnawing action, with the deep masseter passing from the lower jaw through the orbit (eye socket) to the muzzle. Shown here is the skull of a muskrat.

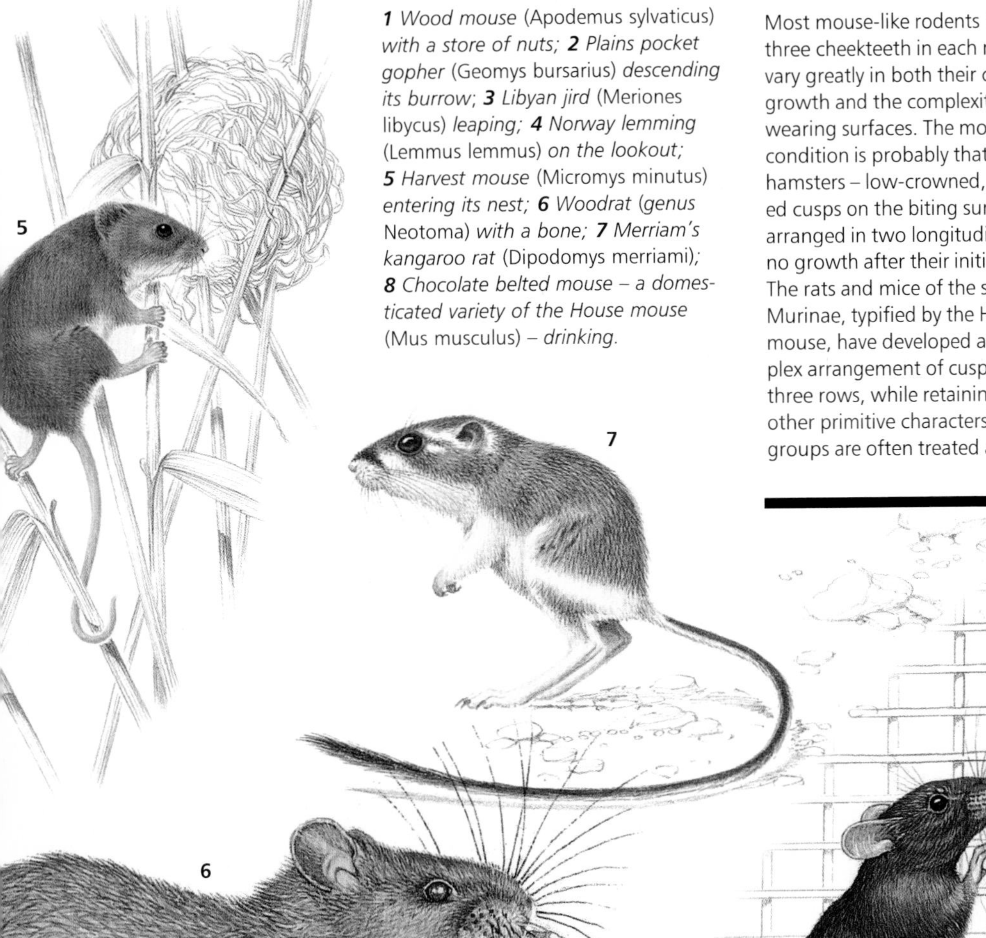

Below *Representative species of mouse-like rodents (not to scale):* **1** *Wood mouse* (Apodemus sylvaticus) *with a store of nuts;* **2** *Plains pocket gopher* (Geomys bursarius) *descending its burrow;* **3** *Libyan jird* (Meriones libycus) *leaping;* **4** *Norway lemming* (Lemmus lemmus) *on the lookout;* **5** *Harvest mouse* (Micromys minutus) *entering its nest;* **6** *Woodrat* (genus Neotoma) *with a bone;* **7** *Merriam's kangaroo rat* (Dipodomys merriami)*;* **8** *Chocolate belted mouse – a domesticated variety of the House mouse* (Mus musculus) *– drinking.*

RATS AND MICE AS PETS

Keeping mice as pets began in the Orient, probably first in China but also in Japan. Early records indicate that mouse breeding was flourishing in the 18th century; in 1787 a Japanese writer published a work on how to breed colored varieties, and domestication must have begun much earlier for such a pamphlet to have been produced. Sailors returning from the Far East probably took pet animals home with them, providing the stimulus for mouse breeding in the west.

Tame mice differ from wild ones in several respects. Apart from their docility, they are larger and have more prominent eyes, larger ears, and longer tails. All are bred to the same body conformation, implying that there are no breeds as such, only varieties.

Keeping rats as pets is a relatively recent activity. It probably began about 1850, possibly in England. The number of varieties is less than for mice, probably because of the shorter period of domesticity and smaller numbers.

New World Rats and Mice

THE INNOCENT PHRASE "NEW WORLD RATS and mice" hides a plethora of 434 species in 86 genera, found in habitats ranging from the northern forests of Canada to the Americas' southernmost tip. Their adaptations and habits are so diverse that an entire volume could scarcely do them justice.

FACTFILE

New World Rats and Mice

Order: Rodentia

Family: Muridae

Subfamily: Sigmodontinae (Hesperomyinae)

434 species in 86 genera and 15 tribes, combined into 6 groups

HABITAT All terrestrial habitats (including northern forests, tropical forest, and savanna) excluding permanently snow-covered mountain peaks and extreme high Arctic.

SIZE **Head–body length** ranges from 5–8.1cm (2–3.2in) in the Pygmy mouse to 16–28.7cm (6.3–11.3in) in the South American giant water rat; **tail length** from 3.5–5.5cm (1.4–2.2in) to 7.6–16cm (3–6.3in), and **weight** from 7g (0.25oz) to 700g (1.5lb), both in the same two species.

COAT Most New World rat and mouse species have a brown back and white belly, but some exhibit very attractive coat colors; the Chinchilla mouse, for instance, has a strongly contrasting combination of buff-to-gray back and white belly. The color of the back often matches the surrounding soil to provide camouflage against owls and other airborne predators.

DISTRIBUTION N and S America and adjacent offshore islands

DIET Mostly plant material and invertebrates, though some take small vertebrates including fish.

BREEDING Gestation 20–50 days

LONGEVITY Maximum 2 years in the wild; some species live up to 6 years in captivity.

CONSERVATION STATUS 8 species are currently listed as Critically Endangered, and a further 24 as Endangered, including 6 species of woodrats (genus *Neotoma*) and 4 of the 5 species of fish-eating rats (*Neusticomys*).

See tribes table ▷

New World rats and mice are uniformly small: in head–body length, even the largest living species measure less than 30cm (c.12in). The length depends on how much tree-climbing their lifestyle demands; the greater the need to balance in high places, the longer the tail. Tails are usually almost hairless but, as is to be expected in such a large and diverse group, some species such as the aptly named Bushy-tailed woodrat buck the trend. There are, however, very few exceptions to the basic arrangement of teeth common to all highly-evolved rodents, with three molars on each side of the jaw separated by a distinct gap from a pair of incisors, which grow continuously and have enamel on their anterior surfaces, enabling a sharp cutting edge to be maintained.

Natural selection has shaped the New World rats and mice to a multitude of forms for different habitats. Burrowing species have short necks, short ears, short tails, and long claws, while those living in water, such as the marsh rats, often have webbed feet or, like the fish-eaters, a fringe of hair on the hind feet that increases the surface area to form a paddle. In forms even more developed for aquatic life, the external ear is reduced in size or even absent, as is the case with the Ecuadorian fish-eating rat (*Anotomys leander*).

Above the generic level, the classification of New World rats and mice is controversial. The genera can be grouped into 15 tribes, and these in turn can be thought of as belonging to six groups (see table), though these groupings are of varying validity; currently only one of the tribes has been shown convincingly to be monophyletic.

Below *Restricted to salt marshes in the San Francisco Bay area of California, the Salt-marsh harvest mouse is among the rarer New World rodents; it was officially declared endangered by the US government in 1970. As an adaptation to its salty habitat, it has become one of the few mammals able to drink sea water.*

North American Rats and Mice

NEOTOMINE–PEROMYSCINE GROUP

The **white-footed mice and their allies** (tribe *Peromyscini*) are among the most extensively studied murides. In particular, the White-footed mouse itself (*Peromyscus leucopus*) has been used as a model to investigate how males and females compete with members of their own sex to maximize individual reproductive opportunities.

The mating systems of White-footed mice range from polygyny to promiscuity. Males typically occupy home ranges of approximately 500–1,000sq m (600–1,200 sq yds), about twice the size of those occupied by females. During the breeding season, a male may spend most of his time with one primary female, but his home range often overlaps that of two or three others. Males with larger home ranges have access to more potential mates, thereby increasing mating opportunities. Males that mate with the most females pass on the most genes, so males benefit by moving over large areas. At low densities, a male may develop a wandering strategy; he may associate with one primary female until she is pregnant, then wander off to reside with neighboring, or secondary, females while they are in estrus, before returning eventually to his primary female.

Females usually mate with one male, but may mate promiscuously with several, such that a given litter may be sired by as many as three males; however, the actual number of males a

Above *Constantly on the look-out for potential threats, a Deer mouse* (Peromyscus *sp.*) *nurses her two-day-old young in her nest. Like most mouse species, Deer mice are prolific breeders: in the wild they average 3 or 4 litters of up to 9 young a year.*

female mates with could be more. Mating with several males may increase the genetic diversity of the offspring or help ensure fertility, but it also functions to confuse the question of paternity. One reproductive strategy adopted by male White-footed mice, as by many other species of mammals, is to kill offspring that they have not sired. Infanticide both removes the offspring of competitor males and provides a reproductive opportunity for the perpetrator. In White-footed mice, a female that loses her litter stops lactating and is ready to breed again within a few days, so by committing infanticide a male may be able to mate sooner than if the young were permitted to suckle until weaning.

Males do not recognize their own offspring, but they do associate copulation with a given place and time, and will not kill pups within the area where they mated. Hence, a female who confuses paternity by mating with neighboring males also reduces the chances that these males will commit infanticide. Multi-male mating appears to be a behavioral mechanism used by females as a counter-strategy to infanticide by males.

Female White-footed mice also commit infanticide, but as a mechanism to compete for breeding space. Dispersing females that do not have a territory of their own may attempt to kill the offspring of other females as a way of competing to take over their territories. At low densities when space is available for colonization, females space themselves out and are relatively non-aggressive, but when breeding space becomes limited, they aggressively defend their turf to insure the survival of their offspring. At high densities, females defend breeding territories of approximately 300–500sq m (400–600sq yds).

Females provide most of the parental care in White-footed mice, but males do make some contributions. Males often nest with litters, retrieve wandering youngsters, provide warmth for the pups, and deter infanticide by intruding males. Pups are nursed for a period of 21–24 days and are then ready to leave the nest. Leaving involves risks, but if all the offspring were to remain at the natal site, unsustainable competition would ensue for resources and for mates.

An even greater cost to not dispersing is

inbreeding, which is avoided in White-footed mice by sex-biased dispersal. At the time of weaning, young males will leave the natal site if their mothers are present in the territory, while young females depart if their fathers are present. In practice, mothers are almost always present, whereas fathers seldom are, so males disperse much more often than females. If for some reason offspring cannot disperse because all of the neighboring breeding sites are occupied, juveniles remain in their maternal site and form extended families for one to two generations. In these situations inbreeding is avoided by delaying the sexual maturation of older juveniles.

Sometimes mothers will share part of the natal site with their daughters. Mothers and daughters usually raise their young in separate nests, but at very high densities when breeding space is limited, the two will occasionally raise them in the same nest, nursing each others' offspring indiscriminately. In such circumstances the same male may well be the father of both litters, although other males may also have mated with either or both of the females.

During the winter non-breeding season, white-footed mice often aggregate, with 5–8 individuals nesting communally and thereby sharing body heat. In some cases, these communal groups are relatives from the last litters of the breeding season, but occasionally they are unrelated. Food is typically stored during these inclement times. During food shortages, White-footed mice commonly enter daily torpor to conserve energy. Torpor periods vary in length, but typically last less than 12 hours.

White-footed mice are at risk from a variety of aerial and terrestrial predators. Their nocturnal lifestyle protects them from some of these, and they further increase their chances by being less active during full moons and by trying to avoid running on dry fall leaves that crunch underfoot, choosing instead to move over logs or else to forage among branches or shrubs. When active on a noisy substrate such as dry leaves, White-footed mice adjust their gait by placing their hind feet in the path of their forefeet, thus halving the number of foot contacts with the ground. Unsurprisingly, they are most active on cloudy or rainy nights or when the ground is damp.

To have the option of selectively foraging or remaining inactive, the mice must have supplies of food readily on hand. Mice are seasonal hoarders and store non-perishable seeds for consumption in hot summer and cold winter periods. Acorns are a favorite crop that is harvested and stored in underground caches and hollow logs and trees. In fact, the abundance of the acorn crop determines to a large part survival rates and population size among White-footed mice. Following autumns of high mast production, which occur episodically once every 4–5 years, White-footed mice have high winter survival and may even breed all winter. Consequently, the following

Above *Representative species from six tribes of New World rats and mice:* ***1*** *South American climbing rat (genus* Rhipidomys*; tribe Thomasomyini).* ***2*** *Central American vesper rat (genus* Nyctomys*; tribe Nyctomyini).* ***3*** *Central American climbing rat (genus* Tylomys*; tribe Tylomyini).* ***4*** *pygmy mouse (genus* Baiomys*; tribe Baiomyini).* ***5*** *white-footed or deer mouse (genus* Peromyscus*; tribe Peromyscini).* ***6*** *woodrat or pack rat, carrying a bone (genus* Neotoma*; tribe Neotomini).*

Colonizing the Continents

Rats and mice originated in North America from the same kind of primitive rodents as the hamsters of Europe and Asia and the pouched rats of Africa, their nearest Old World relatives today. These ancestors, the so-called "cricetine" rodents, first appeared in the Old World in the Oligocene era about 34 million years ago, and were found in North America by the mid-Oligocene some 5 million years later. They were adapted to living among the treetops in forest environments, but as land dried during the succeeding Miocene era (24–5 million years ago) some became more ground-dwelling and developed into forms recognizable as those of modern New World rats and mice, the Sigmondontinae. In the course of their evolution they adapted to many habitats similar to those also occupied by Old World counterparts; for example, harvest mice of the genus *Reithrodontomys* reflect the Old World *Micromys*, while *Peromyscus* wood mice have counterparts in the Murinae genus *Apodemus*.

The rats and mice of South America developed in a similar way. A land bridge formed between North and South America during the Pliocene era 5–1.8 million years ago, and several stocks of primitive North American rodents moved across it. Equipped for climbing and forest life, they underwent an extensive radiation in the new continent's spacious grasslands. Subsequently they adapted to many other habitats, some of which are not occupied by rodents in any other part of the world. The absence of insectivores and lagomorphs (rabbits and hares) from much of the continent allowed them room for maneuver, and some evolved to resemble shrews, moles, or even rabbits: today's South American species include mole mice, the Shrew mouse, and the Bunny rat.

Above *Like a wolf howling, a Northern grasshopper mouse* (Onychomys leucogaster) *raises its head to utter the shrill cry for which the species is famous. Humans can hear the calls, which can last for a second or more, from as much as 100m (330ft) away. The cries probably serve to warn approaching mice that a patch of land is already tenanted.*

summers, mice are at their highest densities. When mice are abundant, they may selectively prey on larvae such as those of gypsy moths, reducing the numbers that defoliate oak and other deciduous trees in eastern North America. Less beneficially for humans, the mice can host a number of illnesses including Lyme disease and the deadly hantavirus that causes Hantaviral Pulmonary Syndrome.

Closely related to the white-footed mice is the Volcano mouse of Mexico, a burrow-user that is quite terrestrial in its habits. It occurs at elevations of 2,600–4,300m (8,530–14,100ft), and there is a birth peak in July and August

The size of harvest mice varies considerably, from 6–14cm (2–5.7in) in head–body length and from 6.5–9.5cm (2.5–3.7in) in tail length. The North American species tend to be smaller than the Central American species, rarely weighing more than 15g (0.5oz). The Western harvest mouse is typical of the species inhabiting the grassland areas of western North America, emerging at night to eat seeds or grain and living in a globular nest approximately 24cm (9in) off the ground in tall grass.

Grasshopper mice, which are about 10cm (4in) long with short tails, live in arid and semi-arid areas, feeding on insects and small vertebrates. A pair will live together in a burrow during the breeding season, after which the weaned young disperse; it is not known whether the parents then stay together to raise further litters. These rodents are well-known for their high-pitched squeak (usually above 20kHz), which may be used to indicate tenancy of a patch as burrows are widely spaced.

The Golden mouse (*Ochrotomys nuttalli*) is confined to the moderately wet, wooded habitats of the southeastern USA. The distinctive golden-brown color of its back contrasts sharply with its white belly. This is an extremely arboreal form that builds a complex leafy nest in tangles of vines.

The **woodrats and their allies** (tribe Neotomini) are rat-sized rodents, varying in color from dark buff on the back to paler shades on the belly. In general they eat a wide range of foods, but some species are highly adapted for feeding on the green parts of plants; indeed, *Neotoma stephensi* feeds almost entirely on the foliage of juniper trees. Some of the rats take refuge in crevices or cracks in rocky outcrops, while others construct burrows. All like to collect a mound of sticks and other detritus around the nest hole or crack.

The pads of spiny cacti may prove perfect tools for desert species; indeed, the name "pack rats" comes from this habit of transporting materials around their range. Each stick nest tends to be inhabited by a single adult individual, but they do make calls on neighboring nests; in particular, males visit receptive females.

The Magdalena rat occurs in an extremely restricted area of tropical deciduous forest in the states of Jalisco and Colima in western Mexico, where it may have an extended season of reproduction. This small, nocturnal woodrat is an excellent climber.

The Diminutive woodrat (*Nelsonia neotomodon*) is found in the mountainous areas of central and western Mexico, where it is known to shelter in crevices of rock outcroppings at elevations exceeding 2,000m (6,500ft).

The home-range sizes of New World mice vary hugely. **Pygmy mice** and **brown mice** (tribe Baiomyini) are the smallest New World rodents, and they possess correspondingly small home ranges that are often less than 900sq m (9,700sq ft) in extent; in comparison, a larger seed-eating rodent such as *Peromyscus leucopus* has a home range of 1.2–2.8ha (2.9–6.9 acres). Pygmy mice are seed-eaters that inhabit a grass nest, usually under a stone or log. They may be monogamous while pairing and rearing their young. Brown mice are small, subtropical mice with a high-pitched call, probably used to demarcate territory. Males produce this call more frequently than females.

The **Central American climbing rats** (tribe Tylomyini) live in the forest canopy never far from water. They rarely descend from the trees or emerge before sundown, and eat mainly fruits, seeds, and nuts. The Big-eared climbing rat, the smallest of these species, forages both in the trees and on the ground, and is unique among the New World rats in another crucial way: its young are born fully-furred, are few in number, and are very well-developed (precocial), opening their eyes after just 6 days. In contrast, most New World rats and mice produce hairless young whose eyes open only after 10–12 days. Furthermore, the Big-eared climbing rat mother gestates her growing young for 6.5 weeks, while most New World rats and mice gestate for only 3 weeks.

Vesper Rats

NYCTOMYINE GROUP

The tribe Nyctomyini contains just two species of vesper rat. The Central American vesper rat is a specialized, nocturnal, arboreal fruiteater that builds nests in trees and has a long tail and large eyes. Its young are born furred with open eyes.

The Yucatan vesper rat, also highly arboreal, is a relict species in the Yucatan peninsula. It probably was once more broadly distributed under different climatic conditions, but became isolated during one of the drying cycles in the Pleistocene era 1.8 million–10,000 years ago.

Paramo and Rice Rats

THOMASOMYINE–ORYZOMYINE GROUP

The **Paramo rats** and their allies of the tribe Thomasomyini are distributed throughout South America. In the mountains of the Andes, many species are adapted for life at elevations exceeding 4,000m (13,000ft). Otherwise they are almost always confined to forests, or else live along rivers. All that is known about them is that they confine most of their activity to the hours of darkness, and that they eat fruit. Litters of two to four young have been recorded, but in general their reproductive potential is considered to be quite low.

The South American climbing mice are likewise adapted for life in trees. They are also nocturnal, and feed upon fruits, seeds, fungi, and insects. Their litter size is small: for the Long-tailed climbing mouse (*Rhipidomys mastacalis*), two or three young per litter have been recorded.

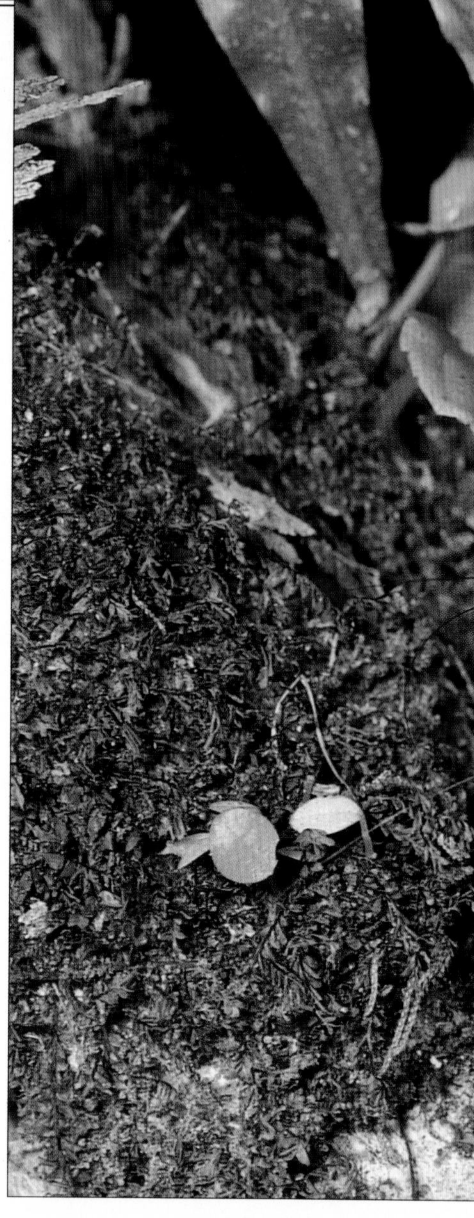

Right *The Vesper rat* (Nyctomys sumichrasti) *is a tree-dweller in the rain forests of central America, where it builds nests rather like those of Red squirrels. Individuals sometimes inadvertently find their way to the USA, concealed in bunches of imported bananas.*

Below *The Cotton mouse owes its name to its distribution in the cotton-growing states of the American south, where it can be found in woodland, swamps, and rocky areas. A nocturnal omnivore, it feeds on insects and other invertebrates as much as on plants.*

Rice rats and their allies (tribe Oryzomini) are an assemblage with three tendencies: they may live in the trees, on the ground, or next to water. The species often replace each other up an altitudinal gradient. In northern Venezuela *Oryzomys albigularis* occurs above elevations of 1,000m (3,300ft), but at lower elevations it is replaced by *O. capito*. Alternatively, when two species occur in the same habitat, one may be more adapted to life in the trees than the other. *O. capito*, a terrestrial species, can happily share territory with *Oecomys bicolor*, a species adapted for climbing.

The South American water rat (one of the three species of *Nectomys*) is semiaquatic and the dominant aquatic rice rat over much of South America. *Oryzomys palustris*, found from northern Mexico up to southern Maryland, is also semiaquatic. This rat has catholic taste in food, though at certain times of year over 40 percent of its diet may consist of snails and crustaceans. It has an extended breeding season (from February to November) across much of its range, and so is able to produce four young every 30 days. As a consequence, it can become a serious agricultural pest.

The small bristly mice, which have a distinctive spiny coat, are nocturnal and eat seeds. *Neacomys tenuipes*, in northern South America, exhibits wide variations in population density.

Over evolutionary time rice rats have excelled at colonizing islands in the Caribbean and the Galapagos group, but many of the genus *Oryzomys* are currently threatened with extinction. The introduction of the Domestic cat and murine rats and mice by humans has had a severe impact on the Galapagos population in particular.

South American Field and Burrowing Mice

AKODONTINE–OXYMYCTERINE GROUP

South American field mice (tribe Akodontini) are adapted for foraging on the ground, and many are also excellent burrowers.

The grass mice of the genus *Akodon* are another group that have excelled at occupying all the available vacancies in the ecosystems they inhabit. In general they are omnivorous, eating green vegetation, fruits, insects, and seeds, and most are adapted to moderate and high elevations. The Northern grass mouse (*A. urichi*) typifies the adaptability of the genus; besides being active both by day and night, it is terrestrial and eats an array of different foods.

Members of the genus *Bolomys* are closely allied to *Akodon* but are more specialized for terrestrial existence. The tail, neck, and ears are shortened for burrowing, and the eyes are also reduced. Members of the genus *Microxus* are similar to *Bolomys* in appearance, but their eyes show yet further reduction in size.

Cane mice (tribe Zygodontomini) are widely distributed in South America, taking the place of *Akodon* at low elevations in grasslands and bushlands. The runways they construct can be visible to a human observer. Cane mice eat a considerable quantity of seeds and do not seem to be specialized for processing green plant food. In grassland habitats subject to seasonal fluctuations in rainfall, the cane mice may show vast oscillations in population density. In the llanos of Venezuela when the grass crop is exceptionally good, enabling them to harvest seeds and increase their production of young, they show population explosions. Densities can vary from a high of 15 per ha (6 per acre) to a low of less than 1 per ha (0.4 per acre), depending on the weather.

The **burrowing mice and their relatives** (tribe Oxymycterini) are closely allied to the South American field mice. These mice, like other rodents that feed on insects, have relatively small molar teeth and an elongated snout used for getting insects out of holes. Long claws help in digging for soil arthropods, larvae, and termites.

Feeding on insects and termites has its downside, however, which may be reflected in the limited number of offspring these groups are capable of producing. Termites are wrapped in protective armor in the form of an exoskeleton made of the protein chitin. The nutritional content of this substance is relatively low compared to other sources of protein or carbohydrate. Consequently, mammals relying on chitin for their daily supper tend to have low metabolic rates to compensate for the low energy return from their food. The relatively small litters of 2–3 young born to *Oxymycterus* species may be a constraint of their poor diet.

The Shrew mouse, with its tiny eyes and ears hidden in its fur, represents an extreme adaptation for a tunneling way of life. It constructs a deep burrow under the litter of the forest floor. Its molar teeth are very reduced in size, a characteristic that indicates adaptation for a diet of insects.

Mole mice of the genera *Chelemys*, *Notiomys*, and *Geoxus* are widely distributed in Argentina and Chile and exhibit an extensive array of adaptations for exploiting both semi-arid steppes and wet forests. Some species are adapted to higher-elevation forests, others to moderate elevations in central Argentina. They have extremely powerful claws, which may exceed 0.7cm (0.3in) in length. The name "mole mice" derives from their habit of spending most of their life underground.

Fish-eating Rats and Mice

ICHTHYOMYINE GROUP

The **fish-eating** or **crab-eating rats and mice** (tribe Ichthyomyini) live the aquatic lives their name suggests, on or near higher-elevation freshwater streams, where they exploit small crustaceans, aquatic arthropods, and fish as their primary food sources. The rest of their biology remains a mystery.

Fish-eating rats of the genus *Ichthyomys* are among the most specialized of the genera. Their fur is short and thick, their eyes and ears are reduced in size, and their whiskers are stout. A fringe of hairs on the toes of the hind feet aids in swimming, and the toes are partially webbed to propel a body about 33cm (9in) in length. They resemble large water shrews, or some of the fish-eating insectivores of West Africa and Madagascar.

Fish-eating rats of the genus *Neusticomys* are similar to *Ichthyomys*, and are distributed disjunctly in the mountain regions of Venezuela, Columbia, and Peru.

Water mice are smaller than the *Ichthyomys* species, rarely exceeding 19cm (7in) in head–body length. They occur in central American mountain streams, and are known to feed on snails, aquatic insects, and possibly fish.

Cotton and Marsh Rats and Allied Species

SIGMODONTINE–PHYLLOTINE–SCAPTEROMYINE GROUP

Cotton rats and marsh rats (tribe Sigmodontini) are united by a common feature, namely folded patterns of enamel on the molars that tend to approximate to an "S" shape when viewed from above. They exhibit a range of adaptations; the species referred to as marsh rats are adapted for a semi-aquatic life, whereas the cotton rats are terrestrial. Both groups, however, feed predominantly on herbaceous vegetation.

The marsh rats, which are web-footed, form the genus *Holochilus*. Two species are broadly distributed in South America, while another is limited to the chaco of Paraguay and northeastern Argentina. The underside of the tail has a fringe of hair that functions as a rudder when swimming. These rats build a grass nest that may exceed 40cm (15.7in) in diameter, locating it near water, sufficiently high up to avoid flooding. In the more southerly parts of their range in temperate South America, breeding tends to be confined to the spring and summer (September–December).

Cotton rats are broadly distributed from the southern USA to northern South America. In line with their adaptations for terrestrial life, tail length is always considerably shorter than head–body length. Cotton rats are active both by day and night; they are omnivores, taking advantage of the fresh growth in herbs and grasses that follows after the onset of rains.

A striking feature of the Hispid cotton rat is that its young are born fully furred; their eyes open within 36 hours of birth. This species has a very high reproductive capacity, and although it produces precocial young, the gestation period is only 27 days. Litter sizes are quite high, ranging from five to eight, with 7.6 as an average. The female is receptive after giving birth and only lactates for 10–15 days. Thus the turn-around time between litters is very brief; a female can produce a litter every month during the breeding season. In agricultural regions this prolific rat can quickly become a serious pest.

Leaf-eared mice and their allies (tribe Phyllotini) are typified by the genera *Phyllotis* and *Calomys*. *Calomys* (vesper mice) includes a variety of species distributed over most of South America. They feed primarily on plant material; arthropods form an insignificant portion of their diet. Most of the species making up the genus *Phyllotis* (the leaf-eared mice) occur at high altitudes. They are often active by day, and may bask in the sun. They feed primarily on seeds and herbaceous plant material.

Above *Large ears have given South America's leaf-eared mice their common name. This Darwin's leaf-eared mouse* (Phyllotis darwini) *is foraging 4,300m (14,000ft) up on the Andean* altiplano.

Below *Representative New World rats and mice:* ***1*** *South American grass mouse (genus* Akodon*; tribe Akodontini) grooming its tail;* ***2*** *cotton rat (genus* Sigmodon*; tribe Sigmodontini) attempting to move an egg;* ***3*** *mole mouse (genus* Chelemys*; tribe Akodontini) in an underground burrow;* ***4*** *South American water rat (genus* Nectomys*; tribe Oryzomini) at the water's edge;* ***5*** *fish-eating rat (genus* Ichthyomys*; tribe Ichthyomyini);* ***6*** *Swamp rat (*Scapteromys tumidus*; tribe Scapteromyini);* ***7*** *leaf-eared mouse (genus* Phyllotis*; tribe Phyllotini).*

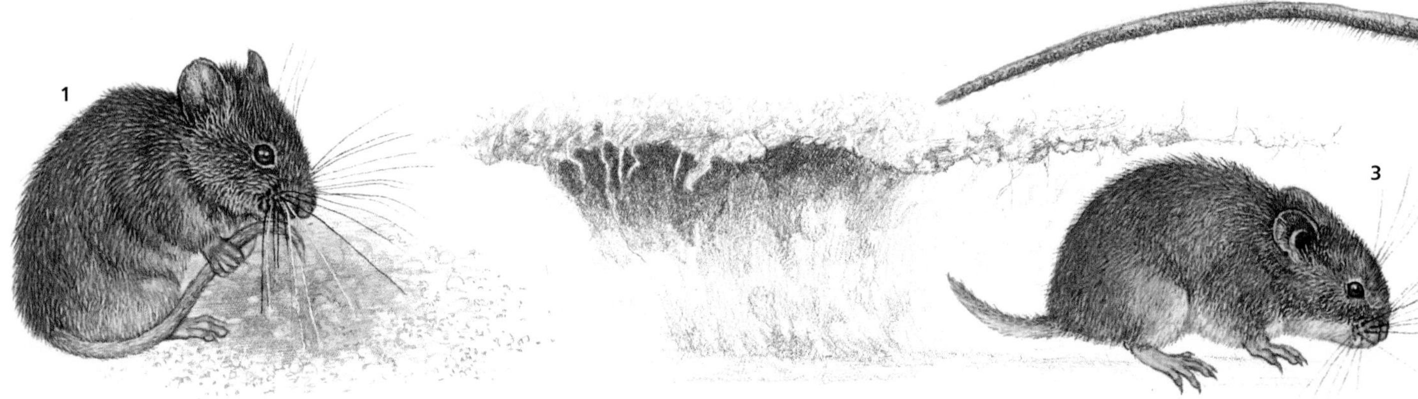

'he variations in form, and the way in which :everal species of different size occur in the same .abitat, are reminiscent of the white-footed mice f the tribe Peromyscini.

The vegetarian, cathemeral Bunny rat (*Rei-ırodon physodes*) is of moderate size and has thick .ır adapted to the open-country plains of temper-te Chile, Argentina, and Uruguay.

The Highland gerbil mouse (*Eligmodontia uerulus*) is one of the few South American odents specialized for semi-arid habitats. Its hind :et are long and slender, resulting in a peculiar, alloping gait in which the forelimbs simultane-usly strike the ground, and then are driven ıpward by a powerful thrust from the hind legs. 'he kidneys of this species are very efficient at ecovering water; it can exist for considerable peri-ods of time without drinking, being able to derive ts water as a by-product of its own metabolism.

Patagonian chinchilla mice are distributed in wooded areas from central Argentina south to Cape Horn. The Puna mouse is found only in the *altiplano* of Peru. This rodent is the most vole-like in bodyform of any South American rodent. It is active both by day and night, and its diet is apparently confined to herbaceous vegetation. The Chilean rat is an inhabitant of humid temperate forests; this extremely arboreal species may be a link between the phyllotines and the oryzomyine rodents or rice rats. The Andean marsh rat occurs at high elevations near streams and appears to occupy a niche appropriate for a vole.

The **southern water rats and their allies** (tribe Scapteromyini) are adapted for burrowing in habitats by or near rivers. The Swamp rat (*Scapteromys tumidus*), also known as the Argentinian water rat, is found near rivers, streams, and marshes. It has extremely long claws and can construct extensive burrow systems.

The giant South American water rats prefer moist habitats and have considerable burrowing ability. The Woolly giant rat (*Kunsia tomentosus*) is one of the largest living New World rats, with a head–body length that may reach 28cm (11in) and a tail length of up to 16cm (6.3in).

Red-nosed rats are small burrowing forms allied to the larger genera. As with so many of the animals in this section, their biology and habits are poorly known. JFE/RB/JOW

New World Rat and Mouse Tribes

THE SHEER NUMBER OF NEW WORLD RAT and mouse species makes it convenient to have some grouping system through which to make sense of their diversity. Yet attempts to combine them at anything between subfamily and genus level (and, in some cases, even at that) have proved taxonomically controversial. The arrangement of groups and tribes suggested here is only provisional; readers should be aware that only one tribe has been convincingly shown to be monophyletic (arising from a single ancestor).

North American Neotomine–Peromyscine Group

White-footed Mice and their Allies
Tribe Peromyscini

10 genera: **White-footed** or **deer mice** (*Peromyscus*, 54 species), from N Canada (except high Arctic) S through Mexico to Panama. **Harvest mice** (*Reithrodontomys*, 19 species), from W Canada and USA S through Mexico to W Panama. **Crested-tailed deer mice** (*Habromys*, 4 species) from C Mexico S to El Salvador. **Florida mouse** (*Podomys floridanus*), Florida peninsula. **Volcano mouse** (*Neotomodon alstoni*), montane areas of C Mexico. **Grasshopper mice** (*Onychomys*, 3 species), SW Canada, NW USA S to north-central Mexico. **Michoacan deer mouse** (*Osgoodomys bandaranus*), W C Mexico. **Isthmus rats** (*Isthmomys*, 2 species), Panama. **Thomas's giant deer mouse** (*Megadontomys thomasi*), C Mexico. **Golden mouse** (*Ochrotomys nuttalli*), SW USA.

Woodrats and their Allies
Tribe Neotomini

4 genera: **Woodrats** (*Neotoma*, 19 species), USA to C Mexico. **Allen's woodrat** (*Hodomys alleni*), W C Mexico. **Magdalena rat** (*Xenomys nelsoni*), W C Mexico. **Diminutive woodrat** (*Nelsonia neotomodon*), C Mexico.

Pygmy Mice and Brown Mice
Tribe Baiomyini

2 genera: **Pygmy mice** (*Baiomys*, 2 species), SW USA S to Nicaragua. **Brown mice** (*Scotinomys*, 2 species), Brazil, Bolivia, Argentina.

Central American Climbing Rats
Tribe Tylomyini

2 genera: **Central American climbing rats** (*Tylomys*, 7 species), S Mexico to W Columbia. **Big-eared climbing rat** (*Ototylomys phyllotis*), Yucatan peninsula of Mexico S to Costa Rica.

Nyctomyine Group

Vesper Rats Tribe Nyctomyini

2 genera: **Central American vesper rat** (*Nyctomys sumichrasti*), S Mexico S to C Panama. **Yucatan vesper rat** (*Otonyctomys hatti*), Yucatan peninsula of Mexico and adjoining areas of Mexico and Guatemala.

Thomasomyine–Oryzomyine Group

Paramo Rats and their Relatives
Tribe Thomasomyini

8 genera: **Paramo rats** (*Thomasomys*, 25 species), Andean areas of high altitude from Colombia S to Bolivia. **Atlantic forest rats** (*Delomys*, 2 species), SE Brazil to adjacent areas of Argentina. **Wilfred's mice** (*Wilfredomys*, 2 species), SE Brazil to NW Argentina and Uruguay. **Brazilian spiny rat** (*Abrawayaomys ruschii*), SE Brazil. **South American climbing rats** (*Rhipidomys*, 14 species), low elevations from extreme E Panama S across northern S America to C Brazil. **Colombian forest mouse** (*Chilomys instans*), high elevations in Andes in W Venezuela S to Colombia and Ecuador. **Montane mice** (*Aepomys*, 2 species), high elevations in Andes in Venezuela, Colombia, Ecuador. **Rio de Janeiro rice rat** (*Phaenomys ferrugineus*), vicinity of Rio de Janeiro.
Note: *Chilomys, Delomys, Wilfredomys, Phaenomys* have uncertain affinities.

Rice Rats and their Allies
Tribe Oryzomini

13 Genera: **Dusky rice rats** (*Melanomys*, 3 species), Central America S to Peru. **Montane dwarf rice rats** (*Microryzomys*, 2 species), mountains of Venezuela S to Bolivia. **Arboreal rice rats** (*Oecomys*, 12 species), lowland tropical forest of C. America to Brazil. **Pygmy rice rats** (*Oligoryzomys*, 15 species), S Mexico to S Brazil. **Rice water rats** (*Sigmodontomys*, 2 species), forests of Costa Rica to Ecuador. **Rice rats** (*Oryzomys*, 36 species), SE USA S through C America and N S America to Bolivia and C Brazil. **Galapagos rice rats** (*Nesoryzomys*, 5 species), Galapagos archipelago of Ecuador. **Spiny mouse** (*Scolomys*, 3 species), Ecuador, Peru. **False rice rats** (*Pseudoryzomys*, 2 species), Bolivia, E Brazil, N Argentina. **Bristly mice** (*Neacomys*, 3 species), E Panama across lowland S America to N Brazil. **South American water rats** (*Nectomys*, 3 species), lowland S America to NE Argentina. (*Amphinectomys savamis*), E Peru. **Brazilian arboreal mouse** (*Rhagomys rufescens*), SE Brazil.

Right *Some desert-dwelling woodrats like this White-throated individual* (Neotoma albigula) *make elaborate dens that may be passed down through successive generations. The spiny materials used in their construction makes it difficult for enemies to enter them, though the rats themselves apparently come and go unscathed.*

Wied's Red-nosed Mouse
Tribe Wiedomyini

Wied's red-nosed mouse (*Wiedomys pyrrhorhinos*), E Brazil. Relationships uncertain.

Akodontine–Oxymycterine Group

South American Field Mice
Tribe Akodontini

11 genera: **South American field mice** (*Akodon*, 45 species), found in most of S America from W Colombia to Argentina. **Bolo mice** (*Bolomys*, 6 species), montane areas of SE Peru S to Paraguay and C Argentina. ***Microxus*** (3 species), montane areas of Colombia, Venezuela, Ecuador, Peru. **Altiplano mice** (*Chroeomys*, 2 species), Andes from Peru to N Argentina. **Mole mice** (*Chelemys*, 2 species), Andes on the Chile–Argentina boundary. **Long-clawed mole mouse** (*Geoxus valdivianus*), S Chile and adjacent Argentina. ***Pearsonomys annectens***, montane-central Chile. **Cerrado mice** (*Thalpomys*, 2 species), cerrado of S Brazil. **Andean rat** (*Lenoxus apicalis*), SE Peru and W Bolivia. **Edwards's long-clawed mouse** (*Notiomys edwardsii*), Argentina and Chile. **Mount Roraima mouse** (*Podoxymys roraimae*), at junction of Brazil, Venezuela, Guyana.

Cane Mice
Tribe Zygodontomini

Cane mice (*Zygodontomys*, 3 species), Costa Rica and N S America. Relationships uncertain.

Burrowing Mice and their Relatives
Tribe Oxymycterini

4 genera: **Burrowing mice** (*Oxymycterus*, 12 species), SE Peru, W Bolivia E over much of Brazil and S to N Argentina. **Brazilian shrew mouse** (*Blarinomys breviceps*), E C Brazil. **Juscelin's mouse** (*Juscelinomys candango*), vicinity of Brasilia. ***Brucepattersonius*** (7 species), S C Brazil and adjacent northeastern Argentina and Uruguay.

Ichthyomyine Group

Fish-eating Rats and Mice
Tribe Ichthyomyini

5 genera: **Fish-eating** or **crab-eating rats** (*Ichthyomys*, 4 species), premontane habitats of Venezuela, Columbia, Ecuador, Peru. **Chibchan water mice** (*Chibchanomys*, 2 species), W Venezuela S in the Andes to Peru. **Ecuador fish-eating rat** (*Anotomys leander*), montane Ecuador. **Fish-eating rats** (*Neusticomys*, 4 species), Andes region of S Colombia and N Ecuador, N Venezuela, W Peru, and French Guyana. **Water mice** (*Rheomys*, 5 species), C Mexico S to Panama.

Sigmodontine–Phyllotine–Scapteromyine Group

Cotton Rats and Marsh Rats
Tribe Sigmodontini

3 genera: **Marsh rats** (*Holochilus*, 4 species), most of lowland S America. **Cotton rats** (*Sigmodon*, 10 species), S USA, Mexico, C America, NE S America as far S as NE Brazil. **Giant water rat** (*Lundomys molitor*), extreme S Brazil and adjacent areas of Uruguay.

Leaf-eared Mice and their Allies
Tribe Phyllotini

17 genera: **Leaf-eared mice** (*Graomys*, 3 species), Andes of Bolivia S to N Argentina and Paraguay. **Chaco mice** (*Andalgalomys*, 3 species), Paraguay and NE Argentina. **Garlepp's mouse** (*Galenomys garleppi*), high altitudes in S Peru, W Bolivia, N Chile. **Big-eared mice** (*Auliscomys*, 3 species), mountains of Bolivia, Peru, Chile, and Argentina. **Puna mouse** (*Punomys lemminus*), montane areas of S Peru. **Bunny rat** (*Reithrodon physodes*), steppe and grasslands of Chile, Argentina, Uruguay. **Vesper mice** (*Calomys*, 9 species), most of lowland S America. **Chinchilla mouse** (*Chinchillula sahamae*), high elevations S Peru, W Bolivia, N Chile, Argentina. **Chilean rat** (*Irenomys tarsalis*), N Argentina, N Chile. **Andean mouse** (*Andinomys edax*), S Peru, N Chile. **Gerbil mice** (*Eligmodontia*, 4 species), S Peru, N Chile, Argentina. **Leaf-eared mice** (*Phyllotis*, 12 species), from NW Peru S to N Argentina and C Chile. **Patagonian chinchilla mice** (*Euneomys*, 4 species), temperate Chile and Argentina. **Andean swamp rat** (*Neotomys ebriosus*), Peru S to NW Argentina. **Bolivian big-eared mouse** (*Maresomys boliviensis*), N Argentina and adjacent Bolivia. **Southern big-eared mouse** (*Loxodontomys micropus*), W Argentina. ***Salinomys delicatus***, known from just 9 specimens found in NW Argentina.

Southern Water Rats and their Allies
Tribe Scapteromyini

3 genera: **Red-nosed rats** (*Bibimys*, 3 species), SE Brazil W to NW Argentina. **Argentinean water rat** (*Scapteromys tumidus*), SE Brazil, Paraguay, E Argentina. **Giant South American water rats** (*Kunsia*, 2 species), N Argentina, Bolivia, SE Brazil.

For full species list see Appendix ▷

Voles and Lemmings

POPULAR LEGEND TELLS OF LEGIONS OF lemmings periodically flinging themselves into rivers or the sea to drown en masse. However, there is no factual evidence of suicidal tendencies either in lemmings or in voles, their partners in the subfamily Microtinae, although it is true that lemmings embark from time to time on mass migrations, in the course of which thousands of individual animals may die.

Lemmings and voles have two features of particular interest. First, their populations expand and contract considerably, in line with cyclical patterns. This has made them the most studied subfamily of rodents (and the basis of much of our understanding about the population dynamics of small mammals). Secondly, though they neither hibernate like such larger mammals as ground squirrels nor can rely on a thick layer of fat like bears, many voles and lemmings live in habitats covered by snow for much of the year. They are able to survive thanks to their ability to tunnel beneath the snow, where they are insulated from extreme cold.

Thickset Bodies and Rounded Muzzles

FORM AND FUNCTION

Voles and lemmings are small, thickset rodents with bluntly rounded muzzles, and tails that are usually less than half the length of their bodies. Only small sections of their limbs are visible. Their eyes and ears tend to be small, and in lemmings the tail is usually very short. Coat colors vary not only between species but often within them. Lemmings' coats are especially adapted for cold temperatures: they are long, thick, and waterproof. The Collared lemming is the only rodent that molts to a completely white coat in winter.

Some species display special anatomical features. The claws of the first digit of the Norway

Above left *A Field vole eating. Although males are quite territorial, the females tend to be less so, with considerable overlapping of territories. Young males will be forced to disperse by older resident males.*

FACTFILE

VOLES AND LEMMINGS

Order: Rodentia

Family: Muridae

Subfamily: Arvicolinae (Microtinae)

143 species in 26 genera. Species include: **Collared lemming** (*Dicrostonyx torquatus*); **Norway lemming** (*Lemmus lemmus*); **Northern mole-vole** (*Ellobius talpinus*); **Bank vole** (*Clethrionomys glareolus*); and **muskrat** (*Ondatra zibethicus*).

DISTRIBUTION N and C America, Eurasia from Arctic S to Himalayas; small relict population in N Africa.

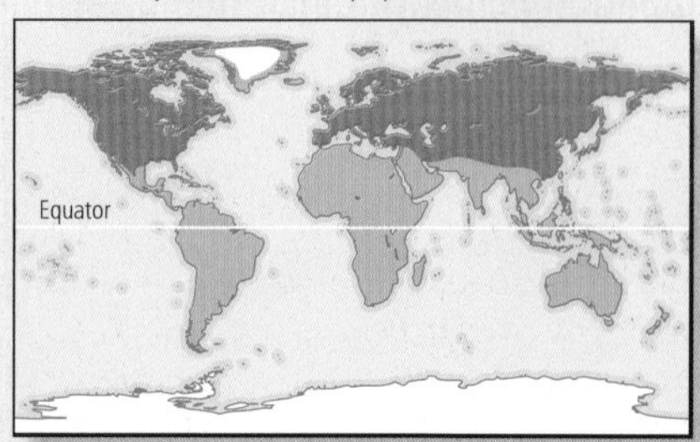

HABITAT Burrowing species are common in tundra, temperate grasslands, steppe, scrub, open forest, rocks; 5 species are aquatic or arboreal.

SIZE In most species, **head–body length** 10–11cm (4–4.5in), **tail length** 3–4cm (1.2–1.6in), **weight** 17–20g (0.6–0.7oz).

COAT Thickly furred in various shades of gray and brown.

DIET Herbivorous; mostly the green parts of plants, though some species eat bulbs, roots, and mosses. Muskrats occasionally eat mussels and snails.

BREEDING Gestation 16–28 days. Litter size varies between species from 1–12; at least 2 litters, and sometimes 4 or more, are produced each year.

LONGEVITY 0.5–2 years

CONSERVATION STATUS 3 species are listed as Critically Endangered and 3 more as Endangered. The Bavarian vole (*Microtus bavaricus*) is now considered Extinct.

See tribes table ▷

Left *Muskrats are well adapted to life around water. They have partial webbing on their back feet and can use their tail, which is slightly flattened, as a rudder. Consequently, muskrats occur in a wide variety of aquatic environments.*

Right *Bank voles such as this* (Clethrionomys glareolus) *live in Europe and Central Asia in woods and scrubs, in banks and swamps, usually within a home range of about 0.8ha (2 acres).*

lemming are flattened and enlarged for digging in the snow, while each fall the Collared lemming grows an extra big claw on the third and fourth digits of its forelegs, shedding them in spring. Muskrats have long tails and small webbing between toes that assist in swimming. The mole lemmings, adapted for digging, have a more cylindrical shape than other species, and their incisors, used for excavating, protrude extremely.

Adult males and females are usually the same color and approximately the same size, though the shade of juveniles' coats may differ from the adults'. Although most adult voles weigh less than 100g (3.5oz), the muskrat grows to over 1,400g (50oz). The size of the brain, in relation to body size, is lower than average for mammals.

Smell and hearing are important, well-developed senses, able to respond, respectively, to the secretions that are used to mark territory boundaries, to indicate social status, and perhaps to aid species recognition, and to vocalizations (each species has a characteristic range of calls). Calls can be used to give the alarm, to threaten, or as part of courtship and mating. Brandt's voles, which live in large colonies, sit up on the surface and whistle like prairie dogs.

Tundra, Grassland, Forest

DISTRIBUTION PATTERNS

Microtines, especially lemmings, are widely distributed in the tundra regions of the northern hemisphere, where they are the dominant small mammal species. Their presence there is the result of recolonization since the retreat of the last glaciers. Voles are also found in temperate grasslands and in the forests of North America and Eurasia.

Because the Pleistocene era (1.8 million–10,000 years ago) has yielded a rich fossil record of microtine skulls, much of the taxonomy of this subfamily is based on the structure and kind of teeth, which are also used to distinguish the microtines from other rodents. All microtine teeth have flattened crowns, and prisms of dentine surrounded by enamel. There are twelve molars (three on each side of the upper and lower jaws) and four incisors. Species are differentiated by the particular pattern of the enamel of the molars. Dentition has not proved sufficient for solving all taxonomic questions, and some difficulties remain both in delimiting species and defining genera. It is often possible to distinguish the species of live animals from the general body size, coat color, and length of tail. The subdivision of species into many subspecies (not listed here) reflects geographic variation in coat color and size in widely distributed species. *Microtus*, the largest genus (accounting for nearly 50 percent of the subfamily), is a heterogeneous collection of species.

Many species have widely overlapping ranges. For example, there are six species – two lemmings and four voles – in the southern tundra and forest of the Yamal Peninsula in Russia. Each can be differentiated by its habitat preference or diet. The ranges of Siberian lemmings and Collared lemmings overlap extensively, but Collared lemmings prefer upland heaths and higher and drier tundra, whereas Siberian lemmings are found in the wetter grass-sedge lowlands. Competition between closely related microtines has been suggested but rarely demonstrated by field experiments.

Busy Grazers and Diggers

DIET

Voles and lemmings are herbivores that usually eat the green parts of plants, though some species prefer bulbs, roots, mosses, or even pine needles. The muskrat occasionally eats mussels and snails. Diet usually changes with the seasons and varies according to location, reflecting local abundances of plants. Species living in moist habitats, such as lemmings and the Tundra vole, prefer grasses and sedges, while those inhabiting drier habitats, such as Collared lemmings, prefer herbs. But animals select their food to some extent; diets do not just simply mimic vegetation composition.

Voles and lemmings can be found foraging both by day and night, although dawn and dusk might be preferred. They obtain food by grazing or digging for roots; grass is often clipped and placed in piles in their runways. Some cache food in summer and fall, but in winter, when the snow cover insulates the animals that nest underneath the surface, food is obtained by burrowing; animals also feed on plants at ground surface. The Northern red-backed vole feeds on (among other items) berries, and in summer has to compete for them with birds; in winter, when the bushes are covered with snow, this vole can burrow to reach the berries, so only during the spring thaw is the animal critically short of food. In winter the Sagebrush vole utilizes the height of snow packs to forage on shrubs it normally cannot reach.

Most microtines – for example the Florida water rat, the Steppe lemming, and all species of *Microtus* – have continuously growing molars, and can chew more (and more abrasive) grasses than species with rooted molars. In tundra and grasslands, voles and lemmings help the recycling of nutrients by eliminating waste products and clipping and then storing food below ground.

A Short, Hectic Life

Social Behavior

The life span of microtines is short. They reach sexual maturity at an early age and are very fertile. Mortality rates, however, are high: during the breeding season, only 70 percent of the animals alive one month will still be alive the next. The age of sexual maturity can vary considerably. The females of some species may become mature only 2–3 weeks after birth. Males take longer, usually 6–8 weeks. The Common vole has an extraordinarily fast development. Females have been observed coming into heat while still suckling, and may be only 5 weeks old at the birth of their first litter. In species that breed only during the summer, young born early will probably breed that summer, while later litters may not become sexually mature until the following spring.

The length of the breeding season is highly variable, but lemmings can breed in both summer and winter. Winter breeding is less common in voles, which tend to breed from late spring to fall. Voles in Mediterranean climates, on the other hand, breed in winter and spring, during the wet season, but not in summer. Within a species, the breeding season often varies widely in different years or in different parts of the range. The muskrat breeds all year in the southern part of its range but only in summer elsewhere. Some meadow vole species breed in winter during the phase of population increase, but not when numbers are falling.

In many species, such as the Montane, Field, and Mexican voles, the presence of a male will induce ovulation in a female. (Physical contact is unnecessary; ovulation is probably induced by social odors produced by the male.) The normal gestation time is 21 days, but it may be less if conditions are good, falling to just 17 days in a Bank vole with optimal nutrition. The gestation period may also lengthen – up to 24 days in the Bank vole's case – if a female has mated within a few hours of giving birth and is therefore pregnant while lactating.

The reproductive cycle is liable to be disrupted. A mated female can sometimes fail to become pregnant, or will abort spontaneously if exposed to a strange male (both phenomena have been observed in Field, Meadow, and Prairie voles in the laboratory; however, attempts to show it in the field have so far failed). Males are under pressure to sire as many litters as possible in the short breeding season. Collared lemming females have

Water Voles in Britain

The changing fortunes of British water vole populations through the 20th century have only recently come to light following the pioneering national surveys carried out by the Vincent Wildlife Trust in 1989–90 and 1996–98. These surveys confirmed that the species had been getting scarcer along waterways since the 1930s due to habitat loss and land use change associated with the intensification of agriculture. In recent years this decline has accelerated due to predation by feral American mink (established as escapees from fur farms).

The decline has now developed into a serious population crash with a further loss of the remaining population by 88 percent in only seven years. The 1996–98 survey reports that the population loss has been most severe in the north and southwest of England, reaching 97 percent of the population in Yorkshire! This makes the water vole Britain's most rapidly declining mammal, and as such it has been given legal protection under the Wildlife and Countryside Act 1981.

The threats to water voles are complex but involve habitat loss and degradation (due to river engineering or agricultural practice), population isolation and fragmentation, fluctuations in water levels, pollution, predation (especially by mink), or indirect persecution through rat control operations.

Saving the water vole is going to take some complex habitat manipulation, and cooperation between many sectors of British society. Habitat restoration creates dispersal corridors for isolated populations to reach each other between restored backwaters, ponds, wet grassland, marshes, reedbeds, and a network of ditches. Riverbanks are central to water vole activity and the possibility of a coordinated scheme to connect whole lengths of river along their banks by the withdrawal of farming would be invaluable. When banks are left free from cultivation, wild plants perfect for water vole food and cover often soon colonize, their seeds carried along the river by water or wind. Where this doesn't happen, planting can speed things considerably. Highly layered bankside vegetation with tall grasses and stands of flowers such as willowherb, loosestrife, and meadowsweet, fringed with thick stands of rushes, sedges, or reed are the best habitat. In winter, the roots and bark of woody species such as willows and sallows are an important part of the water vole's diet, together with rhizomes, bulbs, and roots of herbaceous species. Already underway are projects on the River Cole at Coleshill, Wiltshire, in which 2km of degraded river has been restored to meanders, and on the River Skerne at Darlington, County Durham, where backwater pools are creating small-scale water vole havens. RS

4

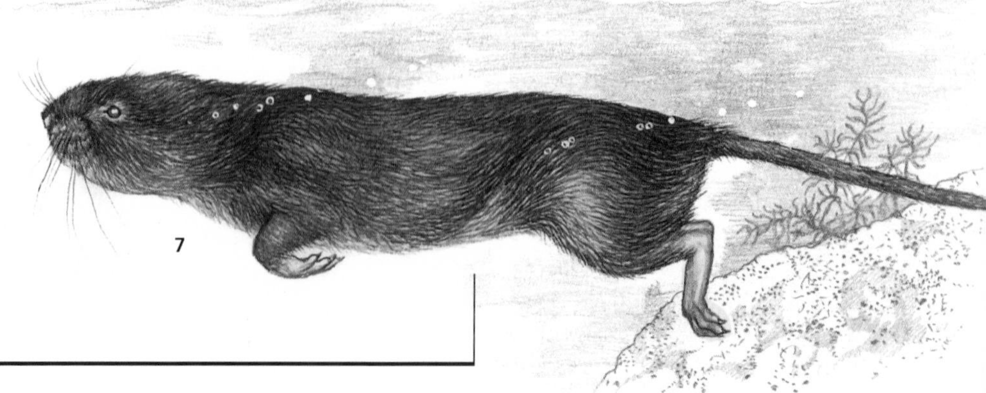

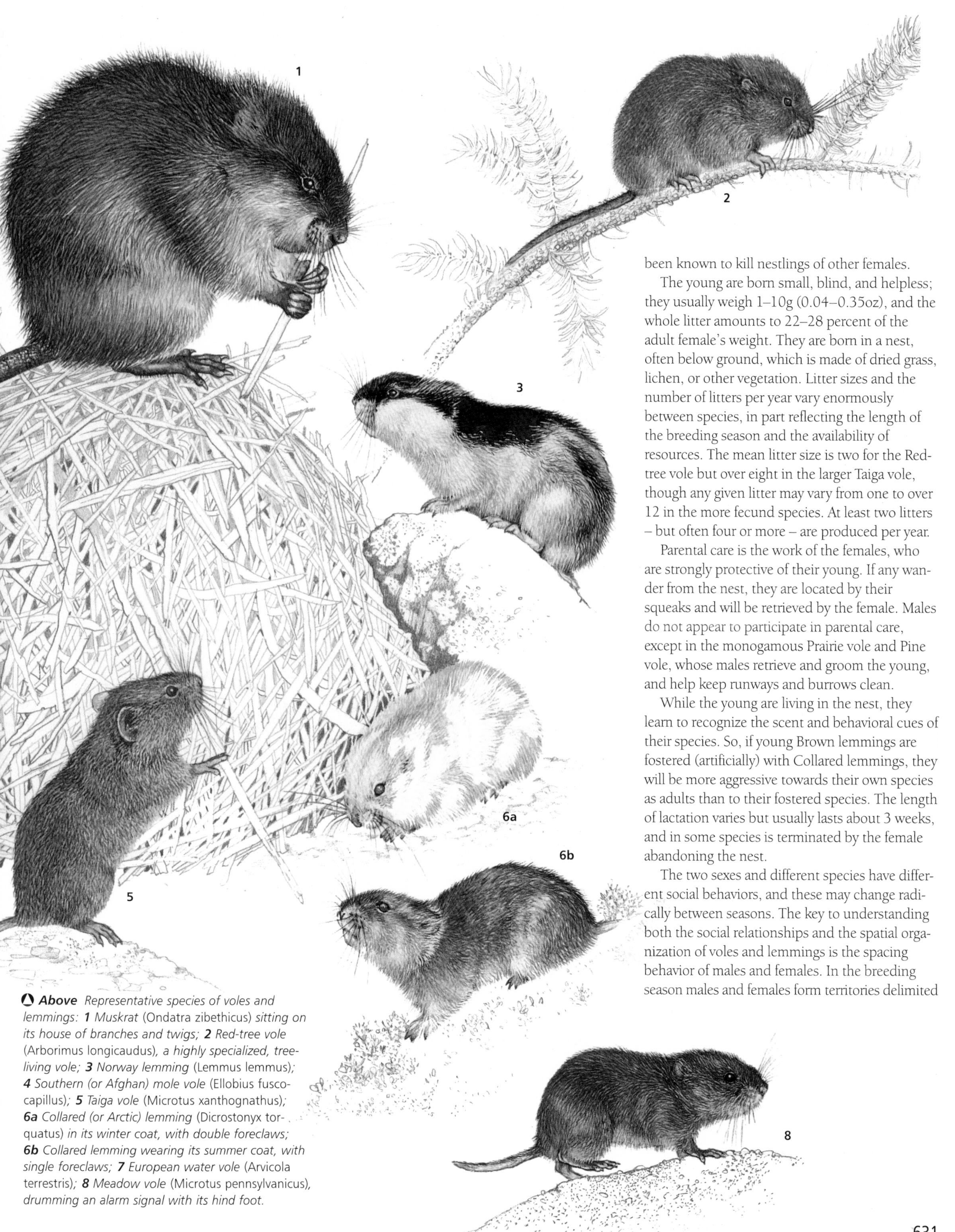

Above *Representative species of voles and lemmings:* **1** *Muskrat* (Ondatra zibethicus) *sitting on its house of branches and twigs;* **2** *Red-tree vole* (Arborimus longicaudus), *a highly specialized, tree-living vole;* **3** *Norway lemming* (Lemmus lemmus)*;* **4** *Southern (or Afghan) mole vole* (Ellobius fuscocapillus)*;* **5** *Taiga vole* (Microtus xanthognathus)*;* **6a** *Collared (or Arctic) lemming* (Dicrostonyx torquatus) *in its winter coat, with double foreclaws;* **6b** *Collared lemming wearing its summer coat, with single foreclaws;* **7** *European water vole* (Arvicola terrestris)*;* **8** *Meadow vole* (Microtus pennsylvanicus), *drumming an alarm signal with its hind foot.*

been known to kill nestlings of other females.

The young are born small, blind, and helpless; they usually weigh 1–10g (0.04–0.35oz), and the whole litter amounts to 22–28 percent of the adult female's weight. They are born in a nest, often below ground, which is made of dried grass, lichen, or other vegetation. Litter sizes and the number of litters per year vary enormously between species, in part reflecting the length of the breeding season and the availability of resources. The mean litter size is two for the Red-tree vole but over eight in the larger Taiga vole, though any given litter may vary from one to over 12 in the more fecund species. At least two litters – but often four or more – are produced per year.

Parental care is the work of the females, who are strongly protective of their young. If any wander from the nest, they are located by their squeaks and will be retrieved by the female. Males do not appear to participate in parental care, except in the monogamous Prairie vole and Pine vole, whose males retrieve and groom the young, and help keep runways and burrows clean.

While the young are living in the nest, they learn to recognize the scent and behavioral cues of their species. So, if young Brown lemmings are fostered (artificially) with Collared lemmings, they will be more aggressive towards their own species as adults than to their fostered species. The length of lactation varies but usually lasts about 3 weeks, and in some species is terminated by the female abandoning the nest.

The two sexes and different species have different social behaviors, and these may change radically between seasons. The key to understanding both the social relationships and the spatial organization of voles and lemmings is the spacing behavior of males and females. In the breeding season males and females form territories delimited by scent marks. Males compete for access to receptive females, while females compete for space that contains high-quality food and serves to protect their litters from infanticide. Male and female territories may be separate (as in the Montane vole and the Collared lemming) or else overlap (as in the European water and Meadow voles); alternatively, several females may live in overlapping ranges within a single male territory (for example, in Taiga and Field voles). Males form hierarchies of dominance – subordinates may be excluded from breeding – and may act to exclude strange males from an area. The Common, Sagebrush, and Brandt's voles live in colonies. The animals build complicated burrows.

Although in most species males are promiscuous, a few are monogamous. In the Prairie vole the males and females form pair bonds, while the Montane vole appears to be monogamous at low density as a result of the spacing of males and females, but with no pair bond. At high density this species is polygynous (males mate with several females). Monogamy is favored when both adults are needed to defend the breeding territory from intruders bent on infanticide.

The social system may vary seasonally. In the Taiga vole, the young animals disperse and their territories break down late in summer. Groups of five to ten unrelated animals then build a communal nest, which is occupied throughout the winter by both sexes. Communal nesting or local aggregations of individuals are also observed in the Meadow, Gray, and Northern red-backed voles. Huddling together reduces energy requirements.

Dispersal (the movement from place of birth to place of breeding) is an important aspect of microtine behavior, and has been the subject of considerable research. It plays an important role in regulating population size, and allows the animals

Vole and Lemming Tribes

Lemmings Tribe Lemmini

17 species in 4 genera. N America and Eurasia, inhabiting tundra, taiga, and spruce woods. Skull broad and massive, tail very short, hair long; 8 mammae.
Collared lemmings (*Dicrostonyx*, 11 species).
Brown lemmings (*Lemmus*, 3 species). Includes Norway lemming (*L. lemmus*).
Bog lemmings (*Synaptomys*, 2 species).
Wood lemming (1 species) *Myopus schisticolor*.

Mole voles Tribe Ellobiini

5 species in 1 genus. C Asia, inhabiting steppe. Form is modified for a subterranean life; coat color varies from ocher sand to browns and blacks; tail short; no ears; incisors protrude forward. Species include Northern mole vole (*Ellobius talpinus*), Southern mole vole (*E. fuscocapillus*).

Voles Tribe Microtini

121 species in 21 genera. N America, Europe, Asia, the Arctic.
Meadow voles (*Microtus*, 61 species). N America, Eurasia, N Africa. Coat and size highly variable, molars rootless, skull weak; 4–8 mammae; burrows on surface and underground. Includes Common vole (*M. arvalis*), Field vole (*M. agrestis*), American pine vole (*M. pinetorum*).
Mountain voles (*Alticola*, 12 species). C Asia. Includes Large-eared vole (*A. macrotis*).
South Asian voles (*Eothenomys*, 9 species) E Asia, includes Père David's vole (*E. melanogaster*).
Red-backed voles (*Clethrionomys*, 7 species). Japan, N Eurasia, N America, inhabiting forest, scrub, and tundra. Back usually red, cheek teeth rooted in adults, skull weak; 8 mammae. Includes Bank vole (*C. glareolus*).
Musser's voles (*Volemys*, 4 species). China.
Tree voles (*Arborimus*, 3 species). W USA.
Brandt's voles (*Lasiopodomys*, 3 species). C Asia.
Snow voles (*Chionmys*, 3 species). S Europe, SW Asia, Turkey.
Japanese voles (*Phaulomys*, 2 species). Japan.
Kashmir voles (*Hyperacrius*, 2 species) Kashmir and the Punjab. Includes True's vole (*H. fertilis*).
Water voles (*Arvicola*, 2 species). N America, N Eurasia. Includes European water vole (*A. terrestris*).
Heather voles (*Phenacomys*, 2 species) W USA and Canada. Includes Red-tree vole (*P. longicaudus*).
Afghan voles (*Blanfordimys*, 2 species). C Asia.
Martino's snow vole (1 species), *Dinaromys bogdanovi*. Yugoslavia.
Duke of Bedford's vole (1 species), *Proedromys bedfordi*. China.
Sagebrush vole (1 species), *Lemmiscus curtatus*. W USA.
Muskrat (1 species), *Ondatra zibethicus*. N America.
Round-tailed muskrat (1 species), *Neofiber alleni*. Florida.
Long-clawed mole vole (1 species), *Prometheomys schaposchnikowi*. Caucasus, Russia.
Steppe lemming (1 species), *Lagurus lagurus*. C Asia .
Yellow steppe lemmings (*Eolagurus*, 2 species). E C Asia.

For full species list see Appendix ▷

LEMMING MIGRATIONS

According to Scandinavian legend, every few years regimented masses of lemmings descend from birch woods and invade upland pastures, where they destroy crops, foul wells, and infect the air with their decomposing bodies. Driven by an irresistible compulsion they press on, not pausing at obstacles, until they end their suicidal march in the sea.

The truth is in fact very different. The life cycle of Norway lemmings is not exceptional. The animals are active throughout the year. During the winter, they tunnel and build nests under the snow, where they can breed in safety from predators except for an occasional attack by a weasel or ermine. With the coming of the spring thaw, the burrows risk collapse, so the animals are forced to move to higher ground, or lower to parts of the birch–willow forest, where they spend the summer months in the safety of cavities in the ground or in burrows dug through shallow layers of soil and vegetation. In the fall, with the freezing of the ground and withering of the sedges, there is a seasonal movement back to sheltered places in the alpine zone. Lemmings are particularly vulnerable at this time: should freezing rain and frost blanket the vegetation with ice before the establishment of snow cover, the difficulties in gathering food can prove fatal.

The mass migrations that have made the Norway lemming famous usually begin in the summer or fall following a period of rapid population growth. The migrations start as a gradual movement from densely-populated areas in mountain heaths down into the willow, birch, and conifer forests. At first the lemmings seem to wander at random, but as a migration continues, groups of animals may be forced to coalesce by local topography, for instance if caught in a funnel between two rivers. In such situations the continuous accumulation of animals becomes so great that a mass panic ensues and the animals take to reckless flight – upward, over rivers, lakes, and glaciers, and occasionally into the sea.

Although the causes of mass migrations are far from certain, it is widely thought that they are triggered by overcrowding. Females can breed in their third week and males as early as their first month; reproduction continues year-round. Litters consist of five to eight young, and may be produced every 3–4 weeks, so in a short period lemmings can produce several generations. Short winters without sudden thaws or freezes, followed by an early spring and late fall, provide favorable conditions for continuous breeding and a rapid increase in population density.

Lemmings are generally intolerant of one another and, apart from brief encounters for mating, lead solitary lives. It is possible that in peak years the number of aggressive interactions increases drastically, and that this triggers the migrations. Supporting evidence comes from reports that up to 80 percent of migrating lemmings are young animals (and thus are likely to have been defeated by larger individuals). Food shortages do not seem to be an important factor, since enough food seems to be available even in areas where the animals are most numerous.

The essential feature of long-distance migrations would therefore seem to be a desire for survival. Lemming species in Alaska and northern Canada also engage in similar, if less spectacular, expeditions. Although countless thousands of Norway lemmings may perish on their long journeys, the idea that such ventures always end in mass suicide is a myth. UWH

to exploit efficiently the highly seasonal, patchy nature of their habitat: strategies change according to population density. Norway lemmings move in summer onto wet, grassy meadows but in the fall move into deep mossy hillsides to overwinter.

Dispersers differ from nondispersing animals in several ways. There is some evidence that, in the Meadow and Prairie voles, animals that disperse are on average genetically different from permanent residents. Juvenile males disperse from their natal site to seek unrelated mating partners and to avoid inbreeding. Additionally, in Field, Prairie, Meadow, Tundra, and European water voles, many young pregnant females will disperse. This is quite unusual for mammals, but it does enable voles to rapidly colonize vacant habitats.

The numbers of juvenile males and females are equal, except in the Wood lemming, which has an unusual feature for mammals: some females are genetically programmed to have only daughters, which is advantageous in an increasing population. In most species, however, there are usually more female adults than males, probably as a consequence of dispersal by males, which are then more susceptible to predators. In the European water vole it is the females who usually disperse, and there is a slight excess of males. In Townsend's vole, the survival and growth of juveniles are higher when the density of adult females is low and adjacent females are relatives.

Territory size varies, but males usually have larger territories than females. In the Bank vole, home ranges were found in one study to be 0.7ha (1.7 acres) for females and 0.8ha (2 acres) for males. Territory size decreases as population density rises. In the Prairie vole, home range length drops from 25m (82ft) at low density to 10m (33ft) at high density.

Potential for Harm

CONSERVATION AND ENVIRONMENT

Many microtines live in areas of little or no agricultural value, or in areas little changed by human habitation. These species are neither persecuted nor endangered. However, species living in temperate areas can be agricultural pests. The American pine vole burrows in winter around the base of apple trees and chews on the roots or girdles the stems, resulting in a loss in apple production. The extensive underground burrows of Brandt's vole can become a danger to grazing stock.

In addition to their status as pests, many species harbor vectors of diseases such as plague (mole voles) and sylvatic plague (Sagebrush vole). In parts of eastern Europe and central Asia, *Microtus* species and water voles carry tularemia, an infectious disease that causes fever, chills, and inflammation of the lymph glands. The clearing of forest has increased both *Microtus* habitat and the incidence of tularemia. In the 1990s, hantavirus disease, which is carried by several species of voles and lemmings, emerged as an important viral ailment that can be fatal in more than 50 percent of cases. As a result, considerable research effort is now being directed to the epidemiology of the rodent-borne hemorrhagic fevers. CJK

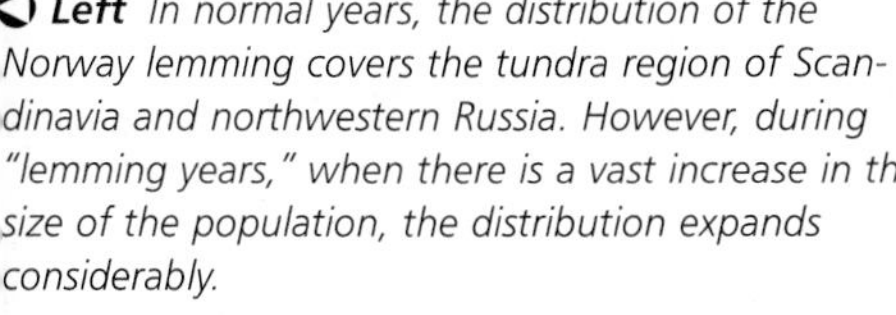

Left *In normal years, the distribution of the Norway lemming covers the tundra region of Scandinavia and northwestern Russia. However, during "lemming years," when there is a vast increase in the size of the population, the distribution expands considerably.*

Below *Norway lemmings are individualistic, intolerant creatures. When their numbers rise so does aggressive behavior, which is well developed. In the conflict shown here, two males box* ***1****; engage in wrestling* ***2****; and adopt a threatening posture toward one another* ***3****.*

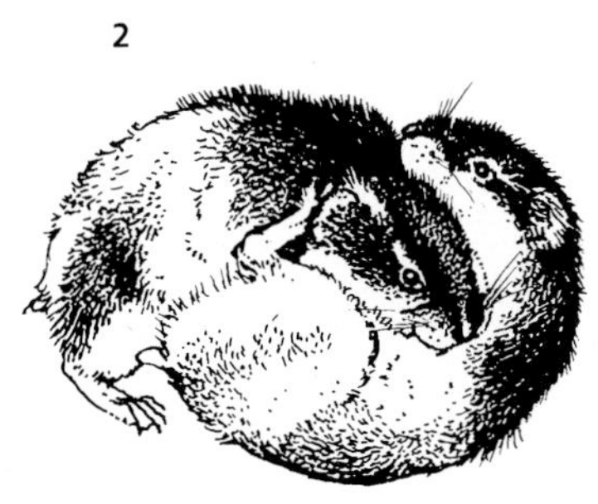

THE SCENT OF DEATH

The effect of weasel odor on vole reproduction

FOR MANY DESERT RODENTS, IT IS BETTER BY FAR to stay hungry but safe for a night or so than to risk death by seeking a meal outside of the burrow during the full moon. In open deserts and other arid areas, birds of prey are a serious threat. They have keen vision, which in many birds extends to seeing UV reflectance. Hunting raptors can even see the scent trails of voles reflecting UV light better than they can the average surroundings.

Things are different in holarctic and boreal areas with a dense cover, whether in forest, meadow, or agricultural areas. Preferred rodent habitats in these areas are characterized by an undercover of grasses, herbs, scrubs, ferns, and bushes. Protection provided by this low canopy of vegetation considerably reduces the risk of airborne predation, and the same applies where there is permanent snow cover that forms a canopy during the winter. However, under the snow's protective canopy, the rodents face the threat of Europe's smallest carnivore, the Least weasel (*Mustela nivalis*). Along with its slightly larger cousin, the stoat (*M. erminea*), it is the only predator of boreal areas that can follow prey into their own world: the burrows and crevices they occupy in old fields and forests, and the subnivean space under the snow.

Small mustelids are specialized vole hunters, and their population fluctuations closely follow those of the boreal voles on which they depend. The Least weasel is committed to a lifestyle based on eating rodents; some specialists reckon that weasel predation is among the forces causing the three-yearly round known as the Northern European vole cycle. During vole population peaks, the Least weasel and stoat can be responsible for 90 percent of total Field and Bank vole mortality.

Any strategy that helps the voles to avoid being killed will be favored in this harsh natural-selection regime. One way of assessing danger is to tap into the communicatory systems of the enemy themselves, and to use the information gained to predict areas of highest risk. Both weasels and stoats secrete a strong odor from their anal glands that is used in territory marking and sexual behavior. So alarming is the scent of weasels to voles that it has been used as a repellent to protect forestry and orchards from browsing. When the presence of scents indicates a lot of predator activity, voles stay in their burrows, becoming less active and avoiding going to distant foraging sites. If they do go out, they are more likely to forage in trees, keeping above the ground and out of the weasels' way.

These behaviors might well decrease the number and intensity of signs left to orient a weasel or stoat lurking somewhere inside the grass thicket or behind stones. Yet there is a price to pay for the enhanced security, in that the voles go hungry. This is especially a problem for females. Breeding is energetically costly, and, during pregnancy and lactation, reproductively active females need two or three times the amount of food required by nonbreeding females.

However, the most intricate tangle of adaptation and counteradaptation concerns the impact of weasel odors on the vole's sex life. Like weasels, voles also communicate by scent, so just as voles can spy on weasel activity by reading their scent messages, so too can weasels use vole scent marks to track them down. More chillingly still, weasels may even be able to distinguish the scents left by voles of different ages and reproductive condition to ensure that, when they do hunt, they are led to the best possible prizes. Female voles that have just given birth make for large, slow, and profitable prey, and the weasel may also benefit from an additional meal – the pups.

Thus if the risk of mortality from weasels is already high, and if it is particularly directed toward breeding females, what can the voles do? Laboratory experiments on the Bank vole and the Field vole have revealed that, in both species, the impact of mustelid odor on breeding was the same: in risky environments over 80 percent of females suppressed breeding, while in safe conditions the same percentage of control-group females bred successfully. Females exposed to mustelid treatment also lost weight, indicating decreased foraging success. Young females were more responsive to predator odor, while older females, closer to death anyway, were more likely to risk breeding.

Field experiments also showed that additional weasel smells suppressed Field vole activity, while removing predators from some areas caused an upturn in reproduction. Yet some parts of this puzzle are still missing. Further experiments are needed in the field to compare rich habitats with dense and thriving rodent populations with the pattern among more marginal, less successful populations. Such studies would clarify the picture by showing what happens when rodents occupying marginal habitats face a strong risk of predation in addition to hunger and cold. Nonetheless, the exciting finding that vole breeding is influenced by the scent of the animals' predators is in itself a clear illustration of the pervasive importance of odor in most mammalian lives. HY

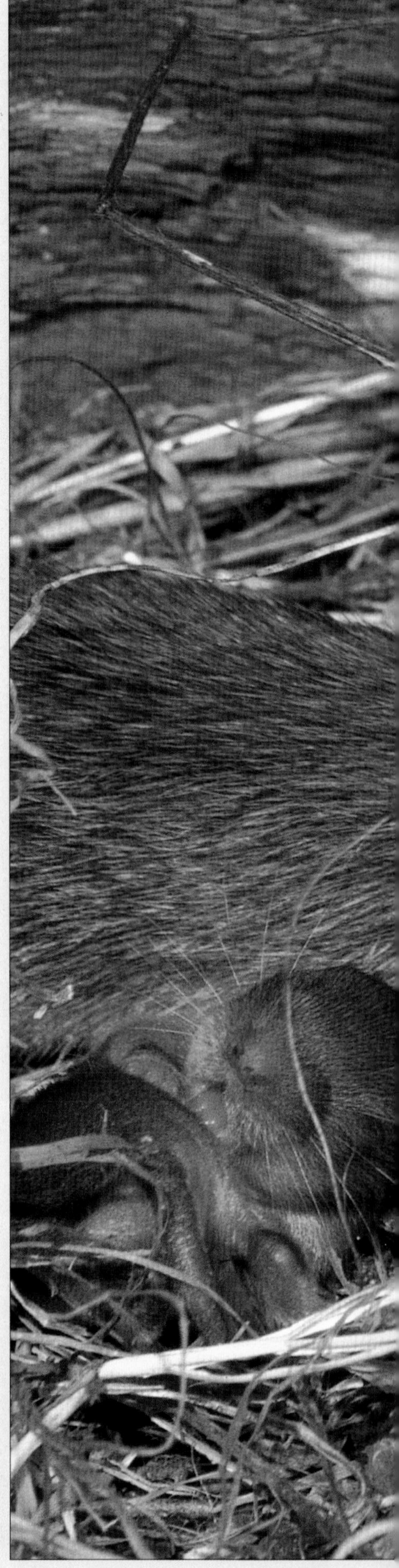

Right *A female Bank vole with young that are just over a week old. Female voles, especially younger ones, will suppress breeding if they detect a powerful presence of weasel odor.*

Old World Rats and Mice

THE OLD WORLD RATS AND MICE, OR MURINAE, include at least 542 species distributed over the major Old World land masses, from immediately south of the Arctic Circle to the tips of the southern continents. If exuberant radiation of species and the ability to survive, multiply, and adapt quickly are criteria of success, then the Old World rats and mice must be regarded as the most successful of all mammals.

The Murinae probably originated in Southeast Asia in the late Oligocene or early Miocene about 25–20 million years ago from a primitive (cricetine) stock. The earliest fossils (*Progonomys*), in a generally poor fossil representation, are known from the late Miocene of Spain, about 8–6 million years ago. Slightly younger fossils (*Leggadina*, *Pseudomys*, and *Zyzomys*) have been discovered recently from the early Pliocene of Australia (5 m.y.a.). Old World rats and mice are primarily a tropical group that have sent a few hardy migrant species into temperate Eurasia.

An Evolutionary Success Story

FORM AND FUNCTION

The murines' success lies in a combination of features probably inherited and adapted from a primitive, mouselike archimurine This is a hypothetical form, but many features of existing species point to such an ancestor, from which they are little modified. The archimurine would have been small, perhaps about 10cm (4in) long in head and body, with a scaly tail of similar length. The appendages would have been of moderate length, thereby facilitating the subsequent development of elongated hind legs in jumping forms and short, robust forelimbs in burrowers. It would have had a full complement of five fingers and five toes. The sensory structures (ears, eyes, whiskers, and olfactory organs) would have been well-developed. Its teeth would have consisted of continuously growing, self-sharpening incisors and three elaborately rasped molar teeth on each side of each upper and lower jaw, with powerful jaw muscles for chewing a wide range of foods and preparing material for nests. The archimurine would have had a short gestation period, would have produced several young per litter, and therefore

Right *Although the tiny Wood mouse* (Apodemus sylvaticus) *is often held responsible for the destruction of young plants, it also plays a role in the dispersal of seeds, which it buries underground. The Wood mouse does not move far from its burrow and may never travel more than 200m (660ft) from its home.*

would have multiplied quickly. With its small size, it could have occupied many different microhabitats. Evolution has produced a wide range of adaptations, but only a few, if highly significant, lines of structural change.

Modifications to the tail have produced organs with a wide range of different capabilities. It has become a long balancing organ, sometimes with a pencil of hairs at its tip (as in the Australian hopping mice) and sometimes without (as in the Wood mouse). In the Harvest mouse it has developed into a grasping organ to help in climbing. In some species including the Greater tree mouse it serves as a sensory organ, with numerous tactile hairs at the end furthest from the animal; in others like the Bushy-tailed cloud rat, it is now thickly furred. In some genera, for example rock rats and spiny mice, as in some lizards, the tail is readily broken, either in its entirety or in part, though unlike lizards' tails it does not regenerate. In species where the proximal part of the tail is dark and the distal part white (for example the Smooth-tailed giant rat), the tail may even serve as an organ of communication

Hands and feet show a similar range of adaptation. In climbing forms, big toes are often opposable, though sometimes relatively small, as in the Palm mouse. The hands and/or feet can be broadened to produce a firmer grip, for instance in the Pencil-tailed tree mouse or Peter's arboreal forest rat. In jumping forms, the hind legs and feet may be much elongated (as in Australian hopping mice), while in species living in wet, marshy conditions (for instance, African swamp rats), the hind feet can be long and slightly splayed, somewhat reminiscent of the webbed feet of ducks. This type of adaptation is at its most pronounced in the Australian and New Guinea water rats and the shrew rats of the Philippines, which possess broadly webbed hind feet.

The claws are also often modified. They may be short and recurved for attaching to bark and other rough surfaces, as in Peter's arboreal forest rat, or large and strong in burrowing forms like those of the Lesser bandicoot rat. In some of the species with a small, opposable digit, the claw of this digit becomes small, flattened, and nail-like (for instance in the Pencil-tailed tree mouse).

Fur is important for insulation. In some species such as spiny mice, some hairs of the back are modified into short, stiff spines, while in others the fur can be bristly (harsh-furred rats), shaggy (African marsh rat), or soft and woolly (African forest rat). The function of spines is not known, although it is speculated that they deter predators.

FACTFILE

OLD WORLD RATS AND MICE

Order: Rodentia

Family: Muridae

Subfamily: Murinae

542 species in 118 genera

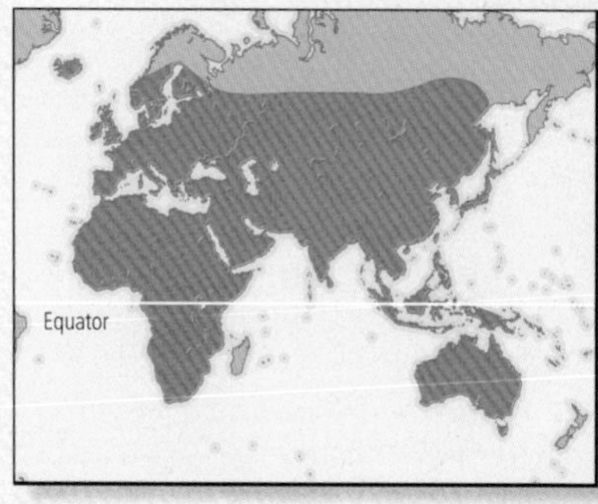

DISTRIBUTION Europe, Asia, Africa (excluding Madagascar), Australasia; also found on many offshore islands.

HABITAT Very varied; mostly terrestrial, but some are arboreal, fossorial, or semi-aquatic.

SIZE **Head–body length** ranges from 4.5–8.2cm (1.7–3.2in) in the Pygmy mouse to 48cm (19in) in Cuming's slender-tailed cloud rat; **tail length** from 2.8–6.5cm (1.1–2.5in) to 20–32cm (8–13in); **weight** from about 6g (0.2oz) to 1.5–2kg (3.3–4.4lb), both in the same two species.

COAT Typically medium to dark brown on the back and flanks, sometimes with a lighter-colored belly. Some species are striped for camouflage.

DIET Mostly plant material and invertebrates, though some take small vertebrates.

BREEDING In most small species, gestation lasts 20–30 days, though longer in species that give birth to precocious young (e.g. 36–40 days in spiny mice), and also in "old endemic" Australian rodents (e.g. 32–50 days in Australian hopping mice). Duration not known for large species.

LONGEVITY Small species live little over 1 year; larger species like the giant naked-tailed rats over 4 years.

CONSERVATION STATUS Although many Murinae species are flourishing, 15 are currently listed as Critically Endangered, including 2 species each of the genera *Zyzomys* (rock rats) and *Pseudomys* (Australian pseudo-mice); 36 are Endangered.

For full species list see Appendix ▷

Ears can range from the large, mobile, and prominent (as in the Rabbit-eared tree-rat) to the small and inconspicuous, well covered by surrounding hair (the African marsh rat). As its common name indicates, the Earless water rat of Papua New Guinea lacks any external ears, an adaptation that helps streamline its body for a life spent in water. In addition, this highly specialized species has a longitudinal fringe of long white hairs on its tail, remarkably similar to that found in the completely unrelated Elegant water shrew (*Nectogale elegans*) of the Himalayas. The tail fringe is thought to act as an effective rudder.

In teeth, there is considerable adaptation among murines in the row of molars. In what is presumed to be the primitive condition, there are three rows of three cusps on each upper molar tooth, but the number of cusps is often much smaller, particularly in the third molar, which is often small. The cusps may also coalesce to form transverse ridges. But the typical rounded cusps, although they wear with age, make excellent structures for chewing a wide variety of foods.

The molars of Australian water rats and their allies, however, show great simplification, lacking as they do the strong cusps or ridges found in most of the murine rodents. In some species the molars are also reduced in number, the extreme being seen in the One-toothed shrew-mouse which has only one small, simple molar in each row. This adaptation is most likely to be related to a diet of fruit or soft-bodied invertebrates.

The shrew rats of the Philippines have slender, protruding incisor teeth like delicate forceps, presumably adapted, as in the true shrews, for capturing insects and other invertebrates. However, the remaining teeth are small and flat-crowned, quite unlike the sharp-cusped batteries of the true shrews. The two species of the genus *Rhynchomys* offer pronounced examples of this adaptation.

These adaptations of teeth have, at the extremes, resulted in the development of robust

Right *The Norway rat – also called the Brown or Common rat – is practically omnivorous (even eating soap) and found mainly in urban areas, where it causes a good deal of damage. The species' particular preference is for animal matter; it has been known to attack poultry and even young lambs.*

a
b
c
d
1
2
3
4
5
6
e
f
g
h

Left *Old World rats and mice: **1** Smooth-tailed giant rat* (Mallomys rothschildi)*; **2** Pencil-tailed tree mouse* (Chiropodomys gliroides)*; **3** African marsh rat* (Dasymys incomtus)*; **4** a spiny mouse (genus* Acomys)*; **5** Natal multimammate rat* (Mastomys natalensis)*; **6** Fawn-colored hopping mouse* (Notomys cervinus)*; **7** a Vlei rat* (Otomys irroratus) *sitting in a grass runway; **8** Brush-furred rat* (Lophuromys sikapusi) *eating an insect; **9** Four-striped grass mouse* (Rhabdomys pumilio)*; **10** Australian water rat* (Hydromys chrysogaster) *diving; **a** tail of a field mouse (genus* Apodemus)*; **b** tail of the Harvest mouse* (Micromys minutus)*; **c** tail of the Greater tree mouse* (Chiruromys forbesi)*; **d** tail of the Bushy-tailed cloud rat* (Crateromys schadenbergi)*; **e** hind foot of the Asiatic long-tailed climbing mouse* (Vandeleuria oleracea)*; first and fifth digits opposable to provide grip for living in trees; **f** hind foot of the Shining thicket rat* (Grammomys rutilans)*; has broad, short digits for providing grip; **g** paw of the Lesser bandicoot rat* (Bandicota bengalensis) *showing long, stout claws; **h** hind foot of an African swamp rat (genus* Malacomys) *showing long, splayed foot with digits adapted for walking in swampy terrain.*

and relatively large teeth (in Rufous-nosed and Nile rats) and in the reduction of the whole tooth row to a relatively small size, as in the Lesser small-toothed rat of New Guinea. The food of this rat probably requires little chewing, possibly consisting of soft fruit or small insects.

Teeming and Ubiquitous

DISTRIBUTION PATTERNS

Murines are found throughout the Old World. There are considerable variations in the numbers of species in different parts of their range, though in examining their natural distribution the House mouse, Roof rat, Norway rat, and Polynesian rat must be discounted, as they have been inadvertently introduced in many parts of the world.

The north temperate region is poor in species. In Europe, countries such as Norway, Great Britain, and Poland have respectively as few as 2, 3, and 4 species each. In Africa, the density of species is low from the north across the Sahara until the savanna is reached, where the richness of species is considerable. Highest densities occur in the tropical rain forest and adjacent regions of the Congo basin. This fact can be shown by reference to selected sites. The desert around Khartoum, the arid savanna at Bandia, Senegal, the moist savanna in Ruwenzori Park, Uganda, and the rain forests of Makokou, Gabon, support 0, 6, 9, and 13 species respectively. The Democratic Republic of Congo boasts 45 species, and Uganda 37.

Moving to the Orient, species are most numerous south of the Himalayas. India and Sri Lanka have about 35 species, Malaysia 22. In the East Indies some islands are remarkably rich: there are about 60 species in New Guinea, 38 in Sulawesi (Celebes), and 35 in the Philippines. Within the Philippines there has been a considerable development of native species, with 10 of the 12 genera and a total of 30 species found only there (in other words, they are endemic); only two species of *Rattus* are found elsewhere. A notable feature is the presence of 10 large species having head–body lengths of about 20cm (8in) or more.

Left *Cane rats are mostly nocturnal creatures but sometimes engage in diurnal activity. As the name suggests, they often inhabit sugar-cane plantations, in which they are capable of wreaking so much damage that large, rodent-eating snakes are occasionally protected, as the lesser of two evils.*

Below *The Plains, or Desert mouse, of Australia inhabits quite shallow burrows. Like many other species native to Australia, its prevalence has been affected by the influx of predators and other animals competing for the same resources that resulted from European colonization and subsequent expansion.*

Right *A Golden-backed tree rat, found in northern parts of Western Australia and the far north of Northern Territory. It is reputed to have a fierce temper and the more annoyed it becomes the louder the noises it makes are. It is known to nest in the Pandanus tree, from which matting is made.*

The largest known murine, Cuming's slender-tailed cloud rat, is found in the Philippines; it grows to over 40cm (16in) in head–body length, and the Pallid slender-tailed rat and the Bushy-tailed cloud rat are only slightly shorter. This high degree of endemism and the tendency to evolve large species is also found in the other island groups. In New Guinea there are 8 species with head–body lengths of more than 30cm (12in), including the Smooth-tailed giant rat, the Rough-tailed eastern white-eared giant rat, the giant naked-tailed rats and the Rock-dwelling giant rat. The most gigantic of all New Guinea's rodents, the Sub-alpine woolly rat, which exceeds 40cm in length and weighs up to 2kg (4.4lb), was discovered only in 1989. There are only two small species, the New Guinea jumping mouse, about the size of the House mouse, and the diminutive Delicate mouse, about half its size again. In Australia there are around 72 species, with the eastern half of the continent having a far greater diversity of species than elsewhere.

Niche Specialists

ECOLOGICAL ADAPTATION

It has proved difficult to give an adequate and comprehensive explanation of the evolution and species richness of the murines. There are some pointers to the course evolution may have followed, based on structural affinities and ecological considerations. The murines are a structurally similar group, and many of their minor modifications are clearly adaptive, so there are few characters that can be used to distinguish between, in terms of evolution, primitive and advanced conditions. In fact only the row of molar teeth has been used in this way: primitive dentition can be recognized in the presence of a large number of well-formed cusps. Divergences from this condition may represent specialization or advancement, while ecological considerations account for a species' abundance and for the types of habitat preference it may show.

From this analysis, two groups of genera have been recognized. The first contains the dominant genera (African soft-furred rats, Oriental spiny rats, Old World rats, giant naked-tailed rats, the *Mus* species, African grass rats and African marsh rats) which have been particularly successful, living in dense populations in the best habitats. These are believed to have evolved slowly, because they display relatively few changes from the primitive dental condition. The second group contains many of the remaining genera which are less successful, living in marginal habitats and often showing a combination of aberrant, primitive, and specialized dental features.

The dominant genera (with the exception of the African marsh rat) contain more species than the peripheral genera and are constantly attempting to extend their range. Considerable numbers of new species have apparently arisen within what is now the range center of a dominant genus (for example, soft-furred rats in central Africa and Old World rats in Southeast Asia). The reasons for this await explanation.

It is quite common for two or more species of murine to occur in the same habitat, particularly in the tropics. One of the more interesting and

important aspects of studies of the animals is to explain the ecological roles assumed by each species in a particular habitat, and then to deduce the patterns of niche occupation and the limits of ecological adaptations by animals with a remarkably uniform basic structure. A particularly favorable habitat, and one amenable to this type of study, is regenerating tropical forest.

In Mayanja Forest, Uganda, 13 species have been found in a small area of about 4sq km (1.5sq mi). Certain species – the Rusty-bellied rat and the Punctated grass mouse – are of savanna origin and are restricted to grassy rides. Of the remaining 11 species, all have forest and scrub as their typical habitat with the exception of the two smallest species, the Pygmy mouse and the Larger pygmy mouse, which are also found in grasslands and cultivated areas. Three species, the Tree rat, the Climbing wood mouse, and Peter's arboreal forest rat, seldom, if ever, come to the ground. The small Climbing wood mouse – often found within the first 60cm (24in) off the ground – prefers a bushy type of habitat. The two other arboreal species are strong branch-runners and are able to exploit the upper and lower levels of trees and bushes. All three species are found alongside a variety of plant species (in the case of the Wood mouse, 37 were captured beside 19 different plants, with *Solanum* among the most favored). All species are herbivorous and nocturnal, and the two larger species construct elaborately woven nests of vegetation.

Two species are found on both the ground and in the vegetation up to 2m (6.5ft) above ground level. Of these, the African forest rat is abundant and the Rufous-nosed rat is much less common. The African forest rat lives and builds its nests in burrows whose entrances are often situated at the bases of trees; it is nocturnal, feeding on a wide range of both insect and plant foods. The Rufous-nosed rat is both nocturnal and diurnal, and constructs nests with downwardly-projecting entrances in the shrub layer, made out of grass, on which this species is known to feed.

Of the 11 forest species, Peter's striped mouse, the Speckled harsh-furred rat, the Long-footed rat, the Pygmy mouse, the Larger pygmy mouse, and Hind's bush rat are all ground-dwellers. Of these, the striped mouse is a vegetarian, preferring the moister parts of the forest. The harsh-furred rat is an abundant species, predominantly predatory, favoring insects but also prepared to eat other types of flesh. The Long-footed rat is found in the vicinity of streams and swamps; it is nocturnal, and includes in its diet insects, slugs, and even toads (a specimen in the laboratory constantly attempted to immerse itself in a bowl of water). The two small mice are omnivores, while Hind's bush rat is a vegetarian species that inhabits scrub.

A further important feature, which could well account for the dietary differences in these species, is their respective sizes. The three mice are in the 5–25g (0.2–0.9oz) range, with the

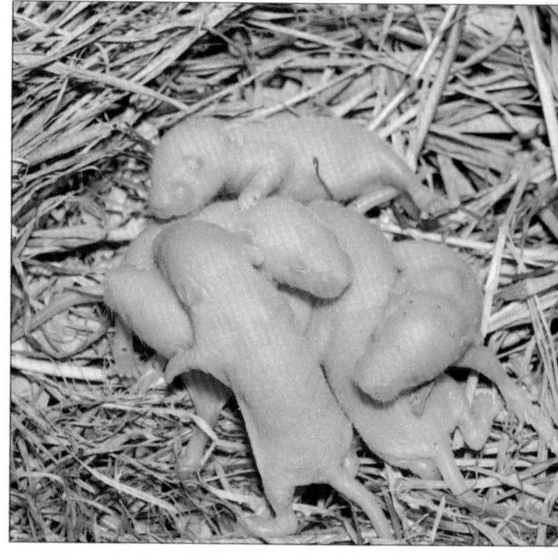

Above *The blind, hairless litter of the House mouse. This species is found almost worldwide, except in locations where it is excluded by either climate or competition from other small mammals.*

Right *A female Brants' whistling rat* (Parotomys brantsii) *with her young clinging to one of her nipples. Rodents of this species seldom venture far from their burrows, to which they return at the slightest provocation after emitting a loud alarm whistle.*

Below *A male Harvest mouse with a non-receptive female. Harvest mice construct nests for the young, usually about 1m (3.3ft) above ground. Litter size is about 4–7; sexual maturity is reached at five weeks.*

Pygmy mouse rather smaller than the other two. The Rufous-nosed rat is in the 70–90g (2.5–3.2oz) range, and the Long-footed rat and Hind's bush rat above this. The remaining species – the Tree rat, Peter's arboreal forest rat, the African forest rat, the Speckled harsh-furred rat, and Peter's striped mouse – have weights between 35g and 60g (1.2–2.1oz).

Within the tropical forests there is a high precipitation, with rain falling in all months of the year. This results in continuous flowering, fruiting, and herbaceous growth, which is reflected in the breeding activity of the rats and mice found there. In Mayanja Forest, the African forest rat and the Speckled harsh-furred rat were the only species obtained in sufficient numbers to permit the monthly examination of reproductive activity. The African forest rat bred throughout the year, while in the Speckled harsh-furred rat the highest frequency of conception coincided with the wetter periods of the year, from March to May and October to December.

Of Mice and Men

CONSERVATION AND ENVIRONMENT

Some Old World rats and mice have a close, detrimental association with humans through consuming or spoiling their food and crops, damaging their property, and carrying disease.

The most important species commensal with humans are the Norway or Brown or Common rat, the Roof rat, and the House mouse; now found worldwide, they originated from around the Caspian Sea, India, and Turkestan respectively. In addition to these cosmopolitan commensals, there are the more localized Multimammate rat in Africa, the Polynesian rat in Asia, and the Lesser bandicoot rat in India.

While the Roof rat and the House mouse have been extending their ranges for centuries, the Norway rat's progress has been much slower; it was unknown in the West before the 11th century, though now it is established in urban and rural situations in temperate regions, and is the rodent most commonly found in sewers; in the tropics, it is mainly found in large cities and ports. The Roof rat is more successful in the tropics, where towns and villages are often infested, though it cannot compete with the indigenous species in the field. In many Pacific, Atlantic, and Caribbean islands, Roof rats are common in agricultural and natural habitats in the absence of competitors.

With even the solitary House mouse capable of causing considerable damage, the scale of mass outbreak damage is difficult to envisage. An Australian farmer recorded 28,000 dead mice on his veranda after one night's poisoning, and 70,000 were killed in a wheat yard in an afternoon.

There are many rodent-borne diseases, transmitted either directly or through an intermediate host. The Roof rat, along with other species, hosts the plague bacterium, which is transmitted through the flea *Xenopsylla cheopis*. The lassa fever virus of West Africa is transmitted through urine and feces of the Multimammate rat. Other diseases in which murines are involved include murine typhus, rat-bite fever, and leptospirosis.

In the past, some Old World rats and mice were persecuted for reasons other than pest control. Notably, the sheer size of the beaver rat of Australia – which weighs 650–1,250g (23–44oz) – once told against it, when it was hunted for its luxurious pelt from the 1930s onward. As a result of hunting restrictions, this species is now on the increase. Most murine species now regarded as threatened are in danger from habitat destruction.

The Old World rats and mice are a remarkably rich and adaptable group of mammals. In spite of their abundance and ubiquity in the Old World, particularly the tropics, the murines remain a poorly studied group. Exceptions include a few species of economic importance and the Wood mouse. Many species are known only from small numbers in museums, supported by the briefest information on their biology. The Bisa rat, for example, is known only from a single skull retrieved from Bisa Island in 1990. There are undoubtedly endless opportunities for future research on this fascinating and accessible group of mammals. GS/CRD/MJD

Old World Rat and Mouse Species

Species include: African forest rat (*Praomys jacksoni*), **African grass rats** (genus *Arvicanthis*), **African marsh rat** (*Dasymys incomtus*), **African soft-furred rats** (genus *Praomys*), **African swamp rats** (genus *Malacomys*), **Asiatic long-tailed climbing mouse** (*Vandeleuria oleracea*), **Australian hopping mice** (genus *Notomys*), **Australian pseudo-mice** (genus *Pseudomys*), **beaver rats** (genus *Hydromys*), **Bisa rat** (as yet undescribed), **Bushy-tailed cloud rat** (*Crateromys schadenbergi*), **Cuming's slender-tailed cloud rat** (*Phloeomys cumingi*), **Delicate mouse** (*Pseudomys delicatulus*), **giant naked-tailed rats** (genus *Uromys*), **Greater tree mouse** (*Chiruromys forbesi*), **Harvest mouse** (*Micromys minutus*), **Hind's bush rat** (*Aethomys hindei*), **House mouse** (*Mus musculus*), **Larger pygmy mouse** (*Mus triton*), **Lesser bandicoot rat** (*Bandicota bengalensis*), **Lesser ranee mouse** (*Haeromys pusillus*), **Lesser small-toothed rat** (*Macruromys elegans*), **Long-footed rat** (*Malacomys longipes*), **Multimammate mouse** (*Mastomys natalensis*), **New Guinea jumping mouse** (*Lorentzimys nouhuysi*), **Nile rat** (*Arvicanthis niloticus*), **Norway, Brown,** or **Common rat** (*Rattus norvegicus*), **Old World rats** (genus *Rattus*), **Oriental spiny rats** (genus *Maxomys*), **Pallid slender-tailed rat** (*Phloeomys pallidus*), **Pencil-tailed tree mouse** (*Chiropodomys gliroides*), **Peter's striped mouse** (*Hybomys univittatus*), **Polynesian rat** (*Rattus exulans*), **Punctated grass mouse** (*Lemniscomys striatus*), **Pygmy mouse** (*Mus minutoides*), **Rabbit-eared tree rat** (*Conilurus penicillatus*), **Rock-dwelling giant rat** (*Xenuromys barbatus*), **rock rats** (genus *Zyzomys*), **Roof rat** (*Rattus rattus*), **Rough-tailed eastern white-eared giant rat** (*Hyomys goliath*), **Rufous-nosed rat** (*Oenomys hypoxanthus*), **Rusty-bellied brush-furred rat** (*Lophuromys sikapusi*), **shrew rats** (genus *Rhynchomys*), **Smooth-tailed giant rat** (*Mallomys rothschildi*), **Yellow-spotted brush-furred rat** (*Lophuromys flavopunctatus*), **spiny mice** (genus *Acomys*), **Subalpine woolly rat** (*Mallomys istapantap*), **thicket rats** (genus *Thamnomys*), **vlei rats** (genus *Otomys*), **whistling rats** (genus *Parotomys*), **Wood mouse** (*Apodemus sylvaticus*).

For full species list see Appendix ▷

A Scent-Based Information Superhighway

Communication patterns among House mice

In the darkness behind our kitchen cupboards or in the house footings, mice live, feeding on the superabundant food that is a by-product of our wasteful lives. As many as 50 may live together in family groups, with several adult females, their offspring, subdominant males, and a dominant male defending the territory. They may be quiet, but be sure that they are communicating, relaying complex and subtle messages of life, death, property rights, sex, and family matters through the medium of their own urine.

Urine – a substance we think of as disposable if not disgusting – is the essence of the mouse information superhighway. In addition to urea and other waste products, their urine contains a complex mixture of chemicals: small-molecular-weight volatile odorants and much higher-weight non-volatile proteins. Together these are the mouse equivalent of visiting cards, providing information on identity, species, sex, social and reproductive status, and state of health. Many genetic differences between individual mice contribute to each individual's unique scent signature, including the highly variable (polymorphic) genes of the major histocompatibility complex (MHC). Because scents used for individual recognition are inherited, mice can recognize relatives they have not met before from the similarity of their scents to that of known family members. The system is much more sophisticated than that used by humans to recognize their long-lost aunts or uncles.

Mice are also capable of producing their scent marks in slow-release capsules to ensure maximum effect. Mouse urine contains a high concentration of small (18–20kDa) lipocalin proteins, termed Major Urinary Proteins (MUPs), that are manufactured in the liver and filtered through the kidneys into the urine. Adult males excrete around 30mg of protein per ml (.004oz per fluidram) of urine per day, while adult females excrete about 40 percent as much. These urinary proteins bind the signalling chemicals inside a central cavity to cause their slow release from scent marks.

Each individual mouse within a group deposits urine in small streaks and spots as they move around their home area, particularly when they encounter an unmarked surface, so that all surfaces become covered with a thin smear. Communication "posts" of dried urine mixed with dust build up like small stalagmites in frequently-used locations, around feeding sites, at entrances to nest areas, or along trails.

Since mice are always surrounded by a familiar mixture of urine marks, they rapidly detect any new objects in their environment – or precipitous edges in the dark – by the absence of the strong, accustomed smell. Scent marks deposited around the territory also help to maintain familiarity and recognition between group members. Intruders can be recognized immediately because their scent contrasts with the background odors, stimulating investigation and attack by resident mice, and especially by the dominant male. In addition, adult males excrete signaling volatiles that are highly attractive to females and induce caution or stimulate aggression from other males.

Dominant males advertise their territory ownership and competitive ability by scentmarking at a much higher rate than other mice. They deposit hundreds of urine marks per hour, compared to only tens of marks by females or subordinates, using hairs on the end of their prepuce as a "wick." Since only a male successfully dominating an area can ensure that it is suffused with his own odor, this is a reliable signal that a territory is owned – by the male. Dominant territory owners seek out and attack any other males that deposit competing scent marks in their territory, and immediately countermark a competitor's scent by depositing their own urine nearby, ensuring that their own scent is always the freshest. Other

Urine donor			
	Unfamiliar adult male	**Non-breeding grouped female**	**Pregnant or lactating female**
Juvenile female	Accelerates puberty	Delays puberty	Accelerates puberty
Adult female	Induces estrus and shortens cycle	Prolongs anestrus or induces pseudopregnancy	Prolongs estrus
Pregnant female	Terminates early pregnancy and induces estrus		

Right As the table shows, the scent of urine has different priming effects on female House mice depending both on the reproductive state of the female and the identity of the donor. The strength of the inhibition caused by urine from non-breeding females increases with the size of the group and length of exposure.

males will usually flee if they meet the owner of the scent marks within his scentmarked area, or will avoid entering it, considerably reducing the need for aggressive defence by the owner. However, if a male is not defending his territory very successfully and other males are able to deposit competing scents, the owner may be challenged.

Females take advantage of these urinary competitions to select the best father for their young. While they will nest in one male's territory, they may well wander off to mate with other males, particularly if their patch is exclusively and freshly marked. They also prefer to mate with animals whose scents are dissimilar to their parents', thus avoiding inbreeding problems. If there is no choice, however, they will mate anyway. Males do not distinguish their own young from those fathered by another male.

If there is no territory available, some males will live within those of dominant males, spreading urine with low concentrations of signaling chemicals in fewer deposits. This allows the dominant male to recognize that they are not a threat, but it also means they are not attractive to females.

Mouse urine can also alter the reproductive state of females according to their opportunities, as spelled out in the table above. A young female exposed to the scent of a new male will come into puberty, and therefore become ready to breed, up to 6 days earlier than normal for first estrus, which typically falls at 36–40 days of age (the familiar odor of a father has no such effects). The urine of an unfamiliar male can also prevent a pregnancy from proceeding in a newly-conceived female who encounters the scent before the embryo has implanted into the wall of her uterus, so that the new male may stand a chance of becoming the father of the next litter. Unfamiliar male urine also induces estrus and shortens the cycle of adult females, thereby synchronizing their estruses.

As female mice live in communes sharing the chores of motherhood, it is unsurprising that a final, major role of mice scents is that of girl talk. Preferring to share their nest sites with other familiar females (usually close relatives), females will come into breeding condition earlier if they are receiving signs that other female kin are already breeding. Overcrowding, however, is a problem, and may inhibit further reproduction. If three or more females live together and are waiting for an opportunity to breed, they produce a scent that can inhibit puberty in other young females by more than 20 days, and that inhibits estrous cycling among other adult females. This behavior delays reproduction in overcrowded conditions. These ingenious responses allow females to reproduce rapidly under favorable conditions, but to delay their reproduction at times of high population density when their offspring would stand little chance of survival. JH

Below At least since the start of cereal-growing around 10,000 years ago, House mice have lived in wary co-existence with humans, using their sensitive noses to sniff out information about their environment and each other.

Other Old World Rats and Mice

1

THROUGHOUT THE OLD WORLD THERE ARE groups of rats and mice that cannot be included in the three major Muridae subfamilies. Here they are grouped in eight subfamilies, placed for convenience in a west–east geographical sequence. Some of these rodents are superficially similar to members of the major groups; for example, the zokors can be considered as specialized voles. Two groups, the blind mole rats and the bamboo rats, are more distinctive and are sometimes treated as separate families.

Blind Mole Rats

SUBFAMILY SPALACINAE

Of all the subterranean rodents, the blind mole rats show the most extreme adaptations to life underground. Their eyes are completely and permanently hidden under their skin, and there are no detectable external ears or tail. The incisors protrude so far that they are permanently outside the mouth and can be used for digging without the mouth having to be opened. A unique feature is the horizontal line of short, stiff, presumably touch-sensitive hairs on each side of the head. Most blind mole rats are about 13–25cm (5–9in) long, but in one species, the Giant mole rat of southern Russia, they can reach 35cm (14in).

Blind mole rats are found in dry but not desert habitats from the Balkans and southern Russia around the eastern Mediterranean to Libya. Apart from being entirely vegetarian they live very much like the true moles (which are predators belonging to the order Insectivora). Each animal makes its own system of tunnels, which may reach as much as 350m (1,150ft) in length, throwing up heaps of soil. They feed especially on fleshy roots, bulbs, and tubers, but also on whole plants. Although originally animals of the steppes, they have adapted well to cultivation and are a considerable pest in crops of roots, grain, and fruit.

Blind mole rats breed in spring. There are usually two or three in a litter, and they disperse away from the mother's tunnel system as soon as they are weaned, at about 3 weeks. There is sometimes a second litter later in the year.

As in many other burrowing mammals, the blind mole rats' limited movement has led to the evolution of many local forms, making the individual species very difficult to define but providing a bonanza for the study of genetics and the processes of evolution and species formation. The number of species that should be recognized, and how they should best be classified, are still very uncertain. Eight species are recognized here, but these have been reduced to three elsewhere.

African Pouched Rats

SUBFAMILY CRICETOMYINAE

The five species of African pouched rats resemble hamsters in having a storage pouch opening from the inside of each cheek. The two short-tailed pouched rats are hamster-like in general appearance also, but the other three species are ratlike, with a long tail and large ears.

The giant pouched rats are among the largest of the murid rodents, reaching 40cm (16in) in head–body length. They are common throughout Africa south of the Sahara, and in some areas are hunted for food. They feed on a wide variety of items, including insects and snails as well as seeds and fruit. In addition to serving to carry food to underground storage chambers, the cheek pouches can be inflated with air as a threat display. The gestation period is about 6 weeks, and litter size usually two or four.

The three large species are associated with peculiar, blind, wingless earwigs of the genus *Hemimerus*, which occur in their fur and in their nests, where they probably share the rats' food.

Left *Species of Other Old World rats and mice:* **1** *Brant's climbing mouse* (Dendromus mesomelas)*;* **2** *Savanna giant pouched rat* (Cricetomys gambianus) *with both pouches full of food;* **3** *East African root rat* (Tachyoryctes splendens) *burrowing with its incisors;* **4** *East Siberian zokor* (Myospalax aspalax)*, kicking back excavated soil with its hind feet;* **5** *Ehrenberg's mole rat* (Nannospalax ehrenbergi)*, showing the broad nose used for ramming soil and tactile hairs on the face – note the absence of external eyes;* **6** *Crested rat* (Lophiomys imhausi) *with mane erect, showing a glandular patch;* **7** Macrotarsomys ingens *(one of the Madagascan rats);* **8** *Spiny dormouse* (Platacanthomys lasiurus).

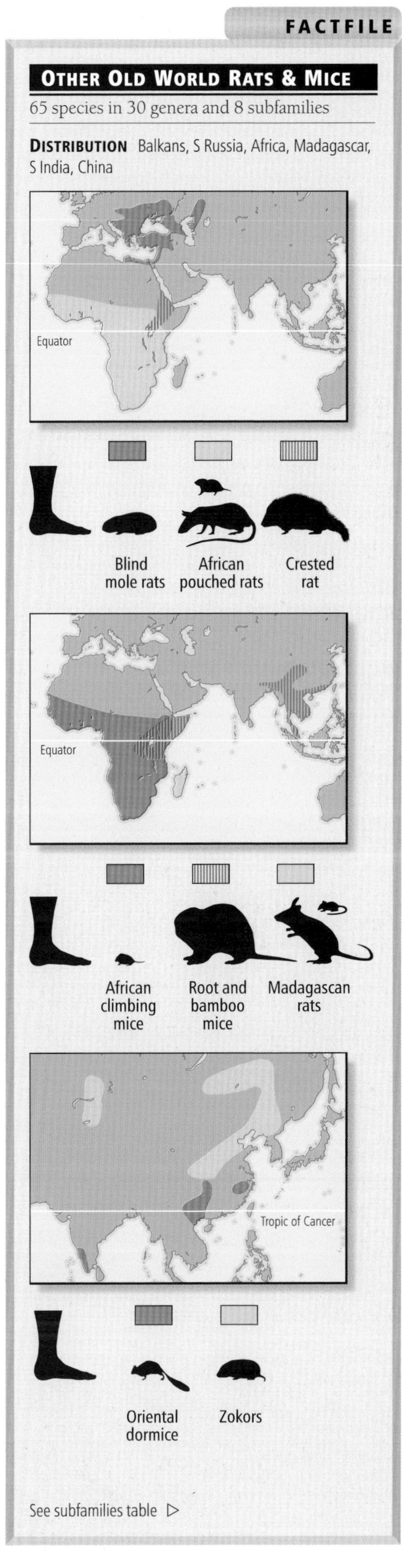

FACTFILE
OTHER OLD WORLD RATS & MICE
65 species in 30 genera and 8 subfamilies
DISTRIBUTION Balkans, S Russia, Africa, Madagascar, S India, China
Equator
Blind mole rats
African pouched rats
Crested rat
Equator
African climbing mice
Root and bamboo mice
Madagascan rats
Tropic of Cancer
Oriental dormice
Zokors
See subfamilies table ▷

Crested Rat

SUBFAMILY LOPHIOMYINAE

The Crested rat has so many peculiarities that it is placed in a subfamily of its own, and it is not at all clear what its nearest relatives are. It is a large, dumpy, shaggy rodent with a bushy tail and tracts of long hair along each side of the back that can be erected. These are associated with specialized scent glands in the skin, and the individual hairs of the crests have a unique, lattice-like structure that probably serves to hold and disseminate the scent. These hair tracts can be suddenly parted to expose the bold, striped pattern beneath, as well as the scent glands themselves.

The skull is also unique in possessing a peculiar granular texture and in having the cavities occupied by the principal, temporal chewing muscles roofed over by bone – a feature not found in any other rodent.

Crested rats are nocturnal and little is known of their way of life. They spend the day in burrows, rock crevices, or hollow trees. They are competent climbers and feed on a variety of vegetable materials. The stomach is unique among rodents in being divided into a number of complex chambers similar to those found in ruminant ungulates such as cattle and deer.

African Climbing Mice

SUBFAMILY DENDROMURINAE

The majority of African climbing mice are small, agile mice with long tails and slender feet, adapted to climbing among trees, shrubs, and long grass. Although they are confined to Africa south of the Sahara, some of them closely resemble mice in other regions that show similar adaptations, such as the Eurasian harvest mouse (subfamily Murinae) and the North American harvest mice (subfamily Sigmodontinae). They are separated from these mainly by a unique pattern of cusps on the molar teeth, and it is on the basis of this feature that some superficially very different rodents have been associated with them in the subfamily Dendromurinae.

Typical dendromurines, for instance those of the genus *Dendromus*, are nocturnal and feed on grass seeds, but are also considerable predators on small insects such as beetles and even on young birds and lizards. Some species in other genera are suspected of being more completely insectivorous. In the genus *Dendromus*, some species make compact, globular nests of grass above ground, for instance in bushes; others nest underground. Breeding is seasonal, with usually three to six naked, blind nestlings in a litter.

Of the other genera, the most unusual are the fat mice. They make extensive burrows and spend long periods underground in a state of torpor during the dry season, after developing thick deposits of fat. Even during their active season fat mice become torpid, with reduced body temperature during the day.

Many species of this subfamily are poorly known. Several distinctive new species have been discovered in the course of the past 40 years, and it is likely that others remain to be found, especially arboreal forest species.

The genera *Petromyscus* and *Delanymys* have sometimes been separated from the others in a subfamily of their own, Petromyscinae.

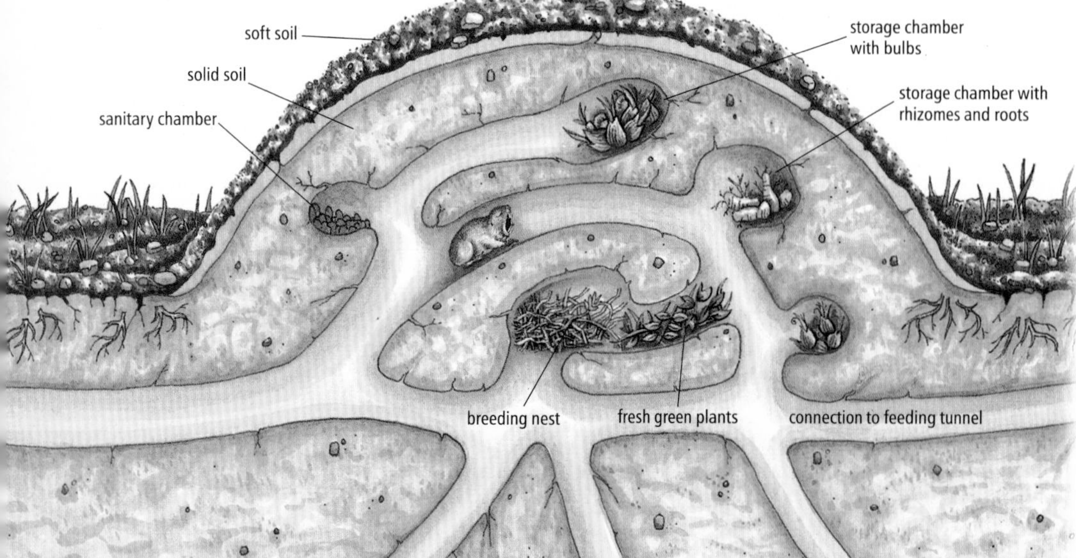

Above *Despite its common name, the Island mouse* (Nesomys rufus) *is ratlike in appearance. It is one of a group of 14 rodents found only on the island of Madagascar, where it lives in complex burrows often located under fallen treetrunks.*

Left *Breeding mound of a blind mole rat, Ehrenberg's mole rat* (Nannospalax ehrenbergi). *Each animal makes its own system of tunnels which may be as much as 350m (1,150ft) long, throwing up heaps of soil. The rats feed especially on fleshy roots, bulbs, and tubers, but also on whole plants, which they pull down into their tunnels by the roots. Their underground food storage chambers have been known to hold as much as 14kg (31lb) of assorted vegetables.*

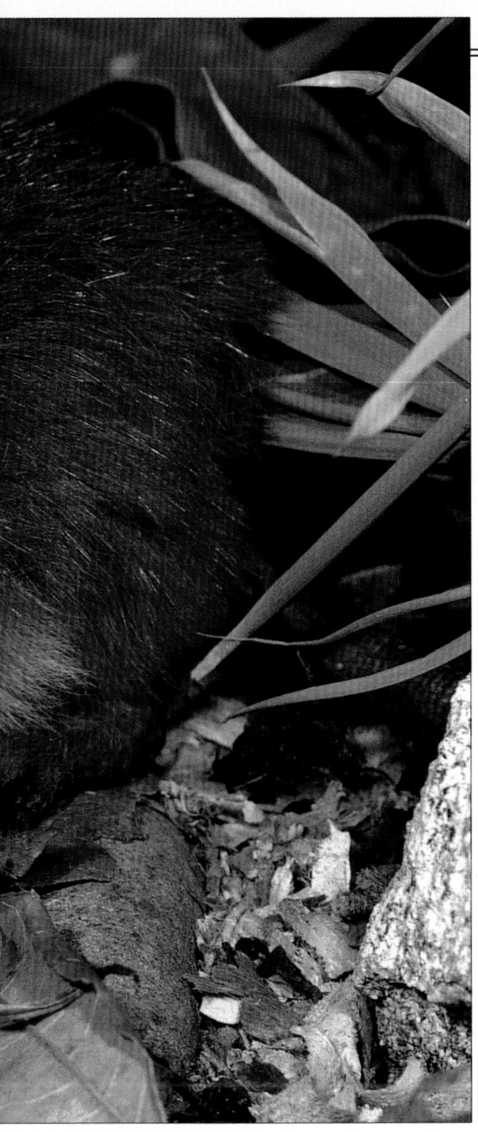

Root and Bamboo Rats

SUBFAMILY RHIZOMYINAE

Root rats are large rats adapted for burrowing, and they show many of the characteristics found in other burrowing rodents – short extremities, small eyes, large, protruding incisor teeth, and powerful neck muscles, reflected in a broad, angular skull. The bamboo rats, found in southeast Asian forests, show all these features in less extreme form than the East African root rats. They make extensive burrows in which they spend the day, emerging at night to do at least some feeding above ground. The principal diet consists of the roots of bamboos and other plants, but above-ground shoots are also eaten. In spite of their size, breeding is similar to the normal murid pattern.

The African root rats are more subterranean than the bamboo rats, but less so than African mole rats or the blind mole rats. They make prominent "mole hills" in open country. Roots and tubers are stored underground. As in most molelike animals, each individual occupies its own tunnel system. The gestation period is unusually long – between 6 and 7 weeks.

Madagascan Rats

SUBFAMILY NESOMYINAE

It has long been debated whether the 14 indigenous rodents of the island of Madagascar form a single interrelated group (implying they have evolved from a single colonizing species), or whether there have been multiple colonizations, meaning that some of the present species may be more closely related to mainland African rodents than to their fellows on Madagascar. The balance of evidence seems to favor the first hypothesis – hence their inclusion here in a single subfamily.

The South African white-tailed rat has also been included – the implication being that it is the sole survivor on the African mainland of the stock that colonized Madagascar – although it actually might be best placed in a separate subfamily, Mystromyinae. The problem arises from the diversity of the Madagascan species, coupled with the fact that none of them match very closely any of the non-Madagascan groups of murid rodents.

The group includes small, agile mice with long tails, long, slender hind feet, and large eyes and ears (e.g. *Macrotarsomys bastardi* and *Eliurus minor*). *Nesomys rufus* is typically ratlike in its proportions, while *Hypogeomys antimena* is rabbit-sized and makes deep burrows, although it forages for food on the surface. The two *Brachyuromys* species are remarkably vole-like in form, dentition, and ecology. They live in wet grassland or marshes, and are apparently adapted to feeding on grass. Externally, they can only be distinguished with difficulty from Eurasian water voles.

Oriental Dormice

SUBFAMILY PLATACANTHOMYINAE

The three species of Oriental dormice have been considered to be closely related to the true dormice (family Gliridae), which they resemble externally and in the similar pattern of transverse ridges on the molar teeth, although there are only three molars on each row, not preceded by a premolar as in the true dormice. More recently, opinion has swung toward treating them as aberrant members of the family Muridae (in its widest sense, as used here). Whatever their affinities, they are distinctive arboreal mice with no close relatives, and little is known of their way of life. The Spiny dormouse, a seedeater, is a pest of pepper crops in numerous parts of southern India.

Zokors

SUBFAMILY MYOSPALACINAE

Zokors are burrowing, vole-like rodents found in steppes and open woodlands in much of China. Although they live almost entirely underground, they are less extremely adapted than the blind mole rats. Both eyes and external ears are clearly visible, although tiny, and the tail is also distinct. Digging is done mainly with the very large claws of the front feet rather than the teeth.

Like the blind mole rats, zokors feed on roots, rhizomes, and bulbs, but they also occasionally collect food such as seeds from the surface. Massive underground stores of food are accumulated, enabling the animals to remain active all winter.

Breeding takes place in spring, when one litter of up to six young is produced. Their social organization is little known, but the young appear to stay with the mother for a considerable time. GBC

Other Old World Rat and Mouse Subfamilies

Blind mole rats Subfamily Spalacinae

8 species in 2 genera: greater blind mole rats (*Spalax*, 5 species); lesser blind mole rats (*Nannospalax*, 3 species). Balkans, S Russia, E Mediterranean, N Africa. 5 species are Vulnerable.

African pouched rats Subfamily Cricetomyinae

5 species in 3 genera: giant pouched rats (*Cricetomys*, 2 species); Lesser pouched rat (*Beamys hindei*); short-tailed pouched rats (*Saccostomus*, 2 species). Africa S of the Sahara. 1 species is Vulnerable.

Crested rat Subfamily Lophiomyinae

1 species, *Lophiomys imhausi*. Kenya, Somalia, Ethiopia, E Sudan, in mountain forests from 1,200m (4,000ft).

African climbing mice Subfamily Dendromurinae

21 species in 10 genera: climbing mice (*Dendromus*, 6 species); Nikolaus's mouse (*Megadendromus nikolausi*); Dollman's tree mouse (*Prionomys batesi*); Link rat (*Deomys ferrugineus*); Velvet climbing mouse (*Dendroprionomys rousseloti*); Groove-toothed forest mouse (*Leimacomys buettneri*); Gerbil mouse (*Malacothrix typica*); fat mice (*Steatomys*, 6 species); rock mice (*Petromyscus*, 2 species); Delany's swamp mouse (*Delanymys brooksi*). Subsaharan Africa. 2 species are Critically Endangered; 1 is Endangered, and 3, including Nikolaus's mouse, are Vulnerable.

Root and bamboo rats Subfamily Rhizomyinae

6 species in 3 genera: bamboo rats (*Rhizomys*, 3 species); Lesser bamboo rat (*Cannomys badius*); root rats (*Tachyoryctes*, 2 species). E Africa and SE Asia.

Madagascan rats Subfamily Nesomyinae

15 species in 8 genera: big-footed mice (*Macrotarsomys*, 2 species); Island mouse (*Nesomys rufus*); White-tailed rat (*Brachytarsomys albicauda*); tufted-tailed rats (*Eliurus*, 6 species); voalavoanala (*Gymnuromys roberti*); Malagasy giant rat (*Hypogeomys antimena*); short-tailed rats (*Brachyuromys*, 2 species); White-tailed mouse (*Mystromys albicaudatus*). Madagascar (1 species in S Africa). 2 species are Critically Endangered, 2 Endangered, and 2 Vulnerable.

Oriental dormice Subfamily Platacanthomyinae

3 species in 2 genera: Spiny dormouse (*Platacanthomys lasiurus*); Chinese pygmy dormice (*Typhlomys*, 2 species). S India (Spiny dormouse); S China and N Vietnam (Chinese dormice). 1 *Typhlomys* species is Critically Endangered.

Zokors Subfamily Myospalacinae

6 species in 1 genus (*Myospalax*). China and Altai Mountains. Underground. 1 species is Vulnerable.

For full species list see Appendix ▷

Hamsters

FAMILIAR IN THE WEST AS CHILDREN'S PETS, hamsters in their natural setting are solitary animals that react with aggression when they encounter other members of their own species. One breed that currently flourishes in captivity virtually all around the world was on the brink of extinction less than a century ago.

FACTFILE

HAMSTERS

Order: Rodentia

Family: Muridae

Subfamily: Cricetinae

24 species in 5 genera: **Mouse-like hamsters** (Genus *Calomyscus*, 5 species; Iran, Pakistan Afghanistan, S Russia,). **Rat-like hamsters** (Genus *Cricetulus*, 11 species; SE Europe, Asia Minor, N Asia). **Common hamster** or **Black-bellied hamster** (*Cricetus cricetus*; C Europe, Russia). **Golden hamsters** (Genus *Mesocricetus*, 4 species; E Europe, Middle East.). **Dwarf hamsters** (Genus *Phodopus*, 3 species; Siberia, Mongolia, N China).

DISTRIBUTION Europe, Middle East, Russia, China.

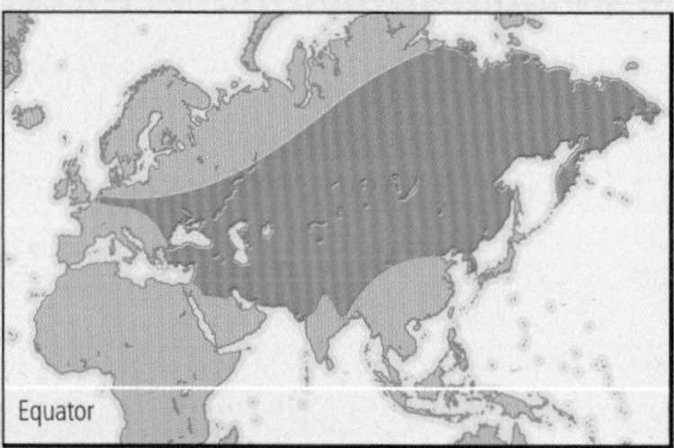

HABITAT Arid or semiarid areas, varying from rocky mountain slopes and steppe to cultivated fields.

SIZE Ranges from **head–body length** 5.3–10.2cm (2–4in) to 20–28cm (7.9–11in), **tail length** 0.7–1.1cm (0.3–0.4in), **weight** 50g (1.8oz) to 900g (32oz)

COAT Soft, thickly-furred in shades of gray and brown

DIET Mostly herbivorous – seeds, shoots, root vegetables – though occasionally hamsters will take insects, lizards, small mammals, and young birds.

BREEDING Gestation ranges from 15 days in the Golden hamster to 37 days in the White-tailed rat.

LONGEVITY 2–3 years

CONSERVATION STATUS *Calomyscus hotsoni* and *Mesocricetus auratus* are listed as Endangered; *Mesocricetus newtoni* is listed as Vulnerable; others not threatened.

For full species list see Appendix ▷

Until the 1930s the Golden hamster was known only from a single specimen found in 1839. However, in 1930 a female with 12 young was collected in Syria and taken to Israel. There the littermates bred, and some descendants were taken to England in 1931 and to the USA in 1938, where they proliferated. Today the Golden hamster is one of the most familiar pets and laboratory animals in the West. The other hamster species are less well-known, though the Common hamster has been familiar for many years.

Pouches for Foraging

FORM AND FUNCTION

Most hamsters have small, compact, rounded bodies with short legs, thick fur, large ears, prominent dark eyes, long whiskers, and sharp claws. Most have cheek pouches that consist of loose folds of skin, starting from between the prominent incisors and premolars and extending along the outside of the lower jaw. When hamsters forage they can push food into the pouches, which then expand, enabling them to carry large quantities of provisions to their underground storage chambers – a useful adaptation for animals that live in habitats where food may occur irregularly but in great abundance. The paws of the front legs are modified hands, giving the animals great dexterity when they manipulate food. Hamsters also use a characteristic forward squeezing movement of the paws as a means of emptying their cheek pouches of food. Common hamsters are reputed to inflate their cheek pouches with air when crossing streams, presumably to create extra buoyancy.

Hamsters are mainly herbivorous. The Common hamster may hunt insects, lizards, frogs, mice, young birds, and even snakes, but such prey contribute only a small amount to its diet. Normally hamsters eat seeds, shoots, and root vegetables, including wheat, barley, millet, soybeans, peas, potatoes, carrots, and beets, as well as leaves and flowers. Small items such as millet seeds are carried to the hamster's burrow in its pouches, larger items like potatoes in its incisors. Food is either stored for the winter, eaten on returning to underground quarters, or, in undisturbed conditions, eaten above ground. One Korean gray rat managed to carry 42 soybeans in its pouches. The record for storage in a burrow probably goes to the Common hamster: chambers of this species have been found to contain as much as 90kg (198lb) of plant material collected by a single hamster. Hamsters spend the winter in hibernation in their burrows, only waking on warmer days to eat food from their stores.

Baby Machines

SOCIAL BEHAVIOR

Though considered docile pets, hamsters in the wild are solitary and exceptionally aggressive toward members of their own species. These characteristics may result from intense competition for patchy but locally abundant food resources, but may also serve to disperse population throughout a particular area. Large species, such as the Korean gray rat, also behave aggressively toward other species, and have been known to attack dogs or even people when threatened. To defend itself from attack, the Korean gray rat may throw itself on its back and utter piercing screams.

Species studied in the laboratory have been shown to have acute hearing. They communicate with ultrasound as well as with squeaks audible to the human ear. Ultrasound appears to be most important between males and females during mating. The sense of smell is also acute; the Golden hamster can recognize individuals, probably from flank gland secretions, and males can detect stages of a female's estrous cycle and recognize a receptive female by odor.

Most hamsters become sexually mature soon after weaning (or even during it). Female Common hamsters become receptive to males at 43 days and can give birth at 59 days. Golden

Above *The Striped dwarf hamster* (Cricetulus barabensis) *has capacious cheek pouches, enabling it to transport large quantities of food quickly to its burrow. This species of ratlike hamster is reported to be very aggressive, especially toward conspecifics.*

Above *The Golden hamster* (Mesocricetus auratus) *has been familiar worldwide for several decades as a pet. However, in the wild it is found only in a small area of northwestern Syria and is classified as Endangered by the IUCN.*

hamsters have slightly slower development and become sexually mature between 56 and 70 days. In the wild they probably breed only once – occasionally twice – per year, during spring and summer, but in captivity they can breed year round.

Courtship is simple and brief, as befits animals that generally meet only to copulate. Odors and restrained movement indicate that the partners are ready and willing to mate. Immature animals or females not in heat will either attack or be attacked by other individuals. After copulation a pair separate and may never meet again. The female builds a nest for the young in her burrow from grass, wool, and feathers, and gives birth after 16–20 days (in the Common hamster). The young – born hairless and blind – are cared for by the female alone. During this time she may live off her food store in another section of the burrow. The young are weaned at about 3 weeks in the Golden hamster. In the slowest-developing species, the Mouse-like hamster, adult coloration and size may not be reached until 6 months old.

One species, *Cricetulus (Tscherskia) triton*, has been extensively studied in the North China Plain. It begins to breed in March and ends by August. While older females can manage 3 litters in a year, hamsters born that year produce only 1–2 litters. Litter size may be between 2 and 22, but the average is 9–10, after a gestation of about 20 days. It takes 2 months for a new-born female hamster to produce a first litter. The interval between two litters for an adult female is also about 2 months. The population oscillates greatly from year to year. Weasels and hawks are the major predators.

Male Dzhungarian hamsters are extraordinary fathers, and even act as midwife to their partner. They help pull the young from the birth canal, and then clean up the newborns, consuming the placenta and licking out the nostrils to allow the little ones to breath. The male then stays close to the mother and young to keep them warm. These behaviors may be mediated by hormonal changes, as just before a birth the male's levels of estrogen and cortisol – the female and stress hormones – rise, to be replaced by testosterone thereafter. Later on, the male babysits while the female goes out to feed.

Pets or Pests?

CONSERVATION AND ENVIRONMENT

Hamsters are considered serious pests to agriculture in some areas – in some countries dogs are trained to kill them. Chinese peasants sometimes catch large hamsters to feed to cats or other pets, and may dig the burrows in autumn to recover stored grain; in addition, the Common hamster is trapped for its skin. Despite these pressures, most hamster species are not endangered, perhaps because most live in inhospitable regions and have high reproduction rates. JF/ZZ

Gerbils

TO MOST PEOPLE, GERBILS ARE ATTRACTIVE pets with large, dark eyes, white bellies and feet, and furry tails. The animal they have in mind, however, is the Mongolian jird, just one of the many species of gerbils, jirds, and sand rats that together make up the world's largest group of rodents adapted to arid environments.

Gerbils are distinctive among rodents, mostly resembling the familiar Mongolian jird in overall appearance but varying in dimensions from the mouse-sized slender gerbils to robust-bodied jirds and sand rats. Within genera, however, the visible differences between species are subtle, often only expressed by small changes in fur and nail color, tail length, and the presence or absence of a tail tuft. Given such complexities, it is impossible to know for sure how many species exist; sometimes it is even hard to identify the genus to which a species belongs without using chromosomal, protein, and molecular comparisons.

Adaptations for Arid Climes

FORM AND FUNCTION

Most gerbils live in arid habitats and harsh climates, having adapted to both in interesting ways. To survive in such inhospitable conditions, an animal must not lose more water than it normally takes in. Water loss usually occurs by evaporation from the skin, in air exhaled from the lungs, and via urination and defecation. The gerbil's predicament is that it has a large body surface compared with its volume, and so has to find ways of minimizing water loss.

As a consequence, gerbils cannot afford to sweat, and indeed cannot survive temperatures of 45°C (113°F) or higher for more than about two hours. Most species are nocturnal; during the day they live underground, often with the burrow entrance blocked, at a depth of about 50cm (20in) where the temperature remains a constant 20–25°C (68–77°F). Only some northern species – for example, Great gerbils and Mongolian jirds – live on the surface in daytime, though some jirds that live farther south also emerge during the day in winter.

In the arid world of the gerbil, the only foods that are often available are dry seeds or leaves. The animal's nocturnal activity enables it to make the most of this poor sustenance. By the time it comes out of its burrow, such foods are permeated with dew, and it can improve the burrow's humidity, already high in relative terms, by taking them back there to eat. The gerbil's digestive system extracts water efficiently from the food, minimizing the water lost in feces, and the kidneys produce only a few drops of concentrated urine.

Other gerbil adaptations reduce the risk of capture by predators. Gerbils take on the color of the ground on which they live; this ability extends even to local populations of a single species living in different habitats, so animals found on dark lava soils are dark brown whereas conspecifics living on red sand are red. The effectiveness of the camouflage is only compromised by the tail, which ends in a tuft of a contrasting color. Even so, the tail is a vital survival aid; it helps with balance during movement; it can be used to twirl sand over a burrow entrance, effectively concealing it; and it may act as a decoy too, distracting predators from the animal's body and coming away either whole or in part if a predator happens to catch hold of it.

Another distinctive feature of gerbil anatomy is a particularly large middle ear, which is at its biggest in species living in open desert habitats; this enables the animals to hear low-frequency sounds such as the beating of an owl's wings. Gerbils also have large eyes, positioned high on the head so that they give the animals a wide field of vision.

The Three Gerbil Zones

DISTRIBUTION PATTERNS

The geographical range of gerbils can be divided into three major regions. The first includes not just the extensive savannas of Africa but also the Namib and Kalahari deserts, where the temperature rarely falls below freezing in winter. The second takes in the "hot" deserts and semidesert regions along the Tropic of Cancer in north Africa and southwest Asia, plus the arid Horn of Africa. The third covers the deserts, semideserts, and steppes of Central Asia, where winter temperatures fall well below freezing. The different gerbil genera fall broadly into groups linked to one or other of these regions. So, with the exception of the Indian gerbil, gerbils of the *Gerbillurus* and *Taterillus* groups occur in the first region, *Ammodillus*, *Gerbillus*, and *Pachyuromys* gerbils live in the second, while only species belonging to the *Rhombomys* group live in the third (though some of these also occur in the second region).

FACTFILE

GERBILS

Order: Rodentia

Family: Muridae

Subfamily: Gerbillinae

95 species in 14 genera

DISTRIBUTION African deserts and savannas; Asian deserts and steppes, from Turkey and Transcaucasia to NE China.

HABITAT Desert, savanna, steppe, rocks, cultivated land.

SIZE **Head–body length** ranges from 6.2–7.5cm (2.4–2.9in) in the Pygmy gerbil to 15–20cm (5.9–7.9in) in the Indian gerbil; **tail length** from 7.2–9.5cm (2.8–3.7in) to 16–22cm (6.2–8.7in) and **weight** from 8–11g (0.3–0.4oz) to 115–190g (4–6.7oz), both in the same two species.

COAT Mouselike in appearance, with soft pale yellow, light brown, or grayish fur. The underparts are customarily white or cream-colored.

DIET Primarily seeds, roots, and other plant matter, although some species also eat insects, snails, and (occasionally) small mammals and reptiles.

BREEDING Gestation 21–28 days

LONGEVITY Usually 1–2 years

CONSERVATION STATUS Of the 95 species, 13 in the genus *Gerbillus* and 1 in the genus *Meriones* (jirds) are listed as Critically Endangered. In addition, 4 *Meriones* species are Endangered, while 3 *Gerbillus* and the sole *Ammodillus* species, *Ammodillus imbellis*, are Vulnerable.

See genera table ▷

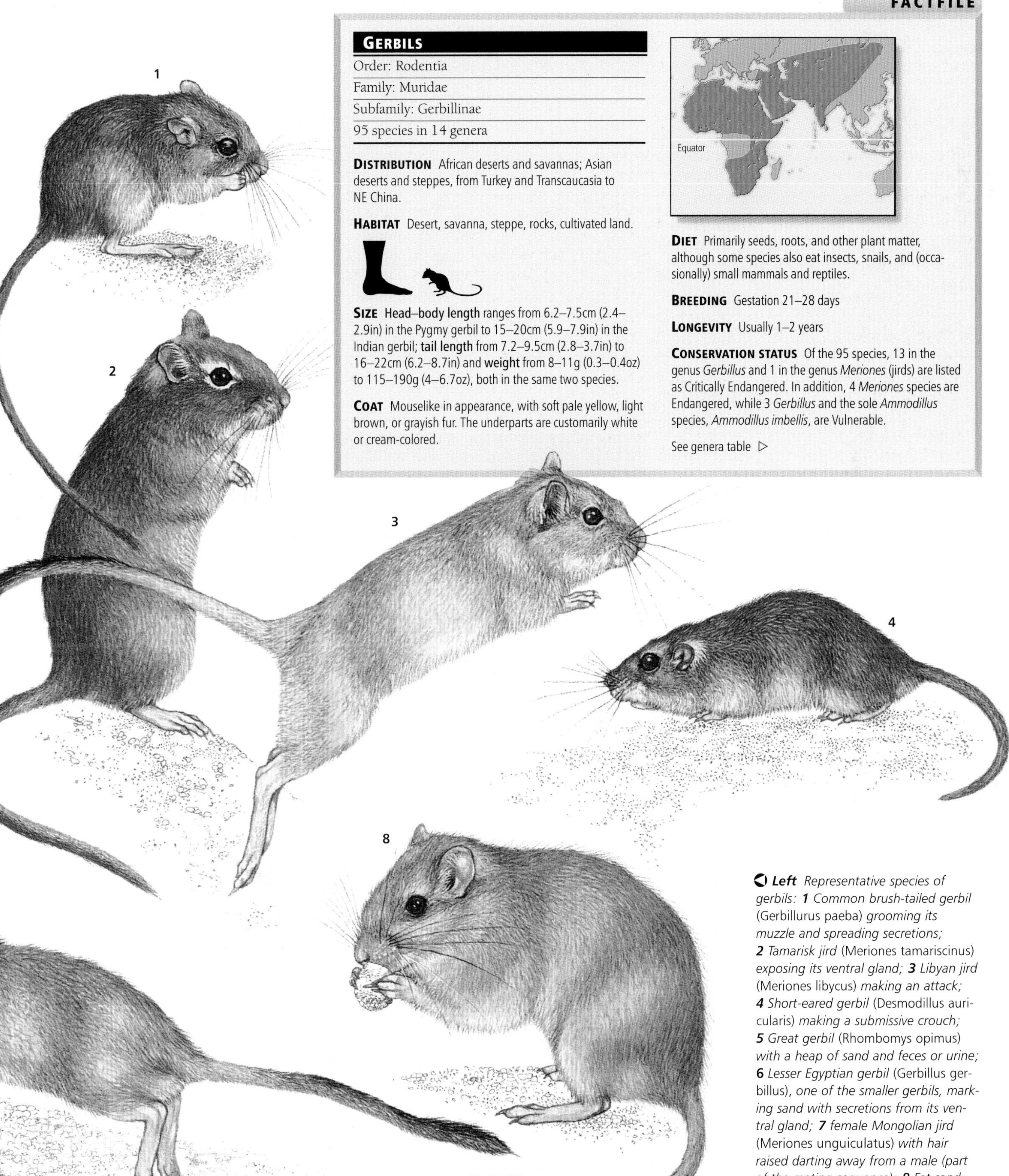

◐ ***Left** Representative species of gerbils: **1** Common brush-tailed gerbil* (Gerbillurus paeba) *grooming its muzzle and spreading secretions; **2** Tamarisk jird* (Meriones tamariscinus) *exposing its ventral gland; **3** Libyan jird* (Meriones libycus) *making an attack; **4** Short-eared gerbil* (Desmodillus auricularis) *making a submissive crouch; **5** Great gerbil* (Rhombomys opimus) *with a heap of sand and feces or urine; **6** Lesser Egyptian gerbil* (Gerbillus gerbillus), *one of the smaller gerbils, marking sand with secretions from its ventral gland; **7** female Mongolian jird* (Meriones unguiculatus) *with hair raised darting away from a male (part of the mating sequence); **8** Fat sand rat* (Psammomys obesus) *holding and sniffing a ball of sand and urine.*

Gerbil Groups

Ammodillus

1 genus: *Ammodillus*
Somali gerbil or **walo** (*Ammodillus imbellis*) Somalia, E Ethiopia, inhabiting savanna and desert.

Gerbillus

3 genera: *Desmodilliscus, Gerbillus, Microdillus*
Pouched pygmy gerbil (*Desmodilliscus braueri*) Senegal and Mauritania E to C Sudan, inhabiting savanna.
Northern pygmy gerbils (*Gerbillus*, 45 species) N Africa, Middle East, Iran, Afghanistan, to NW India, inhabiting desert, semidesert, and coastal plains. Species include Pygmy gerbil (*G. henleyi*), Lesser Egyptian gerbil (*G. gerbillus*), Greater rock gerbil (*G. campestris*), Wagner's gerbil (*G. dasyurus*).
Somali pygmy gerbil (*Microdillus peeli*) Somalia inhabiting dry savanna.

Gerbillurus

2 genera: *Desmodillus, Gerbillurus*
Short-eared gerbil (*Desmodillus auricularis*) S Africa, inhabiting desert, savanna.
Southern pygmy gerbils (*Gerbillurus* 4 species) S Africa, inhabiting savanna and desert. Species include Namib brush-tailed gerbil (*G. setzeri*).

Pachyuromys

1 genus: *Pachyuromys*
Fat-tailed jird (*Pachyuromys duprasi*) Morocco to Egypt inhabiting desert and semidesert.

Rhombomys

5 genera: *Brachiones, Meriones, Psammomys, Rhombomys, Sekeetamys*
Przewalski's gerbil (*Brachiones przewalskii*) N China, inhabiting desert.
Jirds (*Meriones*, 16 species) N Africa, Turkey, SW Asia, Kazakhstan, to Mongolia, N China and NW India inhabiting desert and semidesert. Species include Silky jird (*M. crassus*), Mongolian jird (*M. unguiculatus*).
Sand rats (*Psammomys*, 2 species) N African to Syria and Arabian peninsula inhabiting desert and semidesert. Species include Fat sand rat (*P. obesus*) and Lesser sand rat (*P. vexillaris*).
Great gerbil (*Rhombomys opimus*) Kazakhstan, Iran, Afghanistan, Pakistan to N China and Mongolia inhabiting steppe and desert.
Bushy-tailed jird (*Sekeetamys calurus*) E Egypt, S Israel, Jordan, to C Saudi Arabia inhabiting desert.

Taterillus

2 genera: *Tatera, Taterillus*
Large naked-soled gerbils (*Tatera*, 12 species) S, E, and W Africa, Syria to India, Nepal and Sri Lanka inhabiting savanna and steppe. Species include Indian gerbil (*T. indica*), Black-tailed gerbil (*T. nigricauda*).
Small naked-soled gerbils (*Taterillus*, 8 species) Senegal, Mauritania to S Sudan and S to N Tanzania inhabiting semidesert, savanna and wooded grassland. Species include: Emin's gerbil (*T. emini*), Harrington's gerbil (*T. harringtoni*).

For full species list see Appendix ▷

Omnivorous Vegetarians

DIET

Gerbils are basically vegetarians, eating various parts of plants – seeds, fruits, leaves, stems, roots, and bulbs; many species, however, will eat anything they encounter, including insects, snails, reptiles, and even other small rodents. The nocturnal *Gerbillus* species often search for wind-blown seeds in deserts. Gerbils living in the very dry desert regions of southern Africa are primarily insectivorous.

Some species are very specialized, living on a single type of food. The Fat sand rat, Great gerbil, and Indian gerbil, for example, are all basically herbivorous. Of these, the Fat sand rat only occurs where it can find salty, succulent plants, while the Indian gerbil depends on fresh food all year round, so it tends to occur near irrigated crops. Wagner's gerbil has such a liking for snails that it threatens the existence of local snail populations; big piles of empty shells are found outside this gerbil's burrows.

Most gerbils take the precaution of carrying their food back to their burrows before they consume it. Species that live in areas with cold winters must hoard in order to survive. One Mongolian jird was found to have hidden away 20kg (44lb) of seeds in its burrow. Great gerbils not only hoard plants but also construct stacks outside the burrow that can be 1m (3.3ft) high and 3m (9.8ft) long.

A Link with Climate and Food

SOCIAL BEHAVIOR

The social organization of gerbils is only beginning to be studied. Species that live in authentic deserts, whatever genus or group they belong to, tend to lead solitary lives, sometimes in extensive burrow systems, though the burrows are often close enough not to preclude the existence of colonies. Perhaps because the supply of food cannot be guaranteed in such an environment, each animal fends for itself. In contrast, species from savannas, where food is more abundant, are more social. There have been reports of stable pairs forming, and even of family structures emerging.

The most complex social arrangements of all have been observed in species in the Rhombomys group that live in regions with cold winters. Groups larger than families gather in single, extensive burrows, perhaps to huddle together for warmth but also maybe to guard food supplies. The best-known example is the Great gerbil of the Central Asian steppes, which lives in large colonies composed of numerous subgroups that themselves have developed from male–female pairs. A similar social structure is found among Mongolian jirds, although other jird species in North Africa and Asia are reported to be solitary in hot climates but social in cooler regions.

Above *The characteristically slender form of a northern pygmy gerbil, in this case a Cheesman's gerbil embroiled in clearing sand from the entrance to its burrow in the Wahiba Sands, Oman. Many species of this genus are Endangered.*

In savanna species, reproduction too seems to be linked to climate and food. These gerbils give birth after the rainy season. Species living in areas where fresh food is available may reproduce all the year round, with females giving birth to two or three litters a year. Some desert species, however, reproduce only in the cooler months, although those found in southern African deserts may give birth at any time.

Litter size can vary between 1 and 12, with a mean of 3–5, depending on species. The young are born helpless and hairless with their eyes closed, and are unable to regulate their body

Right *The Bushveld gerbil has a ratlike appearance. These gerbils inhabit sandy plains, savannas, and woodlands. Despite their small stature they can leap vast heights and distances when frightened, fleeing from predators in a series of running bounds. Burrows for this genus may be up to 1m (3ft) underground.*

THE COMMUNAL LIFE OF THE MONGOLIAN JIRD

Mongolian jirds live in sizable social groups that, at their largest in summer, consist of 1–3 adult males and 2–7 adult females, plus numerous subadults and juveniles, all dwelling in a single burrow system. Detailed studies have demonstrated that the animals engage in various group activities, for example collectively hoarding food for the winter and spending the cold months huddled together in the burrow. The integrity of the community seems, under normal circumstances, to be jealously guarded. Strange jirds and other animals are chased off.

Who, then, among the adults in the group are the parents of the subadults and juveniles? Parenthood is not evident from the behavior of the males and females within the community, even though they have been observed to form pairs.

The young might conceivably be the offspring of young adults that have migrated from another burrow, but for many reasons this has never seemed likely to be the case. If animals were to leave their own group late in the summer to establish another burrow community elsewhere, they would be vulnerable to predators and the effects of bad weather, and would also have to contend with other jirds into whose territories they might wander. (When population densities are high, there may be as many as 50 burrows per hectare, or 20 per acre). In addition, they would not have access to the food collected for the winter. The most serious objection, however, is that if the animals traveled in groups, such behavior would perpetuate inbreeding and so produce genetic problems.

The unexpected answer to this conundrum has come from observation of animals in captivity. Such studies have shown that communal groups do remain stable and territorial, but when females are in heat they leave their own territories and visit neighboring communities to mate. The females then return to their own burrows, where their offspring will eventually grow up under the protection and care not of their mother and father but rather of their mother and uncles.

Right *This Mongolian jird is probably two to three weeks old. Litter size is usually 4–6; infants, which open their eyes at 2–3 weeks, are weaned after about 3–4 weeks.*

temperature. For about two weeks they depend entirely on their mother's care, and are nursed constantly. Where there is a breeding season, only those born early within it become sexually mature in time to themselves breed in the same season (when aged about 2 months). Those born later become sexually mature after about 6 months, and breed during the following season.

Contradictory Relations with Humans

CONSERVATION AND ENVIRONMENT

Most gerbils live in largely uninhabited areas of the world. When they do come into contact with humans, especially in the African savannas, the Asian steppes, and India, their activities bring them into conflict with people. When collecting food, especially to hoard for winter, they pilfer from crops. When burrowing, they can cause great damage to pastures, irrigation channels, road and railway embankments, and even to the foundations of buildings. They also carry the fleas that transmit deadly disease, including plague, and are reservoirs of the skin disease leishmaniasis.

Though they serve humankind in medical research and as pets, they become pests when they interfere with peoples' lives. Many gerbils are destroyed by gassing; alternatively, burrow systems may be plowed up, even though they may have been used by generations of gerbils for hundreds of years. In some regions too, the sweet, lightly-colored meat is considered a delicacy and is readily eaten. DAS/GA

Dormice

Right *Worldwide, many dormouse species are at risk; the Japanese dormouse is classified by the IUCN as Endangered.*

Above *Despite its name, the Garden (or Orchard) dormouse* (Eliomys quercinus) *is mostly found in forests across central Europe, though some inhabit shrubs and crevices in rocks.*

THE DORMICE OR MYOXIDAE ORIGINATED AT least as early as the Eocene era, 55–34 million years ago. In the Pleistocene (1.8 million–10,000 years ago), giant forms lived on some Mediterranean islands. Today dormice are the intermediates, in form and behavior, between mice and squirrels.

Key features of dormice are their accumulations of fat and their long hibernation period (about seven months in most European species). The Romans fattened dormice for eating in a special enclosure, the *glirarium*, while the French have a phrase "To sleep like a dormouse," equivalent to the English "To sleep like a log."

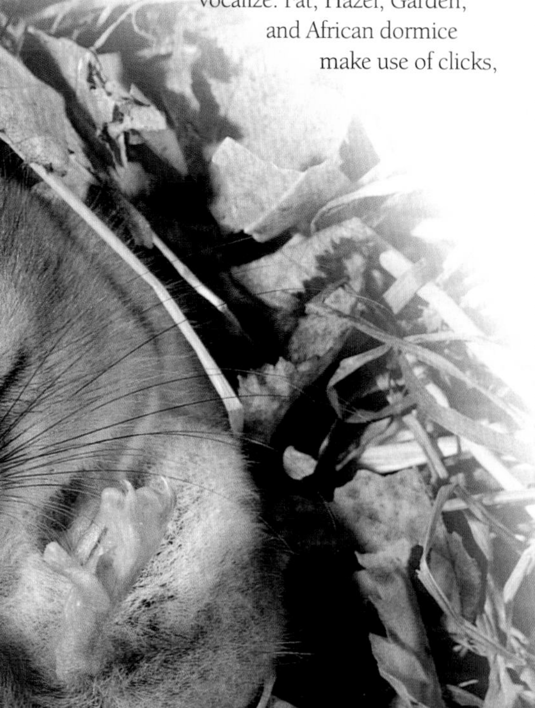

Equipped for Scurrying

FORM AND FUNCTION

Dormice are extremely agile. Most species are adapted to climbing, but some – for example Garden and Forest dormice – also live on the ground; however, the Masked mouse-tailed dormouse is the only species that does so exclusively. The four digits of the forefeet and the five digits of the hind feet have short, curved claws. The underside of each foot is bare with a cushion-like covering. The tail is usually bushy and often long, and in some species such as the Fat, Hazel, Garden, Forest, and African dormice it can detach from the body when it is seized upon by predators – or even other dormice. The sense of hearing is particularly well-developed, as is the ability to vocalize. Fat, Hazel, Garden, and African dormice make use of clicks, whistles, and growling sounds across a broad range of behavior: antagonistic, sexual, explorative, playful.

The Desert dormouse, which is placed in a genus of its own, occurs in deserts to the west and north of Lake Balkhash in eastern Kazakhstan, Central Asia. It has very dense, soft fur, a naked tail, small ears, and sheds the upper layers of skin when it molts. It eats invertebrates such as insects and spiders, and is mostly active at twilight and in the night. It probably hibernates in cold weather.

Feeding Up for Winter

DIET

Dormice are the only rodents that do not have a cecum, which indicates that their diet contains little cellulose. Analysis of the contents of their stomachs has shown that they are omnivores whose diet varies according to season and that there is dietary variation between species according to region. The Fat and Hazel dormice are the most vegetarian, eating quantities of fruits, nuts, seeds, and buds. Garden, Forest, and African dormice are the most carnivorous – their diets include insects, spiders, earthworms, and small vertebrates, but also eggs and fruit.

In France, 40–80 percent of the diet of the Garden dormouse is comprised of insects, according to region and season. However, there is also another factor to be considered. In summer the Garden dormouse eats mainly insects and fruit while in the fall it eats little except fruit, even though the supply of insects is plentiful at this time of year. This change in the content of the diet is part of the preparation for entering hibernation; the intake of protein is reduced, and consequently sleep is induced.

A Long Time Sleeping

SOCIAL BEHAVIOR

In Europe, dormice hibernate from October to April with the precise length of time varying between species and according to region. During the second half of the hibernation period they sometimes wake intermittently – a sign of the onset of the hormone activity that stimulates sexual activity.

Dormice begin to mate as soon as they emerge from hibernation, females giving birth from May onward through to October according to age. (Not all dormice that have recently become sexually mature participate in mating.) The Fat and Garden dormice produce one litter each per year, but Hazel and Forest dormice can produce up to three. Vocalizations play an important part in mating. In the Fat dormouse, the male emits calls as he follows the female; in the Garden dormouse, the female uses whistles to attract the male. The female goes into hiding just before she is due to give birth, and builds a nest, usually globular in shape and located off the ground, for example in a hole in a tree or in the crook of a branch. Materials used include leaves, grass, and moss. The Garden and Fat dormice use hairs and feathers as lining materials. The female Garden dormouse scent-marks the area around the nest and defends it.

Female dormice give birth to between two and nine young, with four being the average litter size in almost all species. The young are born naked and blind. In the first week after birth they become able to discriminate between smells, although an exchange of saliva between mother and young appears to be the means whereby mother and offspring learn to recognize each other. This behavior may also aid the transition from a milk diet to a solid food one. At about 18 days the young become able to hear, and at about the same time their eyes open. They become independent after about 4–6 weeks. Young dormice then grow rapidly until

FACTFILE

DORMICE

Order: Rodentia

Families: Myoxidae

26 species in 8 genera and 3 subfamilies

DISTRIBUTION Europe, Africa, Turkey, Asia, Japan.

HABITAT Wooded and rocky areas, steppe, gardens.

SIZE Head–body length 6.1–19cm (2.4–7.5in); tail length 4–16.5cm (1.6–6.5in); weight 15–200g (0.5–7oz).

COAT Soft-furred and squirrel-like, with bushy tails (except in *Myomimus*).

DIET Omnivorous, including insects, worms, spiders, fruit, seeds, nuts, and eggs.

BREEDING Gestation 21–32 days

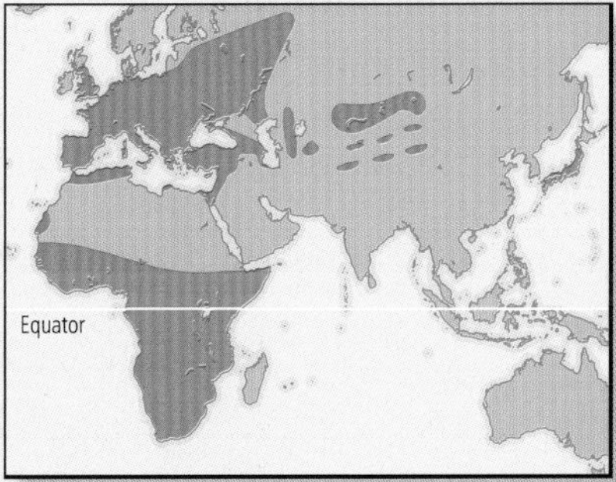

LONGEVITY 3–6 years in wild

CONSERVATION STATUS Half of all dormouse species are listed by the IUCN: 4 as Endangered, 4 as Vulnerable, and 5 as Lower Risk – Near Threatened.

See subfamilies table ▷

Above The richly colored Hazel dormouse lives in thickets and areas of secondary growth in forests; it has a particular fondness for nut trees and is very well-adapted to climbing in them.

Left Curled up tightly, this Hazel dormouse is in hibernation. For its long winter sleep this dormouse resorts to a nest, either in a tree stump, amid debris on the ground, or in a burrow. The length of hibernation is related to climate and can last nine months.

Dormouse Subfamilies

Subfamily Graphiurinae

African Dormice (*Graphiurus*, 14 species). 1 species Vulnerable.

Subfamily Leithiinae

Forest Dormice (*Dryomys*, 3 species). 1 species Endangered.
Garden dormice (*Eliomys*, 2 species). 1 species Vulnerable.
Mouse-tailed dormice (*Myomimus*, 3 species). 1 species Endangered, 2 species Vulnerable.
Desert dormouse (*Selvinia betpakdalaensis*). Endangered.

Subfamily Myoxinae

Japanese dormouse (*Glirulus japonicus*). Endangered.
Hazel dormouse (*Muscardinus avellanarius*).
Fat dormouse (*Myoxus glis*).

For full species list see Appendix ▷

the time for hibernation approaches, when their development slows. Sexual maturity is reached about one year after birth, towards the end of or after the first hibernation.

Dormice populations are usually less dense than those of most other rodents. There are normally between 0.1 and 10 dormice per ha (0.04–4 per acre). They live in small groups, half of which normally consist of juveniles, and each group occupies a home range, the main axis of which can vary from 100m (330ft) in the Garden dormouse to 200m (660ft) in the Fat dormouse. In urban areas the radio-tracking of Garden dormice has indicated that their home range describes an elliptical shape and is related to the availability of food. In the fall the home range is about 1,000sq m (10,800sq ft).

One study of the social organization of the Garden dormouse revealed significant changes in behavior in the active period between hibernations. In the spring, when Garden dormice are emerging from hibernation, males form themselves into groups in which there is a clearcut division between dominant and subordinate animals. As the groups form, some males are forced to disperse. Once this has happened, although groups remain cohesive, behavior within them becomes somewhat more relaxed, so that by the end of the summer the groups have a family character. In the fall social structure includes all categories of age and sex. Despite the high rate of renewal among its members, a colony can continue to exist for many years. CB

Right *Fat dormice have a predilection for fruit and are one of the most vegetarian species. This made them highly suitable for human consumption, hence their alternative name of Edible dormice.*

Jumping Mice, Birchmice, and Jerboas

To a greater or lesser extent, all the members of the Dipodidae family have evolved to move in leaps and bounds. As their names suggest, jumping mice and birchmice are small and mouse-like; the jerboas are mostly rather larger, and are remarkable for their kangaroo-like hind legs.

These small and relatively defenseless rodents seem to have adapted to jumping as an anti-predator strategy. Mostly nocturnal, they are shy and are seldom seen in the wild, not least because some species hibernate for up to 9 months of the year.

Hop, Crawl, and Jump

FORM AND FUNCTION

All jumping mice are equipped for jumping, with long back feet and long tails to help them keep their balance in the air; even so, the most common species, those of the genus *Zapus*, are more likely to crawl under vegetation or make a series of short hops rather than long leaps. And yet the Woodland jumping mouse will often move by bounding 1.5–3m (5–10ft) at a time. Along with the feet and tail, the outstanding characteristics of jumping mice are their colorful fur and grooved upper incisors. The function of the groove is unknown: it may improve the teeth's cutting efficiency, or simply strengthen them.

***Left** A Meadow jumping mouse drinks from a puddle. Like all jumping mice, it will spend more than half the year in hibernation, entering torpor in October and only re-emerging in late April.*

Jumping mice are not burrowers. They live on the surface, although their nests may be underground or in a hollow log or other protected place, and the hibernating nest is often at the end of a burrow in a bank or other raised area. For the most part, though, they hide by day in vegetation. They also usually travel about in thick herbaceous cover, although they will use runways or sometimes other species' burrows when present.

Birchmice differ from jumping mice in having scarcely enlarged hind feet and upper incisors without grooves. Their legs and tail are shorter, yet they too travel by jumping, and climb into bushes using their outer toes to hold on to vegetation and their tails for partial support. Birchmice also dig shallow burrows and make nests of herbaceous vegetation underground.

Jerboas are nocturnal and have large eyes. Their hind limbs are elongated to at least four times the length of their front legs, and in most species the three main foot bones are fused into a single "cannon bone" for greater strength. Jerboas living in sandy areas have tufts of hair on the undersides of the feet that serve as "snowshoes" on soft sand and help them to maintain traction and kick sand backwards when burrowing. These jerboas also have hair tufts to help keep sand out of their ears.

The well-developed jumping ability of jerboas enables them to escape from predators as well as to move about. Only the hind legs are used in moving; the front feet then can be used for gathering food. Jerboas use their long tails as props when standing upright and as balancing organs when jumping. Jumps of 1.5–3m (5–10ft) are used when the animal moves rapidly.

Subfamilies of Jumping Mice, Birchmice, and Jerboas

Jumping mice Subfamily Zapodinae

3 genera: *Zapus, Eozapus, Napaeozapus*

Jumping mice (*Zapus*, 2 species): Meadow jumping mouse (*Zapus hudsonius*) and *Z. princeps*. (The animal sometimes classified separately as *Z. trinotatus* is here considered synonymous with *Z. princeps*). N America, in wooded areas, grassy fields, and alpine meadows.

Woodland jumping mouse (*Napaeozapus insignis*) N America, in forests.

Chinese jumping mouse (*Eozapus setchuanus*) China.

Birchmice Subfamily Sicistinae

1 genus: *Sicista*

Birchmice (*Sicista*, 13 species). Species include *S. betulina* (N Eurasia), *S. caucasica* (W Caucasus and Armenia), *S. subtilis* (Russia and E Europe).

Jerboas 5 subfamilies

Subfamily Cardiocraniinae

2 genera: *Cardiocranius, Salpingotus*

Five-toed pygmy jerboa (*Cardiocranius paradoxus*), W China, Mongolia.

Three-toed pygmy jerboas (*Salpingotus*, 6 species) Asian deserts.

Subfamily Dipodinae

4 genera: *Dipus, Jaculus, Stylodipus, Eremodipus*

Northern three-toed jerboa (*Dipus sagitta*) Russia, Kazakhstan, Turkestan (sandy deserts), C Asia (sandy and sandy-gravel deserts).

Desert jerboas (*Jaculus*, 4 species) N Africa, Iran, Afghanistan, Pakistan, Turkestan, in various desert habitats.

Three-toed jerboas (*Stylodipus*, 3 species) S European Russia, in sandy semidesert; Kazakhstan, Mongolia, China, in clay and gravel deserts.

Lichtenstein's jerboa (*Eremodipus lichtensteini*) Sandy deserts of Turkestan.

Subfamily Paradipodinae

1 genus: *Paradipus*

Comb-toed jerboa (*Paradipus ctenodactylus*) Turkestan, in sandy deserts.

Subfamily Allactaginae

3 genera: *Allactaga, Allactodipus, Pygeretmus*

Four- and five-toed jerboas (*Allactaga*, 11 species) NE Africa (Libyan desert), Middle East, Russia, Kazakhstan, Turkestan, C Asia.

Fat-tailed jerboas (*Pygeretmus*, 3 species) S European Russia, Kazakhstan, Mongolia, China, in salt and clay deserts.

Bobrinski's jerboa (*Allactodipus bobrinskii*) Gravel deserts of Turkestan.

Subfamily Euchoreutinae

1 genus: *Euchoreutes*

Long-eared jerboa (*Euchoreutes naso*) China and Mongolia, in gravel deserts.

For full species list see Appendix ▷

Living off Plants and Insects

DIET

The major animal foods eaten by jumping mice are moth larvae (primarily cutworms) and ground and snout beetles. Also important is the subterranean fungus *Endogone*, which make up about 12 percent of the diet (by volume) in the Meadow jumping mouse and about 35 percent in the Woodland jumping mouse. Meadow jumping mice eat many things, but seeds, especially those from grasses, are the most important food. The seeds eaten change with availability.

Birchmice can eat extremely large amounts of food at one time and can also spend long periods without eating. Their main foods are seeds, berries, and insects.

Jerboas have a wide spectrum of diets. Some species are specialized feeders and use only seeds (*Cardiocranius*), insects (*Euchoreutes*), or fresh leaves and stems of succulent plants (*Pygeretmus*

⮈ *Left* *A desert jerboa (*Jaculus jaculus*) crouches in Saharan sand. The* Jaculus *species are extraordinarily powerful jumpers, able to leap from a standing position to a height of almost 1m (3ft).*

and *Paradipus*). Other species have mixed diets of seeds and insects (*Salpingotus*), seeds and green plants (*Dipus*, *Stylodipus*, *Eremodipus*, *Jaculus* and *Allactodipus*) or equal proportions of seeds, insects, and green and underground plant parts (*Allactaga*). In *Dipus* all individuals in a population emerge for their nightly forays at about the same time, and move by long leaps to their feeding grounds, which may be some distance away. There they feed on plants, especially those with milky juices, but they also smell out underground sprouts and insect larvae in underground galls. Like pocket mice, jerboas do not drink water, but instead manufacture "metabolic water" from food.

The Big Sleep

SOCIAL BEHAVIOR

Jumping mice are profound hibernators, hibernating for 6–9 months of the year according to species, locality, and elevation. The Meadow jumping mouse in the eastern USA usually hibernates from about October to late April. Individuals that hibernate successfully put on 6–10g (0.21–0.35oz) of fat in the two weeks prior to entering hibernation. They do this by sleeping for increasingly longer periods until they attain deep hibernation with their body temperature just a little above freezing. Their heart rate, breathing rate, and all bodily functions drop to low levels. However, the animals wake about every two weeks, perhaps to urinate, then go back to sleep. In the spring the males appear above ground about two weeks before the females. Of the animals active in the fall, only about a third – the larger ones – are apparently able to survive hibernation. The rest – young individuals or those unable to put on adequate fat – perish during the winter retreat.

Jumping mice give birth to their young in a nest of grass or leaves either underground or in some other protected place. Gestation takes 17–18 days, or up to 24 if the female is lactating. Each litter contains 4–7 young. Litters may be produced at any time between May and September, but most enter the world in June and August. Most females probably produce one litter per year.

Like jumping mice, birchmice are active primarily by night. Birchmice hibernate in their underground nests for about half of the year. Gestation probably lasts 18–24 days, and parental care for another four weeks. Studies of *Sicista betulina* in Poland have shown that one litter a year is produced and that any female produces only two litters during her lifetime.

Some jerboas hibernate during the winter, surviving off body fat, and in addition some species enter torpor during hot or dry periods. They are generally quiet, but when handled will sometimes shriek or grunt. Some species have been known to tap with a hind foot when inside their burrows.

In northern species mating first occurs shortly after the emergence from hibernation, but most female jerboas probably breed at least twice in a season, producing litters of 2–6 young.

There are four kinds of burrows used by various jerboas, depending on their habits and habitats: temporary summer day burrows for hiding during the day, temporary summer night burrows for hiding during nightly forays, permanent summer burrows used as living quarters and for producing young, and permanent winter burrows for hibernation. The two temporary burrows are simple tubes, in length respectively 20–50cm (8–20in) and 10–20cm (4–8in).

The permanent summer burrows have one nest chamber and no secondary chamber for food storage, while the permanent winter burrows have the main hibernation chamber 1.5–2.5m (5–8ft) below the surface, and also have secondary chambers 40–70cm (15–28in) down. Permanent burrows have one to three accessory exits. Temporary summer-night burrows have widely open exits with a ground gutter near the entrance, but all other burrow exits are always closed with ground plugs and camouflaged. GIS/JW

FACTFILE

JUMPING MICE, BIRCHMICE, & JERBOAS

Order: Rodentia

Family: Dipodidae

50 species in 15 genera and 7 subfamilies

DISTRIBUTION Jumping mice: N America; 1 species (*Eozapus setchuanus*) in China. **Birchmice:** Eurasia. **Jerboas:** N Africa and Asia.

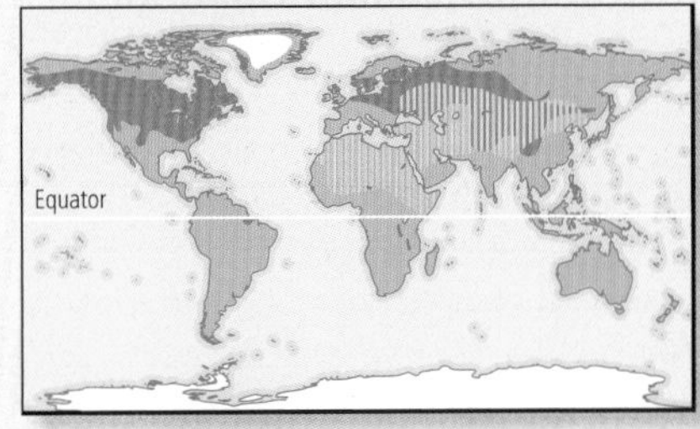

Jumping mice and Birchmice

Jerboas

HABITAT Jumping mice: meadows, moors, steppe, thickets, woods. **Birchmice:** forests, meadows, steppe. **Jerboas:** desert, semidesert, steppe, including patches of bare ground.

SIZE Jumping mice: head–body length 7.6–11cm (3–4.3in), tail length 15–16.5cm (5.9–6.5in), weight up to 29g (1oz). **Birchmice:** head–body length 5–9cm (1.9–3.5in), tail length 6.5–10cm (2.6–3.9in), weight up to 28g (1oz). **Jerboas:** head–body length 4–23cm (1.6–9in), tail length 7–30cm (2.7–11in). hind foot length 2–10cm (0.8–4in).

COAT Coarse in the jumping mice and birchmice, silky in the jerboas. Coloration usually matches the habitat in which the different species live.

DIET Jumping mice: moth larvae, beetles, fungi, and (in Meadow jumping mice) seeds. **Birchmice:** seeds, berries, insects. **Jerboas:** seeds, insects and insect larvae, fresh leaves and stems of succulent plants.

BREEDING Gestation times range from 17–21 days in jumping mice (*Zapus* and *Napaeozapus* species) to 18–24 days in birchmice (*Sicista betulina*) and 25–35 days in jerboas.

LONGEVITY Probably 1–2 years in jumping mice, at least 1.5 years in birchmice, and 2–3 years in jerboas.

CONSERVATION STATUS Among the 4 jumping mouse species, the Chinese jumping mouse (*Eozapus setchuanus*) is listed as Vulnerable. Of 16 birchmouse species, one (*Sicista armenica*) is Critically Endangered and another (*S. caudata*) Endangered. In the jerboas, one *Allactaga* species (*A. firouzi*) is Critically Endangered and another (*A. tetradactyla*) Endangered, as is the Long-eared jerboa (*Euchoreutes naso*). The Five-toed dwarf jerboa (*Cardiocranius paradoxus*) and one of the six Three-toed dwarf jerboa species (*Salpingotus crassicauda*) are Vulnerable.

Pocket Gophers

POCKET GOPHERS ARE ONE OF SEVEN OR EIGHT rodent groups around the world that spend most of their lives below ground in self-dug burrow systems. Native to North and Central America and to northwestern Colombia in South America, they present a paradox: they are extraordinarily diverse, even though all species share a common body plan and a similar life cycle, adapted for a life of digging.

Pocket gophers take their name from the fur-lined cheek pouches that serve as built-in carrier bags both for food items and nesting materials. Unlike those of hamsters and squirrels, the gophers' pouches are external, located on either side of the mouth. They share this feature among mammals only with their close relatives in the family Heteromyidae – the pocket mice, kangaroo mice, and kangaroo rats.

Designed to Dig

FORM AND FUNCTION

Designed for digging, pocket gophers have thickset, tubular bodies, with no apparent neck. Fore and hind limbs are short, powerful, and of approximately equal size. The small, nearly naked tail is particularly sensitive to touch.

The front teeth project through furred lips, an adaptation that allows the gophers to use them for digging or cutting roots without getting dirt in their mouths. Even so, they generally excavate soil with the enlarged claws on their forefeet, and are thus categorized as "claw" or "scratch" diggers, though the incisors are used as helpful adjuncts in some genera. In *Thomomys* species, for example, populations living in harder soils tend to have more forward-pointing incisors that they use as chisels for digging. All pocket gophers push soil from their burrows with rapid movements of their forefeet, chest, and chin, and use an earthen plug to block the entrance. As a result, fresh gopher mounds can be distinguished by their characteristic triangular, or deltaic, shape, with an obvious round plug of soil located at the apex.

Gopher skin fits loosely. It is usually clothed in short, thick fur, interspersed with hairs sensitive to touch. The fur of more tropical species is coarser and less dense, perhaps as an adaptation for warmer climates. The loose skin enables individuals to make tight turns in their constricted burrows. Gophers are very agile and surprisingly fast, capable of rapid movement back or forward on their squat, muscular legs.

The gopher's skull is massive and strongly ridged, with heavy zygomatic arches and a broad temporal region holding powerful jaw muscles. The upper incisors may be either smooth or grooved on their anterior surface, depending on the genus. The four cheek teeth in each quarter of the jaw form a battery for grinding tough and abrasive foodstuffs. Strong and effective, these teeth grow continuously from birth to death, so that the grinding surface is present in all from the youngest to the oldest individuals. In the Valley pocket gopher, both teeth and foreclaws can grow at a rate of 0.5–1mm (0.02-0.04in) per day.

Male pocket gophers are larger than females, though the extent of this dimorphism varies greatly geographically. In the Valley pocket gopher, it seems to depend on habitat quality, and therefore on the density of animals in the population. In high-quality habitat such as agricultural fields, males can have twice the body mass of females, and may be 25 percent larger in skull dimensions, but in poorer habitat such as deserts, the differences shrink to about 15 percent and 6 percent respectively. These variations are a result of both the nutritional quality of the food available to the animals – individuals of both sexes get larger if they eat well, particularly as juveniles – and the cessation of growth in females as they reach reproductive maturity and shift energy from growth to producing young.

Pocket gophers have the oldest fossil record of, and are taxonomically more diverse than, any other group of subterranean rodents. They originated in the late Eocene (36 million years ago) and underwent two pulses of diversification; the latest, beginning in the Pliocene (5 million years ago) led to the living genera. Fossils of both *Geomys* and *Thomomys* species date from Pliocene deposits; the other genera are known either from the early Pleistocene (*Pappogeomys*) or only from the modern record (*Orthogeomys* and *Zygogeomys*). Throughout the history of the family, there has been a relatively stable number of genera and constant extinction and origination rates.

The evolutionary history of the family is characterized by successive attempts to invade the fossorial (digging) niche. Each successive group exhibits better-developed adaptations for subterranean existence, with some modern pocket gophers exhibiting the best adaptations of all. The entire fossil history of the family is contained within North America.

In order to distinguish the genera, biologists examine overall size, details of the skull bones and the cheek teeth, and the presence and number of grooves found on the front surface of the upper incisors. The differences between populations of

Above *In common with many other mammals that lead a subterranean existence, pocket gophers have certain adaptations, such as relatively small eyes and ears, that make it easier to excavate burrows. The Northern pocket gopher* (Thomomys talpoides) *inhabits a wider range of soil types than any other species in the family.*

Below *A Northern pocket gopher feeding with filled pouches. When suitable vegetation is found, the gopher will cut this down into smaller pieces, which can be pushed into the cheek pouches using the fore feet, thus enabling the gopher to transport a relatively large quantity of food back to its burrow.*

FACTFILE

POCKET GOPHERS

Order Rodentia

Family Geomyidae

39 species in 5 genera, divided into 2 tribes

DISTRIBUTION N and C America, from C and SW Canada through the W and SE USA and Mexico to extreme NW Colombia.

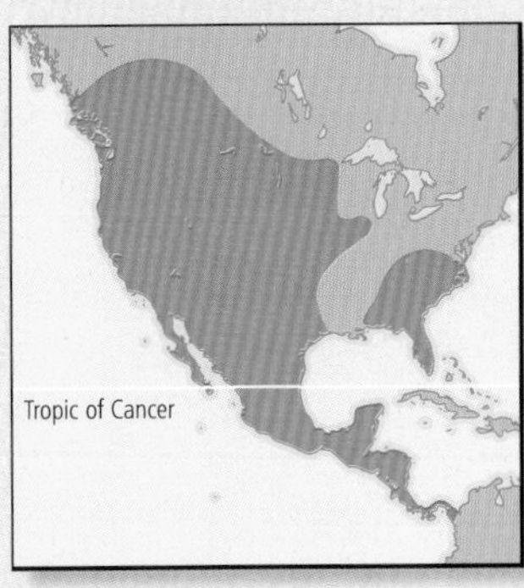

HABITAT Friable soils in desert, scrub, grasslands, montane meadows, and arid tropical lowlands.

SIZE Head–body length ranges from 12–22.5cm (4.7–8.9in) in genus *Thomomys* to 18–30cm (7–11.8in) in genus *Orthogeomys*; **weight** from 45–400g (1.6–14oz) to 300–900g (11–32oz) in the same genera. Males are always larger than females, and up to twice the weight.

COAT Short and thick, in shades of brown and gray.

DIET Plant matter, particularly forbs, grasses, roots, and tubers.

BREEDING Gestation 17–21 days in *Thomomys* and *Geomys* genera.

LONGEVITY Maximum 5 years (6 years recorded in captivity).

CONSERVATION STATUS 2 species are listed as Critically Endangered, one as Endangered, and 3 as Vulnerable. 2 US subspecies are now classed as Extinct.

See tribes table ▷

the same species that live apart can be more striking than the differences that exist between separate species; so the Valley pocket gopher exhibits size, color, and habitat variations that span the entire range exhibited by all the other species in the genus *Thomomys*. As a consequence of this, species limits are, for the most part, poorly known, and a multitude of subspecies have been described – no fewer than 185 in the case of Valley pocket gophers. Some species, including Wyoming, Idaho, and Northern pocket gophers, are distinguished by hidden characteristics, such as chromosome number, rather than by any external features.

Species by Species

DISTRIBUTION PATTERNS

All genera and species are distributed contiguously; except in very narrow zones of overlap, only a single kind of pocket gopher can be found in any particular area, as different species are apparently unable to share the space they occupy underground (the fossorial niche). In areas where several species and/or genera do meet, the pattern of species distribution is mosaic-like. In the mountains of the western USA and Mexico, for example, pocket gopher species replace each other in succession according to altitude.

Within the range of their distribution, pocket gophers are ubiquitous in virtually all habitats that have extensive patches of friable soil. The range of habitats in which they are found can be extreme: for instance, the Valley pocket gopher ranges from desert soils below sea level to alpine meadows that are – at 3,500m (11,500 ft) – well above the timberline. In tropical latitudes populations of the same species may be found in mountain forest meadows and in arid tropical scrub, although few species penetrate true tropical savannas.

Plants are the pocket gophers' dietary mainstay. Above ground, they take leafy vegetation from the vicinity of burrow openings; underground, they devour succulent roots and tubers. They often prefer forbs and grasses, but their diet shifts seasonally according to both availability and the gophers' requirements for nutrition and water. In deserts during the summer, water-laden cactus plants are consumed.

To transport food to caches in the burrow, gophers use their forepaws dextrously to fill their external cheek pouches. Food storage areas are usually sealed off from the main tunnel system.

A Lonely Life Underground

SOCIAL BEHAVIOR

Pocket gophers are solitary creatures. Individuals live in self-excavated burrows that abut those of other gophers. Male burrows tend to be longer and more dendritic than those of females, so that each male burrow might contact those of several females. In the Valley pocket gopher, the maximum territory size is about 250sq m (2,700sq ft), while in the Yellow-faced pocket gopher individual burrow length may exceed 80m (260ft) of actual tunnel. Where habitat is good, the gophers become more crowded and space out evenly across the land; in low-quality habitat, groups of gophers concentrate in the best regions, while other areas remain gopher-free. During breeding, this strict organization relaxes somewhat as males and females briefly cohabit in the same burrow, and females share burrows with their young until they are weaned. Each individual gopher still maintains a tunnel system for their own exclusive use, but adjacent males and females may have both common burrows and deep, shared nesting chambers. Results from genetic studies would seem to suggest that females selected their mate from among the males whose territories are adjacent to their own.

Densities in small species such as the Valley pocket gopher rarely exceed 40 adults per ha (16 per acre), and they sink as low as 7 per ha (3 per acre) in the case of the large-bodied Yellow-faced pocket gopher. In high-quality habitat, individual territories are stable in both size and position, with most individuals living their entire adult lives

Above *Members of the five pocket gopher genera:* ***1*** *male Valley pocket gopher* (Thomomys bottae) *making a mound;* ***2*** *Valley pocket gopher of a different color showing the pouch cheeks in detail;* ***3*** *Plains pocket gopher* (Geomys bursarius) *returning from foraging;* ***4*** *second color variant of Valley pocket gopher, a female;* ***5*** *Large pocket gopher* (Orthogeomys grandis) *in an underground food store making threats;* ***6*** *Michoacan pocket gopher* (Zygogeomys trichopus) *digging with its claws;* ***7*** *Buller's pocket gopher* (Pappogeomys bulleri) *using its incisors for digging.*

in very limited areas; in low-quality environments, they shift throughout the year as animals search for food and mates.

Male and female young disperse from the mother's underground domicile at the same time, still wearing their juvenile coats. In the Valley pocket gopher, female young of the year initially move the farthest to establish territories, although most still seem to settle within 40–50m (130–165ft) of their natal area. If conditions are favorable, they may even breed in the same season as their birth, at just 70 days old. Male young tend to live in shallow systems in marginal and peripheral habitat, until they disperse to establish territories just prior to the next breeding season. As a consequence, their mortality is typically higher than that of females, which results in a prevalence of females in pocket gopher populations. Despite their adaptations for digging, pocket gophers can,

and do, move considerable distances; most dispersal takes place above ground on dark nights, although movements over distances of just a few meters may take advantage of long tunnels just beneath the surface.

Individuals of both sexes are pugnacious and aggressive, and will fight for parcels of land. Males in particular exhibit heavy scarring around the mouth and on the rump, most of which will have been acquired during the breeding season.

As in most animals, reproduction is strongly affected by the changing seasons. In montane regions, breeding follows the melting snow in the late spring and early summer, but in coastal and desert valleys and in temperate grasslands, it coincides with winter rainfall.

Most Valley gopher females have only one litter each season; however, some will have as many as three or four, depending on the quality of habitat – and therefore the nutrition – that is available to the mother. Animals that inhabit irrigated fields may breed nearly year-round while neighboring populations living amid natural vegetation have sharply delimited seasonal breeding. Females of Yellow-faced and Plains pocket gophers typically produce only one or two litters each year. Litter size varies between the species; *Pappogeomys* gophers usually give birth to twins, while *Thomomys* mothers typically produce five young, but may have to deal with as many as 10 per pregnancy. The onset of breeding, the length of the season, and the number of litters are largely controlled by local environmental conditions – primarily temperature, moisture, and the quality of the vegetation.

Pocket gophers are born with both their eyes and cheek pouches closed. The pouches open at 24 days and eyes and ears at 26 days. Additionally, Northern and Valley pocket gophers have a juvenile coat that molts within 100 days of birth.

After a helpless early life spent in the shelter of the mother's burrow, infants are weaned at around 40 days. Within a couple of weeks of weaning the young head off alone, when still less than 2 months old. Unsurprisingly, most of the young that are produced each year do not survive long enough to be able to reproduce. In the Valley pocket gopher, only 6–12 percent of newborns are recruited into the breeding population the following year. The maximum longevity in nature can be as great as 4–5 years, but the average life span of an adult is usually only just over a year. No more than half of the population survive from one year to the next and some years only 15 percent will survive to the next year.

Females live nearly twice as long as males. Among Yellow-faced pocket gophers, they survive on average for 56 weeks, as compared to the average 31-week lifespan of the male; in Valley pocket gophers, females may live for as long as 4.5 years, compared to the maximum expectation of 2.5 years for the male. One possible explanation for this huge difference in longevity is the extreme inter-male aggression that occurs during the breeding season. In addition, males appear to grow continuously throughout their life, while females cease growth when they reach reproductive maturity.

The sex ratio among adults varies geographically in all pocket gophers, a variation that in Valley, Northern, Plains, and Yellow-faced populations has been shown to be related to habitat quality. There are roughly equal numbers of both sexes in poor-quality habitats, such as desert areas, but in agricultural land there is a preponderance of females, with three to four for every adult male. As the population size increases males become less common, which is at least in part due to male mortality during or after fights over territories and access to mates. While almost all the adult females in a population breed each season, many males are not so lucky; some may never breed, while others will monopolize the females and father all the young.

Soil Engineers

CONSERVATION AND ENVIRONMENT

Pocket gophers play a major role in soil dynamics, and in so doing become agriculturalists to suit their own needs. Cattle and other grazers compact soil, but underground, gophers counter this with their constant digging. They cycle soil vertically, which increases porosity, slows the rate of water run-off, and provides increased aeration. Gophers can thus have a profound effect on plant communities, often, through continual soil disturbance, creating conditions that favor growth of the herbaceous plants that they prefer eating.

Voracious eaters, pocket gophers can become agricultural pests; in the western deserts of North America, the annual productivity of irrigated alfalfa fields can be reduced by as much as 50 percent in the year following an invasion by the animals. They also disturb irrigation channels. As a result, millions of dollars have been spent in the USA on programs to control gopher populations, but even so it has proved hard to curb their spread, so perfectly does agriculture suit their expansion.

Even so, some local US populations and subspecies are now considered threatened as a result of habitat loss, while in Mexico the Michoacan pocket gopher is at risk primarily because of its localized geographic distribution. Mostly, though, pocket gophers are among the few animals to have benefited from human development – particularly the replacement of native grass- and shrubland by agriculture. JP

The Two Tribes of Pocket Gophers

Tribe Geomyini

4 genera: *Geomys, Orthogeomys, Pappogeomys, Zygogeomys.*

Eastern pocket gophers (*Geomys*, 9 species), including: Desert pocket gopher (*G. arenarius*); Plains pocket gopher (*G. bursarius*); Texas pocket gopher (*G. personatus*); Southeastern pocket gopher (*G. pinetis*); Tropical pocket gopher (*G. tropicalis*).

Taltuzas (*Orthogeomys*, 11 species) including: Chiriqui pocket gopher (*O. cavator*); Darien pocket gopher (*O. dariensis*); Variable pocket gopher (*O. heterodus*); Big pocket gopher (*O. lanius*).

Yellow and Cinnamon pocket gophers (*Pappogeomys*, 9 species) including: Buller's pocket gopher (*P. bulleri*); Smoky pocket gopher (*P. fumosus*); Queretaro pocket gopher (*P. neglectus*); Zinser's pocket gopher (*P. zinseri*).

Michoacan or Tuza pocket gopher (*Zygogeomys trichopus*, single species).

Tribe Thomomyini

1 genus, *Thomomys*.

Western pocket gophers (Genus *Thomomys*, 9 species) including: Valley pocket gopher (*T. bottae*); Wyoming pocket gopher (*T. clusius*); Northern pocket gopher (*T. talpoides*); Mexican pocket gopher (*T. umbrinus*).

For full species list see Appendix ▷

Pocket Mice and Kangaroo Rats

POCKET MICE AND KANGAROO RATS ARE nocturnal burrow-dwellers that inhabit many different American environments from arid deserts to humid forests. On returning from their nightly forays in search of food, they sometimes plug the burrow entrance with soil for added protection from predators and the weather.

The Heteromyidae family brings together the pocket mice, which take their name from the deep, fur-lined cheek pouches in which they store food, and the kangaroo rats and mice, which as their name suggests are adapted to travel by hopping. They too have cheek pouches, the heteromyids' most distinctive feature; these can be turned inside out for cleaning, then pulled back into place with the help of a special muscle.

Built-in Carrying Pouches

FORM AND FUNCTION

Unusually for burrowers, the heteromyids are thin-skulled, and do most of their digging with their front paws. The most distinctive feature of the head are the two cheek-pouches, opening externally on either side of the mouth and extending back to the shoulders. The mice use their paws to fill the pouches with food and nest material for carrying back to the safety of the burrow.

While the pocket mice travel on all fours, the kangaroo rats and mice, in contrast, have long hind limbs and shrunken forelimbs that are used mainly for feeding; these animals normally move in hops, only lowering themselves onto their front legs to scramble over short distances. The leg muscles are powerful enough to launch kangaroo rats 2m (6.6ft) or more with each leap when hurrying to escape from predators. All heteromyids have long tails; the jumping species rely on them to help keep their balance when traveling and as props to rest on when standing still.

In Search of Seeds

DISTRIBUTION PATTERNS

By day, summer conditions in the deserts of the southwestern USA are formidable. Surface temperatures soar to over 50°C (122°F), sparsely distributed plants are parched and dry, and signs of mammalian life are minimal. As the sun sets, however, the sandy or gravelly desert floor comes alive with rodents. The greatest diversity occurs among the pocket mice, of which five or six species can coexist in the same barren habitat.

Contrast these hot, dry (or in winter, cold and apparently lifeless) conditions with the tropical rain forests of central and northern South America. Rich in vegetation, the rain forest is nearly bare of heteromyid species: only one species, Desmarest's spiny pocket mouse, occurs at most sites.

The most likely explanation for this difference in species richness lies in the diversity and availability of seeds. In North American deserts, seeds of annual species can accumulate in the soil to a depth of 2cm (0.8in) in densities of up to 91,000 seeds per sq m (8,450 per sq ft). Patchily distributed by wind and water currents, small seeds weighing about 1mg tend to accumulate in great numbers under bushes and on the leeward sides of rocks, whereas larger seeds occur in clumps in open areas between vegetation, providing plentiful opportunities for the nocturnal seed-gathering activities of the heteromyid rodents.

Tropical forests are also rich in seeds, but many of those produced by tropical shrubs and trees are protected chemically against predation. This is especially true of large seeds weighing several grams, which considerably reduces the variety of seeds available to rodents. In effect, then, the tropical rain forest is a desert in the eyes of a seed-eating rodent, whereas the actual desert is a "jungle," as far as seed availability is concerned. Most of the 15–20 species of rodents inhabiting New World tropical forests are either omnivorous or exclusively fruiteating.

Above *An Ord's kangaroo rat sets out to forage. Kangaroo rats compete with smaller pocket mice for food where their ranges overlap; experiments in Arizona have shown that pocket mice numbers more than tripled over an 8-month period when kangaroo rats were excluded from their territory.*

Left *The Desert pocket mouse lives in arid areas of the American southwest and northwest Mexico. Like many heteromyids, it emerges from of its burrow at night to exploit the wealth of seeds blown across the desert floor.*

Right *Long tails are a feature of* Dipodomys *species, like this Merriam's kangaroo rat; they serve as counterweights when the rats are traveling and as supports when they are resting.*

FACTFILE

POCKET MICE & KANGAROO RATS

Order: Rodentia

Family: Heteromyidae

59 species in 5 genera

DISTRIBUTION
N, C, and northern S America

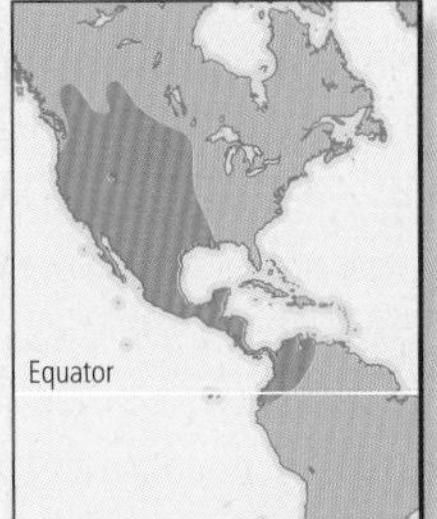

POCKET MICE Genus *Perognathus*
SE Canada, W USA south to C Mexico. Quadrupedal. 24 species, sometimes separated into 2 genera: the silky pocket mice (Genus *Perognathus*) and the coarse-haired pocket mice (genus *Chaetodipus*). HBL from 6–12.5cm (2.4–5in); TL from 4.5–14.5cm (1.8–5.7in); WT from 7–47g (0.25–1.7oz). Longevity: 2 years in the wild; up to 8 in captivity.

SPINY POCKET MICE Genus *Liomys*
Mexico and C America S to C Panama. Mostly in semi-arid country. Quadrupedal. 5 species.

FOREST SPINY POCKET MICE Genus *Heteromys*
Mexico, C America, northern S America. Forests, up to 2,500m (8,200ft). Quadrupedal. 7 species including Desmarest's spiny pocket mouse (*H. desmarestianus*). Conservation status: *H. nelsoni* is Critically Endangered.

KANGAROO RATS Genus *Dipodomys*
SW Canada and USA W of Missouri River to south C Mexico. Arid and semiarid country with some brush or grass. Bipedal (hind legs long, front legs reduced). 21 species. HBL from 10–20cm (4–8in); TL from 10–21.5cm (4–8.5in); WT from 35–180g (1.2–6.3oz). Longevity: up to 9 years in captivity. Conservation status: 3 species – *D. ingens*, *D. insularis*, and *D. margaritae* – are Critically Endangered; 1 more, the San Quintin kangaroo rat (*D. gravipes*) is Endangered, and the Texas kangaroo rat (*D. elator*) is Vulnerable.

KANGAROO MICE Genus *Microdipodops*
USA in S Oregon, Nevada, parts of California and Utah. Near shrubs in gravelly soil or sand dunes. Bipedal (hind legs long, front legs reduced). 2 species.

Abbreviations HBL = head–body length TL = tail length WT = weight

For full species list see Appendix ▷

With large cheek pouches and a keen sense of smell, heteromyid rodents are admirably adapted for gathering seeds. Most of the time that they are active outside their burrow systems is spent collecting seeds within their home ranges. Members of the two tropical genera (*Liomys* and *Heteromys*) search through the soil litter for seeds, some of which will be buried in shallow pits scattered around the home range; others are stored underground in special burrow chambers.

Boom or Bust

SOCIAL BEHAVIOR

Breeding in desert heteromyids is strongly influenced by the flowering activities of winter plants, which germinate only after at least 2.5cm (1in) of rain has fallen between late September and mid-December. In dry years, seeds of these plants fail to germinate, and a new crop of seeds and leaves is not produced by the following April and May. In the face of a reduced food supply heteromyids do not breed, and their populations decline in size. In years following good winter rains most females produce two or more litters of up to five young, and populations increase rapidly.

This "boom or bust" pattern of resource availability also influences heteromyid social structure and levels of competition between species. When seed availability is low, seeds stored in burrow or surface caches become valuable, defended resources. Behavior becomes asocial in most arid-land species (including species of *Liomys*): adults occupy separate burrow systems (except for mothers and their young), and when two members of a species meet away from their burrows they engage in boxing and sand-kicking. In the forest, in contrast, *Heteromys* species are socially more tolerant; individuals have widely overlapping home ranges, share burrow systems, and are less likely to fight .

The diversity and availability of edible seeds is the key to the evolutionary success of heteromyid rodents. Seed availability affects foraging patterns, population dynamics, and social behavior. In North American deserts, because seed production influences levels of competition not just between heteromyids themselves but also with ants and other seedeaters, there is a clear link between resources and the structure of an animal community. Thus the abundance and diversity of seedeaters is directly related to plant productivity. THF

DISTRIBUTION America, Africa, Asia.

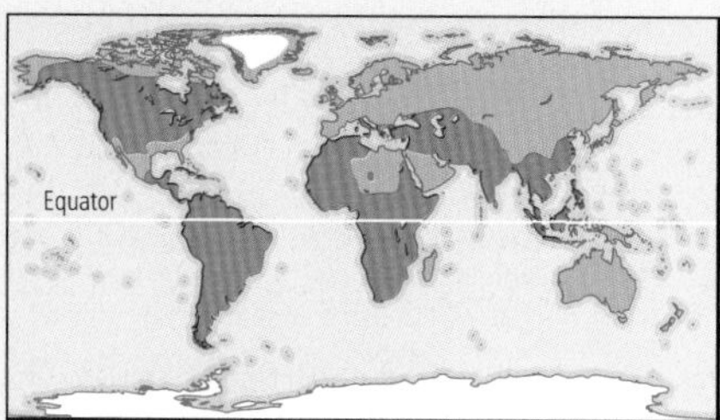

HYSTRICOGNATHS Suborder Hystricognathi
238 species in 63 genera and 18 families

Cavy-like Rodents

THE FAMILIAR GUINEA PIG IS A REPRESENTATIVE of a large group of rodents classified as the suborder Hystricognathi (formerly Caviomorpha). Most are large rodents confined to South and Central America. Although they are extremely diverse in external appearance and are generally classified in separate families, the hystricognaths share sufficient characteristics to make it likely that they constitute a natural, interrelated group.

Externally, many of these rodents have large heads, plump bodies, slender legs, and short tails – as in the guinea pigs, the agoutis, and the giant capybara, the largest of all rodents at over one meter (39in) in length. Others, however – for example, some spiny rats of the family Echimyidae – come very close in general appearance to the common rats and mice.

Internally, the most distinctive character uniting these rodents is the form of the masseter jaw muscles, one branch of which extends forwards through a massive opening in the anterior root of the bony zygomatic arch to attach on the side of the rostrum. At its other end it is attached to a characteristic outward-projecting flange of the lower jaw. Hystricognaths are also characterized by producing small litters after a long gestation period, resulting in well-developed young. Guinea pigs, for instance, usually have two or three young after a gestation of 50–75 days, compared with seven or eight young after only 21–24 days in the Brown (murid) rat.

The modern suborder name Hystricognathi serves to emphasize the close relationship of the Old World porcupines – family Hystricidae – with the South American "caviomorphs." However, despite the fact that both groups share the features described above, there has been considerable debate as to whether such features signify a common ancestry or are merely indicative of convergent evolution. This controversy over systematics is intrinsically linked to the question of whether the caviomorphs reached South America from North America or from Africa (rafting across in the late Eocene, when the continents were far closer together). The African cane rats (family Thryonomyidae) are closely related to the Hystricidae but some other families, namely the gundis (Ctenodactylidae) and the Dassie rat (Petromuridae) are much more doubtfully related, exhibiting only some of the "caviomorph" characters.

Most of the American caviomorphs are terrestrial and herbivorous but a minority, the porcupines (Erethizontidae) are arboreal, and one group, the tuco-tucos (Ctenomyidae) are burrowers. GBC

5

4

SKULLS AND DENTITION

Most cavy-like rodents have rather angular skulls and very strongly developed incisor teeth. The wearing surfaces of the four cheekteeth show enormous variation in pattern and complexity amongst the different species. Those of the coypu are typical of a large group of herbivorous species, including the agoutis and the American porcupines, and are closely paralleled in the Old World porcupines and cane rats. The teeth of the mara and of the capybara. although superficially very different in degree of complexity, resemble each other in being evergrowing as in the unrelated but also grass-eating voles and rabbits. At the other extreme, the tuco-tucos have surprisingly simple cheekteeth considering that they feed mainly on roots and tubers.

One distinguishing feature of cavy-like rodents is the deep masseter muscle in their jaws (see diagram BELOW). This extends forward through an opening in the zygomatic arch to attach to the muzzle and provides the powerful gnawing action characteristic of this suborder. The lateral masseter is only used in closing the jaw.

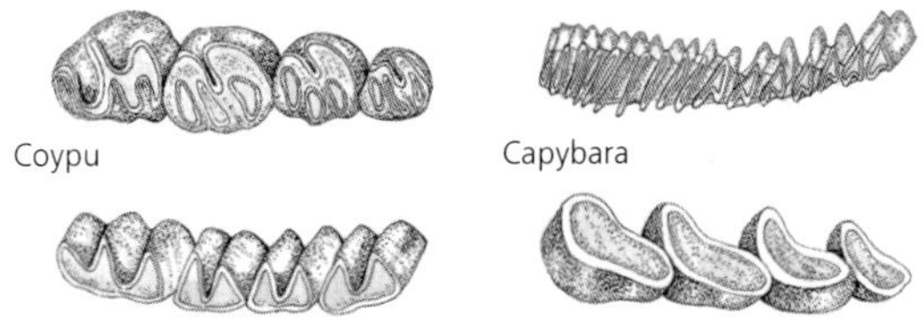

Coypu

Capybara

Mara

Tuco-tuco

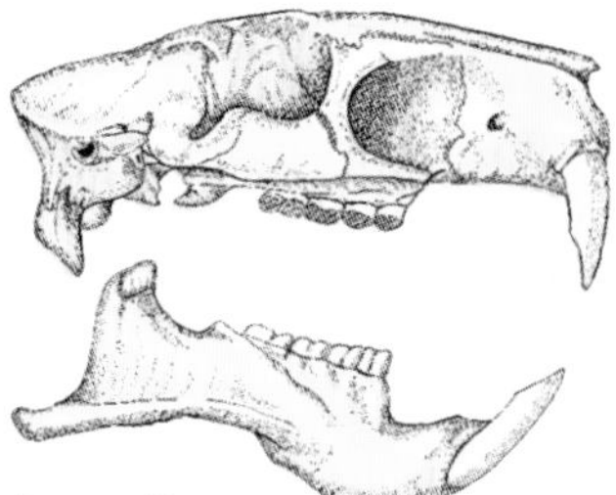

Coypu 13cm

lateral masseter

deep masseter

6

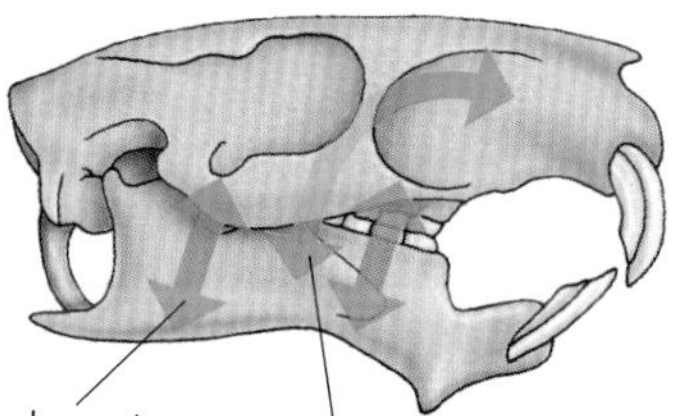

▲ ***Above*** *Representative species of cavy-like rodents (not shown to scale):* ***1*** *North African gundi* (Ctenodactylus gundi)*;* ***2*** *a domesticated form of the cavy, or "guinea pig"* (Cavia porcellus)*;* ***3*** *the distinctively deep, blunt muzzle of the capybara* (Hydrochaeris hydrochaeris)*;* ***4*** *Paca* (Agouti paca) *– this sizable species is particularly prized as a game animal by indigenous peoples of Amazonia;* ***5*** *North American porcupine* (Erethizon dorsatum)*;* ***6*** *Short-tailed chinchilla* (Chinchilla brevicaudata) *– hunted for its soft, valuable fur, this species is now listed as Critically Endangered.*

▼ ***Below*** *Perhaps the most bizarre of the cavy-like rodents are the Naked mole-rats of East Africa – virtually hairless creatures that spend almost all their time underground in extensive colonies. This is a breeding female.*

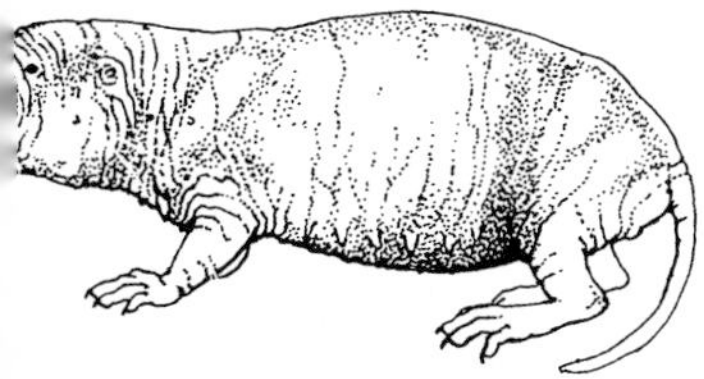

CAVIES AS A FOOD RESOURCE

The true cavy, or "guinea pig," has been highly regarded in South America since ancient times as an important source of meat for human consumption. There is clear evidence that many of the cultures that flourished along the Pacific coast of what is now Peru domesticated the cavy, adding variety to a diet based on fish and cultivated staple crops such as maize and manioc. Cavy bones dating from before 1800BC were discovered in a midden at the coastal site of Culebras. Even the later cultures that arose on the High Andean *altiplano* and farmed llamas and alpacas continued to value the cavy; chroniclers of the 16th century reported that the animal was among sacrificial offerings made by the Inca. Cavies are still widely kept and traded in the region (RIGHT); they make ideal livestock for the smallholder, since they can be kept with little attention in compounds and thrive on greenfoods unpalatable to humans such as brassicas.

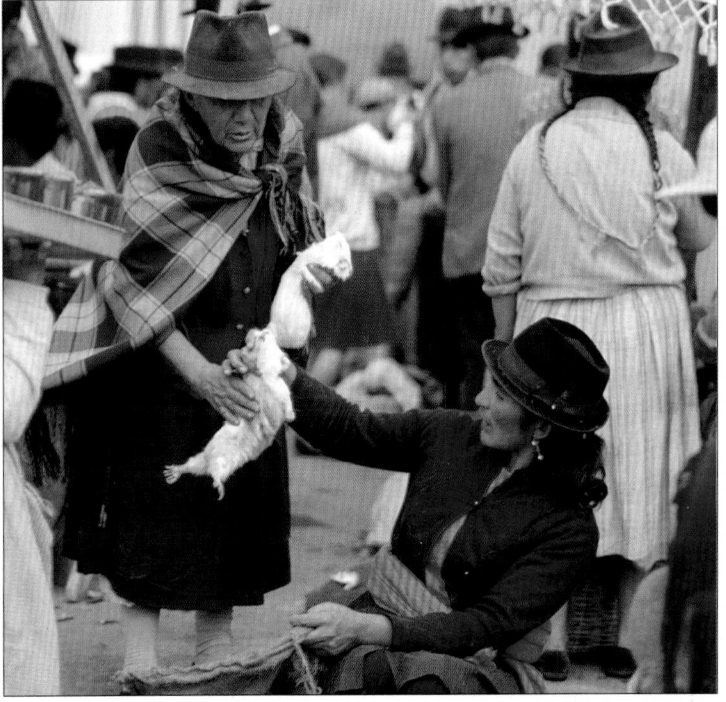

New World Porcupines

New World porcupines bear a strong resemblance to Old World porcupines in both their adaptations and lifestyle. However, while the former are arboreal and have singly embedded quills, the latter are decidedly terrestrial and have quills that are grouped in clusters.

For heavy-bodied animals that can weigh as much as 18kg (40lb), the New World porcupines are excellent climbers, with well-developed claws and unfurred soles on their large feet. The soles consist of pads and creases that increase the gripping power of the feet.

FACTFILE

NEW WORLD PORCUPINES

Order: Rodentia

Family: Erethizontidae

12 species in 4 genera

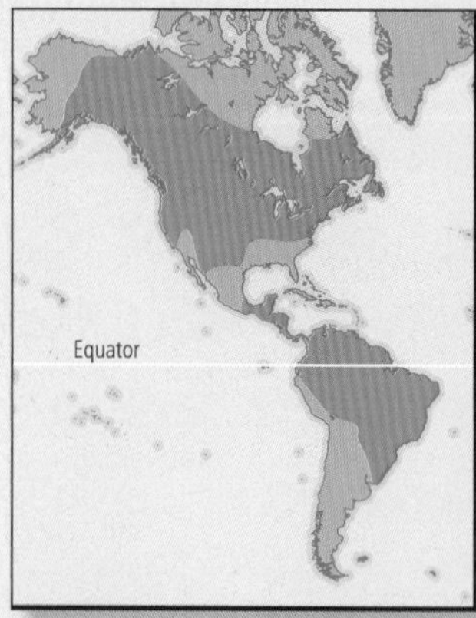

DISTRIBUTION N America (except SE USA), S Mexico, C America, northern S America

Habitat Forest areas, open grasslands, desert, canyon.
Coat Sharp, barbed quills interspersed with long guard hairs cover the upper part of the body. Each quill protrudes individually from the skin, unlike in Old World porcupines in which they are clustered in groups of 4–6.
Diet Bark, leaves, and conifer needles; also roots, stems, berries, fruits, seeds, nuts, grasses, flowers, and, in some species, insects and small reptiles.

PREHENSILE-TAILED PORCUPINES Genus *Coendou*
S Panama, Andes from NW Colombia to N Argentina, NW Brazil. Forest areas. 4 species: Brazilian porcupine (*C. prehensilis*), Bicolor-spined porcupine (*C. bicolor*), Koopman's porcupine (*C. koopmani*), and Rothschild's porcupine (*C. rothschildi*). HBL 30cm (12in); WT 900g (32oz).

HAIRY DWARF PORCUPINES Genus *Sphiggurus*
S Mexico, C America, S America as far S as N Argentina. Forest areas. 6 species, including: Mexican tree porcupine (*S. mexicanus*), South American tree porcupine (*S. spinosus*). Conservation status: one species, *S. vestitus* from Colombia and W Venezuela, is listed as Vulnerable.

NORTH AMERICAN PORCUPINE *Erethizon dorsatum*
Alaska, Canada, USA (except extreme SW, SE, and Gulf coast states), N Mexico. Forest areas. HBL 86cm (34in); WT 18kg (40lb). Breeding: gestation 210 days. Longevity: up to 17 years.

STUMP-TAILED PORCUPINE *Echinoprocta rufescens*
C Colombia. Forest areas.

Abbreviations HBL = head–body length WT = weight
For full species list see Appendix ▷

Equipped for Climbing

FORM AND FUNCTION

Individual genera have further modifications to improve their climbing abilities. The prehensile-tailed porcupines and the hairy dwarf porcupines – the most arboreal genera – have smaller first digits on their hind feet than those in the other genera, but they are incorporated in the footpads, which increases the width and the gripping power of the pads.

The same genera also have long, spineless tails for grasping. Their tips form upward-directed curls and have a hard skin or callus on the upper surface. In the prehensile-tailed porcupines the tail contributes 9 percent of the total body weight; nearly half of the weight of the tail is composed of muscle fibers.

New World porcupines are very near-sighted, but have keen senses of touch, hearing, and smell. They produce a variety of sounds – moans, whines, grunts, coughs, sniffs, shrieks, barks, and wails. All porcupines have large brains and appear to have good memories.

In winter North American porcupines feed on conifer needles and on the bark of a variety of trees, excepting Red maple, White cedar, and hemlock. During the summer these porcupines feed more frequently on the ground and select roots, stems, leaves, berries, seeds, nuts, and flowers. In the spring they frequently come out from forested areas into meadows to feed on grasses in the evening hours. They will eat bark at all times of the year, however, and can be destructive to forest plantations.

Prehensile-tailed and hairy dwarf porcupines feed more on leaves and both have many characteristics of arboreal leaf-eaters. However, they are also reported to feed on tender stems, fruits, seeds, roots, tubers, and insects, and will even consume small reptiles.

On the Move

SOCIAL BEHAVIOR

In habits porcupines range from the North American porcupine, which is semiarboreal, to prehensile-tailed and hairy dwarf porcupines, which are specialized arboreal feeders. All forms spend much of their time in trees, but even tree porcupines are known to come to the ground to feed and to move from one tree to another.

In the North American porcupine, the female reaches sexual maturity when about 18 months old. The estrous cycle is 29 days, and these animals may have more than one period of estrus in a year. They have a vaginal closure membrane, so females form a copulatory plug. The gestation period averages 210 days, and in both North American and prehensile-tailed porcupines usually one young is produced (rarely twins). The weight of the precocial newborn is about 400g

Right *The prehensile-tailed porcupines* (Coendou *sp.) live mainly in the middle and upper layers of forests in Central and South America, only descending to the ground to eat. The tail, which can be coiled around branches, has a callus pad which provides grip.*

Left *Despite its poor eyesight, inability to jump, general clumsiness, and size – large males can weigh in excess of 15kg (33lb) – the North American porcupine frequently climbs to great heights in search of food, such as berries, nuts, and shoots.*

(14oz) in prehensile-tailed porcupines and 600g (21oz) in the North American porcupine. Lactation continues for 56 days, but the animals also feed on their own after the first few days. Porcupine young are born with their eyes open and are able to walk. They exhibit typical defensive reactions, and within a few days are able to climb trees. These characteristics probably explain why infant mortality is very low. Porcupines grow for 3–4 years before they reach adult body size.

The home range of the North American porcupine in summer averages 14.6ha (36 acres). In winter, however, they do not range great distances, instead staying close to their preferred trees and shelters. Prehensile-tailed porcupines can have larger ranges, though these vary from 8 to 38ha (20–94 acres). They are reported to move to a new tree each night, usually 200–400m (660–1,300ft) away, but occasionally up to 700m (2,300ft). Prehensile-tailed porcupines in South Guyana are known to reach densities of 50–100 individuals per sq km (130–260 per sq mi). They have daily rest sites in trees, usually on a horizontal branch 6–10m (20–33ft) above the ground. These porcupines are nocturnal, changing locations each night and occasionally moving on the ground during the day. Male prehensile-tailed porcupines are reported to have ranges up to four times as large as those of females.

Sharing their Fate with the Forests

CONSERVATION AND ENVIRONMENT

Porcupines, in general, are not endangered, and the North American porcupine can in fact be a pest. The fisher (a species of marten) has been reintroduced to some areas of North America to help control porcupines, one of its preferred prey. The fisher is adept at flipping the North American porcupine over so that its soft and generally unquilled chest and belly are exposed. The fisher attacks this area. A study found that porcupines declined by 76 percent in an area of northern Michigan following the introduction of the fisher.

Prehensile-tailed porcupines are often used for biomedical research, which contributes to the problem of conservation, but the main threat is habitat destruction. In Brazil prehensile-tailed porcupines have been affected by the loss of the Atlantic forest, and the Paraguayan hairy dwarf porcupine is on the endangered species list published by the Brazilian Academy of Sciences. One species of porcupine may have become extinct in historic times: *Sphiggurus pallidus*, reported in the mid-19th century in the West Indies, where no porcupines now occur. CAW

Cavies

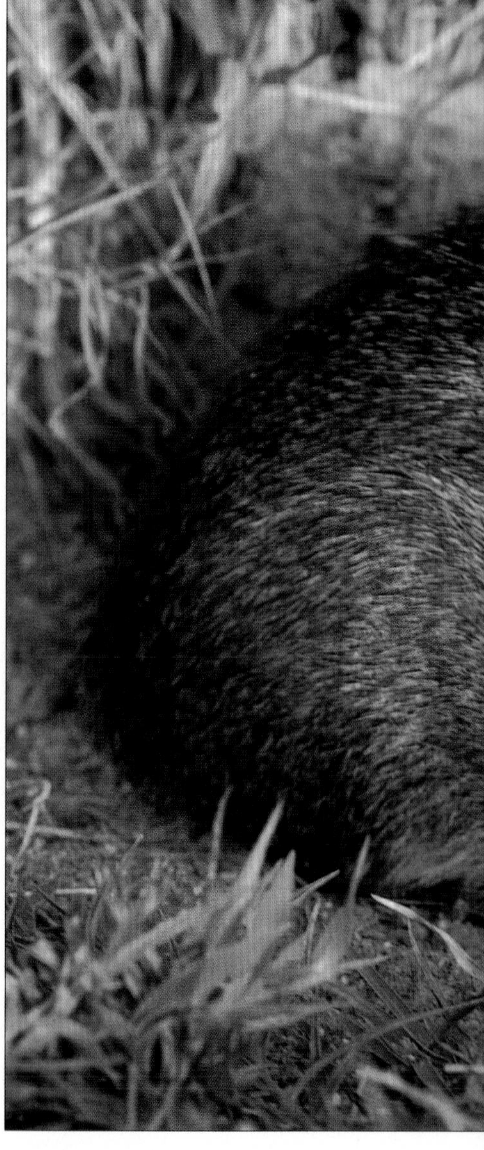

FACTFILE

CAVIES

Order: Rodentia

Suborder: Hystricognathi

Family: Caviidae

14 species in 5 genera

DISTRIBUTION S America (mara in C and S Argentina only)

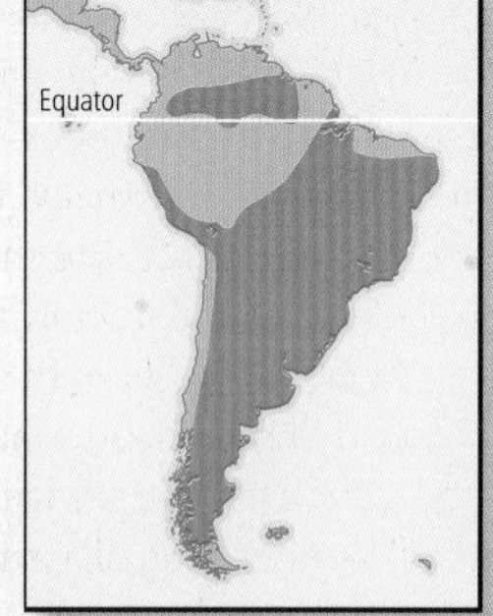

GUINEA PIGS AND CAVIES Genus *Cavia*
S America, in the full range of habitats. Coat: grayish or brownish agouti; domesticated forms vary. 5 species: *C. aperea, C. fulgida, C. magna, C. porcellus* (Domestic guinea pig), *C. tschudii.*

MARA Genus *Dolichotis*
Mara or Patagonian hare or cavy
S America (C and S Argentina), occurring in open scrub desert and grasslands. HBL 50–75cm (19.7–30in), TL 4.5cm (1.8in), WT 8–9kg (17.6–19.8lb). Coat: head and body brown, rump dark (almost black) with prominent white fringe round the base; belly white. Gestation: 90 days. Longevity: up to 15 years. 2 species: *D. patagonum* and *D. salinicola*. Conservation status: Lower Risk – Near Threatened.

YELLOW-TOOTHED CAVIES Genus *Galea*
Yellow-toothed cavies or cuis
S America, in the full range of habitats. Coat: medium to light brown agouti, with grayish-white underparts. Gestation: 50 days. 3 species: *G. flavidens, G. musteloides, G. spixii.*

ROCK CAVY *Kerodon rupestris*
NE Brazil, occurring in rocky outcrops in thorn-scrub. HBL 38cm (15in), WT 1kg (2.2lb). Coat: gray, grizzled with white and black; throat white, belly yellow-white, rump and backs of thighs reddish. Gestation: 75 days.

DESERT CAVIES Genus *Microcavia*
Argentina and Bolivia, in arid regions. HBL 22cm (8.7in), weight 300g (10.7oz). Coat: a coarse dark agouti, brown to grayish. Gestation: 50 days. Longevity: 3–4 years (up to 8 in captivity). 3 species: *M. australis, M. niata, M. shiptoni.*

Abbreviations HBL = head–body length TL = tail length WT = weight

MOST PEOPLE ARE FAMILIAR WITH CAVIES, but under a different and somewhat misleading name: guinea pigs. "Guinea" refers to Guyana, a country where cavies occur in the wild, while "pig" derives from the short, squat body of this rodent (the pork-like quality of the flesh doubtless also played a part in its naming). The Domestic guinea pig was being raised for food by the Incas when the conquistadors arrived in Peru in the 1530s, and is now found the world over, with one exception: it no longer occurs in the wild.

Cavies are among the most abundant and widespread of all South American rodents. They live in a variety of habitats ranging from tropical floodplains through open grasslands and forest edges to rocky meadows 4,000m (13,000ft) up. For a long time, they – or at least their domesticated cousins – were considered very stupid animals; psychologists found it difficult to set up tests in which guinea pigs showed any sign of learning or intelligence. However, it now appears that the tasks set for them may have been inappropriate.

To remedy this and to give cavies a fair chance to display their brainpower, experiments have recently been designed to play to their strengths. The basic prerequisites are a satisfactory laboratory and experiment habituation period; providing the animal with its daily requirements of vitamin C; a choice of task adapted to the animal's natural habits; and, finally, a controlled environment with as little extraneous noise as possible, to prevent the guinea pig from "freezing." In these conditions, guinea pigs have shown that they can learn at a similar rate to other mammalian species, particularly rats. Furthermore, when they were trained to forage under different experimental conditions (which included manipulations of travel time between food sources, food gain rates, and food availability), guinea pigs proved adept at optimizing their foraging efficiency.

Above *The Brazilian guinea pig* (Cavia aperea) *is one of three wild species from which the familiar Domestic guinea pig may have derived. The coats of the wild species are relatively long and coarse.*

Below *Rock cavies' unusual mating behavior entails defending isolated rock piles to which females are drawn in search of shelter; in this way, the males accumulate harems. When approaching a female in estrus, the male circles around her to block her path* ***1****, then passes under her chin* ***2*** *before attempting to mount* ***3****.*

Small, Alert, and Nervous

FORM AND FUNCTION

Cavies are among the most abundant and widespread of all South American rodents. All except the mara share a basic form and structure. The body is short and robust and the head large, contributing about one-third of the total head–body length. The eyes are fairly large and alert, the ears big but close to the head. The fur is coarse and easily shed when the animal is handled. There is no tail. The forefeet are strong and flat, usually with four digits, each equipped with sharp claws; the hind feet, with three clawed digits, are elongated. Cavies walk on their soles, with the heels touching the ground. The incisors are short, and the cheek teeth, which are arranged in rows that converge towards the front of the mouth, have the shape of prisms and are constantly growing. Both sexes are alike, apart from each possessing certain specialized glands.

Cavies are very vocal, making a variety of chirps, squeaks, churrs, and squeals. One genus, *Kerodon*, emits a piercing whistle when frightened. *Galea* species rapidly drum their hind feet on the ground when anxious.

Cavies first appeared in the mid-Miocene era in South America. Since their appearance some 20 million years ago, the family has undergone an extensive adaptive radiation, reaching peak diversity between 5 and 2 million years ago. From a peak of 11 genera during the Pliocene, they fell to their current 5 genera during the Pleistocene, about 1 million years ago.

The Cautious Life of Prey Species

SOCIAL BEHAVIOR

The 12 remaining species of the subfamily Caviinae, which comprises all the cavies except the *Dolichotis* species, are widely distributed throughout South America. All 12 are to a degree specialized for exploiting open habitats. Cavies are found in grasslands and scrub forests from Venezuela to the Straits of Magellan, but each genus has evolved to exist in a slightly different habitat.

Cavia is the genus most restricted to grasslands. In Argentina, *Cavia aperea* is limited to the humid pampas in the northeastern provinces. *Microcavia australis* is the desert specialist, and is found throughout the arid Monte and Patagonian deserts of Argentina. Other *Microcavia* species, *M. niata* and *M. shiptoni*, occur in the arid, high-altitude *puña* (subalpine zone) of Bolivia and Argentina. The specialized genus *Kerodon* is found only in rocky outcrops called *laleiros* that dot the countryside in the thorn-scrub of northeastern Brazil. *Galea* seem to be the "jacks-of-all-trades" of the cavies, found in all the above habitats; it is also the only genus that coexists with other genera.

Regardless of habitat, all cavies are herbivorous. *Galea* and *Cavia* feed on herbs and grasses. *Microcavia* and *Kerodon* seem to prefer leaves; both genera are active climbers, which in the case of the *Kerodon* species is surprising because they lack claws and a tail, two adaptations usually associated with life in trees. The sight of an 800g (28oz) guinea pig hurrying along a pencil-thin branch high in a tree is quite striking.

All cavies become sexually mature early, at 1–3 months. The gestation period is fairly long for rodents, varying from 50–75 days. Litter sizes are small, averaging about 3 for *Galea* and *Microcavia*, 2 for *Cavia*, and 1.5 for *Kerodon*. The young are born highly precocial. Males contribute little obvious parental care, and generally ignore the female and her young once the litter is born.

Three species of cavies have been studied in northeastern Argentina: *Microcavia australis*, *Galea musteloides*, and *Cavia aperea*. *Cavia* and *Microcavia* never occur in the same area, *Cavia* preferring moist grasslands and *Microcavia* more arid habitats. *Galea* occurs with both genera. Competition between *Galea* and *Microcavia* seems to be minimized by the use of different foraging tactics: *Microcavia* is more of a browser, and arboreal. The degree to which *Cavia* and *Galea* interact within the same areas is unknown. Home-range sizes are, on average, 3,200sq m (34,500sq ft) for *Microcavia*, and 1,300sq m (14,000sq ft) for *Galea* and *Cavia*. They are diurnal, and are active mainly during early morning and evening hours.

The genus *Cavia* is the most widely distributed of all cavies, ranging over almost all South America from Colombia to Argentina. *Cavia* species breed year-round, but are less active in the winter. In the Pampas region they can occur in high densities, especially in late autumn, typically inhabiting linear habitats such as field margins and roadsides which have a zone of tall and dense vegetation. They feed in open areas of short vegetation, but return to the borders for protection.

The amount of time cavies spend watching for danger while feeding varies with the chances of being attacked by predators. In the open delta of the Paraná river, Argentina, those feeding in short grass away from cover were found by researchers to look up for danger three times more often than those close to shelter. Almost 50 percent of feeding occurred within 1m (3.3ft) of shelter, and the animals never strayed more than 4m (13ft) from cover. They recognized safety in numbers, staying in the open more than twice as long when in groups than when they were alone.

Cavies have several important predators. The grison – a South American weasel – has been known to almost wipe out a population of cavies over a period of five months. High-density populations also attract several species of raptor, which seem to kill more prey than they can consume, as evidenced by the presence of uneaten cavy carcasses where raptors have been.

Little is known of cavy reproductive behavior in the wild. Captive males are very aggressive, making it almost impossible to keep them together in the presence of females. As a result, only one male is present whenever a female comes into estrus. Females organize themselves in linear dominance hierarchies that are strictly age-dependent.

Cavies are considered a major pest of tree crops in many regions of Argentina. For example, in the Paraná Delta, cavies and Red rats (*Holochilus brasiliensis*) destroyed more than 50 percent of the cultivated salicaceous trees by gnawing a fringe of bark 40cm (16in) up above ground level. Several methods of reducing their impact on forestry have been experimentally introduced, including chemical repellents and covering the stalks of seedlings with polyethylene tubes.

Below *Two Rock cavies* (Kerodon rupestris) *huddle on a branch in northeast Brazil. The animals take their name from the rock piles in which they live and breed, leaving them each evening to forage. Agile climbers, they spend much of their feeding time in trees, where they go in search of their chief staple, tender leaves.*

Galea musteloides can live both at sea level and at altitudes up to 5,000m (16,400ft). Males living in large, mixed-sex groups form hierarchies in which higher-ranking animals sire more young than do subordinates – between 70 and 90 percent of all offspring, according to one observation of captive animals that employed DNA fingerprinting techniques. Aggression directed towards subordinates in the colony by higher-ranking individuals may have the effect of suppressing the losers' sex hormones. Equally intriguing is the fact that females appear to be tolerant of unfamiliar pups while they are breeding, probably as an adaptation to living in reproductive groups of related females with synchronized births and communal suckling.

In *Microcavia* species, aggression between males defines a linear dominance hierarchy within colonies. The stability of the social groups seems to vary between habitats: in deserts they keep strict fidelity to a burrow system to which their

Above A Southern mountain cavy (Microcavia australis) keeps watch over her growing brood on the Valdés Peninsula in Argentina. Three is a typical litter size for this short-lived species; mothers become receptive again immediately after giving birth, and can have up to five litters a year.

group territoriality is restricted, but in less arid habitats they show a lax social organization without permanent groups. *Microcavia* sites its colonial burrows beneath bushes with broad canopies that are low to the ground, presumably as protection from predators. Grazing by *Microcavia* can damage an important number of plants; a study conducted in the Nacuñán Reserve, western Argentina, showed that herbivory by Highland tuco-tucos (*Ctenomys opimus*) and *Microcavia* species affected 35 percent of the total plants of a creosote bush community dominated by *Larrea cuneifolia*.

Two species of cavies coexist in northeastern Brazil: *Kerodon rupestris* and *Galea spixii*. *Galea spixii* is similar to the Argentine *Galea* in ecology, morphology, color, and behavior. The animals inhabit thorn forests, are grazers, and have a noncohesive social organization.

The Rock cavy (*Kerodon rupestris*) is markedly different from all the other small cavy species. It is larger and leaner, and has a face that is almost doglike. All small cavies except *Kerodon* have sharply clawed digits; *Kerodon* has nails growing from under the skin, with a single grooming claw on the inside hind-toes and extensively padded feet. The modifications of the feet facilitate movement on slick rock surfaces. Rock cavies are strikingly agile as they leap from boulder to boulder, executing graceful mid-air twists and turns. They are also exceptional climbers, and forage almost exclusively on leaves in trees. There is little competition for resources with *Galea*.

Perhaps the most interesting difference between *Kerodon* and *Galea* is behavioral. *Galea* species, like the Argentine cavies, live in thornscrub forest and have a promiscuous mating system. Rock cavies inhabit isolated patches of boulders, many of which can be defended by a single male, and single males seem to have exclusive access to two or more females. The boulder piles attract females, so by defending these sites males monopolize the female tenants. This system parallels that of the unrelated hyraxes of eastern Africa (see Subungulates: Hyraxes).

Natural Adaptors in Need of Help

CONSERVATION AND ENVIRONMENT

Most cavy species can adapt to altered and disturbed habitat, and some do well among human settlements. But one species, *Kerodon*, is in trouble. Hunted extensively, these Rock cavies are declining in numbers and are in desperate need of protection. Because they are patchily dispersed throughout their range, large areas will have to be set aside; indeed, two such research reserves have already been established in Brazil. TEL/MC

SPECIAL FEATURE

LIFELONG PARTNERS

Colonial breeding in the monogamous mara

AS DAWN BROKE ACROSS THE PATAGONIAN thorn-scrub a large female rodent, with the long ears of a hare and the body and legs of a small antelope, cautiously approached a den, followed by her mate. They were the first pair to arrive, and so walked directly to the mouth of the den. At the burrow's entrance the female made a shrill, whistling call, and almost immediately eight pups burst out. The youngsters were hungry and all thronged around the female, trying to suckle. Under this onslaught she jumped and twirled to dislodge the melee of unwelcome mouths which sought her nipples. The female sniffed each carefully, chasing off all but her own. Finally, she managed to select her own two offspring from the hoard and led them 10m (33ft) to a site where they would be nursed.

In the meantime her mate sat alert nearby. If another adult pair had approached the den while his female was there, coming to tend their own pups, he would have made a vigorous display directly in front of his mate. If the newcomers had not moved away he would have dashed towards them, with his head held low and neck outstretched, and chased them off. The second pair would then have waited, alert or grazing, at a distance of 20-30m (65-100ft). When the original pair had left the area the new pair would then approach the den, to collect their own pups.

The animal being observed was the mara or Patagonian cavy, an 8kg (17.6lb) hare-like day-active cavy, *Dolichotis patagonum*. (The behavior of the only other member of the subfamily Dolichotinae, the Salt desert cavy, *Dolichotis salinicola*, is unknown in the wild.) A fundamental aspect of the mara's social system is the monogamous pair bond, which in captivity, and probably in the wild, lasts for life. The drive that impels males to bond with females is so strong that it can lead to "cradle snatching" – adult bachelor males attaching themselves to females while the latter are still infants. Contact between paired animals is maintained primarily by the male, who closely follows the female wherever she goes, discouraging approaches from other maras by policing a moving area around her about 30m (100ft) in diameter. In contrast, females appear less concerned about the whereabouts of their mates. While foraging, members of a pair maintain contact by means of a low grumble.

Monogamy is not common in mammals and in the mara several factors probably combine to favor this system. It typically occurs in species where there are opportunities for both parents to care for the young, yet in maras virtually all direct care of the offspring is undertaken by the mother. However, the male does make a considerable indirect investment. Due to the high amount of energy a female uses in bearing and nursing her young, she has to spend a far greater part of the day feeding than the male: time during which her head is lowered and her vigilance for predators impaired. On the other hand, the male spends a larger proportion of each day scanning and is thus able to warn the female and offspring of danger. Also, by defending the female against the approaches of other maras, he ensures uninterrupted time for her to feed and care for his young. Furthermore, female maras are sexually receptive only for a few hours twice a year; in Patagonia females mate in June or July and then come into heat again in September or October, about 5 hours after giving birth, so a male must stay with his female to ensure he is with her when she is receptive.

Mara pairs generally avoid each other and outside the breeding season it is rare to see pairs within 30m (about 100ft) of each other. Then their home ranges are about 40ha (96 acres). Perhaps the avoidance between pairs is an adaptation to the species' eating habits. Maras feed primarily on short grasses and herbs which are sparsely, but quite evenly, distributed in dry scrub desert. So far, detail of their spatial organization is unknown, but there is at least some overlap in the movements of neighboring pairs. Furthermore, there are some circumstances when, if there is an abundance of food, maras will aggregate. In the Patagonian desert there are shallow lakes, 100m to several kilometers in diameter, which contain water for only a few months of the year. When dry, they are sometimes carpeted with short grasses that maras relish. At these times, toward the end of the breeding season, up to about 100

Above *A mother and pups. A female mara will normally give birth at one time to a maximum of three well-developed young. She will nurse them for an hour or more once or twice a day for up to four months.*

Left *Foraging and feeding can be a dangerous time for maras, since it leaves them more vulnerable to attacks from predators. It is therefore essential that one of a pair remains alert.*

maras will congregate. However, individual pairs remain loyal to one another after the congregations have dispersed.

The strikingly cohesive monogamy of maras is noteworthy in its contrast with, and persistence throughout, the breeding season when up to 15 pairs become at least superficially colonial by depositing their young at a communal den. The dens are dug by the females and not subsequently entered by adults. The same den sites are often used for three or more years in a row. Each female gives birth to one to three young at the mouth of the den; the pups soon crawl inside to safety. Although newborn pups are moving about and grazing within 24 hours of birth, they remain in the vicinity of the den for up to four months, and are nursed by their mother once or twice a day during this period. Around the den an uneasy truce prevails amongst the pairs whose visits

Above *In the open grasslands of Patagonia, where cover is scarce, maras spend a large part of the day grazing or basking, but keep their keen sense of hearing attuned to any potential danger.*

coincide. The number of pairs of maras breeding at a den varies from 1 to at least 15 and may depend on habitat. Pairs come and go around the den all day and in general at the larger dens at least one pair is always in attendance there. Even when 20 or more young are kept in a crèche, cohabiting amicably, the monogamous bond remains the salient feature of the social system. Each female sniffs the infants seeking to suckle and they respond by proffering their anal regions to the female's nose. Infants clambering to reach one female may differ by at least one month in age. A female's rejection of a usurper can involve a bite and violent shaking. However, interlopers occasionally secure an illicit drink. Although females may thus be engaging in communal nursing they rarely seem to do so willingly.

The reasons why normally unsociable pairs of maras keep their young in a communal crèche are unknown, but it may be that it lowers the likelihood of them falling victim to a predator; indeed, the more individuals at a den (both adults and young) the more pairs of eyes there are to detect danger. Furthermore, some pairs travel as much as 2km (1.2mi) from their home range to the den, so the opportunities for shared surveillance of the young may diminish the demands on each pair for protracted attendance at the den. The unusual breeding system of the monogamous mara may thus be a compromise, conferring on the pups the benefits of coloniality, in an environment wherein association between pairs is otherwise apparently disadvantageous. AT/DWM

Capybara

Savanna-dwelling social grazers averaging a hefty 50kg (110lb), capybaras are unusual animals. They are found only in South America, where they live in groups near water. Members of the suborder known as caviomorphs, which also includes cavies and chinchillas, they are in fact the largest of all the rodents.

The first European naturalists to visit South America called capybaras "water pigs" or "Orinoco hogs," and the first of those names has carried over into their present scientific designation as hydrochaerids. Yet in truth, they are neither pigs nor totally aquatic; their nearest relatives are actually the cavies. The other hydrochaerids, now all extinct, were larger than present-day capybaras; the biggest were twice as long and probably weighed eight times as much, making them heavier than the largest North American Grizzly.

The Biggest Rodents

Form and Function

Capybaras are ponderous, barrel-shaped animals. They have no tail, and their front legs are shorter than their back legs. Their slightly webbed toes, four on the front feet and three on the back, make them very strong swimmers, able to stay under water for up to 5 minutes. Their skin is extremely tough and covered by long, sparse, bristle-like hairs. The nostrils, eyes, and ears are situated near the top of the large, blunt head, and hence protrude out of the water when the animal swims. Two pairs of large, typically rodent incisors enable capybaras to eat very short grasses, which they grind up with their molar teeth. There are four molars on each side of each lower jaw. The fourth molar is characteristic of the subfamily in being as long as the other three.

Two kinds of scent glands are present in the capybara. One gland, highly developed in males but almost nonexistent in females, is located on top of the snout and is known as the *morrillo* (literally, "hillock" in Spanish). This is a dark, oval-shaped, naked protrusion that secretes a copious, white, sticky fluid. Both sexes also produce odors from two glandular pockets located on either side of the anus. Male anal glands are filled with easily detachable hairs abundantly coated with layers of hard, crystalline calcium salts. Female anal pockets also have hairs, but theirs are not detachable and are coated in a greasy secretion rather than with crystalline layers. The proportions of each chemical present in the secretions of individual capybaras are different, providing a potential for individual recognition via personal "olfactory fingerprints." The snout scent gland also plays a role in signaling dominance status, while the anal gland appears to be important in group membership recognition and perhaps in territoriality.

FACTFILE

CAPYBARA

Hydrochaeris hydrochaeris

Order: Rodentia

Family: Hydrochaeridae

Sole member of genus and of family. 2 populations, one E of the Andes from Venezuela to N Argentina, the other from NW Venezuela through N Colombia up to the Panama Canal. Some authorities regard the two as separate species, named respectively *Hydrochaeris hydrochaeris* and *H. isthmius*.

DISTRIBUTION South America

HABITAT Flooded savanna or grassland next to water holes; also, along ponds and rivers in tropical forest.

SIZE Head–body length 106–134cm (42–53in); shoulder height 50–62cm (20–24in); **weight** male 35–64kg (77–141lb), female 37–66kg (81.6–146lb).

COAT Light brown, consisting of short, abundant hairs in young; adults have long, sparse, bristle-like hairs of variable color, usually brown to reddish.

DIET Mostly grasses, especially aquatic; occasionally water hyacinths and other dicotyledons.

BREEDING Females are sexually mature at 12 months, males at 18 months. Gestation 150 days. 1–8 (average 4) young, born mostly at the end of the wet season, after 150–day gestation. Very precocial: young graze within hours of birth.

LONGEVITY About 6 years (12 in captivity).

CONSERVATION STATUS Lower Risk – Conservation Dependent.

Above *Capybaras are born after a long gestation of over five months. Even though they emerge in a well-developed condition and can eat soon after birth, it is over a year before they reach sexual maturity. Although the young are conceived in water, they are born on land.*

Left *Water is a place of refuge for capybaras. Capybaras live either in groups averaging 10 in number or in temporary larger aggregations, which may contain up to 100 individuals and will be composed of the smaller groups. The situation varies according to the season.*

Capybaras have several distinct vocalizations. Infants and the young constantly emit a guttural purr, probably to maintain contact with their mothers or other group members. This sound is also made by losers in altercations, perhaps to appease their adversary. Another vocalization, the alarm bark, is given when a predator is detected. This coughing sound is often repeated several times, and the reaction of nearby animals may be to stand alert or to rush into the water.

Recycling Cellulose

DIET

Capybaras are exclusively herbivorous, feeding mainly on grasses that grow in or near water. They are very efficient grazers, and can crop the short, dry grasses left at the end of the tropical dry season. Because a large proportion of the grasses they eat consists of cellulose, which is indigestible by any mammal's digestive enzymes, capybaras possess a huge fermentation chamber called the cecum, equivalent to our tiny appendix. However, since the cecum is located between the small and large intestine, the animal cannot absorb the products of the fermentation carried out by microbial symbionts. To solve this problem, capybaras resort to coprophagy – reingestion of feces – in order to be able to take advantage of the work of their symbionts. Thus, for a few hours every morning during their resting period, capybaras recycle what they ate the previous evening and night. Usually they spend the morning resting, then bathe during the hot midday hours; in the late afternoon and early evening they graze. At night they alternate rest periods with feeding bouts. They never sleep for long, instead dozing in short bouts throughout the day.

Gregarious Grazers

SOCIAL BEHAVIOR

Capybaras live in groups of 10–30 animals, apparently depending on the habitat: greener and more homogeneous pastures promote larger groups. Pairs are rarely seen, but a proportion of adult males are solitary or loosely associated to one or more groups. In the dry season groups coalesce around the dwindling pools, forming temporary aggregations of 100 or more animals. When the wet season returns, these large aggregations split up into the original groups that formed them. Thus, capybara social units may last three years, and probably more.

Groups of capybaras are closed social units, in which little variation in core membership is observed. A typical group is composed of a dominant male (often distinguishable by his large *morrillo*), one or more females, several infants and young, and one or more subordinate males. Among the males there is a dominance hierarchy, maintained by aggressive interactions that usually take the form of simple chases. Dominant males repeatedly shepherd their subordinates to the periphery of the group, but fights are rarely seen. Females are much more tolerant of each other, although the precise details of their social relationships, hierarchical or otherwise, are unknown. Each home range is used almost exclusively by one group, and can therefore be regarded as a territory. Territories are defended by all adult members of the group against conspecific intruders. Any animal of either sex may chase an interloper away, irrespective of its sex, as long as the chaser is within its own territory.

Capybaras are found in a wide variety of habitats, ranging from open grasslands to tropical rain forest. Groups may occupy an area varying in size from 2–200ha (4–494 acres), with 10–20ha (24.7–49.4 acres) being most common. Each home range is used mainly, but not exclusively, by one group and can therefore be considered a territory. Territories are defended against conspecific intruders by all adult members of the group. Particularly in the dry season, but at other times as well, two or more groups may be seen grazing side by side. In some areas density may reach 2 individuals per hectare (5 per acre), but lower densities of less than 1 per hectare are more frequent.

Capybaras reach sexual maturity at 18 months. In Venezuela and Colombia they appear to breed year round, with a marked peak at the beginning of the wet season in May. In Brazil, in more temperate areas, they probably breed just once a year. When a female becomes sexually receptive, a male will start a pursuit that may last for an hour or more. The female will walk in and out of the water, repeatedly pausing while the male follows close behind. The mating takes place in the water; the female stops, and the male clambers on her back, sometimes thrusting her under water with his weight. As is usual in rodents, copulation lasts only a few seconds, but each sexual pursuit typically involves several mountings.

150 days later, up to seven babies are born; four is the average litter size. To give birth the female leaves her group and walks to nearby cover. Her young are born a few hours later, and are precocial, able to eat grass within their first week. A few hours after the birth the mother rejoins her group, the young following as soon as they become mobile, which should occur when the babies are still very young. Females seem to share the burden of nursing by allowing infants other than their own to suckle. The young in a group spend most of their time within a tight-knit crèche, moving between nursing females. When active, they constantly emit a churring purr.

Capybara infants tire quickly, and are therefore vulnerable to predators. They have most to fear from vultures and feral or semiferal dogs, which prey on them. Caymans and foxes may also take young capybaras. Jaguar and smaller cats were certainly important predators in the past, though today they are nearly extinct in most of Venezuela and Colombia. In some areas of Brazil, however, jaguars seize capybaras in substantial numbers.

When a predator approaches a group the first animal to detect it will emit an alarm bark. The normal reaction of other group members is to stand alert, but if the danger is very close, or the caller keeps barking, they will all rush into the water, where they form a close aggregation with young in the center and adults facing outward.

Putting a Cap on Hunting

CONSERVATION AND ENVIRONMENT

Capybara populations have dropped so substantially in Colombia that, from 1980 onward, the government prohibited capybara hunting. In Venezuela they have been killed since colonial times in areas that are devoted to cattle ranching. In 1953 hunting became subject to legal regulation and controlled, but to little effect until 1968 when, after a 5-year moratorium, a management plan was devised, based on a study of the species' biology and ecology. Since then, 30–35 percent of the annually censused population in licensed ranches with populations of over 400 animals have been harvested every year. This has apparently resulted in local stabilization of capybara populations. Capybaras are now listed as Lower Risk: Conservation Dependent by the IUCN, in recognition of the fact that control on hunting and harvesting must remain if population levels are to be maintained. DWM/EH

Left A capybara marking its territory. It is instantly recognizable as a male from the prominent, bare lump on top of its snout – the morrillo *sebaceous gland, which contains the animal's highly individual scent.*

Below A South American waterbird – the jacana – searches a capybara's coat for parasitic insects. The capybara spends much of its time in water and can travel long distances submerged; its small eyes and ears are thought to be an adaptation for living in water.

FARMING CAPYBARA

In Venezuela there has been a demand for capybara meat at least since the early 16th century, when Roman Catholic missionary monks classified it, along with terrapin, as legitimate Lenten fare; the amphibious habits of the two species presumably misled the monks into thinking they had an affinity with fish. Today, because of their size and high reproductive rate as well as the tasty meat and valuable leather they can provide, capybaras are candidates for both ranching and intensive husbandry.

It has been calculated that, where the savannas are irrigated to mollify the effects of the dry season, the optimal capybara population for farming is 1.5–3 animals per hectare (or about 1 per acre), yielding 27kg of meat per ha (24lb per acre) per annum. Ranches licensed to harvest the population can sustain yields of about 1 animal per 2 ha in good habitat. An annual cull takes place in February, when reproduction is at a minimum and the animals congregate around waterholes. Horsemen herd them together, and they are then surrounded by a cordon of cowboys on foot. An experienced slaughterman selects adults weighing over 35kg (77lb), excluding pregnant females, and kills them with a blow from a heavy club. The average animal weighs 44.2kg (97.4lb), of which 39 percent (17.3kg; 38lb) is dressed meat. These otherwise unmanaged wild populations thus yield over 8kg of meat per hectare (7lb per acre) annually.

In spite of this yield, farmers have traditionally feared that large populations of capybaras would compete with domestic stock. In fact, however, capybaras selectively graze on short vegetation near water and so do not compete significantly with cattle, which take taller, drier forage, except in wet, low-lying habitats. In these regions, capybaras are actually much more efficient at digesting the plant material than are cattle and horses. So, ranching capybara in their natural habitat appears to be, both biologically and economically, a viable adjunct to cattle ranching.

Other Cavy-like Rodents

THE FAMILIES ASSEMBLED HERE ARE PART OF a disparate group of predominantly South American mammals known as the Caviomorpha. The diversity they show is in marked contrast to the relative homogeneity of other rodent groups, such as the squirrels or the rats and mice.

The group includes small, medium, and very large rodents; some are covered with barbed spines, others have soft, silky fur; nearly all are herbivorous, although a few are not averse to including insects or larger prey in their diet. Many are terrestrial but others live as burrowers or tree-dwellers, or else spend much of their time swimming and feeding in water. They inhabit forests and grasslands, water and rocky deserts, coastal plains and high mountains; some are solitary, others colonial. Some species are common and widespread, others known only from a few specimens in museums; yet others have become extinct, some of them within historical times, often as a result of human activities. Many species are eaten by humans, others are prized for their fur; some are pests, while others carry the diseases of humankind and domestic animals.

The larger species of South American rodents, such as agoutis, pacas, pacaranas, and viscachas, are prey for the large and medium-sized carnivores (jaguars, ocelots, Pampas cats, Maned wolves, Bush dogs, foxes, and others). They are herbivorous, and may be considered the South American equivalents of the vast array of ungulate herbivores that are so important in the African ecosystems. It is thought that these rodents radiated into this role as the primitive native herbivores became extinct, and before the arrival of the new fauna from the north.

Despite their obvious diversity, the cavy-like rodents have many anatomical and other features in common. Particularly striking are the similarities in reproduction, such as the long gestation period exhibited by many species, the small number of young in each litter, and the advanced state of development many show at birth. Especially in many of the medium to large species, the young are born fully furred and with their eyes open; some are able to run within a few hours of birth, and many become independent of their mothers relatively soon afterwards.

Dassie Rat

FAMILY PETROMYIDAE

The Dassie rat is superbly adapted to the dry, rocky hillside country of southern Africa in which it lives. The soft pelage is gray, buff, or tawny in color, making the animal difficult to spot when lying on rocks, and it has a flattened skull and very flexible ribs, enabling it to squeeze into narrow crevices. It forages on the ground or in bushes for leaves, berries, and seeds, and is particularly active at dawn and dusk, resting and sunning itself below projecting rocks so as to avoid predatory birds. Dassie rats are solitary or else found in pairs, yet utter a warning whistling call if alarmed. One or two well-developed young are born once per year, at the start of the rainy season, so the species has a relatively slow reproduction rate for a small tropical mammal.

Cane Rats

FAMILY THRYONOMYIDAE

Cane rats or grasscutters are robust African rodents with a coarse, bristly pelage, the bristles being flattened and grooved along their length. The pelage is brown speckled with yellow or gray above, buff-white below, allowing them to blend well with the grasses and reeds in which they feed and live. The preferred semi-aquatic habitat for *Thryonomys swinderianus* is reed beds, marshes, and the margins of lakes and rivers, while *T. gregorianus* occurs in dryer areas of moist savanna grasslands. Cane rats live in small groups, communicating by calls and stamping their hind feet. They shelter among tall grasses or in rock crevices, or in the abandoned burrows of other animals; they also sometimes excavate their own shallow burrows. As their common name suggests, they feed mainly on grass and cane, but also a variety of other vegetation including bark, nuts, and fruits, and they may be pests of plantations and cultivated land, especially where population densities are high. Cane rats are prey for leopards, mongooses, snakes, birds of prey, and humans. In some parts of their range, they breed all year round, but most have two litters each year with an average of four young per litter. The young are well-developed at birth.

Chinchillas and Viscachas

FAMILY CHINCHILLIDAE

Chinchillids live in relatively barren regions, and all have thick, soft fur, although chinchillas and mountain viscachas, occurring at higher elevations, have denser fur than that of the pampas-dwelling Plains viscacha. They are slender-bodied animals with large ears and tails up to one-third the length of the body, and their long, strong hind legs enable them to run and jump agilely.

All the family are colonial. Chinchillas live in holes and crevices among rocks, and emerge at dusk to forage during the night for any available vegetation. In contrast, mountain viscachas forage for grasses, mosses, and lichens, and sun themselves on rocks during the day; they generally live in family groups that coexist with others to form small to very large colonies. Colonies of Plains viscachas live in extensive and complex burrows, consisting of a central chamber from which radiating tunnels lead to various entrances; they feed mainly on grass and seeds at dawn and dusk.

All species are subject to pressure from human hunting. Chinchillas have been pursued for their valuable fur to near-extinction; mountain and Plains viscachas are prized for both food and fur, and the Plains viscacha competes for grazing with domestic animals. In addition, they destroy pasture with their acidic urine, and so undermine the pampas that men, horses, and cattle are often injured by falling into their concealed tunnels.

Pacarana

FAMILY DINOMYIDAE

The pacarana is solitary or lives in pairs. It has a broad head with short, rounded ears, a robust body, and broad, heavily clawed feet. The pelage is coarse, black or brown with two more or less continuous white stripes on each side. A forest-dwelling species, seldom encountered and little known, this slow-moving, inoffensive herbivore is prey for jaguars, ocelots, and other medium-sized carnivores, and is hunted for food by humans.

FACTFILE

OTHER CAVY-LIKE RODENTS

180 species in 12 families and 40 genera

DISTRIBUTION S America; West Indies; Africa S of the Sahara.

Equator

Coypu | Hutia | Pacarana | Paca

Equator

Agoutis and Acouchis | Chinchilla rat | Chinchillas and Viscachas | Dassie rat

Equator

Spiny rat | Degu | Tuco-tuco | Cane rat

See families table ▷

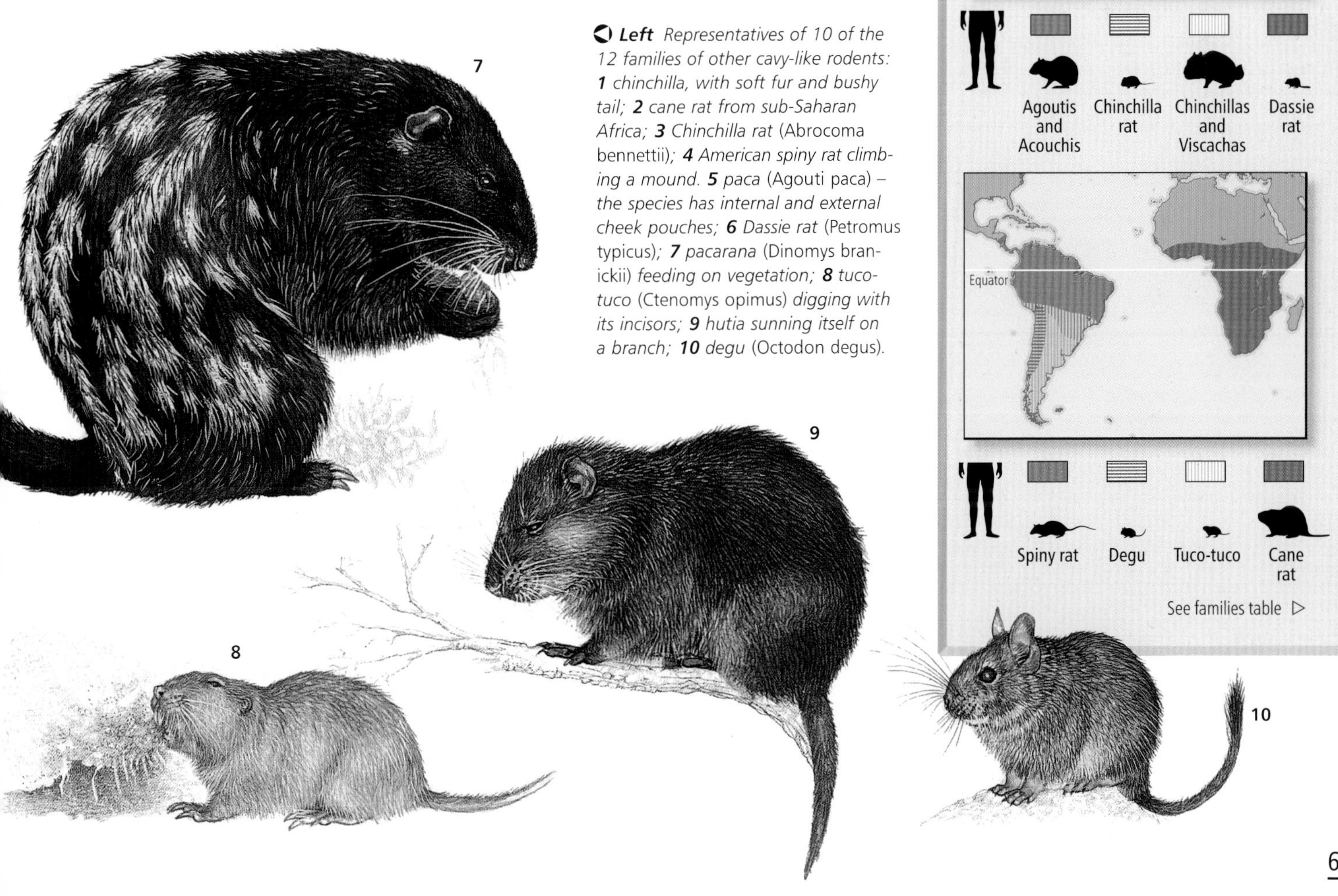

Left *Representatives of 10 of the 12 families of other cavy-like rodents:* **1** *chinchilla, with soft fur and bushy tail;* **2** *cane rat from sub-Saharan Africa;* **3** *Chinchilla rat* (Abrocoma bennettii)*;* **4** *American spiny rat climbing a mound.* **5** *paca* (Agouti paca) *– the species has internal and external cheek pouches;* **6** *Dassie rat* (Petromus typicus)*;* **7** *pacarana* (Dinomys branickii) *feeding on vegetation;* **8** *tuco-tuco* (Ctenomys opimus) *digging with its incisors;* **9** *hutia sunning itself on a branch;* **10** *degu* (Octodon degus).

Agoutis and Acouchis

FAMILY DASYPROCTIDAE

Agoutis and acouchis are relatively large rodents with very short tails and a coarse, usually unpatterned pelage. The long limbs are modifications for running, as is the reduction in the number of toes on the hind feet to three. The animals occur in a variety of habitats including forest, thick brush, and savanna; agoutis in particular are usually found close to water. All species are diurnal but secretive, and in areas of disturbance wait until dusk before emerging to forage for a wide variety of vegetation, especially fruit, nuts, and succulent parts of plants. When food is abundant, agoutis bury some for use in time of dearth, which is an important, if inadvertent, means of dispersing the seeds of many forest trees. In some areas, agoutis are known to breed throughout the year, litters normally consisting of one or two well-developed young that are able to run within an hour of birth. Agoutis and acouchis usually occur in small social groups of an adult male and female with several juveniles. They are preyed upon by a variety of carnivores including humans.

Pacas

FAMILY AGOUTIDAE

Pacas have often been included with the agoutis and acouchis as a separate subfamily of the Dasyproctidae; they are not dissimilar in appearance, but have relatively shorter legs, less reduced digits on the hind feet, and a spotted pelage. To add to the confusion, the scientific name of the paca is *Agouti*, which in common parlance is applied to the *Dasyprocta* species. Pacas usually occur in forested areas near water, often spending the day in burrows excavated by themselves or abandoned by other animals. They emerge at night to feed on leaves, stems, roots, nuts, seeds, and fruit, and may be a major pest of cultivated land. They are hunted by humans in addition to other large carnivores, and have become rare in some areas due to overhunting and habitat loss.

Tuco-tucos

FAMILY CTENOMYIDAE

The body form of tuco-tucos is robust and compact and shows many features associated with their burrowing lifestyle. The head is large, with small eyes and ears. The strong incisors, very prominent with their bright orange enamel, are used to cut through roots when tunneling. The limbs are short and muscular, and the claws on the forefeet in particular are long and strong, serving to dig the extensive burrow systems in which the animals live. The hind feet bear strong bristle fringes, which are used for grooming; the animals' scientific name, meaning "comb-toothed," is derived from the comblike nature of these bristles. The numerous species comprising this genus generally prefer the dry, sandy soils typical of coastal areas, grassy plains, and the *altiplano*, but also of forests. Some species are solitary, others colonial; most individuals occupy single burrows. Both sexes are territorial, with the more aggressive males holding larger territories than the females. Their shallow burrows may have several entrances. Although active mainly during the day, tuco-tucos rarely emerge to forage until after dark. They feed mainly on roots, stems, and grasses. Tuco-tucos are considered pests of cultivated and grazing land in some areas, and their burrows occasionally collapse, injuring people and livestock. As a consequence, some species have been hunted intensively and their numbers greatly reduced.

Degus or Octodonts

FAMILY OCTODONTIDAE

Octodontids occur in southern South America from sea level to about 3,500m (11,500ft). The family name of Octodontidae refers to the worn enamel surface of their teeth, which forms a pattern in the shape of a figure of eight. Most are adapted to digging, particularly rock rats and the coruro, and dig their own burrows, take over burrows abandoned by other animals, or live in rock piles and crevices. The pelage is usually long, thick, and silky. Degus and chozchoris are gray to brown above, creamy yellow or white below. Degus are active during the day; they are colonial and construct extensive burrow systems, with a central section connected to feeding sites by a complex maze of tunnels and surface paths. In contrast, the chozchoz is nocturnal and lives in burrows, rock crevices, and caves, feeding on acacia pods and cactus fruits. Coruros are brown or black, and rock rats dark brown all over; both are adapted to a burrowing lifestyle, with compact bodies, small eyes and ears, muscular forelimbs, and strong incisors, although these are less prominent in rock rats than in coruros. Little is known about the viscacha rats, which are buff above and

Cavy-like Rodent Families

Dassie rat or African rock rat
Family: Petromuridae

1 species (*Petromus typicus*), S Angola, Namibia and NW South Africa

Cane rats or grasscutters
Family: Thryonomyidae

2 species in 1 genus (*Thryonomys*)
Africa S of the Sahara

Chinchillas and viscachas
Family: Chinchillidae

6 species in 3 genera: chinchillas (*Chinchilla*), 2 species; mountain viscachas (*Lagidium*), 3 species; Plains viscacha (*Lagostomus maximus*).
W and S South America. 1 species – the Short-tailed chinchilla – is listed by the IUCN as Critically Endangered, 1 other is Vulnerable.

Pacarana Family: Dinomyidae

1 species (*Dinomys branickii*)
Venezuela, Colombia, Ecuador, Peru, Brazil, and Bolivia. Endangered.

Agoutis and Acouchis
Family: Dasyproctidae

13 species in 2 genera: agoutis (*Dasyprocta*), 11 species; acouchis (*Myoprocta*), 2 species.
S Mexico to S Brazil and Lesser Antilles. 2 agouti species are Endangered, 1 Vulnerable.

Pacas Family: Agoutidae

2 species in 1 genus (*Agouti*)
S Mexico to S Brazil

Tuco-tucos Family: Ctenomyidae

56 species in 1 genus (*Ctenomys*)
Peru S to Tierra del Fuego. 1 species Vulnerable.

Octodonts Family: Octodontidae

10 species in 6 genera:
rock rats (*Aconaemys*), 2 species; degus (*Octodon*), 4 species; Mountain degu or chozchoz (*Octodontomys gliroides*); Viscacha rat (*Octomys mimax*); coruro (*Spalacopus cyanus*); Plains viscacha rat (*Tympanoctomys barrerae*).
Peru, Bolivia, Argentina, Chile. 2 species Vulnerable.

Chinchilla rats or chinchillones
Family: Abrocomidae

3 species in 1 genus (*Abrocoma*), plus 1 other species known only from skeletal remains
Peru, Bolivia, Chile, Argentina. 1 species Vulnerable.

Spiny rats Family: Echimyidae

70 species in 16 genera:
Bristle-spined rat (*Chaetomys subspinosus*); coro-coros (*Dactylomys*), 3 species; Atlantic bamboo rat (*Kannabateomys amblyonyx*); olalla rats (*Olallamys*), 2 species; Arboreal soft-furred spiny rats (*Diplomys*), 3 species; arboreal spiny rats (*Echimys*), 14 species; toros (*Isothrix*), 2 species; Armored spiny rat (*Makalata armata*); Owl's spiny rat (*Carterodon sulcidens*); Lund's spiny rats (*Clyomys*), 2 species; guiara (*Euryzygomatomys spinosus*); Armored rat (*Hoplomys gymnurus*); Tuft-tailed spiny tree rat (*Lonchothrix emiliae*); spiny tree rats (*Mesomys*), 5 species; terrestrial spiny rats (*Proechimys*), 32 species; punare (*Thrichomys apereoides*)
S and C America and West Indies. 5 species are currently listed as Vulnerable; in addition, 4 others have recently been declared Extinct.

Hutias Family: Capromyidae

15 species in 6 genera:
Desmarest's hutia (*Capromys pilorides*); Bahaman and Jamaican hutias (*Geocapromys*), 2 species; small Cuban or sticknest hutias (*Mesocapromys*), 4 species; long-tailed Cuban hutias (*Mysateles*), 5 species; laminar-toothed hutias (*Isolobodon*), 2 species; Hispaniolan hutia (*Plagiodontia aedium*).
W Indies. 6 species are Critically Endangered, and 4 more are Vulnerable; in addition, 6 species have recently been declared Extinct.

Coypu or Nutria
Family: Myocastoridae

1 species, *Myocastor coypus*
S Brazil, Paraguay, Uruguay, Bolivia, Argentina, Chile; introduced into N America, N Asia, E Africa, Europe.

For full species list see Appendix ▷

whitish below with a particularly bushy tail, except that they are nocturnal, burrowing herbivores inhabiting desert scrub. *Tympanoctomys* is similarly poorly known; it is apparently restricted to plains with salt-rich vegetation.

Chinchilla Rats

FAMILY ABROCOMIDAE

As their name suggests, chinchilla rats are soft-furred, rat-like rodents that live in burrows or rock crevices. They are mainly nocturnal and may be colonial. Little information is available about these animals. Their diet includes a wide variety of plant material, although *Abrocoma vaccarum* may be specialized to feed on creosote bush. The pelage is silver gray or brown above, white or brown below, and consists of soft, dense underfur overlain with long, fine guardhairs. The pelts are occasionally sold, but are of much poorer quality than those of true chinchillas.

Spiny Rats

FAMILY ECHIMYIDAE

Spiny rats comprise a diverse group of medium-sized, ratlike, herbivorous rodents, most of which have a spiny or bristly coat, although some are soft-furred. Some are very common and widespread, while others are extremely rare; three genera known only from skeletal remains are probably, and a fourth certainly, extinct. The taxonomy of some genera is poorly understood, and the number of species is only tentative. The body form in this family is correlated with lifestyle. Robust, short-tailed forms (*Clyomys*, *Carterodon*, and *Euryzygomatomys*) are burrowing savanna species; relatively slender, long-tailed forms (*Olallamys*, *Dactylomys*, *Kannabateomys*) are arboreal. Of the intermediate forms, *Proechimys*, *Isothrix*, and *Hoplomys* are more or less terrestrial, and *Mesomys*, *Lonchothrix*, *Echimys*, and *Diplomys* are mostly arboreal. Spiny rats are mainly herbivores: the diet of *Proechimys* and *Echimys* is mainly fruit, *Kannabateomys* and *Dactylomys* eat bamboo and vines, while *Mesomys* eats fruit and other plant material but also insects.

Hutias

FAMILY CAPROMYIDAE

Found only in the West Indies, hutias were once a very diverse group, with different genera grouped into several subfamilies. Approximately half of these genera are now extinct, most within historical times, and a further two include rare and endangered species. The living genera have a robust body, a broad head with relatively small eyes and ears, and short limbs. In size they range from less than 200g (7oz) to 8.5kg (19lb), while the body mass of some extinct species is thought to have exceeded 20kg (44lb). The fur is harsh but with a soft underfur, and is generally brownish or grayish in coloration. Most of the living genera are partially, and the long-tailed Cuban hutias highly, arboreal. They live in forests, plantations, and rocky areas; in addition, small Cuban hutias occur in coastal swamps. Most are nocturnal, although Desmarest's hutia is diurnal in some areas. The diet includes a variety of vegetation but also small animals such as lizards. Hutias are and were preyed on by birds, snakes, introduced domestic dogs, cats, and mongooses, and they have been intensively hunted by humans for food; remains of several of the extinct species have been found in caves and kitchen middens.

Coypu

FAMILY MYOCASTORIDAE

The coypu is a large, robust rodent, well adapted to its semi-aquatic life in marshes, lakes, and streams. The eyes and ears are small, the mouth closes behind the incisors while swimming, the whiskers are long, the limbs relatively short, and the hind feet webbed. The pelage is yellowish or reddish brown; the outer hair is long and coarse and overlays the thick, soft underfur. Coypus live in burrows in river banks and are expert swimmers, able to remain underwater for up to five minutes. Their diet includes a wide range of vegetation, mussels, and snails. They are pests of cane fields and plantations. Coypus are nocturnal and live in pairs or small family groups. Females may have two or three litters a year, with an average of five young to a litter. The young are born fully furred and with their eyes open. Coypus are preyed on by alligators, fish, snakes, and birds of prey, and they are also hunted by humans for their meat. In addition, they are both hunted and farmed for their valuable pelt, known in the fur trade as nutria – a corruption of the Spanish for "otter." They have been introduced for fur farming to many parts of the world, where escaped individuals have formed feral populations (for example, in East Anglia, England), causing extensive damage to watercourses and cultivated land. PJ

Above In Brazil's Amazonas province a Black agouti confronts the problem of opening a large brazil nut. Agoutis often squat on their haunches to consume smaller food items, holding them in their hands squirrel-style to eat.

Right A swimming coypu reveals the bright orange incisors that are unexpected features of this semi-aquatic species. Coypus spend most of their waking hours in water; they live in riverside burrows, and have webbed hind feet.

Old World Porcupines

TO THE HUMAN EYE, PORCUPINES, WITH THEIR array of quills, are among nature's strangest creations, though in terms of bodily protection for the porcupines themselves, the spikes make perfect sense. The two Old World subfamilies are mostly terrestrial, unlike the New World porcupines.

Many people mistakenly believe that porcupines are related to hedgehogs or pigs, but in fact guinea pigs, chinchillas, capybaras, agoutis, viscachas, and cane rats are their closest relatives. Many of these animals have in common an extraordinary appearance, and are well known for their unusual ways of solving the problems of reproduction.

An Armor of Sharp Spines

Function and Form

Old World porcupines belong to two distinct subfamilies, Atherurinae and Hystricinae. The brush-tailed porcupines of the former branch have long, slender tails that end in tufts of stiff, white hairs containing hollow sections that rustle when the tail is shaken. The animals' elongated bodies and short legs are covered with short, chocolate-colored bristles, with a few long quills on the back.

The crested porcupines of the Hystricinae subfamily have short tails surrounded by stout, sharp, cylindrical quills. The tip of the tail is armed with a cluster of hollow, open-ended quills. When the tail is shaken, these produce a rattle that acts as a warning signal that the animal is annoyed. The back of the upper parts and flanks is covered with black and white spines; as a modified form of hair, these are made of keratin.

When threatened, porcupines erect and rattle their quills, stamp their hind feet, and make a grunting noise. If threatened further, they turn their rump to the intruder and run sideways or backwards toward it. If the quills penetrate the skin, they become stuck and detach.

The rest of the body is covered with flat, black bristles. Most species have a crest of erectile hair extending from the top of the head to the shoulders. The head is blunt and exceptionally broad across the nostrils, with small, piglike eyes set far back on either side of the face. The two sexes look alike, though the females have mammary glands that are situated on the side of the body, enabling mothers to suckle lying on their stomachs.

The majority of Old World porcupines are vegetarians, feeding on the roots, bulbs, fruits, and berries of a wide variety of plants. Porcupines play a role in shaping local plant diversity and productivity; for example, the digging sites of Indian porcupines serve as important germination locations for seedlings. In cultivated areas, they will eat such crops as groundnuts, potatoes, and pumpkins. African porcupines are able to feed on plant species that are poisonous to domestic stock. Brush-tailed porcupines are tree-climbers and feed on a variety of fruits.

Porcupines manipulate food with their front feet, pinning it to the ground and gnawing at it. Usually they feed alone, though they will eat in groups of two or three. Bones also appear occasionally in the porcupines' diets, littering their shelters. They are carried there for gnawing, either to sharpen teeth or as a source of phosphates.

Above *The Indonesian porcupine **1** has a dense coat of flat, flexible spines. There are three species – in Borneo, Sumatra, and the Philippines. The African porcupine **2** is one of five crested species. It is very adaptable, inhabiting forest, grassland, and desert.*

Communities on the Cape

Social Behavior

Among Cape porcupines sexual behavior is normally initiated by the female. Approaching a male, she will take up the sexual posture – rump and tail raised and quills pointed away from her partner – who mounts her from behind with his forepaws resting on her back. Intromission only occurs when the female is in heat (every 28–36 days) and the vaginal closure membrane becomes perforated. Sexual behavior without intromission is exhibited during all stages of the sexual cycle.

The young are born in grass-lined chambers that form part of an underground burrow system. At birth they are unusually precocial: fully-furred, they have their eyes open and are covered in

Left *Although lions have been known to eat African porcupines, even they find it very difficult to penetrate the armory of quills. Contrary to folklore, the barbs cannot be projected; they can, however, become embedded and can cause septic, sometimes fatal, wounds.*

bristles that will become quills, although, fortunately for the mother, these only harden after birth. Newborn babies weigh 300–330g (10.6–11.6oz), and start to nibble on solids at 9–14 days. At 4–6 weeks they begin to feed, though they continue to be nursed for 13–19 weeks, by which time they weigh 3.5–4.7kg (7.7–10.4lb). Litter sizes are small; 60 percent of births produce one young and 30 percent produce twins. In the wild porcupines produce only one litter each summer.

Despite foraging alone, Cape porcupines are sociable, living in burrows with as many as 6–8 animals – usually an adult pair and their consecutive litters. Both sexes are aggressive towards strangers, and all colony members protect the young. Only one female in a group reproduces; should a litter be lost, she can conceive again within days. Sexual maturity is attained at 2 years.

Both parents may accompany the young when foraging for up to 6–7 months, although adult males are more frequently encountered with young then adult females. The occurrence of family groups is probably related to the opportunities available for mature offspring to disperse; when a population dips, young individuals are able to take advantage of newly-available territories to reproduce. Thus disturbance can reduce the age of first reproduction from 24 to 12 months.

Porcupines live in groups in order to huddle together for warmth. Newborn offspring do not leave the burrow for the first 9 weeks of their lives, and, when warmed by the bodies of other group members, may be able to allocate more energy to growth. Sharing burrows also reduces porcupines' vulnerability to predation and encourages cooperative rearing.

Population density in the semi-arid regions of South Africa varies from 1–29 individuals per sq km (up to 75 per sq mi). Forty percent of the population are less than 1 year old.

Territories are maintained by scentmarking, using anal glands. Males mark more frequently in preferred feeding patches. Porcupines seem to forage up to 16km (10mi) from their burrows, moving along well-defined tracks, almost exclusively by night. They are catholic in their habitat requirements, provided they have shelter to lie in during the day.

On the Defensive

CONSERVATION AND ENVIRONMENT

Porcupines are often viewed as a threat to crops. In addition, African porcupines carry fleas, which are responsible for the spread of bubonic plague, and ticks, which spread babesiasis, rickettsiasis, and theilerioses. Brush-tailed porcupines are also known to be hosts of the disease organism of malaria, *Plasmodium atheruri*. As a result, they are often persecuted and killed as pests. Indigenous peoples eat their flesh and kill porcupines for recreation.

Nevertheless, the animals occur in great numbers, thanks to the near absence of natural predators over much of their range, and also because of the increase in crop cultivation. At the time of writing, there is no reason to believe that porcupines as a whole are endangered, though certain species and subspecies are now considered at risk. For example, one subspecies of the Asiatic brush-tailed porcupine is currently listed as Endangered by the IUCN, while the Malayan porcupine is classified as Vulnerable.

RJvA

FACTFILE

OLD WORLD PORCUPINES

Order: Rodentia

Family: Hystricidae

11 species in 4 genera and 2 subfamilies

DISTRIBUTION Africa, Asia; some in S Europe.

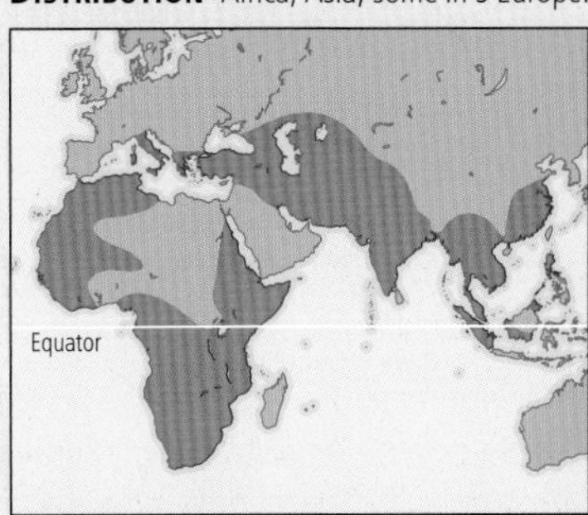

Habitat Varies from dense forest to semi-desert.

Size Ranges from **head–body length** 37–47cm (14.6–18.5in) and **weight** 1.5–3.5kg (3.3–7.7lb) in the brush-tailed porcupines to **head–body length** 60–83cm (23.6–32.7in) and **weight** 13–27kg (28.6–59.4lb) in the crested porcupines.

Coat Head, body, and sometimes the tail, are covered in long, sharp quills – hardened hairs – that are brown or blackish in color, sometimes with white bands.

Diet Roots, tubers, bulbs, fruit, bark, carrion.

Breeding Gestation 90 days for the Indian porcupine, 93–94 days for the Cape porcupine, 100–110 days for the African brush-tailed porcupine, 105 days for the Himalayan porcupine, 112 days for the African porcupine.

Longevity Approximately 21 years recorded for crested porcupines in captivity.

BRUSH-TAILED PORCUPINES Genus *Atherurus*
C Africa and Asia. Forests. Brown to dark brown bristles cover most of the body; some single-color quills on the back. 2 species: **African brush-tailed porcupine** (*A. africanus*), **Asiatic brush-tailed porcupine** (*A. macrourus*).

CRESTED PORCUPINES Genus *Hystrix*
Africa, India, SE Asia, Sumatra, Java and neighboring islands, S Europe; recently introduced to Great Britain. Varied habitats. Hair on back consists of long, stout, cylindrical black and white erectile spines and quills; body covered with black bristles; grayish crest well developed. 5 species: **African porcupine** (*H. cristata*); **Cape porcupine** (*H. africaeaustralis*); **Himalayan porcupine** (*H. hodgsoni*); **Indian porcupine** (*H. indica*); **Malayan porcupine** (*H. brachyura*).

INDONESIAN PORCUPINES Genus *Thecurus*
Indonesia, Philippines. Coat dark brown in front, black on posterior; body densely covered with flattened, flexible spines; quills have a white base and tip, with central parts black; rattling quills on tail are hollow. 3 species: **Bornean porcupine** (*T. crassispinis*); **Philippine porcupine** (*T. pumilis*); **Sumatran porcupine** (*T. sumatrae*).

LONG-TAILED PORCUPINE *Trichys fasciculata*
Malay peninsula, Sumatra, Borneo. Forests. Body covered with brownish, flexible bristles; head and underparts hairy.

Gundis

GUNDIS ARE SMALL, HERBIVOROUS RODENTS of North Africa's mountains and deserts. When the animals first came to the attention of Western naturalists, in Tripoli in 1774, they were given the name of "gundi-mice" (gundi *is the local word). The family name, Ctenodactylidae, means "comb-toes."*

In the mid-19th century, the British explorer John Speke shot gundis in the coastal hills of Somalia, and later French naturalists found three more species; skins and skulls began to arrive in museums. But no attempt was made to study the ecology of the animal. Some authors said gundis were nocturnal, others diurnal; some claimed they dug burrows, others that they made nests; there were reports of the animals whistling, while other sources had them chirping like birds; and there were fantastic tales about them combing themselves with their hind feet in the moonlight. In 1908 two French doctors isolated a protozoan parasite, now known to occur in almost every mammal, from the spleen of a North African gundi and called it *Toxoplasma gondii*.

Powder Puffs in the Desert Sun

FORM AND FUNCTION

Gundis have short legs, short tails, flat ears, big eyes, and long whiskers. Crouched on a rock in the sun with the wind blowing through their soft fur, they look like powder puffs.

The North African and the Desert gundi have tiny, wispy tails, but the other three have fans that they use as balancers. Speke's gundi has the largest and most elaborate fan, which it uses in social displays. Gundis also have rows of stiff bristles – their "combs" – on the two inner toes of each hind foot, and these stand out white against the dark claws. They use the combs for scratching. Sharp claws adapted to gripping rocks would destroy the soft fur coat that insulates them from extremes of heat and cold. The rapid circular scratch of the rump with the combed instep is characteristic of gundis.

The gundi's big eyes convinced some earlier authors that the animal was nocturnal. In fact the gundi is adapted to popping out of sunlight into dark rock shelters. Equally, the gundi can flatten its ribs to squeeze into a crack in the rocks.

Right *An extraordinary feature of the Mzab gundi is that its ears are flat and immovable.*

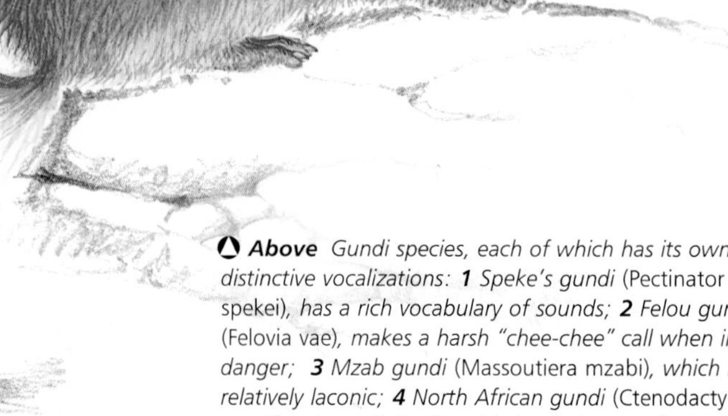

Above *Gundi species, each of which has its own distinctive vocalizations:* **1** *Speke's gundi* (Pectinator spekei), *has a rich vocabulary of sounds;* **2** *Felou gundi* (Felovia vae), *makes a harsh "chee-chee" call when in danger;* **3** *Mzab gundi* (Massoutiera mzabi), *which is relatively laconic;* **4** *North African gundi* (Ctenodactylus gundi), *whose distinctive chirping helps members of this species recognize each other in the desert habitat that they share with the Desert gundi* (Ctenodactylus vali), *a species that whistles.*

FACTFILE

GUNDIS

Order: Rodentia

Family: Ctenodactylidae

5 species in 4 genera

DISTRIBUTION N Africa

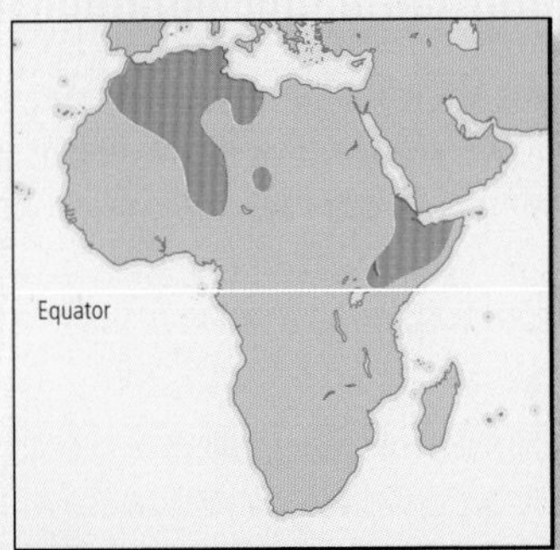

SPEKE'S GUNDI *Pectinator spekei*
Speke's or East African gundi
Ethiopia, Somalia, N Kenya. Arid and semi-arid rock outcrops. HBL 17.2–17.8cm (6.9–7.1in); TL 5.2–5.6cm (2–2.2in); WT 175–180g (6.2–6.3oz). Longevity: Unknown in wild; 10 years recorded in captivity.

FELOU GUNDI *Felovia vae*
SW Mali, Mauritania. Arid and semi-arid rock outcrops. HBL 17–18cm (6.8–7.2in); TL 2.8–3.2cm (1.1–1.3in); WT 178–195g (6.3–6.9oz).

DESERT GUNDI *Ctenodactylus vali*
Desert or Sahara gundi
SE Morocco, NW Algeria, Libya. Desert rock outcrops. Breeding: gestation 56 days.

NORTH AFRICAN GUNDI *Ctenodactylus gundi*
SE Morocco, N Algeria, Tunisia, Libya. Arid rock outcrops. Longevity: 3–4 years.

MZAB GUNDI *Massoutiera mzabi*
Mzab or Lataste's gundi
Algeria, Niger, Chad. Desert and mountain rock outcrops.

Abbreviations HBL = head–body length TL = tail length WT = weight

Foraging in the Cool of the Day

DIET

Gundis are herbivores: they eat the leaves, stalks, flowers, and seeds of almost any desert plant. Their incisors lack the hard orange enamel that is typical of most rodents. Gundis are not, therefore, great gnawers. Food is scarce in the desert, and gundis must forage over long distances – sometimes as much as 1km (0.6mi) a morning. Regular foraging is essential as they do not store food. Home range size varies from a few square meters to 3sq km (1.9sq mi).

Foraging over long distances generates body heat, which can be dangerous on a hot desert day. It is unusual for small desert mammals to be active in daytime, but gundis behave rather like lizards. In the early morning they sunbathe until the temperature rises above 20°C (68°F), and then they forage for food. After a quick feed they flatten themselves again on the warm rocks. Thus they make use of the sun to keep their bodies warm and to speed digestion – an economical way of making the most of scarce food. By the time the temperature has reached 32°C (90°F), the gundis have taken shelter from the sun under the rocks, and they do not come out again until the temperature drops in the afternoon. When long foraging expeditions are necessary, gundis alternate feeding in the sun and cooling off in the shade.

In extreme drought, gundis eat at dawn when plants contain most moisture. They obtain all the water they need from plants; their kidneys have long tubules for absorbing water. Their urine can be concentrated if plants dry out completely, but this emergency response can only be sustained for a limited period.

Family Life in the Colonies

SOCIAL BEHAVIOR

Gundis are gregarious, living in colonies that vary in density from the Mzab gundi's 0.3 per ha (0.12 per acre) to over 100 per ha (40 per acre) for Speke's gundi. Density is related to the food supply and the terrain. Within colonies there are family territories occupied by a male, female, and juveniles or by several females and offspring. Gundis do not make nests, and the "home shelter" is often temporary. Usually a shelter retains the day's heat through a cold night and provides cool draughts on a hot day. In winter, gundis pile on top of one another for warmth, with juveniles shielded from the crush by their mother or draped in the soft fur at the back of her neck.

Each species of gundi has its own repertoire of sounds, varying from the infrequent chirp of the Mzab gundi to the complex chirps, chuckles, and whistles of Speke's gundi. In the dry desert air their low-pitched alert calls carry well. Short, sharp calls warn of predatory birds; gundis within range will hide under the rocks. Longer calls signify ground predators and inform the predator it has been spotted. The Felou gundi's harsh "chee-chee" will last as long as the predator is around.

Long complex chirps and whistles can be a form of greeting or recognition. The *Ctenodactylus* species – whose ranges overlap – produce the most different sounds: the North African gundi chirps, the Desert gundi whistles. Thus members can recognize their own species.

All gundis thump with their hind feet when alarmed. Their flat ears give good all-round hearing and a smooth outline for maneuvering among rocks. The bony ear capsules of the skull are huge, like those of many other desert rodents. The acute hearing is important for picking up the weak, low-frequency sounds of predators – sliding snake or flapping hawk – and for finding parked young. Right from the start, the young are left in rock shelters while the mother forages. They are born fully furred and open-eyed. The noise they set up – a continuous chirruping – helps the mother to home in on the temporary shelter.

The young have few opportunities to suckle: from the mother's first foraging expedition onward they are weaned on chewed leaves. (They are fully weaned after a period of about 4 weeks.) The mother has four nipples – two on her flanks and two on her chest – and the average litter size is two. But a gundi has little milk to spare in the dry heat of the desert. WG

African Mole-Rats

While African mole-rats may not be the most aesthetically pleasing creatures in the animal kingdom, they nonetheless rank among the more interesting members. Discoveries made regarding the habits and behavior of the social species have rocked the scientific community.

The African mole-rat is a ratlike rodent that has assumed a molelike existence and become totally adapted to life underground. It excavates an extensive system of semi-permanent burrows, complete with sleeping and food storage areas; and it pushes the soil it digs out to the surface as "molehills." Whereas most rodents of comparable size grow rapidly and live for only a couple of years, mole-rats take over a year to reach their full adult size and can live for several years. Indeed, the lifespan of captive Naked mole-rats may even exceed 25 years.

Teeth for Tunneling

FORM AND FUNCTION

Mole-rats have cylindrical bodies with short limbs so as to fit as compactly as possible within the diameter of a burrow. Their loose skin helps them to turn within a confined space: a mole-rat can almost somersault within its skin as it turns. Mole-rats can also move rapidly backward with ease, and so they often shunt to and fro without turning round when moving along a burrow.

All genera except the dune mole-rats use chisel-like incisors protruding out of the mouth cavity for digging. To prevent soil from entering the mouth, there are well-haired lip-folds behind the incisors, so that the mouth is closed, so to speak, behind the gnawing teeth. Dune mole-rats dig with the long claws on their forefeet and are less efficient at mining very hard soils; moreover, their body size is larger. These features restrict them to areas with easily dug sandy soil. This difference in digging method is reflected in the animals' teeth: the incisors of tooth-diggers protrude more than those of the dune mole-rats, and the roots extend back behind the row of cheek teeth for strength.

When a mole-rat is tunnelling, it pushes the soil under its body with its forefeet. Then, with the body weight supported by the forefeet, both hind feet are brought forward to collect the soil and kick it behind the animal. Once a pile has accumulated, the mole-rat reverses along the burrow, pushing the soil behind it. Most mole-rats force solid cores of soil out onto the surface, but Naked mole-rats kick a fine spray out of an open hole – an "active" hole looks like an erupting volcano. A number of Naked mole-rats cooperate in digging, one animal excavating, a number transporting soil, and another kicking it out of the hole – this unfortunate individual is particularly vulnerable to predation by snakes. All mole-rats have hind feet that are fringed by stiff hairs, as is the tail in all but Naked mole-rats; both adaptations help hold the soil during digging.

Because mole-rats live in complete darkness for most of their lives, their eyes are small and can only detect light and dark. (Interestingly, the Cape mole-rat, which occasionally travels on the surface, has eyes that are larger than those of other species.) It has been suggested that the surface of the eye may be used to detect air currents that would indicate damage to the burrow system; certainly, if damage occurs, the mole-rats rapidly repair it. Touch is important in finding the way

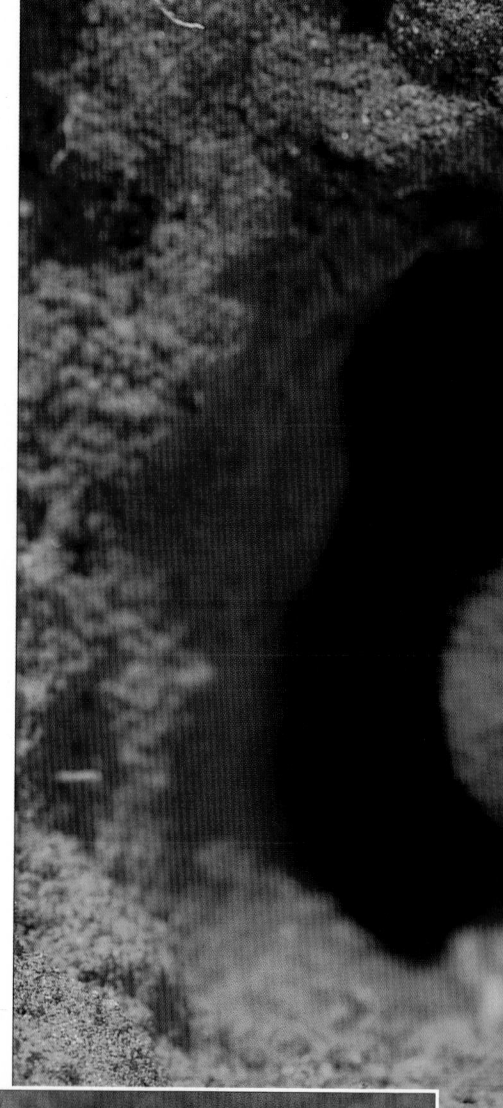

Right *A Damara mole-rat emerging from its tunnel. A tunnel system's depth usually depends on local soil conditions. In areas of loose soil, they lie deeper underground than in areas of hard soil.*

Below *The powerful incisors of Mechow's mole-rat are used for excavation. Clearly, keeping soil out of various orifices is important and its head is well adapted for this: the lips fold behind the incisors, the nostrils can be closed, and the ears and eyes are small.*

FACTFILE

AFRICAN MOLE-RATS

Order: Rodentia

Family: Bathyergidae

At least 15 species in 5 genera: **Dune mole-rats** (genus *Bathyergus*, 2 species), the largest mole-rats,inhabiting sandy coastal soils of S Africa; **common mole-rats** (genus *Cryptomys*, at least 9 species), widespread in W, C, and S Africa; **silvery mole-rats** (genus *Heliophobius*, 2 species), C and E Africa; **Cape mole-rat** (*Georychus capensis*), Cape Province of the Republic of S Africa, along the coast from the SW to the E; and **Naked mole-rat** (*Heterocephalus glaber*), in arid regions of Ethiopia, Somalia, and Kenya.

DISTRIBUTION Africa S of the Sahara

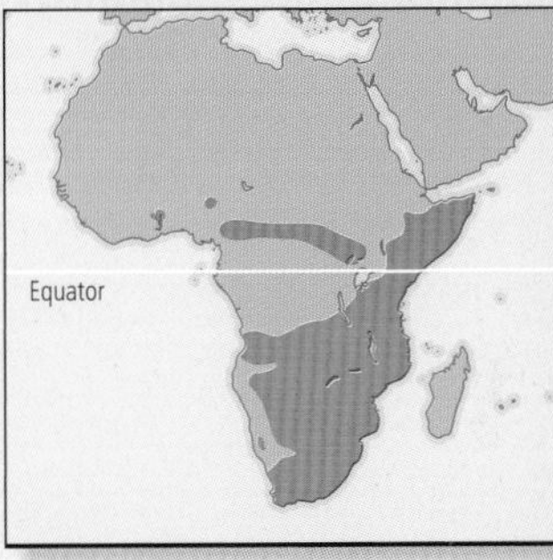

HABITAT Underground in different types of soil and sand.

SIZE Head–body length ranges from 9–12cm (3.5–4.7in) in the Naked mole-rat to 30cm (11.8in) in the genus *Bathyergus*; **weight** from 30–60g (1–2.1oz) to 350–1,800g (26–63oz) in the same species.

COAT Thick, soft, and woolly or velvety in all but the Naked mole-rat, which is almost hairless.

DIET Roots, tubers, geophytes, herbs, and grasses.

BREEDING Gestation 44–111 days (44–48 days in the Cape mole-rat; 66–74 days in the Naked mole-rat; 97–111 days in the Giant Zambian mole-rat).

LONGEVITY Unknown (captive Naked mole-rats have lived for over 25 years and several species of *Cryptomys* for more than 8 years).

CONSERVATION STATUS Four species – *Bathyergus janetta, Cryptomys foxi, C. zechi,* and *Heliophobius argenteocinereus* – are currently listed as Lower Risk: Near Threatened. The other species are not considered threatened.

For full species list see Appendix ▷

around the burrow system; many genera have long, touch-sensitive hairs scattered over their bodies (in the Naked mole-rat, these are the only remaining hairs). The animals' sense of smell and their hearing of low-frequency sounds are good, and their noses and ears are modified on the outside so as to cope with the problems of living in a sandy environment: the nostrils can be closed during digging, while the protruding parts of the ears (pinnae) have been lost.

Subterranean Farmers

DIET

Mole-rats are vegetarians, and obtain their food by digging foraging tunnels. These enable them to find and collect roots, storage organs (geophytes), and even the aboveground portions of plants without having to come to the surface. They appear to blunder into food rather than to detect it – tests failed to reveal any evidence that they could locate food items even half a meter (1.6ft) from the foraging burrow. The large dune mole-rats are not specialist feeders and live on grass, herbs, and geophytes, but Common and Naked mole-rats live entirely on geophytes and roots.

Foraging burrows can be very extensive: one system containing 10 adults and 3 young Common mole-rats was 1km (0.6mi) long, and burrows of Naked mole-rats may exceed 3km (1.9mi) in length. Burrow length depends on the number and ages of mole-rats in a system and on the abundance and distribution of food items.

Apart from providing nest, food storage, and toilet areas, the rest of the burrow system is dug in search of food. Excavating is easiest when the soil is soft and moist. After rain, there is a flurry of digging – indeed, in the first month after rain, colonies of Naked and Damaraland mole-rats can dig 1km (0.6mi) of burrows and throw up more than 2 tonnes of soil as molehills. In the arid regions where the two species live, many months may elapse before it next rains, so it is vitally important to find enough food in these brief periods of optimal digging to see the colony through the drought. Small food items are eaten or stored (by Damaraland, Common, and Cape mole-rats, for example), while larger items are left growing in situ and are gradually hollowed out, thus ensuring a constantly fresh and growing food supply. At a later stage, the hollow is plugged with soil so the tuber will regenerate. This "farming" of geophytes enables colonies to remain resident in the same area for many years.

In some areas where Naked mole-rats occur, tubers may weigh as much as 50kg (110lb). When feeding, the mole-rat holds small items with its forefeet, shakes them free of soil, cuts them into pieces with its incisors, and then chews these with its cheek teeth. In southwestern Cape Province, South Africa, differences in diet, burrow diameter and depth, and perhaps also of social organization enable three genera – *Bathyergus, Georychus*, and *Cryptomys* – to occupy different niches within the same area, and sometimes even the same field. This sympatry is unusual for burrowing mammals, where the normal pattern is for one species to occupy an area exclusively.

Life Below Ground

Social Behavior

The social behavior of three genera (*Bathyergus, Georychus*, and *Heliophobius*) follows the normal pattern for subterranean mammals: they are solitary, and aggressively defend their burrows against conspecifics. They signal to neighboring animals by drumming with their hind feet; at the onset of breeding, *Georychus* males and females drum with a different tempo to attract a mate. Mating is brief, and the male then leaves the female to rear her pups. When about 2 months old, the pups begin to fight, and this is the prelude to dispersing; if forcibly kept together in captivity, siblings will eventually kill each other. In *Georychus*, the dispersing young often burrow away from the parent system and block up the linking burrows; this probably also occurs in the other solitary genera and would ensure that the young are protected from predators during this otherwise very vulnerable phase in their life history.

In the social mole-rats (see box), only a single female breeds in each colony. In all the *Cryptomys* studied there is a strong inhibition against incest, and colonies are founded by a pair of animals originating from different colonies. The rest of the colony is composed of their offspring which, unlike the solitary mole-rats, remain in the natal colony, helping to locate food and rear their siblings until conditions are favorable for dispersing. In fact many never get the chance to breed.

Colony sizes of the Common mole-rat rarely exceed 14 animals, whereas 41 Damaraland mole-rats may occur together (although 14–25 is more common). Naked mole-rat colonies may contain over 300 individuals (the mean is about 80), but here there is no inhibition to incest and the breeders may come from within the colony. The breeding female is the dominant individual in the colony, which she controls through stress-related behaviors such as violent shoving. She has a distinctively elongated body (her vertebrae lengthen during her first few pregnancies), and this serves to better accommodate the large litters typical of this species. Up to three males may mate with her, and multiple paternity of litters can occur. Except for the dispersers, Naked mole-rats are very xenophobic and will kill foreign animals. They recognize colony members by scent, probably through a cocktail of odors that they spread on their bodies in the communal toilet area.

Because they live in a well-protected, relatively safe environment, mole-rats are less exposed to predators than are surface-dwelling rodents. This better life assurance may be why they produce smaller litters, usually of between two and five pups. There are exceptions: the Cape mole-rat produces up to 10 pups, and the Naked mole-rat has as many as 28, although the average litter size is 12. Pups weigh about 1g (0.04oz), while breeding females weigh 65–80g (2.3–2.8oz).

Snakes may occasionally pursue the solitary mole-rats underground, but more often lie in wait for them. Field evidence suggests that the Mole snake (*Pseudaspis cana*) is attracted to the smell of freshly turned soil and will penetrate the burrow system via a new molehill. It usually pushes its head into the burrow and waits for the mole-rat as it reverses with its next load of soil. This may also be true of the Eastern beaked snake (*Rhamphiophis oxythunchus rostrutus*), which has been seen preying on Naked mole-rats as they kick soil out of the burrow. Other predators also take mole-rats: their skulls are not uncommon in the pellets of birds of prey and small carnivores such as jackals, caracals, and zorillas.

In addition to protecting the mole-rat against many predators, the underground environment provides a uniformly humid microclimate. This,

Right *A juvenile Naked mole-rat feeding in an underground chamber. Both the front and rear feet have five digits; the feet are used for digging as well as holding food for eating. Although Naked mole-rats appear to be bald, they do have some hairs scattered about the body and around the feet, which help in sweeping soil back beneath themselves.*

Left *A Naked mole-rat queen suckling several youngsters. The queen is effectively the leader of the colony and is a nonworker, weighing significantly more than the others. Although only a small number of animals are actively involved in the breeding process, the others are not incapable of breeding, merely suppressed; research has shown that if non-breeding females are removed from a colony and paired with a male, they become capable of breeding very quickly.*

THE INSECT-LIKE RODENT

At least two species of mole-rat, the Naked and Damaraland, have a colony structure similar to that of social insects. Within each colony, a single female and 1–3 males breed; the remaining males and females, while not infertile, remain nonreproductive while members of their natal colony. In Naked mole-rat colonies, the more numerous, small-sized, worker mole-rats dig and maintain the foraging burrows and carry food and nesting material to the communal nest. Large-sized individuals spend much of the time in the nest with the breeding female. When the workers give an alarm call, however, they are mobilized to defend the colony.

The young born to the colony are cared for by all the mole-rats but suckled only by the breeding female. Once weaned, they join the worker force, but whereas some individuals apparently remain workers throughout their lives, others eventually grow larger than the rest and become colony defenders. It is from these big, and usually older, individuals that new reproductive Naked mole-rats emerge when a breeder dies.

In Naked as in all the social mole-rats, therefore, a colony is composed of the progeny of a number of closely related litters. As with social insects, this relatedness is probably an important factor in the evolution and maintenance of a social structure in which some individuals in the colony never breed. By caring for closely-related mole-rats that share their genetic make-up, the nonbreeding individuals nonetheless ensure the passing on of their own genetic characteristics.

This system seems to prevail among social mole-rats living in arid regions where for many months conditions are unsuitable for extensive burrowing. When it does rain, a workforce must be mobilized rapidly to find sufficient food to see the colony through the dry months. By joining forces and channeling the energies of colony members along specific avenues (some finding food, some acting as soldiers, and a select few bearing young), these mole-rats can survive in areas where single mole-rats or pairs cannot.

Experiments have shown that nonbreeding mole-rats are not sterile. Suppression is more severe in females than in males, which show some sperm production. Nonbreeders can rapidly become sexually active (within 7–10 days) and can found new colonies or, in the case of Naked mole-rats, can replace the breeding animals if they die. In this latter case, several of the older females initially show signs of sexual activity, and there is often a time of severe (sometimes fatal) fighting, before one female becomes sexually dominant, increases in size, and becomes the new breeder. Occasionally succession occurs without fighting.

Unlike in social insects, pheromones do not seem to be the prime means of control. Yet in many captive colonies, the whole colony is affected by the reproductive state of the breeding female. For example, just before a litter is born all colony members (male and female) develop teats, and some females come close to breeding condition. This strongly suggests that the colony is responding to chemical stimuli produced by the breeding female: in this case, the stimuli seem to prime the colony to receive and care for young that are not their own.

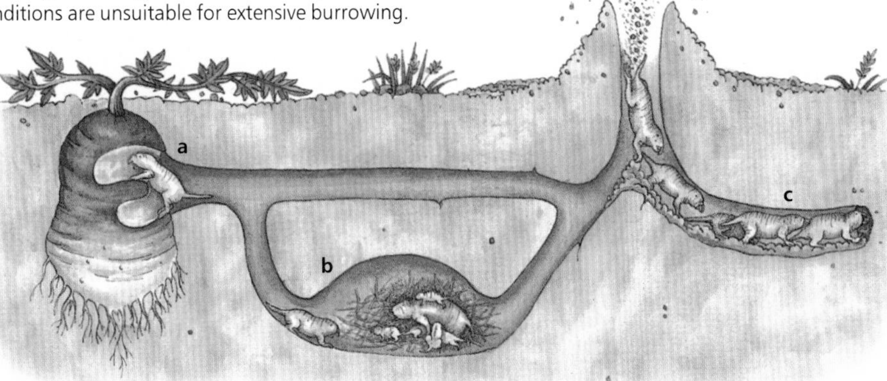

Above *Cross-section of a burrow system. On the left* **a** *Naked mole-rats hollow out a growing tuber; in the center* **b** *is the main chamber, which is occupied by the breeding female, subsidiary adults and young; on the right* **c** *a digging chain is at work.*

Top *A mole-rat digging chain. The lead animal digs with its teeth and pushes the soil backward. The one behind drags the load backward, keeping close to the tunnel floor, and passes it to the animal responsible for dispersal. It then returns to the front, straddling other mole-rats pushing soil away.*

plus the high moisture content of the mole-rat's food, precludes the necessity of having to drink free water. The burrow temperature remains relatively stable throughout the day, often in stark contrast to the surface temperature. In Naked mole-rat country, for example, surface temperatures of over 60°C (140°F) have been recorded while burrows 20cm (8in) below ground registered a steady 28–30°C (82–86°F). In response, Naked mole-rats have almost lost the ability to regulate body temperature, which consequently remains close to that of the burrow. If they need to alter it, they huddle together when cold or bask in surface burrows for warmth; they also take refuge in cooler areas within the system if they overheat, for example after digging near the surface

Helping and Harming

CONSERVATION AND ENVIRONMENT

Though inconspicuous, mole-rats can cause considerable damage to property. Dune mole-rats chew through underground cables, undermine roadways, and sometimes devour root and cereal crops. The molehills they create can damage the blades of harvesting machines, not to mention garden lawns and golf courses. The human response has been to attempt to exterminate all those causing the problems.

Yet mole-rats also have beneficial effects on their environment. They are important agents in soil drainage and soil turnover (a Cape dune mole-rat may throw up as much as 500kg/1,100lb of soil each month). They may play a role in dispersing geophytes (plants with underground storage organs) and they also eat geophytes that are poisonous to livestock. JUMJ

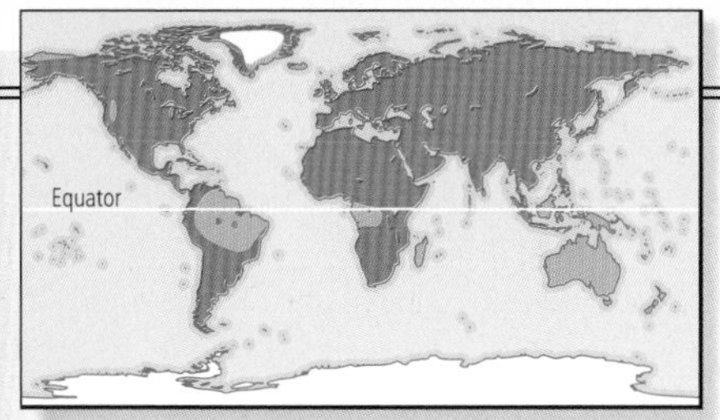

LAGOMORPHS

EITHER AS NATIVE SPECIES OR AS A RESULT OF human introductions, lagomorphs – meaning literally "hare-shaped" – are found all around the globe. The order contains two families: the small, rodent-like pikas, weighing under 0.5kg (1.1lb); and the rabbits and hares, weighing up to 5kg (11lb).

The pikas are thought to have separated from the rabbits and hares in the late Eocene, around 38–35 million years ago. Recent DNA sequence analysis suggests that most rabbit and hare genera arose from a single rapid diversification event approximately 12–16 m.y.a. The first pikas appeared in Asia in the middle Oligocene and spread to North America and Europe in the Pliocene (5–1.8 m.y.a.). Pikas seem to have peaked in distribution and diversity during the Miocene (24–5 m.y.a.) and since declined, while the rabbits and hares have maintained a widespread distribution since the Pliocene.

Rabbits and hares have elongated hind limbs adapted for running at speed over open ground. Their ears are long, the nasal region elongated, and their tail typically has a conspicuous white under-surface. By contrast, pikas are small with short legs, and are well adapted for living in rocky steppe and alpine habitats; the tail is virtually absent, and the ears short and rounded. They are far more vocal than rabbits and hares, but all three groups use scent products from special glands. Other features shared by all lagomorphs include coats of long, soft fur that fully covers the feet, ears that are large relative to body size, and eyes positioned for good broad-field vision.

What Makes a Lagomorph?

DIGESTION, DENTITION, AND DISTRIBUTION

Rabbits, hares, and pikas are all herbivores that feed predominantly on grasses but consume a range of other plant species in different habitats. Their digestive system is adapted for processing large volumes of vegetation, and they re-ingest some of their fecal matter, a behavior known as coprophagy. As herbivores with gnawing incisors, rabbits, hares, and pikas were initially classed by taxonomists within the order Rodentia, but in 1912 J. W. Cridley brought them together within the new order Lagomorpha. Significant among the several distinctive features that set them apart from rodents is the possession of a second pair of small incisors, known as "peg teeth," behind the long, constantly-growing pair in the upper jaw. This gives rabbits and hares a total of 28 teeth and a dental formula of 2/1 incisors, 0/0 canines, 3/2 pre-molars, and 3/3 molars; while pikas have 26 teeth (one fewer upper molar on each side).

Only 2 pika species occur in North America; the rest are distributed across Asia. Rabbits and hares, on the other hand, have diversified to produce species occupying a wide range of habitats, from tropical forest (the Forest rabbit) through swamps, desert, and montane grassland (the North American Marsh rabbit, Black-tailed jackrabbits, and Mexico's Volcano rabbit) to the snow-covered arctic (Snowshoe and Arctic hares).

Phenomenal Numbers

BREEDING PATTERNS

Within their home ecosystems, lagomorphs are often key fodder for a range of mammalian and avian predators. To counter high mortality rates, they breed prodigiously. Most species reach sexual maturity relatively early (after just 3 months in

Right *The European rabbit is a prodigious breeder, capable of producing 10–30 young each year. And yet, in this species, over half of all pregnancies are aborted, with the embryos being resorbed into the female's body.*

female European rabbits). The gestation period is short – 40 days in *Lepus* species, and around 30 days in all other genera – and litter sizes are often large. Other features of lagomorph reproduction that minimize inter-birth intervals for females include the phenomenon of induced ovulation, by which eggs are shed in response to copulation rather than on a cyclic basis; and post-partum estrus, permitting a female to conceive immediately after giving birth. Female lagomorphs are also capable of resorbing embryos under adverse conditions, for example in times of climatic or social stress. There is further evidence that species such as the European hare are capable of conceiving a second litter before the birth of the last young, an amazing feat called "superfetation."

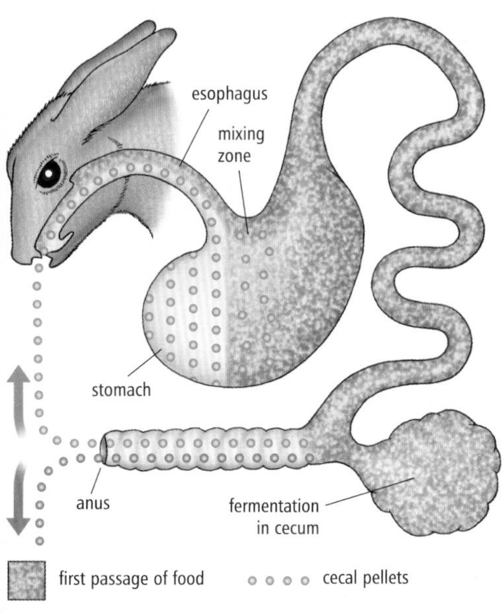

Losses and Gains

CONSERVATION ISSUES

Over 20 percent of lagomorph species are currently listed as threatened. Some are island endemics, like the Mexican Tres Marías cottontail; others, such as Koslov's pika (China), the Riverine rabbit (South Africa), and the Volcano rabbit (Mexico), have a distribution that has been severely reduced by destruction of their highly specialized habitats.

More encouragingly, the Sumatran rabbit (*Nesolagus netscheri*) – representing a unique genus of striped rabbit until recently feared extinct – has been caught on auto-trap camera in its tropical forest habitat. Furthermore, a possible second species of striped rabbit, the Annamite, has also come to light in remote montane forests between Laos and Vietnam. Three freshly-killed specimens were found in a meat market in 1995–96, and live animals have subsequently been photographed. The species' phylogenetic position has yet to be resolved. DB/ATS

Left *The digestive system of lagomorphs is highly modified for coping with large quantities of vegetation. The gut has a large, blind-ending sac (the cecum) between the large and small intestines, which contains bacterial flora to aid the digestion of cellulose. Many products of the digestion in the cecum can pass directly into the blood stream, but others such as the important B vitamins would be lost if lagomorphs did not eat some of the feces (refection) and so pass them through their gut twice. As a result, lagomorphs have two kinds of feces. First, soft black viscous cecal pellets which are produced during the day in nocturnal species and during the night in species active in the daytime. These are usually eaten directly from the anus and stored in the stomach, to be mixed later with further food taken from the alimentary mass. Second, round hard feces which are passed normally.*

Left *Alpine pikas have a fondness for sunning themselves and often bask on rocks, choosing for camouflage those with a similar color to their coat. Although many pikas live in regions with very severe winters, they do not appear to hibernate.*

ORDER: LAGOMORPHA

87 species in 12 genera and 2 families

Distribution Worldwide, except for S South America, the West Indies, Madagascar, and some Southeast Asian islands.

RABBITS AND HARES Family Leporidae p696

58 species in 11 genera

Includes **Riverine rabbit** (*Bunolagus monticularis*); **Hispid hare** (*Caprolagus hispidus*); **European rabbit** (*Oryctolagus cuniculus*); **Amami rabbit** (*Pentalagus furnessi*); **Volcano rabbit** (*Romerolagus diazi*); **Bunyoro rabbit** (*Poelagus marjorita*); **Pygmy rabbit** (*Brachylagus idahoensis*); **Sumatran rabbit** (*Nesolagus netscheri*); **Redrock hare** (*Pronolagus spp*); **Forest rabbit** (*Sylvilagus brasiliensis*); **Eastern cottontail** (*S. floridanus*); **Antelope jackrabbit** (*Lepus alleni*); **Black-tailed jackrabbit** (*L. californicus*); **Snowshoe hare** (*L. americanus*); **European hare** (*L. europaeus*); **Arctic hare** (*L. arcticus*).

PIKAS Family Ochotonidae p710

29 species in the genus *Ochotona*

Includes **Alpine pika** (*O. alpina*); **American pika** (*O. princeps*); **Northern pika** (*O. hyperborea*); **Royle's pika** (*O. roylei*); **Daurian pika** (*O. daurica*); **Gansu pika** (*O. cansus*); **Moupin pika** (*O. thibetana*); **Koslov's pika** (*O. koslowi*); **Plateau pika** (*O. curzoniae*); **Steppe pika** (*O. pusilla*).

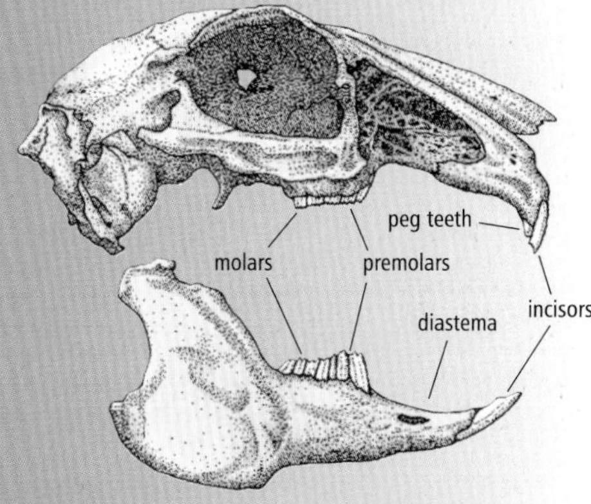

Above *The skull of a rabbit. Lagomorphs have long, constantly growing incisors, as do rodents, but lagomorphs differ in having two pairs of upper incisors, the back non-functional ones being known as peg teeth. There is a gap (diastema) between the incisors and premolars. The dental formula of rabbits and hares (family Leporidae) is I2/1, C0/0, P3/2, M3/3, with the pikas (family Ochotonidae) having one fewer upper molar in each jaw.*

Rabbits and Hares

THERE ARE FEW MAMMALS WHOSE FATE HAS been so intimately intertwined over the centuries with humans' as the European rabbit. Its domestication probably began in North Africa or Italy in Roman times; today there are well over a hundred varieties of domestic rabbit, all selectively bred from this single species. In addition, wild or domesticated offspring of the original stock have, through invasion or deliberate introduction, spread worldwide, with many populations reaching pest proportions. Yet the European rabbit is just one of a family of more than 50 leporid species, some of which lead a far more precarious existence, numbering in the hundreds rather than the tens of millions.

The family Leporidae splits broadly into two groups: the jackrabbits and hares of the genus *Lepus*, and the rabbits in the remaining 10 genera. Just to confuse matters, several species – for example, the African Red rockhare and the endangered Hispid hare – are commonly known as hares, even though behaviorally they are quite clearly rabbits!

Telling Rabbits from Hares

FORM AND FUNCTION

The major differences between the *Lepus* hares and the rabbits relate to differences in the strategies the two groups employ in evading predators and in reproduction. Basically, the longer-legged hares try to outrun their pursuers – some reputedly reaching speeds of 72 km/h (45 mph) in full flight – while the shorter-limbed rabbits run to seek refuge in dense cover or underground burrows. In addition, young hares (leverets) are better developed (precocial) at birth compared to altricial newborn rabbits (kittens). In the non-burrowing hares, the leverets are born after longer gestation periods (37–50 days) with a full covering of fur, their eyes open, and capable of coordinated movement. In contrast, rabbit kittens are born naked or with sparse fur covering after shorter gestation periods (27–30 days), their eyes opening after 4–10 days. Long ears are a conspicuous feature of all leporids, but typically these are at their most magnificent in the jackrabbits, where

Above *Like several other northern species, Snowshoe hares may molt twice a year, donning a white coat for camouflage each winter only to lose it again in the spring. Individuals of the species that live outside areas of continuous snow cover generally do not make the change.*

Left *With its large ears pricked for unfamiliar sounds and its eyes wide open to catch sudden movements, a Savanna hare in Tanzania's Serengeti National Park freezes briefly, alert for danger. In general, hares rely on their exceptional speed to outrun predators, while rabbits make rapidly for the nearest cover.*

they can grow to over 17cm (7in). The eyes of both rabbits and hares are large and adapted to their crepuscular and nocturnal activity patterns. All leporids are herbivorous, but some, like the Mountain and Snowshoe hares, may be more selective in their choice of feeding material than others, including the European rabbit.

Open Terrain or Dense Cover

DISTRIBUTION PATTERNS

Apart from a few forest-dwelling species such as the Snowshoe hare, most hares prefer open habitats with some cover offered by terrain or vegetation. They therefore have a widespread distribution, occurring in habitats ranging from desert to grasslands and tundra. Rabbits, on the other hand, are rarely found far from dense cover or underground tunnels. They occupy a variety of disturbed, successional, and climax habitats, often characterized by grass communities associated with dense cover; in the American cottontails this cover is often provided by plants such as sage brush and bramble. Other species are highly specialized in their habitat requirement, the two striped Sumatran and Annamite rabbits and the Japanese Amami rabbit living in tropical forest cover while the Riverine rabbit and Hispid hare (actually a rabbit!) are restricted respectively to pockets of riverine scrub in the central karoo of South Africa and to tall grassland in the Indian subcontinent. In marked contrast to rabbits, hares tend to use cover for daytime shelter, but will run into the open when confronted with a predator.

Warrens and Absentee Parents

SOCIAL BEHAVIOR

The burrowing lifestyle familiar in the warren-digging European rabbit is actually rather unusual, even among leporid genera. Apart from the European rabbit, only the Pygmy, Amami, and Bunyoro rabbits are reported to dig underground refuges themselves, while a few opportunists (for instance, the Eastern, Desert, and Mountain cottontails) will use burrows dug by other species. A small number of hares are reported to dig burrows to avoid extreme temperatures: the Black-tailed jackrabbit and Cape hare, for example, do so to escape high desert temperatures, while Snowshoe and Arctic hares may burrow into snow. Forms – surface depressions in the ground or vegetation – are more commonly used as resting-up sites by

FACTFILE

RABBITS AND HARES

Order: Lagomorpha

Family: Leporidae

58 species in 11 genera, 7 of them monotypic (containing only a single species): the Riverine rabbit (*Bunolagus monticularis*), Hispid hare (*Caprolagus hispidus*), European rabbit (*Oryctolagus cuniculus*), Amami rabbit (*Pentalagus furnessi*), Volcano rabbit (*Romerolagus diazi*), Bunyoro rabbit (*Poelagus marjorita*), and Pygmy rabbit (*Brachylagus idahoensis*). In addition, there are 2 species of striped rabbit (*Nesolagus*); 3 species of red rockhare (*Pronolagus*); 14 species of cottontail rabbit (*Sylvilagus*), including the Forest rabbit (*S. brasiliensis*) and Eastern cottontail (*S. floridanus*); and 32 species of hare (*Lepus*), including the Antelope jackrabbit (*Lepus alleni*), Black-tailed jackrabbit (*L. californicus*), Snowshoe hare (*L. americanus*), and European hare (*L. europaeus*).

DISTRIBUTION Americas, Europe, Asia, Africa; introduced to Australia, New Zealand, and other islands.

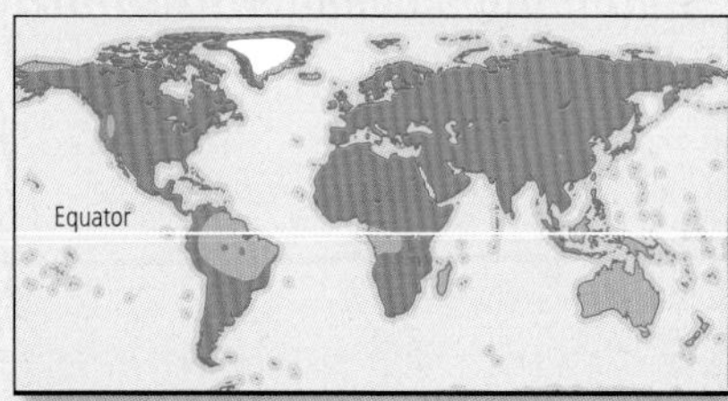

HABITAT Wide-ranging; includes desert, montane forest, tropical rain forest, Arctic tundra, swamp, tall grassland, agricultural landscapes.

SIZE **Head–body length** ranges from 25cm (10in) in the Pygmy rabbit to 75cm (30in) in the European hare; **tail length** from 1.5cm (0.6in) to 12cm (4.7in), and **weight** from around 400g (14oz) to 6kg (13.2lb), both in the same two species. **Ear length** reaches 17cm (7in) in the Antelope jackrabbit. Leporid hind limbs are regularly longer than the forelimbs.

COAT Usually thick and soft, but coarse or woolly in some species (Hispid hare and Woolly hare); hair shorter/sparser on the ears; tail well furred or even bushy (Riverine rabbit and Red rockhares); feet hairy on both surfaces; coloration ranges through reddish brown, brown, buff, and gray, to white; the belly is often covered with lighter or pure white hair. Two species (*Nesolagus*) are striped, and arctic/northern species change into white for winter (Snowshoe, Arctic, Mountain, Japanese hares).

DIET Herbivorous

BREEDING Gestation typically longer in hares (up to 55 days in the Mountain hare) than in rabbits (30 days in the European rabbit). The young are precocial at birth in hares, but altricial in rabbits.

LONGEVITY Average less than 1 year in the wild; maximum of 12 years recorded in European hare and European rabbit.

CONSERVATION STATUS 12 species are listed as threatened (i.e. Critically Endangered, Endangered, or Vulnerable), and another 6 as Near Threatened.

See species table ▷

hares. These may be well-established sites used by successive generations or, alternatively, temporary refuges occupied for only a few hours.

The underground warrens dug by European rabbits form the focus of stable, territorial breeding groups – a social system unknown amongst other leporid species. Most hares and rabbits are non-territorial, moving over individual home ranges of up to 300ha (740 acres) in some hares, with ranges overlapping in favored feeding areas. Temporary feeding aggregations have been seen in a number of leporids, including the Yarkand and Mountain hares, the Black-tailed jackrabbit, and the Brush rabbit. In the European hare these aggregations may be structured, with dominant individuals maintaining priority of access to food patches. Large flocks of Snowshoe hares may be sighted in the winter months, when these animals are in their white camouflage coat. Individual animals are thought to gather together to reduce their chance of being preyed upon.

Apart from "policing," in which males intervene to protect young under attack from adult females, no form of male paternal care has been reported in the leporids, and even maternal care is pretty thin on the ground – a reproductive strategy known as "absentee parentism." In hares, the precocial, fur-covered, fully mobile leverets are born into surface-depression forms, while the poorly developed rabbit kittens are delivered into carefully constructed fur-lined nests built in underground chambers or dense cover (thick clumps of montane bunch-grass in the case of the Volcano rabbit). After birth, a consistent and unusual feature of leporid maternal care is the nursing or suckling of litters for just one brief period, typically less than 5 minutes, once every 24 hours. In fact, the milk is highly nutritious, with a very high fat and protein content, and can be pumped into the youngsters at great speed during the brief lactation period, which lasts for 17–23 days. In the European rabbit, for example, this will be the only contact between mother and young until the kittens are weaned at around 21 days, after which the mother will start preparing for the birth of the litter conceived as she emerged from giving birth to the last one!

As a strategy, this lack of social contact between the mother and her offspring may be designed to reduce the chances of drawing a predator's attention to the highly vulnerable nestlings. In the case of those rabbit species that breed in purpose-dug breeding tunnels, or "stops," the soil entrance will be carefully re-sealed after each short suckling bout. In the surface-breeding hares, the leverets disperse to separate hiding locations about 3 days after birth, but regroup with litter-mates at a specific location at precisely defined intervals (often around sunset) for a frenzied, similarly brief, bout of suckling from mother.

Relationships between climate and reproduction are clearly demonstrated in both the hares and the New World cottontail rabbits. In the cottontails there is a direct correlation between latitude and litter size; species and subspecies in the north produce the largest litters during the shortest breeding season. The Eastern cottontail appears to be the most fecund of the genus, producing up to 35 young a year, whilst the Forest rabbit produces the least, at around 10 per annum. Amongst the *Lepus* hares and jackrabbits, productivity varies from a single litter of 6–8 in the far north to eight litters of 1–2 young at the equator, giving a fairly standard production statistic of about 10 young a year per female. Reproductive output is more variable in the Snowshoe hare (5–18 young per female), a species well-known for its population cycles, which are synchronous over a wide geographical range.

Compared to the more vocal pikas, communicating rabbits and hares appear to rely more heavily on scent than sound for communication. However, some species appear to be more vocal than others, and exceptions like the Volcano rabbit do exist. Most leporids make high-pitched distress squeals when captured by a predator, and five species of rabbit, all comparatively gregarious, give specific alarm calls. The European, Brush, and Desert rabbits are known to thump their hind feet on the ground in response to danger, possibly as a warning to underground nestlings. In addition, many leporids possess a conspicuous white underside to their tail which could serve as a visual warning signal during flight from a predator. Interestingly, those species with tail flags tend to be found in more open habitats than those like the Forest and Volcano rabbits and Hispid hare that have a dark underside to their tails.

All rabbits and hares have scent-secreting glands in the groin and under the chins. These appear to be important in sexual communication; in the gregarious European rabbit, where the activity of the glands is known to be related to testes size and levels of male hormones, they may also signal social status. Dominant male European rabbits are essentially the smelliest, scentmarking at higher frequency with scent-gland secretions and carefully-aimed squirts of pungent urine.

Left *Representative species of rabbits and hares:* ***1*** *Antelope jackrabbit* (Lepus alleni)*;* ***2*** *Amami rabbit* (Pentalagus furnessi) *digging a burrow;* ***3*** *Riverine rabbit* (Bunolagus monticularis) *in an alert posture;* ***4*** *Bunyoro rabbit* (Poelagus marjorita), *hopping;* ***5*** *dominant male European rabbit* (Oryctolagus cuniculus) *scentmarking with its chin;* ***6*** *Sumatran rabbit* (Nesolagus netscheri), *grooming its muzzle and spreading scent;* ***7*** *male Eastern cottontail* (Sylvilagus floridanus) *in an alert posture;* ***8*** *European hare* (Lepus europaeus), *boxing;* ***9*** *Greater red rockhare* (Pronolagus crassicaudatus) *in an alert scanning posture;* ***10*** *Hispid hare* (Caprolagus hispidus), *sitting among cuttings and pellets;* ***11*** *Volcano rabbit* (Romerolagus diazi), *reingesting pellets amid a vegetation of zacatón grasses.*

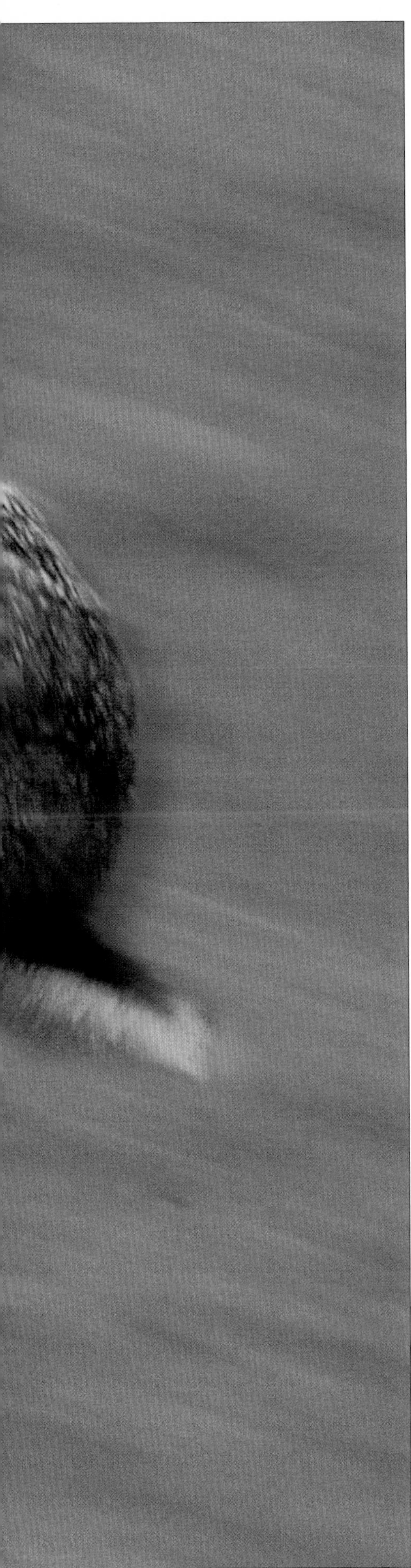

Above *Desert cottontail kittens crowd a shallow nest in California. The mother will not live in the nest with them, but will crouch over it to feed her young. This litter will be only one of several that she will raise in the course of the year, while the kittens themselves will be ready to start breeding within 3 months.*

Left *Built to make quick getaways, a European hare shows a clean pair of heels to a pursuer. Hares are the champion sprinters among the smaller mammals; their long hind legs can drive them forward at speeds of over 70 km/h (45mph).*

UNWELCOME INTRODUCTIONS

Although many rabbit species are now at risk, the European rabbit (*Oryctolagus cuniculus*) is so successful that it has sometimes acquired the status of a major pest. Problems have arisen particularly following the introduction of the species to areas where it was previously unknown – for example Australia and New Zealand in the early 19th century. The result was a population explosion that proved highly costly for the nations' arable and livestock farmers and that was only finally controlled by the deliberate introduction of the rabbit disease myxomatosis in the 1950s

The reasons for the rabbit's dramatic capacity to increase its numbers are built into its biology. Unlike many of its cousins, the European rabbit is not a fussy eater; it can feed off the same plants at many different stages in their growth. It is also almost unique in its habit of living communally in large burrows, a lifestyle that encourages high population densities. Above all, it is hugely prolific. Adult females can bear five litters a year, each containing an average of five or six young. Female offspring will themselves start breeding after just 5–6 months – an unfailing recipe for spiraling demographics.

Relict Species at Risk

CONSERVATION AND ENVIRONMENT

Sadly, the image of rabbits and hares has been somewhat tainted by a few species like the European rabbit and hare that are notorious for the damage they inflict on agricultural crops or forestry plantations. A preoccupation with rabbits' and hares' potential as pests has caused us to neglect the significant positive roles the leporid species play in ecosystems worldwide, both as prey items for small- to medium-sized vertebrate predators and through their grazing activities. Their pestilential reputation can also draw attention away from the larger number of rabbits and hares now listed as highly threatened in international registers.

Typically, those leporids threatened with extinction are primitive relict species, often the only members of their genus and usually the victims of habitat destruction by man. Most are highly specialized in their habitat requirements. The very handsome Riverine rabbit currently clings on in remnants of riverine scrub habitat associated with two seasonal rivers in the central karoo in South Africa: its habitat has been destroyed to promote the irrigation of encroaching agricultural crops. Similarly, the distribution of the primitive Hispid hare is now restricted to isolated fragments of tall thatch grassland habitat located in a few protected areas across the northern Indian subcontinent, while the tiny Mexican Volcano rabbit, or zacatuche, is isolated in pockets of endemic bunch-grass habitat on the slopes of just a few volcanoes around Mexico City, one of the world's largest conurbations. It is time that this group of attractive, long-eared mammals received more attention – and a far more positive press. ATS/DB

Rabbit and Hare Species

GENUS *BRACHYLAGUS*

Pygmy rabbit LR
Brachylagus idahoensis

SW Oregon to EC California, SW Utah, N to SE Montana; isolated populations in WC Washington State. Prefers habitat comprising clumps of dense sagebrush; extensive runways may cross the thickets. Lives in burrows of its own construction.
HBL 21–27cm; TL 1.5–2cm; WT 50–470g. The smallest rabbit.
COAT: reddish, similar to red rockhares with bushy tail.
CONSERVATION STATUS: LR – Near Threatened.

GENUS *BUNOLAGUS*

Riverine rabbit En
Bunolagus monticularis
Riverine rabbit or Bushman hare

Central Cape Province (South Africa). Dense riverine scrub (not the mountainous situations often attributed). Nocturnal, resting by day in hollows on the shady side of of bushes. Now extremely rare.
HBL 34–48cm; TL 7–11cm; WT 1–1.5kg
FORM: coat reddish, similar to red rockhares with bushy tail.

GENUS *CAPROLAGUS*

Hispid hare En
Caprolagus hispidus
Hispid hare, Assam or Bristly rabbit

Uttar Pradesh to Assam; Tripura (India). Mymensingh and Dacca on the W bank of River Brahmaputra (Bangladesh). Sub-Himalayan *sal* forest where grasses grow up to 3.5m in height during the monsoon months; occasionally also in cultivated areas. Inhabits burrows that are not of its own making. Seldom leaves forest shelter.
HBL 48cm; TL 5.3cm; EL 7cm; HFL 10cm; WT 2.5kg.
FORM: coat coarse and bristly; upperside appears brown from intermingling of black and brownish-white hair; underside brownish-white, with chest slightly darker; tail brown throughout, paler below. Claws straight and strong.

GENUS *LEPUS*

Most inhabit open grassy areas, but: Snowshoe hare occurs in boreal forests; European hare occasionally forests; Arctic hare prefers forested areas to open country; Cape hare prefers open areas, occasionally evergreen forests. Instead of seeking cover, hares rely on their well-developed running ability to escape from danger: also on camouflage, by flattening on vegetation. Vocalizations include a deep grumbling, and shrill calls are given when in pain. Usually solitary, but the European hare is more social. Habitat type has a marked effect on home-range size within each species, but differences also occur between species, e.g. from 4–20ha (10–50 acres) in Arctic hares to over 300ha (745 acres) in European hares. Individuals may defend the area within 1–2m of forms, but home ranges generally overlap and feeding areas are often communal. Most live on the surface, but some species, e.g. Snowshoe and Arctic hares, dig burrows, while others may hide in holes or tunnels not of their making.
HBL 40–76cm; TL 3.5–12cm; WT 1.2–5kg.
FORM: coat usually reddish-brown, yellowish-brown or grayish-brown above, lighter or pure white below; ear tips black-edged, with a significant black area on the exterior in most species; in some species the upperside of the tail is black. The Indian hare has a black nape. Species inhabiting snowy winter climes often molt into a white winter coat, while others change from a brownish summer coat into a grayish winter coat.
DIET: usually grasses and herbs, but cultivated plants, twigs, bark of woody plants are the staple food if other alternatives are not available.
BREEDING: breed throughout the year in southern species; northern species produce 2–4 litters during spring and summer. Gestation up to 50 days in Arctic hare, other species shorter. Litter size 1–9.
LONGEVITY: Only a minority of hares survive their first year in the wild, though survivors can reach 5 years; in captivity, hares can live to 6 or 7 years.

Antelope jackrabbit
Lepus alleni

S New Mexico, S Arizona to N Nayarit (Mexico), Tiburon Island. Locally common. Avoids dehydration in hot desert by feeding on cactus and yucca.

Snowshoe hare
Lepus americanus

Alaska, coast of Hudson Bay, Newfoundland, S Appalachians, S Michigan, N Dakota, N New Mexico, Utah, E California. Locally common.

Arctic hare
Lepus arcticus

Greenland and Canadian Arctic Islands south to WC shore of Hudson Bay. Quebec and W maritime provinces of Canada.

Japanese hare
Lepus brachyurus

Honshu, Shikoku, Kyushu (Japan). Locally common.

Black-tailed jackrabbit
Lepus californicus

N Mexico (Baja California), Oregon, Washington, S Idaho, E Colorado, S Dakota, W Missouri, NW Arkansas, Arizona, N Mexico. Locally common.

White-sided jackrabbit LR
Lepus callotis

SE Arizona, SW New Mexico, and Oaxaca (Mexico). Locally common, but declining.
CONSERVATION STATUS: LR – Near Threatened.

Cape hare
Lepus capensis

Sub-Saharan Africa, N Africa through Sinai desert to Arabia, Mongolia, Middle East W of River Euphrates. Locally common.

Broom hare Vu
Lepus castroviejoi

Cantabrian Mountains (N Spain).

Yunnan hare
Lepus comus

Yunnan and W Guizhou (China).

Korean hare
Lepus coreanus

Korean peninsula, S Jilin, S Liaoning, E Heilongjiang (China).

Corsican hare
Lepus corsicanus

Italy (including Sicily); introduced to Corsica (France).

Savanna hare
Lepus crawshayi

S Africa, Kenya, S Sudan; relict populations in NE Sahara. Locally common.

European hare
Lepus europaeus
European or Brown hare

S Scandinavia, S Finland, Great Britain (introduced in Ireland), Europe south to N Iraq and Iran, W Siberia. Locally common but declining.

Ethiopian hare
Lepus fagani

N and W Ethiopia and neighboring SE Sudan south to NW Kenya.

Tehuantepec jackrabbit En
Lepus flavigularis

Restricted to sand-dune forest on shores of saltwater lagoons on N rim of Gulf of Tehuantepec (S Mexico). Nocturnal.

Granada hare
Lepus granatensis
Granada or Iberian hare

Iberian peninsula, Mallorca.

Hainan hare Vu
Lepus hainanus

Hainan (China).

Black jackrabbit LR
Lepus insularis

Espiritu Santo Island (Mexico).
CONSERVATION STATUS: LR – Near Threatened.

Manchurian hare
Lepus mandshuricus

Jilin, Liaoning, Heilongjiang (NE China), far NE Korea, Ussuri region (E Siberia, Russia). Range decreasing.

Indian hare
Lepus nigricollis
Indian or Black-naped hare

Pakistan, India, Sri Lanka (introduced into Java and Mauritius).

Woolly hare
Lepus oiostolus

Tibetan (Xizang) plateau and adjoining areas.

ABBREVIATIONS			
	HBL = head–body length TL = tail length EL = ear length HFL = hind-foot length WT = weight	Ex Extinct	En Endangered
		EW Extinct in the Wild	Vu Vulnerable
	Approximate nonmetric equivalents: 10cm = 4in 30g = 1oz 1kg = 2.2lb	Cr Critically Endangered	LR Lower Risk

Above *The Black-tailed jackrabbit lives in dry, sunny regions, where its huge ears help control heat intake. Even so, it is mostly active at night.*

Alaskan hare
Lepus othus

W and SW Alaska (USA), E Chukotsk (Russia).

Burmese hare
Lepus peguensis

Burma to Indochina and Hainan (China).

Scrub hare
Lepus saxatilis

S Africa, Namibia.

Chinese hare
Lepus sinensis

SE China, Taiwan, S Korea.

Ethiopian highland hare
Lepus starcki

C Ethiopian mountains.

Mountain hare
Lepus timidus
Mountain or Blue hare

Alaska, Labrador, Greenland, Scandinavia, N Russia to Siberia and Sakhalin, Hokkaido (Japan), Sikhoto Alin Mts, Altai, N Tien Shan, N Ukraine, Baltic states. Locally common. Isolated populations in the Alps, Scotland, Wales, and Ireland.

Tolai hare
Lepus tolai

N Caspian Sea S along E shore of Caspian to N Iran, E through Afghanistan, Kazakhstan to Mongolia and W, C, and NE China.

White-tailed jackrabbit
Lepus townsendii

S British Columbia, S Alberta, SW Ontario, SW Wisconsin, Kansas, N New Mexico, Nevada, E California. Locally common.

African savanna hare
Lepus victoriae

Atlantic coast of NW Africa, E across Sahel to Sudan and Ethiopia, S through E Africa to NE Namibia, Botswana, and S Africa.

Malawi hare
Lepus whytei

Malawi. Locally common.

Yarkand hare LR
Lepus yarkandensis

SW Xinjiang (China), margins of Takla Makan desert.
CONSERVATION STATUS: LR – Near Threatened

GENUS *NESOLAGUS*

Sumatran rabbit Cr
Nesolagus netscheri
Sumatran rabbit or Sumatran short-eared hare

W Sumatra (1°–4°S) between 600–1,400m (2,000–4,600ft) in Barisan range. Primary mountain forest. Strictly nocturnal; spends the day in burrows or in holes not of its own making.
HBL 37–39cm; TL 1.7cm; EL 4.3–4.5cm.
FORM: variable; body from buff to gray, the rump bright rusty with broad dark stripes from the muzzle to the tail, from the ear to the chin, curving from the shoulder to the rump, across the upper part of the hind legs, and around the base of the hind foot.
DIET: juicy stalks and leaves.

Annamite rabbit
Nesolagus timminsi

Annamite Mts. between Laos and Vietnam.

GENUS *ORYCTOLAGUS*

European rabbit
Oryctolagus cuniculus
European or Old World Rabbit

Endemic on the Iberian peninsula and in NW Africa; introduced in rest of W Europe 2,000 years ago, and to Australia, New Zealand, S America, and some islands. Opportunistic, having colonized habitats from stony deserts to subalpine valleys; also found in fields, parks, and gardens, rarely reaching altitudes of over 600m (2,000ft). Very common. All strains of domesticated rabbit derived from this

species. Colonial organization associated with warren systems. Utters shrill calls in pain or fear.
HBL 38–50cm; TL 4.5–7.5cm; EL 6.5–8.5cm; HFL 8.5–11cm; WT 1.5–3kg.
FORM: coat grayish with a fine mixture of black and light brown tips of the hair above; nape reddish-yellowish brown; tail white below; underside light gray; inner surface of the legs buff-gray; total black is not rare.
DIET: grass and herbs; roots and the bark of trees and shrubs, cultivated plants.
BREEDING: breeds from February to August/September in N Europe; 3–5 litters with 5–6 young, occasionally up to 12; gestation period 28–33 days; young naked at birth; weight about 40–45g, eyes open when about 10 days old.
LONGEVITY: about 10 years in wild.

GENUS *PENTALAGUS*

Amami rabbit En
Pentalagus furnessi
Amami or Ryukyu rabbit

Two of the Amami Islands (Japan). Dense forests. Nocturnal. Digs burrows.
HBL 43–51cm; EL 4.5cm.
FORM: coat thick and woolly, dark brown above, more reddish below. Claws are unusually long for rabbits at 1–2cm. Eyes small.
BREEDING: 1–3 young are born naked in a short tunnel; two breeding seasons.

GENUS *POELAGUS*

Bunyoro rabbit
Poelagus marjorita
Bunyoro or Central African rabbit, or Uganda grass hare

S Sudan and Chad, NW Uganda, NE Zaire, Central African Republic, Angola. Savanna and forest. Locally common. Nocturnal. While resting, hides in vegetation. Reported to grind teeth when disturbed.
HBL 44–50cm; TL 4.5–5cm; EL 6–6.5cm; WT 2–3kg.
FORM: coat stiffer than that of any other African leporid; grizzled brown and yellowish above, becoming more yellow on the sides and white on the underparts; nape reddish-yellow; tail brownish-yellow above and white below. Ears small; hind legs short.
BREEDING: Young reared in burrows; less precocious than those of true hares.

GENUS *PRONOLAGUS*

Nocturnal, feeding on grass and herbs. Inhabits rocky grassland, shelters in crevices. Utters shrill vocal calls even when not in pain.
HBL 35–50cm; TL 5–10cm; HFL 7.5–10cm; EL 6–10cm; WT 2–2.5kg.
FORM: coat thick and woolly, including that on the feet, reddish.

Greater red rockhare
Pronolagus crassicaudatus
Greater red rockhare or Natal red rock rabbit

SE South Africa, S Mozambique.

Jameson's red rockhare
Pronolagus randensis

NE South Africa, E Botswana, Zimbabwe, W Mozambique, W Namibia.

Smith's red rockhare
Pronolagus rupestris

South Africa to SW Kenya.

GENUS *ROMEROLAGUS*

Volcano rabbit En
Romerolagus diazi
Volcano rabbit, teporingo, or zacatuche

Restricted to two volcanic sierras (Ajusco and Ixtaccihuatl–Popocatepetl ranges) close to Mexico City. Habitat unique "zacatón" (principally *Epicampes*, *Festuca*, and *Muhlenbergia*) grass layer of open pine forest at 2,800–4,000m. Lives in warren-based groups of 2–5 animals. Vocalizations resemble those of picas.
HBL 27–36cm; EL 4–4.4cm; WT 400–500g.
FORM: coat dark brown above, dark brownish-gray below. Smallest leporid; features include short ears, legs, and feet, articulation between collar and breast bones, and no visible tail.
BREEDING: breeding season December to July; gestation 39–40 days; average litter 2. Mainly active in daytime, sometimes at night.

GENUS *SYLVILAGUS*

Most species common. Range extends from S Canada to Argentina and Paraguay, and a great diversity of habitats is occupied. Distributions of some species overlap. Most preferred habitat open or brushy land or scrubby clearings in forest areas, but also cultivated areas or even parks. Various species frequent forests, marshes, swamps, sand beaches, or deserts. All species occupy burrows made by other animals or inhabit available shelter or hide in vegetation. Not colonial, but some species form social hierarchies in breeding groups. Active in daytime or at night. Not territorial; overlapping stable home ranges of a few hectares. Vocalizations rare. Most species are locally common.
HBL 25–45cm; TL 2.5–6cm; WT 0.4–2.3kg.
FORM: coat mostly speckled grayish-brown to reddish-brown above; undersides white or buff-white; tail brown above and white below ("cottontail"); Forest rabbit and Marsh rabbit have dark tails. Molts once a year, except Forest and Marsh rabbits. Ears medium-sized (about 5.5cm) and same color as the upper side; nape often reddish, but may be black.
DIET: mainly herbaceous plants, but in winter also bark and twigs.
LONGEVITY: 10 years (in captivity).

Swamp rabbit
Sylvilagus aquaticus

E Texas, E Oklahoma, Alabama, NW to S Carolina, S Illinois. A strong swimmer.
BREEDING: Gestation period 39–40 days; eyes open at 2–3 days.

Desert cottontail
Sylvilagus audubonii
Desert or Audubon's cottontail

C Montana, SW to N Dakota, NC Utah, C Nevada, and N and C California (USA), and Baja California and C Sinaloa, NE Puebla, W Veracruz (Mexico).

Brush rabbit
Sylvilagus bachmani

W Oregon to Baja California, Cascades to Sierra Nevada ranges.
BREEDING: Average 5 litters per year; gestation 24–30 days; young are covered in hair at birth.

Forest rabbit
Sylvilagus brasiliensis
Forest rabbit or tapiti

S Tamaulipas (Mexico) to Peru, Bolivia, N Argentina, S Brazil, Venezuela.
BREEDING: Average litter size 2; gestation about 42 days.

Mexican cottontail LR
Sylvilagus cunicularius

Sinaloa to Oaxaca and Veracruz (Mexico).
CONSERVATION STATUS: LR – Near Threatened.

Dice's cottontail En
Sylvilagus dicei

C American isthmus: SE Costa Rica to NW Panama.

Eastern cottontail
Sylvilagus floridanus

Venezuela through disjunct parts of C America to NW Arizona, S Saskatchewan, SC Quebec, Michigan, Massachusetts, Florida. Very common.
BREEDING: Gestation period 26–28 days; young naked at birth.

Tres Marías cottontail En
Sylvilagus graysoni

Maria Madre and Maria Magdalena Islands (Tres Marías Islands, Navarit, Mexico).

Omilteme cottontail Cr
Sylvilagus insonus

Sierra Madre del Sur, C Guerrero (Mexico).

Brush rabbit LR
Sylvilagus mansuetus
Brush rabbit or San José brush rabbit

Known only from San José Island, Gulf of California. Often regarded as subspecies of *S. bachmani*.
CONSERVATION STATUS: LR – Near Threatened.

Mountain cottontail
Sylvilagus nuttallii
Mountain or Nuttall's cottontail

Intermountain area of N America from S British Columbia to S Saskatchewan, S to E California, NW Nevada, C Arizona, NW New Mexico.

Appalachian cottontail
Sylvilagus obscurus

New York State (W of Hudson River) to N Alabama along Appalachian Mountain chain.

Marsh rabbit
Sylvilagus palustris

Marsh rabbit or Lower Keys marsh rabbit Florida Keys to S Virginia on the coastal plain. Strong swimmer.

New England cottontail Vu
Sylvilagus transitionalis

S Maine to N Alabama. Distinguished from overlapping Eastern cottontail by presence of gray mottled cheeks, black spot between eyes and absence of black saddle and white forehead. ES

Right As its name suggests, the Marsh rabbit is a habitat specialist, found only in swampy areas of the southern USA. A subspecies from the lower Florida Keys, S. p. hefneri, is considered Endangered.

ABBREVIATIONS
HBL = head–body length TL = tail length EL = ear length
HFL = hind-foot length WT = weight
Approximate nonmetric equivalents: 10cm = 4in 30g = 1oz 1kg = 2.2lb
Ex Extinct
EW Extinct in the Wild
Cr Critically Endangered
En Endangered
Vu Vulnerable
LR Lower Risk

The Social Life of Rabbits

How burrows shape interactions between group members

The European rabbit is unusual among rabbits and hares in constructing its own burrows. These can vary from single-entrance breeding "stops" through to extensive burrow systems, each containing a myriad of interconnected underground tunnels, accessed from as many as 60 entrances and containing a number of potential nest sites. Multiple-entrance burrow systems are generally referred to as "warrens."

The burrowing habit has a number of implications for rabbit ecology and behavior. First, it allows rabbits to live in relatively open habitats, as the burrow affords shelter from predators. Rabbits can also raise large numbers of young by giving birth in the safe confines of the warren.

Female rabbits usually nurse their young in underground nesting chambers situated within pre-existing burrow systems. Alternatively, young are raised in purpose-built stops with only one entrance. The mother animal visits the young only once a day for suckling, which may last as little as 5 minutes. When she leaves, she carefully covers the entrance. Despite these efforts, such stops are prone to being dug out by predators. Hence, nest sites deep in the main warrens are at a premium.

Where space underground is in short supply, natural selection should favor females that successfully defend nest sites. Good evidence of this was obtained from a long-term study of a population in southern England. There, the burrows were clustered together in tight groups in the form of warrens, which were themselves randomly distributed over the down.

Distinct social groups were established, the members of each having exclusive access to one or more burrow systems in their territory. Adult females, which do most of the burrow excavation, rarely attempted more than the expansion of an existing burrow system in the hard chalky soil. Completely new warrens hardly ever appeared, although breeding stops were constructed occasionally, and were at the core of the few new warrens that did appear during the study. Of the disputes between adult females, over 70 percent took place within 5m (16.5ft) of a burrow entrance. These were the most aggressive interactions seen during the course of the study, with the fur sometimes literally flying.

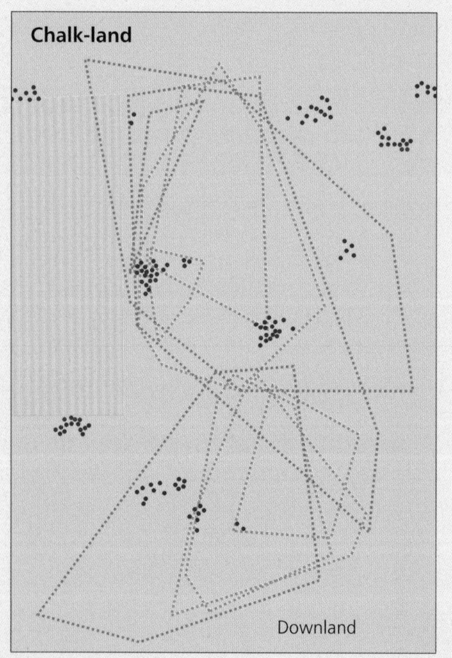

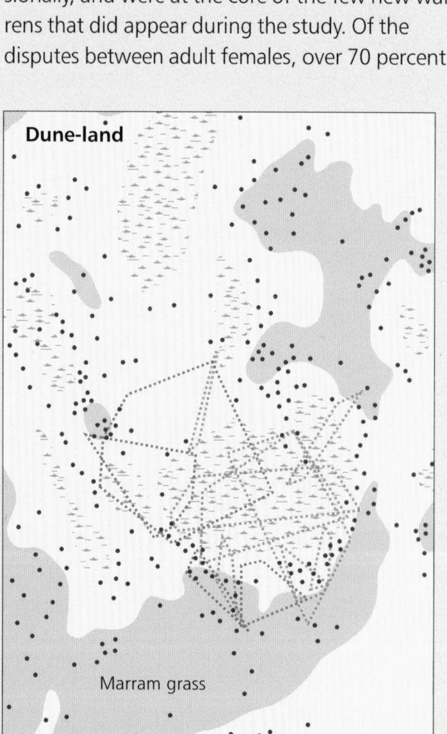

Male territories · Female territories · Burrow entrances
Agricultural field · Dune slacks · 0 — 50m Scale

Above. *Maps comparing group-living and social behavior in European rabbits on chalk- and dune-land respectively. On chalk-land rabbits have clustered burrows with females living as reluctant partners around each cluster. Fights often break out between females and their home ranges overlap considerably within each group but not with those of adjacent groups.*

On dune-land rabbit burrows are not clustered and are randomly distributed, although they do not occur in the slacks, which are prone to flooding. Females move freely between burrows and there is little fighting between individuals. Home ranges overlap less than on chalk-land. In both habitats males have larger territories which overlap those of several females.

There was also a direct relationship between the size of a warren and the number of adult females taking refuge in it: the larger the warren, the more females that lived there. Thus a group of females sharing one or more warrens and feeding in extensively overlapping ranges around them are best regarded as reluctant partners in an uneasy alliance. This reluctance reflects the fact that only costs seem to accrue to group-living females, especially in terms of the survival of their offspring, which can be impaired by increases in both disease and predation. Solitary females, with exclusive access to a warren and attendant nest sites, do best in terms of reproductive success.

The pattern of dispersal also gave some insight into the importance of burrow availability. Overall, both adult males and females were rather sedentary, with only 20 percent and 5 percent respectively moving breeding groups between years. Dispersal was much more common among juvenile males, two-thirds of which bred in a different group from that into which they had been born, as compared with only one-third of juvenile

Above *Post-coital grooming between a mating pair. Females enter estrus again shortly after giving birth. With a gestation period of only about 30 days, females can produce perhaps 4–5 litters per year.*

females. Interestingly, however, those juvenile females that did disperse entered breeding groups that were significantly less crowded in terms of burrow availability.

What of the males? Male rabbits do not contribute directly to parental care. Consequently, their reproductive success will reflect how many matings they have achieved. Females come into heat for 12–24 hours about every seventh day, or soon after giving birth. Males apparently monitor female condition closely: adult does were "escorted" by single males for about a quarter of their time above ground in the breeding season. Male home ranges were on average about twice the size of those of neighboring females, with which they had an extensive overlap. Consequently, these bucks could have been acquiring information, perhaps largely on the basis of scent rather than direct encounters, about the reproductive state of numerous females. The frequent aggressive interactions observed between males, whether or not they were escorting does, may be best interpreted as attempts to curtail each other's use of space and access to females. The behavior of bucks following females around can be regarded as "mate-guarding." Each female is usually accompanied by only one male.

Despite bucks' efforts to monopolize matings with does, the females are promiscuous. An Australian study involved the genetic typing (using blood proteins) of all potential parents in a population, together with their weaned young. The resulting analysis showed that at least 16 percent of the young were not fathered by the male known to be the usual escort of their mother.

Long-term studies of a natural population of European wild rabbits living in eastern England have provided insights into relationships between social organization and the genetic structure of populations. Here, the rabbits live in highly territorial breeding groups that defend areas around the warrens. These breeding groups typically comprise 1–4 males and 1–9 females. Analyses of genetic relatedness between colony members confirmed predictions made from patterns of juvenile dispersal where sons disperse while daughters stay on to breed. As a consequence, females within breeding groups tend to be very closely related, with several generations living together. Indeed, the genetic relatedness of females in breeding groups is at least twice the average relatedness of males. This pattern is reinforced by very infrequent movement of individual rabbits between breeding groups once they reach adulthood.

Between 1987 and 1990, the number of adults in the population more than doubled, from 22 to 45. This resulted in an increase in the number of adults per social group rather than in the formation of new groups. Groups did seem to split when the number of female kin members expanded above six individuals; one or two females would restrict their activities to peripheral burrows within the group territory, and new males would start defending these does to form a new breeding group.

We can now recognize that the burrow shapes both the ecology and behavior of the wild rabbit. It is at the heart of its success as an invasive species throughout the world, so perhaps it is not surprising that it is also at the core of much conflict in rabbit societies. DPC/DB

Above *A European rabbit at a burrow entrance in southern England. Differences in soil type play a major role in social systems. Locations in which warrens can be dug quickly and easily cause rabbit populations to spread out, whereas in hard-soil areas populations center on long-established warren systems.*

SPECIAL FEATURE

THE TEN-YEAR CYCLE

Population fluctuations in the Snowshoe hare

ANIMAL POPULATIONS RARELY, IF EVER, REMAIN constant from year to year. Populations of the great majority of species fluctuate irregularly or unpredictably. An exception is the Snowshoe hare, whose populations in the boreal forest of North America undergo remarkably regular fluctuations that peak every 8–11 years.

These cycles were first analyzed quantitatively when wildlife biologists began to plot the fur-trading records of the Hudson's Bay Company during the early 1900s. Established in 1671, the company kept meticulous account of the numbers of furs traded from different posts across Canada. The most famous time series drawn from the records was that of the Canadian lynx published by Charles Elton and Mary Nicholson in 1942. The lynx is a specialist predator of Snowshoe hares, and the 9–10 year rise and fall in lynx numbers turned out to mirror, with a slight time lag, the rise and fall of Snowshoe hare populations (see graph).

The spectacular cycles of Snowshoe hares and their predators seem to violate the implicit assumption of many ecologists that there is a balance in nature; anyone living in the boreal forest would be hard pressed to recognize any equilibrium in the boom-and-bust pattern of nature's economy. The challenge to biologists has been to understand the mechanisms behind these cycles. Over the last 40 years ecologists working in Alberta, the Yukon Territory, and Alaska have put together an array of studies that have resolved most, but not all, of the enigmas underlying them.

The demographic pattern of the hare cycle is remarkably clear and consistent. The key finding is that both reproduction and survival begin to decay in the increase phase of the cycle, two years before peak densities are reached. Maximal reproduction and highest survival rates occur early in this phase; as it progresses, reproduction slows and survival rates of both adults and juveniles fall. Both reproduction and survival rates continue to decline for 2–3 years after the cycle has peaked, then start to recover over the low phase. This time lag in changes in reproduction and mortality is the proximate cause of the variation in density that makes up the hare cycle, which can see populations rise or fall 30- or even 100-fold.

What causes the changes? The three main factors involved seem to be food, predation, and social interactions. There are two variants of the food hypothesis: first, that the hares may simply run out of food in winter and starve, or that the quality of the food available to them may decline. Yet feeding experiments have failed to change the pattern of the cycle or prevent the cyclic decline, so this hypothesis has now been rejected. Food by itself does not seem to be the primary limitation on hare numbers.

Predation is the next most obvious explanation. Studies with radio-collared hares have in fact shown that the immediate cause of death of 95 percent of adult hares is predation by a variety of animals, the main ones in Canada being lynx, coyotes, goshawks, and Great horned owls. For leverets the figure is 81 percent, most of these being killed by various small raptors or else by Red or Arctic ground squirrels. Few animals die of malnutrition. Obviously such huge losses must play an important role in the cycle.

As for the predators themselves, the evidence indicates that all show strong numerical changes that lag behind the hare cycle by 1–2 years. In addition, both lynx and coyotes kill more hares per day in the peak and decline phases than during the increase. These kill rates have turned out to be well above previous estimates, and are also in excess of energetic demands. Surplus killing seems to be a characteristic of these predators.

By constructing electric fences around 1sq km (0.39sq mi) blocks of boreal forest, researchers

Left *A Snowshoe hare in its summer coat. Female hares show remarkable variations in litter size at different stages of the population cycle, bearing more than twice as many young when numbers are low.*

Right *A lynx carries off its prey. When hare numbers decline, the shortage of alternative food sources leads to a rise in mortality among lynx kittens as well as to a drop in the number of young that are born.*

have been able to test the impact of excluding mammalian predators from hares. The main effects were to increase survival rates and to temporarily stop the cycle inside the fence, indicating that predation is indeed the immediate cause of the mortality changes over the cycle.

Yet if predation causes the changes in mortality, what about the varying rates of reproduction that accompany the cycle? Two possible explanations for these have been suggested. Lower-quality food in the time of high density may reduce reproductive output; alternatively, predators may cause the decline by stressing hares through repeated, unsuccessful attacks. Chronic stress has many direct detrimental effects on mammals, one of which is reduction in reproductive rate. Stress effects may also be indirect and long-term, affecting offspring viability.

Ecologists are now close to understanding the Snowshoe hare cycle. They believe that it results from the interaction between predation and food; but of these two factors, predation is clearly the main one. The impact of food shortages is felt largely in winter and is indirect; hares do not die directly of starvation or malnutrition. But food quality and quantity may nonetheless play a part by affecting the hares' body condition, and so predisposing them to increased parasite loads and higher levels of chronic stress, which in turn probably cause reduced reproductive output.

Hares in peak and declining populations must trade off safety for food. The result is a time lag in both the direct and indirect effects of predation that causes the cyclical pattern. CJK

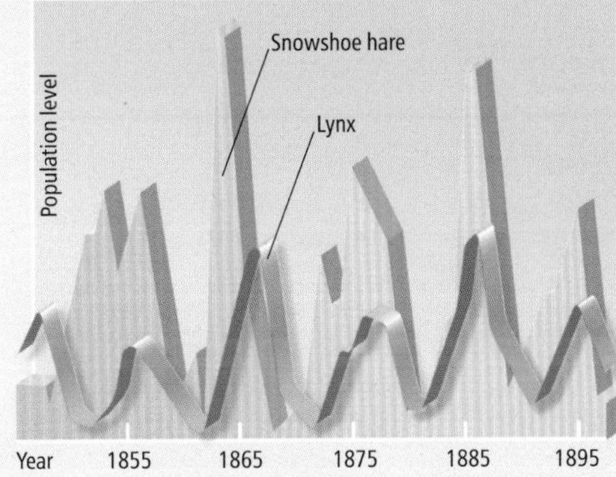

Above *The Snowshoe hare makes up 80–90 percent of the Canadian lynx's diet. Populations of the two mammals closely mirror each other; when food shortages and predation bring about a decline in hare numbers, the lynx population drops sharply soon after; conversely, the recovery of hare populations presages an upturn in lynx numbers. This parallel fluctuation in the animals' fortunes is plotted on the graph, which derives from the records of the Hudson's Bay Company.*

STORY

Mad World of the European Hare

❶ *With no underground sanctuary to provide refuge, the hare relies on its superb senses – and its long legs – for survival. Its sensory equipment includes eyes on the side of the head for allround vision, huge ears, and a sensitive nose.*

1

2

3

❷❸ *"Mad as a March hare," people say, recalling the seemingly wild behavior of hares in the mating season (January–August). At this time does are receptive for just a few hours on one day in each of their six-weekly cycles, perhaps six well-spaced days in all. Local bucks then compete for their favors; the dominant male strives to keep all others at bay, while the doe herself will fight off any that approach before she is ready. "Mad" behavior becomes visible in March only because the nights, which hares prefer for their activities, become shorter, forcing them to enter the daylight arena.*

❹ *Females do not pull their punches when beating off over-eager suitors, as the scarred ears of many bucks testify. When she is ready, the doe will start a wild chase over the countryside, shaking off the following bucks until only one, probably the fittest, remains. Then at last she will stop and allow him to mate.*

4

5 *A young hare shelters in its form, or daytime resting place; unlike their rabbit cousins, leverets enter the world fully coated, sighted, and mobile. Around sunset it will move cautiously to the spot where it was born a few days previously, where it will be joined by its littermates to await the mother. She will arrive about 45 minutes after sunset to suckle them for perhaps 5 minutes, and will then depart again, not to return for another 24 hours. By 4–5 weeks of age the leverets will be consuming vegetation and the doe's visits will cease.* *TH*

Pikas

PIKAS ARE SMALL, EGG-SHAPED LAGOMORPHS with relatively large, rounded ears, short limbs, and a barely visible tail. They are lively and agile, but often sit hunched up on a rock or in an alpine meadow, their long, silky fur making them resemble balls of fluff.

Pikas' generic name, *Ochotona*, is derived from the Mongolian term for the animals (*ogdoi*), while the word "pika" itself evolved from a vernacular term used by the Tungus of Siberia in attempting to mimic the call ("peeka...") of the Northern species. To this day, most pikas are denizens of high, remote mountains and wild country, serving as symbols of untamed nature.

Miniature Haymakers

FORM AND FUNCTION

Pikas are primarily active by day; only the Steppe pika is predominantly nocturnal. They do not hibernate, and are well-adapted to the cold alpine environments that they inhabit. They have a high body temperature, yet can perish in even moderately warm environmental conditions (American pikas have been known to die following a 30-minute exposure to 25°C/77°F). Thus they have little margin for error in their exposure to heat, and most species are active only during cool times of the day. High-altitude American pikas can be active all day, whereas populations at low altitudes (where it is hotter) emerge from their shelters only at dawn and dusk. Himalayan species demonstrate the same trend; Royle's pika is active in the morning and evening, while the Large-eared pika, which lives at cooler elevations above 4,000m (13,000ft), basks in the midday sun.

Pikas either live among rocks or else dig burrows in open meadow–steppe environments. Afghan and Pallas's pikas are intermediate in their use of habitat (they sometimes live in rocks, but also burrow), but their life-history closely parallels those of the burrowing pikas. Almost every facet of the biology of pikas is sharply divided between rock-dwelling and burrowing forms. Rock-dwelling pikas have very low reproductive rates due to the combination of small litter size and few litters per year; for example, most American pikas successfully wean only two young from one litter annually. In contrast, female burrowing pikas are baby machines; some species have as many as five litters containing up to 13 young.

Pikas can utilize whatever plants are available near their burrows or at the edge of their rocky scree territories, although they prefer those plants highest in protein or other important chemicals. They cannot grasp plants with their forepaws, so they eat grasses, leaves, and flowering stalks with a side-to-side motion of their jaws. During summer and fall, most species devote considerable time to harvesting mouthfuls of vegetation that they carry back to their dens to store for winter consumption in caches that resemble large piles of hay. American pikas may devote over 30 percent of their active time to haying, dashing back and forth with mouthfuls of vegetation. Haypiles rarely run out – pikas tend to overharvest, and there are often midden-like remains from the previous year.

In winter pikas also make tunnels in the snow to harvest nearby vegetation. Some species, such as Royle's, Large-eared, and Plateau pikas, live where winter snows are uncommon, and consequently do not construct haypiles – instead they continue to forage throughout the winter.

Pikas, like other lagomorphs, produce two different types of feces: small, spherical pellets resembling pepper seeds, and a soft, dark-green viscous excrement. The soft feces have high energy value (particularly in B vitamins) and are re-ingested either directly from the anus or after being dropped.

Gregarious or Reclusive?

SOCIAL BEHAVIOR

The dramatic differences between rock-dwelling and burrowing pikas are most apparent in their social behavior. Rock-dwelling pikas defend large territories, either as individuals (in North American species) or in pairs (Asian species). The resulting population density is low (about 2–10/ha, or 5–25/acre) and fairly stable over time. Rock-dwelling pikas rarely interact, and when they do it is usually to repel an intruding neighbor. Even the Asian forms that contribute to a shared haypile spend most of the day living solitary lives. The apparent lack of social activity can be somewhat misleading, however, as these animals are clearly aware of all the goings-on across the talus (see special feature: Securing a Vacancy).

In sharp contrast, burrowing pikas are among the most social of mammals. Family groups occupy communal dens, and local densities can exceed 300 animals per hectare (750 per acre) at the end of the breeding season, though the figures can fluctuate wildly, both seasonally and annually. During the breeding season family groups are composed of many siblings of different ages, and social interactions may occur as frequently as once

FACTFILE

PIKAS

Order: Lagomorpha

Family: Ochotonidae

29 species of the genus *Ochotona*. Rock-dwelling species include: Alpine pika (*O. alpina*); North American pika (*O. princeps*); Northern pika (*O. hyperborea*); Collared pika (*O. collaris*). Burrowing pikas comprise: Daurian pika (*O. dauurica*); Gansu pika (*O. cansus*); Koslov's pika (*O. koslowi*); Moupin pika (*O. thibetana*); Plateau pika (*O. curzoniae*); Steppe pika (*O. pusilla*).

DISTRIBUTION Mountains of W North America; across much of Asia N of the Himalayas, from the Middle East and the Ural Mountains E to the N Pacific Rim; from sea level to 6,130 m (20,100 ft).

HABITAT Rock-dwellers: talus (rocky scree) on mountains or occasional piles of fallen logs. Burrowing forms: alpine meadow, steppe, or semi-desert.

SIZE Head–body length 120–285mm (4.7–11.2in); weight 50–350g (1.8–12.3oz); tail length barely visible at about 5mm (0.2in); ear length 12–36mm (0.5–1.4in).

Equator

COAT Dense and soft; grayish-brown in most species (though one is reddish), usually darker above than below.

DIET Generalized herbivorous

BREEDING Rock-dwellers: litter size 1–5; 2 litters a year, but generally only 1 successfully weaned; gestation approximately 1 month. Burrowing forms: litter size 1–13; up to 5 litters a year; gestation approximately 3 weeks.

LONGEVITY Rock-dwellers: up to 7 years. Burrowing forms: up to 3 years, but most live only 1 year.

CONSERVATION STATUS 6 species and 13 subspecies are listed as threatened in some measure. The Helan Shan pika (*O. helanshanensis*) is Critically Endangered.

For full species list see Appendix ▷

Left The collared pika is found mainly on rocky outcrops in Alaska and northwestern Canada. Its name derives from the grayish patches below its cheek and around its neck. It spends roughly half its time above ground, sitting on prominent rocks

Above A characteristic activity of pikas in late summer is the gathering of vegetation to store in haypiles, in part to serve as food during the winter. Most of these stores of food are kept under overhanging rocks.

a minute. Pikas sit in contact, rub noses, socially groom, and play-box together. Young line up behind an adult – generally their father – and follow him like a miniature train. Nearly all these friendly social interactions occur within family groups, while interactions involving animals from different groups are normally aggressive – most notably the long chases of adult males.

Communication styles also differ between rock-dwelling and burrowing pikas. Most rock-dwellers have only two characteristic vocalizations: a short call used to announce their presence on the talus or to warn others of approaching predators, and a long call (or song) uttered by adult males during the breeding season. Some rock-dwelling species (Large-eared and Royle's pikas, for instance) rarely utter even weak sounds. Burrowing pikas, on the other hand, have a vast repertoire: predator alarm calls (short, soft, and rapidly repeated); long calls (given by adult males); and also whines, trills, muffle calls, and transition calls, these last two usually uttered by young pikas and serving to promote cohesion among siblings.

Burrowing pikas also have an unusually flexible mating system. In adjoining Plateau pika burrows, one can observe monogamous, polygynous, polyandrous, and complex (multiple male and female) adult associations side-by-side. Polyandry is extremely rare in mammals, yet two males from the same burrow may be seen alternately mating with the resident female and then sitting side by side or grooming one another, even while the female is in estrus – apparently an adaptation to maximize reproductive rates in face of harsh environmental conditions.

Not Just a Pest

CONSERVATION AND ENVIRONMENT

Several species and subspecies of pika are globally listed as threatened. In general, these are forms confined to restricted rocky habitats or found in isolated locations of central Asia. In reality, the status of many pika species is difficult to determine because they inhabit such remote areas.

On the other hand, some of the burrowing pikas are treated as pests because they reach such high densities and are believed to cause rangeland degradation. For example, the Plateau pika has been poisoned across 200,000sq km (77,000sq mi) in Qinghai province of China alone – an area half the size of California.

However, others consider the Plateau pika a keystone species for biodiversity on the Tibetan plateau because its burrows contribute to increased plant diversity and are the primary homes for a wide variety of birds and lizards. In addition, it serves as the principal prey for many predator species; and it contributes positively to ecosystem-level dynamics by recycling nutrients and minimizing erosion. TK/ATS

Right A pika at its burrow entrance in Ladakh, northeastern India. Only some pika species live in burrows and even these are not especially well adapted for digging. One of the main distinctions between burrowing and rock-dwelling pikas is that the former tend to be far less social than the latter.

SECURING A VACANCY

The social organization of the North American pika

TWO NORTH AMERICAN PIKAS DARTED INTO AND out of sight on a rock-strewn slope (known as a talus). The second, a resident male, was in aggressive pursuit of the first, an immigrant male. The chase continued onto an adjoining meadow, before turning into the dense cover of a nearby spruce forest. When next seen, dashing back toward the talus, the pair were being chased by a weasel. The pursuing pika was caught, and death followed swiftly less than 1m (3.3ft) from the safety of the talus. Immediately, all the pikas in the vicinity, with one exception, broke into a chorus of consecutive short calls, the sounds that pikas utter when they are alarmed by the presence of predators. The dead pika had initiated the chase, but the object of his aggression had managed to escape the weasel. Perched in silence on a prominent rock, he now surveyed his new domain.

Most accounts of the natural history of pikas have emphasized their individual territoriality;

Above *A pika sits amid its pile of winter provisions. Many pikas have a tendency to overprovision and collect far more leaves and grass than they will consume over the cold season.*

Left *Pikas have two characteristic vocalizations: the short call and the long call (or song). Long calls (a series of squeaks lasting up to 30 seconds) are given by males primarily during the breeding season. Short calls normally contain one-to-two note squeaks and may be given from promontories either before or after movement, in response to calls from another pika, while chasing or being chased, or when predators are active.*

however, studies conducted in the Rocky Mountains of Colorado have helped add detail to this basic insight. For example, adjacent territories are normally occupied by pikas of opposite sexes. Male and female neighbors overlap each other's home ranges more, and have centers of activity that are closer to one another's, than the ranges or activity centers of same-sex neighbors. The possession and juxtapositions of territories tend to be stable from year to year, consequently – as North American pikas can live for up to 6 years – the appearance and whereabouts of vacant territories on the talus are unpredictable. For a pika, therefore, trying to secure a vacancy on the talus is like entering a lottery in which an animal's sex in part determines whether or not it will have a winning ticket, for territories are almost always claimed by a member of the same sex as the previous occupant.

The behaviors that sustain this pattern of occupancy are apparently a compromise between the contrasting aggressive and affiliative tendencies of the pika. Although all pikas are pugnacious when they are involved in defending territories, females are less aggressive toward neighboring males, and conversely more aggressive to proximate females. Male residents rarely exhibit aggression toward each other because they do not come into contact with each other very frequently, apparently avoiding one another by the use of both scentmarking and vocalizations. Resident males will, however, vigorously attack unfamiliar (immigrant) males. The pika described in the account above had forayed from his home territory to chase an unfamiliar, immigrant adult male.

Affiliative behavior is seen in pairs of neighboring males and females, who are not only frequently tolerant of each other but also engage in duets of short calls. Such behavior is rarely seen between neighbors of the same sex or between non-neighbor heterosexual pairs.

Adults treat their offspring in the same way as neighbors of the opposite sex. Some aggression is directed toward juveniles, but there are also frequent expressions of social tolerance. Most juveniles will remain on the home ranges of their parents throughout their first summer before subsequently dispersing.

Ecological constraints have apparently led to a monogamous mating system in rock-dwelling pikas. Although males do not contribute directly to the raising of their offspring, they still primarily associate with a single neighboring female. Polygyny evolves when males can either monopolize sufficient resources to attract several females, or when they can directly defend several females. But in pikas the essentially linear reach of vegetation at the base of the talus precludes resource-defense polygyny, while males cannot defend groups of females because the females are dispersed and kept apart by their mutual antagonism, which thus precludes female-defense polygyny.

Juveniles of both sexes are likely to be repelled should they disperse and attempt to colonize an occupied talus. As a result, they normally settle close to their place of birth ("philopatry"). This pattern of settlement may lead to incestuous matings, which contributes to the low genetic variability that is found in pika populations.

The close association among male–female pairs and the near relatedness of neighbors may actually underlie the evolution of cooperative behavior patterns in pikas. First, attacks on intruders by residents may be an expression of indirect parental care: if the adults can successfully repel immigrants, they may increase the probability of their own offspring obtaining a territory should a local site subsequently become available for colonization. Second – to return to the opening account – the alarm calls given by both sexes when the weasel struck the resident pika served as a warning to the close kin – note that the unrelated immigrant was the only pika that did not call out. Uncontested, the newcomer immediately moved across the talus to claim the slain pika's territory, a half-completed haypile, and access to a neighboring female. ATS

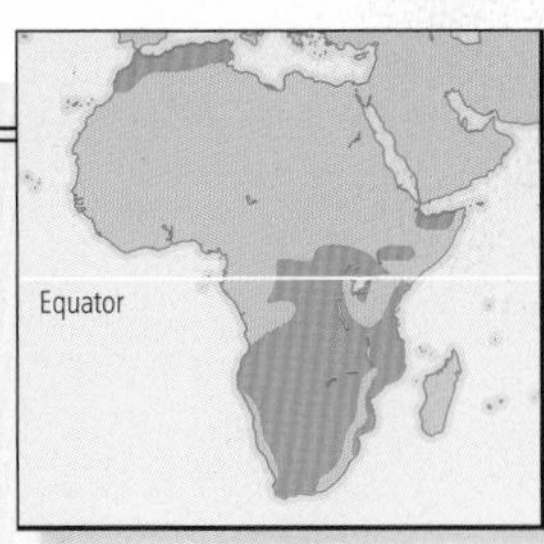

ELEPHANT SHREWS

Appearing much like giant versions of true shrews, elephant shrews were not described in the scientific literature until the mid-19th century, partly because they are cryptic, difficult to trap, and confined to Africa. Almost another century passed before a few short notes on their natural history appeared in print. Knowledge of them has expanded greatly over the last 50 years, revealing for the first time just how unique these animals really are.

In the past, the animals were sometimes referred to as "jumping shrews," but this is something of a misnomer since only the smaller species make pronounced leaps when alarmed; the normal method of locomotion is to walk on all fours. The name "elephant shrew," bestowed by field naturalists in Africa, alludes to their long snouts. With large eyes, a trunklike nose, high-crowned cheek teeth, and a large cecum similar to that found in herbivores, long legs like those of small antelopes, and a long, ratlike tail, elephant shrews sometimes seem like walking anthologies of other animals.

Anteaters on Stilts

FORM AND FUNCTION

The elephant shrews found in Africa today give little insight into the family's long and diverse evolutionary history. Fossil sengis first appeared in the early Eocene 50 million years ago, but they reached their maximum diversity by the Miocene (24 million years ago), when they comprised six subfamilies. One included a small, herbivorous form (*Mylomygale*) weighing about 50g (1.8oz) that resembled a grass-eating rodent; another a large planteater (*Myohyrax*) ten times that weight that was so ungulate-like it was initially thought to be a hyrax. Today, all that remains from these ancient forms are representatives of two well-defined, insectivorous subfamilies, the giant elephant shrews (Rhynchocyoninae) and the soft-furred elephant shrews (Macroscelidinae). The other four subfamilies mysteriously died out by the Pleistocene, 1.5 million years ago.

Taxonomically, elephant shrews have long been a source of controversy. At first, biologists included them with other insect-eaters in the Insectivora. Then they were briefly thought to be distantly related to ungulates. Next there was a scheme to include them with the tree shrews in a new grouping, the Menotyphla. More recently, they have been associated with rabbits and hares. Most biologists now agree that elephant shrews belong in their own order, the Macroscelidea. Perhaps to avoid the old association with true shrews, it has been recently suggested that elephant shrews should also be known as sengis, a name derived from several African Bantu languages.

So what is their exact phylogenetic relationship with other mammals? With the advent of molecular techniques to unravel evolutionary relationships, there is a growing consensus that the Macroscelidea belong to an ancient radiation of African mammals that today share few obvious morphological similarities. The latest proposal includes the elephant shrews in the superorder Afrotheria, which also includes elephants, hyraxes, sea cows (the Paenungulata), aardvarks, golden moles, and tenrecs.

Widespread, but Not Common

DISTRIBUTION PATTERNS

Elephant shrews are widespread in Africa, occupying very diverse habitats. For example, the distribution of the Short-eared elephant shrew (*Macroscelides proboscideus*) includes the Namib Desert in southwestern Africa as well as gravelly thornbush plains in South Africa's Cape Province, while the two rock elephant shrew species (*Elephantulus myurus* and *E. rupestris*) are largely restricted to rocky outcrops and boulder fields in southern Africa. Most other species of *Elephantulus* live in the vast steppes and savannas of southern and eastern Africa. The three giant elephant shrews of the *Rhynchocyon* genus and the Four-toed elephant shrew (*Petrodromus tetradactylus*) are restricted to lowland and mountain forests and associated thickets in central and eastern Africa. *Elephantulus rozeti* is found in semi-arid, mountain habitats in extreme northwestern Africa, isolated from all other species by the Sahara. The absence of sengis from western Africa has never been adequately explained. Nowhere are elephant shrews particularly common, and despite being highly terrestrial and mostly active above ground during the day and in the evening, they often escape detection because of their swift locomotion and secretive habits.

Golden-rumped elephant shrews spend up to 80 percent of their active hours searching for invertebrates, which they track down in the leaf litter on the forest floor by using their long, flexible noses as probes, in the manner of coatis or pigs. *Rhynchocyon* species also use their forefeet, which have three long claws, to excavate small, conical holes in the soil. Important prey include beetles, centipedes, termites, spiders, and earthworms. The soft-furred species spend only half as much time foraging. They normally glean small invertebrates, especially termites and ants, from leaves, twigs, and the soil's surface, but they also eat plant matter, especially small, fleshy fruits and seeds. All elephant shrews have long tongues that extend well beyond the tips of their noses and are used to flick small food items into their mouths.

Monogamy and Trail-Clearing

SOCIAL BEHAVIOR

While elephant-shrew species look diverse and live in vastly differing habitats, they all have similar sex lives. Individuals of the Golden-rumped, Four-toed, Short-eared, Rufous, and Western rock species live as monogamous pairs, but there appears to be little affection between partners.

Rufous elephant shrews that inhabit Kenya's densely wooded savannas are distributed as male–female pairs on territories that vary in size from 1,600 to 4,500sq m (0.4–1.1 acres). The same pattern is found in Golden-rumped elephant shrews in coastal forests of Kenya, although the territory sizes are larger, averaging 1.7ha (4.2 acres). Although monogamous, individuals of both species spend

ORDER: MACROSCELIDEA

Elephant shrews or Sengis
Family: Macroscelididae
2 subfamilies; 4 genera; 15 species

Distribution N Africa, E, C and S Africa, absent from W Africa and Sahara

Habitat Varied, including montane and lowland forest, savanna, steppe, desert.

Size Ranges from Short-eared elephant shrew, with **head-body length** 10.4–11.5cm (4.1–4.5in), **tail length** 11.5–13cm (4.5–5in), **weight** about 45g (1.6oz), to the Golden-rumped elephant shrew with a **head-body length** of 27–29.4cm (11–12in), **tail length** 23–25.5cm (9.5–10.5in), **weight** about 540g (19 oz).

Coat Soft, in various shades of gray and brown.

Diet Beetles, spiders, centipedes, earthworms, ants, termites and other small invertebrates; also fruits and seeds.

Breeding Gestation 57–65 days in the Rufous elephant shrew, about 42 days in the Golden-rumped elephant shrew.

Longevity 2½ years in the Rufous elephant shrew (5½ in captivity), 4 years in the Golden-rumped elephant shrew.

Conservation status *R. chrysopygus, R. petersi,* and *E. revoili* are classed as Endangered; *R. cirnei, M. proboscideus, E. edwardii,* and *E. rupestris* as Vulnerable.

GENUS *RHYNCHOCYON*

Golden-rumped elephant shrew (*Rhynchocyon chrysopygus*), **Black and rufous elephant shrew** (*R. petersi*), **Checkered elephant shrew** (*R. cirnei*).

GENUS *PETRODROMUS*

Four-toed elephant shrew (*Petrodromus tetradactylus*)

GENUS *MACROSCELIDES*

Short-eared elephant shrew (*Macroscelides proboscideus*)

GENUS *ELEPHANTULUS*

Short-nosed elephant shrew (*Elephantulus brachyrhynchus*), **Cape elephant shrew** (*E. edwardii*), **Dusky-footed elephant shrew** (*E. fuscipes*), **Dusky elephant shrew** (*E. fuscus*), **Bushveld elephant shrew** (*E. intufi*), **Eastern rock elephant shrew** (*E. myurus*), **Somali elephant shrew** (*E. revoili*), **North African elephant shrew** (*E. rozeti*), **Rufous elephant shrew** (*E. rufescens*), **Western rock elephant shrew** (*E. rupestris*).

Note: Elephant shrews may shortly be reassigned, with other endemic African placentals (e.g. golden moles, tenrecs, and elephants), to a new grouping, the "Afrotheria" – see p.723.

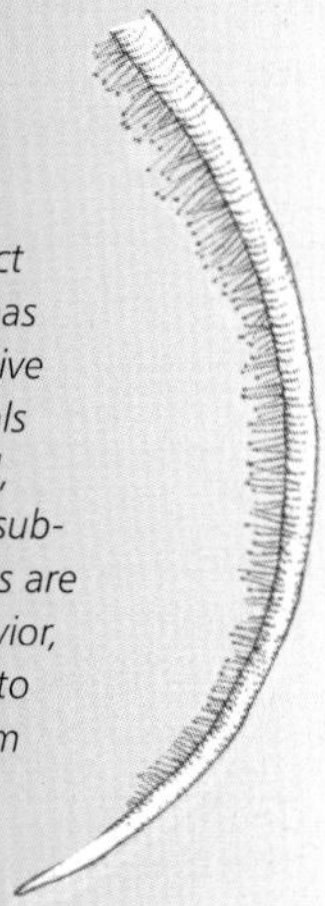

3

4

5

6

7

Right *Representative species of elephant shrews:* ***1*** *Checkered elephant shrew* (Rhynchocyon cirnei) *scentmarking with its anal glands;* ***2*** *Rufous elephant shrew* (Elephantulus rufescens) *foraging for insects;* ***3*** *Short-eared elephant shrew* (Macroscelides proboscideus) *clearing a trail;* ***4*** *North African elephant shrew* (Elephantulus rozeti) *washing its face at a burrow entrance;* ***5*** *Four-toed elephant shrew* (Petrodromus tetradactylus) *extruding its tongue after insects;* ***6*** *Black and rufous elephant shrew* (Rhynchocyon petersi) *tearing at prey with its teeth and claws;* ***7*** *Golden-rumped elephant shrew* (R. chrysopygus) *stalking before a chase.*

Right *Elephant shrew tails – the one shown is from the Four-toed elephant shrew – are lined with knobbed bristles. Their exact function is controversial, but it has been noted that, during aggressive and sexual encounters, individuals lash their tails across the ground, dragging the bristles across the substrate. It may be that the animals are scentmarking through this behavior, with the knobs acting as swabs to spread scent-bearing sebum from large glands on the tail's undersurface.*

Right *Rufous elephant shrews visibly mark their territories by creating small piles of dung in areas where the paths of two adjoining pairs meet. Occasionally aggressive encounters occur in these territorial arenas. In these situations, two animals of the same sex face one another and, while slowly walking in opposite directions, stand high on their long legs and accentuate their white feet, much like small mechanical toys. If neither of the animals then retreats, a fight usually develops and the loser is routed from the area.*

Below *A perfect ball of fur but for the protruding nose, a Short-eared elephant shrew basks in the sun. The species – the only representative of the* Macroscelides *genus – is limited to a region of about 20,000sq km (7,700sq mi) of southern African plain and desert, and is listed as Vulnerable by the IUCN.*

little time together. The male and female share precisely the same territory, but defend this area individually, with females seeing off other females and males evicting intruding males. This system of monogamy, characterized by limited cooperation between the sexes, is also found in several small antelopes, such as the dikdik and klipspringer. As in most monogamous mammals, the sexes are similar in size and appearance, but male giant elephant shrews have larger canine teeth.

In territorial encounters, visual signals are important, but elephant shrews also bring to bear their scent glands to mark out their land. These are located on the bottom of the tail in several *Elephantulus* species, on the soles of the feet in the Rufous elephant shrew, on the chest of the Dusky-footed, Rufous, and Somali elephant shrews, and just behind the anus in the giant elephant shrews of the *Rhynchocyon* genus. Vocal communication is unimportant, although the Four-toed elephant shrew and some species of *Elephantulus* create sounds by drumming their rear feet on the ground, while *Rhynchocyon* species slap their tails on the leaf litter. When captured, several elephant-shrew species emit sharp, high-pitched screams, although all are surprisingly gentle when handled and rarely attempt to bite despite their well-developed teeth.

In several respects elephant shrews are similar to small ungulates, especially in their avoidance of predators. Initially they rely on camouflage to elude detection, but if this fails they use their long legs to swiftly outdistance pursuing snakes and carnivores. This is no mean feat for a creature standing only 6cm (2.4in) high at the shoulder and weighing 58g (2oz); the trick is achieved by utilizing a system of trails to rival the road network of a city like London.

Even so, this explanation begs a question about the purpose of monogamy in the case of the Golden-rumped elephant shrew, which does not clear trails, apparently leaving the male of the pair jobless. The answer in its case lies in the forest habitat it inhabits. The tropical climate allows these particular sengis to breed continuously throughout the year, and their food resources are relatively evenly and widely distributed. Under these circumstances, the most productive strategy for a male may be to remain with one female, ensuring that he fathers her young, rather than to wander over huge expanses of forest trying to keep track of the reproductive condition of several females, and thereby running the very real risk of missing opportunities to mate. This resource-based explanation for monogamy is also thought to explain the paired sex lives of several small antelopes, such as the dikdik and some duikers.

In contrast to the path-using Four-toed and Rufous elephant shrews, several other sengis, including the Short-eared, Western rock, and Bushveld species, dig short, shallow burrows in sandy substrates for shelter. Where the ground is too hard, these species will use abandoned rodent burrows. But even the burrow-using sengis do not incorporate nesting material in their shelters, as do most rodents. The giant elephant shrews are more typical of small mammals, in that they spend each night in a leaf nest on the forest floor. To deny predators the reward of a meal in each nest that they find and tear open, the elephant shrews build several nests and then sleep alone in a different nest every few nights.

On the Trail of the Rufous Elephant Shrew

Near Tsavo National Park in Kenya, the Rufous elephant shrew lives in dense thickets in which each pair builds, maintains, and defends a complex network of criss-crossing trails. To enable the sengis to run at full speed along these paths, the trails must be kept immaculately clean. Just a single twig could break an elephant shrew's flight from a fast-moving predator with disastrous consequences, so the sengis regularly go road-sweeping. Every day, individuals of a pair spend 20–40 percent of the daylight hours separately traversing much of their trail network, removing accumulated leaves and twigs with swift sidestrokes of their fore feet. Little-used paths consist merely of a series of small, bare, oval patches on the sandy soil on which the sengi lands as it bounds along the trail; those that are heavily used form continuous bare channels through the litter.

The trails of Rufous elephant shrews, and also those of *Petrodromus*, are exceptionally important because neither species nests or lives in burrows or shelters. They spend their entire lives relatively exposed, as would small antelopes. Their distinct black-and-white facial pattern probably serves to disrupt the contour of their large black eyes, thus camouflaging them from predators while they are exposed on the trails.

The Rufous elephant shrew produces only 1 or 2 highly precocial and independent young per litter. Since the female alone can nurse her young, the male of the pair can do little to assist. This begs the question of why the animals should be monogamous in the first place.

In the Rufous elephant shrew, part of the answer apparently relates to the system of paths. Males spend nearly twice as much time trail-cleaning as females do – a rather similar arrangement to that of mara couples, which are also monogamous and in which the males put their effort into vigilance, freeing the female to graze. Although this sort of indirect help is not as obvious as the direct cooperation of wolf and marmoset pairs in raising their altricial young, it is just as vital to the elephant shrew's reproductive success, for without paths, its ungulate-like habits would be completely ineffective.

Above *Looking somewhat like a miniature anteater, a melanistic variant of the Checkered elephant shrew* (Rhynchocyon cirnei) *combs the forest floor for insects.*

Sengis in the tropics produce several litters throughout the year, but at higher latitudes reproduction becomes seasonal, usually in association with the wet season. Litters normally contain one or two young, but the North African elephant shrew and Checkered elephant shrew may produce three young per litter. Although all elephant shrews are born in a well-developed state with a coat pattern similar to that of adults, the young of giant elephant shrews are not as precocial as those of the soft-furred species, and thus they are confined to the nest for several days before they accompany their mother.

Giant elephant shrews are exceedingly difficult to keep in captivity, and they have never been bred. In contrast, the Rufous and Short-eared elephant shrews have been successfully exhibited and bred in several zoos, which has resulted in numerous laboratory studies of their biology. For example, physiological studies of Short-eared elephant shrews have shown that they can go into torpor, with body temperatures dropping from about 37°C (98.6°F) to as low as 9.5°C (49°F) for short periods when food resources are limited. This is thought to be an adaptation to conserve energy. Research on captive Rufous elephant shrews has shown that they can recognize the identities of family members and neighbors from scentmarks alone, so individuals can presumably closely monitor the use of their large territories by smell as well as by sight.

Disappearing Forests

CONSERVATION AND ENVIRONMENT

Generally, sengis are of little economic importance to man, although Golden-rumped and Four-toed elephant shrews are snared and eaten along the Kenya coast. This subsistence trapping is illegal, but is so far thought to be sustainable. A bigger problem for these forest-dwellers is severe habitat depletion, especially for those species occupying small, isolated patches of woodland in eastern Africa that are being degraded by tree-cutting for the woodcarving trade, or else being destroyed outright to make way for subsistence farming, exotic tree plantations, or urban developments. It would be a dreadful loss if these unique, colorful mammals were to disappear after more than 50 million years just because their dwindling patches of forest could not be adequately protected. GBR

ESCAPE AND PROTECTION

The tactics and adaptations of the Golden-rumped elephant shrew

THE AFRICAN SUN WAS JUST STARTING TO SET when a Golden-rumped elephant shrew made its way up to an indistinct pile of leaves about 1m (3ft) wide on the forest floor. The animal paused at the edge of the low mound for 15 seconds, sniffing, listening, and watching for the least irregularity. Sensing nothing unusual, it quietly slipped under the leaves. The leaf nest shuddered for a few seconds as the elephant shrew arranged itself for the night, then everything was still.

At about the same time the animal's mate was retreating for the night into a similar nest located on the other side of the pair's home range. As this elephant shrew prepared to enter its nest, a twig snapped somewhere. The animal froze, and then quietly left the area for a third nest, which it eventually entered, but not before dusk had fallen.

Every evening, within a few minutes of sunset, pairs of elephant shrews like this one separately approach and cautiously enter any one of a dozen or more nests they have constructed throughout their home range. They use a different nest each evening to discourage forest predators such as leopards and eagle-owls from ascertaining exactly where they can be found.

Changing nests is just one of several stratagems Golden-rumped elephant shrews regularly use to avoid predators. The problem they face is considerable. During the day they spend over 75 percent of their time exposed while foraging in leaf litter on the forest floor, where they fall prey to Black mambas, Forest cobras, and harrier eagles. To prevent capture by such enemies the animals have developed tactics that involve not only the ability to run fast but also a distinctive coat pattern that is notable for its flashy coloration.

Extraordinarily, Golden-rumped elephant shrews can bound across open forest floor at speeds above 25km/h (16mph)—about as fast as an average person can run. Because they are relatively small, they can also pass easily through patches of undergrowth, leaving larger terrestrial and aerial predators behind as they do so. Despite their speed and agility, however, they still remain vulnerable to ambush by sit-and-wait predators, such as the Southern banded harrier eagle. Most small terrestrial mammals have cryptic coloration on their coats or skins to serve as camouflage. However, the forest floor along the coast of Kenya where the Golden-rumped elephant shrew lives is relatively open, so any defense against predation that relied on camouflage would be ineffective.

Instead, the elephant shrew's tactic is to actively invite predators to take notice of it. It has a rump patch that is so visible that a waiting predator will discover a foraging shrew while it is too far away to make a successful ambush. The predator's initial reaction to the sight, such as rapidly turning its head or shifting its weight from one leg to another, may be enough to reveal its presence. By inducing the predator to disclose prematurely its intent to attack, a surprise ambush can be averted.

An elephant shrew that discovers a predator while still outside its flight distance does not bound away; instead it pauses and then repeatedly slaps the leaf litter with its tail at intervals of a few seconds. The sharp sound produced probably conveys a message to the predator: "I know you are there, but you are outside my flight distance, and I can probably outrun you if you attack." Through experience, the predator learns that when it hears this signal it is generally futile to attempt a pursuit,

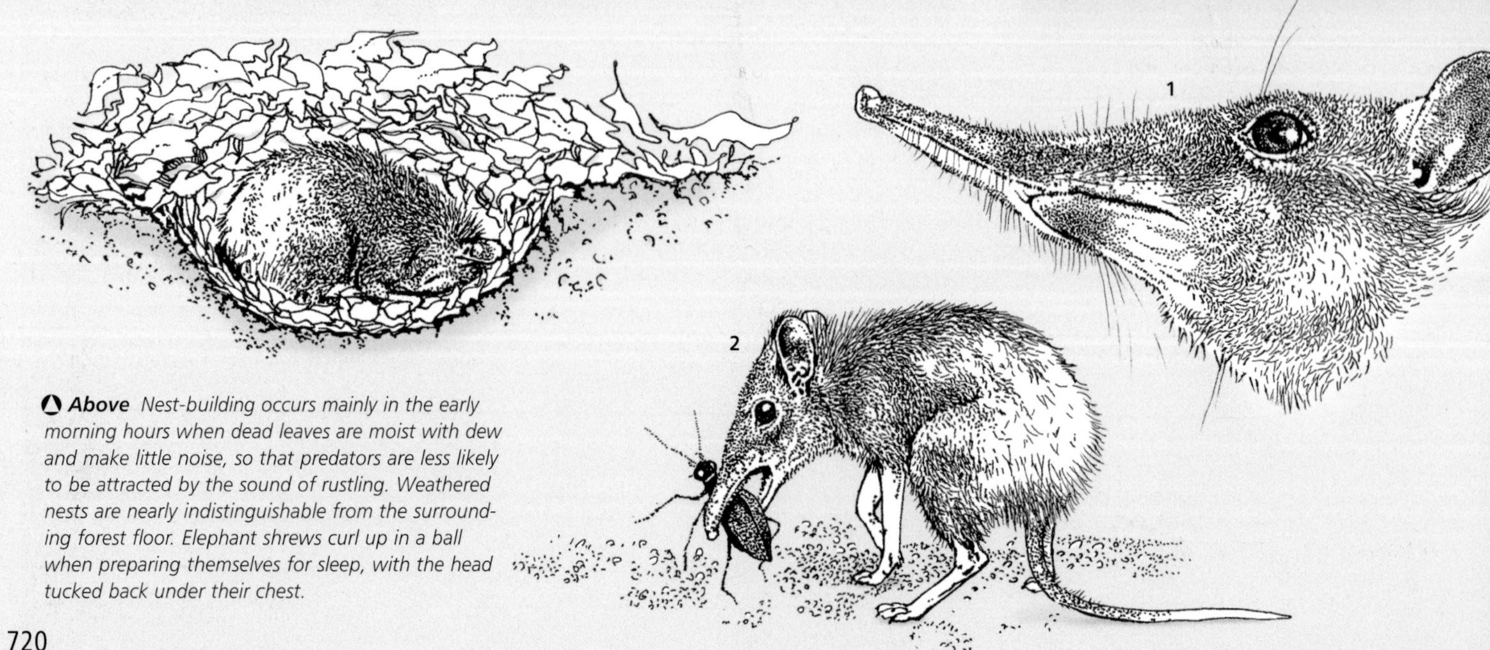

Above *Nest-building occurs mainly in the early morning hours when dead leaves are moist with dew and make little noise, so that predators are less likely to be attracted by the sound of rustling. Weathered nests are nearly indistinguishable from the surrounding forest floor. Elephant shrews curl up in a ball when preparing themselves for sleep, with the head tucked back under their chest.*

towards the nearest cover as the bird swoops to make its kill, noisily pounding the leaf litter with its rear legs as it bounds away. Only speed and agility can save it in such a situation.

The Golden-rumped elephant shrew is monogamous, but pairs spend only about 20 percent of their time in visual contact with each other; the remainder is spent resting or foraging alone. So for most of the time they must communicate via scent or sound. The distinct sound of an elephant shrew tail-slapping or bounding across the forest floor can be heard over a large part of a pair's 1.5ha (3.7 acre) territory. These sounds not only signal to the predator that it has been discovered, but also communicate to the elephant shrew's mate and young that an intruder has been detected.

Each pair of elephant shrews defends its territorial boundaries against neighbors and wandering subadults in search of their own territories. During an aggressive encounter a resident will pursue an intruder on a high-speed chase through the forest. If the intruder is not fast enough, it will be gashed by the long canines of the resident.

These conflicts between elephant shrews can be thought of as a special type of predator–prey interaction, revealing yet another way in which the animal's coloration may serve to avoid successful predation. The skin under the animal's rump patch is up to three times thicker than that on the middle of its back. The golden color of the rump probably serves as a target, diverting attacks on such vital parts of the body as the head and flanks to an area of the body that is better suited to take assaults.

Deflective marks are common in invertebrates, and have been shown to be effective in foiling predators; for example, the distinctive eye spots on the wings of some butterflies attract the predatory attacks of birds, allowing the insects to escape relatively unscathed. The yellow rump and the white tip on the black tail of the elephant shrew may serve a similar function by attracting the talons of an eagle or the fangs of a striking snake, thus improving the animal's chances of making a successful escape. GBR

Above and Left *Foraging in the leaf litter on the forest floor. The Golden-rumped elephant shrew has a small mouth* **1** *located far behind the top of its snout, which makes it difficult to ingest large prey items. Small invertebrates are eaten by flicking them into the mouth with a long, extensible tongue.* **2** *In the Arabuko-Sokoke forest of coastal Kenya, elephant shrews feed mainly on beetles, centipedes, termites, cockroaches, ants, spiders, and earthworms, in decreasing order of importance.*

because the animal is on guard and can easily make its escape back to a place of refuge.

The situation is very different when an elephant shrew becomes aware of a predator – say, for example, an eagle – so close that a safe escape cannot be guaranteed. In those circumstances, the animal will take flight across the forest floor

Right *Elephant shrews chase intruders from their territory using a half-bounding gait.*

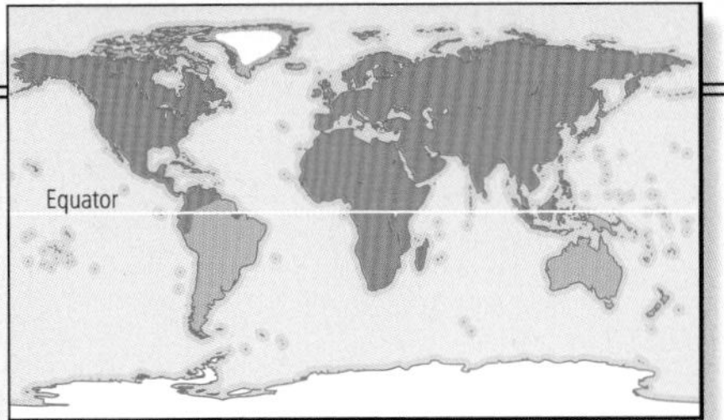

INSECTIVORES

Although among the largest orders of mammals, the Insectivora (or Lipotyphla) is still one of the least well-studied. All insectivores are small animals (none larger than rabbits) with long, narrow snouts that are usually very mobile. Most move by walking or running, although some swim and/or burrow. Body shapes vary widely, from the streamlined form of the otter shrews to the short, fat body of hedgehogs and moles. All walk with their soles and heels on the ground (plantigrade gait) and most have short limbs, with five digits on each foot. Eyes and ears are sometimes so small as not to be visible.

Above *An Alpine shrew* (Sorex alpinus) *in a Bavarian forest. Shrews are by far the most speciose family of the order Insectivora.*

The insectivores are often divided into three suborders to emphasize the relationships between the families. The Tenrecomorpha comprises tenrecs and golden moles; the hedgehogs and moonrats (the latter considered the most primitive of the living insectivores) are placed in the Erinaceomorpha; and the Soricomorpha consists of shrews, moles, and solenodons.

However, recent DNA analysis suggests that the golden moles and tenrecs should be assigned to a new order – the Afrosoricida – which is part of the supraordinal assemblage that make up the Afrotheria (see panel opposite and the introductory essay What is a Mammal?). Correspondingly in this new scheme, the shrews, moles, and hedgehogs become a new order, the Eulipotyphla. While anticipating these changes, this section retains traditional insectivore systematics until a consensus is confirmed.

While the order as a whole is very widely distributed, only three families can be said to be widespread. These are the Erinaceidae (hedgehogs and moonrats), Talpidae (moles and desmans), and Soricidae (shrews), which between them account for almost all of the worldwide distribution. The other three families have very limited distributions indeed! The Solenodontidae (solenodons) are found only on the Caribbean islands of Hispaniola and Cuba. The Tenrecidae (tenrecs) are also found mainly on islands – Madagascar and the Comoros in the Indian Ocean – with some members of the family (the otter shrews) occurring only in the wet regions of Central Africa. Because of the differences in their distribution, lifestyle, and habitat, the otter shrews were at various times considered to be in a separate family, the Potamogalidae, although their teeth indicate that they are true tenrecs; they are treated in the following pages as a subfamily of the tenrecs. The golden moles occur only in the drier parts of southern Africa.

Primitive Placentals

Insectivore Systematics

As a group, the insectivores are generally considered to be the most primitive of living placental mammals and therefore representative of the ancestral mammals from which modern mammals are derived. This was not the original purpose of the grouping. The term "insectivore" was first used in a system of classification produced in 1816 to describe hedgehogs, shrews, and Old-World moles (all primarily insect-eaters). The order soon became a "rag bag" into which any animal was placed that could not be neatly assigned elsewhere. In 1817, the naturalist Georges Cuvier added the American moles, tenrecs, golden moles, and desmans. Forty years later, tree shrews, elephant shrews, and colugos were included. All were new discoveries in need of classification but none looked much like any other members of the group.

Confronted in 1866 by an order Insectivora containing a number of very different animals, the taxonomist Ernst Haeckel subdivided it into two distinct groups that he called Menotyphla and Lipotyphla. Menotyphlans (tree shrews, elephant shrews, and colugos) were distinguished by the presence of a cecum (the human appendix) at the beginning of the large intestine; lipotyphlans (moles, golden moles, tenrecs, and shrews) by its absence. Menotyphlans also differ greatly from

Below *In contrast to shrews and moles, hedgehogs are not generally territorial and appear to wander at random on their foraging expeditions.*

lipotyphlans in external appearance: large eyes and long legs are only two of the more obvious characters. The colugos are so different that the new order Dermoptera was created for them as early as 1872. In 1926 the anatomist Le Gros Clarke suggested that the tree shrews are more similar to lemur-like primates than to insectivores, but the most modern view is that tree shrews comprise a separate order, the Scandentia. The elephant shrews also cannot be readily assigned to any existing order, so they have become the sole family in the new order Macroscelidea. Modern phylogenetic analyses also present conflicting results concerning the origins of remaining families in the Insectivora. Morphological analyses suggest that lipotyphlan members of the group are probably descended from a common ancestor, but molecular evidence indicates that this is unlikely. Multiple origins may also be assumed for the fossil members of the Insectivora, which includes a vast assortment of early mammals and remains very much a "waste-basket" group. Many of these early forms are known only from fossil fragments and teeth; they are assigned to the Insectivora largely as a matter of convenience, having insectivore affinities and no clear links with anything else.

Interestingly, the phylogenetic placement of afrotherians (and edentates – the armadillos) – see below – at the base of the placental mammal evolutionary tree suggests that the earliest placental mammals were habitually terrestrial, not arboreal, as has been argued in the past.

Primitive and Derived Characters

EVOLUTIONARY BIOLOGY

Not all of the insectivores are primitive mammals. Most living species have evolved specializations of form and behavior that mask some of their truly primitive characters (namely, those features that probably would have been found in their ancestors). These are contrasted with "derived" (or advanced) characters, found in animals that have developed structures and habits not found in their ancestors. The cecum is a primitive character, and its lack is therefore a derived character, a feature of the Insectivora as it now stands. There are, however, a number of characters considered to be primitive which are more commonly found in the

ORDER: INSECTIVORA (LIPOTYPHLA)
6 families; 67 genera; 424 species

TENRECS Family Tenrecidae p728

24 species in 10 genera
Includes **Aquatic tenrec** (*Limnogale mergulus*), **Common tenrec** (*Tenrec ecaudatus*), **Giant otter shrew** (*Potamogale velox*).

SOLENODONS Family Solenodontidae p732

2 species in a single genus
Hispaniola solenodon (*Solenodon paradoxus*), **Cuban solenodon** (*Solenodon cubanus*).

HEDGEHOGS Family Erinaceidae p734

23 species in 7 genera
Includes **Western European hedgehog** (*Erinaceus europaeus*), **Greater moonrat** (*Echinosorex gymnura*), **Lesser gymnure** (*Hylomys suillus*).

SHREWS Family Soricidae p742

312 species in 23 genera
Includes **Eurasian water shrew** (*Neomys fodiens*), **European common shrew** (*Sorex araneus*), **Vagrant shrew** (*Sorex vagrans*).

GOLDEN MOLES Family Chrysochloridae p748

21 species in 9 genera
Includes **Giant golden mole** (*Chrysospalax trevelyani*).

MOLES AND DESMANS Family Talpidae p750

42 species in 17 genera
Includes **European mole** (*Talpa europaea*), **Pyrenean desman** (*Galemys pyrenaicus*), **Russian desman** (*Desmana moschata*), **Star-nosed mole** (*Condylura cristata*).

THE AFROTHERIA – A NEW GROUPING

The Insectivora or Lipotyphla may be totally re-aligned as a result of molecular findings that point to an ecologically diverse assemblage of endemic African placentals – the "Afrotheria."

Placental mammals began to diversify in the later Cretaceous, after continental drift had isolated the Afro-Arabian landmass. Because primitive placentals were present in northern continents (Laurasia) at this time, many paleontologists discounted Afro-Arabia as a major center for early placental diversification. Yet it now appears that the southern continents (Gondwana) in fact played a seminal role, with Afrotheria possibly as the very first branch on the placental family tree. Certain characters, such as a very basic male reproductive system and a poorly-developed, almost "reptilian" thermoregulatory system, show the afrotherians to be more primitive than other placentals.

The recognition of Afrotheria as a distinct clade reveals a remarkable pattern of parallel adaptation in the independent evolutionary histories of Gondwanan and Laurasian placentals. Similar selection pressures acting on a similar skeletal body plan produced ricochetal herbivorous forms (e.g., Gondwanan elephant shrews vs. Laurasian rabbits), burrowing forms (golden moles vs. true moles), habitually aquatic forms (manatees vs. whales), and dedicated herbivores (hyraxes vs. perissodactyls). ES

Insectivora than in other mammalian orders. These include relatively small brains, with few wrinkles to increase the surface area, primitive teeth, with incisors, canines, and molars easily distinguishable, and primitive features of the auditory bones and collar bones. Other primitive characteristics shared by some or all insectivores are intra-abdominal testes (i.e. no scrotal sac), a plantigrade gait, and possession of a cloaca, a common chamber into which the genital, urinary and fecal passages empty. Some of these primitive features, such as the cloaca and abdominal testes, are also characteristic of the marsupials, but insectivores, like all Eutherian (placental) mammals, are distinguished by the possession of the chorio-allantoic placenta, which permits the young to develop fully within the womb.

Many insectivores have acquired extremely specialized features, such as the spines of the hedgehogs and tenrecs, the poisonous saliva of the solenodons and some shrews, and the adaptations for burrowing found in many insectivore families. A number of shrew and tenrec species are thought to have developed a system of echolocation similar to that used by bats.

If all these derived characters are disregarded, it is possible to sketch a very general impression of early mammals. They would have been shy animals, running along the ground in the leaf litter but capable of climbing trees or shrubs. Small and active, about the size of a modern mouse or shrew (the largest known fossil is about the size of a Eurasian badger), they probably fed mainly on insects; some may have been scavengers. They would have looked much like modern shrews, with small eyes and a long, pointed snout with perhaps a few long sensory hairs or true whiskers. A dense coat of short fur would have covered all of the body except the ears and soles of the paws. It is speculated that they may have had a dun-colored coat, with a stripe of darker color running through the eye and along the side of the body – a common pattern, found even on reptiles and amphibians. The development of the ability to regulate body temperature, combined with the warm mammalian coat, meant that the early mammals could be active at night when the dinosaurs (their competitors and predators) were largely inactive due to lower air temperatures.

From this basic stock, two slightly different forms are believed to have developed, known today only from teeth and fragments of bone dating from the late Cretaceous (80 million years ago). These two groups are characterized mainly by very different teeth. It appears that one group, the Paleaoryctoidea, eventually gave rise to the creodonts, a type of early carnivore, while the Leptictoidea were once thought to have produced the modern insectivores. Recent research on leptictoid fossils suggests instead that most of them were less closely related to the Insectivora, and were perhaps "dead-end" offshoots from the main branch of insectivore evolution.

Below *Hedgehogs will feed on birds' eggs and young when the opportunity arises. Their extremely varied diet also includes all manner of invertebrates, such as earthworms, beetles, and slugs.*

Insectivore Body Plan

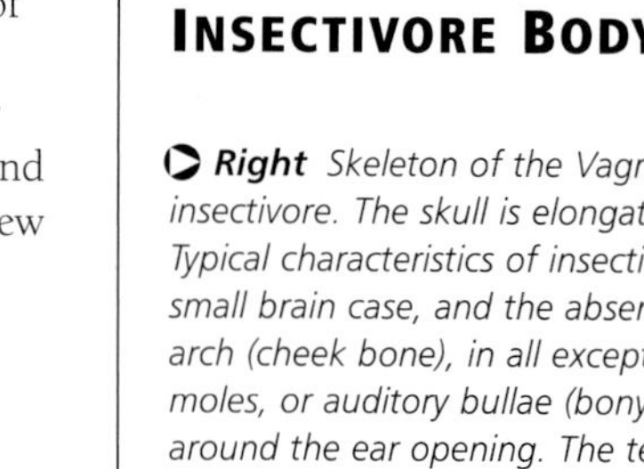

Right *Skeleton of the Vagrant shrew, a typical insectivore. The skull is elongate and flattened. Typical characteristics of insectivores include a small brain case, and the absence of a zygomatic arch (cheek bone), in all except hedgehogs and moles, or auditory bullae (bony) prominences around the ear opening. The teeth of shrews are well differentiated into molars, premolars, and canines, with pincer-like front incisors. The dental formula of the Vagrant shrew is I3/1, C1/1, P3/1, M3/3 = 32. The teeth are partially colored by a brownish-red pigment.*

Below *Skulls of insectivores. Unlike shrews, the cheek bone of hedgehogs is fully formed. The front incisors are enlarged and the molars are adapted to an omnivorous rather than an insectivorous diet. The dental formula is I2–3/3, C1/1, P3–4/2–4, M3/3 = 36–44. The Common tenrec has a long, tapered snout and, in the adult male, long canines, the tips of the bottom pair fitting into pits in front of the upper ones. The dental formula is I2/3, C1/1, P2/3, M3/3 = 36. Solenodons have an unusual cartilaginous snout which articulates with the skull via a "ball-and-socket" joint. Solenodons produce a toxic saliva which is released from a gland at the base of the second lower incisor. The dental formula is I3/3, C1/1, P3/3, M3/3 = 40.*

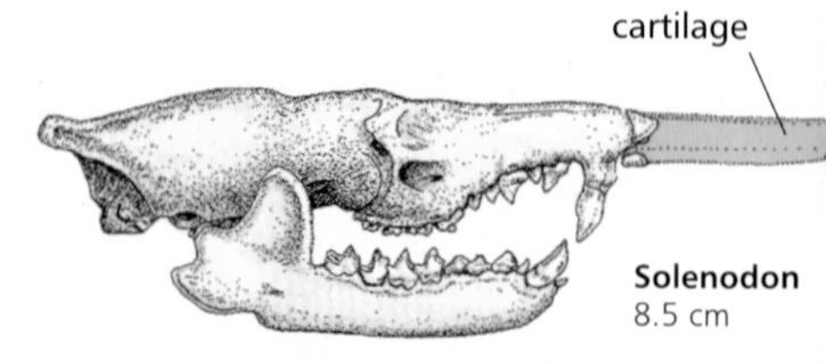

Diversify and Survive

Adaptive Success

Despite the evolutionary relationships between the families, insectivores have little in common other than their apparent primitiveness. It is perhaps this diversity that is responsible for the success of the three larger families. The family Talpidae contains both the true moles, which live mainly in subterranean burrows, and the desmans, which spend much of their time in the water and construct burrows in stream banks only

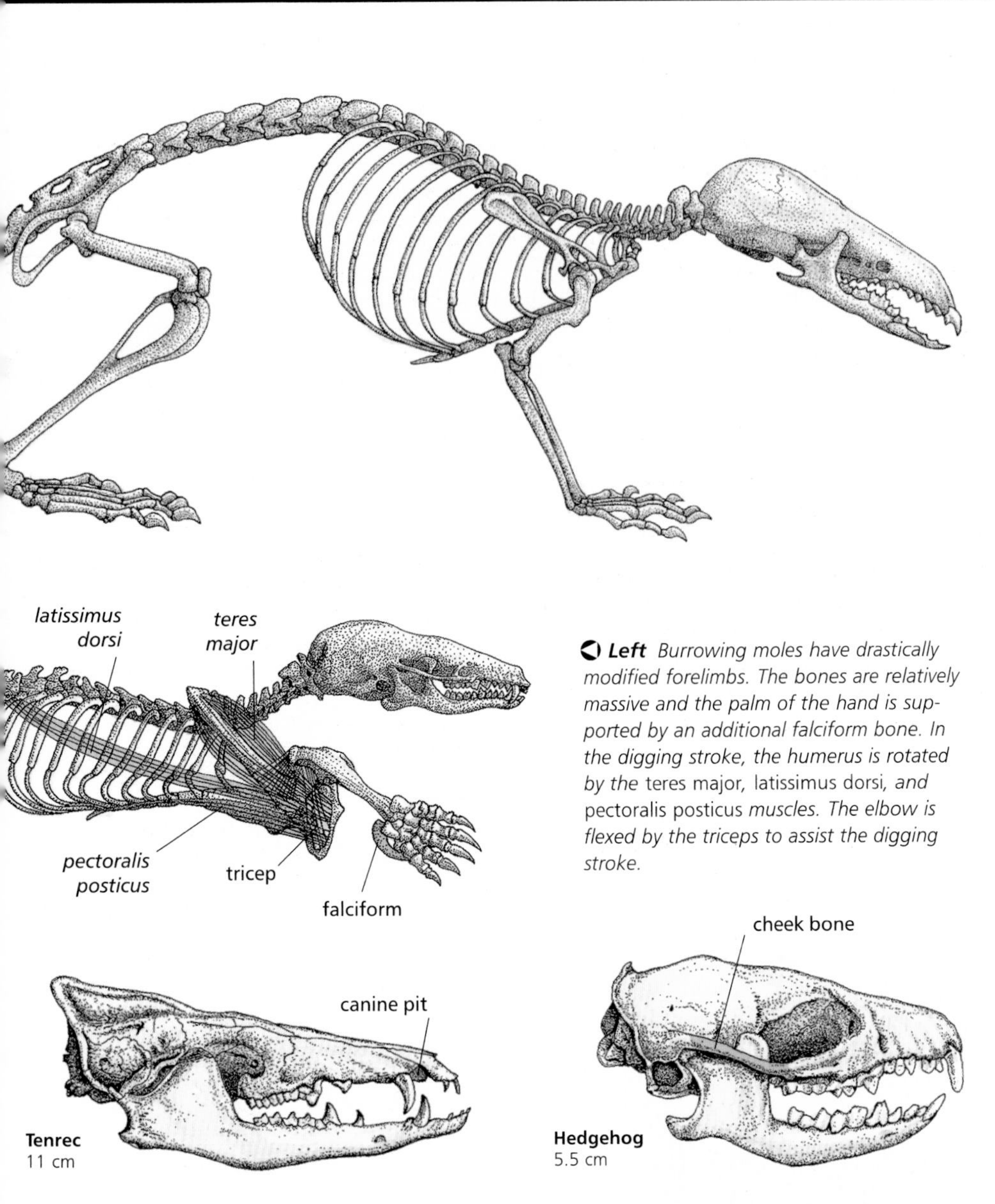

◐ ***Left*** *Burrowing moles have drastically modified forelimbs. The bones are relatively massive and the palm of the hand is supported by an additional falciform bone. In the digging stroke, the humerus is rotated by the* teres major, latissimus dorsi, *and* pectoralis posticus *muscles. The elbow is flexed by the triceps to assist the digging stroke.*

for shelter. Similarly, the mainly terrestrial family Tenrecidae also includes the rice tenrecs, which burrow in the banks of rice paddies, while the Aquatic tenrec leads a semi-aquatic life. The Soricidae show adaptations to every type of habitat – as well as the "standard" shrew, running along the ground, there are also species, such as the American short-tailed shrew, that burrow like miniature moles, and those like the European water shrew that swim like the tenrecid otter shrews.

The order Insectivora is rich in examples of convergent evolution. Moles and golden moles, for example, are not closely related within the order yet both have adopted similar burrowing lifestyles and even look much alike. According to the fossil record, moles developed from an animal resembling a shrew, whereas golden moles appear to be more closely related to the tenrecs. The similarity between golden moles and the marsupial moles of Australia is even more remarkable. In this case, the lineages have been separate for 70 million years; one is placental, the other pouched – yet the golden mole is more similar in appearance to the marsupial moles (even in the texture of its fur) than to the true mole. The eyes of both marsupial and golden moles are covered by skin – the eyelids have fused, whereas the minute eyes of the true moles are still functional.

The ability to curl up, combined with a dense coat of spines (as in hedgehogs and some species of tenrecs), is an obvious deterrent to predators. Some non-spiny species, moles and shrews for example, have strongly distasteful secretions from skin glands which may have a similar effect. Many of the other specializations are more likely to be the result of competition for food. If two or more species compete for food or some other resource, then either the worst competitor will become (at least locally) extinct, or all will evolve to specialize on different aspects of the resource and reduce the competition. Moles, for example, may have developed their burrowing lifestyle to avoid both predation and competition with surface-dwelling insectivores. Some species of both shrews and moles have become semi-aquatic, probably to exploit a different source of invertebrate food.

The tenrecs are thought to have been one of the first mammalian groups to arrive on Madagascar. Like the Australian marsupials, they provide a fascinating example of adaptive radiation, having evolved, in the absence of competition with an established fauna, a variety of forms which use most of the available habitats.

Insectivores rely heavily on their sense of smell to locate their prey, as would be expected from the relatively large center of smell in the brain. Invertebrates are the main food – most insectivores are thought to feed on insects and earthworms, although some of those associated with water also eat mollusks and possibly fish. Where food habits have been closely studied, the animals appear to eat almost anything organic they find or can catch, and many will attempt to kill prey which are substantially larger than themselves: hedgehogs, for example, can kill chickens; water shrews can kill frogs. The poisonous saliva of solenodons and some shrews may have evolved to enable these animals to catch larger prey than their body size would normally permit. The poison acts mainly on the nervous system, paralyzing the victim.

Shrews, in particular, have a reputation for gluttony, consuming more than their own bodyweight daily, but recent studies indicate that their food is less nutritious than the dry seeds eaten by rodents of a similar size, so they need to consume a correspondingly greater bulk. High food requirements also result from the high metabolic rates of the shrews, which in the case of some northern shrews may be an adaptation to a highly seasonal climate. The respiration rate of a shrew can be more than 5,000cu cm (305cu in) of air per kilogram of body weight per hour, compared with 200cu cm (12cu in) for humans.

Reported litter sizes vary widely in those species that have been studied. Females of the genus *Tenrec* have been found to have up to 32 developing embryos, of which probably only 12–16 survive to birth. Other tenrec species, along with moles, hedgehogs, and shrews, have litters of 2–10; solenodons have only one or two young per litter. Young are typically altricial or underdeveloped at birth; newborn shrews of the genus *Sorex* may weigh as little as 0.25g. The offspring develop rapidly, and, in large litters, mortality can result from high levels of competition. The timing of the breeding season is governed mainly by food availability and thus the young of animals living in arid areas are usually born in the rainy season (winter), while temperate species breed in the spring and summer. Some tropical species (the moonrats, for example) are thought to breed throughout the year.

Solitary but Promiscuous

SOCIAL BIOLOGY

Because most insectivores are shy, nocturnal, secretive creatures, they do not lend themselves to field study. Thus, relatively little is known about their social biology. Those few species that have been the subject of research (mainly shrews, plus some hedgehogs and tenrecs) are solitary, with little communication between adults except at breeding times. Studies of European hedgehogs indicate that individuals do not defend territories, although shrews do.

Shrews are territorial in that their home ranges – about 500sq m (5,400sq ft) in the European common shrew – are mutually exclusive. When shrews meet they generally act aggressively. Home ranges are often maintained by juveniles. Reproductive females may extend their ranges when breeding, which can result in some overlap of their movements. Mating systems have been studied for several species of shrews. Promiscuity is probably the most common mating pattern and multiple paternity is known in some species. In the European common shrew, litters may be sired by up to six different males, and there is considerable overlap of male movements during the breeding season. Monogamy is also known to occur in at least one shrew – the European house shrew. Unlike the European common shrew, males of this species defend exclusive access to females and may invest in care of their offspring. Moles are even less sociable and, once the juveniles have been expelled from a female's burrow system, she will not normally tolerate another mole in her home burrow except for a few hours in the spring when she is ready to mate. It is likely that scent plays an important role in keeping moles apart, and they are probably circumspect about entering unfamiliar places smelling of other moles. This pattern of mutual avoidance is probably common among some other insectivore groups too. Mole home ranges are essentially linear, being constrained by tunnel walls. The burrow system acts as a pitfall trap which collects soil invertebrates as prey and in poor soils, longer burrow systems are needed to supply sufficient food. Thus, a mole's linear home range may vary from 30–120m (100–400ft) or more, depending on soil type and food density. Surface activity, especially among dispersing juveniles may extend this home range considerably. Hedgehogs, such as the European hedgehog, which forage over an area from 1–5ha (2.5–12.5 acres) in a night, may use a home range of up to 30ha (74 acres) in a season; males travel further and have larger ranges than females. Unlike some small rodents, insectivores do not normally seem to make use of a three-dimensional home range by climbing into bushes and trees. In desmans, the home range is linear, along a stream edge; in tenrecs it is an area around the burrow or den; but for many species there is no detailed ecological information at all.

Friends and Enemies

CONSERVATION ISSUES

A large proportion of insectivore species is found only in tropical countries, where the primary emphasis is on development and exploitation of resources rather than wildlife conservation. Being in the main unprepossessing animals, threats to the survival of insectivores do not usually receive wide publicity.

Habitat destruction is undoubtedly the single most significant threat. In Western Europe, increasingly intensive farming methods and the drainage of wetlands are having a major impact on insectivore numbers and distribution. Worldwide, many species are in great peril. Thirteen Eurasian species are currently classified as Critically Endangered. Alarmingly, the lack of field data may conceal an even more desperate situation. Human introduction of new competitors and predators also poses a serious threat, particularly to the tenrecs and solenodons. These competitors, which are effective generalists and can thrive almost anywhere, also have high rates of reproduction, and so are able to overwhelm native species before they have a chance to adapt to the changed situation. For example, solenodons were the principal small carnivores in the Antilles until the 17th century, when the Spaniards brought with them dogs, cats, and rats. Mongooses were introduced in the late 19th century to combat a proliferation of Black rats. The latter two proved highly effective as competitors for food and all apparently added solenodons to their diet. Predation and forest clearance continue to put intense pressure on the remaining solenodon species, which are heavily dependent on conservation programs for their survival.

Left and Below Moles are solitary animals and usually do not trespass into each others' tunnel systems. However, they will quickly take over a territory if its occupant is removed. The extensive damage that moles can cause to pasture is evident from the line of spoil heaps in this English field.

The remnant Russian desman population is now protected by law, though numbers continue to decline. This species was once a staple of the fur trade, with tens of thousands of skins exported annually to western Europe, but hunting, combined with pollution of its aquatic habitat and competition from introduced coypus and muskrats, saw numbers fall drastically. The small population of Pyrenean desmans in the Pyrenees is threatened by water pollution and escaped mink. Several species of golden mole are also threatened by various changes in their habitat.

Few attempts have been made to domesticate insectivores, although hedgehogs are growing in popularity as pets, particularly in the United States. Hedgehogs have generally had a favorable relationship with humans. The folklore of both Europe and Asia abounds in tales of hedgehogs and although they are regarded as pests by some – notably gamekeepers and the greenkeepers on golf courses – householders in the urban areas of western Europe regularly provide them with bowls of food. The Asian musk shrew has apparently adapted well to use as a laboratory animal; in contrast, many other insectivores are difficult to breed in captivity. Elsewhere in the world, the larger insectivores are occasionally eaten (especially in Madagascar and the West Indies) and the smaller ones are probably rarely noticed. Most insectivores are too small and too scarce to be any use for food, and equally are unlikely to be serious economic pests. Yet some are extremely abundant, and it has been suggested that if all shrews were to disappear suddenly, the number of insect pests in fields and gardens would increase noticeably. Other insectivores are more likely to become accidental victims of human activities than to be persecuted as pests or fostered as allies.

Some species have been exploited as an economic resource. The Russian desmans have a dense lustrous coat and, in the past, considerable numbers of these animals were caught in nets and traps. They were also chased from their burrows and shot or clubbed. A century ago, a fortunate hunter might obtain 40 desman skins during the month-long hunting season at the time of the spring thaw. Despite their value, desman pelts were never a major commodity and the "industry" seems to have been opportunistic.

Moleskins were used extensively for garments and accessories from Roman times up to the 20th century. So heavy was the demand in Germany from the 17th century onward that fears were raised for the mole's survival there; however, changing fashions saw the market slump by 1914. However, just as demand was falling in Europe, a new market opened up in North America; even as late as the 1950s, a million moles a year were being trapped in Britain alone for skinning and export. Moleskins have now largely been supplanted by cheaper, more durable synthetic substitutes. AW/PS

Tenrecs

Tenrecs and otter shrews are remarkable in having a greater diversity of shape and form than any other living family of insectivores. Yet this has been achieved in virtual isolation, since tenrecs themselves are confined to Madagascar, which they were one of the first mammals to colonize, and the otter shrews (Potamogalinae; sometimes regarded as a separate family) to West and central Africa.

Tenrecs retain characters that were perhaps more widespread among early placental mammals. These conservative features include a low and variable body temperature, retention of a common opening for the urogenital and anal tracts (the cloaca), and undescended testes in the male. Some of the family comprise a more conspicuous part of the native fauna than temperate zone insectivores, being either an important source of food, relatively large and bold, or conspicuously colored. Even though detailed studies of the family are few and restricted to a handful of species they provide an excellent basis for the elucidation of mammalian evolution. The earliest fossils date from Kenyan Miocene deposits (about 24 million years ago), but by then the Tenrecidae were well differentiated, and had probably long been part of the African fauna.

Diversity in Isolation

FORM AND FUNCTION

Among the largely nocturnal tenrecs and otter shrews, eyesight is generally poor, but the whiskers are sensitive, and smell and hearing are well developed. Vocalizations range from hissing and grunting to twittering and echolocation clicks. The brain is relatively small and the number of teeth ranges from 32–42.

The aquatic Tenrecidae are active creatures of streams, rivers, lakes, and swamps. The Giant otter shrew and Mount Nimba least otter shrew are confined to forest, but the Ruwenzori least otter shrew and the Aquatic tenrec are

Above *Species of tenrecs:* **1** *Aquatic tenrec* (Limnogale mergulus); **2** *Giant otter shrew* (Potamogale velox); **3** *Ruwenzori least otter shrew* (Micropotamogale ruwenzorii); **4** *Streaked tenrec* (Hemicentetes semispinosus nigriceps); **5** *Common tenrec* (Tenrec ecaudatus); **6** *Lesser hedgehog tenrec* (Echinops telfairi); **7** *Long-tailed tenrec* (Microgale melanorrachis); **8** *Greater hedgehog tenrec* (Setifer setosus); **9** *Four-toed rice tenrec* (Oryzorictes tetradactylus).

less restricted. All four species have a sleek, elegant body form with a distinctive, flattened head that allows the ears, eyes, and nostrils to project above the surface while most of the body remains submerged. Stout whiskers radiate from around the muzzle, providing a means of locating prey. The fur is dense and soft; frequent grooming ensures that it is waterproof and traps insulating air during dives. Grooming is accomplished by means of the two fused toes on each hind foot, which act as combs. All otter shrews and the Aquatic tenrec have a chocolate-brown back; the Aquatic tenrec has a gray belly and otter shrews have white bellies. The Madagascan Aquatic tenrec shows strong convergence with the least otter shrews, with a rat-size body and a tail approximately the same length. The Ruwenzori least otter shrew and the Aquatic tenrec have webbed feet, which probably provide most of the propulsion in the water, and their tails are slightly compressed laterally, providing each with an effective rudder and additional propulsion.

The Mount Nimba least otter shrew is probably the least aquatic, having no webbing and a rounded tail. However, all are probably agile both in water and on land. The Giant otter shrew, which is among the most specialized of the aquatic insectivores, often figures as part fish and part mammal in African folklore, giving rise to such names as "transformed fish." The unmistakable deep, laterally flattened tail that tapers to a point is the main source of such beliefs, even though it is covered by fine, short hair, but the animal's proficiency in water no doubt also plays a part. Sinuous thrusts of the powerful tail extending up the lower part of the body provide the propulsion for swimming, allowing a startling turn of speed and great agility. Although most of the active hours are spent in the water the agility also extends to foraging on dry land. The Boulou of southern Cameroon call the Giant otter shrew the *jes*: a person is said to be like a *jes* if he flares up in anger but calms down again just as rapidly.

The long-tailed and Large-eared tenrecs are shrew-like, and the former have the least modified body plan within the Tenrecidae. Evergreen forest and wetter areas of the central plateau of Madagascar are the primary habitats for long-tailed tenrecs, with only one species extending into the deciduous forests of the drier western region. These tenrecs have filled semi-arboreal and terrestrial niches. The longest-tailed species, with relatively long hind legs, can climb and probably spring among branches; jumpers and runners live on the ground, together with short-legged semi-burrowing species. The Large-eared tenrec is also semi-burrowing in its western woodland habitat and is apparently closely related to one of the oldest fossil species.

The rice tenrecs, with their mole-like velvet fur, reduced ears and eyes, and relatively large forefeet, fill Madagascar's burrowing insectivore niche. In undisturbed areas of northern and western Madagascar, these tenrecs burrow through the humus layers in a manner similar to the North American shrew mole, but the extensive cultivation of rice provides new habitats for them.

The subfamily Tenrecinae contains some of the most fascinating and bizarre insectivores. The tail has been lost or greatly reduced and varying degrees of spininess are linked with elaborate and striking defensive strategies. Both the Greater hedgehog tenrec and its smaller semi-arboreal counterpart, the Lesser hedgehog tenrec, can form a nearly impregnable spiny ball when threatened, closely resembling the Old World hedgehogs. Continued provocation may also lead to them advancing, gaping, hissing, and head-bucking, the latter being common to all Tenrecinae. The brown adult Common tenrec, which is among the largest living insectivores, is the least spiny species, but it

FACTFILE

TENRECS

Order: Insectivora

Family: Tenrecidae

24 species in 10 genera and 4 subfamilies

DISTRIBUTION Madagascar, with one species introduced to the Comoros, Réunion, and the Seychelles; W and C Africa.

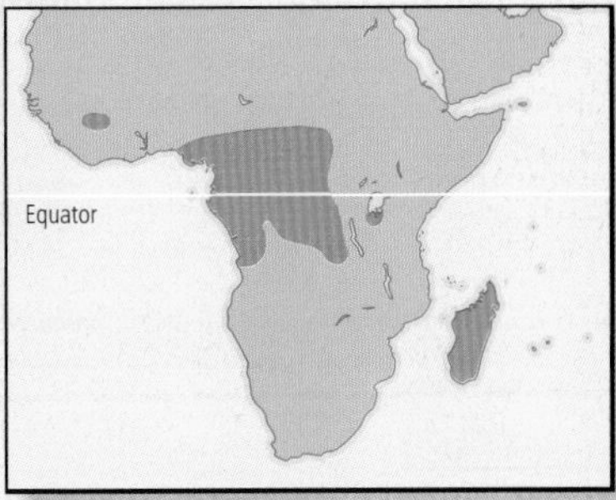

HABITAT Wide-ranging, from semi-arid to rain forest, including mountains, rivers, and human settlements.

SIZE Head–body length ranges from 4.3cm (1.7in) in the Pygmy shrew tenrec to 25–39cm (10–15in) in the Common tenrec; **tail length** from 5–10mm (0.2–0.4in) to 4.5cm (1.8in), and **weight** from 5g (0.18oz) to 500–1,500g (18–53oz), both in the same two species.

COAT Soft-furred to spiny; coloration ranges from brown or gray to contrasted streaks.

DIET Tenrecs and otter shrews are opportunistic feeders, taking a wide variety of invertebrates as well as some vertebrates and vegetable matter. Rice tenrecs mostly live off invertebrate prey, but also consume vegetable matter. Fruit supplements the invertebrate diet of the more omnivorous species such as the Common and hedgehog tenrecs. Common tenrecs are also large enough to take reptiles, amphibians, and even small mammals. The Streaked tenrec feeds on earthworms.

BREEDING Gestation relatively aseasonal within the Oryzorictinae, Geogalinae, and Tenrecinae where known (50–64 days); unknown in Potamogalinae.

LONGEVITY Up to 6 years.

CONSERVATION STATUS Nine species are considered Vulnerable or Endangered by the IUCN (6 Oryzorictinae, 3 Potamogalinae), while one – the Tree shrew tenrec (*Microgale dryas*) – is Critically Endangered.

See subfamilies table ▷

combines a lateral, open-mouthed slashing bite with head-bucking that can drive spines concentrated on the neck into an assailant. A fully-grown male with a gape of 10cm (4in) has canines that can measure up to 1.5cm (0.6in), and the bite is powered by the massively developed masseter (jaw) muscles. A pad of thickened skin on the male's mid-back provides some additional protection. The black-and-white striped offspring relies less on biting but uses numerous barbed, detachable spines to great effect in head-bucking. Common tenrecs have better eyesight than most other species in the family but may also detect disturbances through long, sensitive hairs on the back. When disturbed, the young can communicate their alarm through stridulation, which involves rubbing together stiff quills on the mid-back to produce an audible signal. Streaked tenrecs are remarkably similar to juvenile Common tenrecs in coloration, size, and possession of a stridulating organ. Like juvenile Common tenrecs, they forage in groups, and their principal defense involves scattering and hiding under cover. If they are cornered, they advance, bucking violently, with their spines bristling.

Seeking Prey by Land and Water

DIET

Tenrecs and otter shrews are opportunistic feeders, taking a wide variety of invertebrates as well as some vertebrates and vegetable matter. Otter shrews scour the water, stream bed, and banks with their sensitive whiskers, snapping up prey and carrying it up to the bank if caught in the water. Crustaceans are the main prey, including crabs of up to 5–7cm (2–3in) across the carapace. Rice tenrecs probably encounter most of their invertebrate prey in underground burrows or surface runs, but also consume vegetable matter. Fruit supplements the invertebrate diet of the more omnivorous species such as the Common and hedgehog tenrecs. Common tenrecs are also large enough to take reptiles, amphibians, and even small mammals. Prey are detected by sweeping whiskers from side to side, and by smell and sound. Similarly, semi-arboreal Lesser hedgehog tenrecs and long-tailed tenrecs perhaps encounter and eat lizards and nestling birds. The Streaked tenrec, which is active during daytime, has delicate teeth and an elongated, fine snout for feeding on earthworms.

Complex Multigenerational Groupings

SOCIAL BEHAVIOR

Tenrec reproduction is diverse and includes several features peculiar to the family. Where known, ovarian processes differ from those in other mammals in that no fluid-filled cavity, or antrum, develops in the maturing ovarian follicle. Spermatozoa also penetrate developing follicles and fertilize the egg before ovulation; this is known in only one other mammal, the Short-tailed shrew.

Tenrec and Otter Shrew Subfamilies

Tenrecs

Subfamily Oryzorictinae

Sixteen species in 3 Madagascan genera, including **Aquatic tenrec** (*Limnogale mergulus*); **rice tenrecs** (*Oryzorictes*, 3 species), **long-tailed tenrecs** (*Microgale*, 12 species). The Tree shrew tenrec (*Microgale dryas*) is Critically Endangered; the Pygmy shrew tenrec (*M. parvula*), Greater Long-tailed shrew tenrec (*M. principula*), and Aquatic tenrec are Endangered; while the Gracile shrew tenrec (*M. gracilis*), the Dark pygmy shrew tenrec (*M. pulla*), and Thomas's shrew tenrec (*M. thomasi*) are Vulnerable.

Subfamily Geogalinae

One species in one Madagascan genus: **Large-eared tenrec** (*Geogale aurita*).

Subfamily Tenrecinae

Four species in 4 Madagascan genera: **Greater hedgehog tenrec** (*Setifer setosus*); **Lesser hedgehog tenrec** (*Echinops telfairi*); **Common tenrec** (*Tenrec ecaudatus*), introduced to Réunion, Seychelles, and Mauritius; **Streaked tenrec** (*Hemicentetes semispinosus*).

Otter shrews

Subfamily Potamogalinae

Three species in 2 African genera: **Giant otter shrew** (*Potamogale velox*), Nigeria to W Kenya and Angola; **Ruwenzori least otter shrew** (*Micropotamogale ruwenzorii*), Uganda, DRC; Mount Nimba; **Least otter shrew** (*M. lamottei*), Guinea, Liberia, and Côte d'Ivoire. All species are Endangered.

For full species list see Appendix ▷

TENREC BODY TEMPERATURE

Body temperature is relatively low among tenrecs, with a range of 30–35°C (86–95°F) during activity. The Large-eared tenrec and members of the Tenrecinae enter seasonal hypothermia, or torpor, during dry or cool periods of the year, which ranges from irregular spells of a few days to continuous periods lasting six months; then it is integral to the animal's physiological and behavioral cycles. So finely arranged are the cycles of hypothermia, activity, and reproduction that the Common tenrec must complete such physiological changes as activation of the testis or ovary while still torpid, since breeding begins within days of commencing activity.

The Giant otter shrew, some Oryzorictinae, and the Tenrecinae save energy at any time of year because body temperature falls close to air temperature during daily rest. Interactions between these fluctuations in body temperature and reproduction in the Tenrecidae are unique. During comparable periods of activity in the Common tenrec, body temperatures of breeding males are on average 0.6°C (1.1°F) lower than those of nonbreeding males. This is because sperm production or storage can only occur below normal body temperature. Other mammals either have a mechanism for cooling reproductive organs or, rarely, tolerate high temperatures. Normally, thermoregulation improves during pregnancy, but female Common tenrecs continue with their regular fluctuations in temperature dependent on activity or rest, regardless of pregnancy. This probably accounts for variations in gestation lengths, as the fetuses could not develop at a constant rate if so cooled during maternal rest. Although torpor during pregnancy occurs among bats, it is well regulated, and the type found in tenrecs is not known elsewhere.

Left *The Streaked tenrec* (Hemicentetes semispinosum semispinosum) *has brown markings and a stripe that runs from the nose to the back of the head; stripes also run lengthwise on the body. The hairs on the underbelly are spiny, in contrast to the softer ones of* Hemicentetes semispinosum nigriceps.

Most births occur in the wet season, coinciding with maximum invertebrate numbers, and the offspring are born in a relatively undeveloped state. Litter size varies from two in the Giant otter shrew and some Oryzorictinae to an extraordinary maximum of 32 in the Common tenrec. This seems to be related to survival rates, which in turn are conditioned by the stability of the environment. For example, oryzorictines in the comparatively stable high rainforest regions seem to be long-lived and bear small litters. Average litter size of Common tenrecs in relatively seasonal woodland/savanna regions with fluctuating climatic conditions is 20, compared to 15 in rainforest regions, and 10 in Seychelles rain forests within 5° of the Equator. Weight variation within the litter can reach 200–275 percent in Common and hedgehog tenrecs.

A recent study of Large-eared tenrecs revealed that four out of ten breeding females exhibited postpartum estrus, the first time this phenomenon has been recorded within the Tenrecidae. Gestation length is around 57 days, confirming that all tenrecs have a uniformly slow fetal development rate. Pregnant females may enter torpor, which contributes to variety in length of gestation. Litter size ranged from 2–5 neonates.

The Common tenrec feeds her offspring from up to 29 nipples, the most recorded among mammals. Nutritional demands of lactation are so great in this species that the mother and offspring must extend foraging beyond their normal nocturnal regime into the relatively dangerous daylight hours. This accounts for the striped camouflage coloration of juveniles, which only become more strictly nocturnal at the approach of the molt to the adult coat. Moreover, adult females have a darker brown coat than adult males, presumably because it affords better protection for daylight feeding throughout their brief spell of lactation. The striking similarity between juvenile Common tenrecs and adult Streaked tenrecs suggests that a striped coat associated with daylight foraging has been an important factor in the evolution toward modern Streaked tenrecs.

Rainforest Streaked tenrecs form multigenerational family groups comprising the most complex social groupings among insectivores. Young mature rapidly and can breed at 35 days after birth, so that each group may produce several litters in a season. The group, of up to 18 animals, probably consists of three related generations. They forage together, in subgroups or alone, but when together they stridulate almost continuously. Stridulation seems to be primarily a device to keep mother and young together as they search for prey.

The primary means of communication among the Tenrecidae is through scent. Otter shrews regularly deposit feces either in or near their burrows and under sheltered banks. Marking by tenrecs includes cloacal dragging, rubbing secretions from eye glands, and manual depositing of neck-gland secretions. Common tenrecs cover 0.5ha (1.2–5 acres) per night, although receptive females reduce this to about 200sq m (2,150sq ft) in order to facilitate location by males. Giant otter shrews may range along 800m (0.5mi) of their streams in a night.

Mixed Fortunes

CONSERVATION AND ENVIRONMENT

Common tenrecs have been a source of food since ancient times, but are not endangered by this traditional hunting. Undoubtedly, some rainforest tenrecs are under threat as Madagascar is rapidly being deforested, but some species thrive around human settlements. Tourism may be a threat to other species. At the mid-altitude rainforest reserve of Analamazaotra, Madagascar, which is used for tourism, seven endemic tenrec species, and three endemic rodent species were recently found. Of the sites surveyed, most biological diversity was demonstrated at the most undisturbed site, though individual species abundance was reduced. Forest subjected to infrequent logging by local people exhibited an intermediate level of species richness, and it seems apt to conclude that core areas of the reserve should be left undisturbed in order to preserve small mammal species diversity. Forest destruction is also reducing the range of the Giant otter shrew and perhaps also of the Mount Nimba least otter shrew and the Ruwenzori least otter shrew. MEN

Solenodons

Left *When hunting, solenodons use their long sharp claws to upturn stones and tear off bark from fallen branches in search of prey. Although able to climb, solenodons spend most of their time foraging on the forest floor.*

THE EXTRAORDINARY SOLENODONS OF CUBA and Hispaniola face a real and immediate threat to their survival. They are so rare and restricted that the key to their conservation lies in prompt government action to set aside suitable, well-managed forest reserves in remote mountainous regions. Without such efforts, these distinctive, ancient, Antillean insectivores are likely to follow their relatives, the West Indian shrews (Nesophontidae), into extinction.

In addition to the living genus, solenodons are known from North American middle and late Oligocene deposits (about 32–26 million years ago). Their affinities are difficult to ascertain owing to their long isolation, but their closest allies are probably the true shrews (Soricidae), or the Afro-Madagascan tenrecs and otter shrews. Some mammalogists have also considered the extinct West Indian nesophontid shrews to be within the Solenodontidae. Solenodons were among the dominant carnivores on Cuba and Hispaniola before Europeans arrived with their alien predators, and were probably only occasionally eaten themselves by boas and birds of prey.

Poisoning Prey

FORM AND FUNCTION

The solenodons are among the largest living insectivores, resembling, to some extent, large, well-built shrews. Their most distinctive feature is the elongated snout, extending well beyond the length of the jaw. In the Hispaniola solenodon, the remarkable flexibility and mobility of the snout results from a unique ball-and-socket joint attaching it to the skull. The snout of the Cuban solenodon is also highly flexible but lacks the round articulating bone. Solenodons have 40 teeth, and the front upper incisors project below the upper lip. The Hispaniola solenodon secretes toxic saliva, and this probably occurs also in the Cuban species. Each limb has five toes, and the forelimbs are particularly well developed, bearing long, stout, sharp claws. Only the hind feet are employed in self-cleaning; they can reach most of the body surface, thanks to flexible hip joints. Only the rump and the base of the tail cannot be reached, but because these areas are hairless they require little attention. The tail is stiff and muscular and may play a role in balancing.

As in most nocturnal terrestrial insectivores, brain size is relatively small, and the sense of touch is highly developed, while smell and hearing are also important. Vocalizations include puffs, twitters, chirps, squeaks, and clicks; the clicks comprise pure high-frequency tones similar to those found among shrews, and probably provide a crude means of echolocation. Scentmarking is probably important, as evidenced by the presence of anal scent glands, while contact is thought to play a role in some situations.

Soil- and litter-dwelling invertebrates make up a large part of the solenodons' diet, including beetles, crickets, and various insect larvae, together with millipedes, earthworms, and termites. Vertebrate remains that have appeared in feces may be the result of scavenging carrion, but solenodons are large enough to take small vertebrates such as amphibians, reptiles, and perhaps small birds. Solenodons are capable of climbing near-vertical surfaces, but spend most time foraging on the ground. The snout is used to investigate cracks and crevices, while the massive claws are used to expose the prey under rocks, bark, and soil. A solenodon may lunge at prey and pin it to the ground with the claws and toes of the forefeet, while simultaneously scooping up the prey with the lower jaw. Occasionally the prey is pinned to the ground only by the nose, and must be held there as the solenodon advances. These advances take the form of rapid bursts to prevent the prey's escape, and a maneuvering of the lower jaw into a scoop position. Once it is caught, the prey is presumably immobilized by the toxic saliva.

Slow Breeders

SOCIAL BEHAVIOR

The natural history of solenodons is characterized by a long life span and low reproductive rate, features resulting from its position as one of the dominant predators in pre-Columbian times. The frequency and timing of reproduction in the wild is not known, but receptivity lasts less than one day and recurs at approximately 10-day intervals. Events leading up to mating involve scentmarking by both sexes, soft calling, and frequent body contacts. In captivity, the scentmarking involves marking projections in the female's cage with anal drags and also defecating and urinating in locations previously used by the female. The young (one, rarely two) are born in a nesting burrow, and they remain with the mother for an extended

FACTFILE

SOLENODONS

Order: Insectivora

Family: Solenodontidae

2 species in a single genus

DISTRIBUTION
Cuba and Hispaniola

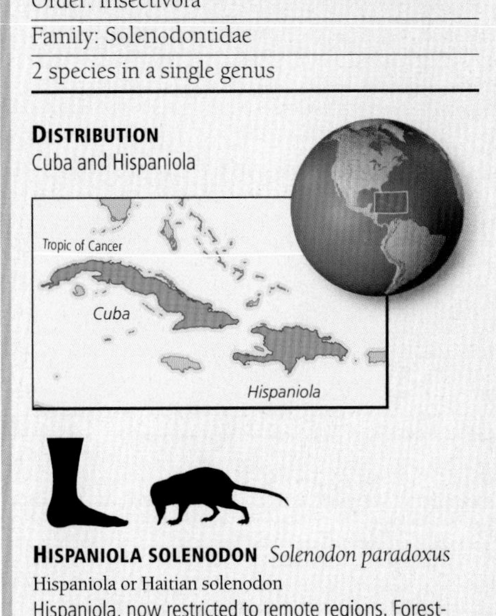

HISPANIOLA SOLENODON *Solenodon paradoxus*
Hispaniola or Haitian solenodon
Hispaniola, now restricted to remote regions. Forest-dwelling, nocturnal. HBL 28.4–32.8cm (11–13in); TL 22.2–25.5cm (8.5–10in); WT 700–1,000g (25–35oz). Coat: forehead black, back grizzled gray-brown, white spot on the nape, yellowish flanks; tail gray except for white at base and tip. Breeding: single young (rarely two), weighing 40–55g (1.4–1.9oz). Longevity: 11 years in captivity; unknown in wild. Conservation status: Endangered.

CUBAN SOLENODON *Solenodon cubanus*
Cuba. Forest-dwelling; nocturnal. HBL 28.4–32.8cm (11–13in); TL 17.5–25.5cm (7–10in); WT 700–1,000g (25–35oz). Coat: finer and longer than in the Hispaniola solenodon, dark gray except for pale yellow head and mid-belly. Breeding: as for Hispaniola solenodon. Longevity: 6.5 years in captivity; unknown in wild. Conservation status: Endangered.

Abbreviations HBL = head–body length TL = tail length HT = height WT = weight

Above *The solenodon's snout is a unique feature, in that it is a cartilaginous (not osseous) appendage that extends well beyond the jaw. In the Hispaniola solenodon the snout is connected to the skull via a ball-and-socket joint, making it very flexible.*

period of several months, which is exceptionally long among insectivores. During the first two months, each young solenodon may accompany the mother on foraging excursions by hanging onto her greatly elongated teats by the mouth. Solenodons are the only insectivores that practice teat transport, and carrying the offspring in the mouth is more widespread within this order. Initially, the offspring are simply dragged along, but as they grow they are able to walk with the mother, pausing when she stops. Teat transport would undoubtedly be useful if nursing solenodons change burrow sites regularly. More advanced offspring continue to follow the mother, learning food preferences from her by licking her mouth as she feeds, and getting to know routes around the nest burrow. The mother–offspring tie is the only enduring social grouping among solenodons; adults are otherwise solitary.

Threatened by Progress

CONSERVATION AND ENVIRONMENT

Skeletal remains discovered on Hispaniola and Cuba indicate that there were formerly two further species of solenodon – *S. arredondoi* (on Cuba) and *S. marcanoi* (on Hispaniola); the former is thought to have been half the size again of extant species and twice as heavy, while the latter was slightly smaller than living solenodons. They may have gone extinct after European encroachment on the islands.

There are no accurate estimates of solenodon numbers on Cuba or Hispaniola. The Cuban species appears to be the rarer, though small, formerly unknown populations were found in the east of the island in the 1990s. The low reproductive rate is one factor in the decline in solenodon abundance, but more significant reasons for their rarity are habitat destruction and predation by introduced carnivores, against which solenodons have no defense. Mongooses and feral cats are the main predators on Cuba, whereas dogs decimate solenodon populations in the vicinity of settlements on Hispaniola. There is little hope for the Hispaniola solenodon in Haiti, the nation comprising the western half of Hispaniola, but protected areas of dense forest now exist in remote regions of the neighboring Dominican Republic, and on Cuba. These require prompt, efficient management to ensure the solenodon's survival. Such is the pressure for new land accompanying the human population explosion on these islands that the solenodons' survival may ultimately depend upon the efforts of zoos. MEN

Hedgehogs and Moonrats

The hedgehog is one of the most familiar wild mammals seen in the European landscape and also among the most thoroughly studied in the field. One reason for this familiarity is the hedgehog's intriguing defensive adaptation against predators. Having spines reduces the requirement for hedgehogs to run for cover, which means that they are relatively easy to spot in gardens, prime habitat in which to amble about on lawns looking for tasty beetles, worms, and other invertebrate prey. However, the habit of not running has ill-served them in the age of the motor vehicle and nowadays they are as common a sight – dead or dying – by the roadside as they are in the garden.

Not all hedgehogs, though, live in close proximity to humans in densely-populated Europe. Some species range the dry steppes and deserts of Africa and the Middle East. The rather poorly known moonrats and gymnures, hedgehog relatives that are lacking spines, inhabit the humid forests of south Asia and exhibit behavior more akin to that of the elusive shrews.

Spiny and Spineless

FORM AND FUNCTION

Hedgehogs, moonrats, and gymnures are plantigrade animals, which means that at each step the entire sole of the foot makes contact with the ground. They have an elongated head and snout, a small braincase, and well-developed eyes and ears. Males and females generally look alike, but the distance between the anus and genitals is larger in the males. Hedgehogs have 2–3 incisors, 1 canine, 2–4 premolars, and 3 molar teeth on each half dental arch, and the first incisor is usually larger than the others.

The spiny coat covering the back and the crown of the head makes hedgehogs unmistakable (see box). The spines have sharp tips, and incorporate many small internal cavities to reduce weight. The basal portion is flexible and works as a shock absorber if the hedgehog is hit hard. Spines normally lie flat along the back: each is erected by a single muscle (as is true of hairs in all mammals), and when lifted they crisscross and support each other. Hedgehogs' underparts are furry rather than spiny, which prevents the animals from spiking themselves when they roll up!

Below *Hedgehogs eat a wide variety of foods and will consume almost any invertebrate prey. Here, a Western European hedgehog feasts on snails.*

Hedgehogs have other features that distinguish them from moonrats and gymnures, including a larger number of mammae (4–5 as opposed to 2–4); they also lack the well-developed anal glands that, in moonrats and gymnures, produce an unpleasant odor that presumably assists in deterring predators. Hedgehogs have powerful front limbs and strong claws that they can use to dig when they are searching for food or constructing nest burrows; they cannot run very quickly – the maximum speed reported is around 10km/h (6mph) – but they can easily climb over such obstacles as wire-netting fences. Moonrats and gymnures move much more speedily, in the manner of large shrews; however, they are also much less efficient at digging.

While spines may protect the hedgehog from larger animals that seek to prey on their flesh, they positively encourage the presence of small bloodsuckers, since their impenetrability makes it difficult for hedgehogs to groom themselves. Fleas, ticks, mites, and fungal infections of the skin can reach a very high density on hedgehogs; some individuals may carry in excess of 1,000 fleas.

Left A Greater moonrat foraging. This species is generally black with a whitish head and shoulders, but some animals, like this individual, are white all over.

Right A Western European hedgehog curls up on a pile of dead leaves, revealing its relatively vulnerable underside. The tighter the ball that a hedgehog curls itself up into, the spinier it becomes.

FACTFILE

Hedgehogs and Moonrats

Order: Insectivora

Family: Erinaceidae

23 species in 7 genera

DISTRIBUTION Africa, Europe, and Asia N to the limits of deciduous forest; SE Asian islands. Absent from Madagascar, Sri Lanka, and Japan. European hedgehog introduced to New Zealand.

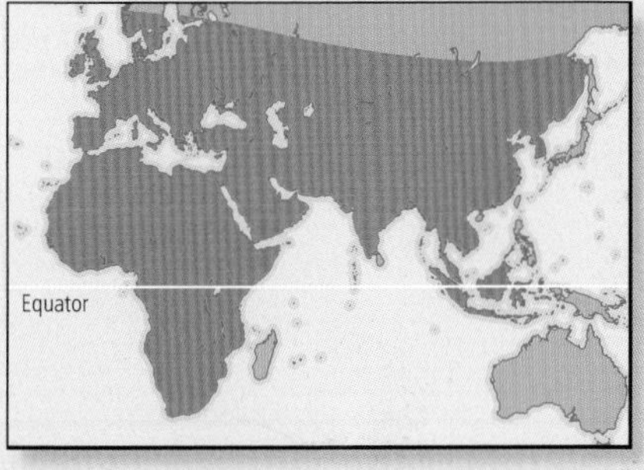

HABITAT Woodland, grassland, urban areas, dry steppe, desert, lowland forest, mangroves, tropical forest, montane areas.

SIZE Head–body length ranges from 10–15cm (4–6in) in the Lesser and Dwarf gymnures to 27–45cm (11–18in) in the Greater moonrat; **tail length** from 1–3cm (0.4–1.2in) to approximately 20cm (8in), and **weight** from 15–80g (0.5–3oz) to 1–2kg (35–70oz), both in the same species.

COAT Hedgehogs are distinctively covered in sharp spines – modified hairs that are typically 2–3cm (0.8–1.2in) long. Moonrats and gymnures have fur in place of spines.

DIET Typically invertebrates including beetles, earthworms, caterpillars, earwigs, slugs, grasshoppers, plus some carrion. African and Asian hedgehogs eat more vertebrates than their European counterparts – up to 40 percent of total food intake in the case of the Collared hedgehog.

BREEDING Gestation period ranges from 30–32 days in the Long-eared hedgehog to 40–48 days in the Algerian hedgehog.

LONGEVITY Up to 7 years, both in the wild and in captivity.

CONSERVATION STATUS 5 of the 7 species of moonrats and gymnures are listed by the IUCN, along with 1 of 16 species of hedgehogs. The Dwarf gymnure is Critically Endangered, and the Hainan gymnure and Dinagat and Mindanao moonrats are both Endangered due to the rapid habitat destruction and fragmentation occurring in their already narrow range; the Chinese Hugh's hedgehog is Vulnerable for the same reason.

See subfamilies box ▷

They also play host to a variety of internal parasites as well as bacterial diseases, including leptospirosis, which they can transmit to humans through their urine.

Reports of rabid hedgehogs foaming at the mouth, however, probably stem from the animals' unique habit of self-anointing, which involves spreading a huge amount of foamy saliva from the mouth all over their own back. This practice, which requires a surprising amount of agility on the hedgehog's part, may be exhibited in response to either strong-smelling or novel food, or the presence of other hedgehogs, foxes, or glue. Only tentative explanations exist for self-anointing: it may serve to clean the spines and act as an insecticide, however, it may also prove to have a part to play in courtship.

Hedgehogs are primarily nocturnal creatures. Their eyes are good enough to enable them to distinguish objects, but they probably only have monochromatic vision. Like many other nocturnal animals they rely mainly on their senses of smell and hearing to relate to the external world. The olfactory lobes of the brain are accordingly well-developed, and are augmented by a Jacobson's organ in the palate, which also serves an olfactory function. This supplementary sensory organ is found in a number of vertebrate species, in which it is associated with functions as different as prey detection and mating: the functions with which it is associated in hedgehogs still need to be thoroughly investigated.

A variety of hedgehog glands produce potentially odorous secretions. There are sexual scent-marking glands present in males, lubricating glands in the vagina of females, and sebaceous glands in the corners of the mouth. The sense of hearing has only been well studied in the Long-eared hedgehog, a species in which hearing is thought to be especially sensitive. Long-eared hedgehogs seem to perceive high-frequency sounds up to 45KHz, in comparison with a human range of only up to about 18–20KHz; this sense probably helps hedgehogs to locate underground invertebrate prey making high-pitched noises as they move in the soil and leaf litter. In contrast, low-frequency sound perception is relatively weak. Little is known about the aural senses of moonrats and gymnures, but the general considerations outlined for hedgehogs should also apply to them.

The genetic relationships of hedgehog species are controversial, although recent studies of mitochondrial DNA and karyotype are shedding some light on the situation; the DNA research helps to determine the genetic relatedness between populations on the basis of differences in the DNA sequence, while karyology studies account for differences in the number of chromosomes – 48 in all hedgehog species analysed so far – and their appearance. Western and Eastern European hedgehogs are capable of interbreeding, which would normally link them within a single species; however, mitochondrial DNA analyses have suggested that they are in fact sufficiently different to qualify as two separate species. Individuals from islands, such as Great Britain and Crete, are smaller than their Eastern counterparts; in addition, Western individuals are paler on the back and darker on the underparts. Karyotypes of African and European hedgehogs have confirmed that species that are assigned to the same genus on the basis of similar appearance are indeed genetically more closely related than those assigned to different genera.

Prey and Predator

DIET

Although hedgehogs, moonrats, and gymnures live in diverse habitats and pursue different lifestyles, most include beetles and earthworms in their diet, and also show a penchant for other invertebrate prey including caterpillars, earwigs, slugs, crickets, and grasshoppers. European hedgehog populations foraging in house gardens and on playing fields feed mainly on earthworms and slugs (but not on large snails, since they seem to be incapable of breaking their thick shells), while those foraging in bushes and maquis (a dense shrub vegetation found in wild, dry Mediterranean areas) rely on other invertebrates. The stomachs and droppings of European hedgehogs also commonly contain the remains of vertebrate prey; frogs, bird chicks, mice, shrews, moles, voles, lizards, and snakes have all been detected there. With the exception of chicks, which some hedgehogs might occasionally kill, these are probably not the results of predatory behavior but were more likely consumed as carrion. Generally, meat does not make up a significant proportion of the hedgehogs' diet. Dietary changes with age have been investigated, showing that European hedgehogs apparently learn to forage more efficiently as they grow older. Young hedgehogs eat prey of a wide variety of size, while older ones focus on the biggest insects; this may reduce competition between different age classes.

African, Long-eared, Collared, and Asian hedgehogs feed on a higher proportion and variety of vertebrates than the European hedgehog, including frogs, toads, sand and spiny lizards, snakes, and small rodents. A study of the Collared hedgehog found that vertebrates (amphibians and mammals in particular) made up 40 percent by dry weight of the stomach contents. The very few studies that have been conducted on the diets of moonrats and gymnures living in montane or tropical forests indicate that they forage on invertebrates, and possibly also on some vegetable matter. The only exception is the Greater moonrat, which also inhabits mangroves and lowland plantations and has been reported to enter the water to prey on crabs, mollusks, and fish.

A Solitary Life behind the Prickles

SOCIAL BEHAVIOR

Thanks to their spiny coats, hedgehogs have few natural enemies. Dogs and foxes can only occasionally overcome their defenses, and even lions have sometimes been observed abandoning a rolled-up African hedgehog after painful attempts to bite it. The only specialized hedgehog predators are large owls, which are able to "shell" the hedgehog with their claws, and badgers. These last can sneak their muzzles into the tightly-clenched opening of a rolled-up hedgehog, and eat it, leaving behind an empty, spiny coat. Badgers as hedgehog predators are, in fact, an interesting ecological case study, since both species feed on

Right *Some representative species of hedgehogs and moonrats:* ***1*** *Desert hedgehog* (Hemiechinus aethiopicus) *eating a beetle;* ***2*** *North African hedgehog* (Atelerix algirus)*;* ***3*** *Shrew gymnure* (Hylomys sinensis)*;* ***4*** *Long-eared hedgehog* (Hemiechinus auritus)*;* ***5*** *Short-tailed gymnure* (Hylomys suillus)*;* ***6*** *Greater moonrat* (Echinosorex gymnura)*;* ***7*** *Mindanao moonrat* (Podogymnura truei)*;* ***8*** *Hainan gymnure* (Hylomys hainanensis).

the same prey (earthworms) and badgers feed opportunistically on hedgehogs whenever they encounter them. This can affect the distribution pattern of hedgehogs, since they can be excluded from areas rich in food and cover that would otherwise be suitable habitat for them solely because of predation by badgers. Experiments conducted over the last 10 years suggest that hedgehog densities can be ten times higher in badger-free areas than where badgers are present (less than 0.5 as opposed to 2–3 hedgehogs per hectare). Captive and free-ranging hedgehogs both tend to avoid foraging areas tainted by badger odor. Moreover, the presence of badgers can increase mortality rates and the dispersal of hedgehog populations.

When it gets cold hedgehogs enter hibernation,

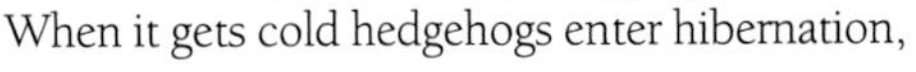

Hedgehog Subfamilies

Hedgehogs
Subfamily Erinaceinae

16 species in 4 genera, including **Western European hedgehog** (*Erinaceus europaeus*), W Europe and British Isles; **Eastern European hedgehog** (*E. concolor*), E Europe; **North African hedgehog** (*Atelerix algirus*), SW Europe and N Africa; **Four-toed hedgehog** (*A. albiventris*), N and C Africa; and **Indian hedgehog** (*Hemiechinus micropus*), Pakistan and NW India. The **Bare-bellied hedgehog** (*H. nudiventris*) and **Hugh's hedgehog** (*Mesechinus hughi*) are both Vulnerable.

Moonrats and Gymnures
Subfamily Hylomyinae

7 species in 3 genera, including **Greater moonrat** (*Echinosorex gymnura*), SE Asia; **Short-tailed gymnure** (*Hylomys suillus*), SE Asia; **Hainan gymnure** (*H. hainanensis*), SE Asia; **Dinagat moonrat** (*Podogymnura aureospinula*), Philippines; and **Mindanao moonrat** (*P. truei*), Philippines. The **Dwarf gymnure** (*Hylomys parvus*) is Critically Endangered, the Hainan, Dinagat, and Mindanao gymnures all Endangered.

For full species list see Appendix ▷

drastically reducing the energy they normally expend on remaining active and maintaining body temperature. While hibernation does increase survival rates, approximately half of the individuals that enter it nonetheless die before the next spring. Hedgehogs hibernate when and where they need to in response to adverse climatic conditions, rather than rigidly every season. In North Africa, for example, the African hedgehog is active throughout the year; those living in southern Europe have a short, 0–4 month hibernation period during winter, while the populations inhabiting the coldest South African regions go into hibernation from June to August. On average European hedgehogs hibernate from October to April in the northern parts of the continent, while those living in temperate climes undergo dormancy only during the coldest winters. Hedgehogs in captivity never hibernate, provided that they are kept warm and are given sufficient food.

Hibernating requires accumulating resources to survive long periods without eating. This is accomplished by depositing subcutaneous and abdominal white fat, and axillary, thorax, neck, and spinal brown fat, in late summer. White fat (the commonest type) acts as an insulating layer and is used for ordinary metabolic activity. Brown fat is characteristic of hibernators and is used to kickstart the heating needed to arouse individuals from hibernation.

Even with large fat deposits hedgehogs need to stop hibernation for 1–2 days every week or two in order to forage, urinate, and defecate, thereby avoiding accumulating catabolites, the organic and inorganic by-products of the organism's biochemical cycles, which could otherwise poison the hedgehog as they concentrate. During hibernation hedgehog body temperature falls from around 35°C to 15–20°C, and the heart rate decreases from about 250 beats/min to around 10 beats/min. The respiration rate similarly decreases, and can even cease entirely for up to 2 hours. Depending on external conditions, the metabolic rate of a hedgehog can slow down a hundredfold during hibernation.

Besides external cues, hormonal circannual cycles play a key role in the onset of dormancy in male hedgehogs. Melatonin (secreted by the pineal gland) and testosterone (secreted by gonads) have opposite effects on the animals' hibernation and sexual activity. The lengthening of the hours of darkness in winter causes melatonin

Spines and Curling in Hedgehogs

The most distinctive features of hedgehogs are their spines. An average adult carries around 5,000, each about 2–3cm (1in) long, with a needle-sharp point. Each creamy white spine usually has a subterminal band of black or brown. Spines are actually modified hair, and along the animal's sides where spines give way to true hair, thin spines or thick, stiff hairs can often be found, which may show the transition from one to the other. To minimize weight without losing strength, each spine is filled with many small, air-filled chambers, separated by thin plates. Towards its base, each spine narrows to a thin, angled, flexible neck, and then widens again into a small ball that is embedded in the skin. This arrangement transforms any pressure exerted along the spine (from a blow or a fall, for example) into a bending of the thin, flexible part rather than driving the base of the spine into the hedgehog's body. Connected to the base of each spine is a small muscle that is used to pull it erect. Normally, the muscles are relaxed and the spines laid flat along the back. If threatened, a hedgehog will often not immediately roll up but will first simply erect the spines and wait for the danger to pass. When erected, the spines stick out at a variety of different angles, criss-crossing over one another and supporting each other to create a virtually impenetrable barrier.

Hedgehogs are additionally protected by their ability to curl up into a ball. This is achieved by the presence of rather more skin than is necessary to cover the body, beneath which lies a powerful muscle (the *panniculus carnosus*) covering the back. The skin musculature is more strongly developed around its edges than at the center (where it forms a circular band, the *orbicularis* muscle) and is only very loosely connected to the body beneath. When the orbicularis contracts, it acts like the drawstring around the opening of a bag, forcing the contents deeper into the bag as the string is drawn tighter.

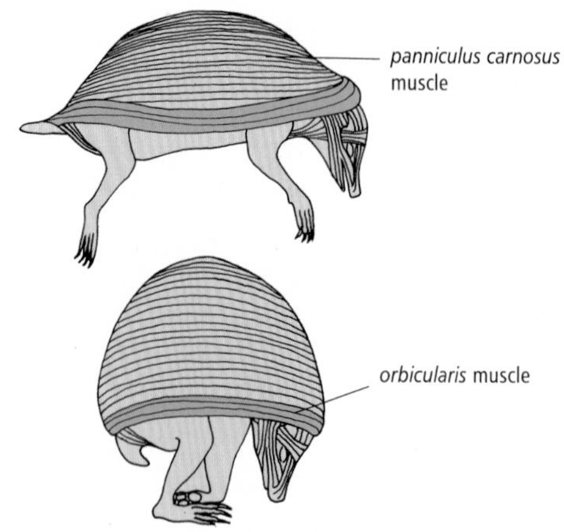

When a hedgehog starts to curl up, two small muscles first pull the skin and underlying circular muscle forward over the head and down over the rump. Then the circular *orbicularis* muscle contracts, the head and hindquarters are forced together, and the spine-covered skin of the back and sides is drawn tightly over the unprotected underparts. So effective is this stratagem that, on a fully-curled hedgehog, the spines that formerly covered its flanks and the top of the head are brought together to block the small hole (smaller than the width of a finger) corresponding to the opening of the bag. As the skin is pulled tightly over the body, the muscles that erect the spines are automatically stretched and the spines erected, so that the tighter the hedgehog curls, the spinier it becomes. AW

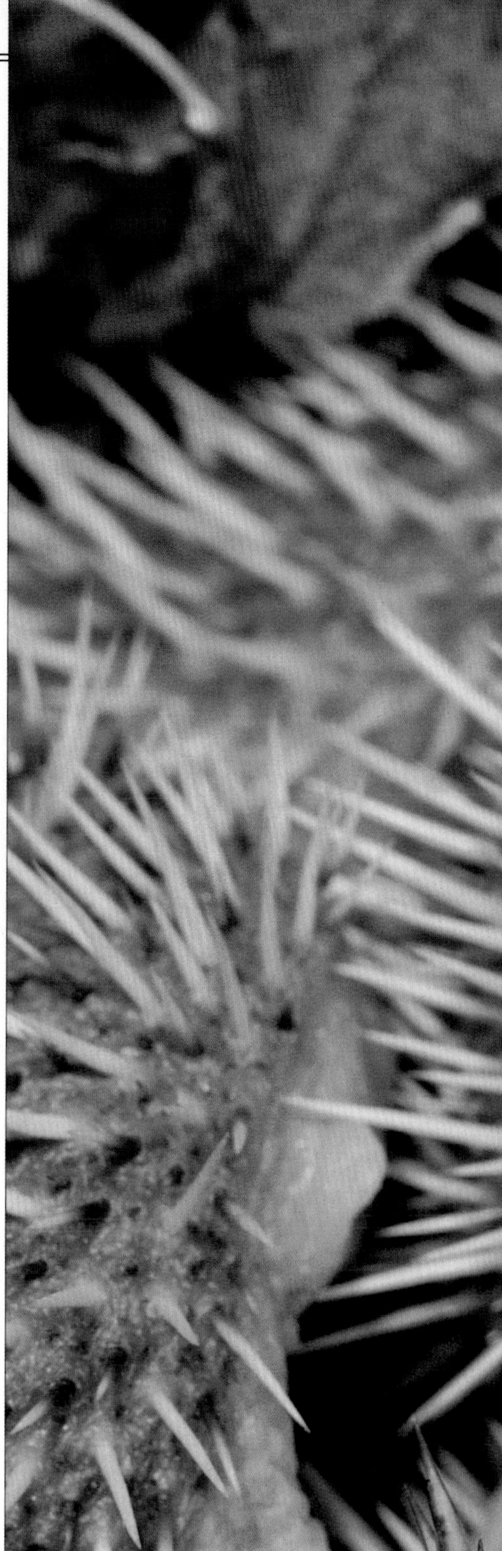

Above *Hedgehogs are born with the spines present, however, they are located just beneath the skin in order to avoid damaging the mother's birth canal. These spines, which are white, emerge from beneath the skin in a matter of hours. Within a couple of days of birth several additional, darker pigmented spines will also have sprouted. As well as being born naked, hedgehogs also leave the womb deaf and blind; yet within a few weeks, the young are generally quite functional and are fully weaned at about 6 weeks.*

levels to rise, reducing the activity of the sexual organs and the level of blood testosterone and inducing dormancy. Conversely, testosterone promotes gonadal activity and inhibits hibernation. Experiments involving the administration of hormones to female hedgehogs have demonstrated that hibernation in females is, in contrast, more environment-driven

During the daytime, hedgehogs rest in hidden nests lined with leaves, grass, and twigs. If the climate is warm enough, they may sleep under wood piles, thick bushes, pine needles, or simply foliage. A hedgehog uses many day nests in a season, and each nest is usually frequented by many individuals. Although hedgehogs in captivity often sleep together, no report exists of the simultaneous sharing of a nest in the wild. Breeding nests, where females give birth to their litter, have the same structure as day nests, but wintering nests are normally more robust since they have to last for months of continuous use. Long-eared, Collared, and Desert hedgehogs, all of which live in relatively arid regions, use burrows more frequently than the European hedgehog.

Every night a hedgehog moves around its home range at an average speed of 100–200m (110–220yds) per hour. Unlike many other mammals, European hedgehogs do not own and defend an exclusive territory: the area they use for foraging, resting, and breeding is normally shared with others of both sexes. This type of spatial organization is likely to arise when resources are so dispersed, or so unpredictably distributed in space and/or time, that a territory would be too large to be defended efficiently. This may be the case for hedgehogs, as earthworms and other invertebrates are patchily distributed and their availability fluctuates over time. Hedgehogs can forage in close contact with others if the feeding patch is rich enough to allow it: otherwise they stay apart, but no aggressive interaction seems to occur in order to achieve a more even spacing.

A European hedgehog's yearly home range is generally less than 40ha (100 acres) in area, and often much smaller if resources are abundant: suburban areas with gardens allow hedgehogs to reach very high densities (many individuals per hectare), each occupying an area of just 5–10ha (12–25 acres). On average males have larger home ranges than females – up to twice the size – and accordingly move longer distances in one activity cycle. Little is known about the home ranges of other hedgehogs, moonrats, and gymnures, although desert hedgehogs probably have larger yearly home ranges than European ones. The Greater moonrat is solitary and territorial.

Male hedgehogs emerge from hibernation before females, and start foraging to gain weight. When females also emerge, 3–4 weeks later, the males widen their ranging area looking for mates. Mating lasts from May to the end of August, with limited differences between species. Courtship is a conspicuous behavior and occurs frequently, but only in a few cases (less than 10 percent according to the data available hitherto) does it end in mating. When approached, the female reacts aggressively, lowering her spiny forehead and snorting loudly. The male circles her (generally silently, although the male African hedgehog makes a characteristically high-pitched courtship vocalization), and from time to time tries to mount. The male often gives up during this phase, in which case the pair will separate.

Above *A hedgehog in aggressive posture. Because they rely on their spines for protection, hedgehogs often wander in areas with little or no cover. If disturbed they usually freeze and erect their spines.*

Right *Although they are primarily terrestrial, hedgehogs are quite adept at swimming. Here, a Western European hedgehog is seen crossing a pond.*

Hedgehogs and Adders

For centuries folktales passed on knowledge of an adaptation peculiar to hedgehogs that was long neglected by the scientific community: the animals' resistance to adder venom. This resistance, probably incomplete and individually variable, is conferred by an anti-hemorrhagic factor called erinacin, a protein obtained from hedgehog muscle extracts that inhibits the venom's hemorrhagic and proteolytic activity. In combination with the defensive spines, erinacin actually allows hedgehogs to attack (BELOW) and eat snakes, although this does not happen frequently.

Venom resistance is not restricted to hedgehogs – other animals as unrelated as opossums and mongooses also exhibit this quality. Among European insectivores, extracts from shrew and mole muscles also have an anti-hemorrhagic effect, although this is less marked; those of such other mammals as mice, rats, and rabbits do not.

Mating starts if the female accepts the male. Contrary to Aristotle's opinion, hedgehogs do not mate face to face, but from the rear like most other mammals. To facilitate male penetration, the female will press her belly to the ground, exposing the vagina, which is located very close to the anus. The male penis lies forward under the belly, which assists him in copulating without being pricked. While mounting, the male grips the female's shoulder spines with his teeth to hold her in position.

At the end of mating the pair separates: the male does not guard the female to protect her from other males, nor does he help her build a nest or to rear the offspring. Rather than try to invest in the survival of his progeny by any one female, the male practices a lottery policy, gambling that the more females he can impregnate the greater his chances of leaving surviving young. Females too can mate more than once before starting gestation, but it is unknown if they can in some way select a preferred male's sperm, or if a single litter is fathered by more than one male. In mild climates females try to maximize reproductive success by raising a second litter later in the season, even though few individuals from it will grow large enough to survive hibernation.

After 35 days of gestation, four or five baby hedgehogs are born, each 7cm (3in) in length and weighing 10–25g (0.3–0.9oz). At birth the spines are hidden underneath the skin in a space filled with fluid, to avoid hurting the mother during parturition. The fluid is absorbed within 24 hours and the spines emerge, to be supplemented and then replaced by pigmented adult spines within 2–3 weeks. Baby hedgehogs are capable of erecting spines at the age of 2–3 days and can roll up into a ball after 2–3 weeks. As their milk teeth erupt, around the third week, they start to venture out of the nest with their mother. Weaning occurs around 6 weeks of age, after which the young forage intensively to accumulate fat reserves and search for a suitable wintering nest of their own. They will be sexually mature by the next spring. Living in tropical climates, the Greater and Lesser moonrat and the Dwarf gymnure breed throughout the year (the latter probably has no more than two litters annually, while nothing is known about the former), generating 1–3 offspring per litter.

Coexisting with Humans

CONSERVATION AND ENVIRONMENT

Concern has been raised about a possible decline in the numbers of European hedgehogs in the last 20 years. This decline can mainly be attributed to the huge number of hedgehogs killed on roads, where the roll-up defensive strategy that has proved so effective against animal predators often dooms them to be crushed by vehicles.

One suspected effect of roads on hedgehog decline is habitat fragmentation as a result of human activity, a key issue in conservation biology that has been the main cause of animal extinctions in the last two centuries. Habitat fragmentation divides large populations into many smaller subpopulations, which are no longer connected to each other. As individuals cannot disperse from one subpopulation to another, genetic mixing is prevented, and the probability of subpopulation survival in the long term is reduced. One study has demonstrated that hedgehog populations living as close as 15km (10mi) to each other have a different genetic composition, meaning that dispersal between populations rarely occurs. Since no correlation between geographic and genetic distance was found, it is possible that roads could act as barriers to hedgehog dispersal. However, a viability analysis on another group of hedgehog populations has concluded that fencing roads to avoid hedgehog roadkills would make them impenetrable, definitely isolating each one.

Roads are a major cause of hedgehog mortality, accounting for around half of all deaths except those occurring during hibernation. Yet roads can sometimes be successfully crossed, and their verges serve as corridors for dispersal. Hedgehogs following road verges often end up in urban areas, where they tend to prefer foraging in fields and gardens. Since urban areas support higher hedgehog densities than the surrounding woods and arable land, the abundance of food probably overcompensates for the added risk of roadkill.

Above *The head of the Desert hedgehog. Increasing desertification is leading to the fragmentation of populations of this species.*

Different lifestyles and geographic distribution will most likely ensure different destinies for the members of the Erinaceidae family. The animals' wide distribution and tolerance of people will probably see them survive in densely-populated areas of Europe and Asia, while African hedgehogs seem secure in the unpopulated desert and steppes. By contrast, the survival of moonrats and gymnures, like that of many other poorly known tropical forest species, is by no means assured, being intrinsically linked to the fate of the remnant forest itself. CR

Shrews

IN 1607 THE ENGLISH NATURALIST EDWARD *Topsell wrote one of the first known descriptions of the Eurasian common shrew, and it was not flattering. "It is a ravening beast," he stated, "feigning itself gentle and tame, but being touched it biteth deep, and poisoneth deadly. It beareth a cruel mind, desiring to hurt anything, neither is there any creature it loveth."*

This negative attitude toward the shrew has even established itself in the English language; the words "shrewd," "shrewish," and "shrew" were all coined to describe cunning, ill-tempered, or villainous people, although the meaning of "shrewd" at least has become less pejorative over time. Elsewhere, shrews have been more appreciated; the ancient Egyptians, for example, mummified – and are thought to have deified – the African giant and Egyptian pygmy shrews.

The Red and the White

FORM AND FUNCTION

Shrews are small, secretive mammals, superficially rather mouse-like but with characteristically long, pointed noses. They are typically terrestrial, foraging in and under the litter in woods and in the vegetation mat beneath herbage. Some are able to climb trees, others live underground, while a number of species are aquatic. A highly successful group of small insectivores, shrews comprise the third most speciose family of mammals, occurring over much of the globe except Australasia and the major part of South America.

The eyes of shrews are small, sometimes hidden in the fur, and vision seems to be poor. Hearing and smell, however, are acute. Even so, the external ears are reduced and difficult to discern in some species. The species that shows the greatest reduction in eye and ear size is the burrowing Mole shrew, which closely resembles a mole in external appearance, except that the forelimbs are barely modified from the typical shrew condition. There is a more dramatic change to the feet in the Tibetan water shrew, which has webbing between the digits. Other "aquatic" species, like the American water shrew, have their feet, digits, and tail all fringed with stiff hairs, which increase the surface area, thereby aiding propulsion underwater. The hairs also trap air, allowing the shrews to "run" on the surface of the water. Fringing of hair on the feet is additionally found in the Piebald shrew, in its case to aid running on sand.

The first set of teeth is shed or resorbed during embryonic development, so that shrews are born with their final set. The teeth are very important taxonomically. Some species can only easily be distinguished from close relatives by differences in tooth shape. At a higher taxonomic level, shrews are divided into two subfamilies on the basis of whether they have red tips to their teeth (the so-called "red-toothed shrews," or Soricinae) or not (the "white-toothed shrews," or Crocidurinae). The redness reflects the presence of a deposition of iron in the enamel.

A skeletal feature found in the Armored shrew – and in no other mammal – is the possession of interlocking lateral, dorsal, and ventral spines on the vertebrae. Along with the large number of facets for articulation, the spines create an exceptionally sturdy vertebral column. There are reliable reports of the Armored shrew surviving the pressure of a full-grown man standing on it.

The most conspicuous vocalizations of shrews are the high-pitched screams and twitterings used in disputes with members of their own species. One species, the House shrew, makes a sound like jangling coins as it moves around buildings and is consequently known as the "money shrew." Some species, at least of the genera *Sorex* and *Blarina*, may use ultrasound; this seems to be generated in the larynx and may be used to provide a crude form of echolocation.

Shrews are often viewed as "primitive" forms, and the earliest mammals are often portrayed as shrew-like. In fact, shrews are a clearly modern family of eutherians with origins within the Tertiary. The earliest shrew fossils have been found in North America in the middle Eocene (45 million years ago). Eurasian fossils date back to the early Oligocene (34 million years ago), and African shrews are known from the middle Miocene (14 million years ago). Those features that shrews probably do have in common with the earliest mammals, for example their plantigrade gait, put them at no disadvantage for their way of life as small terrestrial mammals. Shrews are in fact highly successful forms, supremely adapted to survive as miniature insectivores.

One of the most striking features of shrews is their extraordinary chromosomal variation. Species of shrew often differ from close relatives in the number and morphology of their chromosomes. Indeed the chromosomal differences may, in some cases, have contributed to the reproductive isolation necessary for speciation. In this context, the within-species variation of chromosomes

Left *The smallest terrestrial mammal is the Pygmy white-toothed shrew which, at 2g (0.07oz), is similar in size to the tiniest bats and hummingbirds.*

Far right *Plastered with air bubbles, a Eurasian water shrew plunges vertically downward to forage underwater. These shrews feed on aquatic invertebrates, small fish, and amphibians.*

Below *The white-toothed shrews are by far the most numerous of the various shrew groups. This is the Lesser white-toothed shrew* (Crocidura suaveolens).

FACTFILE

Shrews

Order: Insectivora

Family: Soricidae

312 species in 23 genera and 2 subfamilies

Distribution Eurasia, Africa, N America, northern S America.

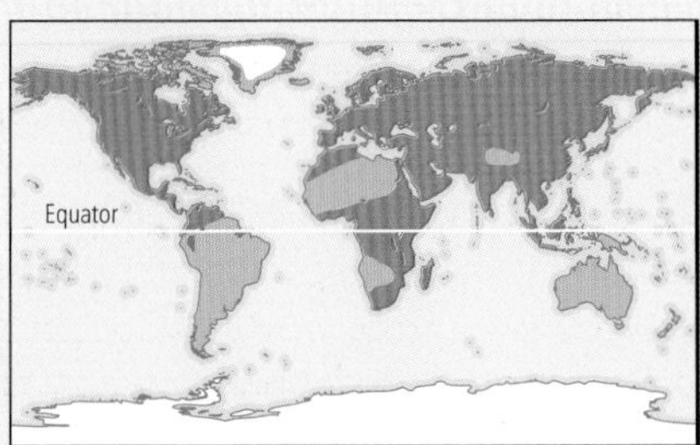

Habitat Forest, woodland, grassland, desert, terrestrial; but some species partially aquatic.

Size **Head–body length** from 3.5cm (1.4in) in the Pygmy white-toothed shrew, the smallest living terrestrial mammal, to 15cm (5.9in) in the House shrew; **weight** from 2g (0.07oz) to 106g (3.7oz), in the same two species.

Coat Small, mouse-like bodies are covered in short, thick fur, mostly in shades of gray or brown.

Diet Depending on size, seeds, nuts, and other plant matter; invertebrate and vertebrate prey, including earthworms, lizards, newts, frogs, and fish.

Breeding Gestation period 17–32 days

Longevity 12–30 months

Conservation status 29 species of shrew are classed as Critically Endangered, 30 as Endangered, and 56 as Vulnerable.

See subfamilies box ▷

of some species of shrews is of particular interest. Some of the "races" within species that have distinctive chromosomes may themselves be progenitors of new species. The Eurasian common shrew is particularly impressive as regards its chromosomal variation; altogether, over 60 different chromosome races with chromosome numbers ranging from 20 to 33 have been described. Eurasian common shrews are also unusual among mammals in that males and females of a particular race have different chromosome numbers. Females have two sex chromosomes (XX) while males have three (XY1Y2).

Tenacious Survivors

DIET

Shrews are very active and consume large amounts of food for their size. Some species cannot survive more than an hour or two without food. Their high metabolic rate is associated with other extraordinary features; for example, heartbeats of over a thousand beats per minute have been recorded (for an adult human, the resting rate is closer to 70 beats per minute). In several northern species, including the Eurasian common shrew, the skull, skeleton, and certain internal organs shrink during the winter in order to reduce energy demands.

Shrews cope with their high requirement for food and water primarily by living in habitats where these are abundant. However, their generally small size enables them to utilize thermally-protected microhabitats. At least one species of red-toothed shrew, the Desert shrew, has mastered the physiological problems of living in a hot, arid climate by lowering its metabolic rate in a similar way to other desert-dwelling mammals. Several species, mainly among the white-toothed shrews but also including the Desert shrew, are capable of going into torpor at times of day when they are not able to obtain food. Hibernation, however, has not been demonstrated for any shrew.

Interestingly, the red-toothed shrews have higher metabolic rates than the white-toothed. This difference may be related to their different origins, northern and tropical respectively. The iron-rich red tips to the teeth may increase resistance to wear, of importance given the substantial throughput of hard and gritty invertebrate food required by these animals. Despite this, the teeth may wear down substantially, rendering them ineffective. Adult specimens of the Eurasian common shrew are often found dead in the open during the fall. These are old individuals that have bred the previous summer and whose teeth are worn. The apparent cause of death is starvation, and the carcasses remain uneaten owing to the presence of scent glands on the flanks, whose secretions make shrews unpalatable to most carnivores.

Many shrews are opportunists in their feeding habits, showing little specialization. The Eurasian common shrew, for instance, will eat almost every invertebrate that it comes across. It spends its time scurrying along rodent runs or through vegetation, coming across prey in a haphazard way.

Above *A Greater white-toothed shrew dines on a grasshopper. Shrews have voracious appetites – some species need to eat every couple of hours to survive.*

Other species are more adventurous in their feeding habits. The aquatic species will make dives of over 30 seconds to find small fish or other prey. The Piebald shrew feeds on lizards.

The bite of some shrews is venomous. The salivary glands of the American short-tailed shrew, for instance, produce enough poison to kill about 200 mice by intravenous injection. The poison acts to kill or paralyze before ingestion, and may be important in helping to subdue large vertebrate prey. The poison may also immobilize insects for eating at a later time. Food-caching is an activity known in several species of shrews, including the American short-tailed, and seems to be a means of securing short-term food supplies when competition is fierce or food is temporarily very abundant.

A number of shrew species – very possibly all – practice refection (the reingestion of excreted food). In the Eurasian common shrew, the animal curls up and begins to lick its anus, sometimes gripping the hind limbs with the forefeet to maintain position. After a few seconds, abdominal contractions cause the rectum to extrude and the end is then nibbled and licked for some minutes before being withdrawn. It appears that refection does not start until the intestine is free of feces. If the shrew is killed, the stomach and first few

Right A young Eurasian water shrew swallows a worm. The animals' small size belies their status as fierce predators with omnivorous eating habits.

Shrew Subfamilies

White-toothed shrews
Subfamily Crocidurinae

White-toothed shrews Genus *Crocidura*
151 species in Africa and S Eurasia, including African giant shrew (*C. olivieri*), Egyptian pygmy shrew (*C. religiosa*), Greater white-toothed shrew (*C. russula*), Lesser white-toothed shrew (*C. suaveolens*); forest to semidesert. Seventeen species are classified as Critically Endangered.
Mouse shrews Genus *Myosorex*
12 species in central and southern Africa, including Dark-footed forest shrew (*M. cafer*); forest. The Rumpi mouse shrew (*M. rumpii*) and Schaller's mouse shrew (*M. schalleri*) are Critically Endangered.
African shrews Genus *Paracrocidura*
3 species in central Africa, including Lesser large-headed shrew (*P. schoutedeni*); forest. Grauer's shrew (*P. graueri*) is Critically Endangered.
Pygmy and Dwarf shrews Genus *Suncus*
16 species in Africa and southern Eurasia, including Pygmy white-toothed shrew (*S. etruscus*), House shrew (*S. murinus*), Lesser dwarf shrew (*S. varilla*); forest, scrub, savanna; commensal. The Black shrew (*S. ater*), Flores shrew (*S. mertensi*), and Gabon dwarf shrew (*S. remyi*) are Critically Endangered.
Kenyan shrews Genus *Surdisorex*
2 species in Kenya: Mount Kenya mole shrew (*S. polulus*); montane; Aberdare shrew (*S. norae*); Aberdare range. Both species are Vulnerable.
Forest musk shrews Genus *Sylvisorex*
10 species in C Africa, including Climbing shrew (*S. megalura*); forest, grassland. The Arrogant shrew (*S. morio*) is Endangered.
Congolese shrew (*Congosorex polli*)
S Democratic Republic of Congo (DRC); forest. Critically Endangered.
Piebald shrew (*Diplomesodon pulchellum*)
W C Asia; desert.
Kelaart's long-clawed shrew (*Feroculus feroculus*). India and Sri Lanka; montane forest. Endangered.
Ruwenzori shrew (*Ruwenzorisorex suncoides*). C Africa; montane forest. Vulnerable.
Armored shrew (*Scutisorex somereni*)
C Africa; forest.
Pearson's long-clawed shrew (*Solisorex pearsoni*). C Sri Lanka; montane. Endangered.

Red-toothed shrews
Subfamily Soricinae

American short-tailed shrews Genus *Blarina*. 3 species in E North America, including Northern short-tailed shrew (*B. brevicauda*); forest grassland.
Asiatic short-tailed shrews Genus *Blarinella*. 2 species in China and Burma: Chinese short-tailed shrew (*B. quadraticauda*), montane forest; Ward's short-tailed shrew (*B. wardi*), Burma, Yunnan.
Oriental water shrews Genus *Chimarrogale*. 6 species in S and E Asia, including Himalayan water shrew (*C. himalayica*); montane streams. The Malayan (*C. hantu*) and Sumatra water shrews (*C. sumatrana*) are Critically Endangered.
Small-eared shrews Genus *Cryptotis*
14 species in E North America, C America, N South America, including American least shrew (*C. parva*); forest, grassland. Enders's small-eared shrew (*C. endersi*) is Endangered.
Old World water shrews Genus *Neomys*
3 species in N Eurasia, including Eurasian water shrew (*N. fodiens*); woodland, grassland, streams, wetlands.
Holarctic shrews Genus *Sorex*
70 species in N and C Eurasia and N America, including Eurasian common shrew (*S. araneus*), Masked shrew (*S. cinereus*), Smoky shrew (*S. fumeus*), Eurasian pygmy shrew (*S. minutus*), American water shrew (*S. palustris*), Trowbridge's shrew (*S. trowbridgii*); tundra, grassland, woodland. Kozlov's shrew (*S. kozlovi*) and the Gansu shrew (*S. cansulus*) are Critically Endangered.
Genus *Soriculus*. 10 species in S Central Asia from Pakistan to China and Vietnam, including Long-tailed mountain shrew (*S. macrurus*); montane forest. Salenski's shrew (*S. salenskii*) is Critically Endangered.
Mole shrew (*Anourosorex squamipes*)
S Asia; montane forest.
Mexican giant shrew (*Megasorex gigas*)
SW Mexico; forest, semi-arid areas.
Tibetan water shrew (*Nectogale elegans*)
Sikkim to Shenshi; montane streams.
Desert shrew (*Notiosorex crawfordi*)
S USA and Mexico; semidesert scrub, montane.

For full species list see Appendix ▷

centimeters of the intestine can be seen to be filled with a milky fluid containing fat globules and partially digested food. It is thought that shrews may obtain trace elements and vitamins B and K in this way.

Caravans and Tunnels

SOCIAL BEHAVIOR

In species occupying temperate and arctic zones, breeding is seasonal, while tropical species may breed continuously. An example of a northern temperate species is the Eurasian common shrew. Individuals of this species generally do not breed until the year following birth. The breeding season starts in April, and females can potentially produce up to 4 or 5 litters of as many as 11 young. Usually, however, adult females produce one or two litters of 4–8 young, and then die. Likewise, there is mortality of adult males, so by late summer the population is dominated by immature individuals. The gestation period is 20 days in this species and lactation lasts for about 23 days. The first estrus of the breeding season is synchronized within a population, with the result that almost all females are mated within a few days. These must be frantic periods, for the estrus of each female lasts no more than a few hours. It has been demonstrated by DNA fingerprinting of offspring, their mothers, and potential fathers, that a single female may mate with as many as six different males during one estrus. Mating itself has been observed in captivity. The male grips the female on the nape and it is possible to identify mated females by the bite mark , which develops into a patch of white hair.

Among the tropical species, the best studied is the House shrew. Like the Eurasian common shrew, it is promiscuous and has been observed to accept as many as eight males and to copulate 278 times within a period of 2 hours. House shrews have a gestation period of 30 days and a litter size of usually 2–4 young. As with other shrews, the young are born naked and blind, but develop fast.

The House shrew is one of a number of species of *Suncus* and *Crocidura* whose young exhibit "caravanning" behaviour. When mature enough to leave the nest the young form a line. Each animal grips the rump of the one in front and the foremost grips that of the mother. The grip is quite tenacious and the whole caravan can be lifted off the ground intact by picking up the mother.

Most shrews are solitary and it is likely that the promiscuous mating system found in the Eurasian common shrew and the House shrew is frequent. Because of high food requirements, individuals of some species defend territories for all or part of their lives. This is seen particularly clearly in the Eurasian common shrew during the winter months. The only time many shrews come into close contact is during fighting, mating, and at the nestling stage. However, at least one species, the Lesser dwarf shrew, has been shown to form very long-lasting pair bonds. Other species are even more social. The American least shrew is believed to be more or less permanently colonial, and individuals of the Greater white-toothed shrew huddle together in groups during the winter, presumably to help maintain body temperature.

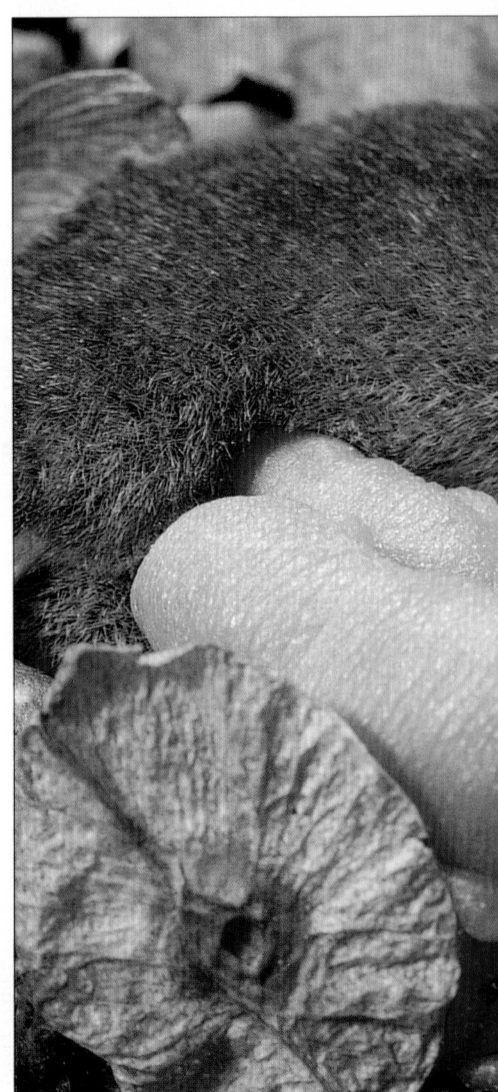

Several species dig tunnel systems and these may be the focal point of defended territories. The American short-tailed shrew, the American least shrew, the Eurasian common shrew, and the Eurasian water shrew have all been observed digging. Observations in captivity of the Smoky shrew, the Masked shrew, Trowbridge's shrew, and the Eurasian pygmy shrew suggest that they do not burrow. In the Eurasian water shrew, the tunnel system is important in squeezing water from the fur, and it also seems to be significant in maintaining fur condition in the Eurasian common shrew. In captivity, Eurasian common shrews often cache food in their tunnels. Tunnel systems may also be important in avoiding predators and usually have more than one entrance. Nests of grass and other plant material are usually built in a chamber off the tunnel system, and shrews spend most of their sleeping and resting time there.

Endemics in Danger

CONSERVATION AND ENVIRONMENT

It might be thought that shrews, as a highly successful group of small, fast-breeding mammals with great reproductive potential, would be rather resistant to human-induced extinction. Sadly, nothing could be further from the truth. Many tropical species are extremely restricted in their distribution and occur at low densities. At current rates of destruction of the tropical forest, it is certain that many of these species will go extinct.

Above *Caravanning: young shrews find their way around their terrain by forming chains, led by their mother. They continue to hang on even if the mother is picked up off the ground.*

Right *Two Northern short-tailed shrews* (Blarina brevicauda) *confront one another. This species, like several others, digs tunnel systems that form the basis of defended territories.*

Left *Refection: some species, like this Eurasian pygmy shrew* (Sorex minutus), *obtain vital nutrients by licking the rectum, which is extruded from the anus to facilitate access.*

Below *A Common Eurasian shrew* (Sorex araneus) *suckling her infants. Males make no contribution to raising the young.*

In contrast, some shrews appear to have flourished alongside humans. The House shrew is an extreme example: it is a tropical, commensal species that feeds on insects found around people's houses. House shrews have been unwittingly transported with humans to many parts of the tropics. Unfortunately, such passive transport of human-tolerant shrews can be very dangerous to endemic species of shrews on islands. Thus, it is thought that the genus *Nesiotites*, known from Mediterranean islands such as Corsica, Sardinia, and the Balearics, became extinct following the accidental introduction of Greater, Lesser, and Pygmy white-toothed shrews by humans. This is an instance of competitively superior shrews leading to the demise of other species.

It is not just very local species of shrews that are vulnerable to extinction. Recent evidence from Britain suggests that the Eurasian common shrew is declining there. This is particularly alarming, since this species would otherwise have appeared to be the ultimate survivor. It can utilize marginal habitats such as roadside verges, and appears to have unusual resistance to some pollutants such as heavy metals. Shrews, as much as tigers and rhinos, need to be monitored and cared for during the coming years of threat to the world's mammalian fauna. JBS/CJB

Golden Moles

GOLDEN MOLES ARE SO CALLED FROM THEIR family name, Chrysochloridae, deriving from the Greek terms for "gold" and "pale green." The appellation refers to the iridescent sheen of coppery green, blue, purple, or bronze on the animals' fur rather than to the color of the fur itself, which is usually a shade of brown.

Golden moles are known from as far back as the late Eocene, about 40 million years ago. Climatic modifications may be responsible for their present discontinuous distribution, but they have special adaptations for a burrowing mode of life in a wide geographic range of subterrestrial habitats.

FACTFILE

GOLDEN MOLES

Order: Insectivora

Family: Chrysochloridae

21 species in 9 genera

DISTRIBUTION Sub-Saharan Africa, including Somalia.

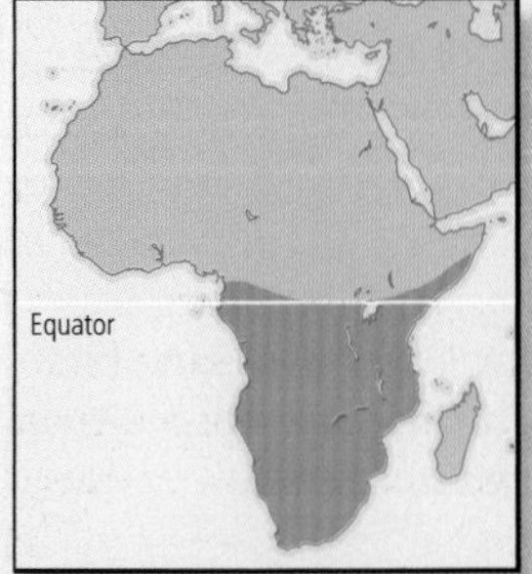

Habitat Almost exclusively burrowing.

Size Head–body length ranges from 7–8.5cm (2.7–3.3in) in Grant's desert golden mole to 19.8–23.5cm (7.8–9in) in the Giant golden mole.

LARGE GOLDEN MOLES Genus *Chrysospalax*
The 2 largest species: **Giant golden mole** (*C. trevelyani*), forests in E Cape Province, Endangered; and **Rough-haired golden mole** (*C. villosus*), grasslands and swamp in E South Africa, Vulnerable.

SECRETIVE GOLDEN MOLES Genus *Cryptochloris*
2 species, both found in arid regions of Little Namaqualand, W Cape: **De Winton's golden mole** (*C. wintoni*), Vulnerable; and **Van Zyl's golden mole** (*C. zyli*), Critically Endangered.

CAPE GOLDEN MOLES Genus *Chrysochloris*
3 species: **Stuhlmann's golden mole** (*C. stuhlmanni*), mountains in C and E Africa; **Cape golden mole** (*C. asiatica*), W Cape to Little Namaqualand; and **Visagie's golden mole** (*C. visagiei*), succulent karoo of W Cape, Critically Endangered.

AFRICAN GOLDEN MOLES Genus *Chlorotalpa*
2 species: **Sclater's golden mole** (*C. sclateri*), high-altitude grasslands and scrub in W Cape, Lesotho, E Free State, and Mpumalanga, Vulnerable; and **Duthie's golden mole** (*C. duthieae*), coastal forests in W and E Cape, Vulnerable.

SOUTH AFRICAN GOLDEN MOLES Genus *Amblysomus*
5 species, including the **Hottentot golden mole** (*A. hottentotus*), all found in grasslands and forests in W and E Cape, KwaZulu–Natal, NE Free State, Mpumalanga, Swaziland.

GENUS *NEAMBLYSOMUS*
2 species: **Gunning's golden mole** (*N. gunningi*), forests in Northern Province, Vulnerable; and **Juliana's golden mole** (*N. julianae*), sandy soils in savannas of Gauteng and Mpumalanga, Critically Endangered.

GENUS *CALCOCHLORIS*
3 species: **Yellow golden mole** (*C. obtusirostris*), Zululand to Mozambique and SE Zimbabwe; **Congo golden mole** (*C. leucorhinus*), forests of W and C Africa; **Somali golden mole** (*C. tytonis*), NE Somalia.

GRANT'S GOLDEN MOLE *Eremitalpa granti*
Sandy desert and semidesert of W Cape, Little Namaqualand, Namib Desert. Vulnerable.

AREND'S GOLDEN MOLE *Carpitalpa arendsi*
Forests and adjacent grasslands in E Zimbabwe and NE Mozambique.

For full species list see Appendix ▷

Equipped for Extremes

FORM AND FUNCTION

Golden moles are solitary, burrowing insectivores with compact, streamlined bodies, short limbs, and no visible tail. The backward-set fur is moisture-repellent, remaining sleek and dry in muddy situations; a dense, woolly undercoat provides insulation. The skin is thick and tough, particularly on the head. The eyes are vestigial and covered with hairy skin, and the optic nerve is degenerate. The ear openings are covered by fur, and the nostrils are protected by a leathery pad that assists in soil excavations; in some species such as the Yellow golden mole, the nostrils also have foliaceous projections that prevent sand from entering the nose during burrowing. The wedge-shaped head and extremely muscular shoulders push and pack the soil, whereas the strong forelimbs are equipped with curved, picklike digging claws.Of the four claws, the third is extremely powerful, while the first and fourth are usually rudimentary. The hind feet are webbed with five digits, each bearing a small claw, and are used to shovel loose soil backward along tunnels.

The key to the evolutionary success of golden moles lies in their unique physiology. Despite a high thermal conductance, they have a low basal metabolic rate and do not thermoregulate when at rest, thereby considerably reducing their energy requirements. All species enter torpor, either daily or in response to cold. Body temperature in the thermal neutral zone is lower than in other similarly-sized mammals. Lowered metabolism and efficient renal function effectively reduce water requirements to the extent that most species do not need drinking water. Far from being "primitive" characteristics, such physiological specializations allow the moles to survive in habitats where temperatures are extreme and food is scarce.

The chief senses in golden moles are hearing, touch, and smell. The ear ossicles of some species are disproportionately large, giving great sensitivity to vibrations, which trigger rapid locomotion either toward prey or unerringly toward an open burrow entrance (on the surface) or a bolthole (when underground). Those species without enlarged ear ossicles have a well-developed hyoid apparatus that may transmit low-frequency sounds to the inner ear. Golden moles also have an extraordinary ability to orientate themselves underground; when parts of burrow systems are damaged, the new tunnels that repair them always link up precisely with the existing tunnels.

Golden moles forage in subsurface tunnels, visible from above as soil ridges. Desert-dwelling species "swim" through the sand just below the surface, leaving U-shaped ridges. Most species also excavate deeper burrows connecting grass-lined nests, defecation chambers, and spiraling boltholes, depositing excess soil in small mounds on the surface. Average sustained burrowing lasts for about 44 minutes, separated by inactive periods of about 2.6 hours. Non-random surface locomotion minimizes foraging costs.

The Underground Larder

DIET

Golden moles are opportunistic foragers, preying predominantly on earthworms and insect larvae. Grant's golden moles feed mainly on soft-bodied termites, a sedentary prey occurring in patches of high concentration. De Winton's golden moles also prey on legless lizards, using their long, slender claws to hold these reptiles. Giant golden moles feed mainly on oniscomorph millipedes, which abound in leaf litter, but also take giant earthworms (*Microchaetus* spp.); they probably also consume any small vertebrates they may stumble across. Captive moles take a wide spectrum of terrestrial invertebrates except for mollusks, and can be trained to eat ground beef.

Contacts between the Burrows

SOCIAL BEHAVIOR

Territorial behavior is influenced by the availability of food. Hottentot golden mole burrow systems are more numerous in the summer when food is more abundant, and a certain amount of home range overlap is tolerated. The systems are larger and more aggressively defended in less fertile areas; a neighboring burrow may be taken over by an individual as an extension of its home range. Occupancy is detected by scrutiny of tunnel walls by smell. Fighting occurs between individuals of the same sex, and sometimes between male and female. Hottentot golden moles tolerate herbivorous mole rats in the same burrow systems, and in the Drakensberg range golden mole burrows open into those of Sloggett's vlei rats (*Otomys sloggetti*).

Courtship in Hottentot golden moles involves much chirruping vocalization, head-bobbing, and foot stamping in the male, and grasshopper-like rasping and prolonged squeals with the mouth wide open in the female. Both sexes have a single external urogenital opening. Females display aseasonal polyestry. Litters comprise 1–3 (usually 2) naked young with a head–body length of 4.7cm (1.9in) and a weight of 4.5g (0.16oz). Birth and lactation take place in grass-lined nests, and eviction from the maternal burrow system occurs once the young weigh 35–45g (1.2–1.6oz).

Sadly, eleven species of golden moles are now threatened with extinction owing to habitat degradation induced by human activities. These threats include urbanization, the mining of alluvial sands for diamonds and building materials, poor agricultural practices, and predation by domestic dogs and cats. GB/MP

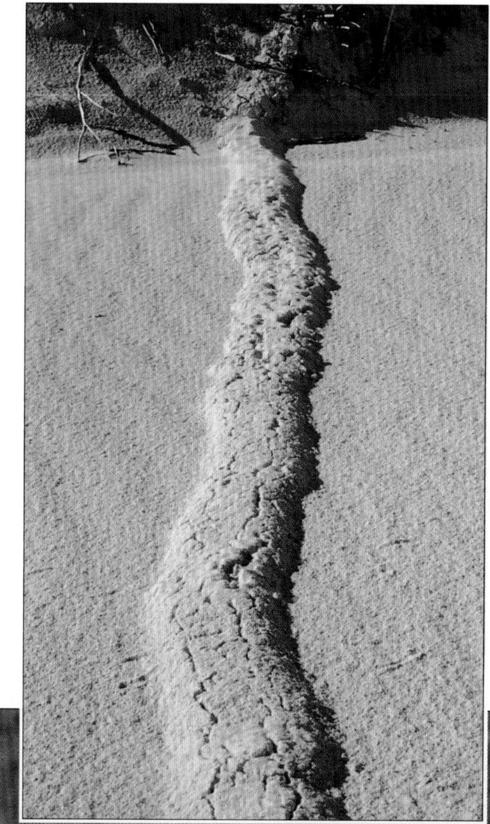

Right *A telltale ridge of earth indicates the location of a golden mole foraging tunnel. Most species also dig deeper nesting burrows.*

Below *In the Namib Desert in southern Africa, a Grant's golden mole feeds on a locust – a change from its usual diet of soft-bodied termites.*

Moles and Desmans

THE FAMILY TALPIDAE CONTAINS AS QUEER A set of bedfellows as one could hope to meet; on the one hand the moles and shrew-moles, which are more or less fossorial (diggers), and, on the other, the desmans, which are semi-aquatic (swimmers). Today there are 40 species of moles placed in 15 genera and just two species of desmans, each in its own genus.

Most people in Europe and North America have seen molehills, but very few have seen moles. Their subterranean way of life makes these animals difficult to study, and it is only in the last three decades that detailed information has been gathered. The other major group within the family, the desmans, are also little-known, due to their inaccessibility in the mountain streams of the Pyrenees and in the vastness of Russia.

Adaptations for Digging and Swimming

FORM AND FUNCTION

Moles and desmans have elongated, cylindrical bodies. The muzzle is long, tubular, and naked, apart from sensory whiskers. It is highly mobile and extends beyond the lower lip. In the Star-nosed mole the nose divides at the end into a naked fringe of 22 mobile and touch-sensitive fleshy tentacles, used to detect prey. The penis is directed to the rear and there is no scrotum.

The eyes – structurally complete but minute – are largely hidden within the fur, or, in the case of the Mediterranean mole, covered by protective skin. The eyes are sensitive to changes in light level but have little visual acuity. Moles have no external ears except in the Asiatic shrew moles. Both moles and desmans rely to a great extent on touch. The muzzle is richly endowed with projections called Eimer's organs, which are probably touch-sensitive. In addition, various parts of the body including the muzzle, tail, and also the legs of desmans, have sensory whiskers.

Desmans are adapted for swimming: they have a long, flat tail like a rudder, broadened by a fringe of stiff hairs. The legs and feet are proportionally long and powerful, ending in webbed toes and half-webbed fingers, both also fringed with stiff hairs. The nostrils and ears are opened and closed by valves. When swimming, the hind legs provide the main propulsive force.

In moles, the forelimbs are adapted for digging. The hands are turned permanently outward, rather like a pair of oars. The hands are large, almost circular, and equipped with five large and strong claws. The teeth are unspecialized and typical of the Insectivora.

The moles and desmans originated in Europe, where their fossil record extends back some 45 million years into the mid-Eocene. Today, moles are spread throughout Europe, Asia, and North America, but are absent from Africa, where their niche as underground hunters of invertebrates is occupied by the golden moles (family Chrysochloridae). Recent analyses suggest that the ancestors of *Condylura*, *Neurotrichus*, and

Below The Star-nosed mole has a unique nose that is divided into a number of fleshy tentacles – almost coral-like in appearance – which are used for detecting prey.

Left *Representative species of moles and desmans:* ***1*** *European mole* (Talpa europaea)*;* ***2*** *American shrew-mole* (Neurotrichus gibbsi)*;* ***3*** *Lesser Japanese shrew-mole* (Urotrichus pilirostris)*;* ***4*** *Star-nosed mole* (Condylura cristata)*;* ***5*** *Pyrenean desman* (Galemys pyrenaicus).

Parascalops dispersed to Europe and North America separately from the common ancestor of *Scalopus* and *Scapanus*.

During the Pleistocene, the Russian desman, *Desmana moschata*, was found across Europe in a broad band from southern Britain to the Caspian Sea. Since then it has become progressively restricted in distribution, and today is confined to the southern part of European Russia, in the basins of the rivers Dnieper, Don, Kama, Ural, and Volga. Fossils of *Galemys pyrenaicus*, the Pyrenean desman, are known only from the Pyrenees and their foothills, and today the species is restricted to the northern half of the Iberian Peninsula and to the French side of the Pyrenees.

Diets of Worms

DIET

Moles dig permanent tunnels and obtain most of their food from soil invertebrates that fall into them. When digging new tunnels, they brace themselves with their hind feet and dig with the fore feet, which are thrust alternately into the soil and moved sideways and backward. Periodically, they dig a vertical shaft to the surface and push up the soil to make a molehill. Tunnels range in depth from a few centimeters to 1m (3.3ft). Probably 90 percent of the diet is foraged from the permanent tunnels; moles eat whatever is locally available but favor earthworms, beetle and fly larvae, and slugs. The European mole stores worms with their heads bitten off near its nest in October and November.

Desmans obtain nearly all their food from water, especially aquatic insects such as stonefly and caddis-fly larvae, freshwater shrimps, and snails. The Russian desman also takes larger prey such as fish and amphibians.

FACTFILE

MOLES AND DESMANS

Order: Insectivora

Family: Talpidae

42 species in 17 genera and 3 subfamilies. Subfamily Desmaninae contains 2 monotypic genera of desmans; subfamily Talpinae 14 genera of moles, including the Old World moles (*Talpa* spp.); subfamily Uropsilinae 1 genus of shrew mole (*Uropsilus* spp.)

DISTRIBUTION Europe, Asia, and N America

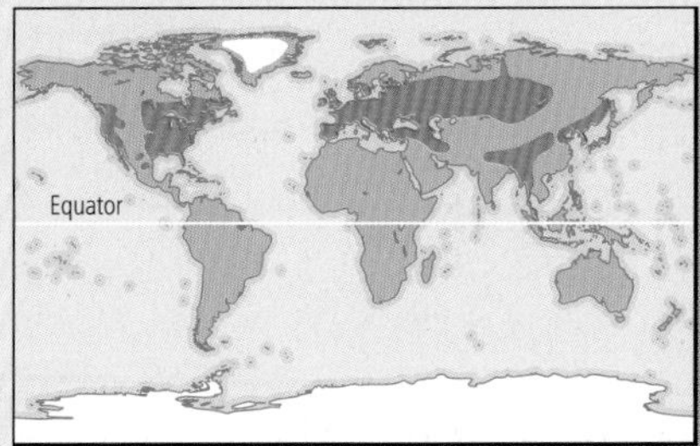

HABITAT Moles are largely subterranean, usually living under forests and grasslands but also under heaths. Desmans are aquatic in lakes and rivers. Shrew moles construct tunnels, but forage in the litter layer.

SIZE **Head–body length** ranges from 2.4–7.5cm (1–3in) in the shrew moles to 18–21.5cm (7–8.5in) in the Russian desman; **tail length** from 2.4–7.5cm (1–3in) to 17–21.5cm (6.5–8.5in), and **weight** from under 12g (0.4oz) to about 550g (19.5oz), both in the same species.

COAT Desman fur is double-layered, with a short, dense waterproof underfur and oily guard hairs; stiff hairs enlarge the paws and tail for swimming. Moles have short fur of uniform length, which lies in any direction during tunneling. Shrew moles have guard hairs and underfur directed backward. Moles are usually uniformly brownish black or gray, desmans brown or reddish on the back, merging to gray below.

DIET Moles live mainly on earthworms, beetle and fly larvae, and slugs; desmans on larvae, freshwater shrimps, snails, small fish, and amphibians.

BREEDING Gestation unknown (but greater than 15 days) in shrew moles and desmans; 30 days in the European and 42 days in the Eastern American mole.

LONGEVITY Up to 4 years in the European and Eastern American mole, 4–5 years in the Hairy-tailed mole.

CONSERVATION STATUS Two species – the Small-toothed and Persian moles – are Critically Endangered, and a further 5 species Endangered. The Japanese mountain mole and both species of desman are listed as Vulnerable.

For full species list see Appendix ▷

From Tunnel to Tunnel

SOCIAL BEHAVIOR

Little is known about the population density and social organization of desmans beyond the following sparse details: Pyrenean desmans appear to be solitary and inhabit small permanent home ranges, which they scentmark with latrines. There is some evidence that Russian desmans are at times nomadic as a result of unpredictable water levels. They may be social, since as many as eight adults have been found together in one burrow.

There is much more information on the moles, although it relates in the main to the European mole. There are equal numbers of males and females in populations of European moles, and populations appear to be relatively stable, unlike those of small rodents. Most mole species are solitary and territorial, with individuals defending all or the greater part of their home range. The Star-nosed mole is exceptional in that, for reasons that are poorly understood, a male and female may live together during the winter. In other species, males and females meet only briefly, to mate.

Population densities vary from species to species and from habitat to habitat. Between 5–25 European moles may be found per hectare; the equivalent figures for Eastern American moles are 5–12, for Star-nosed moles 5–25, and for Hairy-tailed moles 5–28.

Radio-tracking studies have shown that territory sizes in the European mole vary between the sexes, from habitat to habitat, and from season to season. The habitat and sex differences reflect differences in food supply and the greater energy demands of the larger male. Males increase territory size dramatically in the spring, as they seek out receptive partners. Normally, moles remain within their territories for their whole lives. However, radio-tracking has revealed that during hot, dry weather some moles leave their range and travel as far as 1km (0.6mi) to drink at streams. This entails crossing the territories of up to 10 other moles!

Moles' social organization is better understood than their ranges. European moles are territorial and spend almost their whole lives underground. Radio-tracking studies have shown that although neighboring moles each inhabit their own tunnel systems, territories do overlap to a small extent. Whether tunnels in overlapping areas are shared or separate, running between each other in the soil column, is not known. There is evidence, however, that moles are aware of the presence and activities of their neighbors. For example, during any particular activity period, neighbors forage in non-adjacent parts of their territories, avoiding contact. It is a testimony to the efficiency of this system of avoidance that conflict between established neighbors has not been recorded.

When a mole dies, or is trapped and removed, neighbors quickly detect its absence and invade the vacated area. For example, after several weeks of observation one radio-tagged animal was trapped and removed: within 12 hours a neighboring mole was spending its morning activity period foraging in the vacated territory, and his afternoon shift in his own territory.

In other cases, a vacated territory may be shared among neighbors. Such was the case with a group of four moles that occupied neighboring territories until one was removed. Within a matter of hours, the others had enlarged their territories to incorporate the vacated area. The enlarged territories were retained for at least several weeks.

Moles probably advertise their presence and their tenure of an area by scentmarking. Both sexes possess preputial glands that produce a highly odorous secretion that accumulates on the fur of the abdomen; this is deposited on the floor of tunnels, as well as at latrines. The scent is highly volatile and must be renewed regularly if the mole is to maintain its claim to ownership of the territory. In the absence of such a scent, the territory is quickly invaded.

Desmans are primarily nocturnal, but they often also have a short active period during the day. In contrast, most of the moles are active both by day and night. Until recently it was thought

that European moles always had three active periods per day, alternating with rest periods. Recent studies of moles fitted with radio transmitters reveal a more complex picture. In the winter, males and females do show three activity periods, each of about four hours, separated by a rest of four hours in the nest. At this time they almost always leave the nest at sunrise. Females maintain this pattern for the rest of the year, except for a period in summer when they are lactating. Then they return to the nest more often in order to feed their young.

Males are less predictable. In spring they start to seek out receptive females and remain away from the nest for days at a time, snatching catnaps in their tunnels. In the summer they return to their winter routine, but in September they display just two activity periods per day.

Life-cycle details are known only for a few species, again the European mole in particular; little is known about the desmans and shrew moles. In general, moles have a short breeding season, and produce a single litter of 2–7 young each year. Lactation lasts for about a month. The males take no part in the care of the young, and the young normally breed in the year following their birth.

In a number of species including the European mole, it is difficult to tell the two sexes apart. This is because the external sexual features are similar, a problem exacerbated by the fact that the testes do not descend into an external scrotum. Females of these species are exceptional among mammals in that they possess ovotestes. The ovotestis contains a morphologically normal ovarian component that develops during the spring and that produces eggs, but also a testicular region that enlarges during the autumn and produces high levels of testosterone. This may be the reason why the sexes are externally similar, and also why female moles are just as aggressive to other moles as males. Whether this physiological system conveys real advantages to the mole, or is just an evolutionary quirk, is unknown, and for the moment remains an unresolved mystery.

Above right The fate of many moles is to be killed by farmers. For centuries moles have been trapped, to reduce damage to fields and for their skins. Eradication campaigns are waged when they become serious pests, but moles' pelts no longer have any real commercial value.

Left Breaking cover, a European mole emerges from the middle of a molehill. The relatively huge, powerful, outward-facing forelimbs are clearly visible. Molehills– the most obvious sign of the presence of these animals – are often all that most people see of this elusive animal.

In Britain, European moles mate in March to May, and the young are born in May or June. The time of breeding varies with latitude: the same species is pregnant in mid-February in northern Italy, but not until May or June in northeast Scotland. This variation suggests that the length of daylight controls breeding, which may seem strange for a subterranean animal. However, moles do come to the surface to collect grass and other materials for their nests – one nest, located close to a licensed hotel, was constructed entirely of discarded potato-chip bags!

The average litter size in Britain is 3.7, but it can be higher in continental Europe; an average of 5.7 has been reported for Russia. One litter per year is the norm, but there are records of animals that were pregnant twice in one year (the second pregnancy coming in the fall). Sperm production lasts for only two months, but sufficient spermatozoa may be stored to allow the insemination of females coming into this late or second period of heat.

The young are born in the nest. They are naked at birth, have fur at 14 days, and open their eyes at 22 days. Lactation lasts for 4–5 weeks, and the young leave the nest weighing 60g (2.1oz) or more, some 35 days after birth.

After leaving the nest, the young leave their mother's territory and move overground to seek an unoccupied area. At this time many are killed by predators and by cars. With an average litter size of about four, the numbers of animals present in May must be reduced by 66 percent by death or emigration if the population is to remain stable.

Martyrs to Moleskin

CONSERVATION AND ENVIRONMENT

The Russian desman is widely trapped for its lustrous fur, and most of the skins in Western museums have come from the fur trade. To increase production the species was introduced to the River Dnieper.

All moles, particularly North American and European, are regarded as pests by farmers, gardeners, and greenkeepers. Moles disturb the roots of young plants, causing them to wilt and die; in addition, soil from molehills contaminates silage, and the stones they bring to the surface cause damage to cutting machinery. On the positive side, they have on occasion brought worked flints and Roman tesserae to the surface.

Nowadays, moles are usually controlled by poison. In Britain the most widely used agent is strychnine, which has the disadvantage of being unacceptably cruel and also endangers other animals: although available only by government permit, a disturbing amount is diverted to killing other forms of "vermin," including birds of prey.

In the past, professional trappers and agricultural workers trapped European moles on a massive scale and sold the skins to be made into breeches, waistcoats, and ladies' coats. There was a similar trade in moleskins in North America, particularly from the Townsend mole. This trade largely collapsed after the First World War.

The two extant species of desmans are both endangered: the Russian because of over-hunting for its fur, and the Pyrenean from overzealous scientific collecting and from habitat destruction, such as the damming of mountain streams. The formulation of management plans is hampered by a lack of knowledge of the basic biology of both of these species. MLG

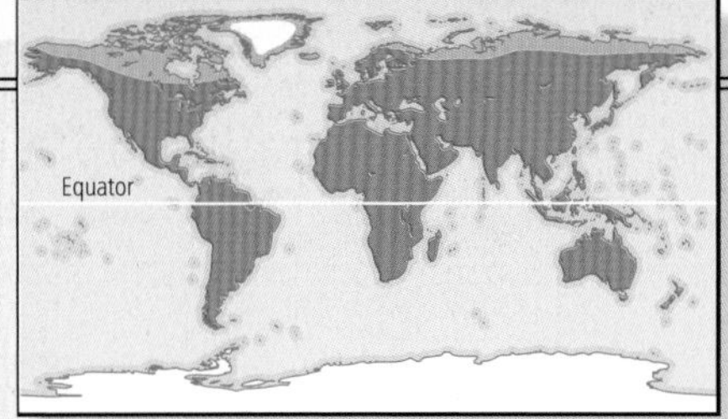

BATS

BATS ARE EXTRAORDINARY CREATURES. THEY *are exemplary mothers, and some also care for each other's young; they can fly at speeds of up to 50km/h (30mph) through complete darkness, thanks to sophisticated orientation systems; and one of the few convincing examples of altruistic behavior in the animal kingdom is exhibited by bats. Bats can show specialized reproductive adaptations, including sperm storage, delayed fertilization, and delayed implantation. They are heterothermic; their body temperature may vary from up to 41°C (106°F) in flight to under 2°C (36°F) during hibernation. They can form aggregations of 20 million animals – the largest known in vertebrates. Their wonderful diversity and specializations have inspired important programs for their conservation worldwide.*

Even so, bats rarely appear on people's top-ten lists of favorite animals. Perhaps their lack of public charisma stems from the same characteristics that make them so fascinating and unique. About one quarter of all living mammal species are bats. Houses, caves, mines, and trees in leaf all provide them with roosting sites, while some species make their own tents from plant parts. Most are largely nocturnal – small bats fly at night to avoid predators, while it is thought large megachiropterans do so to avoid overheating during the day. They are found everywhere except on the highest mountains and isolated oceanic islands, and have even been discovered breeding north of the Arctic Circle; on some islands such as the Azores, Hawaii, and New Zealand, they are the only native land mammals. Flight and echolocation have especially contributed to their diversity and cosmopolitan nature, permitting them to effectively exploit a food resource for which they have no animal competitors – insects that fly at night.

Mega and Micro Species

CLASSIFICATION AND EVOLUTION

Bats (order Chiroptera) are classified in two suborders, the Megachiroptera and Microchiroptera. Megachiropterans comprise a single family, the flying foxes (Pteropodidae), and are restricted to the tropics and subtropics of the Old World, where they feed mainly on fruit and nectar. The smallest megachiropterans weigh only 15g (0.5oz), while the heaviest weigh over 1.3kg (2.9lb). Flying foxes usually have a claw on their second fingers as well as on their thumbs, and their simple ears lack the tragus – a lobe in the ear – seen in many microchiropterans. Megachiropterans have well-developed vision and make little use of echolocation; the bats of only one Megachiroptera genus – the rousettes – use an echolocatory system, a relatively crude one employing tongue movements to produce largely ultrasonic double clicks of a few milliseconds' duration to aid orientation in dark caves. In addition, the Dawn bat slaps its wings in flight, perhaps to create echoes for similar reasons.

Above *An epauletted fruit bat* (Epomophorus *genus) rests at a roost in the Okavango Delta, Botswana. Its folded wings reveal not only the clawlike thumb possessed by nearly all bat species but also a longer claw on the second finger; this feature, found in most Megachiroptera genera, is never found in the Microchiroptera, the other division of the bat order.*

⮞ ***Right*** *The huge ears of this Bechstein's bat* (Myotis bechsteinii) *help in the hunt for insects. The 84* Myotis *species are found everywhere except the polar regions and some oceanic islands, and are reckoned to have the widest natural distribution of any group of terrestrial mammals other than the human race.*

Extraordinarily, some scientists have proposed that the Megachiroptera may in fact be more closely related to primates than to the Microchiroptera. As well as having no vocal echolocation, megachiropterans share advanced features of their vision with primates, and also have similar ratios between the lengths of their finger bones. This "flying-primate" hypothesis, which would require wings to have evolved independently in each of the two suborders, has, however, received little support from molecular studies, which instead emphasize the similarity between the bat groups, in that both have high levels of adenine and thymine bases in their genes. Biologists who support the flying-primate theory argue against this that the genetic resemblance may simply be a consequence of both suborders needing to consume high levels of energy in flight; in their view, the genomes of the two groups are similar because evolution has shaped them separately in similar ways rather than because they shared a recent common ancestor. The debate continues, but it remains true to say that most molecular studies now support a monophyletic origin for bats, by which flying foxes and microchiropterans would share a common ancestor.

The molecular evidence also suggests that horseshoe bats, along with such near relatives as the Old World false vampires, are more closely related to the Megachiroptera than they are to other Microchiroptera. Confusingly, however, horseshoe bats have some of the most sophisticated echolocation systems seen in the animal kingdom, while the Megachiroptera species have almost none. This hypothesis would therefore suggest either that horseshoe bats evolved echolocation independently of the other microchiropteran species, or alternatively that the various Megachiroptera species once had similar sensory capacities that they subsequently lost. Neither of these propositions seems very likely, and a third possibility is that the molecular evidence placing horseshoe bats close to the Megachiroptera may simply be misleading.

The earliest known fossil bat is the 50-million-year-old *Icaronycteris index*, which was found in Wyoming, USA. Superb fossil bats also dating from the Eocene have been found in the Messel beds in Germany, their stomachs containing identifiable fossils insects. Many of these early bats resemble living Microchiroptera, so earlier stages in their evolution are difficult to reconstruct, though the structure of the cochlea (inner ear) suggests that they almost certainly echolocated. The earliest fossils resembling megachiropteran

ORDER: CHIROPTERA

18 families, 174 genera, more than 900 species; figures change constantly as new species are described and taxonomic revisions are made.

Distribution Worldwide except for the highest mountains, some isolated islands, and extreme Polar regions (though some bats breed N of the Arctic Circle).

Size The smallest species is the Hog-nosed bat, with a body mass of 1.9g (0.07oz) and a wingspan of 16 cm (6.3in). Some flying foxes (*Pteropus* spp.) may exceed 1.3kg (2.9lb) in weight and have wingspans of 1.7m (5.6ft).

Diet In 70 percent of bat species, primarily insects and other small arthropods; the rest mainly subsist on fruit, nectar, and pollen. Some tropical species are carnivorous, and three species of vampire bats feed on blood.

Coat Variable; mostly browns, grays, and blacks.

Gestation Variable; can range from 3–10 months in a single species with delayed implantation.

Longevity Maximum 33 years; probably averages 4–5 years in many species.

SUBORDER MEGACHIROPTERA

FLYING FOXES Family Pteropodidae

41 genera and 164 species in the Old World, including: **Straw-colored flying fox** (*Eidolon helvum*); **rousettes** (*Rousettus* spp.); **Rodriguez flying fox** (*Pteropus rodricensis*); **Mariana flying fox** (*Pteropus mariannus*); **Large flying fox** (*P. vampyrus*); **Hammer-headed bat** (*Hypsignathus monstrosus*); **Dawn bat** (*Eonycteris spelaea*); **Short-nosed fruit bat** (*Cyanopterus sphinx*); **Long-tongued fruit bats** (*Macroglossus* spp.).

SUBORDER MICROCHIROPTERA

MOUSE-TAILED BATS Family Rhinopomatidae

1 genus and 3 species in the Old World, including: **Greater mouse-tailed bat** (*Rhinopoma microphyllum*).

HOG-NOSED BAT
Family Craseonycteridae

1 species in the Old World: **Hog-nosed bat** (*Craseonycteris thonglongyai*).

SHEATH-TAILED BATS
Family Emballonuridae

12 genera and 47 species in Old and New Worlds, including: **Greater white-lined bat** (*Saccopteryx bilineata*).

SLIT-FACED BATS Family Nycteridae

1 genus and 13 species in the Old World, including: **Large slit-faced bat** (*Nycteris grandis*).

OLD WORLD FALSE VAMPIRE BATS
Family Megadermatidae

4 genera and 5 species in the Old World, including: **Greater false vampire bat** (*Megaderma lyra*); **Yellow-winged bat** (*Lavia frons*).

HORSESHOE AND OLD WORLD LEAF-NOSED BATS
Family Rhinolophidae

10 genera and 129 species in the Old World, including: **Greater horseshoe bat** (*Rhinolophus ferrumequinum*); **Diadem leaf-nosed bat** (*Hipposideros diadema*); **Short-eared trident bat** (*Cloeotis percivali*).

NEW ZEALAND SHORT-TAILED BATS
Family Mystacinidae

1 species in the Old World: **New Zealand lesser short-tailed bat** (*Mystacina tuberculata*).

BULLDOG BATS
Family Noctilionidae

1 genus and 2 species in the New World: **Greater bulldog bat** (*Noctilio leporinus*); **Lesser bulldog bat** (*N. albiventris*).

SPECTACLED BATS Family Mormoopidae

2 genera and 8 species in the New World, including: **Parnell's moustached bat** (*Pteronotus parnellii*).

NEW WORLD LEAF-NOSED BATS
Family Phyllostomidae

48 genera and 139 species in the New World, including: **California leaf-nosed bat** (*Macrotus californicus*); **Greater spear-nosed bat** (*Phyllostomus hastatus*); **Fringe-lipped bat** (*Trachops cirrhosus*); **Seba's short-tailed bat** (*Carollia perspicillata*); **Tent-building bat** (*Uroderma bilobatum*); **Common vampire bat** (*Desmodus rotundus*).

FUNNEL-EARED BATS Family Natalidae

1 genus (*Natalus*) and 5 species in the New World.

THUMBLESS BATS Family Furipteridae

2 genera and 2 species in New World: **Smoky bat** (*Amorphichilus schnablii*) and **Thumbless bat** (*Furipterus horrens*).

DISK-WINGED BATS Family Thyropteridae

1 genus and 2 species in the New World: **Peter's disk-winged bat** (*Thyroptera discifera*) and **Spix's disk-winged bat** (*T. tricolor*).

OLD WORLD SUCKER-FOOTED BAT
Family Myzopodidae

1 species in the Old World: **Sucker-footed bat** (*Myzopoda aurita*)

VESPERTILIONID BATS Family Vespertilionidae

34 genera and 308 species in the Old and New Worlds, including: **Large mouse-eared bat** (*Myotis myotis*); **Daubenton's bat** (*M. daubentonii*); **Little brown bat** (*M. lucifugus*); **Common pipistrelle** (*Pipistrellus pipistrellus*); **noctule** (*Nyctalus noctula*); **Big brown bat** (*Eptesicus fuscus*); **Silver-haired bat** (*Lasionycteris noctivagans*); **Bamboo bat** (*Tylonycteris pachypus*).

PALLID BAT Family Antrozoidae

1 species in the New World: *Antrozous pallidus*

FREE-TAILED BATS Family Molossidae

12 genera and 77 species in the Old and New Worlds, including: **Brazilian free-tailed bat** (*Tadarida brasiliensis*); **European free-tailed bat** (*T. teniotis*); **Black mastiff bat** (*Molossus ater*); **Hairless bat** (*Cheiromeles torquatus*).

See family table ▷

species date to about 35 million years ago.

There are three main hypotheses to explain the evolution of flight and echolocation in bats. The first suggests that flying and finding food by biosonar are inextricably linked. Echolocation calls can contain considerable acoustic power, and are often energetically expensive to produce when a bat is at rest. When flying, however, many bats use the same muscular processes both for powering their wings and for emitting echolocation calls, which are thus produced at no extra cost. The link between flying and calling explains why many bats produce one intense call per wingbeat when searching for prey.

Other researchers have, however, suggested that flapping flight evolved before echolocation, or vice versa. Under either of these hypotheses, the initial function of echolocation was orientation, and it only later became refined for the more demanding tasks involved in the detection and localization of prey.

Comprising 17 families, the microchiropterans are ecologically diverse, particularly in the tropics. They include the smallest bat species, the Hog-nosed bat, which at about 1.9g (0.07oz) is the tiniest of all mammals.

The largest microchiropteran species are the New World false vampire bat (a phyllostomid, unrelated to the Old World false vampires), and the Hairless bat of Indo-Malaysia, which weigh in at 175g (6oz). All microchiropterans echolocate, some just to find their way about. The microchiropterans' small size may help them to produce the high-pitched frequencies needed for finding small insects; because ultrasonic echolocation is a short-range process, large bats may be unable to maneuver quickly enough to respond to echoes that can only be detected at close range. Moreover, large bats flap their wings slowly and so, because echolocation and wingbeat are coupled, produce calls at a rate that may be too drawn-out for efficient prey detection.

As biologists try to establish the exact relationships between the various types of bats, the number of ascribed species changes constantly. Molecular techniques have revealed several "cryptic" species that look very similar to other species but show large genetic differences. In Europe, the common and easily observed pipistrelle bat has recently been found to comprise two cryptic species that emit echolocation calls at different frequencies (see feature on Hidden Biodiversity).

Wings Fitted for the Job

Functional Morphology and Flight

Like other mammals, bats have heterodont (complex) dentition, including incisors, canines, premolars, and molars. Small insectivorous bats may have 38 teeth, while vampire bats have only 20 since they have no need to chew. Among insectivorous bats, species that feed on hard-bodied prey

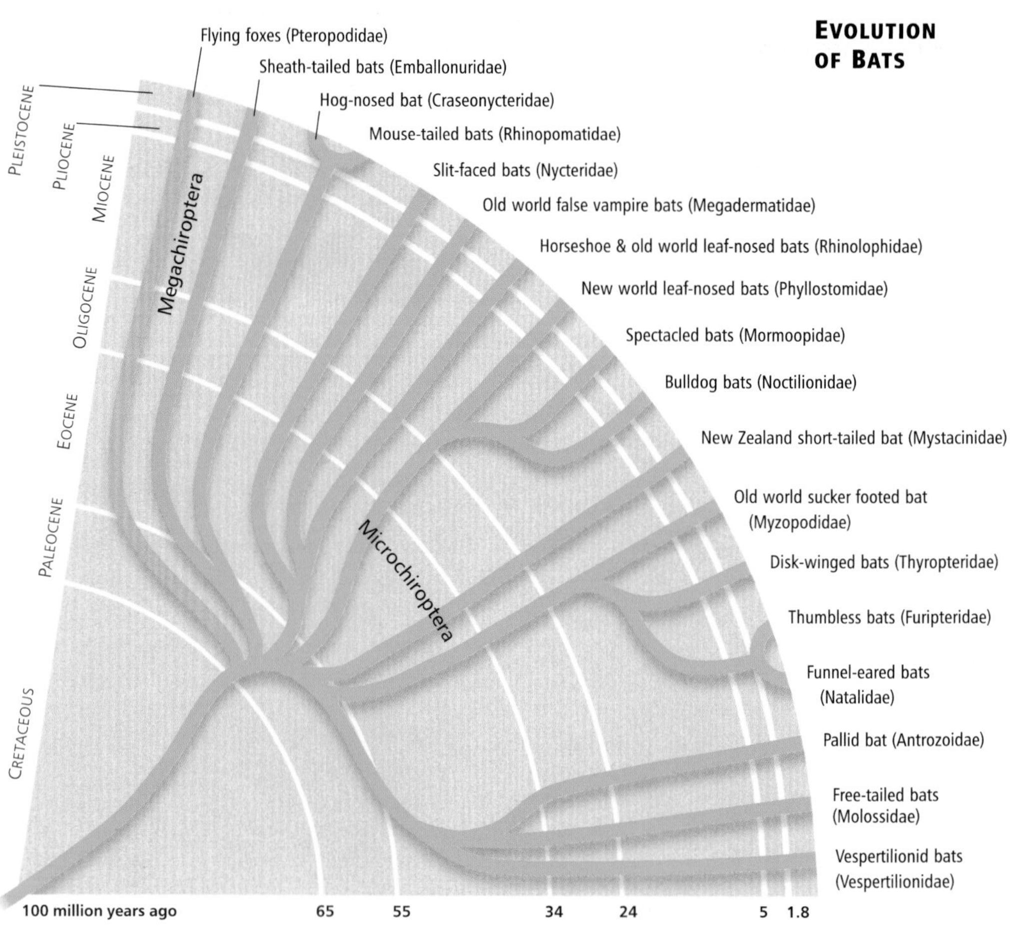

Above *The evolutionary relationships between bat families are controversial as the fossil record is poor. In particular, there are no known early flying foxes (Megachiroptera), so it is not known when the two suborders split, although most taxonomists now accept that they did once share a common ancestor.*

Below *The Australian false vampire* (Macroderma gigas) *is an oddity among bats in that it is carnivorous. After sunset it flies from its roost, usually in a cave or abandoned mine shaft, to a feeding site on a branch. There it hangs vertically, waiting to drop on mice and other small animals that pass underneath.*

BAT BODY PLAN

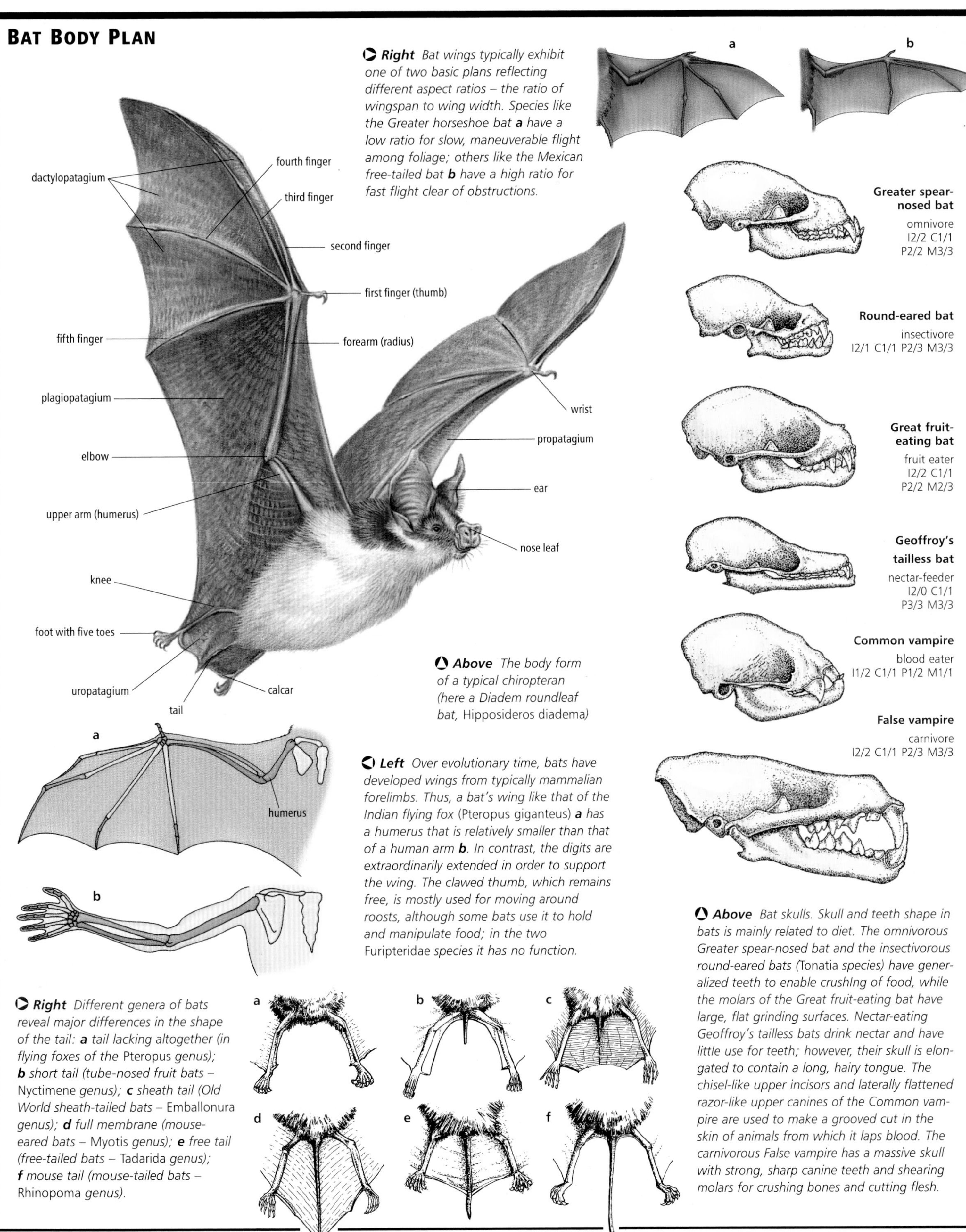

Right *Bat wings typically exhibit one of two basic plans reflecting different aspect ratios – the ratio of wingspan to wing width. Species like the Greater horseshoe bat* ***a*** *have a low ratio for slow, maneuverable flight among foliage; others like the Mexican free-tailed bat* ***b*** *have a high ratio for fast flight clear of obstructions.*

Above *The body form of a typical chiropteran (here a Diadem roundleaf bat,* Hipposideros diadema*)*

Left *Over evolutionary time, bats have developed wings from typically mammalian forelimbs. Thus, a bat's wing like that of the Indian flying fox (*Pteropus giganteus*)* ***a*** *has a humerus that is relatively smaller than that of a human arm* ***b****. In contrast, the digits are extraordinarily extended in order to support the wing. The clawed thumb, which remains free, is mostly used for moving around roosts, although some bats use it to hold and manipulate food; in the two* Furipteridae *species it has no function.*

Above *Bat skulls. Skull and teeth shape in bats is mainly related to diet. The omnivorous Greater spear-nosed bat and the insectivorous round-eared bats (*Tonatia *species) have generalized teeth to enable crushing of food, while the molars of the Great fruit-eating bat have large, flat grinding surfaces. Nectar-eating Geoffroy's tailless bats drink nectar and have little use for teeth; however, their skull is elongated to contain a long, hairy tongue. The chisel-like upper incisors and laterally flattened razor-like upper canines of the Common vampire are used to make a grooved cut in the skin of animals from which it laps blood. The carnivorous False vampire has a massive skull with strong, sharp canine teeth and shearing molars for crushing bones and cutting flesh.*

Right *Different genera of bats reveal major differences in the shape of the tail:* ***a*** *tail lacking altogether (in flying foxes of the* Pteropus *genus);* ***b*** *short tail (tube-nosed fruit bats –* Nyctimene *genus);* ***c*** *sheath tail (Old World sheath-tailed bats –* Emballonura *genus);* ***d*** *full membrane (mouse-eared bats –* Myotis *genus);* ***e*** *free tail (free-tailed bats –* Tadarida *genus);* ***f*** *mouse tail (mouse-tailed bats –* Rhinopoma *genus).*

tend to have larger and fewer teeth, with more robust mandibles and longer canines, than those that feed on soft-bodied insects. Nectar-eating microchiropterans have long snouts, large canines, and small cheek teeth, while frugivorous microchiropterans have modified cusps on their cheek teeth that operate like mortars and pestles to crush fruit.

Bats are the only mammals capable of powered flight. Although flight is energetically expensive in terms of units of time, the energy expenditure per unit of distance covered is low. Bats can therefore fly considerable distances and exploit food over a wide range, allowing them to migrate and exploit farflung parts of the globe.

The finger and arm bones in mammals have evolved into many different types of tools, but none is perhaps so unusual as the bat's wing. The thumb is free, while the fifth digit spans the entire width of the wing. The other three digits support the area of the wing between the thumb and fifth digit, which is known as the dactylopatagium, or "hand wing." The upper arm bone (the humerus) is shorter than the major forearm bone (the radius), and the wing area supported by these bones is called the plagiopatagium, or "arm wing." During flight, inertial forces are greatest on the arm wing.

Many species of microchiropteran also have a tail membrane. This feature, which is absent in *Pteropus* species and reduced in the mouse-tailed bats, is most developed in species such as the slit-faced bats (Nycteridae) that use it to scoop prey from surfaces. The tail membrane is also supported by the legs. Fishing bats and species such as Daubenton's bat that trawl insects from the water's surface often capture prey with the claws on their feet.

Bats' legs project sideways and backward, and the knee bends back rather than forward as in other mammals. The legs are adapted for pulling rather than pushing; the lower leg is formed of a single bone, the tibia. Resting bats hang with their weight suspended on the toes and their well-developed claws. Most bats have a tendon-locking mechanism that keeps the claws bent without the need for muscular contraction. Hanging upside down allows bats to take flight rapidly from a resting position.

Some species such as Common vampires and the New Zealand short-tailed bat crawl on all fours, a feat of which horseshoe bats and others are incapable of accomplishing. Common vampires often approach prey when on the ground, while the New Zealand short-tailed bat evolved in the absence of small terrestrial mammals and with few predators, and so may fill a more terrestrial niche than other species.

The wingbeat of bats is configured primarily to generate thrust, most of it from the hand-wing. Visualizations of the airflow behind flying bats shows that the downstroke generates lift at all speeds, while the upstroke becomes active only at high speed when the bat changes gait.

Wing-shape profoundly influences bats' flying performance. Two aerodynamic properties are especially significant – wing loading and aspect ratio. Wing loading describes the ratio of weight to wing area; a high loading – in other words, small wings for a given weight – means fast flight but limited maneuverability. Aspect ratio is a measure of the wing's relative width, and is calculated as wingspan squared divided by wing area; a high aspect ratio wing is long and narrow and experiences little drag. High aspect ratio wings are therefore efficient, and are often associated with high wing loading and fast flight.

The shape of the wings helps determine where different species can survive. Those that fly in obstacle-rich habitats such as woodlands need to be maneuverable, and therefore have low wing loadings. Species like the noctules and Hoary bats that operate in open areas, however, need to fly

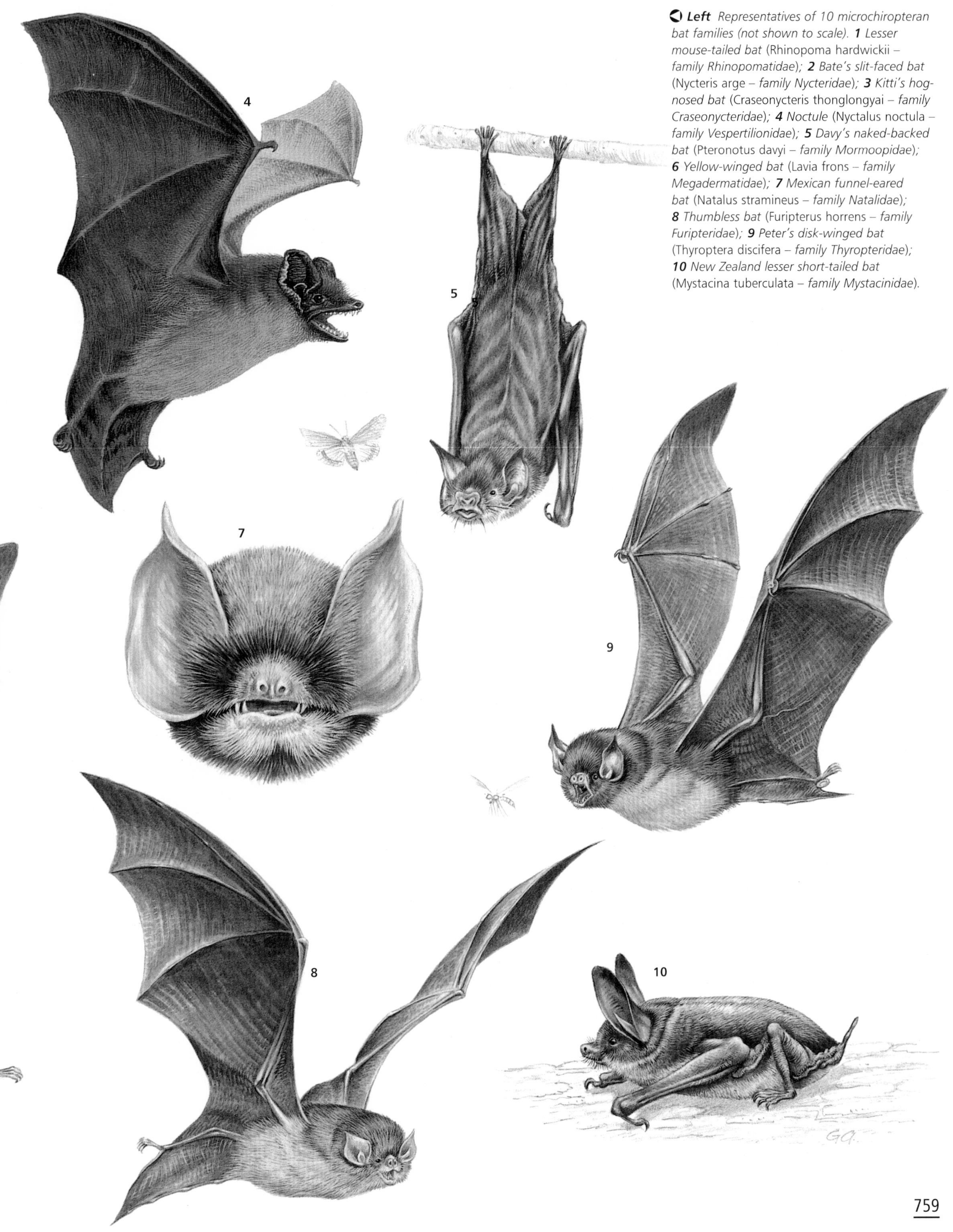

◐ ***Left*** *Representatives of 10 microchiropteran bat families (not shown to scale).* ***1*** *Lesser mouse-tailed bat* (Rhinopoma hardwickii – *family Rhinopomatidae);* ***2*** *Bate's slit-faced bat* (Nycteris arge – *family Nycteridae);* ***3*** *Kitti's hog-nosed bat* (Craseonycteris thonglongyai – *family Craseonycteridae);* ***4*** *Noctule* (Nyctalus noctula – *family Vespertilionidae);* ***5*** *Davy's naked-backed bat* (Pteronotus davyi – *family Mormoopidae);* ***6*** *Yellow-winged bat* (Lavia frons – *family Megadermatidae);* ***7*** *Mexican funnel-eared bat* (Natalus stramineus – *family Natalidae);* ***8*** *Thumbless bat* (Furipterus horrens – *family Furipteridae);* ***9*** *Peter's disk-winged bat* (Thyroptera discifera – *family Thyropteridae);* ***10*** *New Zealand lesser short-tailed bat* (Mystacina tuberculata – *family Mystacinidae).*

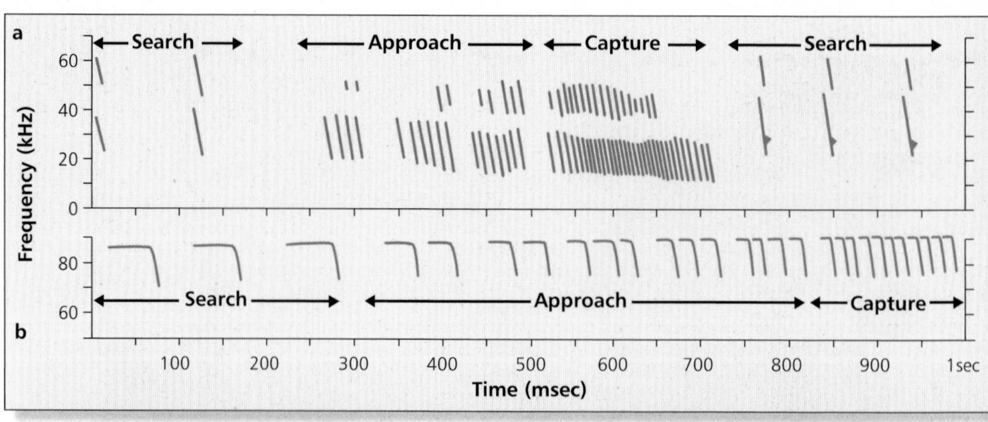

Right Sonograms show the search and capture phases of hunting activity in two bat species that use echolocation to track down insect prey. The North American Big brown bat **a** produces broadband calls (see text) that sweep steeply from 65–20kHz at a rate of 5–6 per second while the bat is foraging. When an insect is located, the rate accelerates rapidly, up to a peak of about 200 pulses a second. In contrast, horseshoe bats **b** produce higher-frequency narrowband calls – well suited for detecting prey but not so accurate when homing in on a target – at a rate of about 10 per second during the search phase. Their calls also speed up on approach, although less dramatically; in addition, they rise slightly in frequency.

fast and efficiently, and so have high aspect ratio wings that confer high wing loadings. The flight speed of small- to medium-sized bats searching for prey varies between 3–15m (10–50ft) per second, and species with the highest wing loadings fly the fastest. Migratory species also have high aspect ratios; Brazilian free-tailed bats may migrate over more than 1,000km (620mi) from the southeastern United States to overwinter in Mexico. In contrast, some nectar-feeding bats have such low loading that they can hover.

Steering by Sound

SENSES

Bats are not blind. Megachiropterans use their large eyes to locate food and to orientate themselves; in low light, flying foxes can see better than humans. A few microchiropteran species also see well; for example, the Californian leaf-nosed bat flexibly switches off echolocation in adequate light and instead uses vision to locate prey. Most microchiropterans, however, do not see well, and their echolocation range is usually more limited than small mammals' vision; a medium-sized bat may detect a beetle at about 5m (16.5ft). Large landmarks may be detected at greater range – perhaps 20m (65ft). Bats are thus susceptible to predation in the daytime, a risk that may explain why most echolocating bats are nocturnal.

Bat have also not lost the sense of smell that is so vital to most mammals. New World leaf-nosed bats locate ripe pepper-plant fruits first by using olfaction, relying on echolocation only at close range. Large mouse-eared bats and New Zealand short-tailed bats may sniff out prey buried in leaf litter. Olfaction is also used in communication; for example, Soprano pipistrelles use smell to distinguish other bats in their maternity roosts from strangers, and female Brazilian free-tailed bats sometimes use olfactory cues to identify their offspring in crèches. Male Greater white-lined bats store secretions from their genital region and gular (throat) glands, together with urine and saliva, in glands in the wing. When courting, they hover in front of females, seducing them with the odor, and females gather about the males that perform the greatest number of these dances. What exactly attracts the females to the smell is unknown.

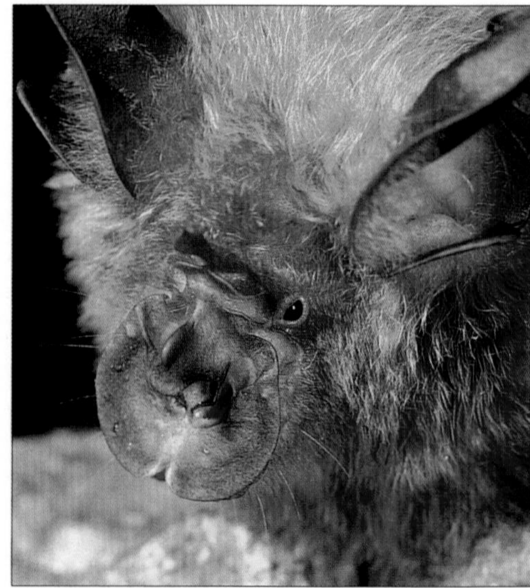

Above and right *Bats rely on their echolocating abilities to succeed as nocturnal insect-hunters. Many have large ears, such as those that give the Brown big-eared bat (above) its name, and curiously shaped noses, as on this Eastern horseshoe bat (right), which are used to focus the ultrasonic squeak generated in the larynx.*

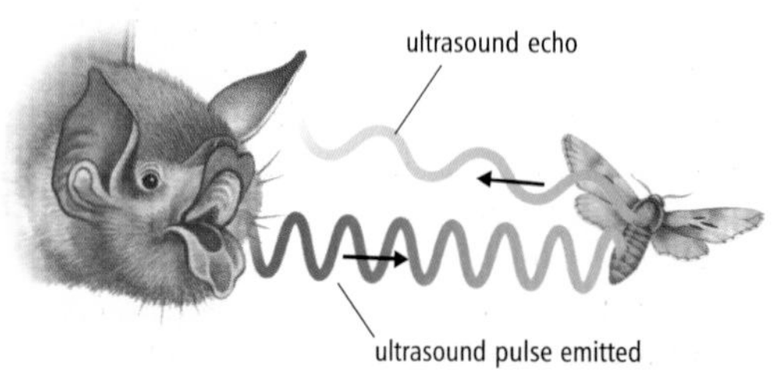

Below *By analyzing the ultrasound waves bouncing back off a flying insect, the bat can pinpoint it in total darkness, and detect its direction and size.*

Bats' auditory powers are exceptional; some are capable of hearing an insect walking on a leaf. Old World false vampire bats have the most sensitive hearing yet discovered in mammals, particularly in the frequencies between 10 and 20kHz produced by prey rustling through vegetation. In cluttered environments such as forests in leaf, listening for prey-generated sounds is a more effective tool than echolocation, so many gleaning bats rely on their ears to find food.

Bats also converse. Social calls may serve to attract mates, defend feeding patches, attract conspecifics, perhaps aid in repelling predators. These calls are often pitched at low frequencies (many are audible to humans) so they can travel over considerable distances, and they may be individually distinctive; for example, infant bats utter "isolation calls" when separated from their mothers, which can apparently identify their own infants from their vocalizations. Unsurprisingly, many microchiropterans are most sensitive to the sound frequencies they use in echolocation.

Above *Like many microchiropterans, Natterer's bat* (Myotis nattereri) *hunts at night and rests by day. Such bats rely chiefly on their sophisticated sense of hearing to navigate, using the ultrasonic echoes bounced back from high-frequency calls to draw a mental map of their continually changing surroundings.*

In 1793 an Italian, Lazzaro Spallanzani, discovered that bats lost their sense of direction when deafened, but could still find their way about when blinded. Spallanzani never discovered how bats oriented themselves, but he did propose that they used their ears to hunt insects. Details of how microchiropterans orientate and detect prey remained unresolved until 1938, when Donald Griffin of Harvard University listened to bats with a microphone that could detect ultrasound.

Bats emit sounds from their larynxes, like humans, and these bounce back off their surroundings; the echoes are then used to create a mental map of the area through which the bat is traveling. Bat echolocation calls are usually ultrasonic – in other words, above the human range of hearing. Among other feats, bats can determine the location of an object to within 2–5° in the horizontal and vertical planes, and can detect objects as small as 1mm (0.04in) in dimension, as well as distinguishing between two identical objects 12mm (0.5in) apart. In addition, Greater horseshoe bats can determine the wingbeat rate of flying insects with a high degree of accuracy.

The ornate faces of bats reflect the importance of sound in their lives. Many bats have a lobe in

the ear (the tragus) that helps them to pinpoint echoes in the vertical plane. Although most bat species emit calls through their mouths, some, like the horseshoe bats, the Old World false vampires, and the Old and New World leaf-nosed bats, transmit sound through their nostrils, in effect calling through their noses. These bats have elaborate ornamentation around their nostrils to direct and focus sound.

From a bat's perspective human responses would seem sluggish, for bats process information at a phenomenal rate, varying the structure of their calls according to the tasks required of them. Search calls are intense, and are often emitted at a rate of one call per wingbeat, which equates to 5–15 calls a second for many species. Once a target insect has been located, the calling gets faster; homing in, the bat passes from the approach phase of echolocation to the terminal phase, in which the shortest calls of all are produced, at rates of up to 200 per second. After the moment of capture there is a pause as the prey is handled, although bats can if necessary continue echolocating while chewing food.

Bats have a "duty cycle" when echolocating that represents the proportion of time actually spent generating sound. Some bats spend only 20 percent of their time echolocating, because they cannot process outgoing pulses and returning echoes at the same time; to do so, they would run the risk of deafening themselves, for echolocation calls can have a volume equivalent in bat terms to that exerted by a pneumatic drill on human ears. To counter the danger, the middle-ear muscles contract, acting as an effective earplug when calls are being made. These bats must leave long gaps between the sounds they emit, during which they process the returning echoes; they also shorten the length of the pulse as they approach targets, so that outgoing pulses do not overlap with echoes that come back all the sooner as objects are approached. To separate the outgoing pulse and the returning echo in time to avoid overlap, their calls are usually shorter than 30 milliseconds in duration. Thus bats shout and listen, then shout and listen again, at a rate faster than humans can well imagine – a fact that explains

Above *A foraging Greater horseshoe bat homes in on a moth – a staple item of its diet. Studies using radio-tracking devices have revealed much about the behavior and roost selection of this species. For example, two of its key habitats are woodland and adjacent old pasture, both of which are rich in insect life, yet which are in decline.*

KEEPING AN EAR OPEN FOR BATS

How hearing helps moths evade echolocating predators

OBSERVED UNDER AN ISOLATED STREET-LIGHT, a flying moth spirals toward the light. Suddenly the moth changes its flight behavior and dives for the ground, just before a bat zooms through the air along the moth's original flight path. The simple explanation is that the moth heard the bat coming and took evasive action; more specifically, it responded to the echolocation or biosonar calls emitted by the bat. The change in the moth's flight behavior in response to the bat´s imminent attack is a revelation in the remarkable world of the ongoing war between predators and prey.

Prey species from a variety of groups including many moths, some lacewings, and a selection of crickets, katydids, mantids, and beetles can detect echolocation calls, indicating that bat-detecting ears have evolved independently many times in insects. In most cases, bat sounds appear to be the only ones that many of these insects heed. Calling insects, however, such as crickets, katydids, and some moths, not only listen to their enemies' echolocation calls but also to the communication signals of their own kinds.

Bat-detecting ears derive from pre-existing sense organs – the so-called chordotonal organs – and occur in different parts of the insects' bodies, from the head, thorax, and abdomen to the wings and legs. Most insect ears are elaborate enough to allow the hearer to distinguish strong from weak signals, and insects use the strength of impulses on the auditory nerve to assess the proximity of the bat and the threat it poses. Moths can hear bats from about 20–40m (65–130ft) away, although bats can normally only detect moths within about 5m (16.5ft). Moths that simply change their flight direction in response to a distant bat will launch themselves into an unpredictable repertoire of spirals, loops, and dives that usually terminate on the ground when the bat is in close range. The ultrasonic echolocation calls used by bats attenuate rapidly in air, and since the sounds detected by the bats are echoes that have traveled two ways instead of one, the insect will nearly always be at an advantage. Insects with pairs of ears show a very general left/right directional response to a bat's echolocation calls; mantids, however, have only a single ear, and so cannot determine the direction of the sound.

Advances in the study of bat echolocation and moth hearing indicate that there is more to the bat–insect story than first meets the ear. On the bat side, echolocators can use very strong or very weak calls, and different species use calls dominated by different acoustic frequencies. Bats using strong calls are aerial feeders hunting flying insects, which they detect, track, and evaluate through the use of echolocation. The intensity of the calls is impressive, ranging up to 125dB measured 10cm (4in) from the bat´s mouth. Bats that employ such strident calls include noctules, serotines, pipistrelles, Big and Little brown bats, the Red and Hoary bat, and many others. On the other hand, some insect-eating bats that feed near and within vegetation use low-intensity (60–80dB) calls. These "whispering" bats – which include slit-faced bats, false-vampire bats, many mouse-eared bats, long-eared bats, and New World leaf-nosed bats – detect and track their prey by the sounds of fluttering wings or insect footfalls rather than by echolocation.

On the insect side, moths do not hear all sounds equally well. For example, the quiet echolocation calls of whispering bats are not picked up until the bat is within 1m (3.3ft) or so – too late for the moth to take evasive action. Most moths hear best at frequencies of 20–60kHz, the frequency band used by most bats. But some aerial-feeding species, including many horseshoe bats, Old World leaf-nosed bats, and a few free-tailed bats, use echolocation calls dominated by frequencies either above 60kHz or below 20kHz. Although these calls may be intense, they are not conspicuous to moths because they contain frequencies that the moths cannot detect easily.

One study showed that moths with ears were 40 percent less likely to be caught by bats than moths that had been experimentally deafened. Within the diets of aerial-feeding bats, moths are more commonly food for those species whose calls are not conspicuous. This generalization requires qualification, however, for some bats with conspicuous echolocation calls eat a lot of moths. Red and Hoary bats, for example, often hunt around street-lights, which not only attract insects but also make them easy to catch by interfering with their defensive responses when they are looking into the light. Likewise, the barbastelle seems to specialize on deaf species of moths.

Some moths have escalated the sound-based arms race with bats. Many species of tiger moths have an unpleasant taste, because their caterpillars eat plants containing secondary compounds serving to protect them from herbivores. These moths are often brightly colored and conspicuous, defensive signals apparently directed at predators like birds, which learn to associate the bright colors with an unappetizing taste. Some of these moths also have noisemakers, and the clicks they produce are often directed at bats. Hunting Red and Hoary bats will normally approach, but then veer away from, flying Painted lichen moths; if the moths' noisemakers are damaged, however, the bats will seize them, only to quickly release them unharmed, presumably because they recognize the smell or taste immediately. So the clicks emitted by the moths apparently serve as warnings to the bats that the moths are inedible. In time and frequency parameters, the clicks of tiger moths match the intensity, frequency pattern, and duration of echo of the bats' calls, and are therefore well designed to get an echolocating bat's attention. It has also been suggested that the clicks work by startling the bat or else interfere with its information-processing by disturbing the echoes.

Some insects use ears for listening to sounds other than bat echolocation calls, and there are also many nocturnal insects that lack bat-detectors. However, these too have various means of evasion. Some deaf moths practice "acoustic concealment" by hiding among the massive echoes from the vegetation on which they feed. Others are particularly fast and swift flyers, and use an elevated body temperature to achieve the same end. Many dipterans and mayflies hide in dense swarms of conspecifics.

The most efficient of all means to avoid bats is, of course, to fly by day, when bats do not hunt. This may in fact explain why some moths long ago turned into butterflies. Moths apparently did indeed evolve many new traits during the Eocene and Oligocene eras 30–50 million years ago, the period when the first echolocating bats evolved.

Meanwhile, male moths flying toward a female releasing a plume of pheromones will disappear rapidly when presented with bat-like sounds. Bats are clearly enemies to taken seriously if, for some moths at least, avoiding them even takes precedence over sex!

JR/BF

Below *Moths evade bats in several ways. Some zig-zag* ***1*** *to escape attack; others drop to the ground, either passively* ***2*** *or in a powered dive* ***3****, where they may be difficult to detect among plants and stones. Certain unpleasant-tasting tiger moths emit clicks* ***4*** *that apparently warn attackers that they are not prey species, causing the bats to veer away.*

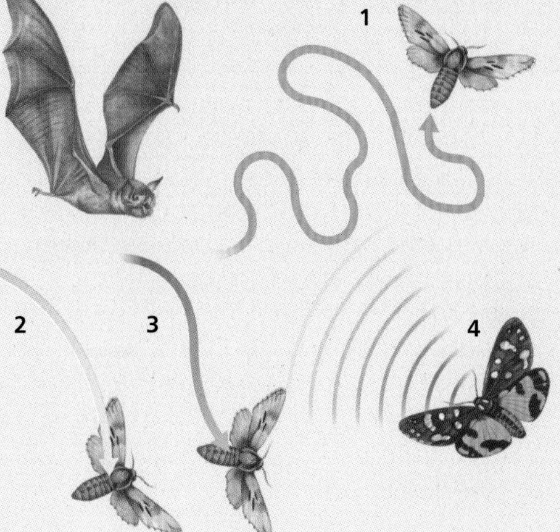

1

2

3

5

why they can catch flies in their mouths while most people have difficulty even swatting them with their hands.

Horseshoe bat calls can be 70 milliseconds long, and yet these bats can forage in the forest canopy, where they are bombarded with echoes from objects at close range. Long pulses are well suited for the detection and classification of targets, but most bats cannot benefit from this advantage as they must take time to listen in between calls. Horseshoe bats, however, which echolocate for more than 50 percent of the time, manage to do so by separating call and echo by frequency rather than in time. They can do so because their ears pick up sounds at an optimal pitch slightly lower than that at which they generate the calls. Their auditory system has evolved to be maximally sensitive to frequencies close to those emitted when the bat is resting – about 83kHz for a Greater horseshoe bat.

In effect, horseshoe bats can "shout" at the same time as listening. They are able to do so through a specialization known as "Doppler shift compensation." Humans can perceive Doppler shifts as an increase in the frequency of, say, an ambulance siren as the vehicle approaches and a decrease as it travels away. Sound waves compress more the faster a bat flies, so frequency increases; in turn, the bats lower the frequency of each call in relation to their flight speed in order to compensate for Doppler shifts. Echoes always return at about 83kHz, but because the emitted call frequency is lowered, sending and receiving are performed on different bandwidths. Thus call and echo are separated in frequency, and the bats can use long calls.

Echolocation calls function to detect, localize, and classify targets, and bats are able to vary from frequency modulated (FM, or broadband) to constant frequency (CF) mode to carry out these different tasks. Only rhinolophid bats and Parnell's mustached bat emit CF calls that are associated with Doppler shift compensation. Calls that are almost constant in frequency are termed "narrowband." Broadband calls sweep through a wide range of frequencies in a short time (almost 100–20kHz in under 2.5 milliseconds in a Natterer's bat), and are well suited for the localization of targets but are not so good at target detection (finding prey in the wider landscape). Narrowband and constant frequency calls, in contrast, are ideal for the detection of targets but are poorly suited for localization. Hence some bat species such as the pipistrelles use narrowband calls to detect insects but switch to broadband calls to locate them. Horseshoe bats add broadband components to their CF calls and extend the bandwidth of the terminal broadband component during localization.

Long CF calls are also used to classify targets. Insect wingbeats produce small modulations in the frequency and amplitude of echoes (glints). Because small insects flap their wings more rapidly than large ones, glint frequency relates to insect size. Horseshoe bats can distinguish profitable prey items from unprofitable ones by glint characteristics, and eat unprofitable items such as small dung flies only when profitable prey such as moths and large beetles are scarce.

Most bat species call by using frequencies in the 20–60kHz range. Frequencies lower than 20kHz have wavelengths larger than most insects, and so travel round them rather than bouncing back to the bat. Frequencies higher than 60kHz attenuate rapidly in air, which restricts their useful range, although for slow-flying bats that can react more easily at close range they are ideal. The Short-eared trident bat emits the highest-frequency call of all, at 212kHz, whereas some free-tailed species, such as the Spotted bat, echolocate at frequencies as low as 11kHz. Another advantage of using exceptionally high or low frequencies is that these are poorly detected by insect prey, allowing bats that call at these frequencies to catch large numbers of insects that are otherwise equipped to hear ultrasound.

An insect's life depends on whether or not it survives a predation event, so it is hardly surprising that insects have evolved defenses against echolocating bats. Some of these are behavioral; for example, male Ghost swift moths, which display at leks, do so early in the evening, when predation risks from both bats and diurnal birds are minimized. The moths display by flying close to the ground, where their acoustic profiles are masked by ground echoes; if they fly above the highest grass panicles, they lose their acoustic crypsis and are readily caught by bats.

Ghost swifts are primitive moths that lack hearing organs, but sensitivity to ultrasound has evolved in orthopterans, lacewings, beetles, flies,

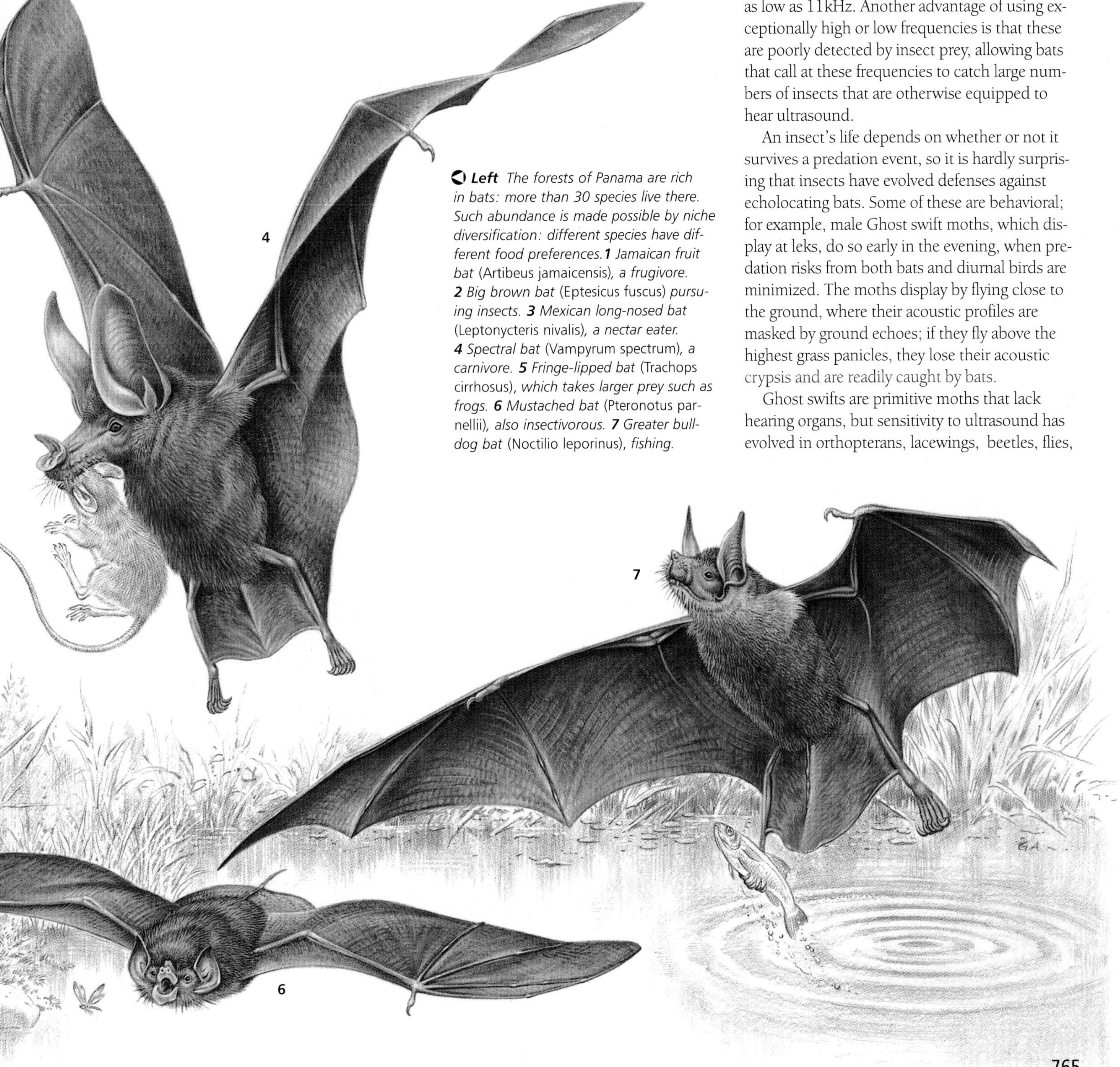

Left *The forests of Panama are rich in bats: more than 30 species live there. Such abundance is made possible by niche diversification: different species have different food preferences.* ***1*** *Jamaican fruit bat* (Artibeus jamaicensis), *a frugivore.* ***2*** *Big brown bat* (Eptesicus fuscus) *pursuing insects.* ***3*** *Mexican long-nosed bat* (Leptonycteris nivalis), *a nectar eater.* ***4*** *Spectral bat* (Vampyrum spectrum), *a carnivore.* ***5*** *Fringe-lipped bat* (Trachops cirrhosus), *which takes larger prey such as frogs.* ***6*** *Mustached bat* (Pteronotus parnellii), *also insectivorous.* ***7*** *Greater bulldog bat* (Noctilio leporinus), *fishing.*

Bat Blood Donors

Feeding and sharing in vampire bat colonies

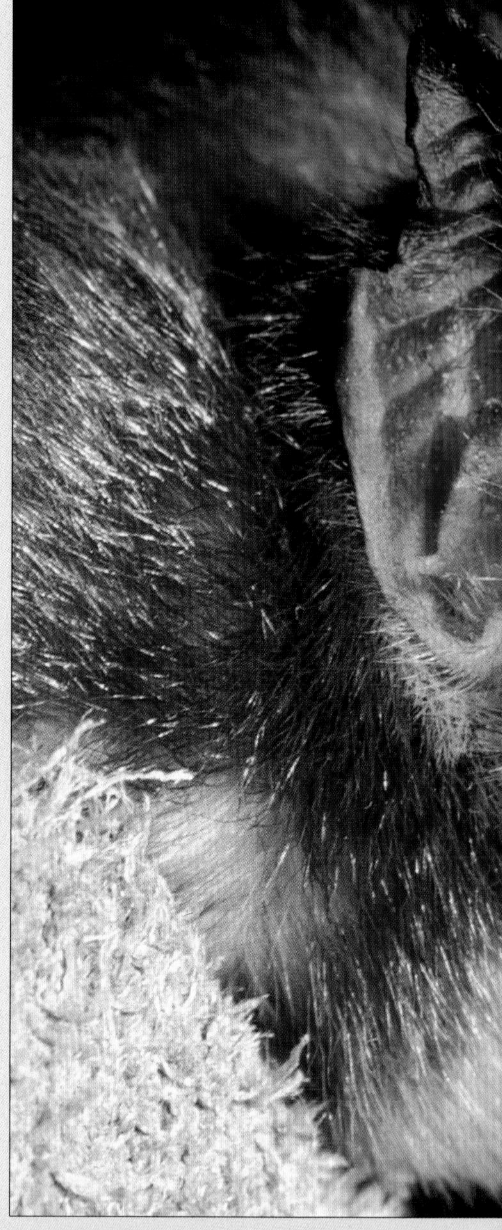

No species have contributed more to the misunderstanding and fear of bats than the vampires. Three species of vampire bats occur in Central and South America, where, as their name implies, they feed exclusively on blood. Hairy-legged (*Diphylla ecaudata*) and White-winged vampire bats (*Diaemus youngi*) favor bird blood, and are adept at climbing tree branches to feed on roosting chicks. Common vampires (*Desmodus rotundus*) prefer mammal blood, and are usually found where cattle, horses, and other livestock are common. If livestock are absent, they feed instead on tapirs, deer, peccaries, agoutis, and sea lions.

People have some reason to fear vampire bats. Attacks on humans do sometimes occur, often following the removal of livestock from an area. Vampire bites are not painful, but they can be dangerous, for they are known to transmit paralytic rabies. Because the bats themselves are also susceptible to the virus, their populations undergo periodic crashes in response to rabies epidemics. The remarkable blood-sharing behavior of the bats – by which well-fed bats regurgitate blood to hungry companions – almost certainly facilitates the transmission of saliva-borne viruses, including the one responsible for rabies.

Vampire blood-sharing is a rare example among animals of reciprocity (the "You scratch my back, I'll scratch yours" principle). Understanding why bats would share food even at the risk of infection requires an appreciation of the social organization and life history of these extraordinary creatures.

Common vampire bats often use caves, tunnels, or hollow trees as day roosts. While some roosts may contain more than 2,000 individuals, colonies most often contain 20–100 bats. Within a colony, groups of 10–20 females often roost together for years. Some of these females are related, because female offspring remain with their mothers after reaching sexual maturity in their second year. However, female groups also contain unrelated animals, as a consequence of adult females occasionally switching day roosts. Groups of up to ten males also occur, but they are not related and do not remain together for extended periods. Young males from 12–18 months of age disperse independently, often after fighting with an adult male in their natal group. Typically one adult male roosts with females and their young, while others hang nearby and periodically fight viciously to gain access to the females. On average, a male in a preferred spot fathers half of the pups born to females in that group, retaining control of his position for about 2 years. A roosting group of Common vampire bats typically consists, therefore, of a few unrelated adult males and of sets of females with young related through different matrilines.

For their size, vampires expend more time caring for their young than any other bat. Females give birth to single pups weighing nearly 20 percent of their body weight (30–35g/1–1.2oz). Although the pups are active at birth, they grow slowly and continue to receive milk from their mothers for over 6 months. Females supplement their diet with regurgitated blood shortly after birth, and also periodically during the first year. Pups begin to fly after 6 months, but do not reach adult weight until they are 1 year old.

Common vampires use smell and sound to find prey. Females from a roosting group hunt in adjacent areas, and will defend bite sites by chasing other bats away. Even when prey are abundant, successfully obtaining a blood meal can be difficult. To make a bite, a bat must first locate a warm spot where blood vessels are near the surface of the skin, using heat receptors located on its nose pad. It then uses its razor-sharp incisors to remove a small piece of skin, rather like a golf divot. Anticoagulants in the bat's saliva ensure that the blood flows freely as the vampire laps up the blood with its tongue. Bats' feeding skills improve with age: those aged 1–2 fail to feed one night out of three on average, but those over 2 years old are unsuccessful only one time in ten. Failure results from the wariness of the animals under attack, which will sometimes try to brush off the feeding bats. Not surprisingly, young bats sometimes feed simultaneously or sequentially from the same wound site as their mothers, and individuals may return to the same wound site on consecutive nights.

If a bat fails to obtain a meal, it will return to the roost and beg blood from a roost mate by licking its lips. The likelihood that a bat will regurgitate and share blood depends on its association and kinship with the hungry bat. Bats do not share blood unless they have roosted together for more than 60 percent of the time. Some, but not all, blood-sharing events involve related bats.

Failing to feed is risky, as bats that go hungry starve to death within 3 days. Because starving bats lose weight more slowly than recently-fed bats, the transfer of blood buys the recipient more survival time than the donor loses. Reciprocal blood-sharing therefore results in a net benefit to participating bats. In the absence of reciprocity, annual mortality should exceed 80 percent, yet female vampire bats are known sometimes to survive for more than 15 years in the wild.

One problem for the donor is knowing how to ensure that the recipient is a genuine reciprocator and not a cheat who will receive blood without giving it in return. One way bats can at least assess each other's hunger levels is during episodes of mutual grooming. Bats that have successfully fed typically ingest over half their body weight in blood in a 30-minute period, which causes their stomach to bulge. This gut distention is likely to be noticed by another bat in the course of grooming, which frequently occurs just prior to blood regurgitation. Since both mutual grooming and blood sharing only occur between individuals that have reliably roosted together, partner fidelity appears to be essential for the persistence of this amazing reciprocal-exchange system. JW

Above *A Common vampire bat* (Desmodus rotundus) *reveals the razor-sharp incisors that make the puncture marks through which the bat laps its prey's blood. The truncated muzzle allows the bat to press its mouth close up against the flesh of the animal on which it is feeding.*

Left *In Trinidad, a vampire bat takes blood from a resting donkey. To feed, the bat first of all chooses a site on an animal's skin where a blood vessel is close to the surface. It then licks the spot with its tongue before using its teeth to shear away protective hairs or feathers. Finally it removes an almost circular patch of skin to access the blood beneath.*

Right *Bats cluster in a roost within a crevice. Although vampires sometimes gather in colonies up to 2,000 strong, these large agglomerations subdivide into smaller units, typically composed of groups of 10–20 females and their young. Blood-sharing is normally limited to these close-knit groupings.*

and mantids, as well as several times in other moth species. Many insects show startle responses on hearing ultrasound pulses, flying away from those exhibiting low sound intensities (probably corresponding to distant predators), but sometimes making dives or spiral flights when alerted by intense sounds. Some arctiid moths and tiger beetles click in response to ultrasound pulses, perhaps in order to startle the predator, to interfere with echo processing, or else to warn the bat that they are distasteful prey.

Many insects clearly evolved ears as bat detectors. Whether bats have in turn responded by changing their calls to increase their chances of catching these insects remains controversial. Some bats, such as those that use very high- or low-frequency echolocation calls, catch many prey that can hear ultrasound. Exploitation of these insects could simply be a by-product of other advantages associated with these calls, however, rather than reflecting specific adaptations to facilitate predation on hearing insects.

Helping Out with the Young

BREEDING AND REPRODUCTION

Many tropical bats are polyestrous, undergoing several reproductive cycles in a year. Female Common vampire bats, for instance, may experience four reproductive cycles in a year, giving them four opportunities for breeding. Most temperate bat species are monestrous, however, with only one reproductive cycle annually. Hibernation constrains the timing of reproduction, and spermatogenesis in males and estrus in females typically occur in late summer or autumn. Females may store sperm overwinter for as long as 7 months in their uterus and oviduct, and ovulation and fertilization are delayed until a few days after their arousal from hibernation. Male bats can also store sperm in their epididymes (tubules at the back of the testes) or in the vas deferens (the sperm duct) for long periods. Bats in three families – Pteropodidae, Rhinolophidae, and Vespertilionidae – have also shown delayed implantation, and several species including the Jamaican fruit-eating bat show diapause (delays) during embryonic development. Lactation is energetically costly, and parturition is synchronized to coincide with times of high food abundance.

Most bats produce one offspring, but some species produce twins (which may have different fathers, as in noctules) and occasionally triplets or quadruplets (as in Red bats). Infants are relatively large at birth, weighing as much as 40 percent of the mother's body mass, and grow rapidly. Lactation lasts about 45 days in Greater horseshoe bats. During their first foraging trips (at between 28 and 30 days old), young Greater horseshoe bats forage close to the roost independently of their mothers, which may travel as far as 2–3km (1.2–1.9mi) from the roost. When the young are past their first winter, however, they may share foraging grounds with their mothers and other close relatives. Some species including the Lesser bulldog bat may provide maternal tuition by guiding the young to suitable foraging areas.

Parental care is usually given exclusively by mothers, although fathers of the monogamous Yellow-winged bat may provide assistance by defending foraging territories where the young are learning to feed. In some bat species, individuals other than the parents help raise the babies. For example, Evening bats undertake allosuckling, by which females (31 percent in one study) suckle infants (24 percent, and mostly female) to which

Above *Newborn Brazilian free-tailed bats* (Tadarida brasiliensis) *crowd a nursery colony in Texas. No other bat or mammal species comes together in larger numbers than this one: an estimated 25–50 million individuals – mainly mothers and young – are recorded to have inhabited a single cave.*

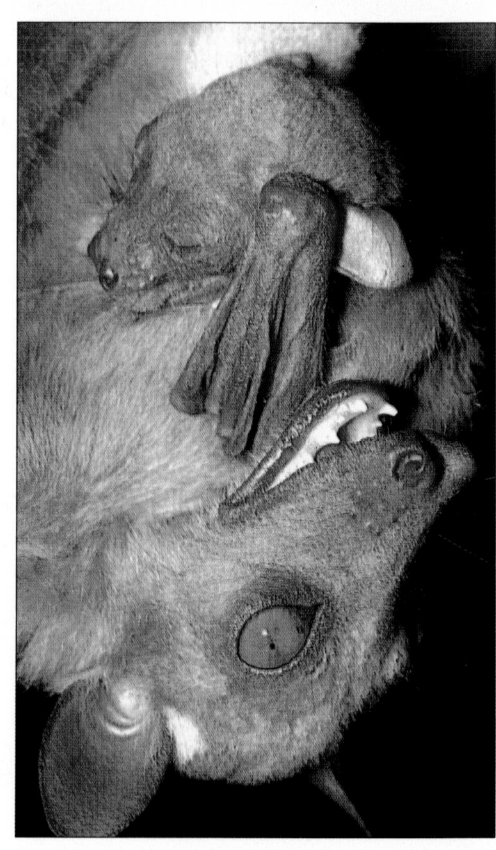

Right *A female flying fox nurses her offspring. Mothers normally bear one baby at a time, typically carrying their young with them for anything from 1 to 6 months. In the nursery colonies they may receive help with their maternal duties; other females have been seen to assist in grooming baby bats and also in maneuverng them into a suckling position.*

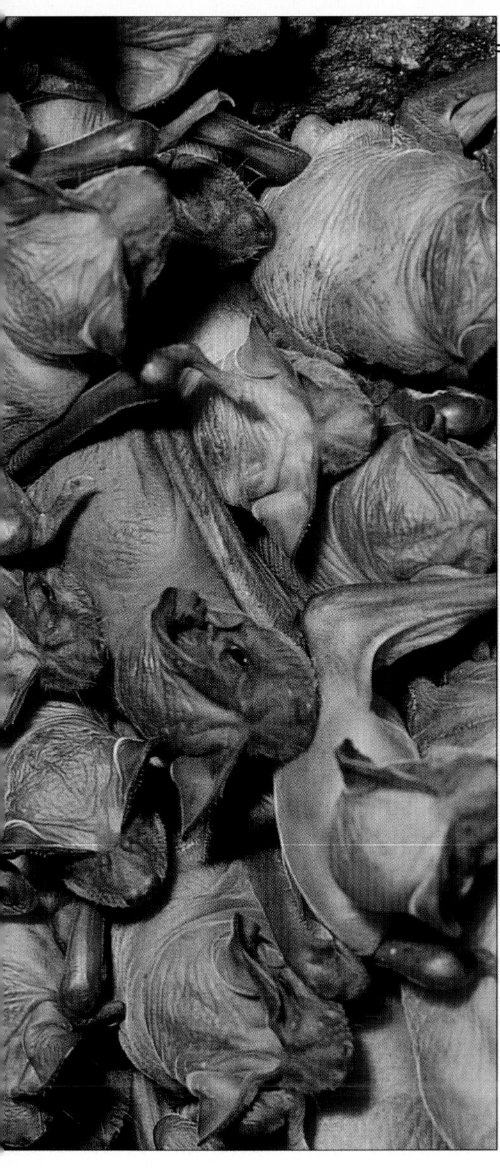

Above *A highly unusual roosting technique is employed by the tent-making bats* (Uroderma bilobatum) *of Central and South America. They chew through structural parts of the leaves of palms and other plants, causing them to collapse into tent-like shelters.*

they are not related. Because females are usually philopatric, returning to their natal colony to breed, allosuckling may have evolved in this species to increase future colony size and to improve the potential for transferring information about feeding sites.

Mating Habits and Mutual Aid

SOCIAL ORGANIZATION

Bats show considerable variety in social organization. A few species, such as the Yellow-winged bat, are monogamous. Promiscuity (with both males and females mating with several individuals) may occur in certain species, especially when males mate with torpid females during hibernation, although in some of these – the Little brown bat for one – molecular analysis of parentage now suggests that mating is biased towards particular males rather than being purely promiscuous as originally believed. Most bats are polygynous, with some males mating with more than one female and many failing to mate at all. However, females may also mate polyandrously – with several males – in one breeding cycle, as has been shown by DNA fingerprinting of twins

Many bats including pipistrelles and noctules show "resource defense polygyny," by which a male defends a resource (for example a roost site) that is important to females. The reproductive success of Greater white-lined bats that defend roost sites is three times higher than that of peripheral males that linger around the edges of the roosts. Males are unable to monopolize access to females, however, and on average 29 percent of the young born to females in a male's mating group are fathered by other males. Greater horseshoe bats defend roost sites and/or groups of females during the autumn, while pipistrelle males perform songflights to attract females.

Some female Greater horseshoe bats rear young sired by the same male in consecutive years. Their reasons for doing so can only be speculated upon – it may simply be that the males concerned are of high quality – but it is possible that the aim is to increase the co-ancestry of the offspring; if they are closely related, their behavior toward each other is more likely to be altruistic, which may bring benefits. After copulating, some of the male's ejaculate produces a "vaginal plug" that probably functions to safeguard paternity.

Certain species of bats even congregate to mate in the strange system known as lekking, by which males gather at recognized locations where they defend a small territory that has no benefit to the females other than as a place to come for sex. Hammerheaded bat males aggregate in groups and display vigorously at these mating arenas, producing ritualized calls that sound like glass being rapped hard on a porcelain sink. A tiny number of males (6 percent) are responsible for 79 percent of the copulations, since females seem to have

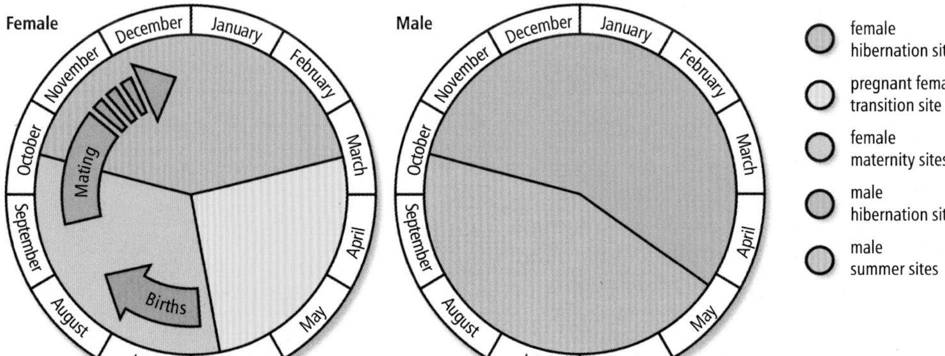

Left *A year in the life of the Greater horseshoe bat. Adult males and females generally lead separate lives except during the mating season, when females will usually go in search of a partner at a traditional male roost, although males occasionally visit females at the maternity roost instead. Juveniles are seen more with mature females than males. The female cycle shows that, on leaving hibernation sites (at which the oldest females tend to be solitary), pregnant females gather at transition roosts before moving on to the maternity sites where birth and mating occurs. Most juveniles remain in the maternity roost during the summer. In contrast, males tend to occupy the same traditional sites both during winter hibernation and in the summer.*

strong preferences for particular sites at the lek.

In temperate regions, female bats gang together during the summer in maternity colonies in which they give birth and rear offspring. Aggregation keeps the growing young warm and allows mothers to indicate the best foraging sites to one another. Maternity roosts may be large – one colony of Brazilian free-tailed bats holds the record as the largest aggregation of vertebrates ever known, containing over 20 million individuals. Males are usually more dispersed in the summer.

Common vampire bats show some of the best evidence of reciprocal altruism – when an individual chooses to act altruistically to another that acted thus in the past – in the animal kingdom. Reciprocity can evolve when the recipient gains more than the donor loses, and when at some future time the two are likely to swap roles.

Another essential element for the evolution of such behavior is that the animals involved should be able to recognize individuals who cheat the system by failing to provide help in their turn, in order to prevent them from benefiting from the system at no expense. These criteria appear to be fulfilled in the regurgitation of blood between individuals within colonies of Common vampire bats. Bats that go hungry will beg blood from other bats in the colony and are often rewarded, but only if they have a family connection or have proved themselves good friends in the past (see Bat Blood Donors).

Natural Energy-saving

TORPOR AND HIBERNATION

In order to get through the many rainy and cold and lean days of winter, many temperate bat species shut down body systems, lower their body temperature close to ambient levels, and enter torpor. During hibernation, a bat's heart rate may drop from 400 beats a minute to just 11–25 beats. Bat hibernacula can vary in temperature from –10° to 21°C (14°–70°F), with a mode of 6°C (43°F) for vespertilionids and 11°C (52°F) for horseshoe bats. A Brown long-eared bat in a torpid state at 5°C (41°F) expends only 0.7 percent as much energy as when fully active at the same temperature. Insectivorous bats may become torpid even during lactation, when both the young and their mothers may enter the state in periods of severe food shortage. Torpor has been studied in bats from the Vespertilionidae, Rhinolophidae, Molossidae, and Mystacinidae families. Some small megachiropterans may also enter shallow torpor for short periods.

In southwest England, Greater horseshoe bats enter torpor for periods of up to 12 days, waking up at dusk on mild winter nights to forage. The energy to wake up comes initially from brown adipose tissue, fat specialized for heat production; later in arousal, bats shiver in order to generate more heat. On arousing from 3.5°C (38°F), a Brown long-eared bat takes about 40 minutes to

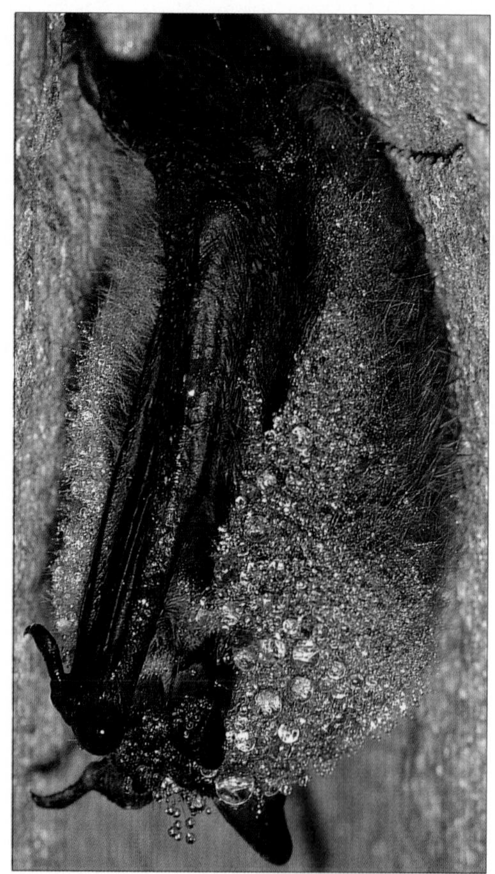

Above *Dew droplets caught in the fur of a hibernating Daubenton's bat* (Myotis daubentonii) *indicate that its body temperature has fallen to that of its very humid surroundings in a cold cave. Other hibernation sites favored by this species include cellars, mines, and disused military bunkers.*

reach a body temperature suitable for flight. Arousals must be crucial to survival, since they are costly, accounting for 85 percent of overwinter fat depletion in Little brown bats, even though the bat is fully homeothermic for only 2–4 percent of the time. Arousal may allow bats to drink, to restore metabolic imbalances, to resist disease, and even to make up for sleep deprivation, because the low brain temperatures during torpor inhibit the normal restorative functions of sleep.

Fruit Bats and Insect-eaters

FEEDING BEHAVIOR

About 70 percent of bat species eat mainly insects and other small arthropods, usually on the wing. Millimeter-long midges and beetles measuring 2.5cm (1in) are at opposite ends of the bat insect intake spectrum. Flies, beetles, moths, orthopterans, bugs, termites, hymenopterans, mayflies, and caddis flies are among the preferred insect prey.

The movements of millions of Brazilian free-tailed bats from caves in Texas have been studied by radar. Aggregations of bats can be detected moving from the caves and heading towards concentrations of billions of Corn earworm moths over croplands. Corn earworm moths are the major agricultural pest in the USA, and free-tailed bats from the largest roosts in Texas may eat 1,000 tonnes of insects in a night. The bats can detect insects by echolocation at altitudes as high as 750m (2,500ft) or more, and they may fly as high as 3,000m (10,000ft).

Some slow-flying species such as the Brown long-eared bat glean insects from vegetation. Others, including many horseshoe bats, hunt from perches. Some species including Daubenton's bat and the Lesser bulldog bat trawl prey from the water's surface. The Greater bulldog bat catches fish, which it finds by detecting irregularities or movements on the water's surface by means of echolocation, or by raking randomly in areas where it has previously captured prey. Some tropical bats are carnivorous, and three species of vampire bats feed on blood.

One third of the world's bat species eat fruit, nectar, and pollen. In the Old World tropics the bats in the family Pteropodidae (Megachiroptera) occupy these dietary niches, which are filled in the New World tropics by the Phyllostomidae (Microchiroptera). Plant-visiting bats pollinate or disperse many economically important plants including balsa, mango, durian, and wild-banana trees. Phyllostomids feed on nectar from the maguey plants that are used to produce tequila. At least 443 products used by humans derive from 163 different plant species that rely to some extent on pollination or dispersal by bats.

Sixty-two percent of pteropodids live on islands, where they can be keystone species for fruit dispersal; on many they are the only seed dispersers and pollinators, filling niches occupied by sunbirds, hummingbirds, primates, and other organisms elsewhere in the tropics. In the Philippines, large pteropodids tend to eat fruit in the tree of origin, while smaller bats carry fruits to other trees. Many Australian eucalyptus species may largely depend on flying foxes for their pollination. In West Africa, Straw-colored flying foxes are the only known dispersers of the iroko tree, whose timber has great commercial value.

Some plants have evolved flowers and fruits that increase their conspicuousness to bats. Flowers of bat-pollinated plants are often large, strong, inconspicuously colored, and heavily scented; they are also often exposed and open at night. They produce copious nectar, and their pollen is high in protein. Plant-visiting bats may feed singly on plants with few fruits, but forage in groups at superabundant food sources such as trees with many fruits. Female Greater spear-nosed bats emit low-frequency "screech calls" to attract members

Right *One of the smallest of all Megachiroptera species, the tiny Southern blossom bat* (Syconycteris australis) *is a nectar-eater averaging just 6cm (2.5in) in head–body length. Its exceptionally long tongue is equipped with brushlike projections that pick up nectar and pollen from the flowers on which it feeds.*

Left *Nearly three-quarters of all bat species rely mainly on insects for their food. Here a Striped hairy-nosed bat* (Mimon crenulatum) *makes a meal of a grasshopper in Ecuador's Yasuni National Park.*

Below *Another principal source of food for bats is fruit, as demonstrated by this Hairy big-eyed bat* (Chiroderma villosum) *carrying a fig. Bats of this genus, which is part of the family Phyllostomidae, inhabit Central and South America and the Caribbean.*

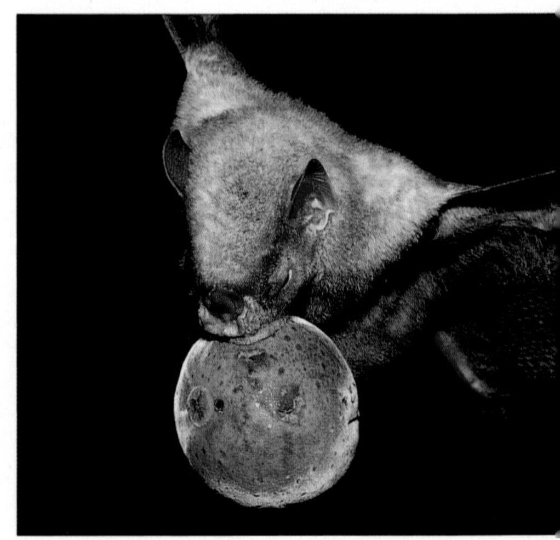

HIDDEN BIODIVERSITY

Pipistrelles are the most abundant bat species in Britain, with perhaps 2 million bats feeding in a range of habitats from urban areas to rivers and lakes. Until 1993 they were considered as a single species, *Pipistrellus pipistrellus*. Research on their echolocation calls, however, showed that some echolocated close to 45kHz while others did so nearer 55kHz. Bats from any one maternity colony consisted of only one phonic type, and differences in the surroundings could not explain the differences in call frequency. Subsequently it was found that the two phonic types also differ in habitat use (the 55kHz bats are more dependent on riparian habitats) and in diet, with the 55kHz bat eating more biting and nonbiting midges while the 45kHz bats prefer owl- and window-midges and dung flies. In addition, the maternity roosts of 55kHz pipistrelles usually contain more bats than those of the 45kHz bats, sometimes attracting 1,000 individuals. it has also been discovered that the two phonic types emit subtly different "social calls" to warn off conspecifics at feeding sites when food is scarce, and males utter different mate-attraction calls during songflights – perhaps a mechanism to facilitate reproductive isolation, as males only associate with females of the same phonic type as themselves at mating roosts.

Overwhelming evidence that the two phonic types are indeed different species also came from molecular studies. The cytochrome-b gene of mitochondrial DNA evolves rapidly, and is often used by evolutionary biologists interested in determining genetic differences between species. A portion of this gene was found to differ in its sequence by 11 percent between the two phonic types – a larger gap than that separating other bat species that appear quite different to the naked eye.

The two phonic types are in fact what biologists call "cryptic species": ones whose distinct identity is camouflaged by similar appearance. Although no one morphological character has yet been found to distinguish them, the 55kHz bats are usually lighter in color, and the 45kHz bats often have a black face mask. Presumably there has been stronger selection for the species to differ in their acoustic characteristics than in looks.

The two species are sympatric over much of Europe. As the first known representation of *Pipistrellus pipistrellus* is found in a book published in 1774 that clearly shows a bat with a black face mask, that name has been retained for the 45kHz phonic type, while the 55kHz species has been proposed as *P. pygmaeus*. Because the existence of *P. pygmaeus* was first inferred from its high-pitched call, the title "Soprano pipistrelle" has been proposed as its vernacular name.

Left *Caves make favorite roosting sites for many species of bats, providing shelter from the elements and usually also from human interference. Here a mixed flock of funnel-eared, spear-nosed, and mustached bats takes to flight in Tamana Cave, Trinidad.*

of their social group to rich foraging sites, often flowering or fruiting trees. Flying foxes find fruit by a mixture of vision and olfaction. Phyllostomids detect fruit at a distance by olfaction, then used echolocation for localization. One bat-pollinated vine in the Neotropics – *Mucuna holtonii* – directs bats to its nectar with a specialized petal that reflects most of an echolocation call's energy directly back at the bat; if this acoustic guide is removed, the visitation rate of bats to the flowers decreases dramatically.

Gathering-places on a Grand Scale

ROOSTS

Bats may spend over half their lives in roosts, where they cluster together for warmth and improved protection from predators. Sometimes over a million flying foxes will roost together, exposed in forest "camps." Roost sites include crevices in rocks and trees, tree cavities and foliage, mines, and caves (which are also sometimes used for breeding). Vespertilionid bats in British Columbia are very particular, preferring tall trees that are close to others but with little surrounding foliage. Silver-haired bats in Oregon roost in trees that protrude above the forest canopy, which perhaps act as signposts to guide in bats that fly past. Roosting in the canopy can be dangerous, however, and many foliage-roosting species have coloring that breaks up their outline, such as the two white lines on the back of the Greater white-lined bat. The Golden-tipped bat roosts in the abandoned, dome-shaped nests of birds, while in West Africa woolly bats of the *Kerivoula* genus have been found roosting in the large webs of colonial spiders.

Bats in three families (Thyropteridae, Myzopodidae, and Vespertilionidae) have cling-on pads: modified thumb and foot pads or disks to facilitate attachment on moist surfaces such as leaves or the insides of bamboo stems. At least one megachiropteran (the Short-nosed fruit bat) and 16 species of phyllostomids make tents from leaves and other plant parts. A single male and several females usually occupy the tents, sometimes with their young. Many bat species now roost in manmade structures such as mines and buildings; where their natural roosts have been removed with the harvesting of mature trees, they can be encouraged into artificial boxes.

Many bats change roost sites frequently. Most (70 percent) of the roosts used by the New Zealand long-tailed bat are used for only a single night, and females of this species often carry their young between roosts, perhaps in order to reduce predation or else to move them to more favorable microclimates. Pallid bats that move roost frequently have fewer ectoparasites than more sedentary individuals, so roost-switching may function to reduce parasite load.

Threats and Promises

CONSERVATION AND ENVIRONMENT

Throughout the world the populations of many bat species are declining. In Britain, the numbers of pipistrelles emerging from maternity colonies declined by 62 percent between 1978 and 1986. Lesser horseshoe bats are now almost extinct over much of northwest Europe.

Flying foxes are threatened mainly by habitat destruction, by conflict with fruit growers, and by hunting. A looming threat to flying foxes on oceanic islands comes from global warming, bringing with it the risk of higher sea levels and changing weather patterns. On the island of Anjouan in the Comoros group, which holds the world's largest population of Livingston's flying foxes, forest cover has decreased by 70 percent in the last 20 years; educational programs and a captive-breeding project have been set up to protect the bats. In Israel, cave roosts of rousettes have been fumigated in an attempt to restrict damage to fruit crops, resulting in the deaths not just of many rousettes but also of other, nontargeted

insectivorous bats; because the caves were sprayed with the persistent organochlorine lindane, they will remain toxic for years to come.

The native people of Guam have hunted flying foxes in small numbers for food for at least a millennium. After World War I the introduction of guns resulted in the Mariana flying fox becoming endangered. By 1975 native flying foxes were becoming difficult to find, so about 230,000 bats of at least 10 species were imported onto the island for food between 1975 and 1990. Since 1989 all pteropodids, and also related species in the genus *Acerodon*, have been protected under the Convention on International Trade in Endangered Species (CITES). Yet enforcing CITES regulations to protect flying foxes has proved difficult; perhaps the greatest hope for Guam's flying foxes comes from changing dietary habits on the island: a recent study suggested that most of the people who eat flying foxes are more than 55 years old.

Other problems face insectivorous bats in temperate regions. Roost sites such as old barns have been renovated or destroyed, and disturbances at hibernacula have probably contributed to population declines. The loss of natural and seminatural habitats has sometimes occurred at alarming rates. Hedgerows provide commuting routes and feeding sites for many bat species, and the drastic reduction in their numbers has probably affected bats significantly in western Europe. The intensification of farming practices, with increased emphasis on the application of pesticides, has probably reduced the availability of quality foraging areas for bats. Antiparasitic drugs that cure livestock of worms also kill dung beetles and other invertebrates that live in cowpats, thereby reducing the availability of an important prey species for Greater horseshoe bats.

Many bats have been killed as a result of coming into contact with the chlorinated hydrocarbon insecticides and fungicides that were formerly used to protect roof timbers against wood-boring beetles and timber-rotting fungi. In many countries bats have also accumulated toxic residues from agricultural pesticides such as DDT; fortunately, the dangers associated with these substances are now recognized, and their use is often restricted. Water pollution, with its associated effects on aquatic insects, and the accumulation of metals such as cadmium and methyl mercury may also have contributed to population declines among pipistrelles in industrialized areas.

Above *Thousands of Mexican free-tailed bats emerge from their cave. Such impressive sights may become increasingly rare, as bat species around the world suffer habitat loss and poisoning by agricultural chemicals.*

Bats' reputations have not been helped by the negative mythological connotations associated with their name. They are frequently linked to witchcraft and magic; for instance, "wool of bat" features in the potion prepared by the three witches in Shakespeare's play *Macbeth*. There is also, of course, an extensive folklore about vampires – the transformed bodies or souls of the dead, who suck the blood of humans at night. The first time that vampires took the form of bats, however, was in Bram Stoker's novel *Dracula*, which appeared in 1897, although it is not clear whether Stoker knew of the existence of genuine vampire bats at the time. Interestingly, bats are viewed more positively in Chinese culture, where they are symbols of joy; stylized bats appear in many Chinese decorations and artforms. Bats also suffer from stories of their getting tangled in people's hair, although the sophistication of their echolocation equipment make such an occurrence unlikely.

Bats are also often identified as vectors of disease. A small number of bat species do indeed carry rabies, although this disease now accounts for only one or two human deaths per year in the USA (in comparison, about 20 people are killed in dog attacks annually). Flying foxes in Australasia may be the natural host of the *Hendra* virus that is thought to cause death in humans and horses. Public reaction to rabies and *Hendra* viruses in bats has been hysterical at times, although humans can easily avoid infection not just through vaccination but also by taking standard hygiene precautions and avoiding the handling of bats.

Given the bad publicity that often attaches to bats, it is encouraging to note that highly successful programs exist for their conservation. Many important hibernacula have been fitted with grilles at their entrances to restrict human access, a measure that can increase population numbers dramatically. Bat boxes may encourage bats to roost in habitats that otherwise would be unsuitable, and artificial hibernacula have also been constructed with some success.

Recently a major campaign undertaken to protect the single most important hibernation site in Europe ended in success. Thirty thousand bats of 12 different species annually hibernate in underground fortifications that were built in western Poland to combat the German invasion during World War II. The Polish government initially planned to store radioactive waste at the site, but extensive campaigning by bat conservationists has resulted in its protection as a bat reserve.

In Britain, the public's attitude to bats has warmed dramatically over the past 20 years. The International Convention on Biological Diversity, established in Rio de Janeiro in 1992, has triggered the formulation of action plans to conserve five species of British bats, and a network of about 90 groups now works to promote bat conservation. Research on the foraging ecology of Greater horseshoe bats has led to recommendations for the protection of their maternity roosts, and landowners can now obtain government grants to enhance this species' favored habitats.

National and regional policies to protect bats have now been developed in many countries, including Australia, Brazil, and the United States. Migratory bats are protected in the European Union, and European law also confers some protection on foraging sites. Bat Conservation International exists to promote the conservation cause. In addition, many researchers working on these fascinating creatures have become so impressed by their amazing adaptations that they have also become committed to their preservation. It is not hard to see why. GJ

Above *Three representative endangered species suggest the range of problems now facing bats around the world. The Ryukyu flying fox* ***1*** *is only found on the Ryukyu Islands at Japan's southern tip; long hunted for its meat, it is now also threatened by the felling of the trees on which it feeds and roosts. The Gray bat* ***2*** *from the southeastern USA is at risk because the caves where it hibernates are increasingly disturbed by tourists and quarryers. The Sucker-footed bat* ***3*** *is restricted to Madagascar, and like many of the island's mammalian species is threatened as the growing human population encroaches on the forests where it lives.*

Bat Families

NEW BAT SPECIES ARE BEING IDENTIFIED continuously, often as a consequence of molecular studies, and various authorities recognize differing numbers of families, genera, and species. This categorization lists diagnostic features of the families as recognized by the most recent, phylogeny-based approach to bat relationships, based on molecular and morphological (i.e. "total-evidence") research.

SUBORDER MEGACHIROPTERA

Flying foxes
Family Pteropodidae
Flying foxes or Old World fruit bats

164 species in 41 genera. Old World tropics and subtropics from Africa to E Asia and Australia, including many islands in the Indian and Pacific Oceans. A few species reach warmer temperate regions N to Turkey and Syria and to the extreme S of Africa and SE Australia. Species include: **Indian flying fox** (*P. giganteus*); **Leschenault's rousette** (*Rousettus leschenaulti*); **Blanford's fruit bat** (*Sphaerias blanfordi*); **Southern blossom bat** (*Syconycteris australis*); **Sulawesi fruit bat** (*Acerodon celebensis*); **Lesser naked-backed fruit bat** (*Dobsonia minor*); **Common tube-nosed fruit bat** (*Nyctimene albiventer*).
SIZE: Ranges from fairly small to the largest of all bats, with wing spans of 1.7m; HBL 50–400 mm; tail short or absent except in one species where it is half head–body length; FL 37–220mm; WT 15–1,500g. Male Hammer-headed bats (*Hypsignathus monstrosus*) show the most extreme sexual dimorphism in bats; males are almost twice the mass of females, and have swollen muzzles with flaring lip flaps. A huge larynx fills more than half of the body cavity, and is used to produce calls at lek sites. Males of other species have tufts of hair or glandular patches on their shoulders that are used in display.
COAT: Usually brown, but a few species are brightly colored, e.g. the Rodriguez flying fox (*Pteropus rodricensis*) varies from black to silver, yellow, orange, and red. Tube-nosed bats (genus *Nyctimene*) can be brightly colored, with speckled membranes and a dorsal stripe; cryptic coloration may camouflage bats roosting in foliage.
FORM: Most species have doglike faces, large eyes, and conspicuous, widely separated simple ears that lack a tragus. The second finger retains independence and, like the thumb, usually bears a claw. Most species orientate visually, except bats in the genus *Rousettus*, which echolocate by clicking the tongue to detect obstacles in caves. Dawn bats (*Eonycteris spelaea*) clap their wings when flying in caves, and this might function as a rudimentary form of echolocation.
DIET: Mainly plant material, largely ripe fruit but also flowers, nectar, and leaves. Some species may supplement their diet with a small numbers of insects. Nectar-feeding species (e.g. blossom bats in genus *Macroglossus*) have long tongues with bristle-like papillae. Many species are essential for pollination and dispersal of economically important plants.
BEHAVIOR: Some of the larger species may form "camps" in forest patches, sometimes containing over a million bats. The largest colonies of all are formed in caves by rousettes (*Rousettus* spp); they may contain several million bats, which may make round trips of 40–50km in a night to forage. Straw-colored fruit bats (*Eidolon helvum*) are migratory, and may make round trips in E Africa of up to 2,500km.
CONSERVATION STATUS: 8 species are now considered Extinct by the IUCN; in addition, 14 species (including all 4 *Pteralopex* spp) are Critically Endangered, 7 are Endangered, and 39 Vulnerable.

SUBORDER MICROCHIROPTERA

Mouse-tailed bats
Family Rhinopomatidae
Mouse-tailed, rat-tailed, or long-tailed bats

3 species in 1 genus (*Rhinopoma*). N Africa to the S Sudan, Middle East, India, and SE Asia; Sumatra. Arid or semi-arid areas, also agricultural and disturbed habitats. Species are: **Lesser mouse-tailed bat** (*Rhinopoma hardwickii*); **Greater mouse-tailed bat** (*R. microphyllum*); **Small mouse-tailed bat** (*R. muscatellum*).
SIZE: Generally small. HBL 50–90mm; TL 40–80mm; FL 45–75mm; WT 6–14g.
FORM: Tail exceptional, being almost as long as head–body length and entirely free of membrane. Ears joined by inner margins, tragus well developed; snout bears a small, simple noseleaf like a pig's muzzle. Least derived of Microchiroptera.
BEHAVIOR: Fast-flying aerial insectivores that emit long, narrowband, multiharmonic echolocation calls. Roost in caves and buildings, including Egyptian pyramids where they have occurred for 3,000 years or more. May enter torpor when insects are scarce, and then utilize fat stored in the abdominal region. Roosts may comprise thousands of bats.
CONSERVATION STATUS: 1 species (*Rhinopoma hardwickii*) is Vulnerable.

Hog-nosed bat
Family Craseonycteridae
Craseonycteris thonglongyai
Bumblebee or Butterfly bat

1 species, first described in 1974. W. Thailand. Bamboo forests and teak plantations (much of the natural vegetation in area removed). SIZE: The world's smallest bat (and mammal). HBL 29–33mm; tail absent; FL 22–26mm; WT 2g.
COAT: Upperparts brown to reddish gray, underside paler, wings darker.
FORM: Muzzle piglike in appearance. Glandular swelling on underside of throat in males.
BEHAVIOR: Aerial insectivore, feeding on small beetles and other insects. Hunts in open areas for 30–45 mins after sunset and for a short period before dawn. Echolocation calls narrowband and multiharmonic. Forms small colonies in caves.
CONSERVATION STATUS: Endangered.

Sheath-tailed bats
Family Emballonuridae
Sheath-tailed or sac-winged bats

47 species in 12 genera. Africa, S and SE Asia, Australia, tropical America, many islands in the Indian and Pacific Oceans. Species include: **Shaggy bat** (*Centronycteris maximiliani*); **Greater ghost bat** (*Diclidurus ingens*); **Peter's sheath-tailed bat** (*Emballonura atrata*); **Lesser doglike bat** (*Peropterxy macrotis*); **Proboscis bat** (*Rhynchonycteris naso*); **Egyptian tomb bat** (*Taphozous perforatus*).
SIZE: Varies from small to relatively large. HBL 37–157mm; TL 6–36mm; FL 37–97mm; WT 5–105g.
COAT: Mostly drab or brown, although some species have cryptic patterns and tufts of hair. Bats in one genus, *Diclidurus* (ghost bats), are largely white.
FORM: Ears often joined, tragus present. Short tail pierces the tail membrane, so its tip appears exposed on membrane's dorsal surface. Many species have glandular wing sacs (larger in males) that open on the upper surface of the wing; in the Greater white-lined bat (*Saccopteryx bilineata*), these are used for odor storage. Some species have throat glands that secrete odorous compounds.
BEHAVIOR: Many are fast-flying aerial insectivores that emit long, narrowband, multiharmonic echolocation calls. Roost in a wide range of sites from hollow trees to buildings and caves. Some species highly colonial (e.g. the tomb bats, *Taphozous* spp). Most are tropical, but some species in cooler regions use torpor and may hibernate.
CONSERVATION STATUS: 2 species are Critically Endangered, 2 Endangered, and 10 Vulnerable.

Right A Pallas's tube-nosed fruit bat (Nyctimene cephalotes) *rests with its wings folded at a roost in northern Sulawesi, Indonesia. The* Nyctimene *species makes up one of the 41 genera of the Pteropodidae – the flying-fox family of Old World fruit bats.*

Slit-faced bats
Family Nycteridae
Slit-faced, hollow-faced, or hispid bats

13 species in 1 genus. Africa, SW and SE Asia. Arid areas as well as rain forests. Species include: **Large slit-faced bat** (*Nycteris grandis*); **Hairy slit-faced bat** (*N. hispida*); **Dwarf slit-faced bat** (*N.nana*).
SIZE: Medium-sized. HBL 40–93mm; TL 43–75mm, FL 32–60mm, WT 10–30g.
COAT: Usually long, rich brown to gray.
FORM: Furrow in muzzle from nostrils to pit between eyes; noseleaves and fur sometimes conceal furrow externally. Ears large, tragus small. Broad wings. Tail T-shaped at tip, uniquely among mammals.
DIET: Orthopterans, spiders, caterpillars, scorpions. The Large slit-faced bat (*Nycteris grandis*) eats fish, frogs, birds, and other bats.
BEHAVIOR: The bats often carry prey to a perch for eating. Echolocation calls faint, broadband, and multiharmonic. Roost often in small groups in a variety of sites including hollow trees, culverts, and even aardvark burrows.
CONSERVATION STATUS: 2 species are Vulnerable.

Old World false vampire bats
Family Megadermatidae
Old World false vampire bats, false vampire bats, yellow-winged bats

5 species in 4 genera. Old World tropics and subtropics of Africa, SC and SE Asia, and Australasia. Species include: **Heart-nosed bat** (*Cardioderma cor*); **Greater false vampire bat** (*Megaderma lyra*).
SIZE: Among the largest microchiropterans, with the Australian false vampire bat (*Macroderma gigas*) the largest. HBL 65–140mm; FL 50–115mm; WT 37–123g.
COAT: Drab but variable, from gray to brown, even whitish. Yellow-winged bat (*Lavia frons*) very colorful, with ears and wings yellow-orange and fur gray to olive.

ABBREVIATIONS HBL = head–body length TL = tail length FL = forearm length WT = weight
Approximate nonmetric equivalents: 25mm = 1in 10cm = 4in; 100g =3.5oz

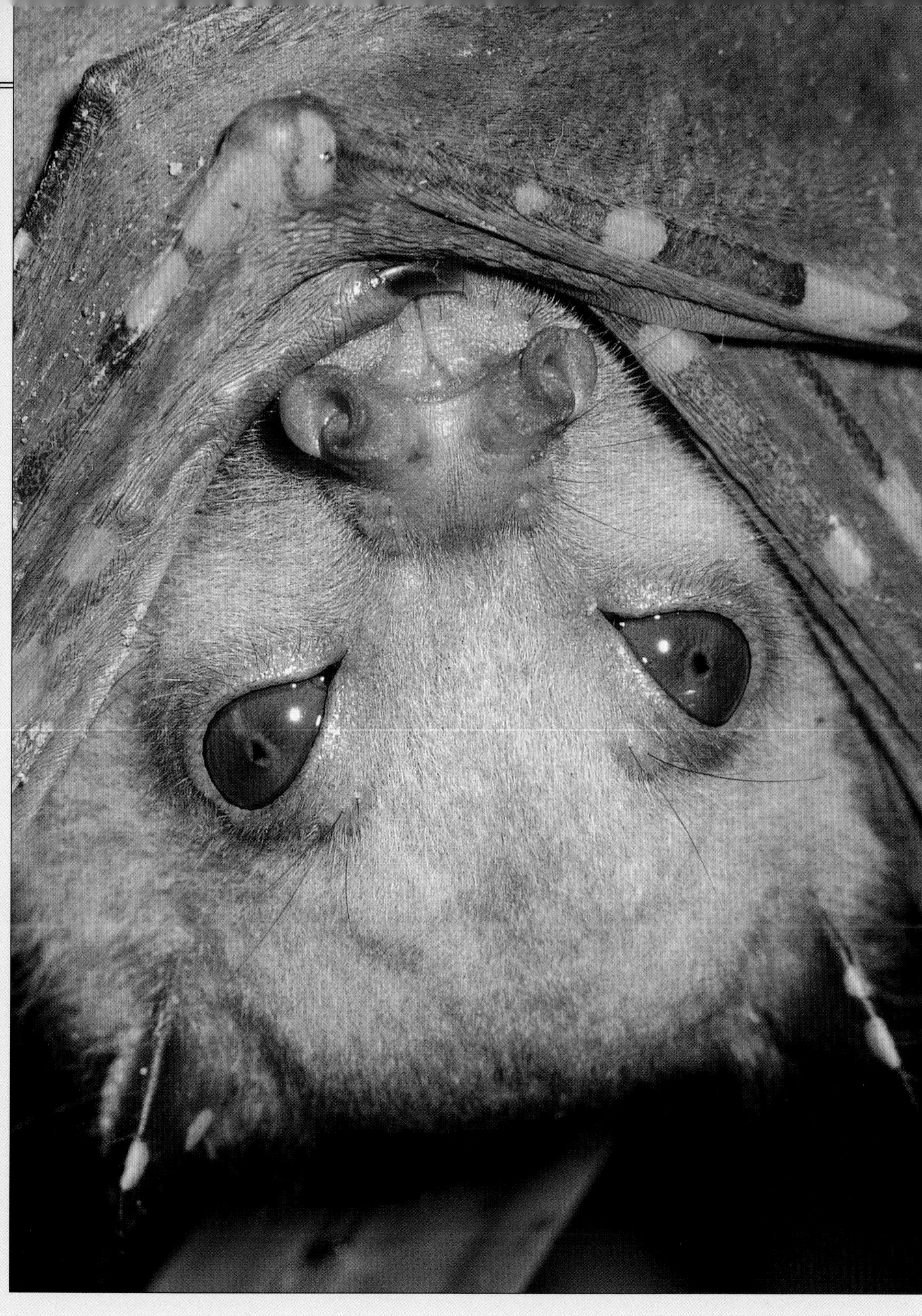

Dusky roundleaf bat (*Hipposideros ater*).

SIZE: Mostly relatively small. HBL 35–110mm; TL 15–70mm; FL 30–105mm; WT 4–180g.
COAT: Variable, including yellows, browns, grays, blacks; some species have whitish fur patches. Some species occur in two color phases.
FORM: Prominent noseleaves; in horseshoe bats the front part of the noseleaf is horseshoe-shaped, and a sella projects forward with a generally pointed lancet running lengthways. Old World leaf-nosed bats have a noseleaf lacking a well-defined horseshoe, and the lancet is a transverse leaf often with three points; there is no sella in the center of the noseleaf. Ears generally large and pointed, and always lack a tragus. Heads face downward. Hind legs poorly developed; the bats are unable to walk quadrupedally. Dummy teats on female abdomens for attachment of young.
BEHAVIOR: Broad wings confer slow, maneuverable flight, often within dense vegetation. The bats catch prey (usually insects) by aerial hawking and gleaning, and large species often hunt from and eat prey at perches. Sophisticated echolocation involves nasal emission of relatively long constant-frequency pulses, usually as second harmonic of call. Horseshoe bats echolocate by compensating for Doppler shifts induced by their flight speed; Old World leaf-nosed bats compensate partially. All roost in caves, mines, trees, buildings. Some species may be monogamous, others polygynous. At roosts, fold their wings around themselves. Can hibernate.
CONSERVATION STATUS: 2 of the 64 horseshoe bats are Endangered, and 8 are Vulnerable. Of the 65 Old World leaf-nosed bats, 2 are Critically Endangered, 1 Endangered, and 15 Vulnerable.

FORM: Long, erect, simple noseleaves. Ears very large and fused at base, each with large tragus. Broad tail membrane, but tail itself vestigial or absent. Eyes large.
DIET: Some of the most carnivorous bats, eating small vertebrates, other bats, reptiles, amphibians, fish, and arthropods. Often hunt from and eat prey at perches.
BEHAVIOR: Broad wings confer slow, maneuverable flight and strong take-off from ground. The Yellow-winged bat is often active by day and lives in territorial, monogamous pairs, which remain with young for about 30 days after weaning. Hearing very sensitive to rustling sounds made by prey moving in vegetation. These bats often hunt by listening for prey-generated sounds, although they also detect prey by echolocation in non-cluttered environments. Echolocation calls faint, broadband, and multiharmonic. Roost in caves, buildings, and hollow trees, often in small groups. The African false vampire bat (*Cardioderma cor*) "sings" to establish defended foraging areas.
CONSERVATION STATUS: The Australian false vampire bat (*Macroderma gigas*) is Vulnerable, and exists in isolated populations that show strong genetic divergence.

Horseshoe and Old World leaf-nosed bats

Family Rhinolophidae

129 species in 10 genera. Many authors treat the horseshoe bats and Old World leaf-nosed bats as separate families (Rhinolophidae and Hipposideridae respectively), although modern classifications consider both as belonging to one family. Africa, S Eurasia, SE Asia, Australia, especially in tropical regions. Species include: **Thomas's horseshoe bat** (*Rhinolophus thomasi*); **Trident leaf-nosed bat** (*Asellia tridens*); **Flower-faced bat** (*Anthops ornatus*);

New Zealand short-tailed bat

Family Mystacinidae
Mystacina tuberculata
New Zealand short-tailed or Lesser short-tailed bat

1 species. New Zealand and adjacent islands. Another putative species, *M. robusta*, is now almost certainly extinct.
SIZE: Small. HBL 60mm; TL 18mm; FL 40–46mm; WT 13–22g.
COAT: Fur brown-gray, short and velvety.
FORM: Thumb and toe claws have extra projection or talon, uniquely among bats; wings can be rolled tightly against body; tail perforates upperside of tail membrane, as in the Emballonuridae; ears simple, separate, and large, with long tragus; tongue with papillae at tip.
DIET: Broad diet includes aerial insects, terrestrial arthropods, nectar, pollen, fruit.

BEHAVIOR: Agile on the ground; evolved in the absence of terrestrial mammals and perhaps with few predators, facilitating terrestrial habits. Molecular studies suggest affinities to Mormoopidae, Noctilionidae, and Phyllostomidae families, suggesting Gondwanaland ancestry. Echolocation calls broadband and multiharmonic. The bat catches aerial prey by echolocation; listens for prey-generated sounds and uses olfaction to locate prey in leaf litter; roosts in tree holes, often in large numbers. Males "sing" at aggregations of tree roosts, probably to attract females. Large foraging range. Often hosts a wingless fly that feeds on fungi that grow in its guano. Can use torpor and probably hibernates. This and a vespertilionid bat are the only indigenous nonmarine mammals in New Zealand.
CONSERVATION STATUS: Vulnerable

Bulldog bats
Family Noctilionidae
Bulldog or fisherman bats

2 species in 1 genus (*Noctilio*). Tropical C and S America, Caribbean islands.
SIZE: **Greater bulldog bat**, or **Fisherman bat** (*Noctilio leporinus*): HBL 98–132mm; FL 70–92mm; WT 54–90g. **Lesser bulldog bat** (*N. albiventris*): HBL 57–85mm; FL 54–70mm; WT 18–44g.
COAT: Fisherman bat has short yellow or orange fur that sheds water easily.
BEHAVIOR: The fisherman bat is the only bat to specialize in eating fish; also eats insects and crabs. Fish up to 8cm (3in) long are detected by echolocation when they protrude or form ripples on the water's surface; huge feet with long, sharp claws are used to catch fish from just under the surface. Catches are stored in cheek pouches. Roosts often in sea caves, where colonies can be located by their musky odor. Insectivorous Lesser bulldog bat captures insects in midair or from water's surface, and may be preadapted for piscivory. Echolocation calls include narrowband and broadband components to detect fish and insects that often break the water's surface.
CONSERVATION STATUS: Not threatened.

Spectacled bats
Family Mormoopidae
Spectacled bats, naked-backed bats, leaf-chinned bats, mustached bats

8 species in 2 genera. Extreme SW USA, through C America and Caribbean S to central S Brazil. Species include: **Ghost-faced bat** (*Mormoops megalophylla*); **Sooty mustached bat** (*Pteronotus quadridens*).
SIZE: Small to medium-sized. HBL 40–77 mm; TL 15–30 mm; FL 35–65 mm; WT 3.5–20g.
COAT: Fur short and dense, reddish through brown and gray.
FORM: Lack a noseleaf but can funnel lips to create a dish shape. Several species have naked backs because wing membranes join on dorsal midline of body. Ears small with tragus. Tail projects slightly beyond tail membrane.
BEHAVIOR: Insectivorous. Can form large colonies in caves. Echolocation calls include narrowband and broadband components. Parnell's mustached bat (*Pteronotus parnellii*) evolved echolocation with Doppler shift compensation independently of the Rhinolophidae.
CONSERVATION STATUS: 1 species (*Pteronotus macleayii*) is Vulnerable.

New World leaf-nosed bats
Family Phyllostomidae
New World leaf-nosed bats or spear-nosed bats

About 140 species in 48 genera. New World from extreme SW USA throughout C America and Caribbean south to N Argentina. Species include: **California leaf-nosed bat** (*Macrotus californicus*); **Golden bat** (*Mimon bennettii*); **Pale-faced bat** (*Phylloderma stenops*); **Greater spear-nosed bat** (*Phyllostomus hastatus*); **Greater round-eared bat** (*Tonatia bidens*); **Western nectar bat** (*Lonchophylla hesperia*); **Lesser long-tailed bat** (*Choeroniscus minor*); **Tent-making bat** (*Uroderma bilobatum*); **Large fruit-eating bat** (*Artibeus amplus*); **Jamaican flower bat** (*Phyllonycteris aphylla*); **Hairy-legged vampire bat** (*Diphylla ecaudata*).
SIZE: Variable. HBL 40–135mm; TL absent or 4–55mm; FL 31–105mm, WT 7–200g. Includes the microchiropteran with the longest forearm length, the carnivorous Spectral bat (*Vampyrum spectrum*).
COAT: Apart from one almost white bat, the Honduran white bat (*Ectophylla alba*), others are brown, gray, or black, occasionally with red or white hair tufts. Several species have whitish lines on the face and/or body.
FORM: Most species have a spear–shaped noseleaf, but 5 have none, or a more complex shape. Ears usually simple, can be large; tragus present. Vampire bats' muzzles are swollen and glandular, resembling a noseleaf, and they have specialized teeth, with a reduction in tooth number and sharp incisors and canines.
DIET: Show virtually every type of food habit known for bats. Many species feed on fruit, pollen, or nectar; others are insectivorous, and a few are omnivorous or even carnivorous. Three species of vampire bats in C and S America feed on blood, whose flow is increased by anti-coagulants in the saliva. The Common vampire bat (*Desmodus rotundus*) feeds mainly on mammalian blood, the other two principally on bird blood.
BEHAVIOR: Phyllostomids detect food by echolocation, olfaction, and probably vision. Echolocation calls usually faint, broadband, and multiharmonic. One species, the Fringe-lipped bat (*Trachops cirrhosus*), hunts frogs, which it locates by listening for their calls. The bats roost in a variety of sites including caves, culverts, hollow trees, and among foliage. Some species make "tents" by biting through leaf ribs. None hibernate. Vampire bats are agile on the ground.
CONSERVATION STATUS: The Puerto Rican flower bat (*Phyllonycteris major*) is now thought to be extinct; 4 other species are Endangered, and 25 are Vulnerable.

Funnel-eared bats
Family Natalidae
Funnel-eared or long-legged bats

5 species in 1 genus (*Natalus*). N Mexico through C America to Brazil, also Caribbean islands. Species include **Mexican funnel-eared bat** (*N. stramineus*).
SIZE: Small and delicate. HBL 35–55mm; TL 50–60mm; FL 27–41mm; WT 4–10g.
COAT: Often reddish or brown.
FORM: Second digit of wing reduced to metacarpal, phalanx lost. Lightly built with long, slender wings and legs; tails longer than head and body. Ears funnel-shaped and large, with a short, triangular tragus; nose simple.
BEHAVIOR: Poorly studied. Often roost in caves, also in tree hollows. Slow-flying and maneuverable. Insectivorous. Adult males have a "natalid organ" on face or muzzle, of unknown function.
CONSERVATION STATUS: 1 species – *N. tumidifrons* – is Vulnerable.

Thumbless bats
Family Furipteridae
Thumbless or smoky bats

2 species in 2 genera: **Thumbless bat** (*Furipterus horrens*) and **Smoky bat** (*Amorphochilus schnablii*). Costa Rica to N Chile and SE Brazil; Trinidad.
SIZE: Small bats. HBL 33–58mm; TL 24–36mm; FL 30–40mm; WT 3–5g.
COAT: Brown to gray.
FORM: Small, functionless thumbs; ears funnel-shaped with small tragus. Wings relatively long; crown elevated above face. Truncated snout ends in disklike structure. Females have one pair of abdominal mammae.
BEHAVIOR: Little is known about these bats, except that they are insectivorous and roost in caves, buildings, tunnels, among boulders, or in tree hollows.
CONSERVATION STATUS: The Smoky bat is Vulnerable.

Disk-winged bats
Family Thyropteridae
Disk-winged bats or New World sucker-footed bats

2 species in 1 genus: **Peter's disk-winged bat** (*Thyroptera discifera*) and **Spix's disk-winged bat** (*T. tricolor*). S Mexico to N Bolivia and SE Brazil; Trinidad.
SIZE: Small bats. HBL 34–52mm; TL 25–33mm; FL 27–38mm; WT 4–5g.
COAT: Red-brown or blackish, white below.
FORM: Wrists, ankles, and functional thumbs have disk suckers borne on small stalks. One sucker can support the bat's weight as it roosts in smooth, furled leaves, such as bananas or those of *Heliconia*. Ears funnel-shaped, tragus present. As in Smoky bats, third and fourth toes are joined.
BEHAVIOR: Insectivorous, roost in small groups in moist, evergreen forest. Echolocation calls low in intensity, variable, including both broadband, multiharmonic, and narrowband calls.

Old World sucker-footed bat
Family Myzopodidae
Myzopoda aurita

1 species. Restricted to Madagascar.
SIZE: Small to medium-sized. HBL 57mm; TL 48mm; FL 46–50mm.
FORM: Suction disks present on wrists and ankles differ in structure from those in *Thyroptera*, and probably evolved independently. Large ear, with tragus and mushroom-shaped structure at base. Long tail projects beyond tail membrane.
BEHAVIOR: Rare and little studied. Roosts in palm leaves; one captured bat had fed on microlepidoptera. Emits unusual echolocation calls that are long, multiharmonic, and consist of several elements. Similar bats found in E Africa in Pleistocene.
CONSERVATION STATUS: Vulnerable.

Vespertilionid bats
Family Vespertilionidae

At least 308 species in 34 genera. The second largest mammalian family after the Muridae (Old World rats and mice). Worldwide except for extreme Polar regions and remote islands. Species include: **Smith's woolly bat** (*Kerivoula smithii*); Pied bat (*Chalinolobus superbus*); **Cape serotine** (*Eptesicus capensis*); **Large myotis** (*Myotis chinensis*); **noctule** (*Nyctalus noctula*); **Mouselike pipistrelle** (*Pipistrellus musciculus*); Robust yellow bat (*Scotophilus robustus*); **Brown tube-nosed bat** (*Murina suilla*); **Greater bamboo bat** (*Tylonycteris robustula*); Peter's tube-nosed bat (*Murina grisea*).
SIZE: Mostly small, but a few are medium to large. HBL 32–105mm; TL 25–75mm; FL 22–75mm; WT 4–50g.
COAT: Usually browns and grays, undersides often paler. Some species yellow, red, or orange. Painted bat (*Kerivoula picta*) has scarlet or orange fur, black membranes with orange finger bones. The cryptic coloration may provide camouflage in flowers and foliage; small clusters resemble the

hanging mud nests of wasps. Butterfly bats (*Glauconycteris* spp.) usually have white spots and stripes on their pelage. These bats often roost in foliage, and may show cryptic or disruptive color patterns.
FORM: Most have simple muzzles, though bats in 2 genera have a slight noseleaf, and tubular nostrils are also present in 2 genera. Ears normally separate, with tragus, and vary from small to enormous, especially in the long-eared bats (*Plecotus* spp.) where they approach head–body length. Bats in 5 genera have wing and/or foot discs to aid gripping smooth leaves or bamboo.
BEHAVIOR: Mainly insectivorous, with insects captured by aerial hawking, gleaning, or even trawling from water's surface (e.g. *Myotis daubentonii*). At least one species, the Fish-eating bat (*Myotis vivesi*) sometimes eats fish. Echolocation calls vary from broadband in species that feed in obstacle-rich environments to narrowband in more open habitats. Species that feed in clutter often locate prey by listening for prey-generated sounds rather than by echolocation. Roost sites include tree hollows, under bark, rock crevices, birds' nests, in foliage, and in buildings and caves. Bamboo bats (*Tylonycteris* spp.) have flat heads and suckers on their thumbs and feet, and roost in bamboo stems. Roost sizes vary from a few individuals to over 1 million. Some species migrate, though temperate species generally hibernate.
CONSERVATION STATUS: Of the 308 species, 6 are considered Critically Endangered, 19 Endangered, and 50 Vulnerable. 1 pipistrelle – *Pipistrellus sturdeei* – is listed as Extinct.

Pallid bat
Family Antrozoidae
Antrozous pallidus

1 species. Recent phylogenetic studies based on morphology place the Pallid bat closer to free-tailed bats than to vespertilionids, and suggest that it should be placed in a family of its own, as here. N America to C Mexico; Cuba.
SIZE: Medium-sized. HBL 60–85mm; TL 35–57mm; FL 45–60mm; WT 17–28g.
COAT: Cream , yellowish, or light brown, paler ventrally.
FORM: large ears with tragus; small, horseshoe-shaped ridge on square muzzle, with nostrils underneath. Broad wings confer maneuverable flight.
DIET: Eats beetles, orthopterans, moths, spiders, scorpions, centipedes, and even small vertebrates, which are often captured on the ground.
BEHAVIOR: Often inhabits desert areas, though also occurs in forests. Locates prey on the ground by listening for rustling sounds. Highly social, with a rich repertoire of communication calls.

Right *Proboscis bats (*Rhynchonycteris naso*) form a line on a tree stump. These bats may roost on branches during the day. Pale lines on their backs disrupt their shape; this, plus their habit of tilting their heads backward, makes the roosting bats resemble the curved edges of lichen.*

Free-tailed bats
Family Molossidae

At least 77 species in 12 genera. Warm or tropical areas from central USA south to C Argentina; S Europe, Africa, E to Korea, the Solomons, and Australasia. Species include: **Spotted free-tailed bat** (*Chaerephon bivittata*); **Hairless bat** (*Cheiromeles torquatus*); **Malayan free-tailed bat** (*Mops mops*); **Cinnamon dog-faced bat** (*Molossops abrasus*).
SIZE: Most are small- to medium-sized, though the Hairless bat (*Cheiromeles torquatus*) is the heaviest microchiropteran. HBL 40–130mm; TL 14–80mm; FL 27–85mm; WT 8–180g.
COAT: Fur usually short and sleek, brown, gray, or black, but some are reddish with whitish patches, e.g. in *Mops spurrelli*.
FORM: Robust bats with a large proportion of the thick tail projecting beyond the tail membrane. Membranes leathery, wings long and narrow, conferring fast, unmaneuverable flight. Ears small to relatively long, usually joined across forehead and directed forward in small-eared species. Tragus present. Several species in the genus *Chaerephon* have tufts of glandular hairs on crown. The large Hairless bats have very thick skin, with wings joining at the midline of the back, and large skin pouches laterally into which the wing tips are tucked when at rest. They are largely naked, with a few hairs around a throat sac. They emit a pungent odor.
BEHAVIOR: Molossids are usually aerial hawkers that emit long-duration, narrowband calls. They feed mainly on moths and beetles captured in open habitats. Roost sites include caves, rock crevices, hollow trees, and buildings. Nursery colonies of the Brazilian free-tailed bat (*Tadarida brasiliensis*) in SW USA form the largest aggregations of vertebrates known, with up to 20 million bats in a colony. Sodium nitrate extracted from guano at these large roosts was used to produce gunpowder. Some species (e.g. *T. brasiliensis*) may migrate, others (e.g. the European free-tailed bat, *Tadarida teniotis*) can hibernate in rock crevices, emerging to feed on mild winter nights.
CONSERVATION STATUS: 3 species are Critically Endangered, 1 Endangered, and 16 Vulnerable. Brazilian free-tailed bat declines are partly attributable to poisoning by breakdown products of DDT. GJ

For full species list see Appendix ▷

SPECIAL FEATURE

Bats and Columnar Cacti

A symbiotic relationship between pollinator and pollinated

FLOWERS AND BATS MAY NOT SEEM THE MOST likely bedfellows, but they have formed mutually beneficial partnerships throughout the tropics worldwide. In the Old World, six genera and 14 species of pteropodid bats – the flying foxes and their allies – feed from flowers; they apparently evolved to do so independently from the 15 genera and 38 species of New World Phyllostomidae (leaf-nosed bats) that do the same. Together, these bats are the major, if not exclusive, pollinators of hundreds of species of tropical and subtropical trees, shrubs, lianas, and epiphytes, including such commercially important species as kapok, balsa, wild bananas, and durian.

While a majority of bat-pollinated plants occur within the wet tropics, one ecologically conspicuous group of neotropical plants, the columnar cacti, live in semi-arid to very arid habitats and depend on a specialized group of phyllostomid bats for their pollination (and, often, for seed dispersal). Containing about 100 genera and 1,500 species classified in three or four subfamilies, the Cactaceae form one of the New World's most distinctive plant families. Whereas most species are relatively small in stature and are pollinated by insects, the giants of this family, especially those of the Pachycereeae and Leptocereeae tribes, are often pollinated by bats.

Columnar cacti such as saguaros, cardóns, organ pipes, and their relatives attain the dizzy heights of 10–15m (33–50ft), and tend to be the dominant species in the deserts of the southwestern United States, Mexico, northern South America, and the Andes. Located high above the desert floor at or near the tips of long branches or stems, their robust, cream-colored flowers open up at night to produce substantial amounts of energy-rich nectar (up to 2ml and 8.8kJ) and pollen (up to 0.5g). Unlike insect-pollinated flowers, which typically remain open for several days at a time, most bat-pollinated cactus flowers close within 12 hours of opening.

Although a variety of phyllostomid bats opportunistically visit the flowers of columnar cacti, most of the pollination is accomplished by just four species, all specialized members of the subfamily Glossophaginae. Three of these – *Leptonycteris curasoae*, *L. nivalis*, and *Choeronycteris mexicana* – occur in Mexico and the southwestern United States; one (*Platalina genovensium*) is exclusive to South America, where *L. curasoae* is also found. In this group of bats, certain adaptations – a greatly elongated rostrum and tongue, long, relatively narrow wings, and large size (for the subfamily) – have evolved independently three times to aid their floral way of life. Nectar and pollen can be rapidly extracted from large cactus flowers, and the bats can fly long distances quickly and efficiently. Physiologically, they can extract all of the nutrients contained within the pollen grains in their stomachs and use the amino acids for protein synthesis, which means that they have to eat fewer insects to make up their diet, which in some seasons and locations consists almost entirely of nectar and pollen. Finally, these bats are migratory and often move seasonally among habitats in search of rich patches of flowering plants.

Each spring, pregnant female Lesser long-nosed bats (*L. curasoae*) migrate nearly 1,200km (750mi) from the tropical dry forests of south-central Mexico to the Sonoran Desert of northwestern Mexico and Arizona, where they form large maternity colonies containing tens of thousands of individuals. Although details of its migration routes, or "nectar corridors," are unknown, *P. genovensium* also probably migrates over long distances among cactus-rich habitats in the Peruvian Andes, especially in years affected by the El Niño weather phenomenon, when heavy rains disrupt the normal flowering schedules of Andean columnar cacti.

When the bats were experimentally kept away from night-blooming columnar cacti, the results revealed an interesting geographic pattern. In the arid regions of central Mexico and northern Venezuela, including the island of Curaçao, nectar-feeding bats were found to be the only pollinators of these cacti, and flowers from which bats were excluded invariably aborted. North and south of these areas, however, diurnal flower visitors such as birds and bees become important pollinators. In the Sonoran Desert, for example, fruit set is strongly dependent on Lesser long-nosed bats only in cardón (*Pachycereus pringlei*); white-winged doves are the major vertebrate pollinators of saguaro (*Carnegiea gigantea*), and hummingbirds of organ pipe (*Stenocereus thurberi*). In the Andes of southern Peru, the cactus *Weberbauerocereus weberbaueri* is effectively pollinated by the bat *P. genovensium* at night and by the Andean giant hummingbird (*Patagona gigas*) during the day. In the Sonoran Desert, changes in the time of flower closing and the duration of stigma receptivity, rather than changes in flower morphology, have permitted diurnal flower visitors to become effective pollinators. In the Peruvian Andes, *W. weberbaueri* opens its blooms in the later afternoon, and considerable variation in the size, shape, and color

▲ ***Above*** *Lesser long-nosed bats queue up to feed from a cactus flower. Because the bats cannot hover, they spend only a fraction of a second feeding at each pass. They extract the nectar with their extraordinary tongues, which extend for almost the length of their bodies and are tipped with fleshy bristles.*

of its flowers indicates that selection for both bat and hummingbird pollination has been strong in this species. Thus the scarcity of migratory bats has led to the evolution of more generalized pollination systems in bat-pollinated columnar cacti at both the northern and southern limits of their geographical distribution.

In both the the Sonoran Desert and in Venezuela, the blooming seasons of bat-pollinated cacti tend to be seasonally displaced in order to avoid competition for nocturnal pollinators. In Sonora, for example, flowering peaks of cardón and saguaro occur in late April, whereas that of organ pipe occurs in mid-June; saguaro avoids competing for pollinators with cardón by staying open longer during the day. In central Mexico, however, co-occurring columnar cacti tend to bloom at the same time of year (in April–June); how these species avoid competing for bat pollinators is currently unknown.

In another example of bat–cactus coevolution, the form of cardón's breeding system in the Sonoran Desert varies geographically, in part as a result of geographic variation in the abundance of Lesser long-nosed bats. In areas near maternity roosts, cardón populations contain separate male, female, and hermaphrodite individuals (a breeding system technically known as "trioecy"). Away from these roosts, populations contain only females and hermaphrodites, which produce both pollen and ovules in their flowers (this system is known as "gynodioecy"). Males are present in cardón populations only in areas where the nocturnal pollinators are abundant.

Finally, as an unusual method of reducing the impact of competition for bat pollinators, some organ pipe cacti flower early, and in the absence of their own pollen use cardón or saguaro pollen to stimulate the asexual production of fruits bearing mature seeds. In this way, organ pipe flowers that receive the "wrong" pollen, which in the Sonoran Desert happens frequently in April and early May, are not wasted via abortion. Thus, bats serve as the bees of the desert. THF

SHOWING OFF TO THE FEMALES

How lek mating works among Hammerheaded bats

THE HAMMERHEADED FRUIT BAT IS ONE OF A small number of mammalian species that practice lek mating. A lek is an aggregation of displaying males to which females come solely for the purpose of mating. Females usually visit a lek, examine a number of males, and then select one with which to mate. Females are remarkably consistent in their choice, so that only a few males do all the mating. Females undertake all the parental care in lek species. Other forms exhibiting lek behavior include the Uganda kob, the topi, and the Fallow deer among mammals, as well as some birds, frogs, fishes, and insects.

The bat (*Hypsignathus monstrosus*) occurs in tropical forests from Senegal through the Congo basin to western Uganda. In the core of the range in Gabon, the bats form leks at traditional locations and mate during the two annual dry seasons, from January to early March and from June to August respectively. Populations in more peripheral parts of the range may show more dispersed and less stable distributions of displaying males.

A Gabon Hammerheaded bat lek nearly always borders a waterway, and varies from 0.7–1.5km (0.4–0.9mi) in length. Males are spaced about 10–15m (33–49ft) apart along the site, and the array is usually about two males deep, the full range being 1–4. There are calling males on the sites for about 15 weeks each dry season. Early and late in the season, only a few males call. The number increases rapidly to a peak in February and July, and then declines more slowly as the season draws to an end.

Each night at sunset during the mating period, males leave their day roosts and fly directly to the traditional lek sites. At the lek they hang in the foliage at the canopy's edge and emit a loud, metallic call while flapping their wings at twice the call rate. Early in the mating season, there is usually some fighting between males for calling territories. By the time females begin visiting the lek – and they start to do so before they are ready to mate – males are settled in their territories and there is little subsequent interference between them. This is in marked contrast to most ungulate leks, in which male turnover is very high even during the season and males frequently interfere with each others' courtship efforts.

Typical Hammerheaded bat leks contain 30–150 displaying males, each one calling 1–4 times a second and flapping its wings furiously. Females fly along the ranks and periodically hover before a chosen male. This causes the male to perform a staccato variation of its call and to tuck its wings close against its body. Females will make repeated visits on the same night to a decreasing number of males, each time eliciting a "staccato buzz." Finally, selection is complete, the female lands by the male of her choice, and mating is accomplished in 20–30 seconds. Females usually terminate mating with several squeals, and then fly off.

The importance of display in enabling a male to breed has obviously favored a heavy investment in the equipment the bats uses to advertise themselves. Males are twice as large as females, weighing 425g (15oz) compared to 250g (8.8oz); they have an enormous bony larynx that fills their chest cavity, and a bizarre head with enlarged cheek pouches, inflated nasal cavities, and a funnel-like mouth. The larynx and associated head structures are all specializations for producing the loud call.

Females can mate at 6 months, and reach adult size at about 9 months. They can thus produce their first offspring (only one young is born at a time) as yearlings. Females come into heat immediately after birth (post-partum estrus), and thus can produce two successive young each year. In fact, many of the females mating during any dry season are carrying newborn young conceived at the last mating or lek period. As with many lek species, males mature later than females.

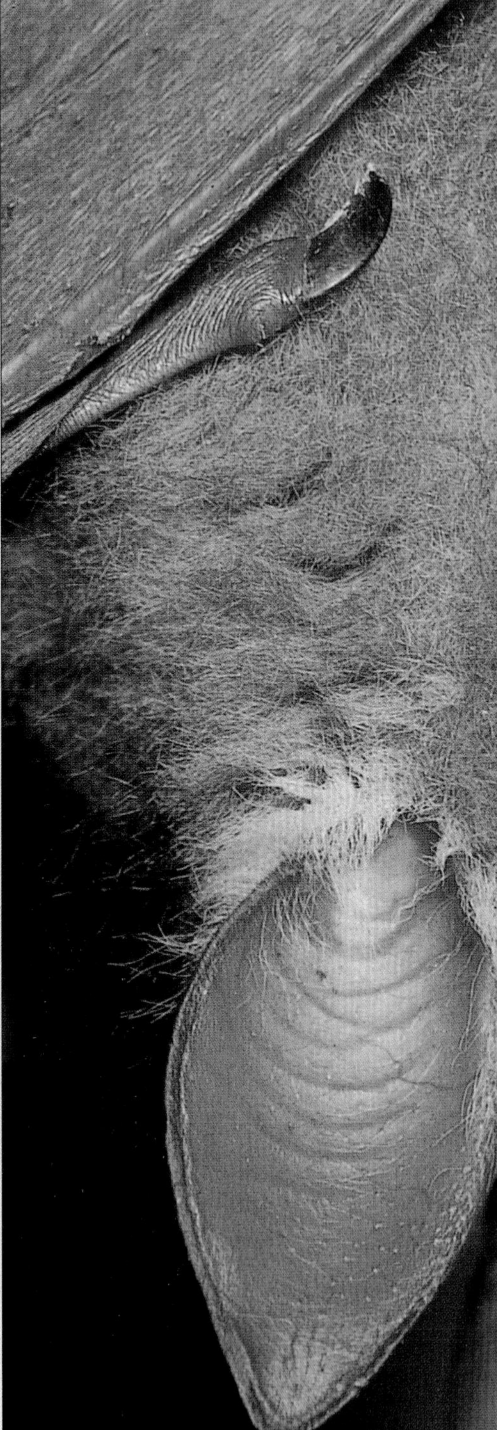

Above *Male Hammerheaded bats owe their bizarre head shape to the competitive demands of lek mating. The broad muzzle encloses inflatable air sacs and a hugely enlarged larynx that extends back to fill much of the chest cavity. The continuous croaking produced with its aid apparently serves to attract potential mates. One naturalist has compared the noise of a lek in full throat to that of "a pondful of noisy American wood-frogs, greatly magnified and transported to the tree-tops."*

Left *Female Hammerheads are more conventional in their appearance, and are also much smaller, with a body length little more than half that of the males.*

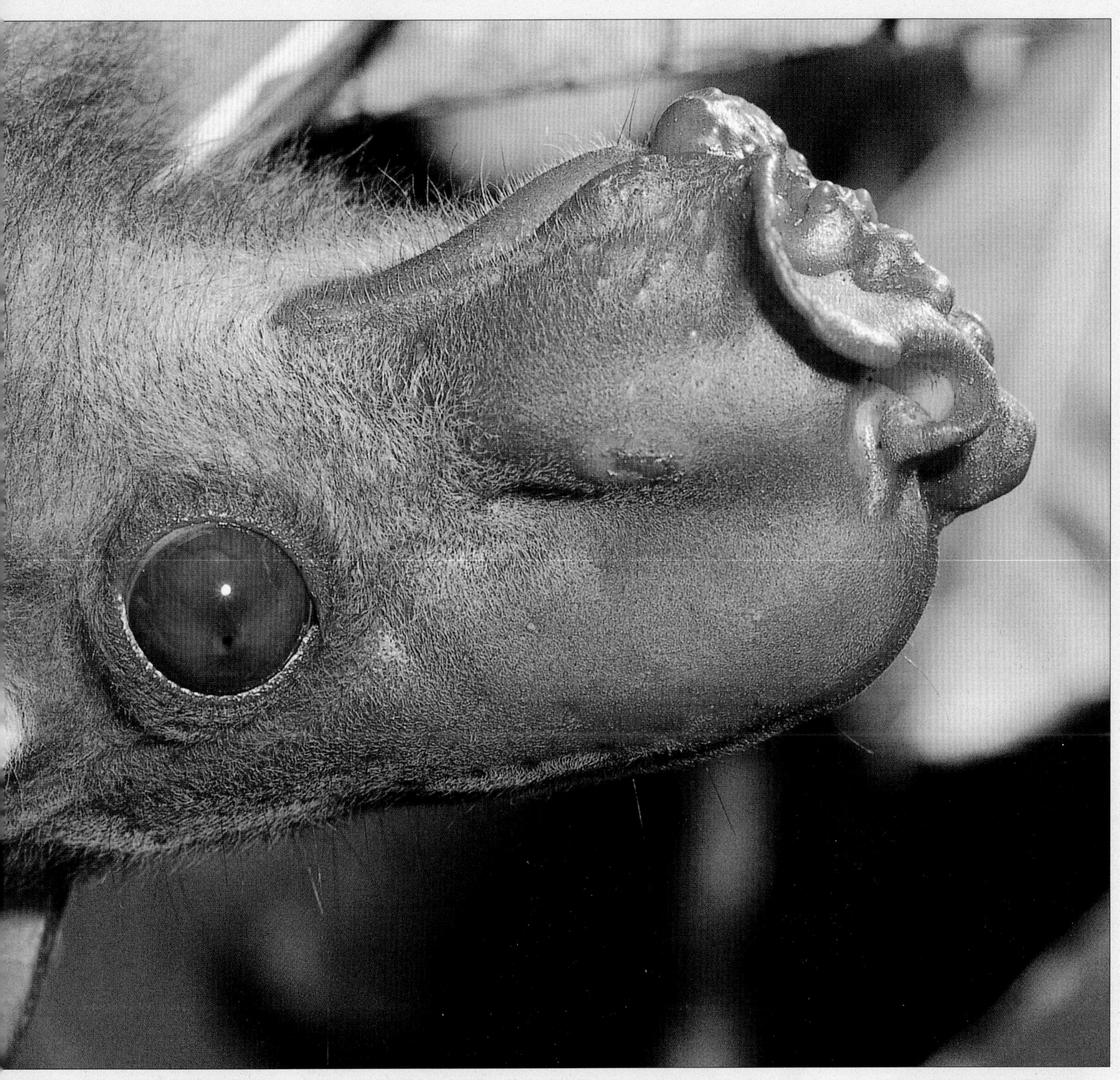

Despite early reports to the contrary. these bats are exclusively frugivorous. Figs and the fruit of several species of *Anthocleista* form the major part of their diets in Gabon. Females and juvenile males appear to feed more on the easily located *Anthocleista*, which is generally found closer to the roost (within 1–4km/0.6–2.5mi) but is also less profitable, for its fruits ripen slowly and a few at a time. By contrast, adult males fly 10km (6.2mi) or more to find the less predictable but more profitable patches of ripe figs, where large numbers of fruits are available on a single trip, but only for a short while. This extra effort by the males presumably pays off by providing more energy for vigorous mating displays. It has the cost, however, that the males may risk starvation if they are unsuccessful. The effects of variable food levels and the high energetic outlays during display may explain the higher parasitic loads (primarily hemosporidians in the blood cells) found in adult males, and also their higher mortality rates. They are also reflected in the abandonment of display by all males following days of colder than average weather, even though females may be visiting for mating.

Lek mating is often considered a "default" mating system, adopted when males cannot provide parental care, protect resources that females require, or defend groups of females. It is easy to understand that the expensive and chancy business of self-advertisement in competition with other males might be undesirable for males if not absolutely necessary. But Hammerheaded bats do not appear to fit these generalizations: there is little males could do to assist itinerant females with young, females rarely form groups (and those formed are at best transient aggregations), and neither roosts nor food sources are defensible, the latter because fig trees are widely dispersed and come randomly and unpredictably into fruit. The costs of display are certainly significant, yet even so males are committed both physically and behaviorally to the system. JWB

PHOTO STORY

A Second to Live

1

2

3

❶❷ *The Greater false vampire bat* (Megaderma lyra), *which is native to South and Southeast Asia, is an expert night hunter that skims low through trees and undergrowth in search of its prey. Using a combination of acute hearing, echolocation, and keen eyesight, this individual homes in on a mouse.*

❸ *At the moment of impact, the bat delivers a bite to the mouse's neck, killing it instantly. Rodents are just part of this truly carnivorous species' diet, which also includes birds, frogs, lizards, spiders, insects, and other bats.*

❹ *Captured prey may be eaten on the wing or carried back to the bat's roost – in a cave, crevice, or hollow tree. False vampire bats get their name from the mistaken notion that they draw blood from their prey, like true vampires. In truth, while they will sometimes drink their victim's blood before eating its flesh, they never feed solely off the blood of other animals.*

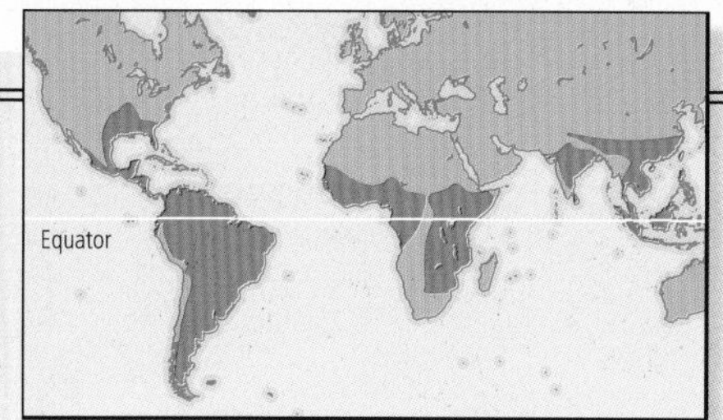

EDENTATES

THE DESIGNATION "EDENTATES" IS A misnomer. Firstly, it contradicts the fact that most species treated here are not toothless, but have at least vestigial teeth. Secondly, it has no taxonomic validity; the order Edentata, which once embraced anteaters, sloths, armadillos, pangolins (and even the aardvark), is now defunct. The perceived similarities that occasioned such a grouping are now thought to be the result of convergent evolution, and not of any phylogenetic connection. "Edentate" is thus retained only as a convenient umbrella term for this most diverse group of mammals.

By the early Tertiary 60 million years ago (the beginning of the "Age of Mammals"), the ancestral edentates had already diverged into two quite distinct lines. The first, comprising small, armorless animals of the suborder Palaeanodonta, rapidly became extinct (there is now a broad consensus that this suborder was ancestral to the Pholidota only, and not to the Xenarthra). The other line, comprising the xenarthrans, was on the brink of a spectacular radiation that was later to produce some of the most distinctive and bizarre of all the New World mammals.

Sluggish Specialists

ANATOMY AND EVOLUTION

The living and recently extinct members of the order Xenarthra are distinguished from all other mammals – including pangolins – by additional articulations between the lumbar vertebrae, which are called xenarthrales (or xenarthrous vertebrae). These bony elements provide lumbar reinforcement for digging, and are especially important for the armadillos. The living xenarthrans also differ from most mammals in having a double posterior vena cava vein (single in other mammals), which returns blood to the heart from the hindquarters of the body. Females have a primitive, divided womb only a step removed from the double womb of marsupials, and a common urinary and genital duct, while males have internal testes, and a small penis with no glans.

Despite these unifying characteristics, the extinct xenarthrans differed greatly in size and appearance from their modern relatives and, in terms of numbers of genera, were more than ten times as diverse. The rise and fall of these early forms is closely linked to the fact that throughout the Tertiary South America was a huge, isolated island. At the beginning of this epoch, ancestral xenarthrans shared the continent only with early marsupials and other primitive mammals, and flourished in the virtual absence of competition. By the late Eocene (38 million years ago), three families of giant ground sloths had emerged, with some species growing to the size of modern elephants. In their heyday during the late Miocene, 30 million years later, ground sloths appeared in the West Indies and southern North America, apparently having rafted across the sea barriers as waif immigrants. Four families of armored, armadillo-like xenarthrans were contemporary with the ground sloths for much of the Oligocene. The largest species, *Glyptodon*, achieved a length of 5m (16.5ft) and carried a rigid 3m (10ft) shell on its back, while the related *Doedicurus* had a massive tail with the tip armored like a medieval mace. Although *Glyptodon* and the giant ground sloths survived until historical times – and are spoken of in the legends of the Tehuelche and Araucan Indians of

Right *Prehistoric edentates. The edentates produced three major groups: "shelled" forms (Loricata), including the extinct glyptodonts and living armadillos; "hairy" forms (Pilosa), including the extinct Giant ground sloth and living tree sloths; and the anteaters (Vermilingua).* **1** *The Giant ground sloth* (Megatherium), *from the Pleistocene of South America, was up to 6m (20ft) long.* **2** Eomanis waldi, *a small armored pangolin from the Eocene of Germany.* **3** Glyptodon panochthus, *a giant shelled form from the Pleistocene of South America.* **4** *Giant anteater* (Scelidotherium), *from the Pleistocene of South America.*

Patagonia – only the smaller tree sloths, anteaters, and armadillos persisted to the present day.

The extinct xenarthrans are believed to have been ponderous, unspecialized herbivores that inhabited scrubby savannas. They were probably out-competed and preyed upon by the new and sophisticated northern invaders. In contrast, the success of the living xenarthrans was due to their occupation of relatively narrow niches, which allowed little space for the less specialized newcomers. The anteaters and leaf-eating sloths, for example, have very specialized diets. To cope with the low energy contents of their foods, both groups evolved metabolic rates that are only 33–60 percent of those expected for their bodyweights, and variable but low (32.7–35°C/91–95°F) body temperatures that burn fewer kilojoules. Armadillos eat a wide range of foods, but are specialized for a partly subterranean way of life; they also have low metabolic rates and body temperatures (33–35.5°C/91.5–96°F) to avoid overheating in their closed burrows. Lacking similarly sluggish metabolisms, the invading mammals were not able fully to exploit these habitats, so competition was probably minimal.

Living in Niches

SOCIAL BEHAVIOR

As consequences of specializing and slowing their metabolisms, the sloths and anteaters use energy frugally, and generally move slowly over small home ranges. Females attain sexual maturity at 2–3 years of age and breed only once a year thereafter. They produce small, precocious litters (usually one young), and invest much time and energy in weaning and post-weaning care. Defense against predators is passive and primarily dependent on cryptic camouflage. While anteaters, and occasionally sloths, may try to flee from an assailant, they more often stand their ground and strike out with their claws. Sloths are reputedly able to survive the most severe injuries; bite wounds and deep scars rarely become infected, and heal completely within weeks. Armadillos show similar trends toward economizing their use of energy, but these are not as marked as in their ant- and plant-eating relatives. The armadillos are less constrained because of their more varied and energy-rich diets, and the ability (at least of some species) to store fat and enter torpor.

The social lives of xenarthrans and pangolins are probably dominated by the sense of smell. All species produce odoriferous secretions from anal glands, which are used to mark paths, trees, or conspicuous objects; these probably advertise the presence, status, and possibly the sexual condition of the marking individual. Scentmarks may also serve as territorial markers, and allow individuals priority of access to scarce resources, such as food. Pangolins can employ their anal secretions as a form of defense, squirting a jet of foul-smelling fluid at an aggressor.

With their lack of teeth, long, sticky tongues, and taste for ants, the pangolins exploit a niche equivalent to that occupied by the South American anteaters. These similarities suggest that selection for the anteating habit has acted in parallel in both the Old and New World. CRD

ORDERS: XENARTHRA AND PHOLIDOTA

36 species in 14 genera and 5 families

ORDER XENARTHRA

ANTEATERS Family Myrmecophagidae p788

4 species in 3 genera. **Silky anteater** (*Cyclopes didactylus*), **Giant anteater** (*Myrmecophaga tridactyla*), **Northern tamandua** (*Tamandua mexicana*), and **Southern tamandua** (*T. tetradactyla*).

SLOTHS

Families Megalonychidae and Bradypodidae p792

5 species in 2 genera. **Brown-throated three-toed sloth** (*Bradypus variegatus*), **Pale-throated three-toed sloth** (*B. tridactylus*), **Maned three-toed sloth** (*B. torquatus*), **Hoffmann's two-toed sloth** (*Choloepus hoffmanni*), and **Southern two-toed sloth** (*C. didactylus*).

ARMADILLOS Family Dasypodidae p796

20 or 21 species in 8 genera. Includes **Nine-banded** or **Common armadillo** (*Dasypus novemcinctus*), **Southern naked-tailed armadillo** (*Cabassous unicinctus*), **Larger hairy armadillo** (*Chaetophractus villosus*), **Brazilian three-banded armadillo** (*Tolypeutes tricinctus*), **Greater fairy armadillo** (*Chlamyphorus retusus*), **Yellow armadillo** (*Euphractus sexcinctus*), **Pichi** (*Zaedyus pichiy*), and the **Giant armadillo** (*Priodontes maximus*).

ORDER PHOLIDOTA

PANGOLINS Family Manidae p800

7 species in 1 genus. **Indian pangolin** (*Manis crassicaudata*), **Giant pangolin** (*M. gigantea*), **Malayan pangolin** (*M. javanica*), **Chinese pangolin** (*M. pentadactyla*), **Ground pangolin** (*M. temminckii*), **Long-tailed pangolin** (*M. tetradactyla*), and **Tree pangolin** (*M. tricuspis*).

SKULLS AND DENTITION

Edentates have the least complex skulls of all mammals; for example, pangolins have smooth, conical skulls with a simple, bladelike structure for their lower jaw. Although the name Edentata means "without teeth," it is in fact only the pangolins, and among the xenarthrans only the anteaters (such as the Southern tamandua) that are completely toothless. Both sloths and armadillos are equipped with a series of uniform, peg-shaped cheek or grinding teeth (premolars and molars). These lack an enamel covering and have a single so-called "open root" that allows continuous growth of the teeth throughout life. True incisor and canine teeth are absent in all edentates, but sloths have enlarged, canine-like premolars.

The diet of edentates ranges from an almost total reliance on ants and termites in the anteaters and pangolins, through a wide range of insects, tubers, and carrion in the armadillos, to plants in the sloths.

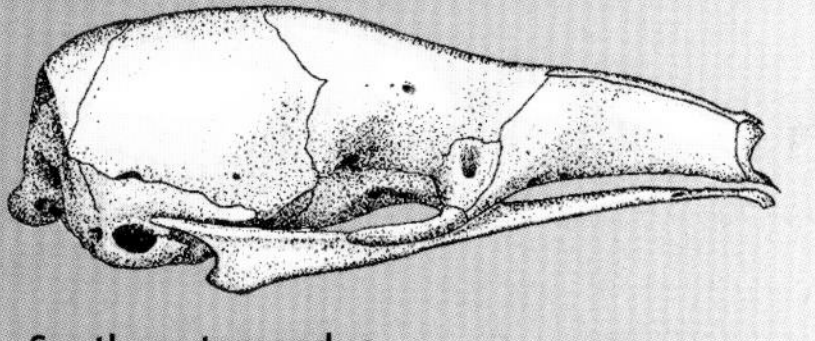

Southern tamandua
12 cm

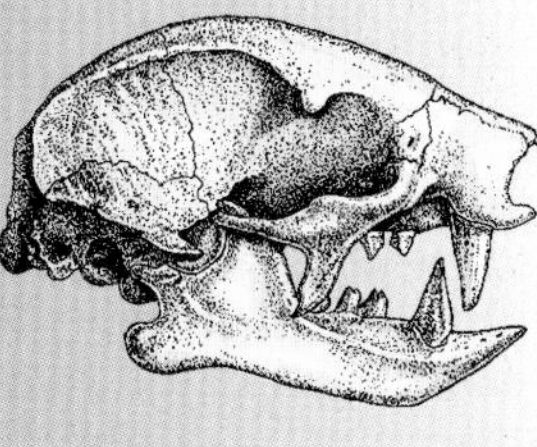

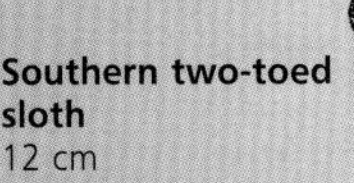

Southern two-toed sloth
12 cm

Anteaters

ANTEATERS FEED EXCLUSIVELY ON SOCIAL *insects, primarily ants and termites. Their adaptations to this diet affect not only masticatory and digestive structures but behavior, metabolic rate, and locomotion. Anteaters are solitary, except that a mother may carry her young on her back for up to a year, until it is nearly adult in size.*

***Right** The Southern tamandua has strong claws and a powerful tail, which is used to gain additional purchase when climbing in trees. The tail can also act as a prop, enabling the animal to rear up on its back legs.*

While the different anteater species do not overlap greatly in distribution, they nonetheless operate at different times and in different strata: the Giant anteater feeds mostly by day (although it becomes nocturnal when it is disturbed by people), whereas tamanduas are variably active both by day and by night and the Silky anteater is strictly nocturnal. Similarly, Giant anteaters are terrestrial, tamanduas partially arboreal, and Silky anteaters almost exclusively arboreal. All anteaters can both dig and climb, as well as walk on the ground. However, the Giant anteater rarely climbs and the Silky anteater descends to the ground only infrequently. There is further niche separation in diet, with Giant anteaters eating the largest-bodied ants and termites, tamanduas the medium-sized insects and Silky anteaters the smallest.

Toothless Insect-eaters

FORM AND FUNCTION

The anteaters share membership in the Order Xenarthra with the sloths, the armadillos, and the extinct glyptodonts, but are the only toothless (edentulous) members of the Order. Mouths in all species are small and only open to a small oval. Anteater snouts are disproportionately long, the Giant anteater's head appearing to be almost tubular and over 30cm (12in) in length. Their narrow, rounded tongues are even longer than their heads; the tongues of tamanduas protrude some 40cm (16in), while that of the Giant anteater can extend up to 61cm (24in). In all anteaters, the tongues are covered in minute, posteriorly directed spines and coated with a thick, sticky saliva secreted from salivary glands relatively larger than those of any other animals. Anteater stomachs are unusual in not secreting hydrochloric acid, but depend instead on the formic acid content of the ants they eat to assist with digestion.

The only natural predators of Giant anteaters are pumas and jaguars – if threatened they rear up on their hind legs, slashing with claws that can be up to 10cm (4in) in length. They have even been known to embrace and crush an attacker. The largest claws are on fingers two and three in Giant and Silky anteaters, but digits two, three, and four

in tamanduas. All have five fingers and four or five toes, although some fingers are reduced in size and enclosed within the skin of the hand. The fifth finger of the Giant anteater and the first, fourth and fifth fingers of the Silky anteater are the reduced digits. Anteaters move with the fingers of the forefeet flexed and turned inward to keep the sharp claw tips from contacting the ground. Sometimes they walk on the sides of their hind feet, turning the claws inward, much as did some of the extinct ground sloths to which they are related. Climbing in trees the tamanduas and Silky anteaters use their prehensile tails, and claws that may be up to 400mm (16in) in length to grip branches. When threatened, a tamandua on the ground balances on hind feet and tail, swiping ferociously with the foreclaws. The defensive posture of Silky anteaters also uses the prehensile tail and hind feet to grasp a supporting branch, but initially the forefeet are raised to the level of the shoulders with the claws aimed forward and inward. Amazingly, Silky anteaters can stretch out horizontally from the supporting branch, an unusual feat (shared with tree sloths) made possible by additional (xenarthrous) articulations between vertebrae. Furthermore, an additional (and unique) joint in the sole of the foot allows the claws to be turned back under the foot to enhance the grasp. The most common predators of arboreal anteaters include Harpy eagles, hawk eagles and the Spectacled owl. These hunters fly above the canopy and search visually for prey; thus, the coat of the Silky anteater, which closely resembles the massive balls of silvery fluff that make up the seed pods of the silk-cotton Ceiba tree, may serve as protective coloration. Silky anteaters are frequently found in these trees. None of the anteaters is particularly vocal, but Giant anteaters bellow when threatened. If separated from the mother, young animals produce short, high-pitched whistles.

Digging for Dinner

DIET

Anteaters detect prey mainly by smell, but their vision is probably poor. Giant anteaters feed on large-bodied colonial ants and termites. Anteaters feed rapidly. Typically they dig a small hole in the nest, and lick up worker ants as they emerge, and with tongue movements as rapid as 150 times a minute take larvae and cocoons as well. Insects trapped on the sticky saliva-coated tongue are crushed against the hard palate prior to swallowing. Anteaters avoid large-jawed ant and termite

FACTFILE

ANTEATERS

Order: Xenarthra

Family: Myrmecophagidae

4 species in 3 genera

DISTRIBUTION S Mexico, C & S America S to Paraguay and N Argentina; Trinidad.

GIANT ANTEATER *Myrmecophaga tridactyla*
C America; S America E of the Andes to Uruguay and NW Argentina. Grassland, swamp, lowland tropical forest. HBL 1–1.3m (3.3–4.2ft); TL 65–90cm (25.5–35.5in); WT 22–39kg (48–86lb); male anteaters are 10–20 percent heavier than females. Coat: coarse, stiff, dense; coloration gray with black-and-white shoulder stripe. Breeding: 1 young born in spring after a gestation of 190 days. Longevity: unknown in the wild, but up to 26 years in captivity. Conservation status: Vulnerable.

NORTHERN TAMANDUA *Tamandua mexicana*
Northern tamandua, Northern collared or lesser anteater
S Mexico to NW Venezuela and NW Peru. Savanna, thorn scrub, wet and dry forest. HBL 52.5–57cm (21–22in); TL 52.5–55cm (21–21.5in); WT 3.2–5.4kg (7–12lb). Coat: light fawn to dark brown with variable patches of black or reddish-brown from shoulders to rump. Breeding: Gestation 130–150 days. Longevity: unknown in the wild but to at least 9 years in captivity.

SOUTHERN TAMANDUA *Tamandua tetradactyla*
Southern tamandua, Southern collared or lesser anteater
S America E of the Andes from Venezuela to N Argentina; Trinidad. HBL 58–61cm (23–24in); TL 50–52.5cm (19.5–21.5in); WT 3.4–7kg (7.5–15.5lb). Coat: as for Northern tamandua, but black "vest" is only present in specimens from SE portion of range.

SILKY ANTEATER *Cyclopes didactylus*
C and S America, from S Mexico to the Amazon basin and N Peru. Tropical forest. HBL 18–20cm (7–8in); TL 18–26cm (7–10in); WT 375–410g (13.2–14.4oz). Coat: soft, silky gray to yellowish-orange, with darker mid-dorsal stripe.

Abbreviations HBL = head–body length TL = tail length WT = weight

*◐ **Left** A Giant anteater with offspring in Brazil. The young Giant anteater may continue to ride around on its mother's back for up to a year, well past the weaning stage, which occurs at about six months.*

soldiers. Even though the skin on their muzzles is thick it is evidently not impervious to the bites of insect soldiers. Because they utilize each nest for only a short period, and take as few as 140 insects (only about 0.5 percent of their daily food requirement) per feeding bout, anteaters cause little permanent damage to nests. Their density appears to depend on the number of nests that are available in a given area; many must be visited daily to get sufficient nutrition (which may amount to 35,000 ants a day). Beetle larvae are also taken. Water requirements are generally met by their food.

The way anteaters eat is unique among mammals. They contract their chewing (temporal and masseter) muscles to roll the two halves of the lower jaw towards the middle, thereby separating the anterior tips to open the mouth. The mouth is closed by the pterygoid muscles that pull the lower rear (posteroventral) edges of the two lower jaw bones inward (medially), raising the anterior tips to close the mouth. The result is simplified and minimal jaw movement which, when coupled with movements in and out of the tongue and nearly continuous swallowing, maximize the rate of food intake. The extraordinary movements of the tongue are controlled by a sternoglossus muscle that attaches to the base of the sternum.

Tamanduas specialize in smaller-bodied termites and ants than do the Giant anteater, and also avoid the soldier castes. They also refrain from eating ant and termite species that have chemical defenses, and will eat bees and honey. A tamandua will typically consume 9,000 ants in a day. The average length of arboreal ants and termites eaten by Silky anteaters is 4mm (0.15in) as opposed to the 8mm (0.3in) or larger prey of Giant anteaters.

Precocious Young

SOCIAL BEHAVIOR

All species of anteater are usually solitary. Home ranges in Giant anteaters may be as small as 0.5sq km in areas of high food availability, such as the tropical forests of Barro Colorado Island, Panama, or the southeastern highlands of Brazil. In habitats that support fewer ant and termite colonies such as the mixed deciduous forests and semi-arid

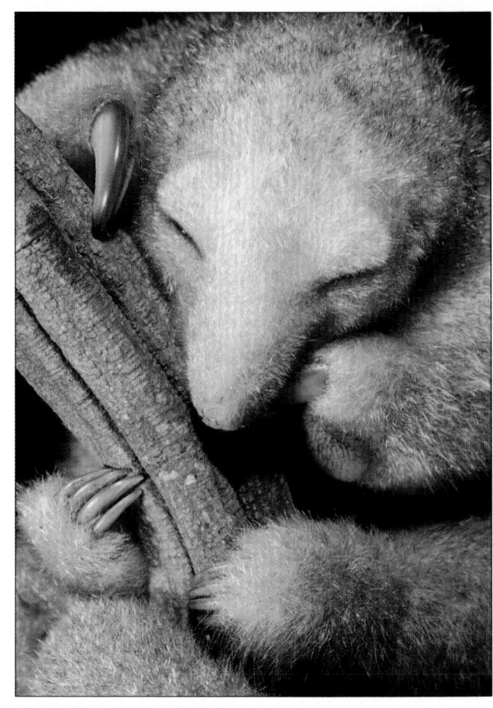

Above and below *Silky anteaters taking a break: above, curled in defensive posture with the claws in front of the face; below, suspended from a branch. An adaptation in the foot of Silky anteaters allows the claws to be turned back under the foot to improve grip.*

llanos of Venezuela one individual Giant anteater may require as much as 2,480ha (6,200 acres). The ranges of female Giant anteaters may overlap by as much as 30 percent, while those of males typically overlap by less than 5 percent. Tamanduas are less than half the size of the Giant anteater and in favorable habitats such as Barro Colorado, occupy home ranges of 50–140ha (124–346 acres). In the open llanos one animal may require as much as 340–400ha (840–988 acres). Silky anteater females on Barro Colorado have a home range that averages 2.8ha (7 acres) while that of a single male is approximately 11ha (27 acres). The home range of this male overlapped those of two females, but not the range of an adjacent male. Although the geographic distribution of the four anteater species differ, when they occur in the same habitat, home ranges of one do not appear to be affected by the presence of another.

Giant anteaters and tamanduas mate in the fall, and the single young is born in the spring. The Giant anteater gives birth standing, using her tail as a third support. The young are precocious and have sharp claws that allow them to crawl to the mother's back shortly after being born. Twins occur rarely and the young are suckled for approximately six months but may remain with the mother up to the age of two years, by which time they will have reached sexual maturity. Young Giant anteaters can gallop by about a month after birth, but generally either move slowly or are carried on the mother's back. Tamanduas may place the young one on a branch near a preferred feeding location, or leave them for a short period of time alone in a leaf nest, a practice that is shared with Silky anteaters. The young of the Silky anteater is fed semidigested ants that are regurgitated by both parents, and the infant may be carried by either parent. Giant anteater young are miniatures copies of their parents; however, tamandua infants do not resemble the parents and range in color from white to black.

Giant anteaters do not actually burrow, but instead scoop out shallow depressions in which they rest for up to 15 hours a day. They remain cryptic by covering their bodies with the great fanlike tail. Tamanduas generally rest in hollows in trees while Silky anteaters sleep during the day curled up on a branch with the tail wrapped around the feet. They generally do not spend more than one day in a single tree.

Giant anteaters and tamanduas can produce strong-smelling secretions from their anal glands. Silky anteaters have a facial gland; however, its purpose is unknown. Giant anteaters can also distinguish the scent of their own saliva, although it is not known whether they use salivary secretions to communicate.

All anteaters have low metabolic rates: Giant anteaters have the lowest recorded body temperature for a placental mammal, 32.7°C (90.9°F), and the tamanduas' and Silky anteater's body temperatures are not especially higher. Daily activity periods generally do not average more than eight hours for the Giant anteater and tamandua and about four hours for the Silky anteater.

Primarily resulting from slight color pattern differences, Giant anteaters have been divided into three subspecies, and *T. mexicana* into five. Color variation in *T. mexicana* depends on the size and darkness of the black vest, although all individuals of this species show some degree of this marking. In contrast, *T. tetradactyla* shows great variation in this trait. In animals in the northern part of the range the coat is a uniform light color, while those in the southern part may have striking vest development. The species differences are most striking where the geographic ranges abut, and may be an excellent example of character displacement. This coat color variation probably explains why this species has been divided into thirteen subspecies. Differences in coat color probably also explain the naming of seven subspecies of Silky anteater. In the northern regions the animal is a uniformly golden color, or has a darker dorsal stripe, but it becomes progressively grayer, and the mid-dorsal stripe darker, in the south.

Above *Inserting its long snout into a hollow log, a Giant anteater feeds on the insects inside. Anteaters are choosy about the type of ants they consume, taking care to avoid the aggressive soldiers.*

Prey for Trophy Hunters

CONSERVATION AND ENVIRONMENT

With the exception of small-scale use of tamandua skin in local leather industries, anteaters have little commercial value and are seldom hunted for food. However, the Giant anteater has disappeared from most of its historic range in Central America as a result of habitat loss and human encroachment. In South America, it is frequently hunted as a trophy or captured by animal dealers. It has been extirpated in some parts of Peru and Brazil. Tamanduas also suffer when they occur near human habitations. They put on spectacular defenses and may be hunted with dogs for sport, or are often killed on the roads around areas of human settlement. In the Venezuelan llanos, young individuals may be tamed and prove to be popular as pets. However, the most serious threat to these creatures is the loss of habitat and the destruction of the limited number of prey species upon which they feed. VN

Sloths

ALTHOUGH SLOTHS ARE RENOWNED FOR their almost glacial slowness of movement, they are the most spectacularly successful large mammals in Central and tropical South America. On Barro Colorado Island, Panama, two species – the Brown-throated three-toed sloth and Hoffmann's two-toed sloth – account for two-thirds of the biomass and half of the energy consumption of all terrestrial mammals, while in Surinam they comprise at least a quarter of the total mammalian biomass. Success has come from specializing in an arboreal, leaf-eating way of life to such a remarkable extent that the effects of competitors and predators are scarcely perceptible.

Oviedo y Valdes, one of the first Spanish chroniclers of the Central American region in the 16th century, wrote that he had never seen an uglier or more useless creature than the sloth. Fortunately, little commercial value has since been attached to these animals, although large numbers, especially of two-toed sloths, are hunted locally for their meat in many parts of South America. Beauty, however, is in the eye of the beholder, and modern-day tourists will pay to have their photograph taken with sloths stolen from the forests and touted on the streets of South American cities. The Maned three-toed sloth of southeastern Brazil is considered Endangered due to the destruction of its coastal rainforest habitat, and the fortunes of all five species are inextricably bound up with the future of the tropical forests.

A Walking Ecosystem

FORM AND FUNCTION

Sloths have rounded heads and flattened faces, with small ears hidden in the fur; they are distinguished from other tree-dwelling mammals by their simple teeth (five upper molars, four lower), and their highly modified hands and feet which terminate in curved claws 8–10cm (3–4in) long.

Below *At home in the trees, a Brown-throated three-toed sloth takes its ease in Bolivia's Gran Chaco National Park. The three-toed sloths spend almost all their lives in the branches, only descending to the ground once or twice a week to defecate.*

Their general appearance is extraordinary, but most remarkable of all is the fact that sloths are green; they possess a short, fine underfur and an overcoat of longer and coarser hairs which, in moist conditions, turn green, owing to the presence of two species of blue-green algae that grow in longitudinal grooves in the hairs. This helps to camouflage animals in the tree canopy. The ecology of sloth fur does not end there, for it also harbors animals, including moths (*Cryptoses* spp), ticks (*Amblyomma varium*, *Boophilus* spp), and beetles (*Trichilium* spp). All species have extremely large, multi-compartmented stomachs, which contain cellulose-digesting bacteria. A full stomach may account for almost a third of the body weight of a sloth, and meals may be digested there for more than a month before passing completely into the relatively short intestine. Feces and urine are passed only once a week, at habitual sites at the bases of trees.

The sloths are grouped into two distinct genera and families, which can be distinguished most easily by the numbers of fingers: those of genus *Choloepus* have two fingers and those of genus *Bradypus* have three. Misleadingly, despite the fact that both genera have three toes, the two-fingered forms are known as two-toed and the three-fingered forms as three-toed sloths.

Both two- and three-toed sloths maintain low but variable body temperatures, from 30–34°C (86–93°F), which fall during the cooler hours of the night, during wet weather, and whenever the animals are inactive. Such labile body temperatures help to conserve energy: sloths have metabolic rates that are only 40–45 percent of those expected for their body weights as well as reduced muscles (about half the relative weight for most terrestrial mammals), and so cannot afford to keep warm by shivering. Both species frequent trees with exposed crowns and regulate their body temperatures by moving in and out of the sun.

Sharing the Forest

DISTRIBUTION PATTERNS

While representatives of both sloth families occur together in tropical forests through much of Central and South America, sloths within the same genus occupy more or less exclusive geographical ranges. These closely related species differ little in body weight (staying within a 10 percent range), and have such similar habits that they are apparently unable to coexist.

Where two- and three-toed sloths occur together, the two-toed form is 25 percent heavier than its relative and it uses the forest in different ways. In lowland tropical forest on Barro Colorado Island in Panama's Canal Zone, the Brown-throated three-toed sloth achieves a density of 8.5 animals per ha (3.5 per acre), over three times that of the larger Hoffmann's two-toed sloth. The smaller species is sporadically active for over 10 hours out of 24, compared with just 7.6 hours for the two-toed sloth and, unlike its nocturnal relative, it is active both by day and by night. Three-toed sloths maintain overlapping home ranges averaging 6.6ha (16.3 acres), three times those of the larger species. Despite their apparent alacrity, however, only 11 percent of three-toed sloths travel further than 38m (125ft) in a day, and some 40 percent remain in the same tree

Left A Southern two-toed sloth rests on the fork of a branch. Two-toed sloths spend much of their lives hanging upside down, supported by their hooked claws; they even sleep and give birth in that position.

Below This Brown-throated three-toed sloth clinging to a tree in Panama owes its green coloration to an algal growth. The algae provide camouflage and possibly also a source of nutrition, either absorbed through the skin or licked directly from the hair.

FACTFILE

SLOTHS

Order: Xenarthra

Families: Megalonychidae (two-toed sloths) and Bradypodidae (three-toed sloths)

5 species in 2 genera

DISTRIBUTION C and S America

Habitat Lowland and upland tropical forest; montane forest to 2,100m/7,000ft (Hoffmann's two-toed sloth only).

Coat Stiff, coarse, grayish-brown to beige, with a greenish cast provided by the growth of blue-green algae on the hairs; dark hair on face and neck, lighter fur on shoulders; hair grows to 6cm (2.4in) on three-toed sloths and to 15cm (6in) on two-toed sloths.

Breeding Gestation period 6 months (Southern two-toed sloth, three-toed sloths); 11.5 months (Hoffmann's two-toed sloth).

Longevity 12 years (up to at least 31 in captivity).

TWO-TOED SLOTHS Genus *Choloepus*
From Nicaragua S through C American isthmus to Colombia, Venezuela, Surinam, Guyana, French Guiana, NC Brazil, and N Peru. 2 species: **Hoffmann's two-toed sloth** (*C. hoffmanni*); **Southern** or **Linné's two-toed sloth** (*C. didactylus*). HBL 58–70cm (23–28in), WT 4–8kg (8.8–17.6lb), tail absent.

THREE-TOED SLOTHS Genus *Bradypus*
From Honduras S through C American isthmus to Colombia, Venezuela, Surinam, Guyana, and French Guiana; coastal Ecuador, Bolivia, Paraguay, and N Argentina. 3 species: **Brown-throated three-toed sloth** (*B. variegatus*), **Pale-throated three-toed sloth** (*B. tridactylus*), **Maned three-toed sloth** (*B. torquatus*). HBL 56–60cm (22–24in), TL 6–7cm (2.4–2.8in), WT 3.5–4.5kg (7.7–9.4lb). The Maned three-toed sloth is classed as Endangered.

Abbreviations HBL = head–body length TL = tail length WT = weight

on two consecutive nights; the three-toed sloths, by contrast, change trees four times as often.

Three Maned three-toed sloths in an Atlantic forest reserve of south-eastern Brazil were observed to eat 99 percent leaves, with tree leaves (83 percent) preferred to liana leaves (16 percent). Moreover, young leaves (68 percent) were favored over mature ones (7 percent) throughout the year. Their diet included a total of 21 plant species (16 tree and 5 liana), but each individual made up its diet from an even smaller number of plant species (7–12). The sloths consumed only a tiny fraction of the species available to them, and those they ate were not particularly abundant. It seems likely that sloths have evolved resistance to the defensive poisons produced by certain plants and so eat predominantly those. Their metabolism is extremely slow, which may allow their gut to neutralize the plant toxins as they pass through, contributing to the sloth's success as the ultimate plant-eater.

Inheriting the Mother's Domain

SOCIAL BEHAVIOR

Sloths are believed to breed throughout the year, but in Guyana births of the Pale-throated three-toed sloth occur only after the rainy season, between July and September. Reproduction in the Maned three-toed sloth is aseasonal. The single young, weighing 300–400g (10.5–14oz), is born above ground and is helped to a teat by the mother. The young of all species cease nursing at about 1 month, but may begin to take leaves even earlier. They are carried by the mother alone for 6–9 months and feed on leaves they can reach from this position; they utter bleats or pure-toned whistles if separated. After weaning, the young inherit a portion of the home range left vacant by the mother, as well as her taste for leaves. A consequence of inheriting preferences for different tree species is that several sloths can occupy a similar home range without competing for food or space; this will tend to maximize their numbers at the expense of howler monkeys and other leaf-eating rivals in the forest canopy. Two-toed sloths may not reach sexual maturity until the age of 3 years (females) or 4–5 years (males).

Adult sloths are usually solitary, and patterns of communication are poorly known. However, males are thought to advertise their presence by wiping secretions from an anal gland onto branches, and the pungent-smelling dung middens conceivably act as trysting places. Three-toed sloths produce shrill "ai-ai" whistles through the nostrils, while two-toed sloths hiss if disturbed. CRD

Left *A large adolescent Pale-throated three-toed sloth clings protectively to its mother in the rain forest of Brazil's Manaus province. Even though they are weaned at 4 weeks, the young usually stay with their mothers for at least another 5 months, relying on them for transport through the trees.*

Armadillos

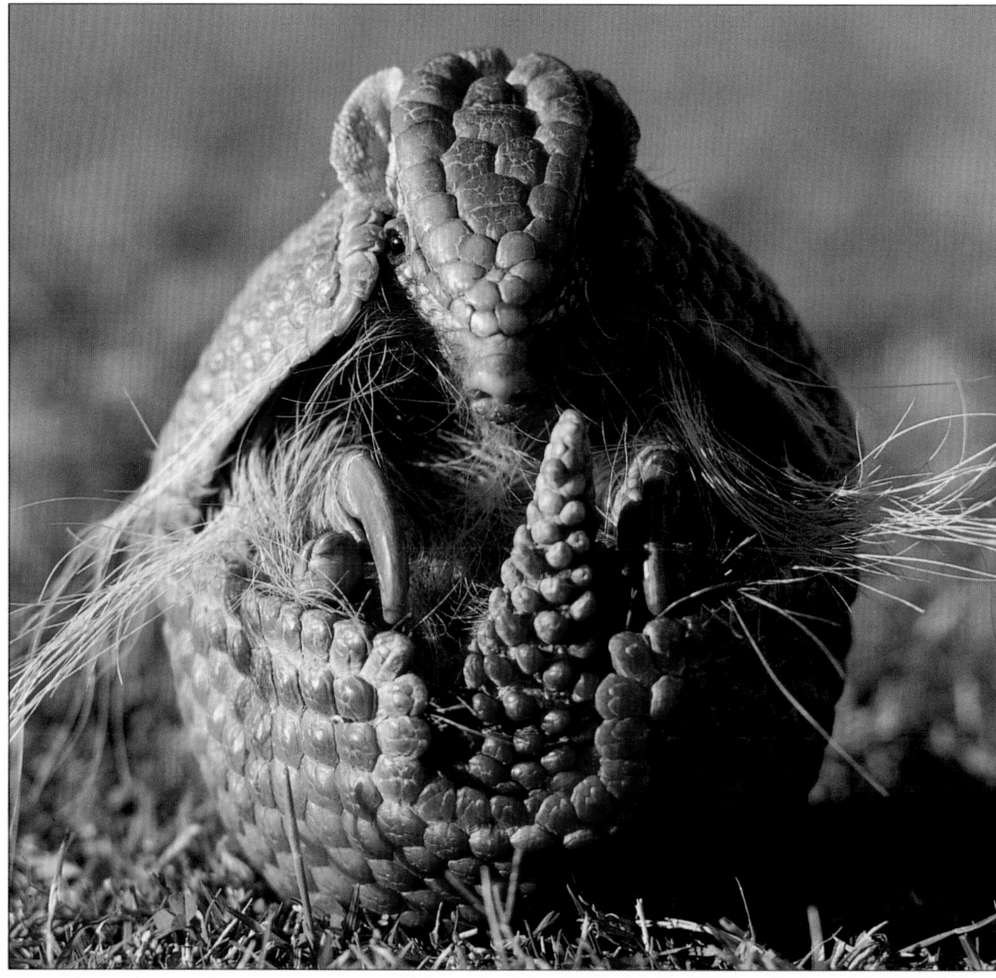

ARMADILLOS ARE ONE OF THE OLDEST, AND *oddest, groups of mammals. Because of the tough protective carapace they all possess, early zoologists often linked them with shelled vertebrates like turtles. Modern taxonomists put them in the order Xenarthra with the anteaters and sloths.*

Fossil evidence suggests that the armadillo lineage forms one of the earlier branches in the evolution of placental mammals, arising about 65–80 million years ago. However, recent molecular genetic studies have indicated armadillos may be closely related to such later-evolving groups as the ferroungulates, which include the carnivores, cetaceans, and artiodactyls.

Presently classified in the order Xenarthra, armadillos were long included in the now-obsolete order Edentata, which means "without teeth." This was always spurious, as they all possess rudimentary, peglike teeth that are undifferentiated (in other words, not divided into incisors, canines, or molars) and that serve to mash up their food. Most species have 14–18 teeth in each jaw, but the Giant armadillo, with 80–100, has more than almost any other mammal. In the long-nosed species of the genus *Dasypus*, the jaws do not open very wide, so, as with the anteaters, they capture prey with their long tongues.

Though hardly lightweights, modern armadillos are puny compared to their ancestors. The largest extant species, the Giant armadillo (*Priodontes maximus*), weighs 30–60kg (66–132lb), but the extinct glyptodonts were far more massive, with weights estimated at 100kg (220lb) or more. Some of these fossil forms were so large that their carapaces (up to 3m/10ft long) were used as roofs or tombs by early South American Indians.

Our knowledge of living armadillos is extremely sparse. Many species have not been studied extensively in the wild, and attempts to breed armadillos for study in captivity have been largely unsuccessful. The only well-known species currently is the Nine-banded armadillo, which has been the subject of a few long-term field studies.

Insect-eaters in Armor

FORM AND FUNCTION

All armadillos possess a number of distinctive features, most notably a tough carapace that covers some portion of the upper surface of their bodies. This shell probably provides some protection from predators and minimizes damage from the thorny vegetation that armadillos frequently pass through. The carapace develops from the skin, and is composed of strong bony plates, or scutes, overlaid by horny skin. There are usually broad and rigid shields over the shoulders and hips, and a variable number of bands (from 3–13) over the middle of the back that are connected to the flexible skin beneath. The tail, the top of the head, and the outer surfaces of the limbs are also usually armored (although the tail is not covered in the genus *Cabassous*), but the undersurface is just soft, hairy skin. To protect this vulnerable area, most species are able to withdraw the limbs under the hip and shoulder shields and sit tight on the ground, while some, such as the three-banded armadillos of the genus *Tolypeutes*, can roll up into a ball. While this strategy may prove effective against most predators, it has unfortunately made these species easy prey for human hunters, possibly contributing to their current listing as Endangered species.

While apparently well-protected, armadillos are not invulnerable to predation. Juvenile mortality can be twice that of adults, much of it due to coyotes, bobcats, mountain lions, some raptors, and even domestic dogs. Juveniles may be more vulnerable to predators because of their small size

Above *A Three-banded armadillo rolling itself up into a defensive ball. When fully curled, it has the appearance of a puzzle ball, leaving no chinks for natural predators to attempt to prize it open. Yet this defense has afforded little help against predation by humans, who have exploited armadillos as a source of food for centuries.*

ARMADILLOS AND LEPROSY

In the 1960s, Eleanor Storrs made the remarkable discovery that armadillos inoculated with the leprosy bacillus can develop the disfiguring human disease, and in the 1970s the condition was found in wild populations. In wild Nine-banded armadillos, its occurrence varies regionally: Floridian armadillos lack leprosy, while as many as 20 percent of the animals from populations in Texas and Louisiana may be infected.

Unlike humans, armadillos exhibit no external symptoms until the disease has progressed sufficiently to fatally damage the internal organs. It is not yet known whether people can contract leprosy from armadillos, but the risk may depend on where they are from: unlike individuals from more tropical regions, people of northern European descent are relatively immune to the malady.

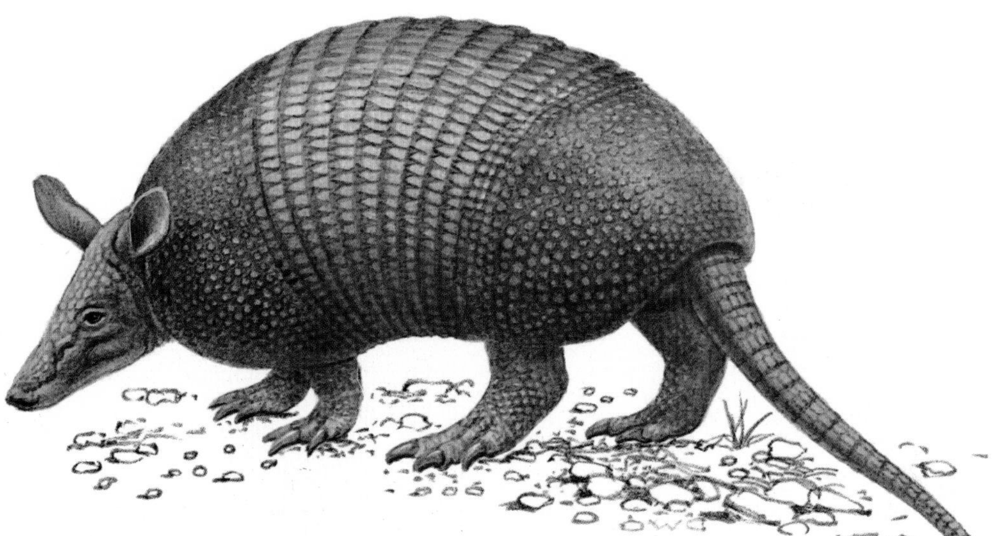

and softer carapace, but adult carcasses have also been found in the guts of large predators like jaguars, alligators, and black bears.

Most armadillos find prey by digging in the soil. In addition, many species excavate burrows that are used as refuges, resting-places, and nest sites for rearing young. Consequently, most armadillos have muscular fore and hind limbs, ending in large, sharp claws that facilitate digging. While the hind limbs always bear five-clawed digits, the fore limbs may have from three to five digits with curved claws, depending on the species. In some species such as the Naked-tailed armadillos and the Giant armadillo the front claws are greatly enlarged, perhaps to facilitate opening ant and termite mounds for foraging. As a result, however, these species are unable to run quickly when danger appears.

Because they are usually active at night, most armadillos have poor eyesight. They seem, however, to have well-developed senses of hearing and smell, which may be used in the detection of both predators and prey. Olfaction may also be employed to determine the identity (and, during the breeding season, the reproductive condition) of other armadillos. The Yellow armadillo and species in the genus *Chaetophractus* have 3–4 gland pits located on the back carapace. The long-nosed armadillos of the genus *Dasypus* have glands on the ears, eyelids, and soles of the feet, as well as a bean-shaped pair of anal glands that produce a yellowish secretion. These glands may be important in chemical communication, as armadillos are frequently observed sniffing this area when they encounter one another; they may also rub the glands on the ground in spots along the periphery of their home range. Although their hearing seems fairly acute, most armadillos are silent. What sounds they do produce are usually just low grunts or squeals, so most communication is probably chemical.

While there are reports of modest differences in body size between the sexes, the males being larger than the females, most armadillos exhibit no obvious sexual dimorphism. One interesting anatomical feature is the male penis, one of the longest among mammals, extending two-thirds of the body length in some species. Armadillos were at one time thought to be the only mammals other than humans to copulate face to face, though this is now no longer believed to be the case; it seems instead that the males mount females from behind, as in most other mammals. If so, then the long penis may be necessary to permit intromission, given the necessity to extend beyond the

Above *The Nine-banded armadillo, unlike the other species, is able to traverse water by inflating its stomach and intestine with air for buoyancy. Since it can hold its breath for several minutes, it can cross smaller streams underwater.*

Below *The configuration of the armored shell varies markedly between species of armadillo:* ***1*** *the Southern three-banded armadillo* (Tolypeutes matacus)*;* ***2*** *the pichi* (Zaedyus pichiy)*;* ***3*** *the Lesser fairy armadillo* (Chlamyphorus truncatus).

FACTFILE

ARMADILLOS

Order: Xenarthra

Family: Dasypodidae

20 or 21 species in 8 genera

DISTRIBUTION Florida (except Everglades), Georgia, and South Carolina W to Kansas; E Mexico, C and S America to Straits of Magellan; Trinidad and Tobago, Grenada, Margarita.

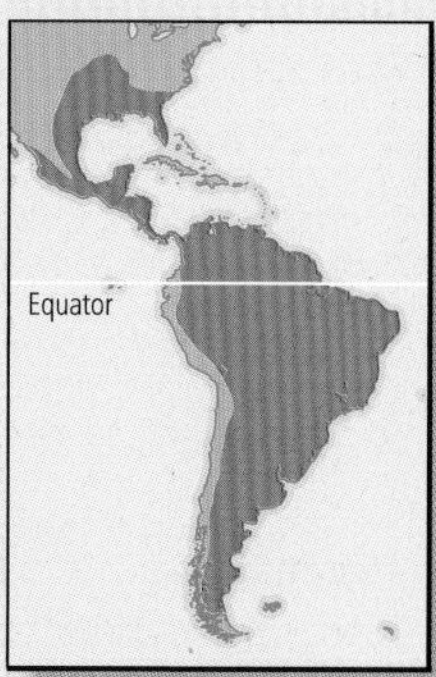

HABITAT Savanna, pampas, arid desert, thorn scrub, and deciduous, cloud, and rain forest.

SIZE **Head–body length** ranges from 12.5–15cm (5–6in) in the Lesser fairy armadillo to 75–100cm (30–39in) in the Giant armadillo; **tail length** ranges from 2.5–3cm (1–1.2in) to 45–50cm (18–20in) and **weight** from 80–100g (2.8–3.5oz) to 30–60kg (66–132lb), both in the same two species.

COAT Broad shield of pale pink or yellowish dark brown armor (scute plates) over shoulders and pelvis, with varying numbers of flexible half-rings over middle of back; some species have white to dark brown hairs between the scute plates.

DIET Soil invertebrates, especially ants and termites. Will also feed on some plant matter, and are occasionally observed scavenging vertebrate carcasses.

BREEDING Both sexes are sexually mature at about 1 year. Breeding can occur year-round, but is most frequent during summer. Gestation varies from 60–65 days (in the Yellow and hairy armadillos) to 120 days (prolonged by delayed implantation) in the Nine-banded armadillo. Litter size usually 1–4, but reaching 8–12 in some species.

LONGEVITY Unknown in the wild, but may be about 8–12 years (up to 20 in captivity).

CONSERVATION STATUS At present, 6 species are considered Vulnerable or Endangered by the IUCN, and 2 more are listed as Lower Risk: Near Threatened. Four species are classed as Data Deficient.

See genera box ▷

1

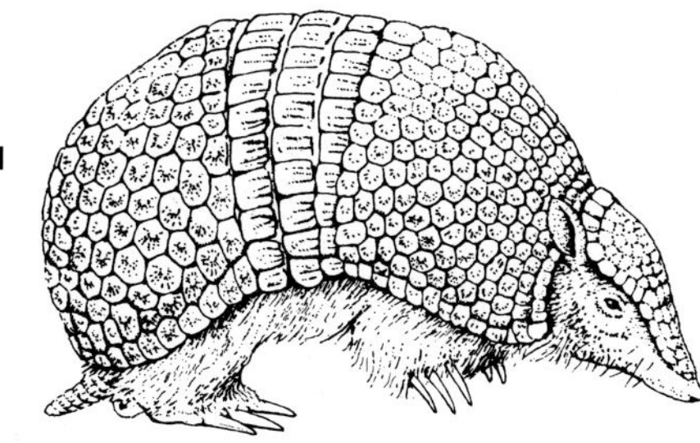

2

3

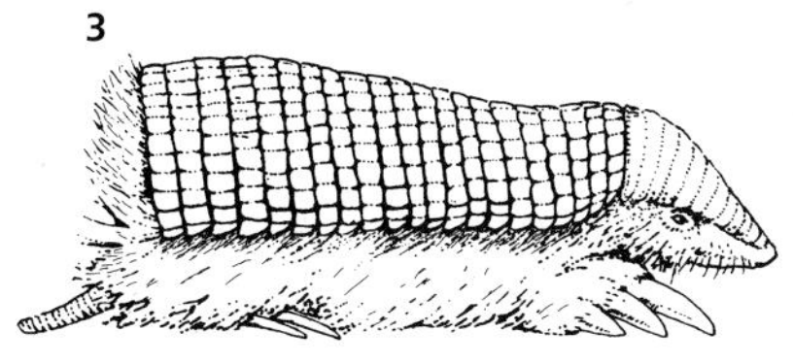

armored carapace to reach the vaginal opening. The musculature of the penis consists of a series of longitudinal and circumferential supporting fibers, an arrangement that is thought to be common to all mammals.

New World Burrowers

DISTRIBUTION PATTERNS

Armadillos are strictly New World species; the majority of fossil forms come from South America, suggesting that this was where the group originally evolved. They subsequently colonized North America, where glyptodont fossils are found as far north as Nebraska, during periods when a land bridge connected the two continents. These fossil forms eventually went extinct, leaving no armadillos in North America until recent times, when, from the late 1800s onward, the Nine-banded armadillo (*Dasypus novemcinctus*) rapidly expanded its range from northern Mexico to include much of the southern USA, with current sightings as far north as Nebraska, southern Missouri, and southwestern Tennessee. In Florida, several armadillos escaped from zoos or private owners in the 1920s, and these also established wild populations, which have slowly spread northward and westward. Florida-derived and Texas-derived populations of armadillos have probably made contact by now, possibly in Alabama or Mississippi.

Although normally associated with moist, tropical habitats, armadillos can be found almost anywhere in the New World. For example, the pichi is found in the Patagonian region of Argentina all the way south to the Straits of Magellan; the Hairy long-nosed armadillo is known only from high-altitude regions of Peru from 2,400–3,200m (7,900–10,500ft); while the Greater long-nosed armadillo occurs only in the rain forests of the Orinoco and Amazon basins.

Species of armadillo vary dramatically in their abundance. The Nine-banded armadillo, also aptly known as the Common long-nosed armadillo, can reach population densities of 50 per sq km (130 per sq mi) in the coastal prairies of Texas and elsewhere. However, the maximum density estimated for the Southern naked-tailed armadillo in the Venezuelan llanos is only 1.2 per sq km (3.1 per sq mi), and the Giant armadillo, even in optimum lowland forest habitat in Surinam, was only found to reach half that figure. Yellow armadillos have been found in Brazilian savanna and forest at densities of up to 2.9 animals per sq km (7.5 per sq mi), while Southern three-banded armadillos in southern Brazil may reach densities of 7 animals per sq km (18 per sq mi).

Even if no live animals are observed, a good sign of the presence of armadillos in a habitat is a burrow. Armadillos dig between 1 and 20 burrows, each 1.5–3m (5–10ft) long, in their home ranges, occupying a given burrow for anything from 1–29 consecutive days. Because armadillos use multiple burrows, counting burrows in a habitat may not provide a reliable estimate of actual population density. Burrows are generally not very long and tend to run horizontally under the surface rather than to extend vertically down into the ground. Most burrows usually have just one or two entrances.

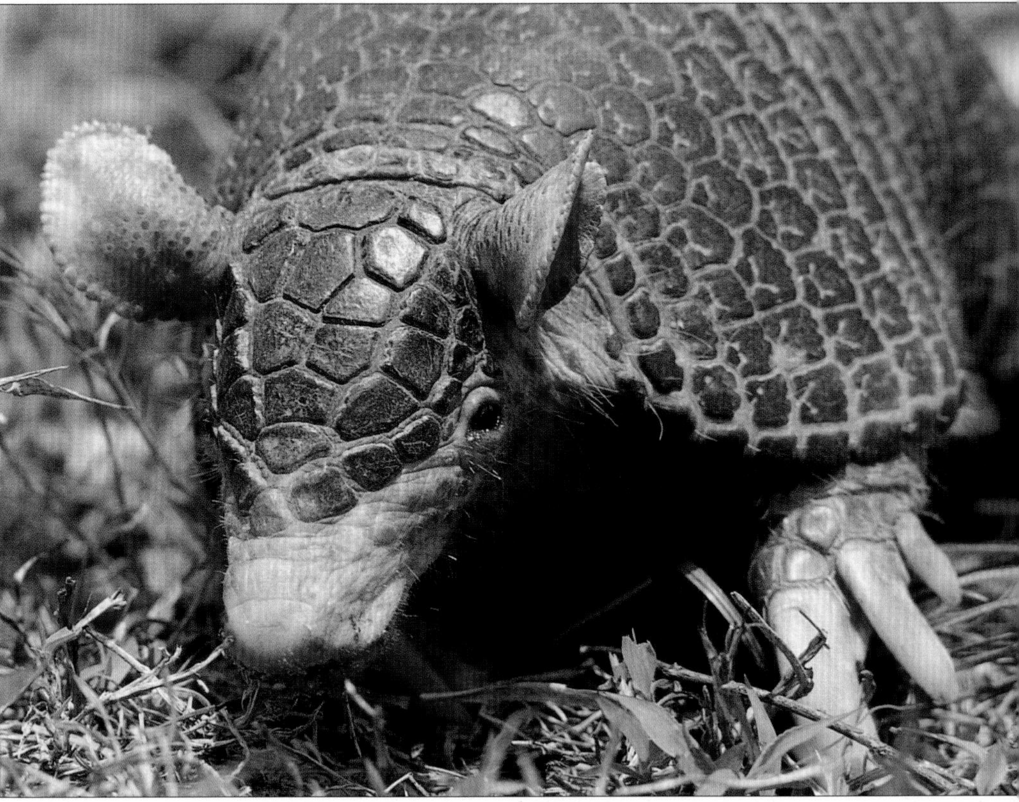

***Above** The Southern naked-tailed armadillo has five curved claws on its forefeet, the middle one of which is especially powerful. Its gait is unusual – it walks on the soles of its hind feet (plantigrade), but on the tips of the claws on its fore feet (digitigrade).*

Snuffling through the Leaf Litter

DIET

Perhaps because of their armor, armadillos are often fairly conspicuous, making a considerable racket as they snuffle along searching for food in dry vegetation. For armadillos, foraging consists of moving slowly along with the nose in the soil and leaf litter, then digging up material with the foreclaws. The animals may also use their claws to rip open rotting logs.

With their large front claws, Giant and naked-tailed armadillos seem specialized for ripping open ant and termite mounds. Other species have a more catholic diet; analyses of gut contents have revealed a mix of invertebrates, mostly beetles and ants. The Nine-banded armadillo is one of the few species to have been observed bravely feeding on fire ants (*Solenopsis geminata*), enduring their painful stings to dig open the nest and eat the larvae within. In addition to invertebrates, armadillos are also known to feed on some vegetable matter, including persimmons and other fruits, and, rarely, on certain vertebrates, such as snakes and small lizards; they may also scavenge carrion and the eggs of ground-nesting birds.

NATURE'S CLONES

Armadillos in the genus *Dasypus* are the only vertebrates known to exhibit obligate polyembryony, in which a female produces one fertilized egg that divides into multiple embryos. Because of this, all the offspring produced are genetically identical to one another. Genetic uniformity of siblings in the Nine-banded armadillo (BELOW) has been confirmed using modern molecular techniques.

One proposed reason for such a strange system may be to encourage offspring to help each other out. Altruistic behavior reaps more evolutionary dividends when directed towards relatives, as they share a higher proportion of genes. Because of polyembryony, armadillo littermates are clones, and thus might be predicted to be particularly helpful to each other. As it turns out, however, armadillos get very little opportunity to mix with their sibs as adults, and it seems more likely that polyembryony represents an ingenious way of countering a physical restriction on reproduction, helping to overcome a constraint in the female's reproductive system that leaves space for only one egg prior to implantation.

Armadillo Genera

Long-nosed armadillos
Genus *Dasypus*

S USA, Mexico, C America, Colombia, Venezuela, Guiana, Surinam, Brazil, Paraguay, Argentina, Ecuador, Peru; also Grenada and Trinidad and Tobago. 6 or 7 species: the **Nine-banded** or **Common long-nosed** (*D. novemcinctus*), **Seven-banded** or **Brazilian lesser** (*D. septemcinctus*), **Greater** (*D. kappleri*), **Southern lesser** (*D. hybridus*), **Hairy** (*D. pilosus*), and **Northern** (*D. sabanicola*) **long-nosed armadillos**; and, possibly, *D. yepesi* (from northeastern Argentina). The Hairy long-nosed armadillo is classed as Vulnerable.

Naked-tailed armadillos
Genus *Cabassous*

C and S America E of the Andes from S Mexico and Colombia to Paraguay, Uruguay, and N Argentina. 4 species: the **Southern** (*C. unicinctus*), **Northern** (*C. centralis*), **Chacoan** (*C. chacoensis*), and **Greater** (*C. tatouay*) **naked-tailed armadillos**. The Greater naked-tailed armadillo is classed as Lower Risk: Near Threatened.

Yellow armadillo
Euphractus sexcinctus

Yellow or **Six-banded armadillo**. S Surinam and adjacent areas of Brazil; also E Brazil to Bolivia, Paraguay, N Argentina, Uruguay.

Giant armadillo *Priodontes maximus*

S America E of the Andes, from N Venezuela and the Guianas to Paraguay and N Argentina. Endangered.

Hairy armadillos
Genus *Chaetophractus*

Bolivia, Paraguay, Argentina, Chile. 3 species: the **Andean** (*C. nationi*), **Screaming** (*C. vellerosus*), and **Larger** (*C. villosus*) **hairy armadillos**. The Andean hairy armadillo is classed as Vulnerable.

Three-banded armadillos
Genus *Tolypeutes*

Brazil and E Bolivia S through the Gran Chaco of Paraguay to Buenos Aires (Argentina). 2 species: the **Southern** (*T. matacus*) and **Brazilian** (*T. tricinctus*) **three-banded armadillos**. The Brazilian three-banded armadillo is classed as Vulnerable, while the Southern three-banded armadillo is Lower Risk: Near Threatened.

Fairy armadillos
Genus *Chlamyphorus*

Argentina, Paraguay, Bolivia. 2 species: the **Greater** or **Chacoan** (*C. retusus*) and **Lesser** or **Pink** (*C. truncatus*) **fairy armadillos**. The Lesser fairy armadillo is classed as Endangered, while the Greater fairy armadillo is Vulnerable.

Pichi *Zaedyus pichiy*

C and S Argentina, and E Chile S to the Straits of Magellan.

Life beneath the Carapace

SOCIAL BEHAVIOR

Most armadillos are relatively solitary, and with a few exceptions most species are active at night, although this can vary with age – juveniles are often active in the late morning or early afternoon – and with the time of year: there is more diurnal activity when the weather is colder.

Among adults, most social interactions usually occur during breeding. For most species, breeding is seasonal, with matings occurring primarily during the summer months, although captive animals may breed year-round. Prior to mating, males and females may engage in extended bouts of courtship in which the males avidly follow the females. After mating, most species initiate embryonic development right away, but in northern populations of the Nine-banded armadillo (*Dasypus novemcinctus*), possibly as an adaptation to the different timing of the seasons, implantation of the fertilized egg may be delayed by 3–4 months or more; one captive female of this species reportedly gave birth at least three years after the last date at which she could possibly have been inseminated. *D. novemcinctus* may be polygynous, with males mating with 2–3 females per breeding season, while females typically mate with just a single male. Yet not all individuals are reproductively successful: as many as one third of the females in a population may not reproduce in a given year, and, in a 4-year study, only about a third of all adults were identified as parents of at least one litter. Litter sizes are usually small, varying from 1–4 young per litter for most species, with a maximum of 8–12 reported for captive *D. hybridus*. The majority of species only reproduce once per year.

In *D. novemcinctus*, males occasionally chase, kick, and scratch each other with their claws either in defence of their home ranges or when competing for access to females. Although female home ranges may overlap quite extensively, pregnant or lactating adult females can also be quite aggressive, primarily targeting other females and younger individuals of both sexes.

D. novemcinctus home ranges vary from about 1.5ha (4 acres) for a population in Oklahoma to over 10ha (25 acres) for a population in Florida. Home range sizes for other species are largely unknown, but the low population densities of most species suggest they may range widely.

Conserving the Hoover Hog

CONSERVATION AND ENVIRONMENT

For centuries, armadillos have been exploited by humans for their meat, and they continue to be a favored food item in many areas of Latin America. In North America people partake of armadillo meat less frequently; however, during the Great Depression of the 1930s, destitute southern sharecroppers came to rely on armadillos for food, and the animals were nicknamed "Hoover hogs," a wry allusion to US President Herbert Hoover.

Habitat loss from deforestation, agriculture, and other sources is another significant cause of declining populations, as is the eradication of digging armadillos from both agricultural areas and the well-manicured lawns of suburban communities. On the bright side, the Brazilian three-banded armadillo, which was formerly classed as Extinct in the Wild, has been rediscovered recently in several areas of Brazil. CMM/WJL

Right The Larger hairy armadillo has been known to burrow under animal carcasses in order to feed on the maggots and other insects that accumulate there. This species relies on insects, rodents, and lizards in summer, but half of its winter diet is vegetation.

Pangolins

APPEARING LIKE TILES ON A HOUSE ROOF, *the overlapping, horny scales on pangolins' backs distinguish these animals from all other Old World mammals. The scales, which grow from the thick underlying skin, protect every part of the body except the underside and inner surfaces of the limbs, and are shed and replaced periodically.*

In Africa, large numbers of pangolins are caught up in the bushmeat trade and are killed for their flesh, while the scales are used in traditional medicine. As a result, the Cape pangolin is threatened. In Asia, powdered scales are believed to have medicinal and aphrodisiacal qualities, and the animals are hunted indiscriminately. Unless controlled, the population densities and ranges of the three Asian pangolins will continue to dwindle.

A Suit of Armor

FORM AND FUNCTION

Pangolins specialize in eating ants and termites and, like South American anteaters, they probe the nests of their prey with long, narrow tongues. In the largest species, the Giant pangolin, the strap-like tongue can be extended for 40cm (16in), in total it is 70cm (27.5in) long, and is housed in a sheath that extends to an attachment point on the pelvis. Viscous saliva is secreted onto the tongue by an enormous salivary gland, 360-400cu cm (22–24.5cu in) in capacity, which sits in a recess in the chest. The simple skull lacks teeth and chewing muscles; captured ants are ground up in the specialized, horny stomach. Pangolins have a small, conical head with a reduced or absent outer ear, and an elongate body that tapers to a stout tail. Thick lids protect their eyes from the bites of ants, and special muscles close the nostrils during feeding. The limbs are short but powerful and terminate in five clawed digits; the three middle claws on the forefoot are 55–75mm (2.2–2.9in) long and curved.

In Africa, two of the four pangolin species are principally arboreal, inhabiting the rain-forest belt from Senegal to the Great Rift Valley. While the common Small-scaled tree pangolin occupies home ranges of 20–30ha (49–74 acres) in the lower strata of the forest, the smaller Long-tailed pangolin is more restricted to the forest canopy. Here, moving often by day to avoid its larger relative, it seeks the soft, hanging nests of ants and termites (preferring arboreal species), or attacks the columns that move among the leaves. Both species have a slender but strongly prehensile tail (the 46 or 47 tail vertebrae of the Long-tailed pangolin are a mammalian record), with a bare patch at the tip containing a sensory pad. They scale vertical tree trunks by gaining a purchase with the foreclaws and then drawing up the hindfeet just behind them; the jagged edges of the tail-scales provide additional support. These arboreal pangolins sleep aloft, curled up among epiphytes (plants growing on trees) or in forked branches.

The terrestrial African pangolins are larger than their tree-dwelling relatives, and occur in a spectrum of habitats from forest to open savanna. They sleep in the burrows of other digging animals. As protection against predators pangolins

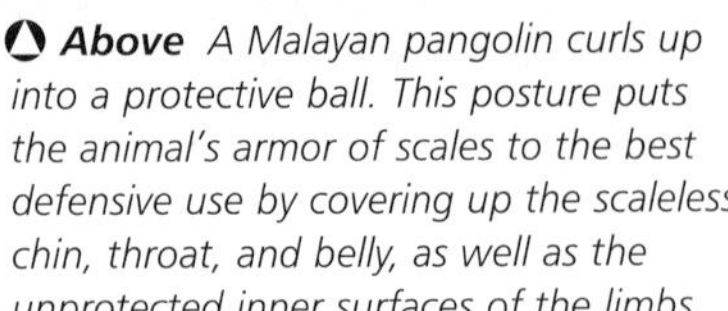

Above *A Malayan pangolin curls up into a protective ball. This posture puts the animal's armor of scales to the best defensive use by covering up the scaleless chin, throat, and belly, as well as the unprotected inner surfaces of the limbs.*

Left *A Tree pangolin* (Manis tricuspis) *hangs from a branch with the aid of its long, prehensile tail. The tail has a bare patch at its tip, equipped with a sensory pad that makes it easier for the animal to find and secure a good grip.*

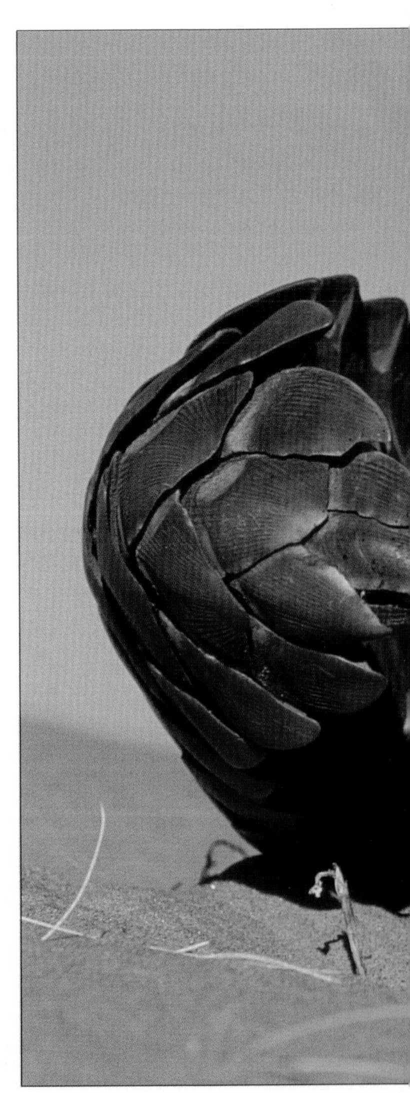

curl tightly into a ball, their scales forming a shield that is impregnable to all but the larger cats and hyenas. They use their powerful claws to demolish the nests of ground termites and ants. The Giant pangolin may take 200,000 ants a night, weighing over 700g (25oz). Because of their huge digging claws, the terrestrial pangolins must walk slowly on the outer edges of their forefeet with the claws tucked up underneath; the sight of this plated animal with its curious, shuffling gait has been likened to that of a perambulating artichoke! However, all species can move more swiftly up to 5km/h (3mph) by rearing up and running on their hind legs, using the tail as a brace.

The three Asian pangolins are less well known than their African counterparts, and are distinguished from them by the presence of hair at the bases of the body-scales. Intermediate in size between the African species, the Asian pangolins are nocturnal and usually terrestrial, but can climb with great agility. They inhabit grasslands, subtropical thorn forest, rain forest, and barren hilly areas almost devoid of vegetation, but are nowhere abundant. The geographical range of the Chinese pangolin is said to approximate that of its preferred prey species, the subterranean termites.

Socializing by Smell

SOCIAL BEHAVIOR

Although pangolins are usually solitary, their social life is dominated by the sense of smell. Individuals advertise their presence by scattering feces along the tracks of their home ranges, and by marking trees with urine and a pungent secretion from an anal gland. These odors may communicate dominance and sexual status, and possibly facilitate individual recognition. The vocal expressions of pangolins are limited to puffs and hisses; however, these are not known to serve any particular social function.

Pangolins usually bear one young weighing 200–500g (7–18oz), although two and even three young have been reported in the Asian species. In the arboreal species, the young clings to the mother's tail soon after birth, and may be carried in this fashion until weaned at the age of 3 months. When alarmed, mothers protect the infants by curling up around them. Young of the terrestrial species are born underground with small, soft scales, and are first carried outside on the mother's tail at 2–4 weeks. In all species, births usually occur between November and March; sexual maturity is reached at 2 years. CRD/RAR

FACTFILE

PANGOLINS

Order: Pholidota

Family: Manidae

7 species of the genus *Manis*. Four African species: **Giant pangolin** (*M. gigantea*), **Cape** or **Ground pangolin** (*M. temminckii*), **Tree** or **Small-scaled tree pangolin** (*M. tricuspis*), **Long-tailed pangolin** (*M. tetradactyla*); three Asian species: **Indian** or **Thick-tailed pangolin** (*M. crassicaudata*), **Chinese pangolin** (*M. pentadactyla*), **Malayan pangolin** (*M. javanica*).

DISTRIBUTION Senegal to Uganda, Angola, W Kenya, S to Zambia and N Mozambique; Sudan, Chad, Ethiopia to Namibia and South Africa; India, Sri Lanka, Nepal, and S China to Taiwan and Hainan; S through Thailand, Myanmar (Burma), Laos, Malaysia, Java, Sumatra, Kalimantan, and offshore islands.

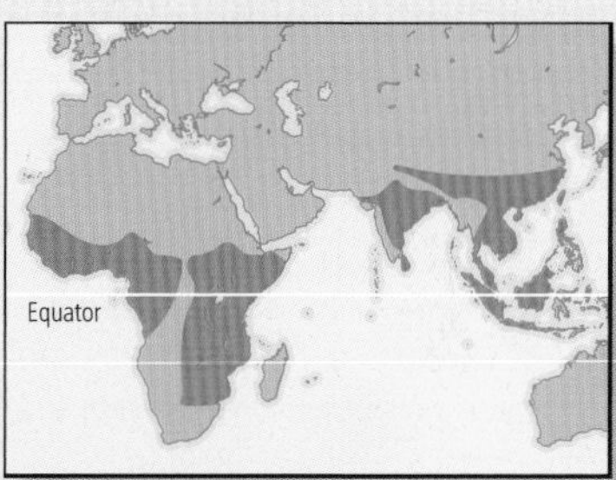

HABITAT Forest to open savanna

SIZE **Head–body length** ranges from 30–35cm (12–14in) in the Long-tailed pangolin to 75–85cm (30–33in) in the Giant pangolin, **tail length** from 55–65cm (22–26in) to 65–80cm (26–31in), and **weight** from 1.2–2.0kg (2.6–4.4lb) to 25–33kg (55–73lb), both in the same two species. In most species, males are 10–50 percent heavier than females, but up to 90 percent heavier in the Indian pangolin.

COAT Horny, overlapping scales on head, body, outer surfaces of limbs, and tail, varying in color from light yellowish-brown through olive to dark brown. The scales of young Chinese pangolin are purplish-brown. Undersurface hairs are white to dark brown.

DIET Termites and ants

BREEDING Gestation period from 65–70 days (Indian pangolin) to 139 days (Cape and Tree pangolins); litter size one, rarely two.

LONGEVITY At least 13 years in captivity (Indian pangolin); unknown in wild.

CONSERVATION STATUS The Cape, Indian, Chinese, and Malayan pangolins are all classed as Lower Risk: Near Threatened.

Left *Its scales the shape of artichoke leaves, a Ground pangolin advances across sand in southern Africa. As the name suggests, this species is terrestrial, living in burrows that may be several meters deep and that terminate in circular chambers that are sometimes big enough for a man to stand up in.*

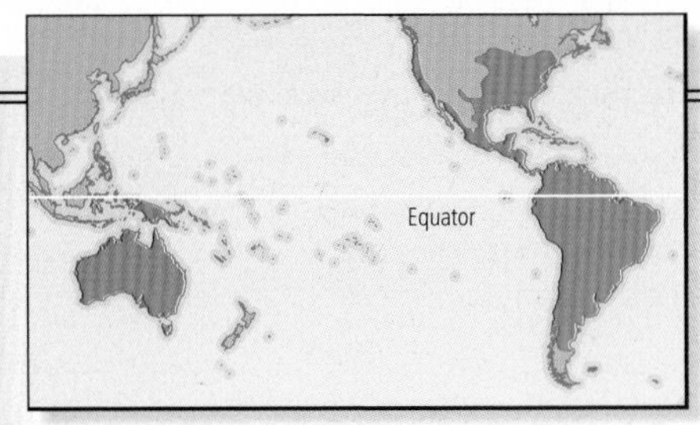

MARSUPIALS

Although marsupials had long been familiar to the indigenous peoples of the Americas and Australasia, they remained unknown to the rest of the world until the 16th century. The first marsupial seen in Europe was a Brazilian opossum that was brought back in 1500 as a gift to Queen Isabella and King Ferdinand of Spain by the explorer Vicente Yáñez Pinzón. Yet it was three centuries before zoologists ascertained that marsupials were not aberrant rodents but a distinctive natural group of mammals united by a unique mode of reproduction.

From North to South

Early Beginnings

The most ancient marsupial-like fossil, *Kokopellia juddi*, hails from early Cretaceous deposits in Utah aged at least 100 million years, but its position as a true marsupial is contentious. The earliest undisputed marsupial, from the now-extinct family Stagodontidae, has been dated at about 80 million years from other fossil deposits in Utah. At least 20 other genera of slightly younger Cretaceous marsupials are known from other North American sites, suggesting that the group originated in that region. Despite their early ascendancy in the north, marsupials soon dwindled there as placental mammals increased in diversity, and they became extinct in North America by 15–20 million years ago. The one presently extant marsupial in North America, the Virginia opossum, recolonized less than 1 million years ago from the south.

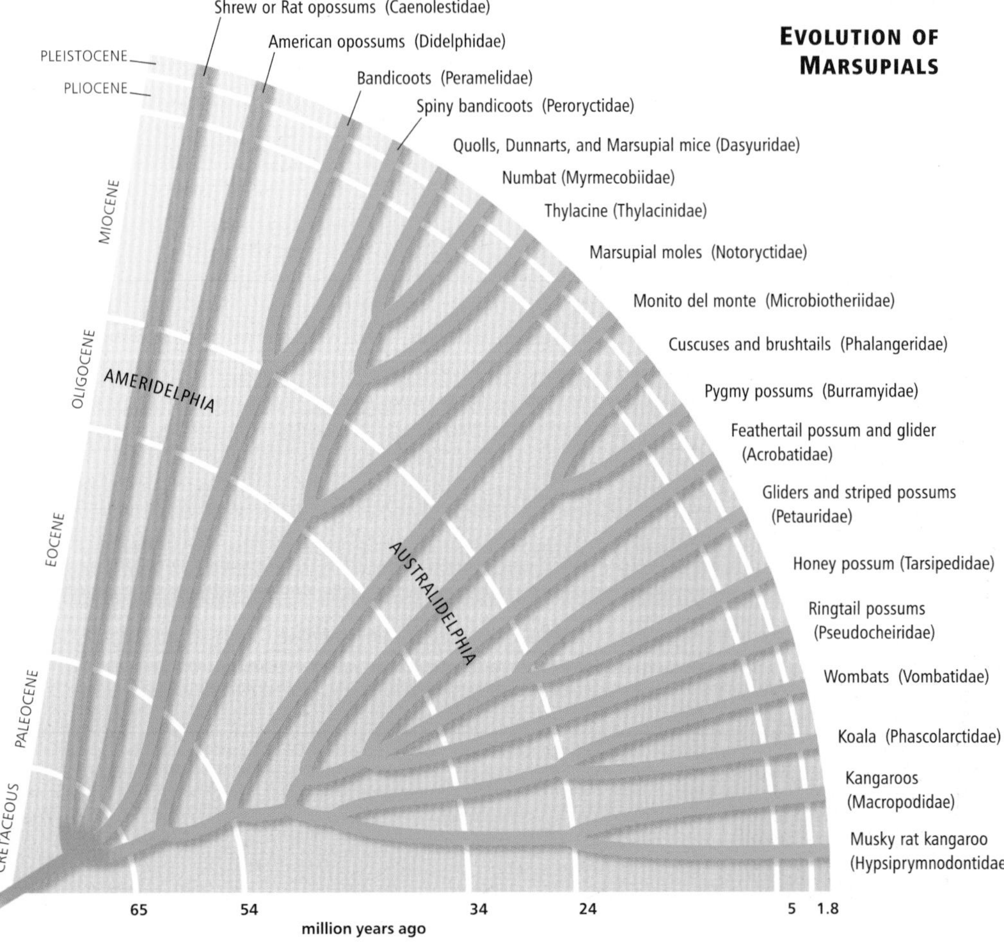

Above In spite of its distinctive appearance, the Long-footed potoroo of Australia was only identified as a separate species as recently as 1980.

Right At birth a koala weighs only one-fiftieth of an ounce. Weaning begins after five months when the mother provides partly digested leaves for the infant.

Shortly after their appearance and radiation in North America, ancestral marsupials dispersed across land bridges to South America and also to Europe. Marsupial diversity in South America remained high from the middle Paleocene (60 million years ago) until the Pliocene (5–1.8 million years ago) during a period of "splendid isolation." In the Pliocene, a land bridge to North America reformed, allowing invasion of the southern continent by placental mammals including raccoons, bears, cats, and other carnivores. In the face of this onslaught, all South America's large (i.e. over 5kg/11lb) carnivorous marsupials disappeared. Its present marsupial fauna is dominated by omnivores

weighing less than 1kg (2.2lb), with a minority of species specializing on insects or small vertebrates. The marsupial emigrants to Europe arrived by the early Eocene (52 m.y.a.) and stayed for perhaps 35 million years. Although this radiation was not spectacular, comprising only six described fossil genera in the family Didelphidae, it extended to many parts of Europe, North Africa, and east to Thailand and eastern China.

The origin of Australasian marsupials is unclear, but the most likely scenario is an invasion of taxa with South American affinities sometime in the early Paleocene, some 65–60 million years ago. The earliest Australian marsupial fossils date to the early Eocene 55 million years ago and, intriguingly, two genera appear similar to marsupials of the same age from Argentina. During the Paleocene Australia, Antarctica, and South America were united as the last surviving remnants of the old supercontinent Gondwana. Australia severed its geological umbilical cord between 46 and 35 million years ago, while Antarctica and South America parted company 5 million years later. The climate of the early Eocene was warm and humid, with beech forest covering much of the Antarctic. That it was conducive to dispersal is suggested by the discovery of two genera of polydolopid marsupials from the middle Eocene of Antarctica.

Following its separation from Antarctica, the Australian ark drifted northward towards the equator, allowing its marsupial cargo to incubate in isolation. There is, frustratingly, a 30-million year "dark age" lasting until the Miocene 24 million years ago that has yielded no fossils. However, by the early Miocene a spectacularly rich forest fauna had appeared, with many arboreal and browsing terrestrial marsupials represented. Climatic oscillations in the middle Miocene 15

SUPERCOHORT: MARSUPIALIA
289 species in 83 genera, 18 families, and 7 orders.

million years ago were associated with losses of entire marsupial families, including the enigmatic miralinids, pilkipildrids, and wynyardiids. With the end of greenhouse conditions, forest contracted to coastal areas and savanna woodland and grassland dominated the interior. These shifts favored grazing marsupials, and set the scene for an explosive radiation of grazing kangaroos.

At the same time that the Miocene climate deteriorated, the Australian and Southeast Asian crustal plates collided, creating the highlands of New Guinea. These new mountainous areas, and parts of northeastern Queensland, allowed the continuation of lush conditions that had prevailed earlier in the Miocene, and provided opportunities for colonization for forest taxa. Many of the marsupials in New Guinea's rain forests now bear a striking resemblance to extinct taxa known only from fossils from the Miocene of central Australia. In contrast to the Australian marsupial fauna, in which only 30 of the 155 species are arboreal, at least 50 of New Guinea's 83 present marsupial species are restricted to the tree top environment.

Modern Radiation

DISTRIBUTION PATTERNS

Although today's marsupials comprise only some 7 percent of the world's mammals, they occur widely in the Americas and predominate in Australasia. There is no evidence that marsupials ever dispersed naturally to New Zealand (the Common brushtail possum was introduced to establish a fur trade in 1858, and a further six species of wallabies were introduced shortly after), but two species of cuscus occur on Sulawesi, one occurs throughout the Solomons, and others are scattered as island endemics in the Banda, Timor, Arafura, Coral, and Solomon seas. Australasian marsupials are ecologically more diverse than their American counterparts, and often occupy similar niches to placental mammals elsewhere. Thus, there are insectivores, carnivores, herbivores, and omnivores, and a honey possum that is unusual in specializing on nectar and pollen. All terrestrial habitats are exploited, from deserts to rain forests and high alpine areas. The smallest species, the Long-tailed planigale, weighs no more than 4.5g, some 20,000 times less than male Red kangaroos.

There is much that is still to be learned about the marsupials. Since the 1980s there has been a steady stream of discoveries of new species, for example the distinctive Long-footed potoroo in Australia in 1980, and two spectacular species of tree kangaroos in New Guinea in the early 1990s, the Dingiso and Tenkile.

The Amazing Journey

REPRODUCTIVE STRATEGIES

Despite their ecological and morphological diversity, it is the mode of reproduction that unites marsupials and sets them apart from other mammals. In its form and early development in the

MARSUPIAL BODY PLAN

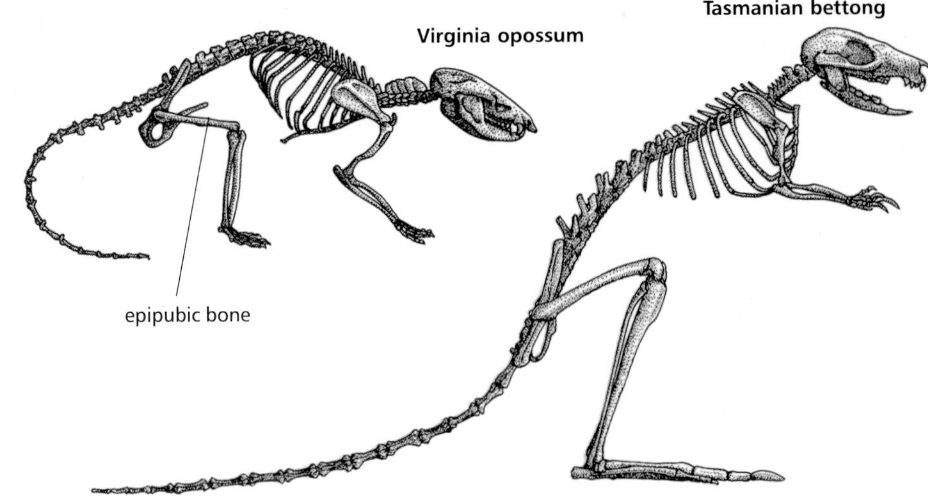

Above Skeletons of the Virginia opossum and Tasmanian bettong. The Virginia opossum is medium-sized, with unspecialized features shared with its marsupial ancestors. These include the presence of all digits in an unreduced state, all with claws. The skull and teeth are those of a "generalist," the long tail is prehensile, acting as a "fifth hand," and there are epipubic or "marsupial bones" that project forward from the pelvis and help support the pouch. The hindlimbs in this quadruped are only slightly longer than the forelimbs. The larger rat-kangaroo has small forelimbs, and larger hindlimbs for leaping. The hindfoot is narrowed and lengthened (hence macro-podoid, "large-footed"), and the digits are unequal. Stance is more, or completely, upright, and the tail is long, not prehensile but used as an extra prop or foot.

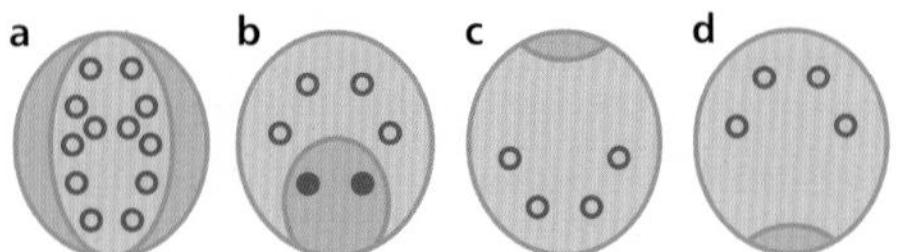

Above Pouches (marsupia) occur in females of most marsupials. Some small terrestrial species have no pouch. Sometimes a rudimentary pouch **a** is formed by a fold of skin on either side of the nipple area that helps protect the attached young (eg mouse opossums, antechinuses, quolls). In **b** the arrangement is more of a pouch (e.g., Virginia and Southern opossums, Tasmanian devil, dunnarts). Many of the deepest pouches, completely enclosing the teats, belong to the more active climbers, leapers or diggers. Some, opening forward **c**, are typical of species with smaller litters of 1–4 (e.g., possums, kangaroos). Others **d** open backward and are typical of digging and burrowing species (e.g., bandicoots, wombats).

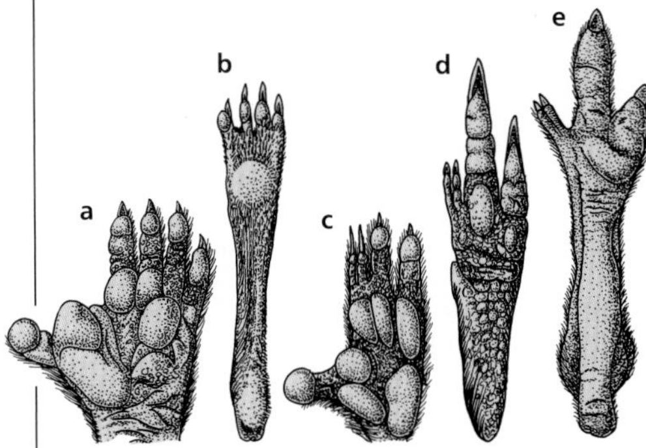

Above Feet of marsupials: **a** opposable first digit in foot of the tree-dwelling Virginia opossum; **b** long narrow foot, lacking a first digit, of the kultarr, a species of inland Australia with a bounding gait: both these species have the second and third digits separate (didactylous); in many marsupials (eg kangaroos and bandicoot) these digits are fused (syndactylous), forming a grooming "comb"; **c** opposable first digit and sharp claws for landing on trees in Feathertail glider; **d** first digit much reduced In long foot of terrestrial Short-nosed bandicoot – fourth digit forms axis of foot; **e** first digit entirely absent in foot of kangaroo.

Below Anatomy of reproduction, and its physiology, set marsupials apart. In the female, eggs are shed into a separate (lateral) uterus, to be fertilized. The two lateral vaginae are often matched in the male by a two-lobed penis. Implantation of the egg may be delayed, and the true placenta of other mammals is absent. The young are typically born through a third, central, canal; this is formed before each birth in most marsupials, such as American opossums: in the Honey possum and kangaroos the birth canal is permanent after the first birth.

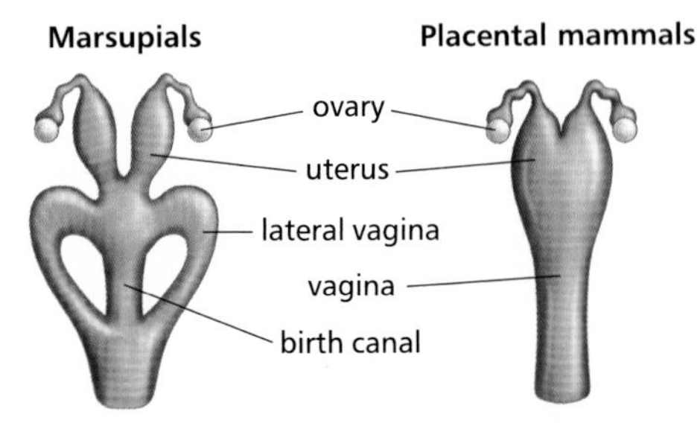

Below *Marsupial skulls generally have a large face area and a small brain-case. There is often a sagittal crest for the attachment of the temporal muscles that close the jaws, and the eye socket and opening for the temporal muscles run together, as in most primitive mammals. There are usually holes in the palate, between the upper molars. The rear part of the lower jaw is usually turned inward, unlike placental mammals.*

Many marsupials have more teeth than placental mammals. American opossums for instance have 50. There are usually three premolars and four molars on each side in both upper and lower jaws.

Marsupials with four or more lower incisors are termed polyprotodont. The Eastern quoll (Dasyurus viverrinus) *has six. Its chiefly insectivorous and partly carnivorous diet is reflected in the relatively small cheek teeth, each with two or more sharp cusps, and the large canines with a cutting edge. Its dental formula is I4/3, C1/1, P2/2, M4/4 = 42. The largely insectivorous bandicoot* (Perameles) *has small teeth of even size with sharp cusps for crushing the insects which it seeks out with its long pointed snout (I4–5/3, C1/1, P3/3, M4/4 = 46–48).*

Diprotodont marsupials have only two lower incisors, which are usually large and forward-pointing. The broad, flattened skull of the leaf-eating Brush-tailed possum contains reduced incisors, canines, and premolars, with simple low-crowned molars (I3/2, C1/0, P2/l, M4/4 = 34). The large wombat has rodent-like teeth and only 24 of them (I1/1, C0/0, P1/1, M4/4), all rootless and ever-growing to compensate for wear in chewing tough, fibrous grasses.

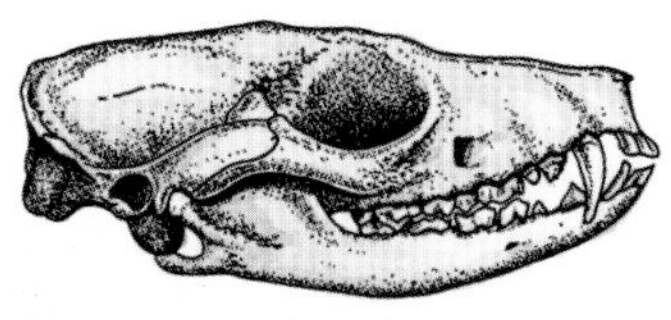

Eastern quoll
7 cm

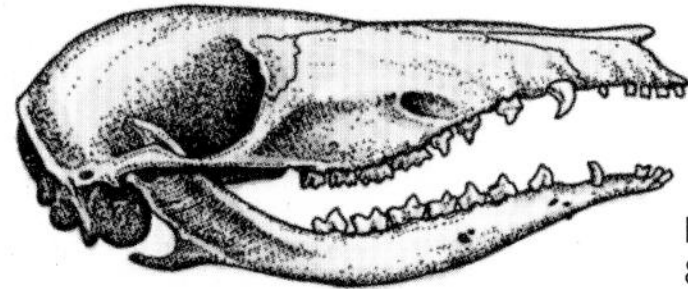

Bandicoot
8 cm

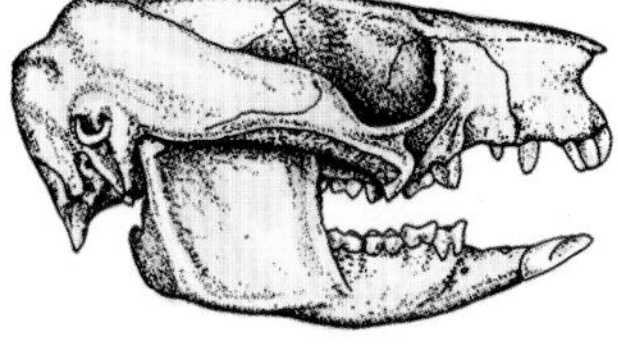

Brushtail possum
8 cm

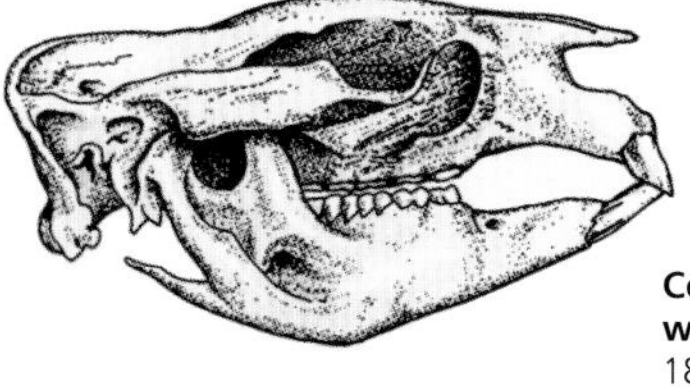

Common wombat
18 cm

Above *A tiny, hairless, infant Common brushtail possum attached to the mother's nipple in the pouch. It will be weaned at about six months.*

uterus, the marsupial egg is like that of reptiles and birds and quite unlike the egg of placental mammals (eutherians). Whereas placental young undergo most of their development and considerable growth inside the female, marsupial young are born very early in development. For example, a female Eastern gray kangaroo of about 30kg (66lb) gives birth after 36 days' gestation to an offspring that weighs about 0.8g (under 0.03oz). This young is then carried in a pouch on the abdomen of its mother where it suckles her milk, develops and grows until after about 300 days it weighs 5kg (11lb) and is no longer carried in the pouch. After it leaves the pouch the young follows its mother closely and continues to suckle until about 18 months old.

Female young are ready to breed just after weaning, but males do not mature until aged two years or more. An extraordinary adaptation in kangaroos and all other marsupials studied is that the amount and quality of the milk changes during suckling, matching the needs of the growing young. Carbohydrates typically make up over half the solid fraction of milk for the first 6 months of lactation, and protein and fat the remainder. By 8–9 months carbohydrates have almost disappeared, and lipids constitute up to two-thirds of the solid bulk of the milk.

Immediately after birth, the newborn young makes an amazing journey from the opening of the birth canal to the area of the nipples. Forelimbs and head develop far in advance of the rest of the body, and the young is able to move with swimming movements of its forelimbs. Although it is quite blind, it moves against gravity, locates (how is not yet known) a nipple and sucks it into its circular mouth. The end of the nipple enlarges to fit depressions and ridges in the mouth, and the young remains firmly attached to the nipple for 1–2 months until, with further development of its jaws, it is able to open its mouth and let go. In many marsupials the young are protected by a fold of skin that covers the nipple area, forming the pouch (see diagram).

As the main emphasis in the nourishment of the young in marsupials is on lactation, most development and growth occurs outside the uterus. For the short time that the embryo is in the uterus, it is nourished by transfer of nutrients from inside the uterus across the wall of the yolksac that makes only loose contact with the uterine wall. In eutherians, nourishment of the young during its prolonged internal gestation occurs by way of the placenta, in which the membranes surrounding the embryo make close contact with the uterine wall, become very vascular and act as the means of transport of material between maternal circulation and embryo. Although it is commonly believed that marsupials do not have a placentation system at all, the yolksac has a placental function; in bandicoots and the Koala there is also development of a functional, albeit short-lived, placenta.

A further peculiarity of reproduction, similar to the delayed implantation found in some other mammals, occurs in all kangaroos (except the Western gray) and a few species in other marsupial families. Pregnancy in these marsupials occupies more or less the full length of the estrous cycle but does not affect the cycle, so that at about the time a female gives birth, she also becomes receptive and mates. Embryos produced at this mating develop only as far as a hollow ball of cells (the blastocyst) and then become quiescent, entering a state of suspended animation or "embryonic diapause." The hormonal signal (prolactin) that blocks further development of the blastocyst is produced in response to the sucking stimulus from the young in the pouch. When sucking decreases as the young begins to eat other food and to leave the pouch, or if the young is lost from the pouch, the quiescent blastocyst resumes development, the embryo is born, and the cycle begins again. In some species that do not breed all year round, such as the Tammar wallaby, the period of quiescence of the blastocyst is extended by seasonal variables such as changes in day length. The origin of embryonic diapause may have been to prevent a second young being born while the pouch was already occupied, but it has other advantages, allowing rapid replacement of young which are lost, even in the absence of a male.

Four Strategies

SOCIAL ORGANIZATION

While large marsupials such as kangaroos and wombats usually produce single young, smaller species are not so constrained. Litter sizes in arboreal browsers, such as the Common ringtail possum range from 1–3, while litters of 3–4 can be produced twice a year by the nectar-feeding Honey possum, pygmy possums, and Feathertail glider. Similar productivity occurs in omnivores such as bandicoots and some didelphid species. Small insectivores such as antechinuses produce 8–12 young in a single bout of reproduction each year. Hensel's short-tailed opossum of southern Brazil has up to 25 nipples and probably rears the largest litters of any marsupial.

In species that produce litters of several young, individual young weigh less than 0.01g, by far the smallest of any mammalian species. At weaning, between 50 days in some didelphids to over 100 days in antechinuses, the combined weight of the young may be two to three times that of the mother. This represents an extraordinary investment of energy by females, and is often accompanied by high mortality of mothers. Not surprisingly, small marsupials tend to be shorter-lived than larger species and achieve reproductive maturity earlier at 5–11 months. Larger litter sizes or more frequent bouts of reproduction compensate for their shorter life spans, with some species being capable of breeding year round. The shortest-lived marsupials are some male dasyurids that die after mating aged 12 months. In contrast, the Mountain pygmy possum is the longest-lived of any small mammal; wild females may live over 11 years. Kangaroos live up to 25 years in the wild.

In some dasyurid marsupials, males disperse upon weaning up to several hundred meters from their mother's home range and then associate with unrelated females. Daughters remain at home, with one usually inheriting the maternal range after the death or dispersal of the mother. This pattern of dispersal reduces the chance of breeding among kin. Intriguingly, in species where females breed in two or more seasons, the sex ratio of the first litter is often skewed toward sons by a ratio of about 6:4, but returns to parity or less thereafter. The over-production of sons by young mothers probably reduces the chance of future

Below *Red kangaroos watering. This species forms small social groups, known as mobs, which may contain between 2–10 members. These mobs are not temporary aggregations, but are quite organized.*

MARSUPIAL MOLES

The two species of Marsupial moles are the only Australian mammals that have become specialized for a burrowing (fossorial) life. Others, including small native rodents, have failed to exploit this niche. except for a few species that nest in burrows.

Because of their extensive and distinct modifications, Marsupial moles (*Notoryctes typhlops* and *N. caurinus*) are placed in their own order, the Notoryctemorphia. Their limbs are short stubs. The hands are modified for digging, with rudimentary digits and greatly enlarged flat claws on the third and fourth digits. Excavated soil is pushed back behind the animal with the hind limbs, which also give forward thrust to the body and, like the hands, are flattened with reduced digits and three small flat claws on the second, third, and fourth digits. The naked skin (rhinarium) on the tip of the snout has been extended into a horny shield over the front of the head, apparently to help push through the soil. The coat is pale yellow and silky. The nostrils are small slits, there are no functional eyes or external ears, and the ear openings are concealed by fur. The neck vertebrae are fused together, presumably to provide rigidity for thrusting motions. Females have a rear-opening pouch with two teats. The tail, reduced to a stub, is said to be used sometimes as a prop when burrowing. Head–body length is approximately 13–14.5cm (5.1–5.7in), tail length 2–2.5cm (1in), and weight 40g (1.4oz). Dentition is I4/3, C1/1, P2/3, M4/4 = 44.

Little is known of these moles in the wild. They occur in the central deserts, using sandy soils in river-flat country and sandy spinifex grasslands. For food, they favor small reptiles and insects, particularly burrowing larvae of beetles of the family Scarabaeidae. In captivity Marsupial moles will seek out insect larvae buried in the soil and consume them underground. They also feed on the surface. They are not known to make permanent burrows, the soil caving in behind them as they move forward, and in this respect they are most unusual among fossorial mammals. Captive animals have been seen to sleep in a small cavity which collapses after they leave.

Compared to other burrowing animals, Marsupial moles show differences of detail in the adaptive route they have followed. The head shield is much more extensive than in many others, the eyes are more rudimentary than in most, and the rigid head/neck region with fused vertebrae appears to be specific to the Marsupial moles. GG

conflict for resources with stay-at-home daughters. By contrast, in species where females usually breed just once, litter sex ratios are variable. Well-fed mothers produce more sons than daughters, perhaps providing the sons with an edge in growth that will allow them to compete successfully with other males and father more young themselves. Mothers in poor condition produce more daughters. Experimental provision of food allows female Agile antechinuses, Black-eared and American opossums to bias their litters in favor of sons, indicating the importance of body condition. The mechanism producing bias remains elusive, but appears to operate prior to fertilization.

Communication between marsupials is primarily through hearing and smell. Arboreal species in particular use sound to communicate over distances up to several hundred meters. Vocalizations range from chirps and squeaks in small possums to full-blooded bellows in the koala. Olfactory communication occurs by passive deposition of urine and feces, but all species mark actively using secretions from glands in the skin. The secretions may be used to self-anoint, to mark other animals, or to lay claim to nests or other key sites. The Sugar glider recognizes strangers on the basis of scent alone; it is likely that many other species share this ability. Most marsupials are nocturnal, so vision is relatively unimportant.

Four broad types of marsupial social organization can be identified. In the first, the social unit is an individual whose range overlaps with that of several others. Males have large ranges that take in those of several females, and mating is promiscuous. Marsupials with this social system often exploit dispersed food resources, and include small didelphids, dasyurids, and the Honey possum. In a second type of social system, the unit is an individual with limited range overlap. Typically a male's home range overlaps that of one or two females who share exclusive mating rights. Arboreal folivores such as koalas and Brushtail possums exemplify this type of system. A third type of social system is founded on cohesive family units that share a common, and often defended, home range. Groups may contain monogamous pairs and their offspring, or dominant males with several adult females and young. Group-living marsupials include sap- and exudate-feeding gliders and possums that may need force of numbers to defend focal food sources and shared dens. Finally, larger members of the kangaroo family exemplify the fourth type of social organization, in which the social unit is a flexible "mob" of gregarious individuals. Mobs have a promiscuous mating system, in which access to females is based on size and dominance.

Growing Human Encroachment

CONSERVATION ISSUES

Pressures on marsupial habitats have intensified in the last 200 years, with European colonization. Forests and woodland have been cleared for agriculture, while exotic herbivores and predators are widespread. Overhunting occurs in New Guinea. No marsupials have become extinct recently in the Americas or New Guinea, but 10 species and 6 subspecies have disappeared from Australia. A further 55 species in the Australasian region and 22 species from Central and South America are considered Vulnerable or Endangered.

Protection of habitat continues to be an important conservation measure. In New Guinea, where most of the land is owned by indigenous peoples, community support is crucial for conservation, but is sometimes subverted by the interests of rapacious multinationals. In southwestern Australia, a program of poisoning the Red fox has aided the recovery of at least six marsupial species. Habitat protection and better ecological understanding will be the key to effective future conservation of marsupials. CRD/EMR

American Opossums

WHEN MARSUPIALS WERE INTRODUCED TO Europeans for the first time in 1500AD, it was in the shape of a female Southern opossum from Brazil, presented by the explorer Vicente Pinzón to Spain's rulers, Ferdinand and Isabella. The monarchs examined the female with young in her pouch and dubbed her an "incredible mother."

Despite this royal introduction, the popular image of the opossum has never been a lofty one; the animals are often portrayed as foul-smelling and rather slow-witted. Although not as diverse as Australia's marsupials, the American opossums are in fact a successful group incorporating a variety of different species, ranging from the highly specialized tree-dwelling woolly opossums to generalists like the Southern and Virginia opossums.

Generalists and Specialists

FORM AND FUNCTION

American opossums range from cat- to mouse-sized. The nose is long and pointed, and has long, tactile hairs (vibrissae). Eyesight is generally well-developed; in many species, the eyes are round and somewhat protruding. When an opossum is aroused it will often threaten the intruder with mouth open and lips curled back, revealing its 50 sharp teeth. Hearing is acute, and the naked ears are often in constant motion as an animal tracks different sounds. Most opossums are proficient climbers, with hands and feet well adapted for grasping. Each foot has five digits, and the big toe on the hind foot is opposable. The round tail is generally furred at the base, with the remainder either naked or sparsely haired. Most opossums have prehensile tails used as grasping organs when animals climb or feed in trees. Not all female opossums have a well-developed pouch; in some species the pouch is absent altogether, while in others there are simply two lateral folds of skin on the abdomen. In males the penis is forked and the pendant scrotum often distinctly colored.

The Virginia or Common opossum of North and Central America, the Southern opossum of Central and South America, and the White-eared opossum of higher elevations in South America are generalized species, occurring in a variety of habitats from grasslands to forests. They have cat-sized bodies, but are heavier than cats and have shorter legs. Although primarily terrestrial, these opossums are capable climbers. In tropical grasslands, the Southern opossum becomes highly arboreal during the rainy season when the ground is flooded. Opportunistic feeders, they eat fruit, insects, small vertebrates, carrion, and garbage, varying their diets with seasonal availability. In the tropical forests of southeastern Peru, the Southern opossum climbs to heights of 25m (80ft) to feed on flowers and nectar during the dry season.

Above *A Gray-bellied slender mouse opossum foraging on the forest floor in Venezuela. Animals of this species are mainly nocturnal and, in common with most other mouse opossums, feed primarily on insects and fruit.*

FACTFILE

AMERICAN OPOSSUMS

Order: Didelphimorphia

Family: Didelphidae

63 species in 15 genera

HABITAT Wide-ranging, including temperate deciduous forests, tropical forests, grasslands, mountains, and human settlements. Terrestrial, arboreal, and semi-aquatic.

SIZE **Head–body length** ranges from 6.8cm (2.7in) in the Formosan mouse opossum to 33–55cm (13–19.7in) in the Virginia opossum; **tail length** from 4.2cm (1.7in) in the Pygmy short-tailed opossum to 25–54cm (9.8–21.3in) also in the Virginia opossum, whose **weight** is 2–5.5kg (4.4–12.1lb).

COAT Either short, dense, and fine, or woolly, or a combination of short underfur with longer guard hairs. Color varies from dark to light grays and browns or golden; some species have facial masks or stripes.

DISTRIBUTION Throughout most of S and C America, N through E North America to Ontario, Canada; Virginia opossum introduced to the Pacific coast.

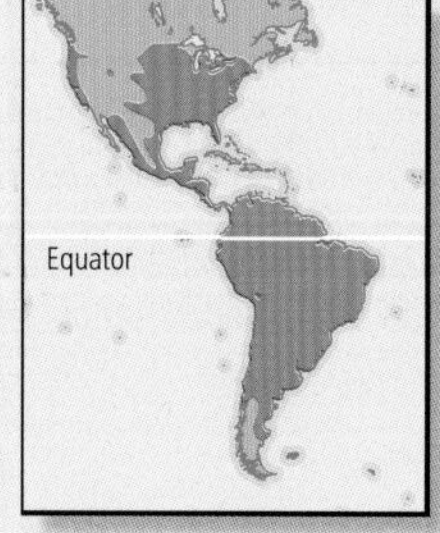

DIET Insectivorous, carnivorous, or (most often) omnivorous.

BREEDING Gestation 12–14 days

LONGEVITY 1–3 years (to about 8 in captivity)

CONSERVATION STATUS 44 of the 63 Didelphidae species are currently listed by the IUCN. Of these, 3 are Critically Endangered, 3 are Endangered, and 15 are Vulnerable; most of the others are ranked as Lower Risk: Near Threatened.

See subfamilies box ▷

The four-eyed opossums from the forests of Central and northern South America are also mostly generalist species. They are smaller than the Virginia opossum, with more slender bodies and distinct white spots above each eye, from which their common name is derived. These opossums are adept climbers, but the degree to which they climb seems to vary between habitats. The four-eyed opossums are also opportunistic feeders; earthworms, fruit, insects, and small vertebrates are all eaten.

The yapok, or Water opossum, is the only marsupial highly adapted to an aquatic lifestyle. The hind feet of this striking species are webbed, making the big toe less opposable than in other didelphids. When swimming, the hind feet alternate strokes, while the forefeet are extended in front, allowing the animals to either feel for prey or carry food items. Yapoks are primarily carnivorous, feeding on crustaceans, fish, and frogs as well as on insects. Although they can climb, they rarely do so, and the long, round tail is not very prehensile. Both male and female yapoks possess a pouch, which opens to the rear. During a dive, the female's pouch becomes a watertight chamber; fatty secretions and long hairs lining its lips form a seal, and strong sphincter muscles close it. In males, the scrotum can be pulled into the pouch when the animal is swimming or moving swiftly.

The Lutrine or Little water opossum is also a good swimmer, although it lacks the specializations of the yapok. Unlike the yapok, which is found primarily in forests, Lutrine opossums often inhabit open grasslands. Known as the *comadreja* ("weasel") in South America, this opossum has a long, low body with short, stout legs. The tail is densely furred and very thick at the base. Lutrine opossums are able predators, being excellent swimmers and climbers and also agile on the ground. They feed on a variety of prey including small mammals, birds, reptiles, frogs, and insects.

The mouse, or murine, opossums are a diverse group, with individual species varying greatly in size, climbing ability, and habitat. All are rather opportunistic feeders. The largest species, the Ashy mouse opossum, is one of the most arboreal, whereas others such as *Marmosa fuscata* are more terrestrial. The tail in most species is long, slender, and very prehensile, but in some species including the Elegant mouse opossum it can become swollen at the base for fat storage. The large, thin ears may become crinkled when the animal is aroused. The females lack a pouch, and the number and arrangement of mammae vary between species. Mouse opossums inhabit most habitats from Mexico through South America; they are absent only from the high Andean *paramo* and *puna* zones, the Chilean desert, and Patagonia, where they are replaced by another small species, the Patagonian opossum, which has the most southerly distribution of any didelphid.

This opossum broadly resembles the mouse opossums, though the muzzle is shorter, which allows for greater biting power; in the light of this adaptation, and because insects and fruit are rare in their habitat, Patagonian opossums are believed to be more carnivorous than mouse opossums. The feet are stronger than in mouse opossums and possess longer claws, suggesting fossorial (burrowing) habits. As in some of the mouse opossums, the tail of the Patagonian opossum can become swollen with fat.

The short-tailed opossums are small didelphids inhabiting forests and grasslands from eastern Panama through most of northern South America east of the Andes. The tail of these shrew-like animals is short and naked; their eyes are smaller than

THE MONITO DEL MONTE

The monito del monte or colocolo lives in the forests of south-central Chile. Once thought to belong to the same family as the American opossums, this small marsupial is now considered the only living member of an otherwise extinct order, the Microbiotheria.

Monitos ("little monkeys") have small bodies with short muzzles, round ears, and thick tails. Head–body length is 8–13cm (3–5in), and they weigh just 16–31g (0.6–1.1oz). They are found in cool, humid forests, especially in bamboo thickets. Conditions are often harsh in these environments, and monitos exhibit various adaptations to the cold. The dense body fur and small, well-furred ears both help prevent heat loss. During winter months when food (mostly insects and other small invertebrates) is scarce, they hibernate. Before hibernation, the base of the tail becomes swollen with fat deposits.

There are various local superstitions about these harmless animals. One is that their bite is venomous and produces convulsions. Another maintains that it is bad luck to see a monito; some people have even reportedly burned their houses to the ground after spotting one in their homes.

Right *A large American opossum (Didelphis sp.) in Minnesota, near the northerly limit of its range. Since the 19th century these opossums have spread rapidly northward through the USA. However, they are poorly adapted to cold weather conditions. During severe spells they may remain inactive in their nests for days on end, although they do not hibernate.*

in most didelphids and not as protruding. As these anatomical features suggest, short-tailed opossums are primarily terrestrial, but they can climb. As in mouse opossums, the females lack a pouch, and the number of mammae varies between species. Short-tailed opossums are omnivorous, feeding on insects, earthworms, carrion, and fruit, among other foods. Often they will inhabit human dwellings, where they are a welcome predator on insects and small rodents.

The three species of woolly opossum, along with the Black-shouldered and Bushy-tailed opossums, are placed in a separate subfamily – the Caluromyinae – from other didelphids on the basis of differences in blood proteins, the anatomy of the females' urogenital system, and the males' spermatozoa. The woolly opossums and the Black-shouldered opossum are among the most specialized of all didelphids. Highly arboreal, they have large, protruding eyes that are directed somewhat forward, making their faces reminiscent of those of primates. Inhabitants of humid tropical forests, these opossums climb through the upper tree canopy in search of fruit. During the dry season, they also feed on the nectar of flowering trees, and serve as pollinators for the trees they visit. While feeding, they can hang by their long prehensile tails to reach fruit or flowers.

Although the Bushy-tailed opossum resembles mouse opossums in its general appearance and proportions, dental characteristics, such as the size and shape of the molars, indicate that it is actually more closely related to the woolly and Black-shouldered opossums. This species is known only from a few museum specimens, all of them taken from humid tropical forests.

Back to North America

DISTRIBUTION PATTERNS

North American fossil deposits from 70–80 million years ago are rich in didelphid remains, and it was probably from North America that didelphids entered South America and Europe. Yet by 10–20 million years ago they had become extinct in both North America and Europe. When South America again became joined to Central America, about 2–5 million years ago, many South American marsupials became extinct in the face of competition from the northern placental mammals that took the opportunity to spread into their territory. Didelphids persisted, however, and even moved north into Central and North America.

During historical times, the early European settlers of North America found no marsupials north of the modern states of Virginia and Ohio. Since then, in the eastern USA, the Virginia opossum has steadily extended its range as far as the Great Lakes. Moreover, following introductions on the Pacific coast in 1890, this species has also spread from southern California to southern Canada. These expansions are most probably related to human impact on the environment.

Left *A Virginia opossum feigning death to deter a potential aggressor. The condition, which brings on symptoms akin to fainting, can last for less than a minute or as long as six hours.*

Playing Possum

SOCIAL BEHAVIOR

Reproduction in didelphids is typical of marsupials: gestation is short and does not interrupt the estrous cycle. The young are poorly developed at birth, and most development takes place during lactation. In the past there was a popular misconception that opossums copulated through the nose and that the young were later blown out through the nostrils into the pouch! The male's bifurcated penis, the tendency for females to lick the pouch area before birth, and the small size of the young at birth (they are just 1cm/0.4in long, and weigh 0.13g/0.005oz) all probably contributed to this notion.

Most opossums appear to have seasonal reproduction. Breeding is timed so that the first young leave the pouch when resources are most abundant; for example, the Virginia opossum breeds during the winter in North America, and the young leave the pouch in the spring. Opossums in the seasonal tropics breed during the dry season, and the first young leave the pouch at the start of the rainy season. Up to three litters can be produced in one season, but the last litter often overlaps the beginning of the period of food scarcity, and these young frequently die in the pouch. Opossums in aseasonal tropical forests may reproduce throughout the year, as may the White-eared opossum in the arid region of northeast Brazil.

There are no elaborate courtship displays nor long-term pair-bonds. The male typically initiates contact, approaching the female while making a clicking vocalization. A non-receptive female will avoid contact or be aggressive, but a female in estrus will allow the male to mount. In some species courtship behavior involves active pursuit of the female. Copulation can be very prolonged – for up to six hours in Robinson's mouse opossums (*Marmosa robinsoni*).

Many of the newborn young die, as many never attach to a teat. A female will often produce more

American Opossum Subfamilies

Subfamily Didelphinae

58 species in 12 genera. S and C America, E USA to Ontario. Virginia opossum introduced to the Pacific coast.

Large American opossums (*Didelphis*, 4 species), including the Virginia opossum (*Didelphis virginiana*), Southern opossum (*D. marsupialis*), and White-eared opossum (*D. albiventris*)

Mouse opossums (*Marmosa*, 9 species), including the Murine mouse opossum (*M. murina*) and Robinson's mouse opossum (*M. robinsoni*). 1 species is Critically Endangered, another Endangered.

Gracile mouse opossums (*Gracilinanus*, 6 species), including the Wood spirit gracile mouse opossum (*G. dryas*). 1 species is Critically Endangered, 2 are Vulnerable.

Slender mouse opossums (*Marmosops*, 9 species), including the Gray-bellied slender mouse opossum (*M. fuscatus*). 1 species is Critically Endangered, 1 Endangered, 1 Vulnerable.

Woolly mouse opossums (*Micoureus*, 4 species), including the Pale-bellied woolly mouse opossum (*M. constantiae*).

Short-tailed opossums (*Monodelphis*, 15 species), including the Gray short-tailed opossum (*M. domestica*) and the Pygmy short-tailed opossum (*M. kunsi*). 1 species is Endangered, 8 are Vulnerable.

Gray and Black four-eyed opossums (*Philander*, 2 species), comprising the Gray four-eyed opossum (*P. opossum*) and Black four-eyed opossum (*P. andersoni*).

Fat-tailed opossums (*Thylamys*, 5 species), including the Elegant fat-tailed opossum (*T. elegans*)

Brown four-eyed opossum (*Metachirus nudicaudatus*)

Water opossum (*Chironectes minimus*)

Lutrine opossum (*Lutreolina crassicaudata*)

Patagonian opossum (*Lestodelphys halli*). Vulnerable.

Subfamily Caluromyinae

5 species in 3 genera. S Mexico through C America and most of northern S America.

Woolly opossums (*Caluromys*, 3 species). 1 species Vulnerable.

Black-shouldered opossum (*Caluromysiops irrupta*). Vulnerable.

Bushy-tailed opossum (*Glironia venusta*). Vulnerable.

For full species list see Appendix ▷

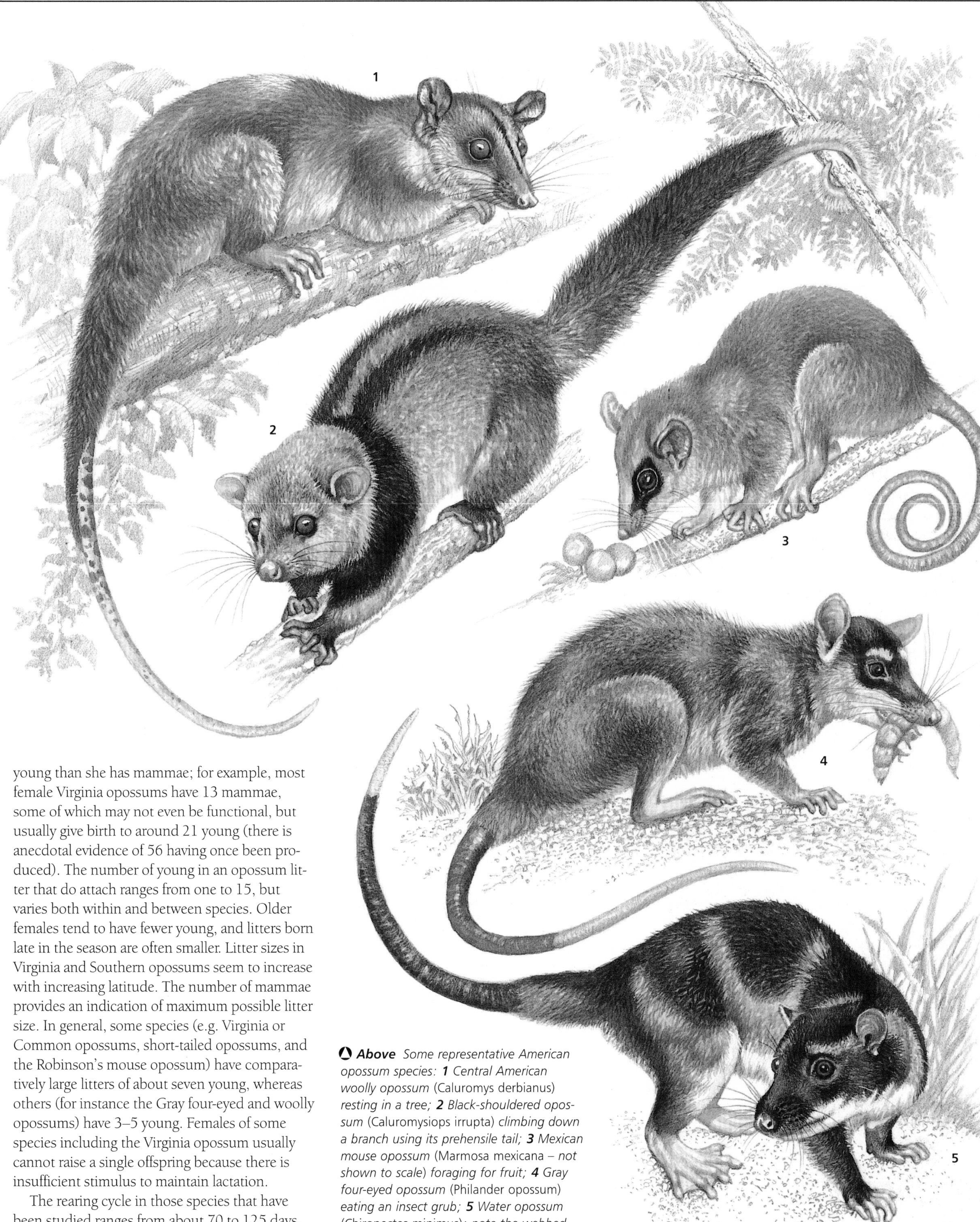

young than she has mammae; for example, most female Virginia opossums have 13 mammae, some of which may not even be functional, but usually give birth to around 21 young (there is anecdotal evidence of 56 having once been produced). The number of young in an opossum litter that do attach ranges from one to 15, but varies both within and between species. Older females tend to have fewer young, and litters born late in the season are often smaller. Litter sizes in Virginia and Southern opossums seem to increase with increasing latitude. The number of mammae provides an indication of maximum possible litter size. In general, some species (e.g. Virginia or Common opossums, short-tailed opossums, and the Robinson's mouse opossum) have comparatively large litters of about seven young, whereas others (for instance the Gray four-eyed and woolly opossums) have 3–5 young. Females of some species including the Virginia opossum usually cannot raise a single offspring because there is insufficient stimulus to maintain lactation.

The rearing cycle in those species that have been studied ranges from about 70 to 125 days. For example, the Gray four-eyed opossum and

Above *Some representative American opossum species:* ***1*** *Central American woolly opossum* (Caluromys derbianus) *resting in a tree;* ***2*** *Black-shouldered opossum* (Caluromysiops irrupta) *climbing down a branch using its prehensile tail;* ***3*** *Mexican mouse opossum* (Marmosa mexicana *– not shown to scale) foraging for fruit;* ***4*** *Gray four-eyed opossum* (Philander opossum) *eating an insect grub;* ***5*** *Water opossum* (Chironectes minimus)*; note the webbed hind feet.*

woolly opossums are similar in size (usually about 400g/14oz), but the time from birth to weaning is 68–75 days in the former and 110–125 days in the latter. Initially the young remain attached to the mother's teats, but later they begin to crawl about her body. Toward the end of lactation, they begin to follow her when she leaves the nest to forage. Female Robinson's mouse opossums will retrieve detached young within a few days of birth; in contrast, Virginia opossum mothers do not respond to their distress calls until after they have left the pouch, at about 70 days.

Although individual vocalizations and odors allow for some mother–infant recognition, maternal care in opossums does not appear to be restricted solely to a female's own offspring. Female Robinson's mouse opossums will retrieve young other than their own, and Virginia and woolly opossums have been observed carrying other females' young in their pouches. Toward the end of lactation, females cease any maternal care, and dispersal is rapid. Sexual maturity is attained within 6–10 months. Age at sexual maturity is not related directly to body size.

In general, opossums are not long-lived. Few Virginia opossums survive beyond 2 years in the wild, and the smaller mouse opossums may not live much beyond one reproductive season. Although animals kept in captivity may survive longer, females are generally not able to reproduce after 2 years. Thus, among many of these didelphids, there is a trend towards the production of a few large litters during a limited reproductive life; indeed, a female Robinson's mouse opossum may typically reproduce only once in a lifetime.

American opossums appear to be locally nomadic, and seemingly do not defend territories. Radio-tracking studies reveal that individual animals occupy home ranges, but do not exclude others of the same species (conspecifics). The length of time for which a home range is occupied varies both between and within species. In the forests of French Guiana, for example, some woolly opossums have been observed to remain for up to a year in the same home range, whereas others shifted home range repeatedly; Gray four-eyed opossums were more likely to shift home range.

In contrast to some other mammals, didelphids do not appear to explore their entire home range on a regular basis. Movements primarily involve feeding and travel to and from a nest site, and are highly variable depending upon food resources and/or reproductive condition. Thus home range estimates for Virginia opossums in the central United States vary from 12.5–38.8 hectares (31–96 acres). An individual woolly opossum's home range may vary from 0.3–1ha (0.75–2.5 acres) from one day to the next. In general, the more carnivorous species have greater movements than similar-sized species that feed more on fruit. Male didelphids become more active during the breeding season, whereas reproductive females generally become more sedentary.

Most opossums use several to many nest sites within their range. Nests are often used alternately with conspecifics in the area. Virginia and Southern opossums use a variety of nest sites, both terrestrial and arboreal, but hollow trees are a common location. Four-eyed opossums also nest in holes and open limbs of trees, and additionally on the ground, in rock crevices, or under tree roots or fallen palm fronds. Mouse opossums nest either on the ground under logs and tree roots or in trees, using holes or abandoned birds' nests.

Shrew Opossums of the High Andes

Five small, shrew-like marsupial species are found in the Andean region of western South America from southern Venezuela to southern Chile. Known sometimes as shrew (or rat) opossums, they are unique among American marsupials in having a reduced number of incisors, the lower middle two of which are large and project forward. The South American group represents a distinct line of evolution that diverged from ancestral stock before the Australian forms did, and its members are now placed not just in their own family, the Caenolestidae, but are also assigned an order, the Paucituberculata, of which they are the only representatives.

Fossil evidence indicates that about 20 million years ago seven genera of shrew opossums occurred in South America. Today the family is represented by only three genera and five species. There are three species of *Caenolestes*: the Gray-bellied (*C. caniventer*), Blackish (*C. convelatus*), and Silky (*C. fuliginosus*); the Incan (*Lestoros inca*) and Chilean (*Rhyncholestes raphanurus*) shrew opossums are placed in separate genera. Known head–body lengths of these small marsupials are in the range of 9–14cm (3.5–5.5in), tail lengths mostly 10–14cm (3.9–5.5in), and weights from 14–41g (0.5–1.4oz).

The animals have elongated snouts equipped with numerous tactile whiskers. The eyes are small and vision is poor. Well-developed ears project above the fur. The ratlike tails are about the same length as the body (rather less in the Chilean shrew opossum) and are covered with stiff, short hairs. The fur on the body is soft and thick and is uniformly dark brown in most species. Females lack a pouch, and most species have four teats (five in the Chilean shrew opossum). Shrew opossums are active during the early evening and/or night, when they forage for insects, earthworms, and other small invertebrates, and also small vertebrates. They are able predators, using their large incisors to kill prey.

Shrew opossums travel about on well-marked ground trails or runways. More than one individual will use a particular trail or runway. When moving slowly, they have a typically symmetrical gait, but when moving faster Incan shrew opossums, and possibly other species, will bound, allowing them to clear obstacles.

The Blackish and Silky shrew opossums are distributed at high elevations in the Andes of western Venezuela, Colombia, and Ecuador. The Gray-bellied shrew opossums of southern Ecuador occur at lower elevations. The Incan shrew opossum is found high in the Peruvian Andes, but in drier habitats than that of the other species; it has been trapped in areas with low trees, bushes, and grasses. The Chilean shrew opossum inhabits the forests of southern Chile. As winter approaches, the tail of this species becomes swollen with fat deposits.

Very little is known about the biology of these elusive marsupials. They inhabit inaccessible and (for humans) rather inhospitable areas, which makes them difficult to study. They have always been considered rare, but recent collecting trips have suggested that some species may be more common than was previously thought.

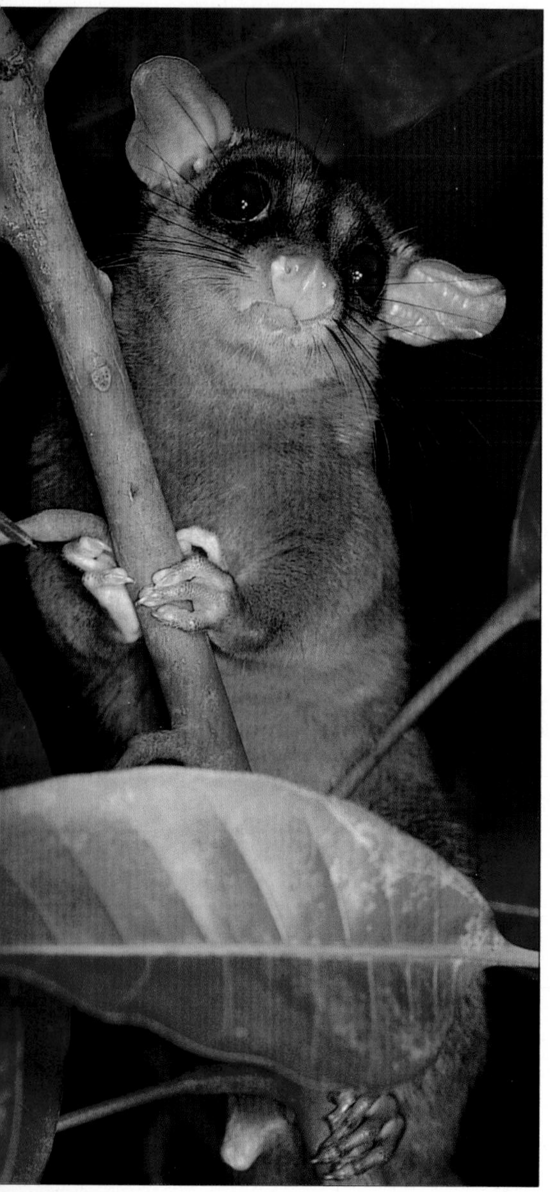

Left *A Bare-tailed woolly opossum perches on a branch in French Guiana. These opossums, which take their name from their long, thick fur, are agile climbers, using the five digits of all four limbs to help them clamber through the forests in which they live.*

Above *Hanging from a branch, a short-tailed opossum mother in Brazil shows off her seven newborn young, each attached to a separate teat. Pouches are not developed in these species.*

Right *An infant Virginia opossum hitches a ride on its mother's back. For a month or two between leaving the pouch and achieving independence, this is the usual mode of transport for the young.*

Occasionally mouse opossums make nests in banana stalks, and more than once animals have been shipped to grocery stores in the United States and Europe! In the open grasslands, the Lutrine opossum constructs globular nests of leaves or uses abandoned armadillo burrows. In more forested areas, these opossums may use tree holes. Unlike other didelphids, yapoks construct more permanent nests; their underground nesting chambers are located near the waterline and are reached through holes dug into stream banks.

Opossums are solitary animals. Although many may congregate at common food sources during periods of food scarcity, there is no interaction unless individuals get too close. Typically, when two animals do meet, they threaten each other with open-mouth threats and hissing and then continue on their way. If aggression does persist (usually between males), the hissing changes into a growl and then to a screech. Communication by smell is very important; many species have well-developed scent glands on the chest. In addition, male Virginia opossums, Gray four-eyed opossums, and Gray short-tailed opossums have all been observed marking objects with saliva. Marking behavior is carried out primarily by males, and is thought to advertise their presence in an area.

In tropical forests up to seven species of didelphids may be found at the same locality. Competition between these species is avoided through differences in body size and varying tendencies to climb. For example, woolly opossums and the Ashy mouse opossum will generally inhabit the tree canopy, the Common mouse opossum mostly haunts the lower branches, Southern and Gray four-eyed opossums are found either on the ground or in the lower branches, while short-tailed opossums live solely on the ground. Some species appear to vary their tendency to climb depending upon the presence of similar-sized opossums. For example, in a Brazilian forest where both Gray and Brown four-eyed opossums were found together, the former was more arboreal than the latter, yet in a forest in French Guiana where only the Gray was present, it was primarily terrestrial, and elsewhere the Brown four-eyed opossum is mostly arboreal.

Despite being hunted for food and pelts, the Virginia opossum thrives both on farms and in towns and even cities. Elsewhere in the Americas, man's impact on the environment has been detrimental to didelphids. Destruction of humid tropical forests results in loss of habitat for the more specialized species. MAO'C

Large Marsupial Carnivores

THE LARGE MARSUPIAL CARNIVORES CONSIST of six species of quolls, sometimes referred to as marsupial cats, and the Tasmanian devil, a single species assigned to its own genus. Tasmanian devils are perhaps best known to the general public in the form of "Taz," the Warner Brothers cartoon character. But the comic image of a tail-spinning terror devouring everything in its path belies the true nature of the devil, as does its reputation in Tasmania itself as an odious scavenger. While belligerent towards other devils, these predators and specialized scavengers responsive, intelligent, and full of character.

Vertebrate prey is a major part of the diet of large marsupial carnivores. Current diversity is low, but fossil evidence suggests many more species existed in the past. At least nine thylacinids have roamed Australia; their heyday was during the Miocene era (24–5 million years ago), while during the Pleistocene (1.8 million–10,000 years ago), several species of giant dasyurids (quolls) existed. The last thylacine, also known as the Tasmanian wolf because of its dog-like appearance, became extinct only in the 1930s.

Marsupial carnivores are not exclusively Australasian. Three species of South American marsupials are highly carnivorous, and thylacine-like borhyaenids and sabretooth-cat-like marsupial thylacosmilids roamed there in prehistoric times.

A Classic Case of Convergence

FORM AND FUNCTION

At least superficially, quolls look like mongooses, thylacines like dogs, while devils resemble small hyenas, even down to the sloping hindquarters and rolling gait. The remarkable similarities between marsupial and placental carnivores that have evolved on different continents can give the impression of parallel universes.

The observed similarities between the two groups are in fact related to these species leading similar lives. Devils, like hyenas, have highly carnivorous dentition and adaptations for bone consumption, including robust premolar or molar teeth used for cracking bones and a relatively short snout with massive jaw-closing muscles, giving the animals a strong, crushing bite.They are able to consume all parts of a carcass, including thick skin and all but the largest bones. With teeth that are adapted for crushing invertebrates as well as for slicing meat, quolls group with the mongoose and stoat families to fill the role of predator of small to medium-sized mammals and invertebrates. The thylacine, on the other hand, was probably ecologically closer to smaller canids like the coyote than it ever was to the wolf. The animal's extremely long snout, very low rates of canine tooth wear and fracture, and limb ratios typical of slow runners, suggest that it hunted prey such as wallabies that were smaller than itself, and did not use long, fast pursuits.

Below *The Tasmanian devil will take living prey, including lambs and poultry, but has a preference for carrion. It can even chew and swallow sheep bones.*

FACTFILE

LARGE MARSUPIAL CARNIVORES

Order: Dasyuromorphia

Family: Dasyuridae

7 extant species in 2 genera; 1 species (*Thylacinus cynocephalus*) in the related Thylacinidae family recently extinct

DISTRIBUTION Australia and New Guinea

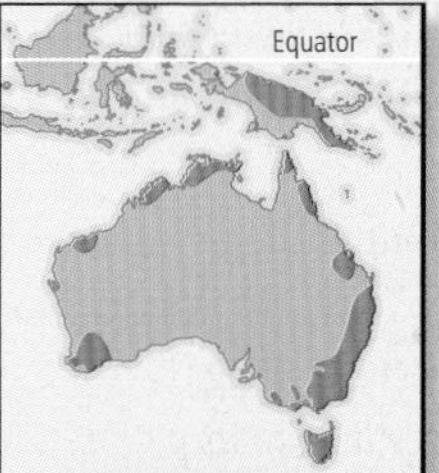

HABITAT Mostly in rain forest or woodland

SIZE **Head–body length** ranges from 12.3–31cm (5–12in) in the male Northern quoll to 50.5–62.5cm (20–25in) in the Tasmanian devil; **weight** ranges from 0.3–0.9kg (0.7–2lb) to 4.4–13kg (10–30lb) in the same two species.

COAT Short-furred; upper parts mostly gray or brown, with white spots or blotches. The Tasmanian devil has thicker fur, and usually only one or two white patches.

DIET Smaller species eat mainly insects, but the Spotted-tailed quoll and the Tasmanian devil will take mammals as large as wallabies.

BREEDING Mostly 4–8 young, carried in pouch for 8–10 weeks in smaller species but for 4–5 months in Tasmanian devil.

LONGEVITY Ranges from 12 months in the smaller species to 6 years in the Tasmanian devil.

CONSERVATION STATUS Four of the 7 extant species are currently listed as Vulnerable, two as Lower Risk: Near Threatened.

See species table ▷

Above *Four-month old Eastern quolls resting in a grass-lined den. Young are usually deposited in a den by the mother about ten weeks after birth, having spent their time up until then in their mother's pouch.*

Marsupial carnivore species exhibit variation in the strength of their canine teeth and temporal (jaw) muscles and take prey of different sizes, which consequently minimizes the competition for food. The shape of the canines is oval, being intermediate between the narrow canine teeth of the dog family and the more rounded canines of the cats. Marsupial carnivores kill their prey by use of a generalized crushing bite, which is applied to the skull or nape.

Two morphological features suggest that Dasyuromorphs and Carnivorans may simply have evolved different solutions to the same problems. Marsupials have a longer snout than the placental carnivores, suggesting that they may possess a weaker bite, but this comparative deficiency appears to be compensated for by extra space for larger jaw muscles, which is derived from their smaller brain cases. Limb–bone ratios suggest that devils, Spotted-tailed quolls, and thylacines, but not the smaller quolls, are slow runners compared with their placental counterparts. The granular foot pads, which are limited to the digits in placental carnivores, extend to the ankle and wrist joints in the marsupial carnivores, which sometimes walk and rest on their heels, suggesting differences in their locomotory function.

There is a quoll adapted to every environment in Australia and New Guinea: tropical rain forest, monsoonal savanna, woodlands, deserts, grasslands, and temperate forests. The degree of carnivory increases with body size, from the quite insectivorous smaller quolls to the completely carnivorous Spotted-tailed quoll, devil, and thylacine. The largest remaining guild, in Tasmania, had four species until the recent extinction of the thylacine: the Eastern quoll, a small (1kg/2.2lb) ground-dwelling insectivore/carnivore; the Spotted-tailed quoll, a middle-sized (2–4kg/4.5–9lb) tree-climbing predator; the devil, a larger (4.4–13 kg/10–30lb) ground-dwelling carnivore and specialist scavenger; and the predatory thylacine (15–35kg/33–77lb). Spotted-tailed quolls are naturally rare, perhaps because they are specialists on forest habitat and vertebrate food, and because they may compete for food with Eastern quolls, devils, and introduced cats. Their tree-climbing abilities may allow them to exploit the tree-dwelling possum prey resource, however, which is less available to other quolls and devils. Devils are the most common species; their larger size gives them an advantage in stealing carcasses from quolls and monopolizing them.

Solitary but Social

SOCIAL BEHAVIOR

Early settlers can be excused for giving the devil its alarming name. The sight of a jet-black animal with large white teeth and bright red ears (the ears of devils blush red when they are stirred up) creating mayhem in a poultry coop, combined with spine-chilling screeches in the dark, must have been more than enough to convince them that they had indeed "raised the Devil." Devils make a wide range of sounds, including growls, whines, soft barks, snorts, sniffs, and whimpers. Devils are solitary creatures but highly social, feeding together at carcasses and developing quite affectionate relationships for several days while mating. Females and males have very different agendas, however. A female loves and leaves several males in succession over a period of a week, perhaps in an attempt to ensure that her young are sired by the best males available. Intent on protecting his paternity, the male devil indulges in "cave-man" tactics, charging the escaping female and then using a neck-bite to drag or lead her back to the den, where he will copulate with her for prolonged periods. During the mating season female devils develop a swelling on the back of their neck that may serve to protect them from injury. Males fight for females, sustaining deep bites and gouges to the head and rump that may sometimes be life-threatening.

Quolls, like devils, are solitary and rest underground or in hollow log dens during the day. Neither group is territorial, although females sometimes maintain an exclusive part of their home range, especially when they have young in a den. Male young are more likely than females to disperse away from the home range of their mother. Young devils disperse at 12–18 months of age, immediately after weaning, moving some 10–30 km (6–19mi) from their birth site. Scent is important in communication, with latrines possibly assuming the role of community noticeboards in some species. Devils frequently scentmark by dragging their cloacae on the ground. Quolls, unlike devils, call infrequently, making a variety of coughing and hissing sounds or else abrupt, piercing screams; in the case of the Spotted-tailed quoll, these cries have been likened to the sound made by a circular saw.

The Threat from Placentals

CONSERVATION AND ENVIRONMENT

All Australian species have declined in range and abundance as a result of human presence. Concern has also been expressed for the quolls in New Guinea, although their conservation status is poorly known.

Devils and quolls, like many placental carnivores, are persecuted for attacking livestock. Both groups take insecurely-penned poultry, and newly-born lambs and sick sheep are also vulnerable. As a result, devils and quolls are trapped, shot, and poisoned in the tens and occasionally in the hundreds. Similar persecution, combined with loss of habitat and prey as a result of more sheep farming, was a main cause of the thylacine's decline, leading ultimately to its extinction.

The greatest threats to marsupial carnivores, however, are loss of habitat and killings by introduced placental carnivores. These often go hand in hand. Like many carnivores, devils and quolls live at low density (one individual to 1–10sq km/ 0.4–4sq mi, depending on habitat, in Spotted-tailed quolls) and require lots of space (home ranges of male devils average 30sq km/12sq mi). Extensive areas of habitat have been lost through

THE TASMANIAN WOLF

Up to the time of its extinction, the thylacine was the largest of recent marsupial carnivores. Fossil thylacines are widely scattered in Australia and New Guinea, but the living animal was confined in historical times to Tasmania.

Superficially, the thylacine resembled a dog. It stood about 60cm (24in) high at the shoulders, head–body length averaged 80cm (31.5in), and weight 15–35kg (33–77lb). The head was dog-like with a short neck, and the body sloped away from the shoulders. The legs were also short, as in large dasyurids. The features that clearly distinguished the thylacine from dogs were a long (50cm/20in), stiff tail, which was thick at the base, and a coat pattern of black or brown stripes on a sandy yellow ground across the back.

Most of the information available on the behavior of the thylacine is either anecdotal or has been obtained from old film. It ran with diagonally opposing limbs moving alternately, could sit upright on its hindlimbs and tail rather like a kangaroo, and could leap 2–3m (6.5–10ft) with great agility. Thylacines appear to have hunted alone or in pairs, and before Europeans settled in Tasmania they probably fed upon wallabies, possums, bandicoots, rodents, and birds. It is suggested that they caught prey by stealth rather than by chase.

At the time of European settlement, the thylacine appears to have been widespread in Tasmania, and was particularly common where settled areas adjoined dense forest. It was thought to rest during the day on hilly terrain in dense forest, emerging at night to feed in grassland and woodland.

From the early days of European settlement, the thylacine developed a reputation for killing sheep. As early as 1830, bounties were offered for killing thylacines, and the consequent destruction led to fears for the species' survival as early as 1850. Even so, the Tasmanian government introduced its own bounty scheme in 1888, and over the next 21 years, before the last bounty was paid, 2,268 animals were officially killed. The number of bounties paid had declined sharply by the end of this period, and it is thought that epidemic disease combined with hunting to bring about the thylacine' s final disappearance.

The last thylacine to be captured was taken in western Tasmania in 1933; it died in Hobart zoo in 1936. Since then the island has been searched thoroughly on a number of occasions, and even though occasional sightings continue to be reported to this day, the most recent survey concluded that there has been no positive evidence of thylacines since that time. In 1999, the Australian Museum in Sydney decided to explore the possibility of cloning a thylacine, using DNA from a pup preserved in alcohol in 1866, although it admitted that to do so successfully would require substantial advances in biogenetic techniques. AKL

◐ *Left* *Rare, if not extinct, in some regions that it traditionally occupied, the Spotted-tailed quoll is vulnerable to predation by placental carnivores in many areas.*

◑ *Right* *The Spotted-tailed quoll is an active hunter, killing its prey – which includes gliders, small wallabies, reptiles, and birds – with a bite to the back of the head.*

degradation, in addition to direct losses to agriculture and intensive forestry, which in turn can limit the availability of prey, especially for the forest-dependent Spotted-tailed quoll, which disappears if more than 50 percent of the canopy is removed. The decline in the number of old-growth, hollow-bearing trees is a cause of particular concern; they support populations of hollow-nesting possums, which are major prey species for Spotted-tailed quolls. The smaller quolls become increasingly vulnerable to predation by introduced predators if vegetative cover is removed by livestock grazing or frequent fires. Northern, Western, and Eastern quolls have lost large tracts of habitat in this way.

Dingoes, cats, and foxes were all introduced to Australia by humans, respectively 3,500–4,000 years ago, as early as the 17th century, and in 1871. Introduced carnivores kill marsupial carnivores for food or to reduce competition. Thylacines and devils declined and went extinct on mainland Australia at the same time as the dingo arrived. Now foxes and cats seem to be major causes of death for the smaller species of quolls; certainly, Northern quoll death rates from predation rise after fires in savanna habitats, but not in rocky habitats where shelter from predators is available; similarly, populations increase when foxes are controlled. On a local scale, quolls tend to become extinct wherever Red foxes live.

Fortunately, there is also some good news to report. The Western quoll – once considered endangered – will shortly be removed from threatened-species lists, following the success of a recovery program that has translocated 200 captive-bred animals to four new sites since the early 1990s, although their continued survival will remain dependent on ongoing fox and cat control.

The Spotted-tailed and Northern quolls are currently focuses for concern. The populations of both species are fragmented and declining, and their survival and population recovery will require a nationally-coordinated approach, with deleterious processes, habitat loss, and predator control all needing to be addressed. MJ

Northern quoll
Dasyurus hallucatus

N Australia. Formerly found in broad band across wet–dry tropics; now only in lowland savanna woodland and rocky terrain.
HBL male: 12.3–31cm (5–12in), female: 12.5–30cm (5–12in); TL male: 12.7–30.8cm (5–12in), female: 20–30cm (8–12in); WT male: 0.4–0.9kg (0.8–2lb), female: 0.3–0.5kg (0.7–1.1lb).
COAT: white spots on brown body, striated foot pads, clawless hallux (1st toe) on 5-toed hind foot, tail long-haired.
BREEDING: mates in June at 12 months old, usually up to 8 young (variation 5–10) carried in pouch for 8–10 weeks, weaned at 6 months.
LONGEVITY: male 12–17 months, female 12–37 months.
CONSERVATION STATUS: Lower Risk, Near Threatened; in extensive decline, with just six fragmented populations remaining.

New Guinea quoll
Dasyurus albopunctatus

New Guinea. Widespread in rainforest habitats above 1,000m (3,300ft).
HBL male: 22.8–35cm (9–14in), female: 24.1–27.5cm (9.5–10.5in); TL male: 21.2–29cm (8–11in), female: 22.1–28cm (9–11in); WT male: 0.6–0.7kg (1.5lb), female: 0.5kg (1.1lb).
COAT: white spots on reddish-brown body, well-developed hallux, tail short-haired.
BREEDING: not seasonal, 4–6 young.
CONSERVATION STATUS: Vulnerable.

Bronze quoll
Dasyurus spartacus

SW New Guinea. Lowland savanna woodlands. Known from only a few specimens; first reported 1979, recognized as a separate species from *D. geoffroii* in 1988.
HBL male: 34.5–38cm (14–15in), female: 30.5cm (12in); TL male: 28.5cm (11in), female: 25cm (10in); WT male: 1.0kg (2.2lb), female: 0.7kg (1.5lb).
COAT: white spots on brown body, small hallux, tail long-haired.
BREEDING: may be seasonal.
CONSERVATION STATUS: Vulnerable.

Western quoll
Dasyurus geoffroii
Western quoll or chuditch

SW Australia. Formerly over two-thirds of the county, from desert to forest, but now restricted to SW alone.
HBL male: 31–40cm (12–16in), female: 26–36cm (10–14in); TL male: 25–35cm (10–14in), female: 21–31cm (8–12in); WT male: 0.7–2.2kg (1.5–4.9lb), female: 0.6–1.1kg (1.3–2.4lb).
COAT: white spots on brown body, white belly, hallux, tail long-haired.
BREEDING: mates in May–June at 12 months old, up to 6 young carried in pouch for 9 weeks, weaned at 6 months.
LONGEVITY: 3 years.
CONSERVATION STATUS: Vulnerable; currently occupies only 2 percent of its former range.

Eastern quoll
Dasyurus viverrinus

Tasmania; formerly also SE mainland. Grasslands and open forests.
HBL male: 32–45cm (13–18in), female: 28–40cm(11–16in); TL male: 20–28cm (8–11in), female: 17–24cm (7–9.5in); WT male: 0.9–2.0kg (2–4.4lb), female: 0.7–1.1kg (1.5–2.4lb).
COAT: white spots on brown body, white belly, no hallux on 4-toed hind foot, tail long-haired.
BREEDING: mates in May–June at 12 months old, as many as 30 young born in June, up to 6 carried in pouch for 8–9 weeks and weaned at 5.5 months.
LONGEVITY: 3–4 years.
CONSERVATION STATUS: Lower Risk, Near Threatened.

Spotted-tailed quoll
Dasyurus maculatus

Tasmania (separate genetic unit); SE mainland and N Queensland (smaller phenotypic subspecies). Forest-dependent in areas of high rainfall or predictably seasonal rainfall.
HBL male: 45–51cm (18–20in), female: 40.5–43cm (16–17in); TL male: 39–49cm (15–19in), female: 34–44cm (13–17in); WT male: 3.0–7.0kg (6.6–15.4lb), female: 1.6–4.0kg (3.5–8.8lb); all data are for southern subspecies.
COAT: white spots on reddish-brown body and on short-haired tail, cream belly, striated foot pads, well-developed hallux.
BREEDING: mating April–July at 12 months old, up to 6 young carried in pouch for 8 weeks, weaned at 5 months.
LONGEVITY: 3–5 years.
CONSERVATION STATUS: Vulnerable; N Queensland subspecies is Endangered.

Tasmanian devil
Sarcophilus laniarius (harrisii)

Tasmania. Open forest and woodland.
HBL male: 50.5–62.5cm (20–25in), female: 53.5–57cm (21–22.5in); TL male: 23.5–28.5cm (9–11in), female: 21.5–27cm (8.5–10.5in); WT male: 7.7–13.0kg (17–28.7lb), female: 4.5–9.0kg (9.9–19.9lb).
COAT: black with variable white markings on chest, shoulder, and rump, fat store in tail base, no hallux on 4-toed on hind foot.
BREEDING: mates in February–March at 12 months old, up to 4 young carried in pouch for 4–5 months, weaned at 9 months.
LONGEVITY: 6 years.
CONSERVATION STATUS: Not listed.

Abbreviations HBL = head–body length TL = tail length WT = weight.

Small Marsupial Carnivores

THE SMALL MARSUPIAL CARNIVORES INCLUDE some of the smallest mammals on Earth, yet the ferocity they display in hunting belies their size. All show a predilection for live food, preying mainly on insects and other invertebrates, but also taking lizards, fledgling birds, and other small mammals. However, their rapaciousness has sometimes brought them undeserved notoriety – for its attacks on poultry in the Sydney area, the diminutive Brush-tailed phascogale, which weighs just 200g (7oz), was unfairly dubbed a "vampire marsupial" by early European settlers.

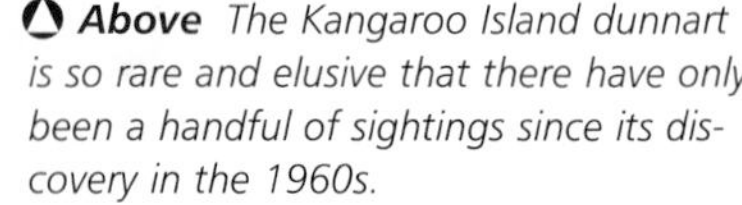

***Above** The Kangaroo Island dunnart is so rare and elusive that there have only been a handful of sightings since its discovery in the 1960s.*

Although other marsupials such as American opossums, bandicoots, and the numbat eat animal flesh, most marsupial carnivores are dasyurids. The majority of species in this family weigh less than 250g (8.8oz) and representatives occur in all terrestrial habitats in Australia and New Guinea. Because of their conservative body form and appearance, dasyurids have long been considered as "primitive" marsupials structurally ancestral to the Australasian radiation. However, recent fossil discoveries show the family to be a recent and specialized addition to the region, with the ancestral dasyurid arising in the early to mid-Miocene, perhaps 16 million years ago.

Coping with Extremes

FORM AND FUNCTION

Despite showing three-thousand-fold variation in body mass, all dasyurids have distinctive pointed snouts with three pairs of similar-sized lower incisors, well-developed canines, and 6–7 sharp cheek teeth. This dentition allows prey to be grasped and quickly killed and then comminuted (chewed into small pieces) before swallowing. Dasyurids are united also in having five toes on the forefeet and 4–5 toes on the hind feet, with all except the hallux (big toe) having sharp claws.

Unlike their larger relatives, the small marsupial carnivores mostly have uniform coat colors that range from shades of brown and gray to black. Three New Guinean dasyurids (the Narrow-striped dasyure, Broad-striped dasyure, and Three-striped dasyure) are unusual in having dark dorsal stripes. All are partly diurnal, the coat pattern helping to camouflage them against the dark background of the rainforest floor. Two further diurnal species, the Speckled dasyure of the New Guinea highlands and the dibbler of southwestern Australia, are also unusual in having grizzled, silvery-gray coats that may again serve a camouflage function. In two species of phascogales, as well as in the mulgara, ampurta, and kowari, the terminal half of the tail is a spectacular black brush that contrasts greatly with the light body fur. The bushy tails are thought to have a signaling function, but there are few observations of these species in the wild to confirm this.

The insectivorous diet of small marsupial carnivores has important implications for their physiology. Insects are rich in protein and fat, but have a free water content of more than 60 percent. Water turnover is relatively high in many dasyurids, with species weighing less than 25g (0.9oz) able to turn over their body weight in water each day. In arid areas, marsupial carnivores can obtain all their water from food for periods of months; juicy prey such as insect larvae and centipedes are preferred; water loss is reduced by the production of concentrated urine. If food is limited, several small marsupial carnivores can reduce their metabolic rates and drop their body temperatures by 10°C (18°F) or more to enter torpor. This can last at least 10 hours, reducing energy expenditure and allowing animals to ride out temporary food shortages. If food is not limited, dasyurids have an extraordinary ability to maintain their body temperatures, producing heat by elevating the metabolic rate 8–9 times its normal resting level. The desert-dwelling kowari maintains its body temperature for at least four hours by this method, even at –10°C (14°F). Alternative means of keeping warm include sun-basking and sharing nests with individuals of the same or different species.

***Left** The nocturnal Fat-tailed pseudantechinus, which stores fat in the base of its tail, is found mainly on rocky hills.*

***Right** Dubbed a "vampire marsupial" by early settlers for its attacks on poultry, the Brush-tailed phascogale can erect the hairs at the end of its tail.*

FACTFILE

SMALL MARSUPIAL CARNIVORES

Order: Dasyuromorphia

Families: Dasyuridae and Myrmecobiidae

64 species in 15 genera

DISTRIBUTION Australia, Papua New Guinea, and Indonesia (Aru Islands).

HABITAT Diverse, from stony desert to forest and alpine heath.

SIZE **Head–body length** ranges from 4.6–5.7cm (1.8–2.2in) in the Pilbara ningaui to 24.5cm (9.7in) in the Numbat; **tail length** ranges from 5.9–7.9cm (2.3–3.1in) to 17.7cm (7in), and **weight** from 2–9.4g (0.07–0.33oz) to 0.5kg (1.1lb), both in the same two species. Males are slightly or much heavier than females.

COAT Varied in color, but mostly short and coarse-furred.

DIET Mainly insects and other small invertebrates.(e.g. beetles, cockroaches, arachnids). Also small mammals and birds, including house mice, lizards, and sparrows.

BREEDING Gestation period ranges from 12.5 days in the Fat-tailed dunnart to 55 days in the Fat-tailed pseudantechinus.

LONGEVITY In the Brown antechinus, males live for 11.5 months, females for 3 years.

CONSERVATION STATUS One subspecies is classed as Critically Endangered, six as Endangered, four as Vulnerable, and five as Lower Risk: Near Threatened.

See families box ▷

Lone Ranger of the Desert

DISTRIBUTION PATTERNS

Small marsupial carnivores occupy all terrestrial habitats in Australia and New Guinea. Up to nine species occur locally in some arid areas and in structurally complex forests, due to the diversity of foraging microhabitats that these environments provide. In contrast, only one or two species usually co-occur in woodland or savanna environments, due to the paucity of opportunities they provide for different species to segregate.

The spectacular success of small dasyurids in the deserts of Australia is perhaps the most remarkable feature of their modern radiation. Many denizens of the arid zone occupy drifting home ranges that appear to track changes in levels of food. The tiny Lesser hairy-footed dunnart, for example, may move 2–3km (1.2–1.9mi) a night in search of food. Few small marsupial carnivores dig their own burrows, and those that inhabit the arid zone exploit soil cracks or abandoned burrows. Some dunnarts and ningauis will move several kilometers from drought-stricken areas toward rain. Lesser hairy-footed dunnarts have been recorded moving 12km (7.5mi) in two weeks, possibly following the scent of wet desert sand on the wind.

A Frenetic Search for Mates

SOCIAL BEHAVIOR

In the 12 species of forest-dwelling antechinuses and phascogales, as well as the Little red kaluta, sexual maturity is reached when males and females are 11 months old. Matings occur over a short period (2–3 weeks), at the same time each year in any locality, with ovulation in females being stimulated by subtle increases in the rate of change of day length in spring. All males die at about 1 year of age within a month of mating, but females can survive and reproduce in a second or occasionally third season. All or almost all females breed annually, producing 6–12 young per litter. Male death is due to increased levels of free corticosteroid (stress) hormones in the blood (see A Once-in-a-Lifetime Breeding Opportunity).

Other marsupial carnivores have more flexible life histories. In the Sandstone antechinus, matings still occur synchronously in winter, but some 70 percent of males survive their first breeding season; about a quarter of individuals of both sexes breed at 2 years of age. Reproductive effort in this species is relatively small, as only 65–88 percent of females breed each year, and litter sizes seldom exceed 4–5 young.

In other species of marsupial carnivores, such as some dunnarts, sexual maturity is achieved at 6–8 months, and females are able to produce 2–3 litters over extended breeding seasons that can last for up to 8 months. Repeated reproduction is

Small Marsupial Carnivore Families

Family Dasyuridae

Antechinuses or Broad-footed marsupial mice Genus *Antechinus*
10 species in Australia, including the Brown antechinus (*A. stuartii*), and Dusky antechinus (*A. swainsonii*). The Atherton (*A. godmani*), Swamp (*A. minimus*), and Cinnamon antechinuses (*A. leo*) are all Lower Risk: Near Threatened. A further 5 species of "antechinus" occur in New Guinea; both the taxonomic and conservation status of these species is uncertain.

Crest-tailed marsupial mice Genus *Dasycercus*
3 species in Australia: the mulgara (*D. cristicauda*), kowari (*D. byrnei*), and ampurta (*D. hillieri*). The ampurta is Endangered, and the mulgara and kowari are Vulnerable.

Long-tailed dasyures Genus *Murexia*
2 species in Indonesia and New Guinea: Short-furred dasyure (*M. longicaudata*) and Broad-striped dasyure (*M. rothschildi*).

Ningauis Genus *Ningaui*
3 species in Australia: the Pilbara ningaui (*N. timealeyi*), the Southern ningaui (*N. yvonneae*), and the Inland or Wongai ningaui (*N. ridei*).

Dibblers Genus *Parantechinus*
2 species in Australia: the Sandstone antechinus (*P. bilarni*) and the dibbler (*P. aplicalis*). The dibbler is Endangered.

Phascogales or Brush-tailed marsupial mice Genus *Phascogale*
2 species in Australia: the Red-tailed phascogale or wambenger (*P. calura*) and the Brush-tailed phascogale (*P. tapoatafa*). The former is Endangered, while the latter is Lower Risk: Near Threatened.

Marsupial shrews Genus *Phascolosorex*
2 species in New Guinea: the Red-bellied (*P. doriae*) and Narrow-striped dasyures (*P. dorsalis*).

Planigales or Flat-skulled marsupial mice Genus *Planigale*
6 species in Australia and New Guinea, including the Paucident planigale (*P. gilesi*) and the Common or Pygmy planigale (*P. maculata*). The Papuan planigale (*P. novaeguineae*) is Vulnerable.

Pseudantechinuses Genus *Pseudantechinus* 5 species in Australia, including the Fat-tailed pseudantechinus (*P. macdonnellensis*), and Woolley's pseudantechinus (*P. woolleyae*).

Dunnarts or Narrow-footed marsupial mice Genus *Sminthopsis*
19 species in Australia and Papua New Guinea, including the Carpentarian dunnart (*S. butleri*) and the Red-cheeked dunnart (*S. virginiae*). The Julia Creek (*S. douglasi*), Kangaroo Island (*S. aitkeni*), and Sandhill dunnarts (*S. psammophila*) are Endangered; and the Carpentarian dunnart is Vulnerable.

Kultarr (*Antechinomys laniger*). Australia.

Little red kaluta (*Dasykaluta rosamondae*). Australia.

Three-striped marsupial mouse (*Myoictis melas*). Indonesia and New Guinea.

Long-clawed marsupial mouse or Speckled dasyure (*Neophascogale lorentzi*). New Guinea.

Family Myrmecobidae

Numbat (*Myrmecobius fasciatus*) Australia. Vulnerable.

For full species list see Appendix ▷

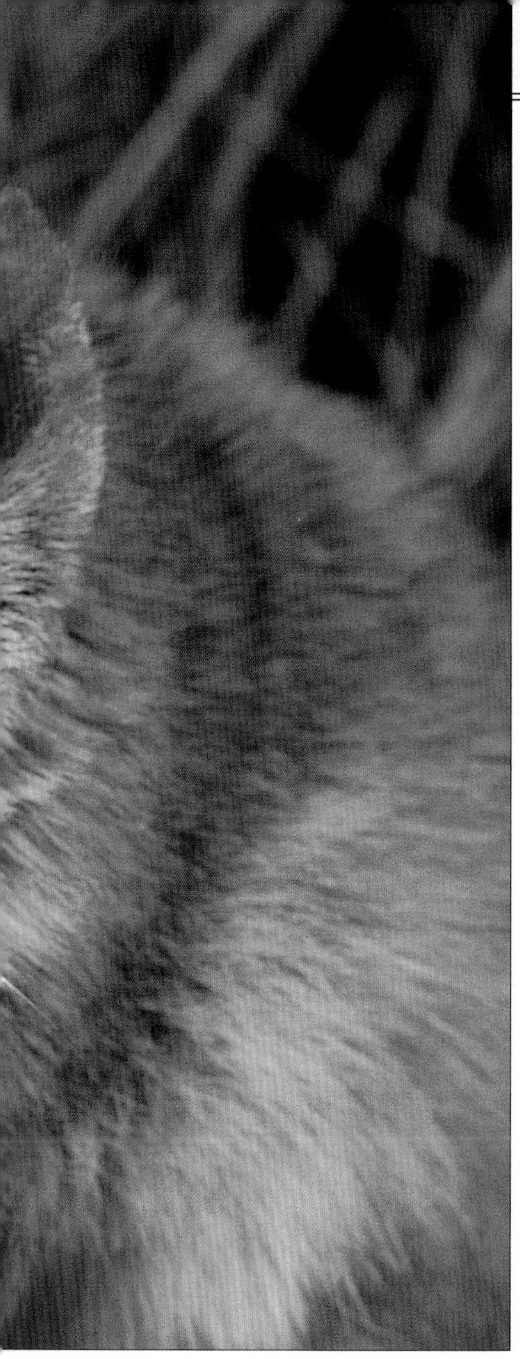

THE NUMBAT – TERMITE-EATER

The numbat (*Myrmecobius fasciatus*), the sole member of the family Myrmecobiidae, is a specialized termite-eater and, perhaps because of the diet, is the only fully day-active Australian marsupial. It sports black-and-white bars across its rump, and a prominent white-bordered dark bar from the base of each ear through the eye to the snout. These distinctive coat markings and its delicate appearance make it one of the most instantly appealing marsupials.

The numbat spends most of its active hours searching for food. It walks, stopping and starting, sniffing at the ground and turning over small pieces of wood in its search for shallow underground termite galleries. On locating a gallery, the numbat squats on its hind feet and digs rapidly with its strong clawed forefeet. Termites are extracted with the extremely long, narrow tongue which darts in and out of the gallery. Some ants are eaten, but it seems that the numbat usually takes these in accidentally while picking up the termites. It does not chew its food, and also swallows grit and soil acquired while feeding.

Numbats are solitary for most of the year, each individual occupying a territory of up to 150ha (370 acres). During the cooler months a male and female may share the same territory, but they are still rarely seen together. Hollow logs are used for shelter and refuge throughout the year, although numbats also dig burrows and often spend the nights in them during the cooler months. The burrows and some logs contain nests of leaves, grass, and sometimes bark. In summer numbats sunbathe on logs.

Four young are born between January and May, and attach themselves to the nipples of the female, which lacks a pouch. In July or August the mother deposits them in a burrow, suckling them at night. By October, the young are half grown and are feeding on termites while remaining in their parents' area. They disperse in early summer (December).

Numbats once occurred across the southern and central parts of Australia, from the west coast to the semi-arid areas of western New South Wales. They are now found only in a few areas of eucalypt forest and woodland in the southwest of Western Australia. Habitat destruction for agriculture and predation by foxes have probably contributed most to this decline. While most of their habitat is now secure, remaining populations are so small that the species is classed as Vulnerable. Efforts are being made to set up a breeding colony from which natural populations may be reestablished. AKL

Left *The mulgara is not a common sight, however it is reported that observed numbers increase when plagues of house mice – a favored food – occur within its range. It eats a mouse from head to tail, inverting the skin of its victim as it goes.*

possible because the gestation period is short (10–13 days) and weaning occurs at 60–70 days. In antechinuses, pseudantechinuses, and phascogales, by contrast, gestation lasts 30–40 days and weaning occurs at least three months after birth. Four marsupial carnivores appear to have no seasonality in their breeding schedules. Northern Australian populations of the Common planigale produce litters of 4–12 young in all months, while three species of antechinus in New Guinea produce smaller litters of 3–4 without any obvious seasonal break.

Differences in life histories of marsupial carnivores have probably arisen in response to variations in the duration and reliability of invertebrate food resources. In the antechinuses and phascogales, peaks in invertebrate abundance occur reliably in spring and summer, and mating is timed so that lactation and weaning coincide with these peaks. The chances of failing to breed at all due to food shortage is thus reduced. In dunnarts, ningauis, and other species where males survive or where females produce two or more litters in a season, food peaks may be smaller or less predictable. The chance of reproductive failure due to food shortage may be high, but the risk can be spread over more than a single litter.

Communication and social organization remain poorly known for most small dasyurids. Most or all species appear solitary except during the breeding season and when the young are dependent for food on the mother. In species that occur usually at low densities or occupy open habitats, such as ningauis, males continuously utter soft clicks or hisses to attract females, while females call in return during periods of receptivity. Loud hissing sounds are made by many species during aggressive encounters or during nest or food defense.

In the Agile antechinus and some other forest-dwelling dasyurids, males disperse from the maternal nest at weaning and reside for periods of days or weeks with unrelated females. Toward the breeding season, males appear to aggregate in tree-top leks where females come to "window shop." Both sexes mate with multiple partners. Females store viable sperm in the reproductive tract for up to two weeks, and produce litters sired by more than one father. For the female, mating with several males causes sperm competition, which may be beneficial in producing genetically diverse offspring. For the male, such competition reduces confidence in paternity and may drive the frenetic search for new mates.

Small Survivors

CONSERVATION AND ENVIRONMENT

In contrast to their larger relatives, the small marsupial carnivores have escaped the worst ravages of European settlement. No species has gone extinct, and only seven have suffered range reductions over 25 percent. Nonetheless, many species have small ranges or sparse populations: 15 Australian species are considered threatened, as is the Papuan planigale. Clearing of vegetation for agriculture and predation by feral cats and foxes are serious threats. Populations of most of the threatened species occur at least partly on protected land. Control of introduced predators has stemmed population declines in the endangered Red-tailed phascogale, and provides hope that it would protect other threatened species if implemented at a broad scale. CRD

A Once-in-a-Lifetime Breeding Opportunity

Sex and Death in the Antechinus

MANY MAMMALS GO TO GREAT LENGTHS FOR sex, but the prize for personal sacrifice has to go to those mammals that reproduce only once in a lifetime, exhausting their own bodies to fuel their reproductive urges. It is the sensible strategy of most mammals to start their reproductive lives with caution, rarely having their largest litter or fattest babies at the first attempt; at least initially, reproductive ability improves with age. In certain species, however, some individuals put all their eggs in one basket. An Alpine vole that has survived the winter often produces only a single litter before it dies, and it will be the offspring of such voles that continue reproduction throughout the spring and summer. And in two groups of carnivorous marsupials, the Australian dasyurids and the American didelphids (opossums), all individuals of certain species commit themselves totally to reproduction at their first attempt. In these species, all the females come into estrus at the same time, once a year. After that all the males die, sometimes over a period of just three or four days. Indeed, all the males can be dead before the females have even ovulated. While the females live on to give birth and suckle their young, they also usually die after rearing only a single litter.

When organisms reproduce only once in their lifetime they are known as semelparous, in contrast

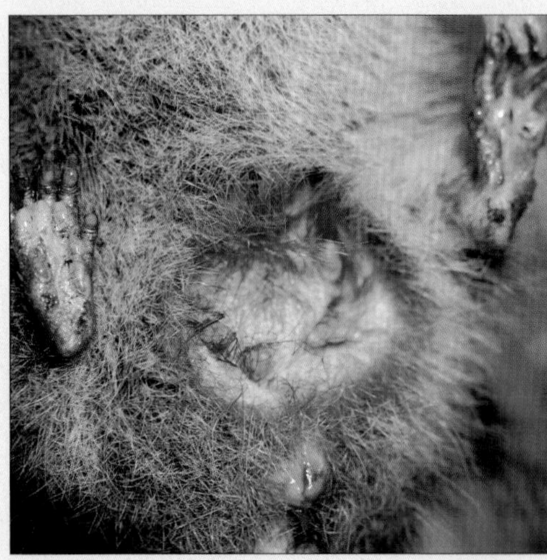

Above *Some species of antechinuses – in this case a Swamp antechinus* (Antechinus minimus) *– lack a pouch as such; instead they have a patch of bare skin from which the mammae or teats protrude.*

Below *Raising litters of as many as ten young places a huge strain on the resources of a female antechinus. To ensure that she can produce enough milk to feed her offspring, the reproductive cycle is timed so that lactation coincides with the period of maximum prey availability, when she is well-fed.*

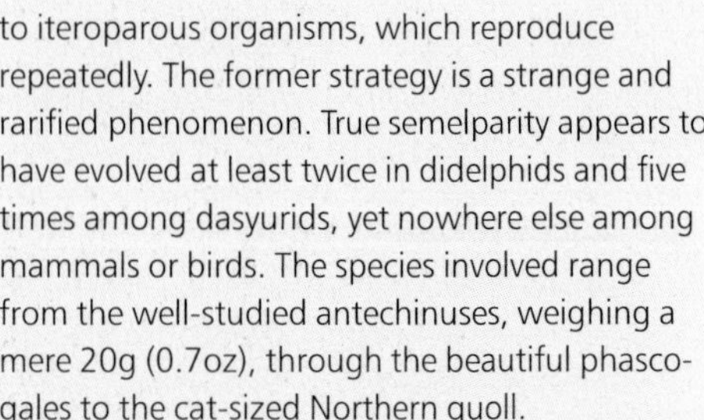

to iteroparous organisms, which reproduce repeatedly. The former strategy is a strange and rarified phenomenon. True semelparity appears to have evolved at least twice in didelphids and five times among dasyurids, yet nowhere else among mammals or birds. The species involved range from the well-studied antechinuses, weighing a mere 20g (0.7oz), through the beautiful phascogales to the cat-sized Northern quoll.

Semelparity only occurs in predictable, highly seasonable environments, and it has its "raison d'être" in the excruciatingly slow reproduction rate of small marsupials. The most famous of the semelparous marsupials, *Antechinus stuartii*, has a four-week pregnancy, at the end of which it gives birth to young weighing only 16mg. The young are deposited in the shallow pouch, where their tiny bodies are unwound by a special swelling on their chest that allows them to attach to a teat, where they continue to develop. After a further 5 weeks of suckling, they have grown to a length of about 1cm (0.4in). They are not weaned until they are 14 weeks of age or more.

Because the mother may be suckling as many as ten young, her metabolic rate in late lactation can be ten to twelve times the basal rate – a mammalian record. Female reproduction is therefore timed to ensure that late lactation coincides with the period of maximum availability of the insect and spider prey on which the species feeds, which falls in late spring or early summer. Females must therefore get pregnant in the winter, when there is little food to eat. But males are put through a further test of endurance by having to congregate in special mating trees where the females come to mate. During the short rut, as many as twenty males may aggregate in the tree cavity, where they are visited by the females.

Males face a threefold dilemma. They have to mate in winter, and at a communal nest as much as 1km (0.6mi) from their own home range; and if they do try to feed at this time, they run the risk of missing the all-important visits by receptive females. It is now that the adaptive advantage of dying becomes apparent. Males resolve the challenge confronting them by exhibiting in an extreme form the stress response mounted by all mammals when presented with an external challenge, or stressor. This response involves the secretion of corticosteroids, of which cortisol is the most potent form in mammals. Cortisol suppresses appetite and promotes gluconeogenesis, the conversion of protein into sugars, which means that reserves other than rapidly-consumed fat can be used to sustain the body during a crisis. Nonetheless, the stress response is a two-edged

Left *A female antechinus returns to her litter of eight-week-old young in the hollow of a tree. Naked and helpless, they will be dependent on their mother's milk until they are weaned at 14 weeks or more.*

weapon. In addition to the benefits it brings, cortisol suppresses the immune and inflammatory responses, exposing stressed animals to a greater risk of disease. In most organisms this eventuality is prevented by corticosteroid-binding globulins that render some of the cortisol inactive, and by negative feedback in the brain that stops cortisol production. Semelparous marsupials, in a suicidal twist, radically reduce the level of binding globulins just as the breeding season starts, and the negative feedback cycle is turned off. As a consequence, males can digest and feed off their own bodies from within, but at the cost of condemning themselves to the most miserable of deaths. The commonest source of mortality is a massive hemorrhage of ulcers in the stomach and intestine, but parasites and other microorganisms that ordinarily have no effect can also often become pathogenic once the immune system fails. Females live on after the males have died, but they need to survive for another 16 months in order to successfully wean a second litter, and they often fail to do so.

This paradoxical solution to the stresses of mating has been arrived at evolutionarily by other organisms that also have to mate in a hostile environment, such as freshwater eels that migrate to the sea to breed or salmon that live in the ocean but spawn in fresh water. They use exactly the same hormonal system to sustain their migration, and they pay a similar price.

The curious life history of these small marsupials is fascinating in its own right, but the extremely simple population structure, producing individuals of identical ages, also provides a unique insight into other intriguing questions affecting mammalian society as a whole. In many mammals, juvenile males disperse, while females remain in the area where they were born. This behavior has been attributed to competition between fathers and sons for mating opportunities. But an alternative explanation, probably relating to incest avoidance, needs to be found for male-biased dispersal in the case of semelparous marsupials. They show extreme male bias in natal dispersal; females continue to live with their mothers after weaning, but all males leave, sometimes traveling many kilometers to a new home-range – and this despite the fact that in these species fathers are dead long before their sons are born. AC

Bandicoots

BANDICOOTS ARE RATLIKE MARSUPIALS, *agile, with long noses and tails. They share a common name and ancestry with the rabbit-eared bandicoots or bilbies, a smaller group adapted for arid environments, and distinguished not just by their long ears, longer limbs, and silkier hair but also by their burrowing habit. One of the two bilby species is now thought extinct, and the other is threatened.*

Bandicoots are notable for having one of the highest reproductive rates among marsupials, exceeded only by that of a single species of dunnart (family Dasyuridae). In this respect, they resemble the rodents in the placental world; like theirs, their life-cycles centre on producing many young with little maternal care. Otherwise, these small insectivores and omnivores fit into an ecological niche similar to that of the shrews and hedgehogs.

Short Necks and Pointed Muzzles

FORM AND FUNCTION

Most bandicoots are rabbit-sized or smaller, have short limbs, a long, pointed muzzle, and a thickset body with a short neck. The ears are normally short, the forefeet have three toes with strong, flattish claws, and the pouch opens backwards. The furthest from this pattern is the recently extinct Pig-footed bandicoot, which had developed longer limbs and hooflike front feet as adaptations to a more cursorial life on open plains. The long-nosed bandicoots of the genus *Perameles* also have longer ears than the other species, but the longest of all are found on the bilbies. Teeth are small and relatively even-sized, and have pointed cusps. Most bandicoots are omnivorous, characteristically obtaining their food from the ground by excavating small, conical pits.

Bandicoots are distinguished from all other marsupials by having fused (syndactylous) toes on the hind feet, forming a comb for grooming, and polyprodont dentition (with more than two well-developed lower incisors). The rear-opening pouch normally has eight teats. It extends forward along the abdomen as the young enlarge, eventually occupying most of the mother's underside, and then contracts again after they have departed. Litter size is normally 2–3.

Bandicoots' sense of smell is well developed. The animals are nocturnal and their eyes are adapted for night vision, although their binocular vision may be limited, perhaps because the elongated nose gets in the way. The Long-nosed bandicoot produces a sharp, squeaky alarm call when disturbed at night, and bandicoots are also sometimes heard to sneeze loudly, probably to clear soil from their noses. They rarely if ever produce loud calls, but a very low, sibilant "huffing" with bared teeth is uttered as a threat by some species.

The dental formula is I5/3, C1/1, P3/3, M4/4 = 48, except in spiny bandicoots, which have four pairs of upper incisors (I4/3). Sexual dimorphism occurs in most species; in some, males may be up to 60 percent heavier and 15 percent longer than females. Males usually have larger canines.

The classification of bandicoots is not fully resolved, although all genera seem to be clearly defined. Currently two families are recognized, one of which (the Peramelidae) contains two distinct subfamilies (see box). Taxonomic relationships within the brown bandicoots and other groups and the taxonomy of most New Guinea forms warrants further study.

Brown bandicoots are stocky, short-eared, plain-colored animals. They inhabit areas of close ground cover, tall grass, or low shrubbery. The Southern brown bandicoot favors heathland, whereas the Northern brown bandicoot occurs in a wide range of habitats from wet forests to open woodland. All have inflated auditory bullae. Dwarfed forms occur, particularly on islands and in more open areas, perhaps as a result of scarcer food resources. Variations in the angle of the ascending rear portion of the lower jaw, and the presence of an extra cusp on the last upper molar, have been used to differentiate between species, but the taxonomy is not satisfactorily understood. The distribution of Southern brown bandicoots does not overlap that of the northern species, except for a single instance. There is, however, overlap between Northern brown bandicoots and the Golden bandicoot.

The long-nosed bandicoots are more lightly built, with a relatively longer skull, small auditory bullae, longer ears, and a preference for areas of open ground cover, although habitat use may be flexible; some species also exhibit barred body markings. The arid-zone species have a wide distribution, from Western Australia to western New South Wales; in contrast, the Long-nosed bandicoot itself is restricted to the eastern coastal areas, and the Eastern barred bandicoot to grasslands and grassy woodlands of the southeast mainland and Tasmania. Important variables between species are ear length and the size of the bullae, which both increase in arid areas, and the positioning of toes on the hind feet. In addition, different species exhibit varying degrees of barring, which is absent in the forest species but conspicuous in grassland species.

Bilbies have lengthened ears, long, narrow rostra, and elongated limbs; other distinctive features include highly developed auditory bullae with twin chambers, long, silky fur, and a long, crested tail. The only burrowing bandicoots, they are an early offshoot from the main bandicoot stock that has become highly specialized for arid areas. Species and populations are differentiated principally by size, coat, and tail coloration, and also by the dimensions of the bullae, which were larger in the Lesser bilby, now thought to be extinct.

The several New Guinea genera (the Peroryctidae) are poorly known. They tend to be little-modified, short-eared, forest bandicoots. The skulls are more cylindrical than in the peramelids, and in the spiny bandicoots and the Seram Island bandicoot the rostrum is long and narrow. The auditory bullae are small. Spiny bandicoots have short tails.

Above The Greater bilby, seen here suckling its young, is a desert species. It now has a much-reduced distribution. Habitat loss and predation have severely reduced its numbers and it is listed as Vulnerable.

Below Behavioral postures of the Northern brown bandicoot. It is nocturnal and frequently sniffs the air **1** to detect any danger. The usual gait is on all fours, but the larger hind limbs are used in an aggressive hop **2** characteristic of males. **3** The Northern brown bandicoot digs out food with its strong fore claws. After the shortest gestation of perhaps any mammal, the newborn young crawl into their mother's rear-opening pouch **4** where they are carried for seven weeks, by which time **5** the pouch is bulging.

FACTFILE

BANDICOOTS

Order: Peramelemorphia

Families: Peramelidae and Peroryctidae

18 extant species in 7 genera

HABITAT All major habitats in Australia and New Guinea from desert to rain forest, including semi-urban areas.

SIZE **Head–body length** ranges from 17–26.5cm (7–10in) in the mouse bandicoots to 50–60cm (20–23in) in the Giant bandicoot; **tail length** from 11–12cm (4.5in) to 15–20cm (6–8in), and **weight** from 140–185g (5–6.5oz) up to 4.8kg (10.5lb), both in the same species. Males of larger species may be up to 50 percent heavier than females.

COAT Mostly short and coarse in bandicoots (stiff and spiny in some New Guinean species); longer and silkier in the Greater bilby.

DIET Insects and other invertebrates, bulbs, roots, tubers.

BREEDING Gestation 12.5 days in the Long-nosed, Eastern Barred, and Northern Brown bandicoots, 14 days in the Greater bilby.

See families box ▷

DISTRIBUTION Australia, Papua New Guinea, West Irian

LONGEVITY About 2–3 years in the Eastern barred, and slightly more in the Northern brown bandicoot.

CONSERVATION STATUS Two bandicoot species – the Pig-footed (*Chaeropus ecaudatus*) and the Desert (*Perameles eremiana*) – have recently been declared Extinct, as has the Lesser bilby (*Macrotis leucura*). Several other species are also at risk; the Western barred bandicoot (*Perameles bougainville*) is listed as Endangered, and three other species are Vulnerable, including the Greater bilby (*Macrotis lagotis*). Little is known about the status of many New Guinea bandicoots; some species have only been collected on one or two occasions.

Above *The Southern brown bandicoot only overlaps with the Northern form in an anomalous population in the Cape York peninsula of northern Australia.*

Bandicoot Families

Family Peramelidae

Subfamily Peramelinae

Long-nosed bandicoots Genus *Perameles*
3 species: Western barred bandicoot (*P. bougainville*); Eastern barred bandicoot (*P. gunnii*); Long-nosed bandicoot (*P. nasuta*). The Desert bandicoot (*P. eremiana*) is now listed as extinct by the IUCN.
Short-nosed or **Brown bandicoots** Genus *Isoodon*
3 species: Golden bandicoot (*I. auratus*); Northern brown bandicoot (*I. macrourus*); Southern brown bandicoot (*I. obesulus*).
Pig-footed bandicoot (*Chaeropus ecaudatus*) is listed in Appendix 1 of CITES, but is now considered extinct by the IUCN.

Subfamily Thylacomyinae

Rabbit-eared bandicoot or **Bilby** Genus *Macrotis*.
Greater bilby (*M. lagotis*). The Lesser bilby (*M. leucura*) is listed in Appendix 1 of CITES, but is now considered extinct by the IUCN.

Family Peroryctidae

New Guinea bandicoots Genus *Peroryctes*
2 species: Giant bandicoot (*P. broadbenti*); Raffray's bandicoot (*P. raffrayana*).
New Guinean mouse bandicoots Genus *Microperoryctes* 3 species: Mouse bandicoot (*M. murina*); Striped bandicoot (*M. longicauda*); Papuan bandicoot (*M. papuensis*).
Spiny bandicoots Genus *Echymipera*
5 species: Rufous spiny bandicoot (*E. rufescens*); Clara's echymipera (*E. clara*); Menzie's echymipera (*E. echinista*); Common echymipera (*E. kalubu*); David's echymipera (*E. davidi*).
Seram Island bandicoot *Rhynchomeles prattorum*

Habitat Specialists

DISTRIBUTION PATTERNS

In recent (Pleistocene) times, bandicoots have to a large extent evolved separately in Australia and New Guinea, as a result both of the intermittent separation of the two land masses and of the marked habitat differences. All but one of the New Guinea genera are endemic; only the Rufous spiny bandicoot extends its range into northern Australia. Conversely, only one Australian species, the Northern brown bandicoot, intrudes into the grassy woodlands of southern New Guinea. This suggests that the main influence of the two land masses on the different bandicoot fauna is habitat, not just the water barrier.

Within New Guinea, different species occur at different altitudes. The Northern brown, Giant, and most species of spiny bandicoots are lowland animals, but some range up to about 2,000m (6,500ft). The Mouse, Striped, and Raffray's bandicoots are all highland species, generally found above 1,000m (3,500ft). The Seram Island bandicoot is only known from high altitudes, at about 1,800m (6,000ft).

Within Australia, there are pronounced climatic influences on the distribution of species, which tend to fall into two groups. Species restricted to semi-arid and arid areas have suffered large population declines since European settlement, and three (the Desert and Pig-footed bandicoots and the Lesser bilby) are now probably extinct. The survivors include the Western Barred and Golden Bandicoots and the Greater bilby.

This pattern is an effect, whether direct or indirect, of rainfall. The Northern brown bandicoot, a coastal species of eastern and northern Australia, is widely distributed as far inland as the 72.5cm (28.5in) isohyet (rainfall line). Beyond this it tends to be largely confined to watercourses, which extend its range much farther inland, almost to the 60cm (23.6in) isohyet. Southern brown bandicoots are more confined to the coast (except in Tasmania), and the Long-nosed bandicoot extends inland beyond the Great Dividing Range only in northeastern Victoria.

Opportunistic and Omnivorous

DIET

Although bandicoots are dentally specialized for feeding on invertebrates, feeding is opportunistic and omnivorous and includes insects, other invertebrates, fruits, seeds of non-woody plants, subterranean fungi, and occasional plant fiber. Diet can also include a high proportion of surface food, and it is likely that bandicoots switch to other food when insects are unavailable. They locate food in the ground by scent and then dig it out with their strong fore claws. The elongated muzzle is presumably used to probe into holes for food.

The Northern brown bandicoot has a characteristic foraging pattern, moving slowly over its whole range. This is an adaptation for finding food that occurs as small, scattered items rather than being concentrated in a few areas. The Eastern barred bandicoot concentrates on areas of increased soil moisture and vegetation diversity, where food species are both more abundant and more readily excavated.

Fast Breeders

SOCIAL BEHAVIOR

Most species are solitary, animals coming together only to mate, and there appears to be no lasting attachment between mother and young. Males are usually larger than females and socially dominant. Dominance between closely-matched males may be established by chases or, rarely, by fights, in which the males approach each other standing on their hind legs.

Male home ranges are larger than females'; for the Northern brown bandicoot 1.7–5.2ha (4.2–12.8 acres) in one study, compared to 0.9–2.1ha (2.2–5.2 acres) for females. Similar values were found for the Eastern barred bandicoot, although these relate to core ranges not entire foraging area. The ranges of both sexes overlap extensively, although core areas may not. Females often dictate local distribution, selecting and perhaps defending high-quality nesting and foraging sites. Males patrol most of the home range each night, perhaps to detect other males or receptive females. In male-biased populations, many males may repeatedly mate with a single female.

Captive Northern brown bandicoots showed intense interest in nests, which consist for them of heaps of raked-up groundlitter with an internal chamber, and dominant males commonly evicted others from them. Nests may therefore be a significant focus of social interactions in the wild in that species. Eastern barred bandicoots make several types of nest, the most complex being a lined, roofed excavation used when females have young. Many species have scent glands present behind the ears; the Northern brown bandicoot uses this gland, which is present in both sexes, to mark the ground or vegetation during aggressive encounters between males. The high reproductive rate of bandicoots means that they are able to recolonize rapidly as habitat recovers from fire or drought.

The reproductive biology of Australian bandicoots has been studied in some detail and is well exemplified by the Northern brown species. The young are gestated for only 12.5 days, less than half the length of time taken by most other marsupials and almost the shortest of any mammal. Development of the embryo is aided by a form of chorioallantoic placentation that is unique to bandicoots among marsupials, in that it resembles the placenta of eutherian mammals. Other marsupials form only a yolksac placenta, whereas bandicoots and eutherians have independently evolved both types of placentation.

At birth the young are about 1cm (0.4in) long and weigh about 0.2g (0.007oz), with well-developed forelimbs. The allantoic stalk anchors the young to the mother whilst the newborn crawls to the pouch, where it attaches to a nipple. The young leave the pouch after 49–50 days and are weaned about 10 days later. In good conditions, sexual maturity may occur at about 90 days, although it is normally attained much later. Females are polyestrous and breed throughout the year in suitable climates; in other conditions they breed seasonally. Mating can occur when the previous litter is near the end of its pouch life. Since the gestation is 12.5 days, the new litter is born at about the time of weaning of the earlier litter. Captive females may have 4–5 litters per year, and may therefore produce about 18 young in a lifetime; in the wild this figure is probably halved. In captivity only about 40 percent of young reach sexual maturity, while in wild populations the survival rate is as low as 11.5 percent, so, despite the high fecundity, recruitment is low.

The reproductive cycle is one of the most distinctive characteristics of bandicoots, setting them apart from all other marsupials. They have become uniquely specialized for a high reproductive rate and reduced parental care. In most bandicoots, this is achieved by accelerated gestation, rapid development of young in the pouch, early sexual maturity, and a rapid succession of litters in the polyestrous females. Female Eastern barred bandicoots may become sexually mature at less than 4 months and, given normal climatic conditions, continue to breed throughout the year for up to 3 years. In one Northern brown bandicoot population with breeding seasons stretching over 6–8-months, females produced an average of 6.4 surviving young in one season, and 9.6 in the next. Litter size, however, while higher than in many marsupial groups, is not exceptional, being smaller than in others, such as dasyurids.

Below *The Eastern barred bandicoot is virtually extinct on mainland Australia, being restricted to a tiny remnant population.*

Above *Young spiny bandicoots* (Echymipera *spp.) alone in a nest. Bandicoots expend little effort on parental care, relying instead on a high birth rate.*

Under Threat of Extinction

CONSERVATION AND ENVIRONMENT

Australian bandicoots have suffered one of the greatest declines of all marsupial groups. All species of the semi-arid and arid zones have suffered massive declines or even become extinct; the survivors are reduced now to a few remnant populations that are still endangered. An important feature of most of the extinctions seems to be grazing by cattle, sheep, or rabbits, and the consequent changes in the nature of ground cover. Some authorities blame introduced predators. Removal of sheep and cattle is an important conservation measure in these areas.

Only a few species that occur in higher rainfall zones, including the Long-nosed and the Northern and Southern brown Bandicoots, can be considered secure, although all have been affected by European settlement. Even these "common" species are under threat of habitat alteration or alienation. For example, the Long-nosed bandicoots have become all but extinct in Sydney, and the same is true of Southern brown bandicoots in the Melbourne metropolitan area.

Conservation of the Eastern barred bandicoot depends on an ongoing reintroduction program, in which captive-bred animals have been released to several protected sites in their former range. Success has been variable, with predation by the introduced Red fox and habitat degradation by grazing herbivores, whether native (kangaroos) or introduced (rabbits), being constraining factors. Control management of these issues is required on a continuing basis, but after more than 10 years of effort some of the reintroduced populations have become successfully established. While the species is more common in Tasmania, the population there is also declining. GG/JHS

Cuscuses and Brushtail Possums

DWELLING IN THE REMOTE OUTBACK AS well as in the suburbs of most Australian cities, the Common brushtail possum is perhaps the most frequently encountered of all Australian mammals and is the most studied of the possums. But most of the remaining 19 phalangerid species are relatively unknown to science, either because of their cryptic behavior in dense rain forest or their restricted distributions; for example, the Telefomin cuscus, from the highlands of central New Guinea, is known from only five museum specimens.

The phalangerids are generally nocturnal, the outstanding exception being the Bear cuscus of Sulawesi (*Ailurops ursinus*), which is the only one with circular pupils, a possible adaptation to diurnal living. The animals are usually arboreal; even the Ground cuscus and Scaly-tailed possum, which habitually rest by day in holes in the ground, spend the night in trees. Cuscuses and possums are careful and deliberate climbers, not given to spectacular leaps; among the adaptations that help them are curved and sharply pointed fore claws, as well as clawless but opposable first hind toes that aid in grasping branches, and prehensile tails with variable amounts of bare skin. Phalangers possess well-developed, forward-opening pouches.

Native Australians

EVOLUTION AND RADIATION

Phalangerids originated in the rain forests of what is now mainland Australia. The earliest fossils of modern genera – *Trichosurus*, *Wyulda*, and *Strigocuscus* – were found in the Miocene rocks of Riversleigh, northern Australia, and date from some 20 million years before the present. *Trichosurus* and *Strigocuscus* were also present in the early Pliocene of southern Australia about 5 million years ago. The genera *Ailurops*, *Phalanger*, and *Spilocuscus* have not appeared in the Australian fossil record and may have originated in New Guinea, possibly from the ancestral phalangerid stock closest to *Ailurops* at a time when New Guinea was connected to Australia during the Miocene or earlier.

Most phalangerid genera, even in the fossil record, are made up of between 1 and 4 species, the only exception being *Phalanger* itself, which numbers 10 species. The stimulus for *Phalanger*'s proliferation was provided by the geographical isolation of its populations, either on islands or on remote mountain ranges in New Guinea, where it is the only genus occurring above 1,200m (4,000ft). *Strigocuscus* is now extinct in Australia, but it has been replaced by two other cuscus genera, the Southern common (*Phalanger intercastellanus*) and Spotted (*Spilocuscus maculatus*) cuscuses. Both are common lowland species in New Guinea that entered Australia less than 2 million years ago over the land bridge that linked the two landmasses during the Pleistocene.

Common and Uncommon

DISTRIBUTION PATTERNS

The Common brushtail possum has the widest distribution of all phalangerids, covering most of Australia in a wide range of habitats from rain forests to semi-arid areas; four subspecies are currently recognized. In temperate Tasmania individuals have thick coats and bushy tails and weigh up to 4.5kg (9.9lb), but there is a general decline in size towards the tropics, with individuals across northern Australia attaining no more than 1.8kg (4lb), having thin coats and little bush to the tail. The predominant color is light gray, but in wetter habitats darker colors are common – black in Tasmania, dark red in northeastern Queensland. The Common brushtail possum's congener, the Mountain brushtail, is geographically much more restricted and not split into subspecies. These possums occupy dense, wet forests in southeastern Australia that are not usually inhabited by the Common brushtail.

Cuscuses are rainforest dwellers, and species often have restricted geographical ranges, either confined to islands or to the highlands of mountain ranges. Most widespread is the Spotted cuscus, found in a wide range of rainforest habitats throughout New Guinea below an elevation of 1,200m (4,000ft), and also on many islands and on Australia's northeastern tip; this species seems able to persist near large centers of human population. Four geographically isolated subspecies are

FACTFILE

CUSCUSES AND BRUSHTAIL POSSUMS

Order: Diprotodontia

Family: Phalangeridae

20 species in 6 genera

DISTRIBUTION Australia, New Guinea, and adjacent islands W to Sulawesi and E to the Solomon Islands. Common brushtail possum introduced to New Zealand; Common and Spotted cuscuses introduced to many of the islands adjacent to New Guinea.

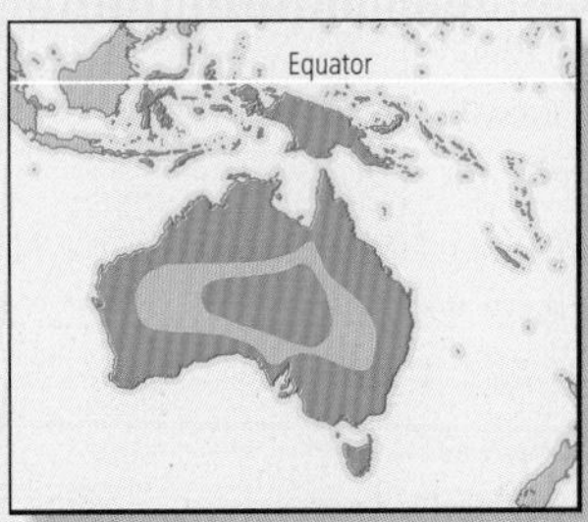

HABITAT All types of forest and woodland: rain forest, moss forest, mangrove, tropical, and temperate eucalypt forest and woodland, arid and alpine woodland.

SIZE Head–body length ranges from 34cm (13.4in) in the Small Sulawesi cuscus to 61cm (24in) in the Bear cuscus; tail length from 34cm (13.4in) to 58cm (22.8in), and weight from about 0.9kg (2lb) to 10kg (22lb), both in the same two species.

COAT Short, dense, gray (Scaly-tailed possum); long, woolly, gray–black (brushtail possums); long, dense, white–black or reddish brown, some species with spots or dorsal stripes (cuscuses).

DIET Leaves, flowers, fruits, seeds, shoots, insects, occasionally small vertebrates and birds' eggs.

BREEDING Gestation lasts 16–17 days in brushtail possums

LONGEVITY Up to 13 years (17 or more in captivity)

CONSERVATION STATUS The Telefomin and Black-spotted cuscuses are currently listed as Endangered, and the Obi and Silky cuscuses as Vulnerable. Three other cuscus and possum species are ranked Lower Risk: Near Threatened.

See species box ▷

currently recognized, and they exhibit considerable variation in color and size. The Spotted cuscus is remarkable in that there is a distinct color dimorphism between males and females: the males have large, irregular, chocolate-brown spots on a creamy white background, whereas the females lack the spots – indeed, in one subspecies they are pure white. Two other members of the genus, the Admiralty and Black-spotted cuscuses, are the only other phalangers with color dimorphism of the sexes.

The Scaly-tailed possum inhabits very rugged, rocky country, with eucalypt forest and rainforest patches, in the remote Kimberley region of northwestern Australia. The last two-thirds of its tail is naked, prehensile, and rasplike, while the hands and feet have greatly enlarged apical pads as an adaptation to life among the rocks.

Of the nine mainland New Guinea cuscuses, the geographical ranges of some species overlap, whereas others fall within more or less exclusive (allopatric) altitudinal zones. These allopatric species all belong to the genus *Phalanger*, are very similar in body size (2.4–3.5kg/ 5.3–7.7lb) and habits, and are apparently unable to coexist. The restriction of Stein's cuscus, for example, to only a narrow altitudinal band of 1,200–1,500m (about 4,000–5,000ft), has been attributed to competition from the Northern and Southern common cuscuses, which occur abundantly below 1,200m (4,000ft), and the Mountain and Silky cuscuses, which occur at altitudes above 1,400m (4,600ft). In localities where the two highland species are absent, Stein's cuscus has been found up to 2,200m (7,200ft).

*⊙ **Above** Brushtail possum and cuscus species: **1** the Gray or Northern common cuscus* (Phalanger orientalis) *lives in New Guinea; **2** Spotted cuscus* (Spilocuscus maculatus)*; **3** the Scaly-tailed possum* (Wyulda squamicaudata) *– only discovered in 1917; **4** Common brushtail possum* (Trichosurus vulpecula).

Where two species do overlap, they usually differ in size or habits. The Ground cuscus, of the genus *Phalanger*, has the widest altitudinal range of all cuscuses, being found from sea level to 2,700m (8,900ft). It is heavier (at 4.8kg /7.0lb) than its congeners, is less arboreal, and has a more frugivorous diet.

Both species of *Spilocuscus* found on the mainland of New Guinea – the Spotted cuscus (6.0kg/8.2lb) and the Black-spotted cuscus (6.6kg/8.8lb) – are heavier than the members of the genus *Phalanger*. They are confined to low altitudes below 1,200m (4,000ft), and cohabit with

◐ **Left** *A Mountain brushtail possum feeding on eucalyptus leaves. This species is both nocturnal and arboreal and nests in tree hollows; the main elements of its diet are leaves, flowers, and young shoots.*

◑ **Right** *A white form of the Spotted cuscus. Mainly inhabiting rain forest, it is active at night and tree-dwelling; the diet of the Spotted cuscus comprises leaves, fruit, and flowers.*

Cuscuses and Brushtail Possums

Cuscuses
4 genera, 17 species

Cuscuses Genus *Phalanger* 10 species: Mountain cuscus (*P. carmelitae*), Ground cuscus (*P. gymnotis*, formerly in *Strigocuscus*), Southern common cuscus (*P. intercastellanus*, formerly in *P. orientalis*), Woodlark cuscus (*P. lullulae*), Telefomin cuscus (*P. matanim*), Gray cuscus or Northern common cuscus (*P. orientalis*), Ornate cuscus (*P. ornatus*), Obi cuscus (*P. rothschildi*), Silky cuscus (*P. sericeus*), Stein's cuscus (*P. vestitus*)
Spotted cuscuses Genus *Spilocuscus* 4 species: Admiralty cuscus (*S. kraemeri*, formerly subspecies of *S. maculatus*), Spotted cuscus (*S. maculatus*); Waigeou cuscus (*S. papuensis*, formerly subspecies of *S. maculatus*), Black-spotted cuscus (*S. rufoniger*)
Plain cuscuses Genus *Strigocuscus* 2 species: Small Sulawesi cuscus (*S. celebensis*), Peleng cuscus (*S. pelengensis* formerly in *Phalanger*)
Bear cuscus (*Ailurops ursinus*)

Brushtail possums
2 genera, 3 species

Brushtail possums Genus *Trichosurus* 2 species: Common brushtail possum (*T. vulpecula*), Mountain brushtail possum (*T. caninus*)
Scaly-tailed possum (*Wyulda squamicaudata*)

both the Ground cuscus and one or other of the common cuscuses. The two *Spilocuscus* species may sometimes be found in the same districts; however, the Black-spotted cuscus inhabits primary forest only, whereas the Spotted cuscus lives in a much broader range of habitats, including secondary forest.

Living Off Leaves

DIET

Most species are nonspecialist leaf-eaters, but their relatively generalized dentition allows them to consume a wide range of foods – fruit or blossom, along with the occasional invertebrate, egg, or small vertebrate. The Common brushtail possum's diet reflects its wide geographical distribution; in some areas up to 95 percent consists of eucalypt leaves, but usually a mix of tree species leaves is taken. In tropical woodland up to 53 percent of the diet may be made up of leaves of the Cooktown ironwood, which are extremely toxic to domestic stock such as cattle. In habitat modified for pasture up to 60 percent of its diet is pasture species, while in suburban gardens it has developed an unwelcome taste for rose buds. The brushtail relies on hindgut microbial activity to extract nutrients from its food, and the large cecum and proximal colon enable food to be retained for relatively long periods. Some cuscuses show evidence of particle-sorting in the cecum, which suggests that they have a more specialized leaf diet than the brushtail.

The Ground cuscus is the most frugivorous phalangerid, with up to 90 percent fruit in the diet of captive animals. Its highly expandable stomach, well-developed pyloric sphincter, and long small intestine are all consistent with delaying the passage of food through the foregut to enable the digestion of lipids from the high fruit diet. Female Ground cuscuses have even been reported to carry fruit back to the den in their pouches by local New Guineans.

A Scent-based Bush Telegraph

SOCIAL BEHAVIOR

Phalangerids are generally solitary, but with a well-organized spatial system based primarily on olfactory communication. The Common brushtail possum actively uses four scent glands. Males, and to a lesser extent females, wipe secretions from mouth and chest glands on the branches and twigs of trees, especially den trees, and deposit sinuous urine trails, containing cells from a pair of paracloacal glands, on branches. These advertise both the presence and the status of the marker to other individuals. When a possum is distressed it produces a sticky, pungent secretion from a second pair of paracloacal glands; this is possibly used as an appeasement signal by low-status individuals. Estrous females produce a copious, gelatinous secretion from the cloaca that becomes smeared on branches and may advertise their readiness to mate.

Little is known about scentmarking in cuscuses and the Scaly-tailed possum, but the sternal and paracloacal glands are generally present, and males of the Spotted cuscus smear the sticky secretion from the paracloacal glands on branches. When distressed, the Spotted cuscus secretes a red–brown substance on the bare skin of its face, particularly round the eyes.

Brushtail possums are one of the most vocal of marsupial genera, and many of their calls are audible to humans at up to 300m (1,000ft). They have about seven basic calls: buccal clicks, agonistic grunts, hisses, loud screeches, alarm chatters, very soft appeasement calls given by the male, and juvenile contact calls; a cartilaginous laryngeal resonance chamber, about the size of a pea and unique to the genus, presumably enhances the repertoire. Cuscuses and the Scaly-tailed possum are not noted for their vocal repertoire, although

buccal clicks, hisses, grunts, and screeches are reported, and the female Spotted cuscus has a call, when in estrus, like the bray of a donkey.

Common brushtails are generally solitary, except when they are breeding and rearing young. By the end of their third or fourth year, individuals establish small exclusive areas centered on one or two den trees within their home ranges, which they defend against individuals of the same sex and social status. Individuals of the opposite sex or lower social status are tolerated within the exclusive areas. Even though the home ranges of males (3–8ha/7.5–20 acres) may completely overlap the ranges of females (1–5ha/2.5–12.4 acres), individuals almost always nest alone, and overt interactions are rare. Territoriality appears to break down in some tropical populations, because Aboriginal hunters may extract up to six individuals from the same hollow tree.

Females defend an individual distance of 1m (3.3ft) against the approach of a male. During courtship a consort male overcomes the female's aggression by repeatedly approaching her and giving soft appeasement calls, similar to those of juveniles. In the absence of a consort male, several males may converge on a female at the time of estrus, and mating is accompanied by considerable agonistic behavior. After mating, the male takes no further interest in the female and is not involved in the raising of the young.

Defence of den trees suggests that preferred nest sites are in short supply. Because few offspring (only 15 percent) die before weaning, relatively large numbers of independent young enter the population each year. These young use small, poor-quality dens, and up to 80 percent of males and 50 percent of females die or disperse within their first year.

Females begin to breed at 1 year and produce 1–2 young annually after a gestation period of 16–18 days. In temperate and subtropical Australian populations, 90 percent of females breed in the fall (March–May), but up to 50 percent may also breed in spring (September–November). In the less seasonal tropics, breeding appears to be continuous, with no seasonal peak of births. Only one young is born at a time, and the annual reproductive rate of females averages 1.4. Population density varies with habitat, from 0.4 animals per hectare (1 per acre) in open forest and woodland to 1.4 per ha (3.5/acre) in suburban gardens and 2.1 per ha (5.2/acre) in grazed open forest.

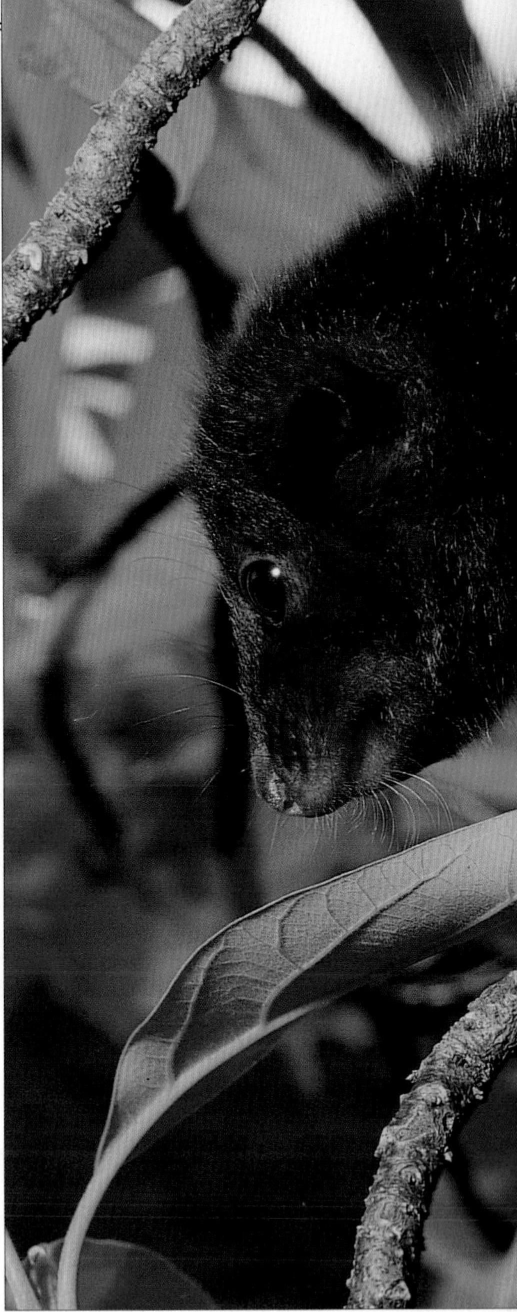

Right The Silky cuscus is found in the mountains of central and eastern New Guinea; they generally inhabit areas of tropical forest at altitudes above 1400m (4600ft). They are heavily built and possess a strong prehensile tail, which assists them in the trees. As is clearly evident here, the end portion of the tail lacks hair and is instead covered with scales. Insects, eggs, and small vertebrates are eaten, but the bulk of their diet is composed of leaves and fruits.

Left A Common brushtail possum carrying young on its back; the young leave the pouch after about five months, with weaning occurring within the next couple of months. Their prospects are not promising, with high numbers being lost during the dispersal period. Common brushtail possums have long been hunted for their fur in Australia and were introduced to New Zealand specifically for this purpose. However, dramatic falls in the price of pelts reduced numbers taken for the fur trade.

The Mountain brushtail has a different strategy, associated to a more stable habitat. Far from being solitary, males and females appear to form long-term pair-bonds. In this species, mortality among the young is greatest before weaning (56 percent); about 80 percent survive each year after becoming independent. Females begin to breed at 2–3 years, produce at most only one young in the fall of each year, and reproduce at an annual rate as low as 0.73. The young are weaned at 8 months, as opposed to 6 for the Common Brushtail, and they disperse at 18–36 months (7–18 for the Common brushtail). Population density is 0.4–1.8 per hectare (1–4.5/acre).

The Scaly-tailed possum bears a single young, and its social strategy is closer to that of the Common brushtail possum. Little is known about the pair relationship in cuscuses, but in the Ground cuscus the male follows the female prior to mating and attempts to sniff her head, flanks, and cloaca; he may also utter soft, short clicks. The only

in Victoria and New South Wales, while in Queensland it frequently raids banana and pecan crops. The Common brushtail also damages pines, and in Tasmania is believed to damage regenerating eucalypt forest.

A potentially much more serious problem is that the Common brushtail may become infected with bovine tuberculosis. This discovery, made in New Zealand in 1970, led to fears that brushtails may reinfect cattle. Although a widespread and costly poisoning program was set up, infected brushtails remain firmly established.

More positively from the economic point of view, the Common brushtail has long been valued for its fur. The rich, dense fur of the Tasmanian form has found special favor, and between 1923 and 1959 over 1 million pelts were exported. Exports from New Zealand have also grown rapidly (see box). In eastern Australia, however, the last open season on possums was in 1963, although in Tasmania the Common brushtail is still subject to control measures in agricultural areas. Although the Common brushtail is considered to be secure, there is a worrying trend of populations crashing in eucalypt woodlands over much of northern and inland Australia.

Cuscuses have long been valued by traditional hunters for their coats and meat, which are sold in local markets. Four cuscuses with restricted ranges or restricted habitats are now considered threatened by overhunting or habitat-clearing. The Black-spotted cuscus is particularly susceptible to hunting with firearms as it sleeps exposed on a branch; for the other three – the Telefomin, Stein's, and Obi cuscuses – habitat-clearing is the major threat. The conservation status of many other cuscuses is not known, but without protective measures the continued survival of the susceptible mainland species and the numerous island forms will be gravely threatened. JW/CRD

phalangerid in which the female is known to take a proactive role in courtship is the Spotted cuscus; at 28-day intervals, assumed to coincide with estrus, she calls throughout the night, which excites males.

Most cuscuses also bear a single young, the exceptions being the two common cuscuses, for which twins are the norm. Long-term pair bonding may only occur in the Bear cuscus, the largest and most diurnal of the phalangerids. Male Spotted cuscuses may use sight in determining territoriality, since they are reputed to use daytime sleeping perches that provide clear views of neighboring rivals.

Pelts or Pests?

CONSERVATION AND ENVIRONMENT

The brushtail possums are of considerable commercial importance, in both a negative and a positive sense. On the downside, the Mountain brushtail causes damage in exotic pine plantations

A MARSUPIAL INVADER: THE COMMON BRUSHTAIL IN NEW ZEALAND

When the first Australian Common brushtail possums were imported to New Zealand around 1840, it was hoped that they would form the basis of a lucrative fur industry. The venture was manifestly successful. Aided by further importations until 1924 and by the freeing of captive-bred animals, populations increased prodigiously, so that sales of pelts became an important source of revenue.

However, the blessings of this marsupial invader are mixed. As well as carrying bovine tuberculosis (see above), the possum has been shown to have subtle but potentially damaging effects on the indigenous vegetation. New Zealand forest trees evolved in the absence of leaf-eating mammals, and, unlike the Australian eucalypts that produce poisonous oils and phenols, the leaves of most species are palatable and lack defenses against predators. When first introduced to particular New Zealand forests, the possums rapidly exploited the new food source, increasing in population density to up to 50 animals per ha (120/acre) – some 25 times more than in Australia. By the time numbers had stabilized at 6–10 per ha (15–25/acre), trees such as ratas and konini had all but disappeared from many areas, and possums were turning their attention to less favored species.

Possums hasten tree death by congregating on individual trees and almost completely defoliating them. These normally solitary creatures evidently abandon their social inhibitions when food is abundant – and, in contrast to their Australian kin, the New Zealand possums occupy small (1–2ha/2.5–5 acre) and extensively overlapping home ranges.

The final verdict on possum damage is unclear. Young individual ratas and other exploited tree species are appearing in many localities, but they now seem to be distasteful to possums. Presumably possums are conferring a selective advantage on unpalatable trees, and so continue, subtly but surely, to alter the structure of the forest. CRD

Ringtails, Pygmy Possums, and Gliders

THE RINGTAIL POSSUMS, GLIDERS, AND PYGMY possums of Australia and New Guinea inhabit a wide range of environments, including forest, shrubby woodland, and even (in the case of the Mountain pygmy possum) alpine upland. Formerly included with the brushtail possums and cuscuses in the family Phalangeridae, they are now divided into four separate families. Although these families may appear superficially similar to one another, the differences between them in external form, internal anatomy, physiology, patterns of genetic variability, and the biochemistry of blood proteins are in fact as great as those between the kangaroos and the koala.

When the Australian continent was invaded some 40–60 million years ago by primitive, possum-like marsupials, it was blanketed in a wet, misty, and humid rain forest. Opening of these forests in the mid to late Tertiary (32–35 million years ago) and their gradual replacement by the marginal eucalypt and acacia forests that now grow there forced this early fauna to seek refuge in the high-altitude regions of northern Queensland and Papua New Guinea, where the ringtail possums radiated to form a diverse family of leaf- and fruit-eating specialists. At the same time, the new nectar-, gum- and insect-rich Australian eucalypt and wattle (*Acacia*) forests provided many niches for the pygmy possums, feeding predominantly on nectar and insects, and the petaurid gliders, which fed on sap and gums. This diversification has led to remarkable convergences of form, function, and behavior with the arboreal lemurs, bush babies, monkeys, and squirrels of other continents.

The Mountain pygmy possum adopted an alternative strategy to the arboreal habits of the other possums and gliders, retreating to a cool, alpine environment, where it became ground-dwelling and inhabited rock deposits formed by periglacial activity. The Mountain pygmy possum is the only Australasian small mammal to undergo deep, seasonal hibernation under snow cover for up to 7 months of the Austral winter.

Fitted for the Forests

FORM AND FUNCTION

The ringtail possums and gliders, and most pygmy possums, are predominantly arboreal, with hand-like feet, an enlarged, opposable big toe on the hind foot, and a range of adaptations suited to moving through wooded environments. In non-gliding species, the tail is prehensile, and may be used for grasping branches and transporting nest material; a naked undersurface effectively increases friction. In gliders (but not the Feathertail glider) the tail is heavily furred and either straight or tapering; it may be used for controlling the direction of flight. Gliding species are specialized for rapid movement in open forest and are thought to have evolved independently in three families during the mid-to-late Tertiary. In the eight species of gliders that survive today, gliding is achieved by use of a thin, furred membrane (patagium) that stretches from fore to hind limbs (wrist to ankle in the Sugar glider, wrist to knee in the Feathertail), increasing surface area in flight to form a large rectangle. It is retracted when not in use and may be seen as a wavy line along the side of the body. The effective surface area has also been increased by a lengthening of the arm and leg bones, and some species cover distances exceeding 100m (330ft) in a single glide, from the top of one tree to the butt or trunk of another. The heavier Greater glider, with a reduced (elbow-to-ankle) gliding membrane, descends steeply with limited control, but the smaller gliders are accomplished acrobats that weave and maneuver gracefully between trees, landing with precision by swooping upwards. What appears to be a gentle landing to the human eye is in fact shown by slow-motion photography to be a high-speed collision. The animals bounce backward after impact and must fasten their long claws into the tree trunk to avoid tumbling to the ground. The fourth and fifth digits of the hand are elongated and have greatly enlarged claws that assist clinging after the landing impact.

Leaves, Insects, Sap, and Nectar

DIET

There are four major dietary groups of possums and gliders – folivores, sapivores and gumivores, insectivores, and nectarivores. All are nocturnal and have large, protruding eyes. Most are also quiet, secretive, and hence rarely seen. The only audible sign of their presence may be the "plop" of gliders landing on tree trunks, the yapping alarm call of the Sugar glider, or the screeching or gurgling call of the Yellow-bellied glider. Ringtail possums are generally quiet but occasionally emit soft twittering calls. The Greater glider is totally silent, emitting only a quiet grumbling sound when being handled. Most possum species (except the Greater glider) make loud screaming and screeching calls when attacked or handled.

Ringtail possums and the Greater glider together form a highly specialized group (Family Pseudocheiridae) of arboreal leaf-eaters (folivores), characterized by an enlargement of the cecum to

Right A Herbert River ringtail displays the tightly-curled tip of the long, prehensile tail that gives these animals their common group name. Increased logging in Queensland is threatening this species' habitat.

Inset Only 8cm (3in) long, a Feathertail glider grooms itself on a branch. Also known as the Pygmy glider, this species is the smallest marsupial capable of gliding, achieving flights of over 50m (165ft).

FACTFILE

RINGTAILS, PYGMY POSSUMS, & GLIDERS

Order: Diprotodontia

Families: Pseudocheiridae, Burramyidae, Petauridae, Acrobatidae

35 species in 13 genera

DISTRIBUTION SE, E, N, and SW Australia, Tasmania, New Guinea, offshore islands of New Guinea.

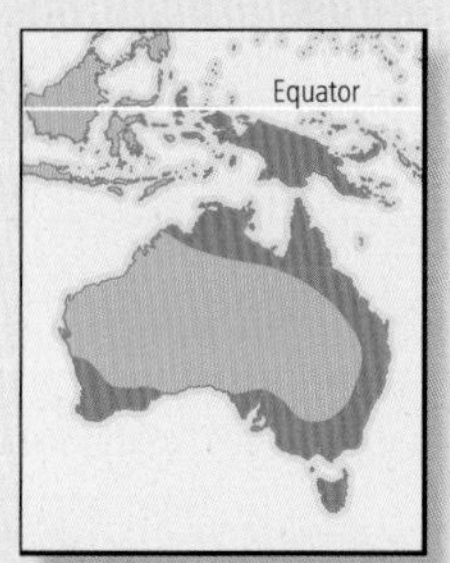

HABITAT Forests, woodland, shrublands, heathland, alpine heathlands

SIZE Head–body length ranges from 6.4cm (2.5in) in the Little pygmy possum to 33–38cm (13–15in) in the Rock ringtail possum; **tail length** from 7.1cm (2.8in) to 20–27cm (7.9–10.6in), and **weight** from 7g (0.2oz) to 1.3–2kg (2.9–4.4lb), both in the same two species.

FORM Coat gray or brown, with paler underside; often darker eye patches or forehead or back stripes (particularly in species feeding on plant gums); tail long, well-furred (in most gliders), prehensile, and part naked, or feather-like.

DIET Ringtails and gliders are primarily folivorous, although they also eat fruit; other species are more omnivorous, also including insects, larvae, spiders, scorpions, and small lizards in their diet.

BREEDING Gestation period 12–50 days; all young weigh less than 1g (0.035oz) at birth.

LONGEVITY 4–15 years (generally shorter in pygmy possums with large litters, and longer in ringtails and large gliders with single young).

CONSERVATION STATUS Four species are classified as Endangered, and four as Vulnerable.

See families box ▷

form a region for microbial fermentation of the cellulose in their highly fibrous diet. Fine grinding of food particles in a battery of well-developed molars with crescent-shaped ridges on the crowns (selenodont molars) enhances digestion. Rates of food intake in these groups are slowed by the time required for cellulose fermentation, and nitrogen and energy is often conserved by slow movement, relatively small litter sizes (averaging 1–1.5 young), coprophagy (reingestion of feces), and adoption of medium to large body size (0.2–2kg/0.4–4.4lb). The preferred diet of the Greater glider of eastern Australia is eucalypt leaves. The quantities of nutrients in these leaves vary substantially between different tree species, and this is a major factor underpinning the patchy patterns of distribution and abundance of the Greater glider through the eastern Australian forests.

The five species of petaurid glider and Leadbeater's possum (all of the Family Petauridae) are specialist plant-exudate (sap and gum) feeders. Arthropods, pollen, and occasionally the green seeds of acacias are also eaten, providing an important source of protein. The petaurid possums and gliders are small to medium in size (70–650g/2.5–21oz). The most primitive member of the group, Leadbeater's possum, is restricted to moist, high-altitude montane eucalypt forests, where it feeds on wattle or acacia gums, insects, and insect exudates. By incising notches in the bark of trees, the possum enhances gum production. Wattle gum is also a principal food of the Sugar glider, and the species may travel hundreds of meters across open pasture to obtain it.

The Sugar glider, which is distributed from Tasmania to northwest Australia and Papua-New Guinea and neighboring islands, also exploits the sap of eucalypts by incising the bark and licking up the sweet, carbohydrate-rich exudates. Such sap-feeding sites are highly prized and may be vigorously defended by chasing and biting intruders. Eucalypt sap also appears in the diet of the rare and highly endangered Mahogany glider in the far northeast of Queensland. Like most other petaurid gliders, this species consumes gum exudates from acacia trees as well as insects. The diet of the Mahogany glider can also include gum tapped from the floral spears of Grass trees. Eucalypt sap feeding has developed to an extreme in the Yellow-bellied glider of eastern Australia, which cuts large notches into the bark of many tree species. The form of these notches varies from deep, V-shaped incisions to long strips of ruffled bark, depending on the eucalypt species that is tapped.

Although a minor component of their diet, pollen and insects are an important protein source for all members of the Petauridae. A high carbohydrate-to-nitrogen ratio in their diet provides additional energy for activity and territorial defense but has limited reproductive potential, and so births are restricted to seasons of insect abundance. The coats of the gum-feeding gliders and possums are characterized by a distinct black dorsal stripe. This is thought to camouflage them when they are feeding – the time when they are most vulnerable to predation by forest owls.

With Leadbeater's possum, the strikingly-colored, black-and-white Striped possum and trioks are the non-gliding members of the Petauridae. In

Ringtail, Pygmy Possum, and Glider Families

Ringtail possums
Family Pseudocheiridae

17 species in 6 genera: 7 species of *Pseudocheirus* (SE, E, N, SW Australia, Tasmania, New Guinea, and West Irian), including Common ringtail possum (*P. peregrinus*) and Western ringtail possum (*P. occidentalis*). The family Pseudocheiridae also includes the Rock ringtail possum (*Petropseudes dahli*) from N Australia; the Greater glider (*Petauroides volans*) from E Australia, and a range of species from the forests of NE Australia such as the Daintree River ringtail possum (*Pseudochirulus cinereus*) and Green ringtail possum (*Pseudochirops archeri*) as well as several poorly-known New Guinea taxa such as the Weyland ringtail possum (*Pseudochirulus caroli*) and Pygmy ringtail possum (*Pseudochirulus mayeri*). The Western ringtail possum, D'Alberti's ringtail possum (*P. albertisii*), and the Plush-coated ringtail possum (*P. corinnae*) are classed as Vulnerable.

Pygmy possums
Family Burramyidae

5 species in 2 genera: pygmy possums (4 species of *Cercartetus*), including the Eastern and Western pygmy possums (*C. nanus* and *C. concinnus*), Tasmania, Kangaroo Island, SE, E, NE, SW Australia, New Guinea; Mountain pygmy possum (*Burramys parvus*), SE Australia. The Mountain pygmy possum is Endangered.

Gliders
Family Petauridae

11 species in 3 genera:. 6 species of *Petaurus* (Tasmania, SE, E, N, NW Australia, New Guinea), including Yellow-bellied or Fluffy glider (*P. australis*), Squirrel glider (*P. norfolcensis*), Mahogany glider (*P. gracilis*), Sugar glider (*P. breviceps*), and two species confined to New Guinea – the Northern glider (*P. abidi*) and *P. biacensis*. The Petauridae also contain the monotypic genus *Gymnobelideus* (Leadbeater's possum, *G. leadbeateri*, Victoria) and four species of *Dactylopsila* (the Striped possum and the trioks). The Striped possum (*Dactylopsila trivirgata*) occurs in both NE coastal Queensland and New Guinea; the remaining three species (*D. megalura*, *D. palpator*, and *D. tatei*) are confined to New Guinea or adjacent offshore islands. The Mahogany glider, Leadbeater's possum, and Tate's triok are classed as Endangered; the Northern glider is Vulnerable.

Feathertail glider and Feathertail possum
Family Acrobatidae

2 species in 2 genera: Feathertail or Pygmy glider or Flying mouse (*Acrobates pygmaeus*), SE to NE Australia; Feathertail possum (*Distoechurus pennatus*), New Guinea.

For full species list see Appendix ▷

a classic case of convergent evolution, the Striped possum and the trioks, like skunks, emit a distinctive, musty odor that is particularly strong in the Long-fingered triok from New Guinea. The four species are medium-sized and are specialized for exploiting social insects, ants, bees, termites, and other wood-boring insects in the tropical lowland rain forests of northern Queensland and New Guinea. A suite of adaptations aids in the noisy extraction of insects from deep within wood crevices – feeding activity may produce a shower of woodchips. These adaptations include an extremely elongated fourth finger (like that of the aye-aye of Madagascar; see Primates: Strepsirhines), an elongated tongue, and enlarged and forward-pointing upper and lower incisors.

Pygmy possums of the genus *Cercartetus* and the Feathertail or Pygmy glider form a fourth group that has diversified in the nectar-rich sclerophyllous Australian heathlands, shrublands, and eucalypt forests. Despite the small size of the Feathertail glider, it is nevertheless highly mobile; the species can glide for distances exceeding 50m (165ft), often spiraling from high in the tree canopy toward the ground like a falling leaf before settling in a flowering shrub. The brush-tipped tongue of the Feathertail glider is used for sipping

Left *Representative species of possums and gliders, exhibiting feeding behavior and movement:* ***1*** *Common ringtail possum* (Pseudocheirus peregrinus) *eating an insect.* ***2*** *Tasmanian pygmy possum* (Cercartetus lepidus) *foraging.* ***3*** *Leadbeater's possum* (Gymnobelideus leadbeateri) *feeding on sap.* ***4*** *Striped possum* (Dactylopsila trivirgata) *on branch.* ***5*** *Feathertail glider* (Acrobates pygmaeus) *in flight.* ***6*** *Mahogany glider* (Petaurus gracilis), *showing the folds on its side where the patagium or flying membrane is stored.* ***7*** *Sugar glider* (P. breviceps) *with the patagium extended in flight.*

nectar from flower capsules, and the small size (under 35g/1.2oz) and extreme mobility of all five species increase nectar harvesting rates. In poor seasons, aggregations of many individuals may be found on isolated flowering trees and shrubs. Most species take insects and the abundant pollen available from flowers to provide protein. The Eastern pygmy possum occasionally eats soft fruits and seeds. The combination of small size and abundant dietary nitrogen permit unusually large litter sizes (4–6), and rapid growth and development rates similar to those of the carnivorous marsupials. The other member of the pygmy possum group, the Feathertail possum of Papua New Guinea, has a tail like that of the Feathertail glider but is larger (50–55g/1.8–1.9oz) and has no gliding membrane. Its diet includes insects, fruit, and possibly plant exudates.

For its spring and summer diet, the Mountain pygmy possum depends largely on Bogong moths (*Agrotis infusa*) and other invertebrates. Huge numbers of these moths migrate to the mountains in spring. As Bogong moths become scarce, fleshy fruits and seeds from heathland plants become increasingly important. The remarkable sectorial premolar tooth is adapted for husking and cracking seeds. Excess seeds may be cached for use during periods of winter or early spring shortage.

Smaller Size, Larger Nesting Groups

SOCIAL BEHAVIOR

Mountain pygmy possums have only one litter of four young per year, following snowmelt. This is an adaptation to the short, alpine summer and the need for both adults and young to gain sufficient fat reserves to enable them to survive the long period of winter hibernation.

Most Australian possums and gliders nest or den in cavities in large, old living or dead trees, although sometimes other types of nest sites are occupied, such as bark strips or fallen logs. The Common ringtail possum can build a stick nest or drey, but in cold subalpine or seasonally hot woodland environments hollow trees are used in favor of dreys. Individuals of all hollow-using possums and gliders have den sites in many different trees and will often swap between them on a regular basis. The entrance to the hollow is typically just large enough to permit the entry of the occupant, but small enough to preclude predators and other species that may attempt to usurp the use of the cavity.

Patterns of social organization and mating behavior in possums and gliders are remarkably diverse, but to some extent predictable from species' body size and diet. The larger folivorous ringtail possums and the Greater glider are often solitary; by day they sleep singly or occasionally in pairs in tree hollows or vegetation clumps, emerging to feed on foliage in home ranges of up to 3ha (7.4 acres) at night. Male home ranges of the Greater glider are generally exclusive but may partially overlap those of one or two females. The occupation of exclusive home ranges by males and of overlapping home ranges by females is associated with a greater mortality of sub-adult males and a consequent female-biased sex ratio.

The tendency toward gregariousness increases with decreasing body size, the Yellow-bellied glider forming nesting groups of up to five individuals, the Common ringtail of eastern Australia up to six, the Sugar glider up to 12, and the Feathertail glider up to 25. Most nesting groups consist of mated pairs with offspring, but the Feathertail glider and the petaurids may form truly mixed groups with up to four or more unrelated adults of both sexes (in the Sugar glider), one male and one or several females (the Yellow-bellied glider), or one female and up to three males (Leadbeater's possum). The chief reason for nesting in groups is thought to be improved energy conservation through huddling during winter. In one species, the Sugar glider, large nesting groups disband into smaller units during summer. The aggregation of females during the winter enables dominant males to monopolize access to up to three females in the petaurid gliders, and a harem defense mating system prevails.

An entirely different mating system occurs in Leadbeater's possum. Individual females occupy large nests in hollow trees and actively defend a surrounding territory of 1–1.5ha (2.5–3.7 acres) from other females. Mating is usually monogamous, and male partners assist females in defense of territories. Additional adult males may be tolerated in family groups by the breeding pairs but adult females are not, and an associated higher female mortality results in a male-biased sex ratio. This pattern appears to be associated with the construction of well-insulated nests, avoiding the necessity for females to huddle together during winter, and with the occupation of dense, highly productive habitats in which food resources are readily defensible and surplus energy is available to meet the cost of territorial defense.

The mating patterns of some possum and glider species can be somewhat flexible, varying spatially and temporally depending on the availability and quality of food and other resources like den sites. For example, small, low-density populations of the Greater glider that occupy eucalypt forests with low levels of foliage nutrients appear to be predominately monogamous. In contrast, higher-density populations in more nutrient-rich forest types maintain a polygamous mating system. The patterns of social organization, group size, and mating systems of Leadbeater's possum and the Yellow-bellied glider may also change over time depending on, for example, year-by-year differences in the availability of food.

Selective pressures exerted during competition for mating partners have led to the prolific development of scentmarking glands in the petaurids, for use in marking other members of the social group. Leadbeater's possum, the most primitive member, shows the least development of special scent glands, and scentmarking between partners involves the mutual transfer of saliva to the tail base with its adjacent anal glands. Sugar glider males, in contrast, possess forehead, chest, and anal glands. Males use their head glands to spread scent on the chest of females, and females in turn spread scent on their heads by rubbing the chest gland of dominant males. Male Yellow-bellied gliders have similar glands, but scent transfer is achieved quite differently, by rubbing the head gland against the female's anal gland. Females in turn rub their heads on the anal gland of the dominant male. Such behavior probably facilitates group cohesion by communicating an individual's social status, sex, group membership, and reproductive position.

Above *Dwarfed by their dinner, a pair of Tasmanian pygmy possums* (Cercartetus lepidus) *prepare to feed on nectar from a* Banksia *flower.*

In contrast to the small gliders, pygmy possums of the genus *Cercartetus* appear mainly solitary. Usually only lactating females share nests with their young, although several males may share a nest, sometimes with a non-lactating female. Mountain pygmy possums seem more social, with sedentary females forming kin clusters in high-quality habitats and sometimes sharing nests with non-dependent, apparently related females. Nest sharing among males is common, and home ranges overlap. Although they are not sexually dimorphic, female Eastern and Mountain pygmy possums may be behaviorally dominant. Mountain pygmy possum males leave the habitat of females after breeding and spend the winter in slightly warmer habitats, with more northerly and westerly aspects and lower elevations. It is still not clear whether the resulting sexual segregation during the non-breeding season is a result of female aggression, or is simply a reproductive strategy.

The optimal temperature for hibernation in males is slightly higher than in females, and they arouse more frequently during winter and finish hibernation earlier in spring. This provides them with a reproductive advantage, because they can undergo spermatogenesis and be ready to breed. Because male survival is often lower than for females, sex ratios are frequently female-biased, especially in high-quality habitats. Pygmy

possums generally have relatively short lifespans (less than 3 years), but Eastern pygmy possums have been known to live for more than 6 years. Mountain pygmy possums are remarkably long-lived, with females living for up to 12 years. This is probably a result of their relatively stable environment, single reproductive effort, larger size, and long periods of hibernation.

Back from the Dead

CONSERVATION AND ENVIRONMENT

The story of the discovery, apparent extinction, and subsequent rediscovery of three widely divergent species typifies the plight of the pygmy possums and gliders. In all cases, human pressure for land use is now once more placing their continued survival under threat.

Just after nightfall one evening in 1961, in the wet, misty mountains just 110km (70mi) from Melbourne, the attention of a fauna survey group from the National Museum of Victoria was caught by a small, bright-eyed, alert gray possum leaping nimbly through the forest undergrowth. Its size at first suggested a Sugar glider, but the absence of a gliding membrane and the narrow, bushy, club-shaped tail led to the exciting conclusion that this was the long-lost Leadbeater's possum. This rare little possum is one of the State of Victoria's faunal emblems, and was first discovered in 1867 in the Bass River Valley. Only six specimens were collected, all prior to 1909, and in 1921 it was concluded that the destruction of the scrub and forest in the area had resulted in the complete extermination of the species. Surveys following the rediscovery, however, led to its detection at some 300 separate sites within a 3,600sq km (1,400sq mi) area. Its preferred habitat is Victoria's Central Highland forests, which are dominated by the majestic Mountain ash (*Eucalyptus regnans*), the world's tallest flowering plant and one of Australia's most valued timber-producing trees. Standing beneath such forest giants provides the most reliable method of catching a glimpse of a Leadbeater's possum, as the animals emerge at dusk from their family retreats in hollow tree trunks to feed. Less than 40 years after its rediscovery, however, the possum is once again threatened with extinction through a combination of inappropriate forest management and natural collapse of the large dead trees that provide nest sites in regrowth forests (in 1999 a fire devastated two-thirds of Victoria's Mountain ash forests).

The Mahogany glider was first described in 1883, but was misidentified at the time as a Squirrel glider. Careful examination of specimens from the Queensland Museum subsequently revealed that gliders from a tiny coastal area in the far north of Queensland were, in fact, different in size, tail length, and a number of other particulars from the Squirrel glider. The species has a highly restricted distribution and is confined to open woodlands and adjacent paperbark swamps. Most potentially suitable habitat has been destroyed, and the survival of the species continues to be threatened by land clearance, particularly for the establishment of sugar cane and banana crops.

The Mountain pygmy possum was described from fossil remains in 1896 and was thought to be extinct until 1966, when one turned up in a ski lodge in the Victorian alps. Since it was believed that all pygmy possums were arboreal and nested in tree-hollows, it was reasoned the animal must have been brought to the alps in a load of firewood. Searches were made in the forests at lower elevations, but to no avail. In 1970, however, another animal was trapped in rocky heath under snow gum woodland at the interface of the sub-alpine and alpine zones on the Kosciuszko plateau of New South Wales. This directed attention back to the ski lodge in Victoria, where trapping in the surrounding rocky heath quickly resulted in the capture of three animals. The following year, 11 animals were trapped well above the tree line on Mt Kosciuszko, the highest mountain on the Australian mainland (2,228m/7,300ft). Since then, intensive research has been conducted on the Mountain pygmy possum. Not coincidentally, because of their requirements for high-elevation sites with good snow cover, the largest local populations all occur within ski-resort concession areas. Here, they are threatened by ski runs and general tourist development. An increasingly apparent and much less easily managed threat is that of increasing temperatures and receding snow cover resulting from global warming.

The survival of Leadbeater's possum, the Mahogany glider, and the Mountain pygmy possum – all of which are nationally endangered – is critically dependent upon effective government action, which is not yet forthcoming. LB/DL

Kangaroos and Wallabies

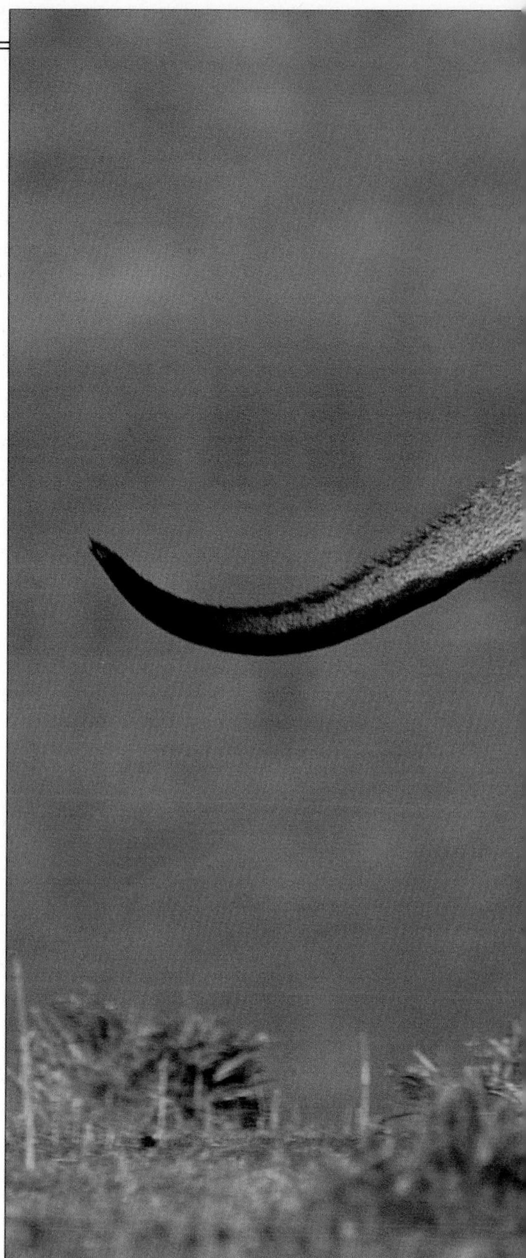

Top left A Swamp wallaby browses on foliage. Despite the name, the species is found in open upland forests as well as in marshy regions and mangroves.

RED KANGAROOS BOUNDING ACROSS THE arid saltbush plains are one of the quintessential images of Australia. Yet the Red is just one among a diverse array of about 68 living species of kangaroos, wallabies, and rat kangaroos that make up the superfamily Macropodoidea. Desert-adapted, grass-eating kangaroos such as the Red have in fact evolved only in the last 5–15 million years. Before then, Australasia was forested, and the ancestors of all macropods were forest-dwelling browsers.

The superfamily Macropodoidea takes its name from *Macropus*, the genus of the Red kangaroo. The word means "big foot" in Latin, and long hind feet do indeed characterize the animals. They are the largest mammals to hop on both feet, a very special gait for a large mammal. Even so, hopping is not the only way that kangaroos get about.

The Mechanics of Hopping

FORM AND FUNCTION

All macropods are furry-coated, long-tailed animals, with thin necks, prominent ears, and strongly developed hindquarters that make the forelimbs and upper body look small. A long, narrow pelvis supports long and muscular thighs; the even more elongate shin bones are not heavily muscled and end in an ankle that is adapted to prevent the foot from rotating sideways (so that the kangaroo cannot twist its ankle while hopping). At rest and in slow motion, the long but narrow sole of the foot bears the animal's weight, making it in effect plantigrade. When hopping, however, macropods rise onto their toes and the "balls" of their hind feet. Only two of the toes, the fourth and fifth, are in fact load-bearing; the second and third are reduced to a single, tiny stump equipped with two claws that are used exclusively for grooming.

The first toe is entirely lost except in the Musky rat kangaroo, the only surviving member of the primitive Hypsiprymnodontidae family. Short-faced kangaroos in the genus *Procoptodon* shed the fifth toe, too. *Procoptodon* went extinct 15–25,000 years ago at the peak of the last ice age, possibly as a result of human-lit fires and hunting, but rock paintings record its distinctive footprints. Limb-lengthening and toe reduction characterize many lineages of mammalian, terrestrial herbivores – they include horses among the perissodactyls and deer, antelopes, and pronghorns among the artiodactyls – that evolved speed to avoid running predators. In these and many other ways, macropods illustrate convergence for lifestyle between eutherian and metatherian mammals.

Macropods do not only hop; they also crawl on all fours when moving slowly, with the pairs of fore- and hind limbs moving together rather than alternately. In medium and large macropodine species, the tail and the forelimbs take the animal's weight while the hind feet are lifted and swung forward. This gait is called "pentapedal" (literally, five-footed). In the larger macropodines the tail is long, thick, and muscular, and the animal sits up by leaning back on it like a sportsman on a shooting stick; it can even briefly be the only means of support during fighting.

In smaller wallabies and rat kangaroos, the tail's major function is for balance and maneuvering, for instance as an aid in abrupt cornering. Most rat kangaroos also use the tail to carry nesting material; grasses or twigs are gathered into a bundle and pushed backward with the hind feet against the underside of the tail, which curls over, holding the bundle against the rat kangaroo's rump as it hops away to the half-built nest.

In contrast to the hind limbs, macropods' forelimbs are relatively small and unspecialized. The forepaws have five equal, strongly-clawed digits set around a short, broad palm. Rat kangaroos dig for their food with elongate second to fourth digits with long claws. Macropods' forepaws can grasp and manipulate food plants; they also serve to grip the skin, hold open the pouch, or scratch the fur while grooming. The larger macropods also use their fore limbs in thermoregulation, licking saliva onto their insides, where it evaporates, cooling blood in a network of vessels that lie just below the skin's surface.

At speeds slower than about 10km/h (6mph), hopping is at best an ungainly way of moving, but above 15–20km/h (9–12mph) it is extremely energy-efficient – more so than four-footed trotting or galloping. At the end of each bound, energy is stored in the tendons of the bent hind legs, contributing to the next driving extension. Like the rider of a spring-loaded pogo-stick, the kangaroo needs only to add a little extra energy in order to keep hopping. Yet hopping probably originated as a way of startling predators by making an explosive burst from cover, a strategy retained by the smaller kangaroos, which conceal themselves in vegetation, and the rat kangaroos, which hide in nests. That explosive burst requires long, strong,

FACTFILE

KANGAROOS AND WALLABIES

Order: Diprotodontia

Families: Macropodidae and Hypsiprymnodontidae

About 68 species in 16 genera

DISTRIBUTION Australia, New Guinea; introduced into Britain, Germany, Hawaii, and New Zealand.

HABITAT Wide-ranging, from deserts to rain forests.

SIZE Head–body length ranges from 28.4cm (11.2in) in the Musky rat kangaroo to 165cm (65in) in the male Red kangaroo; tail length from 14.2cm(5.6in) to 107cm (42in), and weight from 0.5kg (1.2lb) to up to 95kg (200lb), both in the same two species.

COAT Macropod fur, mostly 2–3cm (0.8–1.2in) long, is fine, dense, and not sleek. Colors range from pale gray through various shades of sandy brown to dark brown or black.

DIET Mostly plant foods, including grasses, forbs, leaves, seeds, fruit, tubers, bulbs, and truffles; also some invertebrates, such as insects and beetle larvae.

BREEDING Gestation 30–39 days; newborn attach to a maternal teat within a pouch and remain there for a further 6–11 months.

LONGEVITY Variable according to species and conditions; larger species may attain 12–18 years (28 years in captivity), the smaller rat kangaroos 5–8 years.

CONSERVATION STATUS 2 wallabies and 2 species of hare wallaby (plus 2 subspecies) are now listed as Extinct. A further 7 species of kangaroos and wallabies are Endangered, while 9 more are Vulnerable.

See families box ▷

*▲ **Above** With a baby, or joey, safely secured in her pouch, a female Eastern gray kangaroo goes foraging in an Australian reserve. At full speed, large kangaroos can travel at over 55km/h (35mph).*

*▼ **Below** Red kangaroos fighting. Before a fight two males may engage in a "stiff-legged" walk **1** in the face of the opponent, and in scratching and grooming **2**, **3**, standing upright on extended rear legs. The fight is initiated by locking forearms **4** and attempting to push the opponent backward to the ground **5**.*

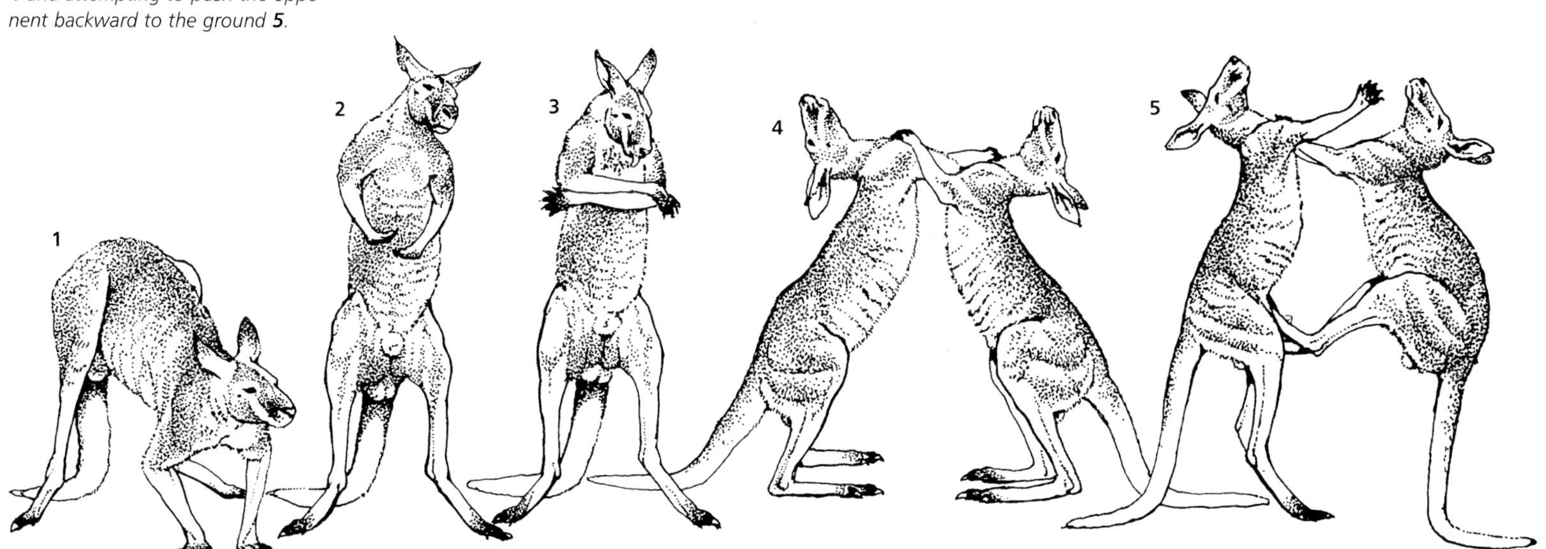

synchronized hind limbs, and initiates a bounding gait. Large kangaroos can sustain hopping speeds faster than 55 km/h (35mph), and small species can manage bursts of over 30 km/h (20mph).

Some macropods are adapted to bounding up narrow ledges on near-vertical cliffs and even to climbing trees. Rock wallabies have rather broad hind feet with nonslip soles, and long, bushy tails to help them keep their balance as they escape. Tree kangaroos find refuge high up in the rain forest. They climb by grasping the trunks with huge, heavily-clawed forepaws on the ends of long, strong arms, and pushing upward on short, broad feet that grip the trunk or branches. Their tails are not prehensile. They descend awkwardly, tail first, but are able to jump short distances from tree to tree. When climbing or even moving along branches, their hind limbs move synchronously – a remnant of their long-gone deerlike gait. Nevertheless, they can move them alternately (as all macropods do when forced to swim or while adjusting their stance).

Macropods' limbs, paws, and feet also function as weapons. Rat kangaroos and small wallabies fight by kicking out as they jump at each other, or else by grappling, rolling on the ground biting and scratching. They sometimes kill opponents with their hind feet. The large kangaroos remain more upright, wrestling with the fore limbs around the opponent's head, shoulders, and neck, and kicking with the powerful hind limbs (the tail briefly taking the animal's weight), driving the large toes hard into the opponent's belly. The shoulders and forearms of males of larger kangaroo species are longer and more muscular than those of the females, and male forepaws are more heavily clawed. Males also grow a shield of thickened skin over the belly – more than twice as thick as on the flanks or shoulders – that helps absorb the impact of kicks to the gut.

Macropods have heads that superficially resemble those of deer or antelopes, with moderately long muzzles, wide-set eyes with some binocular vision, and upright ears that can be rotated to catch sounds from all directions. The upper lip is "split" like that of a hare or a squirrel.

The muzzles, teeth, and tongues of macropods are suited to taking small food items rather than large mouthfuls of food. Most macropods pluck single items, even blades of grass, one at a time. Behind the split upper lip lies an arc of incisor teeth surrounding a fleshy pad at the front of the palate. In macropodine kangaroos and wallabies, the two procumbent (horizontally-set) lower incisors hold leaves against the fleshy pad while they are ripped off along the edge of the upper incisor arc. In potoroine rat kangaroos and the Banded hare wallaby, the only living sthenurine, the lower incisors occlude with the second and third upper incisors, while the central (first) upper incisors protrude and are used for gnawing.

Potoroine rat kangaroos' distinctive premolars form serrated blades to cut the tough and fleshy food that constitutes their diet (mainly plant storage organs or fungal fruiting bodies). Small macropodines have similarly sectorial, persistent premolars; but larger kangaroos shed their unspecialized premolars, opening the way for the molar teeth, which erupt sequentially at the back of the jaw, to migrate forward, being shed as they wear out. This adaptation to an abrasive diet (mainly of grasses) parallels that of elephants. Macropods with persistent premolars do not show molar progression, and all four erupted molars in each half jaw are in wear simultaneously.

Macropods can use their forepaws to handle or dig up food. This ability is most advanced in the potoroines, most of which depend to a large

Right *One of two remaining hare wallaby species, the Rufous hare wallaby is itself listed as Vulnerable, surviving on two islands off Western Australia.*

Below *Representative species of the larger kangaroos and wallabies:* ***1*** *Red kangaroo* (Macropus rufus)*;* ***2*** *Hill wallaroo* (Macropus robustus) *with young in pouch;* ***3*** *Bridled nail-tailed wallaby* (Onychogalea fraenata)*;* ***4*** *Red-legged pademelon* (Thylogale stigmatica) *in pentapedal (literally, five-footed) posture, with tail and all four limbs providing support;* ***5*** *Whiptail or Prettyface wallaby* (Macropus parryi) *in motion;* ***6*** *Goodfellow's tree kangaroo* (Dendrolagus goodfellowi) *resting on a branch.*

extent on digging up underground food items. However, even the largest kangaroos pull plants toward them with their paws, and use their "hands" to remove unwanted plant parts from their mouths.

Macropod digestion is aided by a forestomach enlarged to form a fermentation chamber. Longitudinal and transverse bands of muscles (haustrations) in the stomach wall contract to stir the stomach contents. A moderate-sized small intestine opens into an enlarged cecum and proximal colon, presumably a site for secondary fermentation, beyond which digested food flows through a long, water-resorbing distal colon.

The gut resembles that of some eutherian foregut fermenters, although the macropod stomach has not developed the compartmentalization of form and function to the same extent. The suite of bacteria, ciliate protozoa, and anaerobic fungi found in the stomachs of macropods performs the same functions as in ruminants, but the component species are quite different. Macropods may be less limited by poor-quality food than, for example, sheep. Some macropods recycle urea to help cope with low protein availability.

Macropod fur ranges in color from pale gray to dark brown or black. Many macropods have indistinct dark or pale stripes that visually break their outline: down the spine, across the upper thigh, behind the shoulders, or (most commonly) below or through the eye. The paws, feet, and tail are often darker than the body, and the belly is usually paler, making the animals appear "flat" in the dusk or by moonlight. In a few rock wallabies and tree kangaroos, the tail is longitudinally or transversely striped.

Males of some larger species are more boldly colored than the females; for example, the russet neck and shoulder coloring may be stronger in male Red-necked wallabies. Male Red kangaroos are mostly sandy-red, while the females are blue-gray or sandy gray. But the dimorphism is imperfect; some males may be blue-gray and some females red. The sexes' colors are fixed from the time they first show hair, rather than being acquired under a hormonal surge at or after puberty as in many dimorphic ungulates.

Many macropod males spread scented secretions from the skin of the throat and chest onto trees (especially the tree kangaroos), rocks (rock wallabies), or bushes and tussocks of grass (large kangaroos). They may also rub the scent on females during courtship, indicating to other males their association. Other glands within the cloaca add their scent to the urine or feces.

All over Australia

DISTRIBUTION PATTERNS

The single extant Hypsiprymnodontidae species, the Musky rat kangaroo, is confined to rain forests on the eastern side of Australia's Cape York peninsula. In contrast, the Macropodidae are represented by species of the Macropodinae subfamily all over Australia, in New Guinea, and on offshore islands; but the Potoroinae rat kangaroos (10 recent species) are confined to Australia (including Tasmania and other southern islands), and are rare in the tropical north. Two macropodine genera, *Dorcopsis* and *Dorcopsulus*, are confined to New Guinea; 8 of the 10 *Dendrolagus* tree kangaroos, and one of the four *Thylogale* pademelon species, also occur only there. Just two species, the Red-legged pademelon and the Agile wallaby, occur in both Australia and New Guinea. Feral macropod populations occur in a few countries outside Australasia: Brush-tailed rock wallabies in Hawaii; Red-necked wallabies in England's Pennine hills and in Germany; and both those species plus Tammar and Parma wallabies in New Zealand.

Genera of macropodines that are restricted to the rain forest include *Dorcopsis* and *Dorcopsulus*, and the *Dendrolagus* tree kangaroos. Pademelons

are also associated with wet, dense forests, including eucalypt forests; they occur from New Guinea down the east of Australia to Tasmania. Except for the weakly social tree kangaroos, these forest-dwelling macropodines are solitary.

Hare and nail-tailed wallabies occur in arid and semi-arid habitats including spinifex grassland, shrubland, savanna, and light woodland, and are confined to Australia. So are rock wallabies, which are found in habitats ranging from the arid zone of central, western, and southern Australia to rainforest habitats in the tropics. Yet their habitat always contains boulder piles, rocky hillsides, or clifflines to provide secure diurnal refuges.

The rich *Macropus* genus is almost confined to Australia. Its species occur in habitats ranging from desert to the edges of wet eucalypt forest, all characterized by grasses in the understory. The potoroines are confined to Australia, where they occur in rain forest, wet sclerophyll forest, and scrub, always with dense understory. The bettongs are creatures of the open forest, woodland, and savanna, often with a grassy understory. The Desert rat kangaroo used to occur in lightly-vegetated desert.

Macropod communities in Australia used to contain 5–6 sympatric species in the arid and semi-arid regions, but as many as a dozen in broken woodland and forested country along the Dividing Range. Rainforest communities rarely exceed 4 species.

From Grass to Truffles

DIET

The Musky rat kangaroo eats fleshy fruits and fungi and also regularly takes insects; in addition, it sometimes scatter-hoards seeds, although we do not know how efficient it is at finding them again. Macropodids depend on plants, although some of the smaller species (especially potoroines) will also eat invertebrates such as beetle larvae. Potoroos and bettongs feed largely on the underground storage organs of plants – swollen roots, rhizomes, tubers, and bulbs – and in addition eat the underground fruiting bodies (truffles) of some fungi, playing an important role in dispersing their spores (see A Mutually Beneficial Relationship).

Small macropodine wallabies that occupy dry habitats, including the hare and nail-tailed species, feed selectively on growing leaves of grasses and forbs, augmented with seeds and fruits. In mesic, forested habitats, macropodine diets include more fruits and dicot leaves, and these dominate the diets of tree kangaroos, swamp wallabies, and pademelons. Many macropodines feed opportunistically and seasonally from a large range of plant species and parts. *Macropus* species tend to have diets dominated by grass leaf, and also select seedheads of grasses and other monocots; and the largest kangaroos may rely entirely on grasses.

The smallest macropods tend to be highly selective in their feeding habits, seeking out scattered, high-quality food items, many of which have to be carefully sought and processed. In contrast, the largest species generally tolerate a lower-quality diet taken from a wide range of plant species, selecting mainly leaves but also some higher-value seeds and fruits.

Mobs and Loners

SOCIAL BEHAVIOR

At birth juvenile macropods are tiny, measuring just 5–15mm (0.2–0.6in); they look embryonic, with undeveloped eyes, hind limbs, and tail. Using its strong forelimbs, the newly-born infant will climb unaided up the mother's fur and into her forward-opening pouch. There it clamps its mouth onto one of four teats, remaining attached for many weeks of development – from 150–320 days, depending on species. The pouch provides a warm, humid environment for the juvenile, which cannot yet regulate its own temperature and can lose moisture rapidly through its hairless skin.

Once the juvenile has detached from the teat, the mother in many larger species will allow it out of the pouch for short walkabouts, retrieving it when she moves. She will prevent it from returning to the pouch just before the birth of her next young, but it will continue to follow her about as a dependent young-at-foot, and can put its head into the maternal pouch to suck the teat. The quality of milk provided changes as the joey matures, and a mother suckling a juvenile in the pouch at the same time as a young-at-foot will

Kangaroo and Wallaby Families

Family Hypsiprymnodontidae

Musky rat kangaroo *Hypsiprymnodon moschatus*.

Family Macropodidae

Subfamily Sthenurinae (Sthenurines)

Banded hare wallaby *Lagostrophus fasciatus*. Listed as Vulnerable by the IUCN.

Bettongs and Potoroos
Subfamily Potoroinae

Bettongs and Potoroos are sometimes considered a separate family, the Potoroidae.

Bettongs Genus *Bettongia*, 4 species: Brush-tailed bettong or woylie (*B. penicillata*); Burrowing bettong or boodie *B. lesueur*; Northern bettong (*B. tropica*); and Tasmanian bettong (*B. gaimardi*). All 4 species are listed by the IUCN: the Northern bettong is Endangered, and the Burrowing bettong Vulnerable.

Potoroos Genus *Potorous*, 3 species: Long-nosed potoroo (*P. tridactylus*); Long-footed potoroo (*P. longipes*); and Gilbert's potoroo (*P. gilbertii*). Gilbert's potoroo is Critically Endangered, the Long-footed potoroo Endangered.

Desert rat kangaroo *Caloprymnus campestris* Listed as Extinct by the IUCN.

Rufous rat kangaroo *Aepyprymnus rufescens*.

Kangaroos and Wallabies
Subfamily Macropodinae

Kangaroos, Wallaroos, and Wallabies Genus *Macropus*, 14 species: Red kangaroo (*M. rufus*); Eastern gray kangaroo (*M. giganteus*); Western gray kangaroo (*M. fuliginosus*); Common or Hill wallaroo (*M. robustus*); Black wallaroo (*M. bernardus*); Agile wallaby (*M. agilis*); Antilopine wallaroo (*M. antilopinus*); Red-necked wallaby (*M. rufogriseus*); Black-striped wallaby (*M. dorsalis*); Tammar wallaby (*M. eugenii*); Whiptail or Prettyface wallaby (*M. parryi*); Toolache wallaby (*M. greyi*); Western brush wallaby (*M. irma*); and Parma wallaby (*M. parma*). The Toolache wallaby is now listed as Extinct by the IUCN; the Black wallaroo and the Parma and Western brush wallabies are Lower Risk: Near Threatened.

Tree kangaroos Genus *Dendrolagus*, 10 species: Grizzled tree kangaroo (*D. inustus*); Bennett's tree kangaroo (*D. bennettianus*); Lumholtz's tree kangaroo (*D. lumholtzi*); Matschie's or Huon tree kangaroo (*D.matschiei*); Lowland tree kangaroo (*D. spadix*); Doria's tree kangaroo (*D. dorianus*); Dingiso (*D. mbaiso*); tenkile (*D. scottae*); White-throated tree kangaroo (*D. ursinus*); and Goodfellow's tree kangaroo (*D. goodfellowi*). All 10 *Dendrolagus* species are listed by the IUCN; Goodfellow's, Matschie's, and the Tenkile tree kangaroos are Endangered.

Rock wallabies Genus *Petrogale*, about 15 species including: Yellow-footed rock wallaby (*P. xanthopus*); Brush-tailed rock wallaby (*P. penicillata*); Proserpine rock wallaby (*P. persephone*); Black-footed rock wallaby (*P. lateralis*); Cape York rock wallaby (*P. coenensis*); monjon (*P. burbidgei*); nabarlek (*P. concinna*); and Mount Claro rock wallaby (*P. sharmani*). Seven species are listed, including the Prosperine rock wallaby as Endangered, and the Brush-tailed rock wallaby, which is Vulnerable.

Hare wallabies Genus *Lagorchestes*, 4 species: Spectacled hare wallaby (*L. conspicillatus*); Central hare wallaby (*L. asomatus*); Eastern hare wallaby (*L. leporides*); and Rufous hare wallaby or mala (*L. hirsutus*). The Eastern and Central hare wallabies are now listed as Extinct by the IUCN; the Rufous hare wallaby is Vulnerable , while the Spectacled hare wallaby is Lower Risk: Near Threatened.

Pademelons Genus *Thylogale*, 4 species: Red-necked pademelon (*T. thetis*); Red-legged pademelon (*T. stigmatica*); Tasmanian pademelon (*T. billardierii*); and Dusky pademelon (*T. brunii*). The Dusky pademelon is Vulnerable.

Nail-tailed wallabies Genus *Onychogalea*, 3 species: Bridled nail-tailed wallaby (*O. fraenata*); Northern nail-tailed wallaby (*O. unguifera*); and Crescent nail-tailed wallaby (*O. lunata*). The Crescent nail-tailed wallaby is now listed as Extinct by the IUCN; the Bridled nail-tailed wallaby is Endangered.

Dorcopsises Genus *Dorcopsis*, 3 species: White-striped dorcopsis (*D. hageni*); Gray dorcopsis (*D. luctuosa*); Brown dorcopsis (*D. veterum*).

Forest wallabies Genus *Dorcopsulus*, 2 species: Papuan forest wallaby (*D. macleayi*), and Lesser forest wallaby (*D. vanheurni*). The Papuan forest wallaby is Vulnerable.

Quokka *Setonix brachyurus*. Vulnerable.

Swamp or **Black wallaby** *Wallabia bicolor*.

Below *A Red-necked wallaby and her young relax in the Tasmanian sun. Mothers bear a single young, but the short interval between births means that they often end up rearing an infant in the pouch while still continuing to feed an older offspring that has reached the "young-at-foot" phase.*

produce different qualities of milk from the two teats – a feat achieved by having the mammary glands under separate hormonal control.

The Musky rat kangaroo may give birth to litters of two or even three young, but all macropodids produce only one young at a birth. Few are strictly seasonal breeders; most can conceive and give birth at any time of year. In almost all species gestation lasts a few days short of the length of the estrus cycle – generally 4–5 weeks in macropodines and 3–4 weeks in potoroines.

Giving birth to such small babies is relatively effortless; the female sits with her tail forward between her legs and licks the fur between her cloaca and pouch, producing a path that will keep the climbing neonate moist until it enters the pouch. A few days after giving birth many macropods enter estrus once more. If they are mated and conceive, the new embryo's development halts at an unimplanted blastocyst stage. That "embryonic diapause" lasts until about a month before the current pouch-young is sufficiently developed to quit the pouch. Then the blastocyst implants in the uterus and resumes development. A day or two before birth is due, the mother will exclude the previous young from the pouch, a rebuff that is difficult for it to accept as it has earlier been taught to come when called and to climb back into the pouch. The mother then cleans and prepares the pouch for the next juvenile. Thus many macropod females can simultaneously support a suckling young-at-foot, a suckling pouch-young, and a dormant or developing embryo.

The short interval between births allows females to quickly replace young-at-foot that are killed by predators. It also permits them to easily replace "aborted" pouch-young. A female that is hard-pressed by a dingo may relax the sphincter muscles closing the pouch, dropping her young to get eaten while she escapes. Under the nutritional stress of drought a pouch-young will also die, but will quickly be replaced by the dormant blastocyst, which is stimulated to implant and resume development as soon as the previous pouch-young's suckling stops. At relatively low metabolic cost, a female in drought can maintain a succession of embryos ready to develop as soon as rains break and conditions turn favorable.

The young-at-foot phase comes to an end when the juvenile is weaned; it lasts many months in the large kangaroos, but may be almost absent in small rat kangaroos such as the Rufous bettong. Similarly, large kangaroos grow through a prolonged subadult phase before breeding. Females of the large kangaroos begin breeding at 2–3 years, when they have reached half their full size, and may breed for 8–12 years. Some small rat kangaroos can conceive within a month of weaning, at 4–5 months, but may delay until 10–11 months.

Macropod males may mature physiologically soon after the females, but in larger kangaroos their participation in reproduction is socially inhibited. Female growth decelerates after they begin breeding, but male growth continues strongly, resulting in old males being very much bigger than younger males and females. Indeed, a female Eastern gray or Red kangaroo, in estrus for the first time and weighing as little as 15–20kg (33–44lb), may be courted and mated by a male five or six times her own weight. The large macropods exhibit some of the most exaggerated sexual size dimorphism known for terrestrial mammals, largely because the biggest male in the population gets the majority of the matings. In contrast, males and females of the smaller wallabies and rat kangaroos reach the same adult sizes.

With the exception of females accompanied by dependent young, most macropods are solitary or are found occasionally with one or two others. Potoroines shelter alone during the day in a self-made nest, which a female may share with her unweaned young-at-foot. At night, when she emerges to forage, the female may be found and escorted by a male. In the nights before estrus, several males may attempt to associate with her.

Burrowing bettongs nest in self-dug burrows that form loose colonies, but they too are not truly social. The solitary macropodines that do not use permanent refuges (mostly smaller species living in dense habitat) behave much like potoroines, but association between a female and her most recent offspring may last many weeks beyond weaning. On the day of estrus, a female may be escorted by a chain of ardent males.

Left The Musky rat kangaroo (Hypsiprymnodon moschatus) *is a taxonomic oddity. Unlike all other kangaroos, which belong to the Macropodidae family, it is the sole representative of the Hypsiprymnodontidae. Its most distinctive anatomical feature is the presence of a first toe, lost in all the other species.*

Below *Representative small- and medium-sized kangaroos and wallabies:* ***1*** *Proserpine rock wallaby* (Petrogale persephone)*;* ***2*** *Yellow-footed rock wallaby* (P. xanthopus)*;* ***3*** *Burrowing bettong or boodie* (Bettongia lesueur)*;* ***4*** *quokka* (Setonix brachyurus)*;* ***5*** *Banded hare wallaby* (Lagostrophus fasciatus)*;* ***6*** *Rufous rat kangaroo* (Aepyprymnus rufescens)*. Several of these species are now listed by the IUCN as being at risk: the Proserpine rock wallaby is considered Endangered, and the quokka, Burrowing bettong, and Banded hare wallaby are all Vulnerable.*

Rock wallabies shelter during the day in caves and boulder piles, features that are clustered in the landscape, with the result that colonies of the animals inhabit clusters of daytime refuges. Individuals persistently use the same refuges, and males compete to keep other males away from the refuges of one or more females. In some rock wallaby species, males may consort closely with one or more females during the day, although they will not always forage together. A male tree kangaroo may similarly guard access to the trees used by one or a few females with which he associates.

Some of the largest *Macropus* species form groups (often called "mobs") of 50 or more animals. Membership of these groups is extremely flexible, however, with individuals joining and leaving several times a day. Some sex- and age-classes tend to associate with their peers or with specific other classes. Individual females may also associate with their female kin or with particular unrelated females; this association is frequent and persistent but not permanent. However, the stage of development of a female's young determines her association patterns; females with young about to be excluded from the pouch avoid others at the same stage by retreating to a part of their range generally not much used by other kangaroos, in order to stop the young from becoming confused during this period of rejection.

Males in these species move between groups more frequently than do females, and also move over larger ranges. No males are territorial, nor do any attempt to keep others out of a group of females. Males range widely, inspecting as many females as possible by sniffing the cloaca and urine-tasting. If a male detects a female approaching estrus, he will attempt to consort with her, following her about and mating her when she enters estrus. However, he can be displaced by any larger and more dominant male.

In the medium and large macropods, hierarchical position, based largely on size and thus upon age in these persistently growing animals, is the principal factor in male reproductive success. In the Eastern gray kangaroo, a locally dominant male may obtain up to a half of all matings within his home range. He will usually be able to hold top rank for one year only, however, and may have waited 8–10 years to reach that position. Most males never mate, and very few reach the top of the hierarchy. But those that do may father 20–30 offspring, or even more in dense populations.

In contrast, all females are likely to give birth to about one young a year throughout their adult lives. Among Eastern gray kangaroos, the chances of the young surviving to adulthood are strongly

affected by the number of female relatives a mother has. Those with many female kin are much more likely to rear their own young, and especially their first, successfully through the young-at-foot period, the time of major mortality. In Eastern gray kangaroos, the survival rate is 35 percent for the offspring of mothers whose own mother was still alive at the time of the birth, but only 8 percent if the mother had died. The chances of the first two joeys – baby kangaroos – surviving without a "grandmother" or "aunt" to assist in parenting is just 12 percent, rising to 25 percent when there is a grandmother but no aunts and to 42 percent when there are also one or more aunts.

Extraordinarily, these benefits seem to have translated into a maternal strategy of giving birth to females early in life, while sons are born correspondingly late in a female's breeding career. In Eastern grays the ratio is about 1 daughter to 0.8 sons for the first two offspring, rising to 1 daughter to 1.3 sons in mid career and to 1 daughter to 2.9 sons in the final offspring.

The larger, social macropods all live in open country (grasslands, shrublands, or savanna), and were formerly preyed upon by cursorial and aerial predators such as dingos, Wedge-tailed eagles, and the now-extinct thylacine. Social grouping has conferred the same anti-predator benefits on large kangaroos as on so many other animals, in that dingoes are less able to get close to large groups, which can thus spend more time feeding. Group size of kangaroos relates to their density, the kind of habitat – especially its lateral cover – the time of day, and the weather.

Pests and Prey

CONSERVATION AND ENVIRONMENT

Between 2 and 4 million Red, Eastern, and Western gray kangaroos and Common wallaroos are shot every year in Australia because they are considered pests of pasture and crops. The cull is licensed, regulated, and for the most part humane. These large species were less numerous when Australia was first settled by Europeans, and from 1850–1900 several scientists feared that they might go extinct. Provision of pasture and of well-distributed water for sheep and cattle, together with (particularly) the reduction in numbers of dingoes, their main natural predators, as well as of hunting by Aboriginal peoples, all allowed the kangaroos to flourish.

Kangaroos used to be the main prey for Aboriginal spear-hunters. Smaller wallabies were flushed by fire or driven into nets or else toward lines of hunters armed with spears and throwing-sticks. In New Guinea, they were once pursued with bows and arrows, but are now killed with firearms instead. In some areas commercial hunting is rapidly depleting densities and endangering tree kangaroos and other restricted species.

Above *A kangaroo relaxes with her young in a refreshing pool of water. Large kangaroos mostly keep cool by resting in the heat of the day, coming out to feed at twilight or by night.*

In most of Australia outside the rain forest and wet sclerophyll forest, densities of macropod species with an adult weight of less than 5–6kg (11–13lb) have fallen in the past century or less. On the mainland, several such species are now extinct or very severely limited in range outside the tropics, although some have survived on offshore islands. The extinctions have been caused by a combination of habitat clearing for (or modification by) introduced livestock and, most especially, the impact of foxes. Introduced for sport in Victoria from 1860–80, foxes spread rapidly through sheep country, living primarily on rabbits, but taking as secondary prey bettongs and wallabies, which plummeted in numbers. On fox-free islands these species survived.

Where foxes have been suppressed, such species (if still present) have recovered their former densities. However, foxes, rabbits, and land-clearing are still widespread, and the battle to save small and medium-sized macropods in Australia is not yet won. PJ

A MUTUALLY BENEFICIAL RELATIONSHIP

How rat kangaroos help to cultivate the fungi they feed on

ON THE ROOTS OF MOST VASCULAR PLANTS LIVE fungi. The two symbiotically provide for each other; the plant supplies carbohydrate food, while the fungi enhance the uptake by the plant of nutrients (especially phosphorus) and of moisture from the soil. Associations of this sort between plants and fungi are known as mycorrhizal symbiosis, and they represent a very ancient liaison found in practically all terrestrial ecosystems.

Eucalyptuses typically have very high levels of mycorrhizal association, as Australian soils tend to be poor and deficient in phosphorus. So crucial is this relationship that foresters attempting to recreate natural forest must inoculate the trees with mycorrhizal fungi to ensure success. The fungi that associate with eucalypts are of two kinds. Some are mushrooms that produce fruiting bodies above ground and have spores dispersed by the wind; many others, however, fruit below ground. These hypogeous (subterranean) fungi are similar in many respects to European truffles; like them, they release delicious smells that attract mammals to dig them up and eat the fruiting bodies. The spores survive their passage through the mammals' digestive tracts, and germinate after being deposited on the ground in feces. They may, however, first be carried for quite long distances in the mammals' gut, so this is an effective means of dispersing spores throughout the forest.

For some species of fungus, passage through the digestive tract actually seems to stimulate the germination of spores. The "spore rain" produced by mammals is presumably important in maintaining the mycorrhizal association on mature root systems at a high level, as well as in rapidly establishing mycorrhizal fungi on the root systems of seedlings. Living on trees and dispersed by mammals, the fungi are a central link in the forest ecological web.

There are many willing takers for the proffered fungal delights. In Australia, the rat kangaroos (bettongs and potoroos) that live in eucalypt forests feed almost exclusively on these fungi. These animals will dig as deep as 20cm (8in) into the soil to find truffles, and in captivity they prefer freshly-collected truffles over any other food. There are other species of mammals in eucalypt forests – bandicoots and many rodents – that feed on hypogeous fungi, but the fact that rat kangaroos eat fungi year-round and harvest fruiting bodies in large numbers must make them especially efficient spore-dispersers.

In addition to promoting the reproduction of mycorrhizal fungi, rat kangaroos may also be responsible for protecting their diversity. It is common to find 20 or more different species of fungi living in stands of a single species of eucalypt, many of them on the root system of a single tree. Typically, a small number of these species are very abundant, while most are rare. Rat kangaroos are extremely adept at finding fungi of many different species; for example, a study of the Long-footed potoroo showed that it was eating more than 40 species of hypogeous fungi over an area of only a few hectares, far more than even very experienced human mycologists were able to collect. In the absence of spore dispersal, the fungi would spread by growth of their hyphae – the fungal equivalent of root systems – through the soil, and those with the most vigorous growth would probably eventually displace others from the community. The fact that rat kangaroos are so effective at finding uncommon fungi and broadcasting their spores may be an important factor in keeping them healthy.

The rat kangaroos' role does not end there. Into the already complex but precisely choreographed web of life in an Australian forest comes the element of fire. Eucalypt forests are generally very fire-prone; many eucalypt species have features that encourage its spread as well as adaptations (such as stimulation of seed fall by heat and of seed germination by smoke) that link reproduction to the conflagrations. But if fires kill adult trees, this must lead to the death of their fungal associates soon after, so a mechanism is needed to ensure spore dispersal before the fungi die.

Rat kangaroos are very skilled at surviving fire. One study using radio-tracking to follow Brush-tailed bettongs during a hot fire showed that the animals managed to move around the flames while still remaining within their home ranges. After a fire, the rate of digging for hypogeous fungi by rat kangaroos increases dramatically, as animals in the area increase their feeding on the fungi and others move in from neighboring, unburned areas to feed over the burned ground. The result is that after a fire more fungi get dispersed, both within burned patches and from unburned to burned patches, ensuring that spores are introduced to new seedlings growing in the ashes. Thus the continuity of life is ensured.

There is a beautiful postscript to the story that brings an insect into the equation. Several dung beetle species have evolved specialized relationships with particular species of rat kangaroos. The beetles lay their eggs in dung, which their developing young subsequently eat. Unlike most such beetles, which search for fresh dung on the wing, these species cling to the fur at the base of a rat kangaroo's tail, waiting until feces are passed, when they drop off and bury them. This strategy gives the beetles first access to the feces, and the species that practice it tend to be poor competitors for dung. But the association also aids spore dispersal, providing a mechanism by which fungal spores are immediately transported close to the roots of trees, and so facilitating their germination and the reestablishment of the symbiosis. CJ

Below *Tasmanian bettongs are among the species that eat fungi and disperse their spores. Others include the Northern and Brush-tailed bettongs and all three species of potoroos.*

LIFE IN THE POUCH

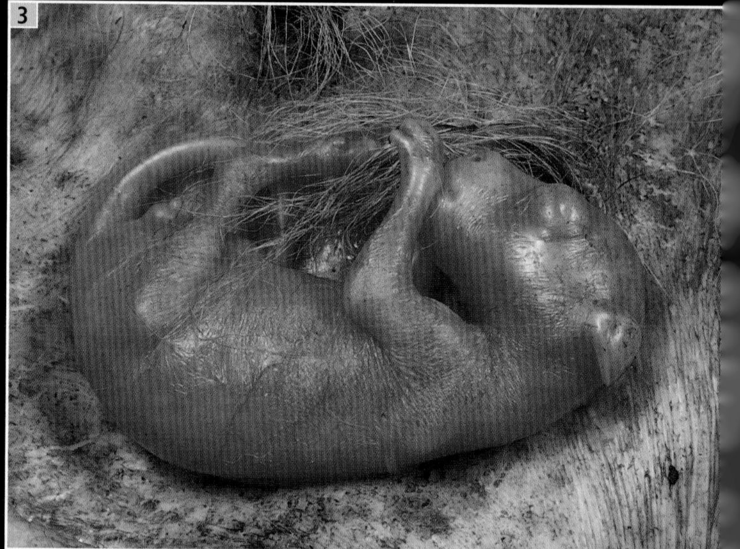

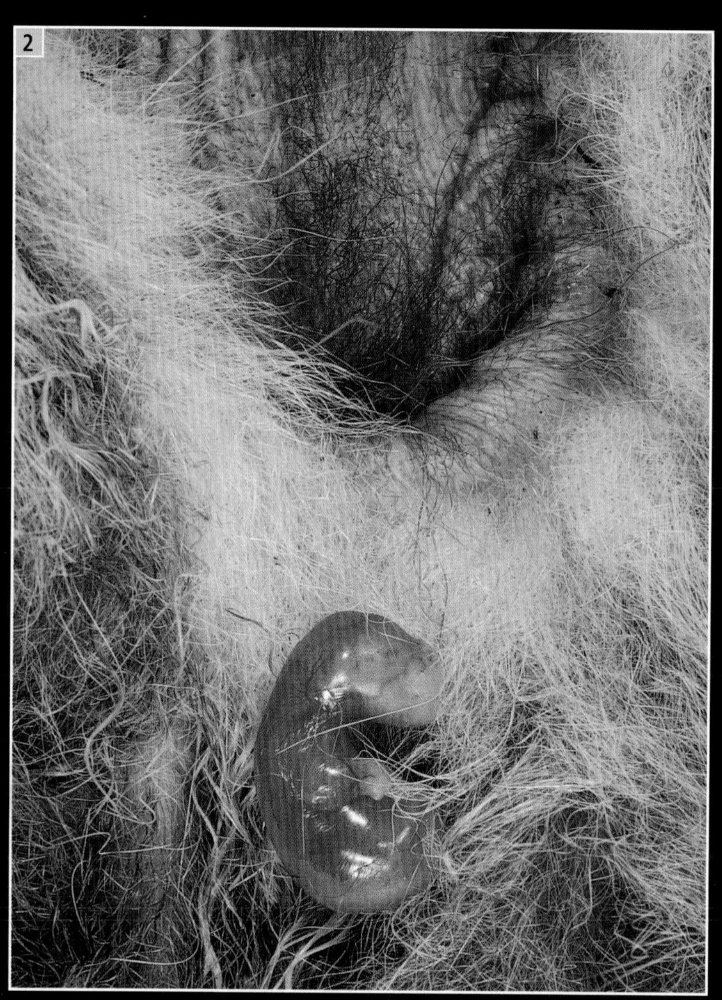

❶❷ *What characterizes all macropods is their highly undeveloped state at birth. In comparison with the newborn of placental mammals, those of marsupials are in an almost embryonic state, with rudimentary hind limbs and tail, ears and eyes closed, and no fur. Yet once the umbilical cord (seen left, below) breaks, the tiny infant is able to propel itself on its relatively strong fore limbs through its mother's fur to reach the safety of her pouch. The climb from birth canal to pouch will take about 2 minutes.*

❸ *The teat (there are four in the pouch) fills the infant's mouth and holds it securely in place; this kangaroo is 4 weeks old. Kangaroos an wallabies suckle for 6–11 months.*

❹ *By 12–14 weeks of age, a kanga roo has grown fast and acquired recognizable features. If a young kangaroo dies or is killed by predators, the mother does not need to mate again to conceive – a second, fertilized egg is immediately implanted in the uterus, and a new embryo begins development.*

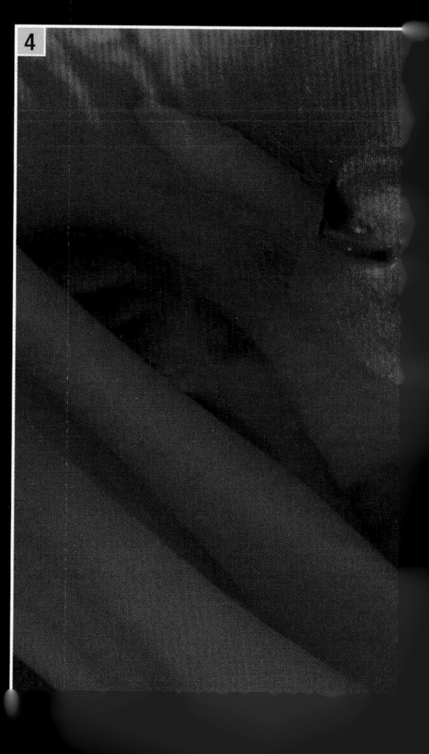

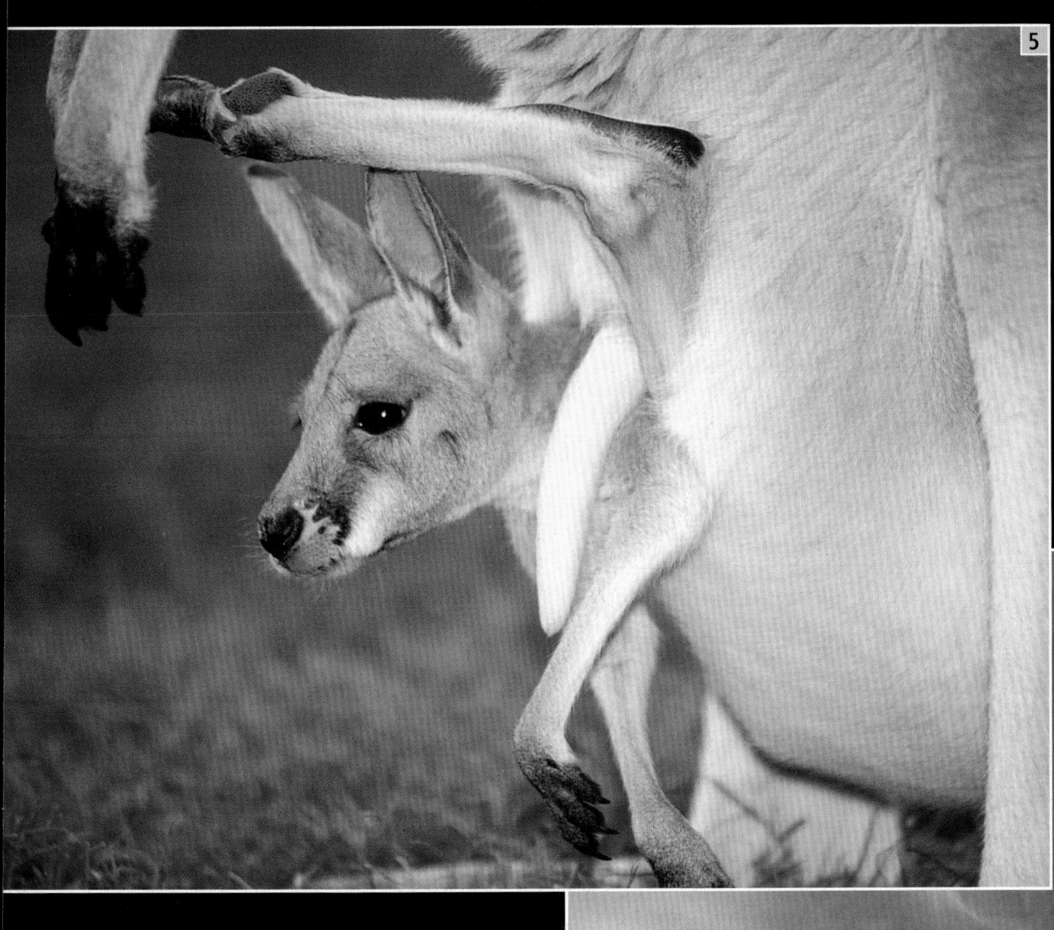

❻ *An Eastern gray kangaroo at the "young at foot" phase. Though it is now denied access to the pouch, it is still not weaned and will suckle for several months to come. To promote rapid growth, the walking young receives fatty milk, while the new infant inside the pouch is provided with fat-free milk. Nurturing two live young simultaneously, with an egg always ready for implantation, has proved an effective reproductive strategy for Australia's large marsupials.*

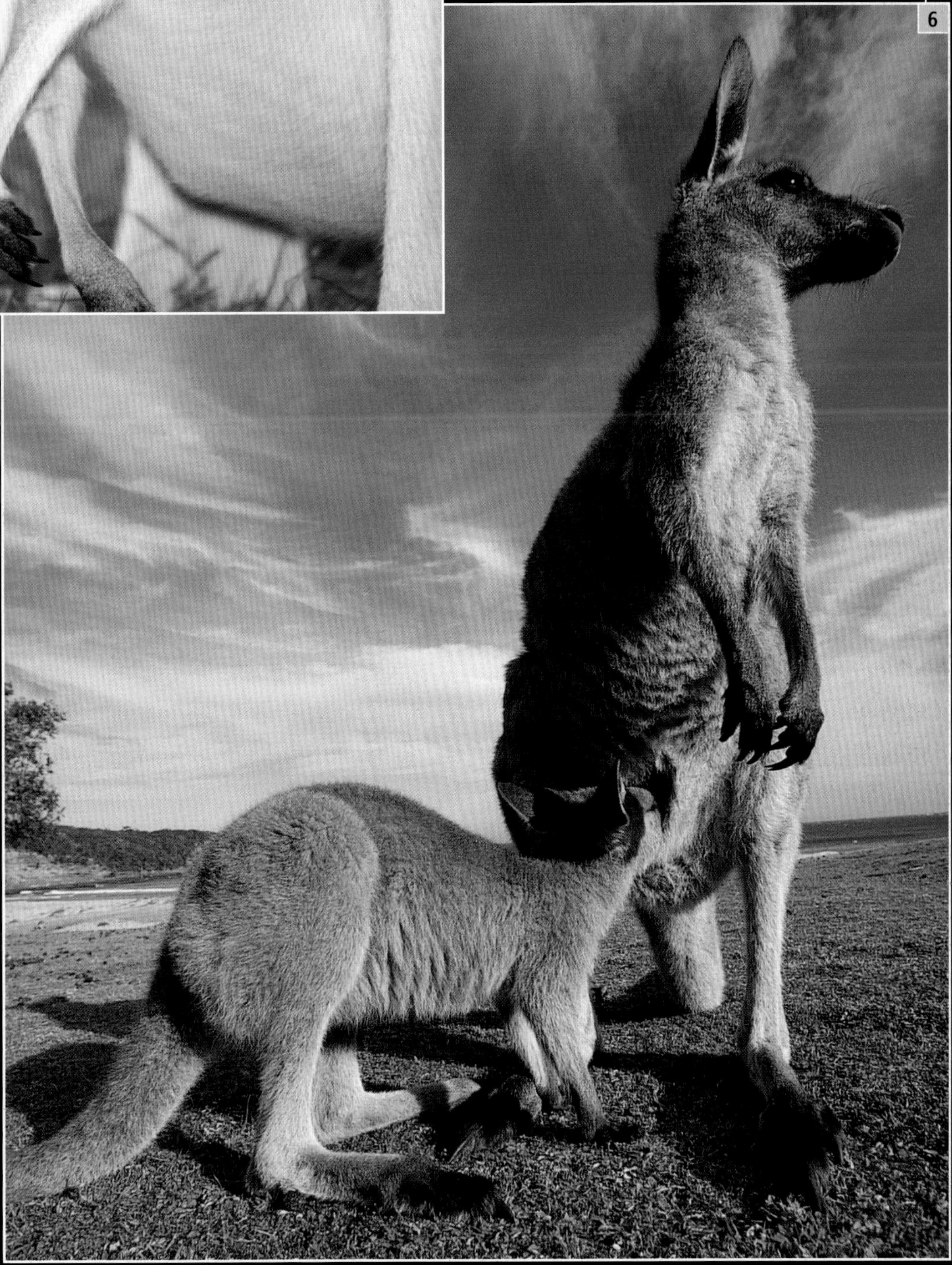

❺ *Almost too large for its accommodation, a well-developed red kangaroo peers from its mother's pouch. This individual is at the stage of semi-independence, where it makes excursions outside but comes back to the pouch to sleep and suckle. It will soon be excluded for good, to make way for a newborn sibling.*

Koala

THE KOALA IS NOW AUSTRALIA'S ANIMAL ICON and one of the world's most charismatic mammals, but this has not always been the case. The first European settlers considered koalas stupid and killed millions for their pelts. Even more serious threats to the animals' survival came from the impact of forest clearance, large-scale forest fires, and the introduction of zoonotic disease, particularly domestic animal strains of chlamydia.

The threat to koalas reached a peak in 1924, when more than 2 million skins were exported. By that time, the species had been exterminated in South Australia, and had largely disappeared from Victoria and New South Wales. As a result of public outcry, bans on hunting were introduced, and intensive management, particularly in the southern populations from 1944 on, has subsequently reversed this decline. Koalas are now once more relatively common in their favored habitat.

***Below** A koala rides on its mother's back. At birth the single young are minuscule, weighing less than 0.5g (0.02oz). After 5 months they start to feed on eucalypt leaves partly predigested by their mothers. At 7 months they leave the pouch, but continue to travel with their mothers for another 4 or 5 months.*

Large Bellies, Small Brains

FORM AND FUNCTION

Trees from the genus *Eucalyptus* are widespread in Australia, and koalas are wedded to them. They spend almost their entire lives in eucalypts. Much of the day is taken up in sleeping (which occupies more than 80 percent of their time); less than 10 percent is required for feeding, and the rest is mainly spent just sitting.

Koalas display numerous adaptations for this relatively inactive, arboreal lifestyle. As they use neither dens nor shelters, their tailless, bearlike bodies are well-insulated with a dense covering of fur. Their large paws are equipped with strongly recurved, needle-sharp claws on most digits, and these make the koala a most accomplished climber, able to ascend the largest smooth-barked eucalypts with ease. To climb, they grip onto the trunk with their claws and use their powerful forearms to heave upwards, while simultaneously bringing the hind limbs up in a bounding motion. The forcipate structure of the forepaws (the first and second digits are opposable to the other three) enables them to grip smaller branches and climb into the outer canopy. They are less agile on the ground, but move frequently between trees, using a slow, quadrupedal walk.

The teeth of the koala are adapted to cope with eucalypt leaves, which are extremely fibrous. Using their cheek teeth, which are reduced to a single premolar and four broad, high-cusped molars on each jaw, they chew the leaves into a very fine paste. This digesta then undergoes microbial fermentation in the cecum, which, at 1.8–2.5m (5.9–8.2ft), is the largest of any mammal in proportion to body size, stretching three times the koala's body length or more.

The small brain of the koala may also be an adaptation for a low-energy diet. Energetically, brains are expensive organs to run as they consume a disproportionate amount of the body's total energy budget. Relative to its body size, the koala's brain is one of the smallest found in marsupials. The brain of a southern koala of average size (9.6kg/21lb)) weighs only about 17g/0.6oz (0.2 percent of body weight).

Male koalas are 50 percent heavier than females, and have a broader face, comparatively smaller ears, and a large, odoriferous sternal (chest) gland. The principal secondary sex characteristic of the females is the pouch, which contains two teats and opens to the rear.

Koalas have a broadly polygynous mating system in which some males do most of the mating, but precise details on the distribution of matings

FACTFILE

KOALA

Phascolarctos cinereus

Order: Diprotodontia

Family: Phascolarctidae

Sole memb\er of genus

DISTRIBUTION Disjunct in E Australia S of latitude 17°

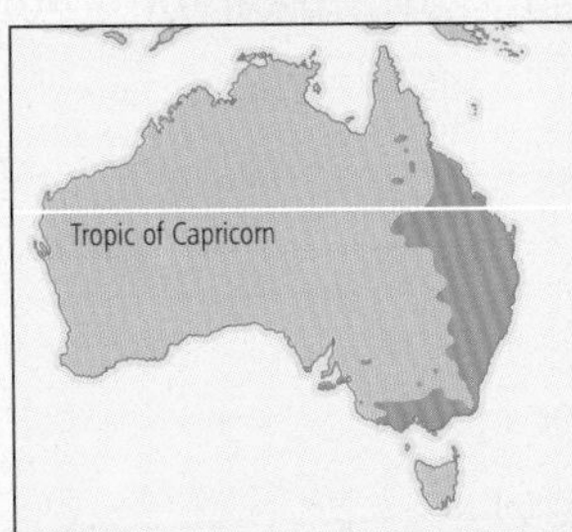

HABITAT Eucalypt forests and woodlands

SIZE **Head–body length** male 78cm (30.7in), female 72cm (28.3in); **weight** male 11.8kg (26lb), female 7.9kg (17.4lb). Animals from N of the range are significantly smaller, averaging only 6.5kg (14.3lb) for males and 5.1kg (11.2lb) for females.

COAT Gray to tawny; white on chin, chest, and inner side of forelimbs; ears fringed with long white hairs; rump dappled with white patches; coat shorter and lighter in N of range.

DIET Foliage, mainly from a limited range of eucalypt species, although leaves from some non-eucalypts including *Acacia*, *Leptospermum*, and *Melaleuca* are also browsed.

BREEDING Females sexually mature at 21–24 months; single young born in summer months (Nov–March) after gestation of about 35 days. Young become independent after 12 months. Females are capable of breeding in successive years.

LONGEVITY Up to 18 years

CONSERVATION STATUS Lower Risk: Near Threatened. Common where habitat is intact, particularly in the S of its range, rarer in the N. Large-scale clearing of woodlands is threatening the northern populations.

between dominant and subdominant animals have not been comprehensively researched and await elucidation. Female koalas are sexually mature and commence breeding at 2 years of age. Males are fertile at the same age, but their mating success is usually poor until they are older, at about 4–5 years, when they are large enough to compete successfully for females.

Above *Traveling between trees, a koala moves cautiously across the forest floor. The animals usually walk sedately, but can bound forward in emergencies.*

Following the Forest

DISTRIBUTION PATTERNS

The koala is often thought of as a fragile and rare species, but in reality it tolerates a wide range of environmental conditions. The eucalypt forests on which koalas depend are widespread but fragmented, and the distribution of the animals now reflects the state of the forest.

Koala populations are often widely separated from each other, usually by extensive tracts of cleared land. Even so, they still occur across several hundred thousands of square kilometers, stretching in a broad swathe across eastern Australia from the edge of the Atherton Tablelands in North Queensland to Cape Otway at the southernmost tip of Victoria.

Koalas occupy a surprisingly diverse range of habitats across this range. These encompass wet montane forests in the south, vine thickets in the tropical north, and woodlands in the semi-arid west of their range. Their abundance varies markedly with the productivity of the habitat. In fertile, high-rainfall country in the south, abundances as high as 8 animals per hectare (more than 3 to an acre) are not uncommon, while in the semi-arid zone 100 hectares (250 acres) may be required to support a single animal.

An Unpromising Food

DIET

As evergreen plants, eucalypts are a constantly available resource for leaf-eating animals. An adult koala eats about 500g (1.1lb) of fresh leaf daily, yet while there are more than 600 species of eucalypts to choose from, koalas feed from only 30 or so of these. Preferences differ between populations, with animals usually focusing on species growing in the wetter, more productive habitats. In the south *Eucalyptus viminalis* and *E. ovata* are preferred, while the northern populations feed predominantly on *E. camaldulensis*, *E. microcorys*, *E. propinqua*, *E. punctata*, and *E. tereticornis*.

Such a diet at first sight might seem unpromising. Eucalypt leaf is inedible, if not downright toxic, to most herbivores. It is low in essential nutrients, including nitrogen and phosphorus; it contains high concentrations of indigestible structural materials such as cellulose and lignin; and it is laced with poisonous phenolics and terpenes (essential oils). Recent research has shown that these last compounds may hold the key to koalas' preferences, as the acceptability of browse species has been found to correlate inversely with the concentration of certain highly toxic phenol–terpene hybrids.

The koala shows a number of adaptations that enable it to cope with such inauspicious food. Some leaves they obviously avoid altogether. Toxic components in others are detoxified in the liver and excreted. Coping with the low available energy provided by such a diet, however, requires behavioral adjustments, and koalas sleep a lot, for up to 20 hours a day. This has given rise to the popular myth that they are drugged by the eucalypt compounds they ingest. Koalas also exhibit very tight water economy and, except in the hottest weather, obtain all of their water requirements from the leaves.

Left *Firmly wedged in a fork in a tree, a koala indulges in the species' favorite pastime: sleeping. The animals spend as much as 80 percent of their time asleep, and a substantial proportion of their waking hours is passed in resting. Their inactive lifestyle is linked to a low-energy diet made up almost entirely of eucalypt leaves, which are short of essential nutrients, including nitrogen and phosphorus.*

Right *Foraging in the branches, a koala strikes an unlikely King Kong pose as it stretches for a eucalypt sprig. Eucalypts are poisonous to most herbivores, but koalas' livers have adapted to be able to cope with at least some of the toxins they contain. Evolution thereby opened up for the animals a relatively uncontested food resource available all the year round.*

Solitary and Sedentary

SOCIAL BEHAVIOR

Koalas are solitary animals. They are also sedentary, with adults occupying fixed home ranges. The size of these ranges is related to the productivity of the environment. In the more prolific forests of the south the ranges are comparatively small, with males occupying only 1.5–3ha (3.7–7.4 acres) and females 0.5–1ha (1.2–2.5 acres). In semi-arid areas, however, they are much larger, and males occupy 100ha (250 acres) or more. The home range of socially dominant males overlap the ranges of up to nine females, as well as those of subadult and subordinate males.

Koalas are principally nocturnal, and in the breeding season adult males move around a great deal in the summer nights. Fights usually occur if they meet up with other adult males and matings if they encounter a receptive (estrous) female. Copulation is brief, usually lasting less than 2 minutes, and occurs in a tree. The male mounts the female from behind, usually holding her between himself and the branch while mating.

Females give birth to a single young, with the majority of births occurring in midsummer (December–February). The newborn animal weighs less than 0.5g (0.02oz) and climbs unaided from the urogenital opening to the pouch, where it firmly attaches to one of the two teats there. Over the next 6 months of pouch life, the young grows and develops while suckling this same teat. Weaning commences after 5 months, and is initiated by the young feeding on partially digested leaf material produced from the female's anus. The mother is able to clear the normally hard fecal pellets from the lower bowel before producing this soft material, and may be stimulated into doing so by the young nuzzling the region. The high concentration of microorganisms in this pap is thought to inoculate the gut of the young with the microbes it needs to digest eucalypt leaf. Growth is rapid from this time onward. The young leaves the pouch after 7 months to travel around clinging to the mother's back. It becomes independent at around 11 months of age, but usually continues to live close to the mother for several months afterward.

Males bellow incessantly through the early months of the breeding season, and these calls, which consist of a series of harsh inhalations each followed by a resonant growling expiration, appear to serve both as an advertisement to potential mates as well as a warning threat to competing males. The call of one male usually elicits a response from all the adult males in the area. The only loud vocalization heard from females is a wailing distress call, usually given when they are being harassed by an adult male. At this time of the year males are often seen scentmarking by rubbing their sternal gland against the trunk of trees, but the precise role of this behavior has yet to be elucidated.

When Clusters Become Crowds

CONSERVATION AND ENVIRONMENT

So far there has been no official attempt to enumerate the total population of wild koalas, but unofficial estimates range from 40,000 to more than 1 million. Genetic studies indicate strong differentiation between northern and southern populations, and suggest that there might be a number of distinct subpopulations in the north. No such pattern has been detected among the southern populations, a fact that is thought to reflect the homogenizing effect of the extensive translocation program that has occurred there.

Habitat loss is threatening the viability of many koala populations, particularly in the northern part of their range. Urban and tourist developments are claiming important areas of habitat in coastal regions, but the situation is especially serious in the semi-arid woodlands of central Queensland, where around 400,000 hectares (1 million acres) are being cleared annually for pastoral and other agricultural purposes. While environmentalists are attempting to stop this clearance, it is a politically difficult issue in the conservative farming areas of central Queensland.

Management problems in the south make koala conservation there an even more complex issue. Historically the koala has been a rare species on the mainland, but overpopulation has been a persistent problem in a few koala colonies established on offshore islands in the late 19th century. This issue has been addressed via translocations, and over the last 75 years more than 10,000 koalas have been relocated back to the mainland.

While the translocation program has reestablished the koala throughout most of its southern range, it has also transferred the problems of overpopulation and habitat degradation to many forest remnants on the mainland. Culling to control overpopulation is widely unpopular, and contraceptive methods are now being tried as a method of capping the growth of these populations. RM

Wombats

THERE ARE ONLY THREE LIVING SPECIES OF wombats, but they represent, together with the koala, one of the two great lineages of marsupial herbivores. The suborder Vombatiformes – which comprises the koala and wombats – diverged from the other marsupial herbivores at least 40 million years ago, and since then wombats have evolved a way of life that is unique among mammals.

The wombats are large-bodied, burrowing herbivores. This combination of traits is extraordinary: the few other large mammals that burrow are either carnivores or specialized insectivores, while the numerous other burrowing herbivores are all small. Much of the interest in the biology of wombats comes from the fact that they defy the general rule which dictates that burrowing and large body size among herbivores are mutually exclusive. The time and effort required to dig a burrow are usually precluded by the need to feed constantly, in order to gain sufficient nutrition from low-energy grass or browse.

Equipment for Burrowing and Grazing

FORM AND FUNCTION

Wombats are stocky, with short tails and limbs. The pectoral girdle is heavy and strong, and the humerus is very broad relative to its length, making the shoulders and forelimbs exceptionally powerful. The forepaws are massive, and bear long, heavy claws. Wombats burrow by scratch-digging with the forepaws, throwing soil behind them with the hind feet and using their ample rumps to bulldoze it clear of the burrow entrance. The wombat skeleton has many detailed features that increase the power of the limbs. For example, the posterior angle of the scapula (shoulder blade) is extended to increase the lever arm for the *teres major* muscle. The male and female are similar in all species, but male Northern hairy-nosed wombats have shorter bodies than females, along with thicker necks and heavier shoulders. The significance of this is unknown – perhaps their thick necks equip males for head-to-head confrontations down burrows.

Wombats feed primarily on grasses that are high in fiber and (especially on sandy soils) high in abrasive silica. The skull, teeth, and digestive tract are specially adapted to this diet. The skull is massive, broad, and flattened, allowing the jaws to exert great compressive force that is used to grind their coarse food. This compensates for the lack of a deeply furrowed rasping surface on the teeth, and enables wombats to grind their food to a very small particle size (about half that achieved by kangaroos). The teeth grow throughout life to compensate for tooth wear. The stomach and small intestine are small and simple and there is almost no cecum, but the colon is expanded and elongated, forming about 80 percent of total gut volume. Microbial fermentation takes place primarily in the colon, contributing about one-third of the total energy assimilated by the animal. Food passes slowly through the gut: particles remain there for an average of 70 hours and solutes for about 50 hours, far longer than in other similarly-sized herbivores.

The Low-Maintenance Marsupial

SOCIAL BEHAVIOR

A common feature of all of the three species of wombats is their ability to maintain high population densities even in very unproductive habitats. Common wombats occur in alpine environments up to and above the snowline, as well as in sandy coastal environments where they may reach very high densities. Hairy-nosed wombats occur in dry habitats, where soil fertility may be too low to support grazing by domestic livestock.

Wombats thrive in such environments because they have extremely reduced energy requirements. The basal metabolic rates of wombats are very low – the Southern hairy-nosed wombat has a rate only 44 percent of that predicted for eutherian mammals of similar mass – and their maintenance energy requirements are the lowest known for marsupials. The combination of low maintenance needs and efficient digestion of high-fiber diets means that the daily food intake of wombats is also very low – only about half that of similarly-sized kangaroos. A wombat spends much less time foraging than might be expected for a herbivore of its size. Total feeding times of as little as two hours per day have been recorded for Northern hairy-nosed wombats in good seasons, and feeding ranges are only about 10 percent of those used by kangaroos in similar habitat. This extreme conservatism allows wombats to spend most of their time underground, and this behavior in turn contributes further to energy conservation by protecting them from unfavorable weather conditions. Although they do not enter torpor, wombats may spend several days at a stretch in their burrows, during which time their energy expenditure must be very low indeed.

Because wombats are nocturnal, secretive, and spend so much time underground, little is known of their behavior. Burrows may be 30m (98ft) or more long, and often have several entrances, side tunnels, and resting chambers. The Southern

Above *A Common wombat mother and her offspring use a fallen tree trunk as a table for a meal of foliage. After leaving the pouch at the age of 6 months, the young may continue to follow their mothers about for as much as a year.*

Above left *A Southern hairy-nosed wombat displays the furry muzzle that distinguishes the two* Lasiorhinus *species from the Common wombat.*

hairy-nosed wombat inhabits large warrens that remain in constant use for decades. Groups of up to ten animals may live there, but interactions are rare and they feed solitarily. The family structure of these groups is unknown.

Among Northern hairy-nosed wombats, burrows are arranged in loose clusters, with entrances about 10m (33ft) apart; different clusters are typically several hundred meters distant. Genetic relatedness is higher for animals in the same cluster than between neighboring clusters. There may be up to ten animals in a group, with equal numbers of males and females. This species displays an unusual pattern of dispersal, in which both young males and females remain in their home burrow cluster, but adult females may disperse to a different cluster after rearing an offspring. Dispersal distances of up to 3km (1.8mi) have been recorded. Common wombats appear to be more solitary than Hairy-nosed wombats.

Cause for Concern

CONSERVATION AND ENVIRONMENT

Over the past 200 years both the Common wombat and the Southern hairy-nosed wombat have suffered range reductions of 10–50 percent as a result of habitat clearance and competition with rabbits, but they remain secure across much of their original range. *Vombatus ursinus ursinus* has gone extinct from all the Bass Strait islands except Flinders Island. In parts of Victoria, Common wombats are considered pests because of the damage they do to rabbit-proof fences, and some local control is carried out.

The Northern hairy-nosed wombat is now one of the rarest mammals in the world. This species has only ever been confirmed in three localities, and it went extinct from two of these early last century as a result of competition with cattle and sheep, habitat change, and poisoning campaigns directed at rabbits. The last population was not finally protected until the 3,000-ha (7,400-acre) Epping Forest National Park was declared over its entire range in 1974 and cattle were excluded in 1980. At that time the population probably consisted of just 35 individuals. By 1995, capture–mark–recapture studies estimated a population of around 70 animals.

To date, the management of this population has focussed on protection of its habitat, especially the control of fire and exclusion of cattle, while minimizing direct interference with the animals. This appears to have allowed the species to begin its recovery, but its fate remains precariously balanced, and plans are afoot to begin captive breeding and to establish other wild populations. CJ

FACTFILE

WOMBATS

Order: Diprotodontia

Family: Vombatidae

3 species in 2 genera

DISTRIBUTION SE Australia

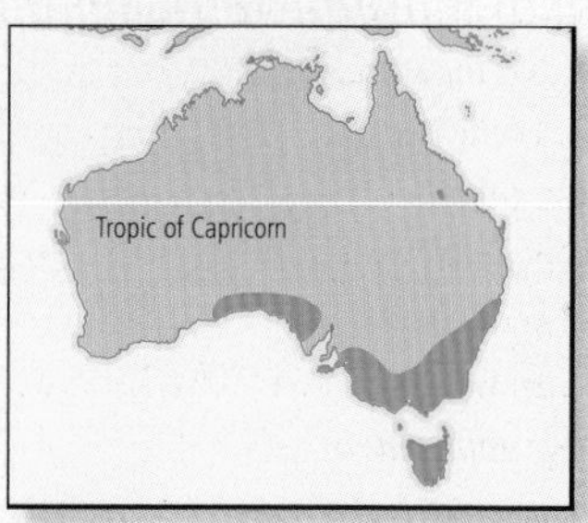

COMMON WOMBAT *Vombatus ursinus*

Three subspecies: *Vombatus ursinus ursinus* (Flinders Island); *V. u. tasmaniensis* (Tasmania); *V. u. hirsutus* (mainland Australia). Temperate forests and woodlands, heaths, and alpine habitats throughout SE Australia, including Flinders Island and Tasmania. HBL 90–115cm (35–45in); TL c.2.5cm (1 in); HT c.36cm (14in); WT 22–39kg (48.5–86lb). Form: coat coarse, black or brown to gray; bare muzzle; short, rounded ears. Diet: primarily grasses, but also sedges, rushes, and the roots of shrubs and trees. Breeding: one offspring, may be born at any time of the year. Pouch life is about 6 months, and the young remains at heel for about another year; sexual maturity is at 2 years of age. Longevity: unknown in the wild, up to 26 years in captivity. Conservation status: Vulnerable.

SOUTHERN HAIRY-NOSED WOMBAT
Lasiorhinus latifrons

Central southern Australia; semi-arid and arid woodlands, grasslands, and shrub steppes. HBL 77–94cm (30–37in); TL c.2.5cm (1 in); HT c.36cm (14in); WT 19–32kg (42–70lb). Form: coat fine, gray to brown, with lighter patches; hairy muzzle; long, pointed ears. Diet: grasses, including forbs and foliage of woody shrubs during drought. Breeding: single young, born in spring or early summer, remains in the pouch for 6–9 months; weaning occurs at approximately 1 year, and sexual maturity at 3 years. Longevity: unknown in the wild, more than 20 years in captivity.

NORTHERN HAIRY-NOSED WOMBAT
Lasiorhinus krefftii

Sole population in Epping Forest National Park, near Clermont in central Queensland; semi-arid woodland. HBL male 102cm (40in), female 107cm (42in); HT c.40cm (16in); WT male 30kg (66lb), female 32.5kg (72lb). Coat: silver gray, dark rings around the eyes. Diet: grasses, plus some sedges and forbs. Breeding: one young born in spring or summer; pouch life c.10 months; weaning age unknown; females breed on average twice every 3 years. Longevity: unknown in the wild; one captive animal lived at least 30 years. Conservation status: Critically Endangered.

Abbreviations HBL = head–body length TL = tail length HT = height WT = weight

Honey Possum

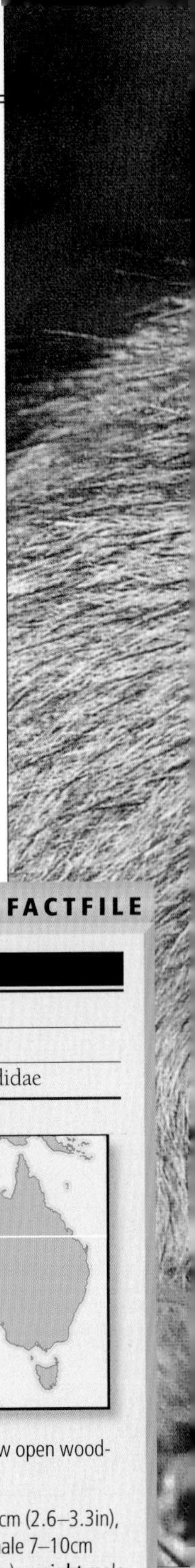

Left Restricted to the southwestern corner of Australia, the Honey possum has a tail that is longer than its head and body combined, and a tongue that can be extended 2.5cm (1in) beyond its nose. Bristles on the tongue's tip brush pollen from flowers.

Right Nectar and pollen make up the entire diet of the Honey possum, shown here feeding on Banksia. The pointed snout is a specialization for this food, used to probe deep into individual flowers in search of nectar. With its grasping hands, feet, and tail, the animal can feed even on small terminal flowers on all but the most slender branches.

THE HONEY POSSUM IS THE SOLE MEMBER of a line of marsupials that diverged very early from possum–kangaroo stock. Although its entire fossil history is contained within the last 35,000 years, it probably evolved about 20 million years ago, when heathlands were widespread.

Today, Honey possums are most abundant in heathlands on coastal sandplains in southwestern Australia, one of 25 biodiversity hotspots worldwide, where about 2,000 species of plants ensure a yearlong supply of the flowers upon which the animals depend for their food.

Foragers among the Flowers

FORM AND FUNCTION

The Honey possum's teeth are reduced in number and size, with a dental formula of I2/1, C1/0, P1/0, M3/3 = 22, but the molars are merely tiny cones. They use a long, protrusive tongue with a brush surface to lick nectar from flowers. Combs on the roof of the mouth remove pollen grains from the brush tongue. The contents of these pollen grains are digested during a rapid, six-hour transit through a simple intestine to provide the Honey possum's sole source of nutrients; nectar (20 percent sugar solution) provides only energy and water. Unusual kidneys allow the animals to excrete up to their own weight in water daily. They feed mainly on *Banksias* and dryandras, plants with large blossoms containing from 250 to 2,500 flowers. The food plants they favor have drably-colored flowers that are concealed either inside bushes or else close to the ground.

These tiny, shrewlike mammals use their long, pointed snouts to probe the flowers. The first digit of the hind foot is opposable to the others for gripping branches, and all digits have rough pads on the tips, not claws. Honey possums run fast on the ground and clamber with great agility over dense heathland vegetation. A long, partially prehensile tail provides balance and support for climbing, and frees the grasping hands to grip branches and manipulate flowers while feeding.

Nonstop Motherhood

SOCIAL BEHAVIOR

Honey possums communicate through a small repertoire of visual postures and high-pitched squeaks, a reflection of their mainly nocturnal activity. Smell appears very important in their social behavior and also helps them locate the flowers of their food plants.

The short lifespan of Honey possums is balanced by their continuous reproduction. A female carries young in her pouch for almost all of her adult life. Both sexes mature at around the age of 6 months. Many females breed for the first time while not yet fully grown, just 3 or 4 months after leaving the pouch in which they have spent the first 2 months of their lives. Births occur throughout the year, but reach very low levels when food is scarce. Population sizes are larger, body condition better, and births more common when nectar is most abundant. The timing of these cycles varies seasonally and between years in relation to rainfall, as well as geographically with differences in plant assemblages. Females appear to breed opportunistically whenever food is abundant, irrespective of the later consequences for their young.

A second litter is often born very soon after the first leaves the pouch or is weaned, since the Honey possum exhibits embryonic diapause (the temporary cessation of development in an embryo). In good times, some females can give

FACTFILE

HONEY POSSUM

Tarsipes rostratus

Order: Diprotodontia

Sole member of the family Tarsipedidae

DISTRIBUTION SW Australia

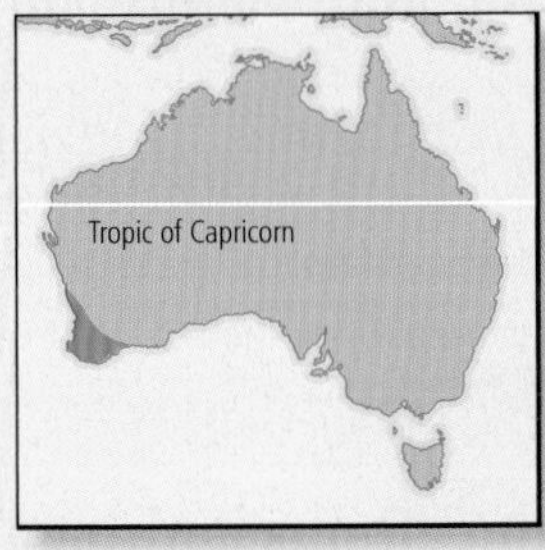

HABITAT Heathland, shrubland, and low open woodland with heath understory.

SIZE **Head–body length** male 6.5–8.5cm (2.6–3.3in), female 7–9cm (2.8–3.5in); **tail length** male 7–10cm (2.8–3.9in), female 7.5–10.5cm (3–4.1in); **weight** male 7–11g (0.3oz), female 8–16g (0.3–0.6oz), averaging 12g (0.4oz). Females are one-third heavier than males, and are larger in the southern part of their distribution than in the northern.

COAT Grizzled grayish-brown above, with orange tinge on flanks and shoulder, next to cream undersurface. Three back stripes: a distinct dark brown stripe from the back of the head to the base of the tail, with a less distinct, lighter brown stripe on each side.

DIET Nectar and pollen

BREEDING Gestation uncertain, but about 28 days

LONGEVITY Typically 1 year, never more than 2

CONSERVATION STATUS Not listed by the IUCN, but dependent on a continued supply of *Eucalyptus*, *Banksia*, and *Callistemon* blossoms.

birth to four litters in a year. At birth, the young are the smallest mammals known, weighing only about 0.0005g (0.00002oz). Their subsequent development is typical of young marsupials. The deep pouch has four teats and, although litters of four occur, two or three are normal. No nest is constructed; instead, mothers carry their young in the pouch as they forage. The small litter size and slow growth of the young, which spend about 60 days in the pouch, indicate the difficulties mothers experience in harvesting enough pollen grains to supply the sucklings with milk.

The young leave the pouch weighing about 2.5g (0.09oz), covered in fur and with eyes open. They follow their mother around as she forages, suckling occasionally, and may even ride on her back. At this time a litter of four young weigh as much as their mother. They disperse to live independently within a week or two of leaving the pouch. Honey possums, especially juveniles, sometimes huddle together to save energy. Their unusually high body temperature and metabolic rate is offset by short-term periods of deep torpor in cold weather when food is short. At such times the body temperature may remain as low as 5°C (41°F) for 10 hours before reviving spontaneously.

Both sexes are solitary and sedentary, living in overlapping home ranges that average 700sq meters (0.17 acres) for females and 1,280sq meters (0.32 acres) for males. In captivity, females are dominant to males as well as to juveniles. In the wild, females that have large young appear to monopolize areas rich in food.

The larger home ranges of males reflect not just their wider search for food, but also their quest for females approaching estrus. Honey possum testes weigh 4.2 percent of the animal's total body weight, the largest proportion for any mammal. Their sperm is also the longest known among mammals (0.36mm). These features imply intense competition between males to father offspring. Courtship is brief; males follow a female nearing estrus, but are only able to mount her when the larger female allows this. DNA microsatellite profiling has shown that two or more males are responsible for fathering each litter.

The Honey possum is still locally abundant in some areas, although its already restricted distribution continues to shrink. Clearance of habitat for agriculture has largely ceased, but within reserves plant diseases and introduced predators (cats and foxes) still pose threats. In addition, management burning can kill the food plants on which the Honey possum depends. RDW/EMR

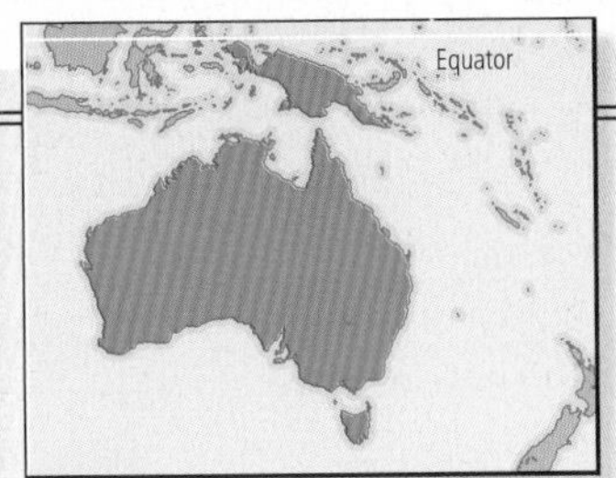

MONOTREMES

WHILE THE DESCRIPTION OF MONOTREMES as "the egg-laying mammals" distinguishes them from other living animals, it exaggerates the significance of egg-laying in this group. The overall pattern of reproduction is mammalian, with only a brief, vestigial period of development of the young within the egg. The soft-shelled eggs hatch after about 10 days, whereupon the young remain (in a pouch in echidnas) dependent on the mother's milk for up to six months in echidnas.

The platypus is confined to eastern Australia and Tasmania, the Long-beaked echidna occurs only in New Guinea, while the Short-beaked echidna is found in all of these regions, in almost all habitats. However, these distributions are relatively recent. There are Pleistocene fossils of Long-beaked echidnas at numerous sites in mainland Australia and Tasmania. Fossil monotremes from the Pleistocene epoch (which began 1.8 million years ago) are much the same as the living types. A platypus fossil from the mid-Miocene (10 million years ago) has been found, which appears very similar to living platypus except that it had fully developed, functional teeth as an adult. In 1991 paleontologists found isolated teeth in early Paleocene (60 million years ago) beds in Patagonia which are almost a perfect match for those of the mid-Miocene platypus. This, however, is not the oldest monotreme fossil; that distinction goes to a partially opalized jaw from the early Cretaceous (120 million years ago). These findings show that monotremes have undergone very conservative evolution, with little change over 100 million years. However, the fossils do not provide any evidence of the origins of monotremes and their ancestral relationships and how they relate to marsupial and placental mammals remains an enigma.

Strange Specialists

PHYSIOLOGY AND DISTRIBUTION

The term "egg-laying mammal" has long been synonymous with "reptile-like" or "primitive mammal," despite the fact that monotremes possess all the major mammalian features: a well-developed fur coat, mammary glands, a single bone in the lower jaw, and three bones (incus, stapes, and malleus) in the middle ear. Monotremes are also endothermic: their body temperature, although variable in echidnas, remains constant regardless of environmental temperatures.

Monotremes have separate uteri entering a common urino-genital passage joined to a cloaca, into which the gut and excretory systems also enter. The one common opening to the outside of the body gave the name to the group that includes the platypus and echidnas – the order Monotremata ("one-holed creatures").

Monotremes are highly specialized feeders. The semiaquatic, carnivorous platypus feeds on invertebrates living on the bottom of freshwater streams. Echidnas are terrestrial carnivores, specializing in ants and termites (Short-beaked or Common echidna) or noncolonial insects and earthworms (Long-beaked echidna). Such diets require grinding rather than cutting or tearing, and, as adults, monotremes lack teeth. In the platypus, teeth actually start to develop and may even serve as grinding surfaces in the very young, but the teeth never fully develop. Rather, they regress and are replaced by horny grinding plates at the back of the jaws. Reduction of teeth is common among ant-eating mammals, and echidnas never develop teeth, nor are their grinding surfaces part of the jaw. In the Long-beaked echidna, a pad of horny spines is located in a groove on the back of the tongue, running from the tip about one-third of the way back. Earthworms are hooked by these spines when the long tongue is extended. The food is broken up as the tongue spines grind against similar spines on the palate. The Long-beaked echidna takes worms into its

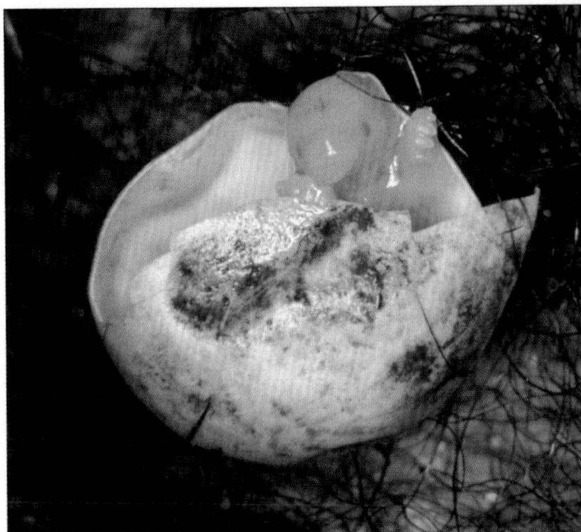

long snout by either the head or tail, and if necessary the forepaws are used to hold the worm while the beak is positioned. In the platypus. the elongation of the front of the skull and the lower jaw to form a bill-like structure is also a foraging specialization. The bill is covered with shiny black skin. Echidnas have a snout that is based on exactly the same modifications of the skull and jaws but is relatively smaller and cylindrical. The mouth is at the tip of this snout and can only be opened enough to allow passage of the cylindrical tongue. MLA

THE POISONOUS SPUR

Monotremes are one of only two groups of venomous mammals (the other includes certain shrews). In the two species of echidnas, the structures that produce and deliver the venom are present but not functional. It is only the male platypus that actually secretes and can deliver the venom. The venom-producing gland is located behind the knee and is connected by a duct to a horny spur on the back of the ankle. This spur is completely lost in the female platypus, but in the male it is hollow and full of venom, which is injected with forceful jabs of the hind limbs. The poison causes agonizing pain in humans and can kill a dog. Because the venom gland enlarges at the beginning of the breeding season, it has been assumed to have a connection with mating behavior. The marked increase in aggressive use of the spurs observed between males in the breeding season may serve to decide spatial relationships in the limited river habitat. Yet this does not explain why, in echidnas, the system is present but non-functional. The spur in male echidnas makes it possible to distinguish them from females, which is otherwise difficult in monotremes since the testes never descend from the abdomen. However, the echidnas' venom duct and gland are degenerate, and the male cannot erect the spur. If it is pushed from under its protective sheath of skin, few echidnas can even retract the spur. It may be that the venom system in monotremes originated as a defense against a predator that has long since gone extinct. Today adult monotremes have few, if any, predators. Dingos occasionally prey on echidnas, but dingos are themselves a relatively recent arrival in Australia.

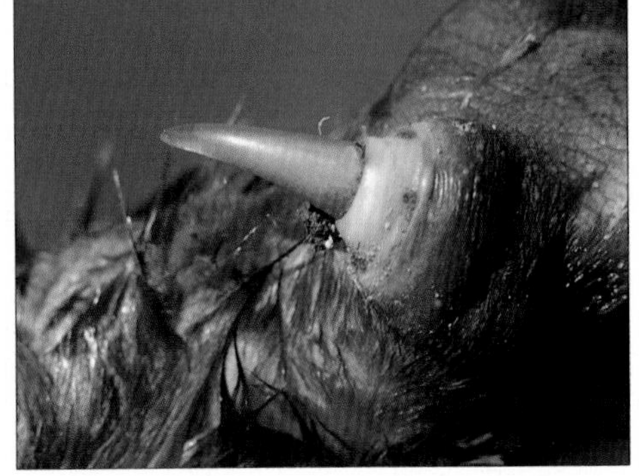

Left *The platypus's spur, which is curved and hollow, is connected by ducts to the venom glands.*

ORDER: MONOTREMATA
2 families; 3 genera; 3 species

Left *A Short-beaked echidna hatches from its soft-shelled egg. Following hatching the tiny newborn will increase in weight at an astonishing rate, perhaps 100–200 times in just a few weeks*

Below *The mouth of the echidna is positioned at the end of its snout, from which a long tongue is extended to catch termites and ants. The Short-beaked echidna is the most widely dispersed monotreme.*

Platypus

EVER SINCE THE FIRST PLATYPUS SPECIMEN (a dried skin) was sent to Britain from the Australian colonies in around 1798, this animal has been surrounded by controversy. At first it was thought to be a fake, stitched together by a taxidermist from the beak of a duck and body parts of a mammal!

Even when the specimen was found to be real, the species was not accepted as a mammal. Although it had fur, it also had a reproductive tract similar to that of birds and reptiles. This led researchers to conclude (correctly) that the platypus laid eggs, and (incorrectly) that it could therefore not be a mammal; all mammals known at the time were viviparous (i.e. gave birth to live young). Ultimately, however, the platypus was recognized as a mammal when it was found to possess the essential characteristic from which the class Mammalia takes its name – mammary glands.

The Sleek Swimmer

FORM AND FUNCTION

At just under 1.7kg (3.7lb), the platypus is smaller than most people imagine. Females are smaller than males and the young are about 85 percent of adult size when they first become independent. The animal is streamlined, with a covering of dense, waterproof fur over its entire body except the feet and the bill. The bill looks superficially like that of a duck, with the nostrils on top, set immediately behind the tip, but it is soft and pliable. Its surface is covered with an array of sensory receptors, which have been found to respond to both electrical and tactile stimuli, and is used by the animal to locate food and find its way around underwater. The eyes, ears, and nostrils are closed when diving. Behind the bill are two internal cheek pouches opening from the mouth. These contain horny ridges that functionally replace the teeth lost by the young soon after they emerge from the burrows. The pouches are used to store food while it is being chewed and sorted.

The limbs are very short and held close to the body. The hind feet are only partially webbed, being used in water only as rudders while the fore feet have large webs and are the main mode of propulsion. The webs of the front feet are turned back to expose large, broad nails when the animal is walking or burrowing. The rear ankles of the males bear a horny spur that is hollow and connected by a duct to a venom gland in the thigh. The venom causes extreme pain in humans; at least one component of the venom has recently been found to act directly on pain receptors, while other components produce inflammation and swelling. The tail is broad and flat and is employed as a fat-storage area.

Below *The prominent bill of the platypus is pliable and touch-sensitive. Underwater, it is the animal's main sensory organ for navigation and locating food.*

FACTFILE

PLATYPUS

Ornithorhynchus anatinus

Duckbill

Order: Monotremata

Family: Ornithorhynchidae

Sole species of the genus

DISTRIBUTION E Australia from Cooktown in Queensland to Tasmania. Introduced in Kangaroo Island, S Australia.

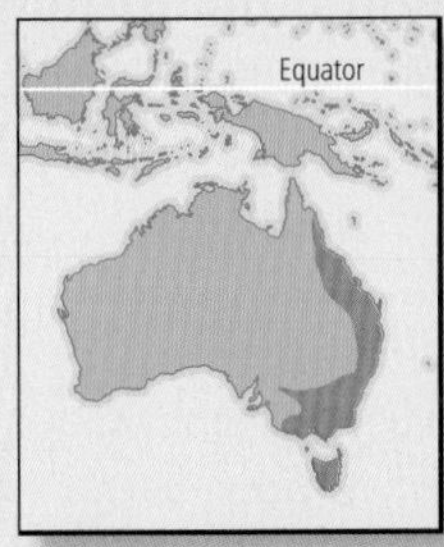

HABITAT Inhabits most streams, rivers, and some lakes that have permanent water and banks suitable for burrows.

SIZE Lengths and weights vary from area to area, and weights change with season. **Head–body length** male 45–60cm (17.7–23.6in) female 39–55cm (15.4–21.7in); **bill length** average male 5.8cm (2.3in) female 5.2cm (2in); **tail length** male 10.5–15.2cm (4.1–6in) female 8.5–13cm (3.3–5.1in); **weight** male 1–2.4kg (2.2–5.3lb) female 0.7–1.6kg (1.5–3.5lb).

COAT Dark brown back, silver to light brown underside with rust-brown midline, especially in young animals, which have the lightest fur. Short, dense fur (about 1cm/0.4in depth). Light patch below eye/ear groove.

BREEDING Gestation period not known (probably 2–3 weeks). Incubation not known (probably about 10 days).

LONGEVITY 10 or more years (17 or more in captivity).

CONSERVATION STATUS Not under threat.

Effective Care of the Young

SOCIAL BEHAVIOR

The platypus is mainly nocturnal in its foraging for prey items, which are almost entirely made up of bottom-dwelling invertebrates, particularly the young stages (larvae) of insects. Normal home ranges of platypuses vary with river systems, ranging from less than 1km (0.6 miles) to over 7km (4.3 miles), with many individuals foraging over 3–4km (1.8–2.5 miles) of stream within a 24-hour period. Two non-native species of trout feed on the same sort of food and are possible competitors of the platypus. Despite this dietary overlap, the platypus is common in many rivers into which these species have been introduced. A study in

one river system showed that the trout ate more of the swimming species of invertebrates, while the platypus fed almost exclusively on those inhabiting the bottom of the river. Waterfowl may also overlap in their diets with platypuses, but most also consume plant material which does not appear to be eaten by the platypus.

Certain areas occupied by platypuses experience water temperatures close to, and air temperatures well below, freezing in winter. When the platypus is exposed to such cold conditions it can increase its metabolic rate to produce sufficient heat to maintain its body temperature around its normal level of 32°C (89.6°F). Good fur and tissue insulation, including well-developed countercurrent blood flow, help the animal to conserve body heat, and its burrows also provide a microclimate that moderates the extremes of outside temperature in both winter and summer.

Although mating is reputed to occur earlier in northern Australia than in the south, it occurs sometime during late winter to spring (between July and October). Mating takes place in water and involves chasing and grasping of the tail of the female by the male. Two (occasionally one or three) eggs measuring 1.7 by 1.5cm (0.7 by 0.5in) are laid. When hatched the young are fed on milk, which they suck from the fur of the mother around the ventral openings of the mammary glands (there is no pouch) for 3–4 months while they are confined to a special breeding burrow. This burrow is normally longer and more complex than the burrows inhabited for resting. Such nesting burrows are reported to be up to 30m (100ft) long and be branched with one or more nesting chambers. The young emerge from these burrows in summer (late January–early March). It is not known how long they continue to take milk from their mothers after leaving the burrow, although they do feed on benthic organisms from the time they enter the water. Individual animals will use a number of resting burrows in an area but it is thought that there is attachment by breeding females to nesting burrows.

Although normally two eggs are laid, it is not known how many young are successfully weaned each year. Not all females breed each year, and new recruits to the population do not breed until they are at least two years of age. In spite of this low reproductive rate, the platypus has returned from near extinction in certain areas since its protection and the cessation of hunting around 1900. This indicates that the reproductive strategy of having only a few young, but looking after them well, is effective in this long-lived species.

The platypus owes its success to its occupation of an ecological niche which has been a perennial one, even in the driest continent in the world. By the same token, because the platypus is such a highly specialized mammal it is extremely susceptible to the effects of changes in its habitat. Changes wrought by humans in Australia, particularly since European settlement began in the late 18th century, have brought about localized reductions and fragmentation of platypus populations. Care and consideration for the environment will have to be rigorously maintained if this unique species is to survive. TRG

Above *The streamlined surface presented by the long guard hairs conceals the thick, dry underfur that insulates the platypus's body in cold water.*

Echidnas

ECHIDNAS ARE READILY RECOGNIZED BY their covering of long spines, which are shorter in the Long-beaked species. There is fur present between the spines as well as on the head, legs and ventral surfaces where there is an absence of spines. In both the Long-beaked echidna and the Tasmanian form of the Short-beaked echidna the fur may be longer than the spines.

Both genera are further distinguished by their elongated, tubular snout, which in the case of the Long-beaked echidna curves somewhat downward and accounts for two-thirds the length of the head. Echidnas are generally solitary creatures that are seen relatively infrequently, despite the fact that the Short-beaked echidna is in fact quite common across its geographic range; the Long-Beaked echidna, however, is only found in the mountainous areas of New Guinea.

Spiny Anteaters

FORM AND FUNCTION

The echidna's coat of spines (it is sometimes called the "spiny anteater") provides an excellent defense. If surprised on hard ground, an echidna curls up into a ball; on soft soil it may rapidly dig straight down, rather like a sinking ship, until all that can be seen of it are the spines of its well-protected back. By using its powerful limbs and erecting all its spines, an echidna can wedge itself securely in a rock crevice or hollow log.

Echidna spines are individual hairs that are anchored in a thick layer of muscle (*panniculus carnosus*) in the skin. The spines obscure the short, blunt tail and the rather large ear openings, which are vertical slits just behind the eyes. The snout is naked and the small mouth and relatively large nostrils are located at the tip. Echidnas walk with a distinctive rolling gait, although the body is held well above the ground.

Males can be distinguished from females by the presence of a horny spur on the ankle of the hind limb. Males are larger than females within a given population. Yearling Short-beaked echidnas usually weigh less than 1kg (2.2lb), but beyond that there is no way of determining age.

Echidnas have small, bulging eyes. Although they appear to be competent at telling objects apart in laboratory studies, in most natural habitats vision is probably not important in detecting food or danger; their hearing, however is very good. In locating prey, usually by rooting through the forest litter or undergrowth, they use their well-developed sense of smell. When food items are detected, they are rapidly taken in by the long, thin, highly flexible tongue, which Short-beaked echidnas can extend up to 18cm (7in) from the tip of the snout. The tongue is lubricated by a sticky secretion produced by the very large salivary glands. Ants and termites form the bulk of

Below *Echidnas' sense of smell is particularly acute. They use their long snout to probe the undergrowth or leaf litter, and have even been seen to use it as a snorkel when crossing water.*

FACTFILE

ECHIDNAS

Order: Monotremata

Family: Tachyglossidae

2 species in 2 genera

DISTRIBUTION Mainland Australia, Tasmania, New Guinea (Short-beaked Echidna); mountains of New Guinea (Long-beaked Echidna).

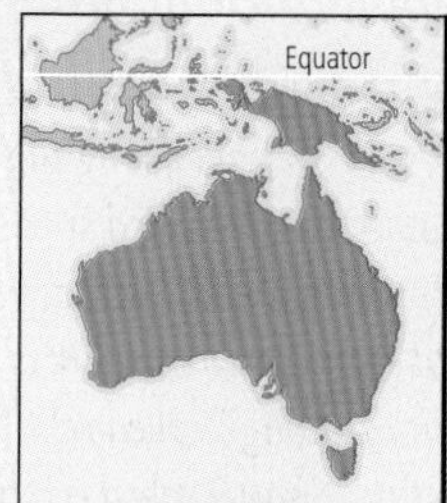

SHORT-BEAKED ECHIDNA *Tachyglossus aculeatus*
Short-beaked or Common echidna (or spiny anteater)
Australia, Tasmania, New Guinea. In almost all types of habitat, from semi-arid to alpine. HBL 30–45cm (12–18in); WT 2.5–8kg (5.5–17.6lb). Males 25 percent larger than females. Coat: black to light brown, with spines on back and sides; long narrow snout without hair. Breeding: gestation about 14 days. Longevity: not known in wild (extremely long-lived in captivity – up to 49 years). Conservation status: generally not threatened, though the Kangaroo Island subspecies (*T. a. multiaculeatus*) is classed as Lower Risk: Near Threatened.

LONG-BEAKED ECHIDNA *Zaglossus bruijni*
Long-beaked or Long-nosed echidna (or spiny anteater)
New Guinea, in mountainous terrain. HBL 45–90cm (18–35in); WT 5–10kg (11–22lb). Coat: brown or black; spines present but usually hidden by fur except on sides; spines shorter and fewer than Short-beaked echidna; very long snout, curved downward. Breeding: gestation period unknown. Longevity: not known in wild (up to 30 years in captivity). Conservation status: Conservation status: Endangered.

Abbreviations HBL = head–body length; WT = weight

Above *The Long-beaked echidna feeds principally on earthworms. Echidnas can survive several weeks with no nourishment at all.*

the Short-beaked echidna's diet. Around August and September Short-beaked echidnas attack the mounds of the Meat ant (*Iridomyrmex detectus*) to feed on the fat-laden females; this is done in the face of spirited defense by the stinging worker ants, although the mounds are prudently avoided for the rest of the year.

Males in Tow

SOCIAL BEHAVIOR

Short-beaked echidnas are essentially solitary animals, inhabiting a home range the size of which varies according to the environment. In wet areas with abundant food it covers some 50ha (124 acres). The home range appears to change little, and within it there is no fixed shelter site. When inactive, echidnas take shelter in hollow logs, under piles of rubble and brush, or in thick clumps of vegetation. Occasionally they dig shallow burrows up to 1.2m (4ft) in length, which may be reused. A female incubating an egg or suckling young has a fixed burrow. The home ranges of several individuals overlap.

During the mating season the female leaves a scent track by everting the cloaca, the wall of which contains numerous glands. This presumably attracts males in overlapping ranges. At this time echidnas break their normally solitary habits to form "trains," in which a female is followed by as many as six males in a line. In captivity, echidnas that are kept in spacious accommodation do not form any sort of groups but are mutually tolerant. By contrast, if they are kept in confined, overcrowded quarters they may form a size-related dominance order, but this does not seem to be a natural behavior.

The chief periods of activity are related to environmental temperature. Short-beaked echidnas are usually active at dusk and dawn, but in the hot summer are nocturnal. During cold spells they may be active in the middle of the day. They avoid rain and will remain inactive for days if it persists. In cold parts of their range, such as the Snowy Mountains of eastern Australia, echidnas hibernate during winter, with body temperatures as low as 5°C (41°F). Inactive echidnas may enter torpor, with body temperatures as low as 18°C (64.4°F), under less rigorous environmental temperatures.

The female pouch is barely detectable for most of the year. Before the start of the breeding season, folds of skin and muscle on each side of the abdomen enlarge to form an incomplete pouch with milk patches at the front end. There are no teats. The single egg is laid into the pouch by extension of the cloaca while the female lies on her back. After about 10 days the young hatches, using an egg tooth and a horny carbuncle at the tip of the snout. The young remains in the pouch until about the time spines begin to erupt. There are rare reports of females found moving about freely with an egg or young in the pouch. However, for most of the suckling period (up to six months), the offspring is left behind in the nursery burrow while the mother forages, often for days at a time. The young become independent and move out to occupy their own home ranges at about 1year old.

The Short-beaked echidna is widespread and common on mainland Australia and Tasmania. However, its status in New Guinea is uncertain. The Long-beaked echidna and several other similar genera were once distributed throughout Australia but disappeared by the late Pleistocene. Today there is only one species, which is restricted to the New Guinea highlands. It is likely that the disappearance of Long-beaked echidnas from Australia is related to climatic changes that have taken place there. Both the Short-beaked and the Long-beaked echidna are hunted for food; most zoologists now consider the latter species to be under severe threat of extinction. MLA

Appendix: Species list

Including all families, genera, and species of Rodents, Lagomorphs, Elephant Shrews, Insectivores, Bats, Edentates, and Marsupials.

ORDER RODENTIA

RODENTS

SUBORDER SCIUROGNATHI

FAMILY APLODONTIDAE

Aplodontia
A. rufa **Mountain beaver or Sewellel** SW British Columbia (Canada) to C California (USA)

FAMILY SCIURIDAE

Squirrels, Chipmunks, Marmots, and Prairie dogs

SUBFAMILY SCIURINAE

Ammospermophilus Antelope Squirrels
A. harrisii **Harris' antelope squirrel** Arizona to SW New Mexico (SW USA) and adj. Sonora (Mexico)
A. insularis **Espirito Santo Island antelope squirrel** Espirito Santo Is, Gulf of California (S Baja California, Mexico)
A. interpres **Texas antelope squirrel** New Mexico and W Texas (SW USA) to Durango (Mexico)
A. leucurus **White-tailed antelope squirrel** SW USA to S Baja California (Mexico)
A. nelsoni **Nelson's antelope squirrel** San Joaquin Valley, S California (USA)
Atlantoxerus
A. getulus **Barbary ground squirrel** Morocco, NW Algeria
Callosciurus Oriental Tree Squirrels
C. adamsi **Ear-spot squirrel** Sabah and Sarawak (N Borneo)
C. albescens **Kloss squirrel** Sumatra (Indonesia)
C. baluensis **Kinabalu squirrel** Sabah and Sarawak (N Borneo)
C. caniceps **Gray-bellied squirrel** Burma Pen., Thailand, Malyasia Pen.
C. erythraeus **Pallas' squirrel** India, Burma, SE and S China, Taiwan, Thailand, Malay Pen., Indochina
C. finlaysonii **Finlayson's or Variable squirrel** SC Burma, Thailand, Laos, Cambodia, Vietnam
C. inornatus **Inornate squirrel** S Yunnan (China), N Vietnam, Laos
C. melanogaster **Mentawai squirrel** Mantawi Is (Indonesia)
C. nigrovittatus **Black-striped squirrel** Thailand, S Vietnam, Malaysia, Sumatra and Java (Indonesia), Borneo
C. notatus **Plantain squirrel** Thailand and Malay Pen. to Java, Bali, Lombok and Salayar (Indonesia), Borneo
C. orestes **Borneo black-banded squirrel** Malaysia, Sabah and Sarawak (N Borneo)
C. phayrei **Phayre's squirrel** S Burma
C. prevostii **Prevost's squirrel** Parts of Indo-Malayan region (excl. Java)
C. pygerythrus **Irrawaddy squirrel** NE India and Nepal to Burma, Yunnan (China), and N Vietnam
C. quinquestriatus **Anderson's squirrel** NE Burma, Yunnan (China)
Cynomys Prairie Dogs
C. gunnisoni **Gunnison's prairie dog** SE Utah, SW Colorado, NW New Mexico, NE Arizona (USA)
C. leucurus **White-tailed prairie dog** SC Montana, W and C Wyoming, NW Colorado, NE Utah (USA)
C. ludovicianus **Black-tailed prairie dog** Saskatchewan (Canada), Montana through C and SC USA to Sonora and Chihuahua (Mexico)
C. mexicanus **Mexican prairie dog** Coahuila and San Luis Potosi (NC Mexico)
C. parvidens **Utah prairie dog** SC Utah (USA)
Dremomys Red-cheeked Squirrels
D. everetti **Bornean mountain ground squirrel** N and W Borneo
D. lokriah **Orange-bellied Himalayan squirrel** E India, C Nepal, Xizang (China) and Bhutan east to Salween R (Burma)
D. pernyi **Perny's long-nosed squirrel** NE India, N Burma, China, Taiwan, N Vietnam
D. pyrrhomerus **Red-hipped squirrel** C and S China, N Vietnam
D. rufigenis **Asian red-cheeked squirrel** NE India, N and C Burma, Yunnan (China), Laos, Vietnam, Thailand, Malay Pen.
Epixerus African Palm Squirrels
E. ebii **Western palm squirrel** Sierra Leone, Liberia, Ivory Coast, Ghana
E. wilsoni **Biafran palm squirrel** Cameroon, Rio Muni (Equatorial Guinea), Gabon
Exilisciurus Pygmy Squirrels
E. concinnus **Philippine pygmy squirrel** Mindanao, Basilan, Biliran, Bohol, Dinagat, Leyte, and Samar Is (Philippines)
E. exilis **Least pygmy squirrel** Borneo, Banggi Is (Malaysia)
E. whiteheadi **Tufted or Whitehead's pygmy squirrel** Sabah, Sarawak, W Kalimantan (Borneo)
Funambulus Asiatic Palm Squirrels
F. layardi **Layard's palm squirrel** S India, S and C Sri Lanka
F. palmarum **Indian palm squirrel** C and S India, Sri Lanka
F. pennantii **Northern palm squirrel** SE Iran to Pakistan, Nepal and N and C India
F. sublineatus **Dusky palm squirrel** SW India, C Sri Lanka
F. tristriatus **Jungle palm squirrel** Coastal W India
Funisciurus Rope Squirrels
F. anerythrus **Thomas' rope squirrel** SW Nigeria, Cameroon, CAR, DRC, Uganda
F. bayonii **Lunda or Bayon's tree squirrel** SW DRC, NE Angola
F. carruthersi **Carruther's mountain squirrel** Ruwenzori Mts (S Uganda), Rwanda, Burundi
F. congicus **Congo rope squirrel** DRC, Angola, Namibia
F. isabella **Lady Burton's rope squirrel** Cameroon, CAR, Congo
F. lemniscatus **Ribboned rope squirrel** Cameroon, CAR, DRC
F. leucogenys **Red-cheeked rope squirrel** Ghana, Togo, Benin, Nigeria, Cameroon, CAR, Equatorial Guinea
F. pyrropus **Fire-footed rope squirrel** W and C Africa to Angola
F. substriatus **Kintampo rope squirrel** Ivory Coast, Ghana, Togo, benin, SE Nigeria
Glyphotes
G. simus **Sculptor squirrel** N Borneo
Heliosciurus Sun Squirrels
H. gambianus **Gambian sun squirrel** Senegal and Gambia to Sudan and Ethiopia south to Angola, Zambia and Zimbabwe
H. mutabilis **Mutable sun squirrel** SE Zimbabwe, N Mozambique, Malawi, Tanzania
H. punctatus **Small sun squirrel** E Liberia, S Ivory Coast, S Ghana
H. rufobrachium **Red-legged sun squirrel** Senegal and W Gambia to SE Sudan and Kenya south to Malawi, Mozambique and E Zimbabwe
H. ruwenzorii **Ruwenzori sun squirrel** Ruwenzori Mts of E DRC, Rwanda, Burundi, SW Uganda
H. undulatus **Zanj sun squirrel** SE Kenya, NE Tanzania (incl. Mafia and Zanzibar Is)
Hyosciurus Sulawesi Long-nosed Squirrels
H. heinrichi **Montane long-nosed squirrel** C Sulawesi (Indonesia)
H. ileile **Lowland long-nosed squirrel** N Sulawesi (Indonesia)
Lariscus Striped Ground Squirrels
L. hosei **Four-striped ground squirrel** Sarawak and Sabah (N Borneo)
L. insignis **Three-striped ground squirrel** Thailand, Malay Pen., Sumatra and Java (Indonesia), Borneo
L. niobe **Niobe ground squirrel** Sumatra, Java and Mentawi Is (Indonesia)
L. obscurus **Mentawai three-striped squirrel** Sumatra (Indonesia)
Marmota Marmots
M. baibacina **Gray marmot** Altai Mts and SW Siberia (Russia), SE Kazakhstan, Kyrgyzstan, Mongolia, Xinjiang (China)
M. bobak **Bobak marmot** E Europe to Ukraine and Russia to N and C Kazakhstan
M. broweri **Alaska marmot** N Alaska (USA)
M. caligata **Hoary marmot** C Alaska (USA), Yukon and Northwest Territories (Canada) south to NW USA
M. camtschatica **Black-capped marmot** E Siberia (Russia)
M. caudata **Long-tailed marmot** W Tien Shan Mts to the Pamirs (Tajikistan and Kyrgyzstan) to Hindu Kush (Afghanistan), Pakistan, Kashmir, and W Xinjiang and Xizang (China)
M. flaviventris **Yellow-bellied marmot** SC British Columbia and S Alberta (SW Canada) south to SW USA
M. himalayana **Himalayan marmot** N India, Nepal, W China
M. marmota **Alpine marmot** French, Swiss, and Italian Alps, S Germany, W Austria, Carpathians (Romania), and Tatra Mts (Czech Rep. and Poland); introd. to Pyrenees, E Austria, and Yugoslavia
M. menzbieri **Menzbier's or Tien Shan marmot** W Tien Shan Mts (S Kazakhstan and Kyrgyzstan)
M. monax **Woodchuck** Alaska (USA) through S Canada to NE and SC USA and south in Rocky Mts
M. olympus **Olympic marmot** Olympics Mts (W Washington, NW USA)
M. sibirica **Tarbagan marmot** SW Siberia (Russia) to N and W Mongolia, Heilongjiang and Inner Mongolia (China)
M. vancouverensis **Vancouver marmot** Vancouver Is (British Columbia, Canada)
Menetes
M. berdmorei **Indochinese ground squirrel** C Burma to S Yunnan (China), Thailand, S Laos, S Vietnam, Cambodia
Microsciurus Dwarf Squirrels
M. alfari **Central American dwarf squirrel** Nicaragua, Costa Rica, Panama, Colombia
M. flaviventer **Amazon dwarf squirrel** Amazon Basin of Colombia, Ecuador, Peru, and W Brazil
M. mimulus **Western dwarf squirrel** Panama, N Colombia, NW Ecuador
M. santanderensis **Santander dwarf squirrel** Colombia
Myosciurus
M. pumilio **African pygmy squirrel** SE Nigeria, Cameroon, Gabon, Bioko (Equatorial Guinea)
Nannosciurus
N. melanotis **Black-eared squirrel** Sumatra and Java (Indonesia), Borneo
Paraxerus Bush Squirrels
P. alexandri **Alexander's bush squirrel** NE DRC, Uganda
P. boehmi **Boehm's bush squirrel** S Sudan, DRC, Uganda, W Kenya, NW Tanzania, N Zambia
P. cepapi **Smith's bush squirrel** SW Tanzania to N South Africa
P. cooperi **Cooper's mountain squirrel** Cameroon
P. flavovittis **Striped bush squirrel** S Kenya, Tanzania, Mozambique
P. lucifer **Black and red bush squirrel** SW Tanzania, N Malawi, E Zambia
P. ochraceus **Ochre bush squirrel** S Sudan, Kenya, Tanzania
P. palliatus **Red bush squirrel** S Somalia to E South Africa
P. poensis **Green bush squirrel** Sierra Leone to S Nigeria and Cameroon, Bioko (Equatorial Guinea), Congo, W DRC
P. vexillarius **Swynnerton's bush squirrel** C and E Tanzania
P. vincenti **Vincent's bush squirrel** N Mozambique
Prosciurillus Sulawesi Dwarf Squirrels
P. abstrusus **Secretive dwarf squirrel** SE Sulawesi (Indonesia)
P. leucomus **Whitish dwarf squirrel** Sulawesi, Buton, and Sangihe Is (Indonesia)
P. murinus **Sulawesi dwarf squirrel** NE and C Sulawesi (Indonesia)
P. weberi **Weber's dwarf squirrel** C Sulawesi (Indonesia)
Protoxerus African Giant Squirrels
P. aubinnii **Slender-tailed squirrel** Liberia, Ivory Coast, Ghana
P. stangeri **Forest giant or Stanger's squirrel** Sierra Leone to S Sudan and W Kenya south to N Tanzania, DRC and N Angola
Ratufa Oriental Giant Squirrels
R. affinis **Pale giant squirrel** S Vietnam, Malaysia, Indonesia (excl. Java), Philippines (excl. SW)
R. bicolor **Black giant squirrel** Assam (India), E Nepal, SE Xizang to S Yunnan and Hainan (China), Burma, Thailand, Laos, Vietnam, Cambodia, Malay Pen. to Java and Bali (Indonesia)
R. indica **Indian giant squirrel** C and S India
R. macroura **Sri Lankan giant squirrel** S India, Sri Lanka
Rheithrosciurus
R. macrotis **Tufted ground squirrel** Borneo
Rhinosciurus
R. laticaudatus **Shrew-faced squirrel** Malaysia, Indonesia (excl. Java), Philippines (excl. SW)
Rubrisciurus
R. rubriventer **Sulawesi giant squirrel** N, C, and SE Sulawesi (Indonesia)
Sciurillus
S. pusillus **Neotropical pygmy squirrel** Surinam, Fr. Guiana, Peru, Brazil
Sciurotamias Asian Rock Squirrels
S. davidianus **Pere David's rock squirrel** Hebei to Sichuan and Hubei, Guizhou (China)
S. forresti **Forrest's rock squirrel** Yunnan and Sichuan (China)
Sciurus Tree Squirrels
S. aberti **Abert's squirrel** Wyoming, Utah, Colorado, New Mexico, Arizona (USA) to Durango (Mexico)
S. aestuans **Guianan squirrel** Venezuela, Guyana, Surinam, Fr. Guiana, Brazil
S. alleni **Allen's squirrel** Mexico
S. anomalus **Caucasian squirrel** Turkey and Caucasus, N and W Iran, Syria, Lebanon, Israel
S. arizonensis **Arizona gray squirrel** SE and C Arizona, WC New Mexico (USA), Sonora (Mexico)
S. aureogaster **Red-bellied squirrel** Mexico, Guatemala
S. carolinensis **Eastern gray squirrel** E Texas (USA) to Saskatchewan (Canada) east to Atl. Oc.; introd. to Britain and South Africa
S. colliaei **Collie's squirrel** W and WC Mexico
S. deppei **Deppe's squirrel** Tamaulipas (Mexico) to Costa Rica
S. flammifer **Fiery squirrel** Venezuela
S. gilvigularis **Yellow-throated squirrel** Venezuela, Guyana, N Brazil
S. granatensis **Red-tailed squirrel** Costa Rica, Panama, Colombia, Venezuela (incl. Margarita Is), Ecuador, Trinidad and Tobago
S. griseus **Western gray squirrel** W USA to Baja California (Mexico)
S. ignitus **Bolivian squirrel** Peru, Brazil, Bolivia, Argentina
S. igniventris **Northern Amazon red squirrel** Colombia, Venezuela, Ecuador, Peru, Brazil
S. lis **Japanese squirrel** Honshu, Shikoku, and Kyushu (Japan)
S. nayaritensis **Mexican fox squirrel** Jalisco (Mexico) north to SE Arizona (USA)
S. niger **Eastern fox squirrel** Texas (USA) and adj. Mexico north to Manitoba (SC Canada) east to Atl. Oc.
S. oculatus **Peters' squirrel** Mexico
S. pucheranii **Andean squirrel** Andes of Colombia
S. pyrrhinus **Junin red squirrel** E slopes of Andes of Peru
S. richmondi **Richmond's squirrel** Nicaragua
S. sanborni **Sanborn's squirrel** Madre de Dios Dept. (Peru)
S. spadiceus **Southern Amazon red squirrel** Colombia, Ecuador, Peru, Brazil, Bolivia
S. stramineus **Guayaquil squirrel** SE Ecuador, NE Peru
S. variegatoides **Variegated squirrel** S Chiapas (Mexico) to Panama
S. vulgaris **Eurasian red squirrel** Forested regions of Palearctic from Iberian Pen. and Britain east to Kamchatka Pen. and Sakhalin Is (Russia) south to Mediterranean and Black Seas, N Mongolia, NE China, Korea

S. yucatanensis **Yucatan squirrel** Yucatan Pen. (Mexico), N and SW Belize, N Guatemala
Spermophilopsis
S. leptodactylus **Long-clawed ground squirrel** SE Kazakhstan, Uzbekistan, Turkmenistan, NE Iran, NW Afghanistan, Tajikistan
Spermophilus Ground Squirrels
S. adocetus **Tropical ground squirrel** Jalisco, Michoacan, Guerrero (Mexico)
S. alashanicus **Alashan ground squirrel** SC Mongolia, N China
S. annulatus **Ring-tailed ground squirrel** Nayarit to N Guerrero (Mexico)
S. armatus **Uinta ground squirrel** Utah, Wyoming, Idaho, Montana (NW USA)
S. atricapillus **Baja California rock squirrel** Baja California (Mexico)
S. beecheyi **California ground squirrel** Washington (USA) to Baja California (Mexico)
S. beldingi **Belding's ground squirrel** W USA
S. brunneus **Idaho ground squirrel** WC Idaho (USA)
S. canus **Merriam's ground squirrel** E Oregon, NW Nevada, WC Idaho (USA)
S. citellus **European ground squirrel** SE Germany, SW Poland, Czech Rep. to W Ukraine, Moldova, and W Turkey
S. columbianus **Columbian ground squirrel** SE British Columbia and W Alberta (SW Canada) to NE Oregon, C Idaho, and C Montana (NW USA)
S. dauricus **Daurian ground squirrel** Russia, Mongolia, N China
S. elegans **Wyoming ground squirrel** NE Nevada, SE Oregon, S Idaho and SW Montana to C Colorado and W Nebraska
S. erythrogenys **Red-cheeked ground squirrel** E Kazakhstan, SW Siberia (Russia), and Xinjiang (China); also Mongolia and Inner Mongolia (China)
S. franklinii **Franklin's ground squirrel** Great Plains of Canada to Alberta, Saskatchewan and Manitoba (Canada) south to Kansas, Illinois and Indiana
S. fulvus **Yellow ground squirrel** Kazakhstan, Uzbekistan, W Tajikistan, Turkmenistan, NE Iran, Afghanistan, W Xinjiang (China)
S. lateralis **Golden-mantled ground squirrel** Montana, W North America, from C British Columbia (Canada) south to S New Mexico, Nevada and California (USA)
S. madrensis **Sierra Madre ground squirrel** SW Chihuahua (Mexico)
S. major **Russet ground squirrel** Steppe between Volga and Irtysh R (Russia, N Kazakhstan)
S. mexicanus **Mexican ground squirrel** S New Mexico and W Texas (USA) to Jalisco and S Puebla (Mexico)
S. mohavensis **Mohave ground squirrel** S California (USA)
S. mollis **Piute ground squirrel** SE Oregon and Snake River Valley (Idaho) south to Nevada, W Utah and EC California (USA); also NW Washington
S. musicus **Caucasian Mountain ground squirrel** N Caucasus Mts (Georgia)
S. parryii **Arctic ground squirrel** Alaska (USA), NW Canada, NE Russia
S. perotensis **Perote ground squirrel** Veracruz and Puebla (Mexico)
S. pygmaeus **Little ground squirrel** SW Ukraine, S Ural Mts to Crimea (Ukraine), Kazakhstan, Uzbekistan, Dagestan (Georgia)
S. relictus **Tien Shan ground squirrel** Tien Shan Mts in SE Kazakhstan and Kyrgyzstan
S. richardsonii **Richardson's ground squirrel** Great Plains of Canada from S Alberta, S Saskatchewan, and S Manitoba (Canada) south to Montana, Dakotas, W Minnesota, and NW Iowa (USA)
S. saturatus **Cascade golden-mantled ground squirrel** Cascade Mts of SW British Columbia (SW Canada) and Washington (NW USA)
S. spilosoma **Spotted ground squirrel** C Mexico to S and SC USA
S. suslicus **Spectacled ground squirrel** Steppes of E and S Europe, incl. Poland, E Romania, Ukraine north to Oka R and east to Volga R (Russia)
S. tereticaudus **Round-tailed ground squirrel** SE California, S Nevada, W Arizona (SW USA), NE Baja California and Sonora (Mexico)
S. townsendii **Townsend's ground squirrel** SE Washington (NW USA)
S. tridecemlineatus **Thirteen-lined ground squirrel** Great Plains from C Texas to E Utah and Ohio (USA) and SC Canada
S. undulatus **Long-tailed ground squirrel** E Kazakhstan, S Siberia and SE Russia, N Mongolia, Heilongjiang and Xinjiang (China)
S. variegatus **Rock squirrel** S Nevada and Utah to SW Texas (USA) to Puebla (Mexico)
S. washingtoni **Washington ground squirrel** SE Washington, NE Oregon (USA)
S. xanthoprymnus **Asia Minor ground squirrel** Caucasus, Turkey, Syria, Israel
Sundasciurus Sunda Squirrels
S. brookei **Brooke's squirrel** Borneo
S. davensis **Davao squirrel** Mindanao (Philippines)
S. fraterculus **Fraternal squirrel** Sipora, Siberut, and N Pagi Is (Sumatra, Indonesia)
S. hippurus **Horse-tailed squirrel** S Vietnam, Malaysia, Indonesia (excl. Java), Philippines (excl. SW)
S. hoogstraali **Busuanga squirrel** Busuanga Is (Philippines)
S. jentinki **Jentink's squirrel** N Borneo
S. juvencus **Northern Palawan tree squirrel** N Palawan (Philippines)
S. lowii **Low's squirrel** Malaysia, Indonesia (excl. Java), Philippines (excl. SW)
S. mindanensis **Mindanao squirrel** Mindanao (Philippines)
S. moellendorffi **Culion tree squirrel** Calamien Is (Philippines)
S. philippinensis **Philippine tree squirrel** S and W Mindanao and Basilan (Philippines)
S. rabori **Palawan montane squirrel** Palawan (Philippines)
S. samarensis **Samar squirrel** Samar and Leyte (Philippines)
S. steerii **Southern Palawan tree squirrel** S Palawan and Balabac Is (Philippines)
S. tenuis **Slender squirrel** Malaysia, Indonesia (excl. Java), Philippines (excl. SW)
Syntheosciurus
S. brochus **Bangs' mountain squirrel** Costa Rica to N Panama
Tamias Chipmunks
T. alpinus **Alpine chipmunk** Sierra Nevada Mts (EC California, USA)
T. amoenus **Yellow-pine chipmunk** C British Columbia (SW Canada) south to California east to C Montana and W Wyoming (USA)
T. bulleri **Buller's chipmunk** Sierra Madre in S Durango, W Zacetecas and N Jalisco (Mexico)
T. canipes **Gray-footed chipmunk** SE New Mexico and W Texas (SW USA)
T. cinereicollis **Gray-collared chipmunk** C and E Arizona and C and SW New Mexico (SW USA)
T. dorsalis **Cliff chipmunk** WC and SW USA to Mexico
T. durangae **Durango chipmunk** Mexico
T. merriami **Merriam's chipmunk** California (USA) to N Baja California (Mexico)
T. minimus **Least chipmunk** Canada, USA
T. obscurus **California chipmunk** S California (USA) to C Baja California (Mexico)
T. ochrogenys **Yellow-cheeked chipmunk** Coastal N California (USA)
T. palmeri **Palmer's chipmunk** Charleston Mts (S Nevada, USA)
T. panamintinus **Panamint chipmunk** SE California and SW Nevada (USA)
T. quadrimaculatus **Long-eared chipmunk** Sierra Nevada Mts of EC California, C and WC Nevada (USA)
T. quadrivittatus **Colorado chipmunk** Colorado and E Utah south to NE Arizona and S New Mexico (USA)
T. ruficaudus **Red-tailed chipmunk** NE Washington to W Montana (USA), SE British Columbia (Canada)
T. rufus **Hopi chipmunk** E and S Utah, W Colorado, NE Arizona (USA)
T. senex **Allen's chipmunk** California, WC Nevada, Oregon (USA)
T. sibiricus **Siberian chipmunk** N Europe and Siberia to Sakhalin and S Kurile Is (Russia), E Kazakhstan to N Mongolia, China, Korea and Hokkaido (Japan)
T. siskiyou **Siskiyou chipmunk** Siskiyou Mts and coastal N California to C Oregon (USA)
T. sonomae **Sonoma chipmunk** NW California (USA)
T. speciosus **Lodgepole chipmunk** California, W Nevada (USA)
T. striatus **Eastern chipmunk** S Manitoba and Nova Scotia (Canada) to SE USA
T. townsendii **Townsend's chipmunk** SW British Columbia (Canada), Washington and Oregon (USA)
T. umbrinus **Uinta chipmunk** California and N Arizona to Colorado, Wyoming and SW Montana (USA)
Tamiasciurus Red Squirrels
T. douglasii **Douglas' squirrel** SW British Columbia (Canada) to S California (USA)
T. hudsonicus **American red squirrel** Alaska (USA), Canada, south of tundra (incl. Vancouver Is), W USA and NE USA south to South Carolina
T. mearnsi **Mearns' squirrel** Sierra San Pedro Martir Mts (Baja California, Mexico)
Tamiops Asiatic Striped Squirrels
T. macclellandi **Himalayan striped squirrel** Assam (India) to E Nepal, N and C Burma, Yunnan (China), Thailand, Vietnam, Laos, Cambodia, Malay Pen.
T. maritimus **Maritime striped squirrel** China to S Vietnam and Laos, Taiwan
T. rodolphei **Cambodian striped squirrel** E Thailand, S Vietnam, S Laos, Cambodia
T. swinhoei **Swinhoe's striped squirrel** SW Gansu through Xizang, Sichuan and Yunnan (China) to N Burma and N Vietnam
Xerus African Ground Squirrels
X. erythropus **Striped ground squirrel** Morocco, Mauritania, Senegal and Gambia east to Sudan, W Ethiopia, W Kenya and N Tanzania
X. inauris **South African ground squirrel** South Africa, W Zimbabwe, Botswana, Namibia, S Angola
X. princeps **Damara ground squirrel** W Namibia, S Angola
X. rutilus **Unstriped ground squirrel** NE and E Africa

SUBFAMILY PETAURISTINAE
Aeretes
A. melanopterus **North Chinese flying squirrel** Hebei and Sichuan (China)
Aeromys Large Black Flying Squirrels
A. tephromelas **Black flying squirrel** Malaysia, Indonesia (excl. Java), Philippines (excl. SW)
A. thomasi **Thomas' flying squirrel** Borneo (excl. SE)
Belomys
B. pearsonii **Hairy-footed flying squirrel** Sikkim and Assam (India) to Hunnan, Sichuan, Yunnan, Guizhou, Hainan (China), Taiwan, Bhutan, N Burma, Indochina
Biswamoyopterus
B. biswasi **Namdapha flying squirrel** Arunachal Pradesh (India)
Eupetaurus
E. cinereus **Woolly flying squirrel** N Pakistan and Kashmir to Sikkim (India) to Xizang (China)
Glaucomys New World Flying Squirrels
G. sabrinus **Northern flying squirrel** Alaska (USA), Canada, NW USA to S California and South Dakota and NE USA to S Appalachian Mts
G. volans **Southern flying squirrel** Nova Scotia (Canada) to E and C USA, NW Mexico to Honduras
Hylopetes Arrow-tailed Flying Squirrels
H. alboniger **Particolored flying squirrel** Nepal and Assam (India) to Sichuan, Yunnan, and Hainan (China), Indochina
H. baberi **Afghan flying squirrel** EC and NW Afghanistan, Kashmir
H. bartelsi **Bartel's flying squirrel** Java (Indonesia)
H. fimbriatus **Kashmir flying squirrel** Punjab (India), Kashmir
H. lepidus **Gray-cheeked flying squirrel** S Vietnam, Thailand to Java (Indonesia), Borneo
H. nigripes **Palawan flying squirrel** Palawan and Bancalan Is (Philippines)
H. phayrei **Indochinese or Phayre's flying squirrel** Fujian and Hainan (China), Burma, Thailand, Laos, S Vietnam
H. sipora **Sipora flying squirrel** Sipora Is (Indonesia)
H. spadiceus **Red-cheeked flying squirrel** Burma, Malaysia, Thailand, S Vietnam, Sumatra (Indonesia)
H. winstoni **Sumatran flying squirrel** N Sumatra (Indonesia)
Iomys Horsfield's Flying Squirrels
I. horsfieldi **Javanese flying squirrel** Malay Pen. to Java (Indonesia), Borneo
I. sipora **Mentawi flying squirrel** Mentawi Is (Indonesia)
Petaurillus Pygmy Flying Squirrels
P. emiliae **Lesser pygmy flying squirrel** Sarawak (N Borneo)
P. hosei **Hose's pygmy flying squirrel** Sarawak (N Borneo)
P. kinlochii **Selangor pygmy flying squirrel** Selangor (Malay Pen.)
Petaurista Giant Flying Squirrels
P. alborufus **Red and white giant flying squirrel** S and C China, Taiwan
P. elegans **Spotted giant flying squirrel** Nepal, Sikkim (India), Sichuan and Yunnan (China), N and W Burma, Laos, Vietnam, Malay Pen., Sumatra and Java (Indonesia), Borneo
P. leucogenys **Japanese giant flying squirrel** Gansu, Sichuan, Yunnan (China), Japan (excl. Hokkaido)
P. magnificus **Hodgson's giant flying squirrel** Sikkim (India), Nepal, Xizang (China), Bhutan
P. nobilis **Bhutan giant flying squirrel** Sikkim (India), C Nepal, Bhutan
P. petaurista **Red giant flying squirrel** E Afghanistan, Kashmir, Punjab to Assam (India), Yunnan and Sichuan (China), Burma, Thailand, Indochina, Malaysia, Sumatra and Java (Indonesia), Borneo
P. philippensis **Indian giant flying squirrel** India, Sri lanka, S China, Taiwan, Burma, Thailand, Indonesia
P. xanthotis **Chinese giant flying squirrel** Mountains of W China
Petinomys Dwarf Flying Squirrels
P. crinitus **Mindanao flying squirrel** Basilan, Dinagat, Siargao, and Mindanao (Philippines)
P. fuscocapillus **Travancore flying squirrel** S India, Sri Lanka
P. genibarbis **Whiskered flying squirrel** Malaysia to Sumatra and Java (Indonesia), Borneo
P. hageni **Hagen's flying squirrel** Sumatra (Indonesia), Borneo
P. lugens **Siberut flying squirrel** Siberut and Sipora Is (Sumatra, Indonesia)
P. sagitta **Arrow-tailed flying squirrel** Java (Indonesia)
P. setosus **Temminck's flying squirrel** Burma, Malaysia, Sumatra (Indonesia), S Borneo
P. vordermanni **Vordermann's flying squirrel** S Burma, Malaysia, Borneo
Pteromys Eurasian Flying Squirrels
P. momonga **Japanese flying squirrel** Kyushu and Honshu (Japan)
P. volans **Siberian flying squirrel** N Finland to Chukotskoye Pen. (Russia) south in Ural and Altai Mts, Mongolia, N China, Korea, Hokkaido (Japan)
Pteromyscus
P. pulverulentus **Smoky flying squirrel** S Thailand to Sumatra (Indonesia), Borneo
Trogopterus
T. xanthipes **Complex-toothed flying squirrel** Yunnan to C and E China in montane forests

FAMILY CASTORIDAE **Beavers**
Castor
C. canadensis **American or Canadian beaver** Alaska (USA) to Labrador (Canada) south to N Florida (USA) and Tamaulipas (Mexico); introd. to Europe and Asia
C. fiber **Eurasian beaver** NW and NC Eurasia, from France to L. Baikal and Mongolia

FAMILY GEOMYIDAE **Pocket Gophers**
Geomys Eastern Pocket Gophers
G. arenarius **Desert pocket gopher** W Texas, New Mexico (USA), Chihuahua (Mexico)
G. bursarius **Plains pocket gopher** SC Manitoba (Canada) to Indiana, Louisiana, SC Texas, and New Mexico (USA)
G. personatus **Texas pocket gopher** S Texas, incl. Padre and Mustang Is (USA), NE Tamaulipas (Mexico)
G. pinetis **Southeastern pocket gopher** C Florida to S Georgia and S Alabama (USA)
G. tropicalis **Tropical pocket gopher** SE Tamaulipas (Mexico)
Orthogeomys Giant Pocket Gophers
O. cavator **Chiriqui pocket gopher** C Costa Rica to NW Panama
O. cherriei **Cherrie's pocket gopher** NC Costa Rica
O. cuniculus **Oaxacan pocket gopher** Oaxaca (Mexico)
O. dariensis **Darien pocket gopher** E Panama
O. grandis **Giant pocket gopher** Jalisco (Mexico) to Honduras
O. heterodus **Variable pocket gopher** C Costa Rica
O. hispidus **Hispid pocket gopher** S Tamaulipas and Yucatan Pen. (Mexico), Belize, Guatemala, NW Honduras
O. lanius **Big pocket gopher** Veracruz (Mexico)
O. matagalpae **Nicaraguan pocket gopher** SC Honduras to NC Nicaragua
O. thaeleri **Thaeler's pocket gopher** NW Colombia
O. underwoodi **Underwood's pocket gopher** Pac. Oc. coast of Costa Rica
Pappogeomys Mexican Pocket Gophers
P. alcorni **Alcorn's pocket gopher** S Jalisco (Mexico)
P. bulleri **Buller's pocket gopher** Nayarit, Jalisco, and Colima (Mexico)

P. castanops **Yellow-faced pocket gopher** SE Colorado and SW Kansas (USA) to San Luis Potosi (Mexico)
P. fumosus **Smoky pocket gopher** E Colima (Mexico)
P. gymnurus **Llano pocket gopher** S and C Jalisco and NE Michoacan (Mexico)
P. merriami **Merriam's pocket gopher** S Mexico
P. neglectus **Queretaro pocket gopher** Queretaro (Mexico)
P. tylorhinus **Naked-nosed pocket gopher** Distrito Federal and Hidalgo to C Jalisco (Mexico)
P. zinseri **Zinser's pocket gopher** NE Jalisco (Mexico)
Thomomys **Western Pocket Gophers**
T. bottae **Botta's pocket gopher** W and SW USA south to Sinaloa and Nuevo Leon (Mexico)
T. bulbivorus **Camas pocket gopher** Willamette valley (NW Oregon, USA)
T. clusius **Wyoming pocket gopher** SC Wyoming (USA)
T. idahoensis **Idaho pocket gopher** EC Idaho, adj. Montana, W Wyoming, N Utah (USA)
T. mazama **Western or Mazama pocket gopher** NW Washington to C Oregon to N California (USA)
T. monticola **Mountain pocket gopher** Sierra Nevada Mts of C and N California and WC Nevada (USA)
T. talpoides **Northern pocket gopher** S British Columbia to SW Manitoba (Canada) south to C South Dakota and to N New Mexico, N Arizona, N Nevada and NE California (USA)
T. townsendii **Townsend's pocket gopher** S Idaho to Oregon, NE California, and N Nevada (NW USA)
T. umbrinus **Southern pocket gopher** SC Arizona and SW New Mexico (USA) to Puebla and Veracruz (Mexico)
Zygogeomys
Z. trichopus **Michoacan pocket gopher or Tuza** NC Michoacan (Mexico)

FAMILY HETEROMYIDAE
Pocket Mice, Kangaroo Rats, and Kangaroo Mice

SUBFAMILY DIPODOMYINAE
Dipodomys **Kangaroo Rats**
D. agilis **Agile kangaroo rat** SW and SC California (USA), Baja California (Mexico)
D. californicus **California kangaroo rat** SC Oregon and N California (USA)
D. compactus **Gulf Coast kangaroo rat** S Texas, incl. Padre and Mustang Is (USA) and N Tamaulipas (Mexico)
D. deserti **Desert kangaroo rat** Deserts of E California, W and S Nevada, SW Utah, W and SC Arizona (USA), Sonora and NE Baja California (Mexico)
D. elator **Texas kangaroo rat** SW Oklahoma and NC Texas (USA)
D. elephantinus **Big-eared kangaroo rat** WC California (USA)
D. gravipes **San Quintin kangaroo rat** NW Baja California (USA)
D. heermanni **Heerman's kangaroo rat** C California (USA)
D. ingens **Giant kangaroo rat** WC California (USA)
D. insularis **San Jose Island kangaroo rat** Gulf of California (Baja California, Mexico)
D. margaritae **Margarita Island kangaroo rat** Santa Margarita Is (Baja California, Mexico)
D. merriami **Merriam's kangaroo rat** NE California and NW Nevada to Texas (USA) south to C Mexico (incl. Baja California)
D. microps **Chisel-toothed kangaroo rat** SW USA
D. nelsoni **Nelson's kangaroo rat** NC Mexico
D. nitratoides **Fresno kangaroo rat** WC California (USA)
D. ordii **Ord's kangaroo rat** SW Saskatchewan and SE Alberta (Canada) and SE Washington (USA) south through the Great Plains and W USA to C Mexico
D. panamintinus **Panamint kangaroo rat** E California and W Nevada (USA)
D. phillipsii **Phillips' kangaroo rat** C Durango to N Oaxaca (Mexico)
D. spectabilis **Banner-tailed kangaroo rat** SC Arizona, New Mexico, W Texas (USA) south to N Mexico
D. stephensi **Stephens' kangaroo rat** S California (USA)
D. venustus **Narrow-faced kangaroo rat** WC California (USA)
Microdipodops **Kangaroo Mice**
M. megacephalus **Dark kangaroo mouse** SE Oregon, S Idaho, W Utah, N and C Nevada, and NE California (USA)
M. pallidus **Pale kangaroo mouse** EC California, W and SC Nevada (USA)

SUBFAMILY HETEROMYINAE
Heteromys **Forest Spiny Pocket Mice**
H. anomalus **Trinidad spiny pocket mouse** W and N Colombia to N Venezuela (incl. Margarita Is), Trinidad and Tobago
H. australis **Southern spiny pocket mouse** E Panama to SW Colombia and NW Ecuador
H. desmarestianus **Desmarest's spiny pocket mouse** SE Tabasco (Mexico) south to NW Colombia
H. gaumeri **Gaumer's spiny pocket mouse** Yucatan Pen. (Mexico) and N Guatemala
H. goldmani **Goldman's spiny pocket mouse** S Chiapas (Mexico), W Guatemala
H. nelsoni **Nelson's spiny pocket mouse** Chiapas (Mexico)
H. oresterus **Mountain spiny pocket mouse** Cord. Talamanca (Costa Rica)
Liomys **Spiny Pocket Mice**
L. adspersus **Panamanian spiny pocket mouse** C Panama
L. irroratus **Mexican spiny pocket mouse** S Texas (USA), Mexico
L. pictus **Painted spiny pocket mouse** Mexico, NW Guatemala
L. salvini **Salvin's spiny pocket mouse** E Oaxaca (S Mexico) south to C Costa Rica
L. spectabilis **Jaliscan spiny pocket mouse** SE Jalisco (Mexico)

SUBFAMILY PEROGNATHINAE
Chaetodipus **Coarse-haired Pocket Mice**
C. arenarius **Little desert pocket mouse** Baja California (Mexico)
C. artus **Narrow-skulled pocket mouse** Mexico
C. baileyi **Bailey's pocket mouse** S California, S Arizona, SW New Mexico (USA) to Sinaloa and Baja California (Mexico)
C. californicus **California pocket mouse** C California (USA) to N Baja California (Mexico)
C. fallax **San Diego pocket mouse** SW California (USA) to W Baja California (Mexico)
C. formosus **Long-tailed pocket mouse** SW USA, Baja California (Mexico)
C. goldmani **Goldman's pocket mouse** Mexico
C. hispidus **Hispid pocket mouse** Great Plains from North Dakota to SE Arizona and W Louisiana (USA) south to Tamaulipas and Hidalgo (Mexico)
C. intermedius **Rock pocket mouse** SC Utah and Arizona to W Texas (USA) south to Sonora and Chihuahua (Mexico)
C. lineatus **Lined pocket mouse** San Luis Potosi (Mexico)
C. nelsoni **Nelson's pocket mouse** SE New Mexico and W Texas (USA) to Jalisco and San Luis Patosi (Mexico)
C. penicillatus **Desert pocket mouse** SW USA to C Mexico (incl. NE Baja California)
C. pernix **Sinaloan pocket mouse** S Sonora to N Nayarit (Mexico)
C. spinatus **Spiny pocket mouse** S Nevada and SE California (USA) to Baja California (Mexico)
Perognathus **Silky Pocket Mice**
P. alticola **White-eared pocket mouse** SC California (USA)
P. amplus **Arizona pocket mouse** W and C Arizona (USA) to Sonora (Mexico)
P. fasciatus **Olive-backed pocket mouse** Great Plains from SE Alberta, Saskatchewan and SW Manitoba (Canada) to North Dakota, S Colorado and NE Utah (USA)
P. flavescens **Plains pocket mouse** Great Plains from Minnesota and Utah (USA) to Chihuahua (Mexico)
P. flavus **Silky pocket mouse** SW Great Plains from South Dakota, Wyoming and Utah (USA) to Sonora and Puebla (Mexico)
P. inornatus **San Joaquin pocket mouse** California (USA)
P. longimembris **Little pocket mouse** SE Oregon and W Utah (USA) south to Sonora and Baja California (Mexico)
P. merriami **Merriam's pocket mouse** New Mexico to S Texas (USA) to Chihuahua and Tamaulipas (Mexico)
P. parvus **Great Basin pocket mouse** Great Basin from S British Columbia (Canada) south to E California and east to Wyoming and NW Arizona (USA)
P. xanthanotus **Yellow-eared pocket mouse** Freeman Canyon (California, USA)

FAMILY DIPODIDAE **Jerboas**

SUBFAMILY ALLACTAGINAE
Allactaga **Four- and Five-toed Jerboas**
A. balikunica **Balikun jerboa** Mongolia, NE Xinjiang (China)
A. bullata **Gobi jerboa** S and W Mongolia, Nei Monggol, E Xinjiang, Ningxia, Gansu, N Shaanxi (China)
A. elater **Small five-toed jerboa** Iran, W Pakistan, Afghanistan, NE Turkey, Armenia, Georgia, Azerbaijan, N Caucasus along Caspian Sea to Lower Volga R south to Turkmenistan, east to Kazakhstan, NE Xinjiang, Nei Monggol, N Gansu (China)
A. euphratica **Euphrates jerboa** Turkey, Caucasus, Syria, Jordan, Iraq, N Iran, Afghanistan, N Saudi Arabia, Kuwait
A. firouzi **Iranian jerboa** Isfahan Prov. (Iran)
A. hotsoni **Hotson's jerboa** Persian Baluchistan (SE Iran), W Pakistan, S Afghanistan
A. major **Great jerboa** Caucasus to Moscow (Russia) and Kyyiv (Ukraine) east to Ob R (W Siberia), Kazakhstan, Uzbekistan, Turkmenistan
A. severtzovi **Severtzov's jerboa** Kazakhstan, Uzbekistan, NE Turkmenistan, SW Tajikistan
A. sibirica **Mongolian five-toed jerboa** Lower Ural R (Kazakhstan) and Caspian Sea east through N Turkmenistan to Mongolia, China
A. tetradactyla **Four-toed jerboa** Egypt, E Libya
A. vinogradovi **Vinogradov's jerboa** S Kazakhstan, E Uzbekistan, Kyrgyzstan, Tajikistan
Allactodipus
A. bobrinskii **Bobrinski's jerboa** W and N Turkmenistan, C and W Uzbekistan
Pygeretmus **Fat-tailed Jerboas**
P. platyurus **Lesser fat-tailed jerboa** W, C, and E Kazakhstan
P. pumilio **Dwarf fat-tailed jerboa** Don R (Russia) through Kazakhstan to Irtysh R (Russia) south to NE Iran east to S Mongolia, N Xinjiang, W Nei Monggol, Ningxia (China)
P. shitkovi **Greater fat-tailed jerboa** E Kazakhstan

SUBFAMILY CARDIOCRANIINAE
Cardiocranius
C. paradoxus **Five-toed pygmy jerboa** E Kazakhstan, Mongolia, N Xinjiang, W Nei Monggol, N Ningxia, Gansu (China)
Salpingotus **Three-toed Pygmy Jerboas**
S. crassicauda **Thick-tailed pygmy jerboa** E Kazakhstan, S and SW Mongolia, NW China
S. heptneri **Heptner's pygmy jerboa** Kyzyl Kum Desert (Uzbekistan, S Kazakhstan)
S. kozlovi **Kozlov's pygmy jerboa** S and SE Mongolia, Nei Monggol, Xinjiang, Gansu, Shaanxi, Ningxia (China)
S. michaelis **Baluchistan pygmy jerboa** NW Baluchistan (Pakistan)
S. pallidus **Pallid pygmy jerboa** Deserts of N Aral and S Balqash (Kazakhstan)
S. thomasi **Thomas' pygmy jerboa** Afghanistan

SUBFAMILY DIPODINAE
Dipus
D. sagitta **Northern three-toed jerboa** Don R (Russia), NW coast of Caspian Sea and NE Iran through Turkmenistan, Uzbekistan, Kazakhstan to S Tuva (Russia), Mongolia, China
Eremodipus
E. lichtensteini **Lichtenstein's jerboa** Caspian Sea to L. Balqash (Kazakhstan, Uzbekistan, Turkmenistan)
Jaculus **Desert Jerboas**
J. blanfordi **Blanford's jerboa** E and S Iran, S and W Afghanistan, W Pakistan
J. jaculus **Lesser Egyptian jerboa** NE Nigeria and Niger, SW Mauritania to Morocco to Egypt, Sudan and Somalia, Arabia to SW Iran
J. orientalis **Greater Egyptian jerboa** Deserts of N Africa and Arabia
J. turcmenicus **Turkmen jerboa** SE Coast of Caspian Sea through Turkmenistan to the Kyzyl Kum Desert (C Uzbekistan)
Stylodipus **Three-toed Jerboas**
S. andrewsi **Andrews' three-toed jerboa** NW, S, and C Mongolia, Nei Monggol, Xinjiang, Gansu, Ningxia (China)
S. sungorus **Mongolian three-toed jerboa** SW Mongolia, Xinjiang (China)
S. telum **Thick-tailed three-toed jerboa** E Ukraine, N Caucasus, Kazakhstan, W Uzbekistan, N and W Turkmenistan east to N Xinjiang, N Gansu, Nei Monggol (China)

SUBFAMILY EUCHOREUTINAE
Euchoreutes
E. naso **Long-eared jerboa** S Mongolia, Nei Monggol, Xinjiang, Qinghai, Gansu, Ningxia (China)

SUBFAMILY PARADIPODINAE
Paradipus
P. ctenodactylus **Comb-toed jerboa** Turkmenistan, Uzbekistan, E Aral region (Kazakhstan)

SUBFAMILY SICISTINAE
Sicista **Birch Mice**
S. armenica **Armenian birch mouse** NW Armenia
S. betulina **Northern birch mouse** N and C Europe to SE Siberia (Russia)
S. caucasica **Caucasian birch mouse** NW Caucasus
S. caudata **Long-tailed birch mouse** Ussuri region of NE China, Sakhalin Is (Russia)
S. concolor **Chinese birch mouse** Pakistan, India, Kashmir, Xinjiang, Qinghai, Gansu, Shaanxi, Sichuan (China)
S. kazbegica **Kazbeg birch mouse** Kazbegi (Georgia)
S. kluchorica **Kluchor birch mouse** NW Caucasus
S. napaea **Altai birch mouse** NW Altai Mts (Russia), Kazakhstan
S. pseudonapaea **Gray birch mouse** S Altai Mts (E Kazakhstan)
S. severtzovi **Severtzov's birch mouse** S Russia
S. strandi **Strand's birch mouse** N Caucasus north to Kursk (S Russia)
S. subtilis **Southern birch mouse** E Austria, Hungary, Yugoslavia, Romania east to S Russia, N Kazakhstan, SW Siberia, L. Balqash, Altai Mts, L. Baikal, and NW Xinjiang (China)
S. tianshanica **Tien Shan birch mouse** Tien Shan Mts of Kazakhstan and Xinjiang (China)

SUBFAMILY ZAPODINAE
Eozapus
E. setchuanus **Chinese jumping mouse** Qinghai, Gansu, Ningxia, Shaanxi, Sichuan, Yunnan (China)
Napaeozapus
N. insignis **Woodland jumping mouse** SC and SE Canada, NE and E USA
Zapus **Jumping Mice**
Z. hudsonius **Meadow jumping mouse** Canada, USA
Z. princeps **Western jumping mouse** W North America
Z. trinotatus **Pacific jumping mouse** SW British Columbia (Canada) south along W coast of USA to California

FAMILY MURIDAE
Rats, Mice, Voles, Lemmings, Hamsters, and Gerbils

SUBFAMILY ARVICOLINAE
Alticola **Mountain Voles**
A. albicauda **White-tailed mountain vole** Himalayas of NW India
A. argentatus **Silver mountain vole** Xinjiang (China) through the Tien Shan Mts and Pamirs to N Afghanistan, Pakistan and N India
A. barakshin **Gobi Altai mountain vole** Tuva (S Russia) east to W and S Mongolia
A. lemminus **Lemming vole** NE Siberia (Russia)
A. macrotis **Large-eared vole** Altai and Sayan Mts to N Mongolia, L. Baikal, and Yakutsk (NE Russia)
A. montosa **Central Kashmir vole** Kashmir
A. roylei **Royle's mountain vole** W Himalayas of N India
A. semicanus **Mongolian silver vole** Tuva (S Russia) to N Mongolia
A. stoliczkanus **Stoliczka's mountain vole** N Ladakh (India) and Nepal to W and N Xizang to Gansu (China)
A. stracheyi **Strachey's mountain vole** Himalayas from E Kashmir to S Xizang (China) and N Nepal to Sikkim (India)
A. strelzowi **Flat-headed vole** Quaraghandy (Kazakhstan) east to Altai Mts of NW Mongolia, Siberia (Russia), and Xinjiang (NW China)
A. tuvinicus **Tuva silver vole** Tuva (S Russia) and Altai Mts to L. Baikal and NW Mongolia
Arborimus **Tree Voles**
A. albipes **White-footed vole** Coastal W USA, from Oregon to NW Calfiornia
A. longicaudus **Red-tree vole** Coastal Oregon (USA)
A. pomo **Sonoma tree vole** Coastal NW California
Arvicola **Water Voles**
A. sapidus **Southwestern water vole** Portugal, Spain, W France
A. terrestris **European water vole** Europe east

through Siberia (Russia) south to Israel, Iran, Tien Shan Mts of NW China and L. Baikal
Blanfordimys Afghan Voles
B. afghanus **Afghan vole** S Turkmenistan, Uzbekistan, Tajikistan, Afghanistan
B. bucharicus **Bucharian vole** Mountains of SW Tajikistan
Chionomys Snow Voles
C. gud **Caucasian snow vole** Caucasus, NE Turkey
C. nivalis **European snow vole** Mountains of Europe (incl. Alps, Carpathians, Balkans, Pindus) east to W Caucasus, Turkey, Israel, Lebanon, Syria, and Zagros and Kopet Dagh Mts (Iran)
C. roberti **Robert's snow vole** W Caucasus Mts, NE Turkey
Clethrionomys Red-backed Voles
C. californicus **Western red-backed vole** Coastal W USA from Columbia R south to NW California (USA)
C. centralis **Tien Shan red-backed vole** Tien Shan Mts (Kazakhstan and Kyrgyzstan), Xinjiang (China)
C. gapperi **Southern red-backed vole** Canada (excl. Newfoundland), N USA, Rocky Mts and Appalachians
C. glareolus **Bank vole** W Palearctic from France and Scandinavia to L. Baikal south to N Spain, Italy, the Balkans, W Turkey, N Kazakhstan, Altai and Sayan Mts, Britain, SW Ireland
C. rufocanus **Gray red-backed vole** N Palearctic from Scandinavia through Siberia to Kamchatka Pen. (Russia) south to Ural and Altai Mts, Mongolia, N China, Korea, N Japan
C. rutilus **Northern red-backed vole** N Scandinavia east to Chukotskoye Pen. south to N Kazakhstan, Mongolia, NE China, Korea, Sakhalin Is, Hokkaido (Japan), and St Lawrence Is (Bering Sea, USA); also Alaska (USA) east to Hudson Bay south to N British Columbia and NE Manitoba (Canada)
C. sikotanensis **Sikotan red-backed vole** Sikotan, Daikoku and Rishiri Is (Kurilskiye Is, Russia)
Dicrostonyx Collared Lemmings
D. exsul **St. Lawrence Island collared lemming** St Lawrence Is (Bering Sea, USA)
D. groenlandicus **Northern collared lemming** N Greenland, Queen Eilzabeth Is south to Baffin and Southampton Is (Canada)
D. hudsonius **Ungava collared lemming** Labrador, N Quebec (Canada)
D. kilangmiutak **Victoria collared lemming** Victoria and Banks Is, adj. mainland (Canada)
D. nelsoni **Nelson's collared lemming** Alaska (USA)
D. nunatakensis **Ogilvie mountain collared lemming** Ogilvie Mts, NC Yukon Territory (Canada)
D. richardsoni **Richardson's collared lemming** W Coast of Hudson Bay west to Mackenzie District (Canada)
D. rubricatus **Bering collared lemming** N Alaska (USA)
D. torquatus **Arctic lemming** Palearctic from White Sea (W Russia) to Chukotskoye Pen. and Kamchatka Pen. (NE Russia), Novosibirskiye Is and Novaya Zemlya Is
D. unalascensis **Unalaska collared lemming** Aleutian Arch. (Alaska, USA)
D. vinogradovi **Wrangel lemming** Vrangelya Is (NE Siberia, Russia)
Dinaromys
D. bogdanovi **Martino's snow vole** Mountains of Yugoslavia
Ellobius Mole Voles
E. alaicus **Alai mole vole** Altai Mts (S Kyrgyzstan)
E. fuscocapillus **Southern or Afghan mole vole** E Iran, Afghanistan, W Pakistan, and Kopet Dagh Mts (S Turkmenistan)
E. lutescens **Transcaucasian mole vole** S Caucasus Mts south to E Turkey and NW Iran
E. talpinus **Northern mole-vole** S Ukraine east through Kazakhstan to N L. Balqash, also Turkmenistan
E. tancrei **Zaisan mole vole** NE Turkmenistan and Uzbekistan east through E Kazakhstan to Xinjiang and Nei Monggol (E China) and Mongolia
Eolagurus Yellow Steppe Lemmings
E. luteus **Yellow steppe lemming** N Xinjiang (China), W Mongolia
E. przewalskii **Przewalski's steppe lemming** China, S Mongolia
Eothenomys South Asian Voles
E. chinensis **Pratt's vole** Sichuan and Yunnan (China)
E. custos **Southwest China vole** Sichuan and Yunnan (China)
E. eva **Gansu vole** S Gansu, adj. Shaanxi, Sichuan, Hubei (China)
E. inez **Kolan vole** Shaanxi and Shanxi (China)
E. melanogaster **Pere David's vole** W and S China south to N Thailand and N Burma, Taiwan
E. olitor **Chaotung vole** Yunnan (China)
E. proditor **Yulungshan vole** Sichuan and Yunnan (China)
E. regulus **Royal vole** Korea
E. shanseius **Shansei vole** Shanxi and Hebei (China)
Hyperacrius Kashmir Voles
H. fertilis **True's vole** Kashmir, N Pakistan
H. wynnei **Murree or Punjab vole** N Pakistan
Lagurus
L. lagurus **Steppe lemming** Ukraine through N Kazakhstan to W Mongolia and NW China
Lasiopodomys Brandt's Voles
L. brandtii **Brandt's vole** Mongolia, adj. Russia, Nei Monggol, Heilongjiang, Hebei (China)
L. fuscus **Plateau vole** Qinghai (China)
L. mandarinus **Mandarin vole** N Mongolia, adj. Russia and SE Siberia, C and NE China, Korea
Lemmiscus
L. curtatus **Sagebrush vole** S Alberta and SE Sakatchewan (Canada) south to NW Colorado and EC California
Lemmus Brown Lemmings
L. amurensis **Amur lemming** E Siberia from L. Baikal north to Verkhoyansk, Khrebet Cherskogo, and Kolymskoye Nagorye Mts to Omolon R (NE Russia)
L. lemmus **Norway lemming** Scandinavia
L. sibiricus **Brown lemming** Palearctic from White Sea (W Russia) to Chukotskoye Pen. and Kamchatka Pen., Nunivak and St George Is (Bering Sea); Nearctic from W Alaska (USA) to Hudson Bay south in Rocky Mts to C British Columbia (Canada)
Microtus Meadow Voles
M. abbreviatus **Insular vole** Hall and St Matthew Is (Bering Sea, USA)
M. agrestis **Field vole** Britain, Scandinavia, France east through Europe and Siberia to Lena R (NE Russia) south to Pyrenees (France, Spain), N Yugoslavia, Ural and Altai Mts, Xinjiang (NW China) and L. Baikal
M. arvalis **Common vole** C and N Spain throughout Europe to west coast of Black Sea northeast to Kirov region (W Russia); also Orkney Is (UK), Guernsey (Channel Is, UK), Yeu (France)
M. bavaricus **Bavarian pine vole** Bavarian Alps (Germany)
M. breweri **Beach vole** Muskeget Is (Massachusets, USA)
M. cabrerae **Cabrera's vole** Spain, Portugal, N Pyrenees (S France)
M. californicus **California vole** Coastal W USA, from SW Oregon to California, to N Baja California (Mexico)
M. canicaudus **Gray-tailed vole** Willamette Valley (NW Oregon), adj. Washington (USA)
M. chrotorrhinus **Rock vole** SE Canada to NE Minnesota (USA) and south in Appalachian Mts to E Tennessee and W North Carolina (USA)
M. daghestanicus **Daghestan pine vole** Caucasus Mts (Dagestan, Russia)
M. duodecimcostatus **Mediterranean pine vole** SE France, E and S Spain, Portugal
M. evoronensis **Evorsk vole** Khabarovsk Krai (Russia)
M. felteni **Felten's vole** S Yugoslavia, Greece
M. fortis **Reed vole** SE Russia south through E China to Lower Yangtze Valley
M. gerbei **Gerbe's vole** SW France, Pyrenees (France, Spain)
M. gregalis **Narrow-headed vole** Palearctic from White Sea (W Russia) to NE, and from Urals east to Amur R and NE China, south to Aral Sea, Pamirs, Tien Shan and Altai Mts, N Mongolia, and NW China
M. guatemalensis **Guatemalan vole** C Chiapas (Mexico), Guatemala
M. guentheri **Gunther's vole** S Bulgaria, S Yugoslavia, E Greece, W Turkey
M. hyperboreus **North Siberian vole** NE Siberia (Russia)
M. irani **Persian vole** NE Libya, E Turkey, N Syria, Lebanon, Israel, Jordan, N Iraq, W and N Iran, Kopet Dagh Mts (Turkmenistan)
M. irene **Chinese scrub vole** N Burma, Xizang, Yunnan, Sichuan, Gansu (China)
M. juldaschi **Juniper vole** Tien Shan Mts, Pamirs (Kyrgyzstan, Tajikistan) west to Samarqand (Uzbekistan) south to NE Afghanistan and N Pakistan then east to Xizang (China)
M. kermanensis **Baluchistan vole** Kerman (SE Iran)
M. kirgisorum **Tien Shan vole** S Kazakhstan, Kyrgyzstan, Tajikistan, SE Turkmenistan
M. leucurus **Blyth's vole** S Qinghai, Xizang (China), and west in the Himalayas to Kashmir
M. limnophilus **Lacustrine vole** Qinghai (China) to W Mongolia
M. longicaudus **Long-tailed vole** E Alaska (USA) and Yukon Territory (Canada) south through SE Canada in Rocky Mts to W Colorado and E California; also S California, Arizona and New Mexico (USA)
M. lusitanicus **Lusitanian pine vole** Portugal, NW Spain, SW France
M. majori **Major's pine vole** S Yugoslavia, Mt Olympus (Greece), N and W Turkey, W and N Caucasus
M. maximowiczii **Maximowicz's vole** E shore of L. Baikal to Upper Amur region (SE Russia), E Mongolia, and Heilongjiang (NE China)
M. mexicanus **Mexican vole** S Utah and SW Colorado (USA) south in Sierra Madres to C Oaxaca (Mexico)
M. middendorffi **Middendorff's vole** NC Siberia (Russia)
M. miurus **Singing vole** N and SE Alaska (USA), NW Canada
M. mongolicus **Mongolian vole** Mongolia, NE China, and adj. Russia
M. montanus **Montane vole** SC British Colombia (Canada) to SW USA in Cascade, Sierra Nevada and Rocky Mts
M. montebelli **Japanese grass vole** Honshu, Sado, Kyushu (Japan)
M. mujanensis **Muisk vole** Vitim River Basin (E Russia)
M. multiplex **Alpine pine vole** S Alps and N Apennines (France, Switzerland, Italy, Austria), NW and C Yugoslavia
M. nasarovi **Nasarov's vole** NE Caucasus
M. oaxacensis **Tarabundi vole** NC Oaxaca (Mexico)
M. obscurus **Altai vole** Russia from E Ukraine through Siberia to Upper Yenesei R (Russia) south to Caucasus, N Iran, Altai Mts, L. Balqash, NW Mongolia, Xinjiang (NW China)
M. ochrogaster **Prairie vole** Northern and Central Great Plains of North America
M. oeconomus **Tundra vole** Palearctic from Scandinavia and Netherlands east to Bering Sea, south to E Germany, Ukraine, S Kazakhstan, Mongolia and the Ussuri region, Sakhalin Is (Russia), St Lawrence Is (Bering Sea, USA); Neartic from Alaska (USA) to Yukon Territory, W Northwest Territories and NW British Columbia (Canada)
M. oregoni **Creeping vole** SW British Columbia (Canada) south to NW California (USA)
M. pennsylvanicus **Meadow vole** Canada, USA (incl. Alaska), Chihuahua (Mexico)
M. pinetorum **Woodland vole** E USA, from S Maine to N Florida west to C Wisconsin and E Texas
M. quasiater **Jalapan pine vole** SE San Luis Potosi to N Oaxaca (Mexico)
M. richardsoni **Water vole** Rocky and Cascade Mts from SW Canada to Wyoming, C Utah and Oregon (USA)
M. rossiaemeridionalis **Southern vole** Finland east to Urals south to N Caucasus west to Ukraine, Romania, Bulgaria, Yugoslavia, N Greece, NW Turkey
M. sachalinensis **Sakhalin vole** Sakhalin Is (Russia)
M. savii **Savi's pine vole** SE France, Italy (incl. Sicily and Elba)
M. schelkovnikovi **Schelkovnikov's pine vole** Talysk and Alborz Mts (S Azerbaijan)
M. sikimensis **Sikkim vole** Himalayas from W Nepal through Sikkim (India) to S and E Xizang (China)
M. socialis **Social vole** Palearctic from Dnipro R (Ukraine) east to L. Balqash and NW Xinjiang (China) south through E Turkey to Syria and NE Iran
M. subterraneus **European pine vole** N and C France through C Europe to Ukraine and Don R, south through Yugoslavia to N Greece; also NE Russia
M. tatricus **Tatra pine vole** W and E Carpathian Mts, Tatra Mts, W Ukraine
M. thomasi **Thomas' pine vole** S coastal Yugoslavia to Greece
M. townsendii **Townsend's vole** SW British Columbia (Canada) to NW California (USA), Vancouver Is (Canada)
M. transcaspicus **Transcaspian vole** S Turkmenistan, N Afghanistan, N Iran
M. umbrosus **Zempoaltepec vole** Mt Zempoaltepec (Oaxaca, Mexico)
M. xanthognathus **Taiga vole** EC Alaska (USA) to W Northwest Territories to C Alberta and W coast of Hudson Bay (Canada)
Myopus
M. schisticolor **Wood lemming** Norway and Sweden through Siberia to Kolyma R and Kamchatka Pen. (NE Russia), south to Altai Mts, N Mongolia, Heilongjiang (NE China), and Sikhote Alin Mts; also in Ural Mts
Neofiber
N. alleni **Round-tailed muskrat** Florida to SE Georgia (USA)
Ondatra
O. zibethicus **Muskrat** North America (south of tree-line); introd. to C and N Europe, Ukraine, Russia, adj. China and Mongolia, Honshu Is (Japan), S Argentina
Phaulomys Japanese Voles
P. andersoni **Japanese red-backed vole** Honshu (Japan)
P. smithii **Smith's vole** Dogo, Honshu, Shikoku, Kyushu (Japan)
Phenacomys Heather Voles
P. intermedius **Western heather vole** SW British Columbia, adj. Alberta (SW Canada) south to SW USA
P. ungava **Eastern heather vole** S Yukon to Labrador south to S Alberta and along N Great Lakes and St Lawrence R (Canada)
Proedromys
P. bedfordi **Duke of Bedford's vole** Gansu and Sichuan (China)
Prometheomys
P. schaposchnikowi **Long-clawed mole vole** Caucasus Mts, NE Turkey
Synaptomys Bog Lemmings
S. borealis **Northern bog lemming** Alaska (USA) to N Washington (USA) east to Labrador (Canada); also from Gaspe Pen. (Quebec, Canada) to New Hampshire (USA)
S. cooperi **Southern bog lemming** Midwestern and E USA through SE Canada (incl. Cape Breton Is)
Volemys Musser's Voles
V. clarkei **Clarke's vole** NW Yunnan (China), N Burma
V. kikuchii **Taiwan vole** Highlands of Taiwan
V. millicens **Sichuan vole** E Xizang and Sichuan (China)
V. musseri **Marie's vole** W Sichuan (China)

SUBFAMILY CALOMYSCINAE
Calomyscus Mouse-like Hamsters
C. bailwardi **Mouse-like hamster** Iran
C. baluchi **Baluchi mouse-like hamster** W Pakistan, NC and E Afghanistan
C. hotsoni **Hotson's mouse-like hamster** Baluchistan (Pakistan)
C. mystax **Afghan mouse-like hamster** S Turkmenistan, NC and NE Iran, NW Afghanistan
C. tsolovi **Tsolov's mouse-like hamster** SW Syria
C. urartensis **Urartsk mouse-like hamster** Azerbaijan, NW Iran

SUBFAMILY CRICETINAE
Allocricetulus Mongolian Hamsters
A. curtatus **Mongolian hamster** North of Altai Mts east to Inner Mongolia, Xinjiang, Ningxia, Anhui (China)
A. eversmanni **Eversmann's hamster** N Kazakhstan, steppes from Volga R to Upper Irtysh R
Cansumys
C. canus **Gansu hamster** S Gansu and Shaanxi (China)
Cricetulus Dwarf Hamsters
C. alticola **Tibetan dwarf hamster** W Nepal, Ladak (N India), Kashmir, W Xizang (China)
C. barabensis **Striped dwarf hamster** S Siberia from Irtysh R to Ussuri region (Russia), south to Mongolia, Xinjiang to Nei Monggol (China), and Korea
C. kamensis **Kam dwarf hamster** Xizang (China)
C. longicaudatus **Long-tailed dwarf hamster** Tuva and Altai regions of Russia and Kazakhstan, Xinjiang (NW China), Mongolia, adj. China south to Xizang
C. migratorius **Gray dwarf hamster** SE Europe through Ukraine to Kazakhstan to S Mongolia and N China south to Turkey, Israel, Jordan, Lebanon, Iraq, Iran, Pakistan, and Afghanistan
C. sokolovi **Sokolov's dwarf hamster** W and S Mongolia, C Nei Monggol (China)
Cricetus
C. cricetus **Black-bellied hamster** From Belgium through C Europe, W Siberia and N Kazakhstan to Upper Yenesei R and Altai Mts, and Xinjiang (NW China)

Mesocricetus **Golden Hamsters**
M. auratus **Golden hamster** Aleppo (Syria)
M. brandti **Brandt's hamster** Turkey and Caucasus south to Israel, Lebanon, Syria, N Iraq, NW Iran
M. newtoni **Romanian hamster** E Romania, Bulgaria
M. raddei **Ciscaucasian hamster** Russia, from N Caucasus to Don R and Sea of Azov
Phodopus **Small Desert Hamsters**
P. campbelli **Campbell's hamster** Heilongjiang through Nei Monggol to Xinjiang (China), Mongolia, adj. Russia
P. roborovskii **Desert hamster** Tuva (S Russia) and E Kazakhstan, W and S Mongolia, Heilongjiang to N Xinjiang (China)
P. sungorus **Dzungarian hamster** E Kazakhstan, SW Siberia (Russia)
Tscherskia
T. triton **Greater long-tailed hamster** Shaanxi to Dongbei south to Anhui (China), Korea, north to Ussuri region (Russia)

SUBFAMILY CRICETOMYINAE
Beamys **Long-tailed Pouched Rats**
B. hindei **Long-tailed pouched rat** S Kenya, NE Tanzania
B. major **Greater Long-tailed pouched rat** S Tanzania, Malawi, NE Zambia
Cricetomys **African Giant Pouched Rats**
C. emini **Giant rat** Sierra Leone to S Nigeria, Cameroon and Gabon through Congo and DRC to S Uganda, Bioko (Equatorial Guinea)
C. gambianus **Gambian rat** Senegal and Sierra Leone to S Sudan south to S Angola, S Zambia, and E South Africa
Saccostomus **Pouched Mice**
S. campestris **Pouched mouse** SW Tanzania to C Angola south to S South Africa
S. mearnsi **Mearns' pouched mouse** S Ethiopia, S Somalia, Kenya, E Uganda, NE Tanzania

SUBFAMILY DENDROMURINAE
Dendromus **African Climbing Mice**
D. insignis **Remarkable climbing mouse** Ethiopian highlands, E Kenya, W Uganda south to Rwanda, Mt Kilimanjaro (NE Tanzania)
D. kahuziensis **Mount Kahuzi climbing mouse** Kivu region (E DRC)
D. kivu **Kivu climbing mouse** West and East of the Ruwenzori Mts in W Uganda and E DRC, also Kivu region (DRC)
D. lovati **Lovat's or Ethiopian climbing mouse** Ethiopian plateau
D. melanotis **Gray climbing mouse** South Africa north to Uganda, west to Mt Nimba (Guinea), also Ethiopia
D. mesomelas **Brant's climbing mouse** S, SE, and E South Africa, C Mozambique, N Botswana, NE Zambia, N and S Malawi, SE, NE, and NW DRC, SW and EC Tanzania
D. messorius **Banana climbing mouse** Benin and Nigeria east to DRC, Uganda, Kenya and Sudan
D. mystacalis **Chestnut climbing mouse** South Africa to Ethiopia and S Sudan
D. nyikae **Nyika climbing mouse** E South Africa, E Zimbabwe, Mozambique, Malawi, Zambia, N and C Angola, SC DRC, SW Tanzania
D. oreas **Cameroon climbing mouse** Mt Cameroun, Mt Kupe, and Mt Manenguba (E Cameroon)
D. vernayi **Vernay's climbing mouse** Chitau (EC Angola)
Dendroprionomys
D. rousseloti **Velvet climbing mouse** Brazzaville (Congo)
Deomys
D. ferrugineus **Congo forest mouse or Link rat** Central Africa, from Uganda, Rwanda to S Cameroon, Bioko (Equatorial Guinea)
Leimacomys
L. buettneri **Groove-toothed forest mouse** Bismarckburg (Togo)
Malacothrix
M. typica **Gerbil mouse or Large-eared mouse** South Africa, S Botswana, Namibia, SW Angola
Megadendromus
M. nikolausi **Nikolaus's mouse** Bale Mts (Ethiopia)
Prionomys
P. batesi **Dollman's tree mouse** W and S Cameroon, S CAR
Steatomys **Fat Mice**
S. caurinus **Northwestern fat mouse** W Africa, from Senegal to C Nigeria
S. cuppedius **Dainty fat mouse** W Africa, from Senegal to NC Nigeria and SC Niger
S. jacksoni **Jackson's fat mouse** Ashanti (Ghana), SW Nigeria
S. krebsii **Kreb's fat mouse** South Africa, NE and SE Botswana, Caprivi Strip (NE Namibia), S Angola, W Zambia
S. parvus **Tiny fat mouse** E South Africa, NW Zimbabwe, N Botswana, N Namibia, Angola, Zambia, Tanzania, Uganda, Kenya, Ethiopia, Somalia, S Sudan
S. pratensis **Common fat mouse** South Africa, Swaziland, Zimbabwe, N Botswana, NE and N Namibia, Angola, Zambia, N Malawi, Mozambique, DRC, SW Sudan west to Cameroon

SUBFAMILY GERBILLINAE
Ammodillus
A. imbellis **Ammodile** Somalia, E Ethiopia
Brachiones
B. przewalskii **Przewalski's gerbil** Xinjiang to Gansu (China)
Desmodilliscus
D. braueri **Pouched gerbil** N and C Sudan, N Cameroon, W Niger, N Nigeria, C Mali, Burkina Faso, Senegal, W Mauritania
Desmodillus
D. auricularis **Cape short-eared gerbil** South Africa, S Botswana, Namibia
Gerbillurus **Hairy-footed Gerbils**
G. paeba **South African hairy-footed gerbil** South Africa, Namibia, Botswana, Zimbabwe, Mozambique, SW Angola
G. setzeri **Setzer's hairy-footed gerbil** NW Namibia to SW Angola
G. tytonis **Dune hairy-footed gerbil** Namibia
G. vallinus **Bushy-tailed hairy-footed gerbil** W South Africa through Namibia
Gerbillus **Gerbils**
G. acticola **Berbera gerbil** Somalia
G. agag **Agag gerbil** Mali, Niger and N Nigeria to Chad, Sudan, and Kenya
G. allenbyi **Allenby's gerbil** N of Gaza to Haifa (Israel)
G. amoenus **Pleasant gerbil** Egypt, Libya
G. andersoni **Anderson's gerbil** Nile Delta south to El Faiyum (Egypt)
G. aquilus **Swarthy's gerbil** SE Iran, W Pakistan, S Afghanistan
G. bilensis **Bilen gerbil** Bilen (Ethiopia)
G. bonhotei **Bonhote's gerbil** NE Sinai Pen. (Egypt)
G. bottai **Botta's gerbil** Sudan, Kenya
G. brockmani **Brockman's gerbil** Somalia
G. burtoni **Burton's gerbil** Dharfur (Sudan)
G. campestris **North African gerbil** N Africa, from Morocco to Egypt and Sudan
G. cheesmani **Cheesman's gerbil** SW Iran, Iraq, Kuwait, Saudi Arabia, Oman, Yemen
G. cosensis **Cosens' gerbil** Ngamatak, Kozibiri R (Kenya)
G. dalloni **Dallon's gerbil** Tibesti (Chad)
G. dasyurus **Wagner's gerbil** Arabian Pen., Iraq, Syria, Lebanon, Israel, Sinai (Egypt)
G. diminutus **Diminutive gerbil** Kenya
G. dongolanus **Dongola gerbil** Dongola (Sudan)
G. dunni **Somalia gerbil** Ethiopia, Somalia, Djibouti
G. famulus **Black-tufted gerbil** Yemen
G. floweri **Flower's gerbil** Sinai (Egypt)
G. garamantis **Algerian gerbil** Algeria
G. gerbillus **Lesser Egyptian gerbil** Israel to Egypt and N Sudan to Morocco, also N Chad, N Niger and N Mali
G. gleadowi **Indian hairy-footed gerbil** NW India, Pakistan
G. grobbeni **Grobben's gerbil** Cyrenaica (Libya)
G. harwoodi **Harwood's gerbil** Kenya
G. henleyi **Pygmy gerbil** Algeria to Israel and Jordan, W Saudia Arabia, N Yemen, and Oman, also Burkina Faso, and N Senegal
G. hesperinus **Western gerbil** Coastal Morocco
G. hoogstraali **Hoogstraal's gerbil** Taroudannt (Morocco)
G. jamesi **James' gerbil** Tunisia
G. juliani **Julian's gerbil** Somalia
G. latastei **Lataste's gerbil** Tunisia, Libya
G. lowei **Lowe's gerbil** Jebel Marra (Sudan)
G. mackillingini **Mackillingin's gerbil** S Egypt, adj. Sudan
G. maghrebi **Greater short-eared gerbil** Fes (Morocco)
G. mauritaniae **Mauritanian gerbil** Aouker region (Mauritania)
G. mesopotamiae **Mesopotamian gerbil** SW Iran, Iraq
G. muriculus **Barfur gerbil** Sudan
G. nancillus **Sudan gerbil** El Fasher (Sudan)
G. nanus **Baluchistan gerbil** Morocco through N Africa to Arabian Pen., Jordan, Israel, Iraq, Iran, Pakistan, S Afghanistan to NW India
G. nigeriae **Nigerian gerbil** N Nigeria, Burkina Faso
G. occiduus **Occidental gerbil** Aoreora (Morocco)
G. percivali **Percival's gerbil** Kenya
G. perpallidus **Pale gerbil** N Egypt, west of Nile R
G. poecilops **Large Aden gerbil** Saudi Arabia, Yemen
G. principulus **Principal gerbil** Jebel Meidob (Sudan)
G. pulvinatus **Cushioned gerbil** Ethiopia
G. pusillus **Least gerbil** Kenya, Ethiopia, S Sudan
G. pyramidum **Greater Egyptian gerbil** Nile Delta (Egypt) south to N Sudan
G. quadrimaculatus **Four-spotted gerbil** Nubia (NE Sudan)
G. riggenbachi **Riggenbach's gerbil** Rio de Oro (Western Sahara), N Senegal
G. rosalinda **Rosalinda's gerbil** Sudan
G. ruberrimus **Little red gerbil** Kenya, E Ethiopia, Somalia
G. simoni **Lesser short-tailed gerbil** Egypt, W of Nile Delta, Libya, Tunisia, Algeria
G. somalicus **Somalian gerbil** Somalia
G. stigmonyx **Khartoum gerbil** Sudan
G. syrticus **Sand gerbil** Nofilia (Libya)
G. tarabuli **Tarabul's gerbil** Libya
G. vivax **Vivacious gerbil** Libya
G. watersi **Waters' gerbil** Somalia, Sudan
Meriones **Jirds**
M. arimalius **Arabian jird** Saudia Arabia, Oman
M. chengi **Cheng's jird** N Xinjiang (China)
M. crassus **Sundevall's jird** N Africa, from Morocco through Niger and Sudan to Egypt then to Saudi Arabia, Jordan, Israel, Syria, Iraq, Iran, Aghanistan
M. dahli **Dahl's jird** Armenia
M. hurrianae **Indian desert jird** SE Iran, Pakistan, NW India
M. libycus **Libyan jird** N Africa, from Western Sahara to Egypt, Saudi Arabia, Jordan, Iraq, Syria, Iran, Afghanistan to S Kazakhstan and Xinjiang (W China)
M. meridianus **Midday jird** N Caucasus and Lower Don R to Mongolia and Xinjiang, Qinghai, Shanxi and Hebei (China) south to E Iran and Afghanistan
M. persicus **Persian jird** Iran, adj. regions of Iraq, Turkey, Caucasus, Turkmenistan, Afghanistan, and Pakistan
M. rex **King jird** Mecca to Aden (SW Saudi Arabia)
M. sacramenti **Buxton's jird** S Israel
M. shawi **Shaw's jird** Morocco to N Sinai (Egypt)
M. tamariscinus **Tamarisk jird** N Caucasus and Kazakhstan to Altai Mts and through N Xinjiang and W Gansu (China)
M. tristrami **Tristram's jird** Jordan, Lebanon and Israel to E Turkey, Caucasus, Syria, Iraq, and NW Iran
M. unguiculatus **Mongolian or Clawed jird** Mongolia, adj. regions of Siberia (Russia), N Gansu through Nei Monggol to Heilongjiang (China)
M. vinogradovi **Vinogradov's jird** N Syria, SE Turkey, Armenia, Azerbaijan, N Iran
M. zarudnyi **Zarudny's jird** NE Iran, S Turkmenistan, N Afghanistan
Microdillus
M. peeli **Somali pygmy gerbil** Somalia
Pachyuromys
P. duprasi **Fat-tailed gerbil** N Sahara, from W Morocco to N Egypt
Psammomys **Sand Rats**
P. obesus **Fat sand rat** N Africa, from Algeria to coastal Egypt, Syria, Jordan, Israel, Arabia; also coastal Sudan
P. vexillaris **Thin sand rat or Pale fat sand rat** Algeria, Tunisia, Libya
Rhombomys
R. opimus **Great gerbil** S Mongola through Ningxia, Gansu and Xinjiang (China) to Kazakhstan, Afghanistan, Pakistan, and Iran
Sekeetamys
S. calurus **Bushy-tailed jird** E Egypt to Sinai, S Israel, Jordan, C Saudi Arabia
Tatera **Large Naked-soled Gerbils**
T. afra **Cape gerbil** SW South Africa
T. boehmi **Boehm's gerbil** Angola, Zambia, Malawi, S DRC, Tanzania, Uganda, Kenya
T. brantsii **Highveld gerbil** South Africa, W Zimbabwe, Botswana, C and E Namibia, S Angola, SW Zambia
T. guineae **Guinea gerbil** Senegal and Gambia to Ghana and Burkina Faso
T. inclusa **Gorongoza gerbil** E Zimbabwe, Mozambique, Tanzania
T. indica **Indian gerbil** Syria, Iraq and Kuwait through Iran, Afghanistan and Pakistan to India and north to S Nepal, also Sri Lanka
T. kempi **Kemp's gerbil** Senegal and Guinea to Cameroon
T. leucogaster **Bushveld gerbil** South Africa to S Angola, S DRC, and SW Tanzania
T. nigricauda **Black-tailed gerbil** Tanzania, Kenya, Somalia
T. phillipsi **Phillip's gerbil** Somalia, Rift Valley (Ethiopia, Kenya)
T. robusta **Fringe-tailed gerbil** Chad, Sudan, Ethiopia, Somalia, Kenya, Uganda, Tanzania, also Burkina Faso
T. valida **Savanna gerbil** CAR, Chad east to Sudan and Ethiopia south to SW Tanzania, Zambia, and Angola
Taterillus **Small Naked-soled Gerbils**
T. arenarius **Sahel gerbil** Mauritania to Niger
T. congicus **Congo gerbil** Cameroon, Chad, CAR, DRC, Sudan, Uganda
T. emini **Emin's gerbil** Sudan, W Ethiopia, NW Kenya, Uganda, NE DRC
T. gracilis **Slender gerbil** Senegal and Gambia to Burkina Faso, Niger, and N Nigeria
T. harringtoni **Harrington's gerbil** CAR, Sudan, Ethiopia, Somalia, Kenya, Uganda, Tanzania
T. lacustris **Lake Chad gerbil** NE Nigeria, Cameroon
T. petteri **Petter's gerbil** Burkina Faso, W Niger
T. pygargus **Senegal gerbil** Senegal, Gambia, S Mauritania, W Mali

SUBFAMILY LOPHIOMYINAE
Lophiomys
L. imhausi **Crested or Maned rat** E Sudan, Ethiopia, Somalia, Kenya, Uganda, Tanzania, poss. Arabia

SUBFAMILY MURINAE
Abditomys
A. latidens **Luzon broad-toothed rat** Luzon (Philippines)
Acomys **African Spiny Mice**
A. cahirinus **Cairo spiny mouse** W Sahara to Egypt, N Sudan, N Ethiopia, Jordan, Israel, Lebanon, Syria, Saudi Arabia, Yemen, Oman, S Iraq, Iran, Pakistan
A. cilicicus **Asia Minor spiny mouse** Silifke (W Turkey)
A. cinerasceus **Gray spiny mouse** N Ghana and Burkina Faso to C and S Sudan, N Uganda, and C and S Ethiopia
A. ignitus **Fiery spiny mouse** Usambara Mts (Tanzania), Kenya
A. kempi **Kemp's spiny mouse** S Somalia, Kenya, NE Tanzania
A. louisae **Louise's spiny mouse** Somalia
A. minous **Crete spiny mouse** Crete
A. mullah **Mullah spiny mouse** Ethiopia, Somalia
A. nesiotes **Cyprus spiny mouse** Cyprus
A. percivali **Percival's spiny mouse** S Sudan, Ethiopia, Somalia, Kenya, Uganda
A. russatus **Golden spiny mouse** E Egypt (incl. Sinai), Jordan, Israel, Saudi Arabia
A. spinosissimus **Spiny mouse** NE and EC Tanzania, SE DRC, Zambia, Malawi, C Mozambique, Zimbabwe, E and SE Botswana, N and NW South Africa
A. subspinosus **Cape spiny mouse** SW South Africa
A. wilsoni **Wilson's spiny mouse** S Sudan, S Ethiopia, S Somalia, Kenya to EC Tanzania
Aethomys **African Rock Rats**
A. bocagei **Bocage's rock rat** C and W Angola
A. chrysophilus **Red rock rat** SE Kenya and Tanzania to S Angola, E Botswana, Zimbabwe, Mozambique, and South Africa
A. granti **Grant's rock rat** S South Africa
A. hindei **Hinde's rock rat** N Cameroon, N and NE DRC, S Sudan, SW Ethiopia, Kenya, Uganda, and Tanzania
A. kaiseri **Kaiser's rock rat** S Kenya, Tanzania, SW Uganda, Rwanda, S and E DRC, Malawi, Zambia, and E Angola
A. namaquensis **Namaqua rock rat** South Africa, Botswana, Zimbabwe, S and C Mozambique, S Malawi, SE Zambia, and S Angola
A. nyikae **Nyika rock rat** N Angola, N Zambia, Malawi, S DRC
A. silindensis **Selinda rock rat** Mt Silinda (E Zimbabwe)
A. stannarius **Tinfield's rock rat** N Nigeria to W Cameroon
A. thomasi **Thomas' rock rat** W and C Angola
Anisomys
A. imitator **Squirrel-toothed or Powerful-toothed rat** New Guinea
Anonymomys
A. mindorensis **Mindoro rat** Mindoro (Philippines)
Apodemus **Field Mice**
A. agrarius **Striped field mouse** C Europe to L. Baikal south to Caucasus and Tien Shan Mts, then

from Amur R through Korea to E Xizang, E Yunnan, Sichuan, and Fujiau (China), Taiwan, Cheju do (S. Korea)
A. alpicola **Alpine field mouse** NW Alps (S Germany, Austria, Liechtenstein, Switzerland, N Italy)
A. argenteus **Small Japanese field mouse** Japan
A. arianus **Persian field mouse** N Iran to Lebanon and N Israel
A. chevrieri **Chevrier's field mouse** W China, from Hubei and S Gansu to N and C Yunnan
A. draco **South China field mouse** China, Burma, Assam (India)
A. flavicollis **Yellow-necked mouse** England and Wales, S Scandinavia, France, and N Spain east through Europe to Ural Mts, S Italy, the Balkans, and Syria, Lebanon and Israel
A. fulvipectus **Yellow-breasted field mouse** E Ukraine through S Russia to the Caucasus, N Turkey, N Iran, east to Kopet Dagh Mts
A. gurkha **Himalayan field mouse** Nepal
A. hermonensis **Mount Hermon field mouse** Mt Hermon (Israel)
A. hyrcanicus **Caucasus field mouse** E Caucasus
A. latronum **Sichuan field mouse** Xizang, Sichuan, Yunnan (China), N Burma
A. mystacinus **Broad-toothed field mouse** SE Europe, S Georgia, NW Iran, Iraq, Jordan, Lebanon, Syria, Saudi Arabia, Rodhos and Crete
A. peninsulae **Korean field mouse** Altai Mts and Ussuri region to Korea, E Mongolia, NE China to SW China, Sakhalin Is, Hokkaido (Japan)
A. ponticus **Black Sea field mice** Sea of Azov south through Caucasus to Armenia, E Turkey, Iraq
A. rusiges **Kashmir field mice** Kashmir, N India
A. semotus **Taiwan field mouse** Taiwan
A. speciosus **Large Japanese field mouse** Japan
A. sylvaticus **Long-tailed field mouse or Wood mouse** Europe, north to Scandinavia and east to Ukraine and Belarus, Iceland, Britain, Ireland, most mediterranean islands; also N Africa, from Atlas Mts (Morocco) to Tunisia
A. uralensis **Ural field mouse** E Europe and Turkey, south into the Caucasus, to the Altai Mts and Xinjiang (NW China)
A. wardi **Ward's field mouse** NC Nepal through Kashmir, N Pakistan and Afghanistan to NW Iran
Apomys **Philippine Forest Mice**
A. abrae **Luzon Cordillera forest mouse** Luzon (Philippines)
A. datae **Luzon montane forest mouse** N Luzon (Philippines)
A. hylocoetes **Mount Apo forest mouse** Mindanao (Philippines)
A. insignis **Mindanao montane forest mouse** Mindanao (Philippines)
A. littoralis **Mindanao lowland forest mouse** Mindanao, Bohol, Biliran, Dinagat, and Leyte (Philippines)
A. microdon **Small Luzon forest mouse** S Luzon (Philippines)
A. musculus **Least forest mouse** Luzon and Mindoro (Philippines)
A. sacobianus **Long-nosed Luzon forest mouse** Luzon (Philippines)
Archboldomys
A. luzonensis **Mount Isarog shrew mouse** Mt Isarog (SE Luzon, Philippines)
Arvicanthis **African or Unstriped Grass Rats**
A. abyssinicus **Abyssinian grass rat** Ethiopia
A. blicki **Blick's grass rat** Ethiopia
A. nairobae **Nairobi grass rat** Rift Valley and east, from C Kenya south to Dodoma region (EC Tanzania)
A. niloticus **African grass rat** S Mauritania, Senegal and Gambia east to Sudan, Egypt, and W Ethiopia, south through NE DRC and Tanzania (west of Rift valley), into E Zambia and N Malawi; also SW Arabia
A. newmanni **Newmann's grass rat** N and E Rift Valleys of Ethiopia, Somalia, SE Sudan, south through E Kenya to C and EC Tanzania
Bandicota **Bandicoot Rats**
B. bengalensis **Lesser bandicoot rat** India (incl. Assam), Sri lanka, Pakistan, Kashmir, Nepal, Bangladesh, Burma, Penang Is (Malaysia), Sumatra and Java (Indonesia), Saudi Arabia
B. indica **Greater bandicoot rat** India (incl. Assam), Sri lanka, Nepal, Burma, Yunnan and Hong Kong (China), Taiwan, Thailand, Laos, Vietnam; introd. to Kedah and Perlis regions (Malay Pen.), Sumatra and Java (Indonesia)
B. savilei **Savile's bandicoot rat** E Burma, Thailand, S Laos, Vietnam, Cambodia
Batomys **Hairy-tailed Rats**
B. dentatus **Large-toothed hairy-tailed rat** N Luzon (Philippines)
B. granti **Luzon hairy-tailed rat** Mt Data and Mt Isarog (Luzon, Philippines)
B. salomonseni **Mindanao hairy-tailed rat** Mindanao, Biliran, and Leyte (Philippines)
Berylmys **White-toothed Rats**
B. berdmorei **Small white-toothed rat** S Burma, N and SE Thailand, N Laos, Cambodia, S Vietnam
B. bowersi **Bower's white-toothed rat** NE India, N and C Burma, Yunnan, Guangxi, Fujiau, S Anhui (S China), Thailand, N Laos, N Vietnam, Malay Pen., NW Sumatra (Indonesia)
B. mackenziei **Kenneth's white-toothed rat** Assam (NE India), C and S Burma, Sichuan (China), S Vietnam
B. manipulus **Manipur white-toothed rat** Assam (NE India), N and C Burma
Bullimus **Philippine rats**
B. bagobus **Bagobo rat** Philippines
B. luzonicus **Luzon forest rat** Luzon (Philippines)
Bunomys **Hill Rats**
B. andrewsi **Andrew's hill rat** Sulawesi (Indonesia)
B. chrysocomus **Yellow-haired hill rat** Sulawesi (Indonesia)
B. coelestis **Heavenly hill rat** Mt Lampobatang (SW Sulawesi, Indonesia)
B. fratrorum **Fraternal hill rat** NE Sulawesi (Indonesia)
B. heinrichi **Heinrich's hill rat** Mt Lampobatang (SW Sulawesi, Indonesia)
B. penitus **Inland hill rat** C and SE Sulawesi (Indonesia)
B. prolatus **Long-headed hill rat** Mt Tambusisi (C Sulawesi, Indonesia)
Canariomys
C. tamarani **Canary mouse** Extinct: formerly Canarias Is
Carpomys **Luzon Tree Rats**
C. melanurus **Short-faced Luzon tree rat** Mt Data (N Luzon, Philippines)
C. phaeurus **White-bellied Luzon tree rat** Mt Data and Mt Kapilingan (N Luzon, Philippines)
Celaenomys
C. silaceus **Blazed Luzon shrew rat** N Luzon (Philippines)
Chiromyscus
C. chiropus **Fea's tree rat** E Burma, N Thailand, C Laos, Vietnam
Chiropodomys **Pencil-tailed Tree Mice**
C. calamianensis **Palawan pencil-tailed tree mouse** Busuanga, Palawan, and Balabac Is (Philippines)
C. gliroides **Pencil-tailed tree mouse** Guangxi and Yunnan (China), Assam (NE India), Burma, Thailand, Laos, Vietnam, Malay Pen., Indonesia
C. karlkoopmani **Koopman's pencil-tailed tree mouse** Pagai and Siberut (Mentawai Is, Indonesia)
C. major **Large pencil-tailed tree mouse** Sarawak and Sabah (N Borneo)
C. muroides **Gray-bellied pencil-tailed tree mouse** Mt Kinabulu (Sabah, Borneo), N Kalimantan (Borneo)
C. pusillus **Small pencil-tailed tree mouse** Sabah, Sarawak, S Kalimantan (Borneo)
Chiruromys **Tree Mice**
C. forbesi **Greater tree mouse** SE Papua New Guinea (incl. D'Entrecasteaux Is)
C. lamia **Broad-headed tree mouse** SE Papua New Guinea
C. vates **Lesser tree mouse** Papua New Guinea
Chrotomys **Philippine Striped Rats**
C. gonzalesi **Isarog striped shrew-rat** Mt Isarog (SE Luzon, Philippines)
C. mindorensis **Mindoro striped rat** Luzon and Mindoro (Philippines)
C. whiteheadi **Luzon striped rat** Luzon (Philippines)
Coccymys **Brush Mice**
C. albidens **White-toothed brush mouse** Irian Jaya (New Guinea)
C. ruemmleri **Rummler's brush mouse** New Guinea
Colomys
C. goslingi **African water rat** Liberia, Cameroon, DRC, NE Angola, Uganda, Rwanda, NW Zambia, E Kenya, W Ethiopia, S Sudan
Conilurus **Rabbit Rats**
C. albipes **White-footed rabbit rat** SE Australia
C. penicillatus **Bushy-tailed rabbit rat** Coastal Northern Territory, NE Western Australia (Australia), adj. islands, SC Papua New Guinea
Coryphomys
C. buhleri **Buhler's rat** Timor (Indonesia)
Crateromys **Bushy-tailed Cloud-rats**
C. australis **Dinagat bushy-tailed cloud-rat** Dinagat Is (Philippines)
C. paulus **Ilin bushy-tailed cloud-rat** Mindoro (Philippines)
C. schadenbergi **Luzon bushy-tailed cloud-rat** N Luzon (Philippines)
Cremnomys **Indian Rats**
C. blanfordi **Blanford's rat** India, Sri Lanka
C. cutchicus **Cutch rat** India
C. elvira **Elvira rat** SE India
Crossomys
C. moncktoni **Earless water rat** Papua New Guinea
Crunomys **Philippines Shrew Rats**
C. celebensis **Celebes or Sulawesi shrew rat** C Sulawesi (Indonesia)
C. fallax **Northern Luzon shrew rat** NC Luzon (Philippines)
C. melanius **Mindanao shrew rat** Mindanao (Philippines)
C. rabori **Leyte shrew rat** Leyte (Philippines)
Dacnomys
D. millardi **Millard's rat** NE India, E Nepal, S Yunnan (S China), N Laos
Dasymys **Shaggy African Marsh Rats**
D. foxi **Fox's shaggy rat** Jos plateau (Nigeria)
D. incomtus **African marsh rat** S Africa, Zimbabwe, Angola, Zambia, Malawi, DRC, Tanzania, Uganda, Kenya, Ethiopia, S Sudan
D. montanus **Montane shaggy rat** Ruwenzori Mts (Uganda)
D. nudipes **Angolan marsh rat** N Botswana, NE Namibia, S Angola, SW Zambia
D. rufulus **West African shaggy rat** Sierra Leone, Liberia, Ivory Coast, Ghana, Togo, Benin, E Nigeria, Cameroon
Dephomys **Defua Rats**
D. defua **Defua rat** Sierra Leone, Guinea, Liberia, Ivory Coast, Ghana
D. eburnea **Ivory Coast rat** Ivory Coast, Liberia
Desmomys
D. harringtoni **Harrington's rat** Ethiopian plateau
Diomys
D. crumpi **Crump's or Manipur mouse** NE India, W Nepal
Diplothrix
D. legatus **Ryukyu rat** Japan
Echiothrix
E. leucura **Sulawesi spiny rat** N and C Sulawesi (Indonesia)
Eropeplus
E. canus **Sulawesi soft-furred rat** C Sulawesi (Indonesia)
Golunda
G. ellioti **Indian Bush Rat** SE Iran, Pakistan, N and NE India south to Sri Lanka, Nepal
Grammomys **African Thicket Rats**
G. aridulus **Arid thicket rat** WC Sudan
G. buntingi **Bunting's thicket rat** W Africa, from Sierra Leone and Guinea to Ivory Coast and Liberia
G. caniceps **Gray-headed thicket rat** S Somalia, N Kenya
G. cometes **Mozambique thicket rat** SE and E South Africa, E Zimbabwe, S Mozambique
G. dolichurus **Woodland thicket rat** Nigeria east to SW Ethiopia, south to Kenya, Uganda, Tanzania, N DRC, Angola, Zambia, Malawi, Zimbabwe, South Africa
G. dryas **Forest thicket rat** Ruwenzoris and Kivu region (Uganda, E DRC), NW Burundi
G. gigas **Giant thicket rat** Mt Kenya (Kenya)
G. ibeanus **Ruwenzori thicket rat** NE Zambia, Malawi, to SC Tanzania and Kenya to S Sudan
G. macmillani **Macmillan's thicket rat** Sierra Leone, Liberia, CAR, S Sudan, Ethiopia, Kenya, Uganda, N DRC, Tanzania, Malawi, Mozambique, Zimbabwe
G. minnae **Ethiopian thicket rat** S Ethiopia
G. rutilans **Shining thicket rat** Guinea
Hadromys
H. humei **Manipur bush rat** NE India, W Yunnan (S China)
Haeromys **Ranne Mice**
H. margarettae **Ranee mouse** Sarawak and Sabah (N Borneo)
H. minahassae **Minahassa Ranee mouse** NE and C Sulawesi (Indonesia)
H. pusillus **Lesser Ranee mouse** Borneo, Palawan (Philippines)
Hapalomys **Marmoset Rats**
H. delacouri **Delacour's marmoset rat** S China, N Laos, S Vietnam
H. longicaudatus **Marmoset rat** SE Burma, Thailand, Malay Pen.
Heimyscus
H. fumosus **African smoky mouse** Gabon, S Cameroon, CAR
Hybomys **Striped Mice**
H. basilii **Father Basilio's striped mouse** Bioko (Equatorial Guinea)
H. eisentrauti **Eisentraut's striped mouse** Mt Lefo and Mt Oku (W Cameroon)
H. lunaris **Moon striped mouse** NE and E DRC, W Uganda, Rwanda
H. planifrons **Miller's striped mouse** NE Sierra Leone, Liberia, SE Guinea, W Ivory Coast
H. trivirgatus **Temminck's striped mouse** E Sierra Leone to S Nigeria
H. univittatus **Peter's striped mouse** SE Nigeria, Cameroon, CAR, Equatorial Guinea, Gabon, Congo, DRC, S Uganda and W Rwanda, NW Zambia
Hydromys **Water Rats**
H. chrysogaster **Golden-bellied water rat** Australia (incl. Tasmania), New Guinea, Kei and Aru Is (Indonesia), Obi Is (Maluku, Indonesia)
H. habbema **Mountain water rat** Irian Jaya (New Guinea)
H. hussoni **Western water rat** New Guinea
H. neobrittanicus **New Britain water rat** New Britain Is (Bismarck Arch., Papua New Guinea)
H. shawmayeri **Shaw Mayer's water rat** Papua New Guinea
Hylomyscus **African Wood Mice**
H. aeta **Beaded wood mouse** Equatorial Guinea, Gabon, Cameroon to W Uganda and Burundi
H. alleni **Allen's wood mouse** W Africa, from Guinea to Cameroon, Gabon, and Equatorial Guinea
H. baeri **Baer's wood mouse** Ivory Coast, Ghana
H. carillus **Angolan wood mouse** N Angola
H. denniae **Montane wood mouse** Kenya, Uganda, Rwanda, Tanzania, Tanzania, NE Zambia, E DRC, also WC Angola
H. parvus **Little wood mouse** S Cameroon, S CAR, N and E DRC, N Gabon
H. stella **Stella wood mouse** S Nigeria, Cameroon, CAR, Gabon, N Angola, DRC, S Sudan, W Kenya, Uganda, Rwanda, Burundi, Tanzania
Hyomys **New Guinean Giant Rats**
H. dammermani **Western white-eared giant rat** New Guinea
H. goliath **Eastern white-eared giant rat** Papua New Guinea
Kadarsanomys
K. sodyi **Sody's tree rat** Java (Indonesia)
Komodomys
K. rintjanus **Komodo rat** Nusa Tenggara (Lesser Sunda Is, Indonesia)
Lamottemys
L. okuensis **Mount Oku rat** Mt Oku (W Cameroon)
Leggadina **Australian Native Mice**
L. forresti **Forrest's mouse** Inland Australia
L. lakedownensis **Lakeland Downs mouse** N Queensland, Western Australia (Australia)
Lemniscomys **Striped Grass Mice**
L. barbarus **Barbary striped grass mouse** Tunisia to Western Sahara, Senegal and Gambia to Cameroon, Sudan, Ethiopia, Kenya, N Uganda, Tanzania, E DRC
L. bellieri **Bellier's striped grass mouse** Ivory Coast
L. griselda **Griselda's striped grass mouse** Angola
L. hoogstraali **Hoogstraal's striped grass mouse** N Sudan
L. linulus **Senegal one-striped grass mouse** Sudan, Senegal, Ivory Coast
L. macculus **Buffoon striped grass mouse** S Sudan, Ethiopia, Kenya, Uganda, NE DRC
L. mittendorfi **Mittendorf's striped grass mouse** L. Oku (Cameroon)
L. rosalia **Single-striped grass mouse** South Africa, Swaziland, Zimbabwe, C and N Botswana, N Namibia, Zambia, Malawi, Mozambique to S Kenya
L. roseveari **Rosevear's striped grass mouse** Zambia
L. striatus **Typical striped grass mouse** Burkina Faso and Sierra Leone east to Ethiopia, south to NW Angola, NE Zambia and N Malawi
Lenomys
L. meyeri **Trefoil-toothed giant rat** N, C, and SW Sulawesi (Indonesia)
Lenothrix
L. canus **Gray tree rat** Malay Pen., Penang and Tuangku Is (Malaysia), Sarawak and Sabah (N Borneo)
Leopoldamys **Long-tailed Giant Rats**
L. edwardsi **Edwards long-tailed giant rat** W Bengal, Assam, Sikkim (India), N Burma, S and C China, N Thailand, Laos, Vietnam, Malay Pen., W Sumatra (Indonesia)
L. neilli **Neill's long-tailed giant rat** C and W Thailand
L. sabanus **Long-tailed giant rat** Bangladesh, Thailand, Laos, Vietnam, Cambodia, Malay Pen.,

Sumatra and Java (Indonesia), Borneo
L. siporanus **Mentawai long-tailed giant rat** Mentawai Is (Indonesia)
Leporillus **Australian Stick-nest Rats**
L. apicalis **Lesser stick-nest rat** S Australia
L. conditor **Greater stick-nest rat** Franklin Is, Nuyt's Arch. (W South Australia)
Leptomys **New Guinean Water Rats**
L. elegans **Long-footed water rat** Papua New Guinea
L. ernstmayri **Ernst Mayr's water rat** Papua New Guinea, Arfak Mts (Irian Jaya, New Guinea)
L. signatus **Fly River water rat** W Papua New Guinea
Limnomys
L. sibuanus **Mindanao mountain rat** Mt Apo and Mt Malindang (Mindanao, Philippines)
Lophuromys **Brush-furred Rats**
L. cinereus **Gray brush-furred rat** Parc National du Kauzi-Biege (DRC)
L. flavopunctatus **Yellow-spotted brush-furred rat** NE Angola, N Zambia, Malawi and N Mozambique to DRC, Uganda, and Kenya; also Ethiopia
L. luteogaster **Yellow-bellied brush-furred rat** NE and E DRC
L. medicaudatus **Medium-tailed brush-furred rat** E DRC, Rwanda
L. melanonyx **Black-clawed brush-furred rat** S and C Ethiopia
L. nudicaudus **Fire-bellied brush-furred rat** W Cameroon, Equatorial Guinea (incl. Bioko), Gabon
L. rahmi **Rahm's brush-furred rat** E DRC, Rwanda
L. sikapusi **Rusty-bellied brush-furred rat** Sierra Leone to DRC, Uganda, and W Kenya; also N Angola
L. woosnami **Woosnam's brush-furred rat** E DRC, W Uganda, Rwanda, Burundi
Lorentzimys
L. nouhuysi **New Guinea jumping mouse** New Guinea
Macruromys **New Guinea Rats**
M. elegans **Western small-toothed rat** Mt Kunupi (Irian Jaya, New Guinea)
M. major **Eastern small-toothed rat** New Guinea
Malacomys **African Swamp Rats**
M. cansdalei **Cansdale's swamp rat** E Liberia, S Ivory Coast, S Ghana
M. edwardsi **Edward's swamp rat** Sierra Leone, Guinea, Liberia, S Ivory Coast, S Ghana, S Nigeria
M. longipes **Big-eared swamp rat** Guinea to S Sudan, Uganda, Rwanda south to NW Zambia and NE Angola
M. lukolelae **Lukolela swamp rat** Lukolela (DRC)
M. verschureni **Verschuren's swamp rat** NE DRC
Mallomys **Woolly Rats**
M. aroaensis **De Vis' woolly rat** New Guinea
M. gunung **Alpine woolly rat** Irian Jaya (New Guinea)
M. istapantap **Subalpine woolly rat** New Guinea
M. rothschildi **Rothschild's woolly rat** New Guinea
Malpaisomys
M. insularis **Lava mouse** Extinct: formerly Canarias Is
Margaretamys **Margareta's Rats**
M. beccarii **Beccari's margareta rat** NE and C Sulawesi (Indonesia)
M. elegans **Elegant margareta rat** Mt Nokilalaki (C Sulawesi, Indonesia)
M. parvus **Little margareta rat** Mt Nokilalaki (C Sulawesi, Indonesia)
Mastomys **Multimammate Mice**
M. angolensis **Angolan multimammate mouse** Angola, S DRC
M. coucha **Southern multimammate mouse** South Africa, Zimbabwe, Namibia
M. erythroleucus **Guinea multimammate mouse** Morocco, then from Senegal and Gambia to S Ethiopia and Somalia, south through E Africa to E DRC
M. hildebrandtii **Hildebrandt's multimammate mouse** Senegal and Gambia to CAR and N DRC to Somalia and Djibouti, south to to Kenya and Burundi
M. natalensis **Natal multimammate mouse** South Africa, Zimbabwe, Namibia, Tanzania, Senegal
M. pernanus **Dwarf multimammate mouse** SW Kenya, NW Tanzania, Rwanda
M. shortridgei **Shortridge's multimammate mouse** NW Botswana, NE Namibia
M. verheyeni **Verheyen's multimammate mouse** Nigeria, Cameroon
Maxomys **Oriental Spiny Rats**
M. alticola **Mountain spiny rat** Mt Kinabulu and Mt Trus Madi (Sabah, N Borneo)
M. baeodon **Small spiny rat** Sabah and Sarawak (N Borneo)
M. bartelsii **Bartel's spiny rat** W and C Java (Indonesia)
M. dollmani **Dollman's spiny rat** C and SE Sulawesi (Indonesia)
M. hellwaldii **Hellwald's spiny rat** Sulawesi (Indonesia)
M. hylomyoides **Sumatran spiny rat** W Sumatra (Indonesia)
M. inas **Malayan mountain spiny rat** Malay Pen.
M. inflatus **Fat-nosed spiny rat** W Sumatra (Indonesia)
M. moi **Mo's spiny rat** S Vietnam, S Laos
M. musschenbroekii **Musschenbroek's spiny rat** Sulawesi (Indonesia)
M. ochraceiventer **Chestnut-bellied spiny rat** Sabah, Sarawak, E Kalimantan (Borneo)
M. pagensis **Pagai spiny rat** Mentawai Is (Indonesia)
M. panglima **Palawan spiny rat** Culion, Palawan, Basuanga, and Balabac Is (Philippines)
M. rajah **Rajah spiny rat** S Thailand, Malay Pen., Riau Arch. and Sumatra (Indonesia), Borneo
M. surifer **Red spiny rat** S Burma, Thailand, Laos, Vietnam, Cambodia, S Thailand, Malay Pen., Sumatra and Java (Indonesia), Borneo
M. wattsi **Watts' spiny rat** Mt Tambusisi (C Sulawesi, Indonesia)
M. whiteheadi **Whitehead's spiny rat** S Thailand, Malay Pen., Sumatra (Indonesia), Borneo
Mayermys
M. ellermani **One-toothed shrew mouse** Papua New Guinea
Melasmothrix
M. naso **Sulawesian shrew-rat** Sulawesi (Indonesia)
Melomys **Mosaic-tailed Rats**
M. aerosus **Dusky mosaic-tailed rat** Seram Is (Indonesia)
M. bougainville **Bougainville mosaic-tailed rat** Bougainville Is (Papua New Guinea) and Buka Is (Solomon Is)
M. burtoni **Grassland mosaic-tailed rat** New Guinea, Louisiade Arch. (Papua New Guinea), Australia
M. capensis **Cape York mosaic-tailed rat** Cape York Pen. (Queensland, Australia)
M. cervinipes **Fawn-footed mosaic-tailed rat** E Australian Coast from Cape York Pen. (Queensland) south to Gosford region (New South Wales)
M. fellowsi **Red-bellied mosaic-tailed rat** Papua New Guinea
M. fraterculus **Manusela mosaic-tailed rat** Seram Is (Indonesia)
M. gracilis **Slender mosaic-tailed rat** Papua New Guinea
M. lanosus **Large-scaled mosaic-tailed rat** N New Guinea
M. leucogaster **White-bellied mosaic-tailed rat** New Guinea, islands off coast of Papua New Guinea, Maluku, Seram and Talaud Is (Indonesia)
M. levipes **Long-nosed mosaic-tailed rat** Papua New Guinea
M. lorentzii **Lorentz's mosaic-tailed rat** New Guinea
M. mollis **Thomas' mosaic-tailed rat** New Guinea
M. moncktoni **Monkton's mosaic-tailed rat** Papua New Guinea
M. obiensis **Obi mosaic-tailed rat** Obi Is (Maluku, Indonesia)
M. platyops **Lowland mosaic-tailed rat** Yapen and Biak Is (Indonesia), New Guinea, New Britian (Bismarck Arch., Papua New Guinea), D'Entrecasteaux Is (Papua NG)
M. rattoides **Large mosaic-tailed rat** Yapen Is (Indonesia), New Guinea
M. rubex **Mountain mosaic-tailed rat** New Guinea
M. rubicola **Bramble Cay mosaic-tailed rat** Bramble Cay (Queensland, Australia)
M. rufescens **Black-tailed mosaic-tailed rat** New Guinea (Bismarck Arch., Papua New Guinea)
M. spechti **Specht's mosaic-tailed rat** Buka Is (Solomon Is)
Mesembriomys **Tree Rats**
M. gouldii **Black-footed tree rat** N Western Australia, N Northern Territory, N Queensland, Melville and Bathurst Is (Australia)
M. macrurus **Golden-backed tree rat** N Western Australia, N Northern Territory (Australia)
Microhydromys **Lesser Shrew Mice**
M. musseri **Musser's shrew mouse** Mt Somoro (Papua New Guinea)
M. richardsoni **Groove-toothed shrew mouse** New Guinea
Micromys
M. minutus **Eurasian harvest mouse** Spain through Europe and Siberia to Ussuri region (Russia) and Korea (incl. Quelpart Is) south to N Caucasus and N Mongolia, then from S China west to Assam (NE India); also Britain, Japan, and Taiwan
Millardia **Asian Soft-furred Rats**
M. gleadowi **Sand-colored soft-furred rat** Afghanistan, Pakistan, adj. India
M. kathleenae **Miss Ryley's soft-furred rat** C Burma
M. kondana **Kondana soft-furred rat** S India
M. meltada **Soft-furred rat** India, Sri lanka, E Pakistan, Nepal
Muriculus
M. imberbis **Striped-backed mouse** Ethiopian highlands
Mus **Old World Mice**
M. baoulei **Baoule's mouse** E Guinea, Ivory Coast
M. booduga **Little Indian field mouse** India, Sri Lanka, S Nepal, C Burma
M. bufo **Toad mouse** E DRC, Uganda, Rwanda, Burundi
M. callewaerti **Callewaert's mouse** N and C Angola, S and W DRC
M. caroli **Ryukyu mouse** Ryukyu Is (Japan) to Taiwan, S China, Thailand, Vietnam, Laos, and Cambodia; introd. to Malay Pen., Sumatra, Java, Madura, and Flores (Indonesia)
M. cervicolor **Fawn-colored mouse** Sikkim, Assam (India) to Nepal, Burma, Thailand, Laos, Vietnam, and Cambodia; introd. to Sumatra and Java (Indonesia)
M. cookii **Cook's mouse** India, Nepal, Burma, SW Yunnan (S China), N and C Thailand, Laos, N Vietnam
M. crociduroides **Sumatran shrewlike mouse** W Sumatra (Indonesia)
M. famulus **Servant mouse** Nilgiri Hills (S India)
M. fernandoni **Sri Lankan spiny mouse** Sri Lanka
M. goundae **Gounda mouse** Gounda R (Cameroon)
M. haussa **Hausa mouse** Senegal and S Mauritania to N Nigeria
M. indutus **Desert pygmy mouse** South Africa, W Zimbabwe, Botswana, C and N Namibia
M. kasaicus **Kasai mouse** Kasai (DRC)
M. macedonicus **Macedonian mouse** Yugoslavia, Bulgaria, Turkey, Syria, Jordan, Israel, Iran
M. mahomet **Mahomet mouse** Ethiopian highlands, SW Kenya, SW Uganda
M. mattheyi **Matthey's mouse** Accra (Ghana)
M. mayori **Mayor's mouse** Sri Lanka
M. minutoides **Pygmy mouse** South Africa, northern limits unknown
M. musculoides **Temminck's mouse** SubSaharan Africa south to contact with previous species
M. musculus **House mouse** Cosmopolitan
M. neavei **Neave's mouse** S Tanzania, S DRC, E Zambia, W Mozambique, S Zimbabwe, NE South Africa
M. orangiae **Orange mouse** South Africa
M. oubanguii **Oubangui mouse** CAR
M. pahari **Gairdner's shrew-mouse** Sikkim, Assam (NE India) to Burma, Yunnan (S China), Thailand, Laos, and Vietnam
M. phillipsi **Phillips' mouse** S India
M. platythrix **Flat-haired mouse** S India
M. saxicola **Rock-loving mouse** S India, S Pakistan, S Nepal
M. setulosus **Peter's mouse** Guinea and Sierra Leone to CAR, N DRC, S Sudan, Ethiopia, W Kenya, N Uganda
M. setzeri **Setzer's pygmy mouse** NE Namibia, NW and S Botswana, W Zambia
M. shortridgei **Shortridge's mouse** Burma, Thailand, NW Vietnam, Cambodia
M. sorella **Thomas' pygmy mouse** E Cameroon, NE and SE DRC, Kenya, Uganda, N Tanzania, EC Angola
M. spicilegus **Mound-building mouse** Hungary, Romania, Yugoslavia, Bulgaria, Ukraine
M. spretus **Algerian mouse** S France, Spain, Portugal, N Africa (Morocco to Libya)
M. tenellus **Delicate mouse** Sudan, S Ethiopia, S Somalia, Kenya to C Tanzania
M. terricolor **Earth-colored mouse** India, Nepal, Pakistan; introd. to N Sumatra (Indonesia)
M. triton **Gray-bellied pygmy mouse** Kenya, Uganda, Tanzania, N and E DRC, Angola, Zambia, Malawi, C Mozambique
M. vulcani **Volcano mouse** W Java (Indonesia)
Mylomys
M. dybowskii **African groove-toothed rat** Guinea, Ivory Coast, Ghana, S Cameroon, CAR, Congo, DRC, S Sudan, Kenya, Uganda, Rwanda, Tanzania
Myomys **African Mice**
M. albipes **Ethiopian white-footed mouse** Ethiopian plateau
M. daltoni **Dalton's mouse** Senegal and Gambia to S Chad and CAR to SW Sudan
M. derooi **Deroo's mouse** Ghana, Togo, Benin, W Nigeria
M. fumatus **African rock mouse** EC Tanzania to Somalia, Ethiopia and S Sudan
M. ruppi **Rupp's mouse** SW Ethiopia
M. verreauxii **Verreaux's mouse** SW South Africa
M. yemeni **Yemeni mouse** N Yemen, SW Saudi Arabia
Neohydromys
N. fuscus **Mottled-tailed shrew mouse** Papua New Guinea
Nesokia **Short-tailed Bandicoot Rats**
N. bunnii **Bunn's short-tailed bandicoot rat** SE Iraq
N. indica **Short-tailed bandicoot rat** N India, Bangladesh, Xinjiang (NW China), Tajikistan, Uzbekistan, Turkmenistan, Afghanistan, Pakistan, Iran, Iraq, Syria, Israel, Saudi Arabia, NE Egypt
Niviventer **White-bellied Rats**
N. andersoni **Anderson's white-bellied rat** SE Xizang, Yunnan, Sichuan, Shaanxi (China)
N. brahma **Brahma white-bellied rat** Assam (India), N Burma
N. confucianus **Chinese white-bellied rat** N Burma, China, N Thailand
N. coxingi **Coxing's white-bellied rat** Taiwan
N. cremoriventer **Dark-tailed tree rat** Thailand, Malay Pen., Myeik Kyunzu (Burma), Sumatra, Java, Bali, Anambas, Nias, and Belitung Is (Indonesia), Borneo
N. culturatus **Oldfield white-bellied rat** Taiwan
N. eha **Smoke-bellied rat** Nepal, Sikkim, Assam (NE India), N Burma, N Yunnan (China)
N. excelsior **Large white-bellied rat** Sichuan (China)
N. fulvescens **Chestnut white-bellied rat** Nepal and N India through Bangladesh and S China to S Thailand, Malaysia Pen., Sumatra, Java, and Bali (Indonesia)
N. hinpoon **Limestone rat** Korat plateau (Thailand)
N. langbianis **Lang Bian white-bellied rat** Assam (NE India), Burma, N Thailand, Laos, Vietnam
N. lepturus **Narrow-tailed white-bellied rat** W and C Java (Indonesia)
N. niviventer **White-bellied rat** NE Pakistan, N India, Nepal
N. rapit **Long-tailed mountain rat** Malay Pen., Sumatra (Indonesia), N Borneo
N. tenaster **Tenasserim white-bellied rat** Assam (NE India), Burma, Vietnam
Notomys **Australian Hopping Mice**
N. alexis **Spinifex hopping mouse** Western Australia, Northern Territory, South Australia, W Queensland (Australia)
N. amplus **Short-tailed hopping mouse** Extinct: formerly S Northern Territory, N South Australia (Australia)
N. aquilo **Northern hopping mouse** N Queensland, N Northern Territory (Australia)
N. cervinus **Fawn-colored hopping mouse** SW Queensland, South Australia, S Northern Territory (Australia)
N. fuscus **Dusky hopping mouse** SE Western Australia, S Northern Territory, South Australia, SW Queensland, W New South Wales (Australia)
N. longicaudatus **Long-tailed hopping mouse** Extinct: formerly Western Australia, Northern Territory (Australia)
N. macrotis **Big-eared hopping mouse** Extinct: formerly Moore R (Western Australia, Australia)
N. mitchellii **Mitchell's hopping mouse** S Western Australia, S South Australia, W Victoria (Australia)
N. mordax **Darling Downs hopping mouse** Extinct: formerly Darling Downs (Queensland, Australia)
Oenomys **Rufous-nosed Rats**
O. hypoxanthus **Rufous-nosed rat** S Nigeria south to N Angola east through DRC to Uganda, Rwanda, Burundi, also in S Sudan, SW Ethiopia, Kenya, and W Tanzania
O. ornatus **Ghana rufous-nosed rat** SE Guinea to Ghana
Otomys **African Vlei Rats**
O. anchietae **Angolan vlei rat** W Kenya, SW Tanzania, N Malawi, C Angola
O. angoniensis **Angoni vlei rat** S Kenya to C South Africa
O. denti **Dent's vlei rat** EC Africa, from Ruwenzori Mts (Uganda) to Zambia and the Nyika Pateau (N Malawi) and to the Usambara and Uluguru Mts (EC Tanzania)
O. irroratus **Vlei rat** South Africa, E Zimbabwe, W Mozambique
O. laminatus **Laminate vlei rat** South Africa
O. maximus **Large vlei rat** S Angola, SW Zambia, NW Botswana, Caprivi (NE Namibia)
O. occidentalis **Western vlei rat** Gotel Mts (SE Nigeria), Mt Oku (W Cameroon)

O. saundersiae **Saunders' vlei rat** SW, S, and C South Africa
O. sloggetti **Sloggett's vlei rat or Ice rat** SE and E South Africa
O. tropicalis **Tropical vlei rat** W Kenya, Uganda, Rwanda, Burundi, DRC, Mt Cameroun (Cameroon)
O. typus **Typical vlei rat** E Africa, from Ethiopia south to the Uzungwe Mts (C Tanzania), Nyika plateau (N Malawi), and adj. Zambia
O. unisulcatus **Bush vlei rat** South Africa
Palawanomys
P. furvus **Palawan soft-furred mountain rat** Palawan (Philippines)
Papagomys **Flores Island Giant Tree Rats**
P. armandvillei **Flores giant tree rat** Flores (Indonesia)
P. theodorverhoeveni **Verhoeven's giant tree rat** Flores (Indonesia)
Parahydromys
P. asper **Coarse-haired water rat** New Guinea
Paraleptomys **Montane Water Rats**
P. rufilatus **Northern water rat** NC New Guinea
P. wilhelmina **Short-haired water rat** C New Guinea
Parotomys **Whistling Rats**
P. brantsii **Brants' whistling rat** W South Africa, SW Botswana, SE Namibia
P. littledalei **Littledale's whistling rat** W South Africa, S and W Namibia
Paruromys **Sulawesian Giant Rats**
P. dominator **Sulawesi giant rat** Sulawesi (Indonesia)
P. ursinus **Sulawesi bear rat** Mt Lampobatang (SE Sulawesi, Indonesia)
Paulamys
P. naso **Flores long-nosed rat** Flores (Indonesia)
Pelomys **Groove-toothed Swamp Rats**
P. campanae **Bell groove-toothed swamp rat** WC and N Angola, W DRC
P. fallax **Creek groove-toothed swamp rat** S Kenya, Uganda, Tanzania, DRC south to N Botswana, E and NW Zimbabwe, and Mozambique
P. hopkinsi **Hopkins' groove-toothed swamp rat** SW Kenya, Uganda, Rwanda
P. isseli **Issel's groove-toothed swamp rat** Kome, Bugala, and Bunyama Is (L. Victoria, Uganda)
P. minor **Least groove-toothed swamp rat** W Tanzania, S and E DRC, N Angola, NW Zambia
Phloeomys **Giant Cloud Rats**
P. cumingi **Southern Luzon giant cloud rat** S Luzon, Marinduque, and Catanduanes Is (Philippines)
P. pallidus **Northern Luzon giant cloud rat** N Luzon (Philippines)
Pithecheir **Sunda Tree Rats**
P. melanurus **Red tree rat** Java (Indonesia)
P. parvus **Malayan tree rat** Malay Pen.
Pogonomelomys **New Guinean Brush Mice**
P. bruijni **Lowland brush mouse** New Guinea
P. mayeri **Shaw Mayer's brush mouse** New Guinea
P. sevia **Highland brush mouse** New Guinea
Pogonomys **Prehensile-tailed Tree Mice**
P. championi **Champion's tree mouse** Papua New Guinea
P. loriae **Large tree mouse** New Guinea, D'Entrecasteaux Is (Papua New Guinea)
P. macrourus **Chestnut tree mouse** New Guinea, Yapen Is (Indonesia), New Britain Is (Bismarck Arch., Papua New Guinea)
P. sylvestris **Gray-bellied tree mouse** Papua New Guinea
Praomys **African Soft-furred Mice**
P. delectorum **Delectable soft-furred mouse** NE Zambia, Malawi
P. hartwigi **Hartweg's soft-furred mouse** L. Oku (Cameroon), Gotel Mts (Nigeria)
P. jacksoni **Jackson's soft-furred mouse** C Nigeria to S Sudan, DRC, Uganda, Kenya south to N Angola, N and E Zambia, and E Tanzania
P. minor **Least soft-furred mouse** Lukolela (C DRC)
P. misonnei **Misonne's soft-furred mouse** N and E DRC
P. morio **Cameroon soft-furred mouse** Mt Cameroun (Cameroon)
P. mutoni **Muton's soft-furred mouse** Masako Forest Reserve (N DRC)
P. rostratus **Forest soft-furred mouse** Liberia, Guinea, Ivory Coast
P. tullbergi **Tullberg's soft-furred mouse** Gambia to N and E DRC, also NW Angola, Bioko (Equatorial Guinea)
Pseudohydromys **New Guinea Shrew Mice**
P. murinus **Eastern shrew mouse** Papua New Guinea
P. occidentalis **Western shrew mouse** New Guinea
Pseudomys **Australian Mice**
P. albocinereus **Ash-gray mouse** SW Western Australia, Bernier, Dorre, Shark Bay, and Woody Is (Australia)
P. apodemoides **Silky mouse** SE South Australia, W Victoria (Australia)
P. australis **Plains mouse** S Northern Territory, South Australia, S Queensland, New South Wales (Australia)
P. bolami **Bolam's mouse** S South Australia, S Western Australia (Australia)
P. chapmani **Western pebble-mound mouse** NW Western Australia (Australia)
P. delicatulus **Little native mouse** SC Papua New Guinea, Coastal N Australia
P. desertor **Brown desert mouse** Arid regions of Australia
P. fieldi **Alice Springs mouse** Alice Springs (S Northern Territory, Australia)
P. fumeus **Smoky mouse** Victoria (Australia)
P. fuscus **Broad-toothed mouse** E New South Wales, S Victoria, Tasmania (Australia)
P. glaucus **Blue-gray mouse** S Queensland, New South Wales (Australia)
P. gouldii **Gould's mouse** Extinct: formerly W Western Australia, E South Australia, New South Wales, N Victoria (Australia)
P. gracilicaudatus **Eastern chestnut mouse** Coast of E Australia, from Townsville (N Queensland) to Sydney region (New South Wales)
P. hermannsburgensis **Sandy Inland mouse** Australia
P. higginsi **Long-tailed mouse** Tasmania (Australia)
P. johnsoni **Central pebble-mound mouse** C Northern Territory (Australia)
P. laborifex **Kimberly mouse** N Western Australia, N Northern Territory (Australia)
P. nanus **Western chestnut mouse** W Western Australia, N Northern Territory, NW Queensland (Australia)
P. novaehollandiae **New Holland mouse** E New South Wales, S Victoria, N Tasmania (Australia)
P. occidentalis **Western mouse** SW Western Australia (Australia)
P. oralis **Hastings River mouse** NE New South Wales, SE Queensland (Australia)
P. patrius **Country mouse** Mt Inkerman (Queensland, Australia)
P. pilligaensis **Pilliga mouse** N New South Wales (Australia)
P. praeconis **Shark Bay mouse** Western Australia, Bernier Is (Australia)
P. shortridgei **Heath rat** SW Western Australia, SW Victoria (Australia)
Rattus **Old World Rats**
R. adustus **Sunburned rat** Enggano Is (Indonesia)
R. annandalei **Annandale's rat** Malay Pen., Singapore, E Sumatra, Padang and Rupat Is (Indonesia)
R. argentiventer **Rice-field rat** Thailand, S Vietnam, Cambodia, Malay Pen., Sumatra, Java, Bali, Sulawesi, Kangean Is, Lesser Sunda Is (Indonesia), Mindoro and Mindanao (Philippines), New Guinea
R. baluensis **Summit rat** Mt Kinabulu (Sabah, N Borneo)
R. bontanus **Bonthain rat** Mt Lampobatang (SW Sulawesi, Indonesia)
R. burrus **Nonsense rat** Nicobar Is (Ind. Oc.)
R. colletti **Dusky rat** Coastal Northern Territory (Australia)
R. elaphinus **Sula rat** Sula (Indonesia)
R. enganus **Enggano rat** Enggano Is (Indonesia)
R. everetti **Philippine forest rat** Philippines
R. exulans **Polynesian rat** Bangladesh, Burma, Thailand, Laos, Vietnam, Cambodia, Indonesia, Philippines, New Guinea, New Zealand, Micronesia, Polynesia
R. feliceus **Spiny Seram rat** Seram Is (Indonesia)
R. foramineus **Hole rat** SW Sulawesi (Indonesia)
R. fuscipes **Bush rat** SW Western Australia, South Australia, Victoria, Queensland (Australia)
R. giluwensis **Giluwe rat** Mt Giluwe (Papua New Guinea)
R. hainaldi **Hainald's rat** Flores (Philippines)
R. hoffmanni **Hoffmann's rat** Sulawesi (Indonesia)
R. hoogerwerfi **Hoogerwerf's rat** Mt Leuser (Sumatra, Indonesia)
R. jobiensis **Japen rat** Yapen, Owi, and Biak Is (Indonesia)
R. koopmani **Koopman's rat** Peleng Is (Indonesia)
R. korinchi **Korinch's rat** Mt Kerinci and Mt Talakmau (W Sumatra, Indonesia)
R. leucopus **Cape York rat** New Guinea, Queensland (Australia)
R. losea **Lesser rice-field rat** S China, Taiwan, Thailand (excl. S), Laos, Vietnam, Pescadores Is
R. lugens **Mentawai rat** Mentawai Is (Indonesia)
R. lutreolus **Australian swamp rat** SE South Australia, Victoria, New South Wales, SE and N Queensland, Tasmania (Australia)
R. macleari **MacLear's rat** Extinct: formerly Christmas Is (Australia)
R. marmosurus **Opossum rat** NE Sulawesi (Indonesia)
R. mindorensis **Mindoro black rat** Mindoro (Philippines)
R. mollicomulus **Little soft-furred rat** Mt Lampobatang (Sulawesi, Indonesia)
R. montanus **Nillu rat** Sri Lanka
R. mordax **Eastern rat** Papua new Guinea (incl. D'Entrecasteaux Is and Louisiade Arch.)
R. morotaiensis **Molaccan prehensile-tailed rat** Morotai Is (Maluku Is, Indonesia)
R. nativitatis **Bulldog rat** Extinct: formerly Christmas Is (Australia)
R. nitidus **Himalayan field rat** S China, Vietnam, Laos, N Thailand, Burma, Assam, Bhutan, Sikkim, Kumaun (India), Bangladesh, Nepal; also Sulawesi and Seram (Indonesia), Luzon (Philippines), Irian Jaya (New Guinea), Palau Is
R. norvegicus **Brown or Norway rat** Cosmopolitan
R. novaeguineae **New Guinean rat** Papua New Guinea
R. osgoodi **Osgood's rat** S Vietnam
R. palmarum **Palm rat** Nicobar Is (India)
R. pelurus **Peleng rat** Peleng Is (Indonesia)
R. praetor **Spiny rat** New Guinea, Admirality Is, Bismarck Arch. (Papua New Guinea), Solomon Is
R. ranjiniae **Kerala rat** Kerala (SW India)
R. rattus **House rat** S India; introd. widely and now cosmopolitan
R. sanila **New Ireland rat** New Ireland (Bismarck Arch., Papua New Guinea)
R. sikkimensis **Sikkim rat** NE India, E Nepal, C and N Burma, S China (incl. Hong Kong and Hainan Is), Vietnam, Laos, Cambodia, N Thailand
R. simalurensis **Simalur rat** Simalur, Siumat, Lasia, Babi Is (Indonesia)
R. sordidus **Dusky field rat** New Guinea, Queensland, NE New South Wales (Australia)
R. steini **Stein's rat** New Guinea
R. stoicus **Andaman rat** Andaman Is (India)
R. tanezumi **Tanezumi rat** E Afghanistan through Nepal and N India to S and C China, Korea and Indochina, also SW India (and Andaman and Nicobar Is), Myeik Kyunza (Burma), Taiwan, Japan; prob. introd. to Malay Pen., Indonesia, Philippines, W New Guinea east through Micronesia to Fiji
R. tawitawiensis **Tawi-tawi forest rat** Tawitawi Is (Sulu Arch., Philippines)
R. timorensis **Timor rat** Mt Muti (Timor, Indonesia)
R. tiomanicus **Malayan field rat** S Thailand, Malay Pen., Sumatra, Java, Bali, Enggano Is (Indonesia), Borneo, Palawan (Philippines)
R. tunneyi **Pale field rat** NE and SW Western Australia, Northern Territory, E Queensland, NE New South Wales (Australia)
R. turkestanicus **Turkestan rat** Kyrgyzstan, N and E Afghanistan, N Pakistan, NE Iran, N India, Kashmir, Nepal, Yunnan and Guangdong (S China)
R. villosissimus **Long-haired rat** Australia
R. xanthurus **Yellow-tailed rat** Sulawesi, excl. SW (Indonesia)
Rhabdomys
R. pumilio **Four-striped grass mouse** South Africa to Angola, SE DRC, Uganda, and Kenya
Rhynchomys **Shrew Rats**
R. isarogensis **Isarog shrew rat** Mt Isarog (SE Luzon, Philippines)
R. soricoides **Mount Data shrew rat** N Luzon (Philippines)
Solomys **Naked-tailed Rats**
S. ponceleti **Poncelet's naked-tailed rat** Bougainville Is (Papua New Guinea), Buka and Choiseul Is (Solomon Is)
S. salamonis **Florida naked-tailed rat** Florida Is (Solomon Is)
S. salebrosus **Bougainville naked-tailed rat** Bougainville Is (Papua New Guinea), Buka and Choiseul Is (Solomon Is)
S. sapientis **Isabel naked-tailed rat** Santa Isabel Is (Solomon Is)
S. spriggsarum **Buka naked-tailed rat** Buka Is (Solomon Is)
Spelaeomys
S. florensis **Flores cave rat** Flores (Indonesia)
Srilankamys
S. ohiensis **Ohiya rat** Sri Lanka
Stenocephalemys **Ethiopian Narrow-headed Rats**
S. albocaudata **Ethiopian narrow-headed rat** Ethiopia plateau
S. griseicauda **Gray-tailed narrow-headed rat** Ethiopia highlands
Stenomys **Slender Rats**
S. ceramicus **Seram rat** Seram Is (Indonesia)
S. niobe **Moss-forest rat** New Guinea
S. richardsoni **Glacier rat** Irian Jaya (New Guinea)
S. vandeuseni **Van Deusen's rat** Mt Dayman (Papua New Guinea)
S. verecundus **Slender rat** New Guinea
Stochomys
S. longicaudatus **Target rat** Togo, S Nigeria, CAR, Cameroon, Gabon, DRC, Uganda
Sundamys **Giant Sunda Rats**
S. infraluteus **Mountain giant rat** Mt Kinabulu and Mt Trus Madi (Sabah, N Borneo), W Sumatra (Indonesia)
S. maxi **Bartels' rat** W Java (Indonesia)
S. muelleri **Muller's giant Sunda rat** SW Burma, S Thailand, Malay Pen., Sumatra (Indonesia), Borneo, Palawan (Philippines)
Taeromys **Sulawesi Rats**
T. arcuatus **Salokko rat** SE Sulawesi (Indonesia)
T. callitrichus **Lovely-haired rat** NE, C, and SE Sulawesi (Indonesia)
T. celebensis **Celebes rat** Sulawesi (Indonesia)
T. hamatus **Sulawesi montane rat** C Sulawesi (Indonesia)
T. punicans **Sulawesi forest rat** C and SW Sulawesi (Indonesia)
T. taerae **Tondano rat** NE Sulawesi (Indonesia)
Tarsomys **Long-footed Rats**
T. apoensis **Long-footed rat** Mindanao (Philippines)
T. echinatus **Spiny long-footed rat** Mindanao (Philippines)
Tateomys **Greater Sulawesian Shrew Rats**
T. macrocercus **Long-tailed shrew rat** Mt Nokilalaki (C Sulawesi, Indonesia)
T. rhinogradoides **Tate's shrew rat** Mt Latimodjong, Mt Tokala, Mt Nokilalaki (Sulawesi, Indonesia)
Thallomys **Acacia Rats**
T. loringi **Loring's rat** E Kenya, N and E Tanzania
T. nigricauda **Black-tailed tree rat** South Africa, Namibia, Botswana, Angola, Zambia
T. paedulcus **Acacia rat** South Africa to S Ethiopia and Somalia
T. shortridgei **Shortridge's rat** W South Africa
Thamnomys **Thicket Rats**
T. kempi **Kemp's thicket rat** E DRC, W Uganda, Burundi
T. venustus **Charming thicket rat** E DRC, Uganda, Rwanda
Tokudaia **Ryukyu Spiny Rats**
T. muenninki **Muennink's spiny rat** Okinawa (Ryukyu Is, Japan)
T. osimensis **Ryukyu spiny rat** Amami-oshima Is (Ryukyu Is, japan)
Tryphomys
T. adustus **Luzon short-nosed rat** Luzon (Philippines)
Uranomys
U. ruddi **Rudd's mouse** Senegal, Guinea, Ivory Coast, Togo, N Nigeria, N Cameroon, NE DRC, Uganda, Kenya, C Mozambique, Malawi, SE Zimbabwe
Uromys **Giant Naked-tailed Rats**
U. anak **Giant naked-tailed rat** New Guinea
U. caudimaculatus **Giant white-tailed rat** Aru, Kei, and Waigeo Is (Indonesia), New Guinea, D'Entrecasteaux Is (Papua New Guinea), Queensland (Australia)
U. hadrourus **Masked white-tailed rat** Thornton Peak (NE Queensland, Australia)
U. imperator **Emperor rat** Guadalcanal (Solomon Is)
U. neobritanicus **Bismarck giant rat** New Britain Is (Bismarck Arch., Papua New Guinea)
U. porculus **Guadalcanal rat** Guadalcanal (Solomon Is)
U. rex **King rat** Guadalcanal (Solomon Is)
Vandeleuria **Long-tailed Climbing Mice**
V. nolthenii **Nolthenius' long-tailed climbing mouse** Sri Lanka
V. oleracea **Asiatic long-tailed climbing mouse** India, Sri Lanka, Nepal to Burma, Yunnan (S China), N Thailand, Vietnam
Vernaya
V. fulva **Red climbing mouse** Sichuan and Yunnan (S China), N Burma
Xenuromys
X. barbatus **Rock-dwelling giant rat** New Guinea
Xeromys
X. myoides **False water rat** C and S Queensland, Northern Territory, Melville Is (Australia)
Zelotomys **Broad-heaed Mice**
Z. hildegardeae **Hildegarde's broad-headed mouse**

CAR, DRC, S Sudan, Kenya, W Uganda, Rwanda, Burundi, Tanzania, Malawi, Zambia, Angola
Z. woosnami **Woosnam's broad-headed mouse** S Africa, Botswana, Namibia
Zyzomys **Australian Rock Rats**
Z. argurus **Silver-tailed rock rat** Western Australia to N coastal Queensland (Australia)
Z. maini **Arnhem Land rock rat** Northern Territory (Australia)
Z. palatilis **Carpentarian rock rat** Echo Gorge (Northern Territory, Australia)
Z. pedunculatus **Central rock rat** Northern Territory (Australia)
Z. woodwardi **Kimberly rock rat** Western Australia, Northern Territory (Australia)

SUBFAMILY MYOSPALACINAE
Myospalax Zokors
M. aspalax **False zokor** Upper Amur region (Russia), Nei Monggol (China)
M. epsilanus **Manchurian zokor** NE China, Amur region to Ussuri region (Russia)
M. fontanierii **Common Chinese zokor** China
M. myospalax **Siberian zokor** Russia, Kazakhstan
M. psilurus **Transbaikal zokor** SE Russia (incl. Ussuri region) to E Mongolia and NE and C China
M. rothschildi **Rothschild's zokor** Gansu and Hubei (China)
M. smithii **Smith's zokor** Gansu and Ningxia (China)

SUBFAMILY MYSTROMYINAE
Mystromys
M. albicaudatus **White-tailed mouse** South Africa, Swaziland

SUBFAMILY NESOMYINAE
Brachytarsomys
B. albicauda **White-tailed rat** E Madagascar
Brachyuromys **Short-tailed Rats**
B. betsileoensis **Betsileo short-tailed rat** Madagascar
B. ramirohitra **Gregarious short-tailed rat or Ramirohitra** Madagascar
Eliurus Tufted-tailed Rats
E. majori **Major's tufted-tailed rat** Madagascar
E. minor **Lesser tufted-tailed rat** E Madagascar
E. myoxinus **Dormouse tufted-tailed rat** SW and S Madagascar
E. penicillatus **White-tipped tufted-tailed rat** Fianarantsoa (Madagascar)
E. tanala **Tanala tufted-tailed rat** E Madagascar
E. webbi **Webb's tufted-tailed rat** E Madagascar
Gymnuromys
G. roberti **Voalavoanala** E Madagascar
Hypogeomys
H. antimena **Malagasy giant rat or Votsotsa** WC Madagascar
Macrotarsomys **Big-footed Mice**
M. bastardi **Bastard big-footed mouse** W and S Madagascar
M. ingens **Greater big-footed mouse** Mahajanga (Madagascar)
Nesomys
N. rufus **Island mouse** N, E, and WC Madagascar

SUBFAMILY PETROMYSCINAE
Delanymys
D. brooksi **Delany's swamp mouse** SW Uganda, Kivu region (DRC), Rwanda
Petromyscus **Rock Mice**
P. barbouri **Barbour's rock mouse** Little Namaqualand (W South Africa)
P. collinus **Pygmy rock mouse** W South Africa through Namibia to SW Angola
P. monticularis **Brukkaros pygmy rock mouse** S Namibia, adj. South Africa
P. shortridgei **Shortridge's rock mouse** W and S Angola, N Namibia

SUBFAMILY PLATACANTHOMYINAE
Platacanthomys
P. lasiurus **Malabar spiny dormouse** S India
Typhlomys **Chinese Pygmy Dormouse**
T. chapensis **Chapa pygmy dormouse** Chapa (N Vietnam)
T. cinereus **Chinese pygmy dormouse** Yunnan, Fujian, Guangxi, Anhui (S China)

SUBFAMILY RHIZOMYINAE
Cannomys
C. badius **Lesser bamboo rat** N and NE India, E Nepal, Bengladesh, Burma, Yunnan (S China), Thailand, Cambodia
Rhizomys **Bamboo Rats**
R. pruinosus **Hoary bamboo rat** Assam (NE India), Yunnan, Guangxi, Guangdong (S China), E Burma, Thailand, Laos, Vietnam, Cambodia, Malay Pen.
R. sinensis **Chinese bamboo rat** S and C China, N Burma, Vietnam
R. sumatrensis **Large bamboo rat** Yunnan (S China), Burma, Thailand, Laos, Vietnam, Cambodia, Malay Pen., Sumatra (Indonesia)
Tachyoryctes **Root Rats or African Mole-rats**
T. ankoliae **Ankole mole-rat** S Uganda
T. annectens **Mianzini mole-rat** E of L. Naivasha (Kenya)
T. audax **Audacious mole-rat** Aberdare Mts (Kenya)
T. daemon **Demon mole-rat** N Tanzania
T. macrocephalus **Big-headed mole-rat** S Ethiopian plateau
T. naivashae **Naivasha mole-rat** W and S of L. Naivasha (Kenya)
T. rex **King mole-rat** Mt Kenya (Kenya)
T. ruandae **Ruanda mole-rat** Kivu (E DRC), Rwanda, Burundi
T. ruddi **Rudd's mole-rat** SW Kenya, SE Uganda
T. spalacinus **Embi mole-rat** Plains near Mt Kenya (Kenya)
T. splendens **East African mole-rat** Ethiopia, Somalia, NW Kenya

SUBFAMILY SIGMODONTINAE
Abrawayaomys
A. ruschii **Ruschi's rat** Espirito Santo and Minas Gerais (Brazil), Misiones (Argentina)
Aepeomys **Montane Mice**
A. fuscatus **Dusky montane mouse** W and C Andes of Colombia
A. lugens **Olive montane mouse** Cord. Meridae of Venezuela to Andes of Ecuador
Akodon **Grass Mice**
A. aerosus **Highland grass mouse** SE Ecuador, E Peru, NW Bolivia
A. affinis **Colombian grass mouse** Cord. Occid. of Colombia
A. albiventer **White-bellied grass mouse** SE Peru, WC Bolivia to N Argentina and Chile
A. azarae **Azara's grass mouse** S Brazil, Bolivia, Paraguay, NE Argentina, Uruguay
A. bogotensis **Bogota grass mouse** Andes of E and C Colombia and W Venezuela
A. boliviensis **Bolivian grass mouse** Altiplano of SE Peru and NC Bolivia
A. budini **Budin's grass mouse** Mountains of NW Argentina
A. cursor **Cursor grass mouse** C and SE Brazil, E Paraguay, NE Argentina, Uruguay
A. dayi **Day's grass mouse** C to SC Bolivia
A. dolores **Dolorous grass mouse** Sierra de Cordoba (C Argentina)
A. fumeus **Smoky grass mouse** E Andean slopes of SE Peru and W Bolivia
A. hershkovitzi **Hershkovitz's grass mouse** Outer islands of Chilean Arch.
A. illuteus **Gray grass mouse** NW Argentina
A. iniscatus **Intelligent grass mouse** WC to S Argentina
A. juninensis **Junin grass mouse** E and W Andean slopes of C Peru
A. kempi **Kemp's grass mouse** EC Argentina, adj. Uruguay
A. kofordi **Koford's grass mouse** Cusco and Puno (SE Peru)
A. lanosus **Woolly grass mouse** S Chile, S Argentina
A. latebricola **Ecuadorian grass mouse** Andes of Ecuador
A. lindberghi **Lindbergh's grass mouse** Parque Nacional de Brasilia (Distrito Federal, Brazil)
A. longipilis **Long-haired grass mouse** C to S Chile, Argentina
A. mansoensis **Manso grass mouse** Andes of Rio Negro Prov. (Argentina)
A. markhami **Markham's grass mouse** Wellington Is (Chile)
A. mimus **Thespian grass mouse** E Andean slopes of SE Peru to WC Bolivia
A. molinae **Molina's grass mouse** EC Argentina
A. mollis **Soft grass mouse** Andes of Ecuador to NC Peru
A. neocenus **Neuquen grass mouse** C and S Argentina
A. nigrita **Blackish grass mouse** SE Brazil, E Paraguay, NE Argentina
A. olivaceus **Olive grass mouse** Chile, adj. W Argentina
A. orophilus **El Dorado grass mouse** Mountains of N Peru
A. puer **Altiplano grass mouse** Altiplano of C Peru, through W Bolivia to NW Argentina
A. sanborni **Sanborn's grass mouse** S Chile, adj. Argentina
A. sanctipaulensis **Sao Paulo grass mouse** SE Brazil
A. serrensis **Serrado Mar grass mouse** SE Brazil
A. siberiae **Cochabamba grass mouse** Cochabamba (Bolivia)
A. simulator **Gray-bellied grass mouse** E Andean slopes of SC Bolivia to NW Argentina
A. spegazzinii **Spegazzini's grass mouse** E Andean slopes of NW Argentina
A. subfuscus **Puno grass mouse** W and E Andean slopes of SC Peru to La Paz (NW Bolivia)
A. surdus **Silent grass mouse** Andes of SE Peru
A. sylvanus **Forest grass mouse** NW Argentina
A. toba **Chaco grass mouse** E Bolivia, W Paraguay, and N Argentina
A. torques **Cloud Forest grass mouse** E Andean forest of SE Peru
A. urichi **Northern grass mouse** E Colombia, Venezuela, N Brazil, Trinidad and Tobago
A. varius **Variable grass mouse** E Andean slopes of W Bolivia
A. xanthorhinus **Yellow-nosed grass mouse** S Chile, S Argentina (incl. Tierra del Fuego)
Andalgalomys **Chaco Mice**
A. olrogi **Olrog's Chaco mouse** Catamarca (NW Argentina)
A. pearsoni **Pearson's Chaco mouse** Chaco (W Paraguay), SE Bolivia
Andinomys
A. edax **Andean mouse** Altiplano of Puno (S Peru) and N Chile through W Bolivia to Jujuy and Catamarca (NW Argentina)
Anotomys
A. leander **Ecuador fish-eating rat** N Ecuador
Auliscomys **Big-eared Mice**
A. boliviensis **Bolivian big-eared mouse** Altiplano from Arequipa (S Peru) to N Chile and Potosi (W Bolivia)
A. micropus **Southern big-eared mouse** S Andes of Chile and Argentina
A. pictus **Painted big-eared mouse** Andes from Ancash (C Peru) to La Paz (NW Bolivia)
A. sublimis **Andean big-eared mouse** Altiplano from Ayacucho (S Peru) through SW Bolivia and adj. Chile to NW Argentina
Baiomys **American Pygmy Mice**
B. musculus **Southern pygmy mouse** Nayarit and Veracruz (Mexico) to NW Nicaragua
B. taylori **Northern pygmy mouse** SE Arizona, SW New Mexico, and E Texas (USA) south to Michoacan, Hidalgo, and C Veracruz (Mexico)
Bibimys **Crimson-nosed Rats**
B. chacoensis **Chaco crimson-nosed rat** NE Argentina
B. labiosus **Large-lipped crimson-nosed rat** Minas Gerais (SE Brazil)
B. torresi **Torres' crimson-nosed rat** EC Argentina
Blarinomys
B. breviceps **Brazilian shrew-mouse** Bahia to Minas Gerais and Rio de Janeiro (Brazil)
Bolomys **Bolo Mice**
B. amoenus **Pleasant bolo mouse** SE Peru, WC Bolivia
B. lactens **Rufous-bellied bolo mouse** SC Bolivia, NW Argentina
B. lasiurus **Hairy-tailed bolo mouse** Brazil (south of Amazon R), E Bolivia, Paraguay, N Argentina
B. obscurus **Dark bolo mouse** EC Argentina, S Uruguay
B. punctulatus **Spotted bolo mouse** Ecuador
B. temchuki **Temchuk's bolo mouse** Misiones (NE Argentina)
Calomys **Vesper Mice**
C. boliviae **Bolivian vesper mouse** W Bolivia
C. callidus **Crafty vesper mouse** E Paraguay, EC Argentina
C. callosus **Large vesper mouse** WC to EC Brazil, E Bolivia, W Paraguay, N Argentina
C. hummelincki **Hummelinck's vesper mouse** Llanos of NE Colombia, N Venezuela, Curacao and Aruba (Neth. Antilles)
C. laucha **Small vesper mouse** WC Brazil, SE Bolivia, W Paraguay, N Argentina, Uruguay
C. lepidus **Andean vesper mouse** Altiplano of C Peru through W Bolivia to NE Chile and NW Argentina
C. musculinus **Drylands vesper mouse** E Paraguay, N Argentina
C. sorellus **Peruvian vesper mouse** Andes of Peru
C. tener **Delicate vesper mouse** EC Brazil
Chelemys **Greater Long-clawed Mice**
C. macronyx **Andean long-clawed mouse** S Andes, along Chile-Argentina border
C. megalonyx **Large long-clawed mouse** C and S Chile
Chibchanomys
C. trichotis **Chibchan water mouse** Colombia, W Venezuela, Peru
Chilomys
C. instans **Colombian forest mouse** N Andes, from W Venezuela through N and C Colombia to N Ecuador
Chinchillula
C. sahamae **Altiplano chinchilla mouse** Altiplano of S Peru, W Bolivia, N Chile, NW Argentina
Chroeomys **Altiplano Mice**
C. andinus **Andean Altiplano mouse** Peru, Bolivia, N Chile, N Argentina
C. jelskii **Jelski's Altiplano mouse** Altiplano of S Peru to WC Bolivia and NW Argentina
Delomys **Atlantic forest rats**
D. dorsalis **Striped Atlantic forest rat** Coastal SE Brazil and NE Argentina
D. sublineatus **Pallid Atlantic forest rat** Coastal Espirito Santo to Parana (SE Brazil)
Eligmodontia **Gerbil Mice**
E. moreni **Monte gerbil mouse** E Andean slopes of Argentina, from Salta to Neuquen
E. morgani **Morgan's gerbil mouse** S Argentina, adj. Chile
E. puerulus **Andean gerbil mouse** Altiplano of S Peru through W Bolivia to NE Chile and NW Argentina
E. typus **Highland gerbil mouse** C Argentina, adj. Chile
Euneomys **Chinchilla Mice**
E. chinchilloides **Patagonian chinchilla mouse** S Chile (incl. Tierra del Fuego), nearby islands
E. fossor **Burrowing chinchilla mouse** Salta (Argentina)
E. mordax **Biting chinchilla mouse** WC Argentina, adj. Chile
E. petersoni **Peterson's chinchilla mouse** S Argentina, adj. Chile (excl. Tierra del Fuego)
Galenomys
G. garleppi **Garlepp's mouse** Altiplano of S Peru, Bolivia, N Chile
Geoxus
G. valdivianus **Long-clawed mole mouse** C and S Chile (incl. Chiloe Is), S Argentina
Graomys **Leaf-eared Mice**
G. domorum **Pale leaf-eared mouse** E Andean slopes of SC Bolivia and NW Argentina
G. edithae **Edith's leaf-eared mouse** La Rioja (NW Argentina)
G. griseoflavus **Gray leaf-eared mouse** Brazil, S Bolivia, Paraguay south to Chubut (S Argentina)
Habromys **Crested-tailed Deer Mice**
H. chinanteco **Chinanteco deer mouse** NC Oaxaca (Mexico)
H. lepturus **Slender-tailed deer mouse** NC Oaxaca (Mexico)
H. lophurus **Crested-tailed deer mouse** Chiapas (Mexico), C Guatemala, NW El Salvador
H. simulatus **Jico deer mouse** E slopes of Sierra Madre Orient. (C Veracruz, Mexico)
Hodomys
H. alleni **Allen's woodrat** Mexico
Holochilus **Marsh Rats**
H. brasiliensis **Web-footed marsh rat** SE Brazil, EC Argentina, Uruguay
H. chacarius **Chaco marsh rat** Paraguay, NE Argentina
H. magnus **Greater marsh rat** S Brazil, Uruguay
H. sciureus **Marsh rat** Colombia, E and S Venezuela, Guianas, Ecuador, Peru, N and C Brazil, Bolivia
Ichthyomys **Crab-eating Rats**
I. hydrobates **Crab-eating rat** Colombia, W Venezuela, Ecuador
I. pittieri **Pittier's crab-eating rat** N Venezuela
I. stolzmanni **Stolzmann's crab-eating rat** Ecuador, Peru
I. tweedii **Tweedy's crab-eating rat** C Panama, W Ecuador
Irenomys
I. tarsalis **Chilean climbing mouse** C and S Chile (incl. Chiloe Is), adj. Argentina
Isthmomys **Isthmus Rats**
I. flavidus **Yellow isthmus rat** W Panama
I. pirrensis **Mount Pirri isthmus rat** E Panama, adj. Colombia
Juscelinomys **Juscelin's Mice**
J. candango **Candango mouse** C Brazil
J. vulpinus **Molelike mouse** Minas Gerais (Brazil)
Kunsia **South American Giant Rats**
K. fronto **Fossorial giant rat** EC Brazil, NE Argentina
K. tomentosus **Woolly giant rat** Mato Grosso (WC Brazil), Beni (NE Bolivia)

Lenoxus
L. apicalis **Andean rat** E Andean slopes of SE Peru and W Bolivia
Megadontomys **Giant Deer Mice**
M. cryophilus **Oaxaca giant deer mouse** NC Oaxaca (Mexico)
M. nelsoni **Nelson's giant deer mouse** E slopes of Sierra Madre Orient., from SE Hidalgo to C Veracruz (Mexico)
M. thomasi **Thomas' giant deer mouse** Sierra Madre del Sur of Guerrero (Mexico)
Megalomys **Extinct West Indian Giant Rice Rats**
M. desmarestii **Antillean giant rice rat** Extinct: formerly Martinique Is (Lesser Antilles)
M. luciae **Santa Lucia giant rice rat** Extinct: formerly St Lucia (Lesser Antilles)
Melanomys **Dusky Rice Rats**
M. caliginosus **Dusky Rice Rat** E Honduras through Panama to N and W Colombia to NW Venezuela and to SW Ecuador
M. robustulus **Robust dark rice rat** SE Ecuador
M. zunigae **Zuniga's dark rice rat** WC Peru
Microryzomys **Small Rice Rats**
M. altissimus **Highland small rice rat** Andes of Colombia, Ecuador, and Peru
M. minutus **Forest small rice rat** N Venezuela through Colombia, Ecuador, Peru to WC Bolivia
Neacomys **Bristly Mice**
N. guianae **Guiana bristly mouse** Guianas, S Venezuela, N Brazil
N. pictus **Painted bristly mouse** E Panama
N. spinosus **Bristly mouse** C and W Brazil to Andean foothills of E Colombia, Ecuador, Peru, and E Bolivia
N. tenuipes **Narrow-footed bristly mouse** W and NC Colombia, N Venezuela, E Ecuador, N Brazil
Nectomys **Neotropical Water Rats**
N. palmipes **Trinidad water rat** NE Venezuela, Trinidad
N. parvipes **Small-footed water rat** Comte R (Fr. Guiana)
N. squamipes **South American water rat** NC and E Colombia, Venezuela, Guianas south to SE Brazil, NE Argentina, and Uruguay
Nelsonia **Diminutive Wood Rats**
N. goldmani **Nelson and Goldman's wood rat** Colima and S Jalisco east to N Mexico (Mexico)
N. neotomodon **Diminutive wood rat** Sierra Madre Orient., from S Durango to N Jalisco (Mexico)
Neotoma **Wood Rats**
N. albigula **White-throated wood rat** SW USA south to NE Michoacan and W Hidalgo (Mexico)
N. angustapalata **Tamaulipan wood rat** SW Tamaulipas and adj. San Luis Potosi (Mexico)
N. anthonyi **Anthony's wood rat** Todos Santos Is (Baja California, Mexico)
N. bryanti **Bryant's wood rat** Cedros Is (Baja California, Mexico)
N. bunkeri **Bunker's wood rat** Coronados Is (Baja California, Mexico)
N. chrysomelas **Nicaraguan wood rat** Honduras, NW Nicaragua
N. cinerea **Bushy-tailed wood rat** SE Yukon and W Northwest Territories (NW Canada) to NW USA and south to N New Mexico and Arizona and east to W Dakotas
N. devia **Arizona wood rat** EC and S Utah, W Arizona (SW USA), NW Sonora (Mexico)
N. floridana **Eastern wood rat** SC and E USA, from Colorado to E Texas east to Florida north to Connecticut
N. fuscipes **Dusky-footed wood rat** W Oregon through W and C California (W USA) to N Baja California (Mexico)
N. goldmani **Goldman's wood rat** SE Chihuahua to WC San Luis Potosi (Mexico)
N. lepida **Desert wood rat** SE Oregon and SW Idaho (USA) south to S Baja California (Mexico)
N. martinensis **San Martin Island wood rat** San Martin Is (Baja Califonia, Mexico)
N. mexicana **Mexican wood rat** SE Utah and C Colorado (USA) south through Mexico to Guatemala, El Salvador, and W Honduras
N. micropus **Southern Plains wood rat** SC USA south to N Chihuahua, San Luis Potosi and S Tamaulipas (Mexico)
N. nelsoni **Nelson's wood rat** Perote (Veracruz, Mexico)
N. palatina **Bolanos wood rat** EC Jalisco (Mexico)
N. phenax **Sonoran wood rat** SW Sonora and NW Sinaloa (Mexico)
N. stephensi **Stephens' wood rat** SC Utah, N Arizona, NW New Mexico (USA)
N. varia **Turner Island wood rat** Turner Is (Sonora, Mexico)
Neotomodon
N. alstoni **Mexican volcano mouse** WC Michoacan east to C Veracruz (Mexico)
Neotomys
N. ebriosus **Andean swamp rat** Altiplano of C Peru south through W Bolivia to N Chile and and NW Argentina
Nesoryzomys **Galapagos Mice**
N. darwini **Darwin's Galapagos mouse** Santa Cruz Is (Galapagos Is, Ecuador)
N. fernandinae **Fernandina Galapagos mouse** Fernandina Is (Galapagos Is, Ecuador)
N. indefessus **Indefatigable Galapagos mouse** Santa Cruz, Baltra, and Fernandina Is (Galapagos Is, Ecuador)
N. swarthi **Santiago Galapagos mouse** San Salvador Is (Galapagos Is, Ecuador)
Neusticomys **Fish-eating Rats**
N. monticolus **Montane fish-eating rat** Colombia, Ecuador
N. mussoi **Musso's fish-eating rat** Venezuela
N. oyapocki **Oyapock's fish-eating rat** Oyapock R (Fr. Guiana)
N. peruviensis **Peruvian fish-eating rat** Peru
N. venezuelae **Venezuelan fish-eating rat** S Venezuela, Guyana
Notiomys
N. edwardsii **Edwards' long-clawed mouse** S Argentina
Nyctomys
N. sumichrasti **Vesper rat** S Jalisco and S Veracruz (Mexico) south (excl. Yucatan Pen.) to C Panama
Ochrotomys
O. nuttalli **Golden mouse** SE Missouri east to S Virginia south to E Texas, Gulf of Mexico, and Florida
Oecomys **Arboreal Rice Rats**
O. bicolor **Bicolored arboreal rice rat** E Panama to Colombia, Venezuela, Guianas, Ecuador, Peru, N and C Brazil, E Bolivia
O. cleberi **Cleber's arboreal rice rat** Distrito Federal (Brazil)
O. concolor **Unicolored arboreal rice rat** E Colombia, S Venezuela, NW Brazil, N Bolivia
O. flavicans **Yellow arboreal rice rat** N and W Venezuela west to Sierra Nevada de Santa Marta of NE Colombia
O. mamorae **Mamore arboreal rice rat** WC Brazil, E Bolivia, N Paraguay
O. paricola **Brazilian arboreal rice rat** SE Venezuela, Guianas, N and C Brazil
O. phaeotis **Dusky arboreal rice rat** E Andean slopes of Peru
O. rex **King arboreal rice rat** E Venezuela, Guianas, NE Brazil
O. roberti **Robert's arboreal rice rat** S Venezuela, Guianas, E Peru, N Brazil, N Bolivia
O. rutilus **Red arboreal rice rat** Guyana, Surinam, Fr. Guiana
O. speciosus **Arboreal rice rat** NE Colombia, C and N Venezuela, Trinidad
O. superans **Foothill arboreal rice rat** Andean slopes of E Colombia, Ecuador, Peru
O. trinitatis **Trinidad arboreal rice rat** SW Costa Rica to SE Brazil, incl. Guianas, Trinidad and Tobago; also E Andean slopes of WE Colombia to SC Peru
Oligoryzomys **Pygmy Rice Rats**
O. andinus **Andean pygmy rice rat** W Peru, WC Bolivia
O. arenalis **Sandy pygmy rice rat** Peru
O. chacoensis **Chacoan pygmy rice rat** WC Brazil, SE Bolivia, W Paraguay, N Argentina
O. delticola **Delta pygmy rice rat** S Brazil, EC Argentina, Uruguay
O. destructor **Destructive pygmy rice rat** Andes of S Colombia through Ecuador and Peru to WC Bolivia
O. eliurus **Brazilian pygmy rice rat** C and SE Brazil
O. flavescens **Yellow pygmy rice rat** SE Brazil, Argentina, Uruguay
O. fulvescens **Fulvous pygmy rice rat** S Mexico through Central America to Ecuador, Guianas, and N Brazil
O. griseolus **Grayish pygmy rice rat** W Venezuela, Cord. Orient. of E Colombia
O. longicaudatus **Long-tailed pygmy rice rat** NC to S Andes, of Chile and Argentina
O. magellanicus **Magellanic pygmy rice rat** S Chile, S Argentina (incl. Tierra del Fuego)
O. microtis **Small-eared pygmy rice rat** C Brazil, adj. Peru, Bolivia, and Paraguay, and Argentina
O. nigripes **Black-footed pygmy rice rat** E Paraguay, N Argentina
O. vegetus **Sprightly pygmy rice rat** W Panama
O. victus **St. Vincent pygmy rice rat** St Vincent (Lesser Antilles)
Onychomys **Grasshopper Mice**
O. arenicola **Mearns' grasshopper mouse** SE Arizona, SC New Mexico, and W Texas (SC USA) south to San Luis Potosi and W Tamaulipas (C Mexico)
O. leucogaster **Northern grasshopper mouse** SW Canada south through W USA (Great Plains and Great Basin) to N Tamaulipas (Mexico)
O. torridus **Southern grasshopper mouse** C California, S Nevada, and SW Utah (SW USA) south to N Baja California, W Sonora and N Sinaloa (Mexico)
Oryzomys **Rice Rats**
O. albigularis **Tomes' rice rat** E Panama, Andes of Colombia, N and W Venezuela, Andes of Ecuador to N Peru
O. alfaroi **Alfaro's rice rat** S Tamaulipas and Oaxaca (Mexico) through Central America to W Colombia and Ecuador
O. auriventer **Ecuadorean rice rat** E Ecuador, N Peru
O. balneator **Peruvian rice rat** E and S Ecuador, N Peru
O. bolivaris **Bolivar rice rat** E Honduras through E Nicaragua, Costa Rica, and Panama to W Colombia and W Ecuador
O. buccinatus **Paraguayan rice rat** NE Argentina, E Paraguay
O. capito **Large-headed rice rat** Colombia, S Venezuela, Guianas, Ecuador, Peru, Brazil, and Bolivia
O. chapmani **Chapman's rice rat** Oaxaca, Guerrero, Veracruz to Tamaulipas (Mexico)
O. couesi **Coues' rice rat** S Texas (USA), Mexico (incl. Isla Cozumel) south through Central America to NW Colombia, Jamaica
O. devius **Boquete rice rat** Costa Rica, W Panama
O. dimidiatus **Thomas' rice rat** SE Nicaragua
O. galapagoensis **Galapagos rice rat** San Cristobal and Sante Fe Is (Galapagos Is, Ecuador)
O. gorgasi **Gorgas' rice rat** Antioquia (Colombia)
O. hammondi **Hammond's rice rat** NW Ecuador
O. intectus **Colombian rice rat** NC Colombia
O. intermedius **Intermediate rice rat** SE Brazil, E Paraguay, NE Argentina
O. keaysi **Keays' rice rat** E Andes of Peru
O. kelloggi **Kellogg's rice rat** SE Brazil
O. lamia **Monster rice rat** SE Brazil
O. legatus **Big-headed rice rat** E Andean slopes of SC Bolivia and NW Argentina
O. levipes **Light-footed rice rat** SE Peru to WC Bolivia
O. macconnelli **MacConnell's rice rat** SC Colombia, E Ecuador, Peru east to Venezuela, Guianas and N Brazil
O. melanotis **Black-eared rice rat** S Sinaloa to SW Oaxaca (W Mexico)
O. nelsoni **Nelson's rice rat** Extinct: formerly Maria Madre Is (Nayarit, Mexico)
O. nitidus **Elegant rice rat** E Ecuador, Peru, Mato Grosso (WC Brazil), Bolivia
O. oniscus **Sowbug rice rat** E Brazil
O. palustris **Marsh rice rat** SE USA, from SE Kansas to E Texas east to S New Jersey and Florida
O. polius **Gray rice rat** NC Peru
O. ratticeps **Rat-headed rice rat** E Brazil, Paraguay, NE Argentina
O. rhabdops **Striped rice rat** S Chiapas (Mexico), C Guatemala
O. rostratus **Long-nosed rice rat** C Tamaulipas to Oaxaca (Mexico, incl. Yucatan Pen.) through Guatemala, El Salvador, and Honduras to S Nicaragua
O. saturatior **Cloud Forest rice rat** S Oaxaca and Chiapas (Mexico) through Guatemala, El Salvador, and Honduras to NC Nicaragua
O. subflavus **Terraced rice rat** E Brazil
O. talamancae **Talamancan rice rat** E Costa Rica, Panama, W and NC Colombia, N Venezuela, W Ecuador
O. xantheolus **Yellowish rice rat** SW Ecuador to W Peru
O. yunganus **Yungas rice rat** S Venezuela, Guianas and E Andean foothills from C Colombia to WC Bolivia
Osgoodomys
O. banderanus **Michoacan deer mouse** S Mexico
Otonyctomys
O. hatti **Hatt's vesper rat** Yucatan Pen. (Mexico) south to N Belize and NE Guatemala
Ototylomys
O. phyllotis **Big-eared climbing rat** N Chiapas, S Tabasco, and Yucatan Pen. (Mexico) south to C Costa Rica, also NC Guerrero (Mexico)
Oxymycterus **Hocicudos**
O. akodontius **Argentine hocicudo** NW Argentina
O. angularis **Angular hocicudo** E Brazil
O. delator **Spy hocicudo** E Paraguay
O. hiska **Small hocicudo** Puno (Peru)
O. hispidus **Hispid hocicudo** Misiones (NE Argentina) to Bahia (E Brazil)
O. hucucha **Quechuan hocicudo** Cochabamba (Bolivia)
O. iheringi **Ihering's hocicudo** Misiones (NE Argentina) and SE Brazil
O. inca SC **Incan hocicudo** Peru to WC Bolivia
O. nasutus **Long-nosed hocicudo** SE Brazil, Uruguay
O. paramensis **Paramo hocicudo** E Andes of SE Peru, WC Bolivia, and NW Argentina
O. roberti **Robert's hocicudo** Minas Gerais (Brazil)
O. rufus **Red hocicudo** SE Brazil, EC Argentina, Uruguay
Peromyscus **Deer Mice**
P. attwateri **Texas mouse** NC Texas through E Oklahoma to SE Kansas, SW Missouri, and NW Arkansas (USA)
P. aztecus **Aztec mouse** S Jalisco and C Veracruz (Mexico) to Guatemala, N El Salvador, and Honduras
P. boylii **Brush mouse** California to W Oklahoma (USA) south to Queretaro and W Hidalgo (Mexico)
P. bullatus **Perote mouse** Perote (Veracruz, Mexico)
P. californicus **California mouse** C and S California (USA) to NW Baja California (Mexico)
P. caniceps **Burt's deer mouse** Monserrate Is (Baja California, Mexico)
P. crinitus **Canyon mouse** E Oregon and SW Idaho south through Nevada, Utah, and W Colorado (USA) to Baja California and NW Sonora (Mexico)
P. dickeyi **Dickey's deer mouse** Tortuga Is (Baja California, Mexico)
P. difficilis **Zacatecan deer mouse** W Chihuahua and SE Caohuila south to C Oaxaca (Mexico)
P. eremicus **Cactus mouse** S California east to Texas (USA) south to Baja California, C Sinaloa and N San Luis Potosi (Mexico)
P. eva **Eva's desert mouse** S Baja California and Carmen Is (Mexico)
P. furvus **Blackish deer mouse** E Sierra Madre Orient., from S San Luis Potosi to NW Oaxaca (Mexico)
P. gossypinus **Cotton mouse** SE USA, from SE Oklahoma, S Illinois, and SE Virginia south to Gulf of Mexico and Florida
P. grandis **Big deer mouse** Guatemala
P. gratus **Osgood's mouse** SW New Mexico (USA), W Chihuahua and SE Coahuila to C Oaxaca (Mexico)
P. guardia **Angel Island mouse** Angel dela Guarda, Granito and Mehia Is (Gulf of California, Mexico)
P. guatemalensis **Guatemalan deer mouse** S Chiapas (Mexico), SW Guatemala
P. gymnotis **Naked-eared deer mouse** S Chiapas (Mexico) to S Nicaragua
P. hooperi **Hooper's mouse** C Coahuila to NE Zacatecas (Mexico)
P. interparietalis **San Lorenzo mouse** San Lorenzo and Salsipuedes Is (Gulf of California, Mexico)
P. leucopus **White-footed mouse** S Alberta and S Ontario, Quebec and Nova Scotia (Canada) south through C and E USA (excl. Florida) to N Durango and along Caribbean coast to NW Yucatan Pen. (Mexico)
P. levipes **Nimble-footed mouse** Mexico to Guatemala, El Salvador, and Honduras
P. madrensis **Tres Marias Island mouse** Tres Marias Is (Mexico)
P. maniculatus **Deer mouse** Alaska (USA) and N Canada south through North America (excl. SE and E seaboard) to S Baja California and NC Oaxaca (Mexico)
P. mayensis **Maya mouse** Huehuetenango (Guatemala)
P. megalops **Brown deer mouse** Sierra Madre del Sur of Guerrero and Oaxaca (Mexico)
P. mekisturus **Puebla deer mouse** SE Puebla (Mexico)
P. melanocarpus **Zempoaltepec deer mouse** NC Oaxaca (Mexico)
P. melanophrys **Plateau mouse** S Durango and Coahuila south to Chiapas (Mexico)
P. melanotis **Black-eared mouse** SE Arizona (USA), Mexico
P. melanurus **Black-tailed mouse** W Sierra Madre del Sur of Oaxaca (Mexico)
P. merriami **Mesquite mouse** SC Arizona (USA) south through Sonora to C Sinaloa (Mexico)
P. mexicanus **Mexican deer mouse** S San Luis Potosi

to Isthmus of Tehuantepec and Guerrero/Oaxaca border to Chiapas (Mexico), Guatemala, El Salvador, Honduras, Nicaragua, Costa Rica, W Panama
P. nasutus **Northern rock mouse** C Colorado, SE Utah, New Mexico, and W Texas (USA) to NW Coahuila (Mexico)
P. ochraventer **El Carrizo deer mouse** S Tamaulipas and adj. San Luis Potosi (Mexico)
P. oreas **Columbian mouse** SW British Columbia (SW Canada), W Washington (NW USA)
P. pectoralis **White-ankled mouse** SE New Mexico and C Texas (USA) south to N Jalisco and Hidalgo (Mexico)
P. pembertoni **Pemberton's deer mouse** Extinct: formerly San Pedro Nolasco Is (Sonora, Mexico)
P. perfulvus **Marsh mouse** C Mexico
P. polionotus **Oldfield mouse** SE USA
P. polius **Chihuahuan mouse** WC Chihuahua (Mexico)
P. pseudocrinitus **False canyon mouse** Coronados Is (Baja California, Mexico)
P. sejugis **Santa Cruz mouse** Santa Cruz and San Diego Is (Gulf of California, Mexico)
P. simulus **Nayarit mouse** Nayarit and S Sinaloa (Mexico)
P. sitkensis **Sitka mouse** Alexander Arch. (Alaska, USA)
P. slevini **Slevin's mouse** Santa Catalina Is (Baja California, Mexico)
P. spicilegus **Gleaning mouse** Sierra Madre Occid., from S Sinaloa and SW Durango to WC Michoacan (Mexico)
P. stephani **San Esteban Island mouse** San Esteban Is (Sonora, Mexico)
P. stirtoni **Stirton's deer mouse** SE Guatemala, El Salvador, Honduras, to NC Nicaragua
P. truei **Pinyon mouse** SW Oregon to W and SE Colorado and NC Texas (USA) south to Baja California (Mexico)
P. winkelmanni **Winkelmann's mouse** Michoacan and Guerrero (Mexico)
P. yucatanicus **Yucatan deer mouse** N Yucatan Pen. (Mexico)
P. zarhynchus **Chiapan deer mouse** NC Chiapas (Mexico)
Phaenomys
P. ferrugineus **Rio de Janeiro arboreal rat** Rio de Janeiro (Brazil)
Phyllotis **Leaf-eared Mice**
P. amicus **Friendly leaf-eared mouse** W Peru
P. andium **Andean leaf-eared mouse** E and W Andean slopes from Tungurahua (C Ecuador) to Lima (C Peru)
P. bonaeriensis **Buenos Aires leaf-eared mouse** Buenos Aires (Argentina)
P. caprinus **Capricorn leaf-eared mouse** E Andes from S Bolivia to N Argentina
P. darwini **Darwin's leaf-eared mouse** Junin (C Peru) south through W Bolivia to C Chile and WC Argentina
P. definitus **Definitive leaf-eared mouse** Andes of Ancash (Peru)
P. gerbillus **Gerbil leaf-eared mouse** Sechura Desert (NW Peru)
P. haggardi **Haggard's leaf-eared mouse** Andes of C Ecuador
P. magister **Master leaf-eared mouse** Andes from C Peru to N Chile
P. osgoodi **Osgood's leaf-eared mouse** Altiplano of NE Chile
P. osilae **Bunchgrass leaf-eared mouse** E Andes from Cuzco (SC Peru) through WC Bolivia to Catamarca (N Argentinia)
P. wolffsohni **Wolffsohn's leaf-eared mouse** E Andes of C Bolivia
P. xanthopygus **Yellow-rumped leaf-eared mouse** Catamarca (NW Argentina) and Atacama (C Chile) south to Santa Cruz (Argentina) and adj. Magallanes (Chile)
Podomys
P. floridanus **Florida mouse** Florida Pen. (USA)
Podoxymys
P. roraimae **Roraima mouse** Guyana, poss. adj. Venezuela and Brazil
Pseudoryzomys
P. simplex **Brazilian false rice rat or Ratos-do-Mato** E Brazil to SW Bolivia, W Paraguay, NE Argentina
Punomys
P. lemminus **Puna mouse** Altiplano of S Peru
Reithrodon
R. auritus **Bunny rat** Argentina, Chile, Uruguay
Reithrodontomys **American Harvest Mice**
R. brevirostris **Short-nosed harvest mouse** NC Nicaragua, C Costa Rica
R. burti **Sonoran or Burt's harvest mouse** WC Sonora to C Sinaloa (Mexico)
R. chrysopsis **Volcano harvest mouse** SE Jalisco to WC Veracruz (Mexico)
R. creper **Chiriqui harvest mouse** Costa Rica to Chiriqui region (Panama)
R. darienensis **Darien harvest mouse** E Panama (incl. Azuero Pen.)
R. fulvescens **Fulvous harvest mouse** SC Arizona to SW Missouri to WC Mississippi (USA) south through Mexico (excl. Yucatan Pen.) to W Nicaragua
R. gracilis **Slender harvest mouse** Yucatan Pen. and coastal Chiapas (Mexico) south to NW Costa Rica
R. hirsutus **Hairy harvest mouse** SC Nayarit and NW Jalisco (Mexico)
R. humulis **Eastern harvest mouse** SE USA, from SE Oklahoma and E Texas east to E Coast, from S Maryland to Florida
R. megalotis **Western harvest mouse** SC British Columbia and SE Alberta (SW Canada), W and NC USA south to N Baja California and to C Oaxaca (Mexico)
R. mexicanus **Mexican harvest mouse** S Tamaulipas and WC Michoacan (Mexico) south through Central America to W Panama, and Andes of W Colombia and N Ecuador
R. microdon **Small-toothed harvest mouse** N Michoacan, Distrito Federal, N Oaxaca, and C Chiapas (Mexico), WC Guatemala
R. montanus **Plains harvest mouse** C USA, from W South Dakota and E Wyoming to EC Texas and SE Arizona, NE Sonora and Chihuahua top N Durango (Mexico)
R. paradoxus **Nicaraguan harvest mouse** SW Nicaragua, WC Costa Rica
R. raviventris **Salt marsh harvest mouse** San Francisco Bay (California, USA)
R. rodriguezi **Rodriguez's harvest mouse** Volcan de Irazu (Cartago, Costa Rica)
R. spectabilis **Cozumel harvest mouse** Cozumel Is (Quintano Roo, Mexico)
R. sumichrasti **Sumichrast's harvest mouse** SW Jalisco and S San Luis Potosi to C Guerrero and EC Oaxaca (Mexico); C Chiapas (Mexico) to NC Nicaragua; C Costa Rica to W Panama
R. tenuirostris **Narrow-nosed harvest mouse** S Chiapas (Mexico), C Guatemala
R. zacatecae **Zacatecas harvest mouse** W Chihuahua to WC Michoacan (Mexico)
Rhagomys
R. rufescens **Brazilian arboreal mouse** Rio de Janeiro (Brazil)
Rheomys **Central American Water Mice**
R. mexicanus **Mexican water mouse** Oaxaca (Mexico)
R. raptor **Goldman's water mouse** Costa Rica, Panama
R. thomasi **Thomas' water mouse** S Mexico, Guatemala, El Salvador
R. underwoodi **Underwood's water mouse** C Costa Rica, W Panama
Rhipidomys **American Climbing Mice**
R. austrinus **Southern climbing mouse** E Andean slopes of SC Bolivia and NW Argentina
R. caucensis **Cauca climbing mouse** W Andes of Colombia
R. couesi **Coues' climbing mouse** Colombia, Venezuela, Ecuador, Peru, Trinidad
R. fulviventer **Buff-bellied climbing mouse** Andes of Colombia and W Venezuela
R. latimanus **Broad-footed climbing mouse** C and W Colombia, Ecuador
R. leucodactylus **White-footed climbing mouse** S Venezuela, Guianas, Ecuador, Peru, N Brazil
R. macconnelli **MacConnell's climbing mouse** S Venezuela, poss. N Brazil and adj. Guyana
R. mastacalis **Long-tailed climbing mouse** C and E Brazil
R. nitela **Splendid climbing mouse** S Venezuela, Guianas, NC Brazil
R. ochrogaster **Yellow-bellied climbing mouse** SE Peru
R. scandens **Mount Pirri climbing mouse** E Panama
R. venezuelae **Venezuelan climbing mouse** N and W Venezuela, E Colombia
R. venustus **Charming climbing mouse** N Venezuela
R. wetzeli **Wetzel's climbing mouse** S Venezuela
Scapteromys
S. tumidus **Swamp rat** S Brazil, E Paraguay, NE Argentina, Uruguay
Scolomys **Spiny Mice**
S. melanops **South American spiny mouse** Pastaza (Ecuador)
S. ucayalensis **Ucayali spiny mouse** Loreto (Peru)
Scotinomys **Brown Mice**
S. teguina **Alston's brown mouse** E Oaxaca (Mexico) to W Panama
S. xerampelinus **Chiriqui brown mouse** Costa Rica to Chiriqui region (W Panama)
Sigmodon **Cotton Rats**
S. alleni **Allen's cotton rat** S Sinaloa to S Oaxaca (W Mexico)
S. alstoni **Alston's cotton rat** NE Colombia, N and E Venezuela, Guyana, Surinam, N Brazil
S. arizonae **Arizona cotton rat** SE California and SC Arizona (SW USA) south to Nayarit (W Mexico)
S. fulviventer **Tawny-bellied cotton rat** SE Arizona and WC New Mexico (USA) south to Guanajuato and NW Michoacan (Mexico)
S. hispidus **Hispid cotton rat** SE USA, from S Nebraska to C Virginia south to SE Arizona and Florida, then E Mexico through Central America to N Colombia and N Venezuela
S. inopinatus **Unexpected cotton rat** Andes of Azuay and Chimborazo (Ecuador)
S. leucotis **White-eared cotton rat** SW Chihuahua and S Nuevo Leon to C Oaxaca (Mexico)
S. mascotensis **Jaliscan cotton rat** S Nayarit to E Oaxaca (W Mexico)
S. ochrognathus **Yellow-nosed cotton rat** SE Arizona, SW New Mexico, and W Texas (USA) south to C Durango (Mexico)
S. peruanus **Peruvian cotton rat** W Andean foothills of W Ecuador and NW Peru
Sigmodontomys **Rice Water Rats**
S. alfari **Alfaro's rice water rat** E Honduras to Panama, C and W Colombia to NW Venezuela and NW Ecuador
S. aphrastus **Harris' rice water rat** San Jose (Costa Rica), Chiriqui (W Panama)
Thalpomys **Cerrado Mice**
T. cerradensis **Cerrado mouse** C Brazil
T. lasiotis **Hairy-eared cerrado mouse** C Brazil
Thomasomys **Thomas' Oldfield Mice**
T. aureus **Golden oldfield mouse** C and W Colombia, NC Venezuela, Ecuador, N Peru
T. baeops **Beady-eyed mouse** W Andes of Ecuador
T. bombycinus **Silky oldfield mouse** Cord. Occid. of Colombia
T. cinereiventer **Ashy-bellied oldfield mouse** Andes of Colombia and Ecuador
T. cinereus **Ash-colored oldfield mouse** SW Ecuador, N Peru
T. daphne **Daphne's oldfield mouse** S Peru to C Bolivia
T. eleusis **Peruvian oldfield mouse** NC Peru
T. gracilis **Slender oldfield mouse** Andes of Ecuador to SE Peru
T. hylophilus **Woodland oldfield mouse** Cord. Orient. of Colombia, Cord. Merida of W Venezuela
T. incanus **Inca oldfield mouse** Andes of C Peru
T. ischyurus **Strong-tailed oldfield mouse** N to C Peru
T. kalinowskii **Kalinowski's oldfield mouse** Andes of C Peru
T. ladewi **Ladew's oldfield mouse** Andes of NW Bolivia
T. laniger **Butcher oldfield mouse** Andes of C Colombia and adj. W Venezuela
T. monochromos **Unicolored oldfield mouse** NE Colombia
T. niveipes **Snow-footed oldfield mouse** C Colombia
T. notatus **Distinguished oldfield mouse** SE Peru
T. oreas **Montane oldfield mouse** WC Bolivia
T. paramorum **Paramo oldfield mouse** Andes of Ecuador
T. pyrrhonotus **Thomas' oldfield mouse** Andes of S Ecuador and NW Peru
T. rhoadsi **Rhoads' oldfield mouse** Andes of Ecuador
T. rosalinda **Rosalinda's oldfield mouse** NC Peru
T. silvestris **Forest oldfield mouse** W Andes of Ecuador
T. taczanowskii **Taczanowski's oldfield mouse** NW Peru
T. vestitus **Dressy oldfield mouse** Cord. Merida of W Venezuela
Tylomys **Naked-tailed Climbing Rats**
T. bullaris **Chiapan climbing rat** Tuxtla Gutierrez (Chiapas, Mexico)
T. fulviventer **Fulvous-bellied climbing rat** E Panama
T. mirae **Mira climbing rat** W Colombia, NW Ecuador
T. nudicaudus **Peters' climbing rat** C Guerrero and C Veracruz (Mexico) south (excl. Yucatan Pen.) to S Nicaragua
T. panamensis **Panamanian climbing rat** E Panama
T. tumbalensis **Tumbala climbing rat** Tumbala (Chiapas, Mexico)
T. watsoni **Watson's climbing rat** Costa Rica, W Panama
Wiedomys
W. pyrrhorhinos **Red-nosed mouse** Ceara to Rio Grane do Sul (SE Brazil)
Wilfredomys **Wilfred's Mice**
W. oenax **Greater Wilfred's mouse** SE Brazil to C Uruguay
W. pictipes **Lesser Wilfred's mouse** NE Argentina and SE Brazil
Xenomys
X. nelsoni **Magdalena rat** Colima and W Jalisco (Mexico)
Zygodontomys **Cane Mice**
Z. brevicauda **Short-tailed cane mouse** SE Costa Rica through Panama, Colombia, Venezuela, Guianas, to N Brazil, Trinidad and Tobago
Z. brunneus **Brown cane mouse** N Colombia

SUBFAMILY SPALACINAE
Nannospalax **Lesser Blind Mole-rats**
N. ehrenbergi **Palestine or Ehrenberg's mole-rat** Syria, Lebanon, Jordan, Israel, N Egypt to N Libya
N. leucodon **Lesser mole-rat** Yugoslavia, Hungary, Bulgaria, Greece, and NW Turkey to SW Ukraine
N. nehringi **Nehring's blind mole-rat** Turkey, Armenia, Georgia
Spalax **Greater Blind Mole-rats**
S. arenarius **Sandy mole-rat** S Ukraine
S. giganteus **Giant mole-rat** NW of Caspian Sea, and Kazakhstan betw. Volga, Dnipro, and Ural R and the Caspian Sea
S. graecus **Bukovin mole-rat** Romania, SW Ukraine
S. microphthalmus **Greater mole-rat** Ukraine and S Russia, betw. Dnipro and Volga R
S. zemni **Podolsk mole-rat** SE Poland east to Ukraine betw. Dnetr and Dnipro R and south to Black Sea

FAMILY ANOMALURIDAE
Scaly-tailed Squirrels

SUBFAMILY ANOMALURINAE
Anomalurus **Scaly-tailed Flying Squirrels**
A. beecrofti **Beecroft's scaly-tailed squirrel** Senegal to Uganda and DRC, Bioko (Equatorial Guinea)
A. derbianus **Lord Derby's scaly-tailed squirrel** Sierra Leone to Angola, east to Kenya south to Zambia and Mozambique
A. pelii **Pel's scaly-tailed squirrel** Sierra Leone to Ghana
A. pusillus **Dwarf scaly-tailed squirrel** S Cameroon, Gabon, DRC

SUBFAMILY ZENKERELLINAE
Idiurus **Pygmy Scaly-tailed Flying Squirrels**
I. macrotis **Long-eared scaly-tailed flying squirrel** Sierra Leone to E DRC
I. zenkeri **Pygmy scaly-tailed flying squirrel** S Cameroon to Uganda
Zenkerella
Z. insignis **Cameroon scaly-tail** SW Cameroon, CAR, Gabon, Rio Muni (Equatorial Guinea)

FAMILY PEDETIDAE
Pedetes
P. capensis **Springhare or Springhaas** South Africa to Tanzania, Kenya

FAMILY CTENODACTYLIDAE **Gundis**
Ctenodactylus **Common Gundis**
C. gundi **Gundi** N Morocco to NW Libya
C. vali **Val's or Sahara gundi** S Morocco, W Algeria, NW Libya
Felovia
F. vae **Felou gundi** Senegal, Mauritania, Mali
Massoutiera
M. mzabi **Mzab gundi** SE Algeria, SW Libya, NE Mali, N Niger, N Chad
Pectinator
P. spekei **Pectinator or East African gundi** Ethiopia, Somalia, Djibouti

FAMILY MYOXIDAE **Dormice**

SUBFAMILY GRAPHIURINAE
Graphiurus **African Dormice**
G. christyi **Christy's dormouse** N DRC, S Cameroon
G. crassicaudatus **Jentink's dormouse** Liberia, Ivory Coast, Ghana, Togo, Nigeria, Cameroon
G. hueti **Huet's dormouse** Senegal to Sierra Leone, Liberia, Ivory Coast, Ghana, Nigeria, Cameroon, CAR, Gabon

G. kelleni **Kellen's dormouse** Angola, Zambia, Malawi, Zimbabwe
G. lorraineus **Lorrain dormouse** Sierra Leone, Ivory Coast to Cameroon, Gabon, N Angola, DRC, Uganda, SW Tanzania
G. microtis **Small-eared dormouse** Zambia, Malawi, Tanzania
G. monardi **Monard's dormouse** E Angola, NW Zambia, S DRC
G. murinus **Woodland dormouse** Sudan, Ethiopia, Kenya, Uganda, Tanzania, E DRC, Mozambique, Malawi, Zambia, S Angola, E and N Namibia, Botswana, Zimbabwe, South Africa
G. ocularis **Spectacled dormouse** South Africa
G. olga **Olga's dormouse** N Niger, N Nigeria, NE Cameroon
G. parvus **Savanna dormouse** Sierra Leone, Ivory Coast, Mali, Ghana, Nigeria, Sudan, Ethiopia, Somalia, Kenya, Uganda, Tanzania
G. platyops **Rock dormouse** South Africa to S DRC
G. rupicola **Stone dormouse** Karibib and Mt Brukaros (Namibia) south to Port Nolloth (W coastal South Africa)
G. surdus **Silent dormouse** Rio Muni (Equatorial Guinea), S Cameroon

SUBFAMILY LEITHIINAE
Dryomys Forest Dormice
D. laniger **Woolly dormouse** Toros Doglan (Turkey)
D. nitedula **Forest dormouse** C Europe from Germany, Switzerland, Austria, and Italy east to W Russia south to Turkey, Arabia, Iraq, Iran, Afghanistan, N Pakistan, Tajikistan, Turkmenistan, Uzbekistan, Kyrygstan, C Kazakhstan north to S Altai Mts and E Tien Shan Mts
D. sichuanensis **Chinese dormouse** N Sichuan (China)
Eliomys **Garden Dormice**
E. melanurus **Asiatic garden dormouse** S Turkey, Syria, Iraq, Jordan, Lebanon, Israel, Saudi Arabia, Egypt to Morocco
E. quercinus **Garden or Orchard dormouse** Most of Europe from Portugal, S Spain, France, Belgium, Netherlands east to W Russia
Myomimus Mouse-tailed Dormice
M. personatus **Masked mouse-tailed dormouse** NE Iran, Kopet Dagh and Malyy Balkhan Mts (Turkmenistan), Uzbekistan
M. roachi **Roach's mouse-tailed dormouse** SE Bulgaria, W Turkey
M. setzeri **Setzer's mouse-tailed dormouse** W Iran

FAMILY SELEVINIA
S. betpakdalaensis **Desert dormouse** SE and E Kazakhstan

FAMILY MYOXINAE
Glirulus
G. japonicus **Japanese dormouse** Honshu, Shikoku, Kyushu Is (Japan)
Muscardinus
M. avellanarius **Hazel dormouse** Europe from S England to W Russia south to N Turkey
Myoxus
M. glis **Edible or Fat dormouse** Most of Europe from N Spain, France, Netherlands, Germany east to W Russia south to N Turkey and through Caucasus to N Iran and SW Turkmenistan

SUBORDER HYSTRICOGNATHI

FAMILY BATHYERGIDAE **Blesmols**
Bathyergus Dune Mole-rats
B. janetta **Namaqua dune mole-rat** SW South Africa, S Namibia
B. suillus **Cape dune mole-rat** S South Africa
Cryptomys Common Mole-rats
C. bocagei **Bocage's mole-rat** NW Namibia, C Angola, S DRC
C. damarensis **Damara mole-rat** E Namibia, Botswana, W Zimbabwe, S Zambia, S Angola
C. foxi **Nigerian mole-rat** C Nigeria
C. hottentotus **African or Common mole-rat** South Africa to Namibia, S Zambia, and Tanzania
C. mechowi **Mechow's or Giant Angolan mole-rat** Angola, Zambia, Malawi, S DRC, Tanzania
C. ochraceocinereus **Ochre mole-rat** E Nigeria, CAR, N DRC, S Sudan, NW Uganda
C. zechi **Togo mole-rat** EC Ghana, WC Togo
Georychus
G. capensis **Cape mole-rat** South Africa
Heliophobius
H. argenteocinereus **Silvery mole-rat** E Zambia, N Mozambique to DRC, Tanzania, and Kenya
Heterocephalus
H. glaber **Naked mole-rat** C Somalia, C and E Ethiopia, C and S Kenya

FAMILY HYSTRICIDAE **Old World Porcupines**
Atherurus Brush-tailed Porcupines
A. africanus **African brush-tailed porcupine** Gambia, Sierra Leone, Liberia, Ghana, DRC, S Sudan, Kenya, Uganda
A. macrourus **Asiatic brush-tailed porcupine** E Assam (India), Sichuan, Yunnan, Hupei, and Hainan (China), Burma to Thailand, Laos, Vietnam, Malaysia, Sumatra (Indonesia)
Hystrix Short-tailed Porcupines
H. africaeaustralis **Cape porcupine** South Africa to N Angola, DRC, Uganda, and Kenya
H. brachyura **Malayan porcupine** Nepal, Sikkim and Assam (NE India), C and S China, Burma, Thailand, Indochina, Malaysia, Singapore, Sumatra (Indonesia), Borneo
H. crassispinis **Thick-spined porcupine** N Borneo
H. cristata **Crested porcupine** Morocco to Egypt, Senegal to Ethiopia and N Tanzania, also Italy, Albania, N Greece
H. indica **Indian crested porcupine** Turkey and Caucasus, Israel, Arabia to S Kazakhstan and India, Sri Lanka, Xizang (China)
H. javanica **Sunda porcupine** Indonesia
H. pumila **Philippine porcupine** Palawan and Busuanga (Philippines)
H. Sumatrae **Sumatran porcupine** Sumatra (Indonesia)
Trichys
T. fasciculata **Long-tailed porcupine** Malaysia, Sumatra (Indonesia), Borneo

FAMILY PETROMURIDAE
Petromus
P. typicus **Dassie rat** W South Africa, Namibia to SW Angola

FAMILY THRYONOMYIDAE **Cane Rats**
Thryonomys
T. gregorianus **Lesser canerat** Cameroon, CAR, DRC, S Sudan, Ethiopia, Kenya, Uganda, Tanzania, Mozambique, Malawi, Zambia, Zimbabwe
T. swinderianus **Greater canerat** SubSaharan Africa

FAMILY ERETHIZONTIDAE
New World Porcupines
Coendou Prehensile-tailed Porcupines
C. bicolor **Bicolor-spined porcupine** Colombia, Ecuador, Peru, Bolivia
C. koopmani **Koopman's porcupine** Brazil
C. prehensilis **Brazilian porcupine** E Venezuela, Guianas, C and E Brazil, Bolivia, Trinidad
C. rothschildi **Rothschild's porcupine** Panama
Echinoprocta
E. rufescens **Stump-tailed porcupine** Cord. Orient. of W Colombia
Erethizon
E. dorsatum **North American porcupine** C Alaska (USA) to Labrador and S of Hudson Bay (Canada) south to E Tennessee, C Iowa, and C Texas (USA) to N Mexico and then north to S California (USA)
Sphiggurus Hairy Dwarf Porcupines
S. insidiosus **Bahia hairy dwarf porcupine** Surinam, E and Amazonian Brazil
S. mexicanus **Mexican hairy dwarf porcupine** San Luis Potosi and Yucatan Pen. (Mexico) to W Panama
S. pallidus **Pallid or West Indian hairy dwarf porcupine** Extinct: formerly W. Indies
S. spinosus **Paraguay hairy dwarf porcupine** S and E Brazil, Paraguay, NE Argentina, Uruguay
S. vestitus **Brown hairy dwarf porcupine** Colombia, W Venezuela
S. villosus **Orange-spined hairy dwarf porcupine** Minas Gerais to Rio Grande do Sul (SE Brazil)

FAMILY CHINCHILLIDAE
Viscachas and Chinchillas
Chinchilla Chinchillas
C. brevicaudata **Short-tailed chinchilla** Andes of S Peru, S Bolivia, Chile, and NW Argentina
C. lanigera **Chinchilla** N Chile south to Coquimbo
Lagidium Mountain Viscachas
L. peruanum **Northern viscacha** C and S Peru
L. viscacia **Southern viscacha** S Peru, S and W Bolivia, W Argentina, N Chile
L. wolffsohni **Wolffsohn's viscacha** SW Argentina, adj. Chile
Lagostomus
L. maximus **Plains viscacha** SE Bolivia, S and W Paraguay, N, C, and E Argentina

FAMILY DINOMYIDAE
Dinomys
D. branickii **Pacarana** Colombia, Venezuela, Ecuador, Peru, Brazil, Bolivia

FAMILY CAVIIDAE

SUBFAMILY CAVIINAE
Cavia Guinea-pigs
C. aperea **Brazilian guinea-pig** Colombia, Venezuela, Guianas, Ecuador, Brazil, Paraguay, N Argentina, Uruguay
C. fulgida **Shiny guinea-pig** Minas Gerais to Santa Catarina (E Brazil)
C. magna **Greater guinea-pig** Rocha (Uruguay) to Rio Grande do Sul and Santa Catarina (S Brazil)
C. porcellus **Guinea pig** Domesticated worldwide but poss. feral in South America
C. tschudii **Montane guinea-pig** Peru, S Bolivia, N Chile, NW Argentina
Galea Yellow-toothed Cavies
G. flavidens **Yellow-toothed cavy** E Brazil
G. musteloides **Common yellow-toothed cavy** S Peru, Bolivia, N Chile, Argentina
G. spixii **Spix's yellow-toothed cavy** Brazil, E Bolivia
Kerodon
K. rupestris **Rock cavy** E Brazil
Microcavia Mountain Cavies
M. australis **Southern mountain cavy** S Bolivia, Jujuy to Santa Cruz (Argentina), Aisen (Chile)
M. niata **Andean mountain cavy** Andes of SW Bolivia
M. shiptoni **Shipton's mountain cavy** Tucuman, Catamarca, and Salta (NW Argentina)

SUBFAMILY DOLICHOTINAE
Dolichotis Maras
D. patagonum **Patagonian mara** Argentina
D. salinicola **Chacoan mara** S Bolivia, Chaco (W Paraguay), NW Argentina south to Cordoba

FAMILY HYDROCHAERIDAE
Hydrochaeris
H. hydrochaeris **Capybara** Panama, Colombia, Venezuela, Guianas, Peru, Brazil, Paraguay, NE Argentina, Uruguay

FAMILY DASYPROCTIDAE **Agoutis**
Dasyprocta Agoutis
D. azarae **Azara's agouti** E, C, and S Brazil, Paraguay, NE Argentina
D. coibae **Coiban agouti** Coiba Is (Panama)
D. cristata **Crested agouti** Guianas
D. fuliginosa **Black agouti** Colombia, S Venezuela, Surinam, Peru, N Brazil
D. guamara **Orinoco agouti** Orinoco Delta (Venezuela)
D. kalinowskii **Kalinowski's agouti** SE Peru
D. leporina **Brazilian agouti** Venezuela, Guianas, E and Amazonian Brazil, Lesser Antilles; introd. to Virgin Is
D. mexicana **Mexican agouti** C Veracruz and E Oaxaca (Mexico); introd. to W and E Cuba
D. prymnolopha **Black-rumped agouti** NE Brazil
D. punctata **Central American agouti** Chiapas and Yucatan Pen. (S Mexico) to S Bolivia, SW Brazil, and N Argentina; introd. to W and W Cuba, Cayman Is
D. ruatanica **Ruatan Island agouti** Roatan Is (Honduras)
Myoprocta Acouchis
M. acouchy **Green acouchi** S Colombia, S Venezuela, Guianas, Ecuador, N Peru, Amazonian Brazil
M. exilis **Red acouchi** Colombia, S Venezuela, Guianas, E Ecuador, N Peru, Amazon Basin of Brazil

FAMILY AGOUTIDAE **Pacas**
Agouti
A. paca **Paca** SE San Luis Potosi (Mexico) to Guianas, S Brazil, and Paraguay; introd. to Cuba
A. taczanowskii **Mountain paca** Colombia, NW Venezuela, Ecuador, Peru

FAMILY CTENOMYIDAE **Tuco-tucos**
Ctenomys
C. argentinus **Argentine tuco-tuco** NC Chaco (NE Argentina)
C. australis **Southern tuco-tuco** Buenos Aires (E Argentina)
C. azarae **Azara's tuco-tuco** La Pampa (S Argentina)
C. boliviensis **Bolivian tuco-tuco** C Bolivia, W Paraguay, Formosa (NE Argentina)
C. bonettoi **Bonetto's tuco-tuco** Chaco (NE Argentina)
C. brasiliensis **Brazilian tuco-tuco** E Brazil
C. colburni **Colburn's tuco-tuco** W Santa Cruz (SW Argentina)
C. conoveri **Conover's tuco-tuco** Chaco (W Paraguay), adj. Argentina
C. dorsalis **Chacoan tuco-tuco** N Chaco (W Paraguay)
C. emilianus **Emily's tuco-tuco** Neuquen (S Argentina)
C. frater **Forest tuco-tuco** Jujuy and Salta (NW Argentina), SW Bolivia
C. fulvus **Tawny tuco-tuco** N Chile, NW Argentina
C. haigi **Haig's tuco-tuco** Chubut and Rio Negro (SW Argentina)
C. knighti **Catamarca tuco-tuco** Tucuman and La Rioja to Salta (W Argentina)
C. latro **Mottled tuco-tuco** Tucuman and Salta (NW Argentina)
C. leucodon **White-toothed tuco-tuco** E Peru, W Bolivia
C. lewisi **Lewis' tuco-tuco** S Bolivia
C. magellanicus **Magellanic tuco-tuco** S Chile, S Argentina
C. maulinus **Maule tuco-tuco** Talca to Cautin (SC Chile), Neuquen (S Argentina)
C. mendocinus **Mendoza tuco-tuco** Salta to Chubut (Argentina)
C. minutus **Tiny tuco-tuco** Rio Grande do Sul and Mato Grosso (SW Brazil), NW Argentina, Uruguay
C. nattereri **Natterer's tuco-tuco** Mato Grosso (SW Brazil)
C. occultus **Furtive tuco-tuco** Tucuman and surrounds (NW Argentina)
C. opimus **Highland tuco-tuco** S Peru, SW Bolivia, N Chile, NW Argentina
C. pearsoni **Pearson's tuco-tuco** Soriana, San Jose, and Colonia (Uruguay)
C. perrensis **Goya tuco-tuco** Corrientes, Entre Rios, and Misiones (NE Argentina)
C. peruanus **Peruvian tuco-tuco** Altiplano of S Peru
C. pontifex **San Luis tuco-tuco** San Luis and Mendoza (W Argentina)
C. porteousi **Porteous' tuco-tuco** Buenos Aires and La Pampa (E Argentina)
C. saltarius **Salta tuco-tuco** Salta and Jujuy (NW Argentina)
C. sericeus **Silky tuco-tuco** Santa Cruz, Chubut, and Rio Negro (SW Argentina)
C. sociabilis **Social tuco-tuco** Neuquen (S Argentina)
C. steinbachi **Steinbach's tuco-tuco** E Bolivia
C. talarum **Talas tuco-tuco** Coastal Buenos Aires (E Argentina)
C. torquatus **Collared tuco-tuco** S Brazil, NE Argentina, Uruguay
C. tuconax **Robust tuco-tuco** Tucuman (NW Argentina)
C. tucumanus **Tucuman tuco-tuco** NW Argentina
C. validus **Strong tuco-tuco** Mendoza (W Argentina)

FAMILY OCTODONTIDAE **Octodonts**
Aconaemys Rock Rats
A. fuscus **Chilean rock rat** Andes of Chile, Argentina
A. sagei **Sage's rock rat** Lago Quillen and Lago Hui Hui (Neuquen, S Argentina)
Octodon Degus
O. bridgesi **Bridges' degu** Andes of Chile
O. degus **Degu** W Andes from Vallenar to Curico (Chile)
O. lunatus **Moon-toothed degu** Valparaiso, Aconcagua, and Coquimbo (Chile)
Octodontomys
O. gliroides **Mountain degu** Andes of SW Bolivia, N Chile, and NW Argentina
Octomys
O. mimax **Viscacha rat** W Argentina
Spalacopus
S. cyanus **Coruro** Chile, W of Andes
Tympanoctomys
T. barrerae **Plains viscacha rat** Mendoza (W Argentina)

FAMILY ABROCOMIDAE
Chinchilla Rats
Abrocoma
A. bennetti **Bennett's chinchilla rat** Copiapo to Rio Biobio (Chile)
A. boliviensis **Bolivian chinchilla rat** Santa Cruz (Bolivia)
A. cinerea **Ashy chinchilla rat** SE Peru, W Bolivia, N Chile, NW Argentina

FAMILY ECHIMYIDAE
American Spiny Rats

SUBFAMILY CHAETOMYINAE
Chaetomys
C. subspinosus **Bristle-spined rat** S Bahia and N Espirito Santo (E Brazil)

SUBFAMILY DACTYLOMYINAE
Dactylomys **Neotropical Bamboo Rats**
D. boliviensis **Bolivian bamboo rat** SE Peru, C Bolivia
D. dactylinus **Amazon bamboo rat** Ecuador, Peru, N Brazil
D. peruanus **Peruvian bamboo rat** SE Peru, Bolivia
Kannabateomys
K. amblyonyx **Atlantic bamboo rat** E Brazil, Paraguay, NE Argentina
Olallamys **Olalla Rats**
O. albicauda **White-tailed olalla rat** NW and C Colombia
O. edax **Greedy olalla rat** W Venezuela, adj. Colombia

SUBFAMILY ECHIMYINAE
Diplomys **Arboreal Soft-furred Spiny Rats**
D. caniceps **Arboreal soft-furred spiny rat** W Colombia, N Ecuador
D. labilis **Rufous tree rat** Panama (incl. San Miguel Is), W Colombia, N Ecuador
D. rufodorsalis **Red-crested tree rat** NE Colombia
Echimys **Spiny Tree Rats**
E. blainvillei **Golden Atlantic tree rat** SE Brazil
E. braziliensis **Red-nosed tree rat** S Brazil
E. chrysurus **White-faced tree rat** Guianas to NE Brazil
E. dasythrix **Drab Atlantic tree rat** SE and E Brazil
E. grandis **Giant tree rat** Rio Negro to Ilha Caviana (Amazonian Brazil)
E. lamarum **Pallid Atlantic tree rat** E Brazil
E. macrurus **Long-tailed tree rat** Brazil, S of Amazon R
E. nigrispinus **Black-spined Atlantic tree rat** E Brazil
E. pictus **Painted tree rat** S Bahia (E Brazil)
E. rhipidurus **Peruvian tree rat** C and N Amazonian Peru
E. saturnus **Dark tree rat** Ecuador, N Peru
E. semivillosus **Speckled tree rat** N Colombia, Venezuela (incl. Margarita Is)
E. thomasi **Giant Atlantic tree rat** San Sebastiao Is (Bahia, E Brazil)
E. unicolor **Unicolored tree rat** Brazil
Isothrix **Brush-tailed Rats**
I. bistriata **Yellow-crowned brush-tailed rat** W Colombia, S Venezuela, SW to NC Brazil, Bolivia
I. pagurus **Plain brush-tailed rat** Amazon Basin (C Brazil)
Makalata
M. armata **Armored spiny rat** Andes of N Ecuador and Colombia, Venezuela, Guianas, Amzaon Basin (C Brazil), Trinidad and Tobago

SUBFAMILY EUMYSOPINAE
Carterodon
C. sulcidens **Owl's spiny rat** E Brazil
Clyomys **Lund's Spiny Rats**
C. bishopi **Bishop's fossorial spiny rat** Itapetininga (Sao Paulo, Brazil)
C. laticeps **Broad-headed spiny rat** Minas Gerais to Santa Catarina (E Brazil)
Euryzygomatomys
E. spinosus **Guiara** S and E Brazil, NE Argentina, Paraguay
Hoplomys
H. gymnurus **Armoured rat** EC Honduras to NW Ecuador
Lonchothrix
L. emiliae **Tuft-tailed spiny tree rat** C Brazil, south of Amazon R
Mesomys **Spiny Tree Rats**
M. didelphoides **Brazilian spiny tree rat** Brazil
M. hispidus **Spiny tree rat** E Ecuador, N and E Peru, N Brazil
M. leniceps **Woolly-headed spiny tree rat** Peru
M. obscurus **Dusky spiny tree rat** Brazil
M. stimulax **Surinam spiny tree rat** Surinam, N Brazil
Proechimys **Terrestrial Spiny Rats**
P. albispinus **White-spined spiny rat** Bahia, adj. islands (Brazil)
P. amphichoricus **Venezuelan spiny rat** S Venezuela, adj. Brazil
P. bolivianus **Bolivian spiny rat** Upper Amazon
P. brevicauda **Huallaga spiny rat** S Colombia, E Peru, NW Brazil
P. canicollis **Colombian spiny rat** NC Colombia, Venezuela
P. cayennensis **Cayenne spiny rat** E Colombia, Guianas south to C Brazil
P. chrysaeolus **Boyaca spiny rat** E Colombia
P. cuvieri **Cuvier's spiny rat** Fr. Guiana, Surinam, Guyana
P. decumanus **Pacific spiny rat** SW Ecuador, NW Peru
P. dimidiatus **Atlantic spiny rat** E Brazil
P. goeldii **Goeldi's spiny rat** Amazonian Brazil, betw. Jamunda and Tapajoz R
P. gorgonae **Gorgona spiny rat** Gorgona Is (Colombia)
P. guairae **Guaira spiny rat** NC Venezuela
P. gularis **Ecuadoran spiny rat** E Ecuador
P. hendeei **Hendee's spiny rat** S Colombia to NE Peru
P. hoplomyoides **Guyanan spiny rat** SE Venezuela, adj. Guyana and Brazil
P. iheringi **Ihering's spiny rat** E Brazil
P. longicaudatus **Long-tailed spiny rat** C and E Peru, S Brazil, W Bolivia, Paraguay
P. magdalenae **Magdalena spiny rat** Colombia, W of Rio Magdalena
P. mincae **Minca spiny rat** Sierra Nevada Santa Marta (N Colombia)
P. myosuros **Mouse-tailed spiny rat** Bahia (E Brazil)
P. oconnelli **O'Connell's spiny rat** C Colombia, east of Cord. Orient.
P. oris **Para spiny rat** C Brazil
P. poliopus **Gray-footed spiny rat** NW Venezuela, adj. Colombia
P. quadruplicatus **Napo spiny rat** Amazonian E Ecuador and N Peru
P. semispinosus **Tome's spiny rat** SE Honduras to NE Peru and Amazonian Brazil
P. setosus **Hairy spiny rat** Minas Gerais (E Brazil)
P. simonsi **Simon's spiny rat** S Colombia, E Ecuador, NE Peru
P. steerei **Steere's spiny rat** W Peru, E Brazil
P. trinitatis **Trinidad spiny rat** Trinidad
P. urichi **Sucre spiny rat** N Venezuela
P. warreni **Warren's spiny rat** Surinam, Guiana
Thrichomys
T. apereoides **Punare** E Brazil, Paraguay

SUBFAMILY HETEROPSOMYINAE
Boromys **Cuban Cave Rats**
B. offella **Oriente cave rat** Extinct: formerly Cuba (incl. Juventud Is)
B. torrei **Torre's cave rat** Extinct: formerly Cuba (incl. Juventud Is)
Brotomys **Edible rats**
B. contractus **Haitian edible rat** Extinct: formerly Hispaniola
B. voratus **Hispaniolan edible rat** Extinct: formerly Haiti (incl. Gonave Is), Dominican Rep.
Heteropsomys **Hispaniolan cave rats**
H. antillensis **Antillean cave rat** Extinct: formerly Puerto Rico
H. insulans **Insular cave rat** Extinct: formerly Puerto Rico
Puertoricomys
P. corozalus **Corozal rat** Extinct: formerly Puerto Rico

FAMILY CAPROMYIDAE **Hutias**

SUBFAMILY CAPROMYINAE
Capromys
C. pilorides **Desmarest's hutia** Cuba, many islands and cays in Cuban Arch.
Geocapromys **Bahaman and Jamaican Hutias**
G. brownii **Brown's hutia** Jamaica
G. ingrahami **Bahamian hutia** Plana Cays (Bahamas); introd. to Little Wax Cay and Warderick Well Cay
G. thoracatus **Swan Island hutia** Extinct: formerly Little Swan Is (Gulf of Honduras)
Mesocapromys **Sticknest Hutias**
M. angelcabrerai **Cabrera's hutia** Ciego de Avila (Cuba)
M. auritus **Eared hutia** Las Villas (Cuba)
M. nanus **Dwarf hutia** Matanzas (Cuba)
M. sanfelipensis **San Felipe hutia** Extinct: formerly Pinar del Rio (Cuba)
Mysateles **Long-tailed Cuban Hutias**
M. garridoi **Garrido's hutia** Canarreos Arch. (Cuba)
M. gundlachi **Gundlach's hutia** N Juventud Is (Cuba)
M. melanurus **Black-tailed hutia** E Cuba
M. meridionalis **Southern hutia** Juventud Is (Cuba)
M. prehensilis **Prehensile-tailed hutia** Cuba

SUBFAMILY HEXOLOBODONTINAE
Hexolobodon
H. phenax **Imposter hutia** Extinct: formerly Hispaniola (incl. Gonave Is)

SUBFAMILY ISOLOBOLONTINAE
Isolobodon **Laminar-toothed Hutias**
I. montanus **Montane hutia** Extinct: formerly Hispaniola (incl. Gonave Is)
I. portoricensis **Puerto Rican hutia** Extinct: formerly Haiti and Dominican Rep. (Hispaniola) and offshore islands (incl. La Tortue Is); introd. to Puerto Rico and elsewhere

SUBFAMILY PLAGIODONTINAE
Plagiodontia **Hispaniolan Hutias**
P. aedium **Hispaniolan hutia** Hispaniola (incl. Gonave Is)
P. araeum **San Rafael hutia** Extinct: formerly Hispaniola
P. ipnaeum **Samana hutia** Extinct: formerly Hispaniola
Rhizoplagiodontia
R. lemkei **Lemke's hutia** Extinct: formerly Massif de la Hotte (SW Haiti)

FAMILY HEPTAXODONTIDAE **Key Mice**

SUBFAMILY CLIDOMYINAE
Clidomys **Key Mice**
C. osborni **Osborn's key mouse** Extinct: formerly Jamaica
C. parvus **Small key mouse** Extinct: formerly Jamaica

SUBFAMILY HEPTAXODONTINAE
Amblyrhiza
A. inundata **Blunt-toothed mouse** Extinct: formerly Anguilla (W. Indies)
Elasmodontomys
E. obliquus **Plate-toothed mouse** Extinct: formerly Puerto Rico
Quemisia
Q. gravis **Twisted-toothed mouse** Extinct: formerly Hispaniola

FAMILY MYOCASTORIDAE
Myocastor
M. coypus **Nutria or Coypu** S Brazil, Bolivia, Paraguay, Chile, Argentina, Uruguay

ORDER LAGOMORPHA
RABBITS, HARES, AND PIKAS

FAMILY LEPORIDAE **Hares and Rabbits**
see full species list pp 702–705

FAMILY OCHOTONIDAE **Pikas**
Ochotona
O. alpina **Alpine pika** Altai and Sayan Mts, NW Kazakhstan, S Russia, NW Mongolia, N Gansu-Ningxia border
O. cansus **Gansu pika** Qinghai, Gansu, Sichuan; also Shanxi and Shaanxi (China)
O. collaris **Collared pika** SE Alaska (USA), NW Canada
O. curzoniae **Black-lipped pika** Sikkim (India), E Nepal, Xizang, adj. Gansu, Qinghai, and Sichuan (China)
O. dauurica **Daurian pika** Altai Mts and Tuva south through Mongolia and N China to Qinghai (China)
O. erythrotis **Chinese red pika** E Qinghai, W Gansu, N Sichuan, S Xinjiang, Xizang (China)
O. forresti **Forrest's pika** Assam and Sikkim (India), Bhutan, N Burma, NW Yunnan and SE Xizang (China)
O. gaoligongensis **Gaoligong pika** Mt Gaoligong (NW Yunnan, China)
O. gloveri **Glover's pika** W Sichuan, NW Yunnan, NE Xizang, and SW Qinghai (China)
O. himalayana **Himalayan pika** Mt Jolmolunga (S Xizang, China); poss. adj. Nepal
O. hyperborea **Northern pika** Ural and Sayan Mts east to Chukotskoye and Koryakskoye ranges and Kamchatka Pen., SE Russia (incl. Sakhalin Is), NC Mongolia, NE China, N Korea, Hokkaido (Japan)
O. iliensis **Ili pika** Tien Shan Mts (Xinjiang, China)
O. koslowi **Kozlov's pika** Kunlun Shan Mts (W China)
O. ladacensis **Ladak pika** N Pakistan, Kashmir, SW Xinjiang, Qinghai, E Xizang (China)
O. macrotis **Large-eared pika** W Tien Shan, Pamirs, Hindu Kush and Karakorum Mts in SE Kazakhstan, Kyrgyzstan, Tajikistan, NE Afghanistan, and N Pakistan, the Himalayas from Kashmir and N India through Nepal and Xizang (China) to Bhutan, and mountains of Sichuan and Yunnan (China)
O. muliensis **Muli pika** SE Muli (W Sichuan, China)
O. nubrica **Nubra pika** Ladakh (Kashmir) through E Nepal to E Xizang (China)
O. pallasi **Pallas' pika** E Kazakhstan, Tuva (Russia), Altai Mts, W Mongolia to N Xinjiang and Inner Mongolia (China)
O. princeps **North American pika** C British Columbia (SW Canada) in Rocky, Cascade and Sierra Nevada Mts of W North America to EC California, Utah, and N New Mexico
O. pusilla **Steppe or Small pika** Middle Volga R (Russia), east and south through N Kazakhstan to upper Irtysh R and Chinese border
O. roylei **Royle's pika** Himalayas of NW Pakistan and N India and Kashmir through S Xizang (China) to Nepal
O. rufescens **Afghan pika** Armenia, Iran, Afghanistan, Baluchistan (Pakistan), and SW Turkmenistan
O. rutila **Turkestan Red pika** Pamirs (Tajikistan) to Tien Shan Mts (SE Uzbekistan, Kyrgyzstan, SE Kazakhstan), poss. E Xinjiang (China) and N Afghanistan
O. thibetana **Moupin pika** Sikkim (India), Bhutan, N Burma, SE Qinghai, S Gansu, Shanxi, Shaanxi, W Hubei, W Yunnan, W Sichuan, S Xizang (China)
O. thomasi **Thomas' pika** NE Qinghai, Gansu, and Sichuan (China)
Prolagus
P. sardus **Sardinian pika** Extinct: formerly Corsica (France), Sardinia (Italy), adj. small islands

ORDER MACROSCELIDEA
ELEPHANT SHREWS

FAMILY MACROSCELIDIDAE
Elephantulus Long-eared Elephant Shrews
E. brachyrhynchus **Short-snouted elephant shrew** N South Africa, NE Namibia, Angola, Mozambique and DRC to Uganda and Kenya
E. edwardii **Cape elephant shrew** SW South Africa
E. fuscipes **Dusky-footed elephant shrew** NE DRC, Uganda, S Sudan
E. fuscus **Dusky elephant shrew or Peters' short-snouted elephant shrew** C Mozambique, SE Zambia, S Malawi
E. intufi **Bushveld elephant shrew** W South Africa, Botswana, Namibia, SW Angola
E. myurus **Eastern rock elephant shrew** E and N South Africa, S Zimbabwe, E Botswana, W Mozambique
E. revoili **Somali elephant shrew** N Somalia
E. rozeti **North African elephant shrew** SW Morocco to W Libya
E. rufescens **Rufous elephant shrew** S Sudan, S and E Ethiopia, N and S Somalia, N and SE Kenya, NE Uganda, NC and W Tanzania,
E. rupestris **Western rock elephant shrew** W South Africa, Namibia
Macroscelides
M. proboscideus **Short-eared or Round-eared elephant-shrew** W South Africa, SW Botswana, S and E Namibia
Petrodromus
P. tetradactylus **Four-toed elephant-shrew** South Africa, Zimbabwe, Botswana, NE Angola, Zambia, Malawi, Mozambique, DRC, Congo, S Uganda, Tanzania (incl. Zanzibar and Mafia Is), SE Kenya
Rhynchocyon **Checkered Elephant shrews**
R. chrysopygus **Golden-rumped elephant-shrew** E Kenya
R. cirnei **Checkered elephant-shrew** Mozambique, NE Zambia, Malawi, E DRC, Uganda, S Tanzania
R. petersi **Black and rufous elephant-shrew** E Tanzania (incl. Zanzibar and Mafia Is), SE Kenya

ORDER INSECTIVORA
INSECTIVORES

Note According to a new account of systematics, the tenrecs and golden moles should be assigned to a new order, the Afrosoricida, part of the superorder Afrotheria. For a full explanation of this restructuring see the introductory essay What is a Mammal? and Insectivores p.722. Correspondingly, in this new scheme, the shrews, moles, and hedgehogs become an order in their own right, the Eulipotyphla.

FAMILY TENRECIDAE Tenrecs

SUBFAMILY GEOGALINAE

Geogale
G. aurita **Large-eared tenrec** NE and SW Madagascar

SUBFAMILY ORYZORICTINAE

Limnogale
L. mergulus **Aquatic tenrec** E Madagascar
Microgale Shrew Tenrecs
M. brevicaudata **Short-tailed shrew tenrec** Madagascar
M. cowani **Cowan's shrew tenrec** N, E, and EC Madagascar
M. dobsoni **Dobson's shrew tenrec** E and EC Madagascar
M. dryas **Tree shrew tenrec** NE Madagascar
M. gracilis **Gracile shrew tenrec** E Madagascar
M. longicaudata **Lesser long-tailed shrew tenrec** E and N Madagascar
M. parvula **Pygmy shrew tenrec** N Madagascar
M. principula **Greater long-tailed shrew tenrec** E and SE Madagascar
M. pulla **Dark shrew tenrec** NE Madagascar
M. pusilla **Least shrew tenrec** E and S Madagascar
M. talazaci **Talazac's shrew tenrec** N, E, and EC Madagascar
M. thomasi **Thomas' shrew tenrec** E Madagascar
Oryzorictes Rice Tenrecs
O. hova **Hova rice tenrec** C Madagascar
O. talpoides **Molelike rice tenrec** NW Madagascar
O. tetradactylus **Four-toed rice tenrec** C Madagascar

SUBFAMILY POTAMOGALINAE

Micropotamogale Dwarf Otter Shrews
M. lamottei **Mount Nimba otter shrew** Guinea, Liberia, Ivory Coast
M. ruwenzorii **Ruwenzori otter shrew** Ruwenzori region (Uganda, DRC), NE DRC
Potamogale
P. velox **Giant otter shrew** Tropical Africa, from Nigeria to W Kenya to Angola

SUBFAMILY TENRECINAE

Echinops
E. telfairi **Lesser hedgehog tenrec** S Madagascar
Hemicentetes
H. semispinosus **Streaked tenrec** E Madagascar
Setifer
S. setosus **Greater hedgehog tenrec** C Madagascar
Tenrec
T. ecaudatus **Tailless or Common tenrec** Madagascar, Comoros; introd. to Reunion, Mauritius, and Seychelles

FAMILY SOLENODONTIDAE Solenodons

Solenodon
S. cubanus **Cuban solenodon** Oriente (Cuba)
S. marcanoi **Marcano's solenodon** Extinct: formerly San Rafael (Dominican Rep.)
S. paradoxus **Hispaniola solenodon** Haiti, Dominican Rep.

FAMILY ERINACEIDAE Hedgehogs and Moonrats

SUBFAMILY ERINACEINAE

Atelerix African Hedgehogs
A. albiventris **Four-toed hedgehog** Senegal to Ethiopia south to Zambezi R
A. algirus **North African hedgehog** Western Sahara to Algeria, Tunisia, N Libya; introd. to Canarias and Balearic Is, Malta, S France and Spain
A. frontalis **Southern African hedgehog** South Africa, E Botswana, W Zimbabwe; also Namibia to SW Angola
A. sclateri **Somali hedgehog** N Somalia
Erinaceus Eurasian Hedgehogs
E. amurensis **Amur hedgehog** Russia south through E China to Hunan, Korea
E. concolor **Eastern European hedgehog** E Europe, S Russia and W Siberia to Ob R, Turkey to Israel and Iran, Greek and Adriatic Is
E. europaeus **Western European hedgehog** Spain to Italy and Istra Pen. north to Scandinavia and NW Russia; also Ireland and Britain, Corsica (France), Sardinia and Sicily (Italy) and other islands
Hemiechinus Desert Hedgehogs
H. aethiopicus **Desert hedgehog** Mauritania to Egypt and Ethiopia, Arabia, Djerba Is (Tunisia), Bahrain
H. auritus **Long-eared hedgehog** E Ukraine to Mongolia in north and Libya to W Pakistan in south
H. collaris **Indian long-eared hedgehog** Pakistan, NW India
H. hypomelas **Brandt's hedgehog** Iran and Turkmenistan east to Uzbekistan and Indus R (Pakistan), Oman, Tanb and Kharg Is (Persian Gulf)
H. micropus **Indian hedgehog** Pakistan, NW India
H. nudiventris **Bare-bellied hedgehog** Tamil Nadu and Kerala (S India)
Mesechinus Steppe Hedgehogs
M. dauuricus **Daurian hedgehog** NE Mongolia east to upper Amur basin in Russia, adj. Inner Mongolia and W Manchuria (China)
M. hughi **Hugh's hedgehog** Shaanxi and Shanxi (China)

SUBFAMILY HYLOMYINAE

Echinosorex
E. gymnura **Moonrat** Malay Pen., Sumatra (Indonesia), Borneo
Hylomys Asian Gymnures
H. hainanensis **Hainan gymnure** Hainan (China)
H. sinensis **Shrew or Chinese gymnure** Sichuan and Yunnan (S China), adj. Burma and N Vietnam
H. suillus **Short-tailed gymnure** Malay Pen. to Indochina to Yunnan border (China), Java, Sumatra, and Tioman Is (Indonesia), Borneo
Podogymnura Philippine Gymnures
P. aureospinula **Dinagat gymnure** Dinagat (Philippines)
P. truei **Mindanao gymnure** Mindanao (Philippines)

FAMILY NESOPHONTIDAE

Nesophontes or Extinct West Indian shrews
Nesophontes
N. edithae **Puerto Rican nesophontes** Extinct: formerly Puerto Rico
N. hypomicrus **Atalyae nesophontes** Extinct: formerly Haiti (incl. Gonave Is)
N. longirostris **Slender Cuban nesophontes** Extinct: formerly Cuba
N. major **Greater Cuban nesophontes** Extinct: formerly Cuba
N. micrus **Western Cuban nesophontes** Extinct: formerly Cuba, Haiti, and Pinos Is
N. paramicrus **St. Michael nesophontes** Extinct: formerly Haiti
N. submicrus **Lesser Cuban nesophontes** Extinct: formerly Cuba
N. zamicrus **Haitian nesophontes** Extinct: formerly Haiti

FAMILY SORICIDAE Shrews

SUBFAMILY CROCIDURINAE

Congosorex
C. polli **Poll's shrew** S DRC
Crocidura White-toothed Shrews
C. aleksandrisi **Alexandrian shrew** Cyrenaica (Libya)
C. allex **Highland shrew** SW Kenya, Mt Kilimanjaro, Meru, and Ngorongoro (N Tanzania)
C. andamanensis **Andaman shrew** Andaman Is (India)
C. ansellorum **Ansell's shrew** N Zambia
C. arabica **Arabian shrew** Coastal plains of S Arabian Pen.
C. armenica **Armenian shrew** Armenia
C. attenuata **Indochinese shrew** India, Nepal, Bhutan, Burma, China, Taiwan, Thailand, Vietnam, Malay Pen., Sumatra and Java (Indonesia)
C. attila **Hun shrew** Mt Cameroun (Cameroon) to E DRC
C. baileyi **Bailey's shrew** Ethiopian highlands west of Rift valley
C. batesi **Bates' shrew** S Cameroon, Gabon
C. beatus **Mindanao shrew** Mindanao, Leyte, and Maripipi (Philippines)
C. beccarii **Beccari's shrew** Sumatra (Indonesia)
C. bottegi **Bottego's shrew** Guinea to Ethiopia and N Kenya
C. bottegoides **Bale shrew** Bale Mts and Mt Albasso (Ethiopia)
C. buettikoferi **Buettikofer's shrew** West Africa, from Guinea-Biassau to Liberia, Nigeria
C. caliginea **African foggy shrew** NE DRC
C. canariensis **Canary shrew** E Canarias Is
C. cinderella **Cinderella shrew** Senegal, Gambia, Mali, Niger
C. congobelgica **Congo shrew** NE DRC
C. cossyrensis **Pantellerian shrew** Pantelleria (Italy)
C. crenata **Long-footed shrew** S Cameroon, N Gabon, E DRC
C. crossei **Crosse's shrew** Sierra Leone to W Cameroon
C. cyanea **Reddish-gray musk shrew** South Africa, Namibia, Botswana, Angola, Mozambique
C. denti **Dent's shrew** Cameroon, Gabon, NE DRC
C. desperata **Desperate shrew** Rungwe and Uzungwe Mts (S Tanzania)
C. dhofarensis **Dhofarian shrew** Dhofar (Oman)
C. dolichura **Long-tailed musk shrew** Nigeria, S Cameroon, Gabon, Bioko (Equatorial Guinea), Congo, CAR, DRC, adj. Uganda and Burundi
C. douceti **Doucet's musk shrew** Guinea, Ivory Coast, Nigeria
C. dsinezumi **Dsinezumi shrew** Japan, Quelpart Is (Korea), poss. Taiwan
C. eisentrauti **Eisentraut's shrew** Mt Cameroun (Cameroon)
C. elgonius **Elgon shrew** Mt Elgon (W Kenya), NE Tanzania
C. elongata **Elongated shrew** N and C Sulawesi (Indonesia)
C. erica **Heather shrew** W Angola
C. fischeri **Fischer's shrew** Nguruman (Kenya), Himo (Tanzania)
C. flavescens **Greater red musk shrew** South Africa
C. floweri **Flower's shrew** Egypt
C. foxi **Fox's shrew** Jos plateau (Nigeria)
C. fuliginosa **Southeast Asian shrew** N India, Burma, adj. China, Malay Pen. and adj. islands, poss. Sumatra and Java (Indonesia) and Borneo
C. fulvastra **Savanna shrew** Kenya to Mali
C. fumosa **Smoky white-toothed shrew** Mt Kenya and Aberdares (Kenya)
C. fuscomurina **Tiny musk shrew** Senegal to Ethiopia south to South Africa
C. glassi **Glass' shrew** Ethiopian highlands east of Rift valley
C. goliath **Goliath shrew** S Cameroon, Gabon, DRC
C. gracilipes **Peter's musk shrew** Mt Kilimanjaro (Tanzania)
C. grandiceps **Large-headed shrew** Guinea, Ivory Coast, Ghana, Nigeria
C. grandis **Mount Malindang shrew** Mt Malindang (Mindanao, Philippines)
C. grassei **Grasse's shrew** Belinga (Gabon), Boukoko (CAR), Yaounde (Cameroon)
C. grayi **Luzon shrew** Luzon and Mindoro (Philippines)
C. greenwoodi **Greenwood's shrew** S Somalia
C. gueldenstaedtii **Gueldenstaedt's shrew** Caucasus
C. harenna **Harenna shrew** Bale Mts (Ethiopia)
C. hildegardeae **Hildegarde's shrew** Nigeria, Cameroon, C and E Africa
C. hirta **Lesser red musk shrew** Somalia to DRC to South Africa
C. hispida **Andaman spiny shrew** Andaman Is (India)
C. horsfieldii **Horsfield's shrew** Mysore and Ladak (India), Sri Lanka, Nepal, Yunnan, Fujian, and Hainan (China), Taiwan, Ryukyu Is (Japan), N Thailand to Vietnam
C. jacksoni **Jackson's shrew** E DRC, Uganda, Kenya, N Tanzania
C. jenkinsi **Jenkin's shrew** Andaman Is (india)
C. kivuana **Kivu shrew** Kahuzi-Biega NP (DRC)
C. lamottei **Lamotte's shrew** Senegal to W Cameroon
C. lanosa **Lemara shrew** E DRC, Rwanda
C. lasiura **Ussuri white-toothed shrew** Ussuri region (Russia) and NE China to Korea, also Jiangsu (China)
C. latona **Latona shrew** NE DRC
C. lea **Sulawesi shrew** N and C Sulawesi (Indonesia)
C. leucodon **Bicolored shrew** France to the Volga R and Caucasus, Turkey, south through the Alborz Mts to Israel and Lebanon, Lesvos Is (Greece)
C. levicula **Celebes shrew** C and SE Sulawesi (Indonesia)
C. littoralis **Butiaba naked-tailed shrew** DRC, Uganda, Kenya
C. longipes **Savanna swamp shrew** W Nigeria
C. lucina **Moorland shrew** E Ethiopia
C. ludia **Dramatic shrew** Medje and Tandala (N DRC)
C. luna **Greater gray-brown musk shrew** Zimbabwe, E Angola, Zambia, Malawi, Mozambique, DRC, Uganda, Rwanda, Tanzania, Kenya
C. lusitania **Mauritanian shrew** S Morocco to Senegal, east in Nigeria, Sudan, Ethiopia
C. macarthuri **MacArthur's shrew** Kenya, Somalia
C. macmillani **MacMillan's shrew** Ethiopia
C. macowi **Macow's shrew** N Kenya
C. malayana **Malayan shrew** Malay Pen. and off-shore islands
C. manengubae **Manenguba shrew** Cameroon
C. maquassiensis **Maquassie musk shrew** N South Africa, Zimbabwe
C. mariquensis **Swamp musk shrew** South Africa to Mozambique, W Zimbabwe, and Zambia; also NE Namibia and NW Botswana
C. maurisca **Dark shrew** Uganda, Kenya
C. maxi **Max's shrew** Java, Lesser Sunda Is, Maluku (Indonesia)
C. mindorus **Mindoro shrew** Mt Halcon (Mindoro, Philippines)
C. minuta **Minute shrew** Java (Indonesia)
C. miya **Sri Lankan long-tailed shrew** C Sri Lanka
C. monax **Rombo shrew** W Kenya, N Tanzania
C. monticola **Sunda shrew** Malay Pen., Java (Indonesia), Borneo
C. montis **Montane white-toothed shrew** Mt Ruwenzori (Uganda), Mt Meru (Tanzania), Imatong Mts (Sudan), poss. Kenya
C. muricauda **Mouse-tailed shrew** Guinea to Ghana
C. mutesae **Uganda large-toothed shrew** Uganda
C. nana **Dwarf white-toothed shrew** Somalia, Ethiopia
C. nanilla **Tiny white-toothed shrew** Mauritania to Kenya and Uganda
C. neglecta **Neglected shrew** Sumatra (Indonesia)
C. negrina **Negros shrew** S Negros Is (Philippines)
C. nicobarica **Nicobar shrew** Great Nicobar Is (Nicobar Is, India)
C. nigeriae **Nigerian shrew** Nigeria, Cameroon, Bioko (Equatorial Guinea)
C. nigricans **Black white-toothed shrew** Angola
C. nigripes **Black-footed shrew** N and C Sulawesi (Indonesia)
C. nigrofusca **Tenebrous shrew** S Ethiopia and Sudan through E Africa to DRC, Zambia, Angola
C. nimbae **Nimba shrew** Mt Nimba (Guinea, Liberia), Sierra Leone
C. niobe **Stony shrew** Uganda, DRC
C. obscurior **Obscure white-toothed shrew** Sierra Leone to Ivory Coast
C. olivieri **Olivier's shrew** Egypt, Senegal to Ethiopia, south to N South Africa
C. orii **Amami shrew** Ryukyu Is (Japan)
C. osorio **Osorio shrew** Gran Canaria Is (Canarias Is)
C. palawanensis **Palawan shrew** Palawan (Philippines)
C. paradoxura **Paradox shrew** Sumatra (Indonesia)
C. parvipes **Small-footed shrew** Cameroon to S Sudan, Ethiopia, Kenya, Tanzania, DRC, Zambia, Angola
C. pasha **Pasha shrew** Sudan, Ethiopia
C. pergrisea **Pale gray shrew** Kashmir
C. phaeura **Guramba shrew** Mt Guramba (Ethiopia)
C. picea **Pitch shrew** Cameroon
C. pitmani **Pitman's shrew** C and N Zambia
C. planiceps **Flat-headed shrew** Ethiopia, Sudan, Uganda, DRC, Nigeria
C. poensis **Fraser's musk shrew** Cameroon to Liberia, Bioko (Equatorial Guinea), Principe (Sao Tome & Principe)
C. polia **Fuscous shrew** Medje (DRC)
C. pullata **Dusky shrew** Afghanistan, Pakistan, India, Kashmir, Yunnan (China), Thailand
C. raineyi **Rainey shrew** Mt Garguez (Kenya)
C. religiosa **Egyptian pygmy shrew** Nile valley (Egypt)
C. rhoditis **Temboan shrew** N, C, and SW Sulawesi (Indonesia)
C. roosevelti **Roosevelt's shrew** Angola, Cameroon, CAR, DRC, Uganda, Rwanda, Tanzania
C. russula **White-toothed shrew** S and W Europe, Mediterranean Is (Ibiza, Sardinia), N Africa, from Morocco to Tunisia
C. selina **Moon shrew** Uganda
C. serezkyensis **Serezkaya shrew** Turkey, Azerbaijan, Turkmenistan, Tajikistan, Kazakhstan
C. sibirica **Siberian shrew** L. Ysyk-Kol (Kyrgyzstan) to upper Ob R, L. Baikal, poss. Mongolia and Xinjiang (China)
C. sicula **Sicilian shrew** Sicily, Egadi Is (Italy), Gozo (Malta)
C. silacea **Lesser gray-brown musk shrew** South Africa, Botswana, Zimbabwe, Mozambique
C. smithii **Desert musk shrew** Senegal, Ethiopia, poss. Somalia
C. somalica **Somali shrew** Ethiopia, Sudan, Somalia, Mali
C. stenocephala **Narrow-headed shrew** Mt Kahuzi (E DRC)
C. suaveolens **Lesser shrew** Palearctic from Spain to Korea, many Atlantic Is, Mediterranean Is incl. Corsica (France), Crete (Greece), Menorca (Spain), and Cyprus

C. susiana **Iranian shrew** Dezful (SW Iran)
C. tansaniana **Tanzanian shrew** Usambara Mts (Tanzania)
C. tarella **Ugandan shrew** Uganda
C. tarfayensis **Tarfaya shrew** Atlantic coast from Morocco to Mauritania
C. telfordi **Telford's shrew** Uluguru Mts (Tanzania)
C. tenuis **Thin shrew** Timor
C. thalia **Thalia shrew** Ethiopian highlands
C. theresae **Therese's shrew** Ghana to Guinea
C. thomensis **São Tomé shrew** São Tomé Is (São Tomé & Principe)
C. turba **Tumultuous shrew** Angola, Zambia, Malawi, DRC, Cameroon, Uganda, Tanzania, Kenya
C. ultima **Ultimate shrew** Jombeni Mts (Nyeri, Kenya)
C. usambarae **Usambara shrew** Usambara Mts (Tanzania)
C. viaria **Savanna path shrew** S Morocco to Senegal and east to Sudan, Ethiopia, Kenya
C. voi **Voi shrew** Kenya and Somalia to Ethiopia and Sudan; also Nigeria and Mali
C. whitakeri **Whitaker's shrew** Morocco to Tunisia, coastal Egypt
C. wimmeri **Wimmer's shrew** S Ivory Coast
C. xantippe **Vermiculate shrew** SE Kenya, Usambara Mts (Tanzania)
C. yankariensis **Yankari shrew** Cameroon, Nigeria, Sudan, Ethiopia, Kenya, Somalia
C. zaphiri **Zaphir's shrew** Kaffa (S Ethiopia), Kaimosi and Kisumu (Kenya)
C. zarudnyi **Zarudny's shrew** SE Iran, SE Afghanistan, SW Pakistan
C. zimmeri **Zimmer's shrew** Upemba NP (DRC)
C. zimmermanni **Zimmermann's shrew** Crete (Greece)

Diplomesodon
D. pulchellum **Piebald shrew** W and S Kazakhstan, Uzbekistan, Turkmenistan

Feroculus
F. feroculus **Kelaart's long-clawed shrew** C Sri Lanka

Myosorex Mouse Shrews
M. babaulti **Babault's mouse shrew** E DRC, W Rwanda, W Burundi
M. blarina **Montane mouse shrew** Mt Ruwenzori (Uganda, DRC)
M. cafer **Dark-footed forest shrew** E and NE South Africa, E Zimbabwe, W Mozambique
M. eisentrauti **Eisentraut's mouse shrew** Bioko (Equatorial Guinea)
M. geata **Geata mouse shrew** SW Tanzania
M. longicaudatus **Long-tailed forest shrew** S and SE South Africa
M. okuensis **Oku mouse shrew** Bamenda plateau (Cameroon)
M. rumpii **Rumpi mouse shrew** Rumpi-Hills (Cameroon)
M. schalleri **Schaller's mouse shrew** Itombwe Mts (E DRC)
M. sclateri **Sclater's tiny mouse shrew** KwaZulu-Natal (South Africa)
M. tenuis **Thin mouse shrew** NE South Africa
M. varius **Forest shrew** South Africa, Lesotho

Paracrocidura African Shrews
P. graueri **Grauer's shrew** Itombwe Mts (E DRC)
P. maxima **Greater shrew** DRC, Rwanda, Uganda
P. schoutedeni **Schouteden's shrew** S Cameroon, Gabon, Congo, CAR, DRC

Ruwenzorisorex
R. suncoides **Ruwenzori shrew** W DRC, Uganda, Rwanda, Burundi

Scutisorex
S. somereni **Armored or Hero shrew** DRC, adj. Uganda, Rwanda, Burundi

Solisorex
S. pearsoni **Pearson's long-clawed shrew** C Sri Lanka

Suncus Pygmy and Dwarf Shrews
S. ater **Black shrew** Mt Kinabalu (Sabah, N Borneo)
S. dayi **Day's shrew** S India
S. etruscus **White-toothed pygmy shrew or Etruscan shrew** S Europe, N Africa, from Morocco to Egypt, Arabian Pen. and Turkey to Iraq, Turkmenistan, Afghanistan, Pakistan, India, Sri Lanka, Nepal, Bhutan, Burma, Yunnan (China), and Thailand
S. fellowesgordoni **Sri Lankan shrew** C Sri Lanka
S. hosei **Hose's shrew** Borneo
S. infinitesimus **Least dwarf shrew** Cameroon, CAR, Kenya to South Africa
S. lixus **Greater dwarf shrew** Kenya, Tanzania, DRC, Malawi, Zambia, Angola, Botswana, N South Africa
S. madagascariensis **Madagascan shrew** Madagscar, Comoros
S. malayanus **Malayan pygmy shrew** Malay Pen.
S. mertensi **Flores shrew** Flores (Indonesia)
S. montanus **Sri Lanka highveld shrew** S India, Sri Lanka
S. murinus **Asian house shrew** Afghanistan, Pakistan, India, Sri Lanka, Nepal, Bhutan, Burma, China, Taiwan, Japan, south to Malay Pen.; introd. to coastal NE Africa (Egypt to Tanzania), coastal Arabia, Madagascar, Comoros, Mauritius, other Indian Ocean islands, Philippines, Guam and prob. many other islands
S. remyi **Remy's shrew** Belinga and Makokou (NE Gabon)
S. stoliczkanus **Anderson's shrew** Pakistan, India, Nepal, Bangladesh
S. varilla **Lesser dwarf shrew** South Africa, Zimbabwe, Zambia, Malawi, DRC, Tanzania; also Nigeria
S. zeylanicus **Jungle shrew** Sri Lanka

Surdisorex Kenyan Shrews
S. norae **Aberdare shrew** Aberdares (Kenya)
S. polulus **Mount Kenya shrew** Mt Kenya (Kenya)

Sylvisorex Forest Musk Shrews
S. granti **Grant's shrew** DRC, Uganda, Rwanda, Kenya, Tanzania; also Cameroon
S. howelli **Howell's shrew** Usambara and Uluguru Mts (Tanzania)
S. isabellae **Isabella shrew** Bioko (Equatorial Guinea), Bamenda plateau (Cameroon)
S. johnstoni **Johnston's shrew** Congo, SW Cameroon, Gabon, Bioko (Equatorial Guinea), DRC, Uganda, Burundi, Tanzania
S. lunaris **Crescent shrew** Ruwenzori Mts (Uganda, DRC), Virunga Volcanos (Rwanda), L. Kivu region (Burundi, DRC)
S. megalura **Climbing shrew** Guinea to Ethiopia south to Mozambique and Zimbabwe
S. morio **Arrogant shrew** Mt Cameroun (Cameroon)
S. ollula **Forest musk shrew** S Cameroon, adj. Nigeria, Gabon, S DRC
S. oriundus **Mountain shrew** NE DRC
S. vulcanorum **Volcano shrew** E DRC, Uganda, Rwanda, Burundi

SUBFAMILY SORICINAE

Anourosorex
A. squamipes **Mole-shrew or Sichuan burrowing shrew** Assam (India), Bhutan, N and W Burma, Shaanxi, Hubei to Yunnan (China), Taiwan, Thailand, N Vietnam

Blarina American Short-tailed Shrews
B. brevicauda **Northern short-tailed shrew** Saskatchewan (Canada) east in S and SE Canada to Nebraska and N Virginia (USA)
B. carolinensis **Southern short-tailed shrew** S Illinois east to N Virginia south to E Texas and N Florida
B. hylophaga **Elliot's short-tailed shrew** S Nebraska and Iowa south to S Texas and east to Missouri and NW Arkansas, Oklahoma into Louisiana (USA)

Blarinella Asiatic Short-tailed Shrews
B. quadraticauda **Sichuan short-tailed shrew** Gansu, Shaanxi, Sichuan, and Yunnan (China)
B. wardi **Ward's short-tailed shrew** Burma, Yunnan (China)

Chimarrogale Oriental Water Shrews
C. hantu **Hantu water shrew** Malay Pen.
C. himalayica **Himalayan water shrew** Kashmir through SE Asia to Indochina, C and S China, Taiwan
C. phaeura **Sunda water shrew** Borneo
C. platycephala **Flat-headed water shrew** Japan
C. styani **Styan's water shrew** Shaanxi, Sichuan (China), N Burma
C. Sumatrana **Sumatra water shrew** Sumatra (Indonesia)

Cryptotis Small-eared Shrews
C. avia **Andean small-eared shrew** Cord. Orient. of Colombia
C. endersi **Enders' small-eared shrew** Bocas del Toro (Panama)
C. goldmani **Goldman's small-eared shrew** S Mexico, WC Guatemala
C. goodwini **Goodwin's small-eared shrew** S Mexico, S Guatemala, W El Salvador
C. gracilis **Talamancan small-eared shrew** SE Costa Rica, W Panama
C. hondurensis **Honduran small-eared shrew** Tegucigalpa (Honduras), poss. adj. Guatemala, El Salvador, and Nicaragua
C. magna **Big small-eared shrew** NC Oaxaca (Mexico)
C. meridensis **Merida small-eared shrew** Cord. Merida and Mts near Caracas (W Venezuela)
C. mexicana **Mexican small-eared shrew** Tamaulipas to Chiapas (Mexico)
C. montivaga **Ecuadorean small-eared shrew** Andes of S Ecuador
C. nigrescens **Blackish small-eared shrew** S Mexico (incl. Yucatan Pen.), Guatemala, El Salvador, Honduras, Costa Rica, Panama
C. parva **Least shrew** SE Canada through EC and SW USA, Mexico, Central America to Panama
C. squamipes **Scaly-footed small-eared shrew** Cord. Occid. of W Colombia and Ecuador
C. thomasi **Thomas' small-eared shrew** Cord. Orient. of W Colombia, Ecuador, and N Peru

Megasorex
M. gigas **Mexican shrew** Nayarit to Oaxaca (Mexico)

Nectogale
N. elegans **Elegant water shrew** Nepal and Sikkim (India), Bhutan, N Burma, Shaanxi, Sichuan, Yunnan, and Xizang (China)

Neomys Old World Water Shrews
N. anomalus **Southern water shrew** Portugal to Poland east to Russia
N. fodiens **Eurasian water shrew** Most of Europe (incl. British Is) east to L. Baikal, Yenisei R (Russia), Tien Shan Mts (China), NW Mongolia; also Sakhalin Is (Russia), Jilin (China), and N Korea
N. schelkovnikovi **Transcaucasian water shrew** Caucasus (Armenia, Azerbaijan, Georgia)

Notiosorex
N. crawfordi **Desert shrew** SW and SC USA to Baja California and N and C Mexico

Sorex Holarctic Shrews
S. alaskanus **Glacier Bay water shrew** Glacier Bay (Alaska, USA)
S. alpinus **Alpine shrew** Pyrenees, Jura, Harz, Sudetan, Carpathians, Tatra and other mountains of C Europe
S. araneus **Eurasian shrew** C, E, and N Europe (incl. British Is) east to Siberia (Russia); also in France, Italy, Spain
S. arcticus **Arctic shrew** NW to SE Canada, Dakotas, Minnesota, and Wisconsin (USA)
S. arizonae **Arizona shrew** SE Arizona, SW New Mexico (USA), Chihuahua (N Mexico)
S. asper **Tien Shan shrew** Tien Shan Mts (Kazakhstan, China)
S. bairdii **Baird's shrew** NW Oregon (USA)
S. bedfordiae **Lesser striped shrew** S Gansu and W Shaanxi to Yunnan (China), adj. Burma, Nepal
S. bendirii **Marsh shrew** NW California to Washington (USA), SE British Columbia (Canada)
S. buchariensis **Pamir shrew** Pamirs (Tajikistan)
S. caecutiens **Laxmann's shrew** E Europe to E Siberia and Sakhalin Is (Russia), south to C Ukraine, N Kazakhstan, Altai Mts, Mongolia, Gansu and NE China, Korea
S. camtschatica **Kamchatka shrew** S Kamchatka Pen. (Russia)
S. cansulus **Gansu shrew** Gansu (China)
S. cinereus **Cinereus or Masked shrew** Alaska (USA) and Canada along the Rocky and Appalachian Mts
S. coronatus **Crowned shrew** W Europe, from The Netherlands and NW Germany to France, Switzerland south to N Spain; also in Jersey (Channel Is)
S. cylindricauda **Stripe-backed shrew** N Sichuan (China)
S. daphaenodon **Large-toothed Siberian shrew** Urals to Kolyma R, Kamchatka Pen., Sakhalin Is (Russia), Jilin and Nei Monggol (China)
S. dispar **Long-tailed shrew** S New Brunswick, Nova Scotia (Canada), North Carolina to Maine in Appalachian Mts (USA)
S. emarginatus **Zacatecas shrew** Durango, Zacatecas, and Jalisco (Mexico)
S. excelsus **Lofty shrew** Yunnan, Sichuan (China), poss. Nepal
S. fumeus **Smoky shrew** SE Canada south through Appalachian Mts to NE Georgia (USA)
S. gaspensis **Gaspe shrew** Gaspe Pen., Nova Scotia, N New Brunswick, Cape Breton Is (Canada)
S. gracillimus **Slender shrew** SE Siberia from Sea of Okhotsk to N Korea, Sakhalin Is (Russia), Hokkaido (Japan)
S. granaries **Lagranja shrew** NW Iberian Pen. (Portugal, Spain)
S. haydeni **Prairie shrew** SC Canada, NC USA
S. hosonoi **Azumi shrew** C Honshu (Japan)
S. hoyi **Pygmy shrew** Alaska (USA), Canada, USA south in Appalachian and Rocky Mts
S. hydrodromus **Pribilof Island shrew** Pribilof Is (Bering Sea, USA)
S. isodon **Even-toothed shrew** SE Norway and Finland through Siberia to Pacific coast and Kamchatka Pen, Sakhalin Is, and Kurilskiye Is (Russia), prob. NE China and Korea
S. jacksoni **St. Lawrence Island shrew** St Lawrence Is (Bering Sea, USA)
S. kozlovi **Kozlov's shrew** E Xizang (China)
S. leucogaster **Paramushir shrew** Paramushir Is (Russia)
S. longirostris **Southeastern shrew** SE USA (excl. Florida) west to Louisiana, Arkansas, Missouri, Illinois, Indiana
S. lyelli **Mount Lyell shrew** Sierra Nevada Mts (California, USA)
S. macrodon **Large-toothed shrew** Veracruz and Puebla (Mexico)
S. merriami **Merriam's shrew** EC Washington to N and E California, Arizona to Nebraska and Montana (USA)
S. milleri **Carmen Mountain shrew** Sierra Madre Orient. of Coahuila and Nuevo Leon (NE Mexico)
S. minutissimus **Miniscule shrew** Norway, Sweden, and Estonia to E Siberia, Sakhalin Is (Russia), Hokkaido (Japan), Mongolia, China, S Korea
S. minutus **Eurasian pygmy shrew** Europe to Yenesei R and L. Baikal south to Altai and Tien Shan Mts
S. mirabilis **Ussuri shrew** Ussuri region (Russia), NE China, N Korea
S. monticolus **Montane shrew** Alaska (USA) south through W Canada to California and New Mexico east to Montana, Wyoming, and Colorado (USA), Chihuahua and Durango (Mexico)
S. nanus **Dwarf shrew** Montana to New Mexico in Rocky Mts, South Dakota, Arizona (USA)
S. oreopolus **Mexican long-tailed shrew** Jalisco (Mexico)
S. ornatus **Ornate shrew** California (USA) to Baja California (Mexico), Santa Catalina Is (USA)
S. pacificus **Pacific shrew** Coastal Oregon (USA)
S. palustris **Water shrew** North America, from Alaska (USA) to Sierra Nevada, Rocky and Appalachian Mts
S. planiceps **Kashmir shrew** N Pakistan, Kashmir
S. portenkoi **Portenko's shrew** NE Siberia (Russia)
S. preblei **Preble's shrew** W USA
S. raddei **Radde's shrew** Caucasus, N Turkey
S. roboratus **Flat-skulled shrew** Russia, betw. Ob and Ussuri R, south to Altai Mts and N Mongolia
S. sadonis **Sado shrew** Sado Is (Japan)
S. samniticus **Apennine shrew** Italy
S. satunini **Caucasian shrew** Caucasus, N Turkey
S. saussurei **Saussure's shrew** Coahuila and Durango to Chiapas (Mexico), Guatemala
S. sclateri **Sclater's shrew** Chiapas (Mexico)
S. shinto **Shinto shrew** Honshu, Shikoku, and Hokkaido (Japan)
S. sinalis **Chinese shrew** C and W China
S. sonomae **Fog shrew** Coastal W USA, from Oregon to N California
S. stizodon **San Cristobal shrew** Chiapas (Mexico)
S. tenellus **Inyo shrew** WC Nevada, EC California (USA)
S. thibetanus **Tibetan shrew** Himalayas, NE Xizang (China)
S. trowbridgii **Trowbridge's shrew** Coastal W USA, from Washington to California, SW British Columbia (Canada)
S. tundrensis **Tundra shrew** Siberia, from Pechora R to Chukotskoye Pen. (Russia) south to Altai Mts, Mongolia, and NE China; also Alaska (USA), Yukon and Northwest Territories (NW Canada)
S. ugyunak **Barrend ground shrew** N Alaska (USA), NW Canada
S. unguiculatus **Long-clawed shrew** E Siberian Coast, from Vladivostok to the Amur, Sakhalin Is (Russia), Hokkaido (Japan)
S. vagrans **Vagrant shrew** SW Canada, W USA
S. ventralis **Chestnut-bellied shrew** NW Puebla to Oaxaca (Mexico)
S. veraepacis **Verapaz shrew** Mexico to SW Guatemala
S. volnuchini **Caucasian pygmy shrew** S Ukraine, Caucasus

Soriculus Asiatic Shrews
S. caudatus **Hodgson's brown-toothed shrew** Kashmir to N Burma and SW China
S. fumidus **Taiwan brown-toothed shrew** Taiwan
S. hypsibius **De Winton's shrew** SW and C China
S. lamula **Lamulate shrew** Yunnan, Sichuan, Gansu to Fujian (C China)
S. leucops **Long-tailed brown-toothed shrew** Sikkim and Assam (India), C Nepal, to S China, N Burma, and N Vietnam
S. macrurus **Long-tailed mountain shrew** C Nepal to W and S China, N Burma, and Vietnam

S. nigrescens **Himalayan shrew** Himalayas from Assam (India) and SW China to Xizang (China) and Nepal
S. parca **Lowe's shrew** SW China, N Burma, Thailand, N Vietnam
S. salenskii **Salenski's shrew** N Sichuan (China)
S. smithii **Smith's shrew** C Sichuan to W Shaanxi (China)

FAMILY CHRYSOCHLORIDAE
Golden Moles

Amblysomus South African Golden Moles
A. gunningi **Gunning's golden mole** NE South Africa
A. hottentotus **Hottentot golden mole** South Africa
A. iris **Zulu golden mole** S, SE, and E South Africa
A. julianae **Juliana's golden mole** NC and NE South Africa
Calcochloris
C. obtusirostris **Yellow golden mole** E South Africa, S Zimbabwe, S Mozambique
Chlorotalpa African Golden Moles
C. arendsi **Arend's golden mole** E Zimbabwe, adj. Mozambique
C. duthieae **Duthie's golden mole** S South Africa
C. leucorhina **Congo golden mole** N Angola, DRC, Cameroon, CAR
C. sclateri **Sclater's golden mole** S and C South Africa
C. tytonis **Somali golden mole** Giohar (Somalia)
Chrysochloris Cape Golden Moles
C. asiatica **Cape golden mole** SW South Africa, poss. Namibia
C. stuhlmanni **Stuhlmann's golden mole** Cameroon, N DRC, Uganda, Kenya, Tanzania
C. visagiei **Visagie's golden mole** Gouna (Calvinia, South Africa)
Chrysospalax Large Golden Moles
C. trevelyani **Giant golden mole** King Williams' Town region (SE South Africa)
C. villosus **Rough-haired golden mole** E South Africa
Cryptochloris Secretive Golden Moles
C. wintoni **De Winton's golden mole** Little Namaqualand (South Africa)
C. zyli **Van Zyl's golden mole** Lamberts Bay region (SW South Africa)
Eremitalpa
E. granti **Grant's desert golden mole** SW and W South Africa, W Namibia

FAMILY TALPIDAE
Desmans, Moles, and Shrew Moles

SUBFAMILY DESMANINAE
Desmana
D. moschata **Russian desman** Don, Volga, and Ural R drainages in SW Russia; introd. to Dnipro R (Ukraine) and Ob basin
Galemys
G. pyrenaicus **Pyrenean desman** Pyrenees (S France) and N Iberian Pen. (N Spain, N Portugal)

SUBFAMILY TALPINAE
Condylura
C. cristata **Star-nosed mole** Georgia and NW South Carolina (SE USA) to Nova Scotia and Labrador (SE Canada) and Great Lakes region to S Manitoba
Euroscaptor Oriental Moles
E. grandis **Greater Chinese mole** S China, Vietnam
E. klossi **Kloss' mole** Thailand, Laos, Malay Pen.
E. longirostris **Long-nosed mole** S China
E. micrura **Himalayan mole** Sikkim and Assam (India), Nepal, N Burma, S China
E. mizura **Japanese mountain mole** Honshu (Japan)
E. parvidens **Small-toothed mole** Di Linh and Rakho (Vietnam)
Mogera East Asian Moles
M. etigo **Echigo mole** Echigo Plain (Honshu, Japan)
M. insularis **Insular mole** SE China (incl. Hainan), Taiwan
M. kobeae **Kobe mole** Kyushu, Shikoku, and S Honshu (Japan)
M. minor **Small Japanese mole** Honshu (Japan)
M. robusta **Large mole** Korea to NE China, adj. Siberia (Russia)
M. tokudae **Tokuda's mole** Sado Is (Japan)
M. wogura **Japanese mole** Japan
Nesoscaptor
N. uchidai **Ryukyu or Senkaku mole** Senkaku Is (Ryukyu Is, Japan)
Neurotrichus
N. gibbsii **American shrew mole** SW British Columbia (Canada) south to WC California (USA)
Parascalops
P. breweri **Hairy-tailed mole** SE Canada, NE USA
Parascaptor
P. leucura **White-tailed mole** Assam (India), Burma, Yunnan (China)
Scalopus
S. aquaticus **Eastern mole** Minnesota and Massachusetts through E USA to N Coahuila and N Tamaulipas (NE Mexico)
Scapanulus
S. oweni **Gansu mole** Gansu, Shaanxi, Sichuan (C China)
Scapanus Western American Moles
S. latimanus **Broad-footed mole** SC Oregon (USA) to N Baja California (Mexico)
S. orarius **Coast mole** SW British Columbia (Canada) to WC Idaho, N Oregon, and NW California (USA)
S. townsendii **Townsend's mole** SW British Columbia (Canada) to NW California (USA)
Scaptochirus
S. moschatus **Short-faced mole** Hebei, Shandong, Shaanxi, Shanxi (NE China)
Scaptonyx
S. fusicaudus **Long-tailed mole** Qinghai, Shaanxi, Sichuan, and Yunnan (S China), N Burma
Talpa Old World Moles
T. altaica **Siberian mole** C Siberia betw. Ob and Lena R (Russia), south to N Mongolia
T. caeca **Mediterranean mole** S Europe and Turkey
T. caucasica **Caucasian mole** NW Caucasus
T. europaea **European mole** Temperate Europe (incl. Britain) east to Ob and Irtysh R (Russia)
T. levantis **Levantine mole** Bulgaria, Thrace, N Turkey, adj. Caucasus
T. occidentalis **Iberian mole** W and C Iberian Pen. (Portugal, Spain)
T. romana **Roman mole** Apennines (Italy), SE France
T. stankovici **Stankovic's mole** Stara Planina (Romania, Bulgaria, S Yugoslavia), Greece (incl. Kerkira Is)
T. streeti **Persian mole** N Iran
Urotrichus Japanese Shrew Moles
U. pilirostris **True's shrew-mole** Honshu, Shikoku, and Kyushu (Japan)
U. talpoides **Japanese shrew mole** Honshu, Shikoku, Kyushu, Dogo, and Tsushima Is (Japan)

SUBFAMILY UROPSILINAE
Uropsilus Asiatic Shrew Moles
U. andersoni **Anderson's shrew mole** C Sichuan (China)
U. gracilis **Gracile shrew mole** Sichuan and Yunnan (China), N Burma
U. investigator **Inquisitive shrew mole** Yunnan (China)
U. soricipes **Chinese shrew mole** C Sichuan (China)

ORDER CHIROPTERA
BATS

SUBORDER MEGACHIROPTERA
Fruit-eating bats

FAMILY PTEROPODIDAE
Old World Fruit Bats and Flying Foxes

SUBFAMILY PTEROPODINAE
Acerodon Island Fruit Bats
A. celebensis **Sulawesi fruit bat** Sulawesi, Salayar, and Mangole (Indonesia)
A. humilis **Talaud fruit bat** Talaud Is (Indonesia)
A. jubatus **Golden-capped fruit bat** Philippines
A. leucotis **Palawan fruit bat** Palawan, Busuanga, and Balabac Is (Philippines)
A. lucifer **Panay golden-capped fruit bat** Panay (Philippines)
A. mackloti **Sunda fruit bat** Indonesia
Aethalops
A. alecto **Pygmy fruit bat** Malay Pen., Sumatra, W Java, and Lombok (Indonesia), Borneo
Alionycteris
A. paucidentata **Mindanao pygmy fruit bat** Mindanao (Philippines)
Aproteles
A. bulmerae **Bulmer's fruit bat** New Guinea
Balionycteris
B. maculata **Spotted-winged fruit bat** Thailand, Malay Pen., Riau Arch. (Indonesia), Borneo
Boneia
B. bidens **Manada fruit bat** N Sulawesi (Indonesia)
Casinycteris
C. argynnis **Short-palated fruit bat** Cameroon to E DRC
Chironax
C. melanocephalus **Black-capped fruit bat** Thailand, Malay Pen., Sumatra, Java, and Sulawesi (Indonesia), Borneo
Cynopterus Short-nosed Fruit Bats
C. brachyotis **Lesser short-nosed fruit bat** India (incl. Andaman and Nicobar Is), Sri Lanka, SE Asia, Malaysia, Sumatra and Sulawesi (Indonesia), Borneo, Philippines
C. horsfieldi **Horsfield's fruit bat** Thailand, Malay Pen., Java, Sumatra, and Lesser Sunda Is (Indonesia), Borneo
C. nusatenggara **Nusatenggara short-nosed fruit bat** Lesser Sunda Is (Indonesia)
C. sphinx **Greater short-nosed fruit bat** India, Sri Lanka, S China, SE Asia, Malay Pen., Sumatra (Indonesia), poss. Borneo
C. titthaecheileus **Indonesian short-nosed fruit bat** Sumatra, Java, Lombok, and Timor (Indonesia)
Dobsonia Naked-backed Fruit Bats
D. beauforti **Beaufort's naked-backed fruit bat** Waigeo Is (Indonesia)
D. chapmani **Negros naked-backed fruit bat** Extinct: formerly Philippines
D. emersa **Biak naked-backed fruit bat** Biak and Owii Is (Indonesia)
D. exoleta **Sulawesi naked-backed fruit bat** Sulawesi, adj. islands (Indonesia)
D. inermis **Solomons naked-backed fruit bat** Solomon Is
D. minor **Lesser naked-backed fruit bat** Sulawesi (Indonesia), C and W New Guinea
D. moluccensis **Moluccan naked-backed fruit bat** Aru, Batanta, Mysol, and Maluku (Indonesia), New Guinea, Bismarck Arch. (Papua New Guinea), N Queensland (Australia)
D. pannietensis **Panniet naked-backed fruit bat** Louisiade Arch., D'Entrecasteaux, and Trobriand Is (Papua New Guinea)
D. peroni **Western naked-backed fruit bat** Indonesia
D. praedatrix **New Britain naked-backed fruit bat** Bismarck Arch. (Papua New Guinea)
D. viridis **Greenish naked-backed fruit bat** Sulawesi and Maluku (Indonesia)
Dyacopterus
D. spadiceus **Dyak fruit bat** Malay Pen., Sumatra (Indonesia), Borneo, Luzon and Mindanao (Philippines)
Eidolon Eidolon Fruit Bats
E. dupreanum **Madagascan fruit bat** Madagascar
E. helvum **Straw-colored fruit bat** Senegal to Ethiopia to South Africa, SW Arabia, islands off East Africa
Epomophorus Epauletted Fruit Bats
E. angolensis **Angolan epauletted fruit bat** W Angola, NW Namibia
E. gambianus **Gambian epauletted fruit bat** Senegal to W Ethiopia, S Tanzania to Angola and South Africa
E. grandis **Lesser Angolan epauletted fruit bat** N Angola, S Congo
E. labiatus **Ethiopian epauletted fruit bat** Nigeria to Ethiopia south to Congo and Malawi
E. minimus **East African epauletted fruit bat** Ethiopia to Uganda, Tanzania
E. wahlbergi **Wahlberg's epauletted fruit bat** Cameroon to Somalia south to Angola and South Africa, Pemba and Zanzibar Is (Tanzania)
Epomops African Epauletted Bats
E. buettikoferi **Buettikofer's epauletted bat** Guinea to Nigeria
E. dobsoni **Dobson's fruit bat** N Botswana, Angola to Rwanda, Tanzania, and Malawi
E. franqueti **Franquet's epauletted bat or Singing fruit bat** Ivory Coast to Sudan, Uganda, NW Tanzania, N Zambia, Angola
Haplonycteris
H. fischeri **Philippine or Fischer's pygmy fruit bat** Philippines
Harpyionycteris Harpy Fruit Bats
H. celebensis **Sulawesi Harpy Fruit Bat** Sulawesi (Indonesia)
H. whiteheadi **Harpy fruit bat** Philippines
Hypsignathus
H. monstrosus **Hammer-headed fruit bat** Sierra Leone to W Kenya, south to Zambia and Angola, Bioko (Equatorial Guinea)
Latidens
L. salimalii **Salim Ali's fruit bat** S India
Megaerops Tailless Fruit Bats
M. ecaudatus **Temminck's tailless fruit bat** Thailand, Malay Pen., Sumatra (Indonesia), Borneo
M. kusnotoi **Javan tailless fruit bat** Java (Indonesia)
M. niphanae **Ratanaworabhan's fruit bat** India, Thailand, Vietnam
M. wetmorei **White-collared fruit bat** Malay Pen., Borneo, Philippines
Micropteropus Dwarf Epauletted Fruit Bats
M. intermedius **Hayman's dwarf epauletted fruit bat** N Angola, SE DRC
M. pusillus **Peter's dwarf epauletted fruit bat** Gambia to Ethiopia, south to Tanzania, Burundi, Zambia, Angola
Myonycteris Little Collared Fruit Bats
M. brachycephala **São Tomé collared fruit bat** São Tomé Is (São Tomé & Principe)
M. relicta **East African little collared fruit bat** Shimba Hills (Kenya), Nguru and Usambara Mts (Tanzania)
M. torquata **Little collared fruit bat** Sierra Leone to Uganda, south to Angola and Zambia, Bioko (Equatorial Guinea)
Nanonycteris
N. veldkampi **Veldkamp's bat or Little flying cow** Guinea to CAR
Neopteryx
N. frosti **Small-toothed fruit bat** W and N Sulawesi (Indonesia)
Nyctimene Tube-nosed Fruit Bats
N. aello **Broad-striped tube-nosed fruit bat** New Guinea
N. albiventer **Common tube-nosed fruit bat** Maluku and Kei Is (Indonesia), New Guinea, Bismarck Arch. (Papua New Guinea), Solomon Is, N Queensland (Australia)
N. celaeno **Dark tube-nosed fruit bat** Geelvink Bay (Irian Jaya, New Guinea)
N. cephalotes **Pallas' tube-nosed fruit bat** Sulawesi, Timor, Maluku, and Numfoor (Indonesia), S New Guinea
N. certans **Mountain tube-nosed fruit bat** New Guinea
N. cyclotis **Round-eared tube-nosed fruit bat** New Guinea, New Britain (Bismarck Arch., Papua New Guinea)
N. draconilla **Dragon tube-nosed fruit bat** New Guinea
N. major **Island tube-nosed fruit bat** Bismarck and Louisiade Archs., D'Entrecasteaux and Trobriand Is (Papua New Guinea), Solomon Is, islands off N New Guinea coast
N. malaitensis **Malaita Island tube-nosed fruit bat** Malaita Is (Solomon Is)
N. masalai **Demonic tube-nosed fruit bat** New Ireland (Bismarck Arch., Papua New Guinea)
N. minutus **Lesser tube-nosed fruit bat** Sulawesi, C Maluku (Indonesia)
N. rabori **Philippine tube-nosed fruit bat** Negros Is (Philippines)
N. robinsoni **Queensland tube-nosed fruit bat** E Queensland (Australia)
N. sanctacrucis **Nendo tube-nosed fruit bat** Santa Cruz Is (Solomon Is)
N. vizcaccia **Umboi tube-nosed fruit bat** Umboi Is, Bismarck Arch. (Papua New Guinea), Solomon Is
Otopteropus
O. cartilagonodus **Luzon fruit bat** Luzon (Philippines)
Paranyctimene
P. raptor **Unstriped tube-nosed bat** New Guinea
Penthetor
P. lucasi **Lucas' short-nosed fruit bat** Malay Pen., Riau Arch. (Indonesia), Borneo
Plerotes
P. anchietai **Anchieta's fruit bat** Angola, Zambia, S DRC
Ptenochirus Musky Fruit Bats
P. jagori **Greater musky fruit bat** Philippines
P. minor **Lesser musky fruit bat** Philippines
Pteralopex Monkey-faced Bats
P. acrodonta **Fijian monkey-faced bat** Fiji Is
P. anceps **Bougainville monkey-faced bat** Bougainville Is (Papua New Guinea), Choiseul Is (Solomon Is)
P. atrata **Guadalcanal monkey-faced bat** Santa Isabel Is, Guadalcanal (Solomon Is)
P. pulchra **Montane monkey-faced bat** Guadalcanal (Solomon Is)
Pteropus Flying Foxes
P. admiralitatum **Admiralty flying fox** Bismarck Arch. (Papua New Guinea), Solomon Is
P. aldabrensis **Aldabra flying fox** Aldabra Is (Seychelles)
P. alecto **Black flying fox** Indonesia, S New Guinea, N and E Australia

P. anetianus **Vanuatu flying fox** Vanuatu
P. argentatus **Ambon flying fox** Amboina Is (Maluku, Indonesia)
P. brunneus **Dusky flying fox** Percy Is (Queensland, Australia)
P. caniceps **North Moluccan flying fox** Sulawesi, Halmahera and Sula Is (Indonesia)
P. chrysoproctus **Moluccan flying fox** Indonesia
P. conspicillatus **Spectacled flying fox** Halmahera Is (Indonesia), New Guinea, NE Queensland (Australia)
P. dasymallus **Ryukyu flying fox** Taiwan, Ryukyu, Daito, and S Kyushu (Japan)
P. faunulus **Nicobar flying fox** Nicobar Is (India)
P. fundatus **Banks flying fox** N Vanuatu
P. giganteus **Indian flying fox** Maldives, Pakistan, India (incl. Andaman Is), Sri Lanka, Burma, Qinghai (China)
P. gilliardi **Gilliard's flying fox** New Britain (Bismarck Arch., Papua New Guinea)
P. griseus **Gray flying fox** Indonesia, poss. S Luzon (Philippines)
P. howensis **Ontong Java flying fox** Ontong Java Is (Solomon Is)
P. hypomelanus **Variable flying fox** Maldives, Thailand and Vietnam to Indonesia, New Guinea, Philippines, Solomon Is
P. insularis **Ruck flying fox** Truk Is (C Caroline Is)
P. leucopterus **White-winged flying fox** Luzon and Dinagat (Philippines)
P. livingstonei **Comoro black flying fox** Comoros
P. lombocensis **Lombok flying fox** Lombok, Flores, and Alor (Indonesia)
P. lylei **Lyle's flying fox** Thailand, Vietnam
P. macrotis **Big-eared flying fox** Aru Is (Indonesia), New Guinea
P. mahaganus **Sanborn's flying fox** Bougainville (Papua New Guinea), Santa Isabel Is (Solomon Is)
P. mariannus **Mariana flying fox** Mariana Is (Guam), Caroline Is, Ryukyu Is (Japan)
P. mearnsi **Mearns' flying fox** Mindanao and Basilan (Philippines)
P. melanopogon **Black-bearded flying fox** Indonesia
P. melanotus **Black-eared flying fox** Nicobar and Andaman Is (India), Engano and Nias Is (Indonesia), Christmas Is
P. molossinus **Caroline flying fox** Mortlock and Ponape Is (Caroline Is)
P. neohibernicus **Great or Bismarck flying fox** New Guinea, Bismarck Arch. and Admiralty Is (Papua New Guinea)
P. niger **Greater Mascarene flying fox** Reunion, Mauritius
P. nitendiensis **Temotu flying fox** Ndeni Is (Santa Cruz Is, Solomon Is)
P. ocularis **Seram flying fox** Seram, Buru (Indonesia)
P. ornatus **Ornate flying fox** New Caledonia Is, incl. Loyaute Is (France)
P. personatus **Masked flying fox** Halmahera Is (Indonesia)
P. phaeocephalus **Mortlock flying fox** Mortlock Is (C Caroline Is)
P. pilosus **Large Palau flying fox** Palau Is (Caroline Is)
P. pohlei **Geelvink Bay flying fox** Yapen Is (Indonesia)
P. poliocephalus **Gray-headed flying fox** S Queensland to Victoria (E Australia)
P. pselaphon **Bonin flying fox** Bonin and Volcano Is (Japan)
P. pumilus **Little golden-mantled flying fox** Philippines
P. rayneri **Solomons flying fox** Solomon Is
P. rodricensis **Rodriguez flying fox** Rodrigues and Round Is (Ind. Oc.)
P. rufus **Madagascan flying fox** Madagascar
P. samoensis **Somoan flying fox** Fiji, Western Samoa and American Samoa (Samoa Is)
P. sanctacrucis **Santa Cruz flying fox** Santa Cruz Is (Solomon Is)
P. scapulatus **Little Red flying fox** S New Guinea, Australia; introd. to New Zealand
P. seychellensis **Seychelles flying fox** Seychelles (incl. Aldabra Is), Comoros, and Mafia Is
P. speciosus **Philippine flying fox** Sulu Arch, Basilan and Mindanao (Philippines), islands in Java Sea
P. subniger **Dark flying fox** Reunion, Mauritius
P. temmincki **Temminck's flying fox** Buru, Amboina, and Seram (Indonesia), Bismarck Arch. (Papua New Guinea), adj. small islands
P. tokudae **Guam flying fox** Guam
P. tonganus **Pacific flying fox** Karkar Is (Papua New Guinea), Rennell Is (Solomon Is) to New Caledonia Is (France) and to Cook Is
P. tuberculatus **Vanikoro flying fox** Vanikoro Is (Santa Cruz Is, Solomon Is)
P. vampyrus **Large flying fox** Indochina, Malay Pen., Sumatra, Java, and Lesser Sunda Is (Indonesia), Borneo, Philippines
P. vetulus **New Caledonia flying fox** New Caledonia (France)
P. voeltzkowi **Pemba flying fox** Pemba Is (Tanzania)
P. woodfordi **Dwarf flying fox** Fauro Is to Guadalcanal (Solomon Is)

Rousettus **Rousette Fruit Bats**
R. aegyptiacus **Egyptian rousette** Senegal to Egypt south to South Africa, Cyprus, Turkey, Yemen to Pakistan,
R. amplexicaudatus **Geoffroy's rousette** Cambodia, Thailand, Malay Pen. through Indonesia to New Guinea, Bismarck Arch. (Papua New Guinea), Philippines, Solomon Is
R. angolensis **Angolan rousette** Senegal to Ethiopia south to Angola and Mozambique, Bioko (Equatorial Guinea)
R. celebensis **Sulawesi rousette** Sulawesi and Sangihe Is (Indonesia)
R. lanosus **Ruwenzori long-haired rousette** S Sudan, S Ethiopia, Kenya, Uganda, Tanzania, E DRC
R. leschenaulti **Leschenault's rousette** Sri Lanka, Pakistan to S China and Vietnam, Sumatra, Java, Bali, and Mentawai Is (Indonesia)
R. madagascariensis **Madagascan rousette** Madagascar
R. obliviosus **Comoros rousette** Comoros
R. spinalatus **Bare-backed rousette** Sumatra (Indonesia), Borneo

Scotonycteris **West African Fruit Bats**
S. ophiodon **Pohle's fruit bat** Liberia to Congo
S. zenkeri **Zenker's fruit bat** Liberia to Congo and E DRC, Bioko (Equatorial Guinea)

Sphaerias
S. blanfordi **Blanford's fruit bat** N India, Bhutan, Burma, SW China (incl. Xizang), N Thailand

Styloctenium
S. wallacei **Stripe-faced fruit bat** Sulawesi (Indonesia)

Thoopterus
T. nigrescens **Swift fruit bat** Sulawesi, Maluku, and Sangihe Is (Indonesia)

SUBFAMILY MACROGLOSSINAE

Eonycteris **Dawn Bats**
E. major **Greater dawn bat** Borneo, Philippines
E. spelaea **Lesser dawn bat** N India (also Andaman Is), S China, Burma, Thailand, Malay Pen., Sumatra, Java, Sumba, Sulawesi, and Timor (Indonesia), Borneo, Philippines

Macroglossus **Long-tongued Fruit Bats**
M. minimus **Lesser long-tongued fruit bat** Thailand to Philippines, New Guinea, Bismarck Arch. (Papua New Guinea), Solomon Is, N Australia
M. sobrinus **Greater long-tongued fruit bat** SE Asia, Sumatra and Java (Indonesia)

Megaloglossus
M. woermanni **Woermann's bat or African long-tongued fruit bat** Liberia to Uganda, S DRC, and N Angola, Bioko (Equatorial Guinea)

Melonycteris **Black-bellied Fruit Bats**
M. aurantius **Orange fruit bat** Florida and Choiseul Is (Solomon Is)
M. melanops **Black-bellied fruit bat** Bismarck Arch. (Papua New Guinea)
M. woodfordi **Woodford's fruit bat** Solomon Is

Notopteris
N. macdonaldi **Long-tailed fruit bat** Vanuatu, New Caledonia (France), Fiji Is, Caroline Is

Syconycteris **Blossom Bats**
S. australis **Southern blossom bat** Maluku (Indonesia), New Guinea, Louisiade and Bismarck Archs., D'Entrecasteaux and Trobriand Is (Papua New Guinea), E Queensland and New South Wales (Australia)
S. carolinae **Halmahera blossom bat** Halmahera Is (Maluku, Indonesia)
S. hobbit **Moss-forest blossom bat** C New Guinea

SUBORDER MICROCHIROPTERA
Insect-eating bats

FAMILY RHINOPOMATIDAE
Mouse-tailed Bats

Rhinopoma
R. hardwickei **Lesser mouse-tailed bat** Burma to Morocco, south to Mauritania, Nigeria, and Kenya; also Socotra Is (Yemen)
R. microphyllum **Greater mouse-tailed bat** Morocco and Senegal to Thailand and Sumatra (Indonesia)
R. muscatellum **Small mouse-tailed bat** Oman, W Iran, S Afghanistan

FAMILY CRASEONYCTERIDAE

Craseonycteris
C. thonglongyai **Kitti's hog-nosed bat** Kanchanaburi (Thailand)

FAMILY EMBALLONURIDAE
Sheath-tailed Bats

Balantiopteryx **Least Sac-winged Bats**
B. infusca **Ecuadorian sac-winged bat** W Ecuador
B. io **Thomas' sac-winged bat** S Veracruz and Oaxaca (S Mexico) to EC Guatemala and Belize
B. plicata **Gray sac-winged bat** S Baja California and C Sonora (Mexico) to Costa Rica and N Colombia

Centronycteris
C. maximiliani **Shaggy bat** S Veracruz (Mexico) to Guianas, Peru, and Brazil

Coleura **Peters' Sheath-tailed Bats**
C. afra **African sheath-tailed bat** Guinea-Bissau to Somalia, south to Angola, DRC, and Mozambique; also Yemen
C. seychellensis **Seychelles sheath-tailed bat** Seychelles

Cormura
C. brevirostris **Chestnut sac-winged bat** Nicaragua to Peru, Brazil

Cyttarops
C. alecto **Short-eared bat** Nicaragua, Costa Rica, Guyana, Amazonian Brazil

Diclidurus **Ghost or White Bats**
D. albus **Northern ghost bat** Nayarit (S Mexico) to E Brazil and Trinidad
D. ingens **Greater ghost bat** SE Colombia, Venezuela, Guyana, NW Brazil
D. isabellus **Isabelle's ghost bat** Venezuela, NW Brazil
D. scutatus **Lesser ghost bat** Venezuela, Guyana, Surinam, Peru, Amazonian Brazil

Emballonura **Old World Sheath-tailed Bats**
E. alecto **Small Asian sheath-tailed bat** Sulawesi and Tanimbar (Indonesia), Borneo, Philippines
E. atrata **Peter's sheath-tailed bat** E and C Madagascar
E. beccarii **Beccari's sheath-tailed bat** Kei Is (Indonesia), New Guinea, Trobriand Is (Papua New Guinea)
E. dianae **Large-eared sheath-tailed bat** New Guinea, New Ireland (Bismarck Arch., Papua New Guinea), Rennell and Malaita Is (Solomon Is)
E. furax **New Guinea sheath-tailed bat** New Guinea, Bismarck Arch. (Papua New Guinea)
E. monticola **Lesser sheath-tailed bat** Thailand to Malay Pen., Indonesia, Borneo
E. raffrayana **Raffray's sheath-tailed bat** Seram, Kei Is, and Sulawesi (Indonesia), New Guinea, Bismarck Arch. (Papua New Guinea), Choiseul, Santa Isabel, and Malaita Is (Solomon Is)
E. semicaudata **Polynesian sheath-tailed bat** Mariana and Caroline Is, Vanuatu, Fiji Is, Samoa Is

Mosia
M. nigrescens **Dark sheath-tailed bat** Indonesia, New Guinea, Bismarck Arch. (Papua New Guinea), Solomon Is

Peropteryx **Dog-like Bats**
P. kappleri **Greater dog-like bat** S Veracruz (S Mexico) to Peru and E Brazil
P. leucoptera **White-winged dog-like bat** Colombia, Venezuela, Guyana, Surinam, Fr. Guiana, Peru, N and E Brazil
P. macrotis **Lesser dog-like bat** Guerrero and Yucatan (Mexico) to Peru, S and E Brazil, Paraguay, Aruba Is (Neth. Antilles), Margarita Is (Venezuela), Trinidad and Tobago, Grenada

Rhynchonycteris
R. naso **Proboscis or Sharp-nosed bat** E Oaxaca and C Veracruz (S Mexico) to Guyana, Surinam, Fr. Guiana, Peru, C and E Brazil, Bolivia, Trinidad

Saccolaimus **Pouched Bats**
S. flaviventris **Yellow-bellied pouched bat** SE New Guinea, Australia (excl. Tasmania)
S. mixtus **Troughton's pouched bat** SE New Guinea, NE Queensland (Australia)
S. peli **Pel's pouched bat** Liberia to W Kenya to Angola
S. pluto **Philippine pouched bat** Philippines
S. saccolaimus **Naked-rumped pouched bat** India, Sri Lanka through SE Asia to Sumatra, Java, and Timor (Indonesia), Borneo, New Guinea, Guadalcanal (Solomon Is), NE Queensland (Australia)

Saccopteryx **Sac-winged Bats**
S. bilineata **Greater sac-winged bat** Jalisco and Veracruz (Mexico) to Guianas, E Brazil, Bolivia, Trinidad and Tobago
S. canescens **Frosted sac-winged bat** Colombia, Venezuela, Guianas, Peru, N Brazil
S. gymnura **Amazonian sac-winged bat** Amazonian Brazil, poss. Venezuela
S. leptura **Lesser sac-winged bat** Chiapas and Tabasco (S Mexico) to Guianas, Peru, and E Brazil, Margarita Is (Venezuela), Trinidad and Tobago

Taphozous **Tomb Bats**
T. australis **Coastal tomb bat** SE New Guinea, Torres Str. Is, N Queensland (Australia)
T. georgianus **Sharp-nosed tomb bat** Australia
T. hamiltoni **Hamilton's tomb bat** S Sudan, Chad, Kenya
T. hildegardeae **Hildegarde's tomb bat** Kenya, NE Tanzania (incl. Zanzibar Is)
T. hilli **Hill's tomb bat** Western Australia, South Australia, Northern Territory (Australia)
T. kapalgensis **Arnhem tomb bat** Northern Territory (Australia)
T. longimanus **Long-winged tomb bat** Sri Lanka, India to Cambodia, Malay Pen., Sumatra, Java, Bali, and Flores (Indonesia), Borneo
T. mauritianus **Mauritian tomb bat** Senegal to Sudan, Somalia to South Africa, Madagascar, Mauritius, Reunion, Aldabra Is (Seychelles)
T. melanopogon **Black-bearded tomb bat** Sri Lanka, India, S China, Burma, Thailand, Laos, Vietnam, Malay Pen., Indonesia, Borneo
T. nudiventris **Naked-rumped tomb bat** Mauritania, Senegal, Guinea-Bissau east to Egypt south to Tanzania, Arabian Pen. east to Burma
T. perforatus **Egyptian tomb bat** Senegal to Egypt, Somalia south to Botswana, Mozambique, S Arabia, S Iran, Pakistan, NW India
T. philippinensis **Philippine tomb bat** Philippines
T. theobaldi **Theobald's tomb bat** C India to Vietnam, Java and Sulawesi (Indonesia), Borneo

FAMILY NYCTERIDAE **Slit-faced Bats**

Nycteris
N. arge **Bate's slit-faced bat** Sierra Leone to NE Angola, S and E DRC, W Kenya, SW Sudan, Bioko (Equatorial Guinea)
N. gambiensis **Gambian slit-faced bat** W Africa, from Senegal and Gambia to Benin and Burkina Faso
N. grandis **Large slit-faced bat** Senegal to Kenya, Mozambique, and Zimbabwe, Zanzibar and Pemba Is (Tanzania)
N. hispida **Hairy slit-faced bat** Senegal to Somalia south to Angola and South Africa, Zanzibar (Tanzania), Bioko (Equatorial Guinea)
N. intermedia **Intermediate slit-faced bat** Liberia to W Tanzania south to Angola
N. javanica **Javan slit-faced bat** Java, Bali, and Kagean Is (Indonesia)
N. macrotis **Large-eared slit-faced bat** Senegal to Ethiopia south to Malawi, Mozambique, and Zimbabwe, Madagascar, Zanzibar (Tanzania)
N. major **Ja slit-faced bat** Liberia to Zambia
N. nana **Dwarf slit-faced bat** Ivory Coast to SW Sudan, W Kenya, W Tanzania, southwest to NE Angola
N. thebaica **Egyptian slit-faced bat** Central Arabia, Israel, Sinai, Egypt to Morocco and most of SubSaharan Africa (incl. Zanzibar and Pemba Is)
N. tragata **Malayan slit-faced bat** Burma, Thailand, Malay Pen., Sumatra (Indonesia), Borneo
N. woodi **Wood's slit-faced bat** Ethiopia, Somalia, SW Tanzania south to Zambia and South Africa; also Cameroon

FAMILY MEGADERMATIDAE
False Vampire Bats

Cardioderma
C. cor **Heart-nosed bat** E Africa

Lavia
L. frons **African yellow-winged bat** Senegal to Somalia, south to Zambia and Malawi, Zanzibar (Tanzania)

Macroderma
M. gigas **Australian false vampire bat** N and C Australia

Megaderma
M. lyra **Greater false vampire bat** Afghanistan to S China, south to Sri Lanka and Malay Pen.
M. spasma **Lesser false vampire bat** India, Sri Lanka through SE Asia to Lesser Sunda Is and Maluku (Indonesia) and Philippines

FAMILY RHINOLOPHIDAE
Horseshoe Bats

SUBFAMILY RHINOLOPHINAE
Rhinolophus **Horseshoe Bats**
R. acuminatus **Acuminate horseshoe bat** Thailand, Laos, Cambodia, Sumatra, Java, Lombok, Bali, Nias and Engano Is (Indonesia), Borneo, Palawan (Philippines)
R. adami **Adam's horseshoe bat** Congo
R. affinis **Intermediate horseshoe bat** India (incl. Andaman Is) to S China through Malaysia to Lesser Sunda Is (Indonesia), Borneo
R. alcyone **Halcyon horseshoe bat** Senegal to SW Sudan, N DRC, Uganda; also Gabon and Bioko (Equatorial Guinea)
R. anderseni **Andersen's horseshoe bat** Palawan and Luzon (Philippines)
R. arcuatus **Arcuate horseshoe bat** Indonesia, New Guinea, Philippines
R. blasii **Blasius' horseshoe bat** N South Africa to Ethiopia and Somalia, Morocco to Algeria, Yemen, Israel, Jordan, Syria, Turkey, Yugoslavia, Albania, Bulgaria, Rumania, Italy, Greece, Caucasus, Iran, Turkmenistan, Afghanistan, Pakistan
R. borneensis **Bornean horseshoe bat** Cambodia, Vietnam, Malaysia, Java, Karimata, and South Natuna Is (Indonesia), Borneo
R. canuti **Canut's horseshoe bat** Java, Timor (Indonesia)
R. capensis **Cape horseshoe bat** SW South Africa, Zimbabwe, Mozambique
R. celebensis **Sulawesi horseshoe bat** Indonesia
R. clivosus **Geoffroy's horseshoe bat** SubSaharan Africa, Algeria to Arabia, Turkmenistan to Afghanistan
R. coelophyllus **Croslet horseshoe bat** Burma, Thailand, Malay Pen.
R. cognatus **Andaman horseshoe bat** Andaman Is (India)
R. cornutus **Little Japanese horseshoe bat** Japan
R. creaghi **Creagh's horseshoe bat** Madura, Java, and Timor (Indonesia), Borneo
R. darlingi **Darling's horseshoe bat** N South Africa, Namibia, Botswana, Zimbabwe, S Angola, Malawi, Mozambique, Tanzania
R. deckenii **Decken's horseshoe bat** Kenya, Tanzania (incl. Zanzibar and Pemba Is), Uganda
R. denti **Dent's horseshoe bat** W South Africa, Namibia, Botswana, Zimbabwe, Mozambique; also Ghana, Ivory Coast, Guinea
R. eloquens **Eloquent horseshoe bat** NE DRC, S Sudan, Uganda, Rwanda, Tanzania (incl. Zanzibar and Pemba Is), Kenya, S Somalia
R. euryale **Mediterranean horseshoe bat** Algeria to Morocco, S Europe and Mediterranean islands, Israel to Caucasus, Iran, Turkmenistan
R. euryotis **Broad-eared horseshoe bat** Indonesia, New Guinea, Bismarck Arch. (Papua New Guinea)
R. ferrumequinum **Greater horseshoe bat** S England to Caucasus south to Morocco and Tunisia east through Iran and Himalayas to China and Japan
R. fumigatus **Rüppell's horseshoe bat** SubSaharan Africa
R. guineensis **Guinean horseshoe bat** Guinea, Sierra Leone, Liberia
R. hildebrandti **Hildebrandt's horseshoe bat** South Africa and Mozambique to Ethiopia, S Sudan, and NE DRC
R. hipposideros **Lesser horseshoe bat** S Europe and N Africa (also Sudan and Ethiopia) through Arabia to Kyrgyzstan and Kashmir; also Ireland
R. imaizumii **Imaizumi's horsehose bat** Ryukyu Is (Japan)
R. inops **Philippine Forest horseshoe bat** Mindanao (Philippines)
R. keyensis **Insular horseshoe bat** Halmahera, Seram, Goram, Kei, and Wetter Is (Indonesia)
R. landeri **Lander's horseshoe bat** Senegal to Ethiopia, Somalia south to South Africa, Bioko (Equatorial Guinea), Zanzibar (Tanzania)
R. lepidus **Blyth's horseshoe bat** Afghanistan, N India, Burma, Sichuan and Yunnan (China), Thailand, Malay Pen., Sumatra (Indonesia)
R. luctus **Woolly horseshoe bat** India (incl. Sikkim), Sri Lanka, Nepal, Burma, S China, Taiwan, Vietnam, Laos, Thailand, Malay Pen., Sumatra, Java, and Bali (Indonesia), Borneo
R. maclaudi **Maclaud's horseshoe bat** Guinea, Liberia, E DRC, W Uganda, Rwanda
R. macrotis **Big-eared horseshoe bat** N India to S China, Vietnam, Malay Pen., Sumatra (Indonesia), Philippines
R. malayanus **Malayan horseshoe bat** Vietnam, Laos, Thailand, Malay Pen.
R. marshalli **Marshall's horseshoe bat** Thailand
R. megaphyllus **Smaller horseshoe bat** E New Guinea, Louisiade and Bismarck Archs., D'Entrecasteaux Is (Papua New Guinea), E Queensland, E New South Wales, and E Victoria (Australia)
R. mehelyi **Mehely's horseshoe bat** S Europe (Portgual and Spain to Greece) and N Africa (Morocco to Egypt) to Israel, Turkey, Caucasus, Iran, Afghanistan
R. mitratus **Mitred horseshoe bat** N India
R. monoceros **Formosan horseshoe bat** Taiwan
R. nereis **Neriad horseshoe bat** Anamba and North Natuna Is (Indonesia)
R. osgoodi **Osgood's horseshoe bat** Yunnan (China)
R. paradoxolophus **Bourret's horseshoe bat** Vietnam, Thailand
R. pearsoni **Pearson's horseshoe bat** N India, Burma, Sichuan, Anhui, and Fujian (China), Vietnam, Thailand, Malay Pen.
R. philippinensis **Large-eared horseshoe bat** Sulawesi, Timor, and Kei Is (Indonesia), Borneo, Mindoro, Luzon, Mindanao, and Negros (Philippines), New Guinea, NE Queensland (Australia)
R. pusillus **Least horseshoe bat** India, Thailand, Malay Pen., Java, Mentawai and Lesser Sunda Is (Indonesia)
R. rex **King horseshoe bat** SW China
R. robinsoni **Peninsular or Robinson's horseshoe bat** Thailand, Malay Pen.
R. rouxi **Rufous horseshoe bat** Sri Lanka, India to S China, Vietnam
R. rufus **Large rufus horseshoe bat** Philippines
R. sedulus **Lesser woolly horseshoe bat** Malay Pen., Borneo
R. shameli **Shamel's horseshoe bat** Burma, Cambodia, Thailand, Malay Pen.
R. silvestris **Forest horseshoe bat** Gabon, Congo
R. simplex **Lombok horseshoe bat** Lesser Sunda Is (Indonesia)
R. simulator **Bushveld horseshoe bat** South Africa to S Sudan and Ethiopia; also Cameroon, Nigeria, Guinea
R. stheno **Lesser brown horseshoe bat** Thailand, Malay Pen., Sumatra and Java (Indonesia)
R. subbadius **Little Nepalese horseshoe bat** Assam (India), Nepal, Burma, Vietnam
R. subrufus **Small rufous horseshoe bat** Philippines
R. swinnyi **Swinny's horseshoe bat** South Africa to S DRC and E Africa
R. thomasi **Thomas' horseshoe bat** Yunnan (China), Burma, Vietnam, Thailand
R. trifoliatus **Trefoil horseshoe bat** NE India, Burma, SW Thailand, Malay Pen., Indonesia, Borneo
R. virgo **Yellow-faced horseshoe bat** Philippines
R. yunanensis **Dobson's horseshoe bat** NE India, Yunnan (China), Thailand

SUBFAMILY HIPPOSIDERINAE
Anthops
A. ornatus **Flower-faced bat** Solomon Is
Asellia **Trident Leaf-nosed Bats**
A. patrizii **Patrizi's trident leaf-nosed bat** N Ethiopia, Red Sea islands
A. tridens **Trident leaf-nosed bat** Senegal, Morocco to Sudan, Egypt, Israel, Arabia to Pakistan
Aselliscus **Tate's Trident-nosed Bats**
A. stoliczkanus **Stoliczka's trident bat** S China, Burma, Thailand, Laos, Vietnam, Malay Pen.
A. tricuspidatus **Temminck's trident bat** Maluku (Indonesia), New Guinea, Bismarck Arch. (Papua New Guinea), Solomon Is, Vanuatu
Cloeotis
C. percivali **Percival's trident bat** South Africa, Botswana, Zimbabwe, Mozambique to S DRC, Kenya
Coelops **Tailless Leaf-nosed Bats**
C. frithi **East Asian tailless leaf-nosed bat** NE India to S China, Taiwan, Vietnam, Malay Pen., Java and Bali (Indonesia)
C. hirsutus **Philippine tailless leaf-nosed bat** Mindoro (Philippines)
C. robinsoni **Malayan tailless leaf-nosed bat** Malay Pen., Borneo
Hipposideros **Roundleaf Bats**
H. abae **Aba roundleaf bat** Guinea-Bissau to SW Sudan, Uganda
H. armiger **Great roundleaf bat bat** N India, Nepal, Burma, S China, Taiwan, Vietnam, Laos, Thailand, Malay Pen.
H. ater **Dusky roundleaf bat** Sri Lanka, India to Malay Pen., Indonesia, Philippines, New Guinea to N Queensland, N Northern Territory, N Western Australia (Australia)
H. beatus **Benito roundleaf bat** Guinea-Bissau, Sierra Leone, Liberia, Ghana, Nigeria, Ivory Coast, Cameroon, Rio Muni (Equatorial Guinea), Gabon, N DRC
H. bicolor **Bicolored roundleaf bat** Malaysia to Philippines, Timor (Indonesia)
H. breviceps **Short-headed roundleaf bat** Mentawai Is (Indonesia)
H. caffer **Sundevall's roundleaf bat** SubSaharan Africa, Morocco, SW Arabian Pen.
H. calcaratus **Spurred roundleaf bat** New Guinea, Bismarck Arch. (Papua New Guinea), Solomon Is
H. camerunensis **Greater roundleaf bat** Cameroon, E DRC, W Kenya
H. cervinus **Fawn roundleaf bat** Malay Pen., Sumatra (Indonesia), Philippines to Vanuatu and NE Australia
H. cineraceus **Ashy roundleaf bat** Pakistan to Vietnam and Borneo
H. commersoni **Commerson's roundleaf bat** Gambia to Somalia south to South Africa, Madascar, São Tomé Is (São Tomé & Principe)
H. coronatus **Large Mindanao roundleaf bat** NE Mindanao (Philippines)
H. corynophyllus **Telefomin roundleaf bat** C New Guinea
H. coxi **Cox's roundleaf bat** Sarawak (N Borneo)
H. crumeniferus **Timor roundleaf bat** Timor (Indonesia)
H. curtus **Short-tailed roundleaf bat** Cameroon, Bioko (Equatorial Guinea)
H. cyclops **Cyclops roundleaf bat** S Sudan, Kenya to Senegal and Guinea-Bissau, Bioko (Equatorial Guinea)
H. diadema **Diadem roundleaf bat** Nicobar Is (India), Burma and Vietnam through Thailand, Malay Pen., Indonesia to New Guinea, Bismarck Arch. (Papua New Guina), Solomon Is, Philippines, NE and NC Australia
H. dinops **Fierce roundleaf bat** Peleng and Sulawesi (Indonesia), Bougainville Is (Papua New Guinea), Solomon Is
H. doriae **Borneo roundleaf bat** Borneo
H. dyacorum **Dayak roundleaf bat** Thailand Pen., Borneo
H. fuliginosus **Sooty roundleaf bat** Liberia to Cameroon, DRC and Ethiopia
H. fulvus **Fulvous roundleaf bat** Pakistan to Vietnam south to Sri Lanka
H. galeritus **Cantor's roundleaf bat** Sri Lanka, India through SE Asia to Java (Indonesia), Borneo
H. halophyllus **Thailand roundleaf bat** Thailand
H. inexpectatus **Crested roundleaf bat** N Sulawesi (Indonesia)
H. jonesi **Jones' roundleaf bat** Sierra Leone, Guinea to Mali, Burkina Faso, Nigeria
H. lamottei **Lamotte's roundleaf bat** Mt Nimba (Guinea, Liberia)
H. lankadiva **Indian roundleaf bat** S and C India, Sri Lanka
H. larvatus **Intermediate roundleaf bat** Bangladesh to Vietnam, Yunnan, Guangxi, Hainan (China), Malay Pen. to Sumatra, Java, and Sumba Is (Indonesia), Borneo
H. lekaguli **Large Asian roundleaf bat** Thailand
H. lylei **Shield-faced roundleaf bat** Burma, Thailand, Malay Pen.
H. macrobullatus **Big-eared roundleaf bat** Sulawesi, Seram, and Kangean Is (Indonesia)
H. maggietaylorae **Maggie Taylor's roundleaf bat** New Guinea, Bismarck Arch. (Papua New Guinea)
H. marisae **Aellen's roundleaf bat** Ivory Coast, Liberia, Guinea
H. megalotis **Ethiopian large-eared roundleaf bat** Saudi Arabia, NE Africa, Kenya
H. muscinus **Fly River roundleaf bat** New Guinea
H. nequam **Malayan roundleaf bat** Selangor (Malaysia)
H. obscurus **Philippine Forest roundleaf bat** Philippines
H. papua **Biak roundleaf bat** Maluku and Biak Is (Indonesia), Irian Jaya (New Guinea)
H. pomona **Pomona roundleaf bat** India to S China and Malay Pen.
H. pratti **Pratt's roundleaf bat** S China, Burma, Thailand, Vietnam, Malay Pen.
H. pygmaeus **Philippine pygmy roundleaf bat** Philippines
H. ridleyi **Ridley's roundleaf bat** Malay Pen., Borneo
H. ruber **Noack's roundleaf bat** Senegal to Ethiopia south to Angola, Zambia, Malawi, and Mozambique, Bioko (Equatorial Guinea), São Tomé and Principe
H. sabanus **Least roundleaf bat** Malay Pen., Sumatra (Indonesia), Borneo
H. schistaceus **Split roundleaf bat** S India
H. semoni **Semon's roundleaf bat** E New Guinea, N Queensland (Australia)
H. speoris **Schneider's roundleaf bat** S India, Sri Lanka
H. stenotis **Narrow-eared roundleaf bat** N Western Australia, Northern Territory, N Queensland (Australia)
H. turpis **Lesser roundleaf bat** Thailand, S Ryukyu Is (Japan)
H. wollastoni **Wollaston's roundleaf bat** W and C New Guinea
Paracoelops
P. megalotis **Vietnam leaf-nosed bat** C Vietnam
Rhinonicteris
R. aurantia **Orange leaf-nosed bat** N Western Australia, Northern Territory, NW Queensland (Australia)
Triaenops **Triple Nose-leaf Bats**
T. furculus **Trouessart's trident bat** N and W Madagascar, Aldabra Is (Seychelles)
T. persicus **Persian trident bat** Somalia, Ethiopia south to Mozambique, Congo, Yemen, Oman, SW Iran

FAMILY NOCTILIONIDAE **Bulldog Bats**
Noctilio
N. albiventris **Lesser bulldog bat** S Mexico to Guianas, Peru, E Brazil, N Argentina
N. leporinus **Greater bulldog bat** Sinaloa (S Mexico) to Guianas, Peru, S Brazil, N Argentina, Trinidad, Greater and Lesser Antilles, S Bahamas

FAMILY MORMOOPIDAE
Leaf-chinned Bats
Mormoops **Ghost-faced Bats**
M. blainvillii **Antillean ghost-faced bat or Blainville's leaf-chinned bat** Greater Antilles
M. megalophylla **Ghost-faced bat** S Texas and S Arizona (USA) south to Baja California (Mexico) to N Venezuela (incl. Margarita Is) and NW Peru, Aruba, Curacao, and Bonaire (Neth. Antilles), Trinidad
Pteronotus **Moustached Bats**
P. davyi **Davy's naked-backed bat** S Baja California, S Sonora, and Nuevo Leon (Mexico) to N Venezuela and NW Peru, Trinidad, S Lesser Antilles
P. gymnonotus **Big naked-backed bat** S Veracruz (S Mexico) to Guyana, Peru, and NE Brazil
P. macleayii **MacLeay's moustached bat** Cuba, Jamaica
P. parnellii **Parnell's moustached bat** S Sonora and S Tamaulipas (Mexico) to Venezuela (incl. Margarita Is), Guianas, Peru, Brazil, Cuba, Jamaica, Hispaniola, Puerto Rico, Trinidad and Tobago
P. personatus **Wagner's moustached bat** S Sonora and S Tamaulipas (Mexico) to Colombia, Surinam, Peru, and Brazil
P. quadridens **Sooty moustached bat** Cuba, Jamaica, Hispaniola, Puerto Rico

FAMILY PHYLLOSTOMIDAE
American Leaf-nosed Bats

SUBFAMILY PHYLLOSTOMINAE
Chrotopterus
C. auritus **Big-eared woolly bat** Veracruz (S Mexico) to Guianas, S Brazil, N Argentina
Lonchorhina **Sword-nosed Bats**
L. aurita **Tomes' sword-nosed bat** Oaxaca (Mexico) to Euador, Peru, and SE Brazil, Trinidad
L. fernandezi **Fernandez's sword-nosed bat** S Venezuela
L. marinkellei **Marinkelle's sword-nosed bat** E Colombia to Fr. Guiana
L. orinocensis **Orinoco sword-nosed bat** Colombia, Venezuela
Macrophyllum
M. macrophyllum **Long-legged bat** Tabasco (Mexico) to Peru, SE Brazil, Bolivia, NE Argentina
Macrotus **Leaf-nosed Bats**
M. californicus **California leaf-nosed bat** S California and S Nevada (USA) south to N Mexico (incl. Baja California)
M. waterhousii **Waterhouse's leaf-nosed bat** Sonora and Hidalgo (Mexico) to Guatemala, Cuba, Jamaica, Hispaniola (incl. Beata Is, Dominican Rep.), Bahamas, Cayman Is
Micronycteris **Little Big-eared Bats**
M. behnii **Behni's big-eared bat** S Peru, C Brazil
M. brachyotis **Yellow-throated big-eared bat** Oaxaca

(Mexico) to Fr. Guiana and Brazil, Trinidad
M. daviesi **Davies' big-eared bat** Costa Rica to Fr. Guiana and Peru
M. hirsuta **Hairy big-eared bat** Honduras to Fr. Guiana, Ecuador, Peru, Amazonian Brazil, Trinidad
M. megalotis **Little big-eared bat** Tamaulipas and Jalisco (Mexico) to Peru, Brazil, Bolivia, Margarita Is (Venezuela), Trinidad and Tobago, Grenada
M. minuta **White-bellied big-eared bat** Nicaragua to Guianas, Peru, S Brazil, Bolivia, Trinidad
M. nicefori **Niceforo's big-eared bat** Belize to N Colombia, Venezuela, Guianas, Peru, and Amazonian Brazil
M. pusilla **Least big-eared bat** E Colombia, NW Brazil
M. schmidtorum **Schmidts's big-eared bat** S Mexico to Venezuela, NE Peru, NE Brazil
M. sylvestris **Tri-colored big-eared bat** Nayarit and Veracruz (Mexico) to Peru and SE Brazil, Trinidad
Mimon **Hairy-nosed Bats**
M. bennettii **Golden bat** S Mexico to Colombia, Guianas, SE Brazil
M. crenulatum **Striped hairy-nosed bat** Chiapas and Campeche (Mexico) to Guianas, Ecuador, E Peru, E Brazil, Bolivia, Trinidad
Phylloderma
P. stenops **Pale-faced bat** S Mexico to Peru, SE Brazil, Bolivia
Phyllostomus **Spear-nosed Bats**
P. discolor **Pale spear-nosed bat** Oaxaca and Veracruz (S Mexico) to Guianas, Peru, SE Brazil, Paraguay, N Argentina, Margarita Is (Venezuela), Trinidad
P. elongatus **Lesser spear-nosed bat** Colombia to Guianas, Ecuador, E Peru, E Brazil, Bolivia
P. hastatus **Greater spear-nosed bat** Honduras to Guianas, Peru, E Brazil, Bolivia, Paraguay, N Argentina, Margarita is (Venezuela), Trinidad and Tobago
P. latifolius **Guianan spear-nosed bat** SE Colombia, Guianas
Tonatia **Round-eared Bats**
T. bidens **Greater round-eared bat** Chiapas (Mexico) and Belize to Brazil, Paraguay, and N Argentina, Trinidad
T. brasiliense **Pygmy round-eared bat** Veracruz (Mexico) to NE Brazil and Bolivia, Trinidad
T. carrikeri **Carriker's round-eared bat** Colombia, Venezuela, Surinam, Peru, N Brazil, Bolivia
T. evotis **Davis' round-eared bat** S Mexico, Belize, Guatemala, Honduras
T. schulzi **Schultz's round-eared bat** Guianas, N Brazil
T. silvicola **White-throated round-eared bat** Honduras to Guianas, E Brazil, Bolivia, and NE Argentina
Trachops
T. cirrhosus **Fringe-lipped bat** Oaxaca (Mexico) to Guianas, Ecuador, SE Brazil, Bolivia, Trinidad
Vampyrum
V. spectrum **Spectral bat or Linnaeus's false vampire bat** Veracruz (S Mexico) to Guianas, Ecuador, Peru, and N and SW Brazil, Trinidad

SUBFAMILY LONCHOPHYLLINAE
Lionycteris
L. spurrelli **Chestnut long-tongued bat** E Panama, Colombia, Venezuela, Guianas, Amazonian Peru and Brazil
Lonchophylla **Nectar Bats**
L. bokermanni **Bokermann's nectar bat** SE Brazil
L. dekeyseri **Dekeyser's nectar bat** E Brazil
L. handleyi **Handley's nectar bat** S Colombia, Ecuador, Peru
L. hesperia **Western nectar bat** Ecuador, N Peru
L. mordax **Goldman's nectar bat** Costa Rica to Ecuador, E Brazil
L. robusta **Orange nectar bat** Nicaragua to Venezuela and Ecuador
L. thomasi **Thomas' nectar bat** E Panama, Colombia, Venezuela, Guianas, Peru, Amazonian Brazil, Bolivia
Platalina
P. genovensium **Long-snouted bat** Peru

SUBFAMILY BRACHYPHYLLINAE
Brachyphylla **West Indian Fruit-eating Bats**
B. cavernarum **Antillean fruit-eating bat** Puerto Rico, Virgin Is, Lesser Antilles south to Barbados
B. nana **Cuban fruit-eating bat** Cuba, Hispaniola, Cayman Is, Caicos Is (W. Indies)

SUBFAMILY PHYLLONYCTERINAE
Erophylla
E. sezekorni **Buffy flower bat** Cuba, Jamaica, Hispaniola, Puerto Rico, Bahamas, Cayman Is
Phyllonycteris **Smooth-toothed Flower Bats**
P. aphylla **Jamaican flower bat** Jamaica
P. poeyi **Cuban flower bat** Cuba (Juventud Is), Hispaniola

SUBFAMILY GLOSSOPHAGINAE
Anoura **Tailless Bats**
A. caudifer **Tailed tailless bat** Colombia, Venezuela, Guianas, Ecuador, Peru, Brazil, Bolivia, NW Argentina
A. cultrata **Handley's tailless bat** Costa Rica, Panama, Colombia, Venezuela, Ecuador, Peru, Bolivia
A. geoffroyi **Geoffroy's tailless bat** Sinaloa and Tamaulipas (Mexico) south to Ecuador and Fr. Guiana, Peru, SE Brazil, Bolivia, Trinidad, Grenada
A. latidens **Broad-toothed tailless bat** Colombia, Venezuela, Peru
Choeroniscus **Long-tailed Bats**
C. godmani **Godman's long-tailed bat** Sinaloa (Mexico) to Colombia, Venezuela, Guyana, Surinam
C. intermedius **Intermediate long-tailed bat** Guianas, Peru, Amazonian Brazil, Trinidad
C. minor **Lesser long-tailed bat** C Colombia, Venezuela, Guianas, Ecuador, Peru, N Brazil, Bolivia
C. periosus **Greater long-tailed bat** W Colombia, W Ecuador
Choeronycteris
C. mexicana **Mexican long-tongued bat** SW USA to Honduras and El Salvador
Glossophaga **Long-tongued Bats**
G. commissarisi **Commissaris' long-tongued bat** Sinaloa (Mexico) to Panama, SE Colombia, E Ecuador, E Peru, NW Brazil
G. leachii **Gray long-tongued bat** Costa Rica to S Mexico
G. longirostris **Miller's long-tongued bat** Colombia, Venezuela (incl. Margarita Is), Guyana, N Ecuador, N Brazil, Aruba, Curacao, Bonaire (Neth. Antilles), Trinidad and Tobago, Grenada, St Vincent (Lesser Antilles)
G. morenoi **Western long-tongued bat** Chiapas to Tlaxala (S Mexico)
G. soricina **Pallas' long-tongued bat** S Mexico (incl. Tres Marias Is) to Guianas, Peru, SE Brazil, N Argentina, Margarita Is (Venezuela), Trinidad, Grenada
Hylonycteris
H. underwoodi **Underwood's long-tongued bat** Nayarit and Veracruz (S Mexico) to W Panama
Leptonycteris **Long-nosed Bats**
L. curasoae **Southern long-nosed bat** Aruba, Curacao, and Bonaire (Neth. Antilles)
L. nivalis **Mexican long-nosed bat** SE Arizona, W Texas (SW USA) to S Mexico
Lichonycteris
L. obscura **Dark long-tongued bat** Guatemala, Belize to Amazonian Brazil, Bolivia
Monophyllus **Single Leaf Bats**
M. plethodon **Insular single leaf bat** Anguilla to St Vincent to Barbados (Lesser Antilles)
M. redmani **Leach's single leaf bat** Cuba, Jamaica, Hispaniola, Puerto Rico, S Bahamas
Musonycteris
M. harrisoni **Banana bat or Colima long-nosed bat** Jalisco to Guerrero (S Mexico)
Scleronycteris
S. ega **Ega long-tongued bat** S Venezuela, Amazonian Brazil

SUBFAMILY CAROLLIINAE
Carollia **Short-tailed Bats**
C. brevicauda **Silky short-tailed bat** San Luis Potosi (Mexico) to Peru, E Brazil, Bolivia
C. castanea **Chestnut short-tailed bat** Honduras to Venezuela, Peru, W Brazil, Bolivia
C. perspicillata **Seba's short-tailed bat** S Mexico to Guianas, Peru, SE Brazil, Bolivia, Paraguay, Trinidad and Tobago, Grenada
C. subrufa **Gray short-tailed bat** Jalisco (SW Mexico) to NW Nicaragua, Guyana
Rhinophylla **Little Fruit Bats**
R. alethina **Hairy little fruit bat** W Colombia, W Ecuador
R. fischerae **Fischer's little fruit bat** SE Colombia, Ecuador, Peru, Amazonian Brazil
R. pumilio **Dwarf little fruit bat** Colombia, Guianas, Ecuador, Peru, Brazil, Bolivia

SUBFAMILY STENODERMATINAE
Ametrida
A. centurio **Little white-shouldered bat** Panama, Venezuela, Guianas, Amazonian Brazil, Bonaire (Neth. Antilles), Trinidad
Ardops
A. nichollsi **Tree bat** St Eustatius to St Vincent (Lesser Antilles)
Ariteus
A. flavescens **Jamaican fig-eating bat** Jamaica
Artibeus **Fruit-eating Bats**
A. amplus **Large fruit-eating bat** N Colombia, Venezuela
A. anderseni **Andersen's fruit-eating bat** Ecuador, Peru, W Brazil, Bolivia
A. aztecus **Aztec fruit-eating bat** S Mexico, Guatemala, Honduras, Costa Rica, W Panama
A. cinereus **Gervais' fruit-eating bat** Venezuela, Guianas, N Brazil
A. concolor **Brown fruit-eating bat** Colombia, Venezuela, Guianas, Peru, N Brazil
A. fimbriatus **Fringed fruit-eating bat** S Brazil, Paraguay
A. fraterculus **Fraternal fruit-eating bat** Ecuador, Peru
A. glaucus **Silver fruit-eating bat** S Mexico to S Brazil, Bolivia, Trinidad and Tobago, Grenada
A. hartii **Velvety fruit-eating bat** Arizona (SW USA), Jalisco and Tamaulipas (Mexico) to Venezuela and Bolivia, Trinidad
A. hirsutus **Hairy fruit-eating bat** Sonora to Guerrero (Mexico)
A. inopinatus **Honduran fruit-eating bat** El Salvador, Honduras, Nicaragua
A. jamaicensis **Jamaican fruit-eating bat** Sinaloa and Tamaulipas (Mexico) to Venezuela and Ecuador, Trinidad and Tobago, Greater and Lesser Antilles
A. lituratus **Great fruit-eating bat** Sinaloa and Tamaulipas (Mexico) to S Brazil, Bolivia, N Argentina, Trinidad and Tobago, S Lesser Antilles, Tres Marias Is (Mexico)
A. obscurus **Dark fruit-eating bat** Colombia, Venezuela, Guianas, Ecuador, Peru, Brazil, Bolivia
A. phaeotis **Pygmy fruit-eating bat** Sinaloa and Veracruz (S Mexico) to Guyana and Ecuador
A. planirostris **Flat-faced fruit-eating bat** Colombia, Venezuela to N Argentina and E Brazil
A. toltecus **Toltec fruit-eating bat** S Mexico to Panama
Centurio
C. senex **Wrinkle-faced bat** Tamaulipas and Sinaloa (Mexico) to Venezuela, Trinidad and Tobago
Chiroderma **Big-eyed Bats**
C. doriae **Brazilian big-eyed bat** SE Brazil
C. improvisum **Guadeloupe big-eyed bat** Guadelope and Montserrat (Lesser Antilles)
C. salvini **Salvin's big-eyed bat** N Mexico to Venezuela and Bolivia
C. trinitatum **Little big-eyed bat** Panama to Peru, Amazonian Brazil, Bolivia, Trinidad
C. villosum **Hairy big-eyed bat** Hidalgo (Mexico) to Peru, S Brazil, Bolivia, Trinidad and Tobago
Ectophylla
E. alba **White bat** Honduras to W Panama, W Colombia
Mesophylla
M. macconnelli **MacConnell's bat** Costa Rica to Peru, Amazonian Brazil, Bolivia, Trinidad
Phyllops
P. falcatus **Cuban fig-eating bat** Cuba, Hispaniola
Platyrrhinus **Broad-nosed Bats**
P. aurarius **Eldorado broad-nosed bat** Colombia, S Venezuela, Surinam
P. brachycephalus **Short-headed broad-nosed bat** Colombia to Guianas, Ecuador, Peru, N Brazil, Bolivia
P. chocoensis **Choco broad-nosed bat** W Colombia
P. dorsalis **Thomas' broad-nosed bat** Panama to Peru and Bolivia
P. helleri **Heller's broad-nosed bat** Oaxaca and Veracruz (S Mexico) to Peru, Amazonian Brazil, Bolivia, Trinidad
P. infuscus **Buffy broad-nosed bat** Colombia to Peru, NW Brazil, Bolivia
P. lineatus **White-lined broad-nosed bat** Fr. Guiana, Surinam, Colombia to Peru, S and E Brazil, Bolivia, N Argentina, Uruguay
P. recifinus **Recife broad-nosed bat** E Brazil
P. umbratus **Shadowy broad-nosed bat** Panama, N and W Colombia, N Venezuela
P. vittatus **Greater broad-nosed bat** Costa Rica to Venezuela, Peru, Bolivia
Pygoderma
P. bilabiatum **Ipanema bat** Surinam, S Brazil, Bolivia, Paraguay, N Argentina

Sphaeronycteris
S. toxophyllum **Visored bat** Colombia to Venezuela, Peru, Amazonian Brazil, Bolivia
Stenoderma
S. rufum **Red fruit bat** Puerto Rico, Virgin Is (USA)
Sturnira **Yellow-shouldered Bats**
S. aratathomasi **Aratathomas' yellow-shouldered bat** Colombia, NW Venezuela, Ecuador, Peru
S. bidens **Bidentate yellow-shouldered bat** Colombia, Venezuela, Ecuador, Peru, poss. Amazonian Brazil
S. bogotensis **Bogota yellow-shouldered bat** Colombia, W Venezuela, Ecuador, Peru, Bolivia, NW Argentina
S. erythromos **Hairy yellow-shouldered bat** Venezuela to Bolivia
S. lilium **Little yellow-shouldered bat** Sonora and Tamaulipas (Mexico) to E Brazil, N Argentina, Uruguay, Trinidad and Tobago, Lesser Antilles
S. ludovici **Highland yellow-shouldered bat** Sonora and Tamaulipas (Mexico) to Guyana and Ecuador
S. luisi **Luis' yellow-shouldered bat** Costa Rica to Ecuador, NW Peru
S. magna **Greater yellow-shouldered bat** Colombia, Ecuador, Peru, Bolivia
S. mordax **Talamancan yellow-shouldered bat** Costa Rica, Panama
S. nana **Lesser yellow-shouldered bat** S Peru
S. thomasi **Thomas' yellow-shouldered bat** Guadeloupe (Lesser Antilles)
S. tildae **Tilda yellow-shouldered bat** Colombia, Venezuela, Guianas, Ecuador, Peru, Brazil, Bolivia, Trinidad
Uroderma **Tent-making Bats**
U. bilobatum **Tent-making bat** Veracruz and Oaxaca (Mexico) to Peru, Brazil, Bolivia, Trinidad
U. magnirostrum **Brown tent-making bat** Michoacan (Mexico) to Venezuela, Peru, Brazil, Bolivia
Vampyressa **Yellow-eared Bats**
V. bidens **Bidentate yellow-eared bat** Colombia and Guianas to Peru, Amazonian Brazil, N Bolivia
V. brocki **Brock's yellow-eared bat** Colombia, Guyana, Surinam, Amazonian Brazil
V. melissa **Melissa's yellow-eared bat** S Colombia, Fr. Guiana, Peru
V. nymphaea **Striped yellow-eared bat** Nicaragua to W Ecuador
V. pusilla **Little yellow-eared bat** Oaxaca and Veracruz (Mexico) to Guianas, Brazil, Bolivia, Paraguay
Vampyrodes
V. caraccioli **Great stripe-faced bat** Oaxaca (Mexico) to Peru, N Brazil, Bolivia, Trinidad and Tobago

SUBFAMILY DESMODONTINAE
Desmodus
D. rotundus **Common vampire bat** Mexico to N Chile, N Argentina, Uruguay, Margarita Is (Venezuela), Trinidad
Diaemus
D. youngi **White-winged vampire bat** Tamaulipas (Mexico) to E Brazil and N Argentina, Margarita Is (Venezuela), Trinidad
Diphylla
D. ecaudata **Hairy-legged vampire bat** S Texas (USA) to Venezuela, Peru, E Brazil, Bolivia

FAMILY NATALIDAE
Funnel-eared Bats
Natalus
N. lepidus **Gervais' funnel-eared bat** Cuba, Bahamas
N. micropus **Cuban funnel-eared bat** Cuba, Jamaica, Hispaniola, Providencia Is (Colombia)
N. stramineus **Mexican funnel-eared bat** Mexico (incl. S Baja California) to Brazil, Lesser Antilles, Jamaica, Hispaniola
N. tumidifrons **Bahaman funnel-eared bat** Bahamas
N. tumidirostris **Trinidadian funnel-eared bat** Colombia, Venezuela, Surinam, Curacao and Bonaire (Neth. Antilles), Trinidad and Tobago

FAMILY FURIPTERIDAE **Thumbless Bats**
Amorphochilus
A. schnablii **Smoky bat** W Ecuador (incl. Puna Is), W Peru, N Chile
Furipterus
F. horrens **Thumbless bat** Costa Rica to Peru and E Brazil, Trinidad

FAMILY THYROPTERIDAE
Disk-winged Bats
Thyroptera
T. discifera **Peter's disk-winged bat** Nicaragua, Panama, Colombia to Guianas, Peru, Brazil, Bolivia

T. tricolor **Spix's disk-winged bat** Veracruz (Mexico) to Guianas, Peru, E Brazil, Bolivia, Trinidad

FAMILY MYZOPODIDAE

Myzopoda
M. aurita **Sucker-footed bat** Madagascar

FAMILY VESPERTILIONIDAE **Vesper Bats**

SUBFAMILY KERIVOULINAE

Kerivoula **Woolly Bats**
K. aerosa **Dubious trumpet-eared bat** Poss. South Africa
K. africana **Tanzanian woolly bat** Tanzania
K. agnella **St. Aignan's trumpet-eared bat** Papua New Guinea
K. argentata **Damara woolly bat** S Kenya, Uganda to Angola, Namibia, E South Africa
K. atrox **Groove-toothed bat** S Thailand, Malay Pen., Sumatra (Indonesia), Borneo
K. cuprosa **Copper woolly bat** Kenya, N DRC, S Cameroon
K. eriophora **Ethiopian woolly bat** Ethiopia
K. flora **Flores woolly bat** Lesser Sunda Is (Indonesia), Borneo
K. hardwickei **Hardwicke's woolly bat** India, Sri Lanka, Burma, China, Thailand, Malay Pen., Indonesia, Philippines
K. intermedia **Small woolly bat** Malay Pen., Borneo
K. jagori **Peters' trumpet-eared bat** Indonesia, Borneo, Philippines
K. lanosa **Lesser woolly bat** Liberia to Ethiopia to South Africa
K. minuta **Least woolly bat** S Thailand, Malay Pen., Borneo
K. muscina **Fly River trumpet-eared bat** C New Guinea
K. myrella **Bismarck trumpet-eared bat** Lesser Sunda Is (Indonesia), Bismarck Arch. (Papua New Guinea)
K. papillosa **Papillose woolly bat** NE India, Vietnam, Malay Pen., Sumatra, Java, and Sulawesi (Indonesia), Borneo
K. papuensis **Golden-tipped bat** SE New Guinea, Queensland, New South Wales (Australia)
K. pellucida **Clear-winged woolly bat** Malay Pen., Sumatra and Java (Indonesia), Borneo, Philippines
K. phalaena **Spurrell's woolly bat** Liberia, Ghana, Cameroon, Congo, DRC
K. picta **Painted bat** India, Sri Lanka to S China, Vietnam, Malay Pen., Indonesia, Borneo
K. smithi **Smith's woolly bat** Liberia, Ivory Coast, Nigeria, Cameroon, N and E DRC, Kenya
K. whiteheadi **Whitehead's woolly bat** S Thailand, Malay Pen., Borneo, Philippines

SUBFAMILY VESPERTILIONINAE

Antrozous **Pallid Bats**
A. dubiaquercus **Van Gelder's bat** Mexico (incl. Tres Marias Is), Belize, Honduras, Costa Rica
A. pallidus **Pallid bat** Mexico to Kansas (USA) and British Columbia (SW Canada), Cuba
Barbastella **Barbastelle's**
B. barbastellus **Western barbastelle** England and W Europe to Caucasus, Ukraine, Turkey, Morocco, Mediterranean islands, Canarias Is
B. leucomelas **Eastern barbastelle** Caucasus to Pamir (Tajikistan), N Iran, Afghanistan, India, W China, Honshu and Hokkaido (Japan), Egypt, N Ethiopia
Chalinolobus **Wattled Bats**
C. alboguttatus **Allen's striped bat** DRC, Cameroon
C. argentatus **Silvered bat** Cameroon to Kenya, Tanzania south to Angola
C. beatrix **Beatrix's bat** Ivory Coast, Congo, Kenya
C. dwyeri **Large-eared pied bat** New South Wales, adj. Queensland (Australia)
C. egeria **Bibundi bat** Cameroon, Uganda
C. gleni **Glen's wattled bat** Cameroon, Uganda
C. gouldii **Gould's wattled bat** Australia, Tasmania, New Caledonia, Norfolk Is (Australia)
C. kenyacola **Kenyan wattled bat** Kenya
C. morio **Chocolate wattled bat** S Australia, Tasmania
C. nigrogriseus **Hoary wattled bat** SE New Guinea, N and E Australia
C. picatus **Little pied bat** C and S Queensland, NW New South Wales, South Australia (Australia)
C. poensis **Abo bat** Senegal to Uganda, Bioko (Equatorial Guinea)
C. superbus **Pied bat** Ivory Coast, Ghana, NE DRC
C. tuberculatus **Long-tailed wattled bat** New Zealand
C. variegatus **Butterfly bat** Senegal to Somalia to South Africa
Eptesicus **Serotines**
E. baverstocki **Inland Forest bat** C and S Australia
E. bobrinskii **Bobrinski's serotine** N Caucasus, Turkmenistan, Uzbekistan, Kazakhstan
E. bottae **Botta's serotine** Egypt, Turkey and Yemen east to Pakistan and Mongolia
E. brasiliensis **Brazilian brown bat** Veracruz (S Mexico) to N Argentina, Uruguay, Trinidad and Tobago
E. brunneus **Dark-brown serotine** Liberia to DRC
E. capensis **Cape serotine** Guinea to Ethiopia to South Africa, Madagascar
E. demissus **Surat serotine** Thailand Pen.
E. diminutus **Diminutive serotine** Venezuela, E Brazil, Paraguay, N Argentina, Uruguay
E. douglasorum **Yellow-lipped bat** Kimberley (N Western Australia)
E. flavescens **Yellow serotine** Angola, Burundi
E. floweri **Horn-skinned bat** Sudan, Mali
E. furinalis **Argentine brown-bat** Jalisco and Tamaulipas (Mexico) to Guianas, Brazil, N Argentina
E. fuscus **Big brown bat** S Canada to Colombia, N Brazil, Greater and Lesser Antilles, Bahamas
E. guadeloupensis **Guadeloupe big brown bat** Guadeloupe (Lesser Antilles)
E. guineensis **Tiny serotine** Senegal, Guinea to Ethiopia and NE DRC
E. hottentotus **Long-tailed house bat** South Africa to Angola and Kenya
E. innoxius **Harmless serotine** W Ecuador, NW Peru, Puna Is (Ecuador)
E. kobayashii **Kobayashi's serotine** Korea
E. melckorum **Melck's house bat** S South Africa, Zambia, Mozambique, Tanzania
E. nasutus **Sind bat** Arabia, Iraq, Iran, Afghanistan, Pakistan
E. nilssoni **Northern bat** W and E Europe to E Siberia and Sakhalin Is (Russia) and NW China, Scandinavia south to Bulgaria, Iraq, N Iran, Pamirs (Tajikistan), W China (excl. Xizang), Nepal, Honshu and Hokkaido (Japan)
E. pachyotis **Thick-eared bat** Assam (India), Burma, N Thailand
E. platyops **Lagos serotine** Nigeria, Senegal, Bioko (Equatorial Guinea)
E. pumilus **Eastern Forest bat** N Western Australia, Northern Territory, Queensland, New South Wales, South Australia (Australia)
E. regulus **Southern Forest bat** SW and SE Australia, Tasmania
E. rendalli **Rendall's serotine** Gambia to Somalia to Botswana and Mozambique
E. sagittula **Large Forest bat** SE Australia, Tasmania, Lord Howe Is (Australia)
E. serotinus **Serotine** N and SubSaharan Africa, W Europe to S Russia to Himalayas, China and Thailand to N Korea, Taiwan
E. somalicus **Somali serotine** Guinea-Bissau to Somalia to Namibia and South Africa, Madagascar
E. tatei **Sombre bat** NE India
E. tenuipinnis **White-winged serotine** Senegal to Kenya to DRC and Angola
E. vulturnus **Little Forest bat** SE Australia, Tasmania
Euderma
E. maculatum **Spotted or Pinto bat** SW Canada, Montana (NW USA) south to Queretaro (Mexico)
Eudiscopus
E. denticulus **Disc-footed bat** C Burma, Laos
Glischropus **Thick-thumbed Bats**
G. javanus **Javan thick-thumbed bat** Java (Indonesia)
G. tylopus **Common thick-thumbed bat** Burma, Thailand, Malay Pen., Sumatra and Maluku Is (Indonesia), Borneo, SW Philippines
Hesperoptenus **False Serotines**
H. blanfordi **Blanford's bat** Burma, Thailand, Malay Pen., Borneo
H. doriae **False serotine bat** Malay Pen., Borneo
H. gaskelli **Gaskell's false serotine** Sulawesi (Indonesia)
H. tickelli **Tickell's bat** India (incl. Andaman Is), Sri Lanka, Nepal, Bhutan, Burma, Thailand
H. tomesi **Large false serotine** Malay Pen., Borneo
Histiotus **Big-eared Brown Bats**
H. alienus **Strange big-eared brown bat** SE Brazil, Uruguay
H. macrotus **Big-eared brown bat** S Peru, S Bolivia, Chile, NW Argentina
H. montanus **Small big-eared brown bat** Colombia, Venezuela, Ecuador, Peru, W Bolivia, Chile, Argentina, Uruguay
H. velatus **Tropical big-eared brown bat** E Brazil, Paraguay
Ia
I. io **Great evening bat** NE India, S China, Vietnam, Laos, Thailand
Idionycteris
I. phyllotis **Allen's big-eared bat** Distrito Federal (Mexico) to S Utah and S Nevada (SW USA)
Laephotis **African Long-eared Bats**
L. angolensis **Angolan long-eared bat** Angola, DRC
L. botswanae **Botswanan long-eared bat** N South Africa, Zimbabwe, Botswana, Zambia, Malawi, DRC
L. namibensis **Namib long-eared bat** Namibia
L. wintoni **De Winton's long-eared bat** Ethiopia, Kenya, SW South Africa
Lasionycteris
L. noctivagans **Silver-haired bat** S Canada, USA (incl. S Alaska), NE Mexico, Bermuda
Lasiurus **Hairy-tailed Bats**
L. borealis **Red bat** C Canada to Brazil, Uruguay, Chile, Argentina, Cuba, Jamaica, Hispaniola, Puerto Rico, Bermuda, Bahamas, Trinidad and Tobago, Galapagos (Ecuador)
L. castaneus **Tacarcuna bat** Costa Rica, Panama
L. cinereus **Hoary bat** Canada, USA, Mexico, Guatemala, Colombia, Venezuela to C Chile, C Argentina, Uruguay, Hawaii (USA), Bermuda, Galapagos (Ecuador)
L. ega **Southern yellow bat** SW USA to Brazil, Argentina, Uruguay, Trinidad
L. egregius **Big red bat** Panama, Fr. Guiana, Brazil
L. intermedius **Northern yellow bat** E and S USA to Sinaloa (Mexico) to Honduras, Cuba
L. seminolus **Seminole bat** S and E USA, Cuba
Mimetillus
M. moloneyi **Moloney's flat-headed bat** Sierra Leone to Ethiopia south to Tanzania, Zambia, Angola, Bioko (Equatorial Guinea)
Myotis **Little Brown Bats**
M. abei **Sakhalin myotis** Sakhalin (Russia)
M. adversus **Large-footed bat** Taiwan and Malay Pen. south to New Guinea, Bismarck Arch. (Papua New Guinea), Solomon Is, Vanuatu, N and E coastal Australia
M. aelleni **Southern myotis** SW Argentina
M. albescens **Silver-tipped myotis** Veracruz (Mexico) to N Argentina, Uruguay
M. altarium **Sichuan myotis** Sichuan (China), Thailand
M. annectans **Hairy-faced bat** NE India to Thailand
M. atacamensis **Atacama myotis** S Peru, N Chile
M. auriculus **Southwestern myotis** Arizona and New Mexico (SW USA) to Jalisco and Veracruz (Mexico)
M. australis **Australian myotis** New South Wales (Australia)
M. austroriparius **Southeastern myotis** SE USA north to Indiana
M. bechsteini **Bechstein's bat** Europe to Caucasus, Iran
M. blythii **Lesser mouse-eared bat** Mediterranean Europe and NW Africa, Israel, Crimea (Ukraine), Turkey and Caucasus to Kyrgyzstan, Afghanistan, and Himalayas, Inner Mongolia and Shaanxi (China)
M. bocagei **Rufous mouse-eared bat** Senegal to S Yemen south to Angola, Zambia, N South Africa
M. bombinus **Far eastern myotis** NE China, SE Siberia (Russia), Korea, Japan
M. brandti **Brandt's bat** Britain, Spain, Greece to Kazakhstan, Mongolia, Korea, Japan, Ussuri region and E Siberia (incl. Sakhalin Is, Kamchatka Pen., Kurile Is)
M. californicus **California myotis** S Alaska (USA) to Baja California (Mexico) and Guatemala
M. capaccinii **Long-fingered myotis** Mediterranean Europe and NW Africa, Israel, Iraq, Iran, Turkey, Uzbekistan
M. chiloensis **Chilean myotis** C and S Chile
M. chinensis **Large myotis** Sichuan and Yunnan to Jiangsu (China), Hong Kong (China), Thailand
M. cobanensis **Guatemalan myotis** C Guatemala
M. dasycneme **Pond bat** France and Sweden east to Yenisei R (Russia) south to Ukraine and Kazakhstan, Manchuria (China)
M. daubentoni **Daubenton's bat** Britain, Scandinavia and Europe east to Kamchatka Pen., Vladivostok, Sakhalin and Kurile Is (Russia), Assam (India), Manchuria, E and S China, Korea, Japan
M. dominicensis **Dominican myotis** N Lesser Antilles
M. elegans **Elegant myotis** San Luis Potosi (N Mexico) to Costa Rica
M. emarginatus **Geoffroy's bat** N Africa (Morocco to Tunisia), Israel, Lebanon, S Europe to Netherlands and Poland east to Caucasus, Iran, Afghanistan, Uzbekistan
M. evotis **Long-eared myotis** SW Canada to New Mexico (USA) and Baja California (Mexico)
M. findleyi **Findley's myotis** Tres Marias Is (Mexico)
M. formosus **Hodgson's bat** Afghanistan to Guizhou, Giangsu, Fujian (China), Taiwan, Korea, Tsushima Is (Japan), Sumatra, Java, Sulawesi, Bali (Indonesia), Philippines
M. fortidens **Cinnamon myotis** Sonora and Veracruz (Mexico) to Guatemala
M. frater **Fraternal myotis** Afghanistan, Uzbekistan and S Sbieria to SE Russia, Heilongjiang (NE China), SE China, Korea, Japan
M. goudoti **Malagasy mouse-eared bat** Madagascar, Comoros
M. grisescens **Gray myotis** C and SE USA
M. hasseltii **Lesser large-footed bat** Sri Lanka, Burma, Thailand, Cambodia, Vietnam, Malay Pen., Sumatra, Java, Mentawai Is, and Riau Arch. (Indonesia), Borneo
M. horsfieldii **Horsfield's bat** India, Sri lanka, SE China, Thailand, Malay Pen., Java, Bali, and Sulawesi (Indonesia), Borneo, Philippines
M. hosonoi **Hosono's bat** Honshu (Japan)
M. ikonnikovi **Ikonnikov's bat** Mongolia, NE China, Ussuris (Russia) and N Korea to L. Baikal (Russia), Sakhalin Is (Russia), Hokkaido Is (Japan)
M. insularum **Insular myotis** Samoa
M. keaysi **Hairy-legged myotis** Tamaulipas (Mexico) to Venezuela and N Argentina, Trinidad
M. keenii **Keen's myotis** Alaska (USA) to W Washington (USA), Mackenzie to Prince Edward Is (Canada) south to Florida (USA)
M. leibii **Eastern small-footed myotis** S British Columbia (Canada) south to Mexico (incl. Baja California) east to Maine (USA) and S Quebec (Canada)
M. lesueuri **Lesueur's hairy bat** SW South Africa
M. levis **Yellowish myotis** SE Brazil, Bolivia, Argentina, Uruguay
M. longipes **Kashmir cave bat** Afghanistan, Kashmir
M. lucifugus **Little brown myotis or bat** Alaska (USA) to Labrador and Newfoundland (Canada) south to Distrito Federal (Mexico)
M. macrodactylus **Big-footed myotis** SE Russia, S China, Japan, Kurile Is (Russia)
M. macrotarsus **Pallid large-footed myotis** N Borneo, Philippines
M. martiniquensis **Schwartz's myotis** Lesser Antilles
M. milleri **Miller's myotis** N Baja California (Mexico)
M. montivagus **Burmese whiskered bat** India, Burma, Yunnan to Fujian (China), Malay Pen., Borneo
M. morrisi **Morris' bat** Ethiopia, Nigeria
M. muricola **Whiskered myotis** Afghanistan to Taiwan and New Guinea
M. myotis **Mouse-eared bat** S England, C and S Europe east to Ukraine, Turkey, Lebanon, Israel, Mediterranean islands
M. mystacinus **Whiskered bat** Ireland and Scandinavia to N China south to Morocco, Iran, NW Himalayas, S China
M. nattereri **Natterer's bat** NW Africa, Europe (excl. Scandinavia), Crimea (Ukraine) and Caucasus to Turkmenistan, Turkey, Israel, Iraq
M. nesopolus **Curacao myotis** NE Venezuela, Curacao and Bonaire (Neth. Antilles)
M. nigricans **Black myotis** Nayarit and Tamaulipas (Mexico) to Peru, S Brazil, N Argentina, Trinidad and Tobago, Grenada
M. oreias **Singapore whiskered bat** Singapore
M. oxyotus **Montane myotis** Costa Rica, Panama, Venezuela to Bolivia
M. ozensis **Honshu myotis** Honshu (Japan)
M. peninsularis **Peninsular myotis** S Baja California (Mexico)
M. pequinius **Peking myotis** Hopeh, Shandong, Henan, Jiangsu (China)
M. planiceps **Flat-headed myotis** Mexico
M. pruinosus **Frosted myotis** Honshu and Shikoku (Japan)
M. ricketti **Rickett's big-footed bat** Fujian, Anhui, Jiangsu, Shandong, Yunnan, Hong Kong (China)
M. ridleyi **Ridley's bat** Malay Pen., Sumatra (Indonesia), Borneo
M. riparius **Riparian myotis** Honduras to E Brazil and Uruguay, Trinidad
M. rosseti **Thick-thumbed myotis** Cambodia, Thailand
M. ruber **Red myotis** SE Brazil, Paraguay, NE Argentina
M. schaubi **Schaub's myotis** Extinct: formerly Caucasus, W Iran
M. scotti **Scott's mouse-eared bat** Ethiopia

M. seabrai **Angola hairy bat** NW South Africa, Namibia, Angola
M. sicarius **Mandelli's mouse-eared bat** Sikkim (NE India)
M. siligorensis **Himalayan whiskered bat** N India to S China and Vietnam to Malay Pen., Borneo
M. simus **Velvety myotis** Colombia, Ecuador, Peru, N Brazil, Bolivia, Paraguay, NE Argentina
M. sodalis **Indiana bat** E USA west to Wisconsin and Oklahoma
M. stalkeri **Kei myotis** Maluku (Indonesia)
M. thysanodes **Fringed myotis** Chiapas (Mexico) to SW South Dakota (USA) and S British Columbia (Canada)
M. tricolor **Cape hairy bat** Ethiopia and DRC south to South Africa
M. velifer **Cave myotis** C and SW USA to Honduras
M. vivesi **Fish-eating bat** Coastal Sonora and Baja California (Mexico)
M. volans **Long-legged myotis** Alaska (USA) to Mexico (incl. Baja California) east to South Dakota (USA) and C Alberta (Canada)
M. welwitschii **Welwitch's bat** Ethiopia to South Africa
M. yesoensis **Yoshiyuki's myotis** Hokkaido (Japan)
M. yumanensis **Yuma myotis** Mexico (incl. Baja California) north to British Columbia (Canada) east to Montana and W Texas (USA)
Nyctalus Noctule Bats
N. aviator **Birdlike noctule** E China, Korea, Japan
N. azoreum **Azores noctule** Azores (Portugal)
N. lasiopterus **Giant noctule** W Europe to Urals and Caucasus, Turkey, Kazakhstan, Iran, Morocco, Libya
N. leisleri **Lesser or Leisler's noctule** Britain, Ireland, W Europe to Urals and Caucasus, E Afghanistan, W Himalayas, NW Africa, Madeira Is (Morocco), Azores (Portugal)
N. montanus **Mountain noctule** Afghanistan, Pakistan, N India, Nepal
N. noctula **Noctule** Europe to Urals and Caucasus, Morocco, Algeria, Israel to SE Turkey, W Turkmenistan to SW Siberia (Russia), Himalayas, China, Taiwan, Honshu (Japan), Malay Pen.
Nycticeius Broad-nosed Bats
N. balstoni **Western broad-nosed bat** Australia
N. greyii **Little broad-nosed bat** Australia
N. humeralis **Evening or Twilight bat** N Veracruz (Mexico) to C, E and SE USA
N. rueppellii **Rüppell's broad-nosed bat** E Queensland and E New South Wales (Australia)
N. sanborni **Northern broad-nosed bat** SE New Guinea, NE Queensland, Northern Territory, N Western Australia (Australia)
N. schlieffeni **Schlieffen's bat** SW Arabia, Mauritania and Ghana to NE Africa and Tanzania, Mozambique, Namibia, Botswana, South Africa
Nyctophilus Long-eared Bats
N. arnhemensis **Northern long-eared bat** N Australia
N. geoffroyi **Lesser long-eared bat** Australia (excl. NE), Tasmania
N. gouldi **Gould's long-eared bat** New Guinea, N and E Queensland, Victoria, E New South Wales, N Northern Territory, N and W Western Australia (Australia)
N. heran **Sunda long-eared bat** Lesser Sunda Is (Indonesia)
N. microdon **Small-toothed long-eared bat** EC New Guinea
N. microtis **New Guinea long-eared bat** E New Guinea
N. timoriensis **Greater long-eared bat** E New Guinea, Timor (Indonesia), Australia (incl. Tasmania)
N. walkeri **Pygmy long-eared bat** Northern Territory and N Western Australia (Australia)
Otonycteris
O. hemprichi **Desert or Hemprich's long-eared bat** Morocco and Niger to Egypt and Arabia to Tajikistan, Afghanistan, and Kashmir
Pharotis
P. imogene **New Guinea big-eared bat** SE New Guinea
Philetor
P. brachypterus **Rohu's bat** Nepal, Malay Pen., Sumatra, Java, Sulawesi (Indonesia), Borneo, Philippines, New Guinea, New Britain Is (Bismarck Arch., Papua New Guinea)
Pipistrellus Pipistrelles
P. aegyptius **Egyptian pipistrelle** Egypt, N Sudan, Libya, Algeria, Burkina Faso
P. aero **Mt. Gargyes pipistrelle** NW Kenya
P. affinis **Chocolate pipistrelle** N India, NE Burma, Yunnan (China)
P. anchietai **Anchieta's pipistrelle** S DRC, Angola, Zambia
P. anthonyi **Anthony's pipistrelle** Changyinku (Burma)
P. arabicus **Arabian pipistrelle** Oman
P. ariel **Desert pipistrelle** Egypt, N Sudan
P. babu **Himalayan pipistrelle** Afghanistan, Pakistan, N India, Nepal, Bhutan, Burma, SW China
P. bodenheimeri **Bodenheimer's pipistrelle** Israel, S Yemen, Oman
P. cadornae **Cadorna's pipistrelle** NE India, Burma, Thailand
P. ceylonicus **Kelaart's pipistrelle** Pakistan, India, Sri Lanka, Burma, Guangxi and Hainan (China), Vietnam, Borneo
P. circumdatus **Black gilded pipistrelle** NE India, Burma, SW China, Malay Pen., Java (Indonesia)
P. coromandra **Indian pipistrelle** Afghanistan, Pakistan, India, Sri Lanka, Nepal, Bhutan, Burma, S China, Vietnam, Thailand, Nicobar Is (India)
P. crassulus **Broad-headed pipistrelle** S Sudan, Cameroon, DRC, Angola
P. cuprosus **Coppery pipistrelle** Borneo
P. dormeri **Dormer's pipistrelle** Pakistan, NW, S and E India
P. eisentrauti **Eisentraut's pipistrelle** Liberia to Kenya and Somalia
P. endoi **Endo's pipistrelle** Honshu (Japan)
P. hesperus **Western pipistrelle** Washington to SW Oklahoma (USA), Mexico (incl. Baja California)
P. imbricatus **Brown pipistrelle** Java, Bali, Lesser Sunda Is, Kangean Is (Indonesia), Borneo
P. inexspectatus **Aellen's pipistrelle** Kenya, Uganda, DRC, Cameroon, Benin
P. javanicus **Javan pipistrelle** Ussuris, China, Korea and SE Asia to Lesser Sunda Is (Indonesia), Philippines, Japan
P. joffrei **Joffre's pipistrelle** N Burma
P. kitcheneri **Red-brown pipistrelle** Borneo
P. kuhlii **Kuhl's pipistrelle** S Europe to the Caucasus to Kazakhstan and Pakistan, SW Asia, Africa, Canarias Is
P. lophurus **Burma pipistrelle** Burma Pen.
P. macrotis **Big-eared pipistrelle** Malay Pen., Sumatra and Bali (Indonesia), Borneo
P. maderensis **Madeira pipistrelle** Madeira (Portugal), Canarias Is
P. mimus **Indian pygmy pipistrelle** Afghanistan, Pakistan, India, Sri Lanka, Nepal, Bhutan, Burma, Vietnam, Thailand
P. minahassae **Minahassa pipistrelle** Sulawesi (Indonesia)
P. mordax **Pungent pipistrelle** Java (Indonesia)
P. musciculus **Mouselike pipistrelle** Cameroon, Gabon, DRC
P. nanulus **Tiny pipistrelle** Sierra Leone to Kenya, Bioko (Equatorial Guinea)
P. nanus **Banana pipistrelle** South Africa to Ethiopia and Sudan, Niger, Mali, Senegal, Madagascar
P. nathusii **Nathusius' pipistrelle** W Europe to Urals and Caucasus and W Turkey; also S England
P. paterculus **Mount Popa pipistrelle** N India, Burma, SW China, Thailand
P. peguensis **Pegu pipistrelle** Burma
P. permixtus **Dar-es-Salaam pipistrelle** Tanzania
P. petersi **Peters' pipistrelle** Sulawesi and Maluku (Indonesia), Borneo, Philippines
P. pipistrellus **Common pipistrelle** S Scandinavia and W Europe to Volga R and Caucasus, Morocco, Israel and Turkey to Kazakhstan, Kashmir and Xinjiang (China); also British Isles
P. pulveratus **Chinese pipistrelle** Sichuan, Yunnan, Hunan, Giangsu, Fujian, Hong Kong (China), Thailand
P. rueppelli **Rüppell's pipistrelle** Senegal and Algeria and Egypt and Iraq south to Botswana and N South Africa
P. rusticus **Rusty pipistrelle** Liberia and Ethiopia south to South Africa
P. savii **Savi's pipistrelle** Iberian Pen., Morocco and Canarias Is through the Caucasus to NE China, Korea, and Japan and through Iran and Afghanistan to N India and Burma
P. societatis **Social pipistrelle** Malay Pen.
P. stenopterus **Narrow-winged pipistrelle** Malay Pen., Sumatra and Riau Arch. (Indonesia), N Borneo, Mindanao (Philippines)
P. sturdeei **Sturdee's pipistrelle** Bonin Is (Japan)
P. subflavus **Eastern pipistrelle** SE Canada, E USA, Honduras
P. tasmaniensis **Eastern false or Tasmanian pipistrelle** S Australia (excl. South Australia), Tasmania
P. tenuis **Least pipistrelle** Thailand to New Guinea, Bismarck Arch. (Papua New Guinea), Solomon Is, Vanuatu, N Australia, Cocos Keeling and Christmas Is (Ind. Oc.)
Plecotus Big-eared Bats
P. auritus **Brown big-eared bat** Spain, Ireland, and Norway to Sakhalin Is (Russia), N China, Nepal, Japan
P. austriacus **Gray big-eared bat** Senegal, Spain, and England to Mongolia and W China, Canarias Is, Cape Verde Is (Atl. Oc.)
P. mexicanus **Mexican big-eared bat** Mexico (incl. Cozumel Is)
P. rafinesquii **Rafinesque's big-eared bat** SE USA
P. taivanus **Taiwan big-eared bat** Taiwan
P. teneriffae **Canary big-eared bat** Canarias Is
P. townsendii **Townsend's big-eared bat** S British Columbia (Canada) to W USA to Oaxaca (Mexico) east to Virginia (E USA)
Rhogeessa Little Yellow Bats
R. alleni **Allen's yellow bat** Oaxaca to Zacatecas (Mexico)
R. genowaysi **Genoways' yellow bat** S Chiapas (S Mexico)
R. gracilis **Slender yellow bat** Mexico
R. minutilla **Tiny yellow bat** Colombia, Venezuela (incl. Margarita Is)
R. mira **Least yellow bat** S Michoacan (S Mexico)
R. parvula **Little yellow bat** Oaxaca to Sonoro (Mexico), Tres Marias Is (Mexico)
R. tumida **Black-winged little yellow bat** Tamaulipas (Mexico) to Ecuador, NE Brazil, Bolivia, Trinidad and Tobago
Scotoecus House Bats
S. albofuscus **Light-winged lesser house bat** Gambia to Kenya and Mozambique
S. hirundo **Dark-winged lesser house bat** Senegal to Ethiopia south to Angola, Zambia, Malawi
S. pallidus **Desert yellow bat** Pakistan, N India
Scotomanes Harlequin Bats
S. emarginatus **Emarginate harlequin bat** India
S. ornatus **Harlequin bat** NE India, Burma, S China, Vietnam, Thailand
Scotophilus Yellow Bats
S. borbonicus **Lesser yellow bat** Madagascar, Reunion
S. celebensis **Sulawesi yellow bat** Sulawesi (Indonesia)
S. dinganii **African yellow bat** Senegal and Sierra Leone to Somalia and S Yemen south to Namibia and South Africa
S. heathi **Greater Asiatic yellow bat** Afghanistan to S China south to Sri Lanka and Vietnam
S. kuhlii **Lesser Asiatic yellow bat** Pakistan to Taiwan, south to Sri Lanka and Malay Pen., to Philippines and Aru Is (Indonesia)
S. leucogaster **White-bellied yellow bat** Mauritania and Senegal to Ethiopia and N Kenya
S. nigrita **Schreber's yellow bat** Senegal to Sudan and Kenya to Mozambique
S. nux **Nut-colored yellow bat** Sierra Leone to Kenya
S. robustus **Robust yellow bat** Madagascar
S. viridis **Greenish yellow bat** Senegal to Ethiopia south to Namibia and South Africa
Tylonycteris Bamboo Bats
T. pachypus **Lesser bamboo bat** India (incl. Andaman Is) and S China to Lesser Sunda Is (Indonesia) and Philippines
T. robustula **Greater bamboo bat** S China to Sulawesi and Lesser Sunda Is (Indonesia) and Philippines
Vespertilio Particolored Bats
V. murinus **Particolored bat** Norway and Britain to Ussuris, Korea, Taiwan, Japan
V. superans **Asian particolored bat** China, Ussuris, Korea, Taiwan, Japan

SUBFAMILY MURININAE
Harpiocephalus
H. harpia **Hairy-winged bat** India to Taiwan and Vietnam, south to Indonesia
Murina Tube-nosed Insectivorous Bats
M. aenea **Bronze tube-nosed bat** Malay Pen., Borneo
M. aurata **Little tube-nosed bat** Nepal to SW China and Burma
M. cyclotis **Round-eared tube-nosed bat** India, Sri Lanka to Guangdong and Hainan (China), Vietnam, south to Malay Pen., Lesser Sunda Is (Indonesia), Borneo, Philippines
M. florium **Flores tube-nosed bat** Leser Sunda Is and Sulawesi (Indonesia), New Guinea, NE Australia
M. fusca **Dusky tube-nosed bat** Manchuria (China)
M. grisea **Peters' tube-nosed bat** NW Himalayas
M. huttoni **Hutton's tube-nosed bat** NW India to Vietnam, Fujian (China), Thailand, Malay Pen.
M. leucogaster **Greater tube-nosed bat** E Himalayas, China, Upper Yenisei R (Russia), Altai Mts, Korea, Ussuris, Sakhalin Is (Russia), Japan
M. puta **Taiwan tube-nosed bat** Taiwan
M. rozendaali **Gilded tube-nosed bat** Borneo
M. silvatica **Forest tube-nosed bat** Japan (incl. Tsushima Is)
M. suilla **Brown tube-nosed bat** Malay Pen., Sumatra and Java (Indonesia), Borneo
M. tenebrosa **Gloomy tube-nosed bat** Tsushima Is (Japan)
M. tubinaris **Scully's tube-nosed bat** Pakistan, N India, Burma, Vietnam, Laos, Thailand
M. ussuriensis **Ussuri tube-nosed bat** Ussuris, Kurile Is and Sakhalin Is (Russia), Korea

SUBFAMILY MINIOPTERINAE
Miniopterus **Bent-winged Bats**
M. australis **Little long-fingered bat** Java (Indonesia), Borneo, Philippines, Vanuatu, E Australia
M. fraterculus **Lesser long-fingered bat** South Africa, Mozambique, Malawi, Zambia, Angola
M. fuscus **Southeast Asian long-fingered bat** SE China, Ryukyu Is (Japan), Thailand, Malay Pen., Java and Sulawesi (Indonesia), Borneo, Philippines, New Guinea
M. inflatus **Greater long-fingered bat** W to E Africa
M. magnater **Western bent-winged bat** NE India and SE China to Timor (Indonesia) and New Guinea
M. minor **Least long-fingered bat** Kenya, Tanzania, DRC, Congo, Madagascar, São Tomé and Comoros
M. pusillus **Small bent-winged bat** India to Maluku (Indonesia) and Philippines, Solomon Is to New Caledonia
M. robustior **Loyalty bent-winged bat** Loyaute Is (E of New Caledonia)
M. schreibersi **Schreiber's long-fingered bat** SubSaharan Africa (also Madagascar), Morocco and S Europe to Iran and Caucasus to China and Japan, much of Indo-Malayan region, New Guinea, Bismarck Arch. (Papua New Guinea), Solomon Is, Australia
M. tristis **Great bent-winged bat** Sulawesi (Indonesia), Philippines, New Guinea, Bismarck Arch. (Papua New Guinea), Solomon Is, Vanuatu

SUBFAMILY TOMOPEATINAE
Tomopeas
T. ravus **Blunt-eared bat** W Peru

FAMILY MYSTACINIDAE
New Zealand Short-tailed Bats
Mystacina
M. robusta **New Zealand greater short-tailed bat** Extinct: formerly Big South Cape Is (New Zealand)
M. tuberculata **New Zealand lesser short-tailed bat** New Zealand

FAMILY MOLOSSIDAE
Free-tailed Bats
Chaerephon **Lesser Free-tailed Bats**
C. aloysiisabaudiae **Duke of Abruzzi's free-tailed bat** Ghana, Gabon, DRC, Uganda
C. ansorgei **Ansorge's free-tailed bat** Cameroon to Ethiopia south to Angola and E South Africa
C. bemmeleni **Gland-tailed free-tailed bat** Liberia, Cameroon, DRC, Uganda, Tanzania, Kenya, Sudan
C. bivittata **Spotted free-tailed bat** Sudan, Ethiopia, Kenya, Tanzania, Uganda, Zambia, Mozambique, Zimbabwe
C. chapini **Chapin's free-tailed bat** Ethiopia, Uganda, DRC, Zambia, Angola, Namibia, Botswana, Zimbabwe
C. gallagheri **Gallagher's free-tailed bat** DRC
C. jobensis **Northern mastiff bat** New Guinea, Solomon Is, Vanuatu, Fiji, N and C Australia
C. johorensis **Northern free-tailed bat** W Malaysia, Sumatra (Indonesia)
C. major **Lappet-eared free-tailed bat** Liberia, Mali, Burkina Faso, Ghana, Togo, Niger, Nigeria, NE DRC, Uganda, Sudan, Tanzania
C. nigeriae **Nigerian free-tailed bat** Ghana and Niger to Ethiopia and Saudi Arabia south to Namibia, Botswana and Zimbabwe
C. plicata **Wrinkle-lipped free-tailed bat** India and Sri Lanka to S China and Vietnam, to Lesser Sunda Is (Indonesia), Borneo, Philippines, Cocos Keeling Is (Ind. Oc.)
C. pumila **Little free-tailed bat** Senegal to Yemen south to South Africa, Bioko (Equatorial Guinea), Pemba and Zanzibar Is (Tanzania), Madagascar, Comoros, Seychelles

C. russata **Russet free-tailed bat** Ghana, Cameroon, DRC, Kenya
Cheiromeles
C. torquatus **Hairless bat** Malay Pen., Sumatra, Java and Sulawesi (Indonesia), Borneo, SW Philippines
Eumops Bonneted Bats
E. auripendulus **Black bonneted bat** Oaxaca and Yucatan (Mexico) to Peru, E Brazil, N Argentina, Trinidad
E. bonariensis **Dwarf bonneted bat** Veracruz (Mexico) to NW Peru, Brazil, N Argentina, and Uruguay
E. dabbenei **Big bonneted bat** Colombia, Venezuela, Paraguay, N Argentina
E. glaucinus **Wagner's bonneted bat** Jalisco (Mexico) to Peru, Brazil, and N Argentina, Cuba, Jamaica, Florida (USA)
E. hansae **Sanborn's bonneted bat** Costa Rica, Panama, Venezuela, Guyana, Fr. Guiana, Peru, Brazil, Bolivia
E. maurus **Guianan bonneted bat** Guyana, Surinam
E. perotis **Western bonneted bat** California and Texas (SW USA) to Zacatecas and Hidalgo (Mexico), Colombia to E Brazil and N Argentina, Cuba
E. underwoodi **Underwood's bonneted bat** Arizona (SW USA) to Nicaragua
Molossops Dog-faced Bats
M. abrasus **Cinnamon dog-faced bat** Venezuela, Guyana, Surinam, Peru, Brazil, Bolivia, Paraguay, N Argentina
M. aequatorianus **Equatorial dog-faced bat** Ecuador
M. greenhalli **Greenhall's dog-faced bat** Nayarit (Mexico) to Ecuador and NE Brazil, Trinidad
M. mattogrossensis **Mato Grosso dog-faced bat** Venezuela, Guyana, C and NE Brazil
M. neglectus **Rufous dog-faced bat** Surinam, Peru, Amazonian Brazil
M. planirostris **Southern dog-faced bat** Panama to Surinam, Peru, Brazil, Paraguay, N Argentina
M. temminckii **Dwarf dog-faced bat** Colombia, Venezuela, Peru, S Brazil, Bolivia, Paraguay, N Argentina, Uruguay
Molossus Mastiff Bats
M. ater **Black mastiff bat** Tamaulipas and Sinaloa (Mexico) to Guianas, Peru, Brazil and N Argentina, Trinidad
M. bondae **Bonda mastiff bat** Honduras to Venezuela and Ecuador, Cozumel Is (Mexico)
M. molossus **Pallas' mastiff bat** Sinaloa and Coahuila (Mexico) to Guianas, Peru, Brazil, N Argentina, Uruguay, Margarita Is (Venezuela), Greater and Lesser Antilles (incl. Neth. Antilles), Trinidad and Tobago
M. pretiosus **Miller's mastiff bat** Guerrero and Oaxaca (Mexico), Nicaragua to Colombia, Venezuela and Guyana
M. sinaloae **Sinaloan mastiff bat** Sinaloa (Mexico) to Colombia and Surinam, Trinidad
Mops Greater Free-tailed Bats
M. brachypterus **Sierra Leone free-tailed bat** Gambia to Kenya, Tanzania, Mozambique
M. condylurus **Angolan free-tailed bat** Senegal to Somalia south to Angola, Botswana, and E South Africa
M. congicus **Medje free-tailed bat** Ghana, Nigeria, Cameroon, DRC, Uganda
M. demonstrator **Mongalla free-tailed bat** Sudan, DRC, Uganda, Burkina Faso
M. midas **Midas free-tailed bat** Senegal to Saudi Arabia south to Botswana and N South Africa, Madagascar
M. mops **Malayan free-tailed bat** Malay Pen., Sumatra (Indonesia), Borneo
M. nanulus **Dwarf free-tailed bat** Sierra Leone to Ethiopia and Kenya
M. niangarae **Niangara free-tailed bat** Niangara (DRC)
M. niveiventer **White-bellied free-tailed bat** DRC, Rwanda, Burundi, Tanzania, Mozambique, Zambia, Angola
M. petersoni **Peterson's free-tailed bat** Cameroon and Ghana
M. sarasinorum **Sulawesi free-tailed bat** Sulawesi (Indonesia), Philippines
M. spurrelli **Spurrell's free-tailed bat** Liberia, Ivory Coast, Ghana, Togo, Benin, Equatorial Guinea, DRC
M. thersites **Railer bat** Sierra Leone to Rwanda, Bioko (Equatorial Guinea)
M. trevori **Trevor's free-tailed bat** NW DRC, Uganda
Mormopterus Little Mastiff Bats
M. acetabulosus **Natal free-tailed bat** Madagascar, Réunion, Mauritius; Ethiopia, South Africa
M. beccarii **Beccari's mastiff bat** Maluku (Indonesia), New Guinea, N Australia
M. doriae **Sumatran mastiff bat** Sumatra (Indonesia)
M. jugularis **Peters' wrinkle-lipped bat** Madagascar
M. kalinowskii **Kalinowski's mastiff bat** Peru, N Chile
M. minutus **Little goblin bat** Cuba
M. norfolkensis **Eastern little mastiff bat** SE Queensland, E New South Wales (Australia), Norfolk Is (S Pac. Oc.)
M. petrophilus **Roberts' flat-headed bat** Southern African Subregion
M. phrudus **Incan little mastiff bat** Peru
M. planiceps **Southern free-tailed bat** New Guinea, Australia
M. setiger **Peters' flat-headed bat** S Sudan, Ethiopia, Kenya
Myopterus African Free-tailed Bats
M. daubentonii **Daubenton's free-tailed bat** Senegal, Ivory Coast, NE DRC
M. whitleyi **Bini free-tailed bat** Ghana, Nigeria, Cameroon, DRC, Uganda
Nyctinomops New World Free-tailed Bats
N. aurispinosus **Peale's free-tailed bat** Sonora and Tamaulipas (Mexico) to Peru, Brazil, and Bolivia
N. femorosaccus **Pocketed free-tailed bat** SW USA, Mexico (Incl. Baja California)
N. laticaudatus **Broad-eared bat** Tamaulipas and Jalisco (Mexico) to NW Peru, Brazil and N Argentina, Trinidad, Cuba
N. macrotis **Big free-tailed bat** SW British Columbia (Canada) and Iowa (USA) to Peru, N Argentina and Uruguay, Cuba, Jamaica, Hispaniola
Otomops Big-eared Free-tailed Bats
O. formosus **Java mastiff bat** Java (Indonesia)
O. martiensseni **Large-eared free-tailed bat** Djibouti and CAR to Angola and E South Africa, Madagascar
O. papuensis **Big-eared mastiff bat** SE New Guinea
O. secundus **Mantled mastiff bat** NE New Guinea
O. wroughtoni **Wroughton's free-tailed bat** S India
Promops Crested Mastiff Bats
P. centralis **Big crested mastiff bat** Jalisco and Yucatan (Mexico) to Surinam, Peru, and N Argentina, Trinidad
P. nasutus **Brown mastiff bat** Venezuela, Surinam, Ecuador, Peru, Brazil, Bolivia, Paraguay, N Argentina, Trinidad
Tadarida Free-tailed Bats
T. aegyptiaca **Egyptian free-tailed bat** South Africa to Algeria, Egypt to Yemen and Oman east to India, Sri Lanka
T. australis **White-striped free-tailed bat** New Guinea, S and C Australia
T. brasiliensis **Brazilian free-tailed bat** USA to Chile, Argentina and S Brazil, Greater and Lesser Antilles
T. espiritosantensis **Espirito Santo free-tailed bat** Brazil
T. fulminans **Madagascan large free-tailed bat** N South Africa, Zimbabwe, Zambia, Malawi, Tanzania, Rwanda, DRC, Kenya, Madagascar
T. lobata **Kenyan big-eared free-tailed bat** Kenya, Zimbabwe
T. teniotis **European free-tailed bat** France, Portugal, and Morocco to S China, Taiwan, and Japan, Madeira Is (Portugal), Canarias Is
T. ventralis **African giant free-tailed bat** Ethiopia to South Africa

ORDER XENARTHRA

EDENTATES

FAMILY MYRMECOPHAGIDAE
American Anteaters

Cyclopes
C. didactylus **Silky anteater** Veracruz and Oaxaca (S Mexico) to Colombia then west of Andes to S Ecuador and east of Andes to Venezuela, Guyana, Surinam, Fr. Guiana, Trinidad, and S Colombia and Venezuela south to Santa Cruz (Bolivia) and Brazil
Myrmecophaga
M. tridactyla **Giant anteater** Belize and Guatemala through South America to the Gran Chaco of Bolivia, Paraguay, and Argentina, and Uruguay
Tamandua Tamanduas
T. mexicana **Northern tamandua** Tamaulipas (E Mexico), Central America south to NW Venezuela and NW Peru
T. tetradactyla **Southern tamandua** South America (east of the Andes) from Colombia, Venezuela, Guianas, south to Argentina and Uruguay, Trinidad

FAMILY BRADYPODIDAE
Three-toed Sloths

Bradypus
B. torquatus **Maned three-toed sloth** SE Brazil in coastal forests
B. tridactylus **Pale-throated three-toed sloth** Venezuela, Guyana, Surinam, Fr. Guiana, N Brazil
B. variegatus **Brown-throated three-toed sloth** Honduras to Colombia, W Venezuela, Ecuador, E Peru, Brazil, Bolivia, Paraguay, Argentina

FAMILY MEGALONYCHIDAE
Two-toed Tree Sloths

SUBFAMILY CHOLOEPINAE
Choloepus
C. didactylus **Southern two-toed sloth** Venezuela and Guianas south to Brazil west into upper Amazon Basin of Ecuador and Peru
C. hoffmanni **Hoffmann's two-toed sloth** Nicaragua into South America east to Venezuela and south to Brazil and E Bolivia

FAMILY DASYPODIDAE **Armadillos**

SUBFAMILY CHLAMYPHORINAE
Chlamyphorus Fairy Armadillos
C. retusus **Chacoan fairy armadillo** Gran Chaco of SE Bolivia, W Paraguay, and N Argentina
C. truncatus **Pink fairy armadillo** Argentina

SUBFAMILY DASYPODINAE
Cabassous Naked-tailed Armadillos
C. centralis **Northern naked-tailed armadillo** Chiapas (Mexico) to N Colombia
C. chacoensis **Chacoan naked-tailed armadillo** Gran Chaco of W Paraguay and NW Argentina; also Mato Grasso (Brazil)
C. tatouay **Greater naked-tailed armadillo** S Brazil, SE Paraguay, NE Argentina, Uruguay
C. unicinctus **Southern naked-tailed armadillo** South America (east of the Andes) from Colombia to Mato Grasso (Brazil)
Chaetophractus Hairy Armadillos
C. nationi **Andean hairy armadillo** Cochabamba, Oruro, and La Paz (Bolivia)
C. vellerosus **Screaming hairy armadillo** Chaco Boreal of Bolivia and Paraguay south to C Argentina and west to Tarapaca (Chile)
C. villosus **Large hairy armadillo** Gran Chaco of Bolivia, Paraguay and N Argentina south to Santa Cruz (S Argentina) and Magallanes (S Chile)
Dasypus Long-nosed Armadillos
D. hybridus **Southern long-nosed armadillo** S Brazil, Paraguay, Argentina
D. kappleri **Great long-nosed armadillo** E Colombia, Venezuela (south of Orinoco R), Guyana, Surinam, south through Ecuador, Peru, and the Amazon Basin of Brazil
D. novemcinctus **Nine-banded armadillo** S USA, Mexico, Central and South America to N Argentina, Trinidad and Tobago, Grenada
D. pilosus **Hairy long-nosed armadillo** Andes of Peru
D. sabanicola **Llanos long-nosed armadillo** Llanos of Colombia and Venezuela
D. septemcinctus **Seven-banded armadillo** Lower Amazon Basin of Brazil to the Gran Chaco of Bolivia, Paraguay, and N Argentina
Euphractus
E. sexcinctus **Six-banded armadillo** S Surinam and adj. Brazil; also E Brazil to Bolivia, Paraguay, N Argentina, Uruguay
Priodontes
P. maximus **Giant armadillo** South America (east of the Andes), from N Venezuela and Guianas to Paraguay and N Argentina
Tolypeutes Three-banded Armadillos
T. matacus **Southern three-banded armadillo** SW Brazil and E Bolivia south through Gran Chaco of Paraguay to Buenos Aires (Argentina)
T. tricinctus **Brazilian three-banded armadillo** Bahia, Ceara, and Pernambuco (Brazil)
Zaedyus
Z. pichiy **Pichi** C and S Argentina and E Chile south to the Magellan Str.

ORDER DIDELPHIMORPHIA

AMERICAN OPOSSUMS

FAMILY DIDELPHIDAE

SUBFAMILY CALUROMYINAE
Caluromys Woolly Opossums
C. derbianus **Central American or Derby's woolly opossum** Mexico, Central America, Colombia, Ecuador
C. lanatus **Western or Ecuadorian woolly opossum** N and C Colombia, NW and S Venezuela, E Ecuador, E Peru, W and S Brazil, E Bolivia, E and S Paraguay, Misiones (N Argentina)
C. philander **Bare-tailed woolly opossum** Venezuela (incl. Margarita Is), Guyana, Surinam, Fr. Guiana, Brazil, Trinidad and Tobago
Caluromysiops
C. irrupta **Black-shouldered opossum** SE Peru, W Brazil
Glironia
G. venusta **Bushy-tailed opossum** Ecuador, Peru, Amazonian Brazil, Bolivia

SUBFAMILY DIDELPHINAE
Chironectes
C. minimus **Water opossum or Yapok** Oaxaca and Tabasco (S Mexico) south through Central America to Colombia, Venezuela, Guianas, Ecuador, Peru, Brazil, Paraguay, Argentina
Didelphis Large American Opossums
D. albiventris **White-eared opossum** Colombia, Ecuador, Peru, Brazil, Bolivia, Paraguay, N Argentina, Uruguay; also S Venezuela, SW Surinam, and N Brazil
D. aurita **Big-eared opossum** E Brazil, SE Paraguay, NE Argentina
D. marsupialis **Southern opossum** Tamaulipas (E Mexico) south throughout Central and South America to Peru, Brazil, Bolivia
D. virginiana **Virginia or Common opossum** S Canada, E and C USA, Mexico, Central America south to N Costa Rica
Gracilinanus Gracile Mouse Opossums
G. aceramarcae **Aceramarca gracile mouse opossum** Bolivia
G. agilis **Agile gracile mouse opossum** E Peru, Brazil, E Bolivia, Paraguay, NE Argentina, Uruguay
G. dryas **Wood sprite gracile mouse opossum** Andes of W Venezuela
G. emiliae **Emilia's gracile mouse opossum** NE Brazil
G. marica **Northern or Venezuelan gracile mouse opossum** N Colombia, Venezuela
G. microtarsus **Brazilian or Small-footed gracile mouse opossum** SE Brazil
Lestodelphys
L. halli **Patagonian opossum** Mendoza south to Santa Cruz (Argentina)
Lutreolina
L. crassicaudata **Lutrine opossum or Thick-tailed water opossum** E Colombia, Venezuela, W Guyana; also SE Brazil, E Bolivia, Paraguay, N Argentina, and Uruguay
Marmosa Mouse Opossums
M. andersoni **Anderson's mouse opossum** Cuzco (S Peru)
M. canescens **Grayish mouse opossum** S Sonora to Oaxaca and Yucatan, Tres Marias Is (Mexico)
M. lepida **Little rufous mouse opossum** E Colombia, Surinam, Ecuador, Peru, Bolivia, poss. Brazil
M. mexicana **Mexican mouse opossum** Tamaulipas (Mexico) to W Panama
M. murina **Murine or Common mouse opossum** Colombia, Venezuela, Guianas, Surinam, E Ecuador, E Peru, Brazil, E Bolivia, Trinidad and Tobago
M. robinsoni **Robinson's or Pale-bellied mouse opossum** Belize, Honduras, Panama, Colombia, Venezuela, W Ecuador, NW Peru, Trinidad and Tobago, Grenada
M. rubra **Red mouse opossum** E Ecuador, Peru
M. tyleriana **Tyler's mouse opossum** S Venezuela
M. xerophila **Dryland mouse opossum** NE Colombia, NW Venezuela
Marmosops Slender Mouse Opossums
M. cracens **Slim-faced slender mouse opossum** Falcon (Venezuela)
M. dorothea **Dorothy's slender mouse opossum** Yungas, Beni, and Chaco (Bolivia)
M. fuscatus **Gray-bellied slender mouse opossum** E Andes of Colombia, N Venezuela, Trinidad and Tobago
M. handleyi **Handley's slender mouse opossum** Antioquia (C Colombia)

M. impavidus **Andean slender mouse opossum** W Panama to Colombia, W Venezuela, Ecuador, and Peru; also S Venezuela
M. incanus **Gray slender mouse opossum** Bahia south to Parana (E Brazil)
M. invictus **Slaty slender mouse opossum** Panama
M. noctivagus **White-bellied slender mouse opossum** Ecuador, Peru, Amazonian Brazil, Bolivia
M. parvidens **Delicate slender mouse opossum** Colombia, Venezuela, Guyana, Surinam, Peru, Brazil
Metachirus
M. nudicaudatus **Brown "four-eyed" opossum** Nicaragua to Paraguay and NE Argentina
Micoureus Woolly Mouse Opossums
M. alstoni **Alston's woolly mouse opossum** Belize to Panama
M. constantiae **Pale-bellied woolly mouse opossum** E Bolivia and adj. Brazil south to N Argentina
M. demerarae **Long-furred woolly mouse opossum** Colombia, Venezuela, Guyana, Surinam, Fr. Guiana, Brazil, E Paraguay
M. regina **Short-furred woolly mouse opossum** Colombia, Ecuador, Peru, Bolivia
Monodelphis Short-tailed Opossums
M. adusta **Sepia or Cloudy short-tailed opossum** E Panama, Colombia, Ecuador, N Peru
M. americana **Three-striped short-tailed opossum** Para south to Santa Catarina (E Brazil)
M. brevicaudata **Red-legged short-tailed opossum** Venezuela, Surinam, Fr. Guiana, Amazon Basin of Brazil, Bolivia
M. dimidiata **Southern short-tailed opossum** SE Brazil, NE Argentina, Uruguay
M. domestica **Gray short-tailed opossum** Brazil, Bolivia, Paraguay
M. emiliae **Emilia's short-tailed opossum** Amazon Basin of Peru and Brazil
M. iheringi **Ihering's short-tailed opossum** SE Brazil
M. kunsi **Pygmy or Kuns' short-tailed opossum** Brazil, Bolivia
M. maraxina **Marajo short-tailed opossum** Marajo Is (Para, NE Brazil)
M. osgoodi **Osgood's short-tailed opossum** SE Peru, C Bolivia
M. rubida **Chestnut-striped short-tailed opossum** Goias south to Sao Paulo (E Brazil)
M. scalops **Long-nosed short-tailed opossum** Espirito Santo south to Santa Catarina (SE Brazil)
M. sorex **Shrewish short-tailed opossum** SE Brazil, S Paraguay, NE Argentina
M. theresa **Theresa's short-tailed opossum** Andes of Peru, E Brazil
M. unistriata **One-striped short-tailed opossum** Sao Paulo (Brazil)
Philander Gray and Black "Four-eyed" Opossums
P. andersoni **Black "four-eyed"opossum** E Colombia, S Venezuela, Ecuador, Peru, W Brazil
P. opossum **Gray "four-eyed"opossum** Tamaulipas (Mexico) through Central and South America to Paraguay and NE Argentina
Thylamys Fat-tailed Opossums
T. elegans **Elegant fat-tailed opossum** S Peru and Cochabamba (Bolivia) south to Valdivia (Chile) and Neuquen (Argentina)
T. macrura **Long-tailed fat-tailed opossum** S Brazil, Paraguay
T. pallidior **Pallid fat-tailed opossum** E and S Bolivia, Argentina
T. pusilla **Small fat-tailed opossum** C and S Brazil, SE Bolivia, Paraguay, N Argentina
T. velutinus **Velvety fat-tailed opossum** SE Brazil

ORDER PAUCITUBERCULATA

SHREW OPOSSUMS

FAMILY CAENOLESTIDAE

Caenolestes Northern Shrew Opossums
C. caniventer **Gray-bellied shrew opossum** Andes of SW Ecuador and NW Peru
C. convelatus **Blackish shrew opossum** Andes of W Colombia and NW Ecuador
C. fuliginosus **Silky shrew opossum** Andes of Colombia, NW Venezuela, and Ecuador
Lestoros
L. inca **Incan or Peruvian shrew opossum** Andes of S Peru
Rhyncholestes
R. raphanurus **Chilean shrew opossum** SC Chile (incl. Chiloe Is)

ORDER MICROBIOTHERIA

FAMILY MICROBIOTHERIIDAE

Dromiciops
D. gliroides **Monito del Monte or Colocolos** Chile, adj. WC Argentina

ORDER DASYUROMORPHIA

AUSTRALASIAN CARNIVOROUS MARSUPIALS

FAMILY THYLACINIDAE

Thylacinus
T. cynocephalus **Thylacine or Tasmanian wolf or tiger** Extinct: formerly Tasmania (Australia)

FAMILY MYRMECOBIIDAE

Myrmecobius
M. fasciatus **Numbat or Banded anteater** SW Western Australia (Australia)

FAMILY DASYURIDAE Dasyurids

Antechinus Antechinuses
A. bellus **Fawn antechinus** N Northern Territory (Australia)
A. flavipes **Yellow-footed antechinus** Cape York Pen. (Queensland) to Victoria, SE South Australia, and SW Western Australia (Australia)
A. godmani **Atherton antechinus** NE Queensland (Australia)
A. leo **Cinnamon antechinus** Cape York Pen. (Queensland, Australia)
A. melanurus **Black-tailed antechinus** New Guinea
A. minimus **Swamp antechinus** Coastal SE South Australia to Tasmania (Australia)
A. naso **Long-nosed antechinus** C New Guinea
A. stuartii **Brown antechinus** E Queensland, E New South Wales, Victoria (Australia)
A. swainsonii **Dusky antechinus** SE Queensland, E New South Wales, E and SE Victoria, Tasmania (Australia)
A. wilhelmina **Lesser antechinus** C New Guinea
Dasycercus Crested-tailed Marsupial Mice
D. byrnei **Kowari** Junction of Northern Territory, South Australia, and Queensland (C Australia)
D. cristicauda **Mulgara** NW Western Australia to SW Queensland and N South Australia (Australia)
Dasykaluta
D. rosamondae **Little red kaluta** NW Western Australia (Australia)
Dasyurus Quolls
D. albopunctatus **New Guinean quoll** New Guinea
D. geoffroii **Western quoll** Western Australia (Australia)
D. hallucatus **Northern quoll** N Western Australia, N Northern Territory, N and NE Queensland (Australia)
D. maculatus **Spotted-tailed or Tiger quoll** E Queensland, E New South Wales, E and S Victoria, SE South Australia, Tasmania (Australia)
D. spartacus **Bronze quoll** Papua New Guinea
D. viverrinus **Eastern quoll** Tasmania (Australia), poss. SE Australia
Murexia Long-tailed Dasyures
M. longicaudata **Short-furred dasyure** Aru Is (Indonesia), New Guinea
M. rothschildi **Broad-striped dasyure** SE New Guinea
Myoictis
M. melas **Three-striped dasyure** Salawati and Aru Is (Indonesia), New Guinea
Neophascogale
N. lorentzi **Speckled dasyure** C New Guinea
Ningaui Ningauis
N. ridei **Wongai or Inland ningaui** Western Australia, Northern Territory, South Australia (Australia)
N. timealeyi **Pilbara ningaui** NW Western Australia (Australia)
N. yvonnae **Southern ningaui** Western Australia to New South Wales and Victoria (Australia)
Parantechinus Dibblers
P. apicalis **Southern dibbler** SW Western Australia (Australia)
P. bilarni **Sandstone dibbler** Northern Territory (Australia)
Phascogale Phascogales
P. calura **Red-tailed phascogale or Wambenger** SW Western Australia (Australia)
P. tapoatafa **Brush-tailed phascogale** SW Western Australia, SE South Australia, S Victoria, E New South Wales, Queensland, Northern Territory (Australia)
Phascolosorex Marsupial Shrews
P. doriae **Red-bellied marsupial shrew** W New Guinea
P. dorsalis **Narrow-striped marsupial shrew** W and E New Guinea
Planigale Planigales
P. gilesi **Paucident planigale** NE South Australia, NW New South Wales, SW Queensland (Australia)
P. ingrami **Long-tailed planigale** NE Western Australia, NE Northern Territory, N and E Queensland (Australia)
P. maculata **Pygmy or Common planigale** E Queensland, NE New South Wales, N Northern Territory (Australia)
P. novaeguineae **New Guinean planigale** S New Guinea
P. tenuirostris **Narrow-nosed planigale** NW New South Wales, SC Queensland (Australia)
Pseudantechinus Pseudantechinuses
P. macdonnellensis **Fat-tailed pseudantechinus** N Western Australia, Northern Territory, central deserts (Australia)
P. ningbing **Ningbing pseudantechinus** Kimberley (N Western Australia, Australia)
P. woolleyae **Woolley's pseudantechinus** Western Australia (Australia)
Sarcophilus
S. laniarius **Tasmanian devil** Tasmania, poss. S Victoria (Australia)
Sminthopsis Dunnarts
S. aitkeni **Kangaroo Island dunnart** Kangaroo Is (South Australia, Australia)
S. archeri **Chestnut dunnart** S Papua New Guinea, Queensland (Australia)
S. butleri **Carpentarian dunnart** Kalumburu (Western Australia, Australia), Papua New Guinea
S. crassicaudata **Fat-tailed dunnart** S Western Australia, South Australia, W Victoria, W New South Wales, SW Queensland, SE Northern Territory (Australia)
S. dolichura **Little long-tailed dunnart** Western Australia, South Australia (Australia)
S. douglasi **Julia creek dunnart** NW Queensland (Australia)
S. fuliginosus **Sooty dunnart** SW Western Australia (Australia)
S. gilberti **Gilbert's dunnart** SW Western Australia (Australia)
S. granulipes **White-tailed dunnart** SW Western Australia (Australia)
S. griseoventer **Gray-bellied dunnart** SW Western Australia (Australia)
S. hirtipes **Hairy-footed dunnart** S Northern Territory, WC Western Australia (Australia)
S. laniger **Kultarr** Western Australia, S Northern Territory, N South Australia, N Victoria, W New South Wales, SW Queensland (Australia)
S. leucopus **White-footed dunnart** Tasmania, S and SE Victoria, New South Wales, Queensland (Australia)
S. longicaudata **Long-tailed dunnart** Western Australia (Australia)
S. macroura **Stripe-faced dunnart** NW New South Wales, W Queensland, S Northern Territory, N South Australia, N Western Australia (Australia)
S. murina **Slender-tailed dunnart** SW Western Australia, SE South Australia, Victoria, New South Wales, E Queensland (Australia)
S. ooldea **Ooldea dunnart** SE Western Australia, S South Australia, S Northern Territory (Australia)
S. psammophila **Sandhill dunnart** SW Northern Territory, S South Australia (Australia)
S. virginiae **Red-cheeked dunnart** N Queensland, N Northern Territory (Australia), Aru Is (Indonesia), S New Guinea
S. youngsoni **Lesser hairy-footed dunnart** Western Australia, Northern Territory (Australia)

ORDER PERAMELEMORPHIA

BANDICOOTS AND BILBIES

FAMILY PERAMELIDAE

Australian Bandicoots and Bilbies
Chaeropus
C. ecaudatus **Pig-footed bandicoot** Western Australia, S Northern Territory, N South Australia, SW New South Wales, Victoria (Australia)
Isoodon Short-nosed Bandicoots
I. auratus **Golden bandicoot** NW Western Australia, Barrow Is (Australia)
I. macrourus **Northern brown bandicoot** NE Western Australia, N Northern Territory, E Queensland, NE New South Wales (Australia), S and E New Guinea
I. obesulus **Southern brown bandicoot or Quenda** N Queensland, SE New South Wales, S Victoria, SE South Australia, SW Western Australia, Tasmania (Australia)
Macrotis Bilbies
M. lagotis **Greater bilby** SW Queensland, Northern Territory/Western Australia border and Kimberley (N Western Australia, Australia)
M. leucura **Lesser bilby** C Australia
Perameles Long-nosed Bandicoots
P. bougainville **Western barred bandicoot** Bernier and Dorre Is (off Western Australia, Australia)
P. eremiana **Desert bandicoot** Great Victoria Desert (Western Australia), N South Australia, S Northern Territory (Australia)
P. gunnii **Eastern barred bandicoot** S Victoria, Tasmania (Australia)
P. nasuta **Long-nosed bandicoot** E Queensland, E New South Wales, E Victoria (Australia)

FAMILY PERORYCTIDAE

Rainforest Bandicoots
Echymipera Echymiperas
E. clara **Clara's echymipera** NC New Guinea
E. davidi **David's echymipera** Kiriwina Is (Papua New Guinea)
E. echinista **Menzie's echymipera** Papua New Guinea
E. kalubu **Kalubu echymipera** New Guinea, adj. small islands incl. Bismarck Arch. (Papua New Guinea) and Salawati Is (Indonesia)
E. rufescens **Long-nosed echymipera** Cape York Pen. (Queensland, Australia), New Guinea (incl. small islands off SE coast), Kei and Aru Is (Indonesia)
Microperoryctes Mouse Bandicoots
M. longicauda **Striped bandicoot** New Guinea
M. murina **Mouse bandicoot** W New Guinea
M. papuensis **Papuan bandicoot** SE New Guinea
Peroryctes New Guinean Bandicoots
P. broadbenti **Giant bandicoot** SE New Guinea
P. raffrayana **Raffray's bandicoot** New Guinea
Rhynchomeles
R. prattorum **Seram bandicoot** Seram Is (Indonesia)

ORDER NOTORYCTEMORPHIA

MARSUPIAL MOLES

FAMILY NOTORYCTIDAE

Notoryctes
N. caurinus **Northwestern marsupial mole** NW Western Australia (Australia)
N. typhlops **Marsupial mole** Western Australia, South Australia, S Northern Territory (Australia)

ORDER DIPROTODONTIA

KOALA, WOMBATS, POSSUMS, KANGAROOS, AND RELATIVES

FAMILY PHASCOLARCTIDAE

Phascolarctos
P. cinereus **Koala** SE Queensland, E New South Wales, Victoria, SE South Australia (Australia); introd. elsewhere

FAMILY VOMBATIDAE Wombats

Lasiorhinus Hairy-nosed Wombats
L. krefftii **Northern or Queensland hairy-nosed wombat** SE and E Queensland, Deniliquin (New South Wales, Australia)
L. latifrons **Southern hairy-nosed or Plains wombat** S South Australia, SE Western Australia (Australia)
Vombatus
V. ursinus **Coarse-haired, Common, Forest, or Naked-nosed wombat** SE Queensland, E New South Wales, S Victoria, Tasmania (and islands in Bass Str.), SE South Australia (Australia)

FAMILY PHALANGERIDAE

Cuscuses and Brushtail Possums
Ailurops
A. ursinus **Bear cuscus** Sulawesi, Peleng, Talaud, Togian, Muna, Buton, and Lembeh Is (Indonesia)
Phalanger Cuscuses
P. carmelitae **Mountain cuscus** C New Guinea
P. lullulae **Woodlark Island cuscus** Woodlark Is (Papua New Guinea)
P. matanim **Telefomin cuscus** Telefomin (W Papua New Guinea)
P. orientalis **Gray cuscus** Timor and Seram Is (Indonesia) to N New Guinea and adj. small islands, Bismarck Arch. (Papua New Guinea), Solomon Is, E Cape York Pen. (Queensland, Australia)

P. ornatus **Moluccan cuscus** Halmahera, Ternate, Tidore, Bacan, and Morotai Is (Indonesia)
P. pelengensis **Peleng Island cuscus** Peleng and Sulu Is (Indonesia)
P. rothschildi **Obi Island cuscus** Pulau Is (Maluku, Indonesia)
P. sericeus **Silky cuscus** C and E New Guinea
P. vestitus **Stein's cuscus** C New Guinea
Spilocuscus **Spotted Cuscuses**
S. maculatus **Short-tailed spotted cuscus** Aru, Kei, Seram, Amboina, and Selayar Is (Indonesia), New Guinea and adj. small islands, Cape York Pen. (Queensland, Australia)
S. rufoniger **Black-spotted cuscus** N New Guinea
Strigocuscus **Plain Cuscuses**
S. celebensis **Little Celebes cuscus** Sulawesi, Peleng, Sanghir, Sula, and Obi Is (Indonesia)
S. gymnotis **Ground cuscus** Aru, Wetar, Timor and other Indonesian islands, New Guinea
Trichosurus **Brushtail Possums**
T. arnhemensis **Northern brushtail possum** N Northern Territory, NE Western Australia, Barrow Is (Australia)
T. caninus **Mountain brushtail possum** SE Queensland, E New South Wales, E Victoria (Australia)
T. vulpecula **Silver gray brushtail possum** E Queensland, E New South Wales, Victoria, Tasmania, SE and N South Australia, SW Western Australia (Australia); introd. to New Zealand
Wyulda
W. squamicaudata **Scaly-tailed possum** Kimberley (NE Western Australia, Australia)

FAMILY POTOROIDAE
Bettongs, "Rat" kangaroos, and Potoroos
Aepyprymnus
A. rufescens **Rufous bettong** NE Victoria, E New South Wales, E Queensland (Australia)
Bettongia **Bettongs**
B. gaimardi **Tasmanian or Gaimard's bettong** Tasmania (Australia)
B. lesueur **Burrowing or Lesueur's bettong or Boodie** W Australian Is
B. penicillata **Brush-tailed bettong or Woylie** SW Western Australia, S South Australia (incl. St Francis Is), NW Victoria, C New South Wales, E Queensland (Australia)
Caloprymnus
C. campestris **Desert "rat" kangaroo** South Australia/Queensland border (Australia)
Hypsiprymnodon
H. moschatus **Musky "rat" kangaroo** NE Queensland (Australia)
Potorous **Potoroos**
P. longipes **Long-footed potoroo** NE Victoria (Australia)
P. platyops **Broad-faced potoroo** Extinct: formerly SW Western Australia, Kangaroo Is (Australia)
P. tridactylus **Long-nosed potoroo** SE Queensland, coastal New South Wales, NE Victoria, SE South Australia, SW Western Australia, Tasmania, King Is (Australia)

FAMILY MACROPODIDAE
Kangaroos and Wallabies
Dendrolagus **Tree Kangaroos**
D. bennettianus **Bennett's tree kangaroo** NE Queenland (Australia)
D. dorianus **Doria's, Dusky or Unicolored tree kangaroo** New Guinea
D. goodfellowi **Goodfellow's or Ornate tree kangaroo** E New Guinea
D. inustus **Grizzled tree kangaroo** Yapen Is (Indonesia), N and W New Guinea
D. lumholtzi **Lumholtz's tree kangaroo** NE Queensland (Australia)
D. matschiei **Huon or Matschie's tree kangaroo** NE New Guinea; introd. to Umboi Is
D. scottae **Tenkile tree kangaroo** Torricelli Mts (Papua New Guinea)
D. spadix **Lowland tree kangaroo** S New Guinea
D. ursinus **White-throated tree kangaroo** NW New Guinea
Dorcopsis **Dorcopsises**
D. atrata **Black dorcopsis** Goodenough Is (Papua New Guinea)
D. hageni **White-striped dorcopsis** NC New Guinea
D. luctuosa **Gray dorcopsis** S New Guinea
D. muelleri **Brown dorcopsis** Misool, Salawati, Aru, and Yapen Is (Indonesia), W New Guinea
Dorcopsulus **Forest Wallabies**
D. macleayi **Papuan or Macleay's forest wallaby** SE New Guinea
D. vanheurni **Lesser forest wallaby** New Guinea
Lagorchestes **Hare Wallabies**
L. asomatus **Central hare wallaby** Extinct: formerly L. Mackay (Northern Territory, Australia)
L. conspicillatus **Spectacled hare wallaby** N Western Australia and adj. islands, N Northern Territory, N and W Queensland (Australia)
L. hirsutus **Western or Rufous hare wallaby** Bernier and Dorre Is (Western Australia) and near Alice Springs (Northern Territory, Australia)
L. leporides **Eastern hare wallaby** Extinct: formerly W New South Wales, NW Victoria, E South Australia (Australia)
Lagostrophus
L. fasciatus **Banded hare wallaby or Munning** Bernier and Dorre Is (Western Australia, Australia)
Macropus **Wallabies, Wallaroos, and Kangaroos**
M. agilis **Agile wallaby** NE Western Australia, Northern Territory, Queensland (Australia), S New Guinea, Kiriwina Is and other islands off SE coast of New Guinea
M. antilopinus **Antilopine wallaroo** N Queensland, Northern Territory, NE Western Australia (Australia)
M. bernardus **Black wallaroo** N Northern Territory (Australia)
M. dorsalis **Black-striped wallaby** E Queensland, E New South Wales (Australia)
M. eugenii **Tammar or Scrub wallaby** SW Western Australia, South Australia, Kangaroo and Wallaby Is and others (Australia)
M. fuliginosus **Western gray or Black-faced kangaroo** SW New South Wales, NW Victoria, South Australia, SW Western Australia, Tasmania, King and Kangaroo Is (Australia)
M. giganteus **Eastern gray or Great gray kangaroo** E and C Queensland, New South Wales, Victoria, SE South Australia, Tasmania (Australia)
M. greyi **Toolache wallaby** Extinct: formerly SE South Australia, adj. Victoria (Australia)
M. irma **Western brush wallaby** SW Western Australia (Australia)
M. parma **Parma or White-fronted wallaby** E New South Wales (Australia); introd. to Kawau Is (New Zealand)
M. parryi **Whiptail or Parry's wallaby** E Queensland, NE New South Wales (Australia)
M. robustus **Hill wallaroo or Euro** Western Australia, South Australia, S Northern Territory, Queensland, New South Wales, Barrow Is (Australia)
M. rufogriseus **Red-necked wallaby** SE South Australia, Victoria, E New South Wales, SE Queensland, Tasmania, King Is and adj. islands (Australia); introd. into England
M. rufus **Red kangaroo** Mainland Australia
Onychogalea **Nail-tailed Wallabies**
O. fraenata **Bridled nail-tailed wallaby** Taunton (Queensland, Australia)
O. lunata **Crescent nail-tailed wallaby** SC and SW Western Australia, S Northern Territory (Australia)
O. unguifera **Northern nail-tailed wallaby** Western Australia, Northern Territory, Queensland (Australia)
Petrogale **Rock Wallabies**
P. assimilis **Allied rock wallaby** Queensland (Australia)
P. brachyotis **Short-eared rock wallaby** Coastal NW Australia, N Northern Territory (Australia)
P. burbidgei **Burbridge's rock wallaby or Monjon** Kimberley (NE Western Australia), Bonaparte Arch. and adj. Islands (Australia)
P. concinna **Pygmy rock wallaby or Nabarlek** NE and NW Northern Territory, NE Western Australia (Australia)
P. godmani **Godman's rock wallaby** Cape York Pen. (Queensland, Australia)
P. inornata **Unadorned rock wallaby** N Queensland (Australia)
P. lateralis **Black-footed rock wallaby** Western Australia, South Australia, Northern Territory, W Queensland (Australia)
P. penicillata **Brush-tailed rock wallaby** E Australia
P. persephone **Proserpine rock wallaby** Proserpine (Queensland, Australia)
P. rothschildi **Rothschild's rock wallaby** NW Western Australia (Australia)
P. xanthopus **Yellow-footed rock wallaby** SW Queensland, NW New South Wales, South Australia (Australia)
Setonix
S. brachyurus **Quokka** SW Western Australia, Rottnest and Bald Is (Australia)
Thylogale **Pademelons**
T. billardierii **Tasmanian pademelon** Tasmania (Australia)
T. brunii **Dusky pademelon** C and E New Guinea and adj. small islands, Bismarck Arch. (Papua New Guinea), Aru Is (Indonesia)
T. stigmatica **Red-legged pademelon** E Queensland, E New South Wales (Australia), SC New Guinea
T. thetis **Red-necked pademelon** E Queensland, E New South Wales (Australia)
Wallabia
W. bicolor **Swamp wallaby** E Queensland, E New South Wales, Victoria, SE South Australia, Stradbroke and Fraser Is (Australia)

FAMILY BURRAMYIDAE
Pygmy Possums
Burramys
B. parvus **Mountain pygmy possum** NE Victoria, S New South Wales (Australia)
Cercartetus **Pygmy possums**
C. caudatus **Long-tailed pygmy possum** New Guinea, Fergusson Is (Papua New Guinea), NE Queensland (Australia)
C. concinnus **Western pygmy possum** SW Western Australia, S and SE South Australia, W Victoria, SW New South Wales (Australia)
C. lepidus **Tasmanian pygmy possum** Tasmania, NW Victoria/South Australia border, Kangaroo Is (Australia)
C. nanus **Eastern pygmy possum** SE South Australia, E New South Wales, Victoria, Tasmania (Australia)

FAMILY PSEUDOCHEIRIDAE
Ringtail Possums
Hemibelideus
H. lemuroides **Lemuroid ringtail possum** NE Queensland (Australia)
Petauroides
P. volans **Greater glider** E Australia
Petropseudes
P. dahli **Rock ringtail possum** N Northern Territory, NW Western Australia (Australia)
Pseudocheirus **Ringtails**
P. canescens **Daintree River ringtail** New Guinea, Salawati Is (Indonesia)
P. caroli **Weyland ringtail** WC New Guinea
P. forbesi **Moss-forest ringtail** New Guinea
P. herbertensis **Herbert River ringtail** NE Queensland (Australia)
P. mayeri **Pygmy ringtail** C New Guinea
P. peregrinus **Queensland ringtail** Cape York Pen. (Queensland) to SE South Australia and SW Western Australia, Tasmania, and islands of the Bass Str. (Australia)
P. schlegeli **Arfak ringtail** NW New Guinea
Pseudochirops **Ringtail Possums**
P. albertisii **D'Albertis' ringtail possum** N and W New Guinea, Yapen Is (Indonesia)
P. archeri **Green ringtail possum** NE Queensland (Australia)
P. corinnae **Golden ringtail possum** New Guinea
P. cupreus **Coppery ringtail possum** New Guinea

FAMILY PETAURIDAE
Gliding and Striped Possums
Dactylopsila **Striped Possums**
D. megalura **Great-tailed triok** New Guinea
D. palpator **Long-fingered triok** New Guinea
D. tatei **Tate's triok** Fergusson Is (Papua New Guinea)
D. trivirgata **Striped possum** New Guinea and adj. small islands, Aru Is (Indonesia), NE Queensland (Australia)
Gymnobelideus
G. leadbeateri **Leadbeater's possum** NE Victoria (Australia)
Petaurus **Lesser Gliding Possums**
P. abidi **Northern glider** NC New Guinea
P. australis **Fluffy or Yellow-bellied glider** Coastal Queensland, New South Wales, Victoria (Australia)
P. breviceps **Sugar glider** SE South Australia to Cape York Pen. (Queensland), N Northern Territory, NE Western Australia (Australia), New Guinea and adj. small islands, Aru Is and Maluku (Indonesia); introd. to Tasmania (Australia)
P. gracilis **Mahogany glider** Barrett's lagoon (Queensland, Australia)
P. norfolcensis **Squirrel glider** E Queensland, E New South Wales, E Victoria (Australia)

FAMILY TARSIPEDIDAE
Tarsipes
T. rostratus **Honey possum** SW Western Australia (Australia)

FAMILY ACROBATIDAE
Feathertail Gliders and Possums
Acrobates
A. pygmaeus **Feathertail glider** E Queensland to SE South Australia, inland to Deniliquin (New South Wales, Australia)
Distoechurus
D. pennatus **Feathertail possum** New Guinea

Glossary

Abiotic non-living.
Abomasum the final chamber of the four sections of the ruminant artiodactyl stomach (following the RUMEN, RETICULUM and OMASUM). The abomasum alone corresponds to the stomach "proper" of other mammals and the other three are elaborations of its proximal part.
Adaptation features of an animal which adjust it to its environment. Adaptations may be genetic, produced by evolution and hence not alterable within the animal's lifetime, or they may be phenotypic, produced by adjustment on the behalf of the individual and may be reversible within its lifetime. NATURAL SELECTION favors the survival of individuals whose adaptations adjust them better to their surroundings than other individuals with less successful adaptations.
Adaptive radiation the pattern in which different species develop from a common ancestor (as distinct from CONVERGENT EVOLUTION, a process whereby species from different origins became similar in response to the same selective pressures).
Adult a fully developed and mature individual, capable of breeding, but not necessarily doing so until social and/or ecological conditions allow.
Aerobic deriving energy from processes that require free atmospheric oxygen. (cf **Anaerobic**)
Afrotheria a strongly supported interordinal group of African mammals that includes the golden moles, tenrecs, elephant shrews, aardvark, and the PAENUNGULATA (hyraxes, elephants, sirenians).
Afrotropical see **Ethiopian**
Age structure the proportion of individuals in a population in different age classes.
Aggression behavior in an animal that serves to injure or threaten another animal, but that is not connected with predation.
Agonistic behavior behavior patterns used during conflict with a CONSPECIFIC, including overt aggression, threats, appeasement, or avoidance.
Agouti a grizzled coloration resulting from alternate light and dark barring of each hair. This banding is well exemplified in the eponymous agouti (Family Dasyproctidae: Order Rodentia).
Air sac a side-pouch of the larynx (the upper part of the windpipe), used in some primates and male walruses as resonating chambers in producing calls.
Albinism the heritable condition in which all hairs are white due to the inability to form MELANIN in the hair, skin, or vascular coating of the eyes.
Allantoic stalk a sac-like outgrowth of the hinder part of the gut of the mammalian fetus, containing a rich network of blood vessels. It connects fetal circulation with the placenta, facilitating nutrition of the young, respiration and excretion. (see **Chorioallantoic placentation**)
Alleles one of several alternative forms of a GENE.
Allogrooming grooming performed by one animal upon another animal of the same species. (cf **Autogrooming**)
Alloparent an animal behaving parentally towards infants that are not its own offspring; the shorthand jargon "helper" is most commonly applied to alloparents without any offspring of their own. The term can be misleading if it is used to describe any non-breeding adults associated with infants, but which may or may not be "helping" by promoting their survival. Alloparents may help suckle young (allosuckling).
Allopatry condition in which populations of different species are geographically separated (cf **Sympatry**).
Allosuckling see **Alloparent**.
Allozyme a form of protein that is produced by a given ALLELE at a single gene LOCUS.
Alpine of the Alps or any lofty mountains; usually pertaining to altitudes above 1,500m (4,900ft).
Altricial young that are born at a rudimentary stage of development and require an extended period of nursing by parent(s). (cf **Precocial**)
Altruistic behavior that reduces personal fitness for the benefit of others.
Alveolus a microscopic sac within the lungs providing the surface for gaseous exchange during respiration. Also used to define the socket of the jaw bone into which the tooth fits.
Ambergris a form of excrement of sperm whales.
Amniote a higher vertebrate whose embryo is enclosed in a fluid-filled embryo, the so-called "amnion".
Amphibious able to live on both land and in water.
Amphilestid a family of TRICONODONT mammals that survived for about 50 million years between the mid-Jurassic and the early Cretaceous.
Amphipod a crustacean of the invertebrate order Amphipoda. Includes many freshwater and marine shrimps.
Ampullary glands paired accessory reproductive glands in some male mammals that contribute their products to the semen.
Amynodont a member of the family Amynodontidae, large rhinoceros-like mammals (order Perissodactyla), which became extinct in the Tertiary.
Anaerobic deriving energy from processes that do not require free oxygen. (cf **Aerobic**)
Anal gland (anal sac) a gland opening by a short duct either just inside the anus or on either side of it.
Ancestral stock a group of animals, usually showing primitive characteristics, which is believed to have given rise to later, more specialized forms.
Androgens hormones, secreted by the testes, that regulate the development of male secondary sexual characteristics.
Angle of attack the angle of the wings of a bat relative to the ground.
Anestrus the non-breeding condition of the reproductive cycle of a mammal when sexual organs are quiescent.
Anomodontia a suborder of the THERAPSIDA, which were extinct by the late Triassic.
Antarctic Convergence the region between 50°–55°S where the Antarctic surface water slides beneath the less-dense southward-flowing subantarctic water.
Anthracothere a member of the family Anthracotheriidae (order Artiodactyla), which became extinct in the late Tertiary.
Anthropoid literally "man-like"; a member of the primate suborder Anthropoidea – monkeys, apes and man. In modern classification systems the Anthropoidea (or Simiiformes) comprises the Catarrhini and Platyrrhini and are included with the tarsiers in the suborder Haplorrhini (**Haplorhines**).
Antigen a substance, whether organic or inorganic, that stimulates the production of antibodies when introduced into the body.
Antitragus the lower posterior part of the outer ear, which lies opposite the TRAGUS.
Antlers paired, branched processes found only on the skull of cervids, made of bone, and shed annually. (cf **Horns**)
Antrum a cavity in the body, especially one in the upper jaw hone.
Apomorphic characters that are derived or of more recent origin. The long neck of the giraffe is apomorphic, the short neck of its ancestor is plesiomorphic (primitive). Apomorphic features possessed by a group of biological organisms distinguish these organisms from others descended from the same ancestor.
Aquatic living chiefly in water.
Arboreal living in trees.
Archaeoceti an extinct order of whales that had features intermediate between terrestrial mammals and fully marine species.
Arteriole a small artery (i.e. muscular blood vessel carrying blood from the heart), eventually subdividing into minute capillaries.
Arterio-venous anastomosis (AVA) a connection between the arterioles carrying blood from the heart and the venules back to the heart.
Arthropod the largest phylum in the animal kingdom in number of species, including insects, spiders, crabs etc. Arthropods have hard, jointed exoskeletons and paired, jointed legs.
Artiodactyl a member of the order Artiodactyla, the even-toed ungulates.
Aspect ratio the ratio of the length of a wing to its width; short, wide wings have a low aspect ratio.
Association a mixed-species group (polyspecific association) involving two or more species; relatively common among both Old and New World monkeys, but the most stable associations are found in forest-living guenons.
Astragalus a bone in the ungulate tarsus (ankle) which (due to reorganization of ankle bones following reduction in the number of digits) bears most of the body weight, a task shared by the CALCANEUM bone in most other mammals).
Asymptote a line (usually straight) that is approached progressively by a given curve, but which is never met within a finite distance. An asymptotic population is the maximum sustainable population size for a given habitat or area, i.e. the growth rate of the population tends asymptotically to zero as the population approaches CARRYING CAPACITY (which is the upper asymptote on a graph).
Atlas one of the top two cervical vertebrae that articulate the skull and vertebral column. (see also **Axis**)
Atrophied of a structure or tissue that is diminished or reduced in size.

Auditory bullae see **Bullae**
Australian a geographic region comprising Australia and New Guinea. (see also **Wallace's line**)
Autogrooming grooming of an animal by itself. (cf **Allogrooming**)
Awns the most common guard hairs on mammals. They usually lie in one direction giving PELAGE a distinctive nap.
Axilla the angle between a forelimb and the body (in humans, the armpit).
Axis one of the top two cervical vertebrae that articulate the skull and vertebral column. (see also **Atlas**)

Baculum (**os penis**, **os baculum** or **penis bone**) an elongate bone present in the penis of certain mammals.
Baleen a horny substance, commonly known as whalebone, growing as plates from the upper jaws of whales of the suborder Mysticeti, and forming a fringelike sieve for extraction of plankton from seawater.
Basal metabolic rate the minimum METABOLIC RATE of an animal needed to sustain the life of an organism that is in an environment at a temperature equal to its own.
Bends the colloquial name for caisson disease, a condition produced by pressure changes in the blood as a diving mammal surfaces. Too rapid an ascent results in nitrogen dissolved in the blood forming bubbles which cause excruciating pain.
Benthic the bottom layer of the marine environment.
Bergmann's rule biogeographic rule that races of species from cold climates tend to be composed of individuals physically larger than those from warmer climates.
Bicornuate type of uterus in eutherian mammals characterized by a single cervix and the two uterine horns fused for part of their length. Found in insectivores, most bats, pangolins, primitive primates, most ungulates, elephants, sirenians, and some carnivores.
Bifid of the penis, with the head divided into two parts by a deep cleft.
Bifurcated paired, having two corresponding halves.
Bilophodont cheek teeth having an OCCLUSAL pattern with paired transverse ridges or LOPHS. (see also **Lophodont**)
Binocular form of vision typical of mammals in which the same object is viewed simultaneously by both eyes; the coordination of the two images in the brain permits precise perception of distance.
Binomial nomenclature (binomial classification) a system, introduced by Carolus Linnaeus, for naming all organisms by means of two Latin names, one for the GENUS (first letter capitalized) and one for the SPECIES (and both names written in italics; e.g. lion, *Panthera leo*)
Biodiversity the living plants, animals and other organisms that characterize a particular region, area, country, or even planet.
Biogeography the study of the patterns of distribution of organisms, either living or extinct, their habitats, and the historical and ecological factors that produced the distributions.
Biological control using one species to reduce the population density of another species in an area through parasitism or predation.
Biomass a measure of the abundance of a life-form in terms of its mass, either absolute or per unit area (the population densities of two species may be identical in terms of the number of individuals of each, but due to their different sizes their biomasses may be quite different).
Biome a broad ecosystem characterized by particular plant life, soil type and climatic conditions. They are the largest geographical BIOTIC COMMUNITIES that it is convenient to recognize.
Biotic community a naturally occurring group of plants and animals in the same environment.
Bipartite a type of uterus in eutherian mammals that is almost completely divided along the median line, with a single cervical opening into the vagina. Found in whales and many carnivores.
Bipedal walking on two legs. Only human beings exhibit habitual striding bipedalism. Some primate species may travel bipedally for short distances, and some (e.g. indri, bushbabies, tarsiers) hop bipedally on the ground.
Blastocyst see **Implantation**
Blowhole the opening of the nostril(s) of a whale, situated on the animal's head, from which the "spout" or "blow" is produced.
Blubber a layer of fat beneath the skin, well developed in whales and seals.
Boreal region a zone geographically situated south of the Arctic and north of latitude 50°N; dominated by coniferous forest.
Bovid a member of the cow-like artiodactyl family, Bovidae.
Brachiate to move around in the trees by arm-swinging beneath branches. In a broad sense all apes are brachiators, but only gibbons and siamangs exhibit a freeflight phase between hand-holds.
Brachydont a type of short-crowned teeth whose growth ceases when full-grown, whereupon the pulp cavity in the root closes. Typical of most mammals, but unlike the HYPSODONT teeth of many herbivores.
Bradycardia a condition in which the heart rate is reduced substantially.
Breaching leaping clear of the water.
Brindled having inconspicuous dark streaks or flecks on a gray or tawny background.
Brontothere a member of the family Brontotheriidae (order Perissodactyla), which became extinct in the early Tertiary.
Browser a herbivore which feeds on shoots and leaves of trees, shrubs etc, as distinct from grasses. (cf **Grazer**)
Bruce effect an effect demonstrated in mice where the presence of a strange male or his odor causes a female to abort her pregnancy and become receptive.
Buccal cavity mouth cavity.
Bullae (auditory) globular, bony capsules housing the middle and inner ear structures, situated on the underside of the skull.
Bunodont molar teeth whose cusps form separate, rounded hillocks which crush and grind.
Bursa (pl. **bursae**) a sac-like cavity (e.g. in ear of civets and Madagascan mongooses).

Cache a hidden store of food; also (verb) to hide food for future use.
Calcaneum one of the tarsal (ankle) bones which forms the heel and in many mammalian orders bears the body weight together with the ASTRAGALUS.
Calcar a process that extends medially from the ankle of bats and helps support the UROPATAGIUM.
Callosities hardened, thickened areas on the skin (e.g. the ISCHIAL callosities in some primates).
Camelid a member of the family Camelidae (the camels), of the Artiodactyla.
Cameloid one of the South American camels.
Caniform dog-like.
Canine a unicuspid tooth posterior to the incisors and anterior to the premolars that is usually elongated and single-rooted.
Caniniform canine-shaped.
Cannon bone a bone formed by the fusion of metatarsal bones in the feet of some families.
Canopy a fairly continuous layer in forests produced by the intermingling of branches of trees; may be fully continuous (closed) or broken by gaps (open). The crowns of some trees project above the canopy layer and are known as emergents.
Caprid a member of the bovid tribe Caprini, of the Artiodactyla.
Carnassial (teeth) opposing pair of teeth especially adapted to shear with a cutting (scissor-like) edge; in extant mammals the arrangement is unique to Carnivora and the teeth involved are the fourth upper premolar and first lower molar.
Carnivore any meat-eating organism (alternatively, a member of the order Carnivora, many of whose members are carnivores).
Carotid rete an interwoven network of bloodvessels formed from the carotid artery.
Carpals wrist bones which articulate between the forelimb bones (radius and ulna) and the metacarpals.
Carrion dead animal matter used as a food source by scavengers
Carrying capacity the maximum number of animals that can be supported in a given area or habitat. In a logistic equation, it is represented by the upper ASYMPTOTE.
Catarrhine a "drooping-nosed" monkey, with nostrils relatively close together and open downward; term used for Old World monkeys, gibbons, apes, and man in contrast to PLATYRRHINE monkeys of the New World. The Catarrhini and Platyrrhini (collectively sometimes referred to as the Anthropoidea or Simiiformes) are usually included along with tarsiers in the HAPLORHINES.
Cathemeral applied to an animal that is irregularly active at any time of day or night, according to the prevailing circumstances.
Caudal gland an enlarged skin gland associated with the root of the tail. Subcaudal, placed below the root; supracaudal, above the root.
Cecum a blind sac in the digestive tract, opening out from the junction between the small and large intestines. In herbivorous mammals it is often very large; it is the site of bacterial action on cellulose. The end of the cecum is the appendix; in species with reduced ceca the appendix may retain an antibacterial function.
Cellulose the fundamental constituent of the cell walls of all green plants, and some algae and fungi. It is very tough and fibrous, and can be digested only by the intestinal flora in mammalian guts.
Cementum hard material which coats the roots of mammalian teeth. In some species, cementum is laid down in annual layers which, under a microscope, can be counted to estimate the age of individuals.
Cephalopod a member of an order of molluscs including such marine invertebrates as squid, octopus, and cuttlefish.
Cerebral cortex the surface layer of cells (gray matter) covering the main part of the brain, consisting of the cerebral hemispheres.
Cerrado (central Brazil) a dry savanna region punctuated by patches of sparsely wooded vegetation.
Cetacea mammalian order comprising whales, dolphins and porpoises.
Cervical of, or pertaining to, the neck; of, or pertaining to, the cervix of the uterus.

Cervid a member of the family Cervidae (the deer), of the Artiodactyla.
Cervix the neck of the womb.
Chaco (Bolivia and Paraguay) a lowland plains area containing soils carried down from the Andes; characterized by dry deciduous forest and scrub, transitional between rain forest and pampas grasslands.
Chalicothere a member of the family Chalicotheriidae (order Perissodactyla), which became extinct in the Pleistocene.
Character displacement divergence in the characteristics of two otherwise similar species where their ranges overlap, caused by competition in the area of overlap
Cheek pouch a pouch used for the temporary storage of food, found only in the typical monkeys of the Old World.
Cheek teeth teeth lying behind the canines in mammals, comprising premolars and molars.
Chiropatagium the portion of the wing membrane of a bat that extends between the digits.
Chorioallantoic placentation a system whereby fetal mammals are nourished by the blood supply of the mother. The CHORION is a superficial layer enclosing all the embryonic structures of the fetus, and is in close contact with the maternal blood supply at the PLACENTA. The union of the chorion (with its vascularized ALLANTOIC STALK and yolk sac) with the placenta facilitates the exchange of food substances and gases, and hence the nutrition of the growing fetus.
Chorion the outer cellular layer of the embryonic sac of mammals, birds, and reptiles. In mammals, the outer layer of the chorion forms the PLACENTA that maintains close contact with the maternal tissues.
Chorionic villi finger-like projections from the CHORION that invade the maternal tissues and form the PLACENTA.
Choriovitelline placentation a type of PLACENTA found in all metatherians (except bandicoots) in which there are no CHORIONIC VILLI and there is only a weak connection to the uterus.
Chromatin materials in the CHROMOSOMES of living cells containing the GENES and proteins.
Chromosome a DNA protein thread occurring in the nucleus of the cell. (see also **DNA**)
Circadian rhythms activity patterns with a period of about 24 hours.
Clade a set of species derived from a single common ancestor.
Cladistics a method by which organisms are organized into TAXA on the basis of joint descent from a common ancestor and their shared derived character states.
Cladogram a branching diagram that illustrates hypothetical relationships between TAXA and shows the evolution of lineages of organisms that have diverged from a common ancestor; it does not however represent rates of evolutionary divergence.
Class in taxonomy, a category subordinate to a phylum and superior to an order. (e.g. Class Mammalia)
Clavicle the collar bone.
Cline a gradual change of character states, such as size, within a species across its geographic range.
Clitoris the small erectile body at the anterior angle of the female vulva; HOMOLOGOUS to the penis in the male.
Cloaca terminal part of the gut into which the reproductive and urinary ducts open. There is one opening to the body, the cloacal aperture, instead of a separate anus and UROGENITAL opening.
Closed-rooted teeth teeth that do not grow throughout the life of an individual. (cf **Open-rooted teeth**)
Cloud forest moist, high-altitude forest characterized by dense undergrowth, and abundance of ferns, mosses, orchids and other plants on the trunks and branches of the trees.
Clupeid a bony fish of the family Clupeidae, including herrings and similar fish, with soft fin-rays, a scaly body and four pairs of gills.
Cochlea the portion of the bony labyrinth of the inner ear. In mammals, except monotremes, it is spirally coiled.
Co-evolution complementary evolution of closely related species. (e.g. the related adaptations of flowering plants and their pollinating insects)
Cohort a group of individuals of the same age.
Colon the large intestine of vertebrates, excluding the terminal rectum. It is concerned with the absorption of water from feces.
Colonial living together in colonies. In bats, more usually applied to the communal sleeping habit, in which tens of thousands of individuals may participate.
Colostrum a special type of protein-rich mammalian milk secreted during the first few days before and after parturition. It contains antibodies that confer the mother's immunity to various diseases to the young.
Commensalism a symbiotic relationship in which one organism benefits from the association and the other organism/s is neither helped nor harmed by the relationship. (cf **Mutualism**, **Parasitism**) (see also **Symbiosis**)
Communal pertaining to the co-operation between members of the same generation in nest building but not care for the young.
Concentrate selector a herbivore which feeds on those plant parts (such as shoots and fruits) which are rich in nutrients.
Condylarthra a diverse lineage of Palaeocene herbivores, a generalized ancestral order, from which arose several orders including proboscideans, sirenians, cetaceans, perissodactyls, and artiodactyls.
Condyle a rounded process at the end of a bone, that fits into the socket of an adjacent bone to form an articulating joint. (e.g. occipital condyles provide articulation between the skull and vertebral column)
Congener a member of the same species (or genus).
Coniferous forest forest comprising largely evergreen conifers (firs, pines, spruces etc). typically in climates either too dry or too cold to support deciduous forest. Most frequent in northern latitudes or in mountain ranges.
Consort (consortship) in certain primates (e.g. Rhesus monkey, Savanna baboon, chimpanzees, orang-utan) males form temporary associations (consortships) with the females, ensuring priority of mating at the appropriate time.
Conspecific member of the same species.
Convergent evolution the independent acquisition of similar characters in evolution, as opposed to possession of similarities by virtue of descent from a common ancestor.
Copepod a small marine crustacean of the invertebrate order Copepoda.
Coprophagy the eating of feces or fecal pellets. (see also **Refection**)
Copulatory plug a plug of coagulated semen formed in the vagina after copulation; found only in certain species of mammals (e.g. springhare).
Coracoid a bone in the pectoral girdle of vertebrates between the scapula and the sternum. In mammals, other than monotremes, it is reduced to a small process on the scapula.
Corpus callosum a broad band of nerve fibres that interlinks the right and left cerebral hemispheres in eutherian mammals.
Corpus luteum the progesterone-secreting mass of follicle cells that develops in the ovary after the egg has been released at ovulation.
Coteries small groups of some mammals (e.g. prairie dogs and some squirrels), which occupy communal burrows.
Cotyledonary placenta a type of CHORIOALLANTOIC PLACENTA in which the CHORIONIC VILLI are grouped into tufts or balls separated by regions of smooth CHORION.
Countercurrent heat exchange mechanism an arrangement of blood vessels that allows peripheral cooling, particularly of appendages, and at the same time maintains an adequate blood supply without excessive heat loss.
Cranium the upper portion of the skull including the bones that surround the brain.
Crenulated finely notched.
Crepuscular active in twilight.
Cretaceous geological time period 144–65 million years ago.
Cricetine adjective and noun used to refer to (a) the primitive rodents from which the New World rats and mice, voles and lemmings, hamsters and gerbils are descended, (b) these modern rodents. In some taxonomic classification systems these subfamilies of the family Muridae are classified as members of a separate family called Cricetidae, with members of the Old World rats and mice alone constituting the Muridae.
Crown the portion of a tooth that projects above the gum, composed of enamel and dentine.
Crustaceans members of a class within the phylum Arthropoda typified by five pairs of legs, two pairs of antennae, head and thorax joined, and calcareous deposits in the exoskeleton (eg crayfish, crabs, shrimps).
Crypsis an aspect of the appearance of an organism which camouflages it from the view of others, such as predators or competitors.
Cryptic (coloration or locomotion) protecting through concealment.
Cue a signal, or stimulus (e.g. olfactory) produced by an individual which elicits a response in other individuals.
Cursorial being adapted for running.
Cusp a prominence on a cheek tooth (premolars or molar).
Cuticle the thin, transparent, outer layer of hair. It forms a distinct, scale-like pattern on the surface.
Cyamids amphipod crustaceans of the family Cyamidae that parasitize the skin of the whales (hence the popular name "whale lice").
Cynodontia a diverse group of THERIODONT THERAPSID reptiles from which mammals supposedly evolved.

Deciduous forest temperate and tropical forest with moderate rainfall and marked seasons. Typically, trees shed leaves during either cold or dry periods.
Deciduous placenta a type of PLACENTA in which a portion of the uterine wall is lost at birth.
Deciduous teeth teeth that are replaced usually early in a mammal's life.
Delayed development a type of embryonic development in which the growth rate of the embryo slows following implantation in the uterine lining. Found in some bats.

Delayed fertilization see **Fertilization**
Delayed implantation see **Implantation**
Deme a local population within which breeding occurs more or less at random.
Den a shelter, natural or constructed, used for sleeping, for giving birth and raising young, and/or in winter; also the act of retiring to a den to give birth and raise young, or for winter shelter.
Dendrogram a treelike diagram of the relationships in a PHYLOGENY.
Dental formula a convention for summarizing the dental arrangement whereby the numbers of each type of tooth in each half of the upper and lower jaw are given. The numbers are always presented in the order: incisor (I), canine (C), premolar (P), molar (M). The final figure is the total number of teeth to be found in the skull. A typical example for Carnivora would be I3/3, C1/1, P4/4, M3/3 = 44.
Dentition the arrangement of teeth characteristic of a particular species.
Dentary bone the single bone of the lower jaw or mandible in mammals.
Dentine a bone-like material (containing calcium phosphate) that forms the body of the tooth.
Dermis the layer of skin lying beneath the outer epidermis.
Derived character refers to a character state that is a modified version of that in the ancestral stock. (see also **Apomorphy**)
Desert areas of low rainfall, typically with sparse scrub or grassland vegetation or lacking vegetation altogether.
Diapause temporary cessation in the growth and development of an insect or mammal. (see also **Embryonic diapause**)
Diaphragm the transverse, muscular partition separating the thoracic and abdominal cavities.
Diastema a space between the teeth, usually the incisors and cheek teeth. It is typical of rodents and lagomorphs, though also found in artiodactyls and perissodactyls, and may be used in grooming.
Dicerathere a member of the family Diceratheriidae (order Perissodactyla), which became extinct in the Miocene.
Dichromatic in dichromatic species, males and females exhibit quite different color patterns (e.g. certain day-active lemurs, some New World monkeys, some Old World monkeys, and certain gibbons).
Dicotyledon one of the two classes of flowering plants (the other class comprises monocotyledons), characterized by the presence of two seed leaves in the young plant, and by net-veined, often broad leaves, in mature plants. Includes deciduous trees, roses etc.
Didactylous the condition in metatherians in which the digits are unfused.
Didelphous pertaining to the female reproductive tract of metatherians in which the uterus, oviduct, and vagina are paired.
Digesta digested food or material.
Digit a finger or toe.
Digital glands glands occurring between or on the toes.
Digitigrade method of walking on the toes without the heel touching the ground. (cf **Plantigrade**)
Dilambodont arrangement of the tooth cusps and associated ridges to form a W-shaped pattern.
Dimorphism the existence of two distinct forms (polymorphism several distinct forms); the term "sexual dimorphism" is applied to cases where the male and female of a species differ consistently in, for example, shape, size, coloration and armament.
Dioecious male and female reproductive organs in separate, unisexual individuals.
Diestrus the period between two estrous cycles in a female mammal.
Diphyletic a group whose members are descended from two distinct lineages.
Diphyodont having two sets of teeth during a lifetime, typically a set of deciduous ("milk") teeth and a set of permanent teeth.
Diploid number the total number of paired CHROMOSOME sets in the cell nucleus. The diploid state is expressed as 2n= (cf haploid state of n=). Almost all animal cells are diploid (e.g. human beings have 2n=46).
Diprotodont having the incisors of the lower jaw reduced to one functional pair, as in possums and kangaroos (small, non-functional incisors may also be present). (cf **Polyprotodont**)
distribution geographical distribution of a species that is marked by gaps. Commonly brought about by fragmentation of suitable habitat, especially as a result of human intervention.
Dispersal the movements of animals, often as they reach maturity, away from their previous HOME RANGE (equivalent to EMIGRATION). Distinct from dispersion, however, the pattern in which things (perhaps animals, food supplies, nest sites) are distributed or scattered.
Display any relatively conspicuous pattern of behavior that conveys specific information to others, usually to members of the same species; can involve visual and or vocal elements, as in threat, courtship or "greeting" displays.
Distal far from the point of attachment or origin (e.g. tip of tail).
Diurnal active in daytime.
DNA (Deoxyribonucleic acid) the genetic material of organisms, its sequence of paired bases constituting the genetic code. It is characterized by the presence of a sugar (deoxyribose) and four bases: cytosine, thymine, adenine, and guanine.
Docodonta an extinct order of late Jurassic mammals known only from the remains of complex tooth and jaw fragments.
Domestication selective breeding of animal species by humans in controlled environments in order to accommodate human needs (e.g. cattle).
Dominant see **Hierarchy**
Doppler shift change in sound frequency caused by movement of the source or the receiver.
Dormancy a period of inactivity; many bears, for example, are dormant for a period in winter; this is not true hibernation, as pulse rate and body temperature do not drop markedly.
Dorsal on the upper or top side or surface (e.g. dorsal stripe).
Dryolestidae a family of early omnivorous mammals in the order EUPANTOTHERIA that were extinct by the mid-Cretaceous.
Ductus deferens (or **Vas deferens**) the duct or tube that carries sperm from the epididymus to the urethra or the cloaca in male mammals.
Duplex a type of uterus in which the right and left parts are completely unfused and each has a distinct cervix. Found in lagomorphs, rodents, hyraxes, and aardvark.
Durophagy the eating of hard or chitinous materials, such as shells and hulls.

Echolocation the process of perception, often direction finding, based upon reaction to the pattern of reflected sound waves (echoes).
Ecological succession replacement of populations in a community through a more or less regular progression culminating in a stable climax community.
Ecological zoogeography the study of the relationships between living organisms in relation to their physical and biotic environment.
Ecology the study of plants and animals in relation to their natural environmental setting. Each species may be said to occupy a distinctive ecological niche.
Ecosystem a unit of the environment within which living and nonliving elements interact.
Ecotone an intermediary habitat created by the juxtaposition of distinctly different habitats (e.g. the zone of transition between grassland and woodland).
Ecotype a genetic variety within a single species, adapted for local ecological conditions.
Ectoparasites parasites that occur on or embedded in the surface of their host organism.
Ectothermy maintenance of body temperature by behavioral means (e.g. basking in the sun).
Edentate a term, literally meaning "without teeth," formerly applied to the clade including the Xenarthra (sloths, armadillos, anteaters) and the Pholidota (pangolins).
Eimer's organ a specialized touch receptor located on the snouts of moles and desmans.
Elongate relatively long (e.g. of canine teeth, longer than those of an ancestor, a related animal, or than adjacent teeth).
Emarginate having a notch or notches at the end.
Embryonic diapause the temporary cessation of development of an embryo leg in some bats and kangaroos.
Emigration departure of animal(s), usually at or about the time of reaching adulthood, from the group or place of birth. (see also **Dispersal**)
Enamel a hard crystalline material, similar in composition to bone, which occurs on the outside portion on the crown of a tooth. It is the hardest and heaviest tissue in vertebrates.
Endemic (endemism) a taxon restricted to a limited geographic area and not found anywhere else.
Endogenous originating from within an organism.
Endometrium the inner lining of the uterus in which blastocysts implant during gestation.
Endoparasites parasites that occur inside the body of their host organism.
Endotheliochorial placenta an arrangement of the CHORIOALLANTOIC PLACENTA in which the CHORION of the embryo is in direct contact with the maternal capillaries.
Endothermy maintenance of constant body temperature by means of heat produced by ENDOGENOUS means (e.g. sweating, panting, shivering)
Entelodont a member of the family Entelodontidae, Oligocene artiodactyls which represent an early branch of the pig family, Suidae.
Entoconid a major cusp found in the lingual portion of the TALONID of the lower molars.
Enzootic concerning disease regularly found within an animal population (endemic applies specifically to people) as distinct from EPIZOOTIC.
Eocene geological epoch 55–34 million years ago.
Epidemic a severe outbreak of a particular disease, usually over a widespread area.
Epidermis the outer layer of mammalian skin (and in plants the outer tissue of young stem, leaf, or root).

Epididymus a coiled duct that receives the sperm from the SEMINIFEROUS TUBULES of the testes and then transmits the sperm to the DUCTUS DEFERENS.
Epipubic bones a pair of bones that extend anteriorly from the pubic bones of the pelvis in monotremes, most metatherians, and also reptiles.
Epiphysis the head of a bone, usually bearing a surface for articulation with another bone.
Epitheliochorial placenta an arrangement of the CHORIOALLANTOIC PLACENTA characterized by having six tissue layers separating the fetal and maternal blood supply, and with the CHORIONIC VILLI resting in pockets in the ENDOMETRIUM. This is the least modified placental arrangement.
Epizootic a disease outbreak in an animal population at a specific time (but not persistently, as in ENZOOTIC); if an epizootic wave of infection eventually stabilizes in an area, it becomes enzootic.
Erectile capable of being raised to an erect position (erectile mane).
Esophagus the gullet connecting the mouth with the stomach.
Estivate (noun: estivation) to enter a state of dormancy or TORPOR in seasonal hot, dry weather, when food is scarce.
Estrus (adj.: estrous) the period in the estrous cycle of female mammals at which they are often attractive to males and receptive to mating. The period coincides with the maturation of eggs and ovulation (the release of mature eggs from the ovaries). Animals in estrus are often said to be "on heat" or "in heat." In primates, if the egg is not fertilized the subsequent degeneration of uterine walls (ENDOMETRIUM) leads to menstrual bleeding. In some species ovulation is triggered by copulation and this is called induced ovulation, as distinct from spontaneous ovulation.
Ethiopian a geographical region comprising Africa, south of the Sahara. Sometimes referred to as Afrotropical.
Eucalypt forest Australian forest, dominated by trees of the genus *Eucalyptus*.
Eupantotheria an order of extinct mammals, known from the Jurassic of North America and Europe.
Euphausiids see **Krill**
Eusocial a social system whereby only one female produces offspring and there is reproductive division of labor (castes) and the cooperative rearing of young by members of previous generations (who also help defend and maintain the colony). It is best typified by honey bees, but in mammals is recorded only in two species of mole-rat.
Eutherian a mammal of the subclass Eutheria, the dominant group of mammals. The embryonic young are nourished by an allantoic placenta.
Evaporative cooling loss of heat through the evaporation of sweat or saliva from the skin, or otherwise of water vapour from the nasal mucosa or lungs.
Exogenous originating from outside the organism.
Exotic a species introduced to an area in which it does not occur naturally.
Extant not extinct; still surviving.
External auditory meatus a passageway leading from the base of the pinna or surface of the head to the TYMPANIC MEMBRANE.
Extinction loss of a taxon.
Extirpation the extermination of a population or taxon from a given area.
Exudate natural plant exudates include gums and resins; damage to plants (e.g. by marmosets) can lead to loss of sap as well. Certain primates (e.g. Bush babies) rely heavily on exudates as a food source.

Facultative optional. (cf **Obligate**)
Facultative delayed implantation a form of delayed IMPLANTATION in which a delay results because the female is nursing a large litter or faces harsh environmental conditions.
Falcate curved or hooked.
Family in taxonomy, a division subordinate to an order and superior to a genus (e.g. family Felidae).
Fast ice sea ice which forms in polar regions along the coast, and remains fast, being attached to the shore, to an ice wall, an ice front, or over shoals, generally in the position where it originally formed.
Feces excrement from the bowels; colloquially known as droppings or scats.
Fecundity the number of offspring produced during a certain amount of time.
Female defense polygyny a mating system whereby males control access to females by directly competing or interfering with other males.
Feral living in the wild (of domesticated animals, e.g. cat, dog, pig).
Fermentation the decomposition of organic substances by microorganisms. In some mammals, parts of the digestive tract (e.g. the cecum) may be inhabited by bacteria that break down cellulose and release nutrients.
Fertilization the penetration of an egg by a sperm resulting in the combination of maternal and paternal DNA and formation of a ZYGOTE. Most aquatic mammals achieve fertilization externally. Delayed fertilization occurs following mating when sperm are deposited in the uterine tract of the female but ovulation and fertilization are delayed for several months. The sperm remain viable in the female's reproductive tract during this time.
Fetal development rate the rate of development, or growth, of unborn young.
Fetlock joint above the hooves.
Fetus the mammalian embryo
Filiform thin and threadlike.
Fimbriation a stiff fringe of hairs between the toes that aid some species in locomotion (e.g. shrews).
Fin an organ projecting from the body of aquatic animals and generally used in steering and propulsion.
Fission splitting or parting.
Fissipedia (suborder) name given by some taxonomists to modern terrestrial carnivores to distinguish them from the suborder Pinnipedia which describes the marine carnivores. Here we treat both as full orders, the Carnivora and the Pinnipedia.
Fitness a measure of the ability of an animal (with one genotype or genetic make-up) to leave viable offspring in comparison to other individuals (with different genotypes). The process of natural selection, often called survival of the fittest, determines which characteristics have the greatest fitness, i.e. are most likely to enable their bearers to survive and rear young which will in turn bear those characteristics. (see **Inclusive fitness**, **Natural selection**)
Flehmen German word describing a facial expression in which the lips are pulled back, head often lifted, teeth sometimes clapped rapidly together and nose wrinkled. Often associated with animals (especially males) sniffing scent marks or socially important odors (e.g. scent of estrous female). Possibly involved in transmission of odor to JACOBSON'S ORGAN.
Flense to strip blubber from a whale or seal.
Flipper a limb adapted for swimming.
Floe a sheet of floating ice.
Fluke one of the lobes of a whale's tail; the name refers to their broad, triangular shape.
Folivory consuming mainly leaves.
Follicle a small sac, therefore (a) a mass of ovarian cells that produces an ovum, (b) an indentation in the skin from which hair grows.
Foramen an opening or passage through bone.
Forbs a general term applied to ephemeral or weedy plant species (not grasses). In arid and semi-arid regions they grow abundantly and profusely after rains.
Foregut fermentation see **Ruminant**
Forestomach a specialized part of the stomach consisting of two compartments (presaccus and saccus).
Fossorial burrowing (of life-style or behavior); adapted to a subterranean lifestyle. Animals that are only partially adapted to such a lifestyle are semi-fossorial.
Frequency the number of wave lengths per second, expressed in Hertz (Hz).
Frugivory consuming mainly fruits.
Furbearer term applied to mammals whose pelts have commercial value and form part of the fur harvest.
Fusiform elongated with tapering ends.
Fusion opposite of FISSION; often applied to a fission–fusion social system in some species whereby some group members leave the group and then rejoin it later.

Gadoid cod-like fish of the suborder Gadoidei.
Gait manner of walking.
Gallery forest luxuriant forest lining the banks of watercourses.
Gamete a male or female reproductive cell (ovum or spermatozoon).
Gape the extent to which the mouth can be opened.
Gene the basic unit of heredity; a portion of DNA molecule coding for a given trait and passed, through replication at reproduction, from generation to generation. Genes are expressed as adaptations and consequently are the most fundamental units (more so than individuals) on which natural selection acts.
Generalist an animal whose lifestyle does not involve highly specialized strategems (cf **Specialist**); for example, feeding on a variety of foods which may require different foraging techniques.
Genotype the genetic constitution of an organism, determining all aspects of its appearance, structure and function. (cf **Phenotype**)
Genus (pl. **genera**) in taxonomy, a division superior to species and subordinate to family (e.g. genus *Panthera*).
Gestation the period of development within the uterus; the process of delayed IMPLANTATION can result in the period of pregnancy being longer than the period during which the embryo is actually developing.
Glanils (marking) glandular areas of the skin, used in depositing scent marks.
Glans penis the head or distal portion of the penis.
Gliding aerial locomotion (not powered as in bats) involving the use of a membrane (the PATAGIUM) to provide lift.
Glissant gliding locomotion. Found in flying lemurs, colugos, flying squirrels and some other mammals.
Granivory consuming a diet of seeds or nuts.
Graviportal animals in which the weight is carried by the limbs acting as rigid, extensible struts, powered by extrinsic muscles (e.g. elephants and rhinos).

Grazer a herbivore which feeds upon grasses. (cf **Browser**)
Great call a protracted series of notes, rising to a climax, produced by the female as part of the group song in lesser apes.
Gregarious living in groups or herds.
Grizzled sprinkled or streaked with gray.
Guano bat fecal droppings. They may accumulate in large quantities where colonies roost.
Guard hair an element of the coat of seals consisting of a longer, stiffer, more bristle-like hair which lies outside and supports the warmer, softer underfur.
Guild a group of species that exploits a common resource base in a similar fashion (e.g. the carnivore guild).
Gumivory consuming a diet of gum (plant EXUDATE).

Hallux the first digit of the hind foot.
Haplorhine a member of the primate suborder Haplorrhini that comprises the anthropoids (suborder Simiiformes) and the infraorder Tarsiiformes (tarsiers). In all members the upper lip is whole and the placenta is hemochorial. (cf **Strepsirhine**)
Haplotype a set of genetic determinants located on a single chromosome; also the single species included in a genus at the time of its designation, thereby becoming the type species of the genus.
Harem group a social group consisting of a single adult male, at least two adult females and immature animals; a common pattern of social organization among mammals.
Haulout behavior of sea mammals pulling themselves ashore.
Heath low-growing shrubs with woody stems and narrow leaves (e.g. heather), which often predominate on acidic or upland soils.
Helper jargon for an individual, generally without young of its own, which contributes to the survival of the offspring of another by behaving parentally towards them. (see also **Alloparent**)
Hemochorial placenta the arrangement of the CHORIOALLANTOIC PLACENTA whereby the CHORIONIC VILLI are in direct contact with the maternal blood supply.
Hemoendothelial placenta the arrangement of the CHORIOALLANTOIC PLACENTA in which the fetal capillaries are surrounded by maternal blood. This arrangement shows the last separation between fetal and maternal bloodstreams.
Hemoglobin an iron-containing protein in the red corpuscles which plays a crucial role in oxygen exchange between blood and tissues in mammals.
Herbivore an animal eating mainly plants or parts of plants.
Hermaphrodite an individual that has both male and female reproductive organs.
Heterodont teeth that vary in form and function in different parts of the jaws (e.g. incisors, canines, premolars, and molars in mammals). (cf **Homodont**)
Heterothermy (Poikilothermy) a condition in which the internal temperature of the body follows the temperature of the outside environment. (cf **Homeotherm**)
Hibernaculum the place in which an animal hibernates.
Hibernation a period of winter inactivity during which the normal physiological process is greatly reduced and thus during which the energy requirements of the animal are lowered.
Hierarchy (social or dominance) the existence of divisions within society, based on the outcome of interactions which show some individuals to be consistently dominant to others. Higher-ranking individuals thus have control of aspects (e.g. access to food or mates) of the life and behavior of low-ranking ones. Hierarchies may be branching, but simple linear ones are often called pecking orders (after the behavior of farmyard chickens).
Higher primate one of the more advanced primates (e.g. Chimpanzee).
Hindgut fermenter herbivores among which the bacterial breakdown of plant tissue occurs in the cecum, rather than in the rumen or foregut.
Historical zoogeography the study of the distribution of animal species in terms of their origin, dispersal and extinction.
Holarctic realm a region of the world including North America, Greenland, Europe, and Asia apart from the southwest, southeast and India.
Holotype (type specimen) in taxonomy, the individual specimen chosen as the future representative during the naming and descriptive process of a specific animal. This specimen is housed within a museum collection and marked accordingly.
Home range the area in which an animal normally lives (generally excluding rare excursions or migrations), irrespective of whether or not the area is defended from other animals. (cf **Territory**)
Homodont teeth that do not vary in form or function; they are often peg-like in structure (e.g. in toothed whales). (cf **Heterodont**)
Homoeothermy (homoiothermy) regulation of constant body temperature by physiological means regardless of the external temperature. (cf **Heterothermy**)
Homologous applied to an organ of one animal that is thought to have the same evolutionary origin as the organ of another animal, even though they differ in function (e.g. penis in males and clitoris in females).
Homoplasy the appearance of similar structures in different lineages in the course of evolution (i.e. not inherited from a common ancestor)
Hormones chemical substances, regulatory in function, released into the bloodstream or into bodily fluids from endocrine glands.
Horns cranial processes, found in bovids, formed from an inner core of bone and covered by a sheath of keratinized material and derived from the epidermis. (cf **Antlers**)
Hybrid the offspring of parents of different species.
Hydrophone a waterproof microphone held in position under the sea surface and used to detect the sounds emitted by sea mammals.
Hyoid bones skeletal elements in the throat region, supporting the trachea, larynx and base of the tongue (derived in evolutionary history from the gill arches of ancestral fish).
Hyperthermy a condition in which internal body temperature is above normal. (cf **Hypothermy**)
Hypocone a cusp posterior to the PROTOCONE and lingual (toward the tongue) in upper molars but labial (toward the cheeks) in lower molars (then referred to as a hypoconid).
Hypoconid see **Hypocone**
Hypoconulid a prominent accessory cusp found in the posterior portion of the TALONID of lower molars.
Hypodermis the innermost layer of the integument, consisting of fatty tissue.
Hypothermy a condition in which internal body temperature is below normal. (cf **Hyperthermy**)
Hypsodont high-crowned teeth, which continue to grow when full-sized and whose pulp cavity remains open; typical of herbivorous mammals. (cf **Brachydont**)
Hyracodont a member of the family Hyracodontidae (order Perissodactyla) which became extinct in the Oligocene.
Hystricognathous in rodents, having the angular process of the mandible lateral to the plane of the alveolus of the lower incisor. (cf **Sciurognathous**)
Hystricomorphous in rodents, having a greatly enlarged infraorbital foramen. (cf **Sciuromorphous**)

Ilium the largest and most dorsal of the three pelvic bones.
Imbricate overlapping.
Immigration movement of individuals into a population or given area.
Implantation the process whereby the free-floating BLASTOCYST (early embryo) becomes attached to the uterine wall in mammals. At the point of implantation a complex network of blood vessels develops to link mother and embryo (the placenta). In delayed implantation, the blastocyst remains dormant in the uterus for periods varying, between species, from 12 days to 11 months. Delayed implantation may be obligatory or facultative and is known for some members of the Carnivora and Pinnipedia and others.
Inbreeding mating among related individuals.
Inbreeding depression reduced reproductive success and survival of offspring as a result of INBREEDING.
Incisor a unicuspid tooth in mammals located anterior to the canines.
Inclusive fitness a measure of the animal's fitness which is based on the number of its genes, rather than the number of its offspring, present in subsequent generations. This is a more complete measure of fitness, since it incorporates the effect of, for example, alloparenthood, wherein individuals may help to rear the offspring of their relatives. (see also **Kin selection**, **Alloparent**)
Incus the second of three bones of the middle ear in mammals.
Induced ovulation see **Estrus**
Infanticide the killing of infants. Infanticide has been recorded notably in species in which a bachelor male may take over a harem from its resident male(s) (e.g. lions).
Infraorbital foramen a canal in the maxilla, below and slightly in front of the orbit, through which bloodvessels and nerve fibres pass.
Infrasound sound frequencies below 20 Hz.
Infundibulum a funnel-shaped opening of the oviduct situated near the ovary that receives the oocytes at ovulation.
Inguinal pertaining to the groin.
Innervated having a supply of nerves to and from an organ.
Insectivore an animal eating mainly arthropods (insects, spiders).
Integument the skin.
Interdigital between the digits.
Interfemoral a membrane stretching between the femora, or thigh bones in bats.
Interordinal between different orders.
Interspecific between different species.
Intestinal flora simple plants (e.g. bacteria) which live in the intestines, especially the cecum, of mammals. They produce enzymes which break down the cellulose in the leaves and stems of green plants and convert it to digestible sugars.
Intraspecific between individuals of the same species.
Introduced of a species which has been brought, by man, from lands where it

occurs naturally to lands where it has not previously occurred. Some introductions are accidental (e.g. rats which have travelled unseen on ships), but some are made on purpose for biological control, farming or other economic reasons (e.g. the common brush-tail possum, which was introduced to New Zealand from Australia to establish a fur industry).
Introgression the mixing of gene pools.
Invertebrate an animal which lacks a backbone (e.g. insects, spiders, crustaceans).
Ischial pertaining to the hip.
Ischial callosities specialized, hardened pads of tissue present on the buttocks of some monkeys and apes. Each overlies a flattened projection of the ischium bone of the pelvis. Known also as "sitting pads," they are found in Old World _monkeys and lesser apes. (see also **Callosities**)
Ischium one of the three bones of the pelvis.
Iteroparous the production of offspring on a regular basis by an organism. (cf **Semelparous**)

Jacobson's organ (vomeronasal organ) a structure in a foramen (small opening) in the palate of many vertebrates which appears to be involved in olfactory communication. Molecules of scent may be sampled in these organs.
Joey a young kangaroo that is still nursing but not restricted to the pouch.
Jurassic geological time period 213–144 million years ago.
Juvenile no longer having the characteristics of an infant, but not yet fully adult.

Karyotype the characteristic number and shape of the chromosomes of a cell, individual or species.
Keratin a tough, fibrous material found in epidermal tissues, such as hair and hooves.
Keratinized made of KERATIN
Kin related individuals.
Kin selection a facet of natural selection whereby an animal's fitness is affected by the survival of its relatives or KIN. Kin selection may be the process whereby some alloparental behavior evolved; an individual behaving in a way which promotes the survival of its kin increases its own INCLUSIVE FITNESS, despite the apparent selflessness of its behavior.
Kleptoparasite an animal that steals food from other animals.
Knuckle-walk to walk on all fours with the weight of the front part of the body carried on the knuckles. Found only in gorillas and chimpanzees.
Kopje (koppie) a rocky outcrop, typically on otherwise flat plains of African grasslands.
Krill shrimp-like crustaceans of the genera *Euphausia*, *Meganyctiphanes* etc., occurring in huge numbers in polar seas, particularly of Antarctica, where they form the principal prey of baleen whales.
K-selection selection favouring slow rates of reproduction and growth for maximizing competitive ability as a response to a stable environment. (cf **R-selection**)

Labial of, or pertaining to, the cheek.
Labile (body temperature) an internal body temperature which may be lowered or raised from an average body temperature.
Lactation (verb: lactate) the secretion of milk, from mammary glands.
Lactose a disaccharide sugar that is the principal sugar of milk.
Lambdoidal crest a bony ridge at the rear of the cranium.
Laminae ridges on teeth sometimes with distinct cusps.
Laminar flow streamline flow in a viscous fluid near a solid boundary; the flow of water over the surface of whales is laminar.
Lamoid Llama-like; one of the South American cameloids.
Lanugo the birth-coat of mammals which is shed to be replaced by the adult coat.
Larynx dilated region of upper part of windpipe, containing vocal chords. Vibration of cords produces vocal sounds.
Latrine a place where feces are regularly left (often together with other scent marks); associated with olfactory communication.
Lead a channel of open water between ice floes.
Lek a display ground at which individuals of one sex maintain miniature territories into which they seek to attract potential mates.
Lesser apes the gibbons and siamang.
Liana a climbing plant. In rain forests large numbers of often woody, twisted lianas hang down like ropes from the crowns of trees.
Lingual of, or pertaining to, the tongue.
Lipotyphlan an early insectivore classification; menotyphlan insectivores possess a cecum, lipotyphlans do not. Only lipotyphlans are now classified as Insectivora. Recent phylogenetic reclassification (cf **Afrotheria**) has seen a subdivision of the Lipotyphlans into the orders Afrosoricida and Eulipotyphla.
Llano South American semi-arid savanna country (e.g. of Venezuela).
Lobtailing a whale beating the water with its tail flukes, perhaps to communicate with other whales.
Locus the specific location of a GENE on a CHROMOSOME.
Loph a transverse ridge on the crown of molar teeth.
Lophiodont a member of the family Lophiodontidae (order Perissodactyla) which became extinct in the early Tertiary.
Lophodont molar teeth whose cusps form ridges or lophs.
Lordosis a behavior, performed by females signaling their willingness to mate, in which the lumbar curvature is exaggerated.
Lower critical temperature the temperature at which an animal must increase its METABOLIC RATE in order to balance heat loss. (cf **Upper critical temperature**)
Lower primate one of the more primitive primates (e.g. lorises).
Lumbar a term locating anatomical features in the loin region (e.g. lumbar vertebrae are at the base of the spine).
Luteinizing hormone (LH) a hormone that stimulates development of corpora lutea and progesterone production in females.

m.y.a. abbreviation for million years ago.
Male dominance polygyny a mating system whereby the males maintain a dominance hierarchy thereby influencing their access to females; higher-ranking males obtain more mates.
Mallee a grassy, open woodland habitat characteristic of many semi-arid parts of Australia. "Mallee" also describes the multi-stemmed habit of eucalypt trees which dominate this habitat.
Malleus the first of the three bones of the middle ear in mammals.
Mamma (pl. **mammae**) (mammary glands) the milk-secreting organ of female mammals, probably evolved from sweat glands.
Mammal a member of the class of vertebrate animals (the Mammalia) having mammary glands which produce milk with which they nurse their young.
Mammalogy the study of mammals.
Mammalogist someone who studies mammals.
Mammilla (pl. **mammillae**) nipple, or teat, on the mamma of female mammals; the conduit through which milk is passed from the mother to the young.
Mandible the lower jaw
Mandibular fossa part of the cranium with which the mandible (lower jaw) interacts.
Mandibular ramus one of the major portions of the dentary bone of the lower jaw. The horizontal part holds the teeth while the ascending part articulates with the skull.
Mangrove forest tropical forest developed on sheltered muddy shores of deltas and estuaries exposed to tide. Vegetation is almost entirely woody.
Manus the hindfoot.
Marine living in the sea.
Marsupium the pouch found in many marsupials and in echidnas that encloses the mammary glands and serves as an incubation chamber for the young (see also **pouch**).
Mask colloquial term for the face of a mammal, especially a dog, fox or cat.
Masseter a powerful muscle, subdivided into parts, joining the mandible to the upper jaw. Used to bring jaws together when chewing.
Mastication the act of chewing.
Maternity the state of being the maternal parent (mother) to an offspring.
Matriarchal of a society in which most activity and behavior is centred around the dominant female.
Matriline a related group of animals linked by descent through females alone.
Maxilla one of the paired bones making up the upper jaw and carrying the teeth. Sometimes applied to the whole upper jaw.
Melanin a dark pigment found in the skin.
Melanism darkness of color due to the presence of MELANIN. (cf **Albinism**)
Menstrual cycle an approximately monthly cycle involving alternation of ovulation and menstruation (loss of blood from the vulva at monthly intervals) until pregnancy intervenes; found in humans, great apes, Old World monkeys and, to varying degrees, in New World monkeys.
Menotyphlan see **Lipotyphlan**
Mesic pertaining to conditions of moderate moisture or water-supply; used of animals occupying moist habitats (i.e. mesic-adapted). (cf **Xeric**)
Metabolic rate the rate at which the chemical processes of the body occur.
Metabolism the chemical processes occurring within an organism, including the production of protein from amino acids, the exchange of gasses in respiration, the liberation of energy from foods and innumerable other chemical reactions.
Metacarpal bones of the hand, between the carpals of the wrist and the phalanges of the digits.
Metacone a cusp posterior to the PROTOCONE and labial in upper molars and lingual in lower molars (then called a metaconid).
Metaconid see **Metacone**
Metapodial the proximal element of a digit (contained within the palm or sole). The metapodial bones are metacarpals in the manus and metatarsals in the pes.
Metapopulation a set of semi-isolated populations linked together via dispersal and having some regular gene flow.

Metatarsal bones of the foot articulating between the tarsals of the ankle and the phalanges of the digits.
Metatheria the group of mammals that comprises the marsupials and all extinct relatives.
Metestrus third stage in the estrous cycle, in which the corpora lutea are formed and progesterone levels are high.
Microhabitat the particular parts of the habitat that are encountered by an individual in the course of its activities.
Microsatellite DNA tandem repeats of short sequences of DNA, most often multiples of two to four bases.
Midden a dunghill, or site for the regular deposition of feces by mammals.
Migration movement, usually seasonal, from one region or climate to another for purposes of feeding or breeding.
Miocene geological epoch 24–5 million years ago.
Molar a non-deciduous cheek tooth posterior to the premolar.
Molariform having the shape and appearance of a molar.
Molecular phylogeny a hypothetical representation of the evolutionary history of a group of organisms based on characters defined at the molecular level.
Molting seasonal replacement of hair.
Monestrus having a single estrous period per year or breeding season. (cf **Polyestrus**)
Monogamy a mating system in which individuals have only one mate per breeding season. (cf **Polygamy**)
Monophyletic a group whose members are descended from a common ancestor. (cf **Paraphyletic**)
Monotreme a mammal of the subclass Monotremata (platypus and echidnas). The only egg-laying mammals.
Monotypic a genus comprising a single species.
Monozygotic polyembryony a reproductive process, characteristic of some armadillos, in which a single zygote splits into individual zygotes and forms several identical, same-sex embryos.
Montane pertaining to mountainous country.
Montane forest forest occurring at middle altitudes on the slopes of mountains, below the alpine zone but above the lowland forests.
Morphology (morphological) the structure and shape of an organism.
Moss forest moist forest occurring on higher mountain slopes, e.g. 1,500–3,200m (4,900–10,500ft) in New Guinea. It is characterized by rich growth of mosses and other plants on tree trunks and branches.
Mucosa mucous membrane; a membrane rich in mucous glands such as the lining of the mouth.
Multiparous a female that has had several litters or young.
Murine adjective and noun used to refer to members of the subfamily (of the family Muridae) Murinae, which consists of the Old World rats and mice. In some taxonomic classification systems this subfamily is given the status of a family, Muridae, and the members then are sometimes referred to as murids. (see also **Cricetine**)
Musk scent secreted from scent glands (musk gland) in mustelids and other mammal species.
Musth the period of heightened reproductive activity in male elephants; aggression increases during this period, usually lasting two to three months.
Mutualism a symbiotic relationship that involves a mutually beneficial association between members of two species. (cf. **Commensalism**, **Parasitism**; see also **Symbiosis**)
Mutation a structural change in a gene which can thus give rise to a new heritable characteristic.
Mycophagy consuming a diet of fungi.
Myoglobin a protein related to HEMOGLOBIN in the muscles of vertebrates; like hemoglobin, it is involved in the oxygen exchange processes of respiration.
Myomorphous in rodents, having a slip of the medial masseter muscle pass through an oval or V-shaped infraorbital foramen.
Myopia short-sightedness.
Myrmecophagy consuming a diet of ants and termites.
Mystacial pad the region on the snout from which most facial vibrissae originate.
Mysticete a member of the suborder Mysticeti, whales with baleen plates rather than teeth as their feeding apparatus.

Nares external nostrils.
Nasolacrimal duct a duct or canal between the nostrils and the eye.
Natal (natality) of, or pertaining to, birth.
Natal range the home range into which an individual was born.
Natural selection the process whereby individuals with the most appropriate adaptations are more successful than other individuals, and hence survive to produce more offspring. To the extent that the successful traits are heritable (genetic) they will therefore spread in the population.
Nearctic the geographical region comprising North America south to Mexico.
Nectivory consuming mainly nectar.
Neonate newborn animal.
Neotropical a geographical area comprising Central and South America, as well as the West Indies and the Galapagos.
New World a geographical term for the region including the Nearctic and Neotropical regions. (cf **Old World**)
Niche the role of a species within the community, defined in terms of all aspects of its life-style (e.g. food, competitors, predators, and other resource requirements).
Nicker a vocalization of horses, also called neighing.
Nictitating (nictating) membrane a thin, transparent membrane beneath the eyelid of some vertebrates that can cover and protect the eye.
Nocturnal active at nighttime.
Nomadic among mammals, species that have no clearly defined residence most of the time. Distinct from migratory species, which may be resident except when migrating.
Nonshivering thermogenesis means of heat production in mammals that does not involve muscle contraction. (see also **Thermogenesis**)
Noseleaf characteristically shaped flaps of skin surrounding the nasal passages of horseshoe, or nose-leaf bats (family Rhinolophidae). Ultrasonic cries are uttered through the nostrils, with the nose leaves serving to direct the echolocating pulses forwards.
Nulliparous a female that has never given birth.
Nunatak refugia within ice sheets during periods of glaciation.

Obligate required, binding. (cf **Facultative**)
Obligate delayed implantation a form of delayed IMPLANTATION in which the delay occurs as a normal part of the reproductive cycle (e.g. in armadillos).
Occipital pertaining to the posterior part of the head.
Occlusal the grinding or biting surfaces of a tooth.
Odontocete a member of the suborder Odontoceti, the toothed whales.
Old World a geographical term for the region including the Palearctic, Oriental, Ethiopian, and Australian regions. (cf **New World**)
Olfaction (olfactory) the olfactory sense is the sense of smell, depending on receptors located in the epithelium (surface membrane) lining the nasal cavity.
Oligocene geological epoch 34–24 million years ago.
Omasum third of the four chambers in the ruminant artiodactyl stomach.
Omnivore an animal eating a varied diet including both animal and plant tissue.
Ontogeny the development of an individual from fertilization of the egg to adulthood.
Open-rooted teeth teeth that grow throughout the life of the individual. (cf **Closed-rooted teeth**)
Opposable (of first digit) of the thumb and forefinger in some mammals, which may be brought together in a grasping action, thus enabling objects to be picked up and held.
Opportunist (of feeding) flexible behavior of exploiting circumstances to take a wide range of food items; characteristic of many species. (see **Generalist**, **Specialist**)
Order in taxonomy, a division subordinate to class and superior to family (e.g. order Carnivora).
Oreodont a member of the family Oreodontidae (order Artiodactyla), which became extinct in the late Tertiary.
Oriental the geographical region comprising India and Asia south of the Himalayan-Tibetan barrier, and the Australasian archipelago (excluding New Guinea and Sulawesi).
Os baculum see **Baculum**
Os clitoris a small bone present in the clitoris of some female mammal species. HOMOLOGOUS to the baculum in males.
Os penis see **Baculum**
Os sacrum fused SACRAL vertebrae in mammals.
Ossicles one of the three middle ear bones.
Ossicones short, permanent, unbranched processes of bone forming the horns in giraffes.
Ovaries the site of egg production and maturation in females.
Oviducts (Fallopian tubes) the ducts that carry the eggs from the ovary to the uterus.
Oviparous a method of reproduction involving the laying of eggs.
Ovoviviparous a method of reproduction whereby young hatch from eggs retained within the mother's uterus.
Ovulation (verb: ovulate) the shedding of mature ova (eggs) from the ovaries where they are produced. (see **Estrus**)

Pack ice large blocks of ice formed on the surface of the sea when an ice field has been broken up by wind and waves, and drifted from its original position.
Paenungulata a strongly supported interordinal group of herbivorous African mammals, which includes the hyraxes (Hyracoidea), elephants (Proboscidea), and manatees and dugongs (Sirenia). The Paenungulata are included within the AFROTHERIA.
Pair-bond an association between a male and female, lasting from courtship at least until mating is completed, and in some species, until the death of one partner.
Paleocene geological epoch 65–55 million years ago.
Palearctic a geographical region encompassing Europe and Asia north of the

Himalayas, and Africa north of the Sahara.

Paleothere a member of the family Paleotheriidae (order Perissodactyla), which became extinct in the early Tertiary.

Palmate palm-shaped.

Pampas Argentinian steppe grasslands.

Pandemic a large-scale outbreak of disease over a very wide geographic area.

Panting a thermoregulatory behavior that involves very rapid, shallow breathing in order to increase evaporation of water from the upper respiratory tract.

Papilla (pl. **papillae**) a small, nipple-like projection.

Paracone a cusp that is anterior to the PROTOCONE and labial in upper molars and lingual in lower molars (then called a paraconid).

Paraconid see **Paracone**

Páramo alpine meadow of northern and western South American uplands.

Paraphyletic a taxonomic group in which some, but not all, members are descended from a single common ancestor. (cf **Monophyletic**)

Parasitism a symbiotic relationship whereby one organism benefits from the association and the other is usually harmed. (cf **Commensalism**, **Mutualism**) (see also **Symbiosis**)

Paratype in taxonomy, a specimen (other than the HOLOTYPE) used by the author at the time of the description. It is housed in a museum and labelled accordingly.

Parous a female mammal that is pregnant or shows evidence of previous pregnancies.

Parturition the process of giving birth (hence post-partum – after birth).

Patagium a gliding membrane typically stretching down the sides of the body between the fore- and hindlimbs and perhaps including part of the tail. Found in colugos, flying squirrels, bats etc.

Paternity the state of being the paternal parent (father) to an offspring.

Pecoran a ruminant of the infra-order Pecora, which is characterized by the presence of horns on the forehead.

Pectinate resembling a comb in shape.

Pectoral girdle the bones of the shoulder region providing support for the forelimbs.

Pedicel a bony supporting structure for an antler.

Pelage all the hairs on an individual mammal.

Pelagic the upper part of the open sea, above the BENTHIC zone.

Pelvis a girdle of bones that supports the hindlimbs of vertebrates.

Penis the male copulatory organ.

Pentadactyl having five digits.

Peramuridae a family of Jurassic mammals that probably gave rise to the advanced therians.

Perineal glands glandular tissue occurring between the anus and genitalia.

Perineal swelling a swelling of the naked area of skin around the anus and vulva of a female primate, as in chimpanzees and some Old World monkeys.

Perissodactyl a member of the Perissodactyla (the odd-toed ungulates).

Pes the forefoot.

Phalanges the bones of the digits.

Phenetic similarity based on observable external characteristics.

Phenotype (phenotypic) the sum total of the observable structural and functional properties of an organism. (cf **Genotype**)

Pheromone secretions whose odors act as chemical messengers in animal communication, and which prompt a specific response on behalf of the animal receiving the message. (see also **Scent marking**)

Philopatry living and breeding in the natal area.

Phylogenetic pertaining to evolutionary relationships between groups.

Phylogeny a classification or relationship based on the closeness of evolutionary descent.

Phylogram a tree diagram (not unlike a CLADOGRAM) which shows the degree of genetic divergence among the represented taxa by means of the lengths of the branches and the angles between them.

Phylum in taxonomy, a division comprising a number of classes (e.g. Phylum Chordata)

Physiology study of the processes which go on in living organisms.

Phytoplankton minute plants floating near the surface of aquatic environments. (cf **Zooplankton**)

Piloerection fluffing or erection of the fur or hair.

Pinna (pl. **pinnae**) the projecting cartilaginous portion of the external ear.

Pinnipedia a member of the order of aquatic carnivorous mammals with all four limbs modified into flippers; the true seals, eared seals and walrus. Sometimes classified as a suborder of Carnivora.

Piscivory consuming a diet of fish.

Pituitary gland the main gland of the endocrine system that secretes a range of hormones.

Placenta, placental mammals a structure that connects the fetus and the mother's womb to ensure a supply of nutrients to the fetus and removal of its waste products. Only placental mammals have a well-developed placenta; marsupials have a rudimentary placenta or none and monotremes lay eggs.

Placental scar a pigmented area on the wall of the uterus formed from prior attachment of a fetus. It is therefore indicative of previous pregnancies.

Plankton floating plant and animal life in lakes and oceans.

Plantigrade way of walking on the soles of the feet, including the heels. (cf **Digitigrade**)

Platyrrhine a "flat-nosed" monkey with widely separated nostrils. Term commonly used for all New World monkeys in contrast to CATARRHINE monkeys of the Old World.

Pleistocene geological epoch 1.8 million – 10,000 years ago.

Pliocene geological epoch 5–1.8 million years ago.

Plesiomorphic see **Apomorphic**

Pod a group of individuals, usually applied to whales or dolphins, with some, at least temporary, cohesive social structure.

Pollex the first digit of the forefoot.

Polyandrous see **Polygynous**

Polyestrus having two or more estrous cycles in one breeding season. (cf **Monestrus**)

Polygamous a mating system wherein an individual has more than one mate per breeding season. (cf **Monogamous**)

Polygynous a mating system in which a male mates with several females during one breeding season (as opposed to polyandrous, where one female mates with several males).

Polymorphism occurrence of more than one morphological form of individual in a population. (see also **Sexual dimorphism**)

Polyprotodont having more than three well-developed lower incisor teeth (as in bandicoots and carnivorous marsupials). (cf **Diprodont**)

Population a more or less separate (discrete) group of animals of the same species within a given biotic community.

Postorbital bar a bony strut behind the eye-socket (orbit) in the skull.

Post-partum estrus ovulation and an increase in the sexual receptivity of female mammals, hours or days after the birth of a litter. (see also **Estrus**, **Parturition**)

Pouch a flap of skin on the underbelly of female marsupials which covers the mammillae. The pouch may be a simple open structure as in most carnivorous marsupials, or a more enclosed pocket-like structure as in phalangers and kangaroos. (see also **Marsupium**)

Prairie North American steppe grassland between 30°N and 55°N.

Predator an animal that forages for live prey; hence "anti-predator behavior" describes the evasive actions of the prey.

Precocial of young born at a relatively advanced stage of development, requiring a short period of nursing by parents. (cf **Altricial**)

Prehensile capable of grasping.

Premolar cheek teeth that are anterior to the molars and posterior to the canines. They may be either deciduous or non-deciduous.

Pre-orbital in front of the eye socket.

Preputial pertaining to the prepuce or loose skin covering the penis.

Primary forest forest that has remained undisturbed for a long time and has reached a mature (climax) condition; primary rain forest may take centuries to become established.

Primate a member of the order Primates comprising the apes, monkeys and related forms, including man, tarsiers, as well as the lorises, bushbabies, lemurs, and potto.

Proboscidean a member of the order of primitive ungulates, Proboscidea.

Proboscis a long flexible snout.

Process (anatomical) an outgrowth or protuberance.

Procumbent (incisors) projecting forward more or less horizontally.

Proestrus the first stage of the estrous cycle when estrogen, progesterone and LH levels are at their peak.

Progesterone a steroid hormone, secreted mainly by the CORPUS LUTEUM, which promotes growth of the uterine lining and enables the implantation of the fertilized egg.

Promiscuous a mating system wherein an individual mates more or less indiscriminately.

Pronking (stotting) movement where an animal leaps vertically, on the spot, with all four feet off the ground. Typical of antelopes (e.g. springbok), especially when alarmed.

Propatagium the anterior portion of the PATAGIUM.

Prosimian literally "before the monkeys"; a member of the relatively primitive primate suborder Prosimii (lemurs, lorises, potto, and tarsiers). Modern classification systems include the tarsiers with the anthropoids in the suborder Haplorrhini (**Haplorhines**). The remaining members comprise the suborder Strepsirrhini.

Protein a complex organic compound made of amino acids. Many different kinds of proteins are present in the muscles and tissues of all mammals.

Protein electrophoresis (allozyme analysis) a method that compares the characteristic migration distance of various proteins acting in an electric field to identify and compare individuals.

Protoceratid a member of the family Protoceratidae (order Artiodactyla), which became extinct in the late Tertiary.

Protocone the primary cusp in a molar, lingual in upper molars and labial in lower molars (then called the protoconid).

Protoconid see **Protocone**

Proximal near to the point of attachment or origin (e.g. the base of the tail).
Pseudoallantoic placentation a kind of placenta shown only by the marsupial bandicoots. Compared with the true eutherian kind of plancentation, transfer of food and gas across the chorioallantoic placental interface is inefficient, as contact between the fetal and maternal membranes is never close.
Pseudopregnancy any period characterized by a functional CORPUS LUTEUM and buildup of the uterine layer in the absence of a pregnancy.
Puberty the attainment of sexual maturity. In addition to maturation of the primary sex organs (ovaries, testes), primates may exhibit "secondary sexual characteristics" at puberty. Among higher primates it is usual to find a growth spurt at the time of puberty in males and females.
Pubis one of the three bones of the pelvis.
Puna a treeless tableland or basin of the high Andes.
Purse seine a fishing net, the bottom of which can be closed by cords, operated usually from boats. (cf **Seine**)
Pylorus the region of the stomach at its intestinal end, which is closed by the pyloric sphincter.

Quadrate bone at rear of skull which serves as a point of articulation for lower jaw.
Quadrumanous using both hands and feet for grasping.
Quadrupedal walking on all fours, as opposed to walking on two legs (BIPEDAL) or moving suspended beneath branches in trees (suspensory movements).
Quaternary geological sub-era covering the last two million years and comprising the Pleistocene and Holocene.

Race a taxonomic division subordinate to subspecies but linking populations with similar distinct characteristics.
Radiation see **Adaptive radiation**
Radiotracking a technique used for monitoring an individual's movements remotely; it involves affixing a radio transmitter to the animal and thereafter receiving a signal through directional antennas which enables the subject's position to be plotted. The transmitter is often attached to a collar, hence "radiocollar."
Rain forest tropical and subtropical forest with abundant and year-round rainfall. Typically species rich and diverse.
Range (geographical) area over which an organism is distributed.
Receptive state of a female mammal ready to mate or in ESTRUS.
Reciprocal altruism a situation whereby the short term costs for providing other individuals with some resource are offset when the recipient returns the favour at a later stage.
Reduced (anatomical) of relatively small dimension (e.g. of certain bones, by comparison with those of an ancestor or related animals).
Refection process in which food is excreted and then reingested a second time from the anus to ensure complete digestion (e.g. in some shrews).
Refugium a delimited geographical region that provides temporary shelter or protection.
Regurgitation the reverse movement of food from the stomach to the mouth.
Reingestion process in which food is digested twice, to ensure that the maximum amount of energy is extracted from it. Food may be brought up from the stomach to the mouth for further chewing before reingestion, or an individual may eat its own feces. (see also **Refection**)
Relict a persistent remnant population.
Reproductive rate the rate of production of offspring; the net productive rate may be defined as the average number of female offspring produced by each female during her entire lifetime.
Resident a mammal which normally inhabits a defined area, whether this is a HOME RANGE or a TERRITORY.
Resource defense polygyny a mating system whereby males control access to females indirectly by monopolizing the resources needed by females.
Rete mirabile a complex mass of capillaries which functions mainly as a COUNTERCURRENT HEAT EXCHANGE MECHANISM.
Reticulum second chamber of the ruminant artiodactyl four-chambered stomach. The criss-crossed (reticulated) walls give rise to honeycomb tripe. (see also **Rumen**, **Omasum**, **Abomasum**)
Retractile (of claws) able to be withdrawn into protective sheaths.
Rhinarium a naked area of moist skin surrounding the nostrils in many mammals.
Riparian vegetation or habitat along the banks of a watercourse.
Rodent a member of the order Rodentia, the largest mammalian order, which includes rats and mice, squirrels, porcupines, capybara etc.
Rookery a colony of pinnipeds.
Rorqual one of the eight species of baleen whales of the family *Balaenopteridae*.
Root the portion of tooth below the gum.
Rostrum a forward-directed process at the front of the skull of some whales and dolphins, forming a beak.
Rumen first chamber of the ruminant artiodactyl four-chambered stomach. In the rumen the food is liquefied, kneaded by muscular walls and subjected to fermentation by bacteria. The product, cud, is regurgitated for further chewing; when it is swallowed again it bypasses the rumen and RETICULUM and enters the OMASUM.
Ruminant a mammal with a specialized digestive system typified by the behavior of chewing the cud. Their stomach is modified so that vegetation is stored, regurgitated for further maceration, then broken down by symbiotic bacteria. The process of rumination is an adaptation to digesting the cellulose walls of plant cells.
Rupicaprid a member of the tribe Rupicaprini (the chamois etc.) of the Artiodactyla.
Rut a period of sexual excitement; the mating season.
R-selection selection favoring rapid reproductive rates and growth rates and typical of species found in unstable environments. When favorable conditions occur, the species can rapidly colonize the given area. (cf **K-Selection**)

Sacculated a stomach, characteristic of certain herbivores, whales, and marsupials, having more than one chamber and with microorganisms present in the first chamber for cellulose digestion.
Sacral of, or pertaining to, the vertebrae that are fused to form the sacrum to which the pelvic girdle is attached.
Sagittal crest the bony ridge on the top of the cranium (formed by the temporal ridges).
Saltatorial locomotion that involves jumping or leaping.
Sanguininvory consuming a diet of blood (e.g. vampire bats).
Satellite male an animal excluded from the core of the social system but loosely associated on the periphery, in the sense of being a "hanger-on" or part of the retinue of more dominant individuals.
Savanna (savannah) tropical grasslands of Africa, Central and South America and Australia. Typically on flat plains and plateaux with seasonal pattern of rainfall. Three categories – savanna woodland, savanna parkland and savanna grassland – represent a gradual transition from closed woodland to open grassland.
Scapula the shoulder-blade. Primates typically have a mobile scapula in association with their versatile movements in the trees
Scatterhoarding the storage of food items at various scattered localities within the confines of an animal's territory or home range.
Scent gland an organ secreting odorous material with communicative properties. (see **Scent mark**)
Scent mark a site where the secretions of scent glands, or urine or feces, are deposited and which has communicative significance. Often left regularly at traditional sites which are also visually conspicuous. Also the "chemical message" left by this means; and (verb) to leave such a deposit.
Sciurognathous in rodents, having the angular process of the mandible in line with the alveolus of the incisor. (cf **Hystricognathous**)
Sciuromorphous in rodents, having a relatively small infraorbital foramen. (cf **Hystricomorphous**)
Sclerophyll forest a general term for the hard-leafed eucalypt forest that covers much of Australia.
Scombroid a bony marine fish of the family Scombridae, with two small dorsal fins, small scales and smooth skin (e.g. mackerel and tunny).
Scrotum the bag or pouch containing the testicles in many male mammals.
Scrub a vegetation dominated by shrubs woody plants usually with more than one stem. Naturally occurs most often on the arid side of forest or grassland types, but often artificially created by man as a result of forest destruction.
Scute a bony plate, overlaid by horn, which is derived from the outer layers of the skin. In armadillos, bony scute plates provide armor for all the upper, outer surfaces of the body.
Seasonality (of births) the restriction of births to a particular time of the year.
Sebaceous gland secretory tissue producing oily substances, for example lubricating and waterproofing hair, or specialized to produce odorous secretions.
Secondary forest (or growth) regenerating forest that has not yet reached the climax condition of primary forest.
Secondary sexual character a characteristic of animals which differs between the two sexes, but excluding the sexual organs and associated structures.
Sectorial premolar one of the front lower premolars of Old World monkeys and apes, specially adapted for shearing against the rear edge of the upper canine.
Sedentary pertaining to mammals which occupy relatively small home ranges, and exhibiting weak dispersal or migratory tendencies.
Seine a fishing net with floats at the top and weights at the bottom, used for encircling fish.
Seismic signal a communication signal comprising a series of low frequency vibrations that travel through the ground.
Selective pressure a factor affecting the reproductive success of individuals (whose success will depend on their fitness, i.e. the extent to which they are adapted to thrive under that selective pressure).

Selenodont molar teeth with crescent shaped cusps.
Sella one of the nasal processes of leafnose bats; an upstanding central projection which may form a fluted ridge running backwards from between the nostrils.
Semantic of, or relating to, the meaning of signals.
Semelparous the production of offspring only once in an organisms life. (cf **Iteroparous**)
Semen (seminal fluid) the ejaculatory fluid of the male reproductive system, produced by the testes, and containing spermatazoa and secretions of various glands.
Semi-fossorial see **Fossorial**
Seminal vessicles the portion of the male reproductive tract in which sperm are stored.
Seminiferous tubules the long, convoluted tubules of the testes in which sperm are produced and mature.
Senescence the process of deterioration of an organism with age, eventually culminating in the death of the organism.
Serrate toothed or notched.
Septum a partition separating two parts of an organism. The nasal septum consists of a fleshy part separating the nostrils and a vertical, bony plate dividing the nasal cavity.
Serum blood from which corpuscles and clotting agents have been removed; a clear, almost colorless fluid.
Sexual dimorphism a condition in which males and females of a species differ consistently in form (e.g. size, shape). (see **Dimorphism**, **Polymorphism**)
Serology the study of blood sera; investigates antigen-antibody reactions to elucidate responses to disease organisms and also phylogenetic relationships between species.
Seta a stiff, bristle-like structure.
Sex ratio the ratio of males to females in a population.
Sexual selection the selection of animals in relation to mating. Males may compete for access to females; females may permit particular males mating rights.
Siblicide the killing of siblings by littermates (e.g. in spotted hyenas).
Siblings individuals who share one or both parents. An individual's siblings are its brothers and sisters, regardless of their sex.
Simian (literally "ape-like") a monkey or ape. Often used as a synonym of anthropoid or higher primate (the Simiiformes)
Sinus a cavity in bone or tissue.
Sirenia an order of herbivorous aquatic mammals, comprising the manatees and dugong.
Sister group in phylogenetics, the MONOPHYLETIC group most closely related to another monophyletic group (the two taxa are connected by a single internal node).
Sivathere a member of a giraffe family which became extinct during the last Ice Age.
Social behavior the interactive behavior of two or more individuals all of the same species.
Sociality the tendency to form social groups.
Society a group of individuals of the same species organized in a co-operative manner.
Solitary living on its own, as opposed to social or group-living life-style. (cf **Gregarious**)
Sonar sound used in connection with navigation (sound navigation ranging).
Sounder the collective term for a group of pigs.
Spatulate broad and flattened with a narrow base.
Specialist an animal whose life-style involves highly specialized stratagems (e.g. feeding with one technique on a particular food). (cf **Generalist**)
Speciation the process by which new species arise in evolution. It is widely accepted that it occurs when a single species population is divided by some geographical barrier.
Species a taxonomic division subordinate to genus and superior to subspecies. In general a species is a group of animals similar in structure and which are able to breed and produce viable offspring. (see **Taxonomy**)
Species richness the number of species in an area.
Spermaceti organ an organ found in the head of whales of some toothed whales. The organ contains a waxy fluid which may help heat loss, provide neutral buoyancy, and contribute to the production of sounds.
Spermatogenesis the formation of sperm resulting from a series of cell divisions.
Sperm competition competition between sperm to fertilize female eggs, particularly after a female has copulated with more than one male.
Sphincter a ring of smooth muscle around a pouch, rectum or other hollow organ, which can be contracted to narrow or close the entrance to the organ.
Spinifex a grass which grows in large, distinctive clumps or hummocks in the driest areas of central and Western Australia.
Spontaneous ovulation OVULATION that occurs without copulation. (see also **Estrus**)
Spoor footprints.
Stapes the last of the three middle ear bones found in mammals.
Steppe open grassy plains of the central temperate zone of Eurasia or North America (prairies), characterized by low and sporadic rainfall and a wide annual temperature variation. In cold steppe, temperatures drop well below freezing point in winter, with rainfall concentrated in the summer or evenly distributed throughout year, while in hot steppe, winter temperatures are higher and rainfall concentrated in winter months.
Stotting see **Pronking**
Strepsirhine a member of the primate suborder Strepsirrhini that comprises the lemurs, bushbabies, lorises and potto. All members have a moist RHINARIUM and a cleft upper lip bound to the gum. (cf **Haplorhine**)
Stridulation production of sound by rubbing together modified surfaces of the body. Found in tenrecs.
Subadult no longer an infant or juvenile but not yet fully adult physically and/or socially.
Subfamily in taxonomy, a division of a family.
Subfossil an incompletely fossilized specimen from a recent species.
Sublingua (subtongue) a flap of tissue beneath the tongue in mammals, retained in most primates though vestigial in New World monkeys; particularly in lemurs and lorises.
Suborder in taxonomy, a subdivision of an order.
Subordinate see **Hierarchy**
Subspecies a recognizable subpopulation of a single species, typically with a distinct geographical distribution.
Subunguis the lower or ventral portion of the claw.
Successional habitat a stage in the progressive change in composition of a community of plants, from the original colonization of a bare area towards a largely stable climax.
Suckling taking nourishment or milk from the nipple or teat in mammals.
Suid a member of the family of pigs, Suidae, of the Artiodactyla.
Supernumerary additional teeth in a position where they do not normally occur.
Superordinal above the rank of order.
Supraorbital pertaining to above the eye (eye-socket or orbit).
Surplus killing a phenomenon where more (sometimes very many more) prey are killed than can immediately be consumed by the killer or its companions.
Suspensory movement movement through the trees by hanging and swinging beneath, rather than running along the tops of branches. (see also **Brachiate**)
Suture the contact line between two bones, such as those of the skull.
Sweat gland (eccrine gland) a gland located in the skin that opens on the surface and excretes sweat, the evaporation of which cools the surface.
Symbiosis an interaction between two species in which one benefits, and the other either benefits, is harmed, or is unaffected. (see also **Commensalism**, **Parasitism**, **Mutualism**)
Symmetrodonta an early order of mammals that includes small carnivores and insectivores from the late Triassic.
Sympatry a condition in which the geographical ranges of two or more different species overlap. (cf **Allopatry**)
Symplesiomorphy in pylogenetics, an ancestral (primitive) character shared by two or more taxa. (see also **Plesiomorphic**)
Synapomorphy in phylogenetics, a derived, HOMOLOGOUS character shared by two or more taxa. (see also **Apomorphic**)
Synapsida a subclass of the Reptilia, from which mammals supposedly evolved.
Syndactylous pertaining to the second and third toes of some mammals, which are joined together so that they appear to be a single toe with a split nail. In kangaroos, these syndactyl toes are used as a fur comb. (cf **Didactylous**)
Synonym in taxonomy, a different name for the same species; the earlier name has priority of use.
Synterritorial sharing a territory.
Syntopic present at the same time and place.
Systematics the study of patterns and processes of evolution used to construct phylogenies or classify organisms. It includes TAXONOMY.
Systematist someone who practises SYSTEMATICS.

Taiga northernmost coniferous forest, with open boggy, rocky areas in between.
Talonid the "heel" or posterior part of a lower molar that occludes with the PROTOCONE of an upper molar.
Tandem-marking communal scent-marking, characterized by repeated sniffing and marking (with urine and faeces) of the same (similar) spot or object by several members of a social carnivore. Often along home range boundaries, indicating a territorial function.
Tapetum lucidum a reflecting layer located behind the retina of the eye, commonly found in nocturnal mammals.
Tarsal pertaining to the tarsus bones in the ankle, articulating between the tibia and fibia of the leg and the metatarsals of the foot (pes).
Taxon (pl. **taxa**) a group of organisms of any taxonomic rank.
Taxonomy the science of classifying organisms, grouping together animals

which share common features and are thought to have common descent. Each individual is thus a member of a series of ever-broader categories (individual-species-genus-family-order-class-phylum) and each of these can be further divided where it is convenient (e.g. subspecies, superfamily, or infraorder).
Temporal of, or pertaining to, the side of the skull.
Terrestrial living on land.
Territoriality behavior related to the defence of a TERRITORY against predators.
Territory an area defended from intruders by an individual or group. Originally the term was used where ranges were exclusive and obviously defended at their borders. A more general definition of territoriality allows some overlap between neighbors by defining territoriality as a system of spacing wherein home ranges do not overlap randomly, i.e., the location of one individual's, or group's, home range influences those of others. (see also **Home range**)
Tertiary geological sub-era between 65–1.7 million years ago and comprising the Palaeocene, Eocene, Oligocene, Miocene, and Pliocene.
Testosterone a male hormone synthesized in the testes and responsible for the expression of many male characteristics (contrast the female hormone ESTROGEN produced in the ovaries).
Therapsida an order within the subclass SYNAPSIDA, which supposedly gave rise to the mammals.
Theria a subclass of the Class Mammalia, which includes the marsupials (Metatheria), the placental mammals (Eutheria), and the ancestral mammals (Pantotheria), but not the monotremes (Prototheria).
Theriodontia a suborder within the order THERAPSIDA that were primarily carnivorous.
Thermal conductance heat loss from the skin to the environment.
Thermogenesis generation of heat.
Thermoneutral range (or **Thermal neutral zone**) the range in outside environmental temperature in which a mammal uses the minimum amount of energy to maintain a constant internal body temperature. The limits to the thermoneutral range are the lower and upper critical temperatures, at which points the mammals must use increasing amounts of energy to maintain a constant body temperature. (cf **Heterothermy**) (see also **Lower** and **Upper critical temperature**)
Thermoregulation the regulation and maintenance of a constant internal body temperature in mammals.
Thoracic pertaining to the thorax or chest.
Tine a point or projection on an antler.
Tooth-comb a dental modification in which the incisor teeth form a comb-like structure.
Torpor a temporary physiological state in some mammals, akin to short-term hibernation, in which the body temperature drops and the rate of metabolism is reduced. Torpor is an adaptation for reducing energy expenditure in periods of extreme cold or food shortage.
Tragus a flap, sometimes moveable, situated in front of the opening of the outer ear in bats.
Triassic geological time period 248–213 million years ago.
Tribosphenic molars with three main cusps (the TRIGON) arranged in a triangular pattern.
Triconodont a member of the order that includes the earliest of all mammals, living from the Triassic until the early Cretaceous.
Trigon the three cusps (PROTOCONE, PARACONE, METACONE) of a TRIBOSPHENIC molar.
Trophic of, or pertaining to, food or nutrition.
Trophoblast the superficial layer of the BLASTOCYST in mammals.
Trypanosome a group of protozoa causing sleeping sickness.
Tubercle a small rounded projection or nodule (e.g. of bone).
Tundra barren treeless lands of the far north of Eurasia and North America, on mountain tops and Arctic islands. Vegetation is dominated by low shrubs, herbaceous perennials, mosses, and lichens.
Turbinate (turbinal) bones bones found within the nasal area that provide increased surface area for moisturizing, warming, and filtering inhaled air.
Tylopod a member of the suborder Tylopoda (order Artiodactyla), which includes camels and llamas.
Tympanic membrane (tympanum) the ear drum.

Ultrasound sound frequencies greater than 20 Hz.
Umbilicus navel.
Underfur the thick, soft undercoat fur lying beneath the longer and coarser hair (guard hairs).
Understory the layer of shrubs, herbs and small trees beneath the forest canopy.
Unguiculate having nails or claws instead of hooves.
Unguis the upper or dorsal portion of the claw.
Ungulate a member of the orders Artiodactyla (even-toed ungulates), Perissodactyla (odd-toed ungulates), Proboscidea (elephants), Hyracoidea (hyraxes) and Tubulidentata (aardvark), all of which have their feet modified as hooves of various types (hence the alternative name, hoofed mammals). Most are large and totally herbivorous. Also considered by some to include members of the orders Cetacea (whales and dolphins) and Sirenia (manatees and dugong).
Unguligrade locomotion on the tips of the "fingers" and "toes," the most distal phalanges. A condition associated with reduction in the number of digits to one or two in the perissodactyls and artiodactyls. (cf **Digitigrade**, **Plantigrade**)
Unicuspid teeth having a single cusp.
Upper critical temperature the maximum temperature at which an animal must lose heat in order to maintain a stable internal temperature. (cf **Lower critical temperature**)
Upwelling an upward movement of ocean currents, resulting from convection, causing an upward movement of nutrients and hence an increase in plankton populations.
Urethra the tube through which urine is expelled from the bladder.
Urogenital sinus a common opening from the reproductive and urinary system. In mammals, found in monotremes and marsupials.
Uropatagium the part of the PATAGIUM in some bats that extends between the hindlimbs and the tail.
Uterus the organ in which the embryo develops in female mammals (except monotremes).

Vagina the part of the female reproductive tract that receives the penis during copulation.
Vascular of, or with vessels which conduct blood and other body fluids.
Vector an individual or species which transmits a disease.
Velvet furry skin covering a growing antler.
Ventral on the lower or bottom side or surface; thus ventral or abdominal glands occur on the underside of the abdomen.
Venule a small tributary conveying blood from the capillary bed to a vein. (cf **Arteriole**)
Vertebrate an animal with a backbone; a division of the phylum Chordata which includes animals with notochords (as distinct from invertebrates).
Vestigial a characteristic with little or no contemporary use, but derived from one which was useful and well developed in an ancestral form.
Vibrissae stiff, coarse hairs richly supplied with nerves, found especially around the snout, and with a sensory (tactile) function.
Viviparous giving birth to live young.
Vocalization calls or sounds produced by the vocal cords of a mammal, and uttered through the mouth. Vocalizations differ with the age and sex of mammals but are usually similar within a species.
Volant having powered flight.

Wallace's line an imaginary line passing between the Philippines and the Moluccas in the north and between Sulawesi and Borneo and between Lombok and Bali in the south. It separates the Oriental and Australian zoogeographical regions.
Warren a communal series of burrows used by rabbits or squirrels.
Wavelength the distance from one peak to the next in a sound wave.
Wing loading in bats, the body mass divided by the total surface area of the wings.
Withers ridge between shoulder blades, especially of horses.

Xenarthrales bony elements between the lumbar vertebrae of xenarthran mammals, which provide extra support to the pelvic region for digging, climbing etc.
Xenarthran a member of the order Xenarthra, which comprises the living armadillos, sloths and anteaters.
Xeric having very little moisture; used of animals inhabiting dry regions (i.e. xeric-adapted). (cf **Mesic**)
Xerophytic forest a forest found in areas with relatively low rainfall. Xerophytic plants are adapted to protect themselves against browsing (e.g. well-developed spines) and to limit water loss (e.g. small, leathery leaves, often with a waxy coating).

Yolk sac a sac, usually containing yolk, which hangs from the ventral surface of the vertebrate fetus. In mammals, the yolk sac contains no yolk, but helps to nourish the embryonic young via a network of blood vessels.

Zalambdodont tooth cusps forming a V-shape.
Zoonoses diseases transmitted from vertebrate, non-human mammals to people.
Zooplankton minute animals living near the surface of the sea. (cf. **Phytoplankton**)
Zygomatic arch the bony arch on the side of a mammal skull that surrounds and protects the eye.
Zygote a fertilized egg.

Bibliography

The following list of titles indicates key reference works used in the preparation of this volume and those recommended for further reading. The list is divided into a number of categories: general mammalogy and particular areas of interest related to specific mammal groups.

A full technical bibliography can be found on: http://www.wildcru.org

General

Allen, G. (1939–1940). *The Mammals of China and Mongolia. Vols I & II*. American Museum of Natural History, New York.

Anderson, S. & Knox Jones, J. Jr., (eds) (1984). *Orders and Families of Recent Mammals of the World*. John Wiley and Sons, New York.

Birney, E.C. & Choate, J.R. (eds) (1994). *Seventy-five Years of Mammalogy (1919–1994)*. Special Publ. No. 11, American Society of Mammalogists.

Bjarvall, A. & Ullstrom, S. (1986). *The Mammals of Britain and Europe*. Croom Helm, London.

Bourliere, F. (1970). *The Natural History of Mammals*. Alfred A. Knopf, New York.

Boyle, C. L. (ed) (1981). *The RSPCA Book of British Mammals*. Collins, London.

Chapman, J.A. & Feldhamer, G.A. (eds) (1982). *Wild Mammals of North America: Biology, Management and Economics*. Johns Hopkins University Press, Baltimore.

Corbet, G.B. & Harris, S. (eds). (1991). *The Handbook of British Mammals*. 3rd edn. Blackwell Scientific Publications, Oxford.

Corbet, G.B. & Hill, J.E. (1991). *A World List of Mammalian Species*. 3rd edn. Oxford University Press, Oxford.

Corbet, G.B. & Hill, J.E. (1992). *The Mammals of the Indomalayan Region: a Systematic Review*. Oxford University Press, Oxford.

Cranbrook, G. (1991). *Mammals of South-East Asia*. 2nd edn. Oxford University Press, New York.

Delany, M. J. & Happold, D.C.D. (1979). *Ecology of African Mammals*. Longman, London and New York.

Dorst, J. & Dandelot, P. (1972). *Larger Mammals of Africa*. Collins, London.

Dunstone, N. & Gorman, M.L. (eds) (1993). *Mammals as Predators. Symp. Zool. Soc. Lond*. No. 65. Clarendon Press, Oxford.

Dunstone, N. & Gorman, M.L. (eds) (1998). *Behaviour and Ecology of Riparian Mammals. Symp. Zool. Soc. Lond*. No. 71. Cambridge University Press, Cambridge.

Eisenberg, J.F. (1981). *The Mammalian Radiations: An Analysis of Trends in Evolution, Adaptation and Behavior*. University of Chicago Press, Chicago.

Eisenberg, J.F. (1989). *Mammals of the Neotropics. The Northern Neotropics, Vol. I, Panama, Colombia, Venezuela, Guyana, Suriname, French Guiana*. University of Chicago Press, Chicago.

Eisenberg, J.F. & Redford, K.H. (1999). *The Mammals of the Neotropics. Vol. III. The Central Neotropics: Ecuador, Peru, Bolivia, Brazil*. University of Chicago Press, Chicago

Ellerman, J.R. (1961). *The Fauna of India: Mammalia, Vol III*. Delhi.

Ellerman, J.R. & Morrison Scott, T.C.S. (1951). *Checklist of Palearctic and Indian Mammals, 1758–1946*. British Museum (Natural History), London.

Emmons, L.H. (1997). *Neotropical Rainforest Mammals: a Field Guide*. University of Chicago Press, Chicago.

Estes, R.D. (1991). *The Behavior Guide to African Mammals: Including Hoofed Mammals, Carnivores, Primates*. University of California Press, Berkeley.

Feldhamer, G.A., Drickhamer, L.C., Vessey, S.H. & Merritt, J.F. (1999). *Mammalogy: Adaptation, Diversity, and Ecology*. McGraw-Hill, New York.

Flannery, T.F. (1990). *Mammals of New Guinea*. Robert Brown & Associates, New South Wales.

Forsyth, A. (1999). *Mammals of North America: Temperate and Arctic Regions*. Firefly Books, Willowdale, Ontario.

Garbutt, N. (1999). *The Mammals of Madagascar*. Pica Press, East Sussex.

Hall, E.R. (1981). *The Mammals of North America*. John Wiley and Sons, New York.

Harrison, D.L. & Bates, P.J.J. (1991). *The Mammals of Arabia*. Harrison Zoological Museum, Sevenoaks, Kent.

Harrison Matthews, L. (1969). *The Life of Mammals. Vols I & II*. Weidenfeld & Nicolson, London.

Heptner, V.G., Nasimovich, A.A. & Bannikov, A. (1988). *Mammals of the Soviet Union. Vol. I. Artiodactyla and Perissodactyla* [A Translation of Heptner et al., 1961, Mlekopitayuschie Sovetskovo Soyuza: Parnokopytyne i neparnokopytyne]. Smithsonian Institution Libraries, Washington.

Heptner, V.G. & Sludskii, A.A. (1992). *Mammals of the Soviet Union. Vol. II. Carnivora (Hyaenas and cats)*. [A Translation of Heptner et al., 1972, Mlekopitayushchie Sovetskovo Soyuza. Khishchye (gienyi i koshki)]. Smithsonian Institution Libraries, Washington.

Hilton-Taylor, C. (compiler) (2000). *2000 IUCN Red List of Threatened Species*. IUCN, Gland, Switzerland and Cambridge.

Jewell, P.A. & Maloiy, G. (1989). *The Biology of Large African Mammals in their Environment. Symp. Zool. Soc. Lond*. No. 61, Oxford University Press, Oxford.

King, C.M. (ed). (1995). *The Handbook of New Zealand Mammals*. Oxford University Press, Oxford.

Kingdon, J. (1971–1982). *East African Mammals. Vols I–III*. Academic Press, New York.

Kingdon, J. (1997). *The Kingdon Field Guide to African Mammals*. Academic Press, London.

Lekagul, B. & McNeely, J.A. (1988). *Mammals of Thailand*. Darnsutha Press, Bangkok.

Macdonald, D. & Barrett, P. (1995). *Collins European Mammals: Evolution and Behaviour*. HarperCollins, London.

Macdonald, D.W. & Barrett, P. (1995). *Collins Field Guide to Mammals of Britain and Europe*. HarperCollins, London.

Mares, M.A. & Schmidly, D.J. (eds) (1991). *Latin American Mammalogy, History, Biodiversity, and Conservation*. Univ Oklahoma Press, Norman.

McKenna, M.C. & Bell, S.K. (1997). *Classification of Mammals Above the Species Level*. Columbia University Press, New York.

Meester, J. & Setzer, H. W. (1971–1977) *The Mammals of Africa: an Identification Manual*. Smithsonian Institution, Washington DC.

Mitchell-Jones, A.J., Amori, G., Bogdanowicz, W., Krystufek, B., Reijnders, P.J.H., Spitzenberger, F., Stubbe, M., Thissen, J.B.M., Vohralik, V. & Zima, J. (1999). *The Atlas of European Mammals*. T&AD Poyser Natural History, London.

Morris, D. (1965). *The Mammals*. Hodder & Stoughton, London.

Niethammer, J. & Krapp, F. (1978–1992). *Handbuch der Säugetiere Europas [Handbook of European Mammals]*. Aula-Verlag, Wiesbaden.

Nowak, R.M. (1999). *Walker's Mammals of the World*. Sixth Edition. Johns Hopkins University Press, Baltimore.

Ognev, S.I. (1962–1966). *Mammals of Eastern Europe and Northern Asia* [A Translation of S.I. Ognev, 1928–1950, Zveri vostochnoi Evropy i severnoi Azii]. Israel Program for Scientific Translations, Jerusalem.

Owen-Smith, R.N. (1988). *Megaherbivores: the Influence of Very Large Body Size on Ecology*. Cambridge University Press, New York.

Parker, S.P. (ed) (1990). *Grzimek's Encyclopedia of Mammals*. McGraw-Hill, New York.

Qumsiyeh, M.B. (1996). *Mammals of the Holy Land*. University of Texas Press, Austin.

Redford, K.H. & Eisenberg, J.F. (1992). *Mammals of the Neotropics. Volume II. The Southern Cone: Chile, Argentina, Uruguay and Paraguay*. University of Chicago Press, Chicago

Reid, F.A. (1997). *A Field Guide to the Mammals of Central America and Southeast Mexico*. Oxford University Press, New York.

Roberts, A. (1951). *The Mammals of South Africa*. Trustees of "The mammals of South Africa" book fund.

Roberts, T.J. (1997). *Mammals of Pakistan*. Oxford University Press, Pakistan.

Simpson, G.G. (1945). *The Principles of Classification and a Classification of Mammals. Bul. Amer. Mus. Nat. Hist*. 85: 1–350

Skinner, J.D. & Smithers, R.H.N. (1990). *The Mammals of the Southern African Subregion*. University of Pretoria, Pretoria.

Sokolov, V.E. (1973–1979). *Sistematika mlekopitayushchikh [Systematics of Mammals]*. Vysshaya Shkola, Moscow. (in Russian).

Sokolov, V.E. & Orlov, V.N. (1980). *Guide to the Mammals of the Mongolian People's Republic*. Nauka, Moscow. (in Russian).

Strahan, R. (1998). *The Mammals of Australia*. Reed New Holland, Australia.

Szalay, F.S., Novacek, M.J. & McKenna, M.C. (eds) (1993). *Mammal Phylogeny: Placentals*. Springer-Verlag, New York.

Vaughan, T.A., Ryan, J.M. & Czaplewski, N.J. (2000). *Mammalogy*. Saunders College Publishing, Philadelphia.

Whitaker, J.O. (1996). *National Audubon Society Field Guide to North American Mammals*. Knopf, Canada.

Wilson, D.E. & Cole, F.R. (2000). *Common Names of Mammals of the World*. Smithsonian Institution Press, Washington.

Wilson, D.E. & Reeder, D.M. (eds) (1993). *Mammal Species of the World: a Taxonomic and Geographic Reference*. 2nd edn. Smithsonian Institution Press, Washington.

Wilson, D.E. & Ruff, S. (eds) (1999). *The Smithsonian Book of North American Mammals*. Smithsonian Institution Press, Washington.

Yalden, D. (2000). *The History of British Mammals*. T&AD Poyser Natural History, London.

Young, J.Z. (1975). *The Life of Mammals: their Anatomy and Physiology*. Oxford University Press, Oxford.

Carnivores

Bailey, T.N. (1993). *The African Leopard: Ecology and Behavior of a Solitary Felid*. Columbia University Press, New York.

Bauer, E.A. & Bauer, P. (1998). *Bears: Behaviour, Ecology, Conservation*. Voyageur Press, Stillwater, MN.

Bekoff, M. (1978). *Coyotes: Biology, Behavior and Management*. Academic Press, New York.

Bertram, B. C. (1978). *Pride of Lions*. Charles Scribner, New York.

Busch, R.H. (1996). *The Cougar Almanac: A Complete Natural History of the Mountain Lion*. The Lyons Press, New York, NY.

Busch, R.H. (2000). *The Grizzly Almanac*. The Lyons Press, New York, NY.

Buskirk, S.W., Harestad, A.S., Raphael, M.G. and Powell, R.A. (1994). *Martens, Sables and Fishers: Biology and Conservation*. Cornell University Press, Ithaca, NY.

Caro, T.M. (1994). *Cheetahs of the Serengeti Plains*. University of Chicago Press, Chicago.

Corbett, L. (1995). *The Dingo: in Australia and Asia*. Cornell University Press, Ithaca, NY.

de la Rosa, C.L. & Nocke, C.C. (2000). *A Guide to the Carnivores of Central America:*

Natural History, Ecology, and Conservation. University of Texas Press, Austin.
Dominis, I. & Edey. M. (1968). *The Cats of Africa*. Time-Life, New York.
Eaton, R.L. (1974). *The Cheetah: the Biology, Ecology, and Behavior of an Endangered Species*. Van Nostrand Reinhold, New York.
Ewer, R.F. (1973). *The Carnivores*. Weidenfeld & Nicolson, London.
Fox, M. W. (ed) (1975). *The Wild Canids: their Systematics, Behavioral Ecology, Evolution*. Van Nostrand Reinhold, London and New York.
Frame, G. & Frame, L. (1981). *Swift and Enduring: Cheetahs and Wild Dogs of the Serengeti*. Dutton, New York.
Gittleman, J.L. (ed) (1989, 1996). *Carnivore Behavior, Ecology and Evolution. Vols I & II*. Cornell University Press, Ithaca, NY.
Gittleman, J., Funk, S., Macdonald, D.W & Wayne R (eds) (2001). *Carnivore Conservation*. Cambridge University Press, Cambridge.
Gould, E. & McKay, G. (1998). *Encyclopedia of Mammals. A Comprehensive Illustrated Guide by International Experts*. Academic Press, London.
Griffiths, H.I. (ed) (2000). *Mustelids in a Modern World: Management and Conservation Aspects of Small Carnivore and Human Interactions*. Backhuys, Netherlands.
Gromov, I.M. & Baranova, G.I. (eds) (1981). *Catalog of Mammals of the USSR*. Nauka, Leningrad. (in Russian).
Guggisberg, C.A.W. (1961). *Simba: the Life of the Lion*. Howard Timmins, Cape Town.
Gurung, K.K. & Singh, R. (1998). *Field Guide to the Mammals of the Indian Subcontinent: Where to Watch Mammals in India, Nepal, Bhutan, Bangladesh, Sri Lanka and Pakistan*. Academic Press, London.
Hampton, B. (1997). *The Great American Wolf*. Henry Holt, New York.
Herrero, S. (ed) (1972). *Bears: their Biology and Management*. IUCN Publ. New Series no.23, Morges, Switzerland.
Hinton, H.E. & Dunn, A.M.S. (1967). *Mongooses: their Natural History and Behaviour*. Oliver & Boyd, Edinburgh and London.
Kanchanasakha, B., Simcharoen, S. & Than, U.T. (1998). *Carnivores of Mainland South East Asia*. WWF, Thailand.
King, C. (1989). *The Natural History of Weasels and Stoats*. Christopher Helm, London.
Kitchener, A. (1991). *The Natural History of the Wild Cats*. Christopher Helm, London.
Kruuk, H. (1972). *The Spotted Hyena: a Study of Predation and Social Behavior*. University of Chicago Press, Chicago.
Kruuk, H. (1989). *The Social Badger*. Oxford University Press, Oxford.
Kruuk, H. (1995). *Wild Otters: Predation and Populations*. Oxford University Press, Oxford.
Long, C. & Killingley, C.A. (1983). *The Badgers of the World*. Charles C. Thomas, Springfield, IL.
Macdonald, D.W. (1987). *Running with the Fox*. Unwin Hyman, London.
Macdonald, D.W. (1992). *The Velvet Claw*. BBC Books, London.
Maehr, D.S. (1997). *The Florida Panther: Life and Death of a Vanishing Carnivore*. Island Press, Washington.
Mason, C.F. & MacDonald, S.M. (1986). *Otters: Ecology and Conservation*. Cambridge University Press, Cambridge.
Mech, L.D. (1970). *The Wolf: the Ecology and Behavior of an Endangered Species*. Natural History Press, Garden City, New York.
Mech, L.D. (1995). *The Way of the Wolf*. Voyageur Press, Stillwater, MN.
Mech, L.D. (2000). *The Wolves of Minnesota: Howl in the Heartland*. Voyageur Press, Stillwater, MN.
Meinzer, W. (1996). *Coyote*. University of Texas Press, Austin.
Mills, M.G. (1990). *Kalahari Hyaenas: The Comparative Behavioural Ecology of Two Species*. Chapman & Hall, London.
Mountfort, G. (1981). *Saving the Tiger*. Michael Joseph, London.
Neal, E.G. (1977). *Badgers*. Blandford, Poole, Dorset.
Neal, E. (1986). *The Natural History of Badgers*. Croom Helm, London.
Neal, E. & Cheeseman, C. (1996). *Badgers*. T&AD Poyser Natural History Ltd, London.
Ovsyanikov, N. (1999). *Polar Bears: Living with the White Bear*. Voyageur Press, Stillwater, MN.
Pelton, M. R., Lentfer, I. W & Stokes, G.E. (eds) (1976). *Bears: their Biology and Management*. IUCN Publ. New Series no. 40, Morges, Switzerland.
Powell, R.A. (1993). *The Fisher: Life History, Ecology and Behavior*. University of Minnesota Press, Minneapolis, MN.
Rosevear, D.R. (1974). *The Carnivores of West Africa*. British Museum (Natural History), London.
Schaller, G.B. (1967). *The Deer and the Tiger: a Study of Wildlife in India*. Chicago University Press, Chicago.
Schaller, G.B. (1972). *The Serengeti Lion: a Study of Predator–Prey Relations*. University of Chicago Press, Chicago.
Schaller, G.B. (1993). *The Last Panda*. University of Chicago Press, Chicago.
Seidensticker, J., Christie, S. & Jackson, P. (eds) (1999). *Riding the Tiger: Tiger Conservation in Human-dominated Landscapes*. Cambridge University Press, Cambridge.
Turner, A. & Anton, M. (1997). *The Big Cats and their Fossil Relatives: an Illustrated Guide to their Evolution and Natural History*. Columbia University Press, New York.
van Lawick, H. & van Lawick-Goodall, J. (1970). *The Innocent Killers*. Collins, London.
Verts, B.J. (1967). *The Biology of the Striped Skunk*. University of Illinois Press, Urbana.
Wrogemann, N. (1975). *Cheetah Under the Sun*. McGraw-Hill, Johannesburg.

MARINE MAMMALS

Allen, K. R. (1980). *Conservation and Management of Whales*. Butterworths, London.
Andersen, H.T. (ed). (1969). *The Biology of Marine Mammals*. Academic Press, New York.
Baker, M.L. (1987). *Whales, Dolphins, and Porpoises of the World*. Garden City, New York.
Berta, A. & Sumich, J.L. (1999). *Marine Mammals: Evolutionary Biology*. Academic Press, London.
Bonner, N. (1989). *The Natural History of Seals*. Academic Press, London.
Bonner, N. (1994). *Seals and Sea Lions of the World*. Facts on File Publ., New York.
Bonner, W.N. (1980). *Whales*. Blandford, Poole.
Bonner, W.N. & Berry. R.J. (eds) (1981). *Ecology in the Antarctic*. Academic Press, London.
Boyd, I.L. (ed) (1993). *Marine Mammals: Advances in Behavioural and Population Biology*. Oxford University Press, Oxford.
Bryden, M.M., Marsh, H. & Shaughnessy, P. (1999). *Dugongs, Whales, Dolphins and Seals: A Guide to the Sea Mammals of Australasia*. Allen & Unwin, Sydney.
Carwardine, M. (1998). *Whales and Dolphins*. HarperCollins, New York.
Carwardine, M., Harrison, P. & Bryden, M (eds) (1999). *Whales, Dolphins and Porpoises*. 2nd edn. Checkmark Books, New York.
Carwardine, M., Hoyt, E., Ewan Fordyce, R. & Gill, P. (1998). *Whales, Dolphins and Porpoises*. Time Life Books, New York.
Ellis, R. (1983). *Dolphins and Porpoises*. R. Hale, London.
Evans, P.G.H. (1987). *The Natural History of Whales and Dolphins*. Christopher Helm, London.
Fontaine, P.-H. (1998). *Whales of the North Atlantic: Biology and Ecology*. Editions MultiMondes, Sainte-Foy, Québec.
Gaskin, D.E. (1972). *Whales, Dolphins and Seals*. Heinemann Educational Books, London.
Gaskin, D.E. (1982). *The Ecology of Whales and Dolphins*. Heinemann, London.
Gentry, R.L. (1998). *Behavior and Ecology of the Northern Fur Seal*. Princeton University Press, Princeton, NJ.
Harrison Matthews, L. (1978). *The Natural History of the Whale*. Weidenfeld & Nicolson, London.
Harrison Matthews, L. (1979). *Seals and the Scientists*. P. Owen, London.
Herman, L.M. (1980). *Cetacean Behavior: Mechanisms and Functions*. John Wiley & Sons, Chichester.
King, J.E. (1983). *Seals of the World*. Oxford University Press, Oxford.
Laws, R.M. (ed) (1993). *Antarctic Seals: Research Methods and Techniques*. Cambridge University Press, Cambridge.
Leatherwood, S. & Reeves, R. (1983). *The Sierra Club Handbook of Seals and Sirenians*. Sierra Club, San Francisco, CA.
Le Boeuf, B.J. & Laws, R.M. (1994). *Elephant Seals: Population Ecology, Behavior, and Physiology*. University of California Press, Berkeley.
Mann, J., Connor, R.C., Tyack, P.L. & Whitehead, H. (1999). *Cetacean Societies: Field Studies of Dolphins and Whales*. Chicago University Press, Chicago.
Martin, R.M. (1977). *Mammals of the Seas*. Batsford, London.
Norris, K.S., Würsig, B., Wells, R.S. & Würsig, M. (1994). *The Hawaiian Spinner Dolphin*. California University Press, Berkeley.
Owen, W. (1999). *Whales, Dolphins and Porpoises*. Checkmark Books, New York.
Perrin, W.F., Würsig, B. & Thewissen, J.G.M. (2001). *Encyclopedia of Marine Mammals*. Academic Press, London.
Pryor, K. & Norris, K.S. (1998). *Dolphin Societies: Discoveries and Puzzles*. University of California Press, Berkeley.
Read, A.J., Wiepkema, P.R. & Nachtigall, P.E. (1997). *The Biology of the Harbour Porpoise*. De Spil Publishers, Woerden, Netherlands.
Reeves, R.R., Stewart, B. & Leatherwood, S. (1992). *The Sierra Club Handbook of Seals and Sirenians*. Sierra Club, San Francisco, California.
Renouf, D. (1990). *The Behaviour of Pinnipeds*. Chapman & Hall, London.
Reynolds, J.E. (2000). *The Bottlenose Dolphin: Biology and Conservation*. University of Florida Press, Gainesville.
Reynolds, J.E. & Rommel, S.A. (1999). *Biology of Marine Mammals*. Smithsonian Institution Press, Washington.
Rice, D.W. (1998). *Marine Mammals of the World: Systematics and Distribution*. Allen Press, Lawrence, KS.
Ridgeway, S.H. & Harrison, R.J. (eds) (1981–1998). *The Handbook of Marine Mammals. Vols I – VI*. Academic Press, London.
Riedman, M. (1991). *The Pinnipeds: Seals, Sea Lions, and Walruses*. University of California Press, Berkeley.
Ripple, J. & Perrine, D. (1999). *Manatees and Dugongs of the World*. Voyageur Press, Stillwater, MN.
Slijper, E.J. (1979). *Whales*. Hutchinson, London.
Watson, L. (1981). *Sea Guide to Whales of the World*. Hutchinson, London.
Winn, H.E. & Olla, B.L. (1979). *The Behavior of Marine Mammals. Vol 3. Cetaceans*. Plenum Press, New York.
Würsig, B., Jefferson, T.A. & Schmidly, D.J. (2000). *The Marine Mammals of the Gulf of Mexico*. Texas A & M University Press, College Station.

PRIMATES

Alterman, L., Doyle, G.A. & Izard, M.K. (eds) (1995). *Creatures of the Dark: the Nocturnal Primates*. Plenum Press, New York.
Altmann, J. (1980). *Baboon Mothers and Infants*. Harvard University Press, Cambridge.
Altmann, S.A. & Altmann, J. (1970). *Baboon Ecology*. University of Chicago Press, Chicago.
Barrett, L. (2000). *Baboons: Survivors of the African Continent*. BBC Books, Bristol.
Bramblett, C.A. (1976). *Patterns of Primate Behaviour*. Mayfield Publishing Co., Palo Alto.
Chalmers, N. (1979). *Social Behaviour in Primates*. Edward Arnold, London.
Charles-Dominique, P. (1977). *Ecology and Behaviour of Nocturnal Primates: Prosimians of Equatorial West Africa*. Duckworth, London.
Chivers, D.J. (ed) (1980). *Malayan Forest Primates*. Plenum Press, New York.
Clutton-Brock, T.H. (ed) (1977). *Primate Ecology*. Academic Press, London.

Clutton-Brock, T.H. & Harvey, P.H. (eds) (1978). *Readings in Sociobiology*. W. H. Freeman, Reading.

Coimbra-Filho, A. F. & Mittermeier, R.A. (1981). *Ecology and Behavior of Neotropical Primates*. Academia Brasileira de Ciencias, Rio de Janeiro.

Cowlishaw, G. & Dunbar, R. (2000). *Primate Conservation Biology*. Chicago University Press, Chicago.

Davies, G. & Oates, J. (1994). *Colobine Monkeys: Their Ecology, Behaviour and Evolution*. Cambridge University Press, Cambridge.

Devore, I. (ed) (1965). *Primate Behavior: Field Studies of Monkeys and Apes*. Holt, Rinehart and Winston, New York.

Doyle, G. A. & Martin, R.D. (eds) (1979). *The Study of Prosimian Behavior*. Academic Press, New York.

Fa, J.E. & Lindburg, J.G. (eds) (1996). *Evolution and Ecology of Macaque Societies*. Cambridge University Press, New York.

Fleagle, J.G. (1988). *Primate Adaptation and Evolution*. Academic Press, New York.

Fleagle, J.G. (1999). *Primate Communities*. Cambridge University Press, Cambridge.

Hill, W.C.O. (1953–1974). *Primates: Comparative Anatomy and Taxonomy*. Edinburgh University Press, Edinburgh, and Wiley-Interscience, New York.

Hrdy, S.B. (1977). *The Langurs of Abu: Female and Male Strategies of Reproduction*. Harvard University Press, Cambridge.

Jay, P.C. (1968). *Primates: Studies in Adaptation and Variability*. Holt, Rinehart and Winston, New York.

Jolly, A. (1972). *The Evolution of Primate Behavior*. Macmillan, New York.

Jolly, A. (1966). *Lemur Behavior: A Malagasy Field Study*. University of Chicago Press, Chicago.

Kinzey, W.G. (1997). *New World Primates: Ecology, Evolution and Behaviour*. Aldine de Gruyter, Hawthorne, NY.

Kleiman, D.G. (ed) (1977). *The Biology and Conservation of the Callitrichidae*. Smithsonian Institution Press, Washington.

Kummer, H. (1971). *Primate Societies: Group Techniques of Ecological Adaptation*. Aldine Atherton, Chicago.

Lee, P.C. (1999). *Comparative Primate Socioecology*. Cambridge University Press, New York.

Lindburg, D.G. (ed) (1980). *The Macaques: Studies in Ecology, Behavior and Evolution*. Van Nostrand Reinhold, New York.

Martin, R.D. (1990). *Primate Origins and Evolution*. Princeton University Press, Princeton, NJ.

Martin, R.D., Doyle, G.A. & Walker, A.C. (eds) (1974). *Prosimian Biology*. Duckworth, London.

Michael, R.P. & Crook, J.H. (eds) (1973). *Comparative Ecology and Behaviour of Primates*. Academic Press, London.

Milton, K. (1980). *The Foraging Strategy of Howler Monkeys*. Columbia University Press, New York.

Moynihan, M. (1976). *The New World Primates*. Princeton University Press, Princeton.

Napier, J.R. & Napier, P.H. (1967). *A Handbook of Living Primates*. Academic Press, New York and London.

Napier, J.R. & Napier, P.H. (1970). *Old World Monkeys*. Academic Press, New York.

Rainier III, H.S.H. & Bourne, G.H. (1977). *Primate Conservation*. Academic Press, New York.

Rijksen, H.D. & Meijaard, E. (1999). *Our Vanishing Relative: The Status of Wild Orang-Utans at the Close of the Twentieth Century*. Kluwer Academic Publishers, Boston.

Rylands, A.B. (1993). *Marmosets and Tamarins: Systematics, Behavior, and Ecology*. Oxford University Press, Oxford.

Schaller, G.B. (1963). *The Mountain Gorilla: Ecology and Behaviour*. University of Chicago Press, Chicago and London.

Schwartz, J.H. (1988). *Orang-utan Biology*. Oxford University Press, Oxford.

Short, R.V. & Weir, B.J. (eds) (1980). *The Great Apes of Africa*. Journals of Reproduction and Fertility, Colchester.

Simons, F.L. (1972). *Primate Evolution*. Collier Macmillan, London.

Smuts, B.B., Cheney, D.L., Seyfarth, R.M., Wrangham, R.W. & Struhsaker, T.T. (1986). *Primate Societies*. Chicago University Press, Chicago.

Struhsaker, T.T. (1975). *The Red Colobus Monkey*. University of Chicago Press, Chicago.

Sussman, R.W. (ed) (1979). *Primate Ecology: Problem-orientated Field Studies*. John Wiley, New York.

Szalay, F.S. & Delson, F. (1979). *Evolutionary History of the Primates*. Academic Press, New York.

van Lawick-Goodall, J. (1971). *In the Shadow of Man*. Collins, London.

Whitehead, P.F. & Jolly, C.J. (2000). *Old World Monkeys*. Cambridge University Press, Cambridge.

Wolfheim, J.H. (1983). *Primates of the World: Distribution, Abundance and Conservation*. University of Washington Press, Seattle.

SCANDENTIA

Emmons, L.H. (2000). *A Field Study of Bornean Treeshrews*. University of California Press, Berkeley.

Luckett, W. P. (ed) (1980). *Comparative Biology and Evolutionary Relationships of Tree Shrews*. Plenum Press, New York.

PROBOSCIDEA

Haynes, G. (1993). *Mammoths, Mastodons, and Elephants: Biology, Behaviour and the Fossil Record*. Cambridge University Press, New York.

Moss, C. (2000). *Elephant Memories*. University of Chicago Press, Chicago.

Shoshani, J. (ed) (1992). *Elephants: Majestic Creatures of the Wild*. Rodale Press, Emmaus, PA.

Shoshani, J. & Tassy, P. (eds) (1996). *The Proboscidea: Evolution and Palaeoecology of Elephants and their Relatives*. Oxford University Press, New York.

Spinage, C. (1994). *Elephants*. T&AD Poyser Natural History, London.

Sukumar, R. & Swaminathan, M.S. (1992). *The Asian Elephant: Ecology and Management*. Cambridge University Press, Cambridge.

HOOFED MAMMALS

Andersen, R., Duncan, P. & Linnel, J.D.C. (1998). *The European Roe Deer: The Biology of Success*. Scandinavian University Press, Norway.

Bauer, E.A. & Bauer, P. (1996). *Mule Deer: Behaviour, Ecology, Conservation*. Voyageur Press, Stillwater, MN.

Bauer, E.A. & Bauer, P. (1999). *Elk: Behaviour, Ecology, Conservation*. Voyageur Press, Stillwater, MN.

Boyd, L. & Houpt, K.L. (1994). *Przewalski's Horse: The History and Biology of an Endangered Species*. SUNY Press, Albany, NY.

Bubenik, G.A. & Bubenik, A.B. (eds) (1990). *Horns, Pronghorns and Antlers*. Springer-Verlag, New York.

Byers, J.A. (1998). *American Pronghorn: Social Adaptation and the Ghosts of Predators Past*. University of Chicago Press, Chicago.

Chaplin, R.E. (1977). *Deer*. Blandford, Poole, Dorset, England.

Chapman, D. & Chapman, N. (1975). *Fallow Deer: Their History, Distribution and Biology*. Terrence Dalton, Lavenham, Suffolk, England.

Clutton-Brock, T.H., Guinness, F.F. & Albon, S.D. (1982). *Red Deer: Behaviour and Ecology of Two Sexes*. Edinburgh University Press, Edinburgh.

Dagg, A.I. & Foster, J.B. (1976). *The Giraffe: its Biology, Behavior and Ecology*. Van Nostrand Reinhold, New York.

Danilkin, A. & Hewison, A.J.M. (1996). *Behavioural Ecology of Siberian and European Roe Deer*. Chapman & Hall, London.

Eltringham, S.K. (1982). *Elephants*. Blandford, Poole, Dorset, England.

Eltringham, S.K. (1999). *The Hippos*. T&AD Poyser Natural History Ltd, London.

Franzmann, A.W. (ed) (1998). *Ecology and Management of the North American Moose*. Smithsonian Institution Press, Washington.

Gauthier-Pilters, H. & Dagg. A.I. (1981). *The Camel: its Evolution, Ecology, Behavior and Relationship to Man*. University of Chicago Press, Chicago.

Geist, V. (1971). *Mountain Sheep: a Study in Behavior and Evolution*. University of Chicago Press, Chicago.

Geist, V. (1999). *Deer of the World: Their Evolution, Behavior and Ecology*. Stackpole Books, Mechanicsburg, PA.

Geist, V. & Francis, M.H. (1999). *Moose*. Voyageur Press, Stillwater, MN.

Groves, C.P. (1974). *Horses, Asses and Zebras in the Wild*. David and Charles, Newton Abbot, England.

Habibi, K. (1994). *The Desert Ibex: Life History, Ecology and Behaviour of the Nubian Ibex in Saudi Arabia*. Immel, Saudi Arabia.

Haltenorth, T. & Diller, H. (1980). *A Field Guide to the Mammals of Africa: Including Madagascar*. Collins, London.

Laws, R.M., Parker, I.S.C. & Johnstone, R.C.B. (1975). *Elephants and Their Habitats: the Ecology of Elephants in North Bunyoro, Uganda*. Clarendon Press, Oxford.

Leuthold, W. (1977). *African Ungulates: a Comparative Review of their Ecology and Behavioral Ecology*. Springer-Verlag, Berlin.

Mloszewski, M.J. (1983). *The Behaviour and Ecology of the African Buffalo*. Cambridge University Press, Cambridge.

Moss, C. (1976). *Portraits in the Wild: Animal Behaviour in East Africa*. Hamish Hamilton, London.

Nievergelt, B. (1981). *Ibexes in an African Environment: Ecology and Social System of the Walia Ibex in the Simien Mountains, Ethiopia*. Springer-Verlag, Berlin.

Prins, H.H.T. (1996). *Ecology and Behaviour of the African Buffalo: Social Inequality and Decision Making*. Chapman & Hall, London.

Putman, R. (1988). *The Natural History of Deer*. Cornell University Press, Ithaca, NY.

Schaller, G.B. (1967). *The Deer and the Tiger: a Study of Wildlife in India*. University of Chicago Press, Chicago.

Schaller, G.B. (1977). *Mountain Monarchs: Wild Sheep and Goats of the Himalaya*. University of Chicago Press, Chicago.

Sinclair, A.R.E. (1977). *The African Buffalo*. University of Chicago Press, Chicago.

Spinage, C.A. (1982). *A Territorial Antelope: The Uganda Waterbuck*. Academic Press, London.

Spinage, C.A. (1986). *The Natural History of Antelopes*. Croom Helm, London.

Vrba, E.S. & Schaller, G.B. (eds) (2000). *Antelopes, Deer, and Relatives: Fossil Record, Behavioral Ecology, Systematics, and Conservation*. Yale University Press, New Haven, CT.

Walther, F.R., Mungall. F.C. & Grau, G.A. (1983). *Gazelles and their Relatives: A Study in Territorial Behavior*. Noyes Publications, Park Ridge, New Jersey.

Wemmer, C.M. (1987). *Biology and Management of the Cervidae*. Smithsonian Institution Press, Washington.

Whitehead, G.K. (1972). *Deer of the World*. Constable, London.

Wilson, R.T. (1998). *Camels*. Palgrave, New York.

RODENTS AND LAGOMORPHS

Barash, D. (1989). *Marmots: Social Behaviour and Ecology*. Stanford University Press, Stanford, CA.

Barnett, S.A. (1975). *The Rat: A Study in Behavior*. University of Chicago Press, Chicago and London.

Bennett, N.C. & Faulkes, C.G. (2000). *African Mole-rats: Ecology and Eusociality*. Cambridge University Press, Cambridge.

Berry, R.J. (ed) (1981). *Biology of the House Mouse*. Academic Press, London.

Calhoun, J.B. (1962). *The Ecology and Welfare of the Norway Rat*. US Public Health Service, Baltimore.

Curry-Lindahl, K. (1980). *Der Berglemming*. A. Ziemsen Verlag, Wittenberg.

Delany, M. J. (1975). *The Rodents of Uganda*. British Museum (Natural History), London.

Eisenberg, J.F. (1963). *The Behavior of Heteromyid Rodents*. University of California Publications in Zoology, vol 69, pp 1–100.

Ellerman, J.R. (1940). *The Families and Genera of Living Rodents*. British Museum (Natural History), London.

Elton, C. (1942). *Voles, Mice and Lemmings*. Oxford University Press, Oxford.

Errington, P.L. (1963). *Muskrat Populations*. University of Iowa Press, Iowa City.

Genoways, H.H. & Brown, J.H. (1993). *Biology of the Heteromyidae*. American Society of Mammalogists Spec. Publ. 10, American Society of Mammalogists, USA.

de Graaff, G. (1981). *The Rodents of Southern Africa*. Butterworth, Durban.

Gurnell, J. (1987). *The Natural History of Squirrels*. Croom Helm, London.

Hilfiker, E.L. (1991). *Beavers: Water, Wildlife and History*. Windswept Press, New York.

Hoogland, J.L. (1994). *The Black-tailed Prairie Dog: Social Life of a Burrowing Mammal*. Chicago University Press, Chicago.

King, J. (ed) (1968). *The Biology of Peromyscus*. Special Publication no.2., American Society of Mammalogists, Oswego, N.Y.

Lacey, E.A., Patton, J.L. & Cameron, G.N. (2001). *Life Underground: the Biology of Subterranean Rodents*. Chicago University Press, Chicago.

Laidler, K. (1980). *Squirrels in Britain*. David and Charles, Newton Abbot and North Pomfret, Vermont.

Linsdale, J.M. (1946). *The California Ground Squirrel*. University of California Press, Berkeley.

Linsdale, J.M. & Tevis, L.P. (1951). *The Dusky-footed Woodrat*. University of California Press, Berkeley.

Lockley, R.M. (1976). *The Private Life of the Rabbit*. 2nd edn. Andre Deutsch, London.

Luckett, W.P. & Hartenberger, J-L. (eds) (1985). *Evolutionary Relationships Among Rodents: A Multidisciplinary Analysis*. Plenum Press, New York.

Menzies, J.I. & Dennis, E. (1979). *Handbook of New Guinea Rodents*. Handbook no. 6. Wau Ecology Institute, Wau, New Guinea.

Morgan, L.H. (1868). *The American Beaver and his Works*. Burt Franklin, New York.

Orr, R.T. (1977). *The Little-known Pika*. Collier Macmillan, New York.

Pavlinov, I.J., Dubrovsky, Y.A., Rossolimo, O.L. & Potapova, E.G. (1990). *Gerbils of the World*. Nauka, Moscow.

Prakash, I. & Ghosh, P.K. (eds) (1975). *Rodents in Desert Environments*. Monographae Biologicae, W. Junk, The Hague.

Rosevear, D.R. (1969). *The Rodents of West Africa*. British Museum (Natural History), London.

Rowlands, I. W. & Weir, B. (eds) (1974) *The Biology of Hystricomorph Rodents. Zool. Soc. Symp*. No. 34. Academic Press, London and New York.

Shenbrot, G.Y., Krasnov, B.R. & Rogovin, K.A. (1999). *Spatial Ecology of Desert Rodent Communities: Adaptations of Desert Organisms*. Springer, Heidelberg.

Sherman, P.W., Jarvis, J.U.M. & Alexander, R.D. (eds) (1991). *The Biology of the Naked Mole-rat*. Princeton University Press, Princeton, NJ.

Stenseth, N.C. & Ims, R.A. (eds) (1993). *The Biology of Lemmings*. Academic Press, London.

Strong, P.I.V. (1997). *Beavers: Where Waters Run*. NorthWord Press, Minocqua, WI.

Thompson, H.V. & King, C.M. (1994). *The European Rabbit: History and Biology of a Successful Colonizer*. Oxford University Press, Oxford.

Watts, C.H.S. & Aslin, H.J. (1981). *The Rodents of Australia*. Angus and Robertson, Sydney and London.

INSECTIVORES, EDENTATES AND ALLIES

Churchfield, S. (1990). *The Natural History of Shrews*. Christopher Helm, London.

Crowcroft, P. (1957). *The Life of the Shrew*. Max Reinhart, London.

Dolgov, V.A. (1985). *Burozubki Starovo Sveta [Shrews of the Old World]*. Moscow University Press, Moscow.

Eisenberg, J. F. (1970). *The Tenrecs: a Study in Mammalian Behavior and Evolution*. Smithsonian Institution Press, Washington.

Godfrey, G.K. & Crowcroft, P. (1960). *The Life of the Mole*. Museum Press, London.

Gorman, M.L. & Stone, R.D. (1990). *The Natural History of Moles*. Christopher Helm, London.

Mellanby, K. (1976). *Talpa: Story of a Mole*. Collins, London.

Merritt, J.F., Kirkland, G.L. Jr & Rose, R.K. (1994). *Advances in the Biology of Shrews*. Carnegie Museum Natural History, Pittsburgh.

Montgomery, G.G. (ed) (1978). *The Ecology of Arboreal Folivores*. Smithsonian Institution Press, Washington.

Montgomery, G.G. (ed) (1985). *The Evolution and Ecology of Armadillos, Sloths, and Vermilinguas*. Smithsonian Institution Press, Washington.

Morris, P. (1983). *Hedgehogs*. Whittet Books, London.

Reeve, N. (1994). *The Natural History of Hedgehogs*. T&AD Poyser Natural History, London.

van Zyll de Jong, C.G. (1983). *Handbook of Canadian Mammals. Part I. Marsupials and Insectivores*. National Museum of Natural Sciences, Ottawa.

Wolsan, M. & Wojcik, J.M. (eds) (1998). *Evolution of Shrews*. Polish Academy of Sciences, Bialowieza.

BATS

Allen, G.M. (1939). *Bats*. Harvard University Press, Cambridge.

Altringham, J.D. (1996). *Bats: Biology and Behaviour*. Oxford University Press, New York.

Adams, R.A. & Pederson, S.C. (eds) (2000). *Ontogeny, Functional Ecology and Evolution of Bats*. Cambridge University Press, Cambridge.

Barbour. R.W. & Davis, W.H. (1969). *Bats of America*. University of Kentucky Press, Lexington, Kentucky.

Bates, P.J.J. & Harrison, D.L. (1997). *Bats of the Indian Subcontinent*. Harrison Zoological Museum, Sevenoaks, Kent.

Churchill, S. (1999). *Australian Bats*. New Holland/Struik.

Crichton, E.G. & Kutzsch, P.H. (2000). *Reproductive Biology of Bats*. Academic Press, London.

Fenton, M.B. (1983). *Just Bats*. University of Toronto Press, Toronto.

Fenton, M.B. (1998). *The Bat: Wings in the Night Sky*. Firefly Books, Willowdale, Ontario.

Findley, J.S. (1993). *Bats: a Community Perspective*. Cambridge University Press, Cambridge.

Griffin, D.R. (1958). *Listening in the Dark*. Yale University Press, New Haven.

Hill, J.E. & Smith, J.D. (1984). *Bats: A Natural History*. British Museum (Natural History), London.

Koopman, K.F. (1994). *Chiroptera: Systematics*. De Gruyter, Berlin.

Kunz, T.H. (ed) (1982). *Ecology of Bats*. Plenum Press, New York.

Kunz, T.H. & Racey, P.A. (1998). *Bat Biology and Conservation*. Smithsonian Institution Press, Washington.

Leen, N. & Norvic, A. (1969). *The World of Bats*. Holt, Rinehart and Winston, New York.

Neuweiler, N. (2000). *The Biology of Bats*. Oxford University Press, Oxford.

Racey, P.A. & Swift, S.M. (2001). *Ecology, Evolution and Behaviour of Bats. Symp. Zool. Soc. Lond*. No. 67, Oxford University Press, Oxford.

Ransome, R.D. (1990). *The Natural History of Hibernating Bats*. Christopher Helm, London.

Rosevear, J. R. (1965). *The Bats of West Africa*. British Museum (Natural History), London.

Schober, W. & Grimmberger, E. (1989). *A Guide to Bats of Britain and Europe*. Hamlyn, London.

Stebbings, R.E. (1988). *Conservation of European Bats*. Christopher Helm, London.

Stebbings, R.E. & Griffith, F. (1986). *The Distribution and Status of Bats in Europe*. Institute of Terrestrial Ecology, Abbots Ripton.

Swift, S.M. (1998). *Long-eared Bats*. T&AD Poyser Natural History, London.

Turner, D.E. (1975). *The Vampire Bat: a Field Study in Behaviour and Ecology*. Johns Hopkins University Press, Baltimore.

Tuttle, M.D. (1997). *America's Neighborhood Bats*. University of Texas Press, Austin.

van Zyll de Jong, C.G. (1985). *Handbook of Canadian Mammals. Part II. Bats*. National Museum of Natural Sciences, Ottawa.

Wilson, D.E. & Tuttle, M.D. (1997). *Bats in Question: the Smithsonian Answer Book*. Smithsonian Institution Press, Washington.

Wimsatt, W. A. (ed) (1970, 1977). *Biology of Bats*. Vols I, II & III. Academic Press, New York.

Yalden, D.W. & Morris, P.A. (1975). *The Lives of Bats*. David and Charles, Newton Abbot.

MARSUPIALS AND MONOTREMES

Archer, M. (ed) (1987). *Possums and Opossums: Studies in Evolution. Vols I & II*. Surrey Beatty and Sons, Chipping Norton, New South Wales.

Archer, M. (ed) (1982). *Carnivorous Marsupials*. Royal Zoological Society of New South Wales, Sydney.

Augee, M.L. (ed) (1978). *Monotreme Biology*. Royal Zoological Society of New South Wales, Sydney.

Augee, M.L. (ed) (1992). *Platypus and Echidnas*. Royal Zoological Society, New South Wales.

Dawson, T.J. (1995). *Kangaroos: Biology of the Largest Marsupials*. Cornell University Press, Ithaca, NY.

Flannery, T.F. (1994). *Possums of the World: A Monograph of the Phalangeroidea*. Chatswood, New South Wales.

Flannery, T.F., Martin, R. & Szalay, A. (1996). *Tree Kangaroos: A Curious Natural History*. Reed New Holland, Melbourne.

Fleay, D.M. (1980). *The Paradoxical Platypus*. Jacaranda Press, Brisbane.

Frith, H.J. & Calaby, J.H. (1969). *Kangaroos*. F. W. Cheshire, Melbourne.

Grant, T.R. (1983). *The Platypus*. University of New South Wales Press, Kensington.

Griffiths, M.F. (1978). *The Biology of Monotremes*. Academic Press, New York.

Grigg, G., Jarman, P. & Hume, I. (eds) (1989). *Kangaroos, Wallabies, and Rat-Kangaroos. Vols I & II*. Surrey Beatty and Sons, Chipping Norton, New South Wales.

Hume, I. D. (1999). *Marsupial Nutrition*. Cambridge and New York, Cambridge.

Hunsaker II, D. (ed) (1977). *The Biology of Marsupials*. Academic Press, New York.

Lee, A.K. & Cockburn, A. (1985). *Evolutionary Ecology of Marsupials*. Cambridge University Press, Cambridge.

Mares, M.A. & Genoways, H.H. (eds) (1982). *Mammalian Biology in South America*. University of Pittsburgh, Pennsylvania.

Marlow, B.J. (1965). *Marsupials of Australia*. Jacaranda Press, Brisbane.

Martin, R. & Handasyde, K. (1999). *The Koala: Natural History, Conservation and Management*. 2nd edn. Krieger, Florida.

Mustrangi, M.A. & Patton, J.L. (1997). *Phylogeography and Systematics of the Slender Mouse Opossum Marmosops (Marsupialia, Didelphidae)*. University of California Press, Berkeley.

Ride, W.D.L. (1970). *The Native Mammals of Australia*. Oxford University Press, Melbourne.

Rismiller, P. (1999). *The Echidna: Australia's Enigma*. Hugh Lauter Levin Associates, Southport, CT.

Saunders, N.R. & Hinds, L.A. (1997). *Marsupial Biology: Recent Research, New Perspectives*. University of New South Wales Press, Sydney.

Seebeck, J.H., Brown, P.R., Wallis, R.W. & Kemper, C.M. (eds) (1990). *Bandicoots and Bilbies*. Surrey Beatty and Sons, Chipping Norton, New South Wales.

Smith, A. & Hume, I. (1984). *Possums and Gliders*. Royal Zoological Society of New South Wales, Sydney.

Stonehouse, B. (ed) (1977). *The Biology of Marsupials*. Macmillan, London.

Troughton, E. le G. (1941). *Furred Mammals of Australia*. Angus and Robertson, Sydney.

Tyndale-Biscoe, C.H. (1973). *Life of Marsupials*. Edward Arnold, London.

Wood-Jones, F. (1923–1925). *The Mammals of South Australia. Vols I–III*. Government Printer, Adelaide.

The IUCN (World Conservation Union) produces a number of outstanding Species Action Plans, many of which have become authoritative sources of information. For further details visit the IUCN website http://www.iucn.org

Index

Page numbers in *italic* type refer to picture captions. Page numbers in **bold** indicate extensive coverage or feature treatment of a subject.

A

C

E

F

H

I

J

K

M

N

Q

R

T